THE CENTURY OF SPACE SCIENCE

esa
and
The Century of Space Science

Science is at the origin of countless aspects of modern life. Space science is no exception. We benefit today from pervasive, though barely perceived, space systems.

When Viktor Hess, pursuing measurements at higher altitude, ascended in a balloon and discovered cosmic rays in 1912, few people would have predicted that this phenomenon would become a driver for the initiation of space research forty years later. And almost no one would then have conceived that, at the beginning of the 21st century, humanity would be making daily use of space through communications, navigation and weather satellites.

The relationship between science and applications is symbiotic. Pure science leads to applications that, later on, become a technological base for new scientific endeavours and, in turn, assure future applications. The European Space Agency evolved along a similar line: from science to everyday life. ESRO, the science-oriented European Space Research Organisation founded in 1962, later became the Scientific Programme of ESA, often referred to as ESA's backbone. ESA, though, also carries out now major infrastructure programmes (the development of Ariane launchers, Spacelab and Space Station) and has brought about new application programmes for communications, Earth observation and navigation. TV transmissions via satellite from remote political hot spots or faraway sports events have become a matter of course. So have images of the weather on Earth seen from geostationary orbit; and the taxi driver guided by a global navigation system is becoming the rule rather than the exception.

In the context of Europe, space science has produced more than scientific results and applications: it has contributed to European unification. ESA is a very successful example of the new Europe working together. Like CERN and ESO and the other European International Research Organisations, ESRO came into existence after the middle of the 20th century, following an era where the continent's nation states had been at war now and again. These international organisations helped to smooth the progress toward peaceful co-operation and collaboration among Europeans. It is interesting to note that ESRO and ESA, quite uniquely among their sister organisations, have worked from their inception with an artificial currency, the Accounting Unit (AU), which today is a real currency—the Euro.

We are pleased to present to you this reference work, especially since the three editors belong to ESA's science community. Nonetheless, the content of this work espouses the international spirit of science: the authors of the individual chapters are based in Europe, Japan, and Russia as well as in the USA and have written their articles with a world-wide perspective. Moreover, many of the pioneers and early key players in space science are still active and have described here the early epoch of their field. The two volumes of *The Century of Space Science* therefore not only convey a global view, but also offer a unique, first-hand insight into the development of the subject.

Antonio Rodotà
Director General of ESA

David Southwood
Director of the Scientific Programme

The Century of Space Science

Volume I

Edited by

JOHAN A.M. BLEEKER
SRON – National Institute for Space Research, Utrecht, The Netherlands

JOHANNES GEISS
ISSI – International Space Science Institute, Bern, Switzerland

MARTIN C.E. HUBER
ESA – European Space Agency, Paris, France

History consultant

ARTURO RUSSO
Università degli Studi di Palermo, Italy

KLUWER ACADEMIC PUBLISHERS
DORDRECHT / BOSTON / LONDON

A C.I.P. Catalogue record for this book is available from the Library of Congress.

ISBN 0-7923-7196-8

Published by Kluwer Academic Publishers,
P.O. Box 17, 3300 AA Dordrecht, The Netherlands.

Sold and distributed in North, Central and South America
by Kluwer Academic Publishers,
101 Philip Drive, Norwell, MA 02061, U.S.A.

In all other countries, sold and distributed
by Kluwer Academic Publishers,
P.O. Box 322, 3300 AH Dordrecht, The Netherlands.

Printed on acid-free paper

Cover figure:
The nucleus of Comet Halley (rectangular frame;
copyright Max-Planck-Institut für Aeronomie) and the cosmic
microwave background (oval frame; image credited to NASA and
the Cosmic Background Explorer Team)

www.thecenturyofspacescience.com

All Rights Reserved
© 2001 Kluwer Academic Publishers
No part of this work may be reproduced, stored in a retrieval system, or transmitted in any form or by any
means, electronic, mechanical, photocopying, microfilming, recording, or otherwise,
without written permission from the Publisher, with the exception of any material supplied specifically for the purpose
of being entered and executed on a computer system, for exclusive use by the purchaser of the work.

Printed in The Netherlands.

Contents

Volume I

List of contributors	vii
Foreword	xiii
L Woltjer	

Introduction
1. The century of space science — *JAM Bleeker, J Geiss and MCE Huber* — 3

The Beginnings
2. The space age and the origin of space research — *A Russo* — 25
3. Enabling technology for space transportation — *E Stuhlinger* — 59

The Early Epoch of Space Science
4. The cosmic radiation — *JA Simpson* — 117
5. Magnetospheric physics — *JA Van Allen* — 153
6. Barium cloud experiments in the upper atmosphere — *R Lüst* — 179
7. Alkali metal cloud experiments in the upper atmosphere — *J-E Blamont* — 189
8. Early solar space research — *C de Jager* — 203
9. A history of the solar wind concept — *EN Parker* — 225
10. The terrestrial planets at the dawn of the space age — *WK Hartmann* — 257
11. The Moon before Apollo — *JR Arnold* — 271
12. From the ionosphere to high energy astronomy – a personal experience — *H Friedman* — 277
13. Early ultraviolet spectroscopy from space — *BD Savage* — 287
14. The early days of infrared space astronomy — *M Harwit* — 301

Fundamental Science in Space

Cosmology and Gravitational Physics
15. Verification of general relativity: tests in the Solar System — *K Nordtvedt* — 335
16. Verification of general relativity: strong fields and gravitational waves — *CM Will* — 353
17. The cosmological constants — *GA Tammann* — 373
18. COBE, dark matter and large-scale structure in the Universe — *KM Górski and AJ Banday* — 399
19. The origin of the light elements in the early Universe — *H Reeves* — 423
20. Gravitational lensing — *J Surdej and J-F Claeskens* — 441

Extragalactic Astronomy
21. Clusters of galaxies — *RF Mushotzky* — 473
22. Gamma-ray bursts — *RAMJ Wijers* — 499
23. Quasars — *M Elvis* — 529
24. Blazars — *G Ghisellini* — 549
25. X-ray and infrared properties of normal galaxies — *G Fabbiano and MF Kessler* — 561

The Milky Way
26. The hot part of the interstellar medium — *SL Snowden* — 581
27. Space-borne observations of the life cycle of interstellar gas and dust — *EF van Dishoeck and AGGM Tielens* — 607
28. The interstellar medium of our Galaxy — *PC Frisch* — 647
29. Galactic cosmic rays — *FB McDonald and VS Ptuskin* — 677
30. Stellar populations and dynamics in the Milky Way galaxy — *GF Gilmore* — 699

31. Pulsars and isolated neutron stars *W Becker and G Pavlov*	721	
32. Evolutionary concepts of binaries with compact objects *EPJ van den Heuvel*	759	
33. White dwarf binaries *PA Charles*	791	
34. Low mass X-ray binaries *J van Paradijs and M van der Klis*	811	
35. High-mass X-ray binaries *NE White*	823	
36. Black-hole binaries *Y Tanaka*	839	
37. The formation of stars and protoplanetary disks *C Waelkens*	857	
38. High-energy radiation from outer stellar atmospheres *R Pallavicini*	875	
39. Mass loss from stars *JP Cassinelli*	895	
40. Planetary nebulae *SR Pottasch*	913	
41. Supernovae and supernova remnants *H Tsunemi*	937	

Volume II

The Solar System

42. Acceleration processes of heliospheric particle populations — 963
    G Gloeckler and K-P Wenzel
43. Reconnection — 1007
    G Haerendel
44. The solar interior — 1035
    DO Gough and PH Scherrer
45. The solar atmosphere — 1065
    SK Solanki and R Hammer
46. The active Sun — 1089
    D Alexander and LW Acton
47. The solar wind — 1115
    M Neugebauer and R von Steiger
48. The heliosphere — 1141
    A Balogh and LA Fisk
49. The dusty heliosphere — 1163
    E Grün
50. The interaction of the heliosphere with the interstellar medium — 1191
    R Lallement
51. Comets: coma and beyond — 1217
    K Szegö
52. The morphology of cometary nuclei — 1235
    HU Keller and L Jorda
53. The constituents of cometary nuclei — 1277
    K Altwegg and WT Huntress
54. The Moon and terrestrial planets: geology and geophysics — 1295
    JW Head III
55. Radiometric chronology of the Moon and Mars — 1325
    LE Nyquist, DD Bogard and C-Y Shih
56. Chemical evolution of the Moon and the terrestrial planets — 1377
    H Wänke
57. The atmospheres of the terrestrial planets — 1405
    FW Taylor
58. Jupiter — 1425
    DM Hunten
59. The planets beyond Jupiter — 1431
    T Encrenaz
60. The satellites of the outer planets — 1451
    N Thomas
61. Planetary and lunar magnetism — 1479
    NF Ness

The Earth and its Plasma Environment

62. The magnetosphere as a plasma laboratory — 1495
    RA Treumann and M Scholer
63. Earth's magnetosphere — 1529
    B Hultqvist
64. Earth's ionosphere — 1559
    T Hagfors and K Schlegel
65. Oceanography — 1585
    JA Johannessen, S Sandven and D Durand
66. Satellite geodesy and geosciences — 1623
    F Barlier and M Lefebvre
67. Chemistry and physics of the atmosphere — 1653
    JE Harries

Appendices

A basic chronology of the space age — 1671
A Russo
Catalog of space science launches 1957–2000 — 1677
J McDowell

Indices

Abbreviations and acronyms	1711
Index of cited authors	1719
Name index	1791
Subject index	1801

List of Contributors

Acton, Loren W.
Physics Department
Montana State University
Bozeman MT 59717-0350
USA

Alexander, David
Lockheed Martin Solar & Astrophysics Lab
3251 Hanover Street
Palo Alto CA 94304
USA

Altwegg, Kathrin
Physikalisches Institut
Universität Bern
Sidlerstrasse 5
CH-3012 Bern
Switzerland

Arnold, James R.
California Space Institute
University of California, San Diego
9500 Gilman Drive
La Jolla CA 92093-0524
USA

Balogh, André
Space and Atmospheric Physics Group
The Blackett Laboratory
Imperial College
Prince Consort Road
London SW7 2BZ
United Kingdom

Banday, A. J.
Max-Planck-Institut für Astrophysik
Karl-Schwarzschild-Strasse 1
D-85741 Garching bei München
Germany

Barlier, François
Observatoire de la Côte d'Azur
Avenue Nicolas Copernic
F-06130 Grasse
France

Becker, Werner
Max-Planck-Institut für Extraterrestrische Physik
Giessenbachstrasse
D-85748 Garching bei München
Germany

Blamont, Jacques-Emile
Centre National d'Etudes Spatiales
2, Place Maurice Quentin
F-75039 Paris
France

Bleeker, Johan A. M.
SRON – National Institute for Space Research
Sorbonnelaan 2
NL-3584 CA Utrecht
The Netherlands

Bogard, Donald D.
Planetary Sciences
NASA – Johnson Space Center

2101 NASA Road 1
Houston TX 77058-3696
USA

Cassinelli, Joseph P.
Department of Astronomy
University of Wisconsin
475 N. Charter Street
Madison WI 53706-1582
USA

Charles, Philip A.
Department of Physics and Astronomy
University of Southampton
Highfield
Southampton SO17 1BJ
United Kingdom

Claeskens, Jean-François
Institut d'Astrophysique et de Géophysique
Université de Liège
Allée du 6 Août, 17
B-4000 Liège 1
Belgium

de Jager, Cornelis
SRON – National Institute for Space Research
Sorbonnelaan 2
NL-2584 CA Utrecht
The Netherlands

Durand, Dominique
Nansen Environmental and Remote Sensing Center
Edvard Griegsvei 3a
N-5059 Bergen
Norway

Elvis, Martin S.
Harvard-Smithsonian Center for Astrophysics
60 Garden Street
Cambridge MA 02138
USA

Encrenaz, Thérèse
Départment de Recherche Spatiale
Observatoire de Paris-Meudon
5, Place Jules Janssen
F-92195 Meudon Cedex
France

Fabbiano, Giuseppina
Harvard-Smithsonian Center for Astrophysics
60 Garden Street
Cambridge MA 02138
USA

Fisk, Lennard A.
Department of Atmospheric, Oceanic and Space Science
University of Michigan
2455 Hayward Avenue
Ann Arbor MI 48109-2143
USA

Friedman, Herbert
(author died on 9 September 2000)

Frisch, Priscilla C.
Department of Astronomy and Astrophysics
University of Chicago
5640 South Ellis Avenue
Chicago IL 60637
USA

Geiss, Johannes
International Space Science Institute
Hallerstrasse 6
CH-3012 Bern
Switzerland

Ghisellini, Gabriele
Osservatorio Astronomico di Brera
Via Bianchi 46
I-23807 Merate
Italy

Gilmore, Gerard F.
Institute of Astronomy
University of Cambridge
Madingley Road
Cambridge CB3 0HA
United Kingdom

Gloeckler, George
Institute for Physical Science and Technology
University of Maryland
Computer and Space Sciences Building
College Park MD 20742-4111
USA

Górski, K. M.
European Southern Observatory
Karl-Schwarzschild-Strasse 2
D-85748 Garching bei München
Germany

Gough, Douglas O.
Institute of Astronomy
University of Cambridge
Madingley Road

LIST OF CONTRIBUTORS

Cambridge CB3 0HA
United Kingdom

Grün, Eberhard
Max-Planck-Institut für Kernphysik
Saupfercheckweg 1
D-69117 Heidelberg
Germany

Gursky, Herbert
Space Science Division
Naval Research Laboratory
4555 Overlook Ave., SW
Washington DC 20375
USA

Haerendel, Gerhard
International University Bremen
Campus Ring 1
D-28759 Bremen
Germany

Hagfors, Thor
Max-Planck-Institut für Aeronomie
Max-Planck-Strasse 2
D-37191 Katlenburg-Lindau
Germany

Hammer, Reiner
Kiepenheuer-Institut für Sonnenphysik
Schöneckstrasse 6
D-79104 Freiburg
Germany

Harries, John E.
Space and Atmospheric Physics Group
The Blackett Laboratory
Imperial College
Prince Consort Road
London SW7 2BZ
United Kingdom

Hartmann, William K.
Planetary Science Institute
620 North 6th Avenue
Tucson AZ 85705-8331
USA

Harwit, Martin
511 H Street SW
Washington DC 20024-2725
USA

Head, James William III
Department of Geological Sciences
Brown University
Providence RI 02912
USA

Huber, Martin C.E.
ESA – European Space Agency
8-10, Rue Mario Nikis
F-75738 Paris Cedex 15
France

Hultqvist, Bengt
Swedish Institute of Space Physics
Box 812
S-98128 Kiruna
Sweden

Hunten, Donald M.
Lunar and Planetary Lab
University of Arizona
1629 East University Avenue
Tucson AZ 85721
USA

Huntress, Wesley T., Jr
Geophysical Laboratory
Carnegie Institution of Washington
5251 Broad Branch Road NW
Washington DC 20015-1305
USA

Johannessen, Johnny A.
Nansen Environmental and Remote Sensing Center
Edvard Griegsvei 3a
N-5037 Bergen
Norway

Jorda, Laurent
Max-Planck-Institut für Aeronomie
Max-Planck-Strasse 2
D-37191 Katlenburg-Lindau
Germany

Keller, Horst Uwe
Max-Planck-Institut für Aeronomie
Max-Planck-Strasse 2
D-37191 Katlenburg-Lindau
Germany

Kessler, Martin F.
Research and Scientific Support Department
ESA – European Space Research and Technology Centre

Keplerlaan 1
NL–2200 AG Noordwijk
The Netherlands

Lallement, Rosine
Service d'Aéronomie
CNRS
F-91370 Verrières-le-Buisson
France

Lefebvre, Michel
8, Avenue de Cugnaux
F-31270 Villeneuve Tolosane
France

Lüst, Reimar
Max-Planck-Institut für Meteorologie
Bundesstrasse 55
D-20146 Hamburg
Germany

McDonald, Frank B.
Institute for Physical Science and Technology
University of Maryland
College Park MD 20742-2431
USA

McDowell, Jonathan C.
Harvard-Smithsonian Center for Astrophysics
60 Garden Street
Cambridge MA 02138
USA

McKibben, R. Bruce
Laboratory for Astrophysics and Space Research
Enrico Fermi Institute
University of Chicago
933 East 56th Street
Chicago IL 60637
USA

Mushotzky, Richard F.
Laboratory for High Energy Astrophysics
NASA – Goddard Space Flight Center
Greenbelt MD 20771
USA

Ness, Norman F.
Bartol Research Institute
University of Delaware
217 Sharp Lab
Newark DE 19716-4793
USA

Neugebauer, Marcia
JPL – Jet Propulsion Laboratory
California Institute of Technology
4800 Oak Grove Drive
Pasadena CA 91109
USA

Nordtvedt, Kenneth
118 Sourdough Ridge Road
Bozeman MT 59715
USA

Nyquist, Laurence E.
Astromaterials Research
NASA – Johnson Space Center
Houston TX 77058-3696
USA

Pallavicini, Roberto
Osservatorio Astronomico di Palermo
Piazza del Parlamento 1
I-90134 Palermo
Italy

Parker, Eugene N.
1323 Evergreen Road
Homewood IL 60430
USA

Pavlov, George
Department of Astronomy and Astrophysics
The Pennsylvania State University
University Park PA 16802
USA

Pottasch, Stuart R.
Kapteyn Astronomical Institute
Rijksuniversiteit Groningen
Landleven 12
NL-9747 AD Groningen
The Netherlands

Ptuskin, Vladimir S.
IZMIRAN – Institute for Terrestrial Magnetism,
Ionosphere and Radio Wave Propagation of the
Russian Academy of Science
Troitsk, Moscow Region 142092
Russia

Reeves, Hubert
1, Rue Jacob
F-75006 Paris
France

LIST OF CONTRIBUTORS

Russo, Arturo
Instituto di Fisica
Università di Palermo
Via Archirafi 36
I-90123 Palermo
Italy

Sandven, Stein
Nansen Environmental and Remote Sensing Center
Edvard Griegsvei 3a
N-5037 Bergen
Norway

Savage, Blair D.
Astronomy Department
University of Wisconsin
475 North Charter Street
Madison WI 53706-1582
USA

Scherrer, Philip H.
Hansen Experimental Physics Lab
Stanford University
455 Via Palou
Stanford CA 94305-4055
USA

Schlegel, Kristian
Max-Planck-Institut für Aeronomie
Max-Planck-Strasse 2
37191 Katlenburg-Lindau
Germany

Scholer, Manfred
Max-Plack-Institut für Extraterrestrische Physik
Giessenbachstrasse
D-85748 Garching bei München
Germany

Shih, C.-Y.
Lockheed-Martin Space Operations
2400 Nasa Road 1
Houston TX 77058
USA

Simpson, John A.
(author died on 31 August 2000)

Snowden, Steven L.
Laboratory for High Energy Astrophysics
NASA – Goddard Space Flight Center
Greenbelt MD 20771
USA

Solanki, Sami K.
Max-Planck-Institut für Aeronomie
Max-Planck-Strasse 2
D-37191 Katlenburg-Lindau
Germany

Stuhlinger, Ernst
3106 Rowe Dr SE
Huntsville AL 35801-6151
USA

Surdej, Jean
Institut d'Astrophysique et de Géophysique
Université de Liège
Allée du 6 Août, 17
B-4000 Liège 1
Belgium

Szegö, Karoly
Research Institute for Particle and Nuclear Physics
Hungarian Academy of Sciences
Miklós út 29–33
H-1525 Budapest
Hungary

Tammann, G. A.
Astronomisches Institut der Universität Basel
Venusstrasse 7
CH-4102 Binningen
Switzerland

Tanaka, Yasuo
Max-Planck-Institut für Extraterrestrische Physik
Giessenbachstrasse
D-85748 Garching bei München
Germany

Taylor, Fred W.
Atmospheric, Oceanic and Planetary Physics
Clarendon Laboratory
University of Oxford
Parks Road
Oxford OX1 3PU
United Kingdom

Thomas, Nicolas
Max-Planck-Institut für Aeronomie
Max-Planck-Strasse 2
D-37191 Katlenburg-Lindau
Germany

Tielens, Alexander G.G.M.
Kapteyn Astronomical Institute

Rijksuniversiteit Groningen
Landleven 12
NL-9747 AD Groningen
The Netherlands

Treumann, Rudolf A.
Max-Planck-Institut für Extraterrestrische Physik
Giessenbachstrasse
D-85741 Garching bei München
Germany

Tsunemi, Hiroshi
Department of Earth and Space Science
Osaka University
1-1, Machikaneyama
Toyonaka, Osaka 560-0043
Japan

Van Allen, James A.
Department of Physics and Astronomy
University of Iowa
515 van Allen Hall
Iowa City, IA 52242–1479
USA

van der Heuvel, Ed P.J.
Astronomical Institute Anton Pannekoek
Universiteit van Amsterdam
Kruislaan 403
NL-1098 SJ Amsterdam
The Netherlands

van der Klis, Michiel
Astronomical Institute Anton Pannekoek
Universiteit van Amsterdam
Kruislaan 403
NL-1098 SJ Amsterdam
The Netherlands

van Dishoeck, Ewine F.
Sterrewacht Leiden
Universiteit Leiden
Niels Bohrweg 2
NL-2333 CA Leiden
The Netherlands

Van Paradijs, Jan
(author died on 2 November 1999)

von Steiger, Rudolf
ISSI – International Space Science Institute

Hallerstrasse 6
CH-3012 Bern
Switzerland

Waelkens, Christoffel
Instituut voor Sterrenkunde
Katholieke Universiteit Leuven
Celestijnenlaan 200 B
B-3001 Leuven
Belgium

Wänke, Heinrich
Abteilung Kosmochemie
Max-Planck-Institut für Chemie
Postfach 3060
D-55020 Mainz
Germany

Wenzel, Klaus-Peter
Research and Scientific Support Department
ESA – European Space Research and Technology Centre
Keplerlaan 1
NL-2200 AG Noordwijk
The Netherlands

White, Nicholas
High Energy Astrophysics Science
Archive Research Center
NASA – Goddard Space Flight Center
Greenbelt MD 20771
USA

Wijers, Ralph A.M.J.
Department of Physics & Astronomy
State University of New York at Stony Brook
Stony Brook NY 11794-3800
USA

Will, Clifford M.
McDonnell Center for the Space Sciences
Department of Physics
Washington University
One Brookings Drive
St. Louis MO 63130
USA

Woltjer, Lodewijk
Observatoire de Haute Provence
F-04870 St. Michel l'Observatoire
France

LODEWIJK WOLTJER*

Foreword

Space has become an essential element of the contemporary world. Even though the first satellite was launched only forty-four years ago, we have become dependent on space-based facilities for a wide variety of functions. Telecommunications, television and weather forecasting are unimaginable without satellites. We determine our position on Earth with a few meters accuracy with space-based systems. Observation of the Earth from space allows us to quantitatively and globally ascertain the state of the world's crops and forests, to forecast abundance and scarcity, and to take remedial measures.

Early on scientists realized that much knowledge about the Universe is inaccessible because of the Earth's atmosphere. Not surprisingly, therefore, they were the pioneers in placing instruments in orbit and to develop the necessary technologies for their utilization. Enormous progress resulted. In addition to visible light and radio waves, astronomical objects could be studied in X- and gamma-rays, in the ultraviolet and in the infrared, while also cosmic-ray particles could be studied in detail and their composition determined. Wholly new phenomena were discovered involving black holes and energetic processes, but also the cool gas and dust so important in the formation of stars and planets. In visible light the absence of the atmosphere allowed images to be obtained with superior resolution. Who has not marvelled at the beauty of the world revealed by the Hubble Space Telescope or at the wild scenes shown by SOHO on the surface of the Sun?

The study of the planets in our solar system made perhaps even greater progress. Instead of peering at these from a large distance, they could be visited by spacecraft with sophisticated instruments. Thus, the detailed structure and composition of the surfaces of Mars and the Moon could be analyzed and much could be learned about the other planets, satellites, comets. The early arrival of science in space provided a strong impetus on the generalization of space technology for many different purposes. But it did more. Even in the darkest days of the cold war communication between the world's scientists continued. There was competition, but also collaboration, and a strong feeling provided that space should remain a domain for peace rather than war. One can only hope that this will remain so also in the future.

Mankind is continuously modifying the earth, its atmosphere and its oceans. While many would wish that this were not so, it remains an incontrovertible fact. Only satellites observing the Earth can provide the global knowledge that is needed to ascertain the resulting changes in detail and to indicate the remedial actions that are needed. The observations of the "ozone hole" over antarctica and the subsequent restrictions on the use of the chemical substances responsible is an illustrative example. Our current understanding of the implications of the results of the observations of the earth is still very limited and sometimes controversial. Continuous monitoring and analysis are essential if a prudent management of the Earth System is to be implemented.

Space activities make their own contribution to polluting the planetary environment. Many large and small debris from earlier launches create risks for new satellites. Fortunately, the space agencies have become aware of this and measures are being taken to reduce such problems. Also proposals have been made for advertising in space. Quite apart from the damage this would do to astronomical observations, it would prevent humanity from watching the star covered night sky, which has been an inspiration to prophets and philosophers and to every human from the earliest days.

Space scientists, whose deeds are described in this book, have shown the way into space with its manifold scientific and technological opportunities. In space there are no geographical boundaries and no property rights. So let us ensure that space will remain a heritage for all of mankind for responsible scientific exploration and technological utilization.

---

* Observatoire de Haute Provence, St. Michel l'Observatoire, France

INTRODUCTION

1

JOHAN A.M. BLEEKER*, JOHANNES GEISS** AND MARTIN C.E. HUBER***

The century of space science

In the course of its brilliant evolution through the twentieth century, space science brought spectacular results and led to fundamental scientific findings. In the early quest for higher altitude, cosmic rays were discovered by Viktor Hess during a balloon flight in 1912. Following the second World War, sounding rockets were used, starting in 1946, to study the structure of the terrestrial atmosphere – the threshold of space. Two landmarks of the space age stand out: the launch in 1957 of the first satellite, Sputnik 1, which sensed the near-Earth environment at orbital altitude, and in 1969 the first landing by humans on an extraterrestrial body, the Moon. Today, sophisticated space probes explore distant worlds, and space telescopes look back in time towards the early Universe.

The scientific exploration of space was at the origin and, indeed, the motivation for our first ventures away from the Earth. It is well known, however, that rockets – the enabling tools of space research – were developed with support from the military, and that the 'space race', later on, served as a substitute battleground during the Cold War. Today, fortunately, the adversities of those years have given way to peaceful global collaboration that would have been technologically and ideologically impossible during most of the twentieth century.

The experiences of the astronauts on the Moon, their pictures of the Earth from above, the visits of unmanned spacecraft and robots to the distant planets and their satellites, and the probing of the largest distances and the earliest epochs of our Universe have changed the perception of the world around us, for scientists as well as for the general public. For about a quarter of a century after the end of World War II, in what we might call the early epoch of space science, experiments in space were regarded as daring but rather extravagant. But in the course of the 1970s, the methods and techniques of space science began to enter the mainstream of science, and by the end of the century, space science had become a natural complement, and often an integral part of study in many fields of science.

The present two volumes of *The Century of Space Science* tell the story of space science as it evolved during the twentieth century. The origins of space research, the enabling technology for space transportation and the 'early epoch' of space science mentioned above, are addressed in Chapters 2–14. The chapters in the main body of this Reference Work then describe the development and results of the scientific topics that have benefited extensively from space investigations. A chronology of the space age and a list of space science missions appear as appendices.

We have had to restrict the number of fields covered, and have therefore concentrated on those topics with the most comprehensive record in space, namely the exploration of the Solar System, astronomy and gravitational physics. We note, however, that space science – far from being a coherent discipline itself – has by now become a widely used method that complements in an essential way the investigative tools of an impressive range of initially laboratory- and ground-based research fields.

We have also included three chapters on Earth science. This field is important not only because of its scientific achievements, but also because, from early on in the space

* SRON – National Institute for Space Research, Utrecht, The Netherlands
** ISSI – International Space Science Institute, Bern, Switzerland
*** ESA – European Space Agency, Paris, France

age, it started to raise humanity's awareness of the fragility of their home planet – 'spaceship Earth' – which protects them from the hostile cosmic environment. It is appropriate therefore to include here an image of 'the blue planet in space', photographed during one of the Apollo missions (Figure 1).

Rather than use this chapter to summarize the contents of the rest of the book, we instead highlight a number of results which demonstrate how space science has made major contributions to our knowledge, and sometimes has even forced us to change our concepts of nature. In outlining these results, we start with Earth System science and then proceed – roughly following the historical development – from the terrestrial environment and the Solar System to the Milky Way, and then to extragalactic space and cosmology. (Note, however, that the sequence chosen for the topics presented in the main body of this work is the opposite: it begins with gravitational physics and cosmology and ends with Earth science.)

EARTH SYSTEM SCIENCE

Space-borne observational platforms gave us, for the first time, the means to survey the many features of our planet rapidly and effectively from a global perspective. This global view of the Earth from space – together with the maturation of distinct Earth-science disciplines, such as geophysics of the solid Earth, oceanography, and atmospheric dynamics and chemistry, as well as the recognition of the human role in global change – has stimulated a new approach for studying our home world, what we might call Earth System science. In this approach the Earth System is studied in the context of a related set of interacting processes rather than as a collection of individual components. Interactions among oceans, ice, land-masses, the atmosphere and biological systems are significant but also very complex. The transport of energy and material within and among these subsystems occurs on a global scale across a wide range of time-scales. Observations from space are indispensable for present and future research,

Figure 1 The Blue Planet, with the dry lunar surface in the foreground.

whose ultimate goal is an understanding of the processes responsible for the evolution of the Earth on all time-scales.

The importance of a synoptic view of planet Earth is convincingly demonstrated by the global monitoring of stratospheric ozone (Figure 2). Ozone provides crucial protection against hard ultraviolet radiation from the Sun. The mapping and monitoring of the evolution of the Antarctic ozone hole has helped us to develop and validate representative models for ozone depletion and restoration in the upper atmosphere.

PLASMAS IN SPACE: MAGNETOSPHERE AND HELIOSPHERE

The International Geophysical Year (1957/58) was the first worldwide coordinated effort in space science. Its goal was to understand the aurorae and to map the variation of the Earth's magnetic field. Space science has indeed clarified to a large extent the relation between the two. Sounding rockets on suborbital trajectories, satellites in various orbits together with ground-based observations of geomagnetic field variations were instrumental in revealing the processes that cause auroral phenomena. These early efforts immediately led to the discovery of a magnetically trapped, collisionless particle population, namely the radiation belts and the terrestrial magnetosphere, including its extended magnetotail (Figure 3).

It was found that what is responsible for the aurorae (Figure 4) is a medium-energy particle component being injected – during magnetic substorms – from the magnetotail into the auroral zone, rather than, as had been believed, trapped high-energy particles in the radiation belts.

Space in the terrestrial neighbourhood was also exploited as a natural laboratory: *in situ* measurements led to the discovery of discontinuities in the collisionless space plasma and uncovered the phenomena of magnetic reconnection and shock acceleration, both fundamental processes throughout the cosmos. The terrestrial magnetosphere later became the prototype for modelling other planetary magnetospheres and the magnetospheres of neutron stars, pulsars and rotating black holes.

The heliosphere is formed, because the corona of the Sun is in a state of expansion, forming a continuously flowing solar wind. Because of its expansion into three dimensions, the solar wind pressure is reduced until it can no longer overcome the ambient interstellar plasma pressure. A termination shock, formed at an estimated heliocentric distance of 80–100 AU, converts the supersonic wind into a subsonic flow.

The charged particles in the external interstellar gas cannot penetrate the heliopause, the boundary of the heliosphere. However, interstellar grains and neutral atoms do enter, and

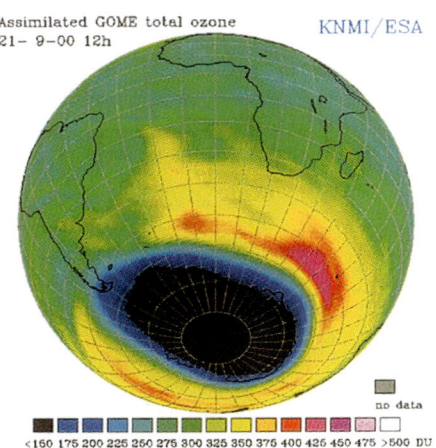

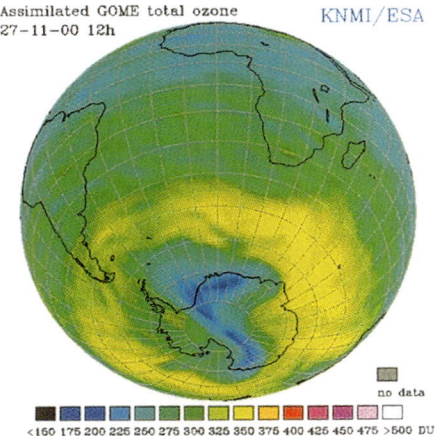

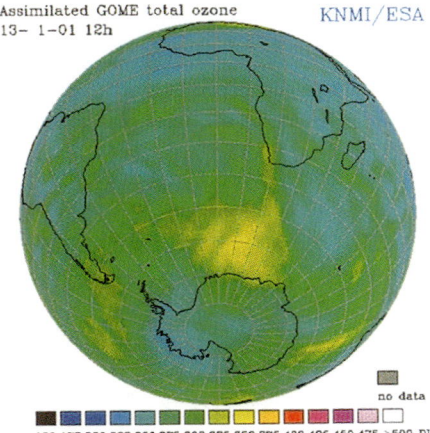

Figure 2 Assimilated total ozone maps showing the evolution of the Antarctic ozone hole between September 2000 and January 2001.

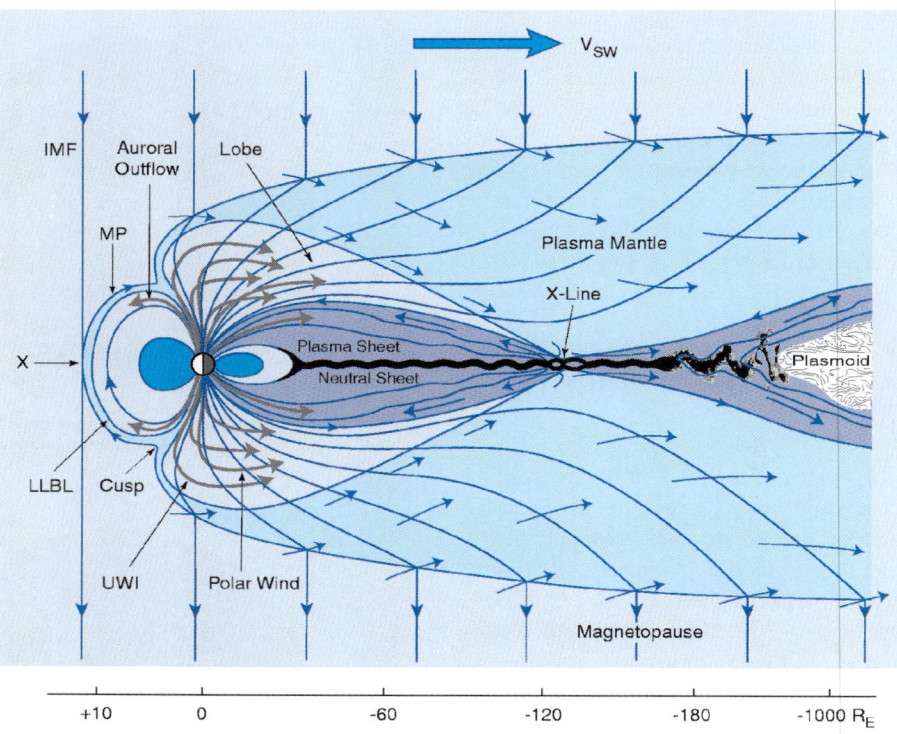

Figure 3 The terrestrial magnetosphere, which results from the interaction of the solar wind with the magnetic field of the Earth, was completely unknown before the advent of the space age. This schematic drawing gives an impression of the complexity of the interaction.

they have been found and investigated by spacecraft in the space between the planets and throughout the heliosphere, allowing a direct analysis of physical, chemical and isotopic properties of the matter in the local interstellar cloud.

A rich variety of heliospheric phenomena have been discovered, and the underlying processes are now generally well understood. And the heliosphere has become a paradigm for the interaction of a main sequence star with the interstellar medium.

THE MOON AND THE TERRESTRIAL PLANETS

The origin of the Earth–Moon system, has long been an enigma. Moons orbit many planets in the Solar System but, with the exception of those of Earth and Pluto, they are tiny in comparison with the planet they accompany. The satellite systems of the giant planets, Jupiter and Saturn, consisting of more than ten moons each were probably formed from circumplanetary disks at the time of the planets' formation. The exceptional relative size of our Moon points to a different genesis. Among the various hypotheses of the Lunar origin, there is only one – the giant impact theory – that explains all the relevant observations (Figure 5).

Evidence gathered from Lunar exploration by manned and unmanned spacecraft shows that water of crystallization and hydrogen-containing minerals are absent from Lunar material, and that other volatile elements, such as carbon and nitrogen, are virtually absent as well (Figure 6). Furthermore, neither *in situ* observations by astronauts nor pictures taken on the lunar surface or from lunar orbit revealed any traces of past water flows or sedimentation (Figure 7). The extremely high temperature of the material ejected by a giant collision can readily explain this complete lack of volatile material.

It is also remarkable that the Moon has much less iron than does the Earth or meteorites. This is indeed what we would expect if the giant impact occurred after the Earth had formed its iron core and if the impact ejected mainly material from the Earth's crust and mantle. Fractionation-corrected abundances of the three isotopes of oxygen also point to a terrestrial origin of the Moon: they are identical in all Lunar and terrestrial samples, yet differ from those of Martian rocks and all classes of meteorites.

The aggregate evidence in favour of the giant impact theory has become so strong that, at the end of the twentieth century, it is generally accepted as the proper explanation for the origin of the Moon.

Figure 4 The aurora borealis as seen from the ground and from the Space Shuttle (note the silhouetted tail fin and engine shrouds), and pictures of aurorae on Jupiter and Saturn, as observed by the Hubble Space Telescope.

Figure 5 All evidence indicates that our Moon was created by a collision between Earth and a Mars-sized object that had formed in a nearby orbit. This painting by William K. Hartmann depicts the situation five hours after the collision.

The Apollo programme and unmanned lunar orbiters and landers have given us a remarkably complete picture of the evolution of the Moon as a geological, geophysical and geochemical entity. Its anorthositic crust indicates complete melting in its early history. Later, between about 600 and 1500 million years after the birth of the Moon, iron- and titanium-rich lava rose from below and filled the large basins that had been excavated a few hundred million years

Figure 6 A thin section of basaltic rock collected by Neil A. Armstrong and Edwin E. Aldrin in the Lunar Mare Tranquillitatis. The Moon is virtually free of hydrogen, carbon, nitrogen and other volatile elements; accordingly, the minerals in this sample of lunar rock do not contain these elements.

Figure 7 The dry Lunar landscape in the Mare Imbrium. Like everywhere else on the Moon, there is no trace of past water flows. The rocks are dry, devoid of hydrogen, carbon and nitrogen, although nearby there is a valley that at times in the past had been considered to be a dry riverbed.

earlier by large impacts during the last stages of accretion in the Solar System. The lava-filled basins are the dark basaltic planes on the Moon that were called maria by Galileo.

The other surface features are younger and are mostly due to more recent impacts, because by about 3 billion years ago the interior of the Moon had cooled to the point where geological activity came to a standstill, and the surface of the Moon was only occasionally changed by impacts. On Earth, the intrinsic geological activity goes on to this day. Impacts have only a small influence on shaping surface features of our planet. However, towards the end of the twentieth century it became more and more apparent that impacts play a major role in the evolution of life.

The space age has also completely transformed planetary science. This field has evolved from a branch of astronomy into a multidisciplinary field that now embraces astronomy, geology, geophysics, geochronology, geochemistry and atmospheric science.

Concepts developed over centuries of Earth science can be extended to those objects not too dissimilar from the Earth, namely the terrestrial planets and the Moon. Although an interplanetary traveller would perceive Earth, Moon and Mars (Figure 8) as entirely different worlds, these members of the Solar System have important similarities: their sizes, chemical compositions and distances from the Sun are rather similar, and so comparative studies can further our understanding of their origin and evolution. Moreover, we note that these three are the only objects in the Universe for which, by the end of the twentieth century, detailed geological surveys had been conducted and of which we possess rock samples for laboratory analysis.

It should be stressed, however, that comparative planetology is not just a matter of transferring experience and information from Earth science to planetology. Quite the contrary: much has been learned about the origin and the early history of the Earth from comparisons with the Moon. The methods used to derive the relative abundances of chemical elements in the Earth and other planetary bodies were developed by comparing results of chemical analyses of rocks from the Earth, Moon and meteorites, including some meteorites from Mars. Comparisons of the molecular and isotopic compositions of the extremely different atmospheres of Venus, Earth, the Moon and Mars have also improved our knowledge of fundamental atmospheric processes such as outgassing and loss by escape, and of factors that determine planetary climates, such as the greenhouse effect. And our

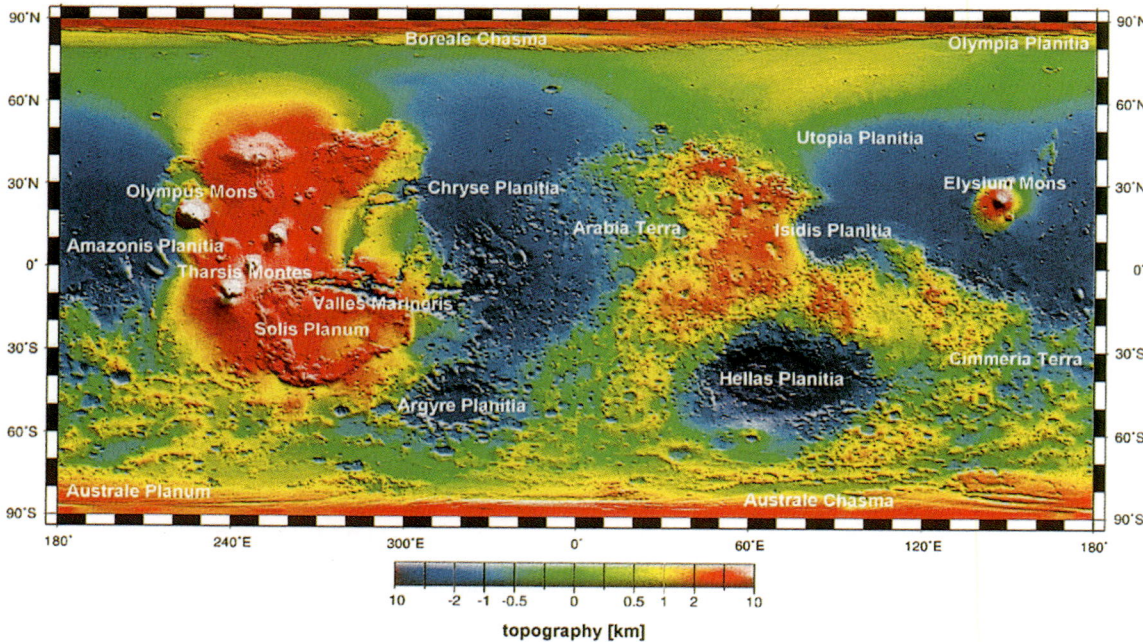

Figure 8 Topographic map of Mars. Low levels are shown in blue, as a reminder that in the past there was probably extensive flooding on Mars.

understanding of the role of impacts in surface geology and in biological evolution on Earth has been greatly advanced by the study of craters and other surface features of the terrestrial planets and the Moon.

The technique of remote sensing from orbit permits us to investigate not only the atmospheres but also the surfaces of the Moon and the planets, including the Earth. Moreover, depending on the thickness of the atmosphere, the surface chemistry and mineralogy of these bodies can be probed by using parts of the electromagnetic spectrum other than the visible, for example gamma rays for Mars and X-rays and gamma rays for the Moon (Figure 9). The atmosphere of Venus, on the other hand, is so thick that surface features can be sensed only by radio and radar techniques (Figure 10). Similarly on Earth, radar can 'see through' the clouds.

Information about the interiors of planets is obtained from seismic data and from measurements of gravity, heat flow and magnetic field. Strong magnetic fields, implying an ongoing dynamo effect inside the Earth, Jupiter and other large planets, have been observed. Magnetic field measurements now also indicate that Mars possessed a dynamo in its early history.

The exploration of Mars gained momentum towards the end of the twentieth century, largely because indications of

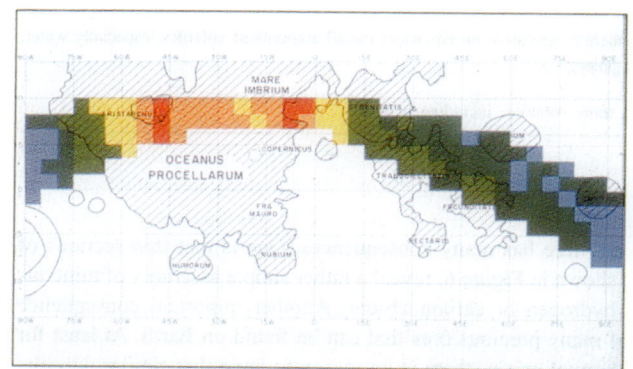 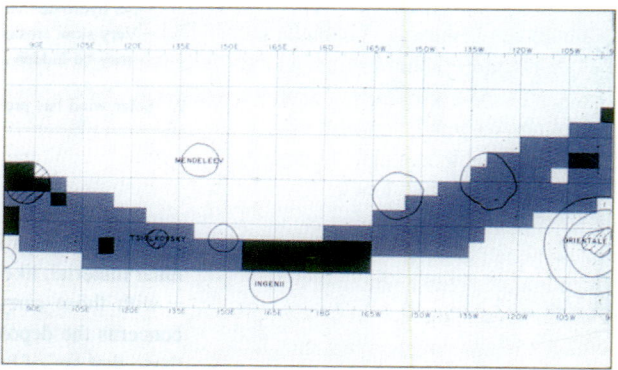

Figure 9 A compositional map of the Lunar surface as determined by the Apollo 15 gamma-ray spectrometer. High concentrations of radioactive elements (red and yellow regions) are found in the maria and other regions on the nearside.

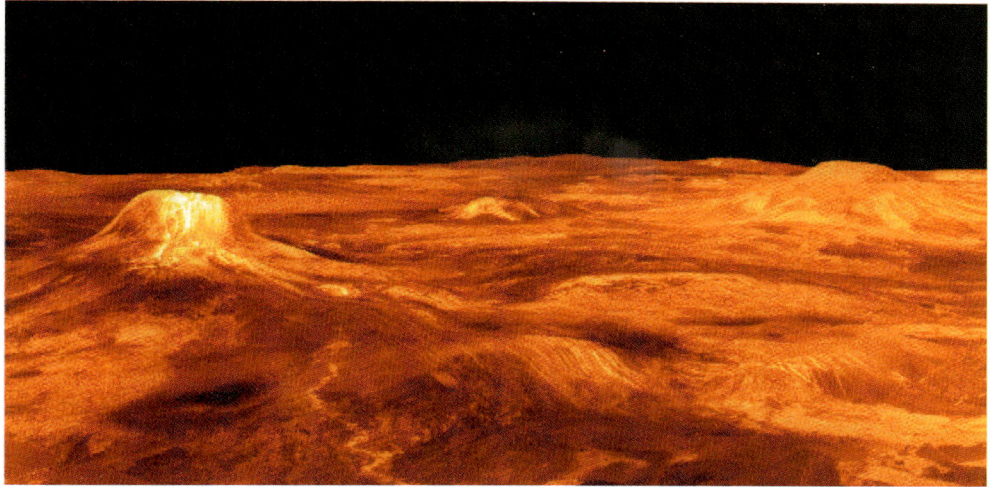

Figure 10 A three-dimensional perspective view of the surface of Venus, showing the western Eistla region with the volcanic peaks of Gula Mons (left, 3 km high) and Sif Mons (right, 2 km high). The image was obtained by use of synthetic aperture radar (SAR) data combined with radar altimetry, both from the US-American Magellan mission. The hues are based on colour images recorded by the Soviet Venera 13 and 14 probes.

extensive water flows at some undetermined earlier epoch were discovered. These observations and findings in meteorites of Martian origin greatly increased interest in the question of whether life, or at least extinct life, could be found on Mars.

THE OUTER SOLAR SYSTEM

Pictures and other data transmitted from probes that have visited the outer planets and their environs, and also observations with space telescopes, have revealed a fantastic variety of surfaces, atmospheric phenomena, ring systems, and Io's plasma torus. Shepherd moons, which constrain thin rings of shards circling planets, were found as well. The larger moons, in particular, showed traces of geologically recent changes on their surfaces and even current activity. The most dramatic example is the ongoing activity of Io, the innermost of Jupiter's Galilean moons (Figure 11). This activity not only shapes the surface of this satellite, it also gives rise to the famous sulphur-rich Io torus, which in turn is a major source of ions for the huge magnetosphere of Jupiter. Given Io's size, which is similar to that of the Earth's Moon, internal energy sources could not possibly drive the strong sulphur volcanism that exists in the present epoch. The answer to the puzzle is the tidal force exerted by Jupiter, which has over 300 times the mass of the Earth.

Europa, though farther from Jupiter than Io, still appears to be affected by this tidal force. Evidence of mobile 'ice rafts' on the surface of Europa indicates fluid flow. Tidal stress may exceed the tensile strength of the ice, lead to cracks in the ice and expose the liquid underneath, which then freezes upon exposure. Changes in the orientation of this satellite's magnetic axis with time are, in fact, consistent with a conductive liquid-water layer less than 100 km below the surface. Europa's icy shell is therefore thought to cover a salt-rich ocean, in which life might exist.

Saturn's largest satellite, Titan, also evoked considerable interest (Figure 12). The temperature structure of its atmosphere and the abundance of nitrogen and hydrocarbons are reminiscent of the early Earth. Titan, however, has not evolved because of the low temperature and the lack of liquid water at 10 AU from the Sun.

THE ORIGIN OF THE SOLAR SYSTEM, METEORITES AND COMETS

The question of the origin of the Solar System remained largely a domain of physicists well into the twentieth century. It was concluded that the process starts with the gravitational collapse of a cloud of gas and dust, and is followed by the formation of a disk from which the Sun and planets evolved. When chemical arguments were introduced, meteorite research became important in studying the origin and evolution of the Solar System. From isotope abundance measurements in terrestrial and meteorite samples, it was found that the age of the Earth and of the Solar System was 4.5 billion years.

In 1986, at the time of the return of Comet Halley, a fleet of spacecraft was despatched to encounter this celestial body and its environment. The Giotto spacecraft, in particular, went deep into the coma, obtaining the first detailed picture of a

Figure 11 Left: the giant planet Jupiter with two of its Galilean moons, Io (reddish) and Europa (white). While these and other moons of Jupiter have nearly the same size as our Moon, they are tiny in comparison with their parent planet. Right: close-ups of Io and Europa, both of which exhibit geological activity that stems from powerful tidal forces exerted by massive Jupiter.

Figure 12 A collage of Saturn and its moons: Dione (foreground) and (from left to right) Thea, Enceladus, Tethys, Mimas and Titan.

Figure 13 The nucleus of Comet Halley imaged by the Halley Multicolor Camera on board Giotto.

cometary nucleus and determining the composition of the gas and dust grains in the coma (Figure 13). This confirmed and considerably refined the so-called dirty snowball hypothesis, and also showed that comets have indeed preserved a wealth of virtually unchanged interstellar material, much like the material from which the Solar System was formed.

THE SUN

The Earth, its climate and the life it supports are all subjected to and governed by solar radiation and its subtle variations. The astronomer, on the other hand, looks upon the Sun as a Rosetta Stone. Being the dominant object in the sky, it is easily accessible to observations and has become a proving ground for advanced observing techniques. Indeed the first space experiments that went beyond investigating the terrestrial atmosphere and ionosphere were devoted to recording solar ultraviolet spectra.

Among the early astronomical satellites were the series of Orbiting Solar Observatories, followed in 1973 by the pioneering Apollo Telescope Mount on Skylab. From this first, albeit short-lived, space station the outer solar atmosphere and the extension of the outer corona into the heliosphere could be investigated by a multiwavelength complement of imaging, spectroscopic and coronagraphic telescopes having focal lengths that enabled observations down to arc-second resolution. (At a time when photoelectric imaging detectors were not readily available, the presence of astronauts onboard Skylab made it possible to use photographic film and return it to Earth for developing.)

With access to a halo orbit around the L_1 Lagrangian point on the Earth–Sun line, it became possible to combine remote-sensing observations with *in-situ* solar-wind measurements. The Solar and Heliospheric Observatory, placed in such an orbit, went a step further. It also enabled space observations by the method of helioseismology, and thus provided a detailed look into the solar interior as well (Figure 14). A complete picture of the structure and dynamics of the solar interior, the processes that maintain the high temperatures in the corona and the processes of solar-wind acceleration is thus emerging from this observatory.

The riddle of whether the 'missing' solar neutrino flux at Earth should be explained by some exotic behaviour of the energy-generating solar core or by adapting the theory of elementary particles is now resolved, since the solar core does not have any unexpected properties. It is currently thought that the electron neutrinos generated in the core change their flavour while passing through the Sun and on their way from the Sun to the Earth. Consequently, since the first large neutrino detectors on Earth were only capable of detecting electron neutrinos, the solar neutrino flux appeared lower since some of them had become μ- or τ-neutrinos.

Measurements by the Solar and Heliospheric Observatory also confirmed that the solar 'constant' varies with the solar cycle. The solar irradiance was found to increase by about 0.1% between solar minimum and maximum.

Moreover, we now know that the Sun's magnetic field supplies the energy that heats the corona. Waves which had previously been thought to contribute to the heating of the corona were identified as the accelerating agent – through the ion-cyclotron mechanism – for the fast solar wind (Figure 15) that emerges from coronal holes (i.e. from open magnetic field configurations).

Coronal mass ejections, a phenomenon discovered by Skylab, can now be followed far into interplanetary space,

Figure 14 Temperature and rotation in the solar interior. The images show the deviations of these properties, as derived from SOHO measurements, from current models.

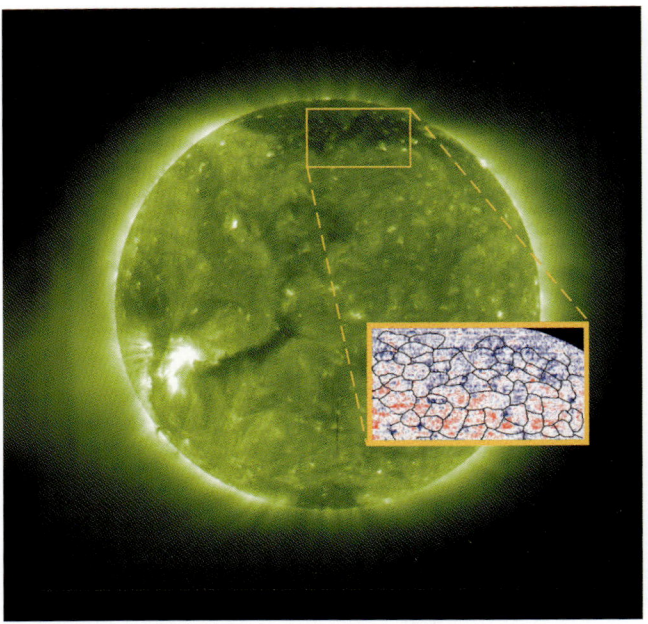
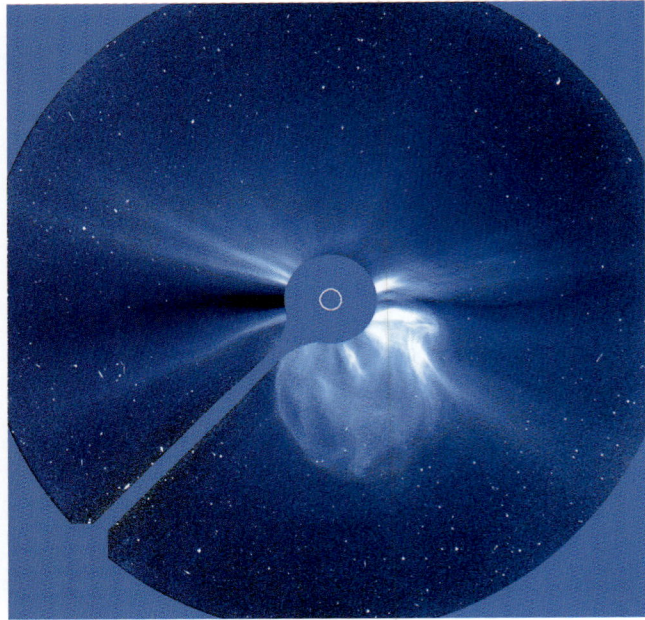

Figure 15 Dopplergram showing the (blueshifted) source regions of the fast solar wind, namely the boundaries and boundary intersections of magnetic network cells in a coronal hole.

Figure 16 A coronal mass ejection propagating into the heliosphere. Stars are visible in this coronagraph image whose field of view is thirty solar diameters wide.

and also as they propagate towards the Earth (Figure 16). This provides a useful means of predicting disturbances in the Earth's environment – a range of phenomena which have been collectively termed 'space weather'. Given the increasing dependence of our civilization on communication, navigation and Earth-observing satellites – all vulnerable to major disturbances in the space environment – space weather forecasting is becoming another important service whose roots lie in space science.

THE MILKY WAY

The impact of the space age on the evolution of, and progress in astronomical research has been huge, since the full breadth of the electromagnetic spectrum and primary cosmic-ray particle population became available as information carriers for cosmic diagnostics. Space observatories exploring the infrared, ultraviolet, X-ray and gamma-ray wavebands have revealed a dynamic and violent Universe harbouring a zoo of exotic objects. Inevitably, we have to limit ourselves here to selecting a few scientific highlights from space-borne astronomy and obviously they reflect a subjective choice.

Although ground-based radio astronomy has unveiled a variety of molecular species, a major asset among the scientific harvest of space-borne spectroscopy in the infrared and far-infrared regions has been the molecular signature of water in circumstellar and interstellar environments. Water possesses a very large number of strong rotational transitions in the far infrared, which suggest that it can act as a major or even dominant coolant in shocks and circumstellar outflows. In particular, the Infrared Space Observatory unveiled a host of water-emitting cosmic sites. Among them are the asymptotic giant branch (AGB) stars, confirming that water molecules are the dominant coolants of the stellar winds emanating from these stars (Figure 17).

Moreover, the Infrared Space Observatory revealed, for the first time in a circumstellar medium, polyacetylenic chains, including C_4H_2 and C_6H_2, and benzene (C_6H_6). From these observations it appears that carbon-rich protoplanetary nebulae are capable of producing prebiotic matter in space (Figure 18).

While our picture of stellar evolution has been established with the aid of ground-based observations, progress in the study of star formation has come to depend increasingly on observations from space. Diagnosing the process of star formation requires penetration into the interiors of protostellar clouds, provided by measurements at infrared and microwave frequencies. Observations from space have played a crucial role in improving our understanding of star formation, notably in the form of spectroscopic measurements from the Infrared Space Observatory, and astrometric measurements and imaging from Hipparcos and the Hubble Space Telescope, respectively.

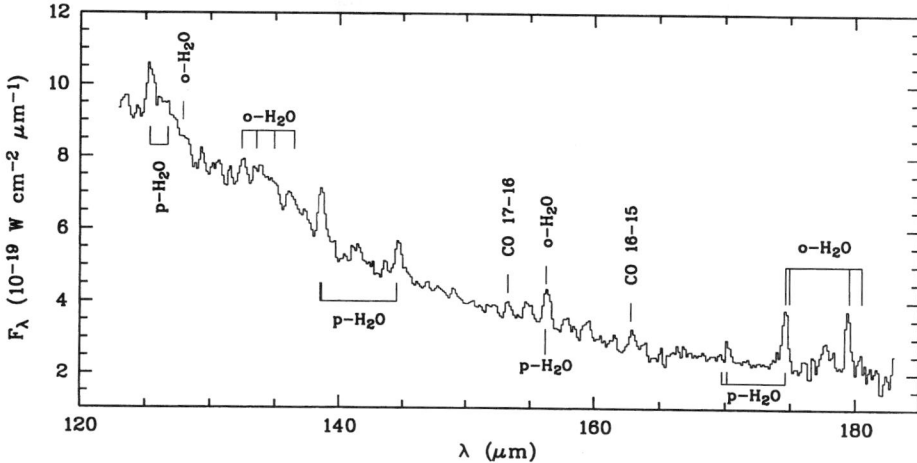

Figure 17 Water acts as a coolant in the stellar winds of AGB stars.

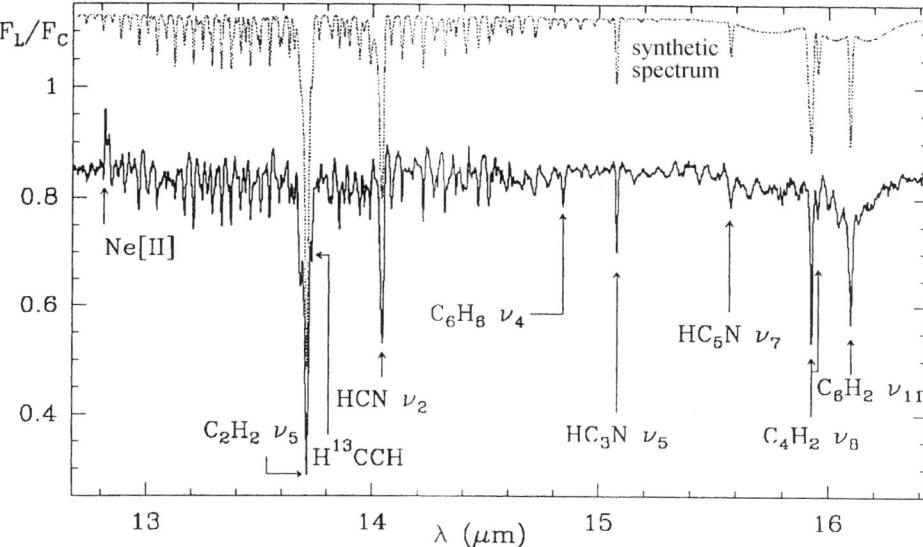

Figure 18 Carbon-rich protoplanetary nebulae as organic chemistry factories in space.

Observationally, it has now been confirmed that practically all low-mass stars like our Sun are born via the formation of a disk, which arises from the angular momentum conservation of the gravitationally collapsing protostellar cloud. The final dimensions of these disks are comparable to the size of our Solar System, and they have a mass approximately ten times the mass of our 'heaviest' planet, Jupiter, which implies the presence of sufficient material to form a planetary system.

Disks of planetary material orbiting newborn stars are called protoplanetary disks. The Hubble Space Telescope has recently obtained high-resolution images of such disks around young stars in the Orion Nebula, where they show up as dark silhouettes in images obtained in visible light (Figure 19). Ground-based observations of exoplanets indicate a great diversity of morphology in planetary systems: Jupiter-like planets are also found much closer to the central star than in our Solar System and they are, most likely, formed first. Earth-like planets are believed to form later, through the aggregation of a large number of planetesimals a few kilometres in size.

Observations from space have not only helped to identify the early stages in the life of a star, but have also revealed the most extreme remnants of stars, namely stellar-mass black holes. The strongest evidence for the existence of such black holes comes from observations of compact X-ray binary sources in which the orbital velocity of the normal companion star can be measured from the Doppler shift of its characteristic spectral lines at optical frequencies.

Using Kepler's laws, a quantity called the mass function, $f(M)$, of the system can be determined from measurements of the orbital period of the compact binary and the semi-amplitude of the radial velocity. This mass function is the observational lower limit of the presumed black hole mass. Fundamental physical arguments show that, among very compact objects, only a black hole can have a mass in excess of 3 solar masses ($M_\odot$), so if mass function values above $3M_\odot$ are found, the compact source has to be a black hole. So far about ten black hole systems (i.e. objects with $f(M)$ greater than $3M_\odot$) have been found in our Galaxy. The largest mass function was found for the recurrent nova V404 Cygni, a low-mass X-ray binary with a 6.5-day periodicity, for which $f(M) = 6.26$ (Figure 20).

A remarkable achievement of space-borne astronomy concerns astrometry, the oldest, most classical subdiscipline of astronomy: measuring precise positions, parallaxes and proper motions of stars to study the Galaxy's structure and kinematics. The Hipparcos satellite has boosted this field by providing positions of nearly 120 000 stars to milli-arc-second accuracy, and established accurate distances for tens of thousands of stars. In fact, Hipparcos pushed the effective range for parallax measurements from 30 pc out to 300 pc. Accurate distances yield accurate luminosities for all kinds of stars, and consequently theories of stellar evolution can be tested much more rigorously (Figure 21). The kinematic data provided by Hipparcos through the measurement of proper motions coupled with accurate distances allows an in-depth assessment of the dynamic evolution of the Milky Way system. Among other things, these dynamical studies will display the interplay between gravity and pressure and the role of instabilities in our Galaxy.

As mentioned in the opening paragraph of this chapter, the discovery of cosmic rays may be regarded as the first result of space research. Although this discovery took place in 1912, the fact that the primary radiation consists of charged particles coming from the depths of space became generally accepted around the middle of the century. Cosmic rays observed at the surface of the Earth are mostly secondary

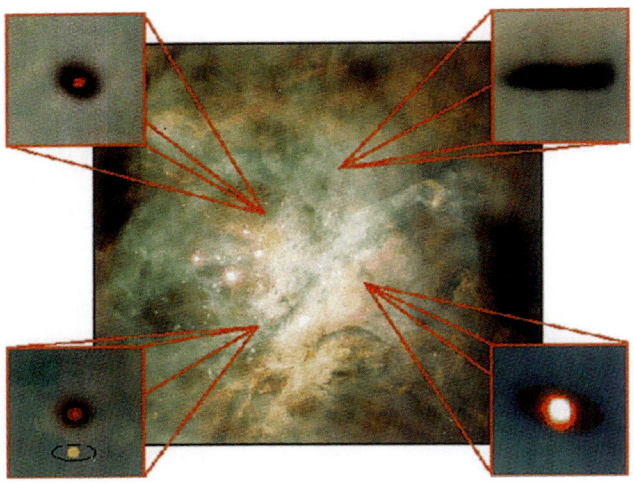

Figure 19 Hubble Space Telescope picture showing disks of planetary material orbiting around young stars in the Orion Nebula. The size of the solar system is indicated by the lower image in the lower left corner.

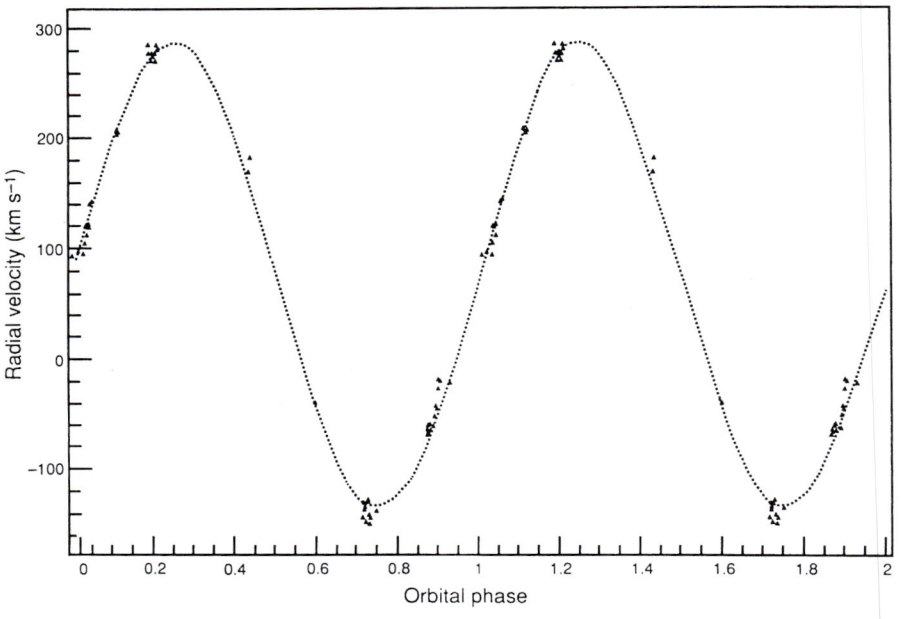

Figure 20 Radial velocity curve of the black hole system V404 Cygni.

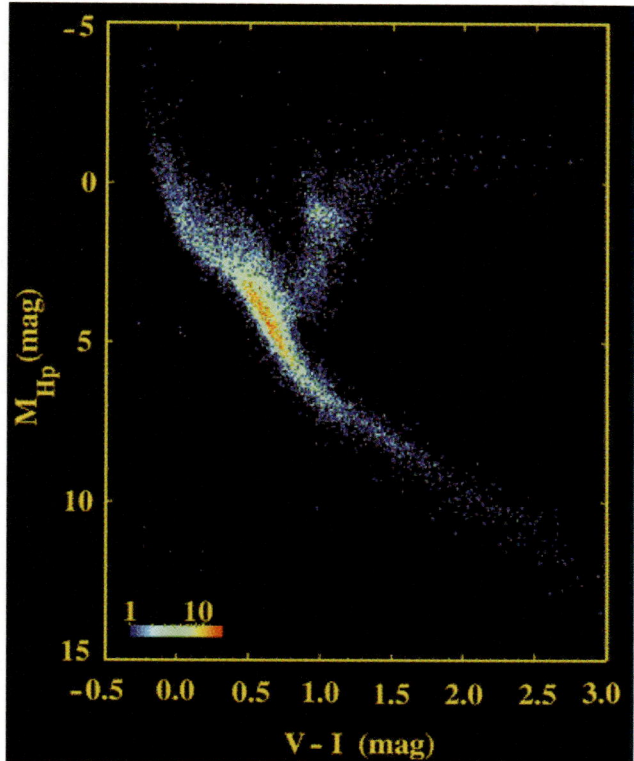

Figure 21 Hertzsprung–Russell diagram, reflecting stellar evolution, for the 16 631 single stars from the Hipparcos Catalogue whose distances and colour magnitudes are known to within 10% and 0.025 mag, respectively. The colour indicates the number of stars in each pixel of the diagram.

particles. Therefore measurements had to be made from space to detect the primary radiation. (In the 1930s and 1940s, cosmic rays were the main source of particles for the emerging field of elementary particle physics. Positrons, muons, pions, kaons and hyperons were discovered as secondary particles produced in the atmosphere by cosmic rays, and their lifetimes and modes of decay were determined.)

Three principal sources of cosmic rays were identified namely (i) high-energy particles originating in the Galaxy, (ii) solar particles accelerated by shocks created by solar flares, coronal mass ejections and the co-rotating interaction regions, and (iii) the so-called anomalous component of cosmic rays. The latter are accelerated pick-up ions produced from the neutral interstellar gas flowing through the heliosphere. Towards the end of the twentieth century, cosmic-ray research, in conjunction with gamma-ray and X-ray astronomy, became more and more important for localizing and studying violent processes in the Galaxy. Supernova remnants were identified as the main source of galactic cosmic rays. However, it is not yet clear to what extent extragalactic sources contribute to cosmic rays of the highest energies.

The modulation of cosmic rays by the reversal of the solar magnetic field in the course of its 22-year cycle has proved to be an important tool for investigating heliospheric processes and dimensions. It has been suggested that the modulation of cosmic rays may also influence – via ionization and nucleation in the troposphere – the formation of low-lying clouds, and may therefore represent a coupling mechanism between solar activity and terrestrial climate variations.

THE GRAVITATING UNIVERSE AND COSMOLOGY

The notion that most of the mass of the Universe is in a form we cannot see is among the most striking discoveries of contemporary science. As early as the 1930s, Fritz Zwicky pointed out that the visible mass we can observe is not sufficient to explain the motions of galaxies in clusters. The nature of this invisible mass, known as dark matter, is under intense discussion and investigation. Despite its invisibility, it can be detected through the effects of its gravitational field, which allows us to probe its distribution in and around galaxies and in galaxy clusters.

Giant clusters of galaxies form in gravitational potential wells dominated by dark matter that extends well beyond the observed galaxy population. However, the dark matter content and distribution can be probed by observing the hot X-ray emitting intracluster gas that is gravitationally bound by the total cluster mass. Space-borne observations by X-ray telescopes play a pivotal role in furthering this research (Figure 22).

An alternative method of probing the large-scale distribution and concentrations of gravitating matter in the Universe is provided by gravitational lenses (Figure 23). Extragalactic gravitational lensing provides us with an 'optical bench' whose length is comparable to the radius of the observable Universe. And here a dissimilarity with traditional astronomical observations emerges most significantly: this method probes all mass – not only 'the visible 10% of the iceberg' (i.e. the luminous content of the Universe).

Confirmation that the enigmatic gamma-ray bursts are objects in the remote Universe, rather than transient sources lying in the close galactic vicinity, came from observations made from space. The location of such events in distant galaxies, through the optical identification of the X-ray afterglows detected by the BeppoSAX X-ray satellite, showed that gamma-ray bursts are the most powerful sources of explosive energy released in the Universe since its very creation in the hot big bang.

The amount of explosive energy release is currently best explained by the so-called fireball model. This postulates a cataclysmic event in which the gamma-ray emission arises from internal shocks generated in a relativistically expanding,

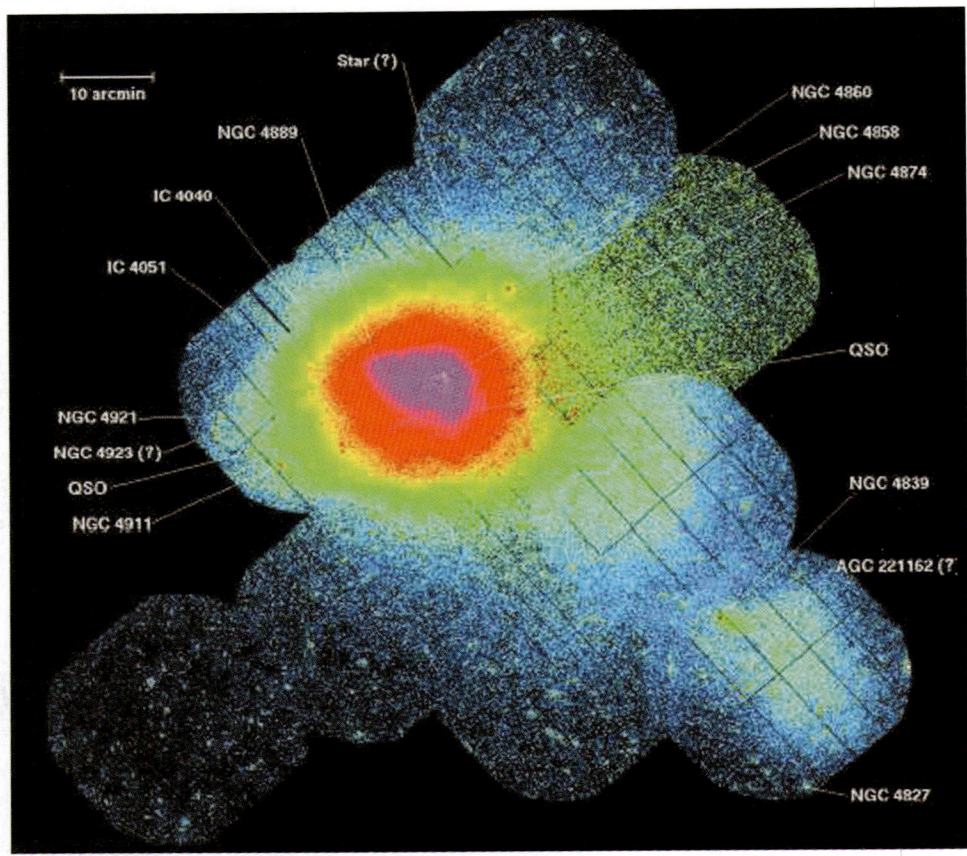

Figure 22 XMM-Newton image of the hot X-ray emitting gas in the Coma Cluster of galaxies.

Figure 23 Although gravitational lenses have been discovered from the ground, the sharper pictures obtained from space permit a refined interpretation of the information provided by the deflected light.

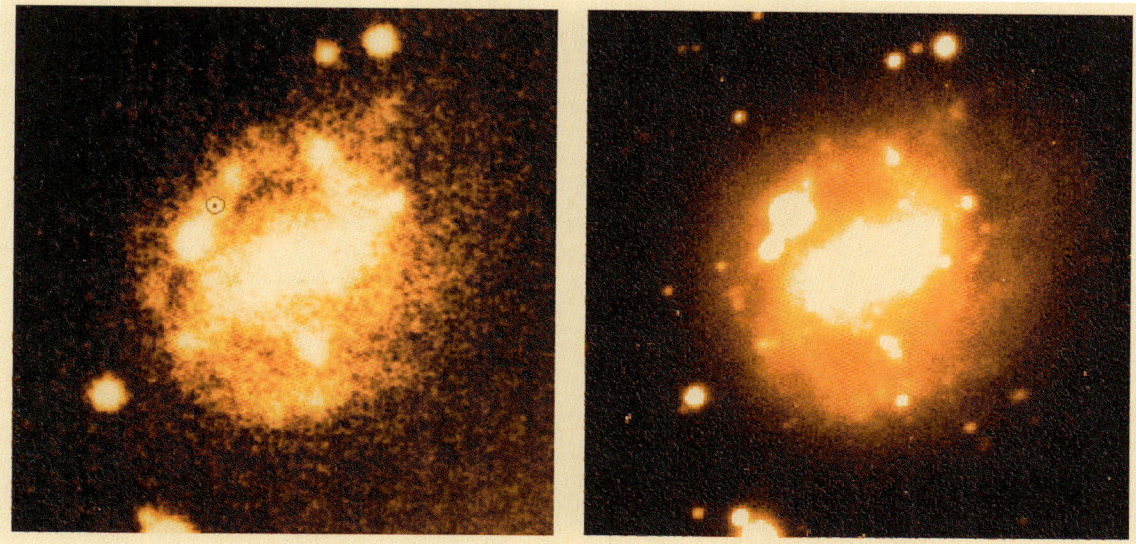

Figure 24   Hypernova 1998bw.

optically thick plasma cloud travelling at more than 99.99% of the velocity of light. Gamma-ray burst sources would thus seem to produce the most extreme form of cosmic acceleration. Some gamma-ray bursts appear to be associated with a peculiar type of highly luminous supernova, a so-called hypernova, implying the explosive death of a very massive star in which a black hole is formed in the gravitationally collapsing core (Figure 24).

The earliest picture of the Universe so far has been obtained by observing in the microwave region. The cosmic microwave background radiation (i.e. the isotropic high-frequency radio emission) was discovered in 1964 by Arno Penzias and Robert Wilson, who used a ground-based antenna. It turned out that the early Universe was very smooth. The radiation in question was last scattered from the universal primordial plasma when the Universe had an age of roughly 300 000 years.

The Cosmic Background Explorer satellite mapped the temperature distribution of the microwave background over the entire sky with an angular resolution of several degrees and revealed tiny spatial fluctuations of the microwave radiation field (Figure 25). This implied the existence of slight density perturbations, the first solid observational evidence confirming the simplest hypothesis for the origin of large-scale structure, namely that it grew out of tiny primordial density perturbations.

The first ideas about the synthesis of chemical elements in the early phase of the expanding Universe were developed in the 1940s. However, the processes and locations that play a significant role in nucleosynthesis were identified only in the 1950s and 1960s. It became clear that the isotopes of hydrogen, helium and lithium were fully or partly

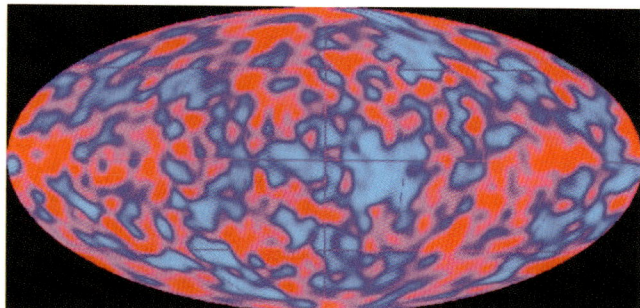

Figure 25   The microwave temperature structure of the sky.

produced in the big bang, but that carbon and all the heavier elements are synthesized in the interiors of stars. Spallation by cosmic rays contributes significantly only for some of the extremely rare elements, such as lithium, beryllium and boron (Figure 26).

The theory of stellar nucleosynthesis was almost exclusively based on element and isotope abundances measured on or from the ground, namely in meteorites and in the solar photosphere. But deriving the primordial abundance of the isotopes of hydrogen and helium, the main products of the big bang, required measurements in space as well.

The primordial helium abundance is one of the most important sources of information about the early Universe. Its value derived from observation agrees, with an uncertainty of only a few per cent, with the theoretical prediction. This confirms that the laws of physics derived from laboratory experiments can be applied without change back to the time when the Universe was as young as one second.

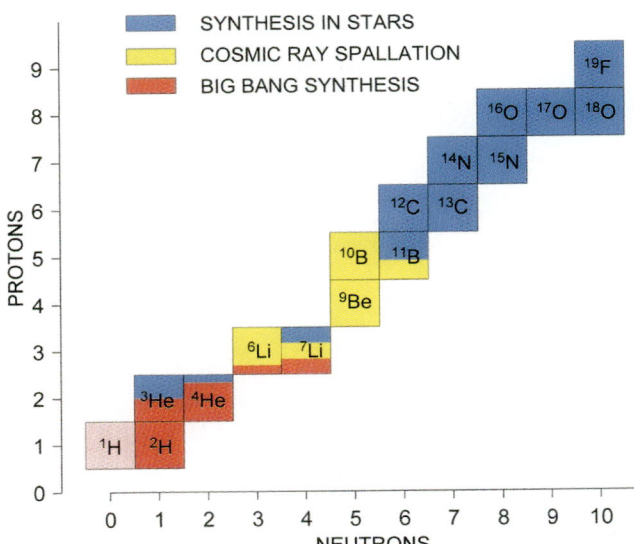

Figure 26 The three principal production sites of nuclei. Carbon and all heavier elements are produced in stars. The big bang yields only the lightest species. For species with mixed origin, such as ^{3}He and ^{7}Li, the relative proportions change with time and location in our Galaxy – and elsewhere.

The average density of the Universe as derived from deuterium and ^{3}He measurements is less than 10% of the critical density, which is the density that would 'close' the Universe and, as mentioned above, it is also less than the matter density that is needed to account for the forces that keep galaxies and clusters of galaxies together. Thus, by the end of the twentieth century it was becoming clear that there exists in the Universe an exotic form of matter that contributes more to the total density than does the visible, baryonic matter. Observations of the subtle inhomogeneities in the cosmic microwave background radiation currently indicate that the exotic matter consists largely of weakly interacting particles that seem to be much more massive than protons.

Perhaps the most striking example of an outcome of the century of space science is the evidence that the cosmological constant, Λ, of Einstein's general theory of relativity is not zero. In 1914, Einstein had to give a non-vanishing value to this integration constant because the observational data available at the time pointed to a static Universe. Consequently, it was necessary to account for the non-collapse of the Universe by giving Λ a finite value. After Edwin Hubble had discovered the expansion of the Universe in 1928, Λ seemed to have lost its meaning. Only towards the end of the twentieth century did space observations of remote supernovae indicate that the Universe had expanded more slowly in an earlier epoch, and therefore that a long-lasting acceleration of the Universe's expansion had taken place, which is described by the cosmological constant Λ or another form of 'dark energy'!

OUTLOOK

Space science has brought us more than new knowledge. Communication and even peaceful collaboration continued between space scientists and engineers living on both sides of the Iron Curtain, even in the darkest times of the Cold War. And space science not only helped to transcend borders between countries and political systems: it also led to intense intellectual exchanges across entrenched confines separating disciplines. As far as everyday life is concerned, we should not forget that reliability considerations and quality control for space hardware – an absolute necessity in view of the unforgiving space environment – has produced a direct payoff in the widespread use of quality assurance in industrial production.

However, the link between basic research and applied science still needs to be strengthened because emphasizing applications without underpinning them by basic research is, in the long run, a dead end. Without science, there is no science to apply.

A case in point is the highly politicized discipline of climate research: understanding the Earth System is required to predict its reaction to changes. Action on the Earth System must not be taken unless it is based on trustworthy predictions. On the other hand, present uncertainties in some of the quantitative predictions must not be used as an excuse for political inaction. The large increase in the quantities of greenhouse gases in the atmosphere caused by human civilization is well documented, and it has undoubtedly begun to affect the climate. Further research is urgently needed, not for recognizing the acute danger – which is obvious – but for predicting the kind of change that will most severely influence life on Earth.

At the beginning of the twenty-first century, the possibilities for science in space are by no means exhausted. Emerging propulsion technology such as solar sailing will expand the range which our spacecraft can explore. And we are just beginning to deploy 'quiet platforms' which provide an essentially acceleration-free environment – often at cryogenic temperatures – and thus enable us to perform tests of the general theory of relativity and, probably within a decade, long-baseline interferometry from free-flying spacecraft. Extending the technique of interferometry into space will widen the diagnostic potential of space science tremendously. With laser interferometry we shall be able to set up giant antennas in space for detecting low-frequency gravitational waves, opening an altogether new window on the Universe and making possible further, more penetrating tests of general relativity.

The technique of "nulling" interferometry, will enable us to observe terrestrial planets that orbit stars other than the Sun, and study the spectra of their atmospheres and thus look for signatures of non-equilibrium that may be caused by life. Interferometry will, of course, also benefit other fields of astronomy. Once available in the X-ray domain, for example, interferometers will allow us to expose the event horizons of black holes and thus probe the limits of the observable Universe in objects that are not at cosmological distances.

Planetary exploration will be a central theme of the space effort in the twenty-first century. How this field develops, and how fast, will depend on available technologies as well as scientific findings. Evolution and breakthroughs in the fields of transportation and robotics will affect the relative roles of human landings and automated exploration, including sample return. And any progress in the search for current or extinct life on other planets or satellites will have an enormous influence on the ways and means of planetary exploration. The search for life in the Universe, and more specifically in the Solar System, is on. And space will play an important role in the new field of astrobiology, which has the potential to lead to a new age of scientific understanding. Astrobiology is also dissolving many boundaries between disciplines because it requires the integration of several disparate fields of science, as for example astronomy, biology, chemistry, Earth sciences and physics.

The outlook for a new century is always uncertain. Yet – excepting a major catastrophe that affects the physical world, or a radical change in the mode of thought of humanity – there is now such a momentum in technology development that the outlook for space science is good. It is advisable not to make any specific predictions: indeed, possibilities often seem unimaginable. As pointed out in Chapter 3, on the enabling technology for space transportation, an author writing at the beginning of the twentieth century would probably not have foreseen the atom bomb, worldwide air traffic, the Global Positioning System, the Internet or, indeed, the supreme reign of the computer. In this sense, we might expect new concepts for access to space to provide new impetus to the overall space effort – including space science.

Acknowledgements

The editors deeply appreciate the enthusiastic collaboration of those who have contributed to this reference work: the pioneers of space research and today's leading space science investigators. We acknowledge the initiative of Peter Binfield at Kluwer, and his safe guiding of the project through the difficulties encountered in the editing phase. We commend Hermine Vloemans' firm and diplomatic leadership during the production process. It is a special pleasure to thank Clare Bingham, Livia Iebba, Lidy Negenborn and Silvia Wenger for their untiring assistance, and Rudolf Treumann for his comprehensive and wise counsel regarding the material reported in this chapter.

FIGURE CREDITS

Figure 1. Photograph by the crew of Apollo-11.

Figure 2. Royal Netherlands Meteorological Institute (KNMI) and ESA. The data were taken by the Global Ozone Monitoring Experiment (GOME) on ESA's ERS-2 satellite.

Figure 3. From: Hultqvist, B., Øieroset, M., Paschmann, G. and Treumann, R., (eds.) (1999). *Magnetospheric Plasma Sources and Losses*. In ISSI Space Science Series Vol. 6, Kluwer, Dordrecht, and *Space Science Reviews*, **88**, Figure 7.1, p. 371.

Figure 4. Upper left, Jan Curtis; upper right, STScI; lower left, NASA; lower right, STScI.

Figure 5. Painting by William K. Hartmann.

Figure 6. NASA/Johnson Space Center.

Figure 7. Photograph by the Apollo 15 crew, NASA Photo AS-15 9212430.

Figure 8. Courtesy NASA and the Mars Orbiter Laser Altimeter (MOLA) Team of the Mars Global Surveyor project. From Kallenbach, R. Geiss, J. and Hartmann, W.K. (eds.) (2001). *Chronology and Evolution of Mars*. In ISSI Space Science Series Vol. 12, Kluwer, Dordrecht and *Space Science Reviews*, **96**, Figure 2, pp. 466–467.

Figure 9. NASA.

Figure 10. Data from the Magellan mission, NASA picture ID P-38724 (excerpt); http://www.jpl.nasa.gov/magellan/gif/p38724.gif

Figure 11. Left: Voyager photograph, 13 February 1979. Upper right: Galileo Orbiter Mission (NASA), Catalog No. PIA 025727 (excerpt); inset: http://www.planetaryexploration.net/jupiter/io/index.html. Lower right:http://photojournal.jpl.nasa.gov/cgi-bin/PIAGenCatalogPage.pl?PIA01127.

Figure 12. http://www.jpl.nasa.gov/saturn/gif/saturn_montage.gif

Figure 13. From: Keller, H.U., Curdt, W., Kramm, J.-R. and Thomas N. (1994). In R. Reinhard, N. Longdon and B. Battrick (eds.) *Images of the nucleus of Comet Halley*. ESA SP-1127, ESTEC, Noordwijk, Vol. 1, Figure 61, p. 63.

Figure 14. D. Gough and P. Scherrer, Chapter 44, and http://sohowww.estec.esa.nl/gallery

Figure 15. From: Hassler, D.M., Dammasch, I.E., Lemaire, Ph., Brekke, P., Curdt, W., Mason, H.E., Vial, J.-Cl. and Wilhelm, K. (1999). Solar Wind Outflow and the Chromospheric Magnetic Network. *Science*, **283**, 810–813.

Figure 16. Courtesy of SOHO/LASCO consortium.

Figure 17. From: Barlow, M.J., Nguyen-Q-Rieu, Truong-Bach, Cernicharo, J., González-Alfonso, E., Liu, X.-W., Cox, P., Sylvester, R.J., Clegg, P.E., Griffin, M.J., Swinyard, B.M., Unger, S.J., Baluteau, J.-P., Caux, E., Cohen, M., Cohen, R.J., Emery, R.J., Fischer, J., Furniss, I.,

Glencross, W.M., Greenhouse, M.A., Gry, C., Joubert, M., Lim, T., Lorenzetti, D., Nisini, B., Omont, A., Orfeil, R., Péquignot, D., Saraceno, P., Serra, G., Skinner, C.J., Smith, H.A., Walker, H.J., Armand, C., Burgdorf, M., Ewart, D., Di Giorgio, A., Molinari, S., Price, M., Sidher, S., Texier, D. and Trams, N. (1996). The rich far-infrared water vapour spectrum of W Hya. *Astron. Astrophys.* **315**, L241–L244.

Figure 18. From: Cernicharo, J., Heras, A.M., Tielens, A.G.G.M., Pardo, J.R., Herpin, F., Guélin, M. and Waters, L.B.F.M. (2001). Infrared Space Observatory's Discovery of C_4H_2, C_6H_2, and Benzene in CRL 618. *Astrophysical Journal Letters*, **546**, L123–L126.

Figure 19. http://antwrp.gsfc.nasa.gov/apod/ap961207.html

Figure 20. From: Casares, J., Charles, P.A., Naylor, T. (1992). A 6.5-day periodicity in the recurrent nova V404 Cygni implying the presence of a black hole. *Nature* **355**, 614–617.

Figure 21. http://astro.estec.esa.nl/Hipparcos/TOUR/hrdiagram2.html

Figure 22. http://sci.esa.int/img/e5/26085.jpg

Figure 23. Hubble Space Telescope Image STScI-PRC00-08, http://astro.hi.is/lens/pict_th.html

Figure 24. From: Baron, E. (1998). How big do stellar explosions get? *Nature*, **395**, 635–636.

Figure 25. Cosmic Background Explorer Team.

Figure 26. From: Geiss, J. and von Steiger, R. (1997). Production of light nuclei in the early Universe. In A. Wilson (ed.) *Fundamental Physics in Space*, ESA SP-420, ESTEC, Noordwijk, pp. 99–106, Figure 1.

THE BEGINNINGS

2

ARTURO RUSSO*

The space age and the origin of space research

INTRODUCTION

The exploration of space is one of the greatest achievements of the twentieth century. In the ideal photo album of this century one could hardly avoid including such pictures as our blue planet seen from space and the human footprint on the Moon's surface; the close-up images of a comet's nucleus and the rough landscape of Mars; the floating of astronauts in the black vacuum around a space station and the rings of Saturn photographed from an approaching spacecraft.

Space technologies have provided us with the possibility of real-time global communications, as well as with unprecedented knowledge of cosmic phenomena. Traditional research fields like astronomy underwent dramatic changes when telescopes could be flown beyond the atmosphere; others emerged brand-new when physicists could put their sophisticated instruments onboard satellites and space-probes. Knowledge of our own planet took a big step forward when it became possible to investigate, from space, the delicate equilibrium of its various components.

This book is devoted to the great advances in scientific knowledge that space technologies have made possible. Scientific research, however, can hardly be considered to be the main driving force in space activities. Political and military considerations, in fact, played the major role in making the old dream of exploring space a technical reality. To the deliberate attempt of the Nazi regime to develop the ultimate weapon to win the war we owe the V2, the prototype of the rockets which helped the United States and the Soviet Union to start successful space programmes in the immediate post-war period. Then the Cold War was to fuel the development of more powerful missiles, capable of delivering a nuclear weapon across the oceans.

These missiles were also capable of launching a satellite into orbit, and a new dimension was thus added to the Cold War confrontation: the demonstration of technical superiority in a field which could greatly affect the public imagination. This was seen as being particularly important in Third World countries, which were, at that epoch, emerging from the colonial era and in search of ideological principles, economic models and international patronage. Being first in space, first on the Moon, became an important political goal for both superpowers, wanting to demonstrate the superiority of their cultural values, the efficiency of their political institutions, the performance of their industrial system, and the strength of their armed forces.

Scientists on both sides of the Iron Curtain were able to take advantage of the opportunity offered by this new dimension, building instruments which could be fitted into the nose-cones of sounding rockets or the payloads of spacecraft, and designing space missions which could solve old and new problems in astronomy and geophysics. A new scientific community emerged from this experience, identified less by their disciplinary affiliation than by the technical means they used.

Space science, in fact, is not a well-defined research field. It includes, in principle, any scientific investigation conducted by the use of rockets, Earth-orbiting satellites and deep-space-probes. In terms of established scientific disciplines, it covers fields as different as atmospheric physics, geophysics, plasma physics, cosmic-ray physics and the various branches of astronomy (solar, stellar and planetary; from radio wavelengths up to the gamma-ray

---
* Università degli Studi di Palermo, Italy

Figure 1 Astronaut's footprint on the Moon, 1969. (Source: NASA.)

region of the electromagnetic spectrum). Space sciences have also revitalized the old subdiscipline of astrometry and have helped make observational cosmology a practical and very fruitful area of research. Moreover, they have made important contributions to material sciences, biology and medicine. Each of these disciplines was dramatically reshaped by the advent of space technologies, both regarding their scientific content (such as *in situ* measurements of ionospheric phenomena, chemical and isotopic investigation of the Moon's soil, astronomical observations in new spectral bands) and with respect to their increasingly interdisciplinary nature (such as the growing importance of physicists in astronomical research).

This chapter presents an historical overview of the origin of space science within the framework of the major political and military events which shaped the early phase of the space age. Given the nature of this overview, it was decided to include a comprehensive bibliography at the end, rather then to insert specific references to the vast literature on the history of space into the body of the text.

DREAMS OF SPACE FLIGHT

A journey to the Moon and beyond has for centuries been the object of popular fantasies, philosophical speculations and poetic dreams. Space travel literature can be traced back to Lucian of Samosota's *Vera Historia* (*True History*), written in the second century, whose heroes are taken to the Moon by a terrible whirlwind that lifts their ship into space. In Lucian's second book, entitled *Icaro-Menippus*, the trip to the Moon is carefully planned and eventually performed by Menippus, the hero, with the help of wings fastened to his shoulders.

At the height of the Italian *Rinascimento*, a voyage to the Moon was conceived in 1516 by the great Italian poet Ludovico Ariosto in his epic poem *Orlando Furioso*. Searching for the lost mind of Orlando, Astolpho is carried to the Moon by Elijah's chariot drawn by four red horses. He visits the Moon, where everything lost on Earth is stored in a valley, and eventually finds the paladin's mind in a phial.

A century later, in Germany, one of the fathers of modern astronomy, Johannes Kepler, told of a Moon flight in his *Somnium* (*Dream*), posthumously published in 1634. In Kepler's dream, conceived as an allegory for astronomy which reveals the reality of celestial bodies, supernatural demons (that is, astronomical knowledge) drive the learned traveller to the Moon along the shadow of the Earth during a lunar eclipse, and then they drive him back along the Moon's shadow during a solar eclipse.

Figure 2 Johannes Kepler. (Source: Istituto e Museo di Storia della Scienza, Florence.)

The same theme was dealt with in those years by two English bishops. Francis Godwin, Bishop of Hereford, wrote *The man in the Moon: or a discourse of a voyage thither* under the pseudonym Domingo Gonsales. The book was published in 1638 and it became very popular all over Europe. The protagonist of the book, Gonsales himself, flies to the Moon by a chair pulled by wild geese. In the same year, John Wilkins, Bishop of Chester, published *The discovery of the man in the Moon: or a discourse tending to prove that 'tis probable there may be another habitable world in that planet*. Wilkins tells nothing about the means of propulsion but he speculates that no effort would be required beyond the point where the influence of terrestrial gravity ceases, which he assumes to be some thirty kilometres above the Earth's surface.

At the court of the king of France we find the most remarkable author of Moon tales of the seventeenth century. This is Savinien de Cyrano Bergerac, dramatist and poet, philosopher and swordsman, whose enormous nose found a place in the history of French literature thanks to Edmond Rostand's famous play of 1897. Cyrano's *L'autre monde, ou les états et empires de la Lune*, posthumously published in 1657, tells of several attempts by himself to reach the Moon. First, he tied a string of vials filled with dew around himself, so that the rays of the morning sun could draw him upward when evaporating the dew, but he only arrived in Canada on that attempt.

He then tried to launch himself from the top of a hill by a special spring-loaded vehicle, but the vehicle crashed down into the valley. After curing the bruises by smearing ox marrow on his body, the hero returned to his vehicle, but he found that some soldiers had taken possession of it and were tying fireworks to it in order to make it appear as a flying fire dragon. He tried to stop the soldier who was setting fire to the fuse, only to find himself trapped in the vehicle while the latter lifted off to the clouds.

Tiers after tiers, all the rockets were rapidly exhausted, but instead of crashing down to Earth with the vehicle, Cyrano continued his ascension to the sky. The waning Moon, he explained, used to suck up the marrow of animals, and that is why his greased body was pulled upwards. A few days later he finally landed on the Moon, where he had many adventures and philosophical discussions with its inhabitants. After returning to Earth, the hero decided to visit the Sun, and the story is told in Cyrano's second book, *Histoire comique des états et empires du Soleil* (1662).

While presenting the first space travel performed by means of rocket propulsion, Cyrano's works do not deal with science, nor even science fiction. They are instead remarkable examples of philosophical novels, a literary device for the author to expound his materialistic ideas and his libertine vision of human affairs. European literature of the seventeenth and eighteenth century counts several other works in which space travel and encounters with alien people serve as witty means for conveying the author's opinions about mundane matters.

But in the century which saw the birth of modern science, other works tried to deal with the old dream of space flight in the light of the new astronomical knowledge (sunspots, Jupiter's satellites, Saturn's rings, mountains on the Moon, and so on), and in accordance with the tenets of the new natural philosophy. A major event was the publication, in 1686, of Bernard de Fontenelle's *Entretiens sur la pluralité des mondes*, a popular astronomy book which was widely read throughout Europe for its fascinating speculations on the nature and habitability of Solar System bodies. Fontenelle claimed that each planet had its own inhabitants and discussed their likely appearance, civilization and habits.

The belief that planets are inhabited by different species of rational beings was also expressed by the great Dutch scientist Christian Huygens in his *Cosmotheros, or conjectures concerning the planetary worlds*, first published in Latin in 1698. However, if the Moon and the planets were no longer considered as metaphorical symbols but as real worlds subject to universal physical laws, tales of space travels had to cope with the problem of escaping the Earth's gravitational pull. In David Russen's *Iter lunare or voyage to the Moon* (1703) a huge spring device is proposed as a means of launching a manned vehicle towards the Moon. Two years later, Daniel Defoe's *The consolidator* tells us that ancient people mastered the art of space flight and used to travel to and from the Moon. Defoe describes in particular an intriguing engine based on the use of hollow bodies filled with 'a certain spirit' which produced a continuous fire. The fire acted on a spring-and-wheel mechanism which kept two large wings in motion.

Many other tales of lunar travel can be found in eighteenth century literature, including Samuel Brunt's *A voyage to Cacklogallinia* (1727), whose heroes are propelled into space by intelligent birds, and Murtagh McDermot's *A trip to the Moon* (1728), where a whirlwind takes the hero to the Moon and gunpowder is used to return him to Earth. In Voltaire's *Micromégas* (1752) we find the first extraterrestrial visitor of literary history: an intelligent giant arriving on Earth from the star Sirius after a long journey through the Milky Way and the Solar System.

In the nineteenth century, science fiction became a well-established literary gense, nourished by the dramatic progress of astronomy and by the growing importance of science and engineering in everyday life. Romantic tales and philosophical speculations were gradually replaced by scientific and technical discussions of the various aspects of space flight. Science fiction writers, however, still lacked the most important element: a credible means to escape the Earth's gravitational field. An antigravity material called *lunarium* was devised by the University of Virginia professor George

Tucker in his novel *A voyage to the Moon with some account of the manners and customs, science and philosophy of the people of Morosofia and other Lunarians*, published in 1827 under the pseudonym of Joseph Atterley. Eight years later, Edgar Allan Poe, a former student of Tucker's, sent his hero Hans Pfaall on a lunar trip in a home-made balloon. In 1865, the Frenchman Achille Eyraud imagined a spaceship powered by a reaction engine based on the ejection of water for his *Voyage à Venus*.

The recognized father of modern science fiction is Jules Verne, whose novels *De la terre à la lune* (1865) and *Autour de la lune* (1870) became very popular all over the world. The former tells of the long and careful preparation of a voyage to the Moon, to be accomplished by three astronauts sitting inside an artillery shell fired into space by a huge cannon, called *Columbiad*. The shell was equipped with a special mechanism for absorbing the shock, a chemical system for air regeneration, and rockets for soft landing on the Moon's surface. The book ends with the successful firing of the cannon and the story is then picked up in the second book. Here we follow the heroes during their space travel. Owing to a large meteorite passing close to their spaceship, the latter is displaced out of the planned trajectory and ends by orbiting around the Moon. An attempt to steer the vehicle towards landing by the use of the rockets fails and the voyagers are pushed towards the Earth, safely descending in the Pacific Ocean where they are rescued by an American ship.

Verne's stories took advantage of the most advanced scientific knowledge of his times, including calculations of escape velocity and flight trajectories, a sound determination of the best geographic location for the launch cannon (in Florida, not far from the present US launch facilities), and a realistic description of weightlessness in outer space. Inspired by the general scientific optimism of the late nineteenth century, Verne and his followers were less interested in writing Moon tales (his heroes, in fact, do not land on the Moon), than in discussing in literary form the possibilities for

Figure 3 Illustration from *De la terre à la lune* by Jules Verne. (Source: AKG.)

Figure 4 Illustration from *Autour de la lune* by Jules Verne. (Source: AKG.)

humankind to make one of its oldest dreams, the conquest of space, a reality.

Spaceships were now imagined as spindle-shaped vehicles, hermetically sealable against the airless condition of outer space; they were provided with life-sustaining equipment, including means for supplying fresh air, heating systems, and sufficient food provisions. Some kind of artificial gravity would also be appreciated by space travellers. Armour protection was required to shield the ship and its crew from meteorites, as well as an airlock and special suits for leaving the spaceship while on some airless world. All these aspects, and several others, were discussed in science fiction novels of the late nineteenth and early twentieth century, which inspired the pioneers of modern astronautics.

Among these works it is worth mentioning what can be considered to be the first proposal for an artificial Earth satellite: Edward E. Hale's *The brick moon*, published in 1869–70 in the New England magazine *Atlantic Monthly*. In this story a group of young Bostonians conceive the idea of launching a satellite into a polar orbit 6000 km above the Earth's surface, where it could be used for navigational aid. The satellite was to be built in bricks because these could withstand the fire generated during the fast motion through the air. The launch system was based on the use of two gigantic fly-wheels, but an accident during the launch operations sent the brick moon off prematurely, with many construction workers and their families aboard it. One year later the satellite was located at an altitude of 8000 km, and accurate observations with a large telescope showed that its unintentional passengers had survived space flight and had managed to organize a pleasant life in their new home planet.

The discovery of the so-called 'canals' on the planet Mars by the Italian astronomer Giovanni Schiaparelli, in 1877, opened a new phase in science fiction literature. Most scientists, including Schiaparelli himself, were convinced that these features were the work of intelligent beings, indeed the first evidence for the existence of extraterrestrial life. The existence of a plurality of inhabited worlds was no longer a matter for philosophical speculations but a concrete scientific reality, and communications with alien species became the ultimate goal of a civilized humankind who had now colonized the whole of planet Earth. Inventors such as Edison and Marconi tried to design systems to establish some kind of communication with Martians. Science fiction writers adapted their stories to the new scientific claims and popular expectations. As it was definitely demonstrated that the Moon was an arid, lifeless desert, Mars became the privileged subject of scientific speculations and space travel tales.

The 'red planet' entered science fiction literature with the novel *Across the Zodiac*, published in 1880 by Percy Gregg, whose heroes travel through space by a huge spaceship similar to an ancient galleon provided with a special antigravitational force called 'apergy'. In France, popular astronomy and science fiction were intertwined in Camille Flammarion's many works, including the two-volume *La planète Mars et ses conditions d'habitabilité*, published in 1892 and 1909, respectively. In the United States, Percival Lowell built an observatory near Flagstaff, Arizona, specially devoted to planetary studies. He published several books on Mars, including *Mars* (1895), *Mars and its canals* (1906), and *Mars as the abode of life* (1908).

Martian invasions were told in Kurd Lasswitz' *Auf zwei Planeten* and Herbert G. Wells' *The war of the worlds*, both published in 1897. In the former, Martian space travellers fly to Earth by means of rockets and antigravity devices and set up a base at the North Pole. Wells' acclaimed book tells the story of a hostile Martian invasion of our planet. After an initial successful conquest of the Earth, the Martians die from terrestrial diseases against which their organisms have no defence. In the opposite direction, epic adventures of humans in Martian settings were described in a very popular series of novels published in 1912 by Edgar Rice Burroughs under the title *Under the moons of Mars*, which was followed by many others after World War I.

In Russia, interplanetary travels and Mars stories were often intertwined with utopian and revolutionary themes. In Alexander Bogdanov's *Krasnoya Zvezda* (Red Star), first published in 1908 but re-edited many times after the Bolshevik Revolution, a Russian Marxist revolutionary is carried to the 'red planet' by its intelligent inhabitants to be shown the communist paradise they have realized there. A Mars story, *Aelita*, was also written by the great Russian author Alexei Tolstoy in 1923, and it provided the subject for a science fiction film in 1927.

Science fiction enjoyed enormous popular success before and after World War I, and many scientists and engineers started a career in rocketry and astronautics inspired by their reading about space travel in popular magazines. Rockets were well known in China since the discovery of gunpowder, in the Middle Ages, and they were widely used in seventeenth and eighteenth century Europe to make firework shows. The British army discovered their military use at the end of the eighteenth century, during their first operations in India, and rocket weapons were eventually introduced in the military arsenals of major European countries. However, the dramatic progress of artillery in the nineteenth century outshone their military potential. It was the dream of space flight which led a few pioneers to reconsider rocket propulsion as a means to escape the Earth and travel through the infinite vacuum of outer space (see the following chapter in this book for a comprehensive history of rocket technology).

THE PIONEERS

In the portrait gallery of the space age the first prominent place should be reserved to Konstantin Eduardovich Tsiolkovsky (1857–1935). Born in the village of Izhevsk, six hundred miles east of Moscow, Tsiolovsky started thinking of space flight in the 1870s, inspired by Verne's space novels and by the philosophical ideas of Nikolai Fyodorov, a librarian who tutored him when he was a student in Moscow. Fyodorov claimed that space colonization was necessary for material and spiritual development of humankind. In 1879 Tsiolkovsky moved to Kaluga, a provincial town about 150 km south of Moscow, where he became a high-school teacher of mathematics and physics. There he pursued his studies in physics and chemistry, and established a small laboratory to conduct experiments in aerodynamics. He designed a metal-made dirigible, and his scientific publications won him admission to the Physical and Chemical Society of St. Petersburg.

The idea of using rocket propulsion for space flight was first expressed by Tsiolkovsky in a 1883 note in his diary. The idea was eventually worked out in all scientific implications in an article published in 1903 in the journal *Nauchnoye Obosreniye* (*Scientific Review*) under the title *Exploration of cosmic space by means of reaction devices*.

In this seminal paper, Tsiolkovsky set out the reasons for using rocket propulsion to escape the terrestrial gravity field and move in empty space, and discussed the advantages of liquid fuels for precise control over spacecraft acceleration. He calculated all the basic theoretical formulae for rocket motion, including the important relation between rocket velocity and exhaust gas velocity, and discussed important technical aspects of rocket design, including the technique of regenerative cooling of the combustion chamber, gyroscopic controls and steerable nozzles. In his subsequent works, Tsiolkovsky discussed other important concepts of space flight, the most remarkable being that of multistage rockets. He also published several science fiction stories which served as a literary training ground for his scientific and technical speculations on astronautics.

Tsiolkovsky's important theoretical work was widely recognized in the new Soviet state and in 1919 he was elected to the Socialist Academy (later the USSR Academy of Sciences). After his death, his house in Kaluga became a museum. His works, however, were not translated into a Western language and he remained practically unknown in the West until the 1920s. While certainly inspiring the rocket pioneers in the USSR, his writings did not influence those who in Germany and the United States were also speculating about the scientific and technical problems of space flight and helped to put the first theoretical ideas on a solid experimental basis.

Hermann Oberth (1894–1989) was one of these. Born in a German family in Sibiu (formerly Hermannstadt), Transylvania, at that time part of the Austro-Hungarian empire and now in Rumania, Oberth studied physics and astronomy at the universities of Cluj (formerly Klausenburg), München, Göttingen and Heidelberg. He developed a strong interest in space travel, reading Verne's novels at the age of twelve. His first designs for a liquid-fuel rocket date back to 1912. Five years later, when Europe was ravaged by World War I, Oberth submitted a report to the German War Department with plans for a gyroscopically controlled long-range missile.

In 1922, Oberth submitted to the university of Heidelberg a dissertation on space flight entitled *Die Rakete zu den Planetenräumen*, but this was rejected because of its poor physical and astronomical content (in fact it was essentially devoted to 'esoteric' rocket technology). He then decided to publish his work as a book, which appeared in 1923 and enjoyed enormous success. In this 90-page book one can find a detailed theoretical discussion of all the main aspects of rocket flight, including liquid-fuel rocket construction, thrust and propulsion, launch operations, inertial guidance and navigation, life-support systems, physiological effects of space flight on human beings, and tracking and recovery techniques. It also included a precise design for a so-called 'Model B' rocket, to be used for scientific investigations

Figure 5 Konstantin Tsiolkovsky (1857–1935). (Source: Novosti.)

Figure 6 Cover of *Die Rakete zu den Planetenräumen* by Hermann Oberth. (Source: Hermann Oberth Raumfahrt Museum.)

in the upper atmosphere. A suitable modification of the Model B rocket, Oberth argued, would make it possible to carry human beings into space, a first step towards the establishment of Earth-orbiting space stations to be used for geographical research, navigational aid, weather forecasting, military reconnaissance, and eventually as a fuelling and launch point for interplanetary travel.

From 1924 to 1938 Oberth was a mathematics and physics professor in the German gymnasium of Mediaş, Transylvania. In this period he continued working on the basic problems of rocketry and space flight, entering into correspondence with other pioneers of astronautics, including Tsiolkovsky, R. Goddard and R. Esnault-Pelterie. Out of this work came the 400-page book *Wege zur Raumschiffart*, published in München in 1929, which Esnault-Pelterie defined as 'the bible of scientific astronautics' when nominating Oberth for a prize of the French Astronomical Society. In this book there is a description of a fictional circumlunar flight by the 'Model E' three-stage rocket *Luna*. This rocket served as a model for the spaceship *Friede* in Fritz Lang's famous science fiction movie *Frau im Mond*. Oberth acted as Lang's scientific advisor during the preparation of this movie, and he also designed a small rocket to be launched on the occasion of its premiere in 1929. The project could not be realized, but the small experimental rocket motor, called *Kegeldüse*, was actually built and successfully tested in July 1930.

In 1938, Oberth went to the Technische Hochschule of Vienna to do rocket research. In 1941, he joined the rocket development programme of the German army in Peenemünde. After the war he spent some years in Switzerland and Italy, where he developed a solid-fuel rocket motor, and published his third important book on space flight, *Menschen im Weltraum* (1954). Finally, in 1955, he joined von Braun's team in Huntsville, Alabama, where he contributed to the American space programme until 1960, when he went back to Germany.

While the work of Tsiolkovsky and Oberth was essentially theoretical, the American Robert Goddard (1882–1945) was the first to undertake an important experimental activity. Born in Worcester, Massachusetts, the young Goddard was a voracious reader of Verne's and Wells' stories. After graduating at Worcester Polytechnic Institute, he got a doctorate in physics in 1911 from Clark University, Worcester, where he eventually became a full professor in 1919. In this period he devoted much thought to the problem of rocket propulsion, which he recognized as the only realistic solution to space flight. By 1910 he had developed a general mathematical theory of liquid-fuel rocket motors and had performed some experiments with gunpowder mixtures in the physics laboratory of Worcester Polytechnic Institute.

Early in 1915 Goddard started an experimental programme with firework rockets, using personal money and the facilities of Clark University's physics laboratories. The results thus obtained won him a research grant from the Smithsonian Institution and an agreement with the US Signal Corps for undertaking secret experiments on rocket weapons at Pasadena, California. Out of this work came a series of rocket projectiles fired from tube launchers, the ancestor of the modern bazooka. After the end of the war, Goddard returned to Clark and in December 1919 he published the results of his work with fireworks in a report to the Smithsonian Institution with the title *A method of reaching extreme altitudes*, a small book which was widely popularized by the press as demonstrating the actual feasibility of Moon travel.

Goddard then turned to experiments with liquid-fuel rocket motors, with continuing support from the Smithsonian Institution. On 16 March 1926, he successfully launched the first liquid-fuel rocket in history from a test area located in a strawberry farm near Auburn, Massachusetts. It was 3.4 m long and weighed 4.6 kg, including 2 kg of fuel consisting of liquid oxygen and gasoline. The flight lasted 2.5 seconds with an average speed of 96 km/hr. The rocket reached an altitude of 12.5 m and fell down 56 m away. A second test flight was successfully achieved on 3 April

Figure 7 Robert Goddard (1882–1945). (Source: NASA.)

1926, and two others, with a different rocket design, on 26 December 1928 and 17 July 1929, respectively. The latter flight was performed for the first time by an instrumented rocket, carrying a barometer, a thermometer and a camera to an altitude of about 30 m before crashing down because of malfunctioning of its parachute.

After these tests, Goddard obtained an important research grant from the Guggenheim Fund for the Promotion of Aeronautics and moved his rocket activity to a remote place in New Mexico, near the town of Roswell. From 1930 to 1941, he developed there an intense experimental programme whose achievements were described in his second report to the Smithsonian Institution, *Liquid-propellant rocket development* (1936), and in the posthumously published book *Rocket development: liquid-fuel rocket research, 1929–1941* (1961).

More than 30 rockets were successfully flight-tested in Roswell and more than 100 static tests were performed, with increasing technical sophistication. On 28 March 1935, a rocket of the A series flew a distance of 4 km from the launch tower, reaching an altitude of 1460 m with an average speed of 880 km/hr. The rocket was 4.5 m long and weighed about 36 kg, and it was provided with a gyroscopic stabilization system which worked perfectly during the 20-second flight. In May of that year, another A series rocket reached an altitude of 2300 m. After a series of static tests of more powerful engines, a new series of rockets, the L series, were tested in 1936–38, while the following period was devoted to the development of increasingly sophisticated pumps and turbines.

Following the US entry into World War II, Goddard offered his services to the Navy and got involved in the development of liquid-fuel rocket boosters for takeoff assistance of heavily loaded seaplanes. He worked on this project until his death, in August 1945.

Unlike his counterparts in the Old Continent, Goddard was never a convinced and active propagandist of space travel. He concentrated all his efforts on the study and technical development of rocket propulsion and did not give in to the temptation of advertising his results as first steps towards space travels. On the contrary, he worked far from the public eye, even in strict secrecy, for much of his lifetime. It was only after his death that his pioneering experimental activity was widely recognized, and eventually the newly created NASA decided to dedicate to him its major centre for space research: the Goddard Space Flight Centre, established in 1959 in Greenbelt, Maryland, between Washington and Baltimore.

Among the founding fathers of modern astronautics we should also include the French engineer Robert Esnault-Pelterie (1881–1957), the only one with experience in aeronautics, both as a designer and as an aviator. His ideas on rocketry were first expressed in a 1912 lecture at the French Physical Society, later published with the rather anodyne title *Considérations sur les résultats d'un allégement indéfini des moteurs* (with reference to continuous mass loss in a rocket because of fuel consumption). After the war, Esnault-Pelterie entered into correspondence with Goddard and Oberth, and promoted the establishment of a prize of 5000 FF for original work in the new science of astronautics (incidentally, the word *astronautique* was first proposed by one of Esnault-Pelterie's converts, the Belgian writer J. Rosny). The prize was funded by his friend the banker André Hirsch, and Oberth was the first recipient in 1929, the amount being increased to 10,000 FF in recognition of his outstanding contribution.

In 1930, Esnault-Pelterie published his major work, entitled *L'astronautique*, the first real textbook on the subject. He then devoted himself to the study of rocket motors with some support from the military authorities. This activity was interrupted following the German invasion at the beginning of World War II.

THE ROCKET SOCIETIES

The work of these pioneers, coupled with the fascination of science fiction in literature and cinema, inspired the formation of several societies of amateur enthusiasts dedicated to the cause of rocketry and space flight. These societies performed a number of important functions. They publicized and legitimated space flight as an actual technical possibility, giving scientific credibility to an enterprise which had previously been seen as the domain of fiction writers and eccentric philosophers. Their members conducted a considerable amount of systematic experimentation, often at great personal risk. And they served as training grounds for some of the most important rocket engineers of this century, notably Werner von Braun (1912–1977) and Sergei Korolev (1906–1966).

The most important rocket society in this period was the German *Verein für Raumschiffahrt* (VfR, Society for Space Travel), established in Breslau (now Wroclaw, Poland) in 1927 by a group of Oberth enthusiasts. Oberth himself joined the Society and served as its president in 1929–30, when membership had reached about 900.

During the first two years of its lifetime the VfR published a journal, called *Die Rakete*, which rapidly became an important forum for discussing ideas and results in rocket development. In 1929, however, it was decided to stop publication in order to save money for experimentation. One year later the Society moved to Berlin, where Oberth was involved in the *Kegeldüse* project, and established a test area in a former military camp on the outskirts of the town. They called it *Raketenflugplatz* (Rocket Proving Ground). It was here that von Braun, a teenager at that time, joined the group.

In May 1931, the VfR successfully launched its first liquid-fuel rocket, called *Repulsor 1*. Not the first in Europe, however, as in March that year one of VfR's founding fathers, Johannes Winkler, had succeeded in launching his *Hückel–Winkler 1* rocket, with funds provided by industrialist Hugo Hückel. By late 1933, when the VfR was finally dissolved because of financial difficulties and internal conflicts, about 100 test flights and about 300 static tests had been made at the *Raketenflugplatz*.

In the United States, humankind's destiny in space was mostly heralded by the American Interplanetary Society (AIS), founded in New York in 1930 by a group of science fiction writers. Its activity was first limited to organizing public lectures and publishing a *Bulletin*. Following a visit to the *Raketenflugplatz* in early 1931, the AIS vice-president G. Edward Pendray resolved to start an experimental activity, and in November 1932 the first AIS liquid-fuel rocket engine was fired on stands in an old farm near Stockton, New Jersey. The first test-flight occurred in May 1933 from Marine Park, Staten Island, followed by a second one in September 1934.

Financial difficulties prevented the group, now renamed the American Rocket Society (ARS), from performing any other flights, and activities were limited to static tests. In the post-war years, the ARS became the largest professional rocket engineering society in the United States. In 1963, it merged with the Institute of Aerospace Sciences forming today's American Institute of Aeronautics and Astronautics.

The Soviet Union was the first government to endorse and support the activity of rocketry amateurs. The earliest group was set up as early as 1924 by a devoted Tsiolkovsky student, the Lithuanian engineer Fridrikh Tsander (1887–1933). At the end of a seminal lecture on interplanetary travel before the Moscow Society of Amateur Astronomers, Tsander proposed establishing a Society for the Study of Interplanetary Travel. This was first constituted as a 'rocket section' of the Zhukovsky Air Force Academy's Military Science Division in Moscow, but eventually evolved into a civilian organization known as the Society for the Study of Interplanetary Communication (OIMS, from its Russian initials).

The ambitious programme of OIMS was short-lived. Lack of funds and the difficult political situation of the young nation in the 1920s made rocket research a luxury which could neither be afforded by the group nor supported by the government. The efforts of the Interplanetary Section of the Association of Inventors, which heralded the organization, in 1927, of the First World Exhibition of Interplanetary Machines and Mechanisms, were eventually also unsuccessful.

New initiatives were undertaken in the USSR at the turn of the decade, linking together astronautics amateurs, academic institutions and the military. In Leningrad, a professor at the Institute of Railway Engineers, Nikolai Rynin, set up in late 1928 a 'Section of Interplanetary Travel', including teachers, engineers and students. Ryinin, a long-standing space enthusiasts, was also the chairman of the Leningrad Institute of Communications Engineers' 'Rocket Research Section' and the author of a nine-volume encyclopedia on '*Interplanetary flight*' (1928–1932).

Much more important was the establishment in Leningrad, also in 1928, of the Gas Dynamics Laboratory (GDL), which took over the activities of a Moscow research group on powder rockets set up in the post-revolution period by the chemical engineer Nikolai Tichomirov under military auspices. The GDL's offices were located in the Admirality building facing the Hermitage palace and a test stand was established on the grounds of the Sts Peter and Paul Fortress across the Neva river. Among the first technical staff was the young engineer Valentin Gloushko (1908–1989), who initiated an important research programme on liquid-fuel rocket engines.

In 1931, a number of Groups for the Study of Reaction Motion (GIRD in Russian initials) were set up in many towns of the Soviet Union, under the general supervision of *Osoaviakhim*, an organization dedicated to the promotion of research of military interest in the fields of aeronautics and chemistry. Although formally established as civil organizations, the GIRDs were in fact integrated in the general effort of the Soviet state to establish a sound scientific basis for the development of its armed forces.

The most important GIRD was that in Moscow, also known as MosGIRD, which soon became the leading organization for rocket research in the USSR. Created upon the initiative of F. Tsander, it was directed by Sergei Korolev, a brilliant graduate in aeronautical engineering from the Moscow Higher Technical School, where he had been a student of Andrei Tupolev, one of the founding fathers of Russian aviation. In November 1933, the MosGIRD activities were taken over by the newly created Jet Scientific Research Institute (RNII), a state institution which also absorbed the activities of the GDL and the Leningrad GIRD (LenGIRD).

The first Soviet liquid-fuel rocket, developed at MosGIRD and called GIRD-09, was launched on 17 August 1933 from a test stand in the woods of Nakhabino, about 30 km west of Moscow. It was followed by GIRD-X on 25 November. The former, designed by Michail Tikhonravov, weighed 18 kg and was powered by a hybrid fuel made of liquid oxygen and solidified gasoline. Its flight lasted less than twenty seconds. The rocket reached an altitude of about 400 m before going on a flat trajectory and hitting the nearby forest. The latter was the first true liquid-fuel Soviet rocket, powered by liquid oxygen and ethanol. Derived from Tsander's OR-2 rocket engine, GIRD-X was 2.2 m high and weighed about 29.5 kg at lift-off. Launched from the same test stand in the Nakhabino woods, it rose to 80 m and fell to the ground at a distance of about 150 m.

Several types of rocket engines were designed in the USSR in subsequent years, but this important research activity suffered from a serious setback during the Stalinist Great Purge of 1937–38. Following the liquidation of Marshall Mikhail Tukachevsky, the main supporter of advanced military technology, many leading rocket engineers were killed or imprisoned in the *gulag* system. The RNII was swept away and Korolev himself was arrested in 1938 and sent to forced labor in a camp in the Kolyma area of far eastern Siberia. Through the intercession of Tupolev, he was released in 1940 and sent to a special penal institution, known as *sharaga*, where prisoners with scientific and technical competence were called to work for the government under the general supervision of the secret police (NKVD).

Korolev was first assigned to the so-called Design Bureau No 29 (KB-29) in Moscow, devoted to aircraft development under Tupolev. Then, in November 1942 he was moved to an aircraft plant in Kazan, where he found Glushko and other rocket specialists from the former GDL. They worked on jet engines for aircraft and ballistic rockets. By the end of the war, the prior convictions against Korolev and his colleagues were expunged and the NKVD handed over its responsibility for the Kazan design bureau to the People's Commissariat of Aviation Industry. Glushko was made chief designer, with Korolev his deputy.

THE V2 ROCKET

The great interest of the military in rocket research in post World War I Germany was motivated by a general concern to rebuild the country's war-fighting capability, coupled with the specific fact that rockets were not subject to the severe restrictions placed on German rearmament in the Treaty of Versailles. An important rocket research programme was launched in 1929 by colonel Karl Emil Becker, chief of the Army's Department of Ballistics and Munitions, under the responsibility of Walter Dornberger. The Reichswehr also established a launch test site at Kummersdorf, in the outskirts of Berlin.

In 1932, Becker and Dornberger visited the VfR's facilities at the *Raketenflugplatz* and were very favourably impressed

Figure 8 Sergei Korolev. (Source: Novosti.)

by von Braun. They offered to support his PhD work and, on 1 November 1932, the 20-year old space enthusiast was recruited by the Army as Dornberger's technical assistant, with the task of developing liquid-fuel rockets at Kummersdorf.

By the end of 1934 von Braun and his team had developed and successfully tested the A-2 (*Aggregat-2*) rocket, thus establishing a firm basis for further development of the German Army's liquid-fuel rocket programme. In the following year, Hermann Göring's powerful Luftwaffe became interested in rocketry, and a joint Army–Air Force rocket research establishment was established in Peenemünde, on the Baltic sea. The staff of Kummersdorf moved to Peenemünde in 1937 and undertook the development of several types of military missiles.

During World War II, several thousand scientists, engineers and other employees worked in Peenemünde on all fundamental problems in rocketry, including aerodynamics, propulsion and guidance systems. Important investments were made to build the necessary facilities, including wind tunnels and test benches for more and more powerful rocket engines. Within this framework, the Army asked von Braun and his team to develop a missile capable of carrying a 1-ton warhead up to a distance of 300 km. This was the A-4 rocket, successfully flown for the first time on 3 October 1942.

Compared to its American, Russian and German forerunners the A-4 was a kind of monster. The rocket was 14 m long with a maximum body diameter of 1.65 m (3.56 m including the fins). It weighed 13 tons at lift-off, including 5 tons of liquid oxygen and 4 tons of ethanol fed by turbo pumps. The engine developed a thrust of 25,000 kg, propelling the rocket, provided with a highly sophisticated guidance system, at a maximum speed of about 5700 km/hr, up to a distance of 300 km and a peak altitude of 90 km, on the fringes of space.

Space, however, was not the goal of Hitler and its army. In fact, they wanted the ultimate weapon to win the war, after the Luftwaffe's strategic bombing of Britain had turned out to a failure, the Soviets had started their counteroffensive in Stalingrad, and western Allies had landed in North Africa. Following the October 1942 test flight, the

Figure 9 Wernher von Braun briefing German officers in 1943 at Peenemünde. (Source: Smithsonian Institution Washington.)

Figure 10 Launch of a V2 rocket, 1944. (Source: AKG.)

Figure 11 The Mittelwerk factory at Thuringia. (Source: Smithsonian Institution Washington.)

Führer awarded top priority to the A-4 programme and the Nazi Propaganda Ministry gave the rocket the ominous name *Vergeltungswaffe 2* (V2), i.e. Vengeance Weapon 2.

Mass production of the missile was organised by the SS using slave labor from Soviet prisoners of war and other concentration camp prisoners. The production plant was first set up in Peenemünde but, after the British Royal Air Force had bombed the rocket centre on 17 August 1943, it was moved to a huge underground factory in Thuringia, near the town of Nordhausen. The facility was called *Mittelwerk* (Central Works), and thousands of Russian, Polish and French slave-prisoners worked (and died) there under miserable conditions. The production rate at Mittelwerk rose to some 900 missiles per months, and a total of about 5800 V2s were eventually built.

The V2 war campaign started on 6 September 1944, when two missiles fired from the Netherlands fell on the outskirts of newly liberated Paris. Two days later the offensive against London and southern England was launched. Between September 1944 and March 1945 more than 1500 missiles reached Britain, with over 500 hitting the capital. The Belgian city of Antwerp and its important harbour were also heavily attacked.

It was too late, however, to change the course of the war. Towards the end of March 1945 the western front collapsed, the allied armies crossed the Rhein and marched into ravaged German territory. The last V2s were fired against Antwerp and London on 27 March. Meanwhile, the Red Army was rapidly advancing on the eastern front, and by mid-January 1945 their tanks had arrived at a few miles from Peenemünde. On 31st of that month an evacuation order finally arrived and von Braun prepared plans to move equipment, technical personnel and archives to central Germany.

Missile production went on until the end of March at the Mittelwerk, whose Dora camp had become the main element of a new concentration camp system called *Mittelbau* (Central Construction), where SS were deporting prisoners evacuated from the east, notably from Auschwitz. Needless to say, prisoners and slave workers were dying by the thousands from disease, starvation and SS brutality. By the best estimates, some 20,000 people died in the Mittelbau–Dora system, about half of the deaths being linked to V2 production.

In April 1945, with the American troops rapidly moving into Thuringia, von Braun and about 500 of his best engineers were driven to Oberammergau, in the Bavarian Alps. Peenemünde's archives were buried in a disused mine northwest of the Mittelwerk. On 2 May, two days after Hitler's suicide, von Braun, Dornberger and their group of veteran Peenemünde rocketeers surrendered to American troops. By that time the Americans had already captured the Mittelwerk plant, whose installations they found intact, with the V2 assembly line left in working condition.

Thuringia was to be part of the Soviet zone of occupation, therefore the US Army Ordnance hurried in seizing parts of V2s and as much important technical equipment as possible. A special 'V2 mission' was set up under the direction of colonel Holger Toftoy, and during the last days of May dozens of trains and trucks left Nordhausen carrying enough hardware to build a hundred V2 missiles. Their destination was the harbour of Antwerp, from where the material was shipped to New Orleans. Toftoy's team also succeeded

Figure 12 von Braun with US Colonel Toftoy, 1945. (Source: Smithsonian Institution Washington.)

in locating the Peenemünde archive; about 14 tons of scientific reports and technical documents were packed and shipped to the United States.

Having secured most of the hardware and software of the German rocketry effort, the Americans wanted to take advantage of the people as well. Within a few weeks, von Braun and some 130 of his best technical staff were hired by Toftoy's Ordnance Rocket Branch to cooperate in the American rocket programme within the framework of the newly established Project Hermes. In autumn 1945 they were moved to the United States and installed at Fort Bliss, near El Paso, Texas. Their first task was to help assemble and launch a number of V2 rockets at White Sands Proving Ground, about 80 miles north, in New Mexico. The first launch occurred on 16 April 1946, followed by 66 others up to 1952.

Soviet troops occupied the Nordhausen area and the Mittelwerk in early July 1945. Among them was a small number of rocket specialists led by Boris Chertok, whose task was to gather as much as possible of the V2 technology, documentation and specialists. Chertok created the so-called Institut Rabe (Raketenbau und Entwicklung, i.e. rocket manufacturing and development), a Russian–German venture which soon managed to employ as many as 1000 people, half of them Russians and the other half Germans, including some 50–60 Peenemünde veterans. The most important among them was Helmut Grottrup, a high-level manager and leading expert in rocket guidance and control. Their task was to restore the V2 design data and

Figure 13 US Army Ordnance units inspecting German supply trains, looking for useful V2 components to ship to America. (Source: Smithsonian Institution Washington.)

Figure 14 German space engineers in the USA, with von Braun in front row. (Source: Smithsonian Institution Washington.)

drawings, set up an assembly line to build rockets from surviving components, and organize a test-launch programme.

By the end of 1945, the Soviet authorities decided to create a more substantial institution, the Nordhausen Institute, which incorporated Rabe and other joint Soviet–German technical institutions. The Institute's director was General Lev Gaidukov, with Korolev as his deputy and chief engineer. They could take advantage of the fact that most of the V2 installations were in the Soviet zone of Germany, including Peenemünde, a missile control plant near Berlin, a design office near Erfurt, the Montania engine assembly plant near Nordhausen, an electric equipment factory in Sonderhausen, and the important engine development facilities and test stands at Lehesten. The latter were in excellent conditions and Glushko was put in charge of their exploitation, as well as of the Montania facilities. Eventually, the assembly line of the Mittelwerk and the whole Lehesten plant were shipped to the USSR.

Grottrup's group eventually numbered some 5000 scientists, engineers and technicians, experienced in rocketry, weapons systems, submarines and aircraft. In October 1946, the whole group and their families were peremptorily shipped to Russia and installed in workplaces near Moscow to be employed within the framework of the Soviet military development programme. By that time, responsibility for ballistic missile development in the Soviet Union had been given to Dimitri Ustinov's Ministry of Armaments and a new rocket research institute had been set up, called NII-88 (Scientific Research Institute-88). Korolev was made responsible for long-range ballistic missiles as the head of NII-88's Special Design Bureau (SKB from its Russian initials). Glushko became chief designer of the OKB-456 rocket engine plant in Khimki, near Moscow.

Most German rocket specialists worked for the NII-88 in a workshop established on the island of Gorodomliya in Lake Seilger, about 200 km from Moscow. Their first task was to help install the A-4 production line and build a number of A-4s to be launched from the newly established test range in Kaputsin Yar, in the Astrakhan area near Volgograd (former Stalingrad). The first launch was performed on 18 October 1947 with only partial success. A second successful one followed two days later. In total, eleven V2 rockets were launched in the Soviet Union. After this, the German rocket specialists were called to contribute to the ongoing

Figure 15 Valentin Glushko. (Source: Novosti.)

Soviet ballistic missile programme but their designs were not implemented. They were eventually shipped back to Germany in November 1953.

The United States and the Soviet Union were not the only beneficiaries of German expertise in rocketry. Britain and France also took a little share. In May 1945, the British launched the so-called Operation Backfire, whose aim was to launch a number of V2s with the help of German specialists. Eight complete missiles were assembled and three of them were launched in October 1945 from a test range in Cuxhaven, on the North Sea near the mouth of the Elbe river. Subsequently, a number of Peenemünde veterans went to the United Kingdom, notably Walter Riedel, head of the design office. The British government, however, did not show much interest in missiles in the immediate post-war period. Other German rocket specialists went to France, where they helped establish a military rocket programme in Vernon, on the Seine north-west of Paris.

THE V2 LEGACY AND THE BIRTH OF SPACE RESEARCH IN AMERICA

The advent of the V2 and the recognition of its performance provided convincing evidence in support of rocket propulsion as a means to achieve space flight. A rocket existed that was capable of lifting a heavy payload to the upper atmosphere, and it was no science fiction to imagine a day when human beings would be carried into space, would build permanent orbiting stations and would eventually journey to the Moon and Mars.

As early as February 1945, with the V2s still hitting London, science fiction writer and amateur scientist Arthur Clarke sent a letter to the British magazine *Wireless World* to suggest that a V2 rocket could be used to launch a communications satellite. In a full article eventually published in October of that year, Clarke proposed for the first time the use of geostationary satellites to achieve a global telecommunications system. A space prophet of the post-war generation, Clarke published in 1951 a best-selling book, *The exploration of space*, which set the stage for a series of articles on the concrete prospects of space flight published by Von Braun in *Collier's* magazine. A new flow of science fiction novels and movies revived space adventures in popular thought and imagination (from Isaac Asimov's *Foundation* cycle of 1942–50 to Irving Pichel's movie *Destination Moon* of 1950).

On the other hand, the development of rockets and spaceships had by now passed irretrievably out of the hands of space flight dreamers, amateur societies or university departments. Such a development required an enormous amount of resources, which only national governments could provide. Moreover, at that time, the immediate aim of any important effort in rocket technology was weapon production rather than the conquest of space. After the war experience, the role of the rocket societies which had flourished in the 1920s and 1930s could only be of an educational nature. They kept advocating the cultural and scientific value of space flight against the ominous prospects of military missiles, and the need for international collaboration in peaceful use of space against the Cold War confrontation between the two superpowers.

In 1949, the Stuttgart-based *Gesellschaft für Weltraumforschung* (GfW) undertook to organize an international meeting of all societies for rocket development, interplanetary travel and space research. The meeting was eventually organized in Paris in 1950 by the *Groupement Astronautique Français*, with representatives from nine astronautical societies. It was ambitiously called 'Premier Congrès International d'Astronautique', as the intention of the participants was to make it the first of a series of annual meetings. The Congress opened with an impressive public meeting in the Richelieu Amphithéatre of the Sorbonne University, attended by an audience of 1000, and was concluded with the decision to set up an international federation of astronautical societies.

The International Astronautical Federation (IAF) was established in London on 4 September 1951, within the framework of the second International Astronautical Congress. Eugen Sänger, an Austrian aeronautical engineer and rocket pioneer, was unanimously elected as its first president. Delegates representing sixteen societies from ten countries attended the London meeting, which included a four-session symposium on artificial satellites. Today, the IAF has 152 members from 45 countries, including learned societies, professional associations, government agencies, aerospace industries and international organizations.

While the IAF can be considered to be the heir of the rocket societies of the interwar period, space research is definitely part of the V2 legacy. If fired vertically, the V2 could reach an altitude of 200 km, i.e. an order of magnitude higher than that achieved by scientific balloons in the 1930s. This made it possible to perform *in situ* measurements on a wide range of phenomena in the upper atmosphere. The military interest of such measurements had already been recognized by von Braun in the Peenemünde days. In fact, in order to develop a ballistic missile capability it was of crucial importance to have reliable information about the altitude profiles of the physical properties of the atmosphere (temperature, pressure, density, chemical composition, etc.). Von Braun contacted the distinguished physicist Erich Regener, a pioneer in scientific ballooning throughout the 1930s, inviting him to provide scientific instruments to be included in the nose-cone of one of the test rockets. Despite his open opposition to the Nazi regime, Regener accepted, and in July 1942 his institute at Friedrichshafen entered into an agreement with the Peenemünde Research Centre to develop a scientific payload for a V2 flight. They designed a protective container that was to be ejected at the peak of the rocket's trajectory and then drifted down to Earth by a special parachute, while a number of barographs and thermographs would record the altitude profile of atmospheric pressure and temperature.

Regener also decided to include in the payload two ultraviolet spectrographs for the study of solar radiation. The first aimed at measuring the increase of intensity during the rocket's ascent in order to calculate the altitude dependence of atmospheric ozone. The second, designed by solar physicist Karl-Otto Kiepenheuer, would investigate the spectral region around the 'Lyman alpha' transition line (1216 Å). This is radiated by hydrogen in the transition from the first excited state to the ground state and it was expected that it should be very strong because of the hydrogen abundance in the Sun. The aim of this investigation was to study the influence of solar radiation on the ionosphere. It fitted within Kiepenheuer's research programme on the correlation between solar activity and short-wave communications for which he had established a network of solar observatories with support from the German Luftwaffe.

Regener's V2 flight was scheduled for spring 1945, but the evacuation of Peenemünde and the German surrender made the event impossible. Regener was interrogated by allied intelligence services and some of his instruments

Figure 16 V2 scheduled for launch at White Sands, New Mexico. (Source: Smithsonian Institution Washington.)

were eventually located among V2 hardware at the Aberdeen Proving Grounds, north of Washington. His plans for ionospheric research by rocket-borne instruments were considered with great interest in the United States, where concrete initiatives were already being taken to establish a scientific agenda for the V2 test flights planned at White Sands. Such an agenda, in fact, was not much different from that elaborated in Peenemünde: an upper atmosphere research programme to be developed by professional scientists under military patronage, whose ultimate aim would be to develop an efficient ballistic missile capability. Solar spectroscopy, cosmic rays and geomagnetic studies fell well within this domain, as well as atmospheric and ionospheric studies.

In January 1946, the US Army Ordnance, responsible for the V2 flights, informed the interested scientific community of the new research opportunity provided by the launch programme. A meeting was convened on 16 January at the Naval Research Laboratory in Washington, attended by some 40 people from 12 institutions, which decided to set up a 'V2 Rocket Panel' to coordinate the scientific use of V2 launches. The panel, eventually renamed 'V2 Upper Atmosphere Research Panel' and later on the 'Upper Atmosphere Rocket Research Panel' (UARRP), met for the first time on 27 February at Princeton University with representatives from the Naval Research Laboratory (NRL) and the Johns Hopkins University's Applied Physics Laboratory (APL); the Army Signal Corps; the universities of Harvard, Michigan and Princeton; and the General Electric Company, contractor for Project Hermes. Ernst Krause, head of the NRL's Communications Security Section was elected chairman. He was succeeded in 1947 by James van Allen, who occupied the post for the next ten years.

Trained at the University of Iowa, van Allen had been interested in geophysical problems during his postgraduate experience at the Carnegie Institution of Washington's Department of Terrestrial Magnetism (DTM). During the war, he was brought by DTM's director Merle Tuve into the newly established Applied Physics Laboratory, where he worked on the development of a successful radio proximity fuse. When the prospects of a V2 research programme emerged, van Allen was called by Tuve to head a High-Altitude Research Group at APL and represent the laboratory in the V2 Panel. In 1949 he moved to the University of Iowa where he continued his studies of cosmic radiation and electrical phenomena in the upper atmosphere.

The V2 Panel's immediate task was to provide advice to the Army Ordnance Department about the allocation of V2s to the various research groups represented in it. Besides van Allen's APL group, the most important was NRL's Rocket-Sonde Research Section, established by Krause in December 1945 and headed since 1947 by Homer Newell. In association with Newell's section worked Richard Tousey, an expert in vacuum ultraviolet techniques who undertook to design V2 instruments to detect the ultraviolet spectrum of the Sun. In 1949, Herbert Friedman joined NRL's upper atmosphere research programme, bringing in his efficient photon counter detectors for studying the solar spectrum in the far ultraviolet and X-ray band. The NRL not only developed a V2 scientific programme, but also provided the other groups with standard nose sections for housing scientific instrumentation and with telemetry services.

A leading position in the research activity coordinated by the V2 Panel was held by the University of Michigan's group headed by William Dow, an electrical engineer with a wartime experience in the development of microwave vacuum tubes at the MIT Radiation Laboratory. Dow's involvement in V2 upper atmosphere research derived from Michigan's Project Wizard, an antiballistic missile programme supported by the Army Air Force. The latter was not initially represented in the V2 Panel, but eventually its Cambridge Field Station, later renamed the Air Force Cambridge Research Laboratories (AFCRL), was included in the membership. The AFCRL's programme in space research was directed by Marcus O'Day, a physicist expert in electronics, and was especially devoted to atmospheric composition and ionospheric radio propagation.

In collaboration with the AFCRL worked a group from the University of Colorado which had developed a fine pointing control system for accurate measurements of the solar far ultraviolet spectrum. This allowed Colorado's William Rense to obtain in 1952 the first good measurement of the Lyman alpha emission line by a grazing incidence spectrograph.

Finally, we should mention the Harvard College Observatory, represented on the V2 Panel by Fred Whipple. Unlike the other charter members of the V2 Panel, Harvard Observatory was not involved in military research and Whipple's interest in upper atmosphere research derived from his earlier studies of meteors.

For more than a decade the V2 Panel occupied a unique position in American science. On the one hand, its members were engaged in scientific and technical studies aimed at solving specific problems of interest to their patrons: their upper atmosphere research work, in fact, was part of a mission whose primary goal was to develop new generation of ballistic missiles, radio communications and radar detection systems, and antiballistic missile systems. On the other hand, they helped establish a new scientific community, i.e. physicists, astronomers, meteorologists, and geophysicists who wanted to seize the unique opportunity offered by space technologies to expand knowledge in their respective field. Raising their instruments to the outer layers of the atmosphere, physicists could study the composition of the primary cosmic radiation before this was altered by nuclear reactions with atmospheric atoms; solar and stellar astronomers could detect that part of the electromagnetic spectrum which is absorbed by the atmosphere; meteorologists gained access to regions of the atmosphere never reached before; geophysicists could finally investigate the complex relations between the ionosphere, the geomagnetic field and the solar radiation by *in situ* measurements.

The importance of the scientific dimension of the V2 Panel's programme was recognized since the very beginning, when it was agreed that it would not be classified. V2 launch schedules and flight information were freely circulated among interested people, and so were the scientific data obtained. Reports on experimental results were presented during panel meetings and a steady flow of papers appeared in the literature that attracted considerable attention as they reported on hitherto impossible measurements. A book by H. Newell (1953) reviewed the results obtained in the first phase of the Panel's work, mainly characterized by V2 launchings.

Besides providing scientific advice to Ordnance on the V2 launch programme, the V2 Panel also set out to take advantage of post-war missile development in the US in order to pursue their upper atmosphere research programme by other sounding rockets. A first possibility was offered by the WAC Corporal rocket, developed under Army Ordnance contract by the Jet Propulsion Laboratory (JPL). The WAC Corporal was successfully launched for the first time in October 1945 from White Sands and it subsequently evolved in improved versions up to the full-scale Corporal E, first launched in May 1947, capable of lifting a 200-kg payload to an altitude of 160 km. The V2 Panel, however, did not consider the Corporal suitable for upper atmospheric sounding, nor was JPL – at that time – particularly interested in research. This rocket, therefore, did not significantly contribute to the upper atmosphere research programme.

Some important results were obtained, however, with the so-called 'Bumper' project, where a WAC Corporal was added as a second stage to a V2 vehicle with the two-fold goal of achieving higher altitudes and testing the technology of two-stage missiles. Eight launches of Bumper were performed between May 1948 and July 1950, but only one was fully successful, on 24 February 1949, when the second stage reached a record altitude of 393 km. The last two test launches are of some historical importance as they were the first performed from the newly established Long Range Proving Ground at Cape Canaveral, Florida.

The most important vehicle for upper atmosphere research in the post-V2 period was the Aerobee, a scaled-up version of the WAC Corporal designed by van Allen and his APL staff and developed with support from the US Navy Bureau of Ordnance. After its first successful test flight, on November 1947, the Aerobee sounding rocket gained favour with all the scientific groups engaged in upper atmosphere research for its good cost-effectiveness. In the following years something like a dozen different versions of the rocket were produced, the last of which was powered by four clustered motors and could reach an altitude of 480 km with a 23-kg payload. Between 1947 and 1985 more than 1000 Aerobee vehicles have been launched, lifting any conceivable payload of scientific interest, including small animals and a special camera which provided in 1954 the first colour photographic image of our blue planet from space.

In parallel with the development of Aerobee, the NRL undertook to develop a large rocket, intended to replace the V2. This was first called *Neptune* but later renamed *Viking*. The first *Viking* was launched on 3 May 1949, but it was soon recognized that it was much too big and expensive for use in atmospheric research. Only 11 Viking vehicles were launched between 1949 and 1954, when the rocket was re-designed to become the first stage of the Vanguard satellite launcher (see below).

In January 1956, in order to celebrate the tenth anniversary of the first meeting of the V2 Panel, the UARRP organized a symposium on the 'scientific uses of earth satellites', whose proceedings were published later that year (van Allen, 1956). After a decade of scientific rocketry, orbiting satellites represented the next logical step in space research, and in fact both the United States and the Soviet Union had undertaken to launch a satellite in the framework the forthcoming International Geophysical Year, planned for 1957–58 (see below). The UARRP itself decided in 1957 to change its name adopting the new Rocket and Satellite Research Panel.

Launching a spacecraft into Earth orbit, however, required a much more powerful rocket than those used for atmospheric sounding. And the management of a scientific satellite programme required a much wider institutional framework than that provided by the UARRP membership. With the creation of NASA in 1958 and the emergence of the National Academy of Sciences' Space Science Board as the major source of scientific advice to the new Agency, the UARRP had definitely lost its momentum. It was in fact disbanded in 1960. As regards the satellite launchers, they were another part of the V2 legacy: the long-range ballistic missiles developed in the framework of the Cold War.

BALLISTIC MISSILES AND SATELLITE LAUNCHERS

Important progress in rocket technology had been made in the United States during the war years, which received a new impulse after the arrival of von Braun. In particular, an important school in applied aerodynamics had been established at the California Institute of Technology (Caltech) by Theodore von Kármán, a professor of aeronautics, who had studied in Germany with Ludwig Prandtl. In 1940, von Kármán and his student Frank Malina developed mathematical techniques that allowed significant progress in solid-fuel rocket technology. Out of this work, and thanks to military support, two important institutions were set up in California before the war's end: the Jet Propulsion Laboratory (JPL), a rocket research centre managed by Caltech, and Aerojet General, a company devoted to battlefield missile production.

In the immediate post-war period, the Army Air Force studied a project for developing an intercontinental ballistic missile, called MX-774, while the Navy launched a V2 rocket from the deck of an aircraft carrier in order to test the eventual use of rocket technology in naval operations. In the aftermath of the first Soviet atomic test (August 1949) and following the outbreak of the Korean war (June 1950), von Braun and his team started a project for developing a mobile missile called *Redstone*, the name of the arsenal near Huntsville, Alabama, where the group had been moved in 1950. The missile, capable of sending a small warhead over a maximum distance of 320 km, was launched for the first time in August 1953 from Cape Canaveral, but it took until 1958 before it entered into service with the American troops in Europe. *Redstone* was largely based on V2 technology and its cost-effectiveness as a nuclear weapon delivery system compared to fighter-bombers was questionable.

In fact, it was only in 1954 that American top policy-makers seriously committed resources to the development of a new generation of rockets which were powerful enough to carry atomic warheads over long distances, and eventually to launch payloads into Earth orbit. Discussions about the development of a ballistic missile capability were affected by three main factors. First was the competition between the various armed services: the Army, the Navy and the newly established autonomous Air Force each pursued its own pet project and claimed control over the missile development programme. Second was the debate about the usefulness of long-range ballistic missiles as weapon delivery systems compared to traditional bomber aircraft. A turning point in this respect was when it became possible to envisage the construction of lightweight nuclear weapons that could be easily fitted in missile nose-cones. Finally, there was the perception of the 'Soviet threat', in particular when intelligence reports indicated the Soviet Union was far ahead of the United States in missile development.

In August 1953, less than one year after the first American H-bomb test (November 1952), the Soviet Union successfully tested a lightweight hydrogen bomb. The US responded by exploding a more powerful device at the Bikini atoll in March 1954 (Bravo test). On May Day that year, at the usual military parade in Moscow, a brand new long-range jet bomber was shown, dubbed *Bison* by NATO. The US long-standing superiority in strategic air forces was directly challenged. All this suggested that intermediate-range and intercontinental ballistic missiles (IRBMs and ICBMs) should be developed soon as delivery systems of nuclear weapons.

In February 1954, a committee chaired by the brilliant mathematician John von Neumann expressed 'grave concern' about the United States' comparative disadvantage in rocket technologies and recommended accelerating ICBM development in order to reach an operational capability in six to eight years. By June 1954, after the Bikini test had demonstrated the feasibility of lightweight nuclear warheads, the US Air Force (USAF) was instructed to reorient and accelerate its *Atlas* ICBM programme 'to the maximum extent that technological development will permit'. One year later, following the report on 'surprise attack' prepared by a Technological Capabilities Panel chaired by MIT president James R. Killian, the USAF was authorized to build a second, more sophisticated ICBM, the *Titan*. Finally, in September 1955, US President Dwight Eisenhower gave the ICBM programme the highest national priority and approved development of an intermediate-range ballistic missile (IRBM) for the European theatre. *Atlas* and *Titan* were test-launched for the first time in 1957 and 1959, respectively, but it took a few years before they became fully operational. Moreover, both *Atlas* and *Titan* were liquid-fuel missiles, requiring a long time for fuelling and launching, and therefore not well suited to respond to a surprise attack.

Following Eisenhower's decision, several IRBM programmes were also initiated. The Army set von Braun and his team to work on developing a missile, called *Jupiter*, derived from the Redstone and capable of delivering a one-ton payload over about 2600 km. Test launched for the first time in March 1957, *Jupiter* became operational one year later and was eventually deployed in Turkey and Italy. An experimental version of this missile, called *Jupiter C*, was launched for the first time on 20 September 1956 from Cape Canaveral, reaching an unprecedented altitude of 1100 km before falling into the sea about 5500 km away from the range.

In direct competition with the Jupiter missile was Air Force's *Thor*, also launched for the first time in 1957. Finally, the Navy, after briefly collaborating on the Jupiter project with the Army, recognized that it needed a solid-fuel IRBM for its submarines, rather than a liquid-fuel missile like the Jupiter and Thor, and in 1957 it was authorized to develop the *Polaris*. That same year, the Pentagon approved Air Force plans for developing a solid–fuel intercontinental missile, lighter and cheaper than Atlas or Titan. The new missile should be capable of delivering a nuclear warhead at 10,000 km distance and should be launched within sixty seconds of an alert. Hence its name: the *Minuteman*.

While becoming obsolete as weapons, Atlas, Titan, Jupiter and Thor represented an efficient and diversified family of boosters for launching military and civil satellites. The foundations had thus been laid for American military power in the decades ahead and for its entry into space.

What of developments behind the Iron Curtain in this period? Contrary to the American doubts and delays, the policymakers in the Soviet Union decided that it was necessary for the country to develop intercontinental delivery systems almost immediately after the war, before they even possessed the atomic bomb. As anticipated above, in the immediate post-war period an important installation for research and development on missile technology, called NII-88, was established in the Moscow suburb of Kaliningrad, with Korolev in charge of long-range missile designs. A major engine development centre was established nearby in Khimki, under the direction of Glushko. And a test range was created at Kapustin Yar, 130 km east of Stalingrad. By the end of the 1940s, some 13 design bureaus and 35 factories were engaged in missile development.

By 1949 Korolev and Glushko were supplying the Red Army with the R-2, a modified version of the V2 having a range of some 600 km, twice its German predecessor. In 1952 they started designing the R-5, an IRBM with a range of 1200 km which flew for the first time in March 1953. At the same time they succeeded in convincing the Soviet government to support a major effort in ICBM development. The programme was approved in 1954, shortly after the successful H-bomb test, and the envisaged missile was designated R-7 or, more familiarly, *Semyorka* (No. 7). A new launch site was created for it in the desert east of the Aral sea, near the town of Tyuratan in Kazakhstan, later known as Baikonur.

Semyorka was designed to carry a 5-ton warhead for 8600 km, but it was also capable of putting a heavy satellite into low earth orbit. Taking into account this possibility, Korolev set to convince the Kremlin and the Academy of Sciences to use it for launching a scientific satellite in the framework of the International Geophysical Year.

By mid-1957, the huge rocket, with its characteristic cluster of four strap-on boosters, was ready for flight. After two unsuccesful attempts which ended in explosions soon after take-off, a third R-7 rocket flew successfully on 21 August to a range of 6000 km, thus becoming the first operational ICBM to fly. A second successful flight occurred a few weeks later, and Korolev was now authorized to modify the rocket in order to make it capable of putting a satellite into orbit as soon as possible. This would be *Sputnik 1*.

In short, by following a very different policy on the development of long-range ballistic missiles, large-rocket technology in the Soviet Union was considerably ahead of that in the USA in the 1950s. This 'technology lag' was one important reason why the Soviets were in space before the Americans and maintained a lead in the space race for almost a decade.

THE INTERNATIONAL GEOPHYSICAL YEAR

In addition to developing missiles for nuclear weapon delivery, the idea of launching spacecraft into orbit around the Earth was also suggested in the USA in the immediate post-war period. As early as May 1946, a report prepared by the Rand Corporation for the Army Air Force analysed the technical aspects of such an undertaking and underlined the great value that such artificial satellites could have for scientific research and national defence. The political and military implications of Earth satellites were again discussed at length in another Rand report in October 1950 which set the stage for the future American satellite programme.

Strategic reconnaissance was the main objective of satellites from the military point of view. In March 1955, while the ICBM programme was receiving its decisive impulse, the Air Force requested industry to submit proposals for a 'strategic reconnaissance satellite weapon system', designated WS-117L. One year later, it started project Pied Piper, foreseeing the development of both a satellite and an upper stage to fit atop the Atlas. This was eventually called *Agena*, and the Atlas–Agena combination played a major role in the future space programme, sending the first American probes to Mars and Venus.

Besides the technical difficulties deriving from the need of using more powerful boosters, reconnaissance satellites posed a severe political problem. A spacecraft overflying foreign territory and gathering photographic data risked being taken as an act of aggression. A vigorous Soviet protest, in particular, was to be expected, with an appeal to international law or even threats against neighbouring states housing American air force bases and tracking stations. Hence, before proceeding with a reconnaissance satellite programme, something like a 'freedom of space principle' had to be established in order to provide a legal right to overfly other countries without becoming snarled up in the complexities of Cold War politics. The International Geophysical Year (IGY), offered a solution to such a problem.

The IGY was originally proposed in 1950 as the Third International Polar Year, a follow up of two Polar Years which had taken place in 1882 and 1932. The first initiative was taken by a small group of British and American scientists, notably James van Allen, Sydney Chapman, a leading expert in atmospheric phyiscs from Oxford University, and Lloyd Berkner, head of the Brookhaven National Laboratory and a Polar Year veteran. The project was endorsed by the International Council of Scientific Unions (ICSU) and, following a wider definition of its scientific objectives in the general field of geophysics, it received the name of International Geophysical Year. It was planned to run for eighteen months, i.e. from 1 July 1957 to 31 December 1958, and one of its main scientific objectives was to gain basic information about upper atmosphere phenomena and cosmic rays during a period of maximum solar activity in 1957–58. In the event, it was an important cooperative scientific venture supported by 66 nations.

To organize the scientific activities of the IGY, the ICSU set up a Special Committee for the International Geophysical Year (CSAGI, from its French initials), with Chapman as president and Berkner as vice president. While the IGY was being organized, the IAF started advocating the launch of a scientific satellite within its framework. Proposed for the first time by the young American physicist Fred Singer at the 1953 IAF Congress in Zurich, the satellite was eventually called MOUSE (Minimum Orbital Unmanned Satellite of the Earth). It was designed to weigh some 45 kg and carry a valuable set of instruments.

The MOUSE concept was eventually endorsed by the International Scientific Radio Union and the International Union of Geodesy and Geophysics, whose representatives sat in the CSAGI. Singer also succeeded in convincing Berkner and the chairman of the US National IGY Committee, Joseph Kaplan, to support the idea. At a CSAGI meeting held in Rome in October 1954, the American delegates recommended that governments should try to launch artificial Earth satellites for scientific purposes during the IGY, in addition to the planned extensive programme of ground-based measurements and rocket observations in the upper atmosphere. The Committee endorsed the proposal and on 29 July 1955, the White House Press Secretary announced that President Eisenhower had agreed to the launch of 'small, Earth-circling satellites' as part of the USA's contribution to the IGY. A few days later, at the sixth IAF congress in Copenhagen, a delegation of Russian scientists, appearing at IAF for the first time, announced that the USSR were planning to do likewise. Leonid Sedov, head of the Soviet delegation and chairman of an Academy of Sciences' Commission on Interplanetary Communications, added that, from a technical point of view, it was possible to launch 'a satellite of larger dimensions than that reported in the newspapers'. As a matter of fact, the Soviet satellite programme was officially approved by the Kremlin in January 1956, shortly after Korolev and Glushko had successfully tested the motor of the R-7 rocket.

The two superpowers took very different policy decisions on the IGY satellite programme. The Soviet Union decided to use a military launcher for the scientific mission envisaged in the IGY framework, and started planning the use of the future R-7 ICBM to launch a heavy satellite. The Eisenhower administration, on the contrary, wanted to stress the scientific image of the venture, partly because they wanted to use their participation in the IGY to establish the freedom of space principle in the scientific domain before probing the Soviet's reaction to military reconnaissance satellites. They thus rejected outright the Air Force's proposal to use the future Atlas rocket to launch the satellite and decided to back a rocket with lesser military connotation.

The choice, in fact, was between von Braun's Project Orbiter and the Naval Research Laboratory's Project Vanguard. The former foresaw a small satellite provided with a simple radio transmitter, launched by an uprated Redstone missile with three upper stages made of *Sergeant* solid-fuel rockets. The NRL's proposal involved a more complex satellite, with better instrumentation and a radio tracking system dubbed *Minitrack*. The satellite would be launched by a three-stage vehicle: the first two stages were based on the existing Viking and Aerobee rockets, respectively; the third stage would be a large, newly designed solid-fuel rocket.

In the event, the Vanguard project was selected. Two major considerations weighed in favour of this choice. Firstly, NRL's project appeared more interesting from a scientific point of view (it had been endorsed by the American Rocket Society and the National IGY Committee) and involved the development of a rocket specially designed as a satellite launch vehicle. Secondly, it was felt that the use of a launcher derived from a military missile and built by Peneemünde veterans in an army arsenal would diminish the purely scientific character of the American IGY satellite project. It was a choice for which the Eisenhower administration was to pay heavily.

SPUTNIK, 'KAPUTNIK' AND EXPLORER

On 4 October 1957 Moscow radio announced that the Soviet Union had successfully launched *Sputnik 1*, the first artificial 'fellow-traveller' of the Earth, according to the meaning of the word 'sputnik', which Russian astronomers had long

Figure 17 Sputnik 1 (1957). (Source: Corbis.)

Figure 18 RD-214 rocket engine that powered the booster of Sputnik 1. (Source: Novosti.)

used to describe hypothetical small natural satellites of the Earth. It was an aluminium sphere of 58 cm diameter, weighing about 84 kg that circled the Earth once every 96.3 minutes on an elliptical orbit with a perigee of 228 km and an apogee of 947 km. Its two radio emitters sent its 'beep-beep' sound into homes all over the world for 92 days. The scientific instruments on board the satellite carried out the first measurements of upper atmosphere density and the first investigations into the transmission of electromagnetic waves through the ionosphere.

The reaction, at least in certain American circles, bordered on the hysterical. A wave of recriminations and self-criticism swept through the country, stimulated by the media. A myriad of explanations were put forward for what *Life* magazine called a 'defeat for the United States': inter-service rivalry between the various sections of the military; underfunding of research and development programmes; a philistine attitude towards 'egghead' scientists; an educational system that was not turning out enough scientists and engineers; and a President who was more interested in golf than in guiding the nation. Indeed the whole American way of life, with its *laisser-faire* approach and its consumerism, was called into question. Perhaps 'totalitarianism', with its ability to mobilize resources and to direct them to a single objective, had some advantages after all.

Not all shared this view, of course. The head of the Strategic Air Command, General Curtis LeMay, for one, disparaged Sputnik as 'just a hunk of iron'. The President also played down its importance, at least in public. Sputnik 'does not raise my apprehensions, not one iota', he claimed in the aftermath of the Soviet announcement. The United States' satellite programme, Eisenhower argued, was intended to reap maximum scientific benefits within the framework of the IGY. For this, an American satellite was to be launched in December, and a fully instrumented satellite was planned for launching in March 1958 (hopefully on board an Atlas rocket). The United States, he told scientists a week later, was not intent on competing with any other nation for first place in a Sputnik race: 'The serving of science, not a high score in an outer space basket ball game, has been and still is our country's goal'.

He was soon forced to revise his public stance. On the morning of 3 November *Sputnik 2* was successfully launched. This satellite, dedicated to the fortieth anniversary of the October Revolution, was over six times heavier than its forerunner (it weighed more than 500 kg), but since the satellite failed to separate from the upper stage of the booster, something like 4 tonnes had gone into orbit. What is more, Sputnik 2 carried a life-support system and the first living being into outer space, the dog Laika, who was wired up for medical and biological studies. Laika was not due to be returned to Earth, as space technology was not developed enough to achieve her safe re-entry into the atmosphere. She lived much less than foreseen, however, as the thermal control system became inoperative and the capsule rapidly overheated.

Against this striking achievement, the US Vanguard project, well known to the public, was due to launch a ten-kilogram satellite and was not yet ready to fly. For all the downplaying of top military and political leaders, the metaphor of the country having suffered something like a new 'Pearl Harbour' became commonplace, with *Life* magazine publishing on 18 November an editorial entitled 'Arguing the case for being panicky'. Space had become a key domestic issue fuelled by the tensions of the Cold War.

The launch of *Sputnik 2* led Eisenhower to increase the pressure on the Vanguard team, who were instructed to bill the first scheduled test flight of their rocket as a full-blown attempt to orbit a satellite. He also instructed von Braun to prepare a modified Jupiter-C rocket to eventually launch a small satellite prepared by the JPL into orbit as a backup to Vanguard. Indeed, in an effort to tranquilize his fellow countrymen and women, a few days after *Sputnik 2* first orbited the Earth the President showed the recovered nose-cone of a Jupiter missile during a radio-TV address, arguing that the nation had achieved missile capability and

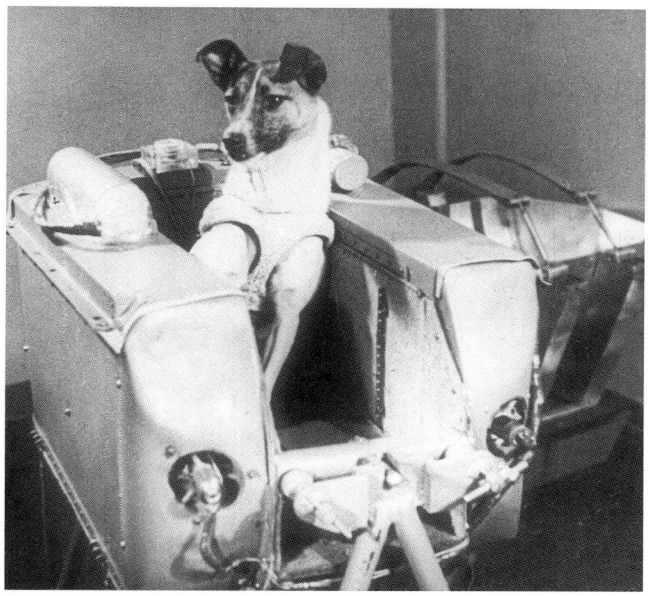

Figure 19 Laika. (Source: Novosti.)

would soon have an American satellite in orbit. He also announced the establishment of a President's Science Advisory Committee (PSAC) and the appointment of James Killian as Special Assistant to the President for Science and Technology.

The competition proved too strong for Vanguard. In early December 1957, reporters from all around the world gathered at Cape Canaveral to witness America's reply to the Soviets. The launch had been prepared in a haste. The first stage had only been tested once a few weeks earlier and the second stage had never flown at all. The satellite had been reduced to a small sphere weighing 1.5 kg with a diameter of less than 15 cm: a grapefruit, it was dubbed. After two days of suspense, the countdown finally reached zero just before noon on 6 December. Vanguard rose four feet off its pad, and slumped back to earth in a ball of thunder and flame.

The programme was tarnished with the image of total failure, and the press was unrelenting. Vanguard was *Kaputnik* and *Flopnik* in London newspapers. 'It seems there is a worm in the grapefruit', wrote the *Paris-Journal*, and the *Ottawa Journal* compared the rocket's 'loss of thrust' to the loss of thrust that 'the Western democracies have been suffering from'. *Time* magazine put Soviet leader Nikita Khrushchev on its cover as Man of the Year, and the editors justified their choice by contrasting the Vanguard failure with the Sputnik successes. Soviet delegates at the United Nations reputedly asked their American counterpart if the US were interested in receiving foreign aid under Moscow's programme of technical assistance to underdeveloped countries.

The Navy desperately prepared for a second launch, but too late. At the end of January 1958, the Army's Jupiter-C carrying the JPL satellite was prepared for launch under conditions of great secrecy at Cape Canaveral. The rocket, rebaptized *Juno 1* to underline its function as a satellite launcher, blasted off successfully on 31 January, and placed its satellite into orbit. *Explorer 1*, as it was called, was essentially the last stage of the Juno rocket. It was cylindrical in shape, a little over 2 m long and 15 cm in diameter. It weighed 14 kg – one sixth the weight of *Sputnik 1* – and was placed in an elliptical orbit with 356 km perigee and 2548 km apogee.

The satellite carried two detectors of micrometeorites and a Geiger counter for studying cosmic-ray particles designed by van Allen. The latter at first seemed to malfunction. In certain regions of the orbit, in fact, the instrument ceased to detect any radiation. A careful analysis by van Allen and his co-workers revealed not a failure but the presence of a very high radiation intensity that saturated the counter. This was confirmed by *Explorer 3*, launched on 26 March after the second in the series had failed three weeks earlier. *Explorer 3* carried the first tape recorder ever launched on a satellite and provided further data on this radiation belt trapped in the Earth's magnetic field. At a conference in the summer of that year, the name 'van Allen radiation belt' was used for the first time.

Meanwhile, after a second launch failure on 5 February, the Navy succeeded in putting the Vanguard satellite into orbit on 17 March. The aluminium 'grapefruit' carried a small set of instruments which revealed the pear-shapedness of the Earth. On 15 May, the Soviet Union launched the 1.3-ton *Sputnik 3*, a spacecraft carrying a large load (968 kg) of scientific instruments to measure micrometeorites, the density of the upper atmosphere, cosmic rays, solar radiation, the possible presence of high-energy particles, and the Earth's own radiation environment. *Sputnik 3*, in particular, could have mapped the van Allen radiation belts systematically before such a mapping was achieved by *Explorer 4*, scheduled for launch by the end of July. Unfortunately, to much disappointment of physicists Roald Sagdeev and Sergei Vernov, the tape recorder onboard the satellite failed to work. The scientific information gathered was thus limited by the area of direct radio visibility and the scientific team could not be sure that the extremely high levels of radiation they were detecting were of planetary significance rather than local.

The year 1958, the first full year of the space age, registered a few other Vanguard failures: the successful flight of *Explorer 4* (launched on 26 July), followed by the failure of *Explorer 5* one month later; three failed attempts by the Soviet Union to launch a lunar probe; and three equally failed attempts of the American *Pioneer* spacecraft to reach the Moon. Finally, on 18 December, an Atlas rocket put

Figure 20 Sputnik 3 (1958). (Source: Novosti.)

Figure 21 Yuri Gagarin with Sergei Korolev, 1961. (Source: Novosti.)

into orbit the 100-kg SCORE (Signal Communication by Orbiting Relay Equipment) satellite, the first experimental communication satellite. It carried a tape recorder with a Christmas message from Eisenhower which was relayed to Earth: 'Through the marvels of scientific advance, my voice is coming to you from a satellite circling in outer space. My message is a simple one. Through this unique means, I convey to you and to all mankind America's wish for peace on Earth and good will towards men everywhere'. The space race between the two Superpowers had definitely started.

THE CREATION OF NASA

In parallel with these developments, the Eisenhower administration began to think about the appropriate institutional framework for America's space programme. The debate over how to do this was intense and vociferous, dominated by the question of the relationship between the civilian and military aspects of space. As all existing space programmes were run separately by the military (the Navy's Vanguard, the Army's Juno-Explorer, and the Air Force's Atlas launching vehicle), the Administration decided in January 1958 to create the Advanced Research Projects Agency (ARPA) within the Department of Defence. Its aim was to run US space programmes on an interim basis under the authority of the Secretary of Defense. Subsequently, a growing consensus emerged that, apart from reconnaissance satellites, the major goals of space flight in the near term were scientific and political. A civilian space agency would therefore best serve American interests by building the image of an open, peaceful programme in contrast to Soviet secrecy. This opinion was supported in particular by the newly established President's Science Advisory Committee. The PSAC also recommended that an existing agency, the National Advisory Committee on Aeronautics (NACA), be expanded and upgraded to take over all aspects of the civilian American space programme.

The NACA had been set up by the Federal Government during World War I to supervise the study of the scientific and technical problems of aeronautics. It had a modest budget until 1940, which by 1945 had increased tenfold to about $40 million. One of the strengths of NACA was that it had some of the best in-house research facilities in the world and maintained close links with an important segment of the scientific community through a university research programme. And while much of its attention was directed to solving aeronautical problems, by the mid-1950s it also did a good deal of advanced research and development in support of missile projects. A few NACA projects eventually led to the development of solid-fuel sounding rockets for upper atmosphere research, such as the successful family derived from the *Nike* anti-aircraft missile. One of the family, the *Nike-Deacon*, was extensively used by van Allen

Figure 22 Vostok 1, 12 April 1961. (Source: Novosti.)

in his 'Rockoon' programme, where a small rocket was launched from a balloon floating in the stratosphere.

On 5 March 1958, Eisenhower approved the recommendation that the leadership of the civil space effort be lodged in a strengthened NACA, and one month later the National Aeronautics and Space Act was submitted to Congress. This was signed into law on 29 July and eventually the National Aeronautics and Space Administration (NASA) came officially into being on 1 October 1958, almost a year to the day after *Sputnik 1* was launched. It inherited NACA's vast organization and facilities and a workforce of some 8000 people. And it expanded rapidly, taking over all

Figure 23 Yuri Gagarin in the cabin of Vostok 1. (Source: Novosti.)

space activities currently under way except those of strict military interest, mainly concentrated in the USAF programmes. In December 1958, the JPL came under NASA's control. In May 1959, the key personnel in the Navy's Project Vanguard were transferred to a new NASA facility at Greenbelt, Maryland, later named the Goddard Space Flight Center in honour of the American rocket pioneer. Finally, NASA acquired von Braun's army team at Huntsville. The centre at Hunstville was renamed the George C. Marshall Space Flight Centre and von Braun was appointed its first director. His specific task was to develop the heavy launchers for the man-in-space programme. By the end of 1960 NASA's staff had doubled to 16,000 and its annual expenditure was over $500 million, three times that of NACA in 1958.

At the beginning of the new decade, both superpowers were firmly engaged in their next step into space, i.e. sending human beings into space. The Soviet Union had their *Vostok* programme, whose experimental capsules had been launched into orbit one after the other since May 1960 with one or two dogs on board. The United States had their *Mercury* programme, with the first manned flight scheduled for 1961. The Russians went first. On 12 April 1961, the Russian cosmonaut Yuri Gagarin onboard Vostok-1 became the first human being to orbit the Earth. Later that year, on 6 August, Guerman Titov performed 17 revolutions around the Earth, remaining more than 24 hours in orbit. The first American into space was Alan Shepard, who went for a suborbital flight onboard the Mercury capsule *Freedom-7* on 5 May 1961, but it was only on 20 February 1962 that John Glenn became the first US astronaut to orbit the Earth.

By that time, the next step had already been planned. On 25 May 1961, in an extraordinary State of the Union message, the newly elected US president John Kennedy announced his committment 'to achieving the goal, before this decade is out, of landing a man on the Moon and returning him safely to the Earth'. The goal was indeed achieved: on 21 July 1969, Neil Armstrong and Buzz Aldrin made Kepler's old dream become a reality.

Figure 24 President John F. Kennedy's speech to Congress, 25 May 1961. (Source: NASA.)

THE ORIGIN OF SPACE RESEARCH IN EUROPE

For both the United States and the Soviet Union space was an important field of political and military confrontation. Both superpowers were prepared to spend huge amounts of money, intellectual resources and industrial capabilities for the pursuit of their respective goals. Nothing comparable to the American and Soviet efforts was of course possible in Europe. However, some of the major European countries had considerable potential for entering space and did develop limited national space programmes in the 1950s which we will rapidly review in this section.

The United Kingdom was the leading country. The British worked on the development of solid-fuel guided rockets during the war, and in October 1945 launched three captured V2s with the help of German specialists in the framework of Operation Backfire. They also obtained the

Figure 25 Edwin E. Aldrin, Jr. walking on the Moon. Photo taken by Neil A. Armstrong after the landing of Apollo 11 in July 1969. (Source: NASA.)

services of one of von Braun's main collaborators, Walter Riedel, who eventually settled at the Rocket Propulsion Establishment in Wescott. In 1946 a rocket research programme was established at the Royal Aircraft Establishment (RAE) near Farnborough, and an agreement was signed between the British and Australian governments to establish a joint launching range at Woomera in South Australia. In the decade that followed, a successful solid-fuel sounding rocket was developed at RAE, initially called CTV5 Series 3 but eventually dubbed *Skylark*, capable of taking a payload of about 50 kg to an altitude of 160 km. In 1955, the United Kingdom also undertook an IRBM programme, aimed at developing the so-called *Blue Streak* missile as a component of the British nuclear deterrent.

The Skylark was first tested at Woomera on 17 February 1957, but as early as 1953 the Ministry of Supply had offered it to the Royal Society for developing an upper atmosphere research programme. The British scientific community was ready to respond positively, taking advantage of a long tradition of ionospheric research, initiated by Edward V. Appleton in the 1920s. In fact, the idea of undertaking a rocket research programme was already in the air. During the war, following a request of the Air Ministry, the Royal Society had asked its Gassiot Committee, responsible for atmospheric studies, to outline a research programme in meteorology and atmospheric physics using rocket-propelled vehicles. In 1950, a leading member of the Gassiot Committee and renowned expert in ionospheric physics and geomagnetism, Sydney Chapman, was invited to attend a meeting of the V2 Panel in the US. He eventually proposed that the Gassiot Committee organize an international conference on rocket exploration of the upper atmosphere. The conference took place in August 1953 in Oxford, with the American V2 Panel being well represented by 15 delegates, including its chairman van Allen. The proceedings of the conference were published, representing an updated review of knowledge in the field (Boyd and Seaton 1954).

Following the Ministry of Supply's offer a Rocket Sub-Committee was established within the Gassiot Committee, under the chairmanship of Harrie Massey. Its task was to advise on the upper atmosphere research programme and to supervise its implementation. The experimental activity started as soon as Skylark was operational and the rocket was widely used during the International Geophysical Year. Since 1957 more than 350 Skylarks have been launched, the rocket undergoing several improvements during this period. Later vehicles, such as the three-stage Skylark 12, were capable of lifting payloads of 350 kg to an altitude of almost 1000 km.

British scientists were also well prepared for seizing the new opportunities offered by satellites. In December 1958, a British National Committee for Space Research (BNCSR) was established. Under the chairmanship of Massey, it

Figure 26 British–NASA Ariel satellite. (Source: Genesis Photo Library.)

included representatives from a wide variety of government departments and scientific societies. When, in March 1959, NASA offered to launch scientific equipment for scientists from other countries, Massey and his colleagues reacted immediately and a cooperative programme was soon agreed on. It foresaw the launch of three satellites with UK instruments on board at roughly yearly intervals. The first in the series, the 60-kilogram *Ariel 1*, was launched from Cape Canaveral atop a *Thor-Delta* rocket on 26 April 1962. It carried out seven experiments built by British scientists to investigate the van Allen radiation belts, the solar radiation and the cosmic rays. A second Ariel satellite, weighing 68 kg and carrying three British experiments, was launched on 27 March 1964 by a *Scout* rocket from Wallops Island, on the east coast of the USA. The third of the series was built in the UK and launched on 5 May 1967 by a Scout rocket from the Vandenberg base, California. It weighed about 90 kg and carried five experiments. Three more Ariel satellites were launched by the UK in 1971, 1974 and 1979, respectively, the first devoted to radio astronomy, the two others to X-ray astronomy.

Second to Britain came France. In fact, a small group of about 40 German V2 experts settled in France in 1946–47 where they formed the nucleus of the first French rocket research effort. In 1949 the French government set up the

Laboratoire de Recherches Balistiques et Aérodynamiques in Vernon, on the Seine north-west of Paris, with the aim of developing ballistic missiles for military use. Though its budget was initially rather small, they did develop an important liquid-fuel sounding rocket, called *Véronique*, modelled on the V2. The first operational flight of *Véronique* occurred in 1954 from the French military base at Hammaguir, in the Algerian desert. In 1957, the Comité d'Action Scientifique de la Défénse Nationale (CASDN) decided to fund the construction of 15 improved *Véronique* rockets to carry out high-altitude atmospheric research in the framework of the IGY. The rocket was subsequently widely used for upper atmosphere research until the 1970s.

Despite these achievements, the French effort tended to limp along until 1958, when it benefited from the happy conjuncture of the launch of *Sputniks* and the arrival of General Charles de Gaulle in power. De Gaulle's determination to develop an independent nuclear capability gave an enormous boost to missile development in the framework of the government's strong support for scientific and technological research. His initial wide-ranging programme included IRBMs, submarine-launched missiles and reconnaissance satellites.

In 1959 the French government set up the SEREB (Société pour l'Etude et la Réalisation d'Engins Balistiques) whose aim it was to develop the in-house knowledge and technology required for the military missile programme. In the same year, it also established a Comité de Recherches Spatiales, chaired by the physicist Pierre Auger, to coordinate scientific research in space. Finally, by a law voted on 19 December 1961, the CNES (Centre National d'Etudes Spatiales) came into being. Its prime task was to develop a satellite launcher, called *Diamant*, based on the military ballistic missiles that France was developing for its strategic deterrent.

The fruits of these investments were soon to be seen. On 26 November 1965, a Diamant rocket rose from the Hammaguir launch pad and placed into orbit the first French satellite, significantly called *Astérix*. France thus became the third space power and confirmed its claim for an independent role in this important strategic field.

Astérix was a 42 kg test satellite whose role was to confirm the rocket's ability to place it into orbit. Ten days later, on 6 December 1965, an American Scout rocket launched France's first scientific satellite, FR-1. This satellite, weighing 60 kg, had been developed by CNES and carried instruments for studying ionization irregularities in the ionosphere and the magnetosphere. It was launched from Vandenberg and placed into a near-circular orbit about 750 km high.

In 1966 and 1967 France launched three small scientific satellites developed by CNES. All of them were launched by Diamant rockets from Hammaguir, and their scientific mission was to make geodetic experiments based on the study of the Doppler effect. At the same time, France

Figure 27 French Astérix satellite. (Source: Genesis Photo Library.)

undertook to establish a new national launching range in Kourou, in French Guyana. The Kourou range became operational on 9 April 1968, with the launch of a Véronique rocket. Two years later, on 10 March 1970, a Diamant B launcher put into orbit the first satellite, a scientific satellite called Dial/Wika, built in collaboration by France and Germany for studying the radiation belts around the Earth.

Italy also had an important national space programme, initially developed at the Turin Polytechnic with military support. In the late 1950s, the Italian Air Force contracted with the newly established SISPRE (Società Italiana per lo Studio della Propulsione a Reazione) the development of a solid-fuel sounding rocket, eventually launched for the first time in 1960. In 1959, the Consiglio Nazionale delle Ricerche (CNR) agreed to support an upper atmosphere research programme under the direction of air force colonel Luigi Broglio, a professor of aeronautical sciences at the University of Rome. A launching range was established at Salto di Quirra, near the small town of Perdasdefogu in Sardinia.

Broglio was quick to react to NASA's offer of collaboration in space research. In 1962, Italy and the USA signed an agreement for the so-called 'San Marco' project, and two years later, on 15 December 1964, the first Italian satellite, *San Marco-1*, was launched by a Scout rocket from

Figure 28 Italian San Marco satellite. (Source: Genesis Photo Library.)

Wallops Island. It was a sphere with a diameter of 66 cm and weighed no less than 115 kg. Built by the University of Rome's Centro di Ricerche Aerospaziali under the direction of Broglio, this was the first all-European satellite to circle the Earth. A second San Marco satellite was launched by Broglio's group in 1967 from a platform anchored in the Indian Ocean off the coast of Kenya, and the programme continued in the following decade.

Finally, we come to Germany, the country which had been in the vanguard of rocketry in Europe in the inter-war period. It was just the history of rocketry in this country that impeded Germany's entry into space. During the Nazi regime, the V2 missile developed by von Braun and his team had been launched on various European cities in the last phase of the war. After the defeat of Germany, the allies imposed strict restrictions on the development of rocket technology in Germany, while the sinister heritage of Peenemünde gave space flight a bad image in West German public opinion. Therefore, although throughout the 1950s internal pressure had been growing for a revival of a national space effort, mainly inspired by scientists and technicians from the Peenemünde project, a number of political obstacles and international constraints had to be overcome before (the Federal Republic of) Germany could take its place alongside the other major European nations in the space effort. Any German involvement in space could only be realized in the framework of an international undertaking, and of course without any military implication.

Germany's re-entry into the field of space occurred along two parallel paths. First was the prompt support to the initiatives undertaken in the early 1960s to establish a joint European space effort (see later in this chapter). A number of scientific groups started developing scientific instruments to be launched within the framework of the forthcoming European sounding rocket and satellite programmes, while industrial effort in rocket technology was also resumed to build the third stage of the envisaged Europa satellite launcher. Secondly, the German government undertook a programme to develop a national scientific satellite, in cooperation with the United States. This was AZUR, a 71 kg satellite for studying the magnetosphere and Earth–Sun relationship which was launched on 8 November 1969 from Vandeberg by a Scout rocket. Five years later, Germany launched the important Helios mission, aimed at studying the Sun from an orbit near the centre of the solar system. The Helios spacecraft was launched on 10 December 1974 from Kennedy Space Centre by the new powerful American rocket Titan IIIE-Centaur.

Besides the national space effort promoted by the governments of the major European countries, a number of European scientists and scientific institutions were also involved in the space programmes of the two superpowers. French scientists were involved in collaborative programmes with their Soviet counterparts within the framework of the USSR space programme; physicists from the University of Graz, Austria, participated in the programmes of Intercosmos, the international space organization of East European countries; and Swiss (University of Bern) experiments were on Apollo 11 and later Apollo missions. The contribution of European scientists (British, German, Swiss, French) to the geochemical and geochronological work on lunar samples returned by the Apollo missions was of high quality.

These national efforts in Europe could not compete in any way with the aggressive space programmes developed in the United States and the Soviet Union. Cold War confrontation, however, was not the only aspect of space. For European scientists, the prospect of launching instruments high in the atmosphere and beyond meant opening new interesting fields of scientific investigation. Moreover, important applications in civilian fields such as telecommunications and meteorology seemed possible in the future. Finally, space appeared to many European industrialists as a key sector for technological development and economic success. In September 1961, European industry created a supranational body called Eurospace, which included all the leading companies in aircraft and missile manufacture. Its aim, according to its statutes, was 'to promote the development of aerospace activities in Western Europe'.

It is against this background that a few initiatives were undertaken in the late 1950s and early 1960s for pooling European resources in a collaborative effort in space. Eventually, these initiatives led to the creation in 1964 of two international space organizations, the European Space Research Organisation (ESRO) and the European Launcher Development Organisations (ELDO).

ESRO was the direct product of an initiative from within the European scientific community. The first steps were taken in 1959 by Edoardo Amaldi and Piere Auger, two physicists who had been among the founding fathers of CERN, the European laboratory for elementary particle physics set up near Geneva in 1954. They hoped to re-create in the space field the positive experience of European scientific collaboration represented by CERN. Following some discussions between them, a number of meetings were organized with fellow scientists from other European countries (notably Harrie Massey, Hendrik van de Hulst, Reimar Lüst and Bengt Hultqvist), and with government representatives. A Preparatory Commission was officially set up in December 1960, with the task of preparing the programme and the legal arrangements of the envisaged organization.

The ESRO Convention was eventually signed in 1962 and entered into force on 20 March 1964, with ten countries joining the new Organisation (Belgium, Denmark, France, Germany, Italy, Netherlands, Spain, Sweden, Switzerland and the United Kingdom). Later that year, ESRO's first sounding rocket campaign started at Salto di Quirra. In 1968, ESRO launched its first three satellites, devoted respectively to the study of the polar ionosphere, to solar astronomy, and to the study of cosmic rays and interplanetary plasma. Three others followed in 1972, among which a large stabilized satellite, called TD-1, carrying various experiments in UV astronomy, cosmic rays and high energy (X- and gamma-ray) astronomy.

ELDO came out from government negotiations prompted by a British initiative. In April 1960, in fact, the British government decided to abandon its Blue Streak IRBM programme. Blue Streak, in fact, was a liquid-fuel, non-mobile missile, which became obsolete before the project was even completed. Rather than cancel the programme outright, the government took the idea of recycling it as the first stage of a civilian satellite launcher. Diplomatic action was undertaken to encourage other European countries to join the UK in a cooperative venture. In the event, this led in 1964 to the establishment of ELDO, including six European countries (Belgium, France, Germany, Italy, Netherlands and the UK) and Australia. Its task was to develop an all-European heavy satellite launcher, called *Europa*, based on Blue Streak as the first stage, a second stage to be developed in France, and a German third stage. Italy was responsible for the development of a test satellite, while Belgium and the Netherlands would provide the guidance station and the telemetry links, respectively. Finally, Australia would make its Woomera base available for the test flights of the rocket.

Both ESRO and ELDO had great difficulty achieving their initial objectives in the following years. Part of the problem was financial: it was soon clear that the money originally set aside by governments for the joint space effort (a small fraction of that being spent by the USA) would be completely inadequate for the agreed programmes. Part of

Figure 29 The ESRO 1 satellite. (Source: ESA.)

Figure 30 TD-1. (Source: ESA.)

Figure 31 ELDO's Europa rocket. (Source: ESA.)

the problem was technical: there were doubts whether the *Europa* rocket was not already obsolete, its technical content of little interest to industry. But above all, there were controversies among member state governments about the general policy to be followed by Europe in space.

Two major issues were hotly debated. Firstly, there was the question of whether Europe should develop more powerful rockets for launching its satellites in order to achieve complete autonomy in the space field, or should rely on the support of the United States. Secondly, with the growing importance of application satellites, notably in the communication field, governments and scientists discussed whether and how Europe should embark in this field, eventually by changing ESRO's charter. The conflicts were aggravated by ELDO's failure to achieve its goal. The Europa rocket, in fact, suffered from an impressive series of launch failures, failing to put a single satellite into orbit. After many discussions, the project was eventually abandoned and ELDO disbanded in 1973.

The resolution of these conflicts was achieved in 1973 with a 'package deal' which saved the ideal of a joint European undertaking in all fields of space (science, applications and launchers), while respecting the different priorities of the participating partners. Out of this compromise, the European Space Agency (ESA) was created in 1975, essentially based on ESRO's technical and managerial structure. In August that year, the new agency boasted the successful launch of its first spacecraft, a scientific satellite called COS-B carrying on board a sophisticated instrument for investigating celestial gamma rays. Fifteen countries are members of ESA today, with Austria, Finland, Ireland, Norway and Portugal added to the former ESRO member states.

Besides inheriting ESRO's successful scientific satellite programme, ESA undertook two applications satellite programmes: an all-European telecommunication satellite programme (OTS, MARECS and ECS satellites), and an aeronautical satellite programme in collaboration with the US which, however, was eventually abandoned. Moreover, ESA undertook the building of a space laboratory to be carried on board the American Space Shuttle (Spacelab) and, mostly important, the development of a heavy satellite launcher called Ariane to be launched from Kourou. Ariane's first flight occurred on Christmas Eve of 1979, finally establishing Europe's full autonomy in space.

In the ESA framework, all programmes except the Science Programme are optional, and each member state is allowed to participate – at the level it deems appropriate – in those programmes in which it has a political and industrial interest.

The Science Programme, on the contrary, is mandatory, and all member states contribute to it in proportion to their gross national product. Science in fact was the main original motivation for European nations to join forces in a space effort, and the European space science community played an important role in keeping alive such a cooperative undertaking against the centrifugal pull of diverging national interests. The intrinsic merit and successful development of ESRO's and ESA's Science Programme indeed represented the strongest element of stability and confidence for Europe in space.

THE ORIGIN OF THE JAPANESE SPACE PROGRAMME

After the defeat of Japan in World War II, the Allied Powers banned Japanese armaments completely. Aircraft-related research was prohibited by the Potsdam Declaration and, for a few years, aeronautical engineers were forced to work on more innocent academic fields. Eventually, they returned to aircraft research after 1952, when the San Francisco Peace Accord came into effect and jet planes were marking the beginning of a new era in aeronautics. Among them was Hideo Itokawa, considered the father of Japanese rocketry.

Born in 1912, Itokawa had graduated in aeronautical engineering at the University of Tokyo. For a few years he worked in the design office of the Nakajima Aircraft Company (today's Nissan Motor Co.), returning to the University of Tokyo just before the end of the war. In 1953 Itokawa spent six months in the United States, where he was fascinated by the prospects of space flight. On his return to Japan, he established a research group on rocket development and undertook to convince industrial leaders to support their activity. Out of their effort came the *Pencil* rocket, a tiny, solid-fuel test rocket, 1.8 cm in diameter and 30 cm in length, launched for the first time in August 1955.

The work of Itokawa and his group caught the attention of Susumi Okano, responsible for coordinating the Japanese contribution to the International Geophysical Year on behalf of the Ministry of Education. Itokawa's rocket group was thus involved in the Japanese IGY research programme, developing the successful K-6 rocket, which could reach 60 km in altitude with the use of solid propellant. This was followed by further improvement of the rocket performance through the development of the K-8 type and others, which opened the way to the routine use of sounding rockets for space research in the 1960s. In 1962, the Kagoshima Space Centre was established in southern Japan and qualified as a launch-site for sounding rockets and scientific satellites.

Meanwhile, following recommendations by the Science Council of Japan, an Institute of Space and Aeronautical

Figure 32 Launch of ESA's COS-B satellite. (Source: ESA.)

Science was established in 1964 at the University of Tokyo to further promote space science in Japan. This was the predecessor of the present Institute of Space and Astronautical Science (ISAS), an inter-university organization created in 1981 under the control of the Ministry of Education and specifically devoted to space research. A second organization was established in 1969 to take charge of application programmes: the National Space Development Agency (NASDA). While ISAS kept responsibility for the Kagoshima centre, a new launch base for geostationary satellites was established nearby in Tanegashima, managed by NASDA.

In 1966, Japanese policy-makers approved a programme aimed at developing the M4-S rocket, capable of launching scientific satellites. The first Japanese satellite, a test satellite called Ohsumi, was launched on 11 February 1970 by a prototype launcher. On 28 September 1971, the first scientific satellite designed by ISAS was successfully launched from Kagoshima by an M-4S-3 rocket. Called Shinsei, it was devoted to the study of solar radio emissions, cosmic rays and the ionosphere. Since then, steady progress has been made in the improvement of the M series rockets, and Japan has been able to put into orbit a fairly large number of scientific satellites.

FURTHER READING

A classical, well-illustrated textbook in the history of rocketry and astronautics has been written by von Braun and Ordway (1975). A more recent version is von Braun, Ordway and Dooling (1985). Ordway and Liebermann (1992) give a good introduction to the history of the space era, with fine illustrations. A concise and comprehensive history of rocketry was published by Winter (1990). Much useful information and illustrations on past and present space activities are in Atlas (1989). Specifically addressed to the general public is a recently published book on the history of the space era: Crouch (1999). On the public image of space in America see McCurdy (1997).

Useful biographies of some of the main pioneers of the space era have been written by Walters (1962), Lehman (1988), Bergaust (1976), Kosmodemyansky (1985), Stuhlinger and Ordway (1994), Rauschenbakh (1994) and Harford (1997).

Winter (1983) presented a thorough history of the rocket societies in the 1920s, and the development of rocketry in Nazi Germany was analysed in great detail by Neufeld (1995). A dramatic insider's account of slave labour for V2 production was given Béon (1985). On von Braun's rocket team, the V2 programme and its post-war extension in the US, see Ordway and Sharpe (1979). The V2 story is also discussed by Kennedy (1983). The American operation to secure German expertise in rocketry in the immediate post-war period is discussed in Lasby (1971) and Gimbel (1986, 1990). For the United Kingdom see Sharpe (1989) and Becklake (1998); for France see Villain (1997).

A thorough discussion of the early history of space research in the USA is given by DeVorkin (1992). Sullivan (1961) deals with the International Geophysical Year. For the history of American space science in the NASA period see Newell (1980). The history of space science in the United Kingdom is told by Massey and Robins (1986).

Contemporary reviews of space science before and after Sputnik 1 were written by Newell (1953), Boyd and Seaton (1954), van Allen (1956) and Berkner and Odishaw (1961). A historical review is that of Hanle and Chamberlain (1981).

The political and military aspects of the space race between the two superpowers are analysed with different approaches by McDougall (1985), Levine (1994), Reeves (1994) and Heppenheimer (1997). The impact of Sputnik on US policy is specifically discussed by Killian (1977), Bulkeley (1991) and Divine (1993). Kennedy's decision to launch the Apollo Moon programme is discussed by Logsdon (1970).

On the history of NACA and NASA see Anderson (1981). A concise history of the US space programme, with a fine anthology of original documents, has been written by Launius (1994), and the history of the Jet Propulsion Laboratory is told by Koppes (1992). The most important documents in the history of the US civil space programme are presented in a series of volumes edited by Logsdon (1995, 1996, 1998, 1999, 2001). For the Soviet space programme see Shelton (1968), Oberg (1981), Harvey (1988) and Siddiqi (2000). The early history of Europe in space is told in detail by Krige and Russo (2000). See also Madders (1997) and Bonnet and Manno (1994), the latter also addresses global collaboration in the space sciences. Useful information on the history of the Japanese space programme is available at the ISAS Web Site *http://www.isas.ac.jp/e/* (English version).

REFERENCES

Anderson, F.W. (1981). *Orders of Magnitude: A History of NACA and NASA, 1915–1980*. Washington: NASA SP-4403.
Atlas (1989). *Le Grand Atlas de l'Espace*. Encyclopedia Universalis France.
Becklake, J. (1998). German engineers: Their contribution to British rocket technology after World War II. In: P. Jung, (ed.), *History of Rocketry and Astronautics*, AAS History Series, vol. 22, San Diego: Univelt.
Béon, Y. (1985). *La Planète Dora*. Paris: Editions du Seuil.
Bergaust, E. (1976). *Wernher von Braun*. Washington: National Space Institute.

Berkner, L.W. and Odishaw, H. (eds) (1961). *Science in Space*. New York: McGraw-Hill.

Bonnet, R. and Manno, V. (1994). *International Cooperation in Space: The Example of the European Space Agency*. Cambridge: Harvard University Press.

Boyd, R.L.F. and Seaton, M.J. (eds) (1954). *Rocket Exploration of the Upper Atmosphere*. Oxford: Pergamon Press; New York: Interscience.

Bulkeley, R. (1991). *The Sputniks Crisis and Early United States Space Policy*. London: MacMillan.

Crouch, T.D. (1999). *Aiming for the Stars: The Dreamers and Doers of the Space Age*. Washington & London: Smithsonian Institution Press.

DeVorkin, D.H. (1992). *Science with a Vengeance*. New York: Springer-Verlag.

Divine, R.A. (1993). *The Sputnik Challenge: Eisenhower's Response to the Soviet Satellite*. New York & Oxford: Oxford University Press.

Gimbel, J. (1986). US policy and German scientists: The early cold war. *Political Science Quarterly*, **101**, 433–451.

Gimbel, J. (1990). German scientists, United States denazification policy, and the '*Paperclip* conspiracy'. *International History Review*, **12**, 441–485.

Hanle, P.A. and Chamberlain, V.D. (eds) (1981). *Space Science Comes of Age*. Washington: Smithsonian Institution Press.

Harford, J. (1997). *Korolev*. New York: John Wiley & Sons.

Harvey, B. (1988). *Race into Space: The Soviet Space Programme*. Chichester: Ellis Horwood.

Heppenheimer, T.A. (1997). *Countdown: A History of Space Flight*. New York: John Wiley & Sons.

Kennedy, G.P. (1983). *Vengeance Weapon 2: The V-2 Guided Missile*. Washington: Smithsonian Institution Press.

Killian, J.R. (1977). *Sputnik, Scientists and Eisenhower*. Cambridge: MIT Press.

Koppes, C.R. (1982). *JPL and the American Space Programme*. New Haven & London: Yale University Press.

Kosmodemyansky, A. (1985). *Konstantin Tsiolkovsky*. Moscow: Nauka.

Krige, J. and Russo, A. (2000). *A History of the European Space Agency, 1958–1987*. Vol. 1: *The Story of ESRO and ELDO from 1958–1973*. Noordwijk: European Space Agency, SP-1235.

Lasby, C.G. (1971). *Project Paperclip: German Scientists and the Cold War*. New York: Atheneum.

Launius, R.D. (1994). *NASA: A History of the US Civil Space Programme*. Malabar: Krieger.

Lehman, M. (1988). *Robert H. Goddard: Pioneer of Space Research* (Reprint of *This High Man*, 1963). New York: Da Capo.

Levine, A.J. (1994). *The Missile and the Space Race*. Westport: Praeger.

Logsdon, J.M. (1970). *The Decision to Go to the Moon: Project Apollo and the National Interest*. Cambridge: MIT Press.

Logsdon, J.M. (1995, 1996, 1998, 1999, 2001). *Exploring the Unknown: Selected Documents in the History of the U.S. Civil Space Program. Volume I, Organizing for Exploration*; *Volume II, External Relationships*; *Volume III, Using Space*; *Volume IV, Accessing Space*; *Volume V, Exploring Cosmos*. Washington: NASA SP-4407.

Madders, K. (1997). *A New Force at a New Frontier*. Cambridge: Cambridge University Press.

Massey, H. and Robins, M.O. (1986). *History of British Space Science*. Cambridge: Cambridge University Press.

McCurdy, H.E. (1997). *Space and the American Imagination*. Washington & London: Smithsonian Institution Press.

Mc Dougall, W.A. (1985). *...the Heavens and the Earth: A Political History of the Space Age*. New York: Basic Books.

Neufeld, M.J. (1995). *The Rocket and the Reich: Peenemünde and the Coming of the Ballistic Missile Era*. New York: the Free Press.

Newell, H.E. (1953). *High Altitude Rocket Research*. New York: Academic Press.

Newell, H.E. (1980). *Beyond the Atmosphere: Early Years of Space Science*. Washington: NASA SP-4211.

Oberg, J.E. (1981). *Red Star in Orbit*. New York: Random House.

Ordway, F.I. and Liebermann, R. (eds) (1992). *Blueprint for Space: Science Fiction to Science Fact*. Washington & London: Smithsonian Institution Press.

Ordway, F.I. and Sharpe, M.R. (1979). *The Rocket Team*. New York: Thomas Y. Crowell.

Rauschenbakh, B.V. (1994). *Herman Oberth, the Father of Space Flight*. Clarence: West-Art.

Reeves, R. (1994). *The Superpowers Space Race*. New York: Plenum.

Sharpe, M.R. (1989). Operation Backfire: England launches the V2. In: F.I. Ordway (ed), *History of Rocketry and Astronautics*. AAS History Series, Vol. 9, San Diego: Univelt.

Shelton, W. (1968). *Soviet Space Exploration: The First Decade*. New York: Washington Square Press.

Siddiqi, A.A. (1998). *Challenge to Apollo: the Soviet Union and the space race, 1945–1974*. Washington: NASA SP-2000-4408.

Stuhlinger, E. and Ordway, F.I. (1993). *Wernher von Braun: Crusader for Space*. Melbourne: Krieger.

Sullivan, W. (1961). *Assault on the Unknown: The International Geophysical Year*. New York: McGraw-Hill.

van Allen, J.A. (ed) (1956). *Scientific Uses of Earth Satellites*. Ann Arbor: University of Michigan Press.

Villain, J. (1997). France and the Penemünde legacy. In: P. Jung (ed.), *History of Rocketry and Astronautics*. AAS History Series, Vol. 21, San Diego: Univelt.

von Braun, W. and Ordway, F.I. (1975). *History of Rocketry and Space Travel* (3rd edition). New York: Thomas Y. Crowell.

von Braun, W, Ordway., F.I. and Dooling, D. (1985). *Space Travel: A History*. New York: Harper & Row.

Walters, H.B. (1962). *Herman Oberth: Father of Space Travel*. New York: Macmillan.

Winter, F.H. (1983). *Prelude to the Space Age. The Rocket Societies: 1924–1940*. Washington: Smithsonian Institution Press.

Winter, F.H. (1990). *Rockets into Space*. Cambridge & London: Harvard University Press.

3

ERNST STUHLINGER*

Enabling technology for space transportation

INTRODUCTION

Although rockets have been in use for a thousand years, it was only 60 years ago when concrete plans were made to transport a package of scientific instruments to high altitudes with a rocket. In 1942, Professor Erich Regener in Stuttgart, Germany, a specialist in atmospheric research, ozone physics, ultraviolet spectroscopy, and cosmic rays, and Wernher von Braun, director of the Peenemünde Rocket Research and Development Center, jointly began to design a large scientific payload for the empty warhead of an A-4 test rocket to be launched into a near-vertical trajectory from the "Greifswalder Oie," a small island near Peenemünde. Regener and von Braun had worked together before on a project to try out radio communication between the ground and a transponder carried aloft with a high-altitude balloon. Instruments for the capsule were built and installed in a large aluminum container, called a "Regener Tonne" (Regener barrel). It contained pressure and density gauges, a UV spectrograph with quartz optics to determine the ozone content of the atmosphere, thermal sensors, containers for atmospheric samples to be taken during the parachute descent of the capsule at various altitudes, instruments to measure ion densities in the atmosphere, drum recorders, radio receivers and transmitters, and a parachute that unfolded under the stiffening force of pressurized tubes. While the quartz spectrograph, built at the Regener Institute, had the primary objective of determining the ozone content of the high atmosphere by measuring the intensity of solar UV radiation, it also promised to furnish the first solar spectrum deep into the ultraviolet, an objective in which solar astronomers, among them K.O. Kiepenheuer, were greatly interested. The capsule was ready for launch late in 1944, but by that time the Nazi government did not permit any activities not directly supportive of the war effort. Then, in January 1945, von Braun received orders to leave Peenemünde and move south; that terminated the Regener Tonne project.

Only one year later, Peenemünde rockets, shipped from Germany to White Sands in New Mexico, were launched with scientific payloads instead of warheads, prepared by scientists from numerous universities and research institutes all over the country. A very active period of high altitude research began which lasted for seven years. Besides its immediate success as a source of scientific results and discoveries, this project proved that rockets represent marvelous tools for scientific research in far-away regions that would not be accessible in any other way.

Stimulated by the successful research projects with the old Peenemünde rockets, several new rockets were developed during the 1940s, among them Van Allen's Aerobee and Aerobee High, and Milton Rosen's Viking, which continued a successful program of scientific space research. They were members of a very small number of rockets developed exclusively for scientific purposes. A large number of rocket projects evolved from about 1950 on in the United States, in Soviet Russia, and in other countries, but most of them were developed and built for military purposes. Some of them, such as the European Ariane and the American Delta II, and also the Space Shuttle, served exclusively, or at least primarily, commercial and other utilitarian uses; the big Saturn-Apollo rocket project arose under the pressure of political circumstances.

On the other hand, all the rockets and rocket technologies developed and built during the past fifty years for various

---

* NASA – Marshall Space Flight Center, Huntsville, AL, USA

purposes have also found applications in purely scientific endeavors. All those spacecraft which were built during the past five decades specifically for astronomical and astrophysical research, for lunar and planetary studies, and for solar and deep space investigations owed their launchings to rockets that had been built primarily for nonscientific purposes. This intimate relationship between technologies created for practical purposes, and the need of scientists who are driven primarily by their quest for knowledge, has been a characteristic feature of rocket and space activities during the past sixty years.

In this essay on enabling technologies for space transportation, the author tries to convey an impression of the multifaceted challenges, problems, failures, and successes the rocket people have encountered on their long way from the dreams of the 1930s to the satellites and the lunar, planetary and space probes that have come to life during the twentieth century, enriching our awareness and our knowledge of the Universe to an extent unforeseen when the century began.

1. EARLY THOUGHTS ABOUT SPACE TRAVEL

Transportation through space, without a firm substrate or a surrounding atmosphere, must rely on principles of propulsion different from those which allow people, and animals, to move from one place to another in our immediate earthly environment. Several ways to produce a propulsive force for travel through space beyond the atmosphere can be envisioned, such as gravitational forces generated by celestial bodies; light pressure; and the rocket principle.

Rockets did not originate from the desire to travel into and through space. History tells us that about one thousand years ago rockets appeared as weapons of war, first in Asia, then in Europe, and later on the American continent. Today, military rockets play major roles in the armaments of all countries that must be prepared for occasional involvement in military conflicts.

Independent of that situation, there have been humans, probably for thousands of years, whose minds dreamed about voyages to the Moon, to planets, and even to stars. Toward the end of the nineteenth century, some of these dreamers began to study whether it would really be possible, on the basis of well-established laws of physics, to travel to the Moon, and perhaps even to one of the nearest planets, with spacecraft built on Earth and propelled by a suitable rocket motor. Today, as the twenty-first century dawns, we realize that those early studies have led to a vast field of activities involving rockets. They gave the twentieth century its most characteristic feature: Mankind's outreach into the unfathomable expanses of space that surround the planet Earth.

This chapter will not present a comprehensive history of rocketry and space projects. Rather, it will highlight those specific events which gave origin to the fabulous evolution of space sciences by providing the technical means to send instruments, and even humans, away from Earth with its many disturbing influences, enabling them to make observations that would never be possible from the surface of the Earth. Also, some thoughts will be offered about the likely development of space flight in the foreseeable future, as seen by those who were actively involved in the development of rockets and spacecraft during past decades.

The first decisive step from dreams to realistic studies was made by a Russian schoolteacher, Konstantin Eduardovitch Tsiolkovskii (1857–1935). Based on Newton's laws of mechanics, he derived in 1895 a simple, but very elegant equation that allows one to calculate the terminal velocity, V_{term}, a rocket will reach in the vacuum of empty space, far away from any noticeable gravity field, when its total initial mass, M_{in}, its mass at burning cutoff, M_{c-o}, and the velocity of its exhaust gases, V_{ex}, are known. He found:

$$V_{term} = V_{ex} \ln [M_{in}/M_{c-o}]$$

Tsiolkovskii's work, remaining largely unknown for many years, did present the proof that rocket behavior is amenable to exacting mathematical analysis.

Around the end of the nineteenth century, a young American physics teacher, Robert Hutchings Goddard (1882–1945), engaged in an extremely active pursuit of rocket and space flight ideas. In 1902, he wrote an essay *The Navigation of Space*, and offered it to *Popular Science*, but his paper was not accepted for publication. Convinced that the key to space travel is a suitable propulsion system, and that space propulsion should be based on rocket principles, he devoted all his further work primarily to the study and development of rocket systems. In 1906, he wrote about "electrified jets" for planetary voyages, but from about 1909 on he concentrated his efforts predominantly on the theoretical and practical development of liquid-propellant rockets.

Beginning in 1914, Goddard applied for and received patents – 214 during the following thirty years – for an astounding variety of subjects: rockets with solid and liquid propellants, multistage rockets (a principle that was mentioned for the first time by Konrad Haas in Romania in 1529!), combustion chambers, nozzles, cooling systems, valves, tanks, propellant feed systems, gyroscopic guidance and control systems, air and jet vanes, and recovery by parachute. Like Tsiolkovskii ten years before him, and independently Hermann Oberth ten years after him, he found that liquid oxygen and liquid hydrogen would make the best propellant combination for high-altitude rockets.

Goddard (1919) described some of his ideas in a paper *A Method of Reaching Extreme Altitudes* which was published by the Smithsonian Institution in 1919. This essay, one of very few Goddard papers published during his lifetime, led to some support from the Navy, and to the permission from Clark University at Worcester, Massachusetts, where he was a professor of physics, to carry out his experiments under the auspices of the university. Goddard's paper, very unfortunately, also evoked some negative comments by the press which ridiculed his contention that humans may one day be able to travel to the Moon. This criticism hurt him deeply. He discontinued publishing his articles, and he put his patents, reports, and diaries in safes where they remained unseen until after his death in 1945. He refused to cooperate, and even to communicate with other rocket scholars, limiting his entire "team" to five persons, including his wife Esther.

Homer E. Newell, NASA Associate Administrator and Chief Scientist during the 1950s, wrote in 1980 (Newell 1980): "By his secrecy, Goddard not only dissipated most of his influence he might have had, he also deprived himself of the engineering expertise that he sorely needed to achieve his dream. The future of rocketry belonged to the team approach. It was, indeed, inextricably tied to the massive funding sources and particular purposes of the national security state... His reluctance to work openly with others deprived Goddard... of the kind of funding support from the military..." It deprived him also of the recognition of peers he would have deserved so abundantly. On March 16, 1926, a Goddard rocket, fueled with kerosene and liquid oxygen, rose to an altitude of 13 meters and covered a distance of 55 meters. It was history's first liquid-propellant rocket. Eleven years and numerous rockets later, Goddard and his co-workers launched a rocket in Roswell, New Mexico, which reached an altitude of 2700 meters (Goddard 1948).

It was several years after Goddard's death that his many prolific notes, diary entries, manuscripts, and patents became available to the public (Goddard *et al.* 1968). There is hardly any area of rocket technology that he did not discuss, and for which he did not propose designs and ideas for further studies. By the time, during the late 1940s and early 1950s, when details of his theoretical and experimental work became known, others had independently arrived at similar ideas, and had even developed them into modern, powerful precision rockets that transformed Goddard's dreams into reality.

Hermann Oberth (1894–1989) learned of Goddard's existence and work in 1922 through a brief note in a newspaper. He wrote Goddard a letter, asking for "a copy of your books." Goddard sent Oberth a copy of his Smithsonian paper. In return, Oberth wrote an appendix to his own book *Die Rakete zu den Planetenräumen* (The Rocket to Planetary Spaces) which was in press at that time (Oberth 1923), with the expression of great admiration and esteem.

For his book, Oberth had independently derived many of the basic formulae of rocketry, including the famous Tsiolkovskii equation. In spite of its small size – 92 pages – Oberth's book is an impressively rich compendium of facts and properties of rockets, most of them presented for the fist time, and also of ideas referring to the use of rockets. Oberth described two types of rockets in great technical detail, the "alcohol rocket," and the "hydrogen rocket." He mentioned the possibility of space stations that orbit the Earth indefinitely "like a little moon" without requiring any sustaining propulsion, providing opportunities to make large-scale weather observations, to communicate through light signals with inaccessible regions on Earth, to investigate the ultraviolet light from stars, to make experiments and study processes under weightlessness, and to carry out observations of strategic interest. "Telescopes of any size could be used in space, since the images of stars do not twinkle...; visiting other celestial bodies would certainly be of utmost scientific value..." And he added very cautiously: "Under certain economic conditions, construction of such machines" – space rockets and orbiting stations – "can be profitable. Such conditions may arise in some decades." How right he was!

2. SYSTEMATIC ROCKET TECHNOLOGY WORK IN GERMANY, 1923–45

Oberth's book of 1923 became the most important and influential book about rocketry ever. Many of the later rocket pioneers quoted it as their initiation into their own rocket career. At the time when it was published, young Wernher von Braun, eleven years old, lived in Berlin. As his mother later wrote, "He involved himself in numerous projects with his friends... they built all kinds of rockets, and they collected pieces of old automobiles from junkyards and built 'new' cars, with and without rocket propulsion..." (Stuhlinger *et al.* 1994). At 14, he wrote an essay *Journey to the Moon: Its Astronomical and Technical Aspects*, published in the *Journal for the German Youth*. One year later, the teenager wrote a letter to Oberth: "I know you believe in the future of rockets. So do I. Hence, I take the liberty of sending you a brief paper on rockets that I wrote recently." This was the beginning of a very close and warm friendship between the two men, based on great and genuine mutual admiration. It lasted until von Braun's death, fifty years later.

Shortly after he had decided to become a space pioneer in 1925, von Braun made another decision, typical of his clarity of thought. If we want to travel to the Moon and to Mars, he said, the most important requirement is a vehicle that can provide proper transportation over the vast distances in space, and that also offers accommodation first for instruments to make scientific observations, and later for human travelers who will live and work on their spacecraft during the voyage, and then on the surface of their celestial target. I will devote my efforts first of all to the development of a powerful precision rocket that is capable of providing an adequate and safe means of transportation through space.

Years later, recalling his early decision, he mentioned to some friends: "Look at the great events of discovery and exploration in our history. Each of them happened when proper means of transportation had become available, good ships for Columbus, horses for the early American pioneers, chuck wagons when the American West was opened for European immigrants, railroads when people began to settle there in numbers, automobiles and airplanes when the continent became a homestead for millions. So, have rocket, will travel through space!"

Von Braun's serious and systematic rocket development work began in 1930 when, as a student at the Technical University of Berlin, he assisted Professor Oberth in his efforts to develop a proper combustion chamber and exhaust nozzle for a gasoline-liquid oxygen rocket motor. This was von Braun's first encounter with a technical enterprise based on careful planning, and on exacting theoretical studies, an approach that was to become a trademark of von Braun's own work in all of his ensuing projects.

Around the same time, von Braun joined a small group of young rocket enthusiasts in Berlin. They called themselves "Verein für Raumschiffahrt" (Society for Space Travel), and they built and tested their liquid-fueled rockets at their "Raketenflugplatz" (Rocket Proving Ground) in nearby Reinickendorf. Although it was a very small team with almost no financial means, they made noticeable progress with their rocket engines. But Rolf Engel, one of the members who later became a prominent rocket expert, remembered: "One day, Wernher told me: 'Look, Rolf, we want to push this thing. But we have no money. The only way how we can get the money, the assistance, and all the means, is the Army'." (Reisig 1997).

It so happened that during 1931–32, the German Army (Deutsche Reichswehr) also had a small rocket development program underway. By some mixture of fate and circumstamces (details are described by Reisig 1997), the Army group made contact with the Society for Space Travel and offered to von Braun a Civil Service contract to develop a rocket that could be used as a weapon replacing heavy artillery that at that time was forbidden to Germany. In 1932, von Braun and most of his few co-workers settled in a modest rocket testing facility in Kummersdorf near Berlin and developed a rocket with alcohol and liquid oxygen as propellants. Their work was successful; in December 1934, they launched Max and Moritz, rocket twins of the A-2 type. Each of them reached a planned altitude of 2500 meters. In 1935, the Army, in a joint enterprise with the Air Force, decided to build a larger rocket development facility at Peenemünde on the Baltic island of Usedom. Two years later, von Braun and his co-workers moved in and began work on a rocket for which the Army had specified a range of about 250 kilometers and a payload of about 1 ton. Von Braun, 25 years old, became the Technical Director of the Peenemünde Rocket Development Station; Colonel (later General) Walter Dornberger was named Commanding Officer. Under the Army's auspices and protection, work at Peenemünde proceeded quickly and successfully, relatively undisturbed by the Nazi government and the Party.

In retrospect, the Peenemünde rocket project stands out as the first step in the realization of von Braun's youthful dream, dreamed ten years earlier, to build a powerful precision rocket whose future descendants one day would enable human voyagers to travel into and through space. The basic features of that first rocket had taken shape in his mind before Peenemünde began: a rocket motor burning alcohol and liquid oxygen, the propellants being fed by centrifugal pumps; the walls of the combustion chamber and nozzle to be cooled by fuel; the attitude of the rocket to be sensed by gyroscopes and controlled by a combination of air and jet vanes; the shape of the rocket to be determined by careful wind tunnel experiments; and its velocity to be measured by a set of accelerometers and integrators.

Only rudimentary knowledge and experience existed in all these areas when work on the Peenemünde rocket started. From the beginning, von Braun organized his fast-growing work force in such a way that intense development work could be conducted simultaneously in all areas. Systems were continuously refined, tested, and further refined. Actually, the Peenemünde rocket grew and approached its definitive form in five distinct steps that had already started in Kummersdorf. Each of the steps utilized the experience gained in its simpler and smaller forerunners; the models were the A-1, A-2, A-3, A-5, and finally the A-4 which was built with the size and capability specified by the Army. Gradually, the A-4 rocket took shape, and after much testing of the instruments in the laboratories, and of the rocket engine with its jet vanes on test stands, flight testing of the A-4 began in the summer of 1942. After the first three launch attempts had failed, a test flight on October 3, 1942 was a full success.

Some improvements were planned and started in Peenemünde, but they were completed only years later when von Braun and part of his team continued their rocket development work in the United States. Among these systems were integral propellant tanks instead of the tank-within-a-shell design, an improvement that became possible after better aluminum alloys and welding techniques had become available; air bearings for the gyroscopes instead of mechanical bearings; a composite gyro-stabilized guidance platform instead of individually mounted gyros; and a unified injector plate with annular spray nozzles instead of individual injector heads. The last two improvements had proceeded to the point where laboratory samples were tested in ground tests, but the war had ended before flight tests could be made.

On a sideline, some of the A-4s readied for flight testing were equipped with wings (A-4b), so that they could glide over some distance at the end of their ballistic trajectories, "... to find out whether the range of the rockets could be extended," as the official explanation read. To some of his close co-workers, von Braun explained that one day, rockets will have to come back from orbital flights, with human passengers. After reentering the atmosphere, he said, they will glide on wings and land like a glider plane. We should find out whether this technique works. Our A-4 provides an excellent opportunity to do that..."

Theoretical considerations showed that if a winged A-4 were put on a powerful booster stage, the A-4 could cover a range of about 4000 kilometers, "just about the distance between Europe and America," one of the team members said. From that time on, the winged A-4b was sometimes called "that America rocket." After the War, an essay writer picked up the story, claiming that Peenemünde was building a rocket that would have been able to bombard and destroy the United States of America. What an absurd idea! Even if a large and powerful booster stage had been available, the winged A-4 would have arrived on a low-level, slow gliding path, carrying an explosive warhead weighing about 25 kg, certainly no match for the American continent! But, barely fifteen years later, Soviet Russia as well as the United States had built their own Intercontinental Ballistic Missiles with nuclear warheads, capable of crossing the Atlantic or the Pacific. The ICBMs have no wings, though, but the Shuttle has.

At the time when test flights with A-4 rockets began at Peenemünde, an opportunity arose to use an A-4 for a first, although modest scientific enterprise. As described briefly in the Introduction, Erich Regener, professor of physics at the technical university of Stuttgart, and von Braun began to make plans for a science payload, called the Regener Tonne, to be carried to an altitude of 60–80 km by an A-4 on a test flight over the Baltic Sea. The fact that the list of investigations included measurements of the pressure, density, and temperature of the atmosphere at high altitudes – data useful also for the analysis of the trajectories of A-4 missiles – made it possible for these studies to survive under the ever-watchful eyes of government officials who wanted to make sure that the Peenemünde project served only military purposes.

By the time the Regener Tonne, filled with scientific instruments and a parachute, was ready for launch in November, 1944, Reichsführer SS Heinrich Himmler had wrenched authority over Peenemünde's test launchings out of von Braun's and Dornberger's hands. That spelled an end to what would have been the first modern rocket launching with a distinctly scientific purpose. Details of the Regener project are described by Reisig (1997).

During the first years of the war, Hitler showed very little interest, and no confidence, in rockets, and the Peenemünde project was able to subsist (in a kind of "splendid isolation") because General von Brauchitsch held his protective hands over it. This changed when Germany lost the air war against England. In 1943, Hitler ordered Himmler to mass-produce the Peeneemünde rocket, and to deploy it against France and England. Dornberger and von Braun, arguing that the A-4, still technically immature, was neither reliable enough nor sufficiently accurate to be an effective weapon, tried to dissuade Hitler and Himmler from producing and deploying their rocket, asking for more time to perfect the technical systems. However, their efforts were in vain. Himmler enforced production in the Mittelwerk underground factory where about one-half of

Figure 1 A-4 (V-2) rocket being erected from its transport vehicle (Meiller Wagen) for a flight test at Peenemünde, 1943. (Source: Corbis.)

the workers were inmates from nearby concentration camps. Several thousand A-4 rockets, named V-2 by Goebbels, were produced at the Mittelwerk between fall 1943 and spring 1945 under absolutely cruel conditions for the inmates, a tragedy that will remain a dark phase in Germany's history for all time.

That high production rate was possible only because design simplifications, changes in materials, reduction of accuracy requirements, and deletion of test procedures were accepted. From the standpoint of rocket technology, the Nazi's frantic effort to mass-produce the A-4 at Mittelwerk made no contribution to the development of the modern precision rocket. Von Braun's systematic work on rocket development in Germany covered a period of 15 years, from 1930 till 1945, while Himmler's crazy orders to mass-produce the A-4 in the Mittelwerk lasted a mere 18 months.

Forty years after the last Peenemünder had left the little island where the modern rocket was born, essays and books began to appear by young writers who claim to portray the "true" history of Peenemünde. They did not know von Braun, and they never worked in a team on the development of a novel technical system. They make no attempt to differentiate between reliable and unreliable documents. They are unaware of the difference between the systematic, careful development and testing of a complex modern rocket system by a large team of engineers and scientists, and – in contrast – the hasty mass production of an immature rocket by slave labor, enforced by a ruthless, totally irresponsible government. These authors are also unfamiliar with the conditions of life and work (and death!) under an absolute dictatorship during times of war. Relying on questionable "sources" that were written for and archived by members of the Nazi government, and declaring as "untruthful" the reports and memoirs written by authors who were active members of the Peenemünde team, they present a "von Braun" who never was, and a "Peenemünde" that never existed.

3. ROCKET DEVELOPMENTS IN OTHER COUNTRIES

Tsiolkovskii's groundbreaking theoretical work toward the end of the nineteenth century, although largely unpublished and therefore unknown, still spawned a surge of rocket activity in Russia (Von Braun *et al.* 1966, Büdeler *et al.* 1979, Harford 1997). In 1924, a Central Bureau for the Study of Rockets was established, and an All-Union Society for the Study of Interplanetary Flight produced an exhibition on rocket technology in 1927. Fridrikh Arturovitch Tsander worked on rockets burning gasoline and liquid oxygen; he started testing them around 1930, achieving thrust levels of up to about 40 kg. During the same period, Valentin Petrovitch Glushko began to experiment with rockets in Leningrad. Both of them wrote books on rocketry, and more and more young engineers were attracted to the art and science of rockets. Nikolai Alexsevitch Rynin, Jakov Isidorovitch Perelman, and others built and launched numerous rockets between 1929 and 1937 with thrust forces between 5 and 600 kg, with liquid and solid propellants; they reached altitudes up to about five kilometers. During the five years from 1928 till 1932, Rynin published a nine-volume work on "Space Flight." The name of one person appears in the records of those times who remained unknown in the West until 30 years later when, after his death, he entered the limelight of recognition and fame as one of the all-time greats in the development of rockets and space flight: Sergei Pavlovitch Korolev (1906–1966) (Harford 1997).

During the war, the Russian military deployed the Katyusha solid propellant rocket, developed by Korolev, in great quantities from salvo launchers; they were called "Stalin organ" by the German soldiers. This author found himself repeatedly at the receiving end of Katyushas in 1942. Plans for a guided missile did not come to fruition in Russia during the war. At war's end, the Russians transported several thousand German engineers, scientists, and technicians to the Soviet Union, and also large amounts of machines and parts of V-2 rockets from Peenemünde, and from the Mittelwerk where mass production of V-2s had been underway. They first produced advanced versions of the V-2 missile, the R-1, R-2, and R-3, and then developed a very successful line of powerful modern military rockets that also found abundant use as satellite and deep-space launchers, among them the famous R-7, Semiorka and Proton rockets.

France also had its own early rocket pioneer: Robert Esnault-Pelterie (1881–1957). He was a theoretician of the art and science of rocketry. Beginning around 1907, he lectured about rockets and space flight, but he had to be very cautious to avoid ridicule by a population that was not yet inclined toward space exploration and travel to the Moon, in spite of Jules Verne! In 1930, Esnault-Pelterie published his famous book *L'Astronautique*. Two years before, he had proposed to the French military the development of "ballistic bombardment missiles," but it took six years before he received a contract to study various designs for potential military rockets. However, no definite project resulted before World War II broke out. Much later, during the 1970s, 1980s and 1990s France began to make very substantial contributions to space transportation within the European space Agency's Ariane Program.

In Great Britain, there was no noticeable rocket development activity during the first decades of the twentieth century, mainly because the Explosives Act of 1875 did not allow rocket testing and launching; besides, the British government did not like rockets anyway in spite of the fact that Sir William Congreve, and also William Hale, during the early and mid 1800s, developed and built rockets that were used in large quantities in various wars during the first half of the nineteenth century. However, the idea of traveling into and through space with rocket-driven spacecraft was wide alive, at least with some Britons. In 1933, young enthusiasts under the leadership of P.E. Cleator founded the British Interplanetary Society; A.V. Cleaver, L.R. Shepherd, K.W. Gatland and L. Carter were early members. From 1934 on, the Society published the *Journal of the British Interplanetary Society*. The JBIS is still one of the most respected space journals today. Cleator's book *Rockets through Space* appeared in 1936.

In 1934, Alwyn Douglas Crow at the Woolwich Royal Arsenal suggested that the British should begin the development of rocket weapons. Two years later, development work on small solid propellant rockets for military uses was begun, and flight testing started first in England, later in Jamaica, but this effort remained on a modest level. An energetic program for the development of modern rockets for military and civilian uses, among them the Blue Streak and the Black Knight, began to develop in England only after the end of the war.

Italian work in rocketry began in 1927 with the development of small powder rockets under the direction of General G.A. Grocco; however, the program was terminated in 1929 after most of the rocket work had switched to liquid-propellant rockets. In spite of reasonable progress, this program, too, was discontinued in 1935. Italian activities in modern rocket and space projects have become quite substantial during recent decades.

Rocket development in the United States had an unusual and unexpected beginning. Robert H. Goddard, one of the three "founding fathers" of the age of rocketry and space flight, did not believe in cooperation between the pioneers, nor in an exchange of results, experience, and plans, nor in open discussions or public presentations. Nor was he inclined to join the American Interplanetary Society which had been founded in March, 1930 by young rocket and space enthusiasts. David Lusser, G. Edward Pendray, Laurence Manning, and half a dozen other young men met regularly, published their Bulletin, and began experimenting. In 1931, Pendray traveled to Germany and met the young rocketeers at Reinickendorf, including Rudolf Nebel and von Braun, a visit that must have been very enjoyable and stimulating for both parties. Upon his return, the activities of the Society, now called the American Rocket Society, intensified, and its membership grew, but progress was slow under the constant pressure of a scant budget and continuing technical problems. Other rocket groups began building and testing rockets, among them H.W. Bull, J.H. Wyld, Robert C. Truax, Ernst Loebell, Leslie Skinner, Willy Ley, Clarence Hickman, and Alfred Africano. Details of their projects, hopes, setbacks, and successes are described by von Braun *et al.* (1966). Although none of these efforts resulted directly in rockets that found use in military or scientific applications, they greatly helped develop and advance our knowledge of rocket engineering and science.

At the Guggenheim Aeronautical Laboratory of the California Institute of Technology (GALCIT), under the directorship of Theodore von Kármán, a group of young men had started a reaction-engine development program in 1936, among them Weld Arnold, Frank Malina, Edward Forman, John Parsons, Hsue-shen Tsien, A.M.O. Smith, and Martin Summerfield. They first built jet-assisted takeoff (JATO) systems for airplanes, and then switched to rockets. In 1943, Th. von Kármán received a classified report about a V-2 rocket that had been built in Peenemünde. His laboratory responded with a proposal to the military that the laboratory, henceforth called Jet Propulsion Laboratory (JPL), should also build a military rocket, although – as von Kármán emphasized – pure scientific objectives should also be kept in mind. JPL started in 1944 to develop first the Private, as a test object, and then the Corporal as a tactical missile with a range of 65–210 km, and with nitric acid and aniline as propellants. Before the Corporal was ready for deployment, Malina asked for permission to develop and build a smaller version of the Corporal, called Wac Corporal, as a high-altitude

Figure 2 Liftoff of a Bumper WAC Corporal two-stage rocket at White Sands, New Mexico, February 24, 1949. (Source: Marshall Space Fligh Center.)

research rocket with a thrust of 680 kg, a payload of 11 kg, and a top altitude of 70 km. This rocket had its first launch in White Sands in September, 1945. It was used successfully for a number of scientific flights between September 1945 and July 1950. Its crowning flight was in combination with a Peenemünde rocket which served as first stage of a two-stage rocket, Project Bumper (Büdeler et al. 1979), an idea that had been brought up by Martin Summerfield and Frank Malina.

Many of the technical problems that plagued the early rocket makers in various countries are forgotten today. Delayed ignition of the propellants caused explosions of the rocket engines; humidity in the air led to a freezing of valves; relays closed or opened prematurely under the heavy vibrations of the rocket motors, disabling the guidance and control systems. It was only through painstaking error analysis, systematic improvements of the subsystems, and tireless testing that these problems were solved, one by one. The early history of Russian rocket technology shows many parallels with the early rocket work in Germany and in the United States, and in other countries as well. Slowly, technologies were improved until they reached a state of efficiency and reliability that made the brilliant rocket and space flight accomplishments possible that had been achieved by the end of the twentieth century.

4. EARLY SPACE SCIENCE PROJECTS IN WHITE SANDS, NEW MEXICO

When the war ended, von Braun and a number of his Peenemünde co-workers were invited by the US Government to come to the United States and to continue their rocket development work in this country. Between September, 1945 and the summer of 1946, 118 of them traveled to El Paso, Texas, where the group was housed in vacant facilities of the Beaumont Army Hospital in Fort Bliss at the outskirts of El Paso. The Armed Forces shipped parts of about 100 Peenemünde V-2 rockets, found in the infamous Mittelwerk underground factory in occupied Germany, first to Antwerp by train, then to New Orleans by ship, and from there by train and flatbed truck to the White Sands Proving Ground in New Mexico, about 130 km north of El Paso (Stuhlinger et al. 1994). The Peenemünde rocket men, assisted by members of the General Electric Company and by enlisted men, assembled 70 complete V-2 rockets. During the seven years from 1946 till 1952, all 70 were taken to the launch pad. Three of them failed, but 67 took off correctly, and 47 reached their expected altitudes between 100 and 213 km. This success rate was considered satisfactory in view of the fact that all of these rockets were several years old, that they had been disassembled and their parts loaded, shipped, unloaded, and then stored openly in the windswept White Sands desert for lack of storage facilities – all by unskilled labor. Finally, they were assembled and tested with very limited tooling, and without the usual cleanroom facilities.

Figure 3 German A-4 (V-2) rocket at White Sands Proving Ground, late 1940s. (Source: Corbis.)

The primary purpose of these V-2 launchings was "to train the Armed Forces, and to give industry an opportunity to learn about rocket technology." The space in the nose-cone originally occupied by the warhead was filled with sand, but there were scientists who realized that those rockets offered a marvelous opportunity of transporting scientific instruments into the highest layers of the atmosphere, and even beyond. These flights would enable projects of scientific research of a kind that had never been possible before.

When Colonel Holger N. Toftoy masterminded and organized the transfer of the German rocketeers to the United States, his primary motive was the transfer of technological knowledge and experience. However, he gladly supported the desire of scientists to put scientific instruments into the rocket nose cones. Gerard Kuiper, professor of astronomy in Tucson, Arizona, was probably the first to express the idea as early as 1944 when he interrogated German war prisoners in France. Arthur C. Clarke had proposed "ionospheric research with V-2s" in February 1945, before the war's end! The National Advisory Committee for

Aeronautics (NACA) suggested the use of V-2 rockets for the collection of atmospheric data up to high altitudes. Professor Leo Goldberg indicated his profound interest in studying the solar spectrum outside the Earth's atmosphere with those rockets. By the end of 1945, two young physicists from the Naval Research Laboratory, Ernst H. Krause and Milton Rosen, decided to act. They founded the Rocket Sonde Research Section at the NRL, enjoying enthusiastic response from scientists not only at NRL, but also at the Johns Hopkins University, and numerous other universities. The list of experimenters who prepared and brought their instruments for a ride on one of the V-2s is most impressive; some of them had already earned their spurs, others were at the beginning of their careers: Fritz Zwicky at CalTech, Jesse Greenstein at Yerkes, Richard Tousey, Herbert Friedman, and Homer E. Newell at NRL, R. Ladenburg and Lyman Spitzer, Jr. at Princeton, M. Tuve, M.H. Nichols, J.J. Hopfield, and James Van Allen at Johns Hopkins, W.B. Klemperer at the Douglas Research Laboratory, Richard W. Porter at General Electric, and J. Allen Hynek at Ohio State University. Krause organized the V-2 Upper Atmosphere Research Panel which included representatives of all the institutions interested at that time in high-altitude research, among them Fred Whipple from Harvard University and G.K. Megarian from General Electric. Homer Newell (1980), in his book *Beyond the Atmosphere*, described many of the details of the high-altitude research at White Sands, and particularly the results of many of those early projects which included, for example, the clarification of the ionization processes in each of the ionized layers of the high atmosphere by H. Friedman, measurements of the Sun's ultraviolet spectrum by R. Tousey, and the discovery of X-rays from the Sun by H. Friedman. Newell wrote: The V-2 Upper Atmosphere Research Panel "became the aegis for the country's first sounding rocket program...the German missile served as an admirable vehicle for high-altitude research. It often carried more than one ton of payload to altitudes better than 160 km." In total, the V-2s in White Sands launched 223 different experiments over a time span of almost seven years!

Originally, this rocket had been conceived as a vehicle for space flight. Then, it was oriented toward military functions and used as a weapon of war. Then, it became the enabling technology for the first comprehensive scientific space research program. Von Braun was very happy about this role change of his brain child. "Finally," he said, "our rocket has met with its real destiny" (Stuhlinger et al. 1994). He followed the events of the High Altitude Research Program with great interest, and in April, 1946, shortly after the project had started, he wrote an essay *Investigation of the Upper Atmosphere with the A-4 (V-2) Rocket Missile*. However, while he kept close contact with the events in White Sands, his mind reached out into the far future. He wrote a book entitled *The Mars Project* (von Braun 1952). Based on the knowledge and experience he had acquired at Peenemünde, he described a manned expedition to the planet Mars, using the A-4 rocket technology as well as the concepts of Martian features available at that time. As a result of the relatively modest capability of the A-4 rocket engine, the Mars rocket came out as a real behemoth. However, it would have been able to make the round trip, as von Braun's flawless equations showed. "I do not recommend," von Braun said, "that such a rocket vehicle should now be built. I just wanted to show that even our present, still fledgling rocket technology would principally be sufficient to enable us to travel to Mars and back. Future work will certainly put us in a position to build a Mars rocket that is smaller, faster, and more efficient than what could be built today on the basis of what was built in Peenemünde as a first step in the evolution of the modern space rocket."

5. ROCKETS FOR RESEARCH AND EXPLORATION

The V-2 launchings in New Mexico, although very successful as a first comprehensive high-altitude and near-space program with a purely scientific exploration and research character, were not very satisfactory from the standpoint of "technology supporting scientific research." To be an ideal research tool, the V-2s were too clumsy, too powerful, and too much designed to military specifications. They were too demanding with respect to launch preparations and activities required from launch crews. Also, the supply of "liberated" V-2s in White Sands began to dwindle. One specific disadvantage of experiment carriers for multiple users – like the big V-2s – is the fact that the various experimenters are never ready for launch at the same time. Some must always wait for others, risking the loss of favorable weather or lighting conditions, or of their own launch readiness.

On the other hand, those V-2 launchings at White Sands, over a time period from 1946 till 1952, proved that rockets offer a very substantial enabling technology for projects of scientific research in space.

It was natural, therefore, that the desire arose among the scientists to have rockets available that were designed and built specifically for the launching of scientific instruments. Dr. James Van Allen from the Applied Physics Laboratory at the Johns Hopkins University, a specialist in cosmic ray research, and an ardent user of V-2 launchings, took action. Supported by the Douglas Aircraft Company and the Aerojet Engineering Corporation, he helped design the Aerobee rocket. With a payload of 65 kg and a peak altitude of 120 km, it was considerably smaller and lighter, and easier and cheaper to launch, than the V-2s. "Also," Van Allen remarked, "I don't have to put up with colleagues who are

Figure 4 Aerobee high-altitude rocket, White Sands Proving Ground, late 1940s. (Source: NASA.)

Figure 5 Scout high-altitude rocket, Wallops Island, 1960 Scout rockets were used frequently for high-altitude and satellite missions during the 1960s and 1970s. (Source: NASA.)

never ready when I am." The first Aerobee was launched in November 1947. About 100 Aerobees were launched in total, financed by the Navy. A successor of the Aerobee, Aerobee High, launched in the mid-1950s, had increased payload (75 kg) and altitude (300 km). Aerobees were mainly used in research concerning the high atmosphere and ionosphere, the magnetosphere, the Sun, and cosmic rays.

Another US rocket designed and built exclusively for scientific research was Viking, the brain-child of Ernst Krause and Milton Rosen, and manufactured by the Glen L. Martin and Reaction Motors companies. Begun in 1946, and using much of the experience gained with the V-2, it became the most advanced US rocket project at that time. Most of the engineering work was done by John Shesta and Edward Neu. Fourteen Vikings were launched between March 1949 and May 1957; payloads varied between 250 and 450 kg, altitudes between 80 and 220 km. They, too, carried mostly instruments for solar, high atmosphere, cosmic-ray, ionosphere, and magnetosphere physics.

An interesting project, using an enabling technology that had been known for more than 400 years, but was demonstrated here in a technically advanced form for the first time, was the Bumper project (von Braun et al. 1966, Büdeler et al. 1979, see also Chapter 4). In 1948, the Jet Propulsion Laboratory (JPL) in Pasadena and the von Braun team in Fort Bliss, Texas, worked together to put a Wac Corporal rocket, developed at JPL in 1944–45, on top of a V-2 as a real two-stage sounding rocket. Eight Bumper-Wac launches were made at White Sands Proving Ground and also at Cape Canaveral in Florida. One of the rockets reached an altitude of 387 km, and a velocity of 8240 km per hour. Bumper-Wac was the first two-stage rocket launched in the United States, demonstrating a technology that was to acquire utmost importance in many subsequent projects that included, for example, the Scout rocket, and the big Saturn V rocket in the lunar landing project.

The decade of the 1950s saw an impressive groundswell of research projects carried aloft by sounding rockets. Besides V-2s, Aerobees, Vikings, and Wac Corporals, there were Van Allen's Rockoons, small rockets attached to balloons and fired when the balloons had reached high altitudes. There was, from 1960 on, the multistage Scout for medium-sized payloads, launched from Wallops Island to high altitudes, amd even into satellite orbits. More than two dozen different kinds of sounding rockets were developed and built in the US, several of them as multistage vehicles. Rocket projects had been underway for a long time in Soviet Russia; they now appeared, or re-appeared, also in many other countries: Canada, Great Britain, France, Italy, Germany, Poland, Japan, Australia, most of them in connection with the International Geophysical Year (IGY), which was organized in 1952 under Sidney Chapman after Lloyd Berkner had suggested, in 1950, that a joint effort on an international basis be undertaken to study the Earth and its environment. The actual IGY covered a period of about 18 months in 1957 and 1958. Under its aegis, America launched 210 sounding rockets, Russia 125, other countries fewer, but still impressive numbers of rockets for scientific research. Also, half a dozen satellites were put in orbit during the IGY. When the 1960s began, rockets had established themselves very solidly as indispensable tools for scientific research in the high layers of the atmosphere, and beyond.

6. A MULTISTAGE ROCKET LAUNCHES THE FIRST AMERICAN SATELLITE

The history of America's first satellite, Explorer 1, illustrates the fact that the availability of an enabling technology is not always sufficient to bring about the birth of a novel technical system.

In 1943, at Peenemünde, where von Braun and his team developed the A-4 rocket, a co-worker asked von Braun one day whether he believed that an artificial little moon could ever be built that traveled around the Earth in an orbit, the way Professor Oberth had predicted it. "Absolutely," was the answer, "if we are lucky enough to survive this present catastrophe, we may even be the ones who build it. But," he added quickly, "let's not talk about this now, it would be far too dangerous under present circumstances" (recollection of the author, who was personally present on that occasion).

Twenty-five years later, when Explorer 1 had reached its orbit, von Braun confessed to a friend: "When I landed on the American continent in September 1945, I had one burning hope: that this step may enable me to contribute to the launching of the first satellite!" In 1929, at 17, von Braun had written a story that described "stations in space" that orbit the Earth (von Braun 1929). For those who knew von Braun well, it was obvious that satellites were orbiting in his overactive mind almost constantly. As soon as his first rocket project in the United States, the Redstone missile, began taking shape during the early 1950s, he started making plans for the building and launching of a little satellite.

Thoughts about satellites gained momentum also in the minds of other young rocketeers at that time. Under their prodding, the US Navy, in the fall of 1945, established the CEFSR (Committee for Evaluating the Feasibility of Space Rocketry), which recommended a satellite project. A study contract was awarded to the Guggenheim Aeronautical Laboratory at the California Institute of Technology, GALCIT. The director of that laboratory, world-famous aerodynamicist Theodore von Kármán, had begun theoretical and experimental work on rockets in 1936 (see Chapter 5). Together with Frank Malina and Hsue-shen Tsien, he proposed a rocket research program in 1943 which, as the authors pointed out, would have "immediate military usefulness." CalTech responded to this proposal by establishing the Jet Propulsion Laboratory (JPL). While its primary duty was the development of guided missiles for the military, von Kármán wished to include "those branches which are more important for peace applications."

A number of studies concerning satellites, for low Earth as well as synchronous orbits, were published during the decade between the mid-1940s and the mid-1950s by authors representing industries, the Army, and the Air Force. In 1947, a Department of Defense committee, chaired by Clark B. Millikan, could not identify any uses for military satellites; this led to a decree by the Secretary of Defense in 1948 to the effect that current satellite activities in the United States should include only studies and component designs, but no satellite projects.

Early in 1950, project work on the Redstone missile had started at the Army's Guided Missile Development Division in Huntsville, Alabama under von Braun's leadership. When the performance figures of the rocket, only unofficially called "Redstone" at that time, had been established, von Braun said to his co-worker Stuhlinger: "With our Redstone, we could do it!" – "Do what?" – "Launch a satellite, of course!" (Stuhlinger *et al.* 1994).

At that time, von Braun requested and received permission from Army Ordnance to make analytical studies of an Earth satellite based on modified V-2 and Redstone missile technology. He had already figured out the details: Take a Redstone rocket; instead of the warhead, put three stages of small solid propellant rockets on top; launch the Redstone on a vertical trajectory, as high as it will go with its available propellant. The front end, including the gyro-control system, the electronics, the three upper stages, and the little satellite, will be separated as one unit from the tank and engine portion of the rocket at burnout. While this front end coasts upward toward the apex of its trajectory, it will be turned slowly by compressed nitrogen jets until it is exactly horizontal. At the moment of apex, a signal from the ground will ignite the first stage of the solid propellant rocket assembly. Stages two and three will follow in quick succession, and they will give our little satelite the necessary velocity of 7.7 km per second in a horizontal direction. "This will certainly not be the most efficient way to put five pounds into orbit," he admitted, "but it will be the only one available this side of the Iron Curtain for the next couple of years."

Figure 6 Army Redstone missile, being prepared for a test flight at White Sands Proving Ground, 1956. (Source: Corbis.)

More proposals for satellite projects were offered by various authors in the United States and in Great Britain. Milton W. Rosen, as chairman of a Space Flight Committee of the American Rocket Society, submitted a report entitled *On the Utility of an Unmanned Earth Satellite* to the National Science Foundation in 1954, but it remained without response. In the meantime, in 1953, two men in Washington decided that a real satellite project could not wait in stagnation any longer. Commander W. Hoover from the Office of Naval Research (ONR), and Frederick C. Durant III, President of the International Astronautical Federation (IAF), deciding that real actions should now be taken, organized an informal meeting with von Braun, Fred L. Whipple from Harvard University, S. Fred Singer, physicist at the University of Maryland, David Young of the Aerojet General Corporation, Alexander Satin from ONR, and some of von Braun's associates from Huntsville.

"Gentlemen," Hoover began the meeting, "the time has come to stop talking and start doing. We will now go ahead and build a satellite" (recollection of the author who was present). Von Braun sketched his proposal which was unanimously accepted, and work assignments were made for the attendees during the following weeks. This author received two assignments: first, to design an "apex predictor" that would determine the exact moment at which the first stage of the solid propellant rocket assembly had to be ignited, and to transmit a radio signal to the system at that moment; and second, to find an "honest-to-goodness" scientist who could be invited – and persuaded – to prepare an instrument for the satellite with which some first-class science could be carried out. Von Braun, together with some of his co-workers, wrote a paper entitled *A Minimum Satellite Vehicle Based on Components Available from Missile Development of the Army Ordnance Corps*, and sent it to Army authorities in September, 1954. He emphasized particularly that all the major components of this project were either already available as tested and proven hardware, or at least presently in the making as parts of the Army's weapon developments. This would guarantee not only a high reliability of the system, but also a low cost of the project. Von Braun also suggested that a satellite project should be a joint enterprise by Army, Navy, and Air Force.

Deciding on a scientist who could be invited to put a scientific instrument onboard the satellite was easy: Dr. James Van Allen at the State University of Iowa had been a researcher in cosmic rays for almost 20 years, well known to this author from his publications on this subject, and also from personal acquaintance during the V-2 launchings in White Sands where Van Allen had been a very eager and successful user of the high-altitude research opportunities offered by the V-2 rockets (Stuhlinger *et al.* 1994).

Von Braun's claim that all the major components of his proposed satellite launcher were virtually available had a very real background. In 1953, the Army had given his Guided Missile Development Division in Huntsville the assignment to develop an intermediate range ballistic missile (IRBM), the Jupiter. This project required the development of a heat protection system for the warhead while it reentered the atmosphere; it became known as an "ablation-type" warhead. For the testing of sample warheads under realistic reentry conditions, some Redstone rockets were equipped with two upper stages of solid propellant rockets. Redstones modified in this fashion, called "Jupiter C" for "composite," were planned and built during the early 1950s. In December 1954, von Braun wrote an update of his earlier satellite proposal which included data from his warhead reentry testing work, and sent it to Washington. He also sent a copy to the Jet Propulsion Laboratory, suggesting a joint program for two projects, a Jupiter C reentry project, and a Jupiter C satellite project. Dr. William H. Pickering, director of JPL, suggested replacing the relatively small Loki solid-propellant rockets for the upper stages with the larger JPL-built Sergeant rockets, and the Navy-built Minitrack transponder with the JPL-developed Microlock transponder, suggestions with which von Braun happily agreed. These changes resulted in a considerably better performance of the satellite and its launcher, later called "Project Orbiter." On that basis, a satellite launch in the summer of 1956 would have been possible. The first successful Jupiter C long-range flight with two upper stages and a model reentry warhead, reaching a maximum altitude of 1091 km and a range of 5440 km, occurred on September 20, 1956.

By that time, a proposal for another satellite launch system had come to life, proposed by Milton W. Rosen at the Naval Research Laboratory (NRL). The launching rocket, named Vanguard, would have three stages, a modified Viking, a modified Aerobee, and a new solid-propellant rocket. It was an attractive and interesting design, but the performance, development time, and cost promised by the project managers appeared too optimistic to the Project Orbiter colleagues in Huntsville who, taking the occasional indications of Russian satellite plans seriously, felt that the United States should try to launch a satellite as soon as possible.

The reports that von Braun sent to Washington did not result in a go-ahead signal for the orbiter project; instead, further committees were installed to study the existing satellite proposals. Only "further planning" was permitted officially to the von Braun team in Huntsville, so all the technical work for the satellite version of the Jupiter C rocket, such as the orientation of the upper portion of the launching vehicle into an exactly horizontal attitude after burnout of the Redstone, the apex predictor, and the satellite itself, was done very quietly by only a few people, and mainly during their free time at home. Von Braun was always very careful and anxious to stay within the limits of his orders from Washington.

President Eisenhower did not wish to agree to a satellite that used a military rocket as the basis for its launcher at a time when he wished to promote a worldwide "open sky" policy. Also, he was not overly excited about the prospect of a satellite orbiting the Earth. So, the Department of Defense called for more committee studies. In August 1955, the Stewart committee decided in favor of the Vanguard Project, against Project Orbiter.

The Vanguard Project proceeded under great efforts of its participants. As is normal for projects of that magnitude and complexity, progress was often interrupted by setbacks, need for redesigns, and launch failures in spite of the dedication and unquestionable capabilities of the Vanguard team members.

Then, on October 4, 1957, Sputnik appeared in its orbit. It was a "shot that was heard and seen around the world," but it took another Russian satellite, the 504 kg Sputnik II with the dog Laika onboard, before the Army team in Huntsville finally received the go-ahead order to launch the first American satellite, Explorer 1. It reached its orbit on January 31, 1958.

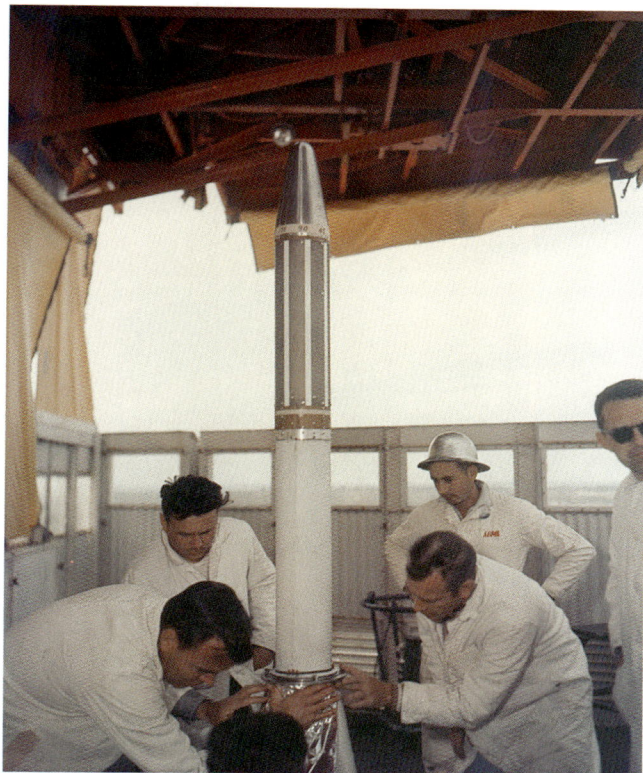

Figure 7 Explorer 1 satellite on Jupiter C rocket, being readied for launch at Cape Canuveral on January 31, 1958. (Source: Marshall Space Flight Center.)

Van Allen's Geiger counters on Explorer 1, and on some of the following Explorers, discovered the famous Van Allen radiation belts, and many of the features of the magnetosphere that surrounds our Earth – highly gratifying events for those Explorer team members in Huntsville and at JPL who insisted from the beginning of the project that America's first satellite should also fulfil a meaningful scientific mission.

The long story of the first American satellite shows a cumbersome, rocky road full of hopes and disappointments, pushed forward by small groups who realized the scientific and technical promises, and also the prestige value, of a satellite project for the nation, but, on the part of the government, delayed by endless committee discussions, by the lack of interest in a new technological achievement, and by the failure to anticipate the psychological impact on the world caused even by a small satellite in the sky.

The history of Explorer 1, which stayed in its orbit for 12.3 years, has been told repeatedly by persons who were deeply involved in its planning, building, and launching (Stuhlinger *et al.* 1994, von Braun *et al.* 1966, Ordway *et al.* 1979). Their accounts show very drastically that it is not only the existence of enabling technology that

Figure 8 Launch of Explorer 1 by a Redstone Jupiter C rocket, Cape Canaveral, Florida, January 31, 1958. (Source: NASA.)

brings a novel idea to life, there is also a need for enabling circumstances: for a consensus and support by powerful officials, noninterference by retarding committees, and – in some cases – a public willingness to provide the necessary funding. This latter point played no role in the Explorer case, because that project needed almost no extra funding, but future space projects may find this requirement formidable.

7. LIQUID HYDROGEN FOR INCREASED ROCKET PERFORMANCE

Each of the serious rocket pioneers, particularly Tsiolkovskii, Ganswindt, Goddard, and Oberth, and also numerous others, had come to the conclusion that the best combination of rocket propellants would be liquid hydrogen as fuel, and liquid oxygen as oxidizer, simply because their chemical reaction delivers more energy, and therefore a higher exhaust velocity (or specific impulse) than any other reaction that might be used in a chemical rocket motor. A disadvantage of liquid hydrogen as fuel for first-stage rocket engines is its low density, requiring huge tank volumes. With hydrogen as first-stage fuel, the cross-section of the hydrogen tank would have to be so large that the advantage of the higher specific impulse would be lost because of the higher air resistance of the large tank while the rocket ascends through the dense layers of the atmosphere at the beginning of its flight. Therefore, rocket systems using hydrogen–oxygen engines in their first stages, such as the Space Shuttle and Ariane V, use powerful solid propellant booster rockets in combination with a less powerful first stage hydrogen–oxygen engine.

During the late 1930s, when rocket development work went into high gear at Peenemünde, the question came up whether liquid hydrogen, in combination with liquid oxygen, should be considered as a potential rocket fuel. Surely, it would be by far the best carrier of energy that could be taken into consideration. However, at that time the technology to produce and handle liquid hydrogen in large quantities had not yet been developed. No work on hydrogen as a fuel for rocket engines was therefore started at Peenemünde.

There was a second reason why von Braun decided against work with hydrogen. Leaking hydrogen presents an extreme danger of explosions because it is very volatile, and inflammable at almost any mixing ratio with air. In all his undertakings, von Braun was always very careful to avoid accidents. Even when he had settled in the United States after the war, and liquid hydrogen became available in large amounts, he still desisted from developing hydrogen–oxygen engines for this very reason. He was afraid that if a major accident occurred with a hydrogen explosion, this would cause a very severe – perhaps even a deadly – blow to the entire rocket and space program.

Dr. Abraham Silverstein, Associate Director at the Lewis Research Center in Cleveland, Ohio, during the 1950s, took a more audacious approach. He started a development program for hydrogen–oxygen rocket motors in the mid-1950s which turned out to be very successful. With support from industrial contractors, particularly the Pratt & Whitney Aircraft Company, a first hydrogen–oxygen engine was developed, the RL 10 engine with a thrust force of about 7000 kg. Test firings began in 1959. A first application was the Centaur, a propulsion stage with two parallel RL 10 engines that was developed by the General Dynamics corporation in San Diego. During its early phases, this project was managed by Krafft Ehricke.

In 1959, when Dr. Silverstein was chairman of a committee that was to judge proposals for rocket vehicles needed for a lunar landing program, he suggested that in the future all stages of space rockets, except the first stage, should use hydrogen and oxygen as propellants; only first stages should continue to use kerosene–oxygen or UDMH–nitrogen tetroxide as propellants. It was a courageous proposal, but it was accepted unanimously by the committee members. This decision led to one of the most momentous

Figure 9 Launch of the HEAO 1 spacecraft with an Atlas-Centaur rocket, March 12, 1977. (Source: Marshall Space Flight Center.)

Figure 11 Launch of the first Saturn rocket, Saturn I, October 1961. (Source: NASA.)

Figure 10 Launch of a Saturn IB rocket with manned capsule. (Source: NASA.)

Figure 12 Production line of hydrogen-oxygen rocket engines J-2, used on Saturn IB and Saturn V rockets. Rocketdyne, Canoga Park, CA, July, 1966. (Source: NASA.)

innovations in the history of rockets during the second half of the twentieth century.

While the Centaur stage was used as a second stage in combination with first-stage Atlas and Titan rockets on missions to Venus and Mars, the RL 10 engine found application also on the first member of the Saturn family, Saturn I. From the fourth Saturn I on, each of seven Saturn I rockets had a second stage powered by a cluster of six RL 10 engines which generated a combined thrust force of 40 000 kg. The Saturn I rocket was built with several objectives in mind: to study the problems of handling large quantities of liquid hydrogen; to develop and test mechanisms to cluster several engines for parallel operation; and to develop engine swivel mechanisms for rocket steering purposes. Besides these prime objectives, Saturn I flights tested stage separation, and also components of the guidance, control, and various communication systems needed for the Saturn V rocket that was to launch the astronauts on their way to the Moon.

The next Saturn family member, Saturn IB, continued this testing program, but it carried a new second-stage hydrogen–oxygen engine, the J-2. The first stage of Saturn I, like the first stage of Saturn IB, had eight kerosene–oxygen engines of the type that powered the Jupiter and Thor missiles. Working in parallel, each of them developed a thrust force of 83 000 kg. Development of the new J-2 engine was begun at the Rocketdyne Division of North American Aviation around 1960. Producing a thrust force of 225 000 kg, the J-2 was flight-tested in 1966. The Saturn IB carried one J-2 engine in its second stage, while the final Moon rocket, Saturn V, had five J-2 engines, clustered in parallel, in its second, and one J-2 engine in its third stage.

Saturn IB rockets found application not only as precursors of the Saturn V, but also for the transfer of three crews to the orbiting Skylab space station in 1973, and for the manned rendezvous and docking mission with the Soviet spacecraft Sojuz in 1975.

Hydrogen rocket technology played an absolutely decisive role not only in the Saturn-Apollo Moon Project, but also in the Shuttle Project that came to life during the 1970s. A new, high performance engine, the Space Shuttle Main Engine (SSME), was developed in a joint effort by the G.C. Marshall Space Flight Center and the Rocketdyne Division of North American Aviation. By that time, hydrogen technology for rocket engines had reached its full maturity.

Figure 13 Launch of the Apollo-Saturn V rocket with Neil A. Armstrong, Edwin E. Aldrin, and Michael Collins on the first manned lunar landing Mission, July 16, 1969, Cape Canaveral, Florida. (Source: NASA.)

Figure 14 Space Shuttle Main Engine, SSME, built by Rocketdyne, North American Aviation Corporation in cooperation with the Marshall Space Flight Center. (Source: NASA.)

Figure 15 Vulcain rocket engine for Ariane 5, built by the Messerschmitt-Bölkow-Blohm Corporation, Munich, Germany. (Source: Courtesy of Helmut Hopmann, from *Schubkraft für die Raumfahrt*, Stedinger Verlag, Lemwerder.)

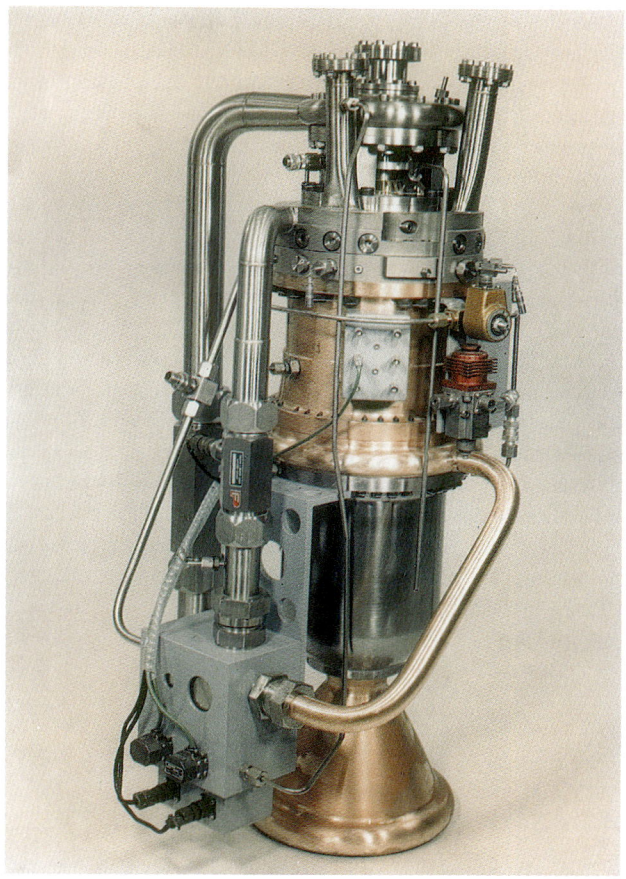

Figure 16 "Closed expander cycle" (Hauptstrom) rocket engine, built by Messerschmitt-Bölkow-Blohm Corporation, Munich, Germany, 1966. (Source: Courtesy of Helmut Hopmann, from *Schubkraft für die Raumfahrt*, Stedinger Verlag, Lemwerder.)

Rocket engines burning hydrogen and oxygen were also developed in several other countries during the past decades. In the mid-1960s, a very energetic development program for hydrogen–oxygen rocket motors began first in Germany by MBB, later to be joined by SEP in France. This work resulted, for example, in the HM 20 motor with a thrust of 20 tons, and then the Vulcain motor (115 tons thrust) which powers the central stage of the three-stage Ariane 5 rocket. The list of other countries that developed hydrogen–oxygen technology includes Russia, Japan, and China.

Among the common goals of these new rocket projects were higher efficiencies and lower weights of the rocket motors, and lower cost of the launch vehicles. The first rockets with ranges above about 400 kilometers that were built after the war, the R-1 in Soviet Russia and the Redstone in the United States, still showed a close relationship with the German A-4. Decisive improvements were soon introduced – some of them had already been started in Peenemünde – and the performance and efficiency of postwar rockets quickly increased. Instead of alcohol, they used kerosene, and in some cases UDMH (unsymmetrical dimethyl hydrazine). The heavy steel double walls of combustion chambers and expansion nozzles were replaced by a system of small, contoured nickel tubes soldered together ("spaghetti-type engine"). The steam generator driving the turbine for the propellant pumps, producing steam by decomposing hydrogen peroxide (H_2O_2), was eliminated by using hot gases from the main combustion system for the turbine. The entire rocket motor was mounted on a universal joint so that it could be swiveled for thrust vector control. The warhead plus instrument compartment of the rocket was designed to separate from the tank and engine section after burning cutoff for higher target accuracy. The numerous small injection nozzles were replaced by one injector plate with a system of several concentric ring slots.

Beginning in 1956, the Bölkow company in Stuttgart, Germany (predecessor of the MBB Corporation), began the development of the "closed expander cycle" technique (Hauptstromtriebwerk) for the pump turbine drive (Hopmann 1999), a technical innovation that led to a quantum jump in engine efficiency.

All previous, and most of the contemporary rocket engines that use hot gas drives for their pump turbines, apply a simple bypass system which channels propellants from the main propulsion systems into a pre-combustion chamber where they burn with a considerable excess of one component (fuel or oxydizer) so that the combustion temperature is low enough for the turbine; from there, the turbine exhaust, still incompletely burned, leaves through an exhaust pipe with expansion nozzle. However, due to the relatively low temperature of the gases, their contribution to the rocket thrust is very modest. In the novel closed expander cycle, the turbine exhaust gases, still rich in one component, are led into the main combustion chamber for complete burning, thus achieving a much improved utilization of the propellants and their combustion energy. Considering that in the SSME engine, the turbine for the oxygen pump develops 25 000 hp, and the turbine for the hydrogen pump 80 000 hp, and that the temperature of the turbine gas is 600 C, while the main combustion chamber temperature is above 3000 C, it may be understood that the afterburning of the turbine exhaust gases makes a substantial difference in engine efficiency.

The closed expander cycle was first used in the P111 oxygen–kerosene rocket engine, built by MBB. It was incorporated in the first hydrogen–oxygen engine developed in the United States, the RL 10. The Space Shuttle Main Engine, SSME, developed during the early 1980s by the Rocketdyne Corporation, uses this cycle. In Russia, work on the closed expander cycle began around 1958. The engines powering the Proton rockets also made use of the closed expander cycle.

8. ROCKETS INTO EARTH ORBITS, TO THE MOON, THE PLANETS, AND TOWARD THE SUN

Beginning a few years after World War II, an impressive activity of rocket launchings to high altitudes, into satellite orbits, and into deep space arose first in Soviet Russia, then followed by developments in the United States and in a number of European countries. Most of these space rockets were derived from military rockets, among them the Russian B-1, Semiorka, and Proton rockets derived from the SS4 and SS6 missiles, and many of the American launch rockets derived from the Redstone, Jupiter, Thor, Atlas, and Titan missiles. The objectives of those nonmilitary rocket launchings were studies of the atmosphere, the ionosphere, and the magnetosphere; of photon and particle radiations from space; and of micrometeoroids. Numerous rockets were launched to furnish pictures and other observational data of the Moon, the planets, and the Sun. Also, many of the rockets were launched for the purpose of Earth observations, geodetic research, weather service, and, increasingly, of developing the novel field of satellite-supported communications.

The characteristic feature of these projects was not primarily the existence of launching rockets as an enabling technology, but the availability of telemeter and radio systems that could transmit observed and incoming data to receiving stations on the ground. It was at Peenemünde during the late 1930s that the development of telemetry began. Based on the technology of radio transmitters of that time, telemeters were large, heavy, and not too reliable. The telemeter for the A-4 rocket in Peenemünde had seven channels for data transmission, and the size of a small

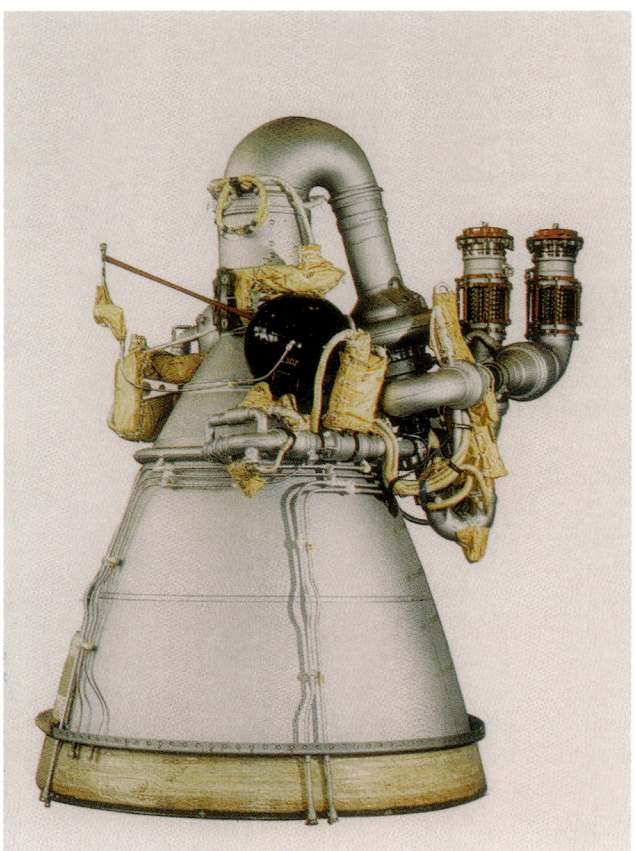

Figure 17 A Russian closed expander cycle rocket engine (RD 0210), built for the second and third stages of the Proton rocket, burning UDMH and nitrogen tetroxide, 1962–65. (Source: Courtesy of Helmut Hopmann, from *Schubkraft für die Raumfahrt*, Stedinger Verlag, Lemwerder.)

coffin. When, during the lengthy A-4 launch preparations, someone asked: "Where are we in the countdown?", the usual answer was: "X minus telemeter."

Early pioneers of rockets and space flight concentrated their studies and efforts on rocket engines, celestial mechanics, and life support systems, but not on the remote acquisition and transmission of performance data. Professor Oberth (1923) may have come closest to the problem of data transmission with his brief remark: "... with suitable mirrors, they can send light signals to Earth stations."

The main reason why this situation changed thoroughly from the 1950s on was the development of semiconductor technology. First the transistor, and then the chip allowed the building of telemeters that had only a small fraction of the weight and the size, and that needed only a very small portion of the power, of their ancient electron-tube predecessors, but they accommodated far more transmission channels – in the thousands – than the old telemeters ever did. During the first two decades after the first Sputnik and Explorer satellites, no less than 2100 satellites for military and civilian purposes were launched, and that number was widely surpassed during the following two decades. All of these satellites, as well as the countless space probes, unmanned and manned, carried electronic systems that allowed extensive two-way communication between them and the Earth.

In the United States, a small number of satellites were launched during the late 1950s by modified Redstone and Jupiter rockets, the Juno I and Juno II launchers, and Vanguard; then, the Thor rocket in various combinations began to appear frequently on the Florida and also the California launch platforms, first with the Discoverer spacecraft, a military reconnaissance satellite whose film capsules were recovered during reentry by airplanes. The Atlas rocket made its debut as a space launcher in 1960, and the multistage Scout rocket became a prolific launcher of smaller satellites. Telstar transmitted the first television program between Europe and the United States in 1962. The Orbiting Solar Observatory (OSO), the British satellite Ariel, the German-French satellites Syncom 1 and 2, and numerous other satellites followed. OGO, the Orbiting Geophysical Observatory, was launched in 1964. It was the tendency of NASA to establish cooperative satellite projects with a growing number of other nations around the world. Heavier satellites were launched with Atlas- and Titan-derived launchers, among them the Orbiting Astronomical Observatory (OAO). More and more "application"-type satellites appeared in orbits, among them the Application Technology Satellites ATS, and the series of Intelsat communication satellites. Among the memorable science-oriented satellites of that time were the German-American solar satellite Helios, and the three High Energy

Figure 18 Helios solar satellite, a joint US–German project for observations of solar particles and fields. (Source: NASA.)

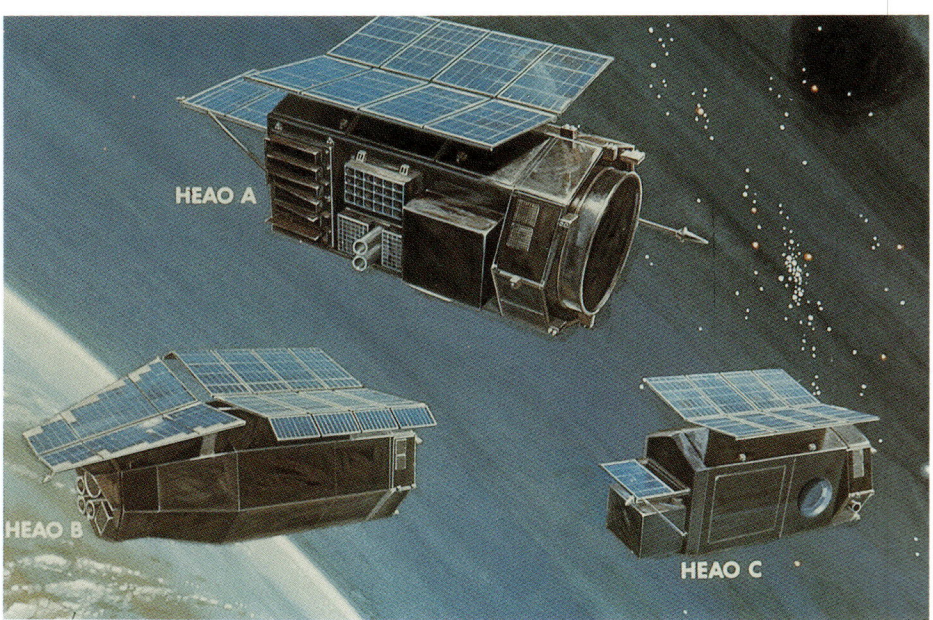

Figure 19 High Energy Astronomical Observatories I, II, and III to study ultraviolet, X-ray, and cosmic radiations, 1977, 1978, and 1979. (Source: NASA.)

Figure 20 Russian Kosmos satellite. On the order of 2000 Kosmos satellites have been launched so far. (Source: Novosti.)

Astronomical Observatories (HEAO). Soviet Russia developed a very active satellite program during the same time period with mixed military and civilian missions, among them the famous Kosmos series. The first Kosmos satellite was launched in 1963; barely fifteen years later, about 1200 Kosmos satellites had been put into their orbits.

By 1978, the list of satellite-launching countries included Russia, the US, France, the European Space Agency, Great Britain, Italy, Japan, India, and China. The science and technology of satellite-launching had reached a high level of perfection by that time. Technologies of guidance and control, orbit determination and orbit changing, sensors, measuring systems, miniturization, solar energy supplies, nuclear isotope power sources, and component reliability enjoyed impressive improvements as these systems were designed, built, and operated. Similar improvements were achieved in the ability of communicating with spacecraft, of giving them orders, of receiving various observational data from them, and of using them as transfer stations for worldwide communication between points on Earth. The evolution of two novel technologies, space transportation and Earth-to-orbit communication, had come together to provide a new dimension to life on Earth.

While a rich program of Earth-orbiting satellites unfolded simultaneously for military purposes and for scientific research, for Earth observations, and for communication, missions to the Moon and to planets were planned and carried out during the same period of time. As early as January 2, 1959, Soviet Russia sent Lunik 1, launched by a Proton rocket, on its way to the Moon. It passed by the Moon at a distance of 5600 km. Lunik 2, launched eight months later, impacted on the lunar surface. Lunik 3

Figure 21 Wernher von Braun, J.R. Casani, and James Van Allen with the lunar probe pioneer IV, launched with a Jupiter rocket Juno II on March 1, 1959. (Source: Marshall Space Flight Center.)

achieved an impressive success in October 1959. It traveled around the Moon, photographed its farside, developed the film by robotic procedures, dried it, scanned it with an electron multiplier tube, and radioed the signals back to Earth where the pictures were reconstituted. The first US lunar mission, Pioneer 4, was launched on March 3, 1959 with a Juno II rocket; it flew by the Moon at a distance of 60 000 km.

The history of unmanned Russian and American lunar probes between 1959 and 1976 shows the great effort extended by both nations to learn more about our neighbor in space, and it also shows the quick increase of the success rate during that period.

In 1959 and 1960, several unsuccessful Pioneer probes were launched with Atlas-based rockets in the United States. A series of nine Ranger spacecraft between 1961 and 1965 was partially successful: three Rangers delivered several thousand photographs of the lunar surface that were very useful when plans for the Saturn-Apollo manned lunar landing project were made. Satisfactory success was achieved by the Surveyor program; of seven Surveyors launched between 1966 and 1968, five accomplished a soft landing on the lunar surface with subsequent soil analysis and photographic scanning of the environment.

Four months before the first American soft landing on the Moon, a Russian spacecraft, Luna 9, landed softly on the lunar surface and transmitted pictures back to Earth. Fifteen more successful Luna mission followed between 1966 and 1976; five of them photographed the surface from orbit, and ten landed softly, taking photographs and analyzing soil samples. Four of them sent probes back to Earth with film and soil samples, and two of them brought roving vehicles to the Moon that traveled many miles over the lunar surface, taking pictures and soil samples, and catapulting them back to Earth.

Besides the Luna program, Russian space exploration activities included the Zond program; Zonds 5, 6, 7, and 8 successfully orbited the Moon, took pictures of the lunar surface, and sent the exposed film back to Earth where the capsules were retrieved after "landing" in the Indian Ocean.

Between 1961 and 1972, the Saturn-Apollo Project was underway in the United States with its extremely high demand on manpower, work, and funds. Unmanned Moon probe projects were mainly undertaken to take pictures, and to perform some chemical and physical analysis of the lunar surface. Consequently, the US probes were primarily designed to furnish photographs (through radio links) of the lunar terrain, and to give an idea of the kind of lunar soil the astronauts would have to expect when they landed on the Moon.

Attempted flights to Venus and Mars during the first 18 years after Sputnik show an impressive degree of activity, but their success rate indicates how demanding and difficult missions to planetary targets really are. Between 1961 and 1978, the USSR started 21 missions to Venus; only four had some partial success, ten suffered failures of the propulsion system, communications failed in three missions. During the same period, the US started six picture-taking missions to Venus; three were successful, two had partial success, while the propulsion system failed in one mission.

In the Mars mission program, the USSR launched 14 probes between 1960 and 1973. None of them enjoyed full success; four probes achieved partial successes with flybys and picture transmission, but six missions suffered from propulsion failures, and four of them from communication problems. The American Mars program was luckier. Of eight launchings between 1964 and 1975, three probes provided pictures taken during flybys from distances between about 10 000 and 3400 km; Mariner 9 in 1971 took pictures and observational data from an orbit around Mars, and the two Mars landers Viking 1 and 2 in 1975 were outstanding successes. Launched with Titan-Centaur rockets, the spacecraft entered into orbits around Mars; from there, landing capsules descended to the surface by retrorockets and parachutes, and eleborate instruments took pictures and analyzed soil samples. Data were transmitted first to the orbiting spacecraft, and from there to Earth. Of particular interest were the observations of instruments that searched for traces of life on the Martian surface. While these instruments did not find any indications of live forms, their sensitivities may not have been sufficient to discover very minute organisms, so that their negative results should not be taken as proof that there are no life forms on Mars. The Viking missions furnished a rich harvest of knowledge concerning the surface conditions, temperatures, atmospheric pressure and composition, winds and weather, colors and

Figure 22 Titan-Centaur rocket being readied for launch. (Source: NASA.)

Figure 23 Russian rocket Vostok with Voskhod spacecraft, early 1960s. (Source: Novosti.)

structure of the surface, solar and cosmic radiations near Mars, and other parameters typical of the Martian environment that had been known only very inaccurately before. Project Viking not only became a decisive source of new knowlege about Mars; it also resulted in several new technologies that formed the basis of future missions of planetary exploration, such as the combination of retrorockets and parachutes to achieve a soft landing, the collection and analysis of soil samples, instruments to study features of the environment, data transmission first from the planetary surface to the orbiting spacecraft, and from there to Earth, and instruments for the search for biological matter.

A number of probes were sent toward the Sun. During the years 1960 to 1968, five US Pioneers circled the Sun in orbits inside the Earth's orbit, measuring the various kinds of solar radiation. The first comprehensive study of the Sun with telescopes from Earth orbit was conducted in 1973 by the space station Skylab (see Chapter 11). In 1974 and 1976, two solar spacecraft approached the Sun to about 48 million km, Helios I and Helios II. Built by Germany and launched by American Titan-Centaur rockets, they were equipped with specially designed reflectors to prevent their overheating. The primary objective of the Helios probes was the study of particle radiations and magnetic fields as close to the Sun as possible. Further solar missions were undertaken to study processes in the outer layers of the Sun.

Lunar, planetary, and solar space missions during the 1960s and 1970s were described in great detail in a number of books (Newell 1980, von Braun *et al*. 1966, Büdeler *et al*. 1979, Ruppe 1982, Eckart 1999). Also, frequent reports on space missions were, and are, given in various space journals, among them the publications of the American Institute of Aeronautics and Astronautics, the American Astronautical Society, the International Astronautical Federation, the *Journal of the British Interplanetary Society*, the Planetary Society, the National Space Society, *Sky and Telescope*, and others.

All the Russian probes were launched by rockets based on the original Russian intercontinental missile R-7. Differing in the number of their thrust units and stages, the R-7 family members included Sputnik, Vostok, Molnija, Luna, Voschod, Luna-Venus, Sojuz, Progress, and Proton (in Russian space projects, the same names are often used for carrier rockets and their spacecraft). American projects used Atlas-Agena and Atlas-Centaur rockets for the Mariner missions to Venus and Mars, and Titan-Centaur launchers for the Mars landers Viking I and II, as well as for Helios I and II and for the Pioneer and Voyager missions to Jupiter and beyond. While the capabilities of these rockets were sufficient for the near planets, their limitations became obvious when plans for missions to Jupiter and the more distant planets were started. That was the time when an old idea was introduced into spaceflight projects. It became known as "gravitational maneuvers," and it made flights as far as Neptune possible, even with bearable lengths of travel times.

The possibility of changing the flight direction, and also the velocity of space probes by encounters with the gravitational fields of planets, and even of moons, was recognized many years ago by astronomers who observed and studied the trajectories of comets. In 1954, Derek Lawden in England presented a theoretical treatment of "gravitational maneuvers," and during the early 1960s members of the Jet Propulsion Laboratory (JPL) worked out details for the application of that method to actual planetary missions. Basically, the principle of such gravitational maneuvers may be compared with events in a soccer ball game when a player intercepts the ball with his head, giving it a new direction and additional speed, and "heading" it – hopefully – into the goal. In a spaceflight gravitational maneuver, the spacecraft's position, velocity, and direction are very precisely controlled so that the spacecraft enters the gravitational field of the planet in such a way that it swings out of that field again in the desired direction, at the desired place, and with the desired velocity. In that process, the spacecraft acquires additional velocity while it is in the grip of the planet due to the planet's own motion around the Sun.

It may be realized how accurately the position, velocity, and direction of both the planet and the spacecraft must be known at any time during the maneuver to carry out such a procedure, and how accurately the spacecraft must be guided and controlled while it approaches the planet. After the celestial mechanics of a gravitational swingby maneuver have been worked out, highest demands are placed upon the ground systems that continuously determine the flight parameters of the spacecraft, upon the communication systems that maintain two-way traffic between ground and spacecraft, and upon the trajectory-correcting propulsion systems on the spacecraft that respond to commands from the ground as well as to signals from onboard planet and star seekers – not to mention the need to know the planet's position and motion at all times with extreme accuracy.

A number of successful planetary swingby maneuvers were carried out during the past decades. The Mercury orbiter reached its target in 1975 after a swing around Venus, Pioneers 10 and 11 reached Jupiter in 1972 and 1973 after swinging around Mars, and the famous Grand Tour missions of Voyagers 1 and 2 in 1977, which visited all other outer planets except Pluto, used several planetary swingby maneuvers around different planets. Galileo, the spacecraft that achieved a remarkably thorough investigation of Jupiter and its moons, arrived at its target after several swingbys around Mars and Earth.

There are numerous ground stations in constant operation around the world, but the most famous among them is the Goldstone Ground Station in California, operated by JPL, which has been in operation now for almost half a century. When "enabling technologies for space transportation" are listed, Goldstone should rank very high on the list.

Gravitational maneuvers have made several deep-space missions possible within acceptable travel times (a few years) that would not have been possible with the traditional Hohmann transfer trajectories. They will certainly retain their importance for many instrumented planetary and interplanetary flights. However, the need for shorter travel times on manned roundtrip flights to planetary targets even as close as Mars, and also for some special unmanned flights to remote targets, will require novel propulsion systems that make direct flights on such missions possible (Chapter 21).

9. A GIANT ROCKET FOR MANNED VOYAGES TO THE MOON

The Saturn rocket, developed for the Saturn-Apollo Project that was to take astronauts to the Moon and back to Earth, was the first large rocket destined for nonmilitary purposes. It would be wrong, though, to assume that the primary motive behind America's Manned Moon Project was solely the quest for scientific exploration and knowledge. Surely, a human voyage to the Moon, and later even to Mars, had been on von Braun's mind ever since he was a teenager, but it took more than a bold dream to make a lunar expedition happen, more than a lifelong systematic effort to develop powerful precision rockets, more than an impressive store of experience in rocket-building, more than teams of able engineers and scientists, and more than the willpower to translate the ideas and hopes into hardware.

All these conditions for a successful Moon project existed at the beginning of the 1960s; besides, there existed also a number of very successful rocket projects that had been accomplished during previous years. But, in addition to these enabling technologies and experiences, there was the need for specific enabling circumstances before a Moon project of that magnitude could come to life. Such circumstances evolved during the early 1960s. There was a Cold War between the superpowers. Soviet Russia surprised the world with impressive rocket and space firsts, the first intercontinental military rocket, the first satellite, the first man in orbit, the first close flyby of the Moon, the first photographs of the farside of the Moon. Also, there was the tragic disaster of the Bay of Pigs in April 1961 – another severe prestige loss for the United States. "Is there any place where we can catch them?" the President asked. Around that time, it became known that the Russians were planning to achieve a manned expedition to the Moon. Under the strong prodding of Vice President Johnson, President Kennedy began to consider the possibility of a manned Moon Project as a way to reestablish America's prestige in the world. The United States, as von Braun explained at length in his famous memorandum to Vice

Figure 24 Goldstone ground station, operated by JPL in California. (Source: NASA.)

President Johnson of April 1961 (von Braun 1961) has "…an excellent chance of beating the Soviets to the first landing of a crew on the Moon (including return capability, of course)…"

On May 25, 1961 Kennedy spoke to Congress: "Now it is time…for this nation to take a clearly leading role in space achievement…," and "I believe that this nation should commit itself to achieving the goal, before this decade is out, of landing a man on the Moon and returning him safely to Earth." These were wonderful words from a great leader, and the American people stood behind him in a rare show of unity.

Years, even decades, before President Kennedy committed the US to the goal of landing a man on the Moon, von Braun had sketched out in his mind the basic features of such a mission: launch two large three-stage rockets – later to be called Saturn V – into Earth orbit; one will haul extra fuel, the other will carry an assembly of a capsule for three astronauts, a propulsion system for travel to the Moon and back, provisions for a lunar landing and subsequent ascent from the lunar surface, and a vehicle to return to Earth, with the capability to land on its surface. After ascent into Earth orbit, the first rocket will fill up the propellant tanks of the second rocket for its long journey to the Moon and back to Earth. The second rocket, rather than descending to the lunar surface upon arriving near the Moon, may remain in orbit while only a landing craft descends to the Moon with one or two astronauts who later ascend to the orbiting vehicle for the return trip to Earth.

Von Braun quoted two important features of this travel mode which make it desirable for a Moon trip. First, the maneuver of refueling the Moon vehicle, and of readying it for its lunar mission, could be done close to Earth, within easy radio contact, and even with rescue possibilities in case of a mishap. Second, this mode would offer a welcome opportunity to develop some orbital maneuver techniques like refueling, checkout procedures, and extravehicular activities. Technologies of this kind, he argued, will certainly be needed in later space projects. This scheme for a lunar expedition, in refined form, became known later as "Earth Orbit Rendezvous mode," or EOR.

While von Braun nurtured his ideas about a lunar voyage, a vigorous effort toward the development of manned spaceflight was underway at the National Advisory Committee for Aeronautics (NACA) under Robert R. Gilruth. During the latter part of the 1950s, the Langley Aeronautical Laboratory, later to become the NASA Langley Research Center, pursued several projects to build and fly rocket-powered airplanes; the best known was the X-15 rocket plane of the 1960s which reached velocities up to Mach 6. In 1958, NACA suggested to Keith Glennan, Administrator of the newly established NASA, that a manned satellite project should be undertaken. A Space Task Group (STG) was formed, headed by Gilruth, and plans for Project Mercury took shape. Alan B. Shepard became the first American astronaut on a ballistic flight with a Redstone rocket in May 1961. In February 1962, one year after Gagarin, John H. Glenn orbited the Earth in a Mercury capsule launched by an Atlas-Centaur rocket. The next step in the program came in 1965 when the two-seater Gemini spacecraft, launched by the powerful Titan-Centaur rockets, began their orbital flights. In the meantime, the Space Task Group under Gilruth had become a full-fledged NASA Center, later to be called the Johnson Space Center, and to settle in Houston, Texas.

The Mercury and Gemini flights proved to be extremely useful, and even indispensable for the layout and design of the Moon rocket. The technology of returning a spacecraft from orbit was developed and tested on the basis of these flights; it turned out that the best reentry body is a cone-shaped structure with a relatively flat front, covered by a material that ablates under the impact of the hot air. Also, a wealth of experience and data was obtained concerning the preparation and the accommodations for the astronauts, as well as their abilities and physiological functions during weightlessness. It was also during some of these manned orbital flights when egress from a space capsule was practiced, and when astronauts gathered extravehicular experience. Without these flights, the Moon travelers would certainly have met with far more problems.

NASA decided to organize the lunar project in such a way that there would be two major centers of activity, one for the development of the powerful Moon rocket and its propulsion systems, the other for the development of the space capsules for the astronauts and their accommodations, including the lunar landing vehicle, and also the selection and training of the astronauts. This decision proved to be logical and correct. The first activity would be the responsibility of the George C. Marshall Space Flight Center in Huntsville under the direction of Wernher von Braun, the second was Robert R. Gilruth's responsibility at the Johnson Space Center in Houston.

At Huntsville, systematic work for a Moon project had started in 1958. Shortly after Secretary of Defense Neil H. McElroy had established the Advanced Research Projects Agency, ARPA, Richard Canright and David Young of that agency visited von Braun to discuss the possibilities of a Moon project, and to find ways how ARPA could help bring such a project to life. Von Braun, convinced that the most important part of this project would be a powerful rocket engine, and that the development of this engine would not only be the pacing element, but in fact the enabling component of the entire Moon project, suggested that developing, building, and testing the big rocket engine should be the first active step of the lunar project.

Figure 25 Production line of thrust chambers for F-1 rocket engines, the most powerful rocket engines ever built, to be used in the first stage of the Saturn V rocket. Rocketdyne, Canoga Park, CA, November, 1963. (Source: NASA.)

Von Braun's concept of a Moon mission, based on the simultaneous launching of two large rockets and a refueling operation in Earth orbit, called the EOR mode, would not have been the most direct way of achieving a manned voyage to the Moon. From the standpoint of necessary space maneuvers, the simplest concept would have been a very large, five-stage rocket, with the astronaut capsule, equipped with landing legs, on top. The first three stages, burning in quick succession, would launch the fourth and fifth stage, and the landing capsule, into a transfer trajectory from Earth directly to the close vicinity of the Moon. The fourth stage would then slow the transfer vehicle down and land it softly on the lunar surface. At the end of their stay on the Moon, the astronauts would go back into their capsule and ignite the fifth stage for lift-off from the Moon and subsequent return all the way to Earth where the Earth's atmosphere would slow the capsule down for a parachute descent into the ocean.

The members of Dr. Gilruth's Space Task Group at NASA's Langley Research Center in Virginia chose this concept, called the "direct mode," as their preferred lunar mission mode because it involved the least amount of in-flight maneuvers. However, as von Braun pointed out, it would require a rocket, tentatively called Nova, about twice the power and the size of the Saturn rocket. He considered the

Figure 26 Test firing of Saturn V first stage, G.C. Marshall Space Flight Center, Huntsville, AL, August 5, 1965. (Source: Marshall Space Flight Center.)

Figure 27 F-1 rocket engine, built by Rocketdyne-North American Aviation for Saturn V, the most powerful rocket engine built so far. Thrust at sea level: 690 tons; in space: 820 tons. Five F-1 engines power Saturn V's first stage. (Source: NASA.)

Saturn, in spite of its large size, as achievable within the time and funding limits established by the Government. At that time, Saturn's development was already underway, including the huge fabrication and testing facilities necessary for its development and fabrication. If a completely new rocket, twice the size of Saturn, with the necessary enormous manufacturing and testing facilities, had to be developed, built, and tested for a Moon project, a lunar mission "before the decade is out" would have been impossible to achieve.

A third mode of travel for a voyage to the Moon and back to Earth, the Lunar Orbit Rendezvous Mode (LOR), would begin with only one rocket of Saturn size. Its first three stages would inject the spacecraft – consisting of the manned capsule (command module), a transfer stage (service module), and a landing stage (lunar excursion module) – into the Earth–Moon transfer trajectory. This entire complex would enter a satellite orbit around the Moon. From there, the lunar lander would descend to the lunar surface with two astronauts, while the third astronaut would remain in the orbiting capsule together with the service module. A few days later, the ascent stage with the two astronauts would return to the orbiting spacecraft, leaving the landing stage of the lunar excursion module behind. After reuniting, the three astronauts with command module, service module, and the lunar ascent stage still attached, would return to Earth, propelled by the service module. During the three-day return trip, the astronauts could use accommodations in the command module as well as in the lunar ascent module. Upon reentering the atmosphere, the lunar ascent module would be detached (it would burn up during its plunge into the atmosphere), while the command module with the astronauts would descend by parachute and finally "land" in the ocean.

Yuri Va. Kondratyuk (1965), one of the early Russian space pioneers, described this way of visiting another celestial body in 1916. Thirty years later, von Braun (1952) suggested the same technique for a mission to Mars in his book *The Mars Project*. Around the same time, the British Interplanetary Society published an article by H.E. Ross (1949) about this landing scheme for a lunar mission, and several more papers on that subject appeared during the late 1950s (Dolan *et al.* 1960). John C. Houbolt (1961), an engineer at the Langley Research Center, suggested using this lunar orbit rendezvous mode in a letter to Robert C. Seamans, NASA Associate Administrator, on May 19, 1961, pointing out that with this mode, a manned lunar landing mission could be achieved with only one Saturn rocket.

Houbolt's appeal for the LOR mode led to a period of frantic activities in committees, study groups, discussions, debates, analyses, and opinions, by knowledgeable rocket and spaceflight experts as well as by individuals of more modest qualifications. The indisputable merit of Houbolt's proposal was fully recognized by the experts, but they also expressed their reservations and concerns. Gilruth's team preferred the direct mode because it required the least number of in-space maneuvers; also, Houbolt's scant payload allowance for astronaut support and accommodations appeared way too low to the engineers responsible for the manned portion of the mission. The main reason why Gilruth was hesitant to embrace the LOR mode was the fact that all those complicated operations to ready the lunar lander for its descent to the lunar surface, and to dispatch it from the orbiting service module, had to be done on the farside of the Moon, without radio connection with the Earth. The idea that these complex functions had to be entrusted entirely to a programmed computer was disturbing to him. Von Braun shared this feeling; also, he felt uneasy without the rescue possibility offered by the EOR mode. Besides these concerns, von Braun wanted the first mission to the Moon to be the beginning of an era of manned space flight which would incorporate various orbital functions, such as rendezvous operations, assembly of spacecraft, refueling, payload transfer, repair capabilities, rescue situations, crew exchange, and others. The Saturn Moon mission, he thought, would offer welcome opportunities to begin such developments. Like Gilruth, von Braun also doubted the very optimistic weight figures Houbolt quoted for spacecraft components and astronaut accommodations.

The waves of excitement and disagreement concerning the lunar mission mode raged high for more than a year. While the development of some of the decisive components of the Moon mission proceeded well, such as the engines at Rocketdyne, the inertial guidance systems at the Marshall Space Flight Center, at the Draper Laboratories at the MIT, at the Honeywell corporation and at other contractors, and the development of the manned reentry capsules at the Johnson Space Center, the binding decision for the actual pathway to the Moon was made as late as November 7, 1962, when Mr. J.E. Webb, NASA's Administrator, finally approved the Lunar Orbit Rendezvous mode for the Moon mission.

The way this decision came about is a classical case of an "enabling technology."

Late in 1961, coming back from a meeting at the Johnson Space Center in Houston, von Braun told his associates that Dr. Gilruth has now completely dropped his concerns about computer-controlled maneuvers on the farside of the Moon, without radio contact with the Earth. "These folks at IBM," Gilruth had said, have now completely convinced him fully that their computers are capable and reliable enough to handle that situation, and that we can accept the LOR mode with confidence. This sweeping success at IBM was mainly due to the efforts and ability of Jim Medlock and his co-workers in Huntsville, Alabama. Without their untiring work, it is doubtful whether American astronauts would have reached the Moon "before the decade is out."

Gilruth's change of mind, based on sound technical progress and very sober facts, also caused von Braun to change his mind, and to support the LOR mode fully. He was even more inclined to do so because so much time had passed during the EOR-LOR-Direct mode debate that any other mode could not have been perfected before the end of the decade. Also, with some of his proverbial foresight, von Braun very fortunately had laid out plans for the Saturn V so that the rocket could accommodate far more payload than what Houbolt had assumed for his mode – three times as much, to be exact – and all of it was needed! Von Braun's hope for a rescue capability in case of a mishap in Earth orbit, and for the development of various orbital operations, had to be postponed for a later time.

Mr. Webb's decision of November 1962 did not convince all the participants in the great lunar mode debate that the LOR mode was the way to go to the Moon, but for the main players in the game, Dr. Seamans and Dr. Mueller in Washington, Dr. Gilruth in Houston, and Dr. von Braun in Huntsville, the marching direction was clear. The Apollo astronauts landed on the Moon and returned safely to Earth half a year before the promised end of the decade.

10. THE RUSSIAN MANNED MOON PROGRAM

With the first Earth satellite, Sputnik, in 1957, and the first astronaut in orbit, Gagarin in 1961, Soviet Russia had clearly established the leading position in space exploration in the early 1960s. It was natural that plans for a manned mission to the Moon were pursued at that time by both the USSR and the US, and that an eager, even fierce spirit of competition between the two space-flight giants developed at that time.

Details about the Russian manned Moon program became known in the West only decades after it had happened (Harford 1997, Vick 1997, Przybilski and Wotzlaw 1996, Hopmann 1999, Michin 1998). It is most impressive, quite complex, and not without drama and human tragedy.

Korolev began making plans for unmanned and also manned Moon missions during the early 1950s. His first successful probe to the Moon, Luna 1, after several probe failures, was launched on January 2, 1959. It missed its target by 6000 km and went into an orbit around the Sun. Luna 3, launched on October 4, 1959 (the second anniversary of Sputnik), photographed the Moon's farside. Later probes, such as Luna 16 (September 12, 1970) sent samples of lunar soil back to Earth. Luna 17, on November 17, 1970, landed Lunokhod 1 on the lunar surface, a robotic roving vehicle that traveled over the lunar landscape for 11 months, taking 20 000 photographs.

James E. Webb, NASA's administrator during the 1960s, mentioned in 1964 that there was evidence that the Russians were working on a manned lunar project. Up to 1964, Russian funding for space projects had increased; appropriations remained about constant until 1970 when they increased again until 1974. After that time, funding for lunar missions was reduced, while emphasis was placed on space stations (Salyut, Mir). Over that same period, the budgets for military rocket work skyrocketed.

From about 1955 on, Korolev had studies made of large booster rockets, heavier than the R-7 rocket which had served as the basis for all the large military and nonmilitary rockets up to that time. A "moon rocket" was designed, the N-1/L-3 rocket (also called Herkules), a four-stage system with the L-3 "lunar module" on top. All the rocket engines were of the highly efficient closed-cycle design, burning kerosene and liquid oxygen. The first stage had thirty engines and four vernier motors, with a diameter of 17 meters! It developed a thrust force of 4620 tons. The complete N-1/L-3 rocket had a height of 105 m and a mass at start of 2750 tons. For comparison, Saturn V's thrust force at takeoff was 3447 tons; it had a mass of 2938 tons, and a height of 110 m. The Russian Moon rocket would have carried two astronauts; one would have remained in lunar orbit, while the other one would have landed on the Moon, and returned to the orbiting capsule for the trip home. In the Saturn-Apollo Project, two astronauts landed on the Moon, while the third remained in the orbiting return vehicle.

Sergei Pavlovich Korolev died in January, 1966. He was succeeded as "General Designer" by Vassily Pavlovich Mishin.

The Russian lunar module, L-3, was flight-tested successfully in Earth orbit flights in 1971 and 1972. The complete N-1 rocket could not be ground-tested because no static test stand accommodating a rocket of that size existed, and funds for its construction were not appropriated. Therefore, five "flight test" vehicles were built, to be launched unmanned. The first test-flight was planned for October 1968. However, hairline cracks were found in its oxygen tank. A flight test with a new first stage took place on February 21, 1969. At an altitude of about 12 km, one rocket engine failed because of a ruptured fuel line. The computer shut off all engines, and the rocket fell back and began to break apart. Range Safety blew the vehicle up.

The second flight test on July 3, 1969 failed because fire broke out in the first stage. The entire launch complex was destroyed in the accident, and a large number of engineers and technicians lost their lives. The third flight test, on June 27, 1971, also ended in failure. When the first-stage engines ignited, a shock wave from the ground developed which broke the vehicle structure. The big rocket traveled about 20 km downrange before it hit the ground. A fourth flight test was undertaken on November 23, 1972. This time, the rocket had a good start, but during the shut-down of the six central engines of the first stage, a fire broke out, so the computer did not permit the second stage to be ignited. The big rocket traveled several hundred miles downrange before it crashed.

That was the end of the N-1/L-3 Moon project, but it was also the beginning of plans for an advanced lunar booster rocket. The rocket was to have upper stages with new, very efficient hydrogen–oxygen engines; the project envisioned two or three astronauts on the lunar surface for periods of 30–120 days. Plans called for the launching of unmanned rockets that would take lunar "cabins," roving vehicles, tools, supplies, and even return rockets to the lunar surface before the astronauts arrived. Flights were expected to begin in the 1980s.

These Moon mission plans did not come to fruition. Around that time, the United States began to make plans for the Shuttle. For reasons that are difficult to understand, planners in Russia believed that the American Shuttle project was primarily a military project to which Russia would have to react with a similar project. A new launcher rocket, Energia, was developed with four LOX–kerosene booster rockets and one LOX–hydrogen central engine for propulsion; it would put the shuttle Buran into a low Earth orbit. Energia made two flights: the first test-flight took place on May 15, 1987, and in November, 1988 Energia launched a Buran shuttle into Earth orbit.

This program, however, was not continued. Beginning in February, 1986, Russia built the great Mir space station, which (at the time of this writing) is looking back over a very successful, and also eventful 14-year life in orbit. With that space station, and with its substantial engagement in the International Space Station (ISS), the Russian space organization cannot afford the continuation of an active program of manned lunar missions (Chapter 13).

For participants in the American and the European space projects, the accomplishments of Russian colleagues in their Moon projects are admirable, and most impressive. Lay people are inclined to speak of "winners" and "losers" in situations like the efforts of Americans and Russians to be first on the Moon. Engineers and scientists who worked on the Saturn-Apollo Project regard their competitors primarily as colleagues who were pursuing similar dreams in space. There can be no doubt that the Russian space farers had to struggle against more severe odds – not of their own making – than the Americans. For their colleagues in the United States, they were not the losers in the race to the Moon – they just came out as the second best. After all, rocket engines originally built for the N-1 and the Energia projects are presently being bought and used for American space projects!

11. SKYLAB, THE FIRST SCIENCE LABORATORY IN SPACE

As the Saturn-Apollo project began to take shape during the early 1960s, thoughts arose about other potential uses of components developed for that project. In one proposed application, astronomers suggested that solar telescopes should be mounted on a command and service module as it had been designed and built for Saturn-Apollo, and that this assembly should be flown with astronauts for several hours in an Earth orbit. In another proposal, worked out by members of the Marshall Space Flight Center, it was suggested that the hydrogen tank of the second stage of the Saturn IB rocket (which also served as the third stage on Saturn V), empty after burnout in orbit, should be modified by three astronauts in space suits who would be launched on a second Saturn IB one day later. Attached to the second stage of the second Saturn IB would be a "telescope mount," together with an ascent stage as used on the lunar excursion module of Saturn-Apollo flights to the Moon, in which the astronauts could eat and sleep during their modification work. Also, this second Saturn IB would bring all the equipment and parts that would have to be installed in the empty hydrogen tank. After modification, the former hydrogen tank would serve as crew quarters and workshop – and also as a real science laboratory – for the astronauts.

This project was called "wet workshop" because the hydrogen tank would be launched full of liquid hydrogen. Plans for the project were offered to NASA headquarters and to Congress in 1966. The project combined the astronomers' wishes to have an assembly of telescopes in orbit, and the wish of the rocket people at the Marshall

Figure 28 Launch of Skylab 1, the orbital workshop with a Saturn V, on May 14, 1973, and of Skylab 2, with three astronauts, with a Saturn IB, on May 25, 1973. (Source: NASA.)

Center to have a manned station and science laboratory in orbit. It must be admitted, though, that the project planners, and particularly the prospective astronauts, did not feel very comfortable when they tried to figure out all the activities that were expected from them in orbit under space conditions. They would have to install and equip the empty hydrogen tank with all kinds of life support systems and accommodations for the future occupants, and also the fast-growing number of experiments and instruments for scientific investigations, brought up in the second Saturn IB that would have to dock somehow with the first Saturn IB, all under weightlessness, and mostly while the astronauts had to wear clumsy space suits.

The relief came on July 22, 1969, two days after the Apollo 11 landing on the Moon. During a Congressional session where the event was celebrated, Dr. George Mueller, Associate Administrator for Manned Space Flight at NASA, announced that the workshop will be launched by Saturn V, and not as a "wet" workshop, but as a "dry" workshop on top of the live second stage of the Saturn V, and that it would be fully equipped and installed on the ground before launching as a space station for extended human occupancy, and with a large array of scientific and technological experiments, including a set of astronomical telescopes.

This change of project details meant a considerable escalation of the original magnitude, and especially of the cost, of the orbital workshop project. However, by making this announcement immediately after the glorious success of the first landing on the Moon by Neil Armstrong and Buzz Aldrin, Mueller could count on an unusual state of exuberance and generosity of all the Congressional members. Indeed, the plans for the dry workshop were unanimously and happily accepted.

Scientists were invited to prepare instruments and projects for scientific observations in space over periods of several months. A large telescope mount was attached to the workshop proper; it carried eight optical instruments to observe the Sun and stellar objects over a wide range of the spectrum from visible wavelengths to X-rays. More than 50 scientists from all over the country, and also from other nations, prepared instruments and experiments for Skylab, and more than 200 individual observations and measurements were made, pertaining to a variety of scientific disciplines, among them solar and stellar astronomy, space physics, Earth observations, life sciences, human physiology, material sciences, and zero-gravity studies. Several experiments were prepared by high-school students. In one of these studies, a student wanted to find out whether a spider (*Araneus diadematus*) would be successful in building a web without the help of gravitational forces. It was: after a few futile attempts, the spider built a beautiful, absolutely perfect web. Unfortunately, there was no fly on Skylab that would have been a reward for the spider.

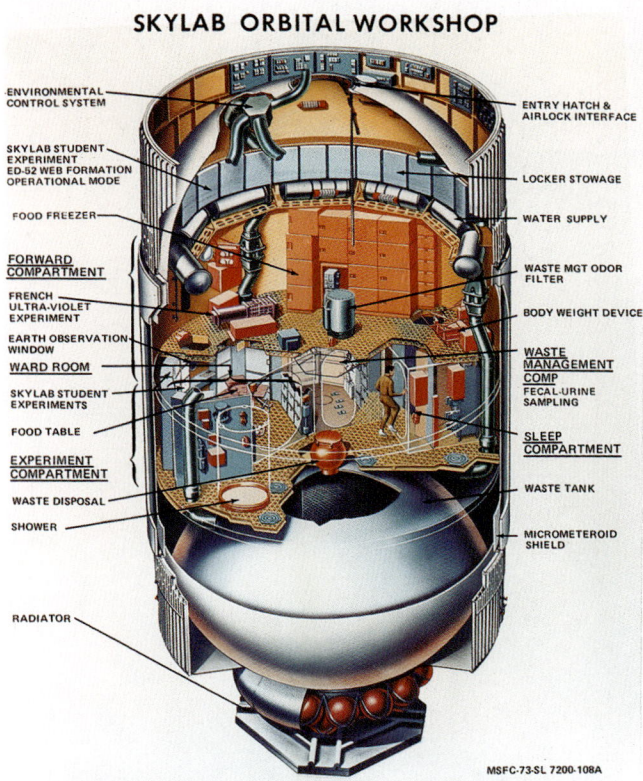

Figure 29 Skylab Orbital Workshop, built from the empty hydrogen tank of the third stage of the Saturn V rocket. (Source: Marshall Space Flight Center.)

Figure 30 Skylab, America's first space station, in orbit. A Saturn Command and Service Module, with a crew of three, has arrived to dock with Skylab. One solar cell wing, and a meteorite shield, were lost during ascent, but the damage could be repaired in orbit. (Source: NASA.)

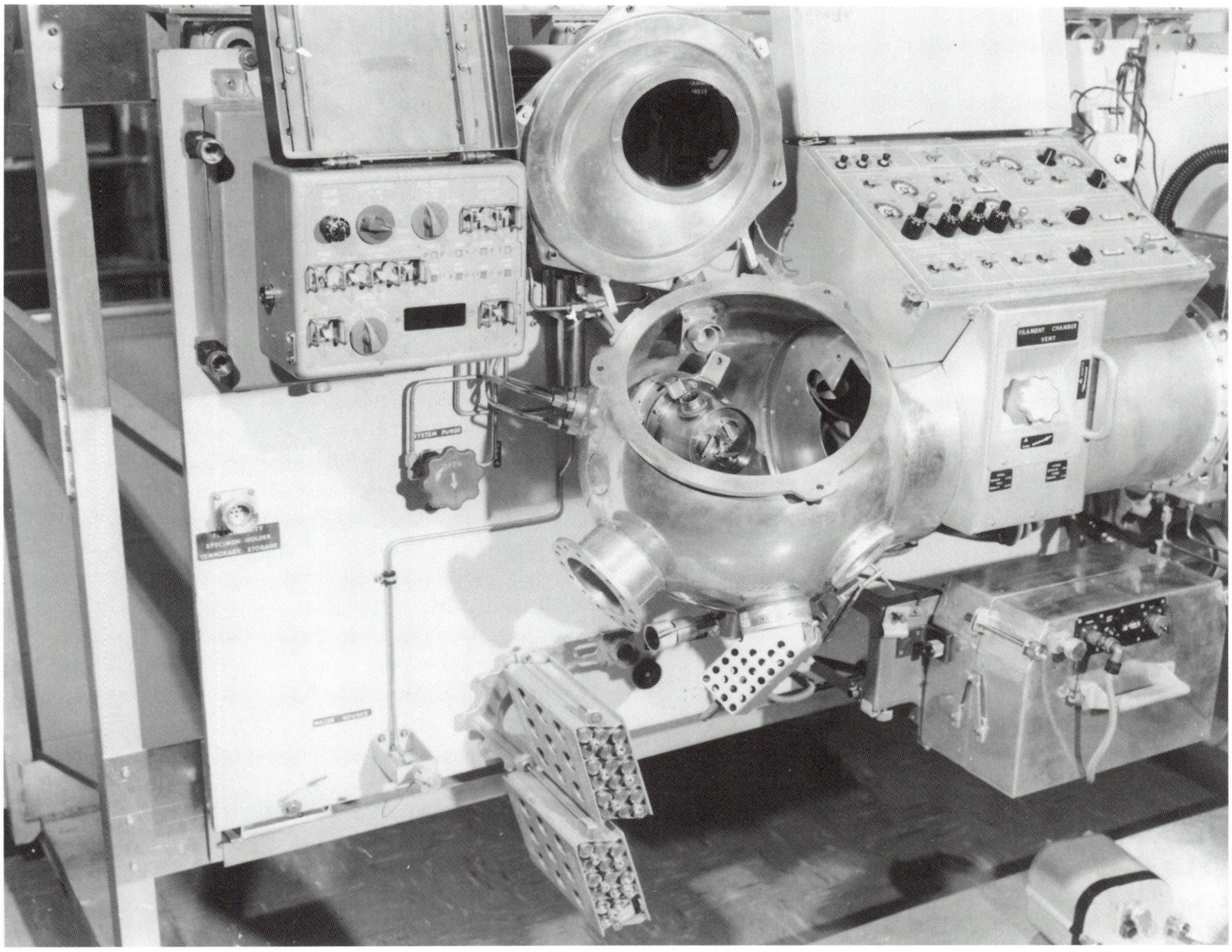

Figure 31 Instrument Unit onboard Skylab for a number of technical and scientific investigations. (Source: United States Space and Rocket Center, Huntsville, Alabama.)

In spite of some initial technical problems that developed during the ascent of Skylab into its orbit, the project became a brilliant success. Three crews occupied Skylab in sequence, each consisting of three astronauts, for a total of 184 man-days in orbit. With a host of new data that could be obtained only in an orbiting laboratory, Skylab proved the outstanding ability of a manned station in orbit to support scientific research under conditions not available on Earth.

The Marshall Center and its contractors prepared three Skylabs with the intent to have the second profit from the experiences of the first, and the third from the experiences of the first and the second Skylab. However, NASA headquarters decided that after Skylab I, which was launched in May, 1973, the program would be terminated. Skylabs 2 and 3 were built, but never launched.

Figure 32 An early web, built under weightlessness, by a spider, *Araneus diadematus* (called "Anita") onboard Skylab; eventually, this spider built a perfect spider web under weightless conditions. (Source: NASA.)

Detailed descriptions of the Skylab project and its experiments were given by (Newell 1980, Belew *et al*. 1973, Büdeler 1973).

An interesting project was carried out on July 15, 1975. In the morning, a Sojuz spacecraft with two Russian cosmonauts was launched from the Baykonur launch site and placed in Earth orbit. Several hours later, a Saturn IB launched an Apollo Command and Service Module (as used in the Saturn-Apollo Moon Project) with three astronauts and a Docking Module to a rendezvous with the Sojuz capsule. After arrival in the correct orbit, the Command and Service Module detached from the launch rocket, turned around, and attached its nose to the Docking Module which it then withdrew from the Saturn. Turning around again and approaching the Sojuz, it docked with the Russian spacecraft. The five-man crew spent two days together on the combined Apollo–Sojuz spacecraft, carrying out joint engineering and science investigations. Then, each crew returned to its home port.

12. SPACELAB OFFERS FURTHER SCIENCE-IN-SPACE FACILITIES

Before Skylab was launched, but after the United States had decided to build a Space Shuttle for repeated manned flights into orbit and back to Earth, the US Government approached the European Space Research Organization of that time, ESRO, with the suggestion that Europe should consider participating in the Shuttle program by building a "sortie laboratory" that could be carried into orbit and back to Earth in the big cargo hold of the Shuttle. In August 1973, the European Research Ministers accepted the invitation (by that time, the ELDO and ESRO organisations had been combined to form the European Space Agency, ESA), and in June, 1974, a contract to develop and build the sortie laboratory, now called Spacelab, was given to the VFW-Fokker/ERNO corporation in Bremen. Germany accepted 53%, Italy 18% of the total budget; Great Britain, France, Spain, The Netherlands, Switzerland, Belgium, Denmark, and Austria shared the rest.

Spacelab was to be launched on the Shuttle together with the "Pallet," an open structure that also fitted into the hold of the Shuttle. The Pallet carried instruments, such as telescopes, that needed an open view of space. Both Spacelab and Pallet would come optionally in various lengths, according to the instruments and experiments to be accommodated on a given flight.

Flight readiness was originally planned for 1979, but continuing changes in technical specifications caused a launch delay until 1983. The Spacelab was transported from Bremen to the Kennedy Space Center in a huge cargo

Figure 33 Launch of a Saturn IB rocket with three astronauts to meet a Russian Sojuz capsule with two cosmonauts in orbit, July 1975. (Source: NASA.)

Figure 34 Apollo–Sojuz orbital rendezvous, July 1975. A Saturn-Apollo Command and Service Module approaches a Sojuz spacecraft in orbit, ready for a docking maneuver. (Source: NASA.)

Figure 35 Spacelab, a science laboratory onboard an orbiting Space Shuttle. (Source: NASA.)

aircraft, the Galaxy C5A; the first launch occurred onboard the Shuttle Columbia, with six astronauts, on December 8, 1983. During the ensuing 15 years, 22 space missions were carried out with Spacelab, 13 on Shuttle Columbia, 3 on Atlantis, and 2 each on Challenger, Discovery, and Endeavor. Altogether, Spacelab spent 231 days in orbit, 149 astronauts worked in it, and 720 scientific experiments were performed under near-weightlessness.

Spacelab experiments and observations fell into numerous categories: astronomical observations of the Sun and stellar objects over a wide range of wavelengths; observations of the Earth, including ocean states and weather situations; biomedical studies; communication experiments; observation of material processes, including crystallization, mixing, welding, alloying; biological observations, including embryonic developments, cellular growth, healing procedures, and physiological processes. Of particular interest were observations concerning human behavior and human abilities under the influence of prolonged microgravity. Experiments that required an extremely vibration-free environment, such as certain astronomical observations, and studies of crystallization processes, were placed on a platform (the Instrument Pointing System, IPS) that was isolated from the rest of the Spacelab structure by magnetic suspension, an arrangement that resulted in an angular stability on the order of one hundredth of one arc second.

The Shuttle proved to be an ideal transporter for the Spacelab into orbit, and back to Earth, not only because it enabled human experimenters to be there continuously to monitor and support the experiments, but also because Spacelab's return to Earth after a number of days in orbit greatly facilitated recalibration and the evaluation of the results of measurements and observations, the preparation of new experiments, and the repair and exchange of instruments, machines and materials for further runs of observations. Both early space laboratories, Skylab and Spacelab, were of great value in the planning of the permanent International Space Station, ISS, for which extended laboratory facilities in orbit are a major objective and ingredient.

Further details about the Spacelab project can be found in Ruppe (1982) and Ottemayer (1976).

13. RUSSIAN SPACE-STATIONS

Soviet Russia's first manned satellite, Yuri Gagarin's Vostok capsule, was launched on April 12, 1961 with a big Semiorka rocket (also designated SS-6 or R-7). At about the same time, Korolev began developing the Proton rocket, with the same multi-engine technology as Semiorka, but with more powerful engines which used the expander cycle system, and worked at increased combustion chamber pressure. They burned the same propellants as their many predecessors: monomethyl hydrazine, nitrogen tetroxide, and nitrogen monoxide. These Proton rockets could transport up to 30 tons of payload into low Earth orbits; they were used for most of the Russian spacecraft launchings from the late 1960s on, and they are also used to carry some of the heavy American components of the International Space Station into orbit.

From 1963 on, Sojuz spacecraft were used for manned flights into orbit, with short stays as satellites. In 1973, Sojuz 13 was launched with P. Klimuk and W. Lebedew, together with a telescope prepared by G. Gursadian. Plans for an orbiting station accommodating cosmonauts for extended periods in space began during the 1960s, resulting in the Salyut Space Station. The first Salyut was launched

Figure 36 MIR, the giant Russian orbital space station. (Source: Science Photo Library.)

unmanned on April 4, 1971. Salyut stations were visited by numerous cosmonauts, one or two at a time, beginning in July, 1974; they used Sojuz capsules for ascent to and descent from Salyut, and stayed on Salyut for up to 175 days. In December 1977, Sojuz 26 docked at one end of Salyut 6, and in January 1978, Sojuz 27 docked at Salyut's other end. Four cosmonauts stayed for six days on this composite station which had a length of 30 meters, a volume of 100 cubic meters, and a mass of 32 tons (Michin 1998).

Salyut stations were visited from time to time by unmanned Progress supply rockets that brought water, oxygen, food, propellant, and other supplies, and then loaded trash and waste from the stations, but were left to burn up in the atmosphere during their return to Earth. Progress supply rockets could carry 2.3 tons of goods into orbit. In June 1985, Salyut 7, launched in 1982, had become inactive. Cosmonauts on Sojuz T-13 were able to repair the ailing station in orbit, and to restore its functionality.

In the meantime, a new and very large space station, Mir, had come into existence. Its first building block, weighing 204 tons, had been assembled in space, brought up in parts by Proton rockets. It was ready to accept further building blocks in 1987; the first was Kvant 1, a component that contained life support systems, attitude and orbit control instruments, an access port for the Progress supply rockets, biomedical experiments, and astrophysical instruments from scientists in several nations. Kvant 2 arrived in November 1989, with a mass of 20 tons, containing an airlock, solar arrays, and more scientific instruments. The Kristall module was added in May 1990, with more solar arrays, scientific and technological instruments, and an airlock that could accommodate Shuttle docking. In March 1995, the Spektr module arrived with still more solar arrays, living quarters, and further scientific instrumentation, including 700 kg of US instruments. The last module to attach to the large Mir Station was Priroda in April 1996; it contained a microgravity laboratory, Earth-observing equipment, and 1000 kg of scientific instruments from scientists all over the world.

During the 1990s, Mir was a very busy place for space planners, project managers, engineers, technicians, scientists, and particularly for cosmonauts and also astronauts. Dozens of them from Russia and also from various other nations, many of them with educational backgrounds in science or medicine, spent weeks and months on Mir, several of them more than one year. The record is held by Valeri Polyakov who lived and worked on the Mir station for 438 days! The station offered research opportunities in a wide variety of fields: astronomy, astrophysics, studies of the high atmosphere, the ionosphere, and the magnetosphere, physico-chemical research, thermophysics, crystallization studies, materials sciences, physiological investigations on plants and animals, biomedical research, human physiological and psychological behavior, research in communication technologies, and other purely technical fields, such as welding and flame propagation.

The life and work of the cosmonauts and astronauts on Mir did not always enjoy smooth sailing. One of the Progress supply craft bumped into Spektr, disabled it, and left a large hole in Mir's wall through which precious breathing air escaped. One day, a fire broke out on Mir, causing moments of extreme danger. The life support system had its times of malfunctioning, and so did the attitude control system. Very fortunately, the cosmonauts could keep all of these mishaps in check; they succeeded in having the station continue in operating condition. On the other hand, wear and tear after a decade of lively activities are beginning to show. After all, Mir's age had already surpassed the station's design lifetime by several years.

In addition to a remarkably rich yield in novel scientific and technological knowledge, the Mir Station has provided its designers, its manufacturers, its operators, and its visitors a wealth of experience about life and work in space. While the design of many of Mir's installations was new territory for its designers, the experience gathered during these past twelve or fourteen years is now available for the design of the International Space Station (ISS). Fortunately, close relations between Russian and US space planners and station designers have developed during recent years, following an official invitation to the Russian Government to be a partner in the ISS project. Present plans (early 2000) call for a first occupancy of the fledgling Space Station early in the new millenium by three crew members, to be transported to the station by a Sojuz capsule. Five months later, a crew of seven will be sent to the station for a stay of three to six months.

The Mir station, in spite of remarkable repair work, continues to show signs of aging and ailing. It is unoccupied at the present time. Russian Mir engineers, understandably, wish to keep the station alive and operating. However, taking care of two large space stations at the same time may just be too much for the Russian space budget. Even if a "Goodbye" to Mir should now become unavoidable, this first large, long-time station in orbit will always have a place of honor in the history of space flight not only as a trail-blazer for space technology, but as an impressive source of scientific knowledge concerning the space environment, and our ability to live and work in it.

MIR was re-occupied in 2000, but early in 2001, it was finally de-activated and put into a controlled descent. It plunged into the Indian Ocean in March 2001, after having orbited the Earth 86 333 times during its lifetime in orbit of more than 15 years.

14. EARTH-TO-ORBIT TRANSPORTATION WITH THE SHUTTLE

All the great rockets of the 1950s and 1960s – Semiorka, Kosmos, Proton, Redstone, Scout, Jupiter, Thor, Atlas, Diamant, Titan, Ariane, and Saturn – proved their abilities to put satellites into orbits and probes into deep space, even with acceptable reliability, but all of them had one decisive shortcoming: each of them was lost forever after one single launching. The quest for reusable launch vehicles, based on the hope that they would make space flight much cheaper, became louder and louder toward the end of the 1960s, particularly from budget planners and congressional leaders. At that time, the directors of the Saturn-Apollo program, George Mueller at NASA headquarters, Robert Gilruth at the Manned Space Center in Houston, and von Braun in Huntsville, began to make project plans for post-Apollo times. Encouraged, optimistic, and unrestrained after the impressive success with the Moon project, some of them came up with a grandiose list of projects: A manned station on the Moon; a big two-stage shuttle between Earth and orbit, with both stages recoverable; a permanent space station in orbit; and a manned expedition to Mars. Not surprisingly, only one project from that list would survive under the impact of fiscal reality. NASA headquarters opted for the Shuttle, still with two stages recoverable. Houston also preferred a recovery of both stages, but von Braun pointed out that only a simple design with one returnable stage, with solid propellant boosters – also recoverable from the ocean after burnout – and with a droppable, non-recoverable external tank, but still with a large payload bay, would be financially bearable.

Shuttle planning intensified in Huntsville and in Houston, but planning activities soon arose also at many other places, not only at NASA and some of its industrial contractors, but also by future scientific customers, by amateur space enthusiasts, by politicians, by military departments, by presidential advisors, and even by members of the Office of Management and Budget, and of Congress. Each of these would-be space experts proposed a different Shuttle design, but all of them agreed on one point: the era of the throw-away launchers must come to an end. George Mueller promised that as soon as the Shuttle is available, all other launchers for satellites and space probes, military and civilian, from small spacecraft to systems with multiperson crews, will be superfluous.

Mueller left NASA in 1969. John M. Logsdon (1978), historian at George Washington University, in an essay *The Space Shuttle Decision: Technology and Political Choice*, wrote: "During the second half of 1971, it was a chaotic exercise... everyone was a Shuttle designer." Finally, in January 1972, President Nixon approved the project, but eager disputes about the best Shuttle design continued for a considerable time.

Slowly, the Shuttle project evolved and took shape, not without detours. A powerful, very advanced rocket engine burning liquid hydrogen and liquid oxygen, the Space Shuttle Main Engine (SSME), was developed at the

Figure 37 Space Shuttle Atlantis during launch at Cape Canaveral, Florida. (Source: Marshall Space Flight Center.)

Rocketdyne Corporation, in close cooperation with the Marshall Space Flight Center. Three of these engines would work in parallel during the ascent of the Shuttle, burning for about nine minutes and developing a joint thrust of 556 tons. A solid propellant booster rocket was developed for the Shuttle by Thiokol in Utah; two boosters would burn in parallel during the first two minutes of the flight, developing a thrust of 2404 tons.

To accommodate a variety of scientific as well as technological experiments, to be attended by science astronauts for periods of days or weeks, the Shuttle can carry the Spacelab within its hold, and also open pallets for instruments that need an open view toward space or toward Earth. It can also accommodate the SPAS, a carrying structure developed by MBB in Munich, that can be released from the orbiting Shuttle into orbits of its own, and will be captured again by the Shuttle after some time for return to Earth. SPAS frames carried telescopes, models of propulsion systems, such as the electric ion thruster RIT, and materials processing experiments that demand an extremely low level of gravitational or man-made accelerations.

Besides providing opportunities to scientists for space-related research, the Shuttle also had to serve as a launcher for large spacecraft, such as ESA's Eureka Platform, the Long-Duration-Exposure-Facility (LDEF), the European-built probe for the joint ESA–NASA out-of-the-ecliptic mission Ulysses, the Hubble Space Telescope, the Chandra X-Ray Telescope, and the Galileo spacecraft to Jupiter and its moons, connected to a large and heavy transfer stage. In all these cases, the scientific payloads profited not only from the considerable payload capability of the Shuttle (30 tons into low Earth orbit), but also from the ability of the Shuttle crew members to give the spacecraft a final hands-on checkup in orbit before the spacecraft had to begin their own space journeys.

The first manned flight of a Shuttle, Columbia, with John W. Young and R.L. Crippen, lasting $2\frac{1}{2}$ days, took place on April 12, 1981, exactly 20 years after Yuri Gagarin's first satellite flight on April 12, 1961. Four Shuttles were built for orbital flights, Columbia, Challenger, Discovery, and Atlantis. More than one hundred Shuttle flights have taken place so far.

One of the Shuttle pilots, coming back from orbit, remarked that the Shuttle is "the most impressive machine ever built by man". Indeed, the Shuttle, within one envelope, works as a multistage rocket, a satellite, an inhabitable space station, a "home away from home" for up to eight crew members, a laboratory for scientists, a workshop for engineers, an orbiting outpost for space observations, a launcher for spacecraft, a test bed for technical systems, a communication center, an occasional nursery for sick astronauts, a reentry vehicle, a supersonic airplane, a glider, and a landing vehicle.

In a long series of successful flight operations, the Shuttle Program had only one serious and very tragic accident, the Challenger disaster in January, 1986. It was caused by a combination of a less than perfect design, extreme weather conditions, time pressure resulting from the launch schedule, and reduced carefulness on the part of decision makers who relied too much on the successes of previous launchings. The Challenger accident taught valuable lessons, although at the price of five human lives. But it certainly helped prevent future accidents.

The most important aspect of the Shuttle is its ability to enable scientists to carry out observations for periods of time of up to about two weeks, under conditions not available on Earth: First, simultaneous observations of large areas of the Earth's surface and its air masses, views of celestial objects without the disturbing influence of the Earth's atmosphere, and studies of interstellar space under the same ideal observational conditions; and second, observations of processes under the near-absence of gravitational forces. A scientist can stay with his or her experiment all the time, observe, take data, and exchange material, components, or experimental setups. Scientists will return to Earth with all their equipment. They can recalibrate, and if desired, modify their instrumentation for the next flight, or continue experimental procedures under one-g conditions.

In spite of the Shuttle's very impressive list of successful flights into orbit and back to Earth, the Shuttle Program is not without critics. When a shuttle project was first offered by space planners around 1968, a cost per pound of payload of less than $50 was quoted. Also, a situation was projected when no "throw-away" rocket launchings would be necessary in the future; all payloads, small and large, would be launched by reusable shuttles. These predictions did not materialize. Thirty years later, and after a hundred Shuttle flights, the cost per pound of payload is still a large multiple of the promised figure, and numerous one-time rockets are launched successfully every year with payloads varying between a few kilograms and many tons.

And yet, the Shuttle has provided a quantum jump in the development of enabling space flight technologies by presenting an opportunity to space scientists to take a very personal part in the execution of their scientific research in space.

15. THE INTERNATIONAL SPACE STATION, A CONTINUOUS OPPORTUNITY FOR SCIENCE IN SPACE

Experimenters on Skylab and Spacelab were limited in their research efforts by the relatively short periods of time during which they could perform their experiments and observations. Desires for a permanent facility in orbit arose early during the first phases of travel into space, first in Soviet Russia, and later in the United States and in Europe.

Plans for a large orbital space station had been underway in the United States since the early 1970s. Skylab was a first opportunity to study some of the technical and human problems of living and working in space over extended periods of time. In 1984, President Reagan approved a space station project, "to be completed within this decade," as he optimistically added. Under the name of Space Station Freedom, planning for a complex, multipurpose station project started. The European Space Agency, ESA, agreed to participate in the project. In 1993, the design was simplified; the new name was Space Station Alpha. Later that year, Russia was invited to participate also, and to contribute the experience the Russian space engineers had gathered with the Mir Station. The name of the station project was changed to International Space Station (ISS).

As a first phase of the project, from 1995 to 1998, a number of Mir–Shuttle rendezvous flights were carried out. Seven US astronauts had the opportunity to spend a total of 975 days on Mir; seven Russian cosmonauts flew on the Shuttle. In 1997, several technical problems were encountered by Mir. On one occasion, an unmanned Progress supply rocket bumped into the Mir station and caused considerable damage, including a significant loss of cabin atmosphere. However, US project managers emphasized that all these technical problems (all of them could be solved without permanently disabling the Mir station!) taught the space engineers valuable lessons which will help them avoid similar problems in their own design and development work.

During the second phase of the International Space Station project, from 1998 to 2000, the Russian module Zarya was launched by a Proton rocket, to be joined promptly by the US module Unity, launched by the Shuttle Endeavour.

The third phase of the ISS project, 2000–2004, will see further growth of the station with components from Russia, Europe, Japan, and Canada. The next large module, the Service Module, to be built by Russia and launched by a Proton rocket, is critical for the further development of the station because it contains living quarters, life support systems, electric power supplies, flight control and data processing systems, and a propulsion system. Its completion and launch dates have been delayed, mainly because of continuing funding problems, but the Service Module, Svezola, was successfully launched and attached to the Space Station on July 26, 2000.

Assembly of the ISS in orbit will require a total of 43 Shuttle flights, five flights of Sojuz, and four flights with Proton rockets. At the beginning of station operation, a

Figure 38 International Space Station, ISS. (Source: NASA.)

Sojuz spacecraft will serve as an emergency return vehicle; later, it will be replaced by an American X-38 spaceplane, and possibly by a new European crew return vehicle to be launched by Ariane 5; the same Ariane launcher is also under consideration as a supply and reboost rocket for the space station.

When completed, the International Space Station will offer very prolific and continuing research facilities in a number of areas: research in human physiology, biotechnology, animal and plant physiology, in materials research, physico-chemical research, crystallography, pharmaceutical research, in Earth monitoring, weather and ocean observations, studies of the high atmosphere and ionosphere, astronomical observations, cosmic ray studies, astrophysics, and probably in novel areas of research not yet known today.

Doing research on the International Space Station, and also using the new station for commercial applications, will certainly not come without several problems, among them the high cost of station operation and maintenance, the relatively rare opportunities for researchers to engage in space research projects, the high risk involved in the participation in any complex technical procedure, high insurance rates, excessive bureaucratic requirements, and others. It will be a challenge to Station project managers and operators to make Space Station research attractive and profitable to members of the scientific community.

On the other hand, a permanent station in orbit, well equipped with various facilities for research in many fields, accessible to scientists either via remote control methods, or even by short personal visits to the orbital station without the need for a lengthy astronautical training, financially affordable to research scientists with limited budgets, with convenient transportation into orbit and back to Earth, and with a satisfactory record of trouble-free operation and of safety, will offer an extremely attractive enabling technology for scientific research of unprecedented dimensions.

16. NOVEL TECHNOLOGIES ENABLE GALILEO'S VOYAGE TO JUPITER

A remarkable step forward in the evolution of propulsion systems for spacecraft was achieved during the 1970s by the Messerschmitt-Bölkow-Blohm Corporation (MBB) in Munich. Up to that time, each of the various (and often numerous!) propulsion modules onboard a spacecraft that are needed for orbit injection, attitude control, orbit correction, orbit changes in altitude and inclination, docking maneuvers, and reentry initiation, had to be equipped with its own tanks for fuel and oxydizer, propellant feed systems, tanking and de-tanking provisions, and the necessary control instruments.

The clumsiness and inefficiency of that situation became so obvious that the decision to develop a better system was almost unavoidable. MBB designed and built a "unified propulsion system," UPS, for satellites and space probes (Hopmann 1999). With a UPS system, all the thrusters onboard a spacecraft, large and small, are fed from the same fuel tank and the same oxidizer tank. A spacecraft would contain one "propulsion module," standardized according to the mission, the size, and the design lifetime of the spacecraft, with one set of pressurized propellant tanks, including feed lines and valves for each of the thrusters on the spacecraft. This simplification would lead to considerable savings not only in the mass and complexity of a spacecraft, but also in the effort needed for its design, manufacturing, testing, launch preparations, operation, and possible repair. The payload capacity of a satellite or space probe using the UPS system would increase by up to 50 percent.

That principle of a unified propulsion system would be equally applicable to all spacecraft, independent of the type of launch vehicles that were to put them either into orbit, or on a deep space trajectory. A UPS module was first applied on the Symphonie communication satellite, an early joint French–German project that was started in 1967 and launched by a Thor Delta rocket in 1974.

The first use of a UPS system on a planetary project came with the Galileo spacecraft that was to travel to Jupiter, send a capsule deep into Jupiter's atmosphere, and then enter into a satellite orbit around the gigantic planet. During the following two years, Galileo was to swing by each of Jupiter's four Galilean moons, Io, Europa, Ganymede, and Callisto. In 1975, NASA invited the European Space Agency to be a partner in this project, and in December 1977, based on earlier bilateral agreements between NASA and the

Figure 39 Spacecraft Galileo on its way to the planet Jupiter. (Source: NASA.)

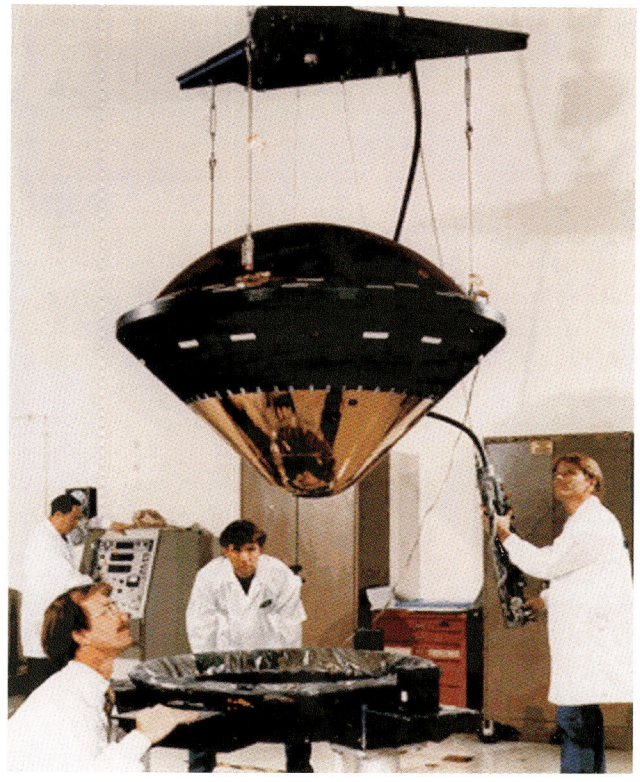

Figure 40 Entry probe, launched from the Galileo spacecraft into the atmosphere of the planet Jupiter. (Source: NASA.)

BMFT (Bundesministerium für Forschung und Technologie, German Ministry for Research and Technology) the DFVLR (Deutsche Versuchsanstalt für Luft- und Raumfahrt, German Agency for Air- and Spaceflight) gave a contractual assignment to MBB in Munich to build the spacecraft, including its propulsion system, and also the necessary accommodations for 17 major experiments to be prepared by scientists in several nations. Spacecraft integration as well as mission control were assigned to JPL in Pasadena, California (Goldman 1999).

It soon became obvious that the propulsion requirements of the probe, which included numerous maneuvers for trajectory corrections and changes, and also the maneuvers to and around the Jupiter moons, could never have been fulfilled with a collection of individual propulsion systems, rather than with the UPS system.

For the launching of the Galileo spacecraft from Earth into orbit, and for its injection into the transfer trajectory from Earth orbit to Jupiter, the combination of a Shuttle and a Centaur upper stage was chosen. As the project began, a launch date in 1982 was envisioned; however, technical problems and necessary redesigns caused this date to slip until May, 1986. Five months earlier, the terrible Challenger accident occurred which caused NASA to rule out the use of the Centaur stage with its hydrogen propellant tank from all Shuttle flights. With the loss of Centaur as a transfer stage, another, weaker solid propellant stage had to be substituted, which made it necessary to apply planetary swing-by maneuvers to the Galileo spacecraft on its way to Jupiter. The propellants used by Galileo's UPS system were monomethyl hydrazine as fuel, and nitrogen tetroxide, mixed with a small amount of nitric oxide, as oxidizer.

Galileo was finally launched on October 18, 1989. Achieving three gravitational maneuvers – at Venus on 2/1990, at Earth on 12/1990, and again at Earth on 12/1992 – it flew by the asteroids Gaspra on 10/1991 and Ida on 8/1993, taking pictures of both. When Galileo had approached Jupiter to a distance of 200 000 kilometers, it released the probe into Jupiter's atmosphere, which promptly sent back data about the planet's atmospheric structure to Galileo, from where the data were then transmitted to JPL. A few days later, Galileo entered into a satellite orbit around Jupiter where it remained for about two years, interrupted by "side trips" to the vicinity of the inner Jovian moons Io, Europa, Ganymede, and Callisto.

At the time of this writing (1999/2000), Galileo is flying away from Jupiter toward an encounter with the spacecraft Cassini which will pass by Jupiter in December, 2000 on its way to the planet Saturn and its moon Titan. Cassini will pick up extra speed by a gravitational maneuver near Jupiter. While Galileo and Cassini are relatively close to each other, they will take pictures and collect data of Jupiter at the same time, but from different locations in space.

The Galileo Project had its problems, too. Without permission to use the Centaur stage, the project engineers had to redesign the project for a weaker transfer stage, and for three gravitational maneuvers. The spacecraft lost its main antenna which could not unfold; therefore, it had to transmit all data at a slower rate through an auxiliary antenna; the tape recorder was sometimes erratic; however, the spacecraft could still be kept working, and it furnished an unbelievably rich treasure of excellent pictures and data about Jupiter, its inner moons, a comet hitting Jupiter, and two asteroids. Galileo owes its outstanding success to the combination of several enabling technologies, each of them contributing decisively to a near-perfect success of an ambitious project: the unified propulsion system; the gravitational swing-by maneuvers; the fabulous electronics at the heart of the guidance and control system, of the cameras, of the programming systems, of the communication systems, and of the receiving and data handling systems on the ground at JPL; and, last but not least, the countless "women and men in the loop" who made and kept the systems working properly.

During 2000 and 2001, Galileo continued to orbit and to observe Jupiter. Also, it took very impressive pictures of Jupiter's moons Io and Europa.

17. MISSIONS TO MARS

The first successful probe to Mars, JPL's Mariner 4, was launched in 1964 with an Atlas Agena. Passing by Mars at a distance of 10 000 kilometers, it took and transmitted 21 pictures of Mars. Numerous further Mars probes followed, launched by Soviet Russia and by the United States, but only with partial success. An exception was Mariner 9, launched with an Atlas Centaur in May 1971, which achieved an orbit around the planet from which it took about 7000 photos. Then, in 1975, the US launched two heavy spacecraft to Mars, Viking I and Viking II, with powerful Titan Centaur rockets. Both Vikings, landing on the surface of Mars, became brilliant successes as scientific probes to our neighbor planet (Chapter 9). However, it would take twenty years before another Mars program was undertaken. In 1996, JPL started an ambitious Mars program, supported by the Lockheed Martin Astronautics Corporation in Denver, Colorado, as developer and builder of the spacecraft.

In this program, called the Mars Surveyor Program, two Mars probes are to be launched every two years. The series began with the launching of the Mars Pathfinder in December 1996. The probe, with an Earth weight of 360 kg, landed without going first into an orbit, on July 4, 1997 in Ares Vallis. An 11 kg roving vehicle, Sojourner, subsequently explored the neighborhood, analyzed rock and soil samples and took pictures. The lander also photographed the landscape, took atmospheric and other data, and communicated with Sojourner and Earth. The landing site had been chosen in an ancient flood plain whose characteristic surface forms could be clearly seen. Pathfinder and Sojourner furnished a wealth of Mars data until September 1997.

Shortly after Pathfinder, but still within the same "launch window," the Mars Global Surveyor was launched. The spacecraft first settled in a highly elliptical orbit; a slow transition into a low, circular, polar orbit was achieved through a number of subsequent aerodynamic braking actions in the tenuous Martian atmosphere. The Surveyor then began to undertake a mapping program of the Martian surface, with a simultaneous close observation of atmospheric conditions. Data are taken 24 hours every day and transmitted to Earth once per day. Also, a short period of live data transmission is inserted on every third day. Data transfer occurs at 40 000 to 80 000 bits per second.

Two years after Pathfinder and Surveyor, the Mars Climate Observer was launched on December 11, 1998 with a Boeing Delta II rocket. Equipped with instruments to study atmospheric and surface conditions, and also to look for traces of water vapor, the satellite first settled in an elliptical orbit, 160 by 40 000 kilometers high. A number of propulsion commands from Earth were to change that orbit into a circular, polar, 50 km orbit. Unfortunately, the spacecraft did not achieve that orbit. Technical information transmitted from the Boeing team in Denver, Colorado to the team at JPL in California was given in the traditional English inch-pound system, while JPL has long been using the international meter-kilogram system, like all the nations on Earth, with the sole exception of Nigeria, Myanmar, and parts of the United States. The consequence of that error was that in September, 1999, the Climate Observer was commanded too deeply into the Martian atmosphere. A crash landing on the surface, with total loss of the spacecraft, was unavoidable – an accident which is hard to understand, and even harder to excuse.

On January 3, 1999, a Delta II rocket launched JPL's Mars Polar Lander. It carried, as part of its payload, two detachable probes, DS 2, which were to separate from the spacecraft as it approached Mars, and to impact on the surface. On each of the two probes, a two meter long, thin sensor with microprobes at its tip was to penetrate into the Martian soil, and to search for traces of water. The probes would transmit observational data to the orbiting Global Surveyor for further transmission to Earth. The landing spots for the probes were selected very near the icy shield around the South Pole of Mars where the surface consists mainly of "layered terrain of ice and dust."

The Polar Lander, also in the near-polar region, would take and analyze soil probes, collect atmospheric data, and

Figure 41 Sojourner roving vehicle as used on the Pathfinder Mars mission, 1997. (Source: NASA.)

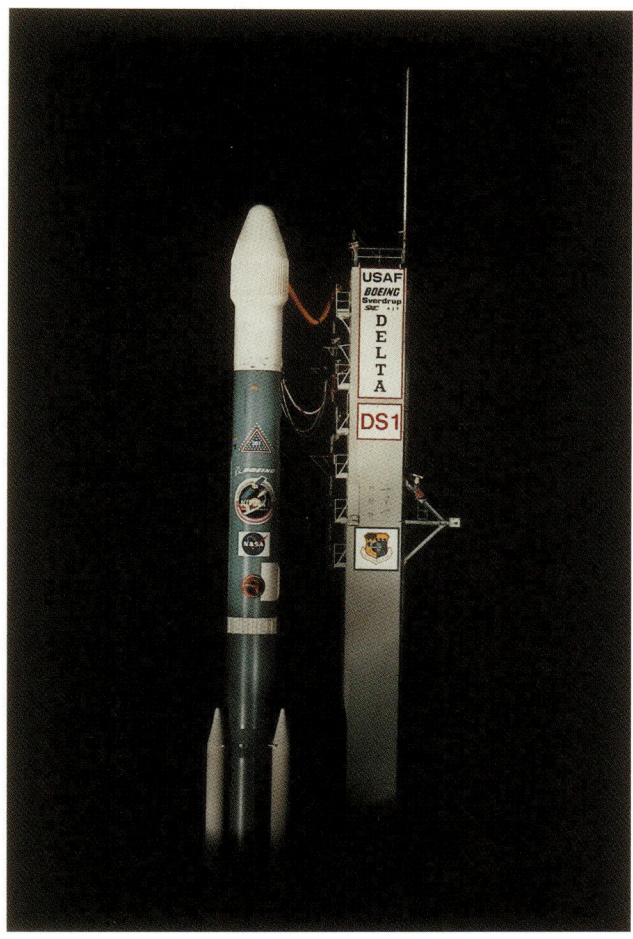

Figure 42 Boeing's Delta II rocket. (Source: NASA.)

look for signs of water. As a second casualty in NASA's Mars program during the 1998–1999 window, the Polar Lander disappeared from the earthbound receiving antennae at the time when it was supposed to touch down on Mars. The two DS 2 probes, likewise, did not send any signals to the Global Surveyor overhead. At the present time (1999–2000), the reason for the death of the Mars Polar Lander is still a matter of speculation.

A number of orbiters and landers for our neighbor planet are planned by NASA in the United States, by ESA in Europe, and by Russia and Japan. Some of the landers will try to return soil samples to Earth. The US plans are, however, being revised in an effort to avoid a recurrence of the past mishaps.

For all these Mars probes, past and future, the most decisive common elements among the enabling technologies are the Earth-to-Mars transportation systems, including aerobraking, the control and communication systems, and the sensors for scientific data. Most of the powerful rockets developed during the past fifty years have also been used, or will be used, to propel spacecraft to Mars: Semiorka, Proton, Atlas-Agena, Atlas-Centaur, Titan-Centaur, Delta II, Ariane, H II, and the Shuttle. The most powerful rocket so far (which also had the most perfect launch record, 100 percent, of all rockets!), Saturn V, was abolished by NASA before it could have been used for a mission to Mars.

In the meantime, the Mars Global Surveyor has provided a huge number of pictures and other data about the Martian surface. The reaction caused by the wealth of new and often bewildering details was expressed by Ken Edgett, in charge of the Surveyor cameras: "We are seeing a Mars you are not going to understand unless you go there!"

18. THE GREAT TELESCOPES

When early space flight pioneers were asked in what way orbiting satellites might be useful to people on Earth, they usually mentioned, as their first choice, space telescopes that could observe the stars and other heavenly objects without the disturbing influence of the Earth's atmosphere. Actually, telescopes were not the first instruments flown on space probes and satellites during the 1950s, simply because they demand larger payloads, and greater pointing accuracy and stability than was available on these first spacecraft. As the Saturn-Apollo program approached its end, astronomers at NASA headquarters suggested that components left over from this program should be used to put telescopes into satellite orbits. These proposals led to Skylab, the first orbiting space station launched in 1973 by the United States (Chapter 11). Its broad program for scientific studies included a set of solar telescopes to observe the Sun in various wavelength regions, including X-rays, and several instruments for stellar astronomy.

Extended opportunities for astronomical and astrophysical observations were offered by the High Energy Astronomical Observatory family of three satellites launched in 1977, 1978, and 1979 by Atlas Centaur rockets. These missions, proposed and managed by the G.C. Marshall Space Flight Center with cooperation from the Goddard Center, were devoted to the study of celestial ultraviolet and X-ray sources, and of cosmic rays. They provided substantial new knowledge about quasars, pulsars, black holes, and other exciting stellar objects (Ruppe 1982).

Plans for a large space telescope were pioneered primarily by Lyman Spitzer, astronomy professor at Princeton University, from the 1950s on. Stimulated and encouraged by his active interest in space telescopes, members of the Marshall Center began in the mid-1960s to make design studies for a large orbiting telescope with a three-meter mirror, and to seek support for such a project among astronomers at universities and observatories, and at

NASA headquarters. Many astronomers were enthusiastic about the prospect of having a large telescope in orbit. Others were reluctant – "you will never be able to control it within a fraction of an arcsecond" they said – and some were downright against it – "how much will it cost? One hundred million dollars? For that money, we can build six Palomars, and that's what we really want!" – "Yessir," members at the Marshall Center answered, "but all six Palomars together cannot give you a resolution better than about 1 arcsecond, and they can let you see only the very nearest ultraviolet!"

Dr. Spitzer set up a model of a gyro-stabilized angular control system for a telescope in the basement of his laboratory at Princeton and proved that the accuracy requirements of a three-meter space telescope can be met, while at the Marshall Center in Huntsville, design work for the "Large Space Telescope" (LST) progressed as an engineering study project, encouraged and supported by Dr. Spitzer. At the Marshall Center, engineers and scientists read the acronym LST as "Lyman–Spitzer Telescope." In 1971, Jesse Mitchell at NASA headquarters assigned the LST formally to the Marshall Center as a development project; the Goddard Space Flight Center in Maryland became a participant in the project. Dr. Charles R. O'Dell, Director of the Yerkes Observatory, agreed to move to Huntsville as chief scientist of the LST project. In 1986, the big telescope was ready to be launched with a mirror diameter of 2.4 meters, a reduction that had become necessary for cost reasons.

Most of the technical requirements for the LST surpassed those of every one of the satellites that had been built so far. Pointing accuracy, long-time attitude stability, freedom from mechanical vibrations, temperature control and stability, and also controllability of the observing instruments, data and picture transmission, and two-way communication – all of these features of the telescope project demanded unusually high degrees of accuracy and reliability.

Launching of the LST had to wait until May 1990 when a Shuttle became available. Soon after deployment in orbit, it turned out that the mirror did not have the correct shape to provide the expected picture quality. A human error had occurred during testing procedures at the Perkin Elmer Corporation where the mirror had been manufactured.

Under the directorship of von Braun, the function of a powerful "quality assurance directorate" had always been a major part of his organization; quality assurance representatives took part in the testing of all the major parts of Marshall Center projects. After von Braun's departure from Marshall in 1970, NASA decided that quality assurance functions by a Center are not necessary, and ordered the discontinuation of the practice.

The error in LST's mirror testing made it necessary to design and build a corrective lens system, to capture the LST in its orbit, to re-attach it to a Shuttle, and to place the corrective lens system into the LST structure, a procedure that required considerable extravehicular activities on the orbiting Shuttle. The operation was successful; the LST came to life and soon began to produce those fabulous pictures of celestial objects that are of equal delight to astronomers and to lay people. Their scientific value is immeasurable. The LST, now called the Hubble Space Telescope, became a huge success, furnishing "about one scientific discovery per week," as one of the happy participating astronomers put it.

The Hubble Space Telescope was not only an impressive accomplishment for the astronomical sciences; it also proved the great value of the Space Shuttle as a launch system, and even as a repair station, that enables a wide variety of outstanding research in space.

The first X-rays in space were discovered during the late 1940s by T.R. Burnight, R. Tousey, and H. Friedman with modest instruments flown to high altitudes by old Peenemünde rockets launched at the White Sands Proving Ground in New Mexico. These X-rays came from the Sun. The first extrasolar X-ray source was discovered in 1962 by Riccardo Giaccconi, Herbert Gursky, Frank Paolini, and Bruno Rossi. More X-ray-emitting sources were found by NASA's first Small Astronomical Satellite (SAS-1), better known under its name "Uhuru," following its launch in 1970 from the Italian San Marco Platform off the coast of Kenya. More X-ray-emitting sources were found by Uhuru; it found the first identifiable black hole in 1972 in Cygnus X-1. Images of stellar X-ray sources were obtained from Skylab in 1973, and from HEAO II in 1978. Imaging telescopes for X-rays can be built with Wolter mirrors which consist of two cylindrical, coaxial mirrors, mounted in tandem, one with a parabolic, the other with a hyperbolic surface, which are designed for grazing incidence, and therefore have a relatively high reflection coefficient, for the impinging X-rays.

A large X-ray telescope for a satellite project, launched in June, 1990 by a Thor Delta rocket, had been proposed by J. Trümper at the Max Planck Institute for Extraterrestrial Research (MPE) in Garching near Munich (Trümper 1999). After extensive studies and technical preparations, the X-ray satellite, named ROSAT and weighing 2.4 tons, was built during the 1980s at the MPE, containing a Wolter mirror manufactured by Zeiss. ROSAT worked until December 1998, for more than eight years. The satellite carried three cameras, two for X-rays, and one for extreme ultraviolet. ROSAT pursued two main objectives, a comprehensive survey of X-ray sources, and a study of individual X-ray objects. In both areas, the satellite was extremely successful, and yielded results that are described in other articles of this book.

Figure 43 The Hubble Space Telescope, built at the Marshall Space Flight Center, shortly after its release from the Shuttle Discovery in orbit on April 25, 1990. The front end of the telescope is still closed by a protective cover. (Source: NASA.)

Figure 44 The Chandra X-Ray Telescope, called AXAF before its launching, built at the Marshall Space Flight Center in cooperation with the Smithsonian Harvard Center, was launched in July, 1999. (Source: NASA.)

ROSAT produced a rich harvest of information and knowledge about cosmic X-rays during the decade of the 1990s. It discovered about 150 000 new X-ray sources, among them more than 500 in the Andromeda galaxy alone, about 200 remnants of supernovae, and 34 neutron stars. No less than 4000 scientists from 26 nations participated in, and profited from the ROSAT Project.

As a successor to ROSAT, an advanced X-ray satellite, ABRIXAS, was developed and built by the OHB Corporation in Bremen. Equipped with seven Wolter mirrors, it was supposed to make an all-sky survey like ROSAT, but with greater accuracy, and with higher resolution. ABRIXAS was launched with a Russian Kosmos rocket in April, 1999 from Kapustin Yar. However, the project suffered from a failing battery shortly after launch; this caused the total loss of the mission.

From the early 1990s on, a large X-ray satellite project, AXAF, had been under development in the United States as a kind of "sister observatory" to the Hubble Space Telescope which observes in the optical part of the spectrum. AXAF has one large X-ray mirror, consisting of four nested Wolter mirrors, which will achieve an angular resolution of the images of better than 1 arcsecond and, at the same time, a high spectral resolution. The Harvard Smithsonian Center is in charge of scientific operations, and of the overall control of the mission. The TRW Corporation in California built the spacecraft. Like in the Hubble Project, responsibility for the technical management of the AXAF Project has been assigned to the Marshall Center in Huntsville; this included the testing of the X-ray mirror and its sensors in Marshall's large X-ray test facility.

AXAF was launched onboard the Shuttle Columbia on July 23, 1999. After being unloaded in orbit, a two-stage solid-propellant rocket system pushed the satellite into a high elliptic orbit. Then, an onboard liquid-propellant propulsion system took over and applied a sequence of perigee and apogee kicks until AXAF had reached a 10 000 km by 140 000 km elliptic orbit. When it had arrived in its final orbit, the satellite was named Chandra in honor of the late Professor Chandrasekhar. By the time of this writing, Chandra had already furnished breathtaking pictures of celestial X-ray sources.

An even more sensitive X-ray satellite, ESA's X-ray Multi-Mirror Mission (XMM), was developed and built by Dornier Satellite Systems in Germany in cooperation with a large number of other companies. The European Space Operations Center (ESOC) in Darmstadt, Germany, is in charge of mission control. Science Operations are centered at ESA's Villafranca Satellite Tracking Station near Madrid. XMM carries three mirrors, each with 58 nested Wolter mirrors that had been produced with a replication procedure and aligned at the Media Lario firm in Italy. The XMM also carries a normal-incidence optical telescope that will allow comparisons between the optical and X-ray images of individual objects. The satellite was launched by the new Ariane 5 rocket from the Kourou launch site into a highly elliptical orbit in December, 1999 (Still 1999).

Figure 45 XMM-Newton Telescope. (Source: ESA.)

Japan's large X-ray satellite, the Astro-E project, unfortunately was lost as a consequence of a launch-vehicle failure.

The third member of the family of great telescopes launched by the Shuttle is the Compton Gamma Ray Observatory, CGRO. Conceived as a successor to the gamma ray satellite COS B, launched in 1975 with a Thor Delta rocket under the auspices of the newly formed European Space Agency (ESA), CGRO was designed to detect, register, and analyze the mysterious gamma-ray bursts that had been observed by COS B and other properly instrumented satellites for years. The CGRO satellite carried several instruments to measure the energy flux, and also the approximate direction of individual bursts. On the average, it recorded about one burst per day. The total energy emitted during each of these bursts is truly unimaginable; some of the bursts obviously emit as much energy within a few seconds as the Sun emits during its entire lifetime. CGRO was developed and built under the direction of the NASA Goddard Space Flight Center, supported by other organizations, including the G.C. Marshall Space Flight Center in Huntsville. The satellite, which had been launched by the Shuttle Atlantis in April, 1991, had to be brought to a controlled atmospheric reentry in June, 2000.

A new gamma ray observatory, ESA's INTEGRAL, is presently in preparation at Alenia Aerospazio in Italy, and at numerous scientific institutes in Europe, Russia, and the United States. INTEGRAL will be launched in 2002 by a Proton rocket from the Baikonur launch site in Kazakstan.

Of all the scientific disciplines that have been touched by the evolution of rocketry and spaceflight technologies during the twentieth century, none was changed as profoundly as astronomy. Indeed, our present understanding of the Universe is very different from our rudimentay knowledge a century ago. It was the creation of novel technologies that made the transition possible. Most decisive was the development of the modern precision rocket which began in Russia, the United States, and Germany during the late nineteenth and early twentieth century, and which was brought to a first stage of perfection at Peenemünde during the late 1930s and early 1940s. After the war, the development continued in Soviet Russia, the United States, and a growing number of other countries. The rocket provided the transportation systems that enabled observations of space objects without the disturbing and limiting effects of the Earth's atmosphere. Parallel to the development of adequate transportation means into space, there was the rapid

Figure 46 The five members of the European Ariane family. (Source: ESA.)

development of novel technologies that made possible the gathering and transmission of data observable in space. It was the development of modern semiconductor electronics that gave us the transistor, the CCD, the chip, microsensors, photovoltaic cells, the miniaturization of measuring and communication devices, and, very particularly, the high-speed computers without which the explosive development of scientific research and knowledge during the second half of the twentieth century would have been unthinkable.

19. EUROPE DEVELOPS THE ARIANE FAMILY

During the 1960s, some European nations tried to implement a joint program for the development of a "Europa rocket" that would provide the member states with the opportunity to participate in the evolving trend of rocketry to become an important element in global commercial enterprises. An intergovernmental organization was established to this end, namely the European Launcher Development Organization (ELDO). This was not successful, however, although a number of industrial enterprises in various nations had started programs to develop components for rockets and spacecraft. Then, mostly under the leadership of the Centre National d'Etudes Spatiales (CNES) in France, several French corporations worked out a plan for a three-stage rocket system, using some components that had been developed during previous years.

The French government suggested that the CNES rocket design be implemented within the framework of a new

Figure 47 Kourou, the European rocket launching facility at Kourou, French Guayana. (Source: ESA.)

European Space Agency (ESA), which would take over the functions of ELDO and its sister organization, the European Space Research Organization (ESRO).

In 1973, ministers of several European nations held a Council Meeting on Ministerial level in Brussels. They accepted France's proposals, and by 1975 the new ESA had been firmly established with eleven member nations. They proposed that a three-stage rocket be developed and built in a joint effort under the leadership of France. The rocket was named Ariane, in memory of the Greek princess who, with her silk thread, helped Theseus find his way out of the chaotic labyrinth of King Minos. Ariane 1, first launched in 1979, was capable of transporting 4.5 tons of payload into low Earth orbits, and 1.8 tons into geostationary orbits. The first stage contained four, the second stage one engine of the Viking type, a rocket motor developed at Vernon in France by H. Bringer, who had been a member of von Braun's Peenemünde team during the war. While Viking engines show traits inherited from the A-4 rocket, they burn UDMH and nitrogen tetroxide (alcohol and liquid oxygen had been used for the A-4). The third Ariane stage was powered by the HM7 engine, a motor burning liquid hydrogen (LH2) and liquid oxygen (LOX). This was the first hydrogen–oxygen rocket engine developed outside of the United States, the product of a joint effort by the French Société Européenne de Propulsion (SEP) and the German Messerschmitt-Bölkow-Blohm (MBB) corporation.

As early as 1964, CNES began building a rocket launch facility at Kourou in French Guayana, 500 km north of the equator. The first rocket launched there was a Diamant in 1970. Ariane's first launch attempt on December 15, 1979 in Kourou did not succeed. Neither did the second try a few days later. Then, on December 24, the project enjoyed a full success, but another failure occurred in 1980. Finally, in 1981, Ariane was ready for use.

Growing rocket payload weights, particularly of communication satellites, made the development of the next Ariane generation necessary, Ariane-4. At the same time, the possibility of adapting the payload capability to a variety of payload weights became desirable. This requirement was met by the optional use of a varying number of clustered engines in the first stage, and by attaching two or four solid propellant boosters to the core of the rocket. The maximum thrust of the first stage was 540 tons. With that initial thrust, Ariane-4 could place payloads of 4.9 tons into geostationary orbits.

Prospects of a growing interest of the communications industry in satellite launchers led to the foundation of Arianespace, a commercial corporation for the manufacturing, marketing, and operation of satellite launchers. In May, 1980 Arianespace signed, as a private enterprise incorporating 36 European companies and 13 banks, its first contract with the French firm Aerospatiale for the production of Ariane rockets.

The Ariane Project became a remarkable success. Between 1981 and 1998, about 120 Ariane rockets were launched for customers from about 20 nations around the world. Arianespace was able to capture more than half of the open market for launch services. Two factors contributed to this impressive growth of the Ariane enterprise: first, the easy adaptability of Ariane-4 to most of the individual needs of customers, and second, several mishaps that occurred to the US launchers Atlas and Titan during that time period. Also, the tragic accident of the Challenger Shuttle cast a dark shadow on the American space program at that time.

While Ariane was primarily a commercial enterprise, there were also several purely scientific projects that profited from Ariane's capabilities, such as the Halley probe Giotto that was launched on an Ariane-1 in 1985. Ariane-4 vehicles brought the German and ESA communication satellites Kopernikus (1989) and Olympus (1989), as well as ESA's astrometry satellite Hipparcos (1989) into orbit. About 70 percent of the Ariane-4 rockets served the telecommunications market, while 30 percent launched satellites and space probes for science projects, and for Earth observations.

In response to the growing demand for Ariane-4 as a launcher for commercial satellites, and also to the continuing rise of the weights of communication satellites, the need for a new generation of Ariane rockets arose during the 1980s. The Ariane-5 program, including the development of a large cryogenic engine, was approved by the ESA member states in 1985. Very interestingly, among a great variety of possible designs, a concept for the proposed large Ariane system was finally adopted that resembled that of the Space Shuttle – not the latter's reusability, though. Ariane-5 uses a hydrogen–oxygen engine, the Vulcain, in the core of its first stage which delivers 88 tons of thrust at take-off; two solid propellant booster rockets with a combined take-off thrust of 1050 tons; and an upper-stage engine, AESTUS, developing 2.7 tons of thrust, burning MMH and N_2O_4. Ariane-5 is capable of delivering 22 tons of payload into a 300 km orbit, 18 tons into 550 km orbits, and 6.8 tons into a transfer trajectory to geostationary altitudes. In comparison, the Shuttle can deliver 30 tons into low Earth orbits. Saturn V was able to put up to 130 tons of payload into low Earth orbits, and 45 tons into a transfer trajectory to the Moon.

ESA has accepted overall responsibility for the Ariane-5 project. Program managemant was assigned to CNES, systems integration to Aerospatiale. Substantial contributions have been made by SEP, DASA (which includes the former MBB), and numerous other European firms from a total of 12 member states of the ESA organization.

The Ariane-5 rocket system was developed and built primarily as a business enterprise aiming at a lucrative worldwide commercial market. Its first demonstration flight, in June 1996, was supposed to bring the first four-satellite cluster system into orbit. A technical failure prevented this; therefore, two, rather than one, demonstration flights followed; the third flight was completely successful. The first user flight then successfully served pure science: On December 10, 1999, an Ariane-5 rocket launched the X-ray Multi-Mirror satellite XMM (now called XMM-Newton) into a transfer trajectory to its 7000 km by 114 000 km orbit.

20. "CLASSICAL" ROCKETRY AT THE TURN OF THE MILLENNIUM

As the science, the art, and the technology of rocket-building evolved during the past century, and as rockets were increasingly recognized as ideal means for space transportation, four distinct lines of activities involving space rockets developed: Rockets for military purposes, rockets for scientific observations and research, rockets for Earth monitoring, and rockets for commercial uses. All four activities profited from the countless technical innovations that continuously improved capabilities, efficiencies, accuracies, and reliability of rocket systems during the past 60 or 70 years.

As military weapons, the modern intercontinental ballistic missiles (ICBMs) with nuclear warheads are credited with the accomplishment, by their mere existence, of having prevented the outbreak of World War III. It is a happy thought for the rocket designers and engineers that not one of the large guided missiles developed and built during the past 55 years has ever been launched in anger.

As instruments of scientific research, space rockets have provided us with an expansion of our knowledge about the structure and history of our Universe never believed possible a hundred years ago.

As outposts for the observation of our Earth, space rockets have given us not only an instantaneous global weather service, but also an appreciation of the fragility and vulnerability of our home planet, and of the urgent need to preserve its integrity.

As elements of commercial transactions, the significance of rockets and satellites is best characterized by a remark in *Time* Magazine in June 1999: "The business volume of satellite-related work in the US amounts to $316 billion per year." In October, another report quoted "more than $500 billion annually."

Fueling that business are a number of activities: Earth observation for a worldwide weather service; agricultural and urban developments; surveys of natural resources; monitoring of ocean states; global positioning systems; communication links for business, private and other nonmilitary users, among them: news services; travel agencies; law enforcement agencies; airline traffic controllers; and other functions. Finally, satellites provide an almost unlimited amount and diversity of information upon the clicking of a few keys.

It is not surprising that in this situation, more and more countries try to make inroads into this fast-growing business by developing rockets capable of launching satellites: Europe with Ariane, Russia with Proton, Japan with H2, China with CZ (Chang Zheng, "Long March"), India with its SLV and PSLV rockets are presently active; Brazil and Korea are potential future members of the club. There is no doubt that several of these rockets are developed with an eye on their military potential, but at least some of them will be serious competitors on the international market of satellite launchers.

Among the urgent needs of present and future users of space transportation systems is a "spaceplane," capable of taking payloads up to about 6 tons, or crews up to about 6 persons, into orbit, or from orbit back to Earth. The spaceplane will be launched from an airplane, such as a Boeing 747, that will take off from any suitable commercial airport and deliver the spaceplane, propelled by a single-stage-to-orbit rocket engine, into a trajectory suitable for its ascent to a specified target in orbit. After a stay of a few days in orbit, the spaceplane will return and land in glider fashion, again on a suitably located large airport. The "Venture Star," a project presently under development in the US, may fill that need.

Another need will arise when very heavy loads, such as large space station components, or systems for a lunar settlement, or for expeditions to the planet Mars, have to be hauled into orbit or directly into a deep space trajectory. At

Figure 48 "Space Express" spaceplane, featuring an air-breathing rocket engine, is testing novel technologies for reausable space vehicles. (Source: Marshall Space Flight Center.)

that time, a worthy successor to Saturn V will have to come into existence.

During recent years, NASA began developing several medium-size Earth-to-orbit rocket systems, all of them substantially smaller and less expensive than the Shuttle, among them the X-33, X-34, X-36, X-38. Innovations of rocket engine technology, such as the "aerospike" engine, are under development. Also, some "new" launch-assist technologies have again come under study that had been considered in the past, but discarded because they did not appear promising, such as an Earth-bound magnetic sled that accelerates the space rocket on a horizontal rail before the rocket engine is ignited, or the use of ramjet or scramjet engines as auxiliary thrusters, or the intake of ambient air as oxidizer for the rocket motors. Although begun with much optimism, progress of these new projects has been slower than hoped for. In fact, some of them (X-33 and X-34) had to be discontinued recently because they did not live up to the expected performance requirements.

It is very likely that the two principal space transportation systems needed in the near future, the Earth-to-orbit space plane, and the post-Saturn (and post-Proton and post-Ariane) heavy duty launch rocket, will be propelled by "classical" rocket motors of refined and advanced designs, but not by systems based on entirely novel propulsion systems.

21. ADVANCED ROCKET SYSTEMS FOR PRESENT AND FUTURE SPACE EXPLORATION

The second half of the twentieth century brought a flood of new knowledge of our cosmic environment, gained by probes sent away from Earth into deep space. The common feature of their discoveries is the way our instruments were transported to their destinations. All of them were sent by chemical rockets, using a technology that has been known and used for a thousand years. During most of the time since the first crude rockets accelerated the arrows of ancient eastern warriors, rockets were of low performance, and of poor accuracy. The turn toward a powerful machine of high precision and reliability came with the Peenemünde rocket during the early 1940s. Mass-produced and deployed as a military weapon while still immature, it enabled, after the war, the first extended program of scientific space research in White Sands, and it also served as a role model for countless postwar rocket projects in many countries.

Rocket technology has improved decisively since the first Peenemünde rocket reached the lower fringes of outer space in 1942; however, the basic principle of chemical rockets has remained the same. It is very likely that the transportation of all kinds of payloads destined for near-Earth space, even as far as the Moon, will continue for a long time with chemical rockets, but missions to more distant targets, among them Sun, planets, comets, asteroids, the Kuiper Belt, and farther out will always encounter restrictions because of the natural limitations of chemical rockets.

There is no shortage of speculations about future deep space propulsion systems. However, when "reality checks" are applied, only very few of them survive.

After the first nuclear reactors had been built and operated, hopes were high for nuclear-powered rockets in which a suitable propellant, preferably hydrogen, would be heated to a high temperature in a reactor and exhausted through a rocket nozzle. During the 1950s, Harold Finger conducted a very impressive development project, ROVER, and showed that a rocket engine with a nuclear reactor as heat source, NERVA, can be built and operated with an exhaust velocity (specific impulse) considerably higher than that of chemical engines. The limit to its performance is dictated by the

Figure 49 The nuclear-thermal rocket motor NERVA in a program during the 1950s to study possibilities of using nuclear thermal reactors for rocket propulsion. (Source: NASA.)

highest temperature that can be accepted by the structural materials of the reactor. Even with the best refractory materials, that temperature was still not high enough to make nuclear engines superior to our best chemical rocket engines, provided that other requirements are also considered that have to be met by nuclear space rockets, such as the heavy shielding needed to protect instruments (and also astronauts!) against the unavoidable strong radiations from the reactor, the need for long-time storage of liquid hydrogen during extended space missions, and the hazard of a nuclear accident occurring during launch operations, or in space.

For these reasons, serious planners of future rocket systems are inclined to exclude nuclear – thermal rockets of the NERVA type from further consideration as possible rocket types that may substitute for chemical rockets on deep space missions. Leslie R. Shepherd, co-founder of the British Interplanetary Society, who wrote one of the first articles about nuclear rockets (*The Atomic Rocket*) in 1948, wrote a paper in 1999 in which he warned: "Future exploration of the solar system on a major scale will require propulsion systems capable of performances far greater than are available with the present generation of rocket engines using chemical propellants." Nuclear-thermal rockets of the NERVA type, he argued, will not offer a significantly higher performance than chemical rockets, and they carry "a heavy burden of radioactivity."

Nuclear rocket systems of the Orion type, which are supposed to use a fast sequence of microbomb explosions behind a "pusher plate" in the rear of the spacecraft, triggered by laser beams, would be capable of producing a considerable propelling thrust. However, no technical base exists for such a system as yet. Also, the problems of harmful radiation affecting the spacecraft would be formidable. Besides, maneuvering near a target, such as a planet, a moon, an asteroid, or a comet, would severely contaminate the target surface.

The best bet for an advanced rocket system that would exceed the performance capabilities of chemical rockets on space missions beyond the Moon, Shepherd concludes, is the nuclear-electric propulsion system, for which a satisfactory theoretical understanding, as well as an extended and very promising technological base, already exist.

22. ION PROPULSION FOR SPACE FLIGHT

Electric rockets were mentioned by the early pioneers of modern rocketry, Tsiolkovskii, Goddard, and Oberth, although without proposals for technical details. A first systematic treatment of the theory of electric rocket systems was published by Stuhlinger (1954), and in a more comprehensive form in 1964. The basic principle of electric space propulsion systems is the same as that of any rocket system: mass is expelled from the spacecraft at high velocity, producing a reactive force which accelerates the spacecraft. Electric rockets eject either ions with an electrostatic field, or a plasma with a combination of electric and magnetic fields.

The physics of electric space propulsion is, however, characteristically different from that of chemical rockets. For instance, while chemical systems should always have an exhaust velocity (specific impulse) as high as possible, electric systems must be designed for an optimum exhaust velocity, determined by the specific power (kilowatt per kilogram) of the onboard power source, and by the length of time the propulsion system has to work on a given mission. The exhaust velocity of ion thrusters is a direct function of the voltage applied to the accelerating grids.

Beginning in the early 1960s, numerous studies of electric thrusters were made, and experimental investigations followed. It soon became evident that the most promising electric propulsion system for long-time, deep space missions would be the ion propulsion system in which a suitable propellant (cesium, mercury, xenon) is first ionized, then the ions are accelerated to a high exhaust velocity by an electrostatic field. A source of electric power is needed onboard to energize this ion beam, either panels of solar-electric cells (solar-electric propulsion), or an electric power plant driven by a nuclear reactor (nuclear-electric propulsion).

In the latter case, the spacecraft will also, like the nuclear-thermal rocket, require an onboard nuclear reactor. The decisive difference is the power level at which each of these two reactors would operate. A nuclear-thermal rocket for large interplanetary spacecraft, unmanned or manned, producing a high thrust for a short time (minutes or hours), will need a reactor producing thermal power on the order of 5000 megawatt at 3000 K for minutes to hours; the reactor will then remain dormant for months or years while the spacecraft coasts toward its target. Then, the reactor is ignited again for a short time while the spacecraft executes approach maneuvers; then, it remains dormant again until it is reignited for the return trip. Only liquid hydrogen has been considered as a propellant for nuclear-thermal rockets so far. Its supply from an onboard propellant tank would have to follow the same pattern.

If the same space mission were driven by a nuclear-electric propulsion system, the reactor would operate continuously over a time period of about one year on manned missions, and up to several years on unmanned missions, producing a low, but persistent thrust force, and at the same time also electric power for onboard use. The thermal power to be generated by the reactor would be 40–50 megawatt at 1500 K, a power level and temperature of the same order at which earthbound commercial power reactors have been operating for many years. Their power level is about 100 times lower than the power level at which the reactor of a nuclear-thermal rocket would have to work.

Two kinds of nuclear-electric power generators appear feasible, the in-core thermionic generator, as used in the Russian Topaz spaceborne power system, developed by G.M. Gryaznov in Moscow (Bober et al. 1992), which was flight tested in 1987 and 1988, and the reactor-heated, turbine-driven generator system as used in earth-bound nuclear power stations, and on ships. In-core nuclear thermionic power generators work without moving parts.

Nuclear power systems are facing the problem of public antagonism because of the potential danger of a radiation catastrophe in case of a reactor accident. However, as described by Stuhlinger (1996), the nuclear reactor on a nuclear-powered ion propulsion space vehicle will be ignited only after the vehicle, launched by a conventional rocket, has left Earth orbit and has reached escape velocity. There will be no chance that a "hot" reactor from a failing spacecraft will ever end up on Earth.

During the second half of the past century, many different kinds of electric thrusters were developed and tested on the ground and in space, among them ion thrusters, plasma thrusters, and arc jet thrusters, and many were successfully applied, particularly on Russian and Japanese spacecraft, but also on European and American satellites. Russian space engineers have been very active in developing and testing electric propulsion systems (Bober et al. 1992).

During the 1960s, scientists and engineers in Russia began using electric thrusters of various designs on spacecraft for attitude and orbit control, and in 1992 a joint venture company was founded with several countries as members: the International Space Technology, Inc. (ISTI), with SEP (Société Européenne de Propulsion), ARC (Atlantic Research Corp.), Space Systems/LORAL, NASA/Glenn Research Center, and Fakel in Russia. The preferred thruster of ISTI is the "Stationary Plasma Thruster" (SPT) which uses the Hall effect (with magnetic fields) to produce a "virtual grid" out of magnetically trapped electrons for the acceleration of the ions. This thruster design has worked on more than one hundred commercial spacecraft successfully during recent years (Voss 2000).

Plasma-dynamic systems, accelerating a mixed beam of ions and electrons with a combination of electric and magnetic fields, permit a simpler and more compact thruster design than pure ion systems, but the latter operate at a considerably higher efficiency, and with longer lifetimes (less erosion). Plasma thrusters are sometimes preferred for satellite control systems where they have to work only for short thrusting maneuvers, and where their limited efficiency is acceptable, but ion systems are superior for main propulsion systems on deep space missions.

Two different designs of ion thrusters, very similar with respect to their performance on long space missions, have emerged as the most successful and promising contenders for electrically propelled space missions of the future. In 1960, Harold Kaufman at the Lewis Flight Laboratory (later NASA Lewis Research Center, and still later NASA Glenn Research Center) published his first paper describing

Figure 50 An electron bombardment ion thruster developed by Harold Kaufman, 1960. (Source: Courtesy of Helmut Hopmann, from *Schubkraft für die Raumfahrt*, Stedinger Verlag, Lemwerder.)

Figure 51 A radio-frequency ion thruster developed by Horst Löb, 1962. (Source: Courtesy of Helmut Hopmann, from *Schubkraft für die Raumfahrt*, Stedinger Verlag, Lemwerder.)

an ion thruster with electron bombardment ionization source, and in 1962, Horst Löb at the Justus Liebig University in Giessen, BRD, presented an ion thruster with a radio-frequency ionization source. Both types have attained remarkable levels of efficiency (above 80 percent) and operational lifetimes well over one year. The Kaufman thruster is manufactured by the Hughes Electron Dynamics Division in California, the Löb thruster by DASA (the former Messerschmitt-Bölkow-Blohm Corporation) in Munich, Germany. A modification of the Löb thruster was recently introduced by replacement of the existing accelerating grid system with a new grid technology developed at AEA Technology Culham in England (Groh et al. 1996). A study of various potential electric propulsion missions to interplanetary and planetary targets was presented by Horst Löb and co-workers in 1996 (Löb et al. 1996).

Although electric space propulsion has been the subject of very intense theoretical and technical work over a period of five decades, and although more than one hundred electric satellite control systems have been operating successfully, it was not before 1996 that planning work for a space mission with an electric propulsion system was begun. Stimulated by NASA Administrator Daniel Goldin as part of his New Millennium Program, the Jet Propulsion Laboratory initiated the Deep Space 1 Mission, an ambitious project that featured 12 different advanced technologies during a voyage through the Asteroid Belt (Rayman et al. 1999). Launched on October 24, 1998 by a Delta II rocket, propelled by a Kaufman–Hughes ion thruster energized by a solar-electric power supply, and managed by Marc D. Rayman at JPL, this project became an outstanding success, providing a rich harvest of novel technological data and experience concerning electric propulsion systems. As planned, the spacecraft passed by asteroid Braille at a distance of only 16 km, taking infrared pictures. On September 22, 2001, almost three years after its launch, Deep Space 1 flew by comet Borrelli, at that time more than 200 million kilometers from Earth. The spacecraft, still accepting control signals from the Jet Propulsion Laboratory in California, took two dozen photographs of the comet with its faint coma of dust from a distance of about 2000 kilometers, at a resolution of about 48 meters, and transmitted the pictures to the California ground station! Borrelli's perihelion distance from the Sun is 1.358 astronomical units, just barely within the Mars orbit; its orbital period is 6.86 years.

Encouraged by the outstanding success of the Deep Space 1 mission, NASA initiated plans for a follow-on ion-propelled space mission, Dawn, to be launched in May, 2006. Dawn will explore the asteroids Vesta and Ceres. Quite generally, ion propulsion systems offer performance features superior to those of chemical rockets on space missions such as flights to planets and their moons, to asteroids and comets, probes close to the Sun, and out of the ecliptic plane. Among their superior performance features are shorter flight times, heavier payloads, a continuous ability to perform trajectory changes or corrections, and more electric power available for onboard instruments. – Electric power sources for ion propulsion systems will be solar-electric or nuclear-electric generators.

Ion propulsion systems will also offer an opportunity for manned roundtrip voyages to Mars that could be achieved within about one year, including a three-week staytime on the Martian surface. On such a mission, the spacecraft would be energized by a nuclear-electric power source. One possible scenario of such a mission was described by Stuhlinger (1996).

Figure 52 Deep Space 1 probe with ion propulsion system, launched in 1998 by JPL. (Source: NASA.)

23. A GLIMPSE INTO THE FUTURE

For all time to come, the twentieth century will be the one in which people from Earth reached out, and even traveled, into the vast spaces beyond the Earth's atmosphere. What will the next century, and future centuries, have in store for us? The adage "have vehicle, will travel" will certainly retain its validity. And there will always be individuals in whom the urge to explore the unknown and to discover new mysteries of our world is so irresistible that they will find ways and means to do it. Also, there will always be others who help them invent and build the enabling technologies that will make voyages deeper and deeper into space possible.

Rockets for military purposes, both defensive and offensive, will certainly remain a part of human civilization, and they will continue to furnish a rich supply of novel technologies for non-military uses, provided that mankind continues to live by the age-old proverb that "war is the father of all things."

In the civilian sector, there are several different activities in space that will continue, and even expand: functions directly useful to people on Earth, such as weather observations, communications, and manufacturing of materials and goods in space; operation of instruments for scientific observations by astrophysicists, astronomers, and cosmologists; one-way missions with robotic systems for scientific exploration; landings on other celestial bodies by unmanned and manned spacecraft, the latter with subsequent return to Earth. The common need of all these activities is transportation from Earth into and through space; some of them also need transportation from space back to the Earth's surface.

All of these missions have several requirements in common: a propulsion system, a source of electric energy, a guidance and control system, and a communication system. Missions with astronauts need a life support system that includes air, water, food, temperature control, protection against radiation from the Sun and from deep space, tools and instruments needed for activities on the celestial targets, provisions to assure the continuing good health of the crew members, including their ability to function as active and proficient researchers in their unusual environment. Thanks to past missions, extended technical bases exist for chemical propulsion systems, solar-electric and nuclear-electric power supplies for relatively modest power levels, guidance and control systems, communication systems, and even for life support systems.

While currently available propulsion systems have proven their usefulness for Earth-to-orbit transportation, for manned lunar landing missions, and for unmanned missions with relatively small payloads to Mars and to all other planets except Pluto, neither our existing propulsion systems, nor our existing electric power sources will be sufficient for many of the demanding near-Earth and deep space missions that appear on the wish lists of present and future space explorers.

There are three different schools of thought among contemporary space mission planners.

Serious scientific experts dig deeply into the teachings of modern astrophysics and cosmology in search of possibilities to overcome the limitations to which we are accustomed in everyday life. Among such possibilities is the utilization of the energies set free when matter reacts with antimatter, such as protons (p^+) with antiprotons (p^-); or the controlled use of the fusion reaction ($H_3 + H_1$ or $He_3 + D_2$), well known from the uncontrolled reaction in the hydrogen bomb; or effects implied by General Relativity Theory; or the mysterious zero-point energy of empty space; or even the "wormholes" that were introduced recently into the debate of modern physics.

Although such considerations appear unrealistic to today's space engineers because there is absolutely no indication yet of any possible engineering approach, it may be appropriate to remember that early in the twentieth century, Ernest Rutherford – in fact in 1911, i.e. nearly hundred years ago – discovered the fact that atoms have a small nucleus and a number of electrons orbiting around the nucleus, that there is much empty space between them, and that atoms, on a small scale, enclose enormous energies; he made the statement that we will never be able to utilize atomic energy on a large scale! Likewise, during the 1940s, when the very first jet engines for airplanes made successful flights, Theodore von Kármán, the guru of modern aerodynamics, asserted that jet engines are so inefficient that they can never be used beyond very special short-time applications on military airplanes because their economy is so poor. Who would dare to say that there will never be space propulsion systems beyond those which we can build and use today?

Who would have foreseen, at the beginning of the twentieth century, the atom bomb, intercontinental ballistic missiles, television in every home, the cellular phone, the microchip, laser beams, the internet, worldwide jetliner traffic, the global positioning system, or the supreme reign of the computer? Probably, no author writing in December of 1899 about events anticipated for the forthcoming century would have mentioned any of these novel technologies as being likely to come, although some adventurous minds may already have been dreaming of them.

The second school of thought is represented by young space planners who share with their older colleagues a great enthusiasm for space flight, but not their long-time exposure to the rigors and pitfalls of planning, designing, building, testing, and launching of rockets and spacecraft, their eternal struggles with the basic laws of nature, their disappointments about failures and setbacks, their fights for budgets, and their continuing learning from past mistakes. Nor do they share with their older colleagues the bitter experience of standing before the public, and before members of Congress, when one of their space missions had failed. Misled by the apparent ease with which the gigantic Saturn-Apollo Project proceeded, beginning with President Kennedy's project initiation, and climaxing with the twelve astronauts on the Moon, including the launching of a total of 32 Saturn rockets with not one single launch failure, these young space planners happily declare that they "take their clues from Star Wars," that they "do their mission planning on the computer," and that "within ten or twenty years, people will fly to the Moon in hours, to Mars in days, and to Alpha Centauri in a year or two." They advise the future Mars astronauts "to travel lightly, live on the land, and produce the propellants for their return vehicle locally."

The third school of thought is represented by engineers and scientists who, during past decades (some of them beginning 65 years ago!) contributed, as active team members, to the evolution of today's space program by planning, developing, building, testing, and launching rockets, satellites, and space probes. The common thought they share with schools 1 and 2 is their firm belief in mankind's destiny in space; but, rather than dreaming about future space travel at near-, and even super-light-velocity in a state of artificial hibernation, they take a careful look at those space transportation modes for which enabling technologies are at least in sight, and assess the realistic prospects of space voyages in the decades to come.

According to their sober predictions, traffic from Earth into orbits, and back to Earth, will continue with the proven rocket systems of past missions. Barely 40 years after the very first artificial satellite began its life in orbit, there are today several thousand active satellites in low Earth and in synchronous orbits. All of them were launched by chemical rockets based on technical principles established a long time ago. Surely, rockets with solid or liquid propellants will retain for near-Earth space transportation a role similar to the role played by gasoline motors, diesel motors, and jet engines for Earth-bound transportation. Incremental technical improvements will continue to improve performance, reliability, and cost-effectiveness of rocket motors, but chemical rocket engines will certainly remain the dominant system for transportation between Earth and orbits for a long time to come, even for injecting spacecraft with advanced propulsion systems into trajectories to more distant targets in planetary space, and beyond.

This phenomenon, the persistence of a fundamental technical principle through a long chain of partial improvements and a tremendous widening of its use, is not unique to the chemical rocket. Michael Faraday's invention of the electric motor and electric generator in 1831 is an example, and so is the automobile, invented and introduced by Otto, Diesel, Benz, Daimler, and Henry Ford around the turn of the 19th to the 20th century. Today, there are about 500 million automobiles in use worldwide, representing one of the most important technical systems in our civilization, and also in our economy. As far as efficiency, reliability, comfort, and technical sophistication are concerned, modern cars are vastly superior to their ancestors. And yet, today's automobile engines are still using cylinders, pistons, valves, and crankshafts, and their power-producing cycles, burning gasoline or oil with air, are reciprocating, just as Otto's and Diesel's engines were 120 years ago.

24. WHEN, AND HOW, WILL WE TRAVEL TO MARS?

This author is convinced that eventually, humans will travel to Mars, land there, explore the ancient river beds and crater fields, search for fossils and live microbes, and return to Earth to report what they found. They will not do it hoping for a thriving business, or for winning political votes, or for military reasons, or to solve specific problems threatening our human existence on Earth – they will do it solely because they, and their sponsors, are driven by the urge to explore the mysteries of the unknown, an urge that has been alive in the human species ever since it has existed, and that will persist as long as there are human beings.

Equally strong is this author's conviction that a manned Mars project will not come about in the fashion of the manned Moon project. In 1961, President Kennedy faced the need "to do something great" in view of the Russian Cold War competition. And the main technologies and components needed for the mission either existed, or were in preparation. Also, the envisioned cost of the project – estimated by von Braun to be $20–25 billion – was not extravagant, and it did not scare the President. (The actual cost of the entire Saturn-Apollo Project, reported by Mr. James E. Webb, turned out to have been $23.5 billion.)

In contrast, none of the major components needed for a voyage to Mars are in existence today, such as a powerful launching rocket of the Saturn V type, or a propulsion system for the Earth–Mars–Earth round-trip, or a long-time life support system, or an adequate electric power supply, or a Mars roving vehicle for human occupancy.

It must not be overlooked that in spite of a host of marvelous novel space technologies developed during the past 60 years, and in spite of many thousand man-hours (and woman-hours!) spent under space conditions during that period of time, human nature with all its physiological and psychological properties has not changed much. People were designed and developed for life under terrestrial conditions; their inborn features will have to be accommodated on an expedition to Mars as far as possible by their space habitat and by the flight plan. Planning for a Martian expedition should be guided by some basic requirements: The astronauts, when arriving on Mars, should be in excellent physical and mental condition, so that they can utilize their time on Mars to the fullest extent for the purpose that brought them to Mars: scientific observations and research. Their return vehicle, its tanks full of propellant, should be ready on Mars and checked out for the return flight at the time the astronauts leave the Earth; it was brought up by robotic procedures. Likewise, a radiation shelter, a supply of air, water and food, and a roving vehicle will be sent to Mars robotically before the astronauts arrive.

Experience on Skylab and Salyut, and the Mir space station, implies that the total time the astronauts have to spend away from Earth should not be longer than about one year. Longer periods of exposure to space conditions seem to affect physiological and psychological abilities in unpredictable ways that may seriously reduce the astronauts' ability to function properly. The only Earth–Mars transfer propulsion system capable of a one-year mission time that may become available in the foreseeable future, and for which a good engineering base exists, is the nuclear-electric ion propulsion system; it could make a Mars expedition, including a three-week stay on Mars, possible within 350 to 400 days round-trip time (Stuhlinger 1996).

It would certainly be wrong to expect that the government would initiate a manned mission to Mars, from scratch to completion, with the commitment for an estimated cost of several hundred billion dollars, in one sweeping decree, in the way President Kennedy did for the Moon Project. Fortunately for the Mars Project, this will not be necessary. A manned Mars mission need not begin with a total commitment for the complete project right at its very beginning. Rather, the manned Mars mission will be one project in a program that already began years ago with the early rocket and space flight pioneers, and which has grown almost like a living organism with the Mariner spacecraft, the Vikings, the Russian Mars probes, Surveyor, and Pathfinder, and which will certainly continue in the future with more unmanned Mars missions.

The four major components needed for a manned Mars mission, a heavy-lift Earth-to-orbit launcher, a nuclear source for electric power, an ion drive for the Earth–Mars–Earth round-trip, and a life support system, will be needed for other space projects that are expected to materialize before a manned Mars mission is initiated. The heavy-lift launcher will be needed to serve and maintain the International Space Station, to launch large deep-space probes, and to establish an astronomical observatory on the farside of the Moon. A nuclear-electric power supply will be needed to serve the lunar astronomical observatory, and also future industrial operations in Earth orbit; long-time life support systems are needed for the Space Station, and for operations on the Moon; and ion propulsion systems with increasing thrust levels will be built for unmanned missions to deep space targets. Once these components have been built, tested, and applied on space missions, it will be relatively easy to adapt them to a mission to Mars, and the manned Mars project will become a manageable, affordable, and even logical step in the continuing growth pattern of space exploration and space science.

REFERENCES

Bassner, H. and Löb, H.W. (1983) Status and Application of the RF-Ion Propulsion System RITA, including Interplanetary Missions, *34th Congress of the International Astronautics Federation*, Budapest, Paper IAF-83-391. Electric thruster Development and Testing in *Schubkraft für die Raumfahrt*, Ref. No 26.

Beatty, J.K. (1999). Galileo, an Image Gallery III, *Sky and Telescope*, July, 40.

Belew, L.F. and Stuhlinger, E. (1973). *Skylab, A Handbook*, US Printing Office, Washington, DC 20402.

Bober, A.S., Burgasov, M.P., Kim, V., Koroteev, A.S., Morosov, A.I., Popov, G.A. and Ryzkov, Y.A. (1992). The Development of Electrical Thrusters in the USSR, Paper RGC-EP 92–03. *First Russian–German Conference on Electric Propulsion Engines and their Technical Applications*, Rauischholzhausen/Giessen University, BRD.

Büdeler, W. (1973). *Skylab, Labor im Weltraum*, Econ Verlag GmBH, Düsseldorf (BRD) und Wien (Austria).

Büdeler, W. and Karamanolis, S. (1979). *Spacelab*, Wilhelm Goldman Publisher, München, BRD.

Daimler (1999). *Twenty-five Years of Spacelab*, Daimler Chrysler Aerospace AG, PO Box 28 61 56, 28361 Bremen, BRD.

Dolan, T.E. and Marshall, E. (1960). *Manned, Modular, Multi-Purpose Space Vehicle*, Vought Astronautics Company Report, Dallas, Texas.

Eckart, P. (1999). *The Lunar Base Handbook*, McGraw-Hill, New York.

Goddard, E.C. and Pendray, G.E. (eds.) (1968). *The Papers of Robert H. Goddard*, 3 vols., McGraw-Hill, New York.

Goddard, R.H. (1919). *A Method of Attaining Extreme Altitude*, Smithsonian Institution, Washington, DC.

Goddard, R.H. (1948). *Rocket Development: Liquid Fuel Rocket Research, 1929–41*, published posthumously by Prentice-Hall, New York.

Goldman, S.J. (1999). Galileo in Retrospect, *Sky and Telescope*, December. 44.

Groh, K.H. and Löb, H.W. (1996). Radio-Frequency Ion Thruster Activities at Giessen University, *20th International Symposium on Space Technology and Science* (ISTS), May 19–25, Gifu, Japan.

Gryaznov, G.M. (1992). Nuclear Power Systems for Spacecraft. Paper RGC-EP-92-22, First Russian–German Conference on Electric Propulsion Engines and Their Technical Applications, Rauischholzhausen/Giessen University, BRD.

Harford, J. (1997). *Korolev*, Wiley, New York.

Hopmann, H. (1999a). *Schubkraft für die Raumfahrt*, Verlag Stedinger, Lemwerder (BDR) PO Box 1165.

Hopmann, H. (1999b). *The Planetary Society Report*, **XIX**, No 4, July/August.

Houbolt, J.C. (1961). Letters to Robert C. Seamans, Jr. of May 19, September 9, and November 15, 1961. NASA Archives, Washington, DC. Technical details about the Saturn Project can be found in Bilstein, R.E. (1980). *Stages to Saturn*, The NASA History Series, NASA SP; 4206, US Government Printing Office, Washington, DC 20402.

Kaufman, H.R. and Reader, P.D. (1960). Experimental Performance of Ion Rockets Employing Electron Bombardment Ion Sources. *American Rocket Society Electrostatic Propulsion Conference*, Monterey, Calif. ARS Preprint 1374–60.

Kondratyuk, Y.V. (1965). In *Pioneers of Rocket Technology: Selected Works*, NASA Technical Translation F-9285, Washington, DC, November 1965.

Löb, H.W. (1962). Ein elektrostatisches Ionentriebwerk mit Hochfrequenzionenquelle, *Astronautica Acta* **VIII**, 1, 49, 1962.

Löb, H.W., Bassner, B., Berry, W., Hechler, M., Leipold, M., Meusemann, H., Noack, E., Schwehm, G. and Seboldt, W. (1996). Electric Propulsion – Ready to Enhance Interplanetary Missions, *Symposium on*

Scientific Satellites Achievements and Prospects in Europe, November 20–22, Paris. (Horst W. Loeb, First Institute of Physics, Justus Liebig University, Giessen, BRD.)

Logsdon, J.M. (1978). The Space Shuttle Decision: Technology and Political Choice, *Journal of Contemporary Business*, **7**, No. 3.

Michin, W.P. (1998). *Sowjetische Mondprojekte*, Informazionno-Isdatelskij Zentr "Inform Snanie", Moscow 1998; German edition by R. Meier (1999). Elbe-Dnjepr Verlag, Bahnhofstrasse 35, 04860 Klitzschen, BRD.

Newell, H.E. (1980). *Beyond the Atmosphere*, NASA SP-4211, Washington, DC.

Oberth, H. (1923) *Die Rakete zu den Planetenräumen*, R. Oldenbourg, München, BRD; Uni Verlag, Feucht-Nürnberg, BRD, 1960, 1964, 1984.

Ordway, F.I., III and Sharpe, M.R. (1979). *The Rocket Team*, Thomas Y. Crowell, New York.

Ottemeyer, D.R. (1976). Das Projekt Spacelab, *Astronautik* **3**, 57, September 1976, Schwetzinger Verlagsdruckerei GmBH, 6830 Schwetzingen, BRD.

Przybilski, O. and Wotzlaw, S. (1996). *N-1 HERKULES*, Schriftenreihe der Deutschen Raumfahrtausstellung e.V. Technische Universität Dresden, Institut für Luft- und Raumfahrttechnik, Germany.

Rayman, M.D., Varghese, P., Lehman, D.H. and Livesay, L.L. (1999). Results from the Deep Space 1 Technology Validation Mission, *50th International Astronautical Congress*, 4–8 October. Amsterdam. IAA-99-IAAA.11.2.01. IAF, 3–5 Rue Mario-Nikis, 75015 Paris.

Reisig, G.H.R. (1997). *Raketenforschung in Deutschland*, Edition Lenser im Profil Verlag, Münster, BRD.

Ross, H.E. (1949). Orbital Bases. *Journal of the British Interplanetary Society*, **VIII**, January.

Ruppe, H.O. (1982). *Die Grenzenlose Dimension*, Vols. 1 and 2, Econ Verlag, Düsseldorf (BRD) and Wien (Austria).

Shepherd, L.R. and Cleaver, A.V. (1948) The Atomic Rocket. *Journal of the British Interplanetary Society*, **7**, 185, and **8**, 23 and 50 (1949).

Shepherd, L.R. (1999). Performance Criteria of Nuclear Space Propulsion Systems. *Journal of the British Interplanetary Society*, **52**, No. 9/10, Sept/Oct.

Sky and Telescope, March 2000, X-Ray Multi-mirror Mission, pg. 27; Roentgensatellit XMM, Luft- und Raumfahrt, Heft 1, Januar-März 2000, 15, Deutsche Gesellschaft für Luft- und Raumfahrt.

Still, M. (1999). X-Ray Astronomy's Golden Age, *Sky and Telescope*, August 56.

Stuhlinger, E. (1954). Possibilities of Electrical Space Ship Propulsion, *Proceedings of the 5th International Astronautical Federation Congress*, Innsbruck, Austria. 100.

Stuhlinger, E. (1964). *Ion Propulsion for Spaceflight*, McGraw-Hill, New York.

Stuhlinger, E. (1976). Skylab Results – Review and Outlook, *COSPAR Space Research* **XVI**, 54, Akademie Verlag, Berlin, BRD.

Stuhlinger, E. and Ordway, F. I., III (1994). *Wernher von Braun, Crusader for Space*, Krieger Publishing Company, Malabar, Florida.

Stuhlinger, E. (1996). Nuclear-Electric Propulsion for Human Mars Missions. In: Carol R. Stoker and Carter Emmart (eds.) *Strategies for Mars*, Science and Technology Series, Vol. 86, American Astronautical Society San Diego, Calif.

Trümper, J. (1999a). Erforschung des Heissen Universums mit ROSAT, *DLR Nachrichten*, N3, 93, 32, Porz-Wahnheide, 51147 Cologne, BRD.

Trümper, J. (1999b). ROSAT und seine Nachfolger, *Physikalische Blaetter* 55, Heft 9, 45, Wiley-VCH Verlag GmBH, PO Box 10 11 69, 69451 Weinheim, BRD.

Vick, C.P. (1997). Why the Soviets Never Beat the US to the Moon: An Interview with Marsha Freeman, *21st Century Science and Technology*, Fall 1997, PO Box 16285, Washington, DC, 20041.

Von Braun, W. (1929). *Lunetta*, published in a Journal for the Youth, *Leben und Arbeit* (Life and Work), edited by the Hermann Lietz Boarding School, Spiekeroog Island, BRD.

Von Braun, W. (1952). *Das Mars Projekt*, Bechtle Verlag, Esslingen, BDR. English Version: *The Mars Project*, University of Illinois, Urbana.

Von Braun, W. (1961) Memorandum to Vice President Lyndon B. Johnson. NASA History Archives, April 1961.

Von Braun, W. and Ordway, F. I., III (1966). *History of Rocketry and Space Travel*, Thomas Y. Crowell, New York.

Voss, L. (2000). New Thrust for US Satellites, *Aerospace America*, **38**, No 2, Feb. American Institute for Aeronautics and Astronautics.

The early epoch of space science

4

JOHN A. SIMPSON[†]

The cosmic radiation

INTRODUCTORY REMARKS

Cosmic ray research has developed as one of the most spectacular and vital contributors to science in the Twentieth Century. From a humble beginning early in this Century, the quest for an understanding of this radiation and the challenges confronting investigators led to the rise of new scientific disciplines, technologies and astrophysical concepts. Particle and high energy physics, radiocarbon dating and magnetic fields and plasmas of astrophysical origin are representative of the many research fields born from cosmic ray research. Examples of technological contributions include radiation instruments and balloon and space flight concepts required to reach the cosmic radiation beyond Earth's atmosphere.

From our concept of a quiet universe at the beginning of the century to the recognition of a violent universe today, the cosmic radiations have – and continue to be – a vital component for understanding these dynamical phenomena on all astrophysical scales. Indeed, cosmic ray particles have become, towards the end of this century, part of what is now called astroparticle physics.

At night, when we look at the Milky Way with naked eyes, we observe an overall glow like that shown in the lower panel in Figure 1. This is the visible light from the excitation of atoms by very low energy processes – i.e., tens of electron volts. On the other hand, recent observations by the EGRET instrument on the Compton Gamma Ray Observatory satellite have revealed a dramatically more energetic view of our galaxy. The light in the upper panel in Figure 1 is from gamma-rays of approximately one-hundred million electron volts.

Nuclear interactions arising from accelerated cosmic rays with atoms in the interstellar medium result in the creation of many fundamental particles, among them pions of zero electrical charge that promptly decay into the gamma rays observed in the upper panel of Figure 1. Thus, it is clear that the cosmic rays propagate throughout the galactic disc.

The early years of cosmic ray research were both exciting and romantic since the investigators worked alone or in small groups world-wide. In the 1930's K.K. Darrow, for many years the secretary of the American Physical Society captured the spirit of these early years by noting:

> [The study of cosmic rays] is unique in modern physics for the minuteness of the phenomena, the delicacy of the observations, the adventurous excursions of the observers, the subtlety of the analysis and the grandeur of the inferences. (Darrow 1932)

This culture of personal involvement of physicists in carrying out experiments and personally recording data on mountains, in aircraft or balloons continued into the 1950's. In return for the personal risks were the rewards of instant discovery.

With the opening of the so-called Space Age it became possible to study directly the acceleration mechanisms and the propagation of charged particles in interplanetary magnetic fields. It also became necessary to include engineers and other technologists to assist in the design, construction and testing of space flight instrumentation.

[†] University of Chicago, Chicago, IL, USA. John Simpson died on August 31, 2000. The editors are indebted to Bruce McKibben for helping them to put the paper into its final form.

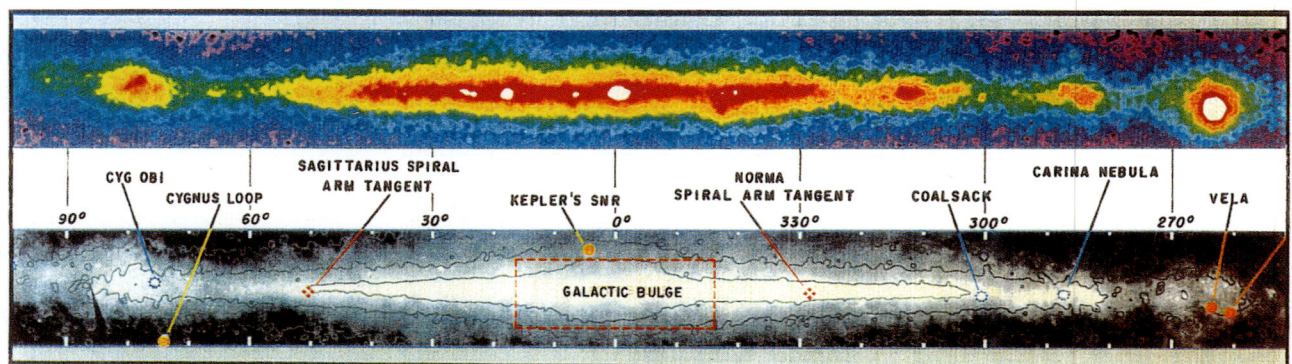

Figure 1 Lower: The galaxy in visible light. Upper: The ~100 MeV Galaxy (NASA).

This cultural change now largely pervades experimental cosmic ray physics. Experimental research in many areas is a big science effort with some young investigators having contact with experiments only by way of computer screens. For many theorists there also has been a cultural shift to massive computer code modeling.

The editors of this publication have requested this author's personal account of the history of cosmic ray research and of his early entry as a cosmic ray researcher into the space era, to be directed to a general scientific readership, and to be carried through to the 1970's or 1980's. Accordingly, this account is divided into two parts: the first part being mainly historical efforts to understand the radiation through the atmosphere; followed in the second part by both the international and the author's research in space beginning in 1958. A guide of references to our present knowledge of cosmic ray physics at the end of the 20th century is also included.

PART A: REACHING TOWARDS SPACE

1. THE DISCOVERY OF EXTRATERRESTRIAL RADIATION

The events leading to the discovery of the cosmic radiation were, in many respects, mysterious and misleading. By the late 1890's the basic instrument for measuring electric charge was the electroscope often used to demonstrate the presence of an electrostatic charge. Aside from leakage of stored charge from imperfect insulators, the electric charge imparted on the electroscope should hold this charge indefinitely. However, even the most highly developed electroscopes slowly lost their charge due to ionization in the atmosphere.

At that time the explanation for this leakage of electric charge appeared obvious – it must arise from radioactivity in the atmosphere ionizing the air around the electroscope.

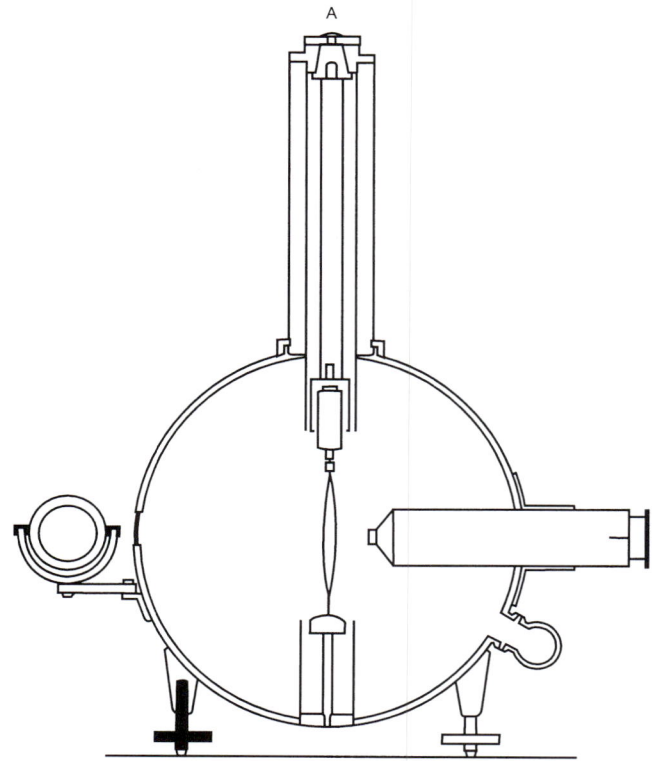

Figure 2 Electroscope and ionization chamber (Th. Wulf 1909).

This explanation had become popular as a result of the discovery of X-ray by Roentgen and discovery by H. Becquerel in 1896 of radioactivity in a compound of Uranium and with the realization that radioactive materials were to be found everywhere on Earth. Even with 5 cm of lead surrounding the electroscope the leakage of charge persisted.

The Dutch physicist, Th. Wulf (1909) had developed a highly stable electroscope and ionization chamber (Figure 2) that he used to test this hypothesis. He carried

Figure 3 Viktor Hess after a balloon landing (1912).

his electroscope up the Eiffel Tower and found a radiation level of only 64% below the value on the Earth's surface for the leakage rate – much lower than he had calculated. He surmised, therefore, that there was either an additional source of radiation from the upper atmosphere, or that the absorption of gamma rays in air was much smaller than had been assumed. At about the same time Gockel (1911) in Switzerland carried a Wulf instrument in a balloon flight to 4500 m. He also questioned from his qualitative measurements whether there might exist a new radiation superposed on the radiation from Earth.

The 28 year old Victor Hess had been following these reports on the source of radiation producing the leakage charge in electroscopes. In a recounting of those years (Figure 3) he stated (Hess 1912, 1940):

> ... At that time in the Spring of 1911, after reading an account of Father Wulf's Eiffel-Tower experiments, I was inclined to believe that a hitherto unknown source of ionization may have been in evidence in all these experiments; and I decided to attack the problem by direct experiments of my own.

> It seemed to me necessary to measure accurately the absorption of gamma rays from radium in air in order to find out how far above the ground gamma rays could act as an ionizing agency.

> The next step was the construction of an air-tight ionization apparatus which could be used during balloon flights and fitted with a sensitive electrometric system which was not influenced by the large fluctuations of temperature occurring in the flights. I used a modification of Th. Wulf's apparatus with walls of zinc, thick enough to withstand the excess pressure of one atmosphere and a temperature compensation for the fiber electrometer. Furthermore, I found it very important always to use two or three of the instruments simultaneously in order to avoid errors from instrumental defects. With such instruments, I made ten balloon ascents: two in 1911, seven in 1912 and one in 1913. Five of them were carried out at night, and some of them continued during the following morning. One flight was made during a solar eclipse, in April 1912.

> By taking successive readings of the ionization with two or three instruments at a time, much more reliable data were obtained. I found that at 500 meters above the ground the ionization was, on the average, about $2\,\mathbf{I}$ ($\mathbf{I}$ = ion pairs-cm^{-3} sec^{-1}) lower than on the ground and that, from about 1800 meters upwards, an increase of ionization is undoubtedly in evidence. At 1500 meters, the ionization increased to the same value as had been found on the ground. At 3500 meters, the increase amounted to no less than $4\,\mathrm{I}$, at 5000 meters to $16\,\mathrm{I}$ above the ground value. No difference between day and night observations was noticed.

> An explanation of the increase of ionization with increasing altitude on account of the action of radioactive substances was impossible....

> The only possible way to interpret my experimental findings was to conclude to the existence of a hitherto unknown and very penetrating radiation, coming mainly from above and being most probably of extra-terrestrial (cosmic) origin ...

Within a year Kolhörster (1913) confirmed Hess's observations and conclusions. Nevertheless, many investigators disputed their measurements and conclusion with arguments concerning the stability of the electroscopes. It was not until after World War I that experiments were resumed. Millikan and Bowen (1923) developed a low mass (~190 g) electrometer and ion chamber for unmanned balloon flights using radiosonar technology developed during World War I. In balloon flights to 15,000 m in Texas they were surprised to find a radiation intensity not more than one-fourth the intensity reported by Hess and Kolhörster. They attributed this difference to a turnover in the intensity at higher altitude, being unaware that a geomagnetic latitude cutoff existed between the measurement in Europe and Texas. Thus, Millikan believed that there was no extraterrestrial radiation until he and Cameron (1926) carried out absorption measurements of the radiation at various depths in snow-fed lakes at high altitudes. Based upon the absorption coefficients and altitude dependence of the radiation, they concluded that the radiation was high energy gamma rays and that "these rays shoot through space equally in all directions" calling them "cosmic rays". They argued that the radiations are "... generated by nuclear changes having energy values not far from [those that they recorded] in nebulous matter in space."

Millikan than proclaimed that this cosmic radiation was the "birth cries of atoms" in our galaxy. His lectures drew considerable attention from, among others, Eddington and Jeans, who struggled unsuccessfully to describe processes that could account for Millikan's claim (cf. De Maria and Russo 1990).

2. COSMIC RAY CHARGED PARTICLES

A key experiment, which would decide whether the incoming radiation was electrically charged or uncharged, was the measurement of the dependence of cosmic ray intensity on geomagnetic latitude. If the radiation was electrically neutral, such as gamma rays, there would be no dependence on the strength of the magnetic field. On the other hand, if the cosmic rays were electrically charged particles – say electrons or protons – their deflection in the Earth's magnetic field would limit access at all but the highest latitudes. Thus the cosmic ray intensity would be less at the equator than at high latitudes.

In 1927, J. Clay (1928) from the Netherlands – by carrying ionization detectors on ships that traveled over a large latitude range – observed a geomagnetic latitude effect in cosmic ray intensity. If confirmed, the radiation must consist, at least in part, of charged particles interacting in the external geomagnetic field. Although Clay's work was disputed by Millikan, A.H. Compton (1933) carried out in 1932 a world-wide survey to settle the dispute. The Earth was divided into nine zones and teams, with all investigators using identical ion chambers. He then reported (1933) that there was a latitude effect, that cosmic rays were charged particles and that Millikan was wrong.

At about the same time, Millikan had sent Victor Neher (his junior colleague from the California Institute of Technology) on another latitude survey to South America. Unknown to the survey party, their electroscope had malfunctioned on the way down and they reported no latitude effect. This was triumphantly proclaimed by Millikan who attacked Compton in a debate at the Christmas meeting of the American Association for the Advancement of Science in 1932.

The electroscope did not fail on the Millikan party's return northward across the equator. This dispute ended in February 1933 when Millikan admitted that there was a latitude effect and that the cosmic rays must be charged particles.

Since Compton and Millikan were supported for their research by the Carnegie Corporation through its Department of Terrestrial Magnetism (DTM), the DTM was eager to settle the acrimonious debate by deciding which instrument was superior – that is, which instrument (Compton's or Millikan's) was the better one to use as a world standard.

In the early 1930's, Auguste and Jean Picard received much publicity for their manned balloon flights in the stratosphere that claimed to study the origin of cosmic rays. In 1933 the promoters of the Chicago "Century of Progress" World's Fair engaged them to make such a flight, with Compton agreeing to arrange all the scientific instrumentation to be carried in the manned gondola. Compton seized on this opportunity to seek a collaboration with Millikan. "It would seem too bad to let an expensive flight of this kind occur without making use of it for some high altitude measurements," he wrote (DeVorkin 1989). He invited Millikan to supply his automatic recording electroscope to be flown with the Chicago instruments in order to reconcile differences in their performance. Millikan, who had claimed his instrument was superior, agreed to meet Compton's challenge.

The Compton-Millikan venture appeared to have no impact on the advancement of cosmic ray physics. However, the development of the Compton-Bennett model-C ionization chamber (Figure 4) (Compton *et al.* 1934) became in the long run the standard adopted by the Carnegie Institution when a worldwide network of stations to search for cosmic ray variations with time was set up with the Chicago-built instruments at world-wide sites of the Institution's Department of Terrestrial Magnetism. The work of Scott E. Forbush, who was responsible for the stations, led to the proof that the observed intensity of cosmic rays within the atmosphere of Earth varied with time (Forbush 1938).

3. CHARGED PARTICLES, + OR −?

In the period 1928–1932 it was clear that cosmic rays included charged particles, but what was the sign of the electrical charge that they carried – plus or minus?

Until about 1930 the only property of cosmic rays that could be measured was their specific ionization (ions $cm^{-3} sec^{-1}$). Fortunately, Geiger and Müller (1928) had invented a cylindrical charged particle detector which made

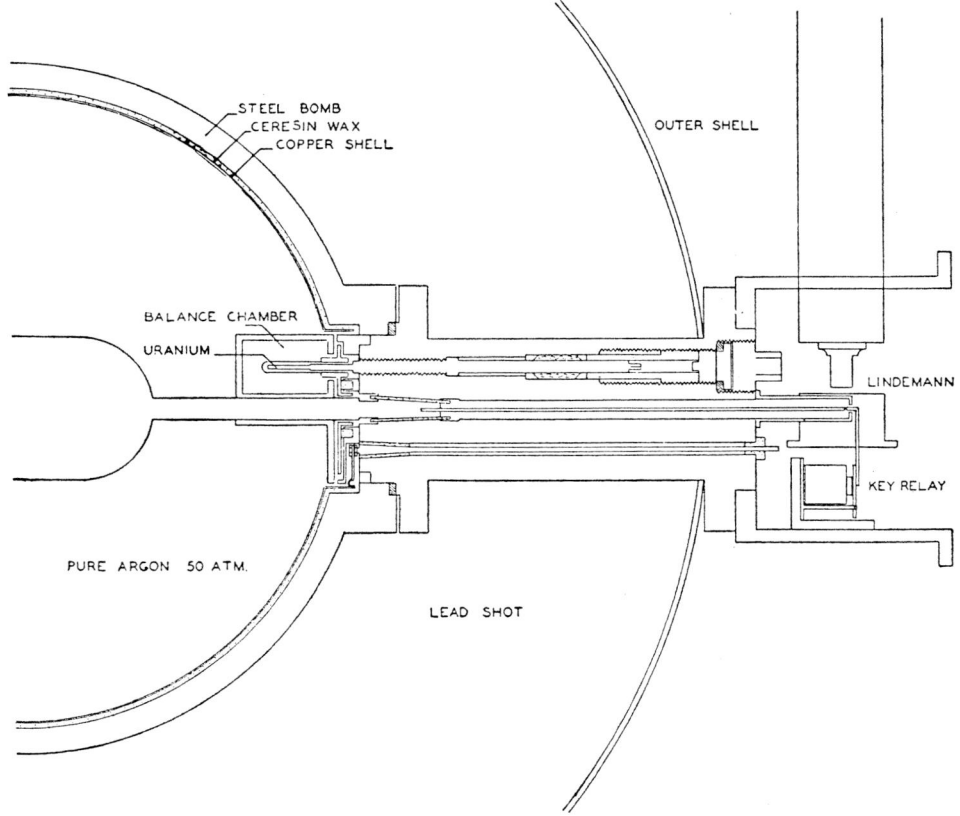

Figure 4 Model C ionization chamber (Compton *et al.* 1934).

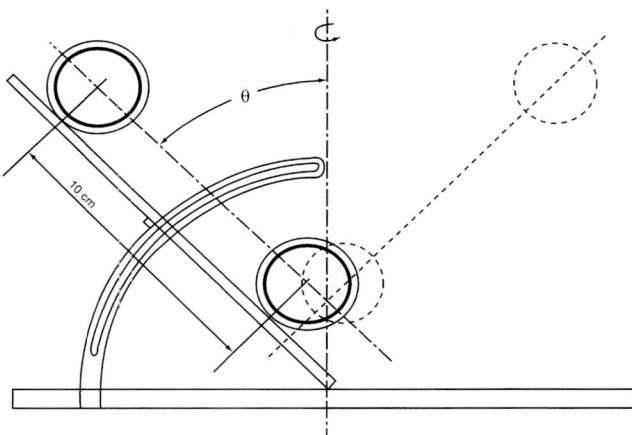

Figure 5 Studying east-west asymmetry of cosmic rays (Alvarez and Compton 1933).

it possible for the first time to determine the direction of the incident cosmic ray particle when two or more of these G-M counters were aligned in parallel, as shown in Figure 5, such as for an east-west experiment. When a charged particle passed through the two counters in Figure 5 the two signal outputs would be coincident. This coincidence method for determining the direction of a particle trajectory was introduced for cosmic ray physics by Bothe and Kolhörster; Rossi (1930a) developed the first vacuum tube electronic coincidence circuit.

Following a suggestion by Mott-Smith (1930) to use a laboratory magnet to deflect cosmic ray particles, Rossi (1930b) turned to consider the use of the geomagnetic field for determining the sign of the incoming charged particles as follows:

If the primary cosmic rays are an electron radiation, then ... the Earth's magnetic field must also be considered. The effect of the Earth's field should be noticeable by an unsymmetric directional distribution of the intensity with respect to the perpendicular. Experiments are being prepared which will test such an effect.

Let us call V the velocity of the electrons in volts, R the Earth's radius, M the magnetic moment of the Earth, λ the magnetic latitude at the point of observation, θ the angle between the plane of the magnetic meridian and the direction of the path of the electrons (positive toward the east), a a constant, whose value is approximately

3×10^2. Then the theory gives the following results. Electrons may impinge at a definite point of the Earth's surface (when $V < aM/R^2$) only if the angle θ satisfies the following inequality:

$$\sin \theta > \frac{aM}{R^2 V} \cos \lambda - \frac{aM}{R^2 V} \zeta^{1/2} \frac{2}{\cos \lambda} \quad (1)$$

For example, at a point where the magnetic latitude is $\lambda = 45°$, if $(aM/R^2V) = 16$, the above condition shows $\sin\theta > 0$. This means that the whole region west of the magnetic meridian is "in shadow". This case comes into consideration for electrons having a velocity of about 4.3×10^9 volt.

If $(aM/R^2V) \cos \lambda - (aM/R^2V)^{1/2} 2/\cos \lambda > 1$ the above inequality has no solution. This means that electrons can never impinge upon the point under consideration.

The race was to decide the question. Vallarta invited Compton at Chicago and Thomas Johnson at the Bartol Foundation to come to the lower latitude and higher altitude of his native Mexico City where, he calculated, the "east-west" effect would be large (see Lemaitre and Vallarta 1933). In Italy Rossi prepared instruments for an expedition to Eritrea with the same objectives.

Compton turned to his graduate student, Luis Alvarez, who had been building G-M counters readily adapted to the problem. Alvarez (1987) recalled working around the clock preparing the equipment (Figure 5).

In Mexico City, the Compton and Johnson groups carried out measurements which displayed more cosmic rays coming from the west than the east. The radiation was at least composed of positively charged particles. Compton gave much of the credit to Alvarez for their joint work. A few months later Rossi also found the east-west effect in Eritrea (Rossi 1985).

Both Johnson (1933) and Alvarez and Compton (1933) published their results back-to-back in the *Physical Review*. Both groups found an excess of flux from the west of approximately 15 percent. Alvarez and Compton also noted that:

> The smallness of the effect confirms our earlier conclusion that most of the rays capable of penetrating the Earth's atmosphere are not sufficiently bent by the Earth's magnetic field to prevent them from reaching the Earth.

Later Alvarez (1987) described these two experiments carried out on the roof of the Hotel Geneva in Mexico City.

The question immediately arose as to which positively charged particles were in the cosmic radiation. Since Dirac had predicted in 1927 the existence of the positron – and Carl Anderson had discovered the positron in 1932 – it was natural to propose, as many physicists did, including Alvarez and Compton, that positrons were the primary particle with positive charge in the cosmic radiation, However, Johnson presented arguments and experiments to suggest that protons were the primary radiation.

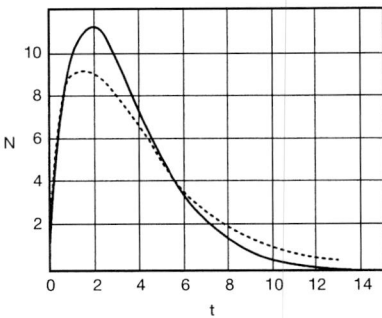

Figure 6 Multiplication curve (–) for an electron of 11×10^9 eV (Serber 1938).

The issue was further confused by an elegant paper by Serber (1938 and earlier references therein) who calculated the multiplication of the soft component in the cosmic rays in the upper atmosphere – showing that the calculated maximum of intensity was in good agreement with the balloon experiments of many earlier investigators (Figure 6). These calculations implied that primary electrons produced the soft component through scattering in the atmosphere, and that "the penetrating component is at least largely of secondary origin". As Serber (1983) later noted:

> This comparison [with experiment] appeared so convincing that it probably delayed recognition of the truth, that the primaries are protons.

That the primary particles were protons was suggested by ingenious balloon flight experiments based on the altitude dependence of the East-West effect by Johnson and Barry (1939). Schein *et al.* (1941) proved that protons were the principal primary.

A cross-section of their G-M counter telescope with interleaved lead absorbers is shown in Figure 7. They found from their measurements that the intensity of the hard component increased continuously to the highest altitudes reached, Curve A, Figure 7. They noted that the particles detected by the telescope could not be electrons since no side counters registered showers generated in the lead by the traversing particles. By comparing their observations with earlier work in Curve B, such as by Pfotzer (1936), the electrons must be of secondary origin. They argued that these electrons were mainly from the decay of mesotrons (present day muons) and knock-on processes. Unfortunately, the significance of this experiment was largely forgotten due to the U.S. entry in World War II.

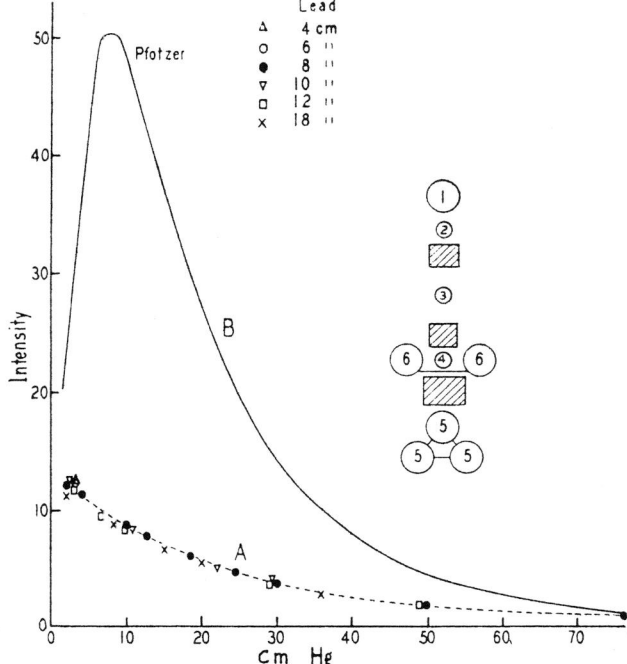

Figure 7 Curve A intensity of the hard component for various lead thicknesses as a function of pressure in the atmosphere (Schein *et al.* 1941).

4. THE RISE OF PARTICLE AND HIGH ENERGY PHYSICS

The reference by Schein, Jesse and Wollan to the secondary particles called mesotrons (now called mesons) had an historical background beginning with the discovery of the positron by Anderson in 1932 and Blackett and Occhialini (1933). These investigators had combined the cloud chamber invented by Wilson (1901) within a strong magnetic field so that the trajectories of charged cosmic ray particles could be determined by curvature. This was a method developed by Skobeltzyn (1929) to provide the first visual means for observing particle processes with higher energies than those from radioactive sources. By controlling the operation of the cloud chamber with G-M. counters, incident cosmic ray particles interacting with atoms in the cloud chamber could be studied. This research is representative of the beginning of the new fields in physics currently identified as particle and high energy physics by leading to the discovery of many new and fundamental particles such a the muon, pion, kaon, etc. Although intimately associated with cosmic radiation, particle physics is beyond the scope of this review. The reader is referred to Serber's excellent review "The birth of particle physics" (1983) which shows the central role played by cosmic ray investigations, especially in the period 1930 to the post-War development of laboratory high energy accelerators.

This convergence of cosmic ray research and both theoretical and experimental high energy particle physics is vividly illustrated by the physicists in the photograph in Figure 8 of participants in an international cosmic ray conference held at the University of Chicago, 1939.

Auger in 1938 and Auger and Maze (1939) reported cosmic ray particle coincidences in G-M counters located in a horizontal plane, first for 150 m separation, and later about 300 m apart. They called these showers from cosmic ray interaction high in the atmosphere "Grandes Gerbes" – that is, extensive showers now called Extensive Air Showers (EAS) composed of electrons and a small portion of muons. Taking account of the high density of the particles and the large area of the EAS, Auger came to the astonishing conclusion that the primary cosmic ray at the top of the atmosphere must have an energy of approximately 10^{15} eV!

After World War II research on the EAS resumed with the establishment by Linsley and Rossi of an array of scintillation counters distributed over 8 km^2 in a New Mexico desert. For example, by sampling 5×10^{10} secondary particles in an atmospheric cascade, Linsley (1963) showed that the energy of the primary particle was 10^{20} eV. This array was the first of a series of giant cosmic ray detectors for the measurement of the cosmic ray spectrum, particle composition and isotropy at the highest energies. Biermann (1997) and Cronin (1999) have reviewed the status of the unresolved quest for the origin and acceleration of these particles. The future of this exciting area of research will be a focus of cosmic ray research in the 21st Century.

5. NEUTRONS AND NUCLEAR INTERACTIONS PRODUCED IN THE ATMOSPHERE

Nuclear interactions with atmospheric atoms had not been seriously investigated until after the discovery of the neutron by Chadwick in 1932. With a half-life of approximately 10 minutes neutrons would only exist as secondary particles as the product of nuclear interactions in the atmosphere by primary cosmic rays. Searches for evidence of nuclear interactions in cloud chambers had been largely unsuccessful. However, with the new photo-emulsion technique developed in Vienna by Marietta Blau (1925, 1927; discussed in Halpern and Shapiro 2001) there began a new series of discoveries based on particle track recognition in thick photo emulsions. Blau and Wambacher (1937) exposed photo emulsion plates at 2300 m on Hafelekar Mountain for five months. After development the emulsions displayed microscopic arrays of particle tracks radiating from a common point, as in Figure 9. This cosmic ray "star" was the breakup of an atomic nucleus by an incident

1. H. Bethe, 2. D. Froman, 3. R. Brode, 4. A.H. Compton, 5. E. Teller, 6. A. Baños, Jr., 7. G. Groetzinger, 8. S. Goudsmit, 9. M.S. Vallarta, 10. L. Nordheim, 11. J.R. Oppenheimer, 12. C.D. Anderson, 13. S. Forbush, 14. Nielsen (of duke U.), 15. V. Hess, 16. V.C. Wilson, 17. B. Rossi, 18. W. Bothe, 19. W. Heisenberg, 20. P. Auger, 21. R. Serber, 22. T. Johnson, 23. J. Clay (Holland), 24. W.F.G. Swann, 25. J.C. Street (Harvard), 26. J. Wheeler, 27. S. Neddermeyer, 28. E. Herzog (?), 29. M. Pomerantz, 30. W. Harkins (U. of C.), 31. H. Beutler, 32. M.M. Shapiro[†], 33. M. Schein*, 34. C. Montgomery (Yale), 35. W. Bostick[†], 36. C. Eckart, 37. A. Code[†], 38. J. Stearns (Denver?), 39. J. Hopfield, 40. E. O. Wollan*, 41. D. Hughes[†], 42. W. Jesse*, 43. B. Hoag, 44. N. Hillberry[†], 45. F. Shonka[†], 46. P.S. Gill[†], 47. A.H. Snell, 48. J. Schremp, 49. A. Haas? (Vienna), 50. E. Dershem, and 51. H. Jones[†] at the Cosmic Ray Conference (Symposium on Cosmic Rays, 1939) convened at the University of Chicago in the summer of 1939. (Courtesy of Maurice M. Shapiro.)

*Then research associate of Compton. [†]Then graduate student of Compton.

Figure 8 The 1939 Conference on the Cosmic Radiation (photo identifications by M. Shapiro).

cosmic ray. They estimated that the total energy was at least one to two hundred MeV and also found that recoil protons from neutron collisions could yield estimates of the incident neutron energy.

Korff (1939) undertook the first systematic measurements of the low energy ("slow") and fast neutrons as a function of altitude in balloons. Since slow neutrons had scattered in the atmosphere, and were moderated in energy,

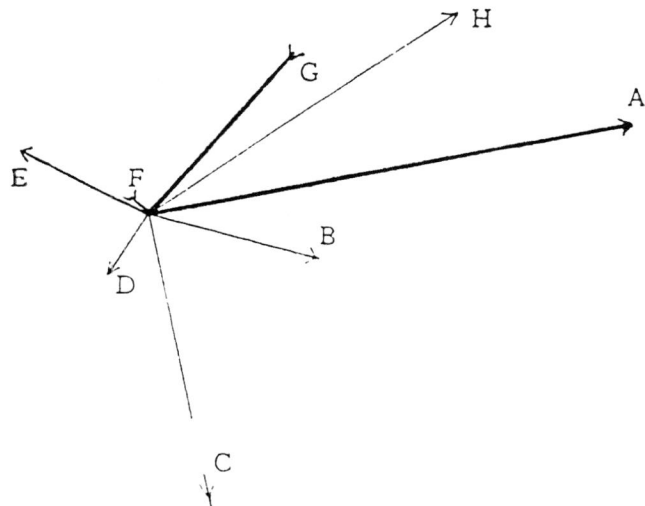

Figure 9 Nuclear "star" event in a photo-emulsion (Blau and Wambacher 1937).

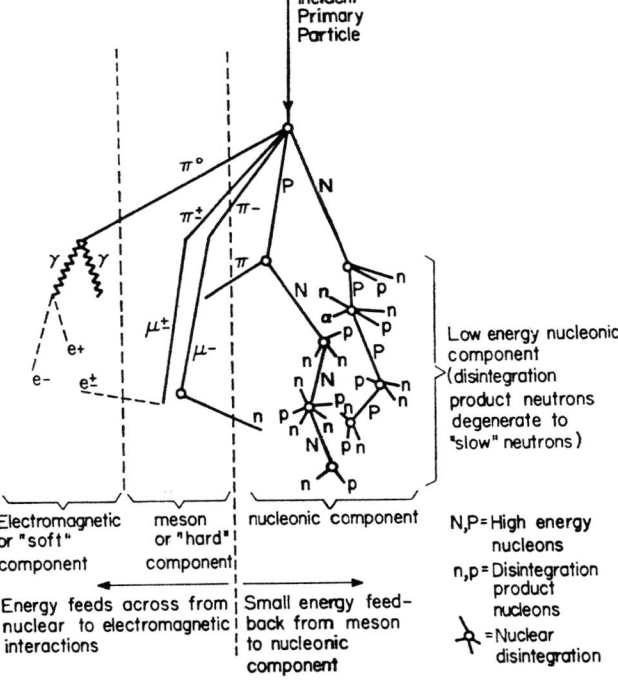

Figure 10 Schematic representation of the typical development of the secondary cosmic radiations within the atmosphere arising from an incident primary particle (from Simpson et al. 1953).

his measurements were in units of neutron density (neutrons-cc^{-1}). Boron trifluoride (BF$_3$) gas proportional counters were developed by Korff for these balloon flights. The reaction for the capture of neutrons by the isotope ^{10}B (20 percent of Boron) is:

$$^{10}B + n = {}^4He^{++} + {}^7Li + Q$$

In the proportional counter the alpha particle pulse was readily detectable against the background of small pulses produced by penetrating cosmic ray charged particles. The production and equilibrium of neutrons in the atmosphere were first analyzed by Bethe et al. (1940). They showed how fast neutrons would scatter and slow down in the atmosphere and the Earth. Since the neutron capture cross-section for nitrogen is inversely proportional to the neutron velocity, more than half the slow neutrons are captured by nitrogen.

The intervention of World War II, with the establishment of the Manhattan (atomic bomb) Project, led to dramatic increases in our knowledge of nuclear physics, especially for nuclear parameters such as capture and scattering cross-sections. Also there were remarkable advances in instrumentation and detectors. For example, the enrichment of the ^{10}B isotope to 96 percent of Boron would contribute to post-War cosmic ray neutron research.

After World War II there was a critical development in the sensitivity of nuclear emulsions that radically advanced discoveries of fundamental particles created by primary cosmic ray interactions in the atmosphere. Berriman (1948) extended for the first time the sensitivity of nuclear emulsion to include minimum ionizing particle tracks. These new emulsions – manufactured by Ilford Ltd. – led to the discovery of the positively charged pion and its decay into a muon and thence a muon into an electron (e.g., Lattes et al. 1947, Serber 1983).

For particle physics, the discoveries resolved the question of which "mesotron" was attributed to Yukawa's 1935 theory for a "mesotron" and led to a comprehensive array of new particles such as the kaon, lambda-zero and other strange particles. Unstable elementary particles continued to be the main focus of attention in the 1940s, mainly through the use of nuclear emulsions and cloud chambers (e.g., Rochester and Butler, 1947). For the high energy physicists, the cosmic ray was their accelerated particle and the atmosphere was their gas target.

For cosmic ray physics the discoveries explained the development and propagation in the atmosphere of the secondary charged and neutral particles. By 1948 there began to emerge a general scenario of the cosmic radiation in the atmosphere, as illustrated in the schematic diagram, Figure 10, of that period.

Although there now was proof of a primary proton component and indications of a helium component in the primary radiation, it came as a surprise in 1948 to discover that the stripped nuclei of elements in the periodic table through iron were also to be found in the primary cosmic radiation (Freier et al. 1948). Figure 11 displays a very

Figure 11 Discovery of high Z stripped nuclei using nuclear emulsions (Freier *et al.* 1948).

heavy stripped nucleon track traced through a stack of nuclear emulsions carried on a high altitude balloon.

It became clear that some of the elements – Li, Be, and B – were overabundant by many orders of magnitude (compared with solar system abundances) and therefore must be secondary nuclei from nuclear spallation interactions during the interstellar propagation of primary cosmic ray nuclei in the galaxy.

To account for these overabundant secondaries, estimates were obtained for the amount of interstellar matter traversed in the galaxy under the assumption that Li, Be, and B were not present in the cosmic-ray sources. Values ranging from ~ 3 to $\sim 5 \, \text{gm-cm}^{-2}$ were obtained, suggesting confinement times in interstellar magnetic fields of order $\sim 10^6$ years. However, Juliusson *et al.* (1972) discovered that the ratio of the galactic secondary nuclei to primary nuclei from sources decreased with increasing energy above $\sim 30 \, \text{GeV} \, \text{nucleon}^{-1}$. Since the measured rigidity dependence is $\propto \mathbf{R}^{-2.6}$, this effect was interpreted as a magnetic rigidity ($\mathbf{R}$) dependence for the escape of cosmic ray nuclei from the galactic disk with a dependence $\propto \mathbf{R}^{-0.6}$, thus assuming an $\mathbf{R}^{-2}$ spectral dependence for the acceleration process in the galaxy.

In 1946–47 very little was known regarding the energy spectrum of the cosmic rays or the development of the nuclear component cascade shown in Figure 10. Consequently, this author decided to investigate the dependence of the nucleonic cascade process on primary cosmic ray particle energy. The tracer, or "tool" for this investigation was to be the fast neutron production in the atmosphere.

To determine the relationship of the disintegration product (fast) neutrons to the cosmic ray proton intensity as a function of energy, he decided to use the geomagnetic field cutoff effect as a function of latitude for cosmic rays.

In the 1930's, Compton and his collaborators had obtained the latitude effect, using ionization chambers, which we now know measured the combined electromagnetic and meson component in the atmosphere. This sea-level latitude effect corresponded to a ≈ 12–15 percent increase between the geomagnetic equator and high latitudes, but there was no knowledge or calculation of what the magnitude of the nucleonic component latitude effect might be. The measurements of fast neutrons were detected by $^{10}\text{BF}_3$ proportional counters surrounded by moderators. Cadmium enclosed the entire assembly in order to capture ambient slow neutrons. Instrumentation also was included to measure the penetrating meson component.

By good fortune, in early 1947 the Office of Naval Research, along with the U.S. Air Force, had established a B-29 bomber base for research purposes in the desert at Inyokern, China Lake, California. The three B-29 aircraft at the base had been modified for Carl Anderson to fly at high altitudes with special propeller controls and for brief periods of time could attain altitudes of $41,000 \, \text{ft.}$ ($\sim 12,500 \, \text{m}$).

The results of these investigations were startling (Simpson 1948, 1951). As shown in Figure 12 the latitude dependence for the nucleonic component was a

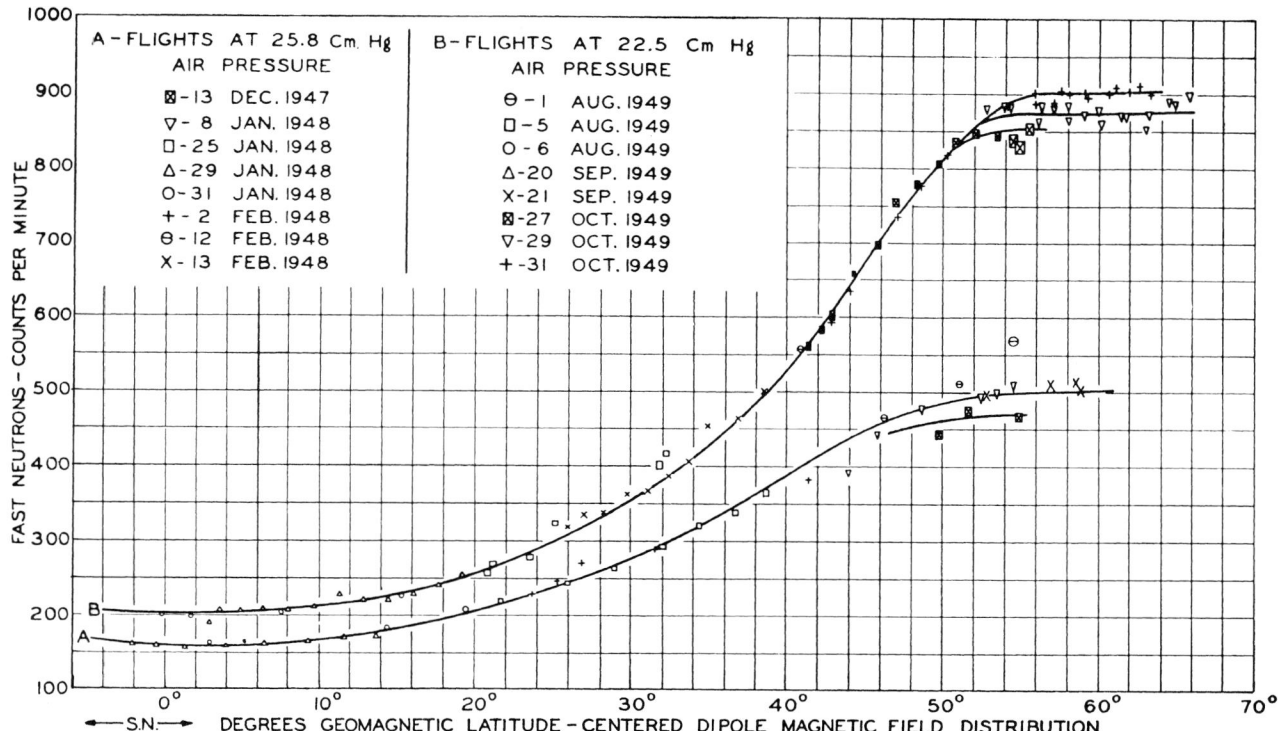

Figure 12 Latitude dependence of fast neutrons at (A) 27,000 feet (8230 m) and (B) 30,000 feet (9150 m) (Simpson 1951).

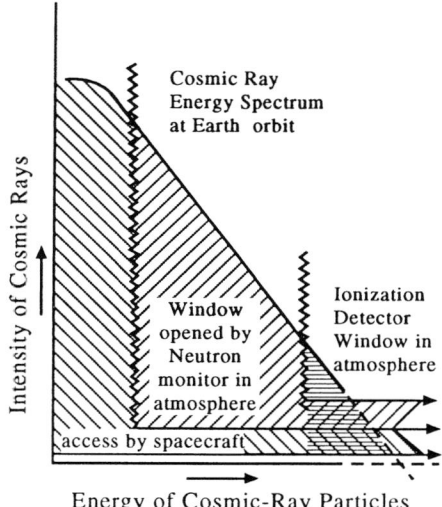

Figure 13 Windows of access to the secondary components inside the atmosphere are compared to the entire range of energies accessible by spacecraft.

300–400 percent effect (depending on atmospheric depth), as compared with a 10–15% effect for the meson component. The fact that the yield of fast neutrons in a column atmosphere from a single primary proton was a rapidly varying function of incident energy down to ≈ 1–$2\,\text{GeV}$ at the top of the atmosphere – and, therefore, that a large latitude effect between 40° and 52° existed – proved that one could extrapolate, from fast neutron measurements deep in the atmosphere, the arrival of primary cosmic ray nuclei at the top of the atmosphere in an energy range inaccessible to ionization detectors. Sketched in Figure 13 is the assumed cosmic ray integral spectrum and the extension to lower energy of the observable nucleonic cascade in the atmosphere.

Following the discovery of the neutrons in the atmosphere and that atmospheric nitrogen has a large neutron capture cross-section, Libby (1946) pointed out the possibility of using the radioactive ^{14}C ($\tau_{1/2} = 5730\,\text{yr.}$) as a tool for dating artifacts that had once lived. But a large discrepancy in the neutron yield between the calculated atmospheric neutron production and the measurements existed until the neutron latitude effect was taken into account (Libby 1955). This technology, derived from cosmic ray physics, has opened new research fields in archeology and anthropology, as shown in the time-line chart.

6. DISCOVERY OF SOLAR MODULATION IN INTERPLANETARY SPACE

Since the Lemaitre and Vallarta (1933) theory for the motion of cosmic ray particles in the dipole magnetic field of Earth showed that a change in dipole strength would only shift the latitude of the "knee" of the latitude curve, why was the nucleonic component intensity changing at high latitude above the geomagnetic cutoff or knee of the latitude curves from flight to flight, as shown in Figure 12? Clearly, the low-energy cosmic ray intensity outside the geomagnetic field (observed above ~50°) appeared to be changing with time. The best way to investigate the origin of this effect was to add automatic recording equipment to the primitive neutron monitor in order to obtain a continuous record. This was installed in one of the B-29 aircraft in 1948 where it operated whether on the ground or in flight.

In 1948 and 1949 with a continuous neutron intensity record over several months, one could select times for undertaking latitude flights to investigate these intensity changes.

Forbush (1938) had studied intensity variations of the electron and meson component from the late 1930's using the worldwide distribution of Compton-Bennett Model-C ionization chambers (Figure 4) [in a program sponsored by the Carnegie Institution of Washington's Department of Terrestrial Magnetism (DTM)]. Through the application of rigorous analytical methods, Forbush discovered all the principal cosmic ray intensity variations that could be measured by ionization chambers. He showed that they were correlated with changes in geomagnetic field intensity, such as occur in geomagnetic storms. He (1938) was the first to observe a solar flare accompanied by high energy particles accelerated by the flare. Perhaps his most widely known correlation was a rapid decrease of cosmic ray intensity correlated with a worldwide change in geomagnetic field intensity (Forbush 1938). It is called a Forbush type decrease, a term widely used today.

The mechanism proposed at that time to account for this Forbush-type decrease was based on increases in the worldwide magnetic field intensity arising from an enhanced equatorial ring current produced by solar ion streams. For incident cosmic ray protons the effect was equivalent to a temporary increase in the Earth's dipole magnetic moment.

To investigate the origin of both the Forbush decrease and the 27-day recurring intensity variations in the cosmic radiation, latitude measurements extending above and below the knee were carried out in order to test the prevailing view that the intensity variations were geophysical in origin. If the cosmic ray intensity variations were due to an enhancement of the geomagnetic cutoff for cosmic rays, the latitude intensity curves for the nucleonic component would shift, as shown in the sketch Figure 14 – whereas if the origin was a change in cosmic ray intensity in the interplanetary medium the latitude curves would display intensity changes above and below the knee.

The experimental results obtained in 1949–1951 were clear for both the Forbush-type decreases (Figure 15) and also for the ~27 day recurring intensity decreases. The

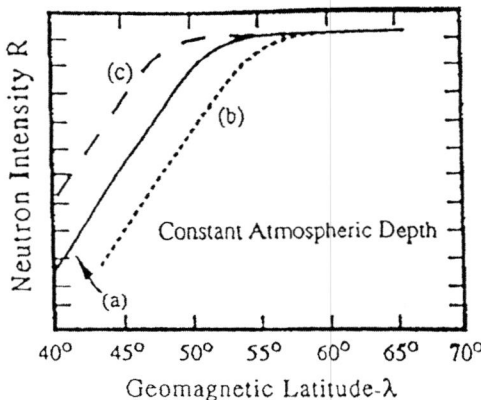

Figure 14 The predicted behavior of neutron intensity as a function of latitude based upon the assumption that the primary cosmic-radiation intensity variations is produced by a geomagnetic field variation (Simpson 2000).

origin of the intensity changes was beyond the range of the geomagnetic field. Since the cosmic ray intensity from the galaxy must be constant over long periods of time, the experimental results pointed to the origin of the Forbush decreases and the ~27-day recurring variations in the intervening interplanetary medium and with control of this medium by the Sun.

By 1952 these investigations proved that interplanetary dynamical processes were the origin of galactic cosmic ray intensity-time variations – a new phenomenon called solar modulation (Simpson 1985 and references therein).

In order to investigate continuously the intensity variations of the nucleonic component while excluding the electron and meson components, a novel detector was developed. It was based on the principle that the nucleonic cascade interacting locally in matter of atomic number **A**, would yield local neutrons whose number was approximately proportional to $A^{2/3}$, as predicted by nuclear physics theory. This idea led to conception of a cosmic ray neutron monitoring system based on measuring the local production of fast neutrons in elements of high atomic weight. The instrument was analogous to a subcritical nuclear reactor in which lead would substitute for uranium as the producer of neutrons, and hydrogenous materials, such as paraffin wax, would substitute for the carbon or heavy water moderator and tamper. Thus the fragmentation of a lead nucleus by an incident high-energy secondary nucleon in the nucleonic cascade of cosmic ray origin would yield a multiplicity (~11) of local fast neutrons, which would then become thermalized in the surrounding paraffin wax and be detected with high efficiency using $^{10}BF_3$ proportional counters embedded in the "pile". With the detection of energetic He^{++} in a proportional counter from slow neutron capture in ^{10}B ($^{10}B + n \rightarrow ^{11}B^* \rightarrow ^{7}Li + ^{4}He^{++} + Q$),

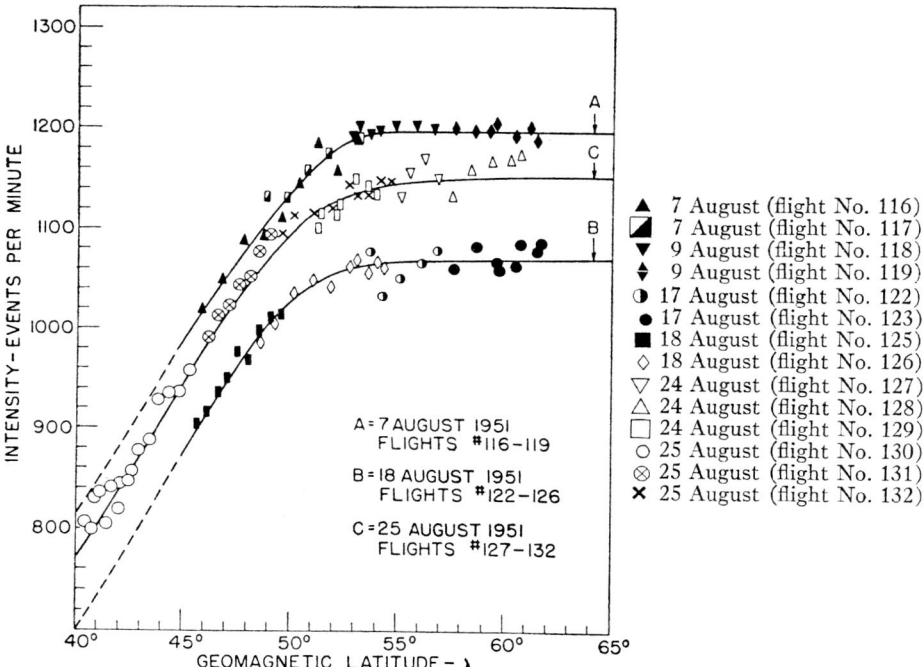

Figure 15 The neutron intensity data used to establish the latitude curves for August 7, 18 and 25, 1951, are shown. The smooth curves (with dashed lines for extrapolations) are used for analysis. The curves are based upon several aircraft flights listed at the right-hand side of the figure (standard deviations are approximately the size of the flight identification symbols) (Adapted from Simpson 1954).

the monitor was immune to both the electromagnetic component and external neutron flux variations due to changing local conditions. The development and use of the cosmic ray neutron monitor has been described by Simpson (1953, 1985, 2000). Figure 16 is a cross-section of the design. The only correction required was the changing atmospheric mass through which the nucleonic cascade passed. This was determined by establishing the barometric coefficient for the neutron monitor counting rate at each site (Simpson et al. 1953).

To relate the observed neutron monitor secondary component in intensity to the incoming primary intensity of charge Ze particles, whether at sea level or at high altitude, a specific yield function S was developed for this purpose. Thus, the observed counting rate R at atmospheric depth x, geomagnetic latitude λ, particle momentum P, and time t is

$$R(\lambda,x,t) = \sum_z \int_{[P/z]_\lambda}^{\omega} S_z J_z\left(\frac{P}{z},t\right) d\left(\frac{P}{z}\right)$$

where P/z_λ is the vertical cutoff magnetic rigidity at λ and $\mathbf{j}_z$ is the differential flux (Simpson et al. 1953).

Although there were extensive high altitude measurements for the determination of the specific yield function S over a wide range of geomagnetic cutoffs, a sea level latitude range of S values was lacking. Consequently, when D.C. Rose of the Canadian National Research Council asked in 1954 for plans to build a neutron monitor in Ottawa, we together worked out a mission whereby we could determine the sea level specific yield function from the Arctic to the Antarctic. He and his research associate, K.B. Fenton, coopted the HMS *Labrador* ice breaker and arranged for a neutron monitor to be installed in a special housing on the deck. After the Arctic survey we met in Boston Harbor to transfer the equipment to the USS *Atka* ice breaker for a 1954–1955 survey into the Antarctic with Fenton in charge. We repeated the surveys in 1955–1956 (USS *Glacier*) and in 1956–1957 (USS *Arneb*) with my graduate student Rochus Vogt in charge. These expeditions covered the period from solar minimum to solar maximum activity and continued into the IGY.

As the intensity level of the approximately 11-year solar cycle declined to solar minimum between 1948 and 1954, neutron monitor and aircraft flight observations proved that an additional flux of low energy primary cosmic ray particles had access to the top of the atmosphere, changing the integral magnetic rigidity spectrum as a function of the $\sim$11 year solar activity level to energies exceeding the geomagnetic equatorial cutoff of $>13\,\text{GV}$.

About this time in 1954, Forbush independently published his important ion chamber measurements from 1937–1952, showing that the approximately 4% intensity variation he observed over each 11-year solar cycle was "negatively correlated with sunspot number" (Forbush 1938). Forbush noted that the long-term ion chamber intensity variations "are not ascribable to transient decreases accompanying some magnetic storms," but he had no other comment on the origin of the negative correlation. In his editorial foreword to the works of Forbush James Van Allen (1993) noted:

> In none of his papers did he propose that either a Forbush decrease or the 11-year intensity cycle was caused by the interplanetary medium; but later with characteristic modesty, he welcomed and embraced this line of interpretation as established by others.

Simpson, by 1955, concluded that the Forbush negative correlation and his own investigations, carried out between 1948 and 1954, provided further evidence for solar modulation of the cosmic rays by interplanetary processes.

In 1951 neutron monitors were in operation in the Peruvian Andes (~0.5° magnetic latitude), Mexico City (+29°), Sacramento Peak, N.M. (+42°), Climax, Co. (+48°) and Chicago, IL (+52°), thus constituting a cosmic ray magnetic spectrometer. During the International Geophysical Year (IGY, SECTION 1B) in 1957–1958 more than 60 neutron monitors were distributed worldwide by many nations. Figure 17 illustrates the current world-wide distribution of neutron monitors, many of which are a larger scale version designed by Carmichael (1964) (the IQSY monitor).

The cosmic ray neutron monitor intensity from 1952 through 1999 is displayed in the lower part of Figure 18, along with the "11" year solar activity cycle and solar magnetic field polarity reversals. The cosmic ray intensity changes at the top of the atmosphere are approximately a factor 2 larger than that shown for the Climax monitor intensity. These investigations proved that the principal cosmic ray intensity variations were due to interplanetary dynamical processes controlled by solar activity and *were not* due to geomagnetic field phenomena. By 1954 this conclusion led to the concept of solar modulation of the galactic cosmic rays.

7. THE ORIGIN OF THE COSMIC RAYS: SOLAR OR GALACTIC?

The author has recounted from the Centennial issue of the *Astrophysical Journal* (20 December, 1999) the critical

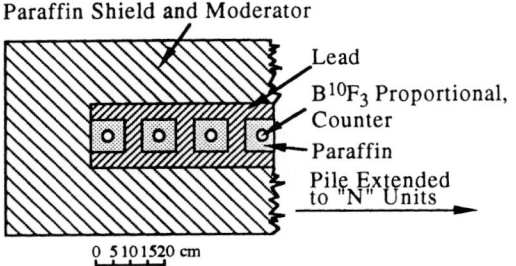

Figure 16 The pile when extended to 12 counters is used as the detector at Chicago and Climax (11,000 feet 3350 m). This pile is composed of 6500 lb. lead plus approximately 3000 lb paraffin. Cross-section view (from Simpson *et al.* 1953).

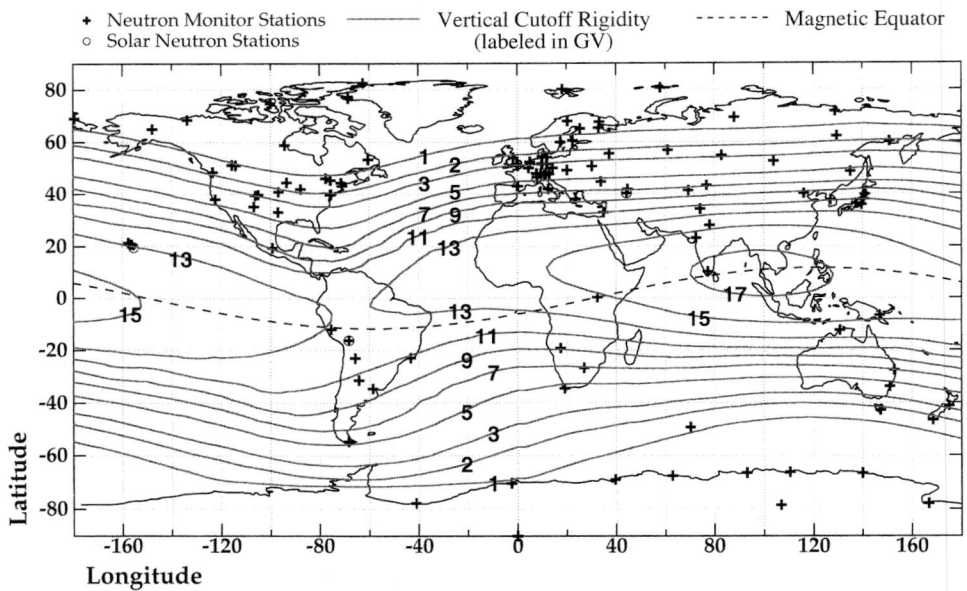

Figure 17 World map of current neutron monitor locations (compiled by C. Lopate).

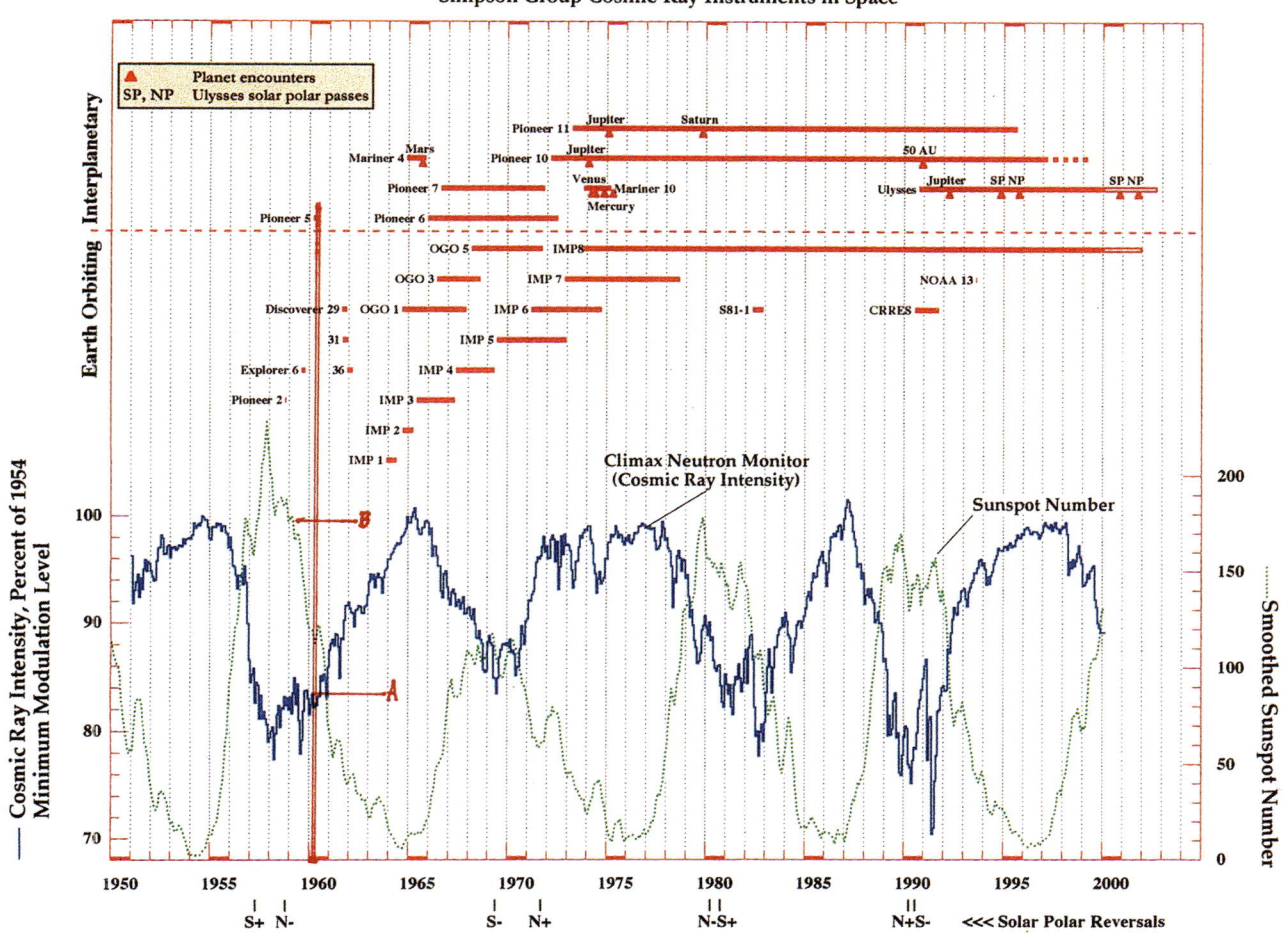

Figure 18 Continuous record of modulated intensity of nucleonic component, 1951–1999.

period after World War II for the convergence of cosmic ray physics and astrophysics.

The debate on the origin of the cosmic rays – were they of solar system or galactic origin – was intensified with the 1948 visit to the University of Chicago by Hannes Alfvén. He had developed further his initial ideas on the motion of an ionized liquid in a magnetic field (Alfvén 1942) that he called "magneto-hydrodynamic" waves to include later the motion of conducting plasmas in magnetic fields on the scale of the solar system. Teller had been arguing for a solar system origin since he claimed that the total energy in the cosmic rays would be too large to be realistic for an origin on a galactic scale.

Alfvén's concepts and arguments for a solar system origin (Alfvén 1949) led in 1948 to a joint paper (Alfvén et al. 1949). In their paper they assumed:

> .... that cosmic rays are generated on or in the neighborhood of the Sun and are kept near the solar system by extended magnetic fields.

On the other hand, Enrico Fermi was fascinated with Alfvén's arguments concerning the motion of magnetic fields imbedded in dilute plasma. Fermi began to consult Teller and me on cosmic ray measurements, composition and spectra.

It was a tradition to have a special dinner near years end at the University Quadrangle Club. In December 1948 my wife and I were at a table with the Fermis. Laura was there but where was Enrico? Dinner was underway when Enrico suddenly appeared with a broad grin saying he may have solved the cosmic ray problem. He had submitted, "On the origin of the cosmic radiation" to the *Physical Review*, (Fermi 1949).

Fermi's abstract says it all:

> A theory of the origin of cosmic radiation is proposed according to which cosmic rays are originated and accelerated primarily in the interstellar space of the galaxy by collisions against moving magnetic fields. One of the features of the theory is that it yields naturally an inverse

power law for the spectral distribution of the cosmic rays. The chief difficulty is that it fails to explain in a straight-forward way the heavy nuclei observed in the primary radiation.

Fermi employed the theory of Alfvén for the velocity, V, of propagation of magneto-elastic waves:

$$V = \frac{H}{(4\pi\rho)^{1/2}} \qquad (3)$$

where, H, is the intensity of the magnetic field and ρ is the density of the interstellar matter. From the above equation he showed that the magnetic field in the dilute interstellar matter should be the order of magnitude 6×10^{-6} gauss.

He illustrated (cf. Figure 19) one of his mechanisms for a charged particle in a magnetic field to gain energy (a) in a head-on collision or to lose energy (b) in an overtaking collision.

The "chief difficulty" according to Fermi was the 1948 discovery by the University of Minnesota and University of Rochester physicists of heavy nuclei in the cosmic radiation consisting of the nuclei of carbon to iron, fully stripped of their electrons (Freier et al. 1948). Fermi claimed "the injection energy of these particles is of several BeV, and it is difficult to imagine a secondary effect of the cosmic radiation on the diffuse interstellar matter which might produce this type of secondary with any appreciable probability." A full understanding of this injection problem remains unsolved today.

The principal hypothesis in Fermi's remarkable theory was the requirement that magnetic fields on a galactic scale existed in the interstellar medium. Did these fields exist?

Apparently Fermi was unaware of concurrent astronomical observations of the polarization of light from distant stars in the interstellar medium in 1948 by Hiltner (1949) at Yerkes and McDonald Observatories; and by Hall (1949) at the U.S. Naval Observatory.

Hiltner's work was undertaken initially to establish observationally the effect pointed out by Chandrasekhar in 1946 that the continuous radiation of early-type stars should be polarized. But Hiltner (1949) found instead "that the measured polarization does not arise in the atmospheres of these stars, but must have been introduced by the intervening interstellar medium." In a later paper Hiltner (1951) proved his hypothesis by reporting polarization measurements from 841 stars, but he did not recognize its importance for the required presence of interstellar magnetic fields. This connection was later established by Davis and Greenstein (1951) among others, by showing that the polarization of starlight was produced by dust grains aligned by interstellar magnetic fields. Davis (1951) appears to be the first to estimate from astronomical data discussed in Davis and Greenstein (1951) that this magnetic field would be $\leq 10^{-4}$ gauss.

Chandrasekhar and Fermi (1953a) joined their expertise in bringing together critical elements from astronomical theory and observation to verify the existence of galactic interstellar magnetic fields, and thus the galactic origin of the cosmic radiation – they changed an hypothesis to a testable theory.

They obtained the value $H \approx 6 \times 10^{-6}$ gauss by two independent methods, and pointed out that Hiltner's observation showed that the magnetic field lines should be described as "wavy". In their first method they used Eq. 3 to determine the velocity of the transverse magneto hydrodynamic wave leading to their equation:

$$H = (4/3 \, \pi p)^{1/2} v/a = 7.2 \times 10^{-6} \, gauss$$

where **v** is the lateral velocity of the lines of force and a (radians) is the mean angular deviation of the plane of polarization from the direction of the spiral arm.

Their second method was based on the requirement of equilibrium of the spiral arm with respect to lateral expansion and contraction. They equated the gravitational pressure in the spiral arm to the sum of the material pressure and the pressure due to the magnetic field. This leads to $H \approx 6 \times 10^{-6}$ gauss, a value close to their first method.

They noted that their values of $H \approx 6 \times 10^{-6}$ gauss is almost two orders of magnitude lower than the value deduced by Davis and Greenstein (1951). Chandra and Enrico in preparing their paper had thereby been unaware of the 1951 note by Davis.

The introduction of their second method obviously led them simultaneously to their second joint effort – namely, "Problems of gravitational stability in the presence of a magnetic field" (1953b) with several of the problems treated "Chandra-style".

Chandra told the author "when you explain to Fermi the results of work on which you have been working for years, only during the first 20 minutes do you know more than Fermi". Their work was one of the great scientific collaborations of this century.

8. EARLY CONCEPTS OF THE INTERPLANETARY MEDIUM AND DISCOVERY OF THE HELIOSPHERE

Initially, ideas concerning phenomena in the interplanetary medium were focused on charged particle plasma streams

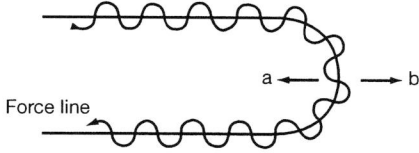

Figure 19 Type B reflection of a cosmic-ray particle (Fermi 1949).

emanating from local active regions on the Sun in order to explain geomagnetic and auroral effects (Chapman and Ferraro 1940); or to accelerate charged particles to support the hypothesis that the bulk of cosmic rays could be of solar system origin (see Section A7). In general, it was assumed that solar particle streams propagated essentially in an interplanetary vacuum. Later Biermann (1952), through the study of the radial distribution of comet tails in space, concluded that there must exist a continuous emission of solar plasma throughout the interplanetary medium. Up to the early 1950's the main focus was on interplanetary medium models restricted to the inner solar system.

Time-intensity variations of the cosmic rays as measured at Earth played an early, leading role in the evolution of models leading to the prediction of a heliosphere. The early research with ionization chambers led to the widely accepted conclusion that the links in these correlations were geocentric, geomagnetic effects (e.g., geomagnetic storm phenomena).

However, the discovery that the low energy primary cosmic rays with access to the atmosphere at high geomagnetic latitudes produced nuclear cascades which could be detected deep in the atmosphere and the introduction of the neutron monitor concept – to observe intensity changes in the nuclear cascade – made it possible to extend the study of the cosmic ray intensity variations in the incident cosmic radiation down to very low energies where they were most sensitive to interplanetary changes in solar plasma and magnetic fields.

Preliminary ideas for a heliosphere began to appear in 1955, which included models for a static magnetic field barrier between the Sun's field and an interstellar field (Figure 20A from Davis 1955), or a sketch of a disordered magnetic field barrier region beyond earth (Figure 20B and adapted from Simpson in 1955). But with a static field the cosmic rays would soon fill the heliosphere and there would be no modulation. Both kinds of models intended to explain how galactic cosmic ray spectra were modified before reaching the observer at Earth (see also Morrison 1956).

Understanding the interplanetary propagation of nuclei accelerated in the flare of 23 February 1956 was a crucial next step. The establishment of a world-wide network of neutron monitors made it possible to show that the accelerated nuclei were emitted from solar flares with a magnetic rigidity spectrum extending to $\leq 20\,GV$ and that these nuclei traveling at relativistic velocities were constrained or "stored" within the inner solar system for many hours (Meyer *et al.* 1956; Figure 21). The energetic charged particles eventually escaped by diffusion through magnetic fields that must be invoked to totally enclose the solar system beyond the orbit of Earth (Figure 22). This was the first direct proof of the existence of a dynamical heliosphere that sustained the modulated galactic cosmic radiation. In Figure 22 the error of an essentially "field free" region between Sun and Earth was corrected by the analysis of Lüst and Simpson (1957).

It was clear that the interplanetary magnetic fields, invoked to account for the February 23, 1956, solar particle propagation, must also be central to the explanation of how galactic cosmic rays are modulated over the long term. The barrier model of Davis (1955) would not work since it was static and would not prevent the full galactic flux from entering the solar cavity within a short time. Parker, who had joined the author's group in 1955 and Morrison (1956), had suggested outward moving magnetic clouds to account for the modulation, but as summarized by Parker (1963) there were serious difficulties with these suggestions. Parker went on to develop the quantitative theory for coronal expansion and extension of the solar wind into interplanetary space carrying solar magnetic fields which formed an archimedian-type spiral field near the equator (Parker 1958, 1963). Parker's work on the solar wind met initially with opposition from the referees: see his attempt to publish in *Astrophysical Journal*, which he described in his acceptance speech for the Bowie Medal (Parker, Chapter 9 in this volume). This dynamical system modulated the incoming galactic cosmic

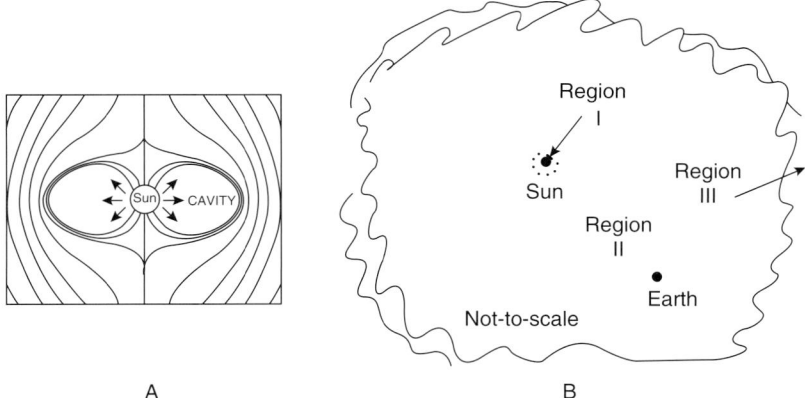

Figure 20 Heliospheric concepts: A (from Davis 1955); B (from Simpson, circa 1956).

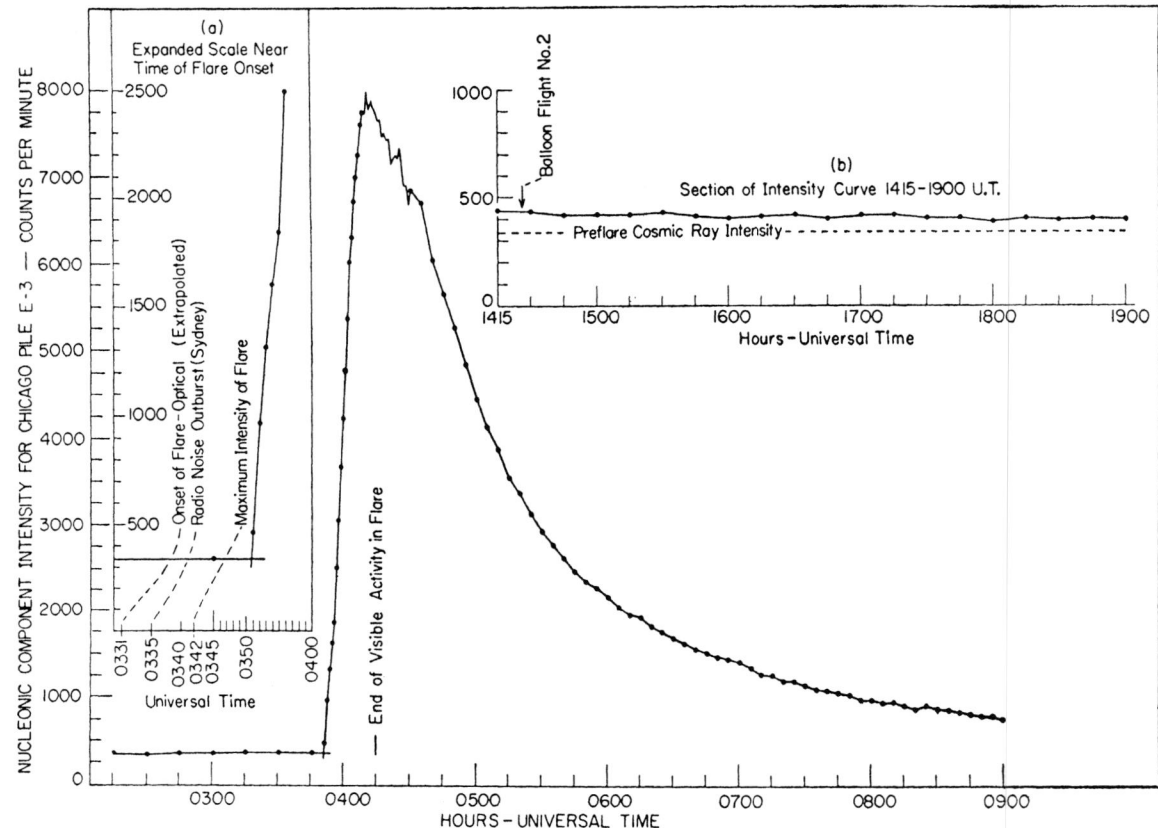

Figure 21 The solar flare of 23 February 1956 recorded in the Chicago Neutron Intensity Monitor (Meyer *et al.* 1956).

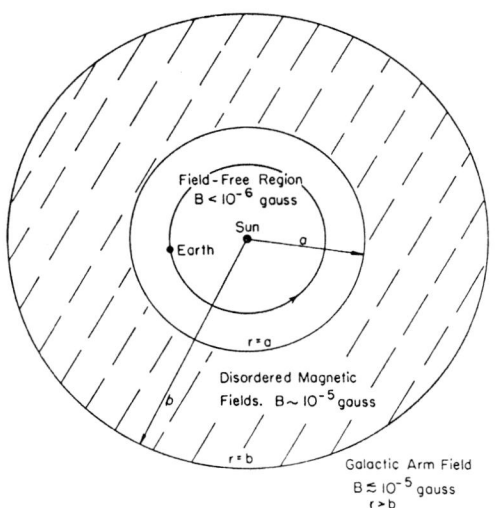

Figure 22 The primitive, dynamic heliospheric model (Meyer *et al.* 1956).

rays by diffusion, convection and adiabatic deceleration to account for the detailed differential energy spectrum of the modulated cosmic rays. Parker (1958, 1963) published important accounts of a basic model for solar modulation that embodied the principles underlying all modulation theories today (see Chapter 9). However, direct experimental tests for the solar wind, interplanetary magnetic fields and modulation of the galactic cosmic rays within the heliosphere had to await the entry of instruments into space.

PART B. IN SPACE

1. THE IGY AND THE BEGINNING OF THE SCIENCE IN SPACE ERA

Scientific conferences after World War II were dominated by particle physicists using the cosmic rays to discover and characterize elementary particles created in atmospheric nuclear interactions (see Section A4). However, this changed rapidly after 1953 – not only due to the post-war construction of high energy accelerators, but also by advances in both recognizing the galactic origin and the intensity variations of cosmic rays in the solar system; namely, the astrophysical and geophysical aspects of the radiation. For example, as reported by Simpson (1985) in his closing remarks for the Third International Cosmic Ray Conference, which was focused on unstable elementary particles, President P.M.S. Blackett state in part:

... the next conference should concern itself mainly with the other aspects of cosmic rays
... under the title of "Geophysical and Astronomical Aspects"
... it will be desirable to invited some of the astronomical colleagues to take part.

Cosmic ray investigations were not included in the initial formulation of plans for the International Geophysical Year (IGY). Simpson (1994) described how cosmic rays as probes of the interplanetary medium became an essential part of IGY research. Marcel Nicolet was Secretary-General of the Comité Spécial de l'Année Géophysique Internationale (CSAGI) (the CSAGI also included Sidney Chapman, President, Lloyd Berkner, Vice President with Vladimir Beloussov [U.S.S.R.] and J. Coulomb [France] to run the IGY with reporters in the 14 disciplines shown in Table 1). Of the 86 nations participating in the various scientific disciplines, more than sixty nations participated in cosmic ray research. The standard IGY instruments distributed world-wide were the meson telescope by Harry Elliot of the U.K. and the neutron monitor by Simpson.

Looking back on the IGY, it is clear the geophysicists of the world brought together a peaceful international enterprise unparalleled in the history of mankind, certainly the most important since the Renaissance or the Copernican revolution. The knowledge of Earth and the space around us derived from most of the disciplines in that 18-month period exceeded the accumulated knowledge of prior centuries of inquiry regarding Earth.

Table 1 Discipline reporters of Comité Spécial de l'Année Géophysique Internationale

Activity	Reporters
World days	A. H. Shapley (United States)
Meteorology	J. Van Mieghem (Belgium)
Geomagnetism	V. Laursen (Denmark)
Aurora and airglow	S. Chapman (United Kingdom) (with F. Roach and C. Elvey, both US)
Ionosphere	W. J. G. Beynon (United Kingdom)
Solar activity	Y. Ohman (Sweden)
Cosmic rays	J. A. Simpson (United States)
Longitudes and latitudes	A. Danjon (France)
Glaciology	J. M. Wordie (United Kingdom)
Oceanography	G. Laclaváre (France)
Rockets and satellites	L. V. Berkner (United States)
Seismology	V. V. Beloussov (U.S.S.R)
Gravimetry	P. Lejay (France)
Nuclear radiation	M. Nicolet (Belgium)

Investigations in the Antarctic were vital for achieving the overall goals of the geophysicists, especially the introduction of cosmic ray research facilities for the first time. This included the development of the Antarctic Treaty to preserve the last continent on Earth for mankind. This treaty was a major achievement and later was a template for discussions leading to a treaty banning atmospheric testing of nuclear weapons.

In this author's opinion, the most important legacy of the IGY may, however, have been the collaborative and collegial character of the enterprise at a time when politically the major powers were facing each other in the Cold War. During the IGY the "doomsday clock" of the Bulletin of the Atomic Scientists hovered at 2 minutes to midnight. However, the exchanges between the geophysics communities of the United State and the U.S.S.R., both scientific and military, for the common goals of the IGY played more than a minor role in keeping the cold war cold.

The IGY also triggered the scientific exploration of space by satellites. At the 1954 CSAGI meeting in Rome, Lloyd Berkner, the reporter for rockets and satellites, had asked the CSAGI to urge the IGY participants to launch satellites. The following year both the U.S.S.R., under Academician Leonid Sedov,[1] and the U.S. White House had announced plans to do so. The CSAGI meeting in Barcelona, September 1956, was largely focused on this issue, with the U.S. committee presenting marvelous color slides and plans for the Vanguard program. On the other hand, I.P. Bardin, who chaired the Soviet IGY committee had little to say and went along with all the resolutions on the subject adopted by the CSAGI. One evening during the sessions, Herbert Friedman and the author had a private dinner with some of the Soviet delegates. We were told that a Soviet satellite would be launched in the fall of 1957 and would transmit a beep signal close to the various wavelengths broadcast by the WWV international time service of the U.S. National Bureau of Standards. This would make it possible for all nations to easily locate the Soviet satellite signals. They also described in general the instrumentation planned for their satellite.

Upon returning to the United States we made contact with our respective channels of intelligence, Friedman through the Naval Research Laboratory and the author through a U.S. security agent. Somewhere up the line of communication to the White House in the fall of 1956 it was decided that the information we provided regarding a Soviet satellite launch in 1957 was ridiculous and not worthy of further action. Our government and the public were surprised a year later with the launch of *Sputnik I* in October 1957.

---

[1] Chairman of the Soviet Academy of Sciences' Commission on Interplanetary Communications.

During 1958 the author visited the U.S.S.R. and many of the laboratories in Moscow and ISMIRAN.[2] K.I. Gringauz, who was later to make the first measurements of the solar wind, showed duplicates of the instruments carried on the Sputniks which looked like equipment installed on WW II aircraft, including shock absorbers! Their Geiger-Muller counters were similar to illustrations in a 1938 book. Later, in Leningrad, this author visited an apparatus shop, found the identical G.M. tubes and bought one. When the authors asked through an interpreter, if the tube worked, his reply was, "It must work!". Clearly, the Soviet scientists and engineers had hurriedly pulled together existing Soviet apparatus with little or no modifications for their first space flight.

On the other hand, the United States had adopted a policy under President Eisenhower that separated military technology from civilian technology for rocket development. This resulted in an unclassified program called "Vanguard" for the U.S. response to the IGY. This program had repeated failures much to the consternation of the scientific community. Finally, with *Sputnik* in orbit the U.S. turned to the military for the Redstone launcher. In a chapter in this volume Van Allen related how the first U.S. satellite was launched with his Geiger counters and his subsequent discovery of a particle radiation belt in the magnetosphere of Earth. Thus, with *Sputnik* and *Explorer I* in orbit the U.S.S.R. and the U.S. opened the era of scientific research in space.

The U.S. National Academy of sciences (NAS) in the Spring of 1958 established the Space Science Board (SSB). Through several scientific committees the Board encouraged U.S. investigators to propose experiments that addressed interplanetary magnetic fields, space plasma (the solar wind) and a wide energy range of charged particles, especially the magnetospheric radiation belts and the cosmic radiation (Berkner and Odishaw 1961).

In the Fall of 1958 the National Aeronautics and Space Administration was established with a charter from the U.S. Congress that included the scientific exploration of space and the planets. The early initiatives of the SSB enabled NASA to attain a fast start for science in space.

2. INTO SPACE

This author's inquiries in the United States regarding the possible satellite payload opportunities for cosmic ray instrumentation led nowhere until after *Sputnik* (Simpson 1985 contains the following personal account). Realizing that the United States could not hold out much longer in providing a wide range of opportunities for space missions, this author went to our University President, Lawrence Klimpton, in December 1957 to request $5,000 support for space experiments to get a part-time engineering group started. We drew upon part-time help from the Laboratories for Applied Sciences, which had been managed by the University during the emergency years of the Korean War. P. Meyer and this author worked with their staff to design and build our first satellite experiments (Simpson 1985, p. 406).

Our initial goal was to prove directly (in space) that the modulation of galactic cosmic rays was not geocentric but heliocentric by determining the radial gradient of cosmic ray intensity – both outward from Earth and away from Earth's orbit – under conditions when the cosmic rays were heavily modulated. The depressed cosmic ray intensity at that phase of the solar cycle in 1958–1960 was ideal for these purposes (see Figure 18). We had designed our experiment by March 1958. The author then went on to Washington to search for opportunities to launch our experiment. H. Odeshaw in the National Research Council called attention to a special meeting of a board on payload selection for missions sponsored by the U.S. Air Force which was meeting the next day. That night I wrote on the backside of Cosmos Club stationery my drawings to display before the committee. As a result of this effort, I came away with approval for half the payload capability for *Pioneer-2*. By June 1 I had invited C.Y. Fan to join P. Meyer and myself to carry forward these experiments. During that month, H. Newell, of the NACA agency (which would soon become NASA) forwarded funding to help us carry through the development. A cross-section of our *Pioneer* detection system, composed of a triple coincidence proportional counter telescope encased in a lead shield, is shown in Figure 23. This design enabled us to distinguish electrons from protons. Being novices on the launching of major rocket-borne experiments we had to learn rapidly many new techniques during the summer of 1958. Without adequate vibration and shock test equipment, we devised a scheme to drop our apparatus out a third floor

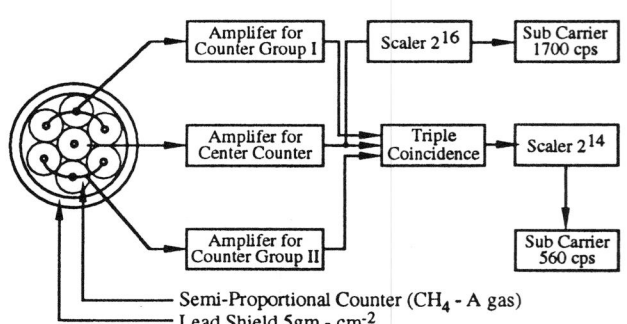

Figure 23 Cross-section of the triple coincidence proportional counter telescope on the first deep space mission, *Pioneer-5* in 1960 (Simpson 1985).

---

[2] Institute of earth magnetism, ionosphere and propagation of radiowaves of the RAS.

window of our laboratory where it would land in a toy sandbox. The test worked and our experiment was launched in *Pioneer-2* on November 8, 1958. Unfortunately, the U.S. Air Force experienced another rocket failure, but not before we discovered that the inner edge of the trapped radiation belt (that Van Allen had discovered and reported in May 1958) contained protons with energies exceeding 75 MeV. At least we had started along the direction of developing a program for conducting experiments in space. In 1959 the satellite *Explorer 6*, carrying another instrument of the same design, showed for the first time that the cosmic ray Forbush decreases of intensity were the same both outside the magnetosphere and at Earth (Fan *et al.* 1960a).

The same type of instrumentation was carried on the first deep space probe, *Pioneer-5*, in March 1960. The 11-year cosmic ray intensity due to modulation was depressed to near minimum flux level. This interplanetary space mission showed that instead of the cosmic ray intensity rising to the level B in Figure 18, it remained close to level **A** at Earth as it traveled inward toward Venus. Thus the principal 11-year modulation was heliocentric in origin, and we noted (Fan *et al.* 1960b, p. 273):

> ...any electromagnetic modulating mechanism required to account for the 11 year intensity variations is located principally outside the orbit of Earth. Such results from *Pioneer-5* place strong constraints upon acceptable heliocentric models for the 11 year cosmic ray intensity variations.

Furthermore, we and the magnetometer group (Coleman *et al.* 1960) discovered the first traveling interplanetary shock from a flare. We proved that a Forbush-type intensity decrease was associated with an interplanetary shock (Fan *et al.* 1960c).

The period 1959 through 1963 was one of frustration and discouragement for me with more than ten launch vehicle failures or mission cancellations (including three *Atlas Ables*, two *Ranger* Missions, *Mariner A* and *B* and others). As a result of these stand-downs I had in my laboratory a dozen instruments qualified for space flight.

Recognizing that all U.S. experiments in the foreseeable future must have low power and small mass in order to compete for space missions, I set up a semi-conductor detector development laboratory in 1959 with A.J. Tuzzolino. In 1959 charged particle solid state detectors were unreliable, especially in a vacuum. Our developments in the 1960's demonstrated that semiconductor devices could replace – with enhanced reliability – the heavier scintillators and photomultiplier tubes required for future charged particle telescopes in space. They were to achieve my goal of measuring the elemental and isotopic composition of the cosmic ray nuclear component.

There was a period in 1959–1962 when NASA was unable to maintain its flight program. However, the U.S. Air Force's Discoverer program came to my rescue by providing "piggy-back" rides on the polar orbits *Discovers 29, 31* and *36* satellites so that my graduate students working with me, including E.C. Stone, could carry out their experimental Ph.D. theses. These *Discoverer* missions became "test beds" for designing several of our more sophisticated solid-state instruments between the 1960's and 1990s.

In 1963 the launch of the first of the series of Interplanetary Monitoring Platform (IMP) satellites, *IMP-1* – with the first semiconductor telescope capable of resolving isotopes, such as ^{2}H (Fan *et al.* 1966) and ^{10}Be (Garcia-Munoz *et al.* 1975) that determined the ~ 10 My lifetime of cosmic rays in the galaxy – was crucial for opening our continuing research opportunities for space missions. IMP-1 was placed into a highly eccentric orbit that provided regular access to the interplanetary medium. Subsequent IMPs, including the still-operating IMP-8 satellite, have provided nearly continuous monitoring of interplanetary conditions since 1963. By setting high standards within NASA and by encouraging University participation for this satellite series, F.B. McDonald played a key role as NASA Program Scientist in the success of these early *IMP* satellites.

3. THE HELIOSPHERE AND MAGNETOSPHERES AS AN ASTROPARTICLE LABORATORY

By the early 1960's it became clear to the author that plasma, magnetic field and energetic particle phenomena – that could not be scaled to Earth-bound laboratories – could, indeed, be investigated on the scales of planetary magnetospheres and, especially, the heliosphere. Furthermore, the interdisciplinary character of the researches formed a framework critical for cosmic ray physics, as shown in Figure 24.

For example, Table 2 delineates the astrophysical sites and scales of charged particle acceleration investigated or predicted by the late 1980's (Simpson 1989). Within the heliosphere the maximum energy and, for most acceleration sites, the particle species and energy spectrum have been investigated. The details of the acceleration mechanisms have been identified along with their theoretical foundations, as reported elsewhere in this publication (see Chapter 48 and 62). These authors have also reported in this volume the recent (1980–1999) research concerning the heliospheric magnetic field, solar wind and the evolution of the heliospheric, wavy current sheet.

These acceleration mechanisms within the heliosphere when extrapolated to a galactic scale, have contributed to an understanding of how cosmic ray particles attain their

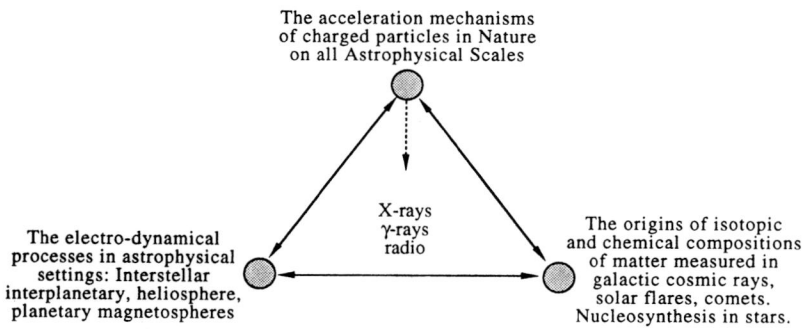

Figure 24 The interdisciplinary character of cosmic ray research.

Table 2 The heirarchy of particle accelerators in astrophysics

Astrophysical Scale	Typical maximum energy (electron volts)	Shocks[a]	Magnetic field reconnection[b]	Stochastic acceleration	Magnetic pumping	Other
Planet Magnetospheres						
– Trapped Radiation	10^7				X	
– Bow Shock	10^5	($\parallel$)				
– Magnetotail	10^6		X			
Solar Flares	10^{10}	X	X			
Interplanetary Medium (Heliosphere)						
– Radial Shock	10^8	($\parallel$) ($\perp$)				
– 27 Day Corotating Shocks	10^7	($\perp$)				
– Anomalous Nuclear Component	10^8	X (?)				
Interstellar Medium						
– Shocks	10^{14}	X				
– Fermi Acceleration	10^{15}			X		
Supernova						
– Explosion	10^{14}	X				
– Remnants	10^{13}	X			X (?)	
Cygnus X-3 Binary with Neutron Star	10^{16}					Neutron-Star
Radio Glaxy	10^{16}	X				?
Intergalactic Medium	10^{20}?			X ?		?

a: ($\parallel$) refers to quasi-parallel shocks; ($\perp$) refers to quasi-perpendicular shocks.
b: See Fermi (1949).

energies. Table 2 also has summarized this extrapolation of these heliospheric acceleration mechanisms.

4. SUMMARY OF PLASMA AND MAGNETIC FIELDS IN INTERPLANETARY SPACE

For understanding the propagation of galactic cosmic radiation through the heliosphere a brief account of the early findings of the solar wind and magnetic fields in the heliosphere is presented here (adapted from Simpson 1989, and references therein).

The magnetic field carried outward by the solar wind was discovered by Wilcox and Ness in 1965 to reverse polarity as the Sun rotated, indicating an azimuthal magnetic field sector structure near the equatorial plane at 1 A.U. These observations were widely interpreted in models in which the sector structures extended north- and south-ward in the interplanetary medium.

Later it was noted by Rosenberg and Coleman (1969) that the observations could also be explained by a three-dimensional field structure in which the interplanetary field was divided into a north polar field and a south polar field extending into the interplanetary medium, but separated by an inclined, neutral current sheet extending far into the outer heliosphere (see, for example, Chapter 48). The experimental proof of this latter concept came from the measurements on *Pioneer-11* (Section B7) as it reached >15° north solar latitude at 3.8–4 A.U. on its journey between Jupiter and Saturn encounters near solar minimum in 1976.

The alternating + and − interplanetary field directions, which indicated the passage of the rotating and inclined current sheet passing over *Pioneer-11* at low latitudes, gave way to an essentially constant + polarity at all solar azimuth angles for measurements made beyond 16° latitude – i.e., *Pioneer-11* was entirely in the northern hemisphere (+) polar field.

In 1986 *Pioneer-11* was again at 15° latitude (near 20 A.U.) The sign of the solar polar field has already reversed in 1980. The wavy, inclined neutral current sheet model and the experimentally-proven Parker spiral magnetic field configuration carried out by the solar wind (Parker 1963), had been simulated by a model of the interplanetary field extending into the outer heliosphere. The *Voyager* and *Pioneer* spacecraft all support these main features of the large-scale interplanetary magnetic field and solar wind near the equatorial plane, including the "tight wrapping" of the Parker spiral field beyond 20–25 A.U.

The solar wind is not uniform over all solar longitudes or latitudes due to underlying dynamical phenomena on the Sun, such as coronal holes which persist for long periods (for a review see, e.g., Zirker 1977) or explosive solar flares (for a review see Chapter 9). Higher than average speed solar wind plasma come from these long-lived active regions on the Sun. These high speed streams overtake the average velocity solar wind in the interplanetary medium to both compress the interplanetary magnetic field and to generate forward (F) and reverse (R) shocks, forming spiral structures in the outer solar system, as sketched in Figure 25. Typically, near solar minimum there are two of these regions, called Corotating Interaction Regions (CIRs) (Smith and Wolfe 1976), which persist for many solar rotations. *Voyager* investigations have demonstrated that these CIRs may merge in the outer heliosphere to become a Merged Interaction Region (MIR), which then continue out beyond 20 A.U. The CIRs (Figure 25) have been shown to be the source of an accelerated nucleon component in the outer heliosphere (Bryant *et al.* 1963; McDonald *et al.* 1975; Barnes and Simpson 1976). The magnetic field, solar wind and accelerated charged particle parameters in Figures 26 and 27 show the discovery of acceleration at the F and R

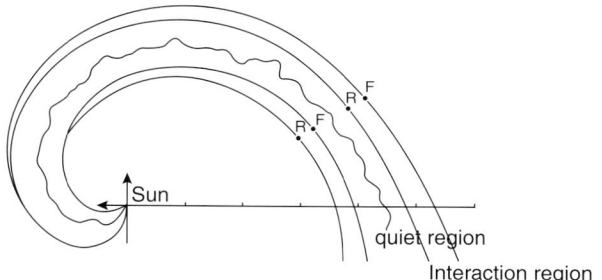

Figure 25 Corotating Interaction Region (CIR) with forward (F) and reverse (R) shocks (Smith and Wolfe 1976).

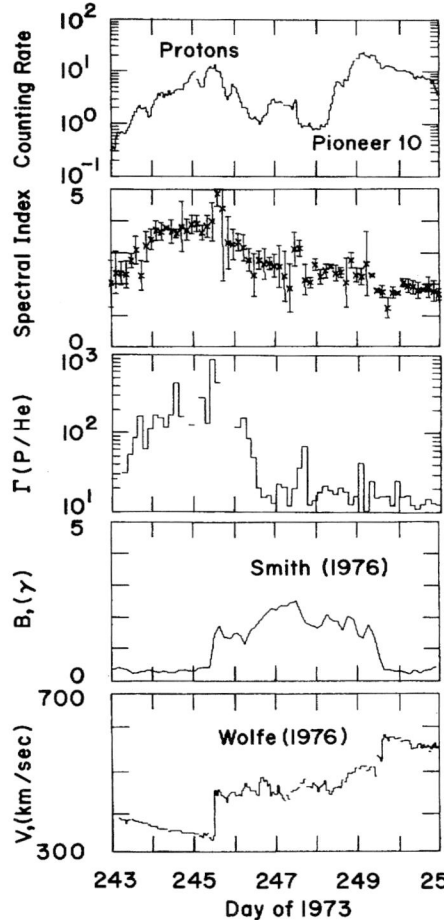

Figure 26 CIR shock acceleration of protons at the F (Day 245) and R (Day 249) shocks (Barnes and Simpson 1976).

shocks (Barnes and Simpson 1976). These and other investigations of magnetic fields and plasmas, along with theoretical developments derived from them, have given the cosmic ray investigator new insights into the large-scale electrodynamical structures surrounding a typical rotating

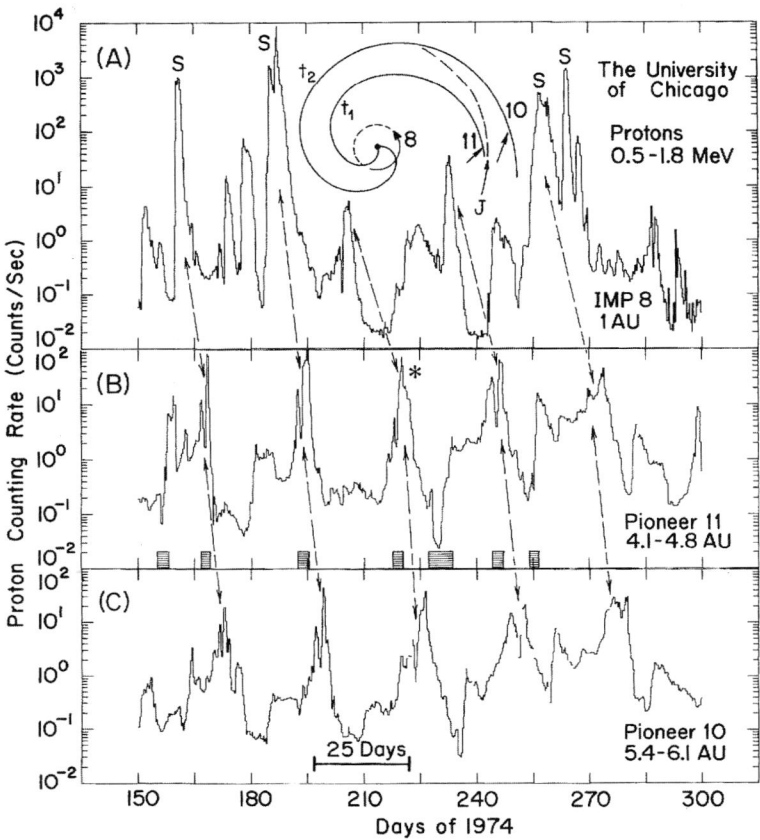

Figure 27 Continuous CIR acceleration of protons from 1 AU to beyond 6 AU (Barnes and Simpson 1976).

magnetic star. The CIR modulation of galactic cosmic rays is discussed in Section B5.

The major physical processes in the heliosphere which determine solar modulation of the galactic cosmic rays are: diffusion, convection, adiabatic deceleration, gradient and curvature drift (Parker 1963). Investigations on the *Pioneer 10/11* and *Voyager 1/2* spacecraft contributed to the resolution of the solar modulation problem – mainly by measurements of radial, latitudinal and longitudinal cosmic ray spectrum gradients in the outer heliosphere.

Following the first radial gradient measurements of galactic cosmic rays inward from the orbit of Earth on *Pioneer-5* in 1960, the *Mariner-4* mission to Mars in 1965 with our experiment, provided the first measurements of intensity gradients beyond the orbit of Earth proving that solar modulation extended far beyond 1.5 A.U. at the time of minimum modulation at Earth. The new era for exploration of the outer heliosphere began with the direct entry of space probes, now extending beyond 70 A.U. by 1998 – namely, *Pioneer-10* and *Pioneer-11* launched in 1972 and 1973, respectively (see Section B7 for details); and *Voyager-1* and *Voyager-2*, launched in 1977. They are on extended journeys that have carried them out of the solar system and towards the heliospheric boundary, as illustrated by their spacecraft trajectories projected into the ecliptic plane in Figure 28.

The experimental investigations with the *Pioneer-10/11* and *Voyager 1/2* spacecraft have had a strong impact on testing current ideas for both cosmic ray propagation and particle acceleration in the outer heliosphere. For example, Jokipii and co-workers (e.g., Jokipii and Kopriva 1979), made important predictions for the behavior of galactic and anomalous components of cosmic rays as tests for determining the role played by gradients and curvature drifts of charged particles in the heliosphere. Figure 29 is an illustration of how positively-charged nuclei from the heliospheric poles would propagate by gradient drift towards the inner solar system at times when the north polar magnetic field is positive. Negative electrons propagate inward near the equator and outward towards the poles. The solar cycle activity dependence of the wavy current sheet – both its tilt and wave amplitude – changes the drift path and, therefore, the modulation level of cosmic rays. When the polar polarity reverses, the inflow of particles of positive charge is along the equatorial zone. (The current sheet shown in Figure 29 has an inclination of $\sim 10°$.)

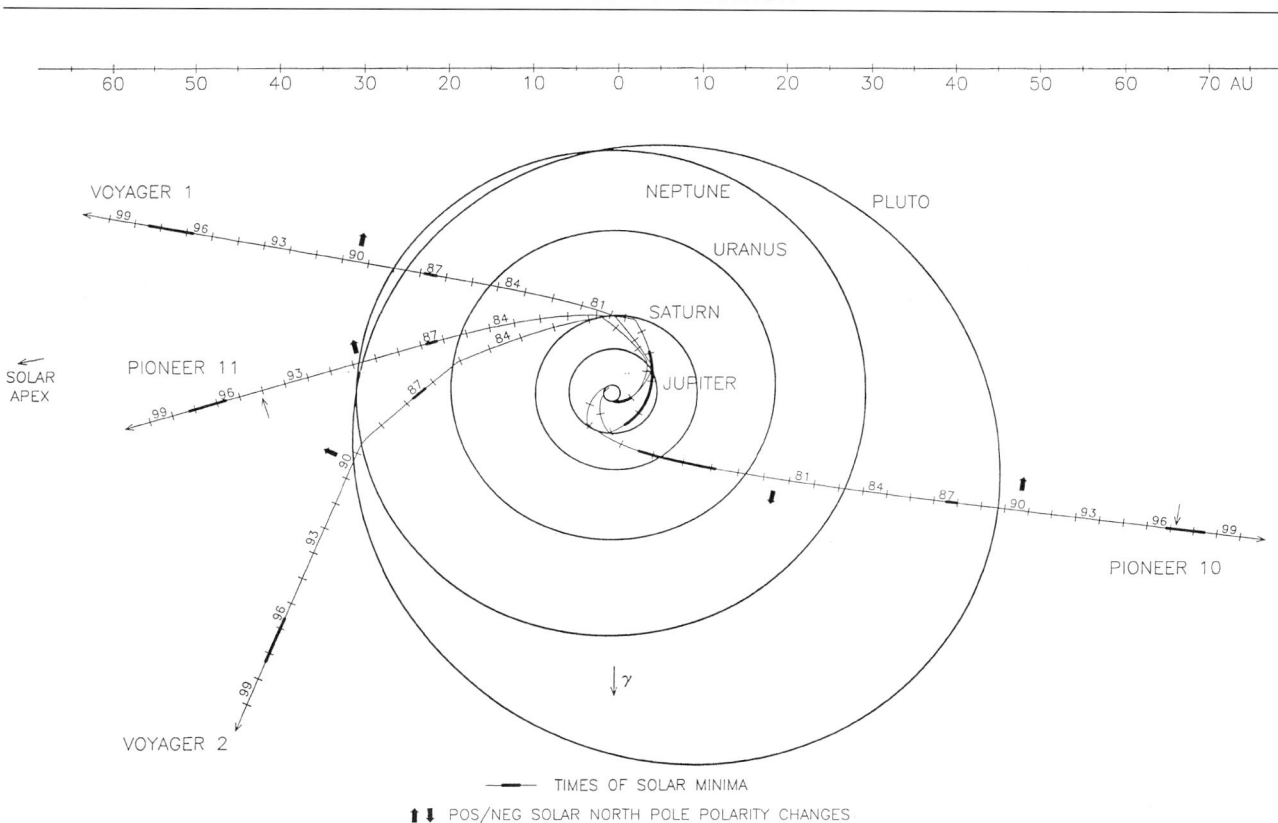

Figure 28 Trajectories of *Pioneer-10*, *Pioneer-11*, *Voyager 1*, and *Voyager 2* projected onto the ecliptic plane (prepared by C. Lopate).

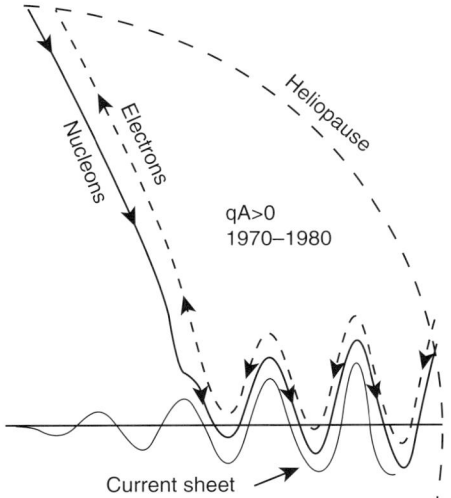

Figure 29 Sketch of propagation of nucleons (—) and electrons (---) with gradient drifts when the solar magnetic field vector potential qA is positive (adapted from Jokipii and Kopriva 1979).

Investigations with *Pioneer* and *Voyager* spacecraft (Figure 28) have demonstrated that drifts are an important factor for solar modulations. For example, the reversals of the sign of the latitude intensity gradients with the prevailing sign reversals of the solar magnetic field are shown in Table 3.

5. RECURRENT SOLAR MODULATION OF THE GALACTIC COSMIC RAYS

The experimental and theoretical investigations on the origin of the 26-day recurrent cosmic ray intensity variations have made vital contributions to solar physics, to the discovery of many dynamical processes in the heliosphere and to the Sun-Earth connection.

Forbush (1938) reported the first conclusive evidence for the existence of a 27 day cosmic ray intensity variation based on measurements with the world-wide network of Compton-C ionization chambers (Figure 4). Since the

Table 3 Summary of latitude gradients 1975–1976 versus 1985–1986

Species	1975–1976 (This paper)		1985–1986 (*Cummings et al. 1987*)	
	Energy, MeV/n	G_θ, %/deg	Energy, MeV/n	G_θ, %/deg
Protons	29–67	0.2±0.1	130–210	−0.9±0.1
	>70	0.01±0.04	>70	−0.34±0.08
Helium	11–20	1.5±0.2	10–22	−2.2±0.7
	29–67	1.2±0.1	30–56	−1.6±0.3

Spacecraft: P-11, IMP 8 (this paper); P-10, Voyager 1, Voyager 2 (Cummings *et al.*).
Latitude, radius: $-7° \leq \theta \leq 16°$, $R \cong 4$ AU (this paper); $0° \leq \theta \leq 27°$, $R \cong 25$ AU (Cummings *et al.*; McKibben 1989).

amplitudes of the 27-day "waves" were much less than a one percent effect in ionization chambers – whether at sea level or mountain altitude – Forbush had to apply elegant statistical analysis methods. He showed that the existence of this world-wide variation was inversely correlated with 27-day variations of the geomagnetic field.

Based on this inverse correlation of ion chamber and geomagnetic field intensities, he concluded that the geomagnetic field variations produced the cosmic ray intensity variations by means of recurrent changes in the geomagnetic cut-off for cosmic rays. He and others suggested that the Bartels recurring M regions on the Sun (Bartels 1940) were the source of the geomagnetic synodic 27 day recurrence. Thus, the concept generally accepted in the 1940's assumed that persistent M regions on the Sun emitted streams of ionized matter (ion streams) traveling in an interplanetary vacuum and interacting with the geomagnetic field.

The discovery of the nucleonic component large scale latitude effect (Section A5) proved that the origin of the 27-day cosmic ray intensity variations was due to dynamical phenomena in the interplanetary medium. The amplitude of these nucleonic component 27-day variations was a factor approximately five times greater than for ion chambers (Figure 30) and increased in amplitude from the geomagnetic equator to high latitudes (Fonger 1953). These and other investigations proved that 27-day changes in the geomagnetic cutoff did not produce the cosmic ray intensity variations – they were interplanetary in origin and correlated with recurring coronal active regions (Simpson 1954). By 1951–1952 the mechanism in the interplanetary medium appeared to be a modulation of the cosmic ray spectrum, but whether by acceleration or deceleration could not yet be decided.

The development by H.W. Babcock and H.D. Babcock of instrumentation utilizing the longitudinal Zeeman effect led to their discovery of recurrent "unipolar" magnetic field regions on the Sun. Based on seven consecutive solar rotations in 1953 it was then discovered (Simpson *et al.* 1955) that these "unipolar" magnetic regions were correlated

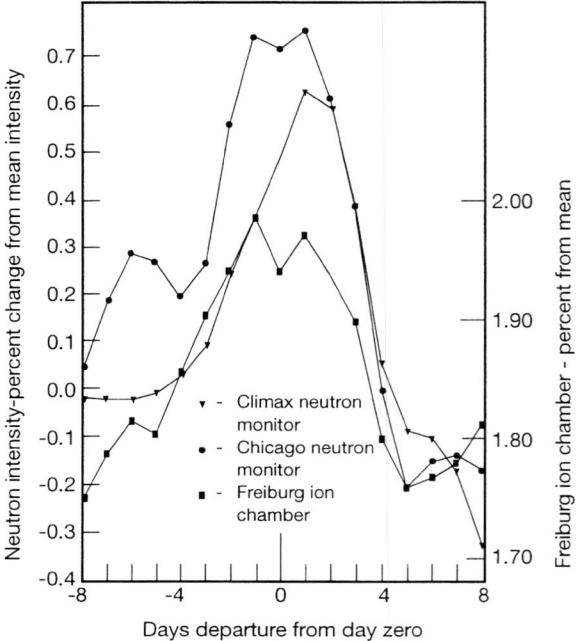

Figure 30 Central meridian pasages of unipolar regions (coronal holes) called Day "0". Neutron monitor intensities compared with ion chamber scaled factor 5 (Simpson *et al.* 1955).

within three to four days by recurrent interplanetary streams with both the increase of cosmic ray 27-day modulation and a recurrent series of geomagnetic storms. (In 1972 Skylab investigators renamed these unipolar regions "coronal holes".) Research in the period 1955–1958 provided further evidence that both the 27-day recurrent intensity variations and the approximately 11 year change of cosmic ray intensity had their origins in interplanetary processes modulating the galactic cosmic ray spectrum.

The experimental searches have been accompanied by extensive theoretical and model investigations that have expanded our understanding of the physical processes of charged particle acceleration and propagation.

Investigations in the 1970's were confined in latitude to ~16° (that is, above the prevailing latitude range of the heliospheric current sheet), by *Pioneer-11* until the 1990's when the two *Voyager* spacecraft were deflected to higher latitudes by planetary encounters (Neptune and Uranus) in the distant heliosphere and, the Ulysses spacecraft carried experiments from solar pole-to-pole in the inner solar system. These *Voyager* and *Ulysses* missions into the third dimension continue to provide new insights for not only the origin of the recurrent modulation of galactic cosmic rays and the anomalous nuclear component, but also for the recurrent interplanetary acceleration of elections and pickup ions.

6. ANOMALOUS "COSMIC RADIATION"

The discovery of a totally new component of nuclear radiation in the heliosphere – now called anomalous cosmic ray (ACR) – has greatly expanded our understanding of pickup ions, particle acceleration and a knowledge of neutral atomic abundances in the local interstellar medium. The story begins with satellite measurements at 1 A.U. of the proton and helium spectra as the heliospheric modulation of galactic cosmic rays declined towards solar minimum condition in 1971–1972 (Figure 18). A spectrum of helium nuclei with energies below ~100 MeV per nucleon appeared which could not be accounted for by either the solar modulation of helium of galactic origin (Garcia-Munoz *et al.* 1973, 1975) or its isotopic composition. The upper graph in Figure 31 illustrates the observations (solid line) at that time, along with the predicted galactic helium component (dotted line).

It soon became clear that this anomalous helium component was also accompanied by anomalous nitrogen and oxygen (Hovestadt *et al.* 1973; McDonald *et al.* 1974). More recently, there has appeared evidence for C, Ne, Ar and H. Garcia-Munoz and collaborators showed that the anomalous component of He is modulated ^{4}He with no spallation ^{3}He and, therefore, must have a "local" origin. Fisk, Koslovsky and Ramaty (1974) proposed the most successful model to account for the anomalous components. As sketched in Figure 32, interstellar neutral atoms with high ionization potentials enter the heliosphere, undergo single ionization by solar ultra-violet radiation in the inner solar system and – by charged particle pickup in the solar wind – are carried outward to an unspecified acceleration region in the distant heliosphere. Some of these accelerated nuclei – now possessing high magnetic rigidities since they are singly charged – propagate inward to undergo solar modulation, along with the galactic low energy cosmic ray nuclei.

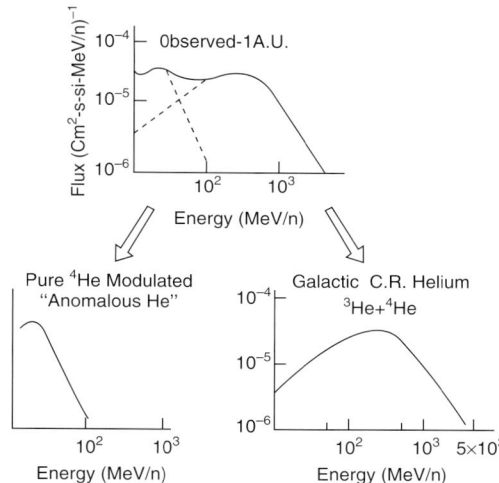

Figure 31 Decomposition of the unusual helium spectrum observed in 1972 into galactic helium and anomalous component helium. Although its true nature was not understood until several years later, this 1972 measurement was the first observation of anomalous component cosmic rays (Garcia-Munoz *et al.* 1973).

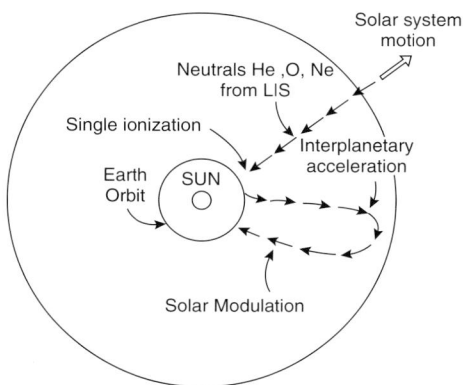

Figure 32 Sketch of concept for the production and acceleration of the anomalous nuclear component – "anomalous cosmic rays" (Fisk *et al.* 1974).

7. ANATOMY OF A DECISIVE SPACE MISSION: PIONEER-10/11

Although the 1964–65 U.S. spacecraft *Mariner IV* in its successful Martian flyby proved that a positive radial gradient of cosmic ray intensity was present to >1.5 A.U. beyond Earth, there remained the challenge of determining – on the larger heliospheric scale – the interplanetary dynamical phenomena that constitutes solar modulation.

This challenge was successfully met by the *Pioneer-10* and *Pioneer-11* mission to Jupiter, Saturn and beyond. Since

Pioneer-10/11 proved that it was possible to reach the planets beyond Mars and the outer interplanetary heliosphere they constituted a major advance for the space sciences, especially cosmic ray physics. In view of the importance of the *Pioneer-10/11* mission, this author will describe in personal detail its evolution and some of the discoveries relevant for charged particle physics as an example of the integration of science, engineering and management.

The possibility for separate missions to Jupiter and a galactic probe was extensively discussed in the 1960's by the Space Science Board and by the NASA centers. It soon became clear that in order to acquire the escape velocity from the solar system a spacecraft would need a velocity increment such as could be provided by a Jovian flyby. Thus studies focused on a Jupiter encounter mission, first at the Goddard Space Flight Center, then at the NASA Ames Research Center where the responsibility for the future *Pioneer-10/11* mission resided after Congressional and NASA approval in 1969.

The TRW systems was awarded a contract to design and fabricate two identical *Pioneer* spacecraft for the mission. A detailed account of mission planning is in NASA Document SP-446 (Fimmel *et al.* 1980) and Simpson (1991).

The hazards to be encountered by the spacecraft mission were daunting – for example:

- Could a spacecraft pass undamaged through the asteroid belt located between the orbits of Mars and Jupiter?
- Would there be radiation damage of the science instruments and/or spacecraft? (It had been discovered by radio telescope in 1955 from synchrotron emissions that Jupiter's magnetic field trapped an intense relativistic electron belt.)
- Could the spacecraft communication be maintained at vast distances where the round-trip radio signal would be 92 minutes for Jupiter and about 170 minutes for Saturn?

Taking into account the time required to fabricate and test the spacecraft, to select investigators and building the scientific instruments, and to organize the management – including tracking and data system support – the first feasible launch opportunity was late February through early March 1972. The second spacecraft could be launched 13 months later. C.F. Hall became Project Manager, with less than 3 years to achieve this 1972 launch objective.

The competition in early 1970 among scientific investigators was fierce for this mission opportunity with more than 150 proposals to NASA for only 10 instruments to be included.

Fortunately, we were prepared for the competition with novel instrument designs that included both cosmic ray studies in the interplanetary medium and the close encounter inside the intense Jovian radiation belts.

Based on our solid-state detector developments in 1959–1962 we included in the 1963 *IMP-1* mission the first multielement, solid-state charged particle telescope. This instrument made possible the first measurements of cosmic ray isotopic composition of the light elements, including the discovery of ^2H (Fan *et al.* 1966).

We also wanted to be prepared for the possibility of a deep space mission in the future for which a radioisotope thermoelectric generator (RTG) would be required. Could our novel telescope perform in the presence of an RTG with its radiation of neutrons, γ-rays, etc.? To answer this question in 1963 we arranged that a U.S. Navy SNAP-type RTG be brought to the GSFC where we exposed our *IMP-1* instrument to the background radiation to validate its performance and to undertake revisions in future designs, including setting detection thresholds above the RTG γ-ray production of electrons (i.e., set about 3 MeV).

These early instrumentation studies and successful *IMP* satellite missions contributed to our being selected by NASA. The final selection of investigations with their Principal Investigators is summarized in Table 4 to include both the interplanetary and planetary encounter phases of the mission.

The University of Chicago instruments included four charged particle detector systems – two primarily devoted to radiation in the interplanetary medium, and two for charged particles in the Jovian magnetosphere and a search for a Saturn magnetosphere with trapped particles. Cross-sections of the basic elements of the four systems are shown in Figures 33A-D. Their novel and critical features include:

a. The MAIN TELESCOPE (MT) for elemental and light element isotopic composition measurements. Our laboratory developed in the 1960's curved, solid-state detectors (D1 and D2) to reduce the average range of trajectories in the detector for an isotropic distribution of nuclei incident within the acceptance cone. To measure electrons with energies above 3 MeV a Cs-I (Tl) crystal, D5, was used as a "calorimeter" to capture the energy of relativistic electrons since they scatter widely in matter. The light output pulse from was sensed by a photodiode D5 coupled to the crystal. This crystal-photo diode detector was an *IMP-1* 1963 development that proved to be successful on many follow-on missions. Differential energy spectra for protons are measured from 6 to 92 MeV – whereas, the mean integral energy for galactic cosmic rays is about 1500 MeV for the proton component penetrating D6. The elemental composition of nuclei from H to Fe was measured by the MT along with the isotopes of H, He and Li.

b. The LOW ENERGY TELESCOPE (LET): This conventional geometry measured protons and helium spectra in the energy range 0.5 to 2 MeV nucleon^{-1}.

c. The ELECTRON CURRENT DETECTOR (ECD): This unique system consists of a single Li-drifted silicon detector (750 μm thick) surrounded by a beryllium absorber sufficient to stop ≤30 MeV protons. Electrons with energies >3 MeV can penetrate the absorber. The ECD normal leakage current was reduced to less than 5×10^{-11} A by cooling to $< -60°C$ by means of surface treatment and insulation from the main body of the spacecraft on a thin-walled titanium pedestal.

d. FISSION CELL for nucleon (FC): In the inner Jovian radiation belt there is great difficulty in identifying high energy protons in the presence of a known, high intensity electron component. To separate these components we proposed to use, for the first time, induced nuclear fission. The ratio of induced fission cross-sections for incident nucleons to electrons is of order 10^3 to 10^5 depending on electron energy. The isotope ^{232}Th was chosen because it is immune to slow neutron fission and has a relatively low level of spontaneous alpha particle emission. The 5-mil foil of ^{232}Th is mounted between two silicon detectors for fission fragments, as illustrated in Figure 33D.

For details on the LET, EDC and the fission cell see Simpson *et al.* (1974b).

Figure 34 is an illustration showing the assembled spacecraft with instruments. To visualize the size of the spacecraft, note that the diameter of the radio-dish is 2.7 m meters. The two RTG assemblies (SNAP-19), with their radiating fins, are shown in flight configuration.

In order to transmit only eight watts of power the spacecraft must constantly face Earth. The spacecraft is dynamically balanced so that it rotates five times a minute. There are gas jets around the rim of the radio dish which keep the dish pointed at Earth at all times so it is able to concentrate all of its transmitting power into a narrow beam. To aid in the pointing, the radio dish is made to wobble slightly as it rotates. This introduces a Doppler effect in the radio

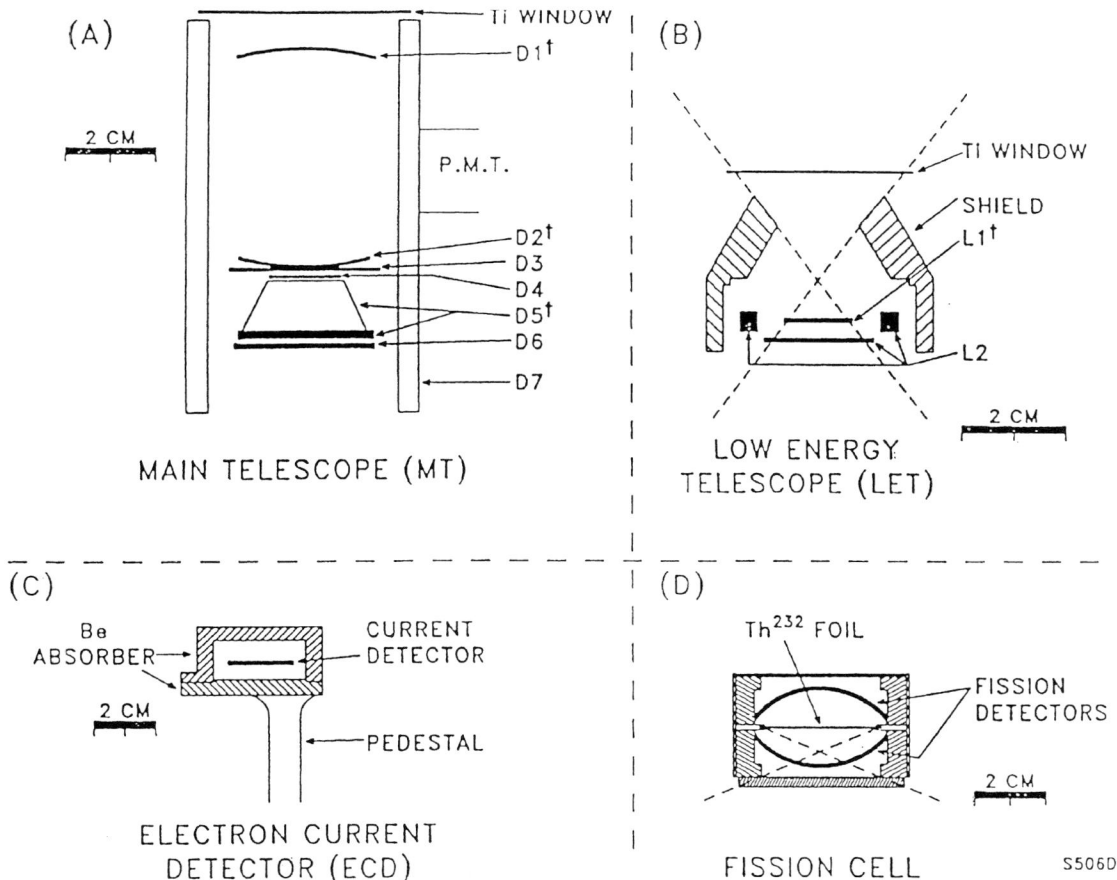

Figure 33 Cross-section of detector systems on *Pioneer-10* and *Pioneer-11* from the University of Chicago.

Table 4 *Pioneer-10* and *Pioneer-11*

Experiments	Principal investigators
Magnetic field	Edward J. Smith
	Jet Propulsion Laboratory
Fluxgate magnetometer	Mario Acuna
(Pioneer-11 Only)	Goddard Space Flight Center
Plasma Analyser	John H. Wolfe (Jupiter, Saturn)
	Aaron Barnes (Beyond Saturn)
	AMES Research Center
Charged Particle Composition	John A. Simpson
	University of Chicago
Energy Spectra of Cosmic	Frank B. McDonald
	Goddgard Space Flight Center
Jovian Charged Particles	James A. Van Allen
	University of Iowa
Jovian Trapped Radiation	R. Walker Fillius
	University of California at San Diego
Asteroid-meteoroid Astronomy	Robert K. Soberman
	General Electric Co. and Drexel University
Meteoroids	William Kinard
	Langely Research Center, NASA
Ultraviolet photometry	Darrell L. Judge
	University of Southern California
Imaging Photopolarmetry	Tom Gehrels
	University of Arizona
Non instument	
Celestial Mechanics	John D. Anderson
	Jet Propulsion Laboratory
Jovian Infrared Thermal Structural	Guido Munch (Jupiter)
	Andrew P. Ingersoll (Saturn)
	California Institute of Technology
S-Band Occultation	Arvydas J. Kliore
	Jet Propulsion Laboratory

emission, so that the engineering staff at Earth can measure the alignment of the radio dish throughout the year as Earth moves about the Sun.

On Earth there are three radio dishes, 64 m in diameter, which are brought into play every twenty-four hours. As the Earth rotates, the dishes come into operation successively to receive the information which is then transmitted to a center where the data for the University are stripped onto a magnetic tape along with all the engineering data, and delivered to our Chicago laboratories, where we do all of our analytic work with computers.

The *Pioneer-10* was launched on 2 March 1972, less than two years after the first meeting of the Principal Investigators at NASA-Ames to organize the science mission. The existence of energetic electron bursts in the interplanetary medium outside the bow shock of Earth led us, with the MT design, to search for similar electron bursts of Jovian origin as *Pioneer* approached Jupiter's magnetosphere. We were not disappointed. At about 1 astronomical unit from Jupiter we found electron bursts in the energy range 6 to 30 MeV in energy, with the average intensity of the bursts increasing as *Pioneer-10* approached Jupiter. We also discovered that the electron energy spectral slope "rocked" or oscillated with a 10 hr. period which was Jupiter's rotation period (9 h 55 rm 30 s). This effect – and the variability of electron intensity – is illustrated in Figure 35, where the intensity ratio ID4/ID5 corresponds to the energy ratio $\langle 6 \rangle Mev/\langle 30 \rangle Mev$. This 10 hr. periodic behavior of the spectrum persisted in phase through the bow shock (BS), magnetopause (MP) and deep inside the Jovian magnetosphere.

Although there were, at that time, many hypotheses proposed to account for this large scale phenomenon that we found, it was not until 1992 – as a result of *Ulysses* passing under Jupiter on its way to the South Poles of the Sun – that

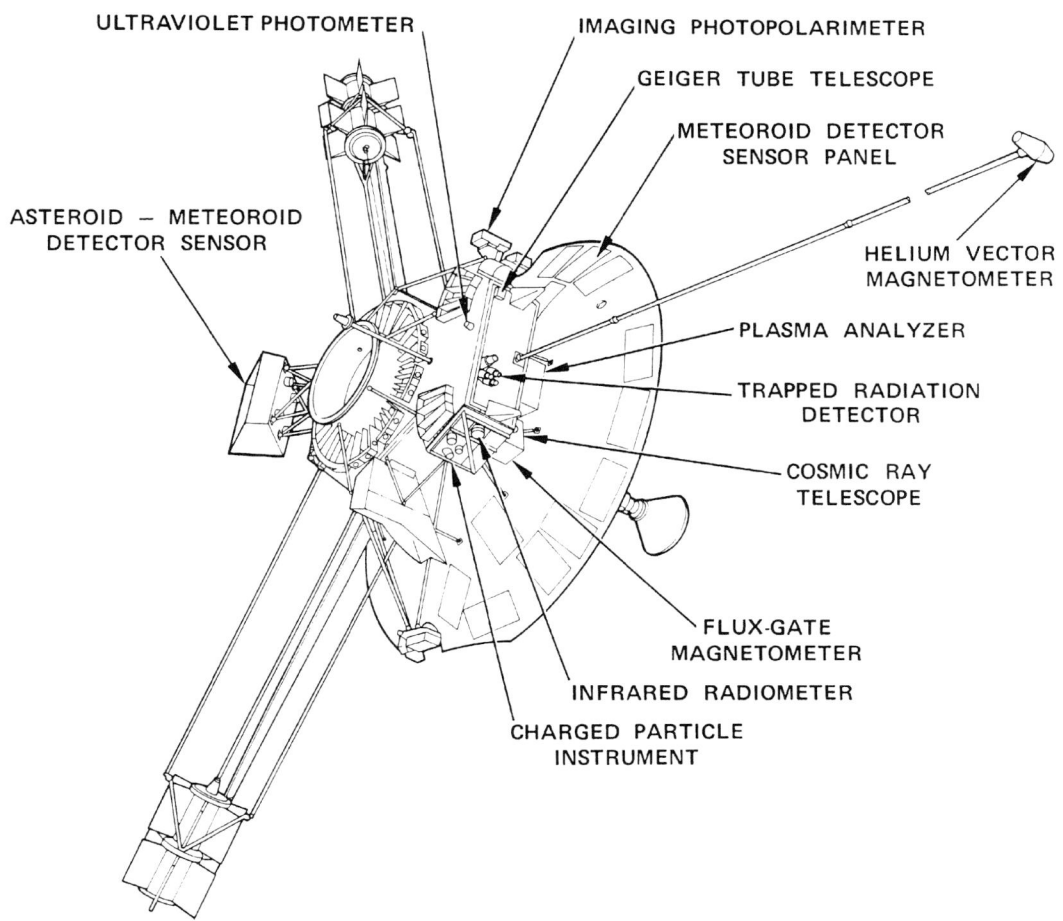

Figure 34 The *Pioneer 10–11* spacecraft with instrument. Only *Pioneer-11* carried the flux-gate magnetometer (NASA).

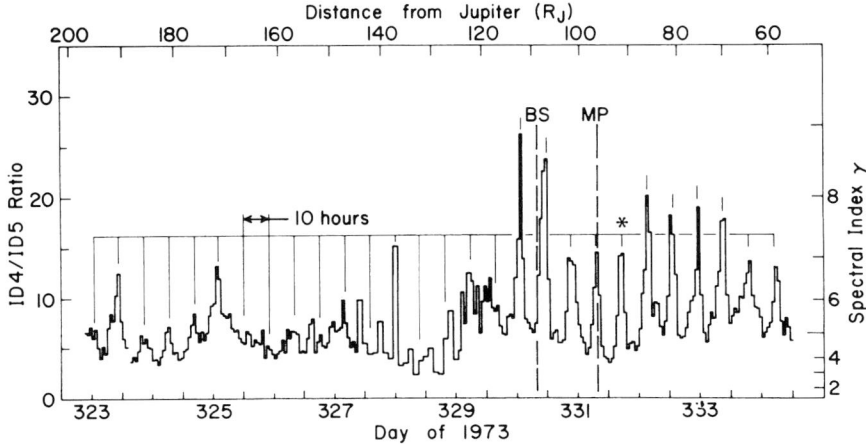

Figure 35 The ratio ID4/ID5 corresponds to the average electron energy ratio $\langle 6 \rangle$ Mev/$\langle 30 \rangle$ Mev. The assumed spectrum is $\propto E^{-\gamma}$ (Chenette et al. 1974)

Figure 36 The *Pioneer-10* and *Pioneer-11* cosmic ray principal investigators F.B. McDonald, J.A. Van Allen and J.A. Simpson.

the mechanism of the 10 hr variation of Jovian electrons and their escape mechanism were discovered (Simpson *et al.* 1993).

The entry into and through the magnetosphere produced many more discoveries reported by the *Pioneer* investigators (cf. Chapter 59). The *Pioneer* encounter with Jupiter brought much public interest. There was a daily summary of the new findings by the Principal Investigators report at News Press Conferences. Three of the cosmic ray and planetary trapped particle investigators are shown in Figure 36 reporting at one of these daily conferences.

Pioneer-10 continued into the distant heliosphere recording the electrons of Jovian origin. It was obvious that Jupiter – as a continual known "point source" of relativistic electrons in the heliosphere – was a new "tool" for investigating charged particle propagation. Indeed, with identical MT and LET charged particle telescopes mounted in the *Mariner-10* spacecraft for encounters with Mercury at 0.3 AU from the Sun, we found the Jovian electron continuously present.

Jovian electrons dominated the heliospheric electron fluxes out to about 50 AU, where we found another electron component whose intensity increased as *Pioneer-10* traveled outward. As *Pioneer-10* approached 73 A.U. in 1999 we concluded that this electron flux is likely of local interstellar origin.

In its transit of the heliosphere there were many discoveries concerning, for example, charged particle acceleration in shocks, solar modulation of galactic cosmic rays, solar flare charged particle propagation, as discussed in Sections B3–B5. All of these phenomenon provided essential knowledge for understanding the acceleratio and propagation mechanisms for cosmic rays in the galaxy.

By the 1990's, the RTG power had decreased to the point where time sharing of instrument operation was required. Except for brief data intervals from time-to-time, the *Pioneer-10* mission ended in 1998 at 73 AU without direct evidence that the solar wind termination shock was nearby. (Even as the 21st century begins *Pioneer-10* still sends a faint signal to Earth.)

The *Pioneer-10* spacecraft established that there was safe passage to the outer heliosphere and opened a new era for the space sciences. For cosmic ray physics this mission demonstrated that, indeed, the heliosphere was a superb venue for studying astroparticle phenomena including acceleration and propagation, as discussed in Sections 3–5.

Turning to *Pioneer-11*, the companion spacecraft of *Pioneer-10*, the launch was in April 1973 long before the fate of *Pioneer-10* in the Jovian radiation belt could be known. *Pioneer-11* later penetrated to within 1.8 Jovian radii and escaped successfully to fly to Saturn on a trajectory direction opposite to *Pioneer-10* (Figure 28), providing high latitude observations of the Jovian magnetosphere (e.g., see Chapter 5).

The post Jupiter trajectory in Figure 28 for the successful transit of Jupiter was planned so that *Pioneer-11* would encounter Saturn in September 1979. There were not only the scientific objectives of a possible planetary magnetic field and trapped radiation, but also the question of spacecraft survival close to or under the rings of Saturn.

This latter problem was a practical one that *Pioneer*-11 was to resolve for follow-on missions to the outer planets beyond Saturn. NASA management considered the two *Pioneer* spacecraft as "precursor missions" for establishing follow-on "Grand Tour" missions to the outer solar system planets – a program fulfilled with two *Voyager* spacecraft.

But should *Pioneer-11* enter deep under the ring plane or play it safe by staying outside the rings? This debate – and the positions taken by the Principal Investigators, by the *Pioneer* Project Office, by the Jet Propulsion Laboratory, Voyager staff and NASA Headquarters – is vividly presented by Wolverton (2000).

In 1979 *Pioneer-11* did pass under the rings far enough to provide investigators with vital planetary discoveries and emerged without damage to proceed toward the "nose" of the heliosphere until 1994 when its RTG power could no longer support a single instrument. *Pioneer-11* opened the "door" for future spacecraft missions to the distant planets.

8. CONCLUDING REMARKS

Access to space has been an essential factor for progress in cosmic ray physics. On the one hand, through direct observations on the scale of the heliosphere it became possible to study directly charged particle acceleration and propagation up to cosmic ray energies. On the other hand, the elemental and isotopic composition of galactic matter from Hydrogen to Uranium – as the only sample of galactic matter directly available to us on Earth – is leading to an understanding of the "seed" matter which becomes the galactic cosmic radiation. An overarching critical factor has been the invention and development of novel particle detectors throughout the 20th Century.

The experiments, observations and theories that evolved during the last two decades of the century support these conclusions, as reported elsewhere in this volume.

> All things keep on, in everlasting motion. Out of the infinite come the particles speeding above, below, in endless dance.
>
> Lucretius
> *De rerum Natura*

Acknowledgments

The author wishes to thank the following publications for permission to use various published illustrations from articles in their Journals: *Astrophysical Journal and Letters; Physical Review and Letters; Journal of Geophysical Research; Space Science Reviews*.

REFERENCES

Alfvén, H. (1942). On the Existence of Electromagnetic-Hydrodynamic Waves. *Arkiv für Matematik, Astronomi och Fysik.*, **29B**, 2.

Alfvén, H. (1949). On the Solar Origin of Cosmic Radiation. *Physical Review, 2nd Series* **75**, No. 1, 1732.

Alfvén, H., Richtmyer, R.D. and Teller, E. (1949). On the Origin of Cosmic Rays. *Physical Review,* **75**, 892.

Alvarez, L.W. (1987). *Adventures of a Physicist*, Basic Books.

Alvarez, L.W. and Compton, A.H. (1933). A Positively Charged Component of Cosmic Rays. *Physical Review,* **43**, 835–836.

Auger, P. and Maze, R. (1939). Extension et Pouvoir Pénétrant des Grandes Gerbes de Rayons Cosmiques. *Compt. Rend.,* **208**, 1641–43.

Barnes, C.W. and Simpson, J.A. (1976). Evidence for Interplantary Acceleration of Nucleons in Corotating Interaction Regions. *Astrophysical Journal,* **210**, L91–L96.

Bartels, J. (1940) in *Geomagnetism*, by Chapman, S. and Bartels, J. (Oxford Univ. Press, U.K.), Vol. **1**, pp. 396–416.

Becquerel, H. (1896). Sur les radiations émises par phosphorescence. *Comptes rendus de l'Académie des Sciences,* **120**, 420–421.

Berkner, L.V. and Odishaw (eds) (1961). *Science in Space*, McGraw-Hill Publ. (see Chapters 11, 13 and 14).

Berriman, R.W. (1948). A Photographic Emulsion Sensitive to Minimum Ionization. *Nature* **162**, 992.

Bethe, H.A., Korff, S.A. and Placzek (1940). On the Interpretation of Neutron Measurements in Cosmic Radiation. *Physical Review,* **54**, 573–587.

Biermann, P. (1997). The Origin of the Highest Energy Cosmic Rays. *J. Phys. G.: Nucl. Part. Phys.,* **23**, 1–27.

Biermann, L. (1952). Über den Schweif des Kometen Halley im Jahre 1910. *Zeitschrift für Naturforschung,* **7a**, 127–136.

Blackett, P.M.S. and Occhialini, G.P.S. (1933). Some Photographs of the Tracks of Penetrating Radiation. *Proc. Roy. Soc.* (London), **A139**, 699–727.

Blau, M. (1925). Über die photographische Wirkung natürlicher H-Strahlen. *Sitzungsberichte Akademie der Wissenschaften in Wien (SBAWW) IIa,* **134**, 427.

Blau, M. (1927). Relative Grain Densities in Tracks of Alpha Particles and Protons. *Sitzungsberichte Akademie der Wissenschaften in Wien (SBAWW) IIa,* **136**, 469.

Blau, M. and Wambacher, H. (1937). Disintegration Processes by Cosmic Rays with the Simultaneous Emission of Several Heavy Particles. *Nature,* **140**, 585.

Bryant, D.A., Cline, T.L., Desai, U.O. and McDonald, F.B. (1963). New Evidence for Long-Lived Solar Streams in Interplanetary Space. *Physical Review Letters,* **11**, 144.

Carmichael, H. (1964). *QSY Instruction Manual 7*, Deep River, Canada.

Chandrasekhar, S. and Fermi, E. (1953a). Magnetic Fields in Spiral Arms. *Astrophysical Journal,* **118**, 113.

Chandrasekhar, S. and Fermi, E. (1953b). Problems of Gravitational Stability in the Presence of a Magnetic Field. *Astrophysical Journal,* **118**, 116.

Chapman, S. and Ferraro, V.C.A. (1940). The Theory of the First Phase of a Geomagnetic Storm. *Terr. Mag.,* **45**, No.3, 245–268.

Chenette, D.L., Conlon, T.F. and Simpson, J.A. (1974). Bursts of Relativistic Electrons from Jupiter Observed in Interplanetary Space with the Time Variation of the Planetary Rotation Period. *Journal of Geophysical Research,* **79**, 3551–3558.

Clay, J. (1928). Penetrating Radiation. *Proc. Amsterdam,* **31**, 1091–1097 (and references therein).

Coleman, P.J., Jr., Sonnett, C.P., Judge, D.L. and Smith, E.J. (1960). Some Preliminary Results of the Pioneer V Magnetometer Experiment. *Journal of Geophysical Research,* **65**, 1856–1857.

Compton, A.H. (1933). A Geographic Study of Cosmic Rays. *Physical Review,* **43**, 387–403.

Compton, A.H., Wollan, E.O. and Bennett, R.D. (1934). A Precision Recording Cosmic Ray Meter. *Rev. Sci. Inst.,* **5**, 415–422.

Cronin, J.W. (1999). Cosmic Rays: The most Energetic Particles in the Universe. *Reviews of Modern Physics,* **71**, S165–S172.

Cummings, A.C., Stone, E.C. and Webber, W.R. (1987). Latitudinal and Radial Gradients of Anomalous and Galactic Cosmic Rays in the Outer Heliosphere. *Geophysical Research Letters,* **14**, 174–177.

Darrow, K.K. (1932). Contemporary Advances in Physics, XXIII. Data and Nature of Cosmic Rays. *Bell System Technical Journal,* **11**, 148–184.

Davis, L. (1951). The Strength of Interstellar Magnetic Field. *Physical Review,* **81**, 890.

Davis, L. and Greenstein, J.L. (1951). The Polarization of Starlight by Aligned Dust Grains. *Astrophysical Journal,* **114**, 206.

Davis, L. (1955). Interplanetary Magnetic Fields and Cosmic Rays. *Physical Review,* **100**, 1440–1444.

De Maria, M., and Russo, A. (1990). Cosmic rays and cosmological speculations in the 1920s. The debate between Jeans and Millikan. In B. Bertotti, R. Balbinot, S. Bergia and A. Messina (eds), *Modern Cosmology in Retrospect,* Cambridge University Press, Cambridge: 401–409.

DeVorkin, D.H. (1989). *Race to the Stratosphere: Manned Scientific Ballooning in America,* Springer-Verlag, New York, 406 pp.

Fan, C.Y., Gloeckler, G. and Simpson, J.A. (1966). Galactic Deuterium and the Energy Spectrum Above 20 MeV Nucleon. *Physical Review, Letters,* **17**, 329.

Fan, C.Y., Meyer, P. and Simpson, J.A. (1960a). Cosmic Radiation Intensity Decreases Observed at the Earth and in the Nearby Planetary Medium. *Physical Review Letters,* **4**, 421.

Fan, C.Y., Meyer, P. and Simpson, J.A. (1960b). Rapid Reduction of Cosmic Ray Intensity Measured in Interplanetary Space. *Physical Review Letters,* **5**, 269–271.

Fan, C.Y., Meyer, P. and Simpson, J.A. (1960c). Experiments on the Eleven Year Change of Cosmic Ray Intensity Using a Space Probe. *Physical Review Letters,* **5**, 272–275.

Fermi, E., 1949, On the Origin of the Cosmic Radiation. *Physical Review,* **75**, 1169.

Fimmel, R.O., Van Allen, J. and Burgers, E. (eds) (1980). *Pioneer: First to Jupiter, Saturn and Beyond,* NASA SP 446, US Government Printing Office, Washington, DC.

Fisk, L.A., Koslovsky, B. and Ramaty, R. (1974). An Interpretation of the Observed Oxygen and Nitrogen Enhancements. *Astrophysical Journal Letters,* **190**, L35–L37.

Fonger, W.H. (1953). Cosmic Radiation Intensity-Time Variations and their Origin: II. Energy Dependence of 27-Day Variations. *Physical Review,* **91**, 351.

Forbush, S.E. (1938). On Cosmic-Ray Effects Associated with Magnetic Storms, *Terr. Mag.,* **43**, 135–150; *Ibid,* 203–218 (More accessible is: in *Cosmic Rays, the Sun and Geomagnetism: The Works of Scott E. Forbush* (Ed. J.A. Van Allen; Pub. Am. Geophys. U., 1993). See discussion: 27-day variations, 362–368. This volume is a collection of some of S.E. Forbush's reprints and lecture notes including the period 1937–1954 on the 27-day variations.

Freier, P., Lofgren, E.J., Ney, E.P., Oppenheimer, F., Bradt, H.L. and Peters, B. (1948). Evidence for Heavy Nuclei in the Cosmic Radiation. *Physical Review,* **74**, 213–217.

Garcia-Munoz, M., Mason, G.M. and Simpson, J.A. (1973). A New Test for Modulation Theory: The May–July 1972 Low Energy Cosmic Ray Proton and Helium Spectra. *Astrophysical Journal,* **182**, L81.

Garcia-Munoz, M., Mason, G.M. and Simpson, J.A. 1975 The Cosmic Ray Age Deduced from the ^{10}Be Abundance. *Astrophysical Journal,* **202**, 265.

Geiger, H. and Müller, W. (1928). Electronenzählrohr zur Messung schwächster Aktivitäten. *Naturwissenschaften,* **16**, 617–618.

Gockel, A. (1911). Messungen der durchdringenden Strahlung bei Ballonfahrten. *Ballonfahrt Zeitschrift,* **12**, 595–597.

Hall, J.S. (1949). Observations of the Polarized Light from Stars. *Science,* **109**, 166.

Halpern, L. and Shapiro, M.M. (2001). Marietta Blau, Pioneer of the Nuclear Emulsion Technique. In N. Byers and G. Williams (eds), *Out of the Shadows: Contributions of 20th Century Women in Physics,* Cambridge University Press, Cambridge, England (in press).

Hess, V.F. (1912). Über Beobachtungen der durchdringenden Strahlung bei sieben Freiballonfarten. *Physikalische Zeitschrift,* **13**, 1084–1091.

Hess, V.F. (1940). The Discovery of Cosmic Radiation. *Thought* (Fordham Quarterly), **15**, 225–236.

Hiltner, W.A. (1949). Polarization of Distant Stars in the Interstellar Medium. *Science,* **109**, 165.

Hiltner, W.A. (1951). Polarization of Stellar Radiation: III The Polarization of 841 Stars. *Astrophysical Journal,* **114**, 241.

Hovestadt, D, Vollmer, O., Gloeckler, G and Fan, C.Y. (1973). Differential Energy Spectra of Low Energy (<8.5 MeV per Nucleon) Heavy Cosmic Rays During Solar Quiet Times. *Physical Review Letters,* **31**, 650–653.

Johnson, T.H. (1933). The Azimuthal Asymmetry of the Cosmic Radiation. *Physical Review,* **43**, 834–835.

Johnson, T.H. and Barry, J.G. (1939). The East-West Symmetry of the Cosmic Radiation at Very High Elevations at the Equator and Evidence that Protons Constitute the Primary Particles of the Hard Component. *Physical Review,* **56**, 219–226.

Jokipii, J.R. and Kopriva, D.A. (1979). Effects of Particle Drift on the Transport of Cosmic Rays. III. Numerical Models of Galactic Cosmic-Ray Modulation. *Astrophysical Journal,* **234**, 384–392.

Juliusson, E., Meyer, P. and Müller, D. (1972). Composition of Cosmic Ray Nuclei at High Energies. *Physical Review Letters,* **29**, 445–448.

Kolhörster, W. (1913). Zur Messung der durchdringenden Strahlung. *Physikalische Zeitschrift,* **14**, 1153–1156.

Korff, S.A. (1939). Fast Neutron Measurements with Recoil Counters. *Physical Review,* **56**, 1241.

Lattes, C.M.G., Occhialini, G.P.S. and Powell, C.F. (1947). Observations on the Tracks of Slow Mesons in Photographic Emulsions. *Nature,* **160**, 453–456, 486–492.

Lemaitre, G. and Vallarta, M.S. (1933). On the Compton's Latitude Effect of Cosmic Radiation. *Physical Review,* **43** 87–91.

Libby, W.F. (1946). Atmospheric Helium Three and Radiocarbon From Cosmic Radiation. *Physical Review,* **69**, 671–672.

Libby, W.F. (1955). *Radiocarbon Dating,* University of Chicago Press.

Linsley, J. (1963). Evidence for a Primary Cosmic Ray Particle With Energy 10^{20} eV. *Physical Review Letters,* **10**, 146–148.

Lüst, R. and Simpson, J.A. (1957). Initial Stages in the Propagation of Cosmic Rays Produced by Solar Flares. *Physical Review,* **108**, 1563–1576.

McDonald, F.B., Teegarden, B.J. and Trainor, J.H. (1974) The Anomalous Abundance of Cosmic-Ray Nitrogen and Oxygen Nuclei at Low Energies. *Astrophysical Journal,* **187**, L105–L108.

McDonald, F.B., Teegarden, B.J., Trainor, J.H., von Rosenvinge, T.T. and Webber, W.R (1975). The Interplanetary Acceleration of Energetic Nucleons. *Astrophysical Journal,* **203**, L149.

McKibben, R.B. (1989). Reanalysis and Confirmation of Positive Latitude Gradients for Anomalous Helium and Galactic Cosmic Rays Measured in 1975–1976 With Pioneer-11. *Journal of Geophysical Research,* **94**, 17,021–17,033.

Meyer, P., Parker, E.N. and Simpson, J.A. (1956). The Solar Cosmic Rays of February 1956 and their Propagation Through Interplanetary Space. *Physical Review,* **104**, 768–783.

Millikan, R.A. and Bowen, I.S. (1923). Penetrating Radiation at High Altitudes. *Physical Review,* **22**, 198.

Millikan, R.A. and Cameron, G.H. (1926). High Frequency Rays of Cosmic Origin. III. Measurements in Snow-fed Lakes at High Altitudes. *Physical Review,* **28**, 851–868.

Morrison, P. (1956). Solar Origin of Cosmic-ray Time Variations. *Physical Review,* **101**, 1397–1404.

Mott-Smith, L.M. (1930). Possibility of Determining the Energy of the Cosmic β-Particles by Magnetic Detections. *Physical Review,* **35**, 1125–1156.

Parker, E.N. (1958). Cosmic Ray Modulation by the Solar Wind. *Physical Review,* **110**, 1445–1449.

Parker, E.N. (1963). *Interplanetary Dynamical Processes,* J. Wiley and Sons, Pub., New York, NY.

Parker, E.N. (2001). A History of Solar Wind Concept. In: *The Century of Space Science,* pp. 225–255.

Pfotzer, G. (1936). Dreifachkoinzidenzen der Ultrastrahlung aus vertikaler Richtung in der Stratosphäre. 1. Messmethode und Ergebnisse, *Zeitschrift für Physik,* **102**, 23–40.

Rochester, G.D. and Butler, C.C. (1947). Evidence for the Existence of New Unstable Elementary Particles. *Nature*, **160**, 855.

Rossi, B. (1930a). Method of Recording Multiple Simultaneous Impulses of Several Geiger's Counters. *Nature*, **125**, 636.

Rossi, B. (1930b). On the Magnetic Deflection of Cosmic Rays. *Physical Review*, **36**, 606–606.

Rossi, B. (1985). Arcetri, 1928–1931. In Y. Sekido and H. Elliot (eds), *Early History of Cosmic Ray Studies*, D. Reidel Publ., Dordrecht, p. 53.

Rosenberg, R.L., and Coleman, P.J. (1969). Heliographic Latitude Dependence of the Dominant Polarity of the Interplanetary Magnetic Field. *Journal of Geophysical Research*, **74**, 5611–5622.

Schein, M., Jesse, W.P. and Wollan, E.O. (1941). The Nature of the Primary Cosmic Radiation and the Origin of the Mesotron. *Physical Review*, **59**, 615.

Sekido, Y. and Elliot, H. (eds) (1985). *Early History of Cosmic Ray Studies*, D. Reidel Publ., Dordrecht.

Serber, R. (1938). Transition Effects of Cosmic Rays in the Atmosphere. *Physical Review*, **54**, 317.

Serber, R. (1983) In L.M. Brown and L. Hodderson (eds), *The Birth of Particle Physics*, Cambridge University Press, p. 215.

Simpson, J.A. (1948). The Latitude Dependence of Neutron Densities in the Atmosphere as a Function of Altitude. *Physical Review*, **73**, 1389–1391.

Simpson, J.A. (1951). Neutrons Produced in the Atmosphere by Cosmic Radiations. *Physical Review*, **83**, 1175–1188.

Simpson, J.A. (1954). Cosmic radiation intensity–time variations and their origin: III. The origin of 27-day variations. *Physical Review*, **94**, 426–440.

Simpson, J.A. (1985). Cosmic Ray Astrophysics at Chicago (1947–1960) In Y. Sekido and H. Elliot (eds), *Early History of Cosmic Ray Studies*, D. Reidel Pub. Co., Dordrecht, 385–409.

Simpson, J.A. (1989). Evolution of Our Knowledge of the Heliosphere. *Advances in Space Research*, **9**, No. 4, 5–20.

Simpson, J.A. (1991). In *Guide to the Simpson Papers*, Dept. of Special Collections, University of Chicago Library.

Simpson, J.A. (1994). A Physicist in the World of Geophysics and Space. *Journal of Geophysical Research*, **99**, 19159–19173.

Simpson, J.A. (2000). The Cosmic Ray Nucleonic Component; The Invention and Scientific Uses of the Neutron Monitor. *Space Science Series of ISSI, Vol. 10 and Space Science Review*, **93**, 11–32.

Simpson, J.A., Fonger, W. and Treiman, S.B. (1953). Cosmic Ray Intensity-Time Variations and Their Origin: Neutron Intensity Variation Method and Meteorological Factors. *Physical Review*, **90**, 934–950.

Simpson, J.A., Babcock, H.W. and Babcock, H.D. (1955). Association of a 'Unipolar' Magnetic Region on the Sun With Changes of Primary Cosmic-Ray Intensity. *Physical Review*, **98**, 1402–1406.

Simpson, J.A. Lentz, G.A. McKibben, R.B., O'Gallagher, J.J., Schroeder, W. and Tussolino, A.J. (1974). National Space Science Data Center, Tech. Ref. File *B21970*, GSFC, Greenbelt, Md.

Simpson, J.A., Smith, D.A., Zhang, M. and Balogh, A. (1993). Jovian Electron Propagation in Three Dimensions of the Heliosphere: The Ulysses Instrument. *Journal of Geophysical Research*, **98, 21**, 128–121.

Skobeltzyn, D. (1929). Uber eine neue Art sehr schneller β-Strahlen. *Zeitschrift für Physik*, **54**, 686–702.

Smith, E.J. and Wolfe, J.H. (1976). Observations of Interaction Regions and Corotating Shocks Between One and Five A.U.: Pioneer 10 and 11. *Geophysical Research Letters*, **3**, 137–140.

Van Allen, J. (1993). Foreword. In reprint of Forbush (1938).

Wilcox, J.M., and Ness, N.F. (1965). Quasi-Stationary Corotating Structure in the Interplanetary Medium. *Journal of Geophysical Research*, **70**, 5793–5805.

Wilson, C.T.R. (1901). On the Ionization of Atmospheric Air. *Proc. Roy. Soc.*, **68**, 151–161.

Wolverton, M. (2000). Pathfinding the Rings: The Pioneer Saturn Trajectory Decision. *Quest: The History of Spaceflight Quarterly*, **7**(4), 5–11.

Wulf, Th. (1909). Über die in der Atmosphäre vorhandene Strahlung von hoher Durchdringungsfähigkeit. *Physikalische Zeitschrift*, **10**, 152–157.

Zirker, J.B. (ed) (1977). *Coronal Holes and High Speed Wind Streams*, Skylab Solar Workshop, Colorado University Press, Boulder, Co.

5

JAMES A. VAN ALLEN*

Magnetospheric physics**

AUTHOR'S PREFACE

How did a Ph.D. in experimental nuclear physics become a planetary astronomer? Well, it wasn't simple! Here is my story.

During the early days of World War II, I helped develop the radio proximity fuze for gun-fired projectiles, working at the Applied Physics Laboratory of Johns Hopkins University. In November 1942, I was commissioned as a line officer in the U.S. Naval Reserve and was dispatched immediately to the South Pacific Fleet to foster the adoption of this innovative, "smart" anti-aircraft fuze. During the next three and a half years, I served a total of 17 months in two tours of duty on combatant ships and at ammunition depots in the South Pacific. In March 1946, now at the exalted rank of lieutenant commander, I was placed on inactive duty and returned as a civilian employee of the APL/JHU.

Then during the period 1946–1950, I developed and oversaw a program of high altitude research in cosmic rays, atmospheric ozone, solar UV spectroscopy, ionospheric currents, and photography of large regions of the Earth's surface. The work was supported by the U.S. Navy Bureau of Naval Ordnance. Our instruments were transported up to altitudes of 160 km by the U.S. Army Ordnance Department's post-World War II test flights of refurbished German V-2 rockets and by American Aerobee rockets. The latter were developed specifically for our investigations of physical phenomena in and above the Earth's atmosphere. Our original Aerobee and its successive upgrades were later adopted by many others. By 1988, over a thousand such vehicles had been flown for a wide variety of atmospheric, solar, astronomical and astrophysical investigations.

In January 1951, I returned to my Ph.D. alma mater as a professor of physics and head of the Department of Physics (later Physics and Astronomy) at the University of Iowa and, with the support of the Research Corporation and the U.S. Office of Naval Research (ONR), initiated and led a student-centered program of similar research using balloons and balloon launched rockets (rockoons) as vehicles. The rockoon technique provided a low cost method of transporting our instruments up to altitudes of 130 km. During the period 1952–1957, we conducted 109 rockoon flights, all from shipboard, over a large range of latitudes from Baffin Bay in the Arctic to the Ross Sea in Antarctica, under the sponsorship of the ONR and, during 1957, by the National Science Foundation as part of the 1957–1958 International Geophysical Year (IGY). Our principal results were a latitude survey of the intensity of the primary galactic cosmic rays, including the heavy nuclei therein, the discovery of X-rays from auroral electrons, and the measurement of ionospheric currents near the equator and at high latitudes.

Beginning in 1956, I participated as a member of several of the committees and panels planning U.S. participation in the IGY and, most importantly, the one on scientific uses of artificial satellites of the Earth. The realistic expectations for investigations with satellite-borne instruments were heavily dependent on the experience and aspirations of a small cadre of us veterans of research with high altitude rockets during the preceding decade.

Following the successful launch of Sputnik I by the U.S.S.R. on 4 October 1957, there was a fast-breaking effort in the United States to speed up our efforts to place artificial satellites in orbit, all within the context of the International Geophysical Year. As part of that response, the

---

* University of Iowa, Iowa City, IA, USA
** This article, except the Author's Preface, was first published in *Icarus*, **122**, 209–232 (1996)

U.S. Army's proven four-stage Juno II vehicle was adapted for the first attempt rather than the previously planned Naval Research Laboratory's new Vanguard vehicle, still plagued by developmental problems.

A cosmic ray instrument developed by George Ludwig and me at the University of Iowa had been selected previously by the relevant IGY panel as one of four payloads slated for early flights on a Vanguard vehicle. Our instrument was essentially ready for flight in late autumn 1957. During the preceding several years, I had followed in detail the relative status of the Juno II and Vanguard vehicles and had decided that it would be wise to design our instrument so that it would be suitable for either. By virtue of this fact and our previously established place on the short list, the Iowa instrument was selected for the first Juno II attempt. We then worked with the Jet Propulsion Laboratory to integrate it into the overall payload for the flight.

The purpose of the investigation was to greatly extend our rocket measurements of cosmic ray intensity above the appreciable atmosphere as a function of latitude, longitude and altitude. Our adopted sensor for this purpose was a single Geiger-Mueller tube.

The Juno II, also known as Jupiter C, delivered the payload (then named Explorer I) into a durable, moderately eccentric orbit on 31 January 1958 (1 February GMT). Telemetered data during and immediately after the launch sequence showed that our instrument was operating properly in free flight. Short segments of real-time data (typically of 2 or 3 minutes duration) gradually flowed into our laboratory from the wide geographical distribution of receiving stations. Some segments showed about the expected cosmic ray counting rates. In others the apparent rate was zero, a physical impossibility for cosmic rays. For several weeks we puzzled over conceivable failure modes, but were much preoccupied by intense preparations for further launches. An upgraded instrument was carried on Explorer II, a Jupiter C launch failure on 5 March. Our third instrument also included Ludwig's magnetic tape recorder for recording data during a complete orbit and then, upon command from a ground station, playing back the record within about 6 seconds. Data were also to be transmitted in real-time. This payload was launched successfully by a Jupiter C on 26 March and called Explorer III. The very first full orbit playback that we received confirmed the existence of expected cosmic ray rates at low altitudes, but then showed rapidly increasing rates as the satellite moved to higher altitudes in its elliptical orbit, and then apparent rates of zero at the highest altitudes. This sequence was repeated in reverse as the satellite descended to lower altitudes. A further series of Explorer III tape playbacks gave repeated examples of similar results and established a coherent pattern. Meanwhile, Carl McIlwain demonstrated by a laboratory test of a similar detector system that the zero rates in Explorer I and Explorer III data were, almost certainly, caused by radiation intensity so great that the GM tube no longer yielded resolvable individual pulses but only a kind of noise level of pulses too small to trigger the counting circuit.

We then returned to the analysis of the real-time Explorer I records, which had much greater dynamic range, and we used a laboratory calibration of a similar system for apparent vs. true (i.e., assuming zero dead-time) counting rates to find true counting rates. The altitude dependence of counting rate was far too rapid and at far too high an altitude to be attributed to any form of electromagnetic radiation (e.g., X-rays or γ-rays) or to any form of corpuscular radiation on direct trajectories from a distant source.

I concluded that the observed "effect" must be attributed to energetic, electrically charged particles trapped mechanically in the Earth's external magnetic field. This was the essence of my "discovery" announcement of the existence of enormous intensities of geomagnetically trapped corpuscular radiation at a joint National Academy of Sciences/American Physical Society meeting in Washington, D.C. on 1 May 1958.

We were immediately given a go-ahead for conducting follow up investigations with detector systems of far greater dynamic range than that of Explorers I, II and III. We worked with the Army Ballistic Missile Agency in Huntsville, Alabama in building payloads for two further launches of upgraded Jupiter C vehicles. The first of these two launches on 26 July 1958 placed Explorer IV in a durable orbit at an inclination of 50° (up from the 33° of the orbits of Explorers I and III). The second attempt on 24 August 1958 was a vehicular failure.

Our multi-fold detector system on Explorer IV yielded immediate and massive confirmation of our earlier results and a substantial advance in particle identification. In addition, it provided the principal observations of artificial radiation belts composed of energetic electrons injected by the radiative decay of the fission products from a series of three high altitude nuclear bomb bursts, called Argus I, II and III. The latter body of observations was classified as secret until public release by the federal government in early 1959. During the following six months, we provided radiation detectors for "moon" shots by Pioneer II, Pioneer III and Pioneer IV. Pioneer III did not achieve escape velocity but reached an apogee of 17 Earth radii (radial) and yielded data during both outbound and inbound legs of its trajectory. These data, combined with the lower altitude data by similar detectors on Explorer IV, clearly established the existence of two major radiation belts, distinguished by the markedly different absorbtivity of the trapped particle population. The very penetrating radiation in the inner belt had been well characterized by Explorers I, III and IV. Meanwhile Sputnik III (May 1958) had provided information on an outer belt. But Pioneer III gave the first complete survey of the radial distribution of trapped particles and established the outer boundary of the trapping region.

Pioneer IV did achieve escape velocity and made a rather remote pass by the Moon as it continued into interplanetary space. Our radiation data from this one-way flight through the Earth's external magnetic field confirmed and extended the basic findings from Pioneer III and gave a generally concurrent determination of the outer boundary of trapping.

Our laboratory then conducted subsequent investigations on the "heavy" IGY satellite Explorer VII of the Army Ballistic Missile Agency and on a series of complete satellites designed, built and instrumented in our small university laboratory. During the period 1961–1974, our six Injun and Hawkeye satellites were placed in a variety of high inclination orbits and provided a wealth of pioneering data on auroral radiation, solar X-rays, VLF radio waves above the ionosphere, galactic cosmic rays, solar energetic particles, and the structure of both radiation belts.

Also during the 1960s and 1970s, we participated in a series of NASA missions, including the Orbiting Geophysical Observatories 1, 2, 3, 4 and 5, Explorers XII and XIV and Explorers 33, 34 and 35.

At one point in time, the University of Iowa had more space flights to its credit than the entire European Space Agency, including individual member states.

During the 1960s, the primary emphasis of the evolving U.S. national space program was on manned missions in low Earth orbit, in preparation of the Apollo manned missions to the Moon. A secondary emphasis was on exploratory (unmanned) missions to the inner planets Venus and Mars. There was a special interest in exploring Mars to establish whether or not the physical conditions on Mars were favorable for the development of some sort of biological activity there, as had been frequently speculated.

Our radiation instruments were carried on Mariner II, the first spacecraft to fly by Venus (1962); on Mariner IV, the first spacecraft to fly by Mars (1964); and Mariner V, the second spacecraft to fly by Venus (1965). We found that neither Mars nor Venus has radiation belts, and we contributed observations establishing significantly small upper limits on the magnitude of their magnetic moments.

As a member of both the Space Science Board of the National Academy of Sciences and NASA's Lunar and Planetary Mission Board, I led special advocacy committees for missions to Jupiter and the other giant gaseous planets Saturn, Uranus and Neptune. It was already known from radio-astronomical evidence that Jupiter had a large radiation belt of relativistic electrons, but there was no credible radio-astronomical evidence of radiation belts around Saturn, Uranus or Neptune.

Our advocacy group provided the scientific rationale for missions to the outer planets. In 1968, NASA adopted two missions designed to pass through the asteroid belt between the orbits of Mars and Jupiter and to pass by Jupiter at a heliocentric distance of 5 AU, beyond any previous missions. Management of the development of the spacecraft was undertaken by NASA's Ames Research Center. Our Iowa package of radiation detectors was one of several selected for these flights.

The missions of the two spacecraft, later named Pioneer 10 and Pioneer 11, were brilliantly successful. Pioneer 10 passed through the asteroid belt unscathed and made the first in situ investigation of Jupiter's huge and intense radiation belt and magnetosphere in November–December 1973. It continued outward in a solar-system escape trajectory. Pioneer 11 made the second and somewhat different passage through Jupiter's magnetosphere a year later. Following the success of Pioneer 10, and at the urging of the scientific investigators, Pioneer 11's encounter trajectory for Jupiter was chosen so that it led to a subsequent encounter with Saturn. Its close flight by Saturn occurred in August–September 1979. We discovered that Saturn is also a highly magnetized planet and has a large radiation belt and magnetosphere, though considerably smaller in magnitude and intensity of trapped particles than that of Jupiter.

There was an intense debate among the scientific investigators and officers of NASA on the best choice of the flight trajectory for Saturn. One ballistic option was a very close passage within the inner edge of Saturn's ring system. Another was an encounter that would lead Pioneer 11 to a subsequent encounter with Uranus. In the end, NASA selected a trajectory that would cross Saturn's ring plane at about the radial distance that was being considered for future Voyager 1 and Voyager 2 encounters, thereby calibrating the survival probability for those missions. After its Saturn encounter, Pioneer 11 also had solar system escape velocity.

Both Pioneers 10 and 11 continued to provide uniquely valuable data as they flew through the outer heliosphere, most notably on the properties of the solar wind and on the cosmic ray intensity over a long time span and over previously unexplored distances from the Sun. Because of a combination of technical limitations, the flow of useful data from Pioneer 11 terminated in January 1995 at a heliocentric radial distance of 42 AU.

As of mid-2000, after over 28 years of flight, Pioneer 10 continues to operate well. Valuable, though rather sparse, data on cosmic ray intensity are still being telemetered reliably from my radiation instrument on Pioneer 10, now at a heliocentric radial distance of over 75 AU (11,220,000,000 km), a truly heroic achievement for NASA's Deep Space Network of receiving stations. (The radiated power of the S/C transmitter is only 8 watts.) We are seeking the boundary of the heliosphere but our most recent data show that the cosmic ray intensity at 75 AU is still influenced by solar activity. Hence, that boundary lies beyond, possibly far beyond, 75 AU.

Meanwhile, my Iowa colleagues are serving as the principal investigators on the currently active missions of Dynamics Explorer I, Geotail, Cluster II, Voyagers 1 and 2, Galileo and Cassini, and they are developing instruments for future missions.

INTRODUCTION

Space science is an eclectic mixture of the traditional disciplines of astronomy, physics, chemistry, geology and biology. The commonalty of its observational component is (a) the use of rocket-propelled vehicles for the delivery of instruments into orbit about the Earth, to the Moon, and through the interplanetary medium to and beyond distant planets, comets and astroids; and (b) the use of radio telemetry as the primary method for transmitting data to terrestrial stations. In addition, physical samples of the solar wind, lunar surface material and meteoric dust have been collected and returned for laboratory study. The return of samples of cometary dust and Martian surface material is planned.

The term space physics designates the sub-field of space science that deals with certain classes of electromagnetic radiations of solar system origin, as well as with electric and magnetic fields, ionized gases (plasma), energetic electrons and energetic ions. The heritage of space physics is well represented by the great monographs of Chapman and Bartels (1940), Störmer (1955), Mitra (1952), and Alfvén (1950).

The modern epoch of space physics began in the late 1940s with the use of high altitude sounding rockets and now encompasses the research efforts of thousands of investigators throughout the civilized world. Sophisticated instruments of superb quality are being flown on all manner of automated, commandable spacecraft. The original data are, for the most part, descriptive of natural physical phenomena, whose interpretation engages a growing cadre of theorists and modelers. In addition, numerous artificial experiments are being conducted in space (Hultqvist and Fälthammar, 1990).

This chapter is a tutorial review of magnetospheric physics, with primary emphasis on the magnetic fields of planets and the energetic particles therein. Aside from this fresh introduction, it is a reproduction of the author's "Kuiper Prize Lecture: Electrons, Protons, and Planets," which was sponsored by the American Astronomical Society, published in the journal *Icarus* (**122**, 209–232, 1996) and reprinted in this volume with the copyright permission of the Academic Press. The relatively short bibliography comprises classical and recent monographs, reviews, major compilations of papers, and a few original papers, but does not cite scores of other original papers.

MAGNETISM OF PLANETARY BODIES

The interior of the Earth has an estimated temperature of several thousand degrees Kelvin, far above the Curie points of all known ferromagnetic substances, i.e., the temperatures above which they lose their ferromagnetic properties. Common evidence for such high temperatures is the flow of molten lava from volcanoes on land surfaces and on the floor of the ocean. Also the permanent magnetization of the cool outer crust is far too weak and fragmentary to be important on a global scale.

It is now almost universally accepted that the general magnetic field of the Earth must be attributed to electromagnetism or the flow of electrical currents in patterns resembling those of laboratory solenoids (Rikitaki, 1966). The consequent magnetic field may be complex because of the likely complexity of the causative current system. However, at distances large compared to the dimensions of the current system, such complexity disappears and the general magnetic field is that of a small current loop or equivalent point magnetic dipole of vector moment **M**. In general, **M** of a planetary body may be offset with respect to the geometric center of the body and tilted with respect to the body's rotational axis.

A first-order approximation to the Earth's external magnetic field is that of a point dipole of moment $M = 7.90 \times 10^{25}$ gauss cm^3 (0.304 gauss $R_\oplus^3$, where the equatorial radius of the Earth, 1.0 $R_\oplus = 6,378$ km) located at the geometric center of the Earth and tilted by 11.5° to its rotational axis. The corresponding geomagnetic poles are at 78.5°N, 69.1°W; and 78.5°S, 110.9°E (epoch 1965). The vector moment **M** points southward.

An improved approximation is obtained by displacing the same vector moment **M**, as in the centered dipole model, by 450 km from the Earth's geometric center toward latitude 17°N and longitude 149°E. This is called the eccentric dipole model. The need for such an improved representation is evident in global charts of the scalar magnitude B as a function of latitude and longitude at zero altitude.

Contemporary measurements establish the secular variation of the Earth's magnetic field and paleomagnetic evidence records reversals of polarity at irregular intervals on a time scale of the order of 10^5–10^6 years (Akasofu and Chapman, 1972), thus testifying to the long-term instability of the internal current system.

Maintenance of the necessary system of electrical currents is attributed to a self-excited dynamo, according to the theory pioneered by Bullard and Elsasser. In this theory, it is visualized that any initial magnetic field, however small, is amplified by the convective flow of electrically conducting material (e.g., molten lava) through that field to induce electrical currents of such strength as to achieve a quasi-steady state between energy input and ohmic and viscous losses. Detailed theories of this effect are complex and have not reached the stage at which quantitative predictions can be made, even given the interior conditions of a planet. But it does appear that two properties of the planet are necessary: (a) an interior that is sufficiently hot to produce a fluid, electrically conducting fluid, and to drive convective flow of that fluid and (b) a rotational rate that is sufficiently rapid to guide the pattern of convective flow. Plausible values of electrical conductivity are derived from laboratory experiments

Table 1 Magnetic moments M and rotational angular momenta Iω

	M (gauss cm³)	Iω (g cm² s⁻¹)
Sun*	3.4 E 32	1.63 E 48
Mercury	4 E 22	9.65 E 36
Venus	<8 E 20	1.83 E 38
Earth	7.95 E 25	5.86 E 40
Moon	<1 E 20	2.34 E 36
Mars	2 E 22	1.98 E 39
Jupiter	1.54 E 30	4.24 E 45
Saturn	4.3 E 28	7.51 E 44
Uranus	3.8 E 27	1.21 E 43
Neptune	2.0 E 27	2.04 E 43
Pluto	—	9.5 E 35
Pulsars[†]	1 E 28 to 1 E 30	5.2 E 44 to 9.6 E 46

* $M_\odot$ assumes equatorial "surface" field of 1.0 gauss.
[†] Data for pulsars derived from Taylor and Stinebring (1986). Notation: $a\,E\,b$ means $a \times 10^b$.

and plausible values of convective speeds are derived from continental drift rates and other geological evidence.

Classical studies of the Earth's surface magnetic field (Chapman and Bartels, 1940) have been extended to large radial distances and in great detail by satellite-borne magnetometers first launched in 1958.

Since 1962, the *in situ* measurement of magnetic fields of other planets has been a central objective of many space missions. Previous to such investigations the only other planet known to be magnetized was Jupiter, based on radioastronomical evidence (synchrotron radiation at microwave frequencies).

There are now comprehensive *in situ* measurements of the magnetic properties of the Moon, as well as of Mercury, Venus, Mars, Jupiter, Saturn, Uranus, and Neptune.

An abridged summary of the dipole representations of the fields of these planets is given in Table 1. Further details concerning individual planets are provided later.

The conditions required for success of a self-excited dynamo are apparently met by the Earth and the four outer, jovian planets, but not by the Moon, Mercury, Venus, and Mars. A rudimentary interpretation of this finding for the latter four objects is that the Moon, Mercury, and Mars probably do not have molten interiors, whereas Venus may have a molten interior that is rotating too slowly (243-day period, retrograde).

Values of magnetic moment M as a function of rotational angular momentum $I\omega$ are plotted in Figure 1. The equation of the straight-line least-squares fit to the data for Mercury, Earth, Mars, Jupiter, Saturn, Uranus, and Neptune, as

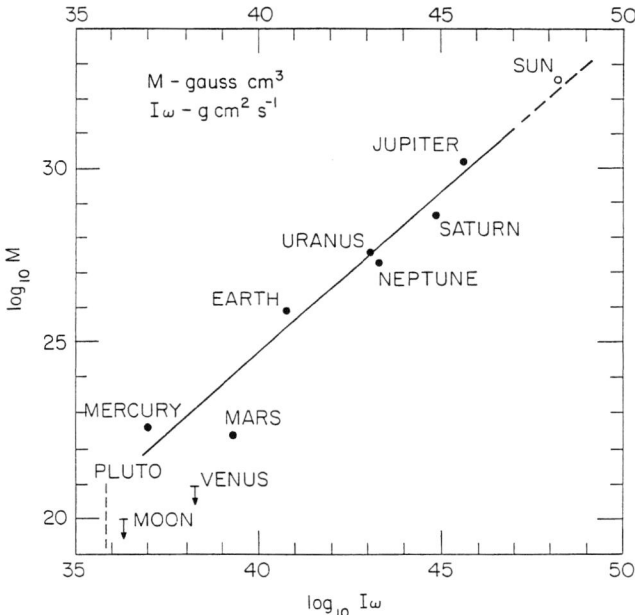

Figure 1 Empirical relationship of the magnetic moments M of the planets to their rotational angular momenta $I\omega$. The least-squares, straight-line fit to seven solid circles is represented by Equation (1) in the text. Upper limits on M for Venus and the Moon are shown and the vertical dashed line for Pluto represents its known angular momentum but unknown magnetic moment. The open circle for the Sun assumes an equatorial surface field of 1.0 gauss.

shown in Figure 1, is

$$\log_{10} M = k_1 + k_2 \log_{10} I\omega \qquad (1)$$

with $k_1 = -12.1$ (± 4.9); $k_2 = 0.92$ (± 0.12). This empirical equation gives a respectable representation of the data over a range of eight orders of magnitude in M. A similar fit to the data for only the five strongly magnetized planets (i.e., omitting Mercury and Mars) yields $k_1 = -8.0$ (± 5.5) and $k_2 = 0.83$ (± 0.13) (not shown in Figure 1). Such relationships are not vectorial ones, of course, but appear to be useful gross summaries of some very complex physical processes.

Despite the absence of an accepted theoretical explanation for these simple relationships, sometimes pejoratively called "magnetic Bode's Laws," the predictive value of a similar 1985 version was confirmed by the *subsequent* measurements of the magnetic moments of Uranus and Neptune.

MAGNETIC STORMS, AURORAE, AND THE IONOSPHERE

A magnetic storm is an episode, typically of several days' duration, during which fluctuations of the surface magnetic field vector are unusually large. There is an enormous literature on magnetic storms and their relationship to solar flares. Long before the epoch of observations in space, such storms were attributed to corpuscular streams of ionized but neutral gas ejected from solar flares on the rotating Sun (Chapman and Bartels, 1940).

As viewed in a nonrotating coordinate system, such a stream was visualized as having the form of an Archimedean spiral (Figure 2), the locus of packets of gas moving radially outward at constant speed from a point on the rotating Sun, like a stream of water from a slowly rotating spigot. The typical delay of two to three days between the occurrence of a noteworthy solar flare and the commencement of a magnetic storm at the Earth corresponds to a propagation speed of about $700 \, \text{km s}^{-1}$.

Each magnetic storm is different in detail, but a simplified generalization identifies two distinct phases. The initial phase begins with a sudden increase in the horizontal component of the field, called a storm sudden commencement or (SSC), with an amplitude of the order of 10 gammas (1 gamma = 1×10^{-5} gauss). The duration of the initial phase is of the order of a few minutes to an hour. It is attributed to the compression of the distant geomagnetic field because of pressure from the impinging (electrically conducting) plasma on the sunward side of the Earth. The subsequent main phase is a more gradual decrease of the field by ≈ 100 gammas, followed by a recovery period of a few days. The main phase is attributed to the magnetic effect of captured plasma from the solar stream to form a westward flowing ring current in the magnetic equatorial plane at a radial distance of 4 to 6 $R_\oplus$.

The other major geophysical effect of such streams is the occurrence of enhanced luminosity in oval-shaped regions of the upper atmosphere which resemble halos encircling both northern and southern magnetic poles at a typical geomagnetic latitude of about 70° and which are called the polar aurora (Störmer, 1955). Such luminosity is observable to a greater or lesser extent nearly every clear night, but has a great range of brightness, structure, and distribution of the geomagnetic latitude associated with magnetic storms. Presence of aurorae during periods of even relatively quiet geomagnetic conditions is evidence for continually present solar corpuscular streams, a conclusion supported by a ubiquitous effect on the ionized gaseous tails (Type I) of comets, which were observed long before *in situ* observations of the interplanetary streams (the solar wind). Because of the dominance of ionized hydrogen in the upper atmosphere of the Sun, the streams were presumed to consist primarily of protons and electrons in the form of a tenuous gas (or plasma) whose bulk electrostatic charge was zero. The speed of propagation corresponded to a directed proton energy of about 1.0 keV, though it was supposed that the Maxwellian temperature of the plasma in the moving frame of reference was of the order of 10^5 K (10 eV).

Knowledge of the partially ionized gas in the Earth's upper atmosphere was contained in the massive field of ionospheric physics (Mitra, 1947, 1952). Charged particles therein have thermal or quasi-thermal energies of the order of a few hundred degrees Kelvin. An important advance in this field came from the study of the desultory occurrence of whistlers, or whistling atmospherics, following atmospheric lightning strikes. Whistlers are electromagnetic waves in the audio frequency range which are guided by the geomagnetic field and whose frequencies f decline in a systematic way with time t during a period of the order of several seconds according to the relationship:

$$f = \left(\frac{D}{t}\right)^2, \tag{2}$$

where the quantity D is called the dispersion. Storey (1953) made important contributions to both the observational and theoretical aspects of this subject. His work was in the context of ionospheric physics but retrospectively might be regarded as discovery of the plasmasphere. Specifically, he found a scale height for the distribution of electrons above the F-layer corresponding to a gas temperature of about 7200 K, and he estimated an electron number density of approximately 400 cm^{-3} at a radial distance of 3 $R_\oplus$ on the equator, values remarkably similar to the ones determined later by *in situ* observations. Also, he suggested that "the source of this ionization is the sun, and that it is some of the

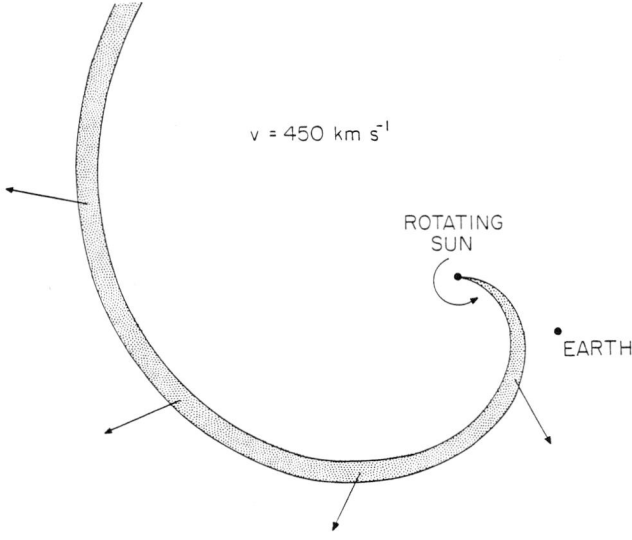

Figure 2 Schematic diagram of the locus of a small segment of a long-lived solar corpuscular stream, as envisioned by Chapman. The radial arrows represent the outward velocities of successive elements of the stream.

material, believed to be principally ionized hydrogen, which is ejected at times of disturbances and is responsible for magnetic storms and the Aurora."

The foregoing is an abridged summary of the author's perception of knowledge about the geophysical roles of electrically charged particles, as it existed before the epoch of high-altitude rockets and satellites. Indeed Alfvén (1950) remarked "Except in problems concerning cosmic rays, there is no problem relating to the terrestrial field where the electronic energy exceeds 10^5 volts." This upper limit was adopted by him as an approximate upper limit to the energy of auroral particles.

Galactic cosmic rays whose energies range upward from about 100 MeV and solar energetic particles of energies much above those characteristic of particles in solar corpuscular streams fall into quite different categories of consideration and are not discussed in this paper.

POST WORLD WAR II ADVANCES

Beginning in 1946, high-performance rockets were adopted for the flight of scientific equipment up to altitudes on the order of 200 km. German V-2 and American Aerobee and Viking rockets became the workhorses for *in situ* measurements in the upper atmosphere and ionosphere, and for observations of solar radiations and cosmic rays above the appreciable atmosphere and over a range of latitudes.

The inexpensive technique of balloon-launched rockets (rockoons) was used to extend measurements of energetic particles and the geomagnetic field to altitudes of about 100 km and over a larger range of latitudes including the Arctic and Antarctic. A significant result of this work in 1953–1957 was the *in situ* detection of auroral bremsstrahlung and the causative primary electrons. These observations yielded direct knowledge of the absolute intensity and spectra of such electrons, which were consistent with the brightness and altitude dependence of auroral luminosity. Beginning in 1957, auroral bremsstrahlung was also detected by balloon-borne detectors at about 30 km altitude and in early 1958 the absolute intensity and spectra of primary auroral electrons and protons were observed directly and more definitively with rocket flights from Fort Churchill, Manitoba. This work provided fresh knowledge about auroral physics and laid both the scientific and technical foundations for planning and conducting scientific research with satellites of the Earth during the 1957–1958 International Geophysical Year.

The first successful American satellite, Explorer I launched in early 1958, yielded the wholly unexpected discovery of enormous numbers of high-energy particles trapped in the Earth's external magnetic field at low latitudes far from the auroral zones (Figures 3, 4, and 5) (Yoshida *et al.*, 1960; Van Allen, 1983; Hess, 1968). In the

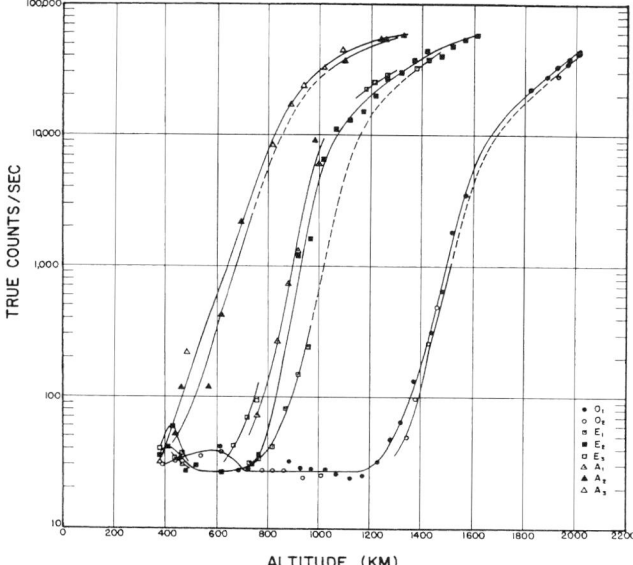

Figure 3 True counting rate (i.e., observed rate after correction for dead time) of the Geiger-Mueller tube on satellite Explorer I as a function of altitude over eight geographic regions, all near the magnetic dip equator. The left-hand two curves are for regions in the approximate geographic longitude 295°E; the central bundle of curves, 5°E; and the right-hand two curves, 105°E. The dependence of these curves on longitude is direct evidence for the 450 km offset of the magnetic center of the Earth from its geometric center (Yoshida *et al.*, 1960).

jargon of the participants, "high energy" meant values distinctively above those identified with aurorae. Intensive follow-up investigations showed the presence of two distinct regions, or radiation belts, and led to great advances in detailed knowledge of the absolute intensities, distributions in space, and energy spectra of the components of the population therein, principally electrons and protons with a minor but significant admixture of heavier ions. Most of the early observations were limited to electrons having energy levels exceeding 40 keV and protons having energy levels exceeding 1 MeV. The low-energy deficiency in spectral coverage was soon remedied by the flight of electrostatic plasma analyzers, whose lower energy limits were extended downward to the eV range.

Magnetometers carried on satellites throughout this period provided knowledge of the large-scale structure of the external magnetic field of the Earth as influenced by the flow of solar corpuscular streams, later called the solar wind, and the diamagnetic effects of the entrapped plasma, including the ring current. The other major area of advance involved the observation of plasma waves, which became recognized as basic elements of the overall physics of the magnetosphere – the larger system within which the geomagnetic field controls the motion of charged particles.

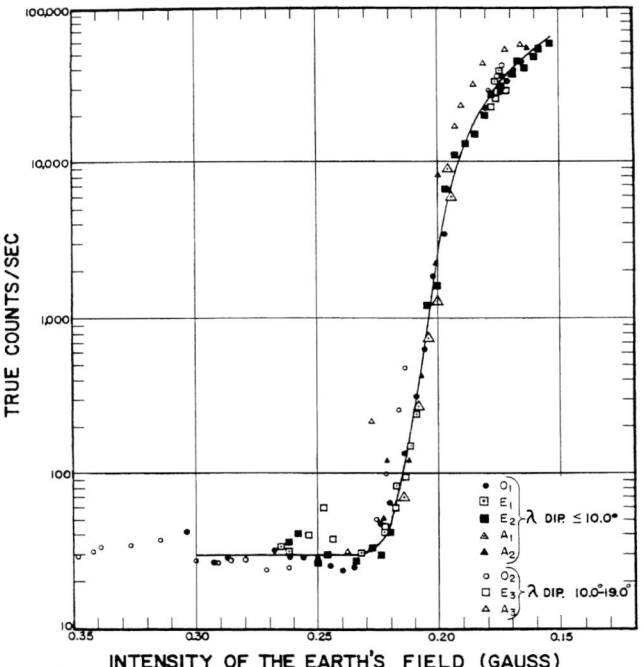

Figure 4 The Explorer I counting rate data of Figure 3 replotted as a function of the intensity B of the Earth's magnetic field at the points of observation (Yoshida et al., 1960).

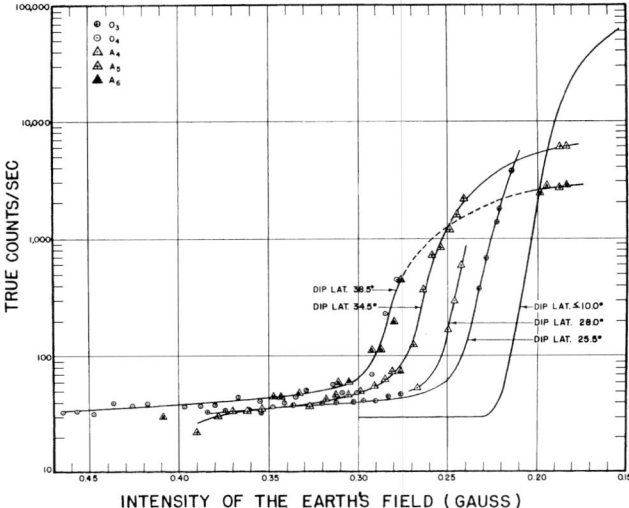

Figure 5 Explorer I counting rate data as a function of B for five different magnetic dip latitudes λ_{dip} as defined in the text (Yoshida et al., 1960).

Thus, although the radiation belts of very energetic particles are a distinctive part of the system, low-energy plasma physical phenomena dominate its detailed morphology and dynamics and, to an important extent, control the energization and distribution of the more energetic particles.

By virtue of its relative convenience, the Earth's magnetosphere has become the prototype for the magnetospheres of other planetary bodies and the reference case for comparison and contrast. At the present time, the principal features of the Earth's magnetosphere are well observed and familiar, but full understanding is elusive and many details remain under active study by hundreds of investigators.

RUDIMENTARY PHYSICS OF A PLANETARY MAGNETOSPHERE (WITH EXAMPLES FOR THE EARTH)

1 A starting point

As a first approximation, a magnetized planet is regarded as simply a current-carrying coil of wire or an equivalent point magnetic dipole of moment **M**, located at distance R from the Sun. In this approximation, the physical body and atmosphere of the planet are omitted.

Such a magnetic dipole may be visualized as an isolated physical entity in the vast interplanetary medium. Its idealized properties for the durable trapping of noninteracting electrically charged particles *in vacuo* are those developed by Störmer (Appendix A). But one of the most basic features of the Störmerian theory is that an electrically charged particle approaching the dipole from infinity has no access to the trapping regions. Its encounter with **M** is simply a scattering or deflection with a total scattering cross section σ given by the approximate formula

$$\sigma = 0.27 \frac{Ze\,M}{pc} \qquad (3)$$

In (3), p is the momentum of the particle, Ze is its electrical charge, c is the speed of light, and the quantity pc/Ze is the magnetic rigidity $\mathcal{R}$ of the particle. Thus, σ is inversely proportional to the magnetic rigidity. Numerical examples for $M = M_\oplus$ are as follows: For a 100 MeV proton, $pc = 4.45 \times 10^8$ electron-volts, $pc/Ze = 1.48 \times 10^6$ erg (e.s.u)$^{-1}$, and $\sigma = 35\,R_\oplus^2$; and for a 1,000 MeV proton, $\sigma = 9.3\,R_\oplus^2$.

In the idealized Störmer model, an external source of noninteracting particles cannot populate the trapping region. Hence, if such a population is found to actually exist, it is necessary to invoke other physical processes.

2 The solar wind and its interaction with a magnetic dipole

The solar wind consists of hot ($\approx 10^5$ K) ionized gas flowing radially outward from the Sun at a speed v, which normally lies in the range 300–700 km s^{-1} and is typically 450 km s^{-1}. The gas consists principally of ionized hydrogen, protons of mass m and number density n, and electrons of equal number density. The solar wind flow is now known by direct measurements with Pioneer 10 to extend outward to at least 62 AU. Hence all known planets of the solar system are exposed to the flow of the solar wind. The speed v is approximately independent of R, and n varies as R^{-2}, having a typical value of 5 cm^{-3} at 1 AU.

In the following discussion, it will be supposed for simplicity that **M** is oriented perpendicular to the radial line from it to the Sun.

As the solar wind, an electrically conducting magnetized fluid, encounters the magnetic field of the dipole, a system of eddy currents is induced in the fluid and the current-carrying plasma is then repelled by **M** and brought to rest at a standoff distance r_s given by the pressure balance equation

$$(n\,m\,v)(v) = k\frac{B^2}{8\pi} = k\left(\frac{M}{r_s^3}\right)^2 \frac{1}{8\pi}$$

or

$$n\,m\,v^2 = \frac{kM^2}{8\pi r_s^6}, \quad (4)$$

where k is a dimensionless factor of the order of, but slightly greater than, 1.0. The standoff distance calculated by (4) is identified as the position of the sunward magnetopause in the framework of fluid dynamics. The fluid then flows around the magnetic obstacle to form a Mach surface, enclosing the magnetic field of the dipole, and a long turbulent wake on the downwind (anti-solar) side (Figure 6).

The resulting plasma physical milieu apparently defies simple analysis but includes a number of suggested effects: turbulence and viscosity, interconnection of the interplanetary magnetic field with that of the dipole, and the development of polar cusps. Some combination of these and perhaps other effects are believed to result in the injection of some small fraction of the solar wind plasma into quasi-bound orbits in the outer fringes of the distorted dipolar field. Some of this plasma may diffuse and/or be convected (by quasi-stationary electric fields) inward to constitute a population of trapped energetic particles. Radial diffusion is visualized as being caused by fluctuating magnetic and electric fields induced by fluctuations in the properties of the solar wind. Such processes, if ones that violate only the third adiabatic invariant (the "weakest" of the three; see Appendices B and C), will raise the original energy of the particles by a factor of the order of 1000 because $p_\perp^2/B = $ const if the first adiabatic invariant is conserved. ($p_\perp$ is the component of the particle's momentum perpendicular to the magnetic field **B** [Appendix B].) For example, an electron having a kinetic energy of 1 keV in the thermalized solar plasma (i.e., after conversion of the bulk energy of the flowing wind to thermal energy as it is brought to rest) will be accelerated to about 600 keV as it diffuses inward. Similarly, a 1 keV proton will be accelerated to about

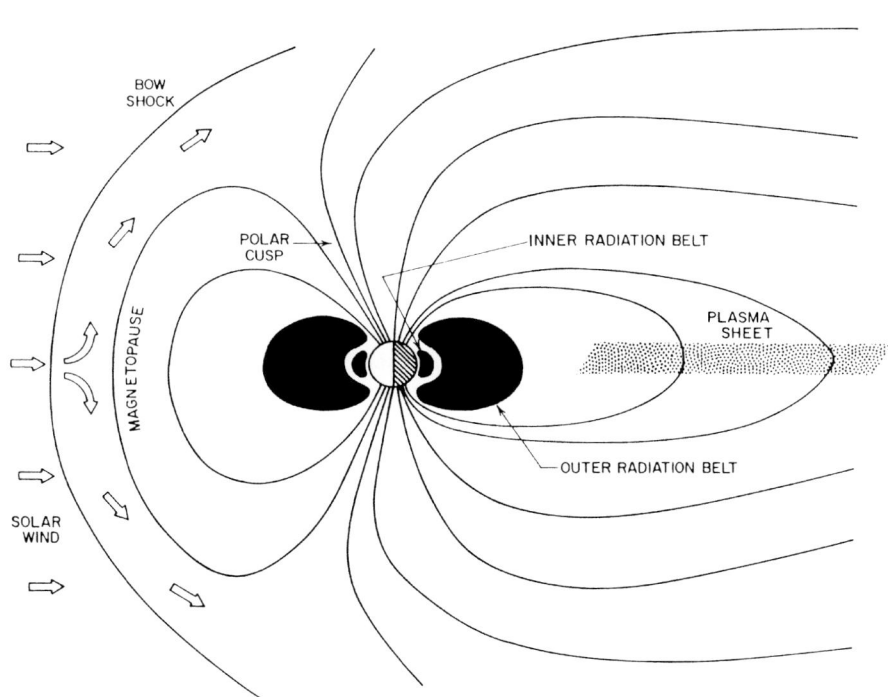

Figure 6 The author's simplified representation of a magnetic meridian-plane cross section of the topology and structure of the Earth's magnetosphere.

1 MeV. Such energies do indeed resemble typical values as observed in the intermediate magnetosphere. This fact supports the belief that the solar wind is an important and probably dominant source of the population of the outer radiation belt of the Earth. More detailed evidence for this belief is provided by the similar proton/alpha particle abundances in the solar wind and the magnetosphere and the corresponding state of ionization (2+) of the helium nuclei. Similar correspondence is also found for the abundances and states of ionization of heavier and much less abundant elements. (See, however, Section 4 below.)

3 Roles of the solid body of a planet

If the solid body of a planet is added to the above conceptual model, at least two additional effects occur.

First, trapped particles can no longer move in Störmerian orbits that reach down to the point dipole but will be prevented from doing so by collisions with the planet. The result is the absence of trapped particles within a "loss cone" around the magnetic vector.

Second, impinging untrapped cosmic rays will produce nuclear reactions in the outer surface of the planet (and in its rings and satellites, if any). Among other products of such reactions are high-energy neutrons, a small fraction of which will fly outward. In turn, a small fraction of these will undergo β-radioactive decay (half life at rest, 12 minutes) within the trapping region and will thereby inject protons and electrons into the magnetic field. It is noted that this process is independent of the existence of the solar wind and is one that can be reduced to relatively straightforward, though complex, calculation. The process is called Crand, cosmic ray albedo neutron decay. The energies of injected electrons range upward to several MeV and of protons up to hundreds of MeV. In the case of the Earth's inner radiation belt, the calculated distribution of energetic protons and their energy spectrum have a convincing resemblance to those observed. The absolute intensities correspond to residence times on the order of years as estimated via the "leaky-bucket model" of equilibrium between sources and sinks. The situation for electrons is less clear, presumably because of an unknown admixture of solar wind electrons that have been accelerated to comparable energies.

4 Roles of the atmosphere, if any, of a planet

An atmosphere increases the efficacy of the Crand process and also broadens the loss cone. It has two other important effects.

First, it acts as an energy-loss medium for trapped particles extending up to altitudes of many scale heights. Thus it establishes an effective altitude limit below which such particles cannot have durable mirror points in their latitudinal oscillations. For the Earth, the lower boundary to the trapped particle population is a function of longitude because of the offset of the geomagnetic center and varies along the geomagnetic equator from about 300 km over the South Atlantic Ocean (the "South Atlantic anomaly") to 1,200 km over Singapore (Figure 3).

Second, an atmosphere and, more importantly, an ionosphere are sources of the thermal or quasi-thermal ($\lesssim 1$ eV) plasma within a magnetosphere. Ionospheric ions and electrons spiral around magnetic lines of force and lose energy to the gravitational field as they move outward. For both of these reasons, they are confined to relatively low altitudes and are not significant sources of high-energy radiation belt particles. However, neutral atmospheric atoms and molecules in the exosphere are not affected by the magnetic field, but only by the gravitational field. For example, a 0.6 eV (7000 K) hydrogen atom has sufficient speed to escape the Earth's gravitational field. Some small fraction of such atoms will be ionized, photoelectrically or otherwise, in the course of their outward flight and become part of the low-energy ion/electron population of the outer reaches of the trapping region. They are then subject to inward diffusion and acceleration, as sketched in Section 2 above, to become much more energetic.

Atmospheres of satellites, sputtered gas from surfaces of satellites and ring material, and volcanic and other emitted gas from satellites are other potentially important internal sources of trapped particles.

5 Losses of trapped particles

Radial diffusion of trapped particles is driven by fluctuating magnetic and electric fields on a global scale. The component of the spectrum of fluctuations that is most effective in a given case is at and near the frequency corresponding to the reciprocal of the longitudinal drift period of the particles under consideration (Appendices B and C and Table C1). Such radial diffusion is, of course, a two-way process, i.e., both outward and inward. Net diffusive flux is in the direction of lesser phase space density, usually but not always inward in planetary magnetospheres. The diffusion coefficient D (not to be confused with the same symbol for the dispersion of a whistler) increases strongly with increasing radius r from the center of a planet because the magnetic field becomes progressively "softer" and hence more subject to perturbations by the fluctuating solar wind. At the Earth $D \propto r^9$. Inward-diffusing particles are eventually lost by collisions with atmospheric gas, if any, or with the solid surface of a planet, its satellites or the particulate matter of rings.

Other loss processes are of a plasma-physical nature, the primary process being pitch-angle scattering (violation of the first adiabatic invariant) (Appendix B) and consequent loss into the atmosphere or surface of the planet by spontaneously generated waves due to instabilities in the ambient plasma. Whistler-mode waves scatter electrons and ion-cyclotron waves scatter protons. The efficacy of such waves is a maximum in the (Doppler-shifted) frequency range corresponding to the gyrofrequency of the spiraling particles. These processes limit the population of trapped particles because the strength of the plasma waves increases as the population of particles increases. Such limits were one of the prominent puzzles of early observational work on the Earth's radiation belts.

The residence times, often called lifetimes, of trapped particles against atmospheric losses are calculable in a reasonably straightforward way, as are losses to satellites and rings. But the losses due to wave-particle interactions are a much more difficult subject. The most definitive, observationally-based residence times are those for ~ 2 MeV electrons from high-altitude nuclear bomb bursts (Figure 7). Otherwise, estimates are based on calculated diffusion coefficients, on the observed intensity of plasma-wave activity, and on the episodic precipitation of previously trapped particles.

6 Temporal variations of trapped particle intensities at the Earth

The intensities of energetic particles in the strong magnetic field in the inner radiation belt are reasonably constant over time, though at low altitudes there is an 11-year solar-activity cycle in intensity, as the scale height of the atmosphere varies, the latter quantity being greater during maximum solar activity and lesser during minimum solar activity. There is the inverse effect on particle lifetimes and intensities.

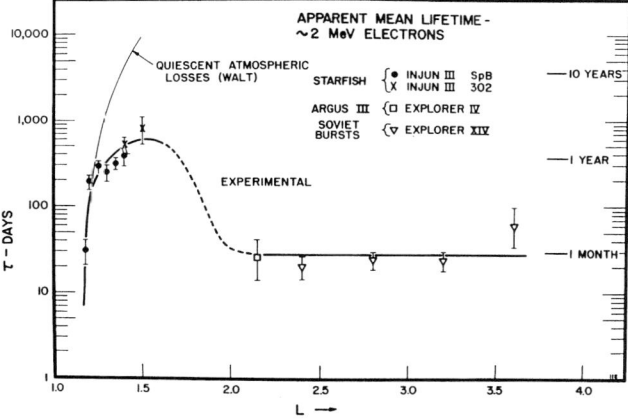

Figure 7 The radial dependence of observed values of the apparent mean lifetime of ≈ 2 MeV electrons from two U.S. and two USSR nuclear bomb bursts at high altitudes. The abscissa is McIlwain's L parameter, which is essentially the equatorial crossing radius, in units of $1.0 R_\oplus$, of a magnetic line of force (Van Allen, 1964).

A few great magnetic storms during the past 38 years have made substantial, though transient, perturbations of the inner belt. Also temporary effects of high-altitude bomb bursts have been well studied.

In contrast, the intensities of energetic particles in the outer radiation belt undergo fluctuations by one or more orders of magnitude in association with magnetic storms. Such fluctuations become more prominent as the radial distance from the Earth increases.

No synoptic observations of other planets yet exist, but corresponding expectations are credible.

7 Absolute intensities

Valuable and widely used models of the absolute intensities of trapped particles within the Earth's magnetosphere, based on dozens of original papers, have been developed by Cladis *et al.* (1973) and by Vette (1991).

A few representative values of absolute integral omnidirectional intensities $J_0(>E)$ of trapped protons and electrons on the geomagnetic equator, as drawn from the above voluminous tables, are given in Table 2. Other presentations of such data are in Figures 8–12.

THE MOON

Several early U.S. space probes – Pioneers 1, 2, 3, and 4 – in 1958–59 were intended to strike the Moon. They contributed importantly to knowledge of the particle populations and extent of the Earth's magnetosphere, but no one of them reached the Moon or approached it closely enough to yield significant information on its magnetic field or particle environment. The Soviet Luna 2 impacted the Moon in 1959, but

Table 2 Some sample equatorial values of trapped particle intensities in the Earth's magnetosphere

Earth radii	Particle species	E MeV	$J_0(>E)^*$ $(cm^2 s)^{-1}$
Inner Belt			
1.4	electron	0.04	5.0 E 8
1.4	electron	0.50	1.0 E 8
1.2	proton	50.	2.1 E 3
1.5	proton	50.	1.6 E 4
1.8	proton	4.0	4.2 E 6
1.8	proton	50.	4.0 E 3
Outer Belt			
4.8	electron	0.04	1.0 E 7
4.8	electron	0.5	2.4 E 6
3.0	proton	4.0	4.3 E 4

* The notation $a E b$ for $J_0(>E)$ means $a \times 10^b$.

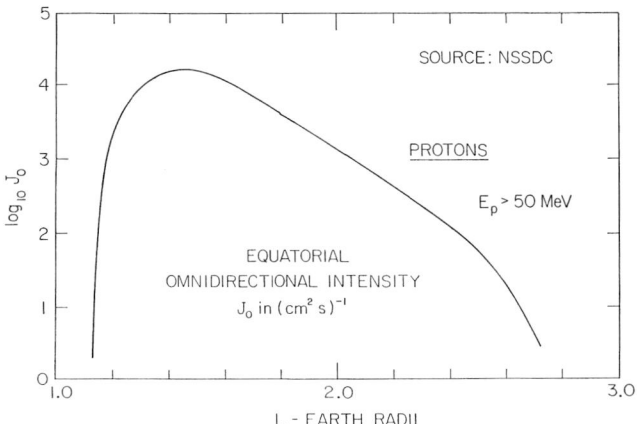

Figure 8 Radial dependence of absolute omnidirectional intensity J_0 in $(cm^2 s)^{-1}$ of protons $E_p > 50$ MeV on the Earth's equator (Vette, 1991).

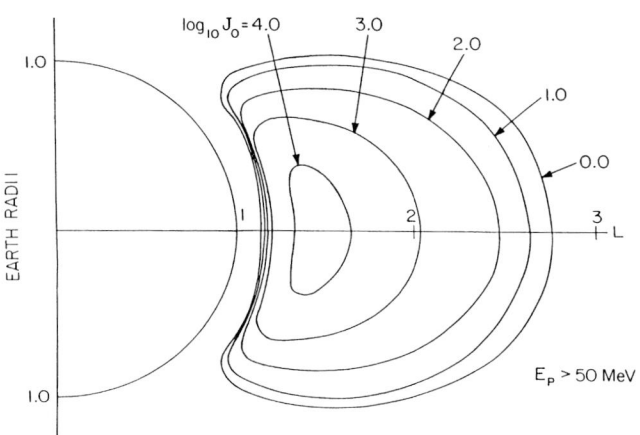

Figure 9 Meridian plane distribution of protons $E_p > 50$ MeV in the Earth's inner magnetosphere (Vette, 1991).

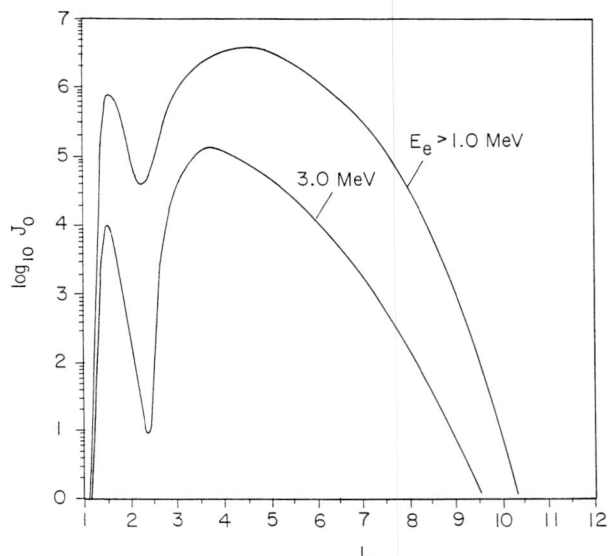

Figure 10 Radial dependence of absolute omnidirectional intensities J_0 in $(cm^2 s)^{-1}$ of electrons $E_e > 1.0$ MeV and $E_e > 3.0$ MeV on the Earth's equator (Vette, 1991).

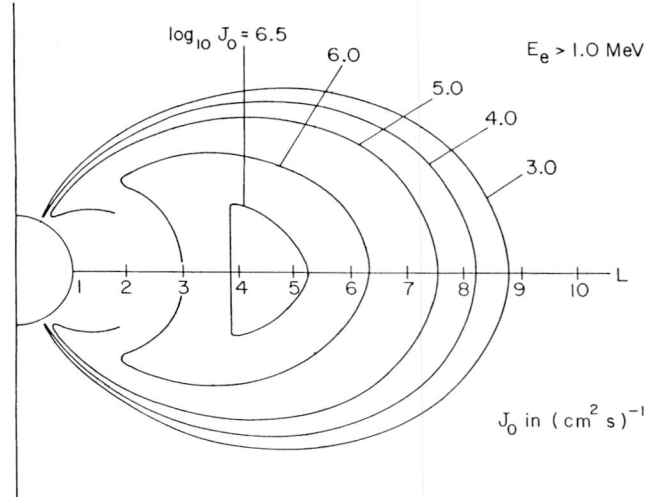

Figure 11 Meridian plane distribution of electrons $E_e > 1.0$ MeV. Note differences of radial scale between Figures 8 and 10 and between Figures 9 and 11 and 12 (Vette, 1991).

its magnetometer registered no magnetic field above the interplanetary level, thereby suggesting an upper limit for the Moon's magnetic moment $M_{\mathbb{C}}$ of $1 \times 10^{-4} M_{\oplus}$.

Explorer 35 was injected into a loose orbit about the Moon in July 1967 with the periapsis at an altitude of 840 km. (Recall that $1.0 R_{\mathbb{C}} = 1{,}738$ km.) Data from the two magnetometers aboard drove down the upper limit substantially. A review of the data suggests that $M_{\mathbb{C}} < 1 \times 10^{20}$ gauss cm^3 or $1.3 \times 10^{-6} M_{\oplus}$. The corresponding upper limit on the equatorial surface field is 1.9 gamma (1 gamma = 1 nanotesla = 1×10^{-5} gauss).

Apollo landers and lunar orbiters have found surface patches with remanent magnetic fields in the range from 6 to 300 gammas. But such patches are small compared to the lunar radius, and the magnitudes and directions of their magnetic vectors appear to be chaotic and do not correspond to a general dipolar field.

The solar wind impacts the sunward face of the Moon (except for periods of a few days each month around the time of full moon when it is in the Earth's magnetotail) and presumably accumulates in a thin ($\approx 1 \times 10^{-5}$ g cm^{-2}) upper layer of surface material. One result is that there is a

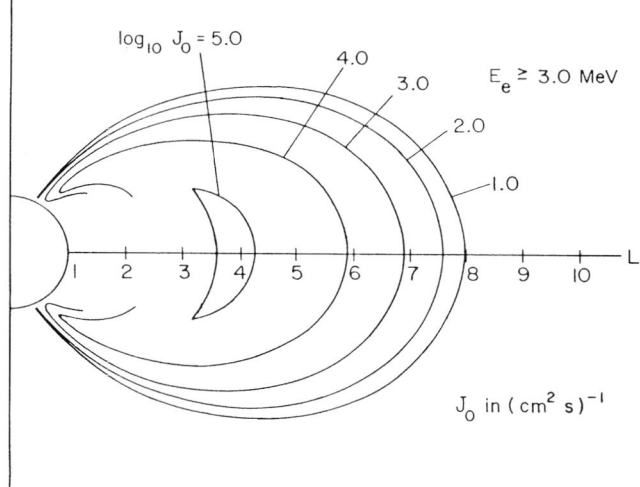

Figure 12 Same as Figure 11, but for electrons $E_e > 3.0$ MeV (Vette, 1991).

cone-shaped void in the solar wind flow on the anti-solar side of the Moon. The half-angle of this cone ($\approx 6°$) is approximately equal to the ratio of the thermal speed of a proton in the solar wind to the bulk speed of the solar wind.

No high-energy particles have been found to be associated with the Moon. Nonetheless, both magnetometers and plasma instruments find that the solar wind induces electrical currents in the solid body of the Moon and in the surrounding medium, and that weak plasma physical phenomena occur both upstream and along the downstream conical wake.

Overall, the Moon has been generally adopted as a well-studied prototype of the interaction of the solar wind with an unmagnetized planetary body.

VENUS

The December 1962 flyby of Venus by Mariner 2 opened the epoch of *in situ* observation of the close environment of other planets. The radius of closest approach (RCA) was 6.77 $R_♀$ (1.0 $R_♀$ = 6052 km) and neither trapped particles nor any magnetic field attributable to the planet was detected. A second and much closer (RCA = 1.68 $R_♀$) trajectory was that of Mariner 5 in October 1967. Again the energetic particle detectors obtained a null result, but both the magnetometer and plasma analyzer observed encounter signatures. By similitude arguments relative to the Earth's magnetosphere, these signatures suggested an upper limit on the magnetic moment $M_♀$ as $1 \times 10^{-3} M_⊕$ and a radial standoff distance of about 1.7 $R_♀$.

The best current upper limit on $M_♀$ is 8.4×10^{20} gauss cm^3 or $1.1 \times 10^{-5} M_⊕$, based on the comprehensive measurements by a magnetometer on the Pioneer Venus Orbiter (Pioneer 12) in an orbit with a periapsis altitude of only a few hundred kilometers.

Nonetheless, many plasma physical phenomena (waves and particle acceleration) are produced by the interaction of the solar wind with the conducting ionosphere of the planet. These phenomena include a strong, well-developed bow shock, an ionopause, and a plasma tail.

The mass and size of Venus are similar to those of the Earth, but its side-real rotational period is 243 days (retrograde), distinctively longer than the periods of all other planets. The latter characteristic is presumably responsible for the absence of a general magnetic field.

See a collection of original papers on Venus (Munch *et al.*, 1980) and later comprehensive reviews in Hunten *et al.* (1983).

MARS

The first spacecraft to encounter Mars was Mariner 4 in July 1965 with an RCA of 3.88 $R_♂$ (1.0 $R_♂$ = 3398 km). Neither the energetic particle instruments nor the magnetometer detected the presence of the planet, though the sensitivity of the instruments was degraded somewhat by an enhanced level of interplanetary activity. Upper limit estimates of $M_♂$ by the Mariner 4 investigators fell in the range 3 to 10×10^{-4} $M_⊕$. A review of the evidence from Mariner 4 and subsequent Soviet missions suggested the value $M_♂ = 1.5$ to 2.2×10^{22} gauss cm^3 ($2.3 \times 10^{-4} M_⊕$) corresponding to an equatorial surface field of ≈ 50 gammas. Measurements on the Phobos orbiting mission cast doubt on the existence of an intrinsic magnetic moment of that order of magnitude. The subject is bewildered by magnetic fields produced by interaction of the solar wind with the (thin) atmosphere and ionosphere of the planet. Hence, the above value, as listed in Table 1, is perhaps best regarded as an upper limit. The observed radial standoff distance of the bow shock is 1.5 $R_♂$. The rotational period of Mars is 24.62 hours.

See review papers in Kieffer *et al.* (1992).

MERCURY

There have been three flybys of Mercury (1.0 $R_☿$ = 2439 km), all by the same spacecraft Mariner 10 – on 29 March 1974, 21 September 1974, and 16 March 1975. The first and third encounters were sufficiently close to yield magnetic and magnetospheric data. Critical reviews of the original work make it appear, in contrast to the essentially null findings for the Moon and Venus, that Mercury does have an intrinsic magnetic moment $M_☿ \approx 4 \times 10^{22}$ gauss cm^3 with its axis approximately parallel to the planet's rotational axis (rotational period 59 days) and the magnetic north pole in the southern hemisphere. Systems of currents on the magnetopause and within the magnetotail, as well as analytical difficulties, place an uncertainty of about a factor of two on the value of the moment.

No durably trapped particles have been identified, but copious quantities of hot plasma ($E_e \lesssim 500\,\mathrm{eV}$) and sporadic bursts of energetic electrons ($E_e > 170\,\mathrm{keV}$) have been detected, suggesting terrestrial-like substorms and energy storage in Mercury's magnetotail. A review of the evidence places Mercury's general magnetic field and its interaction with the solar wind near the low end of the hierarchy of planetary magnetospheres. (See Vilas *et al.*, 1988.)

SUMMARY OF THE TERRESTRIAL PLANETS

Of the four terrestrial planets, Mercury, Venus, Earth, and Mars, only the Earth has a magnetic field of sufficient strength to possess radiation belts and a fully developed magnetosphere, though plasma physical effects are associated with the other three, and with the Moon.

JUPITER

Sporadic bursts of nonthermal radio emission at 22.2 MHz (decametric wavelengths) were first identified with Jupiter in 1955. This subject, which has engaged the interest of many subsequent workers, is reviewed comprehensively in Gehrels (1976). Noteworthy features characterizing studies of the decametric radiation are a precise determination of the physically relevant rotational period of the planet, $9^h\,55^m\,29.7^s$ (System III); identification of the satellite Io as a modulator of the frequency of occurrence of the bursts; and an estimate of the magnetic field in the source region from the upper frequency limit of $f = 40\,\mathrm{MHz}$, interpreted as the gyrofrequency of electrons, i.e., about 14 gauss. The decametric emissions do not appear to require electron energies exceeding a few hundred keV, nor do they suggest the existence of radiation belts.

However, the subsequently discovered nonthermal radio emissions in the decimetric range ($f \gtrsim 200\,\mathrm{MHz}$) are of a quite different nature. See the review by Berge and Gulkis in Gehrels (1976).

(a) The source region is a toroid whose central plane is tilted by about 10° to the planet's equatorial plane. The maximum emissivity occurs at a radial distance of about $1.6\,R_4$.

(b) The radiation is plane-polarized with the electric vector parallel to the central plane of the toroid.

(c) The emission is essentially time-independent in intensity and spectral form.

The accepted interpretation of the decimetric radiation is that it is synchrotron radiation from a huge population of relativistic electrons durably trapped in the planet's dipolar magnetic field. Many authors have proposed models of the energy spectra and distribution of the causative electrons. Such models do not determine these quantities uniquely, lacking knowledge of the magnetic field, but do limit them within a finite region in parametric space. The radio observations do not yield any information on the distributions of protons, subrelativistic electrons, or low-energy plasma.

The observed decimetric radiation from Jupiter, and from no other planet, provided a special motivation for direct observations in its environment.

The Jupiter flybys of Pioneer 10 in 1973 and Pioneer 11 in 1974 opened a new epoch in planetary exploration, namely the *in situ* investigation of the outer planets. The instrumentation on both of these spacecraft emphasized the measurement of the planet's magnetic field and the distribution of energetic particles therein.

Pioneer 10 flew by Jupiter on 4 December 1973 in a prograde trajectory inclined at 14° to the planet's equatorial plane with an RCA of $2.84\,R_4$ ($1.0\,R_4 = 71{,}372\,\mathrm{km}$). Following this successful encounter, the Jupiter flyby trajectory of Pioneer 11 was adjusted so that its subsequent trajectory would lead to Saturn. The resulting encounter with Jupiter on 3 December 1974 was along a retrograde trajectory with an RCA of $1.60\,R_4$ in a plane inclined at 50° to the planet's equatorial plane.

Observations by instruments on the two spacecraft were broadly in agreement and the differences between the two trajectories added importantly to the coverage in latitude and local time. Principal findings included the following:

(a) Jupiter's magnetosphere is huge, with a standoff distance to the bow shock of $7.8 \times 10^6\,\mathrm{km}$ and a transverse dimension of over $2 \times 10^7\,\mathrm{km}$. The latter dimension subtends an angle of 2° to the Earth at times of opposition.

(b) The integral intensity of trapped electrons having energies greater than any specified value is about two orders of magnitude greater than at the Earth (Figures 13 and 14).

(c) The "characteristic" energy of electrons is several times greater than those on the Earth.

(d) The observed distribution of relativistic electrons (Figures 15 and 16) provides a reasonably satisfactory explication of Jupiter's long-known decimetric radiation.

(e) There is a great spun-out magnetodisc, which has no counterpart on the Earth. Observational evidence for this feature is exemplified by Figure 17 and its explanatory caption.

(f) Three of the four Galilean satellites – Io, Europa, Ganymede – and the smaller inner satellite Amalthea strongly reduce the buildup of particle intensities by virtue of their sweeping effects.

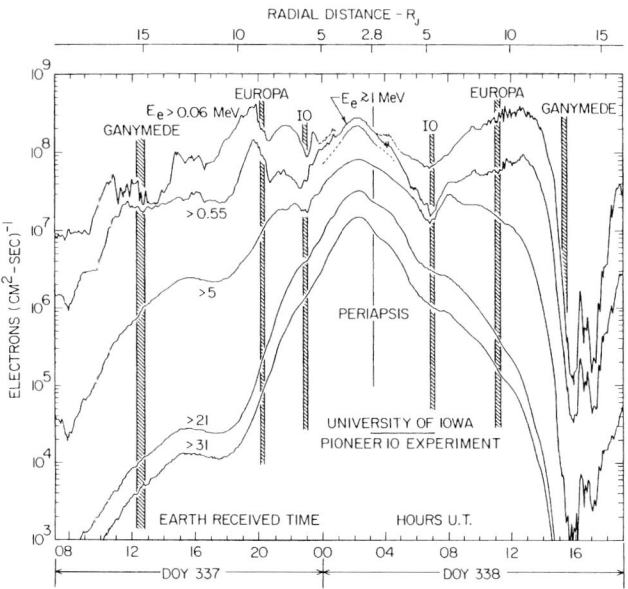

Figure 13 Absolute omnidirectional intensities of electrons in five integral energy ranges as a function of time during Pioneer 10's traversal of Jupiter's inner magnetosphere. The geometrical sweeping regions of the three inner Galilean satellites are shown by vertical shaded bars (Baker and Van Allen, 1977).

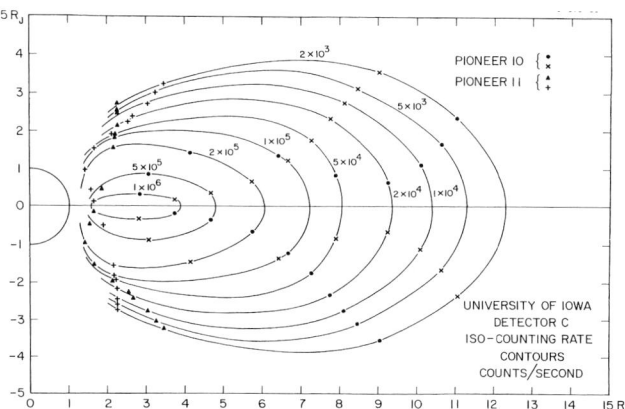

Figure 15 Meridian plane iso-counting rate contours (counts per second) for University of Iowa energetic electron detector C. Omnidirectional intensities J_0 of electrons $E_e > 21$ MeV in Jupiter's magnetosphere are found by multiplying the counting rate by 23. This figure shows combined observations from the Pioneer 10 and Pioneer 11 encounters based on the use of a centered dipole model with a tilt of 9.5° toward System III (1957.0) longitudes of 230° and 233°, respectively. Circles and triangles are observed points for Pioneer 10 and Pioneer 11, respectively; x's and crosses are corresponding reflections in the magnetic equatorial plane (Gehrels, 1976).

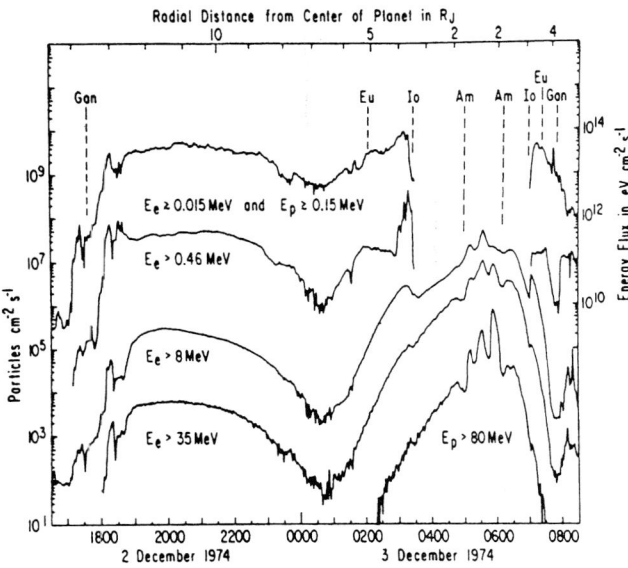

Figure 14 A Pioneer 11 profile of Jupiter's inner radiation belt. The uppermost trace shows the combined energy flux above the specified lower thresholds and below an upper limit of ≈ 0.1 MeV for electrons and several MeV for protons (right-hand scale). The three other traces are referred to the left-hand scale (Fillius and McIlwain, 1974).

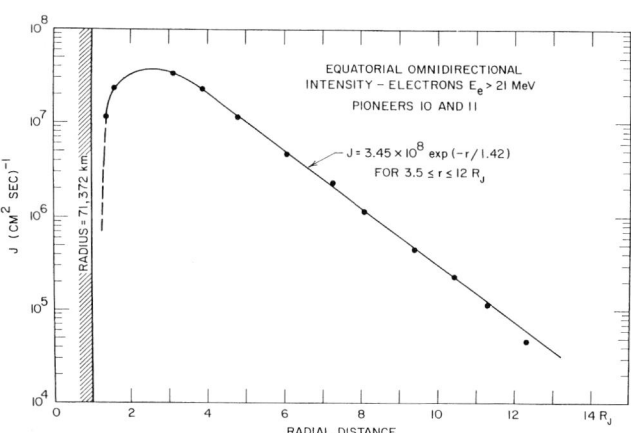

Figure 16 An equatorial profile through Figure 15 (Gehrels, 1976).

(g) High-energy protons ($E_p \gtrsim 80$ MeV) are present in Jupiter's inner magnetosphere (Figure 14).

(h) A marked loss of trapped particles (most clearly protons having $E_p \approx 1$ MeV) at the orbit of Io cannot be attributed to the body of the satellite itself, but must be caused by wave-particle interactions in a much enhanced distribution of plasma associated with this satellite (later called the Io torus).

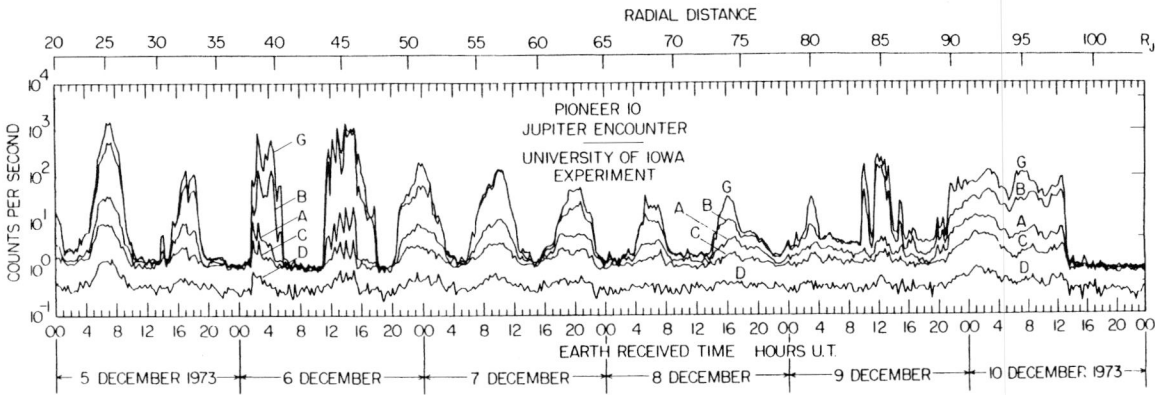

Figure 17 Cyclic variation of the intensities of energetic electrons in five different energy ranges during six days of observation along the outbound leg of the trajectory of Pioneer 10 through the predawn portion of Jupiter's magnetosphere. The modulation pattern is attributed to diurnal wobbling of a tilted magnetodisc past the spacecraft. The nulls in the pattern correspond to regions of open field lines above and below the disc (Gehrels, 1976).

(i) The distribution of energetic particles in the outer magnetosphere is highly time-variable.

(j) The magnetic moment of the planet $M_4 = 1.54 \times 10^{30}$ gauss cm^3 or $1.9 \times 10^4 M_\oplus$. The dipole is offset by about $0.1 R_4$ and tilted at $9.5°$ to the rotational axis, the latter finding being consistent with the radioastronomical evidence cited above. The north magnetic pole is in the northern hemisphere.

(k) The "surface" field strength lies in the range from 4 to 14 gauss.

(l) The planet emits detectable intensities of energetic electrons into interplanetary space at a cyclic rate that is synchronous with its System III rotation.

Three other spacecraft have flown by Jupiter – Voyager 1 (RCA = $4.89 R_4$ on 5 March 1979), Voyager 2 (RCA = $10.11 R_4$ on 9 July 1979), and Ulysses (RCA = $6.31 R_4$ along a retrograde trajectory inclined at $40°$ to the planet's equator on 8 February 1992).

By virtue of their advanced instrumentation for plasma, plasma wave, and imaging observations, these encounters added greatly to knowledge of Jupiter's dynamic magnetosphere, as well as major advances in knowledge of other aspects of the planetary system. In the area of magnetospheric physics, perhaps the most significant additional findings were:

(m) Discovery of volcanic activity on Io and the consequent release of gases that produce the Io plasma torus.

(n) Comprehensive measurements of the distribution and properties of plasmas and the associated plasma waves.

Following its Jupiter encounter, Voyager 2 detected the magnetotail of Jupiter at an anti-solar distance of about 5 AU (7.5×10^8 km).

On the basis of all of the above evidence, it is clear that Jupiter's magnetosphere is one of the grandest phenomena of the solar system.

Extensive reviews are in monographs edited by Gehrels (1976) and by Dessler (1983).

On 7 December 1995, the spacecraft Galileo and its previously released atmospheric entry probe arrived at Jupiter after a six-year flight. The probe transmitted point-by-point data on the temperature, pressure and composition of the planet's atmosphere down to a pressure of 20 bar. The spacecraft obtained data during its traversal of the Io torus and was retarded by a retrorocket so that it was captured into a loose orbit about the planet. During some 11 orbits in the subsequent two years, the planned observations included a detailed study of Jupiter's magnetosphere and the particles and fields phenomena associated with each of the Galilean satellites. The intended data-transmission capabilities of Galileo were drastically reduced by two equipment failures but substantial new findings are still anticipated.

SATURN

Following the successful encounter of Pioneer 10 with Jupiter, the Jupiter-encounter trajectory of Pioneer 11 was adjusted, as noted above, by mid-course maneuvers to one that would lead to a subsequent encounter with Saturn.

In contrast to pre-encounter knowledge of the existence of a radiation belt of relativistic electrons at Jupiter, there was no such prior knowledge of Saturn. From the magnetospheric point of view, Pioneer 11's encounter with Saturn was an event of pure discovery.

Closest approach occurred on 1 September 1979 at an RCA of $1.34 R_\hbar$ ($1.0 R_\hbar = 60,330$ km). The first stage of discovery was the detection of a sequence of three bow-shock crossings near the noon meridian at 24.0, 23.0, and $19.9 R_\hbar$ (average radial distance of 1,345,000 km). These

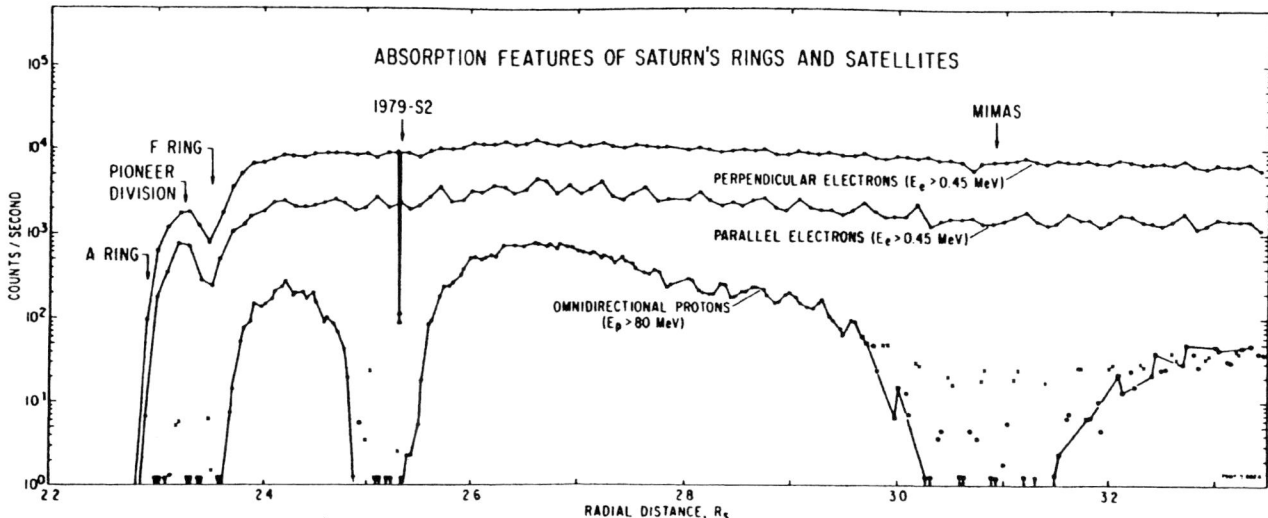

Figure 18 Radical dependences of the specified intensities of electrons $E_e > 0.45$ MeV and protons $E_p > 80$ MeV as observed with Pioneer 11 during its 1979 Saturn encounter (Fillius et al., 1980).

and subsequent observations during the four-day encounter period showed that Saturn's magnetosphere is intermediate between those of the Earth and Jupiter in gross properties: time variability, intensities and spectra of trapped particles, and physical dimensions. The magnetic moment of the planet $\mathbf{M}_{\hbar}$ is (a) 4.6×10^{28} gauss cm^3 or $580 M_{\oplus}$ in magnitude, (b) in the same sense as the planet's angular momentum vector (i.e., opposite to the situation at the Earth), (c) aligned with it to an accuracy of $1°$, and (d) offset northward by about $0.04 R_{\hbar}$. Other distinctive properties of Saturn's magnetosphere based on data from Pioneer 11 and the subsequent encounters of Voyager 1 (RCA = $3.1 R_{\hbar}$ on 12 November 1980) and Voyager 2 (RCA = $2.7 R_{\hbar}$ on 26 August 1981) are related to the absorption of energetic particles by the large inner satellites Rhea, Dione, Tethys, Enceladus, and Mimas and by particulate matter in the system of rings. In the plasma regime, tori of hot ($\approx 10^6$ K) oxygen ions, apparently from satellite out-gassing, are prominent features, as is a comprehensive variety of plasma waves of types previously identified on the Earth. The kilometric radiation from the planet yields an apparent rotational period of $10^h 39^m 24^s$.

As electrons diffuse inward past successive satellites an energy-selection process occurs. Electrons whose longitudinal drift periods are synchronous with the Keplerian orbital period of a satellite diffuse across its orbit as though it were not there. Others collide with the satellite to varying degrees depending on the ratio of the time required to diffuse across the orbital ring occupied by the satellite to the satellite's orbital period. A consequence, unique to Saturn, is that electrons in the inner magnetosphere have a nearly monoenergetic spectrum at about 1 MeV.

As expected, there is a guillotine-like cutoff of the trapped particle population at the outer edge of Ring A at $2.292 R_{\hbar}$. The absence of relativistic electrons in Saturn's strong magnetic field inside this radius precludes a significant level of synchrotron emission and distinguishes Saturn from Jupiter in this important respect.

More generally, the sweeping effects of satellites and rings reduce the population of energetic particles in Saturn's inner magnetosphere by several orders of magnitude below what would otherwise be the case. By the same token, the sweeping effects provide an unusual opportunity for assessing the sources and sinks of particles and for estimating radial diffusion coefficients.

Several bodies of observed data are shown in Figures 18, 19, and 20.

The Crand process is convincingly demonstrated to be the source of inner magnetospheric protons having energies of tens of MeV. Cosmic ray bombardment of the dense rings is the principal source of the requisite neutrons.

Major review papers and a bibliography of original papers through about 1983 are in the monograph edited by Gehrels and Matthews (1984).

The next planned *in situ* investigation of Saturn is in 2004 by the orbiting Cassini spacecraft and its Huygens entry probe for surveying Titan's atmosphere. Launch of this mission is scheduled for 1997.

URANUS

The Saturn flyby trajectory of Voyager 2 was chosen so that its subsequent trajectory led to the first ever flyby

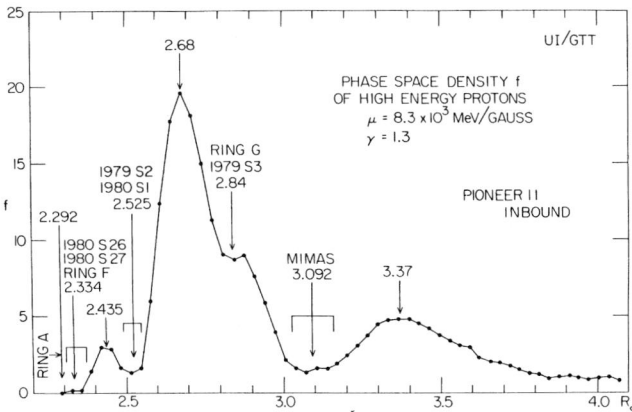

Figure 19 Radical dependence of the phase space density f (in arbitrary units) of high-energy protons in Saturn's magnetosphere at a constant, represesive value of the first adiabatic invariant (adapted from Van Allen et al., 1980a). Absorbing rings and satellites are indicated. (See also Van Allen, 1982.)

of Uranus, on 24 January 1986, in a plane highly inclined to the planet's equatorial plane with RCA = $4.18 R_{c^r}$ ($1.0 R_{c^r} = 25,600$ km).

As was the case before Pioneer 11's flyby of Saturn, the magnetic properties of Uranus and the presence or absence of a well-developed magnetosphere there were matters of conjecture.

It has been long known that the rotational axis of Uranus is tilted at the remarkable angle of 98° to its orbital plane, though the planet's rotational period was not well determined, because of the absence of prominent durable features in its upper atmosphere and the fuzziness of optical images in terrestrial telescopes. Also, there had been no reliable detection of nonthermal radio emission from the planet until they were detected by Voyager 2 a few days before closest approach.

Several authors had suggested that Uranus, if magnetized, would have its magnetic axis approximately aligned with its rotational axis, as is the case for the Earth, Jupiter, Saturn, and possibly Mercury. In such a case, the magnetosphere of Uranus would have a progression of exotic features as the angle between its magnetic axis, and the solar wind flow varied through nearly 360° during each orbital period of 84 years.

At the time of the Voyager 2 encounter, the planet's rotational axis made an angle of 8° to the planet-Sun line. The magnetometer revealed that the physical situation was even more exotic than had been suggested. The eccentric dipole representation of the measured magnetic field is a moment $\mathbf{M}_{c^r}$ of magnitude 3.81×10^{27} gauss cm^3 ($48 M_\oplus$), tilted at an angle of 58.6° to the rotational axis and offset by $0.3 R_{c^r}$.

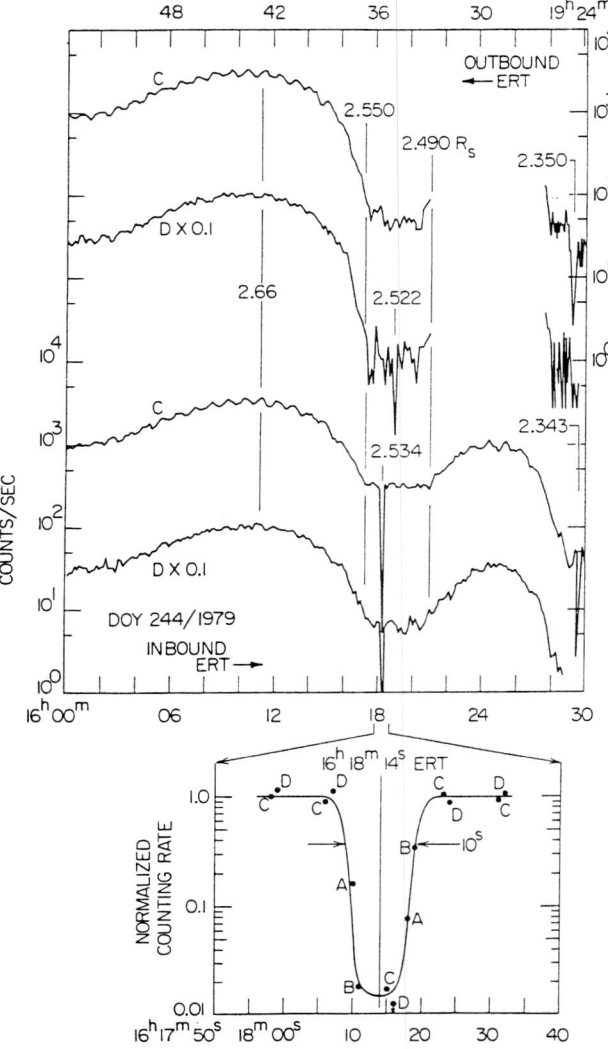

Figure 20 This figure illustrates the discovery of a small satellite in Saturn's inner magnetosphere by its absorption micro-signature of electrons $E_e \approx 1.5$ MeV (Van Allen et al., 1980b). Note also the abrupt cutoff of the particle population at the outer edge of Ring A (right-hand edge of the diagram).

The north magnetic pole is in the planet's northern hemisphere, as defined by the positive sense of its angular momentum vector (departing from the conventional astronomical definition that a planet's geographic north pole is the one on the north side of its orbital plane.)

Corresponding to its state of magnetization, a well-developed magnetosphere is found inside of $18 R_{c^r}$. As is the case for Saturn, the rings and small inner satellites of Uranus have a profound effect in limiting the population of trapped energetic particles. But the situation around Uranus is much more complex, because of the large offset of its magnetic center and, more importantly, the diurnal rocking of the

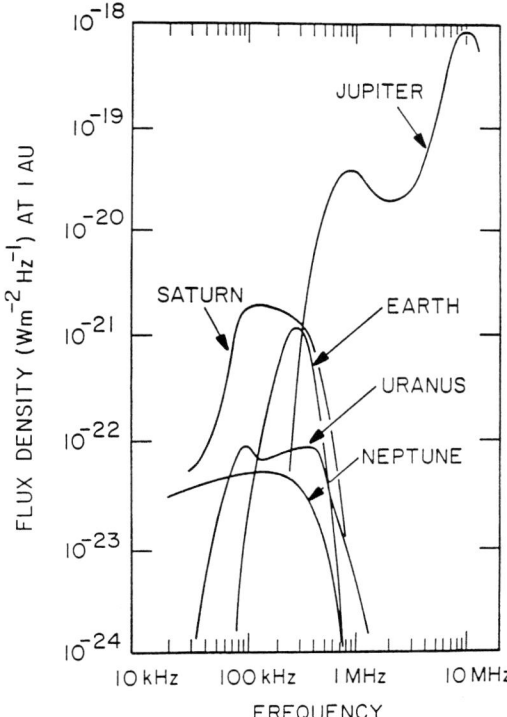

Figure 21 A simplified comparison of the absolute spectra of sporadic radio emissions (all normalized to a distance of 1.0 AU) of the Earth, Jupiter, Saturn, Uranus, and Neptune. The full situations are individually distinctive and far more complex (Gurnett, 1995).

strongly tilted magnetic equator with respect to the rotational equator within which lie the orbits of the rings and satellites. Both the plasma and energetic particle instruments agree in finding no ions heavier than protons. This evidence indicates the dominance of an atmospheric source over a solar wind source. The radio emissions of Uranus are compared with those of the Earth, Saturn, Jupiter, and Neptune in Figure 21.

Periodic variation of the low-frequency radio noise yields a rotational period of 17.24 ± 0.01 hours, the first accurate determination.

Most of the relevant original papers on Voyager 2's Uranus encounter are in *Science* (**233**, 39–109, 1986) and the *Journal of Geophysical Research* (**92**, 14,873–15,375, 1987). Subsequent reviews are in the monograph edited by Bergstralh *et al.* (1991).

NEPTUNE

The next and last planetary target of Voyager 2 was Neptune. Again, as in the cases of Venus, Mars, Mercury, Saturn, and Uranus, there was no prior knowledge of the magnetic and magnetospheric properties of Neptune before the spacecraft passed by the planet.

The encounter trajectory of Voyager 2 with Neptune was in a plane highly inclined to its equatorial plane with an RCA of $1.18 R_\Psi$ ($1.0 R_\Psi = 24,765$ km) on 25 August 1989.

A well-defined bow shock was crossed at $34.9 R_\Psi$ and a poorly defined magnetopause between 26.5 and $23.0 R_\Psi$.

An eccentric dipole model of Neptune's magnetic field as derived from measurements of **B** along the encounter trajectory is as follows: $\mathbf{M}_\Psi$ of magnitude 2.02×10^{27} gauss cm^3 ($26 M_\oplus$), tilted at 47° to the planet's rotational axis and offset from the planet's center by $0.55 R_\Psi$. The north magnetic pole is in the planet's northern hemisphere. (The planet's rotational equator is tilted by 29° to its orbital plane.)

The populations of durably trapped energetic particles are profoundly affected and limited by (a) the large satellite Triton (in a retrograde orbit at $14.3 R_\Psi$), (b) the small inner satellites and rings, (c) the diurnal coning of the magnetic axis relative to the planet-Sun line, (d) the tilt of the magnetic equatorial plane to the rotational equator in which the satellites and rings lie, and (e) the offset of the planet's magnetic center. As a result, the intensities of trapped particles are substantially less than those in the Earth's magnetosphere (Figure 22).

One noteworthy finding is that the composition of the population of energetic ions is expressed by the ratios $H^+/H_2^+/He^{++} = 1300/1/0.1$. The presence of H_2^+ and the paucity of He^{++} favor the atmosphere rather than the solar wind as the dominant source of the magnetospheric population. The spectra of both protons and electrons extend upward to over one MeV.

Voyager 2's radio-astronomical observations in the frequency range from 20 to 865 kHz before, during, and after the encounter establish the planet's rotational period as 16.11 ± 0.05 hours, the first really satisfactory determination.

The primary collection of original papers on Neptune is in *Science* (**246**, 1,417–1,501, 1989). Review articles are in Cruikshank (1995).

PLUTO, ASTEROIDS, AND COMETS

The only planet of the solar system that has not yet been visited by a spacecraft is Pluto. Because of its small size ($1.0 R_P = 1142$ km), low mean density (≈ 2 g cm^{-3}), and slow rotation (period 6.38 days), it seems highly unlikely that it is a magnetized body. Also it seems highly unlikely that any asteroid is magnetized, at least by internal systems of electrical currents. But it is conceivable that a high-density asteroid (none identified yet) might have significant remanent magnetization (cf. Kivelson *et al.*, 1995). Otherwise, the interaction of the solar wind with Pluto (and its satellite Charon) and with asteroids is expected to be Moon-like in nature.

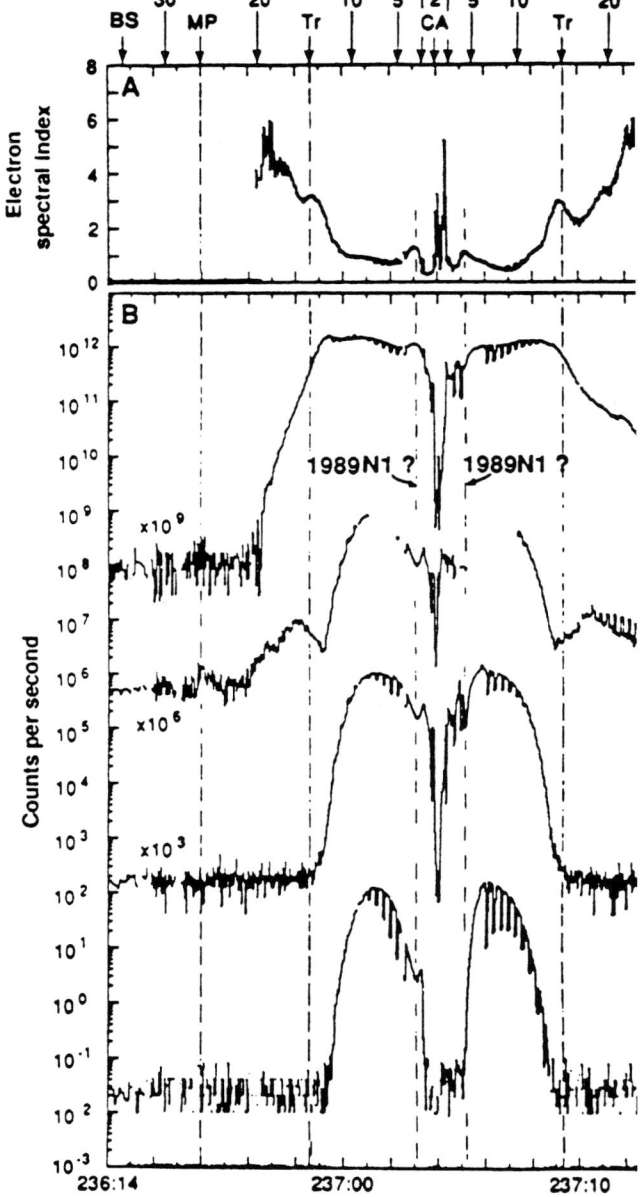

Figure 22 Observed electron intensities during Voyager 2's traversal of Neptune's magnetosphere (Krimigis *et al.*, 1989).

Table 3 Basic sources of energy

(a) Kinetic energy of the flowing plasma constituting the solar wind (or interstellar wind)
(b) Rotational energy of the planet
(c) Orbital energy of satellites

Table 4 Potential sources of energetic particles

(a) The solar wind
(b) Solar energetic particles
(c) Primary cosmic rays
(d) Secondary particles from cosmic-ray interactions in the planet's atmosphere, rings, and satellites
(e) Ionized gas from the planet's ionosphere
(f) Gas sputtered from rings and satellites by particle and photon bombardment
(g) Gas emitted volcanically or out-gassed from rings and satellites

Table 5 Losses of particles and energy

(a) Radial diffusion
  – inward to atmosphere
  – outward to space
(b) Plasma instabilities and pitch angle scattering
  – whistler mode (electrons)
  – ion cyclotron mode (ions)
(c) Collision with satellites and particulate matter in rings
(d) Charge exchange with ambient gas
(e) Gross perturbations of the weak outer magnetic field

Comets are quite different objects by virtue of their emissions of gas, photo-ionization thereof, and collection of the resulting ions by the magnetized solar wind.

CONCLUDING COMMENTS

By virtue of the work that has been sketched above, magnetospheres deserve a place in mainstream planetary astronomy. The fundamental elements of magnetospheric physics have been identified and are now reasonably well understood (Appendices A, B, C, D, and E). But each planet has distinctively different properties and many details remain worthy of fuller investigation and explication, as attested by the contemporary work of hundreds of active investigators and by a variety of current and forthcoming space missions, including active experiments.

A valuable guide to the future is the proceedings of the 1989 Crafoord Symposium of the Royal Swedish Academy of Sciences edited by Hultqvist and Fälthammar (1990).

I may also mention my Nansen Memorial Lecture to the Norwegian Academy of Science and Letters (1990). The following is an excerpt from that lecture.

Very simple questions in magnetospheric physics have proven to be astonishingly difficult to answer. Among these questions are the following:
Where do the observed energetic particles come from?
What is the basic source of energization of the particles and what are the mechanisms for their acceleration to observed energies?
Where do they go?

Is there a natural equilibrium in the population of energetic particles?

What causes the observed distribution of particles within a planetary magnetosphere?

What are the details of the production of aurorae?

What is the physical coupling of magnetospheric phenomena to the solar wind, on the one hand, and to the ionosphere, on the other?

Despite the prevalence of certain basic characteristics, each magnetosphere exhibits a rich variety of distinctive features. The possibility for such distinctiveness rests on the individualized importance of specific processes. "Shopping lists" of broad possibilities are given in Tables 3, 4, and 5. The challenge to planning and interpreting magnetospheric observations lies in assessing the relative importance of these miscellaneous processes in each case. Progress on such matters is substantial but far from definitive. Finally, I may remark that the clarification of natural processes in the Earth's magnetosphere is being greatly enhanced by the conduct of controlled, artificial experiments. These will be of increasing significance in future work.

APPENDIX A

Störmer theory of trapping of an electrically charged particle in a dipolar magnetic field

In this appendix it is assumed that an electrically charged particle moves in a vacuum in the static magnetic field of a point dipole of moment $\mathbf{M}$ and that it has no interaction with any other particle, with any material body, or with any gravitational or electric field. The only force acting on a particle moving in a static magnetic field is the Lorentz force

$$\mathbf{F} = \frac{Ze}{c}(\mathbf{v} \times \mathbf{B}), \tag{A1}$$

where Ze is the particle's electrical charge (in e.s.u.) and may be either positive or negative, $\mathbf{v}$ is its instantaneous velocity vector, $\mathbf{B}$ is the magnetic field intensity at its instantaneous position, and c is the speed of light.

The corresponding differential equation of motion is

$$\frac{d(\mathbf{p}c/Ze)}{dt} = \mathbf{v} \times \mathbf{B}, \tag{A2}$$

where $\mathbf{p}$ is the particle's vector momentum. In (A2), the scalar quantity pc/Ze is called the magnetic rigidity $\mathcal{R}$ of the particle measured in units of electrical potential (ergs per unit charge), B is in gauss and v is in $cm\,s^{-1}$. The value of $\mathcal{R}$ in volts is found by multiplying its value in c.g.s. units by 300.

Inasmuch as the Lorentz force is orthogonal to $\mathbf{v}$ at every point, only the direction of motion is changed and the value of $\mathcal{R}$ and the scalar magnitudes of $\mathbf{v}$ and $\mathbf{p}$ are constants of the motion.

The differential geometric equation of the spatial trajectory of the particle is

$$\frac{\mathcal{R}}{B}\frac{d\hat{\mathbf{v}}}{ds} = \hat{\mathbf{v}} \times \hat{\mathbf{B}}, \tag{A3}$$

where $\hat{\mathbf{v}}$ and $\hat{\mathbf{B}}$ are unit vectors parallel to $\mathbf{v}$ and $\mathbf{B}$, respectively, and s is the arc length measured along the trajectory. By (A3), it is seen that any particle having a given $\mathcal{R}$ (including algebraic sign) and a given $\hat{\mathbf{v}}$ at a given point in the field traces out the same trajectory, though at a rate proportional to its value of v. Also it can be shown from (A3) that the trajectory is reversible, i.e., if $\hat{\mathbf{v}}$ is reversed and the algebraic sign of $\mathcal{R}$ is also reversed, the particle traverses the identical trajectory in the opposite sense. The relationships of $\mathcal{R}$ to other commonly used dynamical quantities for the particle are given by

$$R = \frac{pc}{Ze} = \frac{m_0 c^2}{Ze}\frac{\beta}{\sqrt{1-\beta^2}}$$
$$= \frac{\sqrt{E^2 + 2m_0 c^2 E}}{Ze}, \tag{A4}$$

where E is kinetic energy, m_0 is the rest mass, $m_0 c^2$ is the rest energy, and $\beta = v/c$.

The magnetic field of a point dipole of moment $\mathbf{M}$ is given by

$$\mathbf{B} = -\mathrm{grad}\,\Omega = +\mathrm{grad}\,\frac{M\sin\lambda}{r^2}, \tag{A5}$$

where Ω is the scalar magnetic potential; spherical coordinates, r, λ, and φ (radius, latitude, and longitude) are used; and $\mathbf{M}$ is in the direction of the negative polar axis (for the case of the Earth). In such a field, Eq. (A3) can be integrated numerically to find a trajectory of a particle having a specified value of $\mathcal{R}$ injected at an arbitrary point in the field with an arbitrary $\mathbf{v}$.

However, in a classical contribution to this subject, Störmer (1955) discovered a first integral of (A3) which has been the foundation for most of the subsequent progress in understanding the many classes of trajectories that can occur in a dipolar magnetic field.

One application of this first integral delineates a specific class of trajectories that corresponds to bound, or trapped, motion. For this purpose, the numerical properties of the particle and the field are consolidated in a parameter

$$b \equiv \left(\frac{ZeM}{pc}\right)^{1/2} \tag{A6}$$

whose physical dimension is length, usually called the Störmer unit of length.

The region of bounded motion lies between two surfaces of revolution about the axis of the dipole, namely for

$$r_1 \leq r \leq r_2 \tag{A7}$$

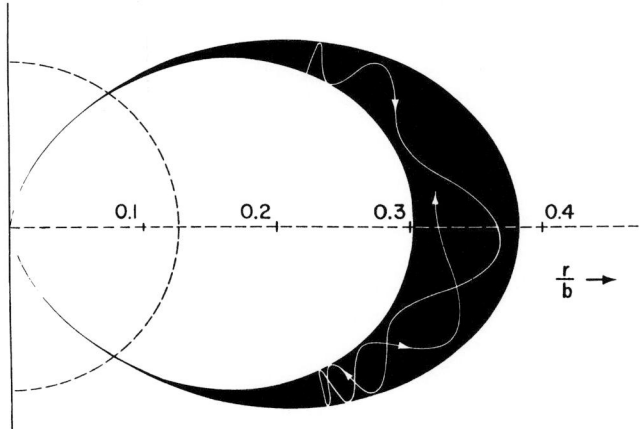

Figure A1 Adaptation of a figure from Störmer (1907) illustrating the meridian-plane projection of the calculated trajectory of a charged particle trapped in a dipolar magnetic field. In this polar plot, r/b is the ratio of the physical radius r to the Störmer unit of length b per Eq. (A6). The dashed semicircle represents the surface of the Earth and the shaded region of bounded motion, $r_1/b \leq r/b \leq r_2/b$, is defined by Eqs. (A7), (A8), and (A9) (Van Allen, 1962).

with

$$\frac{r_1}{b} = \frac{\cos^2 \lambda}{-\frac{\gamma}{b} + \left\{\left(\frac{\gamma}{b}\right)^2 + \cos^3 \lambda\right\}^{1/2}} \quad \text{(A8)}$$

and

$$\frac{r_2}{b} = \frac{\cos^2 \lambda}{-\frac{\gamma}{b} + \left\{\left(\frac{\gamma}{b}\right)^2 - \cos^3 \lambda\right\}^{1/2}}. \quad \text{(A9)}$$

$r_1(\lambda)$ and $r_2(\lambda)$ resemble dipolar lines of force, but are not identical to them, and only approach the dipolar form $r = -(b^2/2\gamma) \cos^2 \lambda$ as $(b/\gamma)^2$ becomes much less than 1.0.

In (A8) and (A9), γ is a negative constant having the dimensions of length and lying between $-\infty$ and -1. Its value is calculable from the injection conditions.

An example of such a trapping region is shown in Figure A1, an adaptation of a figure from Störmer (1907). The curve lying between r_1 and r_2 is the meridian plane projection of a trajectory calculated by one of Störmer's assistants. The spiral motion about a magnetic line of force and the oscillatory motion between "mirror points" in opposite hemispheres is illustrated in this diagram. (See also Appendix B.) The theory also permits calculation of the slower drift of the trajectory in longitude (i.e., into or out of the plane of the diagram), so that the trapping region becomes a figure of revolution about the dipolar axis.

APPENDIX B

The Alfvén adiabatic invariant

Despite the elegance and rigor of the Störmer theory, it provides only a first-order basis for understanding the trapping of particles in the actual geomagnetic field, with all of its departures, both constant and time-variable, from that of an idealized dipole *in vacuo*.

Most discussions of the dynamics of trapped and quasi-trapped particles under actual physical conditions utilize the Alfvén (1950) guiding-center approximation, which may be paraphrased as follows:

For the usual case in which a particle makes many loops whose dimensions are very small compared to those of the magnetic field, each loop is executed in an approximately uniform field. Detailed calculations of the trajectory are both impractical and of relatively little interest. Alfvén proposed to first calculate the motion in a uniform field, then treat the nonuniformity as a perturbation. In a uniform field, the path of a charged particle is a helical one with uniform motion parallel to **B** and circular motion about **B**. The center of the circular motion is called the "guiding center" of the trajectory. Thus the actual helical motion of a particle is reduced to uniform motion of its guiding center parallel to **B** and the spiraling particle is regarded as the equivalent of a current loop or elementary magnetic dipole μ. The scalar value μ is

$$\mu = (\pi \rho^2) \frac{Ze\nu}{c}, \quad \text{(B1)}$$

where ρ is the radius of the loop and ν is the gyrofrequency at which the loop is traversed. With the help of Equation (A1) it follows that

$$\mu = \frac{p_\perp^2 (1 - \beta^2)^{1/2}}{2 m_0 B}, \quad \text{(B2)}$$

with $p_\perp = p \cos \alpha$, the component of **p** perpendicular to **B** with α the pitch angle between **p** and **B**. Equation (B2) gives the relativistically correct value of the "ordinary" magnetic moment of the gyrating particle. In a uniform magnetic field, μ is obviously a constant of the motion.

Alfvén showed that in a nonuniform magnetic field the nonrelativistic value of μ is an adiabatic invariant of the motion if

$$\rho \left| \frac{\text{grad } B}{B} \right| \ll 1. \quad \text{(B3)}$$

Later, Northrop (1963) showed that the relativistically correct invariant is

$$\mu' = \frac{\mu}{(1 - \beta^2)^{1/2}} \quad \text{(B4)}$$

so that

$$\mu' = \frac{p_\perp^2}{2 m_0 B}. \quad \text{(B5)}$$

Inasmuch as p, m_0, and β are rigorous constants, Equation (B2) or (B5) is often written in simplified form

$$\frac{\sin^2 \alpha}{B} = \text{const.} \quad (B6)$$

It can be shown that $\sin^2 \alpha/B$ is a rigorous constant of the motion of a charged particle in the static field of a magnetic unipole, thus adding some physical plausibility to its adiabatic constancy in a dipolar field.

On a given line of force along which the guiding center moves, B has its minimum value B_0 on the magnetic equator ($\lambda = 0°$), and α also has its corresponding minimum value α_0 there. In its latitudinal oscillation (Figure A1), the particle reaches its turning points (or "mirror" points) when $\alpha = \pi/2$ or $3\pi/2$. The value of B at these points is

$$B_m = \frac{B_0}{\sin^2 \alpha_0}. \quad (B7)$$

Additionally, Alfvén showed that the nonrelativistic value of μ is an adiabatic constant in a time-variable magnetic field if

$$\frac{1}{B\nu}\left|\frac{\partial B}{\partial t}\right| \ll 1. \quad (B8)$$

Conditions (B3) and (B8) define the concept of adiabatic invariance and distinguish it from rigorous constancy.

APPENDIX C

Two additional adiabatic invariants and characteristic times in geomagnetic trapping

As discussed in Appendices A and B, the dynamical motion of a trapped particle may be regarded as a composite of three cyclic motions. The three motions are gyration around a magnetic field line, oscillation between mirror points in opposite hemispheres, and drift in longitude. Inasmuch as these three motions have quite different periods, three separate adiabatic invariants can be defined (Northrop, 1963).

A fuller summary of these matters and explicit formulas are given in Van Allen (1962).

A central feature of the guiding-center dynamics of charged particles in a real magnetosphere is the violation of one or more of the adiabatic invariants, whose respective cyclic periods are τ_1, τ_2, and τ_3. For particles of specific species and energy in a specified dipolar field, τ_1 is proportional to L^3, τ_2 is proportional to L, and τ_3 is proportional to L^{-1}, where L is defined in Appendix E. Some illustrative absolute values are listed in Table C1. Note that in each case $\tau_1 \approx \tau_2 \approx \tau_3$.

The power available in the spectrum of fluctuations of magnetic and electric fields near the respective frequencies

Table C1 Sample Values of Cyclic Times τ_1, τ_2, and τ_3 for the Earth's Magnetosphere – on the Geomagnetic Equator at Radial Distance L and for $\alpha_0 = \pi/2$

E MeV	L Earth Radii	τ_1 seconds	τ_2 seconds	τ_3 minutes
Electrons				
0.1	1.5	4.6 E–6	0.17	320
	3.5	5.9 E–5	0.40	137
1.0	1.5	1.14 E–5	0.10	44
	3.5	1.45 E–4	0.23	19
Protons				
1.0	1.5	7.1 E–3	2.05	29
	3.5	9.0 E–2	4.77	13
10.0	1.5	7.2 E–3	0.65	2.9
100.0	1.5	7.9 E–3	0.22	0.31

$1/\tau_1$, $1/\tau_2$, and $1/\tau_3$ determines the rate of violation of the three corresponding invariants. The observed spectrum favors violation of Φ.

APPENDIX D

Liouville's theorem

It should be noted that the theory of Appendices A, B, and C is silent on the subject of particle intensities. That important matter is contained in Liouville's theorem of statistical mechanics. For the purpose of this discussion, the following symbols are adopted:

E	particle kinetic energy
$j_i(E)$	the unidirectional differential intensity of particles of type i having energy in dE at E
r, φ, θ	the geographic polar coordinates of an arbitrary point in the magnetosphere
l, m, n	the direction cosines of the direction in space being considered
t	time.

The raw observational problem is the determination of the function

$$j_i(r, \varphi, \theta, l, m, n, E, t),$$

where i denotes electrons, protons, alpha particles, etc.

A Hamiltonian exists for the motion of an electrically charged particle in any combination of static magnetic and electric fields. Let each particle be represented by a point in six-dimensional momentum/configuration space ("phase space"). The dynamical trajectory of the particle is the trace of its representative point in phase space as time progresses.

Suppose that there is a large swarm of representative points. Nearby points have similar dynamical trajectories. Liouville proved that the density of representative points in phase space is constant along a dynamical trajectory. One consequence of Liouville's theorem is that

$$\frac{j(E)}{p^2} = \text{const}, \tag{D1}$$

along a dynamical trajectory in configuration space. In general, the momentum p may vary, but in a pure static magnetic field p is constant so that

$$j(E) = \text{const}. \tag{D2}$$

Thus, in a quasi-stationary state, $j(\alpha)$ of a given component of the distribution of trapped particles at any point along a given line of force is derivable, in principle, from knowledge of $j(\alpha_0)$ at the equator on that line of force, where α is the pitch angle [=arc cos $(\mathbf{v} \cdot \mathbf{B}/vB)$] at the point in question and α_0 is the equatorial pitch angle. Liouville's theorem is not directly applicable to omnidirectional differential intensity $J = \int_0^{4\pi} j\, dw$, where dw is the element of solid angle, but can be used to derive $J(B)$ along a given line of force from $j(B_0, \alpha_0)$.

APPENDIX E

The B-L coordinate system

Even for a time-stationary state, j_i of Appendix D is a function of three positional and two of the three directional parameters. But there is an enormous reduction in complexity by invoking Liouville's theorem and making two reasonable assumptions, viz.

(a) $j(\alpha) = j(180 - \alpha)$ at any point, where α is the pitch angle of the particle's velocity vector with respect to the local magnetic vector $\mathbf{B}$, and
(b) $j(\alpha)$ is the same on a cone of half-angle α about $\mathbf{B}$.

Further, as noted above, $j(\alpha)$ at any point along the given line of force can be derived from $j(\alpha_0)$, where α_0 is the equatorial pitch angle. Then, if the line of force be labeled by its equatorial crossing distance L (in units of Earth radii), it is clear that

$$j_i(L, \alpha_0)$$

contains all of the information along the magnetic line of force through the point in question.

McIlwain (1961) devised a scheme for transforming the geographic coordinates r, φ, θ to two magnetic shell parameters: B, the scalar value of the geomagnetic field, and L, a magnetic shell parameter for the real magnetic field that is equivalent to the L for a centered dipolar field. The parameters B and L are derived from the spherical harmonic representation of the actual surface geomagnetic field. This coordinate system reduces three-dimensional distributions to two-dimensional ones and is thereby very helpful in assembling observed data into intelligible form. The principal effect of this system is to adjust for the offset and tilt of the dipole. Beyond an L of about 4, more complex models are necessary because of the distortion of the field by the solar wind and by external current systems.

Iso-intensity contours in B-L space can be transformed into a more easily understood polar coordinate system by means of the dipolar relationships

$$r_n = L \cos^2 \lambda_n \tag{E1}$$

and

$$B = \frac{M}{r_n^3}\left(4 - \frac{3r_n}{L}\right)^{1/2}. \tag{E2}$$

In (E1) and (E2), r_n (in Earth radii) and λ_n are called invariant radial distance and invariant latitude, respectively, because McIlwain made use of the first and second adiabatic invariants in deriving the B-L coordinate system. In (E2), B is in gauss and M is in units of gauss $R_\oplus^3$.

The McIlwain procedure has been applied to Jupiter, Saturn, Uranus, and Neptune as well as to the Earth and is almost universally used in organizing radiation belt data for all of these strongly magnetized planets, within the radial range of its applicability.

Explorer I data (Figures 4 and 5) were organized originally by a more primitive but roughly equivalent coordinate system, namely B and magnetic dip latitude λ_{dip} (Yoshida et al., 1960). In that work, B was calculated at each point.

The first of these is the invariant μ' for the gyromotion, as described in Appendix B.

The second is associated with the latitudinal oscillation, first identified by Rosenbluth as applicable to laboratory mirror machines. It is the action integral of the oscillatory motion in latitude, viz.

$$\mathcal{J} = \int_m^{m^*} p_{\parallel} \, ds, \tag{C1}$$

the line integral between mirror point m and its conjugate mirror point m^* in the opposite hemisphere. Inasmuch as p is constant, (C1) can be recast as

$$\mathscr{J} = \frac{\mathcal{J}}{p} = \int_m^{m^*} (1 - B/B_m)^{1/2} \, ds, \tag{C2}$$

upon assuming the constancy of μ'. The quantity $\mathscr{J}$ has the dimension of length, is a property of the magnetic field only, and is assigned to m and m^* as a point function.

The Rosenbluth principle for the constancy of J identifies a unique sequence of segments of magnetic lines of force that constitutes a single-valued three-dimensional surface (or magnetic "shell") on which the guiding center of a trapped particle moves as it oscillates in latitude between specified values of B_m and drifts in longitude. The conservation of J is essential to understanding trapping in the real, irregular geomagnetic field for which the Störmer integral does not exist.

The third of the three invariants is called the flux invariant Φ, the total magnetic flux of **B** through a surface bounded by a magnetic shell whose equatorial crossing radius is r. Thus,

$$\Phi = \frac{2\pi M}{r}. \tag{C3}$$

from the Finch-Leaton spherical harmonic representation of the Earth's magnetic potential, and the magnetic dip latitude λ_{dip} was defined by the dipolar relationship $2 \tan \lambda_{dip} = \tan I$, where the local magnetic dip angle I is the angle between the local vector **B** and the local horizontal.

Acknowledgments

I am indebted to the U.S. Office of Naval Research and to the Ames Research Center of the National Aeronautics and Space Administration for continuing support of my research and of the preparation of this paper.

REFERENCES

Akasofu, S.-I. and Chapman, S. (1972). *Solar-Terrestrial Physics*. Oxford: Clarendon Press.

Alfvén, H. (1950). *Cosmical Electrodynamics*. Oxford: Clarendon Press.

Baker, D.N. and Van Allen, J.A. (1977). Revised Pioneer 10 absolute electron intensities in the inner Jovian magnetosphere. *Journal of Geophysical Research* **82**, 681–683.

Bergstralh, J.T., Miner, E.D. and Matthews, M.S. Eds. (1991). *Uranus*. Tucson: Univ. of Arizona Press.

Chapman, S. and Bartels, J. (1940). *Geomagnetism*. Vols. I and II. Oxford: Clarendon Press.

Cladis, J.B., Davidson, G.T. and Nevkirk, L.L. (1973). *The Trapped Radiation Handbook*, Defense Nuclear Agency, DNA 2524H.

Cruikshank, D.P. (1995). *Neptune and Triton*. Tucson: Univ. of Arizona Press.

Dessler, A.J. Eds. (1983). *Physics of the Jovian Magnetosphere*. Cambridge: Cambridge Univ. Press.

Fillius, R.W. and McIlwain, C.E. (1974). Measurements of the Jovian radiation belts. *Journal of Geophysical Research* **79**, 3 589–3 599.

Fillius, R.W. Ip, W.H. and McIlwain, C.E. (1980). Trapped radiation belts of Saturn: First look. *Science*, **207**, 425–431.

Gehrels, T. Ed. (1976). *Jupiter*. Tucson: Univ. of Arizona Press.

Gehrels, T. and Matthews, M.S. Eds. (1984). *Saturn*. Tucson: Univ. of Arizona Press.

Gurnett, D.A. (1995). Solar system plasma waves, *Radio Science Bull., URSI*, **274**, 4–13.

Hess, W.N. (1968). *The Radiation Belt and Magnetosphere*. Waltham, MA: Blaisdell Publishing Co.

Hultqvist, B. and Fälthammar, C.-G. (1990). *Magnetospheric Physics—Achievements and Prospects*. New York: Plenum Press.

Hunten, D.M., Colin, L., Donahue, T.M. and Moroz, Eds. *Venus*. Tucson: Univ. of Arizona Press.

Journal of Geophysical Research (1987). A compilation of original papers on Uranus, **92**, 14 873–15 375.

Kieffer, H.H., Jakosky, B.M,. Snyder, C.W. and Matthews, M.S., Eds., (1992). *Mars*. Tucson: Univ. of Arizona Press.

Kivelson, M.G., Wang, Z., Joy, S., Khurana, K.K., Polamskey, C., Southwood, D.J. and Walker, R.J. (1995). Solar wind interaction with small bodies: What can Galileo's detection of magnetic rotation tell us about Gaspra and Ida. *Advances in Space Research*, **16**, (4)59–(4)68.

Krimigis, S.M., Armstrong, T.P., Axford, W.I., Bostrom, C.O., Cheng, A. F., Gloeckler, G., Hamilton, D.C., Heath, E.P., Lanzerotti, L.J., Mauk, B.H. and Van Allen, J.A. (1989). Hot plasma and energetic particles in Neptune's magnetosphere. *Science*, **246**, 1 483–1 489.

McIlwain, C.E. (1961). Coordinates for mapping the distribution of magnetically trapped particles. *Journal of Geophysical Research* **66**, 381–391.

Mitra, S.K. (1947). 1952 (2nd ed.) *The Upper Atmosphere*. The Royal Asiatic Society of Bengal, Calcutta.

Munch, T.A. et al. (1980). Original papers on Venus in a special issue of the *Journal of Geophysical Research* **85**, 7 573–8 337.

Northrop, T.G. (1963). *The Adiabatic Motion of Charged Particles*. New York: Interscience Publishers.

Rikitaki, T. (1966). *Electromagnetism and the Earth's Interior*. Amsterdam: Elsevier Publishing Company.

Science (1986). A compilation of original papers on Uranus, **233**, 39–109.

Science (1989). A compilation of original papers on Neptune, **246**, 1 417–1 501.

Storey, L.R.O. (1953). An investigation of whistling atmospherics. *Philos. Trans. R. Soc. London A*, **246**, 113–141.

Störmer, C. (1907). Sur les trajectories des corpuscules électriques dans l'espace sous l'action du magnetism terrestre. *Archives des Sciences Physiques et Naturelles*, **24**, 317–364.

Störmer, C. (1955). *The Polar Aurora*. Oxford: Clarendon Press.

Taylor, J.H. and Stinebring, D.R. (1986). Recent progress in the understanding of pulsars. *Annual Review of Astronomy and Astrophysics*, **24**, 285–327.

Van Allen, J.A. (1962). Dynamics, composition and origin of the geomagnetically trapped corpuscular radiation. *Trans. Int. Astron. Union*, XIB, 99–136.

Van Allen, J.A. (1964). Lifetimes of geomagnetically-trapped electrons of several MeV energy. *Nature*, **203**, 1 006–1 007.

Van Allen, J.A. (1982). Findings on rings and inner satellites of Saturn by Pioneer 11. *Icarus*, **51**, 509–527.

Van Allen, J.A. (1983). *Origins of Magnetospheric Physics*. Washington, D.C.: Smithsonian Institution Press.

Van Allen, J.A. (1990). *The Magnetospheres of Eight Planets and the Moon*. The Norwegian Academy of Science and Letters, Oslo.

Van Allen, J.A., Randall, B.A. and Thomsen, M.F. (1980a). Sources and sinks of energetic electrons and protons in Saturn's magnetosphere. *Journal of Geophysical Research* **85**, 5 679–5 694.

Van Allen, J.A., Thomsen, M.F., Randall, B.A., Rairden, R.L. and Grosskreutz, C.L. (1980b). Saturn's magnetosphere, rings and inner satellites. *Science*, **207**, 415–421.

Vette, J.I. (1991). *The NASA/National Space Science Data Center Trapped Radiation Environment Model Program (1964–1991)*. NASA/Goddard Space Flight Center, Greenbelt, Maryland, and previous reports referenced therein.

Vilas, F., Chapman, C.R. and Matthews, M.S., Eds. (1988). *Mercury*. Tucson: Univ. of Arizona Press.

Yoshida, S., Ludwig, G.H. and Van Allen, J.A. (1960). Distribution of trapped radiation in the geomagnetic field. *Journal of Geophysical Research* **65**, 807–813.

6

REIMAR LÜST*

Barium cloud experiments in the upper atmosphere

1 INTRODUCTION

Space techniques using sounding rockets, satellites and space probes made it possible to send instruments into space not only to measure the physical parameters of the surrounding atmosphere but also to carry out experiments in order to learn about matter and fields in space.

When injecting barium clouds into space, both measurement and experimentation occurs. The barium can be used to trace the movement of atmospheric plasma and thus to measure the electric fields. This is only valid if the artificial plasma cloud does not disturb the surrounding atmosphere too much. By injecting a stronger cloud, it is possible to study the active interaction with the surrounding magnetic field. In this way, one might study interesting general phenomena of a plasma. Experimentation occurs if the pressure of the artificial plasma is much stronger than the pressure of the magnetic field in space.

Experiments with artificial plasma clouds have provided new possibilities for studying the plasma under conditions that cannot be easily set up or may even be impossible to realise in a laboratory.

These experiments are comparable to methods of observing the velocity of a homogeneous fluid. A typical method involves spreading some coloured particles or metallic dust into the fluid. Normally, one uses only very small amounts in order not to disturb the behaviour of the fluid. More than 90 per cent of the cosmic objects are in a plasma state, but are also very dilute and therefore not visible except where concentrated in stars. The cosmic plasma consists mainly of ionized hydrogen and helium molecules which have an extremely small cross section for light-scattering and so, like the even smaller electrons, do not scatter enough light to make their presence visible.

Therefore, it would be interesting to inject into a cosmic plasma a suitable material that with has a cross section large enough for light-scattering to make the motion of cosmic plasma visible. For a plasma with very high electrical conductivity, this is of particular interest, since every motion perpendicular to the magnetic field lines of force can be described as the motion of the magnetic lines force. H. Alfvén used the image of magnetic lines of force frozen the plasma.

2 BIERMANN'S THEORY OF THE INTERACTION OF THE SOLAR WIND WITH THE IONIZED COMETARY TAILS

In 1950 Ludwig Biermann taught a course about comets at the Astronomical Observatory of the University of Göttingen. At that time, I was a PhD student attending these lectures. Biermann was puzzled about why tails with ionized molecules always pointed away from the Sun, while tails consisting of non-ionized molecules and dust were curved toward the Sun? The latter form could be explained by the solar light pressure and the motion of the comets around the Sun.

However, explanations of the light pressure as a force to blow away the ionized tails failed by orders of magnitude. Biermann developed the theory that the corpuscular radiation of the Sun was responsible for the high acceleration observed in the ionized tails. That the Sun sporadically emits a corpuscular radiation was known from the observed perturbation of

---

* Max-Planck-Institut für Meteorologie, Hamburg, Germany

the Earth's magnetic field. Not known at that time was that the Sun emits a corpuscular radiation continuously. This phenomenon was first detected by Russian spacecraft and by the US satellite Explorer X and explained theoretically by Eugene Parker (see Chapter 9).

Biermann published his theory of the interaction of solar corpuscular radiation with the ionized cometary tails in two papers (1951, 1952). In these, he demonstrated the close correlation of events in the tail of comet Whipple-Fedke with the registration of magnetic storms. In a later paper (dedicated to Heisenberg's 50th birthday) he showed that a similar correlation existed for the ionized tails of comet Halley in 1910.

3 PROPOSAL FOR AN ARTIFICIAL COMETARY TAIL

After the successful launches of the first artificial satellites and space probes in the late fifties, I discussed with Biermann at the beginning of 1960 whether German scientists should also get involved actively in space research by using sounding rockets, satellites or space probes and what role the Max Planck Institute for Physics and Astrophysics in Munich could play. In the astrophysical part of the Institute, only theoretical work was then being performed.

During this discussion, the idea of creating an artificial cometary tail in order to understand much better Biermann's theoretical concept of the interaction between the solar wind and the ionized cometary tail was launched.

Of course, using the same molecules observed in a natural cometary tail for such an artificial one would have been most attractive. But the calculation showed that several tons of carbon monoxide (CO) would be needed to create a visible artificial cometary tail (Biermann *et al*. 1961). Therefore other elements or molecules had to be found. In order to keep the cost down and the payload of the sounding rocket as light as possible, it was clear from the outset of the programme that the best energy source for ionizing and exciting the atoms in an artificial cloud was solar radiation. Furthermore, the cloud had to be observable from the surface of the Earth. These conditions led to a number of requirements for suitable elements or molecules: (1) The resonance lines of the ions had to be within the "optical window" of the Earth's atmosphere. (2) The time scales involved in exciting the expected lines of the ions, and photoionizing the neutral atoms had to be sufficiently short, (3) Since a chemical technique for the release was to be used, a low evaporation temperature was highly desirable.

The most promising elements to meet these requirements were some alkaline-earth metals, particularly barium, and probably some of the rare-earth elements, namely europium and ytterbium. We tried strontium and barium, and discovered that visible clouds could be created by using barium. The required quantities were very low, of the order of some 10 to 100 grammes of Ba ions.

4 THE DEVELOPMENT OF BARIUM CLOUD EXPERIMENTS IN THE UPPER ATMOSPHERE

4.1 Development of the technique

In 1961 a small, newly formed group in Garching near Munich – the nucleus of the future MPI for Extraterrestrial Physics – began to develop the necessary technique for an artificial cloud experiment in the Earth's atmosphere with the help of sounding rockets. It was a very lucky coincidence that, during just this period, I met for the first time Jacques Blamont. This was at a meeting at the Royal Society where the first plan for the European Space Research Organisation (ESRO) was discussed. I mentioned to Blamont the idea of creating an artificial cometary tail. He had already used sounding rockets to release neutral sodium (Na) clouds in order to study the upper atmosphere, and invited me to fly a container of barium on one of these Centaure rockets.

Blamont (1983) wrote about the first experiments

> The first operation started badly: Two barium burners (developed by the new group in Garching) were placed on Centaures launched in November 1962 from the French naval base at the Ile du Levant on the Riviera—both rockets failed (the first failures in the whole development history of the rocket...). Two other Centaures were used with complete success in Hammaguir in May 1963. Algeria had then become independent, and the Evian Agreement had authorised the use by the French Government for a further five years, until 1967, of the space complex built in Algeria by the army. The main part of the complex was in the town of Colomb-Béchar and the launch sites were located on a base built ex nihilo 130 km southwest of the town on a flat plateau covered with stones and sand. Our German friends had the responsibility for the burners, which were integrated into the nose cones in Hammaguir, and their photographic equipment was manned by a large team sited at the desert outpost, B-1 Nord', near Colomb-Béchar. I would like to recall the names of this enthusiastic group of fine young engineers, technicians and scientists with whom we spent many tense hours: Gerhard Haerendel, Herbert Bause, Hermann Föppl, Ludwig Heilmeier, Hans Loidl, Friedhelm Melzner, Bernhard Myer and Hans Neuss. It was there (in the bar of the officers' cafeteria!) that I learnt from Prof. Reimar Lüst about the delicate technique of extracting oneself from a sinking submarine.
>
> The results of the May 1963 ejection were decisive: we had to communicate to our German friends the results

given by our spectrographs: their spectacular clouds... contained no barium atoms or barium ions... but useless barium oxides and strong strontium lines (no strontium had been added to the mixture!!...). The principle of these ejections was changed and subsequent firings at the end of 1964 were a complete success. On 22 April 1966, a high-altitude ejection was performed using a Rubis rocket: the cloud which formed at an altitude of 2000 km was also observed from Germany. From then on the barium-cloud technique proved workable and since it has been used extensively in many scientific programmes.

To find an effective way to make barium evaporate, extensive laboratory experiments, as well as theoretical investigations, were carried out. The evaporation of Ba is achieved by a chemical reaction. We tried several different means with the most efficient one being a reaction between atomic Ba and copper oxide (CuO):

$$(n+1)Ba + CuO \rightarrow BaO + Cu + n\,Ba\,(vapor)$$

In this reaction part of the barium is burned and provides the heat necessary for the evaporation of the rest of the barium. An efficiency of about 10 to 20 per cent can finally be achieved: this means that from the volatile barium about 10 to 20 per cent Ba atoms could be observed. The Ba ions are generated by photoionization with a time scale of about 30 seconds. Most effective for this time scale is a metastable energy level of the atomic barium (Föppl *et al.* 1965, 1967).

4.2 The behavior of an artificial plasma cloud

The ionization can be observed not only spectroscopically but also directly with the unaided eye, because the barium cloud changes both colour and shape during the transition phase. The non-ionized cloud radiates in several green, yellow and red lines of the visible spectrum. After the initial, optically thick phase, the radiation in the green (spectrum) is the predominant colour in the neutral cloud. As the cloud becomes fainter because of photoionization, neutral strontium – always present as an impurity – remains. It ionizes much more slowly and radiates in the blue. The ionized barium atoms have spectral lines in the violet, blue and red regions of the spectrum, resulting in a purple colour. Hence the ionized cloud can be easily distinguished from the neutral one because the ionized cloud is purple and the non-ionized one is green and later blue.

A change of shape takes place as well: the neutral cloud is spherical and increases in diameter rather rapidly; the fast expansion is eventually slowed down by collisions of the barium atoms with other atoms and molecules in the Earth's atmosphere; thereafter the neutral cloud increases in size at a much slower rate by diffusion. Meanwhile, the ionized part of the barium cloud undergoes quite different

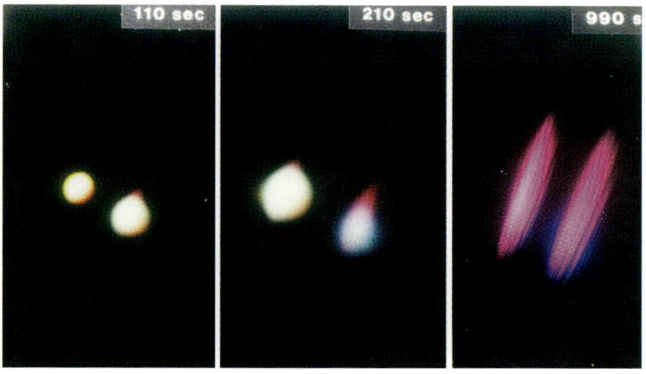

Figure 1 Two barium clouds at an altitude of 260 km. Photoionization by solar UV light is the cause of the changes in color and shape during the first two minutes. The barium clouds are shown at 110, 210 and 990 seconds after release. The neutral clouds are multicolored and spherical and initially show signs of high optical thickness. The ion clouds (purple) become elongated along the magnetic lines of force and show a field-aligned fine structure. (Source: Max-Planck-Institute of Extraterrestrial Physics.)

changes; the positively charged ions and the negatively charged electrons are trapped by the Earth's magnetic field, and they begin to spiral around the lines of magnetic force. For this reason the plasma cloud continues to grow only along the lines of force. The cloud thus becomes cigar-shaped and can be distinguished readily from the spherical non-ionized cloud (see Figure 1). Later, however, considerable distortion of this typical shape can be caused by an inhomogeneous electric field.

4.3 The different sounding-rocket lauchings

During the period (from) 1961 to 1972, the Max Planck Institute for Extraterrestrial Physics participated in 66 sounding-rocket launchings with barium cloud experiments. We used French Centaure, Dragon and Rubis rockets; the English Skylark rocket; the Canadian Black Brant; and the US Javelin, Nike Tomahawk and Nike Apache rockets. Launchings took place as far afield as the Algerian Sahara, at Thumba (India), Sardinia (Italy), Kiruna (Sweden), Andoya (Norway), Fort Churchill (Canada) and Wallops Island (Virginia, USA).

Most of the clouds were released in the ionosphere at altitudes between 150 and 250 km. There were two principal reasons for choosing this range of altitude. The first was that these heights can be reached with small and relatively inexpensive rockets. Second, the motion of the plasma clouds yields information not only about the region of the ionosphere where the clouds are released but also, indirectly, about much higher regions in the magnetosphere.

The experiments were carried out during the twilight period in order to have conditions in which the clouds remained illuminated by the Sun, while the surface of the Earth was dark. The clouds were observed from two or more stations that had to be well separated so that the position of the cloud could be determined by triangulation. The stations were equipped with a variety of cameras, spectrographs and other instruments.

4.4 The drift of the plasma cloud

A magnetic field forces the ions and electrons in a helical path along the lines of magnetic force. It is possible for a particle to encounter two kinds of disturbing forces that will deflect it from this helical path. First, it can collide with another particle. However, in the Earth's atmosphere above 200 km, the frequency of such collisions is very low compared with the frequency of gyration. As a result, the predominant disturbing force is the electric field. In an electric field that is at right angles to the magnetic field, the charged particles of a plasma tend to drift in the direction perpendicular to both fields. An observer moving with the plasma drift velocity will see only the spiralling of the particles around the magnetic lines of force. In other words, in this moving frame of reference, the electric field no longer exists. As mentioned above, Alfvén describes such a case as one in which the lines of force are "frozen into the material". His description is the same as saying that, during its motion, the plasma distributed along certain lines of force or within a flux-tube stays together, as far as motions at right angles to the magnetic field are concerned. Electric fields aperpendicular to the magnetic fields and motions of magnetic field lines are consequently interchangeable notions in many situations.

The observed velocity of a plasma cloud can be expressed in terms of the strength of an electric field: a velocity of 100 m/s perpendicular to a magnetic field of 0.5 gauss (a typical strength for low altitudes) corresponds to an electric field strength of about 5 V/km.

Of course, some care is necessary in interpreting the motion of the plasma cloud in electric fields, since the wind of neutral atoms of the atmosphere might influence the motion of the cloud and, furthermore, the artificial plasma might disturb the surrounding medium by changing the electrical conductivity.

The first effect can be kept small by carrying out the measurements at altitudes where the collision frequency of the barium ions and the neutral particles is small compared with the gyration frequency of the barium ions around the lines of force. At altitudes above 180 to 200 km, this condition is fulfilled and the influence of the neutral wind can be neglected. The perturbation of the cloud in the surrounding medium can be kept small if only small amounts of barium are injected into the atmosphere. Fifteen grammes of barium ions are sufficient for a cloud to be observable for a

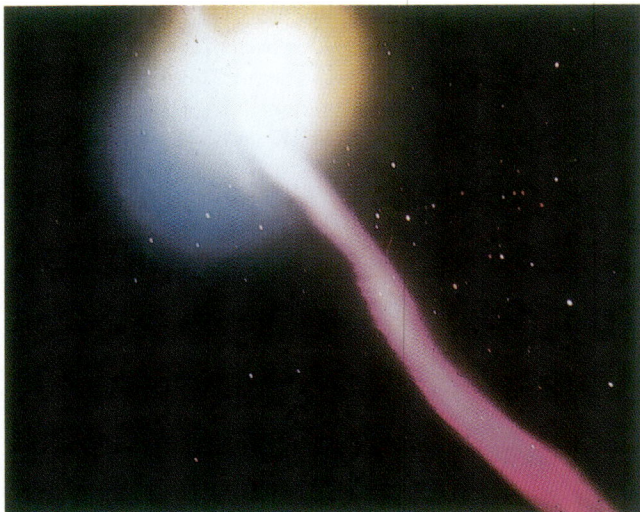

Figure 2 High and inhomogeneous electric fields lead to strong distortions of the ion cloud over hundreds of kilometers. (Source: Max-Planck-Institute of Extraterrestrial Physics.)

reasonable length of time without causing strong perturbations in the atmosphere (Figure 2).

5 MEASUREMENTS IN THE IONOSPHERE

In the following, the results of measurements in the different regions of the ionosphere are summarized (for more details see Haerendel *et al.* 1967, Föppl *et al.* 1968, Lüst 1969 and Haerendel 1987).

5.1 Low and mid-latitudes

To investigate the equatorial electrojet, barium clouds were released over Thumba (India) near the geomagnetic equator. The observed drift motions demonstrated an upward electric field of 1.8 to 5.4 V/km after sunset; the associated horizontal easterly field was weaker by a factor of about 3. These results on the horizontal fields are in agreement with measurements from radar echoes obtained from the incoherent backscatter facility in Jicamarca, although vertical fields could not be measured by this method.

While only a few experiments were done near the geomagnetic equator, quite a number of releases took place at mid-latitudes. In the sixties, all existing low-altitude data on electric fields in this regions were derived from barium cloud experiments.

In analyzing the experimental data one has to consider Ohm's law, which in the ionosphere has the following form:

$$\mathbf{j} = \overleftrightarrow{\sigma} \cdot \left(\mathbf{E} + \frac{1}{c} \mathbf{v}_n \times \mathbf{B} \right)$$

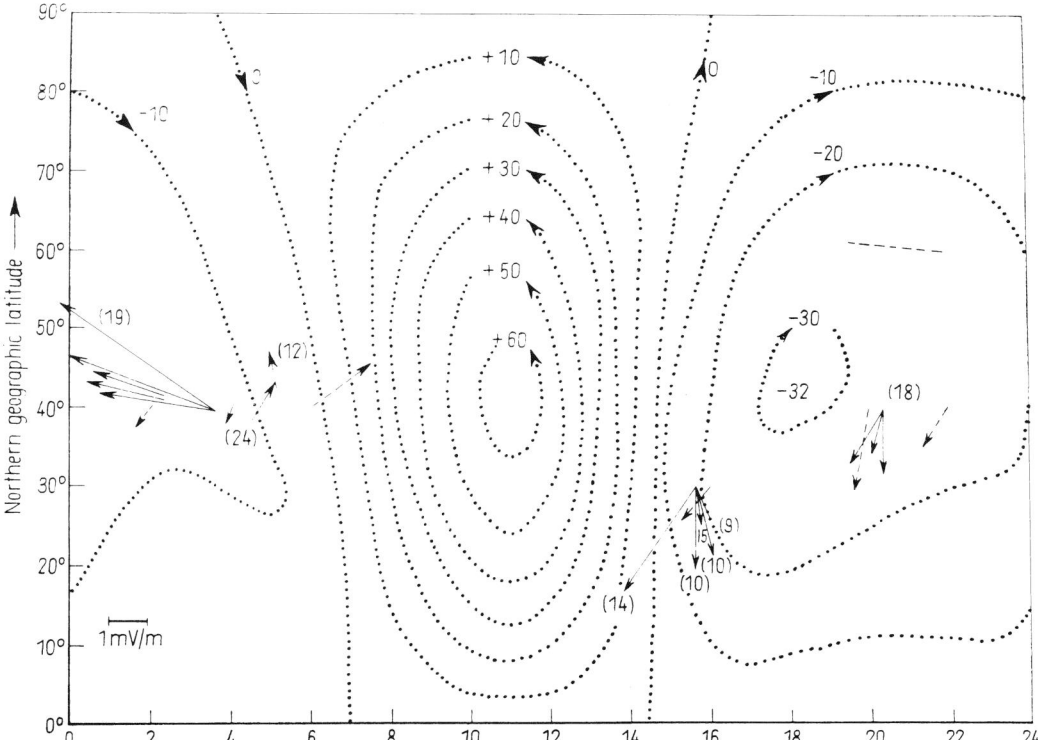

Figure 3 Sq-current pattern, plotted against local time (in hours), for equinoxes according to Chapman and Bartels (between adjacent streamlines [dotted lines] 10,000 amps are flowing), along with theoretical (dashed arrows) and measured (solid arrows) electric fields. (Source: Max-Planck-Institute of Extraterrestrial Physics.)

$\overleftrightarrow{\sigma}$ being the conductivity tensor. The distribution of the conductivities and the neutral air velocity, v_n over the world together with div $\mathbf{j} = 0$ and curl $\mathbf{E} = 0$ determine the distribution of $\mathbf{j}$. Transverse conductivities and currents are confined essentially to the E- and F-layers, but because of the high parallel conductivity, the electric field is projected out into the magnetosphere. Magnetic field lines are approximately electric equipotentials, as long as the magnetic field is sufficiently stiff (low plasma beta).

There are two possibilities for setting up an ionosphere current. Either the dynamo term, $\frac{1}{c}\mathbf{v}_n \times \mathbf{B}$, is driving the current, and $\mathbf{E}$ is adjusted in order to fulfil Ohm's law, or the electric field is impressed from the outer magnetosphere by interaction with the solar wind. If $\mathbf{E}$ and $\mathbf{j}$ are antiparallel, the first case holds.

The reported field measurements clearly support the dynamo theory of the Sq-current system, which, according to Stewart's suggestion, has been well established by many authors. Here Sq stands for 'solar quiet' and means the daily variation of the Earth's magnetic field shown in Figure 3 that is caused by the day-night differences in the ionospheric current system. These currents result from illumination of the dayside ionosphere by the solar UV, which produces excess ionization of the ionosphere, and from the daily tidal and wind motions of the upper atmosphere, which drag the ionization around giving rise to the ionospheric Sq dynamo effect. In certain regions (for instance, at night time and probably in the equatorial electrojet), the dynamo action is very small. Here the electric vector and j should have the same value. The problem is similar to an electric circuit (Figures 3 and 4) consisting of a dynamo with internal resistance, R_i, and external resistance, R_e.

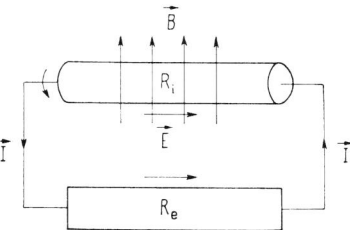

Figure 4 Electric currents as analogue to the current system in the ionosphere. (Source: Max-Planck-Institute of Extraterrestrial Physics.)

Observations of the motion of artificial plasma clouds have demonstrated the existence of such electric fields in the atmosphere, having a strength of one to three V/km at mid-magnetic latitudes. The directions of the electric fields in the twilight zone correspond to the situation inside a dynamo.

Another interesting feature of the barium cloud experiment is the appearance of striations in the clouds aligned with the magnetic field (Figure 5). The striations range in width from half a kilometre to about 10 kilometres. They seem to indicate that the density distribution of the ionized matter in the ionosphere (and probably also in the higher magnetosphere) is not smooth, but rather like a bundle of fibres. The fibres are not constant in space but change within a few minutes. The mechanism for their generation has been explained by Völk and Haerendel (1970).

5.2 High magnetic latitudes

While at mid-latitudes, the magnetic lines of force do not reach far out into the magnetosphere, the situation in the auroral zone is drastically different. Here, the magnetic lines of force meeting the surface of the Earth are linked to the distant part of the magnetosphere and, in particular, to the so-called tail region. The motions set up in the outer magnetosphere by interactions with the solar wind are transported by means of electric fields to lower altitudes. The motions of barium clouds at 200 km altitude, therefore, will also give us information about the state of motion at much higher levels (Figure 6).

Since the twilight period at high latitudes is much longer than at medium latitudes, the drift of the barium clouds can sometimes be observed for several hours.

The electric field component $E_\perp$, *transverse* to the magnetic field, can be determined from the observed *transverse* component of the clouds' velocity $V_\perp$:

$$E = \frac{1}{c} B \times V_\perp$$

Figure 5 The auroral ionosphere is a preferred location for application of the barium cloud technique. The typical striations aligned with the magnetic field become very prominent when the clouds are stretched into long, thin bands and viewed from the side. The striations are modulations of the density profile caused by irregular electric polarization fields. Auroral arcs, although similar in appearance, generally have somewhat different orientations and a dynamic behavior completely different from that of barium clouds. (Source: Max-Planck-Institute of Extraterrestrial Physics.)

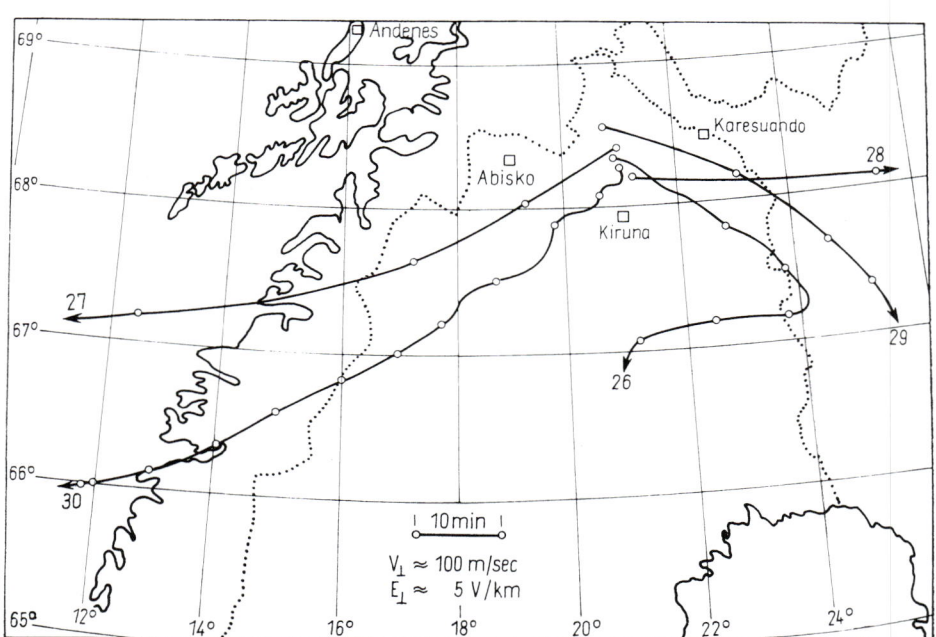

Figure 6 Ground projection of the paths of five ion clouds. The points correspond to fixed time intervals of 10 minutes. The electric field is proportional to the arc length between consecutive points and directed in accordance with the clock-wise trajectory. (Source: Max-Planck-Institute of Extraterrestrial Physics.)

The observed electric field strength is higher in the auroral zone, compared with the medium-latitude region, and may reach values of up to 100 V/km. The parallel component of the electric field is smaller by several orders of magnitude at the experiment heights and cannot be determined by the cloud experiment.

The direction of the fields is generally such that the clouds in the auroral zone drift towards the Sun, which means in the evening to the west and in the morning to the east. Over scales of a few kilometres transverse to the magnetic field, mostly in a north-south direction, or during time intervals of a few minutes, the electric field may change its orientation completely, so that the plasma clouds become severely distorted. We have established a close correlation with magnetic perturbations.

In polar-cap ion-cloud experiments, Wescott *et al.* (1969) showed that the clouds drift in directions opposite to the motion in the auroral zone. The field strength of 15 to 45 V/km is rather smooth there.

Of particular interest are the electric fields in the auroral arcs, since it is not yet clear if the motions of auroral arcs are connected with the motion of magnetic lines of force or with a change in the source of the aurora. Although it has been possible to release several barium clouds so that they intersected the aurora, their behavior was different on different occasions. Some observations showed that, during the time of contact with the aurora, the electric field was considerably depressed. In some cases, the barium clouds moved with the auroral arcs, but on other occasions they did not.

5.3 The magnetosphere

Although it is possible to apply the results about the electric fields in the ionosphere to more distant parts of the magnetosphere by making certain assumptions, it is nevertheless interesting to carry out barium cloud experiments in the distant magnetosphere. As well as the opportunity to measure electric fields, there is the great chance to study the behavior of a very dilute plasma where atomic collisions are no longer of importance.

Two releases of barium clouds into the magnetosphere were carried out by the end of 1971 (Figure 7). In March 1969, it became feasible to release a barium cloud from the ESRO satellite HEOS I at 12.5 Earth radii on the morning side of the magnetosphere. The cloud was observed from many observatories across the whole American continent. The 100 grammes of barium ions that were released formed a cloud with a length of about 5000 km and a width of about 100 km at the end of the observation time-period of 25 minutes (Haerendel and Lüst 1970).

The second release into the magnetosphere was a joint NASA-Max Planck experiment using the NASA Scout vehicle. The release in September 1971 took place at 5.6 Earth radii at a magnetic field line connecting the conjugate

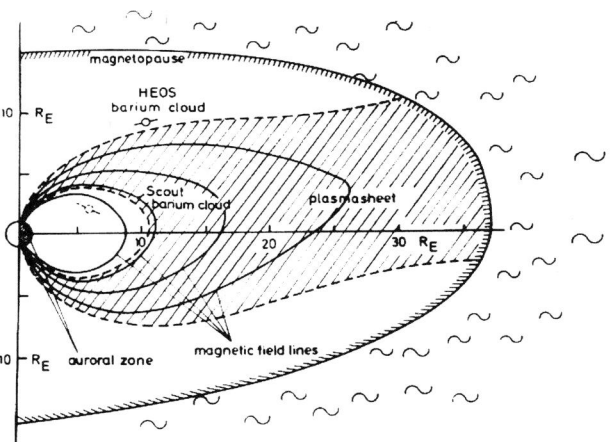

Figure 7 Position of the HEOS and Scout barium cloud. (Source: Max-Planck-Institute of Extraterrestrial Physics.)

magnetic observatories Great Whale (Canada) and Byrd Station (Antarctica), where several geophysical measurements were performed simultaneously. In this release, 1.5 kg of free barium atoms were generated and could be observed for about 1 1/2 hours. Initially, the kinetic pressure of the nearly spherically expanding plasma cloud exceeded by far the magnetic pressure of the ambient field (160γ). Therefore, a magnetic cavity was formed in the central part of the cloud. This was experimentally proven by means of a search coil magnetometer flown on the same rocket. The observed cloud formed striations (Figure 8) due to an instability in the plasma (Brence *et al.* 1973, Haerendel 1973).

6 OUTLOOK

This survey covered the development of a barium cloud technique from 1961 to 1972. The technique was pioneered at the Max Planck Institute of Extraterrestrial Physics in Garching near Munich. But the simplicity of the technique and its wide applicability in studying the dynamics of the near Earth plasma fostered a great many subsequent experiments carried out by about ten groups in Europe and overseas.

My active involvement in the ion-cloud experiments ended at the beginning of 1972, since I left the Max Planck Institute in Garching at that time to become President of the Max Planck Society.

Gerhard Haerendel continued the experiments with plasma clouds with great enthusiasm and new ideas. He had joined the group at the MPI at the very beginning as a PhD student. Since then he has played an essential role in the scientific achievement as well as in the further development of the technique and the necessary logistic of undertaking experiments with it.

A great step forward was made by the application of the shaped-charge technique, which allowed the creation of a

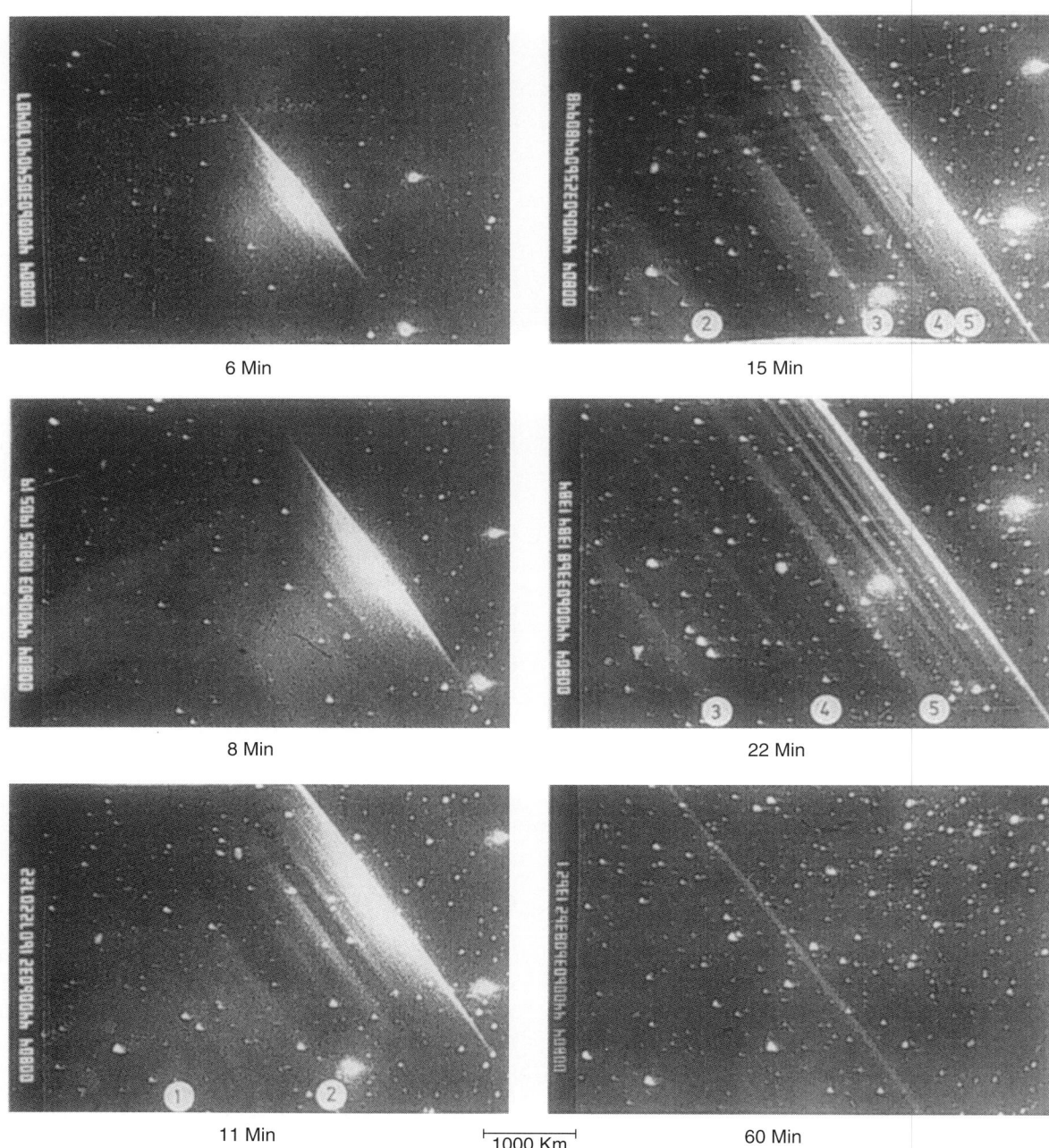

Figure 8 Striations (rays) formed in a barium cloud at $5.6R_E$. (Source: Max-Planck-Institute of Extraterrestrial Physics.)

fast barium jet (up to 14 km/sec.) capable of reaching altitudes of tens of thousand of kilometers and probing the magnetic and electric fields there. This tool was developed independently in England, Germany and the U.S (Haerendel 1987).

Finally Gerhard Haerendel succeeded in achieving the original goal of the ion-cloud experiments: to release an artificial cometary tail outside of the magnetosphere into the solar wind. This remarkable event took place on 27 December 1984 (Haerendel 1983, 1987).

GENERAL REFERENCES

Haerendel, G. and Lüst, R. (1968). Artificial plasma clouds in space. *Scientific American*, **219**, 81–92.

Lüst, R. (1972). Space experiments with barium clouds. *New Scientist*, 154–157.

REFERENCES

Biermann, L. (1951). Kometenschweife und solare Korpuskularstrahlung. *Zeitschrift für Astrophysik*, **29**, 274–286.

Biermann, L. (1952). Über den Schweif des Kometen Halley im Jahre 1910, *Zeitschrift für Naturforschung,* **7a**, 127–136.

Biermann, L., Lüst, R., Lüst, Rh. and Schmidt, H.U. (1961). Zur Untersuchung des interplanetaren Mediums mit Hilfe künstlich eingebrachter Ionenwolken. *Zeitschrift für Astrophysik*, **53**, 226–236.

Blamont, J. (1983). The beginning of space experiments in Munich. In: G. Haerendel and Bruce Battrick (eds.), *Topics in Plasma-, Astro- & Space Physics*, pp. 161–164.

Brence, W.A., Carr, R.E., Gerlach, J.C. and Neuss, Hans (1973). NASA/Max Planck Institute Barium Ion Cloud Project. *Journal of Geophysical Research*, **78**, 5726–5731.

Föppl, H., Haerendel, G., Loidl, J., Lüst, R., Melzner, F., Meyer, B., Neuss, H. and Rieger, E. (1965). Preliminary experiments for the study of the interplanetary medium by the release of metal vapor in the upper atmosphere. *Planetary and Space Science*, **13**, 95–114.

Föppl, H., Haerendel, G., Haser, L., Loidl, J., Lütjens, P., Lüst, R., Melzner, F., Meyer, B., Neuss, H. and Rieger, E. (1967). Artifical strontium and barium clouds in the upper atmosphere. *Planetary and Space Science*, **15**, 357–372.

Föppl, H., Haerendel, G., Haser, L., Lüst, R., Melzner, F., Meyer, B., Neuss, H., Rabben, H.-H., Rieger, E., Stöcker, J., Stoffregen, W. (1968). Preliminary results of electric field measurements in the auroral zone. *Journal of Geophysical Research, Space Physics*, **73**, 21–26.

Haerendel, G., Lüst, R. and Rieger, E. (1967). Motion of artificial ion clouds in the upper atmosphere. *Planetary and Space Science*, **15**, 1–18.[1]

Haerendel, G. and Lüst, R. (1970). Electric fields in the ionosphere and magnetosphere. In: B.M.M. McCormac (ed.), *Electric Fields in the Ionosphere and Magnetosphere*, 213–228.

Haerendel, G. (1973). Results from barium cloud releases in the ionosphere and magnetosphere. In: M.J. Rycroft and S.K. Runcorn (eds.), *Space Research XIII*, **41**, 601–617.

Haerendel, G. (1983). Towards an artificial comet. In: G. Haerendel and Bruce Battrick (eds.), *Topics in Plasma-, Astro-, & Space Physics*, pp. 165–177.

Haerendel, G. (1987). Active plasma experiments. In: S.-I. Akasofu and Y. Kamide (eds.), *The Solar Wind and the Earth*, pp. 215–241.

Lüst, R. (1969). Plasma experiments in space. *Recent Advances in Engineering Science*, **IV**, 39–55.

Völk, H. and Haerendel, G. (1970). Magnetospheric electric fields. In V. Manno and D.E. Page (eds), *Intercorrelated Satellite Observations Related to Solar Events*, D. Reidel Pub. Co., Dordrecht, pp. 280–296.

Wescott, E.M., Stolarik, J.D. and Heppner, J.P. (1969). Electric fields in the vicinity of auroral formes from motions of barium vapor releases. *Journal of Geophysical Research*, **74**, 3469–3487.

---

[1] This is one of the most heavily cited papers in its field, according to data from Science Citation Index.

7

JACQUES BLAMONT*

Alkali metal cloud experiments in the upper atmosphere

Prior to the space age, scientists knew practically nothing about the neutral atmosphere above 30 km. The early sounding rockets greatly advanced our understanding of the physics of this region in a relatively short time. During the period 1954–1970, ejections of sodium atoms by rockets created clouds that remained observable for tens of minutes. Sodium atoms in the atmosphere above 90 km have a relatively long lifetime. They are visible at twilight from optical resonance induced by sunlight at D doublet wavelengths. The first sodium trail experiments gave entirely new insights into the physics, dynamics and structure of the neutral atmosphere from 90 to 400 km. The turbopause was discovered, winds were measured, and quantitative data on eddy diffusion and the first vertical structure of the neutral temperature were obtained.

THE SODIUM TWILIGHT AIRGLOW

Following a suggestion by Otto Struve (Elvey 1950), the word 'airglow' was adopted as a convenient designation for the radiation emitted by the Earth's upper atmosphere, other than that due to aurorae. If it is desired to specify the nocturnal emission alone, 'nightglow' is used. 'Dayglow' and 'twilightglow' are defined analogously.

As discovered independently by Currie and Edwards (1936), Chernaev and Vuks (1937) and Bernard (1938a), the intensity of the nightglow yellow emission line is greatly enhanced at twilight. Bernard (1938b) and, simultaneously, Cabannes *et al.* (1938) identified the emission as

*Centre National d'Etudes Spatiales, Paris, France

the sodium D doublet at $\lambda = 5890$ and $5896\,\text{Å}$. The explanation which immediately suggests itself is that there is resonance scattering of solar radiation by free sodium atoms:

$$\text{Na}\,(^2S) + h\nu\,(\lambda = 5890, 5896\,\text{Å}) \rightarrow \text{Na}\,(^2P) \quad (1)$$

$$\text{Na}\,(^2P) \rightarrow \text{Na}\,(^2S) + h\nu\,(\lambda = 5890, 5896\,\text{Å}) \quad (2)$$

An alternative explanation, advocated by Vegard (1947), was that the excited atoms are released in the photodissociation of some compound of the element.

Bricard and Kastler (1944) studied the absorption of the atmospheric D doublet by a sodium vapour cell kept at various temperatures, and from their results were able to show that the linewidth of the emission is of the order of $10^{-2}\,\text{Å}$, which can be interpreted as indicating a temperature of the emitting atoms in the atmosphere of about $240 \pm 50\,\text{K}$. This favours the resonance scattering theory, since in general the fragments of a molecule that has suffered photodissociation initially have considerable kinetic energy. Any doubt remaining was removed in 1949 when Bricard *et al.* (1949) proved that the $\lambda = 5893\,\text{Å}$ twilight emission is polarized. The extent of the polarization for an observing direction making a right angle with the direction of the Sun was found to be about 9%, as predicted, with resonance scattering assumed. Photodissociation would of course yield no polarization.

The variation is as would be expected. In the case of the morning twilight, for example, the intensity first increases as the Sun approaches the horizon, since the solar radiation can reach more sodium without attenuation in the lower atmosphere; it next reaches a plateau, and finally decreases again (Blamont 1956), this decrease being due to absorption of the solar radiation by sodium on the dayward side (Chamberlain

1956). The behaviour during evening twilight is similar but reversed in sequence.

Sanford (1950) was able to show that the total number of free sodium atoms present at night does not exceed 10^{10} in a column with a 1-cm^2 cross-section that extends from the ground to the top of the atmosphere. Hunten and Shepherd (1954) investigated the distribution. They found that the concentration of free sodium atoms is greatest at 85 ± 3 km, with a maximum around 10^3 cm^{-3}, that below this level it falls off quite rapidly, and that above, up to at least 100 km and probably to 115 km, it falls off with a scale height of 7.5 ± 2 km. Hunten (1956) reported that the seasonal change in this altitude is not more than 1 or 2 km.

The presence of free atoms of sodium near the mesopause level can be explained (Chapman 1939, Bates 1947, Hunten 1954) by an equilibrium between creation processes by reduction of oxides:

$$NaO_2 + O\,(^3P) \rightarrow NaO + O_2 \qquad (3)$$

$$NaO\,(X^2\Pi) + O\,(^3P) \rightarrow Na + O_2 \qquad (4)$$

and destruction processes which are of two kinds, oxidation and ionization. Oxidation occurs through:

$$Na\,(^2S) + O_2\,(X^3\Sigma_g) + M \rightarrow NaO_2 + M \qquad (5)$$

which Bawn and Evans (1937) have shown to be extremely rapid, and through

$$Na\,(^2S) + O_3 \rightarrow NaO + O_2 \qquad (6)$$

Photoionization

$$Na\,(^2S) + h\nu \rightarrow Na^+\,(^1S) + e^- \qquad (7)$$

tends to reduce the number of neutral atoms. It is opposed by several recombination processes, of which

$$Na^+\,(^1S) + O^-\,(^2P) \rightarrow Na + O \qquad (8)$$

is thought to be the most effective in the upper part of the region where sodium can be detected. The rate coefficient for reaction (7) is about 1×10^{-5} s^{-1} (Bates and Seaton 1950), and that for reaction (8) is probably of the order of 10^{-8} cm^3 s^{-1} (Bates and Boyd 1956). Hence, the daytime equilibrium value of the ratio of the concentration of sodium ions to the concentration of sodium atoms is given by

$$\frac{n(Na^+)}{n(Na)} = \frac{10^3}{n(O^-)} \qquad (9)$$

and may thus exceed unity. The two processes concerned are very slow, however, and consequently equilibrium can scarcely be reached.

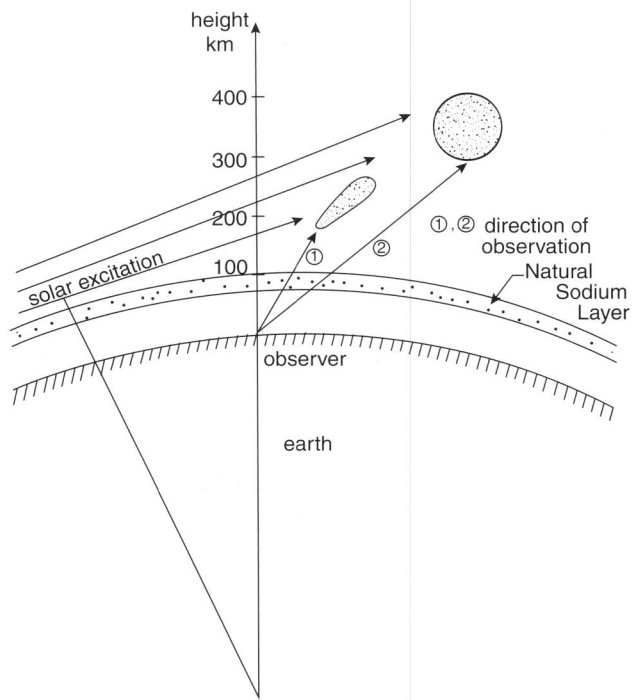

Figure 1 Geometry of a sodium cloud experiment; for example, at Wallops Island, on 13 September 1961, two clouds were formed one around 200 km, the other at 400 km. (From Blamont and Lory 1963b.)

The existence of free atoms of sodium at 85 km, and the proof by Bricard and Kastler that their emission is due to optical resonance excited by solar light, which renders them visible at twilight when the lower atmosphere is in shadow, led D.R. Bates in 1951 to suggest that the ejection of atomic sodium by a rocket in the high atmosphere could create a cloud visible at twilight (Figure 1). The experiment was performed successfully in 1954 by Edwards *et al.* (1956), and a number of Aerobee rockets were devoted to the creation of sodium clouds by the US Air Force Cambridge Research Center (Bedinger *et al.* 1957, 1958; Manring *et al.* 1959). The method was to ignite a mixture of $Al + Fe_2O_3$, called thermite, in which sodium pellets had been dispersed. The oxidation reaction of aluminium by iron oxide is extremely exothermic but stable; the mixture, placed in a container fixed in the rocket nose cone, melts, and sodium vapour is ejected in a steady stream. However, apart from the observation of strong wind shears which quickly deformed the clouds, no scientific results were obtained. The existence of intrinsic atmospheric turbulence was not recognized. Similar results were obtained by Groves (1960) with a Skylark rocket launched on 3 December 1958 at Woomera, Australia.

In 1957, France, having developed and built 15 Véronique rockets to be used for scientific purposes during the

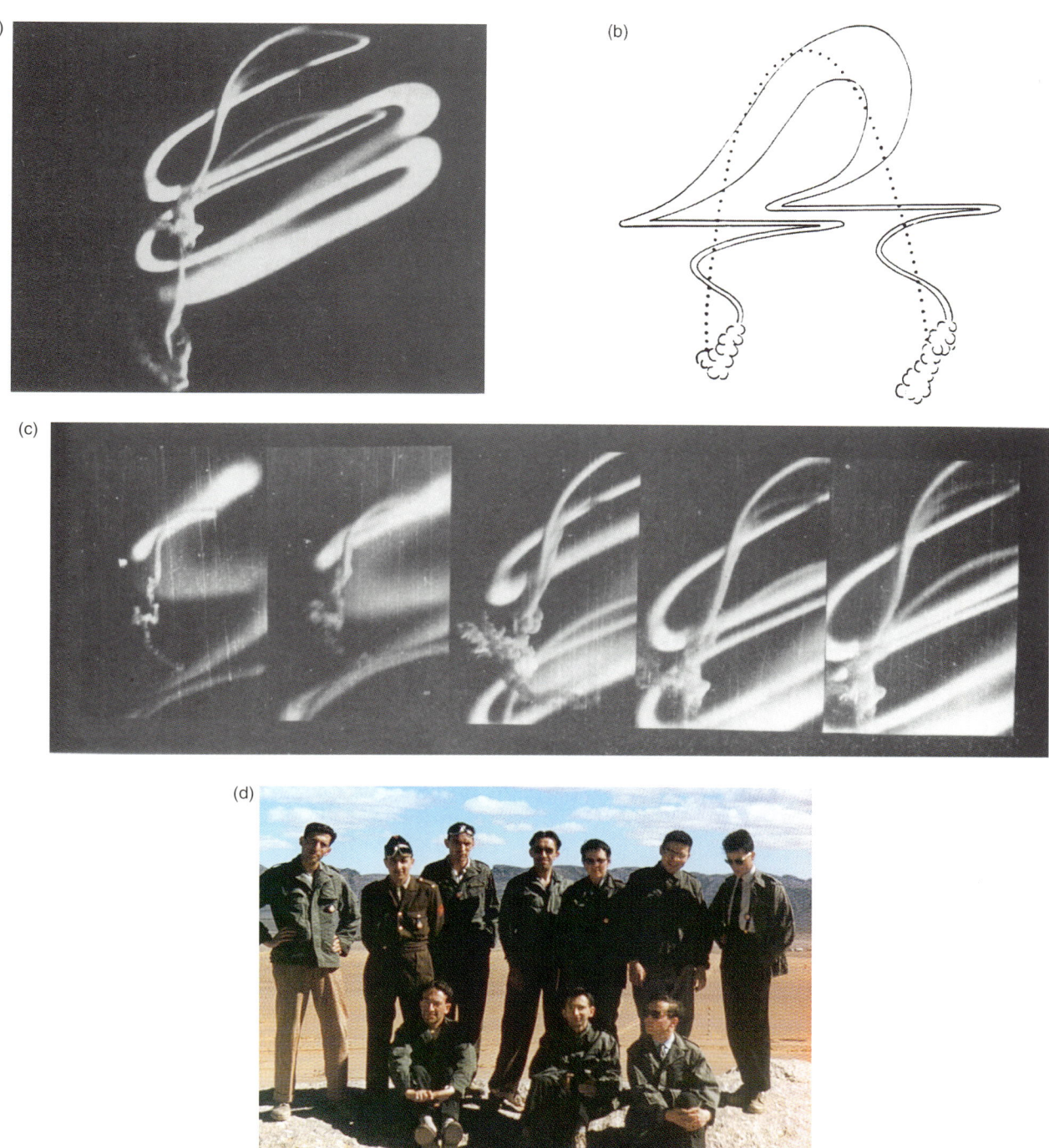

Figure 2 (a) The historic sodium cloud of 10 March 1959. From this picture, obtained at Colomb Bechar, Algeria, the existence of the turbopause was recognized for the first time. In order to understand this image, start at the widest part of the cloud (near the bottom of the picture), which corresponds to the maximum altitude reached by the rocket (130 km). From there, the trail divides into two identical, parallel smooth parts corresponding to the ascent and descent of the rocket. These parts are distorted by the wind. At the lowest altitude (102 km and below) there appears, both at ascent and descent, the sharp separation between the laminar atmosphere (smooth trail) and the turbulent atmosphere (globulous trail). (From Blamont (1966).) (b) Schematic drawing of sodium trail. The dotted line is the path of the rocket. (From Blamont and de Jager 1961.) (c) Evolution in time of the sodium trail obtained on 10 March 1959 at Hammaguir. From left to right: first picture is taken at L (time of launch) + 5 min; second at L + 8 min; third at L + 11 min; fourth at L + 12 min; fifth at L + 15 min. The turbopause already seen in the left-hand picture can be followed through the cloud's development. (From Blamont and Baguette 1961b.) (d) The March 1959 sodium cloud experiment team pictured at Colomb Béchar, Algeria. From left to right: (seated) M. Maguery, F. Roddier and J.Y. Gal; (standing) J. Blamont, C. Cohen-Tannoudji, J.P. Schneider, G. Courtès, M.L. Lory, P. Lena and P. Delache.

International Geophysical Year (IGY), had practically no budget for payloads, no science group, no experimented engineers and no equipment for telemetry, command or tracking. The author proposed to the French IGY Programme Office that a few of the available rockets should be devoted to sodium ejections. The on-board hardware would be extremely simple and cheap, and the rocket was ideally adapted to sodium ejections, since it was designed to carry a 100 kg payload to an altitude of 200 km. The launches would take place on the French Army proving ground of Hammaguir (31°N, 5°W), near Colomb Bechar (in what was then French Algeria) with the help of military resources and personnel.

The scientific objectives would be to study the atmospheric horizontal winds as a function of altitude, and to measure the temperature of the neutral atmosphere as a function of altitude. The observations, all performed on the ground, would be: for dynamical studies, triangulation of the cloud by cameras placed at four sites, each separated from 10 to about 100 km, forming roughly an equilateral triangle; and for temperature measurements, the absorption of the cloud's D line emission by three cells containing sodium vapour at different optical thicknesses.

The first launches took place on 10 and 12 March 1959 from Hammaguir, at dusk and dawn, respectively. On both occasions a sodium cloud was created above 85 km, the first one up to 130 km and the second up to 180 km. They remained visible for 20 minutes (Figure 2). Two major results were obtained from these flights (Blamont 1959, 1960):

1. A spectacular feature of the clouds was an abrupt change in their structure around 102 km: below this altitude, the clouds showed a small-scale motion field consisting of elements with an average diameter of about 0.5 km; above this altitude, the trails displayed a smooth character, not showing the slightest trace of small-scale motions. The region of transition was not thicker than 1 km. This was, as we will see, a major discovery.
2. The temperature of the sodium atoms in the clouds measured with the absorption sodium cell technique were found to be very high (3000 K). This was obviously spurious and due to the multiple scattering of the resonance line inside the cloud: for the first experiments, 10 kg of sodium had been included in the thermite container! It was therefore necessary to reduce greatly the amount of sodium ejected, and also to measure the optical thickness of the cloud in order to make certain that there would be regions in the cloud where the optical thickness would be small enough to allow linewidth measurements.

The potential for scientific purposes of sodium trails generated by rockets had therefore been demonstrated: studies of the dynamical structure of the atmosphere on the one hand, and measurements of the temperature of the neutral atmosphere on the other, could be undertaken. The decision was made by the French IGY Programme Office not only to devote eight Véronique rockets to the ejection of various chemicals, but also a number of the solid-fuel rockets under development, called Centaure (reaching 200 km) and subsequently Dragon (reaching 400 km). The programme was extended to sodium and potassium launches at various latitudes (central Sahara, Argentina, Canada, Sweden, USSR, India), to clouds created by explosives, which were found to exhibit the resonance fluorescence of the oxide AlO, and to barium clouds, in cooperation with R. Lüst's laboratory (Chapter 6). A Panel for Simultaneous Rocket Sounding Launches was created by COSPAR to support the effort. These experiments were carried out during the 1960s.

STUDY OF THE DYNAMICS OF THE ATMOSPHERE BETWEEN 90 AND 200 km

A sodium cloud created by the continuous ejection of vapour by a rocket as it ascends from 80 km to its maximum altitude (around 200 km) and descends is subject to three types of motions (Blamont 1959, Blamont and Barat 1967a):

1. Diffusion of sodium in the medium in a horizontal plane. The diffusion velocity varies inversely with the atmospheric density, and therefore increases with altitude.
2. Horizontal winds with a velocity up to $200\,\mathrm{m\,s^{-1}}$, a randomly distributed direction and horizontal scale above 100 km.
3. Turbulence in the lower part of the cloud (below 102–105 km), which gives it the appearance of a cumulus cloud, and stops abruptly above this altitude.

Diffusion of sodium

The sodium emitted at one point diffuses in two dimensions in a horizontal plane. A determination of the growth rate of the horizontal dimensions of the cloud (i.e. of the diffusion velocity of sodium in the medium) shows the following.

1. On the smooth part of the cloud, above 102 km, measurements show that the law of molecular diffusion is obeyed perfectly. The diffusion of sodium follows the atmospheric diffusion because the molecular (or atomic) masses are nearly identical. Therefore the distribution of sodium atoms at any given time as a function of the distance (r) to the centre of the cloud should be a Gaussian function $\exp(-r^2/L_0^2)$, where L_0 is given by:

$$L_0^2 = 4Dt \qquad (10)$$

where t is the time, D is the atmospheric diffusion coefficient and L_0 can be considered as the width of the cloud. Different authors (Blamont and Baguette 1961a, Manring et al. 1961) have shown that L_0 varies effectively as a function of $t^{1/2}$ and have obtained diffusion coefficients for various altitudes.

2. Below 102 km, the concept of eddy diffusion is needed to explain the rate of growth of the sodium clouds. Small globules originate in many parts of the trail and grow at a rate of $3.3\,\mathrm{m\,s^{-1}}$ (measurements by Blamont and de Jager (1961) on the cloud of 2 March 1960). Similar results have been obtained by Greenhow (1959) using photographs of meteor trails. Figure 3 represents the results of Greenhow and of Blamont and de Jager.

Following Booker (1956) and Booker and Cohen (1956), an element of gas first expands by molecular diffusion according to

$$r^2 = 4Dt \tag{11}$$

as long as it has not travelled a distance equal to the dimension l_0 of the smallest eddies; that is for the time

$$t_1 < l_0 v_0^{-1} \tag{12}$$

where v_0 is the velocity. This remains (Figure 3) the main expansion mechanism for about 30 s. Then, eddy diffusion becomes preponderant with a diffusion coefficient proportional to ϵt^2:

$$r^2 = \frac{4}{3}\epsilon t^3 \tag{13}$$

where ϵ is the energy dissipated as heat per unit mass of fluid, per unit time.

From Figure 3 it can be deduced that $\epsilon = 7 \times 10^{-3}\,\mathrm{W\,kg^{-1}}$. This leads, taking a molecular kinematic viscosity ν of the order of $10\,\mathrm{m^2\,s^{-1}}$ at 95 km, to the following values:

$$l_0 \simeq 30\,\mathrm{m}, \quad t_1 \simeq 30\,\mathrm{s}, \quad v_0 \simeq 1\,\mathrm{m\,s^{-1}} \tag{14}$$

When the trail has reached the dimension L of the large eddies of the field, at a time t_2 after ejection, the relevant coefficient of eddy diffusion is

$$D = vL \tag{15}$$

v and L being, respectively, the velocity and scale of the large-scale turbulence. The law of diffusion then becomes

$$r^2 = 4vLt \tag{16}$$

If we assume that L corresponds to 1 km, the maximum of the measured structure function of the velocity field, the viscosity relevant to this large-scale end of the spectrum is then $2000\,\mathrm{m^2\,s^{-1}}$; this 'eddy viscosity' would then be two hundred times larger than the molecular viscosity.

It can be concluded that the diffusion measurements prove the existence of a transition at around 100 km: above 100 km, molecular diffusion prevails; below 100 km, eddy diffusion dominates.

Hodographs of the horizontal winds

A sodium trail always looks double because the emission continues during ascent and during descent. From the changes with time in the overall shape of the cloud, the horizontal wind vector at each altitude can be deduced. It usually has the same value for the ascending and descending part of the ejection, since the horizontal dimension of a wind system is of the order of hundreds of kilometres. The motions are represented on a diagram, called a hodograph, where the direction and the magnitude of the wind vector are plotted for each altitude.

Data have been published by Kochansky (1964) based on 25 sodium clouds obtained at Wallops Island (38°N, 75.5°W) by E. Manring and A. Bedinger, and by Blamont and Baguette (1961b) based on six clouds obtained by the Service d'Aéronomie du CNRS during the IGY. In the

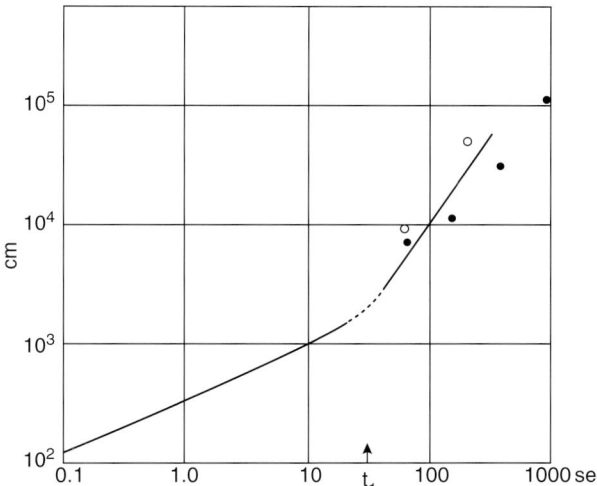

Figure 3 Molecular and eddy diffusion. The solid symbols denote average diameters of meteor trails, as measured by Greenhow (1959); the open symbols the diameters of sodium globules, measured in the trail of 2 March 1960. (After Greenhow 1959 and Blamont and de Jager 1961.)

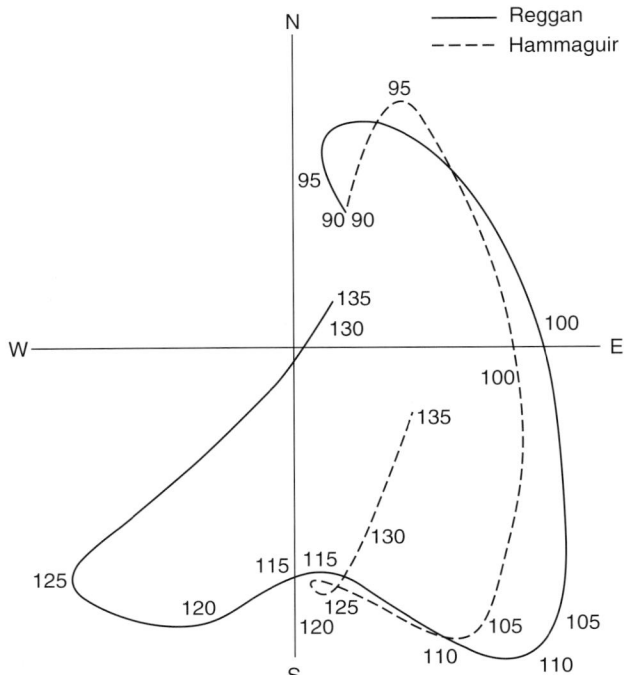

Figure 4 Simultaneous hodographs obtained (Blamont 1966) on 10 December 1961 at Reggan and Hammaguir (separated by 500 km). The numbers indicate the altitude (in km). Scale: 1 cm corresponds to a velocity of $20\,\mathrm{m\,s^{-1}}$.

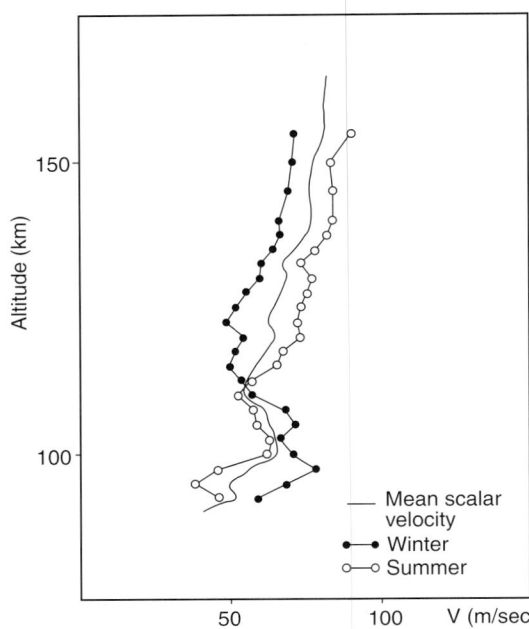

Figure 5 Variation of the scalar velocity as a function of altitude averaged for 31 sodium clouds distributed over various sites. (After Blamont 1966.)

southern hemisphere, data for five clouds obtained at Chamical, Argentina, were analysed by Zaragozza (1965).

In December 1961 data for simultaneous clouds were obtained at Hammaguir (31°N, 5°W) and Reggan (26.7°N, 0°W). In May 1962, May 1963 and November 1964, data for simultaneous clouds were obtained at different launch sites: Île du Levant (43°N, 6°E), Sardinia (39°N, 10°E), Hammaguir and Reggan (Blamont 1996). The data were limited to the altitude range from 95 to 200 km, and to the local times of dusk and dawn. Figure 4 shows typical hodographs.

Below 135 km, the wind vector rotates in the anticyclonic sense when the altitude increases, with usually a maximum between 100 and 110 km. Above 135 km, the wind direction tends to stay in the meridian plane. For 31 clouds studied by Blamont (1966) over sites at different locations, the average velocity gradient is $7.5\,\mathrm{m\,s^{-1}}/10\,\mathrm{km}$ from 100 to 140 km, and $2\,\mathrm{m\,s^{-1}}/10\,\mathrm{km}$ above 140 km (Figure 5).

All the facts point to the existence of some order in the wind structure, which has to be explained in terms of a general circulation, with three major components:

- Short-period oscillations, due to internal gravity waves whose amplitude increases with altitude up to 105 km, with a half wavelength of 3 km at 90 km altitude and 8 km at 125 km. The mean value of the amplitude maximum at 105 km is $90 \pm 30\,\mathrm{m\,s^{-1}}$ at Hammaguir and $48 \pm 23\,\mathrm{m\,s^{-1}}$ at Wallops.
- Atmospheric tides: the observation of the radio echoes of meteor trails has demonstrated the existence of the 12 and 24 h components of planetary tides. The sodium cloud hodographs provide strong evidence for these tides with their anticyclonic rotation with altitude (including the cases observed in the southern hemisphere).
- Seasonal component: when the 'noise' due to gravity and planetary waves is removed, weak zonal winds appear, eastward between 90 and 110 km and westward between 115 and 136 km. We have mentioned the predominance of a meridional component above 135 km.

The large wind gradients that are observed have a major impact on the structure of the ionosphere. In a weakly ionized atmosphere and in the presence of a magnetic field, a neutral horizontal wind can displace the charged particles, which are subject to Laplace forces, along the vertical direction, with an efficiency depending on the angle between the magnetic meridian and the wind vector. In the northern hemisphere a north–south wind will create an ascending motion, and a south–north wind a descending motion. A north–south wind gives a ratio of vertical velocity to horizontal velocity equal to 10^{-3} at 90 km altitude. Vertical wind gradients, as observed with sodium clouds, will therefore create in the lower E region an accumulation or a diminution of the ionization at the points of zero wind. This mechanism explains the occurrence of the sporadic E layer.

The concept of the turbopause

An initial picture

Mass spectrometers mounted on rockets have shown, from measurements of the ratio of argon to nitrogen as a function of altitude, that, if the atmosphere is completely mixed up to 100 km, then above 120 km, on the contrary, the constituents are separated by gravity and their concentration is governed by diffusion equilibria. Between 100 and 120 km there is a transition from complete mixing to diffusion (Meadows and Townsend 1960, Pokhunkov 1961, Istomin and Pokhunkov 1963).

The discontinuity of sodium clouds at 102 km altitude, quantified by the measures of diffusion, led Blamont (1959) to suggest that the sharp transition observed in sodium clouds corresponds to the change from mixing to diffusion. The atmosphere would be turbulent below a certain altitude (around 102 km) and not turbulent above this altitude. This boundary was called the turbopause (probably first by Nicolet) and is considered now to exist on all planets. The sodium clouds, and especially the cloud of 10 March 1959, created a picture of atmospheric turbulence that was very different from the previously accepted one, in which all the atmosphere was considered as turbulent. The clouds demonstrated that the motion field shows a completely different character above and below a sharp transition at around 100 to 105 km. The kinetic energy, stored in big eddies, comes from some organized motion, tidal and gravity waves. It does not pass through the inertial range without dissipation, to be transformed into heat by eddies of size equal to the inner scale, since the laws of isotropic turbulence are not followed.

In the discovery of the turbopause lies the main importance of sodium clouds in the development of space science. The turbopause *does not* coincide with the mesopause.

Evidence for a complex dynamical structure in the 80–110 km region

When pictures are taken of a sodium cloud with a camera of large focal length (600 mm) and immediately after ejection (during the first 30 s), it is found that the simple structure of the field of motion (turbulent and laminar regions separated by a turbopause) is only a very crude first approximation. This configuration can be replaced by the following:

- Below a certain level, around 85 km, where the vertical temperature gradient is negative, the atmosphere is usually turbulent.
- Above a certain level, about 110 km, where the vertical temperature gradient is positive, the atmosphere is in diffusion equilibrium (the field is laminar).
- Between these two limits, in the zone between 85 and 110 km where the vertical gradient is zero or positive, the atmosphere has the vertical structure of a sandwich with alternate regions. At a given altitude the field is turbulent; above or below, at other altitudes, the field is laminar. The vertical span of each region is of the order of a few kilometres. The complete zone between 85 and 110 km can be filled with a turbulent field or with a laminar field, because the vertical location of each region changes with time. The horizontal dimensions of these turbulent regions are much larger (at least one order of magnitude) than the vertical dimensions. This situation is a typical example of turbulent structures in a stably stratified medium, where turbulent and irrotational regions coexist, separated by a sharp dividing boundary.

Figure 6 shows a schematic illustration of the proposed picture. The evidence for this picture comes from the observations of a series of sodium clouds for which a careful analysis of their shape below 110 km has been performed. In all these clouds the sandwich structure is found. The following are three examples from many:

- On 4 November 1964 a sodium cloud was ejected from Hammaguir, Algeria, during evening twilight. The cloud showed the following structure:

  1. From 111 to 104.25 km, the cloud is turbulent. A complete study of all the dynamical characteristics, including a direct measurement of the turbulent energy, has been performed on this region.
  2. From 104.25 to 102 km, the cloud is laminar.
  3. From 102 to 97 km, the cloud is again turbulent.
  4. From 97 to 96 km, the cloud is laminar.
  5. Below 96 km, the cloud is turbulent.

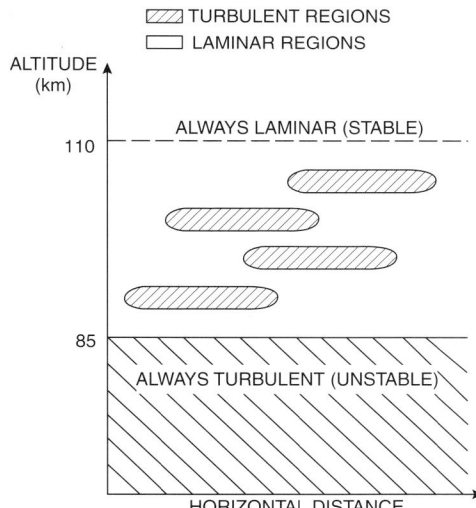

Figure 6 Separation into laminar and turbulent regions as a function of altitude (abscissa and ordinate *not* on the same scale). (From Blamont and Barat 1967a.)

- On 18 September 1965 a sodium cloud was ejected at Surinam during the evening. In the ascending part, the cloud is weakly turbulent between 95 and 90 km and is laminar outside these limits. In the descending part, the cloud is laminar at all altitudes. Thus it is concluded that turbulence, if weak, can differ in different regions situated at the same altitude. Here we are very near the equator and find a situation where the turbulent structure is appearing just marginally.
- On 16 January 1966 a sodium cloud was ejected from Hammaguir, during the evening. The cloud is laminar below 96 km and turbulent from 96 to 102 km.

The phenomenon of a sandwich structure is difficult to observe because the turbulence can be mistaken for a variety of other phenomena, such as irregular wind motions varying quickly with altitude and vertical convective motions of certain parts of the cloud appearing as columns of sodium rising from the body of the cloud.

Origin of the sandwich structure

The best experimental evidence (Blamont and Barat 1967b) available for a discussion of the relation of the turbulent to the laminar region is the hodograph of the Hammaguir cloud of 4 November 1964 (Figure 7). Over the usual anticyclonic rotation is superimposed, between 90 and 110 km, a nearly sinusoidal wave situated in the NE–SW plane. It is easy to separate the wave motion from the slowly varying (anticyclonic rotation) component since, when on the hodograph the velocity gradient as a function of altitude is a maximum (that is, at the nodes of the wave motion), the wave component is zero and the residual velocity is due to the slowly varying component. This component can then be interpolated. Figure 8 shows the projection of the wave on the NE–SW vertical plane and the projection of the slowly varying component. It now appears that:

- the primary motion is a pure sinusoidal wave (dotted line); and
- when the amplitude of this wave increases above a certain limit (which can be related to a critical Reynolds number), the wave is modified by the onset of a turbulence which is generated from the kinetic energy of the wave and disturbs the shape of the wave; the turbulent part of the field can now be obtained by subtracting the real curve from the dotted curve.

These measurements provide a description of the 'sandwich' structure, related to the existence of waves (or organized motions) with a half wavelength of 6–7 km and with amplitude and direction variable with space and time. Since their phase velocity is not horizontal, the altitude of their amplitude maxima will change with time. It is tempting to identify these waves, experimentally observed in sodium clouds, with the internal gravity waves predicted by Hines in 1960, as their observed and corresponding theoretical properties do coincide. Furthermore, wave-induced gradients of air velocities and temperature may lead to local (dynamical) instabilities, thus creating turbulent zones within an

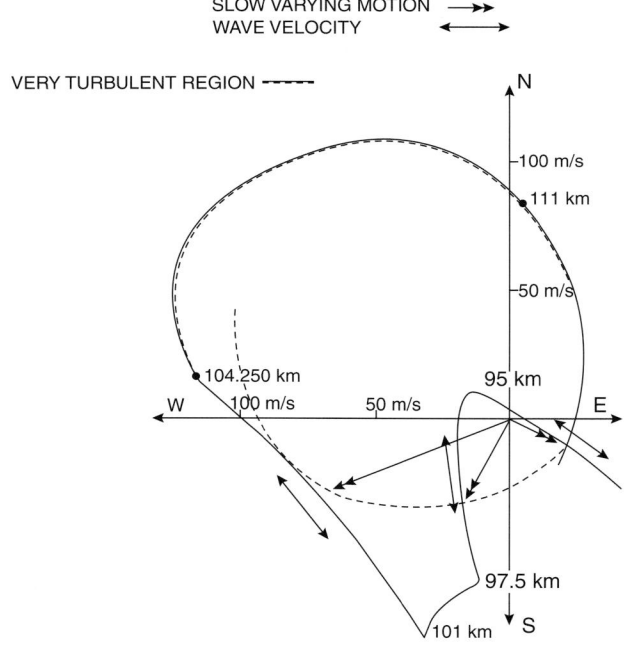

Figure 7 Hodograph (Blamont and Barat 1967b) as a function of altitude from Hammaguir, 4 November 1964.

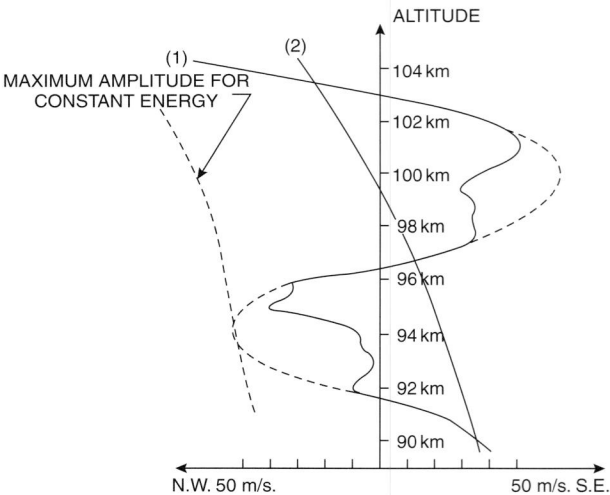

Figure 8 Components (Blamont and Barat 1967b) of velocity due to the wave (1) and to the slow-varying motion (2) as projected into the NE–SW plane for Hammaguir, 4 November 1964.

otherwise stable environment and giving a straightforward explanation of the 'sandwich' structure.

Consequently, in some parts of the wave field:

- the flow is turbulent and the diffusion of constituents is mainly turbulent (mixing); and
- the wave structure appears distorted while, presumably, some wave energy flux divergence feeds the turbulent motions and energy dissipation.

Outside these turbulent regions:

- the flow is laminar and molecular diffusion prevails (diffusion equilibrium); and
- the wave structure is preserved while viscous damping remains negligible.

The 'sandwich' structure cannot exist above 110 km as a result of the dramatic increase of the viscosity and heat conduction with height. There, molecular processes act to reduce velocity and temperature gradients even before they can lead to any instability. Consequently, wave motions are damped and cannot generate any turbulent zone. Neither can the 'sandwich' structure appear in the high mesosphere, below 85 km, as a result of the weak mean temperature gradient: there, convective instabilities most frequently occur, leading to quite different turbulent structures.

Onset of turbulence near the turbopause: The critical Reynolds number

Following Blamont and de Jager (1961), Blamont and Barat (1967b, 1968) introduced a critical Reynolds number which is essentially the experimental value of the ratio Re_{cr} for which the laminar motion becomes turbulent:

$$Re_{cr} = \frac{v_L L}{\nu} \qquad (17)$$

where ν is the kinematic viscosity, and v_L and L are defined from the following considerations. Let a horizontal laminar flow in a vertical plane (z, x) present a distribution $v(z)$ of the fluid velocity. The structure function of the velocities $[v(z_1) - v(z_1 + \Delta z)]^2$, where z_1 is the altitude of an extremum of $v(z)$, has a maximum for a value of Δz corresponding to the following extrema of $v(z)$. Between these two extrema the fluid has an average motion, with dv/dz keeping the same sign. In eqn (10), if the field is laminar, L will be defined as the distance between the two successive extrema. If it is turbulent, L will be the distance between one extremum and the place where the turbulence starts. The parameter v_L will be the difference between the corresponding velocities.

Each stratum of fluid flows as if it were contained in a pipe between extrema of velocities. Barat has measured Re_{cr} at a well-defined transition and found a value of around 600. When the motion of the wave creates shears whose Re_{cr} is higher than 600, the turbulent layer will appear. Using this critical number, defined somewhat arbitrarily, it is possible to predict whether the field will become turbulent when the kinematic properties of the wave have been measured.

Conclusion

The quantitative study of the dynamics of the higher mesosphere and lower thermosphere performed with the help of sodium clouds provided a new picture of the relations between organized and turbulent motions. The physics of the D and E ionospheric regions was clarified by the discovery of the large vertical wind shears.

Apart from the discovery of the turbopause, the main result of this study was to establish that gravity waves are stopped in their ascent by the viscosity of the medium at around 100 km, but that tidal waves propagate into the thermosphere, where they play a major role. The characterization and detailed description of the atmospheric waves in the mesophere have been obtained since 1979 with the powerful lidar technique, developed as a continuation of the research initiated through sodium clouds, by the author and his associates, M.L. Chanin and A. Hauchecorne.

Let us return now to the quantitative explanation of the physicochemistry of atoms at the mesopause. Density profiles of sodium, potassium and lithium atoms in the altitude range (70 to 110 km) were obtained from 1976 onwards by G. Mégie and associates. (Mégie and Blamont 1977; Jegou *et al.* 1985; Granier *et al.* 1989) Their measurements shed new light on the whole chain of reactions, which is much less simple than was previously thought. The atoms originate in meteoroids and micrometeoroids which are vaporized at an altitude of around 87 km. After photoionization the ions form complexes with H_2O molecules: $Na^+(H_2O)_n$, with $1 < n < 3$. The charged complexes can move vertically under the action of the Laplace force generated by their displacement in the terrestrial magnetic field. When they move upwards, they are photodissociated and recycled. If they move downwards, however, they are trapped in the denser atmosphere, oxidized and fall slowly to oblivion. The metallic layer appears not as a closed system, but as a step in the continuous passage from the meteoroidal source to the lower atmospheric sink.

With this new understanding of their behaviour, the properties of the neutral alkali metal atoms in this region of the atmosphere will certainly continue to be used in the near future for the determination of various parameters such as eddy diffusion coefficients, temperature and wave structure.

Furthermore, the use of the sodium layer as a method of creating an artificial star by excitation by a ground-based laser is now an important aid to telescopic observation and may even become essential for the development of high-resolution astronomy. Even astronomers now know of the existence of the atmospheric sodium layer.

MEASUREMENT OF THE TEMPERATURE OF THE ATMOSPHERE

Observations in the 1960s and 1970s with sodium clouds

The temperature of the upper atmosphere as high as 90 km had been found before 1961 by measuring the speed of sound at high altitude in experiments in which grenades were released from rockets and detonated (Stroud et al. 1956, 1958; Newell 1960). The temperature was also inferred from spectrographic observations of the airglow and the aurora at much greater altitudes (Hunten 1960). The very first direct measurements of the temperature were obtained between 100 and 150 km by the observation of sunlight resonantly scattered in a sodium cloud. In such a cloud, the atoms of sodium are thermalized in a few seconds; the lineshape of the resonance light is therefore a pure Doppler one, depending only on the ratio T/M, where T is the temperature of the neutral ambient gas and M the atomic mass of sodium, equal to 23.

The experiment was performed in the evening twilight of 2 March 1960, at Hammaguir, and employed a Véronique rocket which rose to 188 km and released 280 g of sodium (Blamont et al. 1961a). The temperature was measured with a variant of the Bricard and Kastler (1944) technique developed for the airglow. Photographs of selected portions of the cloud of known altitude were obtained repeatedly from a few moments after release of the sodium until the sunlight was extinguished, about half an hour later. These consisted of a series of two images formed side by side in light which passed through two parallel optical systems, identical except for the presence of an absorption cell in one of them. This cell contained sodium vapour at a known temperature T_a and opacity τ. Densitometer measurements of corresponding points in the two photographs were made, and the ratio of the intensities recorded at these points was obtained. For a single hyperfine component and an optically thin cloud the intensity measured without the absorption cell is proportional to

$$I_o = \int \exp(-x^2) dx \qquad (18)$$

where

$$x^2 = Mc^2(\nu-\nu_0)^2/\nu_0^2 RT \qquad (19)$$

since the lineshape is purely Doppler. Here M is the atomic mass of sodium, T the temperature of the sodium cloud and hence of the atmosphere, R the universal gas constant, ν the frequency of light and ν_0 the frequency at the centre of the component. The intensity, for the same point in the cloud, measured through the cell is proportional to

$$I_a = \int \exp(-x^2) \exp[-\tau \exp(-Dx^2)] dx \qquad (20)$$

where

$$D = T/T_a \qquad (21)$$

The ratio I_a/I_o is thus a function of T alone if T_a and τ are known and fixed. Complications arise because the cloud is not always optically thin, because the measured ratios are for the sum of all hyperfine components, and because the cloud is in the shadow of the Earth's own sodium layer during most of the experiment. This last effect produces a serious distortion of the spectrum, but this can be allowed for. The altitude and abundance of the Earth's sodium were measured and the transmission function computed by methods developed for the twilight flash.

It had been recognized earlier that the lineshape of the resonance D lines in the cloud would be distorted by multiple scattering in an optically dense sodium vapour. The optical thickness had therefore to be measured by an interferometric determination of the ratio D_2/D_1 and by the polarization of the scattered resonance lines.

In order to eliminate completely the artefacts of multiple scattering, in subsequent launches a small quantity of potassium was added to the sodium ejectors: the yellow D lines of sodium would be used for localizing the cloud, and the absorption measurements would be made on the invisible red D_1 line of potassium, the D_2 line of potassium is absorbed by a strong molecular absorption band of O_2. The first potassium clouds (Blamont and Lory 1963a) were emitted at Hammaguir with Véronique rockets in 1960. The technique was extended to an altitude of 400 km with the help of two Javelin rockets launched from Wallops Island, Virginia (37°N, 76°W) on 10 December 1960 and 13 September 1961 and provided the first temperature profile of the neutral atmosphere between 100 and 400 km (Blamont et al. 1961b; Blamont and Chanin 1965). Excellent temperature profiles were obtained in a series (Blamont and Lory 1963b) of launches at Hammaguir and at Fort Churchill, Manitoba (68.5°N, 94°W) during 1962 and 1963.

A comparison of all available data showed that the temperature profiles below 150 km altitude exhibit no measurable variation with hour, season or latitude, but display a strong dependence on solar activity from 150 to 400 km (Figure 9).

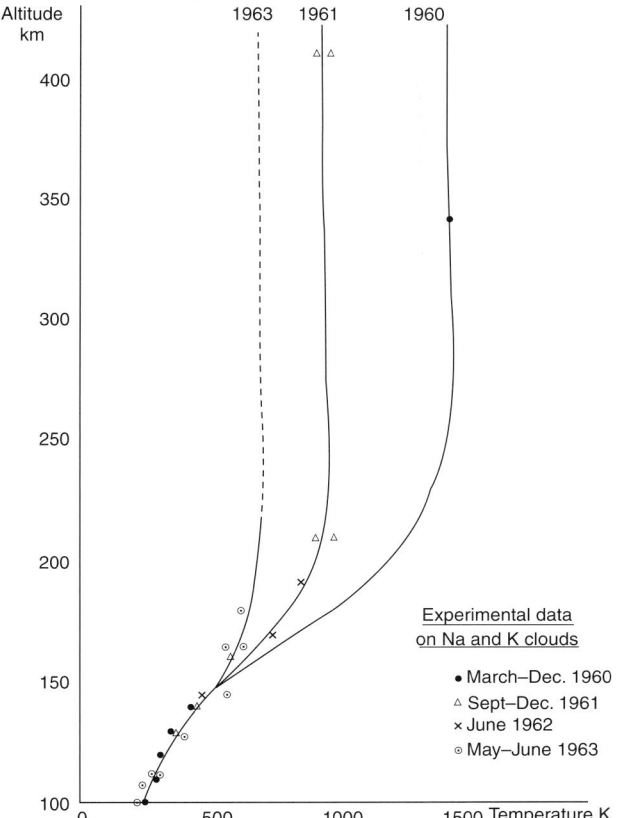

Figure 9 Measured profiles of the neutral temperature (Blamont and Chanin 1965).

The temperature profiles obtained by the sodium–potassium technique were used to build the first COSPAR model of the upper atmosphere.

One of the unexpected consequences of the barium trail experiments (Chapter 6) was the discovery of a strong fluorescence of AlO in the ejected clouds. Such clouds could also provide a way of measuring the temperature. A technique (Authier et al. 1964) was elaborated in 1963 to create highly visible AlO clouds with a mixture of Al and $NaNO_3$. A Centaure launch at Hammaguir on 9 February 1964 produced a bright orange and green cloud, from which a number of rotation and vibration spectra were obtained at an altitude of 177 km. Both types of spectra provided the same result, $T = 750 \pm 50$ K. Another method for creating AlO, the ejection (Authier et al. 1965) of trimethyl aluminium, was also successfully tried in 1964.

The sodium cloud created at Fort Churchill on 22 May 1963 which provided the best measurements ever was visible for 70 minutes, because of the long twilight at high altitude. After 44 minutes, the temperature was observed to rise from 380 ± 10 K to 950 ± 150 K, simultaneously with a sharp increase in the diffusion coefficient. This observation was the first direct detection of a sudden heating of the ionosphere in the auroral zone (Blamont and Chanin-Lory 1964). At the time of the experiment, the magnetic and ionospheric conditions were perturbed but no aurora was visible. In order to study the correlation of the neutral gas temperature with the magnetic parameters, a series of sodium cloud observations was organized in 1967 and 1968 by the Hydrometeorological Service of the USSR at Heyss Island (Franz Joseph Territory, USSR, 80.4°N, 50°E; 72°N magnetic latitude). The rocket used was an M-12 meteorological rocket (Chanin and Tulinov 1979) and the altitude range probed was 120 to 170 km.

Three distinct types of temperature variations were observed at an altitude of 170 km:

- Sudden increases correlated with geomagnetic activity: during March 1968 there was a change from 400–500 K to 950–1100 K during a fast geomagnetic variation (300γ in one hour).
- Slow, regular, continuous decrease of temperature during the night, of the order of 100 ± 50 K.
- Large seasonal variation from the beginning to the end of the polar night (from 1000 K in November 1967 to 500 K in February–March 1968.

The measurements of the neutral temperature in the polar thermosphere were continued at Heyss Island up to 1976 (Poloskov et al. 1969). The programme totalled more than 50 rocket launches. When the heating associated with geomagnetic activity is removed, a large variation of the temperature at 170 km as a function of the solar cycle appears in quiet geomagnetic conditions. A very cold atmosphere associated with a high altitude for the mesopause was found at the maximum of solar activity. The amplitude of variation of the neutral temperature at this altitude was found to be near to 500 K, and the mesopause was observed to move from 80 to 90 km. This variation out of phase by 180° with solar activity has been interpreted (Chanin et al. 1989) by arranging data according to the phases of the quasi-biennial oscillation (QBO) of the equatorial lower atmosphere. The observations of positive or negative correlation with the solar flux, alternating with height, as they have been recorded during winter months from the ground to 30 km by Labitzke and Van Loon (1997), and up to 80 km by Labitzke and Chanin (1988), appear as high as 170 km in the sodium cloud temperature data; this leads in the lower thermosphere to an inverse relationship between temperature and solar activity, in contradiction to current models. The increase with altitude of the temperature dependence of the solar flux and the vertical wavelength of the observed dynamical structures support the idea that this effect is the result of the upward propagation of planetary waves originating in the troposphere.

Present status of research on the temperature of the high atmosphere

A recent review, which has been drawn upon for this section, has been given by Thomas and Chanin (1999). Temperature measurements obtained by studying resonance in clouds were continued up to the end of the 1960s. Providing a precise base for models, they suffered, unfortunately, from a major drawback: they were discontinuous, and the high cost of rocket launches was not compatible with frequent measurements. Other methods were needed. At high altitudes (above 200 km) the linewidth of the airglow 6300–6363 Å doublet of O I, measured aboard NASA's sixth Orbiting Geophysical Observatory with a Fabry–Perot interferometer during 1969–1970 by Blamont and Luton (1972), paved the way for modern techniques. At lower altitudes (below 90 km) the lidar technique, measuring the vertical profile of the Rayleigh backscatter of a laser pulse, operational from 1979 at the Haute-Provence Observatory, has provided permanent and continuous measurements of temperature and density from 30 to 90 km. However, the region between 80 and 150 km is not satisfactorily explored. Nothing has replaced the potassium or aluminium oxide clouds for measuring the neutral gas temperature at this level.

Let us review, as a function of altitude, the progress made in the last 30 years. First, the solar effects, related to the 11-year solar cycle, have to be measured and taken into account. They are not easy to sort out. Series of lidar data obtained at Haute-Provence Observatory reveal a negative response of temperature to solar activity in the upper stratosphere, and a positive response from 50 to 70 km. The response is much stronger in winter than in summer, providing a clue that planetary waves (mostly present in winter) may be involved in amplifying the solar response. The continuation of the data into the mesopause region shows a tendency to a negative response above 80 km. There are other data sets that support the existence of a negative solar response in the mesopause region.

After correction for the solar effects, the data available for the mesosphere (50–180 km) indicate that a dramatic decrease in temperature (up to 7 K per decade) has occurred in the lower part of this region, far greater than can be accounted for by theory (~1 K per decade). Whether or not this change is related to releases of trace gases from anthropogenic sources in the lower atmosphere is clearly of great importance. If sustained, the observed changes in temperature, and resulting hydrostatic contraction, could have future consequences for the ionosphere, atmospheric drag on space vehicles, and possible (and unknown) effects on lower atmospheric climate.

At the mesopause level, the airglow provides information on temperature; the spectra of the atomic oxygen green line at 5577 Å originating at 95 km and of the OH Meinel near-infrared bands originating at 85–89 km have been extensively used for this purpose. The results, somewhat contradictorily, seem to show no real solar dependence of the temperature at the mesopause.

Above 100 km, anthropogenic effects are unlikely to have much impact. The physics of the region is dominated by the variations of the solar flux, more so with increasing altitude, combined with atmospheric dynamics and geomagnetic effects.

In the lower thermosphere, we have noted the anticorrelation of the temperature with solar activity in the range of 110–180 km deduced from sodium cloud measurements. The incoherent scatter technique provides ambiguous results. Whereas the St Santin results show a negative response to solar activity in the 110–125 km region, the Millstone Hill data indicate a positive response at all heights down to 100 km. It is possible that the higher degree of geomagnetic activity at Millstone Hill (magnetic latitude = 56°N, as opposed to 42°N at St Santin) explains the difference.

An indirect but equally potent argument for a negative solar response in the lower thermosphere, at least at 57°N, is the systematic variation in the heights of Perseid meteor trails over the period 1953 to 1965. Since the heights of the endpoints of meteor trails should vary directly with density, and since Perseid meteors all more or less enter the atmosphere with the same velocity, this method provides a semi-quantitative probe of the region around 85 to 95 km. The meteor endpoint heights vary from 85 km at solar minimum to 95 km at solar maximum, roughly equivalent to a factor of two in density. Thus there are three independent data sets, incoherent radar, rocket clouds and echoes from meteors, which suggest strongly that, at least at latitudes above 45°N, the temperature response to increasing solar activity is negative in the 90 to 165 km region.

From the above discussion of the solar cycle response, it is clear that data sets collected over longer times are needed than has been currently available. A period encompassing several solar cycles of data collected by the same, or similar, instrumentation is needed in order to reconcile the discrepancies revealed by analysis of data covering just one 11-year solar cycle. But most of the methods used for determining the temperature are indirect or model-dependent, with the exception of the sodium cloud technique, whose accuracy in the 1960s was not very high. In order to establish on a firm basis the eventual variation of the temperature from the mesosphere to the F layer, it would be necessary to resurrect a sodium cloud programme similar to that of the USSR Hydrometeorological Service, but with a better ground-based instrumentation. We doubt that this will happen. Maybe an improvement of the lidar technique could provide the data from this region.

In the higher thermosphere, a significant positive solar response of temperature has been understood since early in the space age. A convenient measure of the thermospheric response is the exospheric temperature, T_{ex}, constant with altitude above about 300 km. Although many details are yet to be worked out, the large solar cycle variation in T_{ex} is due to a significant solar cycle change in the solar EUV flux. For a 10.7 cm radio flux change from 70 to 230 (in units of 10^{-22} Watt m^{-2}(cycles/sec)$^{-1}$), T_{ex} varies between 700 and 1300 K. Geomagnetic activity is also a strong influence, varying with magnetic latitude. The heat balance of this region is complicated, but the main elements have been identified.

CONCLUSION

The two successive sodium clouds of March 1959 had a decisive effect on the future of space in Europe: they convinced the French Government that the rocket programme was technically sound and that sufficient competence was available for the development of larger boosters. The decision to proceed with the Diamant satellite launcher and the creation of the French Space Agency (CNES) in 1961, which accompanied this strategy, alongside the agreement between the British Prime Minister Harold Macmillan and French President Charles De Gaulle to place a large Véronique on top of the ailing Blue Streak in order to form the Europa-I rocket, plus the creation of ELDO (the European Launcher Development Organisation, a precursor of ESA) as a direct consequence of this concept, can all be traced back to the sodium cloud of 10 March 1959. The Véronique launches in 1959 followed by the successful satellite programme of the European Space Research Organisation (ESRO) created the confidence in Europe to establish the European Space Agency (ESA) and to develop the highly successful Ariane launchers.

REFERENCES

Authier, B., Blamont, J.E. and Carpentier, G. (1964). Mesure de la température de l'ionosphère à partir de la fluorescence crépusculaire du monoxyde d'aluminium. *Annales de Géophysique*, **20**, 342.

Authier, B., Blamont, J.E. and Lory-Chanin, M.L. (1965). Altitude comparison of electron and neutral gas temperatures. In *Space Research VI*, North-Holland, Amsterdam, p. 383.

Bates, D.R. (1947). The equilibrium of atmospheric sodium. *Terrestrial Magnetism and Atmospheric Electricity*, **52**, 71–75.

Bates, D.R. and Boyd, T.J.M. (1956). Inelastic heavy particle collisions involving the crossing of potential energy curves: IV. Unique recombination. *Proceedings of the Physical Society*, **A69**, 910.

Bates, D.R. and Seaton, M.J. (1950). Theoretical considerations regarding the formation of the ionized layer. *Proceedings of the Physical Society*, **B63**, 129–140.

Bawn, C.E.H. and Evans, A.G. (1937). The reaction of sodium atoms with the oxides of nitrogen, nitromethane, ethyl nitrate and amyl nitrate. *Transactions of the Faraday Society*, **33**, 1571.

Bedinger, J.F., Ghosh, S.N. and Manring, E.R. (1957). Emission from sodium ejected from rockets. In M. Zelikoff (ed.), *The Threshold of Space*, Pergamon, London and New York, p. 225.

Bedinger, J.F., Manring, E.R. and Ghosh, S.N. (1958). Study of Na vapor ejected in the atmosphere. *Journal of Geophysical Research*, **63**, 19.

Bernard, R. (1938a). Observation d'un nouveau phénomène de fluorescence dans la haute atmosphère. Présence et variations d'intensité de la radiation λ 5893 Å dans la lumière du ciel au crépuscule. *Comptes Rendues de l'Académie des Sciences*, Paris, **206**, 448.

Bernard, R. (1983b). Etude interférentielle de la radiation jaune λ 5893 Å du ciel crépusculaire et preuve de la présence de sodium dans la haute atmosphère. *Comptes Rendues de l'Académie des Sciences*, Paris, **206**, 928.

Blamont, J.E. (1956). Observations de l'émission atmosphérique des raies D du sodium au moyen d'un appareil à balayage magnétique. In E.B. Amstrong and A. Dalgarno (eds), *The Airglow and the Aurora*, Pergamon, London and New York, p. 99.

Blamont, J.E. (1959). Nuages artificiels de sodium – Vitesse du vent, turbulence et densité de la haute atmosphère. *Comptes Rendues de l'Académie des Sciences*, Paris, **249**, 1248.

Blamont, J.E. (1960). Nuages artificiels de sodium – Diffusion multiple de la lumière de résonnance. *Comptes Rendues de l'Académie des Sciences*, Paris, **250**, 567.

Blamont, J.E. (1963). Turbulence in atmospheric motions between 90 and 130 km of altitude. *Planetary and Space Science*, **10**, 89.

Blamont, J. (1966). Structure horizontale des vents au-dessus de 90 km. In *Les Problèmes météorologiques de la stratosphère et de la troposphère*, Presses Universitaires de France, p. 151.

Blamont, J.E. and Baguette, J.M. (1961a). Quelques résultats déduits de l'étude de la déformation de nuages artificiels de métaux alcalins. *Comptes Rendues de l'Académie des Sciences*, Paris, **252**, 3099.

Blamont, J.E. and Baguette, J.M. (1961b). Mesures déduites des déformations de six nuages de métaux alcalins formés par fusées dans la haute atmosphère. *Annales de Géophysique*, **17**, 319.

Blamont, J.E. and Barat, J. (1967a). Dynamical structure of the atmosphere between 80 and 120 km. In B. MacCormac (ed.), *Aurora and Airglow*, Reinhold, New York, p. 159.

Blamont, J. and Barat, J. (1967b). Introduction d'un modèle pour la structure des mouvements de l'atmosphère entre 85 et 110 km d'altitude. *Annales de Géophysique*, **23**(2), 173.

Blamont, J. and Barat, J. (1968). Variation avec l'altitude de la structure du champ de turbulence atmosphérique. *Annales de Géophysique*, **24**, 1.

Blamont, J.E. and Chanin, M.L. (1965). Temperature measurements in the ionosphere from 100 km to 400 km between 1960 and 1964. *Space Research V*, North-Holland, Amsterdam, p. 1137.

Blamont, J.E. and Chanin-Lory, M.L. (1964). Sudden heating of the ionosphere in the auroral zone. *Nature*, **201**, 593.

Blamont, J.E. and de Jager, C. (1961). Upper atmospheric turbulence near the 100 km level. *Annales de Géophysique*, **17**, 134.

Blamont, J.E., Donahue, T.M. and Lory, M.L. (1961a). Mesure de la température de la haute atmosphère à l'altitude de 370 km. *Physical Review Letters*, **6**, 403.

Blamont, J.E. and Lory, M.L. (1963). Mesure de la température de l'atmosphère au moyen de nuages de potassium créés par des fusées. *Comptes Rendues de l'Académie des Sciences*, Paris, **257**, 1135.

Blamont, J.E. and Lory, M.L. (1963b). New direct measurements of ionospheric temperatures. In H. Wexler and J.E. Caskey (eds), *Proceedings of First International Symposium on Rocket and Satellite Meteorology*, North-Holland, Amsterdam.

Blamont, J.E., Lory, M.L., Schneider, J.P. and Courtès, G. (1961b). Mesure de la température de l'atmosphère à l'altitude de 370 km, *Space Research II*, North-Holland, Amsterdam, p. 974.

Blamont, J.E. and Luton, J.M. (1972). Geomagnetic effects on the neutral temperature of the F-region during the magnetic storm of September 1969. *Journal of Geophysical Research*, **77**, 3534.

Booker, H.G. (1956). Turbulences in the ionosphere with applications to meteor-trails, radiostar scintillation, aurora radar echos, and other phenomena. *Journal of Geophysical Research*, **61**, 673–705.

Booker, H.G. and Cohen, R. (1956). A theory of long-duration meteor echos based on atmospheric turbulence with experimental confirmation. *Journal of Geophysical Research*, **61**, 707–733.

Bricard, J. and Kastler, A. (1944). Recherches sur la radiation D du sodium dans la lumière du ciel crépusculaire et nocturne. *Annales de Géophysique*, **1**, 53–91.

Bricard, J., Kastler, A. and Robley, R. (1949). Polarisation et mécanisme d'excitation de la raie D du sodium dans le ciel crépusculaire. *Comptes Rendues de l'Académie des Sciences*, Paris, **228**, 1601.

Cabannes, J., Dufay, J. and Gauzit, J. (1938). Sur la présence du sodium dans la haute atmosphère. *Comptes Rendues de l'Académie des Sciences*, Paris, **206**, 870.

Chamberlain, J.W. (1956). Resonance scattering by atmospheric sodium: 1. Theory of the intensity plateau in the twilight airglow. *Journal of Atmospheric and Terrestrial Physics*, **9**, 73–89.

Chanin, M.L., Keckhut, P., Hauchecorne, A. and Labitzke, K. (1989). The solar activity – Q B O effect in the lower thermosphere. *Annales de Géophysique*, **7**, 463.

Chanin, M.L. and Tulinov, G.F. (1979). The polar thermosphere temperature behaviour during the 11 year solar cycle. *Journal of Geophysical Research*, **84**, 406.

Chapman, S. (1939). Notes on atmospheric sodium. *Astrophysical Journal*, **90**, 309–316.

Chernaev, V.I. and Vuks, M.F. (1937). The spectrum of the twilight sky. *Doklady Akademii Nauk SSSR*, **14**, 77–80.

Currie, B.W. and Edwards, H.W. (1936). On the auroral spectrograms taken at Chesterfield, Canada, during 1932–33. *Terrestrial Magnetism and Atmospheric Electricity*, **41**, 265–278.

Edwards, H.D., Bedinger, J.F., Manring, E.R. and Cooper, C.D. (1956). Emission from a sodium cloud artificially produced by means of a rocket. In E.B. Armstrong and A. Dalgarno (eds), *The Airglow and the Aurora*, Pergamon, London and New York, pp. 122–134.

Elvey, C.T. (1950). Note on the spectrum of the airglow in the red region. *Astrophysical Journal*, **111**, 432.

Granier, C., Jegou, J.P., Chanin, M.L. and Megie, G. (1985). General theory of the alkali metals present in the earth's upper atmosphere III. Diurnal variations. *Annales Geophysicae*, **3**, 445–450.

Granier, C., Jegou, J.P., Megie, G. (1989). Iron atoms metallic species in the Earth's upper atmosphere. *Geophys. Res. Lett.*, **16**, 243–246.

Greenhow, J.S. (1959). Eddy diffusion and its effect on meteor trails. *Journal of Geophysical Research*, **64**, 2208–2209.

Groves, G.V. (1960). Upper-atmosphere wind studies by Skylark rocket sodium experiment. *Nature*, **187**, 1001.

Hines, C.O. (1960). Internal atmospheric gravity waves at ionospheric heights. *Can. J. Phys.*, **38**, 144.

Hunten, D.M. (1954). A study of sodium in twilight: 1. Theory. *Journal of Atmospheric and Terrestrial Physics*, **5**, 44–56.

Hunten, D.M. (1956). The interpretation of twilight measurements of sodium emission. In E.B. Amstrong and A. Dalgarno (eds), *The Airglow and the Aurora*, Pergamon Press, London and New York, pp. 183–192.

Hunten, D.M. (1960). Measurement of the temperature in the upper atmosphere to 150 km in a rocket experiment. Symposium on Aeronomy, Copenhagen, 1960. *Annales de Géophysique*, **17**, 249.

Hunten, D.M. and Shepherd, G.G. (1954). A study of sodium in twilight: II. Observations on the distribution. *Journal of Atmospheric and Terrestrial Physics*, **5**, 57–62.

Istomin, V.G. and Pokhunkov, A.A. (1963). Mass spectrometer measurements of atmospheric composition in the USSR. In *Space Research III*, North-Holland, Amsterdam, p. 117.

Jegou, J.P., Granier, C., Chanin M.L. and Megie, G. (1985). General theory of the alkali metals present in the Earth's upper atmosphere I. Flux model: chemical and dynamical processes. *Annales Geophysicae*, **3**, 163–176.

Jegou, J.P., Granier, C., Chanin M.L. and Megie, G. (1985). General theory of the alkali metals present in the Earth's upper atmosphere II. Seasonal variations. *Annales Geophysicae*, **3**, 299–312.

Kochansky, A. (1964). Atmospheric motions from sodium cloud drifts. *Journal of Geophysical Research*, **69**, 3651–3662.

Labitzke, K. and Chanin, M.L. (1988). Changes in the middle atmosphere in winter related to the 11 year solar cycle. *Annales de Géophysique*, **6**, 643.

Labitzke, K. and Van Loon, H. (1997). The signal of the 11 year sunspot cycle in the upper-troposphere–lower stratosphere. *Space Science Reviews*, **80**, 393.

Manring, E., Bedinger, J. and Knaflich, H. (1961). Some measurements of winds and of the coefficient of diffusion in the upper atmosphere. In *Space Research II*, North-Holland, Amsterdam, pp. 1107–1124.

Manring, E.R., Bedinger, J.F., Pettit, H.B. and Moore, C.B. (1959). Wind determinations in the upper atmosphere using artificially generated Na clouds. *Journal of Geophysical Research*, **64**, 587.

Meadows, E. and Townsend, J. (1960). IGY rocket measurements of arctic atmospheric composition above 100 km. In *Space Research I*, North-Holland, Amsterdam, pp. 175–198.

Megie, G., Blamont, J.E. (1977). Laser sounding of mesospheric sodium – interpretation in terms of global atmospheric parameters. *Planet. Space Sci.*, **25**, 1093.

Newell, H.E. Jr. (1960). The upper atmosphere studied by rockets and satellites. In J.A. Ratcliffe (ed.), *Physics of the Upper Atmosphere*, Academic Press, New York and London, p. 73.

Pokhunkov, A.A. (1961). Mass-spectrometric investigations of structural parameters of terrestrial atmosphere on heights from 100 to 210 km. *Iskusstvennykh Sputnikov Zemli*, **7**, 89–100.

Poloskov, S.M., Tulinov, G.F., Blamont, J.E., Chanin, M.L. and Maillard, M. (1969). Mesure de la température de la haute atmosphère dans la région polaire. *Annales de Géophysique*, **25**, 393.

Sanford, R.F. (1950). Search for atmospheric D1 on high dispersion stellar spectrograms. *Publications of the Astronomical Society of the Pacific*, **62**, 272.

Stroud, W.G., Bandeen, W.R., Nordberg, W., Bartman, F.L., Otterman, J. and Titus, R. (1958). Atmospheric temperature and winds between 30 to 80 km. In Fifth Assembly of the Comité Spécial pour l'Année Géophysique Internationale, Moscow, *Annals of the International Geophysical Year*, Pergamon, London and New York.

Stroud, W.G., Nordberg, W. and Walsh, J.R. (1956). Atmospheric temperature and winds between 30 to 80 km. *Journal of Geophysical Research*, **61**, 45–58.

Thomas, G.E. and Chanin, M.L. (1999). Global change in the mesosphere and lower thermosphere: temperature and composition. In International Workshop on Long Term Changes and Trends in the Atmosphere, Poona.

Vegard, L. (1947). In *Relations entre les phénomènes solaires et géophysiques*, CNRS, Paris, p. 254.

Zaragoza, A. (1965). Nuages de sodium créés en Argentine. Internal Report, Service d'Aéronomie du CNRS.

8

CORNELIS DE JAGER*

Early solar space research

The Sun, the most luminous celestial object as viewed from Earth, was the first target for space experiments when such possibilities opened up. It is convenient to distinguish between the quasi-permanent and slowly changing solar features, for which incidental rocket observations suffice, and the rapidly changing, transient phenomena, which need satellite monitoring. In the first category are objects such as the photosphere, including the fairly slowly changing features, like centres of activity, the basic features of the chromosphere and of the coronal regions. Solar flares and coronal mass ejections come into the second category.

It is therefore only natural that in the rocket era the emphasis was on the (quasi-)stationary parts of the Sun, while, once satellites were orbiting the Earth, the rapidly varying aspects could be studied. These two different aspects – stationary versus transient – govern the organization of this chapter. That distinction does not lead to a strict chronological description, though we attempt to stick more or less to chronology. Rocket observations persisted until long into the satellite era, while there were occasional flare observations being made in the rocket era; for example, when a rocket was fired accidentally or deliberately during the occurrence of a solar flare.

Like any history, ours has a prelude. During World War II the first ideas for observing the Sun by means of rockets were developed in Germany. These ideas, though at one time very close to their realization, never materialized. The story, though, constitutes an essential part of the history of space research.

This chapter deals with the period between about 1940 and 1970–75, which was dominated by rocket observations

and by the first – mostly fairly simple – satellites. We conclude with a description of the major results of Skylab, but it should be realized that the scientific influence of this space laboratory persisted for a long time after it had been put out of service.

THE WARTIME ERA: KIEPENHEUER AND THE V-2

During World War II German military authorities, under Wernher von Braun, developed the A-IV rocket (later called V2, from *Vergeltung-zwei*, 'Revenge 2'), in order to be able to bombard targets many hundreds of kilometres away. The V2 was able to climb to altitudes of over 100 km. This ability attracted the attention of scientists who realized that – in principle – the V2 could make it possible to study both the structure of the upper atmosphere and solar radiation from wavelengths that are inaccessible from ground-based sites. This fascinating episode is described by Hufbauer (1991, pp. 120–124) and Wolfschmidt (1993).

Wolfschmidt starts her review by quoting Heraclitus: 'War is the father of all things.' This evidently exaggerated statement does indeed apply to many scientific developments that began in that period and would come to dominate the second part of the twentieth century. During that period the technique of radio astronomy was developed, and the foundations of space research were laid. In the USA, Walter Orr Roberts founded, under war pressure, the High Altitude Observatory for solar monitoring, and on the European continent Karl Kiepenheuer founded six solar observatories. The impetus behind these developments was the knowledge that the Sun influences long-distance radio

*SRON – National Institute for Space Research, Utrecht, The Netherlands

communications, and the expectation that understanding the origins of solar variability would lead to greater control over long-distance radio communications. The military importance of the latter is obvious. With some scientific opportunism, solar scientists on either side of the front line thus hoped that military funding would enable them to continue and intensify their research.

The fact that the Sun influences the ionosphere and thus the efficiency of radio propagation was known by the early twentieth century, but the origin of the disturbances was unknown, despite well-organized worldwide solar monitoring since the IAU General Assemblies of 1924 and 1928. At that time a large co-operative project started, to which the names of George Ellery Hale, Sydney Chapman, Charles St John, Giorgio Abetti and others are associated. It had been known for some years that radio disturbances were associated with the occurrence of sunspot groups, and Howard Dellinger (1935) found, on the basis of synoptic solar data (obtained by Henri d'Azambuja) that the just-discovered solar flares were responsible for at least part of the radio disturbances.

As a post-World War I punishment by the victorious allied nations, Germany was excluded from international scientific co-operation during the period between the wars, a humiliation to which decent scientists should never have lent support. German scientific institutions were therefore not included in the campaigns for global solar monitoring. Before the war, Germany had few solar observatories and no modern monitoring equipment. Apparently, in order to cope with that, Grotrian started systematic sunspot observations in Rechlin, some 100 km north of Berlin, at the end of the 1930s. To improve the quality and continuity of solar monitoring, Hans Plendl and Kiepenheuer, the latter then in Göttingen, proposed at the end of 1939 to establish a solar observatory at the Wendelstein mountain.

At about the same time (1940) Kiepenheuer, guided by Grotrian and Otto Heckmann, constructed a spectroheliograph in Göttingen. In co-operation with Plendl he established during wartime a network of solar monitoring stations in Germany and Italy, thus providing those involved in radio communication with reliable solar data. By including observatories in occupied countries in the network (for example, Meudon, Pic du Midi, Ondřejov) he provided these institutes with some protection against military interference with their other scientific work.

Earlier (1936–39), Kiepenheuer had realized the importance of the solar UV flux in controlling the ionization of the ionosphere. In order to obtain information about solar UV radiation, and particularly its variations with time, he made observations at the Jungfraujoch observatory (altitude 3600 m) and (from 1939) with stratospheric balloons (Kiepenheuer 1938). The instrument was a double monochromator based on quartz optics. The Jungfraujoch observations showed him that even at that altitude the solar UV could not be detected, and even using stratospheric balloons did not significantly improve detection. The reason was rapidly found in absorption by terrestrial ozone at heights above a few tens of kilometres.

An immediate challenge resulting from these investigations was therefore to try going higher, above the ozone layer, which means above about 50 km. When the military authorities started developing the A-IV (V2) rocket, the project leader, von Braun, was eager to collect information on the medium through which the rockets were supposed to fly: the upper atmosphere. He succeeded in interesting Erich Regener, who had strong anti-Nazi sentiments, but who was attracted by the prospects of doing pioneering research with the new vehicle. Overcoming his political sentiments, Regener started developing a scientific payload for measuring atmospheric densities and temperatures with the A-IV. Somewhat later, Regener realized the importance of flying a spectrograph in order to study the absorption of solar radiation by the residual atmosphere above the rocket. To that end he asked for the assistance of Paetzold, a specialist in optics, and Paetzold recruited Kiepenheuer into the project. From a letter dated 8 July 1942 we quote:

> The A-IV offers the possibility of using the newly developed methods to carry out measurements in the upper atmosphere. [There follows a description of proposed instrumentation] … above all, an ultraviolet spectrograph. This must be above the absorbing ozone layer, which is to say at an altitude of 50 km or more, if it is to record the solar spectrum. (Author's translation)

At Peenemünde, Paetzold developed a prototype of the spectrograph. A second, more sophisticated flight instrument, conceived by Kiepenheuer and constructed by Paetzold, was a spectrograph based on LiF optics (in order to be able to observe wavelengths less than 2000 Å). In spite of the increasingly difficult war situation, von Braun planned a prototype flight in the winter of 1944–45, to be followed by a scientific flight in the spring. The instrument and nose cone arrived in Peenemünde in January 1945, where they were subsequently integrated and underwent their pre-launch tests. No launch ever took place. Ultimately, the increasing severity of the war prohibited the use of V2 rockets for scientific experiments.

FIRST SOLAR NEAR-UV RADIATION DETECTED BY ROCKETS

After the war the USA took the lead in space research with rockets. The first years after the war set the direction for the subsequent development of solar space research, and showed all the elements that would prove to be so characteristic of the initial phases of various subdisciplines of space research. A slightly humorous aspect was that the scene in the USA was complicated by competition between army, navy and air force.

After the military collapse of Germany in the spring of 1945, a US army intelligence unit under Colonel Holger N. Toftoy rapidly rounded up von Braun's team and captured, almost from under the nose of the approaching Soviet army, sufficient material for assembling around a hundred V2 rockets. Toftoy, realizing the importance of establishing good relations between the military and scientific communities, charged Major James G. Bain with the task of finding scientists who might be interested in assembling payloads for rocket-borne scientific experiments that could be launched from the newly established range at White Sands.

Initially there was great enthusiasm, particularly among the leading scientists, about the new potential for doing research in a spectral region that had so far been inaccessible. Later, though, when confronted with the large technical obstacles and the new and unconventional techniques, along with the timescale for realizing the goals set, scientists showed a tendency to step back and resume their previous modes of research. In the meantime, physicists and engineers who were familiar with the new techniques entered the field from other areas of science or engineering. They seized the initiative and became the main people responsible for the experiments. In later phases, when the scientific results became available and were reaching the level of specifications set by the astronomers, the latter returned. Then, however, they often found that the physicists had in the meantime acquired sufficient knowledge of the field and hardly needed assistance from traditional astronomers. This development could be witnessed in several disciplines, particularly in space research. In addition to solar space physics this same dynamic of community development could also be observed in the disciplines of X-ray and gamma-ray astronomy, where cosmic ray physicists came to dominate this typically astrophysical field. To a lesser degree this development also took place in radio astronomy.

Ultimately, this process appeared to be enriching astronomy, because of the entry of eminent physicists into the field. There were Bernard Lovell, Martin Ryle, Wilbur Christiansen and others in radio astronomy, and a multitude of scientists in space research: Bruno Rossi, Richard Tousey, Herbert Friedman, Sergei Mandel'shtam, Saito Hayakawa, Guiseppe (Beppo) Occhialini, and many others.

We illustrate the situation with an example. In September 1945, confronted for the first time with the possibility of observing the Sun from space, Leo Goldberg wrote to Donald Menzel (Hufbauer 1991, p. 136):

If anyone asked you what technological development could, at one stroke, make obsolete all our textbooks written in astronomy, I am sure your answer and mine would be the same, namely the spectroscopy of the Sun outside the Earth's atmosphere ... I would like nothing more than to be involved in such a project, even if it meant shaving my head and working in a cell for the next ten or fifteen years.

But already in 1948 Goldberg was having doubts when he wrote to Lyman Spitzer (Hufbauer 1991, p. 139):

I have been somewhat disappointed in the development of the rocket project since its inception. It had been my hope and I think also yours, that the rockets would open a new field of solar research ... At least thus far ... they have hardly opened a new field of investigation ... at the moment I would not want to ask for a renewal of the contract solely on the basis of the V2 investigations.

Goldberg's disappointment has been echoed many times in the initial period of development of space research, particularly by theoretically inclined scientists who are essentially interested in the scientific results, and have difficulty in grasping the time and effort needed for solving the technical problems encountered when a new field of research is facing unconventional and often new technological challenges.

In the hectic period immediately after the war, things in the USA went ahead remarkably well, under able leadership. An informal V2 panel set up under Ernest H. Krause identified in a relatively short time institutions and scientists willing and able to develop scientific experiments for the new vehicle. It is remarkable that at that time most of the solar scientists were from two institutions only. One was the Naval Research Laboratory (NRL), directed by Edward O. Hulburt, who had for a long time been deeply interested in the relationship between solar energetic radiation and the ionosphere; the other was the Applied Physics Laboratory of Johns Hopkins University, directed by Merle Tuve.

Richard Tousey's group at the NRL had the first success: on 10 October 1946 a rocket-borne camera took the first photograph of the solar UV spectrum down to 2200 Å. The spectrum was obtained from a height of 55 km.

In 1947 the chairmanship of the V2 panel passed to James Van Allen, while Homer A. Newell became head of the section for rocket research of the NRL. These appointments greatly strengthened the programme.

When, in 1948, the stock of V2s became exhausted, US-made rockets were introduced, first the Aerobees and Vikings. The first Aerobees did not reach sufficient altitude to be scientifically useful, but after increasing the length of the rockets they attained heights of 150 km or more, thus surpassing the greater altitudes reached by the V2s.

None of the people involved in the first rounds of solar experiments had much prior knowledge of solar physics. Richard Tousey's background was in UV spectroscopy, and this made him well suited for the new tasks. But new problems arose. In the first few years, when Sun-pointing devices had still to be developed, methods had to be generated for directing sunlight into the spinning, yawing spectrograph. One remedy was to use large-field spectrographs, but that was not a sufficient solution. Various methods (Tousey 1953) were tried of placing reflecting objects in front of the spectrograph slit. In one of the most successful versions, introduced by the NRL group, two polished spheres of LiF, 2 mm in diameter, were positioned some distance in front of the entrance aperture of the spectrograph. Thus a point-like solar image was produced, at the position usually occupied by the spectrograph slit. A slit was then no longer needed. The introduction of *two* spheres was needed to compensate partly for the rocket's spinning. The drawback was that the 'slit' was smeared out, causing degradation of resolution. Another method, used by the Johns Hopkins group, was based on a diffuse reflector consisting of rough aluminized glass. In that case a slit was necessary. Other variants are described by Tousey (1953) and are shown in Figure 1. In all cases the spectrographs had luminous efficiencies below 1%.

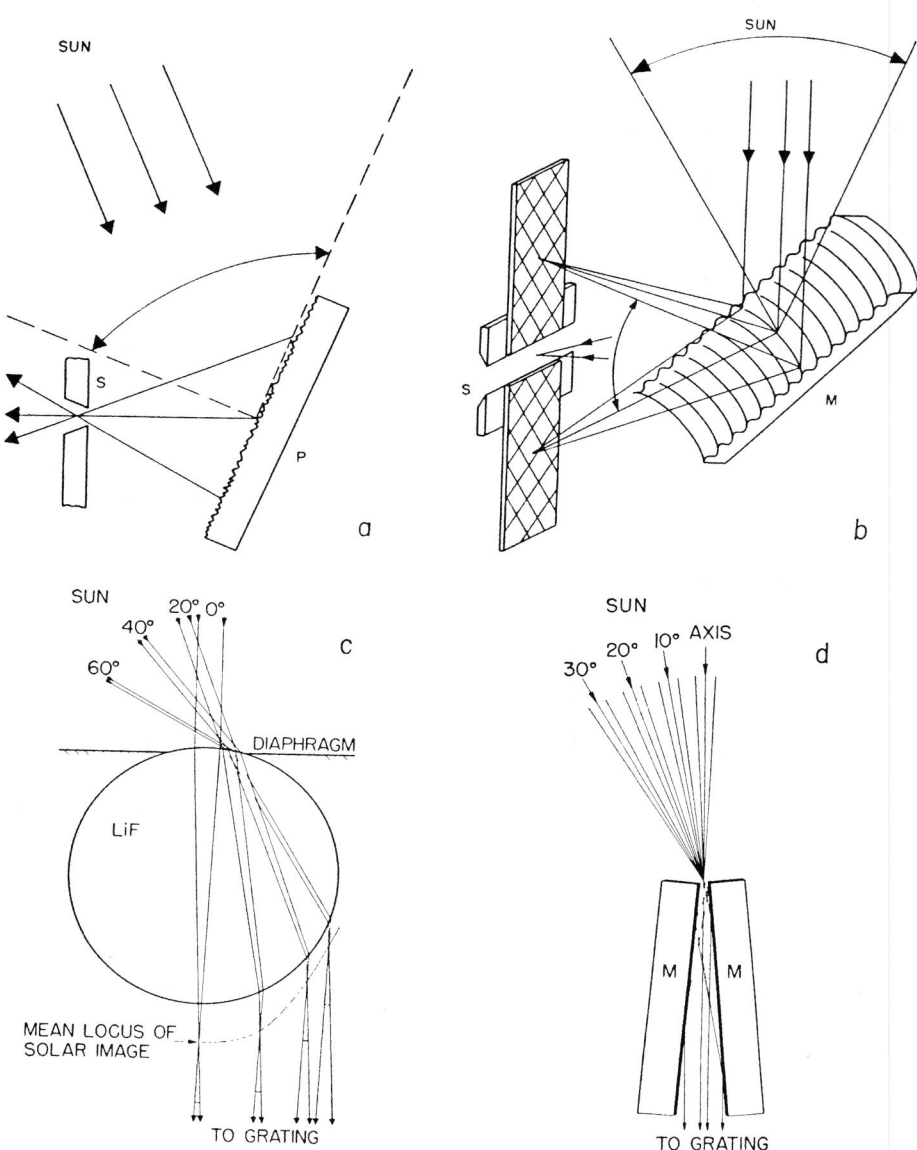

Figure 1 Wide-field-of-view entrance aperture systems for rocket spectrographs: (a) diffuse reflector and slit; (b) corrugated cylindrical mirror and slit; (c) lithium fluoride sphere (diameter 2 mm); (d) mirror-jawed slit. (From Tousey 1953. Reproduced with the permission of the University of Chicago Press.)

The initial observations provided information on a restricted part of the solar spectrum, that above about 2200 Å. The spectrographs were not luminous enough and did not stay high enough for a sufficiently long time to observe the region below 2200 Å. From an astrophysical point of view this meant that the information acquired was about the solar photosphere (see Figure 2, which shows the near-UV energy distribution of the Sun) and hence not to the higher and scientifically more interesting chromosphere. The absorption coefficient of solar matter is such that even at 2200 Å the emitted radiation still originates in the photosphere, albeit in its upper layers.

From data such as those in Figure 2, improved upon in later years, the variation of temperature in the solar atmosphere near the photosphere/chromosphere transition region can be derived. The brightness temperature of the Sun reaches its minimum value of 4700 K at 1400 Å. But what we observe is the result of an integration over the atmosphere and, after correcting for that, a temperature minimum value of 4500 ± 100 K is derived (de Jager 1965, p. 215). That the photospheric region around the temperature minimum emits in this spectral range is also shown by the observation that this spectral region lacks spectral lines. Absorption lines occur at longer wavelengths, emission lines at shorter wavelengths. A third aspect is that in this region the limb darkening (at higher wavelengths) changes into limb brightening at shorter wavelengths (Tousey 1963).

Tousey obtained more spectra on three other V2 flights. Although each flight was welcomed by the scientific community, it was clear at the same time that none of them led to basically new understanding, let alone a breakthrough, because the spectra did not have the resolution needed for astrophysical research. The need for a system for automatic and biaxial pointing of the spectrograph was obvious. It would increase the spectrograph's power and hence its attainable spectral resolution.

The first Sun-follower, constructed by the Applied Physics Laboratory, was launched as early as 29 July 1947. It was based on the aluminized and corrugated mirror shown in Figure 1(b). The mirror was rotated until the slit was illuminated, and the mirror was then kept steady by a servo system based on two phototubes, placed at either side of the slit. Yaw corrections could not yet be applied.

Other, more sophisticated Sun-followers were developed later by the NRL but several of them failed in the initial period, due to difficulties with the rockets. Later, such Sun-followers became essential parts of solar rocket spectrographs and they have been used with great success. While the Sun-followers improved steadily, the spectral resolution increased in parallel. At the same time more luminous spectrographs in rockets, staying longer at high altitudes, enabled the observation of the spectral region below 2200 Å.

THE FAR-UV AND X-RAY SPECTRUM

Exploration of the spectral range below 2000 Å had to wait for the development of biaxial pointing devices. A newly established group, located at the University of Colorado and stimulated by the US Air Force obtained such a spectrum for the first time in 1952. The Sun-follower needed was developed there by F. Wildhusen, and frequently used by Rense and Pietenpol and their associates. The successful result included the first observations of the hydrogen Lyman α line. Biaxial pointing control systems of increasing degrees of sophistication were subsequently used by all research groups at different institutions. Thus, spectra were obtained with steadily improved resolution and richer detail. In 1956 the Colorado group made a further step forward when it obtained the first spectroheliogram in Lyman α. A few years later, another institute entered the field when Hinteregger,

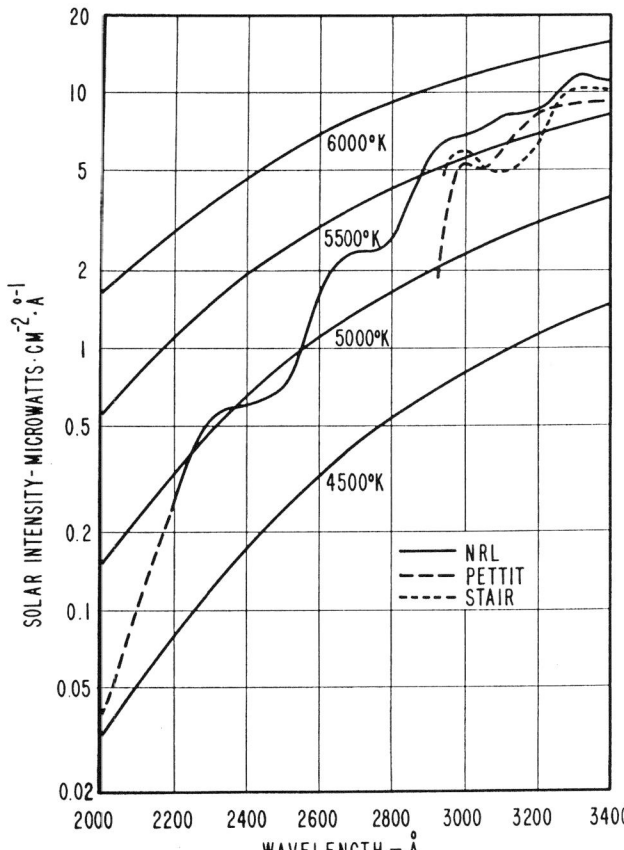

Figure 2 The solar intensity distribution between 2200 and 3400 Å, along with Planckian radiation curves for various temperatures. (From Tousey 1953. Reproduced with the permission of the University of Chicago Press.)

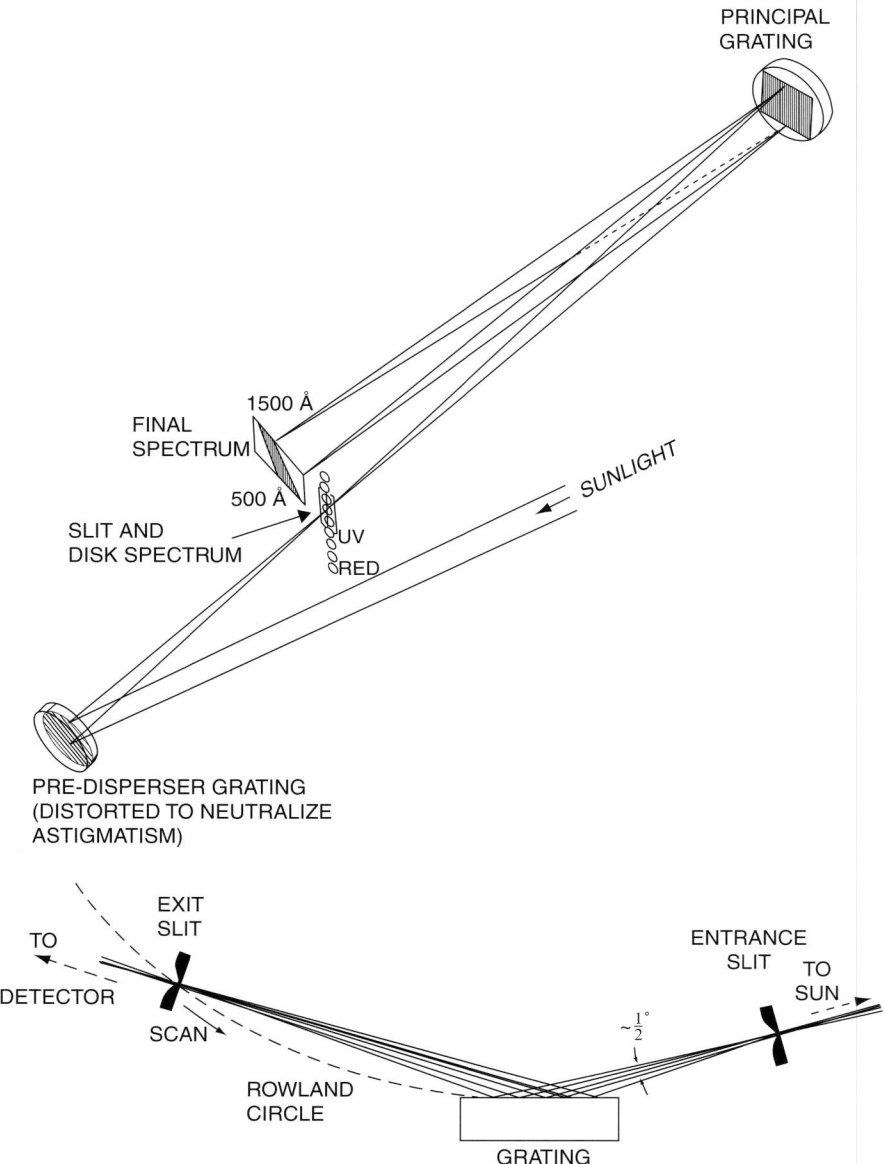

Figure 3 High-dispersion rocket spectrographs used around 1960: (a) double-dispersion spectrograph used by Tousey and coworkers of the NRL group; (b) grazing incidence monochromator used by Hinteregger and others at Goddard Space Flight Center and the Air Force Cambridge Research Laboratory. (Tousey 1963.)

initially with Tousey's group, worked on the same topic at the Air Force Laboratories in Bedford, Massachusetts.

Reviews by Tousey (1963, 1986) (see also interim reports by Tousey (1955, 1958)) and Hinteregger (1965) describes this fascinating period. Figure 3, from Tousey (1963), shows the double-dispersion spectrograph used by Tousey, and Hinteregger's grazing incidence monochromator.

By around 1960, these observations had led to a complete coverage of the short-wavelength spectrum, with observations of increasingly high resolving power, up to 75000. When going to shorter wavelengths, these spectra show first a change from an absorption to an emission spectrum near 1500 Å, as described in the previous section. This is an indication that the excitation temperature is increasing outward above the region where the 1500 Å radiation is emitted. For shorter wavelengths the radiation originates from the low chromosphere. At shorter wavelengths still the degree of ionization of the atoms emitting the lines increases gradually. Near 200–300 Å the observed degree of ionization points to emission from the chromosphere/corona

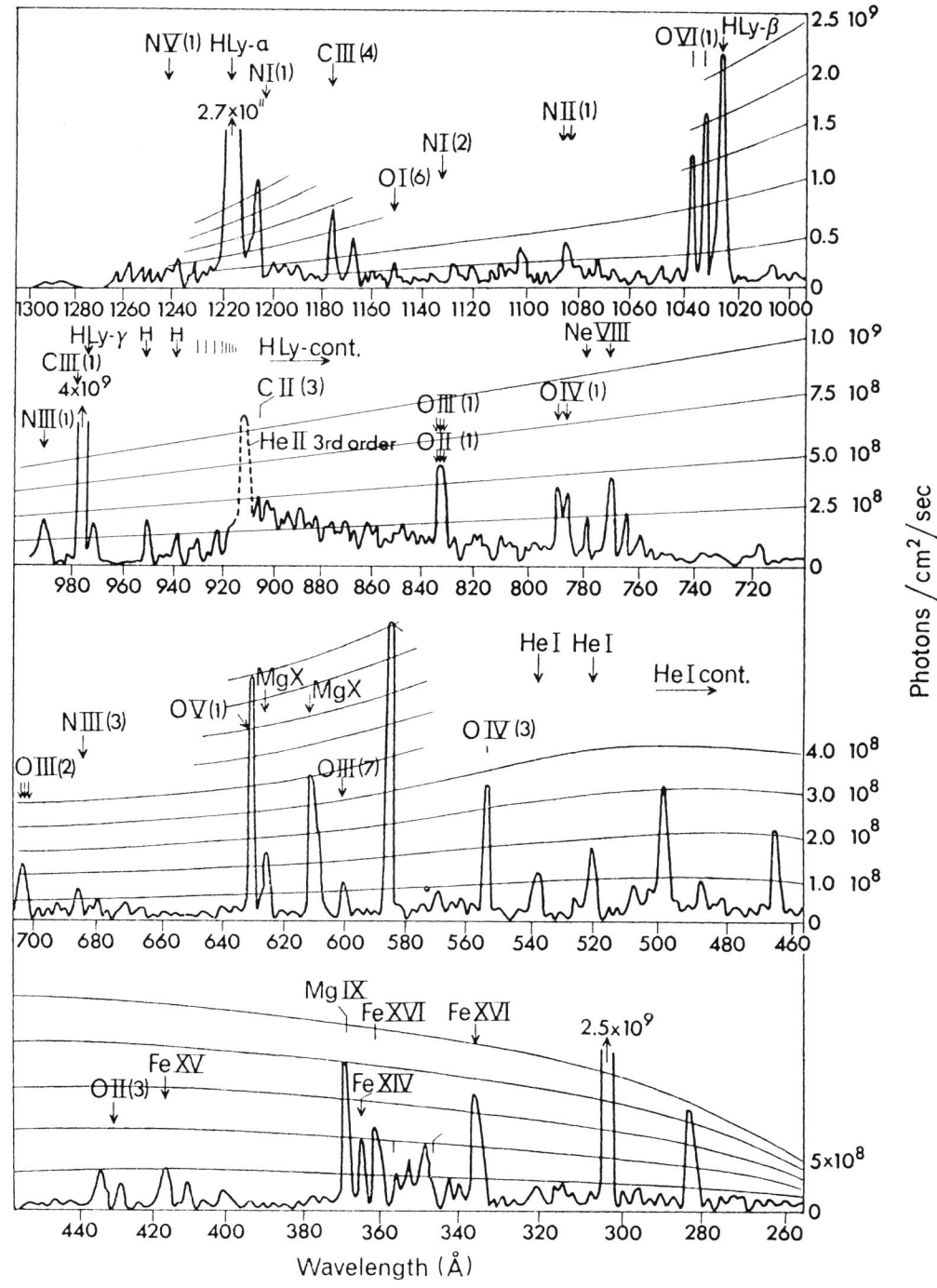

Figure 4 The solar spectrum from 260 to 1300 Å, observed by Hinteregger. (Massey 1964.)

transition region, and for wavelengths of 100 Å and below the emission is purely of coronal origin.

The progress was shown in an impressive way by Hinteregger's (1965) review, in which he presented a number of his 'best' spectra. On 23 August 1961 the spectrum between 260 and 1300 Å was obtained, shown here in Figure 4. A spectrum from 60 to 300 Å was obtained on 2 May 1963. The latter spectra were obtained in the 'satellite' era and they show that in this period rockets were not obsolete if used for special purposes. Much later,

Cushman and Rense (1978) described high-resolution line profiles of the lines at 304 Å (He II) and 303 Å (Si XI), obtained on 30 August 1973 and 20 January 1975.

Since the corona is a highly variable part of the Sun, dominated by its active regions, rocket observations of the Sun below about 250 Å are only useful for obtaining a snapshot, and satellite monitoring is essential for getting insight into solar variability.

CORONAL ROCKET OBSERVATIONS

Solar X-radiation is of coronal origin. Its observation demands an approach that was completely unorthodox to 'classical' astrophysicists: no optics, no conventional spectrographs. Therefore, this region was from the outset the realm of high-energy physicists such as Friedman in the USA and Mandel'shtam in the USSR.

The first detection of solar X-rays was on 5 August 1948, due to Burnight (1949), who sent up photographic plates covered with Be windows 0.8 mm thick. Such Be sheets transmit radiation below 8 Å. Another material that is often used as an X-ray filter is Al, which transmits radiation between 8 and about 15 Å. A related method that has also been used is based on thermolucent phosphors, packed in Al or Be. Their use has the advantage that these phosphors are not sensitive to visible light that might leak through pinholes in the sheets. But both methods have the drawback that they give qualitative results only. In any case, the first observations had already shown that solar X-radiation depends strongly on solar activity. The qualitative character of the information thus obtained was of course unsatisfactory. The need for photon counting devices was obvious.

The first successful measurements based on photon counting were taken on 29 September 1949 by Friedman *et al.* (1951) and independently by Burnight. Burnight's measurements, based on Al windows transmitting in the region between 8 and 15 Å, were successfully repeated on 21 November 1950. The Friedman group used Be windows 0.075 mm thick. Photons were counted when the rocket was above 87 km. At 100 km the intensity was half that at 150 km. The conclusion was that on that day the Sun did not emit X-radiation below about 6 Å. In Burnight's measurements of November 1950 no radiation was detected even below 15 Å. The Sun was extremely quiet that day.

A FIRST COMPARISON WITH THEORETICAL CALCULATIONS

The rocket observations, rudimentary as they were at the beginning of the 1950s, and made mainly to find the solar source of ionospheric ionization, did arouse the interest of astrophysicists. In Tübingen, Elwert (1954, 1956) started elaborate calculations of the expected X-ray spectrum of the solar corona. He had been able to collect the necessary spectroscopic data, such as ionization potentials and oscillator strengths, but the additional problem – that of the physical parameters describing the corona – was less easy. At that time the temperature of the corona (assuming there is *one* temperature!) was not well known, and we now know that it was rather underestimated at the time. The average electron number density and its variation with height were better known (Van de Hulst 1953; de Jager 1959).

Elwert, therefore, made calculations for two assumed coronal temperatures: 0.7 million and 1 million K (too low, as we know now). In addition he also made calculations for a 'coronal condensation' (that is, the coronal part of an activity region) with an assumed temperature of 6 million K. Furthermore, he introduced an arbitrary uncertainty parameter Q'. If needed, the calculated results could be multiplied by Q' in order to improve agreement with observations.

Understandably, the calculated emissions from the coronal condensation, with their higher temperatures, appeared to extend to shorter wavelengths than that of the 'quiet' corona. The condensations emit mainly in the spectral region below 10 Å.

The observational results as they were available in 1955 were compared with Elwert's predictions by the present author (de Jager 1955) and they are presented here in improved form (corrected to 1958) in Figure 5 (de Jager 1959, Figure 82). The conclusions were at that time somewhat bewildering. Roughly, accepting differences up to a factor of 10, there was some agreement, but there were days on which *much harder* radiation was emitted than predicted, suggesting higher coronal temperatures. On other occasions the radiation emitted was *much more intense* than predicted, suggesting a larger extent of the relevant coronal active region.

In the meantime, several well-chosen rocket launchings had shown (some of them fortuitously) that during solar flares energetic regions could temporarily emit hard to very hard X-rays. On 31 August 1959 a high-energy flare was observed from ground and from space. It emitted at energies as high as the 10–100 keV range. Elwert (1961) gave a summarizing overview of the solar spectrum around 1960 (Figure 6). The time was becoming ripe for flare monitoring. Sputnik had been launched, and the satellite era had started.

PRE-SATELLITE OBSERVATIONS OF FLARES

The fascinating account of the various attempts to measure X-rays by means of rockets while flares were occurring is mainly Friedman's story and it has been described in a lucid way in his reminiscences (Friedman 1977). His interest in

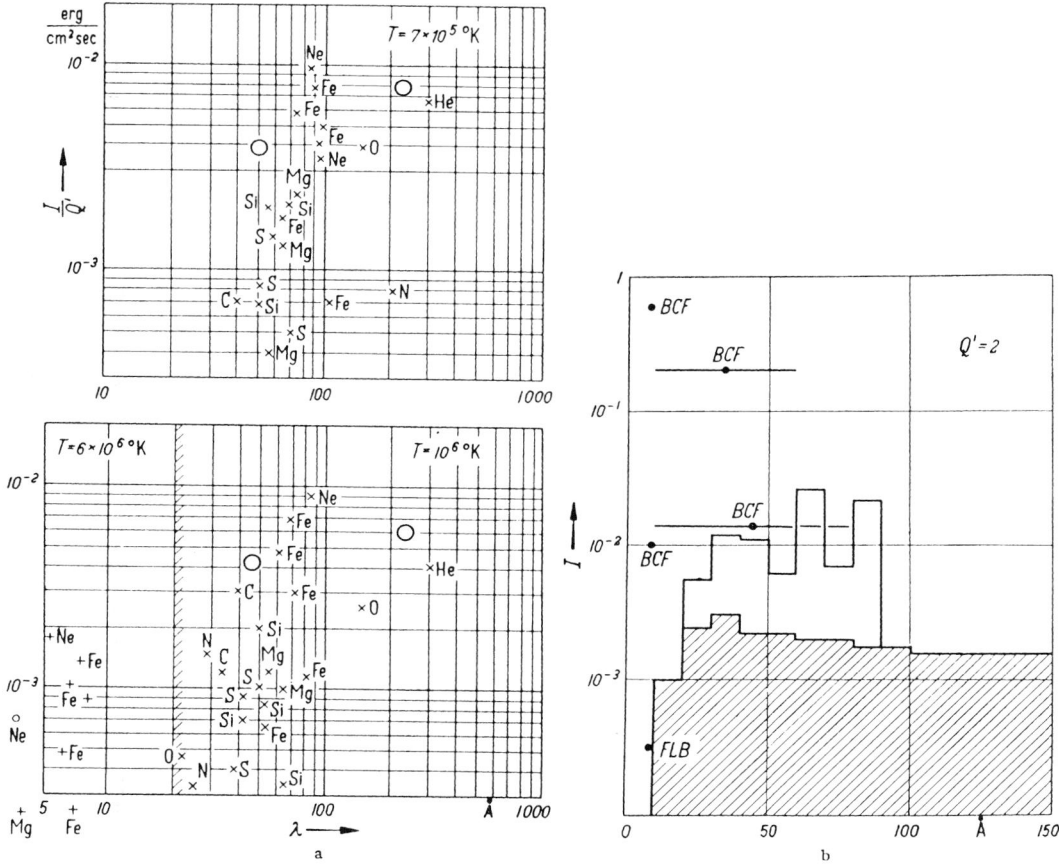

Figure 5 (a) Calculated spectra of the corona for $T = 0.7$ million K, for 1 million K and for a coronal condensation with an assumed temperature of 6 million K (after Elwert 1954, 1956). (b) Comparisons with observations by de Jager (1959, p. 262): BCF, Byram, Chubb and Friedman; FLB, Friedman, Lichtman and Byram. (Reproduced with the permission of Springer-Verlag, Berlin.)

flares was aroused by the enormous particle flux associated with a giant flare of November 1949. The Aerobee, which could attain a height of some 150 km, seemed a suitable vehicle for measuring X-rays during a flare. However, the preparations for launch took a long time, and the Aerobee had to be launched within a few hours of being made ready. Flares, however, arrive at unpredictable instants and do not last longer than a few tens of minutes. In addition, as was already anticipated by some scientists at that time, the first two minutes of a flare's life are essential to an understanding of the flare mechanism. A more or less suitable solution was found in the rockoon, a solid-propellant rocket that could be lifted to 25 km altitude by a balloon, and fired by ground command at the required time. The balloons stayed within reach of the command centre and at altitude during the day, hence the only requirement was that a flare should occur that day. The US Navy arranged for the command centre to be located on the ship from which the balloons were launched. The story of the first expedition to the Pacific is worth reading. There were a lot of improvisations and some amusing failures: flares were absent during the first three days of observations, and when the crew decided to have rest on Sunday, the first two flares occurred. But ultimately a weak flare did present itself, and the observations showed short-wavelength X-rays observable down to an atmospheric altitude of 70 km.

Later the situation was improved when two-stage rockets could be used. Deacon rockets, boosted by a Nike, were launched from the ground (Figure 7). The results essentially confirmed that flares are sources of more energetic X-radiation than the quiet corona.

Along the same lines were the observations of the remaining X-ray intensity of the corona during a solar eclipse, by launching five Nike-Asp rockets from a ship in the Pacific. It was found that the X-radiation is concentrated in the coronal parts of the active regions of the Sun,

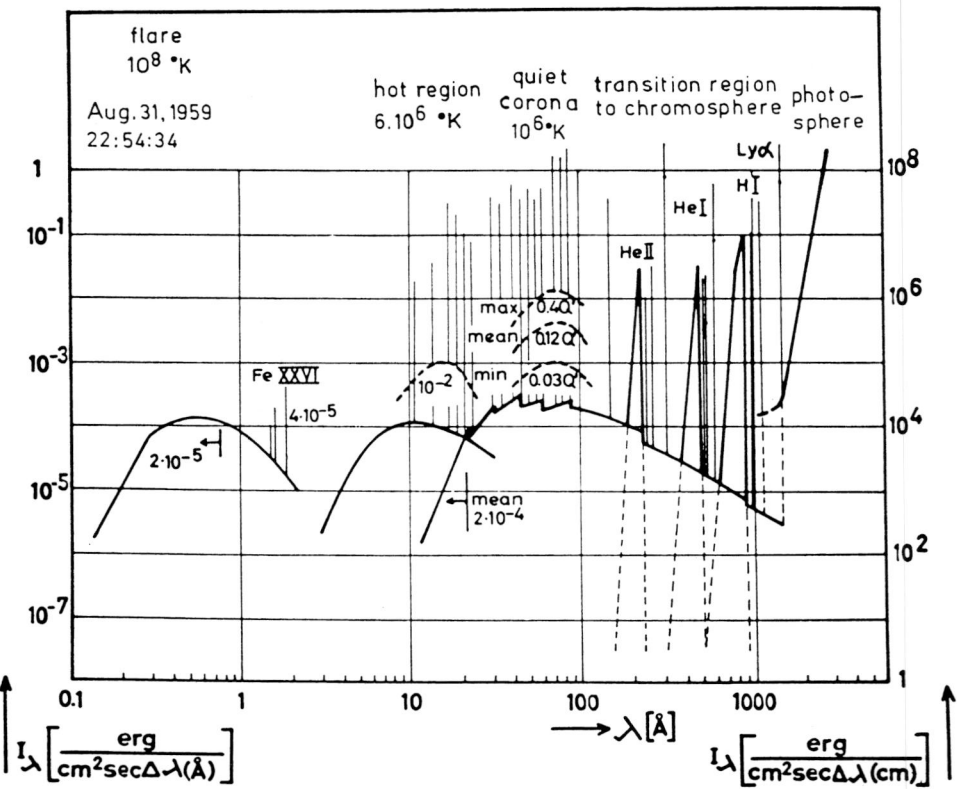

Figure 6 The X-ray spectrum of the Sun as summarized by Elwert (1961). The spectrum of the strong flare of 31 August 1959 is the first indication of occasional very high-energy emission between 10 and 100 keV. (Reproduced with the permission of the *Journal of Geophysical Research*.)

as might be expected, because coronal condensations are denser than the average corona. The emission of a tenuous plasma is proportional with the square of the particle number density.

BROAD-BAND SOFT X-RAY MONITORING OF SOLAR ACTIVITY

The decade after the surprising appearance of Sputnik was marked by the launching of quite a number of generally fairly small satellites with an exploratory task. Although new satellite-launching nations (UK, France) or groups of nations (the Commission Préparatoire Européenne pour la Recherche Spatiale (COPERS), which later became the European Space Research Organisation (ESRO) and still later the European Space Agency (ESA)) prudently entered the scene, the leading roles stayed with the USA and USSR.

In the Soviet Union a rocket programme was started some years after the war. As compared to the US programme there were two differences, and these determined the future direction of research. The first was that biaxial pointing systems were not initially developed to the degree of sophistication reached in the USA. The second aspect was that the scientist leading solar space research (S.L. Mandel'shtam) was a high-energy physicist with experience in counter techniques. This situation naturally defined the direction of research of the Soviet space programme in solar physics. It was centred on counting soft X-rays in the spectral region below 10 Å, with initially not too much emphasis on high spatial resolution. Such a programme evidently yields its best return in flare studies.

An example is the payload of the 'Geophysical Rocket' launched on 21 July 1959. It consisted of Geiger counters with 2 mm mica windows, covered by a 2 μm thick layer of Al. A control counter was turned 15° away from the Sun. Other rockets carried ionization chambers with LiF windows and thermolucent phosphor sensors with photomultipliers for measuring UV and Lyman α emission. The phosphors store energy which is re-radiated at about 5000 Å (Wukelic 1968). Later, this instrument was used several times aboard satellites. In the history of science this

Figure 7 Launching a Nike–Deacon combination for the observation of solar flares from the island of San Nicolas. (Official US Navy photograph.)

instrument was the first solar monitoring instrument to be flown aboard a satellite.

Right at the start of the satellite era, Sputnik 2 (launched on 3 November 1957), as well as Korabl'-Sputnik 2 (launched on 19 August 1960) and Korabl'-Sputnik 3 (1 December 1960), contained elaborate sets of photon counters for measuring radiation between 1 and 120 Å and Lyman α. Similar instrumentation was carried by Elektron 2 (30 January 1964) and Elektron 4 (11 June 1964). Practically all Soviet satellites launched in the three years following Sputnik 1 carried this and similar standard instrumentation for measuring solar energetic radiation. These satellites, along with those launched in the USA in the years thereafter, formed an impressive series of soft X-ray and far-UV monitoring satellites.

In the USA the NRL placed its first instruments aboard Vanguard 3. Later, Friedman and co-workers at the NRL produced a series of Solrad satellites, of which the first was operative in July–November 1960. These spacecraft, monitoring various kinds of energetic radiation, from Lyman α to X-radiation of about 10 keV, were intended for continuous observation of the solar energetic radiation at various wavelengths. In that connection they could be considered a continuation and an important extension of the programme of ground-based and space-borne solar monitoring set up during the International Geophysical Year (1957–58).

The UK Ariel 1 satellite was primarily meant for ionospheric investigations, and solar X-ray monitoring was needed for that purpose. Yet, by pulse-height analysis of the counts secured by proportional photon counters aboard the spacecraft, data on the spectral distribution of radiation were obtained. These results were useful from the viewpoint of solar physics. As an example, Figure 8 presents the X-ray spectra in the range below 10 Å during flares, as compared with the 'quiet' situation. The appearance of very short-wavelength emission, pointing to high temperatures

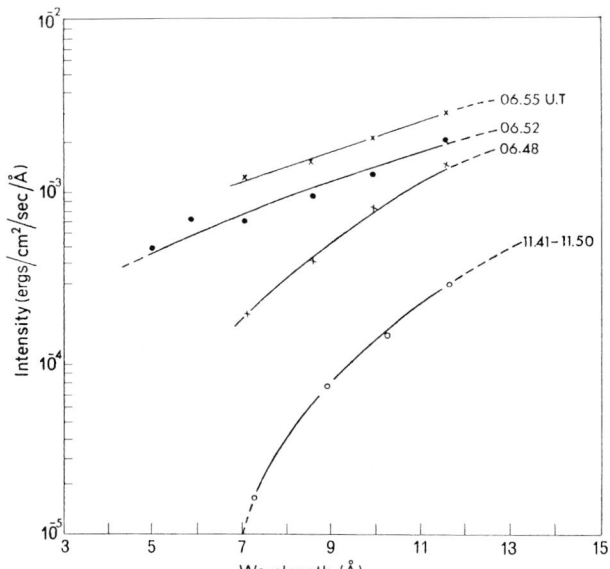

Figure 8 X-ray spectra of a flare on 3 May 1962, obtained by pulse-height analysis of photons observed by proportional counters aboard the Ariel satellite. (From Massey 1964, p. 174, reproduced with the permission of Cambridge University Press.)

or high electron energies, is apparent. In the period between 26 April 1962 and 2 June 1979 six Ariel satellites were launched.

The payload of the ESRO II satellite, the first 'European' satellite, launched in 1968, was for a considerable part similar to Ariel. This shows, among other things, how strongly the British dominated the initial European efforts in space research (later the UK lost its leading position in space research, mainly because of organizational weaknesses). An important addition was that the spacecraft also carried solar instruments developed for an astrophysical goal: monitoring solar variability in X-rays. The restriction to photon energies below some 20 keV was useful for the study of slowly varying phenomena, but appeared to be a limitation when flare physics entered the picture. Attempts to use such observations for studying the structure of the flares, let alone the study of flare dynamics, were doomed to failure because the additional information needed for that study was absent. One would have needed at least data on the flare volume in order to study the radiative properties of the flare, while for the flare dynamics structural information would be needed, along with data on the position of the initial flare kernel(s) with respect to the magnetic field lines. The lack of such data frustrated progress in the study of solar flares. Most important was the absence of high-energy photon data. However, the experiments had their usefulness: they provided a good training period for hosts of young European engineers and physicists who needed to acquaint themselves with this new field. At the same time they provided indications of how to continue the research. One of these was the remarkable observation that the soft X-ray emission did not peak in the very first phase of the flares. 'Something else' apparently started earlier. It is of course obvious that the best approach to solving the flare problem is by searching for the onset of flares and for the physical mechanisms existing at that time, because they determine the subsequent development of the flares.

EVIDENCE FOR HARD X-RAYS ASSOCIATED WITH FLARES: HARD X-RAY MONITORING

During the 1960s there were several indications that flares could sometimes emit hard X-rays in the energy range up to about 100 keV. An example is the flare of 31 August 1959, of which the X-ray spectrum is shown in Figure 6 (Elwert 1961). Another case, serendipitously observed, yielded essential information. On 18 September 1963 J.-P. Legrand launched a balloon from the Arctic to search for high-energy particle fluxes. The instrument package consisted of two Geiger counters that could be used at the same time either individually or as a particle telescope. On that day he observed a strong pulse of hard radiation, correlated in time with a solar flare. The X-ray burst lasted for about 2 minutes (Figure 9).

The burst must have been due to radiation, not particles, since the Geiger telescope showed no response. The X-rays had an energy of ~500 keV, as derived from the relative energy efficiency of the counters. A comparison with the radio data (de Jager 1967) showed that the hard X-ray peak occurred 3 minutes before the peaks in the radio bursts. Precisely at the time of the hard X-ray burst a small knot in the Hα flare occurred above the magnetic inversion line, in a filament connecting the two main bodies of the Hα flare. It disappeared a few minutes later, when also the hard X-ray burst had finished. The analysis yielded the total number of electrons in the gas and the electron density, leading to a volume that, if spherical, would have a diameter of 5000 km.

Studies like this one showed two things. The first is the importance of observing flares at *higher energies* than was customary. The study also suggests that in its very earliest phase a flare might be restricted to small nodules, containing very energetic particles and accelerated at the flare's onset. This leads to the picture of localized flare ignition, resulting in a small volume containing high-energy particles. Hence the importance of *flare images* in hard X-rays. The first demand, that of getting high-energy flare observations, was met by the launch of ESRO's TD1A satellite in 1972. For the second, flare imaging in hard X-rays, one had to wait until 1980, when imaging solar observatories entered the field.

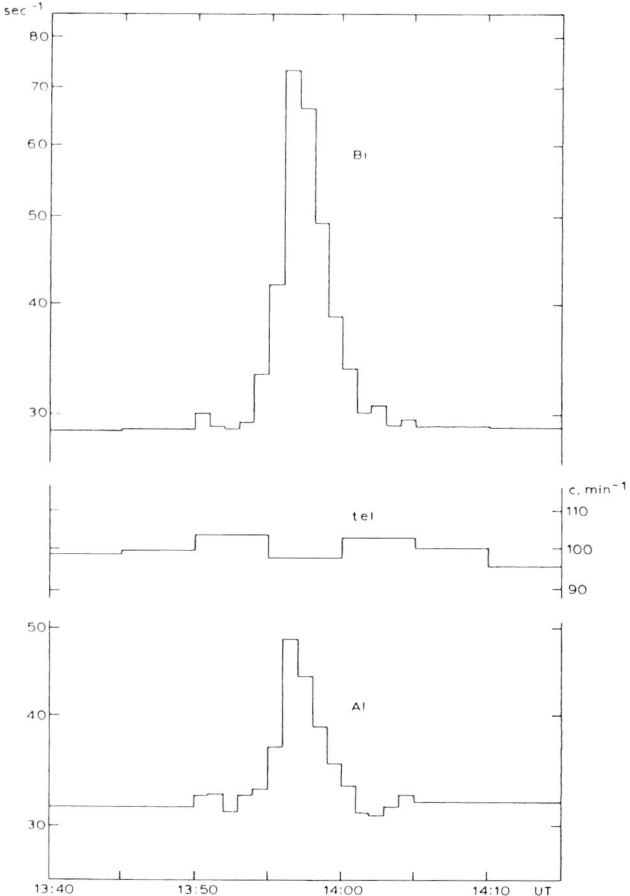

Figure 9 Upper panel: a hard X-ray pulse of solar origin observed by J.-P. Legrand on 18 September 1963 with a Be counter. Middle panel: telescope response. Lower panel: counts with Al window. (de Jager 1967.)

TOWARDS AN INITIAL FLARE MODEL

In the second half of the 1960s the first studies appeared directed at studying the flare physics and finding the flare parameters. In the USSR such data were obtained by the Lebedev Physical Institute, under Mandel'shtam, using rocket and satellite instruments. Their standard instrument was a very high-resolution X-ray spectrograph, operating in the spectral band from 1.85 to 1.87 Å (an example of a spectrum is shown in Figure 14). Apart from getting spectra, the Lebedev group was concerned with measuring spectral polarization in X-rays in flares. The X-ray polarimeter consisted of Be scattering plates surrounded by six photon counters. For flares observed in October and November 1970 the degree and angle of polarization was measured: the degree turned out to be highest in the initial phase, and reducing to zero later on (Tindo et al. 1972). This aspect was clearest in a limb flare observed on 5 November 1970 (Figure 10).

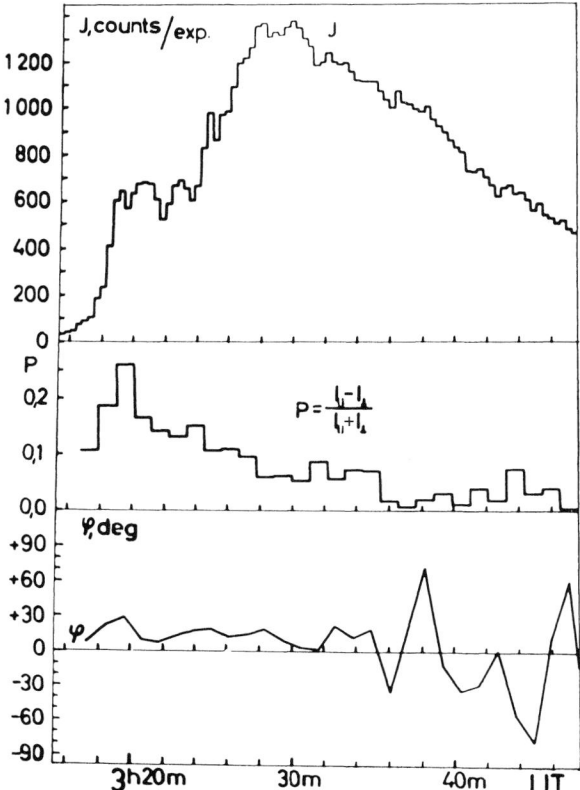

Figure 10 Polarization in the limb flare of 5 November 1970. The diagram shows, from top to bottom: count rates, degree of polarization and polarization angle. (Tindo et al. 1972.)

The observations suggest a flare model with strong unidirectional motion in its early phase. Acceleration in a loop to energies above 10 keV was suggested. The accelerated high-energy electrons would then spiral along the loop. They are part of a hot plasma, for which a temperature of about 20 million K was measured. This scenario finds support in the first X-ray images of a flare (Zhitnik et al. 1967) obtained on Cosmos 166 and Cosmos 230. Although crude, these observations showed for the first time that the flare site appeared to be located at heights between 20,000 and 25,000 km. Moreover it was concluded that the flares must have had a filamentary structure, as follows also from a comparison of the electron density of the emitting plasma with the integrated number of electrons and the flaring volume and from a study of the images.

THE ORBITING SOLAR OBSERVATORIES

At the time of the developments described above, NASA was well ahead in planning a new generation of solar satellites. Ideas for fancy and highly elaborate satellites were drifting around; Homer Newell in particular supported these. At one time Europe too had ambitious plans for very advanced solar observations from space. At a meeting in January 1966 held at Nice, a project for a pointing solar observatory called ELOISE (European Large Orbiting Instrumentation for Solar Experiments) was developed. It did not materialize, being too ambitious. In the USA also, the more ambitious plans failed to materialize. In their place came a project for a realistic and feasible pointing solar observatory, small but sophisticated. That this could happen was mainly due to a fortuitous coincidence of two factors. The main success of the project should be ascribed to a pragmatic and strongly motivated person, John Lindsay. The other positive factor was that a new company, Ball Brothers of Boulder, Colorado, was capable and willing to develop and build the spacecraft. This combination of conditions convinced both the US scientific community and NASA that the construction of an Orbiting Solar Observatory (OSO; Figure 11) was appropriate and timely.

Although OSO was a small satellite, it was a real observatory. Essentially it consisted of a spin-stabilized platform with a Sun-pointing controlled instrument section. Lindsay's achievement and drive were impressive: only three years were needed to get the observatory into orbit – OSO 1 was launched on 7 March 1962. The important novel aspect of the OSO series is that the pointing section made possible high-resolution spectral studies, previously achieved only by rocket-borne spectrographs. For the first time the variation with time of the UV and X-ray spectra of transient phenomena, such as active regions and flares, could be observed. In addition, a simple imaging device yielded positional information on flares and some crude data on their geometry.

The success of the first OSO led to the decision to continue the programme. A start was made with OSO 2, and development work for OSO 3, 4 and 5 went ahead. In the enthusiasm surrounding the initial success a further step forward, the construction of an Advanced Orbiting Solar Observatory, was considered. But from that time (1962) onward everything that could go wrong did go wrong. In April 1964 static electricity led to ignition on board OSO 2, which caused much damage. The rebuilt OSO 2 was later launched into orbit, but electric arcing destroyed several instruments, in particular Goldberg's sophisticated UV spectrometer/spectroheliometer. In August 1965 OSO 3 failed to reach orbit, due to rocket failure. In the same year, Lindsay, the main driver of the project, died of a heart attack, and it was decided to cancel the Advanced OSO project.

After this disastrous series of events things eventually got back on track. In the period to 1975, OSOs 4–8 were successfully launched. The OSO series contributed greatly to the advance of knowledge, both of the quiet Sun and of solar active regions, including flares. As an example, Figure 12 shows the X-ray spectra of a flare on 22 March 1967 (Neupert *et al.* 1969).

THE OSO CONTRIBUTION TO ADVANCING SPECTRAL DIAGNOSTICS

One of the most important contributions of the OSO series was the availability of a continuous series of high-resolution far-UV and X-ray spectra. For a review of the state of knowledge of the emission line EUV spectrum in 1971, see Jordan (1972); the flare X-ray spectrum was reviewed at the same time by Walker (1972). The availability of such spectra and the possibility of comparing them with spectra of hot laboratory plasmas (Fawcett *et al.* 1963) presented the right

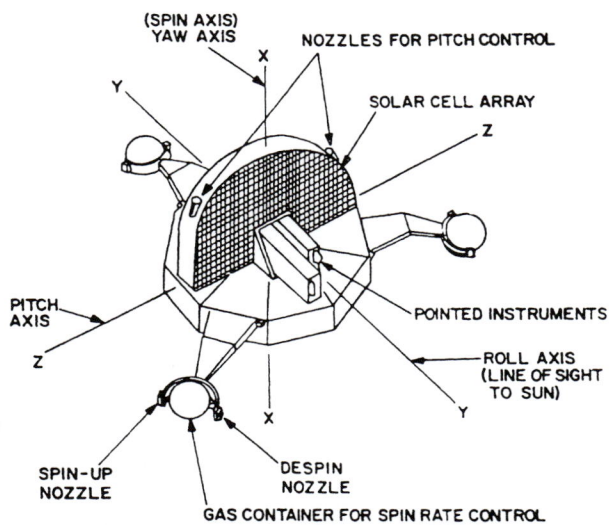

Figure 11 Sketch of the Orbiting Solar Observatory (Hufbauer 1991, p. 171.), above and photograph of OSO-6 in the laboratory (Ball Brothers Corporation).

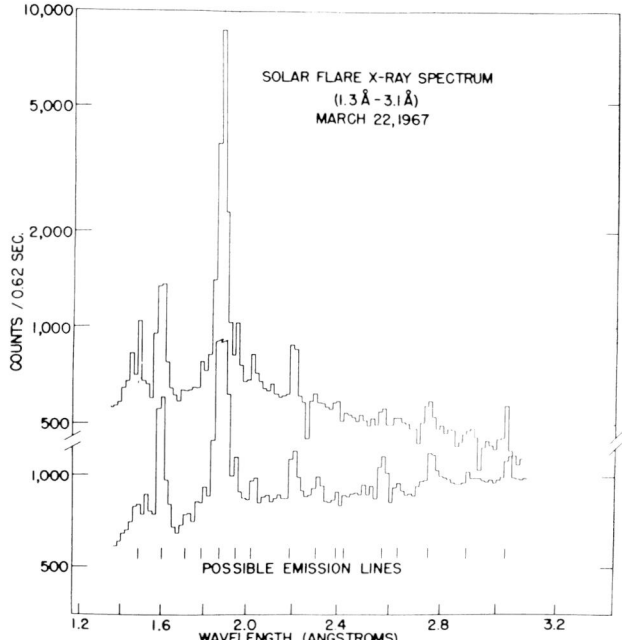

Figure 12 Two spectral scans in the region 1.3 to 3.1 Å during the increasing phase of the flare of 22 March 1967 (Neupert et al. 1969). (Reproduced with the permission of North-Holland.)

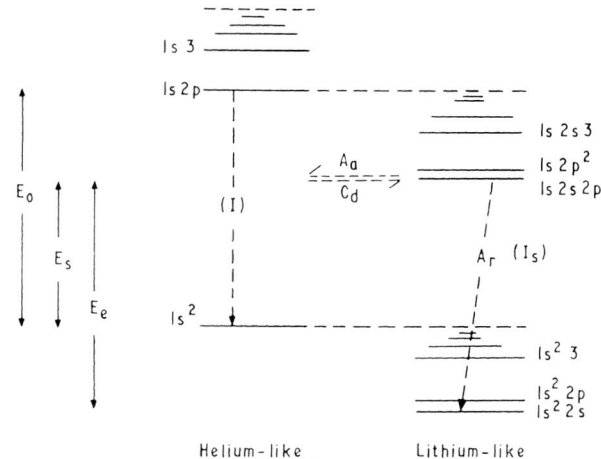

Figure 13 Energy level diagram of the helium- and associated lithium-like ion, showing the helium-like and intercombination transitions (Gabriel 1972a). (Reproduced with the permission of *Space Science Reviews*.)

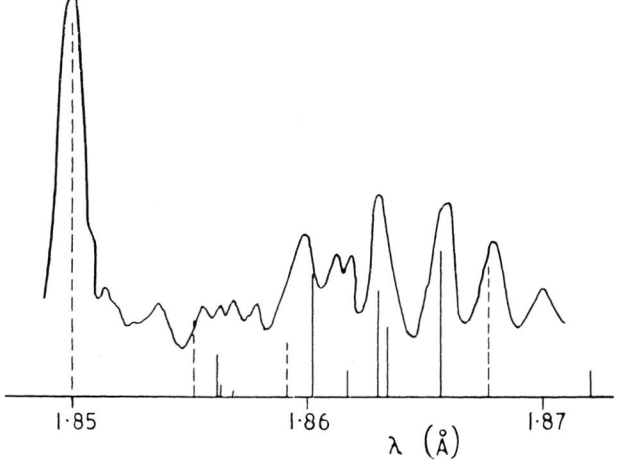

Figure 14 A solar flare spectrum near 1.8 Å observed by the Intercosmos 4 satellite (Grineva et al. 1973). The vertical lines are intensities calculated by Gabriel (1972b). (Reproduced with the permission of *Space Science Reviews*.)

conditions for advancing the technique of spectral diagnosis for these spectral regions.

An important contribution was the identification of the so-called satellite lines in helium-like spectra by Gabriel and Jordan (1969), which were worked out by Gabriel and co-workers (Gabriel et al. 1969; Gabriel and Paget 1972). Edlén and Tyrén (1939) had discovered these lines in laboratory spectra, but their identification remained a mystery for a long time. Gabriel et al. (1969) put forward the hypothesis that the satellite lines are due to dielectronic recombination in helium-like ions, whereby the recombination involves inner shell excitation of the related lithium-like ion, as shown in the energy level scheme of Figure 13. Quantitative verification by Gabriel and Paget (1972) confirmed the identification. Confirmation is shown in Figure 14, where a solar flare spectrum observed by Grineva et al. (1973) is compared with a computed spectrum (Gabriel 1972), showing good agreement.

The importance of these results for spectral diagnostics is that the intensities in the helium-like complexes are mainly density dependent. Temperature dependence is a second-order effect. Therefore their measurement offers a way to determine the density structure of the part of the solar atmosphere studied.

X-RAY SPECTROSCOPY OF FLARES: SOME HIGHLIGHTS

The first X-ray spectrum of a solar flare is due to Neupert (1967) from OSO 3 observations. For a flare on 16 November 1970, Grineva et al. (1973) published detailed spectra showing lines of Fe XXIII to Fe XXV. They found indications for unidirectional macroscopic motion with velocities of the order of 100 km s^{-1}. From data such as those shown in Figures 10, 12 and 14 came the first information on the non-thermal, inhomogeneous aspects of flares. Friedman (1969) determined the temperature parameters for a flare on 1 January 1968:

- kinetic temperature (from the slope of the continuous spectrum): 15–20 million K;
- ionization temperature (from the K spectra): 1–4 million K;
- excitation temperature (from the Lyman α lines): 8 million K.

The hard X-ray observations by the TD1A satellite yielded the result that most of the flares investigated emit in X-rays with energies between about 30 and 100 keV during a relatively short interval of time. Emission lasted on average about 1–2 minutes. It appeared that the emission was in separate bursts that lasted between 4 and 24 seconds (Figure 15) (de Jager and de Jonge 1978). These bursts were coined 'elementary flare bursts' since they could not be decomposed further, given the instrumental time resolution of 1.2 seconds. (Note that later observations with higher resolution, with instruments aboard the Solar Maximum Mission, proved the presence of bursts of shorter duration.) What is important is that for one and the same flare and for the same energy range all elementary flare bursts had the same duration. Hence, for a given energy range it is possible to identify a characteristic duration of the elementary flare burst for every flare investigated. The half-width decreases with increasing photon energy: for 100 keV their duration is about half that at 30 keV. These and similar observations placed clear constraints on flare models.

Such was the situation around 1975. Further progress had to wait for the sophisticated imaging solar satellites of the 1980s and 1990s.

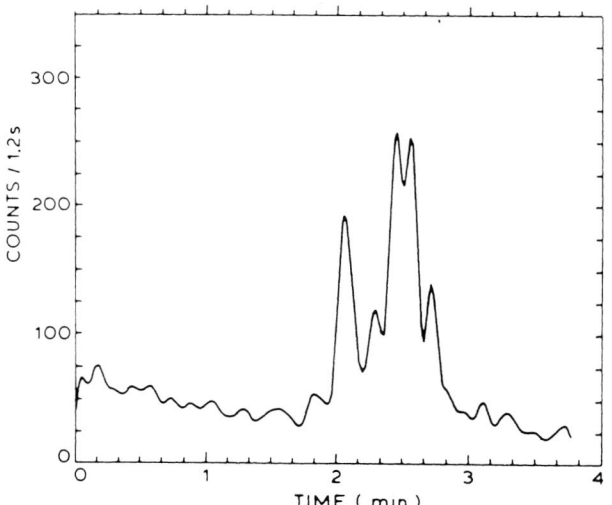

Figure 15 Hard X-ray 'elementary flare bursts' in the range 29–41 keV, during the initial phase of the flare of 2 August 1972 (de Jager and de Jonge 1978). (Reproduced with the permission of *Solar Physics*.)

THE LYMAN-ALPHA CORONA

Following the surprise discovery of the Lyman-α corona (Speer et al. 1970, Figure 16) many years had to pass as well until its detailed investigation could be taken up. During the eclipse of 7 March 1970, a rocket equipped with an extreme-ultraviolet spectrograph was launched into the umbra. A sequence of objective-grating spectra – 'flash' spectra, taken while the lunar disk progressively occulted the solar limb – offered a greatly improved height-resolution (of a few arc sec) in ultraviolet radiation. The extended Lyman-α corona, which appeared on the film that was developed after the recovery of the rocket payload, presented a riddle until it was realised (Gabriel et al. 1971) that the phenomenon was due to fluorescent excitation: neutral hydrogen, whose density in the corona is of the order of 10 atoms cm^{-3} – in spite of the high temperature – is absorbing and re-emitting Lyman-α photons stemming from the chromosphere. The discovery of the Lyman-α corona opened up a new observing technique: ultraviolet spectrometry of the corona by use of coronagraphs. Suitable instruments flown towards the end of the 1970s (Kohl et al. 1980) gave insight into the kinetic temperature first of hydrogen, but later also of minor ions. This led, by the end of the century, to the sophisticated observations of the spectrum of the extended corona by the Ultraviolet Coronagraph Spectrometer (Kohl et al. 1997) on SOHO – the Solar and Heliospheric Observatory – that opened new vistas regarding the acceleration processes of the solar wind.

SKYLAB: THE FIRST CREWED SPACE LABORATORY

No project starts out of nothing. Talking and thinking are involved, often for a considerable time. Ideas for a crewed space laboratory for solar observations had already been put forward by Lindsay before his untimely death. The prospects looked more and more hopeful when the preparations for crewed lunar flights were in full swing. The idea that astronauts would command the instrumentation and react to unexpected events was the main driver. There was the drawback that the stability of the spacecraft would be badly influenced by the presence of people aboard, but this seemed a problem that could be solved.

After the plans for the Advanced OSO had been cancelled, the pressure on NASA for another advanced solar observatory grew. In December 1965 NASA decided to embark on a study of a crewed solar observatory, in the framework of, and parallel to, the Apollo lunar programme. Some of the investigations that had been made for the cancelled Advanced OSO could be used for the new project. In the same way, the contractors initially involved in the Advanced OSO could also be involved in the new mission (Newkirk et al. 1977, Hufbauer

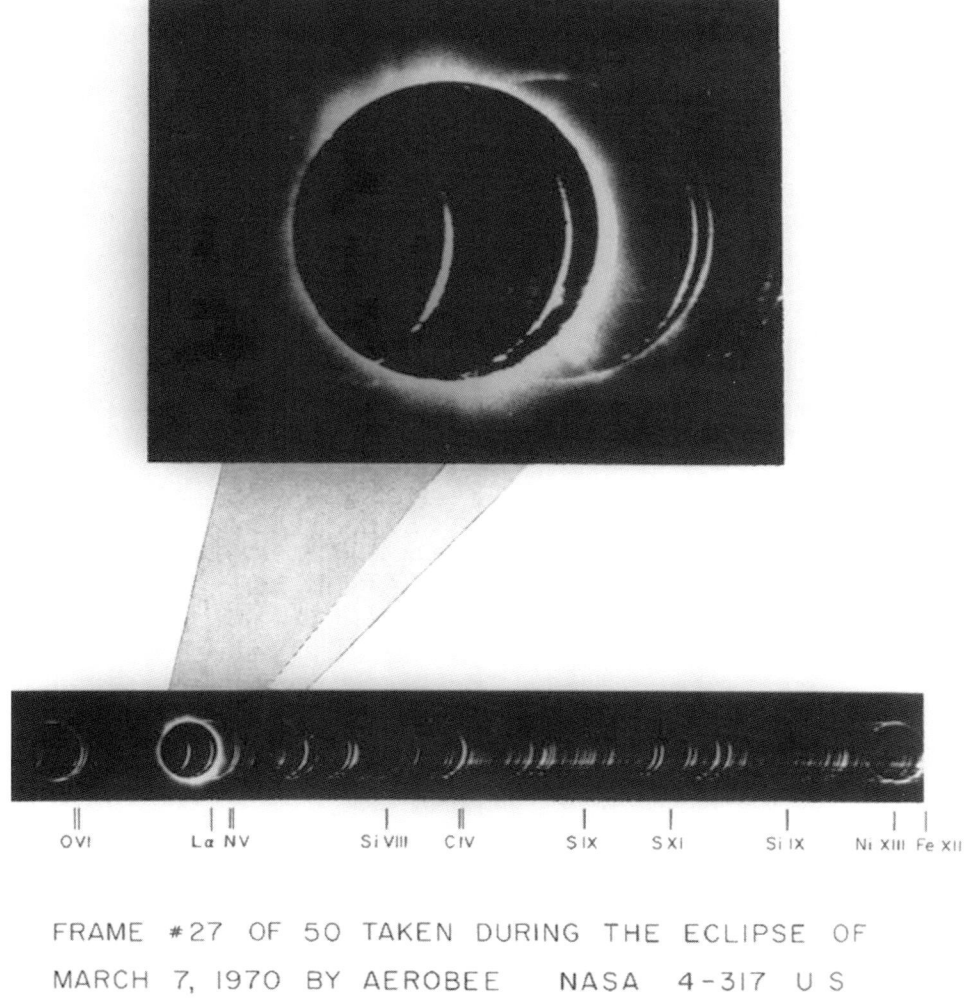

Figure 16 The ultraviolet solar spectrum, taken with an objective grating during the eclipse of 7 March 1970, on which the extended Lyman-α corona was discovered. (Gabriel et al. 1971, courtesy Harvard College observatory.)

1991). In July 1966 the go-ahead was given. The Marshall Space Flight Center (Huntsville, Alabama) would manage the project. Four principal investigators were appointed: Goldberg, Tousey, Newkirk and Giacconi. The instrument, the Apollo Telescope Mount, should be ready for launch at the end of 1968. Evidently that deadline was too close, and almost immediately the principal investigators stated that they would be unable to meet it.

The deadline was extended, also because the mission as initially conceived had been extended in many respects. The most important extension was that more time appeared to be needed for training the astronauts in taking solar observations in an independent way. Specifically, much time appeared to be needed for acquainting the astronauts with the techniques for observing solar flares. The problem of priorities in the day-to-day observing programmes had to be solved. Finally, in 1971 a set of co-ordinated observing routines was agreed upon. Another reason for delaying the project was that the observing period, initially set at two weeks, was extended. In addition, it was decided that not one but three different crews should visit Skylab.

The six instruments aboard Skylab (with their 'home' laboratories and principal scientific investigators) were:

- A white light coronograph (High Altitude Observatory; principal investigator R. McQueen)
- X-ray spectrographic telescope (American Science and Engineering; G. Vaiana)
- X-ray telescope (NASA Marshall Space Flight Center; J. Underwood)
- UV spectrometer/spectroheliometer (Harvard College Observatory; E. Reeves)

- XUV spectroheliograph (Naval Research Laboratory; R. Tousey)
- UV spectrograph (Naval Research Laboratory; R. Tousey)

Skylab, with the Apollo Telescope Mount, was launched on 14 May 1973. The launch did not proceed smoothly. The micrometeorite shield around Skylab had broken off during launch, damaging two solar panels. One was completely out of order. The launch of the first crew was therefore delayed, because the programme of repair had to be studied. On 25 May the crew went up, and the damage was repaired provisionally. A special 'umbrella' that kept the temperature inside the spacecraft within working limits replaced the thermal function of the micrometeorite shield. Also, the damaged solar panel could be repaired. The observatory remained fully operational for eight months. Figure 17 shows Skylab in orbit. The astronauts worked in close co-operation with ground-based observers. An X-ray rocket image taken on the day of the 7 March 1970 eclipse (Van Speybroeck *et al.* 1970) confirmed the existence of coronal holes and showed for the first time the size of one such hole (Figure 18).

CORONAL HOLES AND THE DISCOVERY OF CORONAL MASS EJECTIONS

Two of the major results from Skylab were information on the temporal development of coronal holes and the discovery of coronal mass ejections (CMEs). In the initial publications the main emphasis was on the coronal holes, and the CMEs were mentioned only in passing, and not yet under their present name.

Strictly speaking, the coronal holes had been discovered much earlier, but at that time they received only little attention. Waldmeier (1957, 1975) was the first to mention, from observations with coronagraphs, the existence of extended coronal regions of greatly reduced emission intensity. He called them *Koronale Löcher* ('coronal holes'). Later, it was noted that eclipse observations often showed that certain parts of the corona did not emit any observable radiation. An X-ray rocket image of the Sun taken on 7 March 1970 (Van Speybroeck *et al.* 1970) confirmed the existence of coronal holes and showed for the first time the size of one such hole (Figure 18). Altschuler and Perry (1972) and Munro and Withbroe (1972) confirmed this. The term

Figure 17 Skylab in orbit. (NASA photograph.)

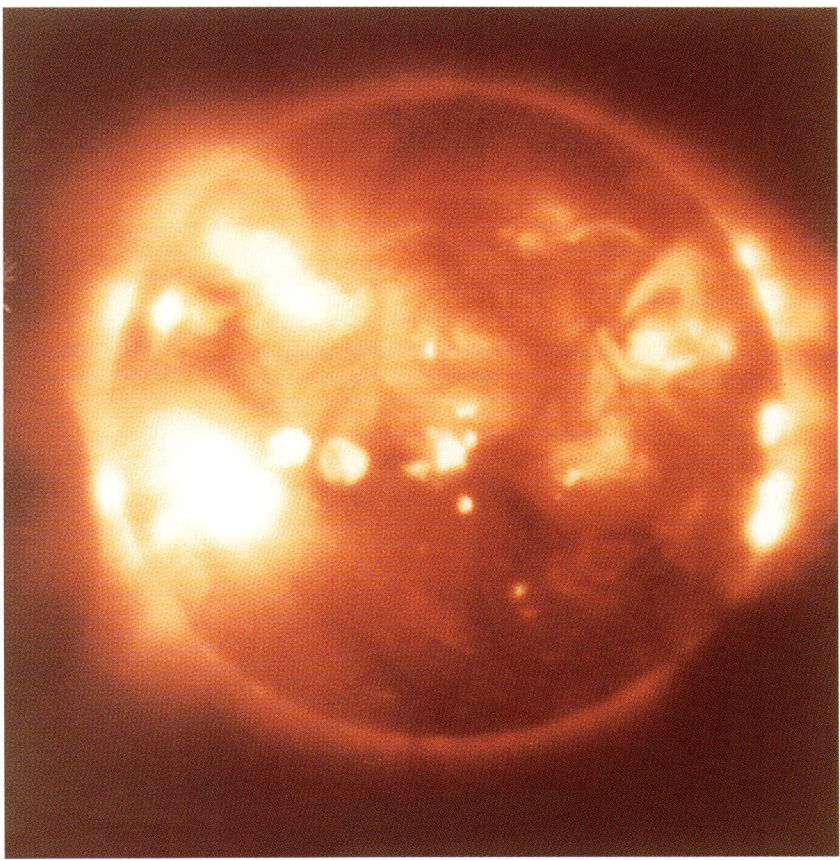

Figure 18 The first X-ray picture of a coronal hole (Van Speybroeck *et al.* 1970). The picture was taken shortly after fourth contact of the eclipse of 7 March 1970; the lunar disk still covers part of the corona seen above the eastern solar limb. (Courtesy American Science and Engineering, Inc.)

'coronal hole' was at that time already in use and has been taken over by others in subsequent years.

An additional aspect was that high-velocity streams in the solar wind might originate from magnetically open regions of the Sun, a hypothesis first put forward by Billings and Roberts (1964). This was confirmed just before the launch of Skylab, when Krieger *et al.* (1973) found that coronal holes are sources of high-speed solar winds. Coronal holes were thus on the agenda for the Skylab crews.

It soon appeared that coronal holes could be observed over a wide range of wavelengths from X-rays to radio wavelengths. They are best 'visible' in X-rays, where their intensity is about one-tenth that of the average adjacent corona (Bohlin 1977). They persist for a fairly long time, and Krieger (1977) published pictures of a coronal hole seen over five successive solar rotations (1 June–18 September 1973). Also, their relation to high-speed solar wind emission was confirmed (Hundhausen 1977).

An unexpected discovery from Skylab was that of the CMEs. We only discuss them briefly here, since they are the subject of a more thorough discussion elsewhere in this volume, and because the bulk of their observations was made with the Solar Maximum Mission (SMM) and later spacecraft such as the ESA–NASA Solar and Heliospheric Observatory (SOHO). Interestingly, the term CME originated only later. Shortly after their discovery they were called 'temporary coronal depletions' (Rust and Hildner 1976).

The CMEs are observed in Thomson-scattered white light observations, and since their discovery have also been observed with K-coronameters on Earth and from other spacecraft such as SMM and SOHO. A few thousand of them have been detected so far. Their average mass is 5 trillion kilograms, and their velocity through the heliosphere ranges between a few hundred and a few thousand kilometres per second. They are often associated with disrupting filaments. The debate on whether CMEs are the cause or the consequence of flares has not yet ended. Anyway, their occurrence is correlated with flares, but not too strictly so. To me it seems that the flare, the disrupting filament and the CME are all three due to a 'grand instability' in an active region (Sakai and de Jager 1996, p. 41). None of the three is the primary cause.

Acknowledgement

The author is obliged to Dr G. Wolfschmidt for useful information and to Professors M.C.E. Huber and A. Russo for helpful suggestions.

REFERENCES

Aleksandrov, S.G. and Fedorov, R.E. (1961). *Sovietskie sputniki i kosmicheskie korabli*. Izd-vo Akademiia Nauk SSSR, Moscow.

Altschuler, M.D., Trotter, D.E. and Orrall, F.Q. (1972). Coronal holes. *Solar Physics*, **26**, 354–365.

Billings, D.E. and Roberts, W.O. (1964). The origin of M-region geomagnetic storms. *Astrophisica Norvegica*, **9**, 147.

Beigman, I.I., Grineva, Yu.I., Mandel'shtam, S.L., Vainstein, L.A. and Zhitnik, I.A. (1969). On the localization, size and structure of the regions of the X-ray flares on the Sun. *Solar Physics*, **9**, 160.

Bohlin, J.D. (1977). An observational definition of coronal holes. In J.B. Zirker (ed.), *Coronal Holes and High-Speed Wind Streams*, Colorado Associated University Press, Boulder, CO, pp. 27–69.

Cushman, G.W. and Rense, W.A. (1978). Solar He II (304 Å) and Si XI (303 Å) line profiles. *Solar Physics*, **58**, 299–305.

de Jager, C. (1955). *Annales de. Géophysique*, **11**, 330.

de Jager, C. (1959). Structure and dynamics of the solar atmosphere. In *Encyclopaedia of Physics*, **52**, 50–362.

de Jager, C. (1967). The hard solar X-ray burst of 18 September 1963. *Solar Physics*, **2**, 327–346.

de Jager, C. and de Jonge, G. (1978). Properties of elementary flare bursts. *Solar Physics*, **58**, 127–137.

Dellinger, J.H. (1935). A new radio transmission phenomenon. *Physical Review*, **48**, 705.

Edlén, B. and Tyrén, F. (1939). Atomic energy states of an unusual type. *Nature*, **143**, 940–941.

Elwert, G. (1954). Die weiche Röntgenstrahlung der ungestörten Sonnenkoronen. *Zeitschrift für Natürforschung*, **9**, 637–653.

Elwert, G. (1956). In W.J.G. Beynon and G.M. Brown (eds), *Solar Eclipses and the Ionosphere*, Pergamon, London, p. 167.

Elwert, G. (1961). Theory of X-ray emission of the Sun. *Journal of Geophysical Research*, **66**, 391–401.

Fawcett, B.C., Gabriel, A.H., Griffin, W.G., Jones, B.B. and Wilson, R. (1963). Observations of the zeta spectrum in the wavelength range 16Å–400Å. *Nature*, **200**, 1303–1304.

Friedman, H., Lichtman, S.W. and Byram, E.T. (1951). Photon counter measurements of solar X-rays and extreme ultraviolet light. *Physical Review*, **83**, 1025–1030.

Friedman, H. (1969). X-ray observations of solar flares. In C. de Jager and Z. Švestka (eds), *Solar Flares and Space Research*, North-Holland, Amsterdam, pp. 87–94.

Friedman, H. (1977). Reminiscences of 30 years of space research. *Naval Research Laboratory Report*, 8113.

Gabriel, A.H. (1972a). Dielectronic satellite spectra in the soft X-ray region. *Space Science Reviews*, **13**, 655–664.

Gabriel, A.H. (1972b). Dielectronic satellite spectra for highly-charged helium-like ion lines. *Monthly Notes of the Royal Astronomical Society*, **160**, 99–119.

Gabriel, A.H., Jordan, C. and Paget, T.M. (1969). In *Proceedings of the 6th International Conference on Physics of Electronic and Atomic Collisions*, MIT Press, MA, p. 558.

Gabriel, A.H., Garton, W.R.S., Goldberg, L., Jones, T.J.L., Jordan, Carole, Morgan, F.J., Nicholls, R.W., Parkinson, W.J., Paxton, H.J.B., Reeves, E.M., Shenton, C.B., Speer, R.J. and Wilson, R. (1971). Rocket Observations of the Ultraviolet Solar Spectrum during the Total Eclipse of 1970 March 7. *Astrophysical Journal*, **169**, 595–614.

Gabriel, A.H. and Paget, T.M. (1972). *Journal of Physics B*, **2**, 1072.

Grineva, Yu.I., Karev, V.I., Korneev, V.V., Krutov, V.V., Mandel'shtam, S.L., Vainstein, L.A., Vasiyev, B.N. and Zhitnik, I.A. (1973). Solar X-ray spectrum observed from the Intercosmos-4 satellite and the Vertical-2 rocket. *Solar Physics*, **29**, 441–446.

Hinteregger, H.E. (1965). In C. de Jager (ed.), *The Solar Spectrum*, Reidel, Dordrecht, p. 179.

Hufbauer, K. (1991). *Exploring the Sun: Solar Science Since Galileo*, Johns Hopkins University Press, Baltimore, MD.

Hundhausen, A.J. (1977). An interplanetary view of coronal holes. In J.B. Zirker (ed.), *Coronal Holes and High-speed Wind Streams*, Colorado Associated University Press, Boulder, CO, pp. 225–329.

Jordan, C. (1972). Identifications of emission lines in the euv solar spectrum. *Space Science Reviews*, **13**, 595–605.

Kiepenheuer, K.O. (1937). Über die Sonnenstrahlung zwischen 2000 und 3000 Å. *Veröffentlichungen der Universitäts-Sternwarte Göttingen*, **57**, 348.

Kohl, J.L., Weiser, H., Withbroe, G.L., Noyes, R.W., Parkinson, W.H., Reeves, E.M., Munro, R.H. and MacQueen, R.M. (1980). Measurements of coronal kinetic temperatures from 1.5 to 3 solar radii. *Astrophysical Journal*, **241**, L117–L121.

Kohl, J.L., Noci, G., Antonucci, E., Tondello, G., Huber, M.C.E., Gardner, L.D., Nicolosi. P., Strachan, L., Fienschi, S., Raymond, J.C., Romoli, M., Spadaro, D., Panasyuk, A., Siegmund, O.H.W., Benna, C., Ciaravella, A., Cranmer, S.R., Giordano, S., Karovska, M., Martin, R., Michels, J., Modigliani, A., Naletto, G., Pernechele, C., Poletto, G. and Smith, P.L. (1997). First Results from the SOHO Ultraviolet Coronagraph Spectrometer. *Solar Physics*, **175**, 613–644.

Korneev, V.V., Krutov, V.V., Mandel'shtam, S.L., Sylwester, B., Tindo, I.P., Uznov, A.M., Valnicek, B. and Zhitnik, I.A. (1980). Solar flare X-ray spectra III. *Solar Physics*, **68**, 381–389.

Korneev, V.V., Krutov, V.V., Mandel'shtam, S.L., Urnov, A.M., Zhitnik, I.A., Kononov, A.Ya., Sylwester, B. and Sylwester, J. (1979). Solar flare X-ray spectra. *Solar Physics*, **63**, 319–327.

Krieger, A.S. (1977). Temporal behavior of coronal holes. In J.B. Zirker (ed.), *Coronal Holes and High-speed Wind Streams*, Colorado Associated University Press, Boulder, CO, pp. 71–102.

Krieger, A.S., Timothy, A.F. and Roelof, E.C. (1973). A coronal hole and its identification as a source of a high velocity solar wind stream. *Solar Physics*, **29**, 505–525.

Massey, H. (1964). *Space Physics*, Cambridge University Press.

Massey, H. and Robins, M.O. (1986). *History of British Space Science*, Cambridge University Press.

Munro, R.H. and Withbroe, G.L. (1972). Properties of a coronal hole derived from extreme-ultraviolet observations. *Astrophysical Journal*, **176**, 511–520.

Neupert, W.M., Gates, W., Swartz, M. and Young, R. (1967). Observation of the solar flare X-ray emission-line spectrum of iron from 1.3 to 20 Å. *Astrophysical Journal*, **149**, L79.

Neupert, W.M., Swartz, M.M., White, W.A. and Young, R.M. (1969). Observations of the solar flare soft X-ray spectrum and comparison with centimeter radio bursts. In C. de Jager and Z. Švestka (eds), *Solar Flares and Space Research*, North Holland, Amsterdam, pp. 95–101.

Newkirk, R.W., Ertel, I.D. and Brooks, C.G. (1977). *Skylab, A Chronology*. NASA SP-4011.

Rust, D.M. and Hildner, E. (1976). Expansion of an X-ray coronal arch into the outer corona. *Solar Physics*, **48**, 381–387.

Speer, R.J., Garton, W.R.S., Goldberg, L., Parkinson, W.H., Reeves, E.M., Morgan, J.F., Nicholls, R.W., Jones, T.J.L., Paxton, J.H.B., Shenton,

D.B. and Wilson, R. (1970). Rocket UV Flash Spectra from the Solar Eclipse of March 7, 1970. *Nature*, **226**, 249–250.

Tousey, R. (1953). Solar work at high altitudes from rockets. In G.P. Kuiper (ed.), *The Sun*, University of Chicago Press, pp. 658–676.

Tousey, R. (1955). Emission lines in the extreme ultra-violet spectrum of the Sun. In P.Th. Oosterhoff (ed.), *Transactions of the IAU*, Vol. 9, Cambridge University Press, pp. 179–180.

Tousey, R. (1958). The extreme ultra-violet spectrum of the Sun. In D.H. Sadler (ed.), *Transactions of the IAU*, Vol. 10, Cambridge University Press, pp. 708–710.

Tousey, R. (1963). The extreme ultra-violet spectrum of the Sun. *Space Science Reviews*, **2**, 3–69.

Tousey, R. (1986). Solar spectroscopy from Rowland to SOT. *Vistas in Astronomy*, **29**, 175–199.

Tindo, I.P., Ivanov, V.D., Valnicek, B. and Livshitz, M.A. (1972). Preliminary interpretation of the polarization measurements performed on 'Intercosmos-4' during three solar X-ray flares. *Solar Physics*, **27**, 426–435.

Van de Hulst, H.C. (1953). The chromosphere and the corona. In G.P. Kuiper (ed.), *The Sun*, University of Chicago Press, pp. 207–321.

Van Speybroeck, L.P., Krieger, A.S. and Vaiana, G.S. (1970). X-ray photographs of the Sun on March 7, 1970. *Nature*, **227**, 818–822.

Waldmeier, M. (1957). *Die Sonnenkorona*, Birkhäuser, Basel.

Waldmeier, M. (1975). The coronal hole at the 7 March 1970 solar eclipse. *Solar Physics*, **40**, 351–358.

Walker, A.B.C. (1972). The coronal X-spectrum: Problems and prospects. *Space Science Reviews*, **13**, 672–730.

Wolfschmidt, G. (1993). Kiepenheuer's Gründung von Sonnenobservatorien im Dritten Reich. In *Wissenschaftliches Jahrbuch 1992/1993*, Deutsches Museum, Munich, pp. 291–315.

Wukelic, G.E. (1968). *Handbook of Soviet Space Science Research*, Gordon & Breach, New York.

9

EUGENE N. PARKER*

A history of the solar wind concept

One presumes that the supersonic solar wind has been blowing since the formation of the Solar System, providing the dynamical plasma and magnetic field throughout interplanetary space. It appears that the early massive solar wind and simultaneous strong solar magnetic fields carried away the initial angular momentum of the young Sun during the first 3×10^8 years or so, leaving the leisurely 25-day equatorial rotation period that we see today (Schatzman 1959, 1973; Biermann 1973). Nowadays the tenuous wind carries no significant angular momentum or mass from the Sun, but it continues to be the dominating condition in interplanetary space, providing the outward sweeping spiral magnetic field, impacting the terrestrial magnetosphere, drastically reducing the intensity of the galactic cosmic rays, and pushing the interstellar gas and galactic magnetic field out beyond the farthest planets. The consequences for the terrestrial environment are profound, but not immediately obvious to us who dwell at the surface of Earth. So the history of the thoughts leading to the recognition of the solar wind extends back more than two millennia. Progress has been paced largely by the development of physics from the days of Gilbert (1544–1603) and Galileo (1562–1642). That is to say, with the exception of the aurora, the effects of the solar wind are not available to our biological senses, but require observations and measurements that can be made only with specially devised scientific instruments. And then it requires an advanced state of physics to infer the implications of the observations. So the history of the solar wind follows both the advances in fundamental classical physics and the studies of the natural phenomena pertinent to the solar wind.

*University of Chicago, IL, USA

The presence of the solar wind is directly indicated by the continuing fluctuations of the geomagnetic field at the surface of Earth, particularly at high latitudes, the continuing aurora at high latitudes, the varying intensity of the galactic cosmic rays, and the antisolar orientation of the gaseous tails of comets. However, these effects, known individually for decades and centuries, are evidence only of some form of external disturbance, and the scientific challenge over the last century has been to work out precisely what that disturbance really is.

Historically there has been no end of ambiguities and distractions. For instance, starting from the classical point of view that space is completely empty, Kelvin proved that the Sun is not capable of such large magnetic variations as would extrapolate to the observed geomagnetic fluctuations at a distance of 1 AU. From this he asserted that there can be no connection between the activity of the Sun and the magnetic fluctuations at Earth. With the same hypothesis that space is completely empty it was thought possible that the observed variations in the cosmic ray intensity are a consequence of large electrostatic potential differences across interplanetary space. The antisolar orientation of gaseous comet tails was attributed to the radiation pressure of sunlight. The implications of the continuing aurora and high-latitude geomagnetic fluctuations were largely ignored, their origin perhaps meteorological.

To go back to the beginning of the concepts on which the solar wind is based, it must be recalled that the concept of space itself is essential and is only a relatively recent development in human thought. In primitive times the sky was not viewed as the window looking out into space. Rather the sky was the abode of gods and spirits, who looked after, or were represented by, the Sun and the Moon and the various planets and stars, which in turn had a powerful influence on our personal destinies. The sky was "that inverted bowl they call the sky whereunder crawling coop'd we live and die" (Omar Khayyam). Knowledge of geography was as limited as the concept of the sky. Some mythologies would have humans descending to Earth from a beginning in the sky, and not so very long ago at that. The concept of the span of time was as stunted as the localized concept of the sky and space. The idea that the whole world was created around us primarily to accommodate us humans is still alive and well today, demonstrating the reluctance of the human mind to relinquish the egocentric world of fantasy.

So it is not surprising that the realization of the vast space in which the tiny Earth resides developed only slowly and sporadically. We owe great respect and admiration to Aristarchus of Samos (circa 275 BC), as the first ancient philosopher-scientist of record to recognize the central position of the Sun in our local system and the vastness of the surrounding space. Armed with a clear understanding of geometry he realized from the converging umbral shadow of Earth during an eclipse of the Moon that the Sun is larger than Earth. From this he recognized that it is more likely that little Earth orbits the large Sun, rather than vice versa. He viewed the Solar System much as Copernicus did 1800 years later, recognizing the great distance to the stars. Needless to say, the contemporary experts found his ideas disturbing, undermining their infallibility, so he was officially rejected. Fortunately he was not entirely ignored, and his penetrating insights, although lost in the original, have come down to us through the writings of others.

Copernicus (1473–1543), beginning in 1530 and finally published in his *De revolutionibus orbium coelestium* in 1543, swept away the Earth-centered crystal spheres of the Ptolemaic geocentric system (first developed by Hipparchus *c.* 150 BC), thereby clearing space of its classical intellectual rubbish and making way for the concept of local terrestrial effects from distant astronomical sources. Equally important, of course, Copernicus, followed by Kepler (1571–1630) and by Galileo with the application of the telescope, began to make sense out of celestial mechanics. This prepared the way for Newton (1642–1727) to develop the Newtonian theory of gravitation and mechanics published in his *Principia* in 1686.

The gradual recognition of the magnetic field of Earth, with the invention of the magnetic compass in China during the Han dynasty about two millennia ago, was another important step on the long road leading to an understanding of the solar wind. The scientific credit here goes particularly to Gilbert (1600) whose ingenious, careful, and quantitative laboratory experiments first established the precise form of the dipole magnetic field around a more or less uniformly magnetized sphere (composed of magnetic iron oxide, magnetite). He recognized that the magnetic field is a special stress system in the space surrounding the magnetized sphere – in other words, that the phenomenon of forces between magnetic objects involves a force-field extending through the space between them. He was the first, then, to recognize the magnetic field as a physical entity throughout the space around a magnet.

The accumulating records of magnetic declination and inclination from extensive ocean voyages were enough for Gilbert to show that Earth itself is a magnetized sphere, and he pointed out that the surrounding space is a special region as a consequence of the magnetic field extending out from Earth. In short, Gilbert recognized the terrestrial magnetosphere, as seen from its lower boundary at the surface of Earth.

GEOMAGNETIC ACTIVITY AND CLASSICAL PHYSICS

It was more than a hundred years after Gilbert's work that Graham (1724) observed a delicately suspended magnetic needle with a microscope and discovered the slight agitation of the needle, implying fluctuations in the magnetic field of Earth. One can appreciate the care that must have been taken to avoid vibrations and air currents to be sure that the agitation was of geomagnetic origin. Celsius (1741) became engaged in similar observations and noticed that the appearance of the aurora was accompanied by enhanced magnetic fluctuations. Celsius and Graham corresponded and soon established that the observed periods of enhanced fluctuation were often simultaneous, thereby showing their large geographic scale.

Wilcke (1777) recognized that the auroral rays are oriented along Gilbert's dipole geomagnetic field. Canton (1759) introduced the first observational connection to the Sun, pointing out that the quiet-time geomagnetic fluctuations are stronger in the summer when the hemisphere is tilted toward the Sun.

It is interesting to note that de Mairan (1754) proposed that the aurora arises from the entry of particles from the Sun into the terrestrial magnetosphere. His idea was based on the contemporary interpretation of the zodiacal light as an extension of the solar corona. Thus Earth orbits through this extended corona, and he proposed that the aurora and the associated geomagnetic fluctuations are the result. It was an inspired conjecture, but he could not do more because the physics of charged particles and magnetic fields was unknown at the time.

Cardan in 1551 had clearly distinguished magnetic and electric effects, and Gilbert emphasized the fundamental difference. In 1733 Du Fay distinguished between positive and negative charge, and a decade later Franklin carried out a number of experiments establishing the conservation of total electric charge. He described positive and negative charge in terms of a surplus or deficit, respectively, of some fundamental electric fluid. We recognize today that electrons make up the mobile electric fluid in most circumstances, so that positive charge actually represents a deficit and negative charge a surplus, but one can see how Franklin advanced the concept.

Priestley discovered the inverse square force law between electric charges in 1767, and Coulomb rediscovered the law in 1785, with the result that today the name Coulomb is associated with the electrostatic inverse square law $1/r^2$. It was not until 1820, however, that Oersted made the experimental connection between electricity and magnetism, demonstrating the magnetic field around an electric current. With this physics in hand, one could understand how electrically charged particles might produce magnetic fluctuations. Only three years later Ampère demonstrated the equivalent magnetic fields of a closed electric circuit and a uniformly magnetized shell whose periphery is bounded by the electric circuit. Today we write Ampère's law in the familiar differential form $c\nabla \times \mathbf{B} = 4\pi \mathbf{j}$, where $\mathbf{j}$ is the electric current density and $\mathbf{B}$ is the magnetic field. By 1831 Faraday had shown that an electric current is driven through a conductor by a time-varying magnetic field, indicating the presence of an electric field whenever $\partial \mathbf{B}/\partial t$ is nonvanishing. The differential form of Faraday's law of induction is the familiar $\partial \mathbf{B}/\partial t = -c\nabla \times \mathbf{E}$, of course. Maxwell completed the electromagnetic equations in 1864, introducing the displacement current $\frac{1}{4\pi}\partial \mathbf{E}/\partial t$ into Ampère's law to obtain $4\pi \mathbf{j} + \partial \mathbf{E}/\partial t = c\nabla \times \mathbf{B}$. Maxwell pointed out that the complete equations predict the electromagnetic wave propagating through vacuum at a speed c given by the ratio of the e.m.u. unit of charge to the c.g.s. unit of charge. Or, as we would say today, the ratio of the units of charge is given by the speed of light, the speed of light being measured directly in the laboratory.

It is an interesting, and not entirely unexpected, historical fact that Maxwell's extension of Ampère's law, based on a direct analogy with an electric current in a dielectric, and the consequent prediction of electromagnetic waves, was largely ignored in England for the next several decades. Kelvin declared that he did not understand Maxwell's electromagnetic theory (which was certainly the case), did not need Maxwell's electromagnetic theory because light and electromagnetic waves could just as well be explained by an elastic aether, and could see no benefit from Maxwell's theory (an immediate consequence of his lack of understanding of the theory). For instance, no reference is given to Maxwell's electromagnetic theory in the extensive articles on electricity, magnetism, and light to be found in the 1894 edition of the *Encyclopaedia Britannica*, except for some sentences in Maxwell's biography where the writer remarks on Maxwell's contributions to electricity and magnetism, citing Maxwell's *Treatise on Electricity and Magnetism* (1873) and noting that the primary test of Maxwell's theory of electromagnetism is whether the velocity of light is accurately given by the ratio of the units of charge in e.m.u. and c.g.s. A few, such as Heaviside, recognized the importance of Maxwell's complete electromagnetic equations, as did Helmholtz and Rowland, and particularly Hertz whose experiments on electromagnetic waves eventually convinced the world of the fundamental importance of Maxwell's equations. Lorentz (1895) demonstrated the implications of Maxwell's equations for transforming $\mathbf{E}$ and $\mathbf{B}$ from one moving coordinate frame to another, which became part of Einstein's (1905) special relativity theory 10 years later.

This is perhaps an appropriate place to note an additional fundamental contribution of Faraday, who, in the course of his epic experiments on magnetic induction, recognized the central importance of the concept of field lines, or magnetic and electric lines of force, as the most direct and effective means for visualizing the form and behavior of a field. The concept is widely applicable today, particularly in magnetohydrodynamics (MHD), wherein the magnetic field is transported bodily with any medium that cannot support a significant electric field in its own moving frame of reference. The magnetic field and, of course, the magnetic energy and stress as well, move with the fluid, as if each field line were a thin line of ink entrained in the fluid, thereby providing a direct visualization of the deformation and transport of the magnetic field.

The theoretical basis for this magnetic transport arises from the researches of Faraday and Lorentz, already mentioned. Beginning with the nonrelativistic Lorentz transformations for the electric field $\mathbf{E}'$ and magnetic field $\mathbf{B}'$ in the frame of reference moving with the fluid velocity $\mathbf{u}$ relative to the laboratory frame, in which the fields are $\mathbf{E}$ and $\mathbf{B}$, we have

$$\mathbf{B}' = \mathbf{B} - \frac{\mathbf{u} \times \mathbf{E}}{c}, \qquad \mathbf{E}' = \mathbf{E} + \frac{\mathbf{u} \times \mathbf{B}}{c}$$

Then since $\mathbf{E}' \cong 0$ in the conducting fluid, it follows that $\mathbf{E} = -\mathbf{u} \times \mathbf{B}/c$, so that $\mathbf{B} = \mathbf{B}'$ upon neglecting terms second order in u/c compared to one, and Faraday's induction equation reduces to the familiar MHD result (Alfvén 1950)

$$\frac{\partial \mathbf{B}}{\partial t} = \nabla \times (\mathbf{u} \times \mathbf{B})$$

Poynting (1884, 1885) showed the mutual compatibility of Newtonian mechanics and Maxwellian electromagnetic theory, with the concept of the electromagnetic stress tensor $-\delta_{ij}(E^2+B^2)/8\pi + (E_i E_j + B_i B_j)/4\pi$ representing the isotropic pressure $(E^2+B^2)/8\pi$ and the tension $E^2/4\pi$ along the electric field lines and the tension $B^2/4\pi$ along the magnetic field lines. Thus, static equilibrium of an electric field involves a balance between the pressure gradient and the tension along curved field lines. The same applies for static equilibrium of a magnetic field. The energy densities of the electric and magnetic fields are $E^2/8\pi$ and $B^2/8\pi$, respectively. At this point in time one could begin to appreciate Gilbert's concept that a magnetic field represents a physically real medium – or stress system (Gauss 1839). The field is not a ponderable medium in that it is not made up of matter, but it is an elastic physical medium with its own energy and, therefore, its own mass density. Thus, for instance, the energy of Gilbert's geomagnetic field above the solid surface of Earth is 8×10^{24} erg, contributing 9 kg to the inertial and gravitational mass of Earth – equivalent to a sack of flour.

The magnetic field becomes a ponderable elastic medium, of course, when a collisionless plasma or equivalent highly conducting medium is present. In particular, Poynting showed that electromagnetic energy is transported with a flux density $c\mathbf{E} \times \mathbf{B}/4\pi$, which becomes $\mathbf{u}_\perp B^2/4\pi$ in MHD, where $\mathbf{u}_\perp$ is the fluid velocity perpendicular to $\mathbf{B}$ and $B^2/4\pi$ represents the magnetic enthalpy. Thus the electromagnetic transport under MHD conditions ($\mathbf{E}' = 0$) is the convective transport of the magnetic field and the enthalpy of that field, representing the convective transport of the magnetic energy density $B^2/8\pi$ plus the work done by $\mathbf{u}_\perp$ pushing the magnetic pressure $B^2/8\pi$ against the fluid ahead. In the nonrelativistic case the electromagnetic momentum density $\mathbf{E} \times \mathbf{B}/4\pi c$ is small to second order in u/c compared to the momentum density in the moving fluid, so that it can be neglected.

We can see how toward the end of the nineteenth century the basic classical physics was falling into place for interpreting the geomagnetic fluctuations. The final piece of physics was the realization that electric charge is not a fluid but consists of discrete particles. The quantized nature of electric charge was indicated by Faraday's experiments with electrolysis around 1831. A small integer times a basic total charge is always associated with the electrolysis of a mole of any substance. Then, noting that the electrolysis experiments showed that matter is composed of discrete particles, Helmholtz in 1881 emphasized that the electric charge, intimately associated with the discrete ions, must also come in discrete units. Johnston Stoney was the first to apply the term "electron" to the basic charge in 1874. In 1897 J.J. Thomson used a Crookes tube (cathode ray tube) to show that the particles emitted by the cathode have a mass that is the fraction 1/1840 of the mass of the hydrogen atom. Thus the idea of the individual lightweight electron and the relatively massive ion as individual free particles was born. The actual structure of the atom was not established until scattering experiments by Hans Geiger and Robert Marsden, using alpha particles from the newly discovered radioactive decay of radium, were performed in 1909, from which Rutherford pointed out in 1911 that the distribution of deflected alpha particles indicated that the atoms from which they scattered contained a massive concentrated central nucleus surrounded by a diffuse electron cloud. Millikan's first measurements of the charge of the electron (4.8×10^{-10} esu) were carried out in the same year.

THE NATURE OF AURORAE AND GEOMAGNETIC ACTIVITY

At this point the physics was ready for interpreting the aurora and the geomagnetic fluctuations, although the scant observational data, limited to instruments at the surface of Earth, held out no brilliant guiding light. Franklin in about 1750, Dalton (1828, 1834), and Gauss (1839) had all come to the view that the aurora is an electrical phenomenon, based on analogy with such electrostatic phenomena as high-voltage coronal streamers and lightning discharges. By about 1890 the shifting ray structure of the aurora reminded physicists of the cathode ray streamers in the partially evacuated Crookes tube, from which Fitzgerald and others suggested that the aurora is a similar electrical discharge. That is to say, the aurora was recognized as being caused by fast particles, and it was presumed that the associated geomagnetic fluctuations were a product of the same particles through Oersted's effect and Ampère's law.

Fortunately the observational studies were moving ahead during the nineteenth century, however slowly. Gauss (1839) became interested in the geomagnetic anomalies and fluctuations, and, beginning in 1832, established the standard magnetic observatory instrumentation. Backed by Humboldt he began organizing the building of magnetic observatories around the world over the next two decades. His instruments used a small mirror attached to a magnetic compass needle, onto which a collimated beam of light was directed. The reflected pencil of light intersected a suitable distant screen, thereby vividly displaying any agitation of the needle without resorting to the cumbersome close-up microscopic scrutiny (see Chapman (1967) for a review of the geomagnetic studies).

Humboldt publicized Schwabe's discovery of the decadal periodicity of sunspots to the scientific world (see von Humboldt 1858). Sabine (1852) pointed out the important fact that geomagnetic disturbances at Toronto are strongly correlated with the number of spots on the Sun. Lamont noted the same correlation for the magnetic variations at München.

The connection of geomagnetic activity with sunspots was reinforced when the four-week variation of the activity was recognized by Broun (1858, 1874), and shown by

Maunder (1904, 1905, 1916) to represent the 27-day equatorial rotation period of the Sun (as observed from the moving Earth). Chree (1912, 1922) and Bartels (1930, 1932, 1934) developed more effective statistical methods for studying the correlation of geomagnetic activity with sunspots, thereby tying the recurring 27-day activity variations to sunspots with increased precision and certainty. It is not surprising, then, that the electrical discharge concept of the aurora led to the idea that the magnetic sunspots were in some way associated with the outward emission of beams of charged particles (referred to as solar corpuscular radiation, abbreviated to SCR in the present writing) from the rotating Sun. Space was regarded as an essentially perfect vacuum, empty of gas and fields except for well defined beams of particles, rotating with the Sun to provide the 27-day recurring geomagnetic activity. This view is not a little curious, because the steadfast zodiacal light clearly filled the space between the Earth and Sun, and it was widely believed that the zodiacal light indicated Thomson scattering from free electrons. Whatever the cause of the zodiacal light, its unvarying presence was simply ignored, presumably on the grounds that the zodiacal light and the associated corona of the Sun are too tenuous to interfere with the outward projected SCR.

In this general context, then, Birkeland (1896) became convinced that Earth and the Sun constitute a celestial Crookes tube, with the Sun as the cathode, emitting electrons which collide with the geomagnetic field and stream in along the magnetic field lines at high latitude to produce the aurora. Birkeland considered the sunspots to be the principal source of electron emission, in view of the close correlation of the terrestrial aurora with the spots on the Sun. This was the first model for the Sun as the emitter of SCR. Birkeland built a remarkable laboratory apparatus – his terrella experiment – consisting of a large vacuum chamber with a hot cathode at one end, simulating the Sun, and a uniformly magnetized sphere as anode to simulate Earth at the other end. An electrostatic potential between the cathode and anode produced a stream of electrons into each hemisphere of the dipole magnetic field of the anode, impacting and fluorescing the surface in a thin band circling the polar regions of the anode. The magnetic fluctuations at the anode resembled the observed geomagnetic field variations often associated with enhanced aurora. One could not ask for a more striking laboratory demonstration of the aurora.

Störmer (1955) was motivated by Birkeland's terrella experiment to devote several decades to the difficult calculations of the trajectories of charged particles incident upon a dipole magnetic field. Neither Birkeland nor Störmer was discouraged by the high magnetic rigidity of 10 MeV or more required for the electrons to enter the atmosphere at auroral latitudes. Such electrons are relativistic with $\gamma \sim 20$, arriving from the Sun only 8 minutes after emission, whereas (see below) the transit time was a couple of days. Then with the discovery of the two or three kilogauss magnetic field of sunspots (Hale 1908, Hale and Nicholson, 1938) the idea of sunspot emission of 10 MeV electrons became problematical, for they could not penetrate out through the sunspot field to escape the Sun. In fact Hale (1913) and Hale *et al.* (1918) thought that they detected a general dipole magnetic field of the Sun, with a polar strength of about 50 G. Such a strong general field would enclose the entire Sun in a magnetic cocoon of some 10^{12} G cm (to be compared with the 10^8 G cm of the geomagnetic field), except for exceptional eruptions of sufficient violence to blow a hole in the field. In retrospect the claim of a general field of such formidable strength was based on a failure to appreciate the noise level in the photographic plates, and the misconception was not corrected until Babcock (1959) and Babcock and Babcock (1955) turned their electronic magnetograph to the Sun more than three decades later.

Both Kelvin and J.J. Thomson adopted a negative view of Birkeland's terrella experiment, and their international influence had the effect of stifling research for a decade or so, until Sydney Chapman and others set it in motion again at about the time of World War I.

Now the flare phenomenon on the Sun was first discovered through a white light flare observed by Carrington (1859) and Hodgson (1859). This remarkable flare was followed a couple of days later by a tremendous geomagnetic disturbance – a magnetic storm. Subsequent studies over the years showed that the phenomenon repeated. The invention of the spectroheliograph, coming into use in 1892, provided an immensely improved instrument for detecting flares far below the intensity required for a flare to appear in white light (Hale and Ellerman 1903, Deslandres 1910). By 1910 the spectrohelioscope was in use (Hale 1929, 1930), and it was firmly established that the more intense flares were often followed one or two days later by enhanced aurora and a magnetic storm. This observational fact indicated that there were bursts of SCR associated with flares in addition to the more or less steadily emitted beams of SCR rotating with the Sun and producing the 27-day recurring periods of geomagnetic activity. It also indicated the very important fact that the burst of SCR associated with the flare, and presumed to be emitted at the time of the flare, travels to Earth in about two days, thereby indicating a speed of the order of 10^3 km^{-1}. Electrons at this speed have a kinetic energy of about 3 eV and cannot penetrate far enough into the geomagnetic field to reach auroral latitudes. Nor would they have enough energy to produce much excitation of the upper atmosphere to provide the visible auroral display. However, protons at the same speed have about 5 keV of energy, plenty to excite the upper atmosphere, but, again, insufficient by far to penetrate to auroral latitudes.

So Birkeland's terrella experiment was spectacular, but evidently irrelevant. However, it set people to thinking, and that is always a useful scientific accomplishment. Schuster (1911) criticized Birkeland's electron beam concept on the simple grounds that the Sun would achieve an enormous positive electric charge and electrostatic potential as a result of the electron emission. For instance, one mole of electrons (0.6×10^{24} electrons) removed from the Sun would leave the Sun (of radius 7.0×10^{10} cm) charged positive to more than 10^6 V. How could the electrons escape from so large an attractive potential? Then one mole of electrons is far too few to produce the aurora, filling a column in interplanetary space with a diameter of, say, 0.1 AU and a length of 1 AU to a number density of only 2×10^{-14} electrons/cm^3. Even at the speed of light this would provide an intensity of only 6×10^{-4} electrons/cm^2 s for a period of only 8 minutes. At the speed of 10^8 cm s^{-1} indicated by the Sun–Earth transit time, the intensity is 2×10^{-6} electrons/cm^2 s, delivering 6×10^{-6} eV cm^{-2} s^{-1} whereas the auroral surface brightness may reach 10^{13} eV/cm^{-2} s^{-1}. Clearly, vastly greater numbers of particles are required to explain the aurora and the magnetic activity. Lindemann (1919) pointed out that the resolution of the dilemma was simply to assume that SCR must be electrically neutral, made up of equal numbers of electrons and protons. Needless to say, this idea of electrically neutral SCR removed the theoretical possibility of the electrostatic acceleration at the Sun.

The chemical composition of the Sun was not clear at that point in time. The lines of such low-excitation atoms as Ca, Si, and Fe dominate the spectrum of the Sun, suggesting that the Sun is composed largely of the vapors of these elements. On the other hand, simple polytropic models of the Sun in hydrostatic equilibrium indicated molecular weights in the neighborhood of one. However, at the surface temperature of 5600 K the H and He lines in the solar spectrum were relatively weak and obscure, and the transparency of hydrogen and helium at 5600 K predicted a fuzzy limb of the Sun rather than the abrupt limb (with characterisitc thickness of the order of only 10^2 km) that is observed. This dilemma was finally resolved with Wildt's (1939) recognition of the negative hydrogen ion (Bethe 1929) as the dominant source of opacity at the surface of the Sun. The detailed calculations of Chandrasekhar (1944, 1945), Chandrasekhar and Münch (1946), Chandrasekhar and Breen (1946), and Chandrasekhar and Elbert (1958) provided the basis for the quantitative model of the limb structure of the Sun (Minnaert 1953). Nonetheless, Lindemann's point was clear enough, whatever the nature of the ions.

The next big step toward understanding the SCR and the consequent geomagnetic activity was the work of Chapman and Ferraro (1931, 1932, 1933, 1940) who showed how a stream of electrically neutral electrons and protons with a speed of 10^8 cm s^{-1} from the Sun would impact on the sunward side of the geomagnetic field to produce the sudden commencement and initial compressive phase of the geomagnetic storm. Their work was the first quantitative theoretical dynamical investigation of the geomagnetic effect of SCR incident on the magnetic dipole field of Earth, pushing the outer regions of the field inward to where the magnetic pressure $B^2/8\pi$ of the compressed dipole field becomes large enough to oppose the impact pressure of the incident SCR across a sharp boundary layer whose thickness is determined by the cyclotron radii of the electrons and protons of the SCR.

The main phase of the geomagnetic storm, following the initial compressive phase, is a weakening of the horizontal component of the magnetic field at the surface of Earth. Chapman and Ferraro noted that, given Ampère's law, the distribution of field reduction with geomagnetic latitude indicated a westward flowing equatorial electric current around Earth somewhere out in space. They did not succeed in showing how the impact of the SCR would produce such a ring current. It was only years later that it was recognized that the main phase represents an outward expansion of the geomagnetic dipole, and that the cause is the internal inflation of the geomagnetic field by keV ions trapped in the field (Dessler and Parker 1959, Dessler et al. 1961, Parker 1968).

The essential point is that Chapman and Ferraro produced the first physical picture of the impact of the SCR on the geomagnetic field, with the basic compressive effect. The everyday time-variations might then be understood as time-varying compression. Still, space continued to be regarded as a hard vacuum except for well-defined beams of SCR, and SCR was believed to be produced in active regions on the Sun, perhaps by some mechanism associated with the strong magnetic fields of the sunspots. It was not clear what that mechanism would be, but no other possibility was known.

COSMIC RAY VARIATIONS

Another line of inquiry leading to SCR was the variation of the intensity of the galactic cosmic rays arriving at Earth. The existence of cosmic rays was first detected by the residual ionization in the air in the laboratory. The laboratory electroscope would not stay charged indefinitely in spite of the essentially perfect electrical insulation employed in its construction. It was shown that the leakage of charge was through the surrounding air. The discovery of radioactivity at the end of the nineteenth century provided an explanation, the α, β, and γ rays all producing ionization in their passage through air. The question was whether the general ionization of the air was entirely a consequence of ambient radioactivity, or whether there was something in addition to the ambient radioactivity, perhaps some hitherto unknown ionizing ray from space. The issue was settled dramatically by Viktor Hess (1912) when he carried an electroscope up in a balloon and found that the higher he rose the more rapidly the electroscope discharged. Clearly the effect was coming from "out there" somewhere. Subsequently the east–west asymmetry of the cosmic rays was discovered and measured, indicating that the cosmic rays are mostly positively charged particles deflected by the magnetic field of Earth. The ionizing particles reaching sea level are mostly mu mesons, with a predominantly positive charge reflecting the primary cosmic ray protons bombarding the upper atmosphere (Rossi 1964, Rossi and Olbert 1970) into which they penetrate a characteristic distance of about $60\,\text{g cm}^{-2}$. Using ionization chambers Forbush (1937) discovered the diurnal variation (after correcting for the diurnal expansion and contraction of the atmosphere). Forbush (1954) announced the worldwide 11-year variation of the galactic cosmic ray intensity by about 4%, anticorrelating with the general level of solar activity. He also pointed out the abrupt decrease at about the time of a sudden commencement of a geomagnetic storm. Subsequently called the Forbush decrease, the decrease had about the same characteristic life as the magnetic storm itself, of the order of 20 hours, except that the Forbush decrease sometimes recovered only part way back to the pre-decrease level, especially during the years of increasing solar activity when the general level of cosmic rays was declining. Somehow the activity of the Sun suppressed the cosmic ray intensity at the orbiting Earth. The occasional outburst of "solar cosmic rays" in association with certain flares on the Sun was an opposite phenomenon, representing an increase in fast particles. Its transient nature and steeply declining energy spectrum clearly stand apart from the Forbush decrease and the 11-year variation whose time integrated effects are much greater.

It was evident that the variations of the galactic cosmic rays were another consequence of SCR, and the energy and momentum spectra of the cosmic ray variations tell us something about electric and magnetic fields and SCR in interplanetary space. The problem was that the mu mesons monitored by the ionization chamber are mainly produced in nuclear collisions of the incoming cosmic ray protons with energies of 15 GeV and more. Particles with such large magnetic rigidity are only slightly affected by the activity of the Sun. On the other hand, the more numerous protons below 15 GeV are primarily responsible for the so-called nucleonic component of the cosmic rays here in the lower atmosphere, mainly neutrons. These lower energy protons are much more strongly influenced (i.e. excluded) by the SCR, as one might expect, but are not detected by the ionization chamber at the surface of Earth. What is more, the lower energy protons are excluded from reaching the atmosphere at low latitudes by deflection in the geomagnetic dipole field. Störmer's (1955) calculations show clearly how the deflection and exclusion come about. Hence the geomagnetic field acts as a magnetic spectrometer, with all energies free to come in at the magnetic poles, whereas vertical incidence on the atmosphere at the equator requires proton energies of about 15 GeV or more. Recognizing the opportunity here to probe the energy spectrum of the reduction of the cosmic ray intensity by the SCR, John Simpson developed the cosmic ray neutron monitor (Simpson et al. 1953). The neutron monitor uses paraffin as a moderator and measures the local intensity of the neutrons produced by the incoming cosmic rays. He set up an array of five neutron monitor stations, from the geomagnetic equator at Huancayo, Peru, where the low-energy cutoff is about 15 GeV for protons, to the University of Chicago where the geomagnetic cutoff is about 2 GeV (Simpson, 1951). The Forbush decreases loom large at these low energies, sometimes decreasing as fast as 5% an hour with a total decrease of as much as 20%.

By this time a number of interplanetary scenarios had been put forth to explain how solar activity and the associated SCR might modulate the galactic cosmic ray intensity observed at Earth. The traditional idea that space is a hard vacuum without significant ambient charged particle populations allowed the possibility of large electrostatic potentials across the solar system. In order to affect the cosmic rays in the 1–15 GeV range the potential differences would have to be of the order of at least 10^9 V.

Davis (1955) recognized the SCR as a plasma filling interplanetary space and sweeping out a giant cavity in the galactic magnetic field. He noted that the sweeping away of the galactic field was essential for the free transit of solar cosmic rays to Earth. He also pointed out that the dimensions of the cavity would vary with the 11-year solar cycle, thereby providing a mechanism for the associated 11-year variation of the cosmic ray intensity discovered by Forbush (1954).

Alfvén (1956) proposed that a well-defined equatorial beam of SCR directed radially outward from the Sun would engage the dipole magnetic field of the Sun, carrying the field outward to large distances from the Sun. A cosmic ray particle in the equatorial plane would be carried along with the magnetic field, of course, and adiabatically decelerated while the number density of the particles diminished in the expanding north–south field within the beam. The result would be a decrease in the cosmic ray intensity at any point moving with the beam for as long as the field continues to decrease. The effect might apply to the Forbush decrease then, but not the sustained 11-year decrease.

Then Morrison (1959) suggested the solar emission of a plasma cloud with an internal tangled magnetic field, so that cosmic ray particles could penetrate into the interior of the cloud only by random scattering in the tangled field, essentially a diffusion process. Thus the interior of the expanding cloud would contain a reduced cosmic ray number density with declining energy per particle as a consequence of the adiabatic expansion. Alternatively, Gold (1959) proposed the emission of a plasma cloud (SCR) from a bipolar active region on the Sun. The ejected cloud would carry the apex of the bipolar field with it, thereby extending a magnetic tongue into space with the plasma cloud at the front. The cosmic ray intensity within the nearly inaccessible interior of the expanding tongue would be greatly reduced through both dilution and adiabatic deceleration.

Parker (1956) proposed the naive idea that Earth is partially shielded from the cosmic rays by the capture of magnetic clouds of plasma by the gravitational field of Earth. The cosmic ray absorption by the terrestrial atmosphere within this shielded volume of the geomagnetic field would provide a reduced cosmic ray intensity at the surface of Earth. Another popular view was that SCR somehow had the effect of deforming the magnetic field of Earth so as to increase the cosmic ray cutoff energy for penetrating through the dipole magnetic field and reaching the atmosphere. The essential point is that the observed cosmic ray variations set a lot of different ideas in motion concerning the nature and consequences of SCR.

Simpson's studies of the energy spectrum of the 11-year variation of the cosmic rays showed that the effect represents a partial removal of the lower energy cosmic rays, with almost complete removal below about 2 GeV at the height of the 11-year sunspot cycle. Thus deceleration alone, by either electrostatic fields or expanding magnetic fields, with the effect of projecting the existing cosmic ray particles to lower energies, was pretty well ruled out, because the depletion of particles is mostly at lower energies (Simpson 1954, Meyer and Simpson 1955). One way or another it appeared that interplanetary space was filled with plasma and magnetic field that somehow partially excluded the galactic cosmic rays from the inner solar system.

The next step came as a total, and very welcome, surprise in the form a large flare on the Sun. The great cosmic ray flare of 23 February 1956 produced an immense burst of relativistic protons with energies up to 20 or 30 GeV. Simpson's cosmic ray neutron monitors, supplemented by a neutron monitor on a ship in the harbor at Wellington, New Zealand, on its way to the Antarctic, together provided an energy spectrum throughout the entire duration of the associated cosmic ray increase. An analysis of the abrupt onset as a function of terrestrial longitude showed a prompt arrival from the direction of the Sun. The intensity passed over a maximum and began to decline after about 20 minutes. The simplest theoretical model fitting the worldwide times of first arrival required direct flight from the Sun, indicating that the magnetic field in interplanetary space is essentially radial between the Sun and Earth. It should be recalled that Davis (1955) had argued that the arrival of solar cosmic rays at Earth is clear evidence that the SCR has swept the galactic magnetic field out of the inner solar system.

Following the maximum the cosmic ray intensity went into a decline of the form $t^{-3/2}$, which eventually became exponential after many hours. This behavior is most simply reproduced by assuming a diffusive barrier encircling the inner Solar System beyond the orbit of Mars. The simplest overall picture was, then, a radial magnetic field out to about the orbit of Mars with a magnetic field beyond Mars with sufficient transverse irregularity to impede the outward passage of the solar cosmic rays. The best fit to the observed decline was obtained if the diffusive magnetic field region terminated at about the orbit of Jupiter (Meyer et al. 1956).

In summary, Simpson's study of the energy dependence of the time variations in the cosmic ray intensity below about 20 GeV per nucleon indicated that interplanetary space is filled with plasma and magnetic field, as distinct from particles emitted into a hard vacuum from special places of origin on the Sun.

COMET TAILS AND BIERMANN'S REVELATION

It was known for a century or more that the gaseous tails of comets point away from the Sun regardless of the direction of motion of the comet. The radiation pressure of sunlight, indicated by the researches of Poynting already mentioned, was the adopted explanation for the effect. Such cometary ions as CO^+ or N_2^+ would be the primary targets for the sunlight. However, by 1950 quantum mechanics had advanced to the point that the absorption cross-sections for these ions could be estimated with confidence. Biermann (1951) pointed out that the cross-sections are far too small to account for the levitation of the comet gases by the radiation pressure of sunlight. In fact, observations of the small-scale features in comet tails indicated an antisolar acceleration of as much as a hundred times the inward acceleration of gravity. The first conclusion was that radiation pressure cannot be the cause of the antisolar comet tails. Then Biermann (1951, 1957) pointed out that the only known alternative to radiation pressure is SCR, noting that the observed time variations in the antisolar acceleration of the tail of Halley's comet in 1910 showed a strong correlation with the known beams of SCR extending out from the Sun at that time. Biermann estimated that a beam of SCR composed of electrons and protons with a number density of the order of 10^3–10^4 cm^{-3} would provide even the extreme observed outward acceleration of the comet tail through Coulomb collisions with the atoms and ions of the comet tail. This number density may seem unreasonably large today, but it must be remembered that at the time the zodiacal light was thought to represent Thomson scattering of sunlight from interplanetary electrons, rather than the dust grains that we now know are the cause. Behr and Siedentopf (1953), Elsässer (1954), Blackwell (1955, 1956), and Blackwell and Ingham (1961) estimated (van de Hulst 1953) that their measurements of the polarized component of the zodiacal light required an electron density of about 700 electrons/cm^3 at 1 AU, not far from Biermann's lower limit of 10^3 electrons/cm^3. So the zodiacal light might have been interpreted as the scattering of sunlight from SCR.

The fundamental point made by Biermann, irrespective of the actual electron and proton density in interplanetary space, was that, if SCR is responsible for the antisolar acceleration of the tails of comets, then the Sun evidently emits SCR in all directions at all times. This follows from the fact that there are occasional comets passing over the poles of the Sun, as well as around low heliographic latitudes, and comets come by as frequently at sunspot minimum as at sunspot maximum. Yet none fail to show an antisolar tail. That is to say, interplanetary space must be completely filled with SCR. There are identifiable denser streams of SCR, indicated by the 27-day recurring geomagnetic events, and there are occasional more violent outbursts of SCR associated with flares and punching their way out through the ambient SCR. But nowhere and at no time is SCR absent. This simple revelation put a whole new face on the problem of SCR.

Kiepenheuer (1953) recognized the implications of Biermann's conjecture and gave some thought to the consequences for the observed solar corona as the SCR streams out through it. Then, as already noted, Davis (1955) investigated the effect of the general outpouring of SCR in sweeping the galactic magnetic field out of the solar system, thereby providing ready access of solar cosmic rays to Earth. Recognizing the SCR as a plasma he described how the galactic magnetic field and the local interstellar gas would be pushed away to distances that he estimated to lie in the range 200–2000 AU, based on the large SCR density of 10^3 ions and electrons per cm^3 and the large galactic magnetic field strength of 10^{-5} G in vogue at that time. Then he noted that the cavity in the galactic magnetic field would vary in size with the 11-year variation of solar activity, which might account for the 4% variation in cosmic ray intensity with the 11-year cycle then recently discovered by Forbush (1954) from ion chambers sensitive mainly to particles above about 15 GeV per unit charge.

Few others took much note of Biermann's proposed solution to the otherwise totally baffling comet tail dilemma. This writer was fortunate in having the opportunity to discuss the conjecture with Biermann during a visit of Biermann to John Simpson's laboratory at the University of Chicago in 1957. The result of the discussion was that, unusual as the conjecture seemed, there was evidently no alternative within known physics.

So once again there was solid evidence that space cannot be considered to be a hard vacuum penetrated only by occasional well-defined beams of SCR. Biermann's conjecture, along with the implications of Simpson's cosmic ray studies, indicated that interplanetary space is filled with a plasma of electrons and protons with densities, it seemed at the time, as large as 10^3 cm^{-3} at the orbit of Earth and outward velocities of the general order of 10^8 cm s^{-1}. But what could be the mechanism of origin at the Sun? Biermann's universal SCR does not require the presence of sunspots or flares, so the acceleration is not basically magnetic as so many of us had vaguely assumed.

THE SOLAR CORONA

Up to this point in the narrative we have ignored the solar corona, that is so awesome during an eclipse of the Sun. The corona plays a central role in the concept of the solar wind, which could not be constructed until the general mystery of the nature and structure of the corona was worked out.

About 99% of the light from the corona is a result of Thomson scattering of photospheric light by free electrons. The photospheric Fraunhofer lines can be identified in the spectrum of the light from the corona, although greatly broadened by the 10^9 cm s^{-1} thermal velocity of the coronal electrons. The zodiacal light merges continuously into the visible corona. About 1% of the light from the corona is emitted by the excited atoms and ions of the coronal gas itself. Spectroscopy during a total eclipse first detected the coronal green emission line in 1869. Over the years other emission lines were detected and their wavelengths measured. Lyot's (1939) invention of the coronagraph was a big step forward and led to the discovery and accurate measurement of a number of emission lines in the inner corona. The problem that arose was that none of the coronal spectral lines could be attributed to any known element. There was talk of a hypothetical element, "coronium", with the expectation that it would soon be discovered in the laboratory, repeating Jansson's and Lockyer's remarkable success with helium, which they first detected by its spectrum in the Sun in 1868. Helium was finally identified in minerals by William Ramsey in 1895. However, the periodic table was eventually filled, leaving no room for coronium.

The broad radial extension of the corona outward from the Sun suggested high kinetic temperatures, of the order of 10^6 K. The ingenious laboratory work of the spectroscopists Grotrian (1933, 1939) and Edlén (1936, 1937, 1942) and then Lyot's (1939) observational determination of several coronal lines and the associated line widths (see discussion in Billings (1966) and in Golub and Pasachoff (1997)) provided the definitive answer to this basic astrophysical question. Extrapolating along isoelectronic series, e.g. Li I, Be II, B III, C IV, N V, etc., they showed that the coronal emission lines are to be identified with forbidden transitions in such highly ionized structures as Fe X, Fe XI, Ca XII, Ca XIII, etc., indicating temperatures of 10^6 K or more. The measured line widths corroborated this astonishing result. The observed scale height of the corona indicated a mean molecular weight comparable to ionized hydrogen. The corona, then, is simply the extended outer atmosphere of the Sun, with elemental abundances of the same general order as to be found at the photosphere, and extended far into space by the incredible temperature (Billings 1966). To give a specific example, the pressure scale height for 2×10^6 K in the low corona is 1×10^{10} cm (the solar radius $R_\odot = 7\times10^{10}$ cm), increasing to 2×10^{10} cm at $r = 1.4 R_\odot$.

This scientific breakthrough is a milestone in astrophysics, uncovering a phenomenon never dreamed of before and today recognized as applying to most stars and even to galaxies and clusters of galaxies. The corona is the basis for the X-ray emission of a star, first detected from the Sun by Burnight (1949). Then, as we shall soon see, the corona is directly connected with SCR.

The obvious question was the source of heat that provides the enormous temperature. A traditional answer was accretion of meteoric particles to maintain the brightness of the Sun as a whole and the accretion of interstellar gas to provide the corona. The antisolar comet tails rule out the accretion of interstellar gas, of course.

The general answer is fundamental to astrophysics, applying to the suprathermal aspects of most stars – that is, some form of mechanical agitation, such as convective overturning, below the visible surface of the Sun (or by an accretion disk, or by an expanding shell in other circumstances) does mechanical work on the fluid and magnetic field. Some part of the mechanical work creates acoustic waves, internal gravity waves, Alfvén waves, or other MHD wave mode, and some goes into magnetic free energy in the slow deformation of bipolar magnetic fields of active regions. The waves are free to propagate up into the chromosphere and into the tenuous corona where they are dissipated by viscosity and thermal conductivity, thereby converting their mechanical energy into heat (Alfvén 1947, Schwarzschild 1948, Biermann 1948, Schatzmann 1949a,b). In addition, the magnetic free energy in the deformed bipolar magnetic fields dissipates vigorously in the flares arising, for example, as a consequence of the spontaneous tangential discontinuities produced by the field topology in static equilibrium Parker (1983, 1988, 1994), or by the collision of one bipolar magnetic field with another (Sweet 1969, Svestka 1976). Shimizu and Tsuneta (1997) found small-scale flickering ("nanoflares") in the X-rays from the active corona observed by Yohkoh. The recent observations by SOHO show how densely the photosphere is covered by active small-scale magnetic fields, flickering and brightening, evidently in response to small bursts of rapid reconnection (flares) (Schrijver et al. 1997, 1999). The heat input required to maintain the quiet corona is estimated (Withbroe and Noyes 1977) to be about 0.5×10^6 erg cm^{-2} s^{-1}, and is a small fraction of the characteristic kinetic energy flux $1/2\rho\langle v^2\rangle^{3/2} \sim 10^8$ erg cm^{-2} s^{-1} in the photospheric convection where $\rho \sim 2\times10^{-7}$ g cm^{-3} and $\langle v^2\rangle^{1/2} \sim 10^5$ cm s^{-1}.

Consider, then, Chapman's fundamental studies of the physical properties of the million degree solar corona. Chapman and Cowling (1958) worked out the kinetic response of a gas disturbed by weak temperature gradients, velocity gradients, and electric fields and current densities, using the method known as the Chapman–Enskog expansion. Chapman (1954) provided the special case of the thermal conductivity κ of fully ionized hydrogen, obtaining $\kappa(T) = 6\times10^{-7} T^{5/2}$ erg cm^{-1} s^{-1} K^{-1}. The conductivity depends upon the density only through a weak logarithmic variation, safely ignored in the present case. Recalling that thermal conductivity is mainly the province of extremely mobile electrons, whose thermal velocities are of the order of 10^9 cm s^{-1} at coronal temperatures, we can appreciate the large thermal conductivity operating in the solar corona.

The thermal radiation loss ε from the optically thin corona is of the order of $\varepsilon \sim 1\times10^{-23} N^2$ erg cm^{-2} s^{-1}. The characteristic number density N is 10^8 atoms/cm^3, with the result that $\varepsilon \sim 10^{-7}$ erg cm^{-3} s^{-1}. Over the characteristic pressure scale height $\Lambda \sim 10^{10}$ cm at 2×10^6 K the total radiative loss is only $W \sim \varepsilon\Lambda \sim 1\times10^3$ erg cm^{-2} s^{-1}. As is shown below the loss by thermal conduction W_t to infinity is a factor of 10 larger than W, so the total outward energy loss from radiation and conduction is essentially W_t.

To appraise the significance of the characteristic outward heat transport, note that Chapman (1959) went on to work out the expected outward radial variation of the coronal temperature beyond where the heat is introduced into the corona. The temperature follows from the heat flow equation

$$\frac{1}{r^2}\frac{d}{dr}\left(r^2\kappa\frac{dT}{dr}\right) = 0$$

Integration shows that the temperature has the form $T = (D+C/r)^{2/7}$ where C and D are arbitrary constants of integration. For all practical purposes the temperature falls to negligible values at $r = \infty$, where $T(\infty) \ll 10^6$ K, so that $D \cong 0$. Thus it follows that

$$T(r) = T(a)(a/r)^{2/7} \qquad (1)$$

for $r \geq a$. With this explicit form for the temperature Chapman turned to the equation

$$dp/dr = -\rho GM_\odot/r^2 \qquad (2)$$

for hydrostatic equilibrium of the corona in the gravitational field of the Sun. The gas pressure p for ionized hydrogen is $2NkT$, so that the gas density $\rho = (M/2kT)p$. The mass of the Sun is denoted by $M_\odot$, G is the gravitational constant, and M is the mass of the hydrogen atom. Integration over r yields

$$p(r) = p(a) \exp\left\{-\frac{7a}{5\Lambda(a)}\left[1 - \left(\frac{a}{r}\right)^{5/7}\right]\right\} \qquad (3)$$

where $\Lambda(a) = 2kT(a)a^2/GM_\odot M$ is the pressure scale height at $r = a$. At large distance the pressure falls asymptotically to the finite value

$$p(r) \sim p(a) \exp\left(-\frac{7a}{5\Lambda(a)}\right) \qquad (4)$$

To get a feeling for the implications, let $a = 10^{11}$ cm $\sim 1.4\, R_\odot$, so that $\Lambda(a) \sim 1.25 \times 10^{10}$ cm for $T(a) = 10^6$ K. The result is $p(r) \sim 1.4 \times 10^{-5} p(a)$. For $T(a) = 2 \times 10^6$ K the result is $p(r) \sim 3.7 \times 10^{-3} p(a)$. Earth lies at $r = 150a$ (so that $(r/a)^{2/7} \sim 4.18$ and $(r/a)^{5/7} \sim 35.6$), and the pressure at 1 AU is close to the final asymptotic value.

The coronal temperature predicted at the orbit of Earth is a factor of 4.18 less than $T(a)$ at the Sun. The gas density is smaller than at $r = a$ by the factor 5.6×10^{-5} for $T(a) = 1 \times 10^6$ K and by 1.55×10^{-2} for $T(a) = 2 \times 10^6$ K. Thus, with 3×10^7 atoms/cm^3 at $r = a$ the coronal gas density at the orbit of Earth would be of the order of 1.7×10^3 atoms/cm^3 and 4.6×10^5 atoms/cm^3, respectively.

Chapman made the startling point that Earth moves through the extended corona of the Sun. This revelation is fundamental to the concept of the solar wind. It is interesting to note, then, that this static corona may have an electron density sufficient to account for the zodiacal light (see Behr and Siedentopf 1953), which takes us all the way back to de Mairan (1754). The essential point is that Chapman showed that the solar corona extends well out into the solar system. Now one can see that the coronal density increases outward beyond the orbit of Earth as the pressure approaches its final asymptotic value and the temperature continues to fall. That part of the model may be hard to believe, in view of its convective instability. But there is no obvious failure in the model out as far as the orbit of Earth.

This author was fortunate in having the opportunity to discuss the model static corona with Chapman in the early phases of the development while visiting the High Altitude Observatory where Chapman was working at the time. The foregoing conclusions, indicated by Chapman's preliminary calculations, were so impressive that the writer repeated them on the journey back to Chicago. It became clear that the resulting finite coronal pressure at infinity did not suit the Sun, surrounded by the relative vacuum of interstellar space. Only by assuming that $T(a)$ is 0.5×10^6 K or less is it possible for the pressure of the corona to fall to typical interstellar values of 10^{-12} dyn cm^{-2}, and observations of the corona near the Sun indicated that the temperature was somewhere in the range $1-2 \times 10^6$ K (Billings 1966) and certainly not as small as 0.5×10^6 K.

It must be appreciated that the million degree corona near the Sun is strongly bound by the gravitational field of the Sun. The gravitational binding energy per hydrogen atom is 2.2×10^{-9} erg at $a = 10^{11}$ cm and 3.1×10^{-9} erg at the base of the corona where $r = 0.7a = R_\odot$, whereas the thermal energy $3kT$ is 0.41×10^{-9} erg at 1×10^6 K and 0.83×10^{-9} erg at 2×10^6 K. So even at 2×10^6 K the gas is strongly bound at $r = a$.

The situation reverses far from the Sun, because the temperature declines only as $(a/r)^{2/7}$, while the gravitational binding energy declines as a/r. Thus, at the orbit of Earth, the thermal energy per atom has declined by factor of 4.2 while the gravitational energy per atom has declined by a factor of 150. So the thermal energy greatly exceeds the gravitational binding energy, from which it follows that the corona cannot be truly static. The outer regions expand away to infinity, allowing the lower corona to expand upward to replace the escaped gas, and departing in turn a little later.

THE BIERMANN–CHAPMAN DILEMMA

The researches of Chapman, indicating that the static corona of the Sun extends far out through interplanetary space, and the researches of Biermann, indicating that interplanetary space is continually swept out by SCR, seemed to be inescapable. One could argue about the precise densities, temperatures, velocities, etc., but there was no evident escape from the qualitative conclusions. However, after some thought on the matter, it became clear that the two were mutually exclusive as a consequence of the two-stream plasma instability (Pierce 1949; Parker 1958b, 1959; Parker and Tidman 1958; Petschek 1958). Two collisionless plasmas cannot stream through each other at supersonic speed without exciting electrostatic plasma oscillations in both plasmas. The growth rate of the instability is characterized by the plasma frequency. The result of the instability is the production of strong plasma oscillations within each plasma, so that the two plasmas interact strongly, quickly arresting any large relative motion, and providing a single plasma with an anisotropic thermal velocity distribution. Shock fronts in a collisionless plasma are a direct manifestation of the effect, in which the supersonic gas flowing into the shock front is decelerated abruptly by the two-stream, or other, instability in the region of interpenetration.

Biermann and Chapman together required two tenuous plasmas to move freely through each other at relative speeds of the order of 10^8 cm s^{-1} for distances of the order of 1 AU or more. That is not possible, and it is also not possible to discredit either of their conclusions. After some thought on the matter, it became clear that Chapman's extended static coronal concept was certainly valid near the Sun, where the corona is strongly bound by gravity, while Biermann's universal SCR was certainly valid at some distance from the Sun where comets commonly pass. The unavoidable expansion of the outer reaches of Chapman's corona suggested some sort of outward motion of the corona in Biermann's more distant domain. So perhaps the two concepts could be reconciled if Chapman's static corona near the Sun somehow went over into Biermann's supersonic SCR at large distance. Otherwise there was a basic contradiction and dilemma.

LARGE-SCALE DYNAMICS OF A TENUOUS PLASMA

The question arose as to how one treats the large-scale dynamics of a tenuous plasma, presumably in the presence of a large-scale magnetic field (Cowling 1953). In the corona near the Sun, where $N \sim 10^8$ cm^{-3}, the mean free path ($\sim 3 \times 10^{16}$ N^{-1} cm^{-1}) of the ions and electrons is substantially shorter than the pressure scale height ($\sim 5 \times 10^9$ cm), so the familiar MHD momentum and induction equations spring to mind. However, farther out in space the coronal gas is a more tenuous plasma, in which collisions are increasingly scarce if the million degree temperature is maintained. To investigate the situation, consider a completely collisionless plasma.

It is well known that the zeroth-, first-, and second-order velocity moments of the collisionless Boltzmann equation provide the equation for conservation of mass, the momentum equation, and the energy equation for the pressure tensor, respectively. These equations are in hydrodynamic form, indicating that the large-scale dynamics of the collisionless plasma can be described by equations in the familiar hydrodynamic form with proper specification of the pressure tensor p_{ij}. In fact, it is not necessary to use the Boltzmann equation because the precise form of the thermal velocity distribution is irrelevant for the large-scale bulk motion. It is only the total number density N and the pressure tensor, rather than the vast pool of information in the velocity distribution function, that are needed in the calculation of the bulk velocity **u**. It is sufficient to consider only the streaming of particles, each with mass M, say, and velocity w_i, through a fixed volume V of scale b, choosing b large enough that $Nb^3 \gg 1$, so that the number of particles in V is large, while b is small compared to the characteristic scale L of the magnetic field and the bulk motion. The Maxwell stress tensor is denoted by M_{ij}, and F_j represents the nonelectromagnetic force per unit volume exerted on the particles in V. The net particle momentum in V is $\Sigma M w_j = \Sigma M u_j$, where u_j is the mean bulk velocity in V, and $w_j = u_j + v_j$, where v_j is the thermal velocity of the individual particle, the sum being taken over the volume V so that $\Sigma M v_j = 0$. The time rate of change of the total particle momentum in V is equal to the total force $V(F_j + \partial M_{jk}/\partial x_k)$ minus the momentum carried out across the surface S of V by the free passage of the particles. The momentum flux of the particles is $M w_j w_k = M(u_j u_k + u_j v_k + v_j u_k + v_j v_k)$, and the sum over the surface S causes the two cross-product terms to vanish. The Reynolds stress tensor is $R_{jk} = \rho u_j u_k$ given by the sum $\Sigma M u_j u_k / V$ and the pressure tensor is given by $p_{jk} = \Sigma M v_j v_k / V$. In these terms the conservation of momentum yields

$$\frac{\partial}{\partial t} \rho u_i = -\frac{\partial R_{ij}}{\partial x_j} - \frac{\partial p_{ij}}{\partial x_j} + \frac{\partial M_{ij}}{\partial x_j} + F_i \qquad (5)$$

However, there are still some nagging questions. The thermal motions in a tenuous (essentially collisionless) plasma consist primarily of motion parallel to the local magnetic field **B** and cyclotron motions perpendicular to **B**. Ampère's law, $4\pi \mathbf{j} = c \nabla \times \mathbf{B}$, must be satisfied in the direction perpendicular to **B** as well as parallel. Parallel to **B** the particles stream relatively freely subject only to the so-called mirror force $(Mv_\perp^2/2B)\partial B/\partial s$ in a converging field, where $B = |\mathbf{B}|$ and s represents distance along the field lines. So it is easy to see how a slight nudge from a weak electric field $\mathbf{E}_\parallel$ supplies the necessary $\mathbf{j}_\parallel$ under ordinary circumstances in the large-scale slowly evolving magnetic field. Generally $\mathbf{E}_\parallel$ can be neglected so far as the slow large-scale variations of **B** are concerned. The important role of $\mathbf{E}_\parallel$ in the special circumstances where it is not negligible has been elaborated by Birn et al. (1989).

In contrast, perpendicular to **B** the particles are completely constrained to cyclotron motion. The introduction of a perpendicular electric field $\mathbf{E}_\perp$ produces only a momentary particle motion and electric current in the direction of $\mathbf{E}_\perp$. The final condition is the electric drift of all the charged particles with the velocity $c\mathbf{E} \times \mathbf{B}/B^2$ perpendicular to **B**, representing no net current in a plasma with overall electrical neutrality. That is to say, application of $\mathbf{E}_\perp$

produces no enduring $\mathbf{j}_\perp$. The electric field $\mathbf{E}'$ in the frame of reference of the electric drift velocity is zero, and Faraday's induction equation shows that the magnetic field $\mathbf{B}$ is carried in that frame of reference. But how is it that Ampère's law can be satisfied?

We should be aware that the equations of Newton and Maxwell are mutually consistent, as demonstrated by Poynting, because they are both fundamental laws of nature, so the natural mechanical motions of the ions and electrons in the plasma must somehow automatically take care of the current required by Ampère. But are there any non MHD features of the dynamics of the collisionless plasma as a consequence of Ampère's requirement? The precise form of the dynamical equations for the large-scale bulk motion of a tenuous plasma was essential for going ahead with the dynamics of tenuous magnetized plasmas in astrophysical settings. In fact there exists a robust confusion on the question even today (Parker 1996), with some authors concerned about the closure of currents and whether Ampères law is properly satisfied, to the exclusion of the dynamics of the system, driven by the momentum of the fluid and the stresses in the elastic magnetic field carried in the fluid. So it is important to understand how to proceed from the equations of Newton and Maxwell to the final dynamical equations for the large-scale bulk velocity $\mathbf{u}$ and magnetic field $\mathbf{B}$.

It occurred to the writer that the well-known guiding center approximation for the motion of an individual ion or electron in a large-scale magnetic field (Alfvén 1950) was applicable to the ions and electrons in a collisionless plasma, allowing the velocity of each particle to be expressed in terms of the local field $\mathbf{B}$ and its gradients. It would then be a simple task, at least in principle, to sum over all particle motions to obtain the total electric current, which ought to satisfy Ampère's law. Fortunately this exercise had been completed (Parker 1957) by the time that the Biermann–Chapman dilemma was appreciated, so that it was possible to proceed immediately to the dynamics of the solar corona, taken up below in the section "Hydrodynamics of the solar corona."

DYNAMICAL EQUATIONS FOR A TENUOUS PLASMA

The basic principle is that the large-scale dynamics of any gas, whether fully or partially ionized, or not ionized at all, and whether collision dominated or collision free, is described by the familiar hydrodynamical momentum equation, for the reasons noted in the previous section. Only the variation of the pressure and the dissipation terms depend upon the detail microscopic structure of the gas. If there are enough free electrons to short circuit and neutralize any electric field in the local frame of reference of the moving gas, then the large-scale magnetic field is carried along bodily with the bulk motion of the gas. The basic concepts follow directly from the equations of Newton, Maxwell, and Lorentz (Parker 1957, 1996). Both Newton's and Maxwell's equations are fundamental laws of physics, so one may be assured that they do not contradict each other under any circumstances. It is instructive to see how this works out in the simple case of the collisionless plasma.

Consider a collisionless plasma composed of N singly charged ions of mass M and electrons of mass m per unit volume, each of which moves freely in the large-scale electric and magnetic fields $\mathbf{E}$ and $\mathbf{B}$. The characteristic scale L of the magnetic field and plasma inhomogeneity is large compared to the characteristic ion cyclotron radius $R = Mv_\perp c/eB$, where $v_\perp$ is the thermal velocity perpendicular to $\mathbf{B}$. The ion cyclotron frequency is $\Omega = v_\perp/R = eB/Mc$. To a first approximation, then, the mean motion perpendicular to $\mathbf{B}$ in the presence of a perpendicular electric field $\mathbf{E}_\perp$ is the electric drift velocity $\mathbf{u}_D = c\mathbf{E} \times \mathbf{B}/B^2$ relative to which the particle gyrates with the cyclotron frequency Ω. For slow changes in B and for motion of the particle along an inhomogenous magnetic field the diamagnetic moment μ of the cyclotron motion, given by $\mu = Mv_\perp^2/2B = Mv^2 \sin^2\theta/2B$, where θ is the pitch angle ($\mathbf{v} \cdot \mathbf{B} = vB\cos\theta$), is invariant. The center of the circular cyclotron orbit is called the *guiding center* and the drift velocity $\mathbf{u}_D$ applies to the motion of the guiding center. It is important to remember that in this drifting frame of reference the electric field $\mathbf{E}'_\perp$ is given by the Lorentz transformation $\mathbf{E}'_\perp = \mathbf{E}_\perp + \mathbf{u}_D \times \mathbf{B}/c = 0$, so the guiding centers of the individual electrons and ions move in the frame of reference in which there is no electric field.

The guiding center approximation goes one step farther, noting that small gradients in $\mathbf{B}$ and small variations in $\mathbf{E}$ produce additional velocities, small to order R/L compared to the thermal velocity v. For instance, if the field magnitude B varies slightly across the cyclotron radius R, then the radius R varies across the cyclotron circle traveled by the particle. Thus the particle travels a little longer on the weak field, large R, side of the cyclotron orbit. The particle velocity is constant around the orbit, so there is a small net drift of the particle in the direction of the particle velocity of the weak field side.

The sum of all these effects can be shown (Parker 1957) to be the total current density perpendicular to $\mathbf{B}$ in the amount

$$\mathbf{j}_\perp = \left(\frac{c}{B^2}\right)\mathbf{B} \times \left\{\nabla p_\perp + \frac{(p_\parallel - p_\perp)}{B^2}(\mathbf{B} \cdot \nabla)\mathbf{B} + NM\frac{d\mathbf{u}_D}{dt}\right\} \quad (6)$$

where $p_\perp$ is the pressure $\Sigma Mv_\perp^2/2$ where the sum is over both electrons and protons in unit volume. The pressure $p_\parallel$ is $\Sigma Mv_\parallel^2$ where the sum is again over both electrons and ions in unit volume.

Substitution into Maxwell's equation gives

$$\frac{\partial \mathbf{E}_\perp}{\partial t} = \frac{4\pi c \mathbf{B}}{B^2} \times \left\{ -NM \frac{d\mathbf{u}_D}{dt} - \nabla\left(p_\perp + \frac{B^2}{8\pi}\right) \right.$$
$$\left. + (\mathbf{B} \cdot \nabla)\mathbf{B}\left[\frac{1}{4\pi} + \left(\frac{p_\perp - p_\parallel}{B^2}\right)\right] \right\} \quad (7)$$

Now, with $\mathbf{u}_D = c\mathbf{E} \times \mathbf{B}/B^2$ it follows that $\mathbf{E}_\perp = -\mathbf{u}_D \times \mathbf{B}/c$. Then since $\partial/\partial t = O(u_D/L)$, it follows that the left-hand side of this relation, $\partial \mathbf{E}_\perp/\partial t$, is small to second order in u_D/c compared to each term on the right-hand side, and so may be set equal to zero, leaving

$$NM \frac{d\mathbf{u}_D}{dt} = -\nabla_\perp\left(p_\perp + \frac{B^2}{8\pi}\right) + \frac{[(\mathbf{B} \cdot \nabla)\mathbf{B}]_\perp}{4\pi}$$
$$\times \left[1 + \frac{4\pi(p_\perp - p_\parallel)}{B^2}\right] \quad (8)$$

for the motion $\mathbf{u}_D$ perpendicular to $\mathbf{B}$. This equation is recognizable as the momentum equation in the presence of the Lorentz force $\partial M_{jk}/\partial x_k$ and the perpendicular pressure gradient, with the additional term in the anisotropy $p_\perp - p_\parallel$. The extra term represents the centrifugal force of the particles streaming along the curved field lines when $p_\parallel - p_\perp > 0$, and the opposite effect when $p_\parallel - p_\perp < 0$. So the result is precisely the equation of motion that we would expect for the bulk motion $\mathbf{u}_D$. That is to say, if the bulk velocity $\mathbf{u}_D$ satisfies Newton's equations of motion, then the thermal gyrations of the ions and electrons in the magnetic field automatically produce the electric current density required by Ampère. It could not be otherwise because the equations of Newton and Maxwell are both fundamental laws of physics. The interested reader is referred to the direct formal derivation of eqn (8) from Newton's Equations of motion provided by Bittencourt (1986).

The particle motion along $\mathbf{B}$ is subject to gradients in $p_\parallel$, of course, and to the so-called mirror force

$$\mathbf{F}_M = -\left(\frac{Mw_\perp^2}{2B^4}\right)\mathbf{B}\{\mathbf{B} \cdot [(\mathbf{B} \cdot \nabla)\mathbf{B}]\} \quad (9)$$

already mentioned, which is just $F_M = -(\frac{1}{2}Mw_\perp^2) \times (1/B)\,\partial B/\partial s = \mu\partial B/\partial s$, where s represents distance along the field line through the guiding center. The mirror force repels particles from regions of strong field. To compute the mean bulk motion along $\mathbf{B}$, consider the distribution of particles $F(s, t, w, \theta)$ at position s and time t with pitch angle θ and speed w. The number of particles per unit volume is then

$$N(s, t) = \int_0^\pi d\theta \int_0^\infty dw\, F(s, t, w, \theta) \quad (10)$$

The mean bulk velocity $\mathbf{u}_\parallel$ along $\mathbf{B}$ is given by

$$N(s, t)\, u_\parallel(s, t) = \int_0^\pi d\theta \int_0^\infty dw\, w \cos\theta\, F(s, t, w, \theta)$$

It is readily shown (write $F = B(s)f$ in eqn (61) of Parker (1957) that F satisfies the collisionless Boltzman equation

$$\frac{\partial F}{\partial t} + w\cos\theta \frac{\partial F}{\partial s} + \frac{w\sin\theta}{2B}\frac{dB}{ds}\frac{\partial F}{\partial \theta} - \frac{w\cos\theta}{2B}\frac{dB}{ds}F = 0 \quad (11)$$

for particles in a stationary ($\partial/\partial t = 0$) field $B(s)$. Thus, for instance, an isotropic distribution $F \sim \sin\theta$ yields $\partial F/\partial t + v\cos\theta\ \partial F/\partial s = 0$, i.e. $dF/dt = 0$, as the particles move along the field.

To obtain the equation for conservation of particles integrate eqn (11) over w and θ, with the result that

$$\frac{\partial N}{\partial t} + \frac{\partial}{\partial s}Nu_\parallel = \frac{Nu_\parallel}{B}\frac{dB}{ds} \quad (12)$$

The right-hand side represents the increase in N as the fluid streams with velocity $u_\parallel$ along a fixed converging field.

To obtain the momentum equation for the motion $\mathbf{u}_\parallel$ along $\mathbf{B}$, multiply eqn (11) by $w\cos\theta$ and integrate over w. Note, then, that the pressure $p_\parallel$ represents the momentum flux of the thermal motions along $\mathbf{B}$, so that $p_\parallel = NM\langle v_\parallel^2\rangle$. Similarly, the pressure $p_\perp$ represents the perpendicular momentum flux transported by $v_\perp$, so that $p_\perp = NM\langle v_\perp^2\rangle/2$. The final result, after integrating $\sin\theta\cos\theta\ \partial F/\partial\theta$ by parts, is the momentum equation

$$\frac{\partial}{\partial t}NMu_\parallel + \frac{\partial}{\partial s}NMu_\parallel^2 + \frac{\partial p_\parallel}{\partial s} - \frac{NMu_\parallel^2}{B}\frac{dB}{ds} + \frac{p_\perp - p_\parallel}{B}\frac{dB}{ds} = 0 \quad (13)$$

Then multiply eqn (12) by $Mu_\parallel$ and subtract it from eqn (13), providing the momentum equation in the form

$$NM\left(\frac{\partial u_\parallel}{\partial t} + u_\parallel \frac{\partial u_\parallel}{\partial s}\right) = -\frac{\partial p_\parallel}{\partial s} + \frac{p_\parallel - p_\perp}{B}\frac{dB}{ds} \quad (14)$$

This is the familiar hydrodynamic equation for a fluid with velocity $u_\parallel$ and pressure $p_\parallel$ with the additional term in the anisotropy $p_\parallel - p_\perp$, representing the net mirror force when the thermal motions are anisotropic.

The essential point is that in an isotropic thermal distribution the mirror force repelling each individual particle from a region of strong field, and thereby decreasing the momentum flux, is exactly compensated by the increase in the momentum flux density as the particles are concentrated by the converging field. That is to say, not surprisingly, the bulk motion $\mathbf{u}$ of a collisionless plasma with statistically isotropic thermal motions is described by the same hydrodynamic equations as a classical fluid. This result is obvious from eqn (5), of course, but we have rederived it here using the guiding center approximation to describe the detailed spiral motion of the individual particles in the inhomogeneous magnetic field. The essential point is that the hydrodynamic equations, usually derived in some restricted context (e.g. the classical fluid) are in fact quite general statements about conservation of particles and momentum, so they cannot be evaded by any special form of internal small-scale motions. The most that a strong anisotropy can do in the presence of a magnetic field is to introduce the mirror force, the last factor in square brackets in eqn (8), and internal instabilities that may soon deplete the anisotropy.

The next question concerns the computation of $p_\parallel$ and $p_\perp$. One can obtain formal equations for their computation in terms of the heat flow tensor, of course, by taking the second-order velocity moments of eqn (11), etc., but such equations are often too complex to be useful. So consider the physics. As already noted, any strong thermal anisotropy is unstable to internal plasma waves, so that a strong anisotropy is not likely to survive for long in the real world unless it is sustained by continuing anisoptropic expansion. For instance, in the slow expansion of the dense corona near the Sun, there can be no more than weak anisotropies. In a shock transition in a tenuous plasma the shock thickness represents the interpenetration distance over which the anisotropy is reduced by the two-stream plasma instability. Within that thickness the anisotropy is large. Then in the large-scale tenuous solar wind far from the Sun there are observed to be significant anisotropies, maintained by the continuing expansion of the plasma in the two directions perpendicular to the radial motion. The large-scale consequences of anisotropy may be judged from the last term in the momentum eqns (8) and (14), and included if necessary.

When thermal anisotropy is not deemed important, the scalar pressure p can be used and computed with the aid of a heat flow equation, or p may be modeled with a polytropic relation $p \sim \rho^\alpha$ for monotonic changes of the density ρ. When it is anticipated that thermal anisotropy may play a role in the dynamics, one may use the Chew–Goldberger–Low (1956) phenomenological approximation, which is based on the transverse invariant $v_\perp^2/B$, the longitudinal invariant $v_\parallel b$, and the invariant Nb/B representing conservation of particles, where b is a characteristic length denoting the scale of variation along **B**. The implicit assumption is that particles do not mix along the field between regions of different temperature and anisotropy. Then, with the fact that $p_\parallel$ varies in proportion to $Nv_\parallel^2$ and $p_\perp$ varies in proportion to $Nv_\perp^2$, it follows that

$$\frac{d}{dt}\left(\frac{p_\perp}{NB}\right) = \frac{d}{dt}\left(\frac{p_\parallel B^2}{N^3}\right) = \frac{d}{dt}\left(\frac{B^5 p_\parallel}{p_\perp^3}\right) = 0 \quad (15)$$

yielding $p_\parallel$ and $p_\perp$ in terms of N and B. The discussion by Bittencourt (1986) is of interest.

HYDRODYNAMICS OF THE SOLAR CORONA

With this formal derivation of the dynamical equations for the large-scale bulk motion of a tenuous plasma, the next step is to apply the equations to the corona of the Sun. The basic physics of the expansion is described in the next section, and the casual reader may wish to jump to that point.

As already noted, the distant solar corona is not fully confined by gravity, however strongly it may be bound by gravity near the Sun. This suggests some sort of coronal outflow. Consider the simple case of a stationary state with radial outflow, as would arise for a Sun with spherical symmetry (Parker 1958e). The effect of an outward flow $u(r)$ is to add an inertial term to Equation (2), so that

$$NMu\frac{du}{dr} = -\frac{dp}{dr} - \frac{GM_\odot MN}{r^2}$$

Conservation of matter requires the mass flux to be independent of r, so that $N(r)u(r)r^2 = N(a)u(a)a^2$, where $r = a$ is some convenient reference level near the Sun. Writing $p = 2NkT$ for ionized hydrogen, the momentum equation can be expressed as

$$\frac{du}{dr}\left(u - \frac{2kT}{Mu}\right) = -\frac{2k}{M}\frac{dT}{dr} + \frac{4kT}{Mr} - \frac{GM_\odot}{r^2} \quad (16)$$

The simplest example of an extended coronal temperature is the isothermal case, $T(r) = T(a)$, and that idealized situation was used for preliminary exploration of the implications. It follows that

$$\frac{du}{dr}\left(u - \frac{2kT}{Mu}\right) = \frac{4kT}{Mr} - \frac{GM_\odot}{r^2} \quad (17)$$

We are interested in a solution $u(r)$ that passes continuously from a strongly bound, quasi-static state ($GM_\odot/a \gg 4kT/M \gg 2u^2$) near the Sun at $r = a$ to a condition of vanishing interstellar pressure at $r = \infty$. Thus at $r = a$ the right-hand side of eqn (17) is negative, and the factor in parentheses on the left-hand side of eqn (17) is also negative. The result is $du/dr > 0$, with the velocity increasing outward.

At infinity the dominant term of the right-hand side is $4kT/Mr$, which declines to zero only as $1/r$. Thus the right-hand side is positive, and we can see from its simple form that in passing from negative to positive it must have crossed zero at the radius $r_c = GM_\odot M/4kT \gg a$. Now if the expansion velocity were to remain subsonic ($u^2 < 2kT/M$) in passing from $r = a$ to $r = \infty$, the factor in parentheses on the left-hand side would remain negative and du/dr would become negative as r passes across r_c. Thus $u(r)$ would decline beyond r_c, and the distant corona would be essentially static with $p(\infty) > 0$ again.

There is no evident way to escape this problem except to suppose that u is nonvanishing at infinity. Figure 1 sketches the topology of the family of solutions $u(r)$ on the (u, r) plane. It is evident that the only solution that passes from small u near the Sun to vanishing pressure, i.e. nonvanishing u, at large r is the solution passing across the critical radius, or sonic point, $r = r_c$ where $u(r_c) = (2kT/M)^{1/2}$. In this way the factor in parentheses $(u - 2kT/Mu)$ vanishes at the same point as the right-hand side, and du/dr and u both pass smoothly across.

With these considerations in mind, consider the integral of Equation (16), which can be written

$$\frac{1}{2}u^2 - \frac{kT}{M}\ln u^2 = \frac{4kT}{M}\ln r + \frac{GM_\odot}{r} + \text{constant} \quad (18)$$

Let $U = (2kT/M)^{1/2}$ represent the characteristic ion thermal velocity and sound speed, and require that $u(r_c) = U$ at the critical radius $r_c = GM_\odot/2U^2$. The result is

$$\frac{u^2}{U^2} - \ln\frac{u^2}{U^2} = 4\ln\frac{r}{r_c} - 3 + 4\frac{r_c}{r} \quad (19)$$

Thus, for $r \ll r_c$, there is slow expansion ($u^2 \ll U^2$) with

$$u(r)^2 \cong U^2 \left(\frac{r_c}{r}\right)^4 \exp\left(-3 - 4\frac{r_c}{r}\right) \quad (20)$$

while for $r \gg r_c$, the velocity is supersonic with

$$\frac{u^2}{U^2} \sim 4\ln\frac{r}{r_c} - 3 + \ln\frac{u^2}{U^2}$$

$$\sim 4\ln\frac{r}{r_c} - 3 + \ln\left(4\ln\frac{r}{r_c} - 3\right) + \cdots \quad (21)$$

The square of the velocity increases logarithmically without bound in the idealized isothermal case because the uniform temperature implies that heat is continually added as the expanding gas flows to infinity.

THE PHYSICS OF SUPERSONIC EXPANSION

It is the continuing addition of heat – by thermal conduction, if nothing else – that allows the coronal gas to climb slowly out of the deep gravitational potential well at the Sun and then accelerate to supersonic velocity with increasing distance ($r > r_c$) from the Sun. This is the crucial difference from ideas for catapulting gas from the surface of the Sun to infinity. For instance, Schlüter's "melon seed" mechanism (Schlüter 1950, 1952) involves a magnetized plasmoid squeezed out along a diverging magnetic field, with the declining magnetic and thermal energy of the expanding plasmoid doing work against the gravitational field and ultimately taking the plasmoid away to infinity with a finite velocity. Sufficient energy for the escape from the Sun must be in the initial thermal and magnetic energy of the plasmoid

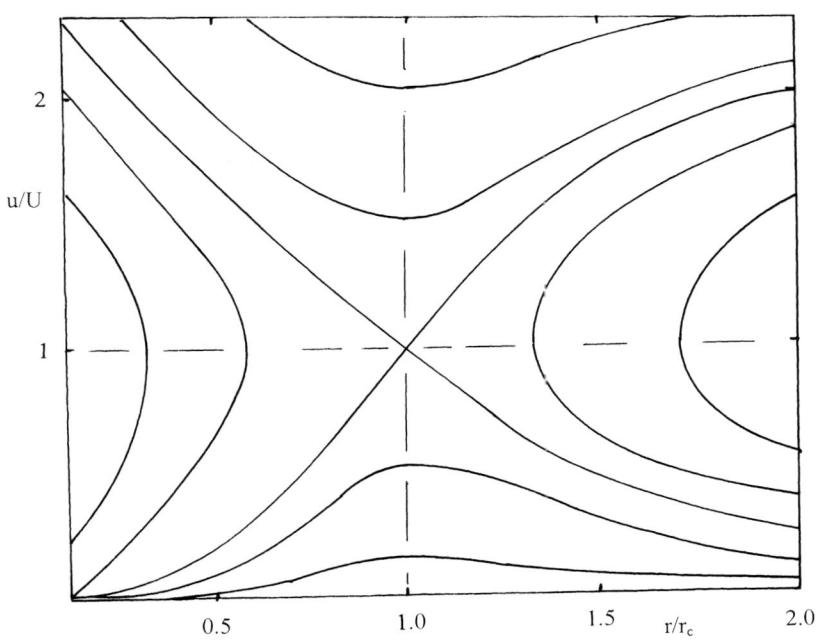

Figure 1 A sketch of the topology of the family of solutions of eqn (25), showing the connection between $u(r)$ at large and small r.

and in the energy of deformation of the surrounding large-scale field by the plasmoid. Once on its way the plasmoid moves too fast to be resupplied with energy along the way. In contrast, the slowly expanding corona lies close to hydrostatic equilibrium with only a modest thermal energy, of perhaps one-fifth of the negative gravitational energy. However, thermal energy is continually supplied to each element of outflowing gas by thermal conduction from the high temperature near the Sun, and perhaps by the dissipation of Alfvén waves, to maintain the temperature as the gas expands outward, eventually putting in many times more energy than the initial resident thermal energy. Thus the gravitationally bound corona near the Sun slowly climbs out of the gravitational potential well and expands away into the vacuum at infinity as thermal energy is continually supplied. This is the essential physics of supersonic coronal expansion.

As a simple illustration, consider a volume V of isothermal ionized gas with number density N starting at the reference level $r = a$ near the Sun with pressure $p(a)$. The initial thermal energy content of the ideal monatomic gas is $\Theta(a) = (3/2)p(a)V(a)$. The product $p(r)V(r)$ remains constant during the isothermal expansion $(\Theta(r) = \Theta(a))$, and the rate at which heat Q is added as the gas streams outward is given by

$$\frac{dQ}{dt} = p\frac{dV}{dt}$$

Integration yields the total heat $Q(r)$ added after departure from $r = a$ as

$$Q(r) = \frac{2}{3}\Theta(a)\ln\left[\frac{N(a)}{N(r)}\right] \quad (22)$$

so that $Q(r)/\Theta(r)$ increases logarithmically as $N(r)$ declines with increasing r.

Note that another elementary approach to the dynamics of the solar corona is the mathematical device of a polytropic law, in which it is assumed that p declines outward from the Sun in proportion to N^α, with $T \sim N^{\alpha-1}$. It is clear that with α equal to the adiabatic value of 5/3 there can be no expansion because the initial thermal energy is insufficient to lift the gas out of the deep gravitational potential well of the Sun. The isothermal corona, $\alpha = 1$, produces a supersonic outflow in the manner already described. The question is the upper limit on α for supersonic expansion. Note, then, that with u at the final asymptotic uniform escape velocity, conservation of matter yields $N \sim 1/r^2$ for the simple radial flow. Hence, $T \sim 1/r^{2(\alpha-1)}$. The requirement for supersonic outflow is that the exponent $2(\alpha-1)$ be less than one, from which it follows that $\alpha < 3/2$ (Parker 1960b, 1963).

The limiting case $\alpha = 3/2$ provides for *subsonic* radial outflow of a corona for the special case $2kT(a) = GM_\odot M/3a$ ($T(a) = 2.6 \times 10^6$ K for hydrogen at $a = 10^{11}$ cm and proportionately higher for heavier molecules) for which $u(r) = u(a)$ for all r. This solution shows that a gas composed of molecules with four degrees of freedom (one of which would be internal), so that adiabatic expansion yields $\alpha = 3/2$, would have a thermal energy and temperature that declines as $N^{1/2}$ or $1/r$ as the gas moves away from the Sun. The thermal energy plus the gravitational potential energy remains equal to zero as the gas expands outward with constant speed. For stronger gravitational binding ($2kT(a) < GM_\odot M/3a$) there is no escape to infinity unless the gas is catapulted out with enough bulk kinetic energy to make up the thermal energy deficit.

The physics of supersonic expansion for an arbitrary extended temperature is nicely illustrated by some simple considerations on the momentum equation and the heat flow equation. For $a < r < r_c$ the inertial term $u\,du/dr$ can be neglected in eqn (16), the approximation breaking down as r approaches r_c. The result is the barometric equation (2), from which we obtain

$$N(r) = N(a)\frac{T(a)}{T(r)}\exp\left[-\int_a^r d\xi \frac{GM_\odot M}{2kT(\xi)\xi^2}\right]$$

Conservation of mass yields

$$u(r) = u(a)\frac{a^2 T(r)}{r^2 T(a)}\exp\left[+\int_a^r d\xi \frac{GM_\odot M}{2kT(\xi)\xi^2}\right] \quad (23)$$

That is to say, the near corona is close to hydrostatic equilibrium and, if $T(r)$ is small enough that the right-hand side of eqn (16) is negative, then the slow outflow $u(r)$ increases outward with declining $N(r)$ so as to conserve matter. As already noted, this simple condition breaks down as r reaches r_c, but, even so, the error in the form of $u(r)$ is not large. The solution is required to pass across the critical point with $u(r_c)^2 = 2kT(r_c)/M$ with r_c defined as the radius at which the right-hand side of eqn (16) vanishes. So, given $T(r)$, r_c and $u(r_c)$ can be calculated without first knowing $u(r)$. The result substituted into eqn (23) gives $u(a)$ and hence $u(r)$ throughout $a < r < r_c$. The process can be iterated, using this first approximation to $u(r)$ to calculate the neglected term $u\,du/dr$. The convergence toward the exact solution is fairly rapid (Parker 1964a,b,c, 1969).

Beyond r_c the opposite approximation becomes useful, neglecting $2kT(r)/u(r)$ compared to $u(r)$. The approximation is not accurate immediately beyond r_c, but the accumulated error in the integrated velocity is not large for declining $T(r)$, and the result is

$$\frac{1}{2}u(r)^2 + \frac{2kT(r)}{M} - \frac{GM_\odot}{r} = \int_{r_c}^r \frac{d\xi}{\xi}\frac{2kT(\xi)}{M} + \frac{3kT(r_c)}{M} - \frac{GM_\odot}{r_c}$$

representing expansion into a vacuum. Again, the calculation can be iterated, using the first estimate of u to calculate the neglected term $2kT(r)/Mu(r)$.

We can see, then, that the supersonic expansion begins as a slow upward flow in the quasi-static corona near the Sun, accelerating with the rapidly decreasing density to the local speed of sound as the critical point is approached, beyond which there is free expansion into a vacuum, retarded only slightly by the residual gravitational attraction of the Sun.

Clauser (1960) pointed out the mathematical analogy of coronal expansion to the expansion of gas through a Laval nozzle to produce a supersonic flow from a steady supply of compressed gas. The nozzle consists of a tube of varying cross-section $A(s)$, where s represents distance along the tube. In the slender tube approximation the steady flow velocity $u(s)$ is assumed to be uniform across $A(s)$ so that conservation of mass requires $u(s)A(s)N(s) = $ constant, where $N(s)$ represents the number density of the gas. For a pressure $p(s) = N(s)kT(s)$, the equation for steady motion is

$$NMu\frac{du}{ds} = -\frac{d}{ds}NkT$$

which becomes

$$\frac{du}{ds}\left(u - \frac{kT}{Mu}\right) = \frac{kT}{MA}\frac{dA}{ds} - \frac{d}{ds}\left(\frac{kT}{M}\right)$$

$$= -A\frac{d}{ds}\left(\frac{kT}{MA}\right)$$

The nozzle begins with a constriction, $dA/ds < 0$, where the subsonic gas flow accelerates with little change in density at first, conserving the mass flux with $A(s)u(s) \sim$ constant. In this stage $u^2 < kT/M$ so that the factor in parentheses on the left-hand side of the equation is negative. Thus $du/ds > 0$, in keeping with the requirement that $A(s)u(s) =$ constant. The flow reaches the speed of sound with $u^2 = kT/M$ in the throat, where $A(s)/T(s)$ is a minimum $(d(A/T)/ds = 0)$. Beyond the throat the tube flares out $(dA/ds > 0)$, and the gas expands freely into the vacuum, or partial vacuum, beyond, with $dT/ds < 0$ so that the right-hand side of the equation is positive at the same time that the factor $(u - kT/Mu)$ on the left is also positive. The sonic point, or critical point, where both $(u - kT/Mu)$ and the right-hand side vanish simultaneously, is simply the location of the minimum in the ratio $A(s)/T(s)$ where the flow goes supersonic.

The analogy with the solar wind is direct, with gravity and the associated strong radial decline in $N(s)$ playing the role of the decline in $A(s)$ to form the throat. The physical difference is only that heat is supplied to the upward moving corona to maintain the temperature as the density declines with height, whereas in the Laval nozzle the density remains nearly constant as the gas approaches the sonic point and no heat needs to be added. That is to say, the expanding corona is doing work against gravity, whereas no such work is involved in expansion through a nozzle.

Note, then, that the afterburner employed in military fighter airplanes, in which fuel is burned in the expanding exhaust channel, is an analogy to the solar wind in which heat is added to the coronal gas after passing the sonic point.

SOLAR WIND SPEEDS

Consider the outward flow velocities that might reasonably be expected at the orbit of Earth ($r = 1$ AU $= 1.5 \times 10^{13}$ cm). The answer depends, of course, on the temperature profile $T(r)$, about which relatively little was known. Observations indicated $T \sim 1$–2×10^6 K near the Sun, but there was no idea as to radial profile of the basic coronal heat source. It was clear that Chapman's very effective heat conduction by electrons would distribute the heat broadly, and the observed outward extension of the corona indicated a broad temperature distribution. Even today, over 40 years later, we do not know the precise run of temperature with radial distance from the Sun. The observational discovery of the higher temperature of the helium and heavier ions in the wind, with the Solar and Heliospheric Observatory (SOHO) measurements of heavy ion temperatures as high as 5×10^6 K (Kohl et al. 1997, Raymond et al. 1997), suggests some form of resonance wave heating. Whatever the detailed form of the heating, it has been suggested by Martin (1984) and by Porter and Moore (1988) that the principal heat input to the expanding (magnetically weak) open corona is the microflaring in the boundaries of the supergranule convective cells and in the general magnetic carpet (Schrijver et al. 1997). As already noted, the necessary heat input is estimated (Withbroe and Noyes 1977) to be approximately 0.5×10^6 erg cm^{-2} s^{-1}, with most of it going into maintaining the temperature of the slowly expanding corona within the first solar radius above the surface of the Sun.

Now it was clear that the corona expands away from the Sun to form the supersonic solar wind only where the magnetic field is not so strong as to prevent the expansion. Thus, for instance, the bipolar magnetic fields of the active regions ($B \sim 10^2$–10^3 G) are not sources of coronal expansion. It was apparent, then, that the coronal expansion from the regions of weak field ($B < 10$ G), now recognized as the coronal holes (Waldmeier 1975, Nolte et al. 1976), may diverge superradially – that is, more rapidly than the idealized radial flow. The equation for conservation of mass is then more like Nur^s with $s > 2$, rather than the simple Nur^2 for the purely radial flow. It followed immediately (Parker 1958e) that this produces a faster expansion for the same temperature profile, with the first term on the right-hand side of eqn (17) replaced by $2skT/Mr$ and the first term on the right-hand side of eqn (19) becoming $2s \ln r/r_c$. The critical radius, or sonic point, moves closer to the Sun, and the wind velocity beyond is larger because the coefficient of the dominant logarithmic term is $2s$ with $s>2$.

However, these details made relatively little qualitative difference in the original idealized isothermal model. Supposing that the effective mean value of T is 1×10^6 K, it is readily shown (see Figure 2) that $U = 130$ km s^{-1}, $r_c = 4.0 \times 10^{11}$ cm $= 5.7\,R_\odot$, and the orbit of Earth lies at $37.5 r_c$. It follows from Figure 3 that $u = 3.7 U = 480$ km s^{-1} at the distance of Earth. A hotter corona, with $T = 2 \times 10^6$ K, provides $U = 182$ km s^{-1}, $r_c = 2.0 \times 10^{11}$ cm $= 2.9\,R_\odot$, with Earth at $75 r_c$ where $u = 4.1 U = 750$ km s^{-1}. The essential point is that the velocities are of the right order of magnitude to provide the SCR indicated by the antisolar comet tails and by the 2–4 day delay in geomagnetic disturbances following special activity on the Sun, indicating 880 km s^{-1} and 440 km s^{-1}, respectively.

The hydrodynamic theory of coronal expansion is not able to estimate the density of the solar wind at the orbit of Earth because of the extreme sensitivity of the density to the temperature near the Sun. The association of the zodiacal light with the extended corona (Behr and Siedentopf 1954, Elsässer 1954, van de Hulst 1954) suggested 500 electrons/cm^3. Biermann suggested comparable or larger densities for the SCR in order to have the Coulomb collisions between the SCR ions and the ions in the comet tail account for the vigorous antisolar acceleration of the comet tail ions. Unsöld and Chapman (1949) proposed that densities might run as high as 10^5 electrons/cm^3 following a large flare on the Sun.

The theoretical problem in predicting the density can be seen from the fact that for $R_\odot < r < r_c$ the inertial effects of the subsonic expansion near the Sun can be neglected to a first approximation and the density of the corona at r_c is given by the static barometric law, from which it follows that $N(r_c) \sim N(a) \exp[2(1 - r_c/a)]$ for the simple isothermal corona. The factor $\exp(2 r_c/R_\odot)$ has a value of 0.95×10^5 for $T = 1 \times 10^6$ and 3×10^2 for 2×10^6 K, differing by a factor of about 300. This gives some idea of the uncertainty in any theoretical predictions of the solar wind density short of precise knowledge of $T(r)$.

Some limits may be placed on the mass flux in the solar wind by considerations of the total heat input to the expanding corona. The energy required to levitate gas from the surface of the Sun to infinity is $GM_\odot/R_\odot = 2 \times 10^{15}$ erg g^{-1}, equivalent to the kinetic energy of the escape velocity of 640 km s^{-1}. Thus, for instance, a wind velocity of 640 km s^{-1} at large radial distance indicates a total energy expenditure of 4×10^{15} erg g^{-1}, and the coronal heat input required to produce a mass flux — Nur^2 gm s^{-1} sr^{-1} is 4×10^{15} — Nur^2 erg s^{-1} sr^{-1}. Lacking any knowledge of the heat input to the corona, N can be either large or small. Only following direct measurements of the solar wind velocity and

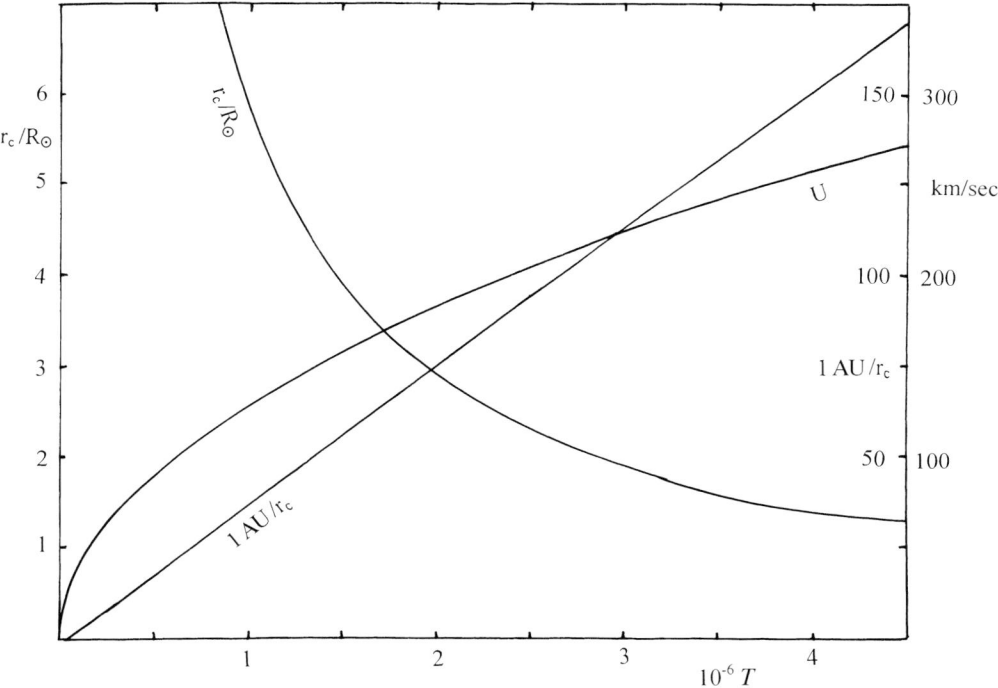

Figure 2 A plot of the thermal velocity $U = (2kT/M)^{1/2}$, the radial distance to the sonic point (critical radius) r_c in units of the solar radius $R_\odot$, and the radial distance to Earth in units of r_c, as a function of the isothermal coronal temperature T.

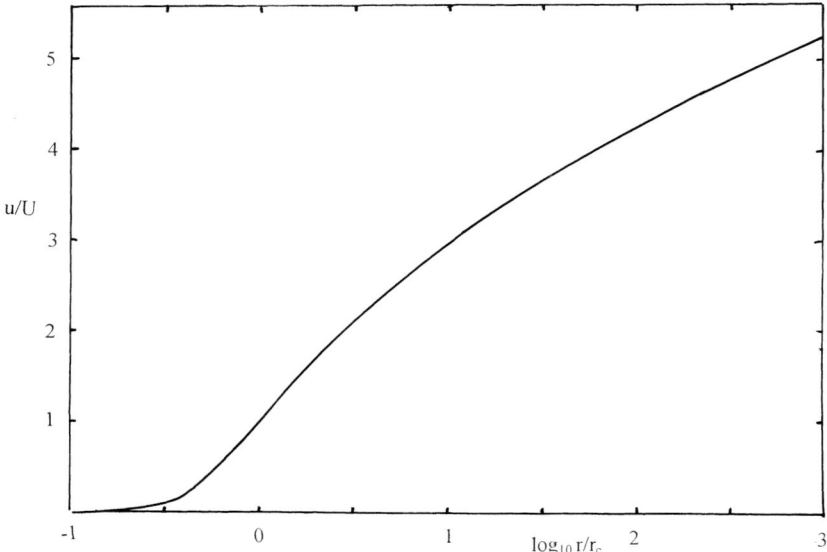

Figure 3 A plot of $u(r)/U$ for the isothermal corona as a function of r/r_c.

density in space was it possible to work back to the Sun to estimate a coronal heat input of 0.5×10^6 erg cm^{-2} s^{-1} (Withbroe and Noyes 1977).

In summary, the first paper (Parker 1958e) laid out the concept of the expansion of the strongly bound corona with a temperature extended by the addition of heat, as distinct from the traditional concept of a static corona limited to evaporation from its exosphere into an otherwise empty space. It showed that, given an extended temperature, there is no obvious alternative to the coronal expansion reaching supersonic velocity at large distance.

INTERPLANETARY MAGNETIC FIELD

Once the mechanics of the solar wind was clear, it was immediately evident that the expanding corona must stretch out any magnetic field that is too weak to hold back the expansion, so that interplanetary space is filled with the magnetic field extended outward from the regions of weak field on the Sun. So for the simple case of radial outflow at some fixed velocity u from a hypothetical Sun that does not rotate, the radial magnetic field $B_r(a, \theta, \varphi)$ at heliocentric polar angle θ and azimuth φ at the reference level $r = a$ is extended to the radial field $B_r(r, \theta, \varphi) = B_r(a, \theta, \varphi)(a/r)^2$ throughout $r > a$. Thus 1 G at a reference level $a = 10^{11}$ cm $= 1.5\,R_\odot$ becomes 0.45×10^{-4} G at 1 AU ($r = 150a$).

The rotation of the Sun makes the extension a little more complicated, twisting the interplanetary field into an Archimedean spiral. Continuing with the idealization of a uniform radial wind velocity u, a steady point source of smoke at some location (a, θ_0, φ_0) rotating with the Sun (so that φ_0 is measured from a reference longitude fixed in the surface of the rotating Sun) produces a thread of smoke in the wind given by $\theta = \theta_0$ and $\varphi = \varphi_0 - \Omega(\theta_0)(r-a)/u$ at the radial distance r, where $\Omega(\theta_0)$ is the angular velocity of the reference surface $r = a$, assumed to rotate with the Sun at $\theta = \theta_0$, and $(r-a)/u$ is the transit time from a to r. The field line from (θ_0, φ_0) threads the smoke and extends out through space along the same path as the smoke. The radial component of the field is transported radially outward in the wind while the Sun rotates through an angle $\Omega(\theta)(r - a)/u$. Thus

$$B_r(r, \theta, \varphi) = B_r[a, \theta, \varphi + \Omega(\theta)(r - a)/u](a/r)^2$$

The smoke thread and the associated field line are inclined to the radial direction by the angle ψ given by

$$\tan \psi = r \sin \theta \, d\varphi/dr$$
$$= \Omega(\theta) r \sin \theta / u$$

Then since $B_\varphi = B_r \tan \psi$, it follows that

$$B_\varphi(r, \theta, \varphi) = B_r[a, \theta, \varphi + \Omega(\theta)(r - a)/u] a^2 \Omega(\theta) \sin \theta / ur$$

Thus, while B_r dominates near the Sun, it declines outward as $(a/r)^2$ so that beyond approximately the orbit of Mars, the azimuthal field B_φ is almost always the stronger component, declining only as a/r for as far as the solar wind extends out into interstellar space (Parker 1958e).

It was noted that the solar wind exerts a retarding torque on the rotation of the Sun, mainly as a consequence of the Maxwell stress (tension) in the spiral magnetic field. It was shown that the present day torque is of no significance for retarding the rotation of the Sun.

It turns out, then, that the radial field between the Sun and Earth allows direct transit of solar cosmic rays to Earth. Beyond the orbit of Mars the field is inclined more than 45° to the radial direction, and the path length along the spiral field increases asymptotically as r^2. Given the expected scattering of fast particles by the small-scale irregularities in the field, one could see that the escape of solar cosmic rays from the inner solar system is substantially impeded, as indicated by the slow decline of the solar cosmic ray intensity following a flare (Meyer *et al.* 1956). In the same way, the entry of galactic cosmic rays into the inner solar system is impeded, so that the outward sweep of the spiral magnetic field substantially reduces the cosmic ray intensity at the orbit of Earth (Parker 1958c, 1963, 1965a).

A blast from the Sun, now recognized as a coronal mass ejection (MacQueen and Holzer 1988, Jackson *et al.* 1988), expands outward through the ambient solar wind and spiral magnetic field, compressing the field into an outward moving magnetic barrier that provides the abrupt sweeping away of the galactic cosmic rays, observed at Earth as the Forbush decrease as the shock front and blast wave pass by (Parker 1961a). The shock front itself produces a fast particle population of its own (Parker 1958d, 1959, 1961c; Parker and Tidman 1958).

The impact of the solar wind and its spatial magnetic field against the geomagnetic dipole is the cause of the geomagnetic variability and the aurora, with a sudden commencement and geomagnetic storm produced by the impact of the occasional blast wave (Parker 1958a, 1961a).

BEHIND THE SCENES

The story behind the scenes provides a lighter note to the formal considerations on the theory of coronal expansion. Thus it was that the original paper presenting the coronal expansion and solar wind concept (Parker 1958e) in the context of the isothermal corona did not contain the term "solar wind" for the simple reason that the author had not yet thought of that term. However, several papers preceding that paper in publication employed the term solar wind, to emphasize the hydrodynamic character of the phenomenon. This temporal inversion came about in the following manner. The original paper was received by the *Astrophysical Journal* for publication on 2 January 1958, having been written in late 1957. The next paper on the solar wind treated the modulation of the galactic cosmic ray intensity by the outward sweeping magnetic irregularities in the solar wind (Parker 1958c) and was received by the *Physical Review* for publication on 19 February 1958. That was the first paper employing the term solar wind. The paper appeared in the 15 June 1958 issue of the *Physical Review*. The study of the impact of the solar wind against the geomagnetic field was next, received on 25 April by the *Physics of Fluids*, revised on 16 May on the basis of some suggestions by the referee, and appeared in the May–June issue (Parker 1958a). Then attention turned to the structure of collisionless shock fronts. That paper (Parker 1959) was received by the *Astrophysical Journal* on 29 July 1958 and appeared in the January 1959 issue.

Now, sometime in the late Spring of 1958 the referee's report on the original paper appeared, with the suggestion that the author familiarize himself with the subject before attempting to write a scientific article on SCR. There was no specific objection to the arguments or calculations in the submitted paper. The author pointed out to the editor the absence of any substantive objections by the referee. The editor, Subrahmanyan Chandrasekhar, sent the paper to a second referee. Sometime in the summer the second referee responded that the paper was misguided and recommended against publication, again with no specific criticism except that the author was obviously unfamiliar with the literature. Again the author's response to the editor was to note that there was no substantive criticism, no specific fault pointed out. Some days later Chandrasekhar appeared in the author's office with the paper in his hand and said something along the lines of, "Now see here, Parker, do you really want to publish this paper?" I replied that I did. Whereupon he said, "I have sent it to two eminent experts in the field and they have both said that the paper is misguided and completely off the mark." I replied that my problem with the referees was that they were clearly displeased, but had nothing more to say. Chandrasekhar was silent for a moment and then he said, "All right, I will publish it." And that is how the paper without the term solar wind finally appeared in the November issue of the *Astrophysical Journal* (Parker, 1958e), substantially after the papers written later and employing the term solar wind. I have sometimes wondered what might have happened had someone else been the editor. In view of the firm assertion by the two eminent referees that the paper was without scientific value, rejection for publication would have been the likely outcome. I would have been obliged to submit the paper to some less prestigious journal, i.e. one that did not use such eminent experts as referees, with further delays, of course.

FURTHER DEVELOPMENT OF THE SOLAR WIND CONCEPT

The next step in the development of the dynamical theory of the supersonic expansion of a strongly bound atmosphere after the isothermal case was the polytrope model in which the temperature declines outward from the Sun in proportion to $N^{\alpha-1}$, or $p \sim N^{\alpha}$, where α is a constant (Parker 1960a,b), showing supersonic expansion for $\alpha < 3/2$, already discussed in terms of energy.

Then came simple models of the outer regions of the solar wind where it impacts on the interstellar medium (Parker 1961b) as the ram pressure of the supersonic wind falls to the level of the interstellar pressure P. Clauser (1960) and Weymann (1960) had both made the point that the solar wind may be presumed to pass through a shock transition to subsonic velocity where the wind meets the interstellar medium, so the shock formation was included in the exploratory models, to show how the subsonic shocked wind might stream away in the local interstellar wind to form a downstream wake. It was not possible at that time to do much in estimating the position of the shock transition. The enormous solar wind densities of 500 atoms/cm^3 at the orbit of Earth, inferred from the belief that the zodiacal light represents Thomson scattering of sunlight from free electrons and that the interaction of the SCR with comet tails is by Coulomb collisions between wind and comet ions, gave rather large distances, of the order of 500 AU or more, along the lines discussed earlier by Davis (1955). So the theoretical study concentrated on the general hydrodynamics of the subsonic solar wind flow beyond the termination shock.

Next was an extension of the coronal expansion and solar wind concept to other stars, with very little to go on in the way of quantitative knowledge of their coronas and stellar winds (Parker 1960b, 1961b). The point was that supersonic outflows were observed from some giant stars, and, if a pedestrian main sequence star like the Sun has an extended million degree corona, then so should most other stars on the lower end of the main sequence. The conclusion was simply that most other stars probably have supersonic winds.

The geomagnetic storm, produced by the impact of a blast wave from the Sun, was another topic opened up by the solar wind concept. Dessler and Parker (1959, 1968) and Parker (1967c) studied the effect of energetic ions trapped in the geomagnetic field and inflating the field to provide the main phase of the storm. It was shown from the mechanics of the inflation that the fractional field reduction $\Delta B/B$ of the magnetic field at low latitudes at the surface of Earth is comparable to the ratio of the kinetic energy of the trapped particles to the energy of the magnetic field above the surface of Earth, some 8×10^{24} erg. It was also pointed out that the principal mechanism for relaxation of the field reduction is by charge exchange of the trapped ions with the ambient thermal neutral atoms, converting each kilovolt trapped ion into a kilovolt neutral atom which freely departs from the region. The large solar wind densities in vogue at the time indicated a ram pressure of the solar wind sufficient to push in the geomagnetic field to within 5 Earth radii of the center of Earth, so it was assumed that some of the kilovolt ions of the solar wind became entangled in the geomagnetic field at about that radius. Then when the pressure of the wind eased, and the initial compressive phase of the magnetic storm came to an end, the inflation of the field by the pressure of the lingering trapped solar wind ions provided the expansive main phase. A position in the vicinity of five Earth radii was about the right location to produce the observed latitudinal profile of the main phase, and it was the right position to account for the subsequent relaxation of the main phase through charge exchange with the ambient neutral hydrogen in a characteristic time of one day.

When there began to be direct observational indications (see the following section) that the solar wind was nowhere near as dense as the 500 atoms cm^{-3} inferred from the zodiacal light and comet tail interaction, it was suggested (Dessler *et al.* 1961) that it is more likely that the solar wind has no means for such deep injection, and it is the ambient atmospheric and magnetospheric ions that are accelerated in some way to kilovolt energies (Parker 1962).

The rise time of the sudden commencement of the initial compressive phase of the geomagnetic storm was calculated (Dessler *et al.* 1960) as the transit time of the interplanetary shock sweeping by the magnetosphere plus the Alfvén wave transit time from each point on the magnetosphere/solar wind boundary – the magnetopause – to the observing point on the surface of Earth. It was pointed out that the observed diffusion and relaxation of the Van Allen trapped particles is primarily a consequence of the fluctuations introduced into the geomagnetic field by the turbulence in the impacting solar wind (Parker 1960c). Then the first calculations of the comet-shaped magnetosphere under the impact of a steady supersonic wind began to appear (Beard 1960, 1962; Ferraro 1960; Davis and Beard 1962; Beard and Jenkins 1962; Spreiter and Briggs 1962a,b; Spreiter and Hyett 1963, Midgley and Davis 1963). The bow shock was introduced into the problem (Axford 1962, Kellogg 1962) together with the effect of the interplanetary magnetic field (Spreiter and Jones 1963), and the geomagnetic picture was beginning to show the form that it has today.

While this activity was underway, J.W. Chamberlain (1960) declared that the supersonic solar wind was a mistake, arising from the improper choice of the constant of integration in the integral of the momentum equation. He subsequently produced a hydrodynamic model (Chamberlain 1961), based on thermal conduction outward from a source of heat at the base of the corona, which was subsonic all the way to infinity. This was accomplished by starting with an enormous gas density ($\sim 10^{10}$ atoms/cm^3) at the Sun, a factor of 100 above the observed density, and by reducing the thermal conductivity to one-eighth of the normal value, thereby greatly reducing the heat supplied to each ion and electron pair. Numerical integration of the radial momentum and heat flow equations led to a subsonic expansion of the corona, closely analogous to the evaporation of gas from an exosphere. His subsonic wind of 10–20 km s^{-1} at 1 AU, was dubbed the solar breeze. Chamberlain felt that this "evaporative" solution was the only proper theoretical treatment of coronal expansion. He justified his suppressed thermal conductivity and high coronal density on the grounds that the calculated electron density at large radial distance agreed with the enormous electron density estimated from the assumption that the zodiacal light represents Thomson scattering from interplanetary electrons (van de Hulst 1953, 1955). The gas pressure fell to zero at infinity, because the suppression of heat transport provided a temperature falling with increasing distance as $1/r$, associated with the vanishing total energy flux in the subsonic solar breeze. The solar breeze velocity declined outward asymptotically as $1/r^{1/2}$ and arrived at infinity with no kinetic, thermal, or gravitational energy, as already noted. Chamberlain gave an invited lecture at the spring 1961 meeting of the National Academy of Sciences, and his audience came away assured that there was no more than the traditional evaporative loss of gas from the solar corona.

Later on, with the iterative method described above for generating analytic solutions for the momentum and heat flow equation (Parker 1964a,b, 1965a,d, 1969), it was possible to treat the various classes of coronal expansion in a general way. A crucial quantity is the total radial energy flux per steradian

$$F = -r^2\kappa(T)\frac{dT}{dr} + N(r)v(r)r^2\left(\frac{1}{2}Mv(r^2) + 5kT(r) - \frac{GM_\odot M}{r}\right)$$

assuming no radiation loss and assuming heat transport only by thermal conduction and bulk convection. The evaporative model corresponds to the condition that $F = 0$, which provides $v \sim 1/r^{1/2}$ and $T \sim 1/r$ at large r. So if the coronal density is sufficiently high and the thermal conductivity sufficiently low, the coronal temperature does not extend far out into space, and there is clearly the possibility of avoiding supersonic expansion (see the exchange of views in Chamberlain (1965) and Parker (1965b,c), thereby leaving the Biermann–Chapman dilemma unresolved.

Noble and Scarf (1962, 1963) and Whang and Chang (1965) provided numerical integrations of the momentum and heat flow equations showing how the thermal conductivity in a corona with realistic densities of 10^8 ions and electrons/cm^3 near the Sun (where the heat is introduced) is sufficient to provide supersonic velocities and wind densities comparable to the observed values.

A complete formal mathematical analysis of the conditions for a conduction corona to provide subsonic or supersonic radial expansion was carried out by Roberts and Soward (1972), and the interested reader is referred to their comprehensive paper for the full mathematical picture. Hundhausen (1972) reviews the many theoretical studies of coronal expansion and the solar wind up to that time.

DIRECT OBSERVATION OF THE SOLAR WIND

It was clear by 1960 that the indirect inferences to which the ground-based observer is limited had gone about as far as they could go. The issue of the solar wind and the density, temperature, velocity, and so on could be addressed only by direct observation with instruments carried out of the geomagnetic field and into the wind on suitable spacecraft. Fortunately the space age arrived in the midst of the debate, with the first direct detection and measurements made by Gringauz et al. (1960, 1961, 1967) on the Luna 2, Luna 3, and Venus 3 spacecraft. The instrument was a plasma-collecting cup shielded by a grid charged 50 V positive, thereby picking up the solar wind electrons on the grid and allowing the solar wind ions with energies above 50 V to pass into the collecting cup to be counted. The device measured a flux of protons of a few times 10^8 cm^{-2} s^{-1}. Noting that a proton with an energy of 50 eV has a velocity of 100 km s^{-1}, the measurements showed that there is a solar wind with velocity not less than 100 km s^{-1} during the times of the flights. Such velocities are supersonic unless the temperature in the wind at the orbit of Earth is assumed to be at least 0.5×10^6 K. Gringauz's measurements were not immediately recognized by his peers for their fundamental importance. A general disbelief was the order of the day, with the unfortunate result that his pioneering scientific measurements received but little support.

The Explorer 10 spacecraft launched in 1961 carried a more sophisticated plasma cup built by the MIT group and able to step the positive voltage on the shielding grid to more than a kilovolt, thereby measuring the proton flux penetrating various retarding potentials (Bonetti et al. 1963). With a cycle of stepwise increases in retarding potential the MIT group found a mean proton velocity of 280 km s^{-1} and a proton flux of $1-2 \times 10^8$ cm^{-2} s^{-1}. This indicated solar wind densities of 4–8 atom/cm^3, essentially a factor of 10^2 lower than the estimates based on the brightness of the zodiacal light and on the strong interaction of the SCR with comet tails. The only problem with the measurements was their intermittent character, with the solar wind appearing for a fraction of an hour alternating with comparable periods in which there was no wind but a magnetic field of 10^{-4} G instead (Heppner et al. 1963). It suggested coronal streamers of solar wind mixed with essentially empty magnetic flux bundles stretched out from the Sun. One wondered whether this peculiar phenomenon had something to do with the fine streamers observed in the solar corona during a total eclipse (Parker 1964c). However, as already noted, quantitative calculations of the dimensions of the comet shape to which the geomagnetic field would be confined by the solar wind were becoming available. The earlier estimates of wind densities of 500 atoms/cm^3 or more had indicated that the sunward boundary of the geomagnetic field should lie somewhere in the vicinity of five Earth radii, well inside the path of Explorer 10 at 10–15 radii. However, the actual low density of the wind revised the estimate to 10–15 radii, depending on location relative to the Sun–Earth line. It was clear from the revised models of the magnetosphere, then, that Explorer 10 had traveled pretty much along the bouncing and flapping magnetopause in the shocked gas behind the bow shock in the solar wind (Bonetti *et al.* 1963). So the plasma detector had alternately sampled the shocked solar wind and the outer geomagnetic field. It showed clearly that the solar wind existed, with supersonic velocities of hundreds of km s^{-1}, temperatures of the order of 10^5 K, and number densities of several protons/cm^3 (Bonetti *et al.* 1963, Scherb 1964). It showed, then, that the zodiacal light was mainly a result of sunlight scattered from interplanetary dust grains, with little or nothing to do with free electrons.

The definitive observations of the solar wind came with the plasma instrument built at the Jet Propulsion Laboratory (JPL) and carried on the Mariner 2 spacecraft launched in late 1962, bound for Venus. The plasma detector was a curved plate electrostatic analyzer, measuring proton flux as a function of proton energy. The instrument showed that the supersonic solar wind was present during the entire voyage of over three months. Flow speeds as slow as 320 km s^{-1} and as large as 770 km s^{-1} were observed, with an average value of 504 km s^{-1}. The proton densities showed a minimum of 0.44/cm^3 and a maximum of 54/cm^3, with an average of 5.4/cm^3 (Snyder and Neugebauer 1964, Neugebauer and Snyder 1966, 1967). Neugebauer and Snyder (1964) went on to show from the data that the proton temperature was sometimes as low as 3×10^4 K and as high as 6×10^5 K, with an average of 1.5×10^5 K. They also pointed out the remarkable fact that the helium ions (He^{2+}) exhibited a temperature that was approximately four times the proton temperature, indicating roughly the same thermal velocities, both of which proved to be about equal to the local Alfvén speed.

This brings us to the interplanetary magnetic field, which is a basic test of the place of origin of the solar wind back at the Sun. Unfortunately the Mariner 2 spacecraft was not magnetically clean, so it was difficult to determine the direction of the magnetic field in the solar wind. However, Coleman *et al.* (1962, 1963) and Davis *et al.* (1964) were able to identify a rapidly fluctuating magnetic field component of 4×10^{-5} G.

In view of the difficulty with the varying spacecraft magnetic fields on Mariner 2, the first of the Interplanetary Monitoring Platforms, IMP 1, was constructed with strict attention of magnetic cleanliness. It must be appreciated that this was no small task. Not only must the spacecraft and its electronics be constructed from entirely nonmagnetic metals, including such fine points as the lead in wires on the transistors, but the electric circuitry has to be designed and laid out so that there are no current loops. Any net circulation of electric current around an area produces a magnetic dipole moment whose magnetic field declines with distance d only as $1/d^3$. The magnetometer was then placed on a long boom to get it as far away from the spacecraft as practicable. The effort paid off and, upon launching IMP 1 out into the solar wind, the magnetometer showed an interplanetary magnetic field fluctuating wildly and rapidly ($\Delta B \sim B$) about a mean value of 6×10^{-5} G and a mean orientation close to the expected 40° or so degrees from the radial direction (Ness *et al.* 1964). A simple radial (inverse square) extrapolation of the 6×10^{-5} G back to the Sun (1 AU = $220 R_\odot$) gave a radial field of about 3 G at the Sun. This order of magnitude was entirely consistent with an origin of the wind in a quiet region at the Sun. In fact, if the wind originates from the corona in the regions of weakest magnetic field at the Sun, so that the corona can push aside the field and escape, then the field flares out more rapidly with radial distance and the measured field indicated fields more like 10 G at the Sun.

There remained the curious discrepancy between the measured solar wind density of the order of 5 protons/cm^3 and the 500 protons/cm^3 or more estimated by Biermann from the vigorous antisolar acceleration of gaseous comet tails. The explanation is that it is the magnetic field carried in the solar wind that interacts with the ions and electrons in the comet tails, rather than the Coulomb collisions and charge exchange of the solar wind ions with the comet ions. The comet ion is picked up by the magnetic field in the wind (i.e. accelerated by the electric field $\mathbf{E} \equiv -\mathbf{u} \times \mathbf{B}/c$ in the comet frame) within a distance of the order of the cyclotron radius of the comet ion computed for the velocity of the ion relative to the solar wind (~ 400 km s^{-1}), which proves to be only a little more than 700 km for a proton at the orbit of Earth where $B \sim 6 \times 10^{-5}$ G. So the coupling of the solar wind to the ions in the comet tail is magnetic, through the general large-scale magnetic field carried in the solar wind. Note, then, that the 5 protons/cm^3 in the solar wind is adequate for the inertia necessary to accelerate the comet tail ions.

So the continuing hydrodynamic coronal expansion to produce the supersonic solar wind was evidently correct, and interplanetary plasma motions and magnetic fields can be understood on that basis. The original proposals by Chapman and Biermann were reconciled as two different aspects of the dynamical corona. The variable reduction of the galactic cosmic ray intensity could be treated on a deductive basis as a consequence of the outward sweep of the magnetic fields carried in the wind. At last we had quantitative observational information on the plasma and magnetic field impacting on the geomagnetic field, and some general idea of the nature of the distant solar wind and its dynamical interaction with the interstellar gas and the galactic magnetic field.

OUTWARD EXTENSION OF THE SOLAR WIND

With the measurements of the solar wind density it was possible to think about the distance to which the solar wind extends into interstellar space, which, until the Mariner 2 observations were made, could only be based on guesswork as to the mean density and velocity of the SCR (Davis 1955, Parker 1961b, 1963). The impact pressure (Reynolds stress density) of a plasma stream of N ions and electrons per unit volume with hydrogen mass M and velocity u is NMu^2. For purely radial motion at constant speed beyond the orbit of Earth ($r > r_E$) the wind density declines as $N(r) = N(r_E)(r_E/r)^2$. The solar wind pushes back the interstellar medium to approximately the distance where the ram pressure NMu^2 falls to the ambient interstellar pressure P, provided by the interstellar gas, magnetic field, and, to some degree, the galactic cosmic rays, whose entry into the solar wind is impeded by the outward sweeping azimuthal field. Now the pressure of the azimuthal magnetic field in the solar wind is negligible compared to the ram pressure. A solar wind with $N = 5$ protons/cm^3 and $u = 500$ km s^{-1} at $r = r_E$ has a ram pressure of about 2×10^{-8} dyn cm^{-2}, declining beyond Earth in proportion to $(r_E/r)^2$. As already noted, the termination of the wind occurs somewhere in the neighborhood of r given by $P = N(r_E)Mu^2(r_E/r)^2$ where the wind passes through a shock transition and becomes subsonic and very hot beyond.

The most uncertain factor in the calculation is P. Forty years ago the conventional numbers were 1 H atom/cm^3 at 10^2 or 10^3 K and a galactic magnetic field estimated at $5-10 \times 10^{-6}$ G. The gas pressure is then small compared to the magnetic pressure of 10^{-12} dyn cm^{-3}. Today it appears that the galactic magnetic field is more like 3×10^{-6} G and the number density of the local interstellar gas is approximately 0.1 ions cm^{-3} at 3×10^3 K. Thus $P \sim B^2 8/\pi \sim 0.4 \times 10^{-12}$ dyn cm^{-3}. The result is a termination of the supersonic wind in the vicinity of $r \sim 250$ AU. A lower mean wind velocity, of the order of 350 km s^{-1}, indicated by more recent studies, yields termination at $r \sim 175$ AU. The impact of the local interstellar wind of 0.1 ions/cm^3 at 20 km/sec provides a ram pressure of about 0.8×10^{-12} dynes/cm^2, pushing the boundary into 180 AU and 120 AU, respectively, on the windward side. So the solar wind pushes back the interstellar gas and fields to some considerable distance, and then streams out behind to form the downstream interstellar wake of tenuous magnetic field and hot tenuous solar gas (Yu 1974).

Somewhere along the line someone introduced the very appropriate term "heliosphere" for the region of interstellar space swept out by the solar wind.

This brings us back to the question of the angular momentum carried away from the Sun by the solar wind and the tension in the spiral interplanetary magnetic field. As already noted, the present day torque is too small to have a sensible effect on the rotation of the Sun (Parker 1958e, Brandt 1966, Brandt et al. 1969, Urch 1969), but the subject is not without scientific interest, particularly when it is recalled that the solar wind torque evidently had the effect of slowing the presumed initial rapid rotation of the Sun during the first 3×10^8 years after formation (Schatzman 1954, 1973). It must also be appreciated that direct observations showed a mean net rotation of the solar wind, with an azimuthal velocity component of about 8 km s^{-1} (in slow winds of 300–350 km s^{-1}) at the orbit of Earth. The theoretical dynamical interaction of the interplanetary magnetic field and the solar wind plasma was worked out by Weber and Davis (1967), Modisette (1967), Alonso-Faus (1968), and Urch (1969). These theoretical investigations predicted an azimuthal velocity of the order of 1–2 km s^{-1}, substantially less than the value inferred from observation. The cause of the discrepancy is not clear.

Another puzzling situation is exhibited by the theoretical demonstration (Parker 1967a,b, Lerche and Parker 1967) that the impact of the solar wind against the outer boundary of the geomagnetic field is not able to provide equilibrium confinement of the geomagnetic field unless certain conditions on the electron temperature are satisfied. The deflection of the obliquely impacting solar wind ions produces a strong rotation of the magnetic field direction in the magnetopause, with the rotated field pushed outward by the passage of the ions. Rapid erosion of magnetic flux from the boundary is the expected result when the conditions for equilibrium are not met, which must be added to the expected rapid reconnection between the geomagnetic and interplanetary magnetic fields at the sunward magnetopause (Dungey 1961). Recent modeling of the impact of the wind on the magnetopause by DeKeyser and Roth (1998) shows the necessary range of conditions for effective confinement, or otherwise, in some detail. Unfortunately, the general dynamical instability of the magnetopause (Parker 1958a, Eviatar and Wolf 1968) makes it difficult to distinguish nonequilibrium and unstable equilibrium by direct observation.

CONCLUSION

With the direct verification and measurement of the solar wind accomplished with Explorer 10, Mariner 2, and IMP 1, it was time to put together the several detailed calculations of solar wind features and effects (already referenced) into a monograph (Parker 1963) including a number of illustrative examples of diffusion of solar cosmic rays and galactic cosmic rays through various interplanetary magnetic field configurations. Besides a detailed exposition of the polytrope model of the expanding corona, the simple model using two polytrope indices was investigated, with an isothermal corona ($\alpha = 1$) from the reference level $r = a$ near the Sun out to some arbitrary distance b, and an adiabatic corona ($\alpha = 5/3$) beyond $r = b$.

In describing the phenomenology of the observed variations in solar wind speed with the rotation of the quiet and active regions on the Sun, it was clear that, by the time the wind reached the orbit of Jupiter at 5 AU, the fast and slow winds, each forming spiral domains ($r = v\varphi/\Omega_E$) in space as a consequence of the rotation of the Sun, interact strongly, with each fast spiral domain ramming into the slow spiral domain ahead. A few simple illustrations were provided. It has been particularly interesting, therefore, to see how subsequent space observations out to Jupiter and beyond have shown the vigor and complexity of this seemingly innocent phenomenon, with forward and backward shocks, particle acceleration, local cosmic ray modulation, and so on (Burlaga 1997).

Then the adiabatic deceleration of cosmic rays in the expanding fields of the solar wind was formulated (Parker 1966) along with the role of the small-scale random walk (stochastic topology) of the field lines of the interplanetary magnetic field and the immediate consequences for the diffusion of cosmic rays from the Galaxy into the heliosphere and the spread of solar cosmic rays outward around the Sun (Jokipii and Parker 1968a,b).

Indeed, with direct observations in space, nearly every aspect of the solar wind has provided unanticipated phenomena. For instance, the helium ions in the solar wind move away from the Sun somewhat faster than the protons. The temperatures of all the heavier ions are substantially higher than the proton temperatures. The wind exhibits fast and slow phases rotating with the Sun (Wilcox and Ness 1965, Ness et al. 1971), with the fast wind now identified with the coronal holes (Nolte et al. 1976). The slow and relatively dense wind is believed to arise around the periphery of active regions on the Sun, perhaps as the expansion of the diffuse X-ray corona. But the picture is not clear at the present time. The heat sources responsible are only vaguely understood in both cases. The large and rapid wave fluctuations and the rotational discontinuities in the interplanetary magnetic field carried in the wind (Burlaga 1968, 1969; Burlaga and Ness 1968, 1969; Belcher and Davis 1971) seem to be telling us something about conditions near the Sun.

In recent years the Ulysses spacecraft (Marsden 1994a,b, Marsden et al. 1994, Simpson et al. 1994) has explored the fast solar wind over the polar coronal holes of the Sun at activity minimum and is in the process now of passing over the poles near activity maximum. The observations have established that the slow solar wind phase is found only at low latitudes ($\leq 30°$), with the high-speed wind everywhere at middle and high latitudes (Phillips et al. 1995a,b, Suess et al. 1998).

The outer heliosphere and the interaction of the solar wind with the local interstellar wind are only beginning to be penetrated by Voyagers 1 and 2. The major interaction of the solar wind with the galactic cosmic rays still lies beyond those spacecraft. The origin of the anomalous component of the cosmic rays (Garcia-Munoz et al. 1973), believed to be created at the termination shock in the wind by the acceleration of pickup ions from infalling interstellar neutral atoms (Fisk et al. 1974, Pesses et al. 1981, Jokipii 1986) remains yet to be checked out by direct observation at the terminal shock. We can be sure the exploration of those distant regions will turn up some surprises.

The Jupiter relativistic electron beacon (Simpson et al. 1974, Chenette et al. 1974) is an astonishing phenomenon, evidently arising from the interaction of the solar wind with the high rotational velocity of the Jovian magnetopause.

The observed alternating flat top and peaked top of successive 11-year cosmic ray maxima has been shown (Jokipii and Kopriva 1979, Morall et al. 1991) to be a consequence of the gradient and curvature drifts of the individual cosmic ray particles in the large-scale interplanetary magnetic field superimposed on the scattering of the individual particles by the small-scale fluctuations in the magnetic field. The direction of drift depends on the sign of the magnetic field, to which general MHD phenomena are oblivious because the Maxwell stresses are quadratic in the field. During one 11-year period the cosmic rays drift into the solar system over the poles of the Sun and outward at low latitudes, reversing the drift direction for the next 11 years, while the outward sweep of the solar wind remains always outward at all heliocentric latitudes. Hence the dissimilarity of successive cosmic ray maxima.

It all shows once again the universal fact that Nature, although rigorously constrained by the equations of Newton, Maxwell, Einstein, Schrödinger, and others, nonetheless is far more complex and imaginative than we are, producing both ourselves and the phenomena that baffle us. It is precisely the open-ended novelty of Nature that makes its scientific pursuit so interesting.

As already emphasized, the history of the concepts leading to our present limited appreciation of the solar wind extends back in time over two thousand years, when space and magnetic fields were first recognized. If that long history has taught us anything, it should be that human familiarity with any special field of physics is not to be taken to delimit that field of physics. Further experimental and observational efforts frequently turn up unanticipated new twists to old principles and familiar phenomena. No matter how hard or how long one has thought about a physical effect or principle, there are usually new facets to be discovered by a fresh look by someone else. That is the fascination of real science. It means that no scientist can become the ultimate expert, and no self-styled expert can operate effectively as a research scientist.

The development of the physics of the solar wind, with the direct studies made possible by scientific space missions since the 1960s, is recorded in the proceedings of the successive solar wind conferences, the first held in 1964 (Mackin and Neugebauer 1966) and continuing through the recent Solar Wind Nine Conference (Habbal et al. 1999). The reader desirous of pursuing the present-day knowledge of the solar wind is referred to such standard monographs as Hundhausen (1972), Burlaga (1997), Golub and Pasachoff (1997), Jokipii et al. (1997), and to the chapter by Neugebauer in the present volume. Present astronomical knowledge of stellar winds is described in the recent treatise by Lamers and Cassinelli (1999). The seemingly universal ejection of matter from diverse active astronomical objects is mentioned in Parker (1997).

REFERENCES

Alfvén, H. (1947). Granulation, magneto-hydrodynamics, and the heating of the solar corona. *Monthly Notices of the Royal Astronomical Society*, **107**, 211–219.

Alfvén, H. (1950). *Cosmical Electrodynamics*, Clarendon Press, Oxford.

Alfvén, H. (1956). The Sun's general magnetic field. *Tellus*, **8**, 1–12.

Alonso-Faus, A. (1968). Rotation of the solar wind plasma. *Planetary and Space Science*, **16**, 1–6.

Axford, W.I. (1962). The interaction between the solar wind and the Earth's magnetosphere. *Journal of Geophysical Research*, **67**, 3791–3796.

Babcock, H.D. (1959). The Sun's polar magnetic field. *Astrophysical Journal*, **130**, 364–365.

Babcock, H.W. and Babcock, H.D. (1955). The Sun's magnetic field. *Astrophysical Journal*, **121**, 349–366.

Bartels, J. (1930). Geophysikalischer Nachweis von Veränderungen der Sonnenstrahlung. *Ergebnisse der Exacten Naturwissenschaften*, **9**, 38–78.

Bartels, J. (1932). Terrestrial magnetic activity and its relation to solar phenomena. *Terrestrial Magnetism and Atmospheric Electriciy*, **37**, 1–32.

Bartels, J. (1934). Twenty seven day recurrences in terrestrial magnetic and solar activity. *Terrestrial Magnetism and Atmospheric Electricity*, **39**, 201–202.

Beard, D. (1960). The interaction of the terrestrial magnetic field with the solar corpuscular radiation. *Journal of Geophysical Research*, **65**, 3559–3568.

Beard, D. (1962). The interaction of the terrestrial magnetic field with the solar corpuscular radiation. 2. Second order approximation. *Journal of Geophysical Research*, **67**, 477–483.

Beard, D.B. and Jenkins, E.B. (1962). The magnetic effects of the magnetospheric surface currents. *Journal of Geophysical Research*, **67**, 3361–3367.

Behr, A. and Siedentopf, H. (1953). Untersuchungen über Zodiakallicht und Gegenschein nach lichtelektrischen Messungen auf dem Jungfraujock. *Zeitschrift für Astrophysik*, **32**, 19–50.

Belcher, J.W. and Davis, L. (1971). Large-amplitude Alfvén waves in the interplanetary medium. 2. *Journal of Geophysical Research*, **76**, 3534–3563.

Bethe, H. (1929). Berechnung der Electronenaffinität des Wasserstoffs. *Zeitschrift für Physik*, **57**, 815–821.

Biermann, L. (1948). Über die Ursache der chromosphärischen Turbulenz und des UV-Exzesses der Sonnenstrahlung. *Zeitschrift für Astrophysik*, **25**, 161–168.

Biermann, L. (1951). Kometenschweife und solare Korpuskularstrahlung. *Zeitschrift für Astrophysik*, **29**, 274–286.

Biermann, L. (1957). Solar corpuscular radiation and the interplanetary gas. *Observatory*, **77**, 109–110.

Biermann, L. (1973). Dynamics of interplanetary material, Part 2. In M.S. Uberoi (ed), *Cosmic Gas Dynamics*, Wiley Interscience, New York, pp. 239–249.

Billings, D.E. (1966) *A Guide to the Solar Corona*, Academic Press, New York.

Birkeland, K. (1896). Un globe magnétique tournant dans un faiseau de rayons cathodiques. *Archives des Sciences Physiques et Naturelles*, Geneva, June 1896 and Sept. 1898 [also in *Kongelige Norske Videnskabers Selskabs Skrifter*, [1] No. 1, **39**, 1901].

Birn, J., Hesse, M. and Schindler, K. (1989). General magnetic reconnection, parallel electric fields, and helicity. *Journal of Geophysical Research*, **93**, 5547–5557.

Bittencourt, J.A. (1986). *Fundamentals of Plasma Physics*, Pergamon Press, Oxford, pp. 312–322.

Blackwell, D.E. (1955). A study of the outer corona from a high altitude aircraft of the eclipse of 1954 June 30. I Observational data. *Monthly Notices of the Royal Astronomical Society*, **115**, 629–649.

Blackwell, D.E. (1956). A study of the outer corona from a high altitude aircraft of the eclipse of 1954 June 30. II Electron densities in the outer corona and zodiacal light region. *Monthly Notices of the Royal Astronomical Society*, **116**, 50–68.

Blackwell, D.E. and Ingham, M.F. (1961). Observations of the zodiacal light from a high altitude station. 1. The average zodiacal light. *Monthly Notices of the Royal Astronomical Society*, **122**, 113–127.

Bonetti, A., Bridge, H.S., Lazarus, A.J., Lyon, E.F., Rossi, B. and Scherb, F. (1963). Explorer 10 plasma measurements. *Journal of Geophysical Research*, **68**, 4017–4063.

Brandt, J.C. (1966). Consequences of the torque exerted on the Sun by the solar wind. *Astrophysical Journal*, **144**, 1221–1222.

Brandt, J.C., Wolff, C. and Cassinelli, J.P. (1969). Interplanetary gas XVI. A calculation of the angular momentum of the solar wind. *Astrophysical Journal*, **156**, 1117–1124.

Broun, J.A. (1858). On certain results of magnetical observations. *Philosophical Magazine*, **16**, 81–99.

Broun, J.A. (1874). Sur les variations barométiques et leurs rapports avec les variations magnétiques. *Comptes Rendus*, **76**, 695–699.

Burlaga, L.F. (1968). Microstructures in the interplanetary medium. *Solar Physics*, **4**, 67–92.

Burlaga, L.F. (1969). Directional discontinuities in the solar wind. *Solar Physics*, **7**, 54–71.

Burlaga, L.F. and Ness, N.F. (1968). *Canadian Journal of physics*, **46**, S962–S965.

Burlaga, L.F. and Ness, N.F. (1969). Tangential discontinuities in the solar wind. *Solar Physics*, **9**, 467–477.

Burlaga, L.F. (1997). *Interplanetary Magnetohydrodynamics*, Oxford University Press, New York.

Burnight, T.R. (1949). Soft X-radiation in the upper atmosphere. *Physical Review*, **76**, 165.

Canton, J. (1759). An attempt to account for the regular diurnal variation of the horizontal magnetic needle. *Philosophical Transactions of the Royal Society of London*, Part 1, **49**, 398–448.

Carrington, R.C. (1859). Description of the peculiar solar event on September 1, 1859. *Monthly Notices of the Royal Astronomical Society*, **20**, 13.

Celsius, A. (1741). Magnetic-nålens misswisning eller afwinkande från norr-streket. *Kongelige Svenska Vetenskaps Academiens Handligar*, **1**, 391 [German translation: 1749, Anmerkungen über die stündlichen Veränderungen der Magnetnadel. *Ibid.*, **2**, 45–48].

Chamberlain, J.W. (1960). Interplanetary gas. II. Expansion of a model solar corona. *Astrophysical Journal*, **131**, 47–56.

Chamberlain, J.W. (1961). Interplanetary gas. III. A hydrodynamic model of the corona. *Astrophysical Journal*, **133**, 675–687.

Chamberlain, J.W. (1965). On the existence of slow solutions in coronal hydrodynamics. *Astrophysical Journal*, **141**, 320–322.

Chandrasekhar, S. (1944). Some remarks on the negative hydrogen ion and its absorption coefficient. *Astrophysical Journal*, **100**, 176–180.

Chandrasekhar, S. (1945). On the continuous absorption coefficient of the negative hydrogen ion. II. *Astrophysical Journal*, **102**, 395–401.

Chandrasekhar, S. and Breen, F.H. (1946). On the continuous absorption coefficient of the negative hydrogen ion. III. *Astrophysical Journal*, **104**, 430–445.

Chandrasekhar, S. and Elbert, D.D. (1958). On the continuous absorption coefficient of the negative hydrogen ion. V. *Astrophysical Journal*, **128**, 633–635.

Chandrasekhar, S. and Münch, G. (1946). The continuous spectrum of the Sun and the stars. *Astrophysical Journal*, **104**, 446–457.

Chapman, S. (1954). The viscosity and thermal conductivity of a completely ioned gas. *Astrophysical Journal*, **120**, 151–155.

Chapman, S. (1959). Interplanetary space and earth's outermost atmosphere. *Proceedings of the Royal Society of London A*, **253**, 462–481.

Chapman, S. (1967). Perspective. In S. Matsushita and W.H. Campbell (eds), *Physics of Geomagnetic Phenomena*, Academic Press, New York, pp. 3–28.

Chapman, S. and Cowling, T.G. (1958). *The Mathematical Theory of Nonuniform Gases*, Cambridge University Press.

Chapman, S. and Ferraro, V.C.A. (1931). A new theory of magnetic storms. *Terrestrial Magnetism and Atmospheric Electricity*, **36**, 77–97, 171–186.

Chapman, S. and Ferraro, V.C.A. (1932). A new theory of magnetic storms. *Terrestrial Magnetism and Atmospheric Electricity*, **37**, 147–156, 421–429.

Chapman, S. and Ferraro, V.C.A. (1933). A new theory of magnetic storms. *Terrestrial Magnetism and Atmospheric Electricity*, **38**, 79–96.

Chapman, S. and Ferraro, V.C.A. (1940). The theory of the first phase of a geomagnetic storm. *Terrestrial Magnetism and Atmospheric Electricity*, **45**, 245–268.

Chenette, D.L., Conlon, T.F. and Simpson, J.A. (1974). Bursts of relativistic electrons from Jupiter observed on interplanetary space with the time variation of the planetary rotation period. *Journal of Geophysical Research*, **79**, 3551–3558.

Chew, G.F., Goldberger, M.L. and Low, F.E. (1956). The Boltzmann equation and the one fluid hydromagnetic equations in the absence of particle collisions. *Proceedings of the Royal Society of London A*, **236**, 112–118.

Chree, C. (1912). Some phenomena of sunspots and terrestrial magnetism at Kew Observatory. *Philosophical Transactions of the Royal Society of London A*, **212**, 75–116.

Chree, C. (1922). The 27-day period (interval) in terrestrial magnetism. *Proceedings of the Royal Society of London A*, **101**, 368–391.

Clauser, F. (1960). In *4th Symposium on Cosmical Gas Dynamics*, Varenna, Italy.

Coleman, P.J., Davis, L., Smith, E.J. and Sonett, C.P. (1962). Interplanetary magnetic field measurements. *Science*, **138**, 1099–1100.

Coleman, P.J., Davis, L., Smith, E.J. and Sonett, C.P. (1963). The magnetic field of Venus. *Science*, **139**, 909–910.

Cowling, T.G. (1953). Solar electrodynamics. In G.P. Kuiper (ed), *The Sun*, University of Chicago Press, pp. 579–583.

Dalton, J. (1828) On the height of the aurora borealis above the surface of the earth: particularly one seen on the 29th of March 1826. *Philosophical Transactions of the Royal Society of London*, Part 1, **118**, 291–302.

Dalton, J. (1834). *Meteorological Observations and Essays*, 2nd edn, London [this adds 27 additional pages on aurorae to the 70 pages on aurorae in the 1st edn (Manchester, 1793)].

Davis, L. (1955). Interplanetary magnetic fields and cosmic rays. *Physical Review*, **100**, 1440–1444.

Davis, L. and Beard, D.B. (1962). A correction to the approximate condition for locating the boundary between a magnetic field and a plasma. *Journal of Geophysical Research*, **67**, 4505–4507.

Davis, L., Smith, E.J., Coleman, R.J. and Sonett, C.P. (1964). Interplanetary magnetic measurements. In R.J. Mackin and M. Neugebauer (eds), *The Solar Wind*, Pergamon Press, New York, pp. 35–49.

De Keyser, J. and Roth, M. (1998). Equilibrium conditions and magnetic field rotation at the tangential discontinuity magnetopause. *Journal of Geophysical Research*, **103**, 6653–6662.

De Marian, J.J. d'Or (1754). Traité physique et historique de l'aurore boréale. *Suite de Mem. Acad. Roy. Sci.*, 1731; Paris, 2nd edition, 1754.

Deslandres, H. (1910). Recherches sur l'atmosphère solaire. *Annales de l'Observatoire de Meudon*, **4**, No. 1.

Dessler, A.J., Francis, W.E. and Parker, E.N. (1960). Geomagnetic storm sudden-commencement rise times. *Journal of Geophysical Research*, **65**, 2715–2719.

Dessler, A.J., Hanson, W.B. and Parker, E.N. (1961). Formation of the geomagnetic storm main phase ring current. *Journal of Geophysical Research*, **66**, 3631–3637.

Dessler, A.J. and Parker, E.N. (1959). Hydromagnetic theory of geomagnetic storms. *Journal of Geophysical Research*, **64**, 2239–2252.

Dessler, A.J. and Parker, E.N. (1968). Corrections to paper by A.J. Dessler and E.N. Parker, "Hydromagnetic theory of geomagnetic storms". *Journal of Geophysical Research*, **73**, 3091–3092.

Dungey, J.W. (1961). Interplanetary magnetic field and the auroral zones. *Physical Review Letters*, **6**, 47–48.

Edlén, J.A. (1936). Mg I-ähnliche Spektren der Elemente Titan bis Cobalt, Ti XI, V XII, Cr XIII, Mn XIV, Fe XV, und Co XVI. *Zeitschrift für Physik*, **103**, 536–541.

Edlén, J.A. (1937). SI-ähnliche Spektren der Elemente Titan bis Eisen, Ti VII, V VIII, Cr IX, Mn X, und Fe XI. *Zeitschrift für Physik*, **104**, 188–193.

Edlén, B. (1942). Die Deutung der Emissionlinien im Spektrum der Sonnenkorona. *Zeitschrift für Astrophysik*, **22**, 30–64.

Einstein, A. (1905). Zur Electrodynamik bewegter Körper. *Annalen der Physik*, **17**, 891–921.

Elsässer, H. (1954). Die räumliche Verteilung der Zodiakallicht Materie. *Zeitschrift für Astrophysik*, **33**, 274–290.

Eviatar, A. and Wolf, R.A. (1968). Transfer processes in the magnetopause. *Journal of Geophysical Research*, **73**, 5562–5576.

Ferraro, V.C.A. (1960). An approximate method of estimating the size and shape of the stationary hollow carved out in a neutral ionized stream of corpuscles impinging on the geomagnetic field. *Journal of Geophysical Research*, **65**, 3951–3953.

Fisk, L.A., Koslovsky, B. and Ramaty, R. (1974). An interpretation of the observed oxygen and nitrogen enhancements in low-energy cosmic rays. *Astrophysical Journal Letters*, **190**, L35–L37.

Forbush, S.E. (1937). On the diurnal variation of cosmic ray intensity. *Terrestrial Magnetism and Atmospheric Electricity*, **42**, 1–16.

Forbush, S.E. (1954). World-wide cosmic ray variations, 1937–1952. *Journal of Geophysical Research*, **59**, 525–542.

Garcia-Munoz, M., Mason, G.M. and Simpson, J.A. (1973). A new test for cosmic ray modulation theory. *Astrophysical Journal Letters*, **182**, L81–L84.

Gauss, C.F. (1839). *Allgemeine Theorie des Erdmagnetismus, Resultate aus den Beobachtungen magnetischen Vereins im Jahre 1838*, pp. 1–59.

Gilbert, W. (1600). *De Magnete*, London [translated from Latin by P.F. Mottelay, Dover Publications, New York 1958].

Gold, T. (1959). Plasma and magnetic fields in the solar system. *Journal of Geophysical Research*, **64**, 1665–1674.

Golub, L. and Pasachoff, J.M. (1997). *The Solar Corona*, Cambridge University Press.

Graham, G. (1724). An account of the observations made of the variations of the horizontal needle at London, in the latter part of the year 1722, and beginning 1723. *Philosophical Transactions of the Royal Society of London*, Part 1, **32**, 96–107.

Gringauz, K.I., Bezrukikh, V.V., Ozerov, V.D. and Rybchinsky, R.E. (1960). A study of interplanetary ionized gas, high energy electrons, and corpuscular radiation from the Sun by means of the three electrode trap for charged particles on the second Soviet cosmic rocket. *Soviet Physics Doklady*, **5**, 361–367.

Gringauz, K.I., Bezrukikh, V.V., Ozerov, V.D. and Rybchinsky, R.E. (1961). Some results of experiments in interplanetary space by means of charged particle traps on Soviet space probes. *Space Research*, **2**, 539–553.

Gringauz, K.I., Ozerov, V.D., Rybchinsky, R.E., Bezrukikh, V.V. and Musatov, S. (1967). Solar wind observations with the *Venus 3* probe. *Cosmic Research*, **5**, 216–222.

Grotrian, W. (1933). Ergebnisse der Potsdamer Expedition zur Beobachtung der Sonnenfinsternis am 9 Mai 1929 in Takengon (Nordsumatra): Über den Intensitäts verlauf und des Intensitätsverhältnis der Koronalinien. *Zeitschrift für Astrophysik*, **7**, 26–45.

Grotrian, W. (1939). Sonnen und Ionosphäre. *Naturwissenshaft*, **27**, 555–563, 569–577.

Habbal, S.R., Esser, R., Holweg, J.V. and Isenberg, P.A. (eds) (1999). *Solar Wind Nine*, American Institute of Physics, New York, AIP Conference Proceedings 471.

Hale, G.E. (1908). On the probable existence of a magnetic field in sunspots. *Astrophysical Journal*, **28**, 100–116, 315–343.

Hale, G.E. (1913). Preliminary results of an attempt to detect the general magnetic field of the Sun. *Astrophysical Journal*, **38**, 27–98.

Hale, G.E. (1929). The spectrohelioscope and its work. *Astrophysical Journal*, **70**, 265–311.

Hale, G.E. (1930). The spectrohelioscope and its work. Part II. The motions of the hydrogen flocculi near sunspots. *Astrophysical Journal*, **71**, 73–101.

Hale, G.E. and Ellerman, F. (1903). The Rumford spectroheliograph. *Publications of the Yerkes Observatory*, **3**, 1–39.

Hale, G.E. and Nicholson, S.B. (1938). *Magnetic Observations of Sunspots, 1917–1924*, Part I, Publications of the Carnegie Institute, No. 498.

Hale, G.E., Seares, F.H., von Maanen, A. and Ellerman, F. (1918). The general magnetic field of the Sun. *Astrophysical Journal*, **47**, 206–254.

Heppner, J.P., Ness, N.F., Scearce, C.S. and Skilman, T.L. (1963). Explorer 10 magnetic measurements. *Journal of Geophysical Research*, **68**, 1–46.

Hess, V.F. (1912). Über Beobachtungen der Durchdringenden Strahlung bei Sieben Freiballonfahrten. *Physikalische Zeitung*, **13**, 1084–1093.

Hodgson, R. (1859). A curious appearance seen on the Sun. *Monthly Notices of the Royal Astronomical Society*, **20**, 15–16.

Hundhausen, A.J. (1972). *Coronal Expansion and the Solar Wind*, Springer-Verlag, Berlin.

Jackson, B.V., Rompolt, B. and Svestka, Z. (1988). Solar wind and interplanetary observations of the mass ejection on 7 May 1979. *Solar Physics*, **115**, 327.

Jokipii, J.R. (1986). Particle acceleration at a termination shock. 1. Application to the solar wind and the anomalous component. *Journal of Geophysical Research*, **91**, 2929–2932.

Jokipii, J.R. and Kopriva, D.A. (1979). Effects of particle drift on the transport of cosmic rays. III. Numerical model of galactic cosmic ray modulation. *Astrophysical Journal*, **234**, 384–392.

Jokipii, J.R. and Parker, E.N. (1968a). Random walk of magnetic lines of force in astrophysics. *Physical Review Letters*, **21**, 44–47.

Jokipii, J.R. and Parker, E.N. (1968b). Stochastic aspects of magnetic lines of force with applications to cosmic ray propagation. *Astrophysical Journal*, **155**, 777–798.

Jokipii, J.R., Sonett, C.P. and Giampapa, M.S. (eds) (1997). *Cosmic Winds and the Heliosphere*, University of Arizona Press, Tucson, AZ.

Kellogg, P.J. (1962). Flow of plasma around Earth. *Journal of Geophysical Research*, **67**, 3805–3811.

Kiepenheuer, K.O. (1953). Solar activity. In G.P. Kuiper (ed), *The Sun*, University of Chicago Press, pp. 437–450.

Kohl, J.L., *et al.* (1997). First results from the SOHO ultraviolet coronagraph spectrometer. *Solar Physics*, **175**, 613–644.

Lamers, H.J.G.L.M. and Cassinelli, J.P. (1999). *Introduction to Stellar Winds*, Cambridge University Press.

Lerche, I. and Parker, E.N. (1967). Nonequilibrium and enhanced mixing at a plasma field interface. *Astrophysical Journal*, **150**, 731–732.

Lindemann, F.A. (1919). Note on the theory of magnetic storms. *Philosophical Magazine*, **38**, 669–684.

Lorentz, H.A. (1895). *Versuch einer Theorie der electrischen und optischen Erscheinungen in bewegten Körpen*. In *Collected Papers, 1935–1939*, **5**, 1–128.

Lyot, B. (1939). A study of the solar corona and prominences without eclipse. *Monthly Notices of the Royal Astronomical Society*, **99**, 580–594.

Mackin, R.J. and Neugebauer, M. (eds) (1966). *The Solar Wind*, Pergamon Press, Oxford.

MacQueen, R.M. and Holzer, T.E. (1988). The coronal mass ejection of 20–21 November 1973. *Solar Physics*, **116**, 349–367.

Marsden, R.G. (ed) (1994a). The high latitude heliosphere. *Space Science Reviews*, **72**(1/2).

Marsden, R.G. (1994b). Surprises and successes from Ulysses' polar pass of the Sun. *Physics World*, **7**(11), 24–25.

Marsden, R.G., Page, D.E. and Smith, E.J. (1994). Ulysses at high solar latitude: An overview. *Transactions of the American Geophysics Union*, **75**, 511.

Martin, S. (1984). In S.L. Keil (ed), *Small-Scale Dynamical Processes in Quiet Stellar Atmospheres*, National Solar Observatory, Sunspot, NM, p. 30.

Maunder, E.W. (1904). Magnetic disturbances at Greenwich, and their association with sunspots. *Monthly Notices of the Royal Astronomical Society, Geophysics Supplement*, **64**, 705–716.

Maunder, E.W. (1905). *Monthly Notices of the Royal Astronomical Society*, **65**, 2–34.

Maunder, E.W. (1916). *Monthly Notices of the Royal Astronomical Society*, **76**, 63–68.

Maxwell, J.C. (1873). *A Treatise on Electricity and Magnetism*, 3rd edn, 1954, Dover, New York.

Meyer, P. and Simpson, J.A. (1955). Changes in the low-energy particle cutoff and primary spectrum of cosmic radiation. *Physical Review*, **99**, 1517–1523.

Meyer, P., Parker, E.N. and Simpson, J.A. (1956). Solar cosmic rays of February 1956 and their propagation through interplanetary space. *Physical Review*, **104**, 768–783.

Midgley, J.E. and Davis, L. (1963). Calculation by a moment technique of the perturbation of the geomagnetic field by the solar wind. *Journal of Geophysical Research*, **68**, 5111–5123.

Minnaert, M. (1953). The photosphere. In G.P. Kuiper (ed), *The Sun*, University of Chicago Press, pp. 88–185.

Modisette, J.L. (1967). Solar wind induced torque on the Sun. *Journal of Geophysical Research*, **72**, 1521–1526.

Moraal, H., Jokipii, J.R. and Mewaldt, R.A. (1991). Heliospheric effects on cosmic ray electrons. *Astrophysical Journal*, **367**, 191–199.

Morrison, P. (1959). Solar origin of cosmic ray time variations. *Physical Review*, **101**, 1397–1404.

Ness, N., Hundhausen, A.J. and Bame, S.J. (1971). Observations of the interplanetary medium: Vela 3 and Imp 3, 1965–1967. *Journal of Geophysical Research*, **76**, 6643–6665.

Ness, N.F., Scearce, C.S. and Seek, J.B. (1964). Initial results of the Imp 1 magnetic field experiment. *Journal of Geophysical Research*, **69**, 3531–3569.

Neugebauer, M. and Snyder, C.W. (1964). Mariner 2 measurements of the solar wind. In R.J. Mackin and M. Neugebauer (eds), *The Solar Wind*, Pergamon Press, Oxford, pp. 3–21.

Neugebauer, M. and Snyder, C.W. (1966). Mariner 2 observations of the solar wind: 1. Average properties. *Journal of Geophysical Research*, **71**, 4469–4484.

Neugebauer, M. and Snyder, C.W. (1967). Mariner 2 observations of the solar wind: 2. Relation of plasma properties to the magnetic field. *Journal of Geophysical Research*, **72**, 1823–1828.

Noble, L.M. and Scarf, F.L. (1962). Hydrodynamic models of the solar wind. *Journal of Geophysical Research*, **67**, 4577–4584.

Noble, L.M. and Scarf, F.L. (1963). Conductive heating of the solar wind. I. *Astrophysical Journal*, **138**, 1169–1181.

Nolte, J.T., Krieger, A.S., Timothy, A.F., Gold, R.E., Roelof, E.C., Vaiana, G., Lazarus, A.J., Sullivan, J.D. and McIntosh, P.S. (1976). Coronal holes as sources of solar wind. *Solar Physics*, **46**, 303–322.

Parker, E.N. (1956). Modulation of the primary cosmic ray intensity. *Physical Review*, **103**, 1518–1533.

Parker, E.N. (1957). Newtonian development of the dynamical properties of ionized gases of low density. *Physical Review*, **107**, 924–933.

Parker, E.N. (1958a). Interaction of the solar wind with the geomagnetic field. *Physics of Fluids*, **1**, 171–187.

Parker, E.N. (1958b). Dynamical instability in an anisotropic ionized gas of low density. *Physical Review*, **109**, 1874–1876.

Parker, E.N. (1958c). Cosmic ray modulation by the solar wind. *Physical Review*, **110**, 1445–1449.

Parker, E.N. (1958d). Suprathermal particles III: Electrons. *Physical Review*, **112**, 1429–1435.

Parker, E.N. (1958e). Dynamics of the interplanetary gas and magnetic fields. *Astrophysical Journal*, **128**, 644–676.

Parker, E.N. (1959). Plasma dynamical determination of shock thickness in an ionized gas. *Astrophysical Journal*, **129**, 217–223; erratum, 860.

Parker, E.N. (1960a). The hydrodynamic treatment of the expanding solar corona. *Astrophysical Journal*, **132**, 175–183.

Parker, E.N. (1960b). The hydrodynamic theory of solar corpuscular radiation and stellar winds. *Astrophysical Journal*, **132**, 821–866.

Parker, E.N. (1960c). Geomagnetic fluctuations and the form of the outer zone of the Van Allen radiation belt. *Journal of Geophysical Research*, **65**, 3117–3130.

Parker, E.N. (1961a). Sudden expansion of the corona following a large solar flare and the attendant magnetic field and cosmic ray effects. *Astrophysical Journal*, **133**, 1014–1033.

Parker, E.N. (1961b). The stellar wind regions. *Astrophysical Journal*, **134**, 20–27.

Parker, E.N. (1961c). Quasi-linear model of plasma shock structure in a longitudinal magnetic field. *Journal of Nuclear Energy C*, **2**, 146–153.

Parker, E.N. (1962). Dynamics of the geomagnetic storm. *Space Science Reviews*, **1**, 62–99.

Parker, E.N. (1963). *Interplanetary Dynamical Processes*, Wiley Interscience, New York.

Parker, E.N. (1964a). Dynamical properties of stellar coronas and stellar winds: I Integration of the momentum equations. *Astrophysical Journal*, **139**, 72–92.

Parker, E.N. (1964b). Dynamical properties of stellar coronas and stellar winds: II Integration of the heat flow equations. *Astrophysical Journal*, **139**, 93–122.

Parker, E.N. (1964c). Dynamical properties of stellar coronas and stellar winds: III The dynamics of coronal streamers. *Astrophysical Journal*, **139**, 690–709.

Parker, E.N. (1965a). The passage of energetic particles through interplanetary space. *Planetary and Space Science*, **13**, 9–49.

Parker, E.N. (1965b). On the existence of slow solutions in coronal hydrodynamics. *Astrophysical Journal*, **141**, 320–324.

Parker, E.N. (1965c). Dynamical properties of stellar coronas and stellar winds: IV The separate existence of subsonic and supersonic velocities. *Astrophysical Journal*, **141**, 1463–1478.

Parker, E.N. (1965d). The dynamical theory of the solar wind. *Space Science Reviews*, **4**, 666–708.

Parker, E.N. (1966). The effect of adiabatic deceleration on the cosmic ray spectrum in the solar system. *Planetary and Space Science*, **14**, 371–380.

Parker, E.N. (1967a). Confinement of a magnetic field by a beam of ions. *Journal of Geophysical Research*, **72**, 2315–2322.

Parker, E.N. (1967b). Small-scale nonequilibrium of the magnetopause and its consequences. *Journal of Geophysical Research*, **72**, 4365–4374.

Parker, E.N. (1967c). Nonlinear inflation of a magnetic dipole. *Journal of Geophysical Research*, **72**, 5287–5293.

Parker, E.N. (1969). Theoretical studies of the solar wind phenomenon. *Space Science Reviews*, **9**, 325–360.

Parker, E.N. (1983). Magnetic neutral sheets in evolving fields: II Formation of the solar corona. *Astrophysical Journal*, **264**, 642–647.

Parker, E.N. (1988). Nanoflares and the solar X-ray corona. *Astrophysical Journal*, **330**, 474–479.

Parker, E.N. (1994). *Spontaneous Current Sheets in Magnetic Fields*, Oxford University Press, New York.

Parker, E.N. (1996). The alternative paradigm for magnetospheric physics *Journal of Geophysical Research*, **101**, 10587–10625.

Parker, E.N. (1997). Mass ejection and a brief history of the solar wind concept. In J.R. Jokipii, C.P. Sonett and M.S. Giampapa (eds), *Cosmic Winds and the Heliosphere*, University of Arizona Press, Tucson, AZ, pp. 3–27.

Parker, E.N. and Tidman, D.A. (1958). Suprathermal particles. II. *Physical Review*, **112**, 1048–1051.

Pesses, M.E., Jokipii, J.R. and Eichler, D. (1981). Cosmic ray drift, shock wave acceleration, and the anomalous component of cosmic rays. *Astrophysical Journal Letters*, **246**, L85–L88.

Petschek, H.E. (1958). Aerodynamic dissipation. *Reviews in Modern Physics*, **30**, 966–972.

Phillips, J.L., Bame, S.I., Goldstein, W.C., Gosling, B.E., Hammond, C.M., McComas, D.J., Neugebauer, M., Scime, E.E. and Seuss, S.T. (1995a). Ulysses solar wind plasma observations at high southerly latitudes. *Science*, **268**, 1030–1032.

Phillips, J.L., Bame, S.J., Barnes, A., Barraclough, B.L., Feldman, W.C., Goldstein, B.E., Gosling, J.T., Hoogeveen, G.W., McComas, D.J., Neugebauer, M. and Seuss, S.T. (1995b). Ulysses solar wind plasma observations from pole to pole. *Geophysical Research Letters*, **22**, 3301–3304.

Pierce, J.R. (1949). Possible fluctuations in electron streams due to ions. *Journal of Applied Physics*, **19**, 231–236.

Porter, J.G. and Moore, R.L. (1988). In R.C. Altrock (ed), *Proceedings of Sacramento Peak Summer Symposium*, National Solar Observatory, Sunspot, NM, p. 30.

Poynting, J.H. (1884). On the transfer of energy in the electromagnetic field. *Philosophical Transactions of the Royal Society of London A*, **175**, 343–361.

Poynting, J.H. (1885). On the connection between electric current and the electric and magnetic inductions in the surrounding field. *Philosophical Transactions of the Royal Society of London A*, **176**, 277–306.

Raymond, J.C., et al. (1997). Composition of coronal streamers from the SOHO ultraviolet coronagraph spectrometer. *Solar Physics*, **175**, 645–665.

Roberts, P.H. and Soward, A.M. (1972). Stellar wind and breezes. *Proceedings of the Royal Society A*, **328**, 185–215.

Rossi, B. (1964). *Cosmic Rays*, McGraw Hill, New York, Chap. 5.

Rossi, B. and Olbert, S. (1970). *Introduction to the Physics of Space*, McGraw Hill, New York, pp. 111, 112.

Sabine, E. (1852). On periodical laws discoverable in the mean effects of the larger magnetic disturbances. *Philosophical Transactions of the Royal Society of London A*, **142**, 103–125; **146**, 357–374.

Schatzman, E. (1949a). Le chauffage de la couronne solaire. *Comptes Rendus*, **228**, 738–739.

Schatzman, E. (1949b). The heating of the solar corona and chromosphere. *Annales d'Astrophysiques*, **12**, 203–228.

Schatzman, E. (1954). *Annales d'Astrophysiques*, **17**, 300–308, 382–389.

Schatzman, A. (1973). Cosmic gas dynamics, Part 1. In M.S. Uberoi (ed), *Cosmic Gas Dynamics*, Wiley–Interscience, New York, pp. 1–176.

Scherb, F. (1964). Velocity distributions of the interplanetary plasma detected by Explorer 10. *Space Research*, **4**, 797–818.

Schlüter, A. (1950). Dynamik des plasma: 1. Grundegleichungen, Plasma in getreuzten Feldern. *Zeitschrift für Naturforschung*, **5a**, 72–78.

Schlüter, A. (1952). Plasma im magnetfeld. *Annalen der Physik*, **10**, 422.

Schrijver, C.J., Title, A.M., Van Ballegooijen, A.A., Hagenar, H.J. and Shine, R.A. (1997). Sustaining the quiet photospheric network: The balance of flux emergence, fragmentation, merging, and cancellation. *Astrophysical Journal*, **487**, 424–436.

Schrijver, C.J., et al. (1999). A new view of the solar outer atmosphere by the Transition Region and Coronal Explorer. *Solar Physics*, **187**, 261.

Schuster, A. (1911). On the origin of magnetic storms. *Proceedings of the Royal Society of London A*, **85**, 44–50.

Schwarzschild, M. (1948). On noise arising from solar granulation. *Astrophysical Journal*, **107**, 1–5.

Shimizu, T. and Tsuneta, S. (1997). Deep survey of solar nanoflares with Yohkoh. *Astrophysical Journal*, **486**, 1045–1057.

Simpson, J.A. (1951). Neutrons produced in the atmosphere by cosmic radiations. *Physical Review*, **83**, 1175–1188.

Simpson, J.A. (1954). Cosmic radiation intensity-time variations and their origins. III. The origin of 27-day variations. *Physical Review*, **94**, 426–440.

Simpson, J.A., Anglin, J.D., Hynds, R.J., Kunow, H., Marsden, R.G., Wenzel, K.P., Bame, S.J. and Balogh, A. (1994). The first solar polar pass of Ulysses: Preliminary findings from cosmic ray and anomalous nuclear component investigations. *Transactions of the American Geophysics Union*, **75**, 513.

Simpson, J.A., Fonger, W. and Treiman, S. (1953). Cosmic radiation intensity-time variations and their origins. I Neutron intensity variation method and meteorological factors. *Physical Review*, **90**, 934–950.

Simpson, J.A., et al. (1974). Protons and electrons in Jupiter's magnetic field: Results from the University of Chicago experiment on Pioneer 10. *Science*, **183**, 306–309.

Snyder, C.W. and Neugebauer, M. (1964). Interplanetary solar wind measurements by Mariner II. *Space Research*, **4**, 89–113.

Spreiter, J.R. and Briggs, B.R. (1962a). Theoretical determination of the form of the boundary of the solar corpuscular stream produced by the interaction with the magnetic dipole field of Earth. *Journal of Geophysical Research*, **67**, 37–51.

Spreiter, J.R. and Briggs, B.R. (1962b). On the choice of condition to apply at the boundary of the geomagnetic field in the steady-state Chapman Ferraro problem. *Journal of Geophysical Research*, **67**, 2983–2985.

Spreiter, J.R. and Hyett, B.J. (1963). The effect of a uniform external pressure on the boundary of the geomagnetic field in a steady wind. *Journal of Geophysical Research*, **68**, 1631–1642.

Spreiter, J.R. and Jones, W.P. (1963). On the effect of a weak interplanetary magnetic field on the interaction between the solar wind and the geomagnetic field. *Journal of Geophysical Research*, **68**, 3555–3568.

Störmer, C. (1955). *The Polar Aurora*, Oxford University Press, Oxford.

Suess, S.T., Phillips, J.L., McComas, D.J., Goldstein, B.E. and Neugebauer, M. (1998). The solar wind-inner heliosphere. *Space Science Reviews*, **83**, 75–86.

Svestka, Z. (1976). *Solar Flares*, Reidel, Dordrecht.

Sweet, P.A. (1969). Mechanisms of solar flares. *Annual Review of Astronomy Astrophysics*, **7**, 149–176.

Unsöld, A. and Chapman, S. (1949). Optical and radio frequency absorption by solar corpuscular bursts. *Observatory*, **69**, 219–221.

Urch, I.H. (1969). A model of the magnetized solar wind. *Solar Physics*, **10**, 219–228.

Van de Hulst (1953). The chromosphere and corona. In G.P. Kuiper (ed), *The Sun*, University of Chicago Press, Chicago, pp. 259–278.

Van de Hulst, H.C. (1955). On the polarization of the zodiacal light. *Mémoires de la Société Royale des Sciences, Liège* [series 4], **15**, 89–95.

Von Humboldt, A. (1858). *Cosmos*, translated by Otte and Dallas, H.G. Bohn, London, pp. 49–87.

Waldmeier, M. (1975). The coronal hole of the 7 March 1970 solar eclipse. *Solar Physics*, **40**, 351–358.

Weber, E.J. and Davis, L. (1967). The angular momentum of the solar wind. *Astrophysical Journal*, **148**, 217–227.

Weyman, R. (1960). Coronal evaporation as a possible mechanism for mass loss from red giants. *Astrophysical Journal*, **132**, 380–403.

Whang, Y.C. and Chang, C.C. (1965). An inviscid model of the solar wind. *Journal of Geophysical Research*, **70**, 4175–4180.

Wilcke, J.C. (1777). Om magnet-nålens årliga och dageliga ändringar i Stockholm. *Kongelige Svenska Vetenskaps Academiens Handlingar*, **38**, 273–300; [German translation: 1782, Über der Magnetnadel jährlichen und täglichen Aenderungen zu Stockholm, Hamburg, **39**, 239–284].

Wilcox, J.M. and Ness, N.F. (1965). Quasi-stationary corotating structure in the interplanetary medium. *Journal of Geophysical Research*, **70**, 5793–5805.

Wildt, R. (1939). Electron affinity in astrophysics. *Astrophysical Journal*, **89**, 295–301.

Withbroe, G.L. and Noyes, R.W. (1977). Mass and energy flow in the solar chromosphere and corona. *Annual Review of Astronomy Astrophysics*, **15**, 363–387.

Yu, G. (1974). The interstellar wake of the solar wind. *Astrophysical Journal*, **194**, 187–202.

10

WILLIAM K. HARTMANN*

The terrestrial planets at the dawn of the space age

It is difficult to overstate the profundity of the change in our perception of the planets during a single generation in the latter part of the twentieth century. The revolution is second only to that in the generation from Copernicus to Kepler at the end the sixteenth century. The very character of the scientific questions changed as the planets went from being astronomical objects to geological objects.

A primary goal of this chapter is to document some of the changes in perception and scientific issues involving terrestrial planets that happened as we went from the purely telescopic to the spacecraft era. While I have documented some technical material, I have allowed myself, as one who lived through this change, to include some anecdotal material. For example, I grew up making backyard telescope observations of planets when such observations could exceed the best photographic imagery; I experienced Sputnik 1 while at college; I was at graduate school under Professor Gerard Kuiper as he supervised a lunar mapping program for the Apollo landings, before a graduate program in planetary sciences was available; I was a co-investigator in the first orbital mapping of Mars; and I attended professional meetings where international friends and colleagues announced the first close-up results from Mercury, Venus, Jupiter, Saturn, Uranus, Neptune, Halley's Comet, Gaspra, Ida, Mathilda, and so on.

An additional aspect of this chapter involves the philosophy of science and exploration. The change from an astronomical to a geological approach is interesting because some of the burning questions of the astronomical era were not so much answered with great fanfare by space probes,

but rather "fizzled out" as scientific interests shifted to new observations. For example, much effort was spent trying to map the markings of Mercury, but this effort virtually died when the first photographs were returned; curiously, it is still unclear to what extent early observers, such as Eugène Antoniadi, were recording real albedo features, and how these features match the geology of Mercury.

This is also an interesting issue from the point of view of pursuing one's own career. For example, major parts of careers and grants were spent on issues such as mapping and interpreting markings, photographing Venusian clouds, trying to estimate surface pressures from inadequate data, and so on. Were these valuable steps toward a goal, or poor choices of research activity? Should the researchers and funding agencies simply have waited for better data – especially once it became clear that probes would be a reality? Often, research time and expenditures on such issues increased dramatically *before* missions, because of interest in what the probe would find, even though the results would soon be rendered obsolete by a single click of a shutter on a spacecraft. With hindsight, we can see that some efforts, such as the first proof of CO_2 on Venus or Mars, became fundamental advances to be cited in many future textbooks, while for others whole subjects were swept away by flyby probes or landers.

A related subject is the subtle abandonment of early "burning questions." Probably the most argued photographic–visual–telescopic question about Mars in the first half of the twentieth century was the fundamental underlying cause of the dark markings. Mariner 9 showed that they involved windblown dust, and the case seemed to be closed. But then the question was, why is the dust not mixed? What is the nature and source of the dark or light material? Do

*Planetary Science Institute, Tucson, AZ, USA

bare rock outcrops exist and are they a source of spectrally fresher and less weathered material, perhaps of lower albedo than the weathered dust? Spectral and thermal observations seem to show that the dark material is coarser, fresher basalt and the light material is finer, oxidized, and weathered material. With this finding, the issue is currently on a backburner. Still unresolved is where the fresh basalt material is coming from (especially since lava flows are more evident in light areas like Tharsis and Amazonis). In that sense, the original, underlying question that drove much of early Mars research is still unanswered. An "answer" was given, based on results obtained with instruments then available, but it did not answer the original question. This example illustrates a slightly unsettling characteristic of science: that the phenomena we measure and debate are not necessarily the "most important" phenomena of nature, identified objectively, but rather are heavily weighted by the particular instruments we happen to have and the subdisciplines that have emerged through socio-scientific processes.

It is unsettling to discover that one's students or younger colleagues don't share certain pivotal experiences that shaped planetary science, and that to them, these experiences are prehistory, or even unknown! Many of the contributors to this book have passed through various moments of shock with various successive waves of students: the ones who don't remember Sputnik, the ones who don't remember Gagarin's flight or President Kennedy's announcement a few weeks later of the Moon landing program, the ones who don't remember the first photographs of Mars, and, finally, the ones who weren't even born when Neil Armstrong stepped onto the Moon. For these reasons, I shall try to capture some of the character of the early developments as experienced at the time.

At the same time, as the twenty-first century begins, one is only too aware that the future progress of planetary exploration and research in general is under threat from anti-intellectual, anti-science, fundamentalist sources. In the USA, the battle over the teaching of biological evolution in schools not only continues, but has expanded into an attack on evolutionary processes, even in physical systems such as stars and galaxies. In 1999, the Kansas state Board of Education, under pressure from fundamentalists who had gained a majority in elections, withdrew not only biological evolution but coverage of the big bang cosmological theory from the subjects covered by the state's educational standards and testing program, effectively excluding them from the curriculum. Generally speaking, the proponents of these views are not well educated about the history of science or the nature of the modern international scientific literature, but take refuge in their own gray literature and national networks of radio programming. Ironically, they wage ghostly echoes of battles already fought and settled nearly two centuries ago by pioneers such as Lamarck and Cuvier – battles subsequently settled by overwhelming and diverse lines of evidence, including radiometric dating. Yet spokespersons for these factions routinely explain in the US media that there is plenty of scientific evidence in their favor. One member of the Kansas Board of Education, a veterinarian who voted to remove evolutionary concepts from the education standards, was described in the *New York Times* and *International Herald Tribune* citing "legitimate scientific doubts about whether the Universe is more than several thousand years old" (Glanz 1999), and this view is widespread in fundamentalist literature. This in spite of the independent and international work by researchers such as Gerling (1942) in Leningrad, Holmes (1946) in Edinburgh, and Houtermans (1947) in Göttingen, who pioneered the modern view of the age of Earth, not to mention the acknowledgment by Pope Pius XII (1951) that the ages of the Earth and Universe must be measured not in thousands, but billions of years. Although the Copernican Revolution started 450 years ago, it has not been won – and this is an indictment of science teachers as well as fundamentalists.

SETTING THE STAGE: THE 1940s AND 1950s

The terrestrial planets were not major objects of study in the 1940s and 1950s, and only a handful of scientists worked professionally on the Solar System. These included astronomer Gerard Kuiper, who discovered the atmosphere of Titan spectroscopically (Kuiper 1944) and trained newly developed infrared technology on the Venusian and Martian atmospheres after World War II (Kuiper 1952); the Nobel laureate geochemist Harold Urey, who published the epoch-marking book *The Planets* (Urey 1952); astronomer Fred Whipple, who published seminal papers on the nature of comets (Whipple 1950, 1951); and Ernst Öpik, who published amazingly prescient papers on a variety of planetary issues, such as impact cratering by interplanetary bodies (Öpik 1951, 1963, 1964) and the surface of Mars (Öpik 1965). Urey and Kuiper both developed far-ranging ideas on the origin of planets and planetary surface features, but rarely saw eye to eye. It is important to realize that although researchers such as Otto Schmidt and, later, Viktor Safronov (1972) developed a strong school of studies on planetary origins, their work was virtually unknown outside the Soviet Union until the 1970s.

The nature of meteorite craters was wholly unappreciated during this period. Much of the literature about the Moon during the preceding century was consumed with arguments over whether the lunar craters were volcanic or impact features. Scattered suggestions that there were many eroded impact features on Earth had about them the aura of fringe science. Arizona's so-called "Meteor Crater" was still listed as "Crater Mound" by the US Board of Geographical Names, even though it was not a mound (Hoyt 1987). This was allegedly because the first director of the survey, G.K. Gilbert, believed it was a volcanic feature and not an impact crater. This was ironic, because Gilbert himself later championed the impact theory for lunar craters. As late as 1945, US Geological Survey (USGS) scientist N.H. Darton delivered a paper decrying any use of the terms "meteor" or "meteorite" in the crater's title, because "I am convinced that no meteorite is present" due to the lack of discovery of an iron mass during drilling (Hoyt 1987, p. 333). This in spite of the fact that the surface around the crater is strewn with small iron meteorites!

In 1949, Ralph Baldwin, a businessman with a PhD in astronomy from the University of Chicago, published *The Face of the Moon* (Baldwin 1949), a highly original work in which he used his experience with bomb craters in World War II to argue methodically that the geometric properties of lunar craters matched impact explosion features, not volcanic features. This was the turning point in the argument about the origin of craters, and led to wide acceptance of the view that the Moon had been peppered with impacts. (The book, from the University of Chicago Press, was not a big seller; I bought my copy on a remainder table of a favorite bookstore in Pittsburgh for $1.49 when I was a boy, attracted by its cover picture of the Moon.)

As for the terrestrial planets, amateurs and professionals alike argued about the cause and significance of the various faint, dusky markings that could be glimpsed on Mars, Mercury, and probably Venus. Frank E. Ross (1928) had shown that the faint, dusky markings of Venus could be photographed with much more clarity and contrast by using ultraviolet filters; they were evidently cloud features, but little more was known about them. Walter Adams and Theodore Dunham (1932) had shown that carbon dioxide was a major atmospheric constituent.

There was a curious flavor to interest in spaceflight or planetary exploration at this time, difficult to recapture today. Popular literature, such as comic books, were full of articles about "post-war marvels." Boys (and a few girls) who read science fiction "knew" that spaceflight was coming, but adults rolled their eyes at such fantasies. In 1952–54, *Collier's* magazine ran a now-famous series of articles by Wernher von Braun, with paintings by the dean of astronomical art, Chesley Bonestell, detailing how we could launch artificial satellites, build a wheel-shaped space station (the design later immortalized in Stanley Kubrick's 1968 film *2001*), and explore the Moon and Mars. In 1956, President Eisenhower surprised many Americans and galvanized science students' imaginations by announcing that the USA, as part of the International Geophysical Year, would attempt to launch an artificial satellite around the Earth. Eisenhower emphasized that the USA would not use existing military rockets, but would develop a new civilian rocket for a fledgling, non-military space program. As is now known publicly, the Soviet Union had also begun development of an artificial satellite project under Sergei P. Korolev, with initial authorization given in January 1956 (Harford 1997, p. 125 ff.).

On 4 October 1957, before the Americans were ready to launch, the Soviet engineers launched Sputnik 1 into Earth orbit. It was followed a month later by the much bigger Sputnik 2 and its booster, which were very prominent in skies around the world. Sputnik 1 was very faint, but the booster of Sputnik 2 was bright and tumbled in its orbit, leading to dramatic variations in brightness with a period of the order of tens of seconds, flashing to about first or second magnitude. This produced profound shock in the USA, leading to massive reorganization of school science programs, and for the first time moved the popular concept of spaceflight from the realm of science fiction to reality. At the same time, the Sputniks galvanized the imagination of young science students in the USA. The spirit of this period is well captured in the film *October Sky*, based on the memoir *Rocket Boys* by Apollo engineer Homer Hickam (1998).

In retrospect, the competing political philosophies of secrecy *v.* scientific openness provide us with important lessons about scientific progress in planetary exploration. The discovery of Earth's radiation belts was credited to James Van Allen from data received by the US Explorer 1 satellite in 1958, but Dessler (1984) reviews how the detectors installed in the earlier Soviet satellite, Sputnik 2, actually measured the radiation belts in 1957. However, according to Dessler, the Soviets chose for reasons of secrecy not to arrange for other countries to detect or decode their satellite signals or pass them on to the Sputnik team. Thus, they did not get enough tracking data from Sputnik 2 to recognize the belts. Dessler remarks that, "Because of their perceived need for secrecy, the Russians missed making one of the most dramatic discoveries in space science."

THE LOW ATMOSPHERIC PRESSURE AND RED COLOR OF MARS: 1947–64

In the twentieth century, the estimated atmospheric pressure and habitability of Mars fell progressively. Percival Lowell's theories at the beginning of the century suggested a thin but Earth-like atmosphere. Although Lowell has been vilified, and even ridiculed, for his theories about civilizations and artificial canals on Mars, he started by propounding underlying ideas that still have validity. He pointed out that smaller planets lose their internal heat faster than larger planets, and are also more subject to the escape of their atmospheres. Thus, Mars was geologically dying and drying, and the atmosphere was being lost to space. But Lowell thought there was evidence for a thicker atmosphere than really exists. He argued that the warm equatorial regions were the most habitable in terms of temperature but that most of the planet's water was locked in the polar ice. Thus, Martian civilizations had built canals to deliver melt water from the polar ice caps to their equatorial settlements. This view of Mars electrified not only scientists, but writers and intellectuals from Alfred Tennyson to H.G. Wells, who combined it with the ideas of Charles Darwin to speculate upon the implications for humanity if we were not alone in the Solar System. This was the Mars that colored much of twentieth-century Martian research, and served as the framework against which to react. In the first decades of the century, researchers began to realize that the planet must be less habitable than Lowell had thought and the surface conditions were likened to a winter day at the top of a mountain on the island of Spitzbergen in the Arctic Ocean.

The changing ideas of Mars were also reflected in popular culture. Edgar Rice Burroughs, in his science fantasy novels, peopled Mars with evil queens and great beasts. Ray Bradbury, in his 1940s classic *The Martian Chronicles*, portrayed the last remnants of Martian civilization, eleing out an existence by the languid canals.

By the middle of the century, the estimated surface pressure had dropped to 100–200 mbar. Kuiper (1947) first detected the carbon dioxide of Mars and erroneously reduced the data to give a surface CO_2 partial pressure of only 0.35 mbar. Because this was so small, most researchers assumed that CO_2 was only a minor constituent in the atmosphere and that nitrogen or some other gas was dominant, with a moderate surface pressure. Gerard de Vaucouleurs (1954), summarizing the available evidence, gave the surface pressure as 85 ± 4 mbar. The erroneously low estimate of CO_2 abundance was a factor in Kuiper's further conclusion, now disproved, that water ice was the main constituent of the polar ice caps.

These views persisted for some years. Horowitz (1986), with some amusement, cites an advisory report to NASA from a prestigious space science board in 1961, stating that "infrared reflection spectra of the polar caps show conclusively that they are not composed of frozen carbon dioxide," but were water ice and frost. They also advised that the total surface pressure was within a factor two of 85 mbar – exactly the value, as Horowitz points out, that Percival Lowell (1910) had published in his book *Mars as the Abode of Life*.

During the first years of infrared studies, another interesting episode occurred. William Sinton (1957) announced confirmation of three faint bands around 3.4 μm wavelength, which he identified with the C–H bond, specifically chlorophyll. In the title of his paper he called this "Spectroscopic evidence for vegetation on Mars" – the long-sought confirmation that the dark markings were caused by simple plant forms. However, a few years later Sinton himself and other observers showed that these bands were due to other causes. This was as close as ground-based telescopic observers ever came to claiming definitive evidence for life on Mars.

Eventually, a clearer view emerged from the ground-based infrared work. Kaplan *et al.* (1964) studied weak, unsaturated CO_2 bands and concluded that the mean surface pressure was 2.515 mbar (Horowitz (1986) refers to this paper as the true beginning of the post-Lowellian understanding of Mars.) In the same year, Owen and Kuiper (1964) used their own spectra and improved laboratory calibration (long-pathlength spectra through low-pressure CO_2) to derive a surface pressure of 1.73 mbar, though they still thought that the atmosphere was likely to be dominated by nitrogen (see also Kuiper 1964).

The same period saw the emergence of the basic understanding of the red color of Mars. By the 1960s, broadband infrared spectroscopy had advanced to the point where it could begin to be used to identify major absorption bands due to minerals on the Martian surface. Two graduate students in Kuiper's infrared spectroscopy laboratory, Binder and Cruikshank (1966), showed that there was a good fit between Mars spectra and red, oxidized basalts found in the desert west of Tucson, Arizona. They proposed, essentially correctly, that the color of Mars comes from oxidized iron minerals that exist as stains on the surface of rocks and dust particles; they emphasized limonite and other iron oxide minerals. This work was generally supported by Adams (1968) and McCord and Adams (1969), who examined additional iron oxide minerals.

Philosophically, work of this sort was an extension of the Copernican Revolution and early ideas about the plurality of worlds. The Copernican Revolution revealed that Earth was not a unique center, but merely one of many planetary worlds. The idea of the plurality of worlds introduced an assumption that the other worlds might be other Earths, with environments similar to our own. William Herschel had believed that Mars, on which he saw clouds and polar caps, was Earth-like. Only in the 1950s and 1960s did solid observations begin to accumulate that allowed these ideas to be assessed. The emerging answer, with profound and still under-appreciated consequences, was that many other planets are indeed worlds with Earth-like minerals and geology, and yet each has a personality of its own.

EARTH: IMPACT CRATERS LEGITIMIZED, c. 1960

Research on planetary geology began to be legitimized around this same time. In the late 1950s geologist Eugene

Shoemaker carried out the most extensive field studies until that time on the Arizona crater, proving that the initially flat-lying beds had been uplifted and bent backwards, like the petals of an opening flower, producing a characteristic hummocky rim and a surrounding ejecta blanket of debris. Shoemaker (1957, 1960, 1963) pointed out that these same features existed on the lunar craters, and used nuclear explosion craters to estimate the energy and mass of the impactor. Shoemaker and Hackman (1962) and Shoemaker et al. (1962) used these ideas in pioneering efforts to show that overlapping stratigraphic relationships could be developed for the Moon, much in the way that the stratigraphic column had been developed for Earth.

At the time that the space program started in earnest in the USA in the 1960s, the success of this work allowed Shoemaker to begin the Astrogeology Branch of the USGS, a group centered in Flagstaff, Arizona, and Menlo Park, California, that was destined to continue the stratigraphic mapping of all planets and satellites using similar principles to those of the present day. Wilhelms (1993) gives an amusing account of the development of this "geological school" of planetary exploration, and its battles with various geochemists, geophysicists, and astronomers, who tended to look down their noses at the geologists as mere descriptive scientists. For example, even as USGS geologists began to train Apollo astronauts, Nobel laureate Harold Urey, who had pioneered the field of cosmochemistry, reportedly argued that geologists had little business in the early exploration of the Moon, which he regarded as more or less as a testbed for broad cosmochemical and geophysical theories of planetary evolution.

I witnessed mild but similar divisions between Kuiper's laboratory and the geology department at the University of Arizona, and indeed throughout the planetary community, during the pre-Apollo lunar explorations of the 1960s, and to some degree it still persists today. Some geophysicists, geochemists, and planetary astronomers tend to criticize geologists as caught up in mapping and analyzing local features, such as lava flows that may reflect accidents of local conditions or topography, without seeing the "big picture" implications for planetary development. Nobel laureate physicist Luis Alvarez, for instance, promoting the impact model of the KT boundary controversies, reportedly dismissed the paleontological/geological opponents as mere "stamp collectors" (Courtillot 1999). Some field geologists, conversely, tend to criticize the "big picture" theorists for missing the complex, real-life, and messy empirical story told in the rocks and craters – a view summarized in the joke about the physicist, who, faced with an agricultural problem on a farm, begins his analysis, "Consider a spherical cow."

Meanwhile, Canadian geologists under C.S. Beals began to recognize from aerial photographs and ground-based field work that the Canadian Shield was dotted with circular impact features. This work was inspired, according to Beals et al. (1963), by the discovery of the 3.2 km New Quebec impact crater around 1950. An early review by Beals et al. (1963) described the discovery of about half a dozen large, ancient, eroded Canadian impact craters.

In the early 1960s, during a mapping program in which lunar photographs were "rectified" by being projected onto a globe, Kuiper's group at Tucson, Arizona, discovered numbers of previously unrecognized, multiple-concentric-ring impact basins on the Moon, including the 1000 km Orientale Basin on the Moon's limb (Hartmann and Kuiper 1962; see also the memoir by Hartmann 1981). This clarified the role of the Imbrium Basin as an impact feature, whose symmetry and origin had been variously interpreted by workers such as Gilbert (1893), Baldwin (1949, 1963), Fielder (1961, 1965), and Urey (1952, 1962). Shoemaker and co-workers immediately recognized that the enormous ejecta blankets of such basins, which spread over sizable fractions of the target planet, could be used as stratigraphic markers in establishing the stratigraphic system of the Moon and presumably of other planets (Shoemaker and Hackman 1962).

Signs of still more ancient, mostly eroded impact scars were being identified. Robert Dietz (1963), in particular, made pioneering studies of what were called "cryptoexplosion" structures. Detecting shatter cones and other features, Dietz and others recognized their impact origin and proposed the name "astroblemes" for these features. This term is now less used, since an entire spectrum of impact structures, of various sizes and in various stages of erosion, is well documented.

The pioneering efforts of Shoemaker, Beals, Dietz, and other field workers led to an important but under-appreciated change in our perception of the terrestrial planets. For the first time, impacts were seen as a potentially important process in planetary development. Probably the last serious argument in favor of volcanic origins of lunar craters came from two books by the geologist Gilbert Fielder (1961, 1965), who, at the same time, contributed valuable work on lunar faults and tectonics.

In this period, the Earth was viewed as much more static than it is today; even "continental drift" was still ridiculed. During the defense of my PhD dissertation, which used the Canadian work to establish the cratering rate and predict that the lunar maria were 3.6 Gyr old (Hartmann 1965), a noted geologist asked with some surprise, "But if this is correct, the Earth would have been hit by many impactors and should have many old craters." Well, yes, but most have been destroyed by plate tectonic movements and subduction – which were not widely accepted until 1968! The point was that they were just beginning to be discovered, and could give us a correct estimate of the ages of other cratered surfaces.

Before the 1950s, conventional wisdom discounted the role of impacts on Earth. One could argue that science was still too much under the ghostly influence of the medieval and pre-Copernican idea that, somehow, "Earth is different" from the rest of the Solar System. But by the 1960s, most workers agreed that Earth and the other terrestrial planets were embedded in a cosmic population of asteroidal and cosmic debris. The data showed that the cratering rate of this debris was extremely high up to about 3.9 Gyr ago, and that, even in later geological time, impacts big enough to affect Earth's global environment could happen on timescales of 100 Myr or so, although no one yet dared to advocate seriously that they might have had effects on biological evolution. Nonetheless, the recognition that Earth and the other planets were linked in a common cratering history was a key step in expanding geological and geophysical thinking to the scale of the Solar System. And in spite of squabbles and "lovers' quarrels," the cross-fertilization of the various disciplines had by the 1960s given birth to a new, interdisciplinary field of planetary science.

THE FIRST REVELATION OF RADICALLY ALIEN CONDITIONS ON VENUS: 1962–67

Venus offers a beautiful example of the revolutionary transformation of our picture of a neighboring world. To early astronomers, the planet seemed potentially Earthlike. Early applications of spectroscopy in the 1890s led to suggestions of water vapor, but the data were erroneous because of terrestrial water vapor.

The recognition that carbon dioxide was a major constituent of the atmosphere, as mentioned earlier, began with the work of Adams and Dunham (1932), who discovered CO_2 absorption bands in the near-infrared spectrum. This work was extended by Kuiper (1947) and Dunham (1948). Even if the atmosphere had a strong CO_2 component, some observers assumed that the whitish clouds of Venus were dense layers of water vapor or high cirrus ice clouds not unlike those of Earth. To Frank E. Ross (1928), who found the ultraviolet cloud patterns, the slightly yellowish tone of Venus suggested dust clouds in the lower atmosphere, seen through breaks in an overlying cirrus level analogous to high Earth-type cirrus.

Simple thermal calculations revealed that if one assumed room temperatures as representative of Earth, and then imagined moving Earth to the position of Venus and adding more cloud cover, the resulting surface temperatures would be around 39–73° C, depending on albedo and cloud cover assumptions, and so a warm, steamy climate was widely proposed. Tenable speculations about Venus ranged from a humid rainforest or a global ocean, to the hot, barren, dust-blown desert of Frank Ross under overcast skies. In the classic 1949 book *The Conquest of Space* (Ley and Bonestell 1949), the scientific artist Chesley Bonestell, having consulted scientific opinion of the day, rendered the latter conception and produced a surprisingly accurate view of the surface. He portrayed eroded rocky outcrops, clear surface air under a thickly overcast, yellowish sky, along with dunes of windblown sand or dust. Meanwhile, more imaginative science fiction of the 1950s portrayed Venus with rainy Cretaceous-style swamps or jungles of dense vegetation (hence the CO_2), dinosaur-like beasts, or perhaps even civilizations hidden beneath the clouds.

New techniques of radiometry in the far infrared showed that something was drastically wrong with the Earthlike picture. Early thermocouple radiometry revealed that the night side was warm, but this was explained (incorrectly) as proving rapid rotation. Rupert Wildt (1934, 1940) took the first steps toward the correct modern understanding of the Venus environment by showing that a thick CO_2 atmosphere could produce a greenhouse effect. He estimated the surface temperature to be over 400 K, or over 127°C, as opposed to the more moderate non-greenhouse temperatures mentioned earlier. Microwave observations in the 1950s showed strong emissions at a wavelength near 3 cm (Mayer et al. 1958), and the infrared–microwave excess was established by the time of the first spacecraft exploration in 1962. But the greenhouse mechanism still seemed speculative, and many researchers looked for other explanations of the microwave emissions, such as ionospheric phenomena and lightning. Carl Sagan (1962), later to become the century's most effective popularizer of planetary exploration, made his earliest major scientific contribution by showing that the thermal infrared spectrum fitted calculations for a strong greenhouse effect due to the dense CO_2 atmosphere. Sagan proposed surface temperatures of 700 K or more and high surface pressure, and is usually credited with establishing the reality of the Venus greenhouse heating.

Venus became the first planet to be observed at close range when Mariner 2 passed at a distance of 38,854 km on 14 December 1962. Microwave and infrared radiometer observations supported the estimates of a dense CO_2 atmosphere and high temperature of at least 698 K in the lower atmosphere on both the day and night side. The Mariner 2 magnetometer revealed a magnetic field less than one-tenth the strength of Earth's field. These results were the first data to put serious boundary conditions on the range of human speculation about what other planets might be like.

Around 1967, infrared spectroscopy gave many indications that the Venusian clouds were very different from the Earth-like clouds that had been assumed. Sill (1967) and Young (1973) showed that they were composed of H_2SO_4 droplets. Connes et al. (1967) discovered hydrogen chloride and hydrogen fluoride in the atmosphere. These results triggered new studies of high-temperature reactions between the surface and the atmosphere.

THE FIRST LOOK AT A PLANETARY SURFACE: MARS IN 1965

The first look at another planetary surface, other than the vague shadings visible in telescopes, came on 15 July 1965 when Mariner 4 sped past Mars at 9600 km above the surface and obtained 21 complete television images in a swath of slightly overlapping footprints, roughly 250 km on a side, across the surface (Figure 2). Many of these pictures were murky at best. (Many modern researchers do not appreciate what the normal haze cover of Mars is like. A generation has been raised on USGS computer-processed photomosaics, which seamlessly link views on cloudless days, showing a (false) crystal-clear desert planet. Thus, the hazes, low contrast, high lighting, and raggedly patchy albedo markings produced nothing very clear in many Mariner 4 frames.) However, the middle, more vertical views, images 5 to 15, revealed about 70 impact craters with diameters from about 5 to 120 km.

The discovery of craters on Mars created something of a sensation. The *New York Times* of 30 July 1965 reported that the existence of the ancient craters showed that Mars

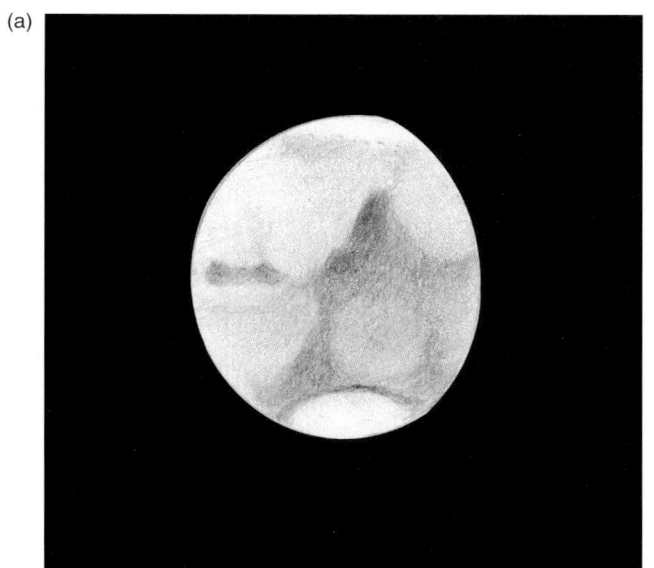

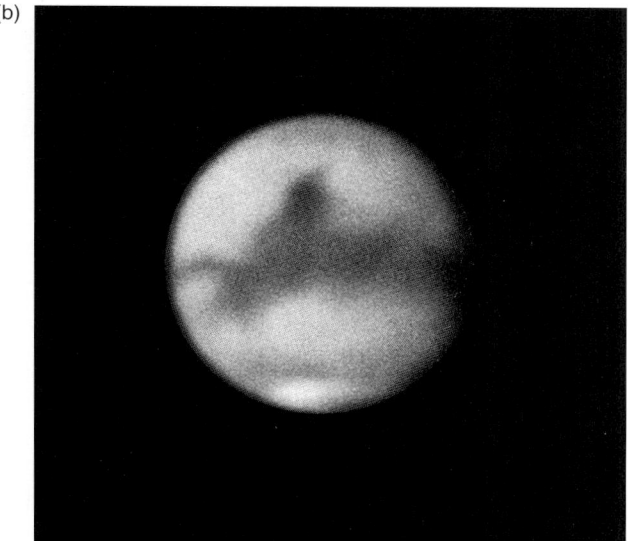

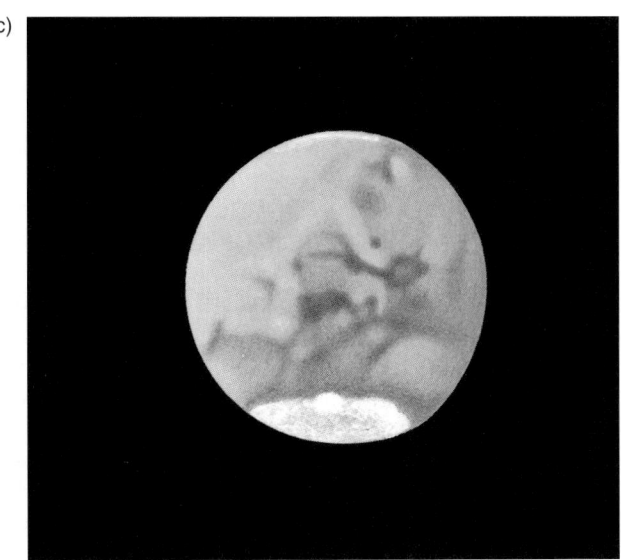

Figure 1 Examples of images of Mars before the first Mariner encounter. (a) Amateurs and professionals equipped with small telescopes were able to monitor cloud patterns, changes in shapes of markings, polar caps, and so on. This drawing by the author, made with the aid of an 8-inch (20-cm) home-made telescope on 28 June 1971 from Tucson, Arizona, shows Syrtis Major and Sabaeus Sinus. (b) A similar view in a photograph by the Lowell Planetary Patrol, taken in red light in 1971. In the era before CCDs, the best photographs required exposures long enough to be blurred by seeing, and visual observers could often record more detail in drawings, though their objectivity was always a question. (c) This drawing by Gerard de Vaucouleurs, made on 11 July 1971, is an example of the best visual observation under the best conditions.

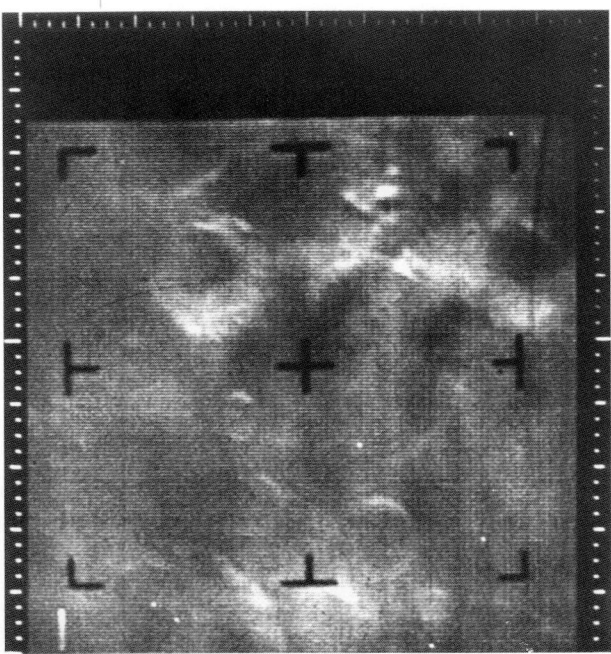

Figure 2 The first close-up photography of Mars, from Mariner 4, in 1965. This is one of the photographs on which craters were first discovered. The image covers an area about 270 × 220 km at 41°S and 152°W in Phaethontis. (NASA/JPL.)

was Moon-like, virtually ruling out the possibility of life there. Leighton *et al.* (1965), calculating a (correct?) characteristic age of the cratered surface as 2 to 5 Gyr, stated that the "remarkable state of preservation of such an ancient surface" precluded an early dense atmosphere or appreciable free water at any time during the history of the surface – ideas contrary to today's understanding.

Looking back from the present, it seems surprising how strong an impact this discovery had. We must remember that the debate about Mars at that time focused on a more Earthlike surface, where the variable dark patches might be vegetation or ash deposits from ongoing volcanic eruptions. Öpik, prescient as usual, had predicted impact craters on Mars and had used asteroid and comet statistics to predict relative impact fluxes on planets (Öpik 1951), but his work had not been widely absorbed. Analyzing the Mariner 4 images independently from Armagh Observatory, Öpik (1965, 1966) correctly deduced that since small craters were under-abundant compared with the lunar crater production function, Mars must have a more active environment in terms of erosion and deposition. However, the dominant response from the Mariner 4 team and other scientists (Leighton *et al.* 1965) was that Mars was *much* more Moonlike than had been supposed. By sheer chance, none of the volcanoes, tectonic canyons, riverbeds, or other "exotic" landforms that we now think of as characterizing the planet were revealed by Mariner 4, whose best pictures were of the old cratered uplands. Thus, while scientific discussions before Mariner 4 were too Lowellian and too unimaginative to appreciate the importance of impact phenomena in shaping planet surfaces, the post-Mariner 4 pendulum of scientific opinion swung too far the other way, picturing Mars as a world completely dead in geological terms, not to mention biological terms. The Mars of Mariner 4 was like the Moon, with just a little air to blow the dust around.

This situation was not to last for long. In 1969 two probes were sent to Mars on flyby missions. Both worked, and both sent back more swaths of close-up pictures as well as somewhat fuzzy far-encounter views that showed a few of the scattered largest craters. This time the photographs revealed tracts of land that were quite unlike either the Moon or the Mariner 4 views. In particular, sharply bounded areas of "higgledy-piggledy" hills were imaged; they came to be called chaotic terrain. As a result, the pendulum began to swing back a bit from the lunar extreme: Mars did have geological activity, and was clearly different from the Moon after all.

EARLY PLANETARY SPACECRAFT IMAGERY, c. 1970: HOW MUCH IS ENOUGH?

The results of Mariners 4, 6, and 7 led to an intriguing problem in designing flyby space missions – a problem that still haunts us. Mission designers facing a first reconnaissance of a new planet, moon, or asteroid, especially on a flyby mission but also on an orbital mapping mission like Mars Global Surveyor, face a choice of resolution in camera design. They can favor either lower resolution in order to sample a large fraction of the surface, or higher resolution in the hope of seeing revealing details, but covering only a small fraction of the surface. So, what is the optimum geological resolution for learning about "important" planetary processes: 10 m, 100 m, 1 km? And what percentage of the planet gives an optimum sample: 100%, 50%, 10%? During the 1960s and 1970s, USGS researchers such as Hal Masursky and Jack McCauley, served on various panels to assess this question. Too much resolution can produce images of intriguing details impossible to interpret without context images of broader surrounding regions. Several later missions, such as Mariner 9, solved the problem with dual bore-sighted low- and high-resolution cameras that took broad "context" frames with a high-resolution "footprint" image in the center. Yet, even today, in the case of Mars, there is a resolution gap between Viking orbiter pictures, which mostly resolve features of 100–500 m in dimension and cover the whole planet, and

Mars Global Surveyor high-resolution images which resolve features 10–20 m (1–2 m/pixel) in size and have footprint widths only around 1 km on a side.

DISCOVERING THE SURFACE ENVIRONMENT OF VENUS: 1967–73

Observations during this same period continued to reveal the unearthly conditions of Venus. The physical rotation of the underlying planet was a long-standing mystery. Kuiper (1954) had correctly deduced from ultraviolet photography that the apparent markings changed rapidly. He inferred a rapid rotation of "no more than a few weeks . . . and definitely not . . . 225 days." Boyer and Guerin (1966) and Smith (1967) correctly measured a retrograde movement of the cloud patterns with periodicity of about four days, which was suspected to mark the planetary rotation. These measures, however, referred to atmospheric circulation, as confirmed in 1974 by Mariner 10, not to the physical planet's spin.

Radar measurements of the actual rotation began to be attempted in 1961. Measurements by 1964 tracked radar-reflective features and established a retrograde, zero-obliquity rotation with a period of about 243 days (Carpenter 1964, Goldstein 1964). Early researchers thought this was locked into resonance with Earth, causing Venus to present the same side to Earth at each inferior conjunction, but it has since been shown that this resonance is not exact (Zohar et al. 1980).

To learn what was beneath the clouds, the Soviet Union began an aggressive program of robotic probes parachuted into the atmosphere. This was the first direct and sustained exploration of another planet, and probably the most spectacular of such explorations, in terms of changing our perceptions of another world. The first success (after roughly ten failed attempts) was on 18 October 1967, five years after the Mariner 2 flyby, when Venera 4 made the first successful exploration of Venus' environment. The probe parachuted into the atmosphere and radioed data back. A major result was to disprove the still widespread suspicion that the atmosphere was largely nitrogen, like Earth's, by revealing the overwhelming preponderance of CO_2. Initial interpretations were that the probe had transmitted data all the way to the surface, where the temperature was estimated at about 550 K. Later analysis showed that the probe had transmitted only from about 55 to 27 km altitude. The Venera 4 bus, flying on past Venus, constrained the magnetic field to less than 1/5000 the strength of Earth's field.

Three years later, on 15 December 1970, Venera 7 made the first successful robotic landing on Venus, sending back more data on the surface atmospheric conditions and confirming the high-temperature and high-pressure conditions.

Finally, on 22 and 25 October 1975, Veneras 9 and 10 sent back the first photographs of the surface (Florensky et al. 1977, Keldysh 1977). The Venera landers featured a clever camera design that scanned a long image, swiveling from one horizon down to the foreground soil at the foot of the lander, and back up to the opposite horizon. This was a safeguard against the possibility of a surface atmosphere so murky that one could see only a few meters away, as in a dense fog. One of the first surprises was that the surface atmosphere was so clear that the horizon was easily visible. Gravel, angular boulders, and platey bedrock textures observed at various sites by several Soviet landers indicated a range of modest weathering processes.

Veneras 8, 9, 10, 13, and 14, along with the later VEGA 1 and 2 landers, made crude composition measurements by various techniques (reviewed in Chapter 54). The first three Venera probes, 8, 9, and 10, made gamma-ray measurements of radioactive potassium, uranium, and thorium in the surface soils. Data from all seven sites together indicate that the surface composition is consistent with basaltic soils, similar to terrestrial oceanic crustal basalts, although the Venera 8 site gave still-controversial indications of a more granitic composition.

These pioneering planetary surface measurements, combined with the 1969–72 return of basaltic lavas from the Moon, and the 1960s' suggestions of weathered basaltic soils on Mars, gave the first direct indications of the universal importance of basaltic volcanism on the terrestrial planets – a subject chosen for an initiative by the Lunar and Planetary Science Institute in the late 1970s and developed at great length in *Basaltic Volcanism on the Terrestrial Planets* (Basaltic Volcanism Team 1981). By the same token, these data gave the first direct confirmation that surface geology of other terrestrial planets could be approached by the same gross petrological and geochemical principles applied in terrestrial crustal geology.

THE FIRST CLOSE-UP GLOBAL MAPPING OF MARS: MARINER 9 IN 1971–72

Following the discovery of chaotic terrain and other non-impact geology on Mars by Mariners 6 and 7 (Figure 3), NASA planned the essentially duplicate Mariners 8 and 9 missions for the 1971 arrival window. Mariner 8 went into the Atlantic just after launch. Mariner 9 went into orbit around Mars on 27 November 1971, during a major global Martian dust storm. Essentially no details were visible on the planet at all, except for four curious black spots in the Tharsis area. Closer imagery revealed volcanic calderas in each black spot – the discovery of the four huge volcanoes of Mars, whose summits, around 23 km high, protruded above the dust storm clouds. Soviet Mars probes arrived at

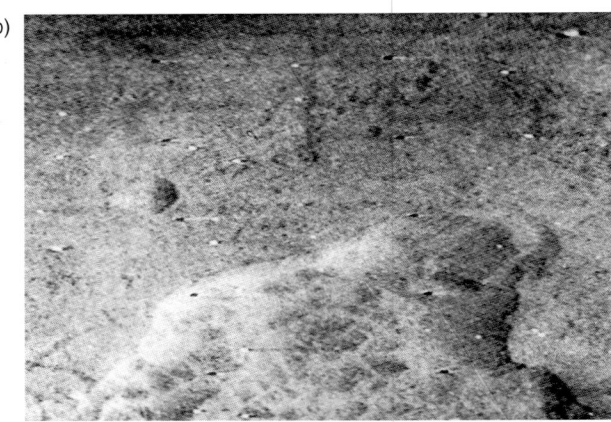

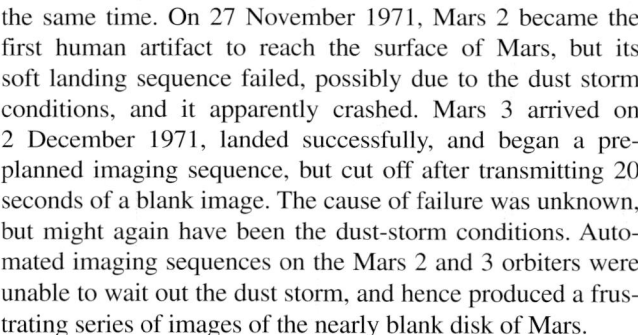

Figure 3 Examples of Mariner 6 and 7 photography of Mars in 1969. (a) Some frames, such as this view from Mariner 6, showed cratered terrain reminiscent of the Moon, as had been photographed by Mariner 4 in 1965. (b) Other frames showed previously unknown types of geological features. This Mariner 6 frame shows a relatively uncratered region with a collapsed depression filled with rough material that came to be known as chaotic terrain. Images such as this were the first to reveal that Mars had regions of geological activity unlike the Moon, and reversed the Mariner 4 conclusion that Mars was a Moonlike, geologically dead world. (c) This Mariner 7 frame shows a sparsely cratered region where the largest crater is muted and pitted, unlike any craters seen on the Moon. (NASA/JPL.)

the same time. On 27 November 1971, Mars 2 became the first human artifact to reach the surface of Mars, but its soft landing sequence failed, possibly due to the dust storm conditions, and it apparently crashed. Mars 3 arrived on 2 December 1971, landed successfully, and began a preplanned imaging sequence, but cut off after transmitting 20 seconds of a blank image. The cause of failure was unknown, but might again have been the dust-storm conditions. Automated imaging sequences on the Mars 2 and 3 orbiters were unable to wait out the dust storm, and hence produced a frustrating series of images of the nearly blank disk of Mars.

During the following weeks, the opaque dust blanket descended to lower altitudes as dust particles of a few micrometers in size settled out of the air. Gradually more and more of the Martian topography was exposed to the Mariner 9 orbiter cameras (Figure 4). Mariner 9 spent 1972 mapping the entire planet for the first time, revealing enormous surprises about the Martian geological features and processes. The most stunning surprise was that Mars – the dry, frozen desert planet – had dry riverbeds, which were named channels (Figure 5) – not to be confused with the "canals" of Schiaparelli and Lowell. The canals turned out to be mostly nonexistent, at least in the extreme straight-line

form championed by Lowell. However, some of the dark markings had an undeniably streaky character, and many of the most famous canals, especially in the low-latitude borders between bright deserts and dark areas, did correspond to streaky, patchy markings, some loosely associated with channels and other elongated features. A huge tectonic canyon complex, similar in scale to the Red Sea rift, occupied the site of the stubby "canal" Coprates, and was named Valles Marineris in honor of the spacecraft.

The attribution of the channels to flowing water caused great debate during the mission and afterwards. In favor of water, it was cited that many channels had tributary systems and grew wider and deeper in the down-slope direction. There were many discussions of whether such features could be carved by wind, glaciers, lava flow, or other fluid compounds unfamiliar on Earth. By the end of the mission, most workers had agreed that the creation of the channels involved flowing water in the form of ephemeral rivers, or perhaps more likely ice-clogged or ice-capped rivers.

There was no sign of life. The albedo markings, which had been seen to change seasonally and were thought to involve possible areas of vegetation, were shown to be dust deposition features. Proof from Mariner 9 came from the

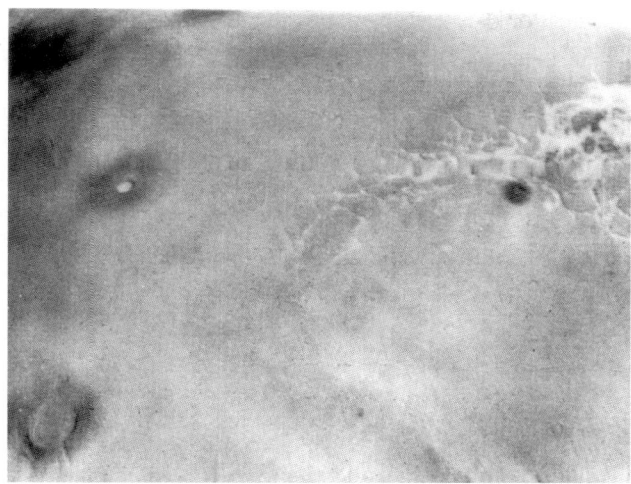

Figure 4 This early Mariner 9 frame, from 17 December 1971, shows the southeast Tharsis region emerging from a global dust storm. At the left, two of the giant Tharsis volcanoes, with central calderas, have partly emerged from the gradually lowering dust pall. In the middle of this view the contrast is greatly reduced because of the dust veil. At the right, the canyon network at the west end of the Valles Marineris is beginning to show through the dust. The valleys are filled with suspended dust; the optical depth through the dust to the canyon floor is much greater than in the surroundings, so that the canyons appear as a bright, spidery network. (NASA/JPL.)

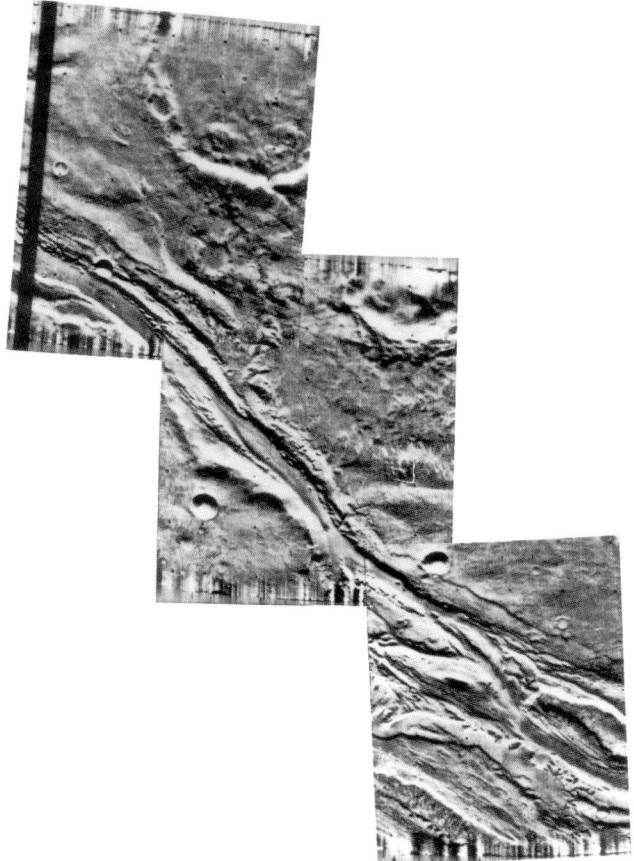

Figure 5 An example of discovery photographs of Martian fluvial channel systems: a mosaic of three Mariner 9 frames from 1972. (NASA/JPL.)

observation of both dark and light streaks, or "tails," on the leeward sides of craters and other topographic obstacles, resolution of dark patches in crater floors into dunes, and the behavior of small-scale changes detected during the course of the mission. The discovery that the dark markings involved windblown dust deposits seemed to solve a centuries-old puzzle, but in reality led to a deeper puzzle that has still not been resolved – and is not even widely debated. If the albedo markings involve winds, why have the darker and lighter materials not been completely mixed? Thermal and spectral observations indicate that the darker markings are usually coarser, less weathered basaltic materials, and the lighter markings are usually more weathered, finer dust. But in that case, what is the source of the fresher basalt? Are fresh lava flows closer to the surface in the dark areas? Seemingly not; curiously, the Tharsis and Amazonis areas, where lava flow morphologies can be seen in moderate-resolution imagery, are bright, not dark, while some famous and classical dark areas such as Syrtis Major do not show lava textures but are heavily cratered.

Mariner 9 also revealed a sea of sand dunes, about equal in area to the Sahara, ringing the north pole. The bright circular features of Hellas and Argyre, mapped generations before by telescopic observers on Earth, turned out to be impact basins.

Many Mariner 9 observations hinted at the important role of water in Mars's past. Polygonal patterns suggested permafrost in upper-latitude plains. Polar data suggested a small, permanent water-ice cap at the north pole under the ephemeral winter CO_2 ice cap, and a possible similar permanent water-ice cap at the south pole as well. The chaotic terrain imaged by Mariners 6 and 7 was found to mark collapsed areas from which water was emitted, perhaps by melting underground permafrost. A number of riverbed channels emanated from areas of collapsed chaotic terrain.

The role of water raised questions about the history and form of the water activity. Some workers proposed that the dendritic tributary systems proved the existence of past rainfall, at least in some areas. Others suggested that the water reached the surface mostly from underground sources, and that the channels were formed by non-rainfall processes such as headward sapping or scouring under a thick, glacial ice cover. These debates continued at least until the Mars Global Surveyor mission at the end of the twentieth century.

The timing of the channel activity was also entirely unclear from the Mariner 9 data. Were the channels associated only with an early thick atmosphere? Did they flow only in the first billion years of Mars history? Was each channel formed in one catastrophic outflow, perhaps by local melting of ground ice? Or were the channels going through multiple periods of activity and inactivity, perhaps separated by hundreds of millions of years? These questions remain unanswered.

The discovery that Martian geology involved very active sculpting of the surface by fluvial activity was another profound change in the evolution of planetary science. It was indeed an incredible reversal from the view propounded only seven years earlier, in 1965, when Mariner 4 scientists pronounced that Mars was a dead, Moon-like world. Meanwhile, the evidence of large volcanoes and lava flow morphologies reinforced the view that basaltic volcanism was a universal phenomenon on terrestrial planets.

MERCURY: 1900–74

Among the terrestrial plants, Mercury has always been the neglected world. The problem began during the era of telescopic observation, for Mercury's proximity to the Sun gives it a greatest solar elongation of only about 28°. Therefore it could be observed at high elevation above the horizon only in the daytime; dawn and dusk observations gave a dark-sky background but high airmass and poor seeing. As a result, several generations of professional and amateur astronomers were able to do no more than record vague, dusky markings on the disk.

So poorly recorded were these markings that there was no definitive map, and no agreement even on the planet's rotation period. Conventional wisdom, based on supposed tracking of the markings, was that Mercury's rotation was very slow and that it kept one face toward the Sun, perhaps due to tidal effects. Some of the markings recorded were undoubtedly real, while others were undoubtedly spurious, produced by the poor conditions. Curiously, no thorough attempt has been made to generate ephemerides of past observations and go back to determine which markings were repeatably observed, or how they relate to the surface features that are still only partially known. On a philosophical level this is all the more ironic, because a number of respected astronomers, such as Antoniadi (1934), devoted years to mapping these features – yet today, even though we have the capability, little interest is shown in finding out how they relate to the lava plains and craters of the planet.

Prior to spaceflight, the most solid geophysical facts were that the reflected visible and radio properties, as well as the thermal infrared properties, were very similar to those of the Moon, suggesting similar igneous rocky types and regolith soil properties, with the regolith containing dark glassy material (McCord and Adams 1972). The planet's density was known to be higher than found for the Moon or Mars, being closer to that of Venus and Earth, at 5.45 g cm^{-3}. The latter value implied a larger iron percentage than for any other planet (Kaula 1968, Reynolds and Summers 1969).

The first actual measurement of the rotation period came with radar astronomy, which was developed in the 1960s. Radar signals bounced off Mercury showed that the rotation period was indeed slow, but instead of matching the 88-day orbital period, it was 59 days (Pettengill and Dyce 1966), and is associated with a 3:2 orbital resonance (Colombo 1966). Murray et al. (1972) re-examined the ground-based telescopic observations with the advantage of hindsight and concluded that they fitted this period, and presented new maps of the dusky shadings.

The first close-up spacecraft data came with the flyby of Mariner 10 on 29 March 1974. The same spacecraft made two more close passes of Mercury, on 21 September 1974 and 16 March 1975, but due to the orbital dynamics, the imagery during the three passes showed nearly the same side of the planet, so that little more than half the planet was mapped. The third encounter was closest by a factor of two, at 327 km, the cameras resolving features of the order of 100 m. Because only about half the planet has been photographed, and because no obvious dark maria are present as on the Moon, the nature of the shadings reported by Antoniadi (1934) and mapped with some care by Murray et al. (1972) remains unknown.

HIATUSES IN PLANETARY EXPLORATION: A GENERAL PHENOMENON?

A curious and little-remarked phenomenon seems to play itself out in many historical phases of exploration. A first burst of exploration occurs, when curiosity and expectations are aroused. Often the resulting discoveries are not quite what was expected, or perhaps we simply learn as much as our imaginations can accept. The thrill of being first is gone, and it is as if a certain social regrouping is required, averaging 2–3 decades, before exploration can continue.

There is a parallel with the Spanish exploration of the New World. While the overall pattern of exploration was relatively continuous after Columbus and does not show the hiatus described above, several regional explorations show the effect perfectly. In Florida, Ponce de Leon made an initial survey in 1512–13, but there was no return until the large, ill-fated Narvaez expedition of 1528. In the American Southwest, the huge Coronado expedition of 1200 people, the Apollo expedition of its day, sought Aztec-like, gold-rich trading cities north of Mexico in 1540. It traveled north

Table 1. Patterns of hiatus following early explorations

Exploration target	Explorations	Hiatus (years)
Florida	Ponce de Leon: 1512–13	15
	Narvaez: 1528	
Southwest USA	First reconnaissance: 1539	28
	Coronado army: 1540–42	
	First return to New Mexico: 1580	
South Pole	Amundsen and Scott 1911–12	18, 36
	Byrd fly-over 1929, 47	
North Pole	Peary near pole 1909	16
	Amundsen, Byrd flyovers 1925–28	
Moon (robotic)	About 20 impactors or landers (1959–76)	20
	Clementine (small probe): 1996	
	Prospector (small probe): 1999	
Moon (human)	Six human expeditions: 1969–74	>26
	None since	
Venus surface	Eight Soviet and one US soft landers with data from surface: 1970–85	>15
Mars surface	Three Soviet and two US landers: 1970–76	21
	Pathfinder lander: 1997	
Mercury	Mariner 10: 1974	>26
	None since	
Jupiter	Voyagers 1 and 2: 1979	16
	Galileo: 1995	
Saturn	Voyagers 1 and 2: 1981	23
	Cassini: 2004	

through Sonora, conquered several pueblos in New Mexico, and then went on as far as Kansas before giving up. So disillusioned were the Spanish that there was no serious return to New Mexico for 28 years, when priests with the Chamascudo expedition returned to New Mexico to settle.

A similar pattern has marked the early phases of planetary exploration. We tend to think of space exploration, especially by robotic orbiters and landers, as a relatively continuous process. However, as shown in Table 1, a hiatus of 16 to 30 years is common for each individual planet after the first burst of exploration of "*terra incognito.*" After Sputnik 1 orbited Earth, newspapers declared that humanity had entered the "space age." In reality, it is as if there are subtle social effects at work, in which it takes societies some years to react to profound new discoveries, and to transform the extraordinary into the ordinary.

REFERENCES

Adams, J.B. (1968). Lunar and Martianz surfaces: Petrologic significance of absorption bands in the near infrared. *Science*, **159**, 1453–1455.

Adams, W.S. and Dunham, T. (1932). Absorption bands in the spectrum of Venus. *Publications of the Astronomical Society of the Pacific*, **44**, 243–247.

Antoniadi, E.M. (1934). *La Planète Mercure et la rotation des satellites*, Gauthier-Villars, Paris reprinted 1974 in English, Keith Reid, London.

Baldwin, R.B. (1949). *The Face of the Moon*, University of Chicago Press.

Baldwin, R.B. (1963). *The Measure of the Moon*, University of Chicago Press.

Basaltic Volcanism Team (1981). *Basaltic Volcanism on the Terrestrial Planets*, Pergamon Press, New York.

Beals, C.S., Innes, M. and Rottenberg, J. (1963). Fossil meteorite craters. In B. Middlehurst and G. Kuiper (eds), *The Moon, Meteorites, and Comets*, University of Chicago Press.

Binder, A.B. and Cruikshank, D.P. (1966). Lithological and mineralogical investigation of the surface of Mars. *Icarus*, **5**, 521.

Boyer, C. and Guérin, P. (1966). Mise en évidence directe, par la photographie d'une rotation rétrograde de Vénus en 4 jours. *Comptes Rendues de l'Académie des Sciences*, **263**, 253–255.

Carpenter, R.L. (1964). Study of Venus by CW radar. *Astronomical Journal*, **69**, 2–11.

Colombo, G. (1966). Rotational period of the planet Mercury. *Nature*, **208**, 575.

Connes, P., Connes, J., Benedict, W. and Kaplan, L. (1967). Traces of HCl and HF in the atmosphere of Venus. *Astrophysical Journal*, **147**, 1230.

Courtillot, V. (1999). On causal links between flood basalts and continental breakup. *Earth and Planetary Science Letters*, **166**, 177–195.

de Vaucouleurs, G. (1954). *Physics of the Planet Mars*, Faber & Faber, London.

Dessler, A.J. (1984). The Vernov radiation belt (almost). *Science*, **226**, (4677), editorial page.

Dietz, R.S. (1963). Astroblemes: Ancient meteorite-impact structures on the Earth. In B. Middlehurst and G. Kuiper (eds), *The Moon, Meteorites, and Comets*, University of Chicago Press.

Dunham, T. (1948). Spectroscopic observations of the planets at Mt. Wilson Observatory. In G.P. Kuiper (ed.), *Atmospheres of the Earth and Planets*, 2nd edn, University of Chicago Press.

Fielder, G. (1961). *Structure of the Moon's Surface*, Pergamon, Elmsford, NY.

Fielder, G. (1965). *Lunar Geology*, Lutterworth, London.

Florensky, C.P., Basilevsky, A. and Ponin, A. (1977). The first panoramas of the Venusian surface. *COSPAR Space Research*, **17**, 645.

Gerling, S.K. (1942). Age of the Earth according to radioactivity data. *Doklady Akademii Nauk*, **34**, 259–261.

Gilbert, G.K. (1893). The Moon's face: A study of the origin of its features. *Bulletin of the Philosophical Society of Washington*, **12**, 241–292.

Glanz, J. (1999). Science vs. Bible wrangle: Debate moves to the cosmos. *International Herald Tribune*, 11 October, p. 2.

Goldstein, R.M. (1964). Venus characteristics by Earth-based radar. *Astronomical Journal*, **75**, 273–284.

Harford, J. (1997). *Korolev*, Wiley, New York.

Hartmann, W.K. (1965). Terrestrial and lunar flux of large meteorites in the last two billion years. *Icarus*, **4**, 157–165.

Hartmann, W.K. (1981). Discovery of multi-ring basins: Gestalt perception in planetary science. In *Multi-Ring Basins*, Proceedings of the Conference on Multi-Ring Basins, Pergamon, New York.

Hartmann, W.K. and Kuiper, G.P. (1962). Concentric systems surrounding lunar basins. *Communications of the Lunar and Planetary Laboratory*, **1**, 51.

Hartmann, W.K., Strom, R., Weidenschilling, S., Blasius, K., Woronow, A., Dence, M., Grieve, R., Diaz, J., Chapman, C., Shoemaker, E. and Jones, K. (1981). Chronology of planetary volcanism by comparative studies of planetary cratering. In *Basaltic Volcanism on the Terrestrial Planets*, Pergamon Press, New York.

Hickam, H. (1998). *Rocket Boys: A Memoir*, Delacorte Press, New York.

Holmes, A. (1946). An estimate of the age of the Earth. *Nature*, **157**, 680–690.

Horowitz, N. (1986). *To Utopia and Back: The Search for Life in the Solar System*, Freeman, New York.

Houtermans, F.G. (1947). Das Alter des Urans. *Zeitschrift für Naturforschung*, **2a**, 322–328.

Hoyt, W.G. (1987). *Coon Mountain Controversies*, University of Arizona Press, Tucson.

Kaplan, L.D., Munch, G. and Spinrad, H. (1964). An analysis of the spectrum of Mars. *Astrophysical Journal*, **139**, 1–15.

Kaula, W.M. (1968). *An Introduction to Planetary Physics: The Terrestrial Planets*, Wiley, New York.

Keldysh, M.V. (1977). Venus exploration with the Venera 9 and 10 spacecraft. *Icarus*, **30**, 605.

Kuiper, G.P. (1944). Titan: A satellite with an atmosphere. *Astrophysical Journal*, **100**, 378–383.

Kuiper, G.P. (1947). Infrared spectra of planets. *Astrophysical Journal*, **106**, 252.

Kuiper, G.P. (1952). Planetary atmospheres and their origin. In G.P. Kuiper (ed.), *Atmospheres of the Earth and Planets*, University of Chicago Press.

Kuiper, G.P. (1954). Determination of the pole of rotation of Venus. *Astrophysical Journal*, **120**, 603–605.

Kuiper, G.P. (1964). Infrared spectra of stars and planets: IV. The spectrum of Mars, 1–2.5 microns, and the structure of its atmosphere. *Communications of the Lunar and Planetary Laboratory*, **2**, 79.

Leighton, R.B., Murray, B., Sharp, R.P., Allen, J.D., Sloan, R.K. (1965). Mariner IV photography of Mars: Initial results. *Science*, **149**, 627–630.

Ley, W.W. and Bonestell, C. (1949). *The Conquest of Space*, Viking Press, New York.

Lowell, P. (1908). *Mars as the Abode of Life*, Macmillan, New York.

Mayer, C.H., Cullough, T. and Slonaker, R. (1958). Observations of Venus at 3.15 cm wavelength. *Astrophysical Journal*, **127**, 1–10.

McCord, T.B. and Adams, J.B. (1969). Spectral reflectivity of Mars. *Science*, **163**, 1058–1060.

McCord, T.B. and Adams, J.B. (1972). Mercury: Interpretations of optical observations. *Icarus*, **17**, 585–588.

Murray, B.B., Dollfus, A. and Smith, B. (1972). Cartography of the surface markings of Mercury. *Icarus*, **17**, 576–584.

Öpik, E.J. (1951). Collision probabilities with the planets and the distribution of interplanetary matter. *Proceedings of the Royal Irish Academy*, **54**, 165.

Öpik, E.J. (1963). Survival of comet nuclei and the asteroids. *Advances in Astronomy and Astrophysics*, **2**, 219.

Öpik, E.J. (1964). Mariner IV and craters on Mars. *Irish Astronomical Journal*, **7**, 92–104.

Öpik, E.J. (1965). The Martian surface. *Science*, **153**, 255–265.

Owen, T.C. and Kuiper, G.P. (1964). A determination of the composition and surface pressure of the Martian atmosphere. *Communications of the Lunar and Planetary Laboratory*, **2**, 113–132.

Pettengill, G.H. and Dyce, R.B. (1966). A radar determination of the rotation of the planet Mercury. *Nature*, **206**, 1240.

Pius XII (1951). Address to the Pontifical Academy of Sciences, 22 November, Tipografia Poliglotta Vaticana.

Reynolds, R.T. and Summers, A.L. (1969). Calculations on the composition of the terrestrial planets. *Journal of Geophysical Research*, **74**, 2494.

Ross, F. (1928). Photographs of Venus. *Astrophysical Journal*, **68**, 57.

Safronov, V.S. (1972). *Evolution of the protoplanetary cloud and formation of the Earth and the Planets*. Nauka, Moscow (NASA TTF-677).

Sagan, C. (1962). Structure of the lower atmosphere of Venus. *Icarus*, **1**, 151.

Shoemaker, E.M. (1957). *Bulletin of the Geological Society of America*, **68**(12), Pt. 2, p. 1846.

Shoemaker, E.M. (1960). US Geological Survey Prof. Paper 400-B, p. 423.

Shoemaker, E.M. (1963). Impact mechanics at Meteor Crater, Arizona. In B. Middlehurst and G. Kuiper (eds), *The Moon, Meteorites, and Comets*, University of Chicago Press.

Shoemaker, E.M. and Hackman, R. (1962). Stratigraphic basis for a lunar timescale. In Z. Kopal and Z. Mikhailov (eds), *The Moon*, Academic Press, London.

Shoemaker, E.M., Hackman, R. and Eggleton, R. (1962). Interplanetary correlation of geologic time. *Advances in Astronautical Sciences*, **8**, 70–89.

Sill, G.T. (1973). Sulfuric acid in the clouds of Venus. *Communications of the Lunar and Planetary Laboratory*, **9**, 191.

Sinton, W.M. (1957). Spectroscopic evidence of vegetation on Mars. *Astrophysical Journal*, **126**, 231–259.

Smith, B.A. (1967). Rotation of Venus, continuing contradictions. *Science*, **158**, 114.

Urey, H.C. (1952). *The Planets: Their Origin and Development*, University of Chicago Press.

Urey, H.C. (1962). Origin and history of the moon. In *Physics and Astronomy of the Moon*, Academic Press, New York.

Wildt, R. (1934). Ozon und Sauerstoff in den Planetenatmosphären. *Nach. Ges. Wiss. Göttingen*, **1**, 1–9.

Wildt, R. (1940). On the chemistry of the atmosphere of Venus. *Astrophysical Journal*, **91**, 266–268.

Wilhelms, D.E. (1993). *To a Rocky Moon*, University of Arizona Press, Tucson.

Whipple, F. (1950). A comet model: I. The acceleration of Comet Encke. *Astrophysical Journal*, **111**, 375–394.

Whipple, F. (1951). A comet model: II. Physical relations for comets and meteors. *Astrophysical Journal*, **113**, 464–474.

Young, A.T. (1973). Are the clouds of Venus sulfuric acid? *Icarus*, **18**, 564.

Zohar, S., Goldstein, R. and Rumsey, H. (1980). A new radar determination of the spin vector of Venus. *Astronomical Journal*, **85**, 1103–1111.

11

JAMES R. ARNOLD*

The Moon before Apollo

When Galileo first turned his telescope toward the sky, he made two major discoveries. One, the four large moons of Jupiter, is justly famous. The other, less familiar, is perhaps more important historically. He saw the Moon with enough resolution to conclude that it is not a 'heavenly body' as that term was understood, made of perfect and everlasting heavenly stuff, but a rough, cratered object more like the Earth – a real world rather than a figment of the human imagination. Its study moved from the field of theology to that of natural philosophy, now called science.

We are approaching the 400th anniversary of that important milestone in intellectual history. I give here a brief account of lunar studies before the first Apollo landing, in 1969. For this period I owe a great debt to Kopal (1969), to which the reader is referred for a more complete and authoritative history. Our knowledge of the Moon up to the modern era divides itself naturally, I believe, into two strands. The first to develop historically was a series of successively refined maps of its visible surface. The second was the growth of understanding of the orbital motion of the Moon, and of its physical nature.

Galileo's first published maps (Kopal 1969, p. 226) were very crude. No known features can be identified on them. The first maps showing features recognizable to the modern student were those of M.F. van Langren in 1649 (Kopal 1969, pp. 228–9), quickly followed by others. The maps of G.B. Riccioli and his colleagues (Kopal 1969, p. 235) show Copernicus and other big craters clearly for the first time. More importantly, most of the feature names given by Riccioli still survive today. He seems to have been responsible for calling the low-lying, dark, relatively smooth areas *maria* or 'seas', names which are still attached to these huge basaltic flows. Although they are low-lying and rather flat, this name was not well chosen. 'Deserts' might have been better, 'lava plains' better still. In this period also the moon's libration (see below), causing somewhat more than half the lunar surface to be visible, was first noticed and used.

The next step in quantitative precision came with Tobias Mayer (Kopal 1969, pp. 241–2), who in 1750 recorded rather accurate coordinates for 23 reference points on the lunar surface. Even so, actual maps using these data and new observations of similar precision did not appear until almost a century later (Beer and Mädler 1837). In Jules Verne's deservedly famous science fiction novels *De la Terre à la Lune* (1865) and *Auture de la Lune* (1870), it was Beer and Mädler's maps that were brought to lunar orbit by his three intrepid astronauts. They remained the standard well into the twentieth century. The only notable advance in this period was in the measurement of altitudes of various features using shadow lengths in early lunar 'morning' and late lunar 'afternoon'.

The second area of research which was within the means of students of the Moon before the space age had to do with its large-scale structure and dynamics. These generally require much more precise quantitative study than the mapping of the surface, and so were slower to develop as areas of research. It is worth noting, however, that Newton's calculation of the gravitational attraction of the Moon to the Earth, by his own account 'agreeing pretty nearly' with an inverse square law of attraction, was a critical early step in his development of his theory of gravitation.

The study of the Moon's orbit may be said to have begun with G.D. Cassini in the mid-seventeenth century. Reasonably precise values of the Moon's orbital motions, including the inclination of its orbital plane to the ecliptic plane (about 5.14°), the inclination I of its axis of rotation to its orbital plane (about 1.53°), the semimajor axis, the eccentricity, and quite a good ephemeris of its motion were determined by Lagrange and Laplace in the late eighteenth century.

---
* University of California San Diego, La Jolla, CA, USA

Darwin (1880) and some of his contemporaries opened up a new scientific frontier by calculating the history of the Moon's orbit, seeking light on the question of its origin. He concluded that the moon has been and is receding from the Earth, due to dissipative tidal forces exerted on the Earth by the Moon. This, he said, should cause the Earth's rotation on its axis to slow over time as a consequence of conservation of angular momentum. Had the Moon ever been as close to the Earth as a few Earth radii, the length of our day would have been only a few hours. This analysis has proven robust.

By the early twentieth century three possible scenarios for the origin of the moon had been proposed and explored to the limited degree possible with the meager data available. What we may call Darwin's (1880) 'fission hypothesis' held that the Moon had separated from the Earth when the Proto-Earth's rapid rotation caused it to deform and become unstable. This idea had been embellished somewhat, and the Pacific Ocean had been suggested as the scar of the fission process. An Earth–Moon 'double planet' scheme, analogous to the formation of double stars, had its advocates. Finally, a capture model, in which the Moon was thought of as a small planet born elsewhere in the Solar System, and later captured in a close passage by the Earth, was seen as a possibility. While many papers on lunar origin were published over the decades, all three of these models continued to be advocated up to and even after the Apollo missions. They were widely considered to be the only three possibilities. It seems now that nature knew better, but that is another story.

While the fact that the Moon's shape is not quite spherical was first deduced by Laplace, the first thorough study was done by Jeffreys (1924). The Moon was shown to have a bulge pointed toward the Earth, accounting for its synchronous rotation. This was consistent at that time with a model in which the bulge was caused by tides raised on a hot early Moon when it was much closer to the Earth. However, the situation is not so simple. Moments of inertia of the Moon around three mutually perpendicular axes can be compared using observations made from Earth: the one called A, pointing toward the center of the Earth; C, close to the rotation axis; and B, perpendicular to the other two. They are all slightly different. $(C-A)/B$ is the best-determined ratio, given by Urey (1952) as 0.000629. The less certain values of $C-A$ and $C-B$ are certainly different from each other, so that to a first approximation the Moon is a triaxial ellipsoid. This was already known by the time of Urey's book as inconsistent with a simple tidal model.

The ellipticity of the lunar orbit, and the inclination of its axis of rotation, combined with the parallax due to the Earth's finite size, give rise to the optical librations, or apparent rocking, of the lunar surface in latitude and longitude. Thus we can see more than half of the Moon's surface, in fact about 59%, at one time or another. There is also a physical libration, or actual rocking of the surface, but it is very much smaller.

We are ready, I think, to move on to the modern era. I choose to start the modern era as Harold Urey, the leading student of the moon 'before Apollo' always did, with the publication of a remarkable book by Ralph Baldwin (1949). In the brief span of 238 pages, Baldwin covered a lot of territory. First he reviewed the history of lunar studies to that point, with an excellent bibliography. Then he gave arguments, clear and convincing to Urey and others able to understand, for some key conclusions about the Moon. Not that he stopped the arguments! As it became apparent that they might soon be settled by direct observation, they in fact tended to grow more intense.

The most interesting of debates was over the mode of origin of the abundant lunar craters, and of the circular maria, such as Mare Imbrium and the 'bull's eye' of Mare Orientale. Baldwin (1949) summarized the nineteenth-century literature on this topic, pointing out that the generally accepted view, then and later, was that these features were volcanic. The other chief idea, that they were formed by the impact of stray bodies, had been proposed quite early, but was not taken seriously – it was even considered a curiosity – by most students of the subject.

The one clear statement of the case for impacts before Baldwin's was given by a geologist famous in his time, G.K. Gilbert (1893). Baldwin (1949) put forward three simple points. First, almost no lunar craters had anything resembling the conical outline familiar in terrestrial volcanoes. There are only a few small, inconspicuous domes. Second, the volume of the crater walls was always comparable, within the rather large errors associated with telescopic observations, to the volume of the central hole; averaged over many craters, the agreement was close. Finally, while the larger craters usually show central peaks, in no case does a central peak rise above the level of the surrounding terrain.

The discussions of this and other questions were interrupted by a major event: the launch of Sputnik 1 on 4 October 1957, followed on 3 November by the larger and more ambitious Sputnik 2. On 31 January 1958 the USA responded by successfully launching the very small but capable Explorer 1, which yielded James Van Allen's discovery of the Earth's radiation belts. It was immediately obvious that great new possibilities for space research were about to be realized. For planetary scientists the Moon was the inevitable first target.

Following the establishment of NASA in 1958, attention was given to the possibility of one or more lunar missions. Urey was particularly effective in advocating such missions. The first NASA committee to explore the subject was appointed early in 1959, with him as its most prominent member (and the author as a new recruit). What might be called 'the first Moon race' then began. The most obvious goal was to produce images of the then still invisible farside.

The mare named Moscoviensis and the large, dark-floored crater named Tsiolkovsky will allow younger readers to guess or remember who won that first race. Indeed, the Soviet Union held the lead for a long time. After two mission failures, their Luna 3 mission flew by the Moon on 8 October 1959 and obtained images of its hidden side, marking the first successful mission to another planetary body. The pictures released were not very clear. This allowed the illusion to persist in the West for a while longer that the Soviet space probes were of poor quality, or even that the images were fraudulent. A scientist working with a US intelligence agency undertook to evaluate the case. He gave a very entertaining talk on the results. His conclusion was that the pictures were retouched 'by experts, as good as ours' before publication. He was sure that the Soviet scientists actually had 'genuine pictures better than the ones they released.' The better pictures obtained later in fact confirmed their results. They could and did name the big features. The most notable difference between the nearside and the farside was the near-absence of maria on the latter.

NASA's plans for lunar missions took some time to develop. They were overtaken by President Kennedy's speech to Congress in May 1961, announcing to the world that he was committing the US Government to 'placing a man on the Moon in this decade and returning him safely to the Earth.' All plans were restudied and, where appropriate, revised. A manned flight on such a schedule required not only the development of advanced capabilities at an accelerated pace, but also their testing by precursor missions of unprecedented scope and complexity.

The new objective led to the announcement of three sets of unmanned NASA missions to the Moon. The first, the Ranger series, was at first largely unchanged from earlier plans. It included a series of impacts on the Moon's visible face, with both scientific and reconnaissance objectives. The second, the Surveyor series, was to land small instrument-bearing spacecraft safely on the surface, again with multiple tasks. Both these programs were to be managed by the Jet Propulsion Laboratory in Pasadena, California. The third, the Lunar Orbiter series, was to produce a photographic map of the Moon, nearside and farside, as complete and detailed as possible. The goal was mainly to support the manned missions, but the data set would also be available for science. Responsibility for this was assigned to the Langley Research Center in Hampton, Virginia.

It was a bumpy road we started down then. There were numerous failures while scientists, engineers, and managers learned their new roles. What is important here is not the race, but what was learned and how that led to the remarkable scientific advances of the Apollo period.

The first NASA lunar mission program was named Ranger. It began badly. Rangers 1 and 2 were engineering test missions – both failed. Then missions 3–5, carrying cameras and scientific instruments and intended to approach and impact the visible face of the moon, also failed (in a different way each time.) However, the experience gained seems to have been useful. Rangers 6–9, launched in 1964–65, were redesigned to carry cameras only. These four succeeded in producing the first close-up images of the Moon, in the few seconds before impact.

These images were clear and in focus. The final frames showed features smaller than one meter, sometimes much smaller. Some of these were craters, not very different in appearance from the larger ones seen through Earth-based telescopes. This was no surprise to those who believed they were caused by impact of stray bodies, small fragments of asteroids or comets. They presented more difficulties to the volcano party.

One group of images displayed parts of bright rays originating from one of the Moon's youngest large craters. These were full of small pits which were elongated in a common direction, appearing quite unlike the large primary craters seen through Earth-based telescopes. There could be little doubt that they were secondary impact features, formed as debris from the major crater-forming event moved out radially from it at subsonic speeds. It was also possible to conclude definitely what had already been inferred from measurements of temperature versus time on the lunar surface (Baldwin 1949, p. 10, Figure 1 and accompanying discussion), namely that the Moon was not covered with bare rock. What the Ranger cameras saw was fine-grained material. In general the surface in all three image sequences was remarkably smooth. Nonetheless, landing a spacecraft without human guidance was seen to have a finite risk of failure due to striking a rock.

Meanwhile the Soviet program was continuing to produce important results and some more 'firsts'. Two long series of spacecraft, Luna and Zond, were launched throughout the 1960s and even later. Perhaps the most important was Luna 9, which was the first to soft-land on the Moon, on 3 February 1966, before the US Surveyor program had begun. It transmitted pictures back to Earth of its landing site in western Oceanus Procellarum. Luna 10, which followed, was the first spacecraft to be put in orbit around the Moon.

This Surveyor series had an ambitious aim. The spacecraft were to land softly and upright on the lunar surface, on three extended feet. They carried not only cameras but also other instruments (described briefly below). Their success rate showed impressive progress since the Ranger series. Of the seven spacecraft launched, only Surveyors 2 and 4 failed to achieve their landings and scientific objectives.

Surveyor 1 landed on 2 June 1966, at a near-equatorial western site not far from the crater Flamsteed in Oceanus Procellarum.* This spacecraft and Surveyor 3 carried vidicon TV cameras capable of producing a black-and-white TV image of 600 lines. The cameras could be rotated 360 degrees, raised and lowered, and zoomed in and out. Over 10,000 pictures of the site (and the sky) were obtained, covering a full lunar day cycle in each case.

A full report of Surveyor 1's findings is given in Jaffe (1967), with an accompanying photographic section. The coverage of surface features was of course far more detailed than that available from the Ranger pictures (or the earlier Soviet ones). One fear was laid to rest by both Luna 9 and Surveyor 1: The two spacecraft did not sink into the lunar surface (confirmed to be finely divided material, now christened 'regolith') as one scientist had warned NASA that it would. Again, the surface was seen to be generally smooth with low slopes, though there were rocks that could present serious obstacles to landing spacecraft. The principal scientific investigator was Eugene Shoemaker, in his first important role in the lunar program.

Surveyor 3 was on the whole a repeat of Surveyor 1, enhancing the database of surface images and strengthening the view that this sort of scene beheld by both craft's cameras might be typical of at least the lunar mare surfaces, which because they showed fewer large craters were preferred for the Apollo manned landings. In addition to the TV camera, Surveyor 3 carried a soil mechanics experiment, essentially a powered, fist-sized bucket with a range of capabilities. In the course of about 18 hours it carried out dozens of tests of soil properties, including bearing strengths and response to impacts. Perhaps most important, it could excavate a trench to a depth of 17 cm and display the sampled material. Shoemaker was again chiefly responsible for these experiments.

Surveyors 5, 6, and 7 landed at new and interesting spots on the nearside, respectively the prominent eastern Mare Tranquillitatis (already emerging as the most likely area for the first manned landing), Sinus Medii near the center of the Moon's face, and the northern rim of the crater Tycho in the south. The best reference for the later Surveyor missions is Jaffe (1969). Camera and soil property instruments were updated for these missions, but the most interesting new development was the inclusion of an alpha-scattering instrument for chemical analysis. This instrument was designed and built at the University of Chicago by a group headed by Anthony Turkevich (1967). It used Rutherford scattering of alpha particles produced by radioactive decay (the same process Ernest Rutherford had used to demonstrate the existence of atomic nuclei). It could identify the nuclei, and hence the elements from which the particles were scattered, using the principle of conservation of momentum. Housed in a box open at the bottom, the instrument was deposited on a flat spot on the surface. It accumulated a spectrum of energies of backscattered particles over many hours of exposure. The concentrations of all major elements (abundance greater than 1%) in the top layer of the soil it sampled were then derived from their scattered energy spectrum.

The Surveyor 5 results clearly identified the material at this mare site as a (ground up) basaltic rock, comparable in general to basalts on earth. However, it was unusual in having a high concentration of the element titanium, 6–8% by weight, which was beyond what geochemists were accustomed to on the Earth. The result, though it evoked a good deal of skepticism at the time, was fully confirmed when the Apollo 11 soil from the same area was analyzed in the laboratory. This was the first chemical analysis of the composition of an extraterrestrial material *in situ*. It was followed by quite a few others, and all the data showed clearly that the Moon is highly differentiated chemically. These and later analyses provided very useful constraints on the origin and history of the Moon.

Meanwhile, the Soviet lunar program had not been idle. A series of Luna missions and one mission called Zond returned important data and images, particularly of the hidden farside of the moon (e.g. Dolginov *et al.* 1967, Lebedinsky *et al.* 1967).

Like the other two US mission series, the Lunar Orbiters were important both as precursors of the Apollo landings, and for the database they provided. The method used to gather the data seems dated now, when CCD cameras and other advances have made to job much easier. But at the time it was ingenious, and above all it worked. So did the spacecraft, all five times. The first launch was in August 1966, the last in August 1967.

Cameras with long rolls of narrow strip film were used to take thousands of pictures of the lunar surface in a pre-planned pattern, both of whole bands of the lunar surface and of particular candidate landing areas at higher magnification. These were developed chemically to bring up the images, and then scanned with beams of light to produce streams of intensity data from which the black-and-white images could be reconstructed back on Earth. Sequences were designed to optimize lighting, so that shadows could help to bring out detail. Many months of exposures were made and recorded. Well over 90% of the lunar surface was eventually covered during the five missions, giving rise to a remarkably valuable database of both the near- and farsides.

Calculations showed that the spacecraft would remain in orbit for a long time after the imaging had been completed,

---

*The convention used then and later reversed that used by astronomers in the preceding centuries, to one in which the direction is that seen from Earth, not as before by an observer on the Moon itself. It was adopted for the convenience of the Apollo program. After Apollo the new convention was adopted by the international astronomical community. Thus, for example, Mare Orientale, which in Latin means 'Eastern Sea' is now at the western edge of the visible hemisphere.

and all data transmitted to Earth. Perhaps some of them are still there.

By the time the Surveyor and Lunar Orbiter programs were complete, the lead in the 'race' had clearly passed to the United States. This is not to say that either the technical or the public relations arena was totally one sided. The most remarkable achievements in the later Luna series were the successful robotic missions, featuring rovers and sample-return capability, in the time period of the Apollo landings.

The increase in scientific knowledge and understanding of the Moon that resulted even from the first Apollo landing mission was of course very large. Still, much of this gain represented confirmation of conclusions already widely shared in the research community. For example, the physical properties and chemical composition of the lunar surface had been basically defined by the Surveyor missions. The farside of the Moon had been mapped in detail. The qualitative jump was mainly in the returned samples and to the studies of them reported from laboratories around the world. The following is the author's list of the six most significant results. The first reports of all of them can be found in *Apollo 11 Lunar Science Conference, Science*, **167**, 449–803, 1970, a special issue of that journal.

1. *Chronology*. The absolute dates of formation of major surface features were only guesswork before Apollo. Radioactive dating of a number of returned rocks revealed that the volcanism that produced Mare Tranquillitatis, and by inference the other large maria, took place between about 3.2 and 4 billion years ago, although the Moon itself was formed much earlier. Like the Earth and the Sun, its original formation occurred around 4.5 billion years ago.
2. *Mineral and chemical composition*. Laboratory studies provided a rich extension of the first Surveyor chemical analyses, down to parts per million or less, and a picture of the complexity of the surface materials. The large percentage of glass among the fine particles was one of many surprises. It was even possible to identify fragments of highland material thrown by impacts onto this mare site, and bits of impacting meteorites.
3. *Solar wind composition*. The lunar regolith was the repository of large amounts of solar wind gases, whose chemical and isotopic signatures became fully available (though meteorites had given earlier information on the subject). Even exposure of an aluminum foil 'window shade' for a few hours allowed measurement of the flux and composition of the solar wind on that date.
4. *Isotopic information*. Isotopic differences from terrestrial materials arose from several processes, and provided insight into them. Besides the solar wind, chemical fractionation, radioactive decay, cosmic ray effects, and the presence of material from projectiles impacting the surface all played a role, and conversely opened up the study of their nature and timing.
5. *Cosmic rays and recent lunar history*. Creation of radioactive nuclides by cosmic ray bombardment had been known for decades, but the lunar samples demonstrated that highly energetic particles from the Sun could produce similar effects. These could be used both to study the rates of turnover, or 'gardening', of the lunar regolith and the history of the radiation itself.
6. *Near-absence of organic matter*. The Apollo project was required to do an immediate study to determine the presence or absence of disease-causing organisms in the returned samples. The decisive test rejecting this improbable risk was the absence of organic compounds at the 1 part per billion level. The Moon does not harbor life (yet).

In the years leading up to the Apollo landings, a number of leading workers in the field made the effort to summarize their views on the origin of the Moon, or of the Earth–Moon system. The goal was to test them against the coming flood of data. In retrospect the outcome was not surprising. Three of those whose ideas had reached a wider public were Harold Urey, Gerard Kuiper, and Eugene Shoemaker. At the first press conference after the preliminary look at the returned samples, a group of us were asked repeatedly by reporters 'Which of the three was right?' The correct answer was, 'None of the above,' though in the presence of these scientists it was a little hard to say so bluntly. It is an inevitable consequence of new scientific knowledge that old ideas require extensive revision. This process takes time. Still, it is wise to pay tribute to these and other leaders who brought our understanding to that time and place.

REFERENCES

Baldwin, R.K. (1948). *The Face of the Moon*, University of Chicago Press.
Beer, W. and Mädler, J.H. (1837). *Der Mond*, Simon Schropp, Berlin.
Darwin, G.H. (1880). *Philosophical Transactions of the Royal Society*, **171**, 713–891.
Dolginov, S.S., Yeroshenko, Y.G., Zhuzgov, L.L. and Zhulin, I.A. (1967). A study of the magnetic field of the Moon. *Geomagnetizm I Aeronomiya*, **7**, 436.
Gilbert, G.K. (1893). The Moon's face, *Bulletin of the Philosophical Society of Washington*, **12**, 241.
Jaffe, L.D. (1967). Lunar surface exploration by Surveyor spacecraft. *Journal of Geophysical Research*, **72**, 773–778.
Jaffe, L.D. (1969). The Surveyor lunar landings. *Science*, **164**, 774–788.
Jeffreys, H. (1924). *The Earth*, Cambridge University Press.
Kopal, Z. (1969). *The Moon*, Reidel, Dordrecht.
Lebedinsky, A.I., Krasnopolsky, V.A., Krysko, A.A., Aleshin, G.M., Iorenas, V.A., Selivanos, A.M. and Zasetsky, V.V. (1967). In A. Dollfus (ed.), *Moon and Planets* (Proc. 7th COSPAR Symp.), North-Holland, Amsterdam, pp. 59–70.
Turkevich, A., Franzgrote, E.J. and Patterson, J.H. (1967). Chemical composition of the lunar surface. *Science*, **158**, 635.
Urey, H. (1952). *The Planets*, Yale University Press, New Haven, CT.

12

HERBERT FRIEDMAN[†]

From the ionosphere to high energy astronomy – a personal experience

In the first half of the 20th century, recognition of the existence of an electrified layer of the upper atmosphere that provided a mode for radio communication over great distances on the Earth grew with little sense of the role of invisible solar radiation in creating the electrical mirror. Solar radiation was thought to be black-body in spectral shape characterized by a temperature of 6000 K with a maximum in the yellow-green, trailing off rapidly in the infrared and ultraviolet. None of this spectrum could produce significant ionization of the major constituents of the atmosphere.

X-rays were discovered by Wilhelm Röntgen shortly before the turn of the century and Guglielmo Marconi, a few years later, demonstrated trans-Atlantic radio communication via a high altitude, natural electrical mirror. Several ingenious physicists and electrical engineers pursued the problems of radio reflection but had few clues to the nature of the solar radiation that keyed the phenomena of ionospheric production and variability. The true nature of solar ionizing radiation remained a baffling puzzle until after WW II when the availability of rockets to carry detectors directly into the ionosphere finally made the studies definitively diagnostic. Captured German V-2 (Vengeance) rockets while being studied by propulsion engineers were also turned from "weapons into plowshares", when they were adapted to ionospheric studies.

To preface the story of the modern era it is interesting to sketch the early ideas of radio propagation science. The perceptions of solar-terrestrial connections had developed slowly over most of the 19th century. Lord Kelvin, one of the most influential physicists of his time, was adamant in rejecting any notion (Kelvin 1892) "that terrestrial magnetic storms are due to the magnetic action of the Sun; or to any dynamic action within the Sun, or in connection with hurricanes in his atmosphere, or anywhere near the Sun outside." Furthermore he held strongly "that the connection between magnetic storms and sunspots is unreal and the seeming agreement between the periods has been mere coincidence." Kelvin's enormous prestige discouraged any dispute and set back solar-terrestrial research for decades.

Toward the end of the 19th century Colonel Sabine of the British army monitored a network of magnetic observatories throughout the empire and noted that by "a most curious coincidence" the magnitude and frequency of magnetic disturbances was synchronized with the appearance and disappearance of sunspots. The direction of the compass needle swung in regular fashion over the diurnal cycle, but at times the movements became more intense and rapid in the auroral zone, giving rise to the name "magnetic storms". It was commonly believed that interplanetary space was a vacuum and that auroras were excited by direct streams of particles from Sun to Earth unimpeded by any interplanetary medium. By timing the appearances of auroras and magnetic storms relative to the visible outbursts of flares on the Sun, the travel speed was calculated to be about 800 km per sec, slower than light, but consistent with concepts of particle streams.

Friedrich Gauss, the great German mathematician-physicist-astronomer, as early as 1839, related fluctuations in the compass needle to the passage of electric currents at high altitudes. In 1882, the Scotsman, Balfour Stewart, defined these currents as a great dynamo of tidal movements of ionized air above 100 km that were driven by solar heating. The vertical movement was only 2 or 3 km, but it was sufficient to generate a great horizontal current sheet of electricity. Observations near the geomagnetic equator indicated circulating systems of electric currents of opposite symmetries in the northern and southern hemispheres. At

---

[†] Naval Research Laboratory, Washington, DC, USA. Herbert Friedman died on 9 September 2000. The editors are indebted to Herbert Gursky for helping them to put the paper and its title into the final form

the equator the currents joined to form a strong flow from west to east at 1100 local time, that Sidney Chapman, in the early 20th century, named the Equatorial Electrojet.

The young Guglielmo Marconi at age 21, in 1895, built a demonstration wireless telegraph on his father's estate near Bologna, Italy. On December 12, 1901 he transmitted a simple Morse code signal from England to Newfoundland, a distance of 2900 km to the astonishment of most scientists who could not fathom how the waves, that were thought to travel in straight lines, could curve over the 160 km high bulge of the surface of the earth. In 1902, Arthur E. Kennelly proposed that the radio waves were ducted around the earth by an electrically conducting layer. Almost simultaneously, Oliver Heaviside reached the same conclusion and the layer came to be called the Heaviside layer. With the above background, the stage was now set for a more focused scientific attack on the nature of a reflecting layer, now called the ionosphere, that eventually came to be associated with solar X-rays. In England, ionospheric research was lead by Edward Appleton and his student Miles Barnett [1925]; in the United States studies were conducted by E.O. Hulburt and A. Hoyt Taylor at the Naval Research Laboratory [1926] and by Gregory Breit and Merle A. Tuve at the Carnegie Institution [1926]. Successful experiments and interpretations were achieved almost simultaneously in England and the United States, both groups working from the ground. When German V-2 rockets were captured by the Americans, they moved ionospheric research into space and outraced the ground-based competition.

Appleton set out to determine the height of reflection of a continuous wave with the cooperation of the British Broadcasting Co. whom he persuaded to provide him with a continuously varying signal from London at the end of the broadcast day so that he could detect the interference pattern of ground and sky waves at Oxford. He observed the elapsed time between emission and reception of the same frequency, as the broadcast frequency was oscillated back and forth. Starting with low frequencies, Appleton probed only the lower portion of the reflecting layer; working later with higher frequencies, he distinguished layered regions of reflection, that he labeled D, E, and F. For these experiments, Appleton later received the Nobel Prize.

Much of the early research by NRL scientists was characterized by an admirable simplicity and economy of means. Hulburt and Taylor cooped the partnership of radio amateurs around the world who used vacuum tubes with power outputs of less than 50 watts and very short radio waves, less than 200 meters, to communicate around the world. Transmissions skipped over a "zone of silence" encircling the transmitter to a distance of 30 to 50 km and at the same time were received out to distances of hundreds of kilometers. Hulburt and Taylor showed that the waves were reflected only when the angle of incidence exceeded a critical value. At smaller angles the waves penetrated the reflecting region and escaped into space. At night, skip distances were greater than during the day and greater in winter than in summer in temperate latitudes. From these simple observations Hulburt calculated the height of reflection and the electron density (about 500,000 electrons and ions per cm^3 at a reflection height of about 150 km). By 1926, Hulburt and Taylor were able to publish a remarkably accurate account of the diurnal variation of ionospheric electron density. The work was almost coincident with Appleton's and Barnett's 1924–25 studies.

In the early years of ionospheric research, theorists speculated about particle radiation as a possible source of ionization of the upper atmosphere. Confronted with an apparent 6000 K solar spectrum it was not possible to model interactions of electromagnetic radiation with atmospheric constituents that would lead to the required ionization. Within the space of a decade after the end of WW II, however, the solar spectrum was revealed from its X-ray limit throughout the ultraviolet with instruments carried on rockets to ionospheric height. Ionizing radiation was observed in X-rays and ultraviolet from the solar corona and chromosphere, where temperatures range from hundreds of thousands to millions of degrees K, and every spectral interval was matched with its absorption at a particular height range.

Throughout this early epoch, some theorists still favored solar particles as the source of ionizing radiation. Confronted with the apparent 6000 K temperature of the solar disk (Figure 1) it wasn't possible to model atmospheric interactions with solar radiation that would lead to the required ionization. At the higher chromospheric and coronal temperatures shorter-wavelength extreme ultraviolet and X-rays would be produced but the particle concentrations

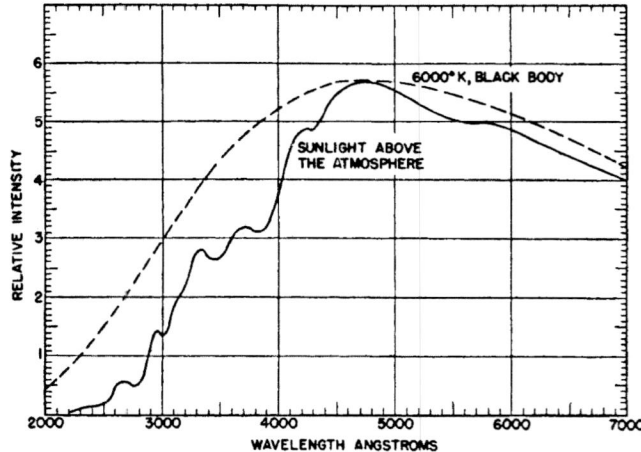

Figure 1 Solar spectral energy distribution from 2000 A to 7000 A compared with 6000 K black body sunlight above the atmosphere.

in the solar atmosphere were estimated to be too thin (emission measure too low) to provide high enough intensities to generate an E or F region. Appleton's early measurements of vertical reflections between midnight and sunrise showed that substantial concentrations of electrons or ions persisted throughout the night, contrary to the prevailing idea that charges would disappear rapidly by attachment once the source of ionization was removed. Radio scientists were thus led to believe that an important portion of the ionization might be produced by corpuscles arriving equally by both day and night. Because a charged particle stream could not readily penetrate the earth's magnetic field, serious thought was given to neutral particle streams.

In 1928, Hulburt proposed that ultraviolet radiation shortward of 1230 Å might be the source of the ionosphere. By 1930 he was intrigued by a possible connection between solar ultraviolet and sunspots and their link to magnetic storms and radio fadeout. In 1935, J.H. Dellinger at the U.S. National Bureau of Standards (NBS), summarized observations of a series of sudden ionospheric disturbances over a period of six months and stressed the importance of understanding the connection with solar activity. He proposed a joint effort of the NBS and the solar observatory on Mt. Wilson. During the same time frame, Robert H. Goddard was developing his rocket to carry instruments to high altitudes for atmospheric research. Correspondence between Hulburt and John Fleming reveals that Hulburt contemplated solar rocket astronomy to understand the basic physics of solar control of the ionization. He noted the theoretical match between atmospheric absorption of solar soft X-rays and the altitude of ionization and suggested that a good test would be to fly photographic film covered with thin aluminum foil or black paper in one of Goddard's rockets to detect X-rays.

Those early glimmerings of high altitude research with rockets were interrupted by the war but the new technologies of the war were soon transferred to peaceful research. In 1942, Ernst Krause in the radio division at NRL undertook to develop a program of guided missiles, specifically a new version of the German V-1 buzz bomb known as the JB-2 and the Lark, a rocket-propelled, guided ship-to-air missile. At the end of the war, Krause pursuaded NRL to commit to a substantial effort in rocket development for high altitude research which led to the resurrection of V-2 rockets late in the 1940s (see Figure 2). The first generation of successful studies of solar X-rays and extreme ultraviolet radiation began in 1949 when my NRL group flew a set of Geiger counters sensitive to a narrow band of X-rays centered at about 8 A, hydrogen Lyman-alpha (1216 A), and the Schumann region, 1425 to 1600 A. As the rocket climbed to an altitude of 150 km, the detectors pointing normal to the spin axis swept the sky repeatedly. X-rays were detected above 80 km with increasing intensity to

Figure 2 A V-2 rocket just prior to launch at the White Sands Proving Ground in New Mexico. About 45 feet tall and 5 feet in diameter it was fueled by 10 tons of alcohol and liquid oxygen. The rocket is shown connected by an umbilical cable to the firing line. To service the rocket and its payload a portable ladder was brought up. Only later was a gantry provided from which each level of the rocket could be reached comfortably. A successful flight could reach 170 km and last for 450 sec. (NRL.)

about 120 km (Figure 3). It appeared that a thermal corona at one to two million deg C made a good fit with the ionization requirement of the E-region.

The Lyman alpha detector and the extreme ultraviolet detector showed how those radiations shaped the bottom of the ionosphere and the upper part of the reflecting E-region. Lyman alpha, originating in the hot solar chromosphere at 10,000 K and higher, contains most of the energy in the extreme ultraviolet and is absorbed between 75 and 90 km but does not interact with any of the major constituents, oxygen or nitrogen, atomic or molecular. Only later on did M. Nicolet, the brilliant Belgian atmospheric scientist point out that it could ionize nitric oxide, a trace constituent present at only 10^8 molecules per cm^{-3}. with almost 100% efficiency and thus have control of D-region. Radiation in the Schumann region produced no ionization but played a very important role in shaping the high ionosphere by dissociating molecular oxygen. By the process of dissociative

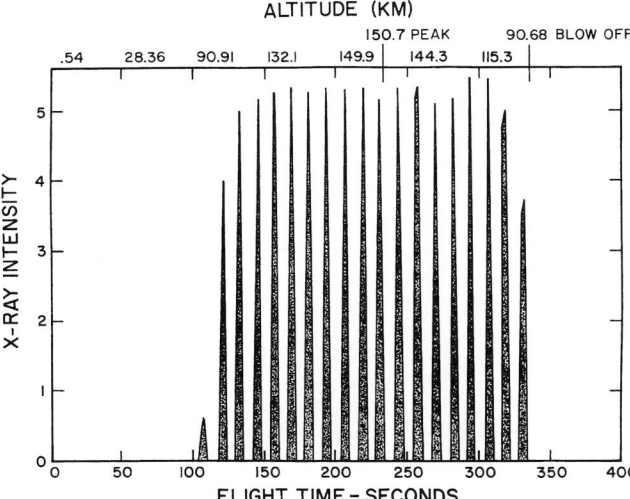

Figure 3 The first measurement of the penetration of solar X-rays into the upper atmosphere made by a V-2 rocket in 1949. The 8 Å X-ray signal was modulated by the spin of the rocket as the Sun came into view once each roll period. X-rays were first detected at about 90 kilometers and reached peak intensity at about 130 km. (U.S. Naval Research Laboratory.)

recombination, molecular oxygen controls the rate of neutralization of F-region electron density much more effectively than atomic oxygen.

The 1949 measurement of harder X-rays (1–8 Å) led to several years of broad band photometry of solar X-rays that extended the range of the spectrum, primarily with the aid of simple filters of beryllium, aluminum, titanium, mylar, formvar, etc. serving as the window materials of the photon counters. It seems in retrospect that the NRL group was almost alone in that decade of pioneering studies of the Sun. The X-ray spectral distribution from 1 Å to 44 Å resembled thermal emission from a thin corona at a temperature of a few million degrees. Successive measurements at intervals of months to years showed flux variations of as much as a factor of 7 for X-rays (8–20 Å) over the sunspot cycle. Such variability was consistent with ionospheric electron density variations in the E-region, supporting a direct connection between solar X-rays and E-region. But the observed variability over a solar cycle made it clear that the concept of X-ray emission from a spherically symmetrical solar corona was very inadequate. Instead it seemed that the corona was structured in condensations, formed over sunspots, that produced enhanced X-ray emission. To resolve the question of spatial origin would require an X-ray scan of the solar disk or an X-ray photograph. Both methods were successfully applied at the end of the decade. The X-ray photograph was obtained with primitive pinhole photography (Figure 6); the scan required a very special combination of a total solar eclipse and an array of rockets launched from the deck of a ship.

SOLAR FLARES

Of all the forms of solar activity, flares are the most spectacular. A solar flare creates a strong impact on the terrestrial environment, producing prompt shortwave radio blackout that may last for two to three hours. The aftereffects may persist for one or two days in the form of great auroral displays, and ionospheric and magnetic storms that seriously degrade shortwave radio communications.

The new arsenal of rockets that became available late in the 1950s made it possible to plan a program of solar flare studies. Although the supply of V-2 rockets was exhausted by 1952 it was replaced by smaller Aerobees and two staged rockets that mated the Deacon with a Nike booster or a Skyhook balloon. The latter combinations were particularly attractive for studies that required a form of instant rocketry. Launch from shipboard at sea offered range safety. By sailing downwind the ship could achieve nearly zero relative wind conditions for inflation and release of the balloon with its suspended rocket. The well deck aboard the U.S.S. Colonial measured 392 feet by 41 feet which we could use to store three trailer-truckloads of helium while the broad helicopter deck above served admirably for the balloon operations. The lumbering ship could make a speed of 15 knots which was slower than the expected drift of the balloon at altitude. To assure radio contact with the balloon payload we were assigned a destroyer, the U.S.S. Perkins that could track the balloon with a speed of 28 knots. Finally a crew of 650 sailors was tasked to man the ship for the naval chase.

Each day as inflation began, the polyethylene balloon, most of it draped in tight folds resembling the stem of an onion, rose 100 feet above the deck, crowned by a 20 foot bulge filled with 5000 cubic feet of helium that would expand further to thirty times that volume at altitude. The 12-feet long Deacon rocket dangled at the end of a 100-foot nylon line (Figure 4). At 80,000 feet, when a flare was observed in visible light, a radio command would fire the rocket and send it upward, piercing the balloon and rushing ahead another 50 or 60 miles through the ionosphere.

An NRL proposal for a naval expedition as part of the International Geophysical Year (IGY) to launch ten Rockoons for solar flare studies with the support of the USS Colonial, an LSD with a large helicopter deck, was approved by the Office of Naval Research. Our ship was a sea-going drydock. The operational plan was to release a Rockoon each morning on ten successive days and allow it to float at 80,000 feet until the onset of a solar flare was detected, when it would be fired. One flare was successfully observed

Figure 4 A Deacon rocket suspended from a Skyhook balloon rises from the helicopter deck of the U.S.S. Colonial. The solid propelllant rocket was 14 feet long and 6 inches in diameter. Floating at 25 km, the rocket could be fired by radio command the moment a flare was detected. No orientation system was carried on the rockoon. During balloon inflation, surface winds had to be less than 10 knots. Cruising downwind the ship could attain nearly zero relative wind condition and the balloon would stand straight up with its suspended payload. (NRL.)

and the X-ray detector confirmed that [1 Å] X-rays produced the required ionization of the base of D-region near 70 km altitude.

Astronomers were disbelieving, so wedded were they to the idea of ultraviolet rays of far higher intensity than the flare X-rays. Because the visible manifestation of a flare is so clear in the hydrogen red line it led most solar physicists to seek a connection between the hydrogen resonance line at 1216 Å, which is responsible for the normal D-region and the increased ionization that is produced by the flare. But a simple analysis of the required radiation enhancements rule out the ultraviolet possibilities. The increase of electron density that accounts for shortwave fadeout during a class 3 flare would require that the ionizing radiation increase by roughly a factor of fifty from the small region of the flare. Such an increase is virtually impossible astrophysically.

Within a year Rockoons were out of style and the preferred launcher was a hybrid two-stage Nike booster plus Deacon rocket. The Nike Booster lifted the Deacon to the stratosphere from where it was fired to rise into the ionosphere. The launch platform was a simple rail that the NRL team set up on San Nicholas Island off Pt. Mugu, California. In 1958 and '59 a series of these rockets was launched at times of solar flares and proved that sudden ionospheric disturbances were triggered by hard X-rays from flares associated with thermal regimes of 150 million K or non-thermal sources of equivalent energies. These flare observations were the first clues to electromagnetic radiations of suprathermal energies from natural sources.

IGY SOLAR ECLIPSE OF 1958

The eclipse of October 10th 1958 presented an opportunity to scan the Sun's distribution of active regions by using the masking effect of the Moon to scan across sections of the solar disk. The eclipse would be total for as long as 6 minutes over the South Seas. In its path across the Pacific Ocean, the eclipse crossed only one piece of land, a group of coral atolls, the Danger Islands, about 400 miles southwest of Samoa. The NRL rocket team was joined by a group of optical astronomers. Our base of operations was the LSD USS Pt. Defiance. The plan was for Jack Evans, leader of the optical astronomers to have his group set up instruments on the Island of Puka Puka and the NRL team to erect six Nike-Asp rockets on the helicopter deck on simple rail launchers. The Navy provided a marine demolition team that worked for weeks to blast a channel through the coral reef that surrounded the island in order to get the heavy astronomical instruments ashore. Aboard ship we had to improvise arrangements to keep our instruments dry and functional. Never before had a barrage of six high altitude rockets been fired from such close spacing aboard ship in the course of an hour (see Figure 5). On eclipse day it rained precisely at totality and the astronomers on Puka Puka were washed out. Just twenty miles away aboard ship we had blue skies and a perfect view of the eclipse from shipboard although it was immaterial to the rockets that rose high above the weather. The rocket observations were excellent and revealed clearly that the X-ray emission came from localized condensations of the corona.

Within two years of the eclipse expedition, the NRL group flew the simplest kind of pinhole camera on an

Figure 5 Six Asp rockets mounted on Nike boosters aboard the helicopter deck of the U.S. Point Defiance in the South Pacific to be fired during the passage of the Moon across the face of the Sun during the eclipse of 1958, observed from Puka Puka in the South Pacific. The observations showed that X-ray emission was concentrated over coronal condensations and that 13% of the X-ray corona was still exposed at totality. (NRL.)

Figure 6 First X-ray photograph of the Sun. The camera was a cylinder the size of a cigarette with a pinhole aperture covered by aluminum foil on the front end and the photographic film mounted at the back end. During the exposure rotation of the rocket smeared the image in an arc of about 60 degrees. Comparison with radioheliograph data showed a one to one correspondence. In spite of the blurring it was clear that about 80% of the X-rays came from localized condensations covering no more that 5% of the disk. (NRL.)

Aerobee rocket from White Sands and obtained the first X-ray photograph of the Sun (see Figure 6). It showed almost all the detail we had deduced from the eclipse experiment with almost trivial effort, but of course without the high adventure of a Polynesian trip. Eighty percent of the X-ray intensity emerged from only five percent of the area of the disk. The brightest coronal condensation was eighty times as intense as the general background of the disk. In spite of the poor resolution of the cameras it could be deduced that the X-rays came from below 40,000 km in the corona. A 9.1 cm radioheliogram, imaged the same day, correlated closely with X-ray details. Within a few years reflecting X-ray telescopes were flown that revealed X-ray detail as fine as optical photographs.

X-RAYS OF NON-SOLAR ORIGIN

Throughout the decade of the 1950s the NRL group was tantalized by the prospect of expanding X-ray astronomy to the galaxy but it appeared we would need detectors millions of times as sensitive as those then available to us. However in the course of rocket flights to extend the range of X-ray flare studies to higher energy X-rays my colleagues and I thought we might have detected a background of X-rays from beyond the solar system. At the time I was aware of only one theoretical prediction that perhaps fit the observation. In 1958, T. Gold and F. Hoyle held to the theory of a steady state universe in which neutrons were continuously created to fill in the void of cosmic expansion. Decay of the neutrons would provide a hot intergalactic gas of electrons and protons at a temperature of the order of a billion K that could recombine to produce a general background of X-ray bremsstrahlung.

On a visit with Fred Hoyle at Cambridge University in 1958, I discussed the possibility that our Rockoon experiments of 1956 had indeed shown evidence of extra-solar system X-rays (Figure 7). We found a discrepancy of more than an order of magnitude between the implied flux from the NRL observations and Hoyle's calculation of an X-ray background. Since we had no alternative at the time for a general X-ray background, I was as disappointed as Hoyle.

At the AGARD conference in Paris in May 1959, I offered under the title "Bremsstrahlung from the Van Allen Belts", a discussion of the results of the 1956 and 1957 attempts to observe background X-radiation above the atmosphere in the energy range from 20 to 300 keV with a scintillation counter. Van Allen cautioned against any suggestion of extra-solar system background and preferred bremsstrahlung from energetic particles in the radiation belts. My NRL colleague, Philip Mange, made a rough calculation of X-rays from the Van Allen Belts and derived a flux about two orders of magnitude smaller than indicated by our rocket observations. I summarized possible interpretations of our observations as follows:

> It is planned to repeat these measurements at night. If the flux of 20–50 kev radiation is missing at night then its source in the daytime must be the sun. If it persists at night, then it may imply a much higher content of lower energy electrons in the Van Allen Belt than was assumed by P. Mange... Still another possible source of the observed radiation may be the bremsstrahlung of hot gas clouds in intergalactic space as described in Hoyle's theory of the condensation of galaxies.

Other speculations about the possibility of detectable X-ray emission from galactic sources appeared as follows:

> Rocket astronomy has not yet undertaken the observation of celestial sources in the X-ray spectrum. Cosmic ray sources like the Crab Nebula, however, have high priority in experiments now being designed and instrumented.

With such comments about the promise of a new decade of the 1960s for X-ray astronomy, rocket astronomers poised for a strong push to develop studies of cosmic X-ray sources and success came rapidly. Giacconi et al. (1962), identified a poorly positioned positive signal from an Aerobee rocket which came from the general vicinity of the galactic center (see Figure 8). The NRL group, within a year, established the position with an accuracy of about 3 deg in the direction of the constellation Scorpius. A weaker signal about 1/10th as strong in the NRL scan was associated with the Crab Nebula. Thus was galactic X-ray astronomy inaugurated. Rossi and Giacconi and their team from American Science and Engineering Co. (AS&E) launched the UHURU X-ray satellite and immediately discovered a sky full of X-ray sources. X-ray astronomy had become a full-fledged component of cosmic studies.

To summarize the first decade of X-ray astronomy we can cite the major achievements of rocket astronomy and the UHURU Satellite. NRL produced a catalogue of about three dozen X-ray sources that included extragalactic detections of the quasar 3C-273 and the giant galaxies, M-87 and NGC1275. These observations proved that X-ray stars and galaxies could be orders of magnitude more luminous than their optical counterparts. Many of the X-ray sources radiated 1000 times as much energy in X-rays as in visible light. UHURU was launched from the San Marco platform, an old oil rig three miles off the coast of Kenya on 12 Dec. 1970. Giacconi led the mission which carried an improved rocket class payload designed by the AS&E group. In somewhat more than two years of operating life it produced a catalogue of 339 discrete sources and led to the discovery

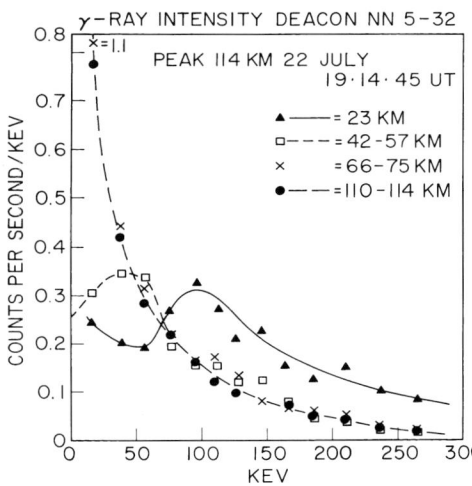

Figure 7 Background X-ray/γ-ray intensity observed as a function of altitude, 22 July 1956. The detector was a scintillation counter flown aboard a Nike Deacon rocket. As the altitude increased a flux of X-rays/γ-rays intensity appeared through the cosmic ray maximum at 23 km and increased steadily up to peak height at 110–114 km. Solar coronal X-rays were below the energy threshold of detection. (Chubb et al. 1957.)

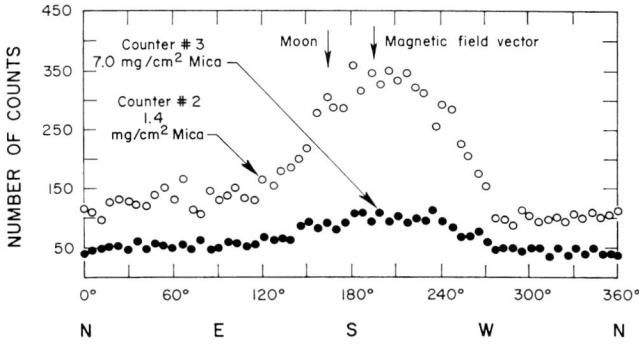

Figure 8 Early evidence (Giacconi et al. 1962) of X-ray emission from the Galaxy obtained in 1962 with proportional counters flown aboard an Aerobee Rocket. Subsequent observations showed that 60 percent of the X-ray intensity came from a single X-ray star, Scorpius X-1.

of periodic variations of intensity, indicating that they belonged to double star systems and that they were powered by gravitational accretion processes. We recognize now that more than 50 percent of the stars in our galaxy belong to double star binaries of which one star is destined to collapse to a compact object – white dwarf, neutron star or black hole – and become a powerful X-ray source by accretion of mass from its companion. UHURU marked the end of "small science" X-ray astronomy, characterized by payloads of a few hundred pounds, and the beginning of an era of "big science" in which truck-sized payloads of 20,000 pounds or more, most notably the NASA Compton Gamma Ray Observatory, were launched from the space shuttle.

GAMMA-RAY ASTRONOMY

Let us now jump to the major emphasis of the 1990s, the distant gamma ray sources. Perhaps the earliest success of gamma ray astronomy came from a secondary instrument package aboard the NASA Orbiting Solar Observatory OSO-3, weighing just a few pounds. It collected only 621 gamma ray photons over its lifetime. The NASA Small Astronomy Satellite SAS-2 launched in 1972 collected 8,000 photons in a flight that lasted 8 months. COS-B, an ESA Satellite flight that lasted 8 years from the time of its launch in 1975 and carried 200 pounds of instruments functioned perfectly for almost 7 years to collect more than 100,000 photons (see Figure 9). For each of the missions cited here, the scientific knowledge was roughly commensurate with the number of photons collected. The spatial definition of COS-B was about 2 deg, expected to be adequate to map the diffuse background but not to resolve discrete gamma-ray sources. In fact it did show that the plane of the Milky Way shines brightly with gamma rays of 100-million eV. These rays derive from interactions of cosmic rays that roam the galaxy, with a dilute hydrogen gas that concentrates towards the plane of the galaxy. The energy conversion occurs by a process of intermediate energy π-meson decay with an ephemeral life ending rapidly in the emission of gamma rays. COS-B demonstrated the direct connection between the density of interstellar matter and the intensity of cosmic gamma rays. More exciting perhaps was the discovery of discrete sources of gamma rays of great intensity, such as the Crab and Vela pulsars, the quasar 3C-273, a giant molecular cloud Rho Ophiuci and Geminga, a mystery source known for 20 years, and subsequently identified with an old X-ray pulsar of 0.237-s period and a 25th-magnitude star, the faintest optical counterpart to a high energy source then known. Virtually all of Geminga's energy is radiated as X-rays and gamma rays.

The most impressive sources in the high energy spectrum have been gamma ray bursts. Their discovery was a fall-out of U.S. military interest in detecting clandestine Soviet tests of nuclear weapons in space after an atmospheric test ban treaty was signed. It was suspected that secretive tests of atomic weapons in deep space were possible. An atomic bomb detonation there would not create the bright fireball of an atmospheric explosion. Its signature more likely would be a brilliant, short-lived X-ray flash. To guarantee concealment, the weapon could even be triggered behind the Moon. Only when the cloud of radioactive debris spread beyond the lunar mask would gamma rays and neutrons be detectable. Accordingly, the Vela Project was implemented by the Air Force to monitor space continuously for X-ray/γ-ray flashes. By 1967, Vela satellites made their first detection of a gamma ray burst but its energy and time profile did not match the theoretical expectation for a nuclear weapon. Since then the search for X-ray/γ-ray bursts was pursued in various spacecraft for scientific studies but few useful clues for the military endeavor were found. When the NASA Compton Gamma Ray Observatory (CGRO) was launched in 1991 it inaugurated a flood of high quality galactic gamma ray information that set high standards for gamma-ray astronomy.

The Burst and Transient Source Experiment (BATSE) on CGRO has recorded about 1000 gamma ray flashes in its first 3 years over the entire sky with energies close to 10-million eV. On the average about once a day a flash of intense gamma rays was detected from some random position in the sky. Individual bursts are often as short as 0.001 sec, but some have been observed to last as long as 80 sec. Most last from 1 to 10 sec. As shown in Figure 10 the distribution of bursts over the sky appears to be totally random with no apparent clustering to the galactic plane. Furthermore, until 1997 no gamma ray burst had been identified with a known astronomical object and no source had repeated at the same position. From the isotropy of the

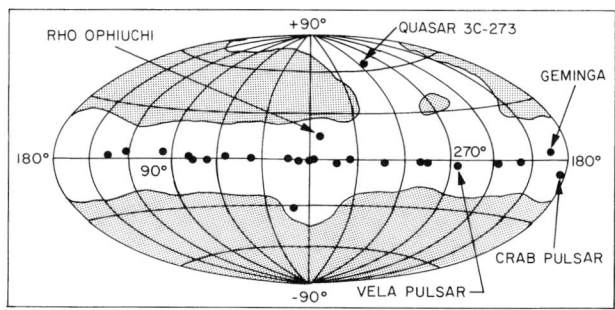

Figure 9 The Gamma-Ray Universe observed by COS-B Satellite. Shaded regions of the sky were not scanned. (Swanenburg et al. 1981.)

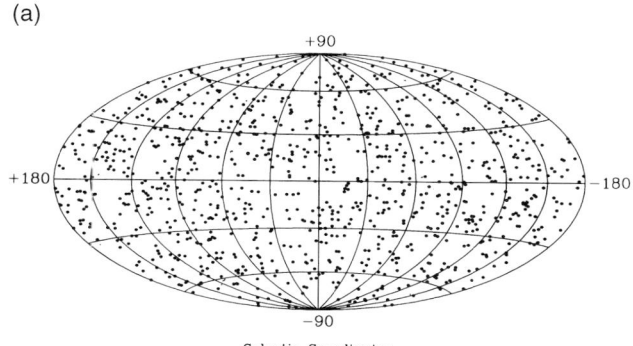

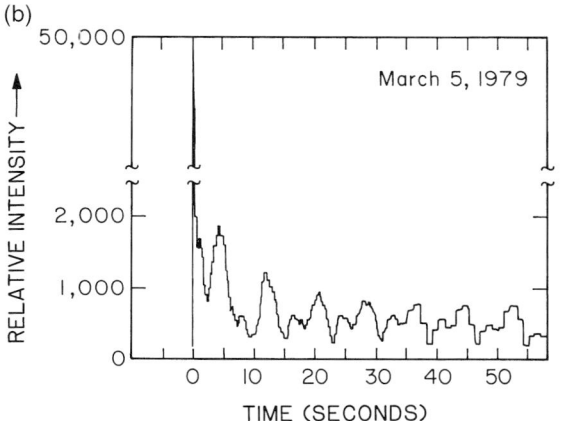

Figure 10 (a) Distribution of gamma ray bursts seen by BATSE, each dot representing a single burst. Note the uniform distribution compared to the sources in Figure 9 that show significant clustering along the milky way. (b) The burst shown in the above figure appeared on 5 March 1979 and was ten times as intense as any previously recorded. It reached its peak in less than 0.0002 sec. As it faded, the intensity oscillated with a period of about 8 sec. The shortness of the burst suggests a neutron star since the diameter of the source cannot exceed 0.0002 light sec or about 60 Km. The oscillation implies a spin period of 8 sec, too fast for any type of star larger than a neutron star. (Meegan et al. 1988.)

distribution of positions the simplest interpretation is that the bursts originate at cosmological distances.

Speculation has been rife about what the gamma ray burst sources might be. A collision between a black hole and a neutron star or between two neutron stars was a popular choice. The neutron-star scenario was especially intriguing because Joseph Taylor and John Hulse in 1974 discovered a binary pulsar (PSR 1913 + 16), one of whose components was a 1.4 solar mass neutron star spinning 17 rotations per second. The orbital period was 7.75 hours, evidence that the two stars were very close. Einstein's General Relativity predicts that the loss of gravitational radiation would cause the orbit to shrink. In over two decades the orbital period decreased by 72 microseconds per year in excellent agreement with Einstein's prediction. As they approach collision, the pair of neutron stars are locked in an ever tightening gravitational grip and radiate a whistle of gravitational waves. Over another 100-million years the frequency would rise from a sub-sonic boom to about 1000 cycles per sec into the death throes of a catastrophic collision. Modeling of the collision is still very hypothetical but almost certainly it would be accompanied by a great gamma ray flash.

The mystery of the location of gamma ray bursts was answered in part in 1997 by the Italian-Dutch "Beppo SAX" Satellite, named in honor of Giuseppi Occhialini by his Italian colleagues. It can accurately locate burst sources on the sky with a wide field X-ray camera through the associated X-ray afterglows. Comparatively long persistence X-ray afterglows, lasting as long as hours to days, are good markers of the bursts. The first convincing optical counterpart was imaged by a telescope in the Canary Islands as 21st magnitude. It faded in a week. A second burst glow was recorded by the Keck telescope in Hawaii. Spectral lines were red-shifted by 0.8, indicating a distance of 1.8 gigaparsecs. The pace of discovery has since been accelerating. A gamma ray burst detected by three spacecraft observatories, CGRO, BeppoSAX, and RXTE, December 14, 1997 claimed a distance record of 12-billion light years. The observers noted that the energy radiated approximated the radiance of the entire Milky Way for a few centuries.

REFERENCES

Bildstein, L. and Strohmeyer, T. (1999). New views of neutron stars. *Physics Today*, **52**, 40–46.

Bowyer, S., Byram, E.T., Chubb, T.A. and Friedman, H. (1964). Lunar occultation of X-ray emission from the Crab Nebula. *Science*, **146**, 913–917.

Bowyer, S., Byram, E.T., Chubb, T.A. and Friedman, H. (1964). X-ray sources in the Galaxy. *Nature*, **201**, 1307–1308.

Chubb, T.A., Friedman, H., Kreplin, R.W., Blake, R.L. and Urizicker, A.E. (1960). Proceedings of the Tenth International Astrophysical Symposium, Liège, 1960, p. 228.

Chubb, T.A., Friedman, H., Kreplin, R.W. and Kupperian, J.E., Jr. (1957). Rocket observation of X-ray emission in a solar flare. *Nature*, **179**, 861–862.

Clark, G.W., Garmire, G.P. and Kraushaar, W.L. (1968). Observation of high-energy cosmic gamma rays. *Astrophysical Journal*, **153**, L203.

Forman, W., Jones, C., Cominsky, L., Julien, P., Murray, S., Peters, G., Tananbaum, H. and Giacconi, R. (1978). The fourth UHURU catalog of X-ray sources. *Astrophysical Journal Supplement*, **38**, 357–412.

Friedman, H. and Byram, E.T. (1967). X-ray sources from 3C273 and M87. *Science*, **158**, 257–259.

Friedman, H., Byram, E.T. and Chubb, T.A. (1967). Distribution and variability of cosmic X-ray sources. *Science*, **156**, 374–378.

Friedman, H., Lichtman, S.W. and Byram, E.T. (1951). Photon counter measurements of solar X-rays and extreme ultraviolet light. *Physical Review*, **83**, 1025–1030.

Giacconi, R., Gursky, H., Paolini, F.R. and Rossi, B. (1962). Evidence for X-rays from sources outside the solar system. *Physical Review Letters*, **9**, 439–443.

Kelvin, W.T. (1892). Presidential address to the Royal Society, November 30, 1892. *Proceedings of the Royal society of London, A*, **52**, 302–310.

Kupperian, J.E., Jr. and Friedman, H. (1958). Proceedings of the 5th CSAGE Assembly, Moscow, U.S.S.R. (July 1958).

Paciesas, W.S., Meegan C.A., Pendleton, G.N., Briggs, M.S., Kouveliotou, C., Koshut, T.M., Lestrade, J.P., McCollough, M.L., Brainerd, J.J., Hakkila, J., Henze, W., Preece, R.D., Connaughton, V., Kippen, R.M., Mallozzi, R.S. and Fishman, G.J. (1988). The 4B BATSE Gamma-Ray.

Swanenburg, B.N., Bennett, K., Bignami, G.F., Buccheri, R., Caraveo, P., Hermsen, W., Kanbach, G., Lichti, G.G., Masnou, J.L., Mayer-Hasselwander, H.A., Paul, J.A., Sacco, B., Scarsi, L. and Wills, R.D. (1981). Second COS-B catalog of high-energy gamma-ray sources. *Astrophysical Journal*, **243**, L69–L73.

Taylor, J.H. and Hulse, R.A. (1998). In H. Friedman (ed.), *The Astronomer's Universe*, W.W. Norton, New York [1990], p. 108.

13

BLAIR D. SAVAGE*

Early ultraviolet spectroscopy from space

The period starting around 1950 and extending into the twenty-first century will probably ultimately be remembered as the golden age of discovery in observational astronomy. This is because of the simultaneous occurrence of three technological advances: the ability to observe the entire electromagnetic spectrum, by combining observations from the ground and from above the absorbing effects of the Earth's atmosphere; the creation of efficient and high-precision detectors of electromagnetic radiation; and the development of computers for processing and manipulating large electronic data sets. The field of ultraviolet (UV) astronomical spectroscopy at wavelengths shorter than 3100 Å has benefited from all three of these technological revolutions and has required access to space for its very existence.

In this chapter I follow the development of the field of UV astronomical spectroscopy from space over the period from approximately 1945 to 1980, with the emphasis on spectroscopy of objects located beyond the Solar System. The discussions concentrate on scientific developments in the wavelength range from ~900 to ~3100 Å. The lower limit of ~900 Å represents the wavelength where bound–free opacity of neutral hydrogen in the interstellar gas becomes large, and the upper limit of 3100 Å is where observations begin to become possible from telescopes situated on the ground. The wavelength regions of 100–900 Å, 900–1200 Å, and 1200–3100 Å are usually referred to as the extreme ultraviolet (EUV), far-UV, and UV regions of the spectrum, respectively. I shall not review the beginnings of EUV spectroscopy from above the atmosphere since those beginnings were delayed by a decade compared with activities in the far-UV and UV bands; for a review of the beginnings of EUV astronomy see Bowyer (1991). This delay resulted mostly from the fear that the interstellar H I opacity in most directions would be so large that only the stars closest to the Sun would be observable. However, the highly irregular distribution of H I in the local interstellar medium and the existence of the local hot bubble of ionized hydrogen rendered these fears invalid. EUV astronomy achieved its first photometric detection of a stellar source during the Apollo–Soyuz mission (Lampton et al. 1976) and the first low-resolution EUV stellar spectra were obtained several years later (Holberg et al. 1980) as part of the Voyager 1 mission.

In preparing this historical overview of developments in the field of UV astronomy from space, I benefited substantially from reference to various papers summarizing the state of UV astronomy over the past 40 years. These include the reviews by Friedman (1959), Wilson and Boksenberg (1969), Wilson (1970), Bless and Code (1972), Code and Savage (1972), Spitzer and Jenkins (1973), Boggess and Wilson (1987), and Brosch (1999).

THE OBSCURING ATMOSPHERE

The attenuation produced by the Earth's atmosphere in the UV as a function of wavelength in angstroms is shown in Figure 1. The solid curve gives the altitude in kilometers at which radiation normal to the atmosphere is reduced in intensity by a factor of $1/e$. The various molecules and atoms mostly responsible for the atmospheric absorption are indicated on the curve. The atmospheric absorption that rapidly sets in near 3100 Å is produced by ozone (O_3) while

*University of Wisconsin, Madison, WI, USA

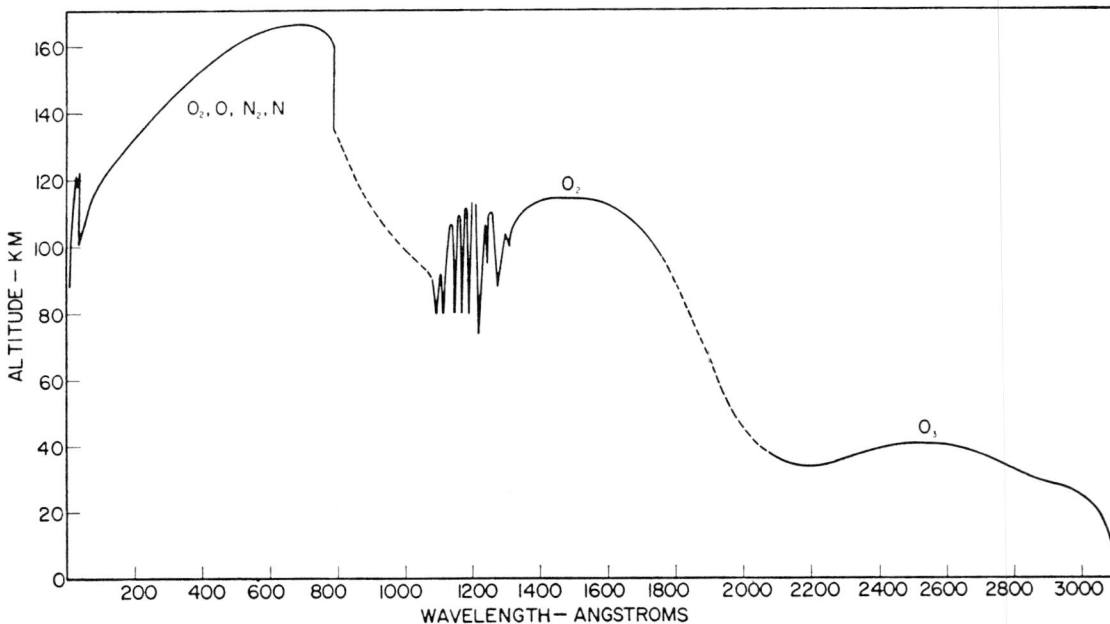

Figure 1 This figure from Friedman (1959) shows the altitude in kilometers at which the fraction of radiation at wavelengths shorter than the visible incident on the Earth's atmosphere is reduced by a factor of $1/e$. Strong atmospheric absorption due to O_3 sets in at about 3100 Å. While in the mid-UV it is possible to carry out observations from high-flying balloons, satellite altitudes are required to observe celestial sources for the UV wavelengths below ~2000 Å. (By permission of the *Journal of Geophysical Research*.)

absorption at shorter wavelengths is mostly from O_2, O, N_2, and N. In the wavelength range ~2000–3000 Å it is possible to carry out UV observations from balloons carried to altitudes of 30–40 km (see later section on Near-UV Spectroscopy from Balloons). However, rocket or satellite altitudes are required to observe at UV wavelengths shorter than ~2000 Å. Even at rocket or satellite altitudes exceeding 300 km, the trace atomic constituents in the atmosphere can interfere with astronomical observations. In particular, the hydrogen geocorona extends many Earth radii from the surface of the Earth, and the H I scatters sunlight in the Lyman series lines producing very strong emission, particularly in the Lyman α, β, and γ lines at 1215.67, 1025.72, and 972.54 Å. Similarly, Earth airglow emission can be strong in various O I transitions. Even at the altitudes of the major modern observatories such as the Hubble Space Telescope, there is enough residual O I in the atmosphere for terrestrial O I absorption lines to be apparent in high-resolution spectra of hot stellar continuum sources.

THE SCIENTIFIC IMPORTANCE OF ACCESS TO SPACE

More than 75 years ago, Oberth (1923) pointed out that an astronomical observatory orbiting above the Earth's atmosphere would have several major advantages over an observatory on the Earth's surface. He noted that at optical wavelengths the orbiting observatory would not suffer the blurring effects of the Earth's atmosphere. Therefore the orbiting observatory could produce images limited only by the quality of the telescope optics and its pointing system. He also pointed out that the orbiting telescope would be capable of observing celestial objects over the entire electromagnetic spectrum, including spectral regions that are not accessible to the ground-based observer. No professional astronomers took these ideas seriously when they were first discussed, and over the subsequent 20 years the only other published speculations about observatories in space that I am aware of appear in the science fiction literature (Richardson 1940). However, everything changed after World War II, in part as a result of the development and capture of the V2 rocket technology and the start of the Cold War between the United States and the Soviet Union.

The first truly serious discussions of the potential scientific importance of placing satellite observatories instrumented for UV spectroscopy above the absorbing atmosphere are found in an internal report entitled "Advantages of an Extra-Terrestrial Observatory," written for the Rand Corporation in 1946 by Lyman Spitzer reprinted in Spitzer (1997). Spitzer's report discusses the science that could be pursued with several types of orbiting observatories of different sizes. In addition to observatories designed to observe the Sun in the UV, he discussed the scientific potential of a modest 0.25 m (10-inch) reflecting telescope designed to obtain UV spectra of stellar sources, and the potential value of a large reflecting telescope operating at optical wavelengths.

The scientific programs Spitzer proposed for this instrument included:

- studies of the composition of planetary atmospheres
- measures of the structure of stellar atmospheres, including the possibility of detecting expanding atmospheres in the strong absorption lines of C, N, and O
- measures of the color temperatures of hot stars
- measures of stellar bolometric magnitudes
- the analysis of eclipsing binary UV light curves to obtain information about stellar masses and atmospheric properties
- using the improved understanding of stellar atmospheric conditions from the UV studies to improve on stellar distance determinations
- measures of the composition of the interstellar gas
- measures of the properties of interstellar absorbing grains
- measures of the UV spectra of supernovae in order to better understand the explosion processes.

For each of these topics Spitzer discussed how observatories above the atmosphere could be used to make major advances in the given science area. For example, in the case of measuring the composition of the interstellar gas, Spitzer pointed out that in interstellar space most atoms and molecules were expected to be found in their ground (lowest-energy) state. Therefore measures of the composition of the gas through absorption line spectroscopy would require an observatory operating at UV wavelengths since most absorption lines out of the ground state of abundant atoms fall in the UV. Slightly more than 25 years after creating this vision for the future of UV astronomy in space, Spitzer pioneered UV studies of the interstellar medium by using the high-resolution UV spectrograph aboard the Copernicus satellite (see the section on the Orbiting Astronomical Observatories).

Shortly after the launch of Sputnik in 1957 and the organization of NASA in 1958, several papers appeared discussing the science that could be pursued with UV spectrometers operating above the Earth's atmosphere, including those of Spitzer and Zabriskie (1959) and Code (1960). I find the Spitzer and Zabriskie paper, entitled "Interstellar research with a spectroscopic satellite," particularly interesting because my professional field of interest is the interstellar medium. I remember referring to that paper in 1965 when taking Professor Spitzer's course on "Physical Processes in the Interstellar Medium." During the course I recall a particularly interesting class assignment. Professor Spitzer asked his students, "What observations would you obtain with a 1 meter satellite observatory equipped with a high-resolution spectrometer operating from 912 to 2000 Å in order to obtain information about the physical properties of the gas between the stars?" In answering the question I was intrigued by the possibility of using a satellite observatory to study aspects of the interstellar gas (the hot phase) that were difficult or impossible to study from the ground. I didn't realize at the time that this assignment was one I would continue to work on for most of my professional career.

TECHNOLOGICAL CHALLENGES

The technological challenges faced by the pioneers of UV spectroscopic space astronomy were considerable. They included gaining reliable access to space, developing small light weight UV spectroscopic optical systems, developing rocket and satellite pointing and control systems, and developing efficient UV sensitive detectors. The most difficult challenge was to design and build complex robotic systems that would actually operate in the space environment.

The progression through these various technological challenges did not always proceed smoothly. As a beginning graduate student in 1964, I witnessed the activities around Princeton University involving the pioneering projects in UV spectroscopy through the efforts of Donald Morton and Lyman Spitzer and in optical diffraction limited imagery through the Stratoscope II balloon-borne telescope program involving Robert Danielson and Martin Schwarzschild. I vividly remember discussions in the halls about rockets blowing up on the launch pad, pointing and control systems that failed to operate, rocket parachute systems that deployed too early, and balloon flights where the 3600 kg Stratoscope II telescope system failed to unlatch from the vertical pointing position, or where the sealed door on the pre-cooled 0.9 m diffraction-limited primary mirror failed to open at altitude. Reflecting on all these problems and wondering how long it might take to complete an experimental thesis in space astronomy, I made an arrangement with Robert Danielson and Martin Schwarzschild to pursue a theoretical PhD thesis and to briefly join the Stratoscope II team following my graduation to continue to explore my experimental interests. The lesson I learned during this period was that the most difficult technical challenge faced by the early pioneers in space astronomy was to achieve a high level of reliability for remotely controlled robotic systems operating in the space environment. Even today, it appears that this is a difficult goal to meet on a regular basis.

SOUNDING ROCKET SPECTROMETERS

UV space astronomy began as an experimental field at about the same time as Spitzer was considering its future.

Using captured German V2 rockets which could carry a 1000 kg of equipment to 150 km altitudes, scientists working at the Naval Research Laboratory (NRL) led by Richard Tousey obtained the first UV spectrum of the Sun to wavelengths as short as 2200 Å (Baum et al. 1946). The first UV stellar photometry followed about ten years later when NRL scientists observed 59 hot stars in a 130 Å wide UV band centered on 1115 Å (Bryam et al. 1957). These early UV stellar observations involved scans of the sky with simple photometer systems from unstabilized sounding rockets.

The flight of a scanning objective grating spectrometer on an unstabilized Aerobee rocket produced the first low-resolution UV stellar spectrophotometry over the 1600–4100 Å region at 50 Å resolution (Stecher and Milligan 1962). In this observation the dispersion direction of the spectrograph was normal to the rocket roll axis. The roll of the rocket produced the spectral scan at the slit with a photomultiplier recording the spectrum. However, to capture enough photons during the short intervals for which the stars were in the field of view, the slit needed to be wide and the resulting resolution of 50 Å was insufficient to resolve discrete stellar absorption or emission lines.

The first moderate-resolution spectroscopic observations of stars required the development of three-axis stabilization systems for pointing rocket-borne instruments at stars for long enough periods of time to record the spectra. The pointing control system developed by the Space General Corporation was a three-axis stabilized gas reaction system that could be used to orient the entire sounding rocket during its free fall. This system could point the payload within 3° of a desired direction and was stabilized to a limit cycle jitter of ±15 arc minutes. In their pioneering experiment, Morton and Spitzer (1966) employed an additional fine stabilization system as part of their instrument package which improved the pointing in the spectrograph dispersion direction to approximately ±16 arc seconds. Their instrument (Figure 2) consisted of two 1200 lines/mm objective plane gratings followed by $f/2$ Schmidt cameras with 100-mm focal lengths and 10° fields. One Schmidt corrector was made of calcium fluoride and transmitted to 1250 Å; the other corrector was quartz and transmitted to 1700 Å. The resulting UV spectra had a dispersion of about 65 Å/mm and were recorded on UV sensitive Kodak Pathe SC5 film. This small pair of UV spectrometers produced spectra with a resolution of approximately 1 Å when the fine stabilization platform achieved its design pointing stability in the dispersion direction.

The pioneering attempts to obtain UV spectroscopic observations of stars did not achieve instant success. During these learning stages there were often problems associated with making the attitude-control pointing systems work properly. For example, quoting from the abstract

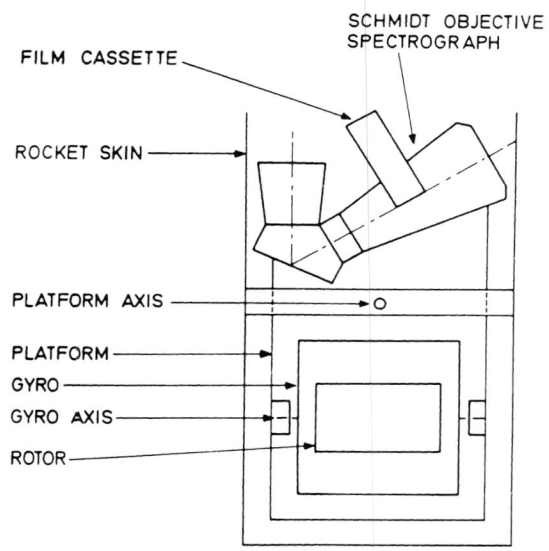

Figure 2 A schematic diagram of the simple UV spectrometer and fine stabilization system flown on an Aerobee rocket by Morton and Spitzer (1966). The fine stabilization in the spectrometer's dispersion direction allowed the instrument to obtain UV spectra of π and δ Sco from 1260 to 1720 Å at a resolution of 1 Å. (By permission of Donald Morton and the *Astrophysical Journal*.)

of the paper by Morton and Spitzer (1966) reporting the observations of first line spectra of stars in the UV,

> On the first two flights the attitude control system failed to stabilize the rocket. On the third flight both coarse and fine systems worked properly, but the parachute failed on re-entry so that the impact damaged the payload beyond repair and admitted light into the film cassettes. Most of the films were totally fogged, but underdeveloping one from the calcium fluoride camera showed wide spectra of the early B-type stars δ and π Sco with a resolution of 1 Å.

I can recall the gloomy mood after this third flight among the returning Princeton scientists and engineers when they discussed their totally blackened images. However, that mood quickly changed to one of joy when the last piece of film was underdeveloped, and revealed clear spectra of π and δ Sco extending from ~1260 to 1720 Å.

One year after the success of obtaining the first moderate-resolution UV spectra of stars in Scorpius, Morton (1967) obtained high-quality spectra of six stars in Orion at 3 Å resolution from 1150 to 1630 Å. Those data provided the first evidence for high-speed mass loss from hot stars in the form of P Cygni profiles of the Si IV and C IV stellar absorption lines (Figure 3). This result must have pleased Lyman Spitzer since in his 1946 Rand Corporation study he remarks that, "In addition the nature of unusual stellar

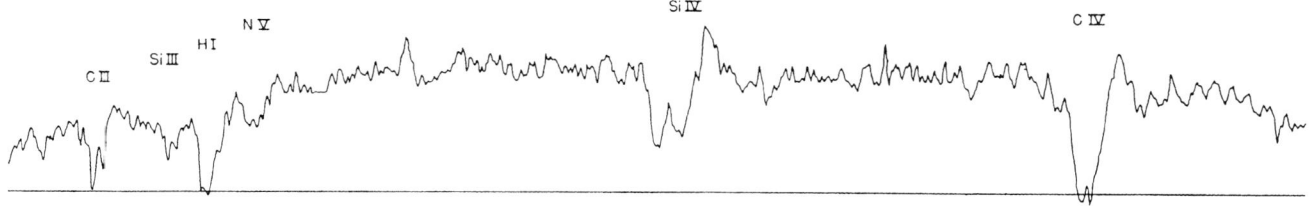

Figure 3 High-speed stellar mass loss from the O9.5 Ib star ζ Ori as revealed through the P Cygni absorption+emission profiles for C IV (1550 Å) and Si IV (1393 and 1403 Å) recorded in this 1160–1600 Å region of the UV spectrum obtained by Morton (1967) with the Princeton UV spectrometer system shown in Figure 2. Wavelength increases from left to right. Wind outflow speeds of 1900 km s^{-1} are implied by the short-wavelength absorption extensions of the C IV and Si IV profiles identified in the spectrum. (From Morton (1969) by permission of D. Reidel Publishing Company.)

atmospheres – expanding, rapidly rotating, etc. – would be more clearly indicated by information on the behavior of such abundant elements as C, N, and O as well as by the behavior of the resonance lines of hydrogen." On subsequent flights the Princeton UV sounding rocket program also produced important information about the basic absorption line properties of hot stellar atmospheres and about the distribution of gas in the interstellar medium (Morton et al. 1968, Jenkins and Morton 1967).

During this period, important work involving various types of UV spectrometers on sounding rockets was also being pursued at the Goddard Space Fight Center by Theodore Stecher, Albert Boggess, and Andrew Smith. Their efforts were directed toward obtaining exploratory low-, medium-, and high-resolution UV spectra of hot stars. The most interesting scientific result emerging from this work was the discovery by Stecher (1965) that the UV interstellar extinction curve contains a broad peak in the middle-UV that could be produced by grains of interstellar graphite (Stecher and Donn 1965). This observation was obtained by an instrument similar to the low-resolution spectrophotometer first flown by Stecher and Milligan (1962). The existence of the 2200 Å extinction bump was convincingly established through higher-quality observations first presented at the IAU Symposium No. 36 in June 1969 by Stecher (1969, 1970) and by Bless and Savage (1970). Stecher's measurements were derived from sounding rocket observations of ζ and ε Per, while the results of Bless and Savage were based on ten star pairs observed by the Orbiting Astronomical Observatory 2 (OAO-2; see the next section).

At NRL, George Carruthers carried out a very ambitious program in UV rocket spectroscopy that resulted in the development of an all-reflecting objective grating spectrograph followed by a very high sensitivity Lallemand (electronographic) image converter detector in which photoelectrons were accelerated to 20 keV and focused onto a nuclear emulsion. The system provided spectral resolutions of 2–3 Å and simultaneous wavelength coverage from 950 to 1400 Å. On the first successful flight the instrument yielded high-quality spectra of γ Vel, ζ Pup, and several stars in Orion (Carruthers 1968). With this sensitive spectrograph it was possible to observe fainter and more reddened stars than those previously observed by less efficient rocket spectrometers. Soon after the first flight, Carruthers (1970) was successful in obtaining a high-quality far-UV spectrum of the reddened star ξ Per (Figure 4). That spectrum clearly revealed the Lyman band absorption produced by interstellar molecular hydrogen. A determination of the column densities of H$_2$ and H I from these observations revealed that approximately half of the hydrogen in the interstellar gas toward ξ Per is in the molecular form. The discovery of the high abundance of H$_2$ in interstellar space (Carruthers 1970) and the discovery of mass loss due to high-speed outflows from hot stars (Morton 1967) must rank as the two most important discoveries made with small rocket-borne UV spectrometers during this period.

THE ORBITING ASTRONOMICAL OBSERVATORIES

Soon after the launch of the USSR Sputnik 1 satellite on 4 October 1957 and the start of the "space race" between the USA and the USSR, a telegram was sent from the Air Force Cambridge Research Laboratory to a number of scientists in the USA seeking expressions of interest in flying astronomical instrumentation on small orbiting satellites. The responses eventually lead to the organization of the Orbiting Solar Observatory (OSO) and Orbiting Astronomical Observatory (OAO) programs under the direction of NASA, which came into existence in late 1958. Experiments proposed by Arthur Code, James E. Kupperien, Lyman Spitzer, and Fred Whipple eventually evolved into the very ambitious OAO program, which was originally

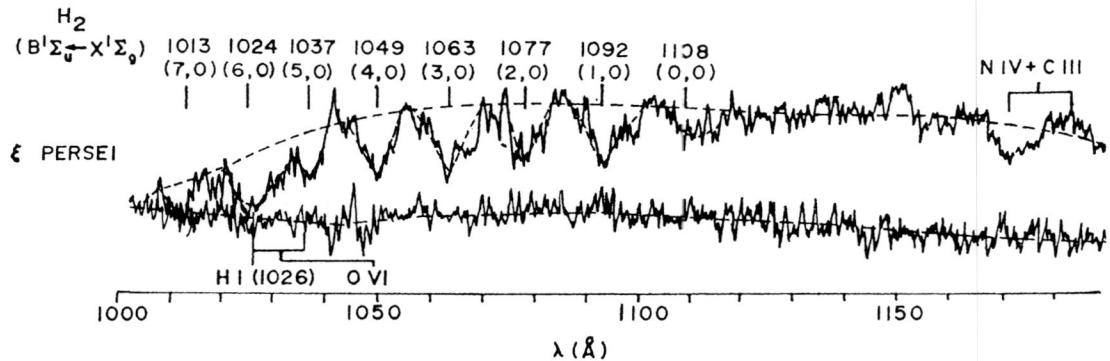

Figure 4 The upper curve is the microdensitometer tracing of the sounding rocket spectrum of ξ Per obtained by Carruthers (1970) that reveals the presence of interstellar molecular hydrogen. The positions of the H_2 Lyman series absorption bands from (0, 0) to (7, 0) are indicated. The lower curve is the background level adjacent to the spectrum. The H_2 absorption band strengths when combined with observations of H I Lyman α absorption implied that ~50% of the hydrogen along the line of sight is molecular. (By permission of George Carruthers and the *Astrophysical Journal*.)

conceived as a series of three UV astronomical satellites, starting with a broadband photometric UV survey satellite (OAO-A), progressing to a moderate-resolution spectroscopic satellite (OAO-B), and finally to a high-resolution spectroscopic satellite (OAO-C). A full description of the OAO program is found in Rogerson (1963). It is noteworthy that the decision to build the ambitious OAO series of satellites was taken about six years before the first moderate-resolution UV spectra of stars were actually obtained with small spectrographic instruments aboard sounding rockets. The OAO satellites were under construction at a time when it was not even possible to stabilize and point small instruments on sounding rockets!

OAO-A, containing broadband UV stellar photometers and low-resolution scanning spectrometers developed at the University of Wisconsin and a γ-ray experiment from MIT, was launched in 1966 and achieved orbit, becoming OAO-1. However, no scientific data were obtained because high-voltage arcing in the attitude control system resulted in the loss of the satellite after only three days. The proper control of high voltages in the low pressure environment of space has always been a difficult problem.

Since back-up components existed for the OAO-A mission, NASA decided to fly the Wisconsin experiment again, along with the Smithsonian Astrophysical Observatory UV imaging experiment known as Celescope. This mission was successfully launched on 7 December 1968, becoming OAO-2. During its four-year lifetime OAO-2 provided broadband UV photometry of ~1150 stars, low-resolution UV spectra of ~330 stars (Code et al. 1970, Bless and Code 1972), and Celescope UV imagery of a substantial fraction of the sky (Davis et al. 1973).

The two scanning spectrographs aboard OAO-2 were simple objective grating instruments consisting of 150 × 200 mm plane reflection gratings followed by 180 × 250 mm parabolic objective mirrors. The dispersed light was focused through a hole in the center of the grating onto an exit slit followed by a photoelectric photometer. Spectrum scanning was achieved by rotating the gratings. One spectrometer provided a resolution of 20 Å and operated from 1800 to 3800 Å; the other spectrometer had a resolution of 12 Å and operated from 1050 to 2000 Å. Both spectrometers used "venetian blind" baffles to restrict the background light. I was extremely fortunate to arrive in Madison, Wisconsin only four months before these spectrometers went into operation in space. As a young member of the OAO-2 science analysis team I marveled at the fact that, only two years after I saw the very first photographic UV spectra of stars, OAO-2 had begun to produce a steady stream of high-quality photoelectric UV spectra of hot O and B stars, and of cooler astronomical objects including planets and comets. With my introduction to interstellar science at Princeton, I immediately started to use the OAO-2 spectroscopic measurements to study UV interstellar extinction (Bless and Savage 1970, 1972) and the distribution of interstellar atomic hydrogen through the very strong Lyman alpha absorption line at 1216 Å (Savage and Code 1970, Savage and Jenkins 1972).

Examples of OAO-2 low-resolution spectra are shown in Figure 5. The two spectra show flux (normalized to the flux at 2400 Å) plotted against wavelength for μ Col (an unreddened O 9.5 V star) and ζ Oph (an O 9.5 V star with a B–V reddening of 0.32). The strong stellar lines of C IV and Si IV and the very strong H I Lyman α interstellar absorption at 1216 Å are evident, even at the relatively low resolution of 12 Å. UV extinction causes the large difference between the two energy distributions and produces the broad depression near 2200 Å in the spectrum of ζ Oph. The OAO-2

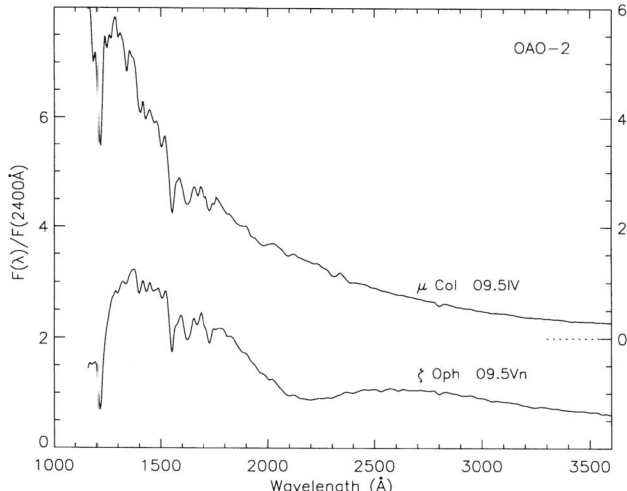

Figure 5 Spectra obtained by the OAO-2 satellite of μ Col (O 9.5 IV, E(B–V) = 0.01) and ζ Oph (O9.5 Vn, E(B–V) = 0.32). Flux (in erg cm^{-1} s^{-1} Å^{-1}) normalized to the flux at 2400 Å is plotted against wavelength, with the scale on the left for ζ Oph and on the right for μ Col. The very different shapes of these two energy distributions is caused by the effects of UV interstellar extinction. The strong stellar absorption lines of C IV (1550 Å) and Si IV (1393 and 1403 Å) are evident even at the relatively low resolution (~12 Å) of these observations. (Spectra from the OAO-2 data archives at the University of Wisconsin.)

spectra yielded fundamental information about the UV continuum and strong absorption line properties of hot stellar atmospheres (Bless et al. 1976), including the effective temperature scale for O and B stars (Code et al. 1976).

With the success of OAO-2, everyone was eagerly awaiting the launch of the more capable OAO-B, which contained a 0.9 m f/5 Ritchey–Chrétien telescope and a UV Ebert–Fastie spectrometer designed to operate from 1050 to 4300 Å. Spectrum scanning was to be achieved by rotating the grating, with the dispersed light falling onto photomultipliers operating with and without entrance slits to provide resolutions from approximately 2–64 Å. Unfortunately, OAO-B, which carried the Goddard Space Flight Center (GSFC) experiment, failed to achieve orbit after its launch on 30 November 1970 because one half of the launch vehicle nose cone did not properly eject. This was a major loss to the very young field of UV spectroscopy and to the GSFC UV science team.

OAO-C became OAO-3 with a successful launch on 21 August 1972, and was soon after named the Copernicus satellite. The experiment from Princeton University consisted of a 0.8 m Cassegrain telescope and a single-grating spectrometer located in a Paschen–Runge mounting configuration between the primary and secondary (Rogerson et al. 1973). Two carriages carrying entrance slits and photomultipliers moved along the spectrograph's Rowland circle in order to scan the spectrum over the wavelength ranges 950–1450 Å with 0.05 Å resolution and 1650–3000 Å with 0.1 Å resolution. LiF optical overcoats of the mirrors and gratings allowed the observatory to efficiently operate in the wavelength range 912–1100 Å, which contains many important atomic and molecular diagnostic lines. The fine guidance system derived error signals from the star being studied and was limited to stars brighter than approximately $V=7$. The Copernicus satellite successfully operated until 1979 and obtained a wealth of information about the physical properties and composition of the interstellar medium (Spitzer and Jenkins 1973) and about the mass outflow from hot stars (Snow and Morton 1976).

In reading the commentary provided by Lyman Spitzer about the Copernicus program in his selected writings (Spitzer 1997), I was amazed to find that on the evening before the satellite launch, while he was reviewing notes from earlier subcontactor meetings, he noticed that an error had been made in the computation of the best focus position for the secondary mirror. After several urgent phone calls he managed to convince the engineers in charge of the project that a focus change should be made before launch. The change was made. This was extremely fortunate, since once the satellite was in orbit the mechanism for moving the secondary mirror was no longer operable. Had the last-minute focus change not been made on the ground, the Copernicus satellite would have been limited to observing only the brightest stars in the sky. By 1972, Professor Spitzer was fully aware from previous experiences that the least reliable spacecraft components are often associated with mechanical systems.

A scientific highlight of the Copernicus mission was the first comprehensive assessment of the physical state and distribution of molecular hydrogen in the interstellar medium (Spitzer et al. 1973). Figure 6 shows the level of spectroscopic detail provided by 0.05 Å resolution Copernicus observations of the H_2 (4, 0) Lyman band near 1150 Å in the spectrum of ζ Oph. H_2 lines from the $J=0$ to 5 rotational levels are detected. A second highlight was the direct detection of the hot phase of the interstellar medium through observations of the absorption produced by the O VI doublet at 1031.93 and 1037.62 Å (York 1974, Jenkins and Meloy 1974, Jenkins 1978). Since O VI requires 113 eV for its production, the ion is likely produced in collisionally ionized hot interstellar gas rather than in warm photoionized gas.

EUROPEAN ROCKET AND SATELLITE PROGRAMS

The UV sounding rocket and satellite program in Europe experienced a slower start than in the USA, where the

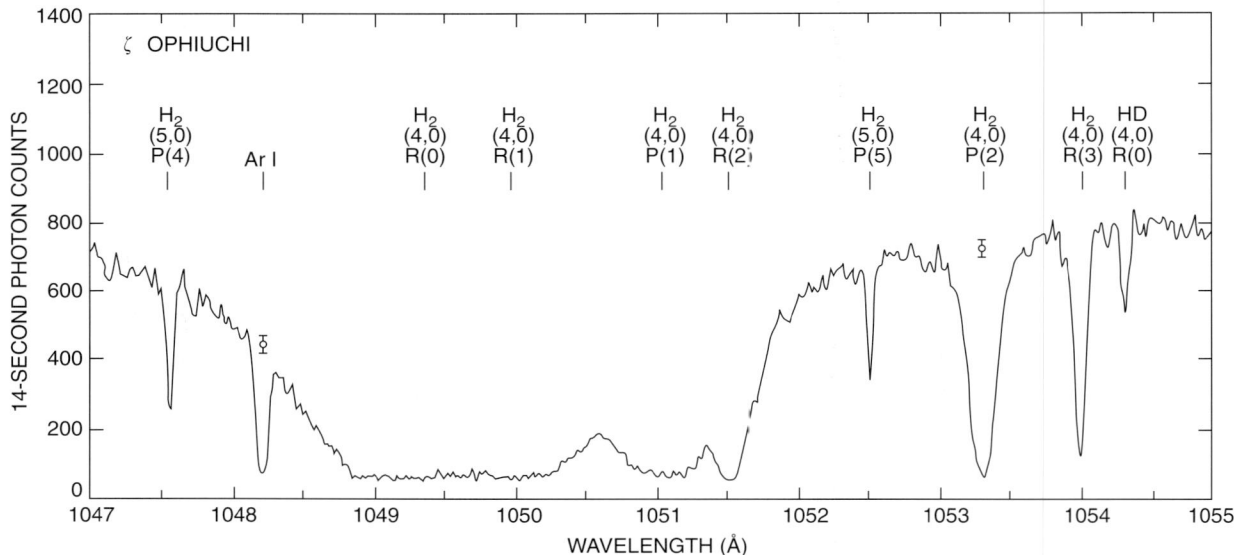

Figure 6 Copernicus Satellite high resolution (0.05 Å) scan of the O 9.5 V star ζ Oph over an 8 Å interval showing the H_2 (4, 0) absorption band near 1050 Å. Absorption from the $J = 0$ to 5 rotational states of the ground vibrational ($v = 0$) level are evident. The R(0), R(1), and P(1) lines are extremely strong and have broad damping wings. The single-channel scanning photomultiplier aboard the spacecraft recorded the spectrum by spending 14 seconds at each scan position. Spectrograph scattered light appears at a level of approximately 50 photon counts per spectrum channel. Of the hydrogen along the line of sight to ζ Oph, 68% is in the form of H_2. (From Spitzer and Jenkins (1973) with permission from the *Annual Reviews of Astronomy and Astrophysics*.)

international political ramifications of the USSR Sputnik satellite helped to create high levels of funding across most fields of space science. The development of a three-axis stabilization system for the British Skylark rocket made it possible to start a British program in UV rocket spectroscopy. Groups at the Royal Observatory Edinburgh and the Astrophysics Research Unit at Culham participated in these ventures. The high-quality, high-resolution (0.3 Å) spectra of γ Vel and ζ Pup obtained by Burton et al. (1973) clearly illustrated the diagnostic potential of high-resolution far-UV spectroscopy.

The first European satellite designed to obtain UV astronomical data was the highly successful TD-1 satellite launched in 1972 by the European Space Research Organisation (ESRO). The satellite contained seven experiments, of which two were devoted to low- and medium-resolution UV spectroscopy. The prime experiment was designed to carry out an all-sky spectrophotometric survey of objects brighter than approximately V=9. It involved the two British UV groups and astronomers from Belgium. The UV Sky Survey Telescope (UVSST) described by Boksenberg et al. (1973) obtained low-resolution (FWHM ~35–40 Å) spectra from 1330 to 2550 Å of 1791 stars and broadband photometric observations of 31,215 objects. The spectra proved valuable for studying stellar spectral energy distributions and UV interstellar extinction. A second UV experiment carried out stellar spectroscopy at moderate (~2 Å) resolution in several bands between 2000 and 3000 Å with an instrument from Utrecht (de Jager et al. 1974, van der Hucht et al. 1976, de Boer and Lamers 1978). These various experiments helped to provide the technical expertise and scientific leadership that enabled European astronomers, the European Space Agency (ESA), and the UK Science and Engineering Research Council (SERC) to play a major role in the International UV Explorer (IUE) satellite project (see below).

NEAR-UV SPECTROSCOPY FROM BALLOONS

The major source of opacity in the Earth's atmosphere in the wavelength range 3200–2200 Å is ozone (Figure 1). Since ozone peaks in density at a relatively low altitude of ~25 km, near-UV astronomical spectroscopy from balloons is possible. A group from Belfast and University College London obtained high-resolution spectra during several balloon flights (Boksenberg et al. 1972). At about the same time, a team from the NASA Manned Spacecraft Center and Houston Baptist College studied chromospheric Mg II emission from cool stars at ~0.3 Å resolution with a 0.4 m balloon-borne telescope feeding an Ebert–Fastie spectrometer and an image dissector tube detector (Kondo et al. 1972). This experiment later evolved into the Balloon Borne UV Stellar Spectrograph (BUSS) with participation of scientists from Utrecht. In the BUSS payload the original spectrometer and detector were replaced with a cross-dispersed echelle spectrograph followed by a secondary-electron conducting (SEC) vidicon detector that could simultaneously and efficiently record a large number of echelle spectral orders. Over a number of flights the BUSS experiment obtained spectra of 56 stars at a resolution of 0.1 Å (Kondo et al. 1979, de Jager et al. 1979).

THE INTERNATIONAL ULTRAVIOLET EXPLORER

The IUE satellite contained a 0.4 m primary mirror followed by echelle spectrographs and SEC vidicon detectors. The IUE was designed for low- and high-resolution spectroscopy over the 1150–3200 Å wavelength region. The IUE revolutionized UV astronomical spectroscopy during its 18-year lifetime from 1978 to 1996.

The history of the origins of the IUE is discussed by Boggess and Wilson (1987), who trace it back to a project that was considered by ESRO in the mid-1960s. The initial ESRO space astronomy program included plans for a major project. After surveying the possibilities, it was decided that the project should be a Large Astronomical Satellite (LAS) devoted to UV spectroscopy of non-solar objects. After releasing a call for proposals, ESRO received three proposals for the LAS instrument and subsequently funded detailed, design studies from all three proposing groups, including one from the UK. The UK instrument design was adopted in 1966 as the instrument package for the LAS. After further study, including detailed cost estimates, ESRO found it necessary to abandon the LAS in 1967 because the costs were too high. However, the UK team, under the direction of Robert Wilson, continued its studies and identified areas where the adoption of new technologies could significantly lower cost with very little impact on the observatory performance. The resulting study produced a design for the Ultraviolet Astronomical Satellite (UVAS) which had more scientific capability than the original LAS design, but was much simpler and less expensive. The UVAS design included a 0.45 m telescope, an echelle spectrometer, and an SEC vidicon detector. The use of the echelle grating helped considerably in relieving the pointing, thermal, and mechanical requirements for the spacecraft and its stabilization system. The spectrograph consisted of a concave collimator, a plane echelle grating, and a concave diffraction grating to separate echelle orders and to bring the spectrum into a two-dimensional format at the SEC vidicon detector. This imaging detector provided a huge increase in multiplexing efficiency over the single-channel scanning photomultipliers planned for use in the OAO series of satellites. The UVAS proposal (Wilson 1968) was submitted to ESRO and received highly favorable reviews, but ultimately was not accepted.

Fortunately, after spending nearly four years on LAS and UVAS and recognizing that the UVAS concept was the correct approach, Robert Wilson decided to press onward and seek help from NASA. He sent a copy of the design report to Leo Goldberg, who was Chairman of NASA's Astronomy Missions Board, and suggested that UVAS might be an excellent project to bridge the time period between the last OAO satellite and the much more ambitious Space Telescope project. The timing was excellent, since NASA and scientists at the GSFC were then considering the possibility of post-OAO UV projects of moderate costs. A visit by Robert Wilson to the GSFC in 1969 resulted in the UVAS concept becoming the IUE. The visit established the spirit of international technical and scientific cooperation in this unique project that continued for 30 years. The main modification of the original UVAS concept was a change to geosynchronous orbit in order to make the IUE operations easier and to allow the observing astronomer to participate in his or her own observing from the scientific operations control room. The 1969 meeting at the GSFC eventually resulted in the formal approval of the IUE as an international project in 1971, involving the UK Science Research Council (SRC), ESRO, and NASA. Scientists in the field of UV astronomy owe Sir Robert Wilson their heartfelt thanks for the scientific, technical, management, and international political roles he played in helping bring the IUE into existence.

I was a member of the IUE Scientific Working Group from 1974 to 1980. I was with the IUE Scientific Commissioning Team from 1977 to 1978, and took on the responsibility, in collaboration with Ralph Bohlin, of overseeing the inflight calibration of the spacecraft. I was therefore able to follow many of the developments associated with assembly, ground testing, and early inflight checkout of the spacecraft.

A geosynchronous orbit is a fine place to locate a satellite from an operations standpoint. However, from a geosynchronous orbit the satellite observatory must carry out many observations in full daylight. Therefore it was necessary to build a telescope baffle that would reject visible scattered sunlight by 16 orders of magnitude so that the observatory attitude control sensor could view and guide on stars with visual magnitudes of ~13. Also, the high orbit meant that satellite weight control was a problem throughout the building of the IUE, and weight-saving measures were often implemented. For example, the 0.45 m primary mirror was made of beryllium to save weight.

Imaging detectors were required to record efficiently the two-dimensional format of the spectra produced by the cross-dispersed echelle spectrograph aboard the IUE. However, at the time there was very little flight experience with UV sensitive imaging detectors. The only plausible candidate was the SEC vidicon, but these tubes were known to have numerous problems. Electrostatically and magnetically focused SEC vidicon tubes were commercially available, but the magnetically focused tubes were expensive and they required a massive magnet which exceeded the detector mass budget. The electrostatically focused tubes had a faceplate that would not transmit UV radiation, so it was necessary to add a proximity focused UV-to-visible light image converter to the front face of the tube. Cesium telluride photocathodes were used for the image converters in order to make the imaging detectors "solar-blind."

The possible effects of the trapped charged particles in the Van Allen belts on the vidicon detectors and their UV-to-visual light converters was a topic of concern throughout the development program. This was further aggravated because adding detector shielding had major satellite weight implications. Close to the launch of the satellite, as part of the launch readiness review, the review team asked for an independent review of the effectiveness of the detector particle radiation shielding. The resulting calculation (which later proved incorrect) suggested that the project estimate was in error by several orders of magnitude and that the detectors might saturate during integrations as short as 30 minutes for substantial portions of an IUE orbit. Since it was too late to add more shielding and the spacecraft was already near its maximum mass limit, the project decided at the last moment to install independent radiation monitors on the spacecraft. The IUE spacecraft was partially disassembled three different times at the launch site: once to add the radiation monitors, again to replace wiring in the camera electronics that had faulty insulation, and the third time to adjust the spectrograph housing after the analysis of laboratory test spectra revealed that the spectrograph structure was probably in mechanical contact with its housing. It is fortunate that all three modifications were permitted.

The IUE was launched on 26 January 1978. The subsequent events were stressful but exciting. The initial checkout went relatively smoothly, and on the third day the first spectra of calibration stars were obtained, revealing that the observatory was not in focus, but that spectroscopic data could be obtained. Two potentially crippling problems were recognized early in the commissioning of the spacecraft. The first images obtained by the short-wavelength prime camera showed extensive noise of unknown origin. The effect was attributed to microphonic noise caused by a source of acoustic interference occurring during the detector readout. It was soon discovered that the source of the disturbance was the panoramic attitude sensor which scanned the sky to detect the position of the Earth. This sensor was needed only during initial attitude acquisition, so switching this scanning sensor off solved the problem, to everyone's relief. A potentially more serious problem was caused by the onboard computer (OBC), which controlled many spacecraft functions including the spacecraft attitude. During the early commissioning, the OBC suddenly commanded the spacecraft to rapidly slew towards the Sun. The presence of the senior spacecraft engineers in the control room at the time may have saved the spacecraft since the attitude control system engineer was able to issue a command putting the spacecraft into a safe mode. Subsequent studies showed that the strange behavior of the OBC was due to overheating. An error in the thermal design of the spacecraft caused the OBC to run much hotter than expected, and when its temperature exceeded 55.8 °C, it often behaved in a highly erratic way. Operating the spacecraft so as to keep the OBC at a temperature below 55 °C solved this particular problem.

With a design lifetime of three years and a goal of five years, the IUE was flown with six operational gyros and was designed to maintain three-axis control as long as three gyros were working. After five years of operation the satellite was down to three operational gyros. Measures were therefore taken to develop alternate attitude control schemes that could work with two gyros and the fine Sun sensor. It was necessary to implement the two-gyro system in 1985. Eleven years later, in March 1996, the IUE was down to one gyro and the mission was terminated on 30 September 1996.

A full description of the IUE mission and its inflight performance is found in Boggess *et al.* (1978a,b). The spacecraft finally flown was very similar to the mission first envisioned by Wilson (1968). The low-resolution modes of the IUE provided spectra with a resolution of 6–9 Å from 1150 to 3100 Å, while the high-resolution modes produced spectra with a resolution of 0.09–0.2 Å, which corresponds to a velocity resolution ($c\Delta\lambda/\lambda$) of ~23 km s^{-1} over the same wavelength range. The imaging SEC vidicon detectors recorded ~800 and 1300 Å of the UV spectrum at short or long wavelengths, respectively.

Following my participation in the IUE commissioning and calibration programs, I eagerly awaited the onset of normal operations, about 60 days after launch. As one of the early guest observers in April 1978 I pursued several exploratory programs, including the acquisition of low-dispersion spectra of HD 38268 (R136a) in the Large Magellanic Cloud (LMC) in collaboration with Klaas de Boer and Jan Koornneef. These spectra were extremely interesting because the interstellar resonance lines of the abundant elements were found to be extremely strong. Also, the required integration time for the low-resolution spectrum was so short (~5 min) it implied that 4 to 6 hour long high-dispersion integrations on bright LMC stars might permit a detailed study of the interstellar gas in the

Milky Way and in the Magellanic Clouds. Later that year Klaas de Boer and I obtained the first high-dispersion UV spectra of the LMC stars R136a and R144. The experience of sitting in the IUE control room and seeing those first spectra successfully obtained by the modest-sized IUE spacecraft ranks as one of the most rewarding observational experiences in my life. Even the very crude spectral extraction techniques available in the IUE observer's control room suggested, in the case of R144, that strong components of interstellar C IV absorption were present at Milky Way and LMC velocities. The first scientific papers based on these spectra soon followed (Savage and de Boer 1979, de Boer et al. 1980). The observations of R144 revealed (Figure 7) very strong Milky Way C IV absorption, and an asymmetric extension to high velocity which allowed us to trace the distribution of highly ionized (hot) gas in the Milky Way halo for the first time. The C IV scale height estimated in that first paper (Savage and de Boer 1979) is very similar to the more accurate value that has been obtained from subsequent IUE and HST observations (Savage et al. 1997). Thirteen years after answering Lyman Spitzer's question about what observations I would obtain to gain more information about the physical state of the gas in the interstellar medium of the Galaxy, I was actually making those observations. The long integrations required to record at high resolution the spectra of twelfth-magnitude LMC stars was possible, since the high-energy particle radiation background effects estimated by the IUE project were more nearly correct than the hundred-fold higher backgrounds estimated by the independent review group assigned during the launch readiness review process.

During the IUE's 18-year lifetime, the 104,000 UV spectra it produced were used to study essentially all classes of astronomical objects in the UV. Some of the scientific highlights include determining measures of the stellar mass-loss phenomena across the Hertzsprung–Russell (HR) diagram; studies of the chromospheres of cool stars; studies of active galactic nucleus (AGN) variability, and nova and supernova evolution; and studies of the composition of comets. In my own areas of interest involving the interstellar medium, the IUE proved extremely valuable in mapping out the distribution of hot gas in the Milky Way halo and in determining the changing nature of interstellar extinction in different interstellar environments.

RECENT DEVELOPMENTS

The experiments described in this brief overview of early UV spectroscopy from space helped to pave the way for

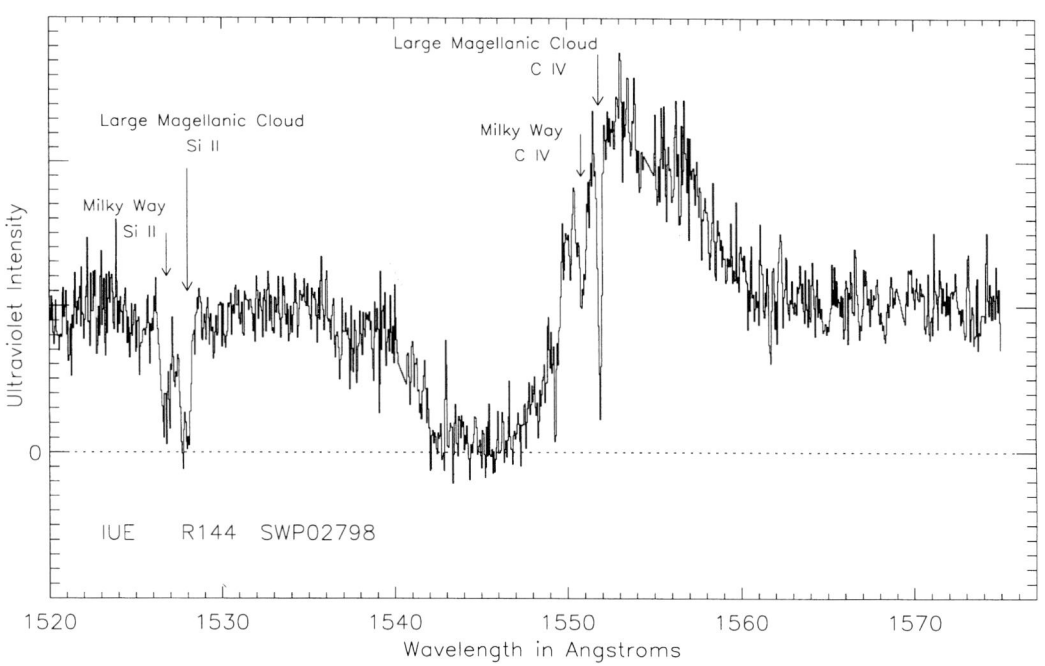

Figure 7 An IUE high-dispersion spectrum of the LMC star R144 in the region of the stellar C IV P Cygni profile near 1550 Å obtained by Savage and de Boer (1979). The spectrum reveals the presence of narrow C IV 1550.77 Å absorption features produced in the gas of the Milky Way halo and in LMC halo and disk gas. The 270 km s^{-1} radial velocity of the LMC separates the Milky Way and the LMC absorption by ~1.4 Å. This spectrum provided the first evidence that the Milky Way has an extended halo of hot ionized gas.

many subsequent exciting UV spectroscopic instruments. These have included the ASTRO 1 and 2 missions, incorporating the Hopkins Ultraviolet Telescope (Davidsen *et al.* 1992) and the Wisconsin Ultraviolet Photopolarimetric Experiment (Nordsieck *et al.* 1994); the Orbiting and Retreivable Far and Extreme Ultraviolet Spectrometers (Grewing *et al.* 1991, Hurwitz *et al.* 1998); the Interstellar Medium Profile Spectrometer (Jenkins *et al.* 1996); the Goddard High Resolution Spectrograph (Brandt *et al.* 1994); the Space Telescope Imaging Spectrograph (STIS, deployed on HST in 1997; Woodgate *et al.* 1998); and the Far Ultraviolet Spectroscopic Explorer (FUSE, launched in 1999; Moos *et al.* 2000). UV spectrographic instruments currently under development include the Cosmic Origins Spectrograph (COS, to be deployed on HST in 2004; Morse *et al.* 1998) and the Galaxy Evolution Explorer (GALEX, to be launched in 2002; Martin *et al.* 1999).

The future of UV spectroscopy in space beyond STIS, FUSE, GALEX, and COS is quite uncertain. In the USA, NASA is considering the Space Ultraviolet-Visible Observatory (SUVO; Shull *et al.* 1999), a large aperture (8-meter class) facility which would provide hundred fold increases in UV spectroscopic throughput and multiplexing efficiency compared to HST+COS. However, the true costs of such an ambitious mission, which would follow sometime after the Next Generation Space Telescope (NGST) – which has no UV capability – are likely very large. This suggests that it may be well into the 2010s before such a facility will exist in space. Hopefully, NASA will create a balanced mix of space mission opportunities so that the field of UV spectroscopy in space can survive after the HST ceases to operate.

Acknowledgments

I thank Marilyn Meade and Robert Bless for comments about a draft version of this paper. Thanks are also extended to Marilyn for preparing Figures 5 and 7.

REFERENCES

Baum, W.A., Johnson, F.S., Oberly, J.J., Rockwood, C.C., Strain, C.V. and Tousey, R. (1946). Solar ultraviolet spectrum to 88 kilometers. *Physical Review*, **70**, 781–782.

Bless, R.C. and Code, A.D. (1972). Ultraviolet astronomy. *Annual Review of Astronomy and Astrophysics*, **10**, 197–226.

Bless, R.C., Code, A.D. and Fairchild, E.T. (1976). Ultraviolet photometry from the Orbiting Astronomical Observatory: XXI. Absolute energy distributions in the ultraviolet. *Astrophysical Journal*, **203**, 410–416.

Bless, R.C. and Savage, B.D. (1970). Observations of interstellar extinction in the ultraviolet with the OAO satellite. In L. Houziaux and H.E. Butler (eds), *IAU Symposium No. 36, Ultraviolet Stellar Spectra and Ground Based Observations (Lunteren, The Netherlands)*, Reidel, Dordrecht, pp. 28–35.

Bless, R.C. and Savage, B.D. (1972). Ultraviolet photometry from the Orbiting Astronomical Observatory: II. Interstellar extinction. *Astrophysical Journal*, **171**, 293–308.

de Boer, K.S., Koornneef, J. and Savage, B.D. (1980). Ultraviolet absorption by interstellar gas near 30 Doradus. *Astrophysical Journal*, **236**, 769–778.

de Boer, K.S and Lamers, H.J.G.L.M. (1978). Interstellar depletion of Fe, Mn and Mg. *Astronomy and Astrophysics*, **69**, 327–332.

Boggess, A., Carr, F.A., Evans, D.C., Fischel, D., Freeman, H.R., Fuechsel, C.F., Klinglesmith, D.A., Krueger, V.L., Longanecker, G.W., Moore, J.V., Pyle, E.J., Rebar, F., Sizemore, K.O., Sparks, W., Underhill, A.B., Vitagliano, H.D., West, D.K., Macchetto, F., Fitton, B., Barker, P.J., Dunford, E, Gondhalekar, P.M., Hall, J.E., Harrison, V.A.W, Oliver, M.B., Sandford, M.C.W., Vaughan, P.A., Ward, A.K., Anderson, B.E., Boksenberg, A., Coleman, C.I., Snijders, M.A.J. and Wilson, R. (1978a). The IUE spacecraft and instrumentation. *Nature*, **275**, 372–377.

Boggess, A., Bohlin, R.C., Evans, D.C., Freeman, H.R., Gull, T.R., Heap, S.R., Klinglesmith, D.A., Longanecker, G.R., Sparks, W., West, D.K., Holm, A.V., Perry, P.M., Schiffer III, F.H., Turnrose, B.E., Wu, C., Lane, A.L., Linsky, J.L., Savage, B.D., Benvenuiti, P., Cassatella, A., Clavel, J., Heck, A., Macchetto, F., Penston, M.V., Selvelli, P.L., Dunford, E., Gondhalekar, P., Oliver, M.B., Sanford, M.C.W., Stickland, D., Boksenberg, A., Coleman, C.I., Snijders, M.A.J and Wilson, R. (1978b). In-flight performance of the IUE. *Nature*, **275**, 377–385.

Boggess, A. and Wilson, R. (1987). The history of IUE. In Y. Konod *et al.* (eds), *Exploring the Universe With the IUE Satellite*, Reidel, Dordrecht, pp. 3–19.

Boksenberg, A., Kirkham, B., Towlson, W.A., Venis, T.E., Bates, B., Courts, G.R. and Ca, P.P.D, (1972). *Nature Physical Science*, **240**, 127–128.

Boksenberg, A., Evans, R.G., Fowler, R.G., Gardner, I.S.K., Houziaux, L., Humphries, C.M., Jamar, C., Macau, D., Malaise, D., Monfils, A., Nandy, K., Thompson, G.I., Wilson, R. and Wroe, H. (1973). The ultraviolet sky-survey telescope in the TD-IA satellite. *Monthy Notices of the Royal Astronomical Society*, **163**, 291–322.

Bowyer, S. (1991). EUV astronomy on the Apollo–Soyuz mission: The first detection of stellar EUV sources and other astrophysical results. In R.F. Mallina and S. Bowyer (eds), *Extreme Ultraviolet Astronomy (Berkeley, California)*, Pergamon, Elmsford, NY, pp. 3–7.

Brandt, J.C., Heap, S.R., Beaver, E.A., Boggess, A., Carpenter, K.G., Ebbets, D.C., Hutchings, J.B., Jura, M., Leckrone, D.S., Linsky, J.L., Maran, S.P., Savage, B.D., Smith, A.M., Trafton, L.M., Walter, F.M., Weymann, R.J., Ake, T.B., Bruhweiler, F., Cardelli, J.A., Lindler, D.J., Malumuth, E., Randall, C.E., Robinson, R., Shore, S.N. and Wahlgren, G. (1994). The Goddard High Resolution Spectrograph: Instrument, goals, and science results, *Publications of the Astronomical Society of the Pacific*, **106**, 890–908.

Bryam, E.T., Chubb, T.A., Friedman, H. and Kupperian, J.E. (1957). In M. Zelikoff (ed.), *The Threshold of Space, Proceedings of the Conference on Chemical Aeronomy*, Pergamon, New York, p. 203.

Brosch, N. (1999), Ultraviolet sky surveys: Instruments, findings, and prospects. *Experimental Astronomy*, **9**, 119–187.

Burton, W.M., Evans, R.G., Griffin W.G., Lewis, C., Paxton, H.J.B., Shenton, D.B., Macchetto, F.D., Boksenberg, A. and Wilson, R. (1973). Ultraviolet spectra of Gamma Velorum and Zeta Puppis. *Nature Physical Science*, **246**, ps37–ps40

Carruthers, G. (1968). Far-ultraviolet spectroscopy and photometry of some early-type stars. *Astrophysical Journal*, **151**, 269–284.

Carruthers, G. (1970). Rocket observation of interstellar molecular hydrogen. *Astrophysical Journal*, **161**, L81–L85.

Code, A.D. (1960). Stellar astronomy from a space vehicle. *Astronomical Journal*, **65**, 278–284.

Code, A.D., Houck, T.E., McNall, J.F., Bless, R.C. and Lillie, C.F. (1970). Ultraviolet photometry from the Orbiting Astronomical Observatory: I. Instrumentation and operation. *Astrophysical Journal*, **161**, 377–388.

Code, A.D. and Savage, B.D. (1972). Orbiting Astronomical Observatory: Review of scientific results. *Nature*, **77**, 213–221.

Code, A.D., Bless, R.C., Davis, J. and Brown, R.H. (1976). Empirical effective temperatures and bolometric corrections for early type stars. *Astrophysical Journal*, **203**, 417–434.

Davidsen, A.F., Long, K.S., Durrance, S.T., Blair, W.P., Bowers, C.W., Conard, S.J., Feldman, P.D., Ferguson, H.C., Fountain, G.H., Kimble, R.A., Kriss, G.A., Moos, H.W. and Potocki, K.A. (1992). The Hopkins Ultraviolet Telescope: Performance and calibration during the Astro-1 mission. *Astrophysical Journal*, **392**, 264–271.

Davis, R.J., Deutschmann, W.A. and Haramundanis, K.L. (1973). *Celescope Catalog of Ultraviolet Magnitudes*, SAO Special Report 350, Smithsonian Astrophysical Observatory, Cambridge, MA.

Friedman, H. (1959). Rocket Spectroscopy. *Journal of Geophysical Research*, **64**, 1751–1764.

Grewing, M., Kramer, G., Appenzeller, I., Eberhard, N., Gringel, W., Kalble, A., Kappelmann, N., Krautter, J., Mandel, H., Ostreicher, R., Petrik, J. and Riegger, J. (1991). The ORFEUS Mission. In R.F. Malina and S. Bowyer (eds), *Extreme Ultraviolet Astronomy*, Pergamon, Elmsford, pp. 437–441.

Holberg, J.B., Sandel, B.R., Forrester, W.T., Broadfoot, A.L., Shipman, H. and Barry, D.C. (1980). Extreme-UV and far-UV observations of the white dwarf HZ 43 from Voyager 2. *Astrophysical Journal Letters*, **242**, L119–L123.

van der Hucht, K.A., Lamers, H.J.G.L.M., Faraffiana, R., Hack, M. and Stalio, R. (1976). Line identification in the near-ultraviolet for nine bright stars. *Astronomy and Astrophysics* Suppl., **25**, 65–128.

Hurwitz, M., Bowyer, S., Bristol, R., Dixon, W.V.D., Dupuis, J., Edelstein, J., Jelinsky, P.S., Timothy P. and Siegmund, O. (1998). Far-ultraviolet performance of the Berkeley Spectrograph during the ORFEUS-SPAS II Mission. *Astrophysical Journal, Letters*, **500**, L1–L7.

de Jager, C., Kondo, Y., Hoekstra, R., van der Hucht, K.A., Kamperman, T.M., Lamers, H.J.G.L.M., Modisette, J.L. and Morgan, T.H. (1979). Balloon-borne Ultraviolet Stellar Spectrograph: II. Highlights of first observational results. *Astrophysical Journal*, **230**, 534–559.

de Jager, C., Hoekstra, R., van der Hucht, K.A., Kamperman, T.M., Lamers, H.J.G.L.M., Hammerschlag, A., Werner, W. and Emming, J.G. (1974). The Orbiting Stellar Ultraviolet Spectrophotometer S59 in ESRO's TD-1A satellite. *Astrophysics and Space Science*, **26**, 207–262.

Jenkins, E.B. and Morton, D.C. (1967). Far ultraviolet spectra of Orion stars. *Nature*, **215**, 1257–1259.

Jenkins, E.B. (1978). Coronal gas in the Galaxy: I. A new survey of interstellar O VI. *Astrophysical Journal*, **219**, 845–860.

Jenkins, E.B. and Meloy, D.A. (1974). A survey with Copernicus of interstellar O VI absorption. *Astrophysical Journal Letters*, **193**, L121–L125.

Jenkins, E.B., Reale, M.A., Zucchino, P.M. and Sofia, U.J. (1996). High resolution spectroscopy in the far-UV: Observations of the interstellar medium by IMAPS on ORFEUS-SPAS. *Astrophysics and Space Science*, **239**, 315–360.

Kondo, Y., Giuli, R.T., Modisette, J.L. and Rydgren, A.E. (1972). Observations of the stellar MG II resonance doublet at 2795 and 2802 Å. *Astrophysical Journal*, **176**, 153–164.

Kondo, Y., de Jager, C., Hoekstra, R., van der Hucht, K.A., Kamperman, T.M., Lamers, H.J.G.L.M., Modisette, J.L. and Morgan, T.H. (1979). Balloon-borne Ultraviolet Stellar Spectrograph. I. Instrumentation and observation. *Astrophysical Journal*, **230**, 526–533.

Lampton, M., Margon, B., Paresce, F., Stern, R. and Bowyer, S. (1976). Discovery of a nonsolar extreme-ultraviolet source. *Astrophysical Journal Letters*, **203**, L71–L74.

Martin, C.D., Bianchi, L., Donas, J., Heckman, T., Madore, B., Malina, R., Milliard, B., Friedman, P., Rich, M., Schiminovich, D., Siegmund, O. and Szalay, A. (1999). The Galaxy Evolution Explorer. In J.A. Morse et al. (eds), *Ultraviolet-Optical Space Astronomy Beyond HST*, ASP Conference Series Vol. 164, pp. 182–193.

Moos, H.W., Cash, W.C., Cowie, L.L., Davidsen, A.F., Dupree, A.K., Feldman, P.D., Friedman, S.D., Green, J.C., Green, R.F., Gry, C., Hutchings, J.B., Jenkins, E.B., Linsky, J.L., Malina, R.F., Michalitsianos, A.G., Savage, B.D., Shull, J.M., Siegmund, O.H.W., Snow, T.P., Sonneborn, G., Vidal-Madjar, A., Willis, A.J.B.E., York, D.G., Ake, T.B., Andersson, B.G., Andrews, J.P., Barkhouser, R.H., Bianchi, L., Blair, W.P., Brownsberger, K.R., Cha, A.N., Chayer, P., Conard, S.J., Fullerton, A.W., Gaines, G.A., Grange, R., Gummin, M.A., Hebrard, G., Kriss, G.A., Kruk, J.W., Mark, D., McCarthy, D.K., Morbey, C.L., Murowinski, R., Murphy, E.M., Oegerle, W.R., Ohl, R.G., Oliveira, C., Osterman, S.N., Sahnow, D.J., Saisse, M., Sembach, K.R., Weaver, H.A., Welsh, B.Y., Wilkinson, E. and Zheng, W. (2000). Overview of the Far Ultraviolet Spectroscopic Explorer mission. *Astrophysical Journal Letters*, **538**, L1–L6.

Morse, J.A., Green, J.C., Ebbets, D.C., Andrews, J.P., Heap, S.R., Leitherer, C., Linsky, J.L., Savage, B.D., Shull, J.M., Snow, T.P., Stern, S.A., Stocke, J.T. and Wilkinson, E. (1998). Performance overview and science goals of the Cosmic Origins Spectrograph for the Hubble Space Telescope, *Proceedings of the Society of Photo-Optical Instrumentation Engineers*, **3356**, 361–368.

Morton, D.C. (1967). The far-ultraviolet spectra of six stars in Orion. *Astrophysical Journal*, **147**, 1017–1024.

Morton, D.C. (1969). Rocket observations of mass loss from hot stars. In M. Hack (ed.), *Mass Loss from Stars*, Reidel, Dordrecht, pp. 36–41.

Morton, D.C., Jenkins, E.B. and Bohlin, R.C. (1968). Rocket observations of Orion stars with an all-reflective ultraviolet spectrograph. *Astrophysical Journal*, **154**, 661–676.

Morton, D.C. and Spitzer, L. (1966). Line spectra of Delta and Pi Scorpii in the far-ultraviolet. *Astrophysical Journal*, **144**, 1–12.

Nordsieck, K.H., Code, A.D., Anderson, C.M., Meade, M.R., Babler, B., Michalski, Donald, E., Pfeifer, R.H. and Jones, T.E. (1994). Exploring ultraviolet astronomical polarimetry: Results from the Wisconsin Ultraviolet Photo-Polarimeter Experiment (WUPPE), *Proceedings of the Society of Photo-Optical Instrumentation Engineers*, **2010**, 2–11.

Oberth, H. (1923). *Die Rakete zu den Planetenraumen*, R. Oldenbourg-Verlag, Munich.

Richardson, R.S. (1940). *Astounding Science Fiction*, Feb., p. 113.

Rogerson, J.B. (1963). The Orbiting Astronomical Observatories. *Space Science Review*, **2**, 621–652.

Rogerson, J.B., Spitzer, L., Drake, J.F., Dressler, K., Jenkins, E.B., Morton, D.C. and York, D.G. (1973). Spectrophotometric results from the Copernicus satellite: I. Instrumentation and performance. *Astrophysical Journal*, **181**, L110–L115.

Savage, B.D. and de Boer, K.S. (1979). Observational evidence for a hot gaseous galactic corona. *Astrophysical Journal Letters*, **230**, L77–L82.

Savage, B.D. and Code, A.D. (1970). Observations of interstellar Lyman-α with the Orbiting Astronomical Observatory. In L. Houziaux and H.E. Butler (eds), *IAU Symposium No. 36, Ultraviolet Stellar Spectra and Ground Based Observations (Lunteren, The Netherlands)*, Reidel, Dordrecht, pp. 302–314.

Savage, B.D. and Jenkins, E.B. (1972). A survey of local interstellar hydrogen from OAO-2 observations of Lyman alpha absorption. *Astrophysical Journal*, **172**, 491–511.

Savage, B.D., Sembach, K.R. and Lu, L. (1997). Absorption by highly ionized interstellar gas along extragalactic and galactic sight lines. *Astronomical Journal*, **113**, 2158–2185.

Shull, J.M., Savage, B.D., Morse, J.A., Neff, S.G., Clarke, J.T., Heckman, T., Kinney, A.L., Jenkins, E.B., Dupree, A.K., Baum, S.A. and Hasan, H.

(1999). The emergence of the modern Universe: Tracing the cosmic web. White Paper of the UV-Optical Working Group, Boulder, University of Colorado, pp. 1–48.

Snow, T.P. and Morton, D.C. (1976). Copernicus ultraviolet observations of mass loss effects in O and B stars. *Astrophysical Journal* Suppl, **32**, 429–465.

Spitzer, L. (1946). Advantages of an extra-terrestrial observatory. Project RAND Report, Douglas Aircraft Co. [Reprinted in Spitzer (1997), pp. 372–380.]

Spitzer, L. (1997). *Dreams, Stars, and Electrons: Selected Writings of Lyman Spitzer, Jr.*, Princeton University Press, pp. 158–168.

Spitzer, L. and Jenkins, E.B. (1973). Ultraviolet studies of the interstellar gas. *Annual Reviews of Astronomy and Astrophysics*, **13**, 133–164.

Spitzer, L. and Zabriskie, F. (1959). Interstellar research with a spectroscopic satellite. *Publications of the Astronomical Society of the Pacific*, **71**, 412–420.

Spitzer, L., Drake, J.F., Jenkins, E.B., Morton, D.C., Rogerson, J.B. and York, D.G. (1973). Spectrophotometric results from the Copernicus satellite: IV. Molecular hydrogen in interstellar space. *Astrophysical Journal*, **181**, L116–L121.

Stecher, T.P. (1965). Interstellar extinction in the ultraviolet. *Astrophysical Journal*, **142**, 1683–1684.

Stecher, T.P. (1969). Interstellar extinction in the ultraviolet: II. *Astrophysical Journal*, **157**, L125–L126.

Stecher, T.P. (1970). Ultraviolet interstellar extinction from a comparison of ε Persei and ζ Persei. In L. Houziaux and H.E. Butler (eds), *IAU Symposium No. 36, Ultraviolet Stellar Spectra and Ground Based Observations (Lunteren, The Netherlands)*, Reidel, Dordrecht, pp. 24–27.

Stecher, T.P. and Donn, B. (1965). On graphite and interstellar extinction. *Astrophysical Journal*, **142**, 1681–1683.

Stecher, T.P. and Milligan, J.E. (1962). Stellar spectrophotometry from above the atmosphere. *Astrophysical Journal*, **136**, 1–13.

Wilson, R. (1968). *Ultraviolet Astronomical Satellite, Final Report*, Vols. 1, 2, and 3. UK Atomic Energy Authority, Culham Laboratory.

Wilson, R. and Boksenberg, A. (1969). Ultraviolet Astronomy. *Annual Review of Astronomy and Astrophysics*, **7**, 421–472.

Wilson, R. (1970). Observations of ultraviolet stellar spectra. In L. Houziaux and H.E. Butler (eds), *IAU Symposium No. 36, Ultraviolet Stellar Spectra and Ground Based Observations (Lunteren, The Netherlands)*, Reidel, Dordrecht, pp. 147–162.

Woodgate, B.E., Kimble, R.A., Bowers, C.W., Kraemer, S., Kaiser, M.E., Danks, A.C., Grady, J.F., Loiacono, J.J., Brumfield, M., Feinberg, L., Gull, T.R., Heap, S.R., Maran, S.P., Lindler, D., Hood, D., Meyer, W., Vanhouten, C., Argabright, V., Franka, S., Bybee, R., Dorn, D., Bottema, M., Woodruff, R., Michika, D., Sullivan, J., Hetlinger, J., Ludtke, C., Stocker, R., Delamere, A., Rose, D., Becker, I., Garner, H., Timothy, J.G., Blouke, M., Joseph, C.L., Hartig, G., Green, R.F., Jenkins, E.B., Linsky, J.L., Hutchings, J.B., Moos, H.W., Boggess, A., Roesler, F. and Weistrop, D. (1998). The Space Telescope Imaging Spectrograph design, *Publications of the Astronomical Society of the Pacific*, **110**, 1183–1204.

York, D.G. (1974). Highly ionized atoms observed with Copernicus. *Astrophysical Journal*, **193**, L127–L125.

14

MARTIN HARWIT*

The early days of infrared space astronomy

INTRODUCTION

Infrared observations can provide a number of unique perspectives on the Universe. This article concerns itself with the men and women who pioneered infrared space astronomy, traces the different techniques they employed, describes the trials and tribulations they had to overcome, and lists some of the gains they achieved. In the process it also examines the social institutions, both academic and military, that were most influential in determining the evolution of the field.

While many infrared and submillimeter observations required going into space, a large number of others were carried out from the ground. The earliest successes of infrared astronomy were largely the work of ground-based astronomers. Going into space required major technological breakthroughs that took time to materialize. A rich texture of sometimes friendly, occasionally aggressive, competition among ground-based, balloon, airborne, rocket, and satellite observers emerged, that persists to this day.

As an active participant in the field, the author makes no claims to objectivity. Nor is he able, in a chapter of twenty or thirty thousand words, to give proper credit to all who made significant contributions. The story told here is overwhelmingly based on personal experiences in the USA. Many of the struggles faced in the USA, however, were duplicated in Europe, Japan, and the Soviet Union. An excellent review written more than 20 years ago by J.E. Beckman and A.F.M. Moorwood (1979) may provide the reader with a complementary European perspective, and D.A. Allen's even earlier book adds an Australian's viewpoint (Allen 1975). The author hopes that contributions such as these may eventually be synthesized to produce a dispassionate, comprehensive, truly international, and perhaps more broadly incisive account.

The cold universe

Much of the Universe is cold. The wavelength, λ, at which stars, galaxies, and interstellar or intergalactic gases radiate is inversely proportional to their temperature, T (Figure 1). The peak emission tends to occur at $\lambda = 3700/T$ micrometers, where T is measured in kelvin (K). One micrometer is a millionth of a meter, traditionally called a micron and abbreviated to μm (or simply μ in the early literature of the field). Observations at infrared wavelengths from 1 to 1000 μm are thus uniquely sensitive to astronomical sources in the temperature range from ~3000 K to 3 K. These include the coolest stars, planets, and interplanetary dust, circumstellar and interstellar matter, and, at the longest wavelengths and coldest temperatures, the earliest known radiation emitted by the entire Universe.

A dusty universe

Interstellar dust – microscopic particles composed of ices, minerals, and common organic and inorganic materials – is abundant throughout the Universe. These particles obstruct a clear view of the cosmos at optical wavelengths. Fortunately, a cloud of dust that may be totally opaque in the visible or ultraviolet can be virtually transparent in the infrared. Thus,

---

*Professor Emeritus of Astronomy, Cornell University, Ithaca, NY, USA; Former Director, National Air and Space Museum, Washington, DC, USA

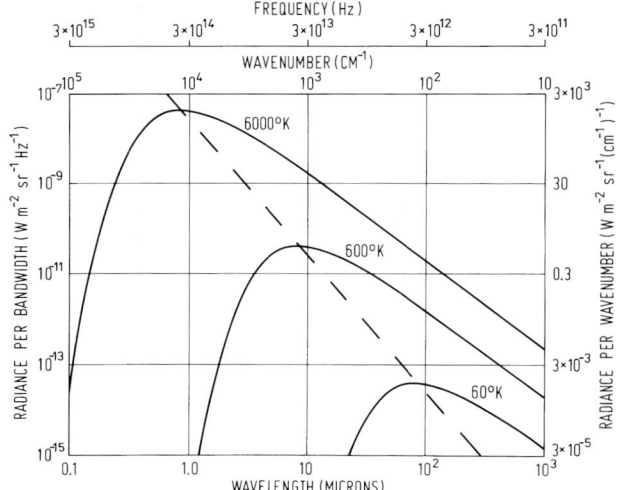

Figure 1 Blackbody radiation spectrum as a function of temperature. As the temperature T decreases by a factor of 10, the wavelength of peak emission increases by the same factor and the total power emitted over the entire wavelength range decreases by a factor of 10^4.

infrared wavelengths can probe regions deep in the core of a dusty galaxy that are inaccessible at shorter wavelengths.

Dust grains are heated by short-wavelength, mainly visible and ultraviolet radiation. They absorb this energy, and re-radiate it at infrared wavelengths. Because of this efficient conversion, the majority of the radiant energy emitted by dense, dusty regions, such as star-forming clouds and occasionally entire galaxies, lies at infrared wavelengths.

Since the transparency of dust clouds increases at longer wavelengths, red light is transmitted more readily than blue. Stars lying behind a tenuous dust cloud thus appear reddened. This reddening can be a useful measure of the amount of dust in the cloud.

The early universe

In the expanding Universe, the more distant an object the greater the velocity with which it recedes from us. This cosmic expansion shifts starlight from distant galaxies into the infrared region; the more distant the source, the farther is its radiation shifted into the infrared. The expansion is characterized by a redshift parameter, z, where $1+z$ gives the ratio of "wavelength reaching the observer" divided by "wavelength actually emitted in the distant source": $1+z = \lambda_{received}/\lambda_{emitted}$.

The most distant quasars and galaxies observed to date have redshifts $z > 5$, so that radiation from the center of the visual band is shifted to beyond 3 μm. Because $1+z$ also represents the factor by which the Universe has expanded since the radiation was emitted, objects at $z = 5$ are seen as they were at an epoch when the Universe was only one-sixth of its present size.

The chemical universe

The infrared band contains the spectral signatures of a variety of atoms, molecules, ions, and solids, at least some of which are found in any astrophysical environment. Infrared spectroscopy allows us to recognize large varieties of substances, ranging from cool ices in the interstellar medium to highly excited ions in active galactic nuclei. By determining the strengths of different spectral features, we obtain important, often unique insights into the chemical and physical conditions in these systems.

INFRARED ASTRONOMICAL CONSTRAINTS

The human eye ceases to respond to radiation beyond the red part of the spectrum. This limit falls at a wavelength of ~ 0.72 μm. The *infrared spectral domain* begins approximately at a wavelength of 0.7–1 μm, and stretches out to ~ 1000 μm (= 1 mm). The wavelength range between ~ 200 μm and 1 mm is often called the *submillimeter domain*.

While the atmosphere tends to be opaque to much of the infrared spectral band, observations can be carried out from high mountain tops in a number of *atmospheric windows*. The lowest panel of Figure 2 shows the approximate widths and depths of the atmospheric windows at an observatory like Mauna Kea, Hawaii, sited at an altitude of 4.2 km. As seen from the upper panels, atmospheric absorption decreases and transmission increases considerably with increasing altitude, and becomes quite high throughout much of the infrared domain at aircraft and balloon altitudes, respectively, at ~ 14 and ~ 28 km.

For the astronomer, good atmospheric transmission is not enough. Low atmospheric emission is also essential. The atmosphere radiates powerfully throughout much of the infrared, and this emission is often orders of magnitude greater than the radiation from astronomical sources. The intensity of atmospheric emission also tends to fluctuate with time. In the near-infrared, at $1 \leq \lambda \leq 4$ μm, emission by OH radicals is highly variable. In the far-infrared, at around 100 μm, variable atmospheric water vapor content produces varying emissivity. Such fluctuations cause the flux, or *signal*, from an astronomical source to be marred by superposed *noise* due to randomly varying atmospheric emission.

Poor atmospheric transmission, and strong atmospheric emission are the prime reasons for undertaking infrared astronomical observations from space. However, in the early 1960s when serious efforts to conduct such observations were beginning, many of these factors were poorly understood. This led to parallel developments in infrared astronomy. Most astronomers worked with ground-based equipment taken to high mountain tops, while a few

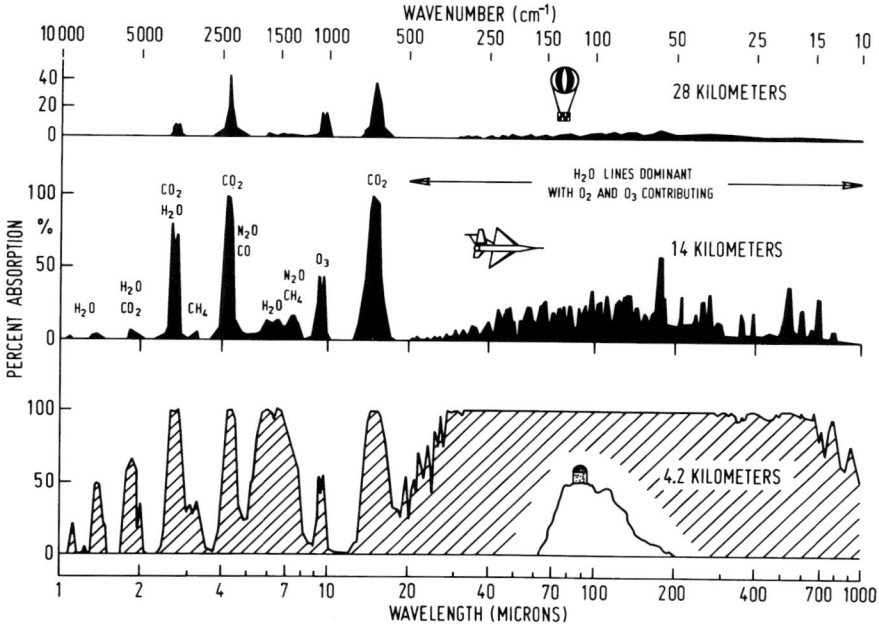

Figure 2 Atmospheric absorption in the infrared at altitudes of balloons, aircraft, and a mountain-top site like Mauna Kea (after Traub and Stier 1976).

attempted to build the instrumentation required to carry out observations from above the atmosphere.

Since rocket payloads were relatively expensive, US ground-based observers feared that the National Aeronautics and Space Administration (NASA) or National Science Foundation (NSF) support for rocket work would drain away funding from their own efforts. Uncertainties about the precise limitations of ground-based techniques often led to acrimonious disputes between ground-based observers and those who wanted to launch sensitive infrared astronomical telescopes above the atmosphere on rockets. Time and again, the more traditional astronomers bluntly recommended that rocket-borne efforts be slashed or pared back to an absolute minimum. Pioneering infrared rocket astronomy was not a happy venture!

To understand the backdrop against which the first successful observations were obtained, one needs to understand not only the technical difficulties that had to be overcome, but also the considerable early successes of ground-based observers who also had to surmount significant instrumental limitations but were able to make headway more quickly in more modest efforts.

EARLY GROUND-BASED EFFORTS

While the earliest attempts to conduct observations of astronomical sources beyond the red part of the spectrum date back to those of Sir William Herschel (1738–1822), who noted the thermal emission of the Sun beyond the visible spectral range, rapid progress did not occur until the 1960s. Two factors limited serious advances: a lack of sufficiently sensitive detectors and a lack of motivation.

In 1838 William Herschel's son John determined the total power emitted by the Sun. His results, obtained by measuring the rise in temperature of water in a blackened vessel on which was focused sunlight, were remarkably accurate given the simplicity of his apparatus. He found that the Sun radiated with enormous power. This posed a problem. Julius Robert Mayer's principle of conservation of energy, enunciated in 1845, led to the realization that the source of this energy had to be huge or the Sun very young, otherwise the Sun's energy would long ago have been depleted. Though many sought to resolve this puzzle, a satisfactory explanation did not emerge until the twentieth century and the development of theories of relativity and nuclear physics.

Thirty years after John Herschel's monumental finding, William Huggins used a thermopile connected to a galvanometer in the far more difficult attempt to measure the heat emitted by stars (Huggins 1869). (A thermocouple is a device consisting of strips of dissimilar metals that generate a voltage when one end of the device is heated.) These measurements were not particularly convincing, but Huggins's contemporary E.J. Stone developed the technique further (Stone 1870). He used two thermopiles back to back, alternating the positioning of a star first on one thermopile and then on the other, in a mode that we would today call "push–pull chopping" (Figure 3). He also calibrated his system against a radiating cube kept at the temperature of boiling water. With these steps he obtained more credible results.

Struggles to perfect increasingly sensitive instrumentation persisted. In the early decades of the twentieth century, pioneering work on better radiometers enabled W.W. Coblentz at the US National Bureau of Standards to report radiometric measurements of 110 stars, and a series of collaborations between Edison Pettit and S.B. Nicholson provided greater insights into planetary and lunar phenomena (Coblentz 1914, Pettit and Nicholson 1922).

The history of infrared detector development, however, was largely guided not by astronomers, but by military needs, such as "night vision" enabling warm objects to be discerned in the dark. World War II brought about the development of lead sulfide (PbS) cells that were far more sensitive than thermopiles. In the post-war era, Peter Felgett in Britain was one of the first to use these devices to measure the fluxes from bright stars (Felgett 1951). Later, the Cold War further accelerated military development of infrared sensors. One particularly successful detector, indium antimonide (InSb), came into use at wavelengths out to 5 μm. At longer wavelengths, the military was developing copper-doped germanium (Ge:Cu) and gallium-doped germanium (Ge:Ga) photoconductors. All these devices worked at their best only over a limited bandwidth, roughly $\Delta\lambda/\lambda \sim \frac{1}{3}$. To obtain continuous coverage over a broader bandwidth a succession of different detector materials had to be employed, each cryogenically cooled to its own optimum operating temperature.

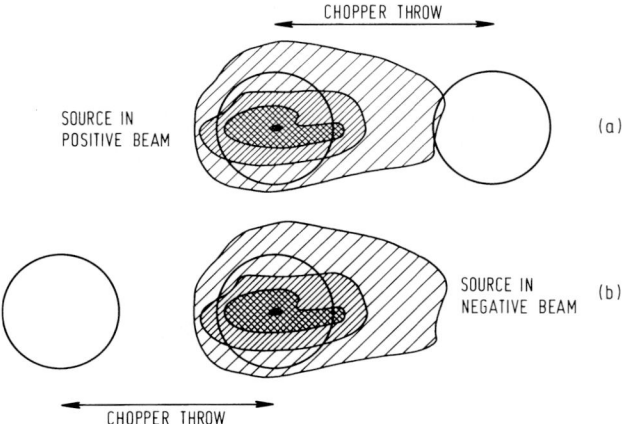

Figure 3 "Push–pull chopping" of a source by switching the field of view of the telescope between an astronomical source and its surroundings on two opposite sides. By subtracting the emission from the surroundings, a first attempt is made to correct for atmospheric emission above the telescope.

If an astronomer could secure a particularly sensitive and stable detector, it was possible to undertake novel infrared measurements. One observer who took good advantage of these detectors was Harold L. Johnson who, with various co-workers, pioneered the Ultraviolet–Blue–Visual–Red–Infrared (UBVRI) photometric system still in wide use. The *I* band lies at 0.9 μm. Johnson also made observations in the *K* band at 2.2 μm and at longer wavelengths. He set out to obtain accurate photometric data on several thousand bright stars and soon realized the importance of reddening and extinction by interstellar dust (Johnson 1965). The general attitude of his time, however, remained that infrared astronomy would continue to be merely an adjunct to optical work. Stars, after all, were known to be visible objects; and the Universe appeared to be largely an aggregate of stars.

Given the still-formidable technical difficulties, most astronomers found little motivation for embarking on a career in infrared astronomy.

This attitude changed with the work of Gerry Neugebauer and Robert W. Leighton at Caltech, who decided to conduct an unbiased survey of the sky at 2.2 μm, where the atmosphere is transparent and PbS is sensitive. Since the amount of observing time that would be required was great, and no telescope was available for the purpose, Leighton and Neugebauer constructed their own. Much of their inexpensive instrumentation was war-surplus, but the 1.57 m (62-inch), *f*/1 parabola they conceived was made from aluminum hollowed out on a lathe. The surface obtained in this fashion was not sufficiently accurate, so they poured a layer of epoxy into the paraboloid, spun it up around its vertical axis and kept it spinning until the epoxy had set. Properly coated, this gave a reflecting surface good enough to produce images of the order of 2 arc minute resolution. This was satisfactory, since their eight-element PbS array pairwise scanning the sky in a push–pull chopping mode would have to cover 30,000 square degrees. This meant that the beam employed would, of necessity, have to be sizeable.

To assure themselves of the performance of their apparatus, Leighton and Neugebauer cross-calibrated against stars already known from Johnson's work. But they also hoped to go further. They were no longer restricted to pointing at visible stars and might find sources that had no optical counterparts.

The final results of their survey were not to be published until the late 1960s. But while the work was in progress the two researchers and one of their students discovered a number of truly remarkable sources. A letter to the *Astrophysical Journal* (Neugebauer et al. 1965) broke the news. It was electrifying!

> The motivation for such a survey is to obtain an unbiased census of objects that emit in the 2.0–2.5 μm atmospheric window [which] might reveal many potentially

interesting objects that would not be included in an *a priori* selection of objects to be measured. In fact, a number of strikingly red stars have already been found... in a few cases the stars seem to be intrinsically extremely red... the value $I-K \approx 7.5$ being a typical color index. Approximately ten such extremely red objects have been discovered so far. Of these only one, TX Cam, has been found listed in a catalog of stars or variable stars... Only the brightest of them has yet been seen visually, and it is a difficult visual object even at the 200-inch telescope... Near-infrared spectra of two of the brightest objects, as well as photometry in the 2.0–2.5 and 3.2–4.1 μ spectral ranges, confirm that a black-body radiation temperature of the order of 1000 °K is appropriate... A 1000 °K black body of apparent magnitude $K=0.0$ should have an angular radius of approximately $0''.04$, or about $10 R_{\odot}$ [ten solar radii] per parsec distance. If the apparent tendency toward occurrence near the plane of the Galaxy is real, this would imply large distances and huge sizes for these objects.

The paper then identified two sources that had been found to have respective K magnitudes of 0 and 1. These immortalized the three authors' names (Neugebauer, Martz, and Leighton) through the designations NML Tau and NML Cyg – still in use today.

Here was a jolting concept: at wavelengths only four times longer than visible light, infrared astronomy had discovered bright sources all but inaccessible in the optical domain, grotesque stars larger than anything known. How many more unbelievably strange objects might we find if we conducted unbiased infrared surveys at wavelengths ten, or a hundred, or a thousand times longer than visible light? Ground-based astronomers began to flock slowly to the infrared.

Four years after this paper, the final catalog was published (Neugebauer and Leighton 1969). It was complete to magnitude 3 at 2.2 μm, and listed 5,612 objects north of declination $-33°$. Famous sources with designations such as Infrared Catalog entry IRC +10216 have graced literally thousands of papers, and continue to attest to the importance of the catalog.

BOLOMETRY

Shortly after World War II, Marcel J.E. Golay continued the development of *bolometric* – wavelength-independent – energy sensors (Golay 1947a,b). He described a pneumatic infrared detector that came into use in laboratory spectroscopy for the identification of infrared spectral features of chemicals. These chemical signatures would, decades later, become important in infrared astronomical spectroscopy.

A truly sensitive bolometer for infrared astronomy, however, did not become available until Frank J. Low entered the field, bringing with him a liquid-helium-cooled germanium bolometer he had invented while working at Texas Instruments Inc. (Low 1961). The difficulties of making observations with liquid-helium-cooled devices at that time are hard to grasp today. Observatories were not equipped with liquid helium, helium transfer lines, vacuum pumps for pumping the helium down to lower temperatures, or most of the necessary electronic instruments. An observer had to arrive bringing all this equipment along. The night assistants in charge of the telescopes, who had never seen anything like this and did not like what they saw, had to be mollified, if not by the observer then by the site director.

Over the years, Low began to extend his observations to increasing wavelengths, to the N band at 10.2 μm and the Q band at 21 μm. By 1965 he and Harold Johnson had observed the quasar 3C 273 in the U, B, V, R, I, K, and N bands (Low and Johnson 1965). Not satisfied with this, Low also began to develop a telescope that could be mounted on a NASA Lear Jet to enable him to observe from above tropospheric water vapor. This permitted him to extend his observations out to 70 μm and beyond.

Starting around 1967, and working first with his student Douglas E. Kleinmann and later with his University of Arizona colleague George H. Rieke, Low conducted studies of galaxies at wavelengths ranging from 2 to 25 μm. To widespread astonishment, Kleinmann and Low found that a number of galaxies emit far more radiation in the infrared than at all other wavelengths combined. Their results were startling (Kleinmann and Low 1970):

> Observations have been presented which support the view that all galaxies emit far-infrared radiation from their nuclei by a common physical process. More than a factor of 10^7 separates the weakest from the most luminous sources, and there are five galaxies which radiate about 1000 times the normal output of an entire spiral galaxy.

While the many difficulties these pioneering observations had to overcome sometimes led to inconsistent results and tendencies to ascribe them to source variability, more reliable measuring techniques were soon developed. As early as 1972, Rieke and Low (1972) reported that:

> The observed range of lower limits to the infrared luminosities for galaxies extends from 10^{32} to 10^{38} watts... A conservative lower limit on the infrared luminosity was calculated using only the energy actually detected at 10 μ. These values, which are tabulated assuming a Hubble constant of $53 \mathrm{~km~s}^{-1} \mathrm{Mpc}^{-1}$, are too low by a factor of 10 for those objects observed out to 25 μ. The undetected energy beyond 25 μ may add significantly to the total infrared luminosity, the determination of which remains an extremely important unsolved problem... Not all QSOs [quasars] emit chiefly in the infrared, but

for 3C 273 the observed infrared luminosity is $\sim 4 \times 10^{39}$ W, nearly 10 times the total emission at all other wavelengths. Assuming cosmological distances for these objects, their infrared luminosities exceed the entire visual output of our Galaxy by more than three orders of magnitude.

These were revolutionary statements. It was one thing, as Neugebauer et al. (1965) had found, that a few stars had infrared luminosities far greater than optical, but to suggest that an entire galaxy could radiate far more powerfully in the infrared than in all other ranges combined was quite another matter. The author remembers puzzling over this, and discussing it with Franco Pacini at Cornell. Eventually, we came up with a model that envisaged epochs during which galaxies underwent bursts of massive star formation in dense clouds of gas and dust. This would produce the large infrared fluxes. Supernova explosions produced by the stars late in their evolution would provide the observed nonthermal radio emission and high gas velocities (Harwit and Pacini 1975). We did not know what might cause these starbursts. That became evident only a decade later when observations with the Infrared Astronomical Satellite (IRAS) identified a sufficiently large number of ultraluminous infrared galaxies (ULIRGs) to show that they often were mergers. Evidently collisions of galaxies produce many of the massive starbursts, but the physics of the process is still not clear.

More recent observations of ULIRGs show that the most powerful among them also have active galactic nuclei that appear to be responsible for their extreme luminosities (Genzel et al. 1998). This too is consonant with Kleinmann and Low's early recognition of the importance of nuclear activity in their sample of sources, the most luminous among which were Seyfert galaxies and quasars.

STRUGGLING FOR ALTITUDE: BALLOON AND AIRBORNE OBSERVATIONS

Efforts to reach above the atmosphere with equipment carried aloft in balloons began under the direction of John Strong at Johns Hopkins with a manned flight in 1959. This was followed by higher-rising, unmanned flights. Bottema et al. (1964) reported a first successful, high-spectral-resolution ($R \sim 5,000$) observation at 1.13 μm of water vapor in the atmosphere of Venus. To be certain that their detections carried out at altitudes of 26.5 km were not due to residual telluric water vapor above the balloon, they used a mercury discharge lamp that provided a fiducial line at 1.1287 μm. They observed a wavelength shift of 0.49 Å in the water vapor line, in good agreement with the expected Doppler shift due to the relative velocity of Earth and Venus.

However, since this was a shift of only one-quarter of a spectral resolution element, an alternative method of checking was deemed necessary. The unmanned system was occasionally pointed at a piece of white cloth that reflected sunlight into the beam. This would show how much of the observed water vapor absorption was terrestrial. The two methods agreed that less than 10% of the water vapor observed was telluric.

Infrared observations were at that time also carried out at Princeton with Stratoscope II, a 0.9 m (36-inch) balloon-borne telescope with which spectral observations were carried out at wavelengths from 1 to 7.5 μm. A first flight provided estimates of the CO_2 content of the Martian atmosphere, and set upper limits on the amount of water vapor in the planet's atmosphere (Danielson et al. 1964). A second flight searched for interstellar absorption due to water ice between Earth and the obscured star μ Cephei. By comparing spectra for α Orionis and μ Cephei at 3.1 μm, Danielson et al. (1965) concluded that not more than one quarter of the interstellar reddening of μ Cephei could be due to water ice.

At longer wavelengths, William F. Hoffmann had the first successful balloon flights with bolometers sensitive in the 100 μm band. Beginning with an observation of the Moon in 1967, he and his co-workers went on to first observe the galactic center at this long wavelength. In a flight on 30 July 1968 they used a 25 mm (one inch) aperture quartz lens to focus a 2.3° diameter field of view on their detector. Their wavelength band spanned a width of ~ 40 μm around an effective wavelength of 100 μm. They observed an intense glow from the Galaxy's center – the first source outside the solar system ever to have been observed in this wavelength range (Hoffmann and Frederick 1969). In a subsequent flight they produced a more detailed map with a 30-cm (12-inch) telescope and a 12 arc minute beam (Hoffmann et al. 1971a). This was followed by a more extended survey along the Galactic plane (Hoffmann et al. 1971b).

By this time, airborne observations had been started through the efforts of Frank Low, who had in the late 1960s persuaded NASA to install a 0.3 m (12-inch) aperture Cassegrain telescope in the fuselage of a Lear Jet. A Cassegrain telescope has a concave parabolic primary mirror, which illuminates an on-axis, convex, hyperbolic secondary mirror. The secondary focuses radiation onto detectors placed in a plane behind the primary. The design is compact and effective.

Low's major problem was to introduce a way of chopping on and off a celestial source that did not also chop the substantial radiation emitted by the warm telescope and the residual atmosphere above the aircraft. On a first flight he found that the signals produced by chopping this foreground were so huge, that no celestial sources could be observed. He recalls removing the telescope from the Lear Jet, taking it back home to Tucson, and going out at night to test it in the parking lot behind the Space Science building at the University of Arizona. There he tried different schemes of chopping that would leave the flux from the telescope and sky largely unmodulated. None of them were successful until he hit on the idea of loosening the mounting of the secondary mirror and moving it back and forth by hand. As he did this he watched the oscilloscope trace of the signal registered by the detector. If he did not tilt the secondary mirror very far, the signal from the sky and telescope was quite steady, even though the radiation from the star in his field of view was moved on and off the detector. With this established, he attached a drive mechanism to the secondary mirror that would rapidly move it back and forth between longer dwell periods at each extreme of the swing. Here was a new effective way of chopping on and off a source.

On his next flight Low was able to test this invention, and remembers, "By golly, it was amazing!" (Low, personal communication, 1999). So successful was the secondary chopper that, over the years, chopping secondaries have gradually been installed on all major ground-based telescopes for infrared observations. Airborne and balloon telescopes have also incorporated them.

With his newly modified airborne instrument, Low and his colleagues concentrated on observations in the 50–300 µm region. In 1970 the group first reported on observations of Jupiter and Saturn but then turned their attention to the galactic center (Aumann and Low 1970). In a second series of flights, they re-examined the galactic center region in greater detail, and were able to measure the emission from the galactic source M17 and the Orion region, as well as the galaxy NGC 1068 (Low and Aumann 1970). Thus, Low and his collaborators could compare their results to those obtained at roughly the same time through balloon observations by Hoffmann and his colleagues. The size of Sgr A that Hoffmann and Frederick (1969) had measured was 38×15 arc minutes, clearly more extended than the 3 arc minute dimension estimated by Low and Aumann (1970), and it emitted a total flux more than three times higher. More detailed observations obtained somewhat later by Low and Aumann (1970) increased their reported source dimension to 8 arc minutes and roughly doubled their measured flux, but still left them short of the values, both for Sgr A and Sgr B2, by a factor of ~ 2, relative to those Hoffmann and his colleagues had obtained with their balloons.

This comparison clearly showed the difficulties, even at aircraft altitudes, of using a chopper throw that was too short. Yet larger chopper throws were problematic too, since they added considerably to the sky noise. The indications were that one needed to go as high above the bulk of the atmosphere as possible to obtain reliable observations. A measure of the true infrared flux of the galactic center, across a broad range of wavelengths, however, was not obtained until the arrival of rocket astronomy.

ROCKET INFRARED ASTRONOMY

In early 1963 the author contacted Herbert Friedman of the US Naval Research Laboratory (NRL) in Washington, DC, to propose the possibility of starting an infrared astronomy program using rocket-borne telescopes. NRL had impressive credentials in ultraviolet and X-ray observations from rockets, and the author had worked with sensitive infrared detection techniques (Harwit 1960). It appeared that a joining of forces would permit the establishment of a new branch of space astronomy that promised ultimately to open the entire infrared spectral domain to sensitive observations.

In a friendly meeting held in the offices of Dr Friedman in Washington, DC, agreement was reached that the author should come to work at NRL for a year, if possible with support of the NSF as the first E.O. Hulburt fellow at NRL. During this year, the author would work closely with NRL scientists and engineers to set up a program in rocket infrared astronomy. At the end of this period, he would return to his home institution, Cornell University, to set up a similar program there, and the two research groups established in this way would thereafter continue to compete in this new field.

The perceived prospects for observations with rockets were:

- A clear view of the sky across the entire infrared spectral range, unencumbered by atmospheric absorption or emission. This contrasted to ground-based, balloon, and airborne observations, which were always restricted by atmospheric opacity.
- An ability to cryogenically cool the entire telescope. This was of fundamental importance. In viewing a faint astronomical source, the detectors would not be burdened with culling out weak signals from the overpowering thermal radiation emitted by an ambient-temperature telescope. Cooling the entire telescope promised huge advantages in observing speed.

For ground-based, balloon, and airborne observers, cooling the telescope made little sense, since the atmosphere through which they viewed the infrared sky was usually as strong an emitter as the telescope. To compensate for atmospheric emission, these observers needed to employ two strategies, neither of which would be required in observations conducted from above the atmosphere. As already indicated, one was to "chop" between source and blank sky. The other was to "integrate" for a long time in order to cull weak signals out of noisy sky mission: since sky emission fluctuated randomly in time but on average was uniform across the sky, long integrations – long-duration observations – with chopping permitted the gradual unveiling of faint celestial sources submerged in a strong, atmospheric, foreground glow.

ASTROPHYSICAL CONSIDERATIONS AND LIMITATIONS

Over the summer of 1963, the author sought to clarify some of the ultimate astrophysical limitations that would have to be faced in a successful program of infrared rocket astronomy with cooled telescopes. In a paper presented at a colloquium held at the University of Liège in late June 1963 he showed that the thermal emission of the zodiacal (interplanetary) dust cloud would dominate the brightness of the infrared sky in the near- and mid-infrared part of the spectrum (Harwit 1964). He wrote:

> One now is in a position to discuss the detrimental effects that zodiacal dust reradiation will have on infrared astronomical observations... [T]he nature of the most promising infrared observations is different from much of the work in the visible region. One often hopes to obtain information about diffuse sources of radiation, so that the zodiacal foreground glow may be an important hindrance... At 42 μ this cloud would radiate of the order of 4×10^{-13} watt/cm^2-sterad-μ at large elongation angles within the plane of the ecliptic.

One needs to recall that, in 1963 the microwave background had not yet been discovered, but the helium content of the Universe was known to account for approximately one quarter of all the atomic matter in the Universe. If the conversion of hydrogen into helium had all taken place in stars, as seemed quite reasonable at this time – the pioneering work of Ralph Alpher and Robert Herman (1949) having been largely forgotten or discounted – then some of the energy liberated in the formation of helium should be observable in the infrared. With this in mind, the author continued:

> [A]n interesting infrared observation concerns the frequently discussed suggestion that the overall cosmic background radiation might amount to as much as 3×10^{-11} watt/cm^2 in the infrared... [T]he cosmic flux could only be detected from the immediate vicinity of the earth, if the radiation were concentrated in a very long wavelength spectral range where interplanetary dust grains are expected to emit inefficiently. (Harwit 1964)

While the limitations produced by a zodiacal glow would, three decades later, turn out to be critically important to our understanding of the diffuse infrared background, its practical significance for other purposes needs to be put in perspective. The background due to the glow of the interplanetary zodiacal dust cloud is some six orders of magnitude fainter than the emission from the atmosphere. This is about the same factor by which the moonless night sky is fainter than the daytime sky, a major reason why optical astronomy is practiced at night, rather than during the day. Flying a cryogenically cooled telescope permits the observer to gain this advantage over the limitations faced by the ground-based observer.

The impact of this million-fold background reduction in space is impossible to overestimate. It produces a thousand-fold increase in sensitivity, or a million-fold increase in the speed of observations. Thus even the first relatively small, cryogenically cooled infrared space telescopes were already able to revolutionize our knowledge of the infrared sky.

EARLY EFFORTS AT THE NAVAL RESEARCH LABORATORY

The Cornell/NRL collaboration started in earnest in September 1963. A large number of technical problems had to be overcome in just 12 months if the work of the first year was to culminate in demonstrable success. NRL provided major resources to the effort. Joining the author were scientists Douglas McNutt, Kandiah Shivanandan, and Blair Zajac, mechanical engineer Henry C. Kondracki, and electronic engineer John M. Reece.

Though the ultimate goal of the group was to construct telescopes cooled to liquid helium temperatures, which would offer unencumbered observations across the entire spectral range from 1 μm out to several hundred microns, the team quickly realized that the design of a liquid-nitrogen-cooled telescope would be considerably simpler. Such a telescope, though not as cold, would still make possible near-infrared observations of great sensitivity, since a telescope cooled to the temperature of liquid nitrogen, $\sim$80 K, would emit negligible thermal radiation at short wavelengths, and the detectors themselves in any case operated optimally at this temperature. Once sufficient experience in the construction of these near-infrared telescopes was gained, the group intended to quickly turn to the technically more difficult task of constructing liquid-helium-cooled telescopes that could be operated at temperatures of 4 K with the helium cryogen at atmospheric pressure, or $\sim$2 K if the helium was pumped down to very low pressure.

The first efforts of the Cornell/NRL collaboration were dedicated to a crash program to take advantage of a free "piggyback" launch that the US Air Force was prepared to provide on an Atlas rocket on a military test mission whose primary objective was not revealed. Time scales were only a few months before the rocket was to be fired from Cape Canaveral, Florida. To take advantage of this opportunity the author and his colleagues built a payload that contained three small liquid-nitrogen-cooled telescopes. These would look out at the sky in three directions, roughly 60° apart. That way a greater amount of data would be simultaneously obtained and telemetered to ground. One advantage of the Atlas rocket was its ability to lift heavy payloads, so weight considerations were minor. The Air Force's requirement for a maximum four-hour hold-time on the launch pad was a more serious restriction. It meant having a sufficiently well heat-shielded payload to prevent evaporation of the small amount of liquid nitrogen to be carried aloft. Eventually, the system was built to hold liquid nitrogen for six hours, considered sufficient for the maximum hold-time plus a roughly 20-minute trajectory aloft.

The payload was completed, on time, under great stress. But then the Atlas launch was delayed and eventually moved to the Van den Bergh Air Force Base in California. This delayed the launch by more than a year.

When the Atlas flight eventually did take place, it was a total fiasco. On the night of the flight, the launch kept getting further and further delayed. When the maximum four-hour hold-time passed the author requested that the payload be removed from the missile, but the Air Force now needed it as ballast and refused. Finally, the launch took place after a six-hour delay on the launch pad. The telemetry signals showed the last of the liquid nitrogen evaporating just as the rocket lifted off. It was heart-breaking. No data were obtained.

The lesson the author learned from this was to never accept an offer for a free "piggyback" flight. It was far better to have a more modest launch on a rocket under the sole control of the scientist, than an opulent flight over which the experimenter had little control.

While waiting for the Atlas launch, the group began work on a smaller payload to be flown on an Aerobee rocket. The early Aerobee flights also encountered technical difficulties, but soon began to yield at least modest results – upper limits to the near-infrared background from the night sky and concrete measures of limb darkening on the Earth's night side (Pipher and Harwit 1969).

EARLY ROCKET EFFORTS AT CORNELL UNIVERSITY

After the first year's efforts at NRL, the author returned to Cornell University in the fall of 1964. Douglas McNutt then took over the direction of the NRL group. Collaboration continued on efforts that had been jointly begun, but as these were completed the two groups began to work independently and compete.

Shortly after the author's return to Cornell, discussions with Nancy Roman, in charge of the astrophysics program at NASA, resulted in grants to Cornell of an initial sum of $250,000 and annual budgets of $100,000 that would constitute sufficient support to conduct a viable research program, initially with Aerobee 150, and later Aerobee 170 rockets. While NASA provided this initial outlay, in some years funding was also obtained from the Air Force Cambridge Research Laboratories (AFCRL).

The Aerobee payloads were launched from the White Sands Missile Range, where the US Navy kept a launch crew. Depending on the mass of the payload, the rockets could reach an altitude of approximately 150 miles (250 km) and provided roughly five minutes of observing time above the atmosphere. Launch support for NASA-funded launches was provided by NASA's Goddard Space Flight Center. For the AFCRL-funded flights, the launches were supported by the Kitt Peak National Observatory, which in the early 1960s had its own rocket support group.

The Cornell group hired Henry C. Kondracki, who left NRL to move to the Ithaca, NY area as full-time mechanical engineer. William Wernsing of Ithaca, NY, joined as electronics engineer. James R. Houck, a graduate student just finishing a PhD at Cornell in solid state physics joined the group after a couple of years, initially as a postdoc. He was soon asked to join the Cornell faculty and a long-lasting collaboration between him and the author ensued.

LIQUID-HELIUM-COOLED TELESCOPES

A number of major engineering problems had to be solved.

1. A cryogenically cooled telescope had to be launched under vacuum. Otherwise atmospheric gases would immediately condense on the optics. But vacuum vessels at that time tended to be constructed with thick steel walls, making them far too massive to be launched on small rockets. A sufficiently lightweight design had to be worked out. Ultimately the problem was solved through the construction of a thin-walled corrugated outer shell that gained strength through its corrugation. A subsidiary problem was the construction of a vacuum-tight telescope cover that could be reliably ejected once the rocket reached altitude so that radiation from the sky would enter the telescope.

2. The small Aerobee rockets that were most economical for these early efforts were only 15 inches (38 cm) in diameter. This defined the maximum outer dimensions of the telescope vacuum jacket. Inside this jacket a number of nested, concentric, cylindrical layers would have to be accommodated. Immediately inside the vacuum jacket, a vacuum layer was needed to isolate the cryogen from the room-temperature vacuum housing (see Figure 4). While the vacuum layer would minimize thermal conduction to the cold inner portions of the telescope, it would not sufficiently diminish the radiative heat load from the outer walls. The vacuum space, therefore, had to be filled with loose, concentrically wound sheets of aluminized milar *superinsulation*. Enveloped in these layers of milar was a cylindrical stainless steel vessel,

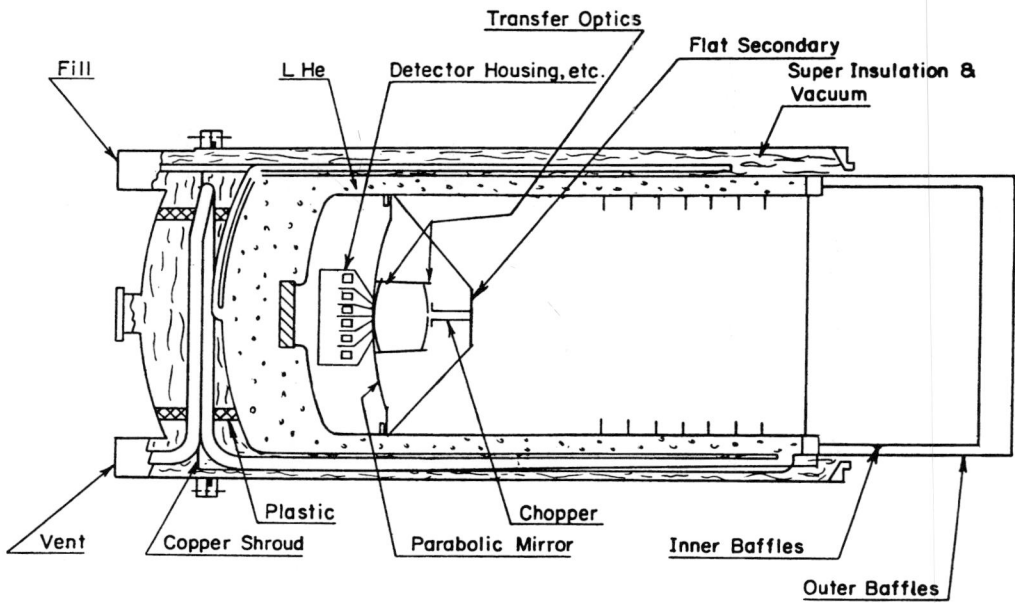

Figure 4 Schematic diagram of the fully liquid-helium-cooled telescope flown by the Cornell University group in their multiwavelength observations of the Galactic center, on 16 July 1971 (see text). Superinsulation refers to sheets of aluminized milar. The inner and outer baffles were made of beryllium–copper stock that opened through spring action when the telescope cover was removed at altitude. The baffles, which were recessed from the detectors' field of view, were painted with an infrared-absorbing black paint on their inner surfaces and kept shiny and reflecting on their outer surfaces (Soifer 1972).

the radius of whose inner wall was only about an inch smaller than that of the outer wall. This cylindrical hollow space was filled with cryogen. The telescope optics and the detectors were housed in a central vacuum space surrounded by the cryogen. The thermal/mechanical design problem of constructing such a telescope, which could survive the vibrations and linear accelerations of launch, and yet have minimal heat-conduction paths to the outer shell at room temperature, was perhaps the most difficult to solve.

3. Since the sensitivity of cryogenically cooled detectors in a cryogenically cooled telescope would be extremely high, observations were possible at high speeds. The bolometers favored by many ground-based observers were deemed too slow reacting to take advantage of this speed. Some of the photoconductors that had been developed for military purposes were far more promising. But it was soon apparent that the very low radiative background a fully cooled telescope provided minimized the photon flux on these detectors, and correspondingly lowered the conductivity of the detector material. The detectors then attained extremely high resistances ranging up to $10^{11}\,\Omega$. Even small capacitive effects would then produce unacceptably slow response times. A major effort had to be undertaken to decrease response times and take full advantage of the detectors' potential sensitivity and speed.

4. With liquid helium a whole set of new techniques had to be pioneered. If there was even the slightest leak from the liquid-helium container into the ambient vacuum jacket, the thermal insulation provided by the vacuum was immediately and totally destroyed. The smallest traces of helium gas in the vacuum spaces could and often did lead to near-explosive boil-off of the cryogen, and presented a genuine danger in working with these telescopes. Heliarc welding techniques, which were relatively new at the time had to be perfected. No welding defects could be tolerated, since they could lead to the opening of tiny pores during the vibrational stresses of launch, threatening leaks of helium into the vacuum space, catastrophic loss of cryogen, and explosive destruction of the payload.

5. The Earth's surface brightness in the infrared was expected to be nine orders of magnitude higher than the basic signals the detectors were able to detect from their $\sim 1°$ fields of view on the night sky. Extreme care had to be taken to baffle the telescope to eliminate any stray light from the Earth's limb that might be scattered or diffracted into the telescope. Eventually this problem was solved by installing two concentrically mounted sets of "pop-out" baffles, beryllium–copper leaf springs that popped out as the telescope cover was removed and in effect provided an extension of the relatively short telescope barrel. Set back

from the primary mirror's field of view, these provided satisfactory additional shielding.

6. Surrounding the telescope with a cryogenic fluid was the simplest way to maintain low temperature. If the cryogen could be contained for a few hours on the launch pad, and for about 15 minutes of flight, there was initially no need to design for longer lifetimes. This eased the design problems somewhat, but the satisfactory design and construction of a liquid-helium-cooled telescope for a successful astronomical flight, nevertheless, required several years of intensive engineering. Once the rocket's powered flight had terminated and the entire payload floated in space under weightless conditions, the liquid cryogen would no longer be gravitationally separated from its boil-off gases. Liquid might then be vented from the telescope together with these gases. If this happened, the telescope could prematurely warm up. Means had to be found to prevent the loss of coolant. The AFCRL group, whose work is described in greater detail below, eventually solved this venting problem elegantly through the use of a porous plug (Price 1988).

Early Cornell designs incorporated a parabolic primary mirror with an 18 cm aperture and focal ratio $f/0.9$. At altitude the entire telescope, except for the entrance aperture, was surrounded by liquid. Three different types of detectors were flown on many of these flights to sample roughly the spectral range from 5 μm to 1.6 mm: copper-doped germanium, gallium-doped germanium, and an n-type indium arsenide hot-electron bolometer (Harwit et al. 1969).

THE INTERPLAY OF THEORY, TECHNOLOGY, AND OBSERVATIONS

An instrumentalist investigating a new spectral range would like to know the most promising part of the sky at which to point a telescope. Throughout the construction of liquid-helium-cooled telescopes, the author tried to anticipate where the most interesting sources might be found.

Neugebauer et al. (1965) had succeeded in identifying a number of luminous, highly evolved stars. The author and Kris Davidson, then a first year graduate student, wondered whether very young stars still surrounded by remnants of the dust cloud from which they had formed would similarly emit powerfully in the infrared (Davidson and Harwit, 1967). They wrote:

> There exist certain stars (e.g., R Monocerotis) which seem to be related to T Tauri stars, and which are associated with very small reflection nebulae... If such stars are really young, the associated dust may be material from the original protostar, which lagged behind during the early collapse. We consider the detectability of very large dust envelopes which might be left behind in the formation of very massive stars... Thus we consider a situation in which a protostar becomes bright and hot before emerging from its envelope. The star's radiation is absorbed by surrounding dust and re-emitted at infrared wavelengths to which the cocoon is transparent. Furthermore, if there is enough time to form an ionized region in the central part of the envelope, radio-frequency radiation also emerges... Both the infrared and the radio emission should be detectable at moderately great distances.

Unknown to the authors, Becklin and Neugebauer (1967) had just discovered a bright star in the Orion Nebula at wavelengths from 1.5 to 13.5 μm, and Kleinmann and Low (1967) had followed up these observations at 22 μm to discover cool nebulosity slightly displaced from the star. While the BNKL region named after these authors did not precisely fit all of the predictions of Davidson and Harwit, the idea of a young star's interaction with the cocoon of dust that originally spawned it has gained wide acceptance.

EARLY ROCKET OBSERVATIONS AND FALSE RESULTS

Great difficulties were encountered on early flights. Rocket exhausts and other gaseous and particulate ejecta formed a diffuse, radiating cloud surrounding the payload. These produced a false signal with all the characteristics of an isotropic flux at the longest wavelengths, from 400 μm to 1.3 mm. The author and his colleagues initially reported these signals as possibly of cosmic origin (Shivanandan et al. 1969). However, as they took increasing care to ascertain the origin of this flux, they became aware that this radiation was not astronomical but was due to more local contaminants and to diffracted Earth shine. Despite these difficulties they were able to set useful upper limits to the background radiation (Pipher et al. 1971).

FIRST SUCCESSES

The first serious successes of these rocket flights involved two quite different types of detections, and resulted from separate flights on 2 December 1970 and on 16 July 1971. The first of these discovered and accurately measured the infrared radiation emitted by the circumsolar zodiacal dust cloud (Soifer et al. 1971). Radiation was detected in three spectral ranges, at 5–6, 12–14, and 16–23 μm. At 70–130 μm initially only an upper limit could be placed. More careful analysis provided a detection even at these long wavelengths (Pipher 1971). Both three- and four-color photometry put the dust temperature at ~280 K. The peak signal measured in the ecliptic plane $\sim 20°$ from the anti-solar direction was, respectively, ~ 3, ~ 6, and $\sim 2.5 \times 10^{-11}\,\mathrm{W\,cm^{-2}\,sr^{-1}\,\mu m^{-1}}$, in the 5–6, 12–14 and 16–23 μm spectral bands, and $\sim 2 \times 10^{-13}\,\mathrm{W\,cm^{-2}\,sr^{-1}\,\mu m^{-1}}$ at 70–130 μm. These results may be compared to the most recent summaries based on Cosmic Background Explorer (COBE) data, which give a dust temperature of 280 K for dust at 1 AU from the Sun (Kelsall et al. 1998). At a solar elongation of 90°, the ecliptic plane emission that these more recent measurements yielded at 12, 25, and 100 μm was, respectively, of the order of 35, 70, and 9 MJy sr^{-1}. For the 12–14, 16–23, and 70–130 μm detectors flown in the early rocket flights the expected fluxes on this basis would have been, respectively, $\sim 6 \times 10^{-11}$, $\sim 2.5 \times 10^{-11}$, and $\sim 2.7 \times 10^{-13}\,\mathrm{W\,cm^{-2}\,sr^{-1}\,\mu m^{-1}}$. The zodiacal emission in the ecliptic plane is not expected to greatly change between solar elongation angles of 90° and 160°, so that this agreement is quite excellent. COBE flew no detectors in the 5–6 μm range. Detectors flown at 3.5 and 4.9 μm measured respective ecliptic plane signals of ~ 0.15 and 0.8 MJy sr^{-1}, indicating a precipitous climb in signal toward long wavelengths. The rocket results of Soifer et al. at 5–6 μm correspond to ~ 2.5 MJy sr^{-1}, showing a continuation of this steep incline. The surprise in these findings was that the dust radiated rather more powerfully than theoretically predicted (Harwit 1964), indicating that the zodiacal dust grains were unexpectedly dark, scattering only a small fraction of the incident light and absorbing an appreciably larger portion.

The second discovery, made with the Cornell liquid-helium-cooled telescope on 16 July 1971, was the magnitude of the total infrared flux emanating from the Galactic center and four other regions in central portions of the Milky Way at 5, 13, 20, and 100 μm (Houck et al. 1971). The 85–115 μm integrated flux over an area of $3° \times 2°$ around the Galactic center was $7 \times 10^{-20}\,\mathrm{W\,m^{-2}\,Hz^{-1}}$, in excellent agreement with the balloon-borne result that had previously been obtained by Hoffmann et al. (1971a). Excellent agreement for this wavelength range was also obtained for the Galactic ionized hydrogen regions M8 and NGC 6357. But the Cornell rocket flight also recorded the previously inaccessible flux from these three regions at 5–6, 12–14, and 16–23 μm (Soifer et al. 1972). Additional results cited by the same authors from an earlier flight provided the 100 μm flux for NGC 1499, a region previously unobserved at this wavelength.

More than a dozen years later, scans of the galactic center by IRAS verified these results (Gautier et al. 1984). Though these authors did not compare their results to any previous work, the 100 μm maps of the galactic center published by the IRAS team gave peak fluxes that, within normal calibration uncertainties, were essentially identical to the Cornell rocket results published a dozen years earlier. Within such uncertainties, the 12 μm IRAS fluxes and the 12–14 μm Cornell detections also showed reasonable agreement.

One significant difference between the rocket and contemporaneous balloon-borne observations was that the latter had been laboriously obtained through differential chopping – on and off source – while the absolute chopping technique employed in the rocket payload, provided by a tuning-fork chopper that alternately opened and shut the detector's view of the sky, gave an absolute sky brightness in this wavelength range. The speed of the rocket observations was also incomparably faster. The total time taken to scan the rocket telescope's beam across a 13° span from galactic longitude 7° to 354°, and back to longitude 3° was about 27 seconds. During this time, three of the five detected sources were detected twice each, and two were traversed once. Moreover, all five sources were observed at four different wavelengths. In contrast, the balloon-borne telescope that had nearly twice the aperture of the rocket instrument required 27 minutes to cover roughly the same swath along the galactic plane, in just one wavelength band, though admittedly over a wider latitude range of $\sim 2.5°$ and with a beam diameter of 12 arc minutes that provided greater structural detail and showed that some of the apparent sources identified in the rocket observations actually were more consistent with groupings of several sources that had all been encompassed within the rocket telescope's 1° field of view. Nevertheless, the advantage of enormously increased sensitivity in observations at rocket or satellite altitudes was clearly apparent.

The importance of the rocket observations was also highlighted by the difference in signal strengths obtained for the galactic center at the three shorter wavelengths. Here the observations from above the atmosphere uniformly yielded higher fluxes than the ground-based observations that had been obtained by Becklin and Neugebauer (1969) and the airborne results of Low and Aumann (1970). Both these groups had used differential chopping techniques and fields of view that were systematically smaller than the extended galactic center region. Becklin and Neugebauer, who obtained measurements at five wavelengths from 1.65 to 19.5 μm, used a chopper throw that ranged from 0.5 to 2 arc minutes away from the source position. They noted a rising flux density out to the largest aperture diameters they employed, 110 arc seconds, but the fraction of the total galactic center emission that could be obtained this way was still only a small part of the radiation seen in the Cornell rocket flight. Aumann and Low also had chosen too small a chopper throw and field of view – choices forced on them by atmospheric constraints.

Observations from within the atmosphere were only possible by selecting small fields of view and small chopper throws. This tended to be laborious and often yielded false results in observations of extended sources. Here, an excessively small chopper throw meant a systematic underestimate of the total source flux because the observer was chopping between a bright region of the source and a less bright region. Rather than measuring the total flux, the observer thus obtained merely a differential flux between the brightest and less bright parts of the source.

This is not to be taken as a criticism of the ground-based, airborne, or balloon observers. They faced great difficulties. The only problem that arose was that these observers did not believe in the possibility of fluxes as high as those observed in the rocket flight. Not only were these higher than the ground-based observations of the galactic center, they also were far higher than any ground-based measurements obtained for the other galactic plane regions the rocket telescope had scanned. Recognizing the validity of the rocket observations would have meant invalidation of most ground-based or airborne observations of extended galactic plane regions, and these more traditional observers were not ready to countenance such a prospect. Not until a dozen years later, when the IRAS results were analyzed, did it become unavoidably clear that the rocket observations both of the ecliptic plane and the Galactic plane regions had been correct. When IRAS was finally launched, the zodiacal light observations made by the Cornell group were fully reproduced. As already mentioned, even more detailed zodiacal light observations carried out in the 1990s with the COBE satellite are in astonishingly good agreement with the rocket results obtained two decades earlier.

The lesson the author learned from this experience is that, while it makes good sense to build powerful new instrumentation, the results obtained with such equipment will not become accepted until at least one other group with similarly powerful apparatus is able to independently verify the results. In this case, the only other rocket group active in the 1970s was the team at AFCRL, which was more interested in shorter wavelength observations, and initially funded the Cornell group to carry out observations at wavelengths at which AFCRL was not prepared to operate. This group unfortunately was not in a position to publish its results until 1975, and by that time NASA had decided to close down the Cornell effort since the agency was going ahead with IRAS that it hoped to launch soon. As recorded below, the Cornell group thus turned to airborne observations, and to obtaining the first mid- and far-infrared spectroscopic results.

Before turning to these, however, we should look at the rocket results obtained by AFCRL.

ROCKET AND SATELLITE EFFORTS AT AFCRL

Writing some years ago, Stephan Price (1988) reported,

In the late 1960s the DoD [US Department of Defense] became actively interested in characterizing the general mid-infrared background by means of a large-scale unbiased survey... Based on their previous experience, NRL was the logical choice to perform the rocket survey. NRL was, however, deeply involved in the[ir] far-infrared experiments and lacked the facilities to handle both programs. The task fell to AFCRL.

Work at AFCRL, later renamed Air Force Geophysical Laboratory (AFGL), was headed by Russell G. Walker, initially assisted and later succeeded by Stephan Price. The Air Force was mainly interested in the extent to which celestial infrared sources might interfere with the detection by infrared means of enemy aircraft, missiles, and spacecraft. This called for a series of sky surveys.

In 1970 the group flew two proof-of-concept experiments – liquid-neon-cooled telescopes with 10 cm apertures, and detectors sensitive to radiation at 12–14 μm. Each of six detectors in a linear array surveyed a 10′ × 10′ field of view. As the Cornell group was also realizing at this time, the sensitivity of even these early telescopes was sufficiently high that a dust grain released from the rocket could be detected out to distances of 100 m.

That same year, the AFCRL group also had two newer survey instruments built by Hughes Aircraft Co., with Advanced Research Project Agency (ARPA) funding, under Air Force Space and Missile Systems Organization (SAMSO) management (Price 1988). Seven Aerobee 170 flights were launched from the White Sands Missile Range between April 1971 and December 1972. The instruments, code-named HISTAR, comprised doubly folded $f/2.2$ Gregorian telescopes with beryllium optics and a 16.5 cm aperture that provided an effective collecting area of ~150 cm^2. Three linear arrays of doped germanium detectors covered the spectral ranges 3–5, 8–14, and 16–24 μm. Each array consisted of eight elements, each with a 3.3 × 10.5 arc minute field of view. The series of flights succeeded in surveying most of the northern celestial sphere. The results of the survey were compiled in a catalog by Walker and Price (1975).

Price (1988) briefly mentions two SAMSO infrared surveys conducted in 1971 – the first two attempts to conduct such a survey from Earth-orbiting satellites. The first of these missions flew a liquid-neon-cooled telescope similar to the one that AFCRL had earlier launched; the second mounted a HISTAR-class instrument with a closed-cycle cooler. Price reports technical problems that these surveys encountered but mentions that the second of these flights did map a considerable portion of the sky in two wavelength bands.

HISTAR rocket launches also continued. Coverage was extended out to 24–30 μm through the use of doped silicon detectors. The sources detected on all these flight series were then compiled as an AFGL catalog (Price and Walker 1976).

The successes of these surveys were considerable. Their legacy is still apparent today in the many sources traditionally referred to by their CRL or their GL or AFGL number, like AFGL 2688, the Egg Nebula. Price also published sky maps based on these data (Price 1981).

R.G. Walker (personal communication, 1999) described the philosophy behind these surveys:

> The Air Force was interested in assessing the "worst case" scenario that the sky could produce in the mid-infrared. The basic parameters of any sky survey are (1) sky coverage, (2) completeness, and (3) reliability. Conducting a survey requires tradeoffs to be made between these parameters. With our limited resources we chose to go for maximum sky coverage (thus little or no re-scan overlap), and emphasize completeness at the expense of reliability. These choices (in the interests of our sponsor) caused no end of grief with the astronomical community. There were many objects we reported that could not be confirmed from the ground. Most turned out to be of an extended nature and were lost to telescopes with small fields and chopping secondary mirrors. However, to be fair, there were some that were spurious, the price paid for choosing completeness at the expense of reliability... In our defense, we did discover many new and interesting sources (e.g., the red rectangle) and set the stage for a satellite to do the job properly.

A further survey conducted jointly by the AFGL and NRL groups, with an instrument carried on an Aries rocket on 22 January 1982, covered nearly a quarter of the sky in four wavelength bands between 20 and 94 μm. A list of the 40 brightest 94 μm sources, mostly brighter than 1000 Jy at 94 μm, was published in December 1983 (Price *et al.* 1983). Three months later, the first results of the IRAS mission became publicly available (Neugebauer *et al.* 1984).

Before discussing IRAS, however, we need to return once more to the early 1970s to record the advances in balloon and airborne observations at that time.

FURTHER ADVANCES IN BALLOON OBSERVATIONS

In attempts to check on the background radiation measurement that had first been reported by the Cornell rocket group and to build better instrumentation to obtain isotropic background measurements, D. Muehlner and R. Weiss at MIT undertook a set of balloon-borne observations. In 1970 they reported results that appeared to bear out the earlier observations on the infrared background that the Cornell group had published. On repeating their observations, however, these two investigators also found their conclusions to have been based on false detections (Muehlner and Weiss 1970, 1973).

It became clear that great care would need to be taken to discriminate against atmospheric emission, stray radiation from the Earth's limb, and any contamination carried aloft.

Paul Richards, and his students J.C. Mather and D.P. Woody at the University of California at Berkeley, tackled the problem using a more discriminating, narrower-band approach based on a liquid-helium-cooled, balloon-borne, Fourier-transform spectrophotometer that permitted them to make best use of atmospheric transparency in the windows between 450 μm and about 3 mm. They obtained great sensitivity through the use of a ^{3}He-cooled composite bolometer, improved rejection of stray radiation by means of a Winston concentrator, and a light-cone concentrator rather than a more conventional focusing telescope. The concentrator defined a clean 7° field of view on the sky (Mather *et al.* 1974, Woody and Richards 1979). With this apparatus they obtained a rough fit to a blackbody spectrum in the wavelength range between 800 μm and 3 mm and a temperature of 2.96 K, slightly higher than the 2.73 K temperature now accepted. At shorter wavelengths down to 450 μm, they obtained significant upper limits consistent with blackbody emission. Equipment of this type was further developed by Richards with his student Andrew Lange and flown as a collaborative effort with Satio Hayakawa, Toshio Matsumoto, and their colleagues on a Japanese rocket (Matsumoto *et al.* 1988). It gave improved cosmic background measurements in broad bands centered at 481 and 709 μm and 1.16 mm, while also probing for interstellar dust in the Galaxy in bands centered at 102, 137, and 262 μm. The instrumentation pioneered for these observations was a direct precursor of apparatus that John Mather and others later developed for COBE, the satellite by means of which the black-body spectrum and temperature of this radiation was finally determined with great precision, as briefly discussed below.

Balloon-borne surveys for discrete sources were also continuing. In Britain, Furniss *et al.* (1972) undertook observations from balloons in the spectral range from 40 to 350 μm. At about the same time, Satio Hayakawa and co-workers also began balloon observations in Japan, initially working at 2.4 μm, but later venturing into the far-infrared (Maihara *et al.* 1978, Hayakawa *et al.* 1981, Okuda 1981). These were precursor efforts that eventually led to the launch of the Japanese Infrared Telescope in Space (IRTS) satellite. Somewhat later, efforts to map the Galaxy at long wavelengths were also proceeding in France, through the work of Gispert *et al.* (1982) and Caux *et al.* (1984, 1985).

THE LEAR JET AND THE KUIPER AIRBORNE OBSERVATORY

While background observations and deep surveys required going above the atmosphere or making use of the exceptional atmospheric transparency that only balloon altitudes were able to offer in the 1970s, far-infrared spectroscopy could be carried out quite effectively from aircraft in major portions of the far-infrared. At Cornell, James R. Houck developed a compact, fully liquid-helium-cooled grating spectrometer for the mid-infrared range that was sufficiently small to be mounted on the 30-cm (12-inch) telescope on the NASA Lear Jet. With this he obtained spectra of Jupiter and the Orion Nebula with a resolving power $R = 30$ across the 16–40 μm range (Houck et al. 1974).

Using a slightly modified copy of this design on the Lear Jet, Dennis B. Ward and the author obtained a first spectrum of the Orion Nebula from ~75 to 100 μm, with a resolving power $R \sim 10$ (Ward and Harwit 1974). Ward, a graduate student at Cornell at the time, then suggested that the 88-μm fine-structure line of doubly ionized oxygen, O^{2+}, might be sought with this instrument. The author was pretty confident that the crude instrument at hand would not detect anything as esoteric as such a line, which had never been observed even in the laboratory at the time. But he did not want to discourage good ideas of young people and went along with the suggestion. This turned out to be a resounding success. The line was soon detected in M17 (Ward et al. 1975), and taught the author how important it is to listen carefully to good graduate students. This one discovery totally changed the direction that his research group was to pursue in subsequent years.

Over the period from 1974 to 1983, the author and his graduate students and post-doctoral associates at Cornell were able to discover the fine-structure lines of O^{2+} at 51.8 and 88.4 μm, neutral atomic oxygen at 63.2 and 145.5 μm, and singly-ionized carbon, C^+, at 157.7 μm. In the meantime, using both NASA's Lear Jet and Kuiper Airborne Observatory (KAO), J.R. Houck and his students, also at Cornell, discovered the fine-structure lines of Ar^{2+} at 21.8 μm, triply ionized oxygen, O^{3+}, at 25.9 μm, Ne^{2+}, at 36.0 μm, and quadruply ionized neon, Ne^{4+} at 24.3 μm. Also using the KAO, A.F.M. Moorwood, J.-P. Baluteau and their colleagues found the lines of S^{2+} at 18.7 and 33.5 μm, and D.T. Jaffe and T.G. Phillips, respectively, found the neutral carbon lines at 370.4 and 609.1 μm (Baluteau et al. 1976, Forrest et al. 1980, Jaffe et al. 1985, Moorwood et al. 1980a, Phillips et al. 1980, Shure et al. 1984). In addition Moorwood et al. (1980b) found the N^{2+} line at 57.3 μm in a set of balloon observations.

The main difficulty in finding these lines was that they had never been detected in the laboratory and that their wavelengths were highly uncertain. Using a spectrometer with a resolving power $R \sim 150$, the author and his co-workers searched for the [C II] line for nearly two weeks in flights on the Lear Jet, carefully scanning the region from 152 to ~157 μm, before almost accidentally hearing that Lawrence Greenberg of the Aerospace Corporation had suggested the line might lie more nearly at 158 μm. With only one flight left in their Lear Jet series for that season, R.W. Russell and the author looked for the line in NGC 2024 and M42 out to 160 μm on the night of 27–28 November 1979, and immediately found it. By begging for an extra flight, to view the calibration source Venus the next day, they established that this was a true astronomical emission line and not a spurious atmospheric feature (Russell et al. 1980).

The importance of these lines can hardly be overstated. Lying at far-infrared wavelengths, few of the lines were appreciably affected by absorption from interstellar dust, and could therefore probe deep into dust-shrouded regions to examine local conditions there. Between them, the lines probed the whole range of interstellar medium conditions – from cold, neutral gas at ~30 K, to highly ionized, extremely hot gases in the $\gtrsim 10^4$ K range. For those species that possess two fine-structure transitions, the local pressure in a region can be determined if the temperature is known or vice versa. Some of the species have relatively low radiative transition probabilities and large collisional cross-sections and tend to be collisionally de-excited even at low ambient hydrogen densities. Others radiate effectively even at high ambient hydrogen densities.

Since atomic and molecular hydrogen have no low-lying energy levels except for the hyperfine atomic transition at 21 cm, which has an extremely low transition probability, the fine-structure lines of these species provide an important means of cooling photodissociation and photon-dominated regions (PDR) associated with massive young stars and star-forming regions.

Far-infrared molecular lines, particularly of carbon monoxide, CO, were also being investigated from aircraft in the late 1970s and early 1980s by Charles H. Townes and his co-workers. These lines turned out to be important tracers of shocked and other warm molecular clouds (Watson et al. 1980).

Far-infrared polarization observations were similarly starting up. A first attempt by the Cornell group to obtain the polarization of the Orion Nebula out to wavelengths as long as 100 μm was made with the Lear Jet and KAO, but only yielded a faint hint at polarization at the 2% level at the longest wavelengths (Gull et al. 1978). Roger Hildebrand and his students at Chicago took up the problem and, in more than a decade of flights aboard the KAO, achieved a series of successes with observations that went out as far as 400 μm (Hildebrand 1996). Their findings gave a general picture of the alignment of interstellar magnetic fields that orient the dust grains and provided additional data on the nature of the grains.

Infrared astronomy was coming into its own as an analytical tool but there was still a great deal that was totally unknown. No unbiased survey of the sky had been made at long infrared wavelengths, so that observers continued to work primarily with sources familiar from visible or radio observations. We had no idea of what else might be found if the methods were at hand. The infrared background radiation also remained a complete mystery. Two satellites were needed to begin dealing with these problems: IRAS and COBE. Before turning to these, however, we need to look first at what was happening in the Netherlands.

EARLY EFFORTS IN THE NETHERLANDS

In 1971 Reinder J. van Duinen, until then primarily interested in ultraviolet observations from space, was placed in charge of the entire space effort at the University of Groningen. This program included an infrared observing program using balloons that had been initiated by Jan Borgman and was led by Jan Wijnbergen. Van Duinen enlarged and encouraged this group, redirecting its efforts toward the design of a telescope with a 60 cm diameter, thin aluminum primary mirror, and a balloon infrared platform (BIRAP) that would have a 1 arc minute pointing accuracy. To this end he brought in the industrial capabilities of Philips Research Laboratories of Eindhoven and Ball Brothers Research Corporation (BBRC) of Boulder, Colorado.

van Duinen, however, was also looking to the future. He recalls (van Duinen personal communication 1999):

> In 1974 the first national astronomical satellite (ANS) had been launched successfully. As principal investigator of the main instrument on ANS, I had been involved in early thinking about a possible follow-up project since 1972. This follow-on project would again combine both scientific interest and the development of industrial space-engineering skills. BBRC, having been responsible for the development of my instrument onboard ANS, and seen as expert on cryogenic cooling, was invited by Dutch industry (Fokker) to participate in the conceptual design. Ball elected to send Dick Herring over for an extended period in 1974. Considerable evolution of the design of ANS-B occurred: from an initially low-orbit Scout-launched 20 cm He-cooled survey instrument to a Delta 2910-launched 60 cm superfluid He-cooled telescope in a 900 km Sun-synchronous orbit. The instrument package contained both a survey instrument and some pointing mode instruments. International collaboration was needed both because the Dutch government required this and obviously also for financial and technical (detectors, cooling) reasons. Encouraged by NASA and by the initial reactions from some of our US colleagues, in the last quarter of 1974 the Dutch authorities formally proposed to NASA a follow-up of the successful ANS collaboration with a proposal which later became known as IRAS. The position for NASA was complicated by the issuing of an earlier Explorer AO [Announcement of Opportunity] in July of that same year, which led to ... some 13 proposals for infrared missions ... In order to help to harmonize the various science objectives and design-inputs I spent a half year in the US at Caltech in the 1975–76 period... The final result was a considerable modification of the original Dutch proposal ... [T]he survey both as a science goal and in terms of its instrument design got a more prominent place in the total concept relative to the original proposal. By the end of 1975 this led to [a] positive letter of intent from NASA to NIVR [The Netherlands Agency for Aerospace Programs – Nederlands Instituut voor Vliegtuigontwikkeling en Ruimtevaart], which in turn allowed NIVR to make a formal proposal to the Dutch government. Meanwhile the British Research Council had been able to decide on a participation "in kind", through assuming the responsibility for satellite operations. Approvals were obtained by the end of 1976.

NASA ANNOUNCEMENT OF OPPORTUNITY FOR AN EXPLORER SATELLITE

In early 1974 NASA circulated an Announcement of Opportunity for Mission Definition for Explorer-class satellite missions. Here was a great opportunity for proposing a first infrared astronomical satellite. Several US groups decided independently to submit proposals. Some were primarily interested in detecting the infrared background. Others proposed unbiased sky surveys.

The infrared community, however, was by no means assured of a mission. The X-ray and ultraviolet astronomical communities had been far more successful up to this point in securing NASA support for satellite projects. Two major hurdles had to be overcome. The first was to demonstrate to NASA that the astronomical community was fully united and wished to see the launch of an infrared survey satellite. The second was to demonstrate, over the doubts of many, the technological readiness for such a satellite.

In its report *Priorities for Space Research 1971–1980* the National Academy of Science's Space Science Board chaired by Charles H. Townes published the views of a distinguished committee of space scientists that had been convened at Woods Hole, Massachusetts, from 27 July to 15 August 1970 (National Academy of Sciences 1971). For astronomy, this report gave its highest priority to High Energy Astronomy Observatory (HEAO; i.e., X-ray and γ-ray) launches, recommended a number of Orbiting Solar Observatories (OSOs), and a continuation of optical astronomical efforts, not from Orbiting Astronomical Observatories (OAOs) as previously recommended by other committees, but from a Large Space Telescope (LST). The report recommended four large and up to ten Small Astronomical Satellites (SASs) for X-ray astronomy, amounting to ~25% of the total budget for the decade. Optical and ultraviolet astronomy were to receive a share amounting to 21%. For infrared astronomy an Orbiting Infrared Observatory (OIRO) was recommended for launch in 1981, at a cost of 4% of the available budget. The rationale was that "to obtain an angular resolution of 1 min of arc at 200 μm, a 70-cm telescope is required, and limitations on the cryogenics payload and pointing accuracy probably rule out the SAS class of spacecraft and will no doubt require a special Infrared Observatory."

This report well demonstrates the political problem infrared astronomers faced. The X-ray astronomy community was far better organized, and the optical/ultraviolet community far more firmly established. Between them, the two groups had positioned themselves to command a 10 times larger share of the astronomical missions and the astronomical budget than the infrared astronomers.

Four years later, the Space Science Board, now chaired by Richard M. Goody, published a report that made more specific recommendations (National Academy of Sciences 1975). The Board was willing to recommend an infrared satellite but only on condition that the infrared astronomical community could clearly demonstrate technological preparedness for such a mission: "In order to clarify the appropriate strategy and timing for such a[n infrared] survey promptly enough for a firm decision on a satellite mission, a study is recommended during the year that would examine the general design of such a mission and the state of development of the required technology and would compare the effectiveness and cost of alternate approaches to infrared astronomy."

This shows the doubts that were still rampant in 1974 about the practicability of an infrared astronomical satellite or the costs entailed.

In responding to the NASA Announcement of Opportunity for Explorers, the issue of technical preparedness was reviewed in depth in a proposal submitted by a group spearheaded by the author. It concentrated on identifying the range of technical problems that a mission definition would have to address, and showed that solutions to these problems were well in hand and the time ripe for a satellite mission. This group included Palmer Dyal and Fred C. Witteborn of NASA Ames, both of whom had been active in getting the KAO successfully constructed; H. Melvin Dyck of the University of Hawaii, who had been active in polarization studies; Robert D. Gehrz and John A. Hackwell who had successfully built a 2.3 m (90-inch) telescope for the University of Wyoming; James R. Houck of Cornell, who had wide-ranging experience in the construction of liquid-helium-cooled telescopes and spectrometers; Richard E. Jennings of University College, London, who had successfully flown infrared astronomical balloon missions and represented British interests in participating on a satellite mission; W.J. Moore of NRL, perhaps the world's leading expert on far-infrared photoconductive detectors at the time; and Judith L. Pipher of the University of Rochester, who had three years earlier obtained her degree at Cornell with a thesis based on far-infrared rocket astronomical observations.

Like van Duinen in the Netherlands, this team worked with Ball Brothers and, through Richard Herring of Ball, established an indirect liaison with the Dutch contingent.

The Cornell-led proposal was highly specific. In detailed discussions among the group's participants, it had become clear that only one design would be effective in providing a satisfactory survey. It would have to entail a highly inclined orbit in which the Earth's equatorial bulge provided precisely the right torque to keep the orbital plane perpendicular to the Sun–Earth radius at all times. In this *Sun-synchronous orbit*, the whole sky would be scanned once every six months as the Earth moved around the Sun. This mode of sky scanning provided enough overlap for the survey to permit the satellite's scans to be arrested from time to time to observe targets of particular interest for a limited period. A set of spectrometers on board could then be used to obtain spectra of sources of particular interest. The lay-out of this proposed instrument is shown in the top panel of Figure 5. The proposal recommended that this design become the basis of the envisaged mission definition study for an infrared astronomical satellite.

The issue of community support was just as squarely addressed in a proposal spearheaded by G. Neugebauer and F. Low. Their proposal listed a large number of co-investigators, many of whom were not involved in the effort once the group's proposals had been submitted, but nevertheless played a persuasive role at this juncture. This was important, because infrared astro-nomers had gained a reputation for contentious infighting and divisiveness, and NASA was not prepared to support a divided community. By gathering a long list of co-investigators, many of whom had impeccable credentials, including membership in the US National Academy of Sciences, this proposal showed the importance that the broader astronomical community now attached to infrared investigations. NASA, like other US governmental agencies is directed by law to consider the advice of the Academy, so the names of so many distinguished academicians on the proposal was a factor that NASA could not well ignore.

Neither Neugebauer nor Low had been involved with cryogenically cooled rocket payloads, but they joined forces with Russell G. Walker of AFGL, who became principal investigator on the proposal. The results of the AFGL efforts had not been published anywhere at that time but rumors were that their efforts were successful, though captive to Air Force secrecy. Also on the proposal was

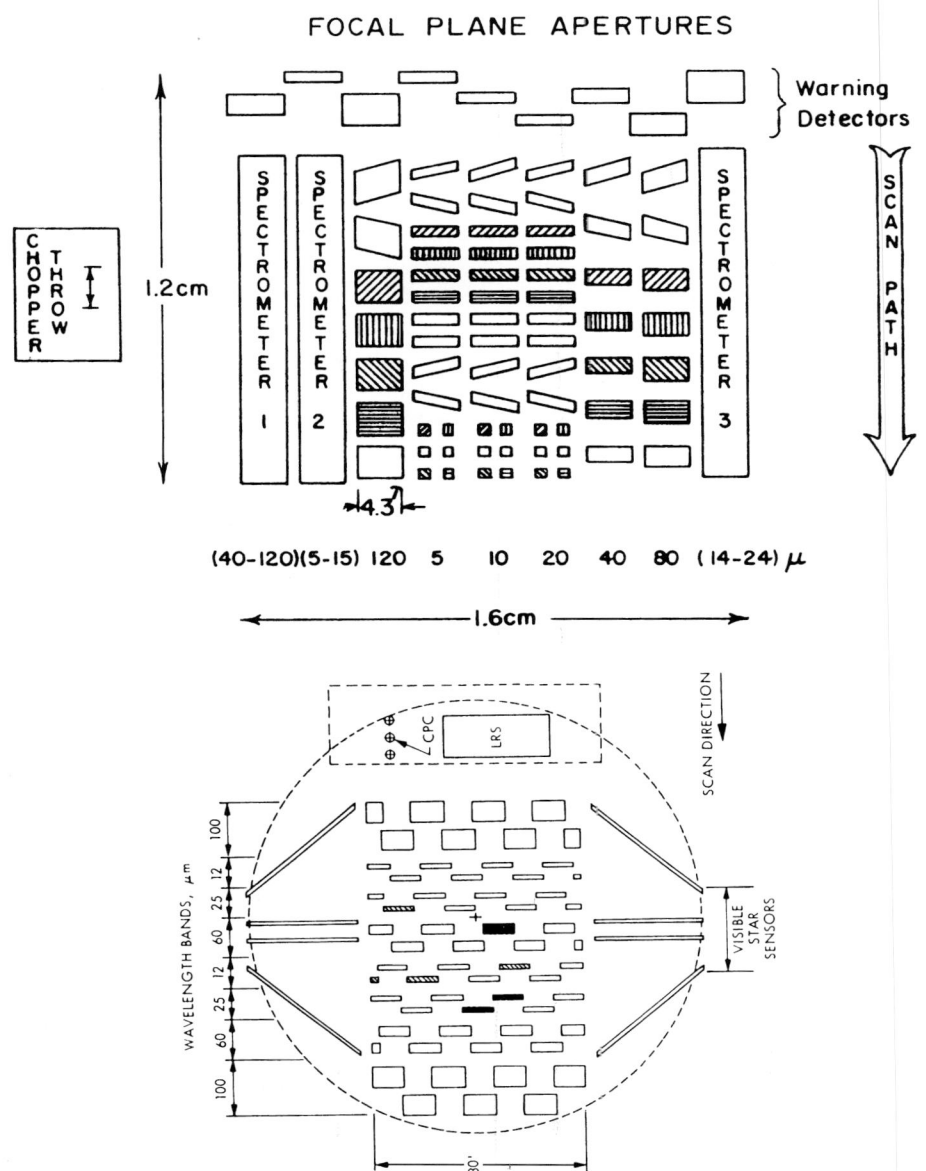

Figure 5 Schematic diagram of two focal plane arrays. Top: the array proposed for a far-infrared astronomical explorer satellite (Harwit et al. 1974). Bottom: array flown on the IRAS mission. LRS is the low resolution spectrometer, CPC the chopped photometric channel (Neugebauer et al. 1984) (see text).

B.T. Soifer, the only person on the team having experience with liquid-helium-cooled telescopes, having just finished his PhD thesis two years earlier with the rocket group at Cornell. Other infrared astro-nomers on the team were H.H. (George) Aumann of the Jet Propulsion Laboratory and Fred C. Gillett of the Kitt Peak National Observatory.

In contrast to the highly specific Cornell-led proposal, the proposal led by Walker undertook to look at a number of different conceivable satellite orbits, uses of different cryogenic techniques, cptical configurations, methods of surveying, etc., but provided few details on the ultimate design for an infrared astronomical satellite that might eventually emerge.

These were to be determined in the mission definition the team hoped to undertake.

NASA awarded the study to Walker and his team, and asked him to lead the mission definition. The study was to be carried out by a ten-member "joint mission definition team" to investigate the various aspects of the problem and recommend an explorer mission concept. Six members of the team came from Walker's original consortium, while four others were to be selected from other groups that had submitted proposals. The author and his Cornell colleague Jim Houck, discussed the situation, and Houck agreed to become the representative of the author's proposal consortium on the joint mission definition team.

Walker (personal communication, 1999) recalls:

> I was the chairman of the study team. It was made up of the persons listed above [Aumann, Gillett, Low, Neugebauer, and Soifer] plus James R. Houck, Michael G. Hauser [of Goddard Space Flight Center, GSFC], Rainer Weiss [of MIT], and John C. Mather [of GSFC]. The length of the study was to be one year. As our study was just getting going, the Dutch proposed to NASA an IR survey satellite with the British as partners... Ten Dutch and British scientists were added to the team and we jointly produced the IRAS mission concept. NASA reviewed our report and decided to proceed with the recommended mission. A new IRAS Science Team was formed with G. Neugebauer and R. van Duinen as co-chairmen. Nancy W. Boggess was the NASA Headquarters scientist for the project. Weiss and Mather withdrew from the team since their main interests were in the cosmic background and the report recommended that one satellite could not do both the stars and diffuse cosmic background. A separate satellite should follow that was optimized for the background measurements [This eventually became COBE]... I remained on the science team and became the IRAS Telescope Scientist (Project Scientist) for NASA Ames.

The Dutch contingent was led by Reinder J. van Duinen joined by his colleagues D.A. Beintema, Jan Borgman, and Stuart R. Pottasch at the Kapteyn Institute in Groningen; Harm J. Habing and George Miley of the Sterrewacht, Leiden; and Teije de Jong of the University of Amsterdam. The British group comprised Richard E. Jennings of University College, London; P.L. Marsden of the University of Leeds; and Peter E. Clegg and Michael Rowan-Robinson of Queen Mary College, London.

When the final design for the satellite emerged from the year's study by the 20 international experts it turned out to be almost precisely what the author's small proposing team had recommended two years earlier. The proposed Sun-synchronous orbit, and the wavelength range to be covered were essentially identical, as were most of the detectors to be used. A grating spectrometer was also incorporated. The focal plane design of the instrument proposed by the Cornell team is shown in the top panel of Figure 5 (Harwit et al. 1974). The IRAS focal plane as flown is shown in the bottom panel (Neugebauer, et al. 1984).

While the author was disappointed that his group's proposal had not won the competition, the lesson he learned from the experience was the importance of establishing community consensus in order to win a major competition.

CONSTRUCTING IRAS

Once the IRAS design had been established, it was still necessary to make it work. Large numbers of complex technical problems had to be solved, and years of careful attention devoted to detail. As actual work on the satellite began, the team of astronomers working on the project gradually expanded. On the US side, the astronomers Charles A. Beichman and T.N. Gautier of the Jet Propulsion Laboratory, and Erick T. Young of the University of Arizona, were brought in to help. For the Europeans, researchers including J.P. Emerson of Queen Mary College, and Paul R. Wesselius of the Kapteyn Institute were added. Other astronomers were brought in as the workload continued to increase.

Walker (personal communication, 1999) describes the division of responsibilities to which the various organizations agreed:

> The NASA Ames Research Center would be responsible for development of the telescope, cryogenics, focal plane array, and electronics... Jet Propulsion Laboratory would be responsible for the data reduction and analysis system, integration and testing of the assembled payload, and overall management of the US portion of the project, with H.H. Aumann the Project Scientist at JPL. The Dutch would develop the spacecraft and all its systems (attitude, data transmission, etc.) as well as operate the satellite on orbit. The Dutch would also develop two focal plane instruments, the Low Resolution Spectrometer (LRS, 7.7 to 28 μm) and the Chopped Photometric Channel (CPC, 50 and 100 μm chopped small-field radiometer). The British would provide the ground station to command the satellite and receive all the downlinked data. They would also provide a "quick look" facility to evaluate rapidly the data quality.

Progress over the years was difficult but successful. Then, just as all the major hurdles seemed to have been overcome and plans were being made for launch, a disaster threatened to derail the entire mission. Jim Houck (personal communication, 1999) recalls:

> IRAS had 62 discrete detectors. Each one had a cryogenic differential JFET [junction field effect transistor] input stage built by Frank Low. Because the total number of wires in the telescope Dewar was nearly 500, wire-to-wire short-testing was impractical... [The] individual wires were checked for shorts to ground and continuity. Unfortunately, one wire which was intentionally set to ground turned out to be shorted to the wire carrying the bias voltage to one of the two modules of Band 2 [25 μm] detectors. One of the JFET inputs on the cryogenic input stage was connected to the detector signal and the other to ground potential. Since the other side of the detector was also connected to ground by the wiring fault, there was no bias on the detector... [This] entire [detector] bank showed no response to infrared signals at all.

During the summer before launch, the combined IRAS Science Working Group met at the Appleton Research Laboratory near Oxford [UK]. By then it was well known that half of the Band 2 20-micron detectors were not working because of [the] shorted bias wire. The NIVR and SERC [the British Science and Engineering Research Council] representatives at the meeting were strongly opposed to launching IRAS without fixing this problem. On the other hand, I knew that the NASA managers had said that the cost of fixing the problem was much too high. I worried that the mission might be delayed by a political stalemate or cancelled outright. I was discouraged and left the meeting early to return to my room in The Gate House Hotel in Oxford. I remembered a comment by Dave Schaack, a graduate student of mine at Cornell, and... tried to find a technical solution to the developing political problem. By dinner time I had found a possible solution, and called Tom Soifer to join me in the bar at the Gate House. I described the idea, and gave him a copy of my notes to give to Gerry Smith, the US project manager. Tom later told me that Gerry bet a case of beer that the solution wouldn't work. By the time I reached Kennedy Airport the next afternoon the engineers at JPL had validated and improved on my solution, and were preparing to test it in the lab. The fix worked well, and was soon installed in the flight hardware. The revived detectors had slightly lower noise than those with the "normal" bias supply, and brief thought was given to rewiring all the detectors in the hope of reducing the system noise. The rapidly approaching launch date put a quick and final end to that idea. The next January IRAS was launched with all detectors working.

THE IRAS MISSION

The IRAS mission was a spectacular success. Approximately two-thirds of the 300-day-long IRAS mission, lasting from January to November 1983, was devoted to an unbiased survey of the sky that succeeded in charting 98% of the celestial sphere in four broad wavelength bands. IRAS was launched into a Sun-synchronous polar orbit at the day–night terminator, which precessed about one degree per day. In this orbit the Earth/Sun/spacecraft geometry varied only slowly so that the survey could be executed with a simple scanning strategy. Observations were carried out with an all-beryllium 57-cm aperture, $f/9.6$ Ritchey–Chrétien telescope whose focal plane was cooled to 3 K and featured a total of ∼60 arsenic-doped silicon (Si:As), antimony-doped silicon (Si:Sb), and gallium-doped germanium (Ge:Ga) discrete photoconductors, each with a separate JFET amplifier readout. The detectors covered, respectively, the 12, 25, 60, and 100 μm bands, with Ge:Ga appropriately filtered to cover the last two of these. A low-resolution spectrometer covered the wavelength range from 7.5 to 23 μm.

A measure of the mission's success was the cataloging of some 250,000 celestial sources, the vast majority of which had never before been detected in the infrared. No area of modern astrophysics remained untouched. The many scientific highlights include:

- The discovery of galaxies emitting up to 50 times more energy at far-infrared wavelengths than in the optical domain, while also emitting from 100 to 1000 times as much total power as the Milky Way. The existence of such highly luminous infrared galaxies came as a major surprise, as did the finding that most of these sources are galaxies in collision.
- The discovery of disks composed of fine dust grains orbiting around a number of stars that, in many ways, are reminiscent of our own Sun. This dust is conjectured to be the remnant of originally far more massive circumstellar clouds of gas and dust from which a system of planets may have already formed. This finding initiated further astronomical searches for signs of planets around these stars. A sharply defined, though far fainter set of dust rings was also found orbiting our own Sun, as were enduring trails of dust left by the passage of comets through the Solar System.
- The successful determination of spectra of planetary nebulae and a variety of other sources at wavelengths previously inaccessible due to telluric absorption.
- The identification of patchy infrared emission from the diffuse interstellar medium, referred to as "infrared cirrus" because of its resemblance to thin, streaky clouds in Earth's atmosphere. Infrared cirrus is important as a tracer of matter high above the disk of our Galaxy, and as a potential source of interference with observations of distant galaxies.
- Perhaps one of the least noted, but most far-reaching achievements of the IRAS mission was its pioneering

success in registering, calibrating, and archiving extremely large and complex databases. The astronomical community had not until then dealt with anything like the wealth of 250,000 sources, observed at different times of the year, in four wavelength ranges, and with scans repeated over ranges of a few hours, days, or months. All this was carefully tracked, artifacts were removed, and a digitized database of great reliability established that is still in wide use today. The efforts required the establishment of a separate Infrared Processing and Analysis Center (IPAC) that has evolved to provide valuable support services to the entire US infrared astronomical community and has served as a model for infrared data processing centers abroad.

THE COSMIC BACKGROUND EXPLORER (COBE)

As Walker pointed out (see above), the joint mission definition team, which had assembled a design for IRAS, had decided that the IRAS survey would not be suitable for also obtaining information on the infrared and microwave background. They recommended that NASA devote a separate satellite mission to this effort. Leading figures in getting this effort started were Michael G. Hauser, John C. Mather, and Rainer Weiss, who had all been on Walker's definition team. They were joined by Samuel Gulkis of JPL, who had great expertise with microwave receivers; George F. Smoot of the University of California, Berkeley, who had previously carried out observations on the anisotropy of the microwave background radiation with instruments carried aloft on a NASA U-2 aircraft; David T. Wilkinson of Princeton University, who had long been involved in short-wavelength radio observations of the background; and Edward L. (Ned) Wright of UCLA, who was interested in cosmological simulations. This group was eventually to be joined by many others who constituted the final science team (Boggess et al. 1992).

COBE was built by NASA's Goddard Space Flight Center (GSFC). In 1989 it was launched into a polar orbit identical to that of IRAS. It carried three instruments, two of which, the Diffuse Microwave Radiometer (DMR) and the Far Infrared Absolute Spectrophotometer (FIRAS), were used to study the cosmic microwave background radiation (CMBR) – the isotropic blackbody radiation with a temperature of ~2.73 K, a relic of the explosive origins of the Universe. The third experiment, the Diffuse Infrared Background Experiment (DIRBE), measured the background at infrared wavelengths from 1 to 200 μm. DIRBE and FIRAS were cooled by liquid helium and repeatedly mapped the entire sky during the ~10 month cryogenic lifetime of COBE. The critical components of the DMR, which was able to operate for about four years, were cooled passively to ~140 K.

The CMBR provides important cosmological information, and DMR and FIRAS succeeded, respectively, in measuring the spatial structure in that radiation field and in establishing its blackbody spectrum to high precision. FIRAS and DMR also measured the foreground radiation from the Milky Way. FIRAS was a polarizing Michelson interferometer instrumented with helium-cooled bolometers as detectors. It registered spectra of the Galactic emission from ~3 μm to ~100 μm with a 7° field of view. Of particular interest was its detection of emission from broadly distributed carbon, nitrogen, oxygen, and carbon monoxide in the Galaxy (Bennett et al. 1994).

DIRBE obtained measurements in ~10 wavelength bands, using a variety of discrete photodiodes and photoconductors from 1 to 100 μm, and helium-cooled bolometers at 140 and 240 μm. All measurements were calibrated against an internal, cold, black reference surface which enabled determination of the absolute sky brightness.

The principal scientific results from DIRBE include (Hauser et al. 1998):

- The temperature and black-body spectrum of the CMBR.
- The anisotropy of this radiation, which can tell us the speed and direction of the Earth's motion through the Universe.
- An improved determination of the distribution of the infrared radiation from the zodiacal dust cloud within the Solar System. This has led to improved models of the dust cloud and its infrared emission. DIRBE also confirmed the discovery by IRAS of a faint enhancement of emission in the Earth-trailing direction, attributed to temporary gravitational trapping by the Earth of zodiacal dust particles that are spiraling inwards towards the Sun.
- Determination of the large-scale distribution of infrared radiation from the Galaxy. This includes both the far-infrared radiation from 25 to 200 μm, which samples the distribution of heated dust, and the near-infrared radiation from 1 to 25 μm, which indicates the large-scale distribution of stars in the Galaxy.
- Evidence for an isotropic background of infrared radiation at 140 and 240 μm, which arises from outside the Galaxy and may represent the integrated emission from star-forming galaxies at redshifts $z \sim 1-2$.

THE INFRARED SPACE OBSERVATORY (ISO)

ISO, built and launched by the European Space Agency (ESA), was the first comprehensive infrared astronomical observatory in space. NASA and the Japanese Institute of Space and Astronautical Sciences (ISAS) provided important technical, operational, and scientific support. ISO mapped celestial sources and analyzed them through spectroscopy, photometry, and linear polarization studies (Kessler et al. 1996).

On the night of 16–17 November 1995 an Ariane 4 rocket launched ISO into a highly elliptical, 24-hour, circumterrestrial orbit, where the observatory operated with great success until its helium cryogen ran out and instruments began warming up in early April 1998. The length of the spacecraft was 5.3 m, its width 2.3 m, and its mass approximately 2,500 kg. At launch it carried a superfluid helium charge of 2,300 liters, which maintained the Ritchey–Chrétien telescope with its optical baffles and scientific instruments at temperatures of 2–8 K. A three-axis-stabilizing system provided an absolute pointing accuracy of a few seconds of arc, with a stability of a fraction of an arc second in both jitter and long-term drift. The diameter of the telescope's fused silica primary mirror was 0.6 m, permitting diffraction-limited observations down to wavelengths of roughly 5 μm.

Four instruments formed the core of the scientific payload.

- A camera containing two 32×32 pixel arrays: indium antimonide (InSb) for the wavelength range 2.5–5.5 μm, and gallium-doped silicon (Si:Ga) for the range 4–17 μm. Each array could be operated with selected filters for broad-band spectrophotometry or continuously variable filters (CVF) for low-resolution ($R \sim 40$) imaging spectroscopy, and could be programmed to view successively sources through three linear polarizers oriented relative to each other at angles of 60°.
- A photometer covered the wavelength range from 2.5 – 240 μm. It employed Si:Ga detectors with peak response at 15 μm, boron-doped silicon (Si:B) detectors with peak response at 25 μm, unstressed gallium-doped germanium (Ge:Ga) detectors with peak response at 100 μm, and stressed Ge:Ga detectors with peak response at 180 μm. Stressed detectors were mounted in a miniature clamp or vise to apply high mechanical pressure to the crystals thereby extending their wavelength response. At 100 and 200 μm, the photometer housed, respectively, 3×3 and 2×2 arrays of unstressed and stressed Ge:Ga to facilitate mapping. Multiple apertures, multiple filters, and polarizers could be used for photometric and photopolarimetric measurements in each range. Scanning and mapping observations were carried out at all wavelengths. Two grating spectrophotometers, each with a 64-element linear Si:Ga detector array, provided low-resolution spectra with resolving power $R \sim 100$ at 2.5–5 and 6–12 μm.
- A short-wavelength spectrometer (SWS) included both grating and Fabry-Perot (FP) instruments. Grating spectra were obtained for the entire wavelength range from 2.38 to 45.2 μm, with resolving power $R \sim 1000$–2000. The FP mode covered the 11.4–44.5 μm portion of this range and gave a resolving power a factor of ~ 20 higher. For the grating mode, the detectors were InSb at 2.38–4.08 μm, Si:Ga at 4.05–12.1 μm, Si:As at 12.0–29.0 μm, and Ge:Be at 29–45.2 μm. For the FP mode, Si:Sb was used at 11.4–26 μm, and Ge:Be at 26–44.5 μm.
- A long-wavelength spectrometer (LWS) permitted coverage from 43 to 196.9 μm. Its grating provided a resolving power $R \sim 150$–200. A FP mode increased the resolving power to $R \sim 6800$–9700. Ten detectors, arranged in a linear array on a curved surface operated in these modes. Detector materials were Ge:Be sensitive at 43–50 μm, unstressed Ge:Ga at 50–100 μm, and stressed Ge:Ga beyond 110 μm.

Among the scientific highlights of ISO were:

- The detection of water vapor in many parts of the interstellar medium of the Galaxy. Before ISO, the infrared emission from interstellar water vapor could not be detected because telluric water vapor absorbs at precisely the emission wavelengths. Water vapor, however, can be one of the primary coolants of interstellar clouds, and the prevalence of this cooling needed to be understood to assess the extent to which it promotes protostellar collapse.
- Detection of polycyclic aromatic hydrocarbons in the spectra of many galaxies. These macromolecules were well known in our Galaxy, but ISO had the sensitivity needed to show that their emission also dominates the 5–12 μm radiation from nearby spiral galaxies. These macromolecules emit through radiative fluorescence: A molecule is excited by absorption of a single ultraviolet or optical photon, and quickly re-radiates this energy through emission of a number of infrared photons as it returns to its ground state.
- Detailed extragalactic source-counts at wavelengths ranging from 4 to 175 μm. These lead to an improved understanding of the origins of the extragalactic diffuse infrared radiation detected by COBE. Many of the randomly observed galaxies appear to be ultraluminous, indicating that they contain extended regions of massive star formation or that they harbor an active galactic nucleus (AGN), possibly surrounding a massive central black hole.

OTHER INFRARED SPACE MISSIONS

For completeness, a number of other infrared space astronomy missions should be noted:

- The Spacelab-2 Infrared Telescope, launched in 1985. A 15-cm diameter helium-cooled telescope was flown on Spacelab 2 and made infrared measurements from 1.7 to 118 μm. It yielded data on the structure of the Galaxy and about the infrared background environment on the Space Shuttle (Kent et al. 1992).
- The Midcourse Space Experiment (MSX), 1995. MSX, a US Department of Defense mission, carried a 35-cm

aperture off-axis telescope and five linear Si:As arrays, which mapped the sky in bands from 4 to 22 μm. Although it was primarily designed to scan the Earth's limb, it carried out a number of astrophysical experiments and produced excellent images of the Galactic plane with ~18 arcsec resolution (Carey et al. 1998). A description of the Galactic plane survey is to be found in Price et al. (2001).

- The Infrared Telescope in Space (IRTS) 1995. IRTS was a predominantly Japanese mission conducted by ISAS with significant NASA participation. IRTS had a 15-cm diameter liquid-helium-cooled telescope and four different focal plane instruments to cover wavelengths from 3 to 800 μm. It was carried on a Japanese satellite called the Space Flyer Unit and surveyed 6% of the sky during its five-week active lifetime (Murakami et al. 1996).
- The Near Infrared Camera and Multi-Object Spectrograph (NICMOS), 1996. NICMOS is a second-generation focal plane instrument installed on the Hubble Space Telescope (HST). It was instrumented with three 256×256 mercury cadmium telluride (HgCdTe) arrays carrying a range of filters covering the 1–2.5 μm spectral band and optimized for high-spatial-resolution imaging. NICMOS was cooled by solid nitrogen and achieved a lifetime of slightly less than two years – shorter than expected because of a partial failure of the cryogenic system in orbit (Thompson et al. 1998).
- The Submillimeter Wave Astronomy Satellite (SWAS), 1998. SWAS was the first space mission to carry heterodyne (radio-type) receivers for spectroscopic exploration. SWAS has a 55×71 cm near-optical-quality off-axis primary mirror and two heterodyne radiometers with Schottky barrier diode mixers and a single acousto-optical spectrometer. SWAS began surveying the Galactic plane in the emission of atomic carbon, C, molecular oxygen, O_2, water vapor, $H_2^{16}O$ and $H_2^{18}O$, and a carbon monoxide, ^{13}CO, transition from rotational level $J = 5$ to 4. All five transitions lie close to either 538 or to 615 μm (Melnick et al. 1998).

WHO WERE THE EARLY INFRARED SPACE ASTRONOMERS?

None of the astronomers who came into the field in the early 1960s arrived with a traditional background in astronomy. At that time, most academic departments taught their graduate students that optical astronomy was the dominant force in the field and that radio astronomy had a future but still had to prove itself.

Radio astronomy was practiced primarily by people who had a background in electrical and radio engineering and knew how to build radio telescopes and receivers. Most, however, lacked an understanding of the more fashionable astronomical problems of the day. They did not need to know a great deal of astronomy to contribute to the field, because every new observation they made was a surprise to optical astronomers. Often the sources observed by radio astronomers were either invisible through optical telescopes or, if found, did not in any obvious way look different from other sources that emitted no radio waves. It was all very puzzling. Theorists were scrambling to get to grips with the way in which radio observations might dovetail with what was known from the visible. They needed more radio observations to tell them how to do this, and the new contingent of radio astronomers were able to oblige without having any deep understanding of traditional optical astronomy (Edge and Mulkay 1976).

To an appreciable extent, the same history repeated itself in the mid- and far-infrared. Those who started the field and had the greatest influence were primarily physicists. William F. Hoffmann, James R. Houck, Frank Low, Gerry Neugebauer, Paul Richards, Charles H. Townes, and the author all had a background in experimental physics. All of them understood optical techniques, and Houck, Low, and Richards all came to infrared astronomy as materials science physicists. Like their US counterparts, British, French, and Japanese colleagues had similar backgrounds. They understood the nature of detectors, and could extract the greatest sensitivity from them. Most of them had to scramble to learn what little traditional astronomy they needed to talk with optical astronomers. But their findings were often so bizarre that initially there was little meaningful conversation between infrared and optical workers. By and large, when the two groups pointed their telescopes at the same source, the optical astronomer would detect emission from the hot regions in the source and the infrared astronomer much colder domains. Sometimes such radiation might arrive from the same field of view but emanate from totally distinct, physically disconnected regions. Explanations for the observed radiation patterns then were hard to come by.

Infrared astronomers initially measured the successes of their efforts in terms of instrumental advances, and this is reflected in the emphasis given to instrumentation in this chapter. At first, there were few really significant new astronomical insights, only a series of successive surprises that produced no coherent theoretical framework. Only later, after many generations of ever more sensitive detection techniques had been developed, was it possible to begin gaining astrophysical insight. If one compares the sensitivity of individual detectors of the early 1960s to those of the 1990s, one has to speak of techniques that earlier were able to reliably cull out signals, under favorable laboratory conditions – and depending on the spectral range – of the order of 10^{-12} to 10^{-14} W in an observation lasting one second.

In contrast, in the early 1990s, detectors could reliably register signals as low as 10^{-17} to 10^{-19} W over a wider range of wavelengths in the same time. In addition, modern arrays coming into use in the late 1990s often contained a million detectors, all of which could simultaneously map a large region. Information could be gathered at billions of times the rates possible earlier and in wavelength ranges that were still inaccessible in 1960.

THE TRANSITION TO "BIG SCIENCE"

Astronomical heritage is handed down from one generation of scientists to the next. Each of the different university groups in which infrared astronomy started has had its own impact on subsequent developments in the field. Those who initiated ground-based efforts, like Gerry Neugebauer, tended to have students who continued along these lines. Others who carried out airborne observations trained students at least some of whom would carry on those traditions. Such dynastic structures pervade infrared space astronomical communities the world over and have played an important role in the transition of infrared space astronomy to "big science."

Starting in the 1980s, infrared space astronomy joined the class of big science projects with costs running into the hundreds of millions, if not billions of dollars. At this stage, the legacy of the groups that had taken an early lead in space work became increasingly influential through the students they had trained. Those early students had by now become the next generation of leaders, occupying most of the positions of influence.

A most striking example of a scientific dynasty and its scions is found in Japan, where the great astrophysicist Satio Hayakawa is considered the single originator of the country's infrared astronomical space efforts. Later leaders in the field, like Haruyuki Okuda, who had been a postdoctoral worker with Hayakawa, and Toshio Matsumoto, who worked with Hayakawa as a graduate student, succeeded each other as heads of the country's leading infrared space group at ISAS.

In Britain, Richard E. Jennings and his students carried out early ballooning observations in the infrared. Among those of Jennings's group who later made major contributions were J.P. Emerson, I. Furniss, K.J. King, and A.F.M. Moorwood. Jennings, probably more than any other individual encouraged Britain to participate in the IRAS project, and brought with him P.E. Clegg, J.P. Emerson, and M. Rowan-Robinson, who together with colleagues like Peter Ade went on to establish a new powerhouse of infrared astronomy at Queen Mary College, London.

At Queen Mary, Peter Clegg later became principal investigator on the Long Wavelength Spectrometer on the ISO satellite. M.J. Griffin who had worked with Clegg on ISO later became principal investigator of the Spectral and Photometric Imaging Receiver (SPIRE) on the Herschel Space Observatory. The now renamed Queen Mary and Westfield College group remained the leading infrared space astronomy group in Britain throughout the 1990s.

In the Netherlands, Jan Borgman and Jan Wijnbergen had been early pioneers in infrared astronomical ballooning. When Borgman took on major administrative responsibilities at the University of Groningen, Reinder J. van Duinen in 1971 took over the group. Van Duinen mentored graduate students, sought closer contacts with industry, and set out to win Dutch participation on IRAS. Joined by colleagues in Groningen's Kapteyn Institute, D.A. Beintema, S.R. Pottasch, P.R. Wesselius, and others, van Duinen became co-principal investigator on IRAS.

Further south in the Netherlands, the European Space Agency's Space Research and Technology Center (ESTEC), at Noordwijk, had also become active. In 1971, at the behest of E.A. Trendelenburg, at the time head of the Space Science Department at ESTEC, the physicist Brian Fitton established an infrared astronomical group. This group, whose first member was Michel Anderegg, dedicated itself to balloon and airborne observations, respectively flown from Palestine, Texas, and on NASA's KAO. In collaboration with Jean-Paul Baluteau at the Observatoire de Meudon, in France, Anderegg began work on an interferometric spectrometer and was soon joined by A.F.M. Moorwood, who by then had completed his studies with Jennings in London.

Thijs de Graauw, fresh out of graduate school at Utrecht, came into this group and initiated a program in heterodyne spectroscopy. For his job interview at ESTEC, de Graauw recalls arriving without a tie, and having to borrow one from his future boss, Fitton, before going in to be interrogated by Trendelenburg. Trendelenburg, who was famously rude, posed a series of blunt, provocative questions, which so outraged the young de Graauw, that he responded in kind. This evidently impressed Trendelenburg, who hired de Graauw straight away (de Graauw, personal communication, 2000).

Once IRAS had been successfully launched, Fitton and van Duinen both pressed ESA to decide on flying ISO, which later became one of ESA's great successes. On ISO, Anderegg served as systems engineer; de Graauw, who had in the meantime left ESTEC to succeed van Duinen in directing the infrared astronomy group at SRON in Groningen, became principal investigator on ISO's SWS; Moorwood, who had by then moved on to the European Southern Observatory (ESO), was selected to serve as one of ISO's mission scientists; and Martin Kessler, who had joined the ESTEC infrared astronomy group straight out of graduate school at Imperial College, London, took over the lead on scientific planning for the mission and eventually became project scientist on ISO.

Launched in late 1995, ISO operated almost flawlessly for 28 months. As the mission on orbit began to wind down, in early 1998, ESA initiated serious steps toward its successor, the Herschel Space Observatory, designed to carry aloft a much larger telescope dedicated exclusively to longer wavelengths. Thijs de Graauw went on to become the principal investigator on the Heterodyne Instrument for the Infrared (HIFI). During the last two decades of the twentieth century, de Graauw effectively served as the principal architect of Dutch infrared space astronomy, and Groningen remained the Netherlands' leading center for the discipline.

In the USA, Gerry Neugebauer's former student Eric Becklin was named director of the Stratospheric Observatory for Infrared Astronomy (SOFIA), while Steven Beckwith, another of his students, was appointed director of the Space Telescope Science Institute.

The most prolific group in the USA may well have been the Cornell University infrared astronomy group, where space infrared astronomy was the goal right from the start. Following the author's early work with infrared rocket telescopes and airborne infrared spectrometers, he worked for many years as the sole US member on ESA's ISO science team, and later also became a member of the science teams of NASA's SWAS and the ESA/NASA Herschel mission.

James R. Houck, who had similarly worked on Cornell's rocket and airborne efforts became a leading member of the science team of IRAS. Later he became principal investigator on the Infrared Spectrometer (IRS), one of three instruments aboard the Space Infrared Telescope Facility (SIRTF) spacecraft.

Many of the graduate students who had obtained their PhD degrees in Cornell's infrared astronomy group went on to become leaders in infrared astronomical space research. Perry Hacking became project scientist on NASA's second astronomical Small Explorer (SMEX) mission, the highly promising Wide-field Infrared Explorer (WIRE), which was to conduct a deep survey of limited portions of the sky, but to unprecedented depths at 12 and 25 μm. This most significant cosmological mission unfortunately failed shortly after launch. Gary Melnick, a co-investigator on Spacelab 2, later became principal investigator on the SWAS mission, the first of NASA's SMEX satellites dedicated to astronomy. Thomas Roellig became the leading US project scientist on the IRTS mission, as well as co-principal investigator on the Mid-Infrared Spectrometer on board. Later he became a facility scientist on SIRTF. Kandiah Shivanandan joined the NRL group that the author had initially started, and later obtained the material for his PhD thesis through collaboration on a flight of a Cornell rocket payload. He went on to direct NRL's infrared collaboration with AFGL on joint AFGL/NRL rocket astronomical surveys. B. Thomas Soifer became professor of physics and astronomy at the California Institute of Technology. For some time he was director of Caltech's IPAC. He later became director of the SIRTF Science Center. Michael W. Werner was project scientist on SIRTF throughout the many years of this mission's initiation and evolution.

Many other former graduates of the Cornell group also were involved in one way or another on these missions, or in developing instrumentation for KAO or SOFIA. Among these were Gordon Stacey, who later joined the Cornell faculty, and Cornell faculty member Terrence Herter, a former student of Judith L. Pipher, professor of physics and astronomy at the University of Rochester, who had been a graduate student at Cornell and the first woman to receive a PhD in infrared rocket astronomy.

At the time of writing, the old university dynasties are rapidly fading. Younger scientists, who have been exposed to a series of post-doctoral posts at different institutions, have begun to sift out the best approaches offered by each of the leading institutions. In addition, the big science projects this new generation of infrared astronomers now heads are generally centered not at universities, but rather at national or international space centers – in the USA, NASA's JPL and GSFC; in Japan, ISAS; in Europe, ESTEC and various other national space centers in different countries. Large armies of technical experts now devote years to an individual mission. Projects take decades to mature from initial concept through construction, launch, data gathering, generation of scientific publications, and completion of a set of final data archives.

In the course of the four decades, from 1960 to the year 2000, the nature of infrared space astronomy has dramatically changed!

THE ROLE OF THE MILITARY

The account of the development of infrared space astronomy given thus far might give the impression that most of the spectacular advances in infrared space astronomy are to be attributed to individual astronomers. Nothing could be more misleading.

The primary advances in the field came about because of technological developments that permitted more incisive observations. The many professional astronomers named in this article most certainly worked extremely hard at adapting basic infrared components to carrying out their observing programs. But they seldom were involved in developing those components most of which were conceived for military purposes. Above, we have seen the involvement of the NRL in the early development of rocket astronomical observations, and the eventual collaboration of this group with the far more extensive efforts initiated by AFCRL, later renamed AFGL and later once again reorganized into the Air Force's Phillips Laboratory.

These US Air Force efforts included the series of rocket flights of the 1970s that have already been mentioned. They provided the largest mid- and far-infrared sky surveys conducted up to that time. In 1982 a collaboration between AFGL and NRL led to the Far Infrared Sky Survey Experiment (FIRSSE), a rocket survey with 36-cm aperture all-beryllium optics and a five-color focal-plane array covering the spectral range from 8 to 120 μm (Price 1988). The instrument was cooled with superfluid liquid helium and used a sintered-nickel porous plug for fluid containment. Containment of cryogen was an important problem that had to be resolved in order to assure that the liquid would not flow out of its reservoir under gravity-free conditions. The results from two such flights, covering 47% of the sky gave rise to a catalog published as an Air Force report (Price and Murdock 1983).

The convention of publishing results in report format, rather than in the traditional astronomical journals, in itself is an indication of why the military's contributions to the field are sometimes underestimated. Only a few specialists in infrared observations might even know of the reports' existence. One example of this is the Air Force's long-term monitoring of long-period stars at 2.7 μm by means of satellites. This survey conducted within 10° of the celestial equator yielded 1,278 sources. The survey had a limiting magnitude 1.5 magnitudes fainter than the Neugebauer and Leighton survey (Neugebauer and Leighton, 1969), though it was only 50–60% complete within its declination range. The results of this survey were never published in the routine astronomical literature, but, according to Price (1988), appeared primarily in the form of technical reports from the Aerospace Corporation, to which few interested astronomers would have access.

Even this, however, hides the true aspect of military involvement, which at enormous cost developed and perfected all of the detector materials and many of the optical materials used in infrared astronomical observations since World War II.

A brief survey of these efforts has been given by John A. Jamieson who, for more than four decades starting in the 1950s, was involved with military infrared developments (Jamieson 1995). Jamieson traces the development of various detector materials from the 1940s to the Vietnam war and concludes:

The two developments of intrinsic HgCdTe [mercury cadmium telluride] and extrinsic silicon, particularly arsenic-doped silicon are, with lead sulphide, the foundations of infrared technology for ballistic missile defense. These detectors individually approach the fundamental limits of detectivity, can be manufactured reproducibly, and can be formed in large arrays... They together with powerful, small, light computers, were to enable the revolution in non-nuclear, hit-to-kill ballistic missile defense which began in the 1970s.

Jamieson outlines the interests of the military as follows:

When the main engines [of an intercontinental ballistic missile] are burning, thrusting the missile onto its trajectory, carbon monoxide and water vapor in the exhausts radiate hundreds [of thousands] to millions of watts per steradian near 2.7 and near 4.3 micrometers. These intense, high contrast, 2500 K signals provide dependable early warning for a hundred seconds. When the acceleration is over and the deadly reentry vehicles are flying on Keplerian midcourse trajectories, they radiate tens of watts per steradian in a broad band around 10 micrometers... At about 290 K their contrast against cold space is enough that they can be sensed at ranges of thousands of kilometers. When they begin to reenter the atmosphere their heat shields glow at hundreds of Kelvins and they radiate several tens of watts in the medium wavelength infrared maximizing around 5 micrometers. These signatures can be detected and tracked with passive sensors and with angular accuracies of tens to hundreds of microradians. To spoof these passive sensors is difficult, particularly if it must be done day or night with a pre-positioned offense.

The range of temperatures cited by Jamieson explains the military's interest primarily in near- and mid-infrared detection. For these spectral ranges, they developed spectacularly sensitive detectors and huge detector arrays.

In 1967 the USA established its Advanced Ballistic Missile Defense Agency (ABMDA) to develop infrared technology for ballistic missile defense (BMD). It developed the Fly-Along Infrared (FAIR) experiment to intercept reentry vehicles with missiles homed in on their targets with cryogenically cooled telescopes and detectors. In 1970 the US Air Force launched the first early warning geosynchronous satellites equipped with infrared sensors. Jamieson (1995) remarks: "If a raid of ICBMs [intercontinental ballistic missiles]... were launched anywhere on earth, sensors on three or four of these satellites would detect them by radiation from their exhausts, track their trajectories and provide sufficient warning to alert our retaliatory forces. This system... was outstandingly successful and has continued with improvements to this day."

The needs for sensitive detectors and detector arrays for defense purposes were clear from the military's perspective, as was the need for an infrared sky survey. Knowledge of the infrared sky map would keep interceptors from locking on celestial sources.

A boost to infrared technology also came from President Ronald Reagan's 23 March 1983 television address, in which he defined plans for a Strategic Defense Initiative (SDI) to "intercept and destroy strategic missiles before they reached our own soil or that of our allies." A study was undertaken to assess the required technology. Headed by James Fletcher, who twice served as NASA Administrator during his career, this study recommended that further advances were needed if such a defensive net was to be viable. Jamieson (1995) states:

> We needed to develop the technology of cryocooled, low-scatter optics which could live in instant readiness for many years. We needed focal plane arrays of detectors for acquisition tracking and kill assessment and for guidance seekers for interceptors. We needed long-life cryocoolers to cool detectors and optics for many years in space. In some cases they needed to cool within 10 Kelvins of absolute zero. We needed very fast signal and data processing to handle, in real time, the enormous volume of signals that would flow from the sensors.

A partial list of the many developments supported by this effort were:

- Measures to radiation-harden detectors against gamma radiation from nuclear debris. This turns out to be particularly important for astronomy from space beyond Earth's radiation belts, where naturally occurring cosmic ray particles and gamma rays produce false signals and can deteriorate detectors. The development of "blocked impurity band" (BIB) detectors were part of this effort.
- Hybrid arrays of detectors bonded through indium bumps to electron readout chips to produce complete multiplexed sensing chips.
- A broad range of materials testing to produce better detectors.
- Increased sizes of detector arrays.
- New polishing techniques to produce very low-scatter optical surfaces, particularly for beryllium optics.
- Development of reflecting optics that could provide excellent spatial resolution over fields of view as wide as 20 to 40°.
- Means for detecting contamination of optics in space and for cleaning.
- Production of lightweight optics and development of economical computer-controlled polishing, figuring, and testing.
- Development of silicon carbide optics.
- Development of mechanical cryocoolers for long-term space missions.

Many of these efforts will undoubtedly continue to have a major impact on infrared space astronomy in the decades ahead.

Jamieson estimates that costs for the whole SDI from 1985 to 1992 amounted to "only about $3.4 billion per year for seven years. The infrared related portions amounted to less than $7.5 billion. (Probably about $1.5 billion of this was spent on infrared technology of optics, focal plane cryocoolers, signal processors and studies, analyses...)". In contrast, the joint expenditures of all civil scientific agencies worldwide on infrared astronomy from space, over the entire history of the field, is likely to have been of the order of half of these $7.5 billion spent in just the years 1985–92.

Given these relatively recent military expenditures, infrared astronomers may expect to inherit even more powerful techniques than those currently available provided they patiently stand by the closed door that normally separates military from academic infrared space astronomy. Occasionally that door opens a crack, and an arm hands out some highly desirable piece of technology.

An obverse of this munificence is also clear. Where the military has had no apparent interests, as in the far-infrared beyond 100 μm and well into the submillimeter range, astronomers have had to develop instrumentation on their own and progress has been far slower. The submillimeter range remains the last hardly-touched frontier, and astronomers still have only the most limited understanding of discoveries they may expect to make in this wavelength range.

Most astronomers tend to remain unaware of the true extent to which military developments have pervaded virtually all aspects of their work. Europeans, in particular, tend to be under the impression that military influences are largely a US phenomenon. But ISO, the most successful infrared astronomical mission of the 1990s, launched by ESA in 1995, is a clear counterexample. The infrared camera on board, constructed primarily in France, had a detector array specially constructed for the mission by the French defense establishment. The Short Wavelength Spectrometer built primarily in the Netherlands, incorporated previously classified infrared detectors provided largely through the efforts of Stephan Price, a co-investigator at the US Air Force (Price, personal communication, 2000). Yet most of the many hundreds of astronomers who have carried out observations with these instruments remain unaware of those contributions and would heatedly deny that this space mission had any military ties.

EPILOGUE

Future historians may well portray the history of infrared space astronomy in quite different terms from those presented here. The traditional review articles published in astronomical journals tend to emphasize the contributions made by academic scientists publishing in traditional academic journals. Moreover, the dynasties created in the academic world tend to propagate (and often exaggerate) stories about the contributions each dynasty has made. These factors undoubtedly have also influenced the author of this article.

A future historian may find that this approach is quite unsatisfactory. Fifty years from now, much of the work carried out by what President Dwight D. Eisenhower called "the military–industrial complex" (Eisenhower 1961) may be declassified. The relative contributions made by defense and academic programs may then become more clearly apparent, provided that the thousands of industrial and defense reports generated over the decades are then still available in industry or defense archives. Unfortunately, such reports are often prematurely destroyed. The archiving of massive series of reports is expensive. Most industries cannot afford to carry such additional expenses, and outdated reports are often discarded rather than maintained once the funding from a government contract expires. With the destruction of these records valuable insights into a true history get lost.

Assuming, however, that the records have been maintained, we may ask, "Where will the true heroes of the field then be found?" Will they be in the academic groups, where much of the work started? Will it be at AFCRL/AGL where, over a period of more than three decades, a stream of rocket and satellite flights was generated? Or will a future historian find that some one ingenious individual on the "black" – the classified – side of the military–academic doorway was the gray eminence of the field, steering it from the shadows, or through a dynasty of younger colleagues that he or she had spawned? The successful academic scientist might then become viewed as merely the tolerated small competitor who at any given time was most adroit at a strategy followed by almost everyone in the field – the begging, borrowing, or buying of the best military hand-me-down equipment available at the time, including detectors, detector arrays, or any associated equipment.

The eminent Columbia University sociologist of science, Robert Merton, long ago formulated his somewhat cynical, yet appropriate *Matthew principle*: "To those who have, more shall be given." By this he meant that scientists' collective memories have a way of distilling and simplifying history to the point where only a small handful of giants is correctly or falsely credited for every conceivable advance – it does not matter which. Those who are once given credit, keep gaining credit even for advances they never made. A kind of comic-strip distillation of history then emerges which hides the true nature and complexity of the scientific enterprise. It is quite likely that those academic scientists who have been credited with advances in infrared space astronomy, in review articles written by other academic scientists – like the one by the present author – will continue to gain increasing recognition and stature, simply through anecdotal repetition of their exploits.

Until a more thorough study of the field is undertaken, a troubling question will persist: "Where are the true heroes of the field of infrared space astronomy?" Despite four decades of work in the field, the author regrets his inability to satisfactorily answer this question.

Acknowledgments

Ther author wishes to thank Drs Michel Anderegg, Thijs de Graauw, James R. Houck, Martin F. Kessler, Stephan D. Price, Russell G. Walker, and Reinder van Duinen for their advice and contributions, and Johan Bleeker and Michelle Little for their careful reading of the manuscript and their comments and corrections. Research leading to this article was funded under several NASA grants for infrared and submillimeter astronomy.

REFERENCES

Allen, D.A. (1975). *Infrared – The New Astronomy*, Keith Reid, Devon.
Alpher, R.A. and Herman, R. (1949). Remarks on the evolution of the expanding universe. *Physical Review*, **75**, 1089–1095.
Aumann, H.H. and Low, F.J. (1970). Far-infrared observations of the galactic center. *Astrophysical Journal*, **159**, L159–L164.
Baluteau, J.-P., Bussoletti, E., Anderegg, M., Moorwood, A.F.M. and Coron, N. (1976). Infrared line emission from the Orion Nebula: Detection of [SIII] (18.71 μ) and [OIII] (88.35 μ). *Astrophysical Journal*, **210**, L45–L48.
Becklin, E.E. and Neugebauer, G. (1967). Observations of an infrared star in the Orion Nebula. *Astrophysical Journal*, **147**, 799–802.
Becklin, E.E. and Neugebauer, G. (1969). 1.65–19.5-Micron observations of the Galactic center. *Astrophysical Journal*, **157**, L31–L36.
Beckman, J.E. and Moorwood, A.F.M. (1979). Infrared astronomy. *Reports on Progress in Physics*, **42**, 87–157.
Bennett, C.L., Fixsen, D.J., Hinshaw, G., Mather, J.C., Moseley, S.H., Wright, E.L., Eplee, R.E. Jr., Gales, J., Hewagama, T., Isaacman, R.B., Shafer, R.A. and Turpie, K. (1994). Morphology of the interstellar cooling lines detected by COBE. *Astrophysical Journal*, **434**, 587–598.
Boggess, N.W. *et al.* (1992). The COBE mission: Its design and performance two years after launch. *Astrophysical Journal*, **397**, 420–429.
Bottema, M., Plummer, W. and Strong, J. (1964). Water vapor in the atmosphere of Venus. *Astrophysical Journal*, **139**, 1021–1022.
Carey, S.J. Clark, F.O., Egan, M.P., Price, S.D., Shipman, R.F. and Kuchar, T.A. (1998). The physical properties of the Midcourse Space Experiment Galactic infrared-dark clouds. *Astrophysical Journal*, **508**, 721–728.
Caux, E. Serra, G., Gispert, R., Coron, N., Puget, J.L. and Ryter, C. (1984). Far-infrared survey of the galactic disk in the southern hemisphere. *Astronomy and Astrophysics*, **137**, 1–5.
Caux, E. Serra, G., Puget, J.L., Gispert, R. and Ryter, C. (1985). Far-infrared survey of the galactic disk II: The sources. *Astronomy and Astrophysics*, **144**, 37–48.
Coblentz, W.W. (1914). A comparison of stellar radiometers and radiometric measurements on 110 stars. *Bulletin of the Bureau of Standards*, **11**, 613–656.
Danielson, R.E., Gausted, J.E., Schwarzschild, M, Weaver, H.F. and Woolf, N.J. (1964). Mars observations from Stratoscope II. *Astronomical Journal*, **69**, 344–352.

Danielson, R.E., Woolf, N.J. and Gausted, J.E. (1965). A search for interstellar ice absorption in the infrared spectrum of Mu Cephei. *Astrophysical Journal*, **141**, 116–125.

Davidson, K. and Harwit, M. (1967). Infrared and radio appearance of cocoon stars. *Astrophysical Journal*, **148**, 443–448.

Edge, D.O. and Mulkay, M.J. (1976). *Astronomy Transformed – The Emergence of Radio Astronomy in Britain*, Wiley, New York.

Eisenhower, D.D. (1961). Televised farewell address to the American public, 17 January 1961.

Felgett, P.B. (1951). An exploration of infrared stellar magnitudes using the photo-conductivity of lead sulphide. *Monthly Notices of the Royal Astronomical Society*, **111**, 537–559.

Forrest, W.J., McCarthy, J.F. and Houck, J.R. (1980). Detection of O [IV] and Ne[V] infrared emission lines from NGC 7027. *Astrophysical Journal*, **240**, L37–L41.

Furniss, I, Jennings, R.E. and Moorwood, A.F.M. (1972). Detection of far-infrared astronomical sources. *Astrophysical Journal*, **176**, L105–L108.

Gautier, T.N. Hauser, M.G., Beichman, C.A., Low, F.J., Neugebauer, G., Rowan-Robinson, M., Aumann, H.H., Boggess, N., Emerson, J.P., Harris, S., Houck, J.R., Jennings, R.E. and Marsden, P.L. (1984). IRAS images of the galactic center. *Astrophysical Journal*, **278**, L57–58 and plates L2–L5.

Genzel, R., Lutz, D., Sturm, E., Egami, E., Kunze, D., Moorwood, A.F.M., Rigopoulou, D., Spoon, H.W.W., Sternberg, A., Tacconi-Garman, L.E., Tacconi, L. and Thatte, N. (1998). What powers ultraluminous IRAS galaxies? *Astrophysical Journal*, **498**, 579–605.

Gispert, R., Puget, J-L. and Serra, G. (1982). Far infrared survey of molecular clouds H II regions complexes along the Galactic plane. *Astronomy and Astrophysics*, **106**, 293–306.

Golay, M.J.E. (1947a). Theoretical consideration in heat and infra-red detection, with particular reference to the pneumatic detector. *Review of Scientific Instrumentation*, **18**, 347.

Golay, M.J.E. (1947b). A pneumatic infra-red detector. *Review of Scientific Instrumentation*, **18**, 357.

Gull, G.E., Houck, J.R., McCarthy, J.F., Forrest, W.J. and Harwit, M. (1978). Far-infrared polarization of the Kleinmann–Low Nebula in Orion. *Astronomical Journal*, **83**, 1440–1444.

Harwit, M. (1960). Measurement of thermal fluctuations in radiation. *Physical Review*, **120**, 1551–1556.

Harwit, M. (1964). Infrared appearance of different Zodiacal cloud models. In *Les Spectres Infrarouges des Astres*, International Colloquium, 24–26 June 1963, Université de Liège, Mémoires de la Société Royale des Sciences de Liège, Series 5, **9**, 506–519.

Harwit, M. (1981). *Cosmic Discovery: The Search, Scope and Heritage of Astronomy*, Basic Books, New York.

Harwit, M., Dyal, P., Dyck, H.M., Gehrz, R.D., Hackwell, J.A., Houck, J.R., Jennings, R.E., Moore, W.J., Pipher, J.L. and Witteborn, F.C. (1974). *A Proposal Submitted Solely to the National Aeronautics and Space Administration for a Far-Infrared Astronomical Explorer Satellite*, Center for Radiophysics and Space Research, Cornell University, NY, p. 12.

Harwit, M., Houck, J.R. and Fuhrmann, K. (1969). Rocket-borne liquid helium cooled telescope. *Applied Optics*, **8**, 473–477.

Harwit, M. and Pacini, F. (1975). Infrared galaxies: Evolutionary stages of massive star formation. *Astrophysical Journal*, **200**, L127–L129.

Hauser, M.G., Arendt, R.G., Kelsall, T., Dwek, E., Odegard, N., Weiland, J.L., Freudenreich, H.T., Reach, W.T., Silverberg, R.F., Moseley, S.H., Pei, Y.C., Lubin, P., Mather, J.C., Shafer, R.A., Smoot, G.F., Weiss, R., Wilkinson, D.T. and Wright, E.L. (1998). The COBE Diffuse Infrared Background Experiment search for the cosmic infrared background: I. Limits and detections. *Astrophysical Journal*, **508**, 25–43.

Hayakawa, S., Matsumoto, T., Murakami, H., Uyama, K., Thomas, J.A. and Yamagami, T. (1981). Distribution of near-infrared sources in the galactic disk. *Astronomy and Astrophysics*, **100**, 116–123.

Hildebrand, R.H. (1996). Problems in far-infrared polarimetry. In W.G. Roberge and D.C.B Whittet (eds), *Polarimetry of the Interstellar Medium*, ASP Conference Series, Vol. 97, pp. 254–268.

Hoffmann, W.F. and Frederick, C.L. (1969). Far-infrared observation of the Galactic-center region at 100 microns. *Astrophysical Journal*, **155**, L9–L14.

Hoffmann, W.F., Frederick, C.L. and Emery R.J. (1971a). 100-Micron map of the galactic center region. *Astrophysical Journal*, **164**, L23–28.

Hoffmann, W.F., Frederick, C.L. and Emery R.J. (1971b). 100-Micron survey of the galactic plane. *Astrophysical Journal*, **170**, L89–L97.

Houck, J.R., Schaack, D. and Reed, R. (1974). 20 to 40 Micron spectroscopy of the Orion Nebula. *Astrophysical Journal*, **193**, L139–L141.

Houck, J.R., Soifer, B.T., Harwit, M. and Pipher, J.L. (1972). The far-infrared and submillimeter background. *Astrophysical Journal*, **178**, L29–L33.

Houck, J.R., Soifer, B.T., Pipher, J.L. and Harwit, M. (1971). Rocket-infrared four-color photometry of the Galaxy's central regions. *Astrophysical Journal*, **169**, L31–L34.

Huggins, W. (1869). Note on the heat of the stars. *Proceeding of the Royal Society (London)*, **17**, 309–312.

Jaffe, D.T., Harris, A.I., Silber, M., Genzel, R. and Betz, A.L. (1985). Detection of the 370 μm $^3P_2$–$^3P_1$ fine-structure line of CI. *Astrophysical Journal*, **290**, L59–L62.

Jamieson, J.A. (1995). Ballistic missile defense and infrared technology – The influence of infrared technology upon the Russian Revolution and the Soviet Empire. *Proceedings of the Infrared Instrumentation Society*, **40**, 13–40.

Johnson, H.L. (1965). Interstellar extinction in the Galaxy. *Astrophysical Journal*, **141**, 923–942.

Kelsall, T., Weiland, J.L., Franz, B.A., Reach, W.T., Arendt, R.G., Dwek, E., Freudenreich, H.T., Hauser, M.G., Moseley, S.H., Odegard, N.P., Silverberg, R.F. and Wright, E.L. (1998). The COBE Diffuse Infrared Background Experiment search for the cosmic infrared background: II. Model of the interplanetary dust cloud. *Astrophysical Journal*, **508**, 44–73.

Kent, S.M., Mink, D., Fazio, G., Koch, D., Melnick, G., Tardiff, A. and Maxson, C. (1992). Galactic structure from the Spacelab Infrared Telescope: I. 2.4 micron map. *Astrophysical Journal* Suppl., **78**, 403–408.

Kessler, M.F., Steinz, J.A., Anderegg, M.E., Clavel, J., Drechsel, G., Estaria, P., Faelker, J., Riedinger, J.R., Robson, A., Taylor, B.G. and Ximenez de Ferran, S. (1996). The Infrared Space Observatory (ISO) mission. *Astronomy and Astrophysics*, **315**, L27–L31.

Kleinmann, D.E. and Low, F.J. (1967). Discovery of an infrared nebula in Orion. *Astrophysical Journal*, **149**, L1–4.

Kleinmann, D.E. and Low, F.J. (1970). Observations of infrared galaxies. *Astrophysical Journal*, **159**, L165–L172.

Low, F.J. (1961). Low-temperature germanium bolometer. *Journal of Optical Society of America*, **51**, 1300–1304.

Low, F.J. and Aumann, H.H. (1970). Observations of galactic and extragalactic sources between 50 and 300 microns. *Astrophysical Journal*, **162**, L79–L85.

Low, F.J. and Johnson, H.L. (1965). The spectrum of 3C 273. *Astrophysical Journal Letters*, **141**, 336–337.

Maihara, T., Oda, N., Sugiyama, T. and Okuda, H. (1978). 2.4–Micron observations of the Galaxy and galactic structure. *Publications of the Astronomical Society of Japan*, **30**, 1–19.

Mather, J.C., Richards, P.L. and Woody, D.P., (1974). Balloon-based measurements of the cosmic background radiation. *IEEE Transactions on Microwave Theory and Technology*, **22**, 1046–1048.

Matsumoto, T., Hayakawa, S., Matsuo, H., Murakami, H., Sato, S., Lange, A.E. and Richards, P.L. (1988). The submillimeter spectrum of the background radiation. *Astrophysical Journal*, **329**, 567–571.

Melnick, G.J. (1998). The Submillimeter-Wave Astronomy Satellite: Science objectives and instrument description. *Proceedings of the International Society for Optical Engineering*, **3357**, 348–358.

Moorwood, A.F.M., Baluteau, J.-P., Anderegg, M., Fitton, B., Coron, N. and Biraud, Y. (1980a). Infrared line emission from H II regions: III. Airborne observations of [S III] (18 μm and 33 μm), [O III] (52 μm and 88 μm), and [N III] (57 μm) on M17. *Astrophysical Journal*, **238**, 565–576.

Moorwood, A.F.M., Salinari, P., Furniss, I., Jennings, R.E. and King, K.J. (1980b). Infrared spectroscopy with a balloon-borne Michelson interferometer: II. Observations of O III, O I and N III fine structure lines in H II regions. *Astronomy and Astrophysics*, **90**, 304–310.

Muehlner, D. and Weiss, R. (1970). Measurement of the isotropic background radiation in the far infrared. *Physical Review Letters*, **24**, 742–746.

Muehlner, D. and Weiss, R. (1973). Further measurements of the submillimeter background at balloon altitudes. *Physical Review Letters*, **30**, 757–760.

Murakami, H., Freund, M.M., Ganga, K., Guo, H., Hirao, T., Hiromoto, N., Kawada, M., Lange, A.E., Makiuti, S., Matsuhara, H., Matsumoto, T., Matsuura, S., Murakami, M., Nakagawa, T., Narita, M., Noda, M., Okuda, H., Okumura, K., Onaka, T., Roellig, T.L., Sato, S., Shibai, H., Smith, B.J., Tanabe, T., Tanaka, M., Watabe, T., Yamamura, I. and Yuen, L. (1996). The IRTS (Infrared Telescope in Space) Mission. *Publications of the Astronomical Society of Japan*, **48**, L41–L46.

National Academy of Sciences (1971). *Priorities for Space Research 1971–1980*, National Academy of Sciences, Washington, DC, pp. 19–20, 73–77, 85–86.

National Academy of Sciences (1975). *Opportunities and Choices in Space Science, 1974*, National Academy of Sciences, Washington, DC, pp. 18–19.

Neugebauer, G. and Leighton R.B. (1969). *Two-micron Sky Survey*, NASA SP-3047.

Neugebauer, G., Martz, D.E. and Leighton, R.B. (1965). Observations of extremely cool stars. *Astronomical Journal*, **142**, 399–401.

Neugebauer, G., Habing, H.J., van Duinen, R., Aumann, H.H., Baud, B., Beichman, C.A., Beintema, D.A., Boggess, N., Clegg, P.E., de Jong, T., Emerson, J.P., Gautier, T.N., Gillett, F.C., Harris, S., Hauser, M.G., Houck, J.R., Jennings, R.E., Low, F.J., Marsden, P.L., Miley, G., Olnon, F.M., Pottasch, S.R., Raimond, E., Rowan-Robinson, M., Soifer, B.T., Walker, R.G., Wesselius, P.R. and Young, E. (1984). The Infrared Astronomical Satellite (IRAS) mission. *Astrophysical Journal*, **278**, L1–L6.

Okuda, H. (1981). The large scale infrared emission in the Galactic plane observations. In *IAU Symposium 96, IR Astronomy*, G.G. Wynn-Williams and D.P. Cruikshank (eds) Reidel, Dordrecht, pp. 247–260.

Pettit, E. and Nicholson, S.B. (1922). The application of thermocouples to problems in astrophysics. *Astrophysical Journal*, **56**, 295–317.

Phillips, T.G., Huggins, P.J., Kuiper, T.B.H. and Miller, R.E. (1980). Detection of the 610 micron (492 GHz) line of interstellar atomic carbon. *Astrophysical Journal*, **238**, L103–L106.

Pipher, J.L. (1971). Rocket submillimeter observations of the Galaxy and background. Astronomy PhD Thesis, Cornell University, pp. 137–143.

Pipher, J.L. and Harwit, M. (1969). Limb darkening on the Earth's night side at wavelengths of 1–3 μ and 3–7.5 μ. *Journal of Atmospheric Sciences*, **26**, 617–619.

Pipher, J.L., Houck, J.R., Jones, B.W. and Harwit, M. (1971). Submillimeter observations of the night sky emission above 120 kilometers. *Nature*, **231**, 375–378.

Price, S.D. (1981). Infrared mapping of the Galactic plane: I. Low-resolution maps between 0° and 320° Longitude. *Astronomical Journal*, **86**, 193–205.

Price, S.D. (1988). The infrared sky: A survey of surveys. *Publications of the Astronomical Society of the Pacific*, **100**, 171–186.

Price, S.D., Egan, M.P., Carey, S.J., Mizuno, D.R. and Kuchar T.A. (2001). Midcourse Space Experiment Survey of the galactic plane. *Astronomical Journal*, **121**, 2819–2842.

Price, S.D. and Murdock, T.L. (1983). *The Revised AFGL Infrared Sky Survey Catalog*, AFGL-TR-83-0161, AD A134 007.

Price, S.D., Shivanandan, K., Murdock, T.L. and Bowers, P.F. (1983). The brighter 94 micron sources observed by the Far-Infrared Sky Survey Experiment. *Astrophysical Journal*, **275**, 125–129.

Price, S.D., and Walker, R.G. (1976). *The AFGL Four Color IR Sky Survey: Catalog of Observations at 4.2, 11.0, 19.8, and 27 μm*. AFGL-TR-76-0208.

Rieke, G.H., and Low, F.J. (1972). Infrared photometry of extragalactic sources. *Astrophysical Journal*, **176**, L95–L100.

Russell, R.W., Melnick, G.J., Gull, G.E. and Harwit, M. (1980). Detection of the 157 micron (1910 GHz) [C II] emission line from the interstellar gas complexes NGC 2024 and M42. *Astrophysical Journal*, **240**, L99–L103.

Shivanandan, K., Houck, J.R. and Harwit, M. (1969). Preliminary observations of the far-infrared night-sky background radiation. *Physical Review Letters*, **21**, 1460–1462.

Shure, M.A., Houck, J.R., Gull, G.E. and Herter, T. (1984). Detection of the [Ne III] 36 micron line in the planetary nebula NGC 6543. *Astrophysical Journal*, **281**, L29–L31.

Soifer, B.T. (1972). Rocket infrared multicolor observations of the interplanetary medium, H II regions, and the Galactic center region. Astronomy PhD Thesis, Cornell University, p. 95.

Soifer, B.T., Houck, J.R. and Harwit, M. (1971). Rocket-infrared observations of the interplanetary medium. *Astrophysical Journal*, **168**, L73–L78.

Soifer, B.T., Pipher, J.L. and Houck, J.R. (1972). Rocket infrared observations of H II regions. *Astrophysical Journal*, **177**, 315–323.

Stone, E.J. (1870). Approximate determinations of the heating powers of Arcturus and α Lyrae. *Proceedings of the Royal Society (London)*, **18**, 159–165.

Thompson, R.I., Rieke, M., Schneider, G., Hines, D.C. and Corbin, M.R. (1998). Initial on-orbit performance of NICMOS. *Astrophysical Journal*, **492**, L95–L100.

Traub, W.A. and Stier, M.T. (1976). Theoretical atmospheric transmission in the mid- and far-infrared at four altitudes. *Applied Optics*, **15**, 364–377.

Walker, R.G. and Price, S.D. (1975). *The AFCRL IR Sky Survey Catalog of Observations at 4, 11 and 20 μm*. AFCRL-TR-0375, AD A016 397.

Ward, D.B., Dennison, B., Gull, G. and Harwit, M. (1975). Detection of the [O III] 88.16 micron line in M17. *Astrophysical Journal*, **202**, L31–L32.

Ward, D.B. and Harwit, M. (1974) Observations of the Orion Nebula at 100 μm. *Nature*, **52**, 27.

Watson, D.M., Storey, J.W.V., Townes, C.H., Haller, E.E. and Hansen, W.L. (1980). Detection of CO $J=21 \rightarrow 20$ (124.2 μm) and $J=22 \rightarrow 21$ (118.6 μm) Emission from the Orion Nebula. *Astrophysical Journal*, **239**, L129–L132.

Woody, D.P. and Richards, P.L. (1979). Spectrum of the cosmic background radiation. *Physical Review Letters*, **42**, 925–929.

FUNDAMENTAL SCIENCE IN SPACE

Cosmology and gravitational physics

15

KENNETH NORDTVEDT*

Verification of general relativity: tests in the Solar System

In 1900, Isaac Newton's worldview of gravity, space, and time still prevailed – that the gravitational force was a universal, direct, and instantaneous action-at-a-distance between the masses of the Universe, that bodies and light rays moved through an "absolute space, in its own nature, without anything external" whose geometric structure was rigidly Euclidean without end, and that the dynamics of physical law unfolded with respect to an "absolute, true, and mathematical time (flowing) equably without relation to anything external" (Newton 1687). Through the twentieth century that edifice was overthrown and replaced by Albert Einstein's general relativity (GR) perspective – that gravity is an interaction transmitted by a causal and dynamic field whose sources are all forms of energy, including its own contributions, and which then acts elsewhere upon the same; and that the metrical relations between the clocks, rulers, and signals throughout the cosmos are dynamic, non-Euclidean, locationally dependent, and established by the fields of gravity. The detailed structure of metric gravitational field components in the Solar System has in all cases been found to match the predictions of GR in a variety of experiments which primarily employed radar and laser ranging between Earth and other planets or spacecraft.

The first half of the twentieth century was occupied mainly by the construction and calculational exploration of Einstein's theory which he built upon the foundations of James Clerk Maxwell's electromagnetic field theory, his own special relativity theory, and then his Equivalence Principle. Further exploration and application of the theory to a variety of *gedanken* experiments, temporarily beyond the reach of experimental test, continued in subsequent decades. But when the post-World War II years brought forth a stream of new technical abilities to launch spacecraft and to send ranging signals out into the Solar System, and other supporting technologies, experiments were carried out with ever-increasing precision to confirm and quantitatively measure the full variety of novel phenomena in the Solar System predicted by GR.

Although the Newtonian model of Solar System dynamics had been quite successful when used to discover the planet Neptune from what seemed to be unexplained perturbations in the observed motions of the planet Uranus, and when it explained the Moon's numerous orbital "irregularities" which result from the competition of the Sun's gravitational influence with that of Earth on that satellite, the Newtonian system still faced problems, such as a robust anomaly found in the observed precession rate of the major axis of Mercury's orbit, and the failure of Albert Michelson and Edward Morley to detect any change in the speed of light passing through their interferometer as the Earth changed its velocity through the cosmos. Newtonian gravity and cosmology also faced theoretical challenges. This was brought into focus in the late nineteenth century by Ernst Mach, physicist and positivist critic of several concepts in physical law, who labeled the notion of absolute time "an idle metaphysical conception" (Mach 1893), and stressed that what is observed in nature and experiment is the behavior of clocks, not the unfolding of "time" as such. He also asserted that it was the relative feature of local motion with respect to the distant "fixed stars" of the Universe that was empirically meaningful, not the notion of absolute motion relative to "space" as such. Einstein acknowledged the influence of Mach's ideas in his formative years. Describing some consequences of his new theory – that the inertia of a

---

* Montana State University, Bozeman, MT, USA

mass was increased by the proximity of other matter, and that accelerated or rotating matter induced corresponding accelerations or rotations of nearby inertial frames, Einstein (1922) pointed out that, "We must see in (these examples) a strong support for Mach's ideas as to the relativity of all inertial actions." A variety of such Machian effects indeed inspired some of the experimental tests of GR and alternative theories which were finally carried out in the late twentieth century. The result of these tests suggested by the Machian perspective is a more comprehensive empirical basis for theory. Einstein still had Mach in mind when he derived from his theory the possibility of a topologically closed Universe without spatial boundaries: "If we think these (Mach's) ideas consistently through to the end we must expect the whole inertia, that is, the whole (gravitational) $g_{\mu\nu}$-field, to be determined by the matter of the Universe, and not mainly by the boundary conditions at infinity."

In taking on the challenge of incorporating gravity into his special-relativity structure of physical law, Einstein worked to replace the instantaneous, action-at-a-distance force of Newtonian gravity with an interaction between separated matter which was carried by a dynamical field. The successful electromagnetic field theory of Maxwell was his guide. This view of the field-mediated interaction has triumphed throughout physical law in the twentieth century. In this paradigm, two bodies interact by a staged, causal process; one body is source of a dynamical field, the field then spreads out in space and time from its origins in accordance with its own dynamical laws, and finally another body located elsewhere is acted upon by the resulting field found at its location.

In the search for a gravitational field, Einstein was profoundly influenced by the empirical fact that gravity accelerates bodies at a rate which does not depend on their chemical composition or other internal properties – the rate is apparently universal. First noted and studied by Galileo and others four hundred years ago, tested by Newton to a one-part-in-a-thousand precision by comparing motions of differently composed pendulums, and tested in Einstein's time by Roland von Eötvös to precisions of a few parts in a billion with torsion balances supporting different substances in Earth's gravity, this strange and apparently exact proportionality between the inertia of bodies and the strength of gravitational force they experience led Einstein, still short of a theory, to make a grand hypothesis. Paraphrasing him: since a local laboratory falling freely in gravity "transforms gravity away" as far as the physics in that laboratory is concerned, the freely falling laboratory must in all phenomenological respects be locally equivalent to an inertial frame (even though it accelerates relative to the distant inertial frame). From this Equivalence Principle, Einstein was able to predict two novel phenomena: that light rays should deflect downward when passing through a gravitational field; and that clocks should tick slower the deeper they are in a gravitational potential, by a given fractional rate which is independent of their internal structure. Measuring these predictions of the Equivalence Principle, and checking the principle's foundational phenomena – the universality of gravitational free-fall rates – to the highest achievable precision continue to be at the core of the experimental program to test GR in the Solar System. In line with the field paradigm of modern physics, if the foundation or predictions of the Equivalence Principle were found to be violated, this would most likely signal the existence of a previously unseen interaction field in physical law which is not "transformed away" in freely falling laboratories. For the purposes of discovering any such inverse-square (or very long Yukawa range) interaction field whose strength of coupling to matter is very weak compared with the gravitational coupling strength, the contemporary space experiments designed to test the Equivalence Principle are unsurpassed instruments.

The universality of free fall also gave Einstein an important clue to the type of field which could be the basis of gravity and to the attribute of matter to which it must couple. From special relativity he learned that a body's inertia equals its total energy content: $E = mc^2$. If the gravitational force on bodies were to be in universal proportion to the bodies' inertial masses, then a gravitational field's coupling strength to bodies must also be proportional to the bodies' energy contents. This led to consideration of a single scalar field ϕ (tensor of rank 0) or a second-rank tensor field $g_{\mu\nu}$ of 10 potentials, being symmetric in its indices μ and ν which range over the four dimensions of space and time, as transmitters of the gravitational interaction; either field could couple rather naturally to the energy content of bodies. (By contrast, the field that transmits the electromagnetic interaction between charges is a first-rank tensor field A_μ of four potentials, and the other interactions of physics, nuclear and weak, have also been found to be based on multiplets of first-rank tensor fields.) Through the decade 1905–15, scalar and tensor gravity were both explored. For a variety of reasons – empirical implications and predictions, theoretical uniqueness, consistency, and completeness – the general theory of relativity emerged based on a pure second-rank tensor gravitational field $g_{\mu\nu}(r, t)$ (Einstein 1916).

Special relativity, by itself, suggests that the motion of bodies interacting gravitationally should deviate in detail from Newtonian form. Since a body's momentum was now known to contain the speed-dependent modifications of special relativity, application of the law of motion in the presence of a force

$$f = \frac{d\mathbf{p}}{dt} = \frac{d}{dt}\left(\frac{m\mathbf{v}}{\sqrt{1 - v^2/c^2}}\right)$$

yields corrections of order v^2/c^2, even before considering the force side of the equation or further modifications to the momentum of the mass. But just like electromagnetism, GR's 10 gravitational potentials produce a total force between bodies which includes corrections from the static situation (with details depending on the motion of one or both of the interacting objects). A further novel feature of GR's interaction is its nonlinearity: the gravitational force due to a sum of sources is not simply the sum of the individual forces; additional forces come into play which are proportional to the product (and higher powers) of the source masses. All of these modifications, fractionally characterized by the factor v^2/c^2 or Gm/c^2r, are very small throughout the Solar System, each amounting to about 2.5×10^{-8} in the case of Mercury's orbit, for example.

General relativity is compatible with the possible existence of additional, very weakly coupled long-range fields which have so far escaped detection. The presence of such fields and their interactions with matter will produce phenomena which either violate the Equivalence Principle and metric foundations of GR, or which diverge from the predictions of pure tensor gravity at the post-Newtonian level. Testing GR can therefore also be viewed as the search for any "new" long-range interaction fields in physical law. This chapter reflects this interpretation of the twentieth century's achievements.

THE EARLY FOUNDATIONAL YEARS

The Equivalence Principle

Soon after formulating his theory of special relativity in 1905, Einstein turned to the task of properly incorporating gravity, the other known force of that time, into physical law. His goal was to replace Newton's instantaneous action-at-a-distance with a causal field theory of gravity, analogous to the electromagnetic theory of interaction between charged particles transmitted by the Maxwell fields. He was also profoundly influenced by the apparent fact that the gravitational forces on bodies were in universal proportion to the bodies' inertial masses. Because of the resulting identity of free-fall rates at a given location, a localised laboratory freely falling in gravity would appear as an inertial frame, though accelerating relative to other distant inertial frames. He elevated this well-confirmed feature of gravity to a grand hypothesis – his *Equivalence Principle*:

> All local phenomena seen in a laboratory freely falling in gravity are equivalent to phenomena in a gravity-free inertial frame, or conversely, phenomena present in a gravity-free but accelerated frame of reference and understood from the basic kinematics of that frame must also be found to occur in a local gravitational field of equivalent acceleration. (Einstein 1907)

As one example of Einstein's ingenious reasoning which followed from this principle, consider two identically constructed clocks freely falling in gravity and separated by a small vertical distance. Light signals triggered by each tick of the higher clock are sent down to the lower clock which, because of its gravitational acceleration and the finite velocity of light, is always receiving the signals at a Doppler-shifted rate which is lower than the transmitted rate. But by the Equivalence Principle, the observed phenomena should be identical to what occurs when two separated clocks at rest in inertial space exchange signals – the frequency of signal transmission recorded by the transmitting clock equals the frequency of signal as recorded by the receiving clock. This will occur only for the gravitationally free-falling clocks if the lower clock ticks slower than the upper clock by an amount needed to compensate for the Doppler shift.

Unlike Newton, who had no underlying theory for the origins and magnitudes of mass (beyond its simple additivity), and who therefore on empirical grounds could simply adopt the equality of inertial and gravitational mass for all objects, Einstein had found from his special theory of relativity that a body's inertial mass was equal to its total energy content, $E = mc^2$, and he was therefore led to seek a theory of gravity in which the gravitational field's coupling strength to a body was also naturally and generally proportional to the body's total energy content. Achieving this took him over a decade. But more immediately, using only his Equivalence Principle, he predicted two novel phenomena. Using the argument outlined in the previous paragraph, he predicted that the rate of any laboratory clock located in a gravitational potential $U(r)$ will differ from the rate of an otherwise identical clock located elsewhere at gravitational potential $U(r')$:

$$\frac{f' - f}{f} \cong \frac{U(r) - U(r')}{c^2}, \qquad U(r) = G \sum_i \frac{m_i}{|r - r_i|} \quad (1)$$

where c is the speed of light, G is Newton's gravitational constant, and m_i are the masses responsible for the gravitational potential. Using similar forms of argument, he also concluded that a light ray's propagation direction $\hat{c}$ will be deflected by the transverse part of any gravitational acceleration field g through which it propagates:

$$\frac{d\hat{c}}{ds} \cong \frac{g - g \cdot \hat{c}\hat{c}}{c^2}, \qquad g(r) = \nabla U(r) \quad (2)$$

These effects are actually closely related. If the locally measured speed of light is to be a universal constant, but clock rates are universally diminished in gravitational potentials, then the globally viewed speed of light must also diminish in those gravitational potentials:

$$c(r) \cong c_\infty(1 - U(r)/c^2)$$

with c_∞ being the speed of light where the gravitational potential of local bodies is negligible. As in any medium with an inhomogeneous speed of wave propagation, this speed of light function then results in the downward deflection of light wavefronts at the rate indicated by eqn (2).

Einstein soon realized that his prediction of light deflection could be experimentally tested. If the path of light between a distant star and an observing telescope on Earth passes the Sun at distance of closest approach D, the integrated deflection angle will be approximately (Einstein 1911)

$$\delta\Theta \cong \frac{2GM}{c^2 D} \cong 0.84 \text{ arcsec for a grazing ray}$$

The angular locations of such star images would therefore move away from the Sun's location and closer to the images from the less distorted regions of the starfield whose light did not pass so close to the Sun. Erwin Freundlich led an expedition to southern Russia in the late summer of 1914 to measure the light deflection by photographing the starfield during an eclipse of the Sun. But World War I was breaking out, and the expedition personnel were arrested and detained by the Russian authorities. Such an experiment was not to be carried out until the occurrence of the next post-war eclipse in 1919, a delay which permitted completion of Einstein's theory and a change in the prediction.

As early as 1907, Einstein was aware of the outstanding anomaly in Newtonian Solar System dynamics, and he sought to account for the discrepancy in his new theory. The accumulated astronomical observations of the previous century had shown that the secular precession rate of the planet Mercury's orbital major axis – the advance of its perihelion – amounted to 574 arcsec/century. This was a quality observation aided by the relatively large eccentricity of Mercury's orbit ($e \cong 0.2$). But only 531 arcsec/century of this precession could be accounted for by the perturbing gravitational accelerations from the other planets in the Solar System, and no other explanations within Newtonian gravity emerged which remained plausible. It seems likely that the failure during the decade 1905–15 of several preliminary versions of a gravitational field theory to naturally account for this excess 43 arcsec/century precession led Einstein to the further considerations from which general relativity was formulated.

General relativity and Mercury's anomalous perihelion advance

Guided by the physical insights gained from his Equivalence Principle and special relativity theory, and with formal help from mathematician and friend Marcel Grossmann, Einstein ultimately built his theory of gravity upon a dynamical second-rank tensor field of 10 potentials – $g_{\mu\nu}(r, t)$ – symmetric in its indices μ and ν, which each range over the four spacetime coordinates. Finding physically acceptable and consistent field equations for these gravitational potentials, including their coupling to matter, consumed several years of labor. In the finished form of the theory, the gravitational potentials fulfill a set of second-order, nonlinear partial differential field equations for which the entire stress–energy–momentum tensor of laboratory matter $T_{\mu\nu}$ is the source of gravity:

$$G_{\mu\nu} = -\frac{8\pi G}{c^4} T_{\mu\nu}$$

where $G_{\mu\nu}$ is a tensor constructed from the ten gravitational potentials, the square of their first partial derivatives, and their second partial derivatives with respect to the space and time coordinates. This tensor is found to be unique by the necessity to be mathematically consistent with the property that its source tensor $T_{\mu\nu}$ fulfills the traditional conservation laws of energy, momentum, and angular momentum under the appropriate local conditions (Einstein 1916). The theory also specified how matter and other fields respond to the gravitational fields; for example, an atom or object of negligible size will move between two spacetime locations on the trajectory $r(t)$, which gives an "extremal" value to the action integral

$$S = -\int \sqrt{g_{\mu\nu}(r,t)\mathrm{d}x^\mu \mathrm{d}x^\nu} \qquad (3)$$

performed along that trajectory; here $\mathrm{d}x^\mu = c\,\mathrm{d}t$, $\mathrm{d}r$ for $\mu = 0, 1, 2, 3$. The specific mass of the atom is absent from this action integral because of the identity of gravitational and inertial mass; a geometrical interpretation of these trajectories as the extremal paths in a curved Riemannian spacetime geometry established by the field potentials $g_{\mu\nu}$ can then straightforwardly follow.

In late 1915, just weeks before arriving at the final form of his theory, Einstein succeeded in using a slightly incomplete version of his gravitational field equations to calculate the static, spherically symmetric tensor gravitational field in the empty space surrounding the Sun; he then used the equation of motion which results from the action integral given by eqn (3) to obtain the relativistic motion of the planet Mercury in that gravity field. Having no free theoretical parameters to adjust, his new theory nevertheless explained the anomalous 43 arcsec/century perihelion advance with good accuracy. Owing to the theory's intrinsic nonlinearity, the Sun's gravitational field included an important correction to the Newtonian inverse-square field: it varied as the inverse third power of distance and was proportional to the square of the solar mass, M^2/R^3. And being a field theory of gravity, there were further corrections to the Newtonian equations of motion proportional to the square of the planet's velocity, and there was a rescaling of Mercury's effective inertia from eqn (3) proportional to the Sun's gravitational potential. These types of correction to the dynamical equations – nonlinear and motional – characterize in fractional terms the size of almost all post-Newtonian effects in the trajectories of Solar System bodies, and for body orbits they are typically of comparable magnitude: for Mercury's orbit,

$$\frac{GM}{c^2R} \cong \frac{V^2}{c^2} \cong 2.5 \times 10^{-8}$$

The total post-Newtonian contribution to the angle of rotation of Mercury's perihelion found by Einstein was

$$\delta\Theta \cong \frac{6\pi\, GM}{(1-e^2)c^2 a} \text{ radians per orbit}$$

which amounts to 42.98 arcsec/century (a and e are the semi-major axis and eccentricity, respectively, of Mercury's orbit, and M is the solar mass). When Einstein obtained this result and found that his new theory had "brought about a full agreement between theory and experience" for this anomaly in Mercury's motion, he found himself "speechless for several days with excitement," as he described to a friend.

Light deflection and the solar eclipse observations

The complete general theory of relativity predicted a deflection of light rays passing close by the Sun which was twice as large as Einstein's earlier Equivalence Principle prediction. Two 1919 expeditions to measure changes in apparent angular positions of stars during a solar eclipse, one to Brazil led by Andrew Crommelin of the Greenwich Observatory, and the other to the island of Principe off the coast of equatorial Africa led by Arthur Eddington (Pais 1982), reported displacements which more closely confirmed the doubled predictions of GR. But because it was difficult to sufficiently calibrate and stabilize the dimensions of the photographic plates which recorded the stellar images, the experimental precisions of these measurements were modest. The reported results, however, caught the public fancy to a much greater degree than did the theory's original and more theoretically comprehensive success in fitting the advance of Mercury's perihelion. This was in part because the theory was published and had its immediate empirical confirmation in the middle of World War I, when people's attention was focused elsewhere, and perhaps it was also because the explanation of the astronomical anomaly had been *a posteriori*.

The response of both the physics community and the public to the light propagation test was also related to the "geometrization of gravity" taking place. Because GR's symmetric, second-rank tensor, the dynamical field $g_{\mu\nu}$, has a structure isomorphic to the metric tensor of a four-dimensional Riemannian geometry, and this field couples in a requisite universal manner with all the particles and other fields of physical law, it permits the interpretation that the gravitational field establishes the metrical properties of the spacetime arena! Because of the distribution of matter, gravity's source, GR's field equations indicate that the metric field, and therefore also the spacetime arena's geometry, deviates both from the globally "flat" four-dimensional Minkowski metric of special relativity, and from the Euclidean geometry of three-dimensional space plus global absolute time in Newtonian physics. The 1919 confirmation of the theory's predicted light deflection dramatically illustrated this revolution in the picture of spacetime geometry – light rays did not propagate in a Euclidean manner! The Sun's deflection of light was now seen as the rays taking a geodesic or "extremal" path through space and time which were both, however, now "curved" by the proximity of matter. In the decades that followed, many influential physicists working in gravitational physics even tended to reverse the ontological connections between gravity and the geometry of events in space and time by promoting the metaphor "gravity *is* spacetime curvature," although the inverse – that the geometry of physical events in the spacetime arena is established by the tensor gravitational field and its coupling to matter – seems to me to be a more appropriate description of the relationship. General relativity includes no "default" geometry for spacetime which is then distorted by the presence of matter and its gravitational fields; the geometry of events is established *in toto* by the metric field of gravity. Experiments later in the century which found neither spatial anisotropy nor violations of local Lorentz invariance in the Solar System's gravitational interactions lent empirical support to there being no "prior" geometric elements in either physical law or its arena.

Predictions of more novel phenomena

Shortly after GR's debut, investigators were finding additional and novel possibilities for tests of the theory. Willem de Sitter (1916) explored the effects of GR on lunar motion. This was a natural testing ground: the Moon's angular position could be observed to high precision by the traditional astronomical techniques, and because of the tidal gravitational accelerations by the Sun on the relatively large lunar orbit, the Moon's motion contained a rich assortment of "irregularities," departures from simple Keplerian motion that had been used since the time of Newton to test gravitational theory. From a perturbative solution to a restricted version of the three-body problem – Sun, Earth, and Moon – de Sitter found a 19.2 milliarcsec/year general relativistic contribution to the Moon's perigee precession rate. Unobservable then, and later, by the angular measurements of astronomy, this relativistic precession eventually becomes one of the important observables measured by lunar laser ranging techniques. Its conceptual importance emerges from the nature of the perturbing lunar acceleration responsible for this orbital precession – it is a Coriolis-like acceleration, indicating that the local inertial frame accompanying the Earth–Moon system in its motion across the Sun's gravitational field actually rotates relative to the more distant inertial space. This becomes another aspect of the spacetime arena's global structure which is established by gravity in GR; all inertial frames are local, and when viewed globally the various local inertial frames are not only freely falling (accelerating) with respect to one another, they are also rotating with respect to one another!

Then calculations by J. Lense and Hans Thirring (1918) found that in GR a rotating celestial body will produce an acceleration field proportional to the angular momentum of its spin, and the resulting field acts noncentrally on other moving bodies. This gravitomagnetic interaction between mass currents, so named by analogy with the magnetic forces between mutually moving charges, is typically even weaker than the interaction discovered by de Sitter, so its experimental confirmation also had to await the space-based techniques of the late twentieth century.

Eddington was one of the first physicists to promote a more comprehensive program for testing gravitational theory, and he took an open-minded stance concerning the interface between theory and experiment (Eddington 1922). Suggesting that Einstein's revolutionary theory asserted more than one independently testable thing about physical law, Eddington provisionally accepted GR's non-Euclidean and dynamical Riemannian spacetime arena formed by a metrical gravitational field, but he then appealed to experiment to map out the specific structure of that arena in the presence of matter. Asking of experiments in gravitation: "Do they verify? Do they suggest? Do they (within certain limitations) compel the laws we adopt?" he pointed out that the 1919 observations of light deflection plus the original explanation of Mercury's anomalous perihelion advance did not provide a comprehensive test of the full theory; this pair of observations measured only two features of the relatively weak, static, and spherically symmetric metric gravitational potentials of the Sun. Making some general and plausible assumptions, Eddington expressed the logically possible metric gravitational field for this special case as a power series in the Sun's mass, and parametrized the leading-order terms with two dimensionless coefficients and γ and β:

$$g_{oo} = 1 - \frac{Gm}{c^2 r} + 2\beta \left(\frac{Gm}{c^2 r}\right)^2 + \cdots$$
$$g_{ab} = -\delta_{ab}\left(1 + 2\gamma \frac{Gm}{c^2 r} + \cdots\right)$$
(4)

where the subscript o indicates the time component and δ_{ab} is the Kronecker delta (equal to 1 for $a = b$ and 0 otherwise), and the spatial subscripts a and b each independently range over the three dimensions. The requirement that metric gravity yield the Newtonian gravity limit fixed the linear term in the potential g_{oo}, and this also then guaranteed the gravitational shifts of clock rates as predicted by the Equivalence Principle. The Eddington coefficients β and γ then quantified, respectively, the nonlinearity of the gravitational field component g_{oo} and the leading-order contribution of nearby matter to the potentials g_{ab} (which establish the geometry of space). The light deflection measurements fixed γ to a modest degree, and Mercury's observed advance of perihelion determined to somewhat higher precision a combination of the two coefficients appearing in eqn (4). Einstein's gravitational field equations specify that both γ and β equal unity, but the structure of the post-Newtonian gravity in other theories would perhaps be different. Eddington's phenomenological focus was to be revived and developed decades later when the experimental possibilities in gravitational physics were rapidly expanding, and a program to map out experimentally the metric field in all its details could be executed.

THE POST-WAR RENAISSANCE OF EXPERIMENTAL GRAVITY

On a brisk, clear dawn in the fall of 1957 I witnessed the beginnings of the space age from atop my university student residence – across the high eastern sky above Boston moved the Russian Sputnik satellite. Soon thereafter, employed on early spacecraft design, enjoying both theoretical and experimental discussions with Stanford gravity physics pioneers Leonard Schiff and William Fairbanks, and briefly working on the Apollo mission to land astronauts on the Moon, I became convinced of both the need and the new opportunities for space experiments in relativistic gravity. The striking fact of the early 1960s was how few in number and modest in precision were the accomplished tests of GR. But a variety of post-war technologies – powerful radar and lasers, atomic clocks, electronic computers, cryogenics, and launching capabilities for spacecraft – were appearing on the scene, and they soon became the ingredients for new, more precise tests of GR. Experimentally oriented theorists were also looking again at GR to find more features of the theory which would have observational consequences.

Gravitational shift of clock rates

During the early 1960s, Robert Pound and co-workers (Pound and Rebka 1960, Pound and Snider 1965) completed the first series of measurements of the gravitational shift of clock rates predicted by the Equivalence Principle. In their experiments, gamma rays were emitted from nonrecoiling (using the Mössbauer effect), excited ^{57}Fe nuclei and were reabsorbed, again recoil-free, by ground state nuclei of ^{57}Fe located at a different height in their laboratory building. To maximize the reabsorption, the emitting and absorbing nuclei were given a small relative velocity of the amount needed to produce a Doppler shift of the gamma-ray frequencies which compensated for gravity's relative shift between the rates of the absorbing and emitting nuclei-as-clocks:

$$\frac{v}{c} \cong \frac{f-f'}{f} \cong \frac{\mathbf{g} \cdot (\mathbf{r'} - \mathbf{r})}{c^2} \cong 2 \times 10^{-15}$$

One percent agreement with this expected shift was obtained, and though the result was consistent with GR, these experiments actually tested broader concepts. If the local laws of nongravitational physics which govern the clocks' internal dynamics are invariant in space and time, then the Equivalence Principle prediction for clock rate shifts in gravity follows straightforwardly from energy conservation. Any metric theory of gravity, a type of theory fulfilling these stated conditions, which also has the Newtonian static limit to the requisite precision will produce the observed shift of clock rates. But this does not diminish the scientific interest in such experiments which measure the effect of proximite matter on the rates of clocks – it only redirects it. They are also a means to search for any new, long-range interaction in physical law which supplements the tensor gravity of general relativity; an additional contribution to the long range interaction energy between the outside world and the various internal states of atomic clocks will generally alter the location dependence of clock rates and thereby violate Equivalence Principle predictions.

In 1976 a hydrogen maser oscillator (clock) was launched aboard a rocket into a suborbital trajectory which included two hours of free-fall duration above the atmosphere. In this Gravity Probe A (GP-A) mission, funded by NASA and developed by Robert Vessot and colleagues of the Smithsonian Astrophysical Observatory, the space clock's phase was continuously and coherently compared to that of another hydrogen maser clock on the ground for over an hour, using a technique involving both one- and two-way radio links. A two-way signal was initiated by the ground clock and transponded back to Earth from the spacecraft, while the one-way signal to Earth was initiated by the spacecraft's clock. Both signals were received and their phases compared by the ground clock. It had been noted by Robert Pound that since the two radio links were at different frequencies, the significant ionospheric dispersion had to be taken into account. Vessot found that a special choice of frequencies would nullify the dispersive effect on the final observables of the experiment. And by subtracting half of the two-way signal phase shift from the one-way signal phase shift, the first-order Doppler shifts caused by the relative motion between the clocks were also canceled. This facilitated measuring the signal, five orders of magnitude smaller, that was of interest – the frequency shift between the clocks resulting from their different locations in the Earth's gravitational potential. The Equivalence Principle's prediction for this underlying frequency shift,

$$\frac{f - f_0}{f_0} \cong \frac{Gm}{c^2 r(t)} - \frac{Gm}{c^2 r_0}$$

where Gm is the Earth's gravitational source strength, was confirmed to 1.4 parts in 10^4 precision (Vessot et al. 1980). $r(t)$ and r_0 are, respectively, the altitudes of the spacecraft clock and of the ground clock.

For over 30 years, an experiment to measure clock rates near the Sun has been periodically studied and proposed as a flight mission. Because of the Sun's strong gravitational potential, clocks sent to within four solar radii of the Sun should run slow by 5 parts in 10^7 relative to clocks on Earth. Modern atomic clocks with frequencies stable to 1 part in 10^{16} could therefore test the Equivalence Principle's prediction to about 1 part in 10^{10}. By sending two or more different kinds of clock on a common spacecraft and comparing their frequencies on board, substantial technical simplifications result. The high-precision, one- and two-way radio Doppler links used in GP-A are not needed, and the spacecraft trajectory does not need to be tracked to the highest precision. Any relative shifts between the different on-board clock frequencies which vary as the clocks quickly sweep by the Sun will directly signal a physical interaction between the Sun and the clocks other than that of metric gravity. This mission remains as unfinished business for the new century.

Interplanetary radar ranging

Radar echoes from the Moon were first recorded in 1946, and by 1959 more powerful radar systems had detected returns from the planet Venus. In 1966 Irwin Shapiro began systematic radar ranging to the planet Mercury in order to map out that planet's orbit and to achieve, among other things, a more precise measurement of the rate of the advance of the orbit's perihelion. For centuries, Solar System orbits could be determined only by using the relative angular bearings these bodies presented in our astronomical observations. In ranging it became the effective distances between these bodies and Earth from which the planetary orbits were determined, and even the earliest ranging measurements, having precisions of the hundreds of meters, were equivalent to angular precisions of tenths of milli-arcseconds, surpassing in significant measure the best of the traditional astronomical measurements. After a decade of ranging to Mercury, GR's prediction for the rate of the advance of the perihelion was confirmed to a precision of half of 1 percent, and by 1990 the observations and theory agreed to a precision of one part in a thousand (Shapiro 1990). The earliest radar ranging echoes were bounced off what ever regions of Mercury's surface happened to be momentarily facing the Earth. The unknown topographical variations of Mercury's surface then produced corresponding uncertainties in the meaning of the range measurements. As specific geographic features were identified on Mercury and the rotational rate of the planet became known from observations, it became possible at certain times to perform repeated ranging to selected sites on that planet whose elevations could be added to the model as solved-for parameters. This eliminated much of the topographical uncertainty as a source of error. Ranging measurements of this higher quality have been accumulated during the 1990s, and further improvements in determination of the historically important rate of the advance of the perihelion and other details of Mercury's orbit were expected at century's end from the analysis of this data.

In the early 1960s, as Shapiro was formulating a general program of interplanetary radar ranging, he made an innovative theoretical discovery which significantly increased the experimental options. He realized that measurement of the round-trip propagation times for radar signals could also be a way to test GR's light propagation equation, finding that when the line of sight between the Earth and the body returning the radar signal passes close by the Sun, there is an anomalous delay in the signal's time of return (Shapiro 1964). This delay, which depends logarithmically on the signal's distance of closest approach to the intervening body's location, is due to GR's reduction, when globally viewed, of the speed of light near gravitating bodies. For a Sun-grazing radio signal between Earth and Venus, for instance, the extra time delay amounts to about 200 μs.

During the late 1960s and through the 1970s, radar rangings to Venus, Mercury, spacecraft, and finally to transponders both on and in orbit around Mars (as part of the Viking missions) were carried out. Preference was given to making these measurements near the times when the targets were on the opposite side of the Sun from the Earth – at (superior) conjunction. Measurement of the Shapiro time delay was achieved with ever-improving precision (Shapiro et al. 1971), especially so in the case of Earth–Mars ranging in which important experimental innovations were introduced. Analysis of the Viking data of 1976–77 confirmed GR's predicted light time delay with a one-part-in-a-thousand precision (Reasenberg et al. 1979). Modeling of the Viking data was greatly improved by employing two different radio frequencies for the transponded return signals. Since the Sun's corona affects the speed of light in a frequency-dependent manner, the radar times of flight at two frequencies allowed the coronal effects on the radar propagation times to be measured in real time, and these effects could then be eliminated to a large extent as a noise or modeling error source in the data. The point-like transponders also eliminated Martian topographic variations as a source of ranging errors. The experiment achieved average range errors of only about 7 m.

Gravitational deflection of light trajectories was suggested as early as the end of the eighteenth century, based simply upon a Newtonian picture of light as a stream of particles traveling at its finite speed through gravity. But such a pre-relativity concept of Newtonian deflection of light also implies that light, like other particles, should then speed up as it falls or deflects, thereby reducing its time of flight when passing close to the gravitating body. Modern confirmation of the Shapiro time delay has therefore been a key conceptual watershed in support of Einstein's revolutionary picture of gravity shaping the spacetime arena's geometry, and light propagating through that arena along the "curved" extremal paths.

The very ranging data which are analyzed to measure the Shapiro time delay and advance of perihelion of orbits also determines, through the same analytical process of fitting the model to the data, every other parameter of the Solar System model, such as the initial conditions of the bodies, and their mass parameters, which are *a priori* unknown or weakly known. In alternative theories of gravity, Newton's gravitational coupling parameter G is generally found to vary in time at a rate comparable to the observed Hubble expansion rate of the universe (see Chapter 17),

$$H \sim \frac{1}{G} \frac{dG}{dt} \cong 5 \times 10^{-11} \, \text{yr}^{-1}$$

When constructing the Solar System dynamical model and then fitting it to the interplanetary radar ranging data, analysts have considered the hypothesis that the Sun's strength of gravitational attraction GM_{sun} varies in this manner. The most observable effects from a changing G would be the corresponding common fractional time evolution in the many frequencies ω_n present in the planetary motions,

$$\frac{1}{\omega_n}\frac{d\omega_n}{dt} \cong 2\frac{1}{G}\frac{dG}{dt}$$

A time variation in G larger than a part in 10^{12} per year has been found to be incompatible with the fits of the radar ranging data, supporting GR in which G remains a true constant.

Interferometric measurements of deflection

By the late 1990s the very long baseline interferometry (VLBI) technique had achieved the lead in precision testing of gravity's effect on light propagation (Eubanks et al. 1997). In VLBI an electromagnetic signal from a distant astronomical source is separately received and recorded as a function of precise local time by two or more geographically separated observatories. The multiple recordings of the signal are then brought together and compared as a function of time delay parameters between the signal copies until optimal matchings of the signals are achieved. If the time delays are determined to a precision τ, and the separation of the observatories is of characteristic distance L, then the angular coordinates of the source relative to the observatory baselines can be determined to a precision of $\delta\theta \cong c\tau/L$. By making more than a million VLBI measurements of the apparent angular locations for extragalactic radio sources located throughout the sky, and then fitting those data with a necessarily complex model, Thomas Eubanks and co-workers found that the deflection of these received signals agreed with the expectation from GR to 3 parts in 10^4 (the "one-sigma" confidence level). Although half of the quality of their fit for the relativistic deflections relied on just a couple of thousand measurements made on sources within 5° of the Sun, where the deflections are greatest, this result was truly a global fit over the whole sky in that the achievable precision diminished only modestly by using the 1.5 million VLBI observations of sources further than 15° away from the Sun.

Gravitational and inertial mass of celestial bodies

My idea for a new space-based test of GR grew out of my interest in the attempts by theorists to adjust (or renormalize) the mass and coupling strengths of particles and bodies because of internal field energies within these objects. At the beginning of the twentieth century, and prior to Einstein's formulation of special relativity, Hendrik Lorentz and others had already discovered that a charged particle's inertial mass was increased by its electromagnetic field energy, although this field-induced inertia diverged as the inverse of the particle's size. Later quantum mechanical treatments of such particles, their fields, and the vacuum softened the divergence of this electromagnetic self-energy (inertia) of a charged electron, for example, to be only logarithmic with size. The electron's coupling strength to an external electromagnetic field was also found to be altered by internal field considerations.

The analogous renormalizations in gravitational theory were considered. Since a celestial body not only interacts with other bodies gravitationally, but also contains internal gravitational forces and consequent binding energy, does the gravity within a body alter both its inertial mass and its gravitational mass (coupling strength) by identical amounts? If not, then celestial bodies should accelerate at different rates in a given gravitational field. Although it was expected that gravitational and inertial mass would remain identical in GR, it was of interest whether yet-untested features of this theory played roles in producing that identity. In the end, this mass ratio for a celestial body was found generally to be different in alternative theories of gravity, and the modification was in proportion to the body's fractional gravitational internal binding energy (Nordtvedt 1968a):

$$\frac{M(G)}{M(I)} = 1 - \eta\frac{G}{2Mc^2}\int\frac{\rho(x)\rho(y)}{|x-y|}d^3x\,d^3y \qquad (5)$$

where η is a combination of dimensionless, theory-dependent coefficients which quantitatively tag the several features of the gravitational interaction contributing to the overall result. Some of those features were yet unmeasured at the time. $\rho(r)$ is the body's mass density. η vanishes in GR by cancellations among many contributions, but this composite factor is nonzero in most other theories of gravity. If configurations of bodies in the Solar System existed in which precise comparisons could be made between the free-fall rates of different celestial bodies in a common gravitational field, then new and comprehensive tests of GR could be developed. The Earth–Moon system, falling in the gravity of the Sun, was soon to serve this purpose.

Complete parametrization of the metric gravitational field

A celestial body's gravitational-to-inertial mass ratio depends on almost every feature of the post-Newtonian, metric gravitational fields produced by the general source of moving and accelerating bodies. To demonstrate that, and to reveal how the outcome varies in alternative theories of gravity, a more complete version of Eddington's phenomenological representation of the metric field, mentioned in the section on "Predictions of more novel phenomena", was required. His special (static and spherically symmetric) case was first augmented by Leonard Schiff in order to describe the theoretical derivation for the precession of a gyroscope in the Earth's gravity. Taking account of the moving matter of the spinning Earth, he added the spatial vector array of three gravitomagnetic potentials, to give

$$g_{oa} = 2(\Delta + \Delta') \frac{G}{c^3 r^3} (\mathbf{J} \times \mathbf{r})_a$$

where $\mathbf{J}$ is the Earth's rotational angular momentum. Schiff suggested determining the new gravitomagnetic parameter $\Delta + \Delta'$ (which is unity in GR) by measuring the Lense–Thirring precession of an orbiting gyroscope which this gravitomagnetic field would produce (Schiff 1960a, b). This suggestion later led to the establishment of NASA's GP-B (Gravity Probe B) program. But the calculation of a body's gravitational-to-inertial mass ratio depends on still further field contributions and on some generalizations of the single source potentials; the total parametrized representation of the metric field needed was found to be (compare Eddington's metric, eqn (4))

$$g_{oo} \cong 1 - 2\frac{U(\mathbf{r},t)}{c^2} + 2\beta \left(\frac{U(\mathbf{r},t)}{c^2}\right)^2$$
$$+ \beta' \frac{G^2}{c^4} \sum_{i,j} \frac{m_i m_j}{|\mathbf{r}-\mathbf{r}_i||\mathbf{r}_i-\mathbf{r}_j|}$$
$$+ \chi \frac{G}{c^4} \sum_i \frac{m_i}{|\mathbf{r}-\mathbf{r}_i|} (\mathbf{r}-\mathbf{r}_i) \cdot \mathbf{a}_i$$
$$+ \text{non-contributing terms} \qquad (6)$$

$$g_{ab} \cong -\delta_{ab}\left(1 + 2\gamma \frac{U(\mathbf{r},t)}{c^2}\right)$$

$$g_{oa} \cong 4\Delta \frac{G}{c^3} \sum_i \frac{m_i}{|\mathbf{r}-\mathbf{r}_i|} (\mathbf{v}_i)_a$$
$$+ 4\Delta' \frac{G}{c^3} \sum_i \frac{m_i}{|\mathbf{r}-\mathbf{r}_i|^3} (\mathbf{r}-\mathbf{r}_i)\cdot \mathbf{v}_i (\mathbf{r}-\mathbf{r}_i)_a$$

where $\mathbf{v}_i$, $\mathbf{a}_i$, and m_i are, respectively, the velocity and acceleration vectors and the mass of each source body i; β' and χ are the two parameter tags for new field contributions supplementing those previously introduced by Eddington and Schiff. The potential of the single source in Eddington's parametrized metric now becomes the linear superposition of generally moving sources,

$$U(\mathbf{r},t) = G \sum_i \frac{m_i}{|\mathbf{r}-\mathbf{r}_i(t)|}$$

The composite coefficient appearing in eqn (5) was found to be

$$\eta = \frac{1}{3}(10\beta + \beta' + 9\gamma + 2 - 24\Delta - 8\Delta' + 2\chi) \qquad (7)$$

(Here η is expressed with a particular choice of local time variable, but its numerical value will be invariant under a change of that coordinate variable in the manner

$$t = t' + \xi \frac{G}{c^4} \sum_i \frac{m_i}{|\mathbf{r}-\mathbf{r}_i|} (\mathbf{r}-\mathbf{r}_i) \cdot \mathbf{v}_i$$

which changes three of the metric field parameters: $4\Delta \to 4\Delta + \xi$, $4\Delta' \to 4\Delta' - \xi$, $\chi \to \chi + 2\xi$. This is an example of the independence of physical predictions from coordinate system choices, an important feature of GR and other metric theories.)

Two relationships among these coefficients were subsequently found to hold in those cases for which gravity preserves the conservation laws for energy, momentum, and angular momentum, and when gravity is locally Lorentz invariant:

$$2\Delta + 2\Delta' = 1 + \gamma \qquad \text{Lorentz invariance}$$
$$\beta' = 2\beta - 1 \qquad \text{conservation laws}$$

In conservative, locally Lorentz-invariant gravity, measurement of the coefficient η in eqn (5) then determines a new combination of the original two Eddington parameters:

$$\eta = 4\beta - \gamma - 3 \qquad (8)$$

This investigative approach of representing GR's predicted post-Newtonian gravitational potentials as a particular case within a broad class of alternative possibilities had the advantage of bringing to light not only the details of this result, but also the basis for GR's many other null experimental predictions. A large number of relativistic effects which are present in alternative theories are effaced in Einstein's theory in the sense that their numerical strengths are zero (albeit a measureable "zero"), and they were consequently more easily overlooked and left undiscovered as tests which either support or refute GR and experimentally discriminate among alternative theories.

It was soon realized that if the parametrized metric field was not assumed to be locally Lorentz invariant, then it must contain additional contributing potentials which made reference to the velocity of the observer's inertial frame with respect to a preferred inertial frame (Will and Nordtvedt 1972). Since the Solar System is moving at a relatively high velocity ($w/c \sim 10^{-3}$) relative to the likely preferred frame – the rest frame of the observed cosmic black-body radiation – these additional preferred-frame metric potentials had important implications for Solar System dynamics and could be tested for with good precision (Nordtvedt and Will 1972, Nordtvedt 1973). In a rearrangement of the original parametrized metric field, and in the case of conservative metric gravity, the new preferred-frame metric potentials were found to require a two-parameter (α_1 and α_2) representation:

$$\delta g_{oo} = (\alpha_1 - \alpha_2) \frac{w^2}{c^4} U(r, t)$$
$$+ \alpha_2 \frac{G}{c^4} \sum_i \frac{m_i}{|r - r_i|^3} (w \cdot (r - r_i))^2$$
$$+ \alpha_1 \frac{G}{c^4} \sum_i \frac{m_i}{|r - r_i|} w \cdot v_i \qquad (9)$$
$$\delta g_{oa} = (\alpha_1/2 - \alpha_2) \frac{1}{c^3} U(r, t) (w)_a$$
$$+ \alpha_2 \frac{G}{c^3} \sum_i \frac{m_i}{|r - r_i|^3} w \cdot (r - r_i)(r - r_i)_a$$

in which $U(r, t)$ is the previously introduced Newtonian potential and w is the velocity of the observer's inertial frame with respect to the preferred inertial frame. In the rearranged parametrization of the metric field, all parameters other than Eddington's γ and β, and the two preferred-frame parameters α_1 and α_2 discussed here, indicate violation of the conservation laws in gravitational physics.

Lunar laser ranging

While still searching for a practical experimental way to measure a celestial body's gravitational-to-inertial mass ratio, I learned of the development of a passive laser reflector by Robert Dicke's gravitational physics group at Princeton University; their goal was to perform laser ranging between Earth and Moon to centimeters-level precision. It was soon realized that a difference in the Sun's acceleration rates of Earth and Moon proportional to the Earth's fractional gravitational binding energy would produce a polarization of the Moon's orbit in the solar direction with about 10 m amplitude (Nordtvedt 1968b). A lunar laser ranging (LLR) signal with monthly periodicity would result whose measurement had the potential to test new features of GR and improve the precision of the existing knowledge about the relativistic gravitational interaction (eqns (5) and (7)).

In anticipation of the new experimental possibilities in space, Dicke and his student Carl Brans (Brans and Dicke 1961) began an in-depth reconsideration of scalar–tensor metric theories of gravity previously studied in Europe (Jordan 1949). This alternative to the pure tensor gravity of GR was particularly viable; the scalar coupling to matter could be constructed so as to be consistent with all predictions of Einstein's Equivalence Principle, and the post-Newtonian modifications of Solar System gravity in this class of theories preserved local Lorentz invariance and all the traditional conservation laws of energy, momentum, and angular momentum. In such theories there is a dynamical scalar field ϕ in addition to the second-rank tensor field $g_{\mu\nu}$, and the total gravitational interaction between matter is a sum of interactions transmitted by each of these two fields. In scalar–tensor *metric* theories a common spacetime arena remains in which all the matter and fields in physical law carry out their dynamics, only now the metric field of that arena is a composite tensor $g^*_{\mu\nu}$ formed as the product of a functional of the scalar field and the original tensor field $g_{\mu\nu}$:

$$g^*_{\mu\nu}(r, t) = S(\phi(r, t)) g_{\mu\nu}(r, t) \qquad (10)$$

One of the Princeton group's chief observational goals was to measure secular changes in the orbits of bodies which result if Newton's coupling parameter G varied in time in proportion to the expansion rate of the Universe; this was predicted to occur in scalar–tensor theories. (By the century's end this measurement became one of the major scientific achievements of LLR!) My calculation of the gravitational-to-inertial mass ratio of bodies revealed another feature of scalar–tensor gravity – the Earth and Moon would accelerate toward the Sun at different rates in such theories. When the relatively large size of this predicted Earth–Moon range signal in scalar–tensor theory became known, added scientific interest was given to the pursuit of a LLR capability.

The US Apollo 11 astronauts placed the first laser reflector on the Moon in July 1969; two later missions left additional reflectors at other lunar sites, and uncrewed Russian landers took French-built reflectors to the Moon. Over the next three decades over 13,000 range measurements, mainly at the McDonald Observatory in Texas and CERGA (the Centre d'Études et des Recherches en Géodynamique et Astrométrie) in Grasse, France, were made between Earth and the lunar reflectors. During a ranging session an observatory typically sends 10 pulses per second toward a selected reflector on the Moon, each directionally focused pulse containing of the order of 10^{19} photons and having a duration of about 100 ps. From a small fraction of the outgoing pulses, the observatory detects and records the round-trip time of flight for a reflected laser photon identified with the help of time, frequency, and directional filtering of the background light. Accumulating returned photons over tens of minutes of observing time, all photon round-trip times of flight are calculationally adjusted to a common fiducial moment, and are then statistically averaged. A "normal point" range measurement results and is delivered to an archive for later distribution to analyst groups throughout the world. Technical advances over the three decades achieved steady improvement in the precision of the individual range measurements – from about 30 centimeters initially down to about a centimeter at century's end.

In order to extract scientific information from the LLR data, an analysis group develops a detailed theoretical model for calculating the time-dependent range between the Earth and Moon sites. The complete $1/c^2$ order relativistic equations of motion for the Solar System N-body system, including rotational motions for key bodies, are formulated with enough flexibility not only to represent the nominal theory – GR – but also to include modifications present in plausible classes of alternative theories. Computer integrations of these equations produce the trajectories of the relevant sites on Earth and Moon, and the round-trip times of flight of laser photons between the trajectories. Corrections to site positions necessitated by tidal deformations of Earth and Moon are made, and factors such as atmospheric delays of the photon propagations are included. Body mass parameters and initial conditions, site positions, and altogether more than a hundred model parameters, are optimally adjusted by traditional least-squares-fit techniques to best match the range data. Additional parameters which quantify theoretical deviations from the underlying gravitational theory can then also be determined by similar procedures (Dickey *et al.* 1994, Samain *et al.* 1998). Such modeling and data analysis techniques as described here are also typical of the procedures used in radar ranging and other space-based tests. Fitting 30 years of LLR data has confirmed GR in several fundamental ways (Williams *et al.* 1996, Müller and Nordtvedt 1998):

1. The Sun's acceleration rates of Earth and Moon are found equal to a precision of 2 parts in 10^{13}. Since the gravitational binding energy of Earth accounts for about 4 parts in 10^{10} of its mass, this confirms GR's predicted nonlinear coupling between the Sun's external gravity and the internal gravity of Earth to parts in 10^4 precision. This equality of free-fall rates also indicates that the Earth's iron–nickel core falls at the same rate as does the Moon's silicate material.
2. The LLR data fits show no evidence for time variation of Newton's G larger than a couple of parts in 10^{12} per year.
3. The de Sitter precession of the lunar orbit's perigee is found to be equal to GR's 19.2 milli-arcsec/year prediction to precision of 0.07 milli-arcsec/year.

The acceleration rates of Earth and Moon toward the dark matter in the Galaxy's center have also been compared using the LLR data. The rates were found equal to a precision of about 10^{-14} cm s^{-2} (Nordtvedt *et al.* 1995; see also Chapter 18).

The performance of the reflectors on the Moon show no signs of deterioration after 30 years on that body's surface.

This experiment should continue in the twenty-first century to provide key measurements of the relativistic dynamics of the Earth–Moon system, and with ever-increasing precision.

Preferred frames in gravity

General relativity and an alternative scalar–tensor theory both predict Lorentz-invariant gravitational fields in the Solar System. This means that the high velocity ($|w| \cong 300$ km s^{-1}) of the Solar System relative to the cosmic rest frame will not affect the measured post-Newtonian perturbations of body and light trajectories. But the metric gravitational field could violate this local Lorentz invariance. To test for this possibility, a variety of "preferred-frame" effects proportional to the cosmic velocity w have been sought in the interplanetary radar ranging data and LLR data. At levels of better than 1 part in 10^4 of the natural size of these effects (which are proportional to the first or second power of w), nothing has been found.

An interesting feature of the Solar System's history and present structure provided one of the most precise tests of gravity's local Lorentz invariance. If this invariance does not hold, then the static gravitational potential energy between two masses in the Solar System may include a small contribution which is directionally dependent and present because of the motion of the Solar System with respect to the cosmic rest frame in our region of the Universe (see eqn (9); the velocity w is the sum of the Sun's motion with respect to the Galaxy and the Galaxy's velocity with respect to the cosmic background radiation's rest frame):

$$\delta V(r_{ab}) = \alpha_2 \frac{Gm_a m_b}{r_{ab}} \frac{(w \cdot r_{ab})^2}{c^2}$$

Because the Sun is a rotating body, its mass distribution is not spherically symmetric. As a result, if $\alpha_2 \neq 0$, then the directionally dependent potential energy between the mass elements of the Sun will produce a body self-torque, and its spin axis will precess about the cosmic velocity vector w. But it is found that after 4.5 billion years of existence the Sun's spin axis is still within 6° of being perpendicular to the orbital plane of the planets. This cannot be understood as a consequence of Newtonian interaction between the distorted rotating Sun and the planets, as this coupling is extremely weak; instead, it indicates that there has been negligible torque acting on the Sun during its several-billion-year existence. A strong limit can therefore be placed upon this form of Lorentz-invariance violation: $\alpha_2 \leq 10^{-7}$ (Nordtvedt 1987). An alternative meaning of these results is that the effective Newtonian coupling parameter G is spatially isotropic to 1 part in 10^{13}.

Gravitomagnetic and inductive forces

A mission to place gyroscopes in orbit around the Earth and measure their precessions with respect to the distant "fixed stars" was conceived and put into development at Stanford University in the 1960s, and funded during subsequent decades by NASA as its Gravity Probe-B (GP-B) program. In Newtonian gravity and cosmology, a freely falling, highly spherical gyroscope's spin axis should remain fixed in orientation with respect to the cosmic inertial frame because no torques are exerted on the device. But in GR the gyroscope's spin axis should precess because of both the Lense–Thirring gravitomagnetic field of the rotating Earth and the de Sitter precession resulting from the gyroscope's orbital motion around the Earth. From a global perspective, the two gravitational interactions studied by Lense and Thirring and by de Sitter represent post-Newtonian torques being applied on the gyroscope through its interaction with the Earth. From a local perspective, both effects represent a nonprecessing gyroscope in the precessing inertial frame which accompanies the gyroscopes in their orbital motion. The geodetic precession occurs in the orbital plane and is 160 times larger than the out-of-orbital-plane gravitomagnetic precession, so both effects are separately measured. A high-performance "drag-free" system slaves the spacecraft to the orbital motion of one of the gyroscopes and thereby further isolates the gyroscopes from disturbances, and electrostatic suspension is then needed only to supply the extremely small, essentially torque-free support forces which confine the other centimeters-size gyroscopes within the spacecraft. The orientations of the gyroscopes with respect to an on-board telescope which is locked onto a "fixed star" (surrogate for distant inertial space) are measured using SQUID circuits which delicately sense changes in magnetic flux when the gyroscopes precess relative to the spacecraft. The gyroscopes, coated with superconducting material and maintained below the critical temperature, will produce London magnetic dipole moments aligned with their spin axes. The cryogenic environment is extended to the telescope and the interconnecting structure to stabilize the tie to the external reference star direction.

The design goal of the experiment is 0.3 milli-arcsec/year measurement precision of the precession rates, this being 1 part in 100 of the expected gravitomagnetic precession and 5 parts in 10^5 of the geodetic precession (Everitt *et al.* 1988). The 16-month GP-B mission is scheduled for launch in late 2002. This experiment was originally conceived as measuring GR's gravitomagnetism, but in the decades since then gravitomagnetism has been confirmed and measured to higher precision in other experiments. In my view, the very important scientific outcome of the GP-B mission will be the higher-precision measurement of the much larger geodetic precession. The de Sitter precession is proportional to the same fundamental metric theory parameter γ that is measured in the light deflection and time delay experiments, in LLR, and in Mercury's advance of perihelion; if GP-B achieves its design goals, it will increase the precision of this fundamental parameter's measurement by an order of magnitude beyond what was achieved from the other twentieth-century experiments.

With the goal of detecting orbital perturbations on Earth satellites due to gravitomagnetism, Ignazio Ciufolini and co-workers (Ciufolini *et al.* 1998) analyzed the laser ranging data to two LAGEOS satellites orbiting near the Earth. These satellites were designed for mapping out the Earth's complex Newtonian multipolar gravitational field; they consist simply of a sphere of very high-density material to achieve a low drag to mass ratio, and several passive corner reflectors mounted on the surfaces (see Chapter 66). These analysts found gravitomagnetic contributions to the precession rates of the two satellites' orbital planes about the Earth's polar axis (precession of the lines of node), and there was also a gravitomagnetic contribution to the perigee precession rate of the second LAGEOS satellite whose measurement was facilitated by this satellite's unexpectedly eccentric orbit. With these three secular observables available from the two satellite orbits, a fit for the presence of GR's gravitomagnetic interaction was possible even in the face of relatively large uncertainties in the strength of the Earth's two leading multipole moments, $J2$ and $J4$ (a combination of the three observables was found which was independent of these two moments).

This discovery underscored the fact that the laser and Global Positioning System (GPS) tracking of near-Earth satellites had reached, by century's end, levels of precision which now requires inclusion of the GR accelerations in the modeling and fitting of the data to an Earth gravity field model. Omitting GR corrections from the orbital dynamics model will bias or alias the fits for the Newtonian multipolar part of that gravity field. In addition to all the practical geophysical reasons for wanting an unbiased gravity field determination for Earth, some of the determined parameters are also of importance for testing GR. For example, the Earth's monopolar gravitational coupling strength GM_{Earth} as measured by fitting Earth satellite orbits can be combined with the corresponding parameter GM_{Moon} measured by the lunar geodesy missions Clementine and Lunar Prospector to construct the total gravitational mass parameter for the Earth–Moon system. The LLR model, which includes this total mass parameter, can then use this externally determined value rather than fit for it using LLR data. The quality of the LLR tests of GR are consequently improved; for example, the formal error of the free-fall test of Earth and Moon is reduced by over 25%. Unbiased measurement of the gravity field from the oceans' tidal bulge using ranging data to Earth satellites is also important because it contributes to the LLR model and its ability to test for time variation of the Newtonian G.

The gravitomagnetic interaction has also played an indispensable role in the successful modeling of the lunar orbit as observed by 30 years of laser ranging data. The equations of motion which are integrated to produce the Earth and Moon trajectories are the GR dynamical equations as formulated with respect to the Solar System's barycenter. In this frame both Earth and Moon are moving bodies, and they exert a gravitomagnetic force on each other. Neglect of this gravitomagnetic interaction in the model's integration of the lunar orbit would result in about a 13 m anomaly in the predicted amplitude of the Earth–Moon range oscillation with monthly period; this key amplitude is measured by LLR to precision of a few millimeters!

While the gravitomagnetic acceleration results from the spatial gradients of the metric field potentials g_{oa} ($a = 1, 2, 3$), the time derivatives of these potentials produce an inductive interaction in which the acceleration of one mass induces an acceleration in nearby masses (in direct analogy with the inductive electric fields produced by accelerated charges in Maxwell's theory):

$$\delta \boldsymbol{a}_i = (4\Delta - \tfrac{1}{2}\chi) \frac{G}{c^2} \sum_{j \neq i} \frac{m_j}{r_{ij}} \boldsymbol{a}_j + (4\Delta' + \tfrac{1}{2}\chi) \frac{G}{c^2} \sum_{j \neq i} \frac{m_j}{r_{ij}^3} \boldsymbol{r}_{ij} \boldsymbol{r}_{ij} \cdot \boldsymbol{a}_j$$

Not only does the induced acceleration of the Moon by the accelerating Earth create a perturbation of the lunar orbit with an amplitude of tens of centimeters, but the inductive accelerations of each element of Earth by all the other accelerating matter of Earth produce together a very large part of the gravitational binding energy's contribution to the Earth's total inertia. This is seen by the strong participation of the parameters Δ and Δ' in eqn (7) in defining the total η parameter which appears in eqn (5). Without these internal inductive forces of inertia the Earth would fall anomalously toward the Sun, and the resulting anomaly in the lunar orbit's polarization amplitude would be tens of meters.

Finite-range interactions

If there is a supplementary interaction between celestial bodies transmitted by a field of extremely small but nonzero quantum mechanical mass $m = \hbar\mu/c$, then that field's potential will not diminish simply as the inverse power of distance from its source, but will show a more rapid exponential weakening with distance. The total static-limit potential energy between the bodies will then include a Yukawa-like contribution from the additional interaction field which supplements gravity:

$$V(r) = -\frac{Gm_i m_j}{r} \pm \frac{k_i k_j}{r} e^{-\mu r} \qquad (11)$$

where k_i and k_j are the bodies' coupling strengths to the additional field which, depending on the nature of the field, may or may not be proportional to the bodies' masses. If the force from the additional interaction deviates from being inverse square, it will contribute to the precession rate of an orbit's major axis. For a low-eccentricity orbit typically encountered in the Solar System, and of radius r, the precession of the major axis per orbital revolution $\delta\theta_p$ is

$$\delta\theta_p \cong \pi \frac{k_i k_j}{Gm_i m_j} (\mu r)^2 e^{-\mu r}$$

Orbits of size comparable to the Yukawa range parameter are the most sensitive probes of the existence of such interactions ($x^2 e^{-x}$ is maximum for $x = 2$). Investigations of the quality of fits of the ranging data to a variety of targets – near-Earth satellites, the Moon, the planets Mercury and Mars – have produced no evidence for any anomalies in the orbit precession rates. Collectively, these results constrain the strength of any supplementary Yukawa-like interaction in physical law having a range anywhere between about an Earth radius and up to the size of Mars's orbit; the product of coupling strengths of the relevant bodies must be smaller than 1 part in 10^{10} of the bodies' corresponding strength of gravitational interaction:

$$\frac{k_i k_j}{Gm_i m_j} \leq 10^{-10}$$

The constraint resulting from LLR for a Yukawa range in the vicinity of $1/\mu \cong 2 \times 10^5$ km is even an order of magnitude lower!

KEY EXPERIMENTS FOR THE NEAR FUTURE

Interplanetary transponded laser ranging

A major improvement in the precision measurements of metric gravity in the Solar System will be possible in the first decade of the twenty-first century using interplanetary laser ranging (IPLR). Deploying laser transponders either on the surfaces of or in orbit around planets, asteroids, or on spacecraft in their own Solar System orbits, and then ranging to these targets with the same centimeter-level precision already achieved in ranging to near-Earth satellites and the Moon, will allow 1 part in 10^6 measurements of the post-Newtonian Solar System dynamics.

A most promising perturbative effect in the orbits of bodies to be measured by IPLR will result if the Sun's gravitational-to-inertial mass ratio differs from unity. If the Sun accelerates anomalously toward Jupiter, which it will if its relatively large fractional gravitational binding energy couples unusually to planetary accelerations, then the inner planets' orbits will be polarized toward Jupiter (Nordtvedt 1970, Anderson *et al.* 1996). This is an analogous effect to the polarization of the lunar orbit toward the Sun which has been measured by LLR over 30 years. Laser ranging between Earth and another body, such as Mars, an asteroid, Mercury, or a spacecraft shows the ability to measure the nonlinearity of GR as parametrized by the Eddington β coefficient with a precision two orders of magnitude higher than today's LLR achievement. And opportunistic measurements of the Shapiro time delay of these laser ranging signals will be possible every time the ranging target passes through (superior) conjunction and its line of sight passes close by the Sun. Because of the inherent higher precision of laser ranging and the fact that the solar coronal effects on the speed of visible light propagation are negligible, the Eddington γ coefficient should be measureable with a precision even better than one part in a million.

Space-based test of the universality of free fall

Testing the universality of free-fall rates of different materials from the periodic table is a method of unsurpassed sensitivity in the search for new long-range interactions in physical law, and it tests the very foundations of the Equivalence Principle and GR. For much of the twentieth century the most precise experiments of this type were performed in the laboratory. But eventually the LLR experiment reached sufficient precision to become competitive, and at century's end the equality of the Sun's acceleration of Earth and Moon to 2 parts in 10^{13} precision could be interpreted as confirmation that the Earth's iron–nickel core and the Moon's silica mantle-like material fall toward the Sun at equal rates with 6 parts in 10^{13} precision (the Earth's core accounts for a third of its mass). Ever since the early years of the space age, proposals have been made to move the laboratory tests of universality of free fall into spacecraft environments. Several advantages would result: the torsion fibers needed to suspend masses in Earth's gravity would be replaced by extremely weak forces of confinement in the nearly gravity-free environment of an orbiting spacecraft; seismic and other Earth laboratory noise sources would be all but eliminated; the full magnitude of the Earth's gravitational acceleration replaces the Sun's acceleration or the Earth's centrifugal acceleration as the experiment's "driving force"; and long time periods of free-fall duration become available in orbit. Experiments under development capable of detecting fractional differences of acceleration as small as 1 part in 10^{18} are under consideration by space agencies in both the USA and Europe, and such a mission ought to be carried out in the first decade of the twenty-first century. Regardless of the outcome of such ultra-precise tests of the universality of free fall, the results will serve as important clues and constraints on the theoretical work seeking to unify gravitation with the rest of physical law.

"Grand fits"

Future performance of combined fits of the data from separate Solar System ranging experiments, so-called grand fits, will result in stronger tests of gravitational theory. In the first place, the masses of the Sun and planets, the initial conditions of the orbits, and the many other model parameters all have specific values in nature; it makes no conceptual sense to allow these quantities of nature to take different values in the separate fits of the different experiments. But perhaps of more importance, sometimes the precision of fit for a scientific parameter in one experiment's data set is limited by the correlation of that parameter's fit with that for another model parameter which is fitted to more modest precision. Doing a grand fit of two or more data sets will in certain cases significantly reduce or eliminate the effect of the offending model parameter's uncertainty on the scientific parameter of interest. Consider the following example.

Fitting the ranging data to Mercury by itself in order to determine that planet's rate of advance of perihelion requires the Earth's absolute sidereal orbital rate to be inferred from the less precisely measured sizes of the orbits of Earth and Mercury. The precision of the perihelion measurement suffers accordingly. But the ranging data to both Mercury and Mars can be incorporated into one grand fit, which eliminates the need for any inferred or otherwise estimated absolute sidereal rate. The "good" observables in planetary ranging are the two frequencies – the relative sidereal (synodic) rate between Earth and the planet $\omega_e - \omega_p$, and the planet's frequency of eccentric (anomalistic) motion ω'_p (if, like Mercury and Mars, the planet's orbit has an appreciable eccentricity). The difference between the perihelion precession rates of Mercury and Mars, after being corrected by all the model's Newtonian perturbative contributions, relates directly to the relativistic features of gravitational theory that one seeks to determine, but observationally this difference can in a grand fit now be measured in terms of the four good frequency observables from the two ranging experiments:

$$\delta\Omega_{\text{Mercury}} - \delta\Omega_{\text{Mars}}$$
$$= (\omega_{\text{Mercury}} - \omega_{\text{Earth}}) - \omega'_{\text{Mercury}} + \omega'_{\text{Mars}} + (\omega_{\text{Earth}} - \omega_{\text{Mars}})$$

By a similar process, the Mercury ranging data and the LLR data can be combined into a grand fit; from the latter data emerges the inferred absolute sidereal rate of Earth, expressed now in terms of the well-measured synodic and eccentric (anomalistic) rates of lunar motion.

SUMMARY

The knowledge gained from the total package of twentieth-century Solar System experiments in gravitational physics can be interpreted in two ways. It collectively tests the GR conceptual revolution which replaced the Newtonian view of a rather rigid space and time existing independently of the rest of physical law with a more integrated picture in which one of nature's force fields – gravity – establishes the Riemannian metric of a dynamically shaped, non-Euclidean spacetime arena. Einstein's metric theory of gravity has (for Solar System purposes) unique field equations for the second-rank tensor field $g_{\mu\nu}$, and therefore for the relativistic corrections to static Newtonian gravity. One way to organize the accumulated experimental results is to ask to what extent the foundational principles of metric gravity are confirmed, and to what precisions several of the post-Newtonian features of metrical gravity are measured. But even if the metric nature of gravity is found to be consistent with observation, the existence of additional very weakly coupled, long-range fields which transmit additional forces between the Solar System's bodies remains possible. The experiments and their results can therefore also be reviewed as a systematic search for such additional interaction fields.

The strongest evidence that gravity is a pure metric phenomenon and that there are no additional long-range interactions of appreciable coupling strength is that no composition dependence of free-fall rates has been observed. Because the Earth has an appreciable iron–nickel core, while the Moon is composed almost entirely of silicate materials, the equality of free fall of these two bodies toward the Sun as measured in LLR supports this composition independence:

$$\frac{|a_{\text{Fe-Ni}} - a_{\text{silicates}}|}{|g_{\text{Sun}}|} \leq 6 \times 10^{-13}$$

This equals or surpasses in precision the best of today's ground-based experiments. Some typical attributes of matter other than total energy to which an Equivalence-Principle-violating field might couple are some combination of lepton and baryon number, or electromagnetic energy content. The ratios of such quantities to total mass of matter vary by about 1 part in 10^3 across the periodic table of elements. So an interaction which couples to one of these attributes and which has a Yukawa range at least as large as an astronomical unit must have a coupling strength less than about 10^{-9} or 10^{-10} that of Newtonian gravity.

If GR were modified in a way that preserves the metric nature of the resulting total gravitational interaction, the strength of the modifying interaction could be much greater. This is because all Equivalence Principle predictions would still be fulfilled, and only the weak corrections to Newtonian gravity will indicate the existence of the additional interaction field. The most theoretically plausible and empirically viable possibility is that metric gravity is a scalar–tensor composite in which a specially coupled scalar field interaction supplements the tensor interaction of GR. In this class of theories, the metric field that produces the post-Newtonian phenomena in the Solar System is Lorentz invariant, leads to dynamics for the system's bodies which fulfills the full complement of conservation laws, and is specified by the two Eddington parameters γ and β (assuming the scalar field is "massless" and has no finite Yukawa range). These parameters each have numerical values of 1 for the pure (rank-two) tensor-based metric gravity. The three most significant types of experiment which measured the Eddington parameters – light deflection and delay, Mercury's advance of perihelion, and LLR-measured free fall of Earth and Moon – have achieved measurements whose precisions now reach

$$|1 - \gamma|_{\text{VLBI}} \leq 6 \times 10^{-4}$$
$$|4\beta - 3 - \gamma|_{\text{LLR}} \leq 5 \times 10^{-4}$$
$$|2\gamma - \beta - 1|_{\text{Mercury}} \leq 3 \times 10^{-3}$$

(different analysts use slightly different definitions of experimental precision; here I have exercised some judgment to put the different numbers on a common basis which approximates a traditional "2-sigma" confidence level). The last published Mercury perihelion precession result above lags the other two experiments in precision, but this gap should close when the Earth–Mercury radar ranging data accumulated in the 1990s are analyzed. $|1 - \gamma|$ is a direct measure of the scalar interaction's fractional participation in producing the total gravitational interaction. That possibility, as well as the possible anomalies in gravity's nonlinearity measured by β, can be no larger than parts in 10^4. This raises a question: how precisely need metric gravity be mapped out before plausible alternatives to pure GR can be either ruled out or discovered?

With Thibault Damour, I studied a general scenario which could naturally explain the very small fractional presence of scalar gravity in the present Universe (Damour and Nordtvedt 1993). We showed that if the functional of the scalar field which couples to particle masses in the metric versions of scalar–tensor gravity (the function $S(\phi)$ in eqn (10)) has a minimum for a particular value of the scalar field ϕ_0, then the cosmological dynamics of the background scalar field in the Universe naturally drives that field toward its value at the functional minimum. The strength of the scalar interaction, being proportional to the square of the slope of the scalar functional, then tends to zero as the minimum is approached, and the pure tensor gravity of Einstein's general relativity emerges as the limiting case. But this is a dynamic process which "turns off" the scalar field, and under plausible assumptions of there being an appreciable fraction of scalar gravity as recently as the end of the radiation-dominated era of the Universe's history, it is found that today in our epoch there should still be a remnant fraction of 10^{-6} or 10^{-7} of scalar interaction in the scalar–tensor mixture. As discussed in the section on "Interplanetary transponded laser ranging", measurements of the Eddington parameters γ and β to this level of precision are within reach by a new generation of interplanetary laser ranging experiments.

A similar mechanism and scenario have been studied for a scalar field which is nonmetrically coupled to matter and is therefore able to produce Equivalence Principle-violating phenomena (Damour and Polyakov 1994). In this case an even stronger dynamical attraction toward the coupling functional minimum was found, and this indicates a much smaller remnant scalar interaction strength with matter. The proposed missions to compare the gravitational free-fall rates of different materials in orbiting spacecraft, which may be able to reach levels of precision as small as 10^{-18} in the measurement of fractional acceleration differences, are presently the most promising chances for testing this nonmetric version of the scenario. Even if such experiments find no violation of the universality of free fall at these much higher levels of precision, that finding will be an important empirical constraint and clue for theoretical work on the unfinished goal of unifying all the pieces of physical law.

Tests of GR in the Solar System have been called weak-field tests in recognition of the small magnitude of the relativistic corrections to Newtonian gravity in that environment. But the effects measured in the Solar System include both the complete inventory of motional corrections to the static interaction and the beginnings of gravity's non-linearity due to its self-coupling. Theoretical investigation of generic scalar–tensor theories of gravity have shown that if the free field energy of the scalar field is positive definite – a very desirable condition for the underlying theory – then deviations from GR must, under all but the most unusual circumstances, begin at the lowest, weak-field, post-Newtonian order; and all higher-order deviations in stronger gravity situations will be proportional to the value of $1 - \gamma$ which is already measured in the Solar System. The quantitatively precise Solar System measurements which have constrained this quantity $|1 - \gamma|$ to be no more than parts in 10^4 give important support to the applications of general relativity in astrophysics and cosmology where stronger gravitational fields are encountered.

REFERENCES

Anderson, J.D., Gross, M., Nordtvedt, K. and Turyshev, S. (1996). The solar test of the equivalence principle. *Astrophysical Journal*, **459**, 365–370.

Brans, C. and Dicke, R.H. (1961). Mach's principle and a relativistic theory of gravity. *Physical Review*, **124**, 925–935.

Ciufolini, I., Pavlis, E., Chieppa, F., Fernandes-Vieira and Perez-Mercader, J. (1998). Test of general relativity and measurement of the Lense–Thirring effect with two Earth satellites. *Science*, **279**, 2100–2103.

Damour, T. and Nordtvedt, K. (1993). General relativity as a cosmological attractor of tensor–scalar theories. *Physical Review Letters*, **70**, 2217–2219.

Damour, T. and Polyakov, A.M. (1994). The string dilaton and a least coupling principle. *Nuclear Physics B*, **423**, 532–558.

deSitter, W. (1916). On Einstein's theory of gravitation and its astronomical consequences. *Monthly Notices of the Royal Astronomical Society*, **77**, 155–184.

Dickey, J.O., Bender, P.L., Faller, J.E., Newhall, X.X., Ricklefs, R.L., Ries, J.G., Shelus, P.J., Veillet, C., Whipple, A.L., Wiant, J.R., Williams, J.G. and Yoder, C.F. (1994). Lunar laser ranging: A continuing legacy of the Apollo program. *Science*, **265**, 482–490.

Eddington, A.S. (1923). *The Mathematical Theory of Relativity*, Cambridge University Press.

Einstein, A. (1907). Über das Relativitätsprinzip und die aus demselben gezogenen Folgerungen. *Jahrbuch der Radioaktivität und Elektronik*, **4**, 411–462.

Einstein, A. (1911). On the influence of gravitation on the propagation of light. In *The Principle of Relativity* (translated by W. Perrett and G.B. Jeffery), Dover Publications, New York, pp. 99–108.

Einstein, A. (1916). The foundation of the general theory of relativity. In *The Principle of Relativity* (translated by W. Perrett and G.B. Jeffery), Dover Publications, New York, pp. 111–162.

Einstein, A. (1922). *The Meaning of Relativity* (original translation by Edwin Plimpton Adams), Princeton University Press, 5th edn, p. 103.

Eubanks, T.M., Matsakis, D.N., Martin, J.O., Archinal, B.A., McCarthy, D.D., Klioner, S.A., Shapiro, S. and Shapiro, I.I. (1997). Advances in Solar System tests of gravity. Abstract K11.05 of paper given at the April 1997 meeting of the American Physical Society.

Everitt, C.W.F., Lipa, J.A., Keiser, G.M., Anderson J.T., Turneaure, J.P., Cornell, E.A., Levine, P.D., van Patten, R.A., Breakwell, J.V. and DeBra, D.B. (1988). The Stanford relativity gyro experiment: History and overview. In J.D. Fairank, B.S. Deaver Jr, C.W.F. Everitt and P.F. Michelson (eds), *Near Zero: New Frontiers of Physics*, Freeman, San Francisco, pp. 587–639.

Jordan, P. (1949). Formation of the stars and development of the Universe. *Nature*, **164**, 637–640.

Lense, J. and Thirring, H. (1918). Über den Einfuss der Eigenrotation der Zentralkörper auf die Bewegung der Planeten und Monde nach der Einsteinschen Gravitationstheorie. *Physikalische Zeitschrift*, **19**, 156–163.

Mach, E. (1912). *The Science of Mechanics: A Critical and Historical Account of Its Development* (English translation by Thomas J. McCormack), Open Court, LaSalle, IL, 1960, Chapter II, Section VI.

Müller, J. and Nordtvedt, K. (1998). Lunar laser ranging and the equivalence principle signal. *Physical Review*, **58**(6), 2001.

Newton, I. (1687). *Principia: Vol. 1 The Motion of Bodies* (Motte's translation, revised by F. Cajori), University of California Press, Berkeley, 1934, Definition VIII.

Nordtvedt, K. (1968a). Equivalence principle for massive bodies: II. Theory. *Physical Review*, **169**, 1017–1025.

Nordtvedt, K. (1968b). Testing relativity with laser ranging to the Moon. *Physical Review*, **170**, 1186–1187.

Nordtvedt, K. (1970). Tests of the equivalence principle and gravitation theory using Solar System bodies. In *Proceedings of the Conference on Experimental Tests of Gravitation Theories*, California Institute of Technology, Pasadena, CA.

Nordtvedt, K. (1973). Post-Newtonian gravitational effects in lunar laser ranging. *Physical Review D*, **7**, 2347–2356.

Nordtvedt, K. (1987). Probing gravity to the second post-Newtonian order and to one part in 10^7 using the spin axis of the Sun. *Astrophysical Journal*, **320**, 871–874.

Nordtvedt, K.L., Müller, J. and Soffel, M. (1995). Cosmic acceleration of the Earth and Moon by dark matter. *Astronomy and Astrophysics*, **293**, L73–L74.

Nordtvedt, K. and Will, C. (1972). Conservation laws and preferred frames in relativistic gravity: II. Experimental evidence to rule out preferred-frame theories of gravity. *Astrophysical Journal*, **177**, 775–792.

Pais, A. (1982). *Subtle is the Lord: The Science and the Life of Albert Einstein*, Oxford University Press, New York, Chapter 16.

Pound, R.V. and Rebka, G.A. Jr. (1960). Apparent weight of photons. *Physical Review Letters*, **4**, 337–341.

Pound, R.V. and Snider, J.L. (1965). Effect of gravity on gamma radiation. *Physical Review*, **140**, B788–B803.

Reasenberg, R.D., Shapiro, I.I., MacNeil, P.E., Goldstein, R.B., Breidenthal, J.C., Brenkle, J.P., Cain, D.L., Kaufman, T.M., Komarek, T.A. and Zygielbaum, A.I. (1979). Viking relativity experiment: Verification of signal retardation by solar gravity. *Astrophysical Journal*, **234**, L219–L221.

Samain, E., Mangin, J.F., Veillet, C., Torre, J.M., Fridelance, P., Chabaudie, J.E., Feraudy, D., Glentzlin, M., Pham Van, J., Furia, M., Journet, A. and Vigouroux, G. (1998). Millimetric lunar laser ranging at OCA (Observatoire de la Cote d'Azur). *Astronomy and Astrophysics, Suppl.*, **130**, 235–244.

Schiff, L.I. (1960). Possible new test of general relativity theory. *Physical Review Letters*, **4**, 215–217.

Schiff, L.I. (1960b). Motion of a gyroscope according to Einstein's theory of gravity. *Proceedings of the National Academy of Sciences*, **46**, 871–872.

Shapiro, I.I. (1964). Fourth test of general relativity. *Physical Review Letters*, **13**, 789–791.

Shapiro, I.I. (1990). In N. Ashby, D.F. Bartlett and W. Wyss (eds), *General Relativity and Gravitation 1989*, Cambridge University Press, p. 313.

Shapiro, I.I., Ash, M.E., Ingalls, R.P., Smith, W.B., Campbell, D.B., Dyce, R.B., Juergens, R.F. and Pettengill, G.H. (1971). Fourth test of general relativity: New radar result. *Physical Review Letters*, **26**, 1132–1135.

Vessot, R.F.C., Levine, M.W., Mattison, E.M., Blomberg, E.L., Hoffman, T.E., Nystrom, G.U., Farrel, B.F., Decher, R., Edy, P.B., Baugher, C.R., Watts, J.W., Teuber, D.L. and Wills, F.O. (1980). Test of relativistic gravitation with a space-borne hydrogen maser. *Physical Review Letters*, **45**, 1081–1084.

Will, C.M. and Nordtvedt, K. (1972). Conservation laws and preferred frames in relativistic gravity: I. Preferred-frame theories and an extended PPN formalism. *Astrophysical Journal*, **177**, 757–774.

Williams, J.G., Newhall, X.X. and Dickey, J.O. (1996). Relativity parameters determined from lunar laser ranging. *Physical Review D*, **53**(12), 6730–6739.

FURTHER READING

Einstein, A. (1955). *The Meaning of Relativity*, 5th edn, Princeton University Press.

Nordtvedt, K. (1996). From Newton's Moon to Einstein's Moon. *Physics Today*, **49**(5), 26–31.

Will, C. (1986). *Was Einstein Right?*, Basic Books, New York.

Will, C. (1993). *Theory and Experiment in Gravitational Physics*, revised edn, Cambridge University Press.

16

CLIFFORD M. WILL*

Verification of general relativity: strong fields and gravitational waves

At the time of the birth of general relativity (GR), experimental confirmation was almost a side issue. Einstein did calculate observable effects of general relativity, such as the deflection of light, which were tested, but compared to the inner consistency and elegance of the theory, he regarded such empirical questions as almost peripheral. But today, experimental gravitation is a major component of the field, characterized by continuing efforts to test the theory's predictions, to search for gravitational imprints of high-energy particle interactions, and to detect gravitational waves from astronomical sources.

The modern history of experimental relativity can be divided roughly into four periods: Genesis, Hibernation, a Golden Era, and the Quest for Strong Gravity. The Genesis (1887–1919) comprises the period of the two great experiments that were the foundation of relativistic physics – the Michelson–Morley experiment and the Eötvös experiment – and the two immediate confirmations of GR – the deflection of light and the perihelion advance of Mercury. Following this was a period of Hibernation (1920–1960) during which relatively few experiments were performed to test GR, and at the same time the field itself became sterile and stagnant, relegated to the backwaters of physics and astronomy.

But beginning around 1960, astronomical discoveries (quasars, pulsars, cosmic background radiation) and new experiments pushed GR to the forefront. Experimental gravitation experienced a Golden Era (1960–1980) during which a systematic, worldwide effort took place to understand the observable predictions of GR, to compare and contrast them with the predictions of alternative theories of gravity, and to perform new experiments to test them. The period began with an experiment to confirm the gravitational frequency shift of light (1960) and ended with the reported decrease in the orbital period of the binary pulsar at a rate consistent with the general relativity prediction of gravity-wave energy loss (1979). The results all supported GR, and most alternative theories of gravity fell by the wayside (for a popular review, see Will 1993a).

Since 1980, the field has moved toward what might be termed a Quest for Strong Gravity. The principal figure of merit that distinguishes strong from weak gravity is the quantity $\epsilon \sim GM/Rc^2$, where G is the Newtonian gravitational constant, M is the characteristic mass scale of the phenomenon, R is the characteristic distance scale, and c is the speed of light. Near the event horizon of a non-rotating black hole, or for the expanding Universe, $\epsilon \sim 0.5$; for neutron stars, $\epsilon \sim 0.2$. These are the regimes of strong gravity. For the solar system, $\epsilon < 10^{-5}$; this is the regime of weak gravity. Figure 1 displays these regimes and various phenomena of interest. Notice that the strong-gravity regime (near the diagonal line corresponding to $\epsilon = 0.5$) encompasses Planck-scale physics where quantum gravity and grand unification of the interactions are important, all the way to the observable universe.

Until the discovery of the binary pulsar, the empirical foundation of general relativity rested on tests in the weak-field regime. In Chapter 15, Kenneth Nordtvedt Jr. surveys experimental tests of general relativity primarily from the point of view of Solar System and laboratory experiments. While most of these focus on weak-gravity effects, some can be viewed as strong-gravity tests, primarily experiments to search for the relics of Planck-scale physics in fundamental interactions, as might show up in tests of the equivalence principle.

*Washington University, St. Louis, MO, USA

Today, much of the focus has shifted to experiments which can probe the effects of strong gravitational fields (see Figure 1). At one extreme are the strong gravitational fields associated with Planck-scale physics. Will unification of the forces, or quantization of gravity at this scale leave observable effects accessible by experiment, even those that themselves are performed under weak-field conditions? Dramatically improved tests of the equivalence principle or of the "inverse square law" are being designed, to search for or to bound the imprinted effects of Planck-scale phenomena (see Chapter 15). At the other extreme are the strong fields associated with compact objects such as black holes or neutron stars or with the universe as a whole. Astrophysical observations and gravitational-wave detectors are being planned to explore and test GR in the strong-field, highly dynamical regime associated with the formation and dynamics of these objects.

In this chapter, we shall focus on tests of general relativity involving strong gravity and gravitational radiation.

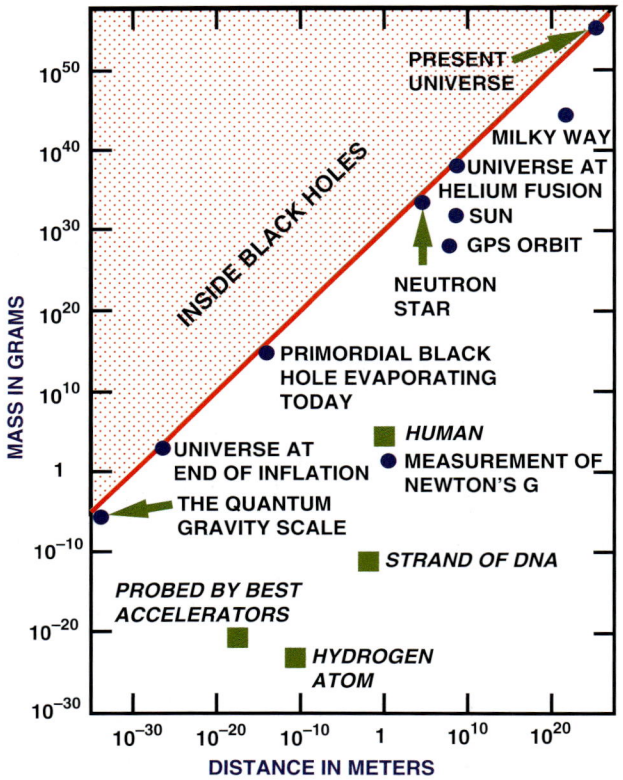

Figure 1 Strong v. weak gravity. Representative phenomena in gravitational physics are indicated by filled circles; other illustrative phenomena where gravitation plays little role are shown by filled squares and italics. Phenomena above the diagonal line are unobservable, because they take place inside black holes. Phenomena close to the diagonal line are in the strong-gravity regime.

These tests mainly involve black holes, neutron stars, and cosmology, and make essential use of gravitational radiation as a carrier of information about strong gravity. Even though the gravitational waves that bathe the Earth are extraordinarily weak (and are themselves describable by the weak-field limit of general relativity), they often carry the imprints of strong-gravity phenomena occurring near black holes, neutron stars, or from the early Universe.

We begin (Section 1) by defining the distinction between strong and weak gravity, and providing (Section 2) an introduction to the general relativistic description of systems involving compact relativistic objects and gravitational radiation. In Section 3 we focus on tests of gravitational theory using stellar binary systems of compact objects, the most famous of these being the binary pulsar PSR 1913+16. Section 4 focuses on the likely direct detection of gravitational radiation by Earth-based antennae in the first decade of the twenty-first century, and the possible tests of gravitational theory that could emerge. In Section 5 we briefly discuss other tests of strong gravity.

We use the standard "geometrized" units of general relativity textbooks (Misner et al. 1973), in which Newton's gravitational constant G and the speed of light c are unity. In these units, mass and energy have units of length, with the basic scale of length set by the solar mass: $M_\odot = 1.476$ km, and velocities are dimensionless, representing fractions of the speed of light. In the few places where indices are used, Greek indices will run over the four spacetime dimensions, $\{0, 1, 2, 3\}$, with the 0 denoting time; while Roman indices will run only over spatial indices $\{1, 2, 3\}$.

Rather than provide complete references to work done in this field, we will refer the reader where possible to the appropriate review articles and monographs, specifically to *Theory and Experiment in Gravitational Physics* (Will 1993b), hereafter referred to as TEGP. References to TEGP will be by chapter or section, e.g. "TEGP 8.9".

1 STRONG-FIELD SYSTEMS IN GENERAL RELATIVITY

1.1 Defining weak and strong gravity

In the Solar System, gravity is weak, in the sense that the Newtonian gravitational potential

$$U(\mathbf{x}, t) \equiv \int \rho(\mathbf{x}', t) |\mathbf{x} - \mathbf{x}'|^{-1} d^3 x' \qquad (1)$$

is much smaller than unity (recall $G = c = 1$) everywhere. So, too, are the quantities v^2 and p/ρ, where v is a typical velocity of bodies in the Solar System, and where p and ρ are typical pressure and density, respectively, inside Solar System bodies. In fact $U \sim v^2 \sim p/\rho \sim \epsilon$. Throughout the Solar System, the metric of spacetime deviates only slightly from its flat-spacetime Minkowski form $\eta_{\mu\nu} = \text{diag}(-1,1,1,1)$. By expanding the metric about this form in powers of the small parameter ϵ, one can obtain the Newtonian limit of general relativity at lowest order, and the first "post-Newtonian", or "1PN", corrections to Newtonian gravity. The metric in such an expansion takes the schematic form

$$g_{00} = -1 + 2U(\mathbf{x}, t) + O(\epsilon^2)$$
$$g_{0i} = O(\epsilon^{3/2}) \qquad (2)$$
$$g_{ij} = \delta_{ij} + O(\epsilon)$$

where the first terms beyond the Minkowski metric and the Newtonian potential at $O(\epsilon^2)$ in g_{00}, $O(\epsilon^{3/2})$ in g_{0i}, and $O(\epsilon)$ in g_{ij}, respectively, are the post-Newtonian terms. It turns out that, in a wide range of alternative theories of gravity, these post-Newtonian terms vary from one theory to the next only in the values of certain numerical coefficients. By inserting arbitrary dimensionless parameters in place of the numerical coefficients, one obtains a parametrized framework that encompasses many theories at once, and is a powerful tool for analyzing Solar System experiments. This "parametrized post-Newtonian" framework is described in more detail in Chapter 15; see also TEGP 4.

In strong-field systems, this simple 1PN approximation is no longer appropriate, for several reasons:

- The system may contain strongly relativistic objects, such as neutron stars or black holes, near and inside which $\epsilon \sim 1$, and the post-Newtonian approximation breaks down. Nevertheless, under some circumstances, the orbital motion may be such that the interbody potential and orbital velocities still satisfy $U \sim v^2 \ll 1$ so that a kind of post-Newtonian approximation for the orbital motion might work; however, the strong-field internal gravity of the bodies could (especially in alternative theories of gravity) leave imprints on the orbital motion.
- The evolution of the system may be affected by the emission of gravitational radiation. The 1PN approximation does not contain the effects of gravitational radiation back-reaction. In the expression for the metric given in eqns (2), radiation back-reaction effects do not occur until $O(\epsilon^{7/2})$ in g_{00}, $O(\epsilon^3)$ in g_{0i}, and $O(\epsilon^{5/2})$ in g_{ij}. Consequently, in order to describe such systems, one must carry out a solution of the equations substantially beyond 1PN order, sufficient to incorporate the leading radiation damping terms at 2.5PN order.
- The system may be highly relativistic in its orbital motion, so that $U \sim v^2 \sim 1$ even for the interbody field and orbital velocity. Systems like this include the late stage of the inspiral of binary systems of neutron stars or black holes, driven by gravitational radiation damping,

prior to a merger and collapse to a final stationary state. Binary inspiral is one of the leading candidate sources for detection by a worldwide network of laser interferometric gravitational-wave observatories nearing completion. A proper description of such systems requires not only equations for the motion of the binary carried to extraordinarily high PN orders (at least 3.5PN), but also requires equations for the far-zone gravitational waveform measured at the detector, that are equally accurate to high PN orders beyond the leading "quadrupole" approximation.

Of course, some systems cannot be properly described by any post-Newtonian approximation because their behavior is fundamentally controlled by strong gravity. These include the imploding cores of supernovae, the final merger of two compact objects, the quasinormal-mode vibrations of neutron stars and black holes, the structure of rapidly rotating neutron stars, and so on. Phenomena such as these must be analyzed using different techniques. Chief among these is the full solution of Einstein's equations via numerical methods. This field of "numerical relativity" is a rapidly growing and evolving branch of gravitational physics, whose description is beyond the scope of this article (see, e.g., the articles in Marck and Lasota 1997).

1.2 Compact bodies and the strong equivalence principle

When dealing with the motion and gravitational-wave generation by orbiting bodies, one finds a remarkable simplification within general relativity. As long as the bodies are sufficiently well separated that one can ignore tidal interactions and other effects that depend upon the finite extent of the bodies (such as their quadrupole and higher multipole moments), then all aspects of their orbital behavior and gravitational-wave generation can be characterized by just two parameters: mass and angular momentum. Whether their internal structure is highly relativistic, as in black holes or neutron stars, or non-relativistic as in the Earth and Sun, only the mass and angular momentum are needed. Furthermore, both quantities are measurable in principle by examining the external gravitational field of the bodies, and make no reference whatsoever to their interiors.

Damour (1987) calls this the "effacement" of the bodies' internal structure. It is a consequence of the strong equivalence principle (SEP), which is satisfied by general relativity, but is violated by almost all other gravitational theories, including scalar–tensor gravity (Brans–Dicke theory and its generalizations). SEP states that (i) all bodies fall in an external gravitational field with the same acceleration (modulo tidal interactions), (ii) the outcome of any local test experiment is independent of the velocity of the (freely falling) apparatus, and (iii) the outcome of any local test experiment is independent of where and when in the Universe it is performed. The term "bodies" here includes everything from laboratory-sized objects to black holes; and the term "experiments" includes everything from electromagnetic experiments to measurements of Newton's G.

General relativity satisfies SEP because it contains one, and only one, gravitational field, the spacetime metric $g_{\mu\nu}$. Consider the motion of a body in a binary system, whose size is small compared to the binary separation. Surround the body by a region that is large compared to the size of the body, yet small compared to the separation. Because of the general covariance of the theory, one can choose a freely falling coordinate system which co-moves with the body, whose spacetime metric takes the Minkowski form at its outer boundary (ignoring tidal effects generated by the companion). There is thus no evidence of the presence of the companion body, and the structure of the chosen body can be obtained using the field equations of GR in this coordinate system. Far from the chosen body, the metric is characterized by the mass and angular momentum (assuming that one ignores quadrupole and higher multipole moments of the body) as measured far from the body using orbiting test particles and gyroscopes. These asymptotically measured quantities are oblivious to the body's internal structure. A black hole of mass m and a planet of mass m would produce identical spacetimes in this outer region.

The geometry of this region surrounding the one body must be matched to the geometry provided by the companion body. Einstein's equations provide consistency conditions for this matching that yield constraints on the motion of the bodies. These are the equations of motion. As a result the motion of two planets of mass and angular momentum m_1, m_2, $\mathbf{J}_1$ and $\mathbf{J}_2$ is identical to that of two black holes of the same mass and angular momentum (again, ignoring tidal effects).

This effacement does not occur in an alternative gravitational theory like scalar–tensor gravity. There, in addition to the spacetime metric, a scalar field Φ is generated by the masses of the bodies, and controls the local value of the gravitational coupling constant (i.e., G is a function of Φ). Now, in the local frame surrounding one of the bodies in our binary system, while the metric can still be made Minkowskian far away, the scalar field will take on a value Φ_0 determined by the companion body. This can affect the value of G inside the chosen body, alter its internal structure (specifically its gravitational binding energy), and hence alter its mass. Effectively, each mass becomes several functions $m_A(\Phi)$ of the value of the scalar field at its location, and several distinct masses come into play; inertial mass, gravitational mass, "radiation" mass, and so on. The precise nature of the functions will depend on the body, specifically on its gravitational binding energy, and as a result, the motion and gravitational radiation may depend on the internal structure of each body. For compact bodies such as neutron stars and black holes, these internal structure effects could be large; for example, the gravitational binding energy of a neutron star can be 40% of its total mass.

At 1PN order, the leading manifestation of this effect is a violation of the equality of acceleration of massive bodies such as the Earth and the Moon. This effect, known as the Nordtvedt effect, has been tested by lunar laser ranging (see Chapter 15 or TEGP 8 for further discussion).

This is how the study of orbiting systems containing compact objects provides strong-field tests of general relativity. Even though the strong-field nature of the bodies is effaced in GR, it is not in other theories, thus any result in agreement with the predictions of GR constitutes a kind of "null" test of strong-field gravity.

2 MOTION AND GRAVITATIONAL RADIATION IN GENERAL RELATIVITY

2.1 Introduction

The motion of bodies and the generation of gravitational radiation are long-standing problems that date back to the first years following the publication of GR, when Einstein calculated the gravitational radiation emitted by a laboratory-scale object using the linearized version of GR, and de Sitter calculated N-body equations of motion for bodies in the 1PN approximation to GR. It has at times been controversial, with disputes over such issues as whether Einstein's equations alone imply equations of motion for bodies (Einstein, Infeld, and Hoffman demonstrated explicitly that they do, using a matching procedure similar to the one described in Section 1.2), whether gravitational waves are real or are artifacts of general covariance (Einstein waffled; Bondi and colleagues proved their reality rigorously in the 1950s), or even over algebraic errors (Einstein erred by a factor of two in his first radiation calculation; Eddington found the mistake). Shortly after the discovery of the binary pulsar PSR 1913 + 16 in 1974, questions were raised about the foundations of the "quadrupole formula" for gravitational radiation damping (and in some quarters, even about its quantitative validity). These questions were answered in part by theoretical work designed to shore up the foundations of the quadrupole approximation, and in part (perhaps mostly) by the agreement between the predictions of the quadrupole formula and the *observed* rate of damping of the pulsar's orbit (see Section 3.1). Damour (1987) gives a thorough review of this subject.

The problem of motion and radiation has received renewed interest since 1990, with proposals for the construction of large-scale laser interferometric gravitational-wave observatories, such as the LIGO project in the USA, VIRGO and GEO600 in Europe, and TAMA in Japan, and the realization that a leading candidate source of detectable waves would be the inspiral, driven by gravitational radiation damping, of a binary system of compact objects (neutron stars or black holes) (Abramovici et al. 1992, Thorne 1995). The analysis of signals from such systems will require theoretical predictions from GR that are extremely accurate, well beyond the leading-order prediction of Newtonian or even post-Newtonian gravity for the orbits, and well beyond the leading-order formulae for gravitational waves.

This presented a major theoretical challenge: to calculate the motion and radiation of systems of compact objects to very high PN order, a formidable algebraic task, while addressing a number of issues of principle that have historically plagued this subject, sufficiently well to ensure that the results were physically meaningful. This challenge is in the process of being met, so that the next few years may see a remarkable convergence between observational data and accurate predictions of gravitational theory that could provide new, strong-field tests of GR.

Here we give a brief overview of the problem of motion and gravitational radiation (see, e.g., Misner et al. 1973).

2.2 Einstein's equations in "relaxed" form

The Einstein equations $G_{\mu\nu} = 8\pi T_{\mu\nu}$ are elegant and deceptively simple, showing geometry (in the form of the Einstein tensor $G_{\mu\nu}$, which is a function of spacetime curvature) being generated by matter (in the form of the material stress-energy tensor $T_{\mu\nu}$). However, this is not the most useful form for actual calculations. For post-Newtonian calculations, a far more useful form is the so-called "relaxed" Einstein equations:

$$\Box h^{\alpha\beta} = -16\pi \tau^{\alpha\beta} \qquad (3)$$

where $\Box \equiv -\partial^2/\partial t^2 + \nabla^2$ is the flat-spacetime wave operator, $h^{\alpha\beta}$ is a "gravitational tensor potential" related to the deviation of the spacetime metric from its Minkowski form by the formula $h^{\alpha\beta} \equiv \eta^{\alpha\beta} - (-g)^{1/2} g^{\alpha\beta}$, g is the determinant of $g_{\alpha\beta}$, and a particular coordinate system has been specified by the deDonder or harmonic gauge condition $\partial h^{\alpha\beta}/\partial x^\beta = 0$ (summation on repeated indices is assumed). This form of Einstein's equations bears a striking similarity to Maxwell's equations for the vector potential A^α in Lorentz gauge: $\Box A^\alpha = -4\pi J^\alpha$, $\partial A^\alpha/\partial x^\alpha = 0$. There is a key difference, however: the source on the right-hand side of eqn (3) is given by the "effective" stress-energy pseudotensor

$$\tau^{\alpha\beta} = (-g)T^{\alpha\beta} + (16\pi)^{-1}\Lambda^{\alpha\beta} \qquad (4)$$

where $\Lambda^{\alpha\beta}$ is the nonlinear "field" contribution given by terms quadratic (and higher) in $h^{\alpha\beta}$ and its derivatives (see Misner et al. (1973), equations (20.20)–(20.21) for formulas). In general relativity, the gravitational field itself generates gravity, a reflection of the nonlinearity of Einstein's equations, and in contrast to the linearity of Maxwell's equations.

Equation (3) is exact, and depends only on the assumption that spacetime can be covered by harmonic coordinates. It is called "relaxed" because it can be solved formally as a functional of source variables without specifying the motion of the source, in the form

$$h^{\alpha\beta}(t, \mathbf{x}) = 4\int_C \frac{\tau^{\alpha\beta}(t - |\mathbf{x} - \mathbf{x}'|, \mathbf{x}')}{|\mathbf{x} - \mathbf{x}'|} d^3 x' \qquad (5)$$

where the integration is over the past flat-spacetime null cone C of the field point $(t, \mathbf{x})$. The motion of the source is then determined either by the equation $\partial \tau^{\alpha\beta}/\partial x^\beta = 0$ (which follows from the harmonic gauge condition), or from the usual covariant equation of motion $T^{\alpha\beta}{}_{;\beta} = 0$, where the subscript $;\beta$ denotes a covariant divergence. This formal solution can then be iterated in a slow-motion ($v < 1$) weak-field ($\|h^{\alpha\beta}\| < 1$) approximation. One begins by substituting $h_0^{\alpha\beta} = 0$ into the source $\tau^{\alpha\beta}$ in eqn (5), and solving for the first iterate $h_1^{\alpha\beta}$, and then repeating the procedure sufficiently many times to achieve a solution of the desired accuracy. For example, to obtain the 1PN equations of motion, *two* iterations are needed (i.e., $h_2^{\alpha\beta}$ must be calculated); likewise, to obtain the leading gravitational waveform for a binary system, two iterations are needed.

At the same time, just as in electromagnetism, the formal integral (5) must be handled differently, depending on whether the field point is in the far zone or the near zone. For field points in the far zone or radiation zone, $|\mathbf{x}| > \lambdabar > |\mathbf{x}'|$ ($\lambdabar$ is the gravitational wavelength/2π), the field can be expanded in inverse powers of $R = |\mathbf{x}|$ in a multipole expansion, evaluated at the "retarded time" $t - R$. The leading term in $1/R$ is the gravitational waveform. For field points in the near zone or induction zone, $|\mathbf{x}| \sim |\mathbf{x}'| < \lambdabar$, the field is expanded in powers of $|\mathbf{x} - \mathbf{x}'|$ about the local time t, yielding instantaneous potentials that go into the equations of motion.

However, because the source $\tau^{\alpha\beta}$ contains $h^{\alpha\beta}$ itself, it is not confined to a compact region, but extends over all spacetime. As a result, there is a danger that the integrals involved in the various expansions will diverge or be ill-defined. This consequence of the nonlinearity of Einstein's equations has bedeviled the subject of gravitational radiation for decades. Numerous approaches have been developed to try to handle this difficulty. The "post-Minkowskian" method of Blanchet and Damour (1986, 1988, 1989, 1992),

Damour and Iyer (1991), and Blanchet (1995) solves Einstein's equations by two different techniques, one in the near zone and one in the far zone, and uses the method of singular asymptotic matching to join the solutions in an overlap region. The method provides a natural "regularization" technique to control potentially divergent integrals. The "direct integration of relaxed Einstein equations" (DIRE) approach of Will and Wiseman (1996) and Pati and Will (2000), retains eqn (5) as the global solution, but splits the integration into one over the near zone and another over the far zone, and uses different integration variables to carry out the explicit integrals over the two zones. In the DIRE method, all integrals are finite and convergent.

These methods assume from the outset that gravity is sufficiently weak that $\|h^{\alpha\beta}\| < 1$ and harmonic coordinates exist everywhere, including inside the bodies. Thus, in order to apply the results to cases where the bodies may be neutron stars or black holes, one relies upon the SEP to argue that, if tidal forces are ignored, and equations are expressed in terms of masses and spins, one can simply extrapolate the results unchanged to the situation where the bodies are ultra-relativistic. While no general proof of this exists, it has been shown to be valid in specific circumstances, such as at 2PN order in the equations of motion, and for black holes moving in a Newtonian background field (Damour 1987).

Methods such as these have resolved most of the issues that led to criticism of the foundations of gravitational radiation theory during the 1970s.

2.3 Equations of motion and gravitational waveform

Among the results of these approaches are formulae for the equations of motion and gravitational waveform of binary systems of compact objects, carried out to high orders in a PN expansion. Here we shall only state the key formulae that will be needed for this chapter. For example, the relative two-body equation of motion has the form

$$\mathbf{a} = \frac{d\mathbf{v}}{dt} = \frac{m}{r^2}\{-\hat{\mathbf{n}} + \mathbf{A}_{1\text{PN}} + \mathbf{A}_{2\text{PN}} + \mathbf{A}_{2.5\text{PN}} + \mathbf{A}_{3\text{PN}} + \mathbf{A}_{3.5\text{PN}} + \cdots\} \quad (6)$$

where $m = m_1 + m_2$ is the total mass, $r = |\mathbf{x}_1 - \mathbf{x}_2|$, $\mathbf{v} = \mathbf{v}_1 - \mathbf{v}_2$, and $\hat{\mathbf{n}} = (\mathbf{x}_1 - \mathbf{x}_2)/r$. The notation $\mathbf{A}_{n\text{PN}}$ indicates that the term is $O(\epsilon^n)$ relative to the Newtonian term $-\hat{\mathbf{n}}$. Explicit formulae for all but the 3PN terms have been calculated by various authors (Damour and Deruelle 1981, Damour 1982, 1987, Grishchuk and Kopeikin 1986, Iyer and Will 1993, 1995, Blanchet 1997, Blanchet et al. 1998, Pati and Will 2000) and as of this writing, various groups were reporting results for 3PN terms; here we quote only the first PN corrections and the leading radiation-reaction terms at 2.5PN order:

$$\mathbf{A}_{1\text{PN}} = \left\{(4 + 2\eta)\frac{m}{r} - (1 + 3\eta)v^2 + \frac{3}{2}\eta\dot{r}^2\right\}\hat{\mathbf{n}} + (4 - 2\eta)\dot{r}\mathbf{v} \quad (7)$$

$$\mathbf{A}_{2.5\text{PN}} = -\frac{8}{15}\eta\frac{m}{r}\left\{\left(9v^2 + 17\frac{m}{r}\right)\dot{r}\hat{\mathbf{n}} - \left(3v^2 + 9\frac{m}{r}\right)\mathbf{v}\right\} \quad (8)$$

where $\eta = m_1 m_2/(m_1 + m_2)^2$. These terms are sufficient to analyze the orbit and evolution of the binary pulsar (Section 3.1). For example, the 1PN terms are responsible for the periastron advance of an eccentric orbit, given by $\dot{\omega} = 6\pi f_b m/a(1 - e^2)$, where a and e are the semimajor axis and eccentricity, respectively, of the orbit, and f_b is the orbital frequency, given to the needed order by Kepler's third law $2\pi f_b = (m/a^3)^{1/2}$.

Another product is a formula for the gravitational field far from the system, written in the form

$$h^{ij} = \frac{2m}{R}\{Q^{ij} + Q^{ij}_{0.5\text{PN}} + Q^{ij}_{1\text{PN}} + Q^{ij}_{1.5\text{PN}} + Q^{ij}_{2\text{PN}} + Q^{ij}_{2.5\text{PN}} + \cdots\} \quad (9)$$

where R is the distance from the source, and the variables are to be evaluated at retarded time $t - R$. The leading term is the so-called quadrupole formula

$$h^{ij}(t, \mathbf{x}) = \frac{2}{R}\ddot{I}^{ij}(t - R) \quad (10)$$

where I^{ij} is the quadrupole moment of the source, and overdots denote time derivatives. For a binary system this leads to

$$Q^{ij} = 2\eta(v^i v^j - m\hat{n}^i \hat{n}^j/r) \quad (11)$$

For binary systems, explicit formulae for all the terms through 2.5PN order have been derived by various authors (Wagoner and Will 1976, Wiseman 1992, Poisson 1993, Blanchet et al. 1995a,b, Will and Wiseman 1996, Blanchet 1996, 1998). Given the gravitational waveform, one can compute the rate at which energy is carried off by the radiation (schematically $\int \dot{h}\dot{h} \, d\Omega$, the gravitational analog of the Poynting flux). The lowest-order quadrupole formula leads to the gravitational-wave energy flux

$$\dot{E} = \frac{8}{15}\eta^2\frac{m^4}{r^4}(12v^2 - 11\dot{r}^2) \quad (12)$$

Formulae for fluxes of angular and linear momentum can also be derived. The 2.5PN radiation-reaction terms in the equation of motion (6) result in a damping of the orbital energy that precisely balances the energy flux (12) determined from the waveform. Averaged over one orbit, this results in a rate of increase of the binary's orbital frequency given by

$$\dot{f}_b = \frac{192\pi}{5} f_b^2 (2\pi \mathcal{M} f_b)^{5/3} F(e) \qquad (13)$$

where $\mathcal{M}$ is the so-called "chirp" mass, given by $\mathcal{M} = \eta^{3/5} m$, and $F(e) = (1 + 73e^2/24 + 37e^4/96)/(1-e^2)^{7/2}$. Notice that by making precise measurements of the phase $\Phi(t) = \int^t f(t') dt'$ of either the orbit or the gravitational waves (for which $f = 2f_b$ for the dominant component) as a function of the frequency, one in effect measures the "chirp" mass of the system.

These formalisms have also been generalized to include the leading effects of spin–orbit and spin–spin coupling between the bodies (Kidder et al. 1993, Kidder 1995).

Another approach to gravitational radiation is applicable to the special limit in which one mass is much smaller than the other. This is the method of black hole perturbation theory. One begins with an exact background spacetime of a black hole, either the non-rotating Schwarzschild or the rotating Kerr solution, and perturbs it according to $g_{\mu\nu} = g_{\mu\nu}^{(0)} + h_{\mu\nu}$. The particle moves on a geodesic of the background spacetime, and a suitably defined source stress-energy tensor for the particle acts as a source for the gravitational perturbation and wave field $h_{\mu\nu}$. This method provides numerical results that are exact in v, as well as analytical results expressed as series in powers of v, both for non-rotating and for rotating black holes. For non-rotating holes, the analytical expansions have been carried to 5.5 PN order, or $\epsilon^{5.5}$ beyond the quadrupole approximation. All results of black hole perturbation agree precisely with the $m_1 \to 0$ limit of the PN results, up to the highest PN order where they can be compared (for a detailed review see Mino et al. 1997).

2.4 Gravitational-wave detection

A gravitational-wave detector can be modelled as a body of mass M at a distance L from a fiducial laboratory point, connected to the point by a spring of resonant frequency ω_0 and quality factor Q. From the equation of geodesic deviation, the infinitesimal displacement ξ of the mass along the line of separation from its equilibrium position satisfies the equation of motion

$$\ddot{\xi} + 2\frac{\omega_0}{Q}\dot{\xi} + \omega_0^2 \xi = \frac{L}{2}\bigg(F_+(\theta,\phi,\psi)\ddot{h}_+(t)$$

$$+ F_\times(\theta,\phi,\psi)\ddot{h}_\times(t)\bigg) \qquad (14)$$

where $F_+(\theta,\phi,\psi)$ and $F_\times(\theta,\phi,\psi)$ are "beam-pattern" factors, that depend on the direction of the source (θ,ϕ), and on a polarization angle ψ, and $h_+(t)$ and $h_\times(t)$ are gravitational waveforms corresponding to the two polarizations of the gravitational wave (for a review see Thorne 1987). In a source coordinate system in which the x–y plane is the plane of the sky and the z-direction points toward the detector, these two modes are given by

$$h_+(t) = \frac{1}{2}(h^{xx}(t) - h^{yy}(t)), \qquad h_\times(t) = h^{xy}(t) \qquad (15)$$

The beam pattern factors depend on the orientation and nature of the detector. For a wave approaching along the laboratory z-direction, and for a mass whose location on the x–y plane makes an angle ϕ with the x-axis, the beam pattern factors are given by $F_+ = \cos 2\phi$ and $F_\times = \sin 2\phi$. For a resonant cylinder oriented along the laboratory z-axis, and for source direction (θ,ϕ), they are given by $F_+ = \sin^2\theta \cos 2\psi$, $F_\times = \sin^2\theta \sin 2\psi$ (the angle ψ measures the relative orientation of the laboratory and source x-axes). For a laser interferometer with one arm along the laboratory x-axis, the other along the y-axis, and with ξ defined as the *differential* displacement along the two arms, the beam pattern functions are

$$F_+ = \frac{1}{2}(1 + \cos^2\theta)\cos 2\phi \cos 2\psi - \cos\theta \sin 2\phi \sin 2\psi$$

and

$$F_\times = \frac{1}{2}(1 + \cos^2\theta)\cos 2\phi \cos 2\psi + \cos\theta \sin 2\phi \cos 2\psi$$

The waveforms $h_+(t)$ and $h_\times(t)$ depend on the nature and evolution of the source. For example, for a binary system in a circular orbit, with an inclination i relative to the plane of the sky, and the x-axis oriented along the major axis of the projected orbit, the quadrupole approximation of eqn (11) gives

$$h_+(t) = -\frac{2\mathcal{M}}{R}(2\pi \mathcal{M} f_b)^{2/3}(1 + \cos^2 i) \cos 2\Phi_b(t)$$

$$h_\times(t) = -\frac{2\mathcal{M}}{R}(2\pi \mathcal{M} f_b)^{2/3} \cos i \cos 2\Phi_b(t) \qquad (16)$$

where $\Phi_b(t) = \int^t f_b(t') dt'$ is the orbital phase.

We now turn to specific ways of testing general relativity using compact objects and gravitational radiation.

3 STELLAR SYSTEM TESTS OF GRAVITATIONAL THEORY

3.1 The binary pulsar and general relativity

The 1974 discovery of the binary pulsar PSR 1913+16 by Joseph Taylor and Russell Hulse during a routine search for new pulsars provided the first possibility of probing new aspects of gravitational theory: the effects of strong relativistic internal gravitational fields on orbital dynamics, and the effects of gravitational radiation reaction. For reviews of the discovery and current status, see the published Nobel Prize lectures by Hulse (1994) and Taylor (1994).

The system consists of a pulsar of nominal period 59 ms in a close binary orbit with an as yet unseen companion. The orbital period is about 7.75 hours, and the eccentricity is 0.617. From detailed analyses of the arrival times of pulses (which amounts to an integrated version of the Doppler-shift methods used in spectroscopic binary systems), extremely accurate orbital and physical parameters for the system have been obtained (Table 1). Because the orbit is so close ($\approx 1 R_\odot$) and because there is no evidence of an eclipse of the pulsar signal or of mass transfer from the companion, it is generally believed that the companion is compact: evolutionary arguments suggest that it is most likely a dead pulsar. Thus the orbital motion is very clean, free from tidal or other complicating effects. Furthermore, the data acquisition is "clean" in the sense that by exploiting the intrinsic stability of the pulsar clock combined with the ability to maintain and transfer atomic time accurately using such devices as the Global Positioning System, the observers can keep track of the pulsar phase with an accuracy of 15 μs, despite extended gaps between observing sessions (including a several-year gap during the middle 1990s upgrade of the Arecibo radio telescope). The pulsar has shown no evidence of "glitches" in its pulse period.

Three factors make this system an arena where relativistic celestial mechanics must be used: the relatively large size of relativistic effects [$v_{\rm orbit} \approx (m/r)^{1/2} \approx 10^{-3}$], a factor of 10 larger than the corresponding values for Solar System orbits; the short orbital period, allowing secular effects to build up rapidly; and the cleanliness of the system, allowing accurate determinations of small effects. Because the orbital separation is large compared to the neutron stars' compact size, tidal effects can be ignored. Just as Newtonian gravity is used as a tool for measuring astrophysical parameters of ordinary binary systems, so GR is used as a tool for measuring astrophysical parameters in the binary pulsar.

The observational parameters that are obtained from a least-squares solution of the arrival-time data fall into three groups:

1. non-orbital parameters, such as the pulsar period and its rate of change (defined at a given epoch), and the position of the pulsar on the sky;
2. five "Keplerian" parameters, most closely related to those appropriate for standard Newtonian binary systems, such as the eccentricity e and the orbital period P_b
3. five "post-Keplerian" parameters.

The five post-Keplerian parameters are $\langle \dot{\omega} \rangle$, the average rate of periastron advance; γ', the amplitude of delays in arrival of pulses caused by the varying effects of the gravitational redshift and time dilation as the pulsar moves in its elliptical orbit at varying distances from the companion and with varying speeds; $\dot{P}_b$, the rate of change of orbital period, caused predominantly by gravitational radiation damping; and r and $s = \sin i$, respectively the "range" and "shape" of the Shapiro time delay of the pulsar signal as it propagates through the curved spacetime region near the companion, where i is the angle of inclination of the orbit relative to the

Table 1 Parameters of the binary pulsar PSR 1913+16

Parameter	Symbol (units)	Value*
(i) "Physical" parameters		
Right ascension	α	$19^{\rm h} 15^{\rm m} 28.^{\rm s}00018(15)$
Declination	δ	$16°06' 27.''4043(3)$
Pulsar period	P_p (ms)	59.029997929613(7)
Derivative of period	$\dot{P}_p$	$8.62713(8) \times 10^{-18}$
(ii) "Keplerian" parameters		
Projected semimajor axis	$a_p \sin i$ (s)	2.3417592(19)
Eccentricity	e	0.6171308(4)
Orbital period	P_b (day)	0.322997462736(7)
Longitude of periastron	ω_0 (°)	226.57528(6)
Julian date of periastron	T_0 (MJD)	46443.99588319(3)
(iii) "Post-Keplerian" parameters		
Mean rate of periastron advance	$\langle \dot{\omega} \rangle$ (° yr^{-1})	4.226621(11)
Redshift/time dilation	γ' (ms)	4.295(2)
Orbital period derivative	$\dot{P}_b$ (10^{-12})	$-2.422(6)$

*Numbers in parentheses denote errors in last digit. Data from http://puppsr8.princeton.edu/psrcat.html

plane of the sky (see Chapter 15 for a discussion of the Shapiro time delay effect).

In GR, these post-Keplerian parameters can be related to the masses of the two bodies and to measured Keplerian parameters by the equations (TEGP 12.1, 14.6(a))

$$\langle \dot{\omega} \rangle = 6\pi f_b (2\pi m f_b)^{2/3}(1-e^2)^{-1}$$
$$\gamma' = e(2\pi f_b)^{-1}(2\pi m f_b)^{2/3}(m_2/m)(1+m_2/m) \quad (17)$$
$$\dot{P}_b = -(192\pi/5)(2\pi \mathcal{M} f_b)^{5/3} F(e)$$
$$s = \sin i$$
$$r = m_2$$

where m_1 and m_2 denote the pulsar and companion masses, respectively. The formula for $\langle \dot{\omega} \rangle$ ignores possible non-relativistic contributions to the periastron shift, such as tidally or rotationally induced effects caused by the companion (for discussion of these effects, see TEGP 12.1(c)). The formula for $\dot{P}_b$ includes only quadrupole gravitational radiation; it ignores other sources of energy loss, such as tidal dissipation (TEGP 12.1(f)). Notice that, by virtue of Kepler's third law, $(2\pi f_b)^2 = m/a^3$, $(2\pi m f_b)^{2/3} \sim m/a \sim \epsilon$, thus the first two post-Keplerian parameters can be seen as $O(\epsilon)$, or 1PN corrections to the underlying variable, while the third is an $O(\epsilon^{5/2})$, or 2.5 PN correction. The current values for Keplerian and post-Keplerian parameters are shown in Table 1. The parameters r and s are not separately measurable with interesting accuracy for PSR 1913+16 because the orbit's 47° inclination does not lead to a substantial Shapiro delay.

Because f_b and e are separately measured parameters, the measurement of the three post-Keplerian parameters provide three constraints on the two unknown masses. The periastron shift measures the total mass of the system, $\dot{P}_b$ measures the chirp mass, and γ' measures a complicated function of the masses. GR passes the test if it provides a consistent solution to these constraints, within the measurement errors.

From the intersection of the $\langle \dot{\omega} \rangle$ and γ' constraints we obtain the values $m_1 = 1.4411 \pm 0.0007 M_\odot$ and $m_2 = 1.3873 \pm 0.0007 M_\odot$. The third of eqns (17) then predicts the value $\dot{P}_b = -2.40243 \pm 0.00005 \times 10^{-12}$. In order to compare the predicted value for $\dot{P}_b$ with the observed value of Table 1, it is necessary to take into account the small effect of a relative acceleration between the binary pulsar system and the Solar System caused by the differential rotation of the galaxy. This effect was previously considered unimportant when $\dot{P}_b$ was known only to 10% accuracy. Damour and Taylor (1992) carried out a careful estimate of this effect using data on the location and proper motion of the pulsar, combined with the best information available on galactic rotation, and found

$$\dot{P}_b^{GAL} \simeq -(1.7 \pm 0.5) \times 10^{-14} \quad (18)$$

Subtracting this from the observed $\dot{P}_b$ (Table 1) gives the residual

$$\dot{P}_b^{CORR} = -(2.408 \pm 0.010[OBS] \pm 0.005[GAL]) \times 10^{-12} \quad (19)$$

which agrees with the prediction, within the errors. In other words,

$$\frac{\dot{P}_b^{GR}}{\dot{P}_b^{CORR}} = 1.0023 \pm 0.0041[OBS] \pm 0.0021[GAL] \quad (20)$$

The consistency among the measurements is displayed in Figure 2, in which the regions allowed by the three most precise constraints have a single common overlap.

A third way to display the agreement with general relativity is by comparing the observed phase of the orbit with a theoretical template phase as a function of time. If f_b varies slowly in time, then to first order in a Taylor expansion, the orbital phase is given by $\Phi_b(t) = 2\pi f_{b0} t + \pi \dot{f}_{b0} t^2$. The time of periastron passage t_P is given by $\Phi(t_P) = 2\pi N$, where N is an integer, and consequently the periastron time will not grow linearly with N. Thus the cumulative difference between periastron time t_P and N/f_{b0}, the quantities actually measured in practice, should vary according to $t_P - N/f_{b0} = -\dot{f}_{b0} N^2 / 2 f_{b0}^3 \approx -(\dot{f}_{b0}/2f_{b0})t^2$. Figure 3 shows the results: the dots are the data points, while the curve is the predicted difference using the measured masses and the quadrupole formula for $\dot{f}_{b0}$ (Weisberg and Taylor 2000).

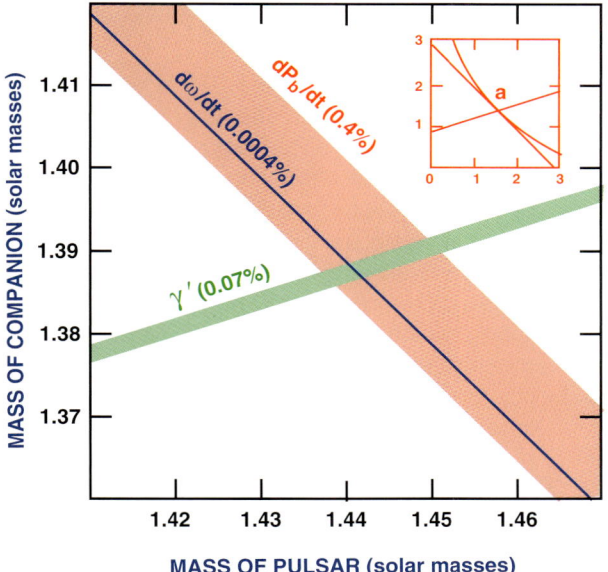

Figure 2 Constraints on masses of pulsar and companion from data on PSR 1913+16, assuming GR to be valid. The width of each strip in the plane reflects observational accuracy, shown as a percentage. The inset shows the three constraints on the full mass plane; intersection region (a) has been magnified 400 times for the full figure.

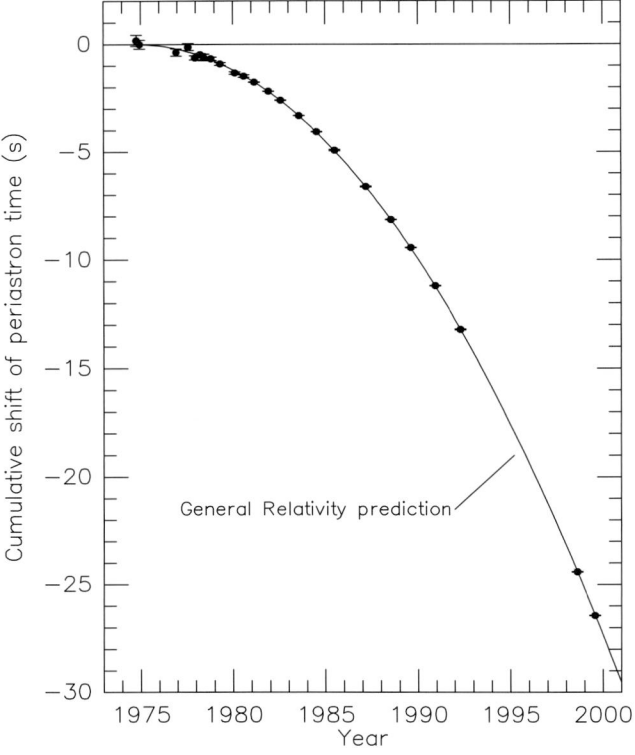

Figure 3 Plot of cumulative shift of the periastron time from 1975 to 2000. Points are data, the curve is the GR prediction. The gap during the middle 1990s was caused by closure of Arecibo for upgrading (J.H. Taylor and J.M. Weisberg, private communication, 2000).

The consistency among the constraints provides a test of the assumption that the two bodies behave as "point" masses, without complicated tidal effects, obeying the general relativistic equations of motion including gravitational radiation. It is also a test of strong gravity, in that the highly relativistic internal structure of the neutron stars does not influence their orbital motion, as predicted by the SEP of GR.

Recent observations (Kramer, 1998, Weisberg and Taylor 2000) indicate variations in the pulse profile, which suggests that the pulsar is undergoing precession as it moves through the curved spacetime generated by its companion, an effect known as geodetic precession. The amount is consistent with GR, assuming that the pulsar's spin is suitably misaligned with the orbital angular momentum. Unfortunately, the evidence suggests that the pulsar beam may precess out of our line of sight by 2020.

3.2 A population of binary pulsars?

Since 1990, several new massive binary pulsars similar to PSR 1913 + 16 have been discovered, leading to the possibility of new or improved tests of GR.

PSR 1534 + 12 is a binary pulsar system in our Galaxy. Its pulses are significantly stronger and narrower than those of PSR 1913 + 16, so timing measurements are more precise, reaching 3 μs accuracy. Its parameters are listed in Table 2 (Stairs *et al.* 1998, 2000). The orbital plane appears to be almost edge-on relative to the line of sight ($i \simeq 80°$); as a result the Shapiro delay is substantial, and separate values of the parameters r and s have been obtained with interesting accuracy. Assuming general relativity, one infers that the two masses are $m_1 = 1.335 \pm 0.002 M_\odot$ and $m_2 = 1.344 \pm 0.002 M_\odot$. The rate of orbit decay $\dot{P}_b$ agrees with GR to about 15%, the precision limited by the poorly known distance to the pulsar, which introduces a significant uncertainty into the subtraction of galactic acceleration.

PSR 2127+11C appears to be a clone of the Hulse–Taylor binary pulsar, with very similar values for orbital period and eccentricity (see Table 2). The inferred total mass of the system is $2.706 \pm 0.011 M_\odot$. Because the system is in the globular cluster M15 (NGC 7078), it suffers Doppler shifts resulting from local accelerations, either by the mean cluster gravitational field or by nearby stars, that are more difficult to estimate than was the case with the galactic system PSR 1913+16. This may make a separate, precision measurement of the relativistic contribution to $\dot{P}_b$ impossible.

PSR 1855+09 is not particularly relativistic, with a long period (12 days) and highly circular orbit. However, because we observe the orbit nearly edge on, the Shapiro delay is large and measurable, as reflected in the post-Keplerian parameters r and s.

PSR 0655+64. This system consists of a pulsar and a white dwarf companion in a nearly circular orbit. Only an upper limit on $\dot{P}_b$ has been placed.

3.3 Binary pulsars and alternative theories

Soon after the discovery of the binary pulsar it was widely hailed as a new testing ground for relativistic gravitational effects. As we have seen in the case of GR, in most respects, the system has lived up to, indeed exceeded, the early expectations.

In another respect, however, the system has only partially lived up to its promise, namely as a direct testing ground for alternative theories of gravity. The origin of this promise was the discovery that alternative theories of gravity generically predict the emission of dipole gravitational radiation from binary star systems. In general relativity, there is no dipole radiation because the "dipole moment" (center of mass) of isolated systems is uniform in time (conservation of momentum), and because the "inertial mass" that determines the dipole moment is the same as the mass that generates gravitational waves (SEP). In other

Table 2 Parameters of other binary pulsars

Parameter	B1534+12	B2127+11C	B1855+09	B0655+64
(i) "Keplerian" Parameters				
$a_p \sin i$ (s)	3.729464(3)	2.520(3)	9.2307802(4)	4.125612(5)
e	0.2736775(5)	0.68141(2)	0.00002168(5)	0.0000075(11)
P_b (day)	0.42073729933(3)	0.335282052(6)	12.3271711905(6)	1.028669703(1)
(ii) "Post-Keplerian" Parameters*				
$\langle \dot\omega \rangle$ (° yr^{-1})	1.755794(19)	4.457(12)		
γ' (ms)	2.071(6)	4.9(1.1)		
$\dot P_b$ (10^{-12})	−0.131(9)			<0.5
r (μs)	6.3(1.3)		1.27(10)	
$s = \sin i$	0.983(8)		0.9992(5)	

*From Wolszczan (1994), Stairs *et al.* (1998, 2000), and http://puppsr8.princeton.edu/psrcat.html

theories, while the inertial dipole moment may remain uniform, the "gravity-wave" dipole moment need not, because the mass that generates gravitational waves depends differently on the internal gravitational binding energy of each body than does the inertial mass (violation of SEP). Schematically, in a coordinate system in which the center of inertial mass is at the origin, so that $m_{I,1}\mathbf{x}_1 + m_{I,2}\mathbf{x}_2 = 0$, the dipole part of the retarded gravitational field would be given by

$$h \sim \frac{1}{R}\frac{d}{dt}(m_{GW,1}\mathbf{x}_1 + m_{GW,2}\mathbf{x}_2)$$

$$\sim \frac{\eta m}{R}\mathbf{v} \times \left(\frac{m_{GW,1}}{m_{I,1}} - \frac{m_{GW,2}}{m_{I,2}}\right) \quad (21)$$

where $\mathbf{v} = \mathbf{v}_1 - \mathbf{v}_2$ and η and m are defined using inertial masses. In theories that violate SEP, the difference between gravitational-wave mass and inertial mass is a function of the internal gravitational binding energy of the bodies. This additional form of gravitational radiation damping could, at least in principle, be significantly stronger than the usual quadrupole damping, because it depends on fewer powers of the orbital velocity v, and it depends on the gravitational binding energy per unit mass of the bodies, which, for neutron stars, could be as large as 40% (see TEGP 10 for further details). As one fulfillment of this promise, Will and Eardley (1977) worked out in detail the effects of dipole gravitational radiation in the bimetric theory of Rosen (1973), and, when the first observation of the decrease of the orbital period was announced in 1979, the Rosen theory suffered a terminal blow. A wide class of alternative theories also fail the binary pulsar test because of dipole gravitational radiation (TEGP 12.3).

The early observations of PSR 1913+16 already indicated that, in GR, the masses of the two bodies were nearly equal, so that, in theories of gravity that are in some sense "close" to GR, dipole gravitational radiation would not be a strong effect, because of the apparent symmetry of the system. The Rosen theory, and others like it, are not "close" to general relativity, except in their predictions for the weak-field, slow-motion regime of the Solar System. When relativistic neutron stars are present, theories like these can predict strong effects on the motion of the bodies resulting from their internal highly relativistic gravitational structure (violations of SEP). As a consequence, the masses inferred from observations of the periastron shift and γ' may be significantly different from those inferred using general relativity, and may be different from each other, leading to strong dipole gravitational radiation damping. By contrast, the Brans–Dicke theory is "close" to GR, roughly speaking within $1/\omega_{BD}$ of the predictions of the latter, for large values of the coupling constant ω_{BD}. Thus, despite the presence of dipole gravitational radiation, the binary pulsar provides at present only a weak test of Brans–Dicke theory, not yet competitive with Solar System tests.

3.4 Binary pulsars and scalar–tensor gravity

Making the usual assumption that both members of the system are neutron stars, and using the methods summarized in TEGP Chapters 10–12, one can obtain formulae for the periastron shift, the gravitational redshift/second-order Doppler shift parameter, and the rate of change of orbital period, analogous to eqn (17), in scalar-tensor theories of gravity. Here we focus on the Brans–Dicke version. These formulae depend on the masses of the two neutron stars, on their self-gravitational binding energy, represented by "sensitivities" s and κ^*, and on the Brans–Dicke coupling constant ω_{BD}. First, there is a modification of Kepler's third law, given by

$$2\pi f_b = (\mathcal{G}m/a^3)^{1/2} \quad (22)$$

Then, the predictions for $\langle\dot{\omega}\rangle$, γ', and $\dot{P}_b$ are

$$\langle\dot{\omega}\rangle = 6\pi f_b(2\pi m f_b)^{2/3}(1-e^2)^{-1}\mathcal{P}\mathcal{G}^{-4/3} \quad (23)$$

$$\gamma' = e(2\pi f_b)^{-1}(2\pi m f_b)^{2/3}(m_2/m)\mathcal{G}^{-1/3}(\alpha_2^* + \mathcal{G} m_2/m + \kappa_1^*\eta_2^*) \quad (24)$$

$$\dot{P}_b = -(192\pi/5)(2\pi\mathcal{M}f_b)^{5/3}F'(e) - 4\pi(2\pi\mu f_b)\xi S^2 G(e) \quad (25)$$

where $\mathcal{M} \equiv \chi^{3/5}\mathcal{G}^{-4/5}\eta^{3/5}m$, and, to first order in $\xi \equiv (2+\omega_{BD})^{-1}$, we have

$$F'(e) = F(e) + \tfrac{5}{144}\xi(\Gamma + 3\Gamma')^2(\tfrac{1}{2}e^2 + \tfrac{1}{8}e^4)(1-e^2)^{-7/2}$$

$$G(e) = (1-e^2)^{-5/2}(1+\tfrac{1}{2}e^2)$$

$$S = s_1 - s_2$$

$$\mathcal{G} = 1 - \xi(s_1 + s_2 - 2s_1 s_2)$$

$$\mathcal{P} = \mathcal{G}\left[1 - \tfrac{2}{3}\xi + \tfrac{1}{3}\xi(s_1 + s_2 - 2s_1 s_2)\right] \quad (26)$$

$$\alpha_2^* = 1 - \xi s_2, \quad \eta_2^* = (1-2s_2)\xi$$

$$\chi = \mathcal{G}^2\left[1 - \tfrac{1}{2}\xi + \tfrac{1}{12}\xi\Gamma^2\right]$$

$$\Gamma = 1 - 2(m_1 s_2 + m_2 s_1)/m, \quad \Gamma' = 1 - s_1 - s_2$$

The quantities s_a and κ_a^* are defined by

$$s_a = -\left(\frac{\partial(\ln m_a)}{\partial(\ln G)}\right)_N, \quad \kappa_a^* = -\left(\frac{\partial(\ln I_a)}{\partial(\ln G)}\right)_N \quad (27)$$

and measure the "sensitivity" of the mass m_a and moment of inertia I_a of each body to changes in the scalar field (reflected in changes in G) for a fixed baryon number N (see TEGP 11, 12 and 14.6(c) for further details). The quantity s_a is related to the gravitational binding energy. Notice how the violation of SEP in Brans–Dicke theory introduces complex structure-dependent effects in everything from the Newtonian limit (modification of the effective coupling constant in Kepler's third law) to gravitational radiation. In the limit $\xi \to 0$, we recover GR, and all structure dependence disappears. The first term in $\dot{P}_b$ (eqn (25)) is the effect of quadrupole and monopole gravitational radiation, while the second term is the effect of dipole radiation.

In order to estimate the sensitivities s_a and κ_a^*, one must adopt an equation of state for the neutron stars. It is sufficient to restrict attention to relatively stiff neutron star equations of state in order to guarantee neutron stars of sufficient mass, approximately $1.4 M_\odot$. The lower limit on ω_{BD} required to give consistency among the constraints on $\langle\dot{\omega}\rangle$, γ, and $\dot{P}_b$ as in Figure 2 is several hundred (Will and Zaglauer 1989). The combination of $\langle\dot{\omega}\rangle$ and γ give a constraint on the masses that is relatively weakly dependent on ξ, thus the constraint on ξ is dominated by $\dot{P}_b$ and is directly proportional to the measurement error in $\dot{P}_b$; in order to achieve a constraint comparable to the Solar System value of 3×10^{-4}, the error in $\dot{P}_b^{OBS}$ would have to be reduced by more than a factor of 10.

Alternatively, a binary pulsar system with dissimilar objects, such as a white dwarf or black hole companion, would provide potentially more promising tests of dipole radiation. Unfortunately, none has been discovered to date; the dissimilar system B0655+64, with a white dwarf companion is in a highly circular orbit, making measurement of the periastron shift meaningless, and is not as relativistic as 1913+16. From the upper limit on $\dot{P}_b$ (Table 2), one can infer at best the weak bound $\omega_{BD} > 100$.

Damour and Esposito-Farèse (1992) have generalized these results to a broad class of scalar–tensor theories. These theories are characterized by a single function $\alpha(\Phi)$ of the scalar field Φ, which mediates the coupling strength of the scalar field. For application to the Solar System or to binary systems, one expands this function about a cosmological background field value Φ_0:

$$\alpha(\Phi) = \alpha_0(\Phi - \Phi_0) + \tfrac{1}{2}\beta_0(\Phi - \Phi_0)^2 + \cdots \quad (28)$$

A purely linear coupling function produces Brans–Dicke theory, with $\alpha_0^2 = 1/(2\omega_{BD} + 3)$. The function $\alpha(\Phi)$ acts as a potential function for the scalar field Φ, and, if $\beta_0 > 0$, during cosmological evolution, the scalar field naturally evolves toward the minimum of the potential – that is, toward $\alpha_0 \approx 0$, $\omega_{BD} \to \infty$, or toward a theory close to,

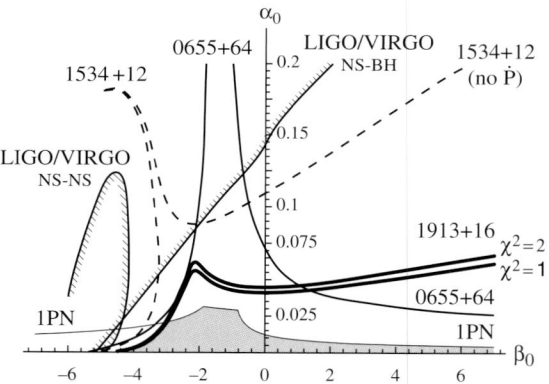

Figure 4 Region of the scalar–tensor theory α_0–β_0 plane allowed by Solar System, binary pulsar, and future gravitational-wave observations. A polytropic equation of state for the neutron stars was assumed. The shaded region is that allowed by all tests. For positive values of β_0, Solar System bounds (labelled 1PN) still are the best. (From Damour and Esposito-Farèse (1998), ©1998 by the American Physical Society, reproduced by permission.)

though not precisely GR (see Damour and Nordtvedt (1993) for further discussion of this cosmological "attractor"). Bounds on the parameters α_0 and β_0 from Solar–System, binary pulsar, and gravitational-wave observations (see Section 4.3) are shown in Figure 4 (Damour and Esposito-Farèse 1998). Negative values of β_0 correspond to an unstable scalar potential; in this case, objects such as neutron stars can experience a "spontaneous scalarization," whereby the interior values of Φ can take on values very different from the exterior values, through nonlinear interactions between strong gravity and the scalar field, dramatically affecting the stars' internal structure and the consequent violations of SEP. On the other hand, $\beta_0 < 0$ is of little practical interest, because, with an unstable Φ potential, cosmological evolution would presumably drive the system away from the peak where $\alpha_0 \approx 0$, toward parameter values that could easily be excluded by Solar System experiments. On the α_0–β_0 plane shown in Figure 4, the α_0 axis corresponds to pure Brans–Dicke theory, while the origin corresponds to pure GR. As discussed above, Solar System bounds still beat the binary pulsars. The bounds labelled "LIGO/VIRGO" are discussed in Section 4.3.

4 GRAVITATIONAL-RADIATION TESTS OF GRAVITATIONAL THEORY

4.1 Gravitational-wave observatories

Some time in this decade, a new opportunity for testing relativistic gravity will be realized, with the commissioning and operation of kilometer-scale, laser interferometric gravitational-wave observatories in the USA (LIGO project), Europe (VIRGO and GEO-600 projects), and Japan (TAMA 300 project). Gravitational-wave searches at these observatories are scheduled to commence around 2002. The LIGO broad-band antennae will have the capability of detecting and measuring the gravitational waveforms from astronomical sources in a frequency band between about 10 Hz (the seismic noise cutoff) and 500 Hz (the photon counting noise cutoff), with a maximum sensitivity to strain at around 100 Hz of $h \sim \Delta l/l \sim 10^{-22}$ (r.m.s.). The most promising source for detection and study of the gravitational-wave signal is the "inspiralling compact binary" – a binary system of neutron stars or black holes (or one of each) in the final minutes of a death dance leading to a violent merger. Such is the fate, for example, of the Hulse–Taylor binary pulsar PSR 1913 + 16 in about 300 million years. Given the expected sensitivity of the "advanced LIGO" (around 2007), which could see such sources out to hundreds of megaparsecs, it has been estimated that from 3 to 100 annual inspiral events could be detectable. Other sources, such as supernova core collapse events, instabilities in rapidly rotating nascent neutron stars, signals from non-axisymmetric pulsars, and a stochastic background of waves, may be detectable (for reviews, see Abramovici et al. (1992) and Thorne (1995); for updates on the status of various projects, see Fritschel (1998) and Brillet (1998)).

A similar network of cryogenic resonant-mass gravitational antennae have been in operation for many years, albeit at lower levels of sensitivity ($h \sim 10^{-19}$). While modest improvements in sensitivity may be expected in the future, these resonant detectors are not expected to be competitive with the large interferometers, unless new designs involving bars of spherical, or nearly spherical shape come to fruition. These systems are primarily sensitive to waves in relatively narrow bands about frequencies in the hundreds to thousands of hertz range (Pallottino 1998, Hamilton 1998, Blair et al. 1998, Prodi et al. 1998).

In addition, plans are being developed for an orbiting laser interferometer space antenna (LISA). Such a system, consisting of three spacecraft separated by 5×10^6 km, would be sensitive primarily in the very low frequency band between 10^{-4} and 10^{-1} Hz, with peak strain sensitivity of order $h \sim 10^{-23}$ (Danzmann 1997).

In addition to opening an altogether new astronomical window, the detailed observation of gravitational waves by such observatories may provide the means to test general relativistic predictions for the polarization and speed of the waves, and for gravitational radiation damping.

4.2 Polarization of gravitational waves

A laser interferometric or resonant bar gravitational-wave detector measures the local components of a symmetric 3×3 tensor which is composed of the "electric" components of the Riemann curvature tensor, R_{0i0j}, via the equation of geodesic deviation, given for a pair of freely falling particles by $\ddot{x}^i = -R_{0i0j}x^j$, where x^i denotes the spatial separation. In general there are six independent components, which can be expressed in terms of polarizations (modes with specific transformation properties under rotations and boosts). Three are transverse to the direction of propagation, with two representing quadrupolar deformations and one representing a monopole "breathing" deformation. Three modes are longitudinal, with one an axially symmetric stretching mode in the propagation direction, and one quadrupolar mode in each of the two orthogonal planes containing the propagation direction. Figure 5 shows the displacements induced on a ring of freely falling test particles by each of these modes. General relativity predicts only the first two transverse quadrupolar modes (a) and (b), independently of the source; these correspond to the waveforms h_+ and $h_\times$ discussed earlier (note the cos 2ϕ and

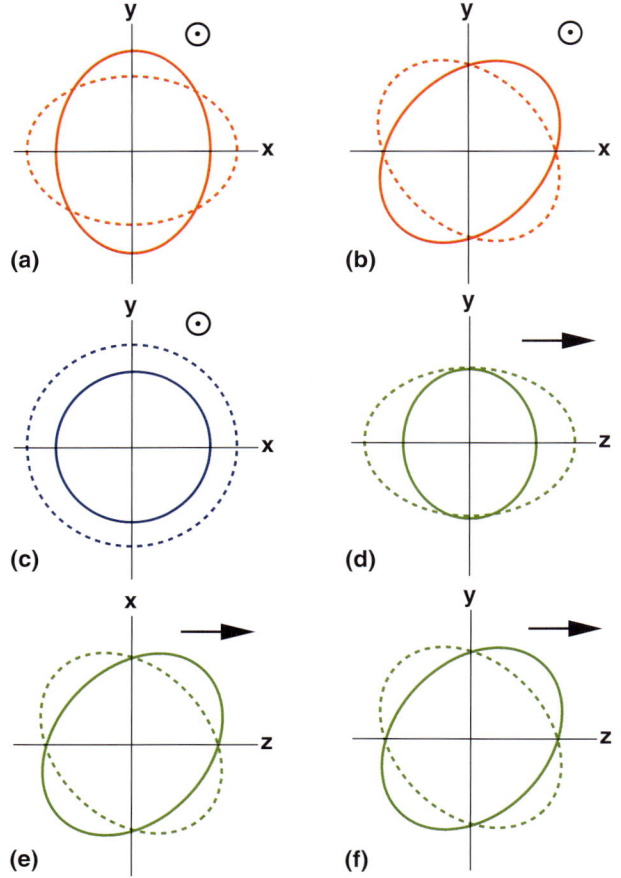

Figure 5 Six polarization modes for gravitational waves permitted in any metric theory of gravity. Shown is the displacement that each mode induces on a ring of test particles. The wave propagates in the +z-direction. There is no displacement out of the plane of the picture. In (a), (b) and (c), the wave propagates out of the plane; in (d), (e), and (f), the wave propagates in the plane. In general relativity, only (a) and (b) are present; in scalar–tensor gravity, (c) may also be present.

$\sin 2\phi$ dependences of the displacements). Scalar–tensor gravitational waves can in addition contain the transverse breathing mode (c). More general metric theories predict additional longitudinal modes, up to the full complement of six (TEGP 10.2).

A suitable array of gravitational antennae could delineate or limit the number of modes present in a given wave. The strategy depends on whether or not the source direction is known. In general there are eight unknowns (six polarizations and two direction cosines), but only six measurables (R_{0i0j}). If the direction can be established by either association of the waves with optical or other observations, or by time-of-flight measurements between separated detectors, then six suitably oriented detectors suffice to determine all six components. If the direction cannot be established, then the system is underdetermined, and no unique solution can be found. However, if one assumes that only transverse waves are present, then there are only three unknowns if the source direction is known, or five unknowns otherwise. Then the corresponding number (three or five) of detectors can determine the polarization. If distinct evidence were found of any mode other than the two transverse quadrupolar modes of GR, the result would be disastrous for GR. On the other hand, the absence of a breathing mode would not necessarily rule out scalar–tensor gravity, because the strength of that mode depends on the nature of the source.

Some of the details of implementing such polarization observations have been worked out for arrays of resonant cylindrical, disk-shaped, spherical and truncated icosahedral detectors (TEGP 10.2; for recent reviews see Lobo (1998) and Wagoner (1997)); initial work has been done to assess whether the ground- or space-based laser interferometers (or combinations of the two types) could perform interesting polarization measurements (Wagoner and Kalligas 1997, Brunetti *et al.* 1999, Maggiore and Nicolis 1999, Gasperini 1999). Unfortunately for this purpose, the two LIGO observatories (in Washington and Louisiana states, respectively) have been constructed to have their respective arms as parallel as possible, apart from the curvature of the Earth; while this maximizes the joint sensitivity of the two detectors to gravitational waves, it minimizes their ability to detect two modes of polarization.

4.3 Gravitational radiation back-reaction

In the binary pulsar, a test of GR was made possible by measuring at least three relativistic effects that depended upon only two unknown masses. The evolution of the orbital phase under the damping effect of gravitational radiation played a crucial role. Another situation in which measurement of orbital phase can lead to tests of GR is that of the inspiralling compact binary system. The key differences are that here gravitational radiation itself is the detected signal, rather than radio pulses, and the phase evolution alone carries all the information. In the binary pulsar, the first derivative of the binary frequency, $\dot{f}_b$, was measured; here the full nonlinear variation of f_b as a function of time is measured.

Broadband laser interferometers are especially sensitive to the phase evolution of the gravitational waves, which carry the information about the orbital phase evolution. The analysis of gravitational-wave data from such sources will involve some form of matched filtering of the noisy detector output against an ensemble of theoretical "template" waveforms which depend on the intrinsic parameters of the inspiralling binary, such as the component masses, spins, and so on, and on its inspiral evolution. How accurate must a template be in order to "match" the waveform from a given source (where by a match we mean maximizing the cross-correlation or the signal-to-noise ratio)? In the total accumulated phase of the wave detected in the sensitive bandwidth, the template must match the signal to a fraction of a cycle. For two inspiralling neutron stars, around 16,000 cycles should be detected during the final few minutes of inspiral; this implies a phasing accuracy of 10^{-5} or better. Since $v \sim 1/10$ during the late inspiral, this means that correction terms in the phasing at the level of v^5 or higher are needed. More formal analyses confirm this intuition (Cutler et al. 1993a, Finn and Chernoff 1993, Cutler and Flanagan 1994, Poisson and Will 1995).

Because it is a slow-motion system ($v \sim 10^{-3}$), the binary pulsar is sensitive only to the lowest-order effects of gravitational radiation as predicted by the quadrupole formula. Nevertheless, the first correction terms of order v and v^2 to the quadrupole formula, were calculated as early as 1976 (Wagoner and Will 1976; see TEGP 10.3).

But for laser interferometric observations of gravitational waves, the bottom line is that, in order to measure the astrophysical parameters of the source and to test the properties of the gravitational waves, it is necessary to derive the gravitational waveform and the resulting radiation back-reaction on the orbit phasing at least to 2PN order beyond the quadrupole approximation, and probably to 3PN order.

For the special case of non-spinning bodies moving on quasi-circular orbits (i.e., circular apart from a slow inspiral), the evolution of the gravitational wave frequency $f = 2f_b$ through 2PN order has the form

$$\dot{f} = \frac{96\pi}{5} f^2 (\pi \mathcal{M} f)^{5/3} \left[1 - \left(\frac{743}{336} + \frac{11}{4} \eta \right) (\pi m f)^{2/3} + 4\pi (\pi m f) + \left(\frac{34103}{18144} + \frac{13661}{2016} \eta + \frac{59}{18} \eta^2 \right) (\pi m f)^{4/3} + O[(\pi m f)^{5/3}] \right] \quad (29)$$

where $\eta = m_1 m_2 / m^2$. The first term is the quadrupole contribution (compare eqn (13)), the second term is the 1PN contribution, the third term, with the coefficient 4π, is the "tail" contribution, and the fourth term is the 2PN contribution first reported jointly by Blanchet et al. (1995a,b) and Will and Wiseman (1996). Calculation of the higher-order contributions is nearing completion.

Similar expressions can be derived for the loss of angular momentum and linear momentum. (For explicit formulas for non-circular orbits, see Gopakumar and Iyer (1997). These losses react back on the orbit to circularize it and cause it to inspiral. The result is that the orbital phase (and consequently the gravitational-wave phase) evolves nonlinearly with time. It is the sensitivity of the broad-band LIGO and VIRGO-type detectors to phase that makes the higher-order contributions to df/dt so observationally relevant. A ready-to-use set of formulae for the 2PN gravitational waveform template, including the nonlinear evolution of the gravitational-wave frequency (not including spin effects) have been published (Blanchet et al. 1996) and incorporated into the Gravitational Radiation Analysis and Simulation Package (GRASP), a publicly available software toolkit.*

If the coefficients of each of the powers of f in eqn (29) can be measured, then one again obtains more than two constraints on the two unknowns m_1 and m_2, leading to the possibility to test GR. For example, Blanchet and Sathyaprakash (1995) have shown that, by observing a source with a sufficiently strong signal, an interesting test of the 4π coefficient of the "tail" term could be performed.

Another possibility involves gravitational waves from a small mass orbiting and inspiralling into a (possibly supermassive) spinning black hole. A general non-circular, non-equatorial orbit will precess around the hole, both in periastron and in orbital plane, leading to a complex gravitational waveform that carries information about the non-spherical, strong-field spacetime around the hole. According to GR, this spacetime must be the Kerr spacetime of a rotating black hole, uniquely specified by its mass and angular momentum, and consequently, observation of the waves could test this fundamental hypothesis of GR (Ryan 1995, Poisson 1996).

Thirdly, the dipole gravitational radiation predicted by scalar–tensor theories will result in a modification of the gravitational-radiation back-reaction, and thereby of the phase evolution. Including only the leading quadrupole and dipole contributions, one obtains, in Brans–Dicke theory,

$$\dot{f} = \frac{96\pi}{5} f^2 (\pi \mathcal{M} f)^{5/3} [1 + b(\pi m f)^{-2/3}] \quad (30)$$

where $\mathcal{M} = (\chi^{3/5} \mathcal{G}^{-4/5}) \eta^{3/5} m$, and b is the coefficient of the dipole term, given by $b = (5/48)(\chi^{-1} \mathcal{G}^{4/3}) \xi \mathcal{S}^2$, where $\chi, \mathcal{G}, \mathcal{S}$ are given by eqns (26), and $\xi = 1/(2+\omega_{BD})$.

The effects are strongest for systems involving a neutron star and a black hole. Double neutron star systems are less promising because the small range of masses available near $1.4 M_\odot$ results in suppression of dipole radiation by symmetry (the sensitivity s turns out to be a relatively weak function of mass near $1.4 M_\odot$, for typical equations of state). For black holes, $s = 0.5$ identically, and so double black hole systems turn out to be observationally identical in the two theories.

---

*http://www.ligo.caltech.edu/LIGO_web/Collaboration/manual.pdf

But for a $1.4 M_\odot$ neutron star and a $10 M_\odot$ ($3 M_\odot$) black hole at 200 Mpc, the bound on ω_{BD} could be 600 (1800) (using advanced LIGO noise curves). The bound increases linearly with signal-to-noise ratio (Will 1994). If one demands that this test be performed annually, thus requiring observation of frequent, and therefore more distant, weaker sources, the bounds on ω_{BD} will be too weak to compete with existing Solar System bounds (this corresponds to the "LIGO/VIRGO" bound on the α_0 axis in Figure 4, which assumes a signal-to-noise ratio of 10). However, if one is prepared to wait 10 years for the lucky observation of a nearby, strong source, the resulting bound could exceed the current Solar System bound. The bounds are illustrated in Figure 6 by the curves marked $N = 1$ and $N = 1/10$. Figure 6 assumes a double neutron star inspiral rate of 10^{-6} per year per galaxy, and a black hole–neutron star rate β times that, where β is highly uncertain.

4.4 Speed of gravitational waves

According to GR, in the limit in which the wavelength of gravitational waves is small compared to the radius of curvature of the background spacetime, the waves propagate along null geodesics of the background spacetime – that is, they have the same speed, c, as light (henceforth, we do not set $c = 1$). In other theories, the speed could differ from c because of coupling of gravitation to "background" gravitational fields. For example, in the Rosen bimetric theory with a flat background metric η, gravitational waves follow null geodesics of η, while light follows null geodesics of $\mathbf{g}$ (TEGP 10.1).

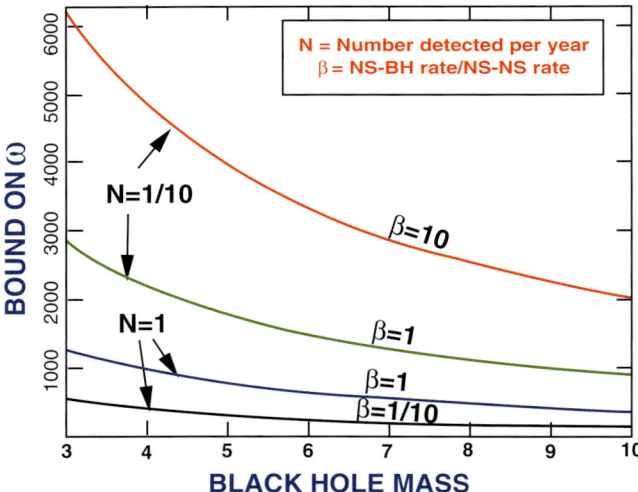

Figure 6 Bounds on the scalar–tensor coupling constant ω_{BD} from gravitational-wave observations of inspiralling black hole–neutron star systems. The Solar System bound is around $\omega_{BD} = 3000$.

Another way in which the speed of gravitational waves could differ from c is if gravitation were propagated by a massive field (a massive graviton), in which case, v_g would be given, in a local inertial frame, by

$$\frac{v_g^2}{c^2} = 1 - \frac{m_g^2 c^4}{E^2} \qquad (31)$$

where m_g and E are the graviton rest mass and energy, respectively. For a recent review of the idea of a massive graviton see Visser (1998).

The most obvious way to test this is to compare the arrival times of a gravitational wave and an electromagnetic wave from the same event, such as a supernova. For a source at a distance D, the resulting value of the difference $1 - v_g/c$ is

$$1 - \frac{v_g}{c} = 5 \times 10^{-17} \left(\frac{200 \text{ Mpc}}{D}\right)\left(\frac{\Delta t}{1 \text{ s}}\right) \qquad (32)$$

where $\Delta t \equiv \Delta t_a - (1 + Z)\Delta t_e$ is the "time difference", where Δt_a and Δt_e are the differences in arrival time and emission time, respectively, of the two signals, and Z is the redshift of the source. In many cases, Δt_e is unknown, so that the best one can do is employ an upper bound on Δt_e based on observation or modelling. The result will then be a bound on $1 - v_g/c$.

For a massive graviton, if the frequency of the gravitational waves is such that $hf \gg m_g c^2$, where h is the Planck constant, then $v_g/c \approx 1 - (1/2)(c/\lambda_g f)^2$, where $\lambda_g = h/(m_g c)$ is the graviton Compton wavelength, and the bound on $1 - v_g/c$ can be converted to a bound on λ_g, given by

$$\lambda_g > 3 \times 10^{12} \text{ km} \left(\frac{D}{200 \text{ Mpc}} \frac{100 \text{ Hz}}{f}\right)^{1/2} \left(\frac{1}{f \Delta t}\right)^{1/2} \qquad (33)$$

The foregoing discussion assumes that the source emits *both* gravitational and electromagnetic radiation in detectable amounts, and that the relative time of emission can be established to sufficient accuracy, or can be shown to be sufficiently small.

However, there is a situation in which a bound on the graviton mass can be set using gravitational radiation alone (Will 1998). That is the case of the inspiralling compact binary. Because the frequency of the gravitational radiation sweeps from low frequency at the initial moment of observation to higher frequency at the final moment, the speed of the gravitons emitted will vary, from lower speeds initially to higher speeds (closer to c) at the end. This will cause a distortion of the observed phasing of the waves and result in a shorter than expected overall time Δt_a of passage of a given number of cycles. Furthermore, through the technique of matched filtering, the parameters of the compact binary can be measured accurately (assuming that GR is a good approximation to the orbital evolution, even in the presence of a massive graviton), and thereby the emission time Δt_e can be determined accurately. Roughly speaking, the "phase interval" $f\Delta t$ in eqn (33) can be measured to an accuracy $1/\rho$, where ρ is the signal-to-noise ratio.

Thus one can estimate the bounds on λ_g achievable for various compact inspiral systems, and for various detectors. For stellar-mass inspiral (neutron stars or black holes) observed by the LIGO/VIRGO class of ground-based interferometers, $D \approx 200$ Mpc, $f \approx 100$ Hz, and $f\Delta t \sim \rho^{-1} \approx 1/10$. The result is $\lambda_g > 10^{13}$ km. For supermassive binary black holes ($10^4 - 10^7 M_\odot$) observed by the proposed LISA, $D \approx 3$ Gpc, $f \approx 10^{-3}$ Hz, and $f\Delta t \sim \rho^{-1} \approx 1/1000$. The result is $\lambda_g > 10^{17}$ km.

A full noise analysis using proposed noise curves for the advanced LIGO and for LISA weakens these crude bounds by factors between 2 and 10. These potential bounds can be compared with the solid bound $\lambda_g > 2.8 \times 10^{12}$ km, derived from Solar System dynamics, which limit the presence of a Yukawa modification of Newtonian gravity of the form $V(r) = (GM/r)\exp(-r/\lambda_g)$, (Talmadge et al. 1988) and with the model-dependent bound $\lambda_g > 6 \times 10^{19}$ km from consideration of galactic and cluster dynamics (Visser 1998).

5 OTHER TESTS

One of the central difficulties of testing general relativity in the strong-field regime is the possibility of contamination by uncertain or complex physics. In the Solar System, weak-field gravitational effects could in most cases be measured cleanly and separately from non-gravitational effects. The remarkable cleanliness of the binary pulsar permitted precise measurements of gravitational phenomena in a strong-field context.

Unfortunately, nature is rarely so kind. Still, under suitable conditions, qualitative and even quantitive strong-field tests of general relativity can be carried out.

One example is in cosmology. From a few seconds after the big bang until the present, the underlying physics of the universe is well understood, although significant uncertainties remain (amount of dark matter, value of the cosmological constant, the number of light neutrino families, etc.). Some alternative theories of gravity that are qualitatively different from GR fail to produce cosmologies that meet even the minimum requirements of agreeing qualitatively with big bang nucleosynthesis (BBN) or the properties of the cosmic microwave background (TEGP 13.2). Others, such as Brans–Dicke theory, are sufficiently close to GR (for large enough ω_{BD}) that they conform to all cosmological observations, given the underlying uncertainties. The generalized scalar–tensor theories, however, could have small ω_{BD} at early times, while evolving through the attractor mechanism to large ω_{BD} today. One way to test such theories is through BBN, since the abundances of the light elements produced when the temperature of the Universe was about 1 MeV are sensitive to the rate of expansion at that epoch, which in turn depends on the strength of interaction between geometry and the scalar field. Because the Universe is radiation-dominated at that epoch, uncertainties in the amount of cold dark matter or of the cosmological constant are unimportant. The nuclear reaction rates are reasonably well understood from laboratory experiments and theory, and the number of light neutrino families (three) conforms to evidence from particle accelerators. Thus, within modest uncertainties, one can assess the quantitive difference between the BBN predictions of GR and scalar–tensor gravity under strong-field conditions and compare with observations. The most sophisticated recent analysis (Damour and Pichon 1999) places bounds on the parameters α_0 and β_0 of the generalized framework of Damour and Esposito-Farèse (see Section 3.4 and Figure 4) that are weaker than Solar System bounds for $\beta_0 < 0.3$, but substantially stronger for $\beta_0 > 0.3$.

Another example is the exploration of the spacetime near black holes via accreting matter. Observations of low-luminosity binary X-ray sources suggest that a form of accretion known as advection-dominated accretion flow (ADAF) may be important. In this kind of flow, the accreting gas is too thin to radiate its energy efficiently, but instead transports (advects) it inward toward the central object. If the central object is a neutron star, the matter hits the surface and radiates the energy away; if it is a black hole, the matter and its advected energy disappear. Systems in which the accreting object is believed to be a black hole from estimates of its mass are indeed observed to be underluminous, compared to systems where the object is believed to be a neutron star. This has been regarded as the first astrophysical evidence for the existence of black hole event horizons (for a review see Menou et al. 1999). While supporting one of the critical strong-field predictions of GR, the observations and models are not likely any time soon to be able to distinguish one gravitational theory from another (except for theories that do not predict black holes at all).

Another example involving accretion purports to explore the strong-field region just outside massive black holes in active galactic nuclei. Here, iron in the inner region of a thin accretion disk is irradiated by X-ray-emitting material above or below the disk, and fluoresces in the $K\alpha$ line. The spectral shape of the line depends on relativistic Doppler and curved spacetime effects as the iron orbits the black hole near the innermost stable circular orbit, and could be used to determine whether the hole is a non-rotating Schwarzschild black hole, or a rotating Kerr black hole.

Because of uncertainties in the detailed models, the results are inconclusive to date, but the combination of higher-resolution observations and better modelling could lead to striking tests of strong-field predictions of GR (see, e.g., Reynolds et al. 1999).

6 CONCLUSIONS

At the end of the twentieth century, we learned that general relativity held up under extensive experimental scrutiny. The question then arises, why bother to continue to test it? One reason is that gravity is a fundamental interaction of nature, and as such requires the most solid empirical underpinning we can provide. Another is that all attempts to quantize gravity and to unify it with the other forces suggest that the standard general relativity of Einstein is not likely to be the last word. A third is that the theory makes important strong-field predictions of the nature of gravitational waves and black holes that have not been tested to date. Furthermore, the predictions of general relativity are fixed; the theory contains no adjustable constants so nothing can be changed. Thus every test of the theory is either a potentially deadly test or a possible probe for new physics. Although it is remarkable that this theory, born 85 years ago out of almost pure thought, has managed to survive every test, the possibility of finding a discrepancy will continue to drive experiments for years to come.

Acknowledgment

This work was supported in part by the National Science Foundation, Grant Number PHY 96-00049.

REFERENCES

Abramovici, A., Althouse, W.E., Drever, R.W.P., Gürsel, Y., Kawamura, S., Raab, F.J., Shoemaker, D., Siewers, L., Spero, R.E., Thorne, K.S., Vogt, R.E., Weiss, R., Whitcomb, S.E. and Zucker., M.E. (1992). LIGO: The laser interferometer gravitational-wave observatory. *Science*, **256**, 325–333.

Blair, D.G., Heng, I.S., Ivanov, E.N. and Tobar, M.E. (1998). Present status of the resonant-mass gravitational-wave antenna NIOBE. In E. Coccia, G. Pizzella and G. Veneziano (eds), *Second Edoardo Amaldi Conference on Gravitational Waves*, World Scientific, Singapore, pp. 127–147.

Blanchet, L. (1995). Second-post-Newtonian generation of gravitational radiation. *Physical Review D*, **51**, 2559–2583 (gr-qc/9501030).

Blanchet, L. (1996). Energy losses by gravitational radiation in inspiralling compact binaries to 5/2 post-Newtonian order. *Physical Review D*, **54**, 1417–1438 (gr-qc/9603048).

Blanchet, L. (1997). Gravitational radiation reaction and balance equations to post-Newtonian order. *Physical Review D*, **55**, 714–732 (gr-qc/9609049).

Blanchet, L. (1998). Quadrupole–quadrupole gravitational waves. *Classical and Quantum Gravity*, **15**, 113–141 (gr-qc/9710038).

Blanchet, L. and Damour, T. (1986). Radiative gravitational fields in general relativity: I. General structure of the field outside the source. *Philosophical Transactions of the Royal Society of London*, **A320**, 379–430.

Blanchet, L. and Damour, T. (1988). Tail-transported temporal correlations in the dynamics of a gravitating system. *Physical Review D*, **37**, 1410–1435.

Blanchet, L. and Damour, T. (1989). Post-Newtonian generation of gravitational waves. *Annales de l'Institut Henri Poincaré (Physique Théorique)*, **50**, 377–408.

Blanchet, L. and Damour, T. (1992). Hereditary effects in gravitational radiation. *Physical Review D*, **46**, 4304–4319.

Blanchet, L., Damour, T. and Iyer, B.R. (1995a). Gravitational waves from inspiralling compact binaries: Energy loss and waveform to second-post-Newtonian order. *Physical Review D*, **51**, 5360–5386 (gr-qc/9501029).

Blanchet, L., Damour, T., Iyer, B.R., Will, C.M. and Wiseman, A.G. (1995b). Gravitational-radiation damping of compact binary systems to second post-Newtonian order. *Physical Review Letters*, **74**, 3515–3518 (gr-qc/9501027).

Blanchet, L., Faye, G. and Ponsot, B. (1998). Gravitational field and equations of motion of compact binaries to 5/2 post-Newtonian order. *Physical Review D*, **58**, 124002 (gr-qc/9804079).

Blanchet, L., Iyer, B.R., Will, C.M. and Wiseman, A.G. (1996). Gravitational waveforms from inspiralling compact binaries to second post-Newtonian order. *Classical and Quantum Gravity*, **13**, 575–584 (gr-qc/9602024).

Blanchet, L. and Sathyaprakash, B.S. (1995). Detecting a tail effect in gravitational-wave experiments. *Physical Review Letters*, **74**, 1067–1070.

Brillet, A. (1998). VIRGO–Status report, November 1997. In E. Coccia, G. Pizzella and G. Veneziano (eds), *Second Edoardo Amaldi Conference on Gravitational Waves*, World Scientific, Singapore, pp. 86–96.

Brunetti, M., Coccia, E., Fafone, V. and Fucito, F. (1999). Gravitational-wave radiation from compact binary systems in the Jordan–Brans–Dicke theory. *Physical Review D*, **59**, 044027 (gr-qc/9805056).

Cutler, C., Apostolatos, T.A., Bildsten, L., Finn, L.S., Flanagan, É.E., Kennefick, D., Marković, D.M., Ori, A., Poisson, E., Sussman, G.J. and Thorne, K.S. (1993a). The last three minutes: Issues in gravitational wave measurements of coalescing compact binaries. *Physical Review Letters*, **70**, 2984–2987 (gr-qc/9208005).

Cutler, C., Finn, L.S., Poisson, E. and Sussman, G.J. (1993b). Gravitational radiation from a particle in circular orbit around a black hole: II. Numerical results for the nonrotating case. *Physical Review D*, **47**, 1511–1518.

Cutler, C. and Flanagan, É.E. (1994). Gravitational waves from merging compact binaries: How accurately can one extract the binary's parameters from the inspiral waveform? *Physical Review D*, **49**, 2658–2697 (gr-qc/9402014).

Damour, T. (1982). Problème des deux corps et freinage de rayonnement en relativité général. *Comptes Rendues de l'Académie des Sciences, Ser. II*, **294**, 1355–1357.

Damour, T. (1987). The problem of motion in Newtonian and Einsteinian gravity. In S.W. Hawking and W. Israel (eds), *300 Years of Gravitation*, Cambridge University Press, Cambridge, pp. 128–198.

Damour, T. and Deruelle, N. (1981). *Physics Letters*, **87A**, 81–84.

Damour, T. and Esposito-Farèse, G. (1992). Tensor–multi-scalar theories of gravitation. *Classical and Quantum Gravity*, **9**, 2093–2176.

Damour, T. and Esposito-Farèse, G. (1998). Gravitational-wave versus binary-pulsar tests of strong-field gravity. *Physical Review D*, **58**, 042001 (gr-qc/9803031).

Damour, T. and Iyer, B.R. (1991). Post-Newtonian generation of gravitational waves: II. The spin moments. *Annales de l'Institut Henri Poincaré (Physique Théorique)*, **54**, 115–164.

Damour, T. and Nordtvedt, K. (1993). Tensor–scalar cosmological models and their relaxation toward general relativity. *Physical Review D*, **48**, 3436–3450.

Damour, T. and Pichon, B. (1999). Big bang nucleosynthesis and tensor–scalar gravity. *Physical Review D*, **59**, 123502 (gr-qc/9807176).

Damour, T. and Taylor, J.H. (1992). Strong-field tests of relativistic gravity and binary pulsars. *Physical Review D*, **45**, 1840–1868.

Danzmann, K. (1997). Lisa – An ESA cornerstone mission for a gravitational-wave observatory. *Classical and Quantum Gravity*, **14**, 1399–1404.

Finn, L.S. and Chernoff, D.F. (1993). Observing binary inspiral in gravitational radiation: One interferometer. *Physical Review D*, **47**, 2198–2219 (gr-qc/9301003).

Fritschel, P. (1998). The LIGO project: Progress and plans. In E. Coccia, G. Pizzella and G. Veneziano (eds), *Second Edoardo Amaldi Conference on Gravitational Waves*. World Scientific, Singapore, pp. 74–85.

Gasperini, M. (1999). On the response of gravitational antennas to dilatonic waves. *Physics Letters B*, **470**, 67–72.

Gopakumar, A. and Iyer, B.R. (1997). Gravitational waves from inspiraling compact binaries: Angular momentum flux, evolution of the orbital elements and the waveform to the second post-Newtonian order. *Physical Review D*, **56**, 7708–7731 (gr-qc/9710075).

Grishchuk, L.P. and Kopeikin, S.M. (1986). Equations of motion for isolated bodies with relativistic corrections including the radiation-reaction force. In J. Kovalevsky and V.A. Brumberg (eds), *Relativity in Celestial Mechanics and Astrometry*, Reidel, Dordrecht; pp. 19–34.

Hamilton, W.O. (1998). The ALLEGRO detector and the future of resonant detectors in the USA. In E. Coccia, G. Pizzella and G. Veneziano (eds), *Second Edoardo Amaldi Conference on Gravitational Waves*, World Scientific, Singapore, pp. 115–126.

Hulse, R.A. (1994). Nobel lecture: The discovery of the binary pulsar. *Reviews of Modern Physics*, **66**, 699.

Iyer, B.R. and Will, C.M. (1993). Post-Newtonian gravitational radiation reaction for two-body systems. *Physical Review Letters*, **70**, 113–116.

Iyer, B.R. and Will, C.M. (1995). Post-Newtonian gravitational radiation reaction for two-body systems: Non-spinning bodies. *Physical Review D*, **52**, 6882–6893.

Kidder, L.E., Will, C.M. and Wiseman, A.G. (1993). Spin effects in the inspiral of coalescing compact binaries. *Physical Review D*, **47**, R4183–4187 (gr-qc/9211025).

Kidder, L.E. (1995). Coalescing binary systems of compact objects to (post)$^{5/2}$-Newtonian order: V. Spin effects. *Physical Review D*, **52**, 821–847 (gr-qc/9506022).

Kramer, M. (1998). Determination of the geometry of the PSR B1913+16 system by geodetic precession. *Astrophysical Journal*, **509**, 856–860 (astro-ph/9808127).

Lobo, J.A. (1998). Spherical GW detectors and geometry. In E. Coccia, G. Pizzella and G. Veneziano, (eds), *Second Edoardo Amaldi Conference on Gravitational Waves*, World Scientific, Singapore, pp. 168–179.

Maggiore, M. and Nicolis, A. (2000). Detection strategies for scalar gravitational waves with interferometers and resonant spheres. *Physical Review D*, **62**, 15 pp. (024004).

Marck, J.-A. and Lasota, J.-P. (eds) (1997). *Relativistic Gravitation and Gravitational Radiation*, Cambridge University Press, Cambridge.

Menou, K., Quataert, E. and Narayan, R. (1999). Astrophysical evidence for black-hole event horizons. In T. Piran and R. Ruffini (eds), *Proceedings of the Eighth Marcel Grossmann Meeting on General Relativity*, World Scientific, Singapore, pp. 204–224.

Mino, Y., Sasaki, M., Shibata, M., Tagoshi, H. and Tanaka, T. (1997). Black hole perturbation. *Progress in Theoretical Physics*, Suppl., **128**, 1–121.

Misner, C.W., Thorne, K.S. and Wheeler, J.A. (1973). *Gravitation*, Freeman, San Francisco.

Pallottino, G.V. (1998). The resonant-mass detectors of the Rome group. In E. Coccia, G. Pizzella and G. Veneziano (eds), *Second Edoardo Amaldi Conference on Gravitational Waves*, World Scientific, Singapore, pp. 105–114.

Pati, M.E. and Will, C.M. (2000). Post-Newtonian gravitational radiation and equations of motion via direct integration of the relaxed Einstein equations. *Physical Review D*, **62**, 28 pp. (124015).

Poisson, E. (1993). Gravitational radiation from a particle in circular orbit around a black hole: I. Analytic results for the nonrotating case. *Physical Review D*, **47**, 1497–1510.

Poisson, E. (1996). Measuring black-hole parameters and testing general relativity using gravitational-wave data from space-based interferometers. *Physical Review D*, **54**, 5939–5953 (gr-qc/9606024).

Poisson, E. and Will, C.M. (1995). Gravitational waves from inspiralling compact binaries: Parameter estimation using second-post-Newtonian waveforms. *Physical Review D*, **52**, 848–855 (gr-qc/9502040).

Prodi, G.A., Conti, L., Mezzena, R., Vitale, S., Taffarello, L., Zendri, J.P., Baggio, L., Cerdonio, M., Colombo, A., Crivelli Visconti, V., Macchietto, R., Falferi, P., Bonaldi, M., Ortolan, A., Vedovato, G., Cavallini, E. and Fortini, P. (1998). Initial operation of the gravitational-wave detector AURIGA. In E. Coccia, G. Pizzella and G. Veneziano (eds), *Second Edoardo Amaldi Conference on Gravitational Waves*, World Scientific, Singapore, pp. 148–158.

Reynolds, C.S., Young, A.J., Begelman, M.C. and Fabian, A.C. (1999). X-ray iron line reverberation from black-hole accretion disks. *Astrophysical Journal*, **514**, 164–179 (astro-ph/9806327).

Rosen, N. (1973). A bi-metric theory of gravitation. *General Relativity and Gravitation*, **4**, 435–477.

Ryan, F.D. (1995). Gravitational waves from the inspiral of a compact object into a massive, axisymmetric body with arbitrary multipole moments. *Physical Review D*, **52**, 5707–5718.

Stairs, I.H., Arzoumanian, Z., Camilio, F., Lyne, A.G., Nice, D.J., Taylor, J.H., Thorsett, S.E. and Wolszczan, A. (1998). Relativistic orbital decay in PSR B1534+12. *Astrophysical Journal*, **505**, 352–357 (astro-ph/9712296).

Stairs, I.H., Nice, D.J., Thorsett, S.E. and Taylor, J.H. (2000). Recent Arecibo timing of the relativistic binary PSR B1534+12. In *Gravitational Waves and Experimental Gravity: XXXIV Rencontres de Moriond* (astro-ph/9903289).

Talmadge, C., Berthias, J.-P., Hellings, R.W. and Standish, E.M. (1988). Model-independent constraints on possible modifications of Newtonian gravity. *Physical Review Letters*, **61**, 1159–1162.

Taylor, J.H. (1994). Nobel Lecture: Binary pulsars and relativistic gravity. *Reviews of Modern Physics*, **66**, 711.

Thorne, K.S. (1987). Gravitational radiation. In S.W. Hawking and W. Israel (eds), *300 Years of Gravitation*. Cambridge University Press, Cambridge; pp. 330–458.

Thorne, K.S. (1995). Gravitational waves. In E.W. Kolb and R. Peccei (eds), *Proceedings of the Snowmass 95 Summer Study on Particle and Nuclear Astrophysics and Cosmology*, World Scientific, Singapore, pp. 398 (gr-qc/9506086).

Visser, M. (1998). Mass for the graviton. *Journal of General and Relativity and Gravitation*, **30**, 1717 (gr-qc/9705051).

Wagoner, R.V. (1997). Resonant-mass detection of tensor and scalar waves. In J.-A. Marck and J.-P. Lasota (eds), *Relativistic Gravitation and Gravitational Radiation*, Cambridge University Press, Cambridge, pp. 419–432.

Wagoner, R.V. and Kalligas, D. (1997). Scalar–tensor theories and gravitational radiation. In J.A. Marck and J.-P. Lasota (eds), *Relativistic Gravitation and Gravitational Radiation*, University Press, Cambridge, pp. 433–446.

Wagoner, R.V. and Will, C.M. (1976). Post-Newtonian gravitational radiation from orbiting point masses. *Astrophysical Journal*, **210**, 764–775.

Weisberg, J.M. and Taylor, J.H. (2000). General relativistic precession of the spin axis of binary pulsar B1913+16: First two-dimensional maps of the emission beam. In M. Kramer, N. Wex and R. Wielebinski (eds), *Pulsar Astronomy – 2000 and Beyond*, p. 127.

Will, C.M. (1993a). *Was Einstein Right?* 2nd edn, Basic Books/Perseus Group, New York.

Will, C.M. (1993b). *Theory and Experiment in Gravitational Physics*, 2nd edn, Cambridge University Press, Cambridge.

Will, C.M. (1994). Testing scalar–tensor gravity with gravitational-wave observations of inspiralling compact binaries. *Physical Review D*, **50**, 6058–6067 (gr-qc/9406022)

Will, C.M. (1998). Bounding the mass of the graviton using gravitational-wave observations of inspiralling compact binaries. *Physical Review D*, **57**, 2061–2068 (gr-qc/9709011).

Will, C.M. and Eardley, D.M. (1977). Dipole gravitational radiation in Rosen's theory of gravity: Observable effects in the binary system PSR 1913+16. *Astrophysical Journal*, **212**, L91–94.

Will, C.M. and Wiseman, A.G. (1996). Gravitational radiation from compact binary systems: Gravitational waveforms and energy loss to second post-Newtonian order. *Physical Review D*, **54**, 4813–4848 (gr-qc/ 9608012).

Will, C.M. and Zaglauer, H.W. (1989). Gravitational radiation, close binary systems and the Brans–Dicke theory of gravity. *Astrophysical Journal*, **346**, 366–377.

Wiseman, A.G. (1992). Coalescing binary systems of compact objects to (post)$^{5/2}$-Newtonian order. II. Higher-order waveforms and radiation recoil. *Physical Review D*, **46**, 1517–1539.

Wolszczan, A. (1994). Binary pulsars and relativistic gravitation. *Classical and Quantum Gravity*, **11**, A227–242.

17

GUSTAV A. TAMMANN*

The cosmological constants

The evolution of the large scale of the Universe is but one chapter of cosmology, yet it is fundamental to all of its aspects. The understanding of the large-scale evolution has once been described as the "Search for two numbers" (Sandage 1970). The first of these numbers is the Hubble constant H_0 which measures the present expansion rate. The second number was thought to account for the gravitational deceleration of the Universe. But since evidence has been found for an additional acceleration, the original deceleration parameter has been split into a decelerating term Ω_m and an accelerating term Ω_Λ. Ω_m is a measure of the gravitating mean matter density of the Universe, while Ω_Λ corresponds to the energy density of Einstein's "cosmological constant" Λ.

The three quantities H_0, Ω_m, and Ω_Λ fix the expansion age T of the Universe and determine also whether the Universe will eventually recollapse or expand forever.

In principle the values of H_0, Ω_m, and Ω_Λ can be measured in the present Universe as well as the age t of the oldest objects which – as an essential test – must be smaller than T.

Yet the cosmological constants are elusive. Even the concerted effort of many of the largest optical and radio telescopes has not led to their satisfactory determination. As a consequence observations from space were added wherever possible. In fact the determination of the cosmological constants was a driving force for some astronomical satellites.

Several satellites will be mentioned in the following which have contributed to our dawning understanding of the cosmological parameters, but none has been as important as the Hubble Space Telescope (HST).

The initial motivation for astronomers to go into space were the wavelengths inaccessible from the ground. The reward was overwhelming with the detection of X-ray and γ-ray sources in the early 1970s (see Chapter 12) and the subsequent results in ultraviolet spectroscopy (see Chapter 13). But optical observations above the atmosphere were also of fundamental interest offering almost 10 times smaller seeing disks and consequently the detection of roughly 100 times fainter objects than from the ground. Individual stars in crowded regions would become accessible and much greater detail would be visible in extended objects, as first pointed out by Hermann Oberth in 1923. However, this enormous gain was offset in part by the size limitations of a telescope in space. Plans for what was to become the 2.4 m HST for imaging and spectroscopy in the optical and near UV developed during the 1960s and were finally approved by the US Congress in 1977. ESA joined the venture with a 15% share, which turned out to be of great benefit to astronomy in Europe. The start of HST, planned for 1983, was delayed by the Space Shuttle accident until 1990. Since then, and particularly since 1993, that is, after the repair mission for the defective mirror, HST has not only brought spectacular progress in astronomical research, but has also stirred an enormous public interest in astronomy.

When plans were made in 1979 (Macchetto et al. 1979; Longair and Warner 1979) for future research with HST, the author was ambivalent about its benefits for the determination of H_0, being convinced that the main uncertainties of the time were due to statistical problems (Malmquist bias; Section 1.3). He was wrong by not foreseeing that standard supernovae of type Ia are so powerful standard candles that they can beat all statistical selection effects, and that HST can provide Cepheid distances to their nearest representatives and thus calibrate their true luminosity. It was clear, however, already in 1979 that the same supernovae can be observed with HST at cosmologically significant redshifts and that this is the route to determine the cosmological parameters q_0 (or Ω_m) and Λ.

---

*Universität Basel, Switzerland

1 THE HUBBLE CONSTANT H_0

Since the Big Bang space has expanded. During the early stages the expansion was governed by complex processes, but after an inflationary period the Universe turned into a constant growth rate carrying with it all matter and the emerging galaxies. As a consequence any observer in the Universe has the impression that all galaxies are being carried away, and the faster they move the larger the distance. Yet strictly speaking the distant galaxies do not recede with any velocity, but space between the observer and the galaxy is stretched. In spite of this the spectra of nearby galaxies are interpreted as if they had recession velocities. This is permissible here because the determination of the *present* expansion rate, that is, the Hubble constant H_0, must be carried out at small distances where the look-back times are short.

If light travels through an expanding space its waves will be stretched and when it arrives at the observer it will appear redder than at the time of emission. Also the spectral lines of the galaxies will be shifted to the red. These lines can be ascribed to certain neutral or ionized elements and each appears in the laboratory at a fixed wavelength λ_0. If that particular line appears in the galaxy spectrum at wavelength λ one defines a redshift z of the galaxy as $z \equiv (\lambda - \lambda_0)/\lambda_0$. The redshift is determined by the stretch factor the Universe has experienced during the light travel time, that is, $z \equiv (R_0/R_e) - 1$, where R_0 is the present radius of the Universe (more exactly its radius of curvature) and R_e is the radius at the time of emission. As long as z is small, say $z \lesssim 0.2$, the redshift can also be interpreted with sufficient accuracy as a Doppler effect. The recession velocity v is in that case simply given by $v = cz$, where c is the velocity of light (in km s^{-1}).

To pin down the expansion rate of the Universe one introduces the Hubble constant H_0 which is defined as

$$H_0 \equiv \dot{R}_0/R_0 = v/r_0 \quad (v \ll c) \tag{1}$$

where v is the "recession velocity" of a galaxy and r_0 is its distance (in Mpc). Thus the units of H_0 are km s^{-1} Mpc^{-1} (the units are not repeated in the following), but its dimension is simply (time)$^{-1}$.

It should be noted that H_0 is constant in the sense that it is the same anywhere in the present Universe, but it has a different value at other times. This is obvious because the distance r is constantly increased by the expansion of the Universe and consequently the Hubble parameter H must decrease with time.

Equation (1) gives the false impression that the determination of the Hubble constant H_0 is simple. One just determines the redshift z of the spectrum of any given galaxy, and hence its recession velocity v, and divides the latter by the galaxy's distance. Indeed the determination of the redshift poses no problem. The crux, however, is introduced by peculiar motions which are superposed on the regular Hubble flow. They are caused by the gravitational pull of galaxies and clusters of galaxies and of any additional dark matter. For instance our Local Group of galaxies is a sufficient density enhancement to have turned its original expansion into a contraction. Our nearest neighbor, the Andromeda galaxy, M31, has thus a *blueshift* and approaches us at about 100 km s^{-1}. The largest peculiar velocity we partake of is a 630 km s^{-1} velocity which manifests itself as a dipole anisotropy of the cosmic microwave background (CMB, see below).

The peculiar velocities cause a dilemma for the determination of H_0. On the one hand one wants to determine it locally to find the *present* value H_0; on the other hand one must determine it at sufficiently large distances, where the recession velocities are significantly larger than any possible peculiar motions. The best compromise is to determine H_0 between recession velocities of 10 000 and 30 000 km s^{-1}. But the determination of the corresponding distances poses formidable problems.

With this requirement in mind a new strategy has developed to determine H_0. The idea is to map the Hubble flow out to the necessary distances by means of *relative* distances of a set of galaxies, for instance from "standard candles" of constant luminosity, and then to determine the true distances of one or a few nearby galaxies of the same set. The latter step has been greatly facilitated by HST. The route is described below.

1.1 The Hubble diagram of standard candles

The expansion of the Universe implies that the recession velocities v increase linearly with the photometric distance. Astronomers measure, in general, photometric distances in terms of the distance modulus $(m - M)$, which is defined as $(m - M) \equiv 5\log r_{\mathrm{Mpc}} + 25$, where m is the apparent magnitude of an object and M its absolute magnitude, that is, the apparent magnitude the object would have if seen from a distance of 10 pc. Objects with (nearly) constant absolute magnitude are called "standard candles." The apparent magnitudes m of standard candles are a measure of their *relative* distances, or more exactly of their log r values.

A plot of log v v. the apparent magnitude m of extragalactic standard candles constitutes the so-called "Hubble diagram." The standard candles concentrate in a linearly expanding Universe along a straight line of slope 0.2. Systematic deviations from this line beyond $\sim$30 000 km s^{-1} are caused by the deceleration or acceleration of the Universe (provided that the apparent magnitudes are corrected for the z-dependent photometric K-term).

Since Hubble and Humason (1931), the Hubble diagram has provided the fundamental evidence for the linear expansion of the Universe. The ever-increasing persuasive power of the Hubble diagram from Humason et al. (1956) to Sandage (1966) and Kristian et al. (1978), who used first-ranked cluster galaxies as standard candles, has shattered all attempts to interpret redshifts as a non-cosmological effect.

1.1.1 The Hubble diagram of supernovae of type Ia

Following an early suggestion of Kowal (1968) it has become increasingly clear that supernovae of type Ia (SNe Ia) are the most powerful standard candles to date. This can be physically rationalized because a SN Ia outburst is caused each time a white dwarf is pushed over its Chandrasekhar limit by a mass-shedding companion.

For 35 SNe Ia in the distance range $1200 < v \lesssim 30\,000$ km s^{-1} good B and V magnitudes at maximum light are known, for 29 of them also $I_{\max}$ magnitudes. The SNe Ia, listed by Parodi et al. (2000), are selected to have $B_{\max} - V_{\max} \lesssim 0.06$, after correction for galactic reddening to guard against absorption within their parent galaxies and to exclude a few intrinsicly red objects with peculiar spectra. Two blue SNe Ia with peculiar spectra at early phases (SN 1991 T, 1995 ac) are also excluded. The selected objects of the sample all have normal spectra as far as is known (Branch et al. 1993).

The 35 SNe Ia define Hubble diagrams in B, V, and I with a very small scatter of $\sigma_M \lesssim 0^m.2$. This means that SNe Ia are standard candles to better than $\pm 20\%$ of their luminosity. In spite of this the residuals about the Hubble line correlate with second parameters. The SNe Ia with slow decline rates Δm_{15} are somewhat brighter, where Δm_{15} is the magnitude decline during the first 15 days after B maximum. The relation between magnitude residual and decline rate is roughly $\delta m \propto 0.45 \times \Delta m_{15}$. Also the intrinsically bluer SNe Ia are somewhat brighter than their redder counterparts by about $\delta m \propto p \times \Delta(B-V)$, where p is ~ 1.2–2.5 for I, V, and B magnitudes, respectively (for the exact relations, see Parodi et al. 2000). Corrected magnitudes $m^{\mathrm{corr}}_{B,V,I}$ are obtained by reducing all SNe Ia to the mean decline rate $\langle \Delta m_{15} \rangle = 1.2$ and to the mean color $\langle B-V \rangle = -0^m.01$.

The corrected magnitudes $m^{\mathrm{corr}}_{B,V,I}$ define Hubble diagrams as shown in Figure 1. *Their tightness is astounding*. The Cerro Tololo collaboration, to whom one owes 70% of the photometry of the fiducial sample, quote a mean observational error of their $m_{\max}$ values of $\sim 0^m.10$ and of their colors $(B-V)$ of $\sim 0^m.05$ (Hamuy et al. 1996). This alone would suffice to explain the observed scatter of $\sigma_m = 0^m.12 - 0^m.13$. An additional error source are the corrections for galactic absorption which were adopted from Schlegel et al. (1998). In fact, if one excludes the nine SNe Ia with large galactic absorption corrections ($A_V > 0^m.2$) the scatter decreases to $0^m.11$ in all three colors. Two important conclusions follow from this. (1) If the total observed scatter of the Hubble diagrams is read vertically as an effect of peculiar motions, a generous upper limit is set of $\Delta v/v = 0.05$, which holds for the range of $3500 \lesssim v \lesssim 30\,000$ km s^{-1}. The (all-sky) distance-dependent variation

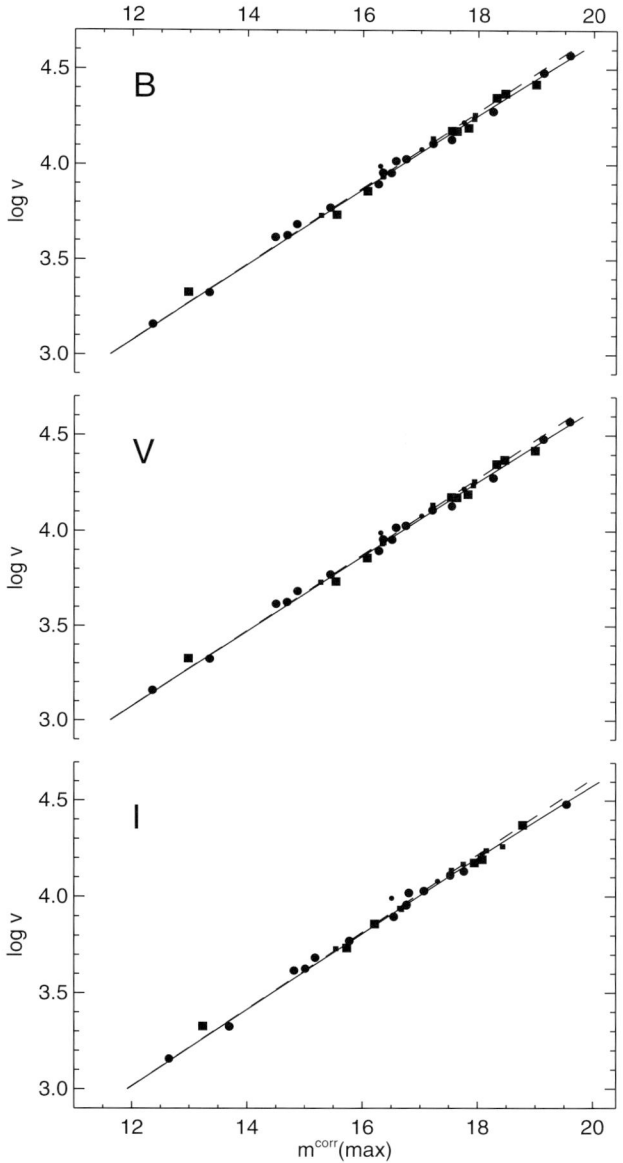

Figure 1 The Hubble diagrams in B, V (and I) for the 35 (29) SNe Ia of the fiducial sample with magnitudes m^{corr}, that is, corrected for decline rate Δm_{15} and color $(B-V)$. Circles are SNe Ia in spirals, squares in E/S0 galaxies. Small symbols are SNe Ia whose observations begin eight days after B maximum or later. Solid lines are fits to the data assuming a flat universe with $\Omega_m = 0.3$ and $\Omega_\Lambda = 0.7$; dashed lines are linear fits with a forced slope of 0.2 (corresponding approximately to $\Omega_m = 1.0$ and $\Omega_\Lambda = 0.0$). (Parodi et al. 2000.)

of H_0 must be even smaller. (2) If, however, the scatter is read horizontally and if allowance is made for the observational errors of the apparent magnitudes (and for any peculiar motions) one must conclude that the luminosity scatter of blue SNe Ia, once they are homogenized in Δm_{15} and color, is smaller than can be measured at present. In other words, they are extremely powerful standard candles.

Linear regressions, which correspond closely to a Universe with $\Omega_m = 1$, $\Omega_\Lambda = 0$, to the data in Figure 1 yield

$$\log v = 0.2 m_\lambda^{\text{corr}} + c_\lambda \qquad (2)$$

with $c_B = 0.676 \pm 0.004$, $c_V = 0.673 \pm 0.004$, and $c_I = 0.616 \pm 0.004$.

It is easy to transform eqns (1) and (2), remembering $(m - M) = 5 \log r_{\text{Mpc}} + 25$, into

$$\log H_0 = 0.2 M_\lambda + c_\lambda + 5 \qquad (3)$$

where M_λ is the *absolute* magnitude of SNe Ia in the appropriate passband. Thus the problem of determining the *large-scale* value of H_0 is reduced, with c_λ now being known, to measuring the absolute magnitude M_λ of one (or a few) *nearby* SNe Ia. It must be stressed that the recession velocity of the nearby calibrator(s) does not enter in any way. The solution for H_0 is therefore independent of any peculiar motions.

1.1.2 Cepheids, HST, and the luminosity calibration of SNe Ia

With the luminosity calibration of a SNe Ia in mind, a small team was formed to observe with HST the Cepheids in IC 4182, home of the well observed SN 1937C. The Cepheids would give a good distance to that inconspicuous galaxy and hence fix the absolute magnitude M in the B and V band. Through eqn (3) this would lead to a high-weight determination of H_0.

The team consisted of Allan Sandage, Abhijit Saha, Lukas Labhardt, Duccio Macchetto, Nino Panagia, and the author. Their application for HST observing time was received with skepticism. Was the project not too easy for HST? Could it not be executed from the ground? It was a happy incidence for the team that the mirror of the most complex and most expensive telescope ever build produced miserable images when it was finally launched in 1990. IC 4182 became an ideal target. In fact the photometry of the extended stellar images on the CCD frames was easier than later, after the repair mission in 1993, when the crisp stellar images became smaller than the pixel size of the WFPC-2 camera, but, of course, offering higher resolution essential for more distant galaxies.

Cepheids are the most uncontroversial distance indicators in extragalactic astronomy. Their pulsational period is a function of their mass which in turn correlates with the luminosity. The corresponding log P–absolute magnitude M relation was discovered in 1912 by Henrietta Leavitt who studied the many Cepheids in the Large Magellanic Cloud (LMC) which all appear at the same distance. The slope of the log P–M relation for different passbands is today well determined from the LMC Cepheids. Remaining uncertainties concern the zeropoint of the relation and the effect of differences of the metallicity of the Cepheids in different galaxies. All authors agree that the metallicity effect is small, but they do not agree on the sign of the corresponding correction. Therefore an uncertainty of $0^m.10$ will be included into the error budget.

At present it is customary to fix the zeropoint by adopting $(m - M)_{\text{LMC}} = 18.50$. There is growing evidence that this value should be increased by $\sim 0^m.05 - 0^m.10$. Based on observations with the International Ultraviolet Explorer (IUE) and HST, Panagia (1998) has derived a *geometrical* distance of $(m - M)_{\text{LMC}} = 18.58 \pm 0.05$ from the fluorescent ring of SN 1987A, a peculiar type II supernova in the LMC. Also the calibration of galactic Cepheids with the astrometric satellite HIPPARCOS tends to give a somewhat larger distance to the LMC (Madore and Freedman 1998; Feast 1999; Pont 1999; Groenewegen and Oudmaijer 2000) in agreement with RR Lyr stars and other distance indicators (Walker 1999; Gratton 2000; Sakai *et al.* 2000; Romaniello *et al.* 2000).

Smaller LMC moduli based on red-clump stars and RR Lyr luminosities, calibrated through statistical parallaxes, have comparatively low weight. Why two eclipsing binaries in the LMC give an unacceptably small modulus of 18.3, including HST observations (Fitzpatrick *et al.* 2000), is not understood at present.

DIVA, the planned next-generation astronomic satellite, is expected to measure very accurate parallaxes of 30–40 galactic Cepheids. This should settle for good the question of the zeropoint. ESA's very ambitious GAIA project would in addition measure the parallaxes of many supergiants and thus make a frontal attack on the distance of the LMC.

For the moment we will retain the customary zeropoint with $(m - M)_{\text{LMC}} = 18.50$ and return to the question in Section 1.1.3.

The HST distance of the Cepheids in IC 4182 provided $M_{B,V}$ of SN 1937C which, inserted into eqn (3), gave a rather low value of $H_0 = 52 \pm 9$ (Saha *et al.* 1994). Essentially the same IC 4182 distance had been derived before from the ground from its brightest stars (Sandage and Tammann 1982), but the result had been criticized. Now, with the HST result the light curve of SN 1937C from archival material came under fire. The obvious step was to calibrate other nearby SNe Ia, which became feasible after the repair of HST.

The five Cepheid distances of SNe Ia host galaxies, which followed subsequently, are compiled in Table 1 together with two such galaxies whose Cepheid distances were observed with HST by other teams. The resulting absolute magnitudes of the eight SNe Ia have different weights because of the different quality of their light curves and of the Cepheid distances whose errors depend heavily on the corrections for internal absorption. But the SNe Ia together determine their mean absolute magnitudes $M_{B,V,I}$ with a small statistical error.

The adopted absolute magnitudes in Table 1 are in very good agreement with model calculations of equally blue SNe Ia by Höflich and Khokhlov (1996).

The HST photometry for the Cepheids in the galaxies listed in Table 1 (except NGC 4414) has been re-analyzed by Gibson et al. (2000). For 114 Cepheids in common with Saha et al. (1994, 1995, 1996a,b, 1997, 1999) they find a brighter photometric zeropoint by $0^m.04 \pm 0^m.02$, which is as satisfactory as can be expected from the subtle photometry with WFPC-2 camera. However, a recent check on the zeropoint by A. Saha suggests that it should become fainter by $0^m.02$. But Gibson et al. (2000) have included in their analysis additional Cepheids which give a distinctly different result. Their reduction of the NGC 5253 modulus by $0^m.39$ is unlikely because it would imply a very faint tip of the red-giant branch. For the remaining galaxies in Table 1 they suggest a mean decrease of the distance moduli by $0^m.11 \pm 0^m.03$. It may be wise to postpone judgement on these results until the arrival of the Advanced Camera on HST when it will be possible to check on some of the Cepheid distances in Table 1 with this powerful instrument.

The slight disagreement, affecting H_0 only at the 5% level, is mentioned here to illustrate two facts. (1) Regardless of what powerful instruments one gives into the hands of scientists, they will work at the limit of the possibilities.

(2) The lead-time of all space equipment is long, causing all instruments in space to be outdated to some extent. This is also the case with the CCD detectors of the WFPC-2 camera installed during the repair mission in 1993. The problem is accentuated by aging effects in space.

The decisive point is that the SNe Ia in Table 1 have the same intrinsic properties as the Hubble diagram SNe Ia which define eqn (2). All spectra known for either set are Branch-normal and their mean intrinsic colors $\langle B-V \rangle$ are identical.

1.1.3 The value of H_0 from SNe Ia

The calibration SNe Ia in Table 1 have been corrected for decline rate Δm_{15} and color $(B-V)$ in exactly the same way as the distant SNe Ia in Figure 1. Their absolute magnitudes can therefore be used to determine H_0 through eqn (3).

With $M_{B,V,I}$ from Table 1 and $c_{B,V,I}$ from eqn (2) one finds from eqn (3) $H_0(B) = 60.3 \pm 2.0$, $H_0(V) = 60.1 \pm 1.8$, and $H_0(I) = 60.0 \pm 2.8$. The close agreement in all three colors provides evidence against any hidden absorption problems. The solution of H_0, including SNe Ia with recession velocities as high as $\sim 30\,000\,\text{km}\,\text{s}^{-1}$, depends slightly on the cosmological model. $\Omega_m = 1.0$, $\Omega_\Lambda = 0$ was assumed; if instead the values favored in Section 3 are adopted, that is, $\Omega_m = 0.3$, $\Omega_\Lambda = 0.7$, the mean value of $H_0(B)$, $H_0(V)$, and $H_0(I)$ are marginally increased to $H_0 = 60.9 \pm 2.0$ (Parodi et al. 2000).

The largest systematic error ($\sim 0^m.10$) may come from the metallicity effect of Cepheids. An upward revision of H_0 of about the same size would be required by the additional Cepheids of Gibson et al. (2000) pending their confirmation. All other systematic errors, discussed in more detail by Parodi et al. (2000), tend to cause an overestimate of H_0, that is, the LMC zeropoint modulus which is probably to be

Table 1 Absolute B, V, and I magnitudes of blue SNe Ia calibrated through Cepheid distances of their parent galaxies

SN	Galaxy	$(m-M)^0$	Reference	M_B^{corr}	M_V^{corr}	M_I^{corr}	Δm_{15}
1937 C	IC 4182	28.36 (09)	1	−19.38 (15)	−19.37 (17)	—	0.87 (10)
1960 F	NGC 4496A	31.04 (10)	3	−19.67 (18)	−19.65 (22)	—	1.06 (12)
1972 E	NGC 5253	28.61 (08)	2	−19.43 (16)	−19.43 (17)	−19.12 (20)	0.87 (10)
1974 G	NGC 4414	31.46 (17)	4	−19.70 (34)	−19.69 (27)	—	1.11 (06)
1981 B	NGC 4536	31.10 (05)	5	−19.48 (14)	−19.46 (10)	—	1.10 (07)
1989 B	NGC 3627	30.22 (12)	6	−19.42 (18)	−19.41 (16)	−19.20 (14)	1.31 (07)
1990 N	NGC 4639	32.03 (22)	7	−19.39 (26)	−19.38 (24)	−19.12 (23)	1.05 (05)
1998 bu	NGC 3368	30.37 (16)	8	−19.56 (31)	−19.55 (26)	−19.21 (21)	1.08 (05)
M (weighted mean)				−19.55 (07)	−19.53 (06)	−19.25 (09)	1.08 (02)
$M^{\text{corr}}(\Delta m_{15} = 1.2; (B-V) = -0.01)$				−19.48 (07)	−19.47 (06)	−19.19 (09)	1.08 (02)

Values in parentheses are the errors in units of $0^m.01$.
References: (1) Saha et al. (1994), (2) Saha et al. (1995), (3) Saha et al. (1996b), (4) Turner et al. (1998), (5) Saha et al. (1996a), (6) Saha et al. (1999), (7) Saha et al. (1997), (8) Tanvir et al. (1995).

increased, photometric blends of the Cepheids in crowded HST frames, incomplete Cepheid sampling near the photometric threshold, and indications for somewhat large adopted corrections for galactic absorption. Adding up all errors in quadrature leads to a correction factor of 0.96 ± 0.08. At the 90% confidence level one obtains (Saha *et al.* 1999; Parodi *et al.* 2000; Tammann *et al.* 2000).

$$H_0 = 58.5 \pm 6.3 \quad (4)$$

1.2 The route through galaxy clusters

1.2.1 The Hubble diagram of clusters of galaxies

The classical Hubble diagram using first-ranked cluster galaxies as standard candles (Sandage and Hardy 1973) is determined by the 72 objects with $3500 < v < 30\,000\,\mathrm{km\,s^{-1}}$ as

$$\log v = 0.2\, m_{1st} + (1.364 \pm 0.007) \quad (5)$$

The scatter of $\sigma_m = 0^m.29$, being considerably larger than for SNe Ia, cannot be caused by peculiar motions, but must be due to intrinsic luminosity dispersion and measuring errors. Still, first-ranked cluster galaxies are useful for the estimate of H_0 if one can determine their absolute magnitude M_{1st} in a few nearby clusters (Section 1.2.2). H_0 is then given by

$$\log H_0 = 0.2\, M_{1st} + (6.364 \pm 0.007) \quad (6)$$

Equation (6) follows from eqn (5) by analogy to eqns (2) and (3).

A variant of the Hubble diagram was introduced by Sandage and Tammann (1990) by plotting $\log v$ not against the apparent magnitude m of some standard candle, but against the cluster distance *relative* to the distance of a nearby cluster. The idea is again to combine the velocity information of clusters out to $\gtrsim 10\,000\,\mathrm{km\,s^{-1}}$ with the distance information of a nearby cluster whose recession velocity may be as much disturbed by peculiar motions as it wants. *Relative* cluster distances have the advantage that it is relatively easy to define comparable galaxy samples in different clusters without the necessity of having complete samples, which are required if one wants to derive absolute distances (Section 1.2.2).

Three examples of Hubble diagrams using relative distances are given here.

(1) Giovanelli *et al.* (1997) and Dale *et al.* (1999) have derived the relative distances of 71 clusters with $3000 < v < 25\,000\,\mathrm{km\,s^{-1}}$ from the Tully–Fisher (TF) relation (Section 1.2.2). If one chooses the Fornax cluster as zeropoint one obtains the Hubble diagram in Figure 2. The clusters define a Hubble line with impressively small scatter of $\sigma_{(m-M)} = 0^m.11$. The Hubble line implies

$$\log H_0 = -0.2\,(m - M)_{\mathrm{Fornax}} + (8.130 \pm 0.003) \quad (7)$$

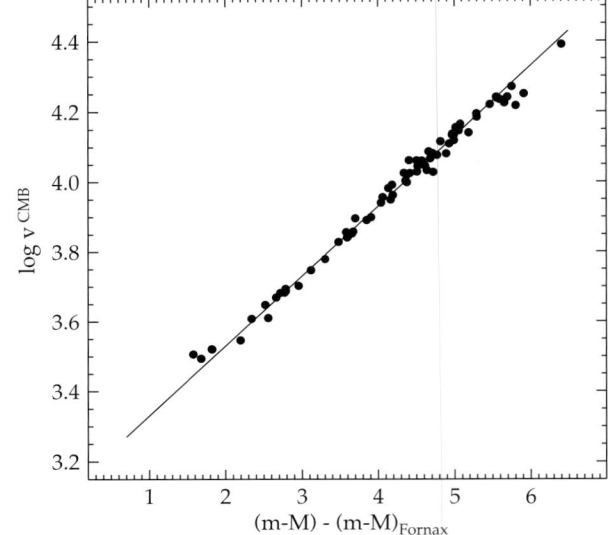

Figure 2 The Hubble diagram of 71 clusters whose distances relative to the Fornax cluster are known from the Tully–Fisher relation. (Data from Giovanelli *et al.* 1997 and Dale *et al.* 1999.)

(2) Federspiel *et al.* (1998) have combined relative cluster distances by Jerjen and Tammann (1993) from various distance indicators and by Giovanelli (1997) from TF distances for 31 clusters within $\sim 10\,000\,\mathrm{km\,s^{-1}}$. They have expressed these relative distances in units of the Virgo cluster distance. The corresponding Hubble diagram has a scatter of $\sigma_{(m-M)} = 0^m.20$; its intercept implies

$$\log H_0 = -0.2\,(m - M)_{\mathrm{Virgo}} + (8.070 \pm 0.007) \quad (8)$$

(3) Relative cluster distances can also be derived from the D_n–σ relation of E/S0 galaxies and of the bulges of spiral galaxies. Their velocity dispersion σ, as measured from the width of the spectral lines, correlates with a normalized linear diameter D_n. The corresponding relative cluster distances by Faber *et al.* (1989) define a useful Hubble diagram. The 10 clusters, for which Kelson *et al.* (2000) have derived relative distances from the D_n–σ method and from the related method of the fundamental plane, have an even smaller scatter of $\sigma_{(m-M)} = 0^m.19$. If their data are expressed relative to the Coma cluster one obtains

$$\log H_0 = -0.2\,(m - M)_{\mathrm{Coma}} + (8.887 \pm 0.012) \quad (9)$$

Before H_0 can be determined from eqns (6)–(9) the *absolute* distances of the Virgo, Fornax, and Coma clusters must be discussed.

1.2.2 The distances of nearby clusters

The distances of the three above-mentioned clusters are determined in the following, where the important role of HST becomes apparent.

The Virgo cluster

Four galaxies with known Cepheid distances from HST lie in the Virgo cluster proper, that is, within the isopleths and the X-ray contours (Binggeli et al. 1993). As can be seen in Table 2 they have widely different distances due to the important depth effect of the cluster. Three of the galaxies have been selected from the atlas of Sandage and Bedke (1988) on the grounds of their exceptionally good resolution; they are therefore *expected* to lie on the near side of the cluster. The fourth galaxy, NGC 4639, which has a *low* recession velocity and can therefore not be assigned to the background, but must be a dynamical member of the cluster, is more distant by almost $1^m.0$ and must lie on the far side of the extended cluster. The relative position of the four galaxies is fully confirmed by their TF distances. The cluster *center* must lie somewhere between the available Cepheid distances, say at $(m-M) = 31.5 \pm 0.3$. Much of the confusion over the extragalactic distance scale comes from the ill-conceived notion that the three highly resolved Virgo galaxies could reflect the *mean* distance of the cluster.

Three well-observed SNe Ia have appeared in Virgo cluster members. Their distances in Table 2, corrected for decline rate Δm_{15} and color $(B-V)$, have been calculated from their individual parameters as compiled by Parodi et al. (2000) and from the HST-based calibration in Table 1.

A special note on the TF relation is appropriate here. It relies on a correlation between the maximum rotation velocity w – as measured from the line width at 21 cm or from any optical spectral line – with the mass and hence with the luminosity of a spiral galaxy. The slope of the TF relation can be well observed from the spiral members of the Virgo cluster. But the zeropoint calibration was uncertain for a long time. A large consortium was therefore formed, first with J. Mould and then with W. Freedman as principal investigators, to measure Cepheid distances, with HST, of spiral galaxies suited for the calibrators of the TF relation. The project was accepted as a Key Project of HST for the determination of H_0.

The Key Project has determined Cepheid distances to 22 spiral galaxies which are sufficiently inclined that a significant component of their rotation velocity is reflected in the line width. The HST Cepheid distance of an additional galaxy, the giant spiral M101 (Freedman et al. 1994), was a valuable confirmation of the previously known distance (Sandage and Tammann 1974), but the galaxy is seen too face-on to be useful for the TF relation. However, the galaxy has two close companions which are important in extending the TF calibration to smaller line widths. In addition there are four Cepheid distances useful for the calibration, which were measured from the ground. The total of 28 spirals with Cepheid distances provides an excellent calibration of the TF relation (Figure 3a).

Table 2 Distances of the Virgo and Fornax clusters

Virgo				Fornax			
Object	$(m-M)^0$	Reference	Remarks	Object	$(m-M)^0$	Reference	Remarks
Cepheids							
NGC 4321	31.04	1	Highly resolved	NGC 1326A	31.49	7	
NGC 4535	31.10	2	Highly resolved	NGC 1365	31.39	8	
NGC 4548	31.04	3	Highly resolved	NGC 1425	31.81	9	
NGC 4639	32.03 (!)	4	Normal resolution				
Mean:	~31.5				31.56 ± 0.13		
SNe Ia							
SN 1984A	31.42	5	In NGC 4419	SN 1980N	31.76	5	In NGC 1316
SN 1990N	32.12	5	In NGC 4639	SN 1981D	31.51	5	In NGC 1316
SN 1994D	31.27	5	In NGC 4526	SN 1992A	31.84	5	In NGC 1380
Mean:	31.60 ± 0.30				31.70 ± 0.10		
Tully–Fisher relation (complete samples)							
Mean ($n=49$):	31.65 ± 0.25	6			(31.25 ± 0.10)	10	
Overall mean:	31.60 ± 0.20			Overall mean:	31.65 ± 0.10		

References: (1) Ferrarese et al. (1996), (2) Macri et al. (1999), (3) Graham et al. (1999), (4) Saha et al. (1997), (5) see text, (6) Federspiel et al. (1998), Federspiel (1999), (7) Prosser et al. (1999), (8) Silbermann et al. (1999), (9) Mould et al. (2000), (10) Schröder 1995; Federspiel 1999.

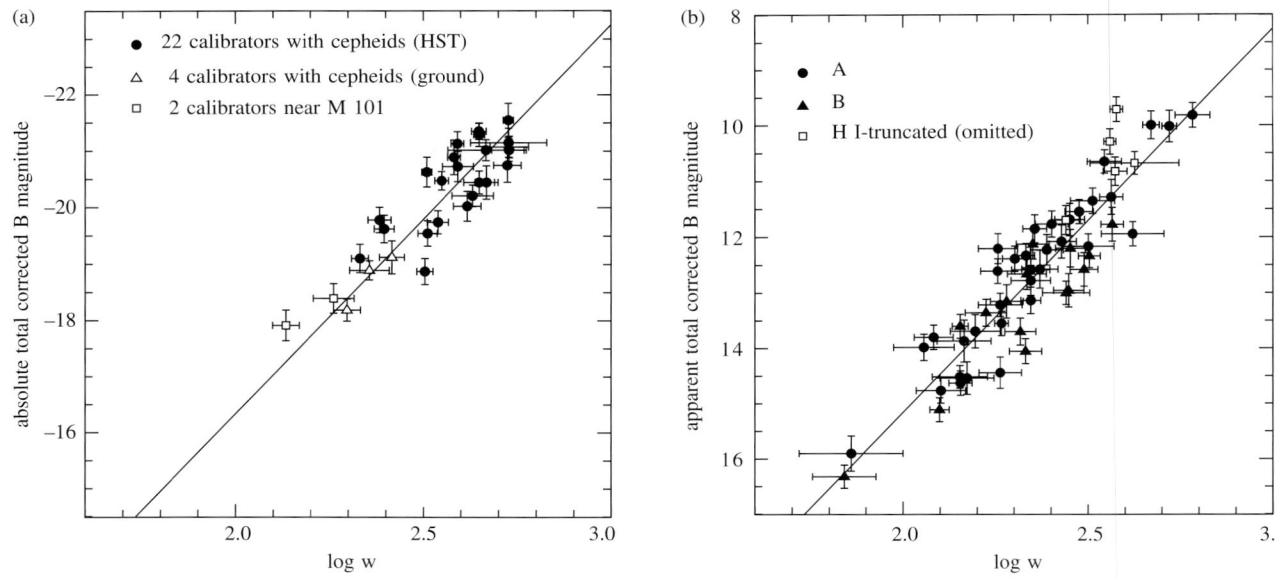

Figure 3 (a) Tully–Fisher relation of 28 galaxies with independently known Cepheid distances mainly from HST. (b) Tully–Fisher relation for a complete sample of 49 Virgo cluster spirals.

Unfortunately the applicability of the TF relation is limited because the scatter of the relation is $\sigma_M \sim 0^m.4$ for all practical purposes (Figure 3a). This invites selection effects that always lead to too small distances. The severeness of the effect is surprising, but it is a constant concern of any astronomer working from existing catalogs of stars and galaxies. The problem is that objects with a non-negligible luminosity scatter, say spiral galaxies of a fixed line width w, have some representatives that are more luminous than average and others that are less luminous. As one goes to larger less luminous ones are apparently too faint to enter the catalog, while the more luminous ones will be easily cataloged. In the case of stars and galaxies this problem is known as the Malmquist bias, and we will return to it in Section 1.3. But it also poses severe problems in the case of galaxy clusters if the galaxy sample is not *complete*. The problem has been formalized by Teerikorpi (1987) (see also Sandage *et al.* 1995; Teerikorpi 1997, and references therein). Any sample of cluster galaxies cut by an apparent-magnitude limit will favor overly bright galaxies (at a fixed line width) and will in addition yield too flat a slope of the TF relation. Both effects cause the resulting distance to be too small.

Spiral galaxies are among the brightest cluster galaxies and reach down seven magnitudes into the cluster population. Fortunately the Virgo cluster has been imaged on $16\,m^2$ of photographic glass plates, and the resulting catalog of Virgo cluster members (Binggeli *et al.* 1985) contains *all* 49 normal spiral members sufficiently inclined to be usable for the TF method. For all these galaxies good magnitudes in the B band and line widths w are known. The resulting TF relation is shown in Figure 3b. Combining Figure 3a and 3b leads to the distance modulus of the Virgo cluster shown in Table 2. The error of the modulus of $\pm 0^m.25$ is relatively large because of the depth of the cluster and some intricacies of the TF method, one being that inclined spiral galaxies must be corrected for internal absorption, which is a rather difficult task.

The complete sample of Virgo cluster spirals is well suited to study the Teerikorpi cluster incompleteness bias. For instance, if one sampled the cluster only down to $12^m.0$ the remaining galaxies would indicate a distance modulus which is too small by $0^m.40$ (20% in linear distance)!

The distance of the Virgo cluster is such an important milestone for the extragalactic distance scale that the value adopted in Table 2 should be compared with other distance indicators. Three distance indicators of early-type galaxies are considered, although their reliability is much less tested. (1) The peak of the luminosity function of globular clusters (LFGC) is interesting because its calibration of the zero-point rests on the Cepheid distance of M31 *and* in excellent agreement with the RR Lyrae distance of galactic globular clusters. (2) Six known novae in Virgo cluster ellipticals can be compared to the novae in M31, whose *apparent* distance modulus can be derived from Cepheids, *or* from galactic novae (Capaccioli *et al.* 1989). Alternatively the semi-theoretical zeropoint of novae can be taken from Livio (1997). (3) The $D_n - \sigma$ relation can be applied to

E/S0 galaxies and to the bulges of S0–Sb galaxies. The mean of the three methods, discussed in more detail by Tammann *et al.* (2000), gives $(m - M)_{\text{Virgo}} = 31.66 \pm 0.17$ and in excellent agreement with the adopted value in Table 2. It is also consistent with the assumption that early-type galaxies and spirals of the Virgo cluster are at the same distance.

The Fornax cluster
Table 2 also shows the distance modulus of the Fornax cluster as derived from the three galaxies with Cepheids and three SNe Ia. The individual distances have much less scatter than in the case of the Virgo cluster; this is obviously a result of the smaller size and depth of the Fornax cluster.

The Fornax cluster, for which also an almost complete spiral sample is available, is not particularly suited for the TF method, because it has only eight inclined full-size spirals and the 19 fainter spirals show considerably more scatter than the Virgo cluster. In spite of this the resulting TF distance comes as a surprise. It suggests that the Fornax cluster is nearer than the Virgo cluster by $0^m.40 \pm 0^m.10$ (Schröder 1995), which is impossible not only in the light of Cepheids and SNe Ia but also from a compilation of 30 authors who found widely different distance ratios of Fornax and Virgo, but suggest on average that Fornax is *more* distant by $\sim 0^m.15 \pm 0^m.10$. At this point one can only speculate about physical differences between Virgo and Fornax spirals.

The Coma cluster
It is noted that the distance of the Coma cluster relative to the Virgo cluster is well determined. The value $\Delta(m - M)_{\text{Coma-Virgo}} = 3.71 \pm 0.08$ (Tammann and Sandage 1999) from brightest cluster galaxies and the TF and D_n–σ methods is quite uncontroversial. With the Virgo modulus in Table 2 one obtains $(m - M)_{\text{Coma}} = 35.31 \pm 0.22$, in good agreement with the globular cluster luminosity functions of two Coma cluster ellipticals, which have been observed with HST (Baum *et al.* 1997; Kavelaars *et al.* 2000), and which give, with the calibration of Tammann and Sandage (1999), $(m - M)_{\text{Coma}} = 35.42 \pm 0.20$. A mean of $(m - M)_{\text{Coma}} = 35.35 \pm 0.20$ is adopted.

1.2.3 The value of H_0 from cluster galaxies

With the Hubble diagrams of clusters in Section 1.2.1 and the absolute cluster distances in Section 1.2.2 it is now straightforward to calibrate H_0.

The apparent V magnitudes of the first-ranked galaxies in the Virgo, Fornax, and Coma clusters as given by Sandage and Hardy (1973) together with the distances of these clusters yield a mean absolute magnitude of the first-ranked galaxies of $\langle M_{1st} \rangle = -23.26 \pm 0.20$. When one inserts this value into eqn (6) one obtains H_0 as shown in Table 3. The other values of H_0 in Table 3 come from eqns (7)–(9) and from inserting the appropriate cluster moduli. The four solutions in Table 3 agree within the statistical errors. The latter are compounded from the error carried by the calibrators plus the scatter of the distance indicators about their Hubble lines. The adopted solution lacks the simplicity and straightforwardness of the method through SNe Ia, but still does remarkably well considering that each cluster route is calibrated by a single cluster (except the first-ranked galaxies whose calibration rests on three clusters), whereas the SNe Ia route is founded on eight calibrators, and that the Hubble diagrams of clusters have considerably more scatter (except the one with TF distances) than the Hubble diagram of SNe Ia. The essential point is, however, that clusters that have not been used for the SNe Ia solution give essentially the same value of H_0 as SNe Ia.

1.3 The value of H_0 from field galaxies

It is desirable not only to measure the expansion rate of clusters of galaxies but also of galaxies outside clusters in the open field. SNe Ia do just this. The majority of all galaxies lie outside clusters and it is therefore not surprising that almost all of the SNe Ia which define the Hubble diagram of Figure 1 have been found "in the field" outside of clusters.

Table 3 Values of H_0 from clusters of galaxies

Hubble diagram from	Eqn	Calibrated through	H_0
72 clusters with first-ranked galaxies	(6)	$\langle M_{1st} \rangle = 23.26 \pm 0.20$	51 ± 6
71 clusters with TF distances	(7)	$(m-M)_{\text{Fornax}} = 31.65 \pm 0.10$	63 ± 5
31 clusters from different methods	(8)	$(m-M)_{\text{Virgo}} = 31.60 \pm 0.20$	56 ± 8
10 clusters with D_n–σ distances	(9)	$(m-M)_{\text{Coma}} = 35.35 \pm 0.20$	66 ± 8
Mean H_0			59 ± 3

The disadvantage of SNe Ia is that they are rare events with only roughly one occurrence per large galaxy per century. One must therefore work with the galaxies that happen to have produced a SN Ia. From this follows that one would like to use other distance indicators which can be used to derive the distances of a preselected sample of field galaxies.

Indeed, it would be a waste not to apply the TF method to field galaxies, after so much HST observing time has gone into its calibration. TF distances have proved to be useful as distance indicators after having provided a tight Hubble diagram of clusters (Figure 2). But the cluster TF distances rest on 15 cluster members on average, and the scatter of an individual galaxy must be much larger. There are several lines of evidence indicating that the intrinsic scatter of the TF method is $\sigma_m = 0^m.3 - 0^m.4$, which implies that it is not possible to determine a single galaxy distance to better than 15–20%.

The problem with the scatter is that it does not only limit the accuracy of individual TF distances, but that it also introduces *systematic* errors whenever one works from *magnitude*-limited catalogs. And astronomers have to work almost always with such catalogs, because ideally suited catalogs which would have to be complete to a fixed *distance* limit are not available. They would require an enormous parent sample with complete redshift information, from which one could select *all* objects within a fixed distance limit. The next step for an application of the TF method would then be to obtain *complete* magnitude and line-width information for all catalog entries. The best thing astronomers have instead are catalogs which are complete to a given apparent-magnitude limit.

The selection effects which magnitude-limited catalogs suffer, have been touched upon already in Section 1.2.2. They were first rigidly treated by Malmquist (1920, 1922), but even half a century later his message has often remained unheard. This is one of the main reasons why one has seen unrealistically high values of H_0 deep into the space age.

The progressive discrimination against underluminous galaxies (at fixed line width w) causes the mean luminosity to increase within every volume element of increasing distance. It is likely that among the faintest galaxies of a catalog is also the most distant one, that is, an ultraluminous "loner" which lies 3σ ($\sim 1^m.0$) above the luminosity its line width w seems to indicate. The great rarity of such an object is compensated for by the very large volume over which it can be sampled. If this object is treated like an average galaxy it suggests a Hubble constant which is too large by a factor of 1.6!

Strategies have been developed to correct for Malmquist bias, particularly for the application of the TF relation but also for other methods, by, for example, Sandage (1988, 1995), Federspiel et al. (1994), Teerikorpi (1984, 1997), Bottinelli et al. (1986, 1995), and Theureau (2000).

A selection of H_0 values, which have been derived from magnitude-limited, yet bias-corrected, samples of field galaxies are compiled in Table 4. The conclusion is

$$H_0 = 55 \pm 3 \ (\pm 6 \text{ for systematic errors}) \qquad (10)$$

from field galaxies, which is valid within various distance ranges up to $\sim 8000\,\text{km\,s}^{-1}$. It should be noted that the different methods use spirals only – except the D_n–σ entry – and their calibration rests only on Cepheids, independent of any adopted distance of the Virgo cluster.

1.4 H_0 from physical methods

All astronomical distance determinations depend on an adopted astronomical distance as zeropoint. Even trigonometric parallaxes depend on the "astronomical unit"

Table 4 Values of H_0 from field galaxies corrected for Malmquist bias

Method	H_0	Range (km s^{-1})	Source
Luminosity classes	56 ± 5	4000	Sandage (1996a)
Morphol. twins of M31, M101	50 ± 5	5000	Sandage (1996b)
TF (using mag. + diam.)	55 ± 5	5000	Theureau et al. (1997)
Galaxy diam.	50–55	5000	Goodwin et al. (1997)
TF	53 ± 5	500	Federspiel (1999)
	58 ± 2	1000	Federspiel (1999)
Luminosity classes	55 ± 3	4000	Sandage (1999)
D_n–σ	52 ± 8	4000	Federspiel (1999)
TF (using mag. + diam.)	55 ± 5	8000	Theureau (2000)
Morphol. twins of M31, M81	60 ± 10	5000	Paturel et al. (1998)
Inverse TF	53 ± 6	8000	Ekholm et al. (1999)
Mean H_0	55 ± 3		

(Earth–Sun distance) which, as the ultimate basis, depends on the radar echo from Venus.

In addition to the astronomical distance determinations there are sometimes favorable conditions where the distance can be directly determined from the geometry or the physics of the object. One example for such a physical distance, namely the ring of SN 1987A, has been mentioned already.

An interesting case are the H_2O masers orbiting around the supermassive nucleus, presumably a black hole, of the active galaxy NGC 4258. Their radial velocities have been measured as well as their tangential motion and acceleration from intercontinental radio interferometry. These motions outline impressively well-defined Keplerian orbits from which the distance can be deduced (Herrnstein et al. 1999). The resulting distance is about $12 \pm 9\%$ smaller than the Cepheid distance of NGC 4258 (Maoz et al. 1999), but it would not be justified to correct the entire distance scale built on Cepheids on the basis of that single measurement. However, there are more than 20 galaxies known at present with H_2O masers, and it seems possible to reach out to distances of $\sim 5000\,\mathrm{km\,s^{-1}}$ with interferometry from space and thus to determine H_0 directly.

An ambitious goal is to observe the light echo of a supernova outburst in the surrounding interstellar medium. Since the feature is of subarc-second size (at a redshift of $\sim 1000\,\mathrm{km\,s^{-1}}$), before its surface brightness becomes too dim the resolving power of HST is required. If the position and the time of maximum polarization can be determined the geometry and hence the distance are fixed (Sparks 1994). A first result of this method is an upper limit of $\lesssim 15\,\mathrm{Mpc}$ for the distance of SN 1991T (Sparks et al. 1999), which compares well with the Cepheid distance of 13.8 ± 0.9 Mpc of its parent galaxy NGC 4527 (Saha et al. 2000). It is, however, questionable whether this method can reach out to $\sim 5000\,\mathrm{km\,s^{-1}}$ to directly measure the large-scale value of H_0.

The Sunyaev–Zeldovich effect is due to the fact that photons of the CMB gain energy by inverse Compton scattering when passing through the hot gas of an X-ray cluster of galaxies. This causes a temperature shift in the Rayleigh–Jeans part of the CMB black-body spectrum by as little as $\Delta T \sim 0.1$–0.3 mK, but which can be measured with modern radio interferometers and single-dish telescopes at centimeter to submillimeter wavelengths in many clusters (Birkinshaw 1999). The strength of the effect is proportional to the depth of the cluster, and since its angular X-ray size is known one can derive a distance assuming sphericity. The method can be applied to a wide range of distances, that is, also at $\sim 10\,000$–$20\,000\,\mathrm{km\,s^{-1}}$ best suited for the present, large-scale value of H_0. The results so far have systematic errors of $\sim 20\%$, mainly due to uncertainties of the temperature structure and the clumpiness of the intracluster gas. The method has become possible through X-ray satellites like Einstein, ROSAT, ASCA, Chandra, and others and will gain from forthcoming data from the XMM mission. Eventually the method will provide competitive distances to better than 10% (Chapter 21).

Cosmological distances can also be obtained from *variable* quasars that are gravitationally lensed by intervening galaxies or clusters producing double or multiple images. The geometrical configuration can be constructed from the redshifts of the quasar and the lens and from the specifications of the images. The configuration can be calibrated in terms of absolute distances by the time delay between the light curves of the individual images. The method appears elegant and attractive, but unfortunately the results are very sensitive to the mass distribution within the lens. Also the time delay, which can vary between a few days and several hundred days, requires careful observations to match the stochastic light curves. It must be hoped that a "golden lens" will be found where the necessary observables can be determined particularly well, possibly involving also observations from space to resolve the gravitational images. An interesting extension of the method has recently been proposed by including the infrared image of the Einstein ring, as observed with the NICMOS camera on HST. The Einstein ring is the lensed image of the host galaxy of the quasar and is believed to strongly constrain the mass distribution of the lens. A first application to the fourfold imaged quasar PG 1115+080 sets an upper limit to the Hubble constant of $H_0 < 60$ (Kochanek et al. 2000). Independently of this Saha (2002) has combined the constraints from several double quasars to restrict the permitted window to $H_0 = 55 \pm 5$. It should be noted that the method does not directly provide the present value of H_0, but the value of H corresponding to the redshift z of the quasar. To obtain H_0 to better than $\sim 10\%$ from quasars with $z \gtrsim 0.4$ certain assumptions on Ω and Λ must be made. Inversely, if H_0 were known with precision the method would allow inferences on Ω and Λ (Chapter 20).

Finally, H_0 leaves its signature in the fluctuation spectrum of the CMB (see below). Present likely solutions, mainly based on the BOOMERANG 98 and MAXIMA 1 balloons, for all parameters influencing the fluctuation spectrum tend to yield values of $H_0 \sim 70$ (Lineweaver 1999; Jaffe et al. 2000). The very preliminary character of this value is underlined by Lahav et al. (2000) who obtained a double solution of H_0 being either ~ 50 or ~ 70. The problem is brought to the point by Jaffe et al. (2000) who write: "Because many of the parameters affect the spectrum in highly correlated (in some cases almost degenerate) ways, limits on any one parameter are necessarily a function of the constraints, implicit or explicit, that one assumes for the other parameters." The difficulty of isolating any one of the cosmological constants from the CMB temperature angular

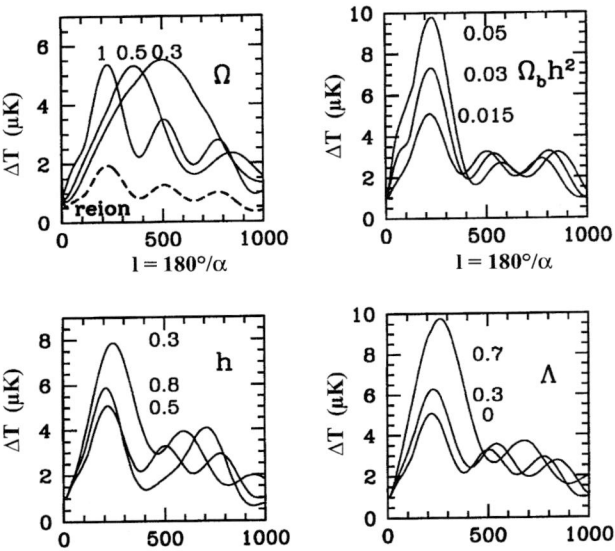

Figure 4 Theoretical model calculations of the angular power spectrum of the temperature fluctuations of the CMB. The underlying model assumes primordial adiabatic perturbations. Each graph shows the effect of varying one of the parameters Ω_m, $\Omega_b h^2$, h (that is, H_0), or Λ. The graph with variable Λ assumes $\Omega_{total} \equiv \Omega_m + \Omega_\Lambda = 1$. (From Kamionkowski and Kosowsky 1999.)

power spectrum is illustrated in Figure 4, where model calculations show the influence of four particular cosmological constants on the observable spectrum. The case of h (that is, H_0 in units of $100 \,\mathrm{km \, s^{-1} \, Mpc^{-1}}$) seems particularly unfortunate, because the height of the primary peak reverses its trend between $h = 0.5$ and $h = 0.8$ and becomes very insensitive for H_0 in the range of interest. The present information on H_0 depends therefore on the just-emerging secondary peak at subdegree scales. For a long time, therefore, it will be important to derive H_0 *independently*, such that it can be used as a prior for the determination of the other parameters.

Summarizing the present evidence on H_0, a value of

$$H_0 = 58 \pm 6 \tag{11}$$

from SNe Ia is consistent with all other determinations albeit of lower weight. The prospects that H_0 will be determined to within 5% during the next 10 years are very bright. This improvement will heavily depend on space research. Astrometric satellites like DIVA, SIM, and GAIA will determine the zeropoint of the log P–M relation of Cepheids. GAIA, reaching also the Cepheids in the LMC and some even in the very metal-poor Small Magellanic Cloud (SMC), will solve the remaining metallicity problem of Cepheids. The Advanced Camera on HST and the New Generation Space Telescope (NGST) will further strengthen the present Cepheid distances of the host galaxies of the SNe Ia, which are used to calibrate the SNe Ia luminosities, and will add the Cepheid-calibrated luminosities of 2–3 nearby blue SNe Ia, which will presumably be discovered during this interval. Supernovae searches from the ground will add 50–100 SNe Ia in the useful distance range, which will not only tighten their already exceptionally tight Hubble diagram, but also give a definitive relation between individual SN Ia luminosities and the light curve decay rate and color. On a parallel track the forthcoming space instrumentation can be used to determine the Cepheid distances of ~20 *randomly* selected Virgo cluster galaxies which will provide an accurate mean cluster distance in spite of its depth effect, which in turn can be used to calibrate other distance indicators with longer range.

2 THE DENSITY PARAMETER Ω

For the global history of the Universe the mean matter density ρ_m is fundamental because it defines the deceleration of the expansion. The customary way to measure the deceleration and hence the mean matter density is to define a deceleration parameter q_0, where

$$q_0 \equiv -\frac{\ddot{R}_0 R_0}{\dot{R}_0} \tag{12}$$

This implies (Sandage 1961)

$$q_0 = \frac{4\pi G \rho_{0m}}{3 H_0^2} \tag{13}$$

For $0 < q_0 < \frac{1}{2}$ the curvature k is -1 and the Universe expands forever; for $q_0 > \frac{1}{2}$ the curvature k is $+1$, the Universe is closed, and will recollapse. The intermediate case $q_0 = \frac{1}{2}$ has flat geometry ($k = 0$) and the Universe expands towards an asymptotic size. This is realized if ρ_m takes the so-called critical value of

$$\rho_{0crit} = \frac{3 H_0^2}{8\pi G} \tag{14}$$

With a rounded value of $H_0 = 60$ (which is used with sufficient accuracy throughout this section) one obtains

$$\rho_{0crit} = 6.76 \times 10^{-30} \,\mathrm{g \, cm^{-3}} \text{ or } 9.99 \times 10^{10} M_\odot \,\mathrm{Mpc^{-3}} \tag{15}$$

For many years attempts were made to determine q_0 directly, either by measuring the deceleration or space curvature. The attacks on q_0 were known as "cosmological tests." They comprised the Hubble diagram with brightest cluster galaxies as standard candles, galaxy counts down to different apparent-magnitude levels or ideally to different redshift limits, or the diameter–redshift relation of galaxies, clusters, or radio sources. An overview of these different methods is given by Sandage (1995). They have all encountered enormous problems due to luminosity or diameter evolution of the objects considered. Only the Hubble diagram, but now with SNe Ia as standard candles, is still vital. We will return to it in Section 3.

Because it has now become probable that the cosmological constant Λ is positive (Section 3), the expansion is not only decelerated by gravitation, but also accelerated by the dark Λ energy, the net effect can still be measured by q_0 (which may become negative), yet the respective contributions of decelerating matter and accelerating cosmological constant Λ are obscured. It has therefore become customary to use the density parameter Ω. The fraction of Ω which is due to matter at time t_0 is defined as

$$\Omega_m(t_0) = \frac{\rho_{0m}}{\rho_{0crit}} \quad (16)$$

Only the value $\Omega_m = 1$ is time independent. For all other values Ω_m converges towards 1 in the very early Universe. This is the puzzling flatness problem of the Universe which is one of the main rationals for inflationary theories.

A brief overview of different methods to determine Ω_m is given in the following.

2.1 The baryon density Ω_b

The primordial nucleosynthesis of the lightest elements ^2D, ^3He, ^4He, and ^7Li determines that part of Ω_m which is due to baryons to be $\Omega_b h^2 = 0.0190 \pm 0.0018$ (Burles et al. 2000; Chapter 19; h being again the Hubble constant measured in units of $100 \, km \, s^{-1} \, Mpc^{-1}$). This corresponds with $H_0 = 60$ to

$$\Omega_b = 0.053 \pm 0.005 \quad (17)$$

An independent ingredient of this value is the D/H abundance as observed with HST (Linsky 1998). It is natural to assume that the intervening hydrogen clouds, which are observed in absorption in quasar spectra out to redshifts of $z = 4$ (the so-called damped Ly α systems), are the progenitors of the present disks of spiral galaxies. This is supported by the baryonic surface brightness being similar for both kinds of objects. One question is raised by the fact that the former have about a 10 times higher co-moving space density than the latter. One of several possible solutions is the current view that many (all?) elliptical galaxies have formed from the merging of spiral galaxies or their progenitors. In any case the damped Ly α systems cannot overrepresent the baryon density.

Observations of damped Ly α systems in seven quasars with redshifts $2.5 \leq z \leq 4.55$ give, in combination with detailed calculations of a standard cold dark matter model (Rauch et al. 1997):

$$\Omega_b \geq 0.066 \left(\frac{H_0}{60}\right)^{1.5} \quad (18)$$

This value could be lower if the H I ionizing flux of quasars is more efficient than assumed. The value may be compared with $\Omega_b = 0.036^{+0.006}_{-0.003}$ ($H_0 = 60$) obtained from HST spectroscopy of intervening He II absorption lines (Wadsley et al. 2000). There may be in addition *diffuse* intergalactic matter. An upper limit for neutral hydrogen of $\Omega_b < 0.03$ is set by the absence of the Lyman break in quasar spectra (Coles and Ellis 1997). An even tighter upper limit of $\Omega_b < 0.02$ is available from an ionized He observation in a high-redshift quasar; the corresponding lower limit is as low as $\Omega_b > 0.002$ ($H_0 = 60$) at $z = 2.9$ (Reimers et al. 1997).

While it thus seems that the baryon density at high redshifts, required by primordial nucleosynthesis, is accounted for by intervening Ly α systems the present whereabouts of baryons in the galactic neighborhood is paradoxically less clear. Luminous stars, H I, and H_2 gas account for less than $\Omega_b = 0.01$. The additional contribution of brown dwarfs in the galactic plane is less than 10% (Rebolo 2001). Microlensing experiments make it increasingly clear that only a fraction of the missing baryons lie in compact, dark halo objects (MACHOs; Milsztajn 2001). Yet the halo mass could be considerably increased by cold gas globules stabilized by MACHO clusters (Gerhard and Silk 1996). The temperature–mass relation of X-ray clusters suggests that the H II gas of the Local Group is not sufficiently heated to light up in X-rays, and that correspondingly an important baryon fraction may be hidden in ionized hydrogen at intermediate temperatures.

The present CMB anisotropy measurements of the BOOMERANG 98 and MAXIMA 1 balloons give typically higher values of the baryon density near $\Omega_b h^2 = 0.03$ (e.g. Jaffe et al. 2000). Such a high value cannot be accommodated by standard Big Bang nucleosynthesis (Burles et al. 2000). If confirmed it may imply new physics (e.g. Hansen and Villante 2000). Yet the confirmation of the discrepancy will only come from polarization measurements with the PLANCK satellite and/or from measurements of the small-scale CMB fluctuations by future balloons and space missions.

2.2 The matter density Ω_m

There are several lines of evidence that the total matter density Ω_m is larger than the baryon density Ω_b. The arguments are presented in the following. The additional mass is thought to reside in weakly interacting particles which are either non-relativistic (cold dark matter, CDM) or relativistic (hot dark matter, HDM). The reality of CDM is also demonstrated by the growth of structures that would remain a mystery if only baryons were involved. The CDM can decouple early on from the primordial radiation field and form the first density enhancements, which are not seen in the CMB, and into which the baryons fall once they have combined with electrons and hence decoupled from the radiation. Model calculations of structure formation with baryons and CDM very successfully reproduce many characteristics of the observed distribution of luminous matter (Bridle *et al*. 1999). But the very largest known structures may still require some HDM. The obvious candidates for HDM are massive neutrinos. If the combined mass of all six kinds of neutrinos is 0.2 eV – a value which is plausible on theoretical grounds (Dixon 1998) – they contribute a neutrino density of $\Omega_\nu \sim 0.007$.

2.2.1 Ω_m from gravitationally bound systems

Structures induce peculiar (virial and other) motions and bind the X-ray gas, they evolve as a function of the mean density, and they bend light. These effects give a handle on their gravitating mass and hence on the mean density Ω_m, if, moreover, the assumption is made that there is no additional dark matter more smoothly distributed than the visible structures.

Several routes arrive at the mean mass-to-light ratio $\mathcal{M}/L$ of the structures. (The mass-to-[blue] light ratio is expressed in solar units, that is, the Sun has $\mathcal{M}_\odot/L_\odot \equiv 1$.) If $\mathcal{M}/L$ is the same in clusters and outside it can be multiplied by the mean universal (blue) luminosity $\mathcal{L}$ of the Universe. $\mathcal{L}$ is found by integrating over the luminosity function of galaxies, which gives (Yahil *et al*. 1980):

$$
\begin{array}{ll}
L \text{ in elliptical (E/S0) galaxies} & 2.6 \times 10^7 \left(\dfrac{H_0}{60}\right) L_{B\odot}\,\text{Mpc}^{-3} \\
\\
L \text{ in spiral galaxies (corrected for internal absorption)} & 1.06 \times 10^8 \left(\dfrac{H_0}{60}\right) L_{B\odot}\,\text{Mpc}^{-3} \quad (19) \\
\\
\text{Sum} & 1.32 \times 10^8 \left(\dfrac{H_0}{60}\right) L_{B\odot}\,\text{Mpc}^{-3}
\end{array}
$$

These values agree to better than factors of 1.2 between different authors, except that some neglect the dust absorption in spirals.

With eqn (19) it takes $\mathcal{M}/\mathcal{L} = 757\,(H_0/60)$ to obtain the critical density of eqn (15). Note that both $\mathcal{L} \times \mathcal{M}/L$ and $\rho_{0\mathrm{crit}}$ depend on H_0^2 and that hence Ω_m is independent of H_0. Ω_m can now be determined by measuring the mean mass-to-light ratio $\mathcal{M}/L$. Different methods have been applied.

Ω_m *in luminous stars*. The mass and light in galaxies within their optical (Holmberg) diameter gives $\mathcal{M}/L \sim 5$ (Persic *et al*. 1996; Kronawitter *et al*. 2000). The *luminous* matter therefore contributes with $\Omega_{\mathrm{lum}} \lesssim 0.007$ very little to Ω_b. Less than 15% of all baryons are luminous. The remaining baryons must reside in form of gas or compact objects. The nature of the unseen baryons is discussed, for example, by Carr (1994).

Ω_m *in the Local Group*. The total mass of the Galaxy and of M31, including their dark halos, is $1.9 \times 10^{12}\,\mathcal{M}_\odot$ (Wilkinson and Evans 2000) and $\sim 8.5 \times 10^{11}\,\mathcal{M}_\odot$ (Evans and Wilkinson 2000), respectively. This is consistent with a total dynamical mass of the Local Group of $4.9 \times 10^{12}\,\mathcal{M}_\odot$ within a radius of ~ 1 Mpc (Lynden-Bell 1999). Dividing the mass by the integrated blue light of the Local Group gives $\mathcal{M}/L = 49$ (independent of H_0), which corresponds, with a critical mass-to-light ratio of 757 ($H_0/60$), to $\Omega_m = 0.06\,(60/H_0)$. This with eqn (17) gives $\Omega_m/\Omega_b \approx 1$. From an arithmetic point of view all matter in the Local Group could hence be baryonic.

The determination of the mass of the Galaxy will be improved by microarc second measurements of its outlying mass tracers with the upcoming astrometric satellites.

Ω_m *in galaxies*. Integrating the mass of elliptical galaxies out to 300 ($60/H_0$) kpc gives a mass-to-light ratio of $\sim 200\,(H_0/60)$ (Bahcall 1997). If the former value is adopted for all types of galaxies one obtains, with eqn (19), $\Omega_m = 0.26$. If, however, one adopts for spirals $\mathcal{M}/L = 40$ as indicated by the Galaxy and M31 (see above) and combines the different mass-to-light ratios with the appropriate luminosity densities in eqn (19) the matter density becomes as low as $\Omega_m = 0.09$. This is only 1.8 $(60/H_0)^2$ times higher than the baryon density in eqn (17). The evidence for non-baryonic matter in galaxies is therefore not overwhelmingly strong.

Ω_m *in clusters*. The mass of galaxy clusters can be determined from the virial theorem and the requirement – assuming hydrostatic equilibrium – that the observed X-ray gas be bound. The latter method is, of course, entirely a child of the space age and its various X-ray satellites. The ensuing mass-to-light ratios level off beyond scales of $\gtrsim 0.4(60/H_0)$ Mpc at a median value of $\mathcal{M}/L \sim 165(H_0/60)$ (David *et al*. 1995; Carlberg *et al*. 1996; Bahcall 1997), corresponding, with a critical mass-to-light ratio of $(\mathcal{M}/L)_{\mathrm{crit}} = 757$, to $\Omega_m = 0.22$. This is an upper limit for two reasons. The value of $\mathcal{M}/L$ in clusters is dominated by E galaxies, and it may be smaller for spirals which dominate the field outside clusters. Moreover the dark matter contributing to $\mathcal{M}/L$ may be specific to the intracluster regions. In that case, however, the high $\mathcal{M}/L$ of individual E galaxies (see above) remains to be explained.

Clusters of galaxies, acting as gravitational lenses, distort the images of background galaxies transforming them into arc-like structures. The properties of these arcs, observed from the ground and with HST, allow a mass determination of the cluster. The well-studied cluster CL0024+16 gives thus a mass-to-light ratio of $\mathcal{M}/L = 190 \pm 20$ $(H_0/60)$ (Broadhurst et al. 2000). This is a lower limit because the cluster mass is integrated out to only 60 kpc and the dark cluster halo could extend to larger radial distances. If this single cluster is taken as representative one arrives at $\Omega_m \geq 0.25$. Also, the statistics of arcs favor low values of Ω_m (Kaufmann and Straumann 2000).

The determination of Ω_m from galaxies and galaxy clusters has been combined by Bahcall et al. (2000). They have determined $\mathcal{M}/L$ for spiral and elliptical galaxies, groups, clusters, and superclusters in regions of low and high density and over a very wide scale length of 20 kpc to $100(60/H_0)$ Mpc. They found an effective mean mass-to-light ratio of $\mathcal{M}/L = (135 \pm 40)$ $(H_0/60)$ and $\Omega_m = 0.16 \pm 0.05$. With the slightly higher luminosity density in eqn (19) one obtains $\Omega_m = 0.18 \pm 0.05$.

Ω_m from Ω_b. The total mass $\mathcal{M}_{total}$ of an X-ray cluster can be determined, as said before, by the virial theorem and by the mass required to bind the X-ray gas. The luminous mass $\mathcal{M}_{lum}$ is simply the sum of the X-ray gas mass and the mass in stars (which contribute only 5–20% of the luminous matter (Roussel et al. 2000). $\mathcal{M}_{lum}$ is due to baryons, but additional baryons may be hidden in compact objects and in form of cool gas. Therefore $\mathcal{M}_b \geq \mathcal{M}_{lum}$. If $f_b = \mathcal{M}_b/\mathcal{M}_{total} \geq \mathcal{M}_{lum}/\mathcal{M}_{total}$ is known, one obtains Ω_m from

$$\Omega_m = \frac{\Omega_b}{f_b} \quad (20)$$

A number of authors have determined $\mathcal{M}_{lum}/\mathcal{M}_{total}$ in a range of X-ray clusters. A typical value is $f_b \geq 0.163 \times (60/H_0)^{3/2}$ as given by Mohr et al. (2000). With the above baryon density of $\Omega_b = 0.053(60/H_0)^2$ from primordial nucleosynthesis (eqn (17)), the matter density becomes $\Omega_m \leq 0.33(60/H_0)^{1/2}$.

This upper limit could still be raised if f_b was higher outside clusters. This is indeed plausible because weakly interacting particles, which at very early epochs formed the first potential wells, but which cannot cool, are in the long run disadvantaged in forming clusters as compared to cooling baryons.

Balloon observations of the CMB suggest $f_b \approx 0.185$ (Jaffe et al. 2000), independent of H_0. This can be taken as confirmation of the above value of f_b. However, the absolute value of Ω_b from the CMB is suspiciously high, as mentioned in Section 2.1, and so is Ω_{CDM}, that is, $\Omega_b = 0.08$ and $\Omega_{CDM} = 0.53$ with $H_0 = 60$.

Rich clusters of galaxies must form in a low-density Universe at early epochs, while they continue to grow in a high-density Universe over extended periods. The observed, only mild evolution between local clusters and clusters at redshift $z \lesssim 1$ requires $\Omega_m = 0.3 \pm 0.1$ (Bahcall et al. 1997). From the temperature evolution of X-ray clusters one obtains $\Omega_m = 0.3$ (Chiu et al. 1997) and $\Omega_m = 0.50 \pm 0.14$ (Henry 1997). But the question as to the cluster formation rate is not settled at present. For other attempts to derive Ω_m from the fluctuation growth rate the reader is referred to Dekel et al. (1997).

2.2.2 Ω_m from unbound systems

Peculiar motions cannot be primordial, but must be induced by density fluctuations. To the extent that visible density fluctuations trace the true density fluctuations, their size correlates with the size of the peculiar motions. This gives a handle on the total matter density.

The cosmic virial theorem
The virial theorem of bound structures can be extended in a statistical sense to field galaxies by combining the two (or three)-point correlation function of these galaxies with their pairwise line-of-sight velocity dispersion.

Because peculiar motions affect the redshifts, but not the angular positions of galaxies, the correlation functions and the power spectra of galaxies must show characteristic distortions when they are viewed in redshift space rather than in real space. Ω_m can therefore be determined from the first and second moments of the velocity distribution function of pairs of galaxies in combination with model simulations (Kaiser 1987). This method has led to $\Omega_m = 0.15$ (Peebles 1993), $\Omega_m = 0.35 \pm 0.28$ (Fisher et al. 1994), and $\Omega_m = 0.74 \pm 0.45$ (Bromley et al. 1997). More exact solutions depend on the question as to whether luminous galaxies trace also the density distribution of dark matter. This seems nearly to be the case judging from model calculations of the formation of structures. In that case the "biasing factor" is close to one. On this assumption the most recent application of the method has given $\Omega_m = 0.35 \pm 0.15$ (Juszkiewicz et al. 2000).

The largest density fluctuations like clusters of galaxies and superclusters cause large-scale streaming motions of the surrounding field galaxies. Determinations of Ω_m from such streaming motions have the advantage of covering scales of up to 200 Mpc, but the disadvantage that the method requires reliable relative distances which must be corrected for severe selection bias occurring over large scales. Present solutions give rather divergent results (Dekel et al. 1997; Willick et al. 1997; da Costa et al. 1998; Sigad et al. 1997).

Ω_m from weak gravitational lensing
Foreground galaxies and clusters distort the images of background galaxies through gravitational lensing. The

Table 5 The density parameter Ω from different methods

	Method	Ω
Ω_b	Primordial nucleosynthesis	0.053
	Ly α absorption lines	≥ 0.066
Ω_{lum}	Integrated star light	≤ 0.007
Ω_v	Consistency arguments	(0.007)
Ω_m	From gravitationally bound systems	
	Galaxies	0.09–0.26
	Clusters	0.18 ± 0.05
	Matter/baryon ratio	$\leq 0.33(60/H_0)^{1/2}$
	Gravitational arcs	>0.25
	Cluster formation epoch	0.3–0.5
Ω_m	From unbound structures	0.35 ± 0.15
Ω_m	Weak lensing	0.16–0.20
Ω_m	Adopted	~ 0.3

effect is very weak, but can be measured from very large samples, that is, hundreds of foreground galaxies and hundreds of thousands of background galaxies. Wilson *et al.* (2001) have obtained in this way $\mathcal{M}/L = 150$ ($H_0/60$), corresponding to $\Omega_m = 0.2$, whereas Smith *et al.* (2000) obtained $\Omega_m = 0.16 \pm 0.03$, and Wittman *et al.* (2000) $\Omega_m = 0.19 \pm 0.03$. The errors do not include systematic effects.

Values of Ω from various methods are listed in Table 5.

3 THE COSMOLOGICAL CONSTANT Λ

Inflation theory strongly favors a flat Euclidean Universe with a *total* density of $\Omega_T = 1$. The most robust result of the BOOMERANG and MAXIMA 1 balloons is that indeed $\Omega_T = 1.0 \pm 0.15$ (Jaffe *et al.* 2000), which follows simply from the position of the first peak of the CMB fluctuation spectrum.

Since Ω_m is most likely smaller than 0.5, one must assume that the difference resides in some form of homogeneous energy. Quantum physics allows for a negative energy density of the vacuum which has driven inflation. Although it is a deep mystery why this energy is not still large in the present Universe, or otherwise why it not has become zero (Straumann 2001), the possibility of a *small* surviving energy density must consequently be accepted as a fact.

Various scalar fields can be thought of which can be associated with some kind of early energy, which has been termed in the most general sense "quintessence." One possible form of quintessence, permitted by general relativity is the cosmological constant Λ.

Einstein's motivation to introduce Λ, which H.P. Robertson called "Einstein's spook in the ether," has long since faded, but the term appears quite naturally in Friedmann's and Lemaître's equations:

$$H^2 = \frac{8\pi G}{3}\rho - \frac{kc^2}{R^2} + \frac{\Lambda c^2}{3} \quad (21)$$

(k is the space curvature $-1, 0, +1$). Λ has in this notation the dimension of (length)$^{-2}$. If the factor c^2 is suppressed, Λ is measured as (time)$^{-2}$. If Λ is positive the expansion of the Universe experiences an acceleration. This property of Λ makes the Universe older than H_0 and Ω_m would suggest (Section 4). A value of Λ of the order of unity has therefore been invoked for some time to accommodate for the globular cluster ages (Sandage and Tammann 1983).

It has become customary to express the Λ energy as a fraction of the critical density by writing

$$\Omega_\Lambda = \frac{\Lambda c^2}{3H^2} \quad (22)$$

While Λ must be constant over much of the history of the Universe, Ω_Λ increases with time because of the decrease of the Hubble constant.

Λ can be measured in principle by the acceleration of the Universe. This expectation has spectacularly been fulfilled using the Hubble diagram of SNe Ia on the assumption that they are *time-independent* standard candles (Figure 5). A positive Λ causes distant SNe Ia to have smaller redshifts than in a Friedmann universe with the same Ω_m and no cosmological constant. The implication is that the Hubble diagram does not actually measure Λ but $\Omega_\Lambda - \Omega_m$ to a first approximation.

In a carefully devised program Perlmutter *et al.* (1999) have analyzed 42 SNe Ia out to a redshift of $z = 0.8$. Their photometry is based in part on HST observations from space. The ensuing light curves are homogenized for a "stretch factor," corresponding to the decline rate discussed in Section 1.1.1. The SNe Ia then define a Hubble diagram with small scatter, which is best fit by $0.8\Omega_m - 0.6\Omega_\Lambda = -0.2 \pm 0.1$. For a flat universe with $\Omega_m + \Omega_\Lambda = 1$ this yields

$$\Omega_\Lambda = 0.7(\pm 0.1) \quad (23)$$

This corresponds to $\Lambda \approx 2 \times 10^{-35} \text{s}^{-2}$. The probability that $\Lambda = 0$ is excluded at the 99% level. The result is confirmed by Riess *et al.* (1998) who used more or less the same local SNe Ia, but another set of 16 distant SNe Ia with $z \lesssim 0.62$.

The result of $\Omega_\Lambda = 0.7$ fits so extraordinarily well with the external evidence of $\Omega_m \approx 0.3$, $\Omega_T = 1$ that it is difficult to think of an artefact. In spite of this the result depends on the assumption that SNe Ia at low z and at high z have the same luminosity. Although they have undistinguishable *observable* parameters, their uniformity is not evident

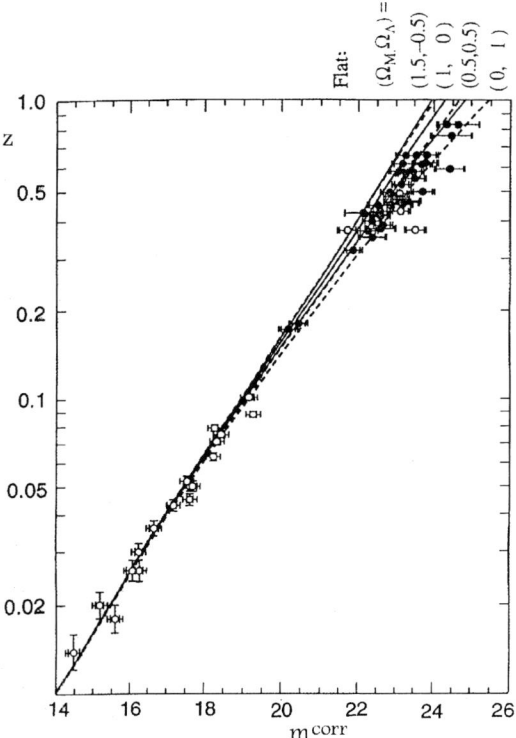

Figure 5 The Hubble diagram of 60 SNe Ia out to $z = 0.83$. The full lines are for $\Lambda = 0$ and $\Omega_m = 2$, 1, and 0 (from left to right). The line for an empty universe ($\Omega_M = 0$, $\Omega_\Lambda = 0$) lies slightly above the points. The dashed lines are for a flat universe ($\Omega_M + \Omega_\Lambda = 1$) and $\Lambda \geq 0$. (Adapted from Perlmutter et al. 1999.)

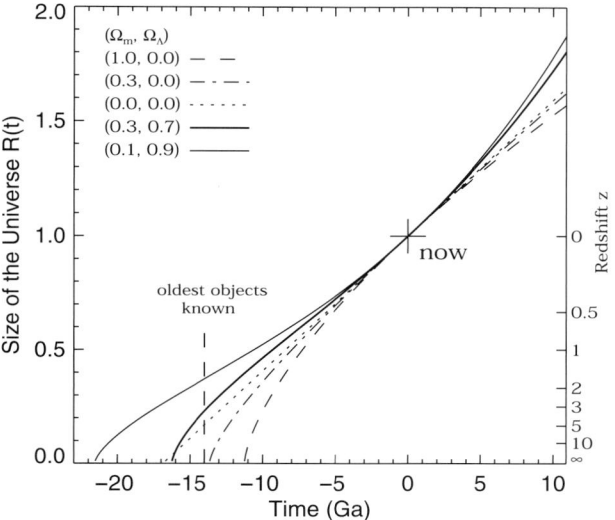

Figure 6 The size of the Universe in units of its present size as a fraction of time for different choices of Ω_m and Ω_Λ. $H_0 = 58$ is adopted.

because very distant SNe Ia may derive from fast-evolving, high-mass stars, which develop different onion shell structures once they have formed white dwarfs, and may lead to different detonation/explosion mechanisms as SNe Ia from low-mass progenitors, which may predominantly feed the present SNIa population (Höflich et al. 1998, 2000).

A space telescope, called SNAP (Supernova Acceleration Probe), is in the planning stage which is expected to gather highly consistent photospectroscopy of several thousand SNe Ia out to redshifts of $z < 1.7$. Going to such high redshifts is not particularly useful for the determination of Λ because at these early epochs its accelerating effect was still small, but in the absence of cosmic luminosity evolution this instrument will put important limits on the nature of Λ or any other form of quintessence.

In the meantime it is expected that automated telescopes on the ground will provide ~100 SNe Ia with $z \lesssim 0.1$ per year over the next years. They are important because in order to determine the curvature of the Hubble line, and hence Ω_m and Ω_Λ, one needs relatively local SNe Ia as well (Figure 1). SNe Ia in the redshift range $0.09 \lesssim z \lesssim 0.12$ will be particularly important, because they allow a "nearby" determination of Λ which strongly eases the problem of SN Ia luminosity evolution. This is because the look-back time to $z = 0.12$ is only 1.8 Gyr as compared to 8.0 Gyr at $z = 0.8$ (with H_0, Ω_m, and Ω_Λ as adopted here). In an $\Omega_m = 0.3$, $\Omega_\Lambda = 0.7$ universe SNe Ia at $z = 0.12$ are fainter by $0^m.12$ than in a flat universe with $\Omega_m = 1$. A hundred SNe Ia – with a cosmic scatter of $\sigma_M = 0^m.11$ after correction for decline rate and color (Section 1.1.1) – at this redshift will allow one to determine the Λ effect with a precision of 10%.

The SN Ia result on Λ receives strong and independent support from the CMB fluctuation spectrum as measured by BOOMERANG and MAXIMA 1. Including constraints of large-scale structure and inserting only a weak prior for H_0, Bond et al. (2000) have determined $\Omega_\Lambda = 0.66 \pm 0.06$.

Several groups have succeeded in measuring the cosmic shear, that is, the eccentricity of distant galaxies due to gravitational lensing of foreground galaxies. The first results on Ω_m are becoming available (Section 2.2.2). Separate values of Λ are not yet available, but the data are entirely consistent with $\Omega_m = 0.3$, $\Omega_\Lambda = 0.7$ (Helbig 1999; Maoli et al. 2000).

The adopted parameters of H_0, Ω_m, and Ω_Λ imply that the present Universe has already entered the phase of exponential expansion (Figure 6). In spite of this it would be imprudent to conclude that the Universe will expand forever. As Λ has decreased for unknown reasons from once gigantic values to its present minute size, it could further decrease at some future epoch and become even negative. In that case it would eventually lead to a re-collapse.

4 THE AGE OF THE UNIVERSE

The determination of the expansion age of the Universe is very easy in principle if H_0, Ω_m, and Ω_Λ are known. If all galaxies have receded from any observer in the Universe since the beginning, each with constant speed, they would have been at zero distance (that is, all matter clumped together) at a time T given by their present distance divided by their recession velocity, that is, $T = 1/H_0$! If the expansion is decelerated by gravity – and it *must* be if $\Omega_m > 0$ – the expansion was faster at earlier times and the galaxies could reach their present distances in a time $T < 1/H_0$. If in addition the expansion is accelerated by Λ, the gravitational deceleration is reduced, balanced, or even overcompensated depending on the relative size of Ω_M and Ω_Λ.

An almost exact formula for the expansion age T as a function of H_0, Ω_m, and Ω_Λ is given by Carroll *et al.* (1992; eqn (17)). With $H_0 = 58$, $\Omega_M = 0.3$, and $\Omega_\Lambda = 0.7$ one obtains

$$1.63 \times 10^{10} \text{ years} = 16.3 \text{ Gigayears (Gyr)} \quad (24)$$

If one changes Ω_m and Ω_Λ by 0.05, but keeping a flat universe with $\Omega_m + \Omega_\Lambda = 1$ as strongly favored by inflation *and* balloon observations of BOOMERANG and MAXIMA 1, T changes by ± 0.8 Gyr. Accounting in addition for a 10% error of H_0 the probable range of T becomes 16.3 ± 2.4 Gyr. For the correlation of the expansion age with Ω_m and Ω_Λ at fixed H_0 see Figure 7.

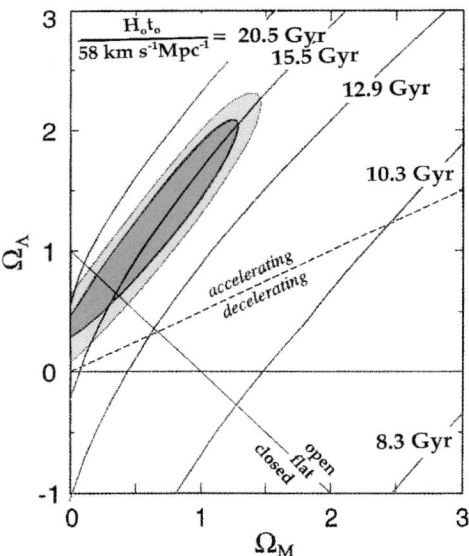

Figure 7 Expansion age isochrones for different values of Ω_m and Ω_Λ adopting $H_0 = 58$. The 68% and 90% confidence contours obtained from distant SNe Ia are shown. (Adapted from Perlmutter *et al.* 1999.)

The CMB fluctuation spectrum, containing information on H_0, Ω_m, Ω_Λ, and other constants, also implies, of course, the expansion age of the Universe. To the extent that these parameters are not yet well determined, the ensuing value of the age is also uncertain. Likely solutions over all parameters tend to give ages 2–4 Gyr smaller than in eqn (24) (Lineweaver 1999; Jaffe *et al.* 2000). A correspondingly low expansion age is not favored by age determinations in our Galaxy, but cannot yet be excluded. The double-valued CMB solution for H_0 of Lahav *et al.* (2000) is a clear indication that additional data from the ground, from balloons, and from the PLANCK satellite especially, as to the polarization are needed, before the CMB will yield an independent age determination.

It may be mentioned here that the concept of the age of the Universe is not any more so clear in modern cosmology as one would naively wish. It was pointed out (e.g. Lineweaver 1999) that the expansion age calculated here measures the time that has elapsed since the early moments when the classical equations of general relativity became valid. Yet models have been proposed with eternal inflation and multiple universes where the age question becomes more complex. Thus the term "age of the Universe" refers here only to the time when our "local" bubble began to expand, and which is taken to be $\sim 16 \pm 2$ Gyr.

The decisive test is whether this age can accommodate the oldest objects known. Some years back, when *HST* was still young and some still propagated values of H_0 of 70 or more, the expansion age was $T < 14.3$ Gyr. But at the same time a value of $\Omega_m = 1$ was widely held – due to a misunderstanding of inflation which actually requires $\Omega_{\text{Total}} = 1$ – which reduced T by another factor of 2/3 to become only 9.5 Gyr. This was clearly younger than the then known age of globular clusters and misled many journalists to lament that the Big Bang was dead (Maddox 1995). The error was simply that H_0 was neither 70 nor was $\Omega_m = 1$.

The age dating of globular clusters (GCs) has been much improved through photometry of *HST* and parallaxes from HIPPARCOS. The method is to plot individual stars of a GC into a color–magnitude diagram (CMD; Figure 8) and to fit them to isochrones, which astrophysicists can construct from the evolutionary tracks of stars of different mass and metallicity. The derived age depends, of course, on the accuracy of the stellar models, but it also depends critically on the distance of a GC and its reddening due to interstellar dust. HST has not only provided two-color photometry of many additional stars in the very crowded regions of GCs, but has also reached to much fainter stars. These faint, low-mass stars are still in the hydrogen–helium burning phase and populate the "main sequence" of the CMD. The latter can be fitted to a fiducial main sequence, which is calibrated in absolute magnitudes, to yield the distance and reddening. The problem is that the position of the main

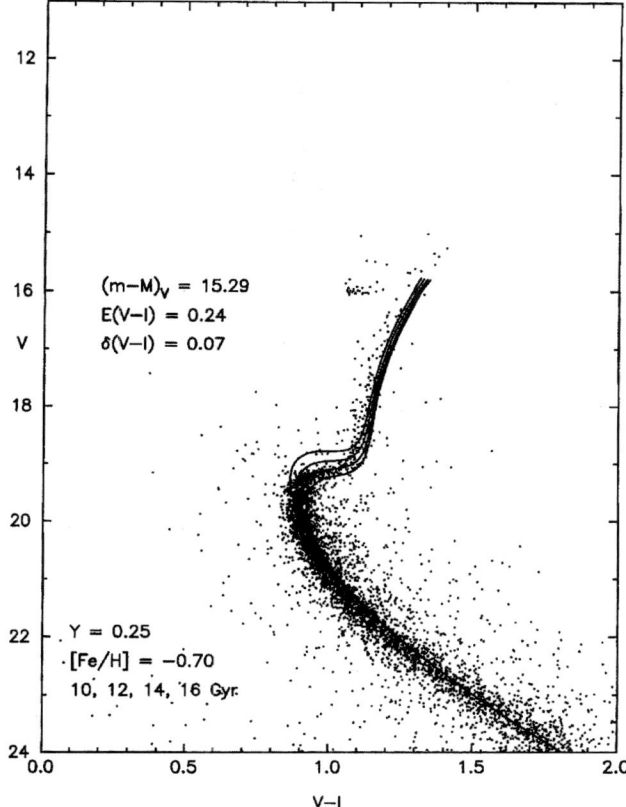

Figure 8 The color–magnitude diagram (CMD) of the globular cluster NGC 6637 as observed with HST. Isochrones from Yale calculations are shown for ages of 10, 12, 14, and 16 Gyr (from top to bottom). An age of 14 Gyr fits very well. (From Heasley et al. 2000.)

sequence depends on the metallicity and that many GCs are so metal poor that it becomes very difficult to calibrate a fiducial main sequence for the appropriate metallicity.

HIPPARCOS has solved the problem in part providing trigonometric parallaxes and hence absolute magnitudes for a number of low-metallicity stars – misnamed "subdwarfs" – but their number is not sufficient to give the calibrated main sequences for the full range of metallicities. The situation is aggravated by the fact that one must guard against the more massive and more luminous (and hence easier) subdwarfs because they may have evolved away from the main sequence. The problem will be solved by the forthcoming astrometric satellites DIVA and GAIA and the astrometric and interferometric satellite SIM. They will also improve the proper motions of GC stars which, combined with their radial velocity measurements and dynamical cluster models, will determine independent cluster distances.

Another possibility to derive GC distances is to match the magnitude of their RR Lyrae stars to the absolute luminosity calibration of these stars which, however, is dependent again on metallicity. Various routes have been used to accomplish this calibration. The definitive calibration will eventually be provided by GAIA.

It is not the place here to discuss all the intricacies of the determination of GC ages. It suffices to quote some results, and the reader can find the information as to which method has been applied and whose stellar evolution models have been used in the references quoted. Being interested only in the age of the *oldest* GCs the question as to their homogeneity or diversity in ages is not addressed here either.

Sandage (1993) determined the mean age of 24 GCs to be $t_{GC} = 14.1 \pm 1.5$ Gyr. This at the time surprisingly small age was due to new stellar models accounting for the excess abundance of oxygen as compared to iron (Bergbusch and VandenBerg 1992) and to his bright RR Lyrae star calibration, which was subsequently vindicated by the HIPPARCOS subdwarfs (Chaboyer et al. 1998). The latter authors found 11.5 ± 1.3 Gyr for the oldest GCs using their own Yale stellar model calculations. Cassisi et al. (1998), again with their own models, found 11.0 ± 1.0 Gyr for metal-poor GCs. Without using any distance Grundahl (1999) obtained 15 Gyr for the GC M15. Also a rather high age of 14 (± 2) Gyr was obtained for several *high*-metallicity GCs by Heasley et al. (2000), but based on a rather faint RR Lyrae star calibration. Chaboyer et al. (2000) derived uniform ages of 11.7 ± 1.6 Gyr for three metal-rich GCs, with a "slight suggestion" for M107 having $t_{GC} = 14.0 \pm 2.2$ Gyr (see also Sandage 1993). The different GC ages can be summarized by a compromise value of 13.0 ± 2.0 Gyr.

A completely independent age determination comes from the radioactive elements in the Solar System. The actinides ^{238}U, ^{235}U, and ^{232}Th are formed by rapid-neutron capture (r-process) presumably in supernovae of type II. If these elements, as found in the Solar System, were formed by a single event, they would require a universal age of ≥ 9 Gyr, but this is a strict lower limit, because there is strong evidence that the r-process elements were formed over an extended period, that is, during the entire star formation history of the Galaxy. The latter is not well specified, and ages of 10 or 20 Gyr cannot be excluded, but the most likely age range is 13–15 Gyr (Cowan et al. 1991). Truran (1999) favors a value at the lower limit of this range: 12.8 ± 3 Gyr.

A more direct way to date the oldest, that is, metal-poorest stars in the Galaxy has recently become feasible from observations from the ground and from HST. The abundance ratio of (radioactive) ^{232}Th and of (stable) Eu has been determined in a few field stars (Westin et al. 2000; Sneden et al. 2000a) and in three stars of the GC M15 (Sneden et al. 2000b). The resulting ages range from 14.3 to 16 Gyr; Sneden et al. (2000b) set a lower limit of 12.3 Gyr. The method has gained additional promise by the detection of ^{238}U in an extremely metal-poor star, giving an age of 12.5 ± 3 Gyr (Cayrel et al. 2001).

Table 6 The age of the Universe

Method	Age (Gyr)	Source
Expansion age	16.3 ± 2.4	This chapter
(1) Globular clusters	14.1 ± 1.5	Sandage (1993)
Metal-poor GCs	11.0 ± 1.0	Cassisi et al. (1998)
Oldest GCs	11.5 ± 1.3	Chaboyer et al. (1998)
Indep. of distance	15	Grundahl (1999)
Metal-rich GCs	11.7 ± 1.6	Chaboyer et al. (2000)
M107	14.0 ± 2.2	Chaboyer et al. (2000)
Metal-rich GCs	14 ± 2	Heasley et al. (2000)
(2) Nucleosynthesis history	14 ± 3	Cowan et al. (1991)
of radioactive actinides	12.8 ± 3	Truran (1999)
(3) Thorium and uranium ages of	15.6 ± 4	Westin et al. (2000)
extremely metal-poor stars	16 ± 4	Sneden et al. (2000a)
	14.3^{+3}_{-2}	Sneden et al. (2000b)
	12.5 ± 3	Cayrel et al. (2001)
(4) White dwarfs	(12)	Ibata and Lewis (1998)
(5) Dynamical age of Local Group	10–17	Lynden-Bell (1999)
Adopted from (1)–(4)	14 ± 2	
Including gestation time (~ 1 Gyr)	15 ± 2	

Lynden-Bell (1999) has determined the dynamical age of the Local Group to lie in the range of 10–17 Gyr depending on the adopted galaxy distances and the orbit eccentricities. An additional possibility to date the oldest stars in the galaxy is to search for the white dwarfs which have most progressed along their evolutionary path. Unfortunately the oldest white dwarfs are not the reddest ones, because molecular opacities cause the evolutionary tracks to turn bluewards at late phases (Hansen 1999; Saumon and Jacobson 1999). Moreover, the theory of cooling white dwarfs is very complex. Ibata and Lewis (1998) have set an uncertain age for an old white dwarf of 12 Gyr. Progress is expected from observations of white dwarfs in nearby GCs with the Advanced Camera of HST (Richer et al. 2000).

The above age determinations are compiled in Table 6. The ages of the oldest known galactic objects carry still relatively large and systematic errors, but by assuming an age of 14 ± 2 Gyr one does not violate any of the age determinations.

So far we have determined only ages of the presumably oldest objects in our Galaxy. But there must be some additional time between the Big Bang and the formation of the first stars. This interval is still poorly known. If one assumes that the first galaxies and stars lit up at redshift $z = 5$ then the interval becomes $\Delta t = 1.6$ Gyr (with H_0, Ω_m, and Ω_Λ as adopted above); if they formed already at redshift $z = 10$ one finds $\Delta t = 0.6$ Gyr. In fact modern model calculations of the formation of the large structures, including CDM and Ω_Λ, indicate that it took ~ 1 Gyr before concentrations of roughly $10^{10} \mathcal{M}_\odot$ could form (Samland 2000) which seems like a necessary prerequisite for the onset of star formation.

It seems indicated therefore to add 1.0 Gyr to the ages of the oldest objects to determine a minimum requirement of $T = 15 \pm 2$ Gyr for the time since the Big Bang. This value is in as good agreement with the expansion age of $T = 16.3 \pm 2.4$ Gyr as can possibly be expected.

The age test has so far led to a triumph. The oldest objects known, including a reasonable gestation time, are indeed younger than the expansion age. Future research with strong involvement of observations from space will further reduce the errors and will make the age test a powerful discriminator between different cosmological models.

Summarizing, it is clear that the determination of the cosmological parameters has gained enormously from space astronomy. It is foreseeable that the interplay of ground-based and space astronomy will continue to be most profitable for astronomy in general. While certain wavelength regions are the exclusive domain of space astronomy, the light-gathering power and relatively low cost of ground-based telescopes – with very large arrays and an optical 100 m telescope now in the planning – will remain competitive for many decades. However, the imaging quality over extended angles is a unique advantage of space astronomy.

REFERENCES

Bahcall, N. (1997). Dark matter. In N. Turok (ed), *Critical Dialogues in Cosmology*, World Scientific, Singapore, p. 221.

Bahcall, N.A., Cen, R., Davé, R., Ostriker, J.P. and Yu, Q. (2000). The mass-to-light function: Antibias and Ω_m. *Astrophysical Journal*, **541**, 1–9.

Bahcall, N.A., Fan, X. and Cen, R. (1997). Constraining Omega with cluster evolution. *Astrophysical Journal*, **485**, L53.

Baraffe, I. and Alibert, Y. (2001). Period–magnitude relationships in BVIJHK-bands for fundamental mode and first overtone Cepheids. *Astronomy and Astrophysics*, **371**, 592–599.

Baum, W.A., Hammergren, M., Thomsen, B., Groth, E.J., Faber, S.M., Grillmair, C.J. and Ajhar, E.A. (1997). Distance to the Coma cluster and a value for H_0 inferred from globular clusters in IC 4051. *Astronomical Journal*, **113**, 1483.

Bergbusch, P.A. and Vandenberg, D.A. (1992). Oxygen-enhanced models for globular cluster stars. II – Isochrones and luminosity functions. *Astrophysical Journal Supplement*, **81**, 163–220.

Binggeli, B., Popescu, C.C. and Tammann, G.A. (1993). The kinematics of the Virgo cluster revisited. *Astronomy and Astrophysics Supplement*, **98**, 275–296.

Binggeli, B., Sandage, A. and Tammann, G.A. (1985). Studies of the Virgo Cluster. II – A catalog of 2096 galaxies in the Virgo Cluster area. *Astronomical Journal*, **90**, 1681.

Birkinshaw, M. (1999). The Sunyaev–Zel'dovich effect. *Physics Reports*, **310**, 97–195.

Bond, J.R. (2000). CMB analysis of boomerang and maxima and the cosmic parameters. In A. Lasenby, A. Wilkinson and A.W. Jones (eds), *New Cosmological Data and the Values of the Fundamental Parameters*, IAU Symposium 201, in press, astro-ph/00 11 378.

Bottinelli, L., Gouguenheim, L., Paturel, G. and Teerikorpi, P. (1986). The Malmquist bias and the value of H_0 from the Tully–Fisher relation. *Astronomy and Astrophysics*, **156**, 157–171.

Bottinelli, L., Gouguenheim, L., Paturel, G. and Teerikorpi, P. (1991). A systematic effect in the use of planetary nebulae as standard candles. *Astronomy and Astrophysics*, **252**, 550–556.

Branch, D., Fisher, A. and Nugent, P. (1993). On the relative frequencies of spectroscopically normal and peculiar type Ia supernovae. *Astronomical Journal*, **106**, 2383–2391.

Bridle, S.L., Eke, V.R., Lahav, O., Lasenby, A.N., Hobson, M.P., Cole, S., Frenk, C.S. and Henry, J.P. (1999). Cosmological parameters from cluster abundances, cosmic microwave background and IRAS, *Monthly Notices of the Royal Astronomical Society*, **310**, 565–570.

Broadhurst, T., Huang, X., Frye, B. and Ellis, R. (2000). A spectroscopic redshift for the Cl 0024+16 multiple arc system: Implications for the central mass distribution. *Astrophysical Journal*, **534**, L15–L18.

Bromley, B.C., Warren, M.S. and Zurek, W.H. (1997). Estimating Omega from galaxy redshifts: Linear flow distortions and nonlinear clustering. *Astrophysical Journal*, **475**, 414.

Burles, S., Nollett, K.M. and Turner, M.S. (2001). What is the big-bang-nucleosynthesis prediction for the baryon density and how reliable is it? *Physical Review D*, **63**, 063512.

Capaccioli, M., della Valle, M., Rosino, L. and D'Onofrio, M. (1989). Properties of the nova population in M31. *Astronomical Journal*, **97**, 1622–1633.

Carlberg, R.G., Yee, H.K.C., Ellingson, E., Abraham, R., Gravel, P., Morris, S. and Pritchet, C.J. (1996). Galaxy cluster virial masses and Omega. *Astrophysical Journal*, **462**, 32.

Carr, B. (1994). Baryonic dark matter. *Annual Review of Astronomy and Astrophysics*, **32**, 531–590.

Carroll, S.M., Press, W.H. and Turner, E.L. (1992). The cosmological constant. *Annual Review of Astronomy and Astrophysics*, **30**, 499–542.

Cassisi, S., Castellani, V., degl'Innocenti, S. and Weiss, A. (1998). An updated theoretical scenario for globular cluster stars, *Astronomy and Astrophysics Supplement*, **129**, 267–279.

Cayrel, R., Hill, V., Beers, T.C., Barbuy, B., Spite, M., Spite, F., Plez, B., Andersen, J., Bonifacio, P., François, P., Molaro, P., Nordström, B. and Primas, F. (2001). Measurement of stellar age from uranium decay. *Nature*, **409**, 691–692.

Chaboyer, B., Demarque, P., Kernan, P.J. and Krauss, L.M. (1998). The age of globular clusters in light of Hipparcos: Resolving the age problem? *Astrophysical Journal*, **494**, 96.

Chaboyer, B., Sarajedini, A. and Armandroff, T.E. (2000). The age of the inner halo globular cluster NGC 6652. *Astronomical Journal*, **120**, 3102–3110.

Chiu, W.A., Ostriker, J.P. and Strauss, M.A. (1998). Using cluster abundances and peculiar velocities to test the Gaussianity of the cosmological density field. *Astrophysical Journal*, **494**, 479–490.

Coles, P. and Ellis, G. (1997). *Is the Universe Open or Closed? The Density of Matter in the Universe*, Cambridge University Press.

Cowan, J.J., Pfeiffer, B., Kratz, K.-L., Thielemann, F.-K., Sneden, C., Burles, S., Tytler, D. and Beers, T.C. (1999). R-process abundances and chronometers in metal-poor stars. *Astrophysical Journal*, **521**, 194–205.

Cowan, J.J., Thielemann, F.-K. and Truran, J.W. (1991). Radioactive dating of the elements. *Annual Review of Astronomy and Astrophysics*, **29**, 447–497.

da Costa, L.N., Nusser, A., Freudling, W., Giovanelli, R., Haynes, M.P., Salzer, J.J. and Wegner, G. (1998). Comparison of the SFI peculiar velocities with the IRAS 1.2-Jy gravity field. *Monthly Notices of the Royal Astronomical Society*, **299**, 425–432.

Dale, D.A., Giovanelli, R., Haynes, M.P., Campusano, L.E. and Hardy, E. (1999). Seeking the local convergence depth. V. Tully–Fisher peculiar velocities for 52 Abell clusters. *Astronomical Journal*, **118**, 1489–1505.

David, L.P., Jones, C. and Forman, W. (1995). Cosmological implications of ROSAT observations of groups and clusters of galaxies. *Astrophysical Journal*, **445**, 578–590.

Dekel, A., Burstein, D. and White, S.D.M. (1997). Measuring Omega. In N. Turok (ed), *Critical Dialogues in Cosmology*, World Scientific, Singapore, p.175.

Dixon, L. (1998). A SLAC physicist responds to the discovery of massive neutrinos. http://www.slac.stanford.edu/slac/announce/9806-japan-neutrino/dixonmore.html

Ekholm, T., Teerikorpi, P., Theureau, G., Hanski, M., Paturel, G., Bottinelli, L. and Gouguenheim, L. (1999). Kinematics of the local Universe. X. H_0 from the inverse B-band Tully–Fisher relation using diameter and magnitude limited samples. *Astronomy and Astrophysics*, **347**, 99–111.

Evans, N.W. and Wilkinson, M.I. (2000). The mass of the Andromeda galaxy. *Monthly Notices of the Royal Astronomical Society*, **316**, 929–942.

Faber, S.M., Wegner, G., Burstein, D., Davies, R.L., Dressler, A., Lynden-Bell, D. and Terlevich, R.J. (1989). Spectroscopy and photometry of elliptical galaxies. VI – Sample selection and data summary. *Astrophysical Journal Supplement*, **69**, 763–808.

Feast, M.W. (1999). Cepheids as distance indicators. *Publications of the Astronomical Society of the Pacific*, **111**, 775–793.

Federspiel, M. (1999). Kinematic parameters of galaxies as distance indicators. Ph.D. thesis, University of Basel.

Federspiel, M., Sandage, A. and Tammann, G.A. (1994). Bias properties of extragalactic distance indicators. III: Analysis of Tully–Fisher distances for the Mathewson–Ford–Buchhorn sample of 1355 galaxies. *Astrophysical Journal*, **430**, 29–52.

Federspiel, M., Tammann, G.A. and Sandage, A. (1998). The Virgo cluster distance from 21 centimeter line widths. *Astrophysical Journal*, **495**, 115.

Ferrarese, L., Freedman, W.L., Hill, R.J., Saha, A., Madore, B.F., Kennicutt, R.C., Jr., Stetson, P.B., Ford, H.C., Graham, J.A., Hoessel, J.G., Han, M., Huchra, J., Hughes, S.M., Illingworth, G.D., Kelson, D., Mould, J.R., Phelps, R., Silbermann, N.A., Sakai, S., Turner, A., Harding, P. and Bresolin, F. (1996). The Extragalactic Distance Scale Key Project. IV. The discovery of Cepheids and a new distance to M100 using the Hubble Space Telescope. *Astrophysical Journal*, **464**, 568.

Fisher, K.B., Davis, M., Strauss, M.A., Yahil, A. and Huchra, J.P. (1994). Clustering in the 1.2-Jy IRAS Galaxy Redshift Survey – II. Redshift distortions and $\xi(r_p,\pi)$. *Monthly Notices of the Royal Astronomical Society*, **267**, 927–948.

Fitzpatrick, E.L., Ribas, I., Guinan, E.F., DeWarf, L.E., Maloney, F.P. and Massa, D.L. (2000). The distance to the Large Magellanic Cloud from eclipsing binaries. American Astronomical Society Meeting 196, 28.06.

Freedman, W.L., Hughes, S.M., Madore, B.F., Mould, J.R., Lee, M.G., Stetson, P., Kennicutt, R.C., Turner, A., Ferrarese, L., Ford, H., Graham, J.A., Hill, R., Hoessel, J.G., Huchra, J. and Illingworth, G.D. (1994). The Hubble Space Telescope Extragalactic Distance Scale Key Project. 1: The discovery of Cepheids and a new distance to M81. *Astrophysical Journal*, **427**, 628–655.

Gerhard, O. and Silk, J. (1996). Baryonic dark halos a model with MACHOS and cold gas globules. In R. Bender and R.L. Davies (eds), *New Light on Galaxy Evolution*, IAU Symposium 171, Kluwer, Dordrecht, pp. 167–178.

Gibson, B.K., Stetson, P.B., Freedman, W.L., Mould, J.R., Kennicutt, R.C., Jr., Huchra, J.P., Sakai, S., Graham, J.A., Fassett, C.I., Kelson, D.D., Ferrarese, L., Hughes, S.M.G., Illingworth, G.D., Macri, L.M., Madore, B.F., Sebo, K.M. and Silbermann, N.A. (2000). The Hubble Space Telescope Key Project on the extragalactic distance scale. XXV. A recalibration of Cepheid distances to Type Ia supernovae and the Value of the Hubble constant. *Astrophysical Journal*, **529**, 723–744.

Giovanelli, R., Haynes, M.P., Herter, T., Vogt, N.P., Wegner, G., Salzer, J.J., da Costa, L.N. and Freudling, W. (1997). The I band Tully–Fisher relation for cluster galaxies: Data presentation. *Astronomical Journal*, **113**, 22–52.

Goodwin, S.P., Gribbin, J. and Hendry, M.A. (1997). A new determination of the Hubble parameter using galaxy linear diameters. *Astronomical Journal*, **114**, 2212.

Graham, J.A., Ferrarese, L., Freedman, W.L., Kennicutt, R.C., Jr., Mould, J.R., Saha, A., Stetson, P.B., Madore, B.F., Bersolin, F., Ford, H.C., Gibson, B.K., Han, M., Hoessel, J.G., Huchra, J., Hughes, S.M., Illingworth, G.D., Kelson, D.D., Macri, L., Phelps, R., Sakai, S., Silbermann, N.A. and Turner, A. (1999). The Hubble Space Telescope Key Project on the extragalactic distance scale. XX. The discovery of Cepheids in the Virgo cluster galaxy NGC 4548. *Astrophysical Journal*, **516**, 626–646.

Gratton, R. (2000). RR Lyrae stars and Cepheids from HIPPARCOS. In É. Aubourg, T. Montmerle, J. Paul and P. Peter (eds), XIXth Texas Symposium on Relativistic Astrophysics and Cosmology, (North-Holland), Mini-symposium 13/03 (CD-Rom).

Groenewegen, M.A.T. and Oudmaijer, R.D. (2000). Multi-colour PL-relations of Cepheids in the bt HIPPARCOS catalogue and the distance to the LMC. *Astronomy and Astrophysics*, **356**, 849–872.

Grundahl, F. (1999). Globular cluster ages and Strömgren CCD photometry. In I. Hubeny, S. Heap and R. Cornett (eds), *Spectrophotometric Dating of Stars and Galaxies*, ASP Conference Series Vol. 192, Astronomical Society of the Pacific, San Francisco, p. 223.

Hamuy, M., Phillips, M.M., Suntzeff, N.B., Schommer, R.A., Maza, J. and Aviles, R. (1996). The Hubble diagram of the Calan/Tololo Type Ia Supernovae and the value of H_0. *Astronomical Journal*, **112**, 2398.

Hansen, B. (1999). Cooling models for old white dwarfs. *Astrophysical Journal*, **520**, 680–695.

Hansen, S.H. and Villante, F.L. (2000). Decaying neutrino and a high cosmological baryon density. *Physics Letters B*, **486**, 1–5.

Heasley, J.N., Janes, K.A., Zinn, R., Demarque, P., Da Costa, G.S. and Christian, C.A. (2000). Hubble Space Telescope photometry of the metal-rich globular clusters NGC 6624 and NGC 6637. *Astronomical Journal*, **120**, 879–893.

Helbig, P. (1999). Gravitational lensing statistics with extragalactic surveys. III. Joint constraints on λ_0 and Ω_0 from lensing statistics and the m-z relation for type Ia supernovae. *Astronomy and Astrophysics*, **350**, 1–8.

Henry, J.P. (1997). A measurement of the density parameter derived from the evolution of cluster X-ray temperatures. *Astrophysical Journal*, **489**, L1.

Herrnstein, J.R., Moran, J.M., Greenhill, L.J., Diamond, P.J., Inoue, M., Nakai, N., Miyoshi, M., Henkel, C. and Riess, A. (1999). A geometric distance to the galaxy NGC 4258 from orbital motions in a nuclear gas disk. *Nature*, **400**, 539–541.

Höflich, P. and Khokhlov, A. (1996). Explosion models for Type Ia supernovae: A comparison with observed light curves, distances, H_0, and q_0. *Astrophysical Journal*, **457**, 500.

Höflich, P., Nomoto, K., Umeda, H. and Wheeler, J.C. (2000). Influence of the stellar population on Type IA supernovae: Consequences for the determination of Ω. *Astrophysical Journal*, **528**, 590–596.

Höflich, P., Wheeler, J.C. and Thielemann, F.K. (1998). Type IA supernovae: Influence of the initial composition on the nucleosynthesis, light curves, and spectra and consequences for the determination of Ω_M and Λ. *Astrophysical Journal*, **495**, 617–629.

Hubble, E.P. and Humason, M.L. (1931). The velocity-distance relation among extra-galactic nebulae. *Astrophysical Journal*, **74**, 43.

Humason, M.L., Mayall, N.U. and Sandage, A.R. (1956). Redshifts and magnitudes of extragalactic nebulae. *Astronomical Journal*, **61**, 97–162.

Ibata, R.A. and Lewis, G.F. (1998). Optimal proper-motion measurements with the Wide Field and Planetary Camera. *Astronomical Journal*, **116**, 2569–2573.

Jaffe, A.H., Ade, P.A.R., Balbi, A., Bock, J.J., Bond, J.R., Borrill, J., Boscaleri, A., Coble, K., Crill, B.P., de Bernardis, P., Farese, P., Ferreira, P.G., Ganga, K., Giacometti, M., Hanany, S., Hivon, E., Hristov, V.V., Iacoangeli, A., Lange, A.E., Lee, A.T., Martinis, L., Masi, S., Mauskopf, P.D., Melchiorri, A., Montroy, T., Netterfield, C.B., Oh, S., Pascale, E., Piacentini, F., Pogosyan, D., Prunet, S., Rabii, B., Rao, S., Richards, P.L., Romeo, G., Ruhl, J.E., Scaramuzzi, F., Sforna, D., Smoot, G.F., Stompor, R., Winant, C.D. and Wu, J.H.P. (2001). Cosmology from Maxima-1, Boomerang and COBE/DMR cosmic microwave background observations. *Physical Review Letters*, **86**, 3457–3479.

Jerjen, H. and Tammann, G.A. (1993). The Local Group motion towards Virgo and the microwave background. *Astronomy and Astrophysics*, **276**, 1.

Juszkiewicz, R., Ferreira, P.G., Feldman, H.A., Jaffe, A.H. and Davis, M. (2000). Evidence for a low-density universe from the relative velocities of galaxies. *Science*, **287**, 109.

Kaiser, N. (1987). Clustering in real space and in redshift space. *Monthly Notices of the Royal Astronomical Society*, **227**, 1–21.

Kamionkowski, M. and Kosowsky, A. (1999). The cosmic microwave background and particle physics. *Annual Review of Nuclear and Particle Science*, **49**, 77–123.

Kaufmann, R. and Straumann, N. (2000). Giant arc statistics and cosmological parameters. *Annalen der Physik*, **9**, 384–394.

Kavelaars, J.J., Harris, W.E., Hanes, D.A., Hesser, J.E. and Pritchet, C.J. (2000). The globular cluster systems in the coma ellipticals. I. The luminosity function in NGC 4874 and implications for Hubble's constant. *Astrophysical Journal*, **533**, 125–136.

Kelson, D.D., Illingworth, G.D., Tonry, J.L., Freedman, W.L., Kennicutt, R.C., Jr., Mould, J.R., Graham, J.A., Huchra, J.P., Macri, L.M., Madore, B.F., Ferrarese, L., Gibson, B.K., Sakai, S., Stetson, P.B., Ajhar, E.A., Blakeslee, J.P., Dressler, A., Ford, H.C., Hughes, S.M.G., Sebo, K.M. and Silbermann, N.A. (2000). The Hubble Space Telescope Key Project on the extragalactic distance scale. XXVII. A derivation of the Hubble constant using the fundamental plane and D_n–σ relations in Leo I, Virgo, and Fornax. *Astrophysical Journal*, **529**, 768–785.

Kochanek, C.S., Keeton, C.R. and McLeod, B.A. (2001). The importance of Einstein rings. *Astrophysical Journal*, **547**, 50–59.

Kowal, C.T. (1968). Absolute magnitudes of supernovae. *Astronomical Journal*, **73**, 1021–1024.

Kristian, J., Sandage, A. and Westphal, J.A. (1978). The extension of the Hubble diagram. II – New redshifts and photometry of very distant galaxy clusters – First indication of a deviation of the Hubble diagram from a straight line. *Astrophysical Journal*, **221**, 383–394.

Kronawitter, A., Saglia, R.P., Gerhard, O. and Bender, R. (2000). Orbital structure and mass distribution in elliptical galaxies. *Astronomy and Astrophysics Supplement*, **144**, 53–84.

Lahav, O., Bridle, S.L., Hobson, M.P., Lasenby, A.N. and Sodré, L. (2000). Bayesian 'hyper-parameters' approach to joint estimation: The Hubble constant from CMB measurements. *Monthly Notices of the Royal Astronomical Society*, **315**, L45–L49.

Lineweaver, C.A. (1999). A younger age for the Universe. *Science*, **284**, 1503.

Linsky, J. (1998). Deuterium abundance in the local ISM and possible spatial variations. In N. Prantzos, M. Tosi and R. von Steiger (eds), *The Primordial Nuclei and their Galactic Evolution*, Kluwer, Dordrecht, pp. 285–296.

Livio, M. (1997). Novae as distance indicators. In M. Livio, M. Donahue and N. Panagia (eds), *The Extragalactic Distance Scale*, Cambridge University Press, pp. 186–196.

Longair, M.S. and Warner, J.W. (eds) (1979). *Scientific Research with the Space Telescope*, IAU Colloquium 54, US Government Printing Office, Washington DC.

Lynden-Bell, D. (1999). Local group dynamics. In P. Whitelock and R. Cannon (eds), *The Stellar Content of Local Group Galaxies*, IAU Symposium 192, Astronomical Society of the Pacific, San Francisco, pp. 39–50.

Macchetto, F., Pacini, F. and Tarenghi, M. (eds) (1979). *Astronomical Uses of the Space Telescope*, ESA/ESO, Geneva.

Macri, L.M., Huchra, J.P., Stetson, P.B., Silbermann, N.A., Freedman, W.L., Kennicutt, R.C., Mould, J.R., Madore, B.F., Bresolin, F., Ferrarese, L., Ford, H.C., Graham, J.A., Gibson, B.K., Han, M., Harding, P., Hill, R.L., Hoessel, J.G., Hughes, S.M.G., Kelson, D.D., Illingworth, G.D., Phelps, R.L., Prosser, C.F., Rawson, D.M., Saha, A., Sakai, S. and Turner, A. (1999). The Extragalactic Distance Scale Key Project. XVIII. The discovery of Cepheids and a new distance to NGC 4535 using the Hubble Space Telescope. *Astrophysical Journal*, **521**, 155–178.

Maddox, J. (1995). Big Bang not yet dead but in decline. *Nature*, **377**, 99.

Madore, B. and Freedman, W.L. (1998). Hipparcos parallaxes and the Cepheid distance scale. *Astrophysical Journal*, **492** 110.

Malmquist, K.G. (1920). A study of the stars of spectral type A. Lund Communications Series II, **22**, 1.

Malmquist, K.G. (1922). On some relations in stellar statistics. Lund Communications Series I, **100**, 1.

Maoli, R., Mellier, Y., van Waerbeke, L., Schneider, P., Jain, B., Erben, T., Bernadeau, F., Fort, B., Bertin, E. and Dantel-Fort, M. (2000). *The Messenger*, **101**, 10.

Maoz, E., Newmann, J.A., Ferrarese, L., Stetson, P.B., Zepf, S.E., Davis, M., Freedman, W.L. and Madore, B.F. (1999). A distance to the galaxy NGC 4258 from observations of Cepheids variable stars. *Nature*, **401**, 351–354.

Milsztajn, A. (2001). In P. Jetzer, K. Pretzl and R. von Steiger (eds), *Matter in the Universe*, ISSI, Bern, in press.

Mohr, J.J., Haiman, Z. and Holder, G.P. (2000). Galaxy cluster baryon fractions, cluster surveys and cosmology. astro-ph/00 04 244.

Mould, J.R., Hughes, S.M.G., Stetson, P.B., Gibson, B.K., Huchra, J.P., Freedman, W.L., Kennicutt, R.C., Jr., Bresolin, F., Ferrarese, L., Ford, H.C., Graham, J.A., Han, M., Hoessel, J.G., Illingworth, G.D., Kelson, D.D., Macri, L.M., Madore, B.F., Phelps, R.L., Prosser, C.F., Rawson, D., Saha, A., Sakai, S., Sebo, K.M., Silbermann, N.A. and Turner, A.M. (2000). The Hubble Space Telescope Key Project on the extragalactic distance scale. XXI. The Cepheid distance to NGC 1425. *Astrophysical Journal*, **528**, 655–676.

Panagia, N. (1998). Distance to SN 1987A and the LMC. In Y.H. Chu, N. Suntzeff, J. Hesser and D. Bohlender (eds), *New Views of the Magellanic Clouds*, IAU Symposium 190, Astronomical Society of the Pacific, San Francisco, pp. 549–553.

Parodi, B.R., Saha, A., Sandage, A. and Tammann, G.A. (2000). Supernova Type Ia luminosities and their dependence on second parameters, and the value of H_0. *Astrophysical Journal*, **540**, 634–651.

Paturel, G., Lanoix, P., Teerikorpi, P., Theureau, G., Bottinelli, L., Gouguenheim, L., Renaud, N. and Witasse, O. (1998). Hubble constant from sosie galaxies and HIPPARCOS geometrical calibration. *Astronomy and Astrophysics*, **339**, 671–677.

Peebles, P.J.E. (1993). *Principles of Physical Cosmology*, Princeton University Press, NJ.

Perlmutter, S., Aldering, G., Goldhaber, G., Knop, R.A., Nugent, P., Castro, P.G., Deustua, S., Fabbro, S., Goobar, A., Groom, D.E., Hook, I.M., Kim, A.G., Kim, M.Y., Lee, J.C., Nunes, N.J., Pain, R., Pennypacker, C.R., Quimby, R., Lidman, C., Ellis, R.S., Irwin, M., McMahon, R.G., Ruiz-Lapuente, P., Walton, N., Schaefer, B., Boyle, B.J., Fillppenko, A.V., Matheson, T., Fruchter, A.S., Panagia, N., Newberg, H.J.M. and Couch, W.J. (1999). Measurements of Omega and Lambda from 42 high-redshift supernovae. *Astrophysical Journal*, **517**, 565–586.

Persic, M., Salucci, P. and Stel, F. (1996). The universal rotation curve of spiral galaxies – I. The dark matter connection. *Monthly Notices of the Royal Astronomical Society*, **281**, 27–47 [and erratum: **283**, 1102].

Pont, F. (1999). The Cepheid distance scale after HIPPARCOS. In D. Egret and A. Heck (eds), *Harmonizing Cosmic Distance Scales in a Post-HIPPARCOS Era*, ASP Conference Series Vol. 167, Astronomical Society of the Pacific, San Francisco, p. 113.

Prosser, C.F., Kennicutt, R.C., Jr., Bresolin, F., Saha, A., Sakai, S., Freedman, W.L., Mould, J.R., Ferrarese, L., Ford, H.C., Gibson, B.K., Graham, J.A., Hoessel, J.G., Huchra, J.P., Hughes, S.M., Illingworth, G.D., Kelson, D.D., Macri, L., Madore, B.F., Silbermann, N.A. and Stetson, P.B. (1999). The Hubble Space Telescope Key Project on the extragalactic distance scale. XXII. The discovery of Cepheids in NGC 1326A. *Astrophysical Journal*, **525**, 80–104.

Rauch, M., Miralda-Escude, J., Sargent, W.L.W., Barlow, T.A., Weinberg, D.H., Hernquist, L., Katz, N., Cen, R. and Ostriker, J.P. (1997). The opacity of the LY alpha forest and implications for Omega B and the ionizing background. *Astrophysical Journal*, **489**, 7.

Rebolo, R. (2001). In P. Jetzer, K. Pretzl and R. von Steiger (eds), *Matter in the Universe*, ISSI, Bern, in press.

Reimers, D., Kohler, S., Wisotzki, L., Groote, D., Rodriguez-Pascual, P. and Wamsteker, W. (1997). Patchy intergalactic He II absorption in HE 2347–4342. II. The possible discovery of the epoch of He-reionization. *Astronomy and Astrophysics*, **327**, 890–900.

Richer, H.B., Hansen, B., Limongi, M., Chieffi, A., Straniero, O. and Fahlman, G.G. (2000). Isochrones and luminosity functions for old white dwarfs. *Astrophysical Journal*, **529**, 318–337.

Riess, A.G., Fillippenko, A.V., Challis, P., Clocchiatti, A., Diercks, A., Garnavich, P.M., Gilliland, R.L., Hogan, C.J., Jha, S., Kirshner, R.P., Leibundgut, B., Phillips, M.M., Reiss, D., Schmidt, B.P., Schommer, R.A., Smith, R.C., Spyromilio, J., Stubbs, C., Suntzeff, N.B. and Tonry, J. (1998). Observational evidence from supernovae for an accelerating universe and a cosmological constant. *Astronomical Journal*, **116**, 1009–1038.

Romaniello, M., Salaris, M., Cassisi, S. and Panagia, N. (2000). Hubble Space Telescope observations of the Large Magellanic Cloud field around SN 1987A: Distance determination with red clump and tip of the red giant branch stars. *Astrophysical Journal*, **530**, 738–743.

Roussel, H., Sadat, R. and Blanchard, A. (2000). The baryon content of groups and clusters of galaxies. *Astronomy and Astrophysics*, **361**, 429–443.

Saha, A., Labhardt, L., Schwengeler, H., Macchetto, F.D., Panagia, N., Sandage, A. and Tammann, G.A. (1994). Discovery of Cepheids in IC 4182: Absolute peak brightness of SN Ia 1937C and the value of H_0. *Astrophysical Journal*, **425**, 14–34.

Saha, A., Sandage, A., Labhardt, L., Schwengeler, H., Tammann, G.A., Panagia, N. and Macchetto, F.D. (1995). Discovery of Cepheids in NGC 5253: Absolute peak brightness of SN Ia 1895B and SN Ia 1972E and the value of H_0. *Astrophysical Journal*, **438**, 8–26.

Saha, A., Sandage, A., Labhardt, L., Tammann, G.A., Macchetto, F.D. and Panagia, N. (1996a). Cepheid calibration of the peak brightness of Type Ia supernovae. V. SN 1981B in NGC 4536. *Astrophysical Journal*, **466**, 55.

Saha, A., Sandage, A., Labhardt, L., Tammann, G.A., Macchetto, F.D. and Panagia, N. (1996b). Cepheid calibration of the peak brightness of Type Ia supernovae. V. SN 1981B in NGC 4536. *Astrophysical Journal Supplement*, **107**, 693.

Saha, A., Sandage, A., Labhardt, L., Tammann, G.A., Macchetto, F.D. and Panagia, N. (1997). Cepheid calibration of the peak brightness of Type Ia supernovae. VIII. SN 1990N in NGC 4639. *Astrophysical Journal*, **486**, 1.

Saha, A., Sandage, A., Tammann, G.A., Labhardt, L., Macchetto, F.D. and Panagia, N. (1999). Cepheid calibration of the peak brightness of Type Ia supernovae. IX. SN 1989B in NGC 3627. *Astrophysical Journal*, **522**, 802–838.

Saha, A., Sandage, A., Thim, F., Tammann, G.A., Labhardt, L., Christensen, J., Macchetto, F.D. and Panagia, N. (2000). Cepheid calibration of the peak brightness of Type Ia supernovae. X. SN1991T in NGC4527. *Astrophysical Journal*, **551**, 973–1015.

Saha, P. (2002). Private communication.

Sakai, S., Zwitsky, D. and Kennicutt, R.C. (2000). The tip of the red giant branch distance to the Large Magellanic Cloud. *Astrophysical Journal*, **119**, 1197–1204.

Samland, M. (2000). Private communication.

Sandage, A. (1961). The ability of the 200-Inch Telescope to discriminate between selected world models. *Astrophysical Journal*, **133**, 355.

Sandage, A. (1966). Radio sources and the expanding universe. In L. Rosino (ed), *Atti del Conv. sulla Cosmologia*, G. Barbera, Firenze, p. 104.

Sandage, A. (1970). Cosmology: A search for two numbers. *Physics Today*, **23**, 34–41.

Sandage, A. (1988). A case for $H_0 = 42$ and $\Omega_0 = 1$ using luminous spiral galaxies and the cosmological time scale test. *Astrophysical Journal*, **331**, 583–604.

Sandage, A. (1993). Globular cluster ages determined from the Oosterhoff period-metallicity effect using oxygen-enhanced isochrones. III. *Astrophysical Journal*, **106**, 719–725.

Sandage, A. (1995). Practical cosmology: Inverting the past. In B. Binggeli and R. Buser (eds), *The Deep Universe, Saas Fee Lectures*, Springer, Berlin, p. 210.

Sandage, A. (1996a). Bias properties of extragalactic distance indicators. V. H_0 from luminosity functions of different spiral types and luminosity classes corrected for bias. *Astrophysical Journal*, **111**, 1.

Sandage, A. (1996b). Bias Properties of extragalactic distance indicators. VI. Luminosity functions of M 31 and M 101 look-alikes listed in the RSA2: H_0 therefrom. *Astrophysical Journal*, **111**, 18.

Sandage, A. (1999). Bias properties of extragalactic distance indicators. VIII. H_0 from distance-limited luminosity class and morphological type-specific luminosity functions for Sb, Sbc, and Sc galaxies calibrated using Cepheids. *Astrophysical Journal*, **527**, 479–487.

Sandage, A. and Bedke, J. (1988). *Atlas of Galaxies useful to Measuring the Cosmological Distance Scale*, NASA, Washington DC.

Sandage, A. and Hardy, E. (1973). The redshift-distance relation. VII. absolute magnitudes of the first three ranked cluster galaxies as functions of cluster richness and Bautz-Morgan cluster type: The effect of q_0. *Astrophysical Journal*, **183**, 743–758.

Sandage, A. and Tammann, G.A. (1974). Steps toward the Hubble constant. III. The distance and stellar content of the M101 group of galaxies. *Astrophysical Journal*, **194**, 223–243.

Sandage, A. and Tammann, G.A. (1982). Steps toward the Hubble constant. VIII. The global value. *Astrophysical Journal*, **256**, 339–345.

Sandage, A. and Tammann, G.A. (1983). In G. Setti and L. Van Howe (eds), *Large-Scale Structure of the Universe, Cosmology and Fundamental Physics*, p. 127.

Sandage, A. and Tammann, G.A. (1990). Steps toward the Hubble constant. IX. The cosmic value of H_0 freed from all local velocity anomalies. *Astrophysical Journal*, **365**, 1–12.

Sandage, A., Tammann, G.A. and Federspiel, M. (1995). Bias properties of extragalactic distance indicators. IV. Demonstration of the population incompleteness bias inherent in the Tully-Fisher method applied to clusters. *Astrophysical Journal*, **452**, 1.

Saumon, D. and Jacobson, S.B. (1999). Pure hydrogen model atmospheres for very cool white dwarfs. *Astrophysical Journal*, **511**, L107–L110.

Schlegel, D., Finkbeiner, D. and Davis, M. (1998). Maps of dust infrared emission for use in estimation of reddening and cosmic microwave background radiation foregrounds. *Astrophysical Journal*, **500**, 525.

Schröder, A. (1995). UBVI photometry of spiral galaxies in the Virgo and Fornax clusters. Ph.D. thesis, University of Basel.

Sigad, Y., Eldar, A., Dekel, A., Strauss, M.A. and Yahil, A. (1997). IRAS versus POTENT density fields on large scales: Biasing parameter and Ω. *Astrophysical Journal*, **495**, 516–532.

Silbermann, N.A., Harding, P., Ferrarese, L., Stetson, P.B., Madore, B.F., Kennicutt, R.C., Jr., Freedman, W.L., Mould, J.R., Bresolin, F., Ford, H., Gibson, B.K., Graham, J.A., Han, M., Hoessel, J.G., Hill, R.J., Huchra, J., Hughes, S.M.G., Illingworth, G.D., Kelson, D., Macri, L., Phelps, R., Rawson, D., Sakai, S. and Turner, A. (1999). The Hubble Space Telescope Key Project on the extragalactic distance scale. XIV. The Cepheids in NGC 1365. *Astrophysical Journal*, **515**, 1–28.

Smith, D.R., Bernstein, G.M., Fisher, P. and Jarvis, R.M. (2001). Weak lensing determination of the mass in galaxy halos. *Astrophysical Journal*, **551**, 643–650.

Sneden, C., Cowan, J.J., Ivans, I.I, Fuller, G.M., Burles, S., Beers, T.C. and Lawler, J.E. (2000). Evidence of multiple R-process sites in the early galaxy: New observations of CS 22892-052. *Astrophysical Journal*, **533**, L139–L142.

Sneden, C., Johnson, J., Kraft, R.P., Smith, G.H., Cowan, J.J. and Bolte, M.S. (2000). Neutron-capture element abundances in the globular cluster M15. *Astrophysical Journal*, **536**, L85–L88.

Sparks, W.B. (1994). A direct way to measure the distances of galaxies. *Astrophysical Journal*, **433**, 19–28.

Sparks, W.B., Macchetto, F., Panagia, N., Boffi, F.R., Branch, D., Hazen, M.L. and della Valle, M. (1999). Evolution of the light echo of SN 1991T, *Astrophysical Journal*, **523**, 585–592.

Straumann, N. (2001). On the mystery of the cosmic vacuum energy density. In H.V. Klapdor-Kleingrothaus (ed), *Dark Matter in Astro- and Particle Physics*, Springer, Berlin, p. 110.

Tammann, G.A. and Sandage, A. (1999). The luminosity function of globular clusters as an extragalactic distance indicator. In D. Egret and A. Heck (eds), *Harmonizing Cosmic Distance Scales in a Post-Hipparcos Era*, ASP Conference Series Vol. 167, Astronomical Society of the Pacific, San Francisco, p. 204.

Tammann, G.A., Sandage, A. and Reindl, B. (2000). The distance of the Virgo cluster. In É. Aubourg, T. Montmerle, J. Paul and P. Peter (eds), *XIXth Texas Symposium on Relativistic Astrophysics and Cosmology*, (North-Holland), Mini-Symposium (CD Rom) 13/11.

Tammann, G.A., Sandage, A. and Saha, A. (2000). H_0 from Type Ia Supernovae. In M. Livio, K. Noll and M. Stiavelli (eds), *A Decade of HST Science*, Cambridge University Press, in press.

Tanvir, N.R., Shanks, T., Ferguson, H.C. and Robinson, D.R.T. (1995). Determination of the Hubble constant from observations of Cepheid variables in the galaxy M 96. *Nature*, **377**, 27.

Teerikorpi, P. (1987). Cluster population incompleteness bias and distances from the Tully–Fisher relation – Theory and numerical examples. *Astronomy and Astrophysics*, **173**, 39–42.

Teerikorpi, P. (1997). Observational selection bias affecting the determination of the extragalactic distance scale. *Annual Review of Astronomy and Astrophysics*, **35**, 101–136.

Theureau, G. (2000). Tully-Fisher distances of field galaxies and the value of H_0. In É. Aubourg, T. Montmerle, J. Paul and P. Peter (eds), *XIXth Texas Symposium on Relativistic Astrophysics and Cosmology*, (North-Holland), Mini-Symposium (CD Rom) 13/12.

Theureau, G., Hanski, M., Ekholm, T., Bottinelli, L., Gouguenheim, L., Paturel, G. and Teerikorpi, P. (1997). Kinematics of the Local Universe. V. The value of H_0 from the Tully-Fisher B and log D_{25} relations for field galaxies. *Astronomy and Astrophysics*, **322**, 730–746.

Truran, J.W. (1999). Dating the universe with r-process radioactivities. In R. Diehl and D. Harmann (eds), *Astronomy with Radioactivities*, MPE Report 274, Max-Planck-Institut für extraterrestrische Physik, Garching, Germany, p. 261.

Turner, A., Ferrarese, L., Saha, A., Bresoliln, F., Kennicutt, R.C., Jr., Stetson, P.B., Mould, J.R., Freedman, W.L., Gibson, B.K., Graham, J.A., Ford, H., Han, M., Harding, P., Hoessel, J.G., Huchra, J.P., Hughes, S.M.G., Illingworth, G.D., Kelson, D.D., Macri, L., Madore, B.F., Phelps, R., Rawson, D., Sakai, S. and Silbermann, N.A. (1998). The Hubble Space Telescope Key Project on the extragalactic distance scale. XI. The Cepheids in NGC 4414. *Astrophysical Journal*, **505**, 207–229.

Wadsley, J., Hogan, C. and Anderson, S. (2000). Measuring Ω_b from the Helium Lyα forest. In A. Mazure, O. Le Fèvre and V. Le Brun (eds), *Clustering at High Redshift*, ASP Conference Series Vol. 200, Astronomical Society of the Pacific, San Francisco, p. 291.

Walker, A.R. (1999). The distances of the Magellanic Clouds. In A. Heck and F. Caputo (eds), *Post-Hipparcos Cosmic Candles*, Kluwer, Dordrecht, pp. 125–144.

Westin, J., Sneden, C., Gustafsson, B. and Cowan, J.J. (2000). The r-process-enriched low-metallicity Giant HD 115444. *Astrophysical Journal*, **530**, 783–799.

Wilkinson, M.I. and Evans, N.W. (1999). The present and future mass of the Milky Way halo. *Monthly Notices of the Royal Astronomical Society*, **310**, 645–662.

Willick, J.A., Strauss, M.A., Dekel, A. and Kolatt, T. (1997). Maximum likelihood comparisons of Tully–Fisher and redshift data: Constraints on Omega and biasing. *Astrophysical Journal*, **486**, 629.

Wilson, G., Kaiser, N. and Luppino, G.A. (2001). Mass and light in the Universe. *Astrophysical Journal*, **556**, 601–618.

Wittman, D., Dell'Antonio, I., Tyson, T., Bernstein, G., Fisher, P. and Smith, D. (2000). The Normal Cluster Weak Lensing Survey: Mass Profiles and *M/L* ratios of eight clusters at $z=0.2$. In F. Durret and D. Gerbal (eds), *Constructing the Universe with Clusters of Galaxies*, astro-ph/00 09 362. (also available on CD-Rom and at http://www.iap.fr/conferences/colloque/coll 2000).

Yahil, A., Sandage, A. and Tammann, G.A. (1980). The velocity field of bright nearby galaxies. III – The distribution in space of galaxies within 80 megaparsecs – The north galactic density anomaly. *Astrophysical Journal*, **242**, 448–468.

18

K.M. GÓRSKI* AND A.J. BANDAY**

COBE, dark matter and large-scale structure in the Universe

PREAMBLE

NASA's first dedicated cosmological space mission, the Cosmic Background Explorer (*COBE*), has provided comprehensive full-sky observations of both microwave and far-infrared frequencies. Each of the three instruments onboard – the Far Infrared Absolute Spectrometer (FIRAS), the Differential Microwave Radiometers (DMR), and the Diffuse Infrared Background Explorer (DIRBE) – has contributed significantly to our fundamental understanding of the universe. FIRAS has measured the Planckian nature of the relic blackbody radiation to unprecedented accuracy, thus establishing stringent constraints on the thermal history of the universe. DMR has discovered full sky structure in the temperature of the blackbody radiation over a range of angular scales down to ~ 10 degrees, thus establishing stringent limits on the evolution of large-scale structure in the universe and providing us with our first glimpse of the initial conditions for structure formation. DIRBE has provided the first evidence for the existence of a cosmic infrared background, thus providing important constraints on the integrated cosmological history of star formation in various pregalactic objects, protogalaxies, and galaxies, and the subsequent conversion of starlight into infrared emission by dust.

Thus *COBE* has become a *tour de force* of modern cosmological studies; it has influenced physical cosmology with its results dramatically, and established a splendid legacy for the community with its unparalleled view and interpretation of the multiwavelength sky.

---

*European Southern Observatory, Garching bei München, Germany; Warsaw University Observatory, Warszawa, Poland
**Max-Planck-Institut für Astrophysik, Garching bei München, Germany

K. M. Górski and A. J. Banday, The Century of Space Science, 399–421
© 2001 *Kluwer Academic Publishers. Printed in The Netherlands.*

1 LARGE-SCALE STRUCTURE OF THE UNIVERSE BEFORE *COBE*

Following closely upon the launch of the *COBE* satellite in November 1989, the January 1990 issue of "Scientific American" presented an article "The Cosmic Background Explorer", which announced the following: "NASA's cosmological satellite will observe a radiative relic of the Big Bang. The resulting wealth of data will be scoured for clues to the evolution of structure in the universe". In this article we describe how the outcome of this remarkable mission did indeed provide important scientific results which helped in a very major way to improve our understanding of the universe. However, to start our story at the beginning, let us first recall what was the status of our knowledge about the universe before *COBE*, and why *COBE* was needed for subsequent exploration of the universe.

It is very well known that our modern understanding of the universe as a physical system rests on three major pillars of observational cosmology:

- **Universal Expansion:** On cosmological distance scales, astronomical objects recede from one another with velocities proportional to their separations – a remarkable discovery by Hubble in the 1920s. As we now know, the universe on the largest scales is very nearly homogeneous and expands isotropically.
- **Primordial Nucleosynthesis:** The observed universal abundances of light elements proved inconsistent with the idea that they were produced in stars. This puzzle was explained in the 1940s by Gamow, Alpher, and Bethe in the context of an initially hot and dense phase in the early

evolution of the expanding universe. This necessarily required that a residual relic thermal radiation permeated the universe, and its temperature was theoretically expected to be ~4–5 K.

- **Microwave Background Radiation:** The relic thermal radiation of temperature ~3 K was serendipitously discovered by Penzias and Wilson (1965) and immediately recognized by Dicke *et al.* (1965) as the "missing link" between the primordial fire-ball of the young Big Bang universe and its present day mature phase dominated by evolved astronomical objects. This discovery delivered a direct proof of the dense and hot early stage of evolution of the universe.

These three phenomenological ingredients laid a solid foundation for the Hot Big Bang model, which became nearly unanimously accepted as the framework for our understanding of the universe at large. The Cosmic Microwave Background (CMB) radiation itself has subsequently become the subject of increasingly vigorous theoretical and observational studies. It was recognised very rapidly following its discovery that the CMB should contain "fossilized" imprints of the processes which could, or even had to, occur in the early universe, and that such signatures of past events should be pristine in comparison with the clues provided to us through the studies of nearby, well evolved astronomical objects. Hence, both the nature of the CMB electromagnetic spectrum, and its possible deviations from a perfectly thermal (Planckian) form, and the CMB anisotropy, or dependence of its temperature on the direction of observation on the sky, very quickly became attractive theoretical and observational research targets, which were pursued relentlessly up until the *COBE* mission, and, indeed, even more so afterward.

It is beyond the scope of this contribution to review systematically the steady progress of observational and theoretical cosmology from the discovery of the CMB until the *COBE* mission, but let us try to mention the essential developments.

- **Large-scale structure of the galaxy distribution:** It has been assessed that galaxies not only form groups and clusters, but their tendency to aggregate in space extends to even larger scales of superclusters (separated by voids), and perhaps beyond. For a long time the extreme scales of detectable inhomogeneity in the 3-D galaxy distribution coincided with the largest scales surveyed, which rendered a determination of the scale of transition to the expected homogeneity of the universe at large somewhat elusive. A rigorous quantitative description of the galaxy distribution in space has been developed, including number counts, correlation functions, and power spectra. During the 1980s, astronomical measurements of the spatial distribution of galaxies matured sufficiently to put strong constraints on theories of galaxy and large-scale structure formation.

- **Dark matter in the universe:** Ever since Zwicky's realisation in the 1930s that clusters of galaxies consisted predominantly of matter in some nonluminous form, the "missing mass" or later "dark matter" problem was one of the most serious puzzles in astronomy. The following observational picture was built over time: astronomical objects of increasing size (galaxies, groups, clusters, superclusters) appear to contain more and more mass that does not manifest itself via luminosity, but can be detected due to its gravitational effects. The amount of this hidden mass is large – perhaps about 20–30% of what is required to render the universe spatially flat. This fraction is sufficiently large that (1) it clearly exceeds the census of baryons in the universe, i.e. the dark matter in known astronomical objects is unlikely to be comprised entirely of ordinary matter, and (2) it invites speculation that perhaps the universe indeed is spatially flat, and therefore that the required dark matter content for the whole universe must still be larger than for individual astronomical objects. This finding tied in with a rich supply of theoretical candidates for weakly interacting massive particles, which could dominate the matter content of the universe, and the idea that perhaps the dynamically detected matter (comprising both luminous, baryonic objects and dark material) is more clumped than the remaining more smoothly distributed dark matter (the so called biasing effect). An alternate speculation involves the idea of a cosmological constant, or vacuum energy density, which was originally introduced by Einstein, and thereafter enjoyed various degrees of popularity with astronomers trying to determine the global properties of the universe. Indeed, the currently available combination of CMB anisotropy, large scale structure, and high-redshift supernovae observations can be interpreted as supportive of these cosmological constant dominated models of the universe.

- **CMB phenomenology:** Immediately after the discovery of the CMB radiation, it was realised that the background radiation should be carefully measured to search for

1. any deviations of its electromagnetic spectrum from thermal, and
2. the expected deviations from isotropy in the angular distribution of its temperature on the sky.

Both effects, if found, would provide invaluable clues to our understanding of physical processes occurring in the early universe. It would be the case regarding spectral distortions because only significant energy releases (e.g. bulk annihilation of exotic particles at some epoch) at very high redshift could measurably perturb the Planckian spectrum of the CMB. It would also be the case regarding CMB anisotropy because it could reveal to us the early predecessors of presently observable structures in the universe, as outlined below.

It was well understood that the primary dynamical factor that must have driven the evolution of structure in the universe was gravitational instability. To understand how the presently existing structure in the universe has arisen, it is necessary to postulate the existence of perturbations in the primordial spatial distribution of matter, which over time were amplified by their self-gravity. The logical appeal of the proposition to use the expected minute CMB anisotropies to study directly the matter distribution in the universe as it just emerged from its embryonic stage was built upon the following considerations:

1. the universe is opaque beyond redshift ~ 1000, thus one may think of CMB photons as having been emitted from the last scattering surface at the epoch corresponding to such a redshift (typically about a few hundred thousand years after the Big Bang); hence the CMB photons, which are influenced very little by the nearly transparent universe at redshifts <1000, bring to us precious information about the conditions as far away and early on in the universe as we can ever probe with electromagnetic radiation;
2. gravitational instability in an expanding universe is a slow process (described by power law functions of time, rather than the familiar Jeans exponentials in a non-expanding medium), hence the amplitudes of large scale inhomogeneities observed today must have then been small but not negligible;
3. the temperature of the CMB radiation photons when they were emitted at the epoch of last scattering was physically related to the perturbations of the matter distribution via both the gravitational potential of inhomogeneities (the so called Sachs-Wolfe effect), the emitting plasma velocity (or the Doppler effect), and the thermal effects in gravitationally compressed or rarified plasma. Thus one is led to conclude that the physical conditions at the time of the last scattering of the CMB photons are necessarily "imprinted" on a sky map of CMB temperature anisotropy.

With such understanding firmly established, all that still remained to be done was to measure these CMB temperature fluctuations.

Surely, the excitement related to the opening of a window in the background radiation through which direct observations of the truly embryonic stages of evolution of the large scale structure of the universe could be attempted was missed by neither experimentalists nor theoreticians in cosmology. But the required measurements proved to be an incredibly hard task, as we can only now appreciate in hindsight. A brief, and somewhat unjust, summary of the experimental CMB anisotropy efforts that were conducted for more than two decades up until the *COBE* mission is simple: there were no detections of cosmological anisotropy, only upper limits. The only exception was the measurement of the CMB dipole temperature pattern, at an amplitude of $\sim 0.1\%$ of the mean, superposed on the isotropic background (Smoot *et al.* 1977). This effect, however, was expected and explained as being dominated by our own motion (i.e. a combination of motions of the solar system in the Galaxy in the Local Group) with respect to the rest frame defined by the CMB radiation, and, hence, a local rather than cosmological effect. Despite this apparent lack of a tangible result, there was ongoing and steady progress stimulated by the increasingly stringent limits on theoretical predictions of expected CMB anisotropy depending on particular scenarios of evolution of structure in the universe. In fact, the search for evidence of anisotropy in the CMB by the experimental community became somewhat akin to the quest for the Holy Grail.

- **Mainstream theoretical cosmology in the 1980s:** The rapid development of theoretical and observational cosmology after the discovery of the CMB radiation resulted in very impressive support for the Hot Big Bang model as a basic paradigm for understanding the universe at large. However, there were still some nagging paradoxes left which could not be answered simply within the framework of the model. Among those were the following:

1. overall homogeneity and isotropy of the universe – why are the very distant regions in the universe, which were never in causal contact (i.e. could not have interacted at any time during the history of the standard Hot Big Bang universe), so similar as suggested by the apparent isotropy of the CMB?
2. flatness of the observed universe – why is the average density of the universe so close (within a factor of a few) to the critical density, which makes the universe spatially flat, just right now, when we (the human race) exist?
3. where did the initial perturbations which seeded structure formation come from?

All these questions could be by-passed with a reference to the initial conditions for the evolution of the universe. This, however, satisfied practically no-one, so when the idea of inflation appeared in the early 1980s, it was rapidly embraced as a compelling explanation for the shortfalls of the standard cosmological model. Inflation postulates that the evolution of the early universe is driven by a scalar field, called the inflaton, which dominates the energy density. Random regions, which get trapped in a state of false vacuum (or a local minimum of the potential energy) of the inflaton, end up expanding very rapidly due to an effectively negative pressure (hence the term inflation). A typical inflating region ends up so big that the currently observed astronomical universe would be just a small fraction thereof. Hence, the explanation of homogeneity and isotropy of the universe that we see. Near flatness is explained because, again, our astronomical universe would be just a small sector of an original arbitrarily curved manifold stretched to enormously large size. Density perturbations which seeded the presently observed structures were generated due to zero-point quantum fluctuations of the inflaton. Even though inflation is not a verified theory its logical appeal is huge, since it is apparently solves a number of seemingly unrelated paradoxes at the price of introducing just one new puzzle – what is the physical nature of the inflaton?

It should be now clear, even with this rather sketchy description of the development of experimental and theoretical cosmology after the discovery of the CMB, that there were important questions about the universe which could be answered with accurate measurements of the attributes of the relic radiation. A hard-learned lesson from early ground-based and balloon-borne experiments was of the enormous difficulties involved with attempts to perform such measurements successfully, and that the better approach would be a well-planned experiment in space. An opportunity to realise such a project was provided by NASA in 1976, when it selected the Cosmic Background Explorer, *COBE*, for a design study.

2 THE *COBE* PROJECT

Perhaps the first issue to address is the simple question, why a satellite? The most succinct response to this can be found in the proceedings of a meeting held in Copenhagen from June 25–29, 1979. "The Universe at Large Redshifts" contained a brief paper by Ray Weiss designed to serve as an introduction to a more complete review of the *COBE* mission (Mather and Kelsall 1980). Nevertheless, the most relevant issues for the reader to understand are indeed recorded there. Based on the integrated experiences of the *COBE* team members with ground-based, balloon-borne and air-borne experiments to measure the background radiation, and more importantly the limitations of these platforms, the arguments in favour of a space-borne experiment were then, and remain today, as follows:

1. Freedom from atmospheric emission and fluctuations in that emission.
2. Full sky coverage with a given single instrument.
3. A benign and controlled thermal environment to reduce systematic errors.
4. The ability to perform absolute primary calibration in flight without the necessity of windows to avoid condensation of the atmosphere on calibrators and instruments.
5. Sufficient time both to peform tests for systematic errors and to gain the increase in sensitivity permitted by extended observation time.

To achieve the full benefit of space observations, the goal of the mission and instrument design was to ensure that the scientific measurements conducted by *COBE* were ultimately limited by the ability to model the various astrophysical components and distinguish them from the cosmological information sought. This goal thus drove the mission strategy, spacecraft and operations design and choice of instruments.

2.1 Satellite overview

The need to minimise, control and measure the impact of systematic errors led to the requirements for an all-sky survey, a minimal survey period of 6 months, and constraints on the amount of interference from local sources of radiation such as the Earth, Sun, Moon and radio emission from the ground. The orbit, spacecraft attitude, and instrument enclosures were therefore carefully selected to avoid direct exposure to the Earth and Sun, and maintain a stable thermal environment for the instruments. For *COBE*, depicted in Figure 1, a 900 km altitude, Sun-synchronous orbit was selected so that the orbit, with a duration of 103 minutes, precessed by 1 degree per day. In order to meet the scientific requirements of the mission (as will be discussed below), the spacecraft must also spin. A rate of 0.8 rpm was adopted, with the spin axis tilted back from the orbital direction by 96 degrees, so that residual atmosphere did not affect the instruments. The attitude of the spacecraft was controlled by inertia wheels and electromagnets, and determined from Sun sensors, Earth sensors and gyroscopes.

2.2 Instrumental design

Three complementary experiments were flown on the *COBE* spacecraft: the Far-Infrared Absolute Spectrometer (FIRAS) designed to measure the frequency spectrum of sky radiation from 100 microns to 1 cm (see Figure 2), the Diffuse Infrared Background Explorer (DIRBE) to map the sky from 1 to 300 microns, and the Differential Microwave Radiometers (DMR) to search for anisotropies in the CMB on scales larger than 7 degrees.

Due to the authors' involvement with the analysis and interpretation of the *COBE*-DMR data, and the aim of this contribution being the assessment of the impact of *COBE*'s results on our understanding of the large-scale structure of the universe, the remaining text is focused on the DMR instrument and the measurements of CMB anisotropy conducted with this instrument. Those readers interested specifically in the results of DIRBE and FIRAS are referred to NASA GSFC www page http://space.gsfc.nasa.gov/astro/cobe/ and references therein.

A differential radiometer is a device which outputs a voltage proportional to the difference in power received by

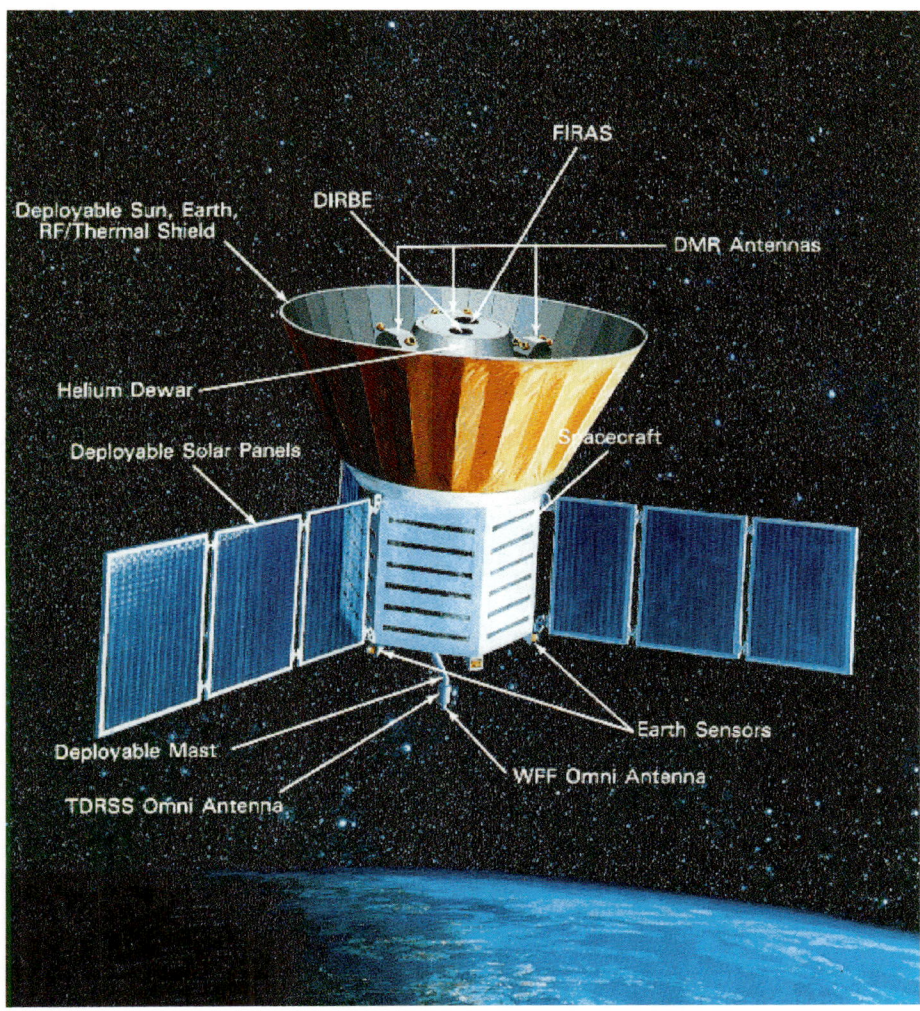

Figure 1 This artist's impression of the *COBE* satellite in orbit around the Earth shows the relative positions of the 3 instruments – DMR, FIRAS and DIRBE. (Courtesy of NASA Goddard Space Flight Center and the *COBE* Science Working Group.)

two antennas. The voltage is calibrated in units of antenna temperature by measuring the voltage change for a known change in temperature of a reference source filling the beam of one antenna. The DMR consisted of six such differential radiometers, two independent channels (A and B) at three frequencies 31.5, 53 and 90 GHz. Each radiometer measured the difference in power between two regions of the sky separated by 60 degrees using a receiver switched between two horn antennas of a beam width of 7 degrees, directed 30 degrees to either side of the spacecraft spin axis. Such a design, in combination with the spacecraft orbital motion and spin, allowed the temperature of any given sky position to be compared to a highly redundant set of all possible temperature differences with this 60 degree spacing over a range of time-periods. Systematic errors, which show specific time-variability, can then easily be distinguished from the true astronomical signal. The multi-frequency measurements then allow a distinction to be made between cosmological emission and that due to more local astrophysical sources, such as our own Milky Way galaxy.

2.3 Data processing

Substantial data processing was required to convert the measured voltage differences into maps of the temperature anisotropy. The DMR telemetry consisted of uncalibrated differential temperatures plus housekeeping data (voltages, temperatures from various devices monitoring the health of the spacecraft and instruments), which are then merged with attitude information. Bad data was flagged, and the measurements calibrated using noise sources of known

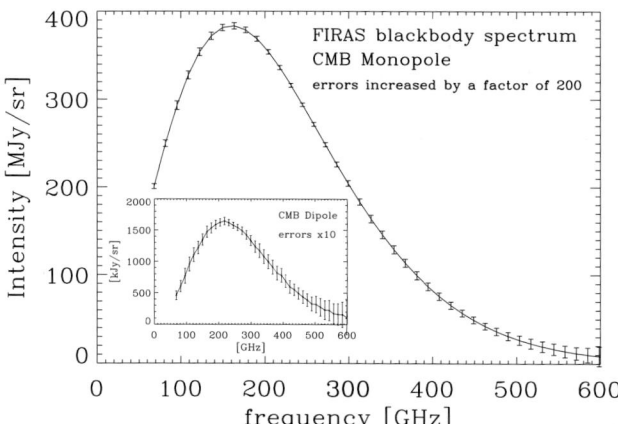

Figure 2 Spectacular measurement of the Planckian energy spectrum of the cosmic microwave background radiation by COBE-FIRAS (Fixsen et al. 1996). An early version of this measurement was the first significant result from COBE (Mather et al. 1990), obtained from only a few minutes of observations in operational mode. This result was deemed sufficiently newsworthy to have made the front page headlines of the "New York Times". Remarkably FIRAS also measured with impressive accuracy the Planckian nature of the spectrum of the dipole anisotropy of the CMB.

antenna temperature. Longer term stability of this calibration was checked by observations of the Moon.

The calibrated data were then corrected for known instrumental systematic effects. Although the instruments were carefully designed to minimise the impact of these parasitic signals, it was inevitable that some contribution remained. An interesting example of such a systematic contaminant resulted from the susceptibility of the instrument to magnetic fields. Since the differential signal was derived from the magnetically driven switching between the two horns, any external magnetic field, either from the spacecraft electromagnets used to control attitude or from the Earth, could modulate the measurement in a time-dependent manner. The latter contribution was removed from the data by using the International Geomagnetic Reference Field model.

2.4 Mission chronology

COBE was initially designed to be launched on a Delta rocket, but soon after the design was underway, the space-shuttle was adopted as the launch vehicle. The Challenger tragedy in 1986 led to a reversion to the Delta launch configuration. Such events demonstrate the peculiar difficulties of carrying out a space mission, which also include the problems of maintaining the project within an assigned budget and the fact that one would be dealing with technology that is 3 to 5 years behind the state-of-the-art (to allow time for space qualification). The latter issue manifested itself earlier in the design phase, when a 23 GHz radiometer channel was removed from the design to allow some additional cooling of the remaining DMR instruments.

COBE was launched on Delta rocket No. 189 at 1434 UT on November 18, 1989 from Vandenberg Air-Force Base in California. The DMR receivers began operating the day after launch. The first full-sky coverage was completed mid-June of 1990. Although one of COBE's gyros failed on the fourth day after launch, no scientific data was lost, and the remaining gyros and spacecraft systems performed flawlessly until the mission neared its end. The cryogenic phase of the mission, essential for the FIRAS instrument, was ended by the depletion of the liquid helium coolant on September 21, 1990. DMR continued to record data into 1994, but none was used beyond December 22, 1993. The combination of additional gyroscope failures beyond July 8, 1993, decreasing scientific returns and rumours of several small objects co-orbiting with the spacecraft (presumably some dislodged thermal shielding) led to the termination of the mission.

3 FROM DISCOVERY TO SCIENTIFIC LEGACY

The defining moment for the COBE-DMR came on April 23, 1992 at a meeting of the American Physical Society in Washington, DC. The DMR team had laboured for many months over the data, convincing themselves that genuine cosmological structure had been observed rather than systematic artifacts, and devising sophisticated schemes to subtract any foreground emission from our own Galaxy, until George Smoot finally announced that anisotropy in the CMB had been discovered. In fact, the announcement was, for some members of the Science Team, somewhat overdue: Ned Wright had been convinced by his analysis of 6 months of data alone that such structure existed. However, the extra months of effort had allowed the claims to be made with considerable certainty and avoided the problems of a previous satellite mission known as RELIKT, which had shown structures in a map more consistent with systematic artifacts. The importance of this discovery to the astronomical community cannot possibly be over-emphasised: Stephen Hawking called it "the scientific discovery of the century, if not all time".[4]

Gradually, the initial euphoria faded, and the general feeling became more skeptical. The first year maps did not contain unique individual structures that were clearly identifiable

---

[4] Whether this was a realistic appraisal is left to the reader, who is also encouraged to read the books "Wrinkles in Time" by George Smoot and Keay Davies, and "The Very First Light" by John Mather and John Boslough for various aspects of the human sociological reaction to such sudden fame and the problems of working in large collaborations.

as cosmological in origin: the detection was merely statistical. As a result, there was an increased vigour in the DMR team to assess the results from two years of data acquisition. In the meantime, data from a balloon-borne instrument had also mapped a reasonable fraction of the sky, and cross-correlation with the DMR data demonstrated statistical consistency (Ganga et al. 1993).

The two-year results were of a substantially improved quality (Bennett et al. 1994, Górski et al. 1994, Wright et al. 1994). From the stand-point of noise, a square-root of two factor of improvement was achieved simply by having twice as much data, but the additional analysis time allowed for significant improvement in the understanding of systematic errors. These sky maps could be interpreted as having well-defined cosmological structure.

Ultimately, four years of data were integrated into the final sky maps released to the general astronomical community (Bennett et al. 1996). These maps are a lasting legacy of the mission and will remain, until the NASA MAP and ESA Planck satellites observe the sky, the only full-sky maps of anisotropy in the CMB. The impact of COBE-DMR on the cosmological community can be easily measured in terms of the number of documents written on COBE-DMR and its implications for Dark Matter. As of April 2000, the Web of Science records 1108.

4 BASIC RESULTS

The basic products of the COBE-DMR mission, shown in Figure 3, are a set of six sky maps, two each at the three frequencies of interest – 31.5, 53 and 90 GHz. Each map represents the temperature deviation from some mean value (unmeasured by the DMR, since it is a differential instrument) recorded on 6144 points on the sphere, or pixels. During the map making process, the known systematic effects are removed from the time ordered data before the sky maps are made. Figure 4 shows a few examples of such

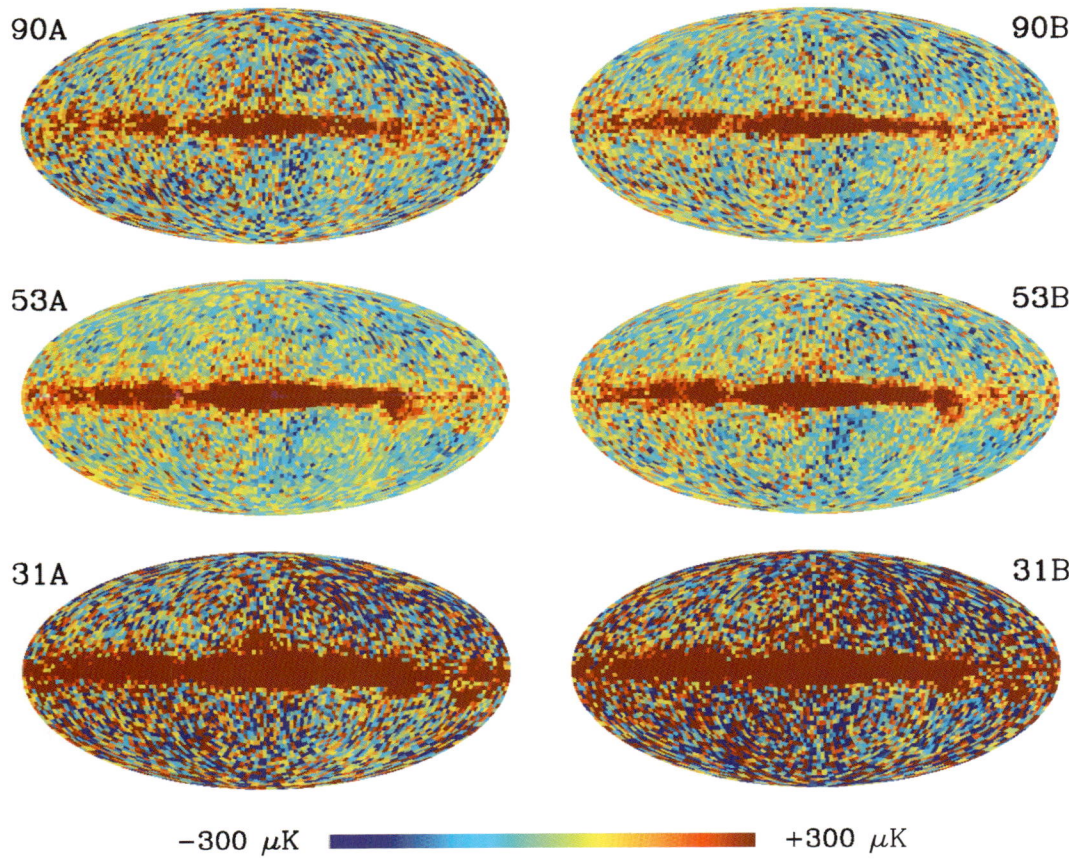

Figure 3 This plot shows the six sky maps derived from 4 years of COBE-DMR data at the three frequencies 31.5, 53 and 90 GHz. Mollweide projection is adopted in these and all other sky maps (unless explicitly stated). The Galactic plane is at the centre of the image. It is easily seen that the least and most noisy of the DMR receivers were those operating at 53 and 31 GHz, respectively.

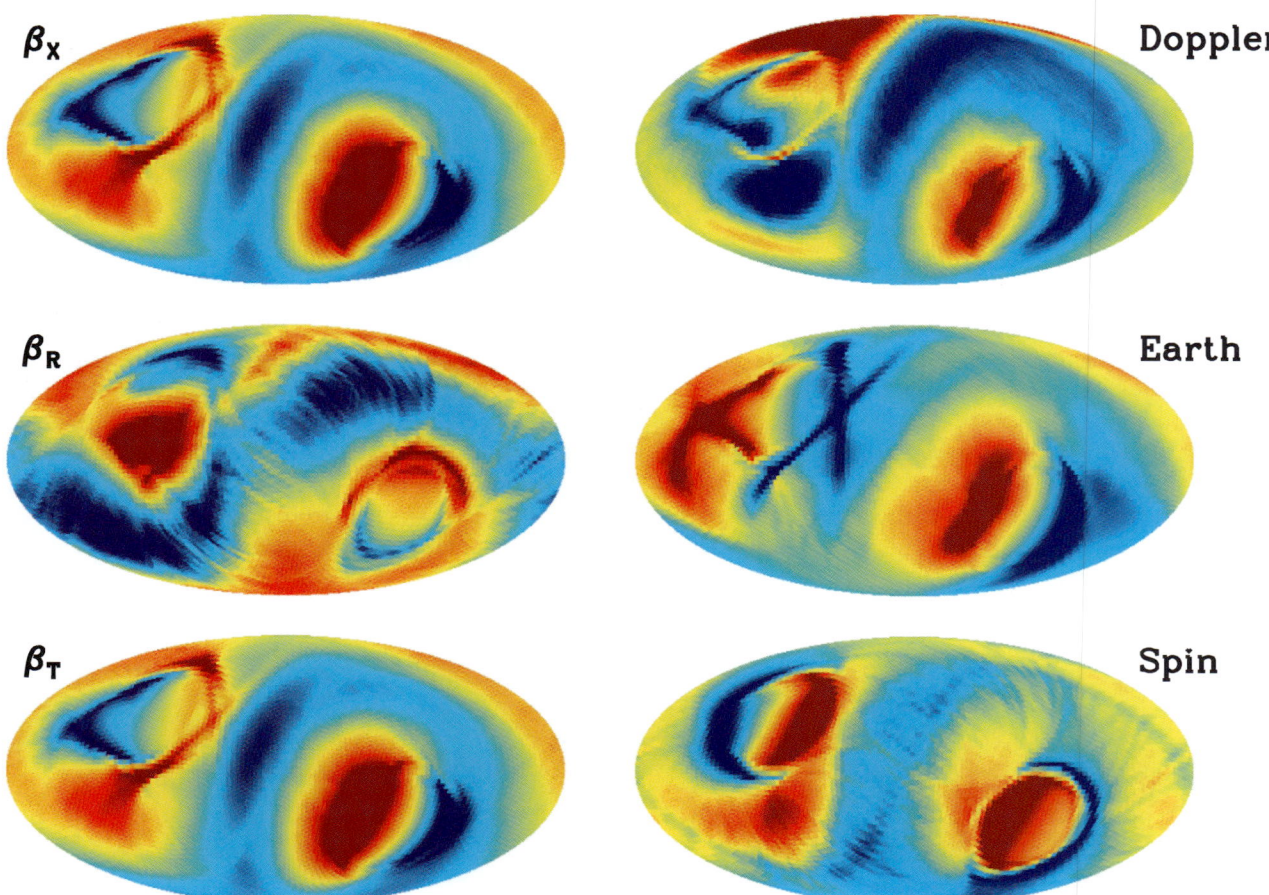

Figure 4 One of the most important aspects of the DMR data reduction involved the modelling and subtraction of potential systematic errors and artifacts due to various instrumental and environmental effects. Here we see the spatial distribution of effects related to the magnetic susceptibility of the radiometers (with the three components β_R, β_T, and β_X); the Doppler correction related to the spacecraft's motion in orbit about the Earth; signal due to the diffraction of the Earth's microwave emission around the shielding; and a contribution due to non-physical variations in signal with a period related to the spin of the satellite. If the time ordered data from the DMR receivers were not corrected for the presence of such known systematic effects, the sky maps produced by the data processing pipeline would indeed contain such patterns as shown in this figure, with amplitudes large enough to be discerned in the subsequent analysis. The ability to model, detect, and remove such effects from the DMR data demonstrated the validity of the mission and instrument concepts, and strengthened confidence in the quantitative analysis of CMB anisotropy.

systematic effects as they would appear on the sky map, if they were not removed from the time ordered data. As a consequence of the map-making process, there is an arbitrary offset on the map, but this has no physical relevance. The major feature of the microwave emission at these wavelengths should be a dipole temperature variation on the sky. However, this feature was discovered many years before *COBE*-DMR, and was removed from the data before map-making. The remaining dominant feature of the map is the bright emission from the Galactic disk. This emission dominates the temperature anisotropy out to about 20 degrees Galactic latitude, with some additional regions, known to exhibit strong emission in the infrared part of the spectrum, seen in Scorpius and Ophiuchus in the northern Galactic hemisphere and Taurus and Orion in the south. The remaining high-latitude part of the sky shows temperature features which are a combination of the true cosmological signal, instrumental noise, small residual systematic artifacts (strongly constrained by extensive analysis, modelling and removal), and potential emission from other local astrophysical sources.

The important aspect of the DMR in having two channels per frequency then manifests itself: by combining them into "sum" and "difference" maps, one now has maps

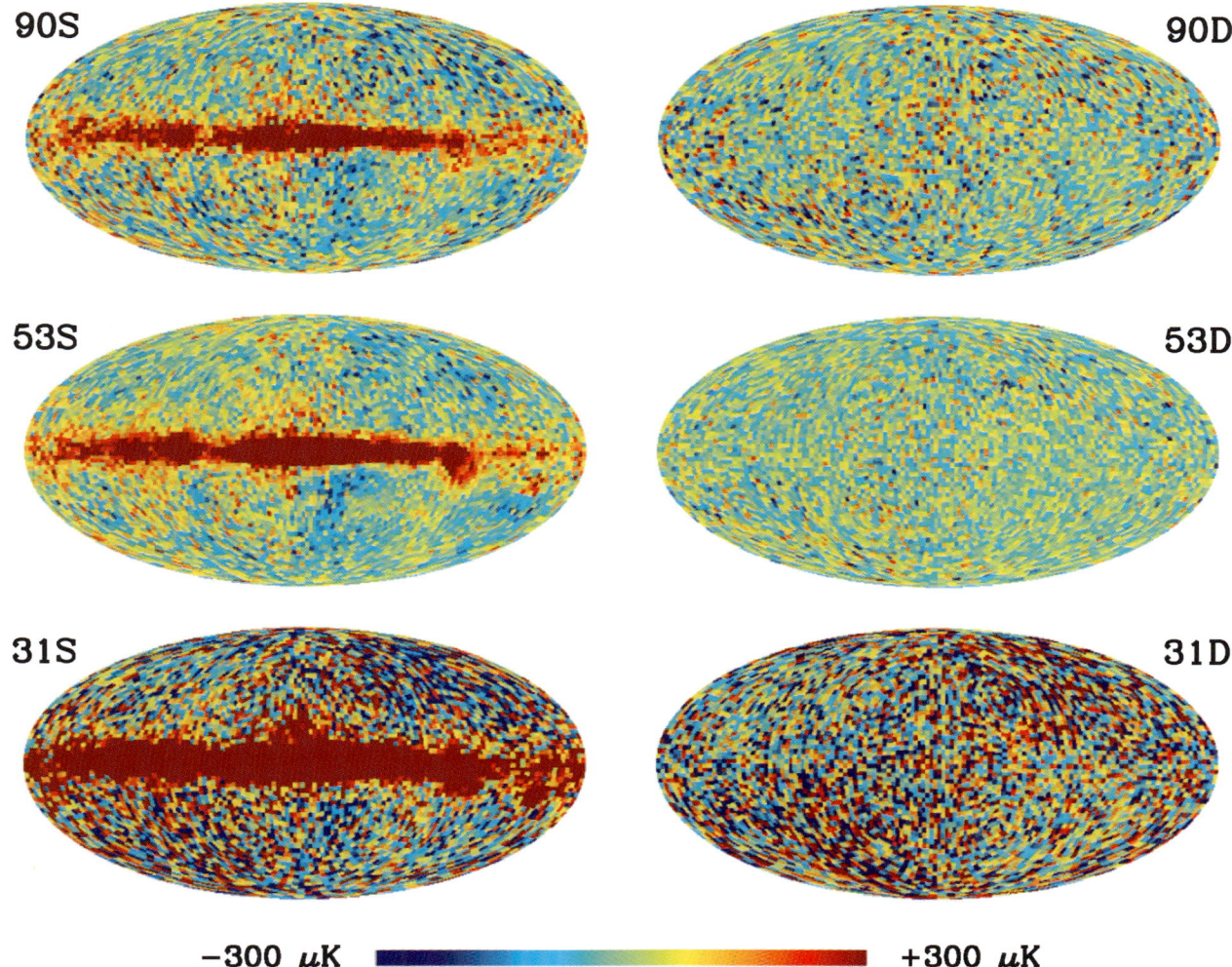

Figure 5 The *sum* and *difference* maps are created by combining the two available channels at each frequency in the combinations (A+B)/2 and (A−B)/2. The respective combinations enhance signal, or create a map which estimates the noise in the signal map. Accurate knowledge of the noise properties of the sky map is a crucial input to sophisticated analyses of its cosmological content. DMR was the first CMB experiment to include multichannel redundancy at each frequency allowing properties of the noise to be extensively tested.

with either enhanced signal and noise only (Figure 5). Understanding the statistical nature of the noise is a key element in interpreting the cosmological content in the map, as we shall discuss later. That the noise maps do indeed only contain noise-like structure is also important in confirming the consistency of the relative calibrations of the two channels; otherwise, residual Galactic plane signals would leak into the difference map. The properties of the noise map should be understandable in terms of the observation pattern for the instrument (Figure 6). Regions of low-noise correspond to those parts of the sky observed the most, close to the ecliptic poles. Other features in the observation pattern reflect the effect of data excision due to systematic effects. For example, the structure seen in the ecliptic plane corresponds to rejection of observations where the Moon was observed in one of the horns of the radiometer.

Of equal importance for deriving cosmological inferences in later analysis is to determine the extent of the presence of foreground emission from local astrophysical sources (Bennett *et al*. 1992). As we have already noted, the sky maps are dominated at low Galactic latitudes by strong emission from the Milky Way. Some weaker emission at high latitudes is also to be expected. This has its origin in three mechanisms: synchrotron emission from relativisitic electrons accelerating in the Galaxy's magnetic

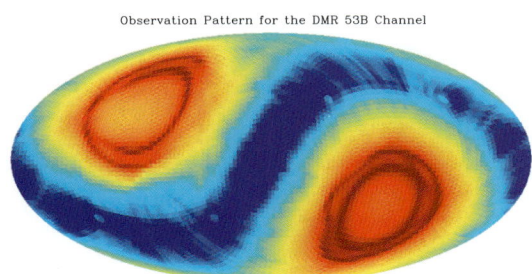

Figure 6 A plot of the number of observations per pixel for the 53 B sky map. This is a histogram-equalised plot with a dynamic range between the minimum ~44 000 and maximum ~164 000 observations per pixel. As a result of the combination of the spacecraft's orbital motion and spin, the highest number of observations lie on rings close to the ecliptic poles. Along the ecliptic plane, low observation count features in the pattern can be discerned which reflect the fact that data has been excised to exclude excessive lunar emission, which otherwise would be present in one of the antenna horns. Since the noise present in the data decreases with the number of observations recorded in a sky pixel (scaling as $\sim\sqrt{1/N_{observations}}$), the final sky maps are least noisy near the ecliptic poles, and most noisy along the ecliptic, as can be seen most easily in the difference maps in Figure 5.

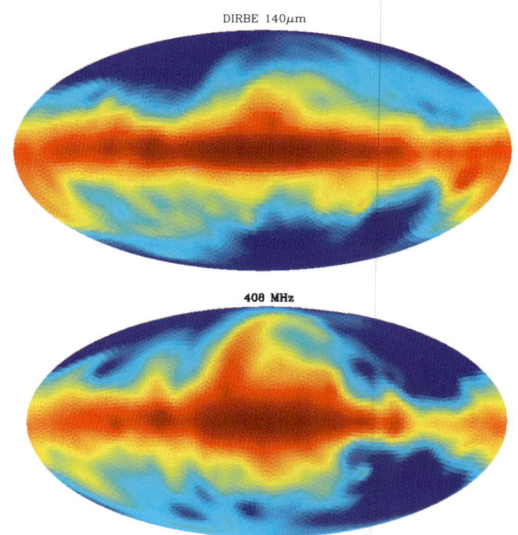

Figure 7 The extent to which Galactic foreground signals can contaminate the DMR sky maps must be estimated. In the upper part of the figure, we see a map of the Galactic dust emission (as observed by the *COBE*-DIRBE instrument but at *COBE*-DMR resolution) at a wavelength of 140 microns; the lower half of the figure represents the Galactic synchrotron emission as observed at 408 MHz. These template maps are cross-correlated with the DMR maps at each frequency to determine the likely level of contamination. Interestingly, at high Galactic latitudes ($|b| \gtrsim 20°$), the dust emission template does correlate detectably with the DMR data, while the one for synchrotron emission does not. Both maps are plotted using a histogram-equalised colour scheme.

field, free-free emission from hot ionised plasma in the interstellar medium, and thermal emission from dust grains. In order to constrain the contribution of such foregrounds to the observed signals in the DMR sky maps, the observed high-latitude emission was compared to maps of the synchrotron and dust emission (Figure 7) made at frequencies where these signals dominate. Further possible contributions could also arise from the emission of discrete extragalactic sources, and from clusters of galaxies. Figure 8 shows maps of the possible distribution of emission from such sources, which are again shown to be insignificant when cross-correlated with the DMR data. The importance of this negative result cannot possibly be overemphasized; it demonstrates that a local origin for the CMB anisotropy detected in the DMR maps, although possible in principle, is unlikely to have been caused by the distribution of known astronomical objects. Hence, after carefully limiting the possibility that the measurable anisotropic structure in the CMB temperature might be related to instrumental effects, our own Galaxy, or the local astrophysical environment, one is finally left with the most exciting option, namely that the signals are cosmological in origin and arise at very high redshift.

The final additional test that must necessarily be satisfied requires that the pattern of perturbations of the average CMB temperature be frequency independent, as it should be for primordial fluctuations.

Demonstration of this requires some quantification of the level of anisotropy present in the maps at the various frequencies. The simplest model-independent measure of statistically significant structure that can be computed from the sky maps is the sky-RMS, or the root mean square deviation of the CMB temperature across the sky. The quantity is determined by consideration of the excess signal present in the sum maps relative to the (noise-only) difference maps for that fraction of the sky not contaminated by the Galactic plane. The typical RMS amplitude is $\sim 35 \pm 2 \, \mu K$. The fact that the sky-RMS temperature fluctuations determined from the *COBE*-DMR four-year sky maps are frequency independent (see Figure 9) and therefore consistent with the Planckian spectrum expected for the Cosmic Microwave Background signal as demonstrated by *COBE*-FIRAS is important evidence that the anisotropy signal is indeed cosmological in origin (Banday *et al.* 1997). Moreover, any residual Galactic emission in the data must necessarily be small.

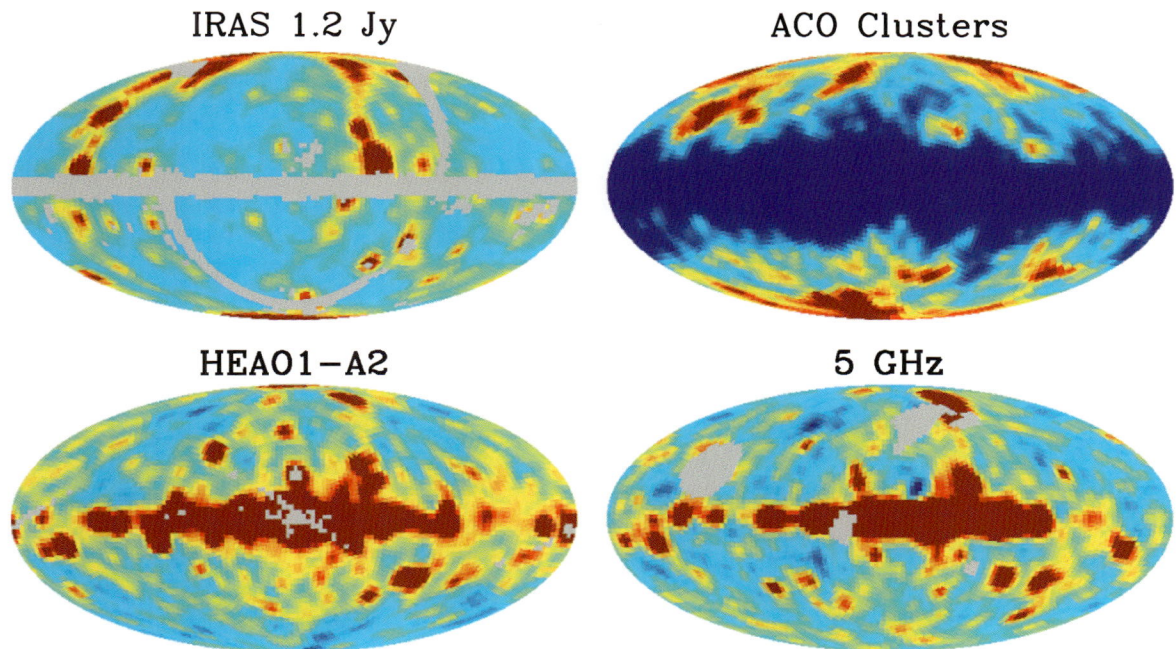

Figure 8 *COBE*-DMR resolution maps of extragalactic emission from known nearly full sky catalogs of the infrared (IRAS), optical (ACO), X-ray (HEAO), and radio (5 GHz) data.

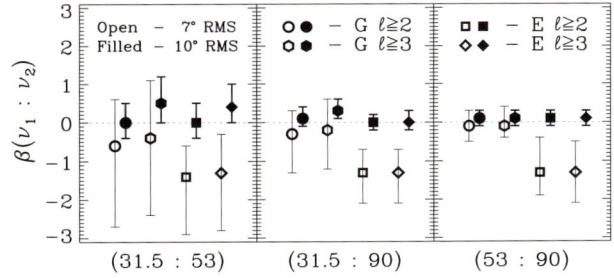

Figure 9 This figure shows the observed sky-RMS values determined from independent sky maps at 31.5, 53 and 90 GHz pixelised in both Galactic and Ecliptic reference frames (G and E symbols), and analysed either including or excluding the quadrupole moment of the CMB anisotropy ($l \geq 2$ and $l \geq 3$ symbols). The data without additional smoothing is referred to as 7° smoothing, since the central lobe of the DMR beam is *approximately* described by a 7° FWHM Gaussian. For determining the sky-RMS at 10°, the data surviving the Galactic cut are smoothed by a 7° FWHM Gaussian kernel with uniform weighting to produce a more closely Gaussian beam of 10° width. The results are consistent with the presence of sky signal in all four years of data. We have computed a frequency dependence in the sky-RMS of the form $\nu^{-\beta}$ between pairs of frequencies. A genuine CMB signal should not exhibit any frequency dependence due to the Planckian nature of its spectrum. The general lack of any frequency dependence in the sky-RMS (expressed in thermodynamic temperature units) without any correction for Galactic emissions is consistent with the dominant signal being cosmic in origin.

5 DETERMINATION OF THE ANGULAR POWER SPECTRUM OF CMB ANISOTROPY FROM THE *COBE*-DMR SKY MAPS

After having established that the high Galactic latitude regions of the *COBE*-DMR sky maps (1) are not compromised by systematic artifacts, (2) are polluted only with relatively small Galactic and extra-galactic foreground emission, (3) contain signals which exceed the instrument generated noise and are independent of frequency, and therefore (4) comprise mostly primordial CMB temperature fluctuations and the DMR instrumental noise, one would like to proceed with a more sophisticated quantitative characterisation of the cosmological signals.

A rudimentary description of the distribution across the sky of the temperature fluctuations – generated by the large-scale structure of matter perturbations in the globally homogeneous and isotropic universe – involves the concept of a statistically isotropic random field on a sphere, say $\delta T(\hat{\gamma})/T_0$, where T_0 is the average CMB temperature, $\delta T = T(\hat{\gamma}) - T_0$, and $\hat{\gamma}$ is a unit vector. Such a field can be naturally expanded in a series of spherical harmonics

$$\frac{\delta T(\hat{\gamma})}{T_0} = \sum_{lm} a_{lm} Y_{lm}(\hat{\gamma}), \quad a_{lm} = \int_{4\pi} d\hat{\gamma} \frac{\delta T(\hat{\gamma})}{T_0} Y^*_{lm}(\hat{\gamma}), \quad (1)$$

where $a_{00} = 0$ by definition, a_{1m} are the coefficients of the dipole moment of the CMB anisotropy induced by our motion with respect to the CMB rest frame, and a_{lm} ($l \geq 2$, $m = -l, ..., l$) are the multipole coefficients of the angular pattern $\delta T(\hat{\gamma})/T_0$. Since there is no possibility of predicting deterministically any specific pattern of primordial CMB temperature fluctuations, one must resort to a statistical description, whereby the moments of the a_{lm}-s are specified by cosmological theories of structure formation. One formally postulates an ensemble of realisations of the full sky distribution $\delta T(\hat{\gamma})/T_0$ (which can be also pictured, in an ergodic sense, as an ensemble of CMB skies as they would be seen by observers at locations other than ours in the globally homogeneous and isotropic universe) for which $\langle a_{lm} \rangle = 0$, and

$$\langle |a_{lm}|^2 \rangle = C_l, \qquad (2)$$

where C_l is the angular power spectrum, independent of m due to the statistical isotropy of the field $\delta T(\hat{\gamma})/T_0$. The angular power spectrum is related to the familiar two point correlation function via the following relation

$$C(\theta) = \frac{1}{4\pi} \sum_l (2l + 1) C_l P_l(\cos \theta), \qquad (3)$$

where P_l are the usual Legendre polynomials. The two-point correlation function was significantly determined already in the *COBE*-DMR data obtained during the first year of the mission and reported in the CMB anisotropy discovery paper (Smoot *et al.* 1992).

While an arbitrary random field requires an infinite set of moments to be prescribed, a simple class of Gaussian random fields is completely specified by the power spectrum C_l, or the correlation function $C(\theta)$. For such fields the one-point probability distribution for $\delta T(\hat{\gamma})/T_0$ is normal with zero mean and variance equal to $C(0) = \Sigma_l (2l + 1) C_l/4\pi$, and all n-point correlations (non-vanishing only for even n) are given in terms of products of the two-point correlation functions. This simple Gaussian model of the random field $\delta T(\hat{\gamma})/T_0$ is a natural choice for application in the analysis and interpretation of the CMB anisotropy data, especially the *COBE*-DMR data, and it facilitates the process of estimation of the power spectrum of the CMB temperature fluctuations. A consistency check requires, however, a careful testing of the hypothesis that the measured parturbations are indeed Gaussian distributed. The angular power spectrum estimated from the experimental data is a considerably compressed form of information about the properties of the measured CMB sky map, since $N_{C_l} \sim \sqrt{N_{a_{lm}}} \lesssim \sqrt{N_{pixels}}$. The angular power spectrum derived from a particular theory of structure formation is usually parameterized with physical quantities that characterise a given theory. Naturally, the theories can be scrutinized by the confrontation of a predicted power spectra with that derived from the data.

Although the above outline of the rudiments of mathematical modeling of the sky distribution of CMB anisotropies suggests a relatively straightforward power spectrum estimation problem, the reality of any attempt to perform such a task on a realistic data set is far from trivial. A proper analysis of a relatively complex data set such as DMR and the correct inference of a parametrisation of the sky distribution of CMB anisotropy requires the simultaneous treatment of several interrelated problems, which include:

1. incomplete sky coverage (i.e. available integration domain in eqn (1) is less than 4π) and the resulting multipole coupling after excision of the Galactic plane region from the analysed maps (necessary since it is impossible to remove the dominant Galactic signal with the required precision;
2. removal of the cosmologically irrelevant low order modes ($l = 0$ – the non-physical offsets in the *COBE*-DMR skymaps resulting from the map-making procedure, and $l = 1$ – the small residuals from inaccurate *a priori* removal of the dipole anisotropy during the map making process), preferably without affecting those higher order modes ($l \geq 2$), which carry cosmological information;
3. modeling of the non-uniform noise distribution;
4. identification and removal of possible diffuse foreground emission in the areas outside the Galactic cut;
5. execution of an unbiased or least-biased parameter inference, preferably using the probability density which describes the tested model of anisotropy in an exact manner.

A program of analysis of the DMR data which addressed all of these points was conducted in Górski (1994, 1997) and Górski *et al.* (1994, 1996). An important feature of this method of analysis is that it employs geometrically constructed orthogonalized spherical harmonics, which are *independent* of any cosmological theories that one might endeavour to test with the DMR data, and that the power spectrum estimation is executed via the maximum likelihood technique, described next, which is bilinear in the data. Other techniques of approaching this problem described in the literature include a pixel-based likelihood method (Tegmark and Bunn 1995, Hinshaw *et al.* 1996), signal-to-noise eigenmode method (Bond 1995), and various quadratic methods of power spectrum estimation (Wright *et al.* 1994, 1996; Tegmark 1995, 1996; Bond *et al.* 1998).

Let us now establish how the method introduced in Górski (1994) can be used to compute an estimator of the full sky angular power spectrum, i.e. the coefficients C_l which can be determined from the data. The pixelised map data are decomposed using the cut-sky orthogonalised spherical harmonics into pseudo-multipole coefficients **c**, which are linear combinations of unknown full-sky multipole coefficients **a**, where bold-face **a**, and **c** is the vector notation for the entire available sets of coefficients a_{lm}, and c_{lm}. Since we work with the Fourier amplitudes **c**, which are linear combinations of pixel temperatures (as they are integrals on the cut-sky of the pixelised maps), their joint probability distribution is Gaussian (see Górski 1994 for all details)

$$P(\mathbf{c})\,d\mathbf{c} = \frac{d\mathbf{c}}{(2\pi)^{N/2}\sqrt{\det(\mathbf{S}+\mathbf{N})}} \exp(-\mathbf{c}^T\cdot(\mathbf{S}+\mathbf{N})^{-1}\cdot\mathbf{c}/2), \quad (4)$$

where N is the dimensionality of the problem, $\mathbf{C} = \mathbf{S} + \mathbf{N}$, and **S** and **N** are covariance matrices of the signal and noise, respectively. The careful design of the differential instrument to measure the CMB anisotropy resulted in a near-perfect pixel-to-pixel independence of the detector noise as projected into pixelised *COBE*-DMR sky maps. The adequacy of the Gaussian model for the noise distribution in both pixel and Fourier space for all DMR maps was tested and confirmed. The Gaussian form of the probability distribution for the physical signal part in the eqn (4) is, of course, a consequence of an *assumption*, requiring further testing, that our model of $\delta T/T_0$ is a multivariate (pixel-space) Gaussian deviate. For the maximum likelihood calculations, we can write the likelihood function defined as

$$F(\mathbf{A}|\hat{\mathbf{c}}) = \det[\mathbf{S}(\mathbf{A})+\mathbf{N}] + \hat{\mathbf{c}}^T\cdot[\mathbf{S}(\mathbf{A})+\mathbf{N}]^{-1}\cdot\hat{\mathbf{c}} \propto -2\ln P(\hat{\mathbf{c}}), \quad (5)$$

where the full sky signal covariance matrix has the trivial diagonal form

$$\mathbf{A} = \mathrm{diag}[\underbrace{C_2, \ldots, C_2}_{5\ \text{times}}, \underbrace{C_3, \ldots, C_3}_{7\ \text{times}}, \ldots, \underbrace{C_{l_{max}}, \ldots, C_{l_{max}}}_{(2l_{max}+1)\ \text{times}}]. \quad (6)$$

The maximum likelihood problem as formulated in eqns (4–6) – seeking the coefficients C_2–$C_{l_{max}}$ given the data vector $\hat{\mathbf{c}}$ – involves the assumption of a Gaussian distribution for both noise and CMB anisotropies, and the statistical isotropy of the field of temperature perturbations (hence only the l dependence of $\langle |a_{lm}|^2\rangle$), but does not imply any relation between C_l-s at different values of l – hence, we are solving for the power spectrum of CMB anisotropy fully *independently* of any models of structure formation, which specify such an l dependence of the angular power spectrum.

A maximum likelihood search for the power spectrum that best matches the distribution of signal on the sky simply entails finding a minimum of the function $F(\mathbf{A})$ in the $l_{max} - 1$-dimensional space. In the present case, it is greatly facilitated by our ability to write down easily the expression for the gradient $\partial F/\partial a_l$, and implement a standard numerical routine for an extremum search with gradient zeroing. The solution to this problem for the DMR 4-year Galactic frame data is shown in Figure 10. Our power spectrum estimator is really an estimator of excess variance of the sky signal over noise in the data. One should note in Figure 10 the asymmetry of the confidence intervals around the maximum likelihood solutions for the C_l-s, and the fact that no derived values go negative. If there is no signal content in the data that can be attributed to a given C_l, an upper limit is established, as e.g. for $l = 16$. The signal-to-noise ratio in Fourier space drops to ~ 1 at $l \sim 15$ in the DMR 4yr data. All the multipoles within that range are determined with great significance. As for the confidence intervals, the plot can not display the correlations between the errors on different multipoles. These correlations are small, and somewhat more significant for multipoles separated by $\Delta l = 2$ than otherwise. Because of the nontrivial (non-Gaussian) structure of confidence interval correlations, it is difficult to implement an exact method of fitting the theoretical model of CMB anisotropy to the DMR data by using the power spectrum, compressed rendition of the data. For such applications, the accurate linear methods, as e.g. Górski *et al.* (1994, 1996) should be preferred.

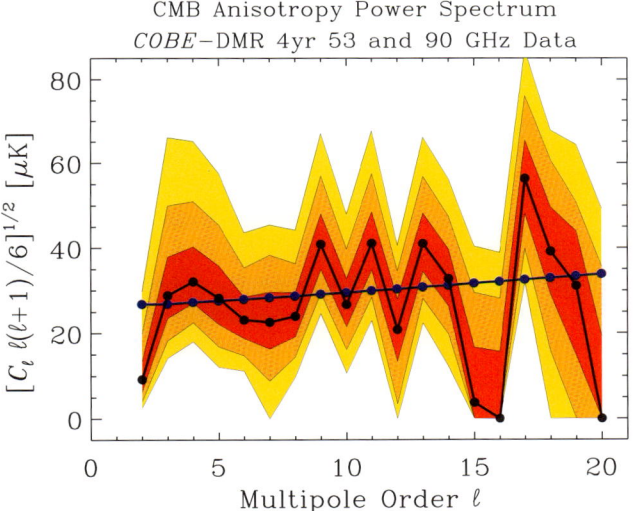

Figure 10 The angular power spectrum (black dots) of the CMB anisotropy derived by the maximum likelihood method with each C_l coefficient treated as a separate parameter. The red, orange and yellow colour bands correspond to 68, 95 and 99% confidence bounds on the derived solution for each multipole amplitude. Noise and incomplete sky coverage induce correlations, impossible to show on the plot, between the derived confidence intervals on each C_l. One should note that a correct estimator of excess variance in signal over noise in the data was applied, hence only positive C_l values were obtained. A low-l tail of the CDM-model CMB anisotropy spectrum (independently fitted to the same data) is shown as a slightly tilted solid blue line.

6 LINEAR FILTERING OF THE COBE-DMR SKY MAPS

The unprocessed *COBE*-DMR 4-year sky maps contain statistically significant signal ($S/N \gtrsim 1$) on angular scales greater than 10 degrees, which correspond to $l \leq 15$. The discretised structure of the maps represented in principle by 6144 pixels allows harmonic decomposition up to $l \sim 70$. Hence, most of the Fourier content of the data comprises noise. Sky signals are filtered by the DMR beam response (7° FWHM), which attenuates their harmonic coefficients. Hence, in order to enhance the visibility of the real sky features one can suppress the noise by removing the high-l harmonic content of the sky maps, which we refer to as cleaning. It has been determined that such low-pass filtering with $l_{max} = 40$ retains nearly unperturbed genuine sky structure (including the sharp features of the Galactic plane) while removing approximately 50% of the pixel noise. The result of such processing of the coadded 53 and 90 GHz 4-year DMR sky maps is shown in Figure 11.

When the statistical properties of the signal (i.e. the power spectrum), polluted by noise in the data, are known, one can apply powerful tools of image processing to improve, oftentimes visually quite dramatically, the presentation of the measurements. The best known device for linear filtering of noisy data is the Wiener filter. Alternatives, usually referred to as sub-optimal filters, do exist, and on occasion provide a preferred means of treating the noisy data. In the case of the DMR raw data, the signal-to-noise ratio is rather poor even in the 4-year data set ($S/N \lesssim 1$ for angular scales $\lesssim 10$ degrees). The Wiener filter, which practically annihilates those components of the data which are assessed to be noise dominated and also significantly attenuates the data well into the $S/N \gtrsim 1$ regime, when applied to data of low quality usually renders the impression of "oversmoothing". In order to avoid that impression, and to save some dynamic range in the filtered *COBE*-DMR skymaps, we choose to apply another device, called the power-equalisation filter, which is constructed as follows.

First let us define the measurement, signal, and noise vectors, and their covariance matrices:

$$\mathbf{m} = \mathbf{s} + \mathbf{n}, \quad \langle \mathbf{s} \cdot \mathbf{s}^T \rangle = \mathbf{S}, \quad \langle \mathbf{n} \cdot \mathbf{n}^T \rangle = \mathbf{N}, \quad \langle \mathbf{m} \cdot \mathbf{m}^T \rangle = \mathbf{S} + \mathbf{N}. \tag{7}$$

Next, define the filtered data vector, and the filter matrix:

$$\mathbf{f} = \mathbf{F} \cdot \mathbf{m} = \mathbf{F} \cdot (\mathbf{s} + \mathbf{n}). \tag{8}$$

If the filter is constructed in accordance with the requirement that the mean square deviation between the filtered and the true signal, $\epsilon = \langle (\mathbf{f} - \mathbf{s})^T \cdot (\mathbf{f} - \mathbf{s}) \rangle = \text{tr}\langle (\mathbf{f} - \mathbf{s}) \cdot (\mathbf{f} - \mathbf{s})^T \rangle$, is minimised, one obtains the Wiener filter

$$\mathbf{F}_W = \mathbf{S} \cdot (\mathbf{S} + \mathbf{N})^{-1}. \tag{9}$$

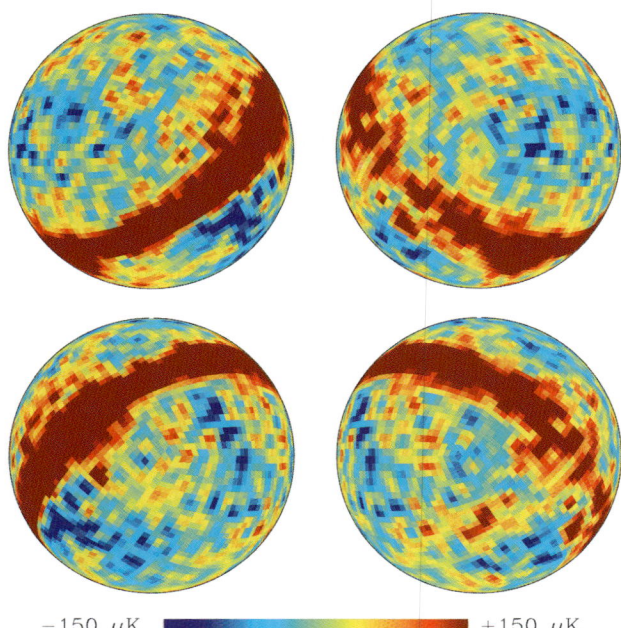

Figure 11 Projection on the sphere of the sky map generated from coadding the 4 maps at 53 and 90 GHz. This combination gives the optimal signal-to-noise for the DMR data. The maps have also been cleaned: that is, a filter has been applied to remove some fraction of the signal associated with noise rather than the true signal from the sky. The map is shown in four orientations: to the left the Galactic center region is most visible, to the right the anti-centre; the upper plots show the view from above onto the northern Galactic cap, the lower plots the view from below onto the southern Galactic cap. The underlying geometry of the quadrilateral cube structure – used for binning the data into a discretised sky map – can also be discerned from the visible pixel distribution.

An alternative possibility for sub-optimal filtering, which produces the power-equalisation filter, arises when $\mathbf{F}$ is chosen such that

$$\langle \mathbf{f} \cdot \mathbf{f}^T \rangle = \mathbf{S}, \quad \text{i.e.} \quad \mathbf{F}_{PE} \cdot (\mathbf{S} + \mathbf{N}) \cdot \mathbf{F}_{PE}^T = \mathbf{S}. \tag{10}$$

Equation (10) means that on average over many applications the power-equalisation filter renders filtered data whose statistical distribution matches that of the underlying signal. Obviously, we are simply using the power-equalisation filter to suppress the small scale power on the maps only just enough to mimic the true signal distribution via the matched power spectra of the filtered and true data, and in any particular realisation, the actual spatial distribution of small scale structure will not match the true underlying signal exactly. In the same regime, the Wiener filter would simply completely remove the small scale structure from the filtered maps. However, as already stated, in the case of poor quality data, the richer visual appearance of the power-equalisation filtered data renders the employment of this technique attractive for graphical display applications, exactly in accord with what interests us here.

Using Choleski decomposition of the relevant matrices, and the following auxiliary matrices, we can construct the power-equalisation filter as follows:

$$\mathbf{S}^{-1} = \mathbf{L}_{\mathbf{S}^{-1}} \cdot \mathbf{L}_{\mathbf{S}^{-1}}^T, \quad \mathbf{\Gamma}_{\mathbf{S}^{-1}} = \mathbf{L}_{\mathbf{S}^{-1}}^{-1}, \quad \mathbf{S} = \mathbf{\Gamma}_{\mathbf{S}^{-1}}^T \cdot \mathbf{\Gamma}_{\mathbf{S}^{-1}}, \quad (11)$$

$$(\mathbf{S} + \mathbf{N})^{-1} = \mathbf{L}_{(\mathbf{S}+\mathbf{N})^{-1}} \cdot \mathbf{L}_{(\mathbf{S}+\mathbf{N})^{-1}}^T,$$
$$\mathbf{\Gamma}_{(\mathbf{S}+\mathbf{N})^{-1}} = \mathbf{L}_{(\mathbf{S}+\mathbf{N})^{-1}}^{-1}, \quad (12)$$
$$\mathbf{S} + \mathbf{N} = \mathbf{\Gamma}_{(\mathbf{S}+\mathbf{N})^{-1}}^T \cdot \mathbf{\Gamma}_{(\mathbf{S}+\mathbf{N})^{-1}},$$

and as a result:

$$\mathbf{F}_{PE} = \mathbf{\Gamma}_{\mathbf{S}^{-1}}^T \cdot \mathbf{L}_{(\mathbf{S}+\mathbf{N})^{-1}}^T \quad (13)$$

This is an upper triangular matrix; hence, the filtering of a given mode in the cut-sky data vector $\hat{\mathbf{c}}$ does not mix in the information from lower order modes, preserving the information ordering imposed by the construction of orthonormal modes (see Górski 1994).

For the presentation purposes of this contribution we constructed separate filters for the 4 year COBE-DMR 53 and 90 GHz maps. The tremendous advantage of DMR is clearly visible here. We can internally compare two sets of decent quality observations at 53 and 90 GHz, and we are able to quantify and visualize the consistency with which the instrument reveals the image of the last scattering surface.

7 THE FIRST IMAGE OF THE LAST SCATTERING SURFACE

It should be realized that the COBE-DMR observations *did* allow us to make a first reliable picture of the last scattering surface. Recognition of this fact is somewhat obscured by the memory of the famous picture of the first DMR all sky map that was circulated in 1992 – a picture which was substantially distorted by noise features.

Four years after that presentation, the quality of DMR data was substantially better, but it is still somewhat difficult to argue about its validity as an image of the last scattering surface, since (at the time of writing) there is simply no other data of comparable sky coverage and quality at microwave frequencies and the same angular scales to allow a direct image to image comparison with DMR. However, since DMR observed the sky at three frequencies, and two of these have rendered sky maps of decent quality, an internal image comparison is possible.

What is presented here is the result of a fairly long chain of events: separate instruments (with different noise properties) were used to observe the sky (looking at different directions at the time of observation), the data were separately analysed and corrected for the systematic effects, the maps were made separately, all the subsequent data reduction was performed separately, and in the end different filters were constructed and applied.

The end product of these operations is visible in two panels on the left side of Figure 12 as the DMR 53 and 90 GHz, power-equalisation filtered sky maps. To facilitate the comparison and assessment of consistency of these maps, we add and subtract them. This is shown in the remaining panels of the color plate in both Mollweide and spherical polar projections.

It is immediately clear that the 53 and 90 GHz maps contain many common features on large angular scales, and (as shown in Banday et al. 1996, Kogut et al. 1996a, Górski et al. 1996), they can not be attributed to foreground emission from either the Galaxy, or nearby extragalactic objects. The epistemological value of this result of the COBE-DMR mission should not be underestimated, as this is the first time ever that we are afforded a reliable picture of the most remote regions of the universe.

The rms anisotropy in the (53+90)/2 power-equalisation filtered map is $35\,\mu\mathrm{K}$, while the rms in the (53–90)/2 map is about $15\,\mu\mathrm{K}$. These numbers define the accuracy to which COBE-DMR has revealed to us a most elusive astronomical realm, and provided a first direct glimpse at the universe in its embryonic state.

8 WHAT DOES COBE-DMR TELL US ABOUT THE UNIVERSE?

Let us now return to the interpretation of the angular power spectrum of the CMB anisotropy derived from the COBE-DMR data (Figure 10), and then proceed to the assessment of the key assumption that was visible in the construction of the maximum likelihood power spectrum estimation tools, namely that of the Gaussianity of the cosmological CMB anisotropies.

Already in the early 1970s, the following arguments were made (see Peebles and Yu 1970, Harrison 1970, Zel'dovich 1972) regarding the scale-dependence of the primordial curvature perturbations that must have seeded the development of the presently observed large scale matter inhomogeneities in the universe. Since we do not observe tiny black holes at the very small scale end of the hierarchy of density perturbations, this seems to suggest that there is more power at long wavelengths of the primordial inhomogeneities than at small. At the same time, the observed near isotropy of the CMB on very large angular scales seems to support the opposite conclusion. The apparent solution to an undesirable run-away in inhomogeneity power at either end of the wavelength range requires a distribution with

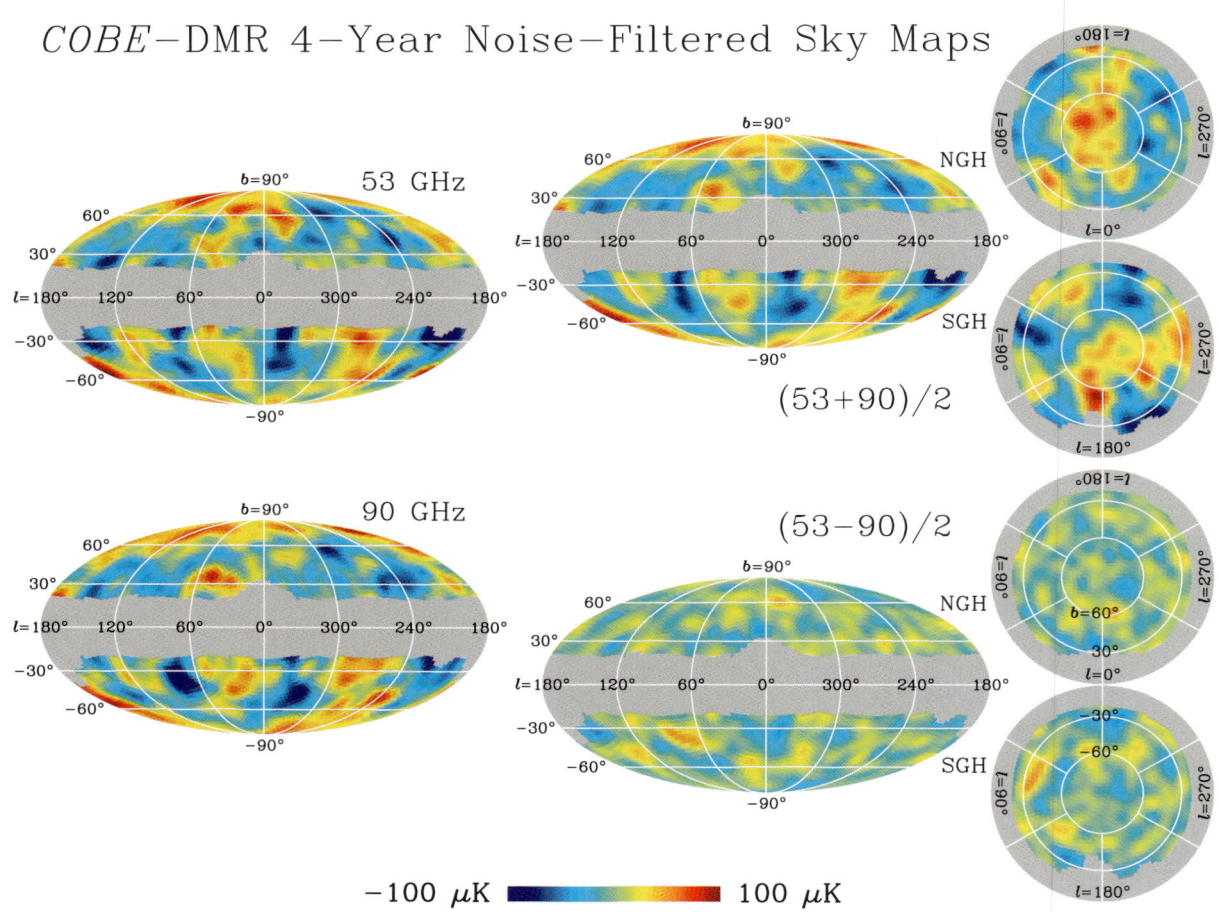

Figure 12 Power-equalisation filtered *COBE*-DMR 4-year sky maps showing the best currently available image of the CMB last scattering surface on large angular scales.

roughly equal power per decade of scale. In the language of 3-D wavenumber k (inverse wavelength), and curvature perturbation Fourier amplitude φ_k, this translates to const $\sim \int d\ln k \, k^3 |\varphi_k|^2$, and hence $|\varphi_k|^2 \propto 1/k^3$. Since the power spectrum of density perturbations $P(k) \propto k^4 |\varphi_k|^2$ (due to the Poisson equation), the "required" form of the power distribution in the density perturbation field is $P(k) \propto k$ – later to be termed the Harrison-Zel'dovich spectrum. At that time, there were no known physical mechanisms which could naturally produce such a scale-invariant power spectrum of fluctuations. After the idea of inflation found its way into cosmology (Starobinsky 1980, Guth 1981), it was soon demonstrated (Bardeen *et al.* 1983) that the mechanism of stretching the microscopically sized zero-point quantum fluctuations (which are then necessarily random phase, or Gaussian distributed) in the inflaton field into present epoch fluctuations on all observable scales and even beyond can actually produce the required nearly Harrison-Zel'dovich distribution of power for the seeds of large-scale structure. Even though everything seemed to have been qualitatively falling neatly together, there were still quantitative problems with the theories failing to predict correctly, in the absence of serious fine-tuning, the amplitude of the power spectrum of inflation generated inhomogeneities. However, at that time, the idea of weakly interacting cold dark matter (CDM) particles (Peebles 1984, Davis *et al.* 1985) as the dominant constituents of an inflationary universe had been proposed and studied via state-of-the-art numerical simulations of structure formation. The results of these simulations matched to the astronomical data and especially the statistical measures of inhomogeneity in the galaxy distribution (correlation function, number counts etc.) allowed calibration of the model.

The scale invariant curvature perturbations generate, via the Sachs-Wolfe effect (Sachs and Wolfe 1967), the CMB anisotropy described by an angular power spectrum $C_l \propto 1/l(l+1)$. The cold dark matter model calibrated angular spectrum of CMB fluctuations defined the target for the CMB anisotropy measurements to be performed by *COBE*-DMR. When these predictions were confronted with the expected performance of the DMR receivers, it appeared that perhaps, unfortunately, *COBE*-DMR would not be sufficiently sensitive to detect the low amplitude theoretically predicted anisotropies. Although not everyone adhered strictly to the apparent constraints imposed by the then very popular CDM model, in general the expectations of the community, that *COBE*-DMR would succeed in determining the longstanding "moving target" of the CMB anisotropy amplitude, thus concluding the game played out between the theorists and experimentalists, were not very high before *COBE* was launched.

Nevertheless, *COBE* was finally launched in 1989, the discovery of anisotropy by the DMR team was reported in 1992, and after four years of integration of signals it was indeed possible to make a very good quality measurement of the angular spectrum of CMB anistropy (Figure 10), and an image of the last scattering surface (Figure 12). Very rapidly after the discovery it was realised (Efstathiou *et al.* 1992) that the amplitude of the detected fluctuations is high compared to the established "standard" CDM scenario predictions. *COBE*-DMR had begun making an impact which we shall illustrate with a few figures comparing the angular power spectrum as derived from the 4-year DMR data (Figure 10) with spectra derived from various theories.

Figure 13 shows the striking discrepancy between the measured signals and the predictions of the low normalisation "standard" CDM model of the late 1980s. Figure 14 shows the same model renormalised to high amplitude, which allows it to fit the *COBE*-DMR data very well, but forces violations with other astronomical constraints – most notably at the scale of galaxies and clusters. Figure 15 shows another theoretical possibility – the model of the universe dominated by a cosmological constant, a model currently enjoying much popularity due to the interpretation of measurements of distant supernovae (Leibundgut 2001, Perlmutter 1998, Riess *et al.* 1998, Riess 2000). A noticeable feature of such models is the upward bend of the C_l spectrum at low multipoles – this "extra" power compared to the flat matter dominated models is due to the so-called integrated Sachs-Wolfe effect, or the influence of the time variation in the gravitational potential of the inhomogeneities at low redshift. This feature actually makes the cosmological constant dominated models fit the *COBE*-DMR data somewhat more poorly than the matter dominated flat models do, but the effect is not significant enough to reject the models purely by confrontation with the DMR data. Figure 16 shows a possible fit for an open cosmological model of the C_l. In this model, the suppression of low order modes is seen when compared to the flat matter dominated cosmology. The effect is driven by attenuation of the power of inhomogeneity at the scales close to the curvature length of the model. The fit to the DMR data is nominally better than that of the flat, matter-dominated models, but the model fits poorly more recent CMB data (e.g. BOOMERanG (de Bernardis *et al.* 2000), MAXIMA

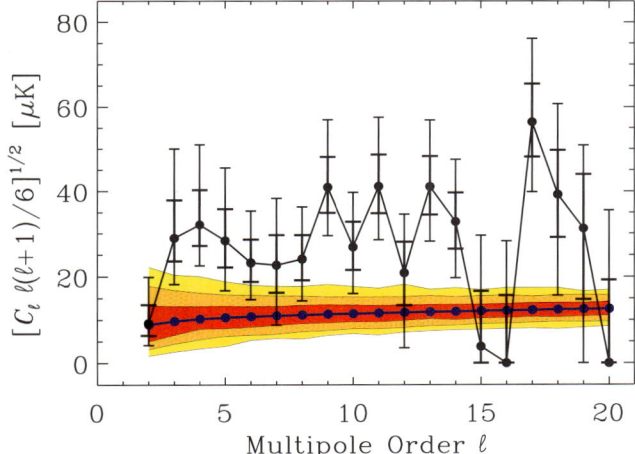

Figure 13 Comparison of theory predictions for the biased cold dark matter model (blue dots), which was the predominant theory of cosmological structure formation in the 1980s, with the *COBE*-DMR power spectrum of CMB anisotropy (black dots with 68, and 95% error bars from Figure 12). The colour bands now show the statistically allowed range of predicted C_l amplitudes for an ensemble of simulated CMB skies, consistent with the theory. The red, orange, and yellow bands correspond to 68, 95, and 99% probability ranges. Clearly the model is dramatically inconsistent with the data.

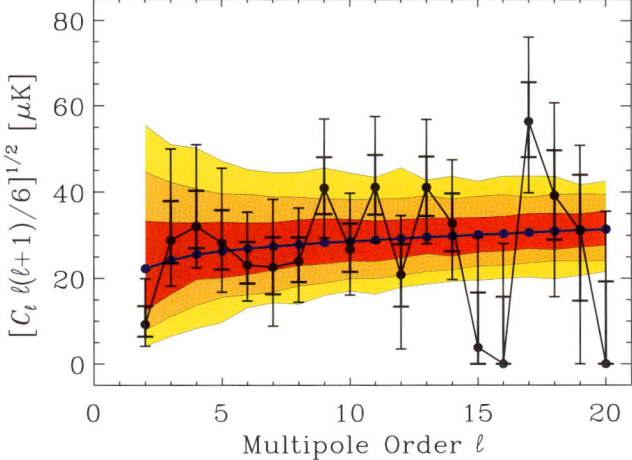

Figure 14 Comparison of theory predictions for the unbiased cold dark matter model with the DMR angular power spectrum of CMB anisotropy. The colour coding is the same as in Figure 13. The model is consistent with the DMR data, but fails in confrontation with other astronomical observations.

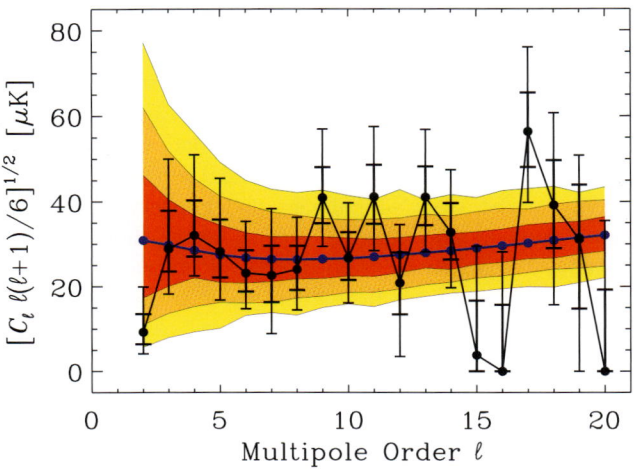

Figure 15 Comparison of theory predictions for the cosmological constant dominated cold dark matter model with the DMR angular power spectrum of CMB anisotropy. The colour coding is the same as in Figure 13. This model is a slightly poorer fit to the DMR data alone than is the unbiased CDM model, but offers greater consistency with other astronomical observations (e.g. the more recent CMB results from BOOMERanG and MAXIMA experiments, and the distant supernova data).

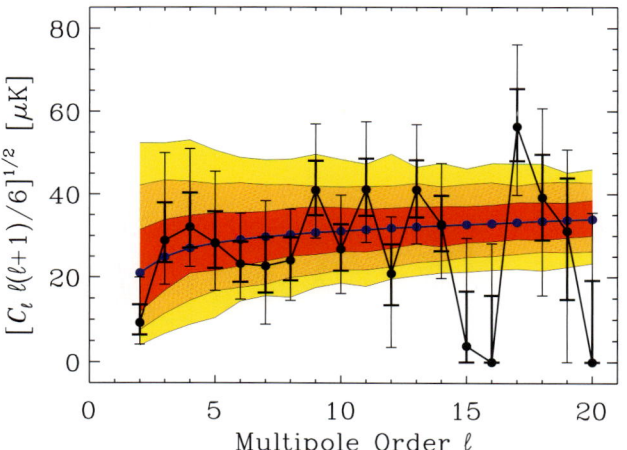

Figure 16 Comparison of theory predictions for the cold dark matter model with sub-critical density with the DMR angular power spectrum of CMB anisotropy. The colour coding is the same as in Figure 13. This model offers a better fit to the DMR data alone than does the one with cosmological constant, but it is currently disfavoured, for example, by the recent CMB results from BOOMERanG and MAXIMA experiments.

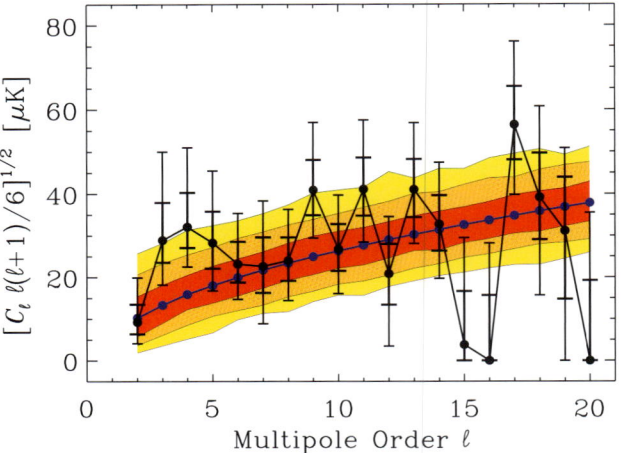

Figure 17 Comparison of theory predictions for the baryon dominated isocurvature perturbation model of structure formation with the DMR angular power spectrum of CMB anisotropy. The colour coding is the same as in Figure 13. The model is significantly inconsistent with the DMR data.

(Hanany et al. 2000) at smaller angular scales than those probed by *COBE*-DMR. Finally, Figure 17 shows a model which fits the DMR data poorly. This is a version of the isocurvature baryon model which was proposed as an alternative to inflation based scale invariant models with curvature perturbations. The C_l spectrum of this model is considerably steeper than scale-invariant (equivalent to $P(k) \propto k^2$ in curvature perturbation parlance).

It is clear then that the measurements performed by the *COBE*-DMR support the nearly scale invariant distribution of power in the matter perturbations that seeded the presently observed large-scale structure of the universe, and detected a sufficiently large amplitude of those perturbations that the structure formation models developed before the DMR data were available had to be significantly readjusted in order to accomodate these new constraints.

We should now return to the issue of the Gaussian nature or otherwise of the primordial CMB anistropies as detected by *COBE*-DMR. Considering that after 4 years of observations the coadded sky 53 and 90 GHz maps show $S/N \gtrsim 1$ on angular scales larger than about 10 degrees, one may argue that DMR has ultimately independently measured the CMB temperature at roughly 700 points on the sky. Further considering that about 30% of the sky is dominated by the Galactic emission, one is left with about $\lesssim 500$ independent temperature measurements with which to assess the statistical properties of the distribution of anisotropy. Although this is not enough for an overwhelming fidelity in results to be achieved, it does allows us to perform a series of useful tests on the fundamentally important hypothesis about the

Gaussianity of the perturbations. Certainly, a detectable violation of this property would be of fundamental importance for our understanding of structure formation in the universe. The first tests of the Gaussian nature of the anisotropy was performed by Smoot et al. (1994) who considered the simplest one-point statistics – the skewness and kurtosis – on the first year DMR data. Given the poor signal-to-noise of the initial data set, it should not be too surprising that little evidence of non-Gaussian features were detected. However, even with the much improved quality of the subsequent data sets, and application of more sophisticated tests, including the 3-point correlation function and excursion set statistics, the Gaussian nature of the detected fluctuations remained uncontested (Kogut et al. 1996b). Interestingly, however, this issue has recently become one of increasing interest again. The application of a new statistical estimator, the angular bispectrum (Ferreira et al. 1998), resulted in the claim that a non-Gaussian signal had indeed been finally detected. Subsequent careful reanalysis of the data sets determined that this was in actuality due to the presence of a systematic artifact which had otherwise eluded detection, but which importantly had no impact on any determination of cosmologically interesting parameters based on the power spectrum (Banday et al. 2000). Other recent analyses based on wavelets statistics (Barreiro et al. 2000) or a Bayesian determination of the skewness of the data simultaneously with the power spectrum (Contaldi et al. 2000) also conclude that the integrity of the COBE-DMR data with respect to the Gaussian nature of the observed anisotropy remained intact.

9 LARGE-SCALE STRUCTURE OF THE UNIVERSE AFTER *COBE* AND FUTURE SATELLITE CMB MISSIONS

The summary of the "Scientific American" article which was referred to in the introduction made the following assertion: "...cosmology will make a leap forward because of *COBE* discoveries". Has this been a correct prediction? In many ways the answer is positive. The DMR instrument, upon which we focused our attention in this contribution, provided us with a wealth of data with which we have been able to

1. discover the cosmological anisotropy of the CMB, which provided the essential proof of existence for those theories purporting that structure in the universe grew slowly from small perturbations, which had to leave traces of their embryonic appearance imprinted on the last scattering surface,
2. confirm that the statistical distribution of cosmological perturbations was most likely Gaussian,
3. measure with good accuracy the angular spectrum of CMB fluctuations, whose shape, very close to scale-invariant, broadly supports the inflationary models of structure formation, but which, due to an unexpectedly high amplitude, required significant revisions and adjustments in the mainstream models of structure formation, and
4. unveil the image of the last scattering surface and thereby finally bring the field of CMB into the domain of astronomy.

Of course right now, *post factum*, it is easy to want more, to debate the fine points of the statistical significance of some of these results, and to lose a little bit of perspective in assessing the impact of *COBE*-DMR in light of all the new results which have appeared since ca. 1996. Nevertheless, anyone who really understands the development of cosmology will always perceive that *COBE*-DMR was, in the truest sense of the word, a *tour de force* of CMB experimentation, and that the considerable costs of building and sending *COBE* to space to conduct its mission were completely justified by the fantastic return of scientific results of fundamental importance.

There was also another way in which *COBE* made an important impact. Its successful mission helped both to generate and renew interest in follow-up CMB observations pursuing the immediately evident new goals of much higher angular resolution and sensitivity. Despite a relatively large number of CMB experiments operating from sub-orbital platforms, both the USA and European CMB communities were successful in making applications to NASA and ESA respectively, for *COBE* follow-up satellite missions. The American mission, the Microwave Anisotropy Probe (MAP), was successfully launched on June 30, 2001, and is scheduled to release maps of its first coverage of the sky by late 2002, while the European next generation mission, Planck, is currently expected to fly in 2007. In Figure 18, we schematically compare the capabilities of *COBE*-DMR and these two future missions. First we see how the combination of 7° FWHM angular resolution, and its receiver noise have both limited, to within $l \lesssim 15$, the window within which, DMR could measure the spectrum of CMB anisotropy. MAP – a conceptual replay of the successful *COBE*-DMR mission concept – is designed to resolve the CMB down to ~12 arcmin at 90 GHz, and reach a noise performance good enough to reveal the structure in the C_l spectrum up to $l \sim 600$, which, in a fiducial flat universe model with adiabatic perturbations, should allow one to measure C_l well beyond the second acoustic peak. Following this, Planck, with its combination of actively cooled, high sensitivity radiometric and bolometric receivers, angular resolution down to ~6 arcmin FWHM in the CMB dominated channels, and unprecedented spectral coverage between 30 and 850 GHz, should be able to measure

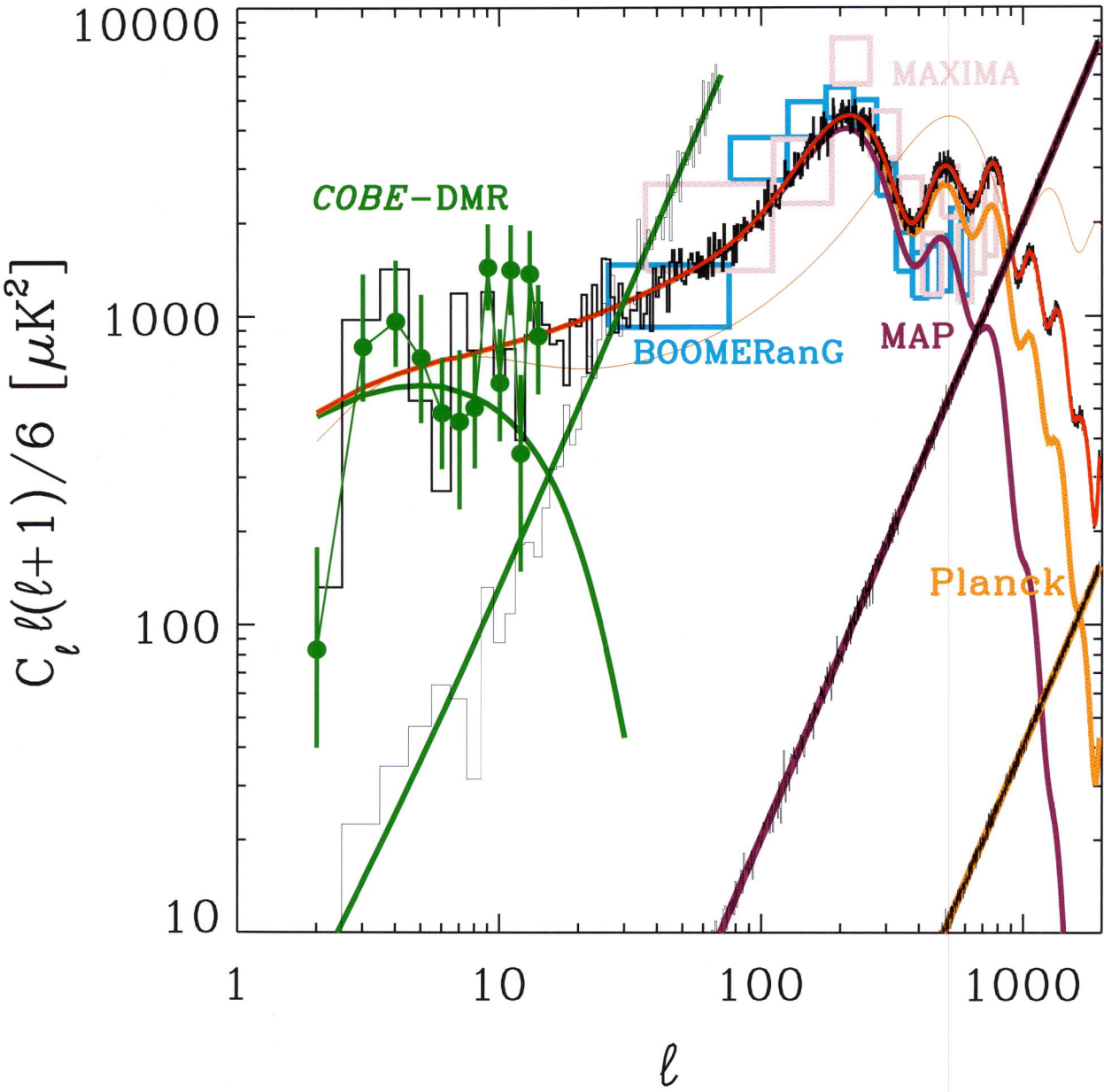

Figure 18 Illustration of the capabilities of the satellite CMB missions *COBE*-DMR, MAP, and Planck. A CDM theory-based average spectrum of CMB anisotropy is shown with a heavy red line. Superposed black histogram illustrates statistical spread of actual C_l values in a particular random realisation of the CMB sky consistent with such a model. The light red line shows, for comparison, the theoretical spectrum for a sub-critical density model of the universe – the first spectrum peak in this model is shifted to higher l values, or smaller angular scales. Current CMB data (most recently from the BOOMERanG and MAXIMA experiments, the results of which are plotted as boxes representing band-power averages) disfavour such models with indications of the first peak structure near $l \sim 250$. At each l the mode of the probability distribution of the C_l is plotted for meaningful comparison with the maximum likelihood solution to the *COBE*-DMR data. Since the C_l-s are χ^2_{2l+1}-distributed, the mode is visibly suppressed at low-l relative to the average value (which if plotted conventionally would appear flat). The high l end of the spectrum is suppressed by the finite antenna resolution of the experiment – the values illustrated here are FWHM = 6 arcmin (Planck), 12 arcmin (MAP), and 7° (DMR). The rising straight lines show the average noise spectra of each experiment.

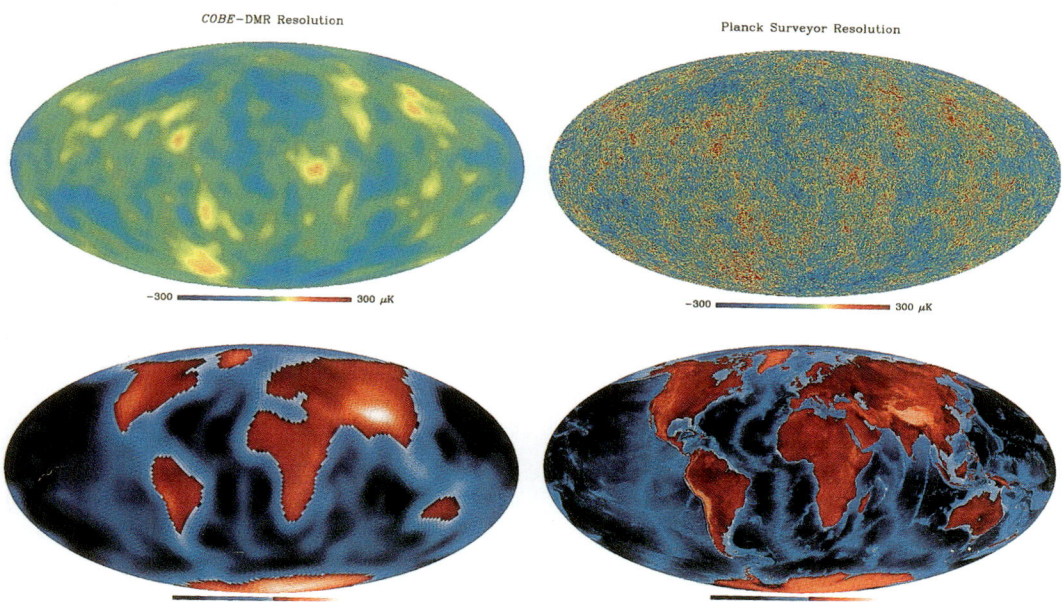

Figure 19 Comparison of the structures which have been resolved by COBE-DMR to those which will be resolved by observations from the MAP and Planck satellites. The top two figures are simulations of the CMB sky for a specific model of the universe dominated by cold dark matter. MAP and Planck will map the sky with an angular resolution typically better than 20 arcminutes and will see a richness of detail in the CMB anisotropy pattern unmeasurable by COBE-DMR, which was only sensitive to structures of angular size in excess of 10°. This dramatic improvement in our picture of the last scattering surface of the universe – to be achieved in only about a decade of development of CMB observations – can be compared to the similar evolution in our understanding of the Earth's topography (depicted in the lower two images) which took a few centuries of geographic exploration.

structure in the C_l up to $l \sim 1500$, and allow us to determine with incredible accuracy a large number of cosmological parameters, which should completely specify the details of a sufficiently adequate model of our universe.

These are wonderful prospects, and it is a truly exciting realm that those involved with MAP and Planck will be exploring. Let us now stop and reflect that the door to that realm was first opened by COBE!

Acknowledgements

We would like to acknowledge the National Aeronautics and Space Administration/Goddard Space Flight Center (NASA/GSFC) which was responsible for the design, development, and operation of COBE, our many colleagues on the DMR project – G. Hinshaw, P. Jackson, E. Kaita, P. Keegstra, A. Kogut, V. Kumar, R. Kummerer, C. Lineweaver, K. Loewenstein, J. Santana, and L. Tenorio – and the Science Working Group – C.L. Bennett, N. Boggess, E. Cheng, E. Dwek, S. Gulkis, M. Hauser, M. Janssen, T. Kelsall, D.T. Leisawitz, P. Lubin, J.C. Mather, S. Meyer, H. Moseley, T.L. Murdock, R. Shafer, R. Silverberg, G. Smoot, R. Weiss, D. Wilkinson and E. Wright. In particular, we thank John Mather for recommending us as authors to the editors of this volume. We gratefully thank Eric Hivon for assistance in the rendition of Figure 19.

REFERENCES

Banday, A.J., Górski, K.M., Bennett, C.L., Hinshaw, G., Kogut, A. and Smoot, G.F. (1996). Noncosmological signal contributions to the COBE DMR 4 year sky maps. *Astrophysical Journal*, **468**, L85–L89.

Banday, A.J., Górski, K.M., Bennett, C.L., Hinshaw, G., Kogut, A., Lineweaver, C., Smoot, G.F. and Tenorio, L. (1997). Root mean square anisotropy in the COBE DMR four-year sky maps. *Astrophysical Journal*, **475**, 393–398.

Banday, A.J., Zaroubi, S. and Górski, K.M. (2000). On the non-Gaussianity observed in the COBE Differential Microwave Radiometer sky maps. *Astrophysical Journal*, **533**, 575–587.

Bardeen, J.M., Steinhardt, P.J. and Turner, M.S. (1983). Spontaneous creation of almost scale-free density perturbations in an inflationary universe. *Physical Review D*, **28**, 679–693.

Barreiro, R.B., Hobson, M.P., Lasenby, A.N., Banday, A.J., Górski, K.M. and Hinshaw, G. (2000). Testing the Gaussianity of the COBE DMR data with spherical wavelets. *Monthly Notices of the Royal Astronomical Society*, **318**, 475–481.

Bennett, C.L., Smoot, G.F., Hinshaw, G., Wright, E.L., Kogut, A., de Amici, G., Meyer, S.S., Weiss, R., Wilkinson, D.T., Gulkis, S., Janssen, M., Boggess, N.W., Cheng, E.S., Hauser, M.G., Kelsall, T., Mather, J.C., Moseley, S.H., Jr., Murdock, T.L. and Silverberg, R.F.

(1992). Preliminary separation of galactic and cosmic microwave emission for the COBE Differential Microwave Radiometer. *Astrophysical Journal*, **396**, L7–L12.

Bennett, C.L., Kogut, A., Hinshaw, G., Banday, A.J., Wright, E.L., Górski, K.M., Wilkinson, D.T., Weiss, R., Smoot, G.F., Meyer, S.S., Mather, J.C., Lubin, P., Loewenstein, K., Lineweaver, C., Keegstra, P., Kaita, E., Jackson, P.D. and Cheng, E.S. (1994). Cosmic temperature fluctuations from two years of COBE differential microwave radiometers observations. *Astrophysical Journal*, **436**, 423–442.

Bennett, C.L., Banday, A.J., Górski, K.M., Hinshaw, G., Jackson, P., Keegstra, P., Kogut, A., Smoot, G.F., Wilkinson, D.T. and Wright, E. L. (1996). Four-year COBE DMR cosmic microwave background observations: Maps and basic results. *Astrophysical Journal*, **464**, L1–L4.

Bond, J.R. (1995). Signal-to-noise eigenmode analysis of the two-year COBE maps. *Physical Review Letters*, **74**, 4369–4372.

Bond, J.R., Jaffe, A.H. and Knox, L. (1998). Estimating the power spectrum of the cosmic microwave background. *Physical Review D*, **57**, 2117–2137.

Contaldi, C.R., Ferreira, P.G., Magueijo, J. and Górski, K.M. (2000). A Bayesian estimate of the skewness of the cosmic microwave background. *Astrophysical Journal*, **534**, 25–28.

Davis, M., Efstathiou, G., Frenk, C.S. and White, S.D.M. (1985). *Astrophysical Journal*, **292**, 371–394.

de Bernardis, P., Ade, P.A.R., Bock, J.J., Bond, J.R., Borrill, J., Boscaleri, A., Coble, K., Crill, B.P., De Gasperis, G., Farese, P.C., Ferreira, P.G., Ganga, K., Giacometti, M., Hivon, E., Hristov, V.V., Iacoangeli, A., Jaffe, A.H., Lange, A.E., Martinis, L., Masi, S., Mason, P.V., Mauskopf, P.D., Melchiorri, A., Miglio, L., Montroy, T., Netterfield, C.B., Pascale, E., Piacentini, F., Pogosyan, D., Prunet, S., Rao, S., Romeo, G., Ruhl, J.E., Scaramuzzi, F., Sforna, D. and Vittorio, N. (2000). A flat Universe from high-resolution maps of the cosmic microwave background radiation. *Nature*, **404**, 955–959.

Dicke, R.H., Peebles, P.J.E., Roll, P.G. and Wilkinson, D.T. (1965). Cosmic black-body radiation. *Astrophysical Journal*, **142**, 414–419.

Efstathiou, G., Bond, J.R. and White, S.D.M. (1992). COBE background radiation anisotropies and large-scale structure in the universe. *Monthly Notices of the Royal Astronomical Society*, **258**, 1P–6P.

Ferreira, P.G., Magueijo, J. and Górski, K.M. (1998). Evidence for non-Gaussianity in the COBE DMR 4 year sky maps. *Astrophysical Journal*, **503**, L1–L4.

Fixsen, D.J., Cheng, E.S., Gales, J.M., Mather, J.C., Shafer, R.A. and Wright, E.L. (1996). The cosmic microwave background spectrum from the full COBE FIRAS data set. *Astrophysical Journal*, **473**, 576–587.

Ganga, K., Cheng, E., Meyer, S. and Page, L. (1993). Cross-correlation between the 170 GHz survey map and the COBE differential microwave radiometer first-year maps. *Astrophysical Journal*, **410**, L57–L60.

Górski, K.M. (1994). On determining the spectrum of primordial inhomogeneity from the COBE DMR sky maps: Method. *Astrophysical Journal*, **430**, L85–L88.

Górski, K.M. Hinshaw, G., Banday, A.J., Bennett, C.L., Wright, E.L., Kogut, A., Smoot, G.F. and Lubin, P. (1994). On determining the spectrum of primordial inhomogeneity from the COBE DMR sky maps: Results of two-year data analysis. *Astrophysical Journal*, **430**, L89–L92.

Górski, K.M., Banday, A.J., Bennett, C.L., Hinshaw, G., Kogut, A., Smoot, G.F. and Wright, E.L. (1996). Power spectrum of primordial inhomogeneity determined from the four-year COBE DMR sky maps. *Astrophysical Journal*, **464**, L11–L15.

Górski, K.M. (1997). In: Bouchet, F.R. (ed), "Microwave Background Anisotropies", *Proceedings of the XVIth Moriond Astrophysics Meeting*, March 1996, Editions Frontieres, p. 77.

Guth, A. (1981). Inflationary universe: A possible solution to the horizon and flatness problems. *Physical Review D*, **23**, 347–356.

Hanany, S., Ade, P., Balbi, A., Bock, J., Borrill, J., Boscaleri, A., de Bernardis, P., Ferreira, P.G., Hristov, V.V., Jaffe, A.H., Lange, A.E., Lee, A.T., Mauskopf, P.D., Netterfield, C.B., Oh, S., Pascale, E., Rabii, B., Richards, P.L., Smoot, G.F., Stompor, R., Winant, C.D. and Wu, J.H.P. (2000). MAXIMA-1: A measurement of the cosmic microwave background anisotropy on angular scales of 10'-5°. *Astrophysical Journal*, **545**, L5–L9.

Harrison, E.R. (1970). Fluctuations at the threshold of classical cosmology. *Physical Review D*, **1**, 2726–2730.

Hinshaw, G., Banday, A.J., Bennett, C.L., Górski, K.M., Kogut, A., Smoot, G.F. and Wright, E.L. (1996). Band power spectra in the COBE DMR four-year anisotropy maps. *Astrophysical Journal*, **464**, L17–L20.

Kogut, A. *et al.* (1996a). Microwave emission at high galactic latitudes in the four-year DMR sky maps. *Astrophysical Journal*, **464**, L5–L9.

Kogut, A., Banday, A.J., Bennett, C.L., Górski, K.M., Hinshaw, G., Smoot, G.F. and Wright, E.L. (1996b). Tests for non-Gaussian statistics in the DMR four-year sky maps. *Astrophysical Journal*, **464**, L29–L33.

Kogut, A., Banday, A.J., Bennett, C.L., Górski, K.M., Hinshaw, G., Jackson, P.D., Keegstra, P., Lineweaver, C., Smoot, G.F., Tenorio, L. and Wright, E.L. (1996c). Calibration and systematic error analysis for the COBE DMR 4 year sky maps. *Astrophysical Journal*, **470**, 653–673.

Leibundgut, B. (2001). Cosmological implications from observations of type Ia supernovae. *Annual Review of Astronomy and Astrophysics*, **39**, 67–98.

Mather, J. and Kelsall, T. (1980). *Physica Scripta*, **21**, 71.

Mather, J.C., Cheng, E.S., Eplee, R.E., Jr., Isaacman, R.B., Meyer, S.S., Shafer, R.A., Weiss, R., Wright, E.L., Bennett, C.L., Boggess, N.W., Dwek, E., Gulkis, S., Hauser, M.G., Janssen, M., Kelsall, T., Lubin, P.M., Moseley, S.H., Jr., Murdock, T.L., Silverberg, R.F., Smoot, G.F. and Wilkinson, D.T. (1990). A preliminary measurement of the cosmic microwave background spectrum by the Cosmic Background Explorer (COBE) satellite. *Astrophysical Journal*, **354**, L37–L40.

Peebles, P.J.E. and Yu, J.T. (1970). Primeval adiabatic perturbation in an expanding universe. *Astrophysical Journal*, **162**, 815–836.

Peebles, P.J.E. (1984). Dark matter and the origin of galaxies and globular star clusters. *Astrophysical Journal*, **277**, 470–477.

Penzias, A.A. and Wilson, R.H. (1965). A measurement of excess antenna temperature at 4080 Mc/s. *Astrophysical Journal*, **142**, 419–421.

Perlmutter, S., Aldering, G., Goldhaber, G., Knop, R.A., Nugent, P., Castro, P.G., Deustua, S., Fabbro, S., Goobar, A., Groom, D.E., Hook, I.M., Kim, A.G., Kim, M.Y., Lee, J.C., Nunes, N.J., Pain, R., Pennypacker, C.R., Quimby, R., Lidman, C., Ellis, R.S., Irwin, M., McMahon, R.G., Ruiz-Lapuente, P., Walton, N., Schaefer, B., Boyle, B.J., Filippenko, A.V., Matheson, T., Fruchter, A.S., Panagia, N., Newberg, H.J.M. and Couch, W.J. The Supernova Cosmology Project (1998). Measurements of Omega and Lambda from 42 high-redshift supernovae. *Astrophysical Journal*, **517**, 565–586.

Riess, A.G., Filippenko, A.V., Challis, P., Clocchiatti, A., Diercks, A., Garnavich, P.M., Gilliland, R.L., Hogan, C.J., Jha, S., Kirshner, R.P., Leibundgut, B., Phillips, M.M., Reiss, D., Schmidt, B.P., Schommer, R.A., Smith, R.C., Spyromilio, J., Stubbs, C., Suntzeff, N.B. and Tonry, J. (1998). Observational evidence from supernovae for an accelerating universe and a cosmological constant. *Astronomical Journal*, **116**, 1009–1038.

Riess, A.G. (2000). The case for an accelerating universe from supernovae. *Publications of the Astronomical Society of the Pacific*, **112**, 1284–1299.

Sachs, R.K. and Wolfe, A.M. (1967). Perturbations of a cosmological model and angular variations of the microwave background. *Astrophysical Journal*, **147**, 73–90.

Smoot, G.F., Gorenstein, M.V. and Muller, R.A. (1977). Detection of anisotropy in the cosmic blackbody radiation. *Physical Review Letters*, **39**, 898–901.

Smoot, G.F., Bennett, C.L., Kogut, A., Wright, E.L., Aymon, J., Boggess, N.W., Cheng, E.S., de Amici, G., Gulkis, S., Hauser, M.G., Hinshaw, G., Jackson, P.D., Janssen, M., Kaita, E., Kelsall, T., Keegstra, P., Lineweaver, C., Loewenstein, K., Lubin, P., Mather, J., Meyer, S.S., Moseley, S.H., Murdock, T., Rokke, L., Silverberg, R.F., Tenorio, L., Weiss, R. and Wilkinson, D.T. (1992). Structure in the COBE differential microwave radiometer first-year maps. *Astrophysical Journal*, **396**, L1–L5.

Smoot, G.F., Tenorio, L., Banday, A.J., Kogut, A., Wright, E.L., Hinshaw, G. and Bennett, C.L. (1994). Statistics and topology of the COBE differential microwave radiometer first-year sky maps. *Astrophysical Journal*, **437**, 1–11.

Starobinsky, A.A. (1980). *Physics Letters B*, **91B**, 99.

Tegmark, M. and Bunn, E.F. (1995). A brute force analysis of the COBE DMR data. *Astrophysical Journal*, **455**, 1–6.

Tegmark, M. (1995). A method for extracting maximum resolution power spectra from microwave sky maps. *Monthly Notices of the Royal Astronomical Society*, **280**, 299–308.

Tegmark, M. (1996). The angular power spectrum of the four-year COBE data. *Astrophysical Journal*, **464**, L35–L38.

Wright, E.L. *et al.* (1994). Angular power spectrum of the microwave background anisotropy seen by the COBE differential microwave radiometer. *Astrophysical Journal*, **436**, 443–451.

Wright, E.L., Bennett, C.L., Górski, K.M., Hinshaw, G. and Smoot, G.F. (1996). Angular power spectrum of the cosmic microwave background anisotropy seen by the COBE DMR. *Astrophysical Journal*, **464**, L21–L24.

Zel'dovich, Y.B. (1972). A hypothesis, unifying the structure and the entropy of the Universe. *Monthly Notices of the Royal Astronomical Society*, **160**, 1P–3P.

19

HUBERT REEVES*

The origin of the light elements in the early Universe

Shortly after World War II, George Gamov and his collaborators (Alpher *et al.* 1948) considered the possibility that all chemical elements might have been generated by a succession of nucleon capture reactions in the cooling primordial Universe. The successful prediction of the existence of a fossil radiation left over from the earliest times of the Universe (Gamov 1946) emerged from the hypothesized formation of helium in the big bang. There are two steps to this argument: (1) in order to have nuclear reactions, the temperature must have been in the MeV range (10^{10} K), and (2) given the fact that a free neutron decays in approximately 15 minutes, the transformation of some but not all of the primordial nucleons into helium requires the mean free path for neutron capture (and hence the cosmic density) to lie in an appropriate range. At too low a density, the Universe would be of pure hydrogen; at too high a density it would be of pure helium. It is, in fact, about one-quarter helium and three-quarters hydrogen (in fractional mass). By combining these numbers, Gamov and his colleagues obtained the first estimate of the ratio of nucleons to photons (of the order of 10^{-10}). From there, they predicted the existence of a fossil radiation in the millimetric range (CMB, cosmic microwave background) and estimated its present temperature (a few kelvin). The theory of big bang nucleosynthesis (BBN) is essentially based on these arguments.

However, the hypothesis of a primordial origin of all the chemical elements soon ran into major difficulties. In the proposed chain of successive nucleon capture, extending from hydrogen all the way to uranium, the nuclear instability of aggregates containing five and eight nucleons constituted a fatal flaw: it could not be seen how the corresponding steps in the chain could be achieved, so the scheme was largely abandoned.

Around the same time, Fred Hoyle and his collaborators (Burbidge *et al.* 1957) proposed a stellar origin for the chemical elements. This scheme turned out to be highly successful in accounting for the chemical elements from carbon to uranium ($12 < A < 238$). However, it ran into problems with the lighter elements. Because of their small electric charges, these nuclides rapidly undergo self-destroying nuclear reactions at the high temperatures of stellar interiors. Other formation mechanisms were clearly needed to account for their presence and abundances in the cosmos.

One requirement for the selection of the formation mechanism of these fragile, light nuclides is that, after their formation by appropriate nuclear phenomena, they should be rapidly evacuated into low-temperature regions. Two different physical processes meet this requirement: (1) BBN, where the rapid cooling of the universe preserves the newly formed nuclei from further thermonuclear reactions; and (2) galactic cosmic-ray (GCR) bombardment of interstellar atoms (mostly C, N and O) ejecting spallation residues (mostly Li, Be and B) into the cold interstellar medium (ISM).

THE CASE FOR BIG BANG NUCLEOSYNTHESIS

The best evidence in favour of the BBN of the ligther nuclides can be seen from Figures 1–3. Figure 1 shows the abundance of ^{4}He as a function of the time-integrated stellar activity, represented by the oxygen abundance in various galaxies. Although an increase in helium from approximately 20 to 30% (fractional mass) is clearly visible,

*Centre National de Recherche Scientifique, Paris, France

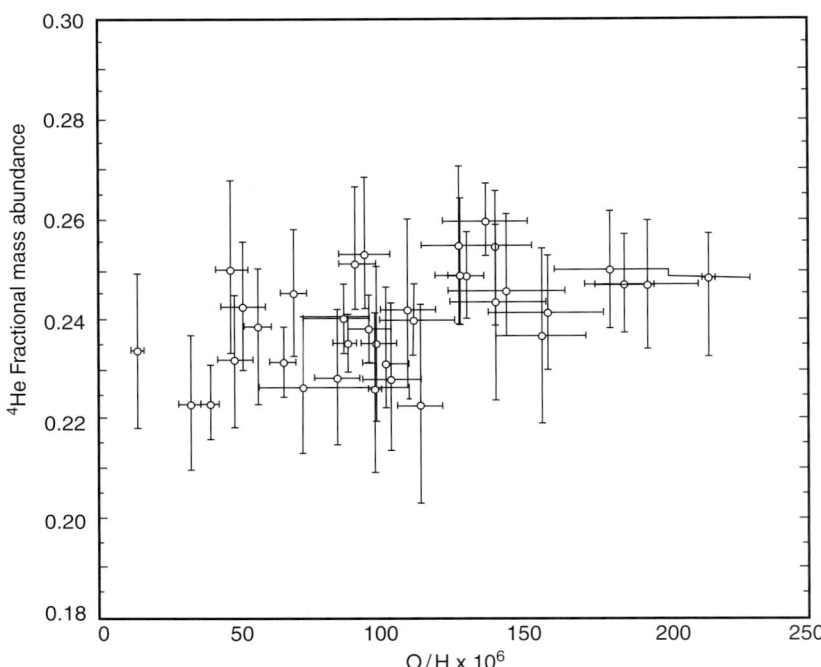

Figure 1 ^{4}He fractional mass abundances as a function of ^{16}O in chemically unevolved galaxies (after Pagel et al. 1992).

manifesting a stellar contribution, no celestial objects are known to have less than 20% helium (10 times less helium than hydrogen, in numbers of atoms).

A similar situation is found for ^{7}Li. As we go backward in time – by observing progressively older stars, representing the state of the ISM at the moment of their formation – the ^{7}Li to hydrogen ratio decreases progressively from 10^{-9} to 10^{-10}, where it levels off for all older stars (Figure 2). In the same fashion, the deuterium to hydrogen ratio appears to increase (Figure 3) with decreasing time.

These observations suggest that these light nuclides were already present in the primordial galactic gas. They point to a common origin: nuclear reactions in the hot big bang. Can this suggestion be made quantitative? Can the observed abundances be accounted for in terms of detailed computations? This is the subject of this chapter. It will appear that the answer is definitely yes, although some aspects are still obscure.

One important difficulty stems from the fact that the available observations (of galaxies, stars, interstellar gas, planets) cannot be readily compared with the computed primordial yields. Many factors have modified the abundances of the nuclides between the big bang and the moment at which they are observed in the history of the cosmos (e.g. the birth of the stars in which they are measured). Extrapolation from the observed data back to BBN has been a difficult task for many decades.

Another problem is the untangling, for each nucleus, of the relative contribution of different production mechanisms: BBN, GCR spallation and stellar nucleosynthesis. This is most important for the isotope ^{7}Li, for which the three mechanisms all contribute. For this reason, the cosmic-ray spallation processes and their resulting abundances will be reviewed in some detail.

NUCLEAR PHYSICS OVERVIEW

It is well known that many features of the universal abundance of chemical elements can be qualitatively understood through a knowledge of their nuclear properties. The iron peak (at $A=56$) corresponds to the most stable nuclear configurations. The secondary peaks correspond to nuclei with so-called magic numbers of neutrons (at $N=50$, 82 and 128), and also to the light nuclei with an integral number of alpha particles (at $A=12$, 16, 20, 24 and 28). In this respect, it is most informative to begin our study of the origin of the light elements with a brief review of their nuclear properties. Since the three elements lithium, beryllium and boron were potentially formed in the big bang, we shall extend this review up to $A<12$.

Two important factors play a crucial role in the nucleosynthesis of the light nuclei: their small electric charges and

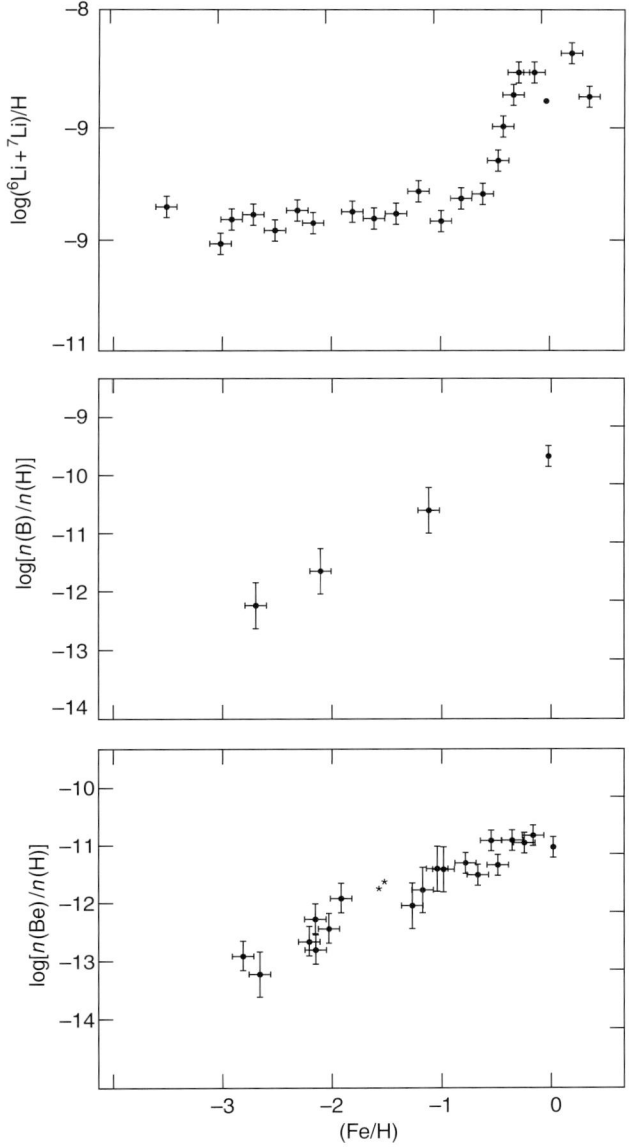

Figure 2 Li/H, Be/H and B/H stellar abundances as a function of Fe/H abundances. The logarithmic scale of the x-axis is in units of fractional solar iron abundance. For more data, see Cunha and Smith (1999) (after Gilmore 1992).

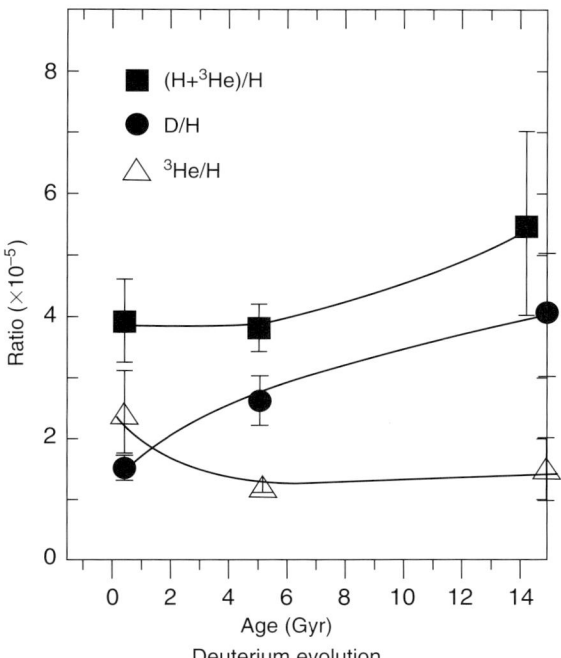

Figure 3 D and ^{3}He abundances as functions of time from the big bang (14 Gyr ago) to the present. The value of ^{3}He at 14 Gyr is not an observation but results from BBN calculations. The apparent variation of this nuclide at later times is well within the uncertainties (adapted from Geiss and Gloeckler 1998).

the large binding energy of alpha particles two relative to other nuclear arrangements.

At BBN or stellar temperatures, the kinetic energy factors are well below the Coulomb barrier. Coulomb penetration factors, proportional to the electric charge, govern the destruction rate of the light nuclei by proton capture. Because of its low Z and low nuclear stability, deuterium is doubly fragile. It disappears at 0.5 million K (MK). Next comes ^{6}Li at 2.0 MK, ^{7}Li at 2.5 MK, ^{9}Be at 3.5 MK, ^{11}B at 5.0 MK and ^{10}B at 5.3 MK. As a result, the light elements (except ^{4}He and, to a certain extent, ^{3}He) cannot resist the heat of stellar interiors. However, in view of the particle instability of ^{4}Li and ^{5}Li, the helium isotopes cannot be destroyed in nucleon-induced reactions, but only by helium-induced reactions, resulting in an appreciably higher Coulomb barrier energy and higher thermal resistance (Figure 4).

The second important nuclear property is the large binding energy of ^{4}He, due to the large pairing effect of nuclear forces when the nucleons are paired four by four: neutron–proton; spin up, spin down. As a result, every nucleus in this mass range is quite unstable toward a rearrangement involving ^{4}He nuclei. No nucleus of mass 5 manages to be stable; the lifetimes are 10^{-21} s. The isotopes ^{8}Li and ^{8}B are beta-unstable with respect to ^{8}Be, which quickly (10^{-16} s) breaks into two alpha particles.

The nuclear stability situation at a mass of 9 is deeply marked by the alpha stability. ^{9}B is unstable against two alphas + p (10^{-19} s), but ^{9}Be barely escapes disintegration (it has a very small binding energy). This weak stability of ^{9}Be is reflected in the fact that the endothermic Q values

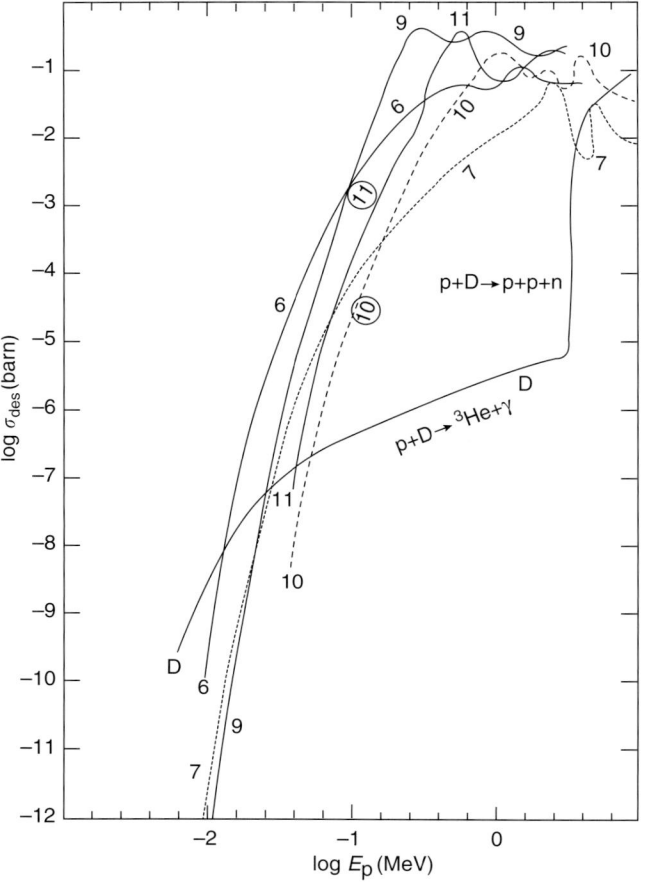

Figure 4 Total destruction cross-sections of light nuclides by protons in barns (10^{-24} cm^{-2}) as a function of energy. The number close to each curve refers to the mass number of the corresponding nuclide (after Reeves 1974).

corresponding to its formation by spallation reactions are remarkably large, and the exothermic Q values corresponding to its destruction are small. It is also reflected in the fact that it has only one 'bound' state; all the excited states are unstable against particle break-up. These facts are instrumental in explaining the low natural abundance of ^{9}Be (one of the lowest in nature). The isotopes ^{6}Li, ^{7}Li, ^{10}B and ^{11}B fare better but all remain comparatively fragile; in high-energy proton-induced reactions they all quickly break down into alpha particles and other products.

The high binding energy of ^{4}He is also responsible for the fact that, at all but the lowest temperatures, hydrogen is transformed rapidly into ^{4}He. The nuclei D and ^{3}He are intermediate steps in this chain of reactions; very small amounts of them remain at the end of the process. These facts dominate the scenario of BBN and also of main sequence stellar energy generation.

The influence of these nuclear properties on the formation rate of the light elements is reflected in their formation cross-sections of the spallation reactions resulting from the bombardment of atoms of C and O by protons (Figure 5). The link is best illustrated by phase-space arguments. In the high-energy region, the break-up of the excited nuclei into a given configuration is proportional to the number of possible channels, which is itself a function of the binding energy, and also of the number of bound excited states for the given configuration. Above 100 MeV or so, the cross-sections reach a plateau which is maintained all the way up to the highest energies. As expected, the cross-sections for the formation of 9Be are the smallest, while for Li and B isotopes they have higher values.

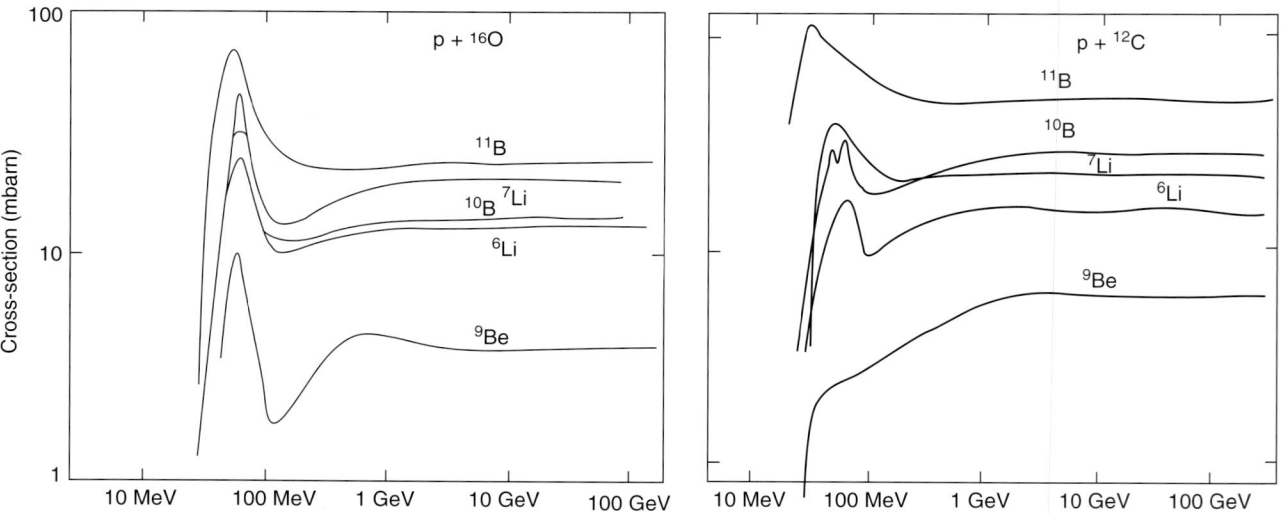

Figure 5 Spallation cross-sections in millibarns (10^{-27} cm^{-2}) for the formation of the Li, Be and B isotopes by proton-induced reactions on ^{12}C and ^{16}O (after Reeves 1974).

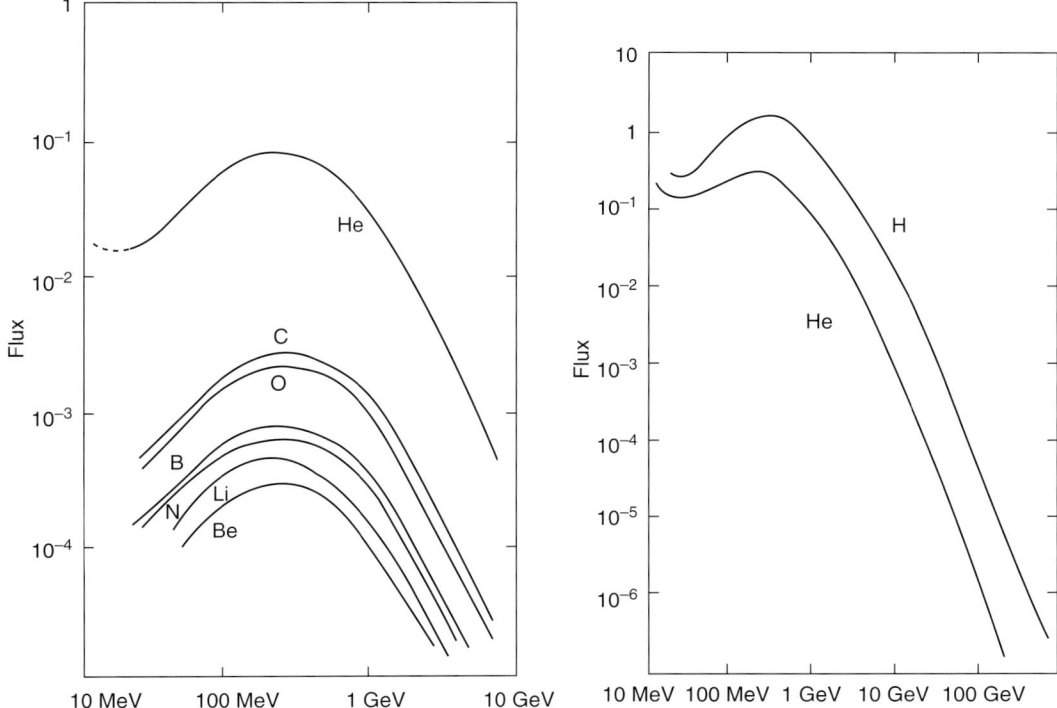

Figure 6 Galactic cosmic-ray fluxes in particles per square centimetre as a function of energy per nucleon (after Reeves 1974).

A similar sequence is observed in the cosmic-ray fluxes of the Li, Be and B isotopes (Figure 6) generated when high-energy C and O collide with interstellar protons. This process, together with its counterpart (fast protons hitting interstellar C and O), is believed to be the main source of ^{6}Li, ^{9}Be, ^{10}B and ^{11}B. The arguments for this view will be summarized later. Let us note for the moment that this process meets our requirement: the light nuclei are not subjected to high temperatures after their formation.

OBSERVATIONS

Deuterium

We have three sets of historical data, pertaining respectively to the present ISM, the protosolar nebula and the early Universe (Figure 3).

Linsky (1998), Linsky and Wood (2000), Vidal-Madjar et al. (1998), Vidal-Madjar (2000) and Sahu (2000) have presented and discussed the D/H ISM values towards bright stars, using UV data returned by the International Ultraviolet Explorer satellite and the Hubble Space Telescope. The estimated value of D/H is 1.5 (± 0.10)$\times 10^{-5}$. Inhomogeneities in the Galaxy may have been detected. The US Far Ultraviolet Spectroscopic Explorer (FUSE) satellite, put into orbit in 1999, is expected to clear up this delicate point.

An argument based on the measurement of the helium isotopic ratio in the solar wind can be used to give an estimate of the D/H ratio in the galactic gas at the birth of the Sun, 4.5 billion years ago on the assumption that the solar wind ^{3}He is the sum of the protosolar D burned to ^{3}He plus the protosolar ^{3}He (Geiss and Reeves 1972). In 1969 Johannes Geiss and his colleagues at Bern measured the ^{3}He/^{4}He ratio in the solar wind from an aluminium sheet brought back from the Moon (Geiss et al. 1970, 1972, Geiss and Gloeckler 1998). Since the solar convective zone is hot enough to burn D + p → ^{3}He, it was clear that this solar wind contained the sum of the primordial D and ^{3}He of the protosolar nebula. But was it possible that during the past history of this zonal material, some ^{3}He had been destroyed in ^{4}He? The presence of ^{9}Be in the solar photosphere at essentially the meteoritic – that is, the protosolar value (Balachandran and Bell 1998) – argues against this possibility. In the outer convective zone of the Sun, the destruction of ^{9}Be occurs at a lower temperature than the ^{3}He burning reactions (Geiss and Reeves 1972). Any ^{3}He depletion would therefore have been accompanied by an even stronger ^{9}Be depletion, and this does not seem to be the case.

The protosolar ^{3}He/^{4}He ratio is obtained from the Jupiter value (discussed later). Subtracting it from the solar wind value one obtains a protosolar D/H=$2.1(\pm0.5)\times10^{-5}$ (Geiss and Gloeckler 1998, Gloeckler and Geiss 2000). This is quite in agreement with the Jupiter D/H value of $2.6(\pm0.7)\times10^{-5}$, measured with the Galileo Entry Probe (Mahaffy et al. 1998). Other Jovian values (Gautier and Owen 1989, Bjoraker et al. 1986, Carlson et al. 1983, Encrenaz et al. 1996) range from 2 to 5×10^{-5}. Because of possible chemical fractionations of hydrogen, one cannot easily obtain the protosolar value from these Jupiter measurements.

Deuterium measurements from the ground by line absorptions on remote intergalactic clouds in front of quasars have led to a period of confusion. A high value, of the order of 10^{-4}, was first obtained by Songaila et al. (1997; discussed by Hogan 1998). A lower value of $3.4(\pm0.25)\times10^{-5}$ was later reported by Burles and Tytler (1998) and Tytler et al. (2000). An analysis of the Tytler data by Levshakov (2000) gives the somewhat higher value of $\sim4.5\times10^{-5}$. There appears to be a consensus for such low values against the earlier high values. The possibility of spatial variations of deuterium in these absorbing clouds is discussed by Vidal-Madjar et al. (1998). Again, the situation should be cleared up with the recently launched FUSE satellite.

Helium-3

A measurement of ^{3}He/^{4}He=$2.48(\pm0.65)\times10^{-4}$ in the ISM near the Solar System has been obtained with the SWICS mass spectrometer on the ESA/NASA spacecraft Ulysses by Gloeckler and Geiss (1996, 1998). Rood et al. (1998) have measured ^{3}He/^{4}He values in ionized hydrogen nebulae (H II regions) with different oxygen abundances situated at various galactocentric distances. These abundances are in approximate agreement with the protosolar and present ISM value and also with the Jupiter value: ^{3}He/^{4}He=$1.66(\pm0.04)\times10^{-4}$ (Mahaffy et al. 1998). With a spread of a factor three in the observed values, there appear to be no systematic trends with galactocentric distance or with metallicity — abundance of elements other than hydrogen and helium. Geiss and Gloeckler (1998) mention the small increase (Figure 3) between the birth of the Solar System and the present gas, which, however, may still be within the uncertainties.

Helium-4

This was generated in the big bang and is also manufactured inside stars. The observed fractional mass abundances range from 24 to 30% (Figure 1), showing the gradual effect of stellar energy generation in galaxies. The search for the primordial yield is conducted in objects which have been least contaminated by stellar synthesis, such as extragalactic H II regions (blue compact galaxies), as is evidenced by their comparatively low abundance of heavier isotopes (Peimbert and Peimbert 2000). One strategy consists of plotting He v. O or C or N, which is tentatively extrapolated to zero for these stellar-generated isotopes. Another estimate is obtained by averaging the He abundances in the objects with the lowest metallicities (Kunth and Sargent 1983, Pagel 1989, Pagel and Simonson 1992 Pagel et al. 1992, Campbell 1992, Fuller et al. 1992, Thuan and Izotov 1998, 2000, Viegas and Gruenwald 2000). A primordial value of the helium mass fraction of 0.23–0.245 is obtained. There is much debate on the uncertainty associated with this value.

Lithium-6 and -7

As mentioned before, one of the best arguments in favour of a big bang contribution to the abundance of ^{7}Li comes from the fact that, for old stars, ^{7}Li/H=1.5×10^{-10} is independent of the metallicity for Fe/H ratios varying from $10^{-3.5}$ to almost 10^{-1} solar (the Spite plateau in Figure 2) (Spite and Spite 1982a,b, Spite et al. 1998, Rebolo et al. 1988, 1992). The 'flatness' of the plateau is discussed by Cayrel (1998), Cayrel and Steffen (2000) and Spite et al. (2000).

To obtain the big bang yield from these observations, one has to estimate the lithium depletion from surface processes in these old stars. This turns out to be a difficult problem, with unsolved questions such as the role of microscopic atomic diffusion (Michaud and Charbonneau 1991, Vauclair 2000, Michaud et al. 2000, Kajino et al. 2000). Nevertheless, the flatness of the plateau (with a dispersion of less than a factor of 2), suggests a rather small amount of surface stellar depletion, less than a factor of 2, at best comparable to the dispersion.

The mean lithium abundance in young stars that have suffered no surface depletion is Li/H=2×10^{-9}, with an uncertainty of a factor of 2 (Pasquini 2000). This is in agreements with the protosolar value, best given by the meteoritic data. (The solar photosphere is depleted by a factor of 100.)

The isotopic composition of lithium

Four cases of stellar ^{6}Li identifications in old population II stars are reported, all at levels of 5% or so (Nissen 2000), which is, in fact, the observational lower limit. ^{6}Li is far more fragile than ^{7}Li, and these detections are often presented as strong arguments for there being no large depletion of ^{7}Li in those stars. One point, however, remains a problem. Given that the theoretically possible range of

^{6}Li/^{7}Li extends up to 100% of the GCR ratio (discussed later), the fact that all four values are at the lower observational limit is somewhat suspicious. Systematic uncertainties may be larger than expected. Finding even one other Population II star at a higher value would ease the situation.

Inhomogeneities in the ^{6}Li/^{7}Li ratios in the interstellar medium ranging from 0.08 (the meteoritic value) to almost 0.5 have been reported (Lemoine *et al.* 1992, Knauth *et al.* 2000).

Beryllium-9

Beryllium has been observed in a large number of young (Population I) stars (Boesgaard and Steigman 1985). The mean value is in good agreement with the solar value, Be/H = 1.4×10^{-11}, measured by Chmielsvski *et al.* (1975). Balachandran (2000) has shown that this value does not differ by more than 20% from the meteoritic (carbonaceous chondrite) value (Grevesse and Noels 1992).

In old (Population II) stars, beryllium shows a systematic decrease with decreasing stellar metallicity (Figure 2, Gilmore *et al.* 1991, 1992, Gilmore 1992, Ryan *et al.* 1991, Ryan 1992, 2000, Rebolo *et al.* 1992). The slope of Be/H v. Fe/H indicates a linear growth of Be with Fe during the lifetime of the Galaxy. Unlike Li, however, Be shows no sign of a plateau at very small metallicity.

Boron-10 and -11

With an abundance ratio of B/Be = 20 to 40, the B/H ratio behaves very much like the Be/H ratio (Boesgaard and Heacox 1978, Primas *et al.* 1999, Primas 2000). It has recently been shown (Cunha and Smith 1999; Cunha 2000) that the solar value is in agreement with the meteoritic value B/H = 7×10^{-10} (Anders and Grevesse 1989), thereby resolving a longstanding reported discrepancy between the two values.

With decreasing metallicity, B/H decreases to 10^{-12} with no sign of a plateau (Figure 2). As for Be, the slope of B/H v. Fe/H is close to 1, showing again a linear growth of B with Fe during the lifetime of the Galaxy.

The boron isotopic ratio in the Solar System is ^{11}B/^{10}B = 4.05 ± 0.05 (Shima and Honda 1962) with variations of a few per cent (Chaussidon and Robert 1998). Measurements of ^{11}B/^{10}B = 3.6 have been made on two stars (Nissen 2000). In the ISM the situation is unclear (Rebull *et al.* 1998).

THE FIRST SECONDS

In order to better understand the physical mechanisms involved in BBN we shall first make a rapid historical survey of the situation. Many crucial events took place during the first seconds of the Universe:

(1) the weak interaction decoupling around 1 MeV, at one second or so;
(2) the electron–positron annihilation around 0.5 MeV, at a few seconds;
(3) the nucleosynthesis of the light nuclei around 0.1 MeV, at a few hundred seconds.

Above 1 MeV, the Universe is in a state of equilibrium with respect to the weak interaction. In other words, neutrinos (ν) are rapidly interacting with nucleons, transforming protons into neutrons and vice versa. Cosmic matter is opaque to neutrinos. The neutron-to-proton ratio is governed by the Boltzmann law: n/p = exp($-\Delta M/kT$), where ΔM is the neutron–proton mass difference (1.293 MeV).

Below 1 MeV, most neutrinos do not have enough energy to keep on interacting with nucleons. The weak equilibrium is broken, and the Universe becomes transparent to neutrinos. More exactly, the mean free paths of these particles is then larger than the cosmic horizon (approximately ct). This event should give rise to a (still undectable) neutrino fossil radiation, analogous to the photon fossil radiation (the CMB).

Above 0.5 MeV (the rest mass of an electron), pair production and annihilation keep the populations of electrons and positrons in equilibrium with the photon gas. Their mutual annihilations at 0.5 MeV create a flux of new photons which increases slightly the photon radiation (more exactly, they lower the cooling rate). Since the neutrino interactions are now very weak, the neutrinos are essentially decoupled from the rest of the Universe and they receive no share of the energy released by these annihilations. As a result, the neutrino fossil radiation is expected to be cooler (1.8 K) than the photon CMB (2.7 K). Present technology does not allow the detection of this neutrino radiation; this is a programme for the future.

Above 0.1 MeV, neutron–proton captures and dissociations are in thermal equilibrium. A (very small) population of deuterons is maintained in the cosmic matter. Below 0.1 MeV the gamma rays (the tail of the Bose–Einstein photon energy distribution) are no longer numerous enough to keep the deuteron population in statistical equilibrium with the nucleons. This is the onset of primordial nucleosynthesis (BBN).

The population of deuterons rises rapidly (Figure 7). As they reach a ratio of D/H of 10^{-3} or so, they undergo further nuclear reactions and are transformed into nuclei of mass 3 (^{3}H and ^{3}He). The population of these nuclei increases, in turn, as the D decreases. A similar fate awaits the nuclei with mass 3, as they are transformed to ^{4}He. As there are no stable nuclei with mass 5 or 8, the ^{4}He suffers essentially no further nuclear depletion. Only a very small fraction of its population gets transformed into ^{7}Li through

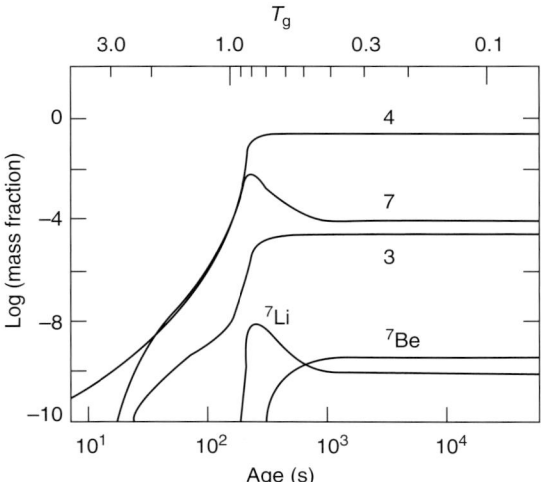

Figure 7 Abundance growth of the light nuclides during BBN. Upper scale: the cosmic temperature in 10^9 K; lower scale: the cosmic age in seconds (after Wagoner et al. 1967).

$^4\text{He} + ^3\text{H} \rightarrow {}^7\text{Li} + \gamma$ and $^4\text{He} + ^3\text{He} \rightarrow {}^7\text{Be} + \gamma$, followed, after many days, by $^7\text{Be} + \text{e} \rightarrow {}^7\text{Li} + \nu$. At temperatures below 0.01 MeV the Universe is too cool to produce any further nuclear transformations. BBN is over.

The relative abundances of the light nuclei generated during this period of cosmic nuclear activity are related to two key parameters of the physical conditions during expansion: the neutron-to-proton ratio and, the nucleon-to-photon ratio.

Let us focus our attention on the fate of the neutrons after weak interaction decoupling at 1 MeV. They may either undergo beta decay into protons (with a lifetime of 890 ± 4 s), later giving rise to hydrogen; or they may interact with protons to form D. The probability of this last event is proportional to the ambient density of nucleons. At low nucleonic density, the neutrons mostly undergo beta decay; at higher densities they undergo nuclear reactions and are essentially all processed into ^{4}He (with very minor formation of light nuclei of masses 2, 3 and 7). The yields of heavier nuclei are negligible.

We have no *a priori* knowledge of the nucleonic density (or, equivalently, of the abundance ratio of nucleons to photons: called η, the baryon number). The problem is thus the following: leaving this quantity as a parameter, can we find a value which would account – after appropriate computations and correct time extrapolations – for the observed abundances?

The yields are computed with the help of reaction networks incorporating a large number of reaction cross-sections for the production and destruction of the various nuclei. They were first computed in a seminal paper by Wagoner et al. (1967). The nuclear data were reviewed by Caughlan and Fowler (1988). The subject has been

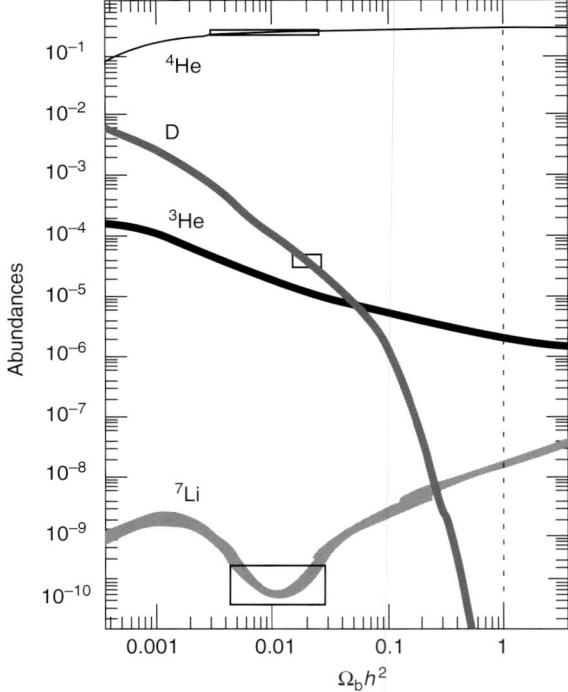

Figure 8 Abundances of the light nuclides at the end of BBN as a function of nucleonic (baryonic) density $\Omega_b h^2$ (the fraction of the critical density times the square of h, the Hubble constant in units of 100 km^{-1}s^{-1}kpc^{-1}). From recent estimates of the Hubble constant H_0, the value of h^2 is close to 0.35. The boxes specify the observed abundances with their uncertainties (after Schramm 1992).

thoroughly re-analysed recently by Smith et al. (1992), Krauss and Romanelli (1990), Thieleman et al. (1991) and Walker et al. (1992). The results are presented in (Figure 8), where the horizontal scale is in units of baryonic density ($\eta = 2.73 \times 10^{-8}\, \Omega_b h^2$; these symbols will be fully defined later). It appears clearly that there is a common range of $\eta = 5 \times 10^{-10}$ which accounts for all three observed nuclides. This fact is by itself already highly suggestive that we are on the right track.

One last point should be considered before we close this section. The 'standard model' used to compute these yields assumes a Universe of homogeneous density at the moment of BBN. Around 1980, progress in particle physics called this assumption into question. Witten (1984) showed that the quark–hadron phase transition, occurring at $T = 150$ MeV, could have generated density inhomogeneities leading, after weak interaction decoupling, to inhomogeneities in the neutron-to-proton ratio during BBN at $T = 0.1$ MeV, and hence possibly to different yields of the light elements (Reeves 1994, Reeves et al. 1990, Kajino et al. 2000, Kurki-Suonio et al. 1990, Kurki-Suonio 1991, 2000).

One key point for this discussion is the order of the quark–hadron transition (or *transitions*, since there are both a chiral and a deconfinement transition, taking place at approximately the same energy). If the transition is of second order, no inhomogeneities are created and the standard model is valid. If it is of first order, the degree of inhomogeneity will depend on a number of parameters, including the duration of the subsequent overcooling period. By varying these parameters, many calculations have been made. None showed significantly different yields with respect to the homogeneous standard model. However, the last word may not have been said, and ongoing experiments on the physics of the transition may alter this conclusion.

How can we relate the baryonic number at BBN, $\eta = n_b/n_\gamma$, to the present nucleonic density? The main point is that we know of no process which could have created new nucleons since BBN. New photons have been created (mostly by stars transforming hydrogen into helium), but the fractional number of newly made photons is less than one-thousandth of the population of photons in the fossil radiation. In consequence, the baryon number has remained practically constant since BBN.

From the present temperature (2.73 K) of the CMB, we know that the number of photons (proportional to T^3) is close to $400\,\text{cm}^{-3}$. Thus, using the baryon number determined from BBN, we may obtain the present nucleonic density as about $3 \times 10^{-31}\,\text{g}\,\text{cm}^{-3}$.

COSMOLOGY

The theory of the big bang has many free parameters (Schramm 1992, Steigman 2000). A crucial one is the mean overall density of the Universe (including all kinds of dark matter), to which the expansion rate is intimately related. Light universes will keep on expanding and cooling indefinitely towards a state called 'big chill', heavy universes will later recollapse and reheat in a 'big crunch'. The critical density is what separates these two fates. Its density is the same as the density of a gas of 10 nucleons per cubic meter. Needless to say, it need not be nucleons.

It is convenient to define Ω_j the fractional density of a component j of matter with respect to the critical density ($\Omega_{\text{crit}} = 1$). The correspondence with the baryonic number η is given by the formula $\eta = 2.73 \times 10^{-8}\,\Omega_b h^2$, where h is the Hubble constant in units of $100\,\text{km}\,\text{s}^{-1}\,\text{Mpc}^{-1}$. Using the recent estimates of the Hubble constant H_0, the value of the h^2 factor is close to 0.35 (see Chapter 17). For nucleonic matter the baryon density Ω_b, as given by BBN, is $\Omega_b = 0.06 \pm 0.15$.

Let us call the 'cosmological window' the range of baryon densities allowed by astronomical observations. The density of luminous matter (Ω_{lum}: photon-emitting matter – stars and hot gases) is evaluated at 0.003. This is clearly a lower limit for the universal nucleon density. The total matter density Ω_{mat}, estimated by its gravitational manifestations (in Einstein's theory of general relativity all forms of matter and energy have gravitational effects) is close to 0.3, which is an upper limit for Ω_b. Thus, the BBN-calculated Ω_b manages to account for the abundances of all the primordial nuclides while being compatible with the cosmological window ($0.003 < \Omega_b < 0.3$). This coherence between the nuclear physics and astronomical domains is another strong argument in favour of the big bang theory.

An evaluation of Ω_b can also be obtained from the analysis of the absorption lines in the spectra of remote quasars. These lines are ascribed to diffuse clouds of matter in intergalactic space. The computations, although somewhat model dependent, again yield values of Ω_b compatible with the BBN value.

We return to the fact that $\Omega_b = 0.06 \pm 0.15$ is well above the density of luminous matter, $\Omega_{\text{lum}} = 0.003$. We conclude that only a small fraction of the nucleons manifest themselves by emitting photons. The rest is called 'nucleonic dark matter'. In what form does it exist? Could it be a population of low-mass stars (brown dwarfs) in the halos of galaxies, too distant to be seen but detectable by microlensing techniques when they are aligned between a bright star and the observer? Although some such stars have been found this way, it seems clearly that their average density is insufficient, and that they represent at most a small fraction of the nucleonic dark matter. Where is the rest?

We note also that the upper limit to Ω_b is well below the estimate of the total gravitating matter detected by its dynamic perturbations ($\Omega_{\text{mat}} = 0.3$). The rest is called 'exotic dark matter' and much research is aimed at finding its nature. Particle physics offers many speculative candidates. Massive particles with very small coupling constants are postulated by theoretical physicists within the realm of 'new physics', such as supersymmetric theories. None have yet been detected unambiguously.

Massive neutrinos would be a nice candidate for this component. If the mean mass of the three varieties of neutrinos were close to 10 eV, they could fit the bill, but unfortunately this does not seem to be the case. Recent experiments (Normile 1998) have given a mass of approximately 0.1 eV for one of the neutrinos. The others are most likely to be even less massive. However, it is interesting to note that the corresponding neutrino cosmic density is $\Omega_\nu = 0.004$, larger than the luminous density. The elusive neutrinos, detected only a few decades ago, have more density than the sum of all the stars and hot gas in the Universe. Quite an astonishing statement!

INFORMATION FROM THE FOSSIL RADIATION

The 3 K background radiation contains very little energy density ($\Omega_{CMB} = 5 \times 10^{-5}$). It is highly isothermal, and its temperature variations are of the order of $\Delta T/T = 10^{-5}$. Much cosmological information can be obtained from the small-angle observations of these inhomogeneities, where a peak has been detected around 1°.

The mass of matter emitting fossil radiation, into a solid angle of one square degree, is approximately 10^{18} solar masses (equivalent to a supercluster of galaxies). This is approximately the mass of matter which, at the moment of CMB emission (300,000 years after the big bang), is entering its 'causal sphere' (no physical process can interact faster than light). The peak density is interpreted as the first manifestation of gravity on a local scale – the birth of the first structures in the Universe. As with a sound wave, the height of the peak should be proportional to the nucleonic density. The exact angular position of the peak should give the geometry of the Universe; the causal sphere will occupy a different solid angle – as seen by the terrestrial observer – if the Universe is flat (Euclidean), spherical or hyperbolic. The geometry is governed by the sum of the total mass density Ω_{mat} plus a term Ω_Λ (if the cosmological constant is not zero). Recent observations (January 2000) seem to favour a flat Universe, implying that $\Omega_{mat} + \Omega_\Lambda = 1$. Forthcoming orbital observatories (e.g. the European spacecraft Planck) should elucidate these points.

BBN AS A TEST FOR COSMOLOGY

The success of BBN in accounting *quantitatively* for the abundances of light nuclides is of great cosmological relevance. It gave the first estimate of the nucleonic (baryonic) density of the Universe, showing that nucleonic matter cannot close the Universe (Reeves 1971). It predicted first the existence of three and only three families of elementary particles in nature (Schramm 1992). It has been a graveyard for many speculative theories of fundamental physics and it has been used to probe various hypotheses such as the possible time variations of the laws of nature.

Most important in this respect is the evaluation of the decoupling temperature of the weak interaction, which, as we saw before, is directly related to the cosmic He/H ratio. Two timescales are involved in the physics of BBN: the macroscopic timescale related to the expansion rate and the microscopic timescale related to the particle reaction rates.

In the early Universe, the Einstein equations yield the simple relation

$$(\dot{R}/R)^2 = 8\pi G_N \rho / 3$$

where R is the scale factor (the distance between two galaxies), $\dot{R}$ its time derivative, ρ the total energy density and G_N is the Newtonian gravitational constant. From dimensional considerations, the expansion time is given by

$$t_{exp} \approx (G_N \rho)^{-1/2}$$

The energy density ρ is dominated by the set of all the relativistic particles i ($kT \gg M_i c^2$) for which ρ_i is proportional to $g_i T_i^4$, where g_i is the statistical multiplicity factor ($g_i = 2$ for the photons). A 'demographic factor' g^* is usually introduced to represent the combined effect of all the particle species, bosons (b) and fermions (f), on the mass density of the early Universe. The multiplicities of each species is given by $g_{b,i}$ and $g_{f,i}$:

$$\rho = \frac{1}{2} g^* \rho_{photons} \approx g^* T_{g\,photons}^4$$

where T_g is the temperature of the photon gas and

$$g^* = \sum g_{b,i} + \frac{7}{8} \sum g_{f,i}$$

The factor $\frac{7}{8}$ reflects the difference between the statistics for bosons (b) and fermions (f). At several MeV, for instance, the 'standard' demography consists of photons, electrons and positrons, three types of neutrinos and their antineutrinos, and thus $g^*(T = \text{several MeV}) = 2 + \frac{7}{8} \times 10 = 10.75$. Note that because it is related to the number of species with $M_i c^2 < kT$, g^* is a function of temperature. The cosmic expansion timescale can thus be written as $t_{exp} \approx (g^* G_N)^{-1/2} T^{-2}$.

For the reaction timescale, let us consider the capture of a neutrino by a neutron to give an electron and a proton. The neutrino capture cross-sections σ are proportional to the square of their energy and also to the square of the Fermi constant G_F. The probability of one capture event is given by the value of the product of the cross-section and the relative velocity: $\langle \sigma v \rangle$, averaged over the velocity distribution of particles at temperature T. This average value is proportional to the strength of the interaction times the square of the the temperature: $\langle \sigma v \rangle \approx G_F^2 T^2$. The probability of reaction per unit volume is proportional to $\langle \sigma v \rangle$ times the number density of capturing particles per unit volume, $n(T)$, which, in the expanding Universe, is $n(T) \approx T^3$. Thus the lifetime t_{reac} for a given neutrino to interact with neutrons (the inverse of the reaction probability) is given by

$$t_{reac}\,(n + v \rightarrow p + e) \sim G_F^{-2} T^{-5}$$

The decoupling temperature of the weak interaction obtained by equating the two timescales

$$T_d \sim (g^*)^{1/6} (G_N)^{1/6} G_F^{2/3}$$

is a function of many fundamental parameters of physics: G_N, G_F and the demography of the Universe (through g^*). It can easily be computed that changing any one of these numbers by more than a few per cent (but excluding the unlikely possibility of a 'conspiracy' between these numbers) would wreak havoc with the concordance between BBN and the observations. Thus we may infer that these numbers have not changed appreciably since the first seconds of the Universe (Irvine 1983, Raesenberg 1988, Scherrer and Spergel 1993).

Becoming somewhat philosophical, we may add a comment to an old question basic to cosmology. Can we really assume that our concept of time based on clocks and rotations of the Earth can be extrapolated so far back into the past? Our pragmatic answer is the following: if we make this hypothesis, we succeed in accounting correctly for the observed abundances of the light nuclides. This, to my mind, is a reasonably good answer.

THE CASE FOR A GALACTIC COSMIC-RAY ORIGIN OF Li, Be AND B

Because of their thermal fragility, the origin of the light elements lithium, beryllium and boron, and also deuterium, has long been a problem (Cameron *et al.* 1964, Burbidge *et al.* 1957). Supernovae explosions were first considered as a possible mechanism. Later on, from measurements of GCR abundances, Bernard Peters (1969, personal communication) suggested a spallation origin related to the destruction of heavy nuclei by fast protons.

Fowler, Greenstein and Hoyle (1961) proposed a local origin of deuterium, lithium, beryllium and boron. They deleveoped a model based on the irradiation of the solar nebula by the Sun in its early T Tauri period. In this model, strong fluxes of fast particles bombarding the planetesimals were held responsible for the generation of the light elements in the Solar System. However, energetic arguments soon showed this view to be untenable (Reeves 1966). Nuclear data on the relevant spallation cross-sections obtained by the Orsay group (Bernas *et al.* 1967) at the CERN synchrocyclotron paved the way for the currently accepted model: bombardment of the interstellar gas by GCRs. Theory and observation are now in agreement concerning the origin of Li, Be and B: ^{6}Li, ^{9}Be and ^{10}B are essentially all produced by GCRs. ^{11}B is partly produced by GCRs and partly by stellar nucleosynthesis. ^{7}Li has three formation sources contributing significantly: the big bang, GCRs and stellar nucleosynthesis.

The origin of beryllium in the Galaxy

Contrary to ^{7}Li, ^{9}Be is not an important secondary product of hydrogen burning, either in stars or during primordial nucleosynthesis. It is a typical case of a nucleus for which we know of no low-energy production mechanism. We have to resort to spallation reactions in interstellar space.

The rate of formation of ^{9}Be by GCRs is given by the product of the flux of high-energy protons ($\sim 16\,\text{cm}^{-2}\,\text{s}^{-1}$) times the cross-sections for ^{9}Be formation by proton collision on the most abundant targets, ^{16}O and ^{12}C ($\sim 5\,\text{mb}$), times the abundance ratio of these targets to hydrogen ($\sim 10^{-3}$) in space. The approximate equality between the formation rate times the age of the Galaxy, and the beryllium-to-hydrogen ratio in recently formed stars ($\sim 10^{-11}$), is the best evidence for a major GCR contribution to ^{9}Be formation and, by the same token, to that of some of the light elements (Reeves *et al.* 1970).

The important parameters are the excitation functions for spallation reactions induced by protons and alpha particles. In principle, all the nuclei with $A > 11$ in interstellar space are target candidates. In practice, only ^{16}O, ^{12}C and, marginally, ^{14}N are abundant enough to contribute appreciably. Reactions between alpha particles may have been important in the early days of the Galaxy (Montmerle 1977, Steigman and Walker 1992).

Owing mostly to the pioneering work at the Orsay laboratory in France, the important excitation functions are now known with sufficient accuracy (for reviews see Reeves (1974) and Read and Viola (1984)). They are shown in Figure 5.

The number of thermalized nuclei added to the interstellar gas per unit time is the sum of two contributions. The first is the spallation of interstellar heavy nuclei by fast protons and alpha particles. In this case, the recoil energy of the light nuclei is small (a few MeV per nucleon) and they suffer no further destruction or loss. The second is the spallation of fast, heavy nuclei by interstellar H and He.

Each of the fast nuclei, created by GCR spallation reactions, is faced with three possible fates. First, it may be destroyed by nuclear collision with interstellar atoms. The total destruction cross-sections have been measured (Figure 4). Second, it may also escape from the Galaxy. The relevant parameter is the 'escape length' L_{esc}: the amount of matter met by a cosmic ray on its way from its source to the border of the galaxy. The value of this parameter is calculated from the abundance ratio of spallated nuclei (Li, Be, B) to the target nuclei (C, N, O); at the energies of interest in the present galactic medium, $L_{\text{esc}} \sim 6\,\text{g}\,\text{cm}^{-2}$. Third, it may be decelerated by collisions with interstellar electrons ('ionization losses') all the way down to interstellar thermal energies. (The energetics of the process have been analysed by Ryter *et al.* (1970), and constitute yet another argument in favour of the GCR origin of the light elements.) These are the atoms that will eventually be incorporated into stars and manifest themselves in stellar spectra.

Calculations are needed in order to evaluate the rate of deceleration of the spallation products and their survival chances against galactic loss and further spallation before thermalization.

Results of the GCR computations

To compare theory and observation, it is convenient to study separately two phases in the lifetime of the Galaxy: first, the 'recent' era (the last 10 billion years or so, corresponding more or less to Population II stars), and second, the early days of the Galaxy (Population II stars).

The recent galactic era

In the cosmic ray sector, we need the flux of H, ^{4}He, ^{12}C, ^{14}N and ^{16}O as functions of energy throughout the Galaxy (Figure 6). Because of solar modulation effects, the low-energy part of these fluxes is damped in the solar neighbourhood. The problem of extrapolating the fluxes outside of the solar cavity is still not entirely solved. Data obtained by the Voyager Probes at 24 AU (farther than Uranus) have been used to obtain the best estimated fluxes and energy spectra (Ferrando et al. 1991, Webber 1998). The extrapolated values become highly uncertain below 100 MeV. Fortunately, most of the contribution to the production rate comes from higher energies, where the fluxes are better known. This may not be the case, however, if reactions between alpha particles contribute significantly to the formation of the lithium isotopes.

Meteoritic data on the production rate of nuclei by spallation reactions show that the GCR fluxes have not varied significantly in the last few billion years (Arnold et al. 1961, Geiss et al. 1961, Voshage and Hintenberger 1961, Lal and Peters 1967, Zanda 1990). The interstellar target abundances of C and O are known to an accuracy of better than a factor of 2 (Anders and Grevesse 1989, Grevesse and Noels 1992). The ratios of the production rates for the various nuclides have been computed (Reeves and Meyer 1978, Walker et al. 1985), normalized to the production rate of ^{9}Be.

Comparing the ratios of the spallation cross-sections of protons on O and C to the ratios of the stellar abundances of Li, Be and B confirms the view that some of the light nuclei are generated by GCRs. The analysis shows that the GCR mechanism satisfactorily accounts for ^{6}Li, ^{9}Be and ^{10}B, and gives a major contribution to the abundance of ^{11}B and a minor (10%) contribution to the abundance of ^{7}Li (Reeves et al. 1970, 1973, Meneguzzi et al. 1971, Austin 1981, Walker et al. 1985, Arnould and Forestini 1989). Another source of ^{7}Li is clearly required. A smaller discrepancy exists also for the boron isotopic ratio, but the solution is highly unlikely to be related to BBN. Low-energy particles (tens of MeV) (Reeves and Meyer 1978) and neutrino-induced reactions in supernovae have both been invoked in order to generate some extra ^{11}B (Woosley et al. 1990).

The early galactic era

The study of the GCR-generated elements in Population II stars may yield important clues as to the physical conditions accompanying the formation of galaxies (Ryan et al. 1991, Ryan 1992, Rebolo et al. 1988, 1992, Spite et al. 1991, Duncan 1997, Gilmore et al. 1991; Smith et al. 1992, Primas 2000, Cunha 2000).

Figure 2 shows the best indication of the different mechanisms involved in the formation of these nuclides. The Be and B abundances are decreasing with decreasing metallicity. At one part in one thousand of the solar metallicity, they have fallen by a factor of more than 100 below their present Population I stellar value, as qualitatively expected from their pure GCR origin. The ^{7}Li abundance first decreases to one-tenth of its present value at one-tenth of the solar iron value (always in the look-back chronology). This is again some of the best evidence in favour of the occurence of BBN.

The growth of the Be/H and B/H ratios as a function of the metallicity (Fe/H) of the Galaxy appears to be linear. The implications of this behaviour are still under discussion. This result suggests that the ratio of fast CNO/H was much larger in the early Galaxy than it is today (Vangioni-Flam and Cassé 2000, Ramaty et al. 2000, Parizot and Drury 2000).

If confirmed, the detection of ^{6}Li metal-poor stars provides the first evidence for the contribution of reactions between alpha particles to the abundance of Li. It would also show that the GCRs did not contribute more than a few per cent to the Population II lithium abundance.

OVERVIEW OF THE INDIVIDUAL NUCLIDES

Deuterium

After protons and neutrons, this is the first nuclide to appear after the big-bang. Its abundance gives the first indication that the Universe is not bound by nucleonic matter. Deuterium is not generated by normal stellar synthesis. The only other source of D is the spallation of ^{4}He by GCRs, which contributes only a part in a thousand of its present abundance.

During galactic evolution, deuterium is systematically destroyed by astration, so we expect a gradual depletion of its BBN abundance as a function of time, as more galactic gas is incorporated in stars and later ejected as D-free gas (Figure 3). This depletion is plausibly tempered by the continuous infall of extragalactic gas on the galactic disk, with presumably pristine D (Lubovich et al. 2000).

Several authors (Tosi 1998, 2000, Prantzos et al. 1992, Prantzos 1996, Matteucci and Romano 2000, Chiappini and Matteucci 2000, Olive 2000, Vangioni-Flam et al. 1994) have presented consistent galactic evolution models incorporating other relevant observations such as the abundance ratios of other elements, galactic mass fractions of stars, and gaseous nebulae as a function of age and radial distance from the centre of the galaxy. These evolution models include up-to-date nuclear and atomic data on the various elements and isotopes of interest. The initial inputs of these models are the rate of star formation, the initial mass function and the rate of infall of extragalactic matter. In this respect the observations by Beers et al. (2000) of very metal-poor stars are of great interest.

The general conclusion is that models which account satisfactorily for the ensemble of observations generate only a modest amount of D depletion (at best by a factor of 3), in good agreement with the data (Figure 3).

Helium-3

This nuclide is produced in significant quantities both by BBN and by stellar synthesis, as an intermediate step of hydrogen burning. In stars, ^{3}He is produced during the main sequence phase by the D + p reaction. In the hot centre of stars, ^{3}He is further transformed into ^{4}He by the ^{3}He + ^{3}He or ^{3}He + ^{4}He reactions. In the outer stellar layers, however, the temperature is high enough to generate ^{3}He but not to destroy it. Stellar models show there an accumulation of ^{3}He, reaching values of ^{3}He/^{4}He = 10^{-2}.

As the star moves toward the red giant branch, a fraction of these nuclei are burned to ^{4}He, while another fraction are convected to the surface and ejected by stellar winds. Other processes in novae or other advanced stages of stellar evolution may further generate significant amounts of ^{3}He.

The final fate of ^{3}He is governed by internal stellar processes from the main sequence to the giant branch and the planetary nebula phases through the effects of various dredge-up and mixings. The conventional stellar models predict an important yield of ^{3}He for stars less than 2 solar masses. Observations of planetary nebulae (PNe) (Rood et al. 1998) have confirmed this prediction, with some detections at the level of ^{3}He/^{4}He $\sim 10^{-3}$. The problem lies with the theoretically expected stellar contribution of these PNe during the life time of the Galaxy. According to conventional models, values of ^{3}He/^{4}He $\sim 10^{-3}$ should be observed in the ISM, in disagreement with the measurements.

Charbonnel (1998) and Charbonnel et al. (2000) have discussed the relation between this problem and the problems raised by the observations of the ^{12}C/^{13}C ratio and of N and O abundances in PNe. The discrepancy could be solved by the operation of a new mixing phase in the red giant branch, which would account for the destruction of ^{3}He and also for the unexpectedly low ^{12}C/^{13}C ratio observed in some PNe.

The large scatter in the observed abundances of ^{3}He makes it difficult to estimate the relative contributions of BBN and stellar evolution to its abundance. The BBN value shown in Figure 3 has been computed within the BBN framework from the baryon number obtained from D, ^{4}He and ^{7}Li. In fact we do not even know whether, after BBN, its galactic abundance increases or decreases with time.

Helium-4

After the BBN production of a helium fractional mass of ~0.24, the main effect of all the main sequence stars has been to increase its galactic value to ~0.30, implying a transformation of ~6% of the hydrogen into ^{4}He, corresponding to an energy release of ~0.4 MeV per nucleon. Compare this value to the present galactic luminosity (Hoyle and Tayler 1964). The standard unit for this discussion is the luminosity per unit mass of the Sun: $2\,\mathrm{erg\,g^{-1}\,s^{-1}}$, or 2×10^{-12} eV/nucleon/second.

Spiral galaxies have a typical luminosity per unit mass of only one-tenth of this solar value. Assuming this luminosity to have been constant in the last 14 billion years, this represents an integrated energy emission of $\sim 9 \times 10^4$ eV, equivalent to the fusion of only 1% of H into He. The helium increase of 6% requires the mean galactic luminosity to have been four or five times larger than the present luminosity.

From the helium increase of 6% we may also compute the radiation density release per nucleon in the cosmos. With the baryonic density obtained from BBN ($r_b \sim 3 \times 10^{-31}$ g cm^{-3}), the radiation density released is ~1 eV/nucleon, quite comparable to the density of the 3 K fossil radiation. In comparison, the stellar radiation density in the galactic plane is 0.4 eV/nucleon (Allen 1973). Because of the universal expansion, the fraction of this radiation coming from intergalactic space is a function of the cosmological model adopted and also of the chronology of helium burning. The interesting conclusion, however, is that a non-negligible fraction of the radiation energy density in the galactic plane comes from helium burned in distant galaxies.

Lithium-6

This nucleus is a virtually pure product of GCRs. It comes from the spallation of interstellar ^{16}O and ^{12}C by GCR protons, and also from reactions between alpha particles. The isotopic ratios (^{6}Li/^{7}Li) observed range from 0.08 for the Solar System (reported values are 0.05 for old stars) to 0.3 and lower for the present ISM close to star-formation regions (Knauth *et al.* 2000). The astrophysical implications of these inhomogeneities are still largely mysterious.

Lithium-7

Uniquely, this nucleus owes its abundance to three different mechanisms, each contributing amounts which differ by less than one order of magnitude.

The BBN contribution dominates the galactic gas abundance in the first several billion years. The flatness of the curve, however, has far-reaching implications for the hydrodynamics of stellar surface layers. The effects of atomic (microscopic) diffusion below the convective zone should result in lithium abundances that vary with the mass and surface temperature of the star in which they are observed, contrary to observations, unless these effects are neutralized by other phenomena. This has been a problem for many years.

Several groups (Delyannis *et al.* 2000, Pinsonneault *et al.* 1997, 2000, Vauclair 2000, Théado and Vauclair 2000) have studied the possible influence of various physical phenomena, such as mixing caused by differential rotation, gravity waves or stellar winds. In all cases the flatness of the curve places important constraints on the efficiency of these processes.

The flatness of the lithium abundance curve as a function of metallicity argues against heavy stellar astration and hence against important D depletion for a metallicity less than one-tenth of the solar value. When the galactic mass fraction of heavy elements became larger than one part in a thousand (one-tenth of the solar value), a new stellar source (Sackman and Boothroyd 2000) managed to increase the abundance of lithium by an extra factor of 10.

Observations of lithium abundances in the surface layers of evolved stars on the asymptotic giant branch (AGB stars) at a value appreciably larger than the Population I value of 10^{-9} suggest that these stars have contributed, by mass ejection, to this enrichment of galactic lithium. Other candidates such as novae and supernovae have also been suggested. Convincing quantitative models of the galactic enrichment are still lacking. The GCR production of ^{7}Li, evaluated through the abundance of ^{9}Be, never dominates the abundance curve.

Lithium has been observed in a large number of stars in various states of evolution. The study of the abundance variations is a very important tool for the elucidation of the physics of stars and galactic evolution.

Beryllium-9

Another pure product of GCR nucleosynthesis, Be is generated only by the bombardment of ^{12}C and ^{16}O. It has increased from Be/H = 10^{-13} to 10^{-11} over the lifetime of the Galaxy. Its unique origin makes it a particularly useful monitor of time-integrated factors of galactic evolution such as the product of particle fluxes and abundance of targets. An early big bang contribution, through the hypothetical effect of the quark–hadron phase transition, appears highly unlikely (the same remarks apply to the two isotopes of boron).

Boron-10 and -11

As for ^{9}Be, these two nuclides are primarily produced by GCRs during the lifetime of the Galaxy. One long-standing problem is the difference between the observed boron isotopic ratio (^{11}B/^{10}B = 4.0 (± 0.1)) in the Solar System and the value 2.3 (± 0.2) expected from the effects of the GCRs, with a mean energy around 1 GeV. Given the shape of the spallation cross-sections leading to these isotopes at low energy, it was proposed that the contribution of lower energy particles, unobservable from the inner Solar System, could account for the difference. However this solution presents serious energy problems (Ramaty *et al.* 1997).

Another solution exists: the neutrino disintegration of ^{12}C, which would generate far more ^{11}B than ^{10}B. The process would take place in exploding stars (supernovae) in which intense fluxes of neutrinos emitted from collapsing cores (Woosley *et al.* 1990) are later reabsorbed by the outer layers, thereby expelling them in the form of remnants.

More isotopic boron ratio measurements in stars would be of great value in this respect, especially since, contrary to the case of the lithium isotiopic ratio, we do not expect differential depletion on the main sequence (except at the very cold end).

REFERENCES

Allen, C.W. 1973 *Astrophysical Quantities*, Athlone Press, London.

Alpher, R.A., Bethe, H. and Gamow, G. (1948). The origin of the chemical elements. *Physical Review*, **73**, 803–804.

Anders, E. and Grevesse, N. (1989). Abundances of the elements: Meteoritic and solar. *Geochimica et Cosmochimica Acta*, **53**, 197–214.

Arnold, J.R., Honda, M. and Lal, D. (1961). Record of cosmic-ray intensity in the meteorites. *Journal of Geophysical Research*, **66**, 3519.

Arnould, M. and Forestini, F. (1989). Synthesis of light elements. *Nuclear Astrophysics; Proceedings of Third International Summer School, La Rabida, Spain, June 19 – July 2, 1988*, Springer, Berlin, pp. 48–85.

Austin, S.M. (1981). The creation of the rare light elements: Cosmic rays and cosmology. *Progress in Particle and Nuclear Physics*, **7**, 1.

Balachandran, S.C. (2000). Beryllium in the Sun: Re-measurement and implications. In L. da Silva, M. Spite and J.R. de Medeiros (eds), *The Light Elements and Their Evolution*, IAU Symposium Vol. 198, Astronomical Society of the Pacific, San Francisco, pp. 383–388.

Balachandran, S.C. and Bell, R.A. (1998). Shallow mixing in the solar photosphere inferred from revised beryllium abundances. *Nature*, **392**, 791–793.

Beers, T.C., Suzuki, T.K. and Yoshii, Y. (2000). The light elements Be and B as stellar chronometers in the early galaxy. In L. da Silva, M. Spite and J.R. de Medeiros (eds), *The Light Elements and Their Evolution*, IAU Symposium Vol. 198, Astronomical Society of the Pacific, San Francisco, pp. 425–431.

Bernas, R., Gradsztajn, E., Reeves, H. and Schatzman, E. (1967). On the nucleosynthesis of lithium, beryllium and boron. *Annals of Physics*, **44**, 426–478.

Bjoraker, G., Larson, H.P. and Kunde, V.G. (1986). The gas composition of Jupiter derived from 5 micron airborne spectroscopic observations. *Icarus*, **66**, 579–609.

Boesgaard, A.M. and Heacox, W.D. (1978). The abundance of boron in B- and A-type stars. *Astrophysical Journal*, **226**, 888–896.

Boesgaard, A.M. and Steigman, G. (1985). Big bang nucleosynthesis: Theories and observations. *Annual Review of Astronomy and Astrophysics*, **23**, 319–378.

Burbidge, E.M., Burbidge, G.R., Fowler, W.A. and Hoyle, F. (1957). Synthesis of the elements in stars. *Reviews of Modern Physics*, **29**, 547–560.

Burles, S. and Tytler, D. (1998). On the measurements of D/H in QSO absorption systems closing in on the primordial abundance of deuterium. *Space Science Reviews*, **84**, 65–75 [and N. Prantzos, M. Tosi and R. von Steiger (eds), *Primordial Nuclei and their Galactic Evolution*, Proceedings of an ISSI Workshop 6–10 May, 1997, Kluwer, Dordrecht, p. 65].

Cameron, A.G.W. *et al.* (1964). *Lecture Notes*. Yale University.

Campbell, A. (1992). On determining the primordial helium abundance from the spectra of H II galaxies. *Astrophysical Journal*, **401**, 157–167.

Carlson, B., Lacis, A.A. and Rossow, W.B. (1993). Tropospheric gas composition and cloud structure of the Jovian North Equatorial Belt. *Journal of Geophysical Research*, **98**, 5251–5290.

Caughlan, G.R. and Fowler, W.A. (1988). *Atomic Data and Nuclear Data Tables*, **40**, 291.

Cayrel, R. (1998). Lithium abundances in low-Z stars. *Space Science Reviews*, **84**, 145–154 [and N. Prantzos, M. Tosi and R. von Steiger (eds), *Primordial Nuclei and their Galactic Evolution*, Proceedings of an ISSI Workshop 6–10 May, 1997, Kluwer, Dordrecht, p. 145].

Cayrel, R. and Steffen, M. (2000). Effects of photospheric temperature inhomogeneities on lithium abundance determinations (2D). In L. da Silva, M. Spite and J.R. de Medeiros (eds), *The Light Elements and Their Evolution*, IAU Symposium Vol. 198, Astronomical Society of the Pacific, San Francisco, pp. 437–447.

Charbonnel, C. (1998). Mixing in stars and the evolution of the ^{3}He abundance. *Space Science Reviews*, **84**, 199–206 [and N. Prantzos, M. Tosi, and R. von Steiger (eds), *Primordial Nuclei and their Galactic Evolution*, Proceedings of an ISSI Workshop 6–10 May, 1997, Kluwer, Dordrecht, p. 199].

Charbonnel, C., Deliyannis, C.P. and Pinsonneault, M. (2000). Sinks of light elements in stars – Part III. In L. da Silva, M. Spite and J.R. de Medeiros (eds), *The Light Elements and Their Evolution*, IAU Symposium Vol. 198, Astronomical Society of the Pacific, San Francisco, pp. 87–97.

Chaussidon, M. (1998). ^{7}Li/^{6}Li and ^{11}B/^{10}B variations in chondrules from the Semarkona unequilibrated chondrite. *Earth and Planetary Science Letters*, **164**, 577–589.

Chiappini, C. and Matteucci, F. (2000). The evolution of ^{3}He, ^{4}He and D in the Galaxy. In L. da Silva, M. Spite and J.R. de Medeiros (eds), *The Light Elements and Their Evolution*, IAU Symposium Vol. 198, Astronomical Society of the Pacific, San Francisco, pp. 540–546.

Chmielewski, Y., Müller, E.A. and Brault, J.W. (1975). The solar beryllium abundance. *Astronomy and Astrophysics*, **42**, 37–46.

Cunha, K. (2000). The abundance of boron in disk-metallicity stars. In L. da Silva, M. Spite and J.R. de Medeiros (eds), *The Light Elements and Their Evolution*, IAU Symposium Vol. 198, Astronomical Society of the Pacific, San Francisco, pp. 415–424.

Cunha, K. and Smith, V. (1999). A determination of the solar photospheric boron abundance. *Astrophysical Journal*, **512**, 1006–1013.

Deliyannis, C.P., Pinsonneault, M.H. and Charbonnel, C. (2000). Sinks of light elements in stars – Part I. In L. da Silva, M. Spite and J.R. de Medeiros (eds), *The Light Elements and Their Evolution*, IAU Symposium Vol. 198, Astronomical Society of the Pacific, San Francisco, pp. 61–73.

Duncan, D. (1998). Key questions for low metallicity stars: Rapporteur summary of the Working Group on Low-Z Stars. *Space Science Reviews*, **84**, 167–174 [and N. Prantzos, M. Tosi and R. von Steiger (eds), *Primordial Nuclei and their Galactic Evolution*, Proceedings of an ISSI Workshop 6–10 May, 1997, Kluwer, Dordrecht, p. 167].

Encrenaz, T., de Graauw, T., Schaeidt, S., Lellouch, E., Feuchtgruber, H., Beintema, D.A., Bezard, B., Drossart, P., Griffin, M., Heras, A., Kessler, M., Leech, K., Morris, P., Roelfsema, P.R., Roos-Serote, M., Salama, A., Vandenbussche, B., Valentijn, E.A., Davis, G.R. and Naylor, D.A. (1996). First results of ISO-SWS observations of Jupiter. *Astronomy and Astrophysics*, **315**, L397–L400.

Ferrando, P., Lal, N., McDonald, F.B. and Webber, W.R. (1991). Studies of low-energy Galactic cosmic-ray composition at 22 AU: I. Secondary/primary ratios. *Astronomy and Astrophysics*, **247**, 163–172.

Fowler, W.A., Greenstein, J. and Hoyle, F. (1961). Deuteronomy: Synthesis of deuterons and the light nuclei during the early history of the Solar System. *American Journal of Physics*, **29**, 393–403.

Fowler, W.A. and Hoyle, F. (1964). Neutrino processes and pair formation in massive stars and supernovae. *Astrophysical Journal*, Suppl., **9**, 201.

Fuller, G.M., Boyd, R.N. and Kalen, J.D. (1991). Primordial He-4 as a test of big bang nucleosynthesis. *Astrophysical Journal*, **371**, L11–L14.

Gautier, D. and Owen, T. (1989). The composition of outer planet atmospheres. In S.K. Atreya and J.B. Pollack (eds), *Origin and Evolution of Planetary and Satellite Atmospheres*, University of Arizona Press, Tucson, AZ, pp. 487–512.

Gamow, G. (1946). Expanding universe and the origin of elements. *Physical Review*, **70**, 572–573.

Geiss, J., Bühler, F., Cerutti, H., Eberhardt, P. and Filleux, C. (1972). *Apollo 16 Preliminary Science Report*, NASA SP-315, Section 14.

Geiss, J., Eberhardt, P., Bühler, F., Meister, J. and Signer, P. (1970). Apollo 11 and 12 solar wind composition experiments: Fluxes of He and Ne isotopes. *Journal of Geophysical Research*, **75**, 5972–5979.

Geiss, J. and Gloecker, G. (1998). Abundances of deuterium and helium-3 in the protosolar cloud. *Space Science Reviews*, **84**, 239–250 [and N. Prantzos, M. Tosi, and R. von Steiger (eds), *Primordial Nuclei and their Galactic Evolution*, Proceedings of an ISSI Workshop 6–10 May, 1997, Kluwer, Dordrecht, p. 239].

Geiss, J., Oeschger, H. and Schwarz, U. (1961). Estratto da Rendiconti della Scuola Internazionale di Fisica E. Fermi, XIX, Corso, p. 247.

Geiss, J. and Reeves, H. (1972). Cosmic and Solar System abundances of deuterium and helium-3. *Astronomy and Astrophysics*, **18**, 126.

Geiss, J. and Reeves, H. (1981). Deuterium in the solar system. *Astronomy and Astrophysics*, **93**, 189–199.

Gilmore, G. (1992). In N. Prantzos, E. Vangioni-Flam and M. Cassé (eds), *On the Origin and Evolution of the Elements*, Cambridge University Press.

Gilmore, G., Edvardsson, B. and Nissen, P.E. (1991). First detection of beryllium in a very metal poor star: A test of the standard big bang model. *Astrophysical Journal*, **378**, 17–21.

Gilmore, G., Gustafsson, B., Edvardsson, B. and Nissen, P. (1992). Is beryllium in metal-poor stars of galactic or cosmological origin? *Nature*, **357**, 379–384.

Gloeckler, G. and Geiss, J. (1996). Abundance of ^{3}He in the local interstellar cloud. *Nature*, **381**, 210–212.

Gloeckler, G. and Geiss, J. (2000). Deuterium and helium-3 in the protosolar cloud. In L. da Silva, M. Spite and J.R. de Medeiros (eds), *The Light Elements and Their Evolution*, IAU Symposium Vol. 198, Astronomical Society of the Pacific, San Francisco, pp. 224–233.

Grevesse, N. and Noels, A. (1992). In N. Prantzos, E. Vangioni-Flam and M. Cassé (eds), *On the Origin and Evolution of the Elements*, Cambridge University Press.

Hogan, C.J. (1998). Extragalactic abundances of hydrogen, deuterium and helium: New steps, missteps and next steps. *Space Science Reviews*, **84**, 127–136 [and N. Prantzos, M. Tosi and R. von Steiger (eds), *Primordial Nuclei and their Galactic Evolution*, Proceedings of an ISSI Workshop 6–10 May, 1997, Kluwer, Dordrecht, p. 127].

Hoyle, F. and Tayler, R.J. (1964). The mystery of the cosmic helium abundance. *Nature*, **203**, 1108–1110.

Irvine, J.M. (1983). Constancy of the laws of physics in the light of prehistoric nuclear reactors. *Contemporary Physics*, **24**, 427–437.

Kajino, T., Suzuki, T.-K., Kawanomoto, S. and Ando, H. (2000). New determination method of primordial Li abundance. In L. da Silva, M. Spite and J.R. de Medeiros (eds), *The Light Elements and Their Evolution*, IAU Symposium Vol. 198, Astronomical Society of the Pacific, San Francisco, pp. 344–349.

Knauth, D.C., Federman, S.R., Lambert, D.L. and Crane, P. (2000). The interstellar lithium isotope ratio toward Per OB2. In L. da Silva, M. Spite and J.R. de Medeiros (eds), *The Light Elements and Their Evolution*, IAU Symposium Vol. 198, Astronomical Society of the Pacific, San Francisco, pp. 338–343.

Krauss, L.M. and Romanelli, P. (1990). Big bang nucleosynthesis: Predictions and uncertainties. *Astrophysical Journal*, **358**, 47–59.

Kunth, D. and Sargent, W. (1983). Spectrophotometry of 12 metal-poor galaxies: Implications for the primordial helium abundance. *Astrophysical Journal*, **273**, 81–98.

Kurki-Suonio, H. (1991). Baryon inhomogeneity from the cosmic quark-hadron phase transition. Paper presented at the Workshop on Strange Quark Matter in Physics and Astrophysics, Aarhus, Denmark.

Kurki-Suonio, H. (2000). Alternative solutions to Big Bang nucleosynthesis. In L. da Silva, M. Spite and J.R. de Medeiros (eds), *The Light Elements and Their Evolution*, IAU Symposium Vol. 198, Astronomical Society of the Pacific, San Francisco, pp. 25–34.

Kurki-Suonio, H., Matzner, R.A., Olive, K.A. and Schramm, D.N. (1990). Big bang nucleosynthesis and the quark–hadron transition. *Astrophysical Journal*, **353**, 406–410.

Lal, D. and Peters, B. (1967). Cosmic ray produced radioactivity on the earth. In K. Sitte (ed.), *Handbuch der Physik*, Vol. 46/2, Springer, Berlin, pp. 551–612.

Lemoine, M., Ferlet, R., Vidal-Madjar, A., Emerich, C., Bertin, P. (1993). Interstellar lithium and the Li-7/Li-6 ratio toward Rho OPH. *Astronomy and Astrophysics*, **269** [1993], 469–476 [also N. Prantzos, E. Vangioni-Flam and M. Cassé (eds), *On the Origin and Evolution of the Elements*, Cambridge University Press].

Levshakov, S.A. (2000). The deuterium abundance in QSO absorption systems: A mesoturbulent approach. In L. da Silva, M. Spite and J.R. de Medeiros (eds), *The Light Elements and Their Evolution*, IAU Symposium Vol. 198, Astronomical Society of the Pacific, San Francisco, pp. 135–140.

Linsky, L.J. (1998). Deuterium abundance in the local ISM and possible spatial variations. *Space Science Reviews*, **84**, 285–296 [and N. Prantzos, M. Tosi and R. von Steiger (eds), *Primordial Nuclei and their Galactic Evolution*, Proceedings of an ISSI Workshop 6–10 May, 1997, Kluwer, Dordrecht, p. 285].

Linsky, J.L. and Wood, B.E. (2000). Deuterium observations in our Galaxy: View A. In L. da Silva, M. Spite and J.R. de Medeiros (eds), *The Light Elements and Their Evolution*, IAU Symposium Vol. 198, Astronomical Society of the Pacific, San Francisco, pp. 141–150.

Lubowich, D.A., Pasachoff, J.M., Galloway, R.P., Balonek, T.J., Tremonti, C., Millar, T. and Roberts, H. (2000). The deuterium abundance in the galactic center 50km/s molecular cloud: Evidence for a cosmological origin of D. In L. da Silva, M. Spite and J.R. de Medeiros (eds), *The Light Elements and Their Evolution*, IAU Symposium Vol. 198, Astronomical Society of the Pacific, San Francisco, pp. 167–175.

Mahaffy, P.R., Donahue, T.M., Atreya, S.K., Owen, T.C. and Nieman, H.B. (1998). Galileo Probe measurements of D/H and ^{3}He/^{4}He in Jupiter's atmosphere. *Space Science Reviews*, **84**, 251–263 [and N. Prantzos, M. Tosi and R. von Steiger (eds), *Primordial Nuclei and their Galactic Evolution*, Proceedings of an ISSI Workshop 6–10 May, 1997, Kluwer, Dordrecht, p. 251].

Matteucci, F. and Romano, D. (2000). Stellar and GCR production of lithium in the Milky Way. In L. da Silva, M. Spite and J.R. de Medeiros (eds), *The Light Elements and Their Evolution*, IAU Symposium Vol. 198, Astronomical Society of the Pacific, San Francisco, pp. 558–562.

Meneguzzi, M., Audouze, J. and Reeves, H. (1971). The production of the elements Li, Be, B by galactic cosmic rays in space and its relation with stellar observations. *Astronomy and Astrophysics*, **15**, 337.

Michaud, G. and Charbonneau, P. (1991). The lithium abundance in stars. *Space Science Reviews*, **57**, 1–58.

Michaud, G., Richer, J. and Richard, O. (2000). Constraints on stellar hydrodynamics from abundance anomalies of LiBeB and metals. In L. da Silva, M. Spite and J.R. de Medeiros (eds), *The Light Elements and Their Evolution*, IAU Symposium Vol. 198, Astronomical Society of the Pacific, San Francisco, pp. 460–469.

Montmerle, T. (1977). Light-element production by cosmological cosmic rays. *Astrophysical Journal*, **217**, 878–882.

Nissen, P.E. (2000). Observations of ^{6}Li in metal poor stars. In L. da Silva, M. Spite and J.R. de Medeiros (eds), *The Light Elements and Their Evolution*, IAU Symposium Vol. 198, Astronomical Society of the Pacific, San Francisco, pp. 259–268.

Normile, D. (1998). Weighing in on neutrino mass. *Science*, **280**, 1689–1690.

Olive, K.A. (2000). The evolution of ^{4}He and LiBeB. In L. da Silva, M. Spite and J.R. de Medeiros (eds), *The Light Elements and Their Evolution*, IAU Symposium Vol. 198, Astronomical Society of the Pacific, San Francisco, pp. 547–557.

Pagel, B.E.J. (1992). In N. Prantzos, E. Vangioni-Flam and M. Cassé (eds), *On the Origin and Evolution of the Elements*, Cambridge University Press.

Pagel, B.E.J. and Simonson, E.A. (1989). Helium in three H II galaxies and the primordial helium abundance. *Revista Mexicana de Astronomía y Astrofísica*, **18**, 153–159.

Pagel, B.E.J., Simonson, E.A., Terlevich, R.J. and Edmunds, M.G. (1992). The primordial helium abundance from observations of extragalactic H II regions. *Monthly Notices of the Royal Astronomical Society*, **255**, 325–345.

Parizot, E. and Drury, L. (2000). The superbubble model for LiBeB production and galactic evolution. In L. da Silva, M. Spite and J.R. de Medeiros (eds), *The Light Elements and Their Evolution*, IAU Symposium Vol. 198, Astronomical Society of the Pacific, San Francisco, pp. 35–50.

Pasquini, L. (2000). Li abundance in pop I stars. In L. da Silva, M. Spite and J.R. de Medeiros (eds), *The Light Elements and Their Evolution*, IAU Symposium Vol. 198, Astronomical Society of the Pacific, San Francisco, pp. 269–278.

Peimbert, M. and Peimbert, A. (2000). The Magellanic Clouds and the primordial helium abundance. In L. da Silva, M. Spite and J.R. de Medeiros (eds), *The Light Elements and Their Evolution*, IAU

Symposium Vol. 198, Astronomical Society of the Pacific, San Francisco, pp. 194–203.

Pinsonneault, M.H., Charbonnel, C. and Deliyannis, C.P. (2000). Sinks of light elements in stars – Part II. In L. da Silva, M. Spite and J.R. de Medeiros (eds), *The Light Elements and Their Evolution*, IAU Symposium Vol. 198, Astronomical Society of the Pacific, San Francisco, pp. 74–86.

Pinsonneault, M.H., Deliyannis, C.P. and Demarque, P. (1992). Evolutionary models of halo stars with rotation: II. Effects of metallicity on lithium depletion, and possible implications for the primordial lithium abundance. *Astrophysical Journal*, Suppl., **78**, 179–203.

Prantzos, N. (1996). The evolution of D and ^{3}He in the Galactic disk. *Astronomy and Astrophysics*, **310**, 106–114.

Prantzos, N., Cassé, M. and Viangoni-Flam, E. (1992). Symposium. In N. Prantzos, E. Vangioni-Flam and M. Cassé (eds), *On the Origin and Evolution of the Elements*, Cambridge University Press.

Primas, F. (2000). The galactic evolution of boron. In L. da Silva, M. Spite and J.R. de Medeiros (eds), *The Light Elements and Their Evolution*, IAU Symposium Vol. 198, Astronomical Society of the Pacific, San Francisco, pp. 405–414.

Primas, F., Duncan, D.K., Peterson, R.C. and Thoburn, J.A. (1999). A new set of HST boron observations: I. Testing light elements stellar depletion. *Astronomy and Astrophysics*, **343**, 545–557.

Ramaty, R., Koslovsky, B., Lingenfelter, R.E. and Reeves, H. (1997). Light elements and cosmic rays in the early Galaxy. *Astrophysical Journal*, **488**, 730.

Ramaty, R., Lingenfelter, R.E. and Kozlovsky, B. (2000). LiBeB evolution: Three models. In L. da Silva, M. Spite and J.R. de Medeiros (eds), *The Light Elements and Their Evolution*, IAU Symposium Vol. 198, Astronomical Society of the Pacific, San Francisco, pp. 51–60.

Read, S.M. and Viola, V.E. Jr (1984). Excitation functions for A ≥ 6 fragments formed in ^{1}H- and ^{4}He-induced reactions on light nuclei. *Atomic Data and Nuclear Data Tables*, **31**, 359.

Reasenberg, R.D. (1988). In V. De Sabbata and V.N. Melnikov (eds), *Gravitational Measurements, Fundamental Metrology and Constants*, Kluwer, Dordrecht, p. 311.

Rebolo, R., Garcia-Lopez, R.J., Martin, E.L., Beckman, J.E., McKeith, C.D., Webb, J.K. and Pavlenko, Y.V. (1992). In N. Prantzos, E. Vangioni-Flam and M. Cassé (eds), *On the Origin and Evolution of the Elements*, Cambridge University Press.

Rebolo, R., Molaro, P. and Beckman, J.E. (1988). Lithium abundances in metal-deficient dwarfs. *Astronomy and Astrophysics*, **192**, 192–205.

Rebull, L., Duncan, D., Johanssen, S. Thorburn, J. and Fields, B. (1998). *Astrophysical Journal*, **507**, 387.

Reeves, H. (1967). High-energy nucleosynthesis. In *High Energy Astrophysics, Vol. II, Summer School of Theoretical Physics, Les Houches, 1966*. Gordon & Breach, New York; Dunod, Paris.

Reeves, H. (1971). American Physical Society Meeting, Porto Rico Dec 1971, p. 256.

Reeves, H. (1974). On the origin of the light elements. *Annual Review of Astronomy and Astrophysics*, **12**, 437–469.

Reeves, H. (1994). On the origin of the light elements. *Review of Modern Physics*, **66**(1), 193–217.

Reeves, H. (1993). The Li/Li-6 ratio and the stellar yield of Li-7. *Astronomy and Astrophysics*, **269**, 166–168 [also N. Prantzos, E. Vangioni-Flam and M. Cassé (eds), *On the Origin and Evolution of the Elements*, Cambridge University Press].

Reeves, H., Audouze, J., Fowler, W.A. and Schramm, D.N. (1973). On the origin of light elements. *Astrophysical Journal*, **177**, 909–930.

Reeves, H., Fowler, W.A. and Hoyle, F. (1970). Galactic cosmic ray origin of Li, Be and B in stars. *Nature*, **226**, 727–729.

Reeves, H. and Meyer, J.P. (1978). Cosmic-ray nucleosynthesis and the infall rate of extragalactic matter in the solar neighborhood. *Astrophysical Journal*, **226**, 613–631.

Reeves, H., Richer, J., Sato, K. and Terasawa, N. (1990). On the origin of lithium. *Astrophysical Journal*, **355**, 18–28.

Rood, R.T., Bania, T.M., Balser, D.S. and Wilson, T.L. (1998). Helium-3: Status and prospects. *Space Science Reviews*, **84**, 185–198 [and N. Prantzos, M. Tosi and R. von Steiger (eds), *Primordial Nuclei and their Galactic Evolution*, Proceedings of an ISSI Workshop 6–10 May, 1997, Kluwer, Dordrecht, p. 185].

Ryan, S.G. (1992). In N. Prantzos, E. Vangioni-Flam and M. Cassé (eds), *On the Origin and Evolution of the Elements*, Cambridge University Press.

Ryan, S.G. (2000). ^{7}Li in metal-poor stars: The spread of the Li plateau. In L. da Silva, M. Spite and J.R. de Medeiros (eds), *The Light Elements and Their Evolution*, IAU Symposium Vol. 198, Astronomical Society of the Pacific, San Francisco, pp. 249–258.

Ryan, S.G., Norris, J.E. and Bessel, M.S. (1991). Subdwarf studies: IV. Abundance ratios in extremely metal-deficient stars. *Astronomical Journal*, **102**, 303–322.

Ryter, C., Reeves, H., Gradstztajn, E. and Audouze, J. (1970). The energetics of L nuclei formation in stellar atmospheres and its relevance to X ray astronomy. *Astronomy and Astrophysics*, **8**, 389.

Sackman, I.-J. and Boothroyd, A.I. (2000). Creation and destruction of ^{7}Li and ^{3}He in RGB and AGB stars. In L. da Silva, M. Spite and J.R. de Medeiros (eds), *The Light Elements and Their Evolution*, IAU Symposium Vol. 198, Astronomical Society of the Pacific, San Francisco, pp. 98–107.

Sahu, M.S. (2000). The D/H ratio in interstellar gas toward the hot, white dwarf G191-B2B. In L. da Silva, M. Spite and J.R. de Medeiros (eds), *The Light Elements and Their Evolution*, IAU Symposium Vol. 198, Astronomical Society of the Pacific, San Francisco, pp. 161–166.

Scherrer, R.J. and Spergel, D.N. (1993). How constant is the Fermi coupling constant? *Physical Review D*, **47**, 4774–4777.

Schramm, D.N. (1992). In N. Prantzos, E. Vangioni-Flam and M. Cassé (eds), *On the Origin and Evolution of the Elements*, Cambridge University Press.

Shima, M. and Honda, M. (1962). Isotopic abundance of meteoric lithium. *Journal of Geophysical Research*, **68**, 2849–2854.

Smith, V.V., Lambert, D.L. and Nissen, P.E. (1992). The 6Li/7Li ratio in the metal-poor halo dwarfs HD 19445 and HD 84937. *Astrophysical Journal*, **408**, 262–276.

Songaila, A., Wampler, E.J. and Cowie, L.L. (1997). A high deuterium abundance in the early universe. *Nature*, **385**, 137–139.

Spite, F. and Spite, M. (1982b). Abundance of lithium in unevolved halo stars and old disk stars: Interpretation and consequences. *Astronomy and Astrophysics*, **115**, 357–366.

Spite, F., Spite, M., Cayrel, R. and Huille, S. (1991). In B. Barbuy and A. Renzini (eds), *The Stellar Populations of Galaxies*, IAU Symposium 149, Kluwer, Dordrecht [1992].

Spite, F., Spite, M. and Hill, V. (1998). Lithium abundance in pop.II stars: A post-HIPPARCOS discussion. *Space Science Reviews*, **84**, 155–160 [and N. Prantzos, M. Tosi, and R. von Steiger (eds), *Primordial Nuclei and their Galactic Evolution*, Proceedings of an ISSI Workshop 6–10 May, 1997, Kluwer, Dordrecht, p. 155].

Spite, M. and Spite, F. (1982a). Lithium abundance at the formation of the galaxy. *Nature*, **297**, 483–485.

Spite, M., Spite, F., Cayrel, R., Hill, V., Depagne, E., Nordström, B. and Beers, T.C. (2000). Lithium depletion in a [Fe/H]=−3.4 star? In L. da Silva, M. Spite and J.R. de Medeiros (eds), *The Light Elements and Their Evolution*, IAU Symposium Vol. 198, Astronomical Society of the Pacific, San Francisco, pp. 356–357.

Steigman, G. (2000). Primordial nucleosynthesis for the new millennium. In L. da Silva, M. Spite and J.R. de Medeiros (eds), *The Light Elements and Their Evolution*, IAU Symposium Vol. 198, Astronomical Society of the Pacific, San Francisco, pp. 13–24.

Steigman, G. and Walker, T.P. (1992). Production of Li, Be, and B in the early galaxy. *Astrophysical Journal, Letters*, **385**, L13–L16.

Théado, S. and Vauclair, S. (2000). Self-regulated hydrodynamical process in halo stars: A possible explanation of the lithium plateau. In L. da Silva, M. Spite and J.R. de Medeiros (eds), *The Light Elements and Their Evolution*, IAU Symposium Vol. 198, Astronomical Society of the Pacific, San Francisco, pp. 520–521.

Thieleman, F.K., Applegate, L., Cowan, J.J. and Wiescher, M. (1991). In H. Oberhummer and C. Rolfs (eds), *Nuclei in the Cosmos*, Springer, Berlin.

Thuan, T.X. and Isotov, Y.I. (1998). The primordial helium-4 abundance from observations of a large sample of blue compact dwarf galaxies. *Space Science Reviews*, **84**, 83–94 [and N. Prantzos, M. Tosi, and R. von Steiger (eds), *Primordial Nuclei and their Galactic Evolution*, Proceedings of an ISSI Workshop 6–10 May, 1997, Kluwer, Dordrecht, p. 83].

Thuan, T.X. and Izotov, Y.I. (2000). Blue compact galaxies and the primordial ^{4}helium abundance. In L. da Silva, M. Spite and J.R. de Medeiros (eds), *The Light Elements and Their Evolution*, IAU Symposium Vol. 198, Astronomical Society of the Pacific, San Francisco, pp. 176–187.

Tosi, M. (1998). Galactic evolution of D and 3He. *Space Science Reviews*, **84**, 207–218 [and N. Prantzos, M. Tosi, and R. von Steiger (eds), *Primordial Nuclei and their Galactic Evolution*, Proceedings of an ISSI Workshop 6–10 May, 1997, Kluwer, Dordrecht, p. 207].

Tosi, M. (2000). Evolution of D and ^{3}He in the Galaxy. In L. da Silva, M. Spite and J.R. de Medeiros (eds), *The Light Elements and Their Evolution*, IAU Symposium Vol. 198, Astronomical Society of the Pacific, San Francisco, pp. 525–534.

Tytler, D., O'Meara, J.M., Suzuki, N., Lubin, D., Burles, S. and Kirkman, D. (2000). Measurements of the primordial D/H abundance towards quasars. In L. da Silva, M. Spite and J.R. de Medeiros (eds), *The Light Elements and Their Evolution*, IAU Symposium Vol. 198, Astronomical Society of the Pacific, San Francisco, pp. 125–134.

Vangioni-Flam, E. and Cassé, M. (2000). LiBeB production and associated astrophysical sites. In L. da Silva, M. Spite and J.R. de Medeiros (eds), *The Light Elements and Their Evolution*, IAU Symposium Vol. 198, Astronomical Society of the Pacific, San Francisco, pp. 41–50.

Vangioni-Flam, E., Olive, K. and Prantzos, N. (1994). On the galactic evolution of D and He-3, *Astrophysical Journal*, **427**, 618–627.

Vauclair, S. (2000). Transport phenomena and light element abundance in the Sun and solar type stars. In L. da Silva, M. Spite and J.R. de Medeiros (eds), *The Light Elements and Their Evolution*, IAU Symposium Vol. 198, Astronomical Society of the Pacific, San Francisco, pp. 470–475.

Vauclair, S. and Charbonnel, C. (1998). Element segregation in low-metallicity stars and the primordial lithium abundance. *Astrophysical Journal*, **502**, 372.

Viegas, S.M. and Gruenwald, R. (2000). Inhomogeneous H II regions and the helium abundance. In L. da Silva, M. Spite and J.R. de Medeiros (eds), *The Light Elements and Their Evolution*, IAU Symposium Vol. 198, Astronomical Society of the Pacific, San Francisco, pp. 188–193.

Vidal-Madjar, A. (2000). Deuterium observation in our Galaxy: View B. In L. da Silva, M. Spite and J.R. de Medeiros (eds), *The Light Elements and Their Evolution*, IAU Symposium Vol. 198, Astronomical Society of the Pacific, San Francisco, pp. 151–160.

Vidal-Madjar, A., Ferlet, R. and Lemoine, M. (1998). *Space Science Reviews*, **84**, 279 [and N. Prantzos, M. Tosi, and R. von Steiger (eds), *Primordial Nuclei and their Galactic Evolution*, Proceedings of an ISSI Workshop 6–10 May, 1997, Kluwer, Dordrecht, p. 279].

Voshage, H. and Hintenberger, H. (1961). Massenspektrometrische Isotopenhäufigkeitsmessungen an Kalium aus Eisenmeteoriten und das Problem der Bestimmung der ^{41}K–^{40}K-Strahlungsalter. *Zeitschrift für Naturforschung*, **16A**, 1042–1053.

Wagoner, R.V., Fowler, W.A. and Hoyle, F. (1967). On the synthesis of elements at very high temperatures. *Astrophysical Journal*, **148**, 3–49.

Walker, T.P., Mathews, G.J. and Viola, V.E. (1985). Astrophysical production rates for Li, Be, and B isotopes from energetic H(1) and He(4) reactions with HeCNO nuclei. *Astrophysical Journal*, **299**, 745–751.

Walker, T.P., Steigman, G., Schramm, D.N., Olive, K.A. and Kang, H.S. (1991). Primordial nucleosynthesis redux. *Astrophysical Journal*, **376**, 51–69.

Webber, W.R. (1998). A new estimate of the local interstellar energy density and ionization rate of galactic cosmic rays. *Astrophysical Journal*, **506**, 329.

Witten, E. (1984). Cosmic separation of phases. *Physical Review D*, **30**, 272–285.

Woosley, S.E., Hartman, D.H., Hoffman, R.D. and Haxton, W.C. (1990). The nu-process. *Astrophysical Journal*, **356**, 272–301.

Zanda, B. (1990). Thèse, Université de Paris.

20

JEAN SURDEJ* AND JEAN-FRANÇOIS CLAESKENS*

Gravitational lensing

In his theory of general relativity (GR), Einstein established that a massive object curves space–time and that photons move along geodesics of this curved space. Einstein's prediction was confirmed during the solar eclipse of 1919. This not only provided a decisive, quantitative proof of the validity of GR but also gave confirmation of the concept that electromagnetic waves do undergo deflections in gravitational fields.

Atmospheric lensing effects may lead to a distortion or to the formation of multiple images of a distant source, resulting in the perception of a 'mirage'. They sometimes literally deform our view of the surrounding areas. Similarly, gravitational lensing may perturb our view of the distant Universe and affect our physical understanding of various classes of extragalactic objects. The great interest in gravitational lensing comes from the fact that this phenomenon can be used as an astrophysical and cosmological tool. Indeed, gravitational lensing may help in deriving:

1. the distance scale of the Universe – via the determination of the Hubble constant H_0 – and possibly the values of other cosmological parameters (the cosmological density Ω_0 and the cosmological constant λ_0),
2. the mass distribution $M(r)$ of the lens,
3. the extinction law in the deflector, usually located at high redshift,
4. the nature and distribution of luminous and dark matter in the Universe,
5. the size and structure of quasars,
6. the size of absorbing intergalactic gas clouds,
7. upper limits on the density of a cosmological population of massive compact objects.

In this chapter we summarize some of the theoretical and observational evidence supporting these claims and show how space observations have significantly contributed in some of the areas listed above.

The chapter is organized as follows. We first review the historical background of gravitational lensing in Section 1. Unlike most other astrophysical discoveries made during the last century, the physics of gravitational lensing was understood well before the first example of a multiply-imaged extragalactic object was found.

In Section 2 we describe the basic principles and concepts of gravitational lensing. We first establish the exact form of the lens equation. By making use of the Einstein deflection angle, we derive the expression for the angular diameter of an Einstein ring, which is also the typical angular separation between multiply-lensed images when the conditions of perfect alignment between the observer, the lens and the source are no longer fulfilled. We then set up a sufficient condition for an observer to see an Einstein ring, or multiple images, of a distant source that is located behind a deflector, and go on to derive, for the case of a point-mass lens model, expressions for the expected image positions and their amplifications. Using wavefront and ray tracing diagrams, we introduce the important concepts of caustics and time delays. We then show how all the lensed image configurations observed in the Universe may be understood in terms of the relative location between the observer and the caustics associated with an asymmetric lens. Several outstanding

*Université de Liège, Liège, Belgium

examples of gravitational lens systems observed with the Hubble Space Telescope (HST) serve as illustrations.

In Section 3, we discuss the most remarkable astrophysical and cosmological applications of gravitational lensing. These include the independent determination of the Hubble constant H_0 based upon the measurement of the time delay Δt between the observed light curves of multiply-imaged quasars. The possibility of weighing the mass of lensing galaxies, galaxy clusters and intervening dark matter from the observation of multiply-imaged distant sources, arcs and arclets is also reviewed. We further discuss the possibility of using gravitational lensing to determine the cosmological density Ω_0 ($=8\pi G\rho_0/3H_0^2$) and the reduced cosmological constant λ_0 ($=\Lambda c^2/3H_0^2$) of the Universe, the cosmological density Ω_L of dark compact lenses, the size of intergalactic absorbing clouds, dust extinction laws in deflecting galaxies and the structure and size of quasars. We also present the status of optical searches for massive astrophysical compact halo objects (MACHOs) based upon microlensing, the formation of giant luminous arcs in galaxy clusters and the observation of 'weak lensing' in the Universe.

The anticipated contributions of ongoing and future space missions such as Chandra, XMM-Newton, MAP, FIRST-Herschel, Planck, NGST and GAIA are presented in Section 4.

1 HISTORICAL BACKGROUND

One of the consequences of the theory of GR is that light rays are deflected in gravitational fields. Although this prediction was made in the twentieth century, speculations that light rays might be bent by gravitation had been proposed much earlier. Indeed, considering that light is composed of elementary constituents, Isaac Newton suggested as early as 1704 that the gravitational field of a massive object could possibly bend light rays, just as it would alter the trajectory of material particles. A century later, Laplace independently made this same suggestion. Furthermore, the astronomer Johann von Soldner (1804) at the Munich Observatory found that, in the framework of Newtonian mechanics, a light ray passing near the limb of the Sun should undergo an angular deflection of 0.875″. However, because the wave description of light prevailed during the eighteenth and nineteenth centuries, neither the conjecture of Newton nor the result of Soldner were ever taken seriously. In 1911 Albert Einstein had re-derived the latter result on the basis of the equivalence principle, unaware of Soldner's work.

In the elaboration of his theory of GR, Einstein predicted that a massive object curves spacetime in its vicinity and that any free particle, massive or not (e.g. photons), will move along geodesics of this curved space. After deriving the full field equations of GR, he predicted in 1915 that a light ray passing near the solar limb should be deflected by an angle given by

$$\hat{\alpha} = 4GM/(c^2R) \ll 1 \qquad (1)$$

where G is the gravitational constant, c is the velocity of light and M and R are the mass and radius of the Sun (or of any other compact lens), respectively. This deflection angle turns out to be exactly twice the value derived by Soldner and by Einstein himself in 1911; the factor of 2 merely reflects the metric curvature. Note that in Newtonian terms the Einstein deflection also follows if one assumes a refractive index n associated with the Newtonian gravitational potential U ($|U| \ll c^2$) via the relation

$$n = 1 - 2U/c^2 \qquad (2)$$

The analogy between atmospheric and gravitational lensing then becomes obvious. Note, however, that the bending of light rays in gravitational fields is predicted to be totally 'achromatic' (see eqn (1)).

Using photographs of a stellar field taken during the solar eclipse in May 1919, and six months apart, Arthur Eddington and his collaborators (Dyson *et al.* 1920) were able to confirm, within a 20–30% uncertainty, the deflection angle predicted by Einstein (Figure 1). This was the second correct prediction of GR – the successful interpretation of the advance of Mercury's perihelion being the first – and marked the full acceptance of the work of Einstein.

It seems that Eddington (1920) was the first to propose the possible formation of multiple images of a background star by the gravitational lensing effect of a foreground one (but see the finding by Renn *et al.* (1997)). Note, however, that Oliver Lodge (1919) had already characterized massive

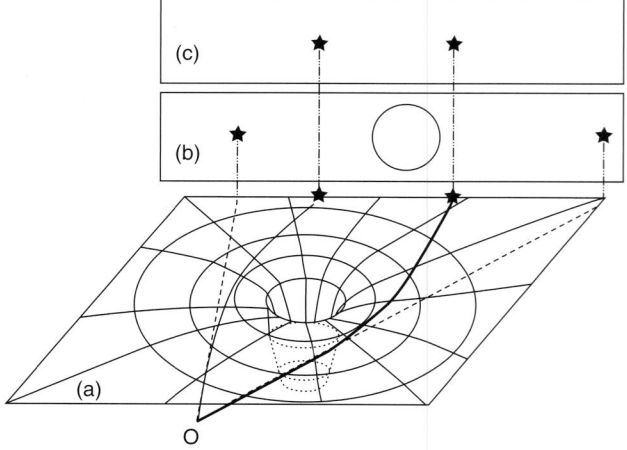

Figure 1 Curved spacetime around the Sun (a) and deflection of light rays from two distant stars as seen by an observer (O) during a solar eclipse (b) and six months apart (c).

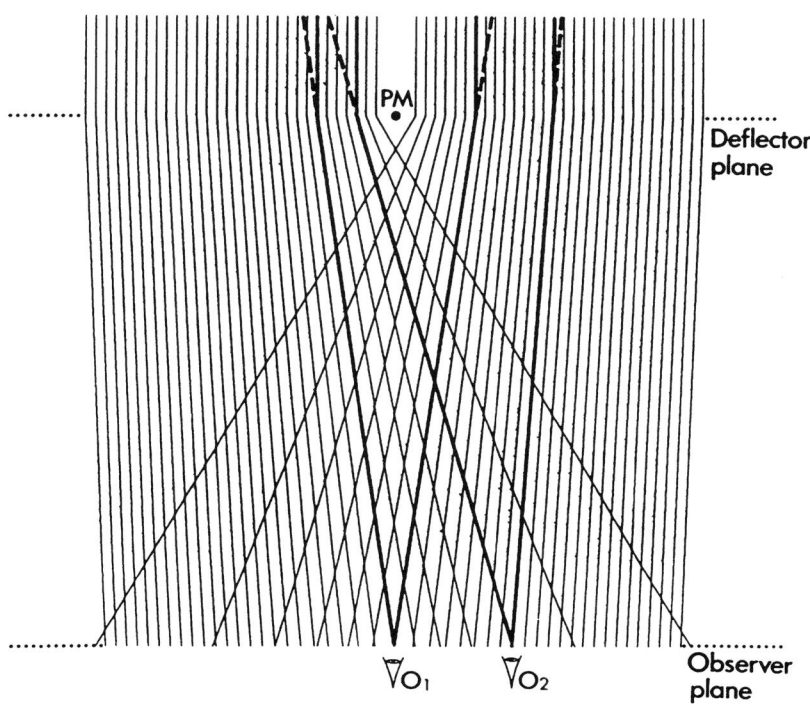

Figure 2 The paths of light rays in a point-mass (PM) lens model. In accordance with eqn (1), a set of parallel rays emitted from a distant source are deflected from their original paths as they cross the deflector plane. Different observer positions are represented from where an Einstein ring (O_1) or a double image (O_2) may be seen.

objects like the Sun as imperfect focusing lenses since they had no real focal length; the light from a background object being mainly concentrated along a focal line of 'infinite' length (Figure 2). In 1923, E.B. Frost, then director of the Yerkes Observatory, initiated a programme to search for multiple images of stars in the Galaxy, but it seems that these observations never really took place. Orest Chwolson (1924) suggested that, in the case of a perfect alignment between an observer and two stars located at different distances, the observer should see a ring-shaped image of the background star around the foreground one.

Independently, at the request of Rudi W. Mandl, a Czech electrical engineer, Einstein (1936) rediscovered the major characteristics (double images, the 'Chwolson' ring usually referred to as the 'Einstein' ring, and so on) of one star lensed by another, but he was very sceptical about the possibility of observing this phenomenon among stars, probably because the expected angular separations between the lensed images were so small. What seems remarkable is that Einstein had first established the whole theory of the formation of multiply-lensed images (the lensing equation, formation of double images by a point-like deflector, amplification of the lensed images, and so on) during the spring of 1912, three years before completing his theory of GR. This finding has recently been reported by Renn *et al.* (1997), who had examined some of Einstein's notebooks. It does therefore seem that by 1936 Einstein had totally forgotten his pioneering gravitational lensing work from 1912, as if he had already convinced himself that the idea was too speculative to have any chance of empirical confirmation.

Fritz Zwicky (1937a,b) was the first to realize the very high probability of identifying a gravitational lens mirage, (i.e. one composed of several distinct images of a single object) among extragalactic sources. He even proposed to use galaxies as natural cosmic telescopes to observe otherwise too faint and distant background objects. He also emphasized the possibility of determining the mass of distant galaxies simply by applying gravitational lens optometry and, in addition, to test the theory of GR. In 1937 Zwicky stated that 'the probability that galactic nebulae which act as gravitational lenses will be found becomes practically a certainty', and he was therefore very much surprised some twenty years later that no such lensing effects had yet been found with the 5 m (200-inch) Hale Telescope at Palomar (Zwicky 1957).

After several decades of little activity, interest in the theory of gravitational lenses was revived by Klimov (1963), Liebes (1964) and Refsdal (1964a,b, 1966a). Some of their proposed applications were particularly promising because of the recent discovery of quasars by Maarten Schmidt

(1963). It would indeed be much easier to prove the lensing origin of multiple QSO images rather than that of extended and diffuse galaxy images, since the former ones consist of very distant, luminous and star-like objects. (For more details on quasars, see Peterson (1997).) In 1964 Sjur Refsdal proposed to apply geometric optics in order to estimate the time delay between the arrival times of two parts of the same distorted wavefront at the observer; this proposal, which formed the second part of his master's thesis, was (erroneously) judged of uncertain quality by his supervisor. Nevertheless he succeeded in publishing his findings (Refsdal 1964a,b); the referee of these papers was Dennis Sciama! Refsdal finally, obtained his Ph.D. in 1970 on the basis of that controversial second part of his master's thesis.

On the basis of the strong similarity between the spectra of quasars and of the nuclei of Class 1 Seyfert galaxies, Barnothy (1965) proposed that high-redshift quasars could actually be the lensed images of distant Class 1 Seyfert galactic nuclei. Sanitt (1971) criticized this view on the basis of statistical arguments.

Theoretical work continued at a low level of activity through the 1970s. Refsdal (1965, 1970) and Press and Gunn (1973) discussed problems on lens statistics, Bourassa and Kantowski (1975) considered extended non-symmetric lenses (Bourassa *et al.* 1973) and Dyer and Roeder (1972) derived a distance–redshift relation for the case of inhomogeneous universes. In spite of clear theoretical predictions, the interest from observers was largely absent, and no systematic search for lenses was initiated.

Forty-two years after Zwicky's prediction, the first example of a distant quasar (Q0957 + 561) doubly imaged by a foreground massive lensing galaxy and its attendant galaxy cluster was serendipitously discovered by Walsh *et al.* (1979). This system consists of two lensed quasar images separated by approximately 6″. Good evidence that Q0957 + 561 A and B correspond to twin lensed images of a single quasar was provided by

1. the similarity between their spectra,
2. a simple lens model which could naturally account for the slight morphological differences detected between the optical and radio images of the quasar (Young *et al.* 1981),
3. the discovery of the lensing galaxy between the twin quasar images, made possible by the use of modern CCD imaging,
4. the observation of a time delay of 1.14 years, first reported by Vanderriest *et al.* (1989), between the light variations of the two quasar components.

In parallel with the discovery of several other multiply-imaged quasars, an even stronger interest in gravitational lensing studies developed with the first identification of giant luminous arcs in 1986, by two independent teams (Section 3.5). Since 1982 a group of French astronomers led by L. Nottale from Meudon Observatory had regularly submitted observing proposals with the Canada–France–Hawaii Telescope (CFHT) to search for such arcs in the centres of rich and compact galaxy clusters, among them Abell 370. However, the observing programme committees were not at all receptive to this kind of proposition. Although these predictions were entirely based on the theory of GR, and had been promoted several decades earlier by Zwicky, they probably still looked too revolutionary to conservative observing programme committee members. Furthermore, during the first international conference that was explicitly dedicated to the field of gravitational lensing, entitled 'Quasars and Gravitational Lenses' (24th Liège International Astrophysical Colloquium, 1983), some of the organizers were very surprised by the high degree of scepticism that was still present among a significant number of the participants. Probably half the audience still then considered gravitational lensing to be an unproved phenomenon, while many others thought that the proposed lensed monsters would remain a mere cosmic curiosity, with no further scientific interest. It also seems that until the 1980s, in some institutes gravitational lensing was not always regarded as an acceptable subject to study. Astronomers who wished to pursue the subject would not have received a research grant for work dedicated to gravitational lensing – nor would they be invited to deliver a seminar on the topic. We know from Refsdal, Nottale and many other pioneers that they had the feeling of being considered somewhat heretic by some of their colleagues.

Following the pioneering detections of multiply-imaged quasars and giant luminous arcs, the levels of observational as well as theoretical activities have increased dramatically. More than fifty multiply-imaged quasars and an even larger number of gravitational arcs have now been discovered, and more than 3100 scientific publications have been written over the past twenty years on the subject of gravitational lensing (Figure 3). Space observations, and in particular

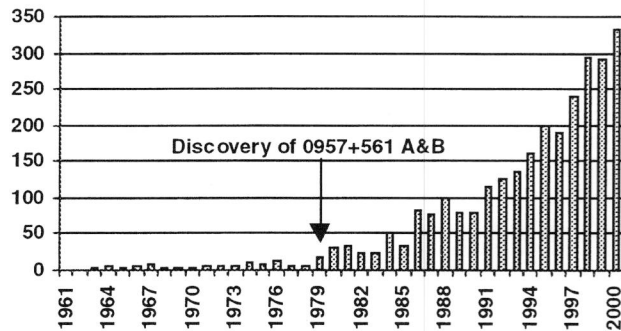

Figure 3 The number of scientific papers published annually on gravitational lensing over the past forty years. (Based on the gravitational lensing bibliography compiled by Pospieszalska-Surdej *et al.* (2000).)

direct imaging with the HST, have contributed decisively to our present understanding of gravitational lensing effects.

Gravitational lensing constitutes now a very important branch of galactic, extragalactic and cosmological astrophysics. There is no doubt that because of the intimate link between the curvature of spacetime and gravitational lensing at all scales and all locations in the Universe, and because of the new technologies and space missions planned for the twenty-first century, our physical understanding of the cosmos will continue to ever increase.

The reader is referred to Schneider *et al.* (1992) for a more detailed account of the history of gravitational lensing, as well as for a more complete and theoretical presentation of this subject.

2 THE PHYSICS OF GRAVITATIONAL LENSING AND SOME IMPORTANT CONCEPTS

The first step in understanding the properties of the images of a lensed source, is to set up the lens equation. We then establish the important relation between the angular diameter of the Einstein ring and the mass of the deflector, and find a sufficient condition for a potential deflector to produce multiple images of a distant source. For a point-mass lens model, we derive expressions for the image positions and amplifications. By means of ray and wavefront tracing diagrams we introduce the concepts of caustics and time delays. Finally, with the help of a gravitational lens simulator, we show how it is possible to reproduce all image configurations that have been observed for the known gravitational lens systems.

2.1 The lens equation and multiple imaging

Along the line of sight to a distant source, there is usually only one mass concentration which acts as a lens, and its size is very much smaller than the distances between the source and the lens, and between the lens and the observer (Figure 4). It is thus reasonable to adopt the 'thin lens approximation', according to which the deflection of a light ray takes place near the deflector at the precise location where the ray crosses the plane perpendicular to the observer–deflector direction. This plane is referred to as the 'lens plane', and the parallel plane containing the source is called the 'source plane'. Furthermore, geometric optics can be used since physical optical effects, induced by optical path differences, are negligible in all realistic situations. Under these assumptions, and given the achromatic property of the gravitational deflection of light, all lensing effects previously described may directly be understood from the lens equation (Figure 4).

Let us define the true position of the source S on the sky by the angle vector $\boldsymbol{\theta}_s$ and the image position(s) by $\boldsymbol{\theta}_i$ ($i = 1, 2, \ldots$). These correspond to the solutions of the lens equation

$$\boldsymbol{\theta} - \boldsymbol{\theta}_s = \boldsymbol{\alpha}(\boldsymbol{\theta}) = (D_{ds}/D_{os})\hat{\boldsymbol{\alpha}}(\boldsymbol{\theta}) \qquad (3)$$

where $\boldsymbol{\alpha}(\boldsymbol{\theta})$ represents the displacement angle and $\hat{\boldsymbol{\alpha}}(\boldsymbol{\theta})$ is the Einstein deflection angle derived in the weak gravitational field approximation (see eqn (1) for the case of a point mass). For an extended lens and within the thin lens approximation, we may simply calculate the effective deflection angle by just summing up all individual deflections due to the projection onto the lens plane of (point) mass elements constituting the lens. Although eqn (3) was originally derived for a static

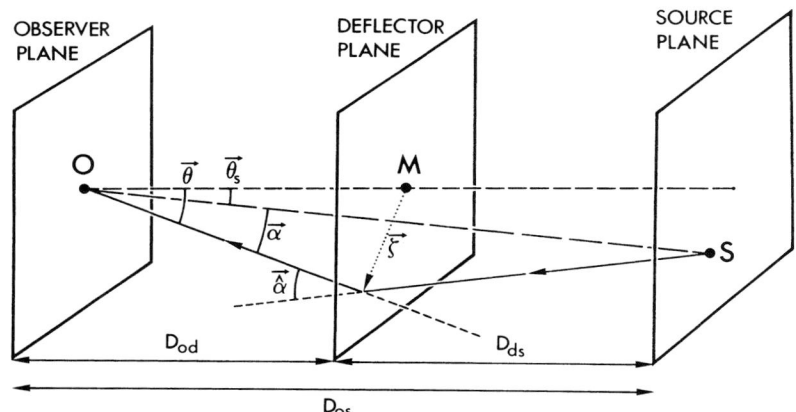

Figure 4 The deflection of a light ray between a source S and an observer O produced by an intervening point-mass lens (M). (For the other quantities, see the text.)

Euclidian space, Refsdal (1966a) showed that it is also valid for Friedmann–Lemaître–Robertson–Walker (FLRW) expanding universe models provided that D_{os}, D_{od} and D_{ds} represent the 'angular size distances' between the observer and the source, the observer and the deflector and the deflector and the source, respectively (e.g. Kayser *et al.* 1997).

We note that a given image position θ always corresponds to a specific source position θ_s, whereas a given source position θ_s may sometimes correspond to several distinct image positions θ. Such cases of multiply-imaged sources constitute the most spectacular and interesting aspect of gravitational lensing.

A typical lens situation is shown in Figure 5, where we have projected onto the plane of the sky the true (but invisible) source position and the resulting apparent image positions (two images in this case). We see again that the image position θ_i is shifted by $\alpha(\theta_i)$ relative to the true source position θ_s; note that $\alpha(\theta_i)$ is usually not constant over an extended lensed image.

Furthermore, Etherington (1933) demonstrated that the specific intensity (or surface brightness) of a light bundle is preserved in curved spaces. Therefore, if the lens is transparent the surface brightness of each lensed image is identical to that of the source, and the resulting amplification is equal to the ratio of the geometrical magnification of the angular size of the corresponding lensed image (e.g. the solid angle subtended by the 'Chwolson' ring) to that of the (unlensed) source. Generally speaking, this magnification μ_i is thus given by the ratio between the solid angle $d\omega_i$ covered by the lensed image and that of the source $d\omega_s$. In a more formal way, the expression of μ_i is given for an infinitesimal source by the inverse Jacobian of the transformation between the source and the image planes:

$$\mu_i = \frac{d\omega_i}{d\omega_s} = \left| \det\left(\frac{\partial \theta_s}{\partial \theta_i}\right) \right|^{-1} \quad (4)$$

If a gravitational lens system is composed of several spatially unresolved images, the total magnification (or amplification) is naturally given by the sum of all individual image magnifications (amplifications).

Finally, since the gravitational lensing amplification may vary over the image of an extended source, and if the latter has a shape that is wavelength dependent, then chromatic effects between the different lensed images can result (note the differently magnified source regions in Figure 5), without contravening the achromatic property stipulated earlier for gravitational lensing (eqn (1)). Furthermore, we shall see in Section 3 how multi-waveband observations of such resolved lensed images may enable us to set very helpful constraints on the lens mass distribution and on the source structure itself.

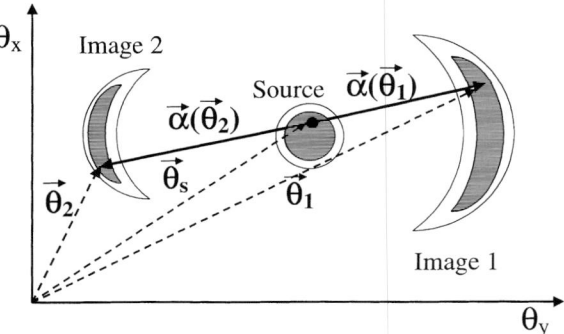

Figure 5 Images of a doubly imaged source as seen projected onto the sky. The white and darker regions represent different emitting areas as seen at two different wavelengths.

2.2 The angular size of the Einstein ring and the condition for multiple imaging

The perfect alignment between the observer, a compact lens characterized by a symmetric mass distribution $M(\xi)$ ($M(\xi)$ representing the mass located within the impact parameter ξ, within the distance of minimum approach between the light ray and the lens) and the source corresponds to $\theta_s = 0$ (Figure 4). It is clear in this axially symmetric case that the observer will actually see a ring (the so-called Einstein ring) of light from a distant source (Figure 6). It is possible to show, for an axially symmetric mass distribution, that the deflection vector angle $\hat{\alpha}(\theta)$ reduces itself to the scalar angle (compare with eqn (1))

$$\hat{\alpha}(\xi) = \frac{4G}{c^2}\frac{M(\xi)}{\xi} \quad (5)$$

Combining this deflection angle and the lens equation (eqn (3)) for $\theta_s = 0$, the angular radius θ_E of this Einstein ring may be expressed conveniently as

$$\theta_E = \sqrt{\left(\frac{4GM(D_{od}\theta_E)D_{ds}}{c^2 D_{od} D_{os}}\right)} \quad (6)$$

where $M(D_{od}\theta_E)$ stands for a mass producing an Einstein ring with radius θ_E, if located at the distance D_{od}. D_{od} represents the distance between the observer and the deflector. Note that when expressing the distances appearing in eqn (6) in terms of the source and lens redshifts, the quantity $D_{ds}/(D_{od}D_{os})$ is found to be directly proportional to H_0.

Listed in Table 1 are typical values of θ_E for different types of deflectors located at various distances, assuming that the distance between the source and the observer is approximately twice that between the deflector and the observer. We see from Table 1 that for a source and a lens located at cosmological distances ($D_{od} \sim 10^9$ pc), the angle θ_E can vary from micro-arcseconds (stellar deflection) to arcsecond (galaxy lensing) scales, and even tens of arcseconds for galaxy cluster lenses.

When there is imperfect alignment between the source, deflector and observer, and for lens mass distributions significantly departing from axial symmetry, the angular separation between multiple lensed images is still about $2\theta_E$, so that eqn (6) may still be used to infer the value of M/D_{od}, or the value of M times the Hubble constant H_0, if the redshift z_d of the lens and the redshift z_s of the source are known (Section 3.4.2).

By means of eqn (6) we may estimate the average surface mass density of the lens within the Einstein ring. We find that

$$\Sigma_c = \frac{c^2 D_{os}}{4\pi G D_{od} D_{ds}} \qquad (7)$$

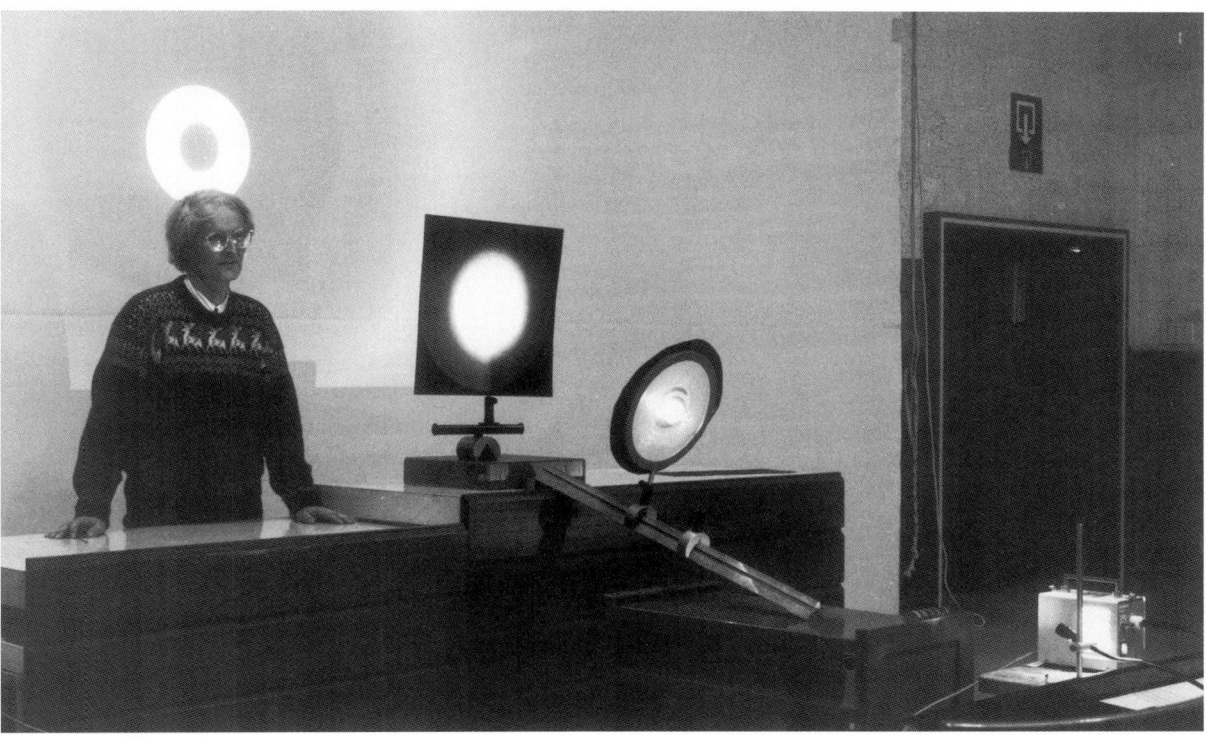

Figure 6 The gravitational lens experiment. When the observer (the pinhole screen visible in this photograph), a symmetric lens (made of Plexiglas, middle right) and a source (light bulb at bottom right) are in perfect alignment, the observer sees an Einstein ring. The simulation of such a ring is seen here projected onto the background screen during an optical experiment prepared by Professor Sjur Refsdal (seen in this photograph) and one of the authors.

Table 1 Angular radius (θ_E) of the Einstein ring produced by a deflector with mass M and linear radius R located at a distance D_{od} and a source at a distance $D_{os} = 2 D_{od}$. The quantity $\Sigma(R)/\Sigma_c$ represents the ratio between the average ($\Sigma(R)$) and critical (Σ_c) surface mass densities of the lens (see text).

Deflectors	Mass ($M_\odot$)	Radius (pc)	D_{od} (pc)	$\theta_E('')$	$\Sigma(R)/\Sigma_c$
Solar planets	10^{-6}–10^{-3}	10^{-10}–10^{-9}	10^{-6}–10^{-3}	2×10^{-3}–2	$< 3\,10^{-4}$
Exoplanets	10^{-6}–10^{-3}	10^{-10}–10^{-9}	1–10	2×10^{-5}–2×10^{-3}	100–10^5
Nearby stars	1–10	10^{-8}–10^{-6}	1–100	6×10^{-3}–0.2	0.1–10^6
Extragalactic stars	1–10	10^{-8}–10^{-6}	10^6–10^9	2×10^{-6}–2×10^{-4}	10^5–10^{13}
Globular clusters	10^5	1–10	10^3–10^5	6×10^{-2}–0.6	10^{-7}–10^{-3}
Local galaxies	10^{10}–10^{12}	10^3–10^4	10^6–10^7	2–64	10^{-5}–1
Distant galaxies	10^{10}–10^{12}	10^3–10^4	10^8–10^9	0.2–6.4	10^{-3}–100
Galaxy clusters	10^{14}–10^{15}	10^5–10^6	10^8–10^9	20–200	10^{-3}–10

irrespective of the mass distribution of the lens. This average surface mass density Σ_c is known in the literature as the critical surface mass density, which depends only on the angular distances between the observer, the lens and the source.

We conclude that the diameter of an Einstein ring adapts itself in such a way that the average surface mass density of the lens inside the ring is always equal to the critical surface mass density Σ_c. It is then straightforward to understand that a sufficient condition for an observer to see a ring, or multiple images, from a distant source is that, somewhere in the lens plane, the surface mass density Σ should exceed the critical value Σ_c, that is $\Sigma \geqslant \Sigma_c$. On the contrary, if we have $\Sigma < \Sigma_c$ everywhere over the lens, then the object is not capable of forming a cosmic mirage. Given typical cosmological distances for the deflector (redshift $z_d \sim 0.5$) and the source ($z_s \sim 2$), $\Sigma_c \sim 1\,\mathrm{g\,cm^{-2}}$ is found to be roughly equivalent to the surface mass density of a 1 cm thick layer of water.

The average surface mass density for a deflector having a typical mass M and radius R is defined as $\Sigma(R) = M/(\pi R^2)$. Listed in Table 1 are values of the ratio $\Sigma(R)/\Sigma_c$ for planets, stars, globular clusters, galaxies and galaxy clusters as possible deflectors. Note that for an axially symmetric lens, the condition $\Sigma(R) \geqslant \Sigma_c$ is equivalent to the condition $\theta_R \leqslant \theta_E$, θ_R ($=R/D_{od}$) being the angular radius of the lens, which automatically ensures that the angular extent of the deflector never covers the lensed images typically separated by an angular distance $2\theta_E$ (actually, $\Sigma(R)/\Sigma_c = (\theta_E/\theta_R)^2$).

We may summarize the important information contained in Table 1 as follows. Because their angular sizes are too large as viewed from the Earth, the solar planets are not capable of producing multiple images of background stars. Depending on the angular resolution of the observations, a slight angular shift of the positions of nearby background stars might, however, be detectable. Exoplanets are found to be excellent gravitational lenses ($\Sigma(R)/\Sigma_c \geqslant 1$) and this explains why several teams of astrophysicists are currently involved in searching for rapid light fluctuations induced by planet-like objects in the light-curves of selected stars (Section 3.3). All nearby and extragalactic stars turn out to be very good potential gravitational lenses. Globular clusters and local galaxies are not sufficiently compact to produce cosmic mirages. Finally, very compact, massive galaxies and/or galaxy clusters located at cosmological distances, for which $\Sigma(R)/\Sigma_c \geqslant 1$, constitute promising 'multiple imaging' deflectors.

2.3 Axially symmetric lens models

In axially symmetric lens models the propagation of light rays reduces to a one-dimensional problem (see eqns (5) and (6)), and the general lens equation (eqn (3)) transforms into the scalar equation

$$\theta^2 - \theta\,\theta_s - \theta_E^2 \frac{M(D_{od}\,\theta)}{M(D_{od}\,\theta_E)} = 0 \qquad (8)$$

The amplification μ of the lensed images (eqn (4)) is then found to be

$$\mu = \frac{\theta\,d\theta}{\theta_s\,d\theta_s} \qquad (9)$$

For the point-mass lens model, we have $M(\xi) = M$, and eqn (8) reduces to a second-degree equation in θ whose solutions are ($i = 1, 2$)

$$\theta_{1,2} = \frac{1}{2}\theta_s \pm \sqrt{\left(\left(\frac{1}{2}\theta_s\right)^2 + \theta_E^2\right)} \qquad (10)$$

and the image amplification is given by

$$\mu_i = \frac{1}{1 - (\theta_E/\theta_i)^4} \qquad (11)$$

a negative value of μ_i indicating that the corresponding lensed image is reversed. The positions of the lensed images being given by eqn (10), the total amplification $\mu_T = \mu_1 + |\mu_2|$ of the source can be found for a given value of θ_s:

$$\mu_T = \frac{\theta_s^2 + 2\theta_E^2}{\theta_s\,\sqrt{(\theta_s^2 + 4\theta_E^2)}} \qquad (12)$$

In particular, we find that when the true position of the source lies inside the imaginary Einstein ring (for $\theta_s \leqslant \theta_E$), the total magnification μ_T of the two images is larger than 1.34. This implies that the cross-section for significant lensing, by convention $\mu_T \geqslant 1.34$, is equal to $\pi\theta_E^2$, which is proportional to the mass M (eqn (6)). We make use of this result when discussing the optical depth for lensing in Section 2.6 and when deriving the expected frequency of multiply-imaged sources to set upper limits on the cosmological density Ω_L of compact objects in the Universe (Section 3.4.2).

When source, lens and observer are in perfect alignment, the observer sees a ring of light due to the axial symmetry (Section 2.2). Figure 6 illustrates a gravitational lens experiment in which a point-mass optical lens, actually made of Plexiglas, reproduces the deflection of light rays by a black hole having one-third of the Earth's mass (the foot of a wine glass may be used as an acceptable lens simulator). In this experiment, a compact light source (representing e.g. a distant quasar) is located on the right-hand side of the photograph. On the left-hand side of the lens is a white screen with a small hole at the centre (a pinhole screen). Further behind, there is a large screen onto which is projected the lensed image(s) of the source (an Einstein ring in this case) as it would be seen if our eyes were located at the position of the pinhole.

Representative examples of simulated lensed images are illustrated in Figure 7. First, the pinhole is set very precisely on the optical axis of the point-mass lens (Figure 7a, see also Figure 6) so that the source, the lens and the observer (the pinhole) are perfectly aligned. As the pinhole is moved slightly away from the axis of symmetry (Figure 7b), the Einstein ring (slightly elliptical in Figure 7h because of projection effects) breaks into two images (Figure 7i), in accordance with eqns (10) and (11). Note that, due to the point-mass lens used in the optical experiment, the distribution of light across the pinhole screen is not uniform: there is a maximum concentration of light near the optical axis. Thus, the maximum amplification is obtained when the pinhole is very precisely set on the optical axis (Figure 7a), corresponding to the formation of an Einstein ring (Figure 7h). Farther from the centre (Figure 7b) the light gets dimmer; the distribution of light does in fact correspond to convergence points due to pairs of light rays deflected by the lens, and the total amplification of the corresponding images tends towards unity with increasing distance from the axis of symmetry. Figures 7o and 7p illustrate corresponding examples of extragalactic multiply-imaged sources. Some of the images shown in Figure 7 were obtained by the WFPC2 (Figures 7p, 7s, 7t) and NICMOS (Figure 7r) cameras aboard the Hubble Space Telescope (HST).

2.4 Asymmetric lens models

As we may expect, symmetric lenses seldom occur in nature; usually the main lens itself is non-symmetric, or some non-symmetric disturbances are induced by the presence of neighbouring masses. In our optical gravitational lens experiment, the effects of a typical non-symmetric (singular) gravitational lens may be simulated by simply tilting the optical lens. The bright focal line along the optical axis in the symmetric configuration (Figure 7a) then becomes a two-dimensional envelope, called a caustic in optics. A section of this caustic is visible as a closed curve having a diamond shape (made of four folds and four cusps) in the pinhole plane (Figures 7c–g). The word 'caustic' in gravitational lensing always refers to this section of the optical two-dimensional caustic (in the symmetric case, the caustic degenerates into a single point, Figure 7a). As a result, the Einstein ring that was observed in the symmetric case (Figure 7h) is now split into four lensed images (Figure 7j). Similar observed examples of gravitational lenses are the multiply-imaged quasars Q2237 + 0305 (known as the Einstein Cross) and H1413 + 117 (the Clover-Leaf; Figure 7q). Such a configuration of four lensed images always arises when the pinhole (observer) is located inside the diamond formed by the caustic.

Figure 7k shows the merging of two of the four images into a single bright image when the pinhole approaches one of the fold caustics (Figure 7d). Some well-known multiply-imaged quasars showing similar image configurations are PG1115 + 080 (the 'triple quasar') and MG0414 + 0534 (the 'dusty lens'). Figure 7r is a direct image of PG1115 + 080 obtained in the near-infrared with the NICMOS camera aboard HST.

Just after the pinhole has passed the fold caustic (Figure 7e), the two merging images totally disappear (Figure 7l). Such a doubly imaged quasar is HE1104-1805 (Figure 7s).

A particularly interesting case occurs when the pinhole (observer) is located very close to one of the cusps (Figure 7f). Three of the four previous lensed images have then merged into one luminous arc, whereas the fourth one appears as a faint counter-image (Figure 7m). Famous luminous arcs have been identified among rich galaxy clusters such as Cl2244−02 and Abell 370 (Figure 7t).

For large sources that cover most of the diamond-shaped caustic (Figure 7g), an almost complete Einstein ring is observed (Figure 7n), although the source, lens and observer are not perfectly aligned and the lens is still tilted. In this last experiment, the increase of the source size has been simulated by significantly enlarging the pinhole radius. Several radio Einstein rings of this sort have been discovered, among them MG1131 + 0456 (Figure 7u) and MG1654 + 1346.

As previously stated, all image configurations illustrated in Figures 7h–n are found among observed gravitational lens systems (Figures 7o–u). If our optical lens had been non-singular in the centre, we would have seen an additional image formed in the central part of the lens. Burke (1981) has in fact demonstrated that a non-singular, transparent (symmetric or asymmetric) lens always produces an odd number of images for a given point source (except when located on a caustic). However, for a singular lens, as in our optical lens experiment, one may obtain an even number of images.

For the known gravitational lenses with an even number of observed images, it may well be that a black hole resides at the centre of the lens. The presence of a compact core could also account for the 'missing' image since the very faint image expected to be seen close to, or through, the core would then be well below the detection limits that are currently achievable. Whereas the formation of multiply-lensed images by non-symmetric gravitational lenses is mathematically well understood (see the pioneering work by Bourassa *et al.* (1973)), the theoretical developments turn out to be very technical and rather tedious. The main results are found to be in very good agreement with the observations simulated in the above optical gravitational lens experiment.

2.5 Propagation of a distorted wavefront and time delays

It is interesting that gravitational fields in the Universe deflect electromagnetic waves in a way similar to the bending of light rays in the Earth's atmosphere. Indeed, as mentioned in

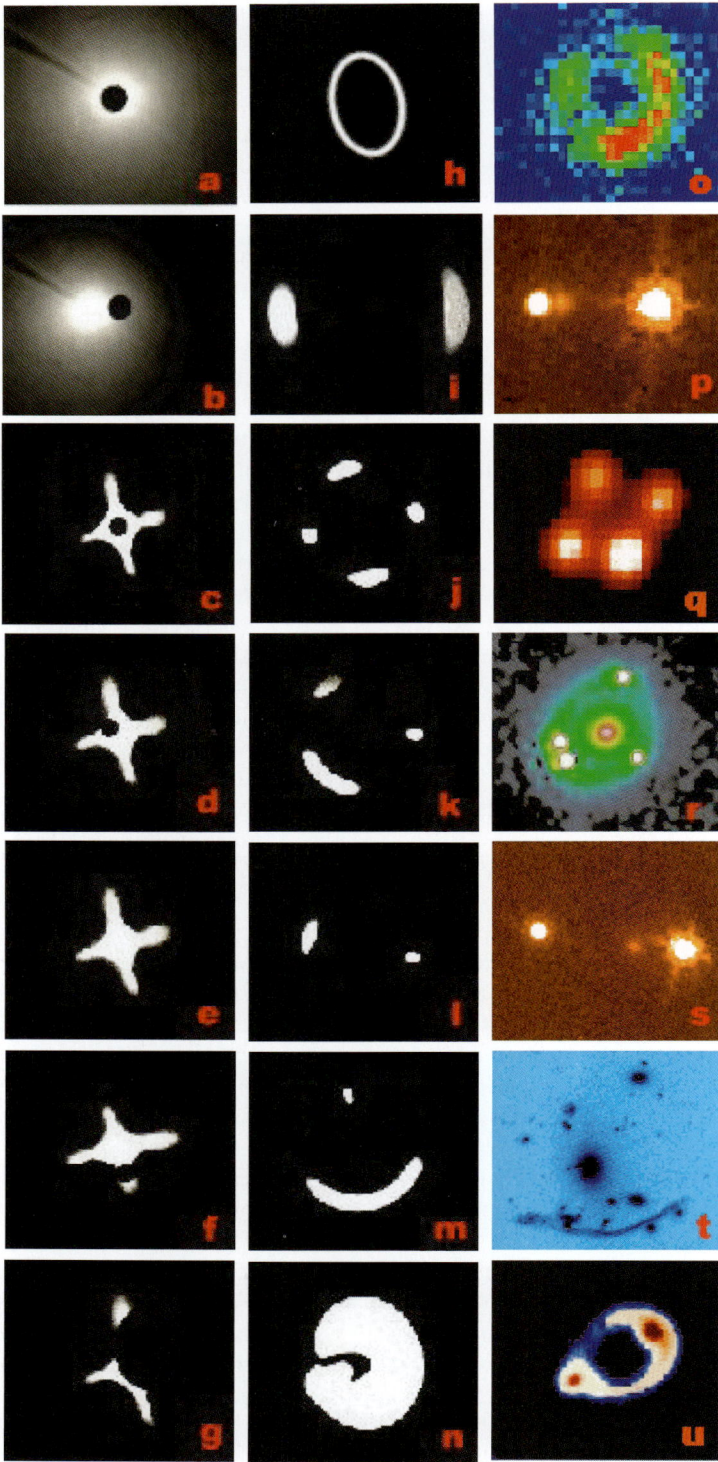

Figure 7 Images in the left-hand column represent the light from a distant source that is redistributed over the pinhole screen (in the experiment shown in Figure 6) by a symmetric (a, b) or a non-symmetric (c–g) optical lens and for various positions of the pinhole (observer). The central column (h–n) illustrates the corresponding lensed images projected onto the large screen located behind the pinhole screen, and the right-hand column (o–u) displays known examples of multiply-imaged sources (0047−28078, 1009−0252, H1413+117, PG1115+080, HE1104−1805, Abell 370 and MG1131+0456). Images (p), (r), (s) and (t) were obtained by the Hubble Space Telescope, and the others by ground-based facilities (ESO and VLA/NRAO). Courtesy of the European Southern Observatory (ESO), the Space Telescope Science Institute (STScI) operated for NASA by AURA, and the Very Large Array (National Radio Astronomy Observatory).

Section 1, we may associate with any point in space an equivalent refractive index n that is entirely defined by the Newtonian potential U of a given deflector (eqn (2)). For the particular case of a deflector characterized by a density distribution of the type $\rho(r) = \rho_0 \exp(-r/r_c)$, r_c being a scaling factor, Figure 8 illustrates the propagation of light rays and of successive wavefronts from a distant source. Close to the source the wavefronts are spherical, but as they approach the lens they become deformed by curvature effects and time retardation ($v = c/n$) and can, in many

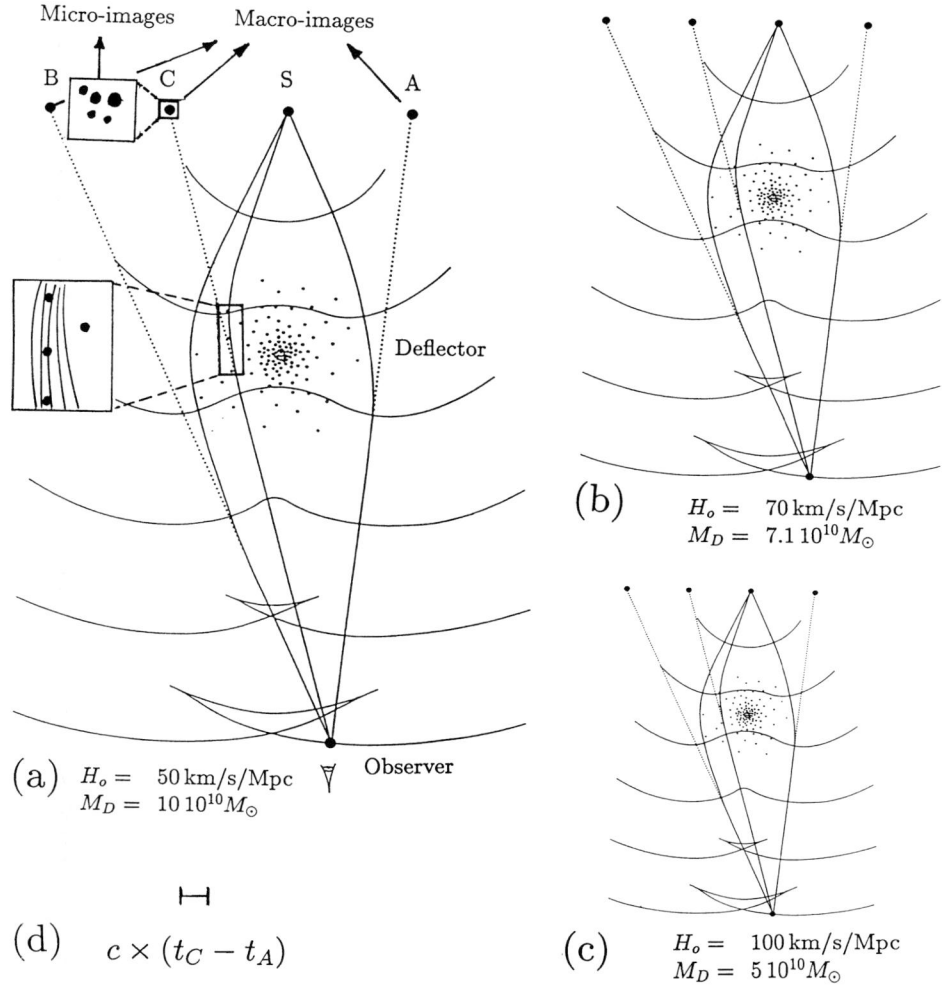

Figure 8 (a) Propagation of light rays and of corresponding wavefronts from a distant source near a deflector characterized by an exponential volume density distribution (see the text). For different values (50, 70 and 100 km s^{-1} Mpc^{-1}) of the Hubble constant H_0 and corresponding values for the deflecting mass M (such that $M H_0 = $ const., see eqn (6) and the note just below), (a), (b) and (c) represent the scaled versions of a gravitational lens system as seen by an observer under a similar angular configuration (all angles measured by the observer on the plane of the sky between the macro-lensed images and the deflector are the same). Given the length of the segment $c (t_C - t_A)$ represented in (d), where c is the velocity of light and $t_C - t_A$ is the observed time delay between the lensed images C and A, it is straightforward to conclude that only one of the proposed models (namely model (b)) is compatible with the measured delay $t_C - t_A$ between the arrival times of the corresponding wavefronts at the observer (compare the length of the segment in (d) with the path differences in (a), (b) and (c) between the wavefronts for images A and C). This is the principle of the method first proposed by Refsdal (1964b) to determine the value of H_0 from the measurement of the time delay. In (a) we have also represented the splitting of macro-image C into several micro-images, with typical angular separations of approximately one micro-arc second, due to gravitational lensing by individual stars located along its line of sight.

cases, self-intersect. This self-intersection is directly related to the formation of multiple images since each crossing of the wavefront at the observer corresponds to the formation of an image in the direction perpendicular to the local wavefront. Note in Figure 8 the complex shape of the distorted wavefront after crossing the deflector plane. In this case, up to three successive parts of a single-folded wavefront (i.e. up to three distinct light rays) can reach an observer, leading to the formation of up to three distinct lensed images. Of course, due to the different geometric lengths of these light trajectories and to the apparently varying speed of light across the deflector, these different parts of a same wavefront reach the observer with time delays. Refsdal (1964b, 1966a) has shown that if it is possible to model the observations of a gravitational lens system successfully, it is straightforward to derive the distance scale of the Universe, or equivalently the value of the Hubble constant H_0, from the measurement of the time delay between a single photometric event detected at two different epochs in the light-curves of two lensed images (Figures 8a, b and c; see Section 3.4.2 for more details).

2.6 Gravitational lensing optical depth

If some luminous sources are *a priori* more likely to be lensed than others, then seeking them out should reveal new cosmic mirages.

Let us assume for simplicity a non-evolving population of lenses with mass M, whose co-moving space density is thus a constant, $n_0(M)$. The optical depth τ for lensing a source located at redshift z_s is then

$$\tau(z_s, M) = V_{\text{eff}}(z_s, M)\, n_0(M) \quad (13)$$

where V_{eff} represents the co-moving volume in which the lens must be located in order to produce an *observable* gravitational mirage (Nemiroff 1989). 'Observable' here means that the angular separation and the magnitude difference between the multiply-lensed images can be resolved with the instrument used for the observations. V_{eff} simply results from the integration of the effective cross-section Σ_{eff} along the line of sight:

$$V_{\text{eff}}(z_s, M) = \int_0^{z_s} (1+z)^3 \Sigma_{\text{eff}}(z, M; z_s, p_i) \frac{cdt}{dz} dz \quad (14)$$

where p_i are the lens parameters.

For Friedmann–Lemaître–Robertson–Walker cosmologies, the line-of-sight derivative element cdt/dz is given as a function of the present matter density Ω_0 and the cosmological constant λ_0 by the following relation (e.g. Kayser *et al.* 1997):

$$\frac{cdt}{dz} = \frac{c}{H_0\sqrt{(1+z)}}$$

$$\times \frac{1}{\sqrt{[(1+z)^3\Omega_0 - (1+z)^2(\Omega_0 + \lambda_0 - 1) + \lambda_0]}} \quad (15)$$

For a point-mass lens, we have seen in Section 2.3 that a typical cross-section for significant gravitational lensing is defined as $\Sigma_{\text{eff}} = \pi\theta_E^2$. Such a cross-section may be generalized for any other type of deflector (e.g. a galaxy lens model). Rigorously, it should of course take into account the limited angular resolution and dynamic range of the instrument used to perform the observations. Given eqns (6) and (15), we see that the expression of the optical depth defined in eqn (13) is independent of the Hubble constant H_0.

The total flux coming from multiply-lensed images is always amplified by a certain factor μ_T with respect to the source flux (Section 2.3). Therefore the probability that a source with redshift z_s and apparent magnitude b_s is a lensed object is not simply $\tau(z_s, M)$, but is given by

$$P(z_s, b_s) = \frac{N_L(z_s, b_s)}{N_s(z_s, b_s)}$$

$$= \frac{\tau(z_s)N_s(z_s, b_s + 2.5\log\mu_T)}{N_s(z_s, b_s)}$$

$$= \tau(z_s)\text{Bias}(z_s, b_s) \quad (16)$$

where the number–magnitude relations $N_L(z_s, b_s)$ and $N_s(z_s, b_s)$ represent respectively the number of multiply-imaged sources and unlensed sources at redshift z_s and blue magnitude b_s per square degree and per magnitude interval; the last equality defines the so-called amplification bias.

Faint sources are usually more numerous than bright ones, so the amplification bias is quite often larger than 1. Moreover, if the number counts of sources increase faster at bright magnitudes, the corresponding proportion of mirages becomes larger. Therefore, eqns (13) and (16) show that the most likely lensed sources are (1) distant (V_{eff} is maximized) and (2) bright objects belonging to a population with a steep number–magnitude relation at the observed wavelength (the amplification bias is then high; Turner 1980).

Highly luminous quasars (HLQs; typically $M_V \leqslant -27.5$ for $H_0 = 50\,\text{km}\,\text{s}^{-1}\,\text{Mpc}^{-1}$) have long been recognized as excellent targets to search for new gravitational lenses because they are both distant and bright sources (Turner *et al.* 1984, Surdej *et al.* 1988). Their number–magnitude relation has a steep bright end with a break around the apparent magnitude $b = 19.2$ (Boyle *et al.* 1988, Hartwick and Schade 1990). They have been selected for search for gravitational lenses by several teams; successful results are presented in Section 3.

3 ASTROPHYSICAL AND COSMOLOGICAL APPLICATIONS

3.1 Introduction

In this section we show how it has been possible to derive unique information on the deflecting objects, on the sources themselves and on the Universe as a whole from the observations and studies of distorted and/or multiple images of distant lensed objects. We look at some of the most important astrophysical and cosmological applications of gravitational lensing with respect to the distance of the lens, going from the Solar System to distant quasars and the very edge of the Universe. During this tour, we shall discover how the disturbing effects of lensing can be transformed into tools to improve our knowledge of the Universe. We also indicate how space observations have helped to establish some of these results.

3.2 Light deflection in the Solar System

Starting our trip from Earth, the closest lens is actually the Earth's atmosphere itself. The refraction of light rays coming from space in the denser layers of the atmosphere slightly changes the apparent positions of stars on the sky. In order to precisely point a telescope towards a given star, corrections for refraction are required. This phenomenon is more pronounced when a star is close to the horizon: a particular example is the deformation of the Sun's image at sunset. Sometimes, multiple images of the Sun can be seen, resulting in the formation of a mirage. Refraction must also be taken into account when the length of the light path in the atmosphere must be precisely known, for example in the context of atmospheric chemistry, either performed from the ground or from spacecraft.

As seen from Earth, gravitational light deflection in the Solar System does not lead to the formation of multiple images of background sources because the apparent size of the deflectors (the Sun or the planets) is larger than their angular Einstein radius (Section 2.2). However, the measurement of the change of the apparent positions of background sources located near the line of sight to the Sun or to Jupiter are very important from the point of view of fundamental physics. We have already seen that the shift in the positions of stars located close to the limb of the Sun observed by Eddington and his collaborators during the total eclipse of 1919 secured the second experimental confirmation of the theory of general relativity. At that time, the relatively poor accuracy (20–30% uncertainty) of the observations was nevertheless just sufficient to distinguish between Newton's prediction and that of Einstein. Thanks to radio interferometric observations of quasi-stellar sources carried out with the VLA, VLBI or VLBA (Fomalont and Sramek 1975a,b; Robertson et al. 1991; Lebach et al. 1995), it has been possible to measure the gravitational deflection of radio waves by the Sun with uncertainties much less than 1%. Because of these high accuracies, selecting among the alternative theories of gravitation has now become feasible (Robertson et al. 1991). General relativity still seems to be preferred. The measurements of the bending of light rays near the Sun at both optical and radio wavelengths do actually constitute the first multi-waveband observations of 'gravitational lensing', verifying at the same time the expected 'achromaticity' of the phenomenon (eqn (1)).

The apparent displacement of the radio source P0201 + 113 by 300 micro-arcseconds due to a close alignment with Jupiter has also been qualitatively measured (Treuhaft and Lowe 1991). Finally, let us note that the light coming from a star located at 90° from the Sun still undergoes a deflection of 4 milli-arcseconds. Therefore the revolution of the Earth around the Sun induces a tiny, annual astrometric shift for all stars in the sky. This has been confirmed with the Hipparcos satellite (Froeschlé et al. 1997). The measurements by the ESA Cornerstone mission GAIA (Global Astrometric Interferometer for Astrophysics, now the Galactic Census Project) should be much more sensitive to the influence of the massive planets Jupiter and Saturn, and will improve the constraint on the alternative theories of gravity by several orders of magnitude (Section 4.1). For more details, see Chapter 15.

3.3 Microlensing in the Milky Way

In order to produce multiple images of a distant star with an angular separation larger than its physical diameter, a deflector like the Sun should be located at a distance from Earth larger than typically 550 AU, or 0.01 light years. This is more than 10 times the size of the Solar System (as defined by Neptune's orbit). So, we are now in our Galaxy. One of the nearest stars, α Centauri, is at a distance of approximately 4.3 light years. If it were used as a deflector, the expected angular separation between the lensed images of background stars would be only about 0.2 arcseconds, and the misalignment should be no worse than about 0.1 arcseconds, otherwise the secondary lensed image will be too faint to be detectable (Section 2.3, eqn (11)), especially in the glare of the foreground star. And the more distant the deflecting star, the smaller the angular separation between the lensed images of background sources, which then falls below the resolution of conventional instruments (eqn (6)). We then talk of *microlensing*. (This is probably why Einstein was so pessimistic about the possibility of observing gravitational lensing among stars.)

However, in an important paper Paczyński (1986) put forward a point concerning gravitational microlensing in the Milky Way: it can be used either to detect or to rule out dark compact objects as a major constituent of the halo of the Galaxy. In 1991, Griest coined the acronym MACHOs for these massive astrophysical compact halo objects. Indeed, everything moves in the Galaxy, so the relative alignments between background stars and potential unseen deflectors continuously change. When the alignment is good, lensed images cannot be resolved but the increase in the total flux can be recorded (eqn (12)). The probability of having an amplification $\mu_T \geqslant 1.34$ (or the optical depth for microlensing; see Section 2.6) is very small (10^{-7}) but this can be compensated for by observing a huge number of stars, located for example in the Magellanic Clouds. The technology was ready for handling such a huge amount of data. Several independent teams (EROS, Expérience de Recherche d'Objets Sombres; MACHO; OGLE, Optical Gravitational Lensing Experiment; and so on) started the ambitious programme of monitoring each night millions of stars located in the Magellanic Clouds, or later towards the galactic centre, waiting for microlensing events – achromatic, symmetric and non-repeating flux variations. The first detections were reported by different groups in 1993, and more than 500 events have been detected so far, mainly towards the galactic bulge (see the review by Paczyński (1996) for further details on the analysis and the first results). Interpreted in the context of microlensing by dark compact objects in the halo of the Milky Way, the current observations of the Large Magellanic Cloud (LMC) imply a MACHO halo fraction of about 20%. The most likely MACHO mass is between $0.15 M_\odot$ and $0.9 M_\odot$; the mass range 10^{-7}–$10^{-2} M_\odot$ is excluded at the 95% confidence level (Aubourg and Palanque-Delabrouille 1999, Alcock et al. 2000).

However, the physical nature of MACHOs is not clear yet. For example, if they are white dwarfs, a strong chemical enrichment of the halo should have been observed. But some of the observed events could be caused by lenses located in the LMC itself (self-lensing): degeneracies between the parameters make it impossible to determine independently the distance of a single lens from the analysis of the light-curve. One solution, discussed by Refsdal (1966b), is to take advantage of the parallax effect: basically, two observers separated by a distance d in the observer plane will see the same light-curve but with a time lag $t = d/v$, where v is the apparent speed of the caustic projected in the observer plane along the line connecting them. (With three observers, the unknown orientation of the caustic can be retrieved.) The parallax induced by the revolution of the Earth around the Sun distorts long-duration light-curves. For short ones, simultaneous observations with two spacecraft located a few astronomical units apart should ideally be performed.

Binary lenses produce different kinds of light-curves. This opens the possibility of detecting planets orbiting the star-lens. The light-curves need to be extremely well sampled. They are usually obtained by means of sustained photometric monitoring campaigns mounted when alerts of ongoing microlensing events are made (e.g. the PLANET – Probing Lensing Anomalies NETwork-collaboration, Albrow et al. 1998).

3.4 Multiply-imaged quasi-stellar objects

Extragalactic gravitational lensing was first mentioned by Zwicky in 1937. He realized that, due to their large mass, external galaxies might lens more distant ones into several *observable* images. However, such mirages are still difficult to identify because galaxies are faint, diffuse and extended objects whose apparent size can be large compared with the Einstein ring of the lens.

In 1963 the first quasar was discovered by Schmidt. Quasars are the beacons of the Universe. Thanks to their huge luminosity, they can be seen at very large distances, so the probability of finding a foreground lensing galaxy close to the line of sight is larger than for normal galaxies. Moreover, multiply-lensed images of quasars can be more easily identified because the source is bright and point-like (Peterson 1997). In 1979, the first cosmic mirage was found by chance as a double image of a radio-loud quasar (Walsh et al. 1979), with identical spectral properties from UV (measured with the International Ultraviolet Explorer satellite by Gondhalekar and Wilson in 1980) to radio through optical and infrared wavelengths. This discovery prompted new interest in gravitational lensing. A large number of applications became possible and motivated several independent teams to find more and more gravitational lenses among selected high-luminosity quasars. HLQs are indeed more likely to be lensed because they are distant and apparently bright: their brightness could be partially due to light amplification by gravitational lensing (the so-called amplification bias; see Section 2.6). However, the probability of being lensed is only about 1%, so large samples have to be observed with ground-based optical or radio telescopes.

Due to the arcsecond scale of the angular separation between the lensed images of QSOs, the HST is often needed to confirm the nature of lens candidates. This is illustrated in the case of the doubly imaged QSO J03.13 A and B ($z_s = 2.55$, $\Delta\theta = 0.84''$), which was discovered at the European Southern Observatory, La Silla, Chile (Figure 9a) to be a good lens candidate, then confirmed with HST to consist of two point-like images (Figure 9b) having identical spectra (Figure 9c).

By the end of the century more than 50 multiply-imaged quasars had been identified (see Claeskens and Surdej (2001) for a complete list). Figures 7 and 9b illustrate several representative ones. In the following we present some of

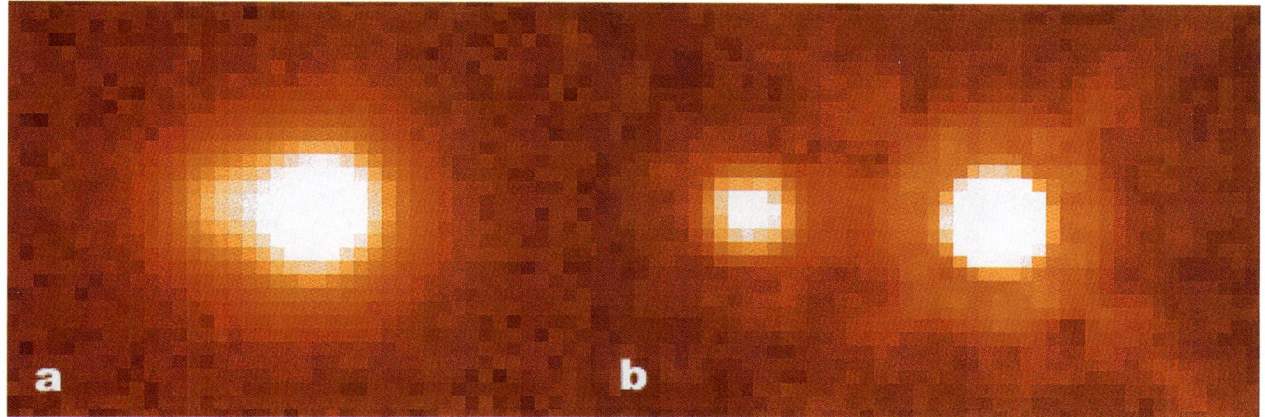

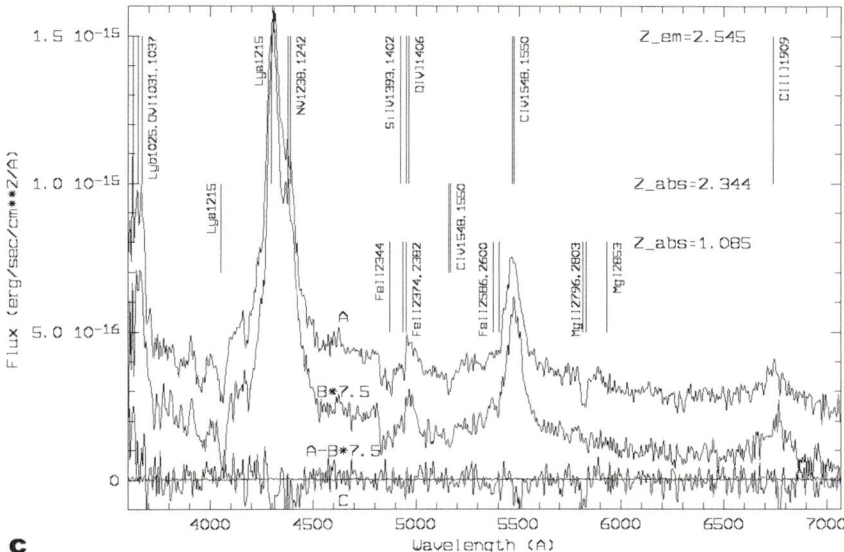

Figure 9 (a) Identification of a gravitational lens candidate based on ground-based direct CCD imagery. On this CCD frame obtained with the New Technology Telescope (NTT) at La Silla (Chile), the image of the QSO J03.13 looks somewhat elongated. The seeing was about 0.7″. Claeskens *et al.* (1996) suspected that J03.13 was composed of two nearby point-like images. (b) Subsequently, the WFPC2 camera abord the HST was used to confirm the image structure of J03.13. This WFPC2 R CCD frame clearly reveals a double image, with an angular separation of 0.84″. (c) For such a small angular separation, it is impossible to obtain individual spectra of the resolved components from the ground. This is an easy task, however, for the HST. Spectra obtained with the Faint Object Spectrograph instrument clearly show that the fingerprints of J03.13 A and B cannot be distinguished from each other, except for a few narrow absorption lines due to distinct intervening clouds located along the two lines of sight. These two spectra show that the quasar is at a redshift $z_s = 2.55$ (Surdej *et al.* 1997). (Courtesy of the European Southern Observatory (ESO) and the Space Telescope Science Institute (STScI) operated for NASA by AURA.)

the most interesting applications and results obtained from gravitational lensing effects in quasar samples.

3.4.1 The lensing galaxies

Multiply-imaged QSOs provide general constraints on the mass distribution of the lensing galaxies. For example, if the value of the Hubble constant H_0 and the lens and source redshifts are known, the total mass can be derived from the angular separation between the QSO lensed images, since $\Delta\theta \sim 2\theta_E$ (eqn (6)). This is the simplest and most direct astrophysical application of gravitational lensing.

If a time delay is also measured, the lensing mass can be obtained independently of the cosmological parameters H_0,

Ω_0, λ_0 and the source redshift (Borgeest 1986, Schneider et al. 1992). The inferred values for the mass are typical of elliptical galaxies (see Claeskens and Surdej (2001) for a summary of mass and mass-to-light ratio estimates).

Furthermore, the absence of observed cosmic mirages with three images was soon recognized as the signature of nearly singular potentials (Wallington and Narayan 1993; Kochanek 1996), in agreement with HST observations of local elliptical galaxies by Gebhardt et al. (1996). Quadruply-imaged quasars can be produced only by asymmetric lenses (Section 2.4; the asymmetry being intrinsic to the lens or due to the perturbing shear of objects located close to the line of sight of the main lens). Generally, the lensed images are better reproduced by lensing isothermal dark matter halos rather than by constant M/L lens models (Maoz and Rix 1993; Kochanek 1996). However, the relative positions and flux ratios of the point-like QSO images yield few constraints on the full mass distribution of the lens. This results in degenerated lens models, as first pointed out by Falco et al. (1985) and Gorenstein et al. (1988), who noted that a given configuration of lensed images can be reproduced with a lighter lensing galaxy inside a galaxy cluster. Thus, additional constraints must be obtained, either from direct observations of the lens environment or from the detection of lensed extended structures, such as the QSO's host galaxy. In the latter case, VLBI or HST high angular resolution is mandatory.

Since the shape and the size of such extended structures may be wavelength dependent, they provide interesting constraints on the mass distribution of the distant deflector and also on the physical modelling of the chromatic structure of the source itself. Existing multi-wavelength observations of multiply-imaged sources, resolved at arcsecond and sub-arcsecond angular scales, have already been successfully inverted. In several cases it has been possible from such analyses to retrieve the mass distribution of the lensing galaxy as well as the multi-wavelength shape of the source (see Wallington et al. (1996) for a discussion of the numerical methods, and Claeskens and Surdej (2001) for a general review of the observations). In order to refine such astrophysical applications, it is very important in the future that not only direct imaging but also imaging spectroscopy, at high angular resolution and with a good dynamic range, become feasible at many different wavelengths, both from space (far-UV, X-ray, γ-ray, near- and mid-infrared, etc.) and from the ground (visible, radio, etc.). An imaging spectroscopy device has been proposed for the Next Generation Space Telescope (NGST). Such an instrument should make it possible to observe very compact gravitational lenses at relatively high angular resolution in the near- and mid-infrared.

The HST plays a crucial role in detecting the main lenses, which are faint, extended high-redshift objects located between bright, small-separation (~1 arcsecond) images of the background quasars. Many lens detections and magnitude measurements have come from the CfA–Arizona Space Telescope LEns Survey (CASTLES), an ongoing HST direct-imaging programme in the H, I and V bands for all the known gravitational lenses as well as published candidates (Muñoz et al. 1999a; Figure 7). So far, spectroscopic redshifts of the lenses have been obtained for 52% of the known multiply-imaged QSOs. The median redshift is 0.47, which is exactly where the lenses are the most efficient since the critical surface mass density is expected to be minimum around $z = 0.5$ for high-redshift sources (eqn (7)).

Finally, multiply-imaged quasars need to be investigated carefully in order to determine the extinction law in the lensing galaxies. Indeed, when the lines of sight corresponding to the different images probe different dust optical depths in the lens, a differential reddening is observed between the lensed images. This is the case with, for example, Q2237+0305 (Nadeau et al. 1991), MG0414+0534 (Angonin-Willaime et al. 1999) and RXJ0911.4+0551 (Burud et al. 1998). Jean and Surdej (1998) have refined this technique to estimate the redshift of the lens, whatever its visibility. Until now only galactic parametric curves have been fitted from photometric data, but high-quality spectrophotometry should allow one to determine directly the extinction law at high redshift and compare it with that observed in the Milky Way, in the Magellanic Clouds and in a few local galaxies.

3.4.2 The cosmological parameters

Extragalactic gravitational lensing provides us with an experiment whose optical bench has a size comparable to that of the observable Universe. Since this size is directly related to the values of the cosmological parameters, it is not surprising that the latter can be constrained from gravitational lensing observations. We refer the reader to Chapter 17 for an extended review of the general determination of the cosmological parameters.

The Hubble constant, H_0

The Hubble constant fixes the actual expansion rate of the Universe. Its determination from the observation of a multiply-imaged extragalactic source is a very original application of gravitational lensing, proposed in 1964 by Refsdal, and is based on the observation of a time delay between flux variations in the lensed images. The major interest of this method is that it avoids the traditional issue of flux calibrations relying on the difficult-to-prove existence of standard candles, like Type Ia supernovae.

Since the angular separation between multiply-lensed images $\Delta\theta \sim 2\theta_E$ is proportional to $\sqrt{(MH_0)}$ (eqn (6) and the note just below it), a larger deflecting mass combined with a lower value of H_0 produces the same angular separation. However, a lower value of H_0 corresponds to larger extragalactic distances and therefore to longer time delays $\Delta t_{i,j}$ between flux variations of two lensed images i and j (Figure 8). More generally, H_0 is inversely proportional to the time delay (e.g. Claeskens and Surdej 2001):

$$\Delta t_{i,j} = H_0^{-1}(1 + z_d) D(z_d, z_s, \Omega_0, \lambda_0) f(\boldsymbol{\theta}_i, \boldsymbol{\theta}_j, \text{lens model}) \quad (17)$$

where z_d is the redshift of the deflector, $D = D_{od}D_{os}/D_{ds}$ and f is a lens-model-dependent function of the observed image positions $\boldsymbol{\theta}_i$ and $\boldsymbol{\theta}_j$.

Therefore, a successful determination of H_0 relies on:

- intensive photometric monitoring of a *variable* multiply-imaged QSO in order to determine precisely $\Delta t_{i,j}$;
- accurate observations of the vector positions $\boldsymbol{\theta}_i$ and $\boldsymbol{\theta}_j$ of the lensed images with respect to the deflector (for example, with HST);
- good knowledge of the lens redshift z_d (the uncertainty due to the lack of accurate values of the other cosmological parameters is negligible as long as the lens redshift is moderate, $z_d \sim 0.5$); and
- a very good estimate of the deflector mass distribution.

As already mentioned in Section 3.4.1, the last item represents the most difficult issue, and complementary observations of the lens and/or the detection of lensed extended structures of the source are required.

Despite the fact that the lens of Q0957 + 561 is complex (denoting the presence of a cluster), it is still probably the best-constrained system, and an accurate value of $\Delta t = 417 \pm 3$ days (Kundić et al. 1997; see also Vanderriest et al. 1989) has been observed for the time delay. Depending on the actual lens model and constraint set, the results currently range, with errors at 2σ, between $H_0 = 77^{+29}_{-24}$ km s^{-1} Mpc^{-1} (Bernstein and Fischer 1999) and $H_0 = 55^{+15}_{-14}$ km s^{-1} Mpc^{-1} (Chae 1999). Other estimates of H_0 are summarized in Claeskens and Surdej (2001).

The cosmological constant, λ_0

The fate of the Universe is intimately related to the values of the cosmological density Ω_0 and the cosmological constant λ_0. Their determination constitutes one of the most important challenges in present-day observational cosmology.

Since the spatial volume of a sphere extending to a given redshift increases for lower values of Ω_0 or for larger values of λ_0, the resulting number of multiply-imaged QSOs in a large sample of HLQs also increases for such values of the cosmological parameters, assuming the co-moving density of galaxies to be constant; this property was first pointed out by Turner (1990) and Fukugita et al. (1990). The influence of λ_0 is much stronger than that of Ω_0: whatever a reasonable value of Ω_0, values of λ_0 close to 1 lead to the prediction of detecting at least twice as many lenses than are actually observed in optical samples. Gravitational lensing statistical studies thus place a natural *upper* limit on the value of the cosmological constant, but no significant constraint on Ω_0.

A statistical analysis based on the concept of lensing optical depth (Section 2.6) in an optical sample of 862 HLQs led Kochanek (1996) to the conclusion that $\lambda_0 < 0.66$ at a 95% confidence level (CL), and Claeskens (1999) derived $\lambda_0 < 0.5$ at the same CL from the number of lenses in an optical sample of 1164 HLQs. Similar analyses have also been performed at radio wavelengths with the Jodrell Bank VLA Astrometric Survey (JVAS) of 2500 flat-radio spectrum sources by Falco et al. (1998) and by Helbig et al. (1999), and with the first part of the Cosmic Lens All Sky Survey (CLASS) by Cooray (1999). They find, respectively, for flat cosmological models, $\lambda_0 < 0.73$, $\lambda_0 < 0.84$ and $\lambda_0 < 0.79$ at 95% CL. Constraints from radio surveys are more compatible with the estimates based on the analysis of high-redshift supernovae (Riess 1998; Perlmutter et al. 1999).

Extinguished secondary lensed images or extremely reddened QSOs could remain undetected in optical samples, resulting in incompleteness in the parent QSO sample and affected statistics. Radio surveys are free from such observational dust-obscuration biases, but the distribution of radio sources as a function of their flux and redshift, intervening in the computation of the amplification bias, is not well known. It must be better studied before further improvements on the statistics can be made.

The cosmological density of dark compact objects

Press and Gunn (1973) noted that counting gravitational lensing events in a sample of distant sources may help in constraining the contribution (Ω_L) to the critical density of the Universe due to a putative cosmological population of dark compact objects with a mass M_L (Section 2.6). Ground-based optical telescopes can reveal the multiple images produced by dark compact objects with a mass M_L between $10^{10.5}$ and $10^{15} M_\odot$. Because of its higher angular resolution, the HST is sensitive to masses down to $10^9 M_\odot$. Radio observations with the VLBI allow one to reach the level of $M_L = 10^6 M_\odot$ (Wilkinson et al. 2001). Below that mass range, the lensed images cannot be resolved and constraints come only from the observed statistical variability of the QSO flux or of the equivalent width of their emission lines due to (micro)lensing by the putative dark compact object population (Canizares 1982, Schneider 1993, Dalcanton et al. 1994, Nemiroff et al. 2001). The available constraints are summarized in Figure 10, and the conclusion is that there is no cosmological population of dark compact objects with mass M_L in the range 10^{-3}–$10^{13} M_\odot$ capable of closing the Universe.

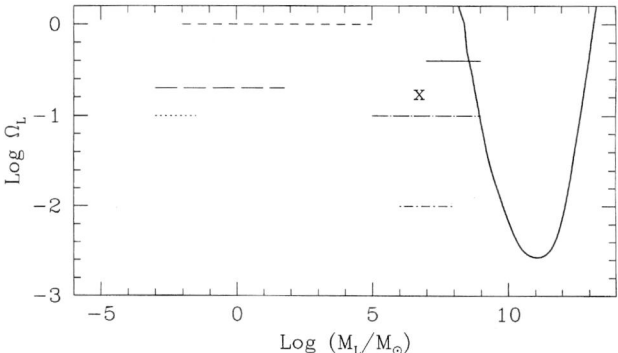

Figure 10 Current constraints from gravitational lens studies on the cosmological density of dark compact objects (in units of the critical density Ω_c) as a function of their mass. Full curve: Claeskens (1999), 99.7% confidence level (CL); straight full line: Kassiola et al. (1991), 99.7% CL; short-dashed line: Canizares (1982); dotted line: Schneider (1993), ~97% CL; long-dashed line: Dalcanton et al. (1994); X: Marani et al. (1999), 90% CL; dot–short-dashed line: Wilkinson et al. (2001), 95% CL; dot–long-dashed line: Nemiroff et al. (2001; $\sim 3\sigma$).

3.4.3 Size of intergalactic clouds

Cold intergalactic clouds, which may be connected with galaxy formation (e.g. Wolfe 1988), are revealed by their Lyman α and/or metallic absorption lines (C IV, Mg II, ...) seen in the spectra of distant quasars. If such a distant QSO is multiply-imaged by gravitational lensing, the different lines of sight probe the clouds at various impact parameters. The large number of coincident absorption lines found in the spectra of the multiply-imaged QSOs UM673 A and B (Smette et al. 1992) and HE1104-1805 A and B (Smette et al. 1995) implies that most of the intervening clouds extend over both lines of sight. A lower limit on their size was derived as 6 and 50 h^{-1} kpc respectively ($h = H_0/100$), in agreement with even larger estimates of about $300h^{-1}$ kpc found from spectroscopic studies of QSO pairs (d'Odorico et al. 1998).

3.4.4 Microlensing in multiply-imaged QSOs

We saw in Section 3.3 that the optical depth for microlensing of a star in the Milky Way by another stellar object is very small. In contrast, since with a multiply-imaged QSO the different lines of sight pass through the halo of the lens, the optical depth for microlensing is much higher, provided the lensing galaxy is made of compact objects (Figure 8a). This was already recognized by Gott (1981). A stellar microlens only quantitatively modifies the flux coming from sources *smaller* than their Einstein radius or smaller than the size of the caustic in the source plane. Thus microlensing preferentially amplifies the continuum emission region, which is the most central part of the QSO ($\sim 10^{-3}$ pc, corresponding to an angular size of $<10^{-6}{''}$) while the broad emission-line flux (emitting region >1 pc) remains unchanged. Therefore microlensing in multiply-imaged QSOs allows one not only to place limits on MACHOs in external galaxies, but also to constrain the size of the continuum-emitting region of the quasars, which is not directly resolvable with current instruments (e.g. Schmidt and Wambsganss 1998). Recall that QSO structures on larger scales can be studied more easily owing to the 'telescope effect' of the whole lensing galaxy on the macro-images (Section 3.4.1).

3.5 Gravitational lensing in galaxy clusters

Giant luminous arcs (GLAs) in galaxy clusters represent the most spectacular effect of gravitational lensing (Figure 7t). They are produced by the merging of two or three lensed images of an extended background galaxy located close to the caustic (Figure 7). However, even after the discovery of the first multiply-imaged QSO in 1979, astronomers were not prepared to recognize the nature of those arcs.

It is worth quoting some comments made soon after the first observations of the arc in the galaxy cluster Abell 370 ($z = 0.373$; Figure 7t). Hoag (1981) reported in the *Bulletin of the American Astronomical Society*:

> Prime focus photographs taken with the Mayall 4-meter telescope at KPNO (Kitt Peak National Observatory) show a filament-like structure in the galaxy cluster Zw 0237.2−0146 (= Abell 370)... If the filament is associated with the cluster, its projected length, judged by the redshift and the diameters of the giant ellipticals, seems to be about 0.2 Mpc. Its width seems not to be resolved.

In 1986, the arc was independently re-observed (together with another one in the cluster Cl2242−02) by Lynds and Petrosian (1986), who wrote:

> We announce the existence of a hitherto unknown type of spatially coherent extragalactic structure having, in the most compelling known examples, the common properties: location in clusters of galaxies, narrow arc-like shape, enormous length, and situation of center of curvature towards both a cD galaxy and the apparent center of gravity of the cluster. The arcs are in excess of 100 kpc in length, have luminosity roughly comparable with those of giant E galaxies, and are distinctly bluer than E galaxies – especially in one case.

In 1987, Soucail and her collaborators reached similar quantitative conclusions, after introducing their observations in the following way (Soucail et al. 1987):

> The data reduction showed a strange ring-like condensation on the R-image very close to a bright elliptical

galaxy near the cluster center. The comparison of several images during the run proved that it was not an artifact.

Many tentative attempts to explain the arcs, in terms of, for example, cooling flows, bow shocks or light echoes, assumed that the arcs belonged to the cluster. Paczyński (1987) first realized that the shape, the orientation, the blue colour and the faint surface brightness of the arcs were a result of gravitational lensing. His prediction was confirmed by Soucail et al. (1988), who showed that the arc in Abell 370 was a strongly distorted view of a faint blue galaxy about twice as distant as the lensing cluster. Thus, after 51 years, Zwicky's prophecy was finally fulfilled!

Besides GLAs, galaxy clusters can also produce multiple images (Figure 11) or single distorted images (called arclets) of background galaxies.

In order to find more arcs, very deep ($B \sim 27$) and high-angular-resolution CCD frames of the richest galaxy clusters have been acquired. The best sample selects clusters with high X-ray luminosity rather than those with many galaxies as seen in the visible domain. For example, 8 arcs have been found by Luppino et al. (1999) among 38 clusters selected from the Einstein Observatory Extended Medium Sensitivity Survey (EMSS). Indeed, it is much more likely that an arc will be observed in a rich cluster than a multiply-imaged QSO will be found in a QSO sample because the population of faint blue galaxies discovered by Tyson (1988) yields a very dense grid of background sources. Furthermore, the angular scale of the arcs is 10 to 100 times larger than that of lensed QSO images.

Astronomers soon realized that gravitational arcs provide an excellent means of probing the central part of the galaxy clusters, including their dark matter content. Once the redshift of an arc is known, the central mass inside the Einstein radius can be approximated from eqn (6), with no assumption about

Figure 11 Multiple images of a well-resolved blue background galaxy lensed by the galaxy cluster Cl0024 + 1654. This is a true-color image built from CCD frames obtained with the HST (Colley et al. 1996). (Courtesy of the Space Telescope Science Institute (STScI) operated for NASA by AURA and the Space Telescope European Coordinating Facility (ST-ECF).)

hydrostatic equilibrium of the gas. Typical masses and mass-to-light ratios derived from gravitational lensing are $M \sim 10^{14} M_\odot$ and $M/L_B > 100$. They confirm that galaxy clusters are dominated by dark matter. The systematic discrepancy between the first values derived for cluster masses and those obtained from X-ray analyses seems to be progressively solved, partially by using more refined and specific lens models (Hattori et al. 2000). Once again, the HST's high angular resolution improved the lens models with new constraints coming from the detection of multiple images of galaxies (Figure 11) or new arcs. Very thin radial spokes have also been detected with the HST in the cluster Abell 370 (Figure 7t, although not visible in this rendition), MS0440+02 and AC118, indicating a central mass density distribution flatter than r^{-2} and a small core radius.

As originally pointed out by Zwicky, lensing clusters may be used as natural telescopes to study faint background sources which would otherwise be out of reach. As mentioned in Section 2, a gravitational lens may help in two different ways: either it amplifies the total flux received from *unresolved* sources, or it improves the effective instrumental resolving power on *extended* objects by increasing their apparent size. The first property has been applied, for example, with the ISOCAM camera aboard the ISO satellite in order to detect mid-infrared sources intrinsically fainter than those directly observable outside the cluster. After correction for the lensing effects, we are left with a significant excess of sources with respect to the predictions, probably because of a higher star formation rate in those distant galaxies (Altieri et al. 1999). An interesting application of the second property has been made by Bunker et al. (2000), who used it in conjunction with the HST's high angular resolution to resolve the stellar populations in a lensed galaxy at $z = 4$; with the same technique, Franx et al. (1997) have discovered a lensed galaxy at $z = 4.92$ behind the cluster Cl1358+62.

The reader interested in gravitational lensing produced by galaxy clusters and its applications will find further details in the reviews by Fort and Mellier (1994) and Hattori et al. (2000).

3.6 Weak lensing

3.6.1 Weak lensing in galaxy clusters

Strong lensing – that is, multiple imaging with the possible formation of giant luminous arcs – is possible only when the misalignment between the source and the lens is typically smaller than the Einstein lens radius. The lens will still slightly distort the shape of background sources at much larger impact parameters: this is the so-called 'weak lensing' regime. The image of a circular source is an ellipse oriented along the local gravitational shear (perpendicular to the direction of the lens centre of mass), with an axis ratio equal to the ratio of the two eigenvalues of the magnification matrix (i.e. it is a function of the derivatives of the deflection angle). It is then easy to understand that, in principle, the mass distribution of the lens can be reconstructed from a dense grid of distorted background sources (Kaiser and Squires 1993, Bartelmann et al. 1996).

However, in practice the observations are difficult. First, a sufficiently dense grid of 40 galaxies per square arcminutes is reached for a limiting B magnitude of about 29 per square arc second (Tyson et al. 1990). Second, the induced ellipticity is only a few per cent, much smaller than the intrinsic ellipticity of galaxies, and a statistical analysis of a large number of (faint) background galaxies over a small area of the sky must be performed to retrieve the local orientation of the shear (atmospheric seeing and instrumental distortion also constitute sources of noise). Third, the redshift of the very faint background population is not known for sources located behind high-redshift lenses.

The first weak lensing effects were observed in the galaxy cluster Abell 370 by Fort et al. (1988), giving another confirmation of the lensed nature of GLAs (Figure 7t). Since then, weak lensing has been studied in more than twenty galaxy clusters, sometimes on scales larger than $1 h^{-1}$ Mpc (as compared with $100 h^{-1}$ kpc for strong lensing). The geometry of the recovered mass distribution is found to be in agreement with that derived from observations in visible light and X-rays, with mass-to-light ratios sometimes as large as $1000 h^{-1}$ (median value of $300 h^{-1}$; Mellier 1999).

3.6.2 Weak lensing in the Universe

The first angular correlation between bright quasars and foreground galaxies on a scale of 10 arc minutes was reported by Fugmann (1990). It has been interpreted by Bartelmann and Schneider (1993) as a consequence of the amplification bias (Section 2.6) due to the large-scale structures possibly associated with galaxies. Since then, other such angular correlations have been confirmed on the basis of independent observations. Moreover, weak shear has been detected around some selected bright quasars (Fort et al. 1996).

Very recently, correlated galaxy ellipticities have been detected by three independent teams in 'empty' large deep fields (where no foreground massive galaxy cluster was known). It is also a consequence of light deflection by the large-scale structures present in the Universe (Van Waerbeke et al. 2000, Wittman et al. 2000, Bacon et al. 2000).

We now reach the end of our trip in the world of lensing, by considering the whole Universe as a thick lens! The general situation is indeed much more complex than in the single-lens plane approximation described in Section 2, which is valid for strong lensing. Light bundles from very distant galaxies and propagating across the Universe are sensitive not only to the expansion and geometry of space-time, but also to the *total* quantity of matter present along their path. This is in contrast with traditional astronomical observations, which probe only the luminous content of the Universe – the visible 10% of the iceberg, as it were. The distortion of light bundles by the large-scale matter distribution present along the trajectory is called the *cosmic shear*. Its signature is the observed statistical correlation of the galaxy ellipticities, while its magnitude and angular scale depend on the mass of the dark matter distribution. It has been detected at a level smaller than 2.5% for angular scales typically less than 30 arcminutes.

The cosmic shear was first theoretically investigated by Blandford *et al.* (1991) and Miralda-Escudé (1991), in the context of the formation of large-scale structures. However, the observations are challenging because of the many systematic effects degrading the signal. More recent progress on the theoretical and observational sides makes it hopeful that this new technique can be applied to map the dark matter in the Universe at intermediate and low redshifts, and to measure the mass power spectrum without any assumption on the connection between dark and luminous matter. Ultimately this should help us in constraining the cosmological parameters, and to identify the intermediate stages between the very faint observed fluctuations of the cosmological microwave background and the presently observed structure of the Universe. Observations need to be extended to a lower shear level, corresponding to a larger angular scale.

Interested readers will find a more extended discussion of weak lensing in the Universe in the review by Mellier (1999).

4 FUTURE SPACE MISSIONS AND THEIR EXPECTED CONTRIBUTIONS TO THE FIELD

Taking advantage of the amplification bias in gravitational lens surveys (Section 2.6) and given the good (sub-arcsecond) angular resolution and high sensitivities of existing large optical telescopes and radio interferometers, we have seen in Section 3 that most of systematic observational searches for lenses had been carried out at optical, near-infrared and radio wavelengths. These brightest multiply-imaged quasars distributed all over the sky offer unique tools for decisive astrophysical and cosmological applications, including constraints on the cosmological parameters Ω_0 and λ_0.

We naturally expect that during the first years of the new century, several large ground-based automated optical surveys will provide additional large samples of multiply-imaged quasars. These surveys include the Sloan Digital Sky Survey (SDSS; Castander 1998), the International Liquid Mirror Telescope (ILMT) project (Surdej and Claeskens 1997) and the Two Degree Field (2dF) survey (Boyle *et al.* 2000). Statistical and individual studies of such gravitational lens systems will enable us to address and/or to improve some of the astrophysical and cosmological applications discussed in Section 3.

4.1 GAIA – the galactic census

The GAIA mission has been selected as a future ESA cornerstone mission and scheduled for launch no later than 2012. It will then perform an on-target, multi-colour imaging and spectroscopic all-sky survey down to the limiting magnitude $V \sim 21$. The aimed astrometric accuracy of this survey is quite challenging: 10 micro-arcseconds at $V = 15$, and 200 milli-arcseconds at $V = 21$. Assuming that the GAIA onboard data analysis and selected data transmission to the ground will enable the identification of quasars showing a complex structure within a field of view of 3″, we have estimated that GAIA observations will lead to the detection of a complete sample of more than 3,500 gravitational lenses among optical quasars down to the limiting magnitude $V = 21$. This number has been derived for the less favorable Universe with $\Omega_0 = 1$, $\lambda_0 = 0$. In a universe with $\Omega_0 = 0$, $\lambda_0 = 1$, the total number of multiply-imaged quasars would be 13 times larger. All these results are, of course, independent of the Hubble constant H_0. If the field of view is further reduced to 2″ or 1″, only 83% or 46% of the gravitational lens systems will be detected, respectively. As expected, the contribution due to lensing galaxies of type E/S0 remains the most important one.

Furthermore, because of its unprecedented astrometric accuracy all over the sky, GAIA will follow the bending of star light by the Sun and the major planets over the entire celestial sphere and therefore directly observe the structure of spacetime. The accuracy of its measurements might lead to the detection of a discrepancy at the second order with the value predicted by GR. Measuring the γ parameter (equal to unity in GR) of the parametrized post-Newtonian formulation of gravitational theories is of great importance to fundamental physics. So far, Hipparcos measurements have confirmed that $\gamma = 1.000 \pm 0.003$. GAIA measurements should improve the accuracy of the determination of γ to the level of 5×10^{-7}, and thus might also test alternative cosmological theories, like the Brans–Dicke theory (Brans and Dicke 1961).

4.2 Submillimetre observations

In addition to the visible spectral range for bright quasars, and to a lesser extent the radio waveband, the amplification bias also turns out to be quite important in the submillimetre waveband, for faint flux-limited samples of galaxies. Indeed, if the surface density of (unlensed) galaxies with an apparent magnitude brighter than m is $n_U(<m)$, an amplification by a factor μ_T due to gravitational lensing will modify the counts as follows (Narayan 1989):

$$n_L(<m) = n_U(<m + 2.5 \log(\mu_T))/\mu_T \qquad (18)$$

The factor $n_U(<m + 2.5\log(\mu_T))$ accounts for the increase in the observed number of galaxies due to the fact that objects as faint as $m + 2.5\log(\mu_T)$ can now become brighter than the limiting magnitude m, as a result of gravitational lensing amplification by a factor of μ_T. However, the galaxies we observe projected in the lens (sky) plane within a solid angle $d\omega$ are actually located within a smaller solid angle $d\omega_s$ (=$d\omega/\mu_T$) in the source plane, so that their observed surface density must be decreased by a factor of $d\omega/d\omega_s$ (eqns (4) and (9)), accounting for the factor $1/\mu_T$ in eqn (18). If s represents the slope of the logarithmic intrinsic number counts of objects as a function of apparent magnitude m, such that

$$s = \frac{d}{dm}(\log(n_U(<m))), \qquad (19)$$

eqn. (18) may then be rewritten as

$$n_L(<m) = n_U(<m)\mu_T^{2.5s-1} \qquad (20)$$

Depending on the value of the slope s (≥ 0.4 or <0.4), we see that the surface density of galaxies $n_L(<m)$ will either be larger or smaller than the original number and the amplification bias $\mu_T^{2.5s-1}$ will either be larger or smaller than 1. The amplification bias was first discussed in the context of bright QSO surveys at optical wavelengths; however, Blain (1997) has pointed out that the number counts of faint galaxies in the submillimetre waveband are extremely steep, with observed values of s even larger than 1.2 (Blain and Longair 1993, 1996). We thus expect the amplification bias induced by galaxy lensing or galaxy cluster lensing in the submillimetre domain to be very high for faint galaxies, and that the largest excess of lenses will occur at flux densities where the submillimetre counts are the steepest. (From Figure 2 in Blain (1997), we find that, typically, the flux is 0.1, 0.2, 0.5, 0.2 and 0.1 Jy at wavelenghts of 1300, 850, 450, 200 and 90 μm, respectively.) Note that the slope s of the counts decreases from longer to shorter wavelengths; the 90 μm counts becoming similar to those at optical wavelengths. Adopting $H_0 = 50 \text{ km s}^{-1} \text{ Mpc}^{-1}$, and various values for the cosmological parameters Ω_0 and λ_0, Blain (1997, 1998a,b) gives estimates for the number counts of faint distant dusty star-forming galaxies and AGNs (active galactic nuclei) at different wavelengths in the submillimetre waveband, under different assumptions (evolving and non-evolving mass distributions of lensing galaxies, different world models, different types of evolution of the source galaxies). As expected, the main conclusion is that the ratio of lensed to unlensed integral counts of faint galaxies in the submillimetre region is highest at $\lambda = 1300$ μm (approximately 3%), then gradually decreases with wavelength (to 1.7% at $\lambda = 850$ μm, 0.8% at λ 450 μm, 0.3% at $\lambda = 200$ μm and only 0.04% at $\lambda = 90$ μm). (For comparison, the ratio of lensed to unlensed HLQs in optical samples is of the order of 1% (Surdej et al. 1993), and approximately 0.1% for radio sources.) Note, however, that above $\lambda = 200$ μm the surface density of both lensed and unlensed galaxies decreases as the wavelength of observation increases, and that a best compromise ought to be sought between a high ratio of lensed to unlensed galaxies and a high total number of lensed galaxies.

4.3 The Herschel Space Observatory and the Planck mission

A clever combination of wide-field surveys obtained by space instruments, such as the forthcoming Herschel Space Observatory (originally named the Far Infrared and Submillimetre Space Telescope (FIRST) satellite) and Planck Surveyor, with the high flux sensitivity and angular resolution of large ground-based submillimetre/millimetre interferometer arrays (such as ALMA, the Atacama Large Millimetre Array), ought to provide a large sample of gravitational lens systems. Several hundred lenses should be identified, and these will be ideally suited to probe the distant Universe as well as to infer the properties, evolution and so on of the faintest and most distant populations of galaxies in the Universe.

From simulations made by Blain (1997, 1998a,b) at wavelengths of 200, 450 and 850 μm, an all-sky survey carried out by Planck should contain approximately 650, 560 and 40 gravitational lenses out of a total population of 240,000, 71,000 and 2,300 unlensed galaxies, respectively. However, the 4′ angular resolution of Planck is not good enough to allow rapid and efficient follow-up observations of those galaxies from the ground. But at $\lambda = 850$ μm, HSO could be used to determine more accurate (sub-arcminute) positions and flux densities of the 2300 galaxies in a relatively short time. These candidates could then be observed in a snapshot mode and/or be further studied with large ground-based submillimetre/millimetre interferometer arrays. However, it is not yet clear how best to select the candidates at shorter wavelengths. As suggested by Blain (1997, 1998a,b), it is likely that 'colour' information provided by the HSO and/or Planck observatories will enable the gravitational lens candidates to be extracted from all the faint submillimetre sources that will be detected. Indeed, because most of the lensed galaxies will be found at large redshifts, one would expect that if the dust temperature in star-forming galaxies is correlated with their luminosity, lensed galaxies will systematically show redder colours than will unlensed galaxies observed at the same flux densities.

Note that optimal targets to search for amplified (Herschel/Planck) and/or magnified (ALMA) submillimetre galaxies are foreground galaxy clusters. The lensed images of faint galaxies are expected to be the brightest sources of submillimetre wave radiation on arc second scales. Although these lensed galaxies could not be resolved by Herschel, they will increase the level of source confusion noise in the direction of the galaxy cluster as compared with observations in the field (Blain 1997, 1998a,b). These could subsequently be observed with large ground-based submillimetre/millimetre interferometric arrays. Herschel observations of catalogued galaxy clusters and many more clusters to be found with the Newton X-ray satellite (orginally known as the X-ray Multimirror Mission, XMM) as well as with Planck could also be used to investigate the gravitational lensing effects of clusters on the population of faint background galaxies emitting large amounts of energy in the submillimetre waveband.

NGST, with its imaging spectroscopy device and direct imaging capabilities in the near- and mid-infrared, will be an ideal telescope to further study and extend the spectral range of studies of gravitational lens candidates that will by then have been identified in the (sub-)millimetre domain. Similarly, the Advanced Camera and/or the WFPC3 camera that will be available aboard HST in the near future should also provide a unique mapping at ultraviolet and optical wavelengths of weak lensing around the galaxy clusters targeted with the Herschel and Planck.

4.4 The MAP mission

From 2001, the Microwave Anisotropy Probe (MAP) satellite will provide a detailed mapping of the microwave background radiation, with a much higher sensitivity than COBE (Cosmic Background Explorer) and an angular resolution approximately 35 times better. (The international Boomerang balloon experiment has provided interesting clues on the detailed structure of the microwave cosmic background over a small sky area that suggest that our Universe is flat (de Bernardis et al. 2000).) Planck's observations should be of still much higher quality, but will not start before 2007. Analysis of the MAP data should not only lead to a very precise determination of the cosmological parameters (H_0, Ω_0 and λ_0) and of the fraction of visible versus dark matter, but also test whether our view of an infinite Universe filled with all its known and unknown constituents could not actually result from a cosmic illusion on the scale of the Universe itself. Could it be that a large fraction of the billions of visible galaxies and quasars are only the repeated images of a much smaller number of galaxies in a finite Universe? Could most of these images just be the result of light emitted from a much smaller number of galaxies, circling the whole finite Universe along different trajectories and seen at different epochs of their cosmic evolution, possibly including images of our own Milky Way? The MAP data will be used by theorists to search for repeated or redundant temperature patterns over the cosmic microwave background which could arise in a finite Universe (Roukema and Luminet 1999). In addition, the specific age of any observed cosmic mirage would depend on the precise size and overall shape of the Universe. Theorists are eagerly awaiting the MAP observations to test this alternative theory.

4.5 X-ray observations

Owing to the low surface density of X-ray sources in the sky and their simple source structure, gravitational lens systems ought to be easily identified from the multiplicity of their X-ray images, provided the angular resolution of the X-ray telescope is good enough. Muñoz et al. (1999b) estimated the expected number of multiply-imaged AGNs in the soft X-ray band (0.3–3.5 kev) as a function of the flux limit. They find that approximately 1, 0.1 and 0.01 multiple X-ray images of AGNs lensed by foreground galaxies should be found per square degree down to fluxes of 10^{-15}, 10^{-14} and 10^{-13} erg s^{-1} cm^{-2}, respectively. Muñoz et al. (1999b) found the maximum value for the ratio of lensed to unlensed X-ray sources to be about 0.2–0.4% in the flux range 10^{-13}–10^{-12} erg s^{-1} cm^{-2}. Taking into account the observational selection function of the HRC (High Resolution Camera) and ACIS (AXAF CCD Imaging Spectrometer) detectors aboard the Chandra X-ray Observatory (CXO, originally known as the Advanced X-ray Astrophysics Facility, AXAF), they found that about 1–3 such lensed sources should be detected serendipitously per year of high-resolution imaging. Such an estimate is almost independent of the flux limit. This low detection rate is essentially due to the small field of view of Chandra. The same authors also determined that if Chandra were used to take deep images of known rich galaxy clusters at intermediate redshifts, one wide-separation, multiply-imaged background X-ray source should be detected down to the above flux limits for every 10, 30 and 300 clusters. Of course, such a programme would occupy a substantial amount of observing time (at least several months).

Because of the wider field of view and high sensitivity of the European XMM-Newton telescope, the prospects for detecting gravitationally lensed images look rather more promising. XMM-Newton's angular resolution is coarser (5″ FWHM) than that of Chandra, but the identification of the multiply-imaged AGNs would rely on a parallel high-angular-resolution optical survey, as planned by the XMM–LSS (Large Scale Structures) Consortium. In this framework, we estimate that the planned XMM–LSS imaging of 64 square degrees with typical exposure times of about 3 hours should lead to the detection of some 20 X-ray sources lensed by foreground galaxies. XMM-Newton would also turn out to be very useful for the detection of lensed objects by intervening clusters. Typically, the XMM–LSS survey should identify approximately 700 galaxy clusters for which some 40 multiply-imaged background X-ray sources ought to be found.

Given the unique high-angular-resolution capabilities of Chandra, this observatory will very likely be used to investigate the photometric and spectroscopic properties of known lensed quasars and/or their individual lensed images with X-ray luminosities that are a factor of μ (their gravitational lens amplification) smaller than unlensed quasars at a comparable flux limit, and to estimate the contribution of such faint quasars to the X-ray cosmic background. Chandra will also certainly play a part in monitoring the fast X-ray variability of known multiply-imaged QSOs with a high time resolution, allowing one to derive much more accurate time delays between the multiple lensed images. Shorter microlensing events could also be probed.

An X-ray observatory with an angular resolution comparable to, or better than that of Chandra and a field of view similar to, or larger than, that of XMM-Newton would be the ideal instrument with which to search the whole sky for a large number of multiple X-ray source images lensed by foreground galaxies and clusters. Lenses would then be directly recognizable as such on the sole basis of the X-ray data.

5 SUMMARY AND CONCLUSIONS

Gravitational lensing creates an intimate relation between the observer, a massive astrophysical object (the gravitational lens), a luminous source and the whole Universe (eqn (3)). We conclude by considering what we have learnt or still hope to learn about each of these four components.

Observers are human beings. The historical part of this chapter has retold an interesting journey into the human mind and psychology, and ultimately into the way of making science. A genius (Einstein!) finds the theoretical location of a gold mine. However, he is very rich and does not feel the need to exploit this mine or even to explore it further. He leaves that to the community. But the mine is located in a wild, unexplored region of human knowledge. Few people want to go there, but even the most adventurous of them are unable to reach the mine. Only theoreticians can dream of it. Most scientists are realistic people who do not believe in gold mines. Many years after the genius's revelation, someone accidentally discovers the mine, as mysterious as a temple in the jungle, but cannot keep its location secret. Soon, pioneers arrive and prepare an easy access to the site, and then many scientists rush to the mine, eager to explore every logical passage. This is the start of the golden age.

Today, the deflection of light by massive bodies, as predicted by Einstein's theory of general relativity, has been measured in the Solar System, in the Milky Way and on extragalactic scales. Thanks to this theory, a large variety of observed effects – image shifts, multiple images, giant arcs, radio rings, time delays, flux variations or weak shear distortions of background sources – can all be qualitatively explained by the gravitational influence of normal stars, galaxies, galaxy clusters or large-scale structures. The deflection angle ranges from micro-arcseconds to tens of arcseconds. It has also been checked to be wavelength independent from the far UV to the radio domain. Since the deflection angle is very small, the basic formalism of gravitational lensing is rather simple, as shown in Section 2.

Gravitational lensing does not only prove general relativity to be a correct description of reality, but it also provides a powerful tool to investigate astrophysical objects and the Universe (Section 3).

Let us now turn to the lenses themselves. Since gravity acts also on light, studying the luminous shapes of background sources yields information on the nature of the foreground lens population. In particular, gravitational lensing has confirmed the presence of large amounts of dark matter in the Universe, associated with external galaxies, galaxy clusters and large-scale structures. The mass-to-light ratio in some clusters can indeed be as high as 1000. However, gravitational lensing shows that if isolated, cosmologically distributed compact dark objects exist, their spatial density is probably not high enough to close the Universe. However, it is not yet clear whether the microlensing events detected in the direction of the LMC are due to dark compact objects located in the halo of the Milky Way.

The multiple image configurations produced by individual lensing galaxies are compatible with those produced by isothermal and singular matter distributions. Gravitational lensing will also enable us in the near future to probe quantitatively the extinction law of those distant galaxies. Galactic microlensing is promising for the detection of planets orbiting the main stellar lens.

Our knowledge of the sources may be improved with gravitational lensing by taking advantage of the so-called 'telescope effect', predicted by Zwicky as early as 1937. Striking examples are the discovery of a very distant galaxy at $z = 4.92$ located behind the galaxy cluster Cl1358 + 62 and the study of stellar populations in a $z = 4$ galaxy. Massive galaxy clusters have also helped in detecting a more distant population of dusty galaxies at submillimetre wavelengths (Smail *et al.* 1997). The continuum size of quasars has been constrained from microlensing observations, and theoretical studies indicate that extragalactic microlensing could even help in resolving the structure of quasar accretion disks (Agol and Krolik 1999). Microlensing in our Galaxy should allow us to study the surface of stars; the limb darkening of a lensed star has indeed already been measured (Sackett 2001, and references therein).

The global properties of the Universe, in which light is travelling from the source to the observer, leave their imprint on the multiply-lensed images of extragalactic sources. The Hubble constant has been derived for several cases from time-delay measurements between flux variations of multiply-imaged QSOs, and the value of the cosmological constant has been constrained from the number of multiply-imaged quasars in optical and radio QSO samples. Gravitational lensing represents an alternative approach to the classical methods and avoids the traditional issue of intermediate standard flux calibrations. However, this technique has its own drawbacks: the determination of H_0 is limited by the poor knowledge of the lens model, and the derived value of λ_0 is sensitive to the incompleteness of the parent QSO sample (selection biases, dust extinction and so on) as well as to the luminosity function of the source population, which is not very well known (especially for radio sources). Homogeneous QSO samples coming from automated surveys (e.g. SDSS, 2dF, ILMT), optical identifications of radio sources analysed by CLASS, and new observations of (new) lenses adequate to constrain the lens model may improve the situation in the near future. We may also believe in our Lucky Star, since Link and Pierce (1998) showed that if we had the chance to observe multiple arcs from sources located behind a given cluster but at different redshifts, the cosmological model could then be determined with a better precision.

We cannot conclude this chapter without coming back to the observers. Observations of gravitational lenses are very demanding in terms of spatial resolution and collecting power, but also in time sampling for flux monitoring or in data handling for microlensing searches in the Milky Way. These difficulties have motivated much work in image processing (e.g. point spread function (PSF) subtraction or deconvolution), in light-curve analysis and in team organization to carry out sustained photometric monitoring campaigns in order to obtain frequent data. In this context, space observations, especially with the HST, have played a crucial role with respect to the angular resolution issue. Most gravitational lenses have been discovered with ground-based telescopes or radio telescopes, but the HST has enlightened us with the confirmation of several cases with small angular separations (like J03.13, Figure 9), with the precise determination of all the lensed image positions, with discoveries of new arcs and the first detection of several lensing galaxies.

Where are we going now? Strong lensing events such as giant arcs or multiply-imaged sources are rare, but automated surveys with (nearly) full-sky coverage (SDSS, CLASS, ILMT, GAIA) will find them more and more systematically. Weak shear distortions are common but difficult to detect; however, the increasing accuracy of both observations and analysis has yielded encouraging results very recently (Section 3.6). Gravitational lenses are no longer rare scientific jewels. Until now, either we have usually assumed a known background source population in order to study the lensing population, or we have estimated the distortion of the apparent properties of a class of sources, assuming a given foreground lens population. In reality, both problems are intrinsically connected and should be solved simultaneously. It will probably be impossible to do so, but there is no doubt that in the future, gravitational lensing will help to build a coherent three-dimensional image of the Universe. We are thus now entering the Iron Age of gravitational lensing!

Acknowledgments

Our research was supported in part by contract P4/05 'Pôle d'Attraction Interuniversitaire' (OSTC, Belgium), by PRODEX (Gravitational lens studies with HST) and by the 'Fonds National de la Recherche Scientifique' (Belgium). We are very grateful to Jean-Pierre Swings for reading the manuscript and to Anna Pospieszalska for her help with some of the figures, the table and the bibliography.

FURTHER READING

Further reading on gravitational lensing includes the textbook by Schneider, Ehlers and Falco (1992), the book in Russian by Bliokh and Minakov (1989) and review papers by Blandford and Narayan (1992), Refsdal and Surdej (1994), Fort and Mellier (1994), Narayan and Bartelmann (1996), Wambsganss (1998), Mellier (1999) and Claeskens and Surdej (2001). We refer the reader to the book by Peterson (1997) for a general introduction to active galactic nuclei and quasars.

REFERENCES

Agol, E. and Krolik, J. (1999). Imaging a quasar accretion disk with microlensing. *Astrophysical Journal*, **524**, 49–64.

Albrow, M., Beaulieu, J-P., Birch, P., Caldwell, J.A.R., Kane, S., Martin, R., Menzies, J., Naber, R.M., Pel, J.-W., Pollard, K., Sackett, P.D., Sahu, K.C., Vreeswijk, P., Williams, A. and Zwaan, M.A. (1998). The 1995 pilot campaign of PLANET: Searching for microlensing anomalies through precise, rapid, round-the-clock monitoring. *Astrophysical Journal*, **509**, 687–702.

Alcock, C., Allsman, R.A., Alves, D.R., Axelrod, T.S., Becker, A.C., Bennett, D.P., Cook, K.H., Dalal, N., Drake, A.J., Freeman, K.C., Geha, M.,

Griest, K., Lehner, M.J., Marshall, S.L., Minniti, D., Nelson, C.A., Peterson, B.A., Popowski, P., Pratt, M.R., Quinn, P.J., Stubbs, C.W., Sutherland, W., Tomaney, A.B., Vandehei, T. and Welch, D. (2000). The MACHO Project: Microlensing results from 5.7 years of LMC observations. *Astrophysical Journal*, **542**, 281–307.

Altieri, B., Metcalfe, L., Kneib, J-P., McBreen, B., Aussel, A., Biviano, A., Delaney, M., Elbaz, D., Leech, K., Lémonon, L., Okumura, K., Pelló, R. and Schulz, B. (1999). An ultra-deep ISOCAM observation through a cluster-lens. *Astronomy and Astrophysics*, **343**, L65–L69.

Angonin-Willaime, M-C., Vanderriest, C., Courbin, F., Burud, I., Magain, P. and Rigaud, F. (1999). About the origin of extinction in the gravitational lens system MG J0414+0534. *Astronomy and Astrophysics*, **347**, 434–441.

Aubourg, E. and Palanque-Delabrouille, N. (1999). A search for galactic dark matter with EROS 2. *New Astronomy*, **4**, 265–273.

Bacon, D., Refregier, A. and Ellis, R. (2000). Detection of weak gravitational lensing by large-scale structure. *Monthly Notices of the Royal Astronomical Society*, **318**, 625–640.

Barnothy, J.M. (1965). Quasars and the gravitational image intensifier. *Astronomical Journal*, **70**, 666.

Bartelmann, M., Narayan, R., Seitz, S. and Schneider, P. (1996). Maximum-likelihood cluster reconstruction. *Astrophysical Journal*, **464**, L115–L118.

Bartelmann, M. and Schneider, P. (1993). Large-scale QSO-galaxy correlations revisited. *Astronomy and Astrophysics*, **271**, 421–424.

de Bernardis, P., Ade, P.A.R., Bock, J.J., Bond, J.R., Borrill, J., Boscaleri, A., Coble, K., Crill, B.P., De Gasperis, G., Farese, P.C., Ferreira, P.G., Ganga, K., Giacometti, M., Hivon, E., Hristov, V.V., Iacoangeli, A., Jaffe, A.H., Lange, A.E., Martinis, L., Masi, S., Mason, P.V., Mauskopf, P.D., Melchiorri, A., Miglio, L., Montroy, T., Netterfield, C.B., Pascale, E., Piacentini, F., Pogosyan, D., Prunet, S., Rao, S., Romeo, G., Ruhl, J.E., Scaramuzzi, F., Sforna, D. and Vittorio, N. (2000). A flat Universe from high-resolution maps of the cosmic microwave background radiation. *Nature*, **404**, 955–959.

Bernstein, G. and Fischer, P. (1999). Values of H_0 from models of the gravitational lens 0957+561. *Astronomical Journal*, **118**, 14–34.

Blain, A.W. (1997). Gravitational lensing by clusters of galaxies in the millimetre/submillimetre waveband. *Monthly Notices of the Royal Astronomical Society*, **290**, 553–565.

Blain, A.W. (1998a). The Planck Surveyor mission and gravitational lenses. *Monthly Notices of the Royal Astronomical Society*, **297**, 511–516.

Blain, A.W. (1998b). Submillimetre-wave gravitational lenses and cosmology. *Monthly Notices of the Royal Astronomical Society*, **295**, 92–98.

Blain, A.W. and Longair, M.S. (1993). Submillimetre cosmology. *Monthly Notices of the Royal Astronomical Society*, **264**, 509–521.

Blain, A.W. and Longair, M.S. (1996). Observing strategies for blank-field surveys in the submillimetre waveband. *Monthly Notices of the Royal Astronomical Society*, **279**, 847–858.

Blandford, R.D. and Narayan, R. (1992). Cosmological applications of gravitational lensing. *Annual Review of Astronomy and Astrophysics*, **30**, 311–358.

Blandford, R.D., Saust, A.B., Brainerd, T.G. and Villumsen, J.V. (1991). The distortion of distant galaxy images by large scale structure. *Monthly Notices of the Royal Astronomical Society*, **251**, 600–627.

Bliokh, P.V. and Minakov, A.A. (1989). *Gravitatsionnye Linzy* [Gravitational Lensing], Naukova Dumka, Kiev.

Borgeest, U. (1986). Determination of galaxy masses by the gravitational lens effect. *Astrophysical Journal*, **309**, 467–471.

Bourassa, R.R. and Kantowski, R. (1975). The theory of transparent gravitational lenses. *Astrophysical Journal*, **195**, 13–21.

Bourassa, R.R., Kantowski, R. and Norton, T.D. (1973). The spheroidal gravitational lens. *Astrophysical Journal*, **185**, 747–756.

Boyle, B.J., Shanks, T., Croom, S.M., Smith, R.J., Miller, L., Loaring, N. and Heymans, C. (2000). The 2dF QSO Redshift Survey – I. The optical QSO luminosity function. *Monthly Notices of the Royal Astronomical Society*, **317**, 1014–1022.

Boyle, B.J., Shanks, T., Peterson, B.A. (1988). The evolution of optically selected QSOs. II. *Monthly Notices of the Royal Astronomical Society*, **235**, 935–948.

Brans, C. and Dicke, R.H. (1961). Mach's principle and a relativistic theory of gravitation. *Physical Review*, **124**, 925–935.

Bunker, A.J., Moustakas, L.A. and Davis, M. (2000). Resolving the stellar populations in a $z=4$ lensed galaxy. *Astrophysical Journal*, **531**, 95–117.

Burke, W.L. (1981). Multiple gravitational imaging by distributed masses. *Astrophysical Journal*, **244**, L1.

Burud, I., Courbin, F., Lidman, C., Jaunsen, A., Hjorth, J., Ostensen, R., Andersen, M.I., Clasen, J.W., Wucknitz, O., Meylan, G., Magain, P., Stabell, R. and Refsdal, S. (1998). High-resolution optical and near-infrared imaging of the quadruple quasar RX J0911.4+0551. *Astrophysical Journal*, **501**, L5–L10.

Canizares, C.R. (1982). Manifestations of a cosmological density of compact objects in quasar light. *Astrophysical Journal*, **263**, 508–517.

Castander, F.J. (1998). The Sloan Digital Sky Survey. *Astrophysics and Space Science*, **263**, 91–94.

Chae, K-Y. (1999). New modeling of the lensing galaxy and cluster of Q0957+561: Implications for the global value of the Hubble constant. *Astrophysical Journal*, **524**, 582–590.

Chwolson, O. (1924). Über eine mögliche Form fiktiver Doppelsterne. *Astronomische Nachrichten*, **221**, 329.

Claeskens, J-F. (1999). Thèse de Doctorat. Aspects statistiques du phénomène de lentille gravitationnelle dans un échantillon de quasars très lumineux. *Societé Royale des Sciences de Liège*, **68**, 1–305.

Claeskens, J-F. and Surdej, J. (2001). Gravitational lensing in quasar samples. *Astronomy and Astrophysics Review*, in the press.

Claeskens, J-F., Surdej, J. and Remy, M. (1996). J03.13A and B: A new multiply imaged QSO candidate. *Astronomy and Astrophysics*, **305**, L9–L12.

Colley, W.N., Tyson, J.A. and Turner, E.L. (1996). Unlensing multiple arcs in 0024+1654: Reconstruction of the source image. *Astrophysical Journal*, **461**, L83–L86.

Cooray, A.R. (1999). Cosmological parameters from statistics of strongly lensed radio sources. *Astronomy and Astrophysics*, **342**, 353–362.

Dalcanton, J.J., Canizares, C.R., Granados, A., Steidel, C.C. and Stocke, J.T. (1994). Observational limits on Omega in stars, brown dwarfs, and stellar remnants from gravitational microlensing. *Astrophysical Journal*, **424**, 550–568.

D'Odorico, V., Cristiani, S., D'Odorico, S., Fontana, A., Giallongo, E. and Shaver, P. (1998). The size and geometry of the Lyα clouds. *Astronomy and Astrophysics*, **339**, 678–686.

Dyer, C.C. and Roeder, R.C. (1972). The distance–redshift relation for universes with no intergalactic medium. *Astrophysical Journal*, **174**, L115–L117.

Dyson, F.W., Eddington, A.S. and Davidson, C.R. (1920). A determination of the deflection of light by the sun's gravitational field, from observations made at the total eclipse of May 29, 1919. *Memoirs of the Royal Astronomical Society*, **62**, 291–333.

Eddington, A.S. (1920). *Space, Time and Gravitation*. Cambridge University Press.

Einstein, A. (1936). Lens-like action of a star by the deviation of light in the gravitational field. *Science*, **84**, 506–507.

Etherington, I.M.H. (1933). On the definition of distance in general relativity. *Philosophical Magazine*, **15**, 761.

Falco, E.E., Gorenstein, M.V. and Shapiro, I.I. (1985). On model dependent bounds on H_0 from gravitational images: Application to Q0957+561 A, B. *Astrophysical Journal*, **289**, L1–L4.

Falco, E.E., Kochanek, C.S. and Muñoz, J.A. (1998). Limits on cosmological models from radio-selected gravitational lenses. *Astrophysical Journal*, **494**, 47–59.

Fomalont, E.B. and Sramek, R.A. (1975a). A confirmation of Einstein's general theory of relativity by measuring the bending of microwave radiation in the gravitational field of the Sun. *Astrophysical Journal*, **199**, 749–755.

Fomalont, E.B. and Sramek, R.A. (1975b). Measurements of the solar gravitational deflection of radio waves in agreement with general relativity. *Physical Review Letters*, **36**, 1475–1478.

Fort, B. and Mellier, Y. (1994). Arc(let)s in clusters of galaxies. *Astronomy and Astrophysics Review*, **5**, 239–292.

Fort, B., Mellier, Y., Dantel-Fort, M., Bonnet, H. and Kneib, J-P. (1996). Observations of weak lensing in the fields of luminous radiosources. *Astronomy and Astrophysics*, **310**, 705–714.

Fort, B., Prieur, J.L., Mathez, G., Mellier, Y. and Soucail, G. (1988). Faint distorted structures in the core of A 370: Are they gravitationally lensed galaxies at $z = 1$? *Astronomy and Astrophysics*, **200**, L17–L20.

Franx, M., Illingworth, G.D., Kelson, D., van Dokkum, P.G. and Tran, K-V. (1997). A pair of lensed galaxies at $z = 4.92$ in the field of Cl1358 + 62. *Astrophysical Journal*, **486**, L75–L78.

Froeschlé, M., Mignard, F. and Arenou, F. (1997). Determination of the PPN parameter γ with the Hipparcos data. In *Proceedings of the ESA Symposium 'Hipparcos Venice' 97'*, ESA SP-402, pp. 49–52.

Fugmann, W. (1990). Statistical gravitational lensing and the Lick Catalogue of galaxies. *Astronomy and Astrophysics*, **240**, 11–21.

Fukugita, M., Futamase, T. and Kasai, M. (1990). A possible test for the cosmological constant with gravitational lenses. *Monthly Notices of the Royal Astronomical Society*, **246**, 24P–27P.

Gebhardt, K., Richstone, D., Ajhar, E.A., Lauer, T.R., Byun, Y-I., Kormendy, J., Dressler, A., Faber, S.M., Grillmair, C. and Tremaine, S. (1996). The centers of early-type galaxies with HST: III. Non-parametric recovery of stellar luminosity distribution. *Astronomical Journal*, **112**, 105–113.

Gondhalekar, P.M. and Wilson, R. (1980). UV spectra of the twin QSOs 0957 + 561 A, B. *Nature*, **285**, 461–463.

Gorenstein, M.V., Falco, E.E., Shapiro, I.I. (1988). Degeneracies in parameter estimates for models of gravitational lens systems. *Astrophysical Journal*, **327**, 693–711.

Gott, J.R. III (1981). Are heavy halos made of low mass stars? A gravitational lens test. *Astrophysical Journal*, **243**, 140–146.

Hartwick, F.D.A. and Schade, D. (1990). The space distribution of quasars. *Annual Review of Astronomy and Astrophysics*, **28**, 437–489.

Hattori, M., Kneib, J-P. and Makino, N. (1999). Gravitational lensing in clusters of galaxies. *Progress of Theoretical Physics*, Suppl., **133**, 1.

Helbig, P., Marlow, D.R., Quast, R., Wilkinson, P.N., Browne, W.A. and Koopmans, L.V.E. (1999). Gravitational lensing statistics with extragalactic surveys: II. Analysis of the Jodrell Bank-VLA Astrometric Survey. *Astronomy and Astrophysics Supplement*, **136**, 297–305.

Hoag, A. (1981). A feature in the galaxy cluster Zw0237.2-0146. *Bulletin of the American Astronomical Society*, **13**, 799.

Jean, C. and Surdej, J. (1998). Redshift estimation of a gravitational lens from the observed reddening of a multiply imaged quasar. *Astronomy and Astrophysics*, **339**, 729–736.

Kaiser, N. and Squires, G. (1993). Mapping the dark matter with weak gravitational lensing. *Astrophysical Journal*, **404**, 441–450.

Kassiola, A., Kovner, I. and Blandford, R.D. (1991). Bounds on intergalatic compact objects from observations of compact radio sources. *Astrophysical Journal*, **381**, 6–13.

Kayser, R., Helbig, P. and Schramm, T. (1997). A general and practical method for calculating cosmological distance. *Astronomy and Astrophysics*, **318**, 680–686.

Kochanek, C.S. (1996). Is there a cosmological constant? *Astrophysical Journal*, **466**, 638–659.

Klimov, Y.G. (1963). The deflection of light rays in the gravitational fields of galaxies. *Soviet Physics Doklady*, **8**, 119–122.

Kundić, T., Turner, E.L., Colley, W.N., Gott, R. III, Rhoads, J.E., Wang, Y., Bergeron, L.E., Gloria, K.A., Long, D.C., Malhotra, S. and Wambsganss, J. (1997). A robust determination of the time delay in 0957 + 561A,B and a measurement of the global value of Hubble's constant. *Astrophysical Journal*, **482**, 75–82.

Lebach, D.E., Corey, B.E., Shapiro, I.I., Ratner, M.I., Webber, J.C., Rogers, A.E.E., Davis, J.L. and Herring, T.A. (1995). Measurement of the solar gravitational deflection of radio waves using very-long-baseline interferometry. *Physical Review D*, **75**, 1439–1442.

Liebes, S. (1964). Gravitational lenses. *Physical Review B*, **133**, 835–844.

Link, R. and Pierce, M.J. (1998). Cosmological parameters from multiple-arc gravitational lensing systems: I. Smooth lensing potentials. *Astrophysical Journal*, **502**, 63–74.

Lodge, O.J. (1919). Gravitation and light. *Nature*, **104**, 354.

Luppino, G.A., Gioia, I.M., Hammer, F., Le Fèvre, O. and Annis, J.A. (1999). A search for gravitational lensing in 38 X-ray selected clusters of galaxies. *Astronomy and Astrophysics*, Suppl., **136**, 117–137.

Lynds, R. and Petrosian, V. (1986). Giant luminous arcs in galaxy clusters. *Bulletin of the American Astronomical Society*, **18**, 1014.

Maoz, D. and Rix, H-W. (1993). Early-type galaxies, dark halos, and gravitational lensing statistics. *Astrophysical Journal*, **416**, 425–443.

Marani, G.F., Nemiroff, R.J., Norris, J-P., Hurley, K. and Bonnell, J.T. (1999). Gravitationally lensed gamma-ray bursts as probes of dark compact objects. *Astrophysical Journal*, **512**, L13–L16.

Mellier, Y. (1999). Probing the Universe with weak lensing. *Annual Review of Astronomy and Astrophysics*, **37**, 127–189.

Miralda-Escudé, J. (1991). The correlation function of galaxy ellipticities produced by gravitational lensing. *Astrophysical Journal*, **380**, 1–8.

Muñoz, J.A., Falco, E.E., Kochanek, C.S., Lehár, J., McLeod, B.A. Impey, C.D., Rix, H.-W., Peng, C.Y. (1999a). The CASTLES project. In: J. Gorgas and J. Zamorano (eds), *Astrophysics and Space Science*, (special issue: Proc. III Scientific Meeting of the SEA), astro-ph/9902131.

Muñoz, J.A., Kochanek, C.S. and Falco, E.E. (1999b). Finding gravitational lenses with X-rays. *Astrophysical Journal*, **521**, L17–L20.

Nadeau, D., Yee, H.K.C., Forrest, W.J., Garnett, J.D., Ninkov, Z. and Pipher, J.L. (1991). Infrared and visible photometry of the gravitational lens system 2237 + 030. *Astrophysical Journal*, **376**, 430–438.

Narayan, R. and Bartelmann, M. (1996). Lectures on gravitational lensing. In A. Dekel and J.P. Ostriker (eds), *Formation of Structure in the Universe*, Proceedings of the 1995 Jerusalem Winter School, Cambridge University Press.

Narayan, R. (1989). Gravitational lensing and quasar–galaxy correlations. *Astrophysical Journal*, **339**, L53–L56.

Nemiroff, R.J. (1989). On the probability of detection of a single gravitational lens. *Astrophysical Journal*, **341**, 579–587.

Nemiroff, R.J., Marani, G.F., Norris, J.P. and Bonnell, J.T. (2001). Limits on the cosmological abundance of supermassive compact objects from a millilensing search in gamma-ray burst data. *Physical Review Letters*, **86**, 580.

Newton, I. (1704). *Opticks, or a treatise of the reflections, refractions, inflections and colours of light*, 2nd edn, London.

Paczyński, B. (1986). Gravitational microlensing by the galactic halo. *Astrophysical Journal*, **304**, 1–5.

Paczyński, B. (1987). Giant luminous arcs discovered in two clusters of galaxies. *Nature*, **325**, 572–573.

Paczyński, B. (1996). Gravitational microlensing in the local group. *Annual Review of Astronomy and Astrophysics*, **34**, 419–460.

Perlmutter, S., Aldering, G., Goldhaber, G., Knop, R.A., Nugent, P., Castro, P.G., Deustua, S., Fabbro, S., Goobar, A., Groom, D.E., Hook, I.M., Kim, A.G., Kim, M.Y., Lee, J.C., Nunes, N.J., Pain, R., Pennypacker, C.R., Quimby, R., Lidman, C., Ellis, R. S., Irwin, M., McMahon, R.G., Ruiz-Lapuente, P., Walton, N., Schaefer, B., Boyle, B.J., Filippenko, A.V., Matheson, T., Fruchter, A.S., Panagia, N.,

Newberg, H.J.M. and Couch, W.J. (1999). Measurements of Omega and Lambda from 42 High-Redshift Supernovae. *Astrophysical Journal*, **517**, 565–586.

Peterson, B.M. (1997). *An Introduction to Active Galactic Nuclei*. Cambridge University Press.

Pospieszalska-Surdej, A., Surdej, J., Detal, A. and Jean C. (2000). The GL bibliography and an interactive database. In T.G. Brainerd and C.S. Kochanek (eds), *Gravitational Lensing: Recent Progress and Future Goals*, in the press.

Press, W.H. and Gunn, J.E. (1973). Method for detecting a cosmological density of condensed objects. *Astrophysical Journal*, **185**, 397–412.

Refsdal, S. (1964a). The gravitational lens effect. *Monthly Notices of the Royal Astronomical Society*, **128**, 295–306.

Refsdal, S. (1964b). On the possibility of determining Hubble's parameter and the masses of galaxies from the gravitational lens effect. *Monthly Notices of the Royal Astronomical Society*, **128**, 307–310.

Refsdal, S. (1965). Proc. Int. Conf. on Relativistic Theories of Gravitation, London.

Refsdal, S. (1966b). On the possibility of determining the distance and masses of stars from the gravitational lens effect. *Monthly Notices of the Royal Astronomical Society*, **134**, 315–319.

Refsdal, S. (1966a). On the possibility of testing cosmological theories from the gravitational lens effect. *Monthly Notices of the Royal Astronomical Society*, **132**, 101–111.

Refsdal, S. (1970). On the propagation of light in universes with inhomogeneous mass distribution. *Astrophysical Journal*, **159**, 357–375.

Refsdal, S. and Surdej, J. (1994). Gravitational lenses. *Reports on Progress in Physics*, **57**, 117–185.

Renn, J., Sauer, T. and Stachel, J. (1997). The origin of gravitational lensing: A postcript to Einstein's 1936 *Science* paper. *Science*, **275**, 184–186.

Riess, A.G., Filippenko, A.V., Challis, P., Clocchiatti, A., Diercks, A., Garnavich, P.M., Gilliland, R.L., Hogan, C.J., Jha, S., Kirshner, R.P., Leibundgut, B., Phillips, M.M., Reiss, D., Schmidt, B.P., Schommer, R.A., Smith, R.C., Spyromilio, J., Stubbs, C., Suntzeff, N.B. and Tonry, J. (1998). Observational evidence from supernovae for an accelerating universe and a cosmological constant. *Astronomical Journal*, **116**, 1009–1038.

Robertson, D.S., Carter, W.E., Dillinger, W.H. (1991). New measurement of solar gravitational deflection of radio signals using VLBI. *Nature*, **349**, 768–770.

Roukema, B.F. and Luminet, J-P. (1999). Constraining curvature parameters via topology. *Astronomy and Astrophysics*, **348**, 8–16.

Sackett, P.D. (2001). Microlensing and the physics of stellar atmospheres. In J.W. Menzies and P.D. Sackett (eds), *Microlensing 2000: A New Era of Microlensing Astrophysics*. ASP Conference Series, Astronomical Society of the Pacific, San Francisco (in press).

Sanitt, N. (1971). Quasi-stellar objects and gravitational lenses. *Nature*, **234**, 199–203.

Schmidt, M. (1963). 3C 273: A star-like object with large red-shift. *Nature*, **197**, 1040.

Schmidt, R. and Wambsganss, J. (1998). Limits on MACHOs from microlensing in the double quasar Q 0957 + 561. *Astronomy and Astrophysics*, **335**, 379–387.

Schneider, P. (1993). Upper bounds on the cosmological density of compact objects with sub-solar masses from the variability of QSOs. *Astronomy and Astrophysics*, **279**, 1–20.

Schneider, P., Ehlers, J. and Falco, E.E. (1992). *Gravitational Lenses*. Astronomy and Astrophysics Library, Springer-Verlag, Berlin.

Smail, I., Ivison, R.J. and Blain, A.W. (1997). A deep submillimeter survey of lensing clusters: A new window on galaxy formation and evolution. *Astrophysical Journal*, **490**, L5–L8.

Smette, A., Robertson, J.G., Shaver, P.A., Reimers, D., Wisotzki, L. and Köhler, Th. (1995). The gravitational lens candidate HE 1104-1805 and the size of absorption systems. *Astronomy and Astrophysics*, Suppl., **113**, 199–236.

Smette, A., Surdej, J., Shaver, P.A., Foltz, C.B., Chaffee, F.H., Weymann, R.J., Williams, R.E. and Magain, P. (1992). A spectroscopic study of UM 673 A and B: On the size of Lyman-α clouds. *Astrophysical Journal*, **389**, 39–62.

Soldner, J. (1804). Über die Ablenkung eines Lichtstrahls von seiner geradlinigen Bewegung durch die Attraktion eines Weltkörpers, an welchem er nahe vorbeigeht. *Berliner Astronomisches Jahrbuch*, 161–172.

Soucail, G., Fort, B., Mellier, Y. and Picat, J-P. (1987). A blue ring-like structure in the center of the A 370 cluster of galaxies. *Astronomy and Astrophysics*, **172**, L14–L16.

Soucail, G., Mellier, Y., Fort, B., Mathez, G. and Cailloux, M. (1988). The giant arc in A 370: Spectroscopic evidence for gravitational lensing from a source at $z = 0.724$. *Astronomy and Astrophysics*, **191**, L19–L21.

Surdej, J., Claeskens, J-F. (1997). Gravitational lens studies with a LMT. In M. Ferrari (ed.), Proceedings of international workshop on 'Science with Liquid Mirror Telescopes', Marseille Observatory, 14–15 April 1997. see the url: http://wood.phy.ulaval.ca/Workshop/Wproceed.html#Surdej

Surdej, J., Claeskens, J-F., Crampton, D., Filippenko, A.V., Hutsemékers, D., Magain, P., Pirenne, B., Vanderriest, C. and Yee, H.K.C. (1993). Gravitational lensing statistics based on a large sample of highly luminous quasars. *Astronomical Journal*, **105**, 2064–2078.

Surdej, J., Claeskens, J-F., Remy, M., Refsdal, S., Pirenne, B., Prieto, A., Vanderriest, Ch. (1997). HST confirmation of the lensed quasar J03.13. *Astronomy and Astrophysics*, **327**, L1–L4.

Surdej, J., Magain, P., Swings, J-P., Remy, M., Borgeest, U., Kayser, R., Refsdal, S., Kühr, H. (1988). Preliminary results from a search for gravitational lensing within a sample of highly luminous quasars. In C. Balkowski and S. Gordon (eds), *Large Scale Structures: Observations and Instruments*, Proceedings of the 1st DAEC [Département d'Astrophysique Extragalactique et de Cosmologie] Workshop, Paris, pp. 97–107.

Treuhaft, R.N. and Lowe, S.T. (1991). A measurement of planetary relativistic deflection. *Astronomical Journal*, **102**, 1879–1888.

Turner, E.L. (1980). The effect of undetected gravitational lenses on statistical measures of quasar evolution. *Astrophysical Journal*, **242**, L135–L139.

Turner, E.L. (1990). Gravitational lensing limits on the cosmological constant in a flat universe. *Astrophysical Journal*, **365**, L43–L46.

Turner, E.L., Ostriker, J.P. and Gott, J.R. III (1984). The statistics of gravitational lenses: The distributions of image angular separations and lens redshifts. *Astrophysical Journal*, **284**, 1–22.

Tyson, J.A. (1988). Deep CCD survey – Galaxy luminosity and color evolution. *Astronomical Journal*, **96**, 1–23.

Tyson, J.A., Wenk, R.A. and Valdes, F. (1990). Detection of systematic gravitational lens galaxy image alignments: Mapping dark matter in galaxy clusters. *Astrophysical Journal*, **349**, L1–L4.

Vanderriest, C., Schneider, J., Herpe, G., Chèvreton, M., Moles, M. and Wlérick, G. (1989). The value of the time delay ΔT (A, B) for the 'double' quasar 0957 + 561 from optical photometric monitoring. *Astronomy and Astrophysics*, **215**, 1–13.

Van Waerbeke, L., Mellier, Y., Erben, T., Cuillandre, J.C., Bernardeau, F., Maoli, R., Bertin, E., McCracken, H.J., Le Fèvre, O., Fort, B., Dantel-Fort, M., Jain, B. and Schneider, P. (2000). Detection of correlated galaxy ellipticities on CFHT data: first evidence for gravitational lensing by large-scale structures. *Astronomy and Astrophysics*, **358**, 30–44.

Wallington, S., Kochanek, C.S. and Narayan, R. (1996). LensMEM: A gravitational lens inversion algorithm using the maximum entropy method. *Astrophysical Journal*, **465**, 64–72.

Wallington, S. and Narayan, R. (1993). The influence of core radius on gravitational lensing by elliptical lenses. *Astrophysical Journal*, **403**, 517–529.

Walsh, D., Carswell, R.F. and Weymann, R.J. (1979). 0957 + 561 A, B: twin quasi-stellar objects or gravitational lens? *Nature*, **279**, 381–384.

Wambsganss, J. (1998). Gravitational lensing in astronomy. *Living Reviews in Relativity*, **1**, No. 1998-12. [This is an electronic journal; the url is http://www.livingreviews.org]

Wilkinson, P.N., Henstock, D.R., Browne, I.W.A., Polatidis, A.G., Augusto, P., Readhead, A.C.S., Pearson, T.J., Xu, W., Taylor, G.B. and Vermeulen, R.C. (2001). Limits on the cosmological abundance of supermassive compact objects from a search for multiple imaging in compact radio sources. *Physical Review Letters*, **86**, 584–587.

Wittman, D.M., Tyson, J.A., Kirkman, D., Dell'Antonio, I. and Bernstein, G. (2000). Detection of large-scale cosmic dark matter structure via gravitational lens distortion. *Nature*, **405**, 143–148.

Wolfe, A.M. (1988). Damped Ly-alpha absorption systems. In: *QSO Absorption Lines: Probing the Universe*; Proceedings of the QSO Absorption Line Meeting, Baltimore, MD, 19–21 May 1987, Cambridge University Press, pp. 297–306.

Young, P., Gunn, J.E., Oke, J.B., Westphal, J.A. and Kristian, J. (1981). Q0957 + 561: detailed models of the gravitational lens effect. *Astrophysical Journal*, **244**, 736–755.

Zwicky, F. (1937a). Nebulae as gravitational lenses. *Physical Review*, **51**, 290.

Zwicky, F. (1937b). On the probability of detecting nebulae which act as gravitational lenses. *Physical Review*, **51**, 679.

Zwicky, F. (1957). *Morphological Astronomy*, Springer, Berlin.

Extragalactic astronomy

21

RICHARD MUSHOTZKY*

Clusters of galaxies

Clusters and groups of galaxies are now recognized to be fundamental constituents of the Universe. According to modern theory, most of the material in the Universe that has collapsed into coherent structures lies in groups of galaxies, while clusters are the only entities in the Universe that should be representative of the Universe as a whole. Thus the study of groups and clusters of galaxies is of tremendous importance. One of the most surprising distinguishing characteristics of these, the largest and most massive objects in the Universe, is that they are filled with hot X-ray-emitting gas whose mass exceeds (often by factors of 3–7) the mass of the galaxies in the groups and clusters. This gas is enriched in heavy elements (e.g., oxygen and iron) and is thus the repository of most of the metals in the Universe.

Clusters are also exceedingly sensitive tracers of the cosmological parameters of the Universe (such as its mass density Ω_m, its age H_0^{-1}, and the cosmological constant Λ) and the nature of the primordial spectrum of density fluctuations. Detailed studies of clusters and their evolution as well as their baryonic content are among the most robust estimators of the cosmological parameters (Bahcall et al. 1999) and at present are one of the strongest arguments for a low-Ω universe. X-ray surveys, at present, are capable of detecting clusters in a uniform robust way to redshifts >1.2 and thus provide the fundamental material for the study of cluster evolution and its cosmological impact.

Cluster mergers are thought to be the prime mechanism of massive cluster formation in a hierarchical universe (White and Frenk 1991) and represent the most energetic events in the Universe since the big bang. These mergers with infall velocities of ~2000 km s^{-1} and total masses of $10^{15} M_\odot$ have a kinetic energy of 10^{65} erg. The shocks and structures generated in the merger have an important influence on cluster shape, luminosity and evolution and may generate large fluxes of relativistic particles. These mergers are best studied with X-ray imaging and spectroscopy studies of the temperature, abundances and gas density, which can reveal the entropy and pressure in these systems. X-ray imaging is also required to distinguish between true mergers and the projections that one expects in a hierarchical universe (Henriksen and Markevitch 1996).

Numerical studies of large-scale structure formation show that groups are the tracers of the cosmic web, being the connection of galaxies to clusters in a hierarchical merger "tree". Thus detailed surveys of groups will provide the best measure of the nature and form of large-scale structure, which in turn will provide detailed cosmological information.

Because detailed measurements of the mass, metallicity, evolution, number density, and structure of groups and clusters of galaxies are best performed with X-ray data, X-ray astronomy has become central to the study of clusters. At present, X-ray data can yield the cluster mass, structure, and iron abundances out to redshifts of ~0.8 (Donahue et al. 1999), measuring the abundances and distribution of other elements at $z \sim 0.1$, finding clusters at $z \sim 1.24$ (Rosati et al. 1999), and determining the nature and form of cluster number and mass evolution (Henry 2000) to $z > 0.5$. Because groups are much less luminous, their study so far has been limited to the local volume of space ($z < 0.05$). Since they are extremely numerous there is hope that detailed X-ray surveys optimized for the study of

---

*NASA – Goddard Space Flight Center, Greenbelt, MD, USA

groups will be able to map the cosmic web seen in detailed numerical simulations (Cen and Ostriker 1999).

The hot gas in the cluster potential well will scatter the microwave background and generate a signal whose strength depends on the path integral of the pressure in the cluster but is independent of distance (Sunyaev and Zel'dovich 1981). The ratio of the X-ray properties to the microwave properties is distance dependent, and provides a direct physical means of determining the angular distance to the objects and thus measuring the Hubble constant and Λ. (Frequently the Hubble constant is given in units of h, which means that it is normalized to a value of 100.) The Sunyaev–Zel'dovich effect provides an independent estimate of the cluster gas mass, which has confirmed the X-ray estimates and reinforces the large baryonic ratios in clusters.

Knowledge of the nature of the galaxies in clusters and their evolution with cosmic time has been vastly enhanced by the imaging capabilities of the Hubble Space Telescope (HST). The detailed Hubble images from the UV to the near-IR have enabled the age dating of elliptical galaxies, the most dominant population of cluster galaxies.

The HST optical data, when combined with the X-ray measurements of cluster evolution, indicates that clusters are very old, even the most massive systems having formation epochs at $z > 1$, and most of the stars in the galaxies forming at $z > 2$. This is contrary to the strong predictions of a $\Omega = 1$ closed universe (Turner 1999). The great age of the stars in cluster galaxies also indicates that most of the metals, which are found in the cluster gas, were formed at high z.

The exquisite Hubble images have also enabled the robust measurement of gravitational lensing of background galaxies by the foreground cluster. While the lenses were discovered from ground-based imaging, one requires the HST images to obtain accurate cluster mass profiles. These data can even place constraints on the fine-grain distribution of dark matter in clusters. In addition to providing measurements of the dark matter, lensing amplifies the light from the lensed galaxies, allowing detailed measurements to be made of the properties of high-redshift galaxies (Steidel et al. 1999).

Detailed numerical simulations of cluster formation and evolution which include the effects of dark matter and gas have reached quite high levels of sophistication (Eke et al. 1998). These calculations indicate that the gas is almost in hydrostatic equilibrium and that the clusters should follow a set of scaling laws. One of the most important of these laws is the relation between gas temperature and the total mass of the cluster, including gas, galaxies, and dark matter. Theory indicates that the total mass should scale as the X-ray temperature to the power 1.5 and that for massive systems the X-ray temperature is an accurate estimate of the cluster mass.

The 0.2–15 keV X-ray emission from most clusters of galaxies is dominated by thermal bremsstrahlung and line radiation from a collisionally dominated plasma with about a third of solar abundances (see Sarazin 1988 for a review). While there may be significant contributions at both lower and higher energies from synchrotron or inverse Compton radiation from relativistic particles (Sarazin 1999), the bulk of the radiated energy and the total energy in the cluster is due to thermal particles. The observed range of temperatures for clusters and groups is 0.2–15 keV, and the luminosities range from 10^{42} to 3×10^{45} erg s^{-1} (Mushotzky and Scharf 1997). There is a strong correlation between cluster luminosity and temperature (Mushotzky 1984). The strongest emission line complex from a plasma of $kT > 3$ keV is the He-like 6.67 triplet from Fe XXV, and the first studies of cluster abundances were only able to measure the flux from that line complex. In more modern work, with the ASCA (Advanced Satellite for Cosmology and Astrophysics) and BeppoSax satellites, the emission from the more abundant elements heavier than oxygen has been measured. At the lower temperatures representative of "poor" clusters and groups the spectrum is line dominated, much of the flux being emitted by a veritable sea of Fe L lines.

There is very little variance in cluster metal abundances for massive systems with $kT > 3$ keV, but at lower temperatures (and masses) there appears to be a wide range of abundances and a variation in the range of Si to Fe abundances, indicative of a different formation of heavy elements as a function of cluster mass. These lower-mass systems also have more entropy per unit particle than is indicated by the numerical simulations, indicating, perhaps, that there has been additional energy injected into the intergalactic medium (IGM) other than by gravity. The amount of energy implied is rather large and should have significant effects on group and cluster formation.

Detailed analysis of the cluster temperature and gas density provide, via the equation of hydrostatic equilibrium, robust estimates of the total cluster mass and gas mass. It became clear at the beginning of the 1990s (White et al. 1993) that the dark matter in clusters accounted for "only" ~80% of the total mass. However, in the closed $\Omega = 1$ Universe popular at the time, dark matter should account for ~94% of the total mass. Since clusters are thought to be fairly representative samples of the Universe, this was a strong indication that either the theory of cluster formation had a serious error or that $\Omega \ll 1$, with a best estimate for Ω being 0.3, for example the ratio of the baryons in the clusters to the dark matter compared to the cosmic ratio.

In the centers of a large fraction of luminous clusters the X-ray-emitting gas has a cooling time, in the absence of any known sources of heat, which is considerably less than the Hubble time, so one anticipates that the gas should be cooling (Fabian 1994). In the most spectacular examples of this phenomenon the gas should be cooling at a rate of $>1000\ M_\odot$ yr^{-1}. X-ray spatial and spectral measurements provide striking confirmation of this idea, but the eventual repository of the cooled material has proven to be elusive. The luminous galaxies that reside at the centers of cooling flows have a wide variety of unique properties in the UV, optical, IR, and radio bands which indicate that the cooling flow phenomenon has important consequences not only for the hot gas but also for galaxy formation and evolution. If the final repository of the cooling gas is stars, as it clearly is for some of the material (O'Connell and McNamara 1988), then this process could be a major factor in massive galaxy formation.

The field has undergone an enormous change, driven primarily by a vast improvement in technical capability since the first discovery of X-ray cluster emission during a rocket flight in 1968 (Friedman and Byram 1967). The development of spacecraft allowed the first all-sky surveys with Uhuru in 1971 and Ariel 5 in 1975 (Gursky et al. 1971, Villa 1976), the development of low-background proportional counters resulted in the first X-ray spectra with Ariel 5 (Mitchell et al. 1976) and OSO 8 (Serlemitsos et al. 1977); and the development of imaging X-ray optics made possible the first X-ray images, with the Einstein Observatory (Jones et al. 1979), and the large number of images and spectra with Rosat, ASCA, and BeppoSAX in the 1990s. The development of X-ray photon-counting CCDs combined with high-throughput thin-foil optics led to the first high-quality spectra of numerous clusters with ASCA. The tremendous improvement in sensitivity, spectral resolution, and spatial resolution allowed by the spectacular increase in angular resolution made possible by the Chandra mirror and the technology breakthrough resulting in the very large collecting area of XMM-Newton will push X-ray cluster studies to new levels of sophistication and detail in the first decade of the new century. The spectacular improvement in spectral resolution that would have been brought about by the calorimeter on Astro-E will have to await a future flight of this technology, due to the failure of Astro-E to reach orbit.

To go beyond these three new missions will require even larger collecting areas and better spectral resolution, as will be possible with the Constellation-X mission under study in the USA. While the progress in the 1990s was truly spectacular, we anticipate even more rapid progress in the next decade.

HISTORICAL CONTEXT

(See Bahcall (1977) for an early review.) Clusters of galaxies are the largest virialized structures in the Universe, with masses of 10^{13}–$3 \times 10^{15} M_\odot$ and sizes of 1–3 Mpc. They were discovered rather early in the history of modern astronomy – by Herschel (as noted by Lundmark 1927) – but it was not until the 1930s (Zwicky 1937, Smith 1936) that they were recognized as very large conglomerations of galaxies at great distances. The first dynamical analysis of clusters showed that there must exist much more gravitational material than was indicated by the stellar content of the galaxies in the cluster. This was probably the first discovery of the preponderance of dark matter in the Universe. However, the field did not advance much until the early 1960s, for what seems, in retrospect, to have been the simple reason that astronomers did not want to accept that most of the material in the Universe was dark.

With ever larger ground-based telescopes, cluster research developed along the lines of measuring the numbers, sizes, luminosity, and spatial and velocity distribution of the galaxies. The development of large catalogs of clusters (Abell 1958, Zwicky and Herzog 1963) based on eye estimates of the number of galaxies per unit solid angle was a major milestone in the field. The relatively strict criteria for the Abell catalog proved to be a good guide to the physical reality of the objects, and it is amazing that 40 years later we are still using this catalog. Extensive optical follow-up studies of objects in these catalogs was more time-consuming, and it was not until the early 1970s (Rood 1974) that the first large samples of estimated cluster mass from measurement of the velocity distribution of the galaxies via the use of the virial theorem were obtained.

By the early 1970s it became clear (see the pioneering paper of Rood et al. (1972), and the detailed study of the Coma Cluster by Kent and Gunn (1982)) that clusters of galaxies were dominated by dark matter, with galaxies representing less than 5% of the total mass, and that there were definitive patterns in their galaxy content (Dressler 1980). Thus the issue of the "missing mass" or "dark matter" became the central theme of cluster research.

The nomenclature of the field was partially defined by Abell's survey of the northern sky for clusters based on a visual search of the Palomar Observatory Sky Survey plates. He defined a set of "richnesses" and distance classes based on the apparent magnitudes of the galaxies and the number of galaxies inside a fixed "metric" radius. Abell used strict criteria to define the physical reality of a cluster.

"Rich" clusters (i.e., those with many galaxies inside a fixed metric (Abell) radius) had a preponderance of "early" type (elliptical and S0) galaxies, while "poorer" clusters had a larger fraction of spiral galaxies. It was clear that many clusters had a rather unusual central galaxy, a cD, or centrally dominant galaxy (Morgan and Osterbrock 1969) which is very seldom, if ever, found outside clusters. There was also an unusual type of radio source found primarily in clusters, a so-called WAT, or wide-angle tailed source (Owen and Rudnick 1976). There were the first indications of cluster evolution (Butcher and Oemler 1978) in which distant clusters at $z \sim 0.2$ tend to have "bluer" galaxies than low-redshift clusters (to an optical astronomer elliptical galaxies have rather "red" colors while spirals tend to be bluer), but the morphology of these galaxies was unknown.

However, the fundamental realization (Binney and Strimpel 1978, Heisler *et al.* 1985) that the mass of the cluster could not be precisely determined from galaxy position and velocity distributions severely hampered further progress. The mass of clusters was parametrized by the so-called mass-to-light ratio (in solar units, so the Sun has a value of 1). This is the ratio of the two quantities that can be well estimated by optical data: the total amount of light in the system and a dynamical estimator of the mass (most often the virial estimator). For most clusters the values were $\sim 400 h^{-1}$ (the mass to light ratio depends on the distance to the cluster since the total light is proportional to flux $\times$ distance^2, while the virial mass estimator is proportional to velocity dispersion $\times$ cluster size). There were also technical difficulties with measuring the effective cluster size and the total amount of light, and with determining the velocity dispersion in the presence of sub-structure in the cluster and foreground and background contamination (see Oegerle *et al.* (1989) for a status report).

Since the mass-to-light ratio of "normal" galaxies is in the range of 2–12, this large value for clusters indicated the presence of large amounts of dark matter. However, mass-to-light ratio is not a basic astrophysical quantity since it depends on the age, initial mass function, dust content, and metallicity of a stellar system as well as on the optical color system in which it is obtained. Sensible values from 1 to 20 can be obtained without the need for dark matter. Thus, while these data strongly indicated the presence of dark, non-baryonic material, its abundance and nature could not be strongly constrained.

One of the most surprising discoveries of the space age has been that most of the baryonic material in clusters is in the form of hot X-ray-emitting gas which is enriched in heavy elements to about a third of solar abundances. This discovery came about in stages (see the next section), and it is now realized that X-ray emission from clusters is a dominant process and that, in some sense, clusters of galaxies are X-ray objects. This discovery was essentially entirely unpredicted and has led to fundamental change in our understanding of the Universe. It is rather surprising to realize not only that most of the material in the Universe is dark and non-baryonic, but that most of the baryons in the Universe do not shine in optical light. The anthropomorphic picture that the Universe is best studied with the light visible to our own eyes is not only seriously in error, it drives science in the wrong directions.

Since the discovery of X-ray emission from clusters, the nature of the optical data has changed radically. The ability of multi-fiber spectrographs to obtain velocities for hundreds of galaxies in the cluster and of CCD imagers to obtain accurate multi-color photometry for many hundreds has vastly improved the quality of the data. At present it is the combination of X-ray, optical, and even radio and IR data that yields the best constraints on the physical parameters of clusters.

The ability of modern instrumentation to observe clusters at all mass scales over a wide range of redshifts has made them one of the prime areas of extragalactic research. Their use as cosmological probes, as large samples of galaxies, and even as "fair samples" of the Universe has made the study of clusters a rather exciting area of research. From a theoretical point of view the rather "simple" physics involved in their formation and evolution makes their properties amenable to "first principles" calculation and allows a direct comparison of theory and observation. A quick look at the main journals shows hundreds of papers a year being published in this area, and the rate of increase is rapid.

THEORETICAL CONTEXT

Physical models of clusters

Modern theories of cosmology and large-scale structure are built on the assumption that cold dark matter (CDM) is the main form of dark matter in the Universe and that, based on inflation, there is an initial power density spectrum of fluctuations. This theory has a set of "free" parameters – the Hubble constant H_0, the mass density in gravitating material Ω_m, the fraction of the closure density in cold dark matter particles, the value of the cosmological constant Λ, the normalization of the power spectrum σ_8, the baryonic fraction (the fraction of the closure density in baryons) Ω_b, and the amount of "hot" dark matter. While this is rather a large set of free parameters, most of them are fairly well determined by observation. In this CDM theory, structure forms in a "bottom-up" scenario in which small masses form first and larger structures such as clusters form later. However, this is a vast oversimplification, and a wide range of mass scales collapse at a wide range of redshifts. For a $\Omega_m = 1$, $\Lambda = 0$ universe (the paradigm before the late 1990s), clusters form at rather low redshifts, and Gott and Gunn (1972) referred to the present as "the epoch of cluster formation."

Because clusters are the largest virialized systems in the Universe they are at the tail end of the distribution of fluctuations, so-called 3σ fluctuations. Their formation and evolution is thus very sensitive to the cosmological parameters (see Bahcall *et al.* 1999 for a recent review). Most cosmological models, in fact, are normalized by the COBE spectrum of fluctuations and the abundance of massive clusters – so-called cluster normalized models.

Before the late 1970s theoretical work focused on mechanisms for "relaxation" (so-called violent relaxation

(Lynden-Bell 1967)) and the formation of the cD galaxy and mechanisms for the differentiation of galaxy types in clusters. The first analytic work with the CDM models (White and Rees 1978, Kaiser 1986) showed broad agreement with observations. There was a major change in the 1970s with the advent of powerful computers and algorithms for the calculation of gravitational forces (Aarseth 1978). The advent of X-ray data in the early 1970s stimulated the first models incorporating gas (Cavaliere and Fusco-Femiano 1978) which predicted relations between optical and X-ray properties (the so-called β model). Until the early 1990s most numerical calculations of cluster formation and evolution (e.g., White 1976) focused on the dark matter distribution. The realization that detailed numerical work can essentially calculate all the parameters of the dark matter distribution has made this a powerful tool in the study of dark matter halos. This work has focused on the merger history of clusters, the physical properties involved in cluster evolution, the calculation of the mass spectrum of clusters and its evolution with redshift, the form of the dark matter distribution, and the correlation function. The progress has been tremendous, with the discovery of a so-called universal form of the dark matter potential (Navarro et al. 1997), and of detailed calculations of the mass spectrum (Bryan et al. 1994) and its evolution rate (Cen et al. 1998). There has also been detailed work on the evolution of cluster mergers (Rottinger et al. 1996).

By the mid-1990s it was realized that, while the dark-matter-only calculations were very powerful, they were difficult to compare with observation since they did not have baryons in them. This led to a large body of work incorporating both the dark matter and the X-ray-emitting hot gas (Evrard 1988). Because the dark matter is "collisionless" the inclusion of the gas component required the addition of a large amount of physics (shocks, thermalization, entropy, etc.). However, there are now a large number of observables: the relation between the temperature of the hot gas and the cluster mass and its evolution with redshift, the gas distribution, the X-ray luminosity, the interaction between cooling and heating, the cluster luminosity function, gas and dark matter mass and its evolution with redshift, the cluster temperature function (the number of clusters per unit volume per unit temperature) and its evolution with redshift, the velocity field of the cluster gas, detailed merger maps, the evolution and distribution of cluster entropy, and even galaxy formation (assumed to proceed through gas cooling). The development of these detailed models and the ability to compare them with data has led to a radical change in our understanding of clusters and our ability to determine their detailed physics (see Eke et al. (1998) for a recent summary).

One of the most powerful statements of the nature of cluster theory is its "relative" simplicity as compared with, for example, galaxy formation. As stressed by Ostriker, much of the physics is straightforward (dark matter evolution, gas hydrodynamics, etc.) and thus comparison of theory and data allows a strong test of the theory.

Models of X-ray emission

(See Sarazin (1988) for a review.) The bulk of the energy is radiated via thermal bremsstrahlung, bound–free radiation, and line emission (Raymond and Smith 1977). Measurement of the gas temperature is obtained by fitting the shape of the continuum and the relative strength of temperature-sensitive line ratios (in particular the ratio of the hydrogenic to helium-like lines of iron, silicon, and oxygen). At $kT > 4$ keV the X-ray spectrum is continuum dominated, while at $kT < 2$ keV the spectrum is line dominated. In particular, at $kT < 1.5$ keV the spectrum is dominated by lines from transitions in the L shell of Fe which, for solar abundance plasmas, can have an equivalent width of >1 keV at an effective energy of 1 keV. The other possible sources of continuum radiation are synchrotron radiation from relativistic particles at $E < 0.5$ keV (Sarazin 1999) and inverse Compton (IC) radiation by scattering of microwave background photons by relativistic particles that produce the radio halos sometimes seen in clusters. With a few singular exceptions, both of these processes are much less important than thermal processes. The IC spectrum is considerably flatter than that of bremsstrahlung, and the detection of IC radiation has been predominately at $E > 10$ keV. Measurement of the IC continuum can constrain the magnetic field in clusters, information on which is difficult to obtain in any other way.

If the gas is not isothermal or is cooling, then the integral spectrum will show a range of effective temperatures which may mimic, with low-quality spectral data, an isothermal spectrum. Detailed spectral analysis can derive cluster gas velocities and turbulence from lineshapes, and the degree of equilibrium from ratios of the helium-like triplet.

EARLY X-RAY OBSERVATIONS

(See Gursky and Schwartz (1977) for an early review.) X-ray emission from clusters of galaxies was not predicted, and its discovery was essentially serendipitous. The first detections of what we now know as cluster X-ray emission was from rocket flights in the 1960s (Friedman and Byram 1967, Bradt et al. 1967), which revealed X-ray emission from the direction of the Virgo Cluster, the closest cluster of galaxies. However, the emission was interpreted as emission from the radio galaxy M87. Four years later another rocket flight detected emission from the direction of the Perseus Cluster. In a paper of remarkable prescience, Felten *et al.* (1966) attributed the claimed detection of X-ray emission from the Coma Cluster to thermal bremsstrahlung. These early rocket results were entirely serendipitous, as no one had any idea that clusters of galaxies should be luminous X-ray sources and like so much else in the field of X-ray astronomy, this was an entirely unexpected discovery.

It was the first all-sky X-ray survey, with the Uhuru satellite (Kellogg *et al.* 1971, Gursky *et al.* 1972) that established X-ray-emitting clusters as a class. While having relatively low angular resolution compared with optical and radio telescopes of the time ($\sim 0.5 \times 0.5$ degrees at best), the Uhuru data had relatively small positional uncertainties (error boxes) of ~ 0.05 square degrees for the brightest sources and ~ 5 square degrees for the weakest "real" objects. The dynamic range of Uhuru, ~ 1000 between the brightest galactic sources and the much dimmer extragalactic objects, was vital to the discovery of cluster emission. The relative rarity of optically selected clusters, ~ 1 per 10 square degrees, and the similarly low areal density of the high-galactic-latitude Uhuru X-ray sources of ~ 1 per 100 square degrees, indicated that the presence of an Abell cluster inside an X-ray error box of size less than 1 square degree was statistically rather unlikely (Bahcall and Bahcall 1975) and allowed a relatively high certainty of identification. Much early effort went into the optical identification of high-galactic-latitude X-ray sources, and by the end of 1976 clusters were the only well-populated class, with ~ 30 suggested identifications (Gursky and Schwartz 1977). The Ariel 5 catalog (Cooke and Maccagni 1976) added about 10 more reasonable identifications. The early results clearly indicated that the luminosity of clusters spanned a range of more than 100, from Virgo at $L_x \sim 2 \times 10^{43}$ erg s^{-1} to Abell 2319 at $\sim 3 \times 10^{45}$ erg s^{-1}. The Uhuru sample allowed the first estimate to be made of the cluster luminosity function, the number of clusters per unit luminosity per unit volume, but it depended heavily on what was the nature of the large number of unidentified high-galactic-latitude X-ray sources.

Like so many other early results of X-ray astronomy, the discovery of X-ray emission from clusters was quickly understood and interpreted and led to extremely rapid progress. For the highest signal-to-noise observations, the Uhuru data were interpreted to show that the three brightest and closest clusters (Virgo, Perseus, and Coma) were extended by $>15'$, corresponding to a physical size of ~ 0.5 Mpc. Thus the emission was probably not from a point source, but was likely from a variety of diffuse sources such as inverse Compton emission or bremsstrahlung from a wide variety of physical processes. These same data provided low-quality spectra which could be characterized by a wide variety of spectral forms, but indicated that luminous clusters had either fairly high temperatures, $kT > 2$ keV, or relatively flat power-law slopes. This first set of observational results generated a large number of theoretical speculations, which included models of emission from infall of primordial material, gas in hydrostatic equilibrium, galactic winds, and emission from IC scattering from relativistic particles present in the radio halos of some clusters. The final analysis of the Uhuru data (Forman *et al.* 1978) contained over fifty potential cluster sources, many with secure identifications. It was already clear from these early results that the X-ray selected clusters were not necessarily the same as those from an optically selected set.

The next major step forward was the discovery (Mitchell *et al.* 1976, Serlemitsos *et al.* 1977) using the first generation of X-ray spectrometers on Ariel 5 and OSO 8, of Fe K emission lines from the three brightest clusters in the sky (Virgo, Coma, and Perseus). The spectrometers on these satellites had an energy resolution of $\sim 16\%$ at 5.9 keV, low background, a bandpass of roughly 2–20 keV, a moderate collecting area (~ 250 cm^2), and reasonable exposure times (often >1 day). This combination of properties allowed good-quality spectra which could detect emission lines of equivalent width >150 eV and constrain the broadband spectral shape accurately enough to distinguish between simple models (power law versus bremsstrahlung or black body).

These data definitely established that the prime mechanism for X-ray cluster emission was thermal bremsstralung and that, most surprisingly, the gas was enriched in heavy elements. The derived X-ray temperatures, 10^7–10^8 K, were in rough agreement with the velocity dispersion of the galaxies and the virial theorem, under the assumption that the gas and the galaxies were both in equilibrium and responded to the same dark matter potential. This, in combination with modeling of the X-ray size of clusters, enabled the first estimates of the gas density ($<\rho> \sim 10^{-3}$ cm^{-3}) and of the mass in the X-ray emitting gas to be made, and they turned out to be rather large: $\sim 10\%$ of the virial mass (Lea *et al.* 1973) in all three clusters with well-estimated sizes. This value was confirmed much later by better data. At the time, this large value was not considered to be a problem. The contribution of IC X-ray emission seemed to be small, which allowed a lower limit to be placed on the average cluster magnetic field of $\sim 0.3 \mu$G in the few clusters with diffuse radio sources.

Analysis of the all-sky survey data from the Uhuru mission (Forman *et al.* 1978) and the somewhat more sensitive Ariel 5 satellite (McHardy *et al.* 1978) showed broad relations between optical properties (richness, velocity dispersion, central galaxy density, spiral fraction) and X-ray luminosity (Bahcall 1977, Jones and Forman 1978). Richer clusters (those with more galaxies per unit surface area) were in general more X-ray luminous, but there was a wide dispersion in the relation, giving little direct correlation between the number of galaxies and the X-ray properties. The X-ray luminosity seemed to be proportional to the velocity dispersion of the cluster galaxies to the fourth power (Solinger and Tucker 1974), $L_x \sim \sigma^4$, and the X-ray luminosity was inversely related to the spiral fraction of the cluster. Thus richer, spiral-poor, high-velocity dispersion clusters were more X-ray luminous. Many of the most luminous clusters had massive centrally dominant (cD) galaxies at their center. The relative lack of spirals in X-ray luminous clusters was consistent with the idea that the hot IGM would remove or severely distort the interstellar medium in normal spirals via ram pressure stripping or conductive heating. At last there was some physical basis for the relative distribution of galaxy types in clusters, but there was very little relation with the radio properties.

However, without temperature or structural information the physical basis of these correlations were difficult to identify. The observed range of X-ray luminosities ranged from a low of $2 \times 10^{43}\,\mathrm{erg\,s^{-1}}$ (Virgo) to $\sim 3 \times 10^{45}\,\mathrm{erg\,s^{-1}}$. With this range of luminosities and estimates of sizes and density distributions from optical data, basic physical parameters could be estimated (Bahcall 1977).

A major step forward was the first large sample of X-ray temperature and abundance measurements (Mushotzky et al. 1978, Mitchell et al. 1979) obtained with OSO 8 and Ariel 5. It was found that luminous clusters were all rather hot with $2 > kT_x > 10$ keV, and that the Fe abundance had a rather narrow range between 0.2 and 0.6 solar abundance. (From here onward I use the Anders and Grevesse (1989) solar Fe abundance of 4.6×10^{-5} by number; while this is now out of date, most of the X-ray cluster literature uses this value and it is easier to change one normalization.) There was no correlation of Fe abundance with any other cluster property.

While there was a good general correlation of X-ray temperature and optical velocity dispersion, there was a broad range in the distribution of kT_x/σ^2 which at the time was not well understood (Smith et al. 1979). There were strong correlations with various optical properties, the strongest being between the central galaxy density and X-ray luminosity and temperature (Smith et al. 1979) and between X-ray luminosity and the optical classification of the cluster structure, such that more centrally condensed clusters were X-ray luminous. But correlations with other optical properties proved to be elusive (Bahcall 1977, Forman and Jones 1978). The best correlation between X-ray properties was between temperature and luminosity $L_x \sim T^3$.

The launch of HEAO 1 in 1977, with considerably larger non-imaging proportional counters (called A1 and A2), a Fourier transform "imager" (A3), and a higher-energy detector (A4), considerably increased the number and quality of cluster X-ray temperature and abundance measurements (Mushotzky 1983) and confirmed and enhanced the previous results. In particular, the cluster Fe abundance distribution was now well defined and many of the correlations with optical properties and X-ray temperature were strengthened. In particular, the strongest single correlation using X-ray data alone was between X-ray luminosity and temperature. Since the temperature is a measure of the total mass of the system and the X-ray luminosity is a convolution of the total gas mass and the structure of the system, this showed that clusters had a rather narrow range of parameters and indicated that hotter, more luminous clusters had a larger fraction of their total mass in gas. There was a strong correlation of X-ray temperature and optical velocity dispersion, expected if the hot gas and the galaxies were test particles in the same potential well, but there was a fair dispersion which was not expected. We now know that much of the dispersion was due to foreground and background contamination of the galaxy catalogs used for the velocity dispersion work. The strongest correlation with optical properties was between X-ray temperature and luminosity and the central galaxy density, the galaxy density being linearly proportional to the X-ray temperature. Since modern theory has the virial mass scaling as $(kT)^{3/2}$, this indicates that the relative fraction of the mass in galaxies is decreasing as the total mass and temperature increase.

The A2 and A1 all-sky surveys yielded the first accurate X-ray luminosity function (Piccinotti et al. 1982, Kowalski et al. 1984) for an X-ray selected sample and the first such measurement for an optically selected sample (McKee et al. 1980). The luminosity function was well fitted at high luminosity ($\log L_x > 43.5\,\mathrm{erg\,s^{-1}}$) by a power law and over the whole luminosity range by a Schechter function (which is an exponential times a power law – we now know that much of the flattening of the luminosity function at low luminosities was an artifact of the inability of the proportional counters on HEAO 1 to detect low-temperature clusters). The space density of X-ray selected clusters was similar to that of Abell clusters, and it became clear that X-ray emission was a ubiquitous feature of clusters of galaxies rather than an unusual property. However, there were several X-ray detected clusters which were not in Abell's catalog, and a large fraction of the Abell clusters were not detected. The broad distribution in X-ray to optical properties was not well understood. The highest redshift in these early surveys was $z \sim 0.09$, and thus there was little or no information on evolution. Because these early surveys relied primarily on sealed proportional counters, only systems with $kT_x > 2$ keV could be detected and thus there was little or no information on lower-mass, cooler systems such as groups.

The HEAO 1 A3 experiment provided some limited information on cluster sizes, in particular for clusters that had strong central surface brightness enhancements associated with either a cooling flow (see below) or a central active galaxy. But in general one relied on optical data for a guide to cluster sizes and shapes. In a few clusters (e.g., Perseus, Virgo, and Cygnus A) there was spectral and spatial evidence for emission from an active galaxy in the cluster which contributed <15% of the total luminosity. The A3 experiment also increased the reliability of the cluster identifications by providing "error boxes" of size <1' for several clusters.

The lack of detection of any clusters at $E > 20$ keV by the A4 high-energy experiment (other than that due to the active galaxies at the center of the Perseus and Virgo Clusters) showed that IC emission in clusters was weak and that the average magnetic field in the cluster was greater than 0.1 μG. These data helped to confirm the thermal origin of most of the cluster X-ray flux and the existence of active galactic nuclei in clusters before the advent of X-ray images.

The science of HEAO 1 was severely compromised by its very short life of 17 months and the small size of the instrument science teams. There was barely time to understand the early data before the mission burned up in the atmosphere, and without a guest observer program or a generally usable archive the number of people working on the data was very limited.

Thus before the launch of the first X-ray imaging mission, the HEAO 2 satellite, the basics of X-ray cluster emission had been established. There were hints from the first images obtained during rocket flights that clusters were not spherically symmetric, and (Gorenstein *et al.* 1977) that there was much to learn. As in any other field of astrophysics one realized that spectroscopy and imaging went hand in hand to understand the physical nature of the objects. Up until 1979 imaging was basically a search for sources, with a small amount of spatial information and spectroscopy was performed with low-resolution but broad-bandpass proportional counters. It is hard to imagine how so much progress could have been made in any other field of astronomy with spatial resolutions of arc minutes (at best) and spectral resolutions (E/dE) of 6!

EARLY X-RAY IMAGES AND SPATIALLY RESOLVED SPECTRA

The Einstein observatory (EO), as HEAO 2 was renamed, provided the first capability of X-ray imaging of a large sample of objects. Launched in 1978, the EO was the first sensitive X-ray imaging mission. With a field of view of ~1°, a bandpass of ~0.2–4 keV, a spatial resolution of ~1' for the most sensitive instrument (the imaging proportional counter, IPC), and a flux limit for good-quality images of extended sources about a hundred times better than the detection threshold of HEAO 1, the phase space for discovery was enormous. In addition to the IPC there was a High Resolution Imager (HRI) with ~5" spatial resolution which was about 10 times less sensitive than the IPC, the first moderate-resolution X-ray spectrometer (the Solid State Spectrometer, SSS), and the first high-resolution spectrometer (the FPCS). All of these instruments were to make important contributions to cluster research. In addition, the EO was the first X-ray mission with a guest observer program, which enabled the community to participate in proposing for and analyzing EO data. Because of the pioneering nature of the observations and the relatively unknown properties of the sources many of the observations were severely underexposed. Given the short life of the EO (about $2\frac{1}{2}$ years) there was very little time to go back and examine in greater detail many of the observed objects. It is somewhat sobering to consider, in retrospect, the very short observational lifetimes of the early X-ray missions. In addition, as with previous and subsequent X-ray missions, many of the most important discoveries were serendipitous.

The first set of cluster images, derived primarily from the IPC (see Forman and Jones (1982) for a review), showed that clusters had a wide variety of morphologies, from round, smooth, centrally condensed objects, highly elliptical and non-centrally condensed objects, to objects with multiple components. There was a broad correlation of the X-ray with the optical morphology such that objects with optically luminous central galaxies (cD systems) tended to be X-ray centrally condensed and luminous, and objects which were "irregular" in their optical morphology (Rood–Sastry class III; Rood and Sastry 1971) showed similar structure in the X-ray images. However, there was not a good one-to-one correspondence between the X-ray and optical morphology, which led Jones and Forman (1984) to develop a terminology for cluster morphology based on the X-ray image.

The surface brightness distribution of most of the objects could be incorporated into the β model in which the X-ray surface brightness can be expressed by the azimuthally symmetric analytic form

$$S(r) \sim S_0[1 + (r/a_x)^2]^{-3\beta_{\text{fit}} + 1/2}$$

where S_0 is a central surface brightness value, β_{fit} is the asymptotic slope of the surface brightness distribution, and a_x is a scale length, the so-called core radius. Typical values of β_{fit} are near $\frac{2}{3}$, and the observed range varied from 0.5 to 0.9. The observed values of a_x varied from ~100 to 400 kpc. However, in these early data there were strong correlations of β and the core radius. While the model fitted the data well, its accuracy could not be tested because of the relatively low surface brightness of clusters at large radii compared with the detector background, and thus the asymptotic form of the surface brightness was poorly known.

The model can be analytically inverted to derive the gas density distribution

$$n(r) = n_0[1 + (r/a_x)^2]^{-3\beta/2}$$

This led to a moderately accurate determination of the total gas density and the first proof that the fraction of the baryons in the gas phase was larger than that in the galaxies. Thus, fundamentally, clusters of galaxies are X-ray objects and the total fraction of the mass that is in hot gas is $\sim 0.15\,h^{-3/2}$. The full impact of this was not realized for eight more years, until theoretical work (White *et al.* 1993) showed that it was unlikely that cluster formation and evolution could concentrate gas and that clusters should be fair samples of the Universe. The fact that clusters were baryon rich, with baryonic fractions of $0.15\,h^{-3/2}$ compared with $0.06\,h^{-3/2}$ for the Universe as a whole, is a strong indication that the value of Ω_m was <1.

Calculation of the cooling time in the central regions (Fabian and Nulsen 1977) based on the cluster average temperature and the observed density indicated that, in many clusters the cooling time, in the absence of any heat sources, was considerably less than the Hubble time. These "cooling flow" clusters all showed surface brightness profiles steeper than that of the β model in their central regions (see Figure 1 for surface brightness distributions of cooling flow and regular clusters) and almost always had a centrally dominant (cD) galaxy at their center. Optical and radio follow-up of these systems often showed that they had extensive systems of optical emission line filaments (Cowie *et al.* 1983) and very high Faraday rotation measures for the central radio sources (Ge and Owen *et al.* 1993). The implied cooling rates of these systems ranged from 10 to 1000 $M_\odot$ yr^{-1}.

Confirmation that the centers of some of these clusters were indeed cooler than the cluster average temperature came with data obtained by the SSS and the FPCS on the EO. These detectors had much better spectral resolution than the imaging detectors on the EO, and went to a lower energy than the older proportional counters, but had relatively coarse spatial resolution. The signature of the cooler gas was X-ray emission lines from iron, oxygen and silicon that could not arise in the hotter cluster gas (Figure 2). However, the limited sensitivity and spatial resolution allowed only a small number of objects to be observed, and did not allow a strong test of the cooling flow theory. Catalogs of cooling flow clusters and estimates of cooling rates were obtained based on IPC and HRI surface brightness data (Arnaud 1988, Edge *et al.* 1992) and the average cluster temperature, and it was found that a large fraction of clusters could harbor cooling flows. Despite this extensive X-ray work and the strong indications from other wavelength bands (see Fabian (1994) and

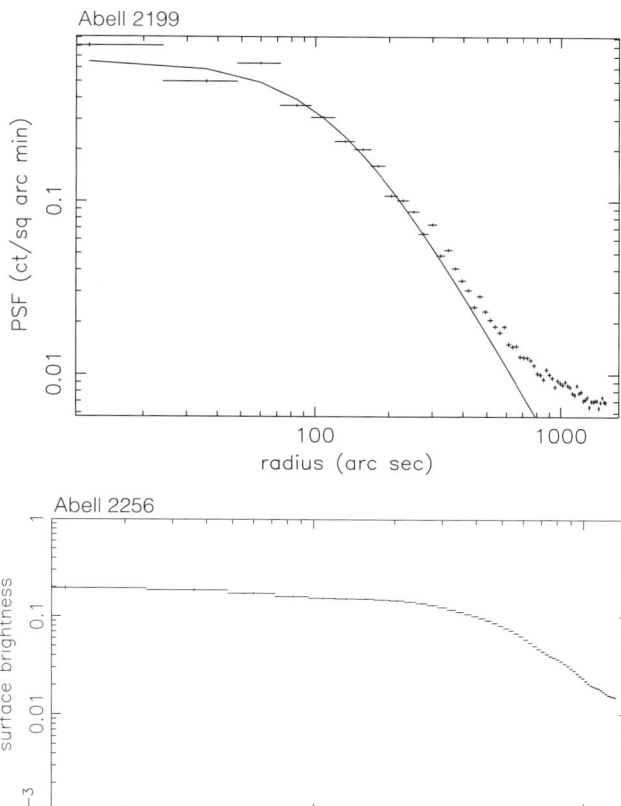

Figure 1 Einstein IPC surface brightness profile for Abell 2199, a cooling flow cluster, and Abell 2256, one without cooling flow. The solid line is the β model fit without background subtraction. Note that the source is only brighter than the background at $r < 5'$ in radius, corresponding to a distance of <500 kpc. Since these are two of the ten brightest clusters in the sky, this shows both the power of the first X-ray images, as well as the limitation of the technique in detecting clusters of large radii. Note the sharp rise in surface brightness in the central regions of Abell 2199 compared to Abell 2256.

Mushotzky and Szymkowiak (1988) for reviews), the eventual reservoir of the cooled material has not yet been found. One of the chief puzzles was the discovery (White *et al.* 1991) that the cooling flows showed a strong indication of absorption in the X-ray band, yet there was little or no indication of this cool material in any other wavelength band.

Cooling flows have a tremendous implication for cosmology and galaxy formation since all theories of structure formation require the cooling of gas from the virial temperature. Since the virial temperature (at least at the present epoch) for all collapsed structures with mass equal to or greater than normal galaxies is greater than 10^6 K, structure formation requires cooling flows. It is a sad commentary

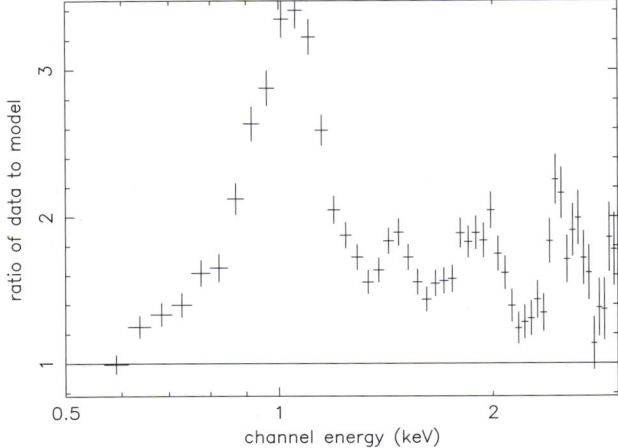

Figure 2 The Einstein SSS spectrum of M87, the central galaxy in the Virgo Cluster. The figure is the ratio of the data to a simple continuum model. This spectrum clearly shows lines from the Fe L blend near 1 keV, helium-like Si near 1.83 keV, and helium-like S near 2.64 keV. The strengths of these lines requires gas of considerably lower temperature than is typical of the outer regions of the Virgo Cluster.

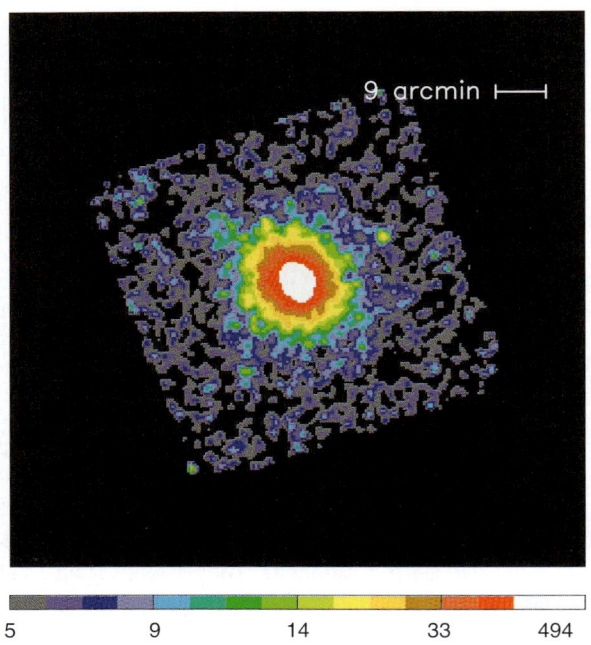

Figure 3 Einstein IPC image of Abell 2199. The exposure time was about 4800 sec, and the field of view is 1°. The colors are chosen to represent the intensity in arbitrary units. The source looks smooth and round.

on the state of scientific research that the mere existence of cooling flows, not to say their fundamental importance, has not been well accepted by the optical astronomy community to this day.

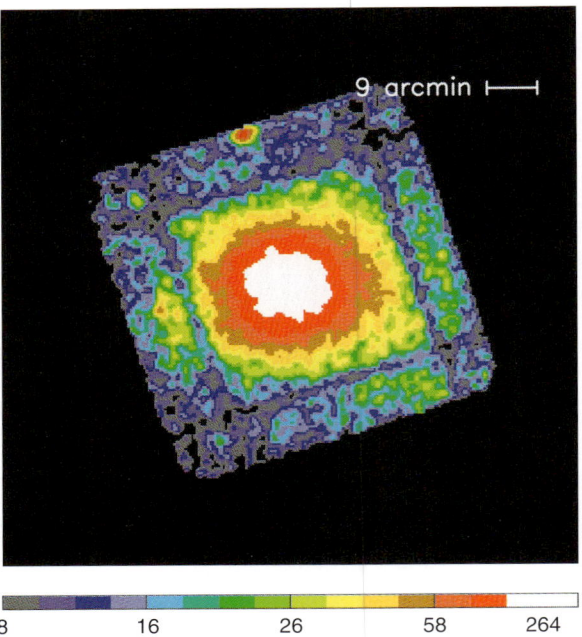

Figure 4 Einstein IPC image of the Coma Cluster (Abell 1656). The cluster is rather elliptical in shape and shows a much flatter surface brightness profile than Abell 2199.

The wide variety of X-ray morphologies (Jones and Forman 1999) can be connected to some other X-ray properties. Luminous systems tend to be either centrally condensed, round, and smooth (a good example is Abell 2199, Figure 3) or elliptical with a large core radius (such as the Coma Cluster, Figure 4). Most of the objects with complex morphologies (such as Abell 194, Figure 5, or the Virgo Cluster, Figure 6) are low-luminosity systems. Some luminous double systems have been seen. An attempt was made (Jones and Forman 1984) to connect the cluster morphology to the evolutionary state of the cluster. The basic idea (see section on "Theoretical context") was that, since clusters are thought to grow hierachically, the round smooth clusters were old relaxed systems which had not undergone a recent merger, the highly elliptical systems were recent merger products, while the clusters with highly clumpy morphology were pre-merger systems soon to collapse. While in broad agreement with the weak theoretical predictions at the time, a lack of homogenous uniform X-ray samples and of a strongly predictive theory prevented detailed comparisons.

The great sensitivity of the EO allowed the first measurements of the X-ray sizes and luminosities of high-redshift ($z > 0.3$) clusters (Henry et al. 1979) and, via the Einstein Extended Medium Sensitivity Survey (EMSS), determination of the luminosity function at redshifts >0.1 and its evolution (Gioia et al. 1990). The EMSS was a

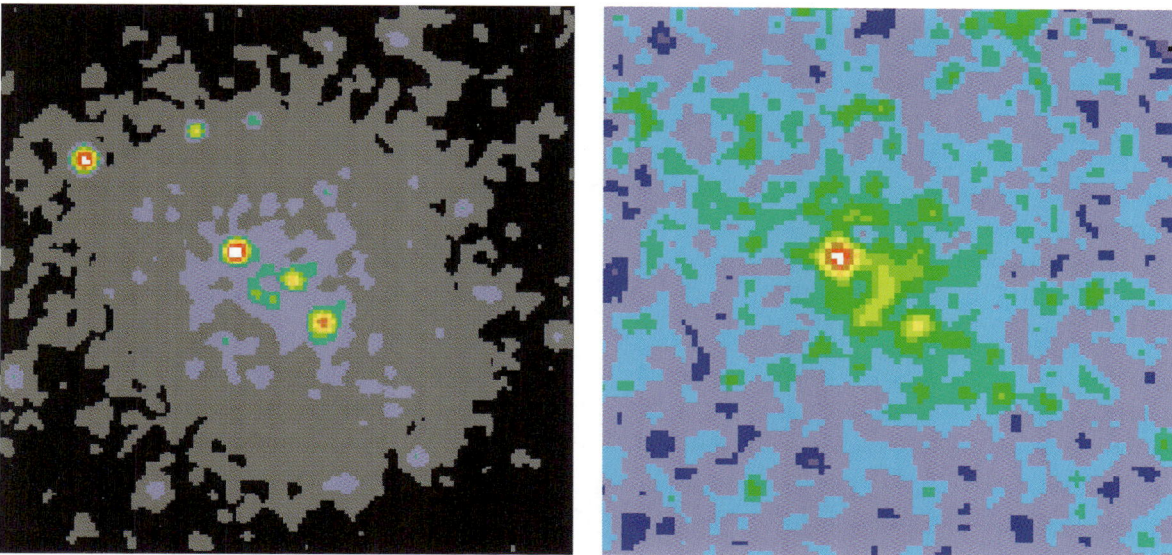

Figure 5 Einstein IPC image of the whole cluster (left) and Rosat PSPC images of the central region (right) of Abell 194. Note the very asymmetric structure, and the evidence for emission from individual galaxies. This object, while also a rich cluster, is rather low in luminosity and temperature.

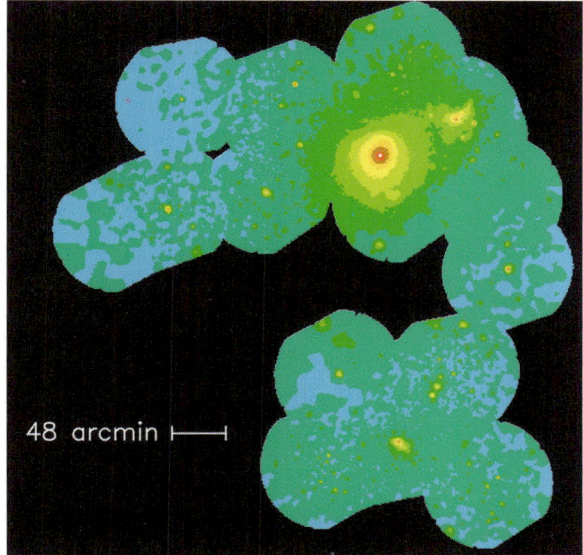

Figure 6 A montage of the Virgo Cluster made from Rosat pointed observations over a solid angle of about 40 square degrees. The wide field, low background, and good imaging qualities of the Rosat PSPC are evident. The wide range of structure, seen in this, the closest of all clusters, is evidence for the formation of clusters via merging.

major effort at identifying serendipitous sources seen in the field of view of the IPC and resulted in the first large catalog of high-redshift X-ray selected objects. Because of the completeness and uniformity of the EMSS sample, these data were the best for over a decade for studying high-redshift clusters in all wavelength bands. In particular they were used for the first study of the velocity dispersion of high-z clusters and estimates of their mass (Carlberg et al. 1997) and a search for strong gravitational lensing (Le Fevre et al. 1994). The early results indicated that there was no strong evolution in the cluster luminosity function with redshift out to $z \sim 0.5$, but there was a statistically significant reduction in luminous clusters at high redshift.

Because of its increased sensitivity and lower energy response than previous experiments, and since $M \sim T^{3/2}$, the EO also allowed the first detection of X-ray emission from much lower-mass systems. Because of the lack of an X-ray survey, the first objects studied were selected from optical catalogs and were primarily elliptical galaxies (Forman et al. 1985 and Trinchieri and Fabbiano 1985) in the Virgo Cluster, but a few unusual objects (such as NGC 3607 and NGC 5846) were detected which hinted at emission from groups. The X-ray emission from ellipticals was often extended beyond the optical image, and if this were caused by hot gas in hydrostatic equilibrium, it implied very extended dark matter halos (Matthews 1978). The lack of good-quality X-ray spectra prevented this conclusion from being confirmed.

The fundamental breakthroughs of the EO were to set a standard for the next decade. The vast improvement in X-ray imaging was not really matched by X-ray spectroscopy, and for a while there was a severe mismatch between the two since the EO had improved the imaging capability by factors of >100 in both angular resolution and sensitivity, but it had much less improvement in X-ray spectral capability.

THE 1980s: EXOSAT, TENMA, GINGA, BBXRT, SPACELAB AND SPARTAN

In the 1980s the field returned to results driven primarily by non-imaging detectors. However, these instruments had the enormous benefit of the EO X-ray images to guide the observations, and thus had a large sample of objects to observe whose nature was known. With their relatively long life and somewhat improved sensitivity over HEAO 1, but with fundamentally similar technology, Exosat and Ginga obtained more precise X-ray temperatures and abundances for a larger sample. This also represented the entry of Europe and Japan into the field. The large archive and guest observer program of Exosat also considerably increased the number of active researchers.

These improvements allowed the construction of well-chosen complete samples and the establishment of robust correlations between various X-ray and optical parameters (Edge and Stewart 1991). The greater statistical accuracy of the Ginga cluster data (Figure 7) confirmed that the Fe abundance had a rather narrow range and showed no or only a weak correlation with other properties. Ginga also provided the first X-ray spectra of objects at $z > 0.1$ (Arnaud et al. 1991) and showed that they were very similar to those of lower-redshift clusters. The 2×1 degree field of view of Ginga allowed a measurement of the temperature and abundance distribution in the Virgo Cluster and established that it was roughly isothermal but had a significant abundance gradient (Takano et al. 1989). Similarly, the 45' field of view of the Exosat ME allowed an abundance and temperature profile of the Coma Cluster (Hughes et al. 1988) to be determined which indicated that, contrary to the Virgo Cluster, the abundance was uniform but the temperature dropped with radius. As before these two clusters and the Perseus cluster were the only clusters large enough and bright enough to allow spatially resolved spectra to be obtained with these instruments, a situation similar to that for X-ray surface brightness with Uhuru 15 years earlier.

The spectral resolution afforded by the Tenma gas scintillation proportional counters was better than that of its predecessors by a factor of 2, which allowed the first accurate measurement of hydrogen as well as helium-like Fe emission from three clusters (Okumura et al. 1988) and hinted at what would be possible with higher-resolution spectroscopy. However, as opposed to other fields of X-ray astronomy (in particular X-ray binaries), these missions did not plow fundamentally new ground in the area of cluster research.

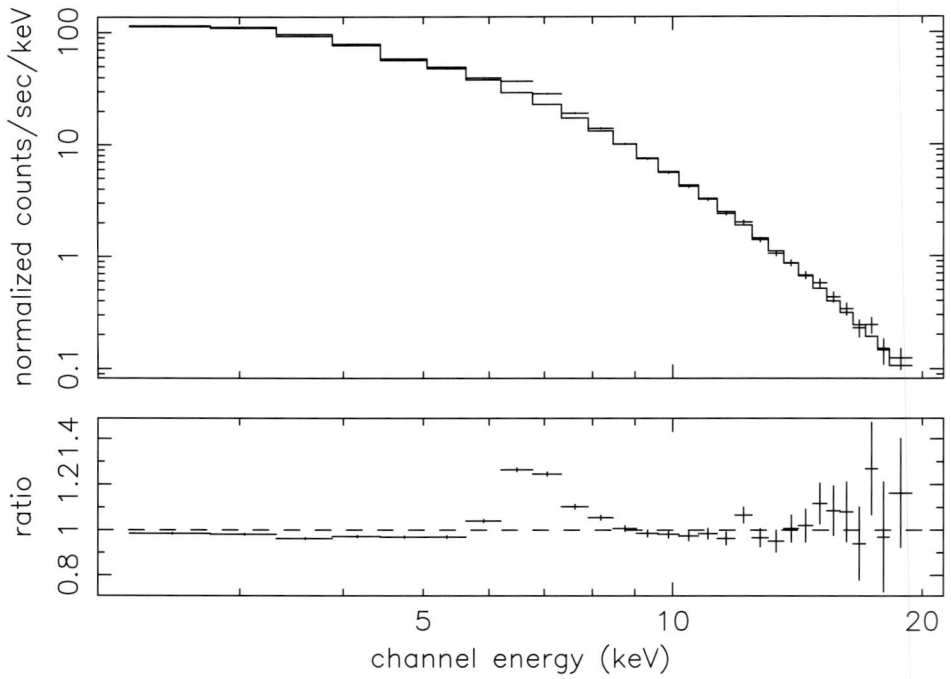

Figure 7 The Ginga spectrum (best-fit model) of the Perseus Cluster. Notice the high signal-to-noise ratio out to ~20 keV. In the bottom panel the ratio of the data to a model without an Fe helium-like K line is shown. The very high significance of this feature allowed accurate determination of the Fe abundance.

The first spectral images of clusters were obtained from the BBXRT, Spacelab, and Spartan experiments flown for short durations on the Space Shuttle. These data, primarily for the Perseus and Coma Clusters, showed the presence of a cooling flow in the spectral domain, confirming the earlier EO results, and provided the first robust temperature maps. These results indicated that clusters were not all alike in their temperature structure, and provided the first accurate estimates of cluster masses. The BBXRT spectra confirmed the apparent presence of absorption discovered by the Einstein SSS in several cooling flows. However, the short observation times and limited number of observations of these shuttle-borne experiments, combined with several operational difficulties, severely limited the scientific usefulness of these experiments.

THE 1990s: ROSAT, BEPPOSAX AND ASCA

The modern era of cluster research began with the development and launch of these three missions, which allowed for the first time true imaging spectroscopy combined with large sample sizes.

Rosat, a German–UK–US collaboration, with roughly three times the sensitivity of the EO, three times the spatial resolution of the imaging proportional counter (called the PSPC), twice the field of view and lower background and better spectral resolution than the IPC, was a giant step forward in X-ray imaging and low-spectral-resolution imaging studies (Figure 4). In addition to a five-year program of pointed observations, Rosat carried out the first imaging all-sky X-ray survey (Trümper 1992) which was roughly 30 times more sensitive for clusters than the HEAO 1 all-sky survey. The all-sky survey, combined with the large number of pointed observations, allowed images of very large objects, such as the Virgo Cluster (Figure 6) to be obtained. The long life of Rosat combined with a vigorous guest observer program resulted in a large database of well-exposed clusters over a wide range of redshift. The soft bandpass (0.2–2 keV) allowed Rosat to be sensitive to the entire temperature range of clusters as opposed to the non-imaging proportional counter surveys. The very high productivity of Rosat in this field is shown by the almost 200 refereed cluster papers using Rosat data.

The large collecting area, moderate spectral resolution ($E/\delta E \sim 45$ at 6 keV), broad bandpass (0.5–10 keV), and adequate spatial resolution of ASCA (~1.5′ radius) (Tanaka et al. 1994) made possible the first detailed X-ray spectroscopy of large numbers of clusters, the determination of elemental abundances other than that of Fe (Figure 8), and the measurement of the temperatures and abundances of groups. Because of the complex point-spread function (PSF) of the telescope, spatially resolved spectra turned out to be difficult, but many robust results have been obtained. ASCA is the spectral match to the Einstein IPC, having similar spatial resolution and a similar sensitivity for spectroscopy that the IPC had for imaging (Figure 9).

The combination of Rosat and ASCA data have made it possible to measure the total mass and mass profile of clusters, the chemical abundance distribution, and the evolution of these quantities to $z \sim 0.5$, providing fundamental cosmological information as well as information on the physics of clusters.

BeppoSAX has a similar field of view but a sharper PSF than ASCA, and broader energy coverage and better spectral resolution than Rosat. For many clusters the BeppoSAX data allow the first accurate determination of abundance and temperature maps and profiles without correction for a complex PSF (DeGrandi and Molendi 1999). The unique capabilities of BeppoSAX for relatively sensitive observations at $E > 10$ keV have opened up a new phase of cluster research, the search for non-thermal emission. However, the lower sensitivity of BeppoSAX than ASCA in the 0.5–8 keV band (by a factor of 4) requires longer observations to obtain similar signal-to-noise ratios and, so far, the results, while interesting, have been somewhat limited.

Cluster evolution, luminosity function, and large-scale structure

The Rosat all-sky cluster surveys based on an optical sample of clusters, XBACs, and purely based on X-ray data, the RBCS (Ebeling et al. 1996, De Grandi and Molendi 1999), have established a robust cluster luminosity function based on over 200 clusters. The X-ray selection technique relied on the fact that at high galactic latitudes clusters are the only extended X-ray sources, and thus, even with the ~1.5′ angular resolution of the Rosat survey data, clusters could be well separated from point sources out to redshifts of ~0.5. Only 70% of the X-ray selected cluster sample are also Abell clusters, consistent with the broad spread in X-ray to optical properties. There are two large peaks in the redshift distribution at $<z> \sim 0.035$ and 0.08 which are due to four superclusters. The all-sky survey data are roughly complete at a flux threshold of 10^{-12} erg cm^{-2} s^{-1}, and for a typical cluster of $L_x \sim 10^{44}$ erg s^{-1} are complete to $z \sim 0.3$. These data represent the largest complete cluster samples ever obtained.

The luminosity function can be well fitted by a power law with a slope of 1.5 over a factor of 300 in luminosity, steepening at $L_x > 3 \times 10^{44}$ erg s^{-1}, with a volume density at 10^{44} erg s^{-1} of $\sim 5 \times 10^{-7}$ Mpc^{-3}. The EMSS indicated (Henry et al. 1992) that there was substantial negative evolution in the cluster luminosity function at the high-luminosity end at $z > 0.3$. The evolution of the cluster luminosity function has been measured by four groups using

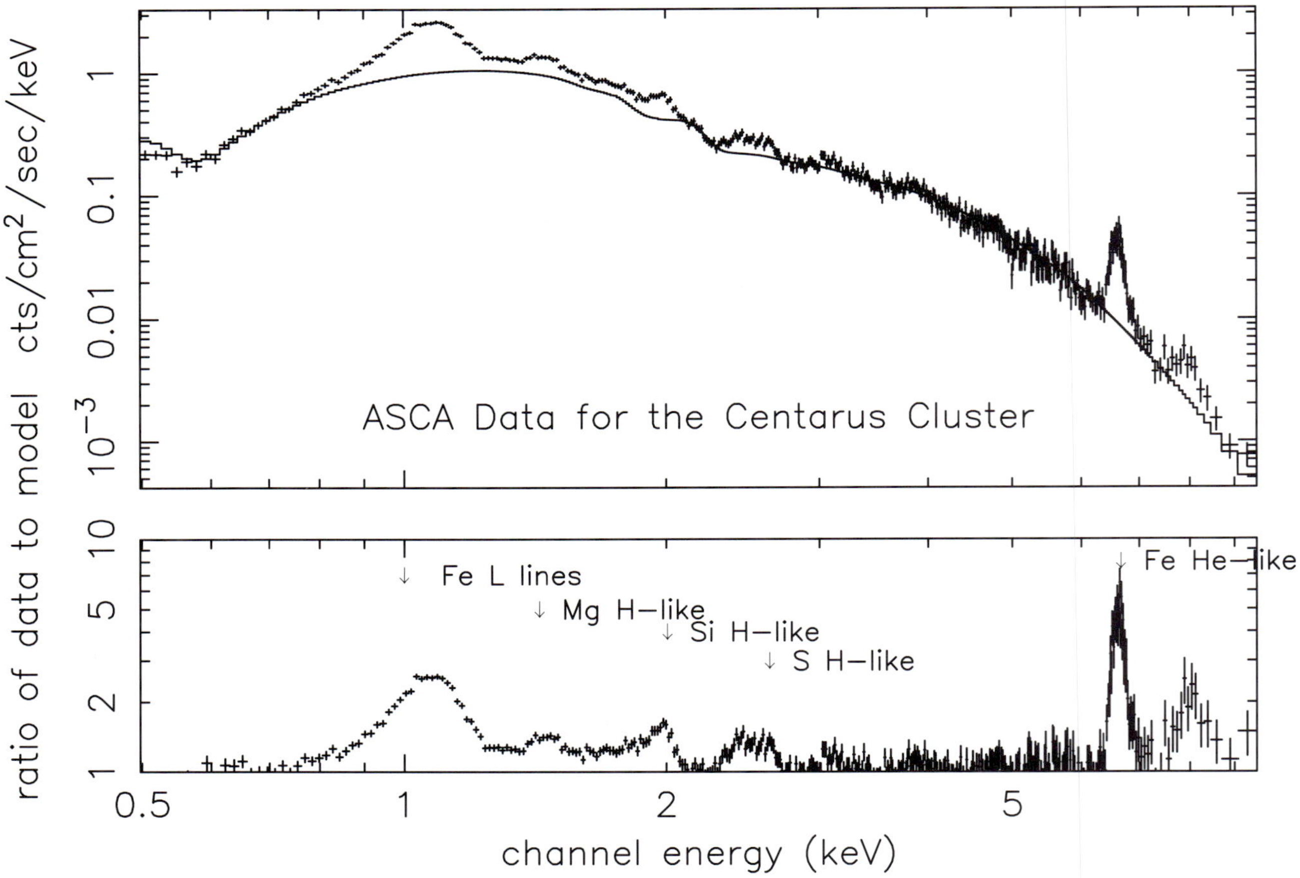

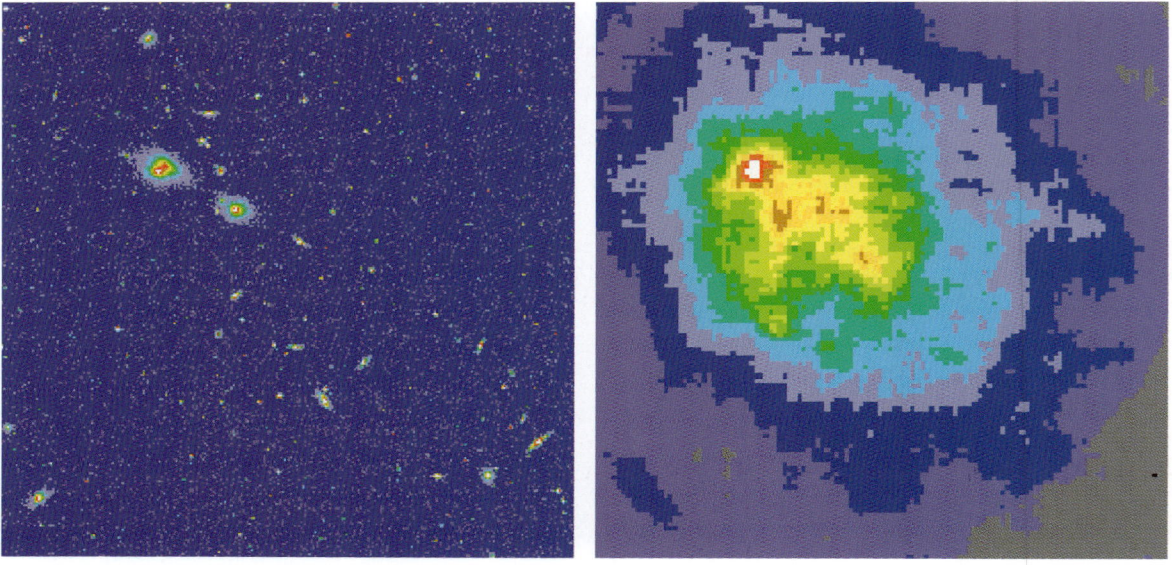

Figure 8 ASCA data for the Centaurus cluster. In the top panel the raw data is shown with a bremsstrahlung continuum fit. In the bottom panel the ratio of the data to this continuum is shown and the strong line emission from Fe L, Mg, Si, S and Fe K lines are labelled.

Figure 9 A comparison of (right) the ASCA GIS image of Abell 194 (Figure 5) and (left) the Palomar Optical Sky Survey data. The relative quality of the ASCA image can be directly compared to the Einstein IPC. Also note the lack of detailed correspondence between the optical and X-ray images, but the overall general similarities.

Rosat pointed data (e.g., Jones et al. 1998), searching for serendipitous clusters in the same general manner as for the EMSS. There is general agreement that the function at log $L_x < 44 \, \mathrm{erg \, s^{-1}}$ evolves only weakly out to a redshift of ~0.5. At higher luminosities there is some disagreement and, if there is any evolution, it is in the sense of fewer luminous clusters at higher redshifts. Comparing the observed cluster luminosity function with theoretical models favors a low-density universe with $\Omega < 0.6$, since in higher-density universes clusters evolve rapidly with redshift.

This work has resulted in the discovery of very-high-redshift clusters (Rosati et al. 2000) (Figure 10), there now being four $z > 1.1$ X-ray selected clusters based on Rosat data. Given the sensitivity and angular resolution of Chandra and XMM-Newton, similar objects could be detected at $z > 1.8$ (depending primarily on cosmological model). The EMSS sample has also been used to find gravitational lenses and arcs and as the basis for detailed optical work. We anticipate that the Rosat samples will extend and enhance this.

The Rosat all-sky data have allowed the first robust detection of large-scale structure using clusters as tracers (the REFLEX survey; Guzzo et al. 1999). Because of the low space density of luminous clusters, a large cluster survey automatically probes very large scales. Since the X-ray properties of the clusters are much more closely connected to mass than the optical properties are, and the selection effects are more easily understood, X-ray surveys are at present probably the best way of determining the form of large-scale structure. The REFLEX data show that the cluster two-point correlation function is extremely similar in form to that of the galaxies, but amplified by a factor of ~10 in power, as predicted if clusters were more strongly biased tracers of mass than galaxies. The power spectrum is robust and shows a turnover at the largest scales, roughly consistent with $\Lambda \sim 0.7$ cosmologies.

X-ray emission from groups

Groups lie at the low-mass end of the cluster spectrum, falling roughly logarithmically between rich clusters at $M \sim 10^{15} M_\odot$ and galaxies at $M \sim 10^{11} M_\odot$. Integration of the theoretical mass functions indicate that most of the mass in collapsed objects in the Universe lies at the mass scale of groups. Since the formation of galaxies is dominated by non-linear processes, groups are the lowest-mass systems that are easily studied theoretically. Also, they are rather

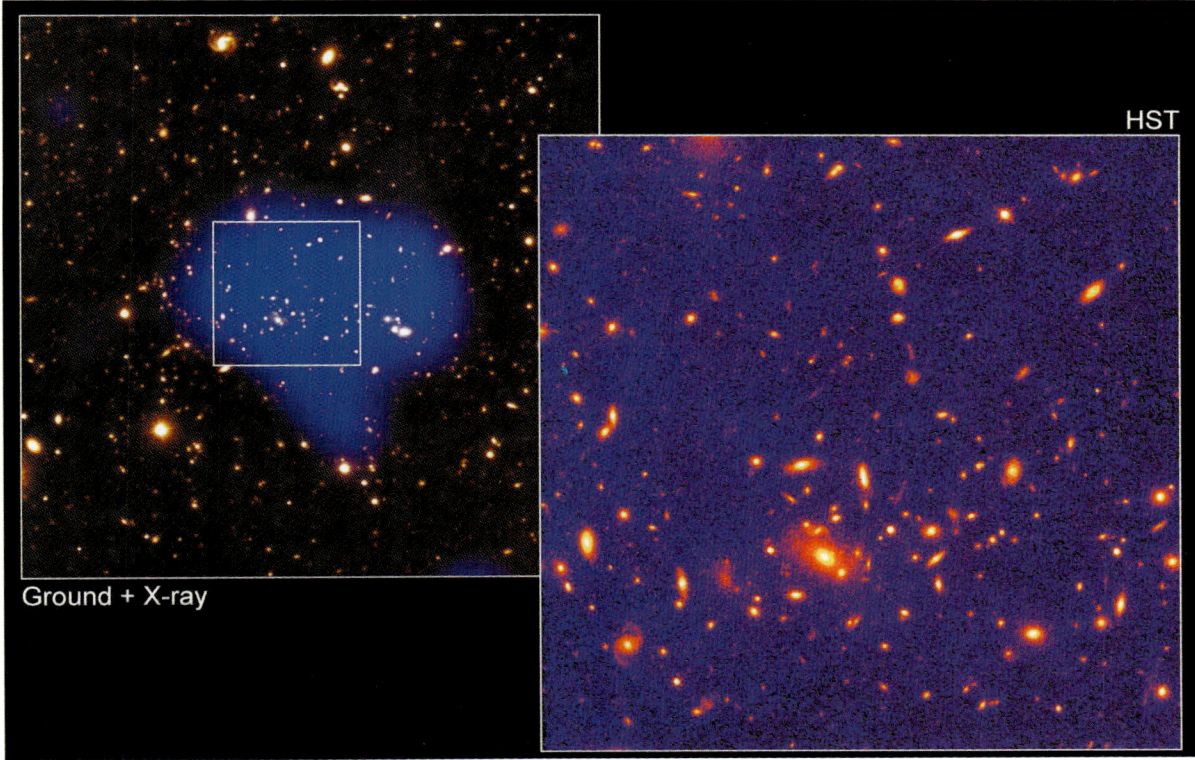

Figure 10 ASCA and ground-based (left) and HST (right) images of a high-redshift $z = 0.83$ cluster, MS1054. The ability of ASCA to measure the temperature and of HST to measure the morphology of the galaxies extends to the highest-redshift clusters so far discovered. (M. Donahue (STScI) and NASA.)

more numerous than rich clusters. However, because of their low galaxy numbers and low velocity dispersion, groups are rather hard to study optically, and even developing a robust group catalog is rather difficult.

There were indications in the early EO data of X-ray emission from groups (Biermann and Kronberg 1983), but the high background and limited sensitivity of the EO IPC prevented much progress from being made in this field. A breakthrough was achieved with Rosat (Mulchaey *et al.* 1993), and there are now over a hundred groups with known X-ray emission (Mulchaey 2000). Consistent with their mass, groups tend to have X-ray temperatures in the $kT \sim 0.3 - 2$ keV range. The X-ray luminosity is only weakly related to the optical properties of the groups, and there are luminous (log $L_x > 43$) systems (Ponman and Bertram 1993) in which there is only one luminous galaxy. The lack of correlation with optical properties is in some way an extension/enhancement of the wide scatter seen in the X-ray to optical properties of rich clusters. Recently, it has been shown that only X-ray luminous groups possess large numbers of low-luminosity galaxies, while optically selected groups which have weak or no X-ray emission are devoid of these objects. The X-ray selected groups are almost always dominated by luminous spheroidal systems with absolute magnitudes $M < -21.5$. There is at present speculation that these galaxies are the result of runaway mergers. However, it is clear from these data that optical group catalogs are very incomplete and represent a poor census of these, the "typical" places in the Universe.

The ability of X-ray imaging spectroscopy to derive the mass of these systems in a fashion similar to that used for rich clusters allowed the first robust estimate of individual group masses to be made, since with only 1–10 luminous galaxies in a group the use of the virial theorem is not justified. The derived masses tend to peak at $\sim 10^{13} M_\odot$, and the range of gas to total mass shows a much wider range than in rich clusters, many of the groups having a significantly smaller baryonic gas fraction, at least in their central regions. Their surface brightness distributions tend to be flatter than that of clusters, but there are systematic uncertainties because of their low surface brightness and the effects of background subtraction. Their low surface brightness makes it rather difficult to measure their gas distribution to large-scale lengths, and it is possible that large amounts of gas exist beyond the well-observed regions. The measured space density is large (Burns *et al.* 1996). Because the X-ray properties of groups are directly related to their mass and because these systems are a thousand times more numerous than rich clusters, an X-ray survey of these systems should be an excellent tracer of large-scale structure.

The ASCA spectra of groups (Davis *et al.* 1999) shows a much wider range in chemical abundances than in clusters, many of the groups having rather low abundances. There may be, on average, a lower Si/Fe ratio than in clusters. The only alternative to the apparent low abundances is if a significant fraction of the emission is produced by a relatively lineless continuum, originating in either much hotter gas (Buote 2000) or a non-thermal continuum. To this reviewer both of these options seem rather unlikely at present, but detailed observations by XMM-Newton and a future replacement for Astro-E will settle the issue.

The relatively low gas content and the low abundances mean that groups tend to have a low ratio of metals to starlight (the metal mass-to-light ratio of Renzini *et al.* (1993)). Since almost all theories of metal formation and stellar evolution indicate that there should be roughly equal amounts of metals formed per unit optical light (for a reasonable range of stellar initial mass functions (Loewenstein and Mushotzky (1996)), this is quite hard to understand. There are several possible explanations. First, the implied abundances are wrong because of the use of Fe L lines to derive the Fe abundance in these systems. Since the Fe L lines are not well resolved in the CCD data and have relatively large atomic physics uncertainties, there is some concern about the robustness of the abundance determination. However, the Si abundances are robust and show the same effects. Second, much of the processed gas might be ejected by galactic winds (David *et al.* 1991). This requires about 1 keV of energy per particle, which is consistent with the energy required in models of cluster evolution (see the section on "The extra energy problem" below). If the ejection idea is correct, most of the metals created in the stars that reside in groups should be in an ionized IGM. Simple calculations (Davis *et al.* 1999) indicate that the IGM would be enriched to >0.04 solar abundance by such a process. The large amount of required energy should have a strong effect on the X-ray surface brightness profile, the gas temperature, and the evolution of groups. All of these effects have been seen to a greater or lesser extent (Lloyd-Davies *et al.* 2000, Loewenstein 2000). The magnitude of the observed effects is roughly consistent with that derived from semi-analytic models of galaxy formation in which feedback from star formation is taken into account. In these models over 70% of all baryons either never collapse into or are ejected from galaxies.

Thus the X-ray observations of groups have the potential for radically changing our view of the Universe from one in which most metals live in stars and gravity totally dominates at all scales to one in which energy injection processes such as supernovae and quasars have an important role and most of the metals in the Universe are in the IGM. These results essentially imply that large-scale structure, metal formation, and tracing of the baryonic Universe require X-ray data. It seems likely that Chandra and XMM-Newton will strongly test these ideas.

The baryonic and total mass of clusters of galaxies

One of the oldest questions in cluster research concerns the amount of and distribution of dark matter in clusters. It was realized in the early 1980s (Fabricant et al. 1980) that if the gas in clusters is in hydrostatic equilibrium with the dark matter potential well, then spatially resolved X-ray spectra could determine the total mass of clusters as well as the gas mass. For a spherical system,

$$M(<R) = -\frac{kT(r)R}{(\mu m_p G)}\left(\frac{d\log T_x}{d\log r} + \frac{d\log \rho}{d\log r}\right)$$

and thus the total gravitational mass depends only on the X-ray observable temperature T_x, its derivative, and the gas density ρ and its derivative. The density distribution is derived directly from the X-ray surface brightness profile. However, it required the spectral capabilities of Rosat and ASCA to obtain spatially resolved spectra.

Because of the relatively poor energy resolution of Rosat and its limited bandpass, the detailed analysis of the temperature profiles is limited to $kT < 4$ keV systems. The first results on the cooling-flow clusters Centarus, Virgo, AWM7, and A0335 + 096 (see Irwin and Sarazin (1995) for a typical analysis) showed a temperature drop in the center and a roughly isothermal profile extending out to $\sim\frac{1}{4}$ of the virial radius (the virial radius is that radius within which the cluster is a bound system, matter outside the virial radius is still falling in or is not bound). For the highest signal-to-noise systems (e.g., the Centarus cluster; Allen and Fabian 1994) the error in temperature was ~15%. Similar results for several groups (WP23, HCG62, and NGC 2300) were obtained. The inferred baryonic fraction (the baryonic fraction is the fraction of the total mass that is in either stars or gas, the rest is in dark matter) was ~15%, with an indication that it was increasing at larger radii. These results strongly supported earlier claims of a large baryonic contribution to the total mass (see next section). However, the limited quality of the Rosat data and the inability to obtain accurate results for high-temperature systems still left some doubts.

ASCA, with its broader bandpass and better spectral resolution, should have obtained much more precise temperature profiles for a wider range of systems. However, the ASCA telescope PSF is complex and has a weak energy dependence, making analysis somewhat difficult. There have been several analyses of the temperature distributions in clusters using ASCA. Markevitch et al. (1998) found a general gradual temperature decrease with radius, while others (e.g., Kikuchi et al. 1999) found more isothermal profiles. The statistical errors are rather small (<10% uncertainty in temperature for a given radial bin) for the 30 brightest objects, so the differences are due entirely to different ways of modeling the PSF. While the effects on the total mass and total mass profile are not large for the massive systems (Nevalainen et al. 1999) being only ~10% at $r \sim r_{500}$ (the radius at which the total density of matter in the cluster is 500 times the mean density of the Universe), there is a marked difference in the fraction of the binding mass that is in gas: the isothermal models show a small rise in gas fraction from the central core radius to ~500 kpc and leveling at larger radii, while the non-isothermal models show a monotonic rise in gas fraction. The BeppoSAX data (de Grandi et al. 1999) have added more confusion, sometimes finding temperature gradients consistent with the Markevitch analysis and sometimes finding isothermal fits. It is fair to say that the situation is not entirely clear at present.

The absolute mass estimates (Horner et al. 1999) scale roughly as predicted theoretically, $M \sim T^{3/2}$, but the absolute mass seems to be about 30% less than predicted at about half the virial radius, a result seen in all the X-ray analyses. Direct comparisons of X-ray determined masses with gravitational lensing masses (see sub-section on "gravitational lensing") are very few at present because of the poor spatial resolution of the X-ray telescopes before Chandra but in a few cases (e.g., A2390; Böhringer et al. 1998) seem to agree rather well.

Theory indicates that if clusters are fair samples of the Universe, their baryonic fraction should be representative of the Universe as a whole and not vary from object to object. However, this applies only to objects that are in hydrostatic equilibrium, and objects that are in the process of merging will add scatter to the relationship. Analysis of relatively large samples of low-redshift objects (Ettori and Fabian 1999, Jones and Forman 1999, Arnaud and Evrard 1999) show mean gas mass fractions of ~15%, with ~25% scatter in the baryonic fraction for massive clusters. This scatter is roughly consistent with that shown by numerical simulations. In addition, the surface brightness profiles show a regular pattern, with small but non-zero dispersion (Neumann and Arnaud 1999) when they are plotted in units of fraction of the virial radius. Since big bang nucleosynthesis indicates that the total baryonic fraction of the Universe is $\sim 0.06\,\Omega_m h^{-2}$, the implied value of Ω_m is ($f_{baryon\ BBN}/f_{baryon\ cluster}$) or $\Omega_m \sim 0.3$ (White et al. 1993, Bahcall et al. 1999). Since the value of f_{gas} is a function of cosmology (Pen 1999) we can use the baryonic fraction as a function of redshift as a cosmology test. While still in its early stages, this test also indicates a low-Ω universe.

Cluster elemental abundances and evolution

The ASCA data gave the first opportunity to obtain abundances of elements other than Fe in regions other than the central cooling flow. The strongest lines from massive clusters of temperature 2–10 keV are the H and He-like lines of Si, S, and Fe. Weaker lines due to Fe L, and the H and He-like lines O, Ne, Ni, Ca, Ar, and Mg are also present (Mushotzky et al. 1996). The general pattern of cluster abundances averaged over the whole cluster (Fukazawa et al. 1998) show that at high temperatures, and thus high masses, the Si/Fe ratio is roughly twice the Anders and Grevesse (1989) solar ratio or consistent with the solar meteoric values. At lower temperatures the Si/Fe ratio monotonically decreases. The S/Fe ratios do not seem to vary and are always consistent with the solar ratios. The lower-Z elements (O, Ne, and Mg) are not as well determined, and their abundance pattern seems to be more consistent with Si rather than Fe. There is real dispersion in the Fe abundance from cluster to cluster, cooling-flow clusters having a higher average abundance of $A \sim 0.35$ (normalized to a solar abundance of unity; Allen and Fabian 1998) than non-cooling flow clusters, for which $A \sim 0.23$ with respect to solar. Groups seem to show a wider range of abundances than do clusters (Davis et al. 1999) (see the sub-section on "X-ray emission from groups").

There seem to be abundance gradients in about of third of all clusters, cooling-flow clusters having a higher probability of having an Fe abundance gradient. In massive clusters there does not seem to be an Si gradient (Fukazawa et al. 2000). There are indications (Dupke and White 2000) of Ni overabundances in the central regions of some clusters. It is not yet clear if the higher average abundances in all cooling-flow clusters are due to the presence of an abundance gradient or are intrinsic to the cluster as a whole. Based on ASCA data (Finoguenov and Ponman 1999), the average cluster Fe abundance at large radii seems to become asymptotic at ~ 0.2 solar, slightly below the cluster average. Combining these data with the stellar masses and abundances, one comes to the conclusion that in massive clusters $\sim 75\%$ of all the metals lie in the gas phase.

The relative abundance of the elements is a strong guide to the processes that created them. In particular the ratio of the "α" elements (those created by rapid burning of oxygen and produced in massive stars) (O, Ne, Mg, Si) which are produced primarily by Type II supernovae while S, Ar, Ca, and Fe can be produced by both Type I and Type II supernovae (Loewenstein and Mushotzky 1996, Gibson et al. 1997). (Type II supernovae are produced by massive stars which have very short lifetimes, while Type I supernovae are produced by accretion onto a white dwarf in a binary system with a rather longer life. Type I supernovae produce a relatively large amount of Fe per explosion while Type IIs produce much more kinetic energy per Fe atom.) The overall cluster trends are consistent with ~ 0.2–0.5 of the Fe being produced by Type I supernovae. The total abundance patterns seem to require a higher Type II to Type I ratio than in the Milky Way.

There is no evidence for evolution in the cluster Fe abundance out to redshifts of ~ 0.5 (Mushotzky and Loewenstein 1997) or perhaps $z \sim 0.8$. There is no sign of massive star formation in clusters out to of $z \sim 0.8$ required to produce the observed metals (see below). The combination of the X-ray and optical data thus indicates that the metals in the cluster gas were created at redshifts > 1.2.

The "extra" energy problem

The observational data for clusters and groups are in general agreement with detailed numerical simulations (Eke et al. 1998). Roughly speaking, clusters forming in a hierarchical dark-matter-dominated universe have the properties that are predicted by numerical simulations. However, there are two major disagreements: the entropy per particle (Ponman et al. 1999) and the gas distribution in groups.

As first noted by Ponman et al., groups have considerably more entropy per particle than predicted by the numerical simulations. Detailed theoretical work by several groups have confirmed this (e.g., Loewenstein 2000) and indicate that about 1 keV of energy per particle is required. Thus some process to inject extra heat, other than gravity, must have been at work. Similarly, groups have considerably more extended X-ray surface brightness distributions than clusters, also consistent with the injection of extra heat (White 1991). This same process can also produce the deviation of the X-ray luminosity versus temperature relation from the theoretically predicted slope of 2 to the observed slope of 3. Recent calculations also show that this amount of extra heat is required to "fine tune" the semi-analytic models of galaxy formation.

Loewenstein (2000) has shown that it is possible to produce this extra heat from supernovae and stellar winds, but it requires very efficient thermalization of the supernova energy. If this is true then most galaxies would, during their rapid star-forming periods, have blown enormous galactic winds, and much of the total material created by nucleosynthesis would have been expelled into the IGM. The injection of extra energy leads to good agreement between the observed cluster luminosity function and theoretical calculations, a serious problem in most numerical calculations.

In addition, as pointed out by Pen (1999), the same indication of energy injection is seen in spiral galaxies. In most hierarchical clustering theories dark matter potential wells either have the average baryon density of the Universe or have concentrated their baryons. Detailed simulations and measurement show that these baryons are not apparent and therefore must have been driven off from spiral and elliptical galaxies.

Since the binding energy of the most massive galaxies is ~ 1 keV per particle, it is clear that if this energy is produced in the stars that lived in galaxies that the whole process of galaxy and group formation and evolution must have been strongly effected by non-gravitational processes. Recent calculations by Valageas and Silk (1999) and others indicate that there is not enough energy in supernovae to produce the extra energy required, and these authors imply that only quasars could produce the needed energy. However, it is difficult to understand the exact mechanism responsible. Also, as pointed out by Pen (1999), most numerical calculations of the formation and evolution of large-scale structure would overproduce the observed soft X-ray background via emission from the gas in groups. If this is true, then the same sort of mechanism is required to drive the gas out of the group potential wells and thus reduce the soft X-ray emission from groups. However, calculations by Davé et al. (2001) do not reproduce this effect. They find that this gas will end up in the IGM, heat it, and enrich it in metals.

Interaction with radio sources and possible non-thermal components

The nature of cluster radio sources has been a long-standing problem. There is a class of radio sources, the so-called wide-angle tailed (WAT) sources, which look as if the radio-emitting plasma is being "pushed" around, like a flag waving in the breeze. In fact this is one of the earliest pieces of evidence for the hot X-ray emitting gas (Rudnick and Owen 1976). Detailed Rosat images of groups and clusters combined with dynamical measurements (Burns et al. 1995), however, have shown that the velocities of the galaxies containing the radio sources are not high enough to produce the ram pressure necessary to bend the radio jets. It is now thought that these radio sources are associated with dynamically complex clusters undergoing merger events, and that the observed morphology is produced in part by the merger-induced bulk motion of the ICM bending the jets.

The central regions of some clusters (Cygnus A, Perseus, 3C 295, e.g. Böhringer et al. 1993) contain extremely luminous radio sources. Rosat HRI images of the Perseus Cluster and Cygmus A show that the radio plasma is distorting the X-ray images, producing "holes" in the X-ray surface brightness. This indicates that the radio plasma sometimes has sufficient pressure to slowly push the X-ray emitting gas so that the gas does not pile up into a shock. Since these luminous radio sources are much more common in the past and always occur in giant elliptical galaxies which preferentially lie in clusters, this might be a very important phenomenon in the high-redshift Universe. Recent Chandra results are providing exciting new information on the interaction of the radio and thermal plasmas.

Very often in cooling-flow clusters there is observational direct evidence for interaction of the relativistic plasma from the radio source and the X-ray emitting gas. The radio emission shows very high Faraday rotation which is almost certainly due to magnetic fields in the high-density cooling gas. The strong association of the optical filaments with the radio sources indicates that much of the heating and cooling may be directly related to the relativistic plasma. In addition, it seems as if recent star formation in the cooling-flow clusters is also associated with the radio emission. It will require Chandra's high angular resolution to make additional progress in this area.

In two radio sources, ASCA and Rosat observations (Feigelson et al. 1995) have found strong evidence for IC emission, where the relativistic particles are Compton scattered off the microwave background to produce hard X-ray emission. The measurement of the hard X-ray image and spectrum allows direct determination of the magnetic field in the radio sources, information vital for the interpretation of the vast amount of radio images obtained to date. The preliminary indications from these two sources are that in one (the emission associated with NGC 1316 in the Fornax Cluster) the magnetic field and relativistic particles are in energy equipartition, an assumption often used to interpret radio data, but in the other, Centuarus B, the particle energy dominates (Tahshiro et al. 1999).

The interaction of thermal and relativistic plasma is also important for the confinement of the radio jets. Detailed Rosat images of several sources (e.g., NGC 1316 and NGC 1399) show that low-luminosity radio sources can indeed be confined by the thermal pressure of the X-ray emitting gas. In at least one case (NGC 3665) the thermal pressure of the hot gas is greater than the jet equipartition pressure of relativistic electrons, implying that the assumptions made in the computation of the internal pressure may not be valid, and that the jet structure can be satisfactorily modeled assuming that the jets are confined by the X-ray emitting atmosphere.

On larger scales in cluster environments, the radio lobes are easily confined by the static pressure of the hot intra-cluster gas and/or the ram pressure generated by the passage of the galaxy through the gas. The radio powers of cluster sources and the optical luminosities of their hosts are both independently correlated with cluster X-ray luminosity. This suggests that radio power is maintained by the confining pressure of the hot gas. Worrall and Birkinshaw (1999) directly compared the X-ray gas pressure against the minimum confining pressure for many Faranoff–Riley Class I (FRI) radio sources, and for the majority of these low-luminosity radio sources the run of gas pressure versus radius matches or exceeds the radio equipartition pressure. The situation for FRIIs is less clear, but they also seem likely to be confined by the hot X-ray emitting gas.

The fact that the radio emission is "confined" by the gas can be inverted to search for high-redshift groups and clusters by looking for extended radio sources. At present (pre-Chandra and XMM-Newton), radio surveys are one of the most sensitive ways of finding high-redshift galaxies.

Recently (Kaastra et al. 1999) there has been a detection of possible non-thermal emission from several clusters of galaxies at both $E > 10$ keV and at $E < 0.2$ keV. For the Coma Cluster the simplest interpretation of the high-energy component is IC scattering of the relativistic particles that produce the diffuse radio halo in this cluster off the microwave background (Felten and Morrison 1966). The relative strength of the radio to the X-ray emission is a measure of the energy density of the photon field (presumably the microwave background) to the magnetic field, and thus detection of this component makes it possible to measure the magnetic field in the cluster. However, for Abell 2199 (Kaastra et al. 1999) there is no diffuse radio source and thus no direct evidence for a population of relativistic particles. The nature of the low-energy component is even more confusing, and at the present time even the flux and spatial extent of this emission are controversial. Results from the Rossi X-ray Timing Explorer (RXTE) and BeppoSAX on several clusters have placed severe upper limits on any hard X-ray emission at the levels seen in Coma and Abell 2199, but this component has been seen in at least one other cluster by BeppoSAX.

Cooling flows

It was hoped that the improved spectral resolution and sensitivity of ASCA would lead to major breakthroughs in our understanding of cooling flows. It is clear from the Rosat and ASCA spectra that there is indeed cooler gas in the central regions of many clusters, and that the amount of cool gas inferred from the X-ray spectral results and the cluster surface brightness models may agree, thus confirming the fundamental assumptions of the cooling-flow model. The measured cooling rates can reach $\sim 1000 \, M_\odot \, \text{yr}^{-1}$. However, the detailed spectra of the highest signal-to-noise cooling-flow clusters do not agree with the original constant-pressure cooling-flow models. Part of the discrepancy can be accounted for if cold material absorbs part of the cooling-flow flux with a typical column density of 3×10^{21} atoms/cm^2 (White et al. 1991). However, other models, while less physically motivated, can give a better fit to the data (Ikebe et al. 1999). Searches for this absorbing material, as well as the massive star formation implied in the optical and radio bands, have proven elusive, but there are indications of such material in the infrared band. Cooling-flow clusters at redshifts of >0.2 have been seen.

THE IMPACT OF THE HUBBLE SPACE TELESCOPE

The other main impact of the space age on the study of clusters and groups has been the HST, the increased angular resolution of which has made possible the study of individual galaxies in clusters out to a redshift of ~ 1, and detailed study of gravitational lenses (Smail et al. 1996).

Study of cluster galaxy populations

At $z < 0.1$, detailed work from the ground has shown a strong correlation between the galaxy population and the cluster properties (Dressler 1980). In particular, rich clusters are dominated in their central regions by elliptical and S0 galaxies, while less rich clusters tend to have more spirals. This is consistent with a general trend of galaxy density versus galaxy type. At larger redshifts, $z \sim 0.2$–0.3, Butcher and Oemler (1978) noticed, from ground-based optical data, that there were more "blue" galaxies than were in lower-redshift systems. The nature of these objects was unclear, but there was some indication that they resulted from a period of enhanced star formation a few billion years earlier. Early Hubble observations indicated that the blue galaxies were both spirals and "unusual" galaxies, and provided convincing evidence of dynamical interactions. The other blue members that display spectral evidence of active or recently completed star formation are predominantly disk-dominated systems whose abundance is greater than that seen in the cores of present-day rich clusters. However, while originally discovered in clusters it is now known that the population of field galaxies as a whole is also undergoing similar evolution, and it is not at all clear at present whether the Butcher–Oemler effect is related to clusters (Poggianti et al. 1999).

Perhaps much more important are the detailed studies of the evolution of the elliptical galaxy population in distant clusters. Low-redshift elliptical galaxies have a set of properties which are highly correlated, and in particular they can be represented by an eigenvector of color, total magnitude, velocity dispersion, size, and surface brightness – the so-called fundamental plane. The HST measurements of the size, shape, and color of the galaxies combined with ground-based measurement of the spectra of clusters out to redshifts of ~ 1 (van Dokkum et al. 1998) show that the galaxies in clusters are evolving "passively." That is, the bulk of the stellar population is consistent with having formed at redshifts >2 (see the discussion of cluster metallicity above); for example, the fundamental plane, at $z \sim 1$, is the same as at $z \sim 0$, if correction is made for the different ages of the stars. Thus the bulk of the stars have not changed since formation and are very old. The properties of galaxies themselves are also consistent with only this "passive" evolution of the stars, for example their sizes, luminosities, surface brightnesses, and so on are consistent with an aging of the stellar population. Thus the luminous massive elliptical galaxies in clusters and the stars in them are very old. What is not clear is whether the clusters themselves are similarly old. There are indications from HST studies of the spatial distribution of $z \sim 3$ galaxies that there is strong clustering in the early Universe, and the spatial density and total masses of these "overdense" regions are consistent with being the progenitors of rich clusters (Steidel *et al.* 1998).

Gravitational lensing

Ever since Einstein predicted the bending of light by a strong mass in the line of sight, it has been clear that the measurement of this signal could be a unique measure of the mass of the deflecting object. It was realized (Refsdal 1964) that massive clusters of galaxies should be capable of distorting the images of background galaxies. The discovery of the first lenses in 1986 (Soucail *et al.* 1987) confirmed this idea. While the theory is not that simple, there is a straightforward connection between the total surface mass density and the nature of the lensing effects. The lens can amplify and distort the images of background galaxies to produce giant arcs (Figure 11) as well as more subtle distortions. What is required is precision measurements of the shape, size, brightness and redshift of the lensed objects. Given such information and a sufficient density of background lensed objects the mass and mass distribution of the cluster can be well determined. This is the only way of "directly" observing the dark matter. (For more detail, see Chapter 20.)

While lenses were discovered by ground-based observations, it soon became clear that the use of ground-based data to infer the masses of clusters was rather difficult. The use of the HST makes for a much more robust result. Perhaps the most exciting results are for the $z = 0.39$ cluster CL0024, for which Tyson *et al.* (1998) were able to show that >98% of the dark matter is not associated with individual galaxies and is distributed in a smooth and symmetric way. Surprisingly, the dark matter has been found to have a soft core, in contrast to predictions from numerical simulations, but outside the core the dark matter distribution is consistent with the theoretical NFW profile (Navarro *et al.* 1997). The dark matter profile in the central regions is more extended than the galaxies, implying a non-constant ratio of gravitational mass to optical light – the so-called mass-to-light ratio. In addition, the data were of sufficiently high quality to constrain the dark matter in individual galaxies. It is now clear that lensing can provide a robust determination of the mass distribution in clusters, given sufficiently good data. However, most clusters do not have such data at present.

The comparison of masses derived from lensing and X-ray techniques is rather complex at the present time. The lensing masses are best determined for $z \sim 0.3$ clusters over an angular scale of $<1'$. The ASCA data used for X-ray temperature measurements do not have sufficient angular resolution to probe this scale and direct comparison is thus difficult. On larger scales strong lensing is not detected and one must derive masses from weak lensing limits. On these scales, for clusters which have smooth, relaxed morphologies, comparison of the X-ray and lensing masses shows good agreement (Allen 1998; Böhringer *et al.* 1998). For objects with more complex morphologies there is a strong tendency for the lensing masses to be a factor of two

Figure 11 HST image of giant arcs in Abell 2218. The blue arcs are the result of gravitational lensing of a background spiral galaxy by the dark matter potential in the cluster. (NASA, A. Fruchter and the ERO Team.)

higher than the best-fitting X-ray models. There are several possible explanations, but the most likely are the existence of structure in the cluster which increases the strength of the lensing signal, but does not effect the X-ray image or temperature significantly, and low spatial resolution in the X-ray data which do not reveal a high-temperature central region, required if the central masses are as high as the lensing signal implies. This should be easily detected by Chandra and XMM-Newton observations.

PROSPECTS

The near future

With the launch of the Chandra Observatory on 23 July 1999 and of XMM-Newton on 10 December 1999, X-ray astronomy is making a spectacular leap. The implications for cluster research are many and important.

Chandra, with its 1″ imaging, is able to image clusters at arbitrary redshift. Early images with Chandra have already shown how easy it is to detect clusters at $z \sim 0.6$ by simple inspection of the images. Given that we know that the cluster luminosity function does not evolve much out to $z \sim 0.8$, it is straightforward to calculate the number of serendipitous clusters per Chandra field and exposure depth. One finds that there should be >2 clusters at $z > 0.8$ per ACIS-I field of view ($17 \times 17'$) in exposures of more than 10 ks. We thus anticipate a revolution in our understanding of the formation and evolution of high-redshift clusters.

The high angular resolution of Chandra will also make the study of cooling flows possible out to very large distances. One of the first Chandra images (3C 295) already shows cooling gas in the cluster center at a redshift of 0.5! Detailed studies at lower redshifts should be extremely exciting. Chandra images of the detailed interaction of the radio-emitting plasma and the hot gas should reveal much of the detailed physics, in particular whether magnetic fields are important in the heating of the gas and how the gas actually confines the radio sources. The detection of other sources with IC emission will provide a large database of magnetic field strengths for comparison with detailed modeling of radio sources.

XMM-Newton, with its very large collecting area and wide field of view, will provide an unprecedented sample of distant clusters and groups. For objects above 10^{-14} erg cm^{-2} s^{-1} spectra will be obtained allowing measurement of the evolution of the cluster temperature function to $z \sim 1$. Detailed studies of individual clusters will allow robust determinations to be made of temperature and abundance gradients and structures, and detailed measurements of the abundances of O and Ni as well as Si, S, and Fe for many objects out to $z \sim 0.3$.

Detailed Chandra images, XMM-Newton temperatures, and ground-based Sunyaev–Zel'dovich effect measurements will derive the scale factor of the Universe and provide estimates of the cosmological parameters such as H_0 and Λ for direct comparison with the microwave background data from the Microwave Anisotropy Probe. Similar comparisons will be made with gravitational lens images for direct comparison of the dark matter distribution. It is possible that serendipitous Chandra and XMM-Newton source catalogs will contain the first clusters and groups to form, and thus allow us essentially to observe the creation of large-scale structure.

Unfortunately the loss of Astro-E means that high spectral resolution (~ 10 eV) images of the central regions of clusters will not be available. Thus robust abundances and the distribution of gas temperatures and emission measures for cooling flows cannot be obtained, and the turbulent velocity of the gas in strong mergers will not be measured. The absence of the Astro-E hard X-ray detector will prevent the measurement of IC high-energy X-ray emission in many clusters and the determination of the magnetic fields in clusters. The Japanese–US satellite Astro-E2, which is a re-build of the original Astro-E satellite, is now scheduled for launch in February 2005 and should allow the lost science of Astro-E to be recovered.

By the end of the next decade the Planck Surveyor mission will have performed an all-sky survey of the Sunyaev–Zel'dovich effect and found many very high-redshift clusters. The addition of the Advanced Camera for Surveys on HST should substantially increase its capabilities for the study of gravitational lenses and cluster evolution.

The distant future

The Constellation-X mission, part of NASA's strategic plan and recommended by the 2000 US National Academy of Sciences decadal survey, will have more collecting area than XMM-Newton and better spectral resolution than Astro-E. This will allow spatially resolved spectra of clusters out to redshifts of ~ 1 and sufficient sensitivity to measure the abundance of essentially all the cosmologically abundant elements (O, Ne, Al, Mg, Si, Ca, Ar, Fe, Ni) to rather high redshifts. This will determine the origin of the elements and the appearance of the first stars. Cooling flows will be mapped in exquisite detail. If clusters exist at $z > 2$ their abundances and temperatures will be measured. Detailed maps of the turbulence and velocity field of the gas will be obtained out to redshifts of ~ 0.5, and the evolution of cluster mergers determined. In the even more distant future XEUS, part of the long-term ESA program, with better spatial resolution and even more collecting area than Constellation-X, will be able to map clusters at the highest conceivable redshifts, find and study groups while they are in the process of formation, and determine how the first structures collapsed and formed. XEUS will also determine the epoch and the mechanism by which the extra heat inferred in the low-z clusters is created.

We anticipate that another X-ray survey at least 10 times more sensitive than Rosat will fly and map the large-scale structure of the Universe by mapping groups at $z < 0.2$ and rich clusters out to $z \sim 1$. This will provide a robust data base for comparison with optical catalogs from the Sloan Digital Sky Survey and allow measurement of fundamental cosmological parameters.

The Next Generation Space Telescope should be flying and determining the velocity field of distant clusters as well as the mass and metallicity of the galaxies therein.

A PERSONAL NOTE

It is rather amazing to look back through the history of space astronomy. In my field of specialization, X-ray astronomy, since 1970 sensitivity has increased by a factor of 10^6, angular resolution by 10^4, spectral resolution by 10^3, and the number of objects available for study by $\sim 10^3$. This explosion in capabilities has been accompanied by an explosion of knowledge. We now confidently discuss the metallicity evolution of objects which were not even known to exist in the mid-1980s. However, the field is still dominated by serendipity. Most of what we study was not predicted by theory, and most of the new discoveries were unanticipated. It is thus very hard to predict where the field will go. It has been a great trip so far – and the future looks even better. However, the rate of improvement has been very uneven. In the first twenty years of the field the progress from simple proportional counters on rockets to precision X-ray imaging was rapid and relatively smooth. However, from 1982 to 1900 there was little, if any, net progress, and even in the 1990s the rate of improvement was slow and uneven. I can only hope that the giant leap represented by Chandra and XMM-Newton has been worth the wait.

Acknowledgments and apologies

I would like to thank my collaborators at the Goddard Space Flight Center (GSFC), especially Mike Loewenstein and Keith Arnaud, for their many years of putting up with me, and I thank present and former postdoctoral fellows and graduate students, especially Caleb Scharf, John Mulchaey and Dave Davis. I thank Don Horner and Wayne Baumgartner for help with preparing this chapter. Most of all I would like to thank the Sciences Directorate at GSFC, whose robust and exciting research atmosphere has sustained and supported me for the 25 years of research represented in this chapter.

I would also like to apologize to all the authors whose work I have not truly represented and whose papers I have not cited. All the errors are mine alone. In a review like this one it is not possible, in such a rich field, to cite all the relevant articles or even to strike a proper balance. There are certain large areas of research such as X-ray emission from elliptical galaxies, cosmology with the Sunyaev–Zel'dovich effect, and detailed studies of cluster mergers that I have been forced to leave out entirely. In addition, because of the emphasis on space observations the theoretical framework for much of the understanding of these objects has only been sketched in.

REFERENCES

Aarseth, S.J. (1978). Computer simulations of galaxy clustering. In M.S. Longair and J. Einasto (eds), *The Large Scale Structure of the Universe*, IAU Symposium 79, Reidel, Dordrecht, pp. 189–194 [Discussion, pp. 194–196].

Abell, G. (1958). The distribution of rich clusters of galaxies. *Astrophysical Journal*, Suppl., **3**, 211–288.

Allen, S.W. (1998). Resolving the discrepancy between X-ray and gravitational lensing mass measurements for clusters of galaxies. *Monthly Notices of the Royal Astronomical Society*, **296**, 392–406.

Allen, S.W. and Fabian, A.C. (1994). A ROSAT PSPC investigation of the Centaurus cluster of galaxies. *Monthly Notices of the Royal Astronomical Society*, **269**, 409–426.

Allen, S.W. and Fabian, A.C. (1998). The relationship between cooling flows and metallicity measurements for X-ray-luminous clusters. *Monthly Notices of the Royal Astronomical Society*, **297**, L63–L68.

Anders, E. and Grevesse, N. (1989). Abundances of the elements – meteoritic and solar. *Geochimica et Cosmochimica Acta*, **53**, 197–214.

Arnaud, K.A. (1988). A catalog(ue) of cooling flows. In A.C. Fabian (ed.), *Cooling Flows in Clusters and Galaxies*, Proceedings of the NATO Advanced Research Workshop, Cambridge, 22–26 June 1987, Kluwer Academic, Dordrecht, pp. 31–40.

Arnaud, M. and Evrard, A.E. (1999). The L–X–T relation and intracluster gas fractions of X-ray clusters. *Monthly Notices of the Royal Astronomical Society*, **305**, 631–640.

Arnaud, M., Lachieze-Rey, M., Rothenflug, R., Yamashita, K. and Hatsukade, I. (1991). GINGA observation of distant galaxy clusters: Temperature and iron abundance of A 2507 and A 483. *Astronomy and Astrophysics*, **243**, 56–66.

Bahcall, J.N. and Bahcall, N.A. (1975). The unidentified high galactic latitude X-ray sources: Bright galaxies or rich clusters. *Astrophysical Journal*, **199**, L89–L92.

Bahcall, N.A. (1977a). Clusters of galaxies. *Annual Review of Astronomy and Astrophysics*, **15**, 505–540.

Bahcall, N.A. (1977b). X-ray clusters of galaxies: Correlation of X-ray luminosity with galactic content. *Astrophysical Journal*, **218**, L93–L95.

Bahcall, N.A., Ostriker, J.P., Perlmutter, S. and Steinhardt, P.J. (1999). The cosmic triangle: Revealing the state of the Universe. *Science*, **284**, 1481–1488.

Biermann, P. and Kronberg, P.P. (1983). Detection of 10 to the 10th solar masses of hot gas in the normal elliptical galaxy NGC 5846 with the Einstein satellite. *Astrophysical Journal*, **268**, L69–L73.

Binney, J. and Strimpel, O. (1978). Predicting the X-ray brightness distributions of cluster sources: 1. Estimating the potentials. *Monthly Notices of the Royal Astronomical Society*, **185**, 473–484.

Böhringer, H., Voges, W., Fabian, A.C. Edge, A.C., Neumann, D.M. (1993). A ROSAT HRI study of the interaction of the X-ray-emitting gas and radio lobes of NGC 1275. *Monthly Notices of the Royal Astronomical Society*, **264**, L25–L28.

Böhringer, H., Tanaka, Y., Mushotzky, R.F., Ikebe, Y. and Hattori, M. (1998). An ASCA-ROSAT study of the distant, lensing cluster A 2390. *Astronomy and Astrophysics*, **334**, 789–798.

Bradt, H., Mayer, W., Naranan, S., Rappaport, S. and Spada, G. (1967). Evidence for X-radiation from the radio galaxy M87. *Astrophysical Journal*, **150**, L199–L206.

Bryan, G.L., Cen, R., Norman, M.L., Ostriker, J.P., Stone, J.M. (1994). X-ray clusters from a high-resolution hydrodynamic PPM simulation of the cold dark matter universe. *Astrophysical Journal*, **428**, 405–418.

Buote, D.A. (2000). X-ray evidence for multiphase hot gas with nearly solar Fe abundances in the brightest groups of galaxies. *Monthly Notices of the Royal Astronomical Society*, **311**, 176–200.

Burns, J.O., Roettiger, K., Pinkney, J., Perley, R.A., Owen, F.N., Voges, W. (1995). Evidence for an on-going cluster/group merger in Abell 2255. *Astrophysical Journal*, **446**, 583–593.

Burns, J.O., Ledlow, M.J., Loken, C., Klypin, A., Voges, W., Bryan, G.L., Norman, M.L. and White, R.A. (1996). The X-ray luminosity function and gas mass function for optically selected poor and rich clusters of galaxies. *Astrophysical Journal*, **467**, L49–L52.

Butcher, H. and Oemler, A. Jr (1978). The evolution of galaxies in clusters: II. The galaxy content of nearby clusters. *Astrophysical Journal*, **226**, 559–565.

Carlberg, R.G., Yee, H.K.C. and Ellingson, E. (1997). The average mass and light profiles of galaxy clusters. *Astrophysical Journal*, **478**, 462–475.

Cavaliere, A. and Fusco-Femiano, R. (1978). The distribution of hot gas in clusters of galaxies. *Astronomy and Astrophysics*, **70**, 677–684.

Cooke, B.A. and Maccagni, D. (1976). New observations of X-ray clusters of galaxies. *Monthly Notices of the Royal Astronomical Society*, **175**, 65P–70P.

Cen, R. (1998). Gaussian peaks and clusters of galaxies: Mass function, correlation function, evolution and beyond. American Astronomical Society Meeting, 192, 42.03.

Cowie, L.L., Hu, E.M., Jenkins, E.B. and York, D.G. (1983). Two-dimensional spectrophotometry of the cores of X-ray luminous clusters. *Astrophysical Journal*, **272**, 29–47.

Davé, R., Cen, R., Ostriker, J.P., Bryan, G.L., Hernquist, L., Katz, N., Weinberg, D.H., Norman, M.L., O'Shea, B. (2001). Baryons in the warm-hot intergalactic medium. *Astrophysical Journal*, **552**, 473–483.

David, L.P., Forman, W. and Jones, C. (1991). Enrichment and heating of the intracluster medium through galactic winds. *Astrophysical Journal*, **380**, 39–48.

Davis, D.S., Mulchaey, J.S. and Mushotzky, R.F. (1999). The enrichment history of hot gas in poor galaxy groups. *Astrophysical Journal*, **511**, 34–40.

De Grandi, S., Böhringer, H., Guzzo, L., Molendi, S., Chincarini, G., Collins, C., Cruddace, R., Neumann, D., Schindler, S., Schuecker, P. and Voges, W. (1999). A flux-limited sample of bright clusters of galaxies from the southern part of the ROSAT All-Sky Survey: The Catalog and LOG N-LOG S. *Astrophysical Journal*, **514**, 148–163.

De Grandi, S. and Molendi, S. (1999). Spatially resolved spectroscopy of the cooling flow cluster PKS 0745-191 with BeppoSAX. *Astronomy and Astrophysics*, **351**, L45–L48.

Donahue, M., Voit, G.M., Scharf, C.A., Gioia, I.M., Mullis, C.R., Hughes, J.P. and Stocke, J.T. (1999). The second most distant cluster of galaxies in the Extended Medium Sensitivity Survey. *Astrophysical Journal*, **527**, 525–534.

Dressler, A. (1980). Galaxy morphology in rich clusters: Implications for the formation and evolution of galaxies. *Astrophysical Journal*, **236**, 351–365.

Dupke, R.A. and White, R.E., III (2000). Constraints on Type IA supernova models from X-ray spectra of galaxy clusters. *Astrophysical Journal*, **528**, 139–144.

Ebeling, H., Voges, W., Böhringer, H., Edge, A.C., Huchra, J.P., Briel, U.G. (1996). Properties of the X-ray-brightest Abell-type clusters of galaxies (XBACs) from ROSAT All-Sky Survey data: I. The sample. *Monthly Notices of the Royal Astronomical Society*, **281**, 799–829. [Erratum: **283**, 1103–1104.]

Edge, A.C. and Stewart, G.C. (1991). EXOSAT observations of clusters of galaxies. Part Two: X-ray to optical correlations. *Monthly Notices of the Royal Astronomical Society*, **252**, 428–441.

Edge, A.C., Stewart, G.C. and Fabian, A.C. (1992). Properties of cooling flows in a flux-limited sample of clusters of galaxies. *Monthly Notices of the Royal Astronomical Society*, **258**, 177–188.

Eke, V.R., Navarro, J.F. and Frenk, C.S. (1998). The evolution of X-ray clusters in a low-density universe. *Astrophysical Journal*, **503**, 569–592.

Ettori, S. and Fabian, A.C. (1999). ROSAT PSPC observations of 36 high-luminosity clusters of galaxies: Constraints on the gas fraction. *Monthly Notices of the Royal Astronomical Society*, **305**, 834–848.

Evrard, A.E. (1988). Beyond N-body: 3D cosmological gas dynamics. *Monthly Notices of the Royal Astronomical Society*, **235**, 911–934.

Fabian, A.C. (1994). Cooling flows in clusters of galaxies. *Annual Review of Astronomy and Astrophysics*, **32**, 277–318.

Fabian, A.C. and Nulsen, P.E.J. (1977). Subsonic accretion of cooling gas in clusters of galaxies. *Monthly Notices of the Royal Astronomical Society*, **180**, 479–484.

Fabricant, D., Lecar, M. and Gorenstein, P. (1980). X-ray measurements of the mass of M87. *Astrophysical Journal*, **241**, 552–560.

Feigelson, E.D., Laurent-Muehleisen, S.A., Kollgaard, R.I., Fomalont, E.B. (1995). Discovery of inverse-Compton X-rays in radio lobes. *Astrophysical Journal*, **449**, L149–L152.

Felten, J.E., Gould, R.J., Stein, W.A. and Woolf, N.J. (1966). X-rays from the Coma Cluster of galaxies. *Astrophysical Journal*, **146**, 955–958.

Felten, J.E. and Morrison, P. (1966). Omnidirectional inverse Compton and synchrotron radiation from cosmic distributions of fast electrons and thermal photons. *Astrophysical Journal*, **146**, 686–708.

Finoguenov, A. and Ponman, T.J. (1999). Constraining the role of Type IA and Type II supernovae in galaxy groups by spatially resolved analysis of ROSAT and ASCA observations. *Monthly Notices of the Royal Astronomical Society*, **305**, 325–337.

Forman, W. and Jones, C. (1982). X-ray-imaging observations of clusters of galaxies. *Annual Review of Astronomy and Astrophysics*, **20**, 547–585.

Forman, W., Jones, C., Cominsky, L., Julien, P., Murray, S., Peters, G., Tananbaum, H. and Giacconi, R. (1978). The fourth Uhuru catalog of X-ray sources. *Astrophysical Journal*, Suppl., **38**, 357–412.

Forman, W., Jones, C. and Tucker, W. (1985). Hot coronae around early-type galaxies. *Astrophysical Journal*, **293**, 102–119.

Friedman, H. and Byram, E.T. (1967). X-rays from the Coma Cluster of galaxies. *Astrophysical Journal*, **147**, 399–401.

Fukazawa, Y., Makishima, K., Tamura, T., Ezawa, H., Xu, H., Ikebe, Y., Kikuchi, K. and Ohashi, T. (1998). ASCA measurements of silicon and iron abundances in the intracluster medium. *Publications of the Astronomical Society of Japan*, **50**, 187–193.

Fukazawa, Y., Makishima, K., Tamura, T., Nakazawa, K., Ezawa, H., Ikebe, Y., Kikuchi, K. and Ohashi, T. (2000). Statistical properties of metal abundances of the intracluster medium in the central region of clusters. *Monthly Notices of the Royal Astronomical Society*, **313**, 21–31.

Ge, J.P. and Owen, F.N. (1993). Faraday rotation in cooling flow clusters of galaxies: I. Radio and X-ray observations of Abell 1795. *Astronomical Journal*, **105**, 778–787.

Gibson, B.K., Loewenstein, M. and Mushotzky, R.F. (1997). Supernovae Types Ia/II and intracluster medium enrichment. *Monthly Notices of the Royal Astronomical Society*, **290**, 623–628.

Gioia, I.M., Henry, J.P., Maccacaro, T., Morris, S.L., Stocke, J.T. and Wolter, A. (1990). The Extended Medium Sensitivity Survey distant cluster sample: X-ray cosmological evolution. *Astrophysical Journal*, **356**, L35–L38.

Gorenstein, P., Fabricant, D., Topka, K., Tucker, W. and Harnden, F.R., Jr. (1977). Structure of the X-ray source in the Virgo Cluster of galaxies. *Astrophysical Journal*, **216**, L95–L99.

Gunn, J.E. and Gott, J.R., III (1972). On the infall of matter into clusters of galaxies and some effects on their evolution. *Astrophysical Journal*, **176**, 1–19.

Gursky, H., Kellogg, E., Murray, S., Leong, C., Tananbaum, H. and Giacconi, R. (1971). A strong X-ray source in the Coma Cluster observed by Uhuru. *Astrophysical Journal*, **167**, L81–L84.

Gursky, H. and Schwartz, D.A. (1977). Extragalactic X-ray sources. *Annual Review of Astronomy and Astrophysics*, **15**, 541–568.

Gursky, H., Solinger, A., Kellogg, E., Murray, S., Tananbaum, H., Giacconi, R. and Cavaliere, A. (1972). X-ray emission from rich clusters of galaxies. *Astrophysical Journal*, **173**, L99–L104.

Guzzo, L., Böhringer, H., Schuecker, P., Collins, C.A., Schindler, S., Neumann, D.M., de Grandi, S., Cruddace, R., Chincarini, G., Edge, A.C., Shaver, P.A. and Voges, W. (1999). The REFLEX Cluster Survey: observing strategy and first results on large-scale structure. *The Messenger*, **95**, 27–32.

Heisler, J., Tremaine, S. and Bahcall, J.N. (1985). Estimating the masses of galaxy groups: Alternatives to the virial theorem. *Astrophysical Journal*, **298**, 8–17.

Henry, J.P. (2000). Measuring cosmological parameters from the evolution of cluster X-ray temperatures. *Astrophysical Journal*, **534**, 565–580.

Henry, J.P., Branduardi, G., Fabricant, D., Feigelson, E., Murray, S., Tananbaum, H., Briel, U. and Soltan, A. (1979). Detection of X-ray emission from distant clusters of galaxies. *Astrophysical Journal*, **234**, L15–L19.

Henry, J.P., Gioia, I.M., Maccacaro, T., Morris, S.L., Stocke, J.T. and Wolter, A. (1992). The extended medium sensitivity survey distant cluster sample: X-ray data and interpretation of the luminosity evolution. *Astrophysical Journal*, **386**, 408–419.

Horner, D.J., Mushotzky, R.F. and Scharf, C.A. (1999). Observational tests of the mass–temperature relation for galaxy clusters. *Astrophysical Journal*, **520**, 78–86.

Hughes, J.P., Yamashita, K., Okumura, Y., Tsunemi, H. and Matsuoka, M. (1988). Isothermality of the gas in the Coma Cluster. *Astrophysical Journal*, **327**, 615–626.

Ikebe, Y., Makishima, K., Fukazawa, Y., Tamura, T., Xu, H., Ohashi, T. and Matsushita, K. (1999). Two-phase intracluster medium in the Centaurus Cluster of galaxies. *Astrophysical Journal*, **525**, 58–79.

Irwin, J.A. and Sarazin, C.L. (1995). ROSAT X-ray observations of the 2A 0335 + 096 cluster of galaxies. *Astrophysical Journal*, **455**, 497–507.

Jones, C. and Forman, W. (1978). X-ray clusters of galaxies and the luminosity–richness relation. *Astrophysical Journal*, **224**, 1–13.

Jones, C. and Forman, W. (1984). The structure of clusters of galaxies observed with Einstein. *Astrophysical Journal*, **276**, 38–55.

Jones, C. and Forman, W. (1999). Einstein Observatory images of clusters of galaxies. *Astrophysical Journal*, **511**, 65–83.

Jones, C., Mandel, E., Schwarz, J., Forman, W., Murray, S.S. and Harnden, F.R., Jr (1979). The structure and evolution of X-ray clusters. *Astrophysical Journal*, **234**, L21–L25.

Jones, L.R., Scharf, C., Ebeling, H., Perlman, E., Wegner, G., Malkan, M. and Horner, D. (1998). *Astrophysical Journal*, **495**, 100–114.

Kaastra, J.S., Lieu, R., Mittaz, J.P.D., Bleeker, J.A.M., Mewe, R., Colafrancesco, S. and Lockman, F.J. (1999). High- and low-energy nonthermal X-ray emission from the Abell 2199 cluster of galaxies. *Astrophysical Journal*, **519**, L119–L122.

Kaiser, N. (1986). Evolution and clustering of rich clusters. *Monthly Notices of the Royal Astronomical Society*, **222**, 323–345.

Kellogg, E., Gursky, H., Leong, C., Schreier, E., Tananbaum, H. and Giacconi, R. (1971). X-ray observations of the Virgo cluster, NGC 5128, and 3c 273 from the UHURU satellite. *Astrophysical Journal*, **165**, L49–L54.

Kent, S.M. and Gunn, J.E. (1982). The dynamics of rich clusters of galaxies: I. The Coma Cluster. *Astronomical Journal*, **87**, 945–971.

Kikuchi, K., Furusho, T., Ezawa, H., Yamasaki, N.Y., Ohashi, T., Fukazawa, Y. and Ikebe, Y. (1999). ASCA measurements of metallicity and temperature distributions in three clusters: A4059, MKW 3s, and 2A 0335 + 096. *Publications of the Astronomical Society of Japan*, **51**, 301–315.

Kowalski, M.P., Cruddace, R.G., Wood, K.S. and Ulmer, M.P. (1984). An X-ray survey of clusters of galaxies: IV. A survey of southern clusters and a compilation of upper limits for both Abell and southern clusters. *Astrophysical Journal*, Suppl., **56**, 403–506.

Lea, S.M., Silk, J., Kellogg, E. and Murray, S. (1973). Thermal-Bremsstrahlung interpretation of cluster X-ray sources. *Astrophysical Journal*, **184**, L105–L111.

Le Fevre, O., Hammer, F., Angonin, M.C., Gioia, I.M., Luppino, G.A. (1994). Imaging of 16 distant EMSS clusters with z greater than or equal to 0.2 and $L_{x,44}$ greater than or equal to 4: New arcs and first consequences. *Astrophysical Journal*, **422**, L5–L8.

Lloyd-Davies, E.J., Ponman, T.J. and Cannon, D.B. (2000). The entropy and energy of intergalactic gas in galaxy clusters. *Monthly Notices of the Royal Astronomical Society*, **315**, 689–702.

Loewenstein, M. (2000). Heating of intergalactic gas and cluster scaling relations. *Astrophysical Journal*, **532**, 17–27.

Loewenstein, M. and Mushotzky, R.F. (1996). Measurement of the elemental abundances in four rich clusters of galaxies: II. The initial mass function and mass loss in elliptical galaxies, enrichment, and energetics in the ICM. *Astrophysical Journal*, **466**, 695–703.

Lynden-Bell, D. (1967). Statistical mechanics of violent relaxation in stellar systems. *Monthly Notices of the Royal Astronomical Society*, **136**, 101.

Mathews, W.G. (1978). The enormous mass of the elliptical galaxy M87 – A model for the extended X-ray source. *Astrophysical Journal*, **219**, 413–423.

McHardy, I. (1978). X-ray sources in clusters of galaxies. *Monthly Notices of the Royal Astronomical Society*, **184**, 783–799.

McKee, J.D., Mushotzky, R.F., Boldt, E.A., Holt, S.S., Marshall, F.E., Pravdo, S.H. and Serlemitsos, P.J. (1980). The HEAO A-2 survey of Abell clusters and the X-ray luminosity function. *Astrophysical Journal*, **242**, 843–856.

Morgan, W.W. and Osterbrock, D.E. (1969). On the classification of the forms and the stellar content of galaxies. *Astronomical Journal*, **74**, 515–524.

Mitchell, R.J., Culhane, J.L., Davison, P.J.N. and Ives, J. (1976). Ariel 5 observations of the X-ray spectrum of the Perseus Cluster. *Monthly Notices of the Royal Astronomical Society*, **175**, 29P–34P.

Mitchell, R.J., Dickens, R.J., Burnell, S. and Culhane, J.L. (1979). The X-ray spectra of clusters of galaxies and their relationship to other cluster properties. *Monthly Notices of the Royal Astronomical Society*, **189**, 329–361.

Mulchaey, J.S., Davis, D.S., Mushotzky, R.F. and Burstein, David (1993). Diffuse X-ray emission from the NGC 2300 group of galaxies: Implications for dark matter and galaxy evolution in small groups. *Astrophysical Journal*, **404**, L9–L12.

Mulchaey, J.S. (2000). X-ray properties of groups of galaxies. *Annual Reviews of Astronomy and Astrophysics*, **38**, 289–335.

Mushotzky, R.F. (1983). X-ray emission from clusters of galaxies. Presented at the Workshop on Hot Astrophysical Plasmas, Nice, 8–10 Nov. 1982 [N83-33826M].

Mushotzky, R.F. and Loewenstein, M. (1997). Lack of evolution in the iron abundance in clusters of galaxies and implications for the global star formation rate at high redshift. *Astrophysical Journal*, **481**, L63–L66.

Mushotzky, R., Loewenstein, M., Arnaud, K.A., Tamura, T., Fukazawa, Y., Matsushita, K., Kikuchi, K. and Hatsukade, I. (1996). Measurement of the elemental abundances in four rich clusters of galaxies: I. Observations. *Astrophysical Journal*, **466**, 686–694.

Mushotzky, R.F. and Scharf, C.A. (1997). The luminosity–temperature relation at z = 0.4 for clusters of galaxies. *Astrophysical Journal*, **482**, L13–L16.

Mushotzky, R.F., Serlemitsos, P.J., Boldt, E.A., Holt, S.S., Smith, B.W. (1978). OSO 8 X-ray spectra of clusters of galaxies: I Observations of twenty clusters: Physical correlations. *Astrophysical Journal*, **225**, 21–39.

Mushotzky, R.F. and Szymkowiak, A.E. (1988). Einstein Observatory solid state detector observations of cooling flows in clusters of galaxies. In *Cooling Flows in Clusters and Galaxies*, Proceedings of the NATO Advanced Research Workshop, Cambridge, England, 22–26 June 1987 (A89-27526 10-90) Kluwer Academic, Dordrecht, Netherlands, pp. 53–62.

Navarro, J.F., Frenk, C.S. and White, S.D.M. (1997). A universal density profile from hierarchical clustering. *Astrophysical Journal*, **490**, 493–508.

Neumann, D.M. and Arnaud, M. (1999). Regularity in the X-ray surface brightness profiles of galaxy clusters and the $M–T$ relation. *Astronomy and Astrophysics*, **348**, 711–727.

Nevalainen, J., Markevitch, M. and Forman, W. (1999). The baryonic and dark matter distributions in Abell 401. *Astrophysical Journal*, **526**, 1–9.

O'Connell, R.W. and McNamara, B.R. (1989). The fate of matter in cooling flows. *Astronomical Journal*, **98**, 180–194.

Oegerle,W., Fichett, M. and Danly, L. (eds) (1990). *Clusters of Galaxies*, Space Telescope Science Institute Symposium 4, Cambridge University Press.

Okumura, Y., Tsunemi, H., Yamashita, K., Matsuoka, M., Koyama, K., Hayakawa, S., Masai, K. and Hughes, J.P. (1988). TENMA observations of the X-ray spectra of the Coma, Ophiuchus, and Perseus clusters of galaxies. *Publications of the Astronomical Society of Japan*, **40**, 639–652.

Pen, U.-L. (1999). Heating of the intergalactic medium. *Astrophysical Journal*, **510**, L1–L5.

Piccinotti, G., Mushotzky, R.F., Boldt, E.A., Holt, S.S., Marshall, F.E., Serlemitsos, P.J., Shafer, R.A. (1982). A complete X-ray sample of the high-latitude/absolute value of B greater than 20 deg/sky from HEAO 1 A-2: Log N-log S and luminosity functions. *Astrophysical Journal*, **253**, 485–503.

Poggianti, B.M., Smail, I., Dressler, A., Couch, W.J., Barger, A.J., Butcher, H., Ellis, R.S. and Oemler, A. Jr (1999). The star formation histories of galaxies in distant clusters. *Astrophysical Journal*, **518**, 576–593.

Ponman, T.J. and Bertram, D. (1993). Hot gas and dark matter in a compact galaxy group. *Nature*, **363**, 51–54.

Ponman, T.J., Cannon, D.B. and Navarro, J.F. (1999). The thermal imprint of galaxy formation on X-ray clusters. *Nature*, **397**, 135–137.

Raymond, J.C. and Smith, B.W. (1977). Soft X-ray spectrum of a hot plasma. *Astrophysical Journal*, Suppl., **35**, 419–439.

Refsdal, S. (1964). The gravitational lens effect. *Monthly Notices of the Royal Astronomical Society*, **128**, 295–306.

Renzini, A., Ciotti, L., D'Ercole, A. and Pellegrini, S. (1993). Production and circulation of iron in elliptical galaxies and clusters of galaxies. *Astrophysical Journal*, **419**, 52–65.

Roettiger, K., Burns, J.O. and Loken, C. (1996). The observational consequences of merging clusters of galaxies. *Astrophysical Journal*, **473**, 651–669.

Rood, H.J., II (1974). Empirical properties of the mass discrepancy in groups and clusters of galaxies. *Astrophysical Journal*, **188**, 451–462.

Rood, H.J., Page, T.L., Kintner, E.C. and King, I.R. (1972). The structure of the Coma Cluster of galaxies. *Astrophysical Journal*, **175**, 627–647.

Rood, H.J. and Sastry, G.N. (1971). 'Tuning fork' classification of rich clusters of galaxies. *Publications of the Astronomical Society of the Pacific*, **83**, 313–319.

Rosati P., Borgani, S, Della Ceca, R., Stanford, S.A., Eisenhardt P.R. and Lidman C. (2000). The most distant X-ray clusters and the evolution of their space density. In *Large Scale Structure in the X-ray Universe*, in the press.

Rudnick, L. and Owen, F.N. (1976). Head–tail radio sources in clusters of galaxies. *Astrophysical Journal*, **203**, L107–L111.

Sarazin, C. (1988). *X-ray Emission from Clusters of Galaxies*. Cambridge Astrophysics Series, Cambridge University Press.

Sarazin, C.L. (1999). The energy spectrum of primary cosmic-ray electrons in clusters of galaxies and inverse Compton emission. *Astrophysical Journal*, **520**, 529–547.

Serlemitsos, P.J., Smith, B.W., Boldt, E.A., Holt, S.S. and Swank, J.H. (1977). X-radiation from clusters of galaxies: Spectral evidence for a hot evolved gas. *Astrophysical Journal*, **211**, L63–L66.

Smail, I., Dressler, A., Kneib, J.-P., Ellis, R.S., Couch, W.J., Sharples, R.M., Oemler, A., Jr (1996). Hubble Space Telescope observations of giant arcs: High-resolution imaging of distant field galaxies. *Astrophysical Journal*, **469**, 508–518.

Smith, B.W., Mushotzky, R.F. and Serlemitsos, P.J. (1979). OSO 8 X-ray spectra of clusters of galaxies: II. Discussion. *Astrophysical Journal*, **227**, 37–51.

Smith, S. (1936). The mass of the Virgo Cluster. *Astrophysical Journal*, **83**, 23–30.

Solinger, A.B. and Tucker, W.H. (1972). Relationship between X-ray luminosity and velocity dispersion in clusters of galaxies. *Astrophysical Journal*, **175**, L107–L111.

Soucail, G., Mellier, Y., Fort, B., Mathez, G. and Hammer, F. (1987). Further data on the blue ring-like structure in A 370. *Astronomy and Astrophysics*, **184**, L7–L9.

Steidel, C., Adelberger, K., Dickinson, M., Giavalisco, M., Pettini, M. and Kellogg, M. (1998). A large structure of galaxies at redshift z approximately 3 and its cosmological implications. *Astrophysical Journal*, **492**, 428–438.

Sunyaev, R.A. and Zeldovich, Ia.B. (1981). Intergalactic gas in clusters of galaxies, the microwave background, and cosmology. *Soviet Scientific Reviews, Section E: Astrophysics and Space Physics Reviews*, **1**, 1–60.

Takano, S., Awaki, H., Koyama, K., Kunieda, H. and Tawara, Y. (1989). Large scale extended X-ray emission from the Virgo Cluster of galaxies. *Nature*, **340**, 289–290.

Tanaka, Y., Inoue, H. and Holt, S.S. (1994). The X-ray astronomy satellite ASCA. *Publications of the Astronomical Society of Japan*, **46**, L37–L41.

Tashiro, M., Iyomoto, N., Kaneda, H., Makishima, K. and Isobe, N. (1999). ASCA measurements of non-thermal pressures in the radio lobes. *Astronomische Nachrichten*, **320**, 217–218.

Trinchieri, G. and Fabbiano, G. (1985). A statistical analysis of the Einstein normal galaxy sample: Part Two. Elliptical and S0 galaxies. *Astrophysical Journal*, **296**, 447–457.

Trümper, J. (1992). ROSAT: A new look at the X-ray sky (The 1991 Grubb-Parsons Lecture). *Quarterly Journal of the Royal Astronomical Society*, **33**, 165–174.

Turner, M.S. (1999). Cosmology solved? Quite possibly! *Publications of the Astronomical Society of the Pacific*, **111**, 264–273.

Tyson, J.A., Kochanski, G.P. and dell'Antonio, I.P. (1998). Detailed mass map of CL 0024 + 1654 from strong lensing. *Astrophysical Journal*, **498**, L107–L110.

Valageas, P. and Silk, J. (1999). The entropy history of the universe. *Astronomy and Astrophysics*, **350**, 725–742.

van Dokkum, P.G., Franx, M., Kelson, D.D., Illingworth, G.D. (1998). Luminosity evolution of early-type galaxies to $z = 0.83$: Constraints on formation epoch and Omega. *Astrophysical Journal*, **504**, L17–L21.

Villa, G., Page, C.G., Turner, M.J.L., Cooke, B.A., Ricketts, M.J. and Adams, D.J. (1976). The Ariel V Sky Survey Instrument and new observations of the Milky Way. *Monthly Notices of the Royal Astronomical Society*, **176**, 609–620.

White, D.A., Fabian, A.C., Johnstone, R.M., Mushotzky, R.F. and Arnaud, K.A. (1991). The discovery of large amounts of cold, X-ray absorbing matter in cooling flows. *Monthly Notices of the Royal Astronomical Society*, **252**, 72–81.

White, R.E., III (1991). The metal abundance and specific energy of intracluster gas. *Astrophysical Journal*, **367**, 69–77.

White, S.D.M. (1976). The dynamics of rich clusters of galaxies. *Monthly Notices of the Royal Astronomical Society*, **177**, 717–733.

White, S.D.M. and Frenk, C.S. (1991). Galaxy formation through hierarchical clustering. *Astrophysical Journal*, **379**, 52–79.

White, S.D.M., Navarro, J.F., Evrard, A.E. and Frenk, C.S. (1993). The baryon content of galaxy clusters: A challenge to cosmological orthodoxy. *Nature*, **366**, 429.

White, S.D.M. and Rees, M.J. (1978). Core condensation in heavy halos: A two-stage theory for galaxy formation and clustering. *Monthly Notices of the Royal Astronomical Society*, **183**, 341–358.

Worrall, D.M. and Birkinshaw, M. (2000). X-ray-emitting atmospheres of B2 radio galaxies. *Astrophysical Journal*, **530**, 719–732.

Zwicky, F. (1937). On the masses of nebulae and of clusters of nebulae. *Astrophysical Journal*, **86**, 217–246.

Zwicky, F. and Herzog, E. (1963). *Catalog of Galaxies and Clusters of Galaxies*. California Institute of Technology, Pasadena, CA.

22

RALPH A.M.J. WIJERS*

Gamma-ray bursts

The only feature that all but one (and perhaps all) of the very many proposed models have in common is that they will not be the explanation of γ-ray bursts. Unfortunately, limitations of time prevent me from telling you which model is the exception. (If I did so, I would suggest Black Hole ridden by Accretion as the favorite in the race with Glitch as a dark horse if only because so many different horses and jockeys are riding under that name.) – Mal Ruderman, Texas Conference, 1974

When Mal so ended the first theory review on gamma-ray bursts (GRBs), I was on the other side of the world, practicing my fractions. But had he spoken these same words at the end of the fifth Huntsville GRB meeting in October 1999, his favorite model (though not his reasoning) would have been instantly recognized by all present as a very plausible and popular contender for GRBs. His dark horse is still easily identified with soft gamma repeaters – *Plus ça change, plus que ça reste le même*. What happened roughly in between the two events is the topic of this tale. We shall see how the field emerged out of the mutual distrust of the Cold War adversaries in the 1960s, and was finally resolved by a determined effort of just about the entire membership of the United Nations in the late 1990s. Its progress, of course, was in fits and starts, interwoven with stagnation and frustration.

Since γ rays are absorbed by less than 1% of the atmosphere, they are quintessentially the domain of space science. The initial motivation for launching satellites with γ-ray capability was not, however, scientific. The purpose of the US Air Force's Vela satellite program was the detection of nuclear explosions in the upper atmosphere or in space, which were prohibited by the first ever test ban treaty.

Like Jansky (radio), Penzias and Wilson (microwave), and Giacconi *et al.* (X-rays), it was what they were *not* looking for that started a new field: occasional flares of gamma rays lasting for just seconds, from extraterrestrial sources. Soon, other satellites confirmed the discovery.

A number of reviews have appeared that summarize the knowledge up to that point. Ruderman's paper (1975) and its observational counterpart by Strong *et al.* (1975) give an account of the initial explosion of discovery and thought. Higdon and Lingenfelter (1990) give a good overview of the pre-BATSE state of the field. The new discoveries made with the Burst and Transient Source Experiment (BATSE) on board the Compton Gamma Ray Observatory (CGRO), and the bewilderment following its overthrow of the galactic disk neutron star paradigm, are described in the review by Fishman and Meegan (1995). The consequences of the discovery of counterparts to GRBs and the state of the art at the end of the 1990s are described in reviews by Piran (1999; emphasis on theory and early stages of the burst) and by van Paradijs *et al.* (2000; emphasis on observations of afterglows – the later part of this chapter is based on parts of that review). To the historically inclined, review papers are only part of the story because they reflect more the final outcome of research than the struggle to reach that outcome. To get a flavor of the latter throughout the history of the subject, I recommend a number of conference proceedings. Theorists' struggles to come to terms with the phenomenon in the early days are reflected in the contributions to the 1974 Texas Symposium (Bergmann *et al.* 1975). Various stages of progress in the next decade are documented in the proceedings of meetings in La Jolla (Lingenfelter *et al.* 1982) and Stanford (Liang and Petrosian 1986). The first conference of the BATSE/CGRO

---
*State University of New York at Stony Brook, NY, USA; present address: Universiteit van Amsterdam, The Netherlands.

era and the accompanying wind of change is documented in the proceedings of the first Huntsville GRB symposium (Paciesas and Fishman 1992). After the discovery of counterparts, a meeting was called on short notice in Elba, Italy (26–27 May 1997). Its atmosphere of excitement mixed with astonishment and confusion is firmly etched in my memory; since the organizers wisely chose not to burden the attendees with writing proceedings papers, this feeling will forever remain the privilege of those who were there. The excitement was still very great during what I think of as the best meeting I ever attended: the fourth Huntsville GRB symposium (Meegan et al. 1998).

At this point I should describe my own path into the field, since it may have an impact on the tone and content of this review. I was in primary school when the discovery of GRBs was announced and remained blissfully unaware of them until some time during my PhD training in Amsterdam. I did not start active research in the field until Bohdan Paczyński's unlimited enthusiasm swept me into it in 1991. Therefore my accounts of the pre-BATSE era are entirely derived from the literature, and from conversations with those researchers I interacted with – by no means an unbiased sample of the field. Reviews are written with the sanitizing censorship of 20/20 hindsight and thus do not fully reflect the bouts of inspiration and confusion that characterize ongoing research: here lies a challenge for those who *were* there to see it. Having come into GRBs by way of Princeton also means that my allegiance in the distance scale debate of the early to mid-1990s was firmly extragalactic. While that point of view did eventually prevail, this outcome cannot have been so clearly predictable as I thought it was then, as is nicely illustrated by the reports of the "Great Debate" on the GRB distance scale (Fishman 1995, Lamb 1995, Paczyński 1995). A good perspective on the merits of such debates can be obtained from Trimble's (1995) discussion of the Shapley–Curtis debate. The field of soft gamma repeaters started as part of GRBs but has slowly split off as a separate discipline; its recent development has been similar in rapidity and magnitude to the GRB revolution, but this has been somewhat overshadowed by the latter. I discuss them only briefly in this chapter (Section 4).

A large part of this chapter is devoted to the events following 28 February 1997, when the first X-ray and optical counterpart to a GRB was discovered. The developments since then have been fast and furious: the first two counterparts settled the distance debate over GRBs firmly in favor of the extragalactic scale, making them the most powerful explosions since the Big Bang. Many issues are still not quite settled, and thus this part of the story has a decidedly less finished feel. Still, the discovery that they are somehow associated with young stars and can be observed out to very high redshifts opens up many new avenues of development.

At the beginning of a new century, this field is likely to explode into a major branch of astrophysics.

The organization of this chapter is mostly historical, though where dictated by logic I complete the discussion of some topics when they first arise, to prevent too much subdivision of the narrative. It begins in the days when every gamma-ray photon had a name, and it ends with the demise of the CGRO, halfway between the popular and the logical end of the twentieth century. I dedicate this account to Jan van Paradijs, who taught me so much. Despite his untimely death on 2 November 1999, his science will last well into the twenty-first century.

1 THE EARLY DAYS

1.1 Coming in from the cold

As described by others in this volume, high-energy astronomy started in the 1950s, because it required high-altitude balloon and satellite technology to get equipment above the atmosphere for long times. Even 1% of the atmosphere absorbs most gamma rays, so being in space is essential (though some observations and new technology tests can be done with balloon experiments). In the history of GRBs, the other important point to note is that the early days of satellite astronomy coincided with the peak of the Cold War. A practical way was needed to verify compliance with the first test ban treaty, which prohibited the detonation of nuclear weapons in space. Since it was possible in principle to hide the initial explosion (by detonation behind the Moon), one type of detector was designed to catch the MeV gamma rays from radioactive decays in the debris cloud that would drift into view some time after the explosion. For this purpose, the US Air Force in collaboration with the Los Alamos and Sandia Laboratories operated the Vela satellites (Figure 1; the name is derived from the Spanish *velar*, which means "to watch over" or "to hold a vigil"). They flew in pairs on opposite sides of a 250 000 km diameter orbit.

Clandestine nuclear explosions were not found, but something else was – short, intense bursts of gamma rays from random directions in the sky. The first one on record dates from 2 July 1967 (Figure 1), but the discovery was not published until 1973 (Klebesadel et al. 1973). It is sometimes stated that this was due to the classified nature of the mission, but this is false: since the objective of the satellites was to deter violations of the treaty, their existence was well advertised (just as shops conspicuously advertise their closed-circuit cameras). However, since only a few people worked on the data, and did so by hand, it took time for them to convince themselves that spikes in the highly variable γ-ray background were real signals and had a cosmic origin. The basic method for doing this is still

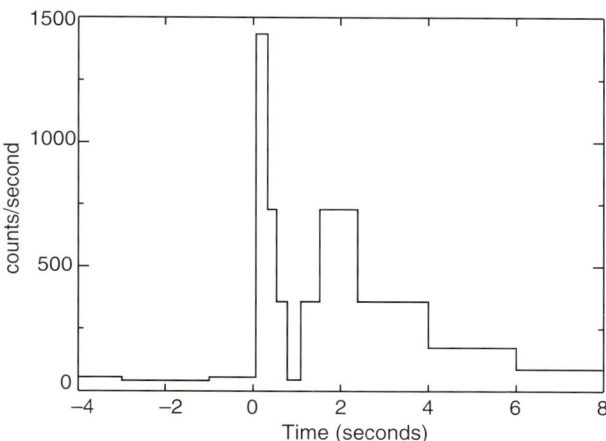

Figure 1 The Vela satellite (here Vela 5b), used by the USA to verify compliance with a ban on nuclear explosions in space (left). The light-curve of the first GRB it ever saw, on 2 July 1967, is also shown (right). (Images courtesy of "Astronomy Picture of the Day," at http://antwrp.gsfc.nasa.gov/apod/astropix.html.)

used in modern instruments: First, one requires more than one detector to see the signal; and second, one uses the difference in arrival time of the burst signal between two widely separated satellites to constrain the arrival direction (Figure 2). In this way, the early researchers eventually convinced themselves that the transient gamma-ray events they saw did not come from the Earth or the Sun.

It is interesting to note that the first observational paper already had a model to discuss. A few years earlier, Colgate (1968) had published his idea that a flash of gamma rays may be produced when the shock powering a Type II supernova breaks out of the surface of a red giant. Because the shock conserves energy, it becomes ever hotter as it moves down the density gradient, and at the point where the (Thomson) optical depth becomes unity, it emits photons with energy comparable to the electron rest mass. Since the medium is moving towards us with a Lorentz factor of about 1500, we see these blueshifted to GeV energies, and the flash time shortened to tens of microseconds. In later papers (Colgate 1973, 1974), he considers cooling before shock break-out and revises his estimates to more nearly the observed burst durations and photon energies. The scanned version of the 1974 paper held in NASA's ADS database (http://adsabs.harvard.edu/abstract_service.html) contains an interesting scribbled note to the conclusion section: the reader comments that such intense flashes in TeV gamma rays should be visible with the TeV telescope at Mount Hopkins. With the kind help of the people at ADS, I discovered that the owner of that copy of the *Astrophysical Journal* was Professor J. Grindlay, who confirmed to me that he had indeed considered the issue seriously. The problem, of course, was the same as with all other counterpart searches: one did not have any means of getting the very early and accurate locations required to aim a telescope at the GRB locations quickly enough.

Klebesadel *et al.* (1973) commented that there have been no detected supernovae near the GRBs they detected, and therefore dismissed the model. Very recent developments have put supernovae very much back into the picture (Section 6.2).

The first few facts about GRBs emerged quickly. Cline *et al.* (1973) used a hard X-ray detector on the IMP 6 satellite, designed for solar flare observations, to confirm the discovery and show that the spectra of the bursts really did peak in gamma rays (thus excluding the possibility that they were just high-energy tails of some type of X-ray event). A telescope with some directional capability on board OSO 7 confirmed the extrasolar origin of one burst

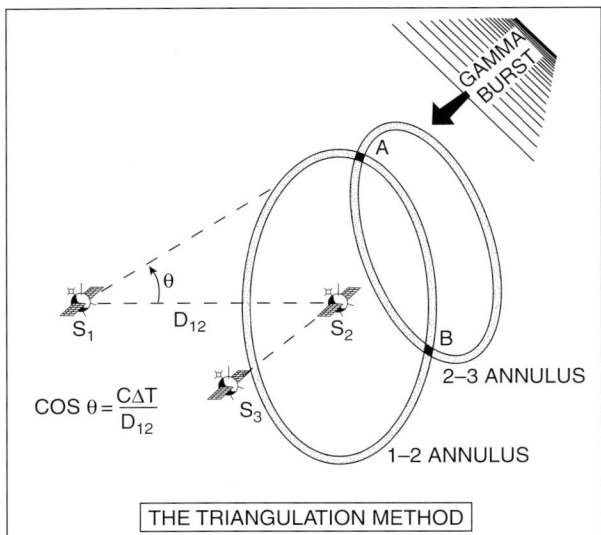

Figure 2 When two or more satellites detect a GRB, the arrival time difference of the signal between each pair of satellites can be used to constrain the location of the burst to a narrow circle on the sky. With three or more satellites, a true error box – sometimes as small as a few arc minutes across – can be constructed. This principle underlies the Interplanetary Network (IPN), still in operation. (Image courtesy of K. Hurley.)

(Wheaton et al. 1973). Incidentally, the first GRBs for which a cosmic origin could be established were observed with the Vela 5 and 6 series satellites, which had enough timing accuracy to pin down the directions to GRBs sufficiently well. The burst of 2 July 1967 was detected with the Vela 3 and 4 series, and is classified as the first GRB only on the basis of its similarity to the later ones, not because the direction to it is known to be inconsistent with a solar or terrestrial origin.

The first published Soviet observation of a GRB, using the Kosmos 461 satellite, dates from 1974, and is by the Leningrad group (Mazets et al. 1974). Igor Mitrofanov has told me that they too knew earlier that they might have something cosmic on their hands, probably in about 1971.

1.2 The earliest observations and theories

From here on the pace of discovery picked up quickly. In the review by Strong et al. (1975), in the proceedings of the seventh Texas Symposium, 34 bursts are listed. Their "sizes" (now called fluence or time-integrated flux) are in the range 10^{-5}–10^{-4} erg cm^{-2}. The spectra, where available, show a power law at low energies and then a bending to a steeper decline at a few hundred keV. At first this bending is fitted with a thermal bremsstrahlung function, that is, an exponential cutoff, but there is already one burst observed with Apollo 16 that has enough signal above 1 MeV to show that the high-energy part is also a power law (Metzger et al. 1974). The time histories are usually quite spiky, with overall durations of 0.1 to 100 s and sharp fluctuations down to 16 ms, the finest time resolution of the Velas. The locations of the bursts seem to show no significant preference for any part of the sky, except that two events are consistent with the location of the well-known accreting source Cyg X-1 (a fact that contributed considerably to Ruderman's pronouncement on theories). It is also noted that the fluence distribution for bright events is $N(>S) \propto S^{-3/2}$, the expectation for sources whose density does not depend on distance (often referred to as the "Euclidean" distribution). Cline (1975), in the same volume, confirms the high-energy end of the spectra to be a power law, and argues that the claimed flattening of the fluence distribution at low fluence is not significant. With hindsight, his balloon detections must have been something other than GRBs, since the fluence distribution is now known to be flatter than Euclidean at the fluence values he reports.

Enter Ruderman, who, also in the same volume, has the task of deciding which of the models to date survive scrutiny. The overview of models later given by Nemiroff (1994) lists 15 models dated 1974 or earlier; most of them are discussed in Ruderman's (1975) paper. He recognizes the limitations of theorists: "Most theoretical astrophysicists function well in only one or two normal modes. Therefore, we often tend to twist rather strenuously to convince ourselves and others that observations of new phenomena fit into our chosen specialties." He finds the brilliant solution of inviting as many as possible to enlist: "For theorists who may wish to enter this broad and growing field, I should point out that there are a considerable number of combinations, for example, comets of antimatter falling onto white holes, not yet claimed."

Ruderman largely dismisses the extragalactic models, in part because Colgate's supernova model is damaged by the non-detection of supernovae coincident with known GRBs. However, he also derives a constraint on distances from a black-body limit to the luminosity, assuming the observed photon energies are comparable to the thermal energies of the emitting particles, which virtually excludes extragalactic models beyond 30 Mpc. He acknowledges this to be wrong for some radiation mechanisms, but nonetheless the constraint appears to have impressed the audience. One of his acknowledged ways around the constraint is synchrotron radiation, since for this mechanism the emitted photons do have much lower energies than the particles that emit them; of course, we now know synchrotron radiation to be the dominant contributor. Among the galactic models he discusses a variety of "conventional" models, namely magnetic flares on various types of object and accretion events in compact objects. He then presents a number of more exotic options; as the later literature shows, none of them had a significant shelf life.

2 BUILDING A FIELD

In the following decade much progress was made in the observational understanding of GRBs. Mostly, the data came from small detectors (3–20 cm^2 of NaI scintillator) flown as "extras" on planetary and other missions, most notably Pioneer Venus Orbiter. A good summary of the state of affairs 10 years after the announcement of GRBs can be found in the proceedings of the Stanford meeting in August 1984 (Liang and Nolan 1984).

2.1 Properties of the gamma-ray emission

2.1.1 Temporal properties

It was clear right from the start that there is no such thing as a typical GRB profile. There were ones shorter than 0.1 s and others longer than 60 s in the initial catalog by Strong *et al.* (1975). Some time histories were simple single rise-and-decay curves, others had a complex, spiky structure, with a dozen or more sub-peaks in their profile. No periodicities were ever detected in any bursts, with the exception of the 8 s "ringing" in the tail of the remarkable 5 March 1979 event (Section 4). Given the prevailing paradigm at the time, that we were seeing events on local neutron stars, the desire to find periods is understandable, but the technical challenges were great: small detector areas and bandwidths implied that not too many photons were collected, and the light-curve was usually available only in binned form. So very short periods were hard to find due to binning and lack of photons; long periods would suffer from the fact that many GRBs are only a few seconds long.

Nonetheless, there is important physical information in the variability. A source of energy of size R cannot vary its output because we see it on a timescale less than the light crossing time R/c across it. (Imagine a sphere of radius R emitting an instantaneous flash of light at $t = 0$. The signal we see is smeared over a time $\delta t = 2R/c$ between when the signal from the point nearest to us arrives and when we see the signal from the farthest point.) Many attempts have been made to invent ways around the argument. For example, one can have a source surrounded by a spherical shell of large radius L. A flash of light from the center energizes it at some time, making it all light up in one go. Then a large portion of the shell near our line of sight can light up on a timescale much shorter than L/c. However, it still takes a time $2L/c$ for the shell to light up. Also, the initial rise of the flash is limited by the rise time of the central energy source, which once again is R/c. Therefore the shortest variation timescale again indicates the true size of the energy source. Similarly, one can consider a source emitting a relativistic shell of matter, with Lorentz factor Γ, which collides with the environment at a large distance L from the source and emits light at that time (a model very relevant to GRBs, as we shall see in Section 5). Due to forward relativistic beaming of the radiation we see only the material within an angle of $1/\Gamma$ from of the line of sight. The light travel time difference between the edge and center of the region we see is $\delta t_{\mathrm{trav}} = L/2\Gamma^2 c$, whereas the size is only $D = L/\Gamma$, so we have $D \sim \Gamma \, c\delta t$, contrary to the stated rule. However, in reality the shell lights up not instantaneously but over an interval $\delta t_{\mathrm{width}}$ equal to the light travel time across the radial width, δR, of the emitted shell. For a relativistic shell the thickness will be related to the time it took the source to emit it, which once again is limited by the source's light crossing time: $\delta t_{\mathrm{width}} \sim \delta R/c \sim R_{\mathrm{source}}/c$. Now, since the variation timescale is the larger of the two, $\delta t_{\mathrm{var}} = \max(\delta t_{\mathrm{trav}}, \delta t_{\mathrm{width}})$, and once again $R_{\mathrm{source}} \lesssim c\delta t_{\mathrm{var}}$.

The bottom line is therefore that since rise times of 0.01 s were seen for spikes within a GRB, it was correctly inferred that the sources are less than 0.01 light seconds, or 3 000 km in size. What is less often appreciated and seldom mentioned is that this also implies a mass and energy limit: in order for us to see any emission, the emitting object must be bigger than its Schwarzschild radius $R_S = 2GM/c^2 = 3(M/M_\odot)$ km. Therefore the size limit from variability implies a mass limit of $10^3 M_\odot$, ruling out, for example, active galactic nuclei. Likewise, an object of mass M is unlikely to store more energy than Mc^2 (assuming its bulk motion is not relativistic), so the energy release in a GRB is therefore constrained to be less than 10^{57} erg. Perhaps these limits were just too far from the physical models that were under consideration at the time to be written up explicitly.

2.1.2 Broadband spectra

The fact that the bulk of the energy emerges in gamma rays was the most distinguishing mark of GRBs right from the start. Detectors were typically a few to a few tens of square centimeters of NaI scintillator, giving a fair energy resolution and sensitivity to detect tens of GRBs per year. The first impression gained from most spectra was of a rapidly dropping power law of photon number with increasing energy, roughly as $n(E) \propto E^{-2.5}$, usually with a steepening at 0.1–1 MeV. The steepening was not well sampled in the vast majority of cases because the steep spectra imply few photons detected at those highest energies. In the literature of the time, one mostly finds the spectra explicitly interpreted as bremsstrahlung spectra, that is, with an exponential cutoff indicating a plasma temperature $kT \gtrsim 200$ keV in the emitter. In some bursts, such as the Apollo 16 event

(Metzger *et al.* 1974, Trombka *et al.* 1974), where enough photons were collected above 1 MeV, the high-energy part seemed to be a power law as well. So whereas it was immediately clear that GRB spectra were much broader than those of black bodies, it was often assumed that they were thermal emission.

The realization that MeV photons do not easily escape from compact regions, and thus that seeing such photons constrains the source, seems to be attributable independently to Cavallo and Rees (1978) and Schmidt (1978). The latter noted that photon–photon scattering (Delbrück scattering) between two MeV photons has a cross-section close to that for Thompson scattering, and thus is quite likely. He found that unless GRBs come from less than about 1 kpc away, the source's optical depth to its own photons would prevent MeV photons from escaping. Cavallo and Rees (1978) formalized the salient source properties into a single compactness parameter, given by

$$\theta_* = \frac{L\sigma_T}{m_p c^3 R} \simeq \frac{F_\gamma d^2 \sigma_T}{m_p c^4 \delta t} \quad (1)$$

which is roughly the photon–photon scattering optical depth of the source if most of the energy is in the MeV range. (L is the luminosity, σ_T the Thompson scattering cross-section, m_p the proton mass, R the source size, d its distance, F_γ its observed flux, and δt its variation timescale.) This number turns out to be about 1 for a distance of 1 kpc, and about 10^{12} for a distance of 1 Gpc, seemingly precluding the larger distance scale. It was shown only much later that relativistic expansion can overcome this limit easily (Paczyński 1986, Goodman 1986), but in the meantime this limit played a significant role in popularizing local models of GRBs.

2.1.3 Lines

A steady stream of reports of lines in GRBs was started by the report by Mazets *et al.* (1981) of annihilation and cyclotron lines in GRBs. These lines fit very well within the already prevailing idea that GRBs originated from confined, hot plasmas near the surface of a nearby neutron star. The annihilation lines were typically observed at energies of a little over 400 keV, making them redshifted by exactly the right amount for the surface of a neutron star. The cyclotron lines were seen around 20–40 keV, the value appropriate for a magnetic field of around 10^{12} G. Two plasmas would be needed: a hot one of ~200 keV to generate the continuum, and a cold one suspended in a magnetic field above the hot plasma to generate the relatively narrow cyclotron lines. Elaborate models of this kind were developed (e.g., Lamb 1981).

The observational reality of the lines is one of the outstanding issues in GRB research to date. The detections of lines with the Ginga satellite seemed rather to strengthen the case for their reality (Murakami *et al.* 1988, Fenimore *et al.* 1988). However, it was often pointed out that the analysis of the lines' significance was not satisfactory, for example in that they seemed to appear only in parts of a burst, and thus arbitrary data cuts were made to optimize their detection. Larger-scale blind searches of samples of GRBs have not turned up significant lines (e.g., Share *et al.* 1993, Briggs 1999, and references therein). At the 29th European Space Laboratory (ESLAB) symposium in Noordwijk, in April 1995, the Russian Payload for High Energy Burst Spectroscopy (PHEBUS) group withdrew all past claims of emission lines in GRB spectra (Denissenko *et al.* 1995, Terekhov *et al.* 1995), apparently including the group's claims based on other experiments as well. The BATSE spectroscopic detectors have never found a convincing instance of a line in a burst spectrum, and the review by Briggs (1999) gives an excellent overview of the low-energy detector response problems and fitting errors that make searching for lines in the 20–40 keV energy range quite hard. Lines in GRBs, whether low-energy absorption or higher-energy emission, appear to be weak and rare at best (they may not exist at all, but this remains to be settled).

2.2 Spatial distribution and locations

One way of saying at least something about the distance of otherwise unknown objects is to look at their sky distribution and see which known distribution it resembles. Since this comparison is for an all-sky pattern, relatively crude positions suffice to do this. The other method is to look at the accurate location of individual objects and to seek counterparts to them at other wavelengths.

Since the detectors employed in most missions had little or no directional sensitivity, the determination of GRB locations was done by timing triangulation (Figure 2). In the beginning, using the Vela satellites' relatively short separation (about one light second) and crude time resolution, the locations were not very good, but it is clear from Strong *et al.* (1975) that the impression of a fairly isotropic distribution on the sky was soon established. At the 1984 Stanford meeting, results of various satellites were shown to be formally consistent with isotropy even with the larger numbers then available. (See the contribution by Hurley *et al.*, especially Figures 1.19–23, in Liang and Nolan 1984.) As noted earlier, a few bursts were attributed to the galactic source Cyg X-1 (Strong *et al.* 1975). A few others were attributed to the class of soft gamma repeaters (Section 4). The large-scale isotropy of the GRB sky distribution was very restrictive in terms of allowed distance scales: it excluded the inner Solar System and even the Oort Cloud (Clarke *et al.* 1994), allowed nearby stars up to less than one scale height of the stellar disk away from us, and then allowed far extragalactic scales. Nearby extragalactic scales were not allowed, since no concentration toward nearby galaxies, the Virgo Cluster, or the supergalactic plane were seen. A later analysis of the two-point correlation function of IPN positions showed that there is no significant correlation down to a scale of one degree, which implies (given the correlation length of galaxies) that typical extragalactic GRBs have to be at least 100 Mpc away (Hartmann and Blumenthal 1989).

A variety of methods were tried to use locations of γ-ray bursts (GRBs) to discover counterparts at other wavelengths, including the following.

1. Archival plate searches for optical transients (OTs) in small GRB error boxes obtained from the IPN network of satellites through triangulation (Hurley *et al.* 2000a,b). Schaefer (1981, 1984) reported the discovery of such transients in the error boxes of several GRBs after extensive searches of the Harvard plate archives. However, Żytkow (1990) and Greiner (1990) have argued, on the basis of an analysis of the three-dimensional distribution of the images in the plate emulsions, that these events are likely to be plate defects (but see Schaefer 1990).
2. Deep imaging observations (optical, X-ray, and radio) of small error boxes. The aim of these observations was to see whether particular objects are found in these error boxes with a statistically significant excess, which might qualify them as possible GRB counterparts. These searches were not successful; that is, they did not lead to a convincing detection of a source population connected to GRBs. For example, some error boxes were conspicuously devoid of host galaxies (Schaefer 1992; Band *et al.* 1999), which was taken as evidence against the extragalactic origin of GRBs or as a problem with specific extragalactic models (Schaefer 1992, Band *et al.* 1999), though correlations with extragalactic objects were also claimed (e.g., Larson *et al.* 1996, Kolatt and Piran 1996) and disputed (Schaefer 1998, Hurley *et al.* 1997).
3. Simultaneous sky coverage using wide-field optical meteor search cameras (see, e.g., Hudec *et al.* 1999, Greiner *et al.* 1993, Krimm *et al.* 1996). In none of the simultaneous photographic images was an optical event detected; the corresponding magnitude limit (for optical flashes lasting as long as the GRB) is about fifth magnitude (McNamara *et al.* 1995).

Extensive summaries of the early attempts to find GRB counterparts have been given by McNamara *et al.* (1995) and Frail (1995). While eventually this would turn out to be the successful method of unraveling the mystery of GRBs (Section 6), we now understand why these early attempts did not succeed: the observations were either too late, or not sensitive enough. The cause of that was purely technical: precise locations could come only from the IPN, but on a timescale longer than it took the counterparts to fade, and the rapid locations required very wide-field follow-up, which could not simultaneously be very deep.

2.3 Models: Arise, local neutron star!

In the proceedings of the 1981 La Jolla meeting, Don Lamb stated that, "Not long ago there were many different tenable ideas about the origin of gamma-ray bursts. It is now widely believed, however, that the bursts come from strongly magnetic neutron stars." No doubt this represented a fairly strong majority opinion at the time, and only with hindsight is it easy to see where the GRB community went astray. However, it should be said that the evidence for cyclotron lines was never very strong, and the local model was only one possible explanation for the uniform and isotropic GRB distribution. So one lesson that can definitely be learnt from this affair is that one should be a bit more critical before getting onto a bandwagon, and a bit more patient with "mavericks," especially those with a proven track record in science. (I make no claim of the uniqueness of this lesson to this area of astronomy. In one of my previous incarnations I experienced the same lesson in the context of the spontaneous magnetic-field decay of neutron stars.) What to my mind is much more mind-boggling is the community's response to the BATSE discovery that excluded the local neutron star hypothesis (Section 3).

3 THE FIRST REVOLUTION: BATSE ON CGRO

The Burst And Transient Source Experiment (BATSE) on the Compton Gamma Ray Observatory (Figure 3), launched in April 1991, was designed to sample the flux distribution of GRBs much more deeply than before. It did so in a revolutionary way – by failing to meet the expectation of most researchers. Given the wide acceptance of the local neutron star model, this expectation was that the faint GRBs would look like the Milky Way: a band in the sky following the galactic plane, with a flux distribution that was inhomogeneous, that is, which did not follow the usual $N(>P) \propto P^{-3/2}$ flux distribution of a distance-independent density of sources.

What BATSE actually found is shown in Figures 4 and 5, albeit with the full nine-year results rather than the six-month ones presented first in Annapolis. It shows a distribution that is isotropic down to the lowest fluxes, and yet inhomogeneous, so the isotropy cannot be caused by our still not seeing more than a scale height away in the local disk model (Meegan *et al.* 1992, Briggs *et al.* 1996). Among the talked-about models this clearly favored a location of GRBs at high redshifts. However, instead of quickly creating a

Figure 3 The Compton Gamma Ray Observatory as it is being released from the Space Shuttle in April 1991. Four of the eight BATSE detectors are visible on the corners of the spacecraft. (Image courtesy of "Astronomy Picture of the Day," at http://antwrp.gsfc.nasa.gov/apod/astropix.html.)

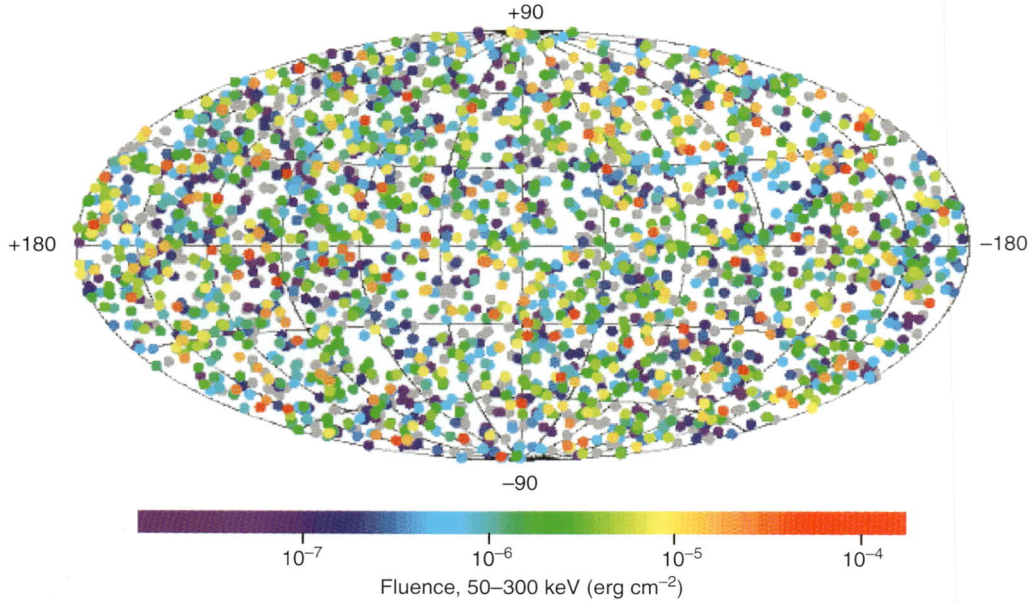

Figure 4 The locations of 2704 BATSE GRBs in galactic coordinates, showing the perfect isotropy of the burst locations. (Courtesy BATSE team, from their web page www.batse.msfc.nasa.gov/batse/.)

new consensus, this result proved tremendously divisive. Roughly half of astronomers (we know the statistics because one strange feature of many meetings in the next few years was counting a "cosmological v. galactic" vote) went for the cosmological distance scale on seeing the data. The other half were split between attempting to question the data (always a reasonable thing to do at first, but it grew rather desperate over the next few years) and producing a novel galactic model of GRBs: neutron stars ejected from the galactic disk at great speed, filling a very large halo around our Galaxy. As the volume of data increased, the radius of this halo did as well, in order to account for the smaller and smaller anisotropy. At this point, I warn the reader again that I entered the field through the good offices of Bohdan

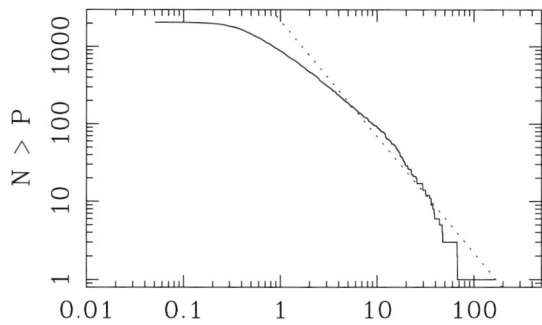

Figure 5 The flux distribution of BATSE GRBs triggered on the 1024 ms timescale. The dotted line is the "Euclidean homogenous" slope of $-3/2$, illustrating that the faint BATSE bursts are not a constant-density local distribution, and perhaps even that the bright end is not either. (Courtesy M. Briggs; from van Paradijs et al. (2000), with permission, from the *Annual Reviews of Astronomy and Astrophysics*, Volume 38, © 2000 www.AnnualReviews.org).

Paczyński, one of the early opponents of this idea. To my mind, this galactic model served no purpose other than "to save the phenomena." It was not called for on any other grounds, and did not even provide a home for the old physical models invented for the local neutron stars, because halo sources required highly super-Eddington luminosities and therefore relativistic outflows just like cosmological GRBs. All it had in common with older models were the words "galactic" and "neutron star." Nonetheless, they had quite a following, and when the American Astronomical Society sponsored a modern-day "Great Debate" between the proponents of the two camps (Fishman 1995, Lamb 1995, Paczyński 1995) the audience remained divided. However, a revolutionary development had definitely been sparked by BATSE: local neutron stars were out.

BATSE also had spectroscopic detectors on board, which were used extensively to search for the cyclotron lines found by Ginga. No convincing detections were ever reported, as discussed above (Section 2.1.3). Still, an entire conference was devoted to high-velocity neutron stars and gamma-ray bursts (Rothschild and Lingenfelter 1996), the last rallying cry for the local model. It would become known as the "*one* meeting on two so *different* topics."

3.1 Further CGRO discoveries

The major discoveries or big improvements by BATSE that we have already discussed are the isotropy and homogeneity of bursts, and the absence or scarcity of spectral lines. Many other discoveries by BATSE and/or the other instruments on CGRO concern the details of bursts. They are harder to place in context, because even now, as we have resolved the distance and energy scales from afterglows, the dynamics and radiative mechanism for the burst proper are not understood very well. As a result, there is no organizing paradigm for the wealth of information collected by the instrument.

Two classes of burst. The BATSE duration distribution of bursts is very wide, ranging from 10 ms to over 1000 s. In addition to this, a scarcity of bursts of around 2 s duration separates two broad peaks centered around 0.3 s and 30 s, the latter containing the majority (~80%) of events (Kouveliotou et al. 1993). However, the short end of the short bursts is no doubt incomplete, because the shortest trigger timescale was 64 ms, so neither the number of short bursts nor the short end of their duration distribution should be considered well established. Furthermore, Kouveliotou et al. (1993) also found that the short bursts are on average harder than the long ones. It has been speculated (e.g., Fryer et al. 1999) that the two could have different physical origins, but no clear evidence of this exists as yet. One problem is that afterglow observations, which have provided the breakthrough in the GRB mystery, have only been made for long bursts.

Terrestrial γ-ray flashes. One of the completely unexpected bonus discoveries from BATSE was the occurrence of short, hard flashes of gamma rays coming from the Earth (Fishman et al. 1994). They appear to correlate fairly well with the direction of thunderstorms, and the leading model attributes them to some variety of "sprites": electric discharges that emerge from the highest thunderclouds in an upward direction, where they may lead to energetic-photon emission under little enough residual atmosphere that the radiation can escape into space.

Hard-to-soft evolution. Most bursts show a hard-to-soft evolution in their spectra as the burst progresses. Single spikes within bursts are harder in their rise than their decline, with the higher-energy signal both rising faster and lasting less long (Fishman and Meegan 1995, and references therein). Generically, this is probably not very constraining, because almost any energetic transient phenomenon that is set off by a sudden injection of energy followed by passive evolution would do this. But the high-fidelity light-curves and spectra collected by BATSE probably pose a considerable challenge to future models for the burst emission and dynamics.

Duration–intensity and hardness–intensity correlation. The hardness–intensity correlation in GRBs was first reported by Golenetskii et al. (1983). This tendency for brighter bursts or burst parts to be harder is not universal but quite frequent (e.g., Kargatis et al. 1994). In the context of cosmological bursts it could be attributed to a redshift effect; the fact that it is also present in the $-3/2$ part of the flux distribution is then unexpected (Dezalay et al. 1992) and may constitute one of the oldest hints that this part of the flux distribution is not simply Euclidean.

Another correlation whose search was motivated by finding signs of redshift in GRBs is that between duration and intensity. Norris et al. (1994, 1995) reported a positive detection of the effect that fainter bursts are on average longer than brighter ones, in a manner consistent with the dimmest ones having a redshift factor $(1+z)$ that is twice that of the brightest ones. This effect now is generally accepted, but because we know that GRBs have a very wide luminosity function, its interpretation as a redshift effect is not. It may instead be at least partly an intrinsic luminosity–duration anticorrelation. Similarly, the discovery that dimmer bursts generally have lower values of the spectral break energy (Mallozzi et al. 1995) may not, after all, be simply a redshift effect.

GeV emission. A number of GRBs have been detected with the EGRET high-energy gamma-ray instrument on board CGRO (Hurley et al. 1994). Since the observed cases are only those where the burst was very bright and near the axis of the EGRET instrument, it is quite possible that the majority of bursts have this component. Remarkably, the high-energy photons last for hours in one case, long after the emission in the MeV range has become undetectable. This indicates that the GeV emission may not simply be the extension of the MeV spectrum, but a good analysis of this will have to await NASA's Gamma-ray Large Area Space Telescope (GLAST) mission.

Millisecond variability. While it has been clear since the beginning that GRBs have a short time structure, the large collecting area of BATSE allows very fine structure to show up significantly. The record is held by a $200\,\mu s$ spike detected in a GRB by Bhat et al. (1992), which indicates that the emitting region is less than 200 light microseconds (60 km) across.

Tails of gamma-ray emission. The detection of long-wavelength afterglow emission in GRBs since 1997 has sparked interest in looking for phenomena that might represent the transition from the burst emission to the afterglow. In a statistical analysis of superposed long bursts, with very careful background subtraction, Connaughton (2000) found strong evidence for tails of emission that decay as power laws in time. This same tail emission appears in some individual bursts as well (Giblin et al. 1999, Burenin et al. 1999). In the particular case of GRB 980923, Giblin et al. found that the tail shares many characteristics of an afterglow emission, and may be the earliest detection of this physical phenomenon. However, a later study reveals that in a sample of about 40 tails to bright bursts, this is not the majority behavior: most tails do show power-law decays with time, but the joint spectro-temporal evolution does not resemble an afterglow (Giblin et al. 2001).

4 SOFT GAMMA-RAY REPEATERS

As stated, I shall not treat soft gamma-ray repeaters (SGRs) in any detail, and the reader is referred to the upcoming review by Kouveliotou (2002) for a paper that does them full justice. However, in the context of the development of GRB astrophysics they have played an important role, so it would be wrong to leave them completely unmentioned. This section briefly mentions some key results.

SGRs first came to notice with the observation of three bursts within two days from SGR 1900 + 14 (Mazets et al. 1979). Observations of a repeating source from the same location as the giant flare of 5 March 1979 (Cline et al. 1980; Golenetskij et al. 1979) added SGR 0525 − 66 to the list. Much later, the third classic SGR was added to the group, starting with a discovery of soft bursts: SGR 1806 − 20 (Laros et al. 1986). In more recent times, two more sources were added: SGR 1627 − 41 and SGR 1801 − 23. Furthermore, an event similar to the 5 March giant flare was detected from SGR 1900 + 14 on 27 August 1998 (Hurley et al. 1999). Their locations were quickly recognized to be in the plane of the Galaxy (or the Large Magellanic Cloud (LMC), for 0525 − 66), and thus there was never much discussion on their distance scale. The fact that the giant flare of 5 March resembled a classic GRB did play a small role in unsuccessful attempts to link the distance scale of SGRs to that of all bursts (e.g., Lamb 1995).

Apart from their repetition, SGRs are recognizable by a number of other features which slowly emerged as class-defining characteristics (e.g., Norris et al. 1991, Kouveliotou et al. 1994). They have soft spectra, with no emission above 100 keV, short durations (typically 0.1 s), and fairly simple, single-pulse time profiles. The association of many SGRs, starting with SGR 0525 − 66, with supernova remnants (e.g., Evans et al. 1980, Kulkarni and Frail 1993) provided early clues to their association with some type of neutron star. The 8 s periodicity and approximate supernova remnant age of SGR 0525 − 66, as well as the very high luminosity of the 5 March event, led to the suggestion that they were in fact very highly magnetized neutron stars, with $B \gtrsim 10^{14}\,G$ (Paczyński 1992, Duncan and Thompson 1992). More recently, similar periodicities have been discovered in other SGRs, and even period derivatives, which imply similarly large magnetic fields (Kouveliotou et al. 1998, 1999), and have led to a fairly broad acceptance of the so-called "magnetar" model for SGRs.

5 THEORY SPEAKS: THE FIREBALL MODEL

Long before the discovery of afterglows, which we discuss next (Section 6), the idea that GRBs were the result of sudden depositions of energy into a small volume was explored (Cavallo and Rees 1978). Rees and Mészáros (1992) revived this idea in a cosmological context, making GRBs the relativistic analogue of a supernova explosion, the radiation generated when the ejecta from an energetic fireball collide with the surrounding medium (Figure 6). Before the first discovery, they predicted that the later evolution of this remnant would lead to longer-wavelength "afterglows" lasting days to months in optical to radio (Mészáros and Rees 1997). The afterglow of GRB 970228 behaved as predicted by their model (Wijers et al. 1997). Since then, this basic dynamical model has been well supported by observations. Of course, there are many possible variations on this theme. The density structure of the ambient medium is important, both on large scales (e.g., a stellar wind with decreasing density v. a uniform interstellar medium) and on smaller scales (e.g., turbulence in a medium might cause extra variability). For reviews on these topics, see Piran (1999) and Van Paradijs et al. (2000).

Briefly, the model works as follows. A large amount of energy E is deposited into a small volume, with a small mass M_0 so that $\eta = E/M_0c^2 \gg 1$. The volume has a very large compactness (eqn (1)) so it cannot radiate and expands adiabatically instead. This converts the hot fireball into a cold shell of matter moving relativistically, with $\Gamma \sim \eta$. When this shell has swept up of the order of Γ^{-1} times its own mass, it begins to decelerate significantly, converting its kinetic energy back into heat, and some of this heat into radiation. This typically happens at large distances from the explosion center, say 10^{16} cm, so that the medium is now optically thin and the radiation escapes. For the most common case of an adiabatic explosion (only a negligible fraction of the total energy is radiated), the Lorentz factor decreases with swept-up mass M as $\Gamma \propto M^{-1/2}$. So far the description looks very much like that of a supernova remnant moving from the ballistic to the adiabatic expansion phase, except that the high Lorentz factors very much reduce the timescales. First, deceleration starts when Γ^{-1} times the mass is swept up, so it happens much earlier for a GRB. Second, a relativistic shell nearly keeps up with the photons it emits, the difference in time between arrival of the photons and the shell being only of

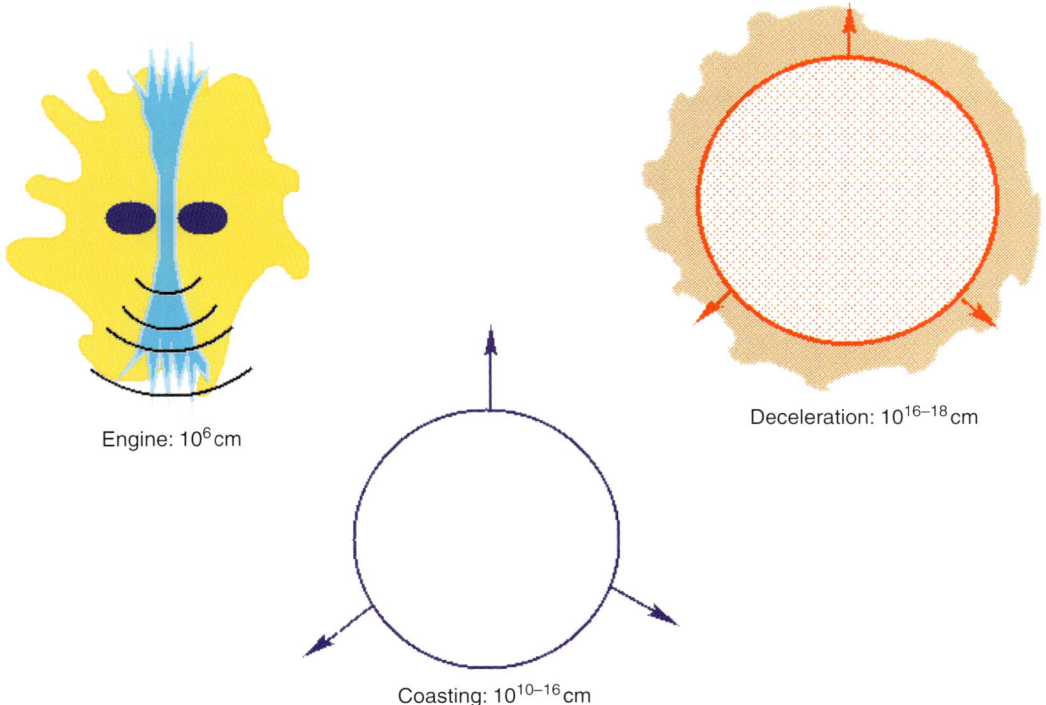

Figure 6 Schematic of the major phases of the fireball/blast-wave model. First, a large amount of energy is deposited into a volume less than a light millisecond across. This "fireball" of energy is opaque, and cools by turning adiabatically into an ultra-relativistic explosion. The ejecta coast along, eventually becoming optically thin, and possibly emit the prompt GRB emission via dissipation in internal shocks. The ejecta are then slowed down by the ambient medium. The kinetic energy is converted into heat of shocked swept-up gas, part of which can be radiated. As the ejecta slow down, the "afterglow" emission becomes softer and fainter with time.

the order of Γ^{-2} times the shell travel time. As a result, the duration of the onset of radiation is reduced by a factor of $\Gamma^{-8/3}$ relative to a supernova remnant, so scaling from a ballistic-phase life time of a typical supernova remnant (a century) with $\Gamma = 1000$, we expect a duration of 30 s for a GRB, which is near the peak of the distribution of durations.

It was expected theoretically, and demonstrated especially well for GRB 970508 (Galama *et al.* 1998d, Figure 15), that the dominant emission mechanism for the afterglow is synchrotron emission. The details of particle acceleration and magnetic field generation in relativistic shocks are not well known, but the simple assumption that constant fractions of the post-shock energy go into magnetic field energy and accelerated electrons works well in explaining observed afterglows. With those assumptions and the above dynamics, the description of the system is complete, and the evolution of the emission at all wavelengths can be calculated. The kinds of complexity that still enter are what the density is as a function of distance from the explosion center (since this determines the evolution of the blast-wave Lorentz factor) and whether the explosion is perhaps not spherical but a collimated, jet-like outflow. For these matters, I refer the reader to the above-mentioned reviews.

6 THE SECOND REVOLUTION: COUNTERPARTS

The idea that searches for counterparts would crack the GRB riddle was not new, of course (Section 2.2), but the Wide Field Cameras (WFC; Jager *et al.* 1995) on board BeppoSAX considerably improved the odds of success: they combined $40° \times 40°$ fields of view and arcminute localization capability with the ability to disseminate the position in hours. Rapid follow-up observations of BeppoSAX error boxes started in earnest in early 1997. The first attempt was made with the burst of 11 January, which came up with some sources but no likely counterpart candidate.

The next opportunity, which would turn out to lead to a breakthrough, arose on 28 February 1997. Pointed BeppoSAX NFI X-ray observations made ~8 hours after the burst revealed the presence of an unknown X-ray point source with a 2–10 keV flux of several 10^{-12} erg cm^{-2} s^{-1} in the WFC error box. Four days later the flux of this X-ray source had decreased by a factor of ~20: the first X-ray afterglow of a GRB had been detected (Costa *et al.* 1997; Figure 7).

In the meantime, GRB 970228 had also become the first GRB for which an optical counterpart was found. From a

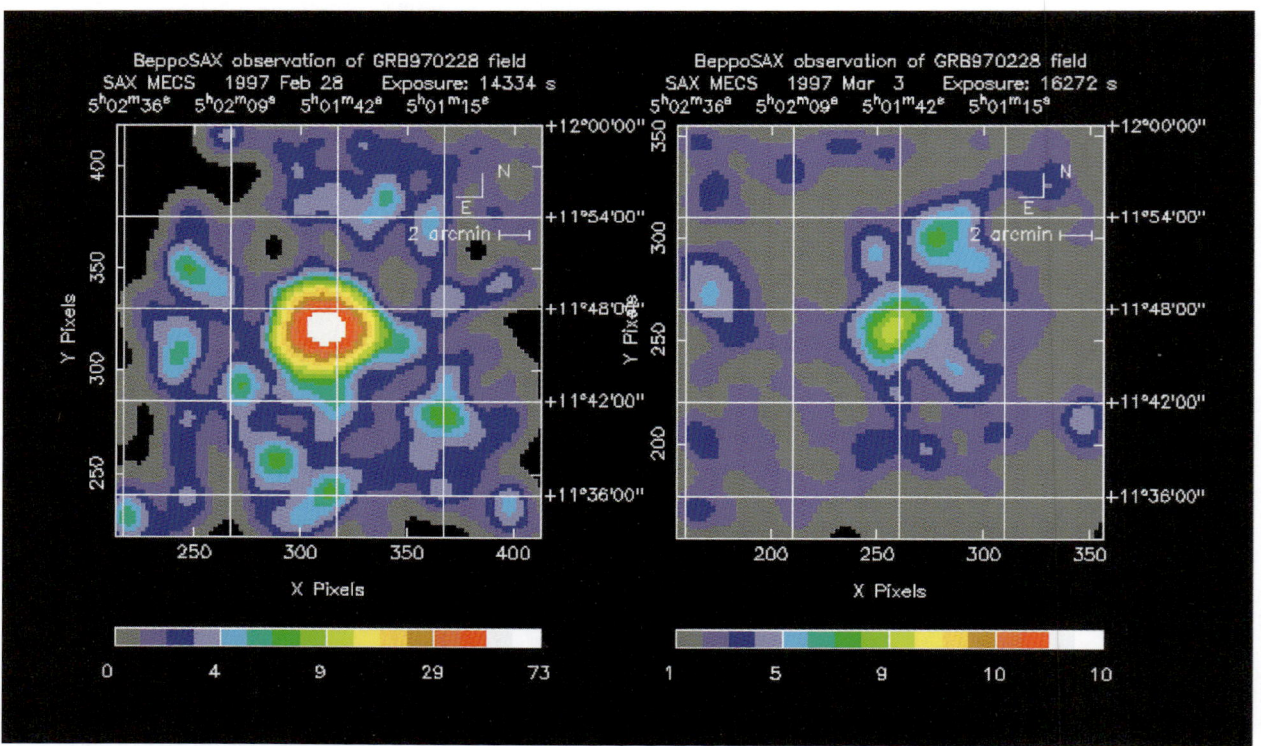

Figure 7 Discovery images of the X-ray afterglow of GRB 970228 with the BeppoSAX Narrow Field Instruments (NFIs). (Reprinted by permission from *Nature* (Costa *et al.* 1997) © 1997 Macmillan Magazines Ltd.)

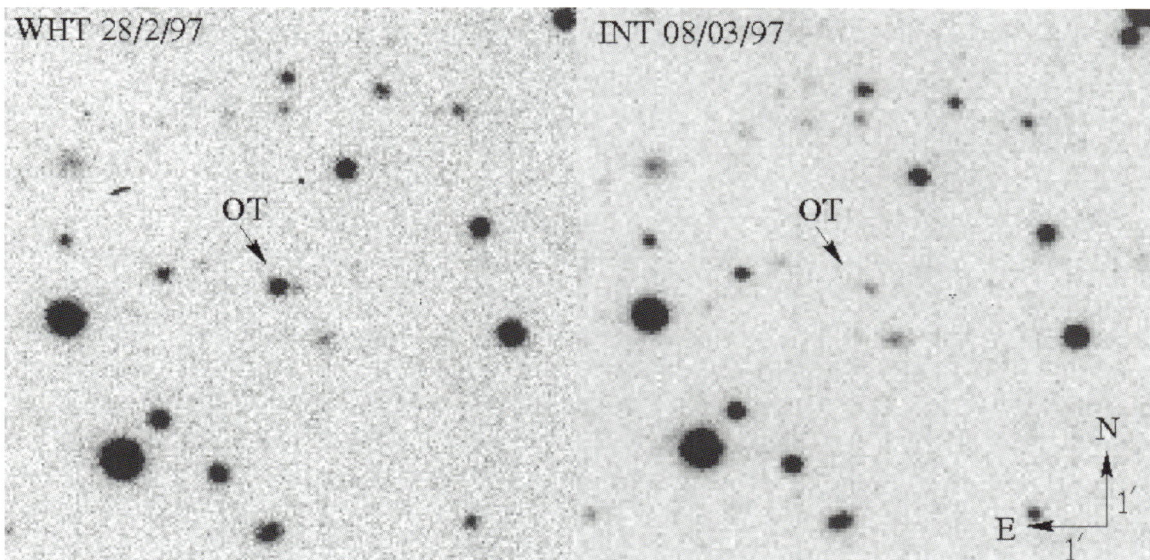

Figure 8 Discovery images of the optical afterglow of GRB 970228 at La Palma ("OT" stands for optical transient). (Reprinted by permission from *Nature* (van Paradijs et al. 1997), © 1997 Macmillan Magazines Ltd.)

comparison of (V and I band) images obtained with the William Herschel and Isaac Newton Telescopes, respectively 21 hours and a week after the burst, Groot et al. (1997b) discovered a decaying 21st-magnitude object at a position consistent with all positional information on the GRB (van Paradijs et al. 1997; Figure 8). Subsequent deep images made with the ESO New Technology Telescope (Groot et al. 1997a) and the Keck Telescope (Metzger et al. 1997c) showed an ~1″ extended object at the location of the optical transient, likely to be the host galaxy of the GRB (implying that the burst came from a distance of the order of Gpc).

Hubble Space Telescope (HST) observations were made in late March and early April 1997 (Sahu et al. 1997) and in September 1997 (Fruchter et al. 1999). These showed the presence of a point source, whose brightness decayed according to a power law, located near the edge of an extended object (diameter ~0.8″; Figure 9). Later observations established its redshift as 0.695 (Djorgovski et al. 1999, Bloom et al. 2001), confirming that the host of GRB 970228 is a distant galaxy.

The optical light-curve of the afterglow of GRB 970228 showed a significant deviation from a pure power law (Galama et al. 1997) about 1–2 weeks after the burst. Recently this has been interpreted as possible evidence for the presence of a supernova component in the light-curve (Reichart 1999, Galama et al. 2000; Figure 10), following an original suggestion for a deviation in GRB 980326 (Bloom et al. 1999a). (See Section 6.2 for further discussion on the connection between GRBs and supernovae.)

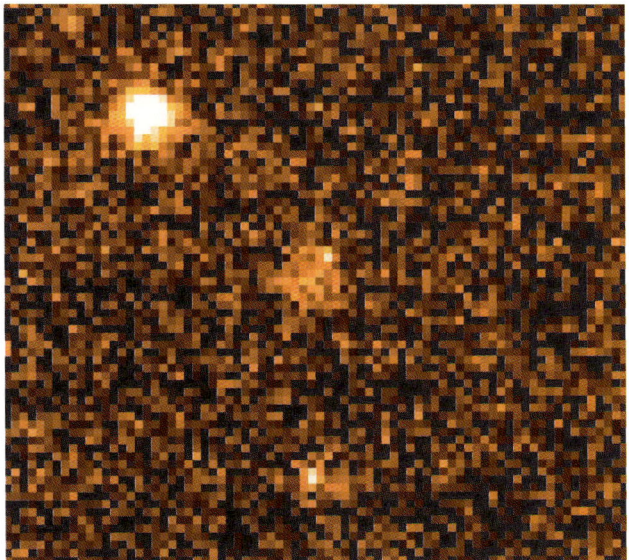

Figure 9 The host galaxy of GRB 970228 (center), imaged with HST. The six-month-old afterglow is still visible (the brightest pixel at the top right edge of the host; Fruchter et al. 1999). (Reproduced by permission of the American Astronomical Society.)

The exponents of the power-law decay X-ray and optical light-curves (Figure 11) were $\alpha_x = -1.33^{+0.12}_{-0.11}$ and $\alpha_{opt} = -1.46 \pm 0.16$. The exponents of the (assumed power-law) spectrum of the afterglow were $\beta_X = -1.06 \pm 0.24$ and $\beta_{X-opt} = -0.78 \pm 0.02$, for the X-ray waveband and

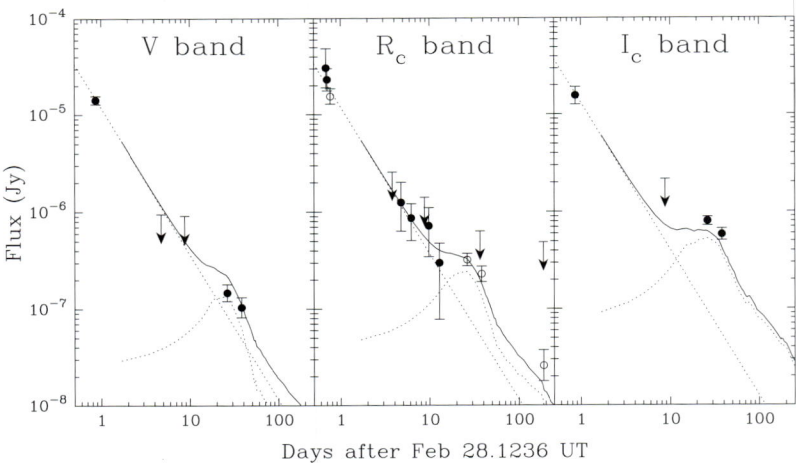

Figure 10 The VRI light curves of the afterglow of GRB 970228, showing evidence of a supernova component superposed on the power-law decline at later times (Galama *et al.* 2000). (Reproduced by permission of the American Astronomical Society.)

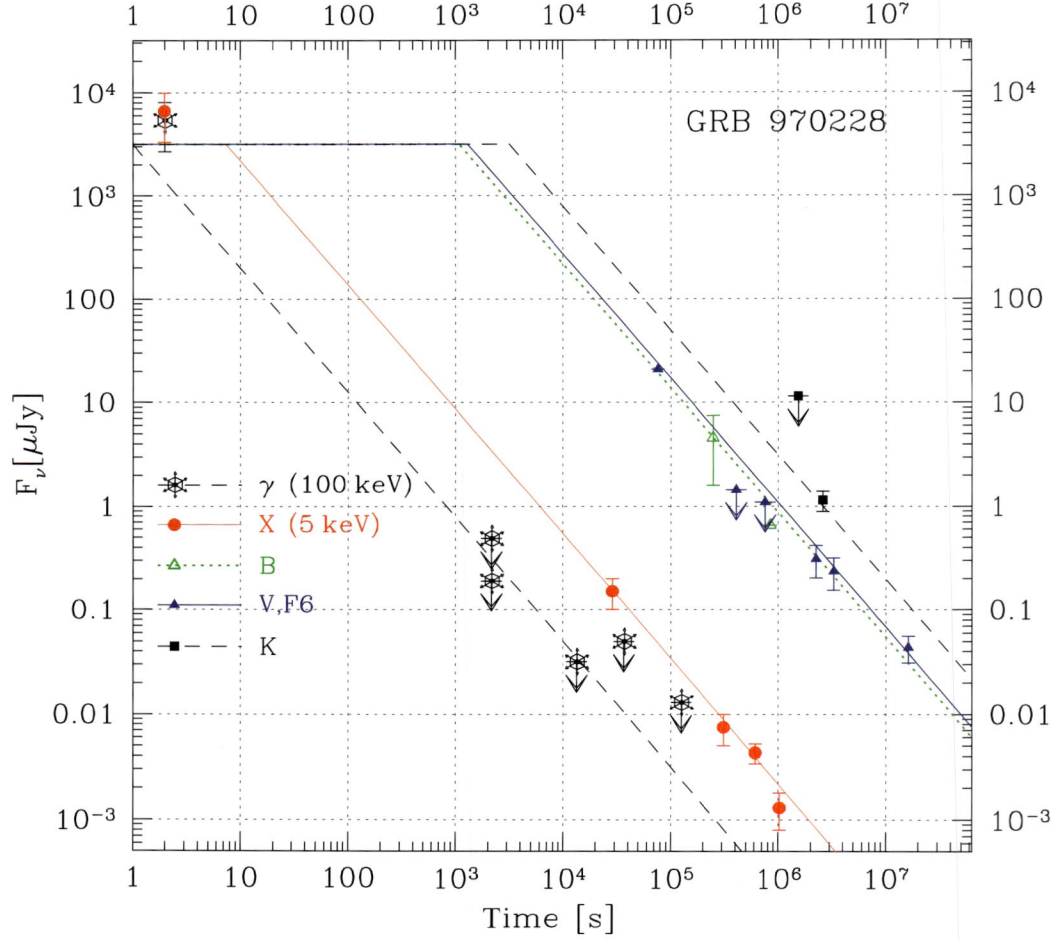

Figure 11 The multi-wavelength afterglow light-curve of GRB 970228. The decay rate in all wavelengths is the same, and the offset between the wavelengths is as given by simple blast-wave models of Rees and Mészáros (1992; Section 5). (From Wijers *et al.* (1997), reproduced by permission of Blackwell Science.)

for the X-ray to R band interval, respectively. The relation between α and β is consistent with that expected from a very simple version of the fireball model (Section 5), and gave initial support to this model (Wijers et al. 1997).

The next big step forward came with GRB 970508, whose optical counterpart (Bond 1997) initially increased in brightness from the first to the second night after the GRB and reached a maximum magnitude $R \simeq 19.8$ two days after the burst (Pedersen et al. 1998). It then started a power-law decline that could be followed for several hundred days (Galama et al. 1998a, Djorgovski et al. 1997, Sokolov et al. 1998, Garcia et al. 1998), and eventually flattened off, revealing the presence of a host (Zharikov et al. 1998, Bloom et al. 1998a, Fruchter et al. 2000).

Optical spectroscopy with the Keck Telescope revealed the presence of absorption lines of Mg II, Fe II, and Mg I (Metzger et al. 1997b; Figure 12), which are often found in quasar spectra (Steidel and Sargent 1992), redshifted by $z = 0.835$ (Metzger et al. 1997b). The subsequent discovery of O II and Ne III emission lines in the spectrum at the same redshift of $z = 0.835$ (Metzger et al. 1997a} established the presence of an underlying host galaxy. HST observations revealed that the OT was at the center of a blue, actively star-forming dwarf galaxy (Pian et al. 1998, Fruchter et al. 2000, Natarajan et al. 1997; Figure 13) of $L_B = 0.12 L_\star$ and a star formation rate of $\leq 1 M_\odot \, \text{yr}^{-1}$ (Bloom et al. 1998a). This result unambiguously established that GRBs originate at "cosmological" distances, and terminated the discussion on the GRB distance scale: it established GRBs as the most luminous photon emitters in the entire Universe, with peak luminosities of the order of $L_\gamma \simeq 10^{52} \, \text{erg s}^{-1}$. The optical afterglow of GRB 970508 reached an absolute magnitude $M_V \simeq -24$; that is, in optical emission it became two orders of magnitude brighter than a Type Ia supernova.

GRB 970508 also has the distinction of being the first GRB for which radio afterglow was detected (Frail and Kulkarni 1997). During the first month the radio flux underwent strong irregular variations, around an average value of

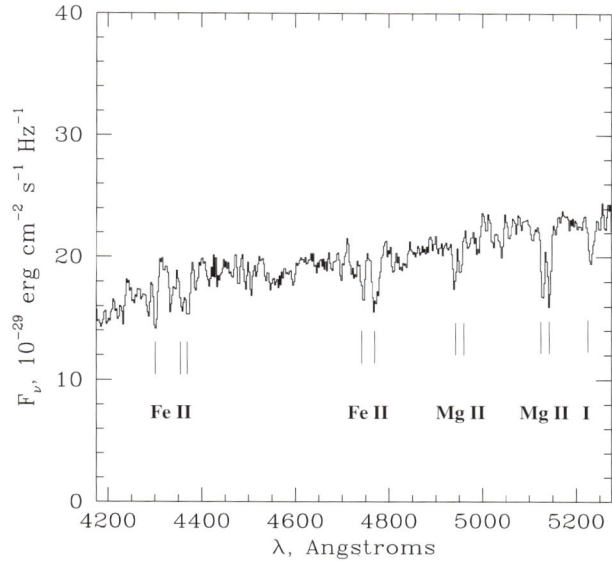

Figure 12 The spectrum of the OT of GRB 970508, showing Fe and Mg absorption lines at $z = 0.835$ and $z = 0.77$. (Reprinted by permission from *Nature* (Metzger et al. 1997b), © 1997 Macmillan Magazines Ltd.)

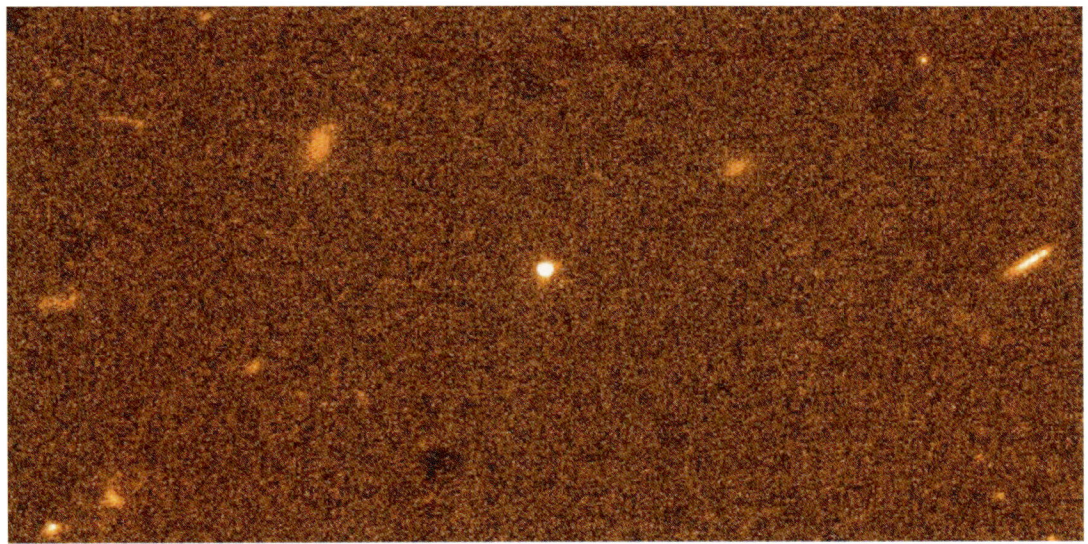

Figure 13 HST/STIS image of the OT of GRB 970508 (Pian et al. 1998). The faint host became visible only after the OT had faded. (Reprinted by permission of the American Astronomical Society.)

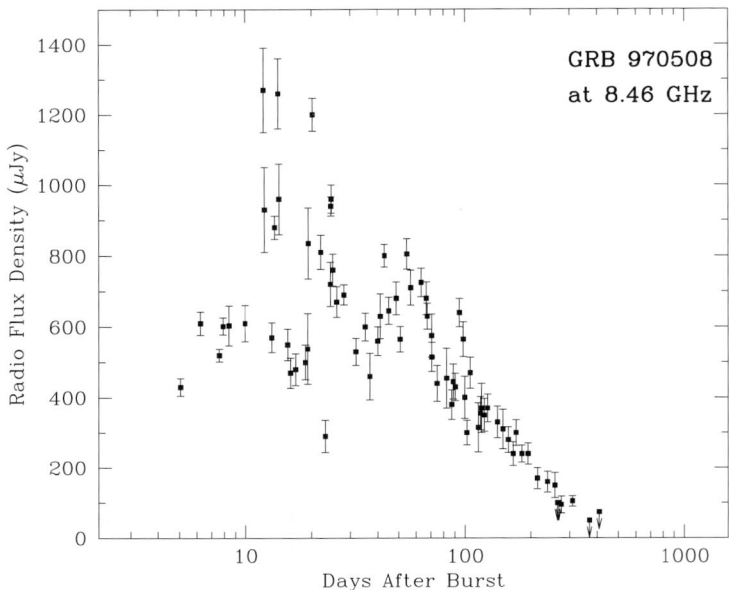

Figure 14 The 8.46 GHz VLA light-curve of the afterglow of GRB 970508. Note the large scintillation fluctuations in the first month and their later absence, indicating that the source expanded (Frail *et al.* 1997, Kulkarni *et al.* 2000). (Reprinted by permission from *Nature* (Frail *et al.* 1997), © 1997 Macmillan Magazines Ltd.)

0.6 mJy (8.5 GHz), which damped out after about a month (Figure 14). The variations are caused by interstellar scintillation in our Galaxy. The damping of the fluctuations reflects the increase in the size of the radio emitter (analogous to the absence of twinkling of planets). From the known source distance and the properties of the interstellar medium along the line of sight, Frail *et al.* (1997) inferred a value for the source size at the time the fluctuations disappeared. They concluded that the radio emitter expanded with a velocity consistent with that of light, which gives strong support to the relativistic fireball model for GRBs and their afterglows.

GRB 970508 was the first GRB for which the afterglow was observed all the way from X-rays, via optical/near-IR, to millimeter and low-frequency radio waves. This allowed Galama *et al.* (1998b) to reconstruct the X-ray to radio spectral energy distribution of the afterglow, as observed 12 days after the GRB (Figure 15). This spectrum consists of piecewise-connected power-law distributions, with three clearly recognizable spectral breaks at the frequencies at which the different power laws are connected. On the low-frequency side we recognize the $\nu^{1/3}$ synchrotron low-frequency limit, which at even lower frequencies turns over due to self-absorption. The high-frequency end of the $\nu^{1/3}$ part of the spectrum connects, at the peak of the spectrum, to a power-law part whose slope depends on the power, p, of the power-law electron energy (or Lorentz factor) distribution. In the near-IR a third break is seen which can be unambiguously identified with the "cooling break," whose frequency corresponds to an electron energy above which the synchrotron loss time is smaller than the flow timescale of the system. The identification is based on two observed facts: (1) the change in spectral slope is 0.5 (Galama *et al.* 1998d; Figure 15); (2) the frequency, ν_c, of the cooling break changed with time since the burst as $\nu_c \propto t^{-1/2}$, as was evident from the progression of the spectral slope change (by 0.5) from the optical to the near-IR passbands (Galama *et al.* 1998a). These results showed that relatively simple versions of the fireball model provide a reasonable description of the observed afterglow spectrum, and provided a strong argument for the idea that GRB afterglows are powered by the synchrotron emission of electrons accelerated in a relativistic shock.

6.1 "Dark" gamma-ray bursts

The relative optical response in the afterglow of a GRB can vary enormously from one burst to another (for the X-ray afterglows the variations are more moderate). A very good example is provided by GRB 970828, a fairly strong gamma-ray event, for which optical observations were made within 4 hours of the burst, and continued for eight consecutive nights after the burst. No potential optical counterpart was found to vary by more than 0.2 mag down to $R = 23.8$ (Groot *et al.* 1998a). Compared with GRB 970508, the peak optical flux in the afterglow of GRB 970828, normalized to the fluence of the GRB, was at least a factor of 10^3 smaller. Absorption in the host galaxy,

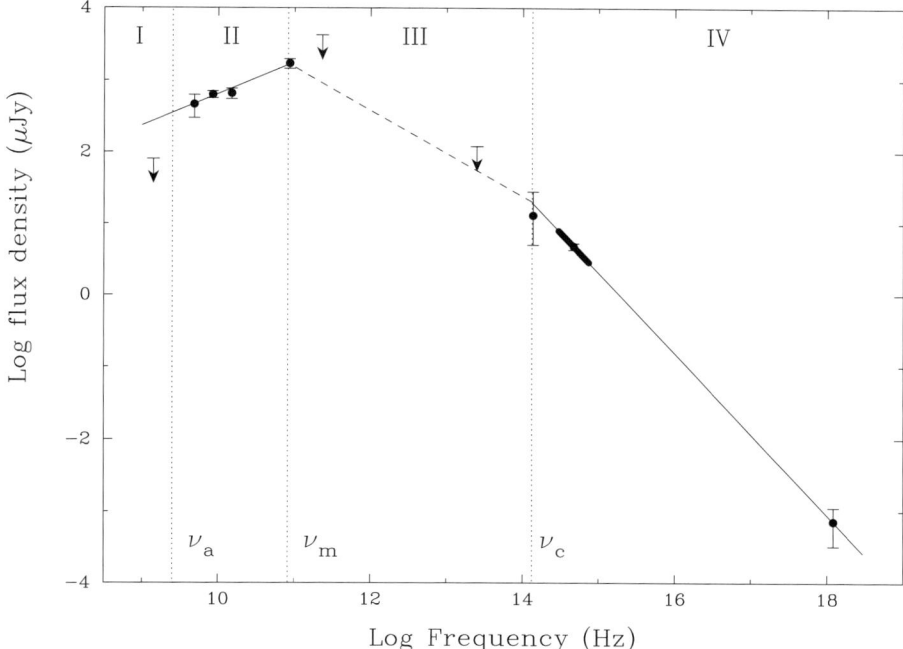

Figure 15 The radio to X-ray spectrum of the afterglow of GRB 970508, 12.1 days after trigger, showing all the characteristics of synchrotron emission (Galama et al. 1998d). (Reprinted by permission of the American Astronomical Society.)

by at least 5 magnitudes in the R band, may explain the large difference in the optical afterglows. Note that for moderate redshifts ($z \sim 1$) the absorbed photons have wavelengths of ~3000Å, at which the interstellar extinction is a factor ~2.5 larger than in the R band. Therefore, with a normal dust-to-gas ratio in the host, a modest column density of $N_H \sim 10^{21}$ cm^{-2} would be sufficient. Recent results support this suggestion: $z = 0.96$ (Djorgovski et al. 2000) and $N_H = 4.1^{+2.1}_{-1.6} \times 10^{21}$ cm^{-2} (as measured in our rest frame by Yoshida et al. 1999). More recent evidence has emerged to suggest that indeed column densities toward GRBs are quite high (Owens et al. 1998, Galama and Wijers 2001), in spite of the generally low optical extinction. This may indicate that the very radiation of GRBs itself destroys the extinction-causing dust (Fruchter et al. 2001, Waxman and Draine 2000), and further supports a strong connection between GRBs and star-forming regions.

6.2 The supernova–GRB connection

The total energy budgets of GRBs, corresponding to their cosmological distances are roughly of the same order of magnitude as those of supernovae, but in GRBs the energy is emitted much more rapidly than in supernovae. The reason for this is that in supernovae the energy (except that in neutrinos) is thermalized by a large amount of mass (several solar masses) and converted into heat and expansion at a speed of typically 10^4 km s^{-1}. In a GRB the energy cannot be shared with more than $\sim 10^{-5} M_\odot$ in baryons, in order not to lose the required high Lorentz factors ($\Gamma \geq 10^2$). The similar amount of total energy (10^{53} erg) involved in both phenomena kept open the possibility of their connection, despite the very large difference in the way the energy is emitted. For instance, in one of the main GRB models, the "collapsar" or "failed supernova" model (Woosley 1993, Paczyński 1998), a GRB is the result of a massive core collapse, with extreme parameters (e.g., mass or rotation of the progenitor). However, until April 1998 direct evidence for a relation between GRBs and supernovae was totally lacking.

It therefore came as something of a surprise when Galama et al. (1998c,e) found that the WFC error box of GRB 980425 contained the supernova SN 1998bw (Figure 16). This supernova is located in a spiral arm of the nearby galaxy ESO 184 − G82, at a redshift of 2550 km s^{-1}, corresponding to a distance of 40 Mpc. On the basis of very conservative assumptions regarding the error box and the time window in which the supernova occurred, Galama et al. (1998c) found that the probability of any supernova with peak optical flux a factor of 10 less than that of SN 1998bw being found in the error box by chance is 10^{-4}, providing strong evidence for a physical relation between the GRB and the supernova. The WFC error box also contains two weak X-ray sources, one of which, 1SAX J1935.0 − 5248,

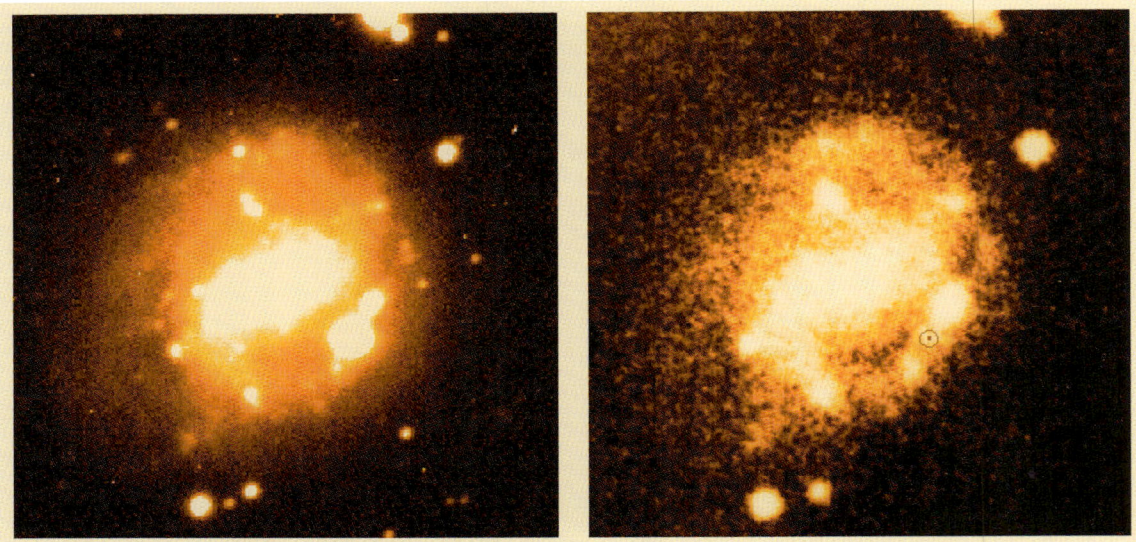

Figure 16 Images of the galaxy ESO 184−G82 with (left) and without (right) SN 1998bw (Galama *et al.* 1998c). (Reproduced by permission from *Nature*, © 1998 Macmillan Magazines Ltd.)

coincides with SN 1998bw and GRB 980425 (Piro *et al.* 1998, Pian *et al.* 1999); the second source initially faded, but was subsequently re-detected at a flux similar to that of the first observation (Pian *et al.* 1999), which excludes it as a viable GRB afterglow. The case for a physical relation between the supernova and the GRB is therefore a strong one.

With respect to its apparent properties (peak flux, duration, burst profile), GRB 980425 was not remarkable. Of course, at its distance of 40 Mpc its total energy (8×10^{47} erg) is some five orders of magnitude smaller than that of "normal" GRBs (Galama *et al.* 1998c). The total energy in GRB 980425 is remarkably close to the value envisaged in Colgate's (1974) model.

Its connection with a GRB aside, SN 1998bw is extraordinary for its very high radio luminosity near the peak of the supernova light-curve (Kulkarni *et al.* 1998b). According to the analysis by Kulkarni *et al.* (1998b), the radio light-curve requires the presence of a mildly relativistic ($\Gamma \sim 2$) outflow, which may account for the GRB emission. However, a sub-relativistic model also explains the data quite well (Waxman and Loeb 1999). An analysis of the optical light-curve (Galama *et al.* 1998c; Figure 17) and its early spectra (Iwamoto *et al.* 1998, Woosley *et al.* 1999, Branch 2001) showed that SN 1998bw was an extremely energetic event (total explosive energy in the range $(2-6) \times 10^{52}$ erg, i.e., a factor of ~30 higher than is typical of a Type Ib/c supernova), in which an extraordinarily large amount of ^{56}Ni ($0.5-0.7 M_\odot$) was ejected. The early expansion speed was as high as $\sim 60{,}000$ km s^{-1}. According to Iwamoto *et al.* (1998; see also Iwamoto 1999a,b), the remnant mass of the core collapse exceeded $3 M_\odot$ and a black hole was likely formed in SN 1998bw. (Note that by allowing for asymmetry in the supernova explosion, Höflich *et al.* (1999) derive somewhat more moderate but still very large values for the energetics of this supernova.) GRB 980425 is the only GRB (out of more than two thousand) for which the evidence of a connection with a supernova appears convincing. Attempts to search for further associations have led to no other strong candidates (Wang and Wheeler 1998, Woosley *et al.* 1999, Bloom *et al.* 1998b, Kippen *et al.* 1998). (I consider the proposed connections between SN 1997cy and GRB 970514 (Germany *et al.* 2000), and between SN 1999eb and GRB 991002 (Terlevich *et al.* 1999), not convincing.)

Since the sampling volume for low-luminosity events like GRB 980425 is smaller than that of "normal" GRBs by a factor $\sim 10^6$, the rate (per galaxy) of the former events may well exceed those of the latter by a large factor. Because of their small distances they are expected to contribute a $P^{-3/2}$ component to the log$N$ ($>P$) distribution. From the absence of a turn-up at the flux limit ($P \simeq 0.2$), Kommers *et al.* (2000) infer that such a Euclidean component can contribute at most 10% to the observed BATSE burst sample (99% confidence limit). With a "normal" GRB rate of $\sim 10^{-8}$ per galaxy per year (Wijers *et al.* 1998), the corresponding limit on SN 1998bw-like events is then of the order of 10^{-4} per galaxy per year. With an observed rate for Type Ib/c supernovae of the order of 10^{-3} per galaxy per year (Van den Bergh and Tammann 1991), this rather weak limit

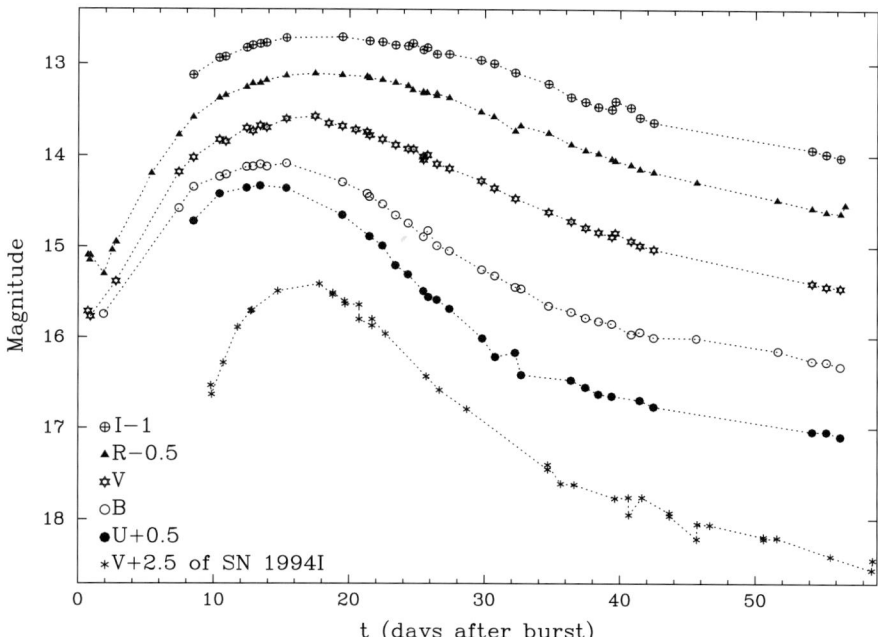

Figure 17 The optical light-curve of SN 1998bw. (Reproduced by permission from *Nature* (Galama *et al.* 1998c), © 1998 Macmillan Magazines Ltd.)

serves to show that at most a fraction of the Type Ib/c supernovae produce GRBs.

The observational basis for a connection between GRBs and supernovae was greatly enriched with the discovery by Bloom *et al.* (1999c) of a late component superposed on the power-law optical light-curve of GRB 980326, which they argue reflects an underlying supernova (Figure 18). A similar interpretation has been proposed by Reichart (1999) and Galama *et al.* (2000) for the long-known deviation from a pure power-law decay of the optical afterglow of GRB 970228 (Galama *et al.* 1997).

The optical light-curve of GRB 980326 showed an initial rapid decay ($\alpha_{opt} = -2.0$; Groot *et al.* 1998b); the light-curve flattened after ~10 days to a constant value $R = 25.5 \pm 0.5$. Such flattening has been seen in the light-curves of other afterglows as well, and has been taken to be the signature of an underlying host galaxy. Observations by Bloom *et al.* (1999c) made ~3 weeks after the burst revealed a surprising brightening of the afterglow, to a flux level 60 times above that expected from an extrapolation of the power-law decay. At the same time the spectral energy distribution became very red. Observations made ~9 months after the burst showed that any host galaxy would have to be fainter than $R = 27.3$. Using the multi-color light-curve of SN 1998bw (Galama *et al.* 1998c) as a template, Bloom *et al.* (1999c) found that they can reproduce the observed optical afterglow light-curve of GRB 980326 by a combination of a power law (exponent -2.0) and a bright supernova at a redshift $z \sim 1$.

Reichart (1999) and Galama *et al.* (2000) make the same decomposition of the optical light-curve of GRB 970228 and find that this provides a good fit to the data. These results support the idea that at least a fraction of the GRBs originate from the collapse of a massive star. This confirms the models proposed by Woosley (1993) and Paczyński (1998), under various names, such as the "failed supernova model" or "collapsar" model, in which it is assumed that a black hole is formed, surrounded by a fairly massive torus. It is as yet unclear what particular circumstances give rise to the GRB (e.g., a very high mass, rapid rotation, a particular evolutionary history), but it seems virtually certain that strong collimation of the outflow is required to accommodate both the extremely high Lorentz factor flow required for the GRB, and the more sluggish flow connected with the supernova photospheric emission.

6.3 Prompt optical emission

Based on the fireball model, prompt optical emission simultaneous with the GRB is expected, with magnitudes ranging between 9 and 18 (Mészáros and Rees 1997, Sari and Piran 1999, Katz 1994). After many years of unsuccessful attempts to catch the optical signal of a GRB in progress

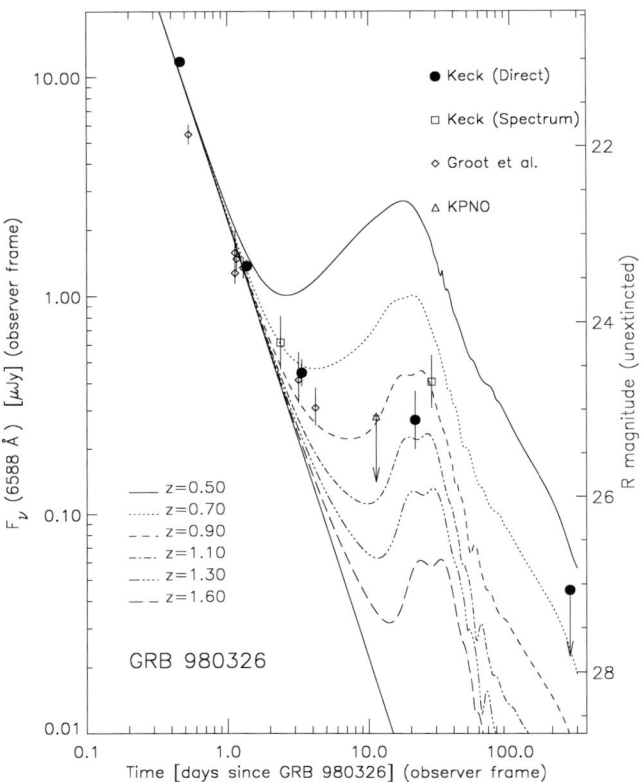

Figure 18 The R light-curve of GRB 980326, with model curves representing the power-law decay of a relativistic afterglow superposed on the light-curve of SN 1998bw, shifted to various redshifts. (Reproduced by permission from *Nature* (Bloom *et al.* 1999a), © 1998 Macmillan Magazines Ltd.)

(e.g., McNamara *et al.* 1995, Krimm *et al.* 1996, Hudec and Soldan 1995, Lee *et al.* 1997, Park *et al.* 1997), the first such simultaneous detection was made of GRB 990123 by Akerlof *et al.* (1999). The robotic camera ROTSE (Robotic Optical Transient Search Experiment), triggered by BATSE, started a sequence of optical images of the BATSE error box 22 s after the start of the burst (Figure 19). An optical transient was detected which, ~50 s after the start of the burst, reached magnitude 8.9, and afterwards decayed to ~15th magnitude in ~10^3 s. The slope of the prompt light-curve changed after ~300 s, and merged smoothly with the later afterglow light-curve (Figure 20; Kulkarni *et al.* 1999a, Galama *et al.* 1999, Castro-Tirado *et al.* 1999). During the peak of the prompt optical emission the GRB reached an absolute magnitude $M_V \approx -36.5$, that is, the event was then for a brief time interval 10 million times brighter than a Type Ia supernova.

The prompt optical emission does not follow the GRB light-curve, nor can it be understood as the low-energy extrapolation of the (variable) GRB spectrum (Galama *et al.* 1999, Briggs *et al.* 1999). This indicates that the prompt optical emission and the gamma rays originate from different regions in the fireball. It has been popular to ascribe the origin of the gamma rays to internal shocks (Rees and Mészáros 1994, Kobayashi *et al.* 1997), and the long-term afterglow to the external shock. (Recent theoretical calculations, however, have reopened the issue: Dermer and Mitman (1999) have presented a plausible external shock model for highly variable prompt gamma-ray emission, and Ramirez-Ruiz and Fenimore (2000) have shown that previous objections to external shock models for the prompt emission can be circumvented.) It has therefore been natural to ascribe the prompt optical emission to the reverse shock (Sari and Piran 1999, Mészáros and Rees 1999), which is observed only during a time interval of the order of the burst duration (i.e. comparable to the time it takes for the reverse shock to travel through the ejecta). The radio afterglow properties of GRB 990123 are peculiar in that radio emission was seen only during a ~1 day interval about a day after the burst (Kulkarni *et al.* 1999b, Galama *et al.* 1999). This brief radio event has been interpreted by Kulkarni *et al.* (1999b) as reverse shock emission. Galama *et al.* (1999) ascribe it to emission from the forward shock and interpret the peculiar nature of the radio emission as the result of a very low value of the synchrotron peak frequency one day after the burst. They suggest that differences in the afterglow properties (peak frequency, cooling frequency) reflect differences in the magnetic field strength in the afterglow emitting regions.

6.4 GRB Polarimetry

Synchrotron radiation is highly polarized, with typical degrees of (linear) polarization for ordered magnetic fields of ~60% (Hughes and Miller 1991); one may therefore expect measurable amounts of polarization in afterglow emission. By analogy to active galactic nuclei, one might expect up to 10–20% polarization if the shock emission takes place in a collimated outflow. The strong intrinsic polarization is lowered by averaging over the unresolved source (Gruzinov and Waxman 1998, Gruzinov 1999, Medvedev and Loeb 1999, Loeb and Perna 1998).

For GRB 990123, Hjorth *et al.* (1999) reported an upper limit to R-band afterglow polarization of 2.3% (95% confidence level). The first positive detection of polarization, however, was made for GRB 990510 (Covino *et al.* 1999, Wijers *et al.* 1999), with polarization $p = 1.7 \pm 0.2\%$, and $1.6 \pm 0.2\%$, 0.77 and 0.86 days after the burst, respectively. An uncertain measurement made 1.8 days after the burst is consistent with these values (Wijers *et al.* 1999). The angle of polarization remained constant during these observations. The rather small observed values of the polarization (compared with the high intrinsic values in the synchrotron process) may

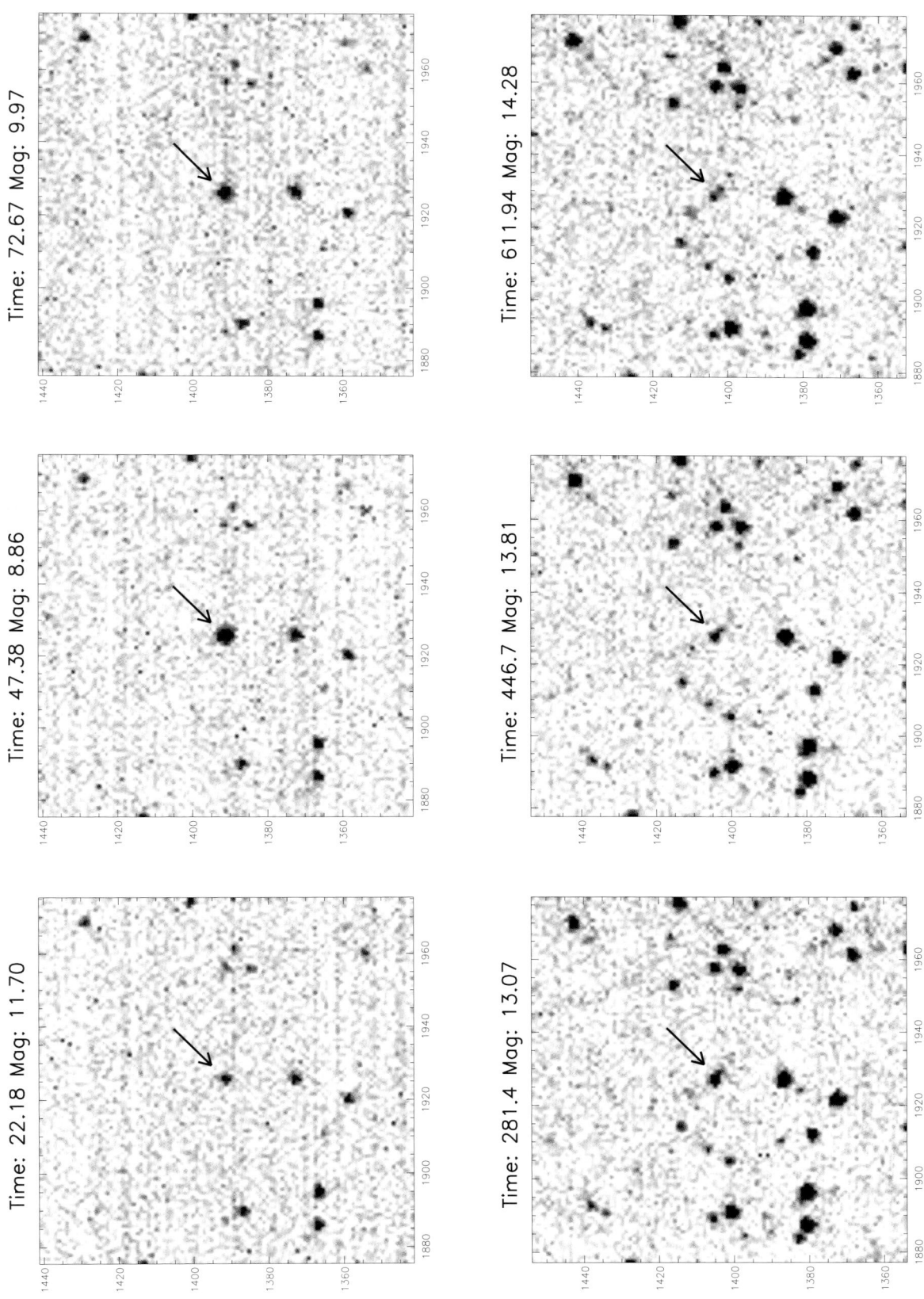

Figure 19 The ROTSE discovery images of the prompt optical emission from GRB 990123, from 22 to 800 s after trigger. (Reproduced by permission from *Nature* (Akerlof et al. 1999), © 1999 Macmillan Magazines Ltd.)

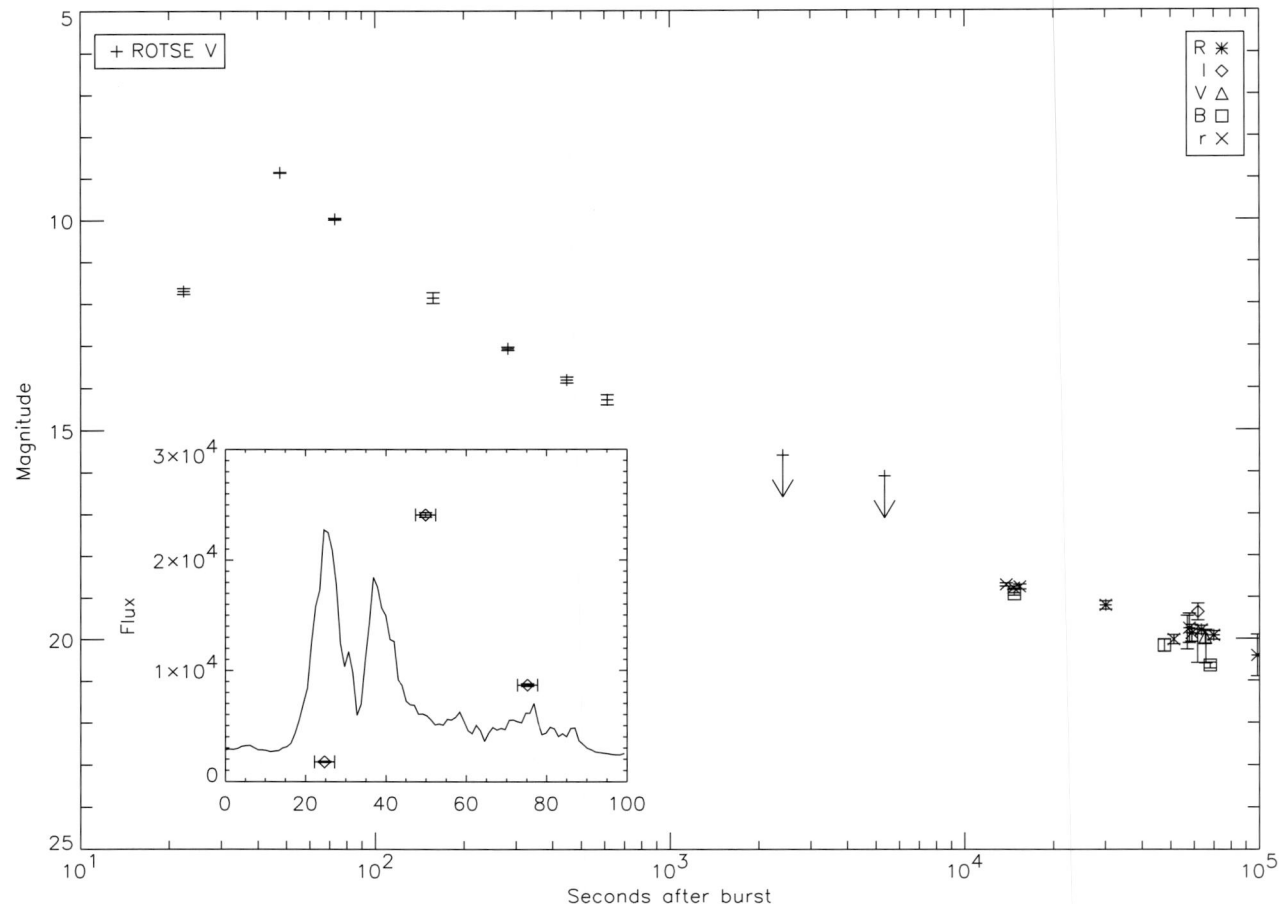

Figure 20 The R light curve of GRB 990123, showing the early emission rising above the back-extrapolated afterglow, and then merging smoothly with the later afterglow light-curve. (Reproduced by permission from *Nature* (Akerlof *et al.* 1999), © 1999 Macmillan Magazines Ltd.)

be the result of a highly tangled structure of the magnetic field, or of very symmetric field geometries. Subsequent observations of GRB 990712 (Rol *et al.* 2000) show three accurate measurements, in which the polarization percentage varies without significant variation of the polarization angle. This is not easily explained by any of the published models.

6.5 Sources, host galaxies, and cosmology

It has become clear that GRBs lie in star-forming galaxies and are associated with supernovae, and that their great brightness allows us in principle to observe them at high redshifts, perhaps up to $z = 20$ (Wijers *et al.* 1998, Ciardi and Loeb 2000, Lamb and Reichart 2000). This has greatly increased the interest in GRBs as tools for cosmology and laboratories of high-energy astrophysics. As the first afterglows were found, it rapidly became clear that almost all detected counterparts lie in a host galaxy. Furthermore, the large energies required also pointed to source models involving stellar collapses and mergers. This prompted a number of attempts to associate the GRB rate in the Universe with the star formation rate (Totani 1997, Wijers *et al.* 1998). It was shown that the observed peak flux distribution of GRBs (Figure 5) is consistent with the assumption that the GRB rate in the Universe is directly proportional to the star formation rate. Recent discoveries of supernovae associated with GRBs (Section 6.2) have lent further support to this conclusion. The increasingly high redshifts of GRBs – for example, $z = 4.5$ for the IPN-located GRB 000131 (Andersen *et al.* 2000) – make them promising future tools of cosmology.

In star formation related models the GRB rate is only 10^{-8} per galaxy per year in the Universe at present, but was much higher at $z \sim 1$, and the characteristic peak luminosity is $10^{52}\,\mathrm{erg\,s^{-1}}$ (Wijers et al. 1998). Typical GRBs at the BATSE threshold would be at $z \simeq 4$. These results are quite different from those of earlier fits to the peak flux distribution, which assumed standard candles and no evolution of the GRB rate; they typically placed the dimmest BATSE bursts at $z \sim 1$ (e.g., Fenimore et al. 1993, Paczyński 1992; but see Fenimore and Bloom 1995).

With close to a dozen redshifts and a somewhat greater number of hosts known, it has become clear that GRB luminosities in all wavelengths range widely, so the results of standard candle fits to the flux distribution should be treated with caution (e.g., Kommers et al. 2000, Krumholz et al. 1998, Schmidt et al. 1999). However, the bursts with known OTs are a much brighter group than the BATSE bursts as a whole, and their median redshift is about 1, making it likely that the dimmest GRBs are very far away indeed. This opens the prospect of using GRBs to study the early Universe, for example by investigating absorption line forests in their spectra, as is done with quasars up to $z = 5$ (Wijers et al. 1998, Lamb and Reichart 2000).

In Figure 21 are assembled images of the known GRB hosts to the middle of 1999. They are rather diverse in nature, but share some important characteristics: all are blue, indicating the presence of abundant amounts of young stars. In virtually all cases the OT does not coincide with the center of the galaxy, but does lie within its detectable light distribution (e.g., Bloom et al. 1999b). Many of the hosts are sub-luminous, but the wide range of values includes $L_\star$ galaxies (e.g., Hogg and Fruchter 1999). Star formation rates in several hosts have been estimated. While they are not particularly high in many cases, the star formation rate per unit luminosity in some is quite substantial (e.g., in the small host of GRB 970508; Natarajan et al. 1997). These average properties support the notion that GRBs occur where massive stars are born and die in the Universe.

The association of GRBs with supernovae and blue host galaxies, as well as the supernova-like energies, clearly suggest an origin of GRBs in some type of stellar death. The most popular among these have been mergers of neutron stars (Paczyński 1986, Goodman et al. 1987, Eichler et al. 1989, Mochkovitch et al. 1993) and massive-star collapses (Woosley 1993, Paczyński 1998). The location of GRB counterparts within the blue parts of galaxies argues against high-velocity progenitors, such as merging neutron stars (Bloom et al. 1999c, Bulik et al. 1999, Bloom et al. 2000). While this is true for the bursts thus far located accurately, it should be noted that those are all long-duration bursts. The short bursts (Kouveliotou et al. 1993) have not yet been followed up, so it is possible that they represent another type of burster (e.g., Fryer et al. 1999).

The energies provided by many possible central engines are quite similar, since all eventually lead to the formation of a rotating compact object surrounded by debris (Mészáros et al. 1999). For some bursts, such as GRB 971214 and GRB 990123, the implied isotropic energy is very great, $\sim 10^{54}$ erg (Kulkarni et al. 1998a, Ramaprakash et al. 1998, Halpern et al. 1998, Kulkarni et al. 1999a, Galama et al. 2000). While they are still within the realm of the possible for the stellar death models, the efficiency of converting the original energy to gamma rays could be low, so some collimation and beaming of the outflows may be necessary (Kumar 1999, Kumar and Piran 2000).

Two mechanisms have been suggested for the extraction of energy from the central engine. Both use a disk-like configuration around a compact object, and therefore naturally lead to some amount of collimation. First, neutrino annihilation can provide a large energy input while the central object is still hot and accreting rapidly. Since it depends strongly on the neutrino luminosity, it is not expected to last for more than a few to ten seconds (e.g., Ruffert and Janka 1999). This suffices to push a jet through a helium star envelope in a collapsar model (MacFadyen and Woosley 1999), but not to power bursts at the long end of the duration distribution (100–1000 s). Second, electromagnetic extraction of rotation energy from a central black hole (Blandford and Znajek 1977) has been proposed. It has the potential of lasting much longer and extracting somewhat higher energies. Its efficacy is not yet universally accepted (Li 2000, Livio et al. 1999), but has recently been discussed in detail by Lee et al. (1999, 2000), who conclude that it is a viable central-engine model.

7 FROM THE PAST, TO THE FUTURE

As the new century begins, so opens the realm of possibilities for this old (by the standards of space science) but recently rejuvenated field. The association of GRBs with dying massive stars has raised the possibility that the very first light-emitting regions of the Universe already created GRBs, and therefore their phenomenal brightness may allow us to probe those remote regions of the Universe's past, if only we knew how to find them. As the chapter draws to a close, the GRB community awaits the first ultra-rapid error boxes from the High Energy Transient Experiment (HETE) 2 satellite, which will enable us to probe the "dark era" between GRB and afterglow within minutes of the burst. This may help us to expand our understanding of the physics from the afterglows to (ironically) the longer-known but still unsolved prompt emission. In the next few years we shall also see the launch of Swift, which will further enhance our ability to localize faint GRBs and get their early light-curves, and of GLAST, which will provide keV to GeV spectra of

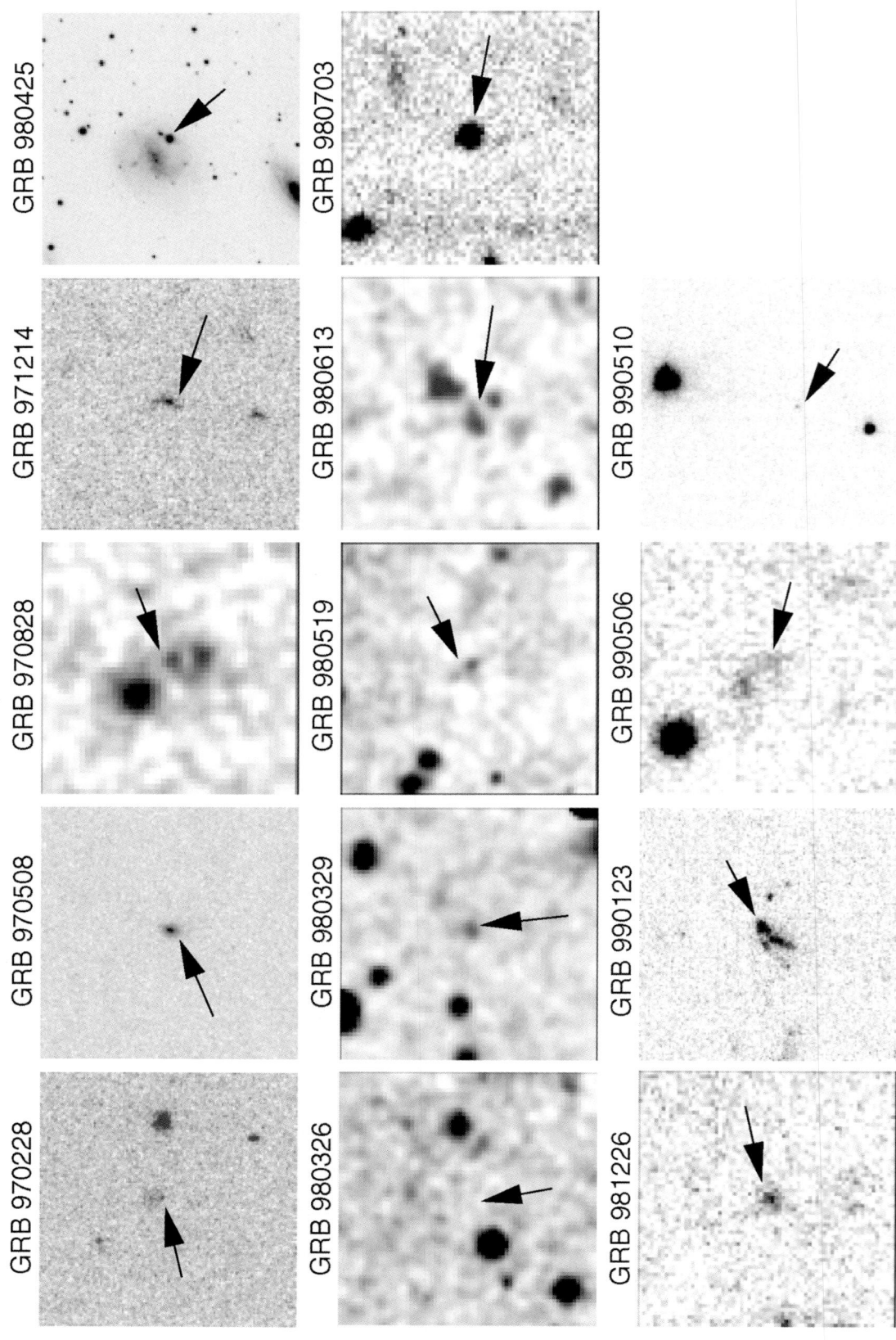

Figure 21 The known host galaxies of GRBs, imaged with HST (0228, 0508, 1214, 0123, 0510), Keck (0828, 0326, 0329, 0519, 0613, 0703, 1226, 0506; courtesy Caltech GRB collaboration), and NTT (0425). Images are 14″ on a side except 0828, 0123, and 1214, which are 7″, and 0425 (2″ on a side). (From Van Paradijs et al. 2000, with permission from the Annual Reviews of Astronomy & Astrophysics, Vol. 38, 2000.)

many GRBs; this may tell us finally what the physics of the prompt emission is.

The twenty-first century, then, may bring to GRBs what was brought to quasars before: a fruitful life as a puzzle for relativists and other theorists and observers, whose main focus remains understanding the beast itself; but also, and perhaps even more importantly, a tool to probe distant, early realms of the Universe, where nothing else casts such a bright light.

REFERENCES

Akerlof, C., Balsano, R., Barthelmy, S., Bloch, J., Butterworth, P., Casperson, D., Cline, T., Fletcher, S., Frontera, F., Gisler, G., Heise, J., Hills, J., Kehoe, R., Lee, B., Marshall, S., McKay, T., Miller, R., Piro, L., Priedhorsky, W., Szymanski, J. and Wren, J. (1999). Observation of contemporaneous optical radiation from a gamma-ray burst, *Nature*, **398**, 400–402.

Andersen, M.I., Hjorth, J., Pedersen, H., Jensen, B.L., Hunt, L.K., Gorosabel, J., Moller, P., Fynbo, J., Kippen, R.M., Thomsen, B., Olsen, L.F., Christensen, L., Vestergaard, M., Masetti, N., Palazzi, E., Hurley, K., Cline, T., Kaper, L. and Jaunsen, A.O. (2000). VLT identification of the optical afterglow of the gamma-ray burst GRB 000131 at $z = 4.50$. *Astronomy and Astrophysics*, **364**, L54–L61.

Band, D.L., Hartmann, D.H. and Schaefer, B.E. (1999). A statistical treatment of the gamma-ray burst "no host galaxy" problem: II. Energies of standard candle bursts. *Astrophysical Journal*, **514**, 862–868.

Bergmann, P.G., Fenyves, E.J. and Motz, L. (eds) (1975). *Seventh Texas Symposium on Relativistic Astrophysics*, Vol. **265** of *Annals of the New York Academy of Sciences*.

Bhat, P.N., Fishman, G.J., Meegan, C.A., Wilson, R.B., Brock, M.N. and Paciesas, W.S. (1992). Evidence of sub-millisecond structure in a gamma-ray burst. *Nature*, **359**, 217–218.

Blandford, R.D. and Znajek, R.L. (1977). Electromagnetic extraction of energy from Kerr black holes. *Monthly Notices of the Royal Astronomical Society*, **179**, 433–456.

Bloom, J.S., Djorgovski, S.G. and Kulkarni, S.R. (2001). The redshift and the ordinary host galaxy of GRB 970228. *Astrophysical Journal*, **554**, 678–683.

Bloom, J.S., Djorgovski, S.G., Kulkarni, S.R. and Frail, D.A. (1998a). The host galaxy of GRB 970508. *Astrophysical Journal*, **507**, L25–L28.

Bloom, J.S., Kulkarni, S.R. and Djorgovski, S.G. (2002). The observed offset distribution of gamma-ray bursts from their host galaxies: A robust clue to the nature of the progenitors. *Astronomical Journal*, **123**, 1111–1148.

Bloom, J.S., Kulkarni, S.R., Djorgovski, S.G., Eichelberger, A.C., Cote, P., Blakeslee, J.P., Odewahn, S.C., Harrison, F.A., Frail, D.A., Filippenko, A.V., Leonard, D.C., Riess, A.G., Spinrad, H., Stern, D., Bunker, A., Dey, A., Grossan, B., Perlmutter, S., Knop, R.A., Hook, I.M. and Feroci, M. (1999a). The unusual afterglow of the gamma-ray burst of 26 March 1998 as evidence for a supernova connection. *Nature*, **401**, 453–456.

Bloom, J.S., Kulkarni, S.R., Harrison, F., Prince, T., Phinney, E.S. and Frail, D.A. (1998b). Expected characteristics of the subclass of supernova gamma-ray bursts. *Astrophysical Journal*, **506**, L105–L108.

Bloom, J.S., Odewahn, S.C., Djorgovski, S.G., Harrison, S.R.K.F.A., Koresko, C., Neugebauer, G., Armus, L., Frail, D.A., Gal, R.R., Sari, R., Squires, G., Illingworth, G., Kelson, D., Chaffee, F., Goodrich, R., Feroci, M., Costa, E., Piro, L., Frontera, F., Mao, S., Akerlof, C. and McKay, T. A. (1999b). The host galaxy of GRB 990123. *Astrophysical Journal*, **518**, L1–L4.

Bloom, J.S., Sigurdsson, S. and Pols, O.R. (1999c). The spatial distribution of coalescing neutron star binaries: Implications for gamma-ray bursts. *Monthly Notices of the Royal Astronomical Society*, **305**, 763–769.

Bond, H.E. (1997). GRB 970508, *IAU Circular*, 6654.

Branch, D. (2001). Direct analysis of spectra of type Ic supernovae. In M. Livio, N. Panagia and K. Sahu (eds), *Supernovae and Gamma Ray Bursts*, Cambridge University Press, Cambridge.

Briggs, M.S. (1999). Gamma-ray burst lines. In J. Poutanen and R. Svensson (eds), *Gamma-Ray Bursts: The First Three Minutes*, Vol. 190 of ASP Conf. Proc., Astronomical Society of the Pacific, San Francisco, pp. 133–150.

Briggs, M.S., Band, D.L., Kippen, R.M., Preece, R.D., Kouveliotou, C., van Paradijs, J., Share, G.H., Murphy, R.J., Matz, S.M., Connors, A., Winkler, C., McConnell, M.L., Ryan, J.M., Williams, O.R., Young, C.A., Dingus, B., Catelli, J.R. and Wijers, R.A.M.J. (1999). Observations of GRB 990123 by the Compton Gamma Ray Observatory *Astrophysical Journal*, **524**, 82–91.

Briggs, M.S., Paciesas, W.S., Pendleton, G.N., Meegan, C.A., Fishman, G.J., Horack, J.M., Brock, M.N., Kouveliotou, C., Hartmann, D.H. and Hakkila, J. (1996). BATSE observations of the large-scale isotropy of gamma-ray bursts. *Astrophysical Journal*, **451**, 40–63.

Bulik, T., Belczynski, K. and Zbijewski, W. (1999). Distribution of compact object mergers around galaxies. *Astronomy and Astrophysics, Suppl.*, **138**, 483–484.

Burenin, R.A., Vikhlinin, A.A., Gilfanov, M.R., Terekhov, O.V., Tkachenko, A.Y., Sazonov, S.Y., Churazov, E.M., Sunyaev, R.A., Goldoni, P., Claret, A., Goldwurm, A., Paul, J., Roques, J.P., Jourdain, E., Pelaez, F. and Vedrenne, G. (1999). GRANAT/SIGMA observation of the early afterglow from GRB 920723 in soft gamma-rays. *Astronomy and Astrophysics*, **344**, L53–L56.

Castro-Tirado, A.J., Rosa Zapatero-Osorio, M., Caon, N., Marina Cairos, L., Hjorth, J., Pedersen, H., Andersen, M.I., Gorosabel, J., Bartolini, C., Guarnieri, A., Piccioni, A., Frontera, F., Masetti, N., Palazzi, E., Pian, E., Greiner, J., Hudec, R., Sagar, R., Pandey, A.K., Mohan, V., Yadav, R.K.S., Nilakshi, N., Bjornsson, G., Jakobsson, P., Burud, I., Courbin, F., Valentini, G., Piersimoni, A., Aceituno, J., Montoya, L.M., Pedraz, S., Gredel, R., Claver, C.F., Rector, T.A., Rhoads, J.E., Walter, F., Ott, J., Hippelein, H., Sanchez-Bejar, V., Gutierrez, C., Oscoz, A., Zhu, J., Chen, J., Zhang, H., Wei, J., Zhou, A., Guziy, S., Shlyapnikov, A., Heise, J., Costa, E., Feroci, M. and Piro, L. (1999). Decay of the GRB 990123 optical afterglow: Implications for the fireball model. *Science*, **283**, 2069–2073.

Cavallo, G. and Rees, M.J. (1978), A qualitative study of cosmic fireballs and γ-ray-bursts. *Monthly Notices of the Royal Astronomical Society*, **183**, 359–365.

Ciardi, B. and Loeb, A. (2000). Expected number and flux distribution of gamma-ray-burst afterglows with high redshifts. *Astrophysical Journal*, **540**, 687–696.

Clarke, T.E., Blaes, O. and Tremaine, S. (1994). Do gamma-ray bursts come from the Oort Cloud? *Astronomical Journal*, **107**, 1873–1878.

Cline, T.L. (1975). Recent observations of cosmic gamma-ray bursts. *Annals of the New York Academy of Sciences*, **262**, 159–163.

Cline, T.L., Desai, U.D., Klebesadel, R.W. and Strong, I.B. (1973). Energy spectra of cosmic gamma-ray bursts. *Astrophysical Journal*, **185**, L1–L5.

Cline, T.L., Desai, U.D., Pizzichini, G., Teegarden, B.J., Evans, W.D., Klebesadel, R.W., Laros, J.G., Hurley, K., Niel, M. and Vedrenne, G. (1980). Detection of a fast, intense and unusual gamma-ray transient. *Astrophysical Journal*, **237**, L1–L5.

Colgate, S.A. (1968). Prompt gamma rays and X-rays from supernovae. *Canadian Journal of Physics*, **46**, S476–S480.

Colgate, S.A. (1973). The production of deuterium in supernova shocks. *Astrophysical Journal*, **181**, L53–L54.

Colgate, S.A. (1974). Early gamma rays from supernovae. *Astrophysical Journal*, **187**, 333–336.

Connaughton, V. (2000). BATSE observations of gamma-ray burst tails. In R.M. Kippen, R.S. Mallozzi and G.J. Fishman (eds), *Gamma-Ray Bursts: 5th Huntsville Symposium*, AIP Conference Proceedings Vol. 526, American Institute of Physics, Melville, NY, pp. 385–389.

Costa, E., Frontera, F., Heise, J., Feroci, M., in 't Zand, J., Fiore, F., Cinti, M.N., Dal Fiume, D., Nicastro, L., Orlandini, M., Palazzi, E. Rapisarda, M., Zavattini, G., Jager, R., Parmar, A., Owens, A., Molendi, S., Cusumano, G., Maccarone, M.C., Giarrusso, S., Coletta, A., Antonelli, L.A., Giommi, P., Muller, J.M., Piro, L. and Butler, R.C. (1997). Discovery of an X-ray afterglow associated with the gamma-ray burst of 28 February 1997. *Nature*, **387**, 783–785.

Covino, S., Lazzati, D., Ghisellini, G., Saracco, P., Campana, S., Chincarini, G., Di Serego, S., Cimatti, A., Vanzi, L., Pasquini, L., Haardt, F., Israel, G.L., Stella, L. and Vietri, M. (1999). GRB 990510: Linearly polarized radiation from a fireball. *Astronomy and Astrophysics*, **348**, L1–L4.

Denissenko, D., Terekhov, O., Sunyaev, R., Barat, C., Dezalay, J.P. and Vedrenne, G. (1995). Spectral properties of grbs observed by phebus. *Astrophysics and Space Science*, **231**, 211–214.

Dermer, C.D. and Mitman, K.E. (1999). Short-timescale variability in the external shock model of gamma-ray bursts. *Astrophysical Journal*, **513**, L5–L8.

Dezalay, J., Barat, C., Talon, R., Sunyaev, R., Terekhov, O. and Kuznetsov, A. (1992). Short cosmic events: A subset of classical GRBs? In W.S. Paciesas and G.J. Fishman (eds), *Gamma-Ray Bursts*, AIP Conference Proceedings Vol. 265, American Institute of Physics, New York, pp. 304–309.

Djorgovski, S.G. et al. (2000). Hosts of GRB 970828 and GRB 980329, in preparation.

Djorgovski, S.G., Kulkarni, S.R., Bloom, J.S. and Frail, D.A. (1999). GRB 970228: Redshift and properties of the host galaxy. *GRB Coordinates Network Circular*, 289.

Djorgovski, S.G., Metzger, M.R., Kulkarni, S.R., Odewahn, S.C., Gal, R.R., Pahre, M., Frail, D.A., Feroci, M., Costa, E. and Palazzi, E. (1997). The optical counterpart to the gamma-ray burst GRB 970508. *Nature*, **387**, 876–878.

Duncan, R.C. and Thompson, C. (1992). Formation of very strongly magnetized neutron stars: Implications for gamma-ray bursts. *Astrophysical Journal*, **392**, L9–L13.

Eichler, D., Livio, M., Piran, T. and Schramm, D.N. (1989). Nucleosynthesis, neutrino bursts and γ-rays from coalescing neutron stars. *Nature*, **340**, 126–128.

Evans, W.D., Klebesadel, R.W., Laros, J.G., Cline, T.L., Desai, U.D., Teegarden, B.J., Pizzichini, G., Hurley, K., Niel, M. and Vedrenne, G. (1980). Location of the gamma-ray transient event of 1979 March 5. *Astrophysical Journal*, **237**, L7–L9.

Fenimore, E.E. and Bloom, J.S. (1995). Determination of distance from time dilation of cosmological gamma-ray bursts. *Astrophysical Journal*, **453**, 25–36.

Fenimore, E.E., Conner, J.P., Epstein, R.I., Klebesadel, R.W., Laros, J.G., Yoshida, A., Fujii, M., Hayashida, K., Itoh, M., Murakami, T., Nishimura, J., Yamagami, Y., Kondo, I. and Kawai, N. (1988). Interpretations of multiple absorption features in a gamma-ray burst spectrum. *Astrophysical Journal*, **335**, L71–L74.

Fenimore, E.E., Epstein, R.I., Ho, C., Klebesadel, R.W., Lacey, C., Laros, J.G., Meier, M., Strohmayer, T., Pendleton, G., Fishman, G., Kouveliotou, C. and Meegan, C. (1993). The intrinsic luminosity of γ-ray-bursts and their host galaxies. *Nature*, **366**, 40–42.

Fishman, G.J. (1995). Gamma-ray bursts: An overview. *Publications of the Astronomical Society of the Pacific*, **107**, 1145–1151.

Fishman, G.J., Bhat, P.N., Mallozzi, R., Horack, J.M., Koshut, T., Kouveliotou, C., Pendleton, G.N., Meegan, C.A., Wilson, R.B., Paciesas, W.S., Goodman, S.J. and Christian, H.J. (1994). Discovery of intense gamma-ray flashes of atmospheric origin. *Science*, **264**, 1313.

Fishman, G.J. and Meegan, C.A. (1995). Gamma-ray bursts. *Annual Reviews of Astronomy and Astrophysics*, **35**, 415–458.

Frail, D.A. and Kulkarni, S.R. (1995). Radio counterpart searches of gamma-ray bursts. *Astrophysics and Space Science*, **231**, 277–280.

Frail, D.A. and Kulkarni, S.R. (1997). GRB 970508. *IAU Circular*, 6662.

Frail, D.A., Kulkarni, S.R., Nicastro, L., Feroci, M. and Taylor, G.B. (1997). The radio afterglow from the gamma-ray burst of 8 May 1997. *Nature*, **389**, 261–263.

Fruchter, A., Krolik, J. and Rhoads, J. (2001). X-ray destruction of dust along the line of sight to gamma-ray bursters. *Astrophysical Journal*, **563**, 597–610.

Fruchter, A.S., Pian, E., Gibbons, R., Thorsett, S.E., Ferguson, H., Petro, L., Sahu, K., Livio, M., Caraveo, P., Frontera, F., Kouveliotou, C., Macchetto, D., Palazzi, E., Pedersen, H., Tavani, M. and van Paradijs, J. (2000). HST observations of the host galaxy of GRB 970508. *Astrophysical Journal*, **545**, 664–669.

Fruchter, A.S., Pian, E., Thorsett, Stephen E.; Bergeron, L.E., González, R.A., Metzger, M., Goudfrooij, P., Sahu, K.C., Ferguson, H., Livio, M., Mutchler, M., Petro, L., Frontera, F., Galama, T., Groot, P., Hook, R., Kouveliotou, C., Macchetto, D., van Paradijs, J., Palazzi, E., Pedersen, H., Sparks, W. and Tavani, M. (1999). The fading optical counterpart of GRB 970228, 6 months and 1 year later. *Astrophysical Journal*, **516**, 683–692.

Fryer, C.L., Woosley, S.E. and Hartmann, D.H. (1999). Formation rates of black hole accretion disk gamma-ray bursts. *Astrophysical Journal*, **526**, 152–177.

Galama, T., Briggs, M., Wijers, R.A.M.J., Vreeswijk, P.M., Rol, E., Band, D., van Paradijs, J., Kouveliotou, C., Preece, R.D., Bremer, M., Smith, I.A., Tilanus, R.P.J., de Bruyn, A.G., Strom, R.G., Pooley, G., Castro-Tirado, A.J., Tanvir, N., Robinson, C., Hurley, K., Heise, J., Telting, J., Rutten, R.G.M., Packham, C., Swaters, R., Davies, J.K., Fassia, A., Green, S.F., Foster, M.J., Sagar, R., Pandey, A.K., Nilakshi, Yadav, R.K.S., Ofek, E.O., Leibowitz, E., Ibbetson, P., Rhoads, J., Falco, E., Petry, C., Impey, C., Geballe, T.R. and Bhattacharya, D. (1999). The effect of magnetic fields on γ-ray-bursts inferred from multi-wavelength observations of the burst of 23 January 1999. *Nature*, **398**, 394–399.

Galama, T., Groot, P.J., van Paradijs, J., Kouveliotou, C., Robinson, C.R., Fishman, G.J., Meegan, C.A., Sahu, K.C., Livio, M., Petro, L., Macchetto, F.D., Heise, J., in 't Zand, J., Strom, R.G., Telting, J., Rutten, R.G.M., Pettini, M., Tanvir, N. and Bloom, J. (1997). The decay of optical emission from the gamma-ray burst GRB 970228. *Nature*, **387**, 479–481.

Galama, T., Tanvir, N., Vreeswijk, P., Wijers, R., Groot, P., Rol, E., van Paradijs, J., Kouveliotou, C., Fruchter, A., Masetti, N., Pedersen, H., Margon, B., Deutsch, E., Metzger, M., Armus, L., Klose, S. and Stecklum, B. (2000). Evidence for a supernova in reanalyzed optical and near-infrared images of GRB 970228. *Astrophysical Journal*, **536**, 185–194.

Galama, T.J., Groot, P.J., van Paradijs, J., Kouveliotou, C., Strom, R.G., Wijers, R.A.M.J., Tanvir, N., Bloom, J., Centurion, M., Telting, J., Rutten, R.G.M., Smith, P., Mackey, C., Smartt, S., Benn, C., Heise, J. and in 't Zand, J. (1998a). Optical follow-up of GRB 970508. *Astrophysical Journal*, **497**, L13–L16.

Galama, T.J., Vreeswijk, P.M., Pian, E., Frontera, F., Doublier, V. and Gonzalez, J.F. (1998b). GRB 980425. *IAU Circular*, 6895.

Galama, T.J., Vreeswijk, P.M., van Paradijs, J., Kouveliotou, C., Augusteijn, T., Bohnhardt, H., Brewer, J.P., Doublier, V., Gonzalez, J.F., Leibundgut, B., Lidman, C., Hainaut, O.R., Patat, F., Heise, J., In 't Zand, J., Hurley, K., Groot, P.J., Strom, R.G., Mazzali, P.A., Iwamoto, K., Nomoto, K.,

Umeda, H., Nakamura, T., Young, T.R., Suzuki, T., Shigeyama, T., Koshut, T., Kippen, M., Robinson, C., de Wildt, P., Wijers, R.A.M.J., Tanvir, N., Greiner, J., Pian, E., Palazzi, E., Frontera, F., Masetti, N., Nicastro, L., Feroci, M., Costa, E., Piro, L., Peterson, B.A., Tinney, C., Boyle, B., Cannon, R., Stathakis, R., Sadler, E., Begam, M.C. and Ianna, P. (1998c). An unusual supernova in the error box of the gamma-ray burst of 25 April 1998. *Nature*, **395**, 670–672.

Galama, T.J. and Wijers, R.A.M.J. (2001). High column densities and low extinctions of γ-ray-bursts: Evidence for hypernovae and dust destruction. *Astrophysical Journal*, **549**, L209–L213.

Galama, T.J., Wijers, R.A.M.J., Bremer, M., Groot, P.J., Strom, R.G., Kouveliotou, C. and van Paradijs, J. (1998d). The radio to X-ray spectrum of GRB 970508 on May 21.0 UT. *Astrophysical Journal*, **500**, L97–L100.

Garcia, M.R., Callanan, P.J., Moraru, D., McClintock, J.E., Tollestrup, E., Willner, S.P., Hergenrother, C., Robinson, C.R., Kouveliotou, C. and van Paradijs, J. (1998). Power-law decays in the optical counterparts of GRB 970228 and GRB 970508. *Astrophysical Journal*, **500**, L105–L108.

Germany, L.M., Reiss, D.J., Sadler, E.M., Schmidt, B.P. and Stubbs, C.W. (2000). SN1997cy/GRB 970514: a new piece in the GRB puzzle? *Astrophysical Journal*, **533**, 320–328.

Giblin, T., van Paradijs, J., Kouveliotou, C., Connaughton, V., Wijers, R.A.M.J. and Fishman, G.J. (1999). Evidence for an early high-energy afterglow observed with BATSE from GRB 980923. *Astrophysical Journal*, **524**, L41–L50.

Giblin, T., van Paradijs, J., Kouveliotou, C., Connaughton, V., Wijers, R.A.M.J. and Fishman, G.J. (2001). Exploring GRB–afterglow connections using bright BATSE bursts with smooth high-energy tails. *Astrophysical Journal*, in press.

Golenetskii, S.V., Mazets, E.P., Aptekar, R.L. and Ilinskii, V.N. (1983). Correlation between luminosity and temperature in gamma-ray burst sources. *Nature*, **306**, 451–453.

Golenetskii, S.V., Mazets, E.P., Ilinskij, V.N. and Guryan, Y.A. (1979). Recurrent gamma-ray bursts from the flaring source FXP 0520–66. *Soviet Astronomy Letters*, **5**, 340–342.

Goodman, J. (1986). Are gamma-ray bursts optically thick? *Astrophysical Journal*, **308**, L47–L50.

Goodman, J., Dar, A. and Nussinov, S. (1987). Neutrino annihilation in type II supernovae. *Astrophysical Journal*, **314**, L7–L10.

Greiner, J., Wenzel, W. and Degel, W. (1990). Discrimination between star-like defects on photographic plates and possible gamma-ray burst optical counterparts. *Astronomy and Astrophysics*, **234**, 251–261.

Greiner, J., Wenzel, W., Hudec, R., Pravec, P., Rezek, T., Nezvara, M., Moskalenko, E.I., Karnashov, A., Metlov, V., Chernych, N.S., Getman, V.S., Kouveliotou, C., Fishman, G.J., Meegan, C.A., Paciesas, W.S. and Wilson, R.B. (1993). Simultaneous optical/γ observations of GRBs. In M. Friedlander, N. Gehrels and D.J. Macomb (eds), *Compton Gamma-Ray Observatory*, American Institute of Physics, New York, pp. 828–832.

Groot, P.J., Galama, T.J., van Paradijs, J., Kouveliotou, C., Wijers, R.A.M.J., Bloom, J., Tanvir, N., Vanderspek, R., Greiner, J., Castro-Tirado, A.J., Gorosabel, J., von Hippel, T., Lehnert, M., Kuijken, K., Hoekstra, H., Metcalfe, N., Howk, C., Conselice, C., Telting, J., Rutten, R.G.M., Rhoads, J., Cole, A., Pisano, D.J., Naber, R. and Schwarz, R. (1998a). A search for optical afterglow from GRB 970828. *Astrophysical Journal*, **493**, L27.

Groot, P.J., Galama, T.J., van Paradijs, J., Melnick, G., van der Steene, G., Bremer, M., Tanvir, N., Bloom, J., Strom, R.G., Telting, J., Rutten, R.G.M., Kouveliotou, C., in 't Zand, J., Heise, J., Costa, E., Feroci, M., Piro, L., Frontera, F., Zavattini, G., Nicastro, L. and Palazzi, E. (1997a). GRB 970228. *IAU Circular*, 6588.

Groot, P.J., Galama, T.J., van Paradijs, J., Strom, R.G., Telting, J., Rutten, R.G.M., Pettini, M., Tanvir, N., Naber, R., Kouveliotou, C., in 't Zand, J.,

Heise, J., Costa, E., Feroci, M., Piro, L., Frontera, F., Zavattini, G., Nicastro, L. and Palazzi, E. (1997b). GRB 970228. *IAU Circular* 6584.

Groot, P.J., Galama, T.J., Vreeswijk, P.M., Wijers, R.A.M.J., Pian, E., Palazzi, E., van Paradijs, J., Kouveliotou, C., in 't Zand, J.J.M., Heise, J., Robinson, C., Tanvir, N., Lidman, C., Tinney, C., Keane, M., Briggs, M., Hurley, K., Gonzalez, J.F., Hall, P., Smith, M.G., Covarrubias, R., Jonker, P., Casares, J., Frontera, F., Feroci, M., Piro, L., Costa, E., Smith, R., Jones, B., Windridge, D., Bland-Hawthorn, J., Veilleux, S., Garcia, M., Brown, W.R., Stanek, K.Z., Castro-Tirado, A.J., Gorosabel, J., Greiner, J., Jaeger, K., Bohm, A.B. and Fricke, K.J. (1998b). The rapid decay of the optical emission from GRB 980326 and its possible implications. *Astrophysical Journal*, **502**, L123–L126.

Gruzinov, A. (1999). Strongly polarized optical afterglows of gamma-ray bursts. *Astrophysical Journal*, **525**, L29–L31.

Gruzinov, A. and Waxman, E. (1999). Gamma-ray burst afterglow: Polarization and analytic light-curves. *Astrophysical Journal*, **511**, 852–861.

Halpern, J.P., Thorstensen, J.R., Helfand, D.J. and Costa, E. (1998). Optical afterglow of the γ-ray burst GRB 971214. *Nature*, **393**, 41–43.

Hartmann, D. and Blumenthal, G.R. (1989). Angular clustering properties of gamma-ray bursts and quantitative constraints on their distances. *Astrophysical Journal*, **342**, 521–526.

Higdon, J.C. and Lingenfelter, R.E. (1990). Gamma-ray bursts. *Annual Reviews of Astronomy and Astrophysics*, **28**, 401–436.

Höflich, P., Wheeler, J.C. and Wang, L. (1999). Aspherical explosion models for SN 1998bw/GRB 980425. *Astrophysical Journal*, **521**, 179–189.

Hogg, D.W. and Fruchter, A.S. (1999). The faint-galaxy hosts of gamma-ray bursts. *Astrophysical Journal*, **520**, 54–58.

Hudec, R., Ceplecha, Z., Spurny, P., Florián, J., Kolá, A., Bocek, J., Borovicka, J. and Kroll, P. (1999). Real-time and pre-burst optical data for GRBs. *Astronomy and Astrophysics*, Suppl., **138**, 591–592.

Hudec, R. and Soldan, J. (1995). Ground-based optical CCD experiments for GRB and optical transient detection. *Astrophysics and Space Science*, **231**, 311–314.

Hughes, P.A. and Miller, L. (1991). Introduction: Synchrotron and inverse-Compton radiation. In P.A. Hughes (ed.), *Beams and Jets in Astrophysics*, Vol. 19 of Cambridge Astrophysics Series, Cambridge University Press, pp. 1–51.

Hurley, K., Cline, T., Mazets, E., Aptekar, R., Golenetskii, S., Frederiks, D., Frail, D., Kulkarni, S., Trombka, J., McClanahan, T., Starr, R. and Goldsten, J. (2000a). Interplanetary Network localization of GRB 991208 and the discovery of its afterglow. *Astrophysical Journal*, **534**, L23–L25.

Hurley, K., Cline, T., Mazets, E., Barthelmy, S., Butterworth, P., Marshall, F., Palmer, D., Aptekar, R., Golenetskii, S., Il'Inskii, V., Frederiks, D., McTiernan, J., Gold, R. and Trombka, J. (1999). A giant periodic flare from the soft gamma-ray repeater SGR 1900+14. *Nature*, **397**, 41–43.

Hurley, K., Dingus, B., Mukherjee, R., Sreekumar, P., Kouveliotou, C., et al. (1994). Detection of a gamma-ray burst of very long duration and very high energy. *Nature*, **372**, 652–654.

Hurley, K., Feroci, M., Cinti, M., Costa, E., Preger, B., Frontera, F., Dal Fiume, D., Orlandini, M., Amati, L., Nicastro, L., Heise, J., in 't Zand, J. and Cline, T. (2000b). Integrating the BeppoSAX gamma-ray burst monitor into the 3rd Interplanetary Network. *Astrophysical Journal*, **534**, 258–264.

Hurley, K., Hartmann, D., Kouveliotou, C., Fishman, G., Laros, J., Cline, T. and Boer, M. (1997). Are Abell clusters correlated with gamma-ray bursts? *Astrophysical Journal*, **479**, L113–L116.

Iwamoto, K. (1999a). Erratum: On the radio-to-X-ray light-curves of SN 1998bw and GRB 980425. *Astrophysical Journal*, **517**, L67–L67.

Iwamoto, K. (1999b). On the radio-to-X-ray light-curves of SN 1998bw and GRB 980425. *Astrophysical Journal*, **512**, L47–L50.

Iwamoto, K., Mazzali, P.A., Nomoto, K., Umeda, H., Nakamura, T., Patat, F., Danziger, I.J., Young, T.R., Suzuki, T., Shigeyama, T., Augusteijn, T.,

Doublier, V., Gonzalez, J.F., Boehnhardt, H., Brewer, J., Hainaut, O.R., Lidman, C., Leibundgut, B., Cappellaro, E., Turatto, M., Galama, T.J., Vreeswijk, P. M., Kouveliotou, C., van Paradijs, J., Pian, E., Palazzi, E. and Frontera, F. (1998). A hypernova model for the supernova associated with the gamma-ray burst of 25 April 1998. *Nature*, **395**, 672–674.

Jager, R., Heise, J., in 't Zand, J. and Brinkman, A.C. (1995). Wide field cameras for SAX. *Advances in Space Research*, **13**, 315–318.

Kargatis, V.E., Li, H., Liang, E.P., Smith, I.A., Hurley, K., Barat, C. and Niel, M. (1994). Duration versus brightness of gamma-ray bursts: Comparisons between SIGNE and BATSE. *Astrophysical Journal*, **421**, L83–L86.

Katz, J.I. (1994). Low-frequency spectra of gamma-ray bursts. *Astrophysical Journal*, **432**, L107–L109.

Kippen, R.M., Briggs, M.S., Kommers, J.M., Kouveliotou, C., Hurley, K., Robinson, C.R., van Paradijs, J., Hartmann, D.H., Galama, T.J. and Vreeswijk, P.M. (1998). On the association of gamma-ray bursts with supernovae. *Astrophysical Journal*, **506**, L27–L30.

Klebesadel, R.W., Strong, I.B. and Olson, R.A. (1973). Observations of gamma-ray bursts of cosmic origin. *Astrophysical Journal*, **182**, L85–L88.

Kobayashi, S., Piran, T. and Sari, R. (1997). Can internal shocks produce the variability in gamma-ray bursts? *Astrophysical Journal*, **490**, 92–98.

Kolatt, T. and Piran, T. (1996). New evidence for the cosmological origin of gamma-ray bursts. *Astrophysical Journal*, **467**, L41–L44.

Kommers, J.M., Lewin, W.H.G., Kouveliotou, C., van Paradijs, J., Pendleton, G.N., Meegan, C.A. and Fishman, G.J. (2000). The intensity distribution of faint gamma-ray bursts detected with BATSE. *Astrophysical Journal*, **533**, 696–709.

Kouveliotou, C. (2002). Soft gamma repeaters. *Annual Review of Astronomy and Astrophysics*, **40**, in preparation.

Kouveliotou, C., Dieters, S., Strohmayer, T., van Paradijs, J., Meegan, C.A., Hurley, K., Kommers, J., Smith, I., Frail, D. and Murakami, T. (1998). An X-ray pulsar with a superstrong magnetic field in the soft gamma-ray repeater SGR 1806–20. *Nature*, **393**, 235–237.

Kouveliotou, C., Fishman, G.J., Meegan, C.A., Paciesas, W.S., van Paradijs, J., Norris, J.P., Preece, R.D., Briggs, M.S., Horack, J.M., Pendleton, G.H. and Green, D.A. (1994). The rarity of soft gamma-ray repeaters deduced from reactivation of SGR 1806 − 20. *Nature*, **368**, 125–127.

Kouveliotou, C., Meegan, C.A., Fishman, G.J., Bhat, N.P., Briggs, M.S., Koshut, T.M., Paciesas, W.S. and Pendleton, G.N. (1993). Identification of two classes of gamma-ray bursts. *Astrophysical Journal*, **413**, L101–L104.

Kouveliotou, C., Strohmayer, T., Hurley, K., van Paradijs, J., Finger, M.H., Dieters, S., Woods, P., Thompson, C. and Duncan, R.C. (1999). Discovery of a magnetar associated with the soft gamma repeater SGR 1900 + 14, *Astrophysical Journal*, **510**, L115–L118.

Krimm, H.A., Vanderspek, R.K. and Ricker, G.R. (1996). Searches for optical counterparts of BATSE gamma-ray bursts with the Explosive Transient Camera. *Astronomy and Astrophysics*, Suppl., **120**, 251.

Krumholz, M.R., Thorsett, S.E. and Harrison, F.A. (1998). Gamma-ray bursts and the cosmic star formation rate. *Astrophysical Journal*, **506**, L81–L84.

Kulkarni, S.R., Berger, E., Bloom, J.S., Chaffee, F., Diercks, A., Djorgovski, S.G., Frail, D.A., Galama, T.J., Goodrich, R.W., Harrison, F.A., Sari, R. and Yost, S.A. (2000). The afterglows of gamma-ray bursts. In R.M. Kippen, R.S. Mallozzi and G.J. Fishman (eds), *Gamma-Ray Bursts*, AIP Conf. Proc. Ser. Vol. 526, pp. 277–297.

Kulkarni, S.R., Djorgovski, S.G., Odewahn, S.C., Bloom, J.S., Gal, R.R., Koresko, C.D., Harrison, F.A., Lubin, L.M., Armus, L., Sari, R., Illingworth, G.D., Kelson, D.D., Magee, D.K., van Dokkum, P.G., Frail, D.A., Mulchaey, J.S., Malkan, M.A., McClean, I.S., Teplitz, H.I., Koerner, D., Kirkpatrick, D., Kobayashi, N., Yadigaroglu, I.A., Halpern, J., Piran, T., Goodrich, R.W., Chaffee, F.H., Feroci, M. and Costa, E. (1999a). The afterglow, redshift and extreme energetics of the gamma-ray burst of 23 January 1999. *Nature*, **398**, 389–394.

Kulkarni, S.R., Djorgovski, S.G., Ramaprakash, A.N., Goodrich, R., Bloom, J.S., Adelberger, K.L., Kundic, T., Lubin, L., Frail, D.A., Frontera, F., Nicastro, L., Barth, A.J., Davis, M., Filippenko, A.V. and Newman, J. (1998a). Identification of a host galaxy at $z = 3.42$ for the γ-ray-burst of 14 December 1997. *Nature*, **393**, 35–39.

Kulkarni, S.R. and Frail, D.A. (1993). Identification of a supernova remnant coincident with the soft gamma-ray repeater SGR 1806–20. *Nature*, **365**, 33–35.

Kulkarni, S.R., Frail, D.A., Moriarty-Schieven, G.H., Shepherd, D.S., Udomprasert, P., Readhead, A.C. S., Bloom, J.S., Feroci, M. and Costa, E. (1999b). Discovery of a radio flare from GRB 990123. *Astrophysical Journal*, **522**, L97–L100.

Kulkarni, S.R., Frail, D.A., Wieringa, M.H., Ekers, R.D., Sadler, E.M., Wark, R.M., Higdon, J.L., Phinney, E.S. and Bloom, J.S. (1998b). Radio emission from the unusual supernova 1998bw and its association with the gamma-ray burst of 25 April 1998. *Nature*, **395**, 663–669.

Kumar, P. (1999). Gamma-ray burst energetics. *Astrophysical Journal*, **523**, L113–L116.

Kumar, P. and Piran, T. (2000). Energetics and luminosity function of gamma-ray bursts. *Astrophysical Journal*, **535**, 152–157.

Lamb, D.Q. (1995). The distance scale to gamma-ray bursts. *Publications of the Astronomical Society of the Pacific*, **107**, 1152.

Lamb, D.Q. and Reichart, D. (2000). Gamma-ray bursts as a probe of the very high redshift universe. *Astrophysical Journal*, **536**, 1–18.

Lamb, F.K. (1981). Neutron star properties from observations of pulsars and pulsing X-ray sources. In W. Sieber and R. Wielebinski (eds), *Pulsars*, IAU. Symp. 95, Reidel, Dordrecht, pp. 309–319.

Laros, J.G., Fenimore, E.E., Fikani, M.M., Klebesadel, R.W. and Barat, C. (1986). The soft gamma-ray burst GB 790107. *Nature*, **322**, 152.

Larson, S.B., McLean, I.S. and Becklin, E.E. (1996). Luminous galaxies near gamma-ray burst positions. *Astrophysical Journal*, **460**, L95–L98.

Lee, B., Akerlof, C., Band, D., Barthelmy, S., Butterworth, P., Cline, T., Ferguson, D., Gehrels, N. and Hurley, K. (1997). Results from Gamma-Ray Optical Counterpart Search Experiment: A real time search for gamma-ray burst optical counterparts. *Astrophysical Journal*, **482**, L125.

Lee, H.-K., Brown, G.E. and Wijers, R.A.M.J. (2000). Issues in the Blandford–Znajek process for GRB inner engine. *Astrophysical Journal*, **536**, 416–419.

Lee, H.-K., Wijers, R.A.M.J. and Brown, G.E. (1999). The Blandford–Znajek process as a central engine for a gamma-ray burst. *Physics Reports*, **325**, 83.

Li, L.-X. (2000). A toy model for Blandford–Znajek mechanism. *Physical Review D*, **61**, 084016.

Liang, E.P. and Nolan, P.L. (1984). Cygnus, X-1 revisited. *Space Science Reviews*, **38**, 353–384.

Liang, E.P. and Petrosian, V. (eds) (1986). *Gamma-ray Bursts*, Vol. 141 of AIP Conference Proceedings, American Institute of Physics, New York.

Lingenfelter, R.E., Hudson, H.S. and Worrall, D.M. (eds) (1982). *Gamma Ray Transients and Related Astrophysical Phenomena*, Vol. 77 of AIP Conference Proceedings, American Institute of Physics, New York.

Livio, M., Ogilvie, G.I. and Pringle, J.E. (1999). Extracting energy from black holes: The relative importance of the Blandford–Znajek mechanism. *Astrophysical Journal*, **512**, 100–104.

Loeb, A. and Perna, R. (1998). Microlensing of gamma-ray burst afterglows. *Astrophysical Journal*, **495**, 597–603.

MacFadyen, A.I. and Woosley, S.E. (1999). Collapsars: Gamma-ray bursts and explosions in "failed supernovae". *Astrophysical Journal*, **524**, 262–289.

Mallozzi, R.S., Paciesas, W.S., Pendleton, G.N., Briggs, M.S., Preece, R.D., Meegan, C.A. and Fishman, G.J. (1995). The νF_ν peak energy distributions of gamma-ray bursts observed by BATSE. *Astrophysical Journal*, **454**, 597–603.

Mazets, E.P., Golenetskii, S.V., Aptekar, R.L., Gurian, I.A. and I'Linskii, V.N. (1981). Cyclotron and annihilation lines in gamma-ray bursts. *Nature*, **290**, 378–382.

Mazets, E.P., Golenetskii, S.V. and Il'Inskii, V.N. (1974). A cosmic gamma-ray flare observed with Kosmos 461. *JETP Letters*, **19**, 77.

Mazets, E.P., Golenetskij, S.V. and Guryan, Y.A. (1979). Soft gamma-ray bursts from the source B1900+14. *Soviet Astronomy Letters*, **5**, 343.

McNamara, B.J., Harrison, T.E. and Williams, C.L. (1995). Directions for future X-ray, optical and radio follow-up observations of gamma-ray burst counterparts. *Astrophysical Journal*, **452**, L25–L28.

Medvedev, M.V. and Loeb, A. (1999). Generation of magnetic fields in the relativistic shock of gamma-ray burst sources. *Astrophysical Journal*, **526**, 697–706.

Meegan, C.A., Fishman, G.J., Wilson, R.B., Paciesas, W.S., Pendleton, G.N., Horack, J.M., Brock, M.N. and Kouveliotou, C.: (1992), Spatial distribution of GRBs observed by BATSE. *Nature*, **355**, 143–145.

Meegan, C.A., Preece, R.D. and Koshut, T.M. (eds) (1998). *Gamma-Ray Bursts, 4th Huntsville Symposium*, Vol. 428 of AIP Conference Proceedings, American Institute of Physics, New York.

Mészáros, P. and Rees, M.J. (1997). Optical and long-wavelength afterglow from gamma-ray bursts. *Astrophysical Journal*, **476**, 232–237.

Mészáros, P. and Rees, M.J. (1999). GRB 990123: Reverse and internal shock flashes and late afterglow. *Monthly Notices of the Royal Astronomical Society*, **306**, L39–L43.

Mészáros, P., Rees, M.J. and Wijers, R.A.M.J. (1999). Energetics and beaming of gamma-ray burst triggers. *New Astronomy*, **4**, 303–312.

Metzger, A.E., Parker, R.H., Gilman, D., Peterson, L.E. and Trombka, J.I. (1974). Observation of a cosmic gamma-ray burst on Apollo 16: I Temporal variability and energy spectrum. *Astrophysical Journal*, **194**, L19–L25.

Metzger, M.R., Cohen, J.G., Chaffee, F.H. and Blandford, R.D. (1997a). GRB 970508. *IAU Circular*, 6676.

Metzger, M.R., Djorgovski, S.G., Steidel, C.C., Kulkarni, S.R., Adelberger, K.L. and Frail, D.A. (1997b). GRB 970508. *IAU Circular*, 6655.

Metzger, M.R., Kulkarni, S.R., Djorgovski, S.G., Gal, R., Steidel, C.C. and Frail, D.A. (1997c). GRB 970228. *IAU Circular*, 6588.

Mochkovitch, R., Hernanz, M., Isern, J. and Martin, X. (1993). Gamma-ray bursts as collimated jets from neutron star/black hole mergers. *Nature*, **361**, 236–238.

Murakami, T., Fujii, M., Hayashida, K., Itoh, M. and Nishimura, J. (1988). Evidence for cyclotron absorption from spectral features in gamma-ray bursts seen with Ginga. *Nature*, **335**, 234–235.

Natarajan, P., Bloom, J.S., Sigurdsson, S., Johnson, R.A., Tanvir, N.R., Groot, P.J., Galama, T.J., van Paradijs, J. and Kouveliotou, C. (1997). The host to gamma-ray burst 970508: A high-redshift dwarf galaxy? *New Astronomy*, **2**, 471–475.

Nemiroff, R.J. (1994). A century of gamma-ray burst models. *Comments on Astrophysics*, **17**, 189–205.

Norris, J.P., Bonnell, J.T., Nemiroff, R.J., Scargle, J.D., Kouveliotou, C., Paciesas, W.S., Meegan, C.A. and Fishman, G.J. (1995). Duration distributions of bright and dim BATSE gamma-ray bursts. *Astrophysical Journal*, **439**, 542–547.

Norris, J.P., Hertz, P., Wood, K.S. and Kouveliotou, C. (1991). On the nature of soft gamma repeaters. *Astrophysical Journal*, **366**, 240–252.

Norris, J.P., Nemiroff, R.J., Scargle, J.D., Kouveliotou, C., Fishman, G.J., Meegan, C.A., Paciesas, W.S. and Bonnell, J.T. (1994). Detection of signature consistent with cosmological time dilation in gamma-ray bursts. *Astrophysical Journal*, **424**, 540–545.

Owens, A., Guainazzi, M., Oosterbroek, T., Orr, A., Parmar, A.N., Costa, E., Feroci, M., Piro, L., Soffitta, P., dal Fiume, D., Frontera, F., Palazzi, E.,

Pian, E., Heise, J., in 't Zand, J.J.M., Maccarone, M.C. and Nicastro, L. (1998). The absorption properties of gamma-ray burst afterglows measured by BeppoSAX. *Astronomy Astrophysics*, **339**, L37–L40.

Paciesas, W.C. and Fishman, G.J. (eds) (1992). *Gamma-Ray Bursts*, Vol. 265 of AIP Conference Proceedings, American Institute of Physics, New York.

Paczyński, B. (1986). Gamma-ray bursters at cosmological distances. *Astrophysical Journal*, **308**, L43–L46.

Paczyński, B. (1992). Estimating redshifts for gamma-ray bursts. *Nature*, **355**, 521–522.

Paczyński, B. (1992). GB 790305 as a very strongly magnetized neutron star. *Acta Astronomica*, **42**, 145–153.

Paczyński, B. (1995). How far away are gamma-ray bursters? *Publications of the Astronomical Society of the Pacific*, **107**, 1167–1175.

Paczyński, B. (1998). Are gamma-ray bursts in star-forming regions? *Astrophysical Journal*, **494**, L45–L48.

Park, H.S., Ables, E., Band, D.L., Barthelmy, S.D., Bionta, R.M., Butterworth, P.S., Cline, T.L., Ferguson, D.H., Fishman, G.J., Gehrels, N., Hurley, K., Kouveliotou, C., Lee, B.C., Meegan, C.A., Ott, L.L. and Parker, E.L. (1997). Real-time optical flux limits from γ-ray bursts measured by the Gamma-ray Optical Counterpart Search Experiment. *Astrophysical Journal*, **490**, 99–108.

Pedersen, H., Jaunsen, A.O., Grav, T., Ostensen, R. andersen, M.I., Wold, M., Kristen, H., Broeils, A., Naeslund, M., Fransson, C., Lacy, M., Castro-Tirado, A.J., Gorosabel, J., Rodriguez Espinosa, J.M., Perez, A.M., Wolf, C., Fockenbrock, R., Hjorth, J., Muhli, P., Hakala, P., Piro, L., Feroci, M., Costa, E., Nicastro, L., Palazzi, E., Frontera, F., Monaldi, L. and Heise, J. (1998). Evidence for diverse optical emission from gamma-ray burst sources. *Astrophysical Journal*, **496**, 311–315.

Pian, E., Amati, L., Antonelli, L.A., Butler, R.C., Costa, E., Cusumano, G., Danziger, J., Feroci, M., Fiore, F., Frontera, F., Giommi, P., Masetti, N., Muller, J.M., Oosterbroek, T., Owens, A., Palazzi, E., Piro, L., Castro-Tirado, A., Coletta, A., dal Fiume, D., del Sordo, S., Heise, J., Nicastro, L., Orlandini, M., Parmar, A., Soffitta, P., Torroni, V. and in 't Zand, J.J. M. (1999). BeppoSAX detection and follow-up of GRB 980425. *Astronomy and Astrophysics*, Suppl., **138**, 463–464.

Pian, E., Fruchter, A.S., Bergeron, L.E., Thorsett, S.E., Frontera, F., Tavani, M., Costa, E., Feroci, M., Halpern, J., Lucas, R.A., Nicastro, L., Palazzi, E., Piro, L., Sparks, W., Castro-Tirado, A.J., Gull, T., Hurley, K. and Pedersen, H. (1998). Hubble Space Telescope imaging of the optical transient associated with GRB 970508. *Astrophysical Journal*, **492**, L103–L106.

Piran, T. (1999). Gamma-ray bursts and the fireball model. *Physics Reports*, **314**, 575–667.

Piro, L., Butler, R., Fiore, F., Antonelli, A. and Pian, E. (1998). GRB 980425: X-ray revised positions. *GRB Coordinates Network Circular*, 155.

Ramaprakash, A.N., Kulkarni, S.R., Frail, D.A., Koresko, C., Kuchner, M., Goodrich, R., Neugebauer, G., Murphy, T., Eikenberry, S., Bloom, J.S., Djorgovski, S.G., Waxman, E., Frontera, F., Feroci, M. and Nicastro, L. (1998). The energetic afterglow of the γ-ray-burst of 14 December 1997. *Nature*, **393**, 43–46.

Ramirez-Ruiz, E. and Fenimore, E.E. (2000). Pulse width evolution in gamma-ray bursts: evidence for internal shocks. *Astrophysical Journal*, **539**, 712–717.

Rees, M.J. and Mészáros, P. (1992). Relativistic fireballs: Energy conversion and time-scales. *Monthly Notices of the Royal Astronomical Society*, **258**, L41–L43.

Rees, M.J. and Mészáros, P. (1994). Unsteady outflow models for cosmological gamma-ray bursts. *Astrophysical Journal*, **430**, L93–L96.

Reichart, D.E. (1999). GRB 970228 revisited: Evidence for a supernova in the light curve and late spectral energy distribution of the afterglow. *Astrophysical Journal*, **521**, L111–L115.

Rol, E., Wijers, R.A.M.J., Vreeswijk, P.M., Kaper, L., Galama, T.J., van Paradijs, J., Kouveliotou, C., Masetti, N., Pian, E., Palazzi, E., Frontera, F. and van den Heuvel, E.P.J. (2000). GRB 990712: First indication of polarization variability in a gamma-ray burst afterglow. *Astrophysical Journal*, **544**, 707–711.

Rothschild, R.E. and Lingenfelter, R.E. (eds) (1996). *High-Velocity Neutron Stars and Gamma-Ray Bursts*, Vol. 366 of AIP Conference Proceedings, American Institute of Physics, New York.

Ruderman, M. (1975). Theories of gamma-ray bursts. *Annals of the New York Academy of Sciences*, **262**, 164–180.

Ruffert, M. and Janka, H.T. (1999). Gamma-ray bursts from accreting black holes in neutron star mergers. *Astronomy and Astrophysics*, **344**, 573–606.

Sahu, K.C., Livio, M., Petro, L., Macchetto, F.D., van Paradijs, J., Kouveliotou, C., Fishman, G.J., Meegan, C.A., Groot, P.J. and Galama, T. (1997). The optical counterpart to gamma-ray burst GRB 970228 observed using the Hubble Space Telescope. *Nature*, **387**, 476–478.

Sari, R. and Piran, T. (1999). GRB 990123: The optical flash and the fireball model. *Astrophysical Journal*, **517**, L109–L112.

Schaefer, B.E. (1981). Probable optical counterpart of a gamma-ray burster. *Nature*, **294**, 722–724.

Schaefer, B.E. (1990). Gamma-ray burster counterparts: Archival data. *Astrophysical Journal*, **364**, 590–600.

Schaefer, B.E. (1992). Severe new limits on extragalactic models of gamma-ray bursts. In C. Ho, R.I. Epstein and E.E. Fenimore (eds), *Gamma-Ray Bursts – Observations, Analyses and Theories*, Cambridge University Press, pp. 107–112.

Schaefer, B.E. (1998). Severe constraints on possible cosmological models caused by limits on host galaxies. In *Gamma-Ray Bursts*, American Institute of Physics, New York, pp. 595–599.

Schaefer, B.E., Bradt, H.V., Barat, C., Hurley, K., Niel, M., Vedrenne, G., Cline, T.L., Desai, U., Teegarden, B.J., Evans, W.D., Fenimore, E.E., Klebesadel, R.W., Laros, J.G., Estulin, I.V. and Kuznetsov, A.V. (1984). Two probable optical flashes from gamma-ray bursters. *Astrophysical Journal*, **286**, L1–L4.

Schmidt, M. (1999). Luminosities and space densities of γ-ray bursts. *Astrophysical Journal*, **523**, L117–L120.

Schmidt, W.K.H. (1978). Distance limit for a class of model gamma-ray-burst sources. *Nature*, **271**, 525–527.

Share, G.H., Harris, M.J., Leising, M.D. and Messina, D.C. (1993). Search for gamma-ray transients using the smm spectrometer. *Astronomy and Astrophysics*, Suppl., **97**, 341–344.

Sokolov, V.V., Kopylov, A.I., Zharikov, S.V., Feroci, M., Nicastro, L. and Palazzi, E. (1998). $BV R_c I_c$ photometry of GRB 970508 optical remnant: May–August, 1997. *Astronomy and Astrophysics*, **334**, 117–123.

Steidel, C.C. and Sargent, W.L.W. (1992). MgII absorption in the spectra of 103 QSOs: Implications for the evolution of gas in high-redshift galaxies. *Astrophysical Journal*, Suppl., **80**, 1–108.

Strong, I.B., Klebesadel, R.W. and Evans, W.D. (1975). Observations of gamma-ray bursts. *Annals of the New York Academy of Sciences*, **262**, 145–158.

Terekhov, O., Denissenko, D., Sunyaev, R., Sazonov, S., Barat, C., Dezalay, J.P., Vedrenne, G., Lund, N., Castro-Tirado, A.J. and Brandt, S. (1995). Review of GRANAT observations of gamma-ray bursts. *Astrophysics and Space Science*, **231**, 31–34.

Terlevich, R., Fabian, A. and Turatto, M. (1999). Supernova 1999eb and GRB 991002. *IAU Circular*, 7269.

Totani, T. (1997). Cosmological gamma-ray bursts and evolution of galaxies. *Astrophysical Journal*, **486**, L71–L74.

Trimble, V. (1995). The 1920 Shapley–Curtis discussion: Background, issues, and aftermath. *Publications of the Astronomical Society of the Pacific*, **107**, 1133–1144.

Trombka, J.I., Eller, E.L., Schmadebeck, R.L., Adler, I., Metzger, A.E., Gilman, D., Gorenstein, P. and Bjorkholm, P. (1974). Observation of a cosmic gamma-ray burst on Apollo 16: II. X-ray time profile and source location. *Astrophysical Journal*, **194**, L27–L33.

Van den Bergh, S. and Tammann, G.A. (1991). Galactic and extragalactic supernova rates. *Annual Reviews of Astronomy and Astrophysics*, **29**, 363–407.

Van Paradijs, J., Groot, P.J., Galama, T., Kouveliotou, C., Strom, R.G., Telting, J., Rutten, R.G.M., Fishman, G.J., Meegan, C.A., Pettini, M., Tanvir, N., Bloom, J., Pedersen, H., Noerdgaard-Nielsen, H.U., Linden-Voernle, M., Melnick, J., van der Steene, G., Bremer, M., Naber, R., Heise, J., in 't Zand, J., Costa, E., Feroci, M., Piro, L., Frontera, F., Zavattini, G., Nicastro, L., Palazzi, E., Hanlon, L. and Bennett, K. (1997). Discovery of transient optical emission from the error box of the gamma-ray burst of February 28, 1997. *Nature*, **386**, 686–689.

Van Paradijs, J., Kouveliotou, C. and Wijers, R.A.M.J. (2000). Gamma-ray burst afterglows. *Annual Reviews of Astronomy and Astrophysics*, **38**, 381–427.

Wang, L. and Wheeler, J.C. (1998). The supernova–gamma-ray burst connection. *Astrophysical Journal*, **504**, L87–L90.

Waxman and Draine, B.T. (2000). Dust sublimation by GRBs and its implications. *Astrophysical Journal*, **537**, 796–802.

Waxman, E. and Loeb, A. (1999). A subrelativistic shock model for the radio emission of SN 1998bw. *Astrophysical Journal*, **515**, 721–725.

Wheaton, W.A., Ulmer, M.P., Baity, W.A., Datlowe, D.W., Elcan, M.J., Peterson, L.E., Klebesadel, R.W., Strong, I.B., Cline, T.L. and Desai, U.D. (1973). The direction and spectral variability of a cosmic gamma-ray burst. *Astrophysical Journal*, **185**, L57–L61.

Wijers, R.A.M.J., Bloom, J.S., Bagla, J.S. and Natarajan, P. (1998). Gamma-ray bursts from stellar remnants: Probing the universe at high redshift. *Monthly Notices of the Royal Astronomical Society*, **294**, L13–L17.

Wijers, R.A.M.J., Rees, M.J. and Mészáros, P. (1997). Shocked by GRB 970228: The afterglow of a cosmological gamma-ray burst. *Monthly Notices of the Royal Astronomical Society*, **288**, L51–L56.

Wijers, R.A.M.J., Vreeswijk, P.M., Galama, T.J., Rol, E., van Paradijs, J., Kouveliotou, C., Giblin, T., Masetti, N., Palazzi, E., Pian, E., Frontera, F., Nicastro, L., Falomo, R., Soffitta, P. and Piro, L. (1999). Detection of polarization in the afterglow of GRB 990510 with the ESO Very Large Telescope. *Astrophysical Journal*, **523**, L33–L36.

Woosley, S.E. (1993). Gamma-ray bursts from stellar mass accretion disks around black holes. *Astrophysical Journal*, **405**, 273–277.

Woosley, S.E., Eastman, R.G. and Schmidt, B.P. (1999). Gamma-ray bursts and type Ic supernova SN 1998bw. *Astrophysical Journal*, **516**, 788–796.

Yoshida, A., Namiki, M., Otani, C., Kawai, N., Murakami, T., Ueda, Y., Shibata, R. and Uno, S. (1999). What did ASCA see in the GRB 970828 afterglow? *Astronomy and Astrophysics*, Suppl., **138**, 433–434.

Zharikov, S.V., Sokolov, V.V. and Baryshev, Y.V. (1998). $BVR_c I_c$ lightcurves of GRB 970508 optical remnant and colors of underlying host galaxy. *Astronomy and Astrophysics*, **337**, 356–362.

Żytkow, A.N. (1990). Are there optical counterparts to gamma-ray bursts? *Astrophysical Journal*, **359**, 138–154.

23

MARTIN ELVIS*

Quasars

To the naked eye the night sky is filled with stars. But outside of a quite narrow slice of the whole electromagnetic spectrum a radically different sky is seen. Over two huge frequency ranges, each about 10 decades broad, utterly different objects dominate the view. Both at the long, radio wavelengths, and at the high X-ray and γ-ray energies, the brightest things in the sky are *quasars*. Only in the three decades of frequency from the ultraviolet, through the visible band and into the infrared do stars dominate, and in the infrared much of the starlight is seen only indirectly, having been absorbed and re-emitted by dust around the stars. Three decades is much broader than the octave-wide band to which our human eyes are sensitive but, compared with the 20 decades of frequency opened up to astronomy during the space age, it is small.

Quasars[1] are truly space-age astrophysics. Of the two wide bands where quasars dominate, the radio sky can be explored quite well from the ground. To explore the high-energy sky though we have no choice but to use telescopes mounted on spacecraft.

What are these objects that dominate the sky over most of the spectrum? In complete contrast to stars, quasars are extremely distant objects seen back to early cosmic times, and are the most powerful ("luminous") continuous[2] sources of radiation in the Universe. Moreover, the light emitted by quasars, even though it reaches to the highest energy γ-rays we can measure, is not the whole of the quasar power output. They also shoot out enormous jets of fast particles, travelling at 99.999% of the speed of light. These jets stay narrow over huge distances, easily as large as the distances between whole galaxies. Such unusual objects as quasars should surely be found in unusual places, and indeed quasars inhabit the very centers ("nuclei") of galaxies. We believe that the origin of the strange properties of quasars is a black hole with a huge mass, 10^6 to 10^9 times that of the Sun.

Here I give a personal view of quasars across these 20 decades of spectrum, and across four decades of discovery, of which I have only witnessed two myself. After telling the history of quasar discovery, I set out seven key quasar mysteries, describe how space obsevations gradually brought the whole quasar phenomenon into view, and then explain each mystery, so far as we now can. I end with a few speculations on where quasar research will go next.

1 A QUICK QUASAR HISTORY

Understanding the quasar puzzle has been a long-term program. There were hints of strange activity in the nuclei of some galaxies as long ago as 1917 when Slipher found extraordinarily broad emission lines coming from the center of the galaxy NGC 1068, and the next year when Curtis found a jet pointing straight out of the center of the elliptical galaxy M87 in Virgo. Quasar prehistory took another step, 25 years later, when Karl Seyfert's 1943 PhD thesis described a few galaxies with extremely rapid gas motions in their nuclei (as determined from the Doppler-shifted width of their peculiar spectral emission lines; Slipher had rejected this explanation). Just a few years later, using surplus World War II radar equipment, radio astronomy blossomed, and for the first time found a sky not dominated by stars. In 1953 some of these new radio sources turned out to

---
* Harvard-Smithsonian Center for Astrophysics, Cambridge, MA, USA

[1] For simplicity and clarity I will refer to all the types of non-stellar "activity" in galactic nuclei as "quasars," even though the research literature carefully distinguishes high-luminosity "quasars" from lower luminosity "active galactic nuclei" (or AGNs). Evidence for any basic physical difference between these types of active objects has diminished, essentially to the vanishing point.

[2] For a few seconds or minutes a gamma-ray burst can outshine any quasar.

come in pairs, either side of a galaxy, suggesting ejection from the galaxy, but there was little hard evidence for this.

The true history of quasars did not begin for another decade, until 1963. Hard-won precise positions of radio sources enabled the large optical telescopes of the day to take spectra of whatever optical object lay at that position. Some showed the starlight of normal galaxies, but often at large distances (as measured by the redshifts of their spectral lines, using the Hubble relation between redshift and distance that describes the expansion of the Universe) (Chapter 17). The spectra of those with compact, "stellar"-looking, optical objects, though, had a baffling series of strong, broad emission lines. It took three years before Maarten Schmidt realized that one of these "radio stars" was actually at a large redshift, and so – given the Hubble relation between redshift and distance – presumably lay at a large distance from us. This in turn implied a prodigous luminosity.

Schmidt (1990) relates how his key discovery was made:

The puzzle was suddenly resolved in the afternoon of February 5, 1963, while I was writing a brief article about the optical spectrum of 3C 273. Cyril Hazard had written up the occultation results for publication in *Nature* and suggested that the optical observations be published in an adjacent article. While writing the manuscript, I took another look at the spectra. I noticed that four of the six lines in the photographic spectra showed a pattern of decreasing strength and decreasing spacing from red to blue. For some reason, I decided to construct an energy-level diagram based on these lines.

I must have made an error in the process which seemed to contradict the regular spacing pattern. Slightly irritated by that, I decided to check the regular spacing of the lines by taking the ratio of their wavelengths to that of the nearest line of the Balmer series. The first ratio, that of the 5630 line to H-beta, was 1.16. The second ratio was also 1.16. When the third ratio was 1.16 again, it was clear that I was looking at a Balmer spectrum redshifted by 0.16.

I was stunned by this development: stars of magnitude 13 are not supposed to show large redshift! When I saw Jesse Greenstein minutes later in the hallway and told him what had happened, he produced a list of wavelengths of emission lines from a just completed manuscript about the spectrum of 3C 48. Being prepared to look for large redshift, it took us only minutes to derive a redshift of 0.37.

The interpretation of such large redshifts was an extraordinary challenge. Greenstein and I soon found that an explanation in terms of gravitational redshift was essentially impossible on the basis of spectroscopic arguments. We recognized that the alternative explanation in terms of cosmological redshifts, large distances, and enormous luminosities and energies was very speculative but could find no strong arguments against it. The results for 3C 273 and 3C 48 were published six weeks later in four consecutive articles in *Nature* (Hazard, Mackay, and Shimmins 1963; Schmidt 1963; Oke 1963; Greenstein and Matthews 1963).

So these things were not stars, they simply appeared stellar on photographs, and hence they were dubbed "quasi-stellar objects" which quickly became "quasars" (Hazard 1979). Astronomers were not prompt to accept the idea that quasars were at cosmological distances. This was not because this distance seemed too great: after all, there was already the example of the large redshift ($z = 0.46$) of the radio galaxy 3C 295. The problem was that, if the redshifts implied distances according to the normal Hubble law (Chapter 17), then it implied *a second power source must be at work in the Universe, beyond the nuclear burning at the centers of stars*. That is because it is one thing is to find that a steady source is very distant and powerful, and quite another thing to find that a *variable* source is far away, because then it must emit its power from a region smaller than the time it takes light to cross it. In the case of quasars this means the power of a whole galaxy of stars comes from a region smaller than the Solar System! (r(Solar System) $\sim 50 r$(Earth–Sun) = 50 AU = 7.5×10^{14} cm.) Any such source has to be more efficient than nuclear burning can be (see next section). And the first quasars found, 3C 273 and 3C 48, indeed varied quite strongly. It was this variability that had fixed the idea in astronomers' minds that these sources, of stellar optical appearance, must be nearby stars. But the high redshifts made that position untenable. In the 40 years since the first redshift of 0.16, quasars have continued to be found out to larger and larger redshifts (Figure 1), reaching 5.8 in 2000, implying they were already around when the Universe was only 5% of its current age, or less than 1 billion years old.

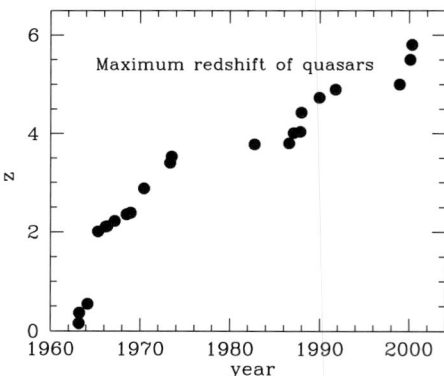

Figure 1 The largest redshift of quasars as a function of time. The current record is $z = 5.80$. (From Fan *et al.* 2001.) For early data see Schneider *et al.* (1992) and Fan *et al.* (1999).

2 SEVEN QUASAR MYSTERIES

The huge research literature on quasars revolves around a quite small number of central questions or mysteries. While these are capable of infinite subdivision I believe there are just seven basic mysteries.

It is remarkable that with the first quasar papers, already four of the seven quasar mysteries were laid out:

1. *Quasar luminosities are enormous, and arise from tiny regions.* The power from quasars covers a wide range ($>10^7$) from object to object: from $<10^{40}\,\text{erg}\,\text{s}^{-1}$ to $>10^{47}\,\text{erg}\,\text{s}^{-1}$, and can compete with, and easily exceed, that of a whole galaxy of 10^{11} stars ($\sim 10^{44}\,\text{erg}\,\text{s}^{-1}$). When Mathews and Sandage (1963) found that quasars could vary their power output by 40% in just a few months, then it seemed that all this luminosity must come from a region similar in size to our Solar System. (From the simple "light travel time argument": a source cannot vary coherently in less time than a light signal takes to cross the source. So if a source varies in a day, it is no bigger than a light-day across (but beware of high velocities; Chapter 24).) Moreover the radio galaxies required that quasars put out enormous power over a long time period since the total energy stored in the giant radio lobes must be $>10^{60}\,\text{ergs}$ (Burbidge 1959). At this point it became a good bet that an energy source quite different from, and much more efficient than, nuclear burning in stars was at work.

2. *Quasar spectra are nothing like that of starlight.* Stars have more or less black-body spectra with well-defined temperatures of a few thousand degrees, so their emission in the radio band is tiny. Instead quasars can have huge radio luminosities with spectra that follow a simple straight-line form in log (flux) vs. log (frequency), i.e., a "power law" from $\sim 100\,\text{MHz}$ to $\sim 100\,\text{GHz}$. So being bright in the radio does not simply mean that they are very cold objects. Their optical spectra have no obvious thermal peak either, but also have a power law shape from $\lambda = 1\,\mu\text{m}$ to $0.1\,\mu\text{m}$ (Figure 3). Confusingly at first this meant that quasars were both "bluer" than stars (i.e., had more short-wavelength emission than expected based on their visual brightness) and "redder" than stars (that is, the same at long wavelengths)! The whole spectrum is not just one power law though, since the radio and optical power laws do not join up. When X-ray spectra first became available, they too had a power-law shape, and again it did not connect with the optical or radio power laws (Figure 3). In this sense quasars have no temperature, and so were called "non-thermal" sources. That does not mean though that no combination of thermal processes underlies the production of the quasar continuum (see Section 5.2). This failure to drop off in power at either long or short wavelengths is the feature that makes quasars dominate the sky over most of the spectrum.

3. *Quasars accelerate material to high velocities.* The widths of the emission lines reach up to 10% of the speed of light ($0.1c$), although most are a smaller, but still considerable, $\sim 0.03c$. Some quasars show absorption lines, and some others show X-ray emission lines, implying velocities up to $0.2c$–$0.3c$. Most extreme of all, to create the observed "superluminal" motions in blazars requires $\delta \sim 10$, i.e., $v \sim 0.999c$ (Chapter 24; Ghisellini et al. 1993). Ways to produce the broad emission and absorption lines without using mass motions are contrived, but not impossible. Large velocities in astronomy imply deep gravitational wells (to accelerate matter, or to produce a general relativistic gravitational redshift), except for short-lived explosive events like supernovae.

4. *Quasars have linear symmetry.* As radio maps became more detailed through the exploitation of interferometry (work which led to a Nobel prize for two pioneers, Martin Ryle and Anthony Hewish) most of the radio sources were found to have a "double lobe" structure (Jennison and Das Gupta 1953) extending well beyond the galaxies they straddled. With the advent of the Very Large Array (VLA) in the 1970s these lobes could be mapped in detail at high dynamic range revealing pairs of large (10^{22}–$10^{24}\,\text{cm}$) highly structured bubbles (Figure 2). In many, but by no means all, cases the central galaxy had a nucleus with a spectrum like that of a quasar. Forty years later the fine detail of Hubble Space Telescope images showed that even the quasars that are not bright radio sources have a linear symmetry. Clear cone-shaped structures were seen stretching over a galaxy scale ($10^{22}\,\text{cm}$). In high-redshift radio-loud quasars (primarily in "radio galaxies") Hubble images showed galaxy-scale optical line emission aligned with the radio jets. A big clue to the inner structure of quasars has to lie in this symmetry.

From these four mysteries an outline of quasars was already becoming clear: a large mass in a small region suggested that the, then outlandish seeming, concept of a black hole might underlie quasar physics, while the symmetry suggested a spin axis, either of the black hole or of a rotating gas disk. A short paper by Donald Lynden-Bell (1969) put this together with extraordinary elegance. He pointed out that energy release from gravitational infall is more efficient than nuclear burning for any mass large enough to produce quasar luminosities. He then outlined how to release this energy as radiation: any material falling onto a black hole from some large distance will have some angular momentum around the black hole and so, unless the incoming matter is perfectly spherically symmetrical, the material will fall into a flattened shape to form an orbiting disk. (This is common in astronomy, recall Saturn's rings, or nascent planetary systems.) If

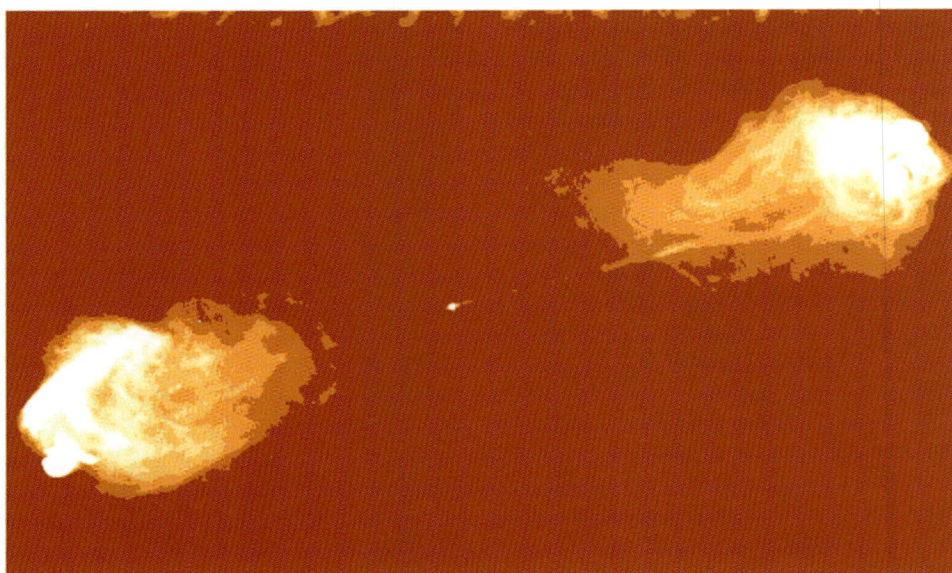

Figure 2 The complex two-sided radio structure of Cygnus A as imaged by the VLA. The large features, or "lobes", are fed by a jet of relativistic particles coming out of the central "core" (white dot), which lies at the nucleus of a large galaxy. (Perley *et al.* 1984; courtesy NRAO.)

enough matter accumulates then some (undetermined) form of viscous friction between adjacent orbits will drag on the faster moving material, moving angular momentum outward, and allowing matter to fall inward, eventuall accreting into the central black hole. As the matter falls inward the gravitational potential energy is released as radiation. At each radius the radiation will be more or less a black body, but since each radius emits a different temperature (getting hotter inward, as the potential gradient of the black hole steepens) so the summed emission from all radii looks nothing like a black body, in fact it can mimic a power law.

These "accretion disks" have since been found in our galaxy around many compact objects: white dwarfs, neutron stars, and stellar-mass black holes (Chapter 32), beginning with the discoveries of the Uhuru satellite (Giacconi and Gursky 1974). These smaller systems in which it was possible to establish that a black hole was almost certainly present made the idea of black holes and accretion disks more familiar and better understood, and they are now a conventional part of astrophysics (Frank *et al.* 1985, 1992). Nikolai Shakura and Rashid Sunyaev (1973) expanded on accretion disk theory by cleverly hiding all the unknown physics of viscosity in a parameter, α, and working out the disk behavior as a function of α.

It took only a few years of searching for fainter and fainter optical counterparts of radio sources up to higher and higher redshifts to discover two further mysteries:

5. *Radio-loud quasars are just the tip of the iceberg.* The blue "non-thermal" spectra of quasars made them easy to pick out from normal stars in large numbers just from a pair of photographs taken through different color filters. This "color selection" was used to find the quasar using the, initially quite uncertain, radio source positions. Hence quite a large area of sky was photographed for each radio source. Alan Sandage (1965) first saw that along with the radio-emitting blue quasar, there were many more blue "interlopers" that had nothing to do with the radio source. Spectra showed that these too were quasars, but ones which were "radio quiet." Radio-quiet quasars turn out to be 10 times more common than the original quasars, which are now called, with impeccable logic, "radio-loud quasars." The difference between radio-loud and radio-quiet quasars is found only in the radio spectrum (and possibly the γ-ray spectrum). The rest of the spectrum, from far-IR to hard X-rays is essentially the same (Elvis *et al.* 1994). (A slight difference of power-law slope in the X-rays being the only exception.) Yet the difference in the radio band is huge, a factor of ~ 100–10^4 in the ratio of radio to optical luminosity separates the two classes.

6. *Quasars are far less common now than they were in the distant past.* This was termed quasar "evolution".* The

---

*I find this an unfortunate choice. All we really mean is "change with time." In biology evolution demands mutation, reproduction, and selection, and is an active, unpredictable process that will produce different outcomes even with identical starting conditions. "Evolution" in astronomy is merely the playing out of physical processes, with no reproduction or competition and with a predictable outcome. Chaotic systems are an exception but still involve no analogs to the biological process of evolution.

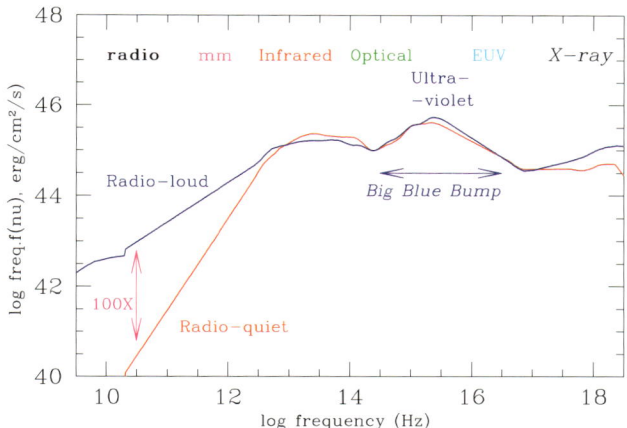

Figure 3 The median radio to X-ray spectra of radio-loud and radio-quiet quasars (Elvis et al. 1994). The two curves are almost the same, except for the factor ~100 divergence in the radio band.

number of quasars kept rising rapidly as fainter and fainter ones were searched for. In a simple Euclidean universe (a good approximation for sources at moderate redshifts), the number of sources ($N(f)$) would go up as the observed flux f was reduced as $N(f) \propto f^{-3/2}$. Instead optical counts of quasars rose much faster, $N(f) \propto f^{-2.2}$. It was Gianni Zamorani and colleagues (Zamorani et al. 1981) using Einstein Observatory X-ray measurements who first showed there had to be a limit, otherwise the total emission from quasars would be brighter than the X-ray emission from the whole sky (the "X-ray background"; Section 4.7.1). Typically, fainter quasars are more distant (though with a large scatter), so the rise in numbers is due to there being more quasars in the distant past. The number of quasars (brighter than some luminosity), $\Phi(L > L_0) \propto (1 + z)^3$ up to $z \sim 2$. So quasars used to be 30 times more common when the Universe was a quarter of its present age.

For a long time quasars were the only objects in the Universe that clearly evolved on a cosmic timescale. As such they were key to the argument between Big Bang and steady-state cosmology. If the past was different from the present, then the steady-state model (Hoyle 1948) was ruled out, since it posited that the Universe was now the same as it ever was. The only escape was to show that the large redshifts of quasars were not due to cosmological expansion but had another origin (Burbidge and Burbidge 1967). This was initially attractive, since if quasar distances are really much smaller than we thought, then the problem with making too much energy from a small region (mystery 1) goes away. There is physics that could produce redshifts from other causes (or "non-cosmological redshifts" as they were called). A strong gravitational field, for example,

can produce a redshift, or rapid ejection from a galaxy could produce a redshift (but also blueshifts, depending on the direction of ejection). This argument continued fiercely until quasars were also found in galaxies for which the spectra of the normal stars in the galaxy showed the same redshift as the quasar emission lines (Stockton 1973). After that the great redshift debate rapidly became a sideshow (Weedman 1976) that is now almost extinguished. We should never forget though, that there can be more than one way to create large redshifts, and some fraction of "quasars" may be something else, more exotic than quasars, that has so far gone unrecognized among the multitudes.

It is extraordinary that inherently "space-age" objects like quasars could present so many of their key mysteries before a single one was detected by a space-based telescope. To solve those mysteries, though, needed a whole fleet of spacecraft equipped with many new types of telescope.

One last quasar mystery did await space-based observations:

7. *Most quasars are heavily obscured by gas and dust.* First, radio-loud quasars were found to be the tip of the iceberg of blue radio-quiet quasars. Now all these quasars have been recognized as themselves just a fragment of the whole quasar population, most of which is hidden by gas and dust. Only infrared (10–100 μm) and X-ray (>2 keV) surveys are immune to this obscuration, and both are space-based astronomies. The infrared is immune because that is where dust, which has temperatures of 1500 K or less, re-radiates the absorbed radiation. The X-rays are immune because, above a few kilovolts energy, only column densities of material that are thick to electron scattering ($N_H > 10^{24}\,\text{cm}^{-2}$) can prevent X-rays emerging directly.

If dust can obscure a quasar so easily, how many quasars are missed by the surveys? The early X-ray work implied that about half were missed. Infrared observations of radio sources led Rachel Webster and colleagues to propose (Webster et al. 1995) that as much as 80% of all quasars were hidden by obscuring dust and gas. Initially a wild-seeming idea, this has moved rapidly into the mainstream because the X-ray background requires it (Section 4.7.1) and ultraluminous infrared galaxies support it. This has major consequences for the total amount of gravitational energy turned into radiation in the Universe (Fabian and Iwasawa 1999).

The arrival of quasars on the astronomical scene created a stir. It was not only evolution that was controversial about quasars. The apparent need for enormous black holes also had to be made compelling. Each of the key mysteries had to be argued over, since none was watertight and the implications were large. "Extraordinary claims require extraordinary evidence" (Sagan 1979). In fact, at least one of the arguments was wrong. As we discussed earlier, the small size of quasars was derived from their variability. We now know that most of the early, extreme variations were not due to small size, but to extremely rapid motion toward us that speeds up all changes (Chapter 24). As it happens, even more rapid variability was soon found in X-rays, and in objects where beaming toward us cannot be the explanation. (Since the luminosity of the emission lines scaled with the X-ray luminosity, and the lines could not be beamed, as they had only relatively modest velocities.) So the quasar sizes deduced have become even smaller, even though the original argument for them was flawed.

The situation just before the advent of space observatories, around 1975, a full 20 years into quasar research proper, was a time of confusion. New data seemed only to complicate things, not clarify them. The subject was in what was, rather dismissively, called its "botanical" phase, where specimens of rare types were collected but little overarching order was established (this is quite unfair to the science of botany; Lawrence 1987). There were then at least two dozen sub-classes of quasar or active galaxy defined and the relationships between them all were vague (Lawrence 1987).

3 SPACE-BASED OBSERVATORIES SHOW "A QUASAR IN FULL"

3.1 Early days

As the armada of spaceborne telescopes set out in the 1970s and 1980s the breadth of what we learned about quasars grew enormously. No longer did we peep through two narrow windows of the spectrum at slivers of the broad quasar spectrum. We could at last see the quasar in full: the whole continuum from high-energy X-rays through the ultraviolet, optical, and infrared, and on into the radio. Here we give a quick outline of space astronomy, as it affected quasars.

The Ariel V survey first established that high-energy X-ray (2–10 keV) emission was a normal feature of active galaxies (this became my thesis work because of a chance list of "Seyfert galaxies" that Richard Bingham gave me during my Master's thesis work at the Royal Greenwich Observatory), as well as that heavy obscuration was common. From 1978 to 1981 the Einstein Observatory, the first imaging X-ray telescope (and as a result some 100 times more sensitive than its predecessors), extended this discovery to include all quasars, up to the highest redshifts then known, but was not sensitive to highly obscured objects because it imaged only lower energy X-rays (0.2–3 keV). Beginning just a year before Einstein the NASA/ESA/UK International Ultraviolet Explorer (IUE) began to give us the first ultraviolet spectra of quasars (0.12–0.3 μm), enabling the recognition of the "Big Blue Bump" (Section 4.2) and, almost in parallel, the 1980 joint Netherlands/UK/NASA IRAS infrared (12–100 μm) all-sky survey found many more "ultraluminous infrared galaxies" (ULIRGs, $L_{IR} > 10^{12}$ L_{Sun} ($L_{Sun} = 3.8 \times 10^{33}$ erg s^{-1}) compared to 10^{11} L_{Sun} for our Galaxy) with implications for quasar numbers (Section 4.3) and evolution (Section 4.6). Meanwhile ground-based radio technology improved, most notably at the NRAO VLA, enabling higher frequency (and so higher angular resolution) work, extending the radio band to over 30 GHz compared with only ~1 GHz before 1970. Rapid X-ray variability and absorbed X-ray spectra were studied intensely by the ESA satellite EXOSAT in the mid-1980s (Lawrence et al. 1987; Turner and Pounds 1989).

The later 1980s was a time of analysis, rather than new missions, largely because of budget problems at NASA and because of the *Challenger* tragedy. A happy exception was the use of the now aging IUE satellite to monitor the ultraviolet spectra of quasars, in order to learn more about the high-velocity gas that produces the broad emission lines. Large amounts of observing time became available as the early IUE research programs reached their limits. The normally competitive quasar research community got together, beginning at one of the critical Santa Cruz workshops (the IAU Symposium No. 134 in 1988), into a few large teams, and secured much of that time for quasar monitoring, adding ground-based optical monitoring to extend the coverage. The idea was to watch for flashes of light from the small quasar continuum source, and then to wait for the "echoes" of this light seen in the strong quasar emission lines. (These lines are produced from the recombination of atoms ionized by the continuum. More continuum ionizes more atoms, making more to recombine and so produce stronger lines.) The delay between seeing the continuum flash and the line flash measures how far the ionized gas is from the continuum source (using the light travel time argument again). In principle this method of "reverberation mapping" can determine the structure of the interior of a quasar in detail. The very first results (Peterson et al. 1991; Clavel et al. 1991) were spectacular: the ionized gas lay 10 times closer to the central continuum that anyone had thought based on the best models then available (Kwan and Krolik 1981), and showed that different ions lay at different distances from the center, so that the geometrically simple "one zone" models then in use had to be discarded.

3.2 The 1990s

Although we can find quasars in optical light from the ground, the launch of the Hubble Space Telescope (HST) in 1990 gave us much more detailed images of nearby active galaxies than ever before. As already noted, this led to the dramatic images of "ionization cones," although they had already been seen, less dramatically, from the ground (Pogge 1988). HST also gave us much better UV spectra than IUE, allowing reverberation mapping at finer scales (Korista *et al.* 1995), and enabling high-resolution studies of absorption lines (Mathur *et al.* 1999). Perhaps though the most important HST contribution has been the use of its high angular resolution to take spectra of starlight close to the nuclei of normal (i.e., non-quasar) galaxies, which has quite accurately shown the presence of large central masses (Kormendy and Richstone 1995) of the size expected for the black holes that are needed to power quasars. Other related HST results tie the galaxies themselves closely to black hole, and so quasar, formation (Section 4.6).

An array of new X-ray satellites beginning in 1990 with ROSAT, followed by ASCA and BeppoSAX (in 1993 and 1996) defined quasar evolution far better and, for the latter two, began the first surveys capable of finding extremely obscured quasars by looking well above 2 kV (Ueda *et al.* 1998; Fiore *et al.* 1999). ASCA produced an astonishing result: a variable emission line due to neutral iron that was broadened by redshifts by up to $0.3c$ (Tanaka *et al.* 1995).

The most dramatic change in our picture of quasar-related activity came from the Compton Gamma Ray Observatory (CGRO) discovery of emission at 10^6 times higher energies than X-rays, making X-rays seem like "medium-energy" astronomy. These strong γ-ray sources are all "blazars" (Fichtel 1994), in which a highly beamed jet points toward us. Several years of study of the γ-ray to radio spectrum have now turned confusion into order. Ghisellini *et al.* (1998) showed that all blazars belong to a simple one-parameter family, which suggests that a simple physics-based solution is at hand (Chapter 24).

Surprising (to this author at least) was the *absence* of γ-ray emission from non-blazar quasars, even at energies only 10 times above the normal X-ray band. This indicates that, outside of their relativistic jets, quasars have a maximum "temperature."

3.3 EUV

The only hole in this "whole view" of quasars was the unavoidable gap in the extreme ultraviolet (EUV). This is a particularly important wavelength range for quasar work, both because the spectrum of quasars seems to peak somewhere in this band, and because EUV photons are particularly good at ionizing hydrogen and other low-atomic-number ions. It is the recombination of these ions which produces the dramatic quasar emission lines. The problem is that, just because of their excellent ionizing properties, these photons are the very ones that are most readily absorbed by the interstellar medium of our own Galaxy. So in the EUV, even in space, we are at the bottom of an atmosphere, that of the Milky Way, as much as we are on Earth. Since we have no way to travel the 300 light years above our Galaxy disk needed to rise above this atmosphere, there is no way to observe quasars, or indeed anything else more than a few light years distant, at wavelengths deep in the EUV. We approach as close as we can though. The EUV begins at the ionization wavelength of hydrogen, 91.2 nm and extends to shorter wavelengths. The Galaxy becomes partially transparent at about 10 nm and two small experiments, the Extreme Ultraviolet Explorer (EUVE) and the ROSAT Wide Field Camera, surveyed the sky at these wavelengths. To most astronomers' surprise they found a substantial population of active galaxies (Fruscione 1996). When corrected for the absorption by our Galaxy these active galaxies turn out to be enormously bright in the EUV, which is an ongoing challenge to accretion disk theories.

3.4 The present

It is fair to say that after 40 years of intense study, including 20 years of space astronomy, the subject of quasars still suffers from the advantage of having a rich phenomenology: we have learned a great deal, but now we know too much detail and not enough structure. This has been true for a long time. Already when Martin Rees reviewed the whole subject at the end of a Cambridge (UK) conference in 1976 it appeared like a spectacular juggling performance: all sorts of assorted facts kept in an orderly arc only with the greatest skill. (Perhaps this is why he now works mainly in cosmology?) While this richness means reseachers in the field have a great deal to do, it leaves outsiders seeing only confusion, an excess of "botany," and so they are not attracted to the field. Although quasars involve 20% of all astronomers, it is currently more fashionable to work on problems that seem more tractable – such as cosmology and galaxy formation! I believe that this is now changing. The complex phenomenology of quasars is just now falling into clear patterns that will soon take the subject onto a higher level of physics explanation. The simplicity of the family of blazars is one example (Chapter 24). Below I give some more examples. For now these are partial clearings in the jungle of observation. They look to be the beginnings of true solutions.

4 PARTIAL ANSWERS TO THE QUASAR MYSTERIES

4.1 Enormous luminosities from small regions: supermassive black holes

In the last few years, convincing estimates for the central mass in quasars have been made (Section 3.2), and these strongly support the idea that a supermassive (10^6–10^9 solar masses) black hole is present.

Reverberation mapping (Section 3.1) tells us about more than the broad emission lines. Since it tells us the radius at which gas moving at a particular velocity is located, it allows us to estimate the mass about which the gas is moving. This requires some assumptions, but they are probably good for a first-order mass estimate. If the gas is simply in a Keplerian orbit then the velocity and distance give the central mass directly. For the moderate luminosity quasars on which this has been tried, masses in the range 10^7 to 10^8 solar masses (10^{40}–10^{41} g) are the result. If these masses are reliable then supermassive black holes must really be at the centers of quasars; nothing else could fit so much mass inside so small a region.

Knowing a quasar's central mass allows many new physical facts to be established: how close to the maximum luminosity (the "Eddington limit") are the quasars emitting? Does the mass of the black hole have a relation to the mass of the host galaxy? The evidence so far is that these quasars are only emitting at about 1% of this limit, which allows them to have been brighter in the past, as some evolution models predict (Section 4.6). There is evidence for a galaxy–black hole mass connection from masses derived by a different technique: measuring stellar motions close to the galaxy nucleus by making use of HST's high angular resolution (Magorrian *et al.* 1998). It seems that the central black hole always has a mass that is about 0.6% of the mass of the surrounding galaxy "bulge" (the almost spherical region in the inner region of a galaxy, that is, the part that is not the flat spiral-armed disk). This implies a strong connection between galaxy formation and the growth of nuclear black holes.

4.2 Not like starlight: the big picture

The first breakthrough resulting from the wall-to-wall view of quasars allowed by space-based telescopes was the realization that they do have a "temperature" after all. Although their emission does not follow a black body, it does have a much broader peak, called the "Big Blue Bump."* Matt Malkan and Wal Sargent (1982) were the first to put together a group of ultraviolet, optical, and infrared spectra after only three years of IUE data. They were following up on an inspired suggestion by Greg Shields (1978) that the whole of the optical and ultraviolet spectrum of a quasar, which appeared to be a straight "power law," was in fact the long-wavelength side of an accretion disk spectrum (formed from adding many black bodies together). The peak of the spectrum defines a maximum temperature, which lies somewhere between 50,000 and 200,000 K, and it is possible that the strong EUV emission in some quasars is the toe of the short wavelength side of the Big Blue Bump nudging into the X-ray band. Qualitatively the Big Blue Bump does seem to fit the expectations of an accretion disk spectrum. But the details do not work so easily. Ski Antonucci, Anne Kinney and their colleagues have fought continuously against a glib interpretation of the Big Blue Bump in this way, pointing out the many difficulties (Antonucci *et al.* 1989; Kinney 1992; Antonucci 1999). The problem has been to find a plausible alternative. Accretion disk models have gradually become more accurate in the physics that they deploy, and some of the initial objections have dissipated, but enough remain that this is not a settled issue. The very broad X-ray emission lines seem to argue strongly for a cool disk close in to the nucleus, but even here there are some puzzling anomalies. One radical alternative is that, as in stars, we do not see the primary energy source in quasars. Suzy Collin and others (Collin-Soufrin *et al.* 1996) have suggested that *all* the emission we see is second-hand: a basic, unseen synchrotron spectrum is reprocessed into the observed form by the same clouds that produce the strong quasar emission lines and by associated warmer gas.

A peculiar thing about the quasar continuum is that it has no clear scaling with luminosity. Stars on the "main sequence" span a similarly wide range ($\sim 10^6$) in luminosity, but the luminous O5 stars are hot ($\sim 40,000$ K) and blue, while low-luminosity M0 dwarf stars are cool (~ 3500 K) and red. Although this is a weak dependence ($T \propto 10^{1/6}$, on average) it has very little scatter. In quasars instead there is no way to tell the luminosity from the continuum shape. There are weak scalings of the strength of the emission lines, both optical and ultraviolet (the "Baldwin effect"; Baldwin 1977), and in X-rays (Iwasawa and Taniguchi 1993), but there is a large scatter of continuum shapes at all luminosities (this is unfortunate, since if we could tell luminosity independent of the redshift, quasars would give us a way to measure the cosmological numbers (Chapter 17) out to the highest redshifts known). An advantage of the accretion disk explanation is that the resulting power-law emission has only a weak scaling with luminosity. The black-body law produces only a weak relation of maximum temperature on central mass: $T_{max} \propto M^{1/4}$. The scale-free nature of the radio and X-ray continuum still has to be dealt with, but this seems like a promising step.

Outside the optical band the several "power-law" shapes are explained quite differently. In the radio, high polarization and surface brightness rule out a thermal origin. Synchrotron emission from relativistic electrons in a magnetic field is undoubtedly at work. In the large radio "lobes" (Figure 2) this emission has a steep spectrum, because the high-energy electrons have already lost their energy and do not get re-accelerated. In the central "core" radio source (Figure 2) the spectra are flat and this is the result of adding many spectra together to get a different shape, as in accretion disks, only in this case each component spectrum is a synchrotron spectrum that is self-absorbed at low frequencies.

---

*I own up to having coined this ugly term during a talk in Santa Cruz, where I was trying to distinguish it from the "Small Blue Bump" which turned out to be due to thousands of blended emission lines (Wills *et al.* 1985). Unfortunately the term stuck, luckily without attribution, but nevertheless an apology is called for.

In the X-ray, instead, the initial "power-law" description is only a rough approximation that worked at first only because of the low signal-to-noise of the spectra. (The first HEAO 1 and EXOSAT spectra had total detection significances of 20–30σ, allowing only two or three parameters to be determined.) As signal-to-noise improved (notably with the Japanese Ginga and ASCA missions) it became clear that there is rarely *any* place in the X-ray band that shows the "underlying" continuum cleanly (Figure 4): there is always some other feature superimposed. At the lowest energies there is a "soft excess" (that leads to EUV detections), then there is the "ionized absorber" (photoelectric absorption and fluorescent emission from 0.5 keV to ~2–3 keV due to partially ionized gas); at higher energies (4–6.7 keV) there is the "relativistically broadened" Fe K line; at 7 keV there is the Fe K absorption edge, and above 7 keV the "Compton hump" reflection feature begins, ending only when the Klein–Nishina reduction in the electron scattering cross-section begins around 40 keV. Above 40 keV the intrinsic spectrum cuts off somewhere in the 60–600 keV range. Given all this, it is not at all clear that the underlying mechanism really produces a perfect power law.

The high energy cut-off hints at a thermal origin for the X-ray "power law," and we have no X-ray polarization measurements of quasars to diagnose a non-thermal origin. A common model (Haardt and Maraschi 1991) currently is of a hot (~10^8 K, from the observed cut-off energies) corona lying above the accretion disk, which electron scatters a fraction of the disk radiation up to X-ray energies.

4.3 Obscured quasars: unified schemes

The first evidence for "hidden quasars" appeared from an early X-ray survey of the sky (the 1974–1978 Ariel V survey, which was the basis of my thesis). This survey found, among the few hundred sources it could pick up, a surprising number where the X-ray position clearly indicated a galaxy as the X-ray source, even though the optical spectrum of the galaxy did not have the broad quasar emission lines (Ward *et al.* 1978). An easy explanation, which turned out to be correct, is that the fairly high-energy (2–10 kilovolt) X-rays that we were detecting were able to penetrate a layer of gas and dust between us and the central continuum radiation source, but that the optical light from the active nucleus was unable to get through. X-ray spectra soon showed that the low-energy X-rays were absorbed, just as they should be in this picture (Maccacaro *et al.* 1982). Infrared photons are less easily absorbed by dust and, as infrared spectra became available, broad quasar-like lines were seen in that part of the spectrum, confirming that obscuration was important. At the time the importance of this discovery was not appreciated.

Evidence of obscuration in many quasars mounted from 1979 on, and showed some link to the host galaxy, but otherwise remained unexplained until 1984. A single observation of the "type 2" active galaxy, NGC 1068, by Ski Antonucci and Joe Miller (1985) then ushered in a whole new era of what were called "unified schemes" or even "grand unified theories." It was a shock to see a single figure in the abstracts "Bulletin" for the American Astronomical Society meeting in 1984. Not only was it the only figure I could remember in the Bulletin, but also a glance showed that it was a breakthrough result.

Astronomers always classify new objects into two types at the first opportunity. With the types of quasars called "Seyfert galaxies" this happened in 1974 when Edward Khachikian and Dan Weedman found that while all Seyfert galaxies had strong emission lines, some had very broad lines, while others had lines only about twice as wide as might be found in a normal galaxy. Since they found few intermediate widths, they called these "type 1" and "type 2" respectively. NGC 1068 was the nearest and brightest type 2. No-one was really sure if type 2 Seyferts really hosted quasars or not.

The discovery that Antonucci and Miller made was that not only did type 2 Seyferts, and NGC 1068 in particular, host a *bona fide* quasar, but that this quasar was completely hidden from our direct view by a "brick wall" of obscuring material. They were able to see the tell-tale broad lines, by looking in *polarized* light. (This is very hard to do, because

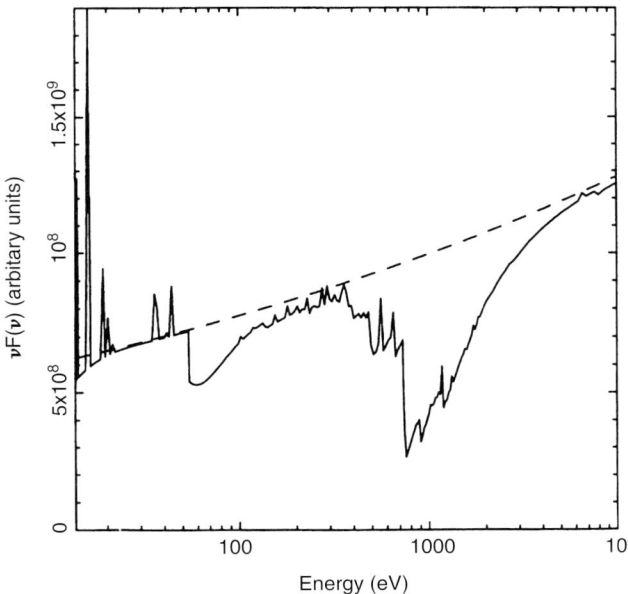

Figure 4 The many features in a quasar X-ray spectrum. From low to high energy: soft excess, ionized absorber, broad iron line, iron edge, Compton hump. (From Reynolds and Fabian 1995.)

even a strong polarization of 10% has to be measured to about 1% to be a good measurement, so it needs 100 times the number of photons that a direct measurement does. (That is, the Lick 3 m diameter telescope they used for polarization was only as effective as an amateur-sized 30 cm telescope doing direct measurements.) Despite this handicap polarization is a powerful tool. Omar Blaes once remarked that the dozen or so polarization papers on quasar emission lines have told us more than the several thousand papers using direct spectroscopy.

In NGC 1068 the polarized light had to be scattered toward us by a gas with many free electrons, and this gas must be in some place that was not "walled off" from the central region of the quasar. There had to be such places, since to induce polarization requires a favored plane, so the obscuring "wall" probably had a flattened, disk- or torus-like geometry. Since it was soon found that even quite high-energy X-rays did not penetrate this "wall" it had to be thick to electron scattering. An extremely strong fluorescent emission line from cold iron was also rapidly found in X-rays, which could only be so strong compared with the continuum if the continuum was heavily absorbed. A spectrum of a highly obscured quasar is shown in Figure 5.

The idea that all quasars (or conservatively all Seyfert galaxies) contained obscuring tori, cold enough to contain molecular material, rapidly caught on for good reasons: several of the confusing classifications of quasar types could be seen as just an accident of the angle at which we happen to view them. The heavy X-ray absorption and strong infrared emission from dust in many quasars could both come from this torus. If the torus covered half or more of the sky (as seen from the central quasar) then it was easy to reproduce the right ratio of type 1 and type 2 objects. Moreover the galaxy-scale cones seen so clearly in HST images could be explained as a shadowing by the torus, so only a cone of the host galaxy was lit up by the quasar continuum. The central idea, that the angle from which we view a quasar determines much of the phenomenology that we see, has been the driving concept behind much of quasar research for the past 15 years.

The result of these "Unified Schemes" has been a great simplification of the many sub-types of quasar; a first clearing away of the confusion by means of a simple viewing-angle parameter. The essence of the scheme is almost certainly correct, though the particulars can be debated.

A close relation between quasars and rapid star formation in galaxy nuclei has been suggested for some time. The IRAS results on quasars (Sanders *et al.* 1989) led to a suggestion of an evolutionary sequence in which quasars and galaxies are born heavily dust-enshrouded, then lose their dust due to heating and radiation pressure from the quasar continuum, emerging as the well-known blue quasars only quite late in their history. At the moment the case for this scenario is rather speculative, but it does give a framework from which to attack the problem.

Some of the ULIRGs found by IRAS turned out to be quasars hidden so thoroughly by dust that it was not possible to tell whether the underlying power source was a quasar, and not an enormous region of rapid star formation, a "starburst." The possibility that a starburst might in fact produce, via supernovae, the broad emission lines characteristic of quasars became an intriguing idea: quasars without black holes (Terlevich and Melnick 1985).

Only the most diligent searches, primarily through far-infrared spectroscopy, have been able to determine the underlying power source in ULIRGs which turn out to be dominated by quasars in some cases and starbursts in others (e.g. Dyson and Lamb 2000).

4.4 High velocities: winds and structure

The broad emission lines that characterize quasars can have widths anywhere from 1000 to 30,000 km s^{-1}, that is, up to $0.1c$. But the simplest knowledge of how the gas is moving to produce these widths – is it dominated by inflow, outflow, rotation, or random motion? – has proved elusive. Eventually greatly improved reverberation mapping, or even interferometric imaging, will provide a definitive answer, but in the meantime we must look for secondary clues.

One clue comes from gas that is seen only by the absorption it produces. Narrow absorption lines are quite commonly superimposed on top of the broad emission lines in ultraviolet spectra. Strikingly, almost all of these lines are shifted toward us, by about 1000 km s^{-1}. Since absorbing gas must lie directly along our line of sight to the continuum source, this gas at least is flowing outward. (For emission lines the gas could be in front, behind, or to the side. The ambiguity arising from this three-dimensional

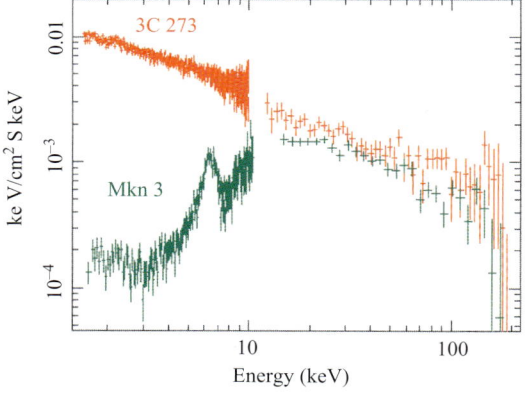

Figure 5 The spectrum of a highly obscured quasar, Markarian 3 (green), compared with an unobscured quasar, 3C 273 (red). (BeppoSAX data.)

integral is a big reason why emission lines have not told us more.) So outflowing winds are a quite normal feature of quasars.

Much more dramatic outflows are seen in about 10% of quasars, the "broad absorption line quasars" (BALs). These BALs show very broad and strong absorption lines that can extend up to $60,000 \text{ km s}^{-1}$ from the emission line peaks, although most are narrower, about $10,000 \text{ km s}^{-1}$. There is a strong suspicion that, although seen in only 10% of quasars, this high-velocity wind is actually present in all quasars, but only in 10% of cases are we looking down the flow. Our viewing angle has complicated the botany.

Has this variety of viewing angles given us a powerful tool? Here is one way of putting together most of these high-velocity gas flows into a single, physically appealing, structure. (Since I proposed this idea, I find it compelling! (Elvis 2000).) The UV and X-ray absorbers seem to require a "shell" of absorbing gas that is much thinner than its distance from the nucleus. Although this gas is outflowing it has not changed its ionization state in decades, against predictions. If the gas lies in a cone-shaped flow crossing our line of sight to the continuum source we can understand this behavior as a steady wind. The same geometry seems to be needed to explain the huge polarization of the BALs, so there may really only be one wind. This wind has to have a funnel-like shape (Figure 6), so that although it starts by rising in a vertical cylinder from a disk, it opens up (perhaps under radiation pressure) as it moves away and accelerates. In this picture the narrow absorption lines are seen by looking through the low end of the funnel, while the broad absorption lines come from looking straight down the fast-moving outer flow. This picture seems to explain many other pieces of the quasar phenomenology.

Even if this idea is wrong in detail, it gives an example of how all the apparently disparate facts about quasars – their absorbers, but also their broad emission lines and the several scattering regions – could be combined into a single structure. This is what gives me great hope for a breakthrough to a new level of understanding of quasars in the near future. Instead of confusing us, the many varieties of quasar can be understood to be telling us about the geometry and kinematics of quasars. Once that is understood the step to a true physical theory seems like a problem that can be tackled.

The highest velocities seen in spectral lines come from the iron K-alpha line which, at rest, is seen at about 6.4 keV in the X-ray band. The ASCA satellite showed that some iron emission lines in quasars are extremely broad, up to $0.3c$ (Tanaka et al. 1995). A simple model using a massive central black hole can explain these lines.

If the X-ray continuum is generated in a corona above the disk then the disk shows the continuum a large solid angle, and provides cool gas that can reprocess the X-ray continuum into the Fe K and Compton hump reflection

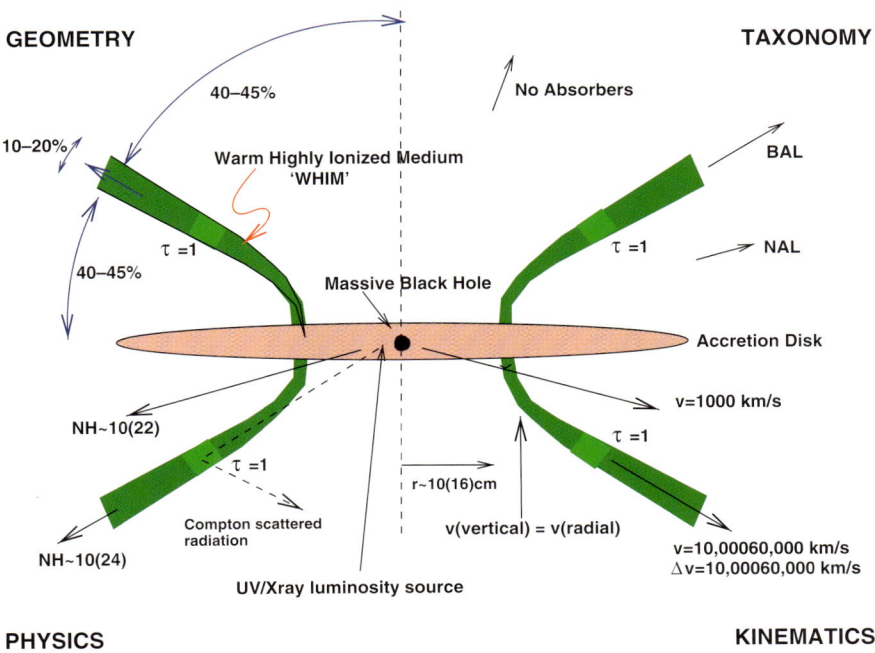

Figure 6 A possible structure for the inner regions of quasars. (From Elvis 2000.)

features. These iron K-alpha lines then come from a rapidly rotating disk at only a few Schwarzchild radii. Doppler broadening, special relativistic boosting, and general relativistic gravitational redshifts will then broaden the line. If this interpretation is correct, then the Fe K line will be the best tool to study black holes to determine at least two of the three properties a black hole can have: mass and spin. (But not, yet, electric charge.)

None of this deals with the highest velocities of all.

4.5 Symmetry and rare radio emission: jets

The highest velocities are evident only in the rare radio-loud quasars (mystery 5). These are all manifestations of relativistically moving jets.

Something had to put energy into the giant radio lobes mapped by early radio observations (Figure 4, mystery 3). Martin Rees suggested jets of relativistic material for this purpose, and predicted that structures in these jets would be seen moving apparently faster than light when viewed end-on, a prediction spectacularly confirmed soon after (Chapter 24). Viewing these jets end-on can produce all the dramatic properties of blazars (Chapter 24). Seen from the side these jets would look like normal, low-power radio galaxies (Urry and Padovani 1995), and better maps show these jets directly in radio galaxies (Figure 4).

So does the presence or absence of a jet decide whether a quasar will be radio loud or not? That is a good approximation, but it cannot be quite so simple. Some radio-quiet quasars show miniature double radio lobes or jets (Blundell and Beasley 1998), and even superluminal motions, although at power outputs 2–3 orders of magnitude fainter. Something must control the efficiency with which power gets into a jet. This might be extrinsic to the particle acceleration region (for example, using low-energy photons to drain energy from the jet electrons by Compton scattering, "Compton drag"; Phinney 1987), or intrinsic to the process (for example, the spin of the black hole regulating the Blandford–Znajek acceleration mechanism). If the spin of the black hole is involved, then why do some black holes spin fast and others not (Wilson and Colbert 1995)? Is this connected with quasar evolution (Section 4.6)?

Blazars and radio galaxies fit into the "Unified Scheme" (Section 4.3) if their jets share the same axis as the obscuring torus. High-redshift radio galaxies showed their own cones of emission lines, which was called the "alignment effect." At first it was assumed that this alignment effect must be due to star formation induced by the shockwave generated by the jet passing through the interstellar medium of the galaxies. Sperello diSerego Alighieri *et al.* (1989) found, however, that this light was highly polarized, and so must be a sign of reflected continuum light from the quasar nucleus. This is very reminiscent of the "Seyfert 2" case (Section 4.3) above. Ski Antonucci (1982) had already shown that the optical light was polarized perpendicular to the radio axis in low-redshift radio galaxies, and so presumably shared the same axis. Radio galaxies (which do not show broad emission lines) were then obscured blazars seen from the side, an appealing bonus for Unified Schemes (Section 4.3).

4.6 Evolution: the co-evolution of galaxies and quasars

Although we have known that quasars were more numerous in the past since 1967 (Schmidt 1967), there is no agreement on why the "age of quasars" is past. Not only is the physics of quasar evolution unknown, with several quite different ideas competing, but we do not even have a good description of how quasars evolve.

There has been a long controversy about why we see more quasars at high redshift ("high" here means from about 1 to 3, that is, when the Universe was a half to a quarter of its present age (for $\Omega = 0$). Is it because the total number was larger, or because the ones we see now were much brighter back then? That is, do they evolve in density or in luminosity? This appeared to have been solved by Brian Boyle and his collaborators back in 1990 in favor of luminosity evolution.

The question matters because, as Sterle Phinney pointed out in his 1983 PhD thesis, the answers have very different implications. If the same quasars have always been present, but have faded, then the massive black holes that quasars trace are rare objects found in a minority of galaxies, and they must by now be extremely massive ($\sim 10^{10}$ solar masses, or 10% of a the stellar mass ($\sim 10^{43}$ g) in a galaxy), to account for the total luminosity of quasars. If instead density evolution applies, then quasars used to be so common that almost every galaxy must harbor a massive black hole, and this allows each black hole to be less giant-sized ($\sim 10^7$ solar masses). Furthermore, if every galaxy has a black hole then perhaps black holes are an inevitable feature of galaxy formation: galaxies and black holes/quasars may co-evolve.

Boyle *et al.* (1988) found a clear break in the distribution of luminosity of the number of quasars, the "quasar luminosity function" as it is called. Above some luminosity the numbers of quasars dropped steeply. They were able to trace how this break in luminosity moved with redshift, and found that it was shifting entirely in luminosity, not number, so luminosity evolution won the day. Moreover, the evolution clearly stopped by a redshift of 3, and Pat Osmer (1982) and others showed that at higher redshifts quasars became rarer again. Some cautionary voices were raised: Emanuale Gialongo and Fausto Vagnetti (1992) showed that the variety of spectra in quasars would bias the evolution calculations, perhaps making a drop in quasar numbers at the highest redshifts unnecessary. Luminosity evolution, though, seemed to fit the radio and X-ray data too, and so was accepted for a decade.

Luminosity evolution does not, however, fit with studies of galaxies. In recent years HST observations have gradually built up measurements of the masses of nuclei in nearby galaxies. These point to there being a "massive dark object" (a carefully non-committal term – everyone has in mind a black hole but this cannot be *proved*) in almost every galaxy. So density evolution ("a black hole in every galaxy") seems favored, contradicting the quasar studies.

Two things have recently changed the picture and these may solve the paradox. First, a new survey that fills some of the redshifts less well covered in 1990, and which uses less restrictive ways to select quasars, has been completed by Lutz Wisotzki (2000). This Hamburg–European Southern Observatory survey finds no break in the luminosity function, and so takes away the marker used to see how the function moved with redshift. Second, the growing realization that as much as 80% of all quasars had been missed because of obscuration by gas and dust means that any results based on the minority 20% could be irrelevant.

In either case, knowing the correct way that the luminosity function moves is just a parameterization of the problem, not a physically based solution. Richard Green (1985) pointed out that population evolution does not equal object evolution. There can be a big gap between the two. A cluster of stars forms a clearly defined Z-like curve on a diagram of color *v*. luminosity (the Hertzprung–Russell diagram). Early on it was assumed, for example by Kelvin, that an individual star moved along the curve (the "main sequence"). We now know that an individual star jumps quickly from one branch of the curve to another as it moves from burning hydrogen to burning helium in its core. The overall curve is produced from having stars of many different masses, and since massive stars burn their fuel faster the stars at one end will all have moved up to a helium-burning state, producing the characteristic bend in the curve. This physics-based explanation is far deeper than the level we are now at with quasars. Progress in cosmology, in particular in the origin of galaxies, and more physically based models of how quasars work, seem likely to combine in the near future to put quasar evolution on a similar physical basis to that of stars.

The HST-based result that every galaxy has a massive black hole at its center is strong, but is not entirely new. Already by 1980 there were strong hints that all galaxies (except perhaps dwarfs) contained at least a weak quasar (Keel 1980). The HST confirmation and extension of this discovery implies a deeper link between the two entities. Massive black holes cannot be just one of the things that might be found in a galaxy nucleus, they are an integral part of what makes a galaxy. An extraordinarily tight relation, with just a few percent scatter, between the mass of the nuclear black hole and the velocity dispersion of the surrounding stars (at radii far beyond where the black hole's gravitational field dominates) has now been found with HST (Ferrarese and Merritt 2000; Gebhardt *et al.* 2000). The galaxy and the quasar co-evolve in some way.

Since quasars are already formed by $z = 5$ it is not clear which came first, the galaxy or the quasar, or whether they both formed together. Only with the emergence of the field of large-scale structure and galaxy formation over the last decade has it become conceivable to form such questions usefully, through increasingly accurate *N*-body simulations.

4.7 High redshifts: applied quasarology

Most of astronomy is forced to search "under the street light," at things that shine, because that is where we can look. The rest is darkness, and how can you study darkness? Quasars give us one way to do so. Quasars can be seen easily across the Universe, currently to redshifts just less than 6 (Fan *et al.* 2001), a time when the Universe was only 5% of its current age (for $\Omega = 1$). The ultraviolet hydrogen Lyman α is seen not at 121.5 nm, but in the far red of the optical spectrum at 850 nm, beyond where the hydrogen Balmer α is seen in the laboratory so quasars can be used as background sources of light to study material that does not make itself obvious by glowing conveniently, particularly cold gas. Anything between us and a quasar acts like bones in a doctor's X-ray: imprinting its presence on the light we receive by absorption. Different elements in different states of ionization will absorb at different wavelengths, and in very different parts of the spectrum. Fortunately quasars emit strongly across the whole spectrum, so with the right choice of telescope we can examine a huge range of conditions in the intergalactic space between us and the distant quasars. Virginia Trimble calls this use of quasars "applied quasarology."

Simply being able to see quasars at redshift 5 in optical light puts strong limits on the density of neutral hydrogen in intergalactic space. Since ultraviolet light shorter than 121.5 nm is absorbed by hydrogen (in the $n = 2 \rightarrow 1$ Lyman α transition), and at redshifts greater than 2.2 this transition enters the optical band, we would see a dramatic cut-off in the spectrum at this (redshifted) wavelength if intergalactic space were filled with even a small density of neutral hydrogen. No such break is universally seen. This is called the Gunn–Peterson effect (Gunn and Peterson 1965) and was used to limit the neutral intergalactic medium (IGM) from 1965. Present limits allow no more than 10^{-5} of the density needed to close the Universe to be in neutral hydrogen (that is, $\Omega_{HI} < 10^{-5}$).

But it is highly unlikely that galaxy formation is so effective that no more than 0.1% of the primordial hydrogen (making the standard assumptions that $\Omega = 1$ and only 10%

of the mass is in normal, baryonic matter) is left in an IGM (Field 1972), so the lack of a Gunn–Peterson effect implies an epoch at which the Universe is re-ionized (after it has become neutral during the Big Bang expansion at $z = 1000$ (the "Dark Ages"). This re-ionization must occur when gravitational contraction has gone far enough to ignite sources of UV photons, either through nuclear burning in stars, or through direct gravitational potential energy release via accretion onto black holes. A huge amount of energy is needed, making the strong ionizing UV continua of early quasars an attractive source. Whether quasars can re-ionize the Universe depends on knowing the true evolution rate of quasars at high z, and on how much of their power comes out at useful wavelengths. Using the standard evolution numbers (Section 4.6) seems not to give enough UV radiation, implying a need for an unobserved first generation of stars. But small observational changes can affect these numbers greatly, as can the true spectral shape to assume in the EUV, which is not readily observed (Section 3.2) and can be quite different, for example, between a stationary accretion disk and a more physical time-evolving disk. Estimates of the *total* power radiated by quasar accretion are easier than estimates of the UV power, and Fabian and Iwasawa (1999) find that, allowing for quasars hidden by dust and gas, the accretion (quasar) luminosity of the Universe is a sizeable, but not equal, fraction of the nuclear (stellar) luminosity. This debate is not yet decided. There is a corresponding helium limit (at 30 nm), which can only be tested from space-based UV observations (and only for fortuitously placed quasars, see below). HST observations by Peter Jakobsen and colleagues (Jacobsen *et al.* 1994) finally found the ionized helium Gunn–Peterson effect in 1994.

In fact though, the IGM had been there in the observations all along, but had not been recognized. The trick is that the IGM is not uniform. Although few quasars show a complete cut-off in their spectra where Ly-α cuts in, all quasars with even a modest ($z > 0.5$) redshift show sharp Ly-α absorption lines, and at high redshifts ($z > 2$) they become so common that they are called the "Lyman α forest" (Figure 7). Since they do not occupy every redshift between us and the quasars, they were thought of as clouds; and since only a wisp of hydrogen (10^{12}–10^{15} cm^{-2}) is needed to produce the observed absorption, then under the simplest assumption – that they are spherical and neutral – then they contain very little mass. These reasonable sounding assumptions were completely wrong. By observing the rare pairs of quasars that happen to lie close to each other (a few arc minutes) on the sky (but at quite different redshifts) Smette and colleagues found (Smette *et al.* 1992) that the same "clouds" were seen in both quasars' spectra, and so were spread over large distances, comparable with clusters of galaxies (10^{24} cm), or other large-scale galaxy structures. Because the "clouds" are so big they have low density, but this means they will be highly ionized by the ambient UV radiation of quasars (and once ionized they will not recombine because of their low density). High ionization in turn implies much larger masses than had been estimated before, a significant fraction of all the baryons in the Universe. The Ly-α forest clouds are now realized to be just moderately high-density fluctuations of the intergalactic medium.

Lyman α absorption lines are seen at larger column densities too. Direct imaging to very faint levels by Jaqueline Bergeron and others (Bergeron *et al.* 1992) showed that some absorption lines (at $N_H \sim 10^{16}$–10^{18} cm^{-2})

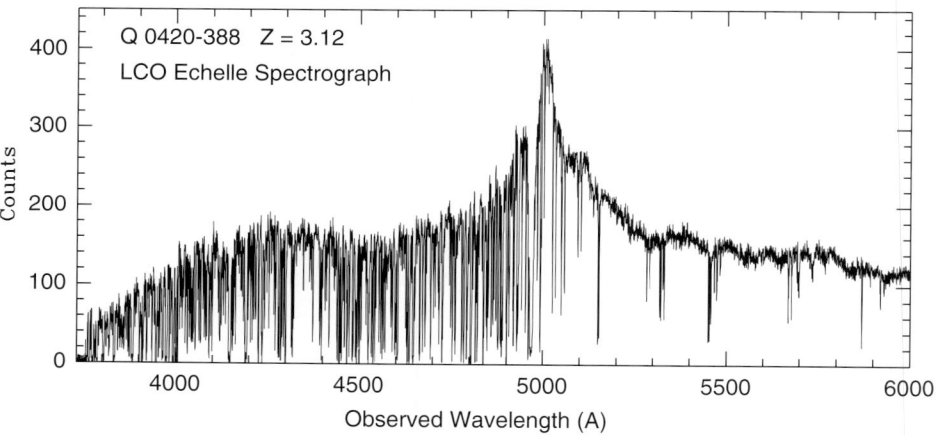

Figure 7 The Ly-α forest of absorption lines. All the gas producing the narrow absorption lines lies to shorter wavelengths than the quasar Ly-α emission line. That is, they are less redshifted, and so closer to us, as they have to be to absorb the quasar light. (J. Bechtold, private communication.)

are due to the outskirts of normal high-redshift galaxies intersecting our line of sight to the quasars. With this discovery we have a tool to study the diffuse gas in galaxies back to the time when most stars were formed (at $Z \sim 2$), and so the history of the recycling of the heavier elements from supernovae back into regions of new star formation can be illuminated using these lines.

More dramatic absorption is seen in a rarer type of broad absorption line (Wolfe 1988). In this case the breadth is not due to Doppler shifts (since absorption lines of other elements at the same redshift are narrow), but are due to the low-probability wings of the Uncertainty Principle widths (called the "damping wings" or the "natural widths" in astronomy). These "damped Ly-α" absorption systems require large column densities ($N_H \sim 10^{19}$–10^{21} cm^{-2}) of hydrogen to make these wings visible, similar to the column densities through the disk of a normal spiral galaxy. The damped Ly-α systems could be early versions of normal galaxies still in formation, "proto-galaxies." Suggestively, the mass in damped Ly-α systems seems to decrease from early times, rather like quasars, until they are rare in the present-day Universe. Back at a redshift of 3, though, these systems held as much mass as is seen in stars and gas found today in galaxies, as though the damped Ly-α systems have been converted into normal galaxies between then and now. This elegant result may be wrong though. Puzzlingly, recent measurements find no decline in dample Ly-α gas toward the present day.

A final use of quasars as background light sources is to examine nearby galaxies, or dark clouds within our own Galaxy. Variable absorption in blazars (Marscher et al. 1993) has been explained by motions of fine-scale structure ($\sim 10^{13}$ cm, roughly the distance from the Earth to the Sun) in an absorbing molecular cloud along our line of sight to the quasar. UV absorption of quasars has had an important role in measuring the state of the outer "atmosphere" (interstellar medium) of the disk of the Milky Way.

4.7.1 The X-ray background

A special case of applied quasarology involves their emission, not absorption.

The first 5 minutes of putting an X-ray detector above the atmosphere (Giacconi et al. 1962) astonished everyone by finding that the sky shone brightly in X-rays. In a sense the whole of the rest of X-ray astronomy has been a quest to find out what could be so hot as to make the whole sky glow. The answer is quasars.

Thanks to ROSAT (Hasinger et al. 1998) and Chandra (Giacconi et al. 2001) we can see that the summed emission from all the quasars in the Universe creates at least 80% of the X-ray background. With low-resolution telescopes all these quasars blur together (there is at least one per square arc minute, or ~ 1000 in an area the size of the Moon), and so they cannot be discerned individually without a telescope that makes arc second images. This blurring together forms the strong X-ray background, against which it is hard to see even much brighter X-ray sources. This reasoning was the push behind X-ray astronomy's constant striving toward higher angular resolution, first Einstein and now Chandra.

The development of X-ray astronomy has paralled that of optical astronomy, but speeded up 20 times. With no source other than the Sun detected before 1962, X-ray astronomy went from a resolution worse than that of the naked eye in 1970 (with the Uhuru satellite, Ariel V, and others) to being comparable with Galileo's telescope of 1610 in 1978 (with the Einstein Observatory and then its successor, ROSAT). (Coincidentially, Einstein had only slightly more effective collecting area than Galileo's telescope, a few tens of square centimeters.) And now, just 20 years on, Chandra has beaten the 1 arc second limit of ground-based telescopes, a step that took optical astronomy 380 years. This final step was essential to answering the question of what made the X-ray background.

The first quasar reported as an X-ray source was 3C273 by the Berkeley group led by Stu Bowyer (Bowyer et al. 1970). (Their motto 'Veritas in Due Sigma' has often proven right.) The first suggestion that quasars produced the X-ray background came from Giancarlo Setti and Lou Woltjer in 1973, using just the X-ray to optical flux ratio of 3C273 and the known number of optical quasars v. brightness. Once Seyfert galaxies were recognized as X-ray sources (Elvis et al. 1978) it was quickly shown (Avni 1978) that this insight was fundamentally correct. However, HEAO 1 results showed that the spectrum of the X-ray background was too flat ($\alpha = 0.4$ ($f_\nu \propto \nu^{-\alpha}$)) compared to the spectra of Seyfert galaxies ($\alpha \sim 0.7$), leading to the "spectral paradox" (Boldt 1988).

Setti and Woltjer (1989) were again the first to suggest the, now conventional, explanation: most quasars are heavily obscured. In X-rays this suppresses the low-energy photons strongly. When summing the contribution of many quasars, some with strong obscuration, some with less and some with none, the result is a spectrum more weighted to high energies, that is, with a flatter slope. Observations with ASCA, BeppoSAX, XMM-Newton, and Chandra seem to bear this out.

5 FUTURE DIRECTIONS

Because quasars dominate the night sky across so many decades of spectrum, it has taken several decades of research to even understand the true scope of the problem. For stars the black-body peak conveniently marched across the one octave of the spectrum that we ourselves see in, and so stars were opened up to understanding over a century ago. Nevertheless the whole process – from the first stellar spectra to a complete physical theory – took almost a century (Lawrence 1987). We are at the point with quasars now that stars had reached around 1910: we can finally see what we are looking at, thanks to the space age. Where do we go from here?

In astronomy theories pass through several stages of abstraction from the data toward a complete "fundamental" theory such as we have for normal stars (Lawrence 1987). Quasars are leaving the descriptive, classification stage, through attempts to unify their confusing phenomenology. This should bring us soon to where we can address the basic quasar mysteries with detailed physics.

5.1 Theory

We think we know the underlying mechanism (gravitational energy release) that powers quasars, but this is a quite superficial understanding, not at all comparable with the precisely calculated chains of nuclear reactions that describe energy generation in stars. Our understanding of stars allows the abundances of the elements to be calculated, and is precise enough that a factor of three discrepancy in the neutrino production of the sun is a real problem.

So the challenge for quasars is to create a physics description of how that gravitational energy is converted into the photons and high-energy particle jets that we observe. Indeed many detailed physics schemes have been worked out for quasars, but they have not cohered. It may be that quasars have been solved, we just do not know it yet, because we cannot decide which is the right solution. We are entering a time when, thanks to unifying schemes, we can describe quasars in terms of physical quantities: mass, accretion rate, specific structures, instead of simple observables (luminosity, spectral shape), which brings us a major step closer to deciding the full physics of quasars.

During the redshift controversy when quasars were first discovered a criticism of the idea that their redshifts were not due to cosmic expansion (that is, the quasars were at "non-cosmological" distances) was that in this case some new physics would be needed to explain their redshifts. This was an effective argument, but it was distinctly unfair and ahistorical. When the main sequence of stars was first delineated there was no physics in existence that could explain their energy production. Quantum mechanics had to be invented first. A great deal though could be understood without a "fundamental" theory of what makes stars shine. The several, complementary, unifying schemes for quasars are at this level. They bring order and comprehension, mostly via geometry and kinematics (for example, viewing angle or relativistic beaming), which allows physics to be applied.

In quasars we imagine that we were lucky to get the new physics we needed – astrophysical black holes – understood only a few years after quasars were discovered. But we may not have the whole story yet. It is even peculiar that such strange objects as massive black holes should be used in current theory only for their potential energy gradients. One mechanism for exploiting their spin energy to power particle jets has been proposed and widely discussed (Blandford and Znajek 1977). Perhaps black holes really do have no other properties to exploit. But perhaps the new excitement in the field of high-energy physics, exploiting many-dimensional structures, will show us that we have been too naive. Certainly the least understood big problem in quasar research is the acceleration of matter to high energies in jets (Chapter 24).

We have seen (Section 4.2) how our view of the quasar continuum was too parochial until space-age observatories opened up the electromagnetic spectrum. The continuum was "too big to be seen." We may have a similar problem with timescales – the "Static Illusion": Quasars are usually treated with disk models that are stationary in time. Quasars certainly appear stationary. With a few notable exceptions (mainly the blazars) quasars flicker only weakly in luminosity. But this is probably deceptive. On timescales of 10^5–10^7 years (much less that the age of the Universe, $\sim 10^{10}$ years) quasar disks should erupt and decay in luminosity by factors of 10^4, just like their low-mass brethren around white dwarfs (Smak 1982). This matters because the structure of an unstable disk is quite different from that of a stable disk, changing the spectrum it emits; with implications for the evolution of quasars that can be quite different. Quasars do not grow overly massive if a disk is only occasionally accreting rapidly.

The other big problem for quasar research is more "astro-" than "physics": what is the relation between quasars and their host galaxies? Here must lie the physical causes of quasar "evolution." The new observational ties between a galaxy and its black hole (Section 4.6), and a clearer structure for quasars to link this to, means that how quasar and host galaxy relate, and how the interaction affects galaxy formation, seem to be set to become a major field in astrophysics. As N-body simulations reach a fine enough scale to treat the processes in newly forming galaxies at a level where black holes can be formed, we can discover how this happens without the process running away, with all the matter flowing into the black hole. This can be re-phrased as "What is the efficiency of black hole formation?" or "How much matter is 'wasted' as galaxy-producing gas?"

5.2 Observation

Space science is nowhere near finished with its contribution to quasars. Here are some quick views of what space observatories will contribute in the next decade or so.

Infrared astronomy has been only weakly represented in this article. The more than 20 year gap between IRAS and NASA's last "great observatory", SIRTF, was filled only with ESA's ISO, which unfortunately was critically less sensitive than expected, and so did not have a great impact on quasar work. SIRTF will have tremendous impact, allowing us to detect faint quasars, study their spectra, and diagnose dust conditions: is there really a torus or not?

Two powerful new X-ray observatories have reached orbit, ESA's large-area XMM, and NASA's high-resolution Chandra. For quasars their greatest contribution is likely be the introduction of real spectroscopy into X-ray astronomy. All the "spectra" up to now have been crude, similar in resolution to a rainbow ($E/\Delta E \sim 10$). These observatories both carry diffraction gratings that give up to 100 times better resolution. This means that X-ray emission, and especially absorption, lines can be studied in quasars at last. The soft X-ray region is rich in atomic transitions, and the impact should be comparable to the introduction of spectroscopy for stars in the late nineteenth century. Early results are living up to this expectation (Kaspi et al. 2000; Kaastra et al. 2000).

However, both missions are small. They have collecting areas no bigger than a good amateur optical telescope, and X-ray photons are much rarer (about 100 times) than optical ones. So until X-ray astronomers can build mirrors that are huge by today's standards, we will always be underexploiting the information coming to us. The ideal would be a mirror with tens of square meters of collecting area (compared to $0.5\,m^2$ for XMM), and better than 1 arc second resolution (in order to see faint sources, and in order to reach the natural spectroscopic limit of the thermally induced widths of the lines). NASA is approaching this ideal by going for large-area ($3\,m^2$) first (but at 15 arc second angular resolution), while ESA is boldly going to very large areas ($30\,m^2$) and good angular resolution (2 arc seconds, still not really good enough). What X-ray astronomy needs is a major investment, not in detectors, but in the making of large, high-resolution mirrors, perhaps even using the active optics techniques pioneered in the optical band. If we could have that capability in hand then the future of X-ray astronomy would be unlimited.

The most ambitious plan of all is to resolve the event horizon around a black hole. This requires micro-arc second or better angular resolution. NASA is also investigating doing this in a truly innovative way, via X-ray interferometry (Cash et al. 2000). For various reasons this micro-arc second resolution cannot be achieved, except at wavelengths shorter than $1\,\mu m$, and is best done in the X-rays because here the mirror separations become "merely" tens of meters. This factor of a million improvement in resolution seems plausible and NASA hopes to test it out in a few years with a pathfinder mission.

6 CONCLUDING REMARKS

It is easy to be pessimistic about quasar research. Quasars as a subject are now 40 years old and, like many of us who reach middle age, have become a little stodgy and a little out of fashion. Hot new topics arise that promise rapid progress, while quasar studies have seemed to offer only elaborations of an already baroque phenomenology. But we have to explain quasars. Astronomy tries to explain what we see in the night sky, and when we throw off our myopic optical attitudes, quasars are what the sky is full of. Fortunately with age comes some small amount of wisdom, which is needed to understand still one of the most puzzling of astronomy's mysteries.

New observatories are now or will soon be delivering radically better data, and plans are being made for another generation of yet grander capabilities across the whole spectrum. "More, better data" is an easy plea to make though. Somewhat more boldly I have asserted that our understanding of quasars is at a breakthrough point, and that a physical understanding of their structure, evolution, and connection with galaxy formation is emerging.

If I am lucky then in a few years these speculations will be seen to be wrong, because quasars, like the rest of the Universe, will continue to surprise us. If I am very lucky, I will have helped to prove myself wrong. Or the mysteries quasars will be solved.

FURTHER READING

These are all review articles or books that make good launching points for further exploration of the quasar literature.

General:

Kembhavi, A.K. and Narlikar, J.V. (1999). *Quasars and Active Galactic Nuclei*, Cambridge University Press.
Krolik, J.H. (1999). *Active Galactic Nuclei*, Princeton University Press, NJ.
Peterson, B. (1998). *An Introduction to Active Galactic Nuclei*, Cambridge University Press.

History:

Hazard, C. (1979). Introduction to quasars and active galactic nuclei. In C. Hazard and S. Mitton (eds), *Active Galactic Nuclei*, Cambridge University Press., p. 1.
Shields, G. (1999). A brief history of active galactic nuclei. *Publications of the Astronomical Society of the Pacific*, **111**, 661–678.
Trimble, V. (1988). *Quarterly Journal of the Royal Astronomical Society*, [possibly: Trimble, V. (1986). Low-luminosity stars. *Quarterly Journal of the Royal Astronomical Society*, **27**, 38–59.]
Weedman, D.E. (1976). Seyfert galaxies, quasars and redshifts. *Quarterly Journal of the Royal Astronomical Society*, **17**, 227–262.

Black hole masses:

Ho, L.C. (1998). Supermassive black holes in galactic nuclei: Observational evidence and some astrophysical consequences. In S.K. Chakrabati (ed),

Observational Evidence for Black Holes in the Universe, Kluwer, Dordrecht.

Kormendy, J. and Richstone, D. (1995). Inward bound – The search for supermassive black holes in galactic nuclei. *Annual Review of Astronomy and Astrophysics*, **33**, 581.

Accretion disks:

Antonucci, R. (1993). Unified models for active galactic nuclei and quasars. *Annual Review of Astronomy and Astrophysics*, **31**, 473–521.

Frank, J., King, A. and Raine D. (1985, 1992). *Accretion Power in Astrophysics*, Cambridge University Press.

Applied quasarology:

Peacock, J.A. (1999). *Cosmological Physics*, Cambridge University Press.

Tytler, D. (1997). (and other articles) In P. Petitjean and S. Charlot (eds), *Structure and Evolution of the Intergalactic Medium from QSO Absorption Line Systems*, Editions Frontieres, Paris.

Unified models:

Urry, C.M. and Padovani, P. (1995). Unified schemes for radio-loud active galactic nuclei. *Publications of the Astronomical Society of the Pacific*, **107**, 803.

Botany:

Lawrence, A. (1987). Classification of active galaxies and the prospect of a unified phenomenology. *Publications of the Astronomical Society of the Pacific*, **99**, 309–334.

REFERENCES

Antonucci, R.R.J. (1982). Optical polarization position angle versus radio source axis in radio galaxies. *Nature*, **299**, 605–606.

Antonucci, R. (1999). Constraints on disks models of the big blue bump from UV/optical/IR observations. In J. Poutanen and R. Svensson (eds), *High Energy Processes in Accreting Black Holes*, ASP Conference Series 161, Astronomical Society of the Pacific, San Francisco, p. 193.

Antonucci, R. and Miller, J.S. (1985). Spectropolarimetry and the nature of NGC 1068. *Astrophysical Journal*, **297**, 621–632.

Antonucci, R., Kinney, A. and Ford, H. (1989). The Lyman edge test of the quasar emission mechanism. *Astrophysical Journal*, **342**, 64–82.

Avni, Y. (1978). Parameter estimation for X-ray sources – Positions. *Astronomy and Astrophysics*, **66**, 307–309.

Baldwin, J.A. (1977). Luminosity indicators in the spectra of quasi-stellar objects. *Astrophysical Journal*, **214**, 679–684.

Bergeron, J., Cristiani, S. and Shaver, P.A. (1992). Discovery of Z of about 1 galaxies causing quasar absorption lines. *Astronomy and Astrophysics*, **257**, 417–424.

Blandford, R.D. and Rees, M.J. (1974). A 'twin-exhaust' model for double radio sources. *Monthly Notices of the Royal Astronomical Society*, **169**, 395–415.

Blandford, R.D. and Znajek R.L. (1977). Electromagnetic extraction of energy from Kerr black holes. *Monthly Notices of the Royal Astronomical Society*, **179**, 433–456.

Blundell, K.M. and Beasley, A.J. (1998). The central engines of radio-quiet quasars. *Monthly Notices of the Royal Astronomical Society*, **299**, 165–170.

Boldt, E. (1988). In Hewitt, A., Burbidge, G. and Fang, L.Z. (eds), *Observational Cosmology*, IAU Symposium 124, Reidel, Dordrecht, p. 611.

Boyle, B.J., Shanks, T. and Peterson, B.A. (1988). The evolution of optically selected QSOs. II. *Monthly Notices of the Royal Astronomical Society*, **235**, 935–948.

Burbidge, G. (1959). Estimates of the total energy in particles and magnetic field in the non-thermal radio sources. *Astrophysical Journal*, **129**, 849–851.

Burbidge, G. and Burbidge, M. (1967). *Quasi-Stellar Objects*, W.H. Freeman, San Francisco.

Cash, W., Shipley, A., Osterman, S. and Joy, M. (2000). Laboratory detection of X-ray fringes with a grazing-incidence interferometer. *Nature*, **407**, 160–162.

Clavel, J., Boksenberg, A., Bromage, G.E., Elvius, A., Penston, M.V., Perola, G.C., Santos-Lleo, M., Snijders, M.A.J. and Ulrich, M.H. (1990). The ultra-compact broad emission line region in NGC4151. *Monthly Notices of the Royal Astronomical Society*, **246**, 668.

Collin-Souffrin, S., Czerny, B., Dumont, A.-M. and Zycki, P.T. (1996). Quasi-spherical accretion of optically thin clouds as a model for the optical/UV/soft X-ray emission of AGN. *Astronomy and Astrophysics*, **314**, 393–413.

Curtis, H.D. (1918). *Publications of Lick Observatory*, 13, 9 (p. 31).

di Serego Alighieri, S., Fosbury, R.A.E., Tadhunter, C.N. and Quinn P.J. (1989). Polarized light in high-redshift radio galaxies. *Nature*, **341**, 307–309.

Dyson and Lamb (2000). *Astrophysics and Space Science*, **266**, Nos. 1–2.

Elvis, M. (2000). A structure for quasars. *Astrophysical Journal*, **545**, 63–76.

Elvis, M., Maccacaro, T., Wilson, A.S., Ward, M.J., Penston, M.V., Fosbury, R.A.E. and Perola, G.C. (1978). Seyfert galaxies as X-ray sources. *Monthly Notices of the Royal Astronomical Society*, **183**, 129–157.

Elvis, M., Wilkes, B.J., McDowell, J.C., Green, R.F., Bechtold, J., Willner, S.P., Oey, M.S., Polomski, E. and Cutri, R. (1994). Atlas of quasar energy distributions. *Astrophysical Journal Supplement*, **95**, 1–68.

Fabian, A.C. and Iwasawa, K. (1999). The mass density in black holes inferred from the X-ray background. *Monthly Notices of the Royal Astronomical Society*, **303**, L34–L36.

Fan X. et al. (1999). High-redshift quasars found in sloan digital sky survey commissioning data. *Astronomical Journal*, **118**, 1–13.

Fan X. et al. (2001). *Astronomical Journal*, **122**, 2833–2849.

Ferrarese, L. and Merritt, D. (2000). A fundamental relation between supermassive black holes and their host galaxies. *Astrophysical Journal*, **539**, L9–L12.

Field, G.B. (1972). Intergalactic matter. *Annual Review of Astronomy and Astrophysics*, **10**, 227.

Fichtel, C. (1994). High-energy gamma-ray observations of active galaxies. *Astrophysical Journal Supplement*, **90**, 917–922.

Fiore, F., La Franca, F., Giommi, P., Elvis, M., Matt, G., Comastri, A., Molendi, S. and Gioia, I. (1999). The contribution of faint active galactic nuclei to the hard X-ray background. *Monthly Notices of the Royal Astronomical Society*, **306**, L55–L60.

Fruscione, A. (1996). X-ray-selected extreme-ultraviolet galaxies: Scraping the bottom of the "invisible" barrel. *Astrophysical Journal*, **459**, 509–519.

Gebhardt, K., Bender, R., Bower, G., Dressler, A., Faber, S.M., Filippenko, A.V., Green, R., Grillmair, Ho, L.C., Kormendy, J., Layer, T.R., Magorrian, J., Pinkney, J., Richstone, D. and Tremaine, S. (2000). A relationship between nuclear black hole mass and galaxy velocity dispersion. *Astrophysical Journal*, **539**, L13–L16.

Giacconi, R., Rosati, P., Tozzi, P., Nonino, M., Hasinger, G., Norman, C., Bergeron, J., Borgani, S., Gilli, R., Gilmozzi, R. and Zheng, W. (2001).

First results from the X-ray and optical survey of the *Chandra* deep field south. *Astrophysical Journal*, **551**, 624–634.

Giacconi, R. and Gursky, H. (eds) (1974). *X-ray Astronomy*, Reidel, Dordrecht.

Giallongo, E. and Vagnetti, F. (1992). The evolution of the QSO luminosity function up to $Z=4$ – The role of K-correction uncertainties. *Astrophysical Journal*, **396**, 411–415.

Ghisellini, G., Padovani, P., Celotti, A. and Maraschi, L. (1993). Relativistic bulk motion in active galactic nuclei. *Astrophysical Journal*, **407**, 65–82.

Ghisellini, G., Celotti, A., Fossati, G., Maraschi, L. and Comastri, A., (1998). A unifying theoretical scheme for gamma-ray bright blazars. *Monthly Notices of the Royal Astronomical Society*, **301**, L51.

Green, R.F. (1986). In G. Swarup and V.K. Kapahi (eds), *Quasars*, IAU Symposium 119, Reidel, Dordrecht, p. 429.

Greenstein, J. and Matthews, T.A. (1963). Redshift of the radio source 3C 48. *Astronomical Journal*, **68**, 279.

Gunn, J.E. and Peterson, B.A. (1965). On the density of neutral hydrogen in intergalactic space. *Astrophysical Journal*, **142**, 1633–1636.

Haardt, F. and Maraschi, L. (1991). A two phase model for the X-ray emission from Seyfert galaxies. *Astrophysical Journal*, **380**, L51–L54.

Hasinger, G., Burg, R., Giacconi, R., Schmidt, M., Truemper, J., Zamorani, G. (1998). The ROSAT Deep Survey. I. X-ray sources in the Lockman Field. *Astronomy and Astrophysics*, **329**, 482–494.

Hazard, C., Mackay, M.B., and Shimmins, A.J. (1963). *Nature*, **197**, 1037.

Hoyle, F. (1948). A new model for the expanding universe. *Monthly Notices of the Royal Astronomical Society*, **108**, 372–382.

Iwasawa, K. and Taniguchi, Y. (1993). The X-ray Baldwin effect. *Astrophysical Journal*, **413**, L15–L18.

Jakobsen, P., Boksenberg, A., Deharveng, J.M., Greenfield, P., Jedrzejewski, R. and Paresce, F. (1994). Detection of intergalactic ionized helium absorption in a high-redshift quasar. *Nature*, **370**, 35–39.

Jennison, R.C. and Das Gupta, M.K. (1953). Fine structure of the extra-terrestrial radio source Cygnus X-1. *Nature*, **172**, 996–997.

Kaastra, J., Mewe, R., Liedahl, D.A., Komossa, S. and Brinkman, A.C. (2000). X-ray absorption lines in the Seyfert 1 galaxy NGC 5548 discovered with Chandra-LETGS. *Astronomy and Astrophysics*, **354**, L83–L86.

Kaspi, S., Brandt, W.N., Netzer, H., Sambruna, R., Chartas, G., Garmire, G.P. and Nousek, J.A. (2000). Discovery of narrow X-ray absorption lines from NGC 3783 with the Chandra High Energy Transmission Grating Spectrometer. *Astrophysical Journal*, **535**, L17–L20.

Keel, W.C. (1980). Inclination effects on the recognition of Seyfert galaxies. *Astronomical Journal*, **85**, 198–203.

Khachikian, E.Ye. and Weedman, D.W. (1974). An atlas of Seyfert galaxies. *Astrophysical Journal*, **192**, 581–589.

Kinney, A. (1992). Lyman edges – Signatures of accretion disks. In S.S. Holt, S.G. Neff and C.M. Urry (eds), *Testing the AGN paradigm*, Proceedings of the 2nd Annual Topical Astrophysics Conference, University of Maryland, College Park, Oct. 14–16, 1991, AIP Conference Series 254, American Institute of Physics, New York, pp. 139–145.

Korista, K., Alloin, D., Barr, P., Clavel, J., Cohen, R.D., Crenshaw, D.M., Evans, I.N., Horne, K., Koratkar, A.P., Kriss, G.A., Krolik, J.H., Malkan, M.A., Morris, S.L., Netzer, H., O'Brien, P.T., Peterson, B.M., Reichert, G.A., Rodriguez-Pascual, P.M., Wamsteker, W., Anderson, K.S.J., Axon, D.J., Benitez, E., Berlind, P., Bertram, R., Blackwell, J.H., Jr., Bochkarev, N.G., Boisson, C., Carini, M., Carrillo, R., Carone, T.E., Cheng, F.-Z., Christensen, J.A., Chuvaev, K.K., Dietrich, M., Dokter, J.J., Doroshenko, V., Dultzin-Hacyan, D., England, M.N., Espey, B.R., Filippenko, A.V., Gaskell, C.M., Goad, M.R., Ho, L.C., Huchra, J.P., Jiang, X.J., Kaspi, S., Kollatschny, W., Laor, A., Luminet, J.-P., MacAlpine, G.M., MacKenty, J.W., Malkov, Yu.F., Maoz, D., Martin, P.G., Matheson, T., McCollum, B., Merkulova, N., Metik, L., Mignoli, M., Miller, H.R., Pastoriza, M.G., Pelat, D., Penfold, J., Perez, M., Perola, G.C., Persaud, J.L., Peters, J., Pitts, R., Pogge, R.W., Pronik, I., Pronik, V.I., Ptak, R.L., Rawley, L., Recondo-Gonzalez, M.C., Rodriguez-Espinosa, J.M., Romanishin, W., Sadun, A.C., Salamanca, I., Santos-Lleo, M., Sekiguchi, K., Sergeev, S.G., Shapovalova, A.I., Shields, J.C., Shrader, C., Shull, J.M., Silbermann, N.A., Sitko, M.L., Skillman, D.R., Smith, H.A., Smith, S.M., Snijders, M.A.J., Sparke, L.S., Stirpe, G.M., Stoner, R.E., Sun, W.-H., Thiele, U., Tokarz, S., Tsvetanov, Z.I., Turnshek, D.A., Veilleux, S., Wagner, R.M., Wagner, S.J., Wanders, I., Wang, T., Welsh, W.F., Weymann, R.J., White, R.J., Wilkes, B.J., Wills, B.J., Winge, C., Wu, H. and Zou, Z.L. (1995). Steps toward determination of the size and structure of the broad-line region in active galatic nuclei. 8: An intensive HST, IUE, and ground-based study of NGC 5548. *Astrophysical Journal Supplement*, **97**, 285–330.

Kormendy, J. and Richstone, D. (1995). Inward bound – The search for supermassive black holes in galactic nuclei. *Annual Review of Astronomy and Astrophysics*, **33**, 581–624.

Kwan, J. and Krolik, J.H. (1981). The formation of emission lines in quasars and Seyfert nuclei. *Astrophysical Journal*, **250**, 478–507.

Lawrence, A. (1987). Classification of active galaxies and the prospect of a unified phenomenology. *Publications of the Astronomical Society of the Pacific*, **99**, 309–334.

Lawrence, A., Watson, M.G., Pounds, K.A. and Elvis, M. (1987). Low-frequency divergent X-ray variability in the Seyfert galaxy NGC 4051. *Nature*, **325**, 694–696.

Lynden-Bell, D. (1969). Galactic nuclei as collapsed old quasars. *Nature*, **223**, 690–694.

Maccacaro, T., Perola, G.C. and Elvis, M. (1982). X-ray observations with the Einstein Observatory of emission-line galaxies. *Astrophysical Journal*, **257**, 47–55.

Magorrian, J., Tremaine, S., Richstone, D., Bender, R., Bower, G., Dressler, A., Faber, S.M., Gebhardt, K., Green, R., Grillmair, C., Kormendy, J. and Lauer, T. (1998). The demography of massive dark objects in galaxy centers. *Astronomical Journal*, **115**, 2285–2305.

Malkan, M. and Sargent, W.L.W. (1982). The ultraviolet excess of Seyfert 1 galaxies and quasars. *Astrophysical Journal*, **254**, 22–37.

Marscher, A.P., Moore, E.M. and Bania, T.M. (1993). Detection of AU-scale structure in molecular clouds. *Astrophysical Journal*, **419**, L101–L104.

Mathur, S., Elvis, M. and Wilkes, B. (1999). Multiple velocity components in the C IV absorption line of NGC 5548. *Astrophysical Journal*, **519**, 605–609.

Matthews, T.A. and Sandage, A.R. (1963). Optical identification of 3c 48, 3c 196, and 3c 286 with stellar objects. *Astrophysical Journal*, **138**, 30–56.

Netzer, H., Wamsteker, W., Wills, B.J. and Wills, D. (1985). The ultraviolet spectra of active galaxies with weak optical Fe II lines. *Astrophysical Journal*, **292**, 143–147.

Oke, B. (1963). Absolute energy distribution in the optical spectrum of 3C 273. *Nature*, **197**, 1040–1041.

Osmer, P.S. (1982). Evidence for a decrease in the space density of quasars at Z more than about 3.5. *Astrophysical Journal*, **253**, 28–37.

Perley, R.A., Dreher, J.W. and Cowan, J.J. (1984). The jet and filaments in Cygnus A. *Astrophysical Journal*, **285**, L35–L38.

Peterson, B.M., Balonek, T.J., Barker, E.S., Bechtold, J., Bertram, R., Bochkarev, N.G., Bolte, M.J., Bond, D., Boroson, T.A., Carini, M.T., Carone, T.E., Christensen, J.A., Clements, S.D., Cochran, A.L., Cohen, R.D., Crampton, D., Dietrich, M., Elvis, M., Ferguson, A., Filippenko, A.V., Fricke, K.J., Gaskell, C.M., Halpern, J.P., Huchra, J.P., Hutchings, J.B., Kollatschny, W., Koratkar, A.P., Korista, K.T., Krolik, J.H., Lame, N.J., Laor, A., Leacock, R.J., MacAlpine, G.M., Malkan, M.A., Maoz, D., Miller, H.R., Morris, S.L., Netzer, H., Oliveira, C.L.M., Penfold, J., Penston, M.V., Perez, E., Pogge, R.W., Richmond, M.W., Romanishin, W., Rosenblatt, E.I., Saddlemyer, L., Sadun, A., Sawyer, S.R., Shields, J.C., Shapovalova, A.I., Smith, A.G., Smith, H.A., Smith, P.S., Sun, W.-H., Thiele, U., Turner, T.J.,

Veilleux, S., Wagner, R.M., Weymann, R.J., Wilkes, B.J., Wills, B.J., Wills, D. and Younger, P.F. (1991). Steps toward determination of the size and structure of the broad-line region in active galactic nuclei. II – An intensive study of NGC 5548 at optical wavelengths. *Astrophysical Journal*, **368**, 119–137.

Phinney, S. (1983). A theory of radio sources. PhD Thesis, University of Cambridge.

Phinney S. (1987). How fast can a blob go? In Zensus, J.A. and Pearson, T.J. (eds), *Superluminal Radio Sources*, Cambridge University Press, pp. 301–305.

Pogge, R.W. (1988). A extended ionizing radiation cone from the nucleus of the Seyfert 2 galaxy NGC 1068. *Astrophysical Journal*, **328**, 519–522.

Reynolds, S. and Fabian, A.C. (1995). *Monthly Notices of the Royal Astronomical Society*, **273**, 1167–1176.

Sagan, C. (1979). *Broca's Brain: Reflections on the Romance of Science*, Balantine, New York, p. 73.

Sandage, A. (1965). The existence of a major new constituent of the Universe: The quasistellar galaxies. *Astrophysical Journal*, **141**, 1560–1578.

Sanders, D.B., Phinney, G., Neugebauer, G., Soifer, B.T. and Matthews, K. (1989). Continuum energy distributions of quasars. *Astrophysical Journal*, **347**, 29–51.

Schinnerer, E., Eckart, A., Tacconi, L.J., Genzel, R. and Downes, D. (2000). Bars and warps traced by the molecular gas in the seyfert 2 galaxy NGC 1068. *Astrophysical Journal*, **533**, 850–868.

Schmidt, M. (1963). 3C 273: A star-like object with large red-shift. *Nature*, **197**, 1040.

Schmidt, M. (1967). Space distribution of quasi-stellar radio sources. *Publications of the Astronomical Society of the Pacific*, **79**, 437.

Schmidt, M. (1990). Discovery of quasars. In R. Bertotti, R. Balbinot, S. Bergia and, A. Messina (eds), *Modern Cosmology in Retrospect*, Cambridge University Press.

Schneider, D.P., van Gorkom, J.H., Schmidt, M. and Gunn, J.E. (1992). Radio properties of optically selected high-redshift quasars I. VLA observations of 22 quasars at 6 cm. *Astronomical Journal*, **103**, 1451–1456.

Setti, G. and Woltjer L. (1989). Active galactic nuclei and the spectrum of the X-ray background. *Astronomy and Astrophysics*, **224**, L21–L23.

Seyfert, C.K. (1943). Nuclear emission in spiral nebulae. *Astrophysical Journal*, **97**, 28–40.

Shakura, N. and Sunyaev, R. (1973). Black holes in binary systems. Observational appearance. *Astronomy and Astrophysics*, **24**, 337–355.

Shields, G.A. (1978). Thermal continuum from accretion disks in quasars. *Nature*, **272**, 706–708.

Smak, J.I. (1982). Accretion in cataclysmic binaries. I – Modified alpha-disks with convection. *Acta Astronomica*, **32**, 199–211.

Smette, A., Shaver, P.A., Foltz, C.B., Chaffee, F.H., Weymann, R.J., Williams, R.E. and Magain, P. (1992). A spectroscopic study of UM 673 A and B – On the size of Lyman-alpha clouds. *Astrophysical Journal*, **389**, 39–62.

Stockton, A. (1973). QSOs – Galaxy associated with 4C37.43. *Nature*, **246**, 25.

Tadhunter, C. and Tsvetanov, Z. (1989). Anisotropic ionizing radiation in NGC 5252. *Nature*, **341**, 422–424.

Tanaka, Y. *et al.* (1995). *Nature*, **375**, 659.

Terlevich, R. and Melnick, J. (1985). Warmers – The missing link between starburst and Seyfert galaxies. *Monthly Notices of the Royal Astronomical Society*, **213**, 841–856.

Turner, T.J. and Pounds, K.A. (1989). The EXOSAT spectral survey of AGN. *Monthly Notices of the Royal Astronomical Society*, **240**, 833–880.

Ueda, Y., *et al.* (1998). *Nature*, **495**, 435.

Urry, C.M. and Padovani, P. (1995). Unified schemes for radio-loud active galactic nuclei. *Publications of the Astronomical Society of the Pacific*, **107**, 803.

Ward, M.J., Wilson, A.S., Penston, M.V., Elvis, M., Maccacaro, T. and Tritton, K.P. (1978). Optical identifications of extragalactic X-ray sources. *Astrophysical Journal*, **223**, 788–797.

Webster, R.L., Francis, P.J., Peterson, B.A., Drinkwater, M.J. and Masci, F.J. (1995). Evidence for a large undetected population of dust-reddened quasars. *Nature*, **375**, 469.

Weedman, D. (1976). Seyfert galaxies, quasars and redshifts. *Quarterly Journal of the Royal Astronomical Society*, **17**, 227–262.

Wilson, A.S. and Colbert, E.J.M. (1995). The difference between radio-loud and radio-quiet active galaxies. *Astrophysical Journal*, **438**, 62–71.

Wisotzki, L. (2000). The bright end of the QSO luminosity function. *Astronomy and Astrophysics*, **353**, 853–860.

Wolfe, A.M. (1988). Damped Ly-alpha absorption systems. In J.C. Blades, D. Turnshek and C.A. Norman (eds), *QSO Absorption Lines: Probing the Universe*, Cambridge University Press, pp. 297–317.

Zamorani, G., Henry, J.P., Maccacaro, T., Tananbaum, H., Soltan, A., Avni, Y., Liebert, J., Stocke, J., Strittmatter, P.A., Weymann, R.J., Smith, M.G. and Condon, J.J. (1981). X-ray studies of quasars with the Einstein Observatory. II. *Astrophysical Journal*, **245**, 357–374.

24

GABRIELE GHISELLINI*

Blazars

Up to the seventeenth century, the unaided eye was the only receiver that humanity could use to observe the Universe. Evolution was able to adapt this "instrument" to be sensitive to the light of the star we happen to be orbiting, the Sun. The invention of the telescope amplified the sensitivity of the eye and its angular resolution, letting humanity discover, less than a century ago, that other galaxies exist, far beyond the Milky way, and that these galaxies are moving apart: the Universe expands. However, all we could discover using the eye and its extension, the optical telescope, regarded a tiny, very tiny, part of the entire electromagnetic spectrum. As soon as technology enabled us to open new windows, we discovered other phenomena, other objects, and could push our knowledge farther out in space and time.

The opening of the radio window made the 1960s the golden decade for astronomy, with the discovery of the microwave background and of pulsars. The third great discovery made in that decade was that of quasars (Chapter 23). The term "quasar" originally stood for "quasi-stellar radio source." In fact when an optical telescope is pointed towards the direction of these radio sources, which can be as extended as minutes of arc in radio maps, the resulting optical plate shows a source which looks like a star, that is, an unextended, or "pointlike" object. This apparently innocuous point is instead a gigantic energy plant, able to produce much more power than an entire galaxy like our own, in a volume which is extremely small, if compared with the dimension of the Galaxy, and comparable with our Solar System. The process that powers the stars, thermonuclear reactions, is not efficient enough to power quasars. We believe that at the core of these sources there is a massive black hole, with a mass between a million and a billion that of the Sun. Matter around the hole is attracted by the black hole gravity, it is compressed, heated, and then it radiates. This was realized quite early (Salpeter 1964; Zeldovich 1964; Lynden-Bell 1969; Shakura and Sunjaev 1973).

Another major advance came with the opening of the X-ray window, first (in the 1960s) with rocket experiments pioneered by Riccardo Giacconi, Bruno Rossi, and others, and then with the first X-ray satellites in the early 1970s. The Uhuru, Ariel 5, HEAO 1 and then Einstein satellites made clear that all kinds of quasars were strong X-ray emitters: at the same time, it started to be believed that quasars were perhaps the major contributors to the cosmic diffuse X-ray background, already discovered in 1962 (Giacconi et al. 1962).

A third qualitative "jump" was the improvements of the interferometric technique in the radio band, again in the early 1970s. Radio telescopes in different continents, looking at the same source, can resolve details as close as a few tenths of a millisecond of arc. This enabled us to discover that some radio-emitting quasars have spots of radio emission which are observed to move. Sometimes this motion corresponds to velocities that exceed the speed of light. Far from challenging special relativity, this "superluminal" motion, as it is now called, was even predicted before it was observed, by Martin Rees in 1966, and corresponds to real fast motion (but slower than the velocity of light!) at an angle close to our line of sight.

The fourth advance came with the Compton Gamma Ray Observatory (CGRO, launched in 1991) which discovered that the subclass of quasars called blazars are very strong γ-ray emitters, producing at energies greater than 100 MeV more power than in the rest of their electromagnetic spectrum. The properties of blazars are very extreme, and we believe that this is partly due to special relativistic effects because their emitting plasma is moving with a bulk motion at large ($v \geq 0.99c$) speeds. Blazars and gamma-ray bursts are indeed the realm of special relativity: effects that

---

* Osservatorio Astronomico di Brera, Merate, Italy

were of academic interest up to 10 years ago suddenly became the keys to understand their behavior.

The discovery of quasars and blazars literally opened up the scale at which we can see the Universe, since they are so powerful that they can be easily seen up to great distances. For 30 years quasars were the most distant known objects in the Universe. And studying these sources we study the effects of both general and special relativity, that is, how space and time change as measured in different frames.

BL LACERTAE OBJECTS AND BLAZARS

The "strange star" BL Lacertae

The star BL in the constellation of Lacertae had been known for many years from its optical variability. When the radio source VRO 42.22.01 was associated with it, people thought they had discovered the first radio star. But unlike other stars, its optical spectrum was completely featureless and this puzzled astronomers. Lots of other things were peculiar about BL Lacertae. The shape of its spectrum did not look thermal, but it was a power law (that is, $F(\nu) \propto \nu^{-\alpha}$). In addition, the optical light was highly polarized, and a high degree of polarization was taken as the signature of synchrotron emission, produced by ultrarelativistic electrons spiraling along magnetic field lines. A lot of observing time was invested in this source to obtain deeper images and spectra. In the end weak absorption lines were observed, from which a redshift of $z = 0.07$ was obtained (Oke and Gunn 1974; Miller and Hawley 1977). This established that BL Lacertae was not a star after all, but a type of quasar.

After the discovery of this prototype, other "stars" were found with similar characteristics, which were named "BL Lacertae objects," or "Lacertids" (or "BL Lacs" for short). At the same time, it was realized that other radio sources, showing the broad emission lines of typical quasars, were also extremely variable in the optical and were named OVVs (optically violent variables) because of this. Their optical spectrum does not generally show the flattening towards the UV indicating the presence of the thermal component due to the blue bump (the thermal emission from the accretion disk) and is therefore non-thermal in origin. Another class of radio-loud quasars, the HPQs (high polarization quasars) showed polarization at a level greater than 3%, and it was finally realized that HPQs were also optically variable, that is, OVVs and HPQs were the same class. It was then found that both BL Lacertae objects and HPQs have a flat radio spectrum, at least at high frequencies, above 1 GHz. These similarities led to the definition of the general class of *blazars* (contraction of BL Lacs and radio quasars). At this time (the 1980s) the LPQs (low polarization quasars) were left out, according to the belief that the polarization characteristics were peculiar and important. But this belief became progressively weaker, up to the discovery of the (often dominant) γ-ray emission in both blazars and LPQs. Then LPQs joined the class of blazars in the 1990s. Linear polarization in the radio band, when observed, is an important tool to diagnose the absence of any cold electron component: if present, these electrons could depolarize the flux due to Faraday rotation of the polarization angle (Wardle 1977). This argument does not apply in the case of electron and positron pairs (they make the polarization angle rotate in opposite directions, and the depolarization effect cancels out).

All blazars so far can be classified as radio-loud objects, but the level of radio emission, with respect to the optical, varies widely. In fact objects selected through X-ray surveys (Extended Medium Sensitivity Survey (EMSS) (Wolter et al. 1991), the Einstein SLEW survey (Perlman et al. 1996), and the Rosat All Sky Survey (RASS)) can be considered as radio weak, since they have a ratio between the radio flux and the X-ray or optical flux much smaller than the blazars found in radio surveys.

To classify a source as a blazar is not an easy task. There is, however, a useful theoretical definition: There are sources characterized by a relativistic jet producing non-thermal radiation. This radiation is Doppler boosted along the velocity direction, coincident with the jet axis. We call *blazar* a source whose non-thermal radiation dominates over the isotropic components (for example, thermal emission from the accretion disk and nonthermal radio emission from extended regions).

Observationally, we have the following properties of blazars:

- Rapid variability – The flux of all blazars varies violently, with large amplitudes in short times. Variations of a factor of two in hours are common at X- and γ-ray energies, slightly smaller variations are seen in the optical in a single night. Some objects have varied by more than two orders of magnitude over a few years. There are many types of variability: there are objects whose flux varies continuously at all frequencies, without any quiescent phase. There are other objects undergoing violent outburst phases from normally quiescent states. Therefore, if an object is not continuously monitored for a long time, it may be difficult to detect its violent variability.
- Emission at all wavelengths – Blazars are active emitters all across the electromagnetic spectrum, from radio to γ-rays. If we plot the quantity νF_ν versus ν their spectrum is characterized by two very broad humps. The peak energies are located in the IR–soft X-ray band and in the MeV–GeV (and even TeV) bands, respectively.

- Non-thermal continuum – Most of the emission (and power) has a power-law shape in restricted energy ranges (a mere decade or so), with weak or absent blue bump.
- Flat radio spectrum – All known blazars, above 1 GHz, have a flat radio spectrum dominated by the core of the jet.

JETS AND FAST MOTIONS

We are now used to seeing beautiful radio maps which can "zoom" in a radio source from the megaparsec to the parsec scale, showing a jet which remains collimated at all scales. Often there are lobes and hot-spots of more intense radio emission at the ends of the jets. There the energy accumulates through million of years, and only a small part of it is radiated away, the rest is used to make the lobe grow and advance in the intergalactic medium.

In the beginning, it was not clear at all how the energy could be transported from the very small center of radio galaxies to regions million of light years away. Then Martin Rees, in his PhD thesis (1966), made the suggestion that this job could be done by matter moving relativistically in a jet. The spectacular prediction he made, out of his idea, was superluminal motion: blobs of radio-emitting matter apparently moving at a velocity greater than the light speed.

Five years later, with the completion of the first Very Long Baseline Interferometry (VLBI), superluminal motion was indeed observed in the blazar 3C 279 (Figure 1; Whitney et al. 1971).

Soon after the discovery of the superluminal motion in 3C 279, 3C 273, and other sources, alternative theories were challenging the idea of Rees. One of the most debated was the so called "Christmas tree" theory: different regions of the source light up at different times, giving us the (false) impression of motion. But observations made it clear that what was observed was always a motion from the center to the outer parts of the source, while the "Christmas tree" model predicts also contraction "motions." Furthermore, if the source is moving relativistically, then its radiation is beamed in the direction of motion, as described below, and we indeed have strong evidence for such beaming.

Superluminal motion

According to the now accepted explanation of superluminal motion, the apparent speed $\beta_{app} = cv_{app}$ of the radio blobs which are observed to move is due to a projection effect. Once the constancy of the speed of light is accepted, superluminal motion can be explained by simple geometry, with no further involvment of special relativity.

The key point is that the source, while emitting photons, is moving towards the direction of the observer. Suppose we take two maps of the same radio source, separated by a time interval Δt_{obs}. When the source emits the photons of the first map, it is located at the position A (Figure 2). After a time Δt_{em}, the source is located at B. The segment AB = $c\Delta t_{em} - \beta c \Delta t_{em}$, and the distance of the two light fronts in the direction of the line of sight is HC = $c\Delta t_{em} - \beta c \Delta t_{em} \cos\theta = c\Delta t_{em}(1 - \beta\cos\theta)$. This translates to a time interval $\Delta t_{obs} = \Delta t_{em}(1 - \beta\cos\theta)$. The two maps show the "blob" in two different positions, whose projected distance, HB, is $\beta c \Delta t_{em} \sin\theta$. We then obtain $\beta_{app} = \beta \sin\theta / (1 - \beta\cos\theta)$. For small viewing angles (a few degrees) and for relativistic speeds ($v \sim c$), β_{app} can be greater than c.

Definition and importance of beaming

It is believed that in blazar jets the plasma is flowing with a bulk speed βc close to the speed of light. In terms of the

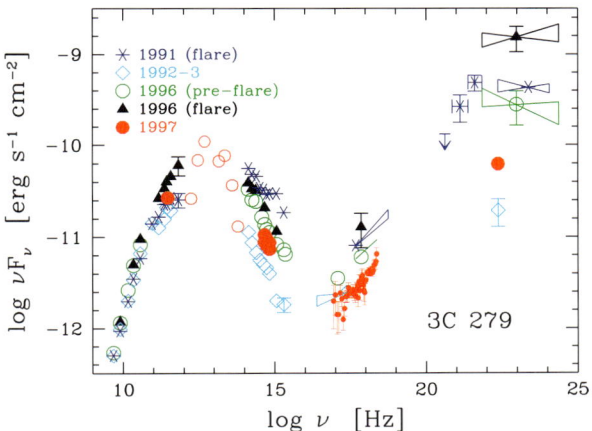

Figure 1 The spectrum of 3C 279 at various epochs. Note the extremely variable flux at all energies, particularly in the CGRO band (0.1–10 GeV). (Adapted from Wehrle et al. 1998 and Maraschi et al. 1998.)

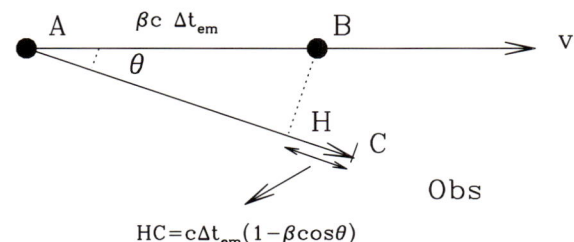

Figure 2 Explanation of the observed superluminal motion. When in A, a "blob" emits a photon. After a time Δt_{em}, the source is in B and emits a second photon. The two photons are separated by HC = $c\Delta t_{em}(1 - \beta\cos\theta)$ and arrive at the observer separated by a time interval $\Delta t_{obs} = \Delta t_{em}(1 - \beta\cos\theta)$.

corresponding Lorentz factor, $\Gamma \equiv (1-\beta^2)^{-1/2}$, we derive values between 5 and 20. Therefore it is essential, in studying blazars, to take into account relativistic effects, the most important of which is called the *beaming* effect. This is the sum of three effects:

(1) Assume that in the frame co-moving with the plasma, the radiation is emitted isotropically. Then half of the photons are emitted in the forward directions, half in the backward directions. An observer, however, sees the plasma moving and so will see that half of the photons are contained in a cone of semi-aperture angle equal to $1/\Gamma$. This is due to the aberration of light.

(2) Since the source of radiation is moving, we must take into account the change of frequency of the observed photons (Doppler effect).

(3) The rate of arrival of photons is not equal to the rate of emission, as measured in the co-moving frame (if we run towards a source, we will collect more photons per second than if we do not run).

These three effects also depend on the angle between our line of sight and the velocity vector of the plasma. They conspire to make the emitting source much brighter if we see it at small viewing angles (photons are concentrated in this small solid angle, they are seen blueshifted, with an enhanced rate of arrival), and much dimmer at large viewing angles. Therefore the (frequency integrated) intensity I of a moving source depends both on its velocity and the viewing angle: $I = \delta^4 I'$ where I' is the intensity seen by a co-moving observer, and $\delta \equiv 1/[\Gamma(1-\Gamma\cos\theta)]$ is the so called Doppler boosting factor. We can see that the intensity can be enhanced by orders of magnitudes when θ is small (a few degrees) and $\beta \sim c$.

Note that relativistic motion greatly affects the variability timescales we measure. Suppose that in the co-moving frame the source flares in a time $\Delta t'$. According to special relativity, this time would be seen, at Earth, longer by a factor Γ. But this is true *only if we do not use photons* to measure this time! If instead we do, then we must take into account that the source has emitted the first photons when it was located in a certain region, and the last photons when it was located *in another position*. Therefore the two bunches of photons traverse paths of different length to reach us. This is the Doppler effect, and contributes a factor $1-\beta\cos\theta$. The two effects combine to give $\Delta t = \Delta t' \Gamma (1-\beta\cos\theta) = \Delta t'/\delta$: for small viewing angles the observed time is *shorter*, not longer!

The enhancement of the observed radiation by beaming can be tested through the following argument. If one assume that the radio emission, as mapped by VLBI techniques, is produced by the synchrotron process in a stationary source, then one can calculate how many electrons are needed to produce the radio emission we see. Knowing directly the size by VLBI, one derives the particle and the photon densities. Therefore one can estimate the probability of interactions between synchrotron photons and relativistic electrons, predicting the flux at high energies, especially in the X-ray band. These estimates often turn out to be orders of magnitude greater than observed. If we now assume that instead the source is moving towards us at relativistic speeds, the fluxes we receive are enhanced, therefore less electrons are needed to produce this emission, and the calculated photon density at the source is lower. The interaction probability is less than before. The estimate of the X-ray flux then depends on the Doppler factor: by matching the predicted X-ray flux with the observed flux we can put an upper limit to the amount of beaming. It is an upper limit because the radio region we are observing at the VLBI may not be the only contributor to the X-ray flux.

Radiation processes

In tenuous plasmas interactions between particles (which depend on their density) are not efficient in sharing the energy between the particles. If the injection of energy is due to the acceleration of few ultrarelativistic particles, then the particle distribution has no chance to become thermal (that is, described by a Maxwellian distribution). In rarefied relativistic plasmas the most efficient ways to produce radiation are the synchrotron and the inverse Compton processes. The first depends on the strength of the magnetic field, the second on the density of the seed photons to be scattered. When these seed photons are produced by the synchrotron process the resulting mechanism is called the synchrotron self-Compton process. For 20 years this was thought to be the main mechanism responsible for the observed radiation, from the radio to the X-ray band, until the high-energy γ-ray emission was discovered (e.g. Ginzburg and Syrovatskii 1965; Rees 1967; Blumenthal and Gould 1970; and Jones et al. 1974a,b). The origin of the high-energy γ-ray emission is still controversial, since the contribution of radiation produced externally to the jet is enhanced by the bulk motion (see below) and can overtake the locally produced synchrotron radiation as a producer of seed photons to be scattered (Maraschi et al. 1992; Dermer and Schlickeiser 1993 Sikora et al. 1994; Blandford and Levinson 1995; Ghisellini and Madau 1996). It is also fair to mention that even the inverse Compton nature of the γ-ray emission is controversial: as Mannheim (1993) suggested, it could be due to the synchrotron process, due to an additional population of ultrarelativistic electrons as a result of photon–meson interactions. Independent of its nature, the high-energy emission (in the X- and γ-ray bands) firmly established the need for relativistic bulk motion.

How fast can a jet go?

There is as yet no general agreement on how jets can originate and be accelerated to relativistic speeds. One of the first ideas to accelerate jets was to use the radiation pressure of the accretion disk. If the matter content of the jet is mainly formed by electron–positron pairs, it is possible to accelerate them even at sub-Eddington rates. If the pairs are hot, it is even possible to use the Compton rocket effect: a hot pair cloud above the accretion disk is illuminated anisotropically, and pairs scatter seed photons preferentially back to the disk, making the cloud recoil (O'Dell 1981). But it was soon realized that the cooling time of these pairs would be very short, and furthermore another effect, called Compton drag, would inevitably limit the maximum bulk Lorentz factor to a few, too small to explain the observations (Phinney 1987). The Compton drag effect can be understood by recalling that, in a moving frame, photons are observed to move along aberrated directions: for increasing Γ more and more radiation comes from the forward hemisphere. When the powers in the forward and backward hemispheres are equal, radiation pressure no longer accelerates the pair cloud, and a limiting Γ can be found.

Blandford and Rees (1974) suggested that jets could be formed by the so called "twin exhaust" model, in which hot gas confined by protons of equal pressure escape through a nozzle along directions of least resistance, which can conceivably coincide with the axis of rotation of the system.

The leading ideas involve rotation and magnetic fields, either braking the rotation of the accretion disk (Blandford and Payne 1982) or the rotation of the black hole itself (Blandford and Znajek 1976).

Another problem concerns the stability of jets which is difficult to understand (e.g. Bodo 1997): how can a jet remain highly collimated over many times its diameter, and not be unstable? However, recent studies and numerical work using relativistic codes are promising (e.g. Aloy *et al.* 1999).

THE "GAMMA-RAY" ERA

Prior to the CGRO there was only one source known to emit γ-rays above 100 MeV: 3C 273, detected by the COS-B satellite. The CGRO discovery that blazars, as a class, emit 90% of their energy in the γ-ray band came as a surprise.

In retrospect, we can hypothesize that researchers in the field were too much influenced by the so called "Compton catastrophe" argument, put forward in the 1960s by Hoyle *et al.* (1966), saying that relativistic electrons emitting more energy by inverse Compton than synchrotron would inevitably be rapidly cooled, decreasing their energy down to the point where inverse Compton and synchrotron losses were equal. But this is not the case, at least during flares, as can be seen in Figure 1, where the spectrum of the first source detected by CGRO, 3C 279, is shown.

What this discovery tells us

The fact that the γ-ray emission is so strong and so strongly variable immediately implies that we are seeing the radiation produced by a plasma moving at relativistic speeds at a small angle with respect to our line of sight. If not, the density of γ-ray photons that we can estimate from the received luminosity (which are assumed to be isotropically emitted) and the variability timescale (limiting the size of the emitting region) are so large that γ-ray photons would interact among themselves and with X-ray photons, producing electron and positron pairs (e.g. Svensson 1987). In this case the γ-rays would disappear.

A source in relativistic motion is observed to emit in a cone around its velocity vector: an observer lying inside this cone receives an increased amount of radiation. This is the lighthouse effect. The observed flux, therefore, corresponds to a decreased amount of emitted power, and this in turn corresponds to a decreased photon density. In this condition γ-ray photons interact less frequently, and can survive. Estimates derived requiring that the X-ray flux is not overproduced and is not absorbing γ-rays are in agreement (in a broad sense, they are both upper limits) with the values derived by superluminal speeds (Ghisellini *et al.* 1993; Dondi and Ghisellini 1995).

Now we ask: where, along the jet, is the observed γ-ray power produced? It cannot be too far away, because at these distances the emitting volume would presumably be too large to account for fast variability. Can it be produced very close to the black hole, at the jet apex? The answer is most likely no. To understand why, we must consider that, besides the jet, there is also an accretion disk, which we do not see because its light is outshone by the jet radiation, but whose existence is indicated by the presence, in flat spectrum radio quasars (FSRQs), of the emission lines. The X-rays produced by the accretion disk interact with the γ-rays produced in the jet, again transforming them into electrons and positrons.

We are then forced to conclude that there must be a preferred distance where γ-rays are produced, and hence there is a preferred distance where the jet produces most of its radiation. This is a few hundred Schwarzchild radii from the black hole, for example, around 10^{17} cm.

Blazars are variable, and even more so in the γ-ray band. Estimates of their contribution to the γ-ray background are therefore uncertain, but they could well be the most important contributor (Padovani *et al.* 1993; Stecker *et al.* 1993), or even explain it all.

More energetic γ-rays can be detected through particle cascades they make in the interaction with the high atmosphere of the Earth. The resulting particles travel at a speed which is larger than the speed of light through air, and then produce optical–UV Cherenkov radiation (see Catanese and Weekes (1999) for a review). Up to now, this technique (which is rather economical, since it needs ground-based telescopes) has discovered four blazars emitting at TeV energies. These observations have a twofold importance: they show how the particle acceleration mechanism works at extreme energies, and directly probe the intergalactic infrared background (Stecker and De Jager 1993), since TeV γ-rays can be absorbed in photon–photon collisions by infrared photons.

UNDERSTANDING BLAZARS?

About 10 years after the discovery of BL Lacertae as a cosmological object, Blandford and Rees (1978) proposed that the extreme properties of these sources could be explained by bulk relativistic motion of their emitting plasma. As a consequence, for each blazar that is pointing at us there must exist $\sim\Gamma^2$ other sources pointing away from us. In other words, this interpretation needs the existence of a *parent population*. Blandford and Rees proposed that this parent population is formed by radio galaxies.

Now there remains little doubt that relativistic bulk motion is the key property of blazars and we understand how beaming affects the observed luminosity function (Urry and Shafer 1984). The blazars–radio galaxies association, investigated in detail in a series of papers (see the review by Urry and Padovani (1995) and references therein), can indeed explain the relative observed number of sources. In the meantime, observational proofs of bulk motion have been accumulated, mainly by observations of superluminal knots on the parsec scale of radio jets and by limits derived by the γ-ray emission. This confirms that the emitting plasma in blazars is moving with a bulk Lorentz factor of between 5 and 20.

Therefore we are confident that the radiation we see comes from the transformation of a fraction of bulk kinetic into random energy and then into beamed radiation. Contrary to what was believed in the past, this transformation does not happen smoothly all along the jet (e.g. Blandford and Königl 1979; Marsher 1980; Königl 1981; Reynolds 1982; Ghisellini et al. 1985), but it has a maximum at a few hundreds of Schwarzschild radii, where most of the radiation we see is produced.

Birth and death of the accelerating jet model

It was in the 1990s that we started to study BL Lac objects systematically, through the definition of complete samples (that is, the complete list of sources, in a given region of sky, which are brighter than some threshold flux in some frequency range).

The first complete sample of blazars was done in the radio band (Stickel et al. 1991), and soon after it was joined by another complete sample in the X-ray band (Wolter et al. 1991). Objects belonging to the two samples were different, but this was in part expected: if we select objects in the radio, we preferentially pick up the brightest objects in this band, and the same is true for the X-ray band. No surprise then that the radio-selected BL Lacs are very bright radio sources, while the X-ray-selected BL Lacs are very bright in X-rays. What was *not* expected was that the X-ray and radio-selected objects had the same X-ray *luminosity*. This means that when we look for BL Lacs in X-rays, we are not discriminating against any particular sub-class, and we should then find the real range of BL Lac behavior. How was it then that the X-ray-selected objects were systematically different from radio-selected ones? Why did not we find any radio-bright object in the X-ray sample of BL Lacs?

A first answer was given by Maraschi and her collaborators (Maraschi et al. 1986; Ghisellini and Maraschi 1989). They propose the "accelerating jet" model, in which: (i) the X-rays are mainly produced by the parts of the jet closer to its apex, while the radio emission is produced at large distances; (ii) the plasma at the base of the jet is smoothly accelerating, until a saturation value is reached. With these assumptions the inner parts of the jet produce quasi-isotropic X-rays, while the outer parts of the jet produce the highly collimated radio fluxes.

Being isotropic (or quasi-isotropic), the X-rays are well visible to all observers, while we see strong radio emission only if we are aligned with the jet. Therefore it is true that in the X-ray band we can discover all kinds of BL Lacs, but the majority of them will be radio weak (the jet is misaligned). In the radio band, on the other hand, one preferentially selects radio-strong sources, the ones whose jet is aligned with the line of sight. The average X-ray luminosity of both kinds of objects is the same. Evidence in favor of this idea came with detailed calculations about the relative numbers of X-ray-selected and radio-selected BL Lacs and FR-I (Fanaroff–Riley-I) radio galaxies, assumed to be the parent population of the entire BL Lac class, due to Padovani and Urry (see the review by Urry and Padovani (1995) and references therein).

But only a few years later Giommi and Padovani (1994) discovered that the overall spectral energy distribution (SED) of the two kinds of BL Lacs, including the new γ-ray data, was difficult to explain on the basis of the "accelerating jet" model. In particular they showed that the peak of the synchrotron emission (in a ν–$\nu F(\nu)$ plot) was at higher energies for X-ray-selected objects than for radio-selected ones. They therefore named the two types of object HBLs (high-frequency peak BL Lacs) and LBLs (low-frequency peak BL Lacs).

Another strong piece of evidence against the accelerating jet model came from the discovery that HBLs are TeV emitters, and that the TeV flux varies with extremely short timescales (doubling time of 20 minutes in Mkn 421; Gaidos *et al.* 1996), which then required a large Lorentz factor even for the parts of the jets emitting γ-rays. Since we observe simultaneous variability in the X-ray and TeV bands, we conclude that the X-ray flux is also highly beamed. This conclusion is important, because it indicates that, while accelerating, the jet is not dissipating much of its bulk kinetic energy.

We are then forced to conclude that the two families of BL Lacs (selected in X-rays and in the radio bands) are not different because we see them under two different angles: their are *intrinsically* different.

The blazar sequence

Figure 3 illustrates the overall averaged SED of blazars. It is characterized by two broadly peaked components: the first, produced by synchrotron emission, peaks in the mm–far-IR band in powerful blazars, and at higher energies for less powerful objects, reaching the EUV–soft X-ray band in HBLs. Correspondingly, also the second most energetic component peaks at higher and higher energies as the total power decreases. This Compton component becomes increasingly dominant (with respect to the synchrotron one) as the total observed power increases. The trends just mentioned have been discussed in detail by Fossati *et al.* (1998), who considered three complete samples of blazars.

Ghisellini *et al.* (1998) studied all blazars detected by EGRET for which we have some γ-ray spectral information, and modeled all sources with a homogeneous synchrotron and inverse Compton model, taking into account, for the scattering process, both the locally produced synchrotron photons and photons produced externally to the jet (Figure 4). The modeling allows one to derive the intrinsic parameters of the sources such as the size, the magnetic field, the energy $\gamma_{peak} m_e c^2$ of the electrons emitting at the peaks, the particle density, and so on.

One of the most important results of this study is the striking correlation found between γ_{peak} and the total (magnetic plus radiative) energy density of the emitting region. Furthermore, γ_{peak} also correlates with the observed power, the amount of external seed photons used for the scattering process and the ratio between the Compton and the synchrotron luminosity. The smaller γ_{peak}, the larger the energy densities and the total power, dominated by the Compton luminosity.

Different blazar subclasses then form a well defined sequence:

- Powerful quasars, both polarized (HPQs) and not (LPQs), are characterized by small values of γ_{peak}, so that their synchrotron spectrum peaks in the mm–far-IR band and their Compton spectrum in the MeV band. The Compton component is stronger than the synchrotron one, and the contribution of photons produced externally to the jet to the scattering process is more important than the synchrotron one.

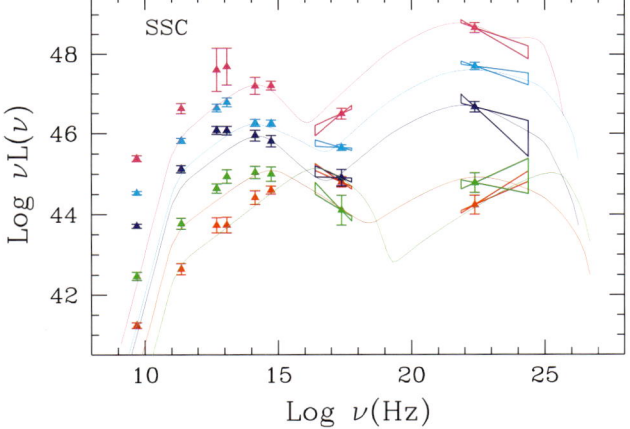

Figure 3 The blazar sequence as proposed by Fossati *et al.* (1998). Low-power objects are characterized by emission peaks located at high energies (X-ray and TeV bands, respectively), with approximately equal luminosities. As the power increases, both peaks shift to lower frequencies, and the high-energy emission becomes more dominant. The lines represent theoretical (synchrotron and inverse Compton one-zone homogeneous) models, as described in Ghisellini *et al.* (1998).

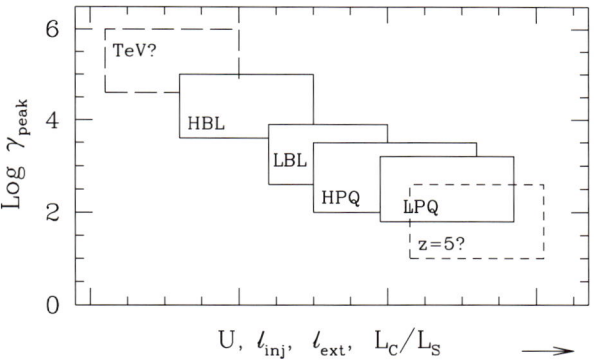

Figure 4 The blazar sequence as proposed by Ghisellini *et al.* (1998). The random Lorentz factor of the electrons emitting at the peaks of the spectral energy distribution is a decreasing function of the energy density U, the total power, and the power contained in the radiation produced externally to the jet, as measured in the co-moving frame. According to this scenario, the upper left of this parameter space should be occupied by low-luminosity sources emitting most of their synchrotron emission in the hard X-ray band, and most of their Compton emission in the TeV band. At the lower right extreme, we should have the most powerful sources, emitting mostly in the far-infrared and in the MeV band.

- At the other extreme, the low-power BL Lacs (the ones peaking at high frequencies, or HBLs) have large values of γ_{peak}, and consequently their spectra peak in the soft X-ray and in the GeV–TeV bands. The Compton emission is roughly as powerful as the synchrotron one. The contribution of externally produced photons is almost negligible (even if photons produced in regions of the jet adjacent to the γ-ray emitting site may be important, as in the case of Mkn 501; Pian *et al.* 1998).

A tentative interpretation of these trends is that in all sources there is a competition between the acceleration and the cooling processes, which determines the relevant value of the electron energy, that is, γ_{peak}. The particular form of the correlations between γ_{peak} and the (co-moving) energy density U, $\gamma_{peak} \propto U^{-0.6}$, implies that the radiative cooling rate ($\dot\gamma \propto \gamma^2 U$) at γ_{peak} is almost the same for all sources. This suggests the existence of a "universal" acceleration process (the same in all sources, and independent of γ and U) which accelerates the electrons up to the energy where their gains balance their losses.

Only in lineless and relatively weak BL Lacs can the electrons attain TeV energies: in these sources the energy densities are relatively small, and correspondingly the balance between heating and cooling rates is achieved at a larger γ_{peak}. On the contrary in powerful blazars with significant emission line luminosity the radiation energy density is large, and the balance between heating and cooling is achieved at a smaller γ_{peak}. In these sources the Compton cooling is enhanced (because of greater radiation energy densities), and this is why the Compton to synchrotron luminosity ratio is larger.

In the proposed picture blazars lie along a sequence: from powerful flat spectrum radio sources to less powerful lineless BL Lacs. The electron cooling rate is the key parameter ruling the value of γ_{peak} and therefore their overall SED. This scenario can then be falsified by the discovery of powerful blazars with relatively strong emission lines and with synchrotron and Compton peaks located at high energies (in the soft X-ray and in the GeV–TeV band). Instead low-luminosity blazars with a SED peaking at low energies (far-IR and MeV–GeV) would not contradict our picture, since they can be slightly misaligned powerful blazars.

PRESENT AND FUTURE

What is the true power of blazars?

It is believed that the observed beamed radiation comes from the transformation of ordered to random kinetic energy in the jet. This transformation cannot be very efficient, however, or otherwise the jet would be left with insufficient bulk kinetic energy for powering the extended radio regions and the lobes of radio sources. Since in these large regions the radiative cooling times are long, we can think of them as big calorimeters, where the energy transported by the jets is stored. Rawlings and Saunders (1991) pointed out that the energy stored in the lobes, divided by their lifetime (that is, the average power that jets had to transport) correlates with the luminosity in the narrow emission lines. Assuming that the narrow line region is illuminated by the accretion disk and that its covering factor is of the order of 1%, one arrives at the very important conclusion that the power in the jet and the power of the accretion disk are roughly equal.

Jets of matter–antimatter?

The matter content of jets is still an open issue, and we recently started to tackle this problem combining theory and observations. The last decade was particularly interesting in this respect, due to the discovery of the conspicuous γ-ray emission of blazars on one hand, and the realization that jets must be *not maximally efficient* at producing radiation on the other hand (otherwise the radio lobes could not be created). The fact that we see γ-rays means that they were not used to make electron–positron pairs; the large power required by the lobes fixes unambiguously the power that the jets must carry. Another severe constraint comes from annihilation suffered by pairs if they are produced right at the start of the jet. With the help of synchrotron theory, we can estimate how many leptons are necessary to account for the synchrotron radiation we see. This puts a lower limit onto the power the jet must carry in bulk kinetic form. These estimates have been done by Celotti and Fabian (1993), who considered the jet at the VLBI scale (the parsec scale), and by Celotti and Ghisellini (2002) on the scale where most of the radiation is produced (the γ-ray region, 0.03–0.3 pc from the black hole). These studies suggest that, for powerful blazars, most of the jet power is carried by protons. Electron–positron pairs could be present, but they are too light to be the only energy carriers. For less powerful blazars (that is, HBL-type BL Lacs), the situation is more uncertain, and a dominant presence of pairs cannot be excluded. Note that these conclusions are still controversial, and recent results about the *circular* polarization in 3C 279 suggest instead that electron–positron pairs could be important also in the jets of powerful quasars (Wardle *et al.* 1998).

Another result concerns the power in Poynting flux which seems not to exceed the power that must be carried by protons. This is in agreement with the fact that, in blazars, the γ-ray emission is due to the inverse Compton process: to be more powerful than the synchrotron process, a relatively weak magnetic field is required.

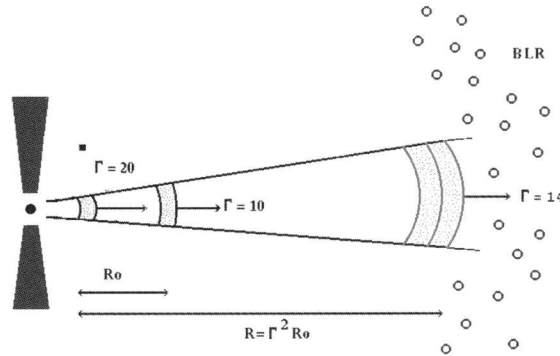

Figure 5. The internal shock scenario for blazars. The plasma flow in the jet is not continuous, but contains different regions moving with different velocities. There is then the possibility that a faster shell of plasma catches up a slower one, producing a shock, which transforms part of the bulk kinetic energy of the fast shell into random energy, and then radiation.

Analogies with gamma-ray bursts

Gamma-ray burst research has the same major problems of understanding how the ordered kinetic energy is transformed into random motions and then to radiation. No wonder then if there are close contacts between researchers in these two related areas. The present understanding of γ-ray burst, summarized in Chapter 22, implies a key role played by a *discontinuos* central engine, and a corresponding inhomogeneous flow, both in density and in velocity. Later but faster parts (or shells) of the flow can catch up earlier and slower parts, producing a shock which converts part of the bulk kinetic energy of the fast shell into random energy. Indeed, this idea was born in the blazar field (Rees 1978), and only later became the leading idea to explain the γ-burst emission.

The schematic of Figure 5 illustrates an example of what can happen in blazars: two shells moving with a bulk Lorentz factors of $\Gamma_1 = 10$ and $\Gamma_2 = 20$, initially separated by a distance R_0, will collide at a distance $R_i \sim \Gamma_1^2 R_0$. If they have equal masses, the Lorentz factor of the merged shells is $\Gamma = 14$. In this specific example, around 6% of the total energy has to be dissipated (to conserve energy and momentum). This model is very promising, since it can explain some basic properties of blazars:

- The efficiency is of the right order: most of the jet power *has not to be dissipated*, in order to power the radio lobes.
- If the initial separation is comparable to a few Schwarzschild radii, that is, $R_0 \sim 10^{15}$ cm, the collision takes place at $R_i \sim 10^{17}$ cm (for $\Gamma_1 \sim 10$), just at the distance where the inverse Compton process off the photons of the broad line region is efficient, where photon–photon collisions are not important, and yet the emitting region is still sufficiently compact to account for the rapid variability.
- There can be a hierarchical structure in shell–shell collisions: pairs of shells can collide once more, at greater distances, where the dominant channel for radiation is synchrotron emission with a smaller value of the magnetic field. Hence there can be a link between the flares at optical and γ-ray energies and the flares in the radio band.

OPEN PROBLEMS

The advance in our understanding can be pictured as an increase in the radius of an imaginary "sphere of knowledge": the volume increases, but also the surface of contact with the "unknown" increases, generating more and more questions. Then, needless to say, there are a number of questions which we hope to tackle in the near future.

Intraday variability – Some blazars (a quarter of the sources in the "1 Jy" sample) vary in the radio band by some percent in a timescale of hours. The implied brightness temperature reaches values up to 10^{20} K, orders of magnitude larger than what can be accommodated for by the incoherent synchrotron self-Compton theory (see Wagner and Witzel 1994 for a review). This phenomenon could be due to an intrinsic effect, or by amplification due to beaming, or by interstellar scintillation. Recently, in a beautifully coordinated experiment on the source PKS 0405−385, done by an Australian and the VLA radio telescopes, it was demonstrated that interstellar scintillation is surely present. But to scintillate, the angular size of the source must be extremely small, which inevitably leads to large brightness temperatures, which in the case of this source turned out to be 10^{15} K (Kedziora-Chudczer *et al.* 1997; Jauncey *et al.* 2000). This value is too large to be explained by beaming (it requires a Doppler factor $\delta \sim 1000$), and suggests that some fraction of the flux is not due to the incoherent synchrotron process (which is one of the pillars on which our knowledge of blazars is based).

Matter content, formation, and acceleration of jets – This issue is still controversial even if a general consensus is growing towards a key role played by magnetic fields and rotation (see Celotti and Blandford 1999 for a review). In the near future, it is likely that our knowledge will be boosted more by observations than by theory. VLBI at high frequencies will resolve scales down to 10^{-2} pc, where jet accelerations is presumably taking place. The Chandra X-ray Observatory (NASA, launched 1999) is already giving us images of jets in X-rays, telling us that emitting leptons can reach energies $\sim 10^8 m_e c^2$. Soon the XMM-Newton X-ray Observatory (ESA, launched 1999) will tell us if jet sources have an accretion disk with similar properties to radio-quiet sources, by looking at the fluorescent K_α Fe line at 6.4 keV. The new generation of γ-ray satellites, such as the Gamma Ray Large Area Telescope (NASA, in preparation), can monitor γ-ray flares from many sources at one time, with at least 20 times the sensitivity of the EGRET instrument aboard the CGRO satellite, and can tell us the fraction of the time a source spends in high flaring states. All this new information will be precious to constrain and advance our ideas about jet physics.

Evolution – HBLs, LBLs, and FSRQs have different evolutionary properties. HBLs show a *negative* evolution, that is, these sources were less numerous or less luminous in the past. In contrast, the radio-strong LBLs, and especially FSRQs, show positive evolution. A closely related issue concerns the relative number density of HBLs and LBLs. A promising idea, to be verified, is that the paucity of HBLs at low fluxes could be the result of luminosity evolution, accompanied by spectral evolution: increasing in luminosity, the SED of a typical blazar changes and become more and more LBL type (Fossati *et al.* 1997). Also the issue of the parent populations of BL Lacs and FSRQs is not completely settled, since some BL Lac objects seem to be hosted in radio galaxies of the FR-II type. Perhaps a more informative line of attack would be to consider FR-I and FR-II together, as the parents of the entire class of blazars, as done by Maraschi and Rovetti (1994). There is finally the issue of the relative evolution of the jet luminosity with respect to the disk luminosity, which can even change accretion mode during its life, as suggested by Cavaliere and Malquori (1999).

Central engine – The importance of the X-ray emission and its spectrum, with respect to the emission in other bands, have changed our picture of the inner parts of the accretion disk, from the standard Shakura and Sunjaev (1973) accretion model to a two-phase and two-temperature model (Haardt and Maraschi 1991), in which a rarefied hot corona (perhaps patchy in space and in time) illuminates a relatively cold and dense accretion disk. This probably implies a key role of magnetic fields and magnetic viscosity, but we still do not understand how, where, and why.

A long debated question is the nature of the disk itself when the mass accretion rate becomes so small to imply inefficient collisions between protons and electrons, letting the disk inflate and become supported by the hot ion pressure. This idea, put forward by Rees *et al.* (1982) to explain radio galaxies, has been recently rejuvenated (and renamed as ADAF: advection-dominated accretion flow by the works of Narayan and collaborators; e.g. Narayan and Yi (1994)), and could play an important role in unifying seemingly different properties of blazars and quasars.

Finally, I would like to draw attention to the still preliminary, but important evidence of rough equality between the powers of jets and accretion disks. Is gravity responsible for both? (that is, is the powering mechanism accretion for both jets and disks?). Or are the two main powering mechanisms (accretion and rotation) strongly linked?

Acknowledgments

It is a pleasure to thank Annalisa Celotti, Andrea Comastri, Andy Fabian, Giovanni Fossati, Paolo Giommi, Francesco Haardt, Laura Maraschi, Silvano Molendi, Paolo Padovani, Roland Svensson, Gianpiero Tagliaferri, Aldo Treves, and Meg Urry for years of fruitful collaborations. I am very grateful to Martin Elvis, who greatly improved this manuscript.

REFERENCES

Aloy, M.A., Ibanez, J.M., Martì, J.M., Gomez, J.-L. and Muller, E. (1999). High-resolution three-dimensional simulations of relativistic jets. *Astrophysical Journal*, **523**, L125–L128.

Begelman, M.C., Blandford, R.D. and Rees, M.J. (1984). Theory of extragalactic radio sources. *Reviews of Modern Physics*, **56**, 255–351.

Blandford, R.D. and Payne, D.G. (1982). Hydromagnetic flows from accretion discs and the production of radio jets. *Monthly Notices of the Royal Astronomical Society*, **199**, 883–903.

Blandford, R.D. and Königl, A. (1979). Relativistic jets as compact radio sources. *Astrophysical Journal*, **232**, 34–48.

Blandford, R.D. and Levinson, A. (1995). Pair cascades in extragalactic jets. 1: Gamma rays. *Astrophysical Journal*, **441**, 79–95.

Blandford, R.D. and Rees, M.J. (1974). A 'twin-exhaust' model for double radio sources. *Monthly Notices of the Royal Astronomical Society*, **169**, 395–415.

Blandford, R.D. and Rees, M.J. (1978). Some comments on radiation mechanisms in Lacertids. In A.M. Wolfe (ed), Pittsburgh Conference of BL Lac objects, Pittsburgh University Press, PA, pp. 328–341.

Blandford, R.D. and Znajek, R.L. (1977). *Monthly Notices of the Royal Astronomical Society*, **176**, 465–481.

Blumenthal, G.R. and Gould, R.J. (1970). Bremsstrahlung, synchrotron radiation, and Compton scattering of high-energy electrons traversing dilute gases. *Reviews of Modern Physics*, **42**, 237.

Bodo, G. (1997). Non-linear instabilities in supersonic jets. In S. Massaglia and G. Bodo (eds), *Astrophysical Jets: Open Problems*, Gordon and Breach, Amsterdam, pp. 161–170.

Catanese, M. and Weekes, T.C. (1999). Very high energy gamma-ray astronomy. *Publications of the Astronomical Society of the Pacific*, 1193–1222.

Celotti, A. and Fabian, A.C. (1993). The kinetic power and luminosity of parsecscale radio jets – An argument for heavy jets. *Monthly Notices of the Royal Astronomical Society*, **264**, 228–236.

Celotti, A. and Blandford, R.D. (2001). On the formation of jets. In L. Kaper, E.P.J. van den Heuvel, P.A. Woudt (eds), *Black Holes in Binaries and Galactic Nuclei*, Springer-Verlag, Berlin, p. 206.

Celotti, A. and Ghisellini, G. (2002). In preparation.

Dermer, C.D. and Schlickeiser, R. (1993). Model for the high-energy emission from blazars. *Astrophysical Journal*, **416**, 458–484.

Dondi, L. and Ghisellini, G. (1995). Gamma-ray-loud blazars and beaming. *Monthly Notices of the Royal Astronomical Society*, **273**, 583–595.

Fossati, G., Maraschi, L., Celotti, A., Comastri, A. and Ghisellini, G. (1998). A unifying view of the spectral energy distributions of blazars. *Monthly Notices of the Royal Astronomical Society*, **299**, 433–448.

Fossati, G., Celotti, A., Ghisellini, G. and Maraschi, L. (1997). Unifying models for X-ray-selected and radio-selected BL Lac objects. *Monthly Notices of the Royal Astronomical Society*, **289**, 136–150.

Gaidos, J.A. et al. (1996). Very rapid and energetic bursts of TeV photons from the active galaxy Markarian 421. *Nature*, **383**, 319–388.

Ghisellini, G., Celotti, A., Fossati, G., Maraschi, L. and Maraschi, L. (1998). A theoretical unifying scheme for gamma-ray bright blazars. *Monthly Notices of the Royal Astronomical Society*, **301**, 451–468.

Ghisellini, G. and Madau, P. (1996). On the origin of the gamma-ray emission in blazars. *Monthly Notices of the Royal Astronomical Society*, **280**, 67–76.

Ghisellini, G., Maraschi, L. and Treves, A. (1985). Inhomogeneous synchrotron-self-Compton models and the problem of relativistic beaming of BL Lac objects. *Astronomy and Astrophysics*, **146**, 204–212.

Ghisellini, G., Padovani, P., Celotti, A. and Maraschi, L. (1993). Relativistic bulk motion in active galactic nuclei. *Astrophysical Journal*, **407**, 65–82.

Giacconi, R., Gursky, H., Paolini, F. and Rossi, B. (1962). Evidence for X-rays from sources outside the Solar System. *Physical Review Letters*, **9**, 439–443.

Giommi, P. and Padovani, P. (1994). BL-lacertae reunification. *Monthly Notices of the Royal Astronomical Society*, **268**, L51–L54.

Ginzburg, V.L. and Syrovatskii, S.I. (1965). Cosmic magnetobremsstrahlung (synchrotron radiation). *Annual Review of Astronomy and Astrophysics*, **3**, 297.

Haardt, F. and Maraschi, L. (1991). A two-phase model for the X-ray emission from Seyfert galaxies. *Astrophysical Journal*, **380**, L51–L54.

Hoyle, F.S.R., Burbidge, G.R. and Sargent, W.L.W. (1966). On the nature of the quasi-stellar sources. *Nature*, **209**, 751–753.

Jauncey, D.L., Kedziora-Chudczer, L.L., Macquart, J.P., Lovell, J.E.J., Perley, R.A., Nicolson, G.D., Rayner, D.P., Reynolds, J.E., Tzioumis, A.K., Wieringa, M.H., Bignall, H.E. (2000). Galaxies and their Constituents at the Highest Angular Resolution. International Astronomical Union. Symposium 205, Manchester, England, August 2000.

Jones, T.W., O'Dell, S.L. and Stein, W.A. (1974a). Physics of compact nonthermal sources. Theory of radiation processes. *Astrophysical Journal*, **188**, 353–368.

Jones, T.W., O'Dell, S.L. and Stein, W.A. (1974b). Physics of compact nonthermal sources. II. Determination of physical parameters. *Astrophysical Journal*, **192**, 261–278.

Kedziora-Chudczer, L., Jauncey, D.L., Wieringa, M.H., Walker, M.A., Nicolson, G.D., Reynolds, J.E. and Tzioumis, A.K. (1997). PKS 0405-385: The smallest radio quasar? *Astrophysical Journal*, **490**, L9–L12.

Königl, A. (1981). Relativistic jets as X-ray and gamma-ray sources. *Astrophysical Journal*, **243**, 700–709.

Lynden-Bell, D. (1969). Galactic nuclei as collapsed old quasars. *Nature*, **223**, 690.

Mannheim, K. (1993). The proton blazar. *Astronomy and Astrophysics*, **269**, 67–76.

Maraschi, L., Celotti, A., Fossati, G., Ghisellini, G., Tagliaferri, G., Pian, E., Treves, A., Raiteri, C., Villata, M., Bassani, L., Cappi, M., Chiappetti, L., Comastri, A., Frontera, F., Giarrusso, S., Grandi, P., Molendi, S., Palumbo, G., Perola, C., Salvati, M., Tanzi, E.G. and Urry, C.M. (1998). Bepposax observations of 3c 279. In L. Scarsi, H. Bradt, P. Giommi and F. Fiore (eds), *The Active X-ray Sky: Results from BeppoSAX and RXTE*, Elsevier, Amsterdam, p. 453 [reprinted from *Nuclear Physics B*, **69**(1–3)].

Maraschi, L., Ghisellini, G. and Celotti, A. (1992). A jet model for the gamma-ray emitting blazar 3C 279. *Astrophysical Journal*, **397**, L5–L9.

Maraschi, L., Ghisellini, G., Tanzi, E.G. and Treves, A. (1986). Spectral properties of blazars. II. An X-ray observed sample. *Astrophysical Journal*, **310**, 325–333.

Maraschi, L. and Rovetti, F. (1994). A unified relativistic beaming model for BL Lacertae objects and flat spectrum radio quasars. *Astrophysical Journal*, **436**, 79–88.

Marscher, A.P. (1980). Relativistic jets and the continuum emission in QSOs. *Astrophysical Journal*, **235**, 386–391.

Miller, J.S. and Hawley, S.A. (1977). The spectrum and redshift of BL Lacertae. *Astrophysical Journal*, **212**, L47.

Narayan, R. and Yi, I. (1994). Advection-dominated accretion: A self-similar solution. *Astrophysical Journal*, **428**, L13–L16.

O'Dell, S.L. (1981). Radiation force on a relativistic plasma and the Eddington limit. *Astrophysical Journal*, **243**, L147–L149.

Oke, J.B. and Gunn, J.E. (1974). The distance of BL Lacertae. *Astrophysical Journal*, **189**, L5.

Padovani, P., Ghisellini, G., Fabian, A.C. and Celotti, A. (1993). Radio-loud AGN and the extragalactic gamma-ray background. *Monthly Notices of the Royal Astronomical Society*, **260**, L21–L24.

Padovani, P. and Giommi, P. (1995). The connection between X-ray- and radio-selected BL Lacertae objects. *Astrophysical Journal*, **444**, 567–581.

Perlman, E.S., Stocke, J.T., Schachter, J.F., Elvis, M., Ellingson, E., Urry, C.M., Potter, M., Impey, C.D. and Kolchinsky, P. (1996). The Einstein Slew Survey sample of BL Lacertae objects. *Astrophysical Journal Supplement*, **104**, 251.

Phinney S. (1987). How fast can a blob go? In J.A. Zensus and T.J. Pearson (eds), *Superluminal Radio Sources*, Cambridge University Press, pp. 301–305.

Pian, E., Vacanti, G., Tagliaferri, G., Ghisellini, G., Maraschi, L., Treves, A., Urry, M., Fiore, F., Giommi, P., Palazzi, E., Chiappetti, L. and Sambruna, R.M. (1998). BeppoSAX observations of unprecedented synchrotron activity in the BL Lacertae object Markarian 501. *Astrophysical Journal*, **492**, L17–L20.

Rawlings, S.G. and Saunders, R.D.E. (1991). Evidence for a common central-engine mechanism in all extragalactic radio sources. *Nature*, **349**, 138–140.

Rees, M.J. (1966). Appearance of relativistically expanding radio sources. *Nature*, **211**, 468–470.

Rees, M.J. (1967). Studies in radio source structure-III. Inverse Compton radiation from radio sources. *Monthly Notices of the Royal Astronomical Society*, **137**, 429–444.

Rees, M.J. (1978). The M87 jet – Internal shocks in a plasma beam. *Monthly Notices of the Royal Astronomical Society*, **184**, 61P–65P.

Rees, M.J., Phinney, E.S., Begelman, M.C. and Blandford, R.D. (1982). Ion-supported tori and the origin of radio jets. *Nature*, **295**, 17–21.

Reynolds, S.P. (1982). Theoretical studies of compact radio sources. I – Synchrotron radiation from relativistic flows. *Astrophysical Journal*, **256**, 13–37.

Salpeter, E.E. (1964). Accretion of interstellar matter by massive objects. *Astrophysical Journal*, **140**, 796–800.

Schneider, D.P., van Gorkom, J.H., Schmidt, M. and Gunn, J.E. (1992). Radio properties of optically selected high-redshift quasars. I – VLA observations of 22 quasars at 6CM. *Astronomical Journal*, **103**, 1451–1456.

Shakura, N.I. and Sunyaev, R.A. (1973). Black holes in binary systems. Observational appearance. *Astronomy and Astrophysics*, **24**, 337–355.

Sikora, M., Begelman, M.C. and Rees, M.J. (1994). Comptonization of diffuse ambient radiation by a relativistic jet: The source of gamma rays from blazars? *Astrophysical Journal*, **421**, 153–162.

Stecker, F.W. and De Jager, O.C. (1993). New upper limits on intergalactic infrared radiation from high-energy astrophysics. *Astrophysical Journal*, **415**, L71.

Stecker, F.W., Salamon, M.H. and Malkan, M.A. (1993). The high-energy diffuse cosmic gamma-ray background radiation from blazars. *Astrophysical Journal*, **410**, L71–L74.

Stern, D., Spinrad, H., Eisenhardt, P., Bunker, A.J., Dawson, S., Stanford, S.A. and Elston, E. (2000). Discovery of a color-selected quasar at $Z = 5.50$. *Astrophysical Journal*, **533**, L75–L78.

Stickel, M., Fried, J.W., Kühr, H., Padovani, P. and Urry, C.M. (1991). The complete sample of 1 Jansky BL Lacertae objects. I – Summary properties. *Astrophysical Journal*, **374**, 431–439.

Svensson, R. (1987). Non-thermal pair production in compact X-ray sources – First-order Compton cascades is soft radiation fields. *Monthly Notices of the Royal Astronomical Society*, **227**, 403–451.

Ulrich, M.-H., Maraschi, L. and Urry, C.M. (1997). Variability of active galactic nuclei. *Annual Review of Astronomy and Astrophysics*, **35**, 445–502.

Urry, C.M. and Padovani, P. (1995). Unified schemes for radio-loud active galactic nuclei. *Publications of the Astronomical Society of the Pacific*, **107**, 803–845.

Urry, C.M. and Shafer, R.A. (1984). Luminosity enhancement in relativistic jets and altered luminosity functions for beamed objects. *Astrophysical Journal*, **280**, 569–573.

Vermeulen, R.C. and Cohen, M.H. (1994). Superluminal motion statistics and cosmology. *Astrophysical Journal*, **430**, 467–494.

Wagner, S.J. and Witzel, A. (1994). Intraday variability in quasars and BL Lac objects. *Annual Review of Astronomy and Astrophysics*, **33**, 163–198.

Wardle, J.F.C. (1977). Upper limits on the Faraday rotation in variable radio sources. *Nature*, **269**, 563–566.

Wardle, J.F.C., Homan, D.C., Ojha, R. and Roberts, D.H. (1998). Electron–positron jets associated with the quasar 3C 279. *Nature*, **395**, 457–461.

Wehrle, A.E., Pian, E., Urry, C.M., Maraschi, L., McHardy, I.M., Lawson, A.J., Ghisellini, G., Hartman, R.C., Madejski, G.M., Makino, F., Marscher, A.P., Wagner, S.J., Webb, J.R., Aldering, G.S., Aller, M.F., Aller, H.D., Backman, D.E., Balonek, T.J., Boltwood, P., Bonnell, J., Caplinger, J., Celotti, A., Collmar, W., Dalton, J., Drucker, A., Falomo, R., Fichtel, C.E., Freudling, W., Gear, W.K., Gonzalez-Perez, N., Hall, P., Inoue, H., Johnson, W.N., Kazanas, D., Kidger, M.R., Kii, T., Kollgaard, R.I., Kondo, Y., Kurfess, J., Lin, Y.C., McCollum, B., McNaron-Brown, K., Nagase, F., Nair, A.D., Penton, S., Pesce, J.E., Pohl, M., Raiteri, C.M., Renda, M., Robson, E.I., Sambruna, R.M., Schirmer, A.F., Shrader, C., Sikora, M., Sillanpaeae, A., Smith, P.S., Stevens, J.A., Stocke, J., Takalo, L.O., Teraesranta, H., Thompson, D.J., Thompson, R., Tornikoski, M., Tosti, G., Treves, A., Turcotte, P., Unwin, S.C., Valtaoja, E., Villata, M., Xu, W., Yamashita, A. and Zook, A. (1998). Multiwavelength observations of a dramatic high-energy flare in the blazar 3C 279. *Astrophysical Journal*, **497**, 178–187.

Whitney, A.R., Shapiro, I.I., Robertson, D.S., Knight, C.A., Clark, T.A., Goldstein, R.M., Marandino, G.E. and Vandenberg, N.R. (1971). Quasars revisited: Rapid time variations observed via very-long-baseline interferometry. *Science*, **173**, 225.

Wolter, A., Gioia, I.M., Maccacaro, T., Morris, S.L. and Stocke, J.T. (1991). The number count distribution for X-ray-selected BL Lacertae objects and constraints on the luminosity function. *Astrophysical Journal*, **369**, 314–319.

Zeldovich, Ya.B. (1964). *Doklady Akademii Nauk SSSR*, **157**, 67.

25

GIUSEPPINA FABBIANO* AND MARTIN F. KESSLER**

X-ray and infrared properties of normal galaxies

WHY GALAXIES FROM SPACE OBSERVATORIES?

Galaxies are key components of the Universe and have been studied in astrophysics for several different and important reasons. Galaxies trace the distribution of matter and can be used as standard candles to derive information on the expansion and large-scale structure of the Universe. From these results constraints can be set on both cosmology and cosmogony scenarios. Galaxies evolve: they form, their stellar population evolves, they interact and merge. Understanding the forces governing galaxy formation and evolution is paramount for our understanding of the properties of galaxies and their components. Galaxies are where primordial matter gives birth to stars, and where stars evolve and form elements, enriching the interstellar medium (ISM) from whence the next stellar generations will be born. Therefore our understanding of stellar evolution and chemical evolution is intimately connected with our understanding of galaxies and their evolution.

Most of the work on galaxies – be it related to cosmology, galaxy formation and evolution, or present-day galaxy properties – has been supported by ground-based, mostly optical observations. While a lot has been learned this way, the optical window is only a small part of the entire emission spectrum of galaxies, and the information it provides is mostly limited to the normal stellar component, whose emission peaks in the optical wavelengths. Moreover, optical light is significantly absorbed by ambient dust, especially in the blue, where the emission of the massive younger stellar population peaks. The importance of widening the observing window on galaxies is made clear by ground-based radio and near- to mid-IR observations, which have provided a way of penetrating deeper into dusty star-forming regions; have allowed the study of the cold ISM and thus the discovery of extended, dark, massive halos in spiral galaxies; and have revealed the properties of both galaxian magnetic fields and highly energetic particles.

With space observatories, the observing window on galaxies has been opened even further, and a whole new set of phenomena has come into view. The waveband coverage has been extended to the UV (International Ultraviolet Explorer, IUE; Hubble Space Telescope, HST; Extreme UltraViolet Explorer, EUVE), mid- and far-IR (Infrared Astronomical Satellite, IRAS; Infrared Space Observatory, ISO), and X-rays (Einstein; Roentgen Satellite, Rosat; Advanced Satellite for Cosmology and Astrophysics, ASCA; BeppoSAX; Chandra; XMM-Newton). Each of these wavebands has given astronomers the means to explore new aspects of galaxies, including states of extreme star formation and nuclear activity, as well as the different phases of the ISM (cold, warm, hot, gaseous, dusty) and the latest stages of stellar evolution. In this chapter we concentrate on the impact of X-ray and IR space observatories. This is not a complete review of all the work that has been done in these areas, but is meant to give a general feel for these fields as they are today.

A BRIEF HISTORY OF X-RAY OBSERVATIONS OF GALAXIES

The first X-ray satellite, Uhuru, was launched from Kenya in December 1970. It was sensitive to photons in the

*Harvard-Smithsonian Center for Astrophysics, Cambridge, MA, USA
**ESA – European Space Research and Technology Centre, Noordwijk, The Netherlands

2–10 keV energy range. Uhuru is not an acronym; it means "freedom" in Swahili, and was so named because the day of the launch was the National Holiday of Kenya. Uhuru had been conceived and built by a team of physicists-turned-astronomers, led by Riccardo Giacconi at AS&E in Cambridge, MA. Uhuru led to the discovery of intense X-ray emission from compact stellar remnants (neutron stars, black holes, white dwarfs) in binary systems in the Milky Way, as well as from active galactic nuclei (AGNs; see Chapter 23), and hot plasmas in clusters of galaxies. X-ray emission from sources in the Magellanic Clouds and from M31 was detected with Uhuru, but other normal galaxies escaped detection because their fluxes were below the Uhuru source detection threshold (see Giacconi and Gursky (1974) for a review of Uhuru results).

To be able to detect and study in X-rays a significant number of normal galaxies, we had to wait for the first true X-ray telescope, the Einstein Observatory, also developed by a team led by Giacconi, including groups at the Smithsonian Astrophysical Observatory, Massachusetts Institute of Technology, Goddard Space Flight Center, and Columbia University. Einstein was launched by NASA in 1978 (Giacconi et al. 1979). With the Einstein grazing-incidence mirrors, X-rays from celestial objects could be focused onto position-sensitive focal plane detectors. Focusing optics resulted in two very significant advantages: the relatively small beam (~5" to ~45", depending on the Einstein focal plane instrument) meant that the background noise in the detection area was much smaller than in any of the previous non-imaging X-ray satellites, resulting in the detection of thousand-fold fainter sources (Figures 1 and 2); and the imaging capability for the first time let us see clearly the morphology of the X-ray sources (Figure 3). These capabilities were crucial for opening up the X-ray window on galaxies. Normal galaxies feature in two papers in the first ever set of Einstein publications. The survey of M31, the Andromeda Galaxy, the large spiral galaxy closest to us, revealed a population of bright point-like sources and gave us for the first time a picture of what our own Milky Way might look like in X-rays to an external observer (Van Speybroeck et al. 1979). The peculiar asymmetric extended emission of M86 in the Virgo Cluster suggested the presence of a hot gaseous halo interacting with the hot cluster gaseous medium, which had been discovered with Uhuru (Formal et al. 1979). The foundations of what we now know about the X-ray emission of galaxies were established with Einstein observations (Fabbiano 1989).

Subsequent X-ray satellites, the Germany–USA Rosat, the Japan–USA ASCA, and the Italy–Netherlands BeppoSAX, have further expanded our knowledge of the X-ray properties of galaxies and widened the observable X-ray window in several ways. Rosat was more sensitive at lower energies than Einstein and had slightly better angular resolution.

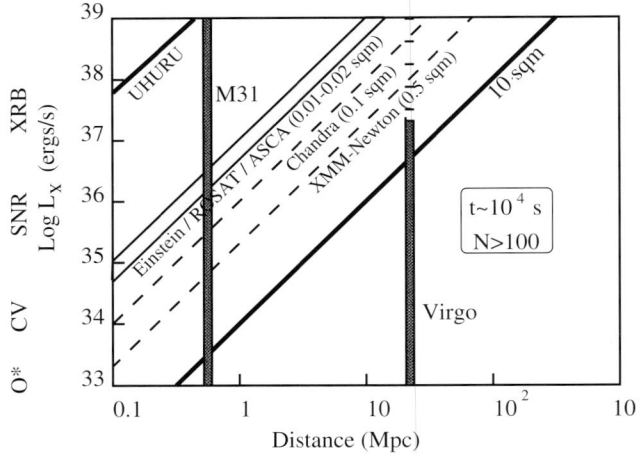

Figure 1 The sensitivity of different X-ray satellites to the detection of 100 counts in 10,000 s observations from typical galactic sources at different distances. For imaging telescopes the collecting area is indicated. Note, however, that unless the telescope has good enough angular resolution (Figure 3) the sources are likely to be confused, and therefore their individual contributions cannot be detected. The y-axis gives X-ray luminosities as well as typical X-ray emitters in given luminosity ranges, from luminous O stars to cataclysmic variables (CV), supernova remnants (SNRs) and X-ray binaries (XRBs). The x-axis is a distance scale. Two benchmark distances, to M31 and the Virgo Cluster of galaxies, are marked by vertical stripes.

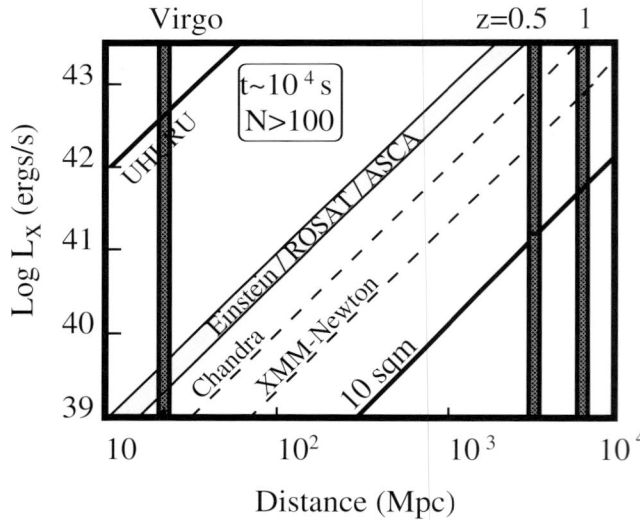

Figure 2 The sensitivity of different X-ray satellites to the detection of global emission from galaxies. Galaxy luminosities are mostly in the 10^{39}–10^{41} erg s^{-1} range, but some E and S0 galaxies can be as luminous as 10^{42}–10^{43} erg s^{-1} (Fabbiano 1989). Note that a significant advance in collecting area (to 10 m^2) will be needed to begin studying normal galaxies at cosmologically relevant distances.

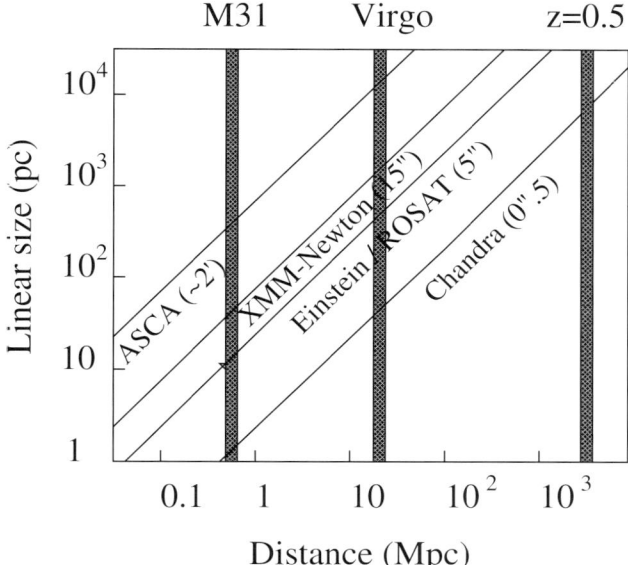

Figure 3 Linear sizes (in parsecs) equivalent to the angular resolution of X-ray telescopes as a function of the galaxy distance. While Chandra's sub-arcsecond resolution is needed to resolve dense stellar regions in M31 (e.g., near the nucleus), Einstein, Rosat, and XMM-Newton can resolve and study most bright X-ray sources. At the distance of the Virgo Cluster, Chandra achieves a linear resolution comparable to that of Einstein and Rosat for M31. The individual source detection sensitivity of previous missions (Einstein and Rosat), as well as that of XMM-Newton, is significantly impaired by confusion within the parent galaxy at the Virgo distance. The angular resolution of ASCA is significantly inferior to that of the other imaging X-ray missions. Note that 10^4 pc – ASCA's linear resolution at Virgo – is a typical galaxy size.

These characteristics have led to a better understanding of the spatial morphology of the X-ray emission of galaxies, and to the detection and study of the soft X-ray-emitting ISM of spiral galaxies. ASCA, was the first X-ray observatory equipped with CCD detectors. These detectors have spectral resolution 5 to 10 times superior than previous instruments, and are able to reveal X-ray emission lines from hot plasmas. Thus for the first time a study of the metal composition of the hot ISM of galaxies could be seriously attempted. ASCA (and BeppoSAX) also extends the observing windows to energies of 10 keV, significantly higher than the 2–4 keV upper reaches of Einstein and Rosat. This allowed the study of different spectral components of the X-ray emission, peaking in different energy ranges. For example, a softer component from a hot ISM and a harder component possibly from a population of X-ray binaries can be detected in the ASCA spectra of elliptical galaxies (Matsushita et al. 1994). However, ASCA's angular resolution (Figure 3) is ~2′, significantly inferior to those of Einstein and Rosat, and as a result detailed studies of individual galaxies cannot be pursued in most cases.

We are now at a very exciting time for X-ray astronomy. In July 1999, NASA successfully launched Chandra, the third Great Observatory, following the Hubble Space Telescope and the Compton Gamma Ray Observatory. The European Space Agency (ESA) successfully followed suit with XMM-Newton in December of the same year. Chandra and XMM-Newton represent a complementary quantum leap in sensitivity, imaging capabilities, and spectral resolution compared with the previous generation of X-ray observatories, and both are going to deepen and perhaps revolutionize the way we look at galaxies in X-rays.

X-RAY PROPERTIES OF GALAXIES

Extrapolating from the results of Uhuru and subsequent non-imaging X-ray missions (the UK Ariel 5; the US–Netherlands Astronomical Netherlands Satellite, ANS; and the US SAS 3 and HEAO 1), one could conclude that not much excitement lay ahead in the X-ray observations of galaxies. This was indeed the feeling of a good number of scientists in the Einstein team during the initial discussions of time allocation to different observing programs. It was thought that normal galaxies would emit because of their population of galactic sources, as happens in the Milky Way, and that the X-ray emission would be totally dominated by the nuclear source in the case of AGNs. While this is true, it is only a small part of the story. Twenty years ago, when X-ray astronomers first pointed Einstein at normal galaxies, they were engaged in a very successful fishing expedition. The Einstein survey of normal galaxies led to the discovery of hot ISM in E and S0 galaxies and of hot gaseous outflows from starburst nuclei, and demonstrated how X-ray emission is not an isolated phenomenon to be studied by X-ray astronomers, but is instead closely connected to the global emission properties of galaxies (see reviews by Fabbiano (1989, 1996a) and the Einstein catalog and atlas of galaxies (Fabbiano et al. 1992)).

Galaxies are key objects in the study of cosmology, the life cycle of matter, and stellar evolution. X-ray observations have given us a new key band for understanding these building blocks of the Universe, with implications ranging from the the study of extreme physical situations, such as can be found in the proximity of black holes, or near the surface of neutron stars; to the interaction of galaxies and their environment; to the measure of parameters of fundamental cosmological importance. In this chapter we discuss the unique contributions to our understanding of the Universe that are provided by X-ray observations of galaxies.

X-ray binaries and supernova remnants

A review of the Einstein and first Rosat results on discrete X-ray sources in external galaxies can be found in Fabbiano (1995a). Although the sample of detected sources has increased since, the basic conclusions are still valid. X-ray observations of the Milky Way and the Magellanic Clouds demonstrate that the majority of bright, individual X-ray sources are binary systems containing one of the compact remnants of stellar evolution: white dwarfes, neutron stars, or black holes. The X-ray emission is the result of gravitational accretion of the atmosphere of the companion star onto this compact object. Thus, the discrete X-ray source population of galaxies gives us a direct view of the end stages of stellar evolution, and of the higher-mass component of the stellar population. While the study of individual sources is best restricted to the Milky Way, because of the much higher fluxes, external galaxies present unique advantages in the study of the population properties of X-ray sources. All sources within a given galaxy are at the same distance from us, so the study of luminosity distributions is much less prone to error than with X-ray sources in our Galaxy, where the distance uncertainties are large. If fairly face-on galaxies are targeted, the effect of differences in interstellar extinction will also be much smaller than in the Milky Way sample, where directional effects due to the position of the Solar System in the galactic plane affect the measurements.

For the same reasons, systematic studies of properties of supernova remnants (SNRs), that can be used to explore their evolution, are best done with external galaxies. Most SNRs are spectacular, extended, expanding shells of hot plasmas and energetic particles, created by the shock waves of the supernova explosion and the interaction of the stellar debris with the surrounding ISM. An example of a study of a well-defined sample of SNRs, for which the distance is known (or all at the same distance within a galaxy), is the diameter–luminosity relation, which can probe both the age of the remnant and the host ISM (e.g., Long and Helfand 1979).

Because high-quality imaging is needed for these studies, Einstein and Rosat have been the two observatories that have contributed most to this field. Observations of nearby spiral galaxies have revealed variable sources with characteristics similar to those of Milky Way sources. The luminosity distributions appear to vary in different galaxies. A notable but not unique case is that of M81, where the X-ray sources appear brighter than in M31 and the Milky Way (Fabbiano 1995a).

While these studies open exciting new avenues, much more sensitivity and resolving power than has been available so far are required to extend them beyond exploratory exercises. Chandra is already making a difference, as can be seen from the images released to the public, but very long exposures will be needed in most cases. A large-area telescope ($\sim 10\,m^2$) with Chandra-like resolution will be needed to sample at significant depths the luminosity function of discrete X-ray sources in galaxies at the distance of the Virgo Cluster and beyond (Figures 1 and 3): The potential of these X-ray population studies is great, with implications for the study of the most massive portion of the stellar mass distribution, that can be detected through remnants in luminous X-ray binaries, and for a systematic study of the interplay of SNRs with the ISM in different galaxies.

Super-Eddington sources: Black hole hunting grounds

A surprising and interesting result of imaging observations of nearby galaxies has been the large number of sources detected with luminosities well above the Eddington luminosity for solar-mass ($1M_\odot$) accreting objects ($\sim 10^{38}\,erg\,s^{-1}$). The Eddington luminosity is the luminosity at which the gravitational pressure of the accreting material is balanced by the radiation pressure, and it represents a natural limit on the power that can be radiated by an accreting object. Given their luminosities, which exceed those of Milky Way sources and sources in M31, these very luminous sources cannot be steadily accreting neutron star binaries. They can be divided into a few categories: nuclear sources; X-ray counterparts of young SNRs; and other unidentified objects. With the exception of the SNR counterparts, and of sources that future higher-resolution observations may reveal as extended complex emission regions, these sources may host intermediate-mass to massive black holes (Fabbiano 1995a, 1998, Makishima *et al.* 2000).

X-ray emission from massive nuclear black holes in nearby galaxies gives us a probe into the accretion mechanisms and has implications for the evolution of AGNs (see Chapter 23). Relatively low-activity AGN and LINER (low-ionization nuclear emission line region) nuclei have been detected as luminous X-ray sources (in the range $\sim 10^{39}-10^{41}\,erg\,s^{-1}$) (Fabbiano 1996b). Although these luminosities would be in the super-Eddington regime for a normal X-ray binary, for nuclear sources due to accretion onto massive nuclear back holes, the accretion must be very inefficient. This is certainly the case for the nuclear source in M104, the Sombrero Galaxy (Fabbiano and Juda 1997), where the dynamics of the nuclear area point to a $10^8 M_\odot$ black hole (Kormendy and Richstone 1995). Given this nuclear mass, the X-ray luminosity of the nucleus of the Sombrero Galaxy is $\sim 4 \times 10^{-7}$ that of the Eddington luminosity.

In M33, a bright and variable nuclear source is also present (e.g., Peres *et al.* 1989), but its nature is more mysterious. It is possible that it may be a super-Eddington accretion binary (Takano *et al.* 1994).

Some at least of the non-nuclear super-Eddington sources may be accretion binaries. Their luminosities suggest that 10–100$M_\odot$ black holes may be present in these sources. The spectral/temporal behavior of at least one of these sources, in the galaxy IC 342, has an uncanny resemblance to those of black hole candidate sources in the Milky Way (Makishima 1994). Recently, a Japanese group (Makishima *et al.* 2000) has compiled and re-examined the ASCA data on a sample of super-Eddington sources. From the X-ray spectra of these sources, they suggest that their X-ray luminosity may be due to accretion onto rotating Kerr black holes.

X-ray observations are the best way – indeed, in most cases the only way – to explore the black hole component of galaxies. However, both high-resolution and large enough telescopes are needed to go beyond the few sources that have been explored in the very local Universe. The sub-arcsecond resolution of Chandra is needed to detect and to firmly associate with galaxian features (i.e. those of other galaxies) luminous sources in the 10^{39} erg s^{-1} range, as well as to establish whether they are point-like or are associated with features of the hot ISM. A good example is given by the Antennae, a merging galaxy pair at a distance of 29 Mpc, where a 5″ resolution Rosat HRI observation led to the detection of 12 candidate super-Eddington sources (Fabbiano *et al.* 1997). A recent Chandra image easily discriminates between point-like sources and diffuse, hot ISM emission. Both Chandra and XMM-Newton can yield spectral information on nearby sources, the limiting factor being collecting area for Chandra and angular resolution for XMM-Newton, which leads to confusion in more distant galaxies (Figure 4). This is a field that would definitely benefit from a future large X-ray telescope with Chandra-like resolution.

Hot plasmas in galaxies

Hot plasmas with temperatures of 1 million to a few tens of millions of degrees emit in the X-ray range. These plasmas are responsible for the X-ray emission of stellar coronae and SNR shells, and may have a role in components of the emission of accretion binaries and AGNs. Extended emission from hot, diffuse plasmas has also been detected in galaxies, although the detections are in some cases at odds with the original expectations. While a luminous hot phase of the ISM was expected in spiral galaxies (Spitzer 1956), this has proven elusive. In E and S0 galaxies, which were thought to be mostly free of ISM, luminous X-ray halos were detected in X-rays. X-ray observations have spurred a rethinking of models of the ISM of spirals (Bregman and Glassgold 1982, McKee 1995), and have opened entirely new avenues in the study of E and S0 galaxies, with implications ranging from cosmology to stellar and chemical evolution.

A lot of energy is released from stars, especially in their supernova stage. If there is a gaseous medium in galaxies that fills the space between the stars, this energy ought to heat the medium to X-ray temperatures, and the luminosity of this hot ISM ought to be larger than the integrated output from a typical population of X-ray binaries. Lyman Spitzer predicted this hot phase of the ISM of spiral galaxies in 1956. However, no significant amount of hot ISM in spirals has yet been detected. Rosat, with its sensitivity at lower energies (~0.1–2.5 keV), has revealed soft emission components localized in bubbles and filaments but with relatively small luminosities (Fabbiano 1996a; see Chapters 26 and 41).

It is now well known that E and S0 galaxies have hot X-ray halos. This is truly an X-ray discovery. Before Einstein was pointed toward the Virgo ellipticals, common wisdom had it that early-type galaxies were mostly devoid of ISM, and certainly optical observations showed no evidence of an ISM in most cases, except for a few dust lanes in some peculiar galaxies (e.g., Cen A/NGC 5128; Fornax A/NGC 1316). IRAS found small amounts of dusty ISM emitting in the IR (Knapp *et al.* 1989), but by and large most of the ISM of early-type galaxies, when present, is an X-ray emitting plasma.

The study of galaxies in the local Universe allows us to establish the astrophysics of hot plasmas. This knowledge can then be used to understand the properties of galaxies at the epoch of formation and their subsequent evolution, both in the field and in clusters (Figure 5). Hot outflows were clearly detected with Einstein and Rosat in two nearby

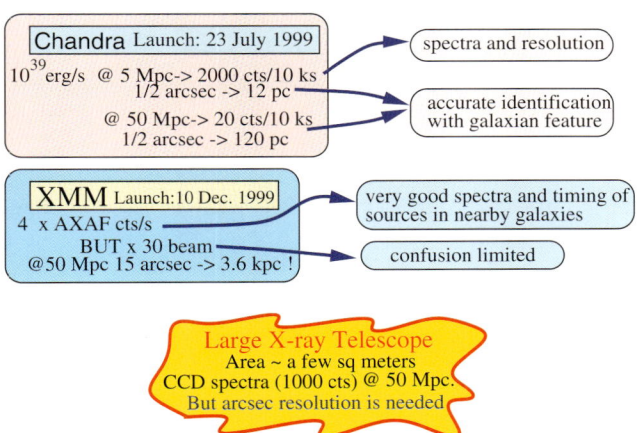

Figure 4 Present and far-future opportunities for studying super-Eddington sources in galaxies. Chandra has expanded the discovery space to nearly 50 Mpc, but detailed spectral studies are feasible only for sources in galaxies within 5 Mpc. XMM-Newton gives very good spectral/timing data for bright sources in nearby galaxies, but its beam makes confusion a serious problem. A future telescope that would allow the bright X-ray galaxian population to be sampled and studied requires a large collecting area and Chandra-like angular resolution.

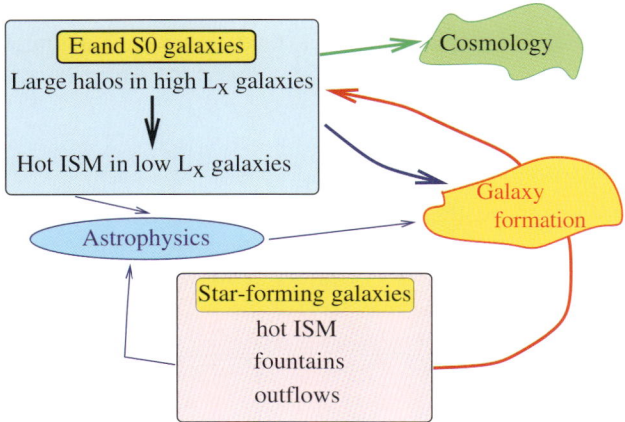

Figure 5 The study of hot plasmas is important for understanding the astrophysical properties of galaxies, their formation, and for constraining cosmological models.

galaxies with intense nuclear star-forming activity (M82 and NGC 253; Fabbiano 1989, 1996a and references therein). Figure 6 shows a Chandra image of M82. Hot outflows are likely to be more widespread, but without higher resolution and more powerful telescopes we cannot explore more distant galaxies. These hot outflows may be a relatively small-scale, nearby example of what was going on in a more widespread and violent fashion during the early phases of galaxy formation. Escape velocities from galaxies, when thermalized, are X-ray temperatures. Hot outflows are instrumental in expelling the elements created by stellar evolution into the circumgalactic medium – witness the elemental abundances found in the intracluster media with X-ray observations (e.g., Fukazawa *et al.* 1998). Galaxy ecology – the study of the cycling of enriched materials from galaxies into their environment – is inherently an X-ray subject.

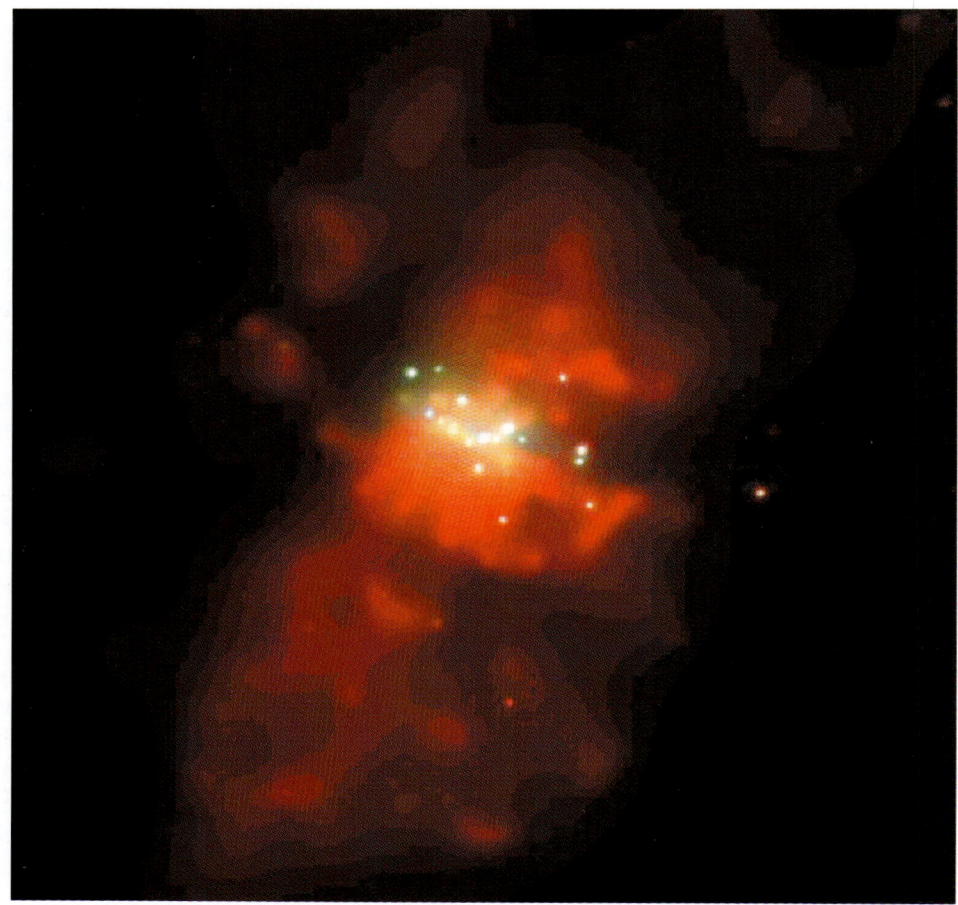

Figure 6 An image of the starburst galaxy M82 obtained with Chandra from the Chandra Archive. The image is color coded with red representing lower energy X-ray emission and blue higher energy X-ray emission. Note the population of bright point-like sources in the central region, as well as the extended gaseous halo to the north and south of the galactic plane.

Hot halos in E and S0 galaxies

Since the early Einstein days, the study of E and S0 galaxies in X-rays has been and continues to be a thriving field, involving many observers as well as theoreticians in lively and at times confrontational debates. The controversies were fueled by the tantalizing yet very limited data and by the data quality, and ranged from establishing the presence and ubiquity of hot halos, to the physical parametrization of the emission, inferring metal abundances of the X-ray plasmas, the ability of using these data for measuring the mass of the parent galaxy, and on to the theoretical modeling of halos and the implications of these models for the evolution of the galaxies.

Hot gaseous halos are uniquely visible in X-rays. Their discovery in E and S0 galaxies has given us a new, potentially very powerful tool for measuring of dark matter in galaxies, as well as for local estimates of Ω. Hot halos were found in a few of the Virgo Cluster galaxies observed early on in the Einstein mission (Forman et al. 1979), but that the phenomenon was more widespread was recognized only when a larger number of galaxies were surveyed (Forman et al. 1985, Trinchieri and Fabbiano 1985; see Fabbiano (1989) for a general review of Einstein results). It had been recognized earlier that if a galaxy has a hot halo in hydrostatic equilibrium, the X-ray observations could be used to measure the mass of the galaxy, once the density and temperature of the emitting plasma are measured as a function of the galactocentric radius. This method was applied to the X-ray halo of M87, the dominant galaxy in the Virgo Cluster, and led to the detection of an extended, dark, massive halo (Fabricant and Gorenstein 1983). With Einstein, mass measurements were attempted for a whole sample of normal galaxies (Forman et al. 1985). For very good reasons, mainly having to do with data quality and analysis methodology (Trinchieri et al. 1986, Fabbiano 1989), these early mass estimates were very controversial, but in principle the method is sound. To obtain accurate mass measurements, however, one needs good spatial and spectral data, both to demonstrate that the emission is from a hot extended halo and to measure accurately the spatial and thermal properties of this halo. Subsequent Rosat and ASCA observations have indeed yielded some convincing measurements of very bright halo galaxies (e.g., Kim and Fabbiano 1995). These results show that in these galaxies the amount of matter present can be more than 10 times that in the visible stellar component, adding to the evidence for dark matter in the Universe. Morever, an odd result – dubbed in some circles the baryon catastrophe – is obtained in some cases: the ratio between baryonic (in stars and X-ray emitting gas) and dark matter masses is incompatible with both nucleosynthesis and an $\Omega_0 = 1$ Universe (e.g., Kim and Fabbiano 1995). Clearly, X-ray observations of hot halos in early-type galaxies can have cosmological implications. Figure 7 shows

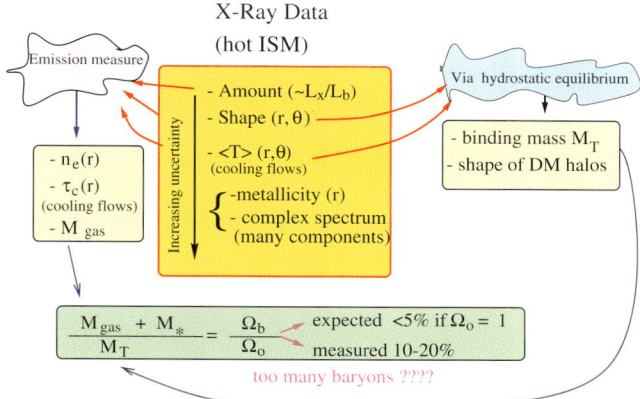

Figure 7 The path from X-ray spectral images of elliptical galaxy halos to the measurement of cosmologically relevant quantities. From the distribution of X-ray surface brightness and temperature, the total mass of the dark matter binding the halo can be derived (Fabricant and Gorenstein 1983), as well as the shape of the dark matter halos (Buote and Canizares 1997). From the X-ray emission measure, the physical parameters of the hot plasma can be derived: density (n_e), cooling time (τ_c), and total gas mass (M_{gas}). Comparison of the visible mass (in X-ray gas and stars) with the total mass M_T results is some cases in a discrepancy with the expected ratio (e.g., Kim and Fabbiano 1995).

schematically the path from X-ray data to this kind of measurement.

Although the existence of hot halos has been established, there are some basic questions that need to be explored further before we understand their evolution and physical state. One question has to do with the presence of the halo itself. Not all E and S0 galaxies harbor large quantities of hot ISM; this was recognized in the earlier Einstein years (Trinchieri and Fabbiano 1985), and it still holds. The other question has to do with the physical properties of the halos, including their shape, thermal properties, and chemical composition. These factors all provide contraints on models of halo formation and evolution, as well as of galaxy stellar and chemical evolution (e.g., Ciotti et al. 1991).

There is a whole range of X-ray emitting halos in E and S0 galaxies (e.g., Fabbiano 1989, 1996a), from halos dominating the X-ray luminosity of the galaxy, to galaxies where, with the present data, no halo emission can be unequivocably distinguished from the integrated emission of X-ray binaries. Although it was suggested as early as the first set of Einstein galaxy survey papers that the emission of a large fraction of E and S0 galaxies is likely to be dominated by their X-ray binary population (Trinchieri and Fabbiano 1985), the issue has been controversial. More recent ASCA and Rosat observations, while confirming the

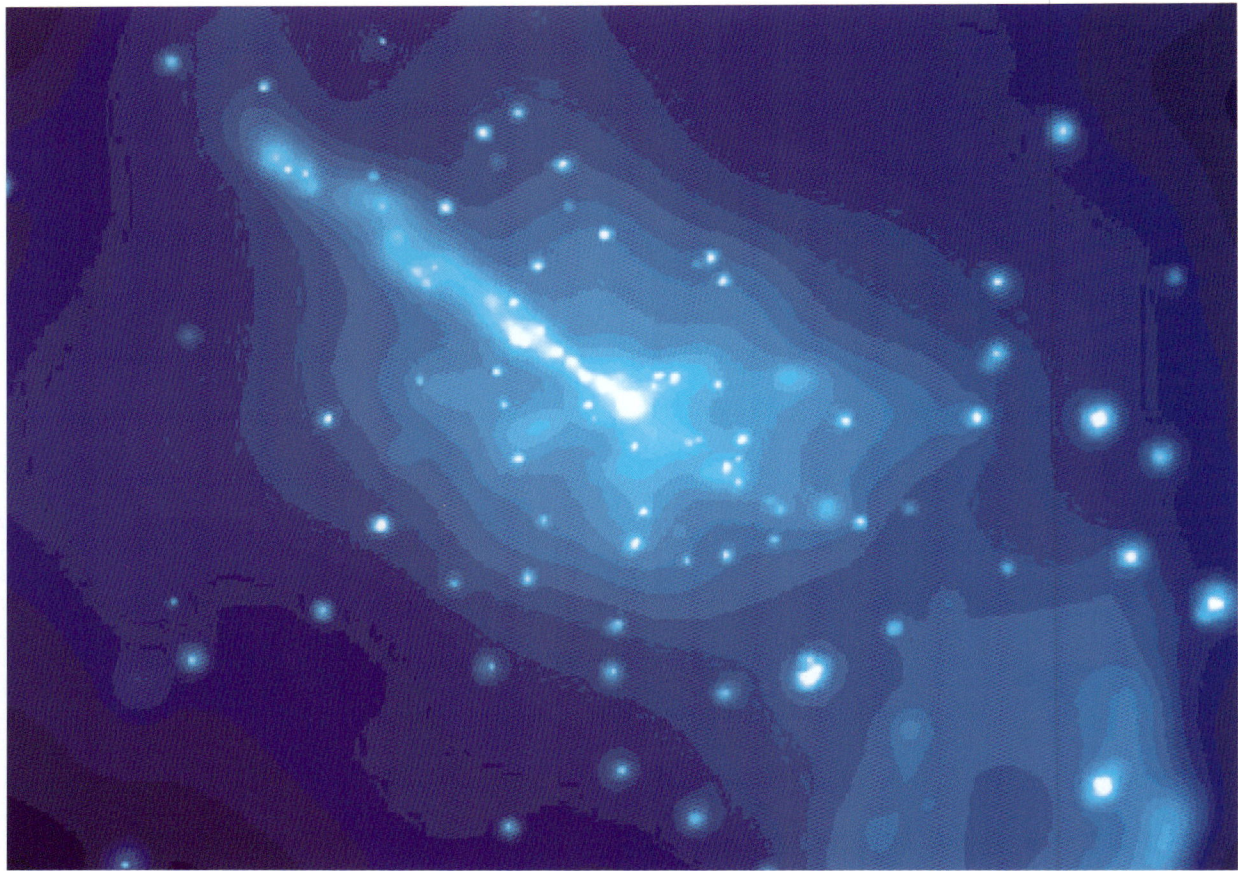

Figure 8 An image of the central region of NGC 5128, the Centaurus A galaxy, as seen with Chandra. Note the population of bright point-like sources, as well as the bright nuclear and jet emission.

presence of an underlaying X-ray binary component because of its spectral signature (Matsushita *et al.* 1994), have shown that even in X-ray faint galaxies there is likely to be a small amount of hot ISM (e.g., Kim *et al.* 1998). But a detailed separation and mapping of gaseous and discrete emission components in galaxies as distant as the Virgo Cluster requires Chandra's sub-arcsecond imaging. The first Chandra image of the nearby S0 radio galaxy Centaurus A clearly revealed a population of ~60 point-like sources, as well as complex extended emission, some of which at least may be of thermal gaseous origin (Kraft *et al.* 2000; Figure 8).

The range of ratios of X-ray to optical luminosities of E and S0 galaxies can vary by up to a factor of 1000 for galaxies with the same optical luminosity (i.e., stellar content). Why is there this range of halo properties? Correlating X-ray properties with other galaxian properties suggests that large, hot halos may be related both to the characteristics of the gravitational potential and to either

Factors favoring halos	Factors opposing halos
Formation	**Environment**
deeper gravitational wells	higher density
rounder/ boxier isophotes	recent merging
presence of central cores	recent star formation
with more recycling of original debris larger (Mg/Fe)	

Figure 9 Factors favoring or opposing hot halo formation.

the galaxy environment or recent evolution (e.g., Eskridge *et al.* 1995a,b, Pellegrini 1999; Figure 9). While the characteristics of the potential, and therefore the original formation mechanisms, are key for the retention of large halos (Figure 9), some environmental effects may have the

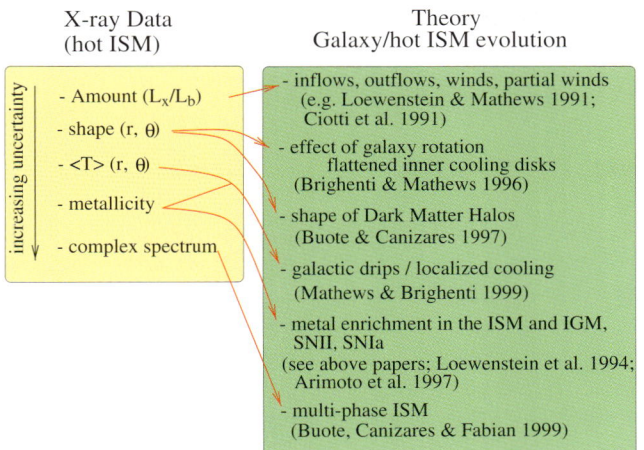

Figure 10 Some of the theoretical work based on the X-ray results.

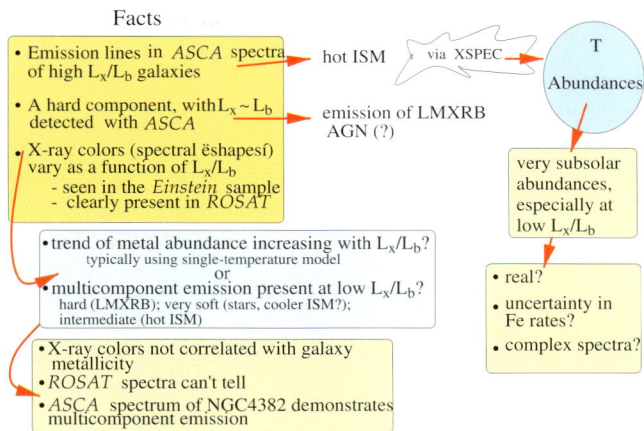

Figure 11 Spectral analysis and issues. While the presence of hot plasmas has been clearly demonstrated by the detection of emission lines in the ASCA spectra of X-ray luminous galaxies (e.g., NGC 4636; Matsushita et al. 1997), the properties of these hot plasmas have been controversial, mostly because of inescapable limitations in both data quality and modeling. Because of the limited spectral resolution, modeling has been done by fitting models convolved with instrumental response, in most cases using software in XSPEC, the X-ray Spectral Analysis Package. (LMXRB = low-mass X-ray binary.)

opposite effect: for example in the case of recent merging and star formation, which may be related to the blowing off of hot ISM (Fabbiano et al. 1997, Mackie and Fabbiano 1997).

A complementary approach is to compare the X-ray results to theoretical predictions (Figure 10). Theoretical work in the first few years after the discovery of hot halos concentrated on exploring the cooling flow model (see review in Fabbiano 1989). In this model the denser inner cores of the halos cause more efficient cooling and result in central inflows. These models predict high X-ray luminosities, as well as typical radial distributions of X-ray surface brightness and temperature. While models can be built to fit the data of the most X-ray luminous galaxies, the spread of X-ray/optical ratios cannot be explained in the cooling flow framework, and requires a more varied approach, ranging from cooling flows to subsonic outflows to winds (e.g., Ciotti et al. 1991). The shape of the halos is also important. Asymmetries as observed in the Virgo Cluster galaxies M86 and NGC 4472 may be indicative of the interaction of the moving galaxy with the intracluster medium (e.g., Fabian et al. 1980). Comparison of the flattening of X-ray and optical isophotes may be used to constrain the mass of the halo (Buote and Canizares 1997), although flattening may also be related to galaxy rotation (Brighenti and Mathews 1996). With the better definition of the X-ray properties of these halos, which will be possible with Chandra and XMM-Newton, some of these comparisons will be more meaningful.

While Einstein and Rosat data showed very clearly that X-ray luminous E and S0 galaxies have softer spectra (peaking below 1 keV) than spiral galaxies (Kim et al. 1992), the final proof of the presence of optically thin plasmas was provided by the ASCA CCD detectors, which revealed the tell-tale signature of X-ray emission lines (e.g., Matsushita et al. 1997). These lines are produced by atomic transitions in elements present in the hot plasma. X-ray spectra can then help in determining the metal abundance of the halos. This is a fundamental question because its answer can place strong constraints on our understanding of the enrichment history, and therefore the stellar evolution history, of the parent galaxy. However, the present answers are controversial because of the relatively poor resolution of the available spectra (e.g., ASCA CCD spectra or, worse, proportional counter spectra), combined with non-unique model fitting (e.g., Fabbiano 1995b, 1996a; Figure 11), that can be solved only with high-quality, high-resolution ($\lambda/\Delta\lambda \sim 1000$), spatially resolved spectra. In particular, the first generation of ASCA spectral analysis produced reports of very subsolar abundances (Mushotzky et al. 1994, Loewenstein et al. 1994). These results were in striking contrast to stellar evolution calculations of metal enrichment (Arimoto et al. 1997), which required a minimum of solar abundances. More recent work (Buote and Fabian 1998, Matsushita et al. 2000) shows that multi-temperature halo models, and a proper weighing of the uncertainties in the Fe plasma emission codes, can result in nearly solar abundances. It is clear that spatially resolved, high-resolution spectroscopy is needed to establish an uncontroversial picture. Chandra and XMM-Newton will enlighten this issue somewhat, but future, larger X-ray telescopes will be needed for truly in-depth studies.

IR OBSERVATIONS OF GALAXIES FROM SPACE

Access to the IR sky from Earth-based telescopes is restricted to a few windows, mainly at the shorter IR wavelengths between 1 and 20 μm. Even in the transparent regions, sensitivities are limited by undesired foreground thermal emission (and associated noise) from the "warm" telescope and from our atmosphere. It is only by going into space and by cooling the telescope and its instrumentation that a clear view of the Universe at IR wavelengths can be achieved.

The US–Dutch–UK IRAS (Neugebauer et al. 1984) operated successfully for nearly 10 months in 1983 before its supply of liquid helium coolant was exhausted. During this time, it surveyed some 96% of the sky in four broad wavelength bands with completeness limits of about 0.5 Jy at 12, 25, and 60 μm, and 1.5 Jy at 100 μm. IRAS gave the first unbiased view of the IR sky. Its Point Source Catalogue contains some 22 000 galaxies, less than half of which had been previously cataloged, with far-IR luminosities ranging from 10^7 to $10^{13} L_\odot$. One of its major discoveries was a new class of galaxies emitting over 95% of their luminosity in the IR (Soifer et al. 1984). The most extreme members of this class – ultraluminous infrared galaxies (ULIRGs or ULIGs) – have bolometric luminosities comparable to those of quasars. The majority of galaxies seen by IRAS are late-type spirals with very few ellipticals and S0s; however, quasars, Seyferts, and early-type galaxies were also detected. Soifer et al. (1987) reviewed extragalactic IRAS results, while Sanders and Mirabel (1996) presented a pre-ISO review of luminous IR galaxies.

The European Space Agency's ISO (Kessler et al. 1996) was the first true spaceborne infrared observatory. It contained four sophisticated scientific instruments, offering a wide range of observing capabilities. Photometry and direct imaging were possible in broad and narrow spectral bands across its entire wavelength range of 2.5 to around 240 μm, with spatial resolutions ranging from 1.5″ (at the shortest wavelengths) to 90″ (at the longer wavelengths). For spectroscopy, resolving powers ranging from 50 to 30 000 were available in the wavelength range from 2.5 to nearly 200 μm. At a wavelength of 12 μm, ISO was a thousand times more sensitive and had a hundred times better angular resolution than IRAS. ISO was launched in November 1995, and during its routine operational phase, which lasted until April 1998 – almost a year longer than specified, made over 10 000 individual extragalactic observations.

Some highlights of ISO studies of normal galaxies are presented below; these examples are extensively drawn from Genzel and Cesarsky (2000), who have presented a full review of ISO extragalactic results, and from Helou (1999), who focused on normal galaxies. Readers are referred to these works for details and full references and may also consult the extragalactic sections of Cox and Kessler (1999), which includes the summary by Genzel (1999).

ISO HIGHLIGHTS

A major theme of ISO's extragalactic work was the investigation of star formation on galactic scales. Normal galaxies, defined as those whose energetics are dominated by nucleosynthesis in stars (Helou 1999), account for most of the star formation in the Universe, and therefore for the bulk of its chemical evolution. ISO results range from detailed studies of nearby systems to distant galaxies with dust-enshrouded star formation, having an impact on our knowledge of the cosmic history of star formation.

The mid-IR spectral region (3–30 μm) marks the transition in the spectrum of galaxies from emission dominated by stars to that dominated by dust re-radiation. ISO's continuous spectral coverage has shown for the first time the details of this transition. There are three components of the dust emission contributing to the mid-IR spectra. First, there is emission from small particles (a few hundred molecules) transiently heated by single photons, and consequently showing large temperature fluctuations. This emission is seen as a family of features centered at 3.3, 6.2, 7.7, 8.6, 11.3, and 12.7 μm and referred to as "unidentified infrared bands" (UIBs) or polycyclic aromatic hydrocarbon (PAH) bands, after the most popular model for their carrier. This mechanism provides the bulk of the mid-IR emission of the interstellar medium. UIB emission is a good tracer of normal and moderately active star formation in spiral and irregular galaxies. Second, in some galaxies or regions of galaxies there is a steeply rising continuum above 10 μm, accompanied by strong fine-structure line emission. This component is characteristic of intense star formation, and the continuum is attributed to very small grains. Third, there is emission from hot dust grains, characteristic of, for example, dust tori in AGNs and showing up at less than 10 μm. Measurements with the ISO camera in two broad bands, centered at wavelengths of 7 and 15 μm, have proven to be an extremely interesting diagnostic of the radiation environment, probing the degree of activity in the ISM of galaxies.

The far-IR emission results from dust re-radiating absorbed short-wavelength radiation as a gray body. ISO photometry at 100–200 μm is beginning to constrain dust temperatures, total IR luminosities, and dust masses in galaxies. For normal spirals, IRAS established that the $\lambda < 100$ μm emission comes from dust grains at a temperature of around 30 K; however, the inferred gas/dust ratios were an order of magnitude different from those in the Milky Way. ISO observations at the longest wavelengths have resolved this problem by establishing the presence of a cooler dust component with a typical temperature of ~20 K. For example, Haas *et al.* (1998) mapped M31 at 175 μm and found that the bulk of the dust has a temperature of only 16 K. They derived a dust mass a factor of 10 higher than was obtained from IRAS results only. Another consequence of the increased dust mass is more extinction in the optical, with implications for the extinction-corrected morphology: for example, turning M31 from an Sb into a ring galaxy (Figure 12).

ISO has been used to search for dark matter in the outer parts of disk galaxies, proposed as an explanation of their rotation curves. Mapping of edge-on spirals with ISO's camera has excluded any significant contribution to massive haloes from hydrogen-burning stars or most brown dwarfs. However, Valentijn and van der Werf (1999) have detected H_2 emission from the extended disk of the edge-on galaxy NGC 891. They trace the emission out to 12 kpc from the nucleus, deriving molecular mass estimates and suggesting that cool H_2 (~80 K) could make a significant contribution to baryonic dark matter. In passing, we note that ISO made the first detections of the lowest-lying pure rotational transitions of H_2 in a wide variety of galactic and extragalactic sources, permitting for the first time a direct probe of the conditions in the dominant constituent of the ISM.

IRAS's detection of mid-IR emission from early-type galaxies left open the question of its origin – whether it was from late-type stars, hot interstellar dust, or an AGN. Comparison of several ISO mid-IR colors has shown that all mechanisms play a role and has permitted a separation of the components.

ISO provided the first opportunity to look at the full 2–200 μm spectra of galaxies at resolving powers of up to 30 000. Figure 13 shows the wealth of spectral features, contained in this region, which can be used as diagnostics of the physical and chemical conditions in galaxies. There are dust continua at different temperatures, solid-state features from dust grains, a multitude of atomic and ionic fine-structure

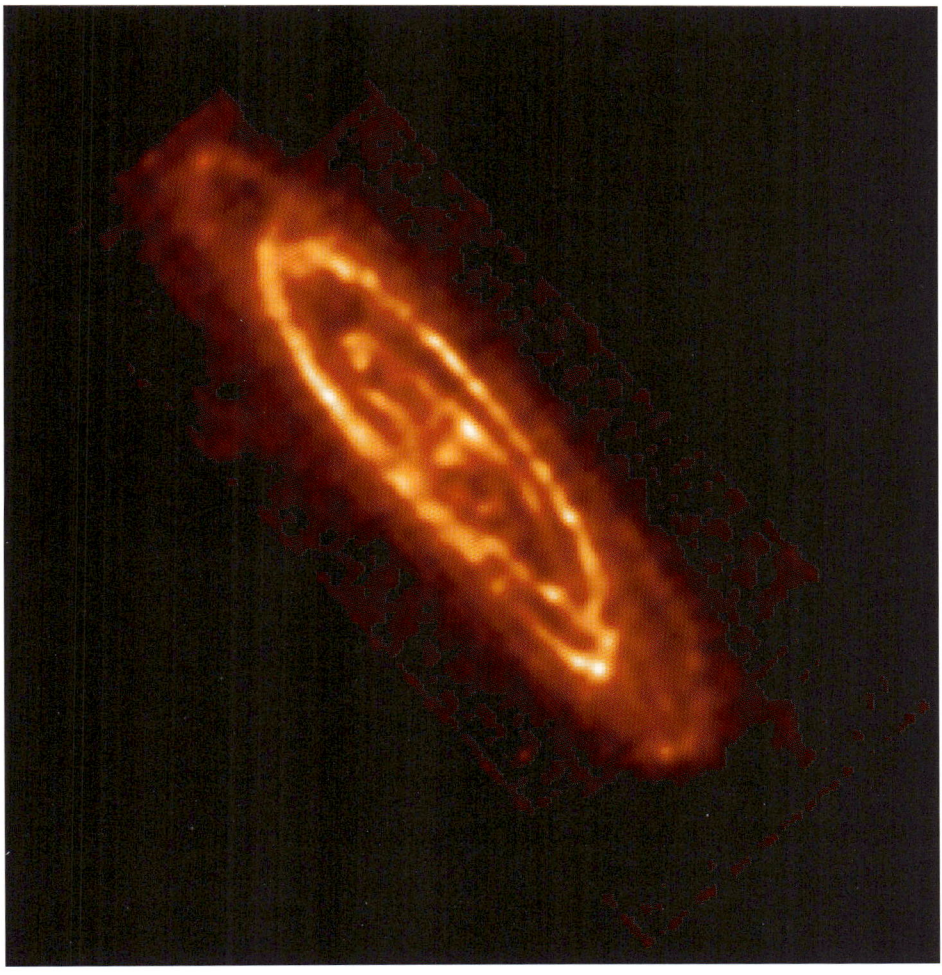

Figure 12 An image of M31 at 175 μm (Haas *et al.* 1998).

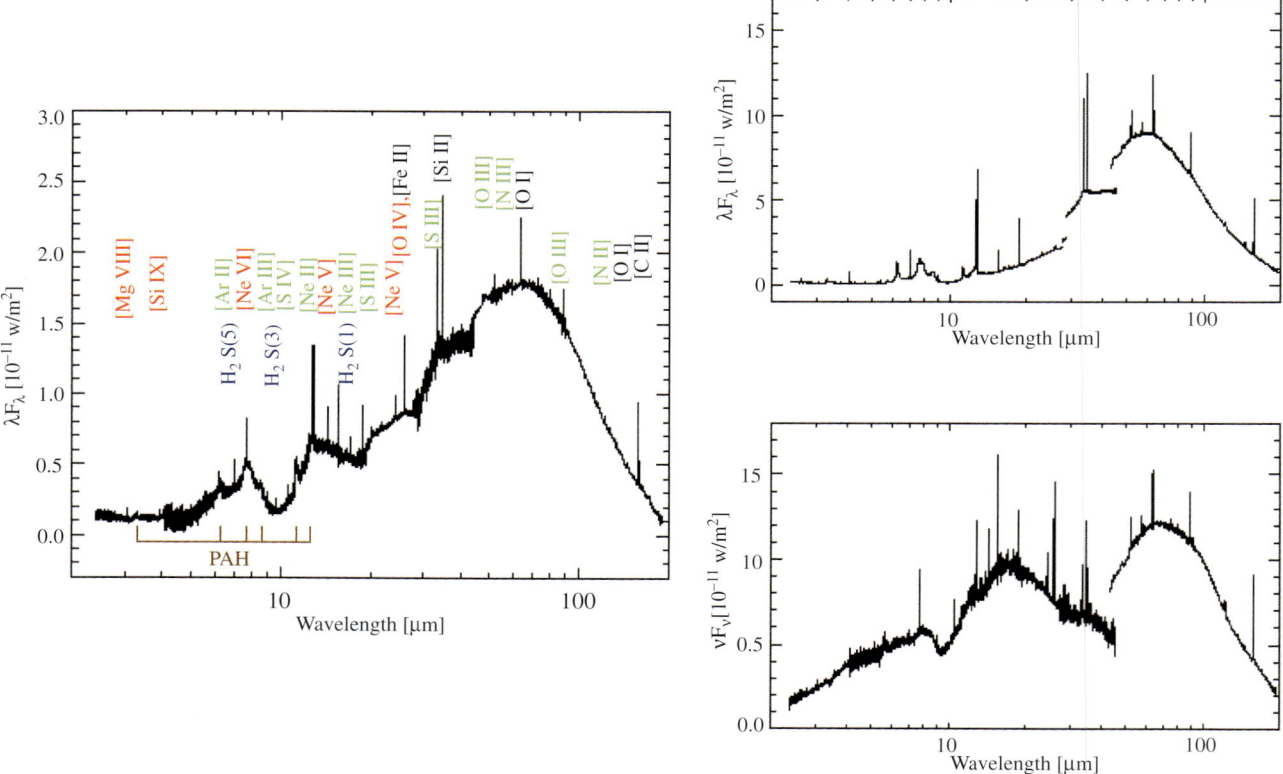

Figure 13 Examples of ISO spectra of galaxies from 2 to 200 μm, showing the Circinus Galaxy, the starburst galaxy M82, and the AGN NGC 1068 (Genzel and Cesarsky 2000, and references therein). Note the wide variety of diagnostic lines, including molecular hydrogen (lettered in blue), low excitation atomic and ionic fine structure lines (black), low-lying (excitation potential<50 eV) ionic fine structure lines (green) and more highly excited ionic lines (red).

lines with a wide range of ionization and excitation states, and molecular lines in emission and absorption for species such as H_2, H_2O, and OH. These lines trace different physical regions, including the molecular ISM, photodissociation regions, H II regions excited by massive stars, shocks, X-ray excited gas, and "coronal" gas photoionized by a central AGN engine. A nice example of ISO confirming a hypothesis is the detection in the near-IR of the transitions responsible for pumping the intense radio "mega-maser" OH emission (Skinner *et al.* 1997). An example where ISO has provided a wealth of data whose interpretation is causing controversy and challenging theoretical models concerns the use of far-IR fine-structure lines, such as [C II] at 158 μm and [O I] at 63 μm, for estimating density and radiation intensity in photodissociation regions. Work by a number of groups (see Helou (1999) for details) has led to a picture in which the ratio of [C II] to far-IR emission rises by an order of magnitude as galaxies move away from complete quiescence, reaches a broad maximum for normal galaxies with active star-forming regions, then decreases again by more than an order of magnitude as galaxies begin to approach the extreme properties of starbursts. However, the reasons for this behavior are not clear.

Dusty starburst galaxies are an important research topic, the key issues being the form of the stellar initial mass function and how the starburst evolves. ISO has contributed to this topic by studying the content of the most massive stars via nebular spectroscopy. It has been shown that H II regions in these galaxies have low excitation, meaning that, on average, hot stars ($T > 40\,000$ K) more massive than $35-40\,M_\odot$ are not present, either because they were never formed or because they have already completed their life cycles and disappeared. The likely explanation is that starbursts are episodic and rapidly aging, with star formation providing strong negative feedback that quenches the burst on the timescale of an OB star's lifetime (10^6–10^7 years). A superb example of ISO's studies of starbursts triggered by interactions concerns the Antennae (Mirabel *et al.* 1998; Figure 14). As shown by the IR contours superimposed on the HST image, the optical/UV data show only part of the picture of star formation on a galactic scale. Note that the most intense IR emission comes from the optically dark region between the two nuclei, where the gas-rich disks of the galaxies are in collision. This region is the site of the most recent star formation in this system.

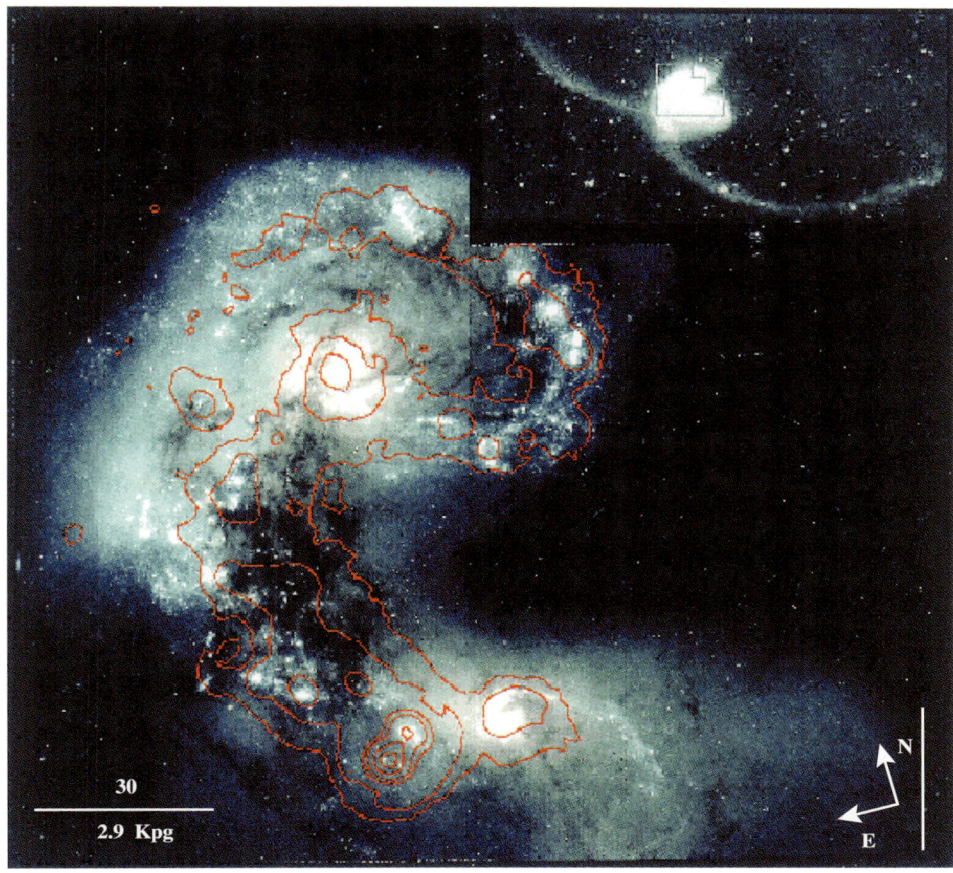

Figure 14 Contours of 15 μm emission measured by ISO superposed on a V/I band HST image (Mirabel 1998) of the Antennae galaxies. The strongest IR emission comes from the region where the two galaxies are colliding (just below the center of the image) and not from the nucleus of NGC 4038 (top right) or NGC 4039 (bottom right).

The ULIRGs, established as a population by IRAS, have quasar-like luminosities emerging almost completely at far-IR wavelengths. They are most often mergers of gas-rich disk galaxies. However, whether the power source of these systems is accretion onto a massive black hole ("monsters") or intense starbursts ("babies") has been a topic of hot debate. ISO spectroscopy, penetrating the obscuring dust and probing the hardness of the intrinsic radiation field (hard for AGNs and soft for starbursts) has allowed a new look at the issues of energetics, dynamics, and evolution of these objects. As seen in Figure 13 the mid-IR spectra of the starburst galaxy M82 is very different from that of the Seyfert NGC 1068. M82 has strong UIB features, prominent low-excitation fine-structure lines, and a weak near-IR continuum, whereas NGC 1068 has almost no UIB features, faint highly excited lines, and a strong mid-IR continuum (Genzel and Cesarsky 2000, and references therein). Thus, "diagnostic diagrams," for example, plotting ratios of high-to-low ionization lines on one axis and UIB strength on the other, have been used by Genzel *et al.* (1998) to distinguish between different energy sources, showing that bright ULIRGs lie between pure AGNs and pure starbursts, and appear to be composite objects with the AGNs and starbursts appearing to be clearly distinguished by the data. Lutz *et al.* (1998) and others extended this work to fainter sources using just the strength of the UIB features (Figure 15). In summary, ISO has shown that for ULIRGs both AGNs and nuclear starbursts play a role, even in the same object, recent massive star formation tending to dominate at the lower luminosities and the AGN fraction increasing significantly at the higher luminosities.

ISO made a number of deep surveys at both short and long IR wavelengths, providing for the first time information on the IR emission of galaxies at $z \geq 0.5$. These measurements address the issue of the cosmic history of star formation, including whether the optical/UV measurements tell the full story or whether dust-enshrouded star formation (as discussed above for the more local Universe) plays a significant role in the early Universe. At wavelengths of 15 μm (and also 7 μm), a range of shallow, deep and

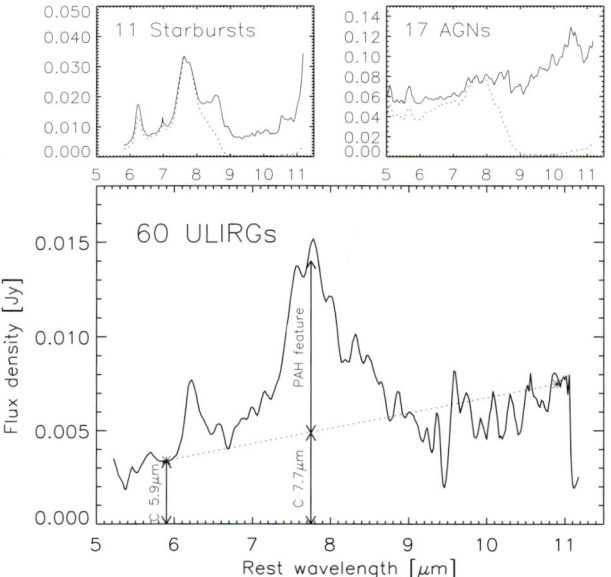

Figure 15 Average mid-IR spectra of ULIRGs, compared to that of starburst galaxies and AGN (Rigopoulou *et al.* 1999).

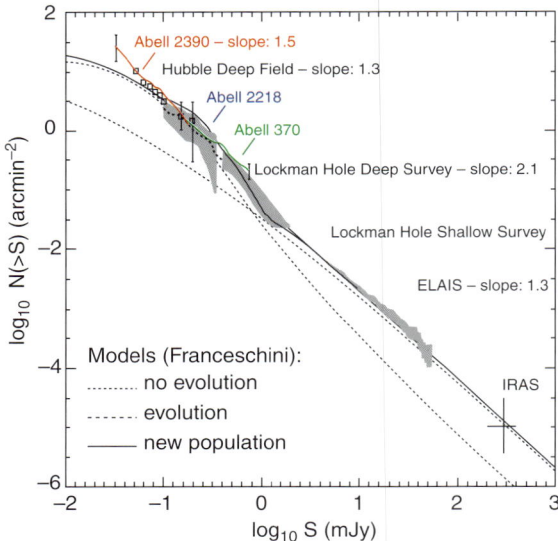

Figure 16 Log–log plot of the number of objects, N, detected at 15 μm above a given flux level, S (mJy) in a number of surveys with ISO, including the European Large-Area Infrared Space Observatory Survey (ELAIS), surveys in various parts of the sky (Lockman Hob and Hubble Deep field) and those using gravitationally-lensing clusters (Abell 370, 2218, 2390). (Metcalfe *et al.* 2000, personal communication; adapted from Elbaz *et al.* 1999 and Metcalfe *et al.* 1999.)

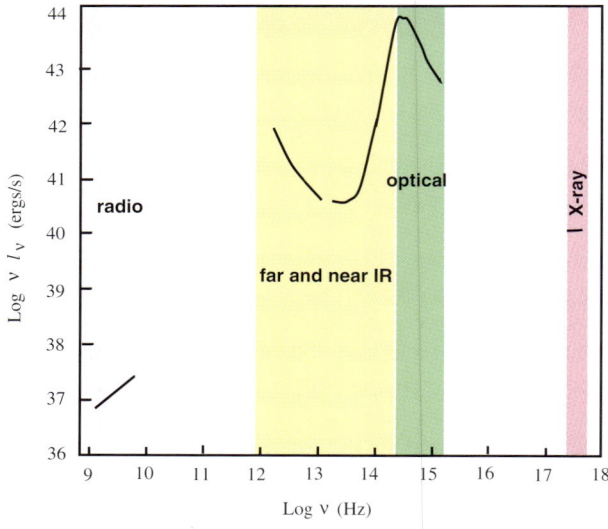

Figure 17 Schematic spectral energy distribution of a normal spiral galaxy, representing the luminosity emitted at each spectral frequency.

ultra-deep surveys have been made (Elbaz *et al.* 1999) trading between depth and sky coverage and reaching down to flux densities of around 30 μJy. Figure 16 summarizes the source counts from these surveys. Counts at the fainter levels rule out models without strong cosmic evolution of the mid-IR emission of galaxies. ISO's deep surveys at long wavelengths, mainly at 175 μm, have been reviewed by Puget and Lagache (1999). Limiting fluxes of ~100 mJy have been achieved, and the counts are consistent with models that include strong evolution. Individual sources detected make up about 10% of the cosmic far-IR background, suggesting that ISO is detecting part of a new population of luminous moderate- to high-redshift galaxies. Fluctuations in this far-IR background have also been reported (Lagache and Puget 2000), giving a glimpse of the properties of fainter populations of discrete sources which will be studied in detail by other missions: XMM-Newton, Chandra, SIRTF, FIRST, and so on.

X-RAY AND IR PROPERTIES IN THE BIG PICTURE

Understanding the structure, formation, and evolution of galaxies is one of the main themes of present-day astrophysics. This quest is made difficult by the complexity of galaxies combined with our limited knowledge of their observational characteristics. Different galactic components (old, new, evolved stars; AGNs; the ISM – gaseous, dusty, hot, cold) contribute in different amounts to the observed emission at different wavelengths, from the radio to X-rays (Gallagher and Fabbiano 1990). Since different emission bands have different sensitivities to absorption, their comparison may also give us some insight into the dust content of the emitting regions. Figure 17 shows a typical spectral energy distribution of normal galaxies. While the emission is dominated by the optical stellar light, important phenomena may be more prominent at other wavelengths. As discussed above, star-formation-related phenomena are best studied in radio and IR, where dust obscuration is less severe and where the energy of young blue stars is re-radiated by hot dust. In X-rays, the resulting hot ISM, SNR shells, and accretion binaries are uniquely visible.

Comparison of global emission properties in a wide range of wavelengths can give us precious insight into the relation between these components, the origin of some parts of the emission spectrum, and the relative importance of different galaxian components throughout the Hubble sequence.

While the study of the statistical properties of samples of galaxies observed at different wavelengths had already been used in comparisons of radio and optical properties, with the advent of space observatories this approach could be extended to encompass a much wider range. From correlation studies including the X-ray band, it is clear that there is a strong connection between X-ray, far-IR, and radio continuum emission in galaxies with pronounced spiral arms, and in irregular galaxies, that points to a common origin through star-formation-related processes (Fabbiano and Shapley 1999). The presence of halo emission in E and S0 galaxies, in excess of the baseline emission of Milky-Way-type X-ray sources, was strongly supported by comparisons of X-ray and B-band correlations in elliptical and spiral galaxies (Trinchieri and Fabbiano 1985). In elliptical galaxies, comparisons among radio continuum, far-IR, optical, and X-ray emission suggest that a hot ISM is a key factor for the generation of the radio continuum emission, and that the nuclei responsible for the radio continuum may be fueled by cooling flows rather than through episodic events of external accretion from close encounters with other galaxies (Eskridge et al. 1995a). These are just examples of what promises to become a fruitful approach. Once larger and better samples are observed, studies can be expanded to take a new look at the present results with increased significance and to explore new territory, including the behavior of these global emission properties with increasing redshift, that can give us novel constraints on models of galaxy evolution.

Multi-wavelength comparisons are also fruitful in the detailed study of individual galaxies. For example, the original conclusion of nuclear bipolar outflow from the starburst galaxy NGC 253 (Fabbiano and Trinchieri 1984) was reinforced by subsequent OH data (Turner 1985). For the Antennae galaxies, comparison of the Rosat HRI (5″ resolution) image with different optical band (I and U), Hα, radio continuum, and CO firmly associates the intense X-ray emission with the star-forming H II regions (Fabbiano et al. 1997), while ISO has shown that the most intense IR emission is associated with the dusty impact region (Figure 14).

FUTURE PROSPECTS

X-ray observations give us a unique view of the hot and massive components of galaxies, and help us to address fundamental questions of galaxy structure and evolution, as well as the physics of extreme gravity.

With Chandra and XMM-Newton we are at an exciting time for new X-ray explorations of galaxies. Chandra is giving us the most detailed ever X-ray images of the sky, and has already demonstrated its power in resolving the images of galaxies in their components. However, galaxies are faint, and Chandra's collecting area will require very long observing times to barely detect X-ray binaries in the relatively nearby Virgo Cluster. To gather spectral information, even longer observations will be required, and may be infeasible. XMM-Newton has a larger area, but also a significantly larger beam, so that crowded regions will not be resolved into their constituents. The poorer angular resolution of XMM-Newton may also be a problem when it comes to deep surveys of the Universe and exploring the X-ray evolution of galaxies. For these types of studies, as well as for very in-depth studies of the populations of X-ray sources in galaxies, we need a significant increase in the telescope area, while retaining a Chandra-like angular resolution.

Planned future X-ray observatories (Constellation-X, under study by NASA, and XEUS, the X-ray Evolving Universe Spectroscopy mission, under study in Europe) will only partially achieve these goals. While the spectra and bandwidth characteristics of these missions will meet our requirements, in both cases the angular resolution is significantly worse than Chandra's. Based on what Chandra has shown and on the characteristics of the objects we want to study – galaxies are complex objects – a Chandra-like resolution is a must to progress in this field.

ISO has already led to major progress in our understanding of normal galaxies, particularly with the use of the mid-IR spectrum as a valuable diagnostic of activity, decomposition of the IR luminosity into its different components, more accurate determinations of dust masses, use of IR spectroscopy to constrain parameters of the active ISM, and distinguishing between various energy sources. These new insights have helped to build a picture of the overall star formation history of the Universe, and are key to understanding the highest-luminosity objects that we must use to probe the early Universe.

The next major space-based IR observatory is NASA's Space Infrared Telescope Facility (SIRTF; e.g., Werner (1999), and references therein). This is due for launch at the end of 2001 and will carry three scientific instruments, the Multiband Imaging Photometer for SIRTF (MIPS), the Infrared Spectrograph (IRS), and the Infrared Array Camera (IRAC). A new, deeper all-sky survey at IR wavelengths from 50 to 200 μm will be carried out by ISAS's

ASTRO-F satellite (e.g., Okuda 1999) due for launch in August 2003. The resulting catalog is expected to contain up to 10 million sources, going around 10–100 times deeper than IRAS, and is certain to contain much interesting information on extragalactic sources. At longer wavelengths, from 80 to 670 μm, the key themes of ESA's Far Infra Red and Submillimetre Telescope (FIRST; e.g., Bonnet et al. 1999), scheduled for launch in 2007, include galaxy formation in the early Universe and large- and small-scale star formation in our own and external galaxies. Looking further ahead in time, it now seems likely that the NASA-led Next Generation Space Telescope (NGST) will have mid-IR capabilities. The scope and sensitivity of these missions and instruments is such that we may confidently expect another leap in our understanding of the topics described above, and more!

REFERENCES

Arimoto, N., Matsushita, K., Ishimaru, Y., Ohashi, T., and Renzini, A. (1997). The iron discrepancy in elliptical galaxies after ASCA. *Astrophysical Journal*, **477**, 128.

Bonnet, R.M. Fridlund, C.V.M., Pilbratt, G.L. and Tauber, J.A. (1999). ESA's future plans in infrared and sub-millimetre astronomy. In P. Cox and M.F. Kessler (eds), *The Universe as Seen by ISO*, ESA SP-427, Noordwijk, pp. 95–101.

Bregman, J.N. and Glassgold, A.E. (1982). X-ray observations to detect hot coronae around galaxies. *Astrophysical Journal*, **263**, 564–570.

Brighenti, F. and Mathews, W.G. (1996). Structure and evolution of interstellar gas in flattened, rotating elliptical galaxies. *Astrophysical Journal*, **470**, 747–761.

Buote, D.A. and Canizares, C.R. (1997). The nature of the X-ray emission and the mass distributions in two early-type galaxies. *Astrophysical Journal*, **474**, 650–658.

Buote, D.A., Canizares, C.R. and Fabian, A.C. (1999). X-ray emission lines and multiphase gas in elliptical galaxies and galaxy clusters. *Monthly Notices of the Royal Astronomical Society*, **310**, 483–510.

Buote, D.A and Fabian A.C. (1998). X-ray spectral analysis of elliptical galaxies from ASCA: The Fe abundance in a multiphase medium. *Monthly Notices of the Royal Astronomical Society*, **296**, 977–994.

Ciotti, L., D'Ercole, A., Pellegrini, S. and Renzini, A. (1991). Winds, outflows, and inflows in X-ray elliptical galaxies. *Astrophysical Journal*, **376**, 380–403.

Cox, P. and Kessler, M.F. (eds) (1999). *The Universe as Seen by ISO*, ESA SP-427, Noordwijk.

Elbaz, D., Aussel, H., Cesarsky, C.J., Desert, F.X., Fadda, D., Franceschini, A., Harwit, M., Puget, J.L. and Starck, J.L. (1999). ISOCAM Extragalactic mid-infrared deep surveys unveiling dust-enshrouded star formation in the universe. In P. Cox and M.F. Kessler (eds), *The Universe as Seen by ISO*, ESA SP-427, Noordwijk, pp. 999–1006.

Eskridge, P.B., Fabbiano, G. and Kim, D.-W. (1995a). A multiparametric analysis of the Einstein sample of early-type galaxies: 1. Luminosity and ISM parameters. *Astrophysical Journal*, Suppl., **97**, 141–184.

Eskridge, P.B., Fabbiano, G. and Kim, D.-W. (1995b). A multiparametric analysis of the Eintein Sample of early-type galaxies: 2. Galaxy formation history and properties of the interstellar medium. *Astrophysical Journal*, **442**, 523–537.

Fabbiano, G. (1989). X-rays from normal galaxies. *Annual Review of Astronomy and Astrophysics*, **27**, 87–138.

Fabbiano, G. (1995a). Normal galaxies and their X-ray binary populations. In W.H.G. Lewin, J. van Paradijs and E.P.J. van den Heuvel (eds), *X-Ray Binaries*, Cambridge University Press, pp. 390–418.

Fabbiano, G. (1995b). X-ray properties of E and S0 galaxies. In A. Buzzoni, A. Renzini, A. Serrano, (eds), *Fresh Views of Elliptical Galaxies*, ASP Conference Series Vol. 86, Astronomical Society of the Pacific, San Francisco, p. 103.

Fabbiano, G. (1996a), The X-ray emission of galaxies. In H.U. Zimmermann, J.E. Truemper and H. Yorke, (eds), *Röntgenstrahlung from the Universe*, MPE Report 263, Max-Planck-Institut für extraterrestrische Physik, Garching, Germany, pp. 347–356.

Fabbiano, G. (1996b). The X-ray properties of LINERs. In M. Eracleous, A. Koratkar, C. Leitherer and L. Ho, (eds). *The Physics of LINERS in View of Recent Observations*, ASP Conference Series Vol. 103, Astronomical Society of the Pacific San Francisco, p. 56.

Fabbiano, G. (1998). Super-Eddington sources in galaxies. In K. Koyama, S. Kitamoto, M. Itoh (eds), *The Hot Universe*, IAU Symposium 188, Kluwer, Dordrecht, pp. 93–96.

Fabbiano, G. and Juda, J.Z. (1997). ROSAT observations of the Sombrero Galaxy: Discovery of an X-ray active nucleus. *Astrophysical Journal*, **476**, 666–676.

Fabbiano, G., Kim, D.-W. and Trinchieri (1992). An X-ray catalog and atlas of galaxies. *Astrophysical Journal*, Suppl., **80**, 531–644.

Fabbiano, G., Schweizer, F. and Mackie, G. (1997). ROSAT HRI observations of NGC 4038/4039, The Antennae Galaxies. *Astrophysical Journal*, **478**, 542–553.

Fabbiano, G. and Shapley, A. (1999). A multi-wavelength look at spiral galaxies. In F. Giovannelli and L. Sabau-Graziati (eds), *Multifrequency Behaviour of High Energy Cosmic Sources*, Memorie della Societa Astronomica Italiana, **70**, 1181–1189.

Fabbiano, G. and Trinchieri, G. (1984). The complex X-ray emission of NGC 253. *Astrophysical Journal*, **286**, 491–497.

Fabian, A.C., Schwarz, J. and Forman, W. (1980). The retention of hot gas in elliptical galaxies. *Monthly Notices of the Royal Astronomical Society*, **192**, 135–142.

Fabricant, D. and Gorenstein, P. (1983). Further evidence for M87's massive, dark halo. *Astrophysical Journal*, **267**, 535–546.

Forman, W., Jones, C. and Tucker, W. (1985). Hot coronae around early-type galaxies. *Astrophysical Journal*, **293**, 102–119.

Forman, W., Schwarz, J., Jones, C., Liller, W. and Fabian, A.C. (1979). X-ray observations of galaxies in the Virgo Cluster. *Astrophysical Journal*, **234**, L27–L31.

Franceschini, A. in "Extragalactic Astronomy in the Infrared", T. Thuan, G. Mamon, J.T.T. Van (eds), XVIIth Moriond Astr. Meeting, Ed. Frontiere, p. 509.

Fukazawa, Y., Makishima, K., Tamura, T., Ezawa, H., Xu, H., Ikebe, Y., Kikuchi, K., and Ohashi, T. (1998). ASCA measurements of silicon and iron abundances in the intracluster medium. *Publications of the Astronomical Society Japan*, **50**, 187–193.

Gallagher, J. Fabbiano, G. (1990). Panchromatic observations of galaxies: Introductory remarks. In G. Fabbiano, J.S. Gallagher, A. and Renzini, (eds), *Windows on Galaxies*, Kluwer, Dordrecht, p. 1.

Genzel, R. (1999). Galaxies as seen by ISO. In P. Cox and M.F. Kessler (eds), *The Universe as Seen by ISO*, ESA SP-427, Noordwijk, pp. 13–20.

Genzel, R. and Cesarsky, C.J. 2000, Extragalactic Results from the Infrared Space Observatory, *Annual Review Astrophysics*, **38**, 761–814.

Genzel, R., Lutz, D., Sturm, E., Egami, E., Kunze, D., Moorwood, A.F.M., Rigopoulou, D., Spoon, H.W.W., Sternberg, A., Tacconi-Garman, L.E., Tacconi, L., and Thatte, N. (1998). What powers ultraluminous IRAS galaxies? *Astrophysical Journal*, **498**, 579–605.

Giacconi, R., Branduardi, G., Briel, U., Epstein, A., Fabricant, D., Feigelson, E., Forman, W., Gorenstein, P., Grindlay, J., Gursky, H., Harnden, F.R., Henry, J.P., Jones, C., Kellogg, E., Koch, D.,

Murray, S., Schreier, E., Seward, F., Tananbaum, H., Topka, K., Van Speybroeck, L., Holt, S.S., Becker, R.H., Boldt, E.A., Serlemitsos, P.J., Clark, G., Canizares, C., Markert, T., Novick, R., Helfand, D. and Long, K. (1976). The Einstein/HEAO 2/X-ray Observatory. *Astrophysical Journal*, **230**, 540–550.

Giacconi, R. and H. Gursky (eds) (1974). *X-ray Astronomy*, Reidel, Dordrecht.

Haas, M., Lemke, D., Stickel, M., Hippelein, H., Kunkel, M., Herbstmeier, U. and Mattila, K. (1998) Cold dust in the Andromeda Galaxy mapped by ISO. *Astronomy and Astrophysics*, **388**, L33–L36.

Helou, G. (1999). The ISO view of star forming galaxies. In P. Cox and M.F. Kessler (eds), *The Universe as Seen by ISO*, ESA SP-427, Noordwijk, pp. 797–803.

Kessler, M.F., Steinz J.A., Anderegg, M.E., Clavel, J., Drechsel, G., Estaria, P., Faelker, J., Riedinger, J.R., Robson, A., Taylor, B.G. and Ximenez de Ferran, S. (1996). The Infrared Space Observatory (ISO) mission. *Astronomy and Astrophysics*, **315**, L27–L31.

Knapp, G.R., Guhathakurta, P., Kim, D.-W. and Jura, M. (1989). Interstellar matter in early-type galaxies: I. IRAS flux densities. *Astrophysical Journal*, Suppl., **70**, 329–387.

Kim, D.-W. and Fabbiano, G. (1995). ROSAT PSPC observations of the early-type galaxies NGC 507 and NGC 499: Central cooling and mass determination. *Astrophysical Journal*, **441**, 182–199.

Kim, D.-W., Fabbiano, G. and Mackie, G. (1998). ROSAT X-ray observations of the radio galaxy NGC 1316 (Fornax A). *Astrophysical Journal*, **497**, 699–712.

Kim, D.-W. and Fabbiano, G. and Trinchieri, G. (1992). The X-ray spectra of galaxies: II. Average spectral properties and emission mechanisms. *Astrophysical Journal*, **393**, 134–148.

Kormendy, J. and Richstone, D. (1995). Inward bound: The search for supermassive black holes in galactic nuclei. *Annual Review of Astronomy and Astrophysics*, **33**, 581–624.

Kraft, R.P., Forman W., Jones, C., Kenfer, A., Murray, S.S., Aldcroft, T.L., Elvis, M.S., Evans, I.N., Fabbiano, G., Isobe, T., Jerius, D., Karovska, M., Kim, D.-W., Prestwich, A.H., Primini, F.A., Schwartz, D.A., Schreier, E.J., Vikhlinin, A.A. (2000). *Astrophysical Journal Letters*, **531**, L9–L12.

Lagache, G. and Puget, J.L. (2000). Detection of the extra-galactic background fluctuations at 170 µm. *Astronomy and Astrophysics*, **355**, 17–22.

Loewenstein, M. and Mathews, W.G. (1991). Hot gas metallicity and the history of supernova activity in elliptical galaxies. *Astrophysical Journal*, **373**, L445–451.

Loewenstein, M., Mushotzky, R.F., Tamura, T., Ikebe, Y., Makishima, K., Matsushita, K., Awaki, H. and Serlemitsos, P.J. (1994). Discovery and implications of very low metal abundances in NGC 1404 and 4374. *Astrophysical Journal*, **436**, L75–L78.

Long, K.S. and Helfand, D.J. (1979). Supernova remnants in the Large Magellanic Cloud. *Astrophysical Journal*, **234**, L77–L81.

Lutz, D. Spoon, H.W.W., Rigopoulou, D., Moorwood, A.F.M. and Genzel, R. (1998). The nature and evolution of ultraluminous infrared galaxies: A mid-infrared spectroscopic survey. *Astrophysical Journal*, **505**, L103–L107.

Mackie, G. and Fabbiano, G. (1997). Environmnetal and internal optical properties and the X-ray content of E and S0s. In M. Arnaboldi, G.S. Da Costa and P. Saha (eds), *The Nature of Elliptical Galaxies*, ASP Conference Series Vol. 116, Astronomical Society of the Pacific, San Francisco, p. 401.

Makishima, K. (1994). In F. Makino and T. Ohashi (eds), *New Horizon of X-ray Astronomy*, Universal Academy Press, Tokyo, p. 171.

Makishima, K. Kubota, A., Mizuno, T., Ohnishi, T., Tashiro, M., Aruga, Y., Asai, K., Dotani,T., Mitsuda, K., Ueda, Y., Uno, S., Yamaoka, K., Ebisawa, K., Kohmura, Y. and Okada, K. (2000). The nature of ultraluminous compact X-ray sources in nearby spiral galaxies. *Astrophysical Journal*, **535**, 632–643.

Mathews, W.G. and Brighenti, F. (1999). Formation of low-mass stars in elliptical galaxy cooling flows. *Astrophysical Journal*, **526**, 114–129.

Matsushita, K., Makishima, k., Awaki, H., Canizares, C.R., Fabian, A.C., Fukazawa, Y., Loewenstein, M., Matsumoto, H., Mihara, T., Mushotzky, R.F., Ohashi, T., Ricker, G.R., Serlemitsos, P.J., Tsuru, T., Tsusaka, Y. and Yamazaki, T. (1994). Detections of hard X-ray emissions from bright early-type galaxies with ASCA. *Astrophysical Journal*, **436**, L41–L45.

Matsushita, K., Makishima, K., Rokutanda, E., Yamasaki, N. and Ohashi, T. (1997). New measurement of metal abundance in the elliptical galaxy NGC 4636 with ASCA. *Astrophysical Journal*, **488**, L125–L128.

Matsushita, K., Ohashi, T., Makishima, K. (2000). *Publications of the Astronomical Society of Japan*, **52**, 685–710.

McKee, C.F. (1995). The multiphase interstellar medium. In A. Ferrara, C.F. McKee, C. Heiles and P.R. Shapiro (eds), *The Physics of the Interstellar Medium and Intergalactic Medium*, ASP Conference Series Vol. 80, Astronomical Society of the Pacific, San Francisco, p. 292.

Metcalfe, L., Altieri, B., McBreen, B., Kneib, J.-P., Delaney, M., Biviano, A., Kessler, M.F., Leech, K., Okumura, K., Schulz, B., Elbaz, D. and Aussel, H. (1999). A deep and ultra-deep ISOCAM cosmological survey through gravitationally lensing clusters of galaxies. In P. Cox and M.F. Kessler (eds), *The Universe as Seen by ISO*, ESA SP-427, Noordwijk, pp. 1019–1022.

Mirabel, I.F., Vigroux, L., Charmandaris, V., Sauvage, M., Gallais, P., Tran, D., Cesarsky, C., Madden, S.C. and Duc, P.-A. (1998). The dark side of star formation in the Antennae galaxies. *Astronomy and Astrophysics*, **333**, L1–L4.

Mushotzky, R.F., Loewenstein, M., Awaki, H., Makishima, K., Matsushita, K. and Matsumoto, H. (1994). ACCA observation of NGC 4636: Dark matter and metallicity gradient. *Astrophysical Journal*, **436**, L79–L82.

Neugebauer, G., Habing, H.J., van Duinen, R., Aumann, H.H., Baud, B., Beichman, C.A., Beintema, D.A., Boggess, N., Clegg, P.E., de Jong, T., Emerson, J.P., Gautier, T.N., Gillett, F.C., Harris, S., Hauser, M.G., Houck, J.R., Jennings, R.E., Low, F.J., Marsden, P.L., Miley, G., Olnon, F.M., Pottasch, S.R., Raimond, E., Rowan-Robinson, M., Soifer, B.T., Walker, R.G., Wesselius, P.R. and Young, E. (1984). The Infrared Astronomical Satellite (IRAS) mission. *Astrophysical Journal*, **278**, L1–L6.

Okuda, H. (1999). From IRTS to IRIS and beyond. In P. Cox and M.F. Kessler (eds), *The Universe as Seen by ISO* ESA SP-427, Noordwijk, pp. 111–114.

Pellegrini, S. (1999). Global X-ray emission and central properties of early-type galaxies, *Astronomy and Astrophysics*, **351**, 487.

Peres, G., Reale, F., Collura, A. and Fabbiano, G. (1989). Time variability of the X-ray sources in M33. *Astrophysical Journal*, **336**, 140–151.

Puget, J.L. and Lagache, G. (1999). Long wavelength cosmological surveys with ISO. In P. Cox and M.F. Kessler (eds), *The Universe as Seen by ISO*, ESA SP-427, Noordwijk, pp. 1007–1010.

Rigopoulou, D., Genzel, R., Lutz, D., Kunze, D., Moorwood, A.F.M., Spoon, H.W.W., Sturm, E., Thornley, M. and Tran, D. (1999). Ultraluminous IRAS galaxies as seen with ISO. In P. Cox and M.F. Kessler (eds), *The Universe as Seen by ISO*, ESA SP-427, Noordwijk, pp. 833–838.

Sanders, D.B. and Mirabel, I.F. (1996). Luminous infrared galaxies. *Annual Review of Astronomy and Astrophysics*, **34**, 749–792.

Skinner, C., Smith, H.A., Sturm, E., Barlow, M.J., Cohen, R.J. and Stacey, G.J. (1997). A starburst origin of the OH-megamaser emission from the galaxy ARP 220. *Nature*, **386**, 472–474.

Soifer, B.T., Neugebauer, G. and Houck, J.R. (1987). The IRAS view of the extragalactic sky. *Annual Review of Astronomy and Astrophysics*, **25**, 187–230.

Soifer, B.T., Rowan-Robinson, M., Houck, J.R., de Jong, T., Neugebauer, G., Aumann, H.H., Beichman, C.A., Boggess, N., Clegg, P.E., Emerson, J.P., Gillett, F.C., Habing, H.J., Hauser, M.G., Low, F.J., Miley, G. and Young, E. (1984). Infrared galaxies in the IRAS minisurvey. *Astrophysical Journal*, **278**, L71–L74.

Spitzer, L. (1956). On a possible interstellar galactic corona. *Astrophysical Journal*, **124**, 20–34.

Takano, M., Mitsuda, K., Fukazawa, Y. and Nagase, F. (1994). Properties of M33 X-8, the nuclear source in the nearby spiral galaxy. *Astrophysical Journal*, **436**, L47–L50.

Trinchieri, G. and Fabbiano, G. (1985). A statistical analysis of the Einstein Normal Galaxy Sample: Part Two – Elliptical and S0 galaxies. *Astrophysical Journal*, **296**, 447–457.

Trinchieri, G., Fabbiano, G. and Canizares, C.R. (1986). The X-ray surface brightness distribution and spectral properties of six early-type galaxies. *Astrophysical Journal*, **310**, 637–659.

Turner, B.E. (1985). VLA observations of OH in galaxies: I. NGC 253 and its nuclear plume. *Astrophysical Journal*, **299**, 312–333.

Valentijn, E.A, and van der Werf, P.P. (1999). First extragalaxtic direct detection of large-scale molecular hydrogen in the disk of NGC 891. *Astrophysical Journal*, **522**, L29–L33.

Van Speybroeck, L., Epstein, A., Forman, W., Giacconi, R., Jones, C., Liller, W. and Smarr, L. (1979). Observations of X-ray sources in M31. *Astrophysical Journal*, **234**, L45–L49.

Werner, M.W. (1999). Scientific Opportunities with SIRFT. In P. Cox and M.F. Kessler (eds), *The Universe as seen by ISO*, ESA SP-427, Noordwijk, pp. 119–122.

The Milky Way

26

STEVEN L. SNOWDEN*

The hot part of the interstellar medium

In the preparation of this chapter there arose issues of semantics. Specifically, what does the "hot part" mean in reference to the interstellar medium (ISM), and just what part of the hot ISM should be considered? As the original task for this chapter referred to the X-ray and EUV (extreme ultraviolet), appropriate temperatures are in the several 10^5 K to several 10^6 K range. In a Universe where the fundamental temperature is currently about 3 K and in a Galaxy where most of the material in the ISM has temperatures less than 10^4 K, any component at 10^6 K must have been produced by fairly extreme and unusually spectacular methods. The most common of these methods in the general galactic disk are supernovae, occasionally aided and abetted by the stellar winds of massive stars. However, as supernova remnants (SNRs) are ably covered elsewhere in this text (see Chapter 41), the topic for this chapter required further refinement. What happens to a SNR when it slips past middle age into its dotage? All of the visual features that allow a SNR to stand out clearly from the background fade: the shock fronts (the interface region between the expanding SNR and the undisturbed ISM) with occasionally spectacular radio, optical, and X-ray emission, and the hot and relatively dense interior with bright X-ray emission. As the SNR fades it also increases in size to become a more significant, or at least a more extensive part of the ISM. In essence it becomes old and boring, and fades into the woodwork; a pale ghost of its former self.

Thus the study of the hot ISM, at least in the galactic disk, is essentially a study of what happens to SNRs that have lost their identity, though there are exceptions. Although the study of hot plasmas in the halo of the galaxy may have cosmological considerations, there are ties to the disk through the galactic gravitational potential and through "chimneys," breakouts of SNRs through the confining cooler ISM of the disk which expel hot plasma into the halo (in this chapter the "halo" of the Galaxy is defined as the region of space above the neutral material of the disk, and includes both ionized and neutral gas associated with the Galaxy and affected by its gravitational potential; this is a fairly loose definition which extends from a couple of hundred parsecs to tens of kiloparsecs, or even farther, above the plane). For the purposes of this chapter, a partially sociological definition of the hot part of the interstellar medium will be used: that part of the ISM which is (1) observed in the 0.05 to 1.0 keV band as the soft X-ray diffuse background (SXRB) and EUV background, (2) associated with the Milky Way disk or halo, and (3) not claimed by others for studies of supernova remnants or stellar wind-blown bubbles. To simplify the presentation and to acknowledge that at least the observed part originates in general from some of the same plasmas, the EUV will be included with $\frac{1}{4}$ keV X-rays, despite the historical separation by NASA of the astrophysics and astronomy programs at 0.1 keV. Furthermore, by and large this chapter will focus on the $\frac{1}{4}$ keV SXRB because it is the dominant galactic background and, at least locally, it is the best understood. Besides, SNRs that are still emitting strongly at energies greater than $\frac{1}{4}$ keV typically still stand out clearly against the general background and fail consideration (3) above.

Despite the observational difficulties in the study of has-been SNRs and their lack of spectacular features (e.g., the wealth of structure revealed by the initial Chandra observations of the more youthful Cas A and Crab Nebula SNRs), a few true believers have pursued the study of the SXRB over the last 30 years with some diligence. Like X-ray astronomy in general, it is a field that has truly required the "Century of Space Science," as photons with energies of 0.05 to 1.0 keV travel at most millimeters in Earth's atmosphere at sea level. The detectors must therefore be placed above the atmosphere at altitudes of $\gtrsim 200$ km. The SXRB

---

*NASA – Goddard Space Flight Center, Greenbelt, MD, USA

required the advent of astronomical research sounding rockets in the late 1950s and early 1960s to provide the facility for its observation. Over three decades of study, the observations of the SXRB have progressed from limited regions of the sky provided by single sounding-rocket flights, to all-sky mosaics provided by multiple flights, to all-sky surveys provided by satellite-borne observatories.

The initial discovery of the $\frac{1}{4}$ keV SXRB in the late 1960s (Bowyer et al. 1968, Henry et al. 1968, Bunner et al. 1969) was interpreted in the context of an extragalactic background. This interpretation was an obvious one suggested by the general negative correlation between the surface brightness of the SXRB and the column density of galactic neutral hydrogen (H I)*, as well as the expectation that there could possibly be extensive X-ray emission from intergalactic space as the low-energy extension of the extragalactic background above 2 keV first observed by Giacconi et al. (1962). The negative correlation observed at $\frac{1}{4}$ keV was assumed to be due to the absorption of an isotropic X-ray flux of distant (extragalactic or galactic halo) origin by the foreground material of the galactic disk. The fact that the required absorption cross-sections were smaller than the theoretical values could be explained by possible clumping of the cooler ISM. The nonzero flux observed in the galactic plane, where the Galaxy is completely opaque to extragalactic photons in this energy range, was attributed to additional background components, of either non-cosmic (e.g., a charged-particle background) or cosmic (e.g., unresolved galactic point sources) origin. Thus, the first big step toward our understanding of the SXRB went, as will be shown, in somewhat the wrong direction. However, also from the first step, the study of the SXRB was correctly linked to the ISM, in this case cooler components of the ISM that can absorb an X-ray flux of distant origin and modulate its apparent surface brightness.

By its nature, the $\frac{1}{4}$ keV SXRB is closely linked to the cooler ISM, whether or not it originates in the galactic disk or as an extragalactic background. The column density of H I (N_H, the amount of material along a line of sight) required for unit optical depth for the absorption of X-rays by the ISM is shown in Figure 1 as a function of energy. In the $\frac{1}{4}$ keV energy range this is $N_H \sim 10^{20}$ cm^{-2}. Even at its minimum high-galactic-latitude column density ($N_H \sim 5 \times 10^{19}$ cm^{-2}), the Milky Way provides roughly half an optical depth of absorption as viewed from Earth for X-rays of extragalactic origin. At lower latitudes with longer path lengths

---

*Astronomy has the quaint convention of using roman numerals to indicate the ionization state of elements. However, a single "I" indicates neutral material. (In fact, the "I" is the label for the "first spectrum" of a given element, which by convention is the spectrum of the neutral overestimate.) Thus, H I refers to neutral hydrogen and H II refers to ionized hydrogen (otherwise known as a proton). The ion O VI referred to later in this chapter is five-times-ionized oxygen, or O^{5+}.

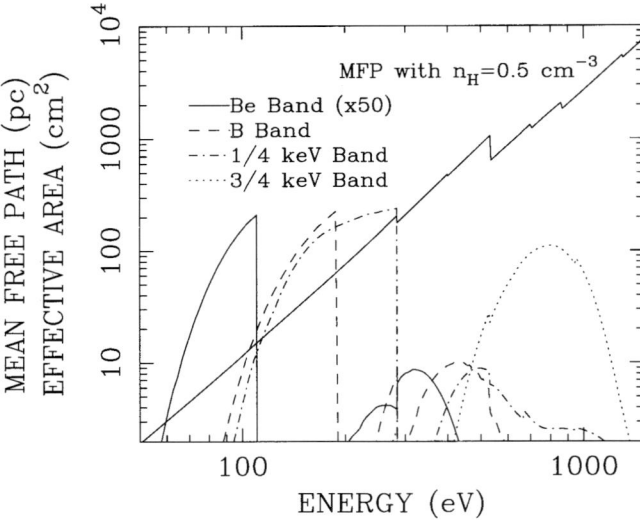

Figure 1 Average mean free path of EUV and soft X-ray photons in the Milky Way as a function of energy, along with typical band-response functions (detector effective areas) for a sample of soft X-ray and EUV detectors. The mean-free-path curve assumes a constant space density of $n_H = 0.5$ cm^{-3} and uses Morrison and McCammon (1983) absorption cross-sections. The Be band response function is from the Wisconsin detector (Bloch et al. 1986), and has been multiplied by a factor of 50 for display purposes. The B band response curve is from the Wisconsin survey (McCammon et al. 1983). The band response functions of the $\frac{1}{4}$ keV and $\frac{3}{4}$ keV bands are from Rosat (Snowden et al. 1995). The band response functions show the sensitivities of the respective instruments as a function of energy (the larger the values, the more able they are to detect X-rays of that energy).

through the Galactic disk, one optical depth is reached after a path length of ~ 100 pc, assuming a smoothly distributed ISM with a space density of 0.5 H I cm^{-3}.

EVOLUTION OF THE DATA

This section will focus on the data collected over the last three decades and how they influenced our understanding of the SXRB. It will not cover all contributions to the field (with apologies to those left out) but will hopefully give a feel for the evolution of the data. While some reference will be made to models for the origin of the SXRB, the detailed discussion of such models will be deferred to later in this chapter.

Initial observations and the early years

As noted above, observations of the SXRB have evolved from the results of a limited number of sounding-rocket flights covering limited regions of the sky to all-sky surveys produced by orbiting observatories. The first published observation was the result of a sounding-rocket scan by Bowyer et al. (1968) which observed the northern Galactic hemisphere between the center and anticenter directions (the anticenter refers to the direction in the Galactic disk which from Earth is 180° from the galactic center), scanning over the galactic pole. They observed a distinct negative correlation between the surface brightness of the SXRB in the $\frac{1}{4}$ keV band and the column density of galactic neutral hydrogen that they attributed to the absorption of an extragalactic background. The fitted absorption cross-section was significantly less than the model prediction from atomic physics, a result which was left as having mostly unspecified strong implications for the ISM after a discussion of a possible incorrect value for the He/H ratio.* Higher X-ray intensities were observed in the direction of the galactic center, which were attributed to an anomalous emission component and interpreted as possible evidence for a galactic corona (a hot galactic halo). The Bowyer et al. results should be considered in the context of the results of Gursky et al. (1963) who had identified the possibility of an extragalactic background contribution in their data at energies above 2 keV, and the suggestion by Gould and Sciama (1964) and others that the intergalactic medium might emit strongly at softer (lower) X-ray energies.

Henry et al. (1968) reported the SXRB observation of another sounding-rocket flight, and like Bowyer et al. (1968) they observed extensive emission at $\frac{1}{4}$ keV. In their analysis, they ruled out galactic corona emission (the density of the corona would need to be unreasonably high) as well as emission from external galaxies and galactic point sources as the origin (because of both spectral considerations and the excessive magnitude of the required total flux), leaving the conclusion that the background arose as free–free emission (radiation from interactions between free electrons and positive ions) from a hot, dense intergalactic medium. Bowyer and Field (1969), when not explaining the differences between the Bowyer et al. (1968) and Henry et al. (1968) results, suggested that the counts observed in the galactic plane were due to an isotropic particle background, and that the apparent absorption cross-sections were reduced from the theoretical values by clumping of the absorbing medium.** The concept of a clumpy, X-ray-absorbing ISM would play a role of some controversy in models of the SXRB for the following 20 years.

Bunner et al. (1969) presented the last of the initial triad of SXRB observations, and reported results which were more-or-less consistent with the previous two. In their analysis they considered three separate alternatives for aspects contributing to the angular structure of the $\frac{1}{4}$ keV SXRB:

1. The small effective absorption cross-sections were due, as in Bowyer and Field (1969), to clumping of the absorbing medium into optically thick clouds. Although they did not find the evidence for the existence of such clouds compelling, clumping of the H I into clouds of $N_H \sim 10^{21}$ cm^{-2} satisfied the intensity-dependence constraint of the $\frac{1}{4}$ keV data with absorption column density but was still consistent with the 0.5–1.0 keV data. However, even with the assumed clumping, an additional soft component was required to explain an excess of emission observed above the extrapolation of the extragalactic power law identified at higher energies.
2. The observed flux in the galactic plane was due to an unknown solar or terrestrial background. This assumption reduced the requirement for clumping of the ISM, but an additional, softer extragalactic emission component was still required.
3. The observed flux in the galactic plane was due to the superposition of unresolved galactic Population II sources (the older, more numerous, and less massive stars in the Galaxy, M dwarfs for example). The required source density, $\sim 10^{-2}$ pc^{-3}, was less than the population of known objects, and so could not be ruled out.

While there was some disagreement about the details of the observational results presented by the three groups, they were all consistent with requiring an emission component in excess of the extrapolation of the extragalactic power law observed above 2 keV to explain the background at $\frac{1}{4}$ keV. The "real" diffuse background was assumed to be extragalactic or galactic halo in origin while any residual flux from the galactic plane was assumed to be either non-cosmic contamination or from unresolved point sources. Diffuse X-ray emission from the interstellar medium of the galactic disk was not yet considered.

---

*While the absorption column density is measured by the column density of neutral hydrogen, N_H, derived from 21 cm measurements of the ISM (21 cm is the wavelength of the neutral hydrogen electron spin-flip transition), the helium associated with the hydrogen provides roughly half of the physical absorption at $\frac{1}{4}$ keV. Therefore if the true He/H ratio is lower than assumed, the model absorption will be overestimated.

**When the ISM is clumped into at least partially optically thick clouds which are not resolved by either the H I or SXRB observations, the effect is to reduce the apparent ability of the material to absorb a diffuse X-ray flux. For example, if the ISM is diffusely distributed then the absorption optical depth should just be the theoretical cross-section multiplied by the observed column density. However, if the ISM is clumped into unresolved "bricks," an X-ray would certainly be stopped by the brick, but the bricks would cover very little of the sky so most X-rays would be unaffected. Once the brick is optically thick it does not absorb more X-rays if you add more material to it, but more material is removed from the ISM, allowing more X-rays to pass between the bricks. The effective absorption cross-section is then reduced. Besides the effect of reduced cross-sections, the absorption becomes energy independent once the bricks are optically thick.

The 1970s

The 1970s were marked by the targeting of observations to attempt to determine the origin of the background and the expanded coverage of the soft X-ray sky. A significant role was played by the decade-long sounding-rocket campaign by the University of Wisconsin–Madison, Space Physics Laboratory under the direction of W.L. Kraushaar. One goal of the campaign was to determine the fraction, if any, of the observed background that was extragalactic in origin; a second goal was to survey the entire sky in the 0.1–10.0 keV band (see below). To address the first issue, the Small Magellanic Cloud (SMC) was scanned to search for the signature of shadowing. The use of "shadowing" here refers to the simple idea that a foreground object such as an H I cloud will absorb part or all of the X-ray emission originating behind it, thus casting a shadow in the light of the background source. This is analogous to the sky on a stormy day when it appears darker in directions where the clouds are thicker, or more starkly, when the Moon passes in front of the Sun during an eclipse. The result of the SMC study was that at least 75% of the observed $\frac{1}{4}$ keV SXRB originated in front of the SMC (McCammon et al. 1971). Another group, Long et al. (1975), searching for detailed negative correlation between the column density of galactic H I and the surface brightness of the SXRB, presented results for a sounding-rocket observation in the direction of the Large Magellanic Cloud. They found that greater than 90% of the observed $\frac{1}{4}$ keV background originated in front of the galactic H I in that direction; a direction which is relatively opaque to $\frac{1}{4}$ keV X-rays of distant origin if the ISM is not clumped.

With increasing sky coverage it became clearer that while there was certainly a general negative correlation between the SXRB surface brightness and the column density of H I (i.e., the plane-to-pole inverse variation of N_H and X-ray intensity), the evidence for detailed negative correlation was in general lacking (i.e., shadows cast by distinct clouds in the ISM). However, it should be remembered that the early detectors were mechanically collimated and typically had fields of view on the order of 10–50 square degrees (the Moon has an apparent diameter of about half a degree). In addition, the observation of the sky during a sounding-rocket flight was limited to around five minutes of useful data collection, so the observations were also photon limited.

An advantage of the Wisconsin data was a second, softer band provided by a detector with a boron filter (the B band, the boron $K\alpha$ absorption edge yields a high-energy cut-off of 0.188 keV, compared to the usual carbon $K\alpha$ absorption-edge cut-off of 0.284 keV for the C or $\frac{1}{4}$ keV band, see Figure 1). The theoretical effective ISM absorption cross-sections for the two bands differ by about a factor of 2; however, the hardness ratio (or X-ray color) of the observed intensities in the two bands remained relatively constant (with some structure) over a factor-of-3 variation in intensity and associated variation in column density (Fried et al. 1980). Two solutions for this discrepancy were that either the absorbing column of galactic H I was significantly clumped, eliminating any differential (in energy) absorption, or that the emission originated in front of the H I (e.g., Sanders et al. 1977). Hayakawa et al. (1978) came to the latter conclusion as well, that a majority of the diffuse Galactic X-ray background originated in a local region surrounding the Sun. They used spectral fitting of sounding-rocket data with the result that little or no interstellar absorption is required.

Another vital aspect of the SXRB became clear in the 1970s: the observed emission most likely had a thermal origin (emission from a hot plasma dominated by collisionally excited emission lines). For example, Williamson et al. (1974) discussed several non-thermal emission mechanisms and found them lacking. Synchrotron production (radiation produced by charged particles spiraling around magnetic field lines) failed because the required X-ray spectral index is much larger than that at radio wavelengths, and the lifetime of the required electrons ($>10^{13}$ eV in a 3 μG magnetic field reasonable for the Galaxy) is less than 10^4 years. The number of ~250 MeV electrons required to produce the $\frac{1}{4}$ keV SXRB by Compton scattering (collisions between high-energy electrons and photons that increases the energy of the photon) off the 3 K background (the fossil radiation left over from the Big Bang) is unreasonably large as well. Thermal bremsstrahlung (free–free) emission from a hot plasma could produce the required spectral contribution but also required too high a space density for the plasmas, and thus excessive pressures, if galactic in origin (e.g., Henry et al. 1968). The inclusion of collisionally excited line emission from a thermal plasma (e.g., Cox and Tucker 1969) provided a much more efficient emission mechanism in the 0.1–1.0 keV energy range, reducing the required density for that plasma to a reasonable level for a galactic origin. Thus thermal emission from extended regions of hot plasma became the most likely candidate for the source of the SXRB. A galactic thermal origin became even more likely with the Copernicus observations of a nearly ubiquitous interstellar O VI absorption line (e.g., Jenkins 1978). The width of the ISM absorption lines in the spectra of Milky Way stars indicated that the O VI could only be produced by collisional excitation by a thermal plasma at $T \gtrsim 10^{5.5}$ K. While a plasma at this temperature is a bit too cool to be responsible for the observed $\frac{1}{4}$ keV emission, it is consistent with what could be expected from H I-cloud interfaces with a hotter plasma. De Korte et al. (1976) presented an analysis of the data from two sounding-rocket flights from the early 1970s. Using data from two detectors with markedly different window thicknesses (which creates a similar but not as strong an effect as having a separate filter, for example the Wisconsin boron band data), their spectral fitting indicated that the diffuse galactic background required at least two thermal emission components: $T \lesssim 10^{6.0}$ K and $T \sim 10^{6.15}$ K $- 10^{6.32}$ K.

A major review of our understanding of the diffuse X-ray background at the end of the 1970s was provided by Tanaka and Bleeker (1977). The authors extensively discussed the data collected to that point (including sources of background contamination), models for both emission and absorption of X-rays in the ISM, and models for the spatial distribution of that emission.

The 1980s

The early 1980s saw the completion and publication of the Wisconsin sounding rocket all-sky survey (McCammon et al. 1983) and the publication of the first satellite-borne survey (from the third in the Small Astronomy Satellite series, SAS 3, Marshall and Clark 1984). The SAS 3 survey of the $\frac{1}{4}$ keV background had the advantage of better photon statistics and a smaller field of view (a FWHM of 2.9° compared to the 7° of the Wisconsin survey), but lacked the spectral (color) information provided by the two soft bands of the Wisconsin survey (B and C bands). The first High Energy Astrophysical Observatory (HEAO 1) A-2 survey of Garmire et al. (1992) should rightfully be placed in the 1980s as well (the data were actually collected in the late 1970s, as were the SAS 3 data); it had a field of view roughly comparable to that of SAS 3, but again improved upon photon statistics. In general, the photon statistics did not allow the full use of the more limited solid angles provided by the satellite experiments, although they did aid in the exclusion of bright point sources. The striking aspect of the three surveys was that they were all in reasonably good agreement with each other in overall normalization and showed the same general diffuse structure in the background. (Figure 2 displays the three all-sky surveys of the $\frac{1}{4}$ keV SXRB listed above, plus that of the Rosat all-sky survey discussed below). This included both the "general" background and the structure of a few distinct features covering large solid angles. Three such objects are the Eridanus X-ray Enhancement (a wind-blown bubble at $l, b \sim 200°$, $-40°$ with an extent of $\sim 20°$ powered by the Orion OB associations), the Monoceros–Gemini Enhancement (Monogem Ring, an older supernova remnant at $l, b \sim 200°$, $5°$ with an extent of $\sim 20°$), and Loop I (a superbubble centered at $l, b \sim 330°$, $20°$ with an extent of $\sim 110°$ powered by the Sco–Cen OB associations). Loop I is best seen at $\frac{3}{4}$ keV (Figure 3a); only the most northerly extent, the North Polar Spur from $l, b \sim 30°, 30°$ to $l, b \sim 290°, 50°$, is clearly seen at $\frac{1}{4}$ keV. These large but distinct features, while interesting in their own right, have typically been separated from the study of the SXRB (following the axiom that if it can be identified, it is not part of the diffuse background).

The 1990s: The age of ROSAT

The Röntgensatellit (Rosat; Trümper 1992) X-ray observatory provided the most significant improvement in SXRB data since the initial observations in the late 1960s. The large effective area and solid angle of its mirrors coupled with the spatial resolution and low background of its detectors opened an entirely new view of the universe. (While the Einstein orbiting observatory flew an imaging proportional counter, the poorer angular resolution, energy resolution, and background rejection reduced its ability to study the SXRB). The Rosat All-Sky Survey (RASS, Snowden et al. 1995, 1997) of the $\frac{1}{4}$ keV SXRB had roughly two orders of magnitude more photons than all previous surveys combined, allowing useful sky-pixel sizes of $12' \times 12'$. Pointed observations with their longer exposures increased the possible resolution to a couple of arc minutes. While the RASS reassuringly agreed well with the previous surveys when coarsely binned on the sky (e.g., Figure 2 and Snowden et al. 1995), the RASS displays a wealth of structure never before observed. Even a casual comparison of the RASS maps with measures of the column density of the galactic ISM, most clearly with the 100 μm maps of the Infrared Astronomy Satellite (IRAS, which observed four bands in the 12 μm to 100 μm wavelength range), shows extensive and detailed negative correlations between the two data sets (see Figures 3 and 4). Essentially every ISM feature in directions of reasonably low total column density (i.e., $N_H \lesssim 3 \times 10^{20}$ H I cm^{-2}) shows a corresponding feature in the $\frac{1}{4}$ keV SXRB. As could be expected, this revolution in the data produced a revolution in our understanding of the SXRB.

MODELS FOR THE $\frac{1}{4}$ keV BACKGROUND

From the initial discovery of the soft X-ray diffuse background in 1968 through to the 1980s, the quality of the data allowed a wide range of models for the relationship between the ISM and the SXRB, and its implication for the galactic disk and halo. Models of the $\frac{1}{4}$ keV SXRB can be loosely separated into three categories ranging from production by an extragalactic source or at least beyond the galactic disk, to the emission being intermixed with the absorbing ISM, to entirely local production. In an ecumenical outcome, the current "best" model is a mixture of all of the above.

The absorption model

As the first observations of the SXRB were from sounding-rocket flights that observed cuts from the galactic pole to plane (not an accidental occurrence), the global general negative correlation between its surface brightness and the column density of galactic H I was clear from the start. The

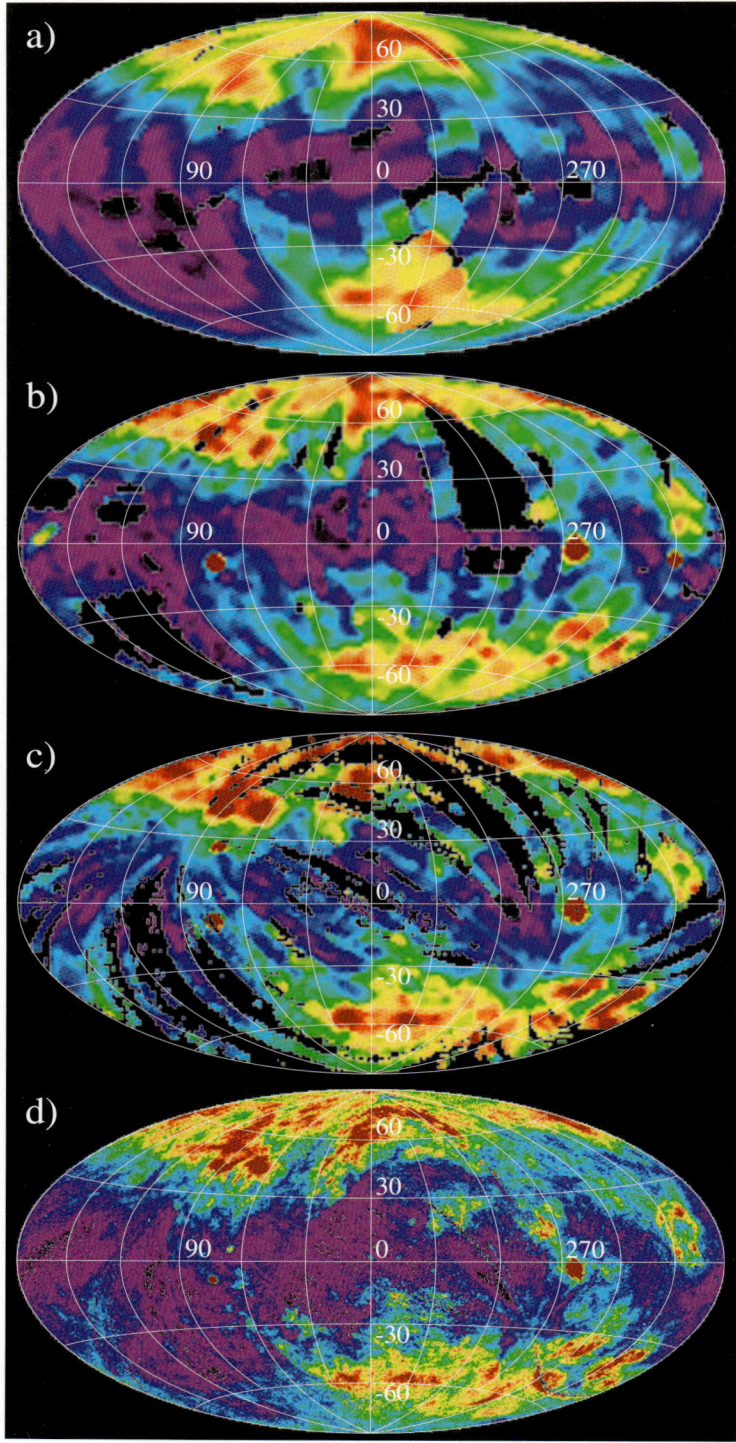

Figure 2 The four all-sky surveys of the $\frac{1}{4}$ keV SXRB in chronological order. (a) Wisconsin (McCammon et al. 1983), (b) SAS-3 (Marshall and Clark 1984), (c) HEAO-1 (Garmire et al. 1992), and (d) Rosat (Snowden et al. 1997). The data are displayed in an Aitoff equal area projection in galactic coordinates with the galactic center at the center of the maps, and with longitude increasing to the left. The data have been linearly scaled so that the maximum to minimum intensity ratio for the displayed range is six for all surveys. The images are color coded, with purple indicating low intensity and red indicating high intensity. Bright point or point-like sources have been removed from the Wisconsin and Rosat maps. Transient sources appear in both the SAS 3 and HEAO 1 maps. Note how the basic features remain the same as the angular resolution of the surveys improves. This improvement in resolution is perhaps most clearly demonstrated by data from the Monogem Ring at $l, b \sim 200°, 10°$, which evolves from a "blob" in the Wisconsin data to a clear ring in the Rosat data.

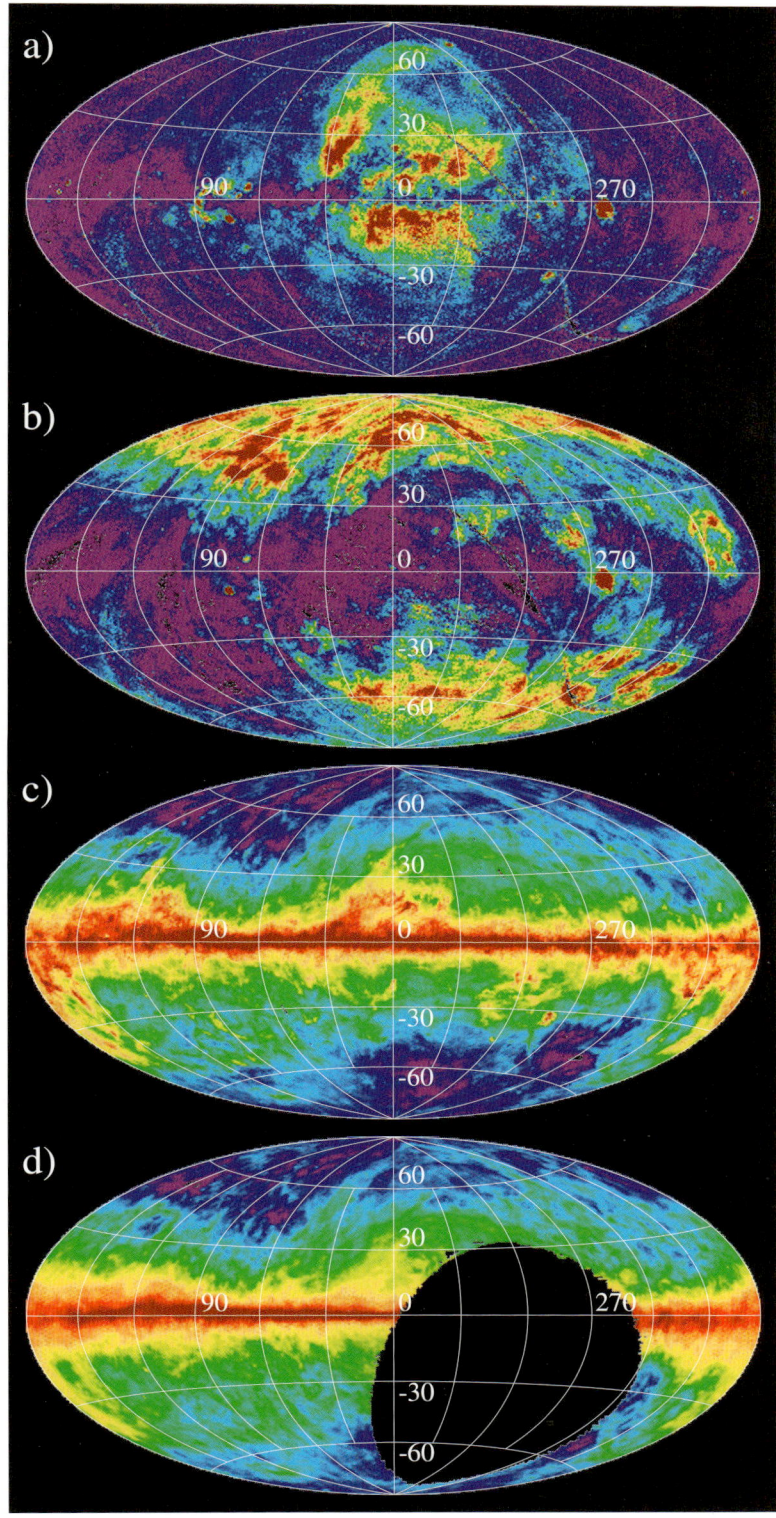

Figure 3 All-sky maps of the (a) Rosat $\frac{3}{4}$ keV band data (Snowden *et al.* 1997), (b) Rosat $\frac{1}{4}$ keV band data (Snowden *et al.* 1997), (c) Schlegel *et al.* (1998) DIRBE-corrected IRAS 100 μm data, and (d) Hartmann and Burton (1997) Leiden–Dwingeloo N_H data. The X-ray data are linearly scaled while the IRAS and N_H data have been logarithmically scaled, all data have been binned into $40'.5 \times 40'.5$ pixels. The images are color coded with purple indicating low intensity and red indicating high intensity. Note the dissimilarity between the $\frac{3}{4}$ keV and $\frac{1}{4}$ keV maps, as well as the clear plane-to-pole general negative correlation between the $\frac{1}{4}$ keV band data and the measures of HI column density.

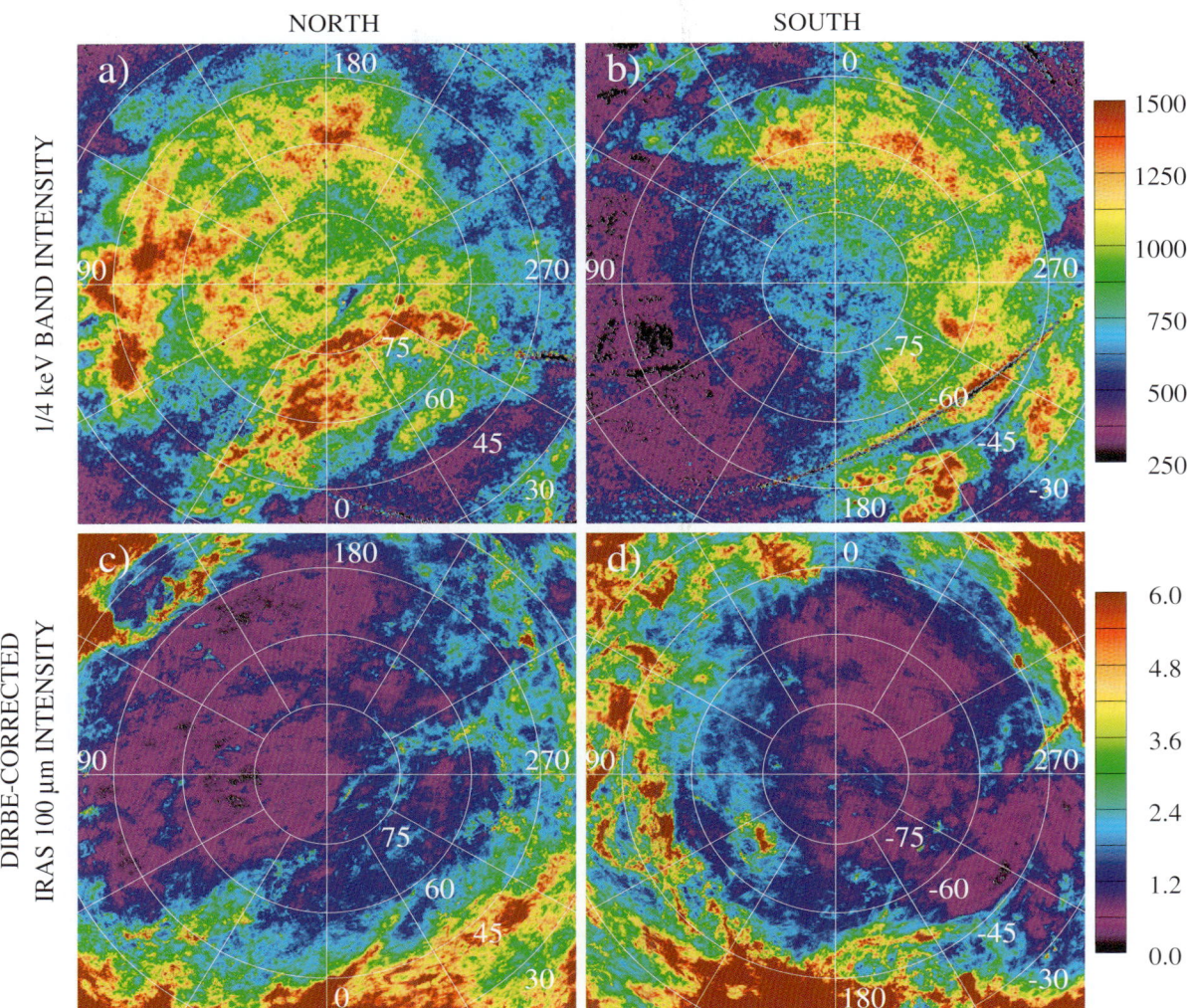

Figure 4 Galactic-pole maps of $\frac{1}{4}$ keV X-ray intensity (Snowden et al. 1997) and IRAS 100 μm intensity (Schlegel et al. 1998). (a) Rosat $\frac{1}{4}$ keV map of the north galactic polar region, (b) Rosat $\frac{1}{4}$ keV map of the south galactic polar region, (c) IRAS 100 μm map of the north galactic polar region, (d) IRAS 100 μm map of the south galactic polar region. All data have been smoothed by a 36'×36' running average for clarity. The units for the X-ray data are 10^{-6} counts s^{-1} arcmin^{-2} and the units of the IRAS 100 μm data are MJy sr^{-1}, and the projection is zenith equal area. The wealth of detailed negative correlation between the two data sets, X-ray intensity, and a measure of the absorbing column density of the galactic ISM, is clearly visible. The Draco Nebula region, the first unambiguous example of a shadowing feature in the general SXRB, is the small area at $l, b \sim 92°, 40°$ (mostly obscured by the $l = 90°$ label).

extragalactic background observed at higher energies was known so it was not unreasonable to postulate that the $\frac{1}{4}$ keV SXRB was just the extension of that power law to lower energies. The angular structure of the SXRB was then produced by the absorption of the galactic ISM, thus the eventual label of absorption model. Minor difficulties such as an inconsistent apparent absorption cross-section (clumping of the absorbing ISM was most successfully invoked to account for the reduced effective cross sections) and a nonzero flux in the galactic plane (unmodeled non-cosmic contamination or the superposition of unresolved galactic point sources were assumed) were explained away by supporters. The excess emission at $\frac{1}{4}$ keV over the extrapolation of the extragalactic power law was attributed to cosmological or galactic halo phenomena. Given the quality of the data (e.g., poor counting statistics and low angular resolution) and preconceived notions concerning the ISM, these assumptions were not unreasonable.

Time and the quality of the observations progressed and the inconsistencies did not fade away. The nonzero flux in the galactic plane particularly required a revision in the model for the SXRB, where a diffuse emission component foreground to the absorbing H I was postulated. The absorption model was modified to include (typically) an isotropic foreground component that took care of the excess emission in the galactic plane as well as the isotropic extragalactic or galactic corona background whose absorption by galactic H I produced the observed structure in the surface brightness of the SXRB. The required distant component was significantly in excess of the low-energy extrapolation of the extragalactic power law observed above 2 keV.

The high-water mark of the absorption model came with the publication of the SAS-3 C band data by Marshall and Clark (1984). In their paper they concentrated on an analysis of the northern hemisphere data from the second galactic quadrant which were fit remarkably well by the modified absorption model. However, the rest of the sky was not as well fit, and in some octants it was quite poorly fit. Their analysis relied on clumping of the ISM to reduce the effective absorption cross-sections to the observed level.

The Interspersed Model

To explain the emission observed at low galactic latitudes, Davidsen et al. (1972) suggested a model where X-ray emission and absorption are intermixed in the galactic disk. The galactic component, when combined with the extragalactic (or galactic corona) component of the absorption model, provided reasonable fits to the coarse data of the time with reasonable parameters (albeit with large uncertainties). Burstein et al. (1977) and Fried et al. (1980) also considered such a model in their analyses of the SXRB.

A theoretical underpinning for an ISM where there is an intermixture of soft X-ray emission and absorption in the galactic disk was provided by McKee and Ostriker (1977). They postulated that the spatial structure of the neutral material of the disk is strongly modified by the effects of, and interaction with, supernova remnants. What may start as diffusely distributed material with occasional clouds (e.g., Field et al. 1969), will be processed into a hot, intercloud medium of large filling factor (the fraction of space taken up by the component by volume) with the neutral material confined to clouds with relatively dense cores and diffuse envelopes.

The high-water mark for the interspersed model was reached with the paper of Jakobsen and Kahn (1986). The results of their analyses required a significantly clumped ISM and scale heights of the X-ray emitting gas which were much greater than those of the absorbing gas. In more simple terms, their model was essentially the revised absorption model without the assumption of a distinct foreground component.

The Displacement Model

Sanders et al. (1977) and Tanaka and Bleeker (1977) presented the displacement model as an alternative for the absorption model that did not suffer from the absorption model's accumulated inconsistencies. While it was still accepted that the extragalactic power law would contribute to the observed $\frac{1}{4}$ keV SXRB, at least in low-column-density directions, the displacement model suggested that most of the observed emission at $\frac{1}{4}$ keV originates in front of the neutral material of the galactic disk. The negative correlation between the SXRB intensity and the H I column density is then due to the hot plasma displacing H I. In any given direction (and most obviously at high or at least constant galactic latitudes away from the galactic disk), if there is a longer path length of X-ray-emitting plasma there is a shorter path length of H I. If both the X-ray emitting plasma and H I are distributed reasonably uniformly in the space where they exist, this leads to a relatively brighter region of X-rays and a relatively lower column density of H I.

A theoretical underpinning for a porous ISM and the existence of regions of hot plasmas in the galactic disk, and thus the displacement model, was provided by Cox and Smith (1974). In their model, supernova remnants in the galactic disk tend to expand until they intersect with a previously formed SNR bubble, at which point the younger remnant will preferentially expand into the preexisting cavity and re-energize it. This leads to interconnecting tunnels and a sponge-like interstellar medium where the cooler ISM forms the matrix. Using H I observations, Knapp (1975, and references therein) had identified a deficit of neutral material in the vicinity of the Sun. Frisch and York (1983) and Paresce (1984) used interstellar absorption-line observations to map out a cavity in the H I of the galactic disk which surrounds the Sun.

The high-water mark for the displacement model came with Cox and Snowden (1986) and Snowden et al. (1990). In their papers they modeled the local interstellar medium as a plane-parallel distribution of H I with a confined cavity filled with plasma at 10^6 K. The shape of the cavity, or more precisely the extent of the X-ray-emitting plasma, the Local Hot Bubble (LHB) as it came to be called, was defined by a linear scaling of the surface brightness of the $\frac{1}{4}$ keV background. The scale factor was determined by the best fit of the predicted-to-observed column density of H I. They showed that the displacement model predicted the negative correlation between the intensity of the $\frac{1}{4}$ keV SXRB and the column density of H I as well as or better than the absorption model, but was free of the latter's inconsistencies. (As one colleague phrased it, the displacement model *was not inconsistent* with any of the observed data.)

When models collide

By the middle of the 1980s there was little controversy concerning the apparent angular structure of the SXRB (e.g., Figure 2). The existing all-sky surveys were in good agreement, at least to the extent that their formal and systematic uncertainties allowed. The interpretation of those data, however, provided some lively discussions at a number of conferences as well as in the literature. The three models for the SXRB had clearly different implications for the ISM, and it is interesting to note that until the mid-1980s our understanding of the ISM was insufficient to distinguish convincingly between a cool ISM which is distributed relatively smoothly or one which is clumped into dense clouds.

A significant downfall for both the absorption and interspersed models was their requirement for extensive clumping of the ISM. Both models required the clumping to reduce the effective absorption cross-sections from the theoretical to the observed values, and the interspersed model was also *based* on the highly-clumped ISM model of McKee and Ostriker (1977) with its large filling factor of hot plasma. However, the clumping of the ISM is an observable phenomena which led to a program to test for it. In an extensive study of a region in Ursa Major now referred to as the Lockman Hole (the region of the minimum column density of galactic H I, in at least the northern hemisphere), significant clumping was ruled out down to the several arc-minute level (Jahoda *et al.* 1986, and references therein). When coupled with the H I emission/absorption measurements of Dickey *et al.* (1979), clumping on all reasonable scales was ruled out. This result was not universally accepted, or even desired. The planned Extreme Ultraviolet Explorer and Rosat Wide Field Camera (WFC) EUV missions needed a clumped ISM to allow the detection of extragalactic and distant galactic sources (in the EUV). While some extragalactic sources were detected by these observatories, they were those which were in directions of low N_H and bright enough to be detected, even after ISM absorption, in the hardest available band passes. For both instruments, these band passes extended into the soft X-ray regime.

The absorption and interspersed models were also problematic in other areas. A clear prediction of both models, and particularly that of the absorption model, is the existence of shadowing of the SXRB by interstellar clouds at high galactic latitudes. Burrows *et al.* (1984) used Wisconsin all-sky survey data to search for evidence for such shadows and placed tight upper limits on the observed flux which could originate beyond the galactic H I.

Burrows (1989) and Burrows and Kraft (1993) examined in detail the requirements and consequences for the radiation transfer of the interspersed model. They concluded that the model cannot predict the correct dependence of the hardness of the $\frac{1}{4}$ keV flux (as determined from the Wisconsin B, C, and Be band data) on absorbing column density when the emission and absorption is relatively mixed. When the emission is more distant than the absorption, the interspersed model resembles the absorption model and fails because of the lack of ISM clumping needed to reduce the effective absorption cross-sections. When the emission is foreground to the absorption, the interspersed model resembles the displacement model and no constraints are placed on the clumping of the ISM. However, few constraints are placed on the more distant emission either, so the interspersed model *becomes* in effect the displacement model. An excellent review of our understanding of the SXRB at the end of the 1980s is provided by McCammon and Sanders (1990). The status of the data existing at that time as well as the perceived strengths and weaknesses of the models are discussed in great detail.

INTO EVERY MODEL A LITTLE RAIN MUST FALL

A shadow is cast

By the end of the 1980s the displacement model had achieved wide, if not universal acceptance. It provided the greatest consistency with all of the available data and had no obvious limitations. The existence of a local cavity in the neutral material of the galactic disk was clearly demonstrated by the ISM absorption-line studies of Frisch and York (1983) and Paresce (1984). Because of the non-zero flux of $\frac{1}{4}$ keV X-rays in the galactic plane, some X-ray emission in the cavity was absolutely required, and so attributing most of the observed flux at high galactic latitudes to the cavity as well was not unreasonable. Since it is likely that SNRs (energetic bubbles of hot plasma) will expand most readily in directions of lower ISM space density, the higher SXRB surface brightness, and therefore the greater required extent of hot plasma, at high galactic latitudes was completely reasonable. The general (i.e., plane to pole) negative correlation of the SXRB surface brightness with H I column density is a natural outgrowth of the extension of the cavity and therefore additional plasma out of the plane of the Galaxy. The appearance of absorption is serendipitous, and the requirement for the clumping of the neutral ISM to produce the reduction in the effective absorption cross-sections from theoretical values is immaterial. Between the paper of Snowden *et al.* (1990) and the reviews of Cox and Reynolds (1987) and McCammon and Sanders (1990), it appeared that the general picture of the local interstellar medium and the SXRB had pretty much been wrapped up.

With all of these convincing arguments, it appeared that only a few details would be required to complete the understanding of the local ISM and the SXRB. The size and shape of the cavity could be mapped by ISM absorption-line studies, the extent (emission measure) of the hot plasma could be mapped by the soon-to-be-launched Rosat observatory with its planned all-sky survey, the ionization states and metallicities of the X-ray emitting plasma could be determined by future experiments with improved spectral resolution, and the global filling factor of such cavities with $\frac{1}{4}$ keV emission could be investigated by using high-angular-resolution observations of external galaxies similar to the Milky Way (because of the limited sampling of the galactic disk due to its high absorption in the $\frac{1}{4}$ keV band). Apparently, however, Nature abhors complacency as well as vacuums.

The design specifications for the Rosat PSPC (Position Sensitive Proportional Counter), were a dream come true for astronomers studying the SXRB. The large effective area, wide field of view, high angular resolution, good background rejection, and long exposure times combined in one instrument would allow studies of any angular structure on much finer scales than ever before attained. The SXRB community eagerly awaited the first "Announcement of Opportunity" for Rosat observations in order to submit proposals. Catalogs of interstellar clouds (e.g., Magnani et al. 1985) were examined for likely candidates for shadowing studies to put the tightest possible upper limits on extragalactic and galactic halo emission at $\frac{1}{4}$ keV. Several of the proposals were accepted and with the launch of Rosat on 1 June 1990 the level of anticipation was very high.

Due to the launch date and the observational constraints of Rosat (Rosat could only observe in a band between 75° and 105° from the Sun), the part of the sky visible during the verification phase of operations included the Draco Nebula, which was observed during both pointed and survey operations. The results of the observations were deep and stunningly beautiful shadows cast by the Draco clouds in the diffuse $\frac{1}{4}$ keV background (Figure 5; see Snowden et al. (1991) and Burrows and Mendenhall (1991) for the initial analyses). The depth of the shadows implied that fully half of the observed flux at $\frac{1}{4}$ keV in that direction originated beyond the Draco Nebula. Since the nebula is several hundred parsecs distant, and at a relatively high Galactic latitude, the distant emission was placed in at least the lower halo. It is not every day that one discovers something so at odds with one's expectations. The clarity of the shadows, even in the raw event images, are an accolade to Rosat, and it is ironic how easy they were to find after some 20 years of various groups searching for just such an effect.

So what happened? Why was it so difficult to distinguish such intense halo emission before Rosat observations made it so painfully obvious? There are a number of different reasons. First, the $\frac{1}{4}$ keV halo in the direction of Draco is, excluding the galactic center/Loop I region, the brightest part of the high-latitude sky, and is relatively limited in angular extent (Snowden et al. 1998). Second, the angular

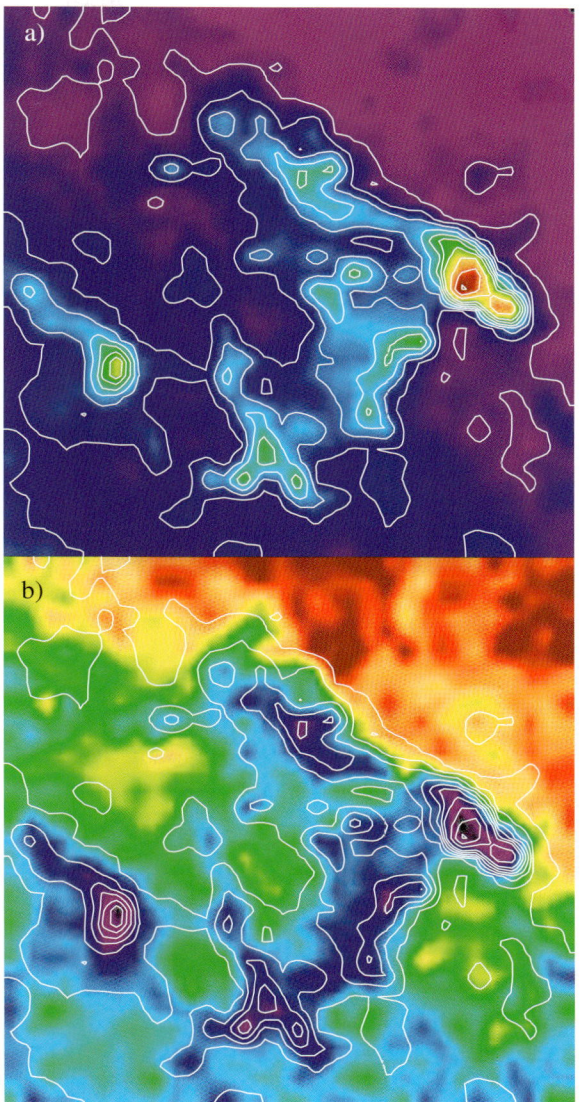

Figure 5 (a) Corrected IRAS 100 μm intensity (Schlegel et al. 1998) and (b) Rosat $\frac{1}{4}$ keV All-Sky Survey (Snowden et al. 1997) images of the Draco Nebula region. Both images have been overlayed with the same corrected 100 μm contours. The field is centered on $l, b \sim 94°.4, 36°.9$ and is $6°.4 \times 6°.4$ in extent. The color coding in both images denotes higher intensity with red and lower intensity with purple. The X-ray surface brightness varies from 400×10^{-6} counts s^{-1} arcmin^{-2} to 1750×10^{-6} counts s^{-1} arcmin^{-2}. The IRAS 100 μm contours are at 1, 1.5, 2, 2.5, 3, 4, and 6.0 MJy sr^{-1}. In both images and for the contours the data have been smoothed for clearer presentation. For the historical record, the Draco shadow paper of Burrows and Mendenhall (1991) used the deep shadow at the right-hand side of the image while Snowden et al. (1991) used the not-quite-so-deep shadow at the left-hand side of the image.

extent of the shadow features were small enough that they would be difficult or impossible to detect in the previous surveys with their coarser angular resolution and limited statistics. The third reason is what could be considered the "spherical cow" syndrome. Basically no one was looking for small angular-scale halo X-ray enhancements casting shadows with even smaller H I or molecular clouds. Because of the poor statistics and angular resolution of the previous X-ray data and limited angular resolution of the N_H data (in the 1980s considerable use was made of the Stark *et al.* (1992) Bell Labs H I survey which had an angular resolution of 2° FWHM), modeling was done over relatively large solid angles. The background flux was assumed to be isotropic or at least slowly varying (e.g., a plane-parallel distribution away from the galactic plane) while the foreground flux was modeled most often as being isotropic. In such situations the clear and distinct shadows as observed by Rosat in Draco would be interpreted as no more than unaccounted-for noise in the large-scale analysis.

A revised perspective, and some not-so-revised results

In the decade since Rosat was launched and $\frac{1}{4}$ keV shadows were discovered, considerable progress has been made on many fronts concerning the SXRB and the interstellar medium. Critical improvements in the data have come in X-rays from both the RASS (Snowden *et al.* 1997) and Rosat pointed observations. The Leiden–Dwingeloo H I survey of Hartman and Burton (1997) improved upon previous surveys both in angular resolution and in baseline correction. Schlegel *et al.* (1998) used Cosmic Background Explorer (COBE, a satellite that observed in the 1.25–240 μm wavelength range) data to clean and correct IRAS 100 μm data providing a high-angular-resolution sampling with a reliable zero level of the column density of galactic material (albeit without the velocity information provided by 21 cm data). When the Rosat data are combined with the improved mappings of the galactic column density of neutral material in the analysis of shadows, an accurate separation of the observed flux into foreground and background components relative to the shadowing object can be obtained. The shadow catalog of Snowden *et al.* (2000) provides a list of 378 such objects plus references to numerous other studies. In addition, the on-going work of Sfeir *et al.* (1999) compiling an extensive catalog of interstellar absorption-line measurements is providing a detailed mapping of the three-dimensional structure of the neutral material in the solar neighborhood.

Snowden *et al.* (1998) used RASS and Schlegel *et al.* (1998) IRAS 100 μm data for a general analysis of the sky to map the foreground and background components of the $\frac{1}{4}$ keV background. Using a shadowing-type analysis after dividing the sky into small regions, they separated the X-ray emission components with the assumption that all of the absorbing material represented by the IRAS data was contained in a region between the two emission components (i.e., there was no intermixture of X-ray emitting and absorbing material). The results for the foreground emission provided an improved view of the LHB while the mapping of the distant emission provided the first view of the angular structure of the halo $\frac{1}{4}$ keV emission. The halo emission shows considerable structure with different appearances in the northern and southern galactic hemispheres. The galactic halo surface brightness ranges from a minimum near zero in the southern hemisphere (assuming that the extragalactic power law continues down in energy to the $\frac{1}{4}$ keV band, and so accounts for most if not all of the observed flux of distant origin) to a maximum of many times that of the LHB in a number of directions.

One of the amusing aspects of working in the same field for a couple of decades is that occasionally one gets to correct one's errors of the past, and this was one such case for the author (although redoing one's PhD thesis is a bit tedious). In Snowden *et al.* (1990) the modeling was based on the assumption that the bulk of the observed $\frac{1}{4}$ keV background arose from the LHB. The word "bulk" was used as it was assumed that the extragalactic power law could provide a limited amount of flux of distant origin at $\frac{1}{4}$ keV in high-galactic-latitude (low N_H) directions. However, since little or no distant emission was required by observations, it was simpler to assume none. In addition, it was not expected that there was much if any other distant emission except for discrete features such as identified supernova remnants. With the observation of shadows demonstrating that there is extensive emission in the galactic halo, this basic assumption was not correct. However, for the purposes of mapping the LHB, the 1990 analysis was reasonably accurate for the local emission, as shown by the tight correlation between the 1990 and 1998 results (Snowden *et al.* 1998). This is because we fortuitously used the Wisconsin survey (McCammon *et al.* 1983) B band data for the modeling to further exclude the contaminating effect of any possible distant emission. As noted above, the B band is slightly softer than a standard $\frac{1}{4}$ keV band (see Figure 1, as represented by the Rosat $\frac{1}{4}$ keV band) as it uses the boron $K\alpha$ absorption edge at 0.188 keV to provide the high-energy cutoff. Because of this cutoff, the effective optical depth of galactic neutral material for the band in a given direction is twice what it is for the standard $\frac{1}{4}$ keV band. The minimum optical depth to the halo is therefore about one with typical high-latitude optical depths of two to three; for the Draco region it is more like four. Thus by using B band data we were indeed effectively sampling the LHB. So we got the LHB right, but as for missing the extensive emission in the galactic halo, well, "You can't win them all."

THE EUV BACKGROUND

Aficionados of the SXRB may have raised issue with my cavalier statement at the begining of this chapter that the EUV background originates from the same source as some of the background at $\frac{1}{4}$ keV. This did not always seem to be the case and there remain some inconsistencies to this date. Cash et al. (1976, sounding-rocket observations) and then Stern and Bowyer (1979, Apollo–Soyuz observations) reported the first unambiguous (if not without contamination) measurements of the diffuse background below the beryllium $K\alpha$ absorption edge at 0.111 keV. Comparing the two data sets, as well as observations at somewhat higher energies, Stern and Bowyer concluded that an emission component with $T \sim 10^{5.5}$ K was required to account for the data. This contrasts with the temperatures of $T \sim 10^{6.0}$ K required by $\frac{1}{4}$ keV band SXRB data implying the existence of multiple emission components contributing to the soft X-ray and EUV backgrounds. Stern and Bowyer also concluded that their results were consistent with the McKee and Ostriker (1977) model for the interstellar medium.

After the flurry of papers in the mid to late 1970s, the study of the EUV background more-or-less languished until the mid 1980s when Bloch et al. (1986) and then Juda et al. (1991) took up the gauntlet with a series of three sounding-rocket flights with beryllium-filter detectors. With significantly better signal-to-noise ratios than the previous experiments, they found that emission in the 0.070–0.111 keV (Be) band range was consistent, almost to a fault, with a constant intensity ratio to the Wisconsin B band data over a sample of directions with a factor-of-three variation in intensity. Furthermore, the model plasma-emission temperature for the required B/Be band ratio was consistent with $T \sim 10^{6.0}$ K, the same temperature as required for the Wisconsin C band to B band ratio. This consistency in the implied emission temperature for the X-ray and EUV data along with the constant B/Be band ratio and nearly constant C/B band ratio was interpreted as strongly supporting the displacement model for the SXRB, and placed a limit of $N_H < 8 \times 10^{18}$ cm^{-2} of neutral material intermixed with the X-ray emission (Juda et al. 1991). The three bands span almost an order of magnitude in their absorption cross-sections, so a constant intensity ratio between the three bands independent of total intensity rules out absorption as a mechanism to produce the observed angular structure, and strongly suggests a single unabsorbed emission source.

The EUVE observatory Lexan/B filter data have provided a more recent view of the EUV background. While the band response is not as soft as Be-filter data, and actually spans the demarcation between the EUV and soft X-ray regimes, the EUVE data do provide a greater sensitivity to emission from the Local Hot Bubble (LHB) than $\frac{1}{4}$ keV band data. Berghöfer et al. (1998, and references therein) used data from the Deep Survey telescope to study several shadows in the EUV/soft X-ray background. Their results are consistent with observations in the $\frac{1}{4}$ keV band, and support the conclusion that there is one diffusely distributed plasma within the local cavity which is responsible for both the EUV and local $\frac{1}{4}$ keV backgrounds.

However, data from the Rosat WFC survey with the S1 filter, a bandpass also intermediate between the Wisconsin Be and B bands, are inconsistent with this pleasant picture. The data provide only an upper limit to the flux from the SXRB (West et al. 1996), an upper limit which is about one-third of the flux expected from the standard thermal equilibrium emission models, the same models which fit the Be and B band data reasonably well. Figure 6 provides a possible explanation why this discrepancy might occur. In Figure 6 are plotted normalized response functions for the Wisconsin Be and B bands, the Rosat WFC S1 band, and Rosat $\frac{1}{4}$ keV band, along with the model spectrum from a 10^6 K thermal equilibrium plasma. Note that the spectrum is dominated by line emission, with the true number of lines obscured by the coarse energy binning. If the line

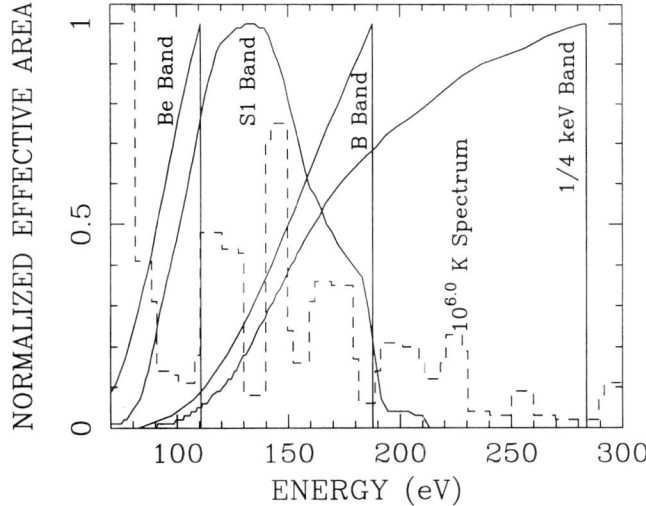

Figure 6 Plot of the normalized (to 1.0) responses of the Wisconsin Be and B bands, the Rosat WFC S1 band, and the Rosat $\frac{1}{4}$ keV band (solid curves). Also shown is the model spectrum from a 10^6 K thermal equilibrium plasma (dashed line, scaled for display).

strengths for certain lines are incorrect, then the different bands will be affected differently. For instance, if the line strengths contributing to the peaks at 120 eV and 145 eV are significantly overestimated, then the model S1 band count rates would also be significantly overestimated compared to those of the other bands.

There is also an inconsistency regarding the existence of spectral features in the EUV band. Spectral models for thermal emission, the same models that predicted the correct ratios for the Wisconsin Be, B, and C bands, also predict significant line emission near 0.070 keV. However, using EUVE spectra, Vallerga and Slavin (1998) found no evidence for such emission.

THE $\frac{3}{4}$ keV BAND

In some respects the general galactic $\frac{3}{4}$ keV background may arise from a more simple distribution of emission. (Again, this is partially an observational effect because when SNRs are still young enough to emit strongly in the $\frac{3}{4}$ keV band, they are typically also young and bright enough to be identified as discrete objects, and so are excluded by the criteria of this chapter.) The brightest extended emission region on the sky is in the direction of the galactic center. While the origin of this emission is confused between Loop I and the galactic center itself, there is strong evidence for the existence of a galactic-center X-ray bulge. Snowden et al. (1997) demonstrated that the 0.5–2.0 keV enhancement in the southern hemisphere with $b \lesssim -10°$ (see Figure 3a) is consistent with an isothermal bulge with a temperature of $\sim 10^{6.6}$ K and a scale height of 1.9 kpc. Park et al. (1997) analyzed a Rosat observation of a deep shadow in the 0.5–2.0 keV background near the galactic center at low latitudes cast by a distant molecular cloud, a cloud which is quite likely part of the molecular ring on the near side of the galactic center. The existence of the shadow indicates, as usual, extensive emission arising from beyond the cloud. The intensity and spectral hardness of the emission extrapolated back to an origin at the galactic center (i.e., de-absorbed by the intervening H I) is consistent with the X-ray bulge of Snowden et al. (1997) extrapolated from high latitudes down to the galactic plane.

There must, however, be an additional diffuse or at least unresolved emission component in the galactic plane. An examination of the Rosat $\frac{3}{4}$ keV map (Figure 3a) reveals that the $\frac{3}{4}$ keV background is nonzero in all directions, even at low latitudes where the galactic disk is optically thick, and there are no obvious emission features. This emission has posed a conundrum since it was first observed: an emission component in the galactic disk which provides sufficient flux to roughly "fill in" for the extragalactic flux absorbed by the ISM. Sanders et al. (1983) modeled the latitude dependence of the Wisconsin all-sky survey $\frac{3}{4}$ keV (M band) data as the sum of four components: the extragalactic power law, a component proportional to the $\frac{1}{4}$ keV band data (which was assumed to arise locally), the superposition of unresolved late-type stars, and the superposition of unresolved SNRs. This somewhat contrived combination of four components each with different latitude dependencies was able, with the "right" scale factors, to model the near-constant value for the latitude dependence of the Wisconsin data crossing the galactic plane in the longitude range $200° < l < 250°$. While their contribution from the superposition of unresolved late-type stars is consistent with more recent calculations (e.g., Schmitt and Snowden 1990), their model emission from the LHB is significantly greater than that detected by Snowden et al. (1993) using a Rosat shadowing observation of the high-latitude molecular cloud MBM 12. This discrepancy requires an even greater contribution of emission from the unresolved SNRs of Sanders et al. (1983). While the higher-resolution Rosat data show additional structure near the galactic plane in the anti-center direction, it is most likely due to the absorption of the extragalactic background, as it shows the pattern of negative correlation with the total column density of galactic H I. There is no significant further detection of distant SNRs which one might expect given the improved quality of the Rosat data, and so the issue of the origin of the diffuse flux from the galactic plane remains, as they say, unresolved.

In a global, high-latitude context, the excess of the observed $\frac{3}{4}$ keV background over the extrapolation of the extragalactic power law has been modeled as an extended galactic halo in hydrostatic equilibrium (e.g., Wang 1998) based on the galactic corona concept of Spitzer (1956). Such a distribution is consistent with the results of Kuntz and Snowden (2000) which show that X-ray emission in the galactic halo is most likely comprised of at least two emission components: a strongly spatially varying component (or components) with temperatures near $10^{6.1}$ K and a more uniform component near $10^{6.5}$ K. The hotter component can be identified with the high-scale-height component discussed by Wang (e.g., a galactic corona). The cooler components(s) may have a more limited spatial extent, and may be linked to the galactic disk through chimneys (e.g., the bright but limited region of $\frac{1}{4}$ keV emission backlighting the Draco Nebula) or be produced *in situ* (e.g., by halo SNR). Pietz et al. (1998) proposed that the $\frac{1}{4}$ keV and $\frac{3}{4}$ keV emission of the galactic halo are produced by a single component with an intermediate temperature; but their model is inconsistent with the spectral energy distribution of the Galactic halo emission and the spatial distribution of the $\frac{1}{4}$ keV emission.

An issue which still remains to be resolved concerns the spectrum of the extragalactic background below 2 keV.

The spectrum is actually the superposition of spectra from a wide variety of unresolved discrete sources, with perhaps some diffuse emission from large-scale cosmological structures thrown in for good measure. The true spectrum may not continue to lower energies as an unbroken power law but turn up or turn down, or both. While not affecting the $\frac{1}{4}$ keV band results extensively as galactic halo emission clearly dominates any extragalactic flux in most directions at that energy, a change in the extragalactic spectrum can significantly affect the amount of the observed $\frac{3}{4}$ keV flux which can be attributed to galactic emission.

A PICTURE OF THE ISM AND THE SXRB

The galactic disk

This is the fun part, at least for the author, where he gets to pontificate about his view of the SXRB and the interstellar medium. Starting locally and working outwards, the solar system is embedded in partially ionized clouds of a few parsecs in extent: the Local Fluff (LF; e.g., Frisch 1995 and references therein). The column density of these clouds is quite low in the grand scheme of things at $N_H \lesssim 5 \times 10^{18}$ cm^{-2}. Beyond and containing the LF is the LHB, the region of low density, $T \sim 10^{6.1}$ K plasma which extends from tens of parsecs in some directions in the galactic plane to 100–200 parsecs at high galactic latitudes. The plasma is distributed somewhat uniformly so that the extent of the LHB is mapped by the surface brightness of the local $\frac{1}{4}$ keV emission. The LHB provides 75% of the observed flux from the general diffuse background in the $\frac{1}{4}$ keV band, nearly all of the diffuse EUV flux, and perhaps $\lesssim$10% of the diffuse flux at $\frac{3}{4}$ keV. While there are possibly other diffuse clouds such as the LF embedded in the LHB, their total column densities must be relatively low (Juda et al. 1991), and/or their filling factor is negligible. The LHB is in turn confined within a cavity in the neutral material of the galactic disk. However, the extent of the cavity in certain directions exceeds the extent of the LHB (although in no direction does the extent of the LHB exceed that of the local cavity, Sfeir et al. 1999). This is most clearly the case in the direction of the Region of Bizarre Emptiness (RBE; Cox 1997) in the third quadrant of the galactic plane. In at least this direction, some of the unaccounted-for path length between the extent of the LHB and that of the cavity is filled with diffuse ionized gas (Gry et al. 1985). While there may be enhanced emission at the boundary layers between the LHB and the cooler material of the cavity wall, there is no evidence to indicate its presence at any significant level of emission in the $\frac{1}{4}$ keV band data (there is, however, the question of the origin of the O VI discussed above, which is most easily attributed to emission from cloud boundary layers).

The exact nature of the neutral material in the galactic disk beyond the wall of the local cavity will be left for others (see, e.g., Chapters 27 and 28); however, it most likely is found in a wide variety of aspects including clouds, sheets, filaments, and perhaps even a truly diffuse component (e.g., Dickey and Lockman 1990).

Figure 7 graphically presents the above picture for the nearest 250 pc of the ISM along a galactic center–anticenter cut. Directions, abbreviations, and distances to many of the objects in Figure 7 are listed in Table 1. Figures 8, 9, and 10 show different views of the nearest 1000 pc of the ISM. The maps primarily show the regions of X-ray-emitting plasmas but also include a number of objects which have been used as shadowing targets, as well as other regions mentioned in this chapter. Before too many protests erupt, there are a number of qualifications and caveats which must be placed on the maps. First, clearly, there is considerably more structure in all components of the ISM than has been drawn. The neutral material is not smoothly distributed in a thick disk with nicely embedded bubbles of hot plasma. The diffuse H I ranges in density from ~ 1 cm^{-3} to regions of $\lesssim 10^{-2}$ cm^{-3} as suggested by the study of the Monogem Ring (Plucinsky et al. 1996). It also does not have a nice, smooth cutoff at high z and there should be structure due to the spiral arms. There are additional clouds of molecular gas as well as H I clouds, sheets, and filaments. However, the maps do present a rough indication of where the X-ray-emitting plasma is located in the solar vicinity in the galactic disk.

Figure 10 shows the "interarm region" of Breitschwerdt et al. (1996), the low-density region between the two thick lines on the plot. In their view the region has been processed by a number of supernovae and stellar-wind activity from the Sco–Cen OB association. The region does contain the nearby X-ray emission regions: Monogem Ring, Eridanus Enhancement, Vela, RX J0852.0-4622, Cygnus Loop, the LHB, Loop I, and probably the Cygnus Superbubble (although it has been suggested that the latter is the superposition of several emission regions rather than a unified object). While the rest of their interarm region is not filled (noticeably) with hot plasma, in the third quadrant it does contain other low-density regions such as the RBE, Gum Nebula, and the volume around the Monogem Ring. Heiles (1998) suggests that this entire region in the third galactic quadrant is the interior of a superbubble, explaining the generally low space densities.

Note that in Figures 7 and 8 the Loop I, LHB, Eridanus, and Monogem regions of hot plasma are not shown as being in communication with each other. In spite of their close proximity, there is little in the way of evidence for the Cox and Smith (1974) model of interconnecting bubbles, at least locally in the galactic disk. Both the Loop I and Eridanus bubbles have emission temperatures significantly higher than those of the LHB and Monogem. In addition,

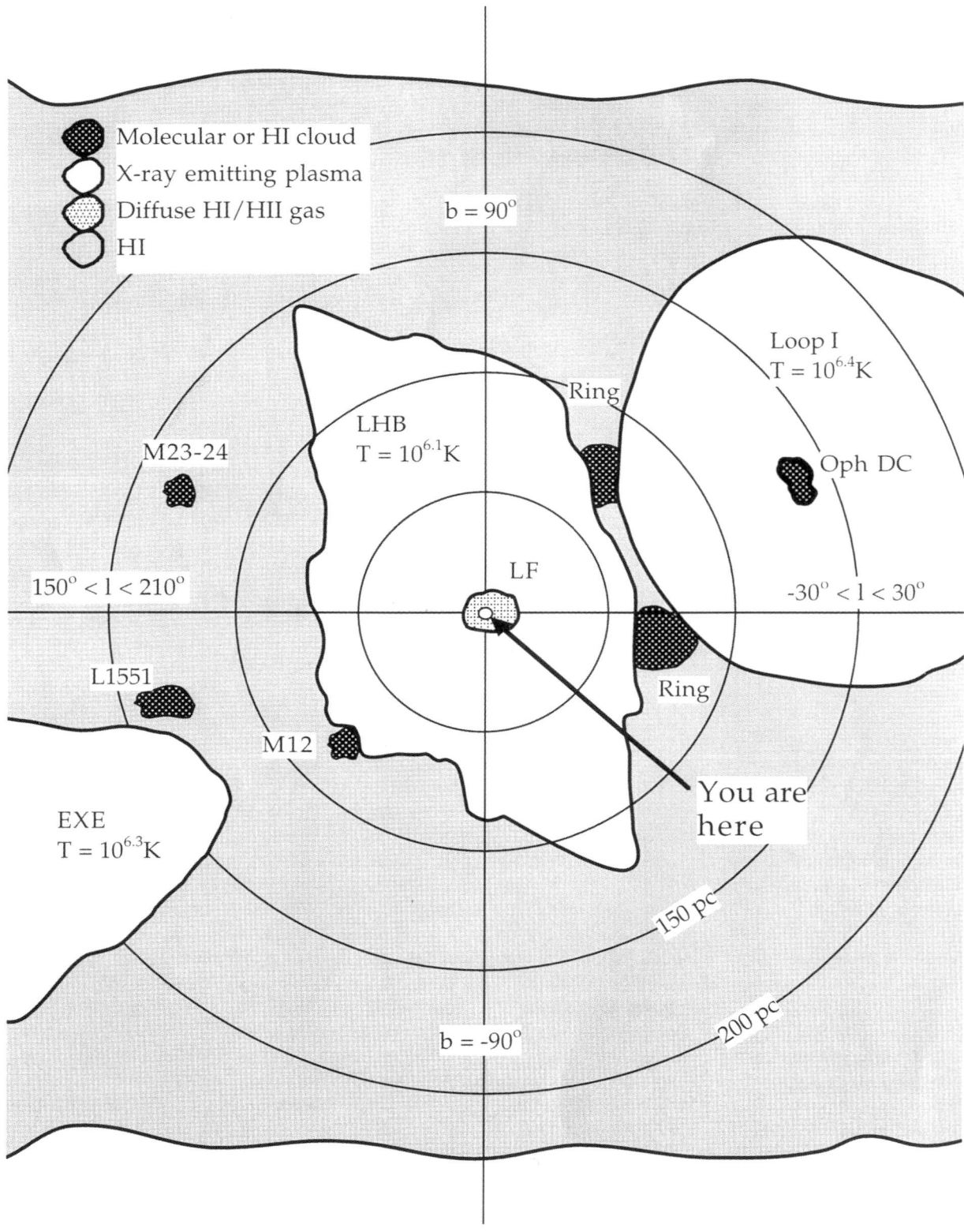

Figure 7 Graphical representation of the nearest 250 pc of the local ISM in a cut along the galactic center–anticenter line. The abbreviations are listed in Table 1.

Table 1 Map objects

Abbreviation	Object	l	b	Distance (pc)	Reference
B92	Barnard 92	13	−01	200	Grant (1999)
Aql R	Aquila Rift	30	05	200	Dame et al. (1987)
Vul R	Vulpecula Rift	58	01	400	Dame et al. (1987)
Cyg R	Cygnus Rift	75	00	700	Dame et al. (1987)
B361	Barnard 361	89	−01	200	Grant (1999)
Draco	Draco Nebula	90	38	540	Grant (1999)
Cyg OB7	Cygnus OB7	93	02	800	Dame et al. (1987)
Ceph Fl	Cepheus Flare	114	10	400	Grant (1999)
Per OB2	Per OB2	162	−15	350	Dame et al. (1987)
M12	MBM 12	159	−34	90	Sfeir et al. (1999)
Taurus	Taurus	170	−15	140	Dame et al. (1987)
M23-24	MBM 23-24	172	27	139	Grant (1999)
B34	Barnard 34	177	02	200	Grant (1999)
L1551	Lynds 1551	179	−20	140	Grant (1999)
EXE	Eridanus Enhancement	185–210	−50 to −15	100–450	Burrows et al. (1993)
MG	Monogem Ring	195–212	−02 to 20	300–500	Plucinsky et al. (1996)
Mon OB1	Monoceros OB1	202	02	800	Dame et al. (1987)
Orion AB	Orion A and B Molecular Clouds	210	−15	450	Burrows et al. (1993)
Mon R2	Monoceros R2	215	−12	830	Dame et al. (1987)
RBE	Region of Bizarre Emptiness	240	−20	100–250	Cox (1997)
Gum	Gum Nebula	240–275	−20 to 20	400–600	Reynoso and Dubner (1997)
Vela SNR	Vela SNR	265	−03	250	Cha et al. (1999)
RXJ	RX J0852.0-4622	266	−01	150	Chen and Gehrels (1999)
Vela Sheet	Vela Sheet	275	02	425	Dame et al. (1987)
Cham I	Chameleon I	297	−16	190	Grant (1999)
Coalsack	Coalsack	303	00	190	Grant (1999)
ρOph	ρ Ophiuchus	358	15	165	Dame (1987)
LHB	Local Hot Bubble	–	–	–	Snowden et al. (1998)
LF	Local Fluff	–	–	–	Frisch (1995)
Ring	Loop I/LHB Ring	–	–	–	Egger and Aschenbach (1995)

there is no evidence whatsoever for an emission region connecting the Monogem and Eridanus bubbles. However, Egger and Aschenbach (1995) have modeled a ring-like enhancement of the local H I in the galactic center direction as the result of the LHB and Loop I bubbles coming into contact with each other. While the two bubbles have considerably different emission temperatures and there is apparently an H I wall remaining between the plasmas, there does appear to be a temperature gradient in the LHB from the galactic center to anti-center direction with a higher temperature in the center direction (Snowden et al. 1993, 1998). This temperature gradient may be the first suggestion of a connection between Loop I and the LHB, allowing the higher temperature and pressure plasma of Loop I to leak into the LHB.

The solar neighborhood in the grand scheme of things

Placing a requirement for an ISM component with a temperature of $\sim 10^6$ K within the nearest 100 pc or so has certain obvious, and not so obvious consequences for models of the ISM. The first tendency is to assume that the solar neighborhood is "typical" of the general ISM; however, from Figures 8–10 this doesn't seem to be the case. While there are other regions of hot plasma, they do not appear to be the dominant component of the ISM. It is very difficult to extrapolate the local geometry of the $\frac{1}{4}$ keV emitting region to a general picture of the galactic ISM. Because of the high optical depth of the neutral ISM for X-rays in this energy range, observations are limited in general to only the nearest few hundred parsecs at low latitudes. The situation is not much better at high galactic latitudes either. If the $\frac{1}{4}$ keV emission is located just above the disk as suggested by its clumpy nature, then only the nearest kiloparsec or so can be sampled.

Of the emission regions indicated in Figure 10, the Loop I superbubble is roughly spherical and is the largest feature on the sky due to its proximity (~ 150 pc distant with a radius of ~ 100 pc), and it is roughly centered on the Sco–Cen OB association. The Eridanus Enhancement is

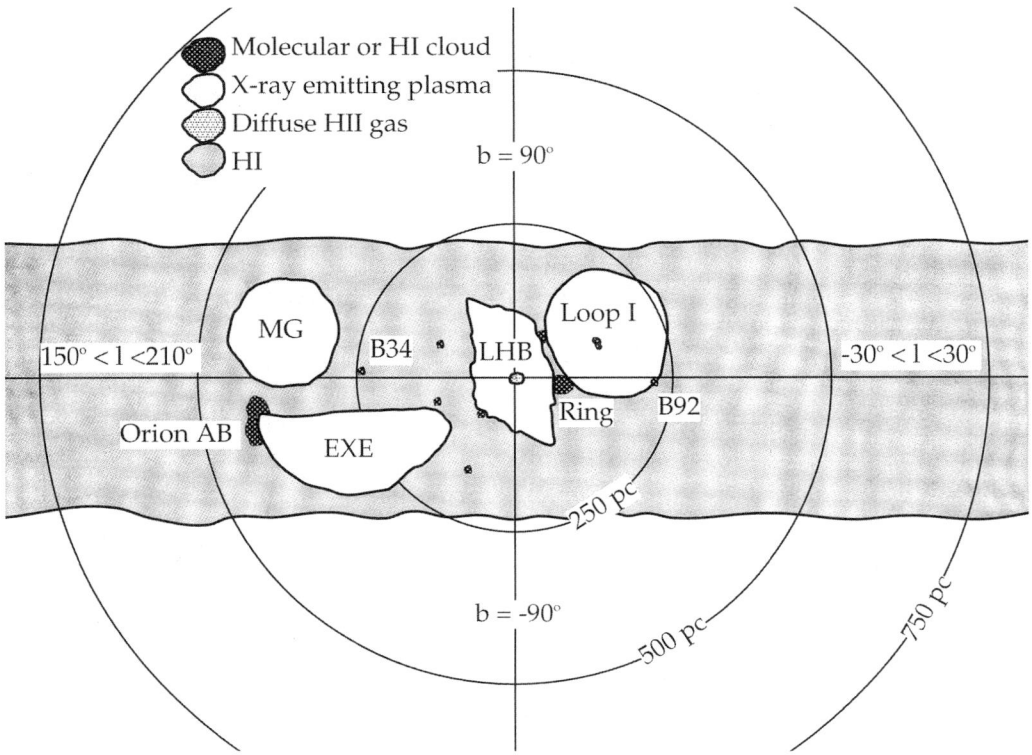

Figure 8 Graphical representation of the nearest 750 pc of the local ISM in a cut along the galactic center–anticenter line. The abbreviations are listed in Table 1.

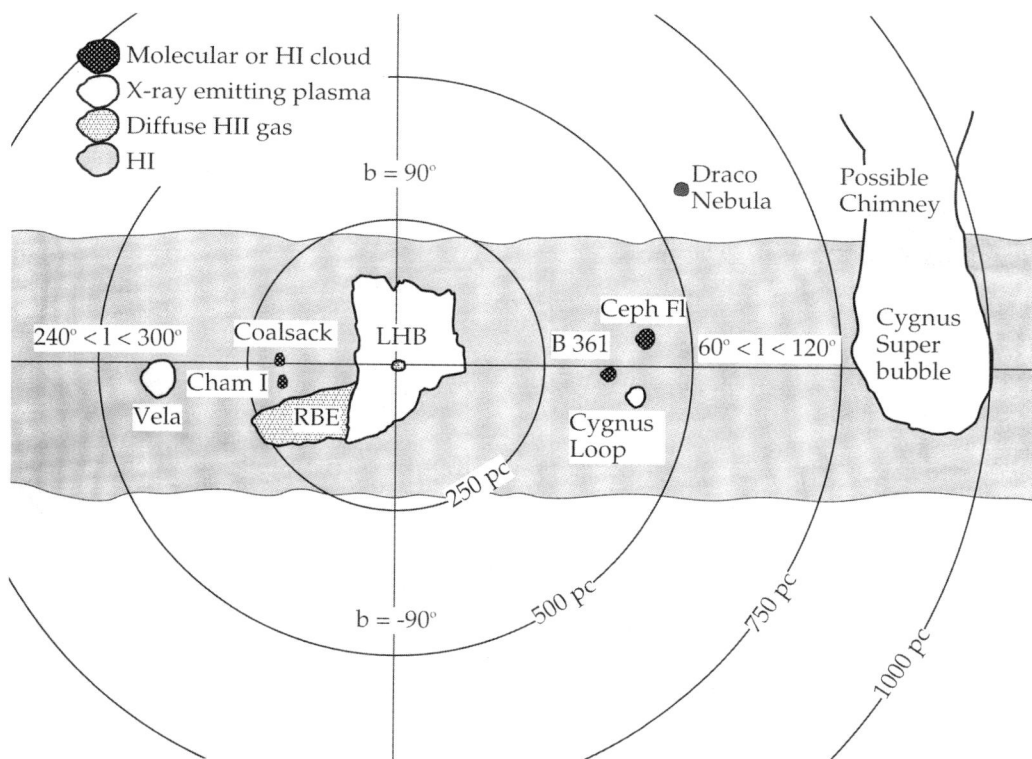

Figure 9 Graphical representation of the nearest 1000 pc of the local ISM in a cut along the $l=90°–l=270°$ line. The abbreviations are listed in Table 1.

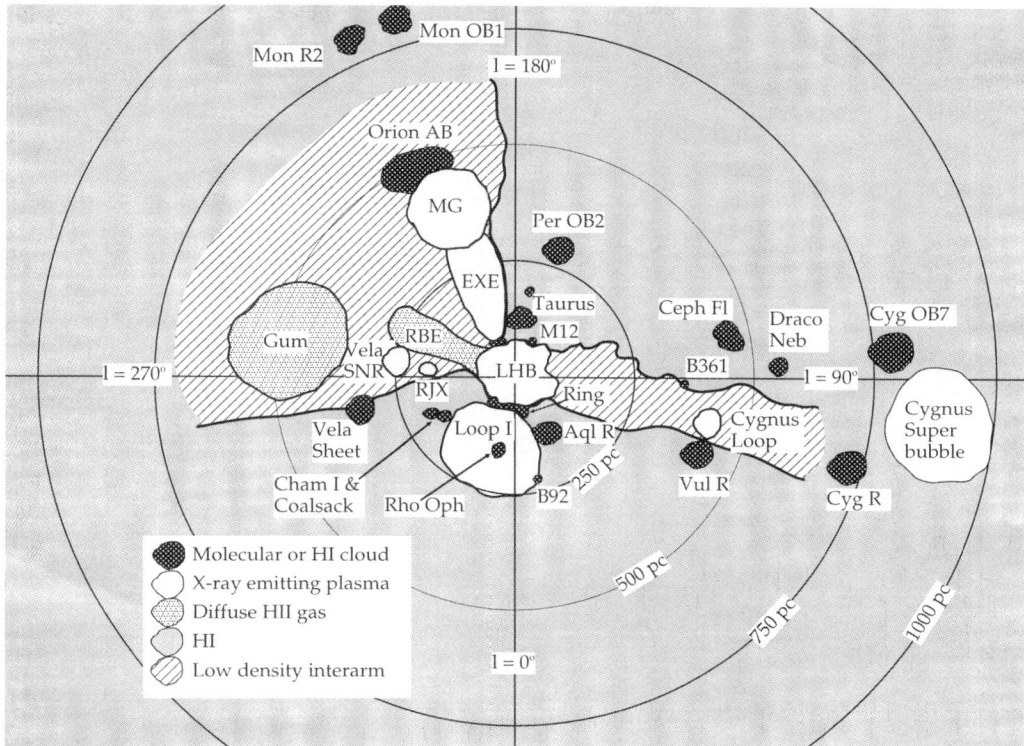

Figure 10 Graphical representation of the nearest 1000 pc of the local ISM compressed to the galactic plane as viewed from above. Additional molecular clouds have been added from Dame et al. (1987). The thick lines outline the low-density inter-arm region of Breitschwerdt et al. (1996). The finite extent of the interarm shading does not mean that it does not continue, just that it is unclear.

powered by the Orion OB associations but instead of being roughly circular it is more cigar-shaped extending from the OB associations to within a couple of hundred parsecs from the Sun (Burrows et al. 1993). The Monoceros–Gemini Enhancement is a likely supernova remnant (Plucinsky et al. 1996) which lies slightly above the plane roughly over the Eridanus Enhancement. Of the three, the Monogem Ring with its cooler plasma and no current power source is the most likely analog to the LHB. The existence of these nearby distinct and discrete bubbles of hot plasma suggest that they are reasonably common in the galactic disk. The fact that all of them appear to be isolated features suggests that the hot plasma does not form a ubiquitous substrate in the Galaxy. The hot plasma is a perturbation on the cooler ISM rather than the cooler H I existing as discrete features (clouds) embedded in the hot plasma as in the McKee and Ostriker (1977) picture.

The question of the existence of a ubiquitous presence of hot plasma in the galactic plane can be revisited by examining the third quadrant, the quadrant containing the RBE (Figure 11). If there is a direction where an extensive distribution of hot plasma should be visible, this would be it as there is very little neutral material out to well over several hundred parsecs. Unfortunately the answer is somewhat ambiguous. In the longitude range $210° < l < 240°$ there is no significant enhancement of the SXRB surface brightness even though the cavity extends at least beyond 150 pc. However, there is a general enhancement in the longitude range $240° < l < 270°$. The high-longitude end of this enhancement is overwhelmed by the coincident emission from the Vela SNR, but by comparing the $\frac{3}{4}$ keV and $\frac{1}{4}$ keV emission maps of the region it is clear that the lower-energy emission is considerably more extensive than that of the Vela SNR. The $\sim 10°$ diameter circular $\frac{1}{4}$ keV enhancement just to the lower-longitude side of the Vela SNR is of particular interest. It is roughly aligned with a direction of maximum galactic-plane extent of the local cavity in Sfeir et al. (1999). This is suggestive of a plasma-filled tunnel in the ISM (perhaps unlikely as it would require the Sun to be in the special location of being able to look down the tunnel), a "hole" into a more distant emission region – as suggested by Heiles (1998) – or a distinct SNR. In any case, since there is only a slight corresponding enhancement in the $\frac{3}{4}$ keV band, the emission is relatively soft indicating a temperature more like those of the LHB and the Monogem Ring, rather than those of Eridanus or Loop I.

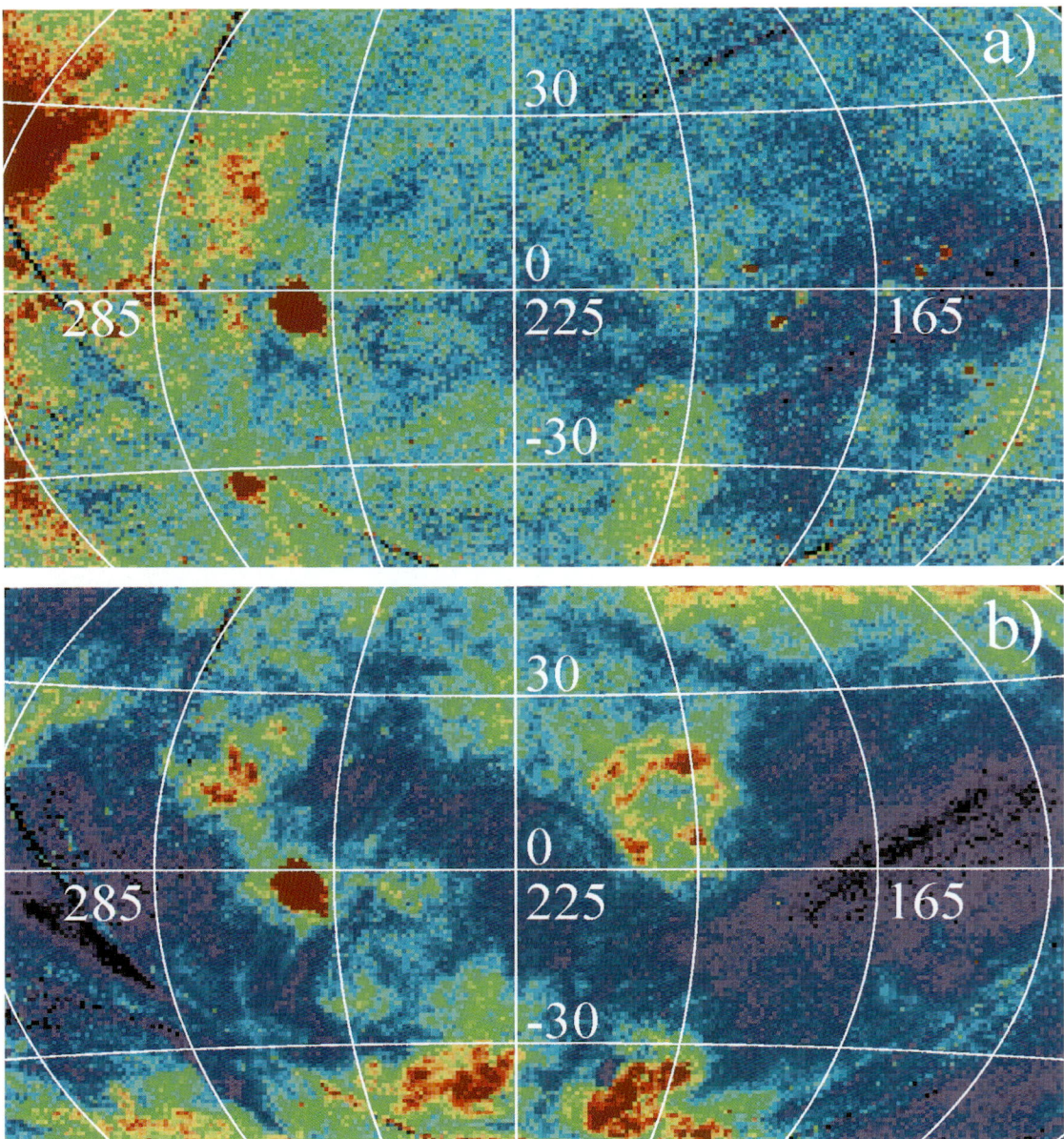

Figure 11 Cut of the galactic plane showing the quadrant containing the RBE (Region of Bizarre Emptiness) in (a) $\frac{3}{4}$ keV X-rays and (b) $\frac{1}{4}$ keV X-rays. The data are from the Rosat All-Sky Survey (Snowden *et al.* 1997) and the projection is in galactic coordinates. The images are color coded, with purple indicating low intensity and red indicating high intensity. The enhancement at $l, b \sim 200°, 7°$ is the Monogem Ring (seen best at $\frac{1}{4}$ keV), the enhancement at $l, b \sim 200°, -30°$ is the Eridanus Enhancement (the full extent is best seen at $\frac{3}{4}$ keV), and the enhancement at $l, b \sim 260°, -3°$ is the Vela SNR. Note the relatively low $\frac{1}{4}$ keV intensities along the plane in the range $210° < l < 240°$ (and likely all the way down to $l < 135°$ if not for emission associated with the Monogem Ring) while the intensity increases over the range $240° < l < 270°$, even excluding the Vela SNR.

Diffuse emission at $\frac{3}{4}$ keV in the galactic disk at the solar radius and greater is likely to be confined to younger SNRs and the stellar wind bubbles and super bubbles surrounding early-type stars and stellar clusters, such as found in Loop I and the Eridanus Enhancement. However, there is still the component which provides the additional emission observed in the galactic plane in the anti-center direction which needs to be identified. Interior to the solar radius, and particularly within the molecular ring, there appears to be quite extensive $\frac{3}{4}$ keV emission tied into the galactic X-ray bulge.

My guess for the general structure of the ISM of the galactic disk is therefore a mixture of neutral material found in sheets, filaments, clouds, and a general diffuse component (Dickey and Lockman 1990), molecular material in the usual dense clouds (e.g., Dame et al. 1987), regions of ionized material at moderate temperatures, and bubbles of hot plasma ranging from older, cooler regions such as the LHB to more active and younger objects such as Loop I and Eridanus, to the very young and active remnants such as Cas A and the Crab.

The galactic halo

The situation is a bit more obscure in the galactic halo. From Snowden et al. (1998, 2000) it is clear that there is extensive $\frac{1}{4}$ keV emission beyond the H I of the galactic disk but that it is rather clumpy. The emission ranges from near zero in certain directions of the southern galactic hemisphere to many times the surface brightness of the LHB in directions such as Draco. The clumpiness of the emission suggests that it is a low-scale-height phenomena and quite likely associated with features of the galactic disk. For instance, Snowden et al. (1997) suggested that the Draco enhancement may be a high-z (distance perpendicular to the Galactic plane) extension of emission associated with the Cygnus Superbubble (a possible chimney or galactic fountain as suggested by Shapiro and Field (1976); see Figure 9). As for the general $\frac{1}{4}$ keV halo, it is likely that it looks similar to what is observed in the solar neighborhood but with increasing amounts of hot gas closer to the galactic center and decreasing amounts at galactic radii greater than that of the Sun. This would be consistent with observations of external galaxies where $\frac{1}{4}$ keV emission is generally found within the effective solar radius of the galaxy and the emission tends to be peaked toward the center (e.g., M101; Snowden and Pietsch 1995).

Except for the possible high-z extension of the Cygnus Superbubble, X-ray emission has not been placed in the halo in Figures 8 and 9. Not that there is not any, it just is not clear where to locate it. There is also considerable H I structure in low-density arcs and high-velocity clouds at high z which is not represented. However, at $\frac{3}{4}$ keV the galactic halo is probably a large-scale-height extended structure in hydrostatic equilibrium such as suggested by Wang (1998), which is tied into the galactic X-ray bulge as the center of the galactic gravitational potential. However, its surface brightness may be relatively low, perhaps 10–20% of the average background at high latitudes.

There is evidence for an extensive distribution of $T \gtrsim 10^5$ K plasma in the halo as demonstrated by observations of the absorption lines of highly ionized atoms (e.g., O VI, N V, and C IV). While the nature of the distribution of the plasma is apparently patchy, its scale height is likely to be $\gtrsim 3$ kpc (Sembach et al. 1999). This material is in competition for space with several other components, with varying scale heights: diffuse H I with a scale height of ~ 500 pc (Lockman 1986), the warm ionized medium (WIM) with a scale height of ~ 1 kpc and temperature of $\sim 10^4$ K (Haffner et al. 1999), intermediate-velocity clouds (IVCs) with heights of hundreds of parsecs and high-velocity clouds (HVCs, at least some of which are a few kiloparsecs above the galactic plane; e.g., Danly et al. 1993), and of course the 10^6 K plasma responsible for the halo SXRB. Above the rest sits the few 10^6 K plasma responsible for the galactic portion of the $\frac{3}{4}$ keV background. This leads to the combination of layer cake and plum-pudding diagram of Figure 12, that the author hopes represents our region of the Galaxy.

Origins

So far this chapter has treated the appearance and the spatial distribution of the emission responsible for the SXRB, and for the most part ignored its physical origin. The diffuse X-ray and EUV emission in the galactic disk generally originates from the hot plasmas and shock fronts produced by supernovae and stellar winds. While there is little if any evidence for enhanced X-ray emission at the boundary layers between the LHB and local cavity, or surrounding any clouds embedded in the LHB, the existence of a ubiquitous ISM O VI line implies the existence of plasmas which should be contributing to the diffuse EUV background. The time scales for heating and cooling, and for the maintenance of hot coherent structures in the ISM is more on the order of tens of millions of years rather than of galactic or cosmological orders. Sound speeds across hot bubbles are of the order of one million years while cooling times are of the order of ten million years (Cox and Snowden 1986). The recurrence time for supernovae in the volume of the LHB is also on the order of ten million years. This all ties the SXRB of the galactic disk closely to the ISM as the reprocessing of the ISM, at least on galactic time scales, must be a near-continuous activity – a point made with different interpretations by both Cox and Smith (1974) and McKee and Ostriker (1977).

The exact origin of the hot plasma of the LHB remains obscure. While the assumption is that the plasma was produced by a supernova, there are no convenient pulsars lurking in the local cavity to prove it. At one point the Geminga pulsar was a likely candidate for the fossil remains of the supernova, but the eventually determined distance of 160 pc placed it outside of the local cavity. Recently, however, evidence for a nearby supernova of the proper age has, literally, surfaced. Knie et al. (1999) presented corroborating evidence for a recent (~ 5 million years ago), nearby

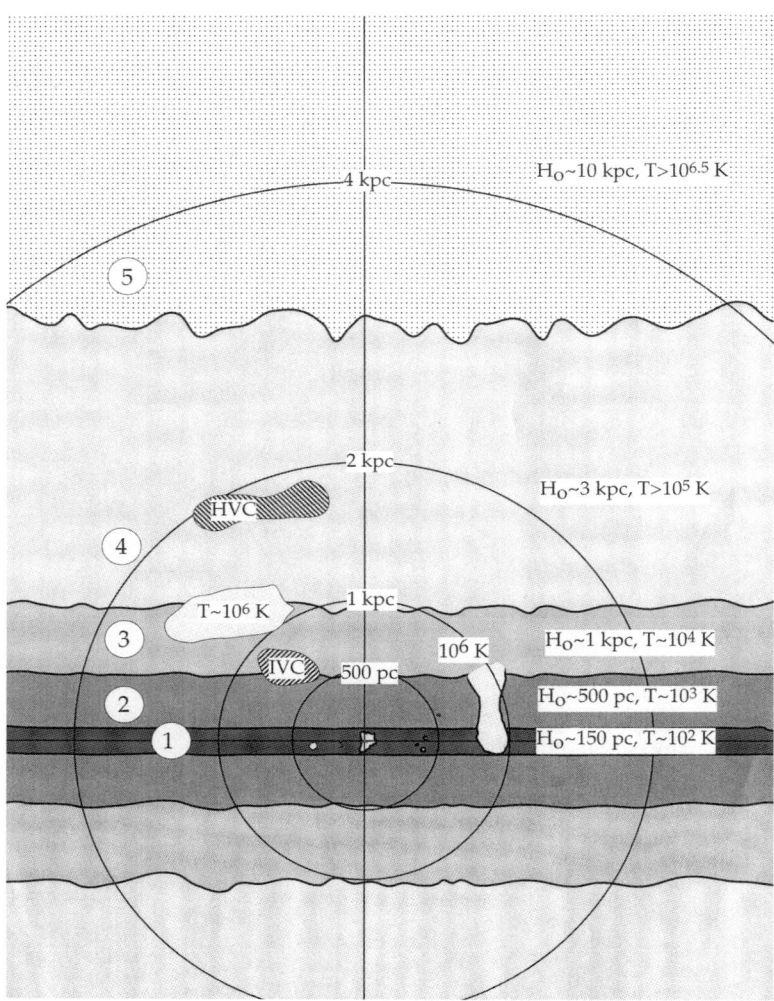

Figure 12 Schematic representing the galactic halo in the solar neighborhood. (1) The galactic disk with a scale height, H_0, of 150 pc. (2) The high-scale-height, $H_0 \sim 500$, diffuse H I of Lockman (1986). (3) The warm ionized medium (WIM) of Haffner et al. (1999) with $H_0 \sim 1$ kpc. (4) The O VI plasma of Sembach et al. (1999) with $H_0 \gtrsim 3$ kpc. (5) The galactic corona with $H_0 \sim 10$ kpc (e.g., Wang 1998). The various components are unlikely to be distributed in such a smooth, plane-parallel manner, but it simplifies the presentation. Embedded in the above components are the intermediate (IVC) and high-velocity (HVC) clouds of predominantly neutral material. There are also patchy regions of 10^6 K plasma.

(~ 30 pc) supernova obtained from ^{60}Fe measurements of a sample of deep-ocean ferromanganese crust. The excess ^{60}Fe detected is assumed to have been produced in the supernova and transported to Earth with the supernova ejecta. The distance of 30 pc places the supernova comfortably within the local cavity irrespective of direction, and the age places the event well within the required cooling time scale.

As for the large-scale structure of the hot galactic halo, it should be considered in terms of the galactic gravitational potential (dark matter and all) and hydrostatic equilibrium (e.g., Wang 1998, Kalberla and Kerp 1998). However, the detailed distribution of material and emission is unlikely to be static or in general smoothly distributed. There is energy expelled into the halo from energetic phenomena in the galactic disk and accretion of material from the intergalactic medium. There are likely to be occasional supernovae from halo stars (e.g., Shelton 1999) and perhaps X-ray and EUV emission from the interaction of HVCs with the halo gas (e.g., by shock heating and/or magnetic reconnection; Kahn and Brett 1996). When the HVCs, IVCs, and diffuse ionized and neutral gas are thrown in, it makes for a complicated situation.

THE FUTURE

So, what is left to be discovered, understood, or corrected (again) about the hot part of the interstellar medium? A fair amount of uncertainty lies in the spectrum of the SXRB, and therefore the physical states of the plasmas. As discussed above, the broad-band data from the Wisconsin Be and B bands are in significant conflict with the Rosat WFC S1 filter data. In addition, while high-spectral-resolution data on the SXRB is limited to the shuttle flight of the Wisconsin Bragg-crystal spectrometer (DXS) and a couple of sounding-rocket flights of the Goddard Space Flight Center/Wisconsin microcalorimeter, it is already clear that thermal equilibrium emission models do not fit these data. Either the models are wrong, the plasmas are not in equilibrium or have non-standard abundances, or most likely all of the above.

There are, as well, models for X-ray emission from the LHB that are radically different from the generally assumed, slowly cooling plasmas. Breitschwerdt and Schmutzler (1994) modeled the emission as delayed recombination from an adiabatically cooled plasma which broke out of a confined region at a higher temperature. Further observations with better statistics and even higher spectral resolution are required to resolve the ionization states and abundances of the various constituents. A recent proposal to address this issue is the Hot Interstellar Medium Spectrometer (HIMS) satellite. This project would use micro calorimeters with $\leq 5\,eV$ resolution to perform an all-sky survey with multiple coverage and $\leq 5'$ angular resolution (for the removal of point-source contamination, statistically useful pixels will be on average significantly larger). HIMS data would allow the identification of ionization states and abundances of the hot plasma in the LHB. With the LHB emission properly modeled it can then be subtracted from high-latitude data, and similar information derived for X-ray emission regions in the halo and beyond. This would provide us with considerably better insights into the physical conditions of the solar neighborhood, the halo, and cosmological structures.

The continuation of the Sfeir et al. (1998) three-dimensional mapping of the local interstellar medium and lower halo by the use of interstellar absorption-line measurements will provide vital information for the determination of the location and structure of the nearby cooler ISM. With the knowledge of the local distribution of H I, the location of the X-ray-emitting plasmas can be more tightly constrained. Currently at high galactic latitudes the choice is whether to place the X-ray emission in front of, or behind the neutral material, with the distances to that material known for just a few directions.

While no major surprises are expected in the evolution of our understanding of the interstellar medium and the EUV/soft X-ray diffuse background, that is what many of us believed in 1990, in the pre-Rosat days. The physical state of the plasmas may prove to be unexpected and the amount of emission arising from boundary layers at the edge of cavities, or at the surface of clouds embedded in the plasmas, may be greater than now seems apparent. Our understanding of the galactic halo, and its multi-component nature, is rapidly increasing which always leaves room for surprises. More data from next-generation observatories with improved angular and spectral resolution, spectral ranges (the $\frac{1}{4}$ keV band is being somewhat ignored by the current generation of X-ray telescopes), and with increased effective areas will be needed to resolve many of the outstanding issues. In short, there are plenty of challenges remaining in this field to take into the next century of space science.

Acknowledgments

The author would like to thank D.N. Burrows, K.D. Kuntz, W.T. Sanders, and S.J. Snowden for their suggestions which helped the clarity and completeness of this chapter.

REFERENCES

Berghöfer, T.W., Bowyer, S., Lieu, R. and Knude, J. (1998). The thermal pressure of the hot interstellar medium derived from cloud shadows in the extreme ultraviolet. *Astrophysical Journal*, **500**, 838–846.

Bloch, J.J., Jahoda, K., Juda, M., McCammon, D., Sanders, W.T. and Snowden, S.L. (1986). Observations of the soft X-ray diffuse background at 0.1 keV. *Astrophysical Journal Letters*, **308**, L59–L62.

Bowyer, C.S., Field, G.B. and Mack, J.E. (1968). Detection of ananisotropic soft X-ray background flux. *Nature*, **217**, 32–34.

Bowyer, C.S. and Field, G.B. (1969). Intensity of the soft X-ray background flux. *Nature*, **223**, 573–575.

Breitschwerdt, D., Egger, R., Freyberg, M.J., Frisch, P.C. and Vallerga, J.V. (1996). The local bubble origin and evolution. *Space Science Reviews*, **78**(1/2), 183–198.

Breitschwerdt, D. and Schmutzler, T. (1994). Delayed recombination as a major source of the soft X-ray background. *Nature*, **371**, 774–777.

Bunner, A.N., Coleman, P.C., Kraushaar, W.L., McCammon, D., Palmieri, T.M., Shilepsky, A. and Ulmer, M. (1969). Soft X-ray background flux. *Nature*, **223**, 1222–1226.

Burrows, D.N. (1989). Observational constraints on an embedded cloud model for the soft X-ray diffuse background. *Astrophysical Journal*, **340**, 775–785.

Burrows, D.N. and Kraft, R.P. (1993). On the source function of the soft X-ray diffuse background. *Astrophysical Journal*, **411**, 685–689.

Burrows, D.N., McCammon, D., Sanders, W.T. and Kraushaar, W.L. (1984). Limits on soft X-ray flux from distant emission regions. *Astrophysical Journal*, **287**, 208–218.

Burrows, D.N. and Mendenhall, J.A. (1991). Soft X-ray shadowing by the Draco cloud. *Nature*, **351**, 629–631.

Burrows, D.N., Singh, K.P., Nousek, J.A., Garmire, G.P. and Good, J. (1993). A multiwavelength study of the Eridanus soft X-ray enhancement. *Astrophysical Journal*, **406**, 97–111.

Burstein, P., Borken, R.J., Kraushaar, W.L. and Sanders, W.T. (1977). Three-band observations of the soft X-ray background and some implications of thermal emission models. *Astrophysical Journal*, **213**, 405–420.

Cash, W., Malina, R. and Stern, R. (1976). An observation of the diffuse soft X-ray/extreme-ultraviolet background. *Astrophysical Journal Letters*, **204**, L7–L11.

Cha, A.N., Sembach, K.R. and Danks, A.C. (1999). The distance to the Vela supernova remnant. *Astrophysical Journal Letters*, **515**, L25–L28.

Chen, W. and Gehrels, N. (1999). The progenitor of the new Comptel/ROSAT supernova remnant in Vela. *Astrophysical Journal Letters*, **514**, L103–L106.

Cox, D.P. (1997). Modeling the Local Bubble. In *The Local Bubble and Beyond*, IAU Colloquium No. 166 (Garching, Germany), Springer-Verlag, Berlin, pp. 121–131.

Cox, D.P. and Reynolds, R.J. (1987). The local interstellar medium. *Annual Review of Astronomy and Astrophysics*, **25**, 303–344.

Cox, D.P. and Smith, B.W. (1974). Large-scale effects of supernova remnants on the Galaxy: Generation and maintenance of a hot network of tunnels. *Astrophysical Journal Letters*, L105–L108.

Cox, D.P. and Snowden, S.L. (1986). Perspective on the local interstellar medium. *Advances in Space Research*, **6**(2), 97–107.

Cox, D.P. and Tucker, W. (1969). Ionization equilibrium and radiative cooling of a low-density plasma. *Astrophysical Journal*, **157**, 1157–1167.

Dame, T.M., Ungerechts, H., Cohen, R.S., de Geus, E.J., Grenier, I.A., May, J., Murphy, D.C., Nyman, L.A. and Thaddeus, P. (1987). A composite CO survey of the entire Milky Way. *Astrophysical Journal*, **322**, 706–720.

Danly, L., Albert, C.E. and Kuntz, K.D. (1993). A determination of the distance to the high-velocity cloud complex M. *Astrophysical Journal Letters*, **416**, L29–L32.

Davidsen, A., Shulman, S., Fritz, G., Meekins, J.F., Henry, R.C. and Friedman, H. (1972). Observations of the soft X-ray background. *Astrophysical Journal*, **177**, 629–642.

de Korte, P.A.J., Bleeker, J.A.M., Deerenberg, A.J.M., Hayakawa, S., Yamashita, K. and Tanaka, Y. (1976). Features in the brightness distribution and spectra of the soft X-ray background. *Astronomy and Astrophysics*, **48**, 235–244.

Dickey, J.M., Salpeter, E.E. and Terzian, Y. (1979). Interpretation of neutral hydrogen absorption. *Astrophysical Journal*, **228**, 465–474.

Dickey, J.M. and Lockman, F.J. (1990). H I in the Galaxy. *Annual Review of Astronomy and Astrophysics*, **28**, 215–261.

Egger, R.J. and Aschenbach, B. (1995). Interaction of the Loop I supershell with the Local Hot Bubble. *Astronomy and Astrophysics Letters*, **294**, L25–L28.

Field, G.B., Goldsmith, D.W. and Habing, H.J. (1969). Cosmic-ray heating of the interstellar gas. *Astrophysical Journal Letters*, **155**, L149–L154.

Fried, P.M., Nousek, J.A., Sanders, W.T. and Kraushaar, W.L. (1980). The soft X-ray background and the structure of the local interstellar medium. *Astrophysical Journal*, **242**, 987–1004.

Frisch, P.C. (1995). Characteristics of nearby interstellar matter. *Space Science Reviews*, **72**(3/4), 499–592.

Frisch, P.C. and York, D.G. (1983). Synthesis maps of ultraviolet observations of neutral interstellar gas. *Astrophysical Journal Letters*, **271**, L59–L63.

Garmire, G.P., Nousek, J.A., Apparao, K.M.V., Burrows, D.N., Fink, R.L. and Kraft, R.P. (1992). The soft X-ray diffuse background observed with the HEAO 1 low-energy detectors. *Astrophysical Journal*, **399**, 694–703.

Giacconi, R., Gursky, H., Paolini, F.P. and Rossi, B.B. (1962). Evidence for X-rays from sources outside the solar system. *Physical Review Letters*, **9**, 439–443.

Gould, R.J. and Sciama, D.W. (1964). Cosmic X- and infrared rays as tools for exploring the large-scale structure of the universe. *Astrophysical Journal Letters*, **140**, 1634–1636.

Grant, C.E. (1999). The three-dimensional structure of the hot interstellar medium. PhD Thesis, The Pennsylvania State University.

Gry, C., York, D.G. and Vidal-Madjar, A. (1985). The exceptionally vacant line of sight to Beta Canis Majoris. *Astrophysical Journal*, **296**, 593–599.

Gursky, H., Giacconi, R., Paolini, F.P. and Rossi, B.B. (1963). Further evidence for the existence of Galactic X-rays. *Physical Review Letters*, **11**, 530–535.

Haffner, L.M., Reynolds, R.J. and Tufte, S.L. (1999). WHAM observations of Hα, [S II] and [N II] toward the Orion and Perseus arms: Probing the physical conditions of the warm ionized medium. *Astrophysical Journal*, **523**, 223–233.

Hartmann, D. and Burton, W.B. (1997). *Atlas of Galactic Neutral Hydrogen*, Cambridge University Press, Cambridge.

Hayakawa, S., Kato, T., Nagase, F., Yamashita, K. and Tanaka, Y. (1978). Distribution and energy spectrum of diffuse soft X-rays. *Astronomy and Astrophysics*, **62**, 21–28.

Heiles, C. (1998). Whence the Local Bubble, Gum, Orion? GSH 238+00+09, a nearby major superbubble toward Galactic longitude 238°. *Astrophysical Journal*, **498**, 689–703.

Henry, R.C., Fritz, G., Meekins, J.F., Friedman, H. and Byram, E.T. (1968). Possible detection of a dense intergalactic plasma. *Astrophysical Journal Letters*, **153**, L11–L18.

Jahoda, K., McCammon, D. and Lockman, F.J. (1986). Small-scale H I structure and the soft X-ray background. *Astrophysical Journal Letters*, **311**, L57–L61.

Jakobsen, P. and Kahn, S.M. (1986). On the interpretation of the soft X-ray background: The effects of an embedded cloud geometry. *Astrophysical Journal*, **309**, 682–693.

Jenkins, E.B. (1978). Coronal gas in the Galaxy: I. A new survey of interstellar O VI. *Astrophysical Journal*, **219**, 845–860.

Juda, M., Bloch, J.J., Edwards, B.C., McCammon, D., Sanders, W.T., Snowden, S.L. and Zhang, J. (1991). Limits on the density of neutral gas within 100 parsecs from observations of the soft X-ray background. *Astrophysical Journal*, **367**, 182–185.

Kahn, F.D. and Brett, L. (1993). Magnetic connection in the disk and halo. *Monthly Notices of the Royal Astronomical Society*, **263**, 37–48.

Kalberla, P.M.W and Kerp, J. (1998). Hydrostatic equilibrium conditions in the Galactic halo. *Astronomy and Astrophysics*, **339**, 745–758.

Knapp, G.R. (1975). Study of galactic gas and dust using observations of elliptical galaxies. *Astronomical Journal*, **80**, 111–116.

Knie, K., Korschinek, G., Paestermann, T., Wallner, C., Scholten, J. and Hillebrandt, W. (1999). Indication for supernova-produced ^{60}Fe activity on Earth. *Physical Review Letters*, **83**, 18–21.

Kuntz, K.D. and Snowden, S.L. (2000). Deconstructing the spectrum of the diffuse X-ray background. *Astrophysical Journal*, **543**, 195.

Lockman, F.J. (1986). 21-centimeter studies of the Galactic halo. In J.N. Bregman and F.J. Lockman (eds), *Gaseous Halos of Galaxies*, NRAO, Greenbank, pp. 63–73.

Long, K., Agrawal, P.C. and Garmire, G. (1975). Soft X-rays from the Large Magellanic Cloud: Implications on the origin of the diffuse X-ray background. *Astrophysical Journal*, **206**, 411–417.

Magnani, L., Blitz, L. and Mundy, L. (1985). Molecular gas at high Galactic latitudes. *Astrophysical Journal*, **295**, 402–421.

Marshall, F.J. and Clark, G.W. (1984). SAS 3 survey of the soft X-ray background. *Astrophysical Journal*, **287**, 633–652.

McCammon, D., Burrows, D.N., Sanders, W.T. and Kraushaar, W.L. (1983). The soft X-ray diffuse background. *Astrophysical Journal*, **269**, 107–135.

McCammon, D., Bunner, A.N., Coleman, P.L. and Kraushaar, W.L. (1976). A search for absorption of the soft X-ray diffuse flux by the Small Magellanic Cloud. *Astrophysical Journal Letters*, **168**, L33–L37.

McCammon, D. and Sanders, W.T. (1990). The soft X-ray background and its origins. *Annual Review of Astronomy and Astrophysics*, **28**, 657–688.

McKee, C.F. and Ostriker, J.P. (1977). A theory of the interstellar medium: Three components regulated by supernova explosions in an inhomogeneous substrate. *Astrophysical Journal*, **218**, 148–169.

Morrison, R. and McCammon, D. (1983). Interstellar photoelectric absorption cross-sections, 0.03–10 keV. *Astrophysical Journal*, **270**, 119–122.

Paresce, F. (1984). On the distribution of interstellar matter around the Sun. *Astronomical Journal*, **89**, 1022–1037.

Park, S., Finley, J.P., Snowden, S.L. and Dame, T.M. (1997). Evidence for an X-ray-emitting Galactic bulge: Shadows cast by distant molecular gas. *Astrophysical Journal Letters*, **476**, L77–L80.

Pietz, J., Kerp, J., Kalberla, P.M.W., Burton, W.B., Hartmann, D. and Mebold, U. (1998). The galactic X-ray halo. *Astronomy and Astrophysics*, **332**, 55–70.

Plucinsky, P.P., Snowden, S.L., Aschenbach, B., Egger, R., Edgar, R.J. and McCammon, D. (1996). ROSAT survey observations of the Monogem Ring. *Astrophysical Journal*, **463**, 224–245.

Reynoso, E.M. and Dubner, G.M. (1997). Cold neutral gas associated with the Gum Nebula. *Astronomy and Astrophysics Suppl.*, **123**, 31–41.

Sanders, W.T., Burrows, D.N., McCammon, D. and Kraushaar, W.L. (1983). Possible contributions of supernova remnants to the soft X-ray diffuse background (0.1–1.0 keV). In J. Danziger and P. Gorenstein (eds), *IAU Symposium 101, Supernova Remnants and their X-ray Emission*, Reidel, Dordrecht, pp. 361–365.

Sanders, W.T., Kraushaar, W.L., Nousek, J.A. and Fried, P.M. (1977). Soft diffuse X-rays in the southern galactic hemisphere. *Astrophysical Journal Letters*, **217**, L87–L91.

Schlegel, D.J., Finkbeiner, D.P. and Davis, M. (1998). Maps of dust infrared emission for use in estimation of reddening and cosmic microwave background radiation foregrounds. *Astrophysical Journal*, **500**, 525–553.

Schmitt, J.H.M.M. and Snowden, S.L. (1990). Contributions of late-type dwarfs to the soft X-ray diffuse background. *Astrophysical Journal*, **361**, 207–214.

Sembach, K.R., Savage, B.D. and Hurwitz, M. (1999). GHRS and ORFEUS II observations of the highly ionized interstellar medium toward ESO 141-055. *Astrophysical Journal*, **524**, 98–111.

Sfeir, D.M., Lallement, R. Crifo, F. and Welsh, B.Y. (1999). Mapping the contours of the local bubble: Preliminary results. *Astronomy and Astrophysics*, **346**, 785–797.

Shapiro P.R. and Field, G.B. (1976). Consequences of a new hot component of the interstellar medium. *Astrophysical Journal*, **205**, 762–765.

Shelton, R.L. (1999). Simulations of supernova remnants in diffuse media: II. Three remnants and their X-ray emission. *Astrophysical Journal*, **521**, 217–233.

Snowden, S.L., Cox, D.P., McCammon, D. and Sanders, W.T. (1990). A model for the distribution of material generating the soft X-ray background. *Astrophysical Journal*, **354**, 211–219.

Snowden, S.L., Egger, R., Finkbeiner, D.P., Freyberg, M.J. and Plucinsky, P.P. (1998). Progress on establishing the spatial distribution of material responsible for the $\frac{1}{4}$ keV soft X-ray diffuse background local and halo components. *Astrophysical Journal*, **493**, 715–729.

Snowden, S.L., Egger, R., Freyberg, M.J., McCammon, D., Plucinsky, P.P., Sanders, W.T., Schmitt, J.H.M.M., Truemper, J. and Voges, W. (1997). ROSAT survey diffuse X-ray background maps II. *Astrophysical Journal*, **485**, 125–135.

Snowden, S.L., Freyberg, M.J., Kuntz, K.D. and Sanders, W.T. (2000). A catalog of soft X-ray shadows and more contemplation of the $\frac{1}{4}$ keV background. *Astrophysical Journal Suppl.*, **128**, 171.

Snowden, S.L., Freyberg, M.J., Plucinsky, P.P., Schmitt, J.H.M.M., Truemper, J., Voges, W., Edgar, R.J., McCammon, D. and Sanders, W.T. (1995). First maps of the soft X-ray diffuse background from the ROSAT XRT/PSPC All-Sky Survey. *Astrophysical Journal*, **454**, 643–653.

Snowden, S.L., McCammon, D. and Verter, F. (1993). The X-ray shadow of the high-latitude molecular cloud MBM 12. *Astrophysical Journal Letters*, **409**, L21–L24.

Snowden, S.L., Mebold, U., Hirth, W., Herbstmeier, U. and Schmitt, J.H.M. (1991). ROSAT detection of an X-ray shadow in the $\frac{1}{4}$ keV diffuse background in the Draco nebula. *Science*, **252**, 1529–1532.

Snowden, S.L. and Pietsch, W. (1995). Diffuse X-ray emission from M101. *Astrophysical Journal*, **452**, 627–633.

Snowden, S.L., Schmitt, J.H.M.M. and Edwards, B.C. (1990). A color gradient in the soft X-ray diffuse background. *Astrophysical Journal*, **364**, 118–122.

Spitzer, L. (1956). On a possible interstellar galactic corona. *Astrophysical Journal*, **124**, 20–34.

Stark, A.A., Gammie, C.F., Wilson, R.W., Bally, J., Linke, R.A., Heiles, C. and Hurwitz, M. (1992). The Bell Laboratories H I survey. *Astrophysical Journal Suppl.*, **79**, 77–104.

Stern, R. and Bowyer, S. (1979). Apollo–Soyuz survey of the extreme-ultraviolet/soft X-ray background. *Astrophysical Journal*, **230**, 755–767.

Tanaka, Y. and Bleeker, J.A.M. (1977). The diffuse soft X-ray sky. *Space Science Reviews*, **20**, 815–888.

Trümper, J. (1992). ROSAT – A new look at the X-ray sky. *Quarterly Journal of the Royal Astronomical Society*, **33**, 165–174.

Vallerga, J. and Slavin, J. (1998). The diffuse extreme ultraviolet background. In D. Breitschwerdt. M.J. Freyberg and J. Trümper (eds), *The Local Bubble and Beyond*, IAU Colloquium 166, Springer-Verlag, Berlin, pp. 79–82.

Wang, Q.D. (1998). The hot galactic corona and the soft X-ray background. In D. Breitschwerdt, M.J. Freyberg and J. Trümper (eds), *The Local Bubble and Beyond*, IAU Colloquium 166, Springer-Verlag, Berlin, pp. 503–512.

West, R.G., Willingale, R., Pye, J.P. and Sumner, T. (1996). A search for the signature of the diffuse soft X-ray background in the ROSAT Wide-Field Camera all-sky survey. In S. Bowyer and R.F. Malina (eds), *Astrophysics in the Extreme Ultraviolet*, IAU Colloquium 152, Kluwer, Dordrecht, pp. 289–293.

Williamson, F.O., Sanders, W.T., Kraushaar, W.L., McCammon, D., Borken, R. and Bunner, A.N. (1974). Observation of features in the soft X-ray background flux. *Astrophysical Journal Letters*, **193**, L133–L137.

27

EWINE F. VAN DISHOECK* AND ALEXANDER G.G.M. TIELENS**

Space-borne observations of the life cycle of interstellar gas and dust

The gas and dust in the interstellar medium (ISM) form an essential part of the evolution of galaxies, the formation of stars and planetary systems, and the synthesis of organic molecules that may lead to the emergence of life elsewhere in the Universe. Over their lifetimes, stars return much of their mass to the ISM through winds and supernova explosions, leading to a slow enrichment in heavy elements and dust that form the building blocks from which future generations of stars and planets are made. Stars also inject energy into the ISM via ultraviolet photons, shocks, and wind-blown stellar bubbles. Cosmic rays, x-rays, and ionizing photons influence the ionization state of the gas, whereas shielding by gas and dust leads to the cold, neutral phases of the ISM where molecules can flourish. As a result, the composition and structure of the ISM is governed by a complex interplay of microscopic and macroscopic processes. Understanding this life cycle of gas and dust in the ISM is a key problem in astrophysics, for understanding not only our own Galaxy but also the much more rapid cycling between the ISM and stars in the earliest star-forming galaxies in the Universe (Figure 1).

Research on the ISM started early in the twentieth century with the ground-based detection of interstellar Na and Ca^+ optical absorption lines (Hartmann 1904, Heger 1919) and the appearance of many dark regions on photographic surveys of the Milky Way (e.g., Barnard 1919). Definite evidence for the presence of interstellar dust came from observations by Trumpler (1930), whereas the first interstellar molecules, CH, CH^+, and CN, were identified between 1937 and 1941 (Swings and Rosenfeld 1937, McKellar 1940, Douglas and Herzberg 1941). Around the same time were detected the diffuse interstellar bands (DIBs; Heger 1922, Merrill 1934), whose identification is still uncertain after more than 75 years. Initially, inspired by the identification of simple diatomic species in the ISM, these bands were attributed to molecular absorbers. However, once interstellar dust was established as an important interstellar component, an origin for the DIBs in absorption by dust grains was taken for granted. Nowadays the pendulum has swung back – and for good reason – to molecules as the prime candidates for the carriers of these absorption features (Snow 1995). The foundation of the theoretical study of the ISM and the physical conditions that may prevail there was put forward by Arthur Eddington (1926) in his famous Bakerian Lecture. It was also he who exclaimed that "atoms are physics but molecules are chemistry" with the scarcely hidden message that astronomers should stay away from molecules. Nowadays, despite Eddington's warning, molecular astrophysics is a thriving field driven to a large extent by the wealth of molecular data that has become available over the last three decades, much of it harvested from space.

With the detection of the 21 cm H I line by Ewen and Purcell (1951), the study of the ISM turned to ground-based radio observations, and data thus gathered still provide a wealth of information on the distribution and kinematics of the neutral, atomic ISM in our Galaxy and other galaxies (e.g., Hartmann and Burton 1997). The detection of the first molecules at radio wavelengths (Weinreb *et al.* 1963, Cheung *et al.* 1968) paved the way for the development of millimeter and submillimeter wave astronomy. The ubiquitous CO molecule was detected by Wilson *et al.* (1970), and a surprisingly large number of other molecules have since been found in dense molecular clouds (see van Dishoeck and Hogerheijde (1999) for a recent overview).

---

* Rijksuniversiteit Leiden, The Netherlands
** Rijksuniversiteit Groningen, The Netherlands

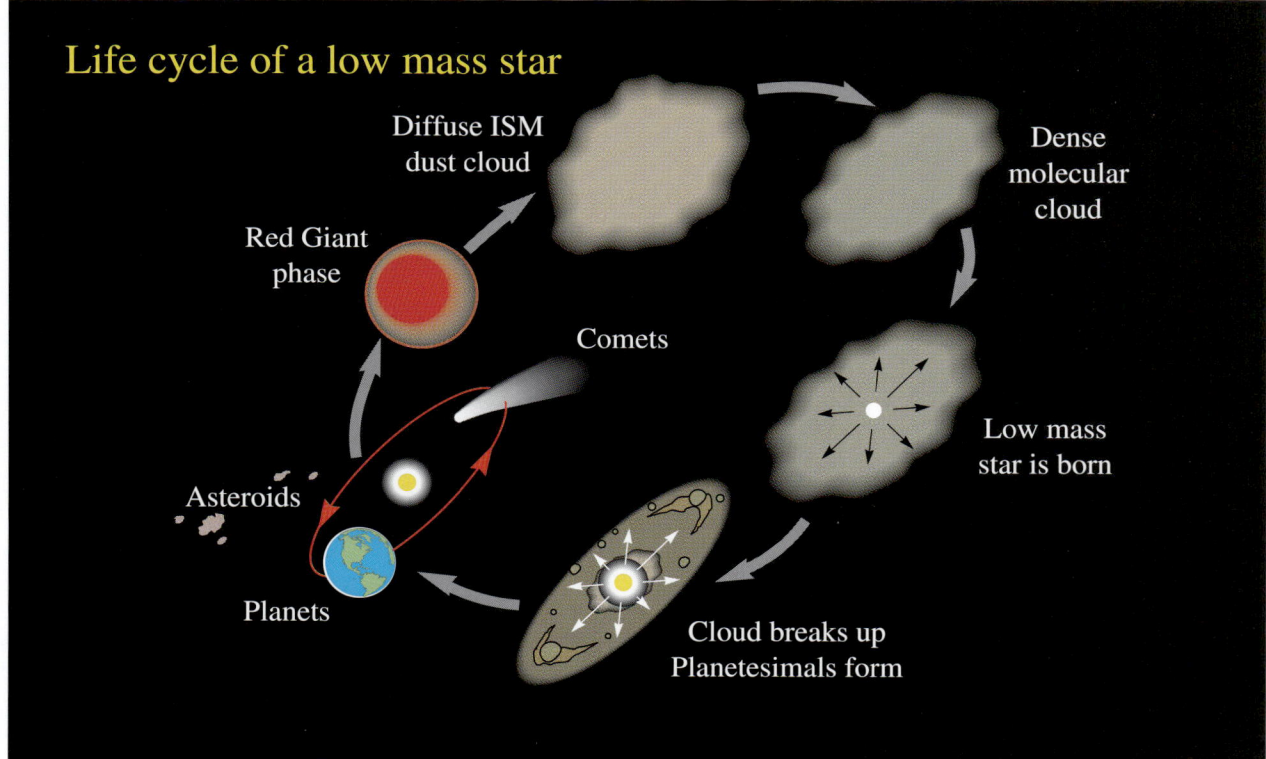

Figure 1 Schematic diagram of the lifecycle of dust and gas from the diffuse ISM, through star formation, to the late stages of stellar evolution. (Pendleton and Cruikshank 1994.)

Space-based ultraviolet observations of the neutral ISM started with small spectrometers flown on rockets, which led to the important detection of H_2 by Carruthers (1970). The later Copernicus satellite (Spitzer and Jenkins 1975) provided extensive new information on H_2 and other interstellar lines, mostly of atoms in various ionization stages. Together with the ground-based H I data, they stimulated the development of the two- and three-phase model of the ISM in which the gas is heated by the action of the photoelectric effect on grains and by large-scale supernova explosions, and is cooled by line radiation from atoms (Field et al. 1969, McKee and Ostriker 1977).

Because the typical temperatures of the cold, neutral ISM range from ~10 to a few hundred K, most of the thermal emission occurs at mid- and far-infrared wavelengths, a part of the spectrum which is largely inaccessible from the ground. It is therefore not surprising that much of our progress over the last 30 years or so has come from space-based observations in the infrared. To our knowledge, the earliest discussion of the importance of the infrared for the study of the ISM goes back to an exchange between Spitzer, Kahn, and Drake at the 3rd Symposium on Cosmical Gas Dynamics (Burges and Thomas 1958). Of course, the first detection of infrared radiation – from the Sun – was by William Herschel in 1800, while his son John Herschel was the first to measure the total power emitted by the Sun, in 1838, and Edward Stone that of stars, in 1869–70 (Harwit 1999). In the late 1870s Thomas Edison observed Arcturus and the solar corona in the infrared from a chicken coop in order to win a bet (Eddy 1972). However, despite these early successes the field of infrared astronomy lay dormant until the late 1960s to early 1970s, when progress in infrared instrumentation truly opened the infrared window of the spectrum.

The 1958 discussion mentioned above went on to consider the observational difficulties of detecting a low-intensity astronomical signal against a 300 K telluric background and emphasized the advantage of balloon or satellite observations. This was elaborated upon in an early discussion of infrared radiation from interstellar grains by Stein (1967), who concluded that it was hopeless even to consider observing it through the mid-infrared telluric windows. Fortunately he was not much impressed with his own pessimistic prediction, and, together with (among others) Fred Gillett, he went on to pioneer the field of mid-infrared spectroscopy of interstellar dust using a circular variable filter-wheel. By the early 1970s it was fully appreciated that the best way to determine the composition of interstellar

Table 1 Selected infrared space missions and instruments relevant to ISM research

Mission	Instrument	Acronym	Wavelength range (μm)	Resolving power	Operational period
IRAS	Photometers		12, 25, 60, 100	3–5	1983
	Low Resolution Spectrometer	LRS	7.7–22.6	20–60	
COBE	Far-InfraRed Absolute Spectrophotometer	FIRAS	100–10 000	100–500	1989–1993
IRTS	Far Infrared Line Mapper	FILM	63, 158	400	1995
ISO	Short Wavelength Spectrometer	SWS	2.5–45	2000, 20 000	1995–1998
	Long Wavelength Spectrometer	LWS	43–197	200, 10 000	
	Camera	CAM	2.5–17	3–20	
	Camera+Circular Variable Filter	CAM-CVF	2.3–16.5	35–50	
	Photometer	PHT	3.3–200	3	
	Photometer-Spectrometer	PHT-S	2.5–5, 6–12	90	
SWAS	Heterodyne Instrument		545, 610	5×10^5	1999–

dust was through infrared spectroscopy (Gaustad 1971). That was an impressive foresight and insight, as well as a change of direction in a short time. It should be remembered that by the late 1960s and early 1970s, rocket-borne ultraviolet studies had revealed the ubiquitous presence of 2200 Å bump in the interstellar extinction curve (Stecher 1965, Bless and Savage 1972), which had been successfully modeled in terms of small graphite grains (Gilra 1972).

Yet, as has become abundantly clear since then, advances in our knowledge of the physics and composition of the dust and gas have followed advances in infrared detector and heterodyne submillimeter receiver technology, and the development of airborne and space-based platforms through pioneers such as Martin Harwit, Gerard Kuiper, Frank J. Low, Charles H. Townes, and many others (e.g., Harwit et al. 1966, Kleinmann and Low 1967). The Lear Jet and the Kuiper Airborne Observatory (KAO) in the USA, and the US–Netherlands–UK Infrared Astronomical Satellite (IRAS), have been pivotal in developing the field, which has culminated with ESA's recent Infrared Space Observatory (ISO) (Table 1). Other important missions include NASA's Cosmic Background Explorer (COBE), balloons, rockets, the Japanese Infrared Telescope in Space (IRTS), the US Air Force project Midcourse Space Experiment (MSX), and NASA's Submillimeter Wave Astronomical Satellite (SWAS). In this respect, the future looks very promising, with the Space Infra Red Telescope Facility (SIRTF), the ASTRO-F mission, the Herschel Space Observatory (formerly known as FIRST), and the Next Generation Space Telescope (NGST) all on the horizon.

This chapter presents an overview of the air- and space-borne observations that over the last 30 years have played a crucial role in our understanding of the cold and dense neutral ISM and the formation of stars. The various phases of the ISM considered here are summarized in Figure 2, and are discussed in order from the diffuse medium to old stars. The richness of the infrared wavelength region is illustrated in Figure 3, which shows the complete ISO Short Wavelength Spectrometer (SWS) spectrum of Orion centered at the IRc2 source (van Dishoeck et al. 1998). Its proximity and extraordinary brightness have made this source the prime target for most of the pioneering observations at infrared wavelengths (e.g., Genzel and Stutzki 1989). Much of the complexity is the result of the disruption of the star-forming environment by powerful outflows from the massive young star, and its intense ultraviolet radiation dissociating and ionizing the gas on the cloud surfaces. Orion is the nearest and best studied region of massive star formation in the Galaxy, and therefore also serves as a template for more distant star-forming regions in other galaxies. The strong continuum is due to thermal emission from warm dust (50–300 K) and peaks around 70 μm. A wealth of superposed lines of atoms, ions, and molecules is seen, which can be used to constrain the physical parameters (see Figure 8 of Genzel 1992) and assess the relative importance of different processes, in particular shocks v. ultraviolet radiation.

Space limitations preclude a comprehensive review of the field, and only topics in which space-based observations have played a crucial role are covered. For example, ground-based submillimeter observations of molecular clouds are not mentioned, but the composition of ices in such clouds, as deduced from mid-infrared spectroscopy is. Ultraviolet observations of the atomic component of the ISM and space-based data on ionized H II regions and continuum observations of young stars are discussed elsewhere in this volume. Each section starts with a brief historical review and an introduction of the basic physics, and goes on to

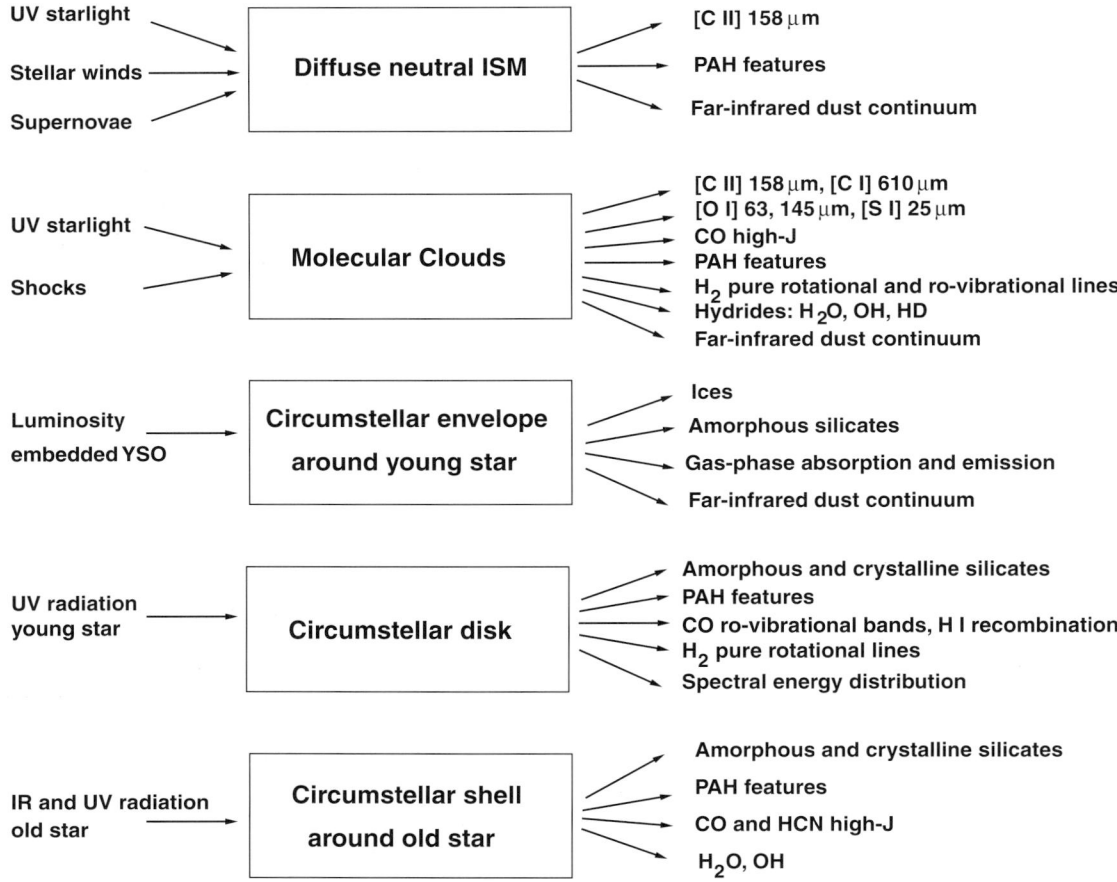

Figure 2 Schematic diagram of the energy input to the various phases of the ISM discussed here, and the resulting diagnostic lines at mid- and far-infrared wavelengths. Most of these features can only be observed from air- and space-borne platforms. PAH stands for polycyclic aromatic hydrocarbons or large carbonaceous molecules in general; YSO for young stellar objects.

focus on recent ISO results to illustrate the current state of knowledge, in particular the spectroscopic data obtained with the ISO's Short Wavelength Spectrometer (SWS, 2.4–45 μm) and Long Wavelength Spectrometer (LWS, 43–197 μm) (see Table 1 for an overview). Excellent previous summaries include the proceedings of the *Airborne Astronomy Symposium* (Haas et al. 1995), which also contains many photographs of key people in the field, and the proceedings of the *The Diffuse Infrared Radiation and the IRTS* (Okuda et al. 1997). Further details of ISO results are contained in *Star Formation with the ISO Satellite* (Yun and Liseau 1998), *ISO's View on Stellar Evolution* (Waters et al. 1998), *Analytical Spectroscopy with ISO* (Heras et al. 1997, 2000), *Solid Interstellar Matter: The ISO Revolution* (d'Hendecourt et al. 1999a), and *The Universe as seen by ISO* (Cox and Kessler 1999). Reviews of various aspects of the neutral ISM include Spitzer (1978), Habing (1988), Genzel (1992), and Hollenbach and Tielens (1999). Recent reviews of astrochemistry can be found in *The Molecular Astrophysics of Stars and Galaxies* (Hartquist and Williams 1998), van Dishoeck and Blake (1998), van Dishoeck and Hogerheijde (1999), Langer et al. (2000), Ehrenfreund and Charnley (2000), and the proceedings of IAU Symposium 197 on *Astrochemistry: From Molecular Clouds to Planetary Systems* (Minh and van Dishoeck 2000). Interstellar dust has been reviewed by, for example, Tielens and Allamandola (1987a,b), Whittet (1992), and Cox et al. (2000).

1 THE DIFFUSE INTERSTELLAR MEDIUM

1.1 Very small grains

The infrared emission from dust in the diffuse, neutral ISM is faint and extended, and observations have been mostly limited to what is possible with cooled telescopes in space. The all-sky maps by IRAS (at 12, 25, 60 and 100 μm) and

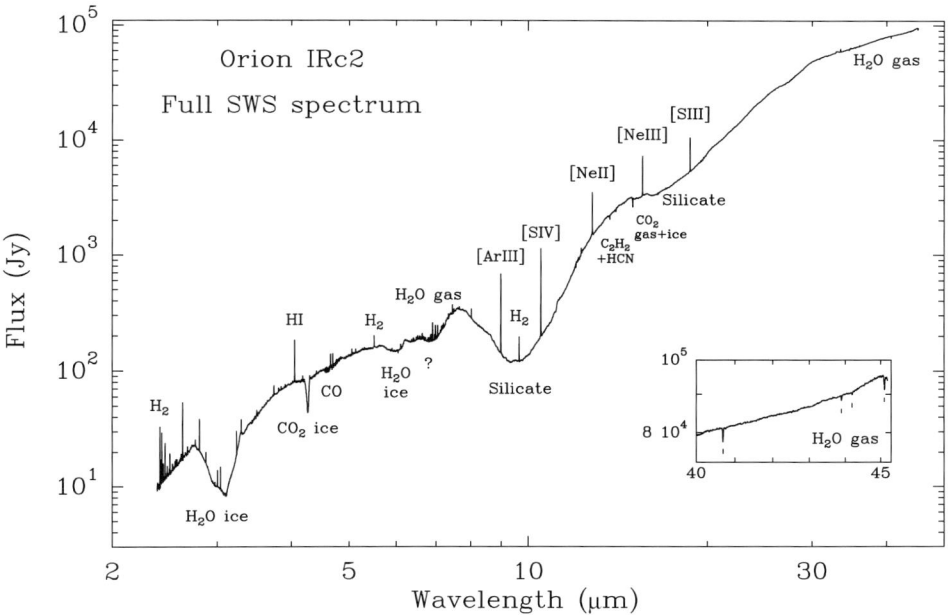

Figure 3 Complete ISO-SWS grating spectrum centered on Orion IRc2 at a resolving power $R = 1300–2500$. The SWS aperture ranges from $14'' \times 20''$ at 2 μm to $20'' \times 27''$ at 45 μm. The principal absorption and emission features are indicated. The inset shows a blow-up of the 40–45 μm region, in which several gas-phase H_2O absorption lines can be seen. (van Dishoeck et al. 1998.)

COBE have been particularly important in revealing the spatial distribution of the dust in the Galaxy as well as its spectral energy distribution from near-infrared to millimeter wavelengths (Sodroski et al. 1994, Reach et al. 1995; Figure 4). The IRAS 100 μm emission is due to thermal emission at $T \approx 18$ K from large (~ 0.1 μm) grains, and correlates well with the ground-based 21 cm H I emission in regions far removed from molecular clouds. This confirms that the dust and gas in these so-called infrared cirrus clouds are well mixed and that the heating of dust is quite homogeneous away from the galactic plane (e.g., Boulanger and Pérault 1988). Polarization observations of dust at optical (Hall 1949, Hiltner 1949) and far-infrared (e.g., Hildebrand et al. 1984, 1995) wavelengths show that the grains are nonspherical and aligned by the magnetic field.

In contrast, the strong and widespread 12 and 25 μm IRAS emission in cirrus clouds was initially a puzzle. Sellgren (1984) proposed that such strong mid-infrared emission is due to very small dust grains, in which the absorptions of individual photons lead temporarily to much higher grain temperatures because of the limited heat capacity of small grains. The presence of small (<10 Å) grains in the ISM had been first proposed in 1956 by Platt to explain optical extinction and polarization, and this was extended by Donn (1968) to ultraviolet extinction. This proposal was revived by Andriesse (1978) in a study of the

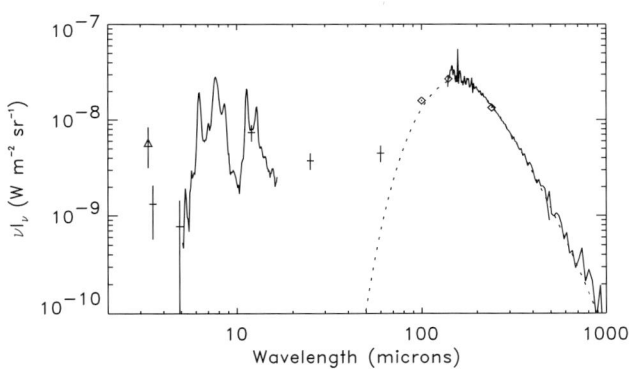

Figure 4 The dust emission spectrum from the near-infrared to millimeter wavelengths normalized for a column density of 10^{20} H atoms per cm^2. This figure combines (1) the 3.3 μm emission of cirrus associated with the molecular ring (Giard et al. 1994), (2) an ISO-CAM spectrum between 5 and 16 μm of the diffuse galactic emission at $l = 28°$ and $b = 0.8°$, and (3) measurements of the high-latitude cirrus emission from the COBE satellite with the Diffuse Infra Red Background Experiment (DIRBE) (Bernard et al. 1994) and the Far Infra Red Absolute Spectrometer (FIRAS) (Boulanger et al. 1996). The dotted line represents the radiation from a modified black body at a temperature $T_{dust} = 17.5$ K with an emissivity law proportional to ν^2. The peak at 158 μm a detection of the [C II] line from high-latitude gas. (Boulanger 1999.)

infrared emission from H II regions. However, none of these insightful suggestions gained wide acceptance because other explanations were favored by Occam's razor: ~ 1000 Å dielectric dust grains for the optical extinction and polarization, ~ 200 Å graphite grains for the ultraviolet extinction, and heating of ~ 100 Å grains by Lyα radiation for the mid-infrared emission from H II regions.

Temperature fluctuations of small grains had been considered by Greenberg and Hong (1974) and Purcell (1976) in connection with the formation (or prevention) of ice mantles, but the possible general importance of the mid-infrared emission from the ISM had not been grasped. After the IRAS discovery of widespread galactic cirrus, this process was studied in more detail by Draine and Anderson (1985), who showed that these temperature fluctuations contribute to the IRAS 60 μm emission as well. The IRAS 12–100 μm data have led to the development of new grain models, such as those by Désert et al. (1990) and Rowan-Robinson (1992). As is discussed in Section 2.2, the very small grains seen by IRAS at 12 and 25 μm form part of a distribution of grain sizes ranging from large molecules, such as polycyclic aromatic hydrocarbons (PAHs), to big grains. ISO-PHOT and ISO-CAM (Table 1) have demonstrated that the spectrum of the mid-infrared cirrus is dominated by the PAH bands (Section 2.2; Mattila et al. 1996; Figure 4). Likewise, spectral imaging studies with ISO-CAM have revealed that the surfaces of dark clouds are set aglow in these PAH bands by impinging far-ultraviolet (FUV) photons (Figure 5).

1.2 Amorphous silicates and hydrocarbon dust

Mid-infrared spectroscopy of the cold dust along the line of sight to bright background sources can provide direct information on the composition of interstellar grains. The absorption spectra are dominated by the broad and strong 10 and 18 μm Si–O stretching and O–Si–O bending features of silicate dust, first detected by Gillett and co-workers (Gillett and Forrest 1973, Gillett et al. 1975). As an example, Figure 6 shows the 10 μm profile observed toward the galactic center, Sgr A, with the ISO-SWS (Vriend 2000). This bright background source suffers some 30 magnitudes of visual extinction and is a prime target for the study of interstellar dust in the diffuse and translucent ISM. The observed profile is remarkably smooth and shows no evidence of the spectral substructure due to crystalline silicate components that would be expected, for example, at 11.3 μm. The crystalline/amorphous silicate ratio in the ISM must be less than a few percent (Demyk et al. 1999). This is in stark contrast with dust around young stellar objects (YSOs), Solar System comets, massive supergiants, or some asymptotic giant branch (AGB) and post-AGB objects, which show up to 75% crystalline silicate material (Section 7.1). The observed 10 μm band is fitted remarkably well by the laboratory spectrum of amorphous olivine with Fe/Mg $\simeq 1$ (FeMgSiO$_4$). However, the profile of the 18 μm band implies that pyroxenes (Fe$_x$Mg$_{1-x}$SiO$_3$) are also present, and this issue has not yet been settled completely. The lack of crystalline material suggests rapid processing of the grains in the ISM, perhaps as a result of bombardment with cosmic rays.

Hydrocarbon grains in the ISM have been detected from the ground through absorption in the C–H stretching mode at 3.4 μm (e.g., Wickramasinghe and Allen 1980, Pendleton et al. 1994). While the origin of this band in the C–H stretching mode of aliphatic hydrocarbon (e.g., –CH$_3$ and –CH$_2$ groups) materials is well established, the precise nature of this carbonaceous material is still uncertain and many candidates have been proposed, including hydrogenated amorphous carbon (HAC, a refractory carbonaceous material analogous to high-temperature soot), quenched carbonaceous carbon (QCC), and ultraviolet processed organic refractory grain mantles, leading to a lively debate (Martin 1989).

This discussion centers largely on the origin and evolution of interstellar dust, and as such is really a continuation of a historical controversy that started in the early 1960s. One camp holds that interstellar dust is formed in stellar ejecta and therefore consists mainly of highly refractory materials such as silicates and graphite (Mathis et al. 1977).

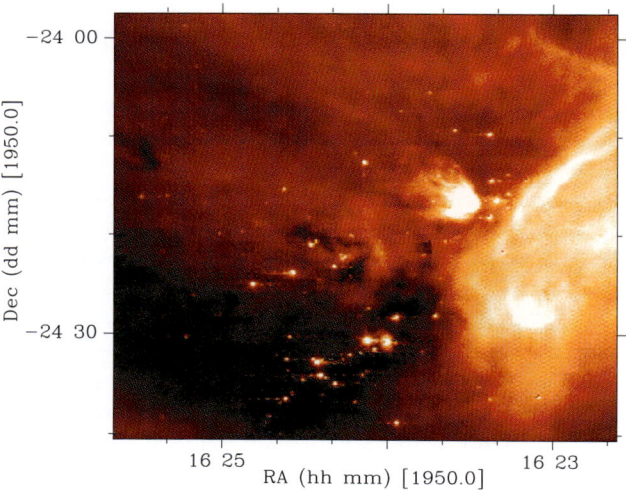

Figure 5 An ISO-CAM composite 6 and 15 μm image of the ρ Ophiuchi molecular cloud. The bright filaments in the west trace hot dust illuminated by the B2V star HD 147889, located just outside the map. The actual dense cloud is to the east. The dark patches are very dense clumps which are optically thick even at 15 μm, whereas the bright spots throughout the image are low- and intermediate-mass protostars. Most of the extended 6 μm emission is due to PAHs. (Abergel et al. 1996.)

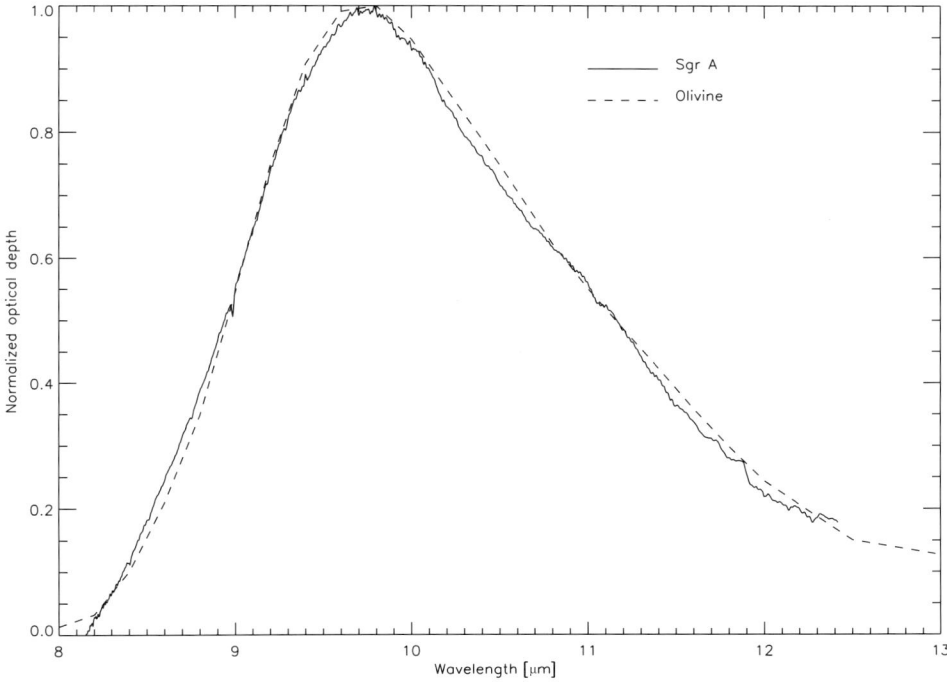

Figure 6 The 10 μm silicate optical depth observed toward the galactic center with the ISO-SWS. The dashed line is the laboratory spectrum of the amorphous olivine FeMgSiO$_4$. The broad feature around 10.5 μm may be due to residual detector memory effects. (Vriend 2000.)

The other side maintains that interstellar dust is also formed by the accretion and processing of icy grain mantles in the ISM itself (Greenberg 1982, Li and Greenberg 1997). Interstellar ice formed by accretion and reaction in the ISM was among the first dust models, originally proposed by van de Hulst (1949) in an entry in a prize essay contest on the origin and growth of solid particles in the ISM organized by Kramers and Oort at Leiden University during the war years (van de Hulst 1997). While this model had attractive features and was generally well received, the problem of nucleation at the low densities of the ISM was recognized. Moreover, by the early 1970s, the expected 3 μm ice band had not been observed in the diffuse ISM (Section 4). The stardust model first saw light when Hoyle and Wickramasinghe (1962) surmounted the nucleation problem by proposing that dust forms at the high densities associated with stellar photospheres and is subsequently driven into space by radiation pressure. They focused on graphite produced in C-rich giants as a likely dust candidate, but it was later concluded that oxides and silicates could form around O-rich giants (Kamijo 1963). By the late 1960s this model was well supported by the discovery of infrared emission from warm, circumstellar dust grains around red giants and by the observed ubiquity of the 2200 Å feature in the interstellar extinction curve, which is well fitted by graphite-like grains. There is no longer any doubt that stardust formed in stellar ejecta is an important component of interstellar dust: genuine stardust grains that can be traced back to AGB stars have been found in meteorites in our laboratories (Anders and Zinner 1993)!

The KAO and ISO have provided further constraints on the composition of interstellar hydrocarbon grains through their study of the 5–8 μm region. This spectral region, which is obscured by telluric absorption, is very characteristic of C=C, C–C, and C=O stretching modes, and C–H deformation modes. The spectra show that the hydrocarbon dust component is largely carbon-based and contains very little oxygen (Tielens et al. 1996, Schutte et al. 1999, Chiar et al. 2000). The observed weak CH deformation modes at 6.85 and 7.3 μm compare well in relative strength and peak position with laboratory spectra of HAC, as originally proposed by Duley and Williams (1983) (Figure 7; Chiar et al. 2000). Moreover, there is little evidence for the strong C=O modes that dominate the spectra of first-generation processed ice mantles (Greenberg and Muñoz-Caro 2000). So, if interstellar mantles play an important role in the ISM, the processing has to convert them almost completely from oxygen-dominated volatile ice mantles (Section 4) to carbon-based refractory mantles, and experimental studies have so far been unable to achieve this in the laboratory. The stardust model is not without problems, either. In particular, stardust proponents still have to come to grips with

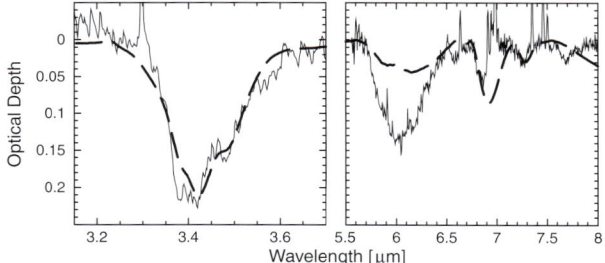

Figure 7 ISO-SWS 3.2–3.7 μm (left) and 5.5–8.0 μm (right) spectra of Sgr A* on an optical depth scale, revealing weak absorption features which are well fitted in peak position and relative strength by the C–H stretching and deformation modes in hydrogenated amorphous carbon (HAC; dashed lines). Most of the feature at 6.0 μm is due to H_2O ice while the 7.7 μm feature is due to CH_4 ice, as discussed in Section 4. (Chiar *et al.* 2000.)

the short dust lifetimes implied both by observations of interstellar depletion variations and by theoretical calculations which suggest an active gas–grain interaction in the ISM (Tielens 1999).

1.3 The [C II] emission from the Milky Way

In addition to studies of dust, the space-based data provide information on the gas in the diffuse ISM. The [C II] emission from the Galaxy has been studied using the FIRAS instrument on COBE and the Japanese Balloon-borne Infrared Carbon Explorer (BICE) instrument (Figure 8; Wright *et al.* 1991, Bennett *et al.* 1994, Nakagawa *et al.* 1998). The [C II] line is the dominant cooling line of the interstellar gas, with an observed luminosity for the entire Milky Way of $5 \times 10^7\ L_\odot$ (see also Figure 4). The [C II]/FIR ratio is typically $\sim 3 \times 10^{-3}$. The [C II] emission from local

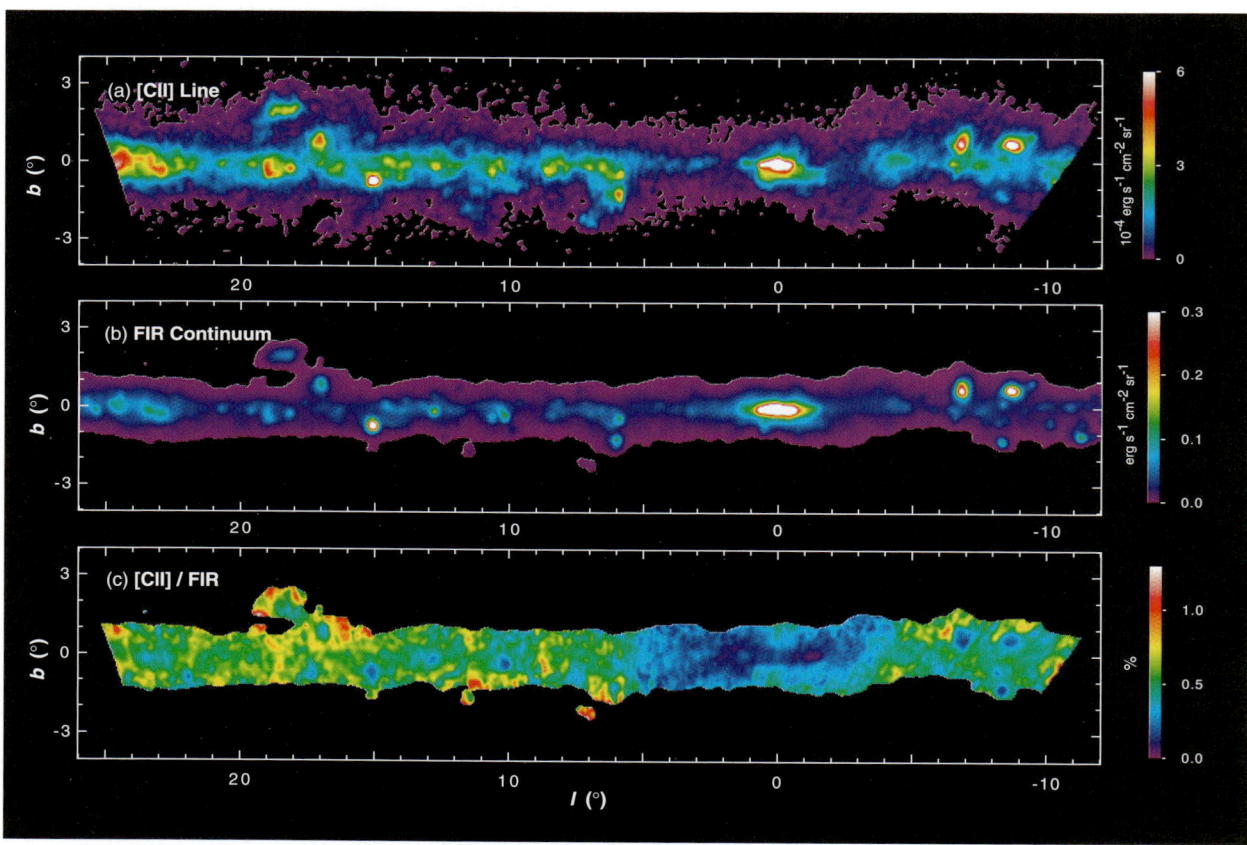

Figure 8 Top: far-infrared [C II] 158 μm line intensity map of the galactic plane obtained by the BICE balloon. The intensity scale runs from 0 (purple) to 6 (red) $\times 10^{-4}$ erg s^{-1} cm^{-2} sr^{-1}. Middle: far-infrared continuum map obtained from IRAS 60 and 100 μm maps. The intensity scale runs from 0 (purple) to 0.3 (red) erg s^{-1} cm^{-2} sr^{-1}. Bottom: the ratio of the far-infrared [C II] line emission to the far-infrared continuum emission. The scale runs from 0.0 (purple) to 1.0% (red). The spatial resolution is about 15'. (Nakagawa *et al.* 1998.)

H I clouds has been measured using small sounding rockets to be $\sim 3\times 10^{-26}\,\mathrm{erg\,s^{-1}}/$ H atom (Bock et al. 1993). Independent estimates of the [C II] emission have been obtained using Copernicus, IUE, and the Hubble Space Telescope (HST) by measuring ultraviolet absorption lines originating in the upper level of this transition, and these give substantially the same value (Pottasch et al. 1979; Gry et al. 1992; Black 1993, using HST data from Cardelli et al. 1993).

The origin of this [C II] emission has been somewhat controversial (Tielens 1995, 1997). Locally, the [C II] emission is dominated by emission from diffuse H I clouds heated by the photoelectric effect. [C II] 158 μm is the principal cooling line, resulting in gas temperatures of $\sim 80\,\mathrm{K}$. The [C II]/FIR ratio in these clouds provides a direct measure of the photoelectric heating efficiency of the dust. Toward the inner Galaxy, where the pressures and densities are higher, a substantial fraction of the [C II] emission may originate from the warm ionized medium (Heiles 1994). Also, much of the H I here is associated with the surfaces of molecular clouds illuminated by the ambient radiation field, and these so-called photodissociation regions (Section 2) can be important contributors to the [C II] emission as well (Stacey et al. 1985, Shibai et al. 1991).

2 PHOTODISSOCIATION REGIONS (PDRS)

Photodissociation regions, also called photon-dominated regions (PDRs), are regions where FUV photons (6–13 eV) dominate the heating of the gas as well as their chemical structure. PDRs can be molecular clouds located close to bright O or B stars, neutral material at the edge of an

Figure 9 A false-color IRAS composite 12 μm (blue), 60 μm (green), and 100 μm (red) image of the constellation Orion, covering a region of 30° × 24° degrees. The bright spot in the bottom part of this image is the sword region of Orion with the great nebula M42. The upper part of the figure is dominated by the ring-like shell surrounding λ Orionis. The bluish dot to the left of this structure is the supergiant Betelgeuse, one of the prominent stars in Orion. The other stars are not very bright in the infrared. (Figure by D. Kester, personal communication.)

H II region, reflection nebulae, the edges of dark clouds, post-AGB stars, or planetary nebulae. The diffuse clouds discussed above are low-density ($n < 1000\,\mathrm{cm}^{-3}$), low incident radiation field (less than 100 times the average interstellar field) versions of PDRs. Indeed, much of the neutral gas in the Galaxy is in the form of PDRs, as is the ISM in other galaxies. Figure 9 shows an IRAS image of molecular clouds in the Orion–Monoceros region. While some of the hot spots trace embedded massive YSOs, much of the diffuse emission is due to the conversion of FUV photons from the ambient O and B stars to infrared photons by the dust in the PDR surfaces.

The study of PDRs was triggered by early observations of strong fine-structure lines of [C II], at 158 μm; [O I], at 63 and 145 μm; and [C I], at 610 μm from dense molecular clouds in regions with nearby O stars (Melnick et al. 1979, Storey et al. 1979, Phillips et al. 1980, Russell et al. 1980, Stacey et al. 1983). These lines were readily observable with the Lear Jet and KAO, and the data provided more accurate wavelengths than were available from laboratory spectroscopy. The strong lines found in the Orion star-forming region stimulated the development of detailed physical and chemical models of PDRs (Tielens and Hollenbach 1985a,b). In addition, the discovery of strong and broad unidentified infrared (UIR) features at mid-infrared wavelengths from the ground and the KAO in the same regions led to the study of PAHs. Together with the wealth of new ISO data, this field has reached a significant level of maturity in a short time, and has provided important insights into the basic physics and chemistry of interstellar gas and dust irradiated by FUV photons. Comprehensive reviews have been given by Hollenbach and Tielens (1997, 1999).

2.1 PDRs: gas phase

The structure of PDRs results from the interaction of FUV photons with the gas and dust through the photoionization of atoms (especially C), photodissociation of molecules (especially H_2 and CO), and the photoelectric effect on grains. Models have been developed by various groups, and calculations presented by Kaufman et al. (1999). The models generally consist of a plane parallel slab of constant density n_H illuminated from one side by an intense radiation field, characterized by an enhancement factor G_0 over the average interstellar radiation field. At the edge most of the gas is in atomic form, but, at an attenuation in the visible of $A_V \approx 1-2$ mag, photodissociation of H_2 is decreased by self-shielding and the transition from H to H_2 occurs. Deeper into the cloud, at $A_V \approx 2-4$ mag, the carbon balance shifts from C^+ to C and eventually to CO (van Dishoeck and Black 1988). The oxygen not locked up in CO remains in atomic form up to $A_V \approx 8$ mag. The gas temperature is high at the edge, ranging from 100 to 1000 K, but drops to less than 20 K deep inside the cloud. Typically about 0.5% of the FUV energy is converted into heating of the gas by the photoelectric effect, whereas the rest is absorbed and re-radiated as far-infrared dust continuum emission.

The high gas temperature at the edge results in bright atomic fine-structure lines of [C II], [O I], [C I], and [Si II] at mid- and far-infrared wavelengths. Since the critical densities and excitation temperatures of the lines span a range of values ($10^3 - 3 \times 10^5$ cm^{-3}, 20–300 K), the observed line ratios are valuable probes of physical parameters, in particular the hydrogen density n_H and the enhancement of the radiation field G_0, which controls the temperature. Figure 10 illustrates how these lines can be used to constrain the PDR environment.

The PDR interpretation was quickly accepted, but the fact that the observed lines persist deeper into the clouds than is predicted by homogeneous models caused much discussion (e.g., Keene et al. 1985, Jaffe et al. 1987, Stutzki et al. 1988, Stacey et al. 1993). Many alternative models were developed, including dynamical cycling of the gas (Boland and de Jong 1983, Xie et al. 1995) and bistabilities in the chemical networks (Le Bourlot et al. 1995). The favored explanation is that the dense ISM is far from homogeneous, and has a "clumpy" structure with a density contrast between the phases of at least a factor of 10.

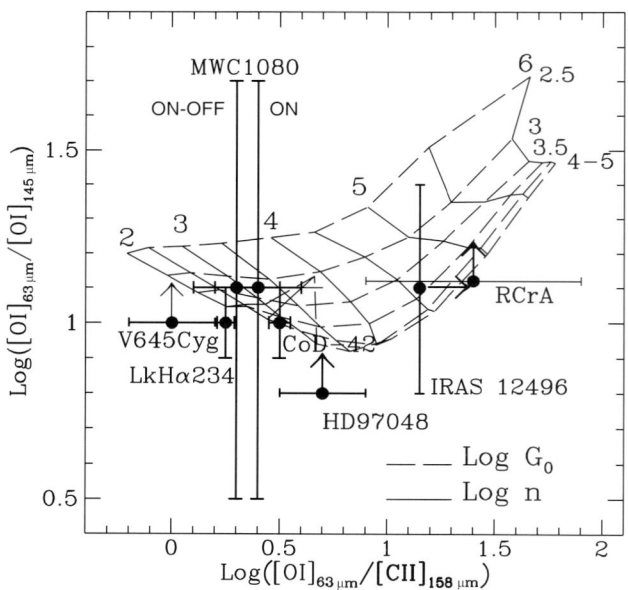

Figure 10 Line ratios of the principal fine-structure lines obtained in PDR models for different density n and radiation field G_0. The observed ratios with the ISO-LWS toward Herbig Ae stars are indicated. (Lorenzetti et al. 1999.)

A new probe of PDRs is provided by the pure-rotational lines of H_2 at 28 μm ($J = 2 \to 0$), 17 μm ($J = 3 \to 1$), 12 μm ($J = 4 \to 2$), and so on. The 12 μm line was first detected by Beck et al. (1979), while the longer-wavelength 17 μm line was first observed by John Lacy and co-workers (Parmar et al. 1991). The ISO-SWS has allowed detection of the fundamental 28 μm line and routine observations of this and other lines in many regions (e.g., Timmermann et al. 1996, Thi et al. 1997, Draine and Bertoldi 1999, Fuente et al. 1999; see also the review by Wright 2000). Published results have been limited mostly to PDRs with strong H_2 lines, corresponding to regions with densities $n_H > 10^4$ cm^{-3} and radiation fields $G_0 > 100$, such as bright reflection nebulae. Maps of the higher-J lines at arcsecond resolution have been made with ISO-CAM (e.g., Cesarsky et al. 2000). The rotational temperatures of ~100–500 K indicated by the $J = 2$–7 lines directly reflect the kinetic temperature structure of the gas, since the lower rotational lines are populated primarily by collisions (Figure 11). The high vibrational temperature of ~2000 K indicated by ISO and ground-based 2 μm observations is due to ultraviolet pumping of the excited states (e.g., Black and van Dishoeck 1987, Sternberg and Dalgarno 1989). Existing PDR models give temperatures which are lower at the edge of the cloud than are inferred from the ISO H_2 data, suggesting either that the photoelectric yield at the cloud edge is underestimated or that H_2 formation on high-temperature grains occurs more rapidly than was previously thought. While this uncertainty currently limits the astrophysical interpretation, it does provide an instructive example of how astrophysical data can constrain basic chemical and physical processes which are difficult to study in a laboratory on Earth.

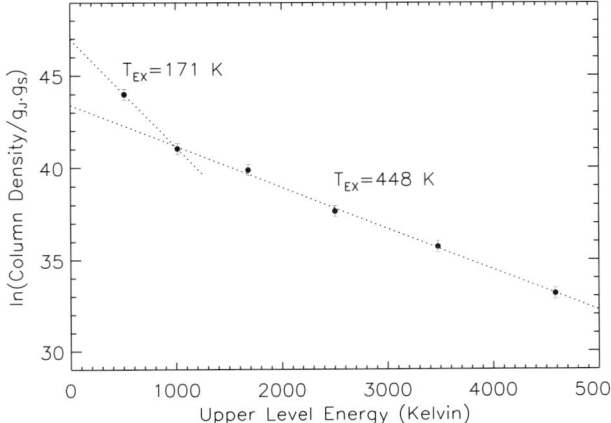

Figure 11 Rotational excitation diagram of H_2 for the PDR IC 1396 based on ISO-SWS observations. The different excitation temperatures derived from the $J = 2$–3 and 3–7 levels indicate the presence of a temperature gradient through the region. (Wright 2000.)

Another exciting new result is provided by the first detection of emission from the HD molecule in the ISM (Figure 12). HD is the main reservoir of deuterium in molecular clouds and can thus be used to constrain the primordial deuterium abundance produced by the Big Bang and its subsequent destruction in stars. Wright et al. (1999) detected the lowest rotational $J = 1 \to 0$ line at 112 μm in the Orion Bar PDR with the ISO-LWS, while Bertoldi et al. (1999) observed the $J = 6 \to 5$ line at 19.4 μm in the Orion shock with the ISO-SWS. Together with the H_2 pure rotational lines, the data give [D]/[H] = $(1.0 \pm 0.3) \times 10^{-5}$ and $(0.51 \pm 0.15) \times 10^{-5}$, respectively, significantly lower than the value of $(1.5 \pm 0.1) \times 10^{-5}$ inferred for the local ISM by Linsky (1996) from ultraviolet absorption line observations. This suggests that deuterium is being destroyed in this active star-forming region.

2.2 Interstellar PAHs

Bright emission features, at 3.3, 6.2, 7.7, 8.6, and 11.3 μm, and a host of weaker ones dominate the mid-infrared spectra of many PDRs. Initially these bands were called UIR features, and one of the first proposed identifications was carbonates, which have a band at 11.3 μm that matched the observed feature rather well (Gillett et al. 1973). This interpretation was supported by the presence of carbonates in meteorites, although it was difficult to envision an interaction between rock, CO_2, and liquid water in an interstellar setting. In any case, this interpretation formed the basis for extensive searches of the 5–8 μm region for the much stronger 6.8 μm carbonate band using the KAO (Russell et al. 1977, Willner et al. 1977). The interstellar spectra in this wavelength region are actually dominated by strong bands at 6.2 and 7.7 μm, and the carbonate suggestion died a quick death. Around the same time, a model was proposed based on infrared fluorescence of simple molecules in an icy grain mantle (Allamandola and Norman 1978), but it was unclear how such simple ices could survive in the harsh FUV light of PDRs. Moreover, the composition of the mixture had to be tailored considerably. In retrospect, it is astonishing that it took almost 10 years to realize that the observed interstellar spectrum is that of PAH compounds (Duley and Williams 1981), but it is now well accepted that the UIR bands are carried by such materials (see Léger and Puget (1984), Allamandola et al. (1985), and Omont (1986) for early discussions, and Tielens et al. (1999, 2000) for recent reviews). The features are bright even far from the illuminating stars; hence, the emission has a nonthermal nature where the absorption of a single FUV photon raises the temperature of a small carrier to some 1000 K, and this species cools through the emission in its vibrational modes (the UIR bands) (Section 1.1). It is more appropriate to describe the carrier as a large molecule rather than a very

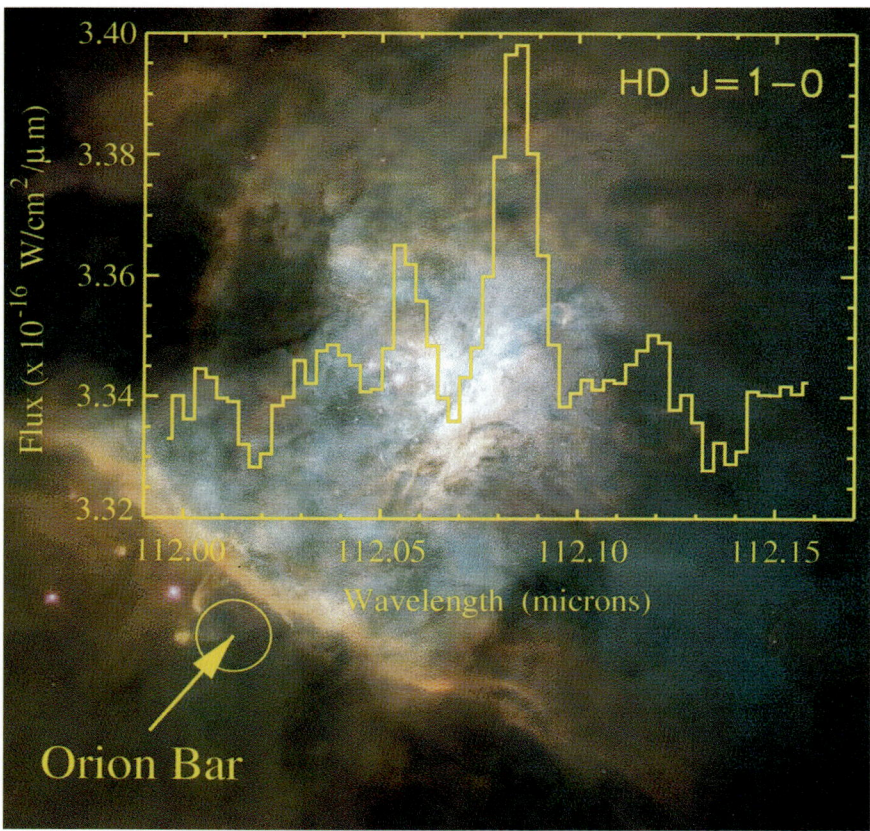

Figure 12 Detection of the HD $J = 1-0$ line at 112 μm with the ISO-LWS Fabry–Perot interferometer aimed in the direction of the Orion Bar. Together with observations of the H_2 pure rotational lines at 28 and 17 μm with the ISO-SWS, these data have been used to infer the [D]/[H] abundance ratio in an active star-forming region. (Wright et al. 1999.)

small grain: borrowing molecular physicists' terminology, the process involves ultraviolet absorption into excited electronic states, which subsequently decay to lower states through a nonradiative process called internal conversion followed by vibrational emission. The observed color temperature corresponds to species containing about 50 C atoms absorbing a ~ 10 eV photon. Thus, the emitters are PAH molecules with about 50 C atoms, such as hexabenzocoronene ($C_{48}H_{24}$) and circumcoronene ($C_{54}H_{18}$). The PAHs also play a significant role in the energy balance, since the excess energy of the electrons produced by photoionization heats the gas (e.g., d'Hendecourt and Léger 1987, Bakes and Tielens 1994).

The ISO-SWS spectra of bright sources have revealed the incredible richness of the interstellar and circumstellar UIR spectrum. As an example, Figure 13 shows three spectra of C-rich, post-AGB objects and planetary nebulae (Hony et al. 2001a). Besides the well-known bands mentioned above, there is a wealth of weaker features: at 3.4, 5.25, 5.65, 6.9, 10.5, and 12.7 μm, to mention a few (Figures 13 and 14). Moreover, many of the prominent bands show satellite bands (e.g., at 6.0 and 11.0 μm) or break up into subfeatures (e.g., at 7.6 and 7.8 μm).

One important characteristic of the UIR spectrum is its spectral variation. All sources show the five well-known UIR bands but their ratios can vary substantially, and the 11.3/7.7 ratios in Figure 13 provide a good example. More dramatically, some H II regions show a 8.6 μm feature that dwarfs the usually much brighter 7.7 μm feature (Roelfsema et al. 1996, Verstraete et al. 1996). Subtler variations are provided by the ratio of the 7.6/7.8 components of the 7.7 μm feature among H II regions and planetary nebulae. Also, the features often vary relative to one another with distance from the illuminating star within a single object (Geballe et al. 1989; Joblin et al. 1996, 2000). These variations make each spectrum unique and imply that the carriers are sensitive to the local physical conditions. It is quite clear from the data amassed that the UIR bands are

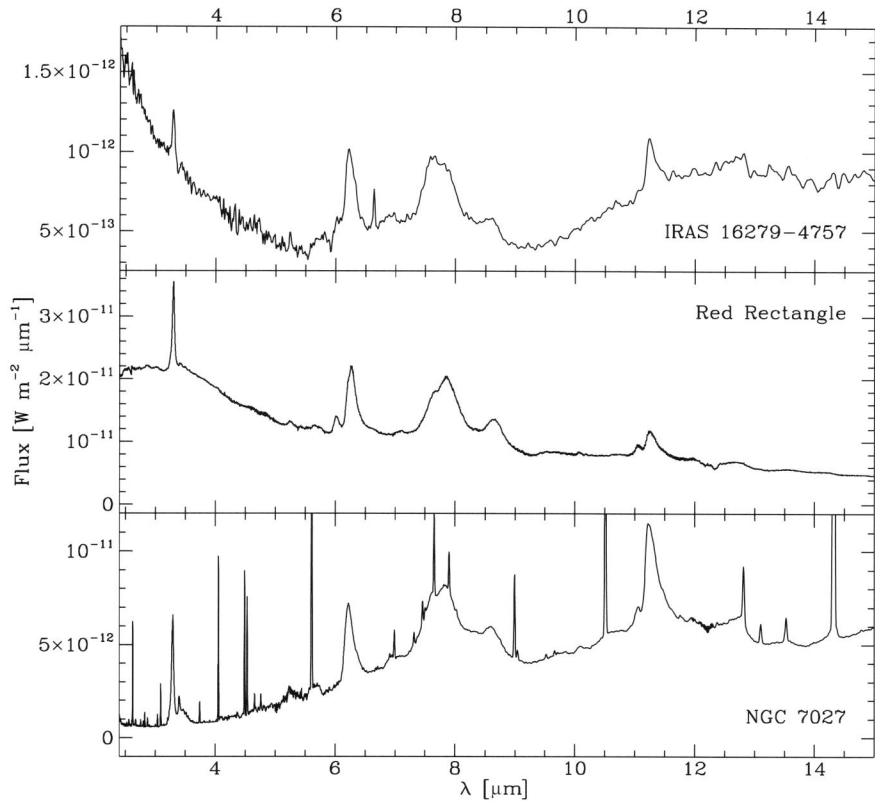

Figure 13 The 3–15 μm emission spectra of the post-AGB objects IRAS 16279–4757 and the Red Rectangle, (BD−10 1476) and the planetary nebula NGC 7027, illustrating the rich and variable circumstellar PAH spectrum. (Hony et al. 2001a.)

carried by a family of related aromatic hydrocarbon compounds. Correlation studies offer the hope that from the observed spectral variations the specific molecular carriers can be identified. Enormous efforts in the laboratory spectroscopy of PAHs (Hudgins and Allamandola 1999, Hudgins et al. 2001, Szczepanski and Vala 1993), supported by theoretical quantum chemical calculations (Langhoff 1996, Bauschlicher 1998), are essential for this purpose.

The 11–15 μm spectral region is characteristic of the C–H out-of-plane deformation mode, and therefore of the edge structure of the emitting aromatic species (e.g., Duley and Williams 1981, Allamandola et al. 1989) (Figure 14). Coupling between the vibrating H atoms means that the exact peak position of these modes depends on the number of directly adjacent H atoms (i.e., those bonded to neighboring C atoms on a ring). Ionization shifts these ranges for the different number of adjacent H atoms to shorter wavelengths. The spectrum of the Red Rectangle (Figure 14), a post-AGB star, suggests a mix of neutral and ionized PAHs, with the strongest bands due to the "1" mode, but with "2" to "4" modes present as well (Hony et al. 2001a).

Finally, an emission plateau in the 15–20 μm region has been discovered in the spectra of two young intermediate-mass stars, so-called Herbig AeBe stars (van Kerckhoven et al. 2000). The dust continuum emission in the spectra of these stars starts at longer wavelengths than in most other sources, making these weak features in the 15–20 μm region more readily recognizable. However, this plateau is present in a variety of objects whose spectra are characterized by the UIR bands. Some spectra show a distinct feature at 16.4 μm superposed on this plateau (Moutou et al. 2000). These longer-wavelength modes have been studied in the laboratory by Moutou et al. (1996) and Hudgins et al. (2001). The emission is due to C–C–C bending modes which cause in-plane and out-of-plane distortions of the C skeleton, and which are more molecule specific than the shorter-wavelength bands. A collection of molecules then gives rise, in a natural way, to a plateau of blended emission features. Spectral substructure, in particular of the 16.4 μm band, is likely characteristic of the smaller PAHs in the interstellar PAH family and may signal the presence of a pendant hexagonal ring. In general this

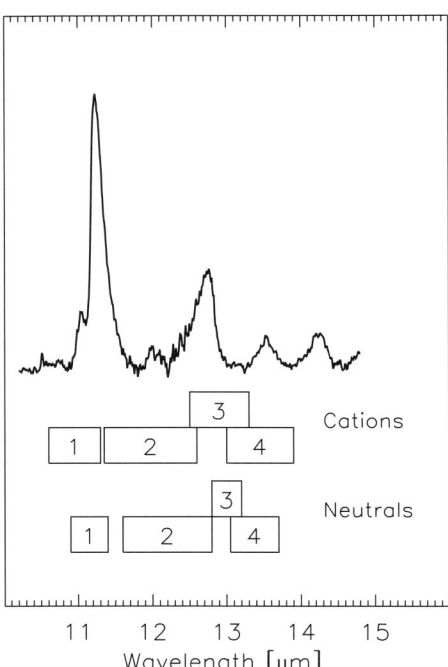

Figure 14 The 10–15 μm spectrum of the Red Rectangle observed with the ISO-SWS, revealing new weak PAH bands (Hony et al. 2001a). Bands in this spectral region are largely due to the C–H out-of-plane bending mode, whose exact peak position is sensitive to the number of adjacent H atoms. The boxes below the spectrum give the ranges in which neutral and cationic PAHs with the indicated number of adjacent H atoms emit (Hudgins and Allamandola 1999). The band at 14.2 μm is attributed to C–C–C bending modes associated with PAHs having three or five adjacent H atoms.

wavelength region has great potential for the identification of molecular structures which dominate the interstellar PAH family.

3 SHOCKS

Energetic input to the ISM is provided not only by ultraviolet photons but also by shocks, which are ubiquitous in the ISM. Shocks can be caused by many astrophysical phenomena, including supernova remnants interacting with the ISM, expanding H II regions, outflows from young stars, cloud–cloud collisions, and accretion onto circumstellar disks. General reviews of shocks have been given by Draine and McKee (1993) and Hollenbach (1997). For fast (>50–$100\,\mathrm{km\,s^{-1}}$) shocks interacting with low-density material, the temperature is instantaneously raised to very high values ($>10^5$ K), and most of the emission appears in atomic lines at ultraviolet wavelengths. These so-called "jump" or J-type shocks dissociate molecules within and ahead of the shock, and the molecules re-form only slowly in a lengthy warm zone in the wake of the shock. For slower shocks impacting on dense molecular clouds with a low ionization fraction, the hydrodynamical variables change more continuously, and the temperatures in these C-type shocks reach peak values of only 2000–3000 K. The emission occurs primarily at infrared wavelengths in atomic fine-structure and molecular lines of H_2, CO, OH, and H_2O. The diagnostic value of these lines has been elaborated by Hollenbach et al. (1989) and Kaufman and Neufeld (1996). Grain cores and mantles are also affected by shocks. In particular, high-velocity J-shocks cause disruption (through shattering) or nonthermal sputtering of the cores, resulting in much enhanced abundances of atomic Si, Mg, Fe, Ti, and Ca in the warm neutral phase of the ISM (Jones et al. 1996).

The best-studied shock in the Galaxy is that associated with the Orion high-velocity outflow. Early KAO observations by several groups (e.g., Stacey et al. 1982, Watson et al. 1985, Melnick et al. 1987, Haas et al. 1991) revealed very strong, broad lines of CO high-J, [S I] 25 μm, OH 119 and 163 μm, and other species, providing the first tests of the model predictions. The subsequent ISO-SWS and LWS data centered at the shock "peak 1" region (in the notation of Beckwith et al. 1978) beautifully illustrate the rich infrared line spectrum (Figure 15; Bertoldi et al. 1999, Rosenthal et al. 2000, Cernicharo et al. 1999). The 2–20 μm region is dominated by a wealth of H_2 lines, which constrain the temperature structure in the shock. The data from Orion and other sources require for their explanation at least two components with excitation temperatures between ~ 700 and ~ 2000 K, respectively. Comparison with the models by Kaufman and Neufeld (1996) suggests a mixture of slow and fast C-shocks, and perhaps a J-shock as well (see Wright (2000) for an overview). Such a mixture of shocks could reflect the fast direct impact and the slower oblique shocks expected for outflows. However, the high population of the $J = 27$ level (which lies 42,500 K above the ground level) as derived from the observed S(25) line toward Orion is still a mystery and suggests that the basic physical structure of shocks is not yet fully understood.

CAM-CVF maps of the higher-J lines at arcsecond resolution have been presented by Cabrit et al. (1999) for YSOs and by Cesarsky et al. (1999) for the supernova remnant IC 443 interacting with the molecular cloud. No continuum emission is found in the latter object, but all of the broadband IRAS 12 and 25 μm emission originally ascribed to warm dust is found to come from pure-rotational H_2 line emission. Similarly, the IRAS 60 μm band may be dominated by the strong [O I] 63 μm line,

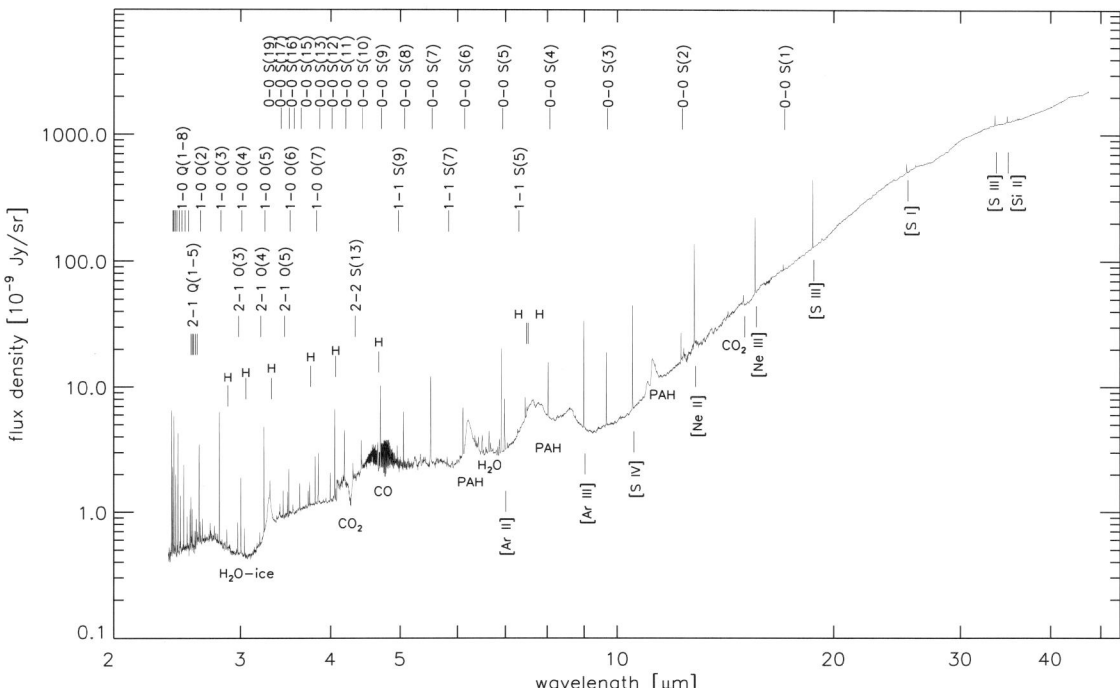

Figure 15 ISO-SWS grating scan of the Orion "Peak 1" shock, showing a rich forest of H_2 lines and other features. (Rosenthal et al. 2000.)

emphasizing the importance of spectroscopy for the interpretation of broadband colors. In general, the [C II] 158 μm/[O I] 63 μm and [O I] 63 μm/145 μm line ratios can be used to distinguish between low-velocity C- and high-velocity J-type shocks, but they often indicate a mix of shock types, like the H_2 lines. The [S I] 25 μm is an excellent diagnostic of shocks v. PDRs.

The power of heterodyne instruments in space is illustrated by spectrally-resolved observations of [C II] and [O I] profiles, as well as high-J lines of OH and CO and their isotopes (Boreiko and Betz 1996a,b; Betz and Boreiko 1989; Keene et al. 1998, 1999), and HCl (Blake et al. 1985). These data place important constraints on the different interpretations, especially in distinguishing shocks and PDRs.

The H_2O molecule is a particularly powerful probe of C-type shocks. H_2O is predicted to be the dominant oxygen-bearing species in warm, dense gas, because most of the available oxygen is driven into H_2O at high temperatures ($T > 230$ K) through the reactions $O + H_2 \rightarrow OH + H$ and $OH + H_2 \rightarrow H_2O + H$, increasing its abundance by orders of magnitude (e.g., Neufeld and Melnick 1987, Ceccarelli et al. 1996, Charnley 1997). In addition, the H_2O level populations are sensitive to collisions and to pumping by far-infrared radiation from warm dust. Finally, H_2O contributes to the cooling and heating of the gas (e.g., Neufeld and Kaufman 1993). Pioneering airborne observations of $H_2^{18}O$ by Zmuidzinas et al. (1995) indicated that the H_2O abundance is low ($<10^{-7}$) in quiescent clouds, but early observations of masing lines by the KAO and from the ground hinted had already at enhanced abundances of up to 10^{-5}–10^{-4} in warm shocked gas (e.g., Waters et al. 1979, Phillips et al. 1980, Melnick et al. 1993, Cernicharo et al. 1994). Ground-based and airborne observations of H_2O are greatly hampered by the Earth's atmosphere, however, and the derivation of abundances from masing lines is tricky. ISO provided the first opportunity to observe a large variety of non-masing H_2O lines in YSOs, including both lower and higher excitation lines. Figure 16 shows a sample of the wealth of pure-rotational H_2O lines detected with ISO toward the Orion shock (Harwit et al. 1998, Cernicharo et al. 1999, Wright et al. 2000). The main conclusion from these data is that the H_2O abundance is indeed high in warm regions and increases up to the order of 10^{-4} with respect to H_2 in the case of Orion, locking up a significant fraction of the available gas-phase oxygen.

The launch in 1998 of the SWAS satellite has allowed observations of spectrally resolved profiles of the lowest ortho-H_2O 1_{10}–1_{01} transition at 538 μm above the Earth's atmosphere (Melnick 2000, Melnick et al. 2000; Figure 17).

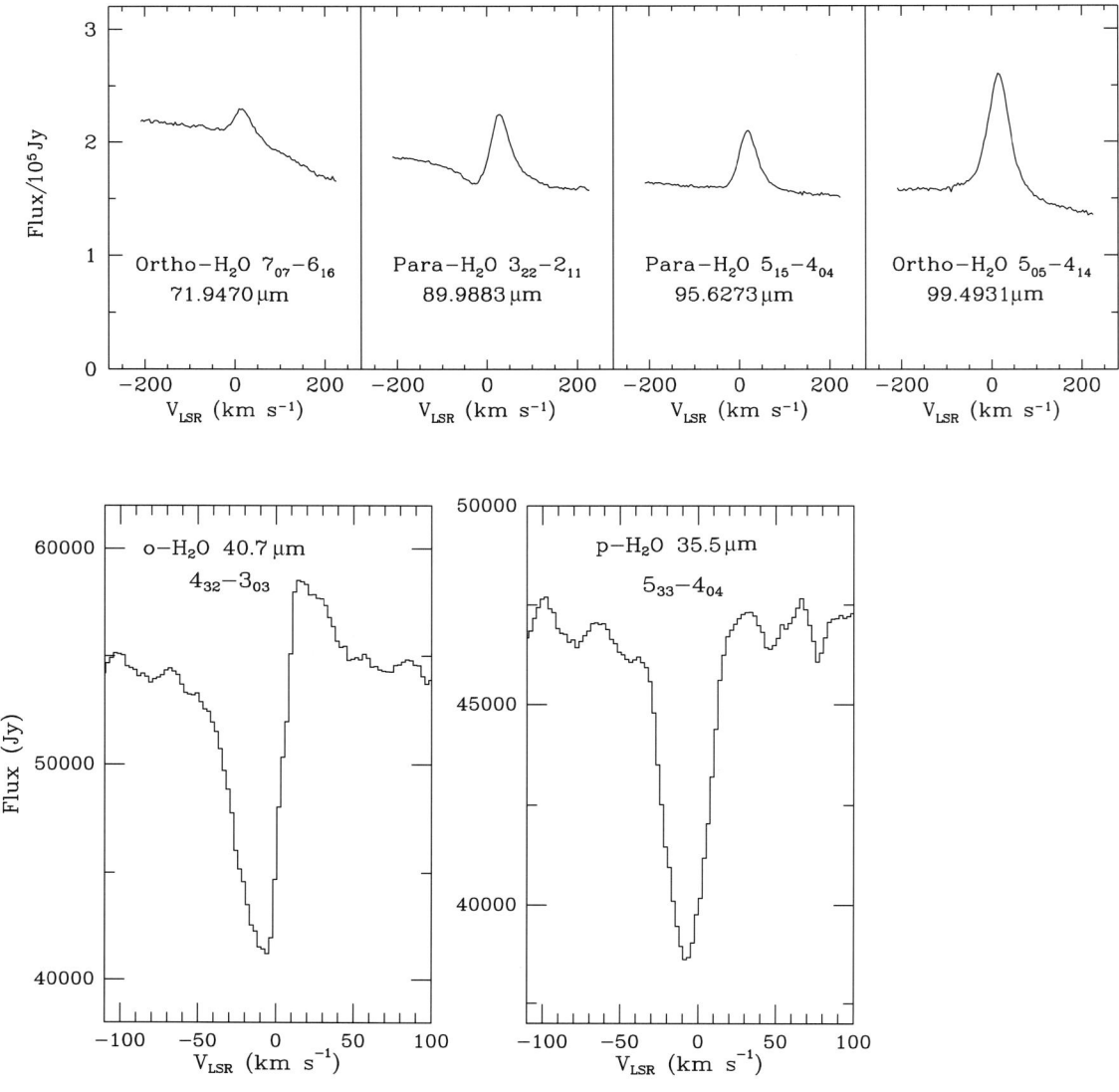

Figure 16 Examples of high-J emission and absorption lines of H_2O as functions of the Local Standard of Rest velocity V_{LSR} for the Orion shock observed with the ISO-LWS (top) and SWS (bottom) Fabry–Perot interferometers, revealing broad and complex line profiles and high abundances. (Harwit et al. 1998, Wright et al. 2000.)

This line can be used to probe the H_2O abundance not only in warm clouds but also in cold gas, to which ISO was insensitive. The first results indicate that the H_2O abundance in quiescent clouds is $\lesssim 10^{-8}$, even lower than previously indicated. Moreover, O_2 has a surprisingly low abundance in molecular clouds, as indicated by SWAS and balloon-borne observations (Olofsson et al. 1998, Goldsmith et al. 2000). Future observations with the Sweden–France–Canada ODIN satellite will allow deeper searches for O_2. The low H_2O and O_2 abundances can be explained by enhanced photodissociation in extended low-density inhomogeneous clouds (Spaans and van Dishoeck 2001) and freeze-out of oxygen in cold, dense clouds (Bergin et al. 2000).

4 COLD ENVELOPES AROUND YOUNG STELLAR OBJECTS: ICES

The gas and dust in the cold outer envelopes around YSOs are best probed through infrared absorption lines toward embedded sources. The young stars heat the dust in their

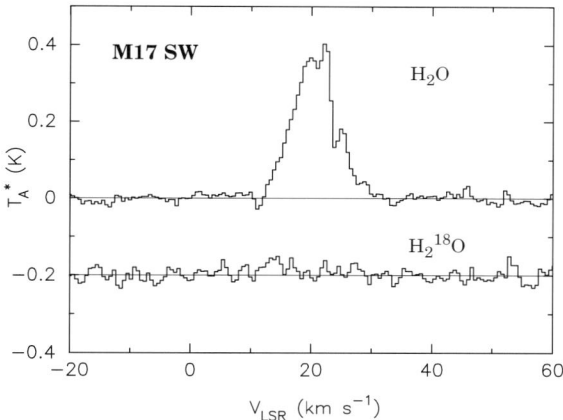

Figure 17 SWAS observations of the 1_{10}–1_{01} ortho-H_2O and $H_2^{18}O$ ground-state line toward the M17 star-forming region. (Melnick 2000, Snell et al. 2000.)

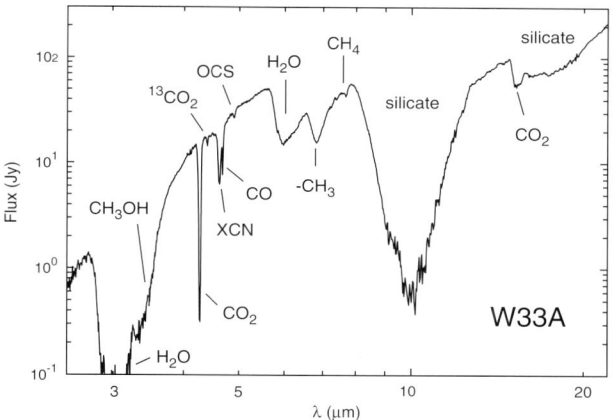

Figure 18 The 2.5–25 μm spectrum of the embedded massive protostar W 33A, obtained with the ISO-SWS at a resolving power of 500–1000. Except for the 10 μm silicate band, all features are due to simple molecules in ice mantles. (Gibb et al. 2000.)

immediate surroundings to a few hundred kelvin, providing a continuum against which the cooler foreground species can be seen in absorption (Section 1.2). The quiescent medium can also be studied by this technique if a bright infrared background star can be found. Such observations provide a powerful complement to submillimeter emission line studies: Only a pencil beam toward the source is probed in absorption, and both the gas and the dust can be observed along the same line of sight.

Since the densities are high and the temperatures low, the timescales for gas-phase atoms and molecules to collide with the cold (~10 K) grains and stick to them are short, less than 10^5 yr. Thus, most gas-phase species are expected to freeze out onto the grain cores and form an icy mantle. While H_2O ice was ruled out as an important dust component in the diffuse ISM by ground-based observations (Gillett et al. 1975), H_2O ice was discovered in the shielded environment of dense interstellar clouds in the 1970s (Gillett and Forrest 1973). The study of ices was subsequently pursued from ground-based and airborne platforms (e.g., Willner et al. 1982, Tielens et al. 1991). However, even from the KAO important regions of the spectrum around 4.2 and 15 μm were made inaccessible by telluric CO_2. The large and systematic database obtained with the ISO-SWS provided the first complete, unbiased census and is particularly useful for studying the composition and evolution of interstellar ices. Recent reviews are given by Tielens and Whittet (1997), d'Hendecourt et al. (1999), and Ehrenfreund and Schutte (2000).

Infrared spectra of YSOs, such as the massive protostar W 33A (Figure 18; Gibb et al. 2000), show a variety of solid-state absorption features due to simple molecules in an icy mixture. Such ices are a major component of interstellar dust inside dense clouds locking up a substantial fraction of the available elements, in abundances quite comparable to gaseous CO. These ices are formed by accretion and reaction on the pre-existing grain cores and their composition allows a direct probe of grain surface chemistry (Tielens and Hagen 1982). In the hot cores around luminous protostars, the evaporation of the ices drives a rich and complex chemistry, mainly by releasing copious amounts of CH_3OH into the gas phase (Section 5). Finally, the agglomeration of interstellar ice grains in the outer regions of the solar nebula is thought to have been the first step in the formation of comets, icy satellites, and planets. Similar processes may play a role in the circumstellar disks around protostars (Section 6).

The spectrum of W 33A shows the spectral signature of H_2O ice (three bands), CO_2 ice (three bands), and solid CO, OCS, CH_4, and CH_3OH (one band each); for a critical discussion of these assignments see Tielens and Whittet (1997) and Ehrenfreund and Schutte (2000). The identification of the bands at 4.62 and 6.85 μm is still somewhat controversial, although a strong case for the identification of the 4.62 μm feature with OCN^- has been made (Schutte and Greenberg 1997, Demyk et al. 1998, Palumbo et al. 2000). Extensive laboratory studies by various groups have been essential in the analysis of these data. While previous low-resolution ground-based and airborne observations had revealed tantalizing hints of the presence of solid CO_2 and CH_4 (d'Hendecourt and de Muizon 1989, Lacy et al. 1991, Boogert et al. 1997), the ISO-SWS made possible the first detailed study of these ice compounds and allowed their potential as diagnostic probes of the environment to be fully developed. Some illustrative examples are discussed below.

4.1 Thermal processing and ice segregation

Solid CO_2 is an important component of interstellar ices in star-forming regions, as evidenced by its strong stretching and bending modes (Figure 18; de Graauw et al. 1996, d'Hendecourt et al. 1996, Gerakines et al. 1999). All of these bands show variations in peak position and profile from source to source, caused by different structures of the ice (Figure 19). The CO_2 bending mode at 15 μm is an especially sensitive diagnostic of the CO_2 ice environment and the degree of thermal processing by the nearby protostar. Extensive laboratory studies of a variety of astrophysically relevant mixtures at Leiden have shown that these variations are caused by the thermal processing of ice mixtures containing CO_2, H_2O, and CH_3OH in a ratio of approximately 1 : 1 : 1 (Ehrenfreund et al. 1998, 1999). The triple structure in the spectra is formed when, upon heating to ∼100 K, the ice matrix rearranges and molecules start to become mobile within the ice matrix. The two narrow peaks are due to "pure" CO_2 clusters in a H_2O/CH_3OH ice environment. The shoulder at longer wavelengths is due to a complex between CO_2 and CH_3OH. The stretching mode of $^{13}CO_2$ ice at 4.38 μm (Figure 18) has also been detected in a large number of sources, and a detailed analysis of its profile corroborates this conclusion (Boogert et al. 2000).

These results provide important evidence for extensive ice segregation and thermal processing in the envelopes around massive protostars.

One of the most prominent, but as yet unidentified, ice bands in the absorption spectra of protostars occurs at 6.85 μm (Schutte et al. 1996, Keane et al. 2001; Figure 20). Because of the limited resolution of earlier airborne observations, the peak position of this feature was assumed to be constant at 6.85 μm (Tielens and Allamandola 1987b). However, the high-resolution ISO-SWS observations have shown that the position varies considerably between the observed lines of sight in a systematic way, similar to that found for solid CO_2. The observed profiles of all sources can be fitted with a combination of two extreme features, suggesting that they reflect different degrees of thermal processing toward these protostars. While the actual carrier of this band is currently unknown, this sensitivity to the physical characteristics of its (ice) environment places additional constraints on the identification, which will assist laboratory identification efforts.

4.2 Grain surface chemistry: solid CH_4

A weak band at 7.6 μm in the spectra of some deeply embedded protostars has been attributed to CH_4 ice

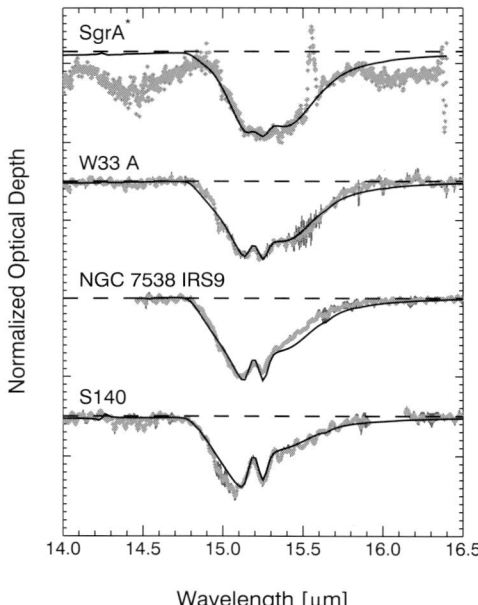

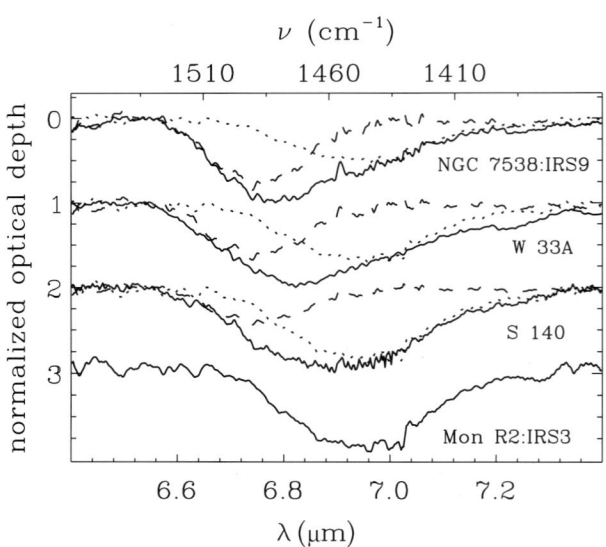

Figure 19 ISO-SWS spectra of the CO_2 bending mode near 15 μm, illustrating the thermal evolution of interstellar ices. The sources are displayed in order of increasing gas temperature. Solid lines are fits based on laboratory data for a H_2O : CH_3OH : CO_2 = 1 : 1 : 1 ice mixture at temperatures ranging from 106 to 118 K in the laboratory (corresponding to 50–60 K under interstellar conditions). (Gerakines et al. 1999.)

Figure 20 The observed ISO-SWS 6.85 μm band in the spectra of luminous protostars, on a normalized optical depth scale. All profiles can be fitted by combinations of two "extreme" profiles (dotted and dashed lines) whose relative strengths reflect the degree of thermal processing of the ice along the line of sight. Note that some profiles show narrow absorption lines due to warm, gaseous water vapor. (Keane et al. 2001.)

(Boogert et al. 1996, 1998), and it is of interest to compare its implications with those derived from the 4.67 μm solid CO band. Studies of the profile of the latter band have revealed that much of the solid CO is not mixed in with polar molecules (i.e., H_2O) (Tielens et al. 1991, Chiar et al. 1998), implying that along most lines of sight there are at least two independent ice components: a water-rich ice dominating the bulk of the interstellar ice, and a second ice component containing only nonpolar molecules such as (most of) the solid CO. The presence of these two independent components may reflect thermal processing (Figure 21) – the volatile CO is outgassed at much lower temperatures than is H_2O – or differences in the chemical history between these two species – H_2O ice is formed on grain surfaces in the presence of atomic H, which also reduces CO to H_2CO and CH_3OH. The importance of thermal processing of interstellar ices is attested to by the observed profiles of the solid CO_2 ice bands (Section 4.1). Solid CH_4 provides another good test for these different scenarios: solid CH_4 is as volatile as solid CO, but chemically CH_4 is formed on grain surfaces by the hydrogenation of accreted atomic C. The observed solid CH_4 band is well fitted by traces of CH_4 in polar, H_2O-rich ices rather than CO-rich ices (Boogert et al. 1996). Hence, for CH_4, chemistry seems to be more important in driving the ice environment than thermal processing.

4.3 The interstellar and cometary ice inventory

Table 2 summarizes the observed composition of interstellar ice toward the massive protostars NGC 7538 IRS 9 and W 33A. H_2O dominates the column density of ice by a large factor. The carbon-bearing species CO, CO_2, and CH_3OH are 10–100 times less abundant, while H_2CO, HCOOH, and CH_4 are at the 1% level compared to H_2O. This dominance of H_2O is directly evident from the complete infrared spectrum of this source (Figure 18). Possibly an exception should be made for the homonuclear molecules O_2, and N_2, which have no infrared active vibrational modes. While the solid matrix will perturb these modes into (very weak) activity, these bands are difficult to detect (Ehrenfreund et al. 1992, Vandenbussche et al. 1999).

This global ice inventory is somewhat misleading since, as emphasized above, analysis of the observed ice profiles have revealed the presence of a number of independent ice components along most lines of sight. These components likely reflect the effects of spatial variations both in the composition of the accreting gas, which sets the stage for grain surface chemistry, and in segregation and outgassing of more volatile ice species driven by heating by the nearby YSO (Figure 21).

The most striking aspect of the infrared spectrum is the apparent simplicity of the composition of interstellar ices. The molecules are highly hydrogenated but also show oxidation. Furthermore, the abundance of these species is much higher in the ice than in the gas phase of dark clouds. The H_2O/CO ratio is quite telling: less than 10^{-3} in the gas in dark clouds (Melnick 2000) compared to 10 in the ice. This preference for saturated species and this abundance difference implies a chemical origin that is different from gas-phase ion–molecule chemistry. Clearly, H_2O ice is a major reservoir of oxygen formed by the hydrogenation of O and/or via O_2 (Tielens and Hagen 1982). Also, given the low abundances of solid NH_3 and CH_4, accretion on the grain mantle must have occured from a molecular gas in which most of the nitrogen and carbon is locked up in N_2 and CO. Furthermore, hydrogenation of CO leads to the formation of the carbon species H_2CO, CH_3OH, and HCOOH, while oxidation results in CO_2. Observations of the deuterated versions of these species in hot cores also support the dominance of reactions on grain surfaces

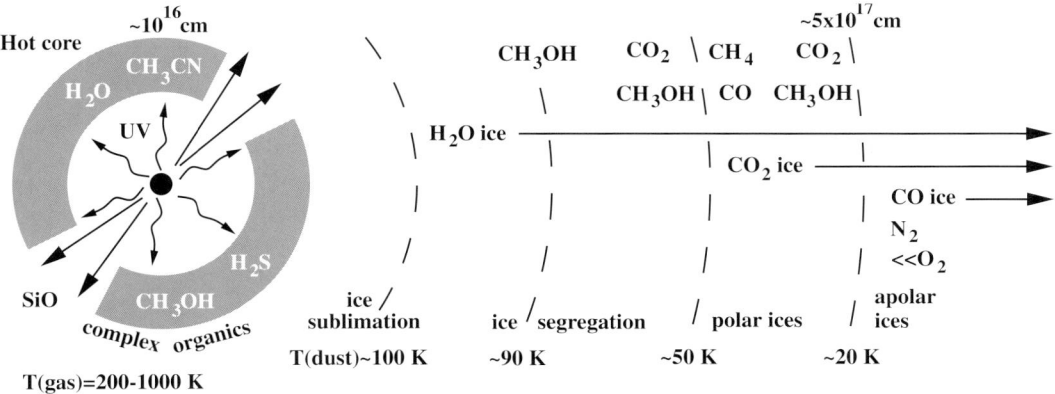

Figure 21 Schematic illustration of the temperature structure of the envelopes of YSOs and the different ice phases due to heating and thermal desorption. In the inner hot core, the ices have evaporated and high-temperature "hot core" chemistry occurs. (van Dishoeck and Hogerheijde 1999.)

(Tielens 1983, Turner 1990, Ceccarelli et al. 1998a). Finally, except for the 4.62 μm XCN band and the 6.85 μm band, whose identifications are still somewhat controversial, no unidentified bands are present. Clearly, grain (surface) chemistry heavily favors simple species and, as a corrollary, the role of surface chemistry and ultraviolet photolysis in producing more complex molecules seems limited to abundances which are at least an order of magnitude lower.

Comets are generally thought to contain a rather pristine record of the interstellar cloud from which the Sun and its planetary system were formed. It is therefore of some interest to compare the interstellar ice abundances with those measured for comets. Table 2 compares observed ice compositions to those of Comet Hale–Bopp (Bockelée-Morvan et al. 2000). It should be understood that a wider array of species has been observed in comets, largely because near the Sun the ices evaporate and simple gaseous species are more readily observed through their rotational spectra at millimeter wavelengths. Nevertheless, the species dominating the infrared spectra of interstellar ices are also abundant in comets. Even their relative abundances are quite comparable (Table 2), lending credence to their evolutionary interrelationship. A "shared" chemical history is also supported by the high degree of deuteration observed in cometary molecules (Irvine et al. 2000).

Table 2 Observed interstellar and cometary ice composition*

Species	NGC 7538 IRS 9	W 33A	Hale–Bopp
H_2O	100	100	100
CO (total)	10	3	23
CO (polar)	3	2	
CO (apolar)	7	1	
CO_2	16	4	6
CO_2 (polar)	1	1	
CO_2 (annealed)	3	2	
CH_3OH	9	10	2
H_2CO	3	2	1
HCOOH	2	0.5	0.1
NH_3	10	4	0.7
CH_4	1	0.4	0.6
OCS	0.1	0.05	0.4
"XCN"	0.8	1	

*Relative to H_2O = 100. Adopted H_2O column densities are 10^{19} and 4×10^{19} cm^{-2} for NGC 7538 IRS 9 and W 33A, respectively.

References. Interstellar ices: H_2O, H_2CO, HCOOH: Keane et al. (2001); CO: Tielens et al. (1991); CO_2: Gerakines et al. (1999); CH_3OH: Allamandola et al. (1992); NH_3: Lacy et al. (1998); CH_4: Boogert et al. (1998); OCS: Palumbo et al. (1997); "XCN": Pendleton et al. (1999). Comet Hale–Bopp: Bockelée-Morvan et al. (2000).

5 WARM INNER ENVELOPES AROUND YOUNG STELLAR OBJECTS

Young stars heat their surrounding envelopes, resulting in a temperature gradient throughout the surrounding gas and dust (Figure 21). For a $\sim 10^5 L_\odot$ source such as W 33A or Orion IRc2, the region in which $T_{dust} \approx T_{gas} > 90$ K extends to $\sim 10^{16}$ cm (~ 1000 AU) from the star, whereas the region with $T_{dust} > 20$ K extends to $\sim 5 \times 10^{17}$ cm (0.1 pc). All ices are evaporated in the inner part of the envelope above 90 K; that is, within a few arcseconds for a typical distance of 1 kpc. In this so-called "hot core" region the ices released can drive a rich chemistry in the gas phase. In particular, CH_3OH is a key ingredient in driving molecular complexity (Charnley et al. 1992, Caselli et al. 1993).

In addition to radiative heating, the interaction of the outflow from the young star with the envelope can result in enhanced temperatures. At the positions of direct impact, the temperature is raised to several thousand kelvin or more, depending on the speed of the shock (Section 3). Turbulent mixing along the outflow walls results in slow entrainment of the gas and dust, with removal of the ice mantles. Finally, once young massive stars clear their surroundings, ultraviolet radiation creates a PDR at the surrounding cloud (Section 2). The different diagnostic probes discussed in the preceeding sections can be used to disentangle their effects, to determine their relative importance, and to establish an evolutionary sequence for the objects (van Dishoeck and van der Tak 2000, van den Ancker et al. 2000a).

The presence of warm gas along the line of sight toward protostars is clearly revealed in their absorption spectra, seen first in ground-based observations of gas-phase CO at 4.7 μm (Mitchell et al. 1990). The ISO-SWS has allowed searches for other gas-phase molecules to be mounted in particular H_2O and CO_2, which cannot be observed from the ground and which are among the dominant O- and C-containing species (see van Dishoeck et al. (1999) for a review). One of the early SWS surprises included the detection of abundant (of the order of 10^{-5} with respect to H_2), hot ($T_{ex} \approx 300$ K) gas-phase H_2O absorption toward a number of YSOs (Helmich et al. 1996, van Dishoeck and Helmich 1996, van Dishoeck 1998; Figure 22). In contrast, the abundance of gas-phase CO_2 was found to be much lower, of the order of 10^{-7} (van Dishoeck et al. 1996, Dartois et al. 1998, Boonman et al. 2000). Gas-phase CH_4 (Boogert et al. 1998), C_2H_2, and HCN (Lacy et al. 1989, Evans et al. 1991, Lahuis and van Dishoeck 2000) have also been detected, with abundances up to 10^{-6} and high excitation temperatures.

The gas-phase results, combined with the solid-state data for the same lines of sight, allow the gas/solid ratios to be determined. Significant trends are found in increasing

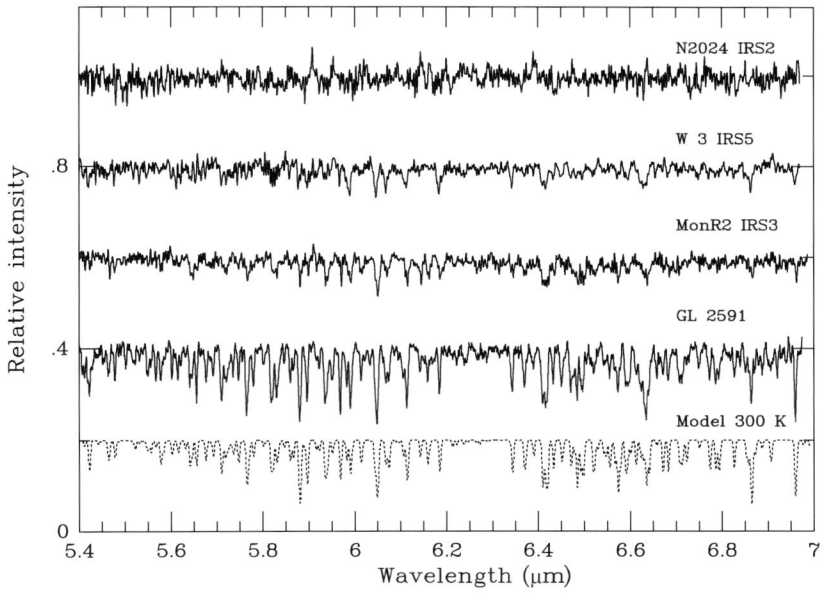

Figure 22 Absorption by gas-phase H_2O in the ν_2 bending mode at 6 μm observed toward massive YSOs with the ISO-SWS. (van Dishoeck and Helmich 1996, Boonman et al. 2000.)

abundances of evaporated molecules and gas/solid ratios with increasing temperature for the different objects. These trends mirror those found for the fraction of heated ices from the solid CO_2 and 6.85 μm bands (Figures 19 and 20) (Boogert et al. 2000). Since these diverse phenomena occur at a range of temperatures, from <100 K to 1000 K, the enhanced temperatures must be communicated to both the inner and outer parts of the envelopes. Van der Tak et al. (2000) have shown that this "global heating" most likely results from the gradual dispersion of the envelopes with time: the more evolved objects have removed a larger fraction of their envelope, resulting in a smaller ratio of envelope mass to stellar mass and thus a higher temperature throughout the envelope. These signatures can therefore be used as a probe of the evolutionary state of the objects.

Emission line observations with the ISO-LWS are also a powerful way of tracing the warm gas surrounding YSOs, and distinguishing between the various mechanisms for heating and clearing of the envelope (see Saraceno et al. (1999) for an overview). The youngest, most deeply embedded, lowest-luminosity ($< 10^3 L_\odot$) sources, so-called "class 0" objects, have very rich LWS spectra with many molecular lines due to H_2O, OH, and CO (e.g., Liseau et al. 1996, Ceccarelli et al. 1998b, Nisini et al. 1999, Spinoglio et al. 2000; Figure 23). The CO excitation indicates high temperatures, of a few hundred kelvin, and high densities over a region of a few arcseconds (500–1000 AU) in size. In objects in later evolutionary stages (classes I and II) the total luminosity in the CO lines, and thus the total cooling, is higher than that in the H_2O or OH lines. By contrast, high-luminosity sources ($> 10^4 L_\odot$) show strong atomic and ionic lines, in particular [C II] 158 μm and [O I] 63 and 145 μm emission, with weaker molecular lines. The main exceptions are the Orion region and massive sources in the galactic center, which exhibit strong H_2O and OH emission and absorption lines (Section 3; Baluteau et al. 1997). One possible explanation for this difference is that heating by radiation plays a larger role in higher-mass objects, whereas heating by shocks is more important in lower-mass objects. Higher-mass objects are also more distant, so that the emission from warm molecular gas suffers more beam dilution.

An interesting example where both heating processes may contribute equally is provided by the intermediate-mass Herbig Ae/Be stars. Van den Ancker et al. (2000a,b) have observed a large sample of these objects with the ISO-SWS, and used various diagnostics to distinguish the dominant excitation mechanism. In these mixed cases the atomic and ionic lines provide good additional diagnostics. Specifically, the [Si II] 34.8 μm and [Fe II] 26.0 μm lines are strong in PDRs, whereas the [S I] 25.2 μm line becomes detectable only in shocks. The same trend is also found from ISO-LWS data for the [O I] 63 μm line: in low-luminosity YSOs the [O I] emission appears to be caused

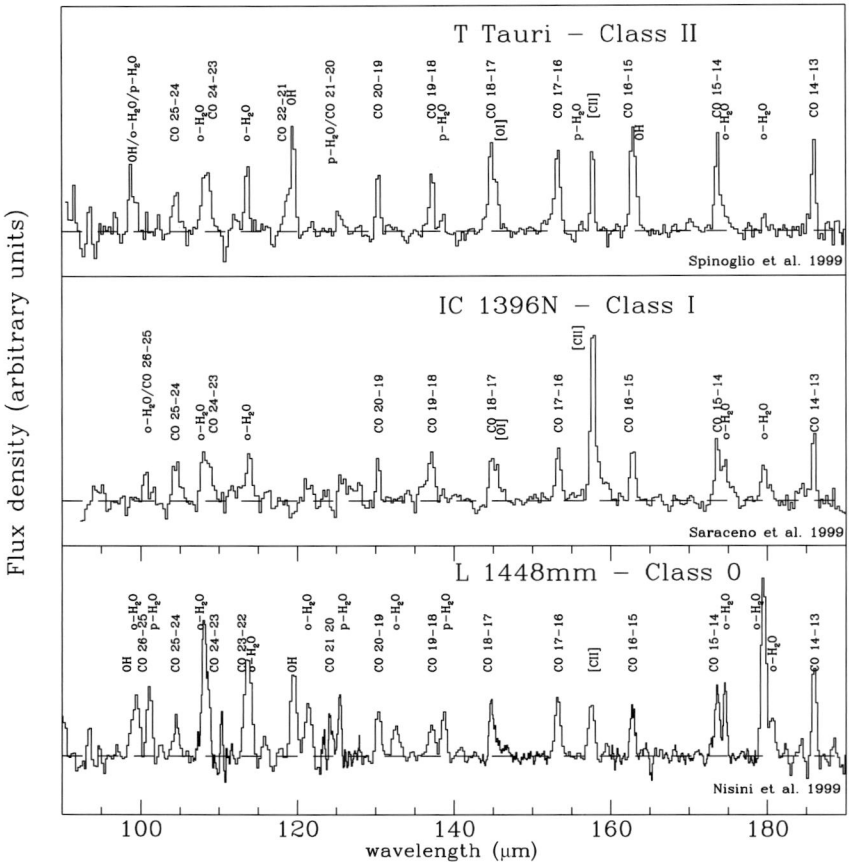

Figure 23 ISO-LWS spectra of low-mass YSOs at different stages of evolution, revealing a wealth of CO, OH, and/or H_2O pure-rotational lines. The continuum has been subtracted. Top: T Tauri (Spinoglio *et al.* 2000); middle: IC 1396N (Saraceno *et al.* 1999); Bottom: L1448mm. (Nisini *et al.* 1999.)

primarily by shocks, whereas higher-mass Herbig Ae/Be objects have additional contributions from PDRs (Figure 10; Lorenzetti *et al.* 1999, Giannini *et al.* 1999).

The evolutionary sequence of intermediate-mass YSOs is beautifully illustrated by their complete mid-infrared spectra (Figure 24; van den Ancker *et al.* 2000a). The ISO-SWS spectrum of the very young, deeply embedded object LkHα 225 reveals a large column of solid water ice and silicates as well as absorption bands due to gas-phase H_2O, CO, and CO_2. This spectrum also contains many emission lines due to molecular hydrogen, [S I], [Si II], and [Fe II], arising in shocks caused by the protostellar outflow flowing into the surrounding material. LkHα 224 is slightly older and its spectrum does not show absorption features due to dust or gas; much of the ice has evaporated and the envelope has dispersed. The absence of dust emission features suggests that the spectrum is dominated by emission from very large grains ($\gg 10\,\mu$m). The spectrum of BD +40 4124, which has already cleared its environment, shows emission bands from PAHs, as well as emission lines from H I, H_2, [Fe II], and [Si II] characteristic of an H II region surrounded by a PDR. The Herbig Ae star AB Aur – a pre-main sequence star surrounded by a circumstellar disk and remnant envelope (Section 6) – shows, like BD +40 4124, the characteristics of ionized gas and a PDR. However, the PAHs are less prominent in this source and silicate emission starts to dominate the mid-infrared dust features. The spectrum of HD 100546, an even more evolved member of this class, reveals the presence of a small fraction of cool crystalline silicates as well as weak PAH features. Except for the H I recombination lines, which originate from the disk, all other spectral lines have disappeared in these latter phases. These spectral variations are related to the relatively rapid (10^6 yr) accretion and/or dispersion of the surrounding molecular cloud material and the slower ($\sim 10^7$–10^8 yr) processing of the dust (agglomeration, settling, and destruction) in the circumstellar disk.

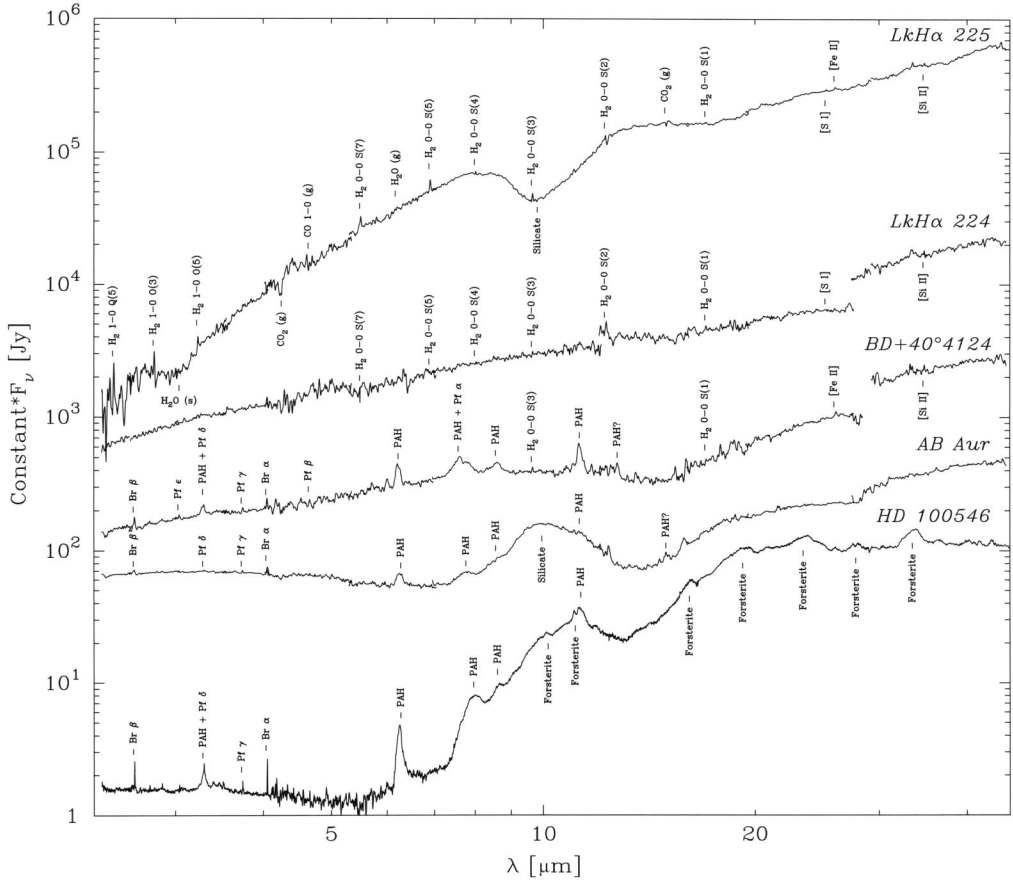

Figure 24 The ISO-SWS mid-infrared spectra of newly formed stars in different stages of formation (van den Ancker et al. 2000a,c, Malfait et al. 1998). From top to bottom – in a rough evolutionary sequence – the spectra change from dominated by solid-state absorption features and gas (shock) emission lines, to featureless, to PAH features and PDR lines, to amorphous and then crystalline silicates with H I recombination lines.

6 CIRCUMSTELLAR DISKS AROUND YOUNG STARS

Circumstellar disks around young stars provide the raw material from which planetesimals and eventually planets may form, and are therefore popular targets for new high-spatial-resolution instrumentation. The presence of these disks was first inferred from ground-based observations at infrared and millimeter wavelengths (see Beckwith and Sargent (1996) for an overview), but they have also been revealed as silhouettes against bright backgrounds through stunning space-based HST images (e.g., McCaughrean and O'Dell 1996). The typical sizes of the disks are $\sim 100–400$ AU, comparable to that of our own Solar System. In order to study the composition and evolution of the dust and gas in such disks, spectroscopy at infrared wavelengths is essential. An overview is presented in Chapter 37, but a few examples are included here to illustrate the cosmic dust cycle.

6.1 Solid-state features

Early ground-based and airborne observations of solid-state features in disks were mostly limited to the 10 μm silicate feature (e.g., Telesco and Knacke 1991, Sitko et al. 1999). This feature has also been observed for a sample of T Tauri stars with ISOPHOT-S (Natta et al. 2000). The full richness of circumstellar disk spectra is revealed only through complete mid- and far-infrared spectroscopy, as provided by the ISO-SWS and LWS. Because of the sensitivity of these instruments, the observations have been possible only for intermediate-mass Herbig Ae stars, and not yet for

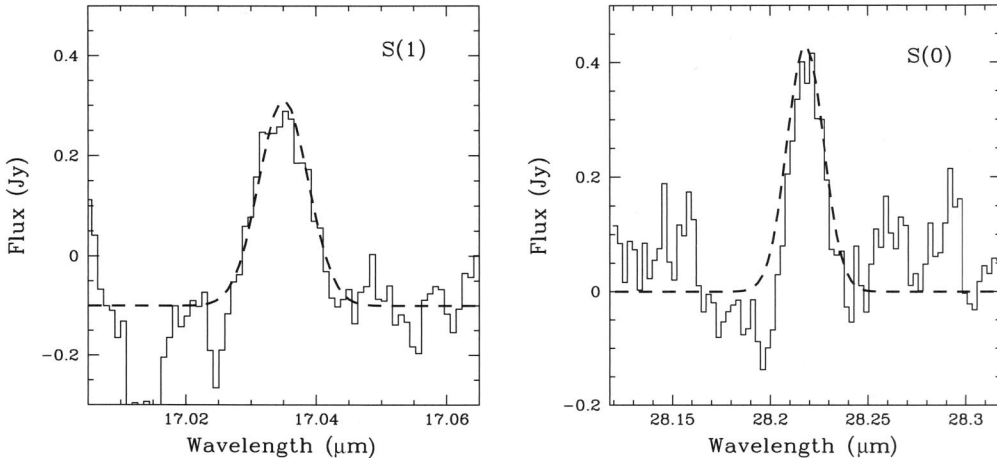

Figure 25 ISO-SWS spectra of the H_2 S(0) and S(1) lines toward the isolated T Tauri object GG Tau. This double binary system is surrounded by a circumstellar dust and gas disk with a size of ~ 800 AU. The H_2 lines provide direct information on the temperature and amount of warm gas. (Thi *et al.* 1999a.)

solar-mass T Tauri stars. A striking example is HD 100546, a ~ 10 Myr old Herbig Ae star, whose disk shows emission from PAHs and crystalline silicates, in particular forsterite (Mg_2SiO_4) (Waelkens *et al.* 1996, Malfait *et al.* 1998; Figure 27). The spectrum of the dust in this disk greatly resembles that of Comet Hale–Bopp (Crovisier *et al.* 1997), suggesting some connection between interstellar and cometary material. In contrast, the dust around AB Aur is dominated by warm amorphous olivines and crystalline iron oxides, with PAH features detected. Crystalline H_2O ice has also been detected in a few cases (van den Ancker *et al.* 2000b, Molinari *et al.* 1999). Thus, the dust spectral characteristics show evidence of significant processing in some sources, which may be partly due to age but which may also reflect local processes caused, for example, by the presence of nearby companions or the possible formation of planetary bodies.

6.2 Gaseous species

A major question with the evolution of disks is the timescale over which the gas disappears from the disk, since this directly affects the formation of gaseous giant planets. Most studies of the gas have been based on ground-based millimeter observations of CO (e.g., Koerner and Sargent 1995, Zuckerman *et al.* 1995, Dutrey *et al.* 1996), but the analysis of these data has been hampered by optical-depth effects in the lines and possible freeze-out of the molecules onto grains in the coldest regions. The lowest H_2 pure-rotational lines provide a powerful complement to CO, since they directly measure the amount and temperature of the gas, provided it is warm enough to emit in these lines (>80 K).

Deep searches for the H_2 S(0) and S(1) lines at 28 and 17 μm have been performed with the ISO-SWS toward a dozen young stars with confirmed circumstellar disks (Thi *et al.* 1999a,b, Stapelfeldt *et al.* 1999, Thi *et al.* 2001a; Figure 25). The sources were chosen because they are isolated objects, away from molecular clouds, so that contamination from surrounding cloud emission is minimal. The S(1) and S(0) lines have been detected in several sources. Analysis of the data suggests that the mass of warm gas at $\sim 100-200$ K is large, ranging from a few percent to a significant fraction of the mass of the disks derived from the dust continuum, assuming a gas/dust ratio of 100. The origin of this warm gas is still a puzzle, although it could be that the surface layers of disks are heated by stellar radiation if they have a flared geometry (e.g., Calvet *et al.* 1991, Chiang and Goldreich 1997).

Surprisingly, evidence for H_2 lines is also found toward three older Vega-type objects, including the debris-disk of β Pictoris (Thi *et al.* 2001b). The data indicate residual primordial gas masses of $0.1-6M_{Jupiter}$, much larger than is inferred from CO, which freezes out onto grains in the disk midplane and is destroyed by photodissociation in the surface layers (Kamp and Bertoldi 2000). For β Pictoris, the gas mass of $0.1M_{Jupiter}$ is insufficient to form new gaseous giant planets, but it does affect the grain dynamics.

7 CIRCUMSTELLAR GAS AND DUST AROUND LATE-TYPE STARS

7.1 The rich spectrum of circumstellar silicates

In the last phases of its life, a star of moderate mass begins to pulsate and lose its outer layers; the ejected material is enriched in elements formed by nucleosynthesis inside the star. In the cooling outflow, solid particles such as silicates, oxides, and carbonaceous material can condense, resulting in very rich mid-infrared spectra. The detection of the infrared emission from this circumstellar dust led to the widespread acceptance of the importance of stardust for the interstellar dust budget (Section 1.2). With the opening up of the telluric mid-infrared windows, the 10 and 18 μm features due to amorphous silicates around O-rich AGB stars and the 12 μm feature due to SiC grains around C-rich AGB stars were quickly identified (Merrill and Stein 1976). That is where the field stalled until the launch of IRAS, which carried a low-resolution spectrometer that took spectra of many of such objects, at $R \approx 15-40$ from about 8 to 23 μm. While most showed these same dust features, some spectra also revealed structure which was attributed to the presence of a small amount of crystalline silicates and to Al_2O_3 dust (e.g., Vardya et al. 1986).

One of the greatest surprises of the ISO mission was the incredible richness of the circumstellar silicate spectra at long wavelengths of AGB stars, post-AGB stars, and planetary nebulae. Invariably, the 15–45 μm spectra of O-rich sources with circumstellar dust show a great number of narrow features, implying a crystalline carrier (Figure 26; Molster 2000). From the laboratory studies of the infrared characteristics of crystalline silicates (Koike et al. 1993, 2000; Jäger et al. 1994, 1998), these features have been identified with crystalline olivine and pyroxene grains; specifically, fosterite (Mg_2SiO_4) and enstatite ($MgSiO_3$), the Mg-rich end members of the olivine and pyroxene families. The measured Fe/Mg abundance ratio is very low, less than 0.05. In contrast, circumstellar and interstellar amorphous silicates contain a substantial fraction of iron with Fe/Mg$\simeq$0.5 (Schutte and Tielens 1989; Section 1.2).

These crystalline silicate features are ubiquitous and, as well as in O-rich AGB stars, post-AGB objects, and planetary nebulae, they have now been observed in such diverse objects as luminous blue variables (i.e., AG Car), Herbig AeBe young stars, nominally C-rich objects such as the Red Rectangle and BD +30°3639, and Comet Hale–Bopp (Figure 27; Voors 1999, Crovisier et al. 1997, Malfait et al. 1998, Waters et al. 1998a,b, Molster 2000). The Hale–Bopp spectrum is of particular interest since it is dominated by forsterite, with very little enstatite present, and hence can serve as a template in the assignment of observed circumstellar silicate features.

Mass-losing, O-rich AGB stars and related objects show evidence for a modest amount (1–10%) of crystalline silicates in their spectra, whose origin is probably to be found in the condensation of the silicates near the star (Sylvester et al. 1999). A group of peculiar objects, however, some of which are confirmed binaries, shows much higher crystalline silicate abundances (above 30%). These objects

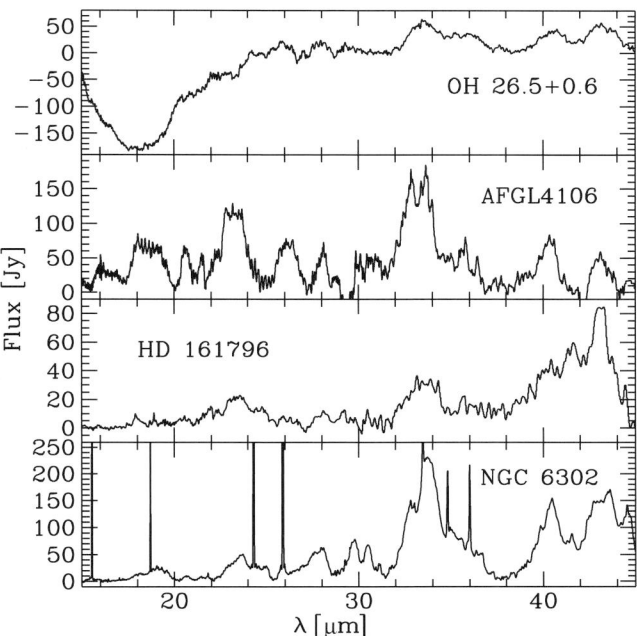

Figure 26 The ISO-SWS 15–45 μm continuum-subtracted spectra of the AGB star OH 26.5 + 0.6 ($\lambda/\Delta\lambda \simeq 250$), the red supergiant GL 4106 ($\lambda/\Delta\lambda \simeq 500$), the protoplanetary nebula HD 161796 ($\lambda/\Delta\lambda \simeq 250$), and the planetary nebula NGC 6302 ($\lambda/\Delta\lambda \simeq 500$), showing a host of features attributed mainly to olivine and pyroxene silicates. These spectra illustrate the incredible richness of the mid-infrared spectra of circumstellar silicates and the detailed variations therein. (Molster 2000.)

have stable, stationary disks and also show evidence of a population of large, cold grains. The large grains are likely the result of coagulation, a process which may eventually lead to cometesimals/planetesimals and eventually planets. The contrast in crystallinity between the amorphous interstellar silicates and the cold crystalline dust in the envelopes around late-type stars, YSOs, and comets is particularly puzzling, the more so since comets are generally thought to represent unprocessed interstellar material. It seems that silicates can anneal even at low temperatures – well below the glass temperature – on the long timescales and under the benign conditions provided by circumstellar disks (Molster et al. 1999).

7.2 Dust around carbon-rich objects

The mid-infrared spectra of C-rich AGB, post-AGB, and planetary nebulae show different dust emission features, reflecting a rich and diverse carbon-based chemistry. Much of this was hinted at by the IRAS-LRS (Willems 1988,

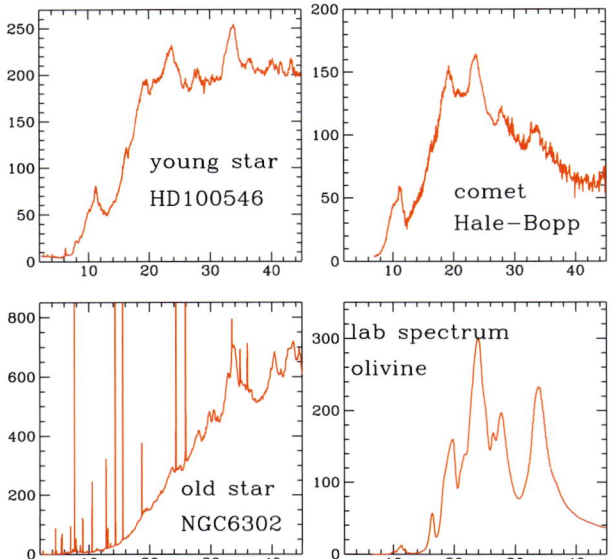

Figure 27 The ISO-SWS mid-infrared spectra of newly formed stars, comets, and planetary nebulae show narrow emission features due to crystalline silicates such as forsterite and enstatite. (Malfait et al. 1998, Molster 2000, Crovisier et al. 1997.)

Kwok et al. 1989, Jourdain de Muizon et al. 1990); the KAO (Buss et al. 1993, Omont et al. 1995) revealed a feature at 21 μm (see below) and a feature near 30 μm in some objects. Figure 28 illustrates the ISO-SWS spectra for various C-rich stars. The AGB stars show a weak emission feature due to SiC at about 11.5 μm, which only for very extreme mass loss rates goes into absorption. Otherwise, besides a feature around 25 μm (see below), the infrared spectra are dominated by a featureless continuum which is generally attributed to amorphous carbon (Martin and Rogers 1987). In the post-AGB phase, broad emission features at 8, 12, 20, and 25 μm radically change the spectral appearance. During this stage, weak PAH features at 6.2, 7.7, and 11.3 μm also appear. In the planetary nebula phase, the PAH emission features dominate the 3–15 μm spectrum (Section 2.2), while a 30 μm feature dominates at longer wavelengths (see below).

The feature near 21 μm has led to the identification of a class of sources known collectively as the "21 μm sources" (Kwok et al. 1989). Visual spectroscopy has shown that these sources are metal-poor, C-rich, red giants with temperatures in the range of 5000–8000 K and enhanced s-process element abundances, supporting their post-AGB nature (Hrivnak 1995, van Winckel and Reynier 2000). These population II stars of around $1 M_\odot$ were formed during the early history of the Milky Way and have now, after some 10^{10} yr, just reached the post-AGB phase which ends the life of all low-mass stars. Hence, in an evolutionary sense, these objects are not directly related to the AGB stars and planetary nebula objects whose spectra are shown in Figure 28. It has recently been shown that the 21 μm feature is well matched by the laboratory spectrum of TiC nanocrystals (Figure 29; von Helden et al. 2000). TiC was exclusively known as a circumstellar dust component through the discovery of small (200 Å) TiC cores inside graphitic stardust (Bernatowicz et al. 1996) – the first identification of this material in an astronomical object. Titanium is a very rare species, and the presence of these dust grains requires very dense conditions for its formation in stellar ejecta. Hence, the 21 μm feature may signal dust formation during the extreme conditions associated with the superwind that terminates the life of all-low mass stars on the AGB and marks the transition to the planetary nebula phase (Renzini 1987).

Traditionally, the strong band between 25 and 40 μm – shown by many C-rich AGB stars, post-AGB objects, and planetary nebulae – has been called the 30 μm feature. This band has been tentatively assigned to solid MgS. Analysis of the SWS spectra shows that the band changes markedly, in both shape and peak position, from the AGB to planetary nebulae phase (Figure 28). Comparison of its shape with laboratory measurements of solid MgS shows that this band cannot be explained by MgS grains alone. It is likely, over time, that the MgS is processed through its exposure to the harsh environment of the ejecta (Hony et al. 2001b, in preparation).

7.3 Gas tracers of the dust condensation zone

A number of O-rich AGB stars, especially those that exhibit the dust emission band at 13 μm ascribed to Al_2O_3, such as EP Aqr, show prominent, narrow emission lines between 13 and 16.5 μm in the ISO-SWS spectra. These have been identified as ro-vibrational transitions of CO_2 and $^{13}CO_2$ (Figure 30; Cami et al. 2000; Justtanont et al. 1996, 1998; Ryde et al. 1999), and this has been confirmed by high-resolution Fabry–Perot observations. Synthetic spectra, assuming optically thin lines formed in local thermal equilibrium (LTE), compare well with the spectrum observed toward EP Aqr. The relatively high temperature (~650 K) confirms the existence of a warm molecular envelope close to this type of star; presumably this is the layer in which dust nucleates and the stellar wind originates.

ISO-SWS studies of the supergiants α Ori and α Sco have revealed the presence of the fine-structure lines of [Si II] and [Fe II] (Justtanont et al. 1999). KAO studies had already shown the presence of intense [O I] 63 μm emission from these objects (Haas and Glassgold 1993). Analysis of the observed intensities shows that the emitting gas is dense ($>10^6 cm^{-3}$), warm (1000 K), and located close to the star ($<20 R_*$). These observations therefore probe the region inside the dust condensation zone for these

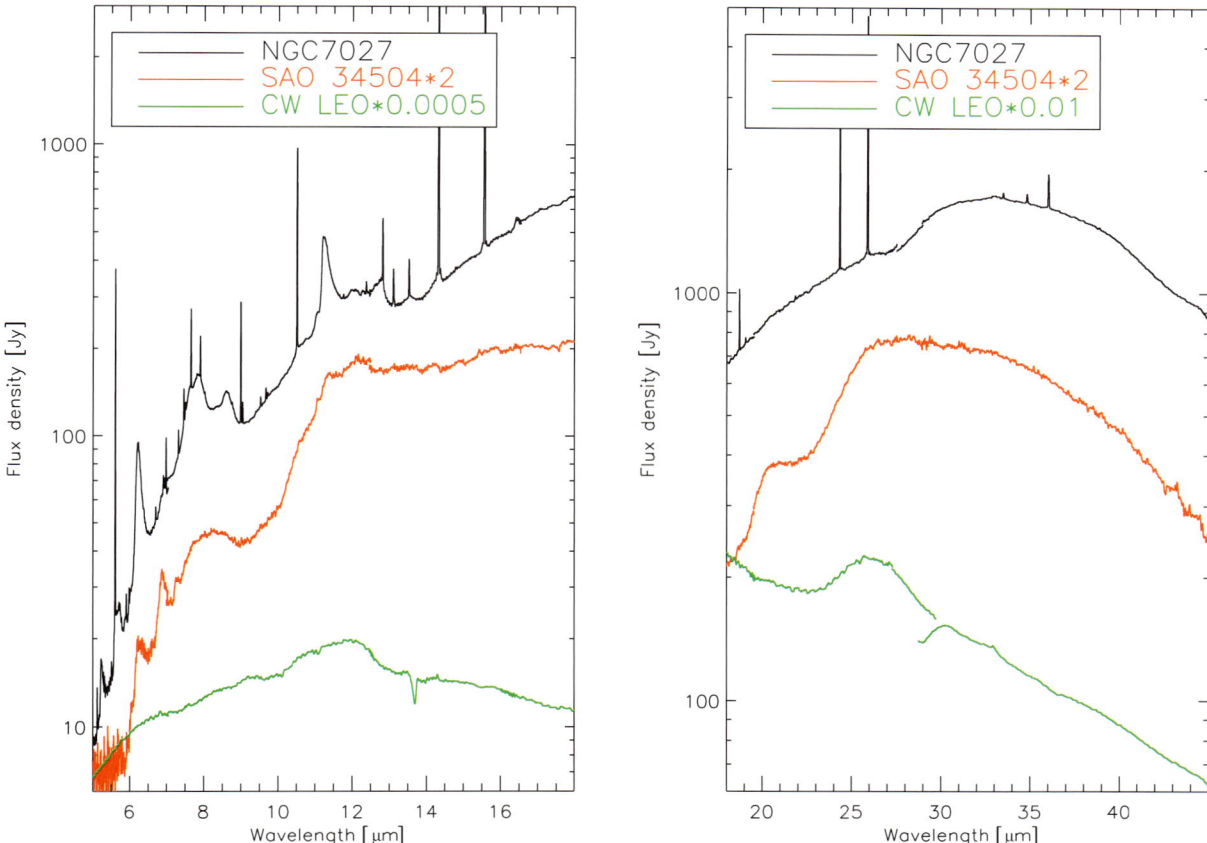

Figure 28 A comparison of the 5–18 μm (left) and 18–45 μm ISO-SWS spectra of three C-rich objects: the AGB star IRC 10216/CW Leo (bottom), the post-AGB object IRAS 22272 (middle) – one of the so-called 21 μm objects (Section 7.2) – and the planetary nebula NGC 7027 (top). Note the strong spectral variation between these objects. (Hony *et al.* 2001b, in preparation.)

supergiants. The observations are in good agreement with theoretical models for the heating and cooling and dynamics of these outflows (Rodgers and Glassgold 1991).

7.4 Molecular emission from AGB winds

AGB stars lose a large fraction of their mass through a cool, slowly expanding stellar wind. The gas in this wind is heated through friction between the gas molecules and the radiatively accelerated dust grains that are driving the wind, and it cools through adiabatic expansion and molecular emission lines. This gas has been probed by rotational emission lines in the submillimeter from the ground and from the KAO. The ISO-LWS and SWS provided the first complete inventory of the dominant cooling lines of such winds. In O-rich objects the cooling is dominated by H_2O (Figure 31; Barlow *et al.* 1996, Neufeld *et al.* 1996, Truong-Bach *et al.* 1999). The lines are highly optically thick and sensitive to the mass loss rate and the H_2O abundance, although, for the shortest-wavelength lines, radiative pumping may also affect the intensity (Barlow 1999). The emission spectrum of C-rich AGB stars, such as IRC+10216, is dominated by lines of CO and HCN (Cernicharo *et al.* 1996). In particular, the HCN spectrum is very rich, showing not only pure-rotational transitions but also ro-vibrational transitions of low-lying vibrational states. Because of this spectral richness, HCN dominates the cooling of these winds. At shorter wavelengths, many absorption bands due to carbon-chain molecules have been detected in absorption, emphasizing a rich chemistry (Cernicharo 2000).

ISO-LWS spectra of C-rich post-AGB objects and planetary nebulae show high-lying pure-rotational emission lines of CO (Cox *et al.* 1996, Liu *et al.* 1996, Justtanont *et al.* 1999). These lines were independently studied with the KAO (Justtanont *et al.* 1997). Detailed modeling shows that

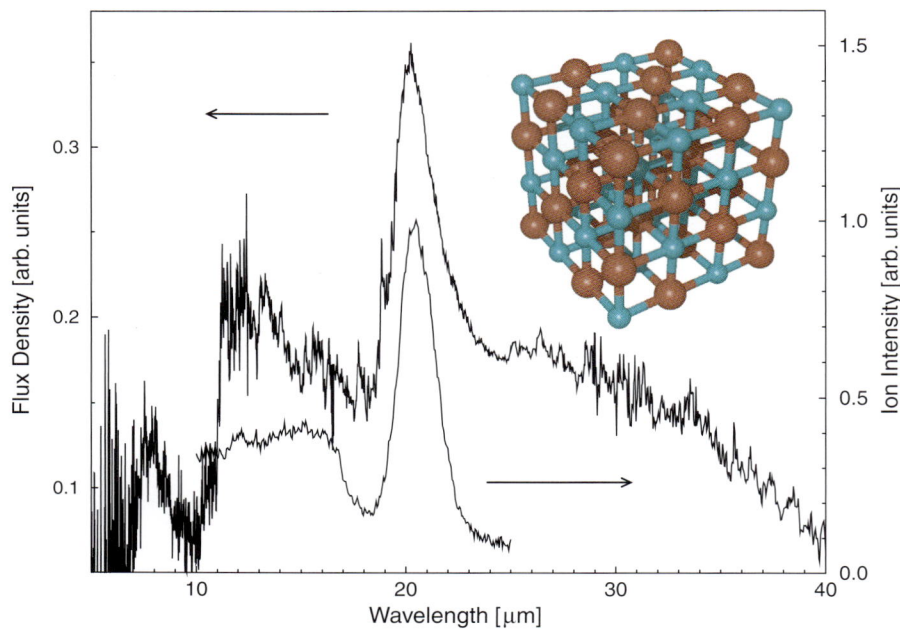

Figure 29 The emission spectrum from the post-AGB object SAO 96709, obtained by the ISO satellite (upper trace, left axis), and the spectra of TiC nanocrystal clusters recorded in the laboratory (lower trace, right axis). Also shown is a pictorial representation of a typical ($4 \times 4 \times 4$ atom) TiC nanocrystal. (von Helden et al. 2000.)

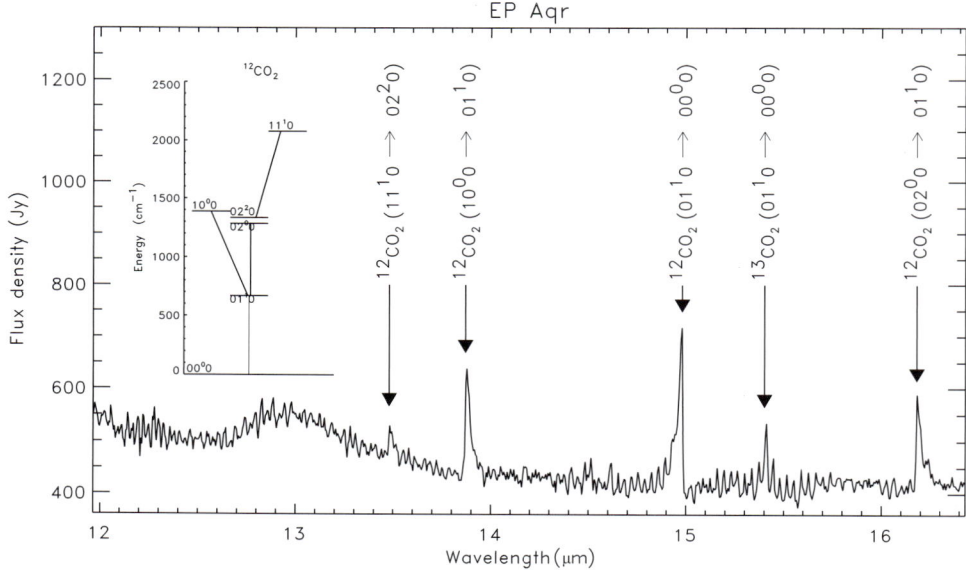

Figure 30 The ISO-SWS 12–16 μm spectrum of the AGB star EP Aq, showing narrow emission features due to the Q-branches of the bending mode and various combination modes of gaseous CO_2 (see the inset). Much of the structure in between is due to P- and R-branch transitions associated with these ro-vibrational modes. (Cami et al. 2000.)

for the planetary nebula NGC 7027, the CO emission originates in a dense PDR created by stellar FUV photons interacting with the previously ejected AGB wind. This warm and dense PDR is also probed by several pure-rotational lines of the CH^+ ion (Cernicharo et al. 1997), some of which were previously ascribed to H_2O. The CO emission in the spectrum of the post-AGB object CRL 2688, in contrast, results from gas heated by the shock driven by the fast

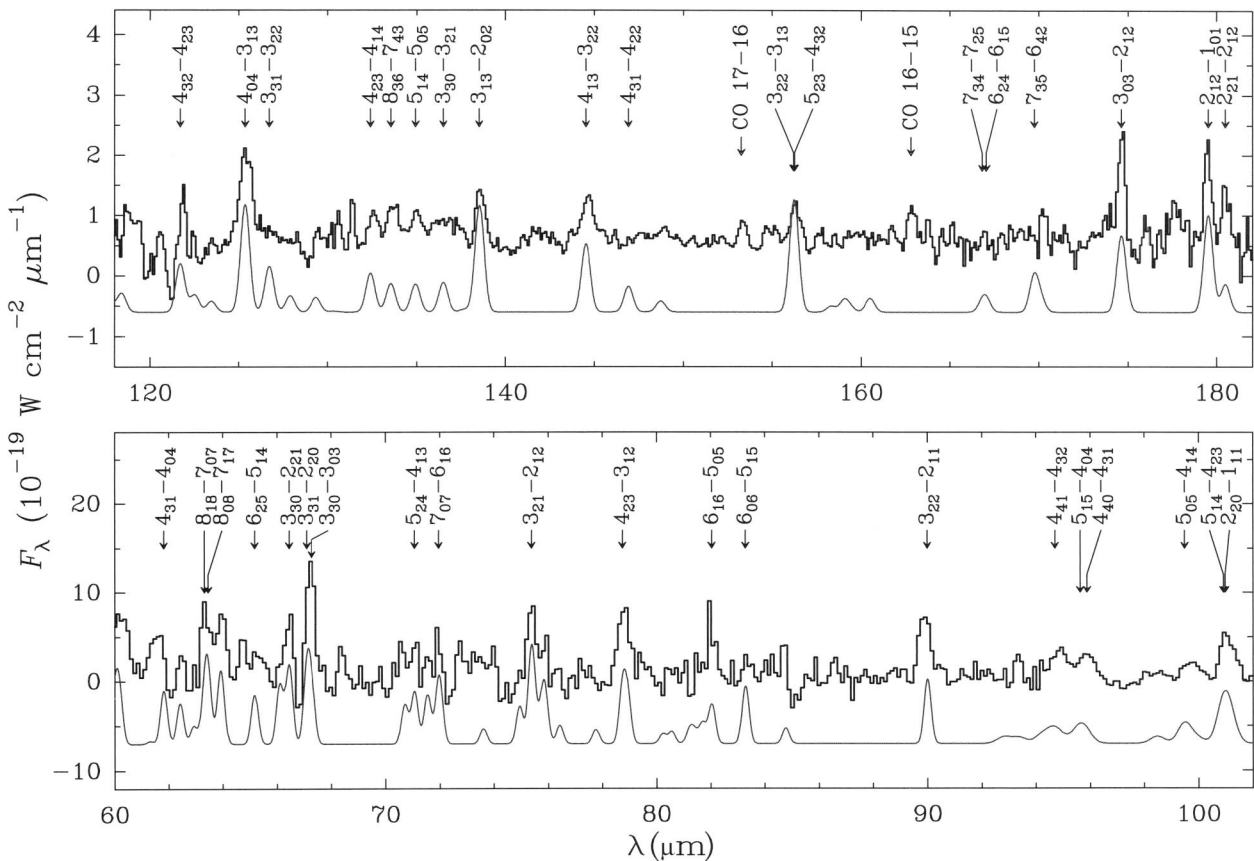

Figure 31 Continuum-subtracted ISO-LWS spectrum of the O-rich AGB star W Hya, showing a rich spectrum of rotational H_2O lines and a few CO lines (top trace in each panel). The dominant transitions are labeled. The bottom trace in each panel shows a model calculation for the water emission. (Barlow et al. 1996.)

wind into the AGB ejecta. Future spectrally resolved line profiles to be obtained with SOFIA and Herschel are needed to distinguish between the mechanisms.

8 CONCLUSIONS AND FUTURE PROSPECTS

The airborne and space-based facilities of the past century have provided essential tools for studying the ISM, star- and planet-forming regions, and the envelopes around dying stars. Together, the spectral features allow a detailed picture of the life cycle of the gas and dust from the quiescent ISM to YSOs to old stars to be built up, and these features can now be used as diagnostic probes of the evolutionary states of objects. Laboratory spectroscopy has played an essential role in the development of the field, and continued support will be needed for future progress.

The ISO has been highly successful in providing the first complete overview of the incredibly rich spectroscopy at infrared wavelengths, but has revealed – almost literally – only the "tip of the iceberg." The interpretation of the ISO data is severely hindered by insufficient spatial and spectral resolution, as well as sensitivity. What are the prospects for making the next leap forward in the coming decades? The enormous progress in mid-infrared detector arrays and instrumentation, means that orders-of-magnitude improvements are still possible, provided suitable air- and space-borne platforms are available. The SOFIA airplane is well placed to pursue follow-up medium- and high-spectral-resolution observations with a wide variety of instruments. The Infra Red Spectrometer (IRS) on SIRTF, operating at $5-40\,\mu m$, will have a sensitivity up to a hundred times better than that of the ISO-SWS, with moderate spectral resolution ($\lambda/\Delta\lambda = 60-600$), and will provide a very powerful probe of the solid-state features in objects which were too faint for ISO. Herschel will open up the terahertz frequency range ($\sim 60-500\,\mu m$), and its High Frequency Instrument (HIFI) will provide spectrally resolved information on H_2O, high-J CO, and other atomic and molecular lines in the ISM and YSOs. The Photoconducting Array Camera and Spectrometer (PACS) instrument will allow efficient mapping of atomic and molecular features at lower spectral resolution. Both SOFIA and Herschel will have considerably better spatial resolution than ISO and SIRTF. The largest gain in sensitivity and spatial resolution at mid-infrared wavelengths will potentially be provided by a camera and spectrometer on board NGST. The ISO has paved the way for the exciting science to be performed with these and other new instruments in the coming century.

Acknowledgments

We would like to salute the dedicated work of all those responsible for building the superb infrared instruments on the various airborne and space-based platforms, including the Kuiper Airborne Observatory and the Infrared Space Observatory, and for getting them operational. We are also grateful to many colleagues and friends for providing preprints and comments, and for permission to reproduce their figures.

REFERENCES

Abergel, A., Bernard, J.P., Boulanger, F., Cesarsky, C., Desert, F.X., Falgarone, E., Lagache, G., Perault, M., Puget, J.-L., Reach, W.T., Nordh, L., Olofsson, G., Huldtgren, M., Kaas, A.A., Andre, P., Bontemps, S., Burgdorf, M., Copet, E., Davies, J., Montmerle, T., Persi, P. and Sibille, F. (1996). ISOCAM mapping of the ρ Ophiuchi main cloud. *Astronomy and Astrophysics*, **315**, L329–L333.

Allamandola, L.J. and Norman, C.A. (1978). Infra-red emission lines from molecules in grain mantles. *Astronomy and Astrophysics*, **63**, L23–L26.

Allamandola, L.J., Sandford, S.A., Tielens, A.G.G.M. and Herbst, T. (1992). Infrared spectroscopy of dense clouds in the C-H stretch region: Methanol and "diamonds", *Astrophysical Journal*, **399**, 134–146.

Allamandola, L.J., Tielens, A.G.G.M. and Barker, J.R. (1985). Polycyclic aromatic hydrocarbons and the unidentified infrared emission bands: Auto exhaust along the Milky Way. *Astrophysical Journal*, **290**, L25–L28.

Allamandola, L.J., Tielens, A.G.G.M. and Barker, J.R. (1989). Interstellar polycyclic aromatic hydrocarbons: The infrared emission bands, the excitation/emission mechanism, and the astrophysical implications. *Astrophysical Journal*, Suppl. **71**, 733–775.

Anders, E. and Zinner, E. (1993). Interstellar grains in primitive meteorites – Diamond, silicon carbide, and graphite. *Meteoritics*, **28**, 490–514.

Andriesse, C.D. (1978). Platt particles in M17. *Astronomy and Astrophysics*, **66**, 169–173.

Bakes, E.L.O. and Tielens, A.G.G.M. (1994). The photoelectric heating mechanism for very small graphitic grains and polycyclic aromatic hydrocarbons. *Astrophysical Journal*, **427**, 822–838.

Baluteau, J.-P., Cox, P., Cernicharo, J., Pequignot, D., Caux, E., Lim, T., Swinyard, B., White, G., Kessler, M., Prusti, T., Barlow, M., Clegg, P. E., Emery, R. J., Furniss, I., Glencross, W., Gry, C., Joubert, M., Liseau, R., Nisini, B., Saraceno, P., Serra, G., Armand, C., Burgdorf, M., Digiorgio, A., Molinari, S., Price, M., Texier, D., Sidher, S. and Trams, N. (1997). Detection of [O I] 63 μm in absorption toward SgrB2. *Astronomy and Astrophysics*, **322**, L33–L36.

Barlow, M.J. (1999). ISO results on circumstellar envelopes. In T. Le Bertre, A. Lebre and C. Waelkens (eds), *Asymptotic Giant Branch Stars*, Astronomical Society of the Pacific, San Francisco, p. 353.

Barlow, M., Nguyen-Q-Rieu, Truong-Bach, Cernicharo, J., Gonzalez-Alfonso, E., Liu, X.-W., Cox, P., Sylvester, R.J., Clegg, P.E., Griffin, M.J., Swinyard, B.M., Unger, S.J., Baluteau, J.-P., Caux, E., Cohen, M., Cohen, R.J., Emery, R.J., Fischer, J., Furniss, I., Glencross, W.M., Greenhouse, M.A., Gry, C., Joubert, M., Lim, T., Lorenzetti, D., Nisini, B., Omont, A., Orfei, R., Pequignot, D., Saraceno, P., Serra, G., Skinner, C.J., Smith, H.A., Walker, H.J., Armand, C., Burgdorf, M., Ewart, D., di Giorgio, A., Molinari, S., Price, M., Sidher, S., Texier, D. and Trams, N. (1996). The rich far-infrared water vapour spectrum of W Hya. *Astronomy and Astrophysics*, **315**, L241–L244.

Barnard, E.E. (1919). On the dark markings of the sky, with a catalogue of 182 such objects. *Astrophysical Journal*, **49**, 1–24.

Bauschlicher, C.W. (1998). The reaction of polycyclic aromatic hydrocarbon cations with hydrogen atoms: The astrophysical implications. *Astrophysical Journal*, **509**, L125–L127.

Beck, S.C., Lacy, J.H. and Geballe, T.R. (1979). Detection of the 12.28 micron rotational line of molecular hydrogen in the Orion molecular cloud. *Astrophysical Journal*, **234**, L213–L216.

Beckwith, S.V.M., Persson, S.E., Neugebauer, G. and Becklin, E.E. (1978). Observations of the molecular hydrogen emission from the Orion nebula. *Astrophysical Journal*, **223**, 464–470.

Beckwith, S.V.W. and Sargent, A.I. (1996). Circumstellar disks and the search for neighbouring planetary systems. *Nature*, **383**, 139–144.

Bergin, E.A. Melnick, G.J., Stauffer, J.R., Ashby, M.L.N., Chin, G., Erickson, N.R., Goldsmith, P.F., Harwit, M., Howe, J.E., Kleiner, S.C., Koch, D.G., Neufeld, D.A., Patten, B.M., Plume, R., Schieder, R., Snell, R.L., Tolls, V., Wang, Z., Winnewisser, G. and Zhang, Y.F. (2000). Implications of Submillimeter Wave Astronomy Satellite observations for interstellar chemistry and star formation. *Astrophysical Journal*, **539**, L133–L136.

Bennett, C.J., Fixsen, D.J., Hinshaw, J.C., Mather, J.C., Moseley, S.H., Wright, E.L., Eplee, R.E., Jr., Gales, J., Hewagama, T., Isaacman, R.B., Shafer, R.A. and Turpie, K. (1994). Morphology of the interstellar cooling lines detected by COBE. *Astrophysical Journal*, **434**, 587–598.

Bernard, J.P., Boulanger, F., Désert, F.X., Giard, M., Helou, G. and Puget, J.L. (1994). Dust emission of galactic cirrus from DIRBE observations. *Astronomy and Astrophysics*, **291**, L5–L8.

Bernatowicz, T.J., Cowsik, R., Gibbons, P.C., Lodders, K., Fegley, B., Amari, S. and Lewis, R.S. (1996). Constraints on stellar grain formation from presolar graphite in the Murchison meteorite. *Astrophysical Journal*, **472**, 760–782.

Bertoldi, F., Timmermann, R., Rosenthal, D., Drapatz, S. and Wright, C.M. (1999). Detection of HD in the Orion molecular outflow. *Astronomy and Astrophysics*, **346**, 267–277.

Betz, A.L. and Boreiko, R.T. (1989). Reversed far-infrared line emission from OH in Orion. *Astrophysical Journal*, **346**, L101–L104.

Black, J.H. (1993). Energy budgets of diffuse clouds. In R.M. Cutri and W.B. Latter (eds), *The First Symposium on the Infrared Cirrus and Diffuse Interstellar Clouds*, ASP Conf. Series, Vol. 58, Astronomical Society of the Pacific, San Francisco, p. 355.

Black, J.H. and van Dishoeck, E.F. (1987). Fluorescent excitation of interstellar H_2. *Astrophysical Journal*, **322**, 412–449.

Blake, G.A., Keene, J. and Phillips, T.G. (1985). Chlorine in dense interstellar clouds: The abundance of HCl in OMC-1. *Astrophysical Journal*, **295**, 501–506.

Bless, R.C. and Savage, B.D. (1972). Ultraviolet photometry from the orbiting Astronomical Observatory: II. Interstellar extinction. *Astrophysical Journal*, **171**, 293–308.

Bock, J.J., Hristov, V.V., Kawada, M. Matsuhara, H., Matsumoto, T., Matsuura, S., Mauskopf, P.D., Richards, P.L., Tanaka, M. and Lange, A.E. (1993). Observation of forbidden C II 158 micron emission from the diffuse interstellar medium at high galactic latitude. *Astrophysical Journal*, **410**, L115–L118.

Bockelée-Morvan, D., Lis, D.C., Wink, J.E. Despois, D., Crovisier, J., Bachiller, R., Benford, D.J., Biver, N., Colom, P., Davies, J. K., Gérard, E., Germain, B., Houde, M., Mehringer, D., Moreno, R., Paubert, G., Phillips, T.G. and Rauer, H. (2000). New molecules found in comet C/1995 O1 (Hale-Bopp). Investigating the link between cometary and interstellar material. *Astronomy and Astrophysics*, **353**, 1101–1114.

Boland, W. and de Jong, T. (1982). Carbon depletion in turbulent molecular cloud cores. *Astrophysical Journal*, **261**, 110–114.

Boreiko, R.T. and Betz, A.L. (1996a). The $^{12}C/^{13}C$ isotopic ratio in photodissociated gas in M42. *Astrophysical Journal*, **467**, L113–L116.

Boreiko, R.T. and Betz, A.L. (1996b). Heterodyne spectroscopy of the 63 micron [O I] line in M42. *Astrophysical Journal*, **464**, L83–L86.

Boogert, A.C.A., Ehrenfreund, P., Gerakines, P., Tielens, A.G.G.M., Whittet, D.C.B., Schutte, W.A., van Dishoeck, E.F., de Graauw, Th., Decin, L. and Prusti, T. (2000), ISO-SWS observations of interstellar solid $^{13}CO_2$: Heated ice and the galactic $^{12}C/^{13}C$ abundance ratio. *Astronomy and Astrophysics*, **353**, 349–362.

Boogert, A.C.A., Helmich, F.P., van Dishoeck, E.F., Schutte, W.A., Tielens, A.G.G.M. and Whittet, D.C.B. (1998). The gas/solid methane abundance ratio toward deeply embedded protostars. *Astronomy and Astrophysics*, **336**, 352–358.

Boogert, A.C.A., Schutte, W.A., Helmich, F.P., Tielens, A.G.G.M. and Wooden, D.H. (1997). Infrared observations and laboratory simulations of interstellar CH_4 and SO_2. *Astronomy and Astrophysics*, **317**, 929–941.

Boogert, A.C.A., Schutte, W.A., Tielens, A.G.G.M., Whittet, D.C.B., Helmich, F.P., Ehrenfreund, P., Wesselius, P.R., de Graauw, T. and Prusti, T. (1996). Solid methane toward deeply embedded protostars. *Astronomy and Astrophysics*, **315**, L377–L380.

Boonman, A.M.S., van Dishoeck, E.F., Lahuis, F., Wright, C.M. and Doty, S.D. (2000). In *Analytical Spectroscopy with the ISO Satellite*, ESA-SP 456, pp. 67–70.

Boulanger, F. (1999). Dust emission and ISM components. In A.R. Taylor, T.L. Landecker and G. Joncas (eds), *New Perspectives on the Interstellar Medium*, ASP Conf. Series, Vol. 168, Astronomical Society of the Pacific, San Francisco, p. 173.

Boulanger, F., Abergel, A., Bernard, J.-P., Burton, W.B., Desert, F.-X., Hartmann, D., Lagache, G. and Puget, J.-L. (1996). The dust/gas correlation at high galactic latitude. *Astronomy and Astrophysics*, **312**, 256–262.

Boulanger, F. and Pérault, M. (1988). Diffuse infrared emission from the galaxy. I – Solar neighborhood. *Astrophysical Journal*, **330**, 964.

Burges, J.M. and Thomas, R.N. (eds) (1958). Proceedings of the 3rd Symposium on cosmical gas dynamics. *Reviews of Modern Physics*, **30**, 1106–1107.

Buss, R.H., Tielens, A.G.G.M., Cohen, M., Werner, M.W., Bregman, J.D. and Witteborn, F.C. (1993). Infrared spectra of transition objects and the composition and evolution of carbon dust. *Astrophysical Journal*, **415**, 250–257.

Cabrit, S., Bontemps, S., Lagage, P.O., Sauvage, M., Boulanger, F., André, P., Nordh, L., Olofsson, G., Cesarsky, C.J., de Boula, O., Sibille, F. and Siebenmorgen, R. (1999). ISOCAM mapping and spectro-imaging of bipolar outflows. In P. Cox and M.F. Kessler (eds), *The Universe as Seen by ISO*, ESA-SP **427**, 449–452.

Calvet, N., Patino, A., Magris, C. and D'Alessio, P. (1991). Irradiation of accretion disks around young objects: I. Near-infrared CO bands. *Astrophysical Journal*, **380**, 617–630.

Cami, J., Yamamura, I., de Jong, T., Tielens, A.G.G.M., Justtanont, K. and Waters, L.B.F.M. (2000). CO_2 emission in EP Aqr: Probing the extended atmosphere. *Astronomy and Astrophysics*, **360**, 562–574.

Cardelli, J.A., Matthis, J.S., Ebbets, D.C. and Savage, B.D. (1993). Abundance of interstellar carbon toward Zeta Ophiuchi. *Astrophysical Journal*, **402**, L17–L20.

Carruthers, G.R. (1970). Rocket observation of interstellar molecular hydrogen. *Astrophysical Journal*, **161**, L81–L85.

Caselli, P. Hasegawa, T.I. and Herbst, E. (1993). Chemical differentiation between star-forming regions: The Orion hot core and compact ridge. *Astrophysical Journal*, **408**, 548–558.

Ceccarelli, C., Castets. A., Loinard, L., Caux, E. and Tielens, A.G.G.M. (1998a). Detection of doubly deuterated formaldehyde towards the low-luminosity protostar IRAS 16293−2422. *Astronomy and Astrophysics*, **338**, L43–L46.

Ceccarelli, C., Caux, E., White, G.J., Molinari, S., Furniss, I., Liseau, R., Nisini, B., Saraceno, P., Spinoglio, L. and Wolfire, M. (1998b). The far infrared line spectrum of the protostar IRAS 16293−2422. *Astronomy and Astrophysics*, **331**, 372–382.

Ceccarelli, C., Hollenbach, D.J. and Tielens, A.G.G.M. (1996). Far-infrared line emission from collapsing protostellar envelopes. *Astrophysical Journal*, **471**, 400–426.

Cernicharo, J. (2000). ISO's view of the molecular content of evolved stars. In Y.C. Minh and E.F. van Dishoeck (eds), *Astrochemistry: From Molecular Clouds to Planetary Systems*, IAU Symposium 197, Astronomical Society of the Pacific, San Francisco, pp. 375–390.

Cernicharo, J., Barlow, M.J., González-Alfonso, E., Cox, P., Clegg, P.E., Nguyen-Q-Rieu, Omont, A., Guelin, M., Liu, X.-W., Sylvester, R.J., Lim, T., Griffin, M.J., Swinyard, B.M., Unger, S.J., Ade, P.A.R., Baluteau, J.-P., Caux, E., Cohen, M., Emery, R.J., Fischer, J., Furniss, I., Glencross, W.M., Greenhouse, M.A., Gry, C., Joubert, M., Lorenzetti, D., Nisini, B., Orfei, R., Pequignot, D., Saraceno, P., Serra, G., Skinner, C.J., Smith, H.A., Towlson, W.A., Walker, H.J., Armand, C., Burgdorf, M., Ewart, D., di Giorgio, A., Molinari, S., Price, M., Sidher, S., Texier, D. and Trams, N. (1996). The ISO/LWS far infrared spectrum of IRC + 10 216. *Astronomy and Astrophysics*, **315**, L201–L204.

Cernicharo, J., González-Alfonso, E., Alcolea, J., Bachiller, R. and John, D. (1994). Widespread water vapor emission in Orion. *Astrophysical Journal*, **432**, L59–L62.

Cernicharo, J., González-Alfonso, E., Sempere, M.J., Leeks, S.J., van Dishoeck, E.F., Wright, C.M., Lim, T., Cox, P. and Perez-Martinez, S. (1999). The water vapour abundance and its spatial distribution in Orion and Sgr B2. In P. Cox and M.F. Kessler (eds), *The Universe as Seen by ISO*, ESA-SP 427, 565–573.

Cernicharo, J., Lim, T., Cox, P., González-Alfonso, E., Caux, E., Swinyard, B.M., Martín-Pintado, J., Baluteau, J. P. and Clegg, P. (1997). Widespread water vapour absorption in Sgr B2. *Astronomy and Astrophysics*, **323**, L25–L28.

Cesarsky, D., Cox, P., Pineau des Forêts, G., van Dishoeck, E.F., Boulanger, F. and Wright, C.M. (1999). ISOCAM spectro-imaging of the H_2 rotational lines in the supernova remnant IC 443. *Astronomy and Astrophysics*, **348**, 945–949.

Cesarsky, D., Lequeux, J., Ryter, C. and Gérin, M. (2000). ISO observations of the reflection nebula Ced 201: Evolution of carbonaceous dust. *Astronomy and Astrophysics*, **354**, L87–L90.

Charnley, S.B. (1997). Sulfuretted molecules in hot cores. *Astrophysical Journal*, **481**, 396–405.

Charnley, S.B., Tielens, A.G.G.M. and Millar, T.J. (1992). On the molecular complexity of the hot cores in Orion A: Grain surface chemistry as "The last refuge of the scoundrel". *Astrophysical Journal*, **399**, L71–L74.

Cheung, A.C., Rank, D.M., Townes, C.H., Thornton, D.D. and Welch, W.J. (1968). Detection of NH_3 molecules in the interstellar medium by their microwave emission. *Physical Review Letters*, **21**, 1701–1705.

Chiang, E.I. and Goldreich, P. (1997). Spectral energy distributions of T Tauri stars with passive circumstellar disks. *Astrophysical Journal*, **490**, 368–376.

Chiang, E.I., Joung, M.K., Creech-Eakman, M.J., Qi, C., Kessler, J.E., Blake, G.A. and van Dishoeck, E.F. (2001). Spectral energy distributions of T Tauri and Herbig Ae disks: Grain mineralogy, parameter dependences, and comparison with ISO LWS observations. *Astrophysical Journal*, **547**, 1077–1089.

Chiar, J.E., Gerakines, P.A., Whittet, D.C.B., Pendleton, Y.L., Tielens, A.G.G.M., Adamson, A.J. and Boogert, A.C.A. (1998). Processing of icy mantles in protostellar envelopes. *Astrophysical Journal*, **498**, 716–727.

Chiar, J.E., Tielens, A.G.G.M., Whittet, D.C.B., Schutte, W.A., Boogert, A.C.A., Lutz, D., van Dishoeck, E.F. and Bernstein, M.P. (2000). The composition and distribution of dust along the line of sight toward the Galactic Center. *Astrophysical Journal*, **537**, 749–762.

Cox, P., Boulanger, F. and Jones, A. (2000). Dust in the interstellar medium, in Les Houches summer school on Infrared Astronomy.

Cox, P., González-Alfonso, E., Barlow, M.J., Liu, X.-W., Lim, T., Swinyard, B.M., Cernicharo, J., Omont, A., Caux, E., Gry, C., Griffin, M.J., Baluteau, J.-P., Clegg, P.E., Sidher, S., Pequignot, D., Nguyen-Q-Rieu, King, K.J., Ade, P.A.R., Towlson, W.A., Emery, R.J., Furniss, I., Joubert, M., Skinner, C.J., Cohen, M., Armand, C., Burgdorf, M., Eward, D., di Giorgio, A., Molinari, S., Texier, D., Trams, N., Unger, S.J., Glencross, W.M., Lorenzetti, D., Nisini, B., Orfei, R., Saraceno, P. and Serra, G. (1996). The ISO/LWS spectrum of the Egg Nebula, AFGL 2688. *Astronomy and Astrophysics*, **315**, L265–L268.

Cox, P. and Kessler, M. (eds) (1999). *The Universe as Seen by ISO*, ESA-SP 427, ESTEC, Noordwijk.

Crovisier, J., Leech, K., Bockelée-Morvan, D., Brooke, T. Y., Hanner, M. S., Altieri, B., Keller, H.U. and Lellouch, E. (1997). The spectrum of Comet Hale–Bopp (C/1995 O1) observed with the Infrared Space Observatory at 2.9 AU from the Sun. *Science*, **275**, 1904–1907.

Dartois, E., d'Hendecourt, L., Boulanger, F., Jourdain de Muizon, M., Breitfellner, M., Puget, J.-L. and Habing, H.J. (1998). Molecular gas phase counterparts to solid state grain mantles features: Implication for gas/grain chemistry. *Astronomy and Astrophysics*, **331**, 651–660.

de Graauw, Th., Whittet, D.C.B., Gerakines, P.A., Bauer, O.H., Beintema, D.A., Boogert, A.C.A., Boxhoorn, D.R., Chiar, J.E., Ehrenfreund, P., Feuchtgruber, H., Helmich, F.P., Heras, A.M., Huygen, R., Kester, D.J.M., Kunze, D., Lahuis, F., Leech, K.J., Lutz, D., Morris, P. W., Prusti, T., Roelfsema, P.R., Salama, A., Schaeidt, S.G., Schutte, W.A., Spoon, H.W.W., Tielens, A.G.G.M., Valentijn, E.A., Vandenbusshe, B., van Dishoeck, E.F., Wesselius, P.R., Wieprecht, E. and Wright, C.M. (1996). SWS observations of solid CO_2 in molecular clouds. *Astronomy and Astrophysics*, **315**, L345–L348.

Demyk, K., Dartois, E., d'Hendecourt, L., Jourdain de Muizon, M., Heras, A.M. and Breitfellner, M. (1998). Laboratory identification of the 4.62 µm solid state absorption band in the ISO-SWS spectrum of RAFGL 7009S. *Astronomy and Astrophysics*, **339**, 553–560.

Demyk, K., Jones, A.P., Dartois, E., Cox, P. and D'Hendecourt, L. (1999). The chemical composition of the silicate dust around RAFGL7009S and IRAS 19110+1045, *Astronomy and Astrophysics*, **349**, 267–275.

Désert, F.-X., Boulanger, F. and Puget, J.L. (1990). Interstellar dust models for extinction and emission. *Astronomy and Astrophysics*, **237**, 215–236.

d'Hendecourt, L.B., Joblin, C. and Jones, A. (eds) (1999a). *Solid Interstellar Matter: The ISO Revolution*, EDP Sciences, Springer, Berlin.

d'Hendecourt, L.B. and Jourdain de Muizon, M. (1989). The discovery of interstellar carbon dioxide. *Astronomy and Astrophysics*, **223**, L5–L8.

d'Hendecourt, L.B., Jourdain de Muizon, M., Dartois, E. Breitfellner, M., Ehrenfreund, P., Benit, J., Boulanger, F., Puget, J.L. and Habing, H.J. (1996). ISO-SWS observations of solid state features towards RAFGL 7009S. *Astronomy and Astrophysics*, **315**, L365–L368.

d'Hendecourt, L., Jourdain de Muizon, M., Dartois, E., Demyk, K., Ehrenfreund, P., Heras, A.M. (1999b). SWS observations of protostars: The relevance of laboratory ices simulations. In P. Cox and M.F. Kessler (eds), *The Universe as seen by ISO*, ESA-SP 427, 589–597.

d'Hendecourt, L.B. and Léger, A. (1987). Effect of photoionization of PAH molecules on the heating of H I interstellar gas. *Astronomy and Astrophysics*, **180**, L9–L12.

Donn, B.D. (1968). Polycyclic hydrocarbons, Platt particles, and interstellar extinction. *Astrophysical Journal*, **152**, L129–L132.

Douglas, A.E. and Herzberg, G. (1941). Note on CH^+ in interstellar space and in the laboratory. *Astrophysical Journal*, **94**, 381.

Draine, B.T. and Anderson, N. (1985). Temperature fluctuations and infrared emission from interstellar grains. *Astrophysical Journal*, **292**, 494–499.

Draine, B.T. and Bertoldi, F. (1999). Heating the gas in photodissociation regions. In P. Cox and M.F. Kessler (eds), *The Universe as seen by ISO*, ESA-SP 427, 553–559.

Draine, B.T. and McKee, C.F. (1993). Theory of interstellar shocks. *Annual Review of Astronomy and Astrophysics*, **31**, 373–432.

Duley, W.W. and Williams, D.A. (1981). The infrared spectrum of interstellar dust: Surface functional groups on carbon. *Monthly Notices of the Royal Astronomical Society*, **196**, 269–274.

Duley, W.W. and Williams, D.A. (1983). A 3.4 micron absorption band in amorphous carbon: Implications for interstellar dust. *Monthly Notices of the Royal Astronomical Society*, **205**, 67P–70P.

Dutrey, A., Guilloteau, S., Duvert, G., Prato, L., Simon, M., Schuster, K. and Menard, F. (1996). Dust and gas distribution around T Tauri stars in Taurus–Auriga: I. Interferometric 2.7 mm continuum and $^{13}CO\ J = 1-0$ observations. *Astronomy and Astrophysics*, **309**, 493–504.

Eddington, A.S. (1926). Diffuse matter in interstellar space. *Proceedings of the Royal Society A*, **111**, 424.

Eddy, J.A. (1972). *Journal of the History of Astronomy*, **3**, 165.

Ehrenfreund, P., Breukers, R., d'Hendecourt, L.B. and Greenberg, J.M. (1992). On the possibility of detecting solid O_2 in interstellar grain mantles. *Astronomy and Astrophysics*, **260**, 431–436.

Ehrenfreund, P. and Charnley, S.B. (2000). Organic molecules in the interstellar medium, comets, and meteorites: A voyage from dark clouds to the early Earth. *Annual Review of Astronomy and Astrophysics*, **38**, 427–483.

Ehrenfreund, P., Dartois E., Demyk K. and d'Hendecourt L. (1998). Ice segregation toward massive protostars. *Astronomy and Astrophysics*, **339**, L17–L20.

Ehrenfreund, P., Kerkhof, O., Schutte, W.A., Boogert, A.C.A., Gerakines, P.A., Dartois, E., d'Hendecourt, L., Tielens, A.G.G.M., van Dishoeck, E.F. and Whittet, D.C.B. (1999). Laboratory studies of thermally processed H_2O–CH_3OH–CO_2 ice mixtures and their astrophysical implication. *Astronomy and Astrophysics*, **350**, 240–253.

Ehrenfreund, P. and Schutte, W.A. (2000). Infrared observations of interstellar ices. In Y.C. Minh and E.F. van Dishoeck (eds), *Astrochemistry: From Molecular Clouds to Planetary Systems*, IAU Symposium 197, Astronomy Society of the Pacific, San Francisco, pp. 135–146.

Elitzur, M., Hollenbach, D.J. and McKee, C.F. (1992). Planar H_2O masers in star-forming regions. *Astrophysical Journal*, **394**, 221–227.

Evans, N.J., Lacy, J.H. and Carr, J.S. (1991), Infrared molecular spectroscopy toward the Orion IRc2 and IRc7 sources: A new probe of physical conditions and abundances in molecular clouds. *Astrophysical Journal*, **383**, 674–692.

Ewen, H.I. and Purcell, E.M. (1951). Radiation from galatic hydrogen at 1,420 Mc./sec. *Nature*, **168**, 356.

Field, G.B., Goldsmith, D.W. and Habing, H.J. (1969). Cosmic-ray heating of the interstellar gas. *Astrophysical Journal*, **155**, L149–L154.

Flower, D.R. and Pineau des Forêts, G. (1999). H_2 emission from shocks in molecular outflows: The significance of departures from a stationary state. *Monthly Notices of the Royal Astronomical Society*, **308**, 271–280.

Fuente, A., Martín-Pintado, J., Rodríguez-Fernandez, N.J., Rodrí-Franco, A., de Vicente, P. and Kunze, D. (1999). Infrared Space Observatory observations toward the reflection nebula NGC 7023: A nonequilibrium *ortho*-to-*para*-H$_2$ ratio. *Astrophysical Journal*, **518**, L45–L48.

Gaustad, J.E. (1971). In B.T. Lynds (ed.), *Dark Nebulae, Globules, and Protostars*, University of Arizona Press, Tucson, AZ, p. 91.

Geballe, T.R., Tielens, A.G.G.M., Allamandola, L.J., Moorhouse, A. and Brand, P.W.J.L. (1989). Spatial variations of the 3 micron emission features within UV-excited nebulae: Photochemical evolution of interstellar polycyclic aromatic hydrocarbons. *Astrophysical Journal*, **341**, 278–287.

Genzel, R. (1992). In D. Pfenniger and P. Bartholdi (eds), *The Galactic Interstellar Medium*, Saas Fee Advanced Course 21, Springer, Berlin, p. 275.

Genzel, R. and Stutzki, J. (1989). The Orion Molecular Cloud and star-forming region *Annual Review of Astronomy and Astrophysics*, **27**, 41–85.

Gerakines, P.A., Whittet, D.C.B., Ehrenfreund, P., Boogert, A.C.A., Tielens, A.G.G.M., Schutte, W.A., Chiar, J.E., van Dishoeck, E.F., Prusti, T., Helmich, F.P. and de Graauw, Th. (1999). Observations of solid carbon dioxide in molecular clouds with the Infrared Space Observatory. *Astrophysical Journal*, **522**, 357; erratum, **526**, 1062.

Giannini, T., Lorenzetti, D., Tommasi, E., Nisini, B., Benedettini, M., Pezzuto, S., Strafella, F., Barlow, M., Clegg, P.E., Cohen, M., di Giorgio, A.M., Liseau, R., Molinari, S., Palla, F., Saraceno, P., Smith, H.A., Spinoglio, L. and White, G.J. (1999). ISO-LWS observations of Herbig Ae/Be stars: II. Molecular lines, *Astronomy and Astrophysics*, **346**, 617–625.

Giard, M., Lamarre, J.M., Pajot, F. and Serra, G. (1994). The large scale distribution of PAHs in the Galaxy. *Astronomy and Astrophysics*, **286**, 203–210.

Gibb, E., Whittet, D.C.B., Gerakines, P.A., Boogert, A.C.A., Chiar, J.E., Ehrenfreund, P., Gerakines, P.A., Keane, J.V., Tielens, A.G.G.M., van Dishoeck, E.F. and, Kerkhof, O. (2000). An inventory of interstellar Ices toward the embedded Protostar W33A. *Astrophysical Journal*, **536**, 347–356.

Gillett, F.C. and Forrest, W.J. (1973). Spectra of the Becklin–Neugebauer point source and the Kleinmann–Low Nebula from 2.8 to 13.5 microns. *Astrophysical Journal*, **179**, 483–491.

Gillett, F.C., Jones, T.W., Merrill, K.M. and Stein, W.A. (1975). Anisotropy of constituents of interstellar grains. *Astronomy and Astrophysics*, **45**, 77–81.

Gilra, D.P. (1972). In A.D. Code (ed.), *The Scientific Results from the Orbiting Astronomical Observatory OAO-2*, NASA SP-310, 297.

Goldsmith, P.F., Melnick, G.J., Bergin, E.A., Howe, J.E., Snell, R.L., Neufeld, D.A., Harwit, M., Patten, B.M., Kleiner, S.C., Plume, R., Stauffer, J.R., Tolls, V., Wang, Z., Zhang, Y.F., Erickson, N.R., Koch, D.G., Schieder, R., Winnewisser, G. and Chin, G. (2000). O$_2$ in interstellar molecular clouds. *Astrophysical Journal*, **539**, L123–L126.

Greenberg, J.M. (1982). Dust in dense clouds: One stage in a cycle. In J.E. Beckman and J.P. Phillips (eds), *Submillimeter Wave Astronomy*, Cambridge University Press, pp. 261–306.

Greenberg, J.M. and Hong, S.S. (1974). The chemical composition and distribution of interstellar grains. In F.J. Kerr and S.C. Simonson (eds), *Galactic Radio Astronomy*, Reidel, Dordrecht, pp. 155–177.

Greenberg, J.M. and Muñoz-Caro, G. (2000). Organics in space: From interstellar dust to comets. In Y.C. Minh and E.F. van Dishoeck (eds), *Astrochemistry: From Molecular Clouds to Planetary Systems*, IAU Symposium 197, Astronomical Society of the Pacific, San Francisco, pp. 331–342.

Gry, C.J., Lequeux, J. and Boulanger, F. (1992). Fine structure lines of C$^+$ and N$^+$ in the Galaxy. *Astronomy and Astrophysics*, **266**, 457–462.

Haas, M.R., Davidson, J.A. and Erickson, E.F. (eds) (1995). *Airborne Astronomy Symposium on the Galactic Ecosystem: From Gas to Stars to Dust*, ASP Conf. Series Vol. 73, Astronomical Society of the Pacific, San Francisco.

Haas, M.R. and Glassgold, A.E. (1993). Detection of [O I] and [Si II] far-infrared fine-structure emission from Alpha Orionis. *Astrophysical Journal*, **410**, L111–L114.

Haas, M.R., Hollenbach, D.J. and Erickson, E.F. (1991). Observations of forbidden [Si II] (35 microns) and [S I] (25 microns) in Orion: Evidence of a wind shock near IRc2. *Astrophysical Journal*, **374**, 555–563.

Habing, H.J. (1968). The interstellar radiation density between 912 Å and 2400 Å. *Bulletin of the Astromical Society of the Netherlands*, **19**, 421.

Habing, H.J. (1989). The interstellar medium. In I. Appenzeller, H.J. Habing and P. Léna (eds), *Evolution of Galaxies and Astronomical Observations*. Proceedings of the First EADN Astrophysics School, pp. 181–242.

Hall, J.S. (1949). Observations of the polarized light from stars. *Science*, **109**, 166–167.

Hartmann, J. (1904). Investigations on the spectrum and orbit of Delta Orionis. *Astrophysical Journal*, **19**, 268–286.

Hartmann, D. and Burton, W.B. (1997). *Atlas of Galactic Neutral Hydrogen*, Cambridge University Press.

Hartquist, T.W. and Williams, D.A. (eds), (1998). *The Molecular Astrophysics of Stars and Galaxies*, Oxford University Press.

Harwit, M. (1999). Neugebauer, Martz and Leighton's observations of extremely cool stars. *Astrophysical Journal*, **525**, 1063–1064.

Harwit, M., McNutt, D.P., Shivanandan, K. and Zajac, B.J. (1966). Results of the first infrared astronomical rocket flight. *Astronomical Journal*, **71**, 1026–1029.

Harwit, M., Neufeld, D.A., Melnick, G.J. and Kaufman, M.J. (1998). Thermal water vapor emission from shocked regions in Orion. *Astrophysical Journal*, **497**, L105–L108.

Heger, M.L. (1919). The occurrence of stationary D lines of sodium in the spectroscopic binaries Beta Scorpii and Delta Orionis. *Lick Observatory Bulletin*, **326**, 59–63.

Heger, M.L. (1922). Further study of the sodium lines in class B stars, the spectra of certain class B stars in the regions 5630 Å –6680 Å and 3280 Å –3380 Å. *Lick Observatory Bulletin*, **337**, 141–148.

Heiles, C. (1994). On the origin of the diffuse C$^+$ 158 micron line emission. *Astrophysical Journal*, **436**, 720–727.

Helmich, F.P., van Dishoeck, E.F., Black, J.H., de Graauw, T., Beintema, D.A., Heras, A.M., Lahuis, F., Morris, P. W. and Valentijn, E.A. (1996). Detection of hot, abundant water toward AFGL 2591. *Astronomy and Astrophysics*, **315**, L173.

Heras, A.M., Leech, K., Trams, N.R. and Perry, M. (eds) (1997). *First ISO Workshop on Analytical Spectroscopy*, ESA-SP 419, ESTEC, Noordwijk.

Hildebrand, R.H., Dotson, J.L., Dowell, C.D., Platt, S.R., Schleuning, David, Davidson, J. A. and Novak, G. (1995). Far-infrared polarimetry. In M. Haas. J.A. Davidson and E.F. Erickson (eds), *Airborne Astronomy Symposium*, ASP Conf. series Vol. 73, Astronomical Society of the Pacific, San Francisco, p. 97.

Hildebrand, R.H., Dragovan, M. and Novak, G. (1984). Detection of submillimeter polarization in the Orion Nebula. *Astrophysical Journal*, **284**, L51–L54.

Hiltner, W.A. (1949). Polarization of light from distant stars by interstellar medium. *Science*, **109**, 165.

Hollenbach, D.J. (1997). The physics of molecular shocks in YSO outflows. in B. Reipurth and C. Bertout (eds), *Herbig–Haro Objects and the Birth of Low-mass Stars*, IAU Symposium 182, Kluwer, Dordrecht, pp. 181–198.

Hollenbach, D.J., Chernoff, D.F. and McKee, C.F. (1989). Infrared diagnostics of interstellar shocks. In B.H. Kaldeich (ed.), *Infrared Spectroscopy in Astronomy*, 22nd ESLAB Symposium, ESA-SP 290, ESTEC, Noordwijk, p. 245–258.

Hollenbach, D.J. and Tielens, A.G.G.M. (1997). Dense photodissociation regions (PDRs). *Annual Review of Astronomy and Astrophysics*, **35**, 179–216.

Hollenbach, D.J. and Tielens, A.G.G.M. (1999). Photodissociation regions in the interstellar medium of galaxies. *Reviews of Modern Physics*, **71**, 173–230.

Hony, S., van Kerckhoven C., Peeters, E., Tielens, A.G.G.M., Hudgins, D.M. and Allamandola, L.J. (2001a). The CH out-of-plane bending modes of PAH molecules in astrophysical environments. *Astronomy and Astrophysics*, **370**, 1030–1043.

Hoyle, F. and Wrickramasinghe, N.C. (1962). On graphite particles as interstellar grains. *Monthly Notices of the Royal Astronomical Society*, **124**, 417–433.

Hrivnak, B.J. (1995). Optical spectroscopy of carbon-rich proto-planetary nebulae, *Astrophysical Journal*, **438**, 341–349.

Hudgins, D.M. and Allamandola, L.J. (1999). Interstellar PAH emission in the 11–14 Micron region: New insights from laboratory data and a tracer of ionized PAHs. *Astrophysical Journal*, **516**, L41–L44.

Hudgins, D.M., Bauschlicher, C.W., Jr. and Allamandola, L.J. (2001). Closed-shell polycyclic aromatic hydrocarbon cations: A new category of interstellar polycyclic aromatic hydrocarbons. *Spectrochimica Acta Part A: Molecular and Biomolecular Spectroscopy*, **57**, 907–930.

Irvine, W.M., Schloerb, F.P., Crovisier, J., Fegley, B., Jr. and Mumma, M.J. (2000). Comets: A link between interstellar and nebular chemistry. In V. Mannings, A.P. Boss and S.S. Russell (eds) *Protostars and Planets IV*, University of Arizona, Tucson, AZ, p. 1159.

Jaffe, D.T., Harris, A.I. and Genzel, R. (1987). Warm dense gas in luminous protostellar regions: A submillimeter and far-infrared CO line study. *Astrophysical Journal*, **316**, 231–242.

Jäger, C., Molster, F., Dorschner, J., Henning, Th., Mutschke, H. and Waters, L.B.F.M. (1998). Steps toward interstellar silicate mineralogy: IV. The crystalline revolution. *Astronomy and Astrophysics*, **339**, 904–916.

Jär, C., Mutschke, H. Begemann, B., Dorschner, J. and Henning, T. (1994). Steps toward interstellar silicate mineralogy: 1. Laboratory results of a silicate glass of mean cosmic composition. *Astronomy and Astrophysics*, **292**, 641–655.

Joblin, C., Bregman J. and Tielens, A.G.G.M. (2000). in preparation.

Joblin, C., Tielens, A.G.G.M., Allamandola, L.J. and Geballe, T.R. (1996). Spatial variation of the 3.29 and 3.40 micron emission bands within reflection nebulae and the photochemical evolution of methylated polycyclic aromatic hydrocarbons. *Astrophysical Journal*, **458**, 610–620.

Jones, A.P., Tielens, A.G.G.M. and Hollenbach, D.J. (1996). Grain shattering in shocks: The interstellar grain size distribution. *Astrophysical Journal*, **469**, 740–764.

Jourdain de Muizon, M., Cox P. and Lequeux, J. (1990). A survey of infrared features in H II regions, planetary nebulae and proto-planetary nebulae from the IRAS-LRS data base. *Astronomy and Astrophysics*, Suppl. **83**, 337–355.

Justtanont, K., de Jong, T., Helmich, F.P., Waters, L.B.F.M., de Graauw, T., Loup, C., Izumiura, H., Yamamura, I., Beintema, D.A., Lahuis, F., Roelfsema, P.R. and Valentijn, E.A. (1996). The ISO-SWS spectrum of NML Cyg. *Astronomy and Astrophysics*, **315**, L217–L220.

Justtanont, K., Feuchtgruber, H., de Jong, T., Cami, J., Waters, L.B.F.M., Yamamura, I. and Onaka, T. (1998). Discovery of CO_2 emission in AGB stars with the 13 μm dust feature. *Astronomy and Astrophysics*, **330**, L17–L20.

Justtanont, K., Tielens, A.G.G.M., Skinner, C.J. and Haas, M.R. (1997). Observations of high rotational CO lines in post-asymptotic giant branch stars and planetary nebulae. *Astrophysical Journal*, **476**, 319–326.

Justtanont, K., Tielens, A.G.G.M., de Jong, T., Cami, J., Waters, L.B.F.M. and Yamamura, I. (1999). Atomic fine-structure lines in the ISO-SWS spectra of the supergiants Alpha Orionis and Alpha Scorpii. *Astronomy and Astrophysics*, **345**, 605–610.

Kamp, I. and Bertoldi, F. (2000). CO in the circumstellar disks of Vega and β Pictoris. *Astronomy and Astrophysics*, **353**, 276–286.

Kamijo, F. (1963). A theoretical study on the long period variable star: III. Formation of solid or liquid particles in the circumstellar envelope. *Publications of the Astronomical Society of Japan*, **15**, 440–448.

Kaufman, M.J. and Neufeld, D.A. (1996). Far-infrared water emission from magnetohydrodynamic shock waves. *Astrophysical Journal*, **456**, 611–630.

Kaufman, M.J., Wolfire, M.G., Hollenbach, D.J. and Luhman, M.L. (1999). Far-infrared and submillimeter emission from galactic and extragalactic photodissociation regions. *Astrophysical Journal*, **527**, 795–813.

Keane, J.V., Tielens, A.G.G.M., Boogert, A.C.A., Schutte, W.A. and Whittet, D.C.B. (2001). Ice absorption features in the 5–8 μm region toward embedded protostars. *Astronomy and Astrophysics*, in press.

Keene, J., Blake, G.A., Phillips, T.G., Huggins, P.J. and Beichman, C.A. (1985). The abundance of atomic carbon near the ionization fronts in M17 and S140. *Astrophysical Journal*, **299**, 967–980.

Keene, J., Lis, D.C., Phillips, T.G., Schilke, P., Werner, M.W. and Zmuidzinas, J. (1999). Atomic oxygen abundance from the dense ISM. In P. Cox and M. Kessler (eds) *The Universe as Seen by ISO*, ESA-SP 427, 687–689.

Keene, J., Schilke, P., Kooi, J., Lis, D.C., Mehringer, D.M. and Phillips, T.G. (1998). Detection of the $^3P_2$–$^3P_1$ submillimeter transition of ^{13}C I in the interstellar medium: Implication for chemical fractionation. *Astrophysical Journal*, **494**, L107–L110.

Kleinmann, D.E. and Low, F.J. (1967). Discovery of an infrared nebula in Orion. *Astrophysical Journal*, **149**, L1–L4.

Koerner, D.W. and Sargent, A.I. (1995). Imaging the small-scale circumstellar gas around T Tauri stars. *Astronomical Jouranl*, **109**, 2138–2145.

Koike, C., Shibai, H. and Tuchiyama, A. (1993). Extinction of olivine and pyroxene in the mid infrared and far infrared. *Monthly Notices of the Royal Astronomical Society*, **264**, 654–658.

Koike, C., Tsuchiyama, A., Shibai, H., Suto, H., Tanabé, T., Chihara, H., Sogawa, H., Mouri, H. and Okada, K. (2000). Absorption spectra of Mg-rich Mg–Fe and Ca pyroxenes in the mid- and far-infrared region. *Astronomy and Astrophysics*, **363**, 1115–1122.

Kwok, S., Volk, K.M. and Hrivnak, B.J. (1989). A 21 micron emission feature in four proto-planetary nebulae. *Astrophysical Journal*, **345**, L51–L54.

Lacy, J.H., Carr, J.S., Evans, N.J., Baas, F., Achtermann, J.M. and Arens, J.F. (1991). Discovery of interstellar methane: Observations of gaseous and solid CH_4 absorption toward young stars in molecular clouds. *Astrophysical Journal*, **376**, 556–560.

Lacy, J.H., Evans, N.J., Achtermann, J.M., Bruce, D.E., Arens, J.F. and Carr, J.S. (1989). Discovery of interstellar acetylene. *Astrophysical Journal*, **342**, L43–L46.

Lacy, J.H., Faraji, H., Sandford, S.A. and Allamandola, L.J. (1998). Unraveling the 10 micron "silicate" feature of protostars: The detection of frozen interstellar ammonia. *Astrophysical Journal*, **501**, L105–L108.

Lahuis, F. and van Dishoeck, E.F. (2000). ISO-SWS spectroscopy of gas-phase C_2H_2 and HCN toward massive young stellar objects. *Astronomy and Astrophysics*, **355**, 699–712.

Langer, W.D., van Dishoeck, E.F., Bergin, E.A., Blake, G.A., Tielens, A.G.G.M., Velusamy, T. and Whittet, D.C.B. (2000). Chemical evolution of protostellar matter. In: V. Mannings, A.P. Boss and S.S. Russell (eds) *Protostars and Planets IV*, University of Arizona, Tucson, AZ, pp. 29–57.

Langhoff, S.R. (1996). Theoretical infrared spectra for polycyclic aromatic hydrocarbon neutrals, cations, and anions. *Journal of Physical Chemistry*, **100**, 2819.

Le Bourlot, J., Pineau des Forêts, G. and Roueff, E. (1995). Complex dynamical behaviour in interstellar clouds. *Astronomy and Astrophysics*, **297**, 251–260.

Léger, A. and Puget, J.L. (1984). Identification of the "unidentified" IR emission features of interstellar dust? *Astronomy and Astrophysics*, **137**, L5–L8.

Li, A. and Greenberg, J.M. (1997). A unified model of interstellar dust. *Astronomy and Astrophysics*, **323**, 566–584.

Linsky, J.L. (1996). GHRS observations of the LISM. *Space Science Reviews*, **78**, 157–164.

Liseau, R., Ceccarelli, C., Larsson, B., Nisini, B., White, G.J., Ade, P., Armand, C., Burgdorf, M., Caux, E., Cerulli, R., Church, S., Clegg, P.E., Digorgio, A., Furniss, I., Giannini, T., Glencross, W., Gry, C., King, K., Lim, T., Lorenzetti, D., Molinari, S., Naylor, D., Orfei, R., Saraceno, P., Sidher, S., Smith, H., Spinoglio, L., Swinyard, B., Texier, D., Tommasi, E., Trams, N. and Unger, S. (1996). Thermal H_2O emission from the Herbig–Haro flow HH 54. *Astronomy and Astrophysics*, **315**, L181–L184.

Liu, X.-W., Barlow, M.J., Nguyen-Q-Rieu, Truong-Bach, Cox, P., Pequignot, D., Clegg, P.E., Swinyard, B.M., Griffin, M.J., Baluteau, J.P., Lim, T., Skinner, C.J., Smith, H.A., Ade, P.A.R., Furniss, I., Towlson, W.A., Unger, S.J., King, K.J., Davis, G.R., Cohen, M., Emery, R.J., Fischer, J., Glencross, W.M., Caux, E., Greenhouse, M.A., Gry, C., Joubert, M., Lorenzetti, D., Nisini, B., Omont, A., Orfei, R., Saraceno, P., Serra, G., Walker, H.J., Armand, C., Burgdorf, M., diGiorgio, A., Molinari, S., Price, M., Texier, D., Sidher, S. and Trams, N. (1996). The ISO LWS grating spectrum of NGC 7027. *Astronomy and Astrophysics*, **315**, L257–L260.

Lorenzetti, D., Tommasi, E., Giannini, T., Nisini, B., Benedettini, M., Pezzuto, S., Strafella, F., Barlow, M., Clegg, P.E., Cohen, M., di Giorgio, A.M., Liseau, R., Molinari, S., Palla, F., Saraceno, P., Smith, H.A., Spinoglio, L. and White, G.J. (1999). ISO-LWS observations of Herbig Ae/Be stars: I. Fine-structure lines. *Astronomy and Astrophysics*, **346**, 604–616.

Malfait, K., Waelkens, C., Bouwman, J., de Koter, A. and Waters, L.B.F.M. (1999). The ISO spectrum of the young star HD 142527. *Astronomy and Astrophysics*, **345**, 181–186.

Malfait, K., Waelkens, C., Waters, L.B.F.M., Vandenbussche, B., Huygen, E. and de Graauw, M.S. (1998). The spectrum of the young star HD 100546 observed with the Infrared Space Observatory. *Astronomy and Astrophysics*, **332**, L25–L28.

Martin, P.G. (1989). Open panel discussion on interstellar grains models. in L.J. Allamandola and A.G.G.M. Tielens (eds), *Interstellar Dust*, Reidel, Dordrecht, p. 375.

Martin, P.G. and Rogers, C. (1987). Carbon grains in the envelope of IRC +10216. *Astrophysical Journal*, **322**, 374–392.

Mathis, J.S., Rumpl, W. and Nordsieck, K.H. (1977). The size distribution of interstellar grains. *Astrophysical Journal*, **217**, 425–433.

Mattila, K., Lemke, D., Haikala, L.K., Laureijs, R.J., Leger, A., Lehtinen, K., Leinert, C. and Mezger, P.G. (1996). Spectrophotometry of UIR bands in the diffuse emission of the galactic disk. *Astronomy and Astrophysics*, **315**, L353–L356.

McCaughrean, M.J. and O'Dell, R.C. (1996). Direct imaging of circumstellar disks in the Orion Nebula. *Astronomical Journal*, **111**, 1977–1986.

McKee, C.F. and Ostriker, J.P. (1977). A theory of the interstellar medium: Three components regulated by supernova explosions in an inhomogeneous substrate. *Astrophysical Journal*, **218**, 148–169.

McKellar, A. (1940). Evidence for the molecular origin of some hitherto unidentified interstellar lines. *Publications of the Astronomical Society of the Pacific*, **52**, 187–192; Wave lengths of the CH band lines, *Publications of the Astronomical Society of the Pacific*, **52**, 312–318.

Melnick, G.J. (2000). First results from the Submillimeter Wave Astronomy Satellite: H_2O and O_2 discoveries. In Y.C. Minh and E.F. van Dishoeck (eds), *Astrochemistry: From Molecular Clouds to Planetary Systems*, IAU Symposium 197, Astronomical Society of the Pacific, San Francisco, pp. 161–174.

Melnick, G., Gull, G.E. and Harwit, M. (1979). Observations of the 63 micron forbidden O I emission line in the Orion and Omega Nebulae. *Astrophysical Journal*, **227**, L29–L33.

Melnick, G.J., Genzel, R. and Lugten, J.B. (1987). Interpretation of rotationally excited far-infrared OH emission in Orion–KL. *Astrophysical Journal*, **321**, 530–542.

Melnick, G.J., Menten, K.M., Phillips, T.G. and Hunter, T. (1993). Discovery of interstellar water lines at 437, 439, and 471 GHz: Strong case for water maser formation behind C-type shocks. *Astrophysical Journal*, **416**, L37–L40.

Melnick, G.J., Stauffer, J.R., Ashby, M.L.N., Bergin, E.A., Chin, G., Erickson, N.R., Goldsmith, P.F., Harwit, M., Howe, J.E., Kleuner, S.C., Koch, D.G., Neufeld, D.A., Patten, B.M., Plume, R., Schieder, R., Snell, R.L., Tolls, V., Wang, Z., Winnewisser, G. and Zhang, Y.F. (2000). The Submillimeter Wave Astronomy Satellite: Science objectives and instrument description. *Astrophysical Journal*, **539**, L77–L85.

Merrill, P.W. (1934). Unidentified interstellar lines. *Publications of the Astronomical Society of the Pacific*, **46**, 206–207.

Merrill, K.M. and Stein, W.A. (1976). 2–14 μm stellar spectrophotometry: II. Stars from the 2 μm Infrared Sky Survey. *Publications of the Astronomical Society of the Pacific*, **88**, 294–307.

Minh, Y.C. and van Dishoeck, E.F. (eds) (2000). *Astrochemistry: From Molecular Clouds to Planetary Systems*, IAU Symposium 197, Astronomical Society of the Pacific, San Francisco.

Mitchell, G.F., Maillard, J.-P., Allen, M., Beer, R. and Belcourt, K. (1990). Hot and cold gas toward young stellar objects. *Astrophysical Journal*, **363**, 554–573.

Molinari, S., Ceccarelli, C., White, G.J., Saraceno, P., Nisini, B., Giannini, T. and Caux, E. (1999). Detection of the 62 μm crystalline H_2O ice feature in emission toward HH 7 with the ISO-LWS. *Astrophysical Journal*, **521**, L71–L74.

Molster, F.J. (2000). Crystalline silicates in circumstellar dust shells. PhD Thesis, Universiteit van Amsterdam.

Molster, F.J., Yamamura, I., Waters, L.B.F.M., Tielens, A.G.G.M., de Graauw, Th., de Jong, T., de Koter, A., Malfait, K., van den Ancker, M.E., Van Winckel, H., Voors, R.H.M. and Waelkens, C. (1999). Low-temperature crystallization of silicate dust in circumstellar disks. *Nature*, **401**, 563.

Moutou, C.C., Léger, A. and d'Hendecourt, L. (1996). Far-infrared emission of PAH molecules (14–40 μm): A preparation for ISO spectroscopy. *Astronomy and Astrophysics*, **310**, 297–308.

Moutou, C., Verstraete, L., Léger, A., Sellgren, K. and Schmidt, W. (2000). New PAH mode at 16.4 μm. *Astronomy and Astrophysics*, **354**, L17–L20.

Nakagawa, T., Yui, Y.Y., Doi, Y., Okuda, H., Shibai, H., Mochizuki, K., Nishimura, T. and Low, F. (1998). Far-infrared [C II] line survey observations of the galactic plane. *Astrophysical Journal*, Suppl. **115**, 259–269.

Natta, A., Meyer, M.R. and Beckwith, S.V.W. (2000). Silicate emission in T Tauri stars: Evidence for disk atmospheres? *Astrophysical Journal*, **534**, 838–845.

Neufeld, D.A., Chen, W., Melnick, G.J., de Graauw, T., Feuchtgruber, H., Haser, L., Lutz, D. and Harwit, M. (1996). Detection of far-infrared rotational lines of water vapour toward W Hydrae. *Astronomy and Astrophysics*, **316**, L237–L240.

Neufeld, D.A. and Kaufman, M.J. (1993). Radiative cooling of warm molecular gas. *Astrophysical Journal*, **418**, 263–272.

Nisini, B., Benedettini, M., Giannini, T., Caux, E., di Giorgio, A.M., Liseau, R., Lorenzetti, D., Molinari, S., Saraceno, P., Smith, H.A., Spinoglio, L. and White, G.J. (1999). Strong H_2O and high-J CO emission towards the Class 0 protostar L1448-mm. *Astronomy and Astrophysics*, **350**, 529–540.

Olofsson, G., Pagani, L., Tauber, J., Febvre, P., Deschamps, A., Encrenaz, P., Floren, H.-G., George, S., Lecomte, B., Ljung, B., Nordh, L., Pardo, J.R.,

Peron, I., Sjoekvist, M., Stegner, K., Stenmark, L. and Ullberg, C. (1998). Low interstellar abundance of O_2 confirmed by the PIROG 8 balloon experiment. *Astronomy and Astrophysics*, **339**, L81–L84.

Okuda, H., Matsumoto, T. and Roellig, T.L. (eds) (1997). *The Diffuse Infrared Radiation and the IRTS*, ASP Conf. Ser. Vol. 124, Astronomical Society of the Pacific, San Francisco.

Omont, A. (1986). Physics and chemistry of interstellar polycyclic aromatic molecules. *Astronomy and Astrophysics*, **164**, 159–178.

Omont, A., Moseley, S.H., Cox, P., Glaccum, W., Casey, S., Forveille, T., Chan, Kin-Wing, Szczerba, R., Loewenstein, R.F., Harvey, P.M. and Kwok, S. (1995). The $30\,\mu m$ emission band in carbon-rich pre-planetary nebulae. *Astrophysical Journal*, **454**, 819–825.

Palumbo, M.E., Geballe, T.R. and Tielens, A.G.G.M. (1997). Solid carbonyl sulfide (OCS) in dense molecular clouds. *Astrophysical Journal*, **479**, 839–844.

Palumbo, M.E., Pendleton, Y.J. and Strazzulla, G. (2000). Hydrogen isotopic substitution studies of the 2165 wavenumber (4.62 micron) "XCN" feature produced by ion bombardment. *Astrophysical Journal*, **542**, 890–893.

Parmar, P.S., Lacy, J.H. and Achtermann, J.M. (1991). Detection of low-J pure-rotational emission from H_2 in the Orion Bar region: Evidence for small-scale clumpiness. *Astrophysical Journal*, **372**, L25–L28.

Peeters, E., Tielens, A.G.G.M., Roelfsema, P.R. and Cox, P. (1999). PAHs in compact H II regions. In P. Cox and M.F. Kessler (eds), *The Universe as Seen by ISO*, ESA, Noordwijk, ESA SP-427, 739–741.

Pendleton, Y.J. and Cruikshank, D.P. (1994). Life from the stars? *Sky and Telescope*, **87**, 36.

Pendleton, Y.J., Sandford, S.A., Allamandola, L.J., Tielens, A.G.G.M. and Sellgren, K. (1994). Near-infrared absorption spectroscopy of interstellar hydrocarbon grains. *Astrophysical Journal*, **437**, 683–696.

Pendleton, Y., Tielens, A.G.G.M., Tokunaga, A.T. and Bernstein, M.P. (1999). The interstellar 4.62 μm band. *Astrophysical Journal*, **513**, 294–304.

Phillips, T.G., Huggins, P.J., Kuiper, T.B.H. and Miller, R.E. (1980). Detection of the 610 μm (492 GHz) line of interstellar atomic carbon. *Astrophysical Journal*, **238**, L103–L106.

Phillips, T.G., Kwan, J. and Huggins, P.J. (1980). Detection of submillimeter lines of CO at 0.65 mm and H_2O at 0.79 mm. In B.H. Andrews (ed.), *interstellar molecules*, IAU Symposium, pp. 21–24. Reidel, Dordrecht.

Phillips, T.G., Scoville, N.Z., Kwan, J., Huggins, P.J. and Wannier, P.G. (1978). Detection of $H_2^{18}O$ and an abundance estimate for interstellar water. *Astrophysical Journal*, **222**, L59–L62.

Platt, J.R. (1956). On the optical properties of interstellar dust. *Astrophysical Journal*, **123**, 486–490.

Pottasch, S.R., Wesselius, P.R. and van Duinen, R.J. (1979). Determination of cooling rates in the interstellar medium. *Astronomy and Astrophysics*, **74**, L15–L17.

Purcell, E.M. (1976). Temperature fluctuations in very small interstellar grains. *Astrophysical Journal*, **206**, 685–690.

Reach, W.T., Dwek, E., Fixsen, D.J., Hewagama, T., Mather, J.C., Shafer, R.A., Banday, A.J., Bennett, C.L., Cheng, E.S., Eplee, R.E., Jr., Leisawitz, D., Lubin, P.M., Read, S.M., Rosen, L.P., Shuman, F.G.D., Smoot, G.F., Sodroski, T.J. and Wright, E.L. (1995). Far-infrared spectral observations of the Galaxy by COBE. *Astrophysical Journal*, **451** 188–199.

Renzini, A. (1987). In I. Iben and A. Renzini (eds), *Physical Processes in Red Giants*, Reidel, Dordrecht, p. 431.

Rodgers, B. and Glassgold, A.E. (1991). The temperature of the circumstellar envelope of Alpha Orionis. *Astrophysical Journal*, **382**, 606–616.

Roelfsema, P., Cox, P., Tielens, A.G.G.M., Allamandola, L.J., Baluteau, J.-P., Barlow, M.J., Beintema, D., Boxhoorn, D.R., Cassinelli, J.P., Caux, E., Churchwell, E., Clegg, P.E., de Graauw, T., Heras, A.M., Huygen, R., van der Hucht, K.A., Hudgins, D.M., Kessler, M.F., Lim, T. and Sandford, S.A. (1996). SWS observations of IR emission features towards compact H II regions. *Astronomy and Astrophysics*, **315**, L289–L292.

Rosenthal, D., Bertoldi, F. and Drapatz, S. (2000). ISO-SWS observations of OMC-1: H_2 and fine structure lines. *Astronomy and Astrophysics*, **356**, 705–723.

Rowan-Robinson, M. (1992). Interstellar dust in galaxies. *Monthly Notices of the Royal Astronomical Society*, **258**, 787–799.

Russell, R.W., Melnick, G., Gull, G.E. and Harwit, M. (1980). Detection of the 157 micron (1910 GHz) [C II] emission line from the interstellar gas complexes NGC 2024 and M42. *Astrophysical Journal*, **240**, L99–L102.

Russell, R.W., Soifer, B.T. and Willner, S.P. (1977). The 4 to 8 micron spectrum of NGC 7027. *Astrophysical Journal*, **217**, L149–L152.

Ryde, N., Eriksson, K. and Gustafsson, B. (1999). The nature of the circumstellar CO_2 emission from M giants. *Astronomy and Astrophysics*, **341**, 579–586.

Salama, A. et al. (eds). (2000). *ISO beyond the Peaks*, Second ISO Workshop on Analytical Spectroscopy, ESA-SP 456, ESTEC, Noordwijk.

Saraceno, P., Benedettini, M., Di Giorgio, A.M., Giannini, T., Nisini, B., Lorenzetti, D., Molinari, S., Pezzuto, S., Spinoglio, L., Tommasi, E., Clegg, P.E., Correia, J.C., Griffin, M.J., Kaufman, M.J., Leeks, S.J., White, G.J., Caux, E., Liseau, R. and, Smith, H. (1999). ISO-LWS study of pre-main sequence sources. In V. Ossenkopf et al. (ed.), *Physics and Chemistry of the Interstellar Medium III*, GCA Verlag, Herdecke, p. 279.

Schutte, W.A., Boogert, A.C.A., Tielens, A.G.G.M., Whittet, D.C.B., Gerakines, P.A., Chiar, J.E., Ehrenfreund, P., Greenberg, J.M., van Dishoeck, E.F. and de Graauw, Th. (1999). Weak ice absorption features at 7.24 and 7.41 μm in the spectrum of the obscured young stellar object W 33A. *Astronomy and Astrophysics*, **343**, 966–976.

Schutte, W.A. and Greenberg, J.M. (1997). Further evidence for the OCN^- assignment to the XCN band in astrophysical ice analogs. *Astronomy and Astrophysics*, **317**, L43–L46.

Schutte, W.A. and Tielens, A.G.G.M. (1989). Theoretical studies of the infrared emission from circumstellar dust shells: The infrared characteristics of circumstellar silicates and the mass-loss rate of oxygen-rich late-type giants. *Astrophysical Journal*, **343**, 369–392.

Schutte, W.A. and Tielens, A.G.G.M., Whittet, D.C.B., Boogert, A., Ehrenfreund, P., de Graauw, T., Prusti, T., van Dishoeck, E.F. and Wesselius, P. (1996). The 6.0 and 6.8 μm absorption features in the spectrum of NGC 7538 : IRS9. *Astronomy and Astrophysics*, **315**, L333–L336.

Sellgren, K. (1984). The near-infrared continuum emission of visual reflection nebulae. *Astrophysical Journal*, **277**, 623–633.

Shibai, H., Okuda, H., Nakagawa, T., Matsuhara, H., Maihara, T., Mizutani, K., Kobayashi, Y., Hiromoto, N., Nishimura, T. and Low, F.J. (1991). Large-scale forbidden [C II] 158 μm emission from the Galaxy. *Astrophysical Journal*, **374**, 522–532.

Sitko, M.L., Grady, C.A., Lynch, D.K., Russell, R.W. and Hanner, M.S. (1999). Cometary dust in the debris disks of HD 31648 and HD 163296: Two "baby" Beta Pictoris stars. *Astrophysical Journal*, **510**, 408–412.

Snell, R.L., Howe, J.E., Ashby, M.L.N., Bergin, E.A., Chin, G., Erickson, N.R., Goldsmith, P.F., Harwit, M., Kleiner, S.C., Koch, D.G., Neufeld, D.A., Patten, B.M., Plume, R., Schieder, R., Stauffer, J.R., Tolls, V., Wang, Z., Winnewisser, G., Zhang, Y.F. and Melnick, G.J. (2000). Water abundance in molecular cloud cores. *Astrophysical Journal*, **539**, L101–L104.

Snow, T.P. (1995). A critique of suggested diffuse band carriers. In A.G.G.M. Tielens and T.P. Snow (eds). *The Diffuse Interstellar Bands*, Kluwer, Dordrecht, p. 379.

Sodroski, T.J., Bennett, C., Boggess, N., Dwek, E., Franz, B.A., Hauser, M.G., Kelsall, T., Moseley, S.H., Odegard, N., Silverberg, R.F. and Weiland, J.L. (1994). Large-scale characteristics of interstellar dust from COBE–DIRBE observations. *Astrophysical Journal*, **428**, 638–646.

Spaans, M. and van Dishoeck, E.F. (2001). The abundance and emission of H_2O and O_2 in clumpy molecular clouds. *Astrophysical Journal*, **548**, L217–L220.

Spinoglio, L., Giannini, T., Nisini, B., van den Ancker, M.E., Caux, E., Di Giorgio, A.M., Lorenzetti, D., Palla, F., Pezzuto, S., Saraceno, P., Smith, H.A. and White, G.J. (2000). Shock excited far-infrared molecular emission around T Tau. *Astronomy and Astrophysics*, **353**, 1055–1064.

Spitzer, L. (1978). *Physicial Processes in the Interstellar Medium*, Wiley, New York.

Spitzer, L. and Jenkins, E.B. (1975). Ultraviolet studies of the interstellar gas. *Annual Review of Astronomy and Astrophysics*, **13**, 133–164.

Stacey, G.J., Jaffe, D.T., Geis, N., Grenzel, R., Harris, A.I., Poglitsch, A., Stutzki, J. and Townes, C.H. (1993). 158 μm forbidden [C II] mapping of the Orion Molecular Cloud. *Astrophysical Journal*, **404**, 219–231.

Stacey, G.J., Kurtz, N.T., Smyers, S.D., Harwit, M., Russell, R.W. and Melnick, G. (1982). The mass of hot, shocked CO in Orion. First observations of the $J=17-J=16$ transition at 153 μm. *Astrophysical Journal*, **257**, L37–L40.

Stacey, G.J., Smyers, S.D., Kurtz, N.T. and Harwit, M. (1983). Observations of the 145.5 μm [O I] forbidden emission line in the Orion Nebula. *Astrophysical Journal*, **265**, L7–L11.

Stacey, G.J., Viscuso, P.J., Fuller, C.E. and Kurtz, N.T. (1985). The 157 μm forbidden [C II] luminosity of the Galaxy: II The presence of knotlike features in the forbidden [C II] emission. *Astrophysical Journal*, **289**, 803–806.

Stapelfeldt, K., Padgett, D. and Brooke, T.Y. (1999). A survey for H$_2$ rotational emission in low-mass pre-main sequence stars. In P. Cox and M.F. Kessler (eds), *The Universe as Seen by ISO*, ESA-SP 427, p. 525.

Stecher, T.P. (1965). Interstellar extinction in the ultraviolet. *Astrophysical Journal*, **142**, 1683–1684.

Stein, W.A. (1967). Infrared radiation from interstellar grains. In J.M. Greenberg and T.P. Roark (eds), *Interstellar Grains*, NASA SP-140, 185.

Sternberg, A. and Dalgarno, A. (1989). The infrared response of molecular hydrogen gas to ultraviolet radiation: High-density regions. *Astrophysical Journal*, **338**, 197–233.

Storey, J.W.V., Watson, D.M. and Townes, C.H. (1979). Observations of far-infrared fine structure lines: Forbidden [O III] 88.35 μm and forbidden [O I] 63.2 μm. *Astrophysical Journal*, **233**, 109–118.

Stutzki, J., Stacey, G.J., Genzel, R., Harris, A.I., Jaffe, D.T. and Lugten, J.B. (1988). Submillimeter and far-infrared line observations of M17 SW: A clumpy molecular cloud penetrated by ultraviolet radiation. *Astrophysical Journal*, **332**, 379–399.

Swings, P. and Rosenfeld, L. (1937). Considerations regarding interstellar molecules. *Astrophysical Journal*, **86**, 483–486.

Sylvester, R.J., Kemper, F., Barlow, M.J., de Jong, T., Waters, L.B.F.M., Tielens, A.G.G.M. and Omont, A. (1999). 2.4–197 μm spectroscopy of OH/IR stars: The IR characteristics of circumstellar dust in O-rich environments. *Astronomy and Astrophysics*, **352**, 587–599.

Szczepanski, J. and Vala, M. (1993). Laboratory evidence for ionized polycyclic aromatic hydrocarbons in the interstellar medium. *Nature*, **363**, 699–701.

Telesco, C.M. and Knacke, R.F. (1991). Detection of silicates in the Beta Pictoris disk. *Astrophysical Journal*, **372**, L29–L31.

Thi, W.F., van Dishoeck, E.F., Jansen, D.J., Spaans, M., Li, W., Evans, N.J. and Jaffe, D.T. (1997). Weak H$_2$ emission from diffuse and translucent clouds. In A.M. Heras, K. Leech, N.R. Trams and M. Perry (eds) *First ISO Workshop on Analytical Spectroscopy*, ESA SP-419, p. 299.

Thi, W.F., van Dishoeck, E.F., Blake, G.A., van Zadelhoff, G.J. and Hogerheijde, M.R. (1999a). Detection of H$_2$ pure rotational line emission from the GG Tauri binary system. *Astrophysical Journal*, **521**, L63–L66.

Thi, W.F., van Dishoeck, E.F., Blake, G.A., van Zadelhoff, G.J., Hogerheijde M.R., Mannings, V., Sargent, A.I., Becklin, E.E., Horn, J., Zuckerman, B., Koerner, D., Mundy, L.G., Waters, L.B.F.M., Wesselius, P.R., van den Ancker, M.E., Natta, A., Waelkens, C. and Malfait, K. (1999b). H$_2$ emission from disks around Herbig Ae and T Tauri stars. In P. Cox and M.F. Kessler (eds), *The Universe as Seen by ISO*, ESA-SP 427, pp. 529–532.

Thi, W.F., Blake, G.A., van Dishoeck, E.F., van Zadelhoff, G.J., Horn, J., Becklin, E.E., Mannings, V., Sargent, A.I., van den Ancker, M.E. and Natta, A. (2001a). Substantial reservoirs of molecular gas in the debris disks around young stars. *Nature*, **409**, 60–63.

Thi, W.F., van Dishoeck, E.F., Blake, G.A., van Zadelhoff, G.J., Horn, J., Becklin, E.E., Mannings, V., Sargent, A.I., van den Ancker, M.E. and Natta, A. (2001b). H$_2$ and CO emission from disks around T Tauri, Herbig Ae, and Vega-like stars: Cold and warm circumstellar gas. *Astrophysical Journal*, in press.

Tielens, A.G.G.M. (1983). Surface chemistry of deuterated molecules. *Astronomy and Astrophysics*, **119**, 177–184.

Tielens, A.G.G.M. (1995). The interstellar medium. In M.R. Haas, J.A. Davidson and E.F. Erickson (eds), *Airborne Astronomy Symposium*, Astronomical Society of the Pacific, San Francisco, pp. 3–22.

Tielens, A.G.G.M. (1997). Heating and cooling of the interstellar medium. In H. Okuda, T. Matsumoto, and T.L. Roellig (eds), *Diffuse Infrared Radiation and the IRTS*, Kluwer, Dordrecht, p. 255.

Tielens, A.G.G.M. (1999). The destruction of interstellar dust. In J.M. Greenberg and A. Li (eds), *Formation and Evolution of Solids in Space*, Kluwer, Dordrecht, p. 331.

Tielens, A.G.G.M. and Allamandola, L.J. (1987a). Composition, structure, and chemistry of interstellar dust, In D.J. Hollenbach and H.A. Thronson (eds), *Interstellar Processes*, Reidel, Dordrecht, pp. 397–469.

Tielens, A.G.G.M. and Allamandola, L.J. (1987b). Evolution of interstellar dust. In G. Morfill and M. Scholer (eds), *Physical Processes in Interstellar Clouds*, Reidel, Dordrecht, pp. 333–376.

Tielens, A.G.G.M. and Hagen, W. (1982). Model calculations of the molecular composition of interstellar grain mantles. *Astronomy and Astrophysics*, **114**, 245–260.

Tielens, A.G.G.M. and Hollenbach, D.J. (1985a). Photodissociation regions: I. Basic model. *Astrophysical Journal*, **291**, 722–746.

Tielens, A.G.G.M. and Hollenbach, D.J. (1985b). Photodissociation regions: II. A model for the Orion photodissociation region. *Astrophysical Journal*, **291**, 747–754.

Tielens, A.G.G.M., Hony, S., van Kerckhoven, C. and Peeters, E. (1999). Interstellar and circumstellar PAHs. In P. Cox and M.F. Kessler (eds), *The Universe as Seen by ISO*, ESA SP-427, 579–587.

Tielens, A.G.G.M., Hony, S., van Kerckhoven, C. and Peeters, E. (2000). Interstellar and circumstellar PAHs. In Y.C. Minh and E.F. van Dishoeck (eds), *Astrochemistry: From Molecular Clouds to Planetary Systems*, IAU Symposium 197, Astronomical Society of the Pacific, San Francisco, pp. 349–362.

Tielens, A.G.G.M., Tokunaga, A.T., Geballe, T.R. and Baas, F. (1991). Interstellar solid CO: Polar and nonpolar interstellar ices. *Astrophysical Journal*, **381**, 181–199.

Tielens, A.G.G.M. and Whittet, D.C.B. (1997). Interstellar ices. In E.F. van Dishoeck (ed.), *Molecules in Astrophysics: Probes and Processes*, IAU Symposium 178, Kluwer, Dordrecht, p. 48.

Tielens, A.G.G.M., Wooden, D.H., Allamandola, L.J., Bregman, J. and Witteborn, F.C. (1996). The infrared spectrum of the galactic center and the composition of interstellar dust. *Astrophysical Journal*, **461**, 210–222.

Timmermann, R., Bertoldi, F., Wright, C.M., Drapatz, S., Draine, B.T., Haser, L. and Sternberg, A. (1996). H$_2$ infrared line emission from S 140: A warm PDR. *Astronomy and Astrophysics*, **315**, L281–L284.

Trumpler, R.J. (1930). Preliminary results on the distances, dimensions and space distribution of open star clusters. *Lick Observatory Bulletin*, **420**, 154–188.

Truong-Bach, R.J., Sylvester, R.J., Barlow, M.J. Nguyen-Q-Rieu, Lim, T., Liu, X.W., Baluteau, J.P., Deguchi, S., Justtanont, K. and Tielens, A.G.G.M. (1999). H$_2$O from R Cas: ISO LWS-SWS observations and detailed modelling. *Astronomy and Astrophysics*, **345**, 925–935.

Turner, B.E. (1990). Detection of doubly deuterated interstellar formaldehyde (D_2CO): An indicator of active grain surface chemistry. *Astrophysical Journal*, **362**, L29–L33.

van de Hulst, H.C. (1949). The solid particles in interstellar space. *Rech. Astr. Obs. Utrecht*, **11**, part 2.

van de Hulst, H.C. (1997). Molecules in astrophysics half a century ago. In E.F. van Dishoeck (ed.) *Molecules in Astrophysics: Probes and Processes*, Kluwer, Dordrecht, p. 13.

van den Ancker, M.E., Bouwman, J., Wesselius, P.R., Waters, L.B.F.M., Dougherty, S.M. and van Dishoeck, E.F. (2000c). ISO spectroscopy of circumstellar dust in the Herbig Ae systems AB Aur and HD 163296. *Astronomy and Astrophysics*, **357**, 325–329.

van den Ancker, M.E., Tielens, A.G.G.M. and Wesselius, P.R. (2000b). ISO spectroscopy of the young bipolar nebulae S106 IR and Cep A East. *Astronomy and Astrophysics*, **358**, 1035–1048.

van den Ancker, M.E., Wesselius, P.R. and Tielens, A.G.G.M. (2000a). ISO spectroscopy of young intermediate-mass stars in the BD +40°4124 group. *Astronomy and Astrophysics*, **355**, 194–210.

Vandenbussche, B., Ehrenfreund, P., Boogert, A.C.A., van Dishoeck, E.F., Schutte, W.A., Gerakines, P.A., Chiar, J., Tielens, A.G.G.M., Keane, J., Whittet, D.C.B., Breitfellner, M. and Burgdorf, M. (1999). Constraints on the abundance of solid O_2 in dense clouds from ISO-SWS and ground-based observations. *Astronomy and Astrophysics*, **346**, L57–L60.

van der Tak, F.F.T., van Dishoeck, E.F., Evans, N.J. and Blake, G.A. (2000). Structure and evolution of the envelopes of deeply embedded massive young stars. *Astrophysical Journal*, **537**, 283–303.

van Dishoeck, E.F. (1998). What can ISO tell us about gas-grain chemistry? *Faraday Discussions*, **109**, 31.

van Dishoeck, E.F. and Black, J.H. (1988). The photodissociation and chemistry of interstellar CO. *Astrophysical Journal*, **334**, 771–802.

van Dishoeck, E.F., Black, J.H., Boogert, A.C.A., Boonman, A.M.S., Ehrenfreund, P., Gerakines, P.A., de Graauw, Th., Helmich, F.P., Keane, J.V., Lahuis, F., Schutte, W.A., Tielens, A.G.G.M., Whittet, D.C.B., Wright, C.M., van den Ancker, M.E., Blake, G.A., Creech-Eakman, M., Waters, L.B.F.M. and Wesselius, P.R. (1999). ISO spectroscopy of young stellar objects. In P. Cox and M.F. Kessler (eds), *The Universe as Seen by ISO*, ESA SP-427, pp. 437–448.

van Dishoeck, E.F. and Blake, G.A. (1998). Chemical evolution of starforming regions. *Annual Review of Astronomy and Astrophysics*, **36**, 317–368.

van Dishoeck, E.F. and Helmich, F.P. (1996). Infrared absorption of H_2O toward massive young stars. *Astronomy and Astrophysics*, **315**, L177–L180.

van Dishoeck, E.F., Helmich, F.P., de Graauw, Th. Black, J.H., Boogert, A.C.A., Ehrenfreund, P., Gerakines, P.A., Lacy, J.H., Millar, T.J., Schutte, W.A., Tielens, A.G.G.M., Whittet, D.C.B., Boxhoorn, D.R., Kester, D.J.M., Leech, K., Roelfsema, P.R., Salama, A. and Vandenbussche, B. (1996). A search for interstellar gas-phase CO_2. Gas : solid state abundance ratios. *Astronomy and Astrophysics*, **315**, L349–L352.

van Dishoeck, E.F. and Hogerheijde, M.R. (1999). Models and observations of the chemistry near young stellar objects. In C.J. Lada and N. Kylafis (eds), *Origins of Stars and Planetary Systems*, Kluwer, Dordrecht, p. 97.

van Dishoeck, E.F. and van der Tak, F.F.S. (2000). Chemistry in the envelopes around massive young stars. In Y.C. Minh and E.F. van Dishoeck (eds), *Astrochemistry: From Molecular Clouds to Planetary Systems*, IAU Symposium 197, Astronomical Society of the Pacific, San Francisco, pp. 97–112.

van Dishoeck, E.F., Wright, C.M., Cernicharo, J., González-Alfonso, E., Helmich, F.P., de Graauw, Th. and Vandenbussche, B. (1998). The ISO-SWS 2.4–45.2 μm spectrum toward Orion IRc2. *Astrophysical Journal*, **502**, L173–L176.

van Kerckhoven, C., Hony, S., Peeters, E., Tielens, A.G.G.M., Allamandola, L.J., Hudgins, D.M., Cox, P., Roelfsema, P.R., Voors, R.H.M., Waelkens, C., Waters, L.B.F.M. and Wesselius, P.R. (2000). The C-C-C bending modes of PAHs: A new emission plateau from 15 to 20 μm. *Astronomy and Astrophysics*, **357**, 1013–1019.

van Winckel, H. and Reyniers, M. (2000). A homogeneous study of the s-process in the 21 μm carbon-rich post-AGB objects. *Astronomy and Astrophysics*, **354**, 135–149.

Vardya, M.S., de Jong, T. and Willems, F.J. (1986). IRAS low-Resolution Spectrograph observations of silicate and molecular SiO emission in Mira variables. *Astrophysical Journal*, **304**, L29–L32.

Verstraete, L., Puget, J.L., Falgarone, E., Drapatz, S., Wright, C.M. and Timmermann, R. (1996). SWS spectroscopy of small grain features across the M17 southwest photodissociation front. *Astronomy and Astrophysics*, **315**, L337–L340.

von Helden, G., Tielens, A.G.G.M., van Heijnsbergen, D., Duncan, M.A., Hony, S., Waters, L.B.F.M. and Meijer, G. (2000). Titanium carbide nanocrystals in circumstellar environments. *Science*, **288**, 313–316.

Voors, R.H.M. (1999). Infrared studies of hot stars with dust. Ph D Thesis, Universiteit Utrecht.

Vriend, W.-J. (2000). Masters Thesis, Rijksuniversiteit Groningen.

Waelkens, C., Waters, L.B.F.M., de Graauw, Th., Huygen, E., Malfait, K., Plets, H., Vandenbussche, B., Beintema, D.A., Boxhoorn, D.R., Habing, H.J., Heras, A.M., Kester, D.J.M., Lahuis, F., Morris, P.W., Roelfsema, P.R., Salama, A., Siebenmorgen, R., Trams, N.R., van der Bliek, N.R., Valentijn, E.A. and Wesselius, P. (1996). SWS observations of young main-sequence stars with dusty circumstellar disks. *Astronomy and Astrophysics*, **315**, L245–L248.

Waters, J.W., Gustincic, J.J., Kakar, R.K., Kuiper, T.B.H., Roscoe, H.K., Swanson, P.N., Rodriguez-Kuiper, E.N., Kerr, A.R. and Thaddeus, P. (1980). Observations of interstellar H_2O emission at 183 GHz. *Astrophysical Journal*, **235**, 57–62.

Waters, L.B.F.M., Beintema, D.A., Zijlstra, A.A., de Koter, A., Molster, F.J., Bouwman, J., de Jong, T., Pottasch, S.R. and de Graauw, Th. (1998a). Crystalline silicates in planetary nebulae with [WC] central stars. *Astronomy and Astrophysics*, **331**, L61–L64.

Waters, L.B.F.M., Cami, J., de Jong, T., Molster, F.J., van Loon, J.Th., Bouwman, J., de Koter, A., Waelkens, C., Van Winckel, H. and Morris, P.W. (1998b). An oxygen-rich dust disk surrounding an evolved star in the Red Rectangle. *Nature*, **391**, 868.

Waters, L.B.F.M., Waelkens, C., van der Hucht, K.A. Zaal, P.A. (eds) (1998) *ISO's View on Stellar Evolution*, Kluwer, Dordrecht.

Watson, D.M., Genzel, R., Townes, C.H. and Storey, J.W.V. (1985). Far-infrared emission lines of CO and OH in the Orion-KL molecular shock. *Astrophysical Journal*, **298**, 316–327.

Weinreb, S., Barrett, A.H., Meeks, M.C. and Henry, J.C. (1963). Radio observations of OH in the interstellar medium. *Nature*, **200**, 829.

Whittet, D.C.B. (1992). *Dust in the Galactic Environment*, Institute of Physics, Bristol.

Willems, F.J. (1988). IRAS low-resolution spectra of cool carbon stars. *Astronomy and Astrophysics*, **203**, 51–70.

Willner, S.P., Gillett, F.C., Herter, T.L., Jones, B., Krassner, J., Merrill, K.M., Pipher, J.L., Puetter, R.C., Rudy, R.J., Russell, R.W. and Soifer, B.T. (1982). Infrared spectra of protostars: Composition of the dust shells. *Astrophysical Journal*, **253**, 174–187.

Willner, S.P., Soifer, B.T., Russell, R.W., Joyce, R.R. and Gillett, F.C. (1977). 2 to 8 micron spectrophotometry of M82. *Astrophysical Journal*, **217**, L121–L124.

Wilson, R.W., Jefferts, K.B. and Penzias, A.A. (1970). Carbon monoxide in the Orion Nebula. *Astrophysical Journal*, **161**, L43–L44.

Wickramasinghe, D.T. and Allen, D.A. (1980). The 3.4-micron interstellar absorption feature. *Nature*, **287**, 518.

Wright, C.M. (2000). Observations of molecular hydrogen in shocks and PDRs with the Infrared Space Observatory. In Y.C. Minh and E.F. van Dishoeck (eds), *Astrochemistry: From Molecular Clouds to Planetary Systems*, IAU Symposium 197, Astronomical Society of the Pacific, San Francisco, pp. 177–190.

Wright, C.M., van Dishoeck, E.F., Cox, P., Sidher, S.D. and Kessler, M.F. (1999). ISO-LWS detection of the $112\,\mu$m HD $J = 1 \rightarrow 0$ line toward the Orion Bar. *Astrophysical Journal*, **515**, L29–L32.

Wright, E.L., Mather, J.C., Bennett, C.L., Cheng, E.S., Shafer, R.A., Fixsen, D.J., Eplee, R.E., Jr., Isaacman, R.B., Read, S.M., Boggess, N.W., Gulkis, S., Hauser, M.G., Janssen, M., Kelsall, T., Lubin, P.M., Meyer, S.S., Moseley, S.H., Jr., Murdock, T.L., Silverberg, R.F., Smoot, G.F., Weiss, R. and Wilkinson, D.T. (1991). Preliminary spectral observations of the Galaxy with a 7 deg beam by the Cosmic Background Explorer (COBE). *Astrophysical Journal*, **381**, 200–209.

Wright, C.M., van Dishoeck, E.F., Black, J.H., Feuchtgruber, H., Cernicharo, J., González-Alfonso, E. and de Graauw, Th. (2000). ISO-SWS observations of pure rotational H_2O absorption lines toward Orion-IRc2. *Astronomy and Astrophysics*, 689–700.

Xie, T., Allen, M. and Langer, W.D. (1995). Turbulent diffusion and its effects on the chemistry of molecular clouds. *Astrophysical Journal*, **440**, 674–685.

Yun, J. and Liseau, R. (eds) (1998). *Star Formation with the Infrared Space Observatory*, Vol. 132, Astronomical Society of the Pacific, San Francisco.

Zmuidzinas, J., Blake, G.A., Carlstrom, J., Keene, Jocelyn, Miller, D., Schilke, P. and Ugras, N.G. (1995). Submillimeter spectroscopy of interstellar hydrides. In M.R. Haas, J.A. Davidson and E.F. Erickson (eds), *Proceedings of the Airborne Astronomy Symposium*, Astronomical Society of the Pacific, San Francisco, pp. 33–40.

Zuckerman, B., Forveille, T. and Kastner, J.H. (1995). Inhibition of giant planet formation by rapid gas depletion around young stars. *Nature*, **373**, 494.

28

PRISCILLA C. FRISCH*

The interstellar medium of our Galaxy

Over the past century our picture of diffuse material in space has grown from a simple model of isolated clouds in thermal equilibrium with stellar radiation fields to one of a richly varied composite of materials with a wide range of physical properties and morphologies. The Solar System interacts with a dynamical interstellar medium. Optical, radio, and UV astronomy allow us to study the clouds which form the galactic environment of the Sun. The composition and distribution of interstellar clouds in the disk and halo tell us about the history of elemental formation in our galaxy, and the past and future environment of the Solar System.

Dark lanes of dusty clouds obscuring portions of the Milky Way are celestial landmarks, but the realization that interstellar gas pervades space is quite recent. The twentieth century opened with the discovery of a "nebulous mass" of interstellar gas in the sightline towards the binary star δ Orionis (Hartmann 1904). A series of over 40 spectra showed that the Ca II K line (3933 Å) absorption was nearly stationary in wavelength, "extraordinarily weak," and "almost perfectly sharp," in contrast to broader variable stellar absorption features. Sharp stationary Na I D1 and D2 lines (5890, 5896 Å) were discovered in δ Ori and β Sco by Mary Lea Heger. An explanation offered was that a stationary absorbing cloud of vapor was present in space between these binary systems and the observer. The Ca II and Na I lines constituted the primary tracer for interstellar gas during the first half of the century.

Interstellar matter (excluding dark matter) provides about 30–40% of the galactic mass density in the solar neighborhood. Trace elements heavier than He, which form the planets, record the chemical evolution of matter in our Galaxy, and provide detailed information on physical conditions in interstellar clouds, represent a small proportion of the interstellar atoms (~0.15%). These same elements trace the metallicity of interstellar gas, and by inference the mineralogy of interstellar grains. A primary goal of interstellar medium studies has been to determine the chemical composition of interstellar clouds compared to, for instance, normal Population I stars such as the Sun. Space observations are required to observe most astronomically interesting elements such as C, N, O, Fe, Mg, and Si, since the resonant ground state transitions of these atoms fall in the ultraviolet (UV, 912–3000 Å).

Gene Parker once asked me "What is an interstellar cloud?" The Rashomon-like answer depends on the context. Early optical data showing velocity components in interstellar absorption lines led to a definition of "clouds" as discrete kinematical units. Alternative descriptions were based on the physical properties of the clouds, for example warm diffuse intercloud material in equilibrium (Stromgren 1948, Field et al. 1969), or hot tenuous coronal gas (Spitzer 1956) to confine the clouds. Ground and space data now show interstellar material (ISM) with densities in the range 10^{-4} atoms cm^{-3} to over 10^3 atoms cm^{-3}, kinetic temperatures $20\,\text{K} < T_\text{k} < 10^6\,\text{K}$, and many levels of ionization. Within 10 pc of the Sun, we see density contrasts of over 400 and temperature contrasts over 100. The distinction between turbulence and "clouds" has been blurred by recent high-spectral resolution data showing that low-resolution spectral data may miss over half of the velocity components in a sightline (Section 3.2), and that ~15% of the mass of cold clouds is contained in tiny (AU-sized) structures (Section 3.1). Are these features "clouds" or manifestations of a turbulent ISM? The discovery of interstellar clouds in the galactic halo (Münch 1957) added new questions about the stratification of ISM in the gravitational potential of the Milky Way Galaxy, and the origin of halo gas.

Surprisingly, interstellar gas constitutes about 98% of the diffuse material inside of the heliosphere, and subparsec spatial variations in interstellar cloud properties near the

---
* University of Chicago, IL, USA

Sun indicate the solar environment could change on time scales $\approx 10^4$ years. Shapley's conjecture in 1921 that interstellar clouds affect planetary climates no longer seems outlandish.

Symbols used here are N (column densities, cm^{-2}), and n_H (total volume density for all forms of H, cm^{-3}). Early optical and 21 cm radio data were insensitive to clouds with column densities $N(\mathrm{H}) < 10^{19}$ cm^{-2}. UV observations of trace elements can detect kinematical objects with $N(\mathrm{H}) \gtrsim 10^{16}$ cm^{-2}. Galactic longitudes and latitudes are quoted in System II. (System II was adopted by the International Astronomical Union in 1958 in order to correct earlier errors in the location of the galactic center. Galactic coordinates published before 1958 are incorrect.) The local standard of rest (LSR) velocity frame represents heliocentric velocities transformed to the rest frame corresponding to the mean motion of comparison stars near the Sun, where the comparison set is selected according to some criterion. (Most LSR interstellar velocities presented in the twentieth century assumed, frequently without explanation, a "standard" solar motion corresponding to a velocity of 19.7 km s^{-1} towards the apex position $l^{II} = 57°$, $b^{II} = +22°$. Recent Hipparcos results give a solar motion of 13.4 km s^{-1} towards the apex direction $l^{II} = 28°$, $b^{II} = +32°$.) Radio data are usually presented in the LSR velocity frame, where this issue is particularly troublesome.

This chapter focuses on the diffuse gas in the space between the stars of our Galaxy. For an eloquent summary of the physical properties of the ISM and data up to the mid-1970s, see Spitzer (1978).

1 DISCOVERING INTERSTELLAR GAS

The observational and theoretical foundations of ISM space studies were formed in the first half of the twentieth century. In 1926 Sir Arthur Stanley Eddington laid out the principles of the ionization equilibrium of atoms in space under the influence of dilute stellar radiation fields (the Bakerian Lecture, Eddington 1926). He evaluated the importance of the short wavelength stellar radiation field ($\lambda < 800$ Å) for cloud heating (although extreme ultraviolet radiation from space was not observable in 1926), and concluded that diffuse clouds are illuminated by a radiation field at a Planck temperature of 10,000 K and have kinetic temperatures $T_k \sim 10,000$ K. He found frequent collisions would establish Maxwellian velocity distributions for electrons and ions in space. Eddington concluded that the material creating the stationary Ca II and Na I absorption lines is uniformly distributed, and argued that stellar dynamics implied $n_\mathrm{H} \lesssim 10$ cm^{-3} for diffuse material. He determined that in space most interstellar Ca is Ca III and most Na is Na II He "reluctantly" concluded that dark nebulae derive their obscuration from "fine solid grains." Eddington noted that radiation with energies greater than ~ 13.6 eV (the ionization potential of hydrogen) would be prevented from entering clouds by abundant hydrogen, and concluded interstellar H$_2$ would be abundant.

In the early part of the century, Harlow Shapley advanced the idea that interstellar clouds were linked to terrestrial climate shifts. He noted that the diffuse luminous and dark nebulae are found throughout space, and that the Sun was receding from the Orion region where dark nebulae are prominent (Shapley 1921). Shapley suggested that a past climate-altering encounter between the Orion molecular clouds and the Solar System, would yield a 20% variation in solar radiation, which if sustained for a period of time would alter Earth's climate. While encounters with dense clouds as envisioned by Shapley are statistically improbable, encounters with clouds of modest density (~ 10 cm^{-3}) are much more likely and would destabilize the heliosphere and modify the interplanetary environment (Zank and Frisch 1999).

1.1 Optical absorption lines

The interstellar nature of the sharp stationary absorption features seen in binary systems was quickly established. Plaskett and Pearce (1930) provided a convincing discussion that the sharp Ca II and Na lines are formed in diffuse space, and labeled these features "interstellar." Observations of 1700 stars with $V < 10.5$ mag by Otto Struve had shown that K line strengths in general increase with increasing magnitude (and thus distance), suggesting an interstellar origin. Plaskett and Pearce measured Ca II K line velocities for ~ 250 OB stars (to within ± 1.8 km s^{-1}), and found they are "almost exactly twice" interstellar Ca II motions expected for the galactic rotation of interstellar clouds located at half the distance to the background star. Peculiar motions within the clouds were found to be equally important for line broadening.

During the late 1930s, Merrill and collaborators surveyed the yellow Na I D1, D2 and blue Ca II H, K lines in over 400 hot, bright stars. The D2/D1 line ratios were seen to increase as total line intensity decreased, providing an early indication of line saturation. The formula relating absorption line equivalent width (W) and column density (N) was found to be faulty for "deep" lines, leading to the development of an empirical "curve of growth" (COG) using the doublet ratio (Ca II H/K, Na I D1/D2) to constrain the functional dependence of W. The COG, formalized later by Stromgren (Section 1.3), has been used to derive column densities from the equivalent widths of interstellar absorption lines for 70 years. Cloud motions were shown to depend on star distance, with velocity–longitude plots showing a double sine pattern with an amplitude at half the expected galactic rotation value (e.g., see review of

Münch 1968). The bulk motions implied by line velocities showed a chaotic or turbulent component, with dispersion $v_{radial} = 5$–$10 \, km \, s^{-1}$, which was interpreted as moving "clouds" or "currents." The Na I line strengths increased with both distance and the amount of interstellar "smoke" (now known as dust) in the sightline (where the dust was determined by the reddening of starlight).

The chemical composition of interstellar clouds was explored by a series of observations at Mt. Wilson, Lick, and the Dominion Astrophysical Observatories. Sharp interstellar absorption lines from Na I ($\lambda\lambda 3302.4, 3303.0, 5895.9, 5890.0$), Ca I ($\lambda 4226.7$), Fe I ($\lambda\lambda 3719.9, 3859.9$), Ti II ($\lambda\lambda 3383.8, 3242.0, 3229.2, 3073.0$), K I ($\lambda\lambda 7664.9, 7699.0$), CH II ($\lambda\lambda 3957.7, 4232.5$), CH I ($\lambda\lambda 4300.3, 3890.2$), and CN I ($\lambda\lambda 3875.8, 3874.0$) were discovered, and upper limits were placed on Al I ($\lambda 3944.0$), Sr II ($\lambda\lambda 4077.7, 4215.5$), and Ba II ($\lambda\lambda 4554.0, 4934.1$) lines (e.g., see review of Münch 1968). The sightlines towards χ^2 Ori and 55 Cyg were shown to be excellent for identifying new interstellar lines (Dunham 1939). Dunham used the weak 3302 Å and strong D1, D2 Na I doublets to construct an empirical COG for the ISM towards $\chi 2$ Ori, avoiding saturation problems with the D lines. Ionization equilibrium calculated for N(Ca I)/N(Ca II) provided the interstellar electron density, $n_e \sim 0.7$–$1.4 \, cm^{-3}$. The anomalous properties of interstellar Ca II lines were apparent in early data. These observations yielded N(Ca II)/N(Na I) $\ll 1.5$, which was the value expected from stellar atmosphere data at that time.

The peculiar behavior of interstellar Ca seen in these early data was a harbinger of the fundamentally different properties of volatiles and refractory elements in the ISM; however, an incomplete understanding of cloud ionization prevented this recognition. When Olin Wilson showed convincingly that Ca II has a velocity distribution which is peculiar in comparison to Na I, he described the results as "unexpected ... and disappointing." The average internal velocity distribution for Ca II was about three times larger than for Na I ($22 \, km \, s^{-1}$ v. $7.5 \, km \, s^{-1}$; Wilson 1939). Although later high-resolution spectroscopy showed that the "lines" observed by Wilson were blends of several components (Section 3.2), the velocity distribution of Ca II was still peculiar. Merrill and Wilson showed that the ratio N(Na I)/N(Ca II) varied strongly from one cloud to another, with values <0.3–10 (e.g., see review of Münch 1968).

Walter S. Adams published an influential survey of the strengths and velocities of Ca II H, K, lines and weak features from CN I, CH I, CH II, Fe I, and Ca I in ~ 300 disk ($|z| < 500 \, pc$) OB stars visible from the northern hemisphere (Adams 1949). The higher resolution of these data ($\sim 6 \, km \, s^{-1}$), compared to earlier photographic data, demonstrated that interstellar gas is concentrated in clouds with velocity separations larger than the atomic velocity dispersion within individual clouds. The velocities in this survey provided a consistent survey of cloud kinematics. Adams estimated the relative strengths of the Ca II lines by visually estimating line strengths, which was a common practice then; the weakest features he detected correspond to equivalent widths of $\sim 3 \, m$Å. Half of the stars in the sample were found to have complex H or K lines, while about six stars showed weak Ca II absorption at 4226 Å. About 25% of the stars had weak blue/near-UV features from CN I, CH I, CH II, Ca I, or Fe I. Adams found 21 Ca II line components with $v_{lsr} > 30 \, km \, s^{-1}$ (the velocities quoted in Adams (1949) should be updated with modern wavelengths), and concluded the velocities represented peculiar cloud motions through the LSR.

Blaauw used Adams' data to determine the form of the velocity distribution of the chaotic component of interstellar cloud velocities (e.g., see review by Spitzer 1968). He compared Ca II component velocities with two possible distributions – a Gaussian and an exponential distribution of component velocities. Blaauw concluded that an exponential form fitted the observed velocity distributions better than a Gaussian form. Munch extended this analysis to halo stars, with observations of Ca II and/or Na I towards 112 stars in the northern hemisphere sky ($l = 50°$–$160°$, $|z| = 1$–$2 \, kpc$; Münch 1957). He confirmed earlier studies showing that bulk cloud motions indicate clouds are aligned with the Orion and Perseus spiral arms, and found blue-shifted absorption components from expanding interstellar gas around O-star associations. Based on the Blaauw results, and reasoning that turbulence is not Gaussian so that the increased energy dissipation by supersonic turbulence would flatten a cloud velocity distribution in comparison to a Gaussian, Münch fitted the observed disk and halo cloud velocities with the form:

$$\Psi(V) = \frac{1}{2\eta} \exp(-|V - V_o|/\eta)$$

where η is the mean radial velocity, found to be $\sim 5 \pm 1 \, km \, s^{-1}$ by Blaauw. With the exponential distribution, a constant value for the velocity dispersion for all Na I D2 line strengths is obtained ($\eta\sqrt{2} \sim 4.6 \, km \, s^{-1}$), providing a better fit to velocity distributions than a Gaussian form. This has created what I view as a conundrum, since the internal velocity distributions of clouds are found to be Gaussian in cold clouds, while the bulk cloud velocities measured at higher resolution show an exponential distribution.

A study of the abundance variations between individual clouds was presented in a classic paper by Routly and Spitzer, in 1952, which confirmed earlier indications that the distribution of Ca II cloud velocities is intrinsically larger than the distribution of Na I velocities. They found both that b(Ca II) $> 1.5 \, b$(Na I) and that Ca II component velocities are larger than for Na I. The systematic decrease of N(Na I)/N(Ca II) with increasing cloud velocity is known as the "Routly–Spitzer (RS) effect", manifested as N(Na I)/N(Ca II) $\lesssim 1$. RS proposed that the process accelerating Ca II also decreased N(Na I)/N(Ca II), either through collisional ionization of Na I during acceleration, or from relative differences between Ca and Na depletions onto dust grains in low- v. high-velocity clouds. The first discovery of ISM within 20 pc of the Sun was possible because of the RS effect. Ca II was detected towards α Oph (14 pc), but not Na I (Munch and Unsold 1962). Recent data give N(Na I)/N(Ca II) ~ 0.1 in the $V_{lsr} \sim -8$ km s^{-1} cloud towards α Oph (Welty et al. 1996). Small-scale ($\sim 1°$) variations in Ca II line strengths suggested the presence of small clouds (<1 pc). These data provided the first evidence of small-scale structure and shocked ISM close to the Sun.

As frequently happens, new techniques proven in first-generation instruments provide limited data, but change the course of science. The first high-spectral-resolution ($R >$ 200,000) observations of optical absorption lines did not achieve the initial goal of observing the Na I D line hyperfine components (separated by ~ 1.0 km s^{-1}), but instead demonstrated cloud velocity structure that is unresolved in photographic plates. High-spectral-resolution ($R = 3-5 \times 10^5$) spectrometers were developed to observe the hyperfine Na I lines towards α Cyg using the long-focal-length McNath solar telescope (Livingston and Lynds 1964), and using a Pepsios spectrometer at Lick Observatory (Hobbs 1965). Much later, after the launch of the Copernicus satellite and after high-resolution optical data were routinely acquired by Hobbs and others (Section 3.2), the Na I D-line hyperfine splitting was discovered towards α Cygnus, using a Michelson interferometer (effective resolution $R \sim$ 500,000; Wayte et al. 1978). Turbulence in cold interstellar clouds was found to be subsonic. The $+1$ km s^{-1} cloud has $b_{Dop} = 0.38$ km s^{-1} and a likely temperature of 70–124 K, indicating turbulent velocities $v_t = 0.26$–0.3 km s^{-1} in comparison to the isothermal sound speed of ~ 0.8 km s^{-1}.

A thorough and now classic study of the ISM in front of ζ Oph (HD 149757) was performed by George Herbig, using the Lick Observatory 120-inch (3 m) Coudé feed spectra with photographic plates (Herbig 1968). The star ζ Oph has a relatively featureless bright continuum (O9.5 Vn, $V =$ 2.6 mag, 140 pc, $v\sin i = 380$ km s^{-1}), and a rich interstellar spectrum ($E(B-V) \sim 0.32$ mag) lending itself to searches for new interstellar species. It is a runaway star from the Scorpius–Centaurus association moving supersonically through the ISM (space velocity ~ 40 km s^{-1}), and with a circumstellar H II region (5° radius). ζ Oph also has a ram-pressure confined stellar wind observed around the star through 60 μm radiation from heated, swept-up interstellar grains. Herbig measured absorption lines from Na I, Ca II, K I, Ti II, Fe I, CH I, CH II, and CN I, and set limits on some 25 additional species at the $W = 1$–2 mÅ level. The ISM is contained in two dominant clouds at -15 km s^{-1} and -29 km s^{-1} (heliocentric velocities). An empirical curve of growth for the main component (heliocentric velocity, $v_{hc} = -15$ km s^{-1}) gives $b_{Dop} = 2.4$ km s^{-1} and the column densities for these elements. The photoionization equilibrium of Na I, which equilibrates photoionization with the temperature-dependent recombination rate of Na$^+$ and an electron (N(Na II)/N(Na I) $\sim n$(Na II)/n(Na I); Section 1.3) gives $n_e = 0.36$–0.54 cm^{-3} if the -15 km s^{-1} cloud is ~ 50 pc from ζ Oph. If photoionization of heavy elements (e.g., C) supplies electrons, densities $n_H =$ 500–900 cm^{-3} are found, indicating the cloud is a thin sheet of thickness ~ 0.15 pc. Later studies show Na is depleted by factors of 4–5, raising densities and reducing cloud thickness by the same factors. Enhanced abundances of Ca II and Ti II in the -29 km s^{-1} cloud identify this material as "intercloud," which Herbig suggested is local (since it is seen in front of many Scorpius stars).

Lewis Hobbs initiated the high-resolution era of optical spectroscopy with a survey of the Na I D lines at resolution ~ 1 km s^{-1} using a Pepsios interferometer at Lick Observatory (Hobbs 1969). Turbulence was found to dominate observed line widths, since cloud temperatures of 100 K yielded $b_{Dop} \sim 0.3$ km s^{-1}, v. observed Doppler widths of $b_{Dop} \sim 1.5$ km s^{-1}. The photoionization balance of Na applied to N(Na I)/N(H I) yielded typical electron densities $n_e \sim 0.008$ cm^{-3} for solar abundances, or $n_H \sim 20$ cm^{-3} if electrons are supplied by the photoionization of metals. High-resolution optical data acquired by Hobbs, his students, and others proved to be a crucial supplement to UV data (Section 3.2).

1.2 Radio astronomy – the first multispectral data

Radio waves provided the first multispectral window on the ISM. Radio emission from the Galaxy was discovered by Karl Jansky in 1932 during an investigation of radio static in a Bell Labs shortwave (~ 15 m) radio antenna. Jansky recognized the interstellar origin of the hiss. Jansky's papers inspired Grote Reber, who, working with a private radio telescope located in his backyard in Wheaton, Illinois, in 1938, detected cosmic static at 2 m and confirmed Jansky's discovery. Over the next several years, Reber used his backyard telescope to map the northern hemisphere radio sky at 160 MHz and 480 MHz. The advent of World War II moved radio astronomy to the forefront of technical interest, both in the USA and Europe, and during the following decade technical advances from the wartime use of radar supported the development of the new radio sciences. Shklovsky (1960) reviews the scientific basis for the developing field of radio astronomy.

The spectral index of the cosmic "radio static" was measured and found to increase towards lower energies, in contrast to the expected increase towards higher energies predicted for a blackbody (or "thermal") source. This puzzle was explained when V.L. Ginzburg calculated the emission from relativistic electrons (~1 GeV–1 TeV) interacting with interstellar magnetic fields in the halo and disk, terming the radiation "magnetobremsstrahlung" in analogy to thermal bremsstrahlung from free–free emission (by free electrons in space plasmas; e.g. Spitzer 1978). Magnetobremsstrahlung is now generally known as synchrotron radiation. In 1952 Shklovsky suggested that a tenuous gaseous envelope must surround the Galaxy in order to explain the geometry of high-latitude, low-frequency non-thermal radio continuum emission. Baldwin concluded that high-latitude, long-wavelength (~3.7 m) emission originated in a galactic halo extending at least ≥ 10 kpc from the plane (Baldwin 1955). Interstellar gas in the halo was considered the carrier for magnetic fields which confine halo cosmic rays (Shklovsky 1960). At low frequencies non-thermal synchrotron emission dominates over optically thick thermal emission, and the ratio of non-thermal to thermal radio emission is $T_{br} \sim \nu^{-1.4}$ (T_{br} is the brightness temperature).

During World War II Dutch astronomers deduced the usefulness of radio emission in probing optically obscured distant portions of the Galaxy, and predicted that the hyperfine transition of the H I ground $^2S_{1/2}$ state would be a useful tracer of H I in interstellar clouds (e.g. van de Hulst 1998). The hyperfine levels are statistically populated, $n_2/n_1 \sim (g_2/g_1)\exp(-E/kT_s)$ (g_2 and g_1 are the statistical weights of the two levels, T_s is the "spin temperature"). Although a single, spontaneous 21 cm hyperfine transition is improbable (~2.85×10^{-15} s^{-1}), this transition is astronomically useful since ~25% of the mass density in the solar neighborhood is H I. Van de Hulst also recognized the possibility of stimulated emission in the 21 cm line through radiative pumping (e.g., see Field 1958).

The discovery of the radio H I 21 cm emission in 1951 at Harvard, in the Netherlands, and in Sydney, provided a new window on the Universe and made it possible to map interstellar gas and the spiral arm structure of the Milky Way Galaxy. Interstellar H I was found to be strongly concentrated in the galactic plane and spiral arms, with emission peaks corresponding to spiral arm velocities. During the 1950s and 1960s, single-beam surveys of H I 21 cm emission at high and low galactic latitudes investigated fundamental cloud properties for clouds with log N(H) $\gtrsim 18.7$ cm^{-2}. H I 21 cm emission was used to reconstruct the basic spiral structure and kinematical rotation curves of our Galaxy (e.g., see review of Burton 1976). The cloud–intercloud nature of the ISM was determined by observations of 21 cm absorption by isolated clouds in front of bright background radio continuum sources such as the Crab Nebula and Cassiopeia A (Hagen *et al.* 1955). In the direction of Cassiopeia A, ~1 kpc away in the Perseus arm, comparisons of on-source v. off-source data showed three strong absorbing components corresponding to clouds in the Orion (velocity dispersion 1.6 km s^{-1}) and Perseus (velocity dispersion ~2.5 km s^{-1}) arms. The first H I shell-like structure in space was discovered in 1958 by Menon, who observed a shell expanding at ~10 km s^{-1} in Orion, centered near the expansion center of the runaway stars AE Aurigae, μ Columbae, and 53 Arietis, and with an age consistent with the formation of these runaways (Menon 1958).

Cold absorbing clouds ($T \leq 80$ K) were found to be common in the galactic plane ($|l^{II}| < 20°$), and broad "intercloud" emission was seen with no associated absorption, indicating the presence of both warm and cold gas (e.g. Clark 1965). The median internal velocity dispersion of 2.0 km s^{-1} is found for cold absorbing clouds towards 12 sources, v. $T \leq 10{,}000$ K for warm emission clouds. About 4 absorbing clouds per kiloparsec are found, while ~7 clouds per kiloparsec are found when interarm regions are excluded. The high-latitude clouds seen optically at high z (z is the distance to the galactic plane) by Munch and Zirin were not detected in 21 cm absorption, which was interpreted as indicating cloud ionization (see review by Münch 1968). Spatial variations in absorption towards Orion A yielded cloud densities of $n_H \sim 680$ cm^{-3}.

Low-column-density, high-velocity infalling gas ($v_{lsr} \sim -100$ km s^{-1}) was discovered at high galactic latitudes and attributed to the galactic halo (see review by Verschuur 1975). High-velocity halo gas included a H I bridge between the south galactic pole and the Large Magellanic Cloud. The high-velocity clouds (HVCs) were recognized as non-normal disk gas, since at $v_{lsr} = -100$ km s^{-1} a $z = 500$ pc cloud would fall into the galactic plane in ~5 Myr. Some HVCs were found to be associated with a gas deficiency at lower velocities, suggesting an origin by the acceleration of low-velocity gas (Verschuur 1975).

The discovery of pulsars in the late 1960s provided a powerful diagnostic of the thermal electron component in the ISM. The wavelength-dependent dispersion of pulsar wave packets is measured by the dispersion measure, DM $\sim \int n(e^-) dL$ cm^{-3} pc (L is pathlength), which gives the mean electron density in the sightline. The dispersing medium was shown to be interstellar from the approximate correlation between DM and pulsar distance over kiloparsec lengths. An early study of the latitude dependence of dispersion measures for 36 pulsars concluded that the thickness of the ionized portion of the galactic disk corresponds to DM ~ 10–40 pc cm^{-3}, yielding $\overline{n_e} = 0.03 - 0.1$ cm^{-3}, in agreement with values derived from thermal absorption of low-frequency, non-thermal radio background by ionized disk gas (Davidson and Terzian 1969).

1.3 Theoretical foundations

Theoretical concepts have played an important role in shaping the picture of the ISM derived from data. Stromgren's explanation for glowing regions of Hα emission ("Stromgren spheres") surrounding hot stars located in spiral arms is famous (Stromgren 1939, O'Dell 1999). Ionized regions which surround hot stars were modeled using a sharp transition between ionized and neutral gas, with no leakage of ionizing photons into general interstellar space.

Competing explanations were presented for the cloud-like nature of the ISM. In 1940 Ambarzumian and Gordeladse used the presence of discrete interstellar clouds defined by dust associated with reflection nebula to construct a picture of the ISM with 6–7 clouds per kiloparsec, each with an average optical depth $\tau \sim 0.25$. The role that supersonic collisions between clouds (relative velocities $\sim 40\,\text{km s}^{-1}$) play in creating ionized layers around clouds and evaporating dust grains was discussed in the 1946 George Darwin Lecture by Oort. He concluded that interstellar clouds must be "continually reshuffled" since grains in both dense and intercloud regions have the same size, and commented that a large Reynolds number ($\sim 10^9$, the ratio of inertial to viscous force) in the ISM suggested turbulence contributes to the reshuffling (Oort 1946). Oort also presented the first, tentative discussion of the interaction between an expanding supernova shell with the surrounding ISM, where he argued that "interstellar friction" heats and eventually brakes (within $\sim 10^4$ years) shell material.

The interaction of a hot star with a patchy ISM was studied in 1948 by Stromgren in a basic theoretical paper that also discussed photoionization equilibrium in a cloudy medium. Based on a clumpy ISM with clouds filling $\sim 14\%$ of space (5–8 clouds per kiloparsec) and the expectation that the space between interstellar clouds contains material of very low density ("intercloud" gas), Stromgren concluded photons would leak away from the immediate vicinity of hot stars, giving a diffuse UV interstellar field ($\lambda > 912\,\text{Å}$) similar to a $T = 25{,}000\,\text{K}$ stellar field diluted by a factor of 2×10^{-17}. This field would, in principle, photoionize low-density ISM and cloud rims. In neutral clouds electrons are contributed by trace element ionization with first ionization potential FIP $< 13.6\,\text{eV}$ (primarily C). For the case where clouds are far from individual stars (say $>100\,\text{pc}$), the general diffuse interstellar radiation field (contributed by many OB stars) would contain sufficient photons with energies greater than the Lyman limit (wavelength $\lambda < 912\,\text{Å}$) to ionize rims around neutral cloud of thickness $\sim 1/n_\text{H}^2$ parsecs. Stromgren predicted emission measure $\mathcal{E} \sim 3\,\text{cm}^{-6}\,\text{pc}$ ($\mathcal{E} = \int n(e^-)^2\,dL$, where the integral is over the pathlength L) for the rims, which was unobservable at that time, contrasting them to $\mathcal{E} = 9 \times 10^5$–$6000\,\text{cm}^{-6}\,\text{pc}$ for O5–B0 star H II regions. Stromgren predicted $n_\text{H} < 0.08\,\text{cm}^{-3}$ for α Vir using Na data, close to the value later found from Copernicus data (see references in Jenkins 1987). Diffuse ionized gas is now known to be widespread, with ionized rims showing $\mathcal{E} = 2$–$10\,\text{cm}^{-6}\,\text{pc}$ present on 10–30% of H I clouds (Reynolds 1995). In the 1948 paper Stromgren presented a picture of intercloud material with density $n_\text{H} \sim 0.1\,\text{cm}^{-3}$ filling the space between interstellar clouds with densities $n_\text{H} \sim 10\,\text{cm}^{-3}$. Stromgren also formalized the "curve of growth" theory, which relates N and W as a function of cloud temperature which describes the statistical distribution of atomic velocities. The key variable is the "Doppler constant" $b_{D_{op}}^2 = 2kT/m + 2v_t^2$, where T is temperature, k is the Boltzmann constant, the atomic mass is m, and v_t is line broadening due to turbulence.

The chemical composition of interstellar clouds can be derived from absorption line data, providing recombination rates and photoionization cross-sections are accurately known. However, atoms may not behave similarly in interstellar clouds and laboratory vacuum, since densities differ by ~ 9 orders of magnitude. Seaton provided the first careful evaluation of recombination and photoionization rates for atoms in space, calculating the ionization equilibrium of H I, C I, Na I, K I, Ca I, He II, Na II, K II, and Ca II (Seaton 1951). Seaton concluded that χ^2 Orionis data are consistent with "mean cosmic abundances," except possibly for calcium which appeared to be deficient. For the temperature-dependent recombination rate (typically $\sim T^{-0.7}$), Seaton used Spitzer's temperatures of $T \sim 100\,\text{K}$ for H I, and $T \sim 10{,}000\,\text{K}$ for H II. Using the ratio $N(\text{Ca I})/N(\text{Ca II})$ from the Dunham observations of χ^2 Orionis (which is not subject to abundance uncertainties) Seaton determined $n_e = 0.066\,\text{cm}^{-3}$. Seaton presented the fundamental principle that both dust and gas must be included to evaluate the composition of interstellar clouds, and suggested that metals are concentrated in dust grains when gas composition departs significantly from cosmic values.

The empirical identification of interstellar clouds by velocity components and the sharp outlines of dust clouds on the Milky Way, suggests clouds are discrete objects. However, in 1952 Chandrasekhar and Münch introduced an alternative picture of the ISM in terms of a continuous distribution of material with statistically distributed density fluctuations (see article accompanying Scalo 1999, S99). The microscale of turbulent eddies was characterized by a fundamental parameter corresponding to the distance where a correlation function falls to $1/e$. The implications and influence of this stochastic model of interstellar clouds in contrast to discrete cloud models is discussed by John Scalo (S99).

The research of Lyman Spitzer, Jr., strongly influenced the study of the ISM for six decades, and ultimately lead to the launch of UV observatories into space. Two oft-cited papers (reprinted in 'Dreams, Stars, and Electrons', along with his personal comments (Spitzer 1997)) published in the 1950s provided proof of the correlation between abundance and velocity anomalies in Ca II (the Routly–Spitzer effect, Section 1.1) and predicted a hydrostatic highly ionized galactic corona. Spitzer summarized the current state of ISM studies in the December 1953 Henry Norris Russell lecture of the American Astronomical Society (Spitzer 1954). Spitzer investigated the putative halo from several approaches. He made an analogy between thermally driven winds in the terrestrial atmosphere and warm heated interstellar gas, and concluded that warm interstellar clouds would be more buoyant than surrounding unheated material and thus would create a "galactic wind" (Spitzer 1954). The discovery at Mt. Palomar of high-latitude halo clouds 1–2 kpc away from the galactic plane, with velocities $\leq 100 \, \text{km s}^{-1}$ (Münch 1957), raised direct questions about the source of the confining pressure. Spitzer made the bold prediction that highly ionized hot "coronal gas" ($n_e \sim 5 \times 10^{-4} \, \text{cm}^{-3}$, $T \sim 10^6 \, \text{K}$) confined these clouds, and also explained high-latitude radio emission (Spitzer 1956). Chandrasekhar and Fermi (1953) had reasoned that the gravitational pressure of ISM in spiral arms was balanced by the sum of magnetic and thermal cloud pressures, for field strengths $B \sim 6 \, \mu\text{G}$. Using a larger thermal pressure, Spitzer concluded thermal and gravitational pressures balanced, so additional interarm pressure from hot coronal gas was required to balance magnetic pressures. The primary tracers of coronal gas are the abundant highly ionized ions Si IV, C IV, N V, and O VI, with UV resonant transitions. At that time the predicted emission measure of coronal gas ($\mathcal{E} \sim 0.1 \, \text{cm}^{-6} \, \text{pc}$) was undetectable in $H\alpha$ photographic surveys. The coronal gas predicted by Spitzer in 1956 was found by Copernicus about 17 years later.

The presence of both cold and warm H I raised theoretical questions about the thermal equilibrium between cold, dense clouds and warm, dilute intercloud material. Thermal stability in a "multiphase medium" was evaluated in a famous paper by Field, Goldsmith, and Habing (Field et al. 1969), using the criterion that heating and cooling balance in thermal stability. (Many excellent reviews of interstellar theory are found in the proceedings of a workshop held in honor of George Field in Elba in 1994 (Ferrara et al. 1995). The Elba volume presents the "scientific genealogy" of George Field, listing his advisor, Lyman Spitzer, Jr. Carl Heiles was Field's first graduate student. Less importantly, I was Heiles' first graduate student.) This requirement means that for cooling $\sim T^\alpha$, α must be ≥ 1 and cooling increases with rising temperature. The dominant cooling mechanism for cool clouds is temperature-dependent fine-structure emission from trace elements (such as the C II $\lambda \, 158 \, \mu\text{m}$ line). FGH assumed temperature-independent cosmic-ray heating of clouds. The result is a thermally unstable phase between two thermally stable phases on the phase-stability $\log(n_H)-\log(T)$ diagram. These three phases were humorously identified as F(ield) for the warm intercloud phase, H(abing) for the cold neutral phase, and G(oldsmith) for the unstable phase (Don Goldsmith now pursues a prominent and stable literary career sharing the wonders of science with the general public). The classic FGH paper for modeling a multiphase ISM, along with the discovery of the Radio Loops, inspired many studies of cloud equilibrium in a supernova-dominated ISM. The FGH two-phase theory has been updated with recent heating and cooling rates, including photoelectric heating by small grains and large molecules (PAHs), and cooling by electron recombination on grain surfaces in warm ($T \geq 8000 \, \text{K}$) gas, and allowing variable input abundances (Wolfire et al. 1995).

1.4 The void in nearby ISM

Our view of nearby interstellar gas is biased by the location of the Sun with respect to Gould's belt and the related local void in interstellar matter now known as the "Local Bubble" (e.g., see Frisch and York 1986). In the nineteenth century, Benjamin A. Gould (founder of the *Astronomical Journal*) realized that nearby bright stars and young stellar OB associations form a "ring" on the sky surrounding the Sun ('Gould's belt'). Up to about 1980, optical and UV ISM studies were heavily biased towards clouds associated with the relatively nearby (<600 pc) massive hot stars of Gould's belt, since these bright, hot, rapidly rotating stars provided suitable stellar continuum for sampling interstellar clouds. Since hot massive stars tend to be associated with the parent cloud from which they formed, the ring-like distribution of the stars in Gould's belt also defines the ring-like spatial distribution of most of the mass of interstellar material within $\sim 600 \, \text{pc}$ of the Sun. The Sun is located inside of this ring and hence inside a void in the distribution of interstellar matter in space. The Local Bubble overlaps the interior of Gould's belt.

The Gum Nebula borders the Local Bubble. It is an extended region of ionized gas originally discovered in $H\alpha$ photographs of the southern Milky Way by Colin Gum in the early 1950s (diameter $> 36°$, $d \sim 250$–$500 \, \text{pc}$, centered near $l^{II} \sim 258°$, $b^{II} \sim 0°$, $\mathcal{E} \sim 1300 \, \text{cm}^{-6} \, \text{pc}$). The Gum Nebula is astronomically rich, with several stellar associations, over six pulsars, several supernova remnants (including Vela), and is bordered by the cluster Collinder 121 which forms one end of the Local Bubble (Heiles 1998). The Vela SNR, $\sim 11,500$ years old, is located at a distance of $\sim 250 \, \text{pc}$ and is on the near side of the Gum Nebula.

Surveys of the reddening ($E(B-V)$) of starlight by interstellar dust in front of nearby B8–A0 stars found that color excesses outline a vacant region surrounding the Sun, extending $\sim 100 \, \text{pc}$ in the directions of the galactic center and anti-center, and $\sim 50 \, \text{pc}$ in the directions of galactic rotation ($l^{II} \sim 90°$) and anti-rotation ($l^{II} \sim 270$; see references in Lucke 1978). An expanding H I ring surrounding the Sun, known as "Lindblad's ring," has morphology and kinematics suggesting association with Gould's belt (e.g. Poeppel *et al.* 1981). The shell expansion age (~60 Myr), elongation, and morphology are consistent with distortion by galactic rotation. Elmegreen argues that both Gould's belt and Lindblad's ring formed during the compression of ISM during the passage of the last spiral arm (about 50 Myr ago; e.g. Elmegreen and Efremov 1998). The intercloud $E(B-V)-N(H)$ relationship determined from Copernicus data (Section 2.5), suggests $E(B-V) = 0.1$ mag, corresponding to $N(H) \sim 5 \times 10^{20}$ cm^{-2}, as the contour boundary approximately defining Local Bubble "walls." A plot of the space-motion of the Sun through the Local Bubble shows that the Sun has been within this low interstellar density region for several million years (Figure 3, and Frisch and York 1986), a fact I do not believe a coincidence given the sensitivity of heliosphere properties to interstellar pressure (Zank and Frisch 1999). By analogy, Sun-like stars with historically stable astrospheres may be the best planetary systems for a search for advanced life forms (Frisch 1997, Frisch 1993).

2 THE COPERNICUS ERA

The dream of flying an ultraviolet spectrometer in space became politically achievable when Cold War politics spawned the space age and the launch of Sputnik (4 October 1957). Before space UV data on the ISM from sounding rockets and early satellites (OAO-2, ESRO, TD-1A), the primary sources of data were optical absorption lines and radio observations which traced primarily cold and warm H I. The ultraviolet, and later extreme ultraviolet (EUV), energy intervals provided a window on hitherto invisible ISM.

The Copernicus satellite changed the course of astronomy with discoveries of deuterium outside of Earth, the non-dominance of highly ionized species in the intercloud medium, the association of H_2 with reddened stars, and variations in the composition of the gas phase of the ISM between intercloud and cloudy regions. These first Copernicus results were published in a series of discovery papers in the May 1973 *Astrophysical Journal*. Several review articles by members of the original science team have summarized the scientific achievements of Copernicus (Spitzer and Jenkins 1975, York 1976, Snow 1976). Copernicus found ~6–9 UV velocity components per ± 20 km s^{-1} velocity interval. The range of ionization, temperature, and densities discovered by Copernicus for diffuse clouds resulted in the slow abandonment of the "intercloud medium" concept. The Copernicus data set remains a unique source of data on disk clouds, and many of the Copernicus targets have never been re-observed in the UV.

The launch of Copernicus coincided with strong growth in radio and optical astronomy. Significant advances occurred simultaneously in many areas of ISM studies. The next satellite with a UV spectrometer was IUE, launched in 1978 as Copernicus was ending its useful scientific lifetime. IUE surpassed expectations by providing extensive data on ISM in the halo and towards faint disk stars ($V \leq 11$ mag). Hubble Space Telescope (HST), launched in 1990, provided a UV observatory with a large collecting area in space. The spectral resolution of Goddard High Resolution Spectrograph (GHRS), gave the physical properties of individual disk and halo clouds. The spectrometers of IUE and HST operated over a broader spectral interval (nominal 1200–3000 Å) than the Copernicus instrument (nominal 912–1400 Å). More recently, IMAPS, EUVE, and FUSE provided data in the 300→1200 Å interval. The understanding we have of ISM in our galaxy as we emerge into the 21st century represents a synthesis of data from diverse sources.

2.1 Cloudy sightlines: ζ Oph

Copernicus conducted in-depth studies of the ISM towards ~20 stars, including moderately reddened stars (ζ Oph, ξ Per, o Per, ζ Per, χ Oph), and unreddened or lightly reddened stars (μ Col, HD 28497, HD 50896, γ Ara, ζ Pup, α Vir, ε Ori, π^5 Ori, 15 Mon, ε Per, δ Per, μ Oph, λ Sco, β CMa; for references see Jenkins 1987). The most influential of these studies was the first thorough study of interstellar UV absorption lines towards the classic ISM target, ζ Oph (Morton 1975). ζ Oph (along with other stars) was also observed by HST, giving a detailed picture of the abundances in warm and cold clouds (Savage *et al.* 1992, SCS92). The ζ Oph sightline is used as an example, because it is consistently selected as a target for improving the quality and breadth of ISM studies.

Copernicus performed a complete scan of the UV spectrum of ζ Oph, and measured ~328 lines of ~40 atoms, >100 H_2 lines, and found ~44 unidentified lines (later attributed to C I, HD, or H_2). The ions Zn II, Ni II, and Cu II were detected for the first time. The -26 km s^{-1} and -14 km s^{-1} cloud blends were later referred to as 'Cloud A' and 'Cloud B', respectively (SCS92). High-resolution optical data (~0.3 km s^{-1} FWHM) show that the two main cloud groups towards ζ Oph represent blends of at least 13 absorption components (Barlow *et al.* 1995). The 10 components common to both the Ca II and Na I absorption profiles show a range of N(Na I)/N(Ca II) ratios, corresponding to 1.1→2.2 for Cloud A (or 0.5 if the -22.0 km s^{-1} component is included), and 5.6→400 for Cloud B (Barlow *et al.* 1995). Therefore the full velocity component structure towards ζ Oph has not yet been resolved in the UV, and the unresolved UV components represent blends of components with different properties.

The chemical abundances of the ISM are found from elemental depletion patterns (Section 2.6). Morton determined the abundances for over 20 elements and found sub-solar abundances, by factors of 3→4,000. Gas-phase abundances are generally described in terms of the "depletion" of the element, δ_X:

$$\delta_X = \frac{N(X)/N(\text{obs})}{N(X)/N(H)_{\text{Sun}}}.$$

The observed and solar abundances for element X with respect to H (or another undepleted element) are $N(X)/N(H)_{\text{obs}}$ and $N(X)/N(H)_{\text{Sun}}$, respectively. HST GHRS data improved abundance precision considerably.

The depletion patterns of Cloud A and Cloud B are distinctly different, and have become a template for defining the properties of warm disk gas and cold disk gas, respectively (e.g. Savage and Sembach 1996). The differences between depletions in Cloud A and Cloud B are shown clearly by the behavior of Ti II and Zn II in Figure 1. Figure 1 shows absorption lines in ζ Oph of, from top to bottom, K I (7698 Å), Na I (5895 Å), Zn II (2026 Å), Ti II (3383 Å), Ca II (3933 Å), Ca I (4226 Å), Fe II (2249 Å), and Cr II (2062 Å). The variations in the Ti II line strengths between Cloud A and B are purely depletion variations, since Ti II is the dominant ionization state. These variations can be compared with the column density differences between the two clouds, shown by the HST Zn II line (since Zn II is undepleted). Cloud A contains ~4% of the ISM mass in the sightline, but about 65% of the Ti II atoms.

In Cloud B, $\log N(H) = 21.13 \text{ cm}^{-2}$. The percentage of H contained in H_2 (~62%) is higher for ζ Oph than for any of the other stars surveyed by Copernicus. H_2 is shielded from optical pumping in Cloud B, giving $T_{ex} = 56$ K from the $J = 1, 0$ levels. Cloud B is sheet-like. Electron densities calculated from C I/C II, Mg I/Mg II, S I/S II and Ca I/Ca II give $n_e \sim 0.7 \text{ cm}^{-3}$ (free from abundance uncertainties). A cloud density and thickness of $n_H = 10^4 \text{ cm}^{-3}$ and $L \leq 0.05$ pc, respectively, follow from the assumption that electrons are from the photoionization of trace elements (primarily C). Cloud A includes contributions from the interstellar gas within 14 pc of the Sun, as is shown by comparative studies of the velocities (Section 3.2) and abundances (Frisch 1981) of α Oph and ζ Oph.

Typical cold cloud temperatures are known from Copernicus H_2 data (45→128 K), H I 21 cm absorption data (typically 25→75 K), and the median Doppler broading of Na I and K I (~80 K). Cloud densities in the range of $n_H = 115$–2700 cm^{-3} and $n_e = 0.15$–0.38 cm^{-3} are found from K I data, while observations of tiny-scale structure in H I 21 cm in the ISM give $n_H \lesssim 10^5 \text{ cm}^{-3}$. The mean H I and median H I + 2H_2 column densities for cold clouds are $N(H\ I) \sim 10^{20} \text{ cm}^{-2}$ (21 cm data, Section 3.1), and $N(H) \sim 2 \times 10^{20} \text{ cm}^{-2}$ (K I, Section 3.2), suggesting that perhaps half of the atoms in cold clouds are in molecular hydrogen.

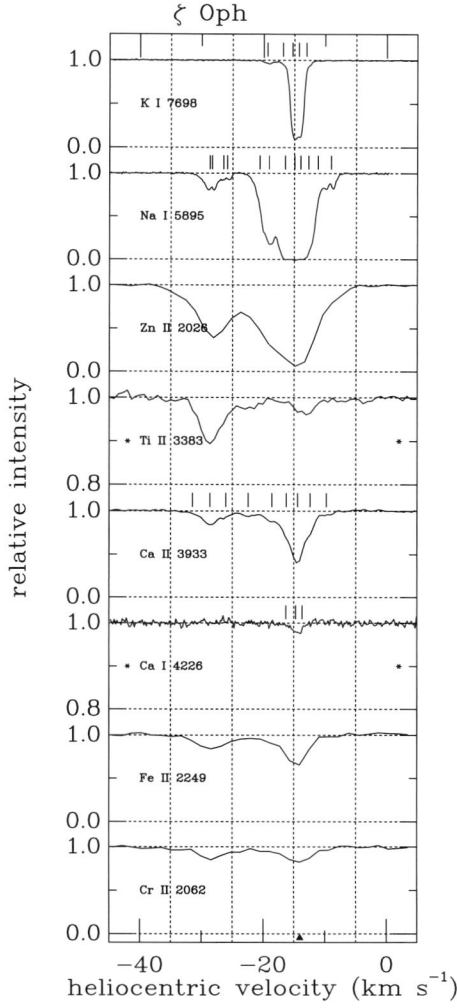

Figure 1 Optical and UV interstellar absorption lines towards ζ Oph. Cloud A and Cloud B are the prominant absorption features at $v_{hc} = -26$ km s^{-1} and -14 km s^{-1}, respectively. Both Zn II and Ti II represent dominant ions. The variations in Ti II line strengths between the two clouds are depletion variations, while the Zn II variations illustrate the differences in the column densities of the two clouds. The tickmarks give the central velocities of components fitted to the line. The triangle on the abscissa represents 0 km s^{-1} in the LSR. (Courtesy of Dan Welty.)

ISM studies at GHRS resolution fail to distinguished over 60% of the component structure (Section 3.2). These uncertainties are partially overcome by modeling UV data using velocity information from high-resolution optical absorption lines, in a method pioneered by Dan Welty to analyze HST observations of 23 Ori (Welty et al. 2000). The three components detected by Copernicus towards 23 Ori resolved into 21 velocity components in the combined optical/UV data (Figure 2). Variations in the spatial distribution between dominant, subordinate, depleted, and

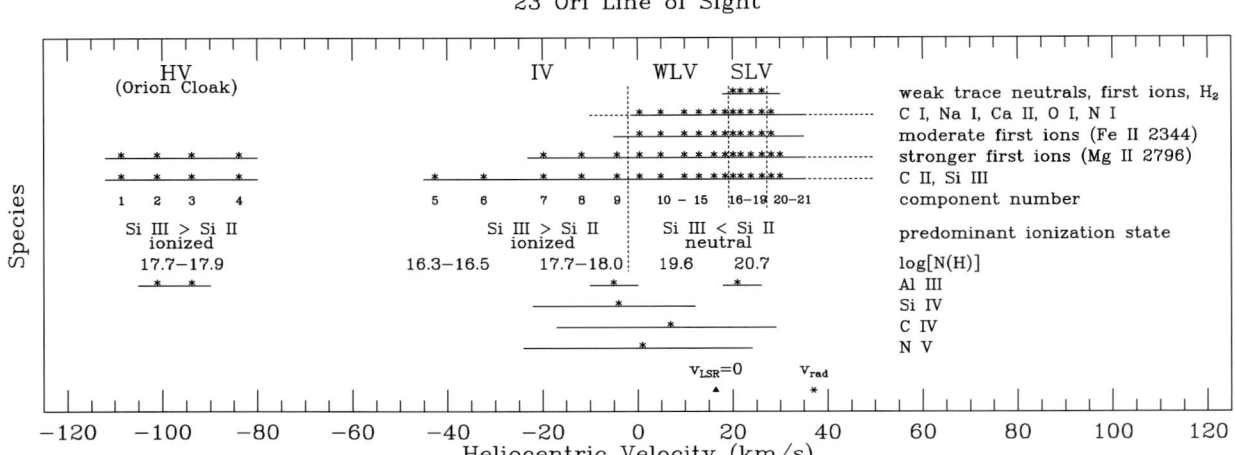

Figure 2 Distribution of interstellar clouds towards 23 Ori, illustrating the distributions of weak trace elements v. strong dominant ions in high-velocity (HV), intermediate-velocity (IV), weak low-velocity (WLV), and strong low-velocity (SLV) absorption components. (From Welty et al. 1999.)

undepleted species prevented the use of a single empirical COG to analyze line strengths and element distributions. Subordinate neutral ionization states sample cold, dense regions (K I, Ca I, Na I in Figure 1). The dominant ionization states of strongly depleted species show the broadest distribution (e.g., Ti II), reflecting enhanced column densities in warm gas, and ionization states of mildly depleted species have intermediate b_{Dop} values. The fine-structure C I lines give densities $n_H \sim 10\text{–}15\,\text{cm}^{-3}$, and a cloud thickness 12–16 pc for the low-velocity gas towards 23 Ori. Copernicus H_2 data gives $T = 65\text{–}150\,\text{K}$, yielding thermal pressure $\log(n_H T) \sim 3.1 \pm 0.1\,\text{cm}^{-3}\,\text{K}$. The inclusion of turbulence increases pressure by less than a factor of two.

The perils of deriving electron densities from a single species are illustrated by comparison of n_e values derived from 12 plasma diagnostics in the cold gas towards 23 Ori (Welty et al. 2000). Photoionization equilibrium calculations for C, Mg, Al, Si, P, S, Ca, Mn, Fe, and Ni, Na I, and K I yield calculated electron densities ranging from $n_e \sim 0.04\,\text{cm}^{-3}$ (S, Mn, Fe), to $n_e \sim 0.95\,\text{cm}^{-3}$ (Ca) and ~ 0.25 (C, Mg). The $N(\text{Ca I})/N(\text{Ca II})$ ratio indicates n_e varies by a factor of 2.6 between the two strongest cloud components. The average properties of the cold, low-velocity components give $n_e \sim 0.15 \pm 0.05\,\text{cm}^{-3}$, and $n_H \sim 10\text{–}15\,\text{cm}^{-3}$, or a fractional ionization of 1%, indicating that H must be partially ionized. The electron density determined from C II*/C II is a factor of 2–3 less than values derived from Na I, perhaps indicating the fine-structure lines are formed in the shielded cloud core.

2.2 Intercloud material

Radio data show that $\sim 60\%$ of interstellar H I atoms are contained in warm clouds ($>500\,\text{K}$, Section 3.1). Possible ionization sources for the intercloud medium include stellar radiation, soft x-rays, or cosmic rays (Dalgarno and McCray 1972). The initial search for intercloud absorption features towards unreddened stars ($E(B-V) < 0.03\,\text{mag}$) did not detect highly ionized species. Observations of bright unreddened stars (λ Sco, α Vir, HD 50896, HD 28497, and μ Col) showed that stars with low mean H I space densities may show foreground clouds that are small and dense ($n_H \sim 10\text{–}10^3\,\text{cm}^{-3}$), in addition to low-density warm gas. Copernicus found that N II is widespread in the ISM. IUE and HST provided data on warm, low-density clouds towards halo stars, and distant disk stars.

The λ Sco sightline (216 pc, $\log N(\text{H}°) \sim 19.5\,\text{cm}^{-2}$, $E(B-V) = 0.03\,\text{mag}$) contains neutral and ionized warm low-density interstellar clouds. λ Sco is in the interior of Loop I (Section 3.1) and samples nearby interstellar gas towards the upwind direction (see references in Jenkins 1987). Five physically distinct absorbing regions are found. A warm neutral cloud ($\log N(\text{H}) = 19.23\,\text{cm}^{-2}$, Component 2 shows typical properties for warm neutral gas, $T \sim 10,000\,\text{K}$, $n_H = 0.3\text{–}9\,\text{cm}^{-3}$, and less than 10% ionization. The cloud thickness is $\leq 5\,\text{pc}$. The spatial distributions of the neutrals H I, N I, O I, and Ar I are similar due to charge exchange (O I, H I) and photoionization (N I, Ar I). Component 3 is an extended density bounded H II region ($n(\text{H}^+) \sim 0.1\text{–}0.3\,\text{cm}^{-3}$) with diameter $\sim 10\text{–}30\,\text{pc}$, $T \sim 10^4\,\text{K}$, and ionization consistent with ionization by the general galactic radiation field with $T_{eff} \sim 24,000\,\text{K}$ (comparable to a B1 star and the diffuse galactic field predicted by Stromgren). The fine-structure populations of C II and N II give $n_e = 0.1\text{–}0.3\,\text{cm}^{-3}$, and Mg I/Mg II implies $n_e \sim 0.5\,\text{cm}^{-3}$. The emission measure of this diffuse H II region towards λ Sco is comparable to the fully ionized H II filaments detected in Hα emission ($\int n(e^-)^2\,dL \sim 1\text{–}3\,\text{cm}^{-6}\,\text{pc}$;

e.g. Reynolds 1995). Depletions in the warm, neutral, and ionized components are similar, log $\delta_{Fe} \sim -1.4$ dex. Mg I formation is enhanced in warm, $T \geq 6{,}000$ K, ionized gas where dielectronic recombination enhances Mg II recombination. Abundances in the H I and H II clouds are similar, including similar depletions for Fe and Si. Component 1 is partially ionized, with similar depletions as the other gas, and may represent a ionized cloud ~ 7 pc thick. Components 4 and 5 are highly ionized (with S IV, and N(Si III)/N(Si II) > 1) and may represent shocked gas.

Low column density (N(H) $< 10^{18}$ cm^{-2}) intermediate-velocity photoionized clouds are a common component of the ISM. For example, HST observations of intermediate velocity gas (IV, $v_{lsr} = 20$–50 km s^{-1}) towards disk (e.g., 23 Ori) and halo (e.g., μ Col) stars show warm (4,000–8,000 K) tenuous components of photoionized gas with N(Si III) $> N$(Si II). Towards 23 Ori, $n_e = n$(H$^+$) = 1.5–5.0 cm^{-3}, and the cloud thickness is ≤ 0.001 pc (~ 200 AU; Welty et al. 2000). Copernicus showed that this intermediate-velocity gas covers Orion, indicating filamentary material.

2.3 Deuterium

The Copernicus discovery that H$_2$ is widespread in interstellar clouds was not a surprise. However, York's discovery of D I was unexpected. As a junior member of the Copernicus staff, Don York was assigned the "uninteresting" stars where no absorption features had been seen optically. However, these were the stars where unsaturated lines permitted the most accurate identification of absorption features. The D I Lymanα line is superimposed on the wings of the H I Lymanα line, and blueshifted by ~ 80 km s^{-1} from the H I Lymanα line center. In the unreddened star β Cen, the D I line is unsaturated, giving the first D/H ratio determined outside of the Earth, N(D I)/N(H I) = 1.4 $\pm$ 0.2 $\times$ 10^{-5} (Rogerson and York 1973). Hydrogen column densities were determined from Lyman $\beta \rightarrow$ Lymanδ lines, minimizing uncertainties due to Lymanα saturation.

Deuterium is produced in Big Bang nucleosynthesis, and $\sim 50\%$ of the primordial D is destroyed by astration in stellar interiors. The early D/H ratio is consistent with ratios derived recently from HST observations of nearby stars (e.g., see papers by Vidal-Madjar, Linsky, and collaborators). However, some studies suggest the observed variations in the D/H ratio between stars are statistically significant and represent real variations in the chemical composition of nearby ISM (Vidal-Madjar and Gry 1984). The high spectral resolution of IMAPS (~ 2.5 km s^{-1}) permitted this question to be revisited (Sonneborn et al. 2000). Significant variations in D/H are still seen, from 0.74 $\times$ 10^{-5} in δ Orion A to 2.2 $\times$ 10^{-5} in γ^2 Vel. These variations, however, do not correlate with N/H, as expected for ISM astration in stellar interiors. Therefore this very important question is unresolved. Unresolved saturation in the D I lines may be the culprit in these discrepant results, and this question may not be resolved until an ultra-high-spectral UV spectrometer is launched.

2.4 Molecular hydrogen

Molecular hydrogen probes cold clouds and provides an interstellar thermometer (see reviews by Spitzer and Jenkins (1975) and Shull and Beckwith (1982)). Theoretical studies predict that most hydrogen is in molecular form in reddened clouds. The formation of H$_2$ on grain surfaces is offset by a two-step dissociation process. Photons, absorbed into the Lyman and Werner bands, cascade from the excited electronic states into unbound vibrational levels. Shielding in cloud cores allows the fraction of hydrogen in H$_2$, $f_{H2} = 2N$(H$_2$)/[N(H I) + 2N(H$_2$)], to approach unity. For N(H$_2$) $> 10^{19}$ cm^{-2}, the H$_2$ $J = 0$ and 1 levels typically have strong damping wings which provide unambiguous column densities. The $J = 0,1$ levels give excitation temperature T_{ex} and $T_{ex} \approx T_k$. Observations of ≈ 61 stars with N(H$_2$) $> 10^{18}$ cm^{-2} give a mean cloud temperature of 77 $\pm$ 17 K, and a range of 45$\rightarrow$128 K.

In unreddened sightlines, self-shielding is no longer able to reduce optical pumping. The detection of complex H$_2$ profiles, with apparent broadening of the H$_2$ lines with increasing J towards many stars (e.g., λ Ori, δ Ori A, μ Col, κ Ori, 15 Mon, 30 CMa, ζ Pup), was explained as due to one or more short-wavelength components that are relatively strong at high J-values and therefore have high T_{ex}. UV pumping models produce the blueward components in dense $n_H >100$ cm^{-3}, sheet-like (<0.02 pc) clouds that may either be close to the background star or associated with compressed neutral post-shock gas or thin expanding shells (Spitzer and Morton 1976).

Molecular hydrogen is widespread in clouds with $E(B - V) > 0.08$ mag, where H makes the transition from primarily atomic ($f_{H_2} < 0.01$) to molecular ($f_{H_2} > 0.01$). For $E(B - V) < 0.15$ mag 90% of the stars have $f_{H_2} < 0.1$, and for $E(B - V) \geq 0.15$ mag 74% have $f_{H_2} \geq 0.1$. The mean gas-to-dust ratio for all sightlines is N(H)/$E(B - V)$ = 5.8 $\times$ 10^{21} atoms cm^{-2} mag, but stars primarily sampling "intercloud" gas towards disk stars ($f_{H_2} < 0.01$) yield [N(H I) + 2N(H$_2$)]/$E(B - V)$ = 5.0 $\times$ 10^{21} atoms cm^{-2} mag (Bohlin et al. 1978). More recent comparisons between high-angular-resolution H I, CO, and 100 μm emission show that ~ 26 infrared cirrus clouds show substantial amounts of H$_2$ for N(H) $> 4 \times 10^{20}$ cm^{-2} (Reach et al. 1994). Recent FUSE results show that H$_2$ is ubiquitous in the galactic halo.

The excited J lines were blended at Copernicus resolution. IMAPS (with a nominal resolution of ~ 2.5 km s^{-1}) measured about 70 absorption lines from the Lyman and Werner H$_2$ bands in π Sco for the rotational levels $J = 0 \rightarrow 5$

(Jenkins et al. 1989). However, systematic shifts in $N(H_2)$ reconstructed from strong v. weak lines indicate that IMAPS did not resolve all of the absorption components. These inconsistent results are resolved if the lines consist of packed unresolved components with pure thermal broadening at $T \sim 80$ K. The high J-level column densities are consistent with population by optical pumping from radiation originating in nearby hot stars.

2.5 The distribution of nearby clouds

Hydrogen is the most abundant element in the ISM, and accurate values for $N(H\ I)$ could be determined from the Lymanα through Lymanϵ lines of H I in the Copernicus bandpass. Accurate $N(H\ I)$ values are found from Lymanα data when the damping wings can be used, typically for $N(H\ I) > 5 \times 10^{18}$ cm^{-2}, and stars hotter than $\sim$B1. A survey of Lymanα and H_2 towards $\sim$100 nearby (<1 kpc) stars provided a detailed picture of the spatial distribution of interstellar hydrogen (e.g. Bohlin et al. 1978). The asymmetric spatial distribution of ISM within $\sim$1 kpc in the disk of the Galaxy, seen originally in reddening data, and again in OAO-2 data, is clearly demonstrated in Copernicus data as a systematic variation of $\overline{n_H}$ as a function of galactic longitude. The Local Bubble, as projected onto the galactic plane, is shown in Figure 3 (also see Paresce 1984). (Different perspectives on the formation of the Local Bubble are presented in Breitschwerdt et al. (1996).)

The mean observed H density is $\overline{n_H} = <n(H^0) + 2n(H_2)> \sim 1.15$ atoms cm^{-3}. The lowest values are in the third quadrant, $\overline{n_H} < 0.008$ cm^{-3}. (The third galactic quadrant corresponds to $l^{II} = 180° \rightarrow 270°$, towards β CMa.) Somewhat larger values are found towards Orion, typically $\overline{n_H} = 0.2 - 0.5$ cm^{-3}. Higher densities are found towards Ophiuchus (e.g., $\overline{n_H} = 5.5$ cm^{-3} towards χ Oph) and Perseus (e.g., $\overline{n_H} = 2.2$ cm^{-3} towards o Per). Lower densities are found in intercloud sightlines ($f_{H_2} < 0.01$), $\overline{n_H} = 0.16$ atoms cm^{-3}. Mean densities for H_2 and H I, respectively, are $\overline{n_{H_2}} = 0.14$ molecules cm^{-3} and $\overline{n(H^0)} \leq 0.86$ atoms cm^{-3} (Bohlin et al. 1978). Figure 3 shows the strong variations in the mean space density of ISM in the nearest 500 pc.

2.6 Chemical composition of disk ISM

Copernicus surveyed H I, H_2, C I, S I, O VI, Mg I, Mg II, P II, Cl I, Cl II, Mn II, Fe II, Cu II, Ni II, and Zn II across a range of interstellar cloud conditions, and characterized both the distribution and composition variations of the ISM within $\sim$500 pc. Together, Copernicus and optical observations measured transitions from two-thirds of the lightest 20 elements in the periodic table. The systematic underabundance of certain elements in neutral gas, compared to solar, could not be attributed to ionization effects. Debates continued on the origin of the dust grains which are the supposed repositories of the atoms missing from the gas phase, as postulated by Seaton. IUE data permitted abundance surveys towards faint disk and halo stars, although moderate resolution compromised the abundances derived from saturated lines. IUE studies of Zn II, for example, were useful (it is a weak feature and is located in a sensitive region of the IUE detectors). HST GHRS data provided extensive data on abundances in individual warm clouds in the disk and halo, and provided the detection of (or limits on) additional rare elements ($N(X)/N(H) < 10^{-8}$) not seen previously (e.g., B, Co, Ga, Ge, As, Se, Rb, Sn, Pb, V, Kr, and Te). The complete picture on elemental abundances in disk clouds provide data that can be used to test grain formation hypotheses and chemical mixing in the ISM. Several excellent reviews discuss ISM abundances, including references not identified here (e.g. Jenkins 1987, Harris 1988, Savage and Sembach 1996).

Several primary trends in abundance patterns were by discovered by Copernicus, and confirmed by IUE and HST data. A. The depletion patterns of elements are grouped, with refractory elements consistently showing greater depletions than volatile elements. B. The most depleted elements also show the largest variations in depletions (which may be

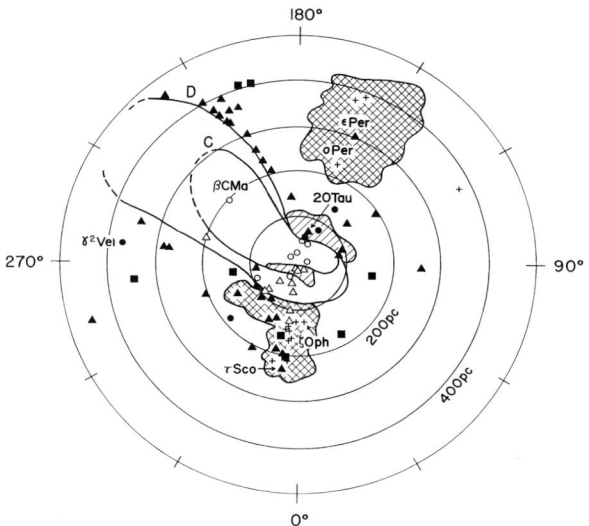

Figure 3 Column densities for interstellar hydrogen ($N(H)$) shown plotted against star position for stars within 500 pc. The inner contour (C) corresponds to $N(H) = 5 \times 10^{18}$ cm^{-2}, and the outer contour (D) to $N(H) = 5 \times 10^{19}$ cm^{-2}, where $N(H) = N(H\ I) + 2N(H_2)$. The apex direction of solar motion, towards $l^{II}, b^{II} = 28°, +32°$ indicates that the Sun has been located in region of space with low interstellar densities for million of years (e.g. Frisch 1995). Note the void around the Sun corresponding to the Local Bubble. (From Frisch and York 1983.)

a tautology, as Don York has pointed out). C. Volatile elements are generally undepleted in most sightlines, and may be undepleted in *all* sightlines if component structures are fully resolved and accurate N(H) values available. D. Depletions generally correlate with the mean density in a sightline, $\overline{n_H}$, although scatter is present. E. Refractory element abundances generally anticorrelate with cloud velocity. These trends are summarized in Figure 4 (trends A and C) and Figure 5 (trends B and C).

Refractory elements (e.g., Fe, Mg, Mn, Ca, Ti, Cr, Ni, Cu, Al, V) are characterized by high condensation temperatures and large depletions (log $\delta \leq 1.0$ dex in the cold clouds). In contrast, volatile elements (Kr, S, Cl, Zn, C, N, O, Na, Cl, Ar) are characterized by low condensation temperatures ($T_{cond} < 1000$ K) and small depletions. The volatiles Zn and S are found at solar abundances in most sightlines (log $\delta < 0.5$ dex), and arguments for small depletions towards reddened stars are unconvincing because of unresolved component structure. This behavior of refractory v. volatile elements is evident in the Routly–Spitzer effect, originally identified for optical Ca II and Na I lines, although Ca II ionization offsets reduced depletions in warm, low-density gas.

A Copernicus survey of Fe II towards 55 early-type stars provided a statistically significant sample which established the anti-correlation between gas-phase abundances of Fe and $\overline{n_H}$ (Savage and Bohlin 1979). Typical values range from log $\delta_{Fe} \sim -1.7$ dex towards Orion stars ($E(B-V) \leq 0.12$ mag) to log $\delta_{Fe} \sim -2.5$ dex towards ρ Oph ($E(B-V) = 0.47$ mag). Ionization is not responsible for these variations, since Fe II is the dominant ionization state in warm neutral and ionized gas. Also, warm ionized and neutral clouds with similar densities and temperatures show similar depletions. The variable $\overline{n_H}$ is also clearly not the best descriptor of depletion variations, since, for example, ζ Oph has two cloud groups exhibiting dramatically different abundance patterns (Figure 1), but the same $\overline{n_H}$ (which is a sightline averaged value). The large variations in gas-phase abundances found for refractory elements is easily understood by noting that $\gtrsim 99\%$ of the Fe atoms, for example, are depleted onto dust grains. The destruction of $\lesssim 1\%$ of the dust grains doubles the gas-phase abundance of Fe.

The true test of ISM metallicity will require accurate abundances for volatile elements in clouds sampling a wide range of physical conditions. B-star abundances are subsolar by a factor of ~ 2, and may instead be the appropriate model for interstellar metallicity (Snow and Witt 1996). Currently, evidence for solar metallicity in the ISM is limited. Copernicus, IUE, and HST consistently show that volatiles (Zn, S, Cl, N, O, Kr) are present is approximately solar abundances in sightlines with low mean densities. Studies of Zn (>200 stars), Cl (>40 stars), and S (>200 stars) show that these elements are depleted by less than factors of two in low-density ($\overline{n_H} < 0.1$ cm^{-3}) sightlines.

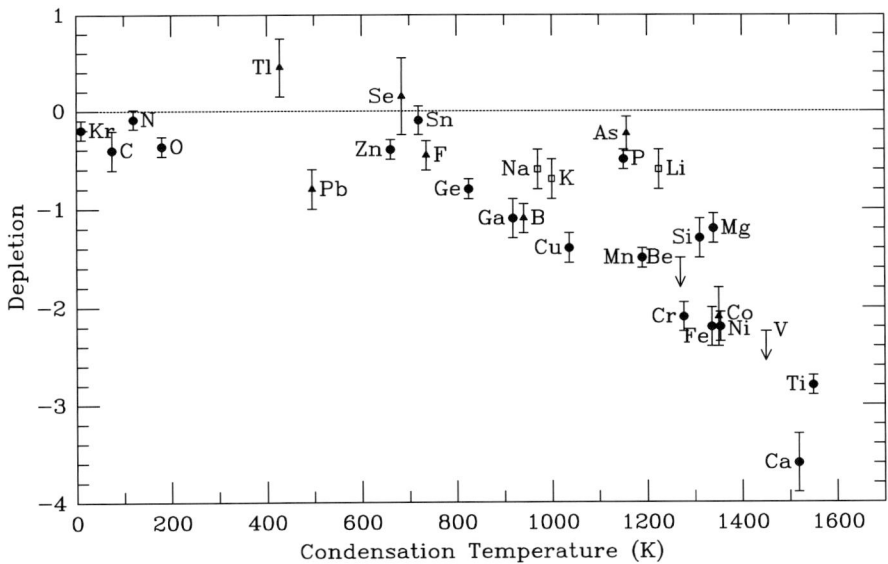

Figure 4 Interstellar depletions (log δ_X = log(X/H) − log(X/H)$_\odot$) in cold disk clouds shown plotted against the condensation temperature (T_{cond}). T_{cond} is the temperature at which 50% of an element has been removed from the gas phase through condensation in a solar composition cooling gas under equilibrium conditions. Variable abundances for a given value of T_{cond} (e.g., Mg v. Fe) reflect the dominant mineralogy for condensation in stellar environments (Ebel 2000). (Courtesy of Dan Welty.)

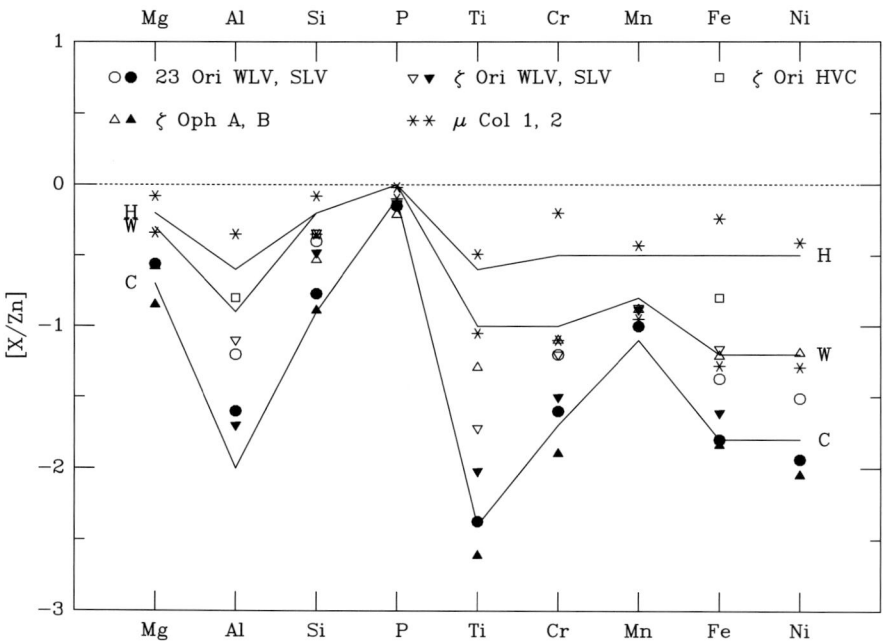

Figure 5 Depletions in disk and halo gas. H, W, and C indicate halo gas, warm disk clouds, and cold disk clouds based on HST data. Depletions from identified clouds towards ζ Oph, 23 Ori, ζ Ori, and μ Col are labeled. WLV, SLV, HVC refer to weak low-velocity, strong low-velocity and high-velocity clouds towards 23 Ori (Welty et al. 2000). (Courtesy of Dan Welty.)

The uniformity of Zn and S abundances at near solar suggest that metallicity is relatively invariant within ~500 pc. Limited results suggesting that Zn depletion increases with increasing f_{H_2} (e.g., log $\delta_{Zn} \sim -0.5$ dex at $f_{H_2} = 0.1$) needs deeper investigation of the velocity component structure. Evidence in favor of subsolar metallicity are observations of Kr, where observations of weak lines suggest abundances that are ~55% solar (Cardelli and Meyer 1997). The noble element Kr should not deplete onto dust grains.

The depletions of volatile elements are generally independent of mean density for $\log N(H^\circ) < 22$ cm^{-2}. Copernicus observations of O I (1356 Å) and N I (1160 Å) towards 53 stars found O and N depletions insensitive to $\overline{n_H}$ with ratios $N(C^0)/N(N^0) \sim 8$ to within 25–50%, and that O and N abundances are 40–70% solar. HST provided C, N, and O abundances with systematically lower errors through observations of weak lines (e.g. Meyer et al. 1998). In clouds with log $N(H\ I) = 20.2$–21.2 cm^{-2}, C, O, and N abundances are, respectively, ~39%, 43%, and 81% of solar values. Sulfur is undepleted in warm clouds, but showed small depletions in cold clouds (~-0.1 dex). Apparent suprasolar abundances are observed for S in cases where $N(H\ I) \sim N(H\ II)$, and ionized gas is not included in estimating $N(H)$. Chlorine is a special case; it is weakly depleted (factors 2–3), but Cl II reacts rapidly with H_2. $N(Cl\ I)/N(Cl\ II)$ is correlated with H_2 for clouds where H_2 is optically thick, while Cl I is unobservable in most other sightlines.

Copernicus studies of the collisionally populated C I fine-structure lines provide an independent diagnostic of cloud pressure (e.g. Jenkins and Shaya 1979). As a trace neutral, $N(C^0) \sim n_H^2$, similar to Na I and K I which correlate tightly with C I. However, the correlations are slightly steeper than linear, possibly indicating that the photoionization rates may vary between the species, possibly from shielding of C-ionizing photons by dust in the thicker clouds. Using the fact that the relative amount of carbon in C I* v. C I** is a function of the local gas pressure and temperature, Jenkins and Shaya examined the parameter range appropriate for neutral clouds. The carbon fine-structure lines towards ~15 stars yielded H I gas pressures $\ll 10^4$ cm^{-3} K, while some stars showed pressures $p/k > 10^4$ cm^{-3} K, exceeding the standard ISM pressure $p/k \sim 1,500$ cm^{-3} K (Spitzer 1978). Possible sources for high-pressure gas include H II regions near hot stars.

The initial discovery of depletion differences between different elements led to discussions as to whether this was an effect of the first ionization potential (FIP) or T_{cond} of the element. Eventually better data resolved this question. Figure 4 shows that T_{cond} is clearly the dominant variable since Zn and Be, for example, have similar FIPs (~9.3 eV), but depletions vary by factors of >15. A second puzzle related to the overall variation of depletions with $\overline{n_H}$. This correlation was interpreted as support for *in situ* grain formation in interstellar clouds (Snow 1975). Spitzer, however, noted that a statistical sampling of a mixture of warm and cold clouds with densities <0.2, 0.7, and >3 cm^{-3} in the sightline would explain the observed variations in the depletion–$\overline{n_H}$ correlation (Spitzer 1985). Spitzer predicted that short pathlengths are more likely to sample intercloud gas, and intercloud contributions are required to explain the better correlation between depletions and $\overline{n_H}$, in comparison with other parameters such as N(H).

The opposing viewpoint was that grains condensed in stellar outflows, with depletions regulated by T_{cond} (Field 1974). The tight envelope of δ–T_{cond} shown in Figure 4 supports this later viewpoint. However, additional grain modification clearly takes place in the ISM, and mantle accumulation may occur. In principle, the mineralogy of interstellar dust grains can be derived from these depletion data. Ebel (2000) compared the condensation sequence of dust grains in a solar composition stellar outflow with observed depletions (Figure 4). He interpreted the natural groupings of elements with similar T_{cond} and depletions as the result of the formation of Ca-aluminates (Ca, Al, Ti), olivine (Si, Mg), and metals (Fe, Ni, Cr, C). The formation of each group corresponds to a phase in the condensation sequence. The ability of condensation calculations to explain the observed depletion groupings, including very different depletions found for elements with similar T_{cond} values (e.g., Mg, Ca), provides strong support for dust grain formation through condensation in stellar outflows.

2.7 The energetic ISM

Highly ionized gas and high-velocity disk clouds are two aspects of an ISM component where collisional ionization is significant. Highly ionized interstellar atoms were expected in the ISM, either through production by X-rays or cosmic rays. The abundances of Si IV, C IV, N V, and O VI peak, respectively, at temperatures ~ 0.6, 1, 2, and 3×10^5 K, in collisionally ionized gas. Stellar evolution supplies energy to the surrounding ISM through supernova explosions and mass loss from massive stars. Routly and Spitzer showed that abundance patterns vary with cloud velocity, an effect attributed to grain destruction in shocks associated with expanding supernova remnants. Such high-velocity gas may also be subject to ionization anomalies.

Broad, shallow, low-velocity O VI lines were found to be widespread. A Copernicus search for lines of O VI, C IV, Si IV, and N V towards five stars ($d \sim 0.1$–1 kpc) found that only O VI was generally seen (York 1974). Combined limits on N(S IV)/N(O VI) and line widths suggest an origin in hot gas with $2 \times 10^5 < T < 2 \times 10^6$ K. A more extensive survey of O VI towards both reddened and unreddened stars found temperature limits $T < 3.6 \times 10^5$ K, and line broadening and ion ratios consistent with formation in hot plasma filling ~ 20–100% of the sightline, and in pressure equilibrium with normal cool clouds (Jenkins 1978). About six O VI absorbing regions per kiloparsec were found, with N(O VI) $\sim 10^{13}$ cm^{-2} per region and a dispersion for bulk O VI cloud velocities of ~ 26 km s^{-1}. The velocity distributions of O VI, Si III, and N II are correlated, indicating the O VI lines are affiliated with less ionized gas (Cowie et al. 1979). The interstellar nature of O VI was shown by the correlation between N(O VI) and stellar distances. Asymmetrical O VI profiles towards low halo stars (HD 28497, μ Col) showed contributions from both local and halo gas.

The detection of truly interstellar C IV and Si IV required long disk sightlines (>1 kpc). The first detections of Si IV and C IV towards disk stars ($d = 0.3$–3 kpc) found the lines strongest towards O and Wolf–Rayet stars connected to nebulosity (Smith et al. 1979, Bruhweiler et al. 1980). Although most of the strong, sharp C IV and Si IV lines towards hot disk stars are formed in nebulosity connected with the stars, broad, shallow C IV lines with no associated Si IV appear to arise in the hot interstellar gas producing O VI (York 1977). HST data showed that a diffusely distributed component of hot gas (N(C IV)/N(Si IV) ~ 4.7 and C IV/N V ~ 3.0) also exists in the galactic disk (Huang et al. 1995). Hot gas towards distant disk stars, $d \lesssim 3$ kpc, shows a strong correlation between C IV, Si IV, and N V line strengths and star distance. The highly ionized and moderately ionized regions are associated, since the widths of high- and low-ionization lines are correlated. In the direction of ζ Oph, three types of highly ionized gas are seen. Narrow Al III components originate in an expanding H II region around ζ Oph, and broad weak Si IV and C IV components are detected at -26 km s^{-1} (Cloud A; Savage et al. 1994). GHRS observations of 23 Ori showed that high ions such as S III, Si IV, C IV, and N V are associated with both intermediate- and low-velocity warm and cold clouds (Figure 2).

The frequency of optical high-velocity clouds (HVCs) and H I 21 cm shells motivated searches for the UV signature of these features. Copernicus determined that HVCs ($|v_{lsr}| > 100$ km s^{-1}) are widespread in the disk, including towards active star-forming regions such as Vela and Orion–Eridanus, and also towards field disk stars. Two out of thirteen nearby associations (Orion and Carina) show high-velocity shells with expansion velocities >100 km s^{-1} and ages $\sim 400{,}000$ years (Cowie et al. 1981). Evidently recombinations lag cooling in these previously shocked clouds. HST observations of HVCs towards 23 Ori, Vela, and μ Col found supraionized gas, showing ionization levels consistent with cloud temperatures of $\sim 25{,}000$ K, but line broadening consistent with kinetic temperatures $T < 12{,}000$ K (e.g. Trapero et al. 1996). The discrepancy is resolved if the HVC represents a cooling radiative shock. The HVC gas is seen in front of most of the Orion stars, and has been denoted "Orion's Cloak." $N(C^{+*})/N(C^{+})$ yields $n_e = n_H = 0.4$–$0.5\,\mathrm{cm}^{-3}$ for $T \sim 8000 \pm 2000\,\mathrm{K}$, or pressure $\log(2n_e T) = 3.7$–$4.0\,\mathrm{cm}^{-3}\,\mathrm{K}$. If carbon is undepleted, the individual HV components have thickness 0.005–$0.12\,\mathrm{pc}$ ($1 - 24 \times 10^3\,\mathrm{Au}$). Several superbubbles were investigated by IUE and GHRS, including the nearby Radio Loops I and IV (e.g. Sembach et al. 1997). The superbubble properties are similar, showing strong Si IV, C IV, and weak N V. Within the bubble interiors, C IV is strongly enhanced compared to N V. Sightlines sampling tangential directions through the shells (e.g., Loop IV) show relatively normal high ion ratios, but broader lines. In many cases, high-ionization gas is kinematically associated with low-ionization species (e.g., N I, S II, Si II, Fe II) which have narrow absorption features at intermediate velocities.

Intermediate-velocity clouds (IVCs, $20 < |v_{lsr}| < 60\,\mathrm{km\,s}^{-1}$) are visible in N I, N II, Si III, Ca II, and Na I (Cowie and York 1978). The ratios N(Fe II)/N(S II) and N(Si II)/N(S II) correlate positively with increasing cloud velocity, indicating grain destruction processes such as produce the Routly–Spitzer effect (Shull et al. 1977).

2.8 The nearest ISM and the local interstellar wind

Prior to Copernicus the only observed features in nearby ISM were Ca II lines in a small region towards α Oph (Section 1.1). Copernicus observations of the nearest stars (1.3–20 pc) showed that low-density ($\overline{n_H} \lesssim 0.1\,\mathrm{atoms\,cm}^{-3}$) interstellar H I is present in front of all known stars, including α Leo (24 pc, $\overline{n_H} \sim 0.02\,\mathrm{cm}^{-3}$; Rogerson et al. 1973), and α Bootis (11 pc, $\overline{n_H} \sim 0.02 - 0.1\,\mathrm{atoms\,cm}^{-3}$; Moos et al. 1974). Contemporary with Copernicus, Boksenberg and co-workers used a balloon-mounted spectrograph to observe Mg I and Mg II in the direction of α Leo, yielding the first estimate of the electron density in nearby interstellar gas ($n_e \lesssim 0.6\,\mathrm{cm}^{-3}$ at $T = 10^4\,\mathrm{K}$; Boksenberg et al. 1975). The nearby ISM was observed in nearby stars of A, B spectral types (e.g., α CMa, α Lyr, α Gru, α Leo; Rogerson et al. 1973, Kondo et al. 1978), and cool (G, K) stars where interstellar Lymanα absorption is superimposed on chromospheric emission features (ϵ Eri, ϵ Ind, α Aur, and α Cen A; e.g. McClintock et al. 1978, Landsman et al. 1984). Interstellar lines superimposed on chromospheric emission require the uncertain analysis step of modeling unattenuated stellar Lymanα emission based on observations of solar Lymanα, which is the only unattenuated stellar Lymanα that can be observed because of the strength of the interstellar Lymanα feature. Nevertheless, a consistent picture emerged showing that the Sun is embedded in a low-density ($n(\mathrm{H}^\circ) \sim 0.1$–$0.15\,\mathrm{cm}^{-3}$) warm ($T \sim 7{,}400\,\mathrm{K}$) interstellar cloud extending to at least $\sim 3.5\,\mathrm{pc}$ from the Sun in all directions, and present inside of the Solar System (Adams and Frisch 1977). Observations of the unreddened stars α Vir and λ Sco found $T \sim 7000\,\mathrm{K}$ for clouds that are 1–10 pc in size and probably embedded in the local ISM (LISM). Nearby ($<30\,\mathrm{pc}$) interstellar gas has an asymmetrical distribution (Figure 3, and Genova et al. 1990). This asymmetry mimics the asymmetry of the Local Bubble in the sense that the lowest column densities through the Local Fluff are towards the third galactic quadrant and north galactic pole, and this asymmetry results from a flow of gas away from Loop I (Frisch 1981).

It interested me that the interstellar H I and He I features discovered within the Solar System showed velocities close to the velocity of interstellar Ca II in α Oph. I concluded that the Sun must be inside of the α Oph cloud, and proposed to use Copernicus to observe the interplanetary Lymanα glow. (However, I did not dare include this motivation in my observing time proposal, since at that time ISM was what was viewed towards distant stars, and not part of the Solar System. Don York, who organized the Copernicus Guest Investigator program, told me that before approving my proposal he had to do some library research to verify that interstellar gas really is observed inside of the Solar System.) The results were the first spectral observations of the Lyα glow, which established the relation between ISM inside and outside of the Solar System (Adams and Frisch 1977). Thus Copernicus obtained the first spectrum of interstellar Lymanα emission within the Solar System during the 1975 solar minimum conditions. We found a velocity for the "local interstellar wind" in good agreement with interstellar velocities towards α Oph, as well as several more distant unreddened stars (e.g., α Vir) where cloud location is less certain.

Copernicus also made the surprising discovery that the H I Lymanα line towards the nearest star, α Cen (1.3 pc), is redshifted by $\sim 8 \pm 2\,\mathrm{km\,s}^{-1}$ with respect to the unsaturated D I line in the same star (e.g. Landsman et al. 1984). This shift was originally interpreted as flagging the existence of two clouds in front of α Cen, with the Sun located close to the boundary between the clouds, and has since been confirmed by IUE and HST data. The heliosheath contribution to this line was recognized later. (e.g. Linsky & Wood 1996, Gayley et al. 1997).

The nearest ISM is nicknamed the "Local Fluff," a term first used by Don Cox at the Local Interstellar Matter COSPAR meeting held in Toulouse, France, in 1986. He was trying to describe the tenuous nature of the interstellar gas surrounding the Sun, and finally waved his hands and said "it's just sort of this local fluff." The Local Fluff represents the ensemble of interstellar clouds (or cloudlets) within $\sim 30\,\mathrm{pc}$ of the Sun, regardless of cloud velocity. (The proceedings of this meeting and a related heliosphere session offer an early look at research into the properties of the LISM.)

The ISM nearest to the Sun is representative of warm, partially ionized diffuse gas. HST studies of the nearest ISM show typical mean densities $\langle n(H°)\rangle \sim 0.1\,\text{cm}^{-3}$ and temperatures $\langle T\rangle \sim 7{,}000\,\text{K}$ are found, and observations of Mg II/Mg I and C II fine-structure lines yield ionization levels of $n_e \sim 0.12\,\text{cm}^{-3}$, which are well modeled by photoionization models which include EUV radiation from an interface between the Local Fluff gas and hot plasma. The LISM gas shows the characteristics of having been shocked, with $\delta_{Fe} \sim -0.67\,\text{dex} \rightarrow -1.34\,\text{dex}$ locally (Frisch et al. 1999). A recent radiative transfer model of the local gas finds $T \sim 7{,}000\,\text{K}$, $n(H°) \sim 0.24\,\text{cm}^{-3}$ and $n_e \sim 0.13\,\text{cm}^{-3}$.

There are several outstanding summary publications discussing nearby ISM (Kondo et al. 1984; Cox and Reynolds 1987; Frisch 1995, 1997; Ferlet 1999).

3 GROUND-BASED ASTRONOMY DURING THE SPACE AGE

The launch of the Copernicus satellite coincided in time with several important advances in ground-based astronomy, including the discoveries of giant shells of radio continuum emission, pervasive components of tenuous neutral gas, diffuse plasma, tiny 'cloudlets', and a highly structured neutral component seen in absorption in both optical and radio data. Ground-based astronomy benefited from large telescopes, long exposure times, and flexibility in detector design, providing an important synergy with UV data.

3.1 The radio sky

Radio astronomy supplies two-dimensional morphological data on the distribution of neutral and ionized gas in space, and provides a key perspective on the structural characteristics of the ISM, including determining the spiral structure of our Galaxy (see review of Burton 1976).

The inhomogeneous H I 21 cm sky led to the search for the lowest column density sightlines out of the Galaxy. In the galactic plane, the direction showing minimum column density ($N(\text{H I}) \sim 4.5 \times 10^{21}\,\text{cm}^{-2}$) is towards the sightline $l^{II} = 245° \pm 6°$ and $b^{II} = 3° \pm 6°$ (Stacy and Jackson 1982). This minimum occurs as a $\sim 6°$ hole in the local gas at $v_{lsr} = -10$ to $+5\,\text{km s}^{-1}$. The "Puppis Window" samples an elongated direction through the third quadrant void (discovered by reddening data) and contains the stars β CMa and ϵ CMa known for their exceptionally low column densities (e.g. Frisch and York 1983, Paresce 1984, Welsh et al. 1997). The Puppis Window appears to result from overlapping shells, including the 0.8 kpc distant supershell GSH 238+00+09. (Another name for the Puppis Window is the "β CMa tunnel.")

The Rosetta Stone for understanding H I shells, anomalous Ca II abundances, the preponderance of negative velocity Ca II components, and the Routly–Spitzer effect was the discovery of large loops of radio continuum emission. Four nearby (<250 pc), high-latitude, non-thermal, radio continuum shells (408 MHz) were found with surface brightness consistent with the brightness–diameter relation found for supernova remnants (SNRs; e.g. Berkhuijsen et al. 1971). The Radio Loops are formed by synchrotron emission from cosmic-ray electrons interacting with the interstellar magnetic fields that have been compressed into giant shells by expanding supernova remnants. In a widely influential talk at an IAU symposium in 1979, Harold Weaver linked the Loop I radio continuum shell with filamentary H I seen in 21 cm emission, and concluded both were formed by a large supershell around the Scorpius–Centaurus Association. The large nearby Radio Loop I (distance $\sim 115\,\text{pc}$, radius $\sim 116\,\text{pc}$), also known as the North Polar Spur (NPS), dominates ISM in the northern hemisphere sky and is associated with a slowly expanding H I shell ($\sim 20\,\text{km s}^{-1}$) and a bright source of x-ray emission. Loop I is a supernova remnant in the radiative phase. The radio continuum, H I, and x-ray emission from Loop I are spatially separated, with ionization decreasing from the interior to exterior of the shell. Comparison of rotation and emission measures, Zeeman splitting, and H I 21 cm data yield best values of $n_e < 0.4\,\text{cm}^{-3}$ $B_{\parallel} \sim 1.2 - 6\,\mu\text{G}$, $n(H°)$ $4\,\text{cm}^{-3}$, and $T \sim 100\,\text{K}$ (21 cm data) in the NPS (e.g. Heiles 1989). Magnetic pressure $B^2/8\pi \sim 4{,}000\,\text{K cm}^{-3}$ dominates the thermal pressure of neutrals, and indicates an unstable gas. This latter point is important because generally magnetic pressure is neglected.

The direction with minimum $N(\text{H I})$ towards the galactic poles is towards $l^{II} = 150°$, $b^{II} = +53°$, where $\log N(H°) = 19.65\,\text{cm}^{-2}$ (the "Lockman Hole"; Dickey and Lockman 1990). Four H I velocity components are seen towards this minimum (0, -10, -50, and $-100\,\text{km s}^{-1}$), with most H I mass at $\sim 0\,\text{km s}^{-1}$. The stratification of H I perpendicular to the galactic plane in this direction indicates that halo gas ($z > 500\,\text{pc}$) contains $\sim 13\%$ of the H I and co-rotates with disk gas. The H I layering can be described as the sum of three distributions. The CNM distribution can be described by a Gaussian with an r.m.s. scale height of $\sim 100\,\text{pc}$. The WNM can be described by the sum of two distributions: a Gaussian with r.m.s. scale height $\sim 250\,\text{pc}$, and an exponential with scale height $\sim 500\,\text{pc}$ (Dickey and Lockman 1990; Table 1).

Observations of H I in absorption v. emission have provided an increasingly precise picture of the relative distributions of cold and warm H I gas. The first large-scale surveys of 21 cm absorption towards bright radio sources yielded a statistical picture of the components of the two-phase ISM (e.g. Hughes et al. 1971, Radhakrishnan and Goss 1972). The equivalent thickness of the galactic plane for cold absorbing clouds was 330 pc, with a midplane density of $0.29\,\text{cm}^{-3}$ for an assumed Gaussian distribution. The WNM extended to larger scale heights, with thickness

Table 1 Stratification of interstellar gas away from the galactic plane

Component[*]		Mid-plane density (cm^{-3})	$h^{\dagger}$ (kpc)	Method[‡]	Reference
H_2		0.14	0.07	UV	Savage et al. (1977)
H I	CNM	0.70	0.07	UV	Bohlin et al. (1978)
	WNM	0.16	0.36	UV	Diplas and Savage (1994)
H I	CNM	0.39	0.11 g	21 cm	Dickey and Lockman (1990)
	WNM	0.11	0.25 g	21 cm	Dickey and Lockman (1990)
	WNM	0.06	0.40	21 cm	Dickey and Lockman (1990)
H II	WIM	0.014	0.07	DM	Reynolds (1995)
	WIM	0.024	0.9	DM	Reynolds (1995)
Fe II			0.5	UV	Edgar and Savage (1989)
Al III	WIM		$1.0^{+0.36}_{-0.24}$	UV	Savage et al. (1990)
Ca II[§]			1	Opt.	Edgar and Savage (1989)
Ti II[§]			>2	Opt.	Edgar and Savage (1989)
O VI	HIM	2.0×10^{-8}	2.7 ± 0.4	UV	Savage et al. (2000)
N V	HIM	2.0×10^{-9}	3.9 ± 1.4	UV	Savage et al. (1997)
C IV	HIM	9.2×10^{-9}	4.4 ± 0.6	UV	Savage et al. (1997)
Si IV	HIM	2.3×10^{-9}	5.1 ± 0.7	UV	Savage et al. (1997)

[*] CNM, WNM, WIM, and HIM are cold neutral, warm neutral, warm ionized, and hot ionized material, respectively. [†] Based on Savage (1995). The scale height h is defined by the density distribution as a function of distance to the galactic plane, z, and generally is described by an exponential distribution: $n(z) = n(0)\exp(-z/h)\,cm^{-3}$, where $n(0)$ is the midplane density and h is the scale height. Distributions labeled by 'g' are Gaussian: $n(z) = n(0)\exp(-z/h)^2/2$. [‡] Methods are: UV = ultraviolet absorption lines; 21 cm = H I hyperfine line; DM = pulsar dispersion measures; Opt. = optical absorption lines. [§] Enhanced abundances of Ca II and Ti II in the WNM yield smoother distributions than for the CNM.

585 ± 100 pc, and lower midplane densities ($0.155\,cm^{-3}$). Cold (60–80 K) absorbing clouds were found to contain 35–45% of the H I, and appear to be confined mainly to spiral arms. The H I absorbing clouds have mean column density $N(H\,I)/T_{spin} = 1.5 \times 10^{19}$ atoms $cm^{-2}\,K^{-1}\,kpc^{-1}$ and a harmonic mean temperature $T = 71° \pm 9°$. Typically over 2.5 cold clouds per kiloparsec are found for a sightline along the galactic plane. The emission is contained in broad features formed in ubiquitous warm intercloud H I. Mean spatial densities for the cold and warm clouds, respectively, are $\overline{n_H} \sim 0.7\,cm^{-3}$ and $\overline{n_H} \sim 0.25\,cm^{-3}$. The temperatures found for the WNM are uncertain, since a single unrecognized cold (~ 80 K) cloud in the sightline containing one-tenth of the warm cloud column density would decrease the WNM temperature-derived T_{spin} to 800 K (Kulkarni and Heiles 1987).

With the discovery of highly-ionized gas, the concept of a two-phase medium fell out of favor, the term "not strongly absorbing" (NSA) gas was introduced to distinguish the kinematically broad H I 21 cm emission components ($T \sim 5000$ K), from narrow cold absorbing components (Dickey et al. 1979). The NSA gas was attributed to either warm surface layers on cold clouds, or independent clouds. NSA clouds have an asymmetric velocity distribution, with $-20\,km\,s^{-1}$ and $+10\,km\,s^{-1}$ components equally frequent, possibly flagging infalling gas since many NSA sightlines are towards the north galactic pole region. The exponential form of the high-velocity tail was similar to the distribution of Ca II components seen optically (Section 1.1). NSA gas is deficient in the interior of the Local Bubble surrounding the Sun, and beyond that forms $\sim 35\%$ of the observed H I (Heiles 1980). Large holes in the NSA H I are common (diameters ≤ 400 pc), evidently corresponding to the interiors of shells surrounding stellar associations. Heiles showed the absence of NSA gas is associated with anomalous velocities and anomalous velocity dispersions for H I, and may represent ISM swept up by expanding shells of gas which have recombined before deceleration.

Recently Heiles (2001) has used high signal-to-noise data on Zeeman splitting of the H I absorption line to produce accurate temperatures for cold clouds. He found that 54% of the CNM clouds (containing 61% of the total CNM column density) have temperatures in the range of $T = 25 \rightarrow 75$ K. Colder clouds, $T = 10 \rightarrow 25$ K, contain $\sim 11\%$ of the mass. About 40% of the WNM components have $T = 500 \rightarrow 5000$ K, containing >47% of the total

WNM column density. About 60% of the H I is WNM with $T > 500$ K. The mean column densities for WNM and CNM, respectively, are $\sim 2.2 \times 10^{20}$ cm^{-2} and $\sim 0.8 \times 10^{20}$ cm^{-2}. The WNM temperature range ($T = 500 \rightarrow 5000$ K) coincides with the thermally unstable range which separates CNM from WNM in the MO theory.

The discovery of radio continuum loops, H I filaments, and H I shells led to a new class of ISM models, based on the injection of energy into the ISM from expanding supernova remnants. Models included overlapping remnants which create tunnels of hot gas in space (Cox and Smith 1974), galactic fountains which release hot buoyant gas into the low halo (Shapiro and Field 1976), models of cold clouds evaporating inside of supernova remnants (McKee and Ostriker 1977, MO), and models of expanding superbubble shells sweeping up the ISM (Mac Low and McCray 1988). The MO multiphase model predicted successive layers of hotter and more ionized gas on the surfaces of cool clouds. In this equilibrium picture, SNR evolution is altered by the evaporation of embedded clouds, and a large fraction of the disk is filled with hot ionized material (HIM, $T = 4.5 \times 10^5$ K, n = 0.0035 cm^{-3}, $\chi = n_e/n_H = 1.0$). MO predicted the outer regions of standard clouds of cold neutral material (CNM, $T = 80$ K, n = 42 cm^{-3}, $\chi = 10^{-3}$) are ionized by diffuse stellar UV and soft x-ray photons, producing layers of warm ionized material (WIM, $T = 8,000$ K, n = 0.37 cm^{-3}, $\chi = 0.15$), and warm neutral material (WNM, $T = 8,000$ K, n = 0.25 cm^{-3}, $\chi = 0.68$) near cloud surfaces. Models of the ISM are summarized nicely in articles by Cox and McKee in the Elba volume (Ferrara et al. 1995).

A seminal advance in understanding high-latitude ISM resulted from the combination of northern and southern hemisphere 21 cm emission data, to create a series of "maps" of the 21 cm sky ($\sim 0.5°$ resolution) at discrete velocity intervals for $|b| > 10°$ (e.g. Colomb et al. 1980). The ISM in the solar neighborhood exhibits rich structural complexity, with most material contained in giant loops, arcs, "worms," and incomplete shell-like features offering little support for the classic Ambarzumian picture of an ISM filled with spherical clouds. Widespread topless shells, capable of supplying hot gas to the halo, are seen throughout the Galaxy. The nearest example of a worm is the NPS. These maps, and subsequent 21 cm studies by Heiles and collaborators, showed that the ISM in the solar neighborhood is superbubble dominated. Over 100 expanding H I shells and superbubbles are detected, with radii <1.2 kpc, masses $2 \times 10^7 M_{sun}$, and expansion velocities <24 km s^{-1} corresponding to energies $\geq 10^{53}$ erg (e.g. Heiles 1982). H I shells are clearly associated with star formation regions (Orion, Carina, Per OB2, Cep OB3, Sco OB3, Loop I), although only $\sim 33\%$ of the radio-emitting supernova remnants also exhibit H I shells. H I gas in shells is significantly cooler ($T = 35–200$ K and $n_H \sim 2$ cm^{-3}) than outside of shells, and the expanding shells associated with Radio Loop I and Eridanus are in statistical equilibrium since shell ages (~ 2 Myr) exceed cooling times for 2 cm^{-3} shell gas (~ 0.4 Myr). These SNRs are also a source of high-velocity gas; $\approx 15\%$ of 100 observed radio continuum SNRs (1.5–9 kpc distant) show high velocity (70–160 km s^{-1}) H I, probably accelerated by SNR blast waves (Koo and Heiles 1991).

The pervasiveness of H I 21 cm high-latitude HVCs raised questions about the origin of this gas, and whether the gas is infalling onto the plane, or ejected from disk supernova remnants. Additionally, warped spiral arms in the outer galaxy are seen at high velocities, leading to the definition that generally gas must deviate by >50 km s^{-1} from predicted galactic rotation velocities to be considered a HVC. Several excellent reviews discuss HVCs (e.g. Verschuur 1975, York 1982, Wakker and van Woerden 1997, Savage 1995). High- and intermediate-velocity clouds are grouped into complexes tens of degrees across on the sky, with predominantly negative velocities at high latitudes. Up to 37% of the sky is covered with H I gas with $|v_{lsr}| > 100$ km s^{-1}. Up to 18% of the sky is covered with HVCs with $|v_{lsr}| > 100$ km s^{-1} and $\log N(H°) > 18.30$ cm^{-2}. Low column density HVCs (17.5–18.4 cm^{-2}) tend to surround higher column density HVCs (>18.4 cm^{-2}), and occupy about 19% of the sky. Up to 10% of the mass of H I in the Galaxy may be in HVCs, and several HVCs have bulk energy comparable to that of supernova (e.g. Verschuur 1975, York 1982, Wakker and van Woerden 1997, Savage 1995).

A new class of tiny dense H I clouds were discovered at Berkeley using VLBI techniques (Dieter et al. 1976), but acceptance of this discovery was delayed for years because it challenged the two-component model of the ISM. The search for H I clumps was motivated by the discovery of tiny ($\sim 10^{-2}$ AU) regions of ionized gas inferred from pulsar scintillation data (e.g. Spitzer 1978). Observations of 3C 147 showed absorption variations on scales of 0.1″. This discovery was later confirmed using VLBI techniques (Diamond et al. 1989), and measurements of time-variable 21 cm absorption line in six transversely moving pulsars (Frail et al. 1994). Over 20% of cold H I is located in small dense clumps with scale sizes 5–100 AU and densities $n(H°) \sim 10^5$ cm^{-3}. The "tiny-scale atomic structures" (TSAS) are a ubiquitous feature of ordinary cold clouds, and are consistent with a composite picture of cold TSAS clouds (~ 15 K, or $b_{DopH} \sim 0.5$ km s^{-1} in the absence of turbulence) interspersed with warmer and less dense "inter-TSAS" gas. The TSAS structures occupy only a few percent of the CNM cloud volume, but contribute 10% of the column density if they have a sheet-like configuration, or 30% if they have a filamentary configuration (Heiles 1997). Absorption lines have been seen to vary in strength between both members of a binary system, and towards the same star observed at multiyear intervals (e.g. Lauroesch et al. 2000). These variations imply structure in the ISM on scale sizes of 10^1–10^3 AU.

3.2 High-resolution optical data

The transition from photographic plates to photon-counting spectrometers (based on Digicon and CCD technology) occurred during the 1970s, allowing high-resolution (≤ 0.5 km s^{-1}), high signal-to-noise ($W < 1$ mÅ) absorption line studies. Although only limited species can be observed optically, ultra-high-spectral-resolution optical data provide a useful template for the component structure and depletion patterns in lower resolution UV data. Early surveys by Hobbs and his students provided the first information about the distributions of Ca II, Ca I, Ti II, Na I, K I, and Li I, and the physical properties of the gas giving rise to these lines. Recent surveys benefit from ultra-high spectral resolutions, high signal-to-noise, and good quantitative analysis tools.

The saturated Na I D lines and the crowding of components in velocity space inspired observations of the weak Na I UV 3302 Å doublet which fall on the linear portion of the COG. The mean widths for the 3302 Å and D-line doublets are $\langle b_{Dop} \rangle = 0.9$ km s^{-1} and 1.5 km s^{-1} (Crutcher 1975), indicating that the weaker features reveal cloud cores, while the D-lines trace more extended regions. After correction for ionization effects, sodium was found to be depleted by a factor of ~ 4–5. During the last decade of the 20th century, a series of high-resolution (~ 0.3–1 km s^{-1}) surveys of optical absorption lines provided a detailed look at the crowding of cloud components in velocity space. A high-resolution study of the Na I D1 line towards 38 stars found 276 absorption components (Welty et al. 1994, WHK). WHK developed a method of fitting complex absorption lines with individual cloud components characterized by component column density (N), velocity (V), and velocity dispersion ($b_{Dop}^2 = 2kT_k/m + 2v_t^2$). The Na I components have median values $b_{Dop} = 0.73$ km s^{-1} and $\log N$(Na I) $= 11.09$ cm^{-2}, and a median separation between adjacent components of ~ 2.0 km s^{-1}. Weaker Na I components exhibit a broader velocity distribution than stronger components, displaced somewhat to negative velocities. The correlation between N(H I) and N(Na I) decreases in quality for N(Na I) $< 10^{11}$ cm^{-2}. For internal cloud temperatures of 80 K, $>38\%$ components have subsonic internal turbulence for assumed Gaussian profiles for individual clouds. WHK showed that component separations for Na I are consistent with a Poisson distribution (Figure 6), and that for this distribution only $\sim 60\%$ of the true total number of cloud components are being detected (given detection limits). Similar results were found later for the distribution of Ca II and K I components.

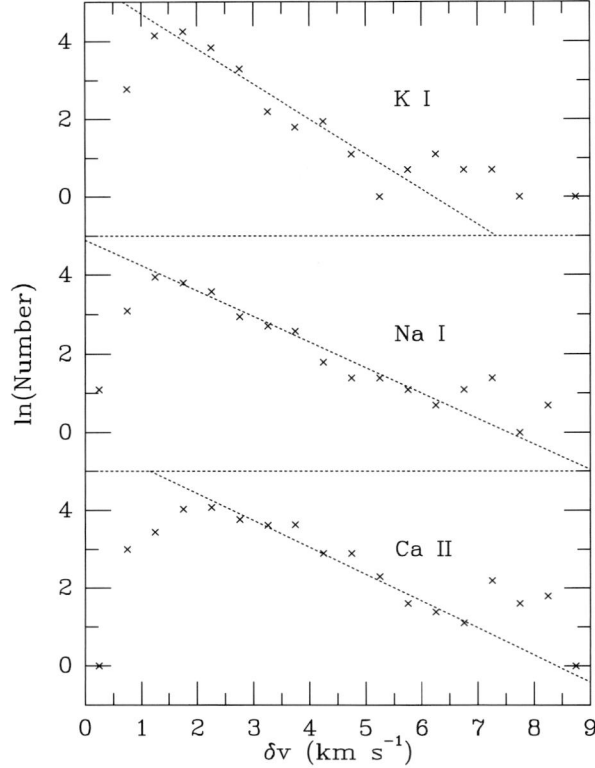

Figure 6 The distribution of velocity separations (δv) between adjacent absorption components for Na I, K I, and Ca II. The dotted line shows the linear fit over the range 2 km s^{-1} $\leq \delta v \leq$ 6.5 km s^{-1}, with slope -0.9 and -0.69, for K I, and Ca II, respectively. (From Welty et al. 1994, 1996; Welty and Hobbs 2001.)

The alkali element Li is observed optically. A survey of the Li I weak 6708 Å (IP $= 5.4$ eV) towards 22 stars found 7 positive detections at the 0.3–3 mÅ level (e.g. White 1986). The depletions of Li, C, Na, and K are correlated, and $\log \delta \sim -0.25$ dex in cold clouds (Figure 4). Recently, the relative isotope ratio ^{7}Li I/^{6}Li I has been observed as a test of the formation of light elements. The ^{7}Li I/^{6}Li I ratio appears to be variable, and for at least one sightline (o Per) cosmic ray spallation appears significant.

In cold clouds Ca II is heavily depleted onto grains, and Ca II is the dominant ionization state, while in warm gas Ca is only moderately depleted but Ca III is the dominant ionization state (see Figure 1). The ratio N(Ca II)/N(Ca I) in individual velocity components yields $n_e = 0.055$–0.57 cm^{-3}, independent of abundance uncertainties (e.g. White 1973). If photoionization of trace elements is the sole source of electrons, densities $n_H \sim 500$–5,000 cm^{-3} are implied. However, H may be partially ionized. Intercloud medium sightlines also show reduced line strengths of Na I (factor of >5) in comparison to sightlines containing clouds.

High-spectral-resolution (0.3–1.2 km s^{-1}) observations of the Ca II K line provide a component-to-component sample of the relative properties of the Na I and Ca II profiles (Welty et al. 1996, WMH). WMH found similar mean LSR velocities for Ca II and Na I (-1.7 km s^{-1}, -0.7 km s^{-1}), but larger velocity dispersions for Ca II (12.3 km s^{-1} v. 8.6 km s^{-1}). The median Ca II column density is 5×10^{10} cm^{-2}, and the median line width is $b_{Dop} = 1.31$ km s^{-1} corresponding to $T \leq 4{,}100$ K. Based on line widths, less than 35% of the components originate in warm gas (6,000–8,000 K). If $T > 500$ K, most components have subsonic internal cloud turbulent velocities. Inside of the same cloud, Ca II is distributed over a broader spatial region than Na I, as shown by b_{Dop} for Ca II and Na I (median values 0.84 and 0.65 km s^{-1} respectively). The ratio N(Na I)/N(Ca II) correlates with Na I (with an offset at larger column densities where Ca II is the dominant ionization state of Ca). Exceptions are sightlines showing the Routly–Spitzer effect (e.g., α Oph). Along low-density sightlines, up to 10% of the total mass in gaseous Ca II is at $V > 10$ km s^{-1}.

The component structure of K I may provide an important clue to tiny cold clouds since potassium is not subject to large depletion variations (Hobbs 1974). Hobbs found that N(Na I), N(K I), $E(B-V)$, and N(H I) (from Copernicus Lymanα data) are mutually correlated to within a factor of two, but N(Ca II) did not correlate with any of these other quantities. Na I and K I show a quadratic dependence on N(H): (N(Na I) $\sim N$(H)$^{2.3}$, N(K I) $\sim N$(H)$^{2.0}$) indicating that n_e/n_H is constant in the cloud and $N \sim n_H^2$ (for photoionization). The differential distribution of K I cloud column densities ($\phi(N)$) follows a power law, $\phi(N(\text{K I})) \sim N(\text{K I})^{-3/2}$, giving $\phi(N(\text{H})) \sim N(\text{H})^{-2}$. About 4.6 K I clouds kpc^{-1} are seen. A recent high-resolution survey (0.4–1.9 km s^{-1}) of K I, with 2σ detection limits of N(K I) $\sim 1.4 \times 10^9$ cm^{-2}, provides the most complete available sample of cold (~ 80 K) diffuse clouds. Observations of 54 stars found 319 components (Welty and Hobbs 2001). The mean component velocity is $v_{lsr} = 1.7$ km s^{-1}, with dispersion $\langle V^2 \rangle^{1/2} = 7.5$ km s^{-1}. Comparisons between K I and other species revealed important trends. N(Na I) and N(K I) show a linear relationship, with similar median b_{Dop} values (0.67 km s^{-1} and 0.73 km s^{-1}, respectively). However, Ca II is formed over a larger velocity interval than Na I or K I, since both the dispersion in bulk velocities, and line broadening, are larger for Ca II components than for Na I and K I components. The linear relationship between N(K I) and N(CH I) and similar profiles suggest these species are closely associated in dense clouds. N(K I) and N(H$^\circ$)+$2N$(H$_2$) are well correlated. Median and maximum values for N(K I) are 4×10^{10} cm^{-2} and 10^{12} cm^{-2}, respectively, which convert to column densities N(H) $= 2 \times 10^{20}$ cm^{-2} and 10^{21} cm^{-2} respectively, the observed N(KI)–N(H) correlation. For $T \sim 80$ K, 35–50% of the components are found to have subsonic turbulence. The K I components are strongly crowded in velocity since $\sim 85\%$ of the component separations (between two adjacent components) are <3 km s^{-1} (Welty and Hobbs 2001, and Figure 5). Few sightlines are dominated by a single cloud.

The ion Ti II is unique among optically visible species because it is the dominant ionization state for Ti in diffuse clouds (IP ~ 13.6 eV). However, Ti has highly variable depletions in the ISM (Figure 1). The Ti II weak 3384 Å line was surveyed towards 68 stars by Stokes (1978). Titanium depletions vary by two orders of magnitude in the solar neighborhood, with mean value $\delta_{Ti} \sim -2$ dex. Comparisons between N(Ti$^+$) and other species show that Ti and Ca correlate more tightly with each other than other species (e.g. Na I, K I, N(H)).

The abundance properties of halo ISM were characterized by a series of studies of Ca II and Na I towards faint blue Feige stars near the galactic poles (Cohen and Meloy 1975). Large N(Ca II)/N(Na I) ratios in halo gas indicate intercloud type gas. Halo gas velocities are skewed to the negative, and the infall requires the replenishment of halo gas on timescales of $\sim 10^8$ years. Observations of halo stars found that Ti II is pervasive in the galactic halo, and Ti abundances increase with increasing $|z|$ (Albert 1983). Albert concluded that two types of gas contributed to the Ti II features: a thick low-velocity disk component ($|z| < 200$ pc), and a second high-velocity weakly depleted gas at high $|z|$ and constituting $\sim 24\%$ of the halo mass. Low-velocity gas shows Ti abundances of $\sim 3\%$ solar, while for $|v_{lsr}| > 10$ km s^{-1} abundances average $\sim 6\%$ solar. An extreme value is seen towards HD 123884, at $z = 8.7$ kpc, where the Ti abundance is 75% solar. The relation between $N\sin|b|$ and $|z|$ for Ti II, Ca II, Fe II, H I, and $E(B-V)$ shows that the Ti II and Ca II gas are smoothly distributed and extend to greater heights than other components (with scale heights >2 kpc, 1 kpc, 0.5 kpc, 0.3 kpc, and 0.1 kpc, respectively, for an exponential distribution (Table 1; Edgar and Savage 1989)).

Time-variable interstellar Ca II and Na I absorption lines have been detected towards sources in the Vela region (e.g. Hobbs et al. 1991). These variable lines are characterized by extremely large widths (>100 km s^{-1}) and ratios N(Ca II)/N(Na I) (1.4 to over 56). Variations are seen on timescales of ~ 7 years towards HD 72127A, located near a bright filament in the Vela supernova remnant.

The velocity of the shocked interstellar gas towards α Oph, compared to the velocity of ISM inside of the Solar System led to the conjecture that Loop I drives a flow of interstellar material past the Sun (Frisch 1981). A series of papers modeling the kinematics of nearby ISM with a linear flow find different flow vectors, depending on the star sample (e.g. Crutcher 1982, Lallement et al. 1986, Vallerga et al. 1993, Frisch 1995). Crawford found that 79% of the components towards a set of stars in Sco–Cen have negative LSR velocities consistent with a shell expanding at $\sim 9\,\mathrm{km\,s^{-1}}$ around the Upper Centaurus–Lupus subgroup of the Sco–Cen Association (Crawford 1991). Although the nearest optical absorption lines show velocities generally consistent with the cloud velocity inferred from observations of Lymanα and He II 584 Å backscattered radiation, nearby stars often show more than one absorption component (e.g., α Aql at 5 pc; Ferlet 1999) and there is a distribution of component velocities about the central flow velocity. The heliocentric velocity vector for the cloud surrounding the solar system ($-25\,\mathrm{km\,s^{-1}}$ from the direction $l^{II} = 3°$, $b^{II} = +16°$,) corresponds to an LSR vector of $-15\,\mathrm{km\,s^{-1}}$ approaching from $l^{II} \sim 344°$, $b^{II} \sim -1°$. Typical line broadening for Ca II K lines towards nearby stars ($b_{\mathrm{Dop}} \sim 2.2\,\mathrm{km\,s^{-1}}$) are consistent with $T_k \sim 7000\,\mathrm{K}$ and $v_t \sim 1\,\mathrm{km\,s^{-1}}$ (Crawford et al. 1998). However, at least two nearby stars (α Pav and δ Vel) have narrow lines, $b_{\mathrm{Dop}} \sim 1\,\mathrm{km\,s^{-1}}$ indicating cold clouds ($<600\,\mathrm{K}$) within 24–50 pc of the Sun.

3.3 Warm ionized material

Classic H II regions surrounding young massive stars in stellar associations delineate the spiral arm structure of the Milky Way Galaxy (Georgelin and Georgelin 1976). However, pulsar data and diffuse optical recombination lines show that diffuse, warm, low-density interstellar plasma is widespread.

A comparisons of pulsar dispersion and rotation measures, for pulsars with an independent distance estimate, show that the layer of free electrons in our Galaxy has a mean density $<n_e> \sim 0.025\,\mathrm{cm^{-3}}$ and scale height $\sim 0.9\,\mathrm{kpc}$ (e.g. Falgarone and Lequeux 1973). Taylor and Cordes (1993) produced a model for the distribution of free electrons in our Galaxy which allowed for a non-uniform distribution of plasma imposed by spiral arm structure and other large-scale variations in the ISM. Several regions of enhanced diffuse n_e are seen, including towards the Gum Nebula ($240° \to 270°$) and the Sagittarius arm ($330° \to 30°$). This model provides a means to estimate pulsar distances from dispersion measures (Taylor and Cordes 1993).

The mapping of diffuse ($n_e \sim 0.1\,\mathrm{cm^{-3}}$) interstellar H$\alpha$ in our Galaxy is largely the result of work by Ron Reynolds and collaborators, who use large-throughput, high-spectral-resolution Fabry–Periot techniques to show that faint Hα and and forbidden optical emission lines closely follow the velocity distribution of H I 21 cm emission visible in spiral arms (Reynolds 1995). The discovery of pulsars in globular clusters allowed the comparison of pulsar dispersion measures with emission measures. Reynolds obtained a midplane density $<n_e> \sim 0.08\,\mathrm{cm^{-3}}$, for $T = 7,000\,\mathrm{K}$ plasma which fills $\sim 20\%$ of a 2 kpc thick layer in the galactic plane (known as the "Reynolds layer"). Most of the Hα luminosity in the ISM arises in traditional H II regions surrounding hot stars; however, most of the interstellar H II mass ($\sim 90\%$) is associated with a diffuse extensive warm ($\sim 8000\,\mathrm{K}$), low-density ($\sim 0.1\,\mathrm{cm^{-3}}$) component. Diffuse H II differs from classic H II regions by large [Si II]$\lambda 6716$/Hα and small [O III]$\lambda 5007$/Hα ratios. The line widths of the metastable lines suggest $T \sim 8,000\,\mathrm{K}$, while the absence of N I indicates the underlying material is ionized ($n_e/n_H > 0.75$). The spectrum of the ionizing source is constrained to be softer than an O-star spectrum by the failure to detect He I $\lambda 5876$ recombination, which indicates that He is mainly neutral ($n(\mathrm{He^+})/n(\mathrm{H°}) < 0.27$) in the ionized WIM.

The local galactic magnetic field has been mapped using the Faraday rotation of pulsar signals, since $\int n(e^-) B_{\|} dL / \int n(e^-) dL$ provides $B_{\|}$ weighted by the electron density (Rand and Kulkarni 1989). Data from ~ 200 pulsars show that RMs which sample the North Polar Spur region are dominated by this feature. When the North Polar Spur contributions are removed from the sample, a local magnetic field of $B_{\|} \sim 1.6 \pm 0.3\,\mu\mathrm{G}$, directed towards $l^{II} = 96° \pm 4°$, is obtained. The galactic field reverses towards the inner Galaxy at $d = 600 \pm 80\,\mathrm{pc}$. Rand and Kulkarni detected a contribution from random magnetic fields, $B \sim 5\,\mu\mathrm{G}$, and cell length 55 pc, which they attributed to fields in unresolved shell structures.

Ionized rims are found on $>10\%$ of interstellar clouds (Reynolds et al. 1995). The distribution of WIM which produces diffuse Hα emission correlates strongly with H I 21 cm clouds: $\sim 30\%$ of the diffuse Hα and ~ 10–30% of 21 cm emission occur in clouds with similar spatial distribution and kinematics, although Hα and 21 cm emission are spatially separated as expected for cloud rims. The clouds with rims typically are low density ($n_H = 0.2$–$0.3\,\mathrm{cm^{-3}}$), and have $N(\mathrm{H\,I}) = 2$–$20 \times 10^{19}\,\mathrm{cm^{-2}}$ and $\mathcal{E} = 2$–$10\,\mathrm{cm^{-6}\,pc}$. The scale heights for clouds with ionized rims ($z = 0.1$–$1\,\mathrm{kpc}$) exceeds the scale heights of cold H I ($\sim 100\,\mathrm{pc}$; Table 1).

4 THE GALACTIC HALO

4.1 Highly ionized gas

The most significant contribution of IUE to ISM studies was the discovery and mapping of highly ionized gas in the galactic halo. Hot gas was discovered in the halo through a remarkable set of IUE observations of faint stars. The C IV 1550 Å and Si IV 1402 Å resonance doublets were detected towards UV-bright O and Wolf–Rayet stars ($V = 9$–$12\,\mathrm{mag}$) in the Large and Small Magellanic Clouds (Savage and de Boer 1981), and ~ 24 faint, high-latitude ($|z| > 1\,\mathrm{kpc}$)

luminous OB stars in the galactic halo (Pettini and West 1982). Hot halo gas extends 10–15 kpc below the galactic plane towards the LMC and SMC, and to velocities ~150 km s^{-1} and ~50 km s^{-1} in the respective directions. Halo gas shares the rotation pattern of the underlying disk gas. The interstellar origin of high-z hot gas was firmly established by a survey of >40 faint ($d > 2$ kpc) disk and halo B stars which were selected to have no obvious association with circumstellar nebulosity (Savage and Massa 1987). Scale heights, h, for the highly ionized species were found to be significantly larger than for neutral and weakly ionized species (Table 1). The unambiguous detection of diffuse halo N V was first made towards halo stars in the inner Galaxy ($|z| = 0.7$–2.2 kpc, $R_g < 6$ kpc). Savage and Massa found line strengths of N V, C IV, and Si IV an order of magnitude larger than expected for a photoionized halo (e.g. Savage and Massa 1987).

A series of HST studies of halo gas, by Blair Savage, Ken Sembach, and collaborators, showed that two types of highly ionized halo gas are present (e.g. Savage and Sembach 1996). In Type I hot gas, N(Si IV):N(C IV):N(N V) ~ 1.0:3.0:0.5 and the narrow Si IV and C IV components are kinematically associated with low-ionization species, possibly originating in interface regions between warm and hot gas. C IV is found together with both the sharp Si IV and broad N V lines. In Type II gas, minimal Si IV is seen, with few low ions, and N(C IV)/N(N V) ~ 1–3. The broad N V features are produced by a widely distributed component, possibly from cooling hot gas in supernova bubbles or galactic fountains.

A classic paper by Savage et al. (1997) combined observations of extragalactic objects (QSOs and Seyfert galaxies) and 15 halo stars to probe the full extent of HIM in the halo. The derived exponential scale heights showed that high ions extend a factor of ~10 further from the plane than H I (Table 1). Ionization process must vary with $|z|$, since the ratios N(C IV)/N(N V) and N(Si IV)/N(N V) are enhanced at mid $|z|$ distances by a factor of 2, compared to lower and higher $|z|$ values. C IV and Si IV correlate well, and both correlate poorly with N V. Halo HIM co-rotates with disk gas out to $|z| \sim 5$ kpc for $R_g > 5$ kpc. HIM halo gas shows typical net inflow velocities of ~20 km s^{-1}, which is about twice that of neutral gas. The turbulent broadening of C IV (~60 km s^{-1}) is inadequate by factors of ~3 to support halo gas, indicating that additional pressure sources such as the galactic magnetic field or cosmic rays are required.

Observations of Al III, Si IV, C IV, and N V showed that profiles of Al III, a tracer of photoionized gas, are generally narrower than those of Si IV, C IV, and N V in halo gas (e.g. Sembach and Savage 1992). The "apparent optical depth technique" was used to derive information on unresolved components or saturation in the IUE profiles. Highly ionized gas extends to greater distances from the galactic plane than Al III, which has a scale height comparable to the Reynolds layer thickness (~1 kpc). N V absorption was detected in 80% of the sightlines, indicating pervasive $T \sim 2 \times 10^5$ K collisionally ionized gas in the galactic halo. These relatively low-resolution IUE data (25 km s^{-1}) show relatively constant mean ratios of N(C IV)/N(Si IV) = 3.6 ± 1.3 and N(C IV)/N(N V) = 4.6 ± 2.7.

Neither IUE nor HST could observe O VI. The far-UV spectrometer on an ORFEUS-SPA mission observed O VI towards galactic and extragalactic sightlines and found O VI to be significantly more confined to the plane than Si IV, C IV, or N V (Hurwitz and Bowyer 1996). The North Polar Spur region (Loop I and Loop IV) are sources of enhanced coronal gas, supplying hot gas to the galactic halo. Variations in the O VI/N V ratio reveals the non-uniform distribution of high-latitude coronal gas. A small O VI excess was tentatively identified towards two stars located behind an intermediate negative velocity cloud ($|v_{lsr}| = 45$–70 km s^{-1}) identified by Danly et al. (1992). Relative line strengths are consistent with either collisionally ionized gas in a galactic fountain, or composite mixture of conductive interfaces and turbulent mixing layers. Recently, FUSE has confirmed many of the overall characteristics of halo O VI found by ORFEUS-SPA. FUSE observations of O VI towards 11 AGNs confirm that O VI is more concentrated towards the galactic plane than Si IV, C IV, or N V, and that the scale height of high ions decreases with increasing ionization (Savage et al. 2000; e.g., see Table 1). N(C IV)/N(O VI) ranges from ~0.15 in the disk to ~0.6 in the halo. Variations in the z-component of N(O VI) (factor ~2.5) confirm that O VI halo gas is highly patchy. HST observations of C IV towards μ Col combined with Copernicus observations of O VI show these features occur at the same velocity, but O VI is twice as broad as C IV (Brandt et al. 1999). The ratio N(C IV)/N(O VI) = 0.11 ± 0.1, and is typical of hot disk material.

The production models for highly ionized gas in the halo are still somewhat uncertain (Spitzer 1996). Possible models include a conductively heated interface, radiative cooling, or turbulent mixing layers. The conductive heating models require contact between cooler material and hot plasma (10^6 K gas), such as found inside supernova remnants or stellar wind bubbles. A representative radiative cooling model would be a Galactic fountain, with the infall of cooling plasma onto the Galactic plane. The turbulent mixing layer model mixes hot and cold gas in a turbulent shear flow of hot gas past cold clouds. Conductive heating models reproduces N(C IV)/N(O VI)~0.15 found at low $|z|$; however for values $\gtrsim 0.6$ a combination of models and multiple ionization mechanisms are required. Smaller scale heights for more highly ionized gas suggests an origin for O VI connected to disk supernova remnants (Savage 1995).

4.2 Abundances in halo gas

When Supernova 1987A (SN1987A) exploded in the LMC, it was quickly observed as a target-of-opportunity by IUE and provided a brief, brilliant opportunity to observe halo gas using a $V \sim 4.5$ mag background source with a featureless continuum (e.g. Blades et al. 1988, Welty et al. 1999). Intermediate-velocity halo gas ($56 \leq v_{lsr} \leq 90 \text{ km s}^{-1}$) was seen to be a mixture of warm and cool gas, with virtually no depletion of the refractory elements Si, Cr, Mn, Fe, Ni, and Fe relative to Zn. The high-velocity halo gas ($109 \leq v_{lsr} \leq 140 \text{ km s}^{-1}$) showed typical abundance characteristics of warm partially ionized halo gas ($T \gtrsim 4500$ K).

A series of HST studies of the ISM towards halo stars by Spitzer, Savage, Fitzpatrick, Sembach, and others show that depletion patterns in halo and disk are similar, except that the least depleted sightlines generally are towards halo stars. Zn depletions in individual clouds towards halo stars range from $\delta_{Zn} \sim -0.4 \rightarrow 0.0$ dex for $\overline{n_H} < 0.1 \text{ cm}^{-3}$ for $|v_{lsr}| < 50 \text{ km s}^{-1}$ (Fitzpatrick 1996). Both warm disk and halo clouds typically show solar abundances of S and at least one IVC (the -70 km s^{-1} component towards HD 93521) shows supersolar abundances for S (Savage and Sembach 1996, Fitzpatrick 1996, Howk et al. 1999). Typically over 10–30 cloud components are seen towards halo stars. In low-velocity clouds ($<20-25 \text{ km s}^{-1}$) there is a clear positive correlation between gas-phase abundances and increasing cloud velocity for refractories (Si, Fe, Cr, Mn, Ti). The strongest variations are found for Fe (~ 1.5 dex) and the smallest variations for Si (0.5 dex). Below $|v_{lsr}| \sim 20 \text{ km s}^{-1}$, the gas-phase abundances of refractory elements (Fe, Mn, Ni, Si, Cr) generally increase with velocity, while above that value depletions vary little with velocity. Fe and Si depletions correlate strongly for both disk and halo stars, with maximum abundances in disk and halo clouds corresponding to $\delta_{Fe} \sim -0.5$ dex, and $\delta_{Fe} \sim -0.15$ dex (with respect to S; Fitzpatrick 1996, Savage and Sembach 1996). For $\overline{n_H} > 0.1 \text{ cm}^{-3}$, Cr/Zn ~ -1.4 dex and is density independent for disk stars. For $\overline{n_H} < 0.1 \text{ cm}^{-3}$, Cr depletions are in the range $\log \delta_{Cr} \sim -1.6$ to -0.8 dex (with respect to H). Warm halo gas components towards HD 116852 ($d = 4.8$ pc, $z = -1.3$ kpc) show that gas-phase abundances increase with $|z|$, and line broadening (velocity dispersion) and ion scale heights increase as the ionization of the gas increases (Sembach and Savage 1996).

4.3 Warm neutral and ionized gas

Comparisons between IUE observations of low and high ions and H I 21 cm data, show halo kinematics, and constrain cloud distances (e.g. Danly 1992). The low halo ($|z| < 1$ kpc) contains approximately equal numbers of negative and positive components, while the distant halo has a preponderance of infalling gas. In the northern hemisphere, low ion (e.g., Si II) lines have component velocities of up to $|v_{lsr}| < 120 \text{ km s}^{-1}$. Intermediate-velocity gas ($>40 \text{ km s}^{-1}$) is formed at $|z| > 1$ kpc, and has an excess of negative velocity components. Low-velocity gas ($<40 \text{ km s}^{-1}$), with an equal number of positive and negative components, tends to be closer to the galactic plane ($|z| < 1$). HST observations of halo stars, which typically sample long sightlines devoid of dense cold clouds, provide some of the best data on warm gas. HST found that up to 30 or more absorption components may be seen towards halo stars with $z \sim 2$ kpc, extending over large velocity intervals ($>100 \text{ km s}^{-1}$). Such sightlines sample both disk and halo gas.

Towards HD 100340, nine regions of intermediate-velocity ($+31 \rightarrow +78 \text{ km s}^{-1}$) partially ionized gas are seen, with column densities in the range $\log N(H^\circ) = 18.38 - 19.36 \text{ cm}^{-2}$. H I 21 cm data show these clouds are warm, neutral gas, $T \sim 7000$ K (from the FWHM values), giving typical densities $n(H^\circ) = 0.022 - 0.10 \text{ cm}^{-3}$. Towards HD 93521 four low-velocity components ($v_{lsr} < 20 \text{ km s}^{-1}$) show $n_e \sim 0.11 \text{ km s}^{-1}$, while intermediate-velocity gas ($v_{lsr} > 25 \text{ km s}^{-1}$) has $n_e \sim 0.04 \text{ km s}^{-1}$, based on C II*/C II (see Fitzpatrick 1996 for references). The 30 components seen towards HD 215733 ($z = 1.7$ kpc), show temperatures $\leq 300 \text{ K} \rightarrow > 1000$ K and electron densities $n_e = 0.02$ to 0.06 cm^{-3}.

In individual warm disk and halo clouds, high and low ionization levels are commonly associated with each other. For example, towards HD 215733 Si IV and C IV lines are at the velocities of three warm clouds, although line widths indicate hot temperatures ($T = 6 \times 10^5$ K and 5×10^4 K, respectively). The neutral, weakly ionized intermediate-velocity ($+70 \text{ km s}^{-1}$) cloud towards HD 20366 is warm (5300–6100 K) with $n_e = 0.15 - 0.34 \text{ cm}^{-3}$; however, highly ionized gas with $N(C \text{ IV})/N(Si \text{ IV}) \sim 4.5$ is present suggesting collisionally ionized cloud interfaces with $T \sim 10^5$ K (e.g. Sembach 1995).

Observations in a number of warm clouds towards halo stars (e.g., HD 215733, HD 154368) show that different diagnostics for n_e in the same cloud yield discrepant results (as also seen for cold clouds, Section 2.6). The ionization equilibrium of C I, Mg I, S I, and Ca II give $\log n_e$ differing by up to 1 dex, with generally $n_e(Mg \text{ I}) > n_e(C \text{ I})$, when the temperature dependence of Mg II and C II recombination rates are assumed similar. These differences are consistent with the trace neutral status of C I, and enhanced Mg I in warm gas ($>5,000$ K) by dielectronic recombination. In addition, Ca II and Ti II (Section 3.2) are distributed relatively smoothly in the halo, and Ti II extends to $z \gg 2$ kpc. Electron densities based on $N(Ca \text{ II})$ vary smoothly in comparison to n_e derived from $N(Na \text{ I})$ and $N(Mg^\circ)/N(Mg^+)$, which sample cloudy regions (Cardelli et al. 1995).

Systematic variations in cloud electron densities as a function of velocity have been detected towards stars in the low and high halo. The stars μ Col and HD 93521, for example, show electron densities which decrease with increasing cloud velocity (e.g. Spitzer and Fitzpatrick 1993). Component n_e values found from both C II*/C II and Ca II show a clear decrease with increasing $|V_{lsr}|$ towards HD 93521. A spread in n_e values of $0.02 \rightarrow 0.05\,\text{cm}^{-3}$ is found for $|V_{lsr}| \sim 50 \rightarrow 70\,\text{km s}^{-1}$. Five warm neutral velocity components are seen towards μ Col ($-29 \rightarrow +41\,\text{km s}^{-1}$; Howk et al. 1999). Warm intermediate-velocity gas ($+31$, $+41\,\text{km s}^{-1}$, $T = 4000 \pm 700\,\text{K}$) is low column density ($\log N(\text{H}) = 17.3$–$17.8\,\text{cm}^{-2}$). Electron densities in component 4, for example, are $n_e = 0.47 \pm 0.14\,\text{cm}^{-3}$ (using $N(\text{C}^{+*})/N(\text{C}^+)$) and $n_e = 0.64 \pm 0.14\,\text{cm}^{-3}$ (using C II and $N(\text{Mg}^\circ)/N(\text{Mg}^+)$).

UV Lymanα data provide the distribution of H I, but generally do not distinguish between cold and warm gas (in contrast to H I 21 cm data where both absorption and emission data are available). H I scale heights were determined from IUE observations of 393 stars ($d = 0.12$–$11\,\text{kpc}$, $|z| < 4\,\text{kpc}$) which are free of stellar Lymanα contamination (Diplas and Savage 1994). Selecting sightlines which avoided obvious clouds yielded scale height $h = 195\,\text{pc}$, and an overall value $\overline{n_H} = 0.23\,\text{atoms cm}^{-3}$ (range 0.017–$8.62\,\text{atoms cm}^{-3}$). Color excess data indicate that dust is patchier than the gas, with dust scale height $h = 152\,\text{pc}$ and midplane value $E(B-V)/d = 0.257\,\text{mag kpc}^{-1}$. Assuming, rather, a two-component exponential model to accommodate gas clumping around the plane gives for the compact component $n_{H,1}(0) = 0.247\,\text{cm}^{-3}$, $h_1 = 73\,\text{pc}$, and for the extended component $n_{H,2}(0) = 0.16\,\text{cm}^{-3}$, $h_2 = 357\,\text{pc}$. These scale heights are comparable to 21 cm results despite different underlying assumptions and sightline lengths. Denser clouds show a larger ratio of $N(\text{H I})/E(B-V)$ than towards interarm sightlines, possibly indicating grain modification in dense clouds. Savage (1995) finds that $\sim 82\%$ of the halo interstellar mass is contributed by neutral gas (both the confined and extended components). An extended component of warm ionized gas provides $\sim 15\%$ of the mass, while the extended hot component contributes $\sim 3\%$ of the halo ISM mass.

5 CLOSING COMMENTS

What does the 21st century hold for ISM studies? Opening up the UV window on our Galaxy initiated the golden age of space astronomy, and revealed 30–40% of the Galactic mass previously invisible. Will the retirement of HST shutter this window? The 21st century may see the first *in situ* measurements of the interstellar cloud surrounding the solar system (Liewer et al. 2000). When we leave our heliosphere and explore interstellar space we will become citizens of the Milky Way Galaxy. However many problems are solvable with earth-orbiting spacecraft. The next instrumental advance for UV astronomy will be to launch an ultrahigh resolution UV space spectrometer (912–3000 Å) with a large dynamic range (V = 1–12 magnitudes) to reveal the physics of cloudlets in the disk and halo.

What are the questions of the future?

1. Where will the solar journey take us, and what are the past and future Galactic environments of the Sun?
2. Is the ISM one grand, turbulent, gravitationally-layered continuum, with locally variable properties sculpted by spiral density waves which initiated epochs of star-formation?
3. Will better data on Galactic halo gas unlock the mysteries of chemical evolution in our Galaxy?
4. Is the ISM composition homogeneous? Is Seaton's postulate that dust grains contain the atoms missing from the ISM gas phase correct?
5. Where does halo ISM originate?
6. Are molecular clouds an ISM reservoir replenishing diffuse gas?
7. What is the answer to Parker's question: "What is an interstellar cloud"? Are turbulence, cloud interface regions, and expanding superbubble shells relevant for the answer to this question?

Finally, we return to Harlow Shapley's idea. Are encounters with interstellar clouds important for evaluating the historical climate of the Earth, and by extrapolation climate stability for extra-solar planets?

Acknowledgments

The author would like to thank Dan Welty for providing many of the figures, and both Dan Welty and Don York for numerous informative scientific discussions. This research has been supported by NASA grants NAG5-6405 and NAG5-7077.

REFERENCES

Adams, T.F. and Frisch, P.C. (1977). High-resolution observations of the Lyman alpha sky background. *Astrophysical Journal*, **212**, 300–308.

Adams, W.S. (1949). Observations of interstellar H and K, molecular lines, and radial velocities in the spectra of 300 O and B stars. *Astrophysical Journal*, **109**, 354–379.

Albert, C.E. (1983). Neutral interstellar gas in the lower galactic halo. *Astrophysical Journal*, **272**, 509–539.

Baldwin, J.E. (1955). The distribution of the galactic radio emission. *Monthly Notices of the Royal Astronomical Society*, **115**, 690–700.

Barlow, M.J., Crawford, I.A., Diego, F., Dryburgh, M., Fish, A.C., Howard, I.D., Spyromilio, J. and Walker, D.D. (1995). First results from the UHRF: Ultra-high-resolution observations of atomic interstellar lines towards zeta Ophiuchi. *Monthly Notices of the Royal Astronomical Society*, **272**, 333–345.

Berkhuijsen, E.M., Haslam, C.G.T. and Salter, C.J. (1971). Are the galactic loops supernova remnants? *Astronomy and Astrophysics*, **14**, 252–386.

Blades, J.C., Wheatley, J.M., Panagia, N., Grewing, M., Pettini, M. and Wamsteker, W. (1988). Abundances in intermediate-velocity clouds toward SN 1987A. *Astrophysical Journal*, **332**, L75–L79.

Bohlin, R.C., Savage, B.D. and Drake, J.F. (1978). A survey of interstellar H I from L-alpha absorption measurements. *Astrophysical Journal*, **224**, 132–142.

Boksenberg, A., Kirkham, B., Pettini, M., Bates, B., Carson, P.P.D., Dufton, P.L. and McKeith, C.D. (1975). Interstellar magnesium absorption in the direction of four unreddened stars. *Astrophysical Journal*, **202**, L91–L95.

Brandt, J.C., Heap, S.R., Beaver, E.A., Boggess, A., Carpenter, K.G., Ebbets, D.C., Hutchings, J.B., Jura, M., Leckrone, D.S., Linsky, J.L., Maran, S.P., Savage, B.D., Smith, A.M., Trafton, L.M., Walter, F.M., Weymann, R.J., Howk, J.C., Snow, M., Ake, T.B. and Sembach, K.R. (1999). Echelle spectroscopy of interstellar absorption toward mu Columbae with the Goddard High Resolution Spectrograph. *Astronomical Journal*, **117**, 400–409.

Breitschwerdt, D., Egger, R., Freyberg, M.J., Frisch, P.C. and Vallerga, J.V. (1996). The Local Bubble origin and evolution. *Space Science Reviews*, **78**, 183–198.

Bruhweiler, F.C., Kondo, Y. and McCluskey, G.E. (1980). Interstellar C IV and Si IV column densities toward early-type stars. *Astrophysical Journal*, **237**, 19–25.

Burton, W.B. (1976). The morphology of hydrogen and of other tracers in the Galaxy. *Annual Review of Astronomy and Astrophysics*, **14**, 275–306.

Cardelli, J.A. and Meyer, D.M. (1997). The abundance of interstellar krypton. *Astrophysical Journal*, **477**, L57–L61.

Cardelli, J.A., Sembach, K.R. and Savage, B.D. (1995). Gas phase abundances and conditions along the sight line to the low-halo, inner galaxy star HD 167756. *Astrophysical Journal*, **440**, 241–253.

Chandrasekhar, S. and Fermi, E. 1953. Magnetic fields in spital arms. *ApJ*, **118**, 113–115.

Clark, B.G. (1965). An interferometer investigation of the 21-centimeter hydrogen-line absorption. *Astrophysical Journal*, **142**, 1398–1422.

Cohen, J.G. and Meloy, D.A. (1975). The interstellar lines of the Feige stars. *Astrophysical Journal*, **198**, 545–550.

Colomb, F.R., Poppel, W.G.L. and Heiles, C. (1980). Galactic H I at $b>10$ deg. II. photographic presentation of the combined southern and northern data. *Astronomy and Astrophysics*, **40**, 47–55.

Cowie, L.L., Hu, E.M., Taylor, W. and York, D.G. (1981). A search for expanding supershells of gas around OB associations. *Astrophysical Journal*, **250**, L25–L29.

Cowie, L.L., Jenkins, E.B., Songaila, A. and York, D.G. (1979). O VI absorption in interstellar cloud surfaces. *Astrophysical Journal*, **232**, 467–472.

Cowie, L.L. and York, D.G. (1978). The velocity distribution of interstellar gas observed in strong UV absorption lines. *Astrophysical Journal*, **223**, 876–883.

Cox, D.P. and Reynolds, R.J. (1987). The local interstellar medium. *Annual Review of Astronomy and Astrophysics*, **25**, 303–344.

Cox, D.P. and Smith, B.W. (1974). Large-scale effects of supernova remnants on the Galaxy: Generation and maintenance of a hot network of tunnels. *Astrophysical Journal*, **189**, L105–L108.

Crawford, I.A. (1991). High resolution observations of interstellar Na I and Ca II towards the Scorpio-Centaurus association. *Astronomy and Astrophysics*, **247**, 183–201.

Crawford, I.A., Lallement, R. and Welsh, B.Y. (1998). Additional ultra-high-resolution observations of Ca^+ ions in the local interstellar medium. *Monthly Notices of the Royal Astronomical Society*, **300**, 1181–1188.

Crutcher, R.M. (1975). Observations of interstellar Na I lines at 3302 Å. *Astrophysical Journal*, **202**, 634–637.

Crutcher, R.M. (1982). The local interstellar medium. *Astrophysical Journal*, **254**, 82–87.

Dalgarno, A. and McCray, R.A. (1972). Heating and ionization of HI regions. *Annual Review of Astronomy and Astrophysics*, **10**, 375–426.

Danly, L. (1992). The ionization conditions in the Milky Way halo – Infalling gas toward the North Galactic Pole. *Publications of the Astronomical Society of the Pacific*, **104**, 819–823.

Davidson, K. and Terzian, Y. (1969). Dispersion measures of pulsars. *Astronomical Journal*, **74**, 849–854.

Diamond, P.J., Goss, W.M., Romney, J.D., Booth, R.S., Kalberla, P.M.W. and Mebold, U. (1989). The structure of the interstellar medium at the 25 Au scale. *Astrophysical Journal*, **347**, 302–306.

Dickey, J.M. and Lockman, F.J. (1990). H I in the galaxy. *Annual Review of Astronomy and Astrophysics*, **28**, 215–261.

Dickey, J.M., Terzian, Y. and Salpeter, E.E. (1979). Interpretation of neutral hydrogen absorption. *Astrophysical Journal*, **228**, 465–474.

Dieter, N.H., Welch, W.J. and Romney, J.D. (1976). A very small interstellar neutral hydrogen cloud observed with VLBI techniques. *Astrophysical Journal*, **206**, L113–L115.

Diplas, A. and Savage, B.D. (1994). An IUE survey of interstellar H I Ly alpha absorption. 2: Interpretations. *Astrophysical Journal*, **427**, 274–287.

Dunham, T. (1939). *Proceedings of the American Philosophical Society*, **81**, 277–293.

Ebel, D.S. (2000). Variations on solar condensation: Sources of interstellar dust nuclei. *Journal of Geophysical Research*, **105**, 10363–10370.

Eddington, A.S. (1926). Diffuse matter in interstellar space. *Proceedings of the Royal Society A.*, **111**, 424–456.

Edgar, R.J. and Savage, B.D. (1989). The density distribution of refractory elements away from the galactic plane. *Astrophysical Journal*, **340**, 762–774.

Elmegreen, B.G. and Efremov, Y.N. (1998). Hierarchy of interstellar and stellar sturctures and the case of the orion star-forming region. [astroph-9801071]

Falgarone, E. and Lequeux, J. (1973). A discussion of the distribution of interstellar matter close to the Sun. *Astronomy and Astrophysics*, **25**, 253–260.

Ferlet, R. (1999). The local interstellar medium. *Astronomy and Astrophysics Review*, **9**, 153–169.

Ferrara, A., McKee, C.F., Heiles, C. and Shapiro, P.R. (ed) (1995). *The Physics of the Interstellar Medium and Intergalactic Medium*. Astronomical Society of the Pacific, San Francisco, 593 pp.

Field, G.B. (1958). Excitation of the hydrogen 21-cm line. *Proceedings of the Institute of Radio Engineers*, **46**, 240–250.

Field, G.B. (1974). Interstellar abundances: Gas and dust. *Astrophysical Journal*, **187**, 453–459.

Field, G.B., Goldsmith, D.W. and Habing, H.J. (1969). Cosmic-ray heating of the interstellar gas. *Astrophysical Journal*, **155**, L149–L154.

Fitzpatrick, E.L. (1996). The composition of the diffuse interstellar medium. *Astrophysical Journal*, **473**, L55–62.

Frail, D.A., Weisberg, J.M., Cordes, J.M. and Mathers, C. (1994). Probing the interstellar medium with pulsars on au scales. *Astrophysical Journal*, **436**, 144–151.

Frisch, P. and York, D.G. (1986). Interstellar clouds near the Sun. In *The Galaxy and the Solar System*, University of Arizona Press, pp. 83–100.

Frisch, P.C. (1981). The nearby interstellar medium. *Nature*, **293**, 377–379.

Frisch, P.C. (1995). Characteristics of nearby interstellar matter. *Space Science Reviews*, **72**, 499–592.

Frisch, P.C. (1997). Journey of the Sun. *http://xxx.lanl.gov/*, page [astroph/9705231].

Frisch, P.C., Dorschner, J.M., Geiss, J., Greenberg, J.M., Grün, E., Landgraf, M., Hoppe, P., Jones, A.P., Krätschmer, W., Linde, T.J., Morfill, G.E., Reach, W., Slavin, J.D., Svestka, J., Witt, A.N. and Zank, G.P. (1999). Dust in the local interstellar wind. *Astrophysical Journal*, **525**, 492–516.

Frisch, P.C. and York, D.G. (1983). Synthesis maps of ultraviolet observations of neutral interstellar gas. *Astrophysical Journal*, **271**, L59–L63.

Gayley, K.G., Zank, G.P., Pauls, H.L., Frisch, P.C., and Welty, D.E. 1997. One- versus two-shock heliosphere: Constraining models with Goddard High Resolution Spectrograph Ly-alpha spectra toward alpha Centauri. *ApJ*, **487**, 259–270.

Genova, R., Beckman, J.E., Molaro, P. and Vladilo, G. (1990). Mg II observed in the local interstellar medium – The local cloud. *Astrophysical Journal*, **355**, 150–158.

Georgelin, Y.M. and Georgelin, Y.P. (1976). The spiral structure of our Galaxy determined from H II regions. *Astronomy and Astrophysics*, **49**, 57–79.

Hagen, J.P., Lilley, A.E. and McClain, E.F. (1955). Absorption of 21-cm radiation by interstellar hydrogen. *Astrophysical Journal*, **122**, 361–375.

Harris, A.W. (1988). Abundances in the diffuse interstellar medium. In *A Decade of UV Astronomy with the IUE Satellite*, Vol. 2, ESA, pp. 3–9.

Hartmann, J. (1904). Investigations on the spectrum and orbit of delta Orionis. *Astrophysical Journal*, **19**, 268–286.

Heiles, C. (1980). Is the intercloud medium pervasive. *Astrophysical Journal*, **235**, 833–839.

Heiles, C. (1982). Temperatures and their variation within interstellar H I structures. *Astrophysical Journal*, **262**, 135–141.

Heiles, C. (1997). Tiny-scale atomic structure and the cold neutral medium. *Astrophysical Journal*, **481**, 193–204.

Heiles, C. (1998). Whence the Local Bubble, Gum, Orion? GSH 238+00+09, a nearby major superbubble toward Galactic longitude 238 degrees. *Astrophysical Journal*, **498**, 689–703.

Heiles, C., Chu, Y., Troland, T.H., Reynolds, R.J. and Yegingil, I. (1980). A new look at the North Polar Spur. *Astrophysical Journal*, **242**, 533–540.

Herbig, G.H. (1968). The interstellar line spectrum of zet Oph. *Zeitschrift Astrophysics*, **68**, 243–277.

Hobbs, L.M. (1965). High-resolution study of an interstellar line. *Astrophysical Journal*, **142**, 160–163.

Hobbs, L.M. (1969). The profiles of the interstellar sodium D-lines. *Astrophysical Journal*, **157**, 165–174.

Hobbs, L.M. (1974). Statistical properties of interstellar clouds. *Astrophysical Journal*, **191**, 395–399.

Hobbs, L.M., Ferlet, R., Welty, D.E. and Wallerstein, G. (1991). Variable interstellar absorption toward HD 72127A. II – 1981–1988. *Astrophysical Journal*, **378**, 586–598.

Howk, J.C., Savage, B.D. and Fabian, D. (1999). Abundances and physical conditions in the warm neutral medium toward mu Columbae. *Astrophysical Journal*, **525**, 253–293.

Huang, J., Songaila, A., Cowie, L.L. and Jenkins, E.B. (1995). Detection of hot gas in the interstellar medium. *Astrophysical Journal*, **450**, 163–178.

Hughes, M.P., Thompson, A.R. and Colvin, R.S. (1971). An absorption-line study of the Galactic neutral hydrogen at 21 centimeters wavelength. *Astrophysical Journal Supplement*, **23**, 323–370.

Hurwitz, M. and Bowyer, S. (1996). Coronal gas in the halo. II. ORFEUS observations of Galactic halo stars. *Astrophysical Journal*, **465**, 296–316.

Jenkins, E.B. (1978). Coronal gas in the Galaxy. II – A statistical analysis of O VI absorptions. *Astrophysical Journal*, **220**, 107–123.

Jenkins, E.B. (1987). Element abundances in the interstellar atomic material. In D.J. Hollenbach and H.A. Thronson (eds), *Interstellar Processes*, Reidel, Dordrecht, pp. 533–559.

Jenkins, E.B., Lees, J.F., van Dishoeck, E.F. and Wilcots, E.M. (1989). Velocities and rotational excitation of interstellar H2 toward pi Scorpii. *Astrophysical Journal*, **343**, 785–810.

Jenkins, E.B. and Shaya, E.J. (1979). A survey of interstellar C I – insights on carbon abundances, UV grain albedos, and pressures in the interstellar medium. *Astrophysical Journal*, **231**, 55–72.

Kondo, Y., Bruhweiler, F.C. and Savage, B.D. (1984). Local interstellar medium. IAU Colloquium no. 81. Technical report, NASA.

Kondo, Y., Talent, D.L., Barker, E.S., Dufour, R.J. and Modisette, J.L. (1978). On the column density of the interstellar Mg II to Sirius and other nearby stars. *Astrophysical Journal*, **220**, L97–L102.

Koo, B. and Heiles, C. (1991). A survey of H I 21 centimeter emission lines toward supernova remnants. *Astrophysical Journal*, **382**, 204–222.

Kulkarni, S.R. and Heiles, C. (1987). The atomic component. In D.G. Hollenbach and H.A. Thronson (eds), *Interstellar Processes*, Reidel, Dordrecht, pp. 87–122.

Lallement, R., Vidal-Madjar, A. and Ferlet, R. (1986). Multi-component velocity structure of the local interstellar medium. *Astronomy and Astrophysics*, **168**, 225–236.

Landsman, W.B., Henry, R.C., Moos, H.W. and Linsky, J.L. (1984). Observations of interstellar hydrogen and deuterium toward Alpha Centauri A. *Astrophysical Journal*, **285**, 801–807.

Lauroesch, J.T., Meyer, D.M. and Blades, J.C. (2000). Evidence of interstellar Na I structure at scales down to 15 Au in low-density gas. *Astrophysical Journal*, **543**, L43–L47.

Liewer, P.C., Mewaldt, R.A., Ayon, J.A. and Wallace, R.A. (2000). NASA's Interstellar Probe mission. pages 911+.

Linsky, J.L. and Wood, B.E. (1996). The alpha Centauri line of sight: D/H ratio, physical properties of local interstellar gas, and measurement of heated hydrogen (the 'hydrogen wall') near the heliopause. *Astrophysical Journal*, **463**, 254–270.

Livingston, W.C. and Lynds, C.R. (1964). Observations of interstellar lines with very high resolution. *Astrophysical Journal*, **140**, 818–820.

Lucke, P.B. (1978). The distribution of color excesses and interstellar reddening material in the solar neighborhood. *Astronomy and Astrophysics*, **64**, 367–377.

Mac Low, M. and McCray, R. (1988). Superbubbles in disk galaxies. *Astrophysical Journal*, **324**, 776–785.

McClintock, W., Henry, R.C., Linsky, J.L. and Moos, H.W. (1978). Ultraviolet observations of cool stars. VII – local interstellar hydrogen and deuterium lyman-alpha. *Astrophysical Journal*, **225**, 465–481.

McKee, C.F. and Ostriker, J.P. (1977). A theory of the interstellar medium – Three components regulated by supernova explosions in an inhomogeneous substrate. *Astrophysical Journal*, **218**, 148–169.

Menon, T.K. (1958). Interstellar structure of the Orion region. I. *Astrophysical Journal*, **127**, 28–47.

Meyer, D.M., Jura, M. and Cardelli, J.A. (1998). The definitive abundance of interstellar oxygen. *Astrophysical Journal*, **493**, 222–229.

Moos, H.W., Linsky, J.L., Henry, R.C. and McClintock, W. (1974). High-spectral measurements of the H I 1216 Å and Mg II 2800 Å emissions from Arcturus. *Astrophysical Journal*, **188**, L93–L96.

Morton, D.C. (1975). Interstellar absorption lines in the spectrum of zeta Ophiuchi. *Astrophysical Journal*, **197**, 85–115.

Münch, G. (1957). Interstellar absorption lines in distant stars. I. Northern Milky Way. *Astrophysical Journal*, **125**, 42–65.

Münch, G. (1968). Interstellar absorption lines. In B.M. Middlehurst and L.H. Aller (eds), *Stars and Stellar Systems*, Vol. 7, pp. 365–402.

Munch, G. and Unsold, A. (1962). Interstellar gas near the Sun. *Astrophysical Journal*, **135**, 711–715.

O'Dell, C.R. (1999). Commentary on Stromgren's 1939 paper. In H.A. Abt (ed), *The Astrophysical Journal: Centennial Issue*, The University of Chicago Press for the American Astronomical Society, pp. 321–323.

Oort, J.H. (1946). Some phenomena connected with interstellar matter (George Darwin Lecture). *Monthly Notices of the Royal Astronomical Society*, **106**, 159–179.

Paresce, F. (1984). On the distribution of interstellar matter around the sun. *Astronomical Journal*, **89**, 1022–1037.

Pettini, M. and West, K.A. (1982). A study of interstellar absorption at high galactic latitudes. I – Highly ionized gas. *Astrophysical Journal*, **260**, 561–578.

Plaskett, J.S. and Pearce, J.A. (1930). The motions and distribution of interstellar matter. *MNRAS*, **90**, 243–268.

Poeppel, W.G., Olano, C.A. and Cappa de Nicolau, C.E. (1981). Neutral hydrogen related to Gould's Belt. *Revista Mexicana de Astronomia y Astrofisica*, **6**, 259–266.

Radhakrishnan, V. and Goss, W.M. (1972). The Parkes survey of 21-centimeter absorption in discrete-source spectra. V. note on the statistics of absorbing H I concentrations in the Galactic disk. *Astrophysical Journal Supplement*, **24**, 161–166.

Rand, R.J. and Kulkarni, S.R. (1989). The local galactic magnetic field. *Astrophysical Journal*, **343**, 760–772.

Reach, W.T., Koo, B.C. and Heiles, C. (1994). Atomic and molecular gas in interstellar cirrus clouds. *Astrophysical Journal*, **429**, 672–693.

Reynolds, R.J. (1995). Diffuse optical emission lines as probes of the interstellar and intergalactic ionizing radiation. In *ASP Conf. Ser. 80: The Physics of the Interstellar Medium and Intergalactic Medium*, pp. 388–726.

Reynolds, R.J., Tufte, S.L., Kung, D.T., McCullough, P.R. and Heiles, C. (1995). A comparison of diffuse ionized and neutral hydrogen away from the Galactic plane: H alpha-emitting H I clouds. *Astrophysical Journal*, **448**, 715.

Rogerson, J.B. and York, D.G. (1973). Interstellar deuterium abundance in the direction of beta Centauri. *Astrophysical Journal*, **186**, L95–L98.

Rogerson, J.B., York, D.G., Drake, J.F., Jenkins, E.B., Morton, D.C. and Spitzer, L. (1973). Spectrophotometric results from the Copernicus satellite. III. Ionization and composition of the intercloud medium. *Astrophysical Journal*, **181**, L110–L115.

Savage, B.D. (1995). The gaseous Galactic corona. In *ASP Conf. Ser. 80: The Physics of the Interstellar Medium and Intergalactic Medium*, pages 233+.

Savage, B.D. and Bohlin, R.C. (1979). The depletion of interstellar gaseous iron. *Astrophysical Journal*, **229**, 136–146.

Savage, B.D., Cardelli, J.A. and Sofia, U.J. (1992). Ultraviolet observations of the gas phase abundances in the diffuse clouds toward zeta Ophiuchi at 3.5 kilometers per second resolution. *Astrophysical Journal*, **401**, 706–723.

Savage, B.D. and de Boer, K.S. (1981). Ultraviolet absorption by interstellar gas at large distances from the galactic plane. *Astrophysical Journal*, **243**, 460–484.

Savage, B.D., Drake, J.F., Budich, W. and Bohlin, R.C. (1977). A survey of interstellar molecular hydrogen. *Astrophysical Journal*, **216**, 291–307.

Savage, B.D., Edgar, R.J. and Diplas, A. (1990). The distribution of interstellar Al III away from the galactic plane. *Astrophysical Journal*, **361**, 107–115.

Savage, B.D. and Massa, D. (1987). Highly ionized interstellar gas located in the galactic disk and halo. *Astrophysical Journal*, **314**, 380–396.

Savage, B.D. and Sembach, K.R. (1996). Interstellar abundances from absorption-line observations with the Hubble Space Telescope. *Annual Review of Astronomy and Astrophysics*, **34**, 279–330.

Savage, B.D., Sembach, K.R. and Cardelli, J.A. (1994). Highly ionized gas absorption in the disk and halo toward HD 167756 at 3.5 kilometers per second resolution. *Astrophysical Journal*, **420**, 183–196.

Savage, B.D., Sembach, K.R., Jenkins, E.B., Shull, J.M., York, D.G., Sonneborn, G., Moos, H.W., Friedman, S.D., Green, J.C., Oegerle, W.R., Blair, W.P., Kruk, J.W. and Murphy, E.M. (2000). Far Ultraviolet Spectroscopic Explorer observations of O VI absorption in the galactic halo. *Astrophysical Journal*, **538**, L27–L30.

Savage, B.D., Sembach, K.R. and Lu, L. (1997). Absorption by highly ionized interstellar gas along extragalactic and galactic sight lines. *Astronomical Journal*, **113**, 2158–2185.

Scalo, J. (1999). Chandrasekhar and Munch's models of fluctuations in brightness of the Milky Way. *Astrophysical Journal*, **525**, C476–C477.

Seaton, M.J. (1951). The chemical composition of the interstellar gas. *Monthly Notices of the Royal Astronomical Society*, **111**, 368–346.

Sembach, K.R. (1995). Properties of the +70 kilometers per second cloud toward HD 203664. *Astrophysical Journal*, **445**, 314–324.

Sembach, K.R. and Savage, B.D. (1992). Observations of highly ionized gas in the galactic halo. *Astrophysical Journal Supplement*, **83**, 147–201.

Sembach, K.R. and Savage, B.D. (1996). The gas and dust abundances of diffuse halo clouds in the Milky Way. *Astrophysical Journal*, **457**, 211–227.

Sembach, K.R., Savage, B.D. and Tripp, T.M. (1997). High-resolution ultraviolet observations of the highly ionized interstellar gas toward Radio Loops I and IV. *Astrophysical Journal*, **480**, 216–234.

Shapiro, P.R. and Field, G.B. (1976). Consequences of a new hot component of the interstellar medium. *Astrophysical Journal*, **205**, 762–765.

Shapley, H. (1921). Note on a possible factor in changes of geological climate. *Journal of Geology*, **29**.

Shklovsky, I.S. (1960). *Cosmic Radio Waves*. Harvard University Press, Cambridge, MA.

Shull, J.M. and Beckwith, S. (1982). Interstellar molecular hydrogen. *Annual Review of Astronomy and Astrophysics*, **20**, 163–190.

Shull, J.M., York, D.G. and Hobbs, L.M. (1977). Abundance variations in high-velocity interstellar gas. *Astrophysical Journal*, **211**, L139–L143.

Smith, L.J., Willis, A.J., Wilson, R., Huber, M.C.E. and Nussbaumer, H. (1979). IUE observations of interstellar Si IV and C IV lines and absorption by hot gas in the ring nebula NGC 6888. In *The First Year of IUE; Proceedings of the Symposium*, University College, London, England, 4–6 April 1979 (A80-16301 04-90), pp. 18–30.

Snow, T.P. (1975). The depletion of interstellar elements and the interaction between gas and dust in space. *Astrophysical Journal*, **202**, L87–L90.

Snow, T.P. (1976). A review of ultraviolet astronomical research with the Copernicus satellite. *Earth and Extraterrestrial Sciences*, **3**, 1–2.

Snow, T.P. and Witt, A.N. (1996). Interstellar depletions updated: Where all the atoms went. *Astrophysical Journal*, **468**, L65–L68.

Sonneborn, G., Tripp, T.M., Ferlet, R., Jenkins, E.B., Sofia, U.J., Vidal-Madjar, A. and Woźniak, P.R. (2000). Spatial variability in the ratio of interstellar atomic deuterium to hydrogen. II. Observations toward gamma2 Velorum and zeta Puppis by the Interstellar Medium Absorption Profile Spectrograph. *Astrophysical Journal*, **545**, 277–289.

Spitzer, L. (1968). Dynamics of interstellar matter and the formation of stars. In B.M. Middlehurst and L.H. Aller (eds), *Stars and Stellar Systems*, Vol. 7, 1–64.

Spitzer, L. (1978). *Physical Processes in the Interstellar Medium*. Wiley, Newark, NJ.

Spitzer, L. (1985). Average density along interstellar lines of sight. *Astrophysical Journal*, **290**, L21–L24.

Spitzer, L. (1997). *Dreams, Stars and Electrons*. Princeton University Press.

Spitzer, L.J. (1996). Highly ionized interstellar atoms – heated, cooled, or mixed? *ApJ*, **458**, L29–L32.

Spitzer, L.J. (1954). Behavior of matter in space. *Astrophysical Journal*, **120**, 1–17.

Spitzer, L.J. (1956). On a possible interstellar galactic corona. *Astrophysical Journal*, **124**, 20–34.

Spitzer, L.J. and Fitzpatrick, E.L. (1993). Composition of interstellar clouds in the disk and halo. I – HD 93521. *Astrophysical Journal*, **409**, 299–318.

Spitzer, L.J. and Jenkins, E.B. (1975). Ultraviolet studies of the interstellar gas. *Annual Review of Astronomy Astrophysics*, **13**, 133–164.

Spitzer, L.J. and Morton, W.A. (1976). Components in interstellar molecular hydrogen. *Astrophysical Journal*, **204**, 731–749.

Stacy, J.G. and Jackson, P.D. (1982). Neutral hydrogen observations towards the Puppis Window of the Milky Way. *Astronomy and Astrophysics*, **50**, 377–422.

Stokes, G.M. (1978). Interstellar titanium. *Astrophysical Journal Supplement*, **36**, 115–141.

Stromgren, B. (1939). The physical state of interstellar hydrogen. *Astrophysical Journal*, **89**, 526–547.

Stromgren, B. (1948). On the density distribution and chemical composition of the interstellar gas. *Astrophysical Journal*, **108**, 242–275.

Taylor, J.H. and Cordes, J.M. (1993). Pulsar distances and the galactic distribution of free electrons. *Astrophysical Journal*, **411**, 674–684.

Trapero, J., Welty, D.E., Hobbs, L.M., Lauroesch, J.T., Morton, D.C., Spitzer, L. and York, D.G. (1996). High-velocity gas in GHRS spectra of three OB stars. *Astrophysical Journal*, **468**, 290–305.

Vallerga, J.V., Vedder, P.W., Craig, N. and Welsh, B.Y. (1993). High-resolution Ca II observations of the local interstellar medium. *Astrophysical Journal*, **411**, 729–749.

van de Hulst, H.C. (1998). Roaming through astrophysics. *Annual Review of Astronomy and Astrophysics*, **36**, 1–16.

Verschuur, G.L. (1975). High-velocity neutral hydrogen. *Annual Review of Astronomy and Astrophysics*, **13**, 257–293.

Vidal-Madjar, A. and Gry, C. (1984). Deuterium, helium and the big-bang nucleosynthesis. *Astronomy and Astrophysics*, **138**, 285–289.

Wakker, B.P. and van Woerden, H. (1997). High-velocity clouds. *Annual Review of Astronomy and Astrophysics*, **35**, 217–266.

Wayte, R.C., Wynne-Jones, I. and Blades, J.C. (1978). Detection of hyperfine structure of interstellar Na I in the alpha Cygni sight-line. *Monthly Notices of the Royal Astronomical Society*, **182**, 5P–10P.

Welsh, B.Y., Sasseen, T., Craig, N., Jelinsky, S. and Albert, C.E. (1997). A minisurvey of interstellar titanium from the Southern Hemisphere. *Astrophysical Journal Supplement*, **112**, 507–526.

Welty, D.E., Frisch, P.C., Sonneborn, G. and York, D.G. (1999). Interstellar abundances in the Magellanic Clouds. II. The line of sight to SN 1987A in the Large Magellanic Cloud. *Astrophysical Journal*, **512**, 636–671.

Welty, D.E. and Hobbs, L.M. (2001). A high-resolution survey of interstellar K I absorption. *Astrophysical Journal*, **133**, 345–393.

Welty, D.E., Hobbs, L.M. and Kulkarni, V.P. (1994). A high-resolution survey of interstellar Na I D1 lines. *Astrophysical Journal*, **436**, 152–175.

Welty, D.E., Hobbs, L.M., Lauroesch, J.T., Morton, D.C., Spitzer, L. and York, D.G. (2000). The diffuse interstellar clouds toward 23 Orionis. *Astrophysical Journal Supplement*, **124**, 465–501.

Welty, D.E., Morton, D.C. and Hobbs, L.M. (1996). A high-resolution survey of interstellar Ca II absorption. *Astrophysical Journal Supplement*, **106**, 533–562.

White, R.E. (1973). Ca I absorption and the electron densities in interstellar clouds. *Astrophysical Journal*, **183**, 81–86.

White, R.E. (1986). Interstellar lithium – differential depletion in diffuse clouds. *Astrophysical Journal*, **307**, 777–786.

Wilson, O.C. (1939). Intercomparison of doublet ratio and line intensity for interstellar sodium and calcium. *Astrophysical Journal*, **90**, 244–248.

Wolfire, M.G., Hollenbach, D., McKee, C.F., Tielens, A.G.G.M. and Bakes, E.L.O. (1995). The neutral atomic phases of the interstellar medium. *Astrophysical Journal*, **443**, 152–168.

York, D.G. (1974). Highly ionized atoms observed with Copernicus. *Astrophysical Journal*, **193**, L127–L131.

York, D.G. (1976). A UV picture of the gas in the interstellar medium. *Memorie della Societa Astronomica Italiana*, **47**, 493–551.

York, D.G. (1977). On the temperature and the interstellar nature of coronal gas observed by Copernicus. *Astrophysical Journal*, **213**, 43–51.

York, D.G. (1982). Gas in the galactic halo. *Annual Review of Astronomy and Astrophysics*, **20**, 221–248.

Zank, G.P. and Frisch, P.C. (1999). Consequences of a change in the galactic environment of the Sun. *Astrophysical Journal*, **518**, 965–973.

29

FRANK B. McDONALD* AND VLADIMIR S. PTUSKIN**

Galactic cosmic rays

Our Galaxy is filled with a relativistic gas of high-energy protons, electrons, and heavy nuclei. The interstellar energy density of these cosmic rays is ~1 eV cm^{-3} – comparable to the energy density of the galactic magnetic field and of the thermal energy of the interstellar medium – with an energy spectra that extends to a maximum energy above 10^{20} eV. High-energy particles are also an important and distinguishing feature of radio galaxies, quasars, and active galactic nuclei. The direct measurement by space and balloon experiments of their charge and mass composition and energy spectra provide information on the source regions within our Galaxy, on injection and acceleration processes, and offer a steadily increasing understanding of cosmic-ray transport through interstellar space. Observations from radio, gamma-ray, and X-ray astronomy define the distribution of energetic particles throughout our Galaxy and establish their presence in extragalactic sources. The interpretation of these observations by allied scientific disciplines is significantly aided by the detailed study of cosmic rays near Earth while our understanding of the sources and of the distribution of galactic cosmic rays is strongly dependent on the data from these other fields.

Space experiments have played a vital role in the development of cosmic ray astrophysics. These experiments have determined the detailed charge and mass composition of cosmic rays for the elements from hydrogen to nickel ($Z = 1$–28) up to energies of ~1 GeV, have mapped the charge distribution of many of the even Z elements to uranium ($Z = 92$) and extended the direct observation of the energy spectra of the more abundant elements up through energies of the order of 10 TeV (10^{13} eV). At high energies ($>$100 TeV), the cosmic-ray intensity is so small that experiments such as large, ground-based air-shower arrays must be used. Throughout the modern era of cosmic rays – defined somewhat arbitrarily as 1946 onward – a vital stimulus has been the guidance and insight from a number of excellent theoretical groups.

Cosmic-ray studies now span an epoch of almost exactly 100 years. In this brief overview we sketch the initial gestation of cosmic-ray astrophysics. A more complete history of the early development of cosmic rays by J. A. Simpson can be found in Chapter 4. At the close of the nineteenth century, scientists using gold-leaf electroscopes to study the conductivity of gases discovered that no matter how carefully they isolated their electroscopes from possible sources of radiation they still discharged at a slow rate. In 1901 two groups investigated this phenomenon, J. Elster and H. Geitel (1901) in Germany, and C. T. R. Wilson (1901) in England. Both groups concluded that some unknown source of ionizing radiation existed. Wilson even suggested that the ionization might be "due to radiation from sources outside our atmosphere, possibly radiation like Röntgen rays or like cathode rays, but of enormously greater penetrating power." A year later two groups in Canada, Ernst Rutherford and H. Lester Cooke (1903) at McGill University, and J. C. McLennan and E. F. Burton (1902), at the University of Toronto showed that 5 cm of lead reduced this mysterious radiation by 30%. An additional 5 tonnes of pig lead failed to further reduce the radiation further.

Nothing significant happened until 1907 when Father Theodore Wulf (1907) of the Institute of Physics of Ignatus College in Valkenburg, Holland, invented a new electroscope. Wulf's electroscope enabled scientists to carry the search for the origin of the mysterious radiation out of the

*University of Maryland, College Park, MD, USA
**IZMIRAN – Institute of Terrestrial Magnetism, Ionosphere and Radio Waves Propagation, Troitsk, Russia

laboratory, into the mountains, atop the Eiffel Tower and, ultimately, aloft in balloons. Assuming that the radiation came from the Earth, they expected to find a rapid decrease in the radiation as they moved away from the surface. They did not find the decrease they expected and in some cases there seemed to be evidence that the radiation actually increased. Intrigued by the conflicting results obtained by Wulf and his colleagues, a young Austrian nuclear physicist, Viktor Hess, obtained support from the Austrian Imperial Academy of Sciences and the Royal Austrian Aero Club to conduct a series of balloon flights to study the radiation. Hess obtained a license to pilot balloons in order to reduce the size of the crew and thereby increase the altitude to which he could carry his electroscopes. On 12 August 1912, using the hydrogen-filled *Böhmen*, Hess reached an altitude of 5,350 m. Carrying two hermetically sealed ion chambers, he found that the ionization rate initially decreased, but then at about 1500 m it began to rise, until at 5,000 m it was over twice the surface rate. Hess (1912) concluded that the results of these observations can best be explained by the assumption that radiation of a very high penetrating power from above enters into the atmosphere and partially causes, even at the lower atmospheric layers, ionization in the enclosed instruments.

On a voyage from Amsterdam to Java, Clay (1927) observed a variation in cosmic-ray intensity with latitude with a lower intensity near the equator, thus establishing that before entering the Earth's magnetic field, the bulk of the primary cosmic rays were charged particles. In 1930 Bruno Rossi, using Störmer theory, showed that if the cosmic rays were predominantly of one charge or the other there should be an east–west effect. In the spring of 1933 two American groups, Thomas H. Johnson (1933) of the Bartol Research Foundation and Luis Alvarez and Arthur H. Compton (1933) of the University of Chicago, simultaneously and independently measured the east–west effect. It showed the cosmic radiation to be predominantly positively charged. In a series of balloon flights in the late 1930s, M. Schein and his coworkers (1941) used Geiger counter telescopes interspersed with lead absorbers to determine that most of the primary particles were not electrons, and hence protons were most plausibly the dominant constituent.

In 1948 research groups from the University of Minnesota and the University of Rochester flew nuclear emulsions and cloud chambers on the same high-altitude Skyhook balloon flight (Freier *et al.* 1948) and discovered the presence of heavy nuclei in the primary cosmic radiation. Further studies by many other groups soon established that essentially all of the elements between H and Fe were present in the cosmic radiation near the top of the atmosphere – including an overabundance of the light elements Li, Be, and B. Then in 1950 it was found that a significant fraction of the cosmic radio emission was synchrotron radiation – indicating the presence of highly relativistic electrons throughout our Galaxy including some discrete sources as well as extragalactic sources (Pacholczyk 1970). However, because of their small abundance (~1% of the intensity of cosmic ray nuclei) electrons were not directly detected in the primary cosmic radiation until 1962 (Earl 1962, Vogt and Meyer 1962). These discoveries made it possible to begin constructing realistic models of the origin and interstellar transport of galactic cosmic rays.

As early as 1934, Baade and Zwicky linked the appearance of supernovae with neutron star formation and cosmic-ray generation. Fermi (1949) regarded cosmic rays as a gas of relativistic charged particles moving in interstellar magnetic fields. His paper laid the groundwork for the modern theory of cosmic ray acceleration and transport. The close link between radioastronomy and cosmic rays was conclusively established at the time of the Paris Symposium on Radioastronomy in 1958 (Paris Symposium 1959). This marked the birth of cosmic-ray astrophysics. The basic model of the origin of galactic cosmic rays was developed by Ginzburg and Syrovatskii (1964).

Over the last three decades the continual evolution of cosmic-ray detector systems for space missions have given us a detailed description of galactic cosmic rays and a growing understanding of their source region, acceleration, and transport.

1. COSMIC-RAY OBSERVATIONS

The elemental composition of cosmic rays provides information on chemical fractionation in the source region as well as some insight into the nature of this region and of the propagation of cosmic rays in interstellar space. Cosmic-ray isotopes probe more deeply the nature of the source region and the timescales of the injection and initial acceleration. Radioactive isotopes such as ^{10}Be, ^{26}Al, and ^{36}Cl reveal the temporal history of cosmic rays in the disk and halo regions. The variation of the charge and mass composition with energy – their energy spectra – can be related to the acceleration process and to particle transport in the Galaxy. When improved measurements are available at ultra-high energies it should be possible to determine whether these particles are of galactic or extragalactic origin. At the highest energies the cosmic-ray arrival direction may also indicate the approximate direction of the most powerful sources.

Over the past four decades of the space age there has been a remarkable improvement in the ability of cosmic ray experiments to measure the elemental and isotopic composition over a steadily increasing range of charges and energy. Space limitations permit discussion of only the more recent results in a given area.

The most remarkable feature of cosmic rays is their energy spectra (Figure 1). From ~10 GeV to $>10^{20}$ eV these spectra, over some 10 orders of magnitude variation in intensity show a relatively featureless power-law distribution. At energies below a few giga-electronvolts the influence of solar modulation becomes important with significant temporal variations at 1 AU related to the 11- and 22-year solar and heliomagnetic cycles. At energies of less than 40 MeV the oxygen spectra in Figure 1 show the presence of so-called anomalous cosmic rays, discussed in Chapter 4. Those are partially ionized interstellar atoms accelerated at the solar wind termination shock. Near 10 MeV there is a highly variable turn-up in the ion spectrum produced by particles of solar/interplanetary origin although the acceleration up to energies more than tens of giga-electronvolts was registered in some solar flare events.

If the differential intensity in Figure 1 is multiplied by $E^{2.7}$, where E is the particle energy, then the structure in the spectrum becomes more apparent at energies >10 Gev (Figure 2). The spectral slope steepens and the energy is proportional to E^{-3} above 10^{14} eV. It decreases back to the form $\sim E^{-2.7}$ above 10^{18} eV, and extends to $\sim 3 \times 10^{20}$ eV. The experimental limitation imposed by the current size of air-shower arrays does not allow measurements of the cosmic-ray intensity at energies greater than 3×10^{20} eV.

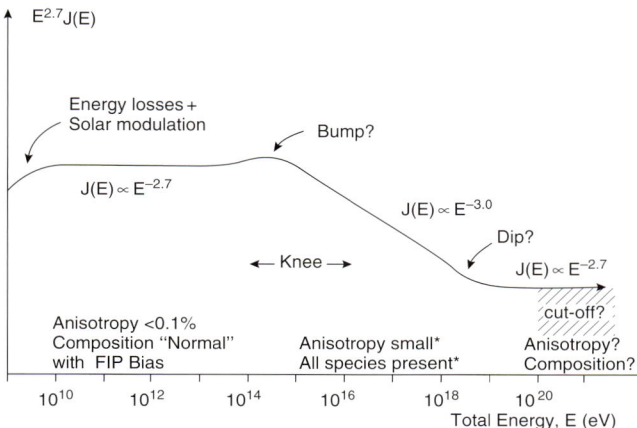

Figure 2 Schematic representation of cosmic ray spectrum (Axford 1994).

Above 10 TeV the composition is not well known; this is the region where acceleration by supernova shocks becomes difficult. It is generally assumed that the cosmic-ray "knee" at $\sim 10^{15}$ eV may reflect a gradual transition in the composition to particles of increasingly higher charge. At energies $>10^{19}$ eV questions of galactic magnetic confinement lead to the assumption that these particles are of extragalactic origin. At energies above 4×10^{19} eV even protons should experience significant deceleration by the 3 K black-body radiation (the Greisen–Zatsepin–Kuzmin effect).

At energies of 10^{12}–10^{14} eV there are small anisotropies of ~0.1% which are thought to be due to local effects. At this time there are no meaningful anisotropies observed at higher energies except the ultra-high energies ~10^{18} eV (see Section 6).

1.1 The elemental composition

Over the charge region $Z = 1–28$ (H–Ni), cosmic-ray experiments in space can resolve the individual elements over an extended energy range. A summary of these data (Figure 3) shows the relative abundance of cosmic rays at ~1 AU along with the Solar System abundance (Simpson 1983) for two different energy regimes, 70–280 MeV/nucleon and 1–2 GeV/nucleon. It can be seen that H and He are the dominant elements, constituting some 98% of the cosmic-ray ions, but are still under-abundant in the cosmic rays relative to the Solar System abundance. There is reasonably good agreement between the cosmic-ray and Solar System abundance data for most of the even elements particularly for C, O, Mg, and Fe. The light elements Li, Be and B, and Sc and V in the sub-iron region are greatly over-abundant when compared to the Solar System abundance – a result of nuclear

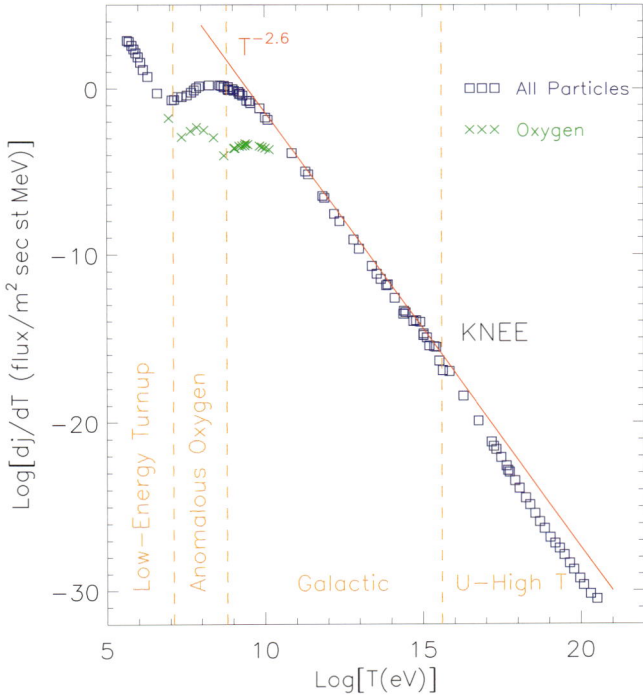

Figure 1 Energy spectrum of cosmic rays measured at the Earth (Jokipii 1989).

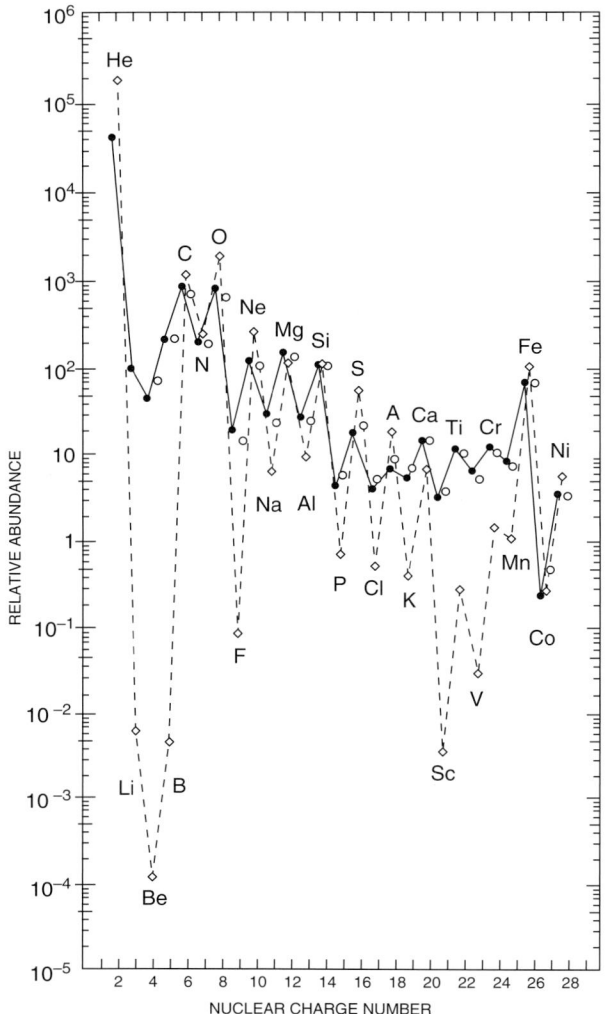

Figure 3 The relative abundances of H to Ni in cosmic rays, solid line and in the Solar System, dashed line. All abundances normalized at Silicam-100 (Simpson 1983).

spallation in interstellar space by nuclei of higher charge. The secondary nuclei generated by these reactions with the interstellar gas will have essentially the same velocity as the incident primary nuclei and hence the same energy per nucleon. Their energy spectra tend to be steeper than those of the primaries due to energy-dependent escape of the higher-energy primaries from the Galaxy.

It is possible to correct the 1 AU observations for the changes produced by nuclear interactions, ionization energy losses, and escape from the Galaxy in the journey of the energetic nuclei through interstellar space and thus obtain an estimate of the source abundance of these elements. When these source abundances are normalized by their Solar System abundances and plotted against the elements' first ionization potential (FIP), a well-defined fractionation effect is observed (Figure 4). Very similar effects are also present in the relative abundance of the solar coronae, solar wind, and gradual solar energetic particle events.

However, particles with low FIP tend to be chemically active and form stable compounds and structures such as dust grains. Meyer *et al.* (1997) have developed in detail the scenario in which galactic cosmic-ray ions originate predominantly from the gas and dust of the interstellar medium. This model gives a natural explanation of the low abundance of H and He. With the presently available data, it is not possible to choose between FIP and volatility. The maximum energy attainable by the shock acceleration of dust grains is of the order of 0.1 MeV/nucleon, so the "FIP *v.* volatility" argument is really about the injection and initial "energization" processes.

1.2 The ultraheavy nuclei ($Z > 30$)

These elements arise from a combination of s-processes (produced in an environment where it is more probable that they undergo beta decay before adding another neutron) and r-processes (produced in a neutron-rich environment where beta decay is less probable). It is expected that r-process elements will dominate in explosive nucleosynthesis such as occurs in supernova explosions. Binns *et al.* (1989) found over the charge range $33 \leq z \leq 60$ that the FIP-corrected observed abundance is similar to that of the Solar System abundance; above $Z > 60$, they found an enhancement of r-process produced nuclei in the source region.

Of particular importance is the actinide region ($Z \geq 88$), where the elements can only be produced via the r-process. These have been studied by an experiment flown on NASA's Long Duration Exposure Facility (LDEF). The experiment consisted of an array of thick, inert, solid-state nuclear track detector stacks – sheets of polycarbonate with a collecting area of 10 m². They were exposed in Earth orbit aboard the LDEF for some 69 months. From this exposure Donnelly *et al.* (1999) measured an actinide/subactinide ratio given by

$$\frac{Z \geq 88}{74 \leq Z \leq 87} = 0.0147 \pm 0.003$$

which is consistent within statistical errors with the abundance ratio =0.013 of propagated primordial Solar System material. Notice that the propagated present-day Solar System material would give 0.0077 for this ratio. Donnelly *et al.* (1999) found that the measured abundance is consistent

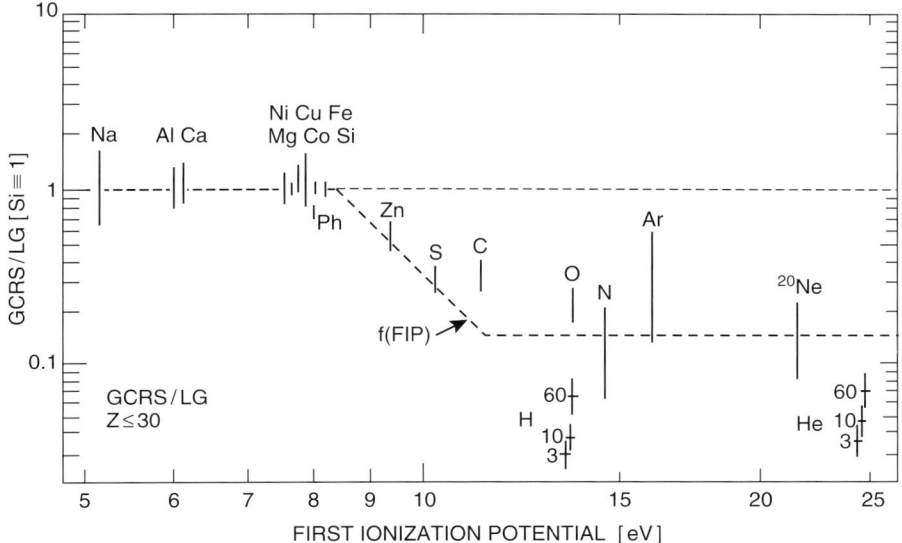

Figure 4 Over-abundance of elements in galactic cosmic ray sources (GCRS) with respect to local galactic composition (LG) v. first ionization potential (Meyer 1985).

with the view that the origin of the cosmic-ray material is predominantly normal interstellar gas and dust.

1.3 Cosmic-ray isotopes

Cosmic-ray experiments on Voyager, Ulysses, and the Advanced Composition Explorer (ACE) represent a significant improvement over previous detector systems and have now resolved the isotopes of essentially all of the primary nuclear components from H ($Z = 1$) to Ni ($Z = 28$) with a relative accuracy that is generally ±10% or better. Resolving isotopes at higher charge or at energies above a few giga-electronvolts becomes increasingly difficult.

The isotopes of the primary cosmic-ray nuclei reflect the properties of the source region. The fractionation process, whether it is FIP or volatility related, is a chemical process operating at the atomic or molecular level and should not affect the relative abundance of the isotopes of a given element. The isotopic abundance of all the primary cosmic-ray elements up to Ni appears to be consistent with the Solar System abundance, ^{22}Ne/^{20}Ne being the only major anomaly. This difficulty can be resolved with a ~3% enhancement of the source regions with ejecta from Wolf–Rayet stars.

Measurements of the radioactive cosmic ray secondaries, which undergo β-decay, can be used to determine the mean confinement lifetime and the mean density of this confinement volume. The cosmic-ray Isotope Spectrometer on the ACE mission has provided high-resolution measurements of ^{10}B, ^{26}Al, ^{36}Cl, and ^{54}Mn over an energy range of 70–400 MeV/nucleon. When treated in the frameworks of the so called "leaky box model" the measured relative abundance for these four species are consistent with a confinement time of 22 million years and a mean interstellar density of the confinement region ranging between 0.15 and 0.4 atoms/cm. (It is worth noting that the interpretation of these measurements is model-dependent. The results for the more physical diffusion model will be given in Section 3.)

1.4 Antiprotons

Antiprotons, $\bar{p}$ are produced in the nuclear interactions of high-energy cosmic rays with the interstellar gas. The kinematics of $\bar{p}$ production results in a unique energy spectrum with a broad peak centered at about 2 GeV, providing a useful probe of interstellar propagation and solar modulation. At 2 GeV the expected ratio of $\bar{p}/p \approx 10^{-4}$ is so small that experimental problems such as background events are severe. However, recent balloon experiments with high-field magnets and multiparameter detector systems have provided excellent measurements over the energy range 0.2–29 GeV (Orito et al. 2000, and references therein). These 1 AU data show the expected peak in the energy spectrum near 2 GeV and that the intensity level and spectral shape are consistent with standard models of cosmic-ray propagation.

While it is not expected that antimatter is a significant constituent of our Universe, the $\bar{p}$ studies also make possible the search for anti-elements of higher charge. The present upper limit on the antihelium/helium flux ratio is $\sim 10^{-6}$.

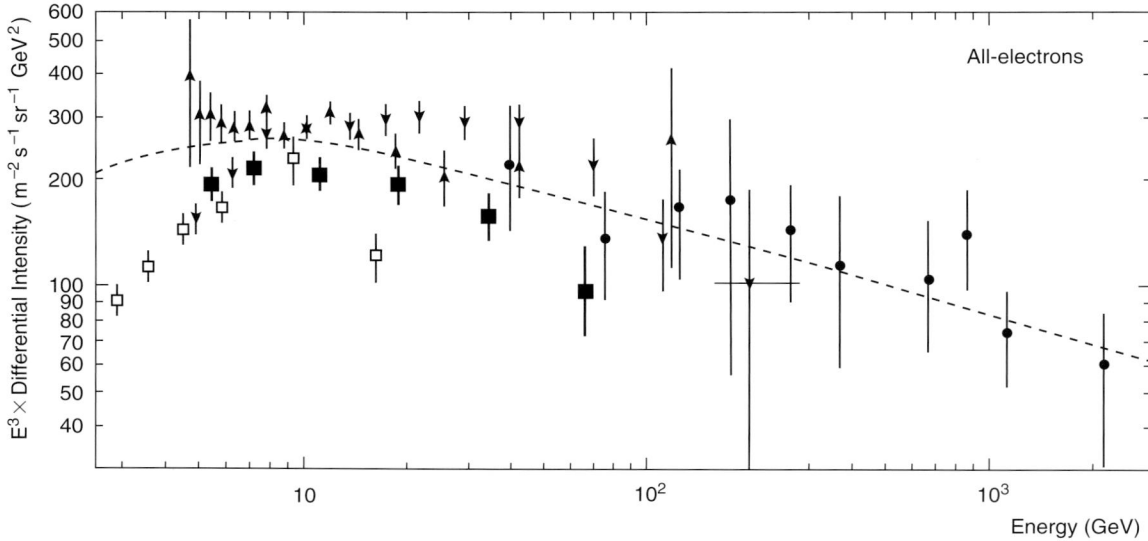

Figure 5 Energy spectrum for all electrons ($e^+ + e^-$) (Barwick *et al.* 1998).

1.5 Electrons and positrons

Electrons and positrons constitute only some 1% of the interstellar cosmic-ray intensity at energies >1 GeV (Figure 5), but provide important information on source regions and on particle transport. Synchrotron radio emission from supernovae identifies these as probable sites of cosmic-ray acceleration. In the interstellar medium, electrons lose energy through inverse Compton scattering and synchrotron emission. Since the rate of loss by electrons through these electromagnetic processes is proportional to the square of their energy, their mean lifetime decreases with increasing energy, resulting in a spectral slope of ~ -3.1 compared with the -2.7 slope of protons. Thus the e/p ratio decreases with energy, and the highest-energy electrons must come from relatively nearby sources.

While primary galactic electrons appear to be generated in supernovae sites, probably along with the nuclear component, there is a secondary electron population produced by nuclear interactions of energetic cosmic-ray ions in the interstellar medium through the production of pions and the subsequent decay chain $\pi^{\pm} \to \mu^{\pm} \to e^{\pm}$. These secondary electrons and positrons are created in approximately equal numbers, and their production spectrum can be calculated from the standard transport models. Thus the positron observations not only provide verification of the standard transport model but also reveal what fraction of the total electron intensity is of primary origin. At 2 GeV the positron fraction $e^+/(e^+ + e^-) \sim 0.1$ and decreases to ~ 0.04 at 10 GeV. These ratios are in agreement with a pure secondary origin of the positron component and indicate that above 2 GeV >80% of the electron component is of primary origin.

1.6 Cosmic ray modulation

The outward flow of the supersonic solar wind carves an enormous cavity in the interstellar medium that reaches far beyond the orbits of the most distant planets of the Solar System. At distances of the order of 100 AU, the pressure of the interstellar medium forces the solar wind to undergo an abrupt transition through the formation of a large standing shock, the termination shock. The hot decelerated solar wind then flows around the termination shock and out along the heliotail while maintaining a separation from the local interstellar medium by another boundary, the heliopause (Holzer 1989).

Galactic cosmic rays entering the Solar System must traverse the region between the heliopause and the termination shock, the heliosheath, before encountering the supersonic solar wind. This termination shock is the source region of a well-defined energetic particle population, the anomalous cosmic-ray component. These predominantly singly charged, low-velocity, high-rigidity ions have their origin as interstellar neutrals, which have been ionized in interplanetary space, convected outward by the solar wind, and accelerated at the termination shock according to the currently accepted paradigm (Fisk *et al.* 1974, Pesses *et al.* 1981).

As anomalous cosmic rays and galactic cosmic rays travel in from the termination shock their intensity decreases with decreasing heliocentric distance. This solar modulation of cosmic rays is most effective at low energies of 0.1–3 GeV/nucleon) and has little influence on very energetic cosmic rays (>20 GeV). The long-term modulation varies with the 11-year solar activity cycle and has the smallest effect when solar activity is at a minimum.

The major physical processes involved in cosmic-ray modulation are: diffusion produced by irregularities in the interplanetary magnetic field embedded in the solar wind; convection and adiabatic energy losses in the outward flowing, expanding solar wind; and gradient and curvature drifts produced by the large-scale structure of the interplanetary magnetic field. Parker (1963) derived a transport equation that includes all of these processes. However, the key problem that remains is the lack of any precise knowledge of the spatial and temporal variations of the diffusion tensor throughout the heliosphere.

The Pioneer and Voyager spacecraft have made precise measurement of the cosmic ray energy spectra at heliocentric distances that now extend to 75 AU up to energies >300 MeV/nucleon. These continuing observations will provide greatly improved estimates of the local interstellar spectra.

The anomalous component is discussed further in Chapter 4. For H, He, O, N, Ne, and Ar these ions constitute a significant background for the study of their low-energy spectra (<60 MeV/nucleon for H and O; <150 MeV for H). However, the simultaneous study of the modulation of anomalous and galactic cosmic rays is a vital means of understanding the transport of energetic particles in the heliosphere.

2. RADIO AND GAMMA-RAY ASTRONOMY: LINKS TO COSMIC RAYS

A feature peculiar to the use of cosmic rays as a tool for astrophysical research is their high isotropy. The long wandering and mixing of energetic charged particles in cosmic magnetic fields washes out information about individual cosmic ray sources and restricts the effectiveness of direct cosmic-ray measurements for astrophysics. However, we can study energetic particles in distant astronomical objects through their electromagnetic radiation.

With the discovery of the synchrotron nature of a significant part of cosmic radio emission it became possible to get information about relativistic electrons in throughout the Galaxy and in various extragalactic sources (Pacholczyk 1970).

In the Galaxy, the diffuse galactic radio emission defines the density distribution of relativistic electrons and their energy spectrum. The maximum in the radiation spectrum of a single electron is at the frequency

$$\nu = 4.6 \left(\frac{B_\perp}{10^{-6} \, \text{G}}\right)\left(\frac{E}{1 \, \text{GeV}}\right)^2 \text{MHz}$$

where $B_\perp$ is the component of magnetic field perpendicular to the velocity of the electron. The electrons (and positrons) with energies 1–30 GeV in a typical interstellar field of 3×10^{-6} G radiate at frequencies of 10 MHz–10 GHz, the frequency interval where the synchrotron radiation of galactic electrons can be observed on Earth. The lower frequency limit is due to the absorption of radiation in regions of interstellar ionized hydrogen, and the upper limit to the increasing influence of the 2.7 K microwave background radiation. The power-law electron spectrum $N(E)dE = KE^{-\gamma} dE$ gives the power-law spectrum of the synchrotron radiation $J(\nu) \propto \nu^{-\alpha}$, $\alpha = \frac{1}{2}(\gamma-1)$; see Ginzburg and Syrovatskii (1964) for detail.

The relativistic electrons in interstellar space experience energy losses by synchrotron radiation and by inverse Compton scattering on background photons (photons of blackbody microwave radiation, photons emitted by stars, etc.). The energy loss rate is given by the equation

$$-\frac{dE}{dt} = bE^2,$$

$$b = 8 \times 10^{-17}\left(w_{\text{ph}} + 6 \times 10^{11}\frac{B^2}{8\pi}\right) \text{GeV s}^{-1} \quad (1)$$

where the quantity w_{ph}, the energy density of photons, is in eV cm^{-3}, B is in Gauss, and the electron energy E is in giga-electronvolts. In the Galaxy, the value of w_{ph} is estimated to be $w_{\text{ph}} = 1$ eV cm^{-3}. The Compton losses are approximately equal to the synchrotron losses at $B = 6 \times 10^{-6}$ G.

The energy losses of relativistic electrons increase with particle energy. Electrons whose energy is sufficiently high (more than about 3 GeV for a of cosmic ray halo of size $H = 5$ kpc) lose their energy so rapidly that they have no time to diffuse from the sources in the disk to the outer halo boundary, so the spatial distribution of diffusing relativistic electrons in the Galaxy is inhomogeneous and energy dependent. The calculation of diffuse galactic synchrotron radiation requires the solution of the transport equation for relativistic electrons using a given distribution of cosmic-ray sources. The calculated radio emission can then be compared with observations. Our incomplete knowledge of the distribution of magnetic fields in the Galaxy introduces significant uncertainty in this procedure. The detailed analysis of galactic radio emission in the diffusion model of cosmic-ray propagation is presented in Berezinskii et al. (1990). The important result that follows from such modeling is the conclusion that the set of parameter values used to explain the radio emission produced by relativistic electrons is consistent with those required for the propagation model of the nucleon component of cosmic rays (the nucleon component is discussed in detail in Section 4). This establishes that the characteristics of cosmic rays obtained from direct observations in the Solar System are typical for the Galaxy as a whole. The source spectrum of electrons needed to fit the radio astronomical observations has the form $E^{-2.2 \pm 0.2}$ (see also Webber 1983).

Another powerful method, beside the radio astronomical one, that gives information on cosmic ray distribution throughout the Galaxy, identifies possible cosmic-ray acceleration sites, and allows one to study very energetic processes in extragalactic sources is offered by gamma-ray astronomy. This branch of modern astronomy is relatively young but is prominent by its outstanding results. The discovery of the gamma-ray blazar class of sources in gamma rays with energies from 100 MeV to the GeV–TeV range, in which active galactic nuclei often radiate more than 10^{48}–10^{49} erg s^{-1}, indicates the extremely efficient acceleration of particles with energies greater than or equal to photon energies (see Mukherjee et al. 1997, Krennrich et al. 1999). The observation of gamma emission from gamma-ray bursts, the most powerful and distant gamma-ray sources with an energy release up to 10^{53} erg (if isotropic) during a fraction of second, proves that ultra-relativistic particles are generated in these events, which are most probably the result of gigantic explosions arising from neutron star or neutron star/black hole mergers (e.g., Piran 1999).

The diffuse gamma-ray continuum emission, which arises from cosmic-ray protons and electrons interacting with the interstellar matter and low-energy photons, is the dominant feature of the gamma-ray sky. This emission in the range of photon energies 50 keV to over 50 GeV was systematically studied by the OSSE, Comptel, and EGRET instruments on the Compton Gamma Ray Observatory (1990–2000). Earlier space experiments on SAS 2 and COS B, launched in the 1970s, also gave a considerable amount of information. Detailed reviews of the results of these observations were presented by Bloemen (1989) and Hunter et al. (1997), see also Dogiel (1996).

Using the EGRET observations, Hunter et al. (1997) analyzed the spatial and spectral distributions of the diffuse gamma-ray emission from the galactic plane. A global galactic model based on a propagation code which allows one to reproduce self-consistently observational data of many kinds related to cosmic rays and gamma rays was developed by Strong and Moskalenko (1998) and Strong et al. (1998).

The most important mechanisms of the generation of diffuse galactic continuum gamma-ray radiation are:

- Production of gamma rays as a result of collisions of energetic protons and nuclei with protons and nuclei of the interstellar gas. These nuclear interactions are accompanied by the production of neutral pions, which decay into two gamma rays. These photons have an energy distribution with a peak at 67.5 MeV, half of the rest mass energy of the π^0 meson.
- Bremsstrahlung of relativistic electrons resulting from their collisions with atoms of the interstellar medium.
- Inverse Compton scattering of relativistic electrons on soft background photons.

The contributions of these mechanisms to the observed diffuse gamma-ray spectrum for the inner Galaxy are shown in Figure 6. The galactic diffuse emission has a narrow

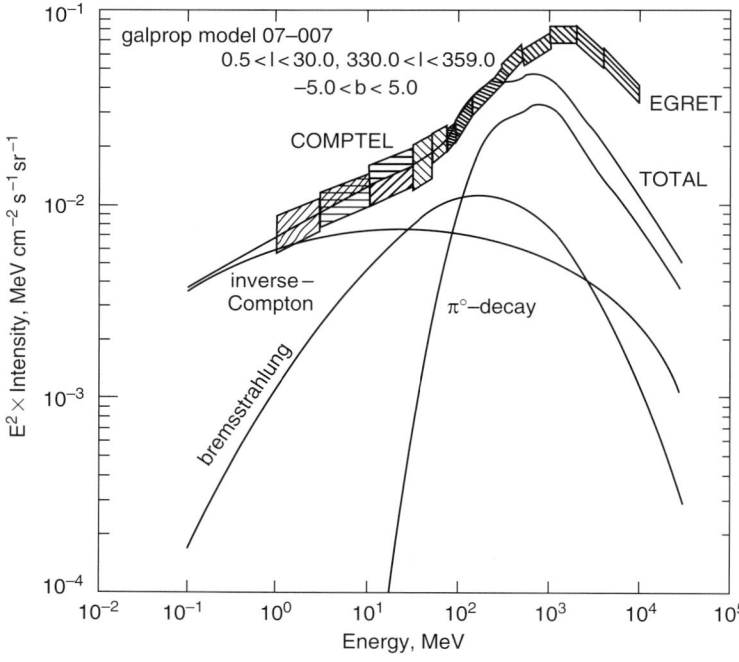

Figure 6 Diffuse gamma-ray spectrum for the inner Galaxy. The curves show the contributions of different processes to the total gamma-ray emissivity (Hunter et al. 1997).

latitude extent along the galactic plane, with enhancements toward the galactic center and at the tangent point of the spiral arms as observed from the Solar System. Astronomical gamma-ray observations in general confirmed the principal features of the galactic model of cosmic-ray origin mainly derived from the direct cosmic-ray observations in the Solar System. However new problems arise after extension of our knowledge about cosmic rays over the entire Galaxy.

From the gamma-ray data, it follows that the density of cosmic-ray protons and electrons has a general galactic gradient decreasing towards the outer Galaxy and perhaps peaking at the Galactic center. This is a natural consequence of the galactic origin of cosmic rays. However, the implied distribution of cosmic-ray sources is much flatter than the supernova distribution although there are large uncertainties in the supernova distribution. The flat cosmic-ray distribution may also reflect some effects of cosmic-ray propagation.

On a smaller few kpc scale, the cosmic-ray density in the galactic disk is correlated with the local surface gas density. This is considered an indication of the dynamical balance between the cosmic rays and the interstellar matter density. The correlation between cosmic-ray density and gas density is not confirmed in the outer Galaxy.

The spectrum of diffuse gamma-ray emission calculated with the use of the local galactic proton and electron spectra at the Earth is, on the whole, in good agreement with observations. The exception is the high-energy excess in the observed gamma-ray emission above about 1 GeV. It may indicate that the cosmic-ray proton and/or electron spectrum observed in the local neighborhood is not representative of the galactic average. A very flat proton spectrum, of the form $E^{-2.4}$–$E^{-2.5}$ is required to explain the excess if it is due to π^0 decay. A flat electron spectrum and increased contribution of the inverse Compton scattering emission is an alternative explanation. It is also not clear how important is the effect of weak unresolved point sources of gamma rays on the interpretation of the diffuse emission.

The methods of gamma-ray astronomy used in addition to traditional methods of radio astronomy are of enormous benefits to cosmic-ray studies. They allow strengthening the cosmic-ray investigations as an instrument of modern astrophysics.

3. COSMIC-RAY PROPAGATION IN THE GALAXY

The relatively large confinement time of cosmic rays in the Galaxy is caused by the interaction of relativistic charged particles with galactic magnetic fields. The diffusion approximation gives an adequate description of cosmic-ray propagation in the Galaxy at energies up to about 10^{17} eV and provides a basis for the interpretation of cosmic-ray data as well as related radio astronomical, X-ray, and gamma-ray observations (Berezinskii et al. 1990, Strong and Moskalenko 1998) (see Figure 7).

Formally, the modeling of cosmic-ray diffusion in the Galaxy includes the solution of a transport equation with a given source distribution and boundary conditions for all cosmic-ray species. The transport equation describes diffusion, convection by the hypothetical galactic wind, and changes of energy (energy losses and possible distributed acceleration). In addition, nuclear collisions with interstellar gas atoms resulting in the production of secondary energetic particles, and the decay of radioactive isotopes should be taken into account when considering the proton–nucleus component. Hundreds of isotopes are included in the calculations of nuclear fragmentation and transformation of energetic nuclei in the course of their interaction with the interstellar gas.

The homogeneous leaky box model is a well-accepted approximation to the diffusion model. The exit of particles from the Galaxy is described in this simple model by some effective escape time T_{lb}, which does not depend on position. The corresponding mean matter thickness traversed by cosmic rays before exit from the Galaxy, the escape length, is defined as $X_{lb} = \rho_{lb} v T_{lb}$, where ρ_{lb} is the mean gas density. Here the subscript lb refers to the leaky box model parameters.

The relation between parameters of the diffusion model with static halo and the leaky box model is given by the following approximate equation, valid for an observer located in the galactic disk (Ginzburg and Ptuskin 1976):

$$X_{lb} \approx \frac{\mu \beta c H}{2D} \qquad (2)$$

where μ is the surface gas density of the galactic disk and H is the height of the galactic cosmic-ray halo.

The cosmic ray halo is flat, so that $R \gg H$, where $R \sim 20$ kpc is the radius of the cylindrical halo. The value of H is defined as the distance from the galactic midplane to the halo boundary, where cosmic rays freely escape from the Galaxy and their number density is negligibly small. It is assumed that cosmic-ray sources are distributed in the relatively thin galactic disk.

The equivalence of the diffusion and the leaky box models with eqn (2) between their parameters holds for stable nuclei with $Z < \sim 30$. In particular, the total nuclear spallation cross-section should be not large: $\sigma \ll mH/(hX)$, where m is the mean mass of an interstellar atom and h is the characteristic half-thickness of the gas disk, $h \ll H$. This limitation on σ ensures that the probability of nuclear fragmentation occurring during one diffusive crossing of the

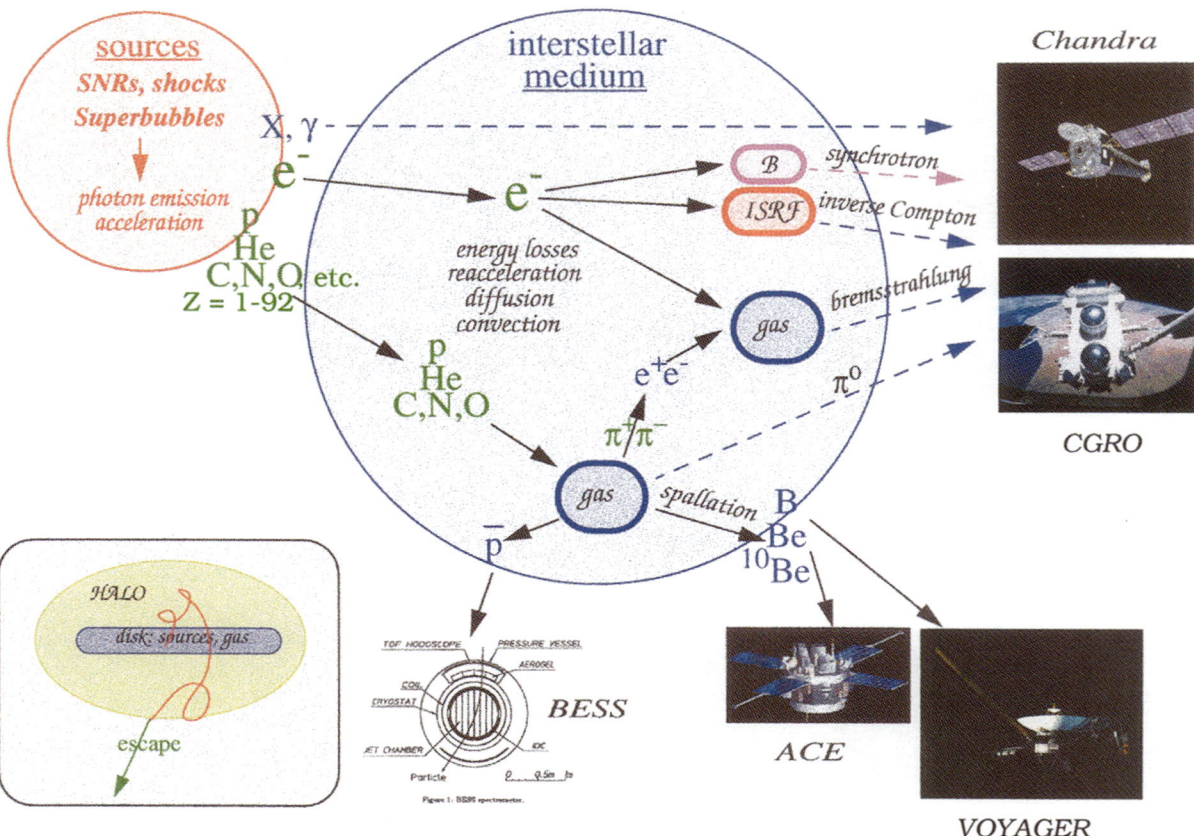

Figure 7 Schematic representation of the cosmic-ray transport process from the source regions through the interstellar medium to their detection in the Solar System. The inset at the lower left shows the thin galactic disk and one possible model of the halo region. The insets along the bottom show some important cosmic ray experiments (BESS) or missions (ACE and Voyager). Along the right-hand side are the Chandra (X-rays) and CGRO (gamma rays) missions (after I. Moskalenko).

gas disk is small. The diffusion and the leaky box models are also not equivalent for those cosmic-ray species which have very inhomogeneous spatial distributions in the Galaxy, for example rapidly decaying radioactive isotopes and the high-energy electrons with strong energy losses.

The data on the abundance of stable secondary nuclei B and Sc + Ti + V in cosmic rays define the value of the escape length. According to Jones *et al.* (2001),

$$X_{lb} = 11.8\beta \text{ g cm}^{-2} \quad \text{at } R < 4.9 \text{ GV},$$
$$X_{lb} = 11.8\beta(R/4.9 \text{ GV})^{-0.54} \text{ g cm}^{-2} \quad \text{at } R \geq 4.9 \text{ GV} \quad (3)$$

This parametrization is valid for particles in the interstellar medium with energies from about 0.4 GeV/nucleon to 300 GeV/nucleon, where data on secondary nuclei are available. The differential source spectrum found from the fit to the observed spectra of primary nuclei C and Fe has a power-law form on rigidity $Q \propto R^{-\gamma_s}$, $\gamma_s = 2.35$.

Equations (2) and (3) imply a constant diffusion coefficient at rigidity $R < 4.9$ GV and give rigidity-dependent diffusion at higher rigidities:

$$D = 2 \times 10^{28}(H/5 \text{ kpc})(R/1 \text{ GV})^{0.54} \text{ cm}^2 \text{ s}^{-1},$$
$$R \geq 4.9 \text{ GV} \quad (4)$$

(we assume here that $\mu = 2.4$ mg cm^{-2}, see Ferriere 1998). Equation (4) shows that the efficiency of cosmic ray confinement is decreasing with particle energy.

The abundance of stable secondary isotopes determines the ratio H/D (eqns (2) and (4)). Additional information can be obtained from the data on radioactive secondary isotopes with beta decay. The isotopes ^{10}Be (the decay lifetime of an ion at rest is $\tau = 2.3 \times 10^6$ yr), ^{26}Al (1.3×10^6 yr), ^{36}Cl (4.3×10^5 yr), and ^{54}Mn (9.1×10^5 yr) are usually employed. The content of these decaying isotopes is determined by the conditions of cosmic-ray propagation in the galactic vicinity $\sim 200\sqrt{[(D/10^{28}\,\text{cm}^2\,\text{s}^{-1})(\tau/10^6\,\text{yr})]}$ pc about an observer position. The surviving fraction of radioactive isotopes derived from low-energy observations gives the value of the diffusion coefficient as about $3 \times 10^{28}\,\text{cm}^2\,\text{s}^{-1}$ at energy 0.4 GeV/ nucleon (see Ptuskin and Soutoul 1998). Equation (4) then gives $H \sim 5$ kpc, and a characteristic time for cosmic-ray diffusion out of the Galaxy of $H^2/2D \sim 10^8$ yr.

This value of H is not in disagreement with radio-astronomical observations (Beuermann et al. 1985) which indicate the presence of a thick nonthermal galactic radio disk with the full equivalent width of the order of 3.5 kpc. The synchrotron radio emission is generated by the electron component of cosmic rays in galactic magnetic field. The density of relativistic stable nuclei is halved at the height about $\frac{1}{2}H \sim 2.5$ kpc above the galactic midplane that gives a somewhat larger full width ~ 5 kpc. This difference is natural since the synchrotron radio emission determined by the density of relativistic electrons decreases with height more rapidly than the density of stable nuclei because of their higher energy losses. Also, the magnetic field that appears in the expression for the synchrotron emissivity should be decreasing with height.

The value of the diffusion coefficient (4) found in the frameworks of the empirical diffusion model can be compared with the predictions of the theory of charged particle motion in random and regular galactic magnetic fields. The typical strength of magnetic field in the galactic disk is estimated to be $B \sim 5\,\mu\text{G}$ (Ruzmaikin et al. 1988, Vallée 1997). The random component of the field exceeds the regular (average) component. The last is predominantly azimuthal and has a value $B_0 \sim 2-3\,\mu\text{G}$. The observed power spectrum of the random field is of the form $W(k)dk \propto k^{-2+a}\,dk$, $a \sim 0.3$–0.6, over a wide range of wavenumbers $1/(10^{20}\,\text{cm}) < k < 1/(10^8\,\text{cm})$. Note that $a = \frac{1}{3}$ corresponds to the Kolmogorov spectrum.

The kinetic theory of cosmic-ray transport in galactic magnetic fields is constructed similarly to the well-studied case of cosmic-ray transport in interplanetary magnetic fields (e.g., Jokipii 1971, Skilling 1975, Berezinskii et al. 1990). A charged particle with Larmor radius r_g ($r_g = pc/ZeB$, where Ze is the particle charge) is mainly scattered by magnetic irregularities of the size $1/k \sim r_g$. This resonant scattering leads to the spatial diffusion of cosmic rays. For the typical parameters of the interstellar magnetic field, the diffusion coefficient can be roughly estimated as

$$D \approx \frac{vr_g}{3}\left(\frac{\delta B_{\text{res}}^2}{B^2}\right)^{-1} \approx 3 \times 10^{28}\left(\frac{R}{1\,\text{GV}}\right)^a \text{cm}^2\,\text{s}^{-1},$$
$$0.3 \leq a \leq 0.6 \qquad (5)$$

Here δB_{res} is the value of random field fluctuations at the resonant scale (the inequality $\delta B_{\text{res}} \lesssim B$ is assumed). The diffusion is anisotropic locally and goes predominantly along the magnetic field, but the large-scale wandering of magnetic field lines makes it close to isotropic on scales $\gtrsim 100$ pc.

The estimate (5) is compatible with the high-energy asymptotics (4) found from the empirical diffusion model. Nonetheless, the resonant scattering by a random field with a power-law spectrum does not provide the energy-independent diffusion at low energies as implied by eqn (2) and scaling $X_{\text{lb}} \propto \beta$ at $R < 4.9$ GV. It was suggested (Jones 1979) that the large-scale convective motion of the interstellar medium may dominate at low energies. The convective transport of cosmic rays may have the form of the galactic wind with velocity 20–30 km s^{-1}, or of turbulent diffusion with a diffusion coefficient of about $4 \times 10^{28}\,\text{cm}^2\,\text{s}^{-1}$. In both cases, however, it is difficult to fit rather sharp peaks in the secondary/primary ratios (e.g., B/C) observed at $R \sim 5$ GV.

An explanation of this peak can be achieved in a model with reacceleration of cosmic rays in the interstellar medium (Simon et al. 1986, Seo and Ptuskin 1994). The distributed reacceleration of cosmic rays after their exit from compact sources (supernova remnants) changes the shape of the particle spectrum. In particular, it leads to an increase in secondary/primary ratios with energy at $R < 5$ GV, where re-acceleration is relatively strong, and to the steep decrease of these ratios at $R > 5$ GV, where the efficiency of re-acceleration is suppressed by the fast escape of particles from the Galaxy. In a minimal model, stochastic re-acceleration occurs as a result of scattering on the same randomly moving magnetohydrodynamic waves which are responsible for the spatial diffusion of cosmic rays. The rate of re-acceleration is determined by the particle diffusion coefficient on momentum, $D_{pp} \sim p^2 V_a^2/D$, where $V_a \approx 20$–40 km s^{-1} is the Alfvén velocity. With such a re-acceleration, the secondary/primary ratios are reproduced with a diffusion coefficient

$$D = 6 \times 10^{28}\beta\left(\frac{H}{5\,\text{kpc}}\right)\left(\frac{R}{1\,\text{GV}}\right)^{0.3}\,\text{cm}^2\,\text{s}^{-1} \qquad (6)$$

This diffusion coefficient corresponds to the scattering on random field with a close to Kolmogorov spectrum at all rigidities.

A weaker dependence of diffusion on energy (6) in the model with re-acceleration compared with the dependence shown in eqn (4) should manifest itself in a weaker energy dependence of the secondary/primary ratios at $E \gg 20$ GeV/nucleon, where re-acceleration is not essential. The data currently available are not adequate to distinguish between these models.

The measurements of secondary K-capture isotopes in low-energy cosmic rays at $E < 1$ GeV/nucleon are also potentially important for testing re-acceleration (Silberberg and Tsao 1990). The isotopes ^{37}Ar, ^{44}Ti, ^{49}V, ^{51}Cr, and others are produced by nuclear interactions during propagation. They rapidly decay by electron capture at low energies, at which energetic ions can have an orbital electron. The probability of there being a bound electron depends strongly on energy, so the surviving fraction of the K-capture isotopes and the abundance of their products are strong functions of energy sensitive to the change of particle energy in the interstellar medium. The first measurements of an energy-dependent decay of the electron-capture nuclei ^{49}V and ^{51}Cr (Niebur et al. 2000) may serve as a test of distributed interstellar re-acceleration.

The continuous distribution in space of cosmic-ray sources was assumed in the preceding discussion of cosmic-ray propagation. This assumption is not valid for very high-energy electrons, which are subject to large energy losses through synchrotron radiation and inverse Compton scattering. The characteristic energy loss time is $t_1 \approx 2 \times 10^5$ (10^{12} eV/E) yr, which allows the electrons with maximum detected energy ~2×10^{12} eV to diffuse to a distance $d \sim (Dt_1)^{1/2} \approx 0.4$ kpc from the source assuming the diffusion coefficient (6). There are only a few close supernova remnants and pulsars that can contribute to the observed flux of high-energy electrons observed at the Earth; these include Vela, Monogem, and Loop I. It can be shown (Nishimura et al. 1997) that it is possible to explain the observed flux of very high-energy electrons, the energetic particles of local galactic origin, under the assumption that typical parameters of the global galactic model are that the diffusion coefficient is given by eqn (6), the electron spectrum at the source has the form $Q \propto R^{-2.4}$, and the source power of relativistic electrons is about 10^{48} erg per SNR.

Anisotropy is another local characteristic of cosmic rays. The amplitude of the first angular harmonic of cosmic-ray intensity in the interstellar medium at the position of the Sun is given approximately by $\delta = (0.5-1.0) \times 10^{-3}$ at energies of $10^{12}-10^{14}$ eV. Possible interpretation of these observations is based on the contribution of local cosmic-ray sources, the supernovae and their remnants (Dorman et al. 1984). The cosmic-ray density is determined by the simultaneous contribution from many galactic sources, but a single nearby source most probably defines the observed anisotropy of cosmic rays. It can be shown that at sufficiently high energies (probably $E > 10^{15}$ eV) the effect of individual local sources is less pronounced and the observed anisotropy is determined by the general streaming of cosmic rays out of the Galaxy.

As was mentioned above, the regular convection of cosmic-ray particles by the galactic wind may work simultaneously with diffusion in a process of cosmic-ray transport in the Galaxy. The wind velocity does not greatly exceed 20 km s^{-1} at a height of 1 kpc above the galactic plane but could reach hundreds of kilometers per second beyond a few dozen kiloparsecs. Different wind models have been considered in the context of cosmic-ray propagation (Jokipii 1976, Berezinskii et al. 1990, Webber et al. 1992, Bloemen et al. 1993). The more advanced version of these models, where cosmic-ray pressure drives the galactic wind, is discussed in Section 5.

However, the mere existence of wind in our Galaxy stands unproved. New cosmic-ray data together with other astronomical observations can clarify the problem; see Berezinskii et al. (1990) for a discussion.

4. GALACTIC SOURCES AND ACCELERATION MECHANISMS

If one uses the cosmic-ray energy requirements and the nonthermal radiation as a guideline, then the most powerful accelerators of relativistic particles in the Galaxy should be supernovae and supernova remnants, pulsars, neutron stars in close binary systems, and winds of young massive stars. The total power L_{cr} needed to maintain the observed energy density of cosmic rays can be estimated as

$$L_{cr} = \frac{w_{cr}V}{T} = \frac{w_{cr}vM_g}{X} \approx 5 \times 10^{40} \text{ erg s}^{-1} \quad (7)$$

Here we have used the relation $X = \rho vT$ between the escape length X, the mean gas density ρ in the volume V occupied by galactic cosmic rays, and the escape time T of cosmic rays from the Galaxy. The quantity $M_g \approx 5 \times 10^{42}$ g is the total mass of the interstellar gas in the Galaxy.

For the acceleration by a supernova, this estimate implies the release of energy in the form of cosmic rays of approximately 5×10^{49} erg per supernova if the supernova rate in the Galaxy is 1 every 30 years. This value comes to about 5% of the kinetic energy of the ejecta that is in agreement with the prediction of the theory of diffusive shock acceleration (e.g., Jones and Ellison 1991). This assumes the acceleration of cosmic rays by the outwardly propagating shock, which results from the supernova explosion and propagates in the interstellar medium or in the wind of the progenitor star.

The rotational energy of a young pulsar with period P that remains after the supernova explosion is estimated to be $2 \times 10^{50}(10 \text{ ms}/P)^2$ erg and may provide an additional energy reservoir for particle acceleration.

The data on the nonthermal radio emission of supernova remnants reinforce the above energy estimates. Assuming an equipartition between cosmic-ray and magnetic field energy densities, and an e/p ratio of 10^{-2}, one finds the total energy of relativistic electrons to be 3×10^{48}–10^{49} erg in each of the remnants Cas A, IC 443, and Cygnus Loop (Lozinskaya 1992). Shell-type SNRs exhibit a broad range of spectral indices, centered roughly on $\alpha \approx -0.5$. This can be related to the electron power-law index through the equation $\gamma = 2\alpha + 1 \approx 2.0$ but the range is from about $\gamma \approx 1.4$ to $\gamma \approx 2.6$. The interpretation of nonthermal radio emission from the galaxy M33 (Duric et al. 1993) showed that SNRs are the sites of the acceleration of relativistic electrons with the same efficiency that is needed to provide the observed intensity of galactic cosmic-ray electrons.

Gamma-ray emission associated with few bright SNRs has been found using the EGRET catalog of gamma-ray sources at $E > 30$ MeV (Sturner and Dermer 1995, Esposito et al. 1996). The gamma-ray fluxes from the two most prominent sources, γ Cygni and IC 443, indicate an energy of about 3×10^{49} erg for relativistic protons and nuclei confined in each envelope, assuming that gamma rays are generated through π^0 decay.

The non-thermal X-ray radiation with a characteristic power-law tail at energies more than few kilo-electronvolts from the bright rims in supernova remnants SN 1006 (Koyama et al. 1995), G347.3–0.5 (Koyama et al. 1997), IC 443 (Keohane et al. 1997), and Cas A (Allen et al. 1997) was detected in ASCA (Advanced Satellite for Cosmology and Astrophysics) and RXTE (Rossi X-ray Timing Explorer) satellite experiments. This detection is considered as evidence of the synchrotron emission of electrons with energies up to 10^{14} eV. The inverse Compton scattering of background photons by these electrons is also the most probable mechanism of the emission in tera-electronvolt gamma rays detected from SN 1006 (Tanimori et al. 1998). A discussion of models of radio, X-ray, and gamma-ray emission from supernova may be found in Reynolds (1998), Gaisser et al. (1987), and Baring et al. (1999).

The correction of observed cosmic-ray composition for nuclear fragmentation in the interstellar medium makes it possible to determine the initial elemental and isotopic composition of accelerated particles, and to clarify the process of cosmic-ray acceleration and the nature of cosmic-ray sources.

The relative abundance of chemical elements in cosmic-ray sources is in general similar to the solar and to the local galactic abundance, but it shows some significant deviations. The elements that appear under-abundant by a factor of about 5 are those elements that are difficult to ionize. The critical ionization potential is ~10 eV, corresponding to ionization at an equilibrium temperature of 10^4 K (characteristic, for example, of the solar photosphere). The correlation of abundance with the first ionization potential is also known for solar energetic particles. So, it is possible that the outer layers of relatively cool stars serve as injectors of the seed particles required for the subsequent acceleration (Meyer 1985).

For most elements the volatility is correlated with the first ionization potential, so volatility is also a possible selection factor for the cosmic-ray population. Then predominant acceleration and breakup of grains, as would be natural for diffusive shock acceleration, could explain the situation (Epstein 1980).

The popular present scenarios include the acceleration of grains together with relatively less abundant volatile ions by shocks in the interstellar medium (Meyer et al. 1997); the acceleration of freshly formed material, particularly grains in young supernova remnants (Lingenfelter et al. 1998); and the acceleration of material in hot superbubbles with multiple supernova remnants (Higdon et al. 1998).

Corrected for the first ionization potential effect, the elemental composition at cosmic-ray sources as compared with the solar and local galactic composition exhibits an under-abundance of H and He by a factor of 10 (if normalized to Fe). It also demonstrates a higher Pt/Pb ratio by a factor of 5 and a higher actinide abundance, the ratio $(Z \geq 88)/(74 \leq Z \leq 87)$ being higher by a factor of 3 (Binns et al. 1989, Thompson et al. 1993). These anomalies may be an indication that cosmic rays arise from supernova material (synthesized in the r-process) mixed with the interstellar gas.

The isotopic composition of cosmic rays has now been measured for all stable isotopes for elements from H through Ni. The isotopic composition of cosmic-ray source material is strikingly similar to the composition found in the Solar System. Thus, the solar-like mix of isotopes of Fe and Ni suggests that Type II and Ia supernovae contribute to cosmic-ray source material in proportions similar to their contributions to the Solar System (Wiedenbeck et al. 1999, Connell and Simpson 1997).

There is one well-established anomaly in the isotopic composition of galactic cosmic rays: the excess of ^{22}Ne. The ratio ^{22}Ne/^{20}Ne is enhanced by a factor of 4 compared with the solar reference value (Lukasiak et al. 1994, Du Vernois et al. 1996). It can be explained only by the special conditions of nucleosynthesis. The enhancement of neutron-rich isotopes would be expected in highly evolved very massive stars in their Wolf–Rayet stage when their surfaces contain large excesses of the products of core helium burning, including ^{22}Ne (Casse and Paul 1982). An increased cosmic-ray C/O ratio by a factor of 2 is also favors Wolf–Rayet stars as a source.

The relatively high and close-to-the-solar value of the ratio ^{59}Co/^{56}Fe (Leske 1993, Mewaldt et al. 1999) testifies that the major part of originally synthesized ^{59}Ni has decayed by the K-capture of an orbital electron into ^{59}Co before the acceleration started; this test was suggested by Casse and Soutoul (1978). The delay between synthesis of this material and acceleration is greater than 10^5 yr. This probably rules out the models where a major portion of the observed mildly relativistic cosmic rays is accelerated in young SNRs.

We come now to the discussion of acceleration mechanisms of galactic cosmic rays. The interpretation of data on the ultra-high-energy particles suggests that acceleration in galactic sources may go at least up to 10^{18}–10^{19} eV (see Section 6). We shall see that an appropriate mechanism in our Galaxy is not easy to identify.

The diffusive shock acceleration at a supernova blast wave, the most generally accepted mechanism for cosmic-ray acceleration in the Galaxy, was suggested by Krymsky (1977), Axford et al. (1977), Bell (1978), and Blandford and Ostriker (1978). The acceleration of a fast particle diffusing near the shock front is a version of Fermi-type acceleration. Fast particles are scattered on inhomogeneities of magnetic field frozen into the background plasma and gain energy each time they cross the shock where plasma is compressed.

The characteristic time of particle acceleration can be estimated as D/u_{sh}^2, where u_{sh} is the shock velocity. The distribution of accelerated particles on momentum has a power-law form $N(p) \propto p^{-(r+2)/(r-1)}$, where r is the gas compression ratio in the shock. For strong shocks $r = 4$ and $N(p) \propto p^{-2}$. In many cases the acceleration is so efficient that the back-reaction of the pressure of accelerated particles changes the shock structure, and cosmic rays cannot be considered as test particles. Additional complication of the acceleration process is caused by plasma instabilities created by energetic particles (see the reviews by Jones and Ellison 1991, Berezhko et al. 1996).

As particle energy increases, the diffusion coefficient D usually increases and the acceleration becomes less efficient. For a spherical shock front with a radius R, the acceleration occurs under the condition $g = Ru_{sh}/D > 1$ (a numerical factor is omitted here). The shock produced by a supernova outburst with initial energy 3×10^{50} erg in the interstellar medium with density 3×10^{-3} cm^{-3} meets the condition $g > 1$ if $D < (1-3) \times 10^{27}$ cm^2 s^{-1} (Prishchep and Ptuskin 1981, Lagage and Cesarsky 1983). The last inequality for D is hardly fulfilled for the cosmic rays with energies $E > 100$ GeV n^{-1} if one takes the typical diffusion coefficient in the interstellar medium (see Section 3).

The cosmic ray stream instability can provide enhanced turbulence in the vicinity of the shock front. This can decrease the diffusion coefficient to the Bohm value $vr_g/3$ (where r_g is the Larmor radius) and increase the maximum energy to about 3×10^{14} eV/nucleon for an interstellar magnetic field of 5×10^{-6} G and a remnant age of 10^3 yr. The maximum energy of accelerated particles can be higher than given by the Bohm limit on the diffusion coefficient if the shock moves strictly perpendicular to the regular magnetic field (Jokipii 1987), but it can hardly exceed the value of 3×10^{15} eV/nucleon, the energy of the "knee" in the observed cosmic-ray spectrum.

Actually the maximum particle energy can be larger if the supernova blast wave propagates in the stellar wind of a pre-supernova star rather than in the interstellar medium.

The magnetic field of the wind is higher and has the form of a Parker spiral. Völk and Biermann (1988) estimated the maximum energy of accelerated particles as about 10^{16} eV/nucleon in the winds of red giants (the surface magnetic field $B = 10$ G at a stellar radius of $r = 2 \times 10^{13}$ cm, where the wind velocity is $u = 10^6$ cm s^{-1}) and Wolf–Rayet stars (50 G, 2×10^{12} cm, 10^8 cm s^{-1}). The initial shock velocity is $(1-3) \times 10^9$ cm s^{-1}. The maximum energy is attained about 1 yr after the explosion and does not change considerably for another few years if the magnetic field decreases with distance as $1/r$ in the wind zone.

The maximum energy which particles can reach in a course of diffusive shock acceleration in the supernova envelope is hardly sufficient to explain the extended cosmic-ray spectrum beyond the knee. Bell (1992) considered the additional acceleration of very-high-energy particles by the regular electric field in a pulsar-driven SNR. For a few millisecond pulsars with a proper (parallel) orientation of magnetic and rotation axes the energy gain can reach $10^{19} Z$ eV. The potential available for acceleration is the same whether the cosmic rays orbit the neutron star near the light cylinder or near the outer edge of the remnant. The scenario might include the diffusive shock acceleration for particles with insufficient energy to enter the nebula and the acceleration by the internal field for high-energy particles, which can drift into the nebula. It is not clear whether this scenario would provide the continuous energy spectrum and proper power-law form of its high-energy part. Some further trajectory calculations are needed to determine the energy distribution of departing particles.

In principle, the maximum particle energy gained at an individual shock wave may be increased after additional acceleration at a different shock. The acceleration can go in two stages (Axford 1994). An individual supernova envelope accelerates particles up to the knee (the first stage). The subsequent acceleration is furnished by the collective action of other SNRs (the second stage). This mechanism is probably the most efficient inside superbubbles. A superbubble may contain an ensemble of shocks generated by the winds of a few tens of massive bright OB stars and the supernova blast waves. The acceleration in a superbubble of size 300 pc can give the particles energies up to $10^{17} Z$ eV if the characteristic turbulent velocity is 3×10^8 cm s^{-1} and the magnetic field strength is as high as 3×10^{-5} G (Bykov and Toptygin 1993).

The largest possible shock that can cause cosmic-ray acceleration in the Galaxy is the (hypothetical) galactic wind termination shock (Jokipii and Morfill 1985). The acceleration up to $3 \times 10^{18} Z$ eV was found for the shock radius of 300 kpc. It remains unclear how to provide a smooth conjugation between the part of the spectrum before the knee (generated in an SNR) and after the knee (generated at the termination shock) in this model.

5. DYNAMICAL EFFECTS OF COSMIC RAYS IN THE GALAXY

As one of the components of interstellar medium, cosmic rays can be considered as a gas of relativistic charged particles. As noted at the beginning of this chapter, an energy density $w_{cr} \approx 1$–2 eV cm^{-3} of cosmic rays is approximately equal to the energy density of the magnetic field and to the energy density of turbulent motions of the interstellar gas. Therefore energetic particles should be an important dynamical factor in the Galaxy.

The presence of a nonthermal component in the interstellar gas leads to collective (plasma) effects (Ginzburg 1966, Lerche 1967) and in particular to stream instabilities that arise when fast charged particles move through background plasma, the problem has been much investigated in plasma physics (e.g., Galeev and Sudan 1983). The instability develops when the cosmic rays move with a bulk velocity exceeding the phase velocity of waves travelling in the same direction.

The contribution of cosmic-ray pressure P_{cr} is probably essential in maintaining the equilibrium distribution of gas and magnetic field above the galactic midplane. The action of cosmic-ray pressure may not only increase the characteristic size of the gaseous halo in the Galaxy, but can produce a non-exponential tail in the gas density distribution at large distances from the galactic plane (Chevalier and Fransson 1984).

The hydrostatic distribution might turn out to be unstable to small perturbations. The characteristic time of the development of instability in an unstable system is of the order of the time of free fall for a gas onto the galactic plane, and ranges from ~10^7 years for the gas disk to ~10^8 years for the gaseous halo of our Galaxy. Parker (1966) found that, as the development of instability progresses, small distortions of the magnetic field lines are enhanced by gas flowing along them, and rising arcs of the magnetic field are formed. It is supposed that the Parker instability is essential in establishing approximate equipartition between the energy densities of cosmic rays and of the interstellar magnetic field acting as a "valve" that releases relativistic particles from the Galaxy when their pressure is too high. It was also suggested (Parker 1992) that production by energetic particle pressure of magnetic arcs and loops above the galactic disk is necessary for the realization the dynamo mechanism of magnetic field generation in the Galaxy.

The interstellar gas at some distance above the galactic plane may not be in static equilibrium, but may undergo large-scale convective motion. In particular, the supersonic wind driven by cosmic rays may exist in the Galaxy. The gas at the galactic disk is not hot enough to drive the thermal wind, but the cosmic-ray pressure can sustain the wind flow of relatively cold gas (Ipavich 1975, Breitschwerdt et al. 1991). In the most general model (Zirakashvili et al. 1996) the production of cosmic rays in a galactic disk largely determines the structure of the wind flow in a rotating Galaxy with a frozen-in, large-scale magnetic field. The stream instability of cosmic rays, moving away from the Galaxy along the spiral magnetic field, creates small-scale Alfvénic turbulence in the system. The equilibrium spectrum of turbulence determined from the nonlinear Landau damping of waves on thermal ions has the form and amplitude that would explain the value and the energy dependence of the cosmic-ray diffusion coefficient as determined from the empirical modeling of cosmic-ray propagation (Ptuskin et al. 1997) (see also Section 3). The wind flow goes through slow, Alfvénic, and fast magnetosonic points at distances about 5 kpc, 7 kpc, and 19 kpc, respectively. The asymptotic value of the wind velocity is close to 450 km s^{-1} and is formally attained at very large distances, of the order of 1 Mpc. In real situations an external pressure can decelerate the galactic wind flow at a smaller distance through a termination shock supposedly around 300 kpc.

While galactic cosmic rays may accelerate the global galactic wind flow, they exert an external pressure on stellar winds and decelerate them (Dorman and Dorman 1967). This effect is included in sophisticated models of modulation of galactic cosmic rays by the solar wind (e.g., le Roux and Fichtner 1997).

6. ULTRA-HIGH ENERGIES

The observed differential power-law spectrum of cosmic rays becomes steeper by about $\Delta\gamma \approx 0.6$ at approximately 3×10^{15} eV. The shape of the spectrum is close to $E^{-3.1}$ at 10^{16} eV $< E < 10^{18}$ eV. The "knee" at ~3×10^{15} eV (Figure 2) is the main reliably established feature in the cosmic-ray spectrum (Kulikov and Khristiansen 1958). In principle, it may reflect either the break at the source spectrum or the more rapid leakage of very high-energy cosmic rays out of the Galaxy. This alternative is not resolved yet.

The required shape of the source spectrum could be obtained in a two-stage model where individual supernova remnants accelerate particles up to the knee and the subsequent energy gain is due to the collective re-acceleration on many shocks produced by other supernovae (Axford 1994) (see also Section 4). In principle, the complicated spectrum could be a superposition of sources of different nature with different spectra (Fichtel and Linsley 1986, Protheroe and Sabo 1992) that implies the surprisingly fine tuning of fluxes from independent sources needed to build the pronounced knee in the total spectrum.

An increased probability of escape from the Galaxy for very high-energy particles may explain the knee in the observed cosmic ray spectrum of a single power-law source spectrum. In a model of this kind (Ptuskin *et al.* 1993), the knee is the result of interplay between the diffusion of cosmic rays along magnetic field lines and the drift (Hall diffusion) perpendicular to the regular, predominantly azimuthal, magnetic field in the Galaxy. Drift has a stronger dependence on energy (which is proportional to the rigidity R) than diffusion (which is proportional to R^a, where $a < 0.6$) and can dominate at high energies, making the cosmic-ray spectrum steeper above the knee.

Most probably, the ultimate explanation of the knee will be obtained when the principal mechanism of cosmic-ray acceleration in the Galaxy has been definitely established. An essential common feature of the processes of particle re-acceleration and diffusion is their dependence on magnetic rigidity. So, one expects an increase in the abundance of heavy nuclei as a function of the total energy per particle at energies above the knee. This prediction has not yet been reliably checked experimentally.

It is customary to consider energies above 10^{17} eV as the region of ultra-high-energy cosmic rays. The boundary of the region is not related to any physics but is determined by the energy threshold at which extensive air showers are detected by the biggest existing detectors. The experimental results that are important for establishing the origin of ultra-high-energy cosmic rays can be summarized as follows (see reviews by Berezinskii *et al.* 1990, Bhattacharjee and Sigl 2000):

- Cosmic rays above the knee have a power-law differential spectrum of the form $E^{-3.1}$ and a flatter highest-energy component starting from $\sim 10^{19}$ eV.
- Cosmic rays appear to have a heavy composition (may be dominated by iron nuclei) at 10^{17}–10^{18} eV; the composition probably becomes lighter at $E > 10^{18}$ eV.
- The cosmic-ray anisotropy is $\sim 4\%$ with a broad cosmic-ray flow from the directions of the galactic center and the Cygnus region at around 10^{18} eV. No significant large-scale anisotropy was found at higher energies. The measurements give an upper limit of $\sim 30\%$ on anisotropy at 10^{19} eV.
- The highest particle energies registered in cosmic rays so far are $\sim 3 \times 10^{20}$ eV.

One may try to explain these data in terms of the two-component empirical model where the heavy galactic component turns down at $E > 3 \times 10^{18}$ eV (Gaisser *et al.* 1993). The extragalactic component, which is assumed to be primarily protons and has a flat spectrum, emerges in the observed spectrum at 3×10^{18}–10^{19} eV and dominates at highest energies. The intensities of these two components are equal at about 10^{19} eV.

The alternative interpretation assumes a purely galactic origin for all particles. To provide anisotropy restrictions, the purely galactic model must have a large magnetic halo in our Galaxy and a heavy cosmic ray composition up to the highest energies. With $Z = 26$, iron is, in principle, easier to accelerate and to confine in the Galaxy. Note that the Larmor radius of a particle with charge Z in the typical galactic halo magnetic field $B \sim 10^{-6}$ G is $r_g = 10 Z^{-1}(E/10^{19}$ eV$)$ $(10^{-6}$ G$/B)$ kpc, which is comparable to the scale of the entire Galaxy, ~ 10 kpc at $E \sim 3 \times 10^{20}$ eV, $Z = 26$.

Present-day data on the most energetic cosmic rays with $E > 1 \times 10^{20}$ eV includes 14 events. There is no indication of a cutoff in the spectrum. The highest energies were registered at the Fly's Eye installation (the energy $E = 3 \times 10^{20}$ eV, coming from galactic coordinates $l = 167°$, $b = 8°$), the Akeno Giant Air Shower Array (AGASA) (2×10^{20} eV, $l = 131°$, $b = -41°$), and the Yakutsk array (2×10^{20} eV, $l = 162°$, $b = 3°$). The cascades of particles (called air-showers) initiated by the ultra-high-energy cosmic rays in the Earth atmosphere are studied in these experiments. The Fly's Eye detector in the high Utah desert measures the fluorescence light produced by air showers. The AGASA installation uses particle detectors spread on the ground over about 100 km^2. The Yakutsk 20 km^2 particle detector array is complemented by the detectors of Cerenkov radiation generated by air showers.

The mere fact that nuclei have energies of $(1–3) \times 10^{20}$ eV limits their age to $T \approx (1–3) \times 10^8$ yr because photo–pion production and nuclear photodisintegration do not allow them to survive longer in the microwave background radiation. This means that the distance to the sources of highest-energy cosmic rays cannot exceed 30–100 Mpc. So, they must come from the local Metagalactic vicinity, say from the Local Supercluster, or even have a galactic origin.

The interaction with the cosmic microwave background in the case of extragalactic sources, homogeneously distributed on Hubble scale $c/H_0 = 3 \times 10^3/h$ Mpc ($H_0 = 100\,h$ km s^{-1} Mpc^{-1}) is the Hubble constant, $h \sim 0.7$), should lead to the Greisen–Zatsepin–Kuzmin cutoff in the cosmic-ray spectrum at about 4×10^{19} eV (Greisen 1966, Zatsepin and Kuzmin 1966). The data (Takeda *et al.* 1998) do not show the cutoff, and this is considered to be a decisive argument against universal extragalactic models of cosmic-ray origin. One has to bear in mind, however, that the natural clustering of galaxies provides source densities that exceed, on a small distance scale of $d < 30$–100 Mpc, the average density for a uniform spatial distribution. This can permit cosmic-ray source distributions commensurate with the observed spectrum, and thus relax the cutoff constraint on extragalactic models.

Let us now estimate the energy budget of extragalactic sources, assuming that they dominate at $E > 3 \times 10^{18}$ eV. The extragalactic cosmic-ray intensity derived from the Fly's Eye experiment is equal to $I_{Mg}(E) = 3 \times 10^{-32}(E/10^{17}$ eV$)^{-2.61}$ (cm^2 s sr eV)$^{-1}$ (Gaisser *et al.* 1993) that gives the energy density $w_{Mg}(>E) = 1.8 \times 10^{-7}(E/10^{17}\text{ eV})^{-0.61}$ eV cm^{-3}. Assuming the steady state balance $n_s L_{cr} = w_{Mg}/T$ (where n_s is the source number density in the Universe, L_{cr} is the source luminosity, and T is the cosmic-ray age), we obtain as the estimate for the minimal source luminosity $L_{cr} \sim 3 \times 10^{39} (d/(3\text{ Mpc}))^3$ erg s^{-1}. Here the integral on energy from 10^9 to 10^{21} eV was taken and the source spectrum of the form $\sim E^{-2}$ was assumed; d is the mean distance between neighboring sources; the age $T(10^{20}\text{ eV}) = 10^8$ yr, limited by the energy losses in microwave background radiation, was assumed.

The estimates inferred from analyzing the total energy release and the spatial density of astronomical objects of different types showed that the galaxies with active nuclei and the Virgo Cluster of galaxies (the central part of the Local Supercluster, approximately 17 Mpc from the Sun), are the most probable sources capable of maintaining the observed intensity of highest-energy cosmic rays (Berezinskii et al. 1990). A possible common origin of ultra-high-energy cosmic rays and cosmological gamma-ray bursts was also pointed out by Waxman (1995) and Vietri (1995).

The most difficult problem for all suggested models of cosmic-ray origin is the necessity to accelerate particles up to the extremely high energy of 3×10^{20} eV. The galactic sources of ultra-high-energy particles were discussed in Section 4; below is a short list of proposed mechanisms of acceleration and types of possible extragalactic sources.

The diffusive shock acceleration at termination shocks produced by sub-relativistic jets from active galactic nuclei can give $E_{max} \sim 3 \times 10^{19} Z$ eV. This estimate was obtained for a typical value of magnetic field $B \sim 3 \times 10^{-5}$ G, although a somewhat stronger field is not excluded (E_{max} is proportional to B). The number of galaxies with active galactic nuclei is 1–3% of the total number of galaxies; the total is ~2500 within a radius of $d < 50$ Mpc, which includes the Local Supercluster.

The acceleration near black holes in the galaxies with active galactic nuclei by unipolar induction in accretion disks, magnetic reconnection, and diffusive shock probably cannot produce the observed cosmic rays since particles with energies above about 10^{19} eV cannot escape from the accretion regions with their large photon density.

The acceleration at the hot spots of powerful Fanaroff–Riley type II radiogalaxies is efficient (Rachen and Biermann 1987) but cannot explain observations beyond the Greisen–Zatsepin–Kuzmin cutoff because all such objects are too far from us, at distances of more than 100 Mpc.

The first-order Fermi acceleration between approaching galactic magnetic coronas or between galactic winds could work in the case of colliding galaxies (Cesarsky and Ptuskin 1993). The acceleration in galaxy collisions may give $E_{max} \sim 3 \times 10^{19} Z$ eV. About 7% of galaxies are presently observed to have disturbances, which indicate an ongoing interaction, so the process is rather common. It is interesting that the triple event found in the AGASA experiment came from the direction of the interacting galaxy Mrk 40 (VV 141, Arp 151; distance ~80 Mpc). (The "triple" means that three events were detected within a limited solid angle of the experimental accuracy.) Based on the distribution of arrival directions of the highest-energy cosmic rays, Al-Dargazelli et al. (1996) claimed that nearby colliding galaxies are most favored as sources.

The acceleration at cluster accretion shocks (Kang et al. 1996) and in cosmological structure formation shock waves (Norman et al. 1995) is too slow and unable to accelerate particles to energies above the Greisen–Zatsepin–Kuzmin cutoff.

The acceleration at the sources of gamma-ray bursts, the product of gigantic explosions in the Universe, can give $\sim 5 \times 10^{20}$ eV (Vietri 1995, Waxman 1995). It is supposed that the fireballs resulting from the merging of neutron stars or black holes expands in the surrounding medium with a Lorenz factor of about 10^3. Shocks, formed in this flow, accelerate particles. It is an open issue as to whether the cosmological local density is sufficiently high to explain observed cosmic-ray events.

The difficulties with the acceleration of cosmic rays to the highest observed energies stimulated interest in the so-called top–down mechanisms. Production of extremely energetic particles with $\sim 10^{21}$ eV and even with $\sim 10^{23}$–10^{25} eV is possible in the course of annihilation or collapse of cosmological topological defects such as monopoles or cosmic strings (Hill et al. 1987, Bhattacharjee et al. 1992). The superheavy gauge and Higgs bosons are produced in these processes. The superheavy bosons decay then into quarks, gluons, and leptons, which materialize into ultra-high-energy nucleons, photons, and neutrinos. In most cases the problem with this mechanism is not the maximal energy, but the fluxes. A dimensional scale for the distance between topological defects is the Hubble distance H_0^{-1}, which is much larger than the attenuation length for particles with energies $E > 10^{20}$ eV. However, the hypothetical superheavy long-lived relic particles produced in the early Universe can accumulate in the big halo around our Galaxy (Berezinsky and Vilenkin 1997). Their decay products could produce ultra-high-energy cosmic rays without the Greisen–Zatsepin–Kuzmin cutoff. The characteristic feature of the model is its prediction of a considerable excess of gamma-ray flux over the nucleon flux in the highest-energy cosmic rays. In this connection, it should be noted that the non-nucleon nature of the main fraction of cosmic-ray events at energies of $E > 10^{19}$ eV is not absolutely excluded by present-day experiments, although the analysis of the highest-energy Fly's Eye event at $\sim 3 \times 10^{20}$ eV does not favor its photon production.

(At present, the Fly's Eye detector is the most effective one in distinguishing between the photon- and proton-induced showers.)

The mystery of the origin of the highest-energy cosmic rays remains unresolved and presents a real challenge to theory and experiment.

7. THE FUTURE

The new initiatives now being planned for cosmic-ray experiments give every promise of extending the great progress that has been made over the past four decades of the space age. Data will continue to flow from the high-resolution detector systems on the Ulysses, ACE and Voyager space missions. It is probable that the Voyagers will operate until 2015 or so, thus reaching out to ~130 AU and possibly providing direct observations of the local interstellar spectra of low- and medium-energy cosmic-ray ions and electrons. A future interstellar probe that will travel out of the Solar System at ~12 AU per year is in the early study phase at NASA.

At this time there are some four balloon-borne, high-resolution magnetic spectrometers that will continue to be flown by various groups over the next few years, extending the $\bar{p}$, e^+, e^- cosmic ray spectra and isotope studies over a greater energy range. These will be followed by flights of large-area ionization calorimeters to extend the detailed composition studies to energies approaching the region of the knee (~10^{14} eV). Some of these will be long-duration balloon flights in the polar regions at an altitude of ~38 km and may last for some 100 days – making them "mini space missions".

In about 2002 the Russian–Italian collaboration will launch the Pamela experiment into a 700 km polar orbit. This permanent magnet spectrometer will measure the $\bar{p}$, e^+, e^- and cosmic-ray spectra of H–C over an energy range of 0.1 to some 200 GeV. This will be followed by the Alpha Magnetic Spectrometer (AMS), which will be placed on the International Space Station in about 2003. This superconducting magnetic spectrometer is being prepared by an extensive international collaboration and will measure the elementary particle spectra and cosmic-ray nuclear spectra into the tera-electronvolt region and will provide a sensitive search for antimatter and the origin of dark matter. A simpler prototype of this instrument using a permanent magnetic spectrometer has been flown on the Space Station.

NASA is currently planning an ionization calorimeter experiment, the Advanced Cosmic Ray Composition Experiment (ACCESS), that will follow AMS on the Space Station. This experiment will extend the composition studies further into the region of the spectral knee.

The Auger experiment will carry out studies of the highest energy cosmic rays at energies $>10^{20}$ eV using two giant air shower arrays, each with an aperture of ~7000 km^2 sr and an energy threshold of 10^{14} eV. The sites are located in Argentina and Utah, and construction of a large-scale prototype is now underway. A follow-on space mission, OWL (Orbiting Array of Wide-Angle Light Collectors), is currently under study. This large, downward-looking space telescope would detect the flourescence from very large electromagnetic showers with a much larger solid angle than would ever be practical from ground-based experiments. A similar program AIRWATCH is under study in Europe. These extensive programs should maintain the role of cosmic rays as a vital part of astrophysics.

REFERENCES

Al-Dargazelli, S.S., Wolfendale, A.W., Smialkowski, A. and Wdowczyk, J.J. (1996). The origin of cosmic rays of the highest energies. *Journal of Physics G*, **22**, 1825–1838.

Allen, G.E., Keohane, J.W., Gotthelf, E.V., Petre, R., Jahoda, K., Rothschild, R.E., Lingenfelter, R.E., Heindl, W.A., Marsden, D., Gruber, D.E., Pelling, M.R. and Blanco, P.R. (1997). Evidence of X-ray synchrotron emission from electrons accelerated to 40 TeV in the supernova remnant Cassiopeia A. *Astrophysical Journal*, **487**, L97–L100.

Alvarez, L. and Compton, A.H. (1933). A positively charged component of cosmic rays. *Physical Review*, **43**, 835–837.

Axford, W.I. (1994). The origins of high-energy cosmic rays. *Astrophysical Journal*, Suppl., **90**, 937–944.

Axford, W.I., Leer, E. and Skadron, G. (1977). *15th International Cosmic Ray Conference*, Plovdiv, **11**, 131.

Baade, W. and Zwicky, F. (1934). Remarks on super-novae and cosmic rays. *Physical Review*, **46**, 76–77.

Baring, M.G., Ellison, D.C., Reynolds, S.P., Grenier, I.A. and Goret, P. (1999). Radio to gamma-ray emission from shell-type supernova remnants: Predictions from nonlinear shock acceleration models. *Astrophysical Journal*, **513**, 311–338.

Barwick, S.W., Beatty, J.J., Bower, C.R., Chaput, C.J., Coutu, S., de Nolfo, G.A., Duvernois, M.A., Ellithorpe, D., Ficenec, D., Knapp, J., Lowder, D.M., McKee, S., Mueller, D., Musser, J.A., Nutter, S.L., Schneider, E., Swordy, S.P., Tarle, G., Tomasch, A.D. and Torbet, E. (1998). The energy spectra and relative abundances of electrons and positrons in the galactic cosmic radiation. *Astrophysical Journal*, **498**, 779–789.

Bell, A.R. (1978). The acceleration of cosmic rays in shock fronts. *Monthly Notices of the Royal Astronomical Society*, **182**, 147–156.

Bell, A.R. (1992). Cosmic ray acceleration in pulsar-driven supernova remnants. *Monthly Notices of the Royal Astronomical Society*, **257**, 493–500.

Berezhko, E.G., Elshin, V.K. and Ksenofontov, L.T. (1996). Cosmic ray acceleration in supernova remnants. *Journal of Experimental and Theoretical Physics*, **82**, 1–21.

Berezinskii, V.S., Bulanov, S.V., Dogiel, V.A., Ginzburg, V.L. and Ptuskin, V.S. (1990). *Astrophysics of Cosmic Rays*, North-Holland, Amsterdam.

Berezinsky, V.S. and Vilenkin, A. (1997). Cosmic necklaces and ultrahigh energy cosmic rays. *Physical Review Letters*, **79**, 5202–5205.

Beuermann, K., Kanbach, G. and Berkhuijsen, E.M. (1985). Radio structure of the Galaxy: Thick disk and thin disk at 408 MHz. *Astronomy and Astrophysics*, **153**, 17–34.

Bhattacharjee, P., Hill, C.T. and Schramm, D.N. (1992). Grand unified theories, topological defects, and ultrahigh-energy cosmic rays. *Physical Review Letters*, **69**, 567–570.

Bhattacharjee, P. and Sigl, G. (2000). Origin and propagation of extremely high-energy cosmic rays. *Physics Reports*, **327**, 109–247.

Binns, W.R., Garrard, T.L., Gibner, P.S., Israel, M.H., Kertzman, M.P., Klarmann, M.H., Newport, B.J., Stone, J. and Waddington, C.J. (1989). Abundances of ultraheavy elements in the cosmic radiation: Results from HEAO 3. *Astrophysical Journal*, **346**, 997–1009.

Blandford, R.D. and Ostriker, J. (1978). Particle acceleration by astrophysical shocks. *Astrophysical Journal*, **221**, L29–L32.

Bloemen, J.B.G.M. (1989). Diffuse Galactic gamma-ray emission. *Annual Review of Astronomy and Astrophysics*, **27**, 469–516.

Bloemen, J.B.G.M., Dogiel, V.A., Dorman, V.L. and Ptuskin, V.S. (1993). Galactic diffusion and wind models of cosmic ray transport. *Astronomy and Astrophysics*, **267**, 372–387.

Breitschwerdt, D., McKenzie, J.F. and Völk, H.J. (1991). Galactic winds: I. Cosmic ray and wave-driven winds from the Galaxy. *Astronomy and Astrophysics*, **245**, 79–97.

Bykov, A.M. and Toptygin, I.N. (1993). Kinetics of particles in the strongly turbulent plasmas. *Physics Uspekhi*, **36**, 1020–1052.

Casse, M. and Paul, J.A. (1982). On the stellar origin of the Ne-22 excess in cosmic rays. *Astrophysical Journal*, **258**, 860–863.

Casse, M. and Soutoul, A. (1978). Time delay between explosive nucleosynthesis and cosmic ray acceleration. *Astrophysical Journal*, **200**, L75–L76.

Cesarsky, C.J. and Ptuskin, V.S. (1993). Acceleration of highest-energy cosmic rays in galaxy collisions. In *Proceedings of the 23rd International Cosmic Ray Conference*, Calgary, Vol. 2, pp. 341–344.

Chevalier, R.A. and Fransson, C. (1984). A cosmic-ray supported corona. *Astrophysical Journal*, **279**, L43–L46.

Clay, J. (1927). Ultra radiation (penetrating radiation III) annual variation and variation with the geomagnetic latitude. *Proc. Nederlandsche Akad. v. Wet* **33**, 711.

Connell, J.J. and Simpson, J.A. (1997). Isotopic abundances of Fe and Ni in galactic cosmic-ray sources. *Astrophysical Journal*, **475**, L61–L64.

Dogiel, V.A. (1996). Gamma-ray direct observations from space. *Nuovo Cimento*, **19C**, 671–700.

Donnelly, J., Thompson, A., Sullivan, D'O, Keane, A.J., Drury, L.O'C. and Wenzel, K.-P. (1999). New results on the relative abundance of actinides in the cosmic radiation. In *26th ICRC*, Vol. 3, Salt Lake City, pp. 109–112.

Dorman, I.V. and Dorman, L.I. (1967). *Journal of Geophysics*, **72**, 1513.

Dorman, L.I., Ghosh, A. and Ptuskin, V.S. (1984). Cosmic ray propagation in the solar neighborhood. *Astronomy Letters*, **10**, 345–347.

DuVernois, M.A., Garcia-Munoz, M., Pyle, K.R., Simpson, J.A. and Thayer, M.R. (1996). The isotopic composition of galactic cosmic-ray elements from carbon to silicon: The combined release and radiation effects satellite investigation. *Astrophysical Journal*, **466**, 457–472.

Earl, J.A. (1961). Cloud chamber observations of primary cosmic-ray electrons, *Physical Review Letters*, **6**, 125.

Elster, J. and Geitel, H. (1901). Weitere versuche über die elektrizitätszerstreuung in abgeschlossenen luftmengen. *Physikalische Zeitschrift*, **22**, 560–563.

Epstein, R.I. (1980). The acceleration of interstellar grains and the composition of the cosmic rays. *Monthly Notices of the Royal Astronomical Society*, **193**, 723–729.

Esposito, J., Hunter, S.D., Kanbach, G. and Sreekumar, P. (1996). EGRET observations of radio-bright supernova remnants. *Astrophysical Journal*, **461**, 820–827.

Fermi, E. (1949). On the origin of the cosmic radiation. *Physical Review*, **75**, 1169–1174.

Ferriere, K. (1998). Global model of the interstellar medium in our Galaxy with new constraints on the hot gas component. *Astrophysical Journal*, **497**, 759–776.

Fichtel, C.E. and Linsley, J. (1986). High-energy and ultra-high-energy cosmic rays. *Astrophysical Journal*, **300**, 474–487.

Fisk, L.A., Kozlovsky, B. and Ramaty, R. (1974). An interpretation of the observed oxygen and nitrogen enhancements in low-energy cosmic rays. *Astrophysical Journal*, **190**, L35–L37.

Frier, P.S., Lofgren, E.J., Ney, E.P., Oppenheimer, F., Bradt, H.L. and Peters, B. (1948). Evidence for heavy nuclei in the primary cosmic radiation. *Physical Review*, **74**, 213–217.

Galeev, A.A. and Sudan, R.N. (eds) (1983). *Basic Plasma Physics*, North-Holland, New York.

Gaisser, T.K., Harding, A.K. and Stanev, T. (1987). Particle acceleration and production of energetic photons in SN1987A. *Nature*, **329**, 314–316.

Gaisser, T.K., Stanev, T., Tilav, S., Corbato, S.C., Dai, H.Y., Dawson, B.R., Elbert, J.W., Emerson, B., Kieda, D.B., Luo, M., Ko, S., Larsen, C., Loh, E.C., Salamon, M.H. and Smith, J.D. (1993). Cosmic-ray composition around 10^{18} eV. *Physical Review D*, **47**, 1919–1932.

Ginzburg, V.L., (1966). Cosmic rays and plasma phenomena in the Galaxy and Metagalaxy. *Astronomische Zeitschrift*, **9**, 877–881.

Ginzburg, L. and Ptuskin, V.S. (1976). On the origin of cosmic rays: Some problems of high-energy astrophysics. *Reviews of Modern Physics*, **48**, 161–189.

Ginzburg, V.L. and Syrovatskii, S.I. (1964). *The Origin of Cosmic Rays*, Pergamon Press, Oxford.

Gleeson, L.J. and Axford, W.I. (1967). Cosmic rays in the interplanetary medium. *Astrophysical Journal*, **149**, 115–118.

Greisen, K. (1966). End to the cosmic-ray spectrum? *Physical Review Letters*, **16**, 748–750.

Hess, V.F. (1911). *Physikalische Zeitschrift*, **12**, 998.

Hess, V.F. (1912). Über Beobachtungen der durchdringenden Strahlung bei sieben Frieballonfahrten. *Physikalische Zeitschrift*, **13**, 1084–1091.

Higdon, J.C., Lingenfelter, R.E. and Ramaty, R. (1998). Cosmic-ray acceleration from supernova ejecta in superbubbles. *Astrophysical Journal*, **509**, L33–L36.

Hill, C.T., Schramm, D.N. and Walker, T.P. (1987). Ultra-high-energy cosmic rays from superconducting cosmic strings. *Physical Review D*, **36**, 1007–1016.

Holzer, T.E. (1989). Interaction between the solar wind and the interstellar medium. *Annual Review of Astronomy and Astrophysics*, **27**, 199–234.

Hunter, S.D., Kinzer, R.L. and Strong, A.W. (1997). Diffuse galactic continuum radiation. In C.D. Dermer *et al.* (eds), *Proceedings of the 4th Compton Symposium*, American Institute of Physics, New York, pp. 192–200.

Ipavich, F. (1975). Galactic winds driven by cosmic rays. *Astrophysical Journal*, **196**, 107–120.

Johnson, T.H. *et al.* (1938). Cosmic ray intensity and geomagnetic effects. *Reviews of Modern Physics*, **10**, 193–244.

Johnson, T.H. (1933). The azimuthal asymmetry of the cosmic radiation. *Physical Review*, **43**, 834–835.

Jokipii, J.R. (1971). Propagation of cosmic rays in the solar wind. *Reviews of Geophysics and Space Physics*, **9**, 27–87.

Jokipii, J.R. (1976). Consequences of a lifetime greater than 10 to the 7th power years for galactic cosmic rays. *Astrophysical Journal*, **208**, 900–902.

Jokipii, J.R. (1989). In W.V. Jones, F.J. Kerr and J.F. Ormes (eds), *Particle Astrophysics. The NASA Cosmic Program for the 1990s and Beyond*. American Institute of Physics, New York, p. 155.

Jokipii, J.R. and Morfill, G.E. (1985). On the origin of high-energy cosmic rays. *Astrophysical Journal*, **290**, L1–L4.

Jones, F.C. (1979). The dynamical halo and the variation of cosmic-ray path length with energy. *Astrophysical Journal*, **229**, 747–752.

Jones, F.C. and Ellison, D.C. (1991). The plasma physics of shock acceleration. *Space Science Reviews*, **58**, 259–346.

Jones, F.C., Lukasiak, A., Ptuskin, V.S. and Webber, W.R. (2001). The modified weighted slab technique: Models and results. *Astrophysical Journal*, **547**, 246–271.

Kang, H., Ryu, D. and Jones, T.W. (1996). Cluster accretion shocks as possible acceleration sites for ultra-high-energy protons below the Greisen cutoff. *Astrophysical Journal*, **456**, 422–427.

Keohane, J.W., Petre, R., Gotthelf, E.V., Ozaki, M. and Koyama, K. (1997). A possible site of cosmic ray acceleration in the supernova remnant IC 443. *Astrophysical Journal*, **484**, 350–359.

Koyama, K., Petre, R., Gotthelf, E.V., Hwang, U., Matsura, M., Ozaki, M. and Holt, S.S. (1995). Evidence for shock acceleration of high-energy electrons in the supernova remnant SN 1006. *Nature*, **378**, 255–258.

Koyama, K., Kinugasa, K., Matsuzaki, K., Nishiuchi, M., Sugizaki, M., Torii, K., Yamauchi, S. and Aschenbach, B. (1997). Discovery of non-thermal X-rays from the northwest shell of the new SNR RX J1713.7-3946: The second SN 1006? *Publications of the Astronomical Society of Japan*, **49**, L7–L11.

Krennrich, F., Biller, S.D., Bond, I.H., Boyle, P.J., Bradbury, S.M., Breslin, A.C., Buckley, J.H., Burdett, A.M., Gordo, J., Bussons, X.X., Carter-Lewis, D.A., Catanese, M., Cawley, M.F., Fegan, D.J., Finley, J.P., Gaidos, J.A., Hall, T., Hillas, A.M., Lamb, R.C., Lessard, R.W., Masterson, C., McEnery, J.E., Mohanty, G., Moriarty, P., Quinn, J., Rodgers, A.J., Rose, H.J., Samuelson, F.W., Sembroski, G.H., Srinivasan, R., Vassiliev, V.V. and Weekes, T.C. (1999). Measurement of the multi-TEV gamma-ray flare spectra of Markarian 421 and Markarian 501. *Astrophysical Journal*, **511**, 149–156.

Krymsky, G.F. (1977). A regular mechanism for the acceleration of charged particles on the front of a shock wave. *Soviet Physics Doklady*, **22**, 327–328.

Kulikov, G.V. and Khristiansen, G.B. (1958). On the size spectrum of extensive air showers. *Soviet Physics JETP*, **35**, 441–447.

Lagage, P.O. and Cesarsky, C.J. (1983). Cosmic-ray shock acceleration in the presence of self-excited waves. *Astronomy and Astrophysics*, **118**, 223–228.

le Roux, J.A. and Fichtner, H. (1997). A self-consistent determination of the heliospheric termination shock structure in the presence of pickup, anomalous, and galactic cosmic ray protons. *Journal of Geophysical Research*, **102**, 17,365–17,380.

Lerche, I. (1967). Unstable magnetosonic waves in a relativistic plasma. *Astrophysical Journal*, **147**, 689–696.

Leske, R.A. (1993). The elemental and isotopic composition of Galactic cosmic-ray nuclei from scandium through nickel. *Astrophysical Journal*, **405**, 567–583.

Lingenfelter, R.E., Ramaty, R. and Kozlovsky, B. (1998). Supernova grains: The source of cosmic-ray metals. *Astrophysical Journal*, **500**, L153–L156.

Lozinskaya, T.A. (1992). *Supernovae and Stellar Wind in the Interstellar Medium*. American Institute of Physics, New York.

Lukasiak, A., Ferrando, P., McDonald, F.B. and Webber, W.R. (1994). Cosmic-ray isotopic composition of C, N, O, Ne, Mg, Si nuclei in the energy range 50–200 MeV per nucleon measured by the Voyager spacecraft during the solar minimum period. *Astrophysical Journal*, **426**, 366–372.

McLennan, J.C. and Burton, E.F. (1902). Some experiments on the electrical conductivity of atmospheric air. *Physical Review*, **16**, 184–192.

Mewaldt, R.A., Wiedenbeck, M.E., Binns, W.R., Christian, E.R., Cummings, A.C., George, J.S., Hink, P.L., Klarmann, J., Leske, R.A., Lijowski, M., Stone, E.C., von Rosenvinge, T.T. and Yanasak, N.E. (1999). The time delay between nucleosynthesis and acceleration based on ACE measurements of primary electron-capture nuclides. In *Proceedings of the 26th International Cosmic Ray Conference*, Vol. 3, Salt Lake City, pp. 45–48.

Meyer, J.P. (1985). Solar-stellar outer atmospheres and energetic particles, and galactic cosmic rays. *Astrophysical Journal*, Suppl., **57**, 173–204.

Meyer, J.P., O'C Drury, L. and Ellison, D.C. (1997). Galactic cosmic rays from supernova remnants. I. A cosmic-ray composition controlled by volatility and mass-to-charge Ratio. *Astrophysical Journal*, **487**, 182–196.

Meyer, P. and Vogt, R. (1961). Electrons in the primary cosmic radiation. *Physical Review Letters*, **6**, 193.

Mukherjee, R., Bertsch, D.L., Bloom, S.D., Dingus, B.L., Esposito, J.A., Fichtel, C.E., Hartman, R.C., Hunter, S.D., Kanbach, G., Kniffen, D.A., Lin, Y.C., Mayer-Hasselwander, H.A., McDonald, L.M., Michelson, P.F., von Montigny, C., Muecke, A., Nolan, P.L., Pohl, M., Reimer, O., Schneid, E., Sreekumar, P. and Thompson, D.J. (1997). EGRET observations of high-energy gamma-ray emission from blazars: An update. *Astrophysical Journal*, **490**, 116–135.

Niebur, S.M., Binns, W.R., Christian, E.R., Cummings, A.C., George, J.S., Hink, P.L., Klarmann, J., Leske, R.A., Lijowski, M., Mewaldt, R.A., Stone, E.S., von Rosenvinge, T.T. and Weidenbeck, M. (2000). Secondary electron-capture-decay isotopes and implication for the propagation of galactic cosmic rays. In R.A. Mewaldt, J.R. Jokipii, M.A. Lee, E. Mobius and T.H. Zurbuchen (eds), *Acceleration and Transport of Energetic Particles Observed in the Heliosphere*. American Institute of Physics, New York, pp. 402–405.

Norman, C.A., Melrose, D.B. and Achterberg, A. (1995). The origin of cosmic rays above $10^{18.5}$ eV. *Astrophysical Journal*, **454**, 60–68.

Nishimura, J., Kobayashi, T., Komori, Y. and Yoshida, K. (1997). Observations of high energy primary electrons and their astrophysical significance. *Advances in Space Research*, **18**, 767–770.

Orito, S., Maeno, T., Matsunaga, H., Abe, K., Anraku, K., Asaoka, Y., Fujikawa, M., Imori, M., Ishino, M., Makida, Y., Matsui, N., Matsumoto, H., Mitchell, J., Mitsui, T., Moiseev, A., Motoki, M., Nishimura, J., Nozaki, M., Ormes, J., Saeki, T., Sanuki, T., Sasaki, M., Seo, E.S., Shikaze, Y., Sonoda, T., Streitmatter, R., Suzuki, J., Tanaka, K., Ueda, I., Yajima, N., Yamagami, T., Yamamoto, A., Yoshida, T. and Yoshimura, K. (2000). Precision measurement of cosmic-ray antiproton spectrum. *Physical Review Letters*, **84**, 1078–1081.

Pacholczyk, A.G. (1970). *Radio Astrophysics*, W.H. Freeman, San Francisco.

Paris Symposium on Radio Astronomy (1959). Stanford University Press.

Parker, E.N. (1963). *Interplanetary Dynamical Processes*, Interscience, New York.

Parker, E.N. (1966). The dynamic state of the interstellar gas and field. *Astrophysical Journal*, **145**, 811–833.

Parker, E.N. (1992). Fast dynamos, cosmic rays, and the Galactic magnetic field. *Astrophysical Journal*, **401**, 137–145.

Pesses, M.E., Eichler, D. and Jokipii, J.R. (1981). Cosmic ray drift, shock wave acceleration, and the anomalous component of cosmic rays. *Astrophysical Journal*, **246**, L85–L88.

Piran, T. (1999). Gamma-ray bursts and the fireball model. *Physics Reports*, **314**, 575–667.

Prishchep, V.L. and Ptuskin, V.S. (1981). The acceleration of fast particles at the front of a spherical shock wave. *Soviet Astronomy*, **58**, 779–789.

Protheroe, R.J. and Sabo, A.P. (1992). High energy cosmic rays from active galactic nuclei. *Physical Review Letters*, **69**, 2885–2888.

Ptuskin, V.S., Rogovaya, S.I., Zirakashvili, V.N., Chuvilgin, L.L., Khristiansen, G.B., Klepach, E.G. and Kulikov, G.V. (1993). Diffusion and drift of very high energy cosmic rays in galactic magnetic fields. *Astronomy and Astrophysics*, **268**, 726–735.

Ptuskin, V.S. and Soutoul, A. (1998). Cosmic ray clocks. *Space Science Reviews*, **85**, 223–236.

Ptuskin, V.S., Völk, H.J., Zirakashvili, V.N. and Breitschwerdt, D. (1997). Transport of relativistic nucleons in a galactic wind driven by cosmic rays. *Astronomy and Astrophysics*, **321**, 434–443.

Rachen, J.P. and Biermann, P.L. (1987). Extragalactic ultra-high energy cosmic-rays: I. Contribution from hot spots in Fr-II radio galaxies. *Astronomy and Astrophysics*, **272**, 161–175.

Reynolds, S.P. (1998). Models of synchrotron X-rays from shell supernova remnants. *Astrophysical Journal*, **493**, 375–396.

Rossi, B. (1930). On the magnetic deflection of cosmic rays. *Physical Review*, **36**, 606.

Rutherford, E. and Cooke, H.L. (1903). A penetrating radiation from the Earth's surface (abstract). *Physical Review*, **14**, 183.

Ruzmaikin, A.A., Sokoloff, D.D. and Shukurov, A.M. (1998). *Magnetic Fields of Galaxies*, Kluwer, Dordrecht.

Seo, E.S. and Ptuskin, V.S. (1994). Stochastic reacceleration of cosmic rays in the interstellar medium. *Astrophysical Journal*, **431**, 705–714.

Schein, M., Jesse, W.P. and Wollan, E.O. (1941). The nature of the primary cosmic radiation and the origin of the mesotron. *Physical Review*, **59**, 615–615.

Silbebrbrg, R. and Tsao, C.H. (1990). Spallation processes and nuclear interaction products of cosmic rays. *Physics Reports*, **191**, 351–410.

Simon, M., Heinrich, W. and Mathis, K.D. (1986). Propagation of injected cosmic rays under distributed reacceleration. *Astrophysical Journal*, **300**, 32–40.

Simpson, J.A. (1983). Elemental and isotopic composition of the Galactic cosmic rays. *Annual Review of Nuclear and Particle Science*, **33**, 323–381.

Skilling, J. (1975). Cosmic ray streaming: I. Effect of Alfvén waves on particles. *Monthly Notices of the Royal Astronomical Society*, **172**, 557–566.

Strong, A.W. and Moskalenko, I.V. (1998). Propagation of cosmic ray nucleons in the Galaxy. *Astrophysical Journal*, **509**, 212–228.

Strong, A.W., Moskalenko, I.V. and Reimer, O. (1998). Diffuse galactic gamma rays, cosmic-ray nucleons and antiprotons. *Astronomy and Astrophysics*, **338**, L75–L78.

Sturner, S.J. and Dermer, C.D. (1995). Association of unidentified, low latitude EGRET sources with supernova remnants. *Astronomy and Astrophysics*, **293**, L17–L20.

Takeda, M., Hayashida, N., Honda, K. et al. (1998). Extension of the cosmic-ray energy spectrum beyond the predicted Greisen–Zatsepin–Kuz'min cutoff. *Physical Review Letters*, **81**, 1163–1166.

Tanimori, T., Hayami, Y., Kamei, S., Dazeley, S.A. et al. (1998). Discovery of TeV gamma rays from SN 1006: Further evidence for the supernova remnant origin of cosmic rays. *Astrophysical Journal*, **497**, L25–L28.

Thompson, A., O'Sullivan, D., Wenzel, K.-P., Bosch, J., Keegan, R., Domingo, C. and Jansen, F. (1993). Some early results from the LDEF ultra heavy cosmic ray experiment. In *Proceedings of the 23rd International Cosmic Ray Conference*, Vol. 1, Calgary, pp. 601–606.

Vallée, J.P. (1997). Observations of the magnetic fields inside and outside the Milky Way, starting with globules (~1 parsec), filaments, clouds, superbubbles, spiral arms, galaxies, superclusters, and ending with the cosmological Universe's background surface (at ~8 Teraparsecs). *Fundamentals of Cosmic Physics*, **19**, 1–89.

Vietri, M. (1995). The acceleration of ultra-high-energy cosmic rays in Gamma-Ray Bursts. *Astrophysical Journal*, **453**, 883–889.

Völk, H.J. and Biermann, P.L. (1988). Maximum energy of cosmic-ray particles accelerated by supernova remnant shocks in stellar wind cavities. *Astrophysical Journal*, **333**, L65–L68.

Waxman, E. (1995). Cosmological gamma-ray bursts and the highest energy cosmic rays. *Physical Review Letters*, **75**, 386–389.

Webber, W.R. (1983). Cosmic ray electrons and positrons: A review of current measurements and some implications. In *Composition and Origin of Cosmic Rays*, Reidel, Dordrecht, pp. 83–100.

Webber, W.R., Lee, M.A. and Gupta, M. (1992). Propagation of cosmic-ray nuclei in a diffusing galaxy with convective halo and thin matter disk. *Astrophysical Journal*, **90**, 96–104.

Wiedenbeck, M.E., Binns, W.R., Christian, E.R., Cummings, A.C., George, J.S., Hink, P.L., Klarmann, J., Leske, R.A., Lijowski, M., Mewaldt, R.A., Stone, E.C., von Rosenvinge, T.T. and Yanasak, N.E. (1999). The isotopic composition of iron, cobalt, and nickel in cosmic ray source material. In *Proceedings of the 26th International Cosmic Ray Conference*, Vol. 3, Salt Lake City, pp. 1–4.

Wulf, Th. (1907). Ein neues Elektrometer für statische Ladungen. *Physikalische Zeitschrift*, **8**, 246–248.

Wilson, C.T.R. (1901). On the ionization of atmospheric air. *Proceedings of the Royal Society A*, **68**, 151–161.

Zatsepin, G.T. and Kuzmin, V.A. (1966). Upper limit of the spectrum of cosmic rays. *Soviet Physics Journal of Experimental and Theoretical Physics Letters*, **4**, 78–80.

Zirakashvili, V.N., Breitschwerdt, D., Ptuskin, V.S. and Völk, H.J. (1996). Magnetohydrodynamic wind driven by cosmic rays in a rotating galaxy. *Astronomy and Astrophysics*, **311**, 113–137.

30

GERARD F. GILMORE*

Stellar populations and dynamics in the Milky Way galaxy

Our Galaxy offers a unique opportunity to deduce the important physics involved in galaxy formation from observations of those old stars that were formed at the time of the formation of the Milky Way, and whose present properties contain some fossil record of the Galaxy's history. Only in the Milky Way galaxy and its immediate neighbour satellite galaxies can one obtain the true three-dimensional stellar spatial density distributions, stellar kinematics and stellar chemical abundances. Knowledge of how stars move and how they are distributed in space measures the Galactic gravitational potential, including its dark matter content, while knowledge of stellar kinematics, ages and chemistry constrains the star formation and gas accretion history.

The scientific goal is to build on the powerful concept of stellar populations, which basically distinguishes old, metal-poor stars (Population II) from younger, more metal-rich stars (Population I), to include the rich complexity of the real Universe.

No single section of the electromagnetic spectrum provides the 'best' view of the Galaxy. Rather, all views are complementary. However, some views are certainly more representative than others. The most fundamental must be a view of the entire contents of the Galaxy. Such a view would require access to a universal property of matter that is independent of the state of that matter. This is provided by gravity, since all matter, by definition, has mass. Mass generates the gravitational potential, which in turns defines the size and the shape of the Galaxy. While the most reliable and comprehensive, such a view is also the hardest to derive. Kinematics and distance data are, however, the closest approach to such a view that is possible. An ideal astrophysicist would have astrometric eyes!

Available information strongly suggests that the Galactic extreme Population II subdwarf system formed early, though the duration of its aggregation into the proto-Galaxy remains unclear. This subdwarf system now forms a flattened, pressure-supported distribution, with axial ratio ~2:1. The thick disk formed close in time to the subdwarf system, with at least the metal-poor tail of the thick disk being comparable in age to the globular cluster system. The thick disk is probably chemically and kinematically discrete from the Galactic old disk – by which we mean those stars of the thin disk with age greater than a few billion years – implying a discontinuous Galactic history. The inner Galaxy is mostly old, almost certainly barred, but as yet remains to be studied in detail, especially in the innermost regions. Importantly, recent dynamical analyses lead to the conclusion that there is no statistically significant amount of non-luminous mass in the solar neighbourhood, and hence no evidence for dissipative dark matter.

We consider in turn the general questions of galaxy formation which the science of stellar population studies aims to address, followed by discussion of star count and infrared studies of Galactic structure, and astrometric and kinematic studies of the Galactic mass distribution. The primary space missions which have contributed to these subjects are the Hubble Space Telescope (HST), whose excellent spatial resolution allows study in crowded regions, whose dark sky background allows the study of very faint objects and, most importantly here, allows reliable image distinction between stars and resolved objects (galaxies); the sequence of infrared observatories, which have allowed observations through the optically thick dust obscuring the inner Galaxy, Infrared

* University of Cambridge, United Kingdom

Astronomical Satellite (IRAS), Diffuse Infrared Background Experiment (DIRBE) on the Cosmic Background Explorer (COBE) and the Infrared Space Observatory (ISO); and the High Precision Parallax Collecting Satellite (Hipparcos), arguably ESA's most substantial contribution to astrophysics in the twentieth century.

THE FORMATION OF DISK GALAXIES

Current understanding of the formation and early evolution of disk galaxies allows a description of the important physical processes at various levels of complexity and generality, and illustrates what we have learned, what we are attempting still to learn, and the relative importances of surveys, targeted research projects and newer space-based studies. At one extreme, one simply considers the global evolution of a gas cloud, and assumes that mean values of relevant parameters suffice for an adequate description of generic properties. Alternatively, one gives up general applicability and instead adopts specific numerical values for those parameters that quantify the important physics, attempting a detailed confrontation of model predictions with observed stellar populations. The relation of any model prediction to detailed observations at a single radius in a specific galaxy clearly needs to be considered with some care. Mindful of this caveat, we outline here the most important timescales and physical processes that are likely to play a role in the determination of the observable properties of galaxies like the Milky Way.

General models of dissipational disk galaxy formation

The existence of cold, thin Galactic disks has strong implications for galaxy formation. To see this, consider a standard picture whereby galaxies form from growing primordial density perturbations, which expand with the background Universe until their self-gravity becomes dominant and they collapse upon themselves. Were there to be no loss of energy in the collapse, and neglecting angular momentum, the transformation of potential energy into thermal (kinetic) energy would lead to an equilibrium system with final radius equal to half its size at maximum expansion, supported by random motions of the constituent particles. Thus an equilibrium, purely gaseous proto-galaxy should have temperature

$$T \equiv T_{\text{virial}} \sim \frac{GMm_p}{kR} \qquad (1a)$$

and a stellar proto-galaxy should equivalently have velocity dispersion

$$\sigma^2 \sim T_{\text{virial}} \frac{k}{m_p} \qquad (1b)$$

where k and m_p are the Boltzmann constant and the mass of the proton, respectively. Numerically, $T_{\text{virial}} \sim 10^6 R_{50}^{-1} M_{12}$ K for gravitational (half-mass) radius, R, in units of 50 kpc, and mass M in units of $10^{12} M_\odot$. Since the disks of spiral galaxies are cold, with $T \ll T_{\text{virial}}$, energy must have been lost. Since this lost energy was in random motions of individual particles, the only possible loss mechanism is through an inelastic collision, leading to the internal excitation of the particles, and subsequent energy loss through radiative de-excitation. Clearly, particles with small cross-section per unit mass for collisions, such as stars, will not dissipate their random kinetic energy efficiently, so that dissipation must occur prior to star formation, while the galaxy is still gaseous. The virial temperature of a typical galaxy-sized potential well is $T_{\text{galaxy}} \sim 10^6$ K, with a corresponding one-dimensional velocity dispersion of ~ 100 km s^{-1}.

The physical conditions in the Universe at the epoch of galaxy formation (redshift ~a few), as deduced from observations of quasar absorption lines (the Gunn–Peterson test for neutral hydrogen), are such that hydrogen is ionized, and correspond to temperatures of the proto-galactic gas of $\sim 10^4$ K, with a sound speed of only ~ 10 km s^{-1}. The collapse of this gas in Galactic potential wells will thus induce supersonic motions, and lead to both thermalization of energy through radiative shocks and subsequent loss of energy by cooling. It is this conversion of potential energy, first to random kinetic energy as described by the virial theorem, and then to radiation via atomic processes, the net result of which is an increase in the binding energy of the system, that is termed dissipation.

The rate at which excited atoms can cool obviously places a fundamental limit on the amount and rate of dissipational energy loss, and hence on the maximum rate at which a gas cloud can radiate its pressure support and collapse. A convenient measure of this timescale is the cooling time of a gas cloud, which is the time for radiative processes to remove the internal energy of the cloud. Defining the cooling rate per unit volume to be $n^2 \Lambda(T)$, where n is the particle number density, and where the functional form of Λ is determined by the relative importances of free–free, bound–free and bound–bound transitions and is thus an implicit function of the chemical abundance, gives

$$t_{\text{cool}} = \frac{3nkT}{n^2 \Lambda(T)} \propto \frac{T}{n\Lambda} \qquad (2)$$

It is usually of most interest to compare this timescale with the global gravitational free-fall collapse time of a system, which is the time it would take to collapse upon itself if there were no pressure support. This timescale depends only on the mean density of the system, and is given by

$$t_{\text{ff}} \sim 2 \times 10^7 n^{-1/2} \text{ yr} \qquad (3)$$

The term 'rapid' is often used to describe evolution that occurs on about a free-fall time.

The important mass scale of any condensations is set by gravitational instability, quantified by the Jeans mass. The Jeans mass is that minimum mass at which gravity overwhelms pressure so that density perturbations of mass $M \gtrsim M_J \sim 10^8 \, T_4^{3/2} \, n^{-1/2} M_\odot$ are unstable and collapse upon themselves, where the numerical factor is for temperature T in units of 10^4 K and number density n in units of particles per cubic centimetre.

The discussion above is based on an extremely idealized model of a proto-galaxy, in that only the *global* cooling and collapse timescales of a *uniform* gas cloud are considered. No analytic descriptions of more plausible models exist.

The above discussion can say nothing about when or how local Jeans-mass condensations actually form stars; the inherent assumption is that cooling is necessary and sufficient for efficient star formation, though the critical distinction between global and local timescales is rarely made explicit. However, it is clear that the existence of gaseous disks requires the star-formation efficiency to be low during the early stages of disk formation. A realistic discussion of galaxy formation must consider the hydrodynamics of the gas in a proto-galaxy. The general conclusion from available studies is that, while it is possible to build models which are somewhat like observations, it is necessary to specify the most sensitive parameters (viscosity and, in effect, the star formation rate) in an *ad hoc* way. Hence the need to be guided by observations.

The angular momenta of galaxies that formed in environments of different density might also be expected to differ. During the build-up of structure, initially strongly bound particles lose both energy and angular momentum, while the weakly bound particles gain energy (become more weakly bound) and also gain angular momentum. There is overall alignment of the angular momentum vector of different shells in binding energy. Disks of spiral galaxies would then form without significant angular momentum transport, provided the baryons remain gaseous until the virialization of the dark halo, and shock-heating, as described earlier, would homogenize the gas. The predictions of these models could be tested in detail if we knew the angular momentum distribution of the outer spheroid of our Galaxy; all we know at present is that the kinematically selected subdwarfs in the solar neighbourhood have a lower specific angular momentum than the disk stars, by roughly a factor of five, and that the metal-poor globular cluster system is consistent with zero net rotation to galactocentric distances of ~30 kpc.

The angular momentum distribution of the material destined to form the disk controls the range of galactocentric radii over which infall occurs at a given epoch, and the duration of the infall at a given location. Thus models of disk chemical and dynamical evolution that appeal to continual infall must also satisfy angular momentum constraints.

Specific models of Milky Way galaxy formation

The most widely referenced model of the formation of our Galaxy is that of Eggen, Lynden-Bell and Sandage (1962; henceforth ELS), which was developed primarily to understand their observations of the kinematics and chemical abundances of stars near the Sun. Certainly the most important effect of the ELS analysis was to emphasize that quantitative conclusions about the epoch of galaxy formation could be derived from observations of stellar abundances and kinematics near the Sun today – a task in which we are still engaged.

The ELS model requires the stellar spheroid to have formed during a period of rapid collapse of the entire proto-galaxy, after which the remaining gas quickly dissipated into a metal-enriched cold disk, in which star formation has continued until the present. The ELS model was designed to provide conditions under which the oldest stars populated radially anisotropic orbits, while stars that formed later had increasingly circular orbits, in accordance with data which implied that the most metal-poor stars, assumed to be the oldest stars, were on more eccentric, lower angular momentum orbits than the more metal-rich stars. This model is based on two crucial assumptions: first, that a pressure-supported, primarily gaseous galaxy (where $T = T_{\text{virial}}$) is stable against star formation, in which case the *global* cooling time is the shortest timescale of interest, and thermal instabilities must be suppressed; and second, that stellar orbits cannot be modified to become more radial after the formation of the star.

If the first assumption were valid, the observed high-velocity stars must have formed from gas clouds that were not in equilibrium in a pressure-supported system. If the second assumption above were valid, these clouds formed stars while on radial orbits at large distances from the Galactic centre. Thus, in this picture, these clouds must have turned around from the background universal expansion, and be collapsing towards the centre of the potential well. Hence, the oldest stars of the Galactic spheroid must have formed as the proto-galaxy coalesced. To determine the rate of the collapse, ELS analysed the evolution of the radial anisotropy of a stellar (or gas cloud) orbit as the Galactic potential changed, and showed it to be approximately conserved during a slow collapse but to become more radially anisotropic in a fast collapse. They argued against a slow collapse on the grounds that such a collapse requires tangentially biased velocities (remember that pressure support has been excluded by assumption), and this tangential bias will be unaffected by the resulting slow changes of the gravitational potential. The observed radial anisotropy of the stellar orbits then implies an initially radially biased velocity ellipsoid, while the calculations of ELS show that such a velocity ellipsoid will have become more radially anisotropic during collapse. Hence, ELS deduced that the gas clouds were in free-fall radial orbits, and that the consequent collapse must have been rapid, with 'rapid' in this sense meaning that the timescale for collapse is comparable to an orbital or a dynamical timescale, which is of the order of 10^8 years. It should be noted that Isobe (1974) came to the opposite conclusion from his analysis of the ELS data, and favoured a slow collapse, while Yoshii and Saio (1982) augmented the ELS sample and also concluded that the halo collapsed over many dynamical times.

Clearly if either of ELS's assumptions were violated, there need be no correlation between the *time* of a star's formation – which they infer from a star's metallicity – and its *present* orbital properties. In a non-rotating pressure-supported system, all stars formed would be on highly radial orbits, as a star has too small a surface area to be pressure supported by the gas. As mentioned above, assumptions about star formation in pressure-supported systems must be treated as *ad hoc* until we understand better the physics of star formation, so conclusions based on such assumptions are at best uncertain. If their second assumption were violated, then the stars that are now the high-velocity stars near the Sun could have originated from more circular orbits inside that of the Sun, and have present orbital properties that depend only on dynamical processes subsequent to their formation. The realization that a forming galaxy undergoes changes in its gravitational potential which are of order the potential itself (violent relaxation) means that stellar orbits can be modified considerably.

Later N-body models (e.g. May and van Albada 1984) for systems in which dissipation does not play a major role show that the final state of the collapsed system depends both on the degree of homogeneity and on the temperature of the initial state: clumps cause angular momentum and energy transport. Violent relaxation never goes to completion, so that final and initial orbital binding energies and angular momenta are correlated, the interior regions becoming more centrally concentrated and the outer regions being puffed up. The typical final steady-state velocity distribution is highly anisotropic outside (roughly) the half-light radius, and more isotropic within that radius. If violent relaxation were completely efficient, all systems would reach the same final state with isotropic velocity distribution. In the Galaxy, the spheroidal half-light radius is ~3 kpc, well inside the Sun's orbit. Thus the expected velocity distribution of old stars near the Sun after virialization of the spheroid is anisotropic, as observed by ELS, even though the dynamical evolution of the system is not as they envisage, and a correlation between kinematics and age is no longer an inevitable conclusion. One might, for example, imagine a situation where later (rapid) collapse of either the disk or the dark halo, or the merger of a few large substructures, could lead to the rapid dynamical evolution of a central spheroidal component which had previously formed on a longer timescale. Models of this type have yet to be studied in detail.

It is the continuing attempt to quantify the metallicity–kinematics distribution function of stars, with sufficient accuracy and large numbers to address these questions, that motivates projects such as Hipparcos, and the use of HST, COBE, IRAS and ISO to quantify the true distribution of stars in space. We now consider that specific continuing challenge.

THE SPATIAL STRUCTURE OF THE MILKY WAY GALAXY

Counting stars is one of the few truly classical scientific techniques used to study high-latitude (and therefore low-obscuration) Galactic structure. Early work in this subject is well reviewed by Paul (1993). The extensive data set and understanding available at that time is reviewed by Blaauw and Schmidt (1965). Relatively little further progress was achieved until the new deep, high-quality data of Ivan King and collaborators at Berkeley became available in the late 1970s. The application of computer modelling to these data by van den Bergh (1979) led to a considerable resurgence of interest, continuing to the present. An accessible overview of the recent advances in the subject is provided by Croswell (1995), while the continuing level of research activity generates several conferences per year. An important specific example, which includes both historical and research articles, is the proceedings of IAU Symposium 164, 'Stellar Populations', held on the 50th anniversary of Walter Baade's publication of the original concept (van der Kruit and Gilmore 1994).

The fundamentals of star-count analyses

The number of stars, N, countable in a given solid angle to a given magnitude limit, m, is given by a simple linear integral equation often known as 'the fundamental equation of stellar statistics':

$$N(m) = \int \Psi(M_v, \boldsymbol{x}) \, D(M_v, \boldsymbol{x}) \mathrm{d}^3 x \qquad (4)$$

where $\Psi(M_v, \boldsymbol{x})$ is the distribution function over absolute magnitude and position, $D(M_v, \boldsymbol{x})$ is the stellar space density distribution, and $\mathrm{d}^3 x$ is a volume element. This (Fredholm) equation is rarely invertible, being ill-conditioned. In general, the luminosity function is too broad to allow any solution for *both* $D(M_v, x)$ and $\Psi(M_v, x)$. The situation can be improved by restricting the data by colour and/or spectral type, which is the technique usually followed. In this case, for an assumed form of the distribution function $\Psi(M_v, x)$ the density function $D(M_v, x)$ is recovered from $N(m, \text{colour})$. This may be done by inverting the data – classical photometric parallax – or computer calculation of the integral with subsequent iterative comparison of data and model (Bahcall 1986). These techniques are clearly entirely equivalent, and should agree. Often they do not.

The fundamental problem with use of the fundamental equation is that both the stellar luminosity function and the stellar density law are functions of many parameters. Few of these are sufficiently well known to be fixed. Consequently, a wide variety of combinations of Ψ and D are allowed mathematically. Other astrophysical constraints are necessary, whose choice has remained subjective until recently due to the lack of adequate observational constraints.

The choice of astrophysical constraints

The technique adopted by almost all workers to date is to fix the very large number of parameters by adopting empirically determined fitting functions for quantities such as the density laws and luminosity functions, and fitting a set of these fitting functions to the observations. The empirical fitting functions are determined primarily from photometric observations of spiral galaxies thought to be similar to the Milky Way, from the Catalogue of Nearby Stars (CNS), and from a small number of well-studied globular and open clusters. As most authors are forced to adopt the same few fitting functions, in the absence of any alternatives, it is not surprising that most conclusions are similar. Very many analyses of this type have been undertaken, leading to the general summary described at the beginning of the chapter.

Very interestingly, extension of the reliable data sets some eight magnitudes fainter, which was made possible by HST, produced no new result! Extension of the data by this huge dynamic range, as for example in the star count studies of the Hubble Deep Field, led to essentially unchanged conclusions. With hindsight one understands that the earlier work had sufficient sensitivity to see luminous giant stars to the effective edge of the Galaxy. Thus, the deeper HST work revealed only intrinsically lower luminosity stars at distances already studied, not new information. The HST data additionally remain limited by a very small sample size, as the HST field of view is tiny.

Some of the constraints adopted have recently been discovered to have been poorly justified. The most important example is the popular impression that the photometric properties of spiral galaxies are adequately described as a sum of a flat exponential (both radially and vertically) disk and a roughly round spheroid with a standard luminosity profile, the '$r^{1/4}$ law'. Recent photometric analyses have shown that such a description is an unacceptably poor description of high-quality surface photometric data in almost all cases. Thus, one should not necessarily expect a star-count model based on a standard spheroid to be an adequate description of the Milky Way galaxy.

A specific point of interest in the relevant star-count literature involved the parameters of the thick disk, with some conflicting claims as to its reality until a few years ago, when the evidence from several kinematic and spectroscopic surveys became overwhelming. This apparent uncertainty in the star-count modelling was due to the extreme sensitivity of computer models to the adopted stellar luminosity function and colour–magnitude relations. Unless careful astrophysical constraints are imposed on these choices, a huge variety of models will reproduce the same data.

These uncertainties illustrate the fundamental limitation of the modelling of star-count data in isolation, which is a severe restriction on its value – too few constraints are usually available to provide a unique model, and too few consistency arguments are usually applied during its use. One requires additional information other than star-count data to constrain the appropriate colour–magnitude relation and, if possible, kinematic data to allow segregation of different stellar populations. The colour–magnitude relation of a stellar population is a function of chemical abundance and age, while the density law depends on both the stellar kinematics and the gravitational potential gradients. In general none of these parameters is known adequately *a priori*.

Analysis of high-latitude star-count data

The most straightforward analysis technique for stellar number–magnitude–colour data is photometric parallax. This involves use of the absolute magnitude–colour relation for an open or globular cluster of appropriate chemical abundance. The absolute magnitude for a field star of measured colour is read directly from this diagram, and combined with the apparent magnitude to give a photometric distance. From a large set of distances, with appropriate Malmquist corrections, a density law is derivable directly. This technique has been extensively applied by many groups, and led to the discovery of the thick disk by Gilmore and Reid (1983). The steep density profile from 2 to 4 kpc was identified by these authors as a Galactic thick disk, with exponential scale height ~1.3 kpc and local normalization ~2% of the old disk stars. They emphasized that a flattened $r^{1/4}$ law was an equally good fit to the data. A density profile of steep exponential form at distances of a few kiloparsecs from the Galactic plane was in fact very well established many years ago (see reviews by Elvius (1965, especially

Figure 2) and Plaut (1965, especially Figure 7b)) though evidently forgotten. Similar results to those of Gilmore and Reid were earlier derived from the Basel surveys by Yoshii (1982), though not widely appreciated at that time. Yoshii showed that the Basel north Galactic pole star-count data required the vertical density profile of the dominant stellar population more than ~1.5 kpc from the Galactic plane to follow an exponential density distribution with scale height a factor of ~6 larger than that of the old disk. The corresponding factor derived by Gilmore and Reid from their exponential fits to the south Galactic pole data was ~4. Rather larger normalizations, implying an even more massive thick disk, are favoured by current analyses.

The alternative analysis technique involves the direct calculation of the integral in the fundamental equation of stellar statistics. This is a straightforward computational exercise. Consequently, many attempts have been made recently to explore parameter space so that the uniqueness of the results from the direct analysis of star-count data can be determined. In relevant form this equation is

$$N(V, \text{colour}) = \text{area} \int \Psi\left(M_v, \left[\frac{\text{Fe}}{\text{H}}\right], \text{age}, r, \ldots\right) D(r, M_v, \text{age}) r^2 \mathrm{d}r \quad (5)$$

The luminosity function Ψ (stars per magnitude class per cubic parsec) has been known for many years to be a function of distance from the Galactic plane (e.g. Bok and MacRae 1942). Similarly, the existence of age–velocity dispersion and age–metallicity relations for old thin-disk stars is well known. This emphasizes the crucial and irreducible limitation of analyses of this type – both the luminosity function and density law are functions of the other phase space parameters. A unique solution of eqn (5) is therefore impossible. Instead, a large number of parameters must be fixed on external astrophysical grounds. Additionally, comparison with observational data requires the adoption of an appropriate absolute magnitude–colour relation, in exactly the way required in the more direct photometric parallax analysis technique described above. It is this non-uniqueness that must be overcome by the use of supplementary chemical abundance and stellar kinematic data. It must be emphasized that star-count analyses *in isolation* are incapable of providing a unique description of the structure of the Galaxy. The results of any such analysis should be viewed as merely indicative of the type of combination of fitting functions that can be used to represent available data. They should not be accepted as a valid description of the stellar populations in the Galaxy until the chemical abundance, luminosity class, age and kinematic assumptions are tested by spectroscopic observations.

Essential supplementary information is available from kinematics, particularly proper motion and radial velocity surveys, and from spectroscopic surveys, which allow dwarf–giant discrimination, and provide chemical abundance data, and hence allow the derivation of reliable distances. Both kinematic and spectroscopic data have become available in large quantities in the last few years, and are being included in star-count modelling. Extension of analyses of this type to the high-precision, high-statistical-weight Hipparcos Catalogue are of considerable interest. Detailed analysis of the Hipparcos data, in spite of its restriction to the immediate solar neighbourhood, is illustrating the way forward here. The considerable complexity of the real distribution of stars in velocity–age–metallicity–position (phase) space which Hipparcos has revealed (e.g. Binney *et al.* 2000) highlights both the kinematic information simply not noticed in previous analyses, and the exciting potential of future space astrometry missions (especially the ESA cornerstone 6 mission, Galactic Astrophysics by Imaging and Astrometry, GAIA). Interpretation and understanding of this complexity has barely started, and remains as a continuing future challenge.

Star counts in the inner Galaxy: on to the infrared

It is widely though erroneously believed that we can 'see' the Milky Way galaxy. In fact, our view of the Milky Way depends more on how we look at it than on what is there to be seen. For reasons related more to human biology than to astrophysics, our eyes are optimized to detect the peak energy output from thermal sources with a surface temperature near 6000 K. Thus, unless the Sun is typical of the entire contents of the Galaxy, there is no reason why we should be able to see by eye a representative part of whatever may be out there. If we had X-ray- or UV-sensitive eyes we would 'see' only hotter objects; if infrared or microwave eyes, only cooler objects.

Complementary and relatively readily available views of much of that part of the ordinary baryonic mass whose state has been identified can be provided by the sum of optical and near-infrared studies. Much of the mass of the Milky Way is in stars, the more massive of which are visually luminous. Lower-mass stars, those objects hiding behind interstellar extinction, and much of the interstellar medium (ISM), are most readily observable in the near infrared. Infrared astronomy provides the closest and greatest complementary match to optical astronomy, while at the same time extending the source temperature range available for study to those lower temperatures at which many known astrophysical sources are to be found, and at which a substantial part of the higher-energy radiation is re-emitted.

The importance of infrared astronomy for the study of obscured objects, a property that is of particular significance when studying the central regions of the Galaxy and regions

Table 1 Optical/IR interstellar extinction

Photometric passband	Wavelength (μm)	Extinction (magnitudes)
U	0.36	1.56
B	0.44	1.33
V	0.55	1.00
R	0.64	0.78
I	0.79	0.59
J	1.25	0.28
H	1.65	0.17
K	2.2	0.11
L	3.5	0.06
M	4.8	0.02
N	10	0.05

of current or recent star formation, is illustrated in Table 1. This gives representative values for the wavelength dependence of extinction, illustrating the relative transparency of interstellar dust at wavelengths just a little longer than those to which our eyes are sensitive. The net effect of this table is clear: if one wishes to observe stars at low latitudes more than a few kiloparsecs distant, and with a finite flux sensitivity limit, then one must operate in crowded low-galactic-latitude fields, in which interstellar extinction is considerable, and one must operate in the near infrared. An order-of-magnitude gain in sensitivity is attained by counting photons at 2 μm instead of 0.5 μm. The important astrophysical questions are buried as much behind interstellar dust as they are buried behind our ignorance. It is not possible to see a very large fraction of the stars, star formation regions, dynamics, or evolution in the Galaxy in a survey restricted to low source density and optical passbands. It is simply not possible to study the astrophysics of the regions of star formation, spiral structure, the inner bulge, the outer disk warp, and so on through the plane of the spiral galaxy in which we live without working in crowded and dusty fields. Such is astrophysics. An ability to gather astrophysically important data under such circumstances is a *sine qua non* for an astrophysically interesting study.

The inner Galaxy

The Milky Way galaxy is believed to be typical of those galaxies that dominate the luminosity in the Universe. As such it provides a cornerstone in our understanding of the structure and evolution of normal galaxies. In particular, the higher surface brightness parts of the Milky Way – the central few kiloparsecs – must be understood before any reliable understanding of the structure, formation and evolution of distant galaxies can be claimed. The central parts of the Milky Way are, however, more than a convenient comparison for studies of other galaxies. The inner disk and bulge are those parts of the Galaxy where most of the stars and ISM, most star formation regions, the greatest number of products of late stages of stellar evolution, most rare and short-lived stages of stellar evolution, exotic objects and the most important effects of spiral and stellar dynamics and large-scale dynamical asymmetries may all be discovered and studied in unique detail.

IRAS and DIRBE/COBE opened the inner Galaxy to study, while raising as many questions as answers (Figure 1).

Figure 1 Composite near-infrared intensity observed by the DIRBE instrument on COBE in the 1.25, 2.2 and 3.5 μm wavelength bands. The images are encoded in the blue, green red colour ranges, respectively. Most of the emission at these wavelengths is from cool, low-mass K stars in the disk and bulge of the Milky Way. Interstellar dust does not strongly obscure emission at these wavelengths. The maps trace emission all the way through the Galaxy, although absorption in the 1.25 μm band is evident in the general direction of the Galactic centre. (Hauser *et al.* 1995.)

The most substantial finding by IRAS was the discovery of very cool late-type stars with substantial mass loss, the asymptotic giant branch (AGB) stars. DIRBE surface brightness mapping discovered the Galactic bar, mapping large-scale structure. Detailed understanding of both these subjects remains a future challenge, however. This is due to a mix of extinction at short wavelengths and poor resolution in available studies at longer wavelengths. It is important to appreciate that effectively no reliable information is available concerning the large-scale spatial structure (inner spiral arms, ring, bar, central massive cluster, bulge, and so on) and the true distribution of reddening in the inner Galaxy. There is no extant description of the distribution of stellar populations in the central Galactic disk and bulge which has more than a crude statistical, and manifestly inadequate, parametrization of Galactic structure.

It is in the central regions that ISO data are providing the most new information, where a vast improvement over pioneering (IRAS, DIRBE, balloon, Midcourse Space Experiment (MSX), ground) data in both spatial resolution and sensitivity is both possible and essential (Figure 2). The ISO data are still under analysis, but have already extended IRAS studies of AGB stars to detect the lowest mass-loss rate phases of stellar evolution. Interestingly, HST NIC-MOS near-IR studies are now also mapping the surprisingly high star formation rate in the very inner Galaxy. A treasure trove is being opened.

The central stellar cluster. The centre of the Galaxy is gravitationally dominated by an extremely dense stellar cluster of unknown origin and history, and poorly known properties. It is not known whether this cluster is a remnant of the core about which the Galaxy grew, is the product of a later merger, is a product of a long-lived bar in the disk feeding gas into continuing star formation at the Galactic centre, or has some other history. Its relationship, if any, to the larger Galactic bulge, halo and disk and to the smaller Galactic non-thermal nucleus is entirely unknown. This cluster, whose density approaches one million solar masses per cubic parsec, or 10 million times that of the solar neighbourhood, is the most extreme dynamical system available for detailed study. In the central regions stellar collisions are expected to be common, while tidal forces are orders of magnitude more extreme than anywhere else in the Galaxy. The central degree or so is also an extended X-ray source, with temperature some 10^8 K, and gas pressures 1000 times those of the normal ISM.

The central bar. At least half of all disk galaxies contain a bar which will dominate the dynamical, gas flow, star formation and central nuclear evolution of its host galaxy. Often, but not always, the presence of a bar correlates with patchy nuclear star formation. Recent studies of IRAS galaxies confirm the existence of patchy near-IR emission, often completely obscured at 1 μm. In addition to this high spatial frequency structure, face-on barred spirals, surprisingly, show inner isophote twists. Modelling of these, in terms of two inner Lindblad resonances is feasible, but only if one can determine the underlying mass distribution. The essential requirement to achieve this is the determination of the surface density distribution of a suitable bulge/inner disk tracer of the dominant mass distribution along the full length of the possible bar. This must be achieved without contamination from reddening and irregular foreground disk structure (spiral arms, patchy dust, star formation regions, and so on) in the

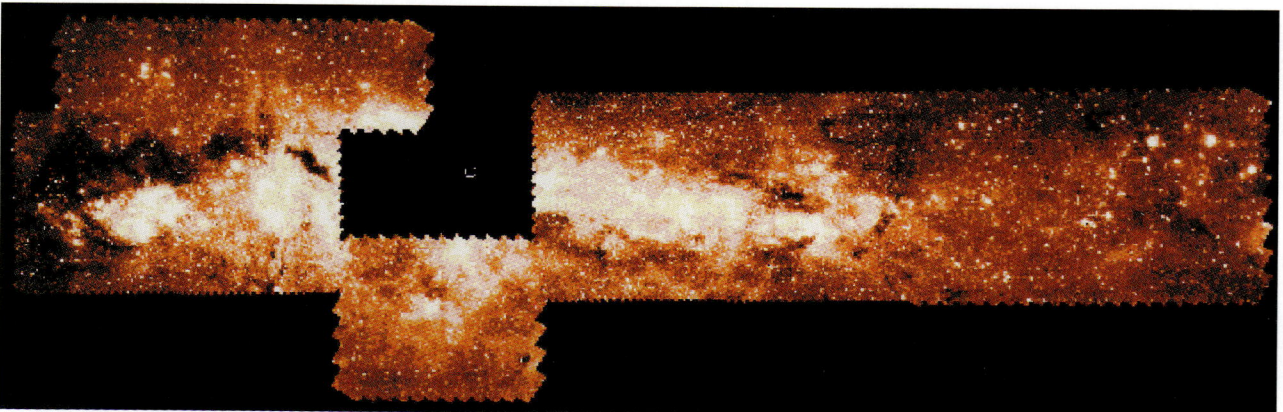

Figure 2 Infrared Space Observatory image of the inner Galaxy at 7 μm. The centre of the Milky Way is the busy core of a metropolis, crowded with huge populations of stars frantically dancing to the rhythm of gravitation. These stars are precious for astronomers: they hold many clues to the past and future history of our Galaxy. The Galactic centre has remaimed a fairly unexplored place so far because of the thick dust covering it. ISO has crossed that dusty barrier and has observed the stellar populations at the Galactic centre with very high resolution during more than 255 hours. The results already show 100,000 stars never seen before. Further analysis of the data could confirm that the Milky Way swallowed neighbouring galaxies in the past. (ESA/ISO, CAM, S. Ganesh, A. Omont and the ISOGAL team.)

plane of the disk, where the perturbation will be greatest, and with sufficient spatial resolution to identify the critical areas where the orbital resonances occur.

Galactic bars are characterized by three key parameters: (1) the pattern speed, (2) the radial density profile of the underlying mass distribution and (3) the strength of the bar, in effect the ratio of the strengths of the quadrupole and monopole contributions to the gravitational field. Numerical dynamical studies demonstrate that steady-state gas flows in bars are intimately connected to the closed orbits allowed, since gas spirals inwards by drifting slowly from one closed orbit to the next deeper in energy. Defining radii for a bar, where observable consequences can be used to test models, are related to the outer Lindblad resonance, where the orbits are roughly circular, and the inner Lindblad resonance(s) (ILR), where orbits become cusped and gas cannot collect. N-body calculations show that the length and pattern speed of bars are related. The existence, number of and importance of the ILRs, and hence the existence of closed circular orbits near the centre, are determined by the inner mass distribution. Current models show that shocks form as gas orbits become self-intersecting, and that these shocks manifest themselves as enhanced dust lanes. If these shocks/dust lanes are associated with an ILR they are parallel to the inner cusped orbit. In a galaxy with a less centrally concentrated mass distribution the dust lanes will be found parallel to the long axis of the bar. Additionally, a strong ILR, by populating elliptical orbits, allows a very high apparent central velocity dispersion for a relatively low central mass. Thus, reliable determination of both the large- and small-scale mass distributions and orbital structures of gas, stars and dust is really a single coherent problem.

In the Galaxy it has long been suspected by those modelling the inner gas dynamics that a bar is required (e.g. Liszt and Burton 1980). The important point for the present is that the best available models have been built from a combination of radio kinematics and space infrared photometry, but are in only moderate agreement with the IRAS and COBE data. Improved understanding of the central regions of the Galaxy awaits much improved determination of the underlying mass distribution from surface density data for old stars.

The inner old disk. In addition to a possible structural asymmetry, the large-scale mass distribution of the inner Galactic disk remains poorly known. Determinations of the radial scale length of an assumed exponential distribution, of the type seen in most galaxies, differ by a factor of nearly 2, as do determinations of the inner scale height. The inner disk is the product of the time-integrated history of star formation, gas flows and mergers in the Galaxy. The disk mass distribution determines the inner gravitational potential, and must be understood to allow modelling of topics as diverse as the dynamics of Galactic spiral structure, the nature of microlensing events and the radial profile of the dark halo.

The molecular ring. CO molecular line observations have long shown that the Galaxy has a molecular ring, some 3–4 kpc from the centre. Stellar rings in grand design spirals are indeed common, and this may be such a feature. IRAS has detected some stellar sources probably in the ring, and there is marginal evidence for some of lower luminosity, with perhaps older stars associated. The true nature of the older stellar population in the ring is extremely difficult to discern without new IR data, as considerable irregular extinction is expected to be associated with the molecular material. IRAS could see only the most luminous sources. ISO data have sufficient sensitivity to detect or constrain any possible underlying older population of M-giant stars, while counting the young OB stars and supergiants. In addition to its intrinsic scientific interest in determining the recent star formation history and dynamical significance of the ring, careful mapping and subtraction of sources and extinction associated with the molecular ring is critically important for a reliable determination of Galactic structure seen through the ring. DIRBE polarimetry has shown substantial $2\,\mu m$ extinction at latitudes up to $4°$ somewhere down the line of sight, giving an entirely spurious 'peanut' shape to the bulge. There is no good *a priori* reason to assume that a molecular ring seen in projection will provide a smooth and symmetric distribution of IR sources on the sky, although this has always been assumed in analyses to date.

Many attempts have been made to determine the properties and the underlying mass distribution in the inner Galaxy, from IRAS, DIRBE and other near-IR data. In general these provide consistent evidence for two features: a central (IR) bulge, which has a vertical scale height less than $2°$, an axial ratio of the order of $b/a = 0.6$, and a possible longitude asymmetry in scale height, with positive Galactic longitudes being somewhat thicker and perhaps brighter (the inner bar?). For comparison, the scale height of the main disk of the Galaxy is two to four times larger than this – the central bulge is much thinner than the disk. It is an extremely small structure, which can be studied properly only by observations at low latitudes. A second feature of present data, which is less well established, is a large-scale asymmetry with radial scale length of some 5–10°. It is suggested in recent analyses that this asymmetry is a bar, and thus the data are consistent with dynamical modelling of the inner gas data.

Between the central 30 arc seconds and the optical windows some degrees away, most extant satellite and balloon data have a spatial resolution of the order of a degree, or provide no useful data. Thus it is not possible reliably to deconvolve the younger and older populations, or the foreground and bulge, which may have very different spatial distributions. The best surveys of the region from a few arc minutes to 2° from the centre able to identify point sources are those of IRAS. The IRAS data, even after Maximum Entropy Method Image Processing (MEM) post-processing by the USAF/MSX group, are seriously affected by crowding, thereby reducing flux sensitivity to some orders of magnitude poorer than achievable by ISO. Ground-based data are seriously restricted by dust extinction and sensitivity. The DIRBE results show that extinction at 2 μm is severe in many places, and is asymmetrically distributed on scales the size of the central cluster, seriously limiting the utility of extant data. In addition, sensitivity limits are such that only the most luminous sources are measurable. Since at least some of these are known to be young, there is no guarantee that their spatial distribution is identical to that of the predominant older star cluster.

Much remains to be done, to build on the substantial scientific impact of IRAS, COBE and the early ISO results.

KINEMATICS AND CHEMISTRY OF OLD STARS

Determination of the locations, kinematics and metallicities for representative samples of stars in the Galaxy is the science case for space astrometry, inaugurated by Hipparcos. We therefore outline the case in a little detail.

Galaxies are today found to have an extremely wide range of masses, sizes, luminosities, central densities, gas contents, chemical abundances and morphologies. Even in the Local Group – our own immediate cosmological neighbourhood, in which only a few tens of galaxies are known – the masses of the known galaxies range over a factor of more than 10^5, the luminosities span a factor of 10^6, the stellar space densities span a range of 10^8, and the mean abundances of the chemical elements vary by a factor of 100, while all possible combinations of stellar age distributions, from exclusively old to predominantly very young, and all possible ranges of star-to-gas mass ratios are seen. Perhaps the only common factor is that all are dominated by the ubiquitous but entirely mysterious dark matter, whose existence is deduced from stellar and gas dynamics.

This diversity is a reflection of the complexity and non-linearity of the many physical processes that are relevant to the formation and evolution of galaxies. It is a considerable observational and theoretical challenge to elucidate the nature of, and the interplay between, these processes. Both detailed theory, incorporating such complex physical phenomena as gravitational and gas dynamics, star formation and the energy balance of the ISM, and sophisticated observations, especially kinematics, and critically near-infrared studies that penetrate the obscuring dust, are required in order to reconcile the huge variety of galaxies with a consistent and reasonably complete theory.

Fortunately, there are many fundamental aspects of the structure of galaxies that can be determined reliably from detailed analysis of the Milky Way galaxy, and only from such analyses. Some of the observable properties of galaxies today retain a fairly direct fossil record of the conditions in the early Universe at the time of galaxy formation (e.g. the present distribution of angular momentum), others are determined by the important physical processes during galaxy formation (e.g. correlations between chemical abundance and kinematics in old stars), while still others are determined by the distribution of the apparently ubiquitous 'missing' mass (e.g. the combination of stellar kinematics and the shape of the stellar system). Thus, detailed analysis of the kinematics, chemical abundances and spatial distribution of the stellar populations and the ISM in a typical large galaxy is a powerful means of studying the important physical processes that have dominated the formation and evolution of galaxies throughout the history of the Universe.

The kinematic properties of stars in the Galaxy are related, through the gravitational potential Φ, to their spatial distribution. The scale length of the spatial distribution is determined by the total energy (kinetic and potential) of the stellar orbits, as well as by the gradient of the potential (i.e. the force on the star). The shape of the spatial distribution depends on the relative populations of the orbits supported by the potential, and on the relative amounts of angular momentum (rotational), and pressure (stellar velocity anisotropy) balance to the potential gradients. The total orbital energy and angular momentum of the gas that will become a star depend on the maximum distance from the centre of the Galaxy it reached before falling out of the background expansion of the Universe, the angular momentum of its orbit at that time, the depth of the potential well through which it fell, the fraction of the total orbital energy which was lost (dissipated) before the gas formed into a star, and the subsequent dynamical evolution of the stellar orbit. That is, the present kinematic properties of old stars in the solar neighbourhood are determined in part by initial conditions in the proto-galaxy at the time of the first star formation, and in part by physical processes during galaxy formation. Hence, local kinematic studies can help to determine both the detailed physics of galaxy formation, and also some of the large-scale structural properties of the Galaxy.

Stellar chemical abundance is determined by the fraction of the available ISM at the time and place of the star's formation that had been processed through the nuclear-burning regions of massive stars. It provides a valuable chronometer for the early evolution of the Galaxy. The chemical abundance of the ISM at any time depends on the local history of formation and evolution of stars sufficiently massive to have created new chemical elements, and the mixing of local gas with more distant material. This more distant gas may or may not itself be enriched, so that the time dependance of the chemical abundance of newly forming stars depends on the local and the global star formation rates, the rate of infall of primordial gas and the efficacy of mixing in the ISM. Thus, while the chemical abundance of newly formed stars is a valuable timepiece, this chronometer need not be a smooth or even a single-valued function of chronological time.

Clearly, however, the distribution function of stellar kinematics, chemistry and age contains a wealth of information on the distribution of proto-galactic gas, the dissipational and star-formation history of that gas, the subsequent dynamical history of the resulting stars and the Galactic gravitational potential.

Observable stellar dynamics

The dynamics of any collisionless system, such as a large number of stars, is governed by the Vlasov equation, which is more commonly referred to as the collisionless Boltzmann equation (CBE):

$$\frac{Df}{Dt} \equiv \frac{\partial f}{\partial t} + \frac{\partial x}{\partial t}\frac{\partial f}{\partial x} + \frac{\partial v}{\partial t}\frac{\partial f}{\partial v} = \frac{\partial f}{\partial t} + v\frac{\partial f}{\partial x} - \nabla\Phi\frac{\partial f}{\partial v} = 0 \quad (6)$$

where f is the phase space density at the point (x,v) in phase space (i.e. there are $f(x, v)\,dx\,dv$ stars in a volume of size dx centred on spatial coordinate x, and with velocity in the volume of size dv about velocity coordinate v). The collisionless nature of stellar interactions allows the substitution of the gradient of the smoothed gravitational potential Φ for the accelerations. The phase space density f does not have to describe the entire Galaxy – we can concentrate on any subsample of stars and apply the CBE (eqn (6)) to it. We shall refer to such subsamples as tracer populations, since we can use their kinematics to trace the gravitational potential of the Galaxy, irrespective of the nature of the mass distribution that generates this potential. For stellar populations whose mass generates the potential as well as traces it, it is necessary to consider joint solutions of both the Boltzmann equation and the Poisson equation. Such self-consistent solutions are discussed below where the local thin disk is discussed.

Since this section is concerned mostly with the old stellar populations, and primarily with high Galactic latitude distributions, we may safely assume that we are concerned with a steady-state tracer population in an axisymmetric time-independent potential, so that time and ϕ derivatives are zero.

It is then convenient to write the CBE in cylindrical polar coordinates (r, ϕ, z) as

$$v_r\frac{\partial f}{\partial r} + v_z\frac{\partial f}{\partial z} + \left(\mathcal{K}_r + \frac{v_\phi^2}{r}\right)\frac{\partial f}{\partial v_r} - \frac{v_r v_\phi}{r}\frac{\partial f}{\partial v_\phi} + \mathcal{K}_z\frac{\partial f}{\partial v_z} = 0 \quad (7)$$

where the accelerations $\dot{v}_r, \dot{v}_\phi, \dot{v}_z$ have been equated to the forces that cause them, ϕ-gradients in f and in the potential have been set to zero, and $\mathcal{K}_r \equiv -\partial\Phi/\partial r$ and $\mathcal{K}_z \equiv -\partial\Phi/\partial z$ are the components of the gravitational force. Clearly, knowledge of a specific $f(x, v)$ allows the force components $\mathcal{K}_r$ and $\mathcal{K}_z$ to be derived. Note, though, that any general function f of two variables need not allow a solution for $\mathcal{K}_r$ and $\mathcal{K}_z$. In view of the intractability of solving the CBE, one proceeds in general by taking velocity moments. Multiplying through by v_z and by v_r and integrating over all velocity space produces the Jeans equations:

$$\nu\mathcal{K}_z = \frac{\partial}{\partial z}(\nu\sigma_{zz}) + \frac{1}{r}\frac{\partial}{\partial r}(r\nu\sigma_{rz}) \quad (8)$$

$$\nu\mathcal{K}_r = \frac{1}{r}\frac{\partial}{\partial r}(r\nu\sigma_{rr}) + \frac{\partial}{\partial z}(\nu\sigma_{rz}) - \frac{\nu\sigma_{\phi\phi}}{r} - \frac{\nu}{r}\langle v_\phi\rangle^2 \quad (9)$$

where $\nu(r, z)$ is the space density of the stars, $\sigma_{ij}(r, z)$ is their velocity dispersion tensor, and the only mean streaming motion is rotation, $\langle v_\phi\rangle$.

Each of the two force components can, in principle, be derived from the measurements of the moments of the velocity distribution and of the spatial density distribution of a tracer stellar population. Such experiments have been carried out for the z-force, and relate the stellar kinematics and space density to the potential of the Galactic disk. They are discussed further below. It is convenient for present purposes to rewrite the radial equation (9) in terms of observables in the Galactic plane ($z = 0$), to get

$$v_c^2 - \langle v_\phi\rangle^2 = \sigma_{\phi\phi} - \sigma_{rr} - \frac{r}{\nu}\frac{\partial(\nu\sigma_{rr})}{\partial r} - r\frac{\partial\sigma_{rz}}{\partial z}$$

$$= \sigma_{rr}\left[\frac{\sigma_{\phi\phi}}{\sigma_{rr}} - 1 - \frac{\partial\ln(\nu\sigma_{rr})}{\partial\ln r} - \frac{r}{\sigma_{rr}}\frac{\partial\sigma_{rz}}{\partial z}\right] \quad (10)$$

In this relation v_c is the circular velocity (i.e. $v_c^2 = r\,\partial\Phi/\partial r = -r\mathcal{K}_r$, where we adopt a locally flat rotation curve with $v_c = 220$ km s^{-1}), $\langle v_\phi\rangle$ is the mean rotation velocity of the relevant sample of tracer stars, which has velocity dispersions $\sqrt{\sigma_{rr}}, \sqrt{\sigma_{\phi\phi}}$ and $\sqrt{\sigma_{rz}}$ and radial spatial density distribution $\nu(r)$, remembering that r is the *planar* radial coordinate. The quantity $v_c - \langle v_\phi\rangle \equiv v_a$ is often called the asymmetric drift of a stellar population. Equation (10) relates measurable local moments of the stellar distribution function to global properties of the Galaxy.

THE MASS DISTRIBUTION IN THE GALACTIC DISK

The distribution of mass in the Galactic disk is characterized by two numbers, its local *volume* density ρ_o and its total *surface* density $\Sigma(\infty)$. They are fundamental parameters for many aspects of Galactic structure, such as chemical evolution (is there a significant population of white dwarf remnants from early episodes of massive star formation?), the physics of star formation (how many brown dwarfs are there?), disk galaxy stability (how important dynamically is the self-gravity of the disk?), the properties of dark matter (does the Galaxy contain *dissipational* dark matter, which may then be fundamentally different in nature from the dark matter assumed to provide flat rotation curves?), non-Newtonian gravity theories (where does a description of galaxies with Newtonian gravity and no dark matter fail?), and so on.

Although $\Sigma(\infty)$ and ρ_0 are different measures of the distribution of mass in the Galactic disk near the Sun, they are related. Of the two, the most widely used and commonly determined measure is the local *volume* mass density – that is, the amount of mass per unit volume near the Sun, which for practical purposes is the same as the volume mass density at the Galactic plane. This quantity has units of $M_\odot \, pc^{-3}$, and its local value is often called the 'Oort limit' in honour of the early attempt at its measurement by Jan Oort (1932). It has taken 70 years to provide a reliable determination of its magnitude, from Hipparcos.

The contribution of identified material to the Oort limit may be determined by summing all local observed matter – an observationally difficult task, which leads to considerable uncertainties. The uncertainties arise in part from the difficulties of detecting very low-luminosity stars, even very near the Sun, in part from uncertainties in the binary fraction among low-mass stars, in part from uncertainties in the stellar mass–luminosity relation, but mostly from uncertainties in determining the volume density of the ISM. This last uncertainty is exacerbated since the physically important quantity (for dynamical purposes) is the mean volume density of the patchily distributed ISM at the solar galactocentric distance. The best available determination of the local mass density in identified material is $\sim 0.1 \, M_\odot \, pc^{-3}$.

The second measure of the distribution of mass in the solar vicinity is the integral surface mass density. This quantity has units of $M_\odot \, pc^{-2}$, and is the total amount of disk mass in a column perpendicular to the Galactic plane. It is this quantity that is required for the deconvolution of rotation curves into disk and halo contributions to the large-scale distribution of mass in galaxies. As an indication of the dynamical significance of this mass density, the contribution of a disk potential generated by this local mass density to the local circular velocity, assuming an exponential disk with the Sun 2.5 radial scale lengths from the Galactic centre, is

$$v_{c,disk} \sim 150 \left(\frac{\Sigma_{local}}{60 \, M_\odot \, pc^{-2}} \right)^{1/2} \, km \, s^{-1} \qquad (11)$$

The local circular velocity is $\sim 220 \, km \, s^{-1}$, and the contributions to this circular velocity from the various components generating the Galactic potential add in quadrature. Thus the Galactic disk is far from dominating the local potential well.

Both these dynamical quantities are derived from a measurement of the vertical Galactic force field, $\mathcal{K}_z(z)$. If one knew both the local *volume* mass density and the integral *surface* mass density of the Galactic disk, one could immediately constrain the scale height of any contribution to the local volume mass density that was not identified. For example, one might suspect that some fraction of the local volume mass density was unidentified (i.e. a local 'missing' mass problem), but also determine a surface density which is effectively fully explained by observed mass. Then the unidentified contribution to the local volume density would have to have a small scale height, if its contribution to the surface density is to be small. In view of the very small scale length on which it must be distributed, it would then be plausible to deduce that any local 'missing' mass unidentified in the volume mass density near the Sun was not the same 'missing' mass as that which dominates the extended outer parts of galaxies.

Determination of the volume mass density and the integral surface mass density near the Sun requires similar observational data, namely distances and velocities for a suitable sample of tracer stars, but rather different analyses.

Measurement of the Galactic potential

All determinations of the mass distribution in the Galactic disk require a solution of the CBE. In view of its intractability, in practice one utilizes its vertical velocity moment, the vertical Jeans equation:

$$\mathcal{K}_z = \frac{1}{\nu} \frac{\partial}{\partial z} (\nu \sigma_{zz}) + \frac{1}{r\nu} \frac{\partial}{\partial r} (r \nu \sigma_{rz}) \qquad (12)$$

where $\nu(r, z)$ is the space density of the stars, and $\sigma_{ij}(r, z)$ their velocity dispersion tensor.

The first term on the right-hand side of this equation is dominant, and contains a logarithmic derivative of the stellar space density, $\nu(r, z)$, and a derivative of the vertical velocity dispersion, σ_{zz}. Since the stellar population in the solar neighbourhood is, within a multiplicative factor of a few, tolerably well described by an isothermal stellar population, the term containing the derivative of the space density dominates the determination of $\mathcal{K}_z(z)$ near the Sun. This point is not often appreciated adequately; it means that one should determine stellar density profiles with even greater care than is required for the velocity dispersions.

The second term in the Jeans equation describes the tilt of the stellar velocity ellipsoid away from the local cylindrical polar coordinate system in which velocity dispersions are measured. One therefore needs the r-gradients of σ_{rz} and of v. There are no general analytical solutions for this term, which must be determined from future astrometric studies.

Given a measurement of the gravitational field $\mathcal{K}(r, z)$ in an axisymmetric galaxy, the total density ρ of gravitating matter follows from Poisson's equation:

$$\nabla \cdot \mathcal{K} = -4\pi G \rho \qquad (13)$$

For a disk galaxy we can express the r-gradient in $\nabla \cdot \mathcal{K}$ in terms of the observed circular velocity at the Sun, v_c, or in terms of the Oort constants of Galactic rotation A and B:

$$\begin{aligned} \rho &= -\frac{1}{4\pi G}\left\{\frac{\partial \mathcal{K}_z}{\partial z} + \frac{1}{r}\frac{\partial}{\partial r}(r\mathcal{K}_r)\right\} \\ &= -\frac{1}{4\pi G}\left\{\frac{\partial \mathcal{K}_z}{\partial z} + \frac{1}{r}\frac{\partial (v_c^2)}{\partial r}\right\} \\ &= -\frac{1}{4\pi G}\left\{\frac{\partial \mathcal{K}_z}{\partial z} + 2(A^2 - B^2)\right\} \end{aligned} \qquad (14)$$

For a disk galaxy with an approximately flat rotation curve the second term is small within a few kiloparsecs of the disk plane (for an exactly flat rotation curve $A^2 - B^2 \equiv 0$ at $z = 0$), so we can integrate in z to obtain the total column density $\Sigma(z)$ between heights $-z$ and z relative to the disk plane $z = 0$:

$$\Sigma(z) = \int_{-|z|}^{|z|} \rho(z)\,dz = \frac{|\mathcal{K}_z|}{2\pi G} - \frac{(A^2 - B^2)}{\pi G}|z| \qquad (15)$$

It is evident from the equations above that determinations of the local volume mass density ρ_o depend on the square of any distance-scale errors in the tracer population, since they are derived from the second derivative of the stellar space density distribution, while determinations of the surface mass density are linearly proportional to the distance scale, being based on the first derivative.

Bahcall (1984) improved the theoretical methods with which to determine the local volume density of matter ρ_o, and re-analysed the available F dwarf and K giant velocity and density-law data with a new, dynamically self-consistent galaxy model, including the dark halo necessary to support a flat rotation curve. The self-consistency means that the matter which generates the gravitational field is required to respond to it via the CBE.

The analytical techniques developed by Bahcall represent a considerable improvement on those applied previously, and for the first time allow a derivation of ρ_o which is limited by the quality of the available observational data, rather than by the approximate nature of the analysis.

Bahcall's analysis is primarily appropriate for determination of ρ_o, and is less suitable for the determination of $\Sigma(\infty)$. Kuijken and Gilmore (1989) therefore developed a new technique for the analysis of stellar kinematic data which is more appropriate for determination of the integral surface mass density of the Galactic disk near the Sun. Their analysis involves maximum likelihood comparison of observed and predicted distribution functions of stellar velocities as a function of distance from the plane. It thus removes the need to describe an array of distance–velocity data by moments, such as the r.m.s. velocity dispersion. It also provides the freedom to include important physical effects (the orientation of the stellar velocity ellipsoid far from the Galactic plane) and constraints (consistency with the Galactic rotation curve) in the modelling.

Determination of the local volume mass density

Determination of the local volume mass density near the Sun – the Oort limit – for many years suggested that perhaps half of the local *volume* mass density near the Sun measured dynamically remained unidentified. The implications of this result are extreme: cold, dissipational dark matter is an important mass contribution even very near the Sun. Was this value robust?

The sensitivity of determinations of the local volume mass density ρ_o to uncertain data lies in the modelling of the stellar velocity distribution near the Galactic plane, and in the determination of the stellar density distribution with distance from this plane. Detailed analysis prior to Hipparcos concluded that available determinations of the volume mass density near the Sun – the Oort limit – remained limited by systematic and random difficulties with the available data.

The space-science breakthrough in this very important field came with the first data which were both accurate and, more importantly, reliably free of systematic errors, from Hipparcos. Creze et al. (1998) utilized the Hipparcos data to determine the local mass density, showing the substantial contribution made by Hipparcos to this subject.

The Hipparcos data provide the first volume-limited and absolute-magnitude-limited homogeneous tracer of stellar density and velocity distributions in the solar neighbourhood. Creze et al. (1998) showed that the density of A stars more luminous than $M_v = 2.5$ can be accurately mapped within a sphere of 125 pc radius, while proper motions in Galactic latitude provide the vertical velocity distribution near the Galactic plane. The potential well across the Galactic plane is then traced directly. The local dynamical density comes out as $\rho_o = 0.076 \pm 0.015\,M_\odot\,\mathrm{pc}^{-3}$, a value below many previous determinations, and in excellent agreement with direct determinations of the observed mass in stars and the ISM. This leaves no room for any disk-shaped component of dark matter. Unfortunately the Hipparcos sample is restricted by satellite limits to only very young stars, about which one has the greatest reservations as tracers of an equilibrium potential. A definitive result awaits future space astrometry missions, especially GAIA. Nonetheless, the Hipparcos answer is a major step forward, and has totally consolidated this field: dissipative dark matter now has zero observational support.

The Hipparcos results noted above severely constrain the nature of dark matter. Space science has contributed to this extremely important problem in other ways which additionally illustrate the power of both HST and ISO. This is an important example of the unique capabilities of space science to resolve long-standing fundamental questions.

Baryonic dark matter: limits from HST and ISO

Recent HST and ISO observations impose very severe limits on any compact baryonic contributions to Galactic (dark) halos. When combined with Milky Way galaxy microlensing results, almost the entire plausible range of massive compact baryonic objects is excluded by direct observation. Deep direct imaging at 7 and 15 μm with ISOCAM on the ISO spacecraft directly exclude hydrogen-burning stars of any mass above the hydrogen-burning limit, and of any chemical abundance, from being the predominant explanation of the dark halos of external spiral galaxies. For the Milky Way galaxy, HST has provided luminosity functions to the hydrogen-burning limit in several globular clusters. The resulting mass functions do not provide any support for dominance by very low-mass stars. This is consistent with field surveys for sub-stellar-mass brown dwarfs, which show such objects to be relatively rare. These results are complemented by very deep HST luminosity functions in the Large Magellanic Cloud, providing strong support for the (near-) universality of the stellar mass function. Very recent HST results are available for the nearby Ursa Minor dwarf spheroidal galaxy. This galaxy, the most dark-matter-dominated object known on kiloparsec scales, has a normal stellar mass function at low masses. The prospects are bright for dark elementary particles.

Background

Low-mass stars and sub-stellar-mass brown dwarfs are the only objects known to exist in potentially interesting numbers that might make up a substantial part of the dark matter required in Galactic halos by dynamical arguments. Although very old white dwarf stellar remnants have been considered, constraints from the combination of an artificially contrived initial mass function, and the inevitable overproduction of all of luminosity, helium and later supernovae by their progenitors, exclude them as astrophysically plausible.

The number of stars as a function of mass, corrected for stellar evolutionary and chemical abundance effects – the initial mass function (IMF) – is one of the two key functions defining the evolution of galaxies and the distribution of (baryonic) matter. (The star formation history is the other.) There is no *ab initio* understanding of either of these two functions. One therefore may speculate on arbitrarily complex variations, but ultimately must be guided by observations. Recent observations indicate an unexpected but now well-established similarity in the IMF over the full range of astrophysical sites in which it can be determined directly or indirectly.

The IMF of high-mass stars – those that would be the progenitors of any putative population of 0.5 $M_\odot$ very old white dwarf stellar remnants – has been determined from HST observations in Local Group galaxies and Hipparcos observations in the solar neighbourhood. Technical details are provided in Gilmore and Howell (1998), and are summarized here.

In spite of the wide range of local chemical abundances and star formation rates, the Local Group and solar neighbourhood IMFs are not distinguishable. In more extreme environments only indirect analyses are possible. Nonetheless, even in extreme starburst galaxies the derived high-mass IMF is consistent with being standard. Thus, the high-mass IMF does not seem to be a (strong) function of the local star formation rate. Since chemical element creation yields from supernovae are progenitor-mass dependent, one may also determine the IMF of the first stars to form in the Milky Way from the element ratio pattern in surviving old low-mass stars. Here again, the evidence is consistent with a standard IMF. Thus, the high-mass IMF does not seem to be a (strong) function of the chemical abundance of the gas from which it forms, over some two to three orders of magnitude.

Given the short lifetimes of high-mass stars, their IMF can be relevant to the dark matter problem only indirectly, through their low-mass remnants. Most of the mass in the observed standard IMF is locked up in long-lived, low-mass stars, all of which ever formed still exist as stars today. Thus, the low-mass IMF can be determined directly, for stars above the minimum mass for hydrogen burning. For even lower-mass stars, which rapidly cool and fade to extremely low luminosity, recent surveys have also provided direct determinations – but only locally, given the intrinsic low luminosities. Nonetheless, recent near- and mid-infrared surveys in regions of recent or continuing star formation determine the low-mass stellar IMF to be adequately described by the standard Kroupa, Tout and Gilmore (KTG) field IMF (Kroupa *et al*. 1993). The KTG IMF, which continues to be uncertainly described, predominantly because of very small sample sizes but possibly due to real small-amplitude variability, is a representation of the local field stars: it thus represents a time-averaged IMF over the history of the local disk, for solar abundance stars. The implication is that the low-mass IMF has not shown (strong) variations at high abundances over cosmic time, in agreement with the conclusions for high-mass stars above. Importantly, the KTG mass function shows a flattening at low masses, below about 0.2 $M_\odot$, such that the total mass is dominated by hydrogen-burning stars. This unimportance of brown dwarfs for the mass budget seems well established at high abundances, being supported by the ongoing field brown dwarf identification programmes from the Deep Near Infrared Sky Survey (DENIS) large-area IR survey.

All these constraints leave open one window of opportunity for compact baryonic dark matter: low masses, low chemical abundances and old ages. Primordial chemical abundance, that is, H + He, is the expected value if the as-yet unidentified baryons required to exist by standard Big Bang nucleosynthesis arguments form compact objects at early times. This requirement is almost unavoidable, for compact objects, since the dark matter seems to be found exclusively in Galactic halos and not Galactic disks. Thus it must have been locked into a non-dissipative form early on. An obvious explanation, that the missing baryons remain distributed in a (possibly chemically enriched) intergalactic medium, is a natural though as-yet unproven explanation for the fate of the baryons not found in galaxies and clusters today if they exist in a non-compact form.

We now consider, in turn, observational limits on the number of compact, low-mass, low-abundance baryonic objects in the Milky Way halo, in other Local Group galaxies, and in more distant galaxies.

The stellar IMF in Milky Way globular clusters

The HST has revolutionized determinations of the relative numbers of stars as a function of luminosity – the luminosity function – in globular clusters of the Milky way. While conversion of an observed luminosity function into an IMF remains very model dependent, differential comparisons of observed luminosity functions between clusters of not dissimilar chemical abundances is a very robust check on possible variations.

How is this possible? In fact, why is the HST so valuable for stellar populations research? The answer is apparent in Figures 3 and 4. Extensive recent results from many research groups justify the general conclusion that, considering the range of chemical abundances, internal cluster structural parameters and differing Galactic orbits, any differences between cluster luminosity functions are certainly not large. In particular, in a beautiful experiment by Piotto *et al.* (1997), four globular clusters with similar chemical abundances were studied with the HST. Three had indistinguishable luminosity functions, while the fourth had a substantial *deficit* of low-mass stars. The deficit cluster is on an orbit close to the Galactic disk, providing an explanation for the lost low-mass stars: they have been tidally removed from the parent, following the normal internal dynamical process of mass segregation inside the cluster. Under this interpretation, all the clusters had a universal luminosity – and hence mass – function, corresponding to the present-day luminosity function of the three similar clusters.

It is worth emphasizing here that the only difference detected between the clusters corresponds to even fewer low-mass stars in one case – that is, even less relevance to baryonic dark matter. Nonetheless, if the IMF can vary down, it could vary up. It is also possible, at least in principle, that the three similar clusters are all stripped of low-mass stars to the same extent, and the true IMF is indeed dominated by now lost low-mass stars. The next step is therefore to confirm the tidal stripping argument for the variable luminosity function, a task yet to be reliably completed either observationally or theoretically.

However, observationally, to at least first order, the metal-poor globular clusters provide evidence for an invariant IMF which has too few low-mass stars to be of relevance to the dark matter problem.

The IMF in globular clusters in other galaxies

The work reviewed above provides clear evidence for an approximately universal IMF at low masses in old globular clusters. Might the IMF change with time, with chemical abundance or with parent galaxy type? To answer these (and related) questions, extensive HST studies are under way to provide information comparable to that available for globular clusters in the Milky Way galaxy on globular clusters in the Large Magellanic Cloud (see Figure 4). These clusters span a much wider range in age and chemical abundance than is available in the Galaxy.

Even from published preliminary data, it is apparent that the luminosity functions are, to first order, similar to one another and to the luminosity function of the Milky Way galaxy. That is, there is no clear evidence to support any variability in the IMF in globular clusters as a function of chemical abundance (over a range of 100), as a function of time (over a range of 12 billion years), or as a function of location (from old halos to modern disks).

The dark halos of spiral galaxies: ISO observations

The strongest direct evidence for local dark-matter dominance is in the outer parts of spiral galaxies, where extended gas rotation curves imply a very extended mass distribution. The various HST studies of field stars, and the gravitational microlensing surveys, study a pencil-beam through this halo in the Milky Way. These observations can be complemented

Figure 3 An illustration of the effects of seeing on spatial resolution, based on HST images of 30 Doradus, a dense young star cluster in the Large Magellanic Cloud. A full-resolution (Planetary Camera) colour composite is shown in the top left, derived from images in the U, B and V passbands. The top right shows the composite after each image is convolved to a seeing of 0.5 arc seconds, representative of a very good ground-based natural seeing image. At bottom left is a more typical ground-based 'survey-type' image, convolved to 1 arc second seeing. The loss of information, and the erroneous apparent information, in poor seeing data is evident.

by direct observations of the halos of external galaxies with known rotation curves. In this case, any observation samples a column through the entire dark halo. It is well known that such studies see no extended luminosity, hence the 'dark'. However, low-mass stars are of very low optical luminosity, so that the direct optical limits are not a strong limit on the mass in such a population.

Low-mass, hydrogen-burning stars are, however, very red, and relatively luminous in the mid-infrared. The ISO provided the ideal mid-infrared imaging system, ISOCAM. Gilmore and Unavane (1998) utilized this to observe four edge-on spiral galaxies, and, after considerable data reduction and some modelling, were able to derive strict limits excluding hydrogen-burning stars of any mass from dominance of the mass in the dark halos of these galaxies.

Summary

Recent HST observations show remarkable evidence that the stellar IMF is surprisingly universal. The IMF shows no dependence on chemical abundance over a factor of 100; no dependence on formation epoch, over 12 billion years; no dependence on type or luminosity of parent galaxy; no dependence on local stellar density; and no dependence on the total local mass content. To an excellent approximation, any old stellar population always has a

Figure 4 An HST colour image of the globular cluster NGC 1850, a young cluster in the Large Magellanic Cloud. In spite of its large distance, the cluster can obviously be studied in detail, with reliable counting of the number of stars to low luminosities, and determination of their colours. Without the HST no such information would be available.

mass-to-light ratio of a few, 2–3 in usual visual light solar units. The observed mass-to-light ratios of a few tens and higher seen in dwarf galaxies, and outer spiral halos, are not generated by a varying number of low-mass stars.

The ISO results effectively exclude hydrogen-burning stars of any mass from generating the dark matter implied by spiral galaxy rotation curves. Their limits may be combined with those of the microlensing groups studying dark matter in our own Galaxy, which exclude very low-mass objects. Clearly, the mass range which is populated by normal compact baryonic objects does not provide the dark matter in normal galaxies.

KINEMATICS: HIPPARCOS AND THE FUTURE

I end this chapter with a prediction for what should be the most important space science achievement in the next few decades: accurate global space astrometry. Everything astrophysicists, physicists, chemists, biologists, and so on study is a form of the roughly 2% of the Universe which is 'ordinary' matter. In the Universe everything we see, know and understand is a tiny and perhaps insignificant perturbation on a different reality, a reality which seems not to know about weak interactions and electromagnetism. Quantifying, mapping, understanding and detecting that reality is the great scientific challenge, comparable in its significance at least to quantum mechanics and relativity, quantum electrodynamics, quantum chromodynamics and the great advances of the twentieth century. How are we to make progress here? There is one property this extra stuff ('dark matter') certainly has, which is its gravity. This tells us the way to go: very precise determinations of gravitational fields on all scales and locations. This is possible, given suitably accurate six-dimensional phase space data – that is, the positions and velocities of tracer particles. Such information is naturally, and uniquely, provided by space astrometry.

As an insurance policy, such data will also address the key problems in what one might term everyday astrophysics. Dynamical analyses, derived primarily from accurate distance and kinematic data, are the key to understanding the wealth of astrophysics in a galaxy. The dynamical evolution of a spiral galaxy is an extremely rich and wide-ranging topic, crucial to all studies of Galactic evolution. Precise astrometric data can uniquely determine the interplay between gravity and pressure, and the role of instabilities, in our Galaxy. Most importantly, though, dynamical studies are the only feasible high-precision experiment that can determine the spatial distribution of the ubiquitous dark matter, which apparently dominates the Universe.

The most dramatic advance in our understanding of Galactic structure and kinematics, and arguably in stellar

Figure 5 The Hipparcos science and operations team: the group who paved the technological path to potentially the biggest scientific advance of the twenty-first century. (Photo by courtesy of M.A.C. Perryman.)

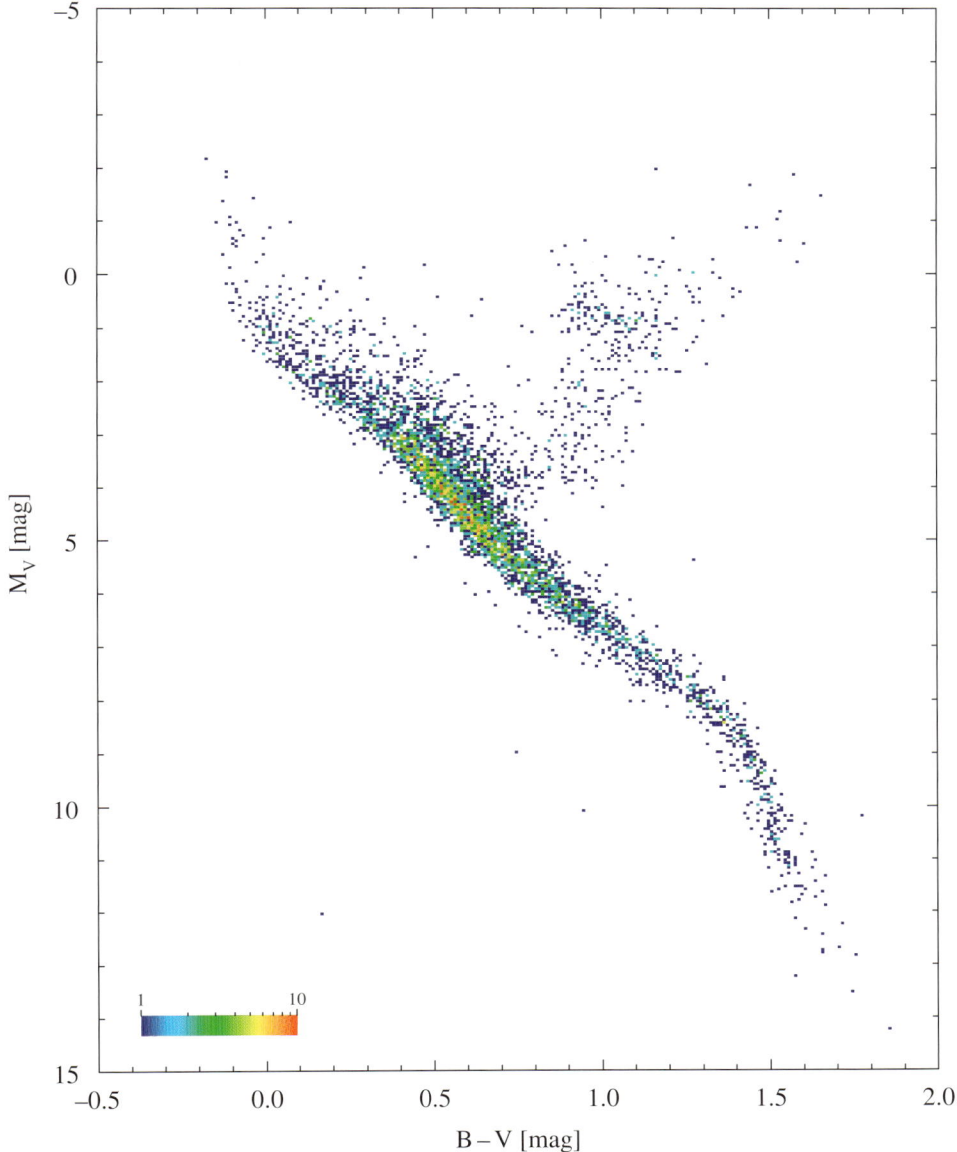

Figure 6 A Hertzsprung–Russell (HR) (M_V, B−V) diagram for the 4902 single stars from the Hipparcos Catalogue with relative distance precision $\sigma_\pi/\pi < 0.05$ and $\sigma(B-V) \leq 0.025$ mag. Colours indicate the number of stars in a cell of 0.01 mag in (B−V) and 0.05 mag in V magnitude (M_V). This is the first ever HR diagram in which the data points are not dominated by observational errors. Its analysis quantifies both stellar evolution and the history of the Milky Way. (From ESA SP-1200, the Hipparcos and Tycho Catalogues.)

evolution, followed a small ESA mission dedicated to global astrometry, Hipparcos. The scientific case for space astrometry had been available for many years before technology, and political will, allowed its implementation. The science case is basically the same as it has been since antiquity – to map and understand the Universe. It is trivial (after a century of devoted effort by huge numbers of people!), given that we now have learned that a typical scale of distances in the Milky Way is several kiloparsecs, to deduce that relevant astrometry must have at least milli-arcsecond precision to address topics in stellar and Galactic astrophysics, and a thousand times better, micro-arcseconds, for gravitational potential mapping. It is somewhat harder to deliver this performance. However, building on a prescient idea of Pierre Lacroute, ESA's Hipparcos mission paved the way.

A very brief history of the project illustrates the challenges of space science, and the two crucial requirements: a brilliant technological step, and a dedicated and hard-working team of people (Figure 5).

A little over 20 years after the original proposal, after much opposition, both political and financial, Hipparcos was launched incorrectly and late. The launch was delayed by a year, after an Ariane launch failure just a few months before the scheduled launch. Eventually, on 8 August 1989, Ariane V33 successfully launched Hipparcos into a geostationary transfer orbit. The apogee boost motor, however, failed to ignite, leaving the satellite in an unplanned highly elliptical orbit, with repeated crossing of the Van Allen radiation belts.

It is a remarkable tribute to the Hipparcos science team, to the ESA project scientist Michael Perryman, who became the Hipparcos 'wet-nurse' for very many years, and the ESOC flight operations team that stable operation was possible for four years. With hindsight, the inadvertent orbit proved mildly advantageous for astrometry, though problematic in every other way. Detailed descriptions of the history, performance and outcome of the Hipparcos mission are available in the first post-mission results presentation (ESA-SP 402), and in the 17-volume data catalogue (ESA-SP 1200) (Figure 6).

The real lesson for the future of Hipparcos comes in two parts: most importantly, ESA and the Hipparcos teams proved that global astrometry to a precision vastly superior to that possible from the ground can be delivered from space. There is an important second lesson, however. Many, arguably a large majority, of the stars observed by Hipparcos are today of specialist or limited interest. Technical requirements forced a selection of targets far in advance of mission operation, inevitably based on scientific knowledge at the time of selection. Astronomy is progressing faster than satellites can be proposed, designed and built: observatory-style missions, based on well-understood and (nearly) complete samples, provide the most science, allowing future scientists maximum opportunity to address timely questions.

The greatest tribute of all to the excellence of Hipparcos is not its immediate scientific return. It is the next generation of satellites being proposed and built to exploit its success: NASA's Space Interferometry Mission SIM, the NASA/USNO/DOD Full-sky Astrometric Mapping Explorer (FAME: of uncertain approval status at the time of writing) and the German national mission Deutsches Interferometer für Vielkanalphotometrie und Astrometrie (DIVA – though no longer an interferometer) and especially GAIA, and their as yet unknown succesors. The first decades of the twenty-first century will look back to the beginnings of space astrometry, proven by Hipparcos, as one of the crucial technological steps in astrophysics.

Space science's unique contributions to the quest to understand stellar populations and our Galaxy include infrared mapping of the Galactic surface brightness (IRAS, COBE, ISO), probing through the optically opaque interstellar dust to the inner Galaxy; improved spatial resolution (HST), allowing the study of dense stellar regions and very low-mass, low-luminosity stars; and accurate astrometry (Hipparcos), quantifying luminosities, velocities and distances directly, certainly among the most important contribution that can and will be made by space science to astrophysics.

REFERENCES

Bahcall, J.N. (1984). K giants and the total amount of matter near the Sun. *Astrophysical Journal*, **287**, 926–944.

Binney, J., Dehnen, W. and Bertelli, G. (2000). The age of the solar neighbourhood. *Monthly Notices of the Royal Astronomical Society*, **318**, 658–664.

Blaauw, A. and Schmidt, M. (eds) (1965). *Galactic Structure*, University of Chicago Press.

Bok, B.J. and MacRae, D.A. (1942). The stellar distribution in high and intermediate latitudes. *Annals of the New York Academy of Sciences*, **42**, 219–258.

Creze, M., Chereul, E., Bienayme, O. and Pichon, C. (1998). The distribution of nearby stars in phase space mapped by Hipparcos: I. The potential well and local dynamical mass. *Astronomy and Astrophysics*, **329**, 920–936.

Croswell, K. (1995). *The Alchemy of the Heavens: Searching for Meaning in the Milky Way*, Anchor Books/Doubleday, New York.

Eggen, O.J., Lynden-Bell, D. and Sandage, A. (1962). Evidence from the motions of old stars that the Galaxy collapsed. *Astrophysical Journal*, **136**, 748.

Elvius, T. (1965). Distribution of common stars in intermediate and high galactic latitudes. In A. Blaauw and M. Schmidt (eds), *Galactic Structure*, University of Chicago Press, pp. 41–60.

Gilmore, G. and Howell, D. (eds) (1998). *The Stellar Initial Mass Function*, ASP Conference Series Vol. 142, Astronomical Society of the Pacific, San Francisco.

Gilmore, G. and Reid, I.N. (1983). New light on faint stars: III. Galactic structure towards the south pole and the galactic thick disc. *Monthly Notices of the Royal Astronomical Society*, **202**, 1025–1047.

Gilmore, G. and Unavane, M. (1998). Dark haloes of spiral galaxies: ISO photometry. *Monthly Notices of the Royal Astronomical Society*, **301**, 813–826.

Hauser, M.G., Kelsall, T., Leisawitz, D. and Weiland, J. (1995). *COBE Diffuse Infrared Background Experiment Explanatory Supplement*, Version 2.0, COBE Ref. Pub. 95-A, NASA/GSFC, Greenbelt, MD.

Isobe, S. (1974). Space motion of subdwarfs and initial contraction of the Galaxy. *Astronomy and Astrophysics*, **36**, 333–340.

Kroupa, P., Tout, C. and Gilmore, G. (1993). The distribution of low-mass stars in the galactic disc. *Monthly Notices of the Royal Astronomical Society*, **262**, 545–587.

Kuijken, K. and Gilmore, G. (1989). The mass distribution in the galactic disc: I. – A technique to determine the integral surface mass density of the disc near the Sun. II – Determination of the surface mass density of the galactic disc near the Sun. III – The local volume mass density. *Monthly Notices of the Royal Astronomical Society*, **239**, 571–603, 605–649, 651–664.

Liszt, H. and Burton, W.B. (1980). The gas distribution in the central region of the Galaxy. III – A barlike model of the inner-Galaxy gas based on improved H I data. *Astrophysical Journal*, **236**, 779–797.

May, A. and van Albada, T.J. (1984). Dynamical constraints during violent relaxation and their effects on the final state. *Monthly Notices of the Royal Astronomical Society*, **209**, 15–23.

Oort, J.H. (1932). The force exerted by the stellar system in the direction perpendicular to the galactic plane and some related problems. *Bulletin of the Astronomical Institute of the Netherlands*, **6**, 249.

Paul, E.R. (1993). *The Milky Way Galaxy and Statistical Cosmology, 1890–1924*, Cambridge University Press.

Piotto, G., Cool, A.M. and King, I.R. (1997). A comparison of deep HST luminosity functions of four globular clusters. *Astronomical Journal*, **113**, 1345–1352.

Plaut, L. (1965). Distribution and motions of variable stars. In A. Blaauw and M. Schmidt (eds), *Galactic Structure*, University of Chicago Press, pp. 267–309.

van den Bergh, S. (1979). Population statistics of faint stellar and non-stellar objects. In M.S. Longair and J.W. Warner (eds), *Scientific Research with the Space Telescope*, IAU Colloquium 54, NASA CP-2111, p. 151–164.

van der Kruit, P.C. and Gilmore, G. (eds) (1994). *Stellar Populations*, IAU Symposium 164, Kluwer, Dordrecht.

Yoshii, Y. (1982). Density distribution of faint stars in the direction of the north galactic pole. *Publications of the Astronomical Society of Japan*, **34**, 365.

Yoshii, Y. and Saio, H. (1979). Kinematics of the old stars and initial contraction of the Galaxy. *Publications of the Astronomical Society of Japan*, **31**, 339–368.

31

WERNER BECKER* AND GEORGE PAVLOV**

Pulsars and isolated neutron stars

HISTORICAL OVERVIEW

The idea of neutron stars can be traced back to the early 1930s, when Subrahmanyan Chandrasekhar, while investigating the physics of stellar evolution, discovered that there is no way for a collapsed stellar core with a mass more than 1.4 times the solar mass, $M_\odot$, to hold itself up against gravity once its nuclear fuel is exhausted (Chandrasekhar 1931). This implies that a star left with $M > 1.4 M_\odot$ (the Chandrasekhar limit) would keep collapsing and eventually disappear from view. After the discovery of the neutron by James Chadwick in 1932, Lev Landau was the first to speculate on the possible existence of a star composed entirely of neutrons (Landau 1932, Rosenfeld 1974). Using the newly established Fermi–Dirac statistics and basic quantum mechanics, he was able to estimate that such a star, consisting of $\sim 10^{57}$ neutrons, would form a giant nucleus with a radius of the order of $R \sim (\hbar/m_n c)(\hbar c/G m_n^2)^{1/2} \sim 3 \times 10^5$ cm in which $\hbar$, c, G and m_n are the rationalized Planck constant, the speed of light, the gravitational constant, and the mass of the neutron. In view of the peculiar stellar parameters, Landau called these objects *unheimliche Sterne* ("weird stars"), expecting that they would never be observed because of their small size and expected low optical luminosity.

Walter Baade and Fritz Zwicky were the first to propose the idea that neutron stars could be formed in supernovae (Baade and Zwicky 1934). The first models for the structure of neutron stars were worked out in 1939 by Robert Oppenheimer and George Volkoff, who calculated an upper limit for the neutron star mass. Using general relativistic equilibrium equations and assuming that the star is entirely described by an ideal (i.e., non-interacting) Fermi gas of neutrons, they found that any star more massive than $3 M_\odot$ (the Oppenheimer–Volkoff limit) will suffer runaway gravitational collapse to form a black hole (Oppenheimer and Volkoff 1939). Unfortunately, their pioneering work did not predict anything astronomers could actually observe, and the idea of neutron stars was not taken seriously by the astronomical community. Neutron stars therefore remained in the realm of imagination for nearly a quarter of a century, until in the 1960s a series of epochal discoveries were made in high-energy and radio astronomy.

X-rays and gamma rays can be observed only from above the Earth's atmosphere (X-rays are absorbed at altitudes of 20–100 km), which requires detectors to operate from high-flying balloons, rockets, or satellites. One of the first X-ray detectors brought to space was launched by Herbert Friedman and his team at the Naval Research Laboratory in order to investigate the influence of solar activity on the propagation of radio signals in the Earth's atmosphere (Chapter 12). Using simple proportional counters put on old V2 (captured in Germany after World War II) and Aerobee rockets, they became the first to detect X-rays from the very hot gas in the solar corona. However, the intensity of this radiation was found to be a factor 10^6 lower than that measured at optical wavelengths. In the late 1950s it was therefore widely believed that all other stars, much more distant than the Sun, should be so faint in X-rays that further observations at that energy range would be hopeless. However, results from high-energy cosmic ray experiments suggested that there exist

---

* Max-Planck-Institut für Extraterrestrische Physik, Garching bei München, Germany
** Penn State University, University Park, PA, USA

celestial objects (e.g., supernova remnants) that produce high-energy cosmic rays in processes which, in turn, may also produce X-rays and gamma rays (Morrison et al. 1954, Morrison 1958). These predictions were confirmed in 1962, when the team led by Bruno Rossi and Riccardo Giacconi accidentally detected X-rays from Sco X-1. With the aim of searching for fluorescent X-ray photons from the Moon*, they launched an Aerobee rocket on 12 June 1962 from White Sands, New Mexico with three Geiger counters as the payload, each having a $\sim 100°$ field of view and an effective collecting area of about $10\,\mathrm{cm}^2$ (Giacconi 1974). The experiment detected X-rays not from the Moon but from a source located in the constellation Scorpius, dubbed Sco X-1, which is now known to be the brightest extrasolar X-ray source in the sky. Evidence for a weaker source in the Cygnus region and the first evidence for the existence of a diffuse isotropic X-ray background were also reported from that experiment (Giacconi et al. 1962). Subsequent flights launched to confirm these first results detected Tau X-1, a source in the constellation Taurus that coincided with the Crab Supernova Remnant (Bowyer et al. 1964). Among the various processes proposed for the generation of the detected X-rays was thermal radiation from the surface of a hot neutron star (Chiu and Salpeter 1964), and searching for this radiation became a strong motivation for the further development of X-ray astronomy. However, the X-ray emission from the Crab Supernova Remnant was found to be of a finite angular size (~ 1 arcmin), whereas a neutron star was expected to appear as a point source. Thus, the early X-ray observations were not sensitive enough to prove the existence of neutron stars. This was done a few years later by radio astronomers.

In 1967, Jocelyn Bell, a graduate student under the supervision of Antony Hewish at Cambridge University, England, came across a series of pulsating radio signals while using a radio telescope specially constructed to look for rapid variations in the radio emission of quasars. These radio pulses, 1.32 seconds apart and with remarkable clock-like regularity, were emitted from an unknown source in the sky at RA $19^\mathrm{h}\,20^\mathrm{m}$, and dec. $+23°$. Further observations refined the pulsating period to 1.33730113 seconds. The extreme precision of the period suggested at first that these signals might be generated by extraterrestrial intelligence. They were subsequently dubbed LGM1, an abbreviation of "Little Green Man 1" (Bell 1977). However, when other similar sources were detected, it became clear that a new kind of celestial object had been discovered. The link between these pulsating radio sources, which were called pulsars, and rapidly spinning neutron stars was provided by Franco Pacini (1967, 1968) and Thomas Gold (1968, 1969). Pacini, then a young postdoctoral worker at Cornell University, had published a paper a few months before the discovery by Bell and Hewish in which he proposed that the rapid rotation of a highly magnetized neutron star could be the source of energy in the Crab Nebula. This prediction was based on the pioneering work of Hoyle et al. (1964), who had proposed that a magnetic field of $10^{10}\,\mathrm{G}$ might exist on a neutron star at the center of the Crab Nebula. The most fundamental ideas on the nature of the pulsating radio sources were published by Gold (1968, 1969) in two seminal *Nature* papers. In these papers Gold introduced the concept of the rotation-powered pulsar which radiates at the expense of its rotational energy – the pulsar spins down as rotational energy is radiated away – and recognized that the rotational energy is lost via electromagnetic radiation from the rotating magnetic dipole and emission of relativistic particles. The particles are accelerated in the pulsar's magnetosphere along the curved magnetic field lines and emit the observed intense curvature and synchrotron radiation. (When a charged relativistic particle moves along a curved magnetic field line, it is accelerated tranversely and radiates. This curvature radiation is closely related to the synchrotron radiation caused by gyration of particles around the magnetic field lines.)

Since those early days of pulsar astronomy more than a thousand radio pulsars have been discovered (see, e.g., the catalog by Taylor et al. (1993), which lists about half of them). The discovery of the first radio pulsar was very soon followed by the discovery of the two most famous pulsars, the fast 33 ms pulsar in the Crab Nebula (Staelin and Reifenstein 1968) and the 89 ms pulsar in the Vela supernova remnant (Large et al. 1968). The fact that these pulsars are located within supernova remnants provided striking confirmation that neutron stars are born in core collapse supernovae from massive main sequence stars. These exciting radio discoveries triggered subsequent pulsar searches at nearly all wavelengths.

Cocke et al. (1969) discovered optical pulses from the Crab Pulsar, whereas its X-ray pulsations in the 1.5–10 keV range were discovered by Friedman's group at the Naval Research Laboratory (Fritz et al. 1969) and by the team at the Massachusetts Institute of Technology (Bradt et al. 1969) three months later. Using a plastic scintillator platform, Hillier et al. (1970) flew a balloon-borne experiment over southern England and detected its pulsed gamma rays at a $\sim 3.5\sigma$ level at energies greater than 0.6 MeV. These early multi-wavelength observations showed that the pulses are all phase aligned, with a pulse profile which was very nearly the same at all wavelengths, suggesting a common emission site for the radiation. Moreover, the power observed at the high photon energies exceeded that in the optical band by more

---

* The Moon was selected as a target because it was expected that a state-of-the-art detector available at that time would not be sensitive enough to detect X-rays from extrasolar sources. "We felt ... that it would be very desirable to consider some intermediate target which could yield concrete results while providing a focus for the development of more advanced instrumentation which ultimately would allow us to detect cosmic X-ray sources" (Giacconi 1974).

than two orders of magnitude, justifying the need for more sensitive satellite-based X-ray and gamma-ray observatories to perform more detailed investigations of the emission mechanism of pulsars and to survey the sky for other X-ray and gamma-ray sources.

The first Earth-orbiting mission dedicated entirely to celestial X-ray astronomy, SAS 1 (Small Astronomy Satellite 1), was launched by NASA in December 1970 from a launch site in Kenya. The observatory, later named Uhuru (which means "freedom" in Swahili), was sensitive in the range 2–20 keV and equipped with two sets of proportional counters having a collecting area of 840 cm^2 (Giacconi et al. 1971). It was designed to operate in survey mode, allowing for the first time a scan of the whole sky with a sensitivity of 1.5×10^{-11} erg s^{-1} cm^{-2}. In over two years of very successful operation, 339 new X-ray sources were detected (Forman et al. 1978): accreting binaries, supernova remnants, Seyfert galaxies, and clusters of galaxies. By far the largest sample of objects was found to belong to the group of accretion-powered pulsars – neutron stars in binary systems accreting matter from a companion star. As the matter spirals onto the neutron star's surface or heats up in an accretion disk, strong X-radiation is emitted (Chapter 32).

The next major step in high-energy astronomy was the launch in November 1972 of SAS 2, the first satellite dedicated exclusively to gamma-ray astronomy (Fichtel et al. 1975). The detector, a spark chamber, was sensitive in the energy range 35–1000 MeV. Although the mission lasted only seven months and was ended by a failure of the low-voltage power supply, its measurements confirmed the existence of the gamma-ray pulses from the Crab (Kniffen et al. 1974) and discovered the gamma-ray pulses from the Vela Pulsar (Thompson et al. 1975), which was found to be the strongest gamma-ray source in the sky. The Vela light-curve was characterized by two relatively sharp peaks, separated by 0.4 in phase (as observed for the Crab) but not phase aligned with the radio and optical pulses.

In addition, a few unidentified gamma-ray sources were detected, among them Geminga, a faint source in the Gemini region from which ~100 gamma-ray photons had been recorded, but which was not finally identified for another 20 years.* Gamma-ray astronomy, from its beginning, was often hampered by the relatively small number of detected photons and large position error boxes, typically ~0.5°–1°. This position uncertainty strongly complicated follow-up observations for optical and X-ray counterparts. Scientific publications describing data analysis techniques optimized for "sparse data," particularly the timing analysis aimed at pulsation searching, were therefore always valued highly.

The first complete and detailed gamma-ray map of the Galaxy was provided by the ESA mission COS-B, launched in August 1975. Developed by a group of European research laboratories known as the Caravane Collaboration (formed of members from MPE-Garching, CEN-Saclay, SRON-Leiden (today Utrecht), IFCAI-Palermo, CNR-Milano, and SSD-ESTEC), the satellite carried two scientific payloads: a digital spark chamber, sensitive in the range 0.03–5 GeV, and a 2–12 keV collimated proportional counter that was used as a pulsar synchronizer. The on-board clock calibration was not very accurate, so the latter instrument was used to ensure the synchronization of the X-ray and gamma-ray pulses from isolated pulsars, like the Crab and Vela Pulsars, and accreting pulsars in X-ray binaries. It was further used to determine pulsar ephemerides from the temporal analysis of X-ray data, independently of the availability of exact radio ephemerides. The high sensitivity of the gamma-ray detector allowed Kanbach et al. (1980) to conduct the first detailed temporal and spectral study of the Vela Pulsar in the range 0.05–3 GeV. The pulsar's spectrum was found to be represented by a power law, $dN/dE \propto E^{-\alpha}$ (with a photon index of $\alpha = 1.89 \pm 0.06$ for the phase-averaged spectrum), but appreciable differences of the photon index were detected for different pulsar phases (e.g., the inter-pulse emission, first detected in the COS-B data, was found to have the hardest spectrum). The COS-B observations of the Crab Pulsar provided much improved photon statistics which resulted in a more accurate pulse profile (Wills et al. 1982) and detailed spectral studies (Clear et al. 1987).

Many radio pulsars had been observed by the mid-1970s, and two of them, the Crab and Vela Pulsars, had been detected at high photon energies. Although the interpretation of both isolated and accreting pulsars as neutron stars with enormous magnetic fields, ~10^{12} G, had been generally accepted, there was no direct evidence for the existence of such huge fields. The evidence came from a remarkable spectral observation of Her X-1, an accreting binary pulsar discovered with Uhuru by Tananbaum et al. (1972). On 3 May 1976, a team of the Max-Planck Institut für extraterrestrische Physik in Garching and the Astronomische Institut of the University of Tübingen, led by Joachim Trümper, launched from Palestine, Texas a balloon equipped with a collimated NaI scintillation counter and a NaI–CsI–phoswich detector, sensitive in the range 15–160 keV. They easily detected the 1.24 s pulsations up to 80 keV (Kendziorra et al. 1977). However, when Bruno Sacco and Wolfgang Pietsch attempted to fit the observed spectrum with usual continuum spectral models, they found that a one-component continuum model could not represent the data – all fits gave unacceptably large residuals at ~40–60 keV. Further data analysis

---

* The source was dubbed Geminga, a pun in Milanese dialect in which *gh'è minga* means "it is not there" or "it does not exist," by Giovanni Bignami – See Bignami and Caraveo (1996) for a comprehensive description of the Geminga story, from the first discovery to the final identification. It is amusing to note that the name inspired Eric Cohez to choose the title of his science fiction book *Geminga: la civilisation perdue*.

confirmed that the spectral feature was not an artifact (e.g., due to incomplete shielding of the in-flight calibration source, ^{241}Am, which emitted a spectral line at $E = 59.5$ keV). It was Joachim Trümper who first recognized that the excess emission at 58 keV (or an absorption feature at 42 keV, depending on interpretation; Figure 1) could be associated with the resonant electron cyclotron emission or absorption in the hot polar plasma of the rotating neutron star. The corresponding magnetic field strength would then be 6×10^{12} or 4×10^{12} G (Trümper et al. 1978). This observation provided the first direct measurement of a neutron star's magnetic field and confirmed the basic theoretical predictions that neutron stars are highly magnetized, rapidly spinning, compact objects.

Beginning in 1977, NASA launched a series of large scientific payloads called High Energy Astrophysical Observatories (the dramatic history of the HEAO project and the experiments aboard HEAO satellites are described in lively fashion by Wallace Tucker (1984)): HEAO 1 (August 1977–January 1979), HEAO 2 (November 1978–April 1981), and HEAO 3 (September 1979–May 1981). Particularly important results on isolated neutron stars, among many other X-ray sources, were obtained with HEAO 2, widely known as the Einstein X-ray observatory (Giacconi et al. 1979), which carried the first imaging X-ray telescope on a satellite. Among four focal plane detectors of Einstein, two proved to be particularly useful for detecting and studying isolated neutron stars. The High Resolution Imager (HRI), a micro-channel plate detector, sensitive in the 0.15–4 keV energy band, with about 5 arc seconds angular resolution, was designed to use the imaging capability of the X-ray telescope. However, it had no energy resolution and its field of view was small, ~ 25 arcmin. The Imaging Proportional Counter (IPC), the workhorse of the observatory, could detect weaker sources than the HRI and had a wider field of view, $\sim 1°$, but its imaging resolution was about 1 arcmin. It was capable of studying spectra with modest energy resolution in the range 0.2–4 keV.

Einstein investigated the soft X-radiation from the previously known Crab and Vela Pulsars and resolved the compact nebula around the Crab Pulsar (Harnden and Seward 1984). It discovered pulsed X-ray emission from two other very young pulsars, PSR B0540−69 in the Large Magellanic Cloud (Seward et al. 1984) and PSR B1509−58 (Seward and Harnden 1982), with periods of 50 ms and 150 ms respectively. Interestingly, these pulsars were the first ones discovered in the X-ray band and only subsequently detected at radio frequencies. No pulsations from the Vela Pulsar were found in the soft X-ray band.

Einstein also detected three middle-aged radio pulsars, PSR B0656+14 (Córdova et al. 1989), B1055−52 (Cheng and Helfand 1983), and B1951+32 (Wang and Seward 1984). Also, X-ray counterparts of two nearby old radio pulsars, PSR B0950+08 and B1929+10, were identified, based on positional coincidence (Seward and Wang 1988). In addition, many supernova remnants were mapped – 47 in our Galaxy (Seward 1990) and 10 in the Magellanic Clouds (Long and Helfand 1979) – and several neutron star candidates were detected as faint, soft point sources close to the centers of the supernova remnants such as RCW 103 (Tuohy and Garmire 1980), PKS 1209−51/52 (Helfand and Becker 1984), Puppis A (Petre et al. 1982), and Kes 73 (Kriss et al. 1985).

Some additional information on isolated neutron stars was obtained by Exosat (European X-ray Observatory Satellite; Taylor et al. 1981), which was equipped with a low-energy detector with imaging capability and grating (0.04–2 keV) and a medium-energy proportional counter (1.5–50 keV). In particular, it measured the soft X-ray spectra of the middle-aged pulsar PSR B1055−52 (Brinkmann and Ögelman 1987) and of a few neutron star candidates in supernova remnants (e.g., PKS 1209−51/52; Kellett et al. 1987).

In spite of the major advance in the field of high-energy astronomy provided by the space observatories (particularly by Einstein) in the 1970s and 1980s, the results obtained

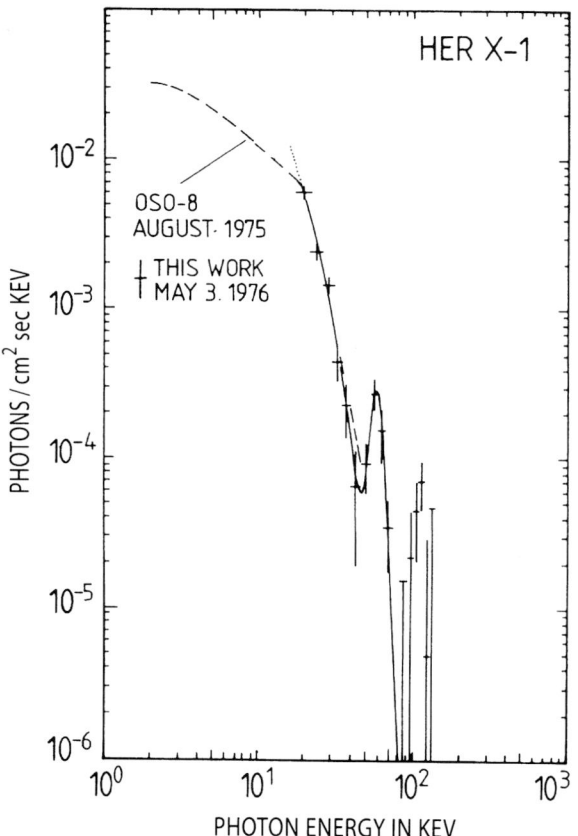

Figure 1 Unfolded X-ray spectrum from Her X-1, showing the first measurement of a cyclotron line in a pulsed spectrum of an accreting neutron star. (From Trümper et al. 1978.)

for isolated neutron stars made it clear that more sensitive instruments and multi-wavelength observations would be required if the spatial, temporal, and spectral emission properties of these objects were to be understood. For instance, Einstein was able to detect X-ray pulses only from the young and powerful Crab-like pulsars, whereas only flux estimates could be obtained for the other detected neutron stars. Only two pulsars, Crab and Vela, were detected in the gamma-ray and optical ranges.

The situation improved drastically in the last decade of the twentieth century, which can be seen as the "decade of space science." The first X-ray satellite in a series of several launched to explore the Universe from space was the German–US–UK mission ROSAT (Röntgen Satellit; Trümper 1983), sensitive in the 0.1–2.4 keV band. Equipped with an imaging X-ray telescope and three detectors, the Position Sensitive Proportional Counter (PSPC), High Resolution Imager (HRI), and EUV Wide-Field Camera, the observatory performed very successful observations of all kinds of astronomical objects in an operational lifetime of more than eight years (June 1990–February 1999). During the first six months of the mission, the ROSAT All-Sky Survey, with a limiting sensitivity of $\sim 3 \times 10^{-12}$ erg s^{-1} cm^{-2}, provided valuable information on fluxes for all the known radio pulsars. This, in particular, made it possible to constrain the neutron star cooling scenarios on a large sample of these objects (Becker et al. 1993, Becker 1995).

The complement to ROSAT, covering the harder X-ray band of 1–10 keV, was the Japanese–US mission ASCA (Advanced Satellite for Cosmology and Astrophysics; Tanaka et al. 1994), launched in 1993. It was the first X-ray observatory equipped with a CCD imager – the Solid-state Imaging Spectrometer (SIS), with a much better spectral resolution than the ROSAT PSPC. The Gas Imaging Spectrometer (GIS), which was operated in parallel, provided timing information in addition. Launched in 1992, EUVE (Extreme Ultraviolet Explorer; Bowyer 1990), sensitive in the range 70–760 Å, has been able to observe several neutron stars at very soft X-rays energies, 0.07–0.2 keV. The contributions to neutron star research by the instruments aboard the Italian–Dutch X-ray mission BeppoSAX (Butler and Scarsi 1990), sensitive in the range of 0.1–200 keV, and the USA's RXTE (Rossi X-ray Timing Explorer; Bradt et al. 1990), both launched in the mid-1990s, were particularly useful for studying X-ray binaries, including accretion-powered pulsars (see Chapter 32).

Further advances in the study of gamma-ray emission from neutron stars was gained from the nine years (1991–2000) of operation of GRO (Compton Gamma Ray Observatory; Kniffen 1990), which has explored the gamma-ray sky in the broad range from 50 keV to 30 GeV with four instruments. Particularly useful for observations of isolated neutron stars was the Energetic Gamma-Ray Experiment Telescope (EGRET), which detected five new gamma-ray pulsars (Thompson et al. 1999) in addition to the previously observed Crab and Vela Pulsars. In particular, the gamma-ray source Geminga, known since 1972, was identified as a pulsar (Bertsch et al. 1992) after the discovery of coherent pulsations in X-rays with ROSAT (Halpern and Holt 1992).

Finally, the outstanding capabilities of the Hubble Space Telescope (HST), launched in 1990, enabled astronomers to observe neutron stars directly, despite their extremely small size, in the IR–optical–UV range (Figure 12), which appeared completely impossible a few decades ago. Of particular interest was the discovery of the (presumably thermal) optical–UV radiation from old neutron stars (Pavlov et al. 1996a, Walter and Matthews 1997).

Our current understanding of the high-energy emission of neutron stars, summarized in the next section, is based largely on the results obtained with these space observatories. Although some of them have completed their service and rest on the ocean floor, new and more powerful X-ray missions have taken their place just before the onset of the twenty-first century – Chandra, with the outstanding imaging capability of its telescope, and XMM-Newton, with its unprecedently high spectral sensitivity and collecting power. It is therefore safe to say that in the very near future a wealth of new X-ray data on various astronomical objects, including isolated neutron stars, will become available and will have a major impact on our current understanding of these objects.

PHYSICS AND ASTROPHYSICS OF ISOLATED NEUTRON STARS

Neutron stars are unique astrophysical laboratories which allow us to explore the properties of matter under the most extreme conditions observable in nature. (Although black holes are even more compact than neutron stars, they can be observed only through interaction with their surroundings.) Studying neutron stars is therefore an interdisciplinary field, in which astronomers and astrophysicist work together with a broad community of physicists. Particle, nuclear, and solid-state physicists are strongly interested in the internal structure of neutron stars, which is determined by the behavior of matter at densities above the nuclear density $\rho_{nuc} = 2.8 \times 10^{14}$ g cm^{-3}. Plasma physicists are modeling pulsar emission mechanisms using electrodynamics and general relativity. It is beyond the scope of this section to describe in detail the current status of the theory of neutron star structure or the magnetospheric emission models. We refer the reader to the literature (Michel 1991, Beskin et al. 1993, Glendenning 1996, Weber 1999) and provide here only the basic theoretical background relevant to the main part of this chapter, which summarizes the observed high-energy emission properties of rotation-powered pulsars and radio-quiet neutron stars.

Rotation-powered pulsars: The magnetic braking model

Following the ideas of Pacini (1967, 1968) and Gold (1968, 1969), the thousand-plus radio pulsars detected so far can be interpreted as rapidly spinning, strongly magnetized neutron stars radiating at the expense of their rotational energy. This very useful concept allows us to obtain a wealth of information on basic neutron star/pulsar parameters just from measuring the Pulsar's period and period derivative. Using the Crab Pulsar as an example will make this more clear. A neutron star with a canonical radius of $R = 10$ km and a mass of $M = 1.4 M_\odot$ has a moment of inertia $I \approx (2/5) MR^2 \approx 10^{45}$ g cm^2. The Crab pulsar spins with a period of $P = 33.403$ ms. The rotational energy of such a star is $E_{\rm rot} = 2\pi^2 I P^{-2} \approx 2 \times 10^{49}$ erg. This is comparable to the energy released in thermonuclear burning by a normal star over many millions of years. Very soon after the discovery of the first radio pulsars it was noticed that their spin periods increase with time. For the Crab Pulsar, the period derivative is $\dot{P} = 4.2 \times 10^{-13}$ s s^{-1}, implying a decrease in the star's rotational energy of $dE_{\rm rot}/dt \equiv \dot{E}_{\rm rot} = -4\pi^2 I \dot{P} P^{-3} \approx 4.5 \times 10^{38}$ erg s^{-1}. Gold (1968) and Pacini (1968) suggested that the pulsar slowdown is due to the braking torque exerted on the neutron star by its magneto-dipole radiation; that yields $\dot{E}_{\rm brak} = -(32\pi^4/3c^3) B_\perp^2 R^6 P^{-4}$ for the energy loss of a rotating magnetic dipole, where $B_\perp$ is the component of the equatorial magnetic field perpendicular to the rotation axis. Equating $\dot{E}_{\rm brak}$ with $\dot{E}_{\rm rot}$, we find $B_\perp = 3.2 \times 10^{19} (P\dot{P})^{1/2}$ G. For the Crab Pulsar this yields $B_\perp = 3.8 \times 10^{12}$ G. From $\dot{E}_{\rm rot} = \dot{E}_{\rm brak}$ we have that $\dot{P} \propto P^{-1}$ for a given $B_\perp$. This relation can be generalized as $\dot{P} = k P^{2-n}$, where k is a constant and n is the so-called braking index ($n = 3$ for the magneto-dipole braking). Assuming that the initial rotation period P_0 at the time t_0 of the neutron star formation was much smaller than today, for $t = t_0 + \tau$ we obtain $\tau = P/[(n-1)\dot{P}]$, or $\tau = P/(2\dot{P})$ for $n = 3$. This quantity is called the characteristic spin-down age. It is a measure of the time required to lose the rotational energy $E_{\rm rot}(t_0) - E_{\rm rot}(t)$ via magneto-dipole radiation. For the Crab Pulsar, $\tau = 1258$ yrs. As the neutron star in the Crab supernova remnant (SNR) is the only pulsar for which its historical age is known (the Crab supernova was observed by Chinese astronomers in AD 1054), we see that the spin-down age exceeds the true age by about 25%. Although the spin-down age is just an estimate for the true age of the pulsar, it is the only one available for pulsars other than the Crab, and it is commonly used in evolutionary studies (e.g., of neutron star cooling).

A plot of observed periods versus period derivatives is shown in Figure 2, using the pulsars from the Princeton Pulsar Catalog (Taylor et al. 1993). Such a P–$\dot{P}$ diagram is extremely useful for classification purposes. The colored symbols represent those 35 pulsars which, by the end of 2000, had been detected at X-ray energies. Among them are the seven gamma-ray pulsars indicated by green symbols. The objects in the upper right corner represent the soft-gamma-ray repeaters (SGRs) and anomalous X-ray pulsars (AXPs) which have been suggested to be magnetars (described in a later section).

Although the magnetic braking model is generally accepted, the *observed* spin-modulated emission, which gave pulsars their name, is found to account for only a small fraction of $\dot{E}$. The efficiencies, $\eta = L/\dot{E}$, observed in the radio and optical bands are typically in the range $\sim 10^{-7} - 10^{-5}$, whereas they are $\sim 10^{-4} - 10^{-3}$ and $\sim 10^{-2} - 10^{-1}$ at X-ray and gamma-ray energies, respectively. It has therefore been a long-standing question how rotation-powered pulsars lose the bulk of their rotational energy.

The fact that the energy loss of rotation-powered pulsars cannot be fully accounted for by the magneto-dipole radiation is known from the investigation of the pulsar braking index, $n = 2 - P\ddot{P}\dot{P}^{-2}$. Pure dipole radiation would imply a braking index of $n = 3$, whereas the values observed so far are $n = 2.515 \pm 0.005$ for the Crab Pulsar (Lyne et al. 1988), $n = (2.8) \pm 0.2$ for PSR B1509$-$58 (Kaspi et al. 1994), $n = 2.28 \pm 0.02$ for PSR B0540$-$69 (Boyd et al. 1995), and $n = 1.4 \pm 0.2$ for the Vela Pulsar (Lyne et al. 1996). The deviation from $n = 3$ is usually taken as evidence that a significant fraction of the pulsar's rotational energy is carried off by a pulsar wind – that is, a mixture of charged particles and electromagnetic fields, which, if the conditions are appropriate, forms a pulsar-wind nebula (PWN) observable at optical, radio, and X-ray energies. Such PWNs (often called plerions or synchrotron nebulae) are known so far only for a few young and powerful (high $\dot{E}$) pulsars and for some center-filled supernova remnants, in which a young neutron star is expected, but only emission from its plerion is detected. The mechanisms of pulsar-wind generation and its interaction with the ambient medium are poorly understood.

Thus, the popular model of magnetic braking provides plausible estimates for the neutron star magnetic field $B_\perp$, its rotational energy loss $\dot{E}$, and characteristic age τ, but it does not provide any detailed information about the physical processes which operate in the pulsar magnetosphere and which are responsible for the broadband spectrum, from the radio to the X-ray and gamma-ray bands (Figure 16). As a consequence, there are several magnetospheric emission models, but no generally accepted theory.

High-energy emission models

Although rotation-powered pulsars are most widely known for their radio emission, the mechanism of the radio emission

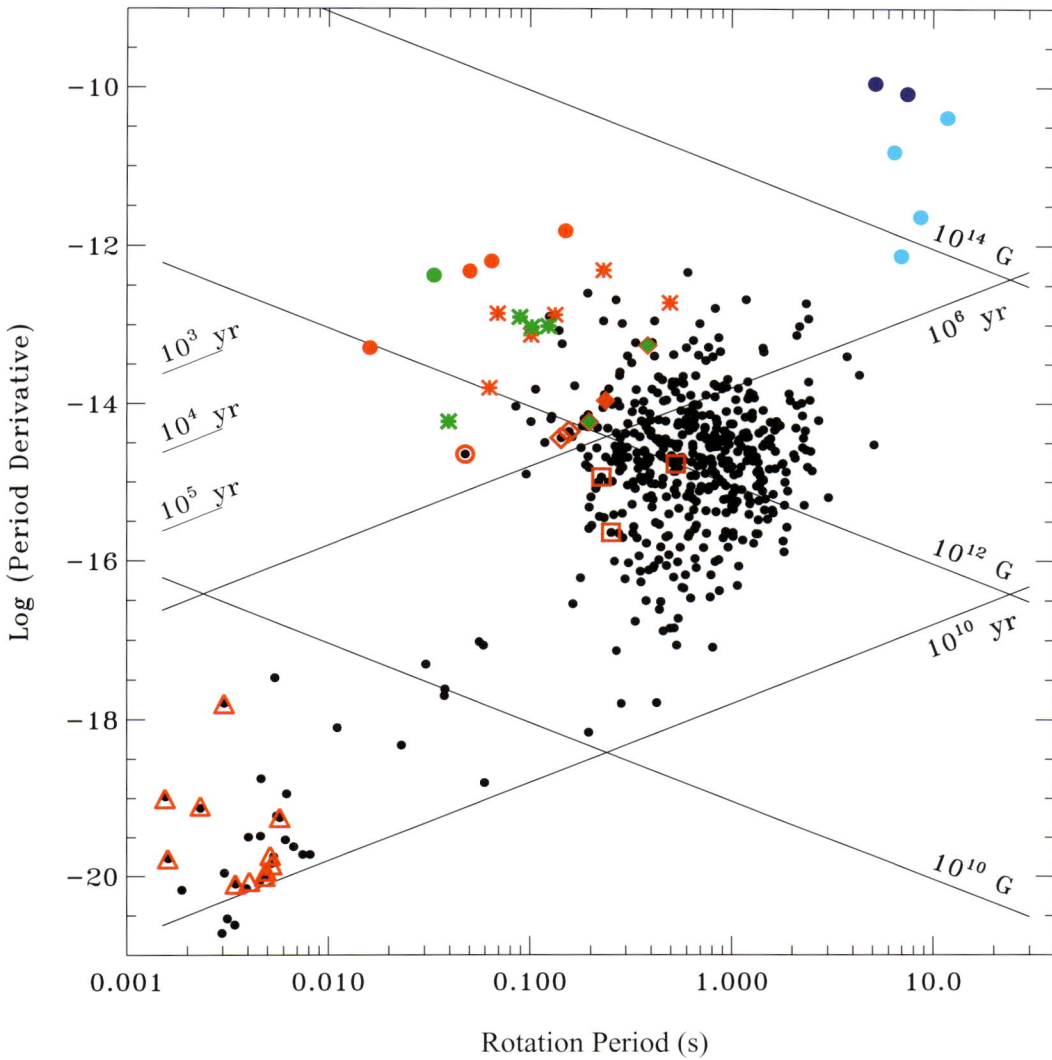

Figure 2 The $P-\dot{P}$ diagram – the distribution of rotation-powered pulsars (small black dots) over their spin parameters. The straight lines correspond to constant ages $\tau = P/(2\dot{P})$ and magnetic field strengths $B_\perp = 3.2 \times 10^{19}\, (\dot{P})$. Separate from the majority of ordinary-field pulsars are the millisecond pulsars in the lower left corner and the putative magnetars – soft gamma-ray repeaters (dark blue) and anomalous X-ray pulsars (light blue) at the upper right. Although magnetars and anomalous X-ray pulsars are not rotation powered, they are included in this plot as it brings out their estimated superstrong magnetic fields. X-ray detected pulsars are indicated by colored symbols. Green symbols indicate gamma-ray pulsars. (From Becker 2000b.)

is poorly understood. However, it is certainly different from those responsible for the high-energy (infrared through gamma-ray) radiation observed with space observatories. It is well known that the radio emission of pulsars is a coherent process, and the coherent curvature radiation has been proposed as the most promising mechanism (Michel 1991, and references therein). However, the optical, X-ray, and gamma-ray emission observed in pulsars must be incoherent. Therefore the fluxes in these energy bands are directly proportional to the densities of the radiating high-energy electrons in the acceleration regions, no matter which radiation process (synchrotron radiation, curvature radiation, or inverse Compton scattering) is at work at a given energy. High-energy observations thus provide the key to understanding the pulsar emission mechanisms. So far, the high-energy radiation detected from rotation-driven pulsars has been attributed to various thermal and non-thermal emission processes, including the following:

- Non-thermal emission from charged relativistic particles accelerated in the pulsar magnetosphere. As the energy distribution of these particles follows a power law, the emission is also characterized by power-law-like spectra in broad energy bands. The emitted radiation can be observed from the optical to the gamma-ray band.

- Extended emission from pulsar-driven synchrotron nebulae. Depending on the local conditions (density of the ambient interstellar medium), these nebulae can be observed from radio to hard X-ray energies.
- X-ray and gamma-ray emission from the interaction of relativistic pulsar winds with a close companion star or with the wind of a companion star, in binary systems (Arons and Tavani 1993).
- Photospheric emission from the hot surface of a cooling neutron star. In this case a modified black-body spectrum and smooth, low-amplitude intensity variations with the rotational period are expected, observable from the optical through the soft X-ray range (Greenstein and Hartke 1983, Romani 1987, Pavlov et al. 1995).
- Thermal soft X-ray emission from the neutron star's polar caps, which are heated by the bombardment of relativistic particles streaming back to the surface from the pulsar magnetosphere (Kundt and Schaaf 1993, Pavlov et al. 1994).

In the following subsections we briefly present the basic ideas on the magnetospheric emission models as well as material relevant to neutron star cooling and thermal emission from the neutron star surface.

Magnetospheric emission models

So far there is no consensus as to where the pulsar high-energy radiation comes from (e.g., Michel 1991, Beskin et al. 1993 and discussion therein). There are two main types of model – the polar cap models, which place the emission zone in the immediate vicinity of the neutron star's polar caps, and the outer gap models, in which this zone is assumed to be close to the pulsar's light cylinder (a virtual cylinder whose radius, $R_L = cP/(2\pi)$, is defined by the condition that the azimuthal velocity of the co-rotating magnetic field lines is equal to the speed of light) to prevent materializing of the photons by one-photon pair creation in the strong magnetic field, according to $\gamma + B \rightarrow e^+ + e^-$ (Figure 3). The gamma-ray emission in the polar cap models (Arons and Scharlemann 1979, Daugherty and Harding 1996, Sturner and Dermer 1994) forms a hollow cone centered on the magnetic pole, producing either double-peaked or single-peaked pulse profiles, depending on the line of sight. The outer gap model was originally proposed to explain the bright gamma-ray emission from the Crab and Vela Pulsars (Cheng et al. 1986a,b) as the

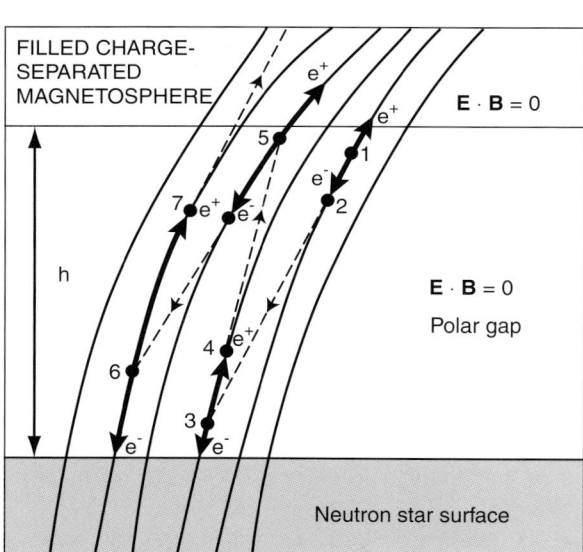 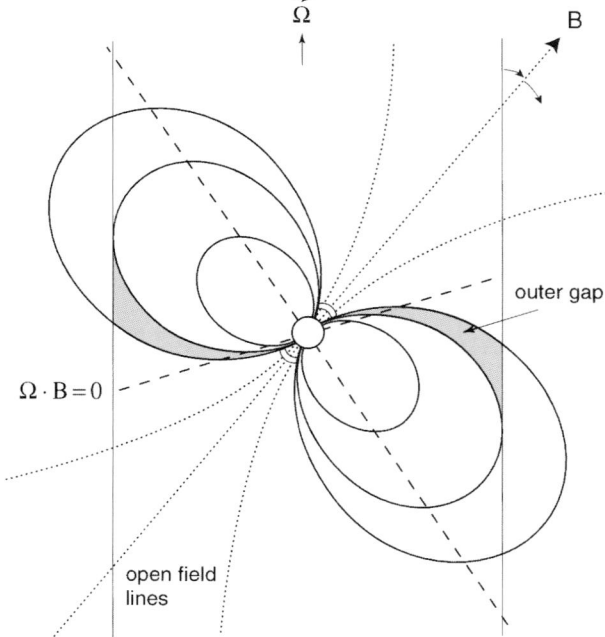

Figure 3 The geometry of the acceleration zones as defined in the polar cap model (left), according to Ruderman and Sutherland (1975), and the outer gap model (right), according to Cheng et al. (1986a,b). The polar cap model predicts "pencil" beams emitted by particles accelerated along the curved magnetic field lines. According to the outer gap model, the pulsar radiation is emitted in "fan" beams. Being broader, the latter can more easily explain two (and more) pulse components observed in several gamma-ray pulsars.

efficiency to get high-energy photons out of the high B-field regions close to the surface is rather small. Placing the gamma-ray emission zone at the light cylinder, where the magnetic field strength is reduced to $B_L = B(R/R_L)^3$, provides higher gamma-ray emissivities which are in somewhat better agreement with the observations. In both types of model, the high-energy radiation is emitted by relativistic particles accelerated in the very strong electric field, $\epsilon \sim (R/cP)B$, generated by the magnetic field co-rotating with the neutron star. These particles are generated in cascade (avalanche) processes in charge-free gaps, located either above the magnetic poles or at the light cylinder. The main photon emission mechanisms are synchrotron/curvature radiation and inverse Compton scattering of soft thermal X-ray photons emitted from the hot neutron star surface.

In recent years the polar cap and outer gap models have been further developed (e.g., Sturner et al. 1995, Harding and Muslimov 1998, Zhang and Harding 2000, Romani and Yadigaroglu 1995, Romani 1996), incorporating the new results on gamma-ray emission from pulsars obtained with the GRO. At present, the observational data can be interpreted within either model, albeit under quite different assumptions about pulsar parameters (e.g., on the direction of the magnetic and rotational axes). The critical observations needed to distinguish between the two models include measuring the relative phases between the peaks of the pulse profiles at different energies. We expect that multiwavelength timing of a large sample of pulsars with the aid of the Chandra, XMM-Newton, Integral, and the HST will resolve this problem in a few years.

Thermal evolution of neutron stars

A neutron star is formed at very high temperatures, $\sim 10^{11}$ K, in the imploding cores of supernova explosions. Much of the initial thermal energy is radiated away from the interior of the star by various processes of neutrino emission (mainly Urca processes and neutrino bremsstrahlung), leaving a day-old neutron star with an internal temperature of about 10^9–10^{10} K. After ~ 100 yr (a typical time for thermal relaxation), the star's interior (at densities $\rho > 10^{10}$ g cm^{-3}) becomes nearly isothermal, and the energy balance of the cooling neutron star is determined by the equation (e.g., Glen and Sutherland 1980)

$$C(T_i)\frac{dT_i}{dt} = -L_\nu(T_i) - L_\gamma(T_s) + \sum_k H_k$$

where T_i and T_s are the internal and surface temperatures, and $C(T_i)$ is the heat capacity of the neutron star. Neutron star cooling thus means a decrease of thermal energy, which is mainly stored in the stellar core, due to energy loss via neutrinos from the interior ($L_\nu = \int Q_\nu dV$, Q_ν is the neutrino emissivity) plus energy loss via thermal photons from the surface ($L_\gamma = 4\pi R^2 \sigma T_s^4$). The relationship between T_s and T_i is determined by the thermal insulation of the outer envelope ($\rho < 10^{10}$ g cm^{-3}), where the temperature gradient is formed. The results of model calculations, assuming that the outer envelope is composed of iron, can be fitted with the simple relation (Gudmundsson et al. 1983)

$$T_s = 3.1 \, (g/10^{14} \, \text{cm s}^{-2})^{1/4} \, (T_i/10^9 \, \text{K})^{0.549} \, \text{MK}$$

where g is the gravitational acceleration at the neutron star's surface, and 1 MK $= 1 \times 10^6$ K. The cooling rate might be reduced by heating mechanisms H_k, such as frictional heating of superfluid neutrons in the inner neutron star's crust or some exothermal nuclear reactions.

Neutrino emission from the neutron star's interior is the dominant cooling process for at least the first 10^5 years. After $\sim 10^6$ years, photon emission from the neutron star surface takes over as the main cooling mechanism. The thermal evolution of a neutron star after ~ 10–100 yr, when the neutron star has cooled to $T_s = 1.5$–3 MK, can follow either of two different scenarios, depending on the still poorly known properties of superdense matter (Figure 4). According to the so-called standard cooling scenario, the temperature decreases gradually, down to ~ 0.3–1 MK, by the end of the neutrino cooling era and then falls down exponentially, becoming lower than ~ 0.1 MK in $\sim 10^7$ yr. In this scenario, the main neutrino generation processes are the modified Urca reactions, $n + N \rightarrow p + N + e + \bar{\nu}_e$ and $p + N + e \rightarrow n + N + \nu_e$ where N is a nucleon (neutron or proton), needed to conserve the momentum of reacting particles. In the accelerated cooling scenarios, associated with higher central densities (up to 10^{15} g cm^{-3}) and/or exotic interior composition (e.g., pion condensation, quark–gluon plasma), a sharp drop of temperature, down to 0.3–0.5 MK, occurs at an age of ~ 10–100 yr, followed by a more gradual decrease, down to the same ~ 0.1 MK at $\sim 10^7$ yr. The faster cooling is caused by the direct Urca reactions, $n \rightarrow p + e + \bar{\nu}_e$ and $p + e \rightarrow n + \nu_e$, allowed at very high densities (Lattimer et al. 1991). Examples of standard and accelerated cooling curves are shown in Figure 4. The neutron star models used in these calculations are based on a "moderate" equation of state which opens the direct Urca process for $M > 1.35 M_\odot$; lower-mass stars undergo the standard cooling. Recent studies have shown that both standard and accelerated cooling can be substantially affected by nucleon superfluidity in the stellar interiors (see Tsuruta 1998 and Yakovlev et al. 1999 for comprehensive reviews). In particular, there can exist many cooling curves intermediate between those of the standard and accelerated scenarios, depending on the properties of nucleon superfluidity, which are also poorly known.

Thus, the thermal evolution of neutron stars between ~ 10 and $\sim 10^6$ yr is very sensitive to the composition and structure of their interiors, in particular to the equation of

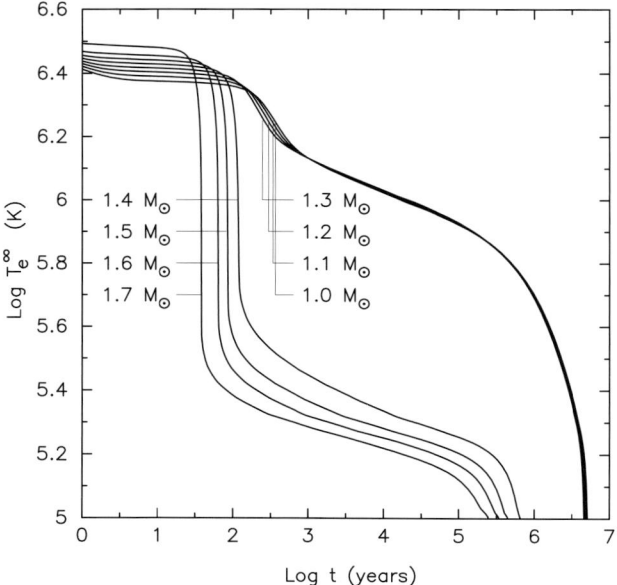

Figure 4 Fast cooling versus standard cooling for neutron stars with different masses. T_e^∞ is the effective surface temperature as observed at infinity (the gravitational redshift is taken into account) and t is the age. Higher-mass stars have a very high core density such that the direct Urca reactions (e.g., direct beta decay) are allowed. This causes a higher neutrino emissivity and hence a faster energy loss (more efficient cooling) by neutrino emission. The sharp temperature drop at an age of 50–100 yr represents the temperature inversion point. Here, the interior of the star, from which the neutrinos have escaped without interaction, is cooler than outer layers, which causes the outer regions to heat up the inner parts of the star. (From Page and Applegate 1992.)

state at super-nuclear densities. Therefore, measuring surface temperatures of neutron stars provides an important means of studying superdense matter. Since typical temperatures of such neutron stars correspond to the extreme UV–soft X-ray range, the thermal radiation from cooling neutron stars can be observed with X-ray detectors sufficiently sensitive at $E \lesssim 1$ keV.

Photospheric emission from cooling neutron stars

Thermal radiation has been observed from about a dozen isolated neutron stars. Much more detailed data on thermal radiation from these and other neutron stars are expected from the Chandra and XMM-Newton X-ray observatories. To interpret these observations, detailed and accurate models for spectra and light-curves of thermal radiation from neutron stars are needed. Properties of the neutron star thermal radiation are determined, as in normal stars, by a thin, partially ionized atmosphere with temperature growing inward. As a result, the neutron star's thermal radiation *may be* substantially different from black-body radiation (Pavlov and Shibanov 1978). Modeling of neutron star atmospheres requires a special approach because neutron stars possess very strong magnetic fields, $B \sim 10^{11}$–10^{13} G. In such fields the electron cyclotron energy, $E_{ce} = 11.6\ (B/10^{12}\,\text{G})$ keV, strongly exceeds the thermal energy, $kT \sim 0.01$–1 keV. As a result, the atmospheres are essentially anisotropic, so that the absorption and emission of photons depend on the direction of the photon wave-vector, and the radiation propagates there as two normal (polarization) modes with nearly orthogonal polarizations and quite different opacities (Gnedin and Pavlov 1974, Bulik and Pavlov 1996). The energy dependences of these opacities are substantially different from each other and from the opacity at $B = 0$. Since the ratio β of the cyclotron energy to the Coulomb energy is very large, (e.g., $\beta = E_{ce}/(Z^2\text{Ry}) = 850\,Z^{-2}(B/10^{12}\,\text{G})$ for one-electron ions; Ze is the ion charge, 1 Ry $= me^4/(2\hbar^2) = 13.6$ eV is the ionization potential of the hydrogen atom), the structure of atoms and ions is distorted by the strong magnetic field, which changes the energies and strengths of spectral features and ionization equilibrium of the atmospheric plasma. As a result, the spectrum, angular distribution, and polarization of the thermal radiation depend on the magnetic field.

Another important effect is that the nonuniform magnetic field leads to a nonuniform surface temperature distribution because of anisotropic heat conduction (Greenstein and Hartke 1983, Shibanov and Yakovlev 1996), which enhances pulsations of thermal radiation due to the neutron star rotation. The high density of the atmospheric matter ~ 1–100 g cm^{-3} at unit optical depth) caused by the immense gravitational acceleration ($g \sim 10^{14}$–10^{15} cm s^{-2}) poses additional complications. In particular, the non-ideality (pressure) effects lead to pressure ionization and smooth out the spectral dependences of the opacities. The huge surface gravity also leads to chemical stratification of neutron star atmospheres, so that the upper layers, which determine the properties of the emitted radiation, are comprised of the lightest element present. This means, in particular, that if a neutron star has accreted some amount of hydrogen (e.g., from the circumstellar medium or from the envelope ejected during the supernova explosion), its radiative properties are determined by the hydrogen atmosphere.

A convenient approach to modeling neutron star atmospheres was described by Pavlov et al. (1995). It includes, as for normal stars, solving of a set of equations for hydrostatic equilibrium, energy balance, ionization equilibrium, and radiative transfer, complemented by calculations of spectral opacities for partially ionized, non-ideal plasma. For atmospheres with strong magnetic fields, two coupled equations of radiative transfer for the intensities of two polarization modes have to be solved. The input parameters for the modeling are the chemical composition, effective temperature T_{eff} (or total radiative flux, $\propto T_{\text{eff}}^4$), magnetic field B (including the field orientation at the radiating neutron star's surface), and gravitational acceleration g (or the star's mass M and radius R).

Low-field neutron star atmospheres

It is commonly accepted that very old neutron stars, such as the 10^8–10^{10} years old millisecond pulsars, have "low" surface magnetic fields, $B \sim 10^8$–10^9 G, which do not affect the X-ray opacities of the atmospheric plasma at temperatures of interest (at $E_{ce} \ll E$, $E_{ce} \ll kT$, and $\beta \ll 1$). The first models of the low-field neutron star atmospheres were developed by Romani (1987). Further work (Rajagopal and Romani 1996, Zavlin et al. 1996) used improved opacities (Iglesias and Rogers 1996) for pure hydrogen, helium, and iron compositions. This work has shown that the spectra of radiation emerging from a light-element (H or He) atmosphere are much harder (less steep) than the black-body spectra at $E \gtrsim kT_{\text{eff}}$ (Figure 5). The reason for such behavior is that the hydrogen and helium opacities decrease with increasing E, so that the radiation of higher energies originates in deeper, hotter layers. As a result, fitting observed spectra with the standard black-body model yields spectral (black-body) temperatures exceeding the true effective temperatures by a factor of 1.5–3, which makes a great difference for the comparison with the models of neutron star cooling.

The spectra emitted from iron atmospheres are much more complex due to numerous spectral features produced by iron ions in various stages of ionization (Figure 5). Some of these features are observable even with moderate-resolution (e.g., CCD) spectrometers. However, when observed with very low energy resolution, the iron atmosphere spectra look very similar to the black-body spectra.

The local specific intensity of radiation decreases with the angle between the neutron surface and the wave-vector, and the shape of the angular distribution depends on photon energy and chemical composition. This (limb-darkening) effect must be taken into account in fitting both the spectra and the pulse profiles if the radiation is emitted from hot spots on the neutron stars's surface, as in millisecond pulsars (Pavlov and Zavlin 1997, Zavlin and Pavlov 1998).

High-field neutron star atmospheres

The first models of magnetic hydrogen atmospheres with $B \sim 10^{11}$–10^{13} G have been constructed recently (Shibanov et al. 1992; Pavlov et al. 1994, 1995; Zavlin et al. 1995a). These models are based on simplified opacities of strongly magnetized, partially ionized hydrogen plasma. These opacities do not include the bound–bound transitions; they neglect the motional Stark effect, and use a simplified model for the ionization equilibrium. Nevertheless, the models provide a qualitatively correct description of the magnetic effects on the emergent radiation, and they are accurate enough for high effective temperatures, $\gtrsim 1$ MK, when the hydrogen is almost completely ionized even in very strong magnetic fields.

Since the magnetic atmospheres are much more transparent in the "extraordinary" polarization mode, whose opacity is strongly reduced by the magnetic field, very deep (hot and dense) layers are responsible for the observed radiation. Their X-ray spectra are harder than the black-body spectrum at the same effective temperature, although not as

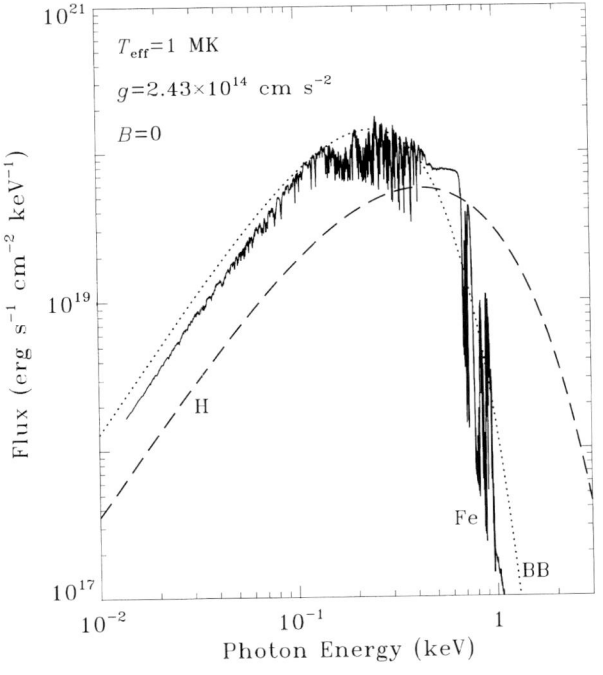
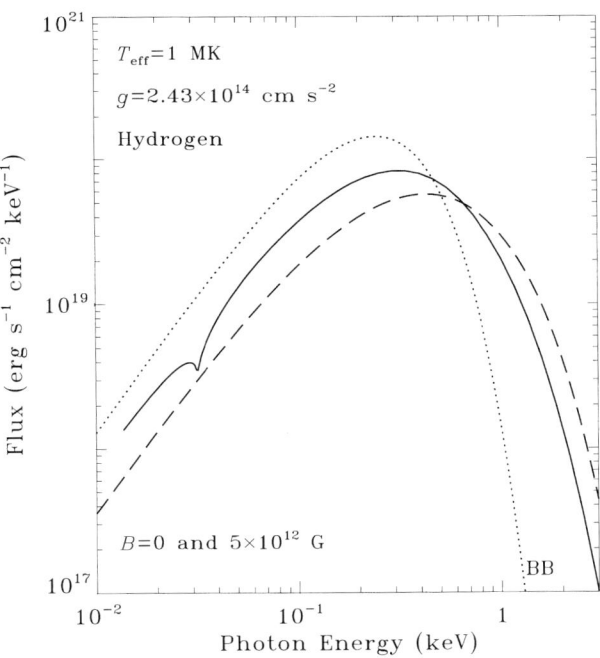

Figure 5 Left: simulated spectra of hydrogen and iron neutron star atmospheres with low magnetic field, together with the black-body (BB) spectrum, for $T_{\text{eff}} = 1$ MK. Right: the simulated spectra of hydrogen neutron star atmospheres with high and low magnetic fields (solid and dashed lines, respectively).

much as the low-field spectra. The only spectral line in the spectra of completely ionized hydrogen atmospheres is the proton cyclotron line at the energy $E_{cp} = (m_e/m_p)E_{ce} = 6.3\,(B/10^{12}\,\text{G})$ keV (Figure 5). The spectra depend not only on the strength, but also on the direction of the magnetic field, which means that the radiative flux emitted by a rotating neutron star is pulsed even if the surface temperature is uniform. The angular distribution of the local intensity shows a sharp peak along the magnetic field and a broader peak at intermediate angles (the "pencil" and "fan" components), the widths of the peaks depending on photon energy. This means that the pulse profiles of radiation emitted from hot polar caps may be much sharper than those emitted from low-field atmospheres. The pulse shape depends strongly on the mass-to-radius ratio due to bending of photon trajectories in the strong gravitational fields (Zavlin et al. 1995b, Shibanov et al. 1995). The radiation emitted from magnetic atmospheres is strongly polarized; the degree of polarization depends on E, B, and M/R (Pavlov and Zavlin 2000).

The first results obtained with the improved hydrogen atmosphere models (Pavlov and Zavlin 2001), which include the bound–bound transitions, show that spectral lines, considerably broadened by the motional Stark effect (Pavlov and Mészáros 1993, Pavlov and Potekhin 1995), become prominent at $T_{\text{eff}} \lesssim 0.5\,\text{MK}$. The strongest line is observed at

$$E \simeq \{75[1 + 0.13\,\ln(B/10^{13}\,\text{G})] + 63\,(B/10^{13}\,\text{G})\}\,\text{ev}$$

Magnetic iron atmosphere models have been considered by Rajagopal et al. (1997). Making use of the so-called adiabatic approximation ($\ln \beta \gg 1$), these authors calculated the energies and wavefunctions of the iron ions and the radiative opacities of the polarization modes. Although these models are inevitably rather crude, they provide a baseline for comparison with the magnetic hydrogen atmosphere models and for future work on heavy-element atmosphere modeling. Similar to the low-field case, the magnetic iron atmosphere spectra are fairly close to the black-body spectra when observed with low-resolution detectors. Developing more accurate iron atmosphere models is important for adequate interpretation of future high-resolution X-ray observations of thermal radiation from neutron stars.

THE CURRENT PICTURE OF HIGH-ENERGY EMISSION PROPERTIES OF ISOLATED NEUTRON STARS

As a result of observations with the satellite observatories ROSAT, EUVE, ASCA, BeppoSAX, RXTE, GRO, HST, Chandra, and XMM-Newton, the number of rotation-powered pulsars seen at X-ray, gamma-ray, and optical energies increased substantially in the last decade of the twentieth century. For the first time it became possible to carry out multi-wavelength studies of the pulsar emission. This is a big advantage as the physical processes that cause the emission in different wavelength bands are obviously related to one another. Although the quality of the data obtained with different instruments is inevitably rather inhomogeneous, and the conclusions drawn from these data are therefore not fully certain in many cases, there is a general consensus that a first big step toward discrimination between different emission scenarios has been made. In this respect, even more is expected from the new observatories, Chandra and XMM-Newton, launched at the twin of the century. Results from the first year of Chandra, which are briefly mentioned in this section, seem to justify these high expectations.

Young neutron stars in supernova remnants

X-ray observations allow us to find both SNRs and the compact objects that may reside within them. In fact, neutron stars and neutron star candidates have been found in a small fraction, <10%, of the 220 known galactic SNRs (Green 1998; Kaspi 2000, and references therein). Less than half of these objects are radio pulsars; the others are radio-silent (or, at least, radio-quiet) neutron stars which are seen only in X-rays (some of them in gamma rays). The young radio pulsars can be divided into two groups, Crab-like and Vela-like pulsars, according to somewhat different observational manifestations apparently associated with the evolution of pulsar properties with age. The radio-silent neutron stars include anomalous X-ray pulsars, soft gamma-ray repeaters, and "quiescent" neutron star candidates. These groups are briefly reviewed below.

Crab-like pulsars

It is well established that magnetospheric emission from charged particles, accelerated in a neutron star's magnetosphere along the curved magnetic field lines, dominates the radiation from young rotation-powered pulsars with ages of $\lesssim 5000$ years (see the previous section "High-energy emission models"). In the Crab Pulsar at least $\sim 75\%$ of the total soft X-ray flux is emitted from the co-rotating magnetosphere (Becker and Trümper 1997). Accordingly, its radiation is characterized by a power-law spectrum (the spectrum of the non-thermal radiation is a power law, $dN/dE \propto E^{-\alpha}$, as the energy distribution of the particles which emit this radiation follows a power law in a broad energy range), and its spin-modulated light-curve exhibits two narrow peaks per period (Figure 6). The Crab Pulsar is also a bright gamma-ray and optical–UV source. Its X-ray, gamma-ray, optical, and radio pulsations are all phase aligned, demonstrating that the emission in these bands is clearly nonthermal and originates

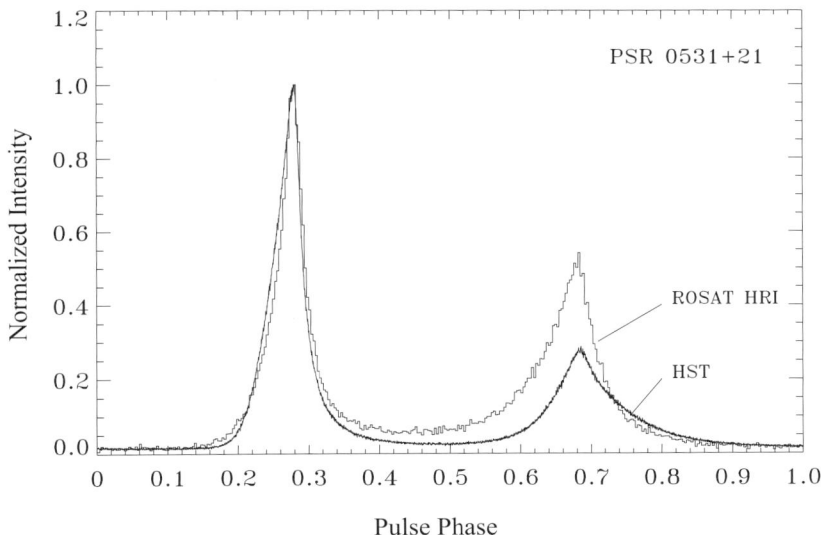

Figure 6 The pulse profile of the Crab Pulsar (PSR 0531+21) as observed with HST and ROSAT in the optical and soft X-ray bands. Its characteristic double-peaked shape is observed at all wavelengths. The phase difference between the first and second peaks shows a weak energy dependence.

from the same site in the pulsar magnetosphere. The slope of its flux spectrum slowly increases with photon energy – the photon index varies from $\alpha = 1.1$ at $E \sim 1\,\mathrm{eV}$ to $\alpha = 2.1$ at $E \sim 10^{10}\,\mathrm{eV}$.

As the Crab Pulsar is the youngest rotation-powered pulsar and thus should be the hottest neutron star, one could expect to observe its thermal surface emission at the off-pulse phases, when the thermal flux is not buried under the powerful magnetospheric emission. However, even the Einstein and ROSAT HRIs, despite their high angular resolution, were not able to completely remove a contribution from the compact synchrotron nebula around the pulsar (Figure 7), so that only an upper limit on the thermal flux has been established from the DC level of the soft X-ray pulse profile. Becker and Aschenbach (1995) found an upper limit of about 2 MK for the surface temperature of the Crab Pulsar from the ROSAT HRI observations, consistent with the predictions of standard cooling models.

The ROSAT HRI data taken from the Crab Nebula have been used to improve our understanding of this object in many respects. Greiveldinger and Aschenbach (1999), using HRI observations spanning a period of more than six years, have shown that the X-ray intensity of the inner synchrotron nebula varies on timescales of years by about 20%. The intensity variations are found to be confined to rather large ($\sim 25'' \times 25''$, or $0.25 \times 0.25\,\mathrm{pc}$) regions in the torus (its radius is $\sim 0.4\,\mathrm{pc}$). Using the instruments aboard Chandra, it will be easy to further investigate these long-term variations. The first images taken with the Advanced CCD Imaging Spectrometer (ACIS) aboard Chandra have already provided spectacular details of the inner nebula structure associated with the pulsar-wind outflow – in addition to the torus ($r \approx 0.38\,\mathrm{pc}$), the inner ring ($r \approx 0.14\,\mathrm{pc}$), jet, and counter-jet have been identified (Weisskopf et al. 2000). To demonstrate the recent progress in X-ray astronomy, images of the Crab Pulsar and its plerion, as seen by the Chandra ACIS and the ROSAT HRI, are shown in Figure 7.

Emission properties similar to those found for the Crab Pulsar are observed from the pulsars B0540−69, J0537−6909 and B1509−58 in the supernova remnants N158A, N157B, and MSH15−52 (the first two are in the Large Magellanic Cloud). In particular, PSR B0540−69 has a compact X-ray nebula strongly resembling that around the Crab Pulsar, and ~40% of the pulsar's soft X-ray photons are pulsed (Seward and Harnden 1994, Gotthelf and Wang 2000). This pulsar has been detected optically (Boyd et al. 1995, Hill et al. 1997) but not in gamma rays. As for the Crab Pulsar, its optical pulse profile is very similar to the profile observed at X-ray energies (Gouiffes et al. 1992, Mineo et al. 1999).

PSR B1509−58 has the highest period derivative of all known pulsars. Optical radiation from this pulsar has not yet been detected, and only upper limits have been obtained for its gamma radiation above ~10–30 MeV, suggesting a break in the gamma-ray spectrum somewhere between 10 and 100 MeV (Kuiper et al. 1999). Its X-ray emission in the ROSAT band is found to have a pulsed fraction of about 65% (Becker and Trümper 1997). The soft X-ray pulse is phase aligned with the hard X-ray and gamma-ray pulses detected by the GRO detectors BATSE, OSSE, and Comptel up to at least 10 MeV (Ulmer et al. 1994, Kuiper et al. 1999). These

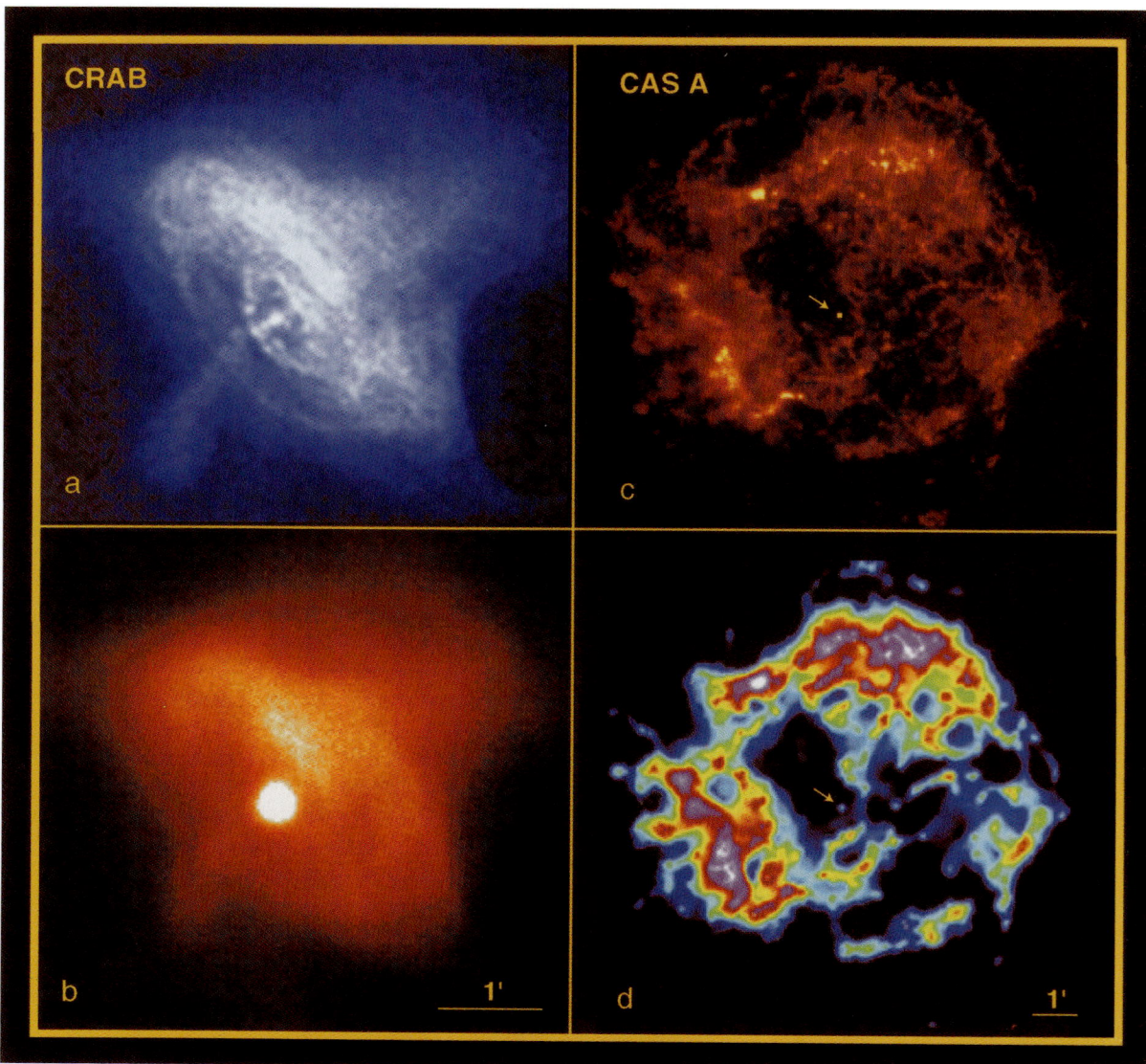

Figure 7 The Crab Pulser as observed with the Chandra ACIs (a) and the ROSAT HRI (b). The images demonstrate an improvement in angular resolution between the two detectors by a factor of 10. In the Chandra image much more detail of the pulsar-driven nebula is visible. Image (c) shows another recent discovery made by Chandra: the point source close to the geometrical center of Cas A, a very young (320 yr) supernova remnant; the coresponding ROSAT HRI image is shown in (d) the unprecedentedly high spatial resolution provided by Chandra enabled the point source, a young neutron star or a black hole to be detected; it was later in earlier deep Einstein HRI and ROSAT HRI images. (From Becker 2000a.)

high-energy pulses appear to be phase shifted by ∼0.3 periods relative to the radio pulse. Based on the ROSAT HRI observations, Brazier and Becker (1997) have proposed that the X-ray nebula surrounding PSR 1509−58 (Figure 8) consists of a torus and a jet, similar to the Crab synchrotron nebula. A nearby region of enhanced X-ray emission, RCW 89, may be caused by the collision of the collimated pulsar wind with the outer shell of the supernova remnant.

PSR J0537−6909 was discovered (Marshall *et al.* 1998) with RXTE and ASCA and later detected with ROSAT (Wang and Gotthelf 1998). It is particularly interesting because of its very short period of 16 ms, the shortest among the "regular" pulsars, despite the fact that it is older ($\tau \sim 5000$ yr) than the other three members of this subclass of pulsars. The ROSAT HRI image shows a bright X-ray nebula whose size (∼2 pc) and cometary shape indicate that the

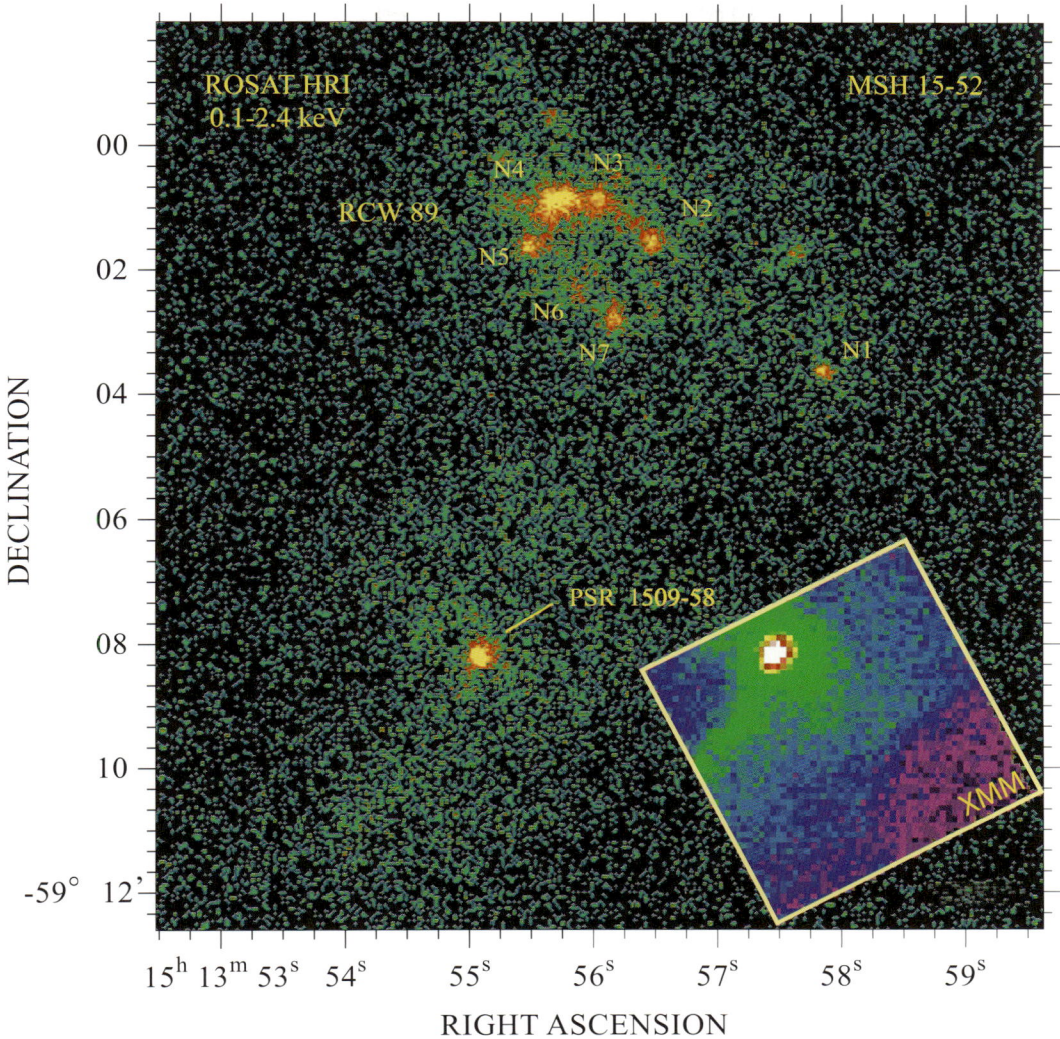

Figure 8 Soft X-ray image of MSH 15−52 and RCW 89 as seen by the ROSAT HRI. The most striking features are the compact knots in RCW 89 (only the brightest seven are labeled, N1–N7), the point source at the location of PSR B1509−58, the synchrotron nebula around PSR B1509−58, and the extended diffuse emission in RCW 89. The inset shows a 4′×4′ area around the pulsar as seen by the PN-camera aboard XMM-Newton. (From Becker 1998.)

pulsar is moving at supersonic velocity, $\sim 1000\,\mathrm{km\,s^{-1}}$, and the X-ray emission of the nebula originates mainly from a bow shock.

Thus all the very young pulsars show strong, nonthermal X-ray emission with an X-ray luminosity $L_x \sim 10^{34}$–$10^{36}\,\mathrm{erg\,s^{-1}}$ in the ROSAT energy range, and are surrounded by pulsar-powered nebulae (plerions) and supernova ejecta. Presumably their magnetospheric emission extends from at least infrared to gamma-ray energies, with typical photon indices varying between about 1 and 2 (about 1.4–1.7 in the soft X-ray range).

Vela-like pulsars

Pulsars with a spin-down age of $\sim 10^4$–10^5 yr are often referred to as Vela-like pulsars because of their apparent similar emission properties. About ten pulsars in this group have been detected in X-rays (Table 1), four of them (the Vela Pulsar B0833−45, PSR B1706−44, B1046−58, and B1951+32) are gamma-ray pulsars, and only the Vela Pulsar has been detected in the optical band. In some respects these objects appear to be different from the Crab-like pulsars. In particular, their pulses at different energies are not phase aligned with one another, their optical radiation

Table 1 Observational properties of X-ray detected radio-quiet neutron star candidates (NSCs). The individual columns are as follows. 1 Object (AXPs anomalous X-ray pulsars, SGRs soft gamma-ray repeaters). 2 Period and pulsed fraction. 3 Period derivative in s s^{-1}. 4 Age estimated as $P/2\dot{P}$ if the period and its derivative are known, or as SNR age if the NSC is associated with SNR. 5 Distance estimate. 6 Hydrogen column density. 7 Temperature and radius for a black-body fit. f_{bb} where present, is the fraction of the black-body flux in a black-body+power-law fit. 8 Photon power-law index. 9 Characteristic X-ray luminosity (persistent luminosity for SGRs. ^{v} indicates a variable source). 10 Optical magnitudes or limits. 11 Host SNRs. 12 References

Object	P/f_p s/%	$\dot{P}$ 10^{-11}	Age kyr	d kpc	N_H 10^{21} cm^{-2}	$T_{bb}/R_{bb}/f_{bb}$ MK/km/%	α	L_x 10^{33} erg s^{-1}	optical	SNR	Ref.
NSCs in SNRs											
CXO J2323+5848	.../<35	...	0.3	3.4	8–15	7/0.3	3.2	2	$R>25$	Cas A	1
1E 1613−5055	21,000/60(?)	...	1–3	3.3	4–10	6/0.5	~4	1–10^v	$R>24$	RCW 103	2
RX J0822−4300	.../<15	...	4	2.2	3–6	4/1	4.8	8	$R>24$	Puppis A	3
1E 1207.4−5209	0.42/10	...	3–20	~2	1–2	3/1.2	5.2	5	$V>23$	G296.5+10.0	4
RX J0002+6246	0.24/10(?)	...	~20	~3	~4	2/3	>4	2	$V>18$	G117.7+0.6(?)	5
AXPs											6
1E 1048−5937	6.45/70	~3	~3	~5	5	7/1/55	2.5	20^v	$V>20$	...	7
1E 1841−045	11.77/35	4	2–4	7	30	–	3.4	400	...	Kes 73	8
1RXS J1708−4009	11.00/30	1.9	9	8	14	5/3/17	2.9	~1000	$R>17$	...	9
1E 2259+586	6.98/30	0.048	3–20	~6	10	5/4/40	3.9	~100^v	$R>25$	G109.1−1.0	10
4U 0142+61	8.69/10	0.2	60	1–5	10	5/2.4/40	~4	80	$V>24$	...	11
AX J1845.0−0300	6.97/50	...	<8	8–15	46	8/1.5	–	50^v	...	G29.6+0.1	12
SGRs											13
SGR 0526−66	8/...	...	~0.5(?)	55	0.5	...	3.5	1000	$V>21$	N49 in LMC	14
SGR 1806−20	7.48/15	~10	1–10	~15	60	...	2.2	400	$K=8.4(?)$	G10.0−0.3(?)	15
SGR 1900+14	5.16/10	~10	~10	~7	10–20	...	1.5	60	$J>18$	G42.8+0.6	16
SGR 1627−41	...	...	~5	11 (?)	80	6/1/20	2.5	50	...	G337.0−0.1(?)	17
Isolated NSCs											18
RX J1856.5−3754	.../<10	...	1000(?)	<0.1	0.1–0.2	0.7/<20	–	<0.5	$V=25.7$	–	19
RX J0720.4−3125	8.39/10	<0.08	...	<0.5	0.1–0.2	0.9/<20	–	<0.6	$B=26.5(?)$	–	20
RX J0420.0−5022	22.7/40(?)	...	...	...	~0.2	0.7/...	–	...	$B>25$	–	21
RX J0806.4−4123	...	...	...	...	0.2–0.3	0.9/...	–	...	$B>24$	–	22
RX J1605.3+3549	...	...	...	...	~0.1	1.2/...	–	...	$B>26$	–	23
RX J1308.6+2127	...	...	...	...	0.1–0.2	1.4/...	–	...	$B>25$	–	24

References: 1 Caraveo et al. (1996); Kaspi (2000); Brazier and Johnston (1999), 2 Pavlov et al. (2000b); Chakrabarty et al. (2001), 3 Gotthelf et al. (1997); Gotthelf et al. (1999a); Garmire et al. (2000), 4 Petre et al. (1996); Zavlin et al. (1999), 5 Mereghetti et al. (1996); Zavlin et al. (2000); 6 Hailey and Craig (1995); 7 Mereghetti and Stella (1995); van Paradijs et al. (1995); Mereghetti (2000), 8 Seward et al. (1986); Oosterbroek et al. (1998); 9 Vasisht and Gotthelf (1997); Gotthelf et al. (1999b), 10 Sugizaki et al. (1997); Israel et al. (1999a); Kaspi et al. (1999), 11 Gregory and Fahlman (1980); Corbet et al. (1995); Rho and Petre (1997); Hulleman et al. (2000), 12 Israel et al. (1994); White et al. (1996); Wilson et al. (1999); Israel et al. (1999b), 13 Torii et al. (1998); Gotthelf and Vasisht (1998); Gaensler et al. (1999), 14 Mazets et al. (1979a); Rothschild et al. (1994); Danner et al. (1998), 15 Kulkarni and Frail (1993); Sonobe et al. (1994); Fenimore et al. (1994); van Kerkwijk et al. (1995); Corbel et al. (1997); Kouveliotou et al. (1998); Hurley et al. (1999a); Mazets et al. (1979b), 16 Hurley et al. (1999b), Murakami et al. (1999); Feroci et al. (1999); Woods et al. (1999a); Marsden et al. (1999); Woods et al. (1999b); Woods et al. (1999c), 17 Woods et al. (1999d); Corbel et al. (1999); Hurley et al. (2000), 18 Neuhäuser and Trümper (1999), 19 Walter et al. (1996); Pavlov et al. (1996b); Walter and Matthews (1997); Walter et al. (2000), 20 Haberl et al. (1997); Motch and Haberl (1998); Kulkarni and Van Kerkwijk (1998), 21 Haberl et al. (1999), 22 Haberl et al. (1998), 23 Motch et al. (1999), 24 Schwope et al. (1999).

is much fainter than that of the very young pulsars, and the overall shape of their high-energy spectra looks different. For instance, the closest ($d \approx 300$ pc) and hence best-investigated Vela Pulsar (Figures 9 and 10) has an optical luminosity four orders of magnitude lower than the Crab Pulsar (Manchester et al. 1978, Nasuti et al. 1997), whereas its rotation energy loss is only a factor of 65 lower. Its light-curve shows two peaks in the gamma-ray range (Kanbach et al. 1994) and at least three peaks in the X-ray range (Strickman et al. 1999, Pavlov et al. 2000a), compared with one peak at radio frequencies, whose phase does not coincide with any of the high-energy pulses. The pulsed fraction in the soft X-ray range, $\sim 12\%$, is much lower than that observed from the Crab-like pulsars.

In contrast to the young Crab-like pulsars, the soft X-ray spectrum of the Vela Pulsar has a substantial thermal contribution with an apparent temperature of ~ 1 MK (Ögelman et al. 1993, Page et al. 1996). However, the spatial structure of the Vela Pulsar's plerion strongly resembles the inner Crab Nebula – it also has a torus-like structure, an inner ring and jets (Figures 7a,b, and 10). The symmetry axis of the nebula, which can be interpreted as the projection of the pulsar's rotation axis onto the sky plane, is co-aligned with the direction of proper motion, exactly as for the Crab Pulsar, which indicates that the "natal kick" of the neutron star occurs along the rotation axis of the neutron star progenitor. The idea of torus configuration formed by a shock-confined pulsar wind was first introduced by Aschenbach and Brinkmann (1975) as a model to explain the shape of the inner Crab Nebula. The discovery of a similar torus-like structure in the Vela synchrotron nebula hints that this model may be applicable to many young pulsars. According to this model, the torus-like structure and its geometrical orientation with respect to the direction of the pulsar's proper motion arise because the interaction of the post-shock plasma with the ambient medium compresses the plasma and amplifies the magnetic field ahead of the moving pulsar. This, in turn, leads to enhanced synchrotron emission with the observed torus-like shape.

Typical sizes of the X-ray nebula structures scale approximately as $\dot{E}^{1/2}$, as one would expect for relativistic pulsar winds shocked by an ambient medium (Rees and Gunn 1974). For instance, the inner ring radii for the Crab and Vela Nebulae are 0.14 pc (for $d = 2$ kpc) and 0.02 pc (for $d = 300$ pc), whereas the full extents of the X-ray nebulae are 1 pc and 0.1 pc, respectively. The X-ray luminosity of the Vela plerion is only 0.04% (0.1–2.4 keV) of the pulsar's spin-down energy loss, versus 5%, 13%, and 1% for the Crab, B0540 − 69, and B1509 − 58 X-ray nebulae, respectively.

Since the other pulsars of this subclass are at much larger distances, it is hard to distinguish them from the putative surrounding X-ray nebulae. What has been observed is therefore mainly emission from a pulsar-powered synchrotron nebula combined with a small contribution of magnetospheric or thermal radiation. The latter is expected to dominate at soft X-ray energies, below 0.5–1 keV, hardly observable in distant pulsars because of interstellar absorption. The relatively small contribution of the pulsar's radiation, perhaps with an intrinsically low pulsed fraction, has precluded the detection of pulsed soft X-ray emission from these objects. Compact X-ray nebulae of physical sizes $\sim 0.3(d/2.4\,\text{kpc})$ pc, $0.4(d/4\,\text{kpc})$ pc, and $0.7(d/2.5\,\text{kpc})$ pc have been observed from PSR B1706 − 44 (Becker et al. 1995, Finley et al. 1998), B1823 − 13 (Finley et al. 1996), and B1951 + 32 (Safi-Harb et al. 1995), respectively. These sizes exceed that of the Vela X-ray nebula, despite close values of $\dot{E}$, which can be tentatively explained by the lower pressure of the ambient medium around these pulsars. It is also possible that future Chandra observations will reveal a fine spatial structure of these nebulae, which would lead to the explanation of the apparent difference from the compact nebula around the prototype Vela Pulsar.

Thus, in spite of the apparent differences between the Crab-like and Vela-like pulsars, the sample of well-investigated objects is still too small to determine whether these differences are caused by a general evolution of pulsar properties during the first millennia of their lives or whether they are due to some incidental properties inherent to the pulsars or their surroundings (e.g., different orientations of the magnetic and rotational axes, or different properties of the ambient medium). Most critical for understanding the nature of these objects will be Chandra observations at high angular resolution, as these observations will allow us to resolve the pulsars from their X-ray plerions.

Radio-silent neutron stars in supernova remnants

X-ray images of some young SNRs show bright point sources which have not been detected in the radio, optical, or gamma-ray bands (Table 2). The youngest among the detected sources of this type is the point source in the very young (320 yr) Cas A supernova remnant (Figure 7). This source was discovered in the first-light Chandra observation (Tananbaum 1999) and subsequently found in archival Einstein HRI and ROSAT HRI images (Aschenbach 1999, Pavlov and Zavlin 1999). The true nature of this source remains elusive (Pavlov et al. 2000b). It shows no long-term (20 yr) or short-term (days, months) variability, and no X-ray pulsations have been detected in the available data. The observed spectra do not have enough counts to distinguish between different simple spectral models (e.g., power law or black body, corresponding to a nonthermal or thermal origin of the detected emission). However, it turns out that the spectrum is much softer than those of young radio pulsars. If the emission occurs from the neutron star surface, the temperature distribution over the surface has to be

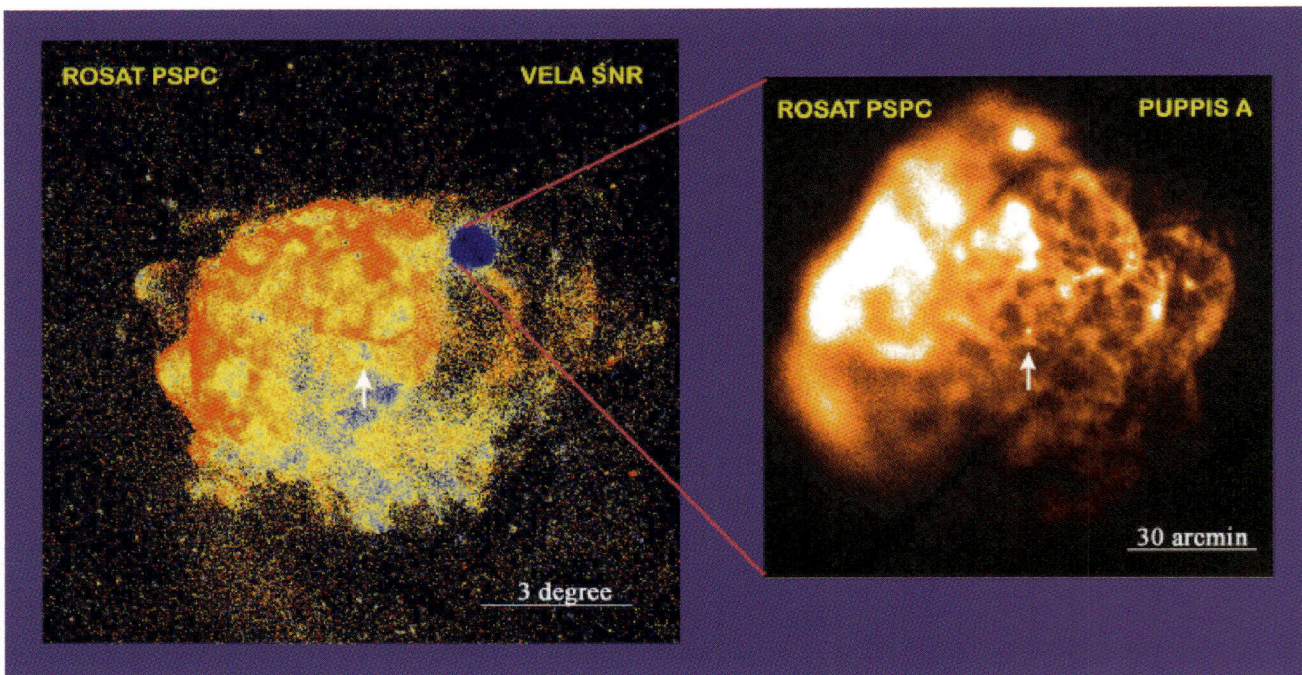

Figure 9 Left: A ROSAT PSPC image of the Vela SNR. Different colors correspond to different energies of X-ray photons, from red (lower energies) to blue (higher energies). The location of the Vela Pulsar is indicated by the arrow. Right: A zoomed image of the Puppis A SNR, located at the northwest edge of the Vela SNR. The arrow indicates the point source RX J0820−4300, a very good candidate for a young cooling neutron star showing photospheric emission (see the next section).

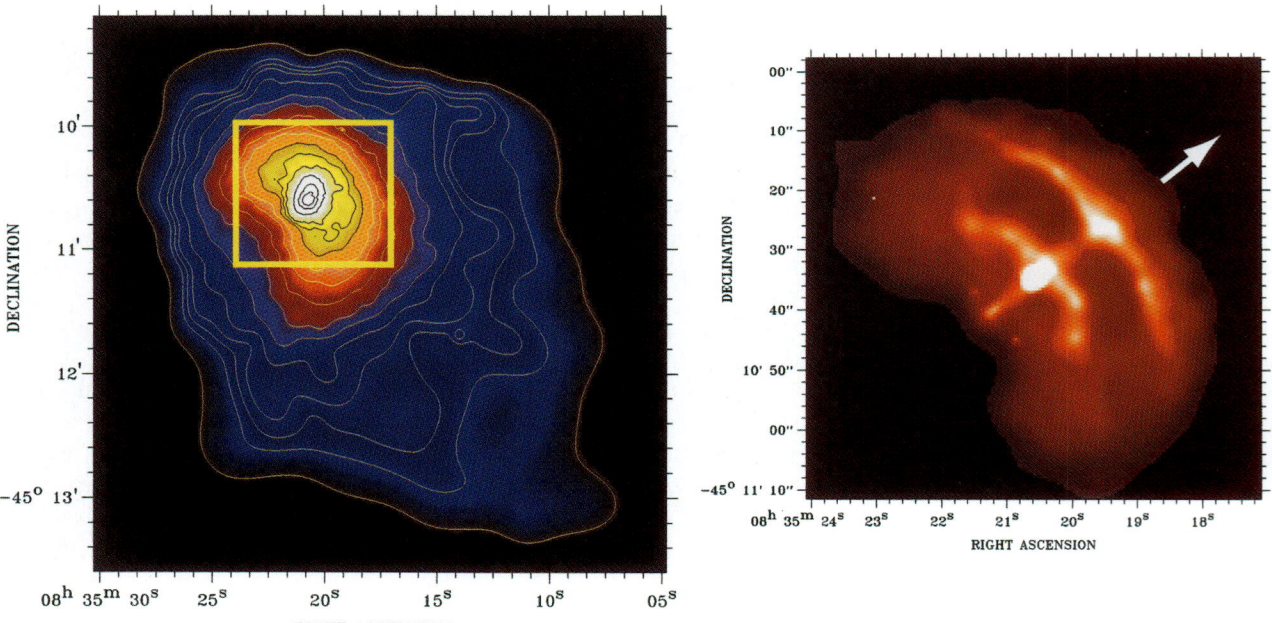

Figure 10 The Vela Pulsar and its X-ray plerion as observed with the ROSAT HRI (left) and Chandra ACIS (right). In both images the pulsar is the brightest source. The Chandra image shows the spatially resolved inner part of the plerion, corresponding to the box in the ROSAT image. The arrow indicates the direction of the pulsar's proper motion, which is aligned with its rotational axis. (From Becker 1999.)

Table 2 List of rotation-powered pulsars that have been detected in the radio, optical, X- and γ-ray wavebands, ordered according to their spin-down flux density at Earth, $\dot{E}/4\pi d^2$. The individual columns are as follows: 1. Pulsar name; 3–8. Energy ranges in which pulsed (p), unpulsed (d) radiation has been detected: R – radio, O – optical, X_s – soft X-rays ($E_\gamma \sim 1$ keV), X_h – hard X-rays ($E_\gamma \sim 10$ keV), γ_s – soft γ-rays ($E_\gamma \sim 1$ MeV), and γ_h – hard γ-rays ($E_\gamma > 100$ MeV). Possible detections are indicated by a question mark. $\dot{E}$ is the pulsar spin-down power/$I\Omega\dot\Omega$; L_x^{tot} the sum of the pulsed and unpulsed X-ray luminosities assuming isotropic emission. L_x^{puls} is the pulsed luminosity; L_x^{pn} is the total X-ray luminosity including synchrotron nebula emission. All luminosities L_x are calculated for the ROSAT energy range 0.1–2.4 keV. For Geminga, PSR 0656+14 and 1055–52 thermal and non-thermal contributions are included. L_{bol}^∞ is the bolometric luminosity. The upper limits have been computed for a neutron star with a medium stiff equation of state (FP – model, $M = 1.4 M_\odot$ and $R = 10.85$ km).

Characteristics of the optical, X-ray and gamma-ray detected rotation-powered pulsars

Pulsar	Comment	R	O	X_s	X_h	γ_s	γ_h	$\dot{E}/(4\pi d^2)$ (erg s^{-1} cm^{-2})	$\log \dot{E}$ (erg s^{-1})	$\log L_x^{tot}$ (erg s^{-1})	$\log L_x^{puls}$ (erg s^{-1})	$\log L_x^{pn}$ (erg s^{-1})	$\log L_{bol}^\infty$ (erg s^{-1})	$\log(P/2\dot{P})$ (yr)	P (ms)	$\dot{P} \times 10^{-15}$ (s s^{-1})	d (kpc)	$\log B_\perp$ (G)
B0531+21	Crab	p	p	p	p	p	p	$9.3 \cdot 10^{-7}$	38.65	35.98	35.85	37.3	≤34.1	3.10	33.40	420.96	2.00	12.58
B0833–45	Vela	p	p	p	p	p	p	$2.3 \cdot 10^{-7}$	36.84	32.25	31.25	32.95	~32.3	4.05	89.29	124.68	0.30	12.53
J0205+6449	in 3C58	—	—	p	—	—	—	$3.2 \cdot 10^{-8}$	37.42	32.20		34.31		3.73	65.67	193.52	2.60	12.55
J2229+6114	in G106.6+2.9	p	—	p	—	—	d	$2.0 \cdot 10^{-8}$	37.34	33.01				4.02	51.62	78.27	≥3	12.30
J1617–5055	near RCW 103	p	—	p	—	—	—	$1.3 \cdot 10^{-8}$	37.20					3.91	69.33	136.05	3.30	12.49
B0633+17	Geminga	p	p	p	—	—	p	$1.1 \cdot 10^{-8}$	34.51	31.10	30.62		31.2	5.53	237.09	10.97	0.16	12.21
B1706–44	in G343.1–02.3	p	—	d	—	—	p	$8.6 \cdot 10^{-9}$	36.53	33.15			≤33.5	4.24	102.45	93.04	1.82	12.49
B1509–58	in MSH 15–52	p	—	d	p	p	—	$7.7 \cdot 10^{-9}$	37.25	34.29	34.10	35.3	≤33.9	3.19	150.23	1540.19	4.30	13.19
B1951+32	in CTB 80	p	—	d	—	—	p	$5.0 \cdot 10^{-9}$	36.57	33.44		34.0	≤33.9	5.03	39.53	5.85	2.50	11.69
J1811–1926	in G11.2–0.3	—	—	p	—	—	—	$2.4 \cdot 10^{-9}$	36.80					4.38	64.67	44.00	≥5	12.23
B1046–58		p	—	d	—	—	p	$1.9 \cdot 10^{-9}$	36.30	≤32.11			≤32.7	4.31	123.65	95.92	2.98	12.54
B1259–63*	Vela twin Be-star/bin	p	—	d	d	d	—	$1.7 \cdot 10^{-9}$	35.92	32.95			≤33.8	5.52	47.76	2.27	2.00	11.51
J0537–6909	in N157B/LMC	—	—	p	—	—	—	$1.6 \cdot 10^{-9}$	38.68					3.71	16.11	51.24	49.4	11.96
J1420–6048	—	p	—	p	—	—	—	$1.4 \cdot 10^{-9}$	37.00	34.46				4.11	68.18	82.85	7.70	12.38
B1823–13	Vela like	p	—	d	—	—	—	$1.4 \cdot 10^{-9}$	36.45	33.39			≤33.9	4.33	101.45	74.95	4.12	12.45
B1800–21	G8.7–0.1	p	—	d	—	—	—	$1.2 \cdot 10^{-9}$	36.35	33.06			≤33.8	4.30	133.61	134.32	3.94	12.63
B1929+10		p	d	p	—	—	—	$1.1 \cdot 10^{-9}$	33.59	30.00	29.5		≤31.4	6.49	226.51	1.16	0.17	11.71
J0437–4715	ms-PSR	p	—	p	—	—	—	$1.0 \cdot 10^{-9}$	33.62	30.86	30.3		≤31.2	9.50	5.75	$2.0 \cdot 10^{-5}$	0.18	8.54
B1937+21	ms-PSR	p	—	p	—	—	—	$7.1 \cdot 10^{-10}$	36.04	≤32.10				8.37	1.55	$1.0 \cdot 10^{-4}$	3.60	8.61
B1821–24	ms-PSR, in M28	p	—	p	p	—	—	$6.2 \cdot 10^{-10}$	36.35	33.24			≤33.7	7.48	3.05	$1.6 \cdot 10^{-3}$	5.50	9.35
B0656+14	cooling NS	p	p	p	—	—	?	$5.5 \cdot 10^{-10}$	34.58	32.98	32.15		32.9	5.05	384.87	55.03	0.76	12.67
J0030+0451	ms-PSR	p	—	p	—	—	—	$5.4 \cdot 10^{-10}$	33.53	30.10				9.88	4.86	$1.0 \cdot 10^{-5}$	0.23	8.32
B0540–69	in N158A, LMC	p	p	p	p	—	—	$5.1 \cdot 10^{-10}$	38.17	36.21	36.1	37.2	≤36.1	3.22	50.37	479.06	49.4	12.70
J2124+3358	ms-PSR	p	—	d	—	—	—	$4.7 \cdot 10^{-10}$	33.15	30.35	29.8		≤31.1	9.86	4.93	$1.08 \cdot 10^{-5}$	0.25	8.36
J1105–6107		p	—	d	—	—	—	$4.3 \cdot 10^{-10}$	36.40					4.80	63.19	15.80	7.00	12.00

Table 2 (Continued)

Charactersitics of the optical, X-ray and gamma-ray detected rotation-powered pulsars

Pulsar	Comment	R	O	X_s	X_h	γ_s	γ_h	$\dot{E}/(4\pi d^2)$ (erg s^{-1} cm^{-2})	$\log \dot{E}$ (erg s^{-1})	$\log L_x^{tot}$ (erg s^{-1})	$\log L_x^{puls}$ (erg s^{-1})	$\log L_x^{pm}$ (erg s^{-1})	$\log L_{bol}^\infty$ (erg s^{-1})	$\log (P/2\dot{P})$ (yr)	P (ms)	$\dot{P} \times 10^{-15}$ (s s^{-1})	d (kpc)	$\log B_\perp$ (G)
B1957+20	ms-PSR	p	—	d	—	—	—	$4.1 \cdot 10^{-10}$	35.06	31.93			≤32.9	9.32	1.60	$1.2 \cdot 10^{-5}$	1.53	8.14
J1024−0719	ms-PSR	p	—	d	—	—	—	$3.6 \cdot 10^{-10}$	33.72	29.48				9.76	5.18	$1.8 \cdot 10^{-5}$	0.35	8.49
B0950+08		p	?	d	—	—	—	$3.3 \cdot 10^{-10}$	32.75	29.35			≤31.0	7.24	253.06	0.23	0.12	11.39
J1744−1134	ms-PSR	p	—	d	—	—	—	$2.4 \cdot 10^{-10}$	33.28	29.30				9.86	4.07	$0.86 \cdot 10^{-5}$	0.26	8.27
B1610−50		p	—	d	—	—	—	$2.5 \cdot 10^{-10}$	36.20					3.87	231.60	492.54	7.26	13.03
J0538+2817	in G180.0−1.7	p	—	d	—	—	—	$1.3 \cdot 10^{-10}$	34.69	32.74			≤33.6	5.79	143.15	3.66	1.50	11.87
J1012+5307	ms-PSR	p	—	d	—	—	—	$1.2 \cdot 10^{-10}$	33.60	30.20			≤30.8	9.76	5.25	$1.4 \cdot 10^{-5}$	0.52	8.45
B1055−52	cooling NS	p	d	p	—	—	p	$1.1 \cdot 10^{-10}$	34.48	33.42	32.57		33.5	5.73	197.10	5.83	1.53	12.03
B0355+54		p	—	d	—	—	—	$8.8 \cdot 10^{-11}$	34.66	31.96			≤33.8	5.75	156.38	4.39	2.07	11.92
B2334+61	G114.3+0.3	p	—	d	—	—	—	$8.6 \cdot 10^{-11}$	34.79	31.86			≤33.4	4.61	495.24	191.91	2.46	12.99
J0218+4232	ms-PSR	p	—	p	—	—	—	$6.5 \cdot 10^{-11}$	35.37	32.75			≤33.1	8.66	2.32	$8.0 \cdot 10^{-5}$	5.85	8.63
B0823+26		p	—	d	—	—	—	$2.6 \cdot 10^{-11}$	32.66	29.83			≤31.0	6.69	530.66	1.72	0.38	11.99
J0751+1807	ms-PSR	p	—	d	—	—	—	$1.5 \cdot 10^{-11}$	33.88	31.60			≤32.2	9.84	3.47	$8.0 \cdot 10^{-6}$	2.02	8.23

* PSR 1259−63 was observed ∼ 13° post-apastron.

Note: For references see Becker and Trümper (1997, 1999). Radio pulsar parameters have been taken from Taylor *et al.* (1995). Proper motion corrected period derivatives have been used for all those pulsars for which this effect is of significance and has been measured. A summary of the optical and γ-ray observations can be found in Caraveo (1995) and Thompson (1996), respectively.

strongly nonuniform. The black-body fit gives a temperature of 7 MK and a radius of the emitting region of 0.3 km. Assuming that there are magnetically confined hydrogen or helium hot polar caps on a cooler iron surface, Pavlov et al. (2000b) obtained 2.8 MK and 1 km for the cap temperature and radius, and 1.7 MK for the surface effective temperature. This anisotropic temperature distribution can cause a spin modulation of the X-ray flux, which remains to be detected in future observations.

A similar point source, 1E 1207−5209, at the center of the ∼7-kyr-old remnant PKS 1209−51/52 (Figure 11), was discovered with HEAO 1 (Tuohy et al. 1979) and studied with Einstein, Exosat, ROSAT, and ASCA (Helfand and Becker 1984, Kellett et al. 1987, Mereghetti et al. 1996). Its X-ray spectrum suggests that the X-radiation is emitted from a hydrogen or helium atmosphere at an effective temperature of 1.2–1.3 MK (Zavlin et al. 1998). The analysis of the Chandra observation of this source has shown that its X-ray flux is modulated with a 424 ms period (Zavlin et al. 2000), which finally proves that it is indeed a neutron star.

Another example of the radio-silent neutron star candidate is 1E 161348−5055 at the center of the SNR RCW 103. This source was discovered with the Einstein Observatory (Tuohy and Garmire 1980) and has an estimated age of 1–3 kyr. Its X-ray spectrum very strongly resembles that of the Cas A central point source. However, comparing two ASCA observations of RCW 103, Gotthelf et al. (1999a) found an order-of-magnitude decrease in its luminosity in four years, which hints that this object may be an accreting source. Even more puzzling is the 6-hour periodicity of its flux reported by Garmire et al. (2000) from the Chandra observations and archival ASCA data. Further investigations of 1E 161348−5055 with Chandra and XMM-Newton are under way, and will resolve the true nature of this source.

Similar to the previous examples is the point source in Puppis A (Figure 9), a SNR located at the edge of the Vela SNR. Puppis A has an age of about 4 kyr and harbors a central, radio-silent X-ray bright source, RX J0822−4300, which is supposed to be a neutron star candidate (Petre et al. 1996). Contrary to the compact stellar remnants in Cas A and RCW 103, its spectrum and luminosity can be interpreted as coming from the entire surface of a neutron star with a 10 km radius and a temperature of 1.6–1.9 MK, assuming that the surface is covered by a hydrogen or helium atmosphere (Zavlin et al. 1999). This temperature, like that inferred for 1E 1207−5209, is compatible with standard neutron star cooling models. It is worth noting that fitting the spectrum with a black-body model gives an improbably small neutron star radius of 1.0–1.5 km and a higher temperature 4–5 MK.

From what we know so far about radio-silent neutron stars in SNRs, we can conclude that such sources are quite different from radio pulsars (in particular, they do not show any activity inherent to radio pulsars). However, it is quite possible that they are more common than radio pulsars, and the relatively small number of known members of this class is due to observational selection – it is much easier to detect and identify active pulsars than these "quiet" sources observable only in the soft X-ray band.

For completeness, we should also mention a number of young SNRs whose central parts show bright, extended

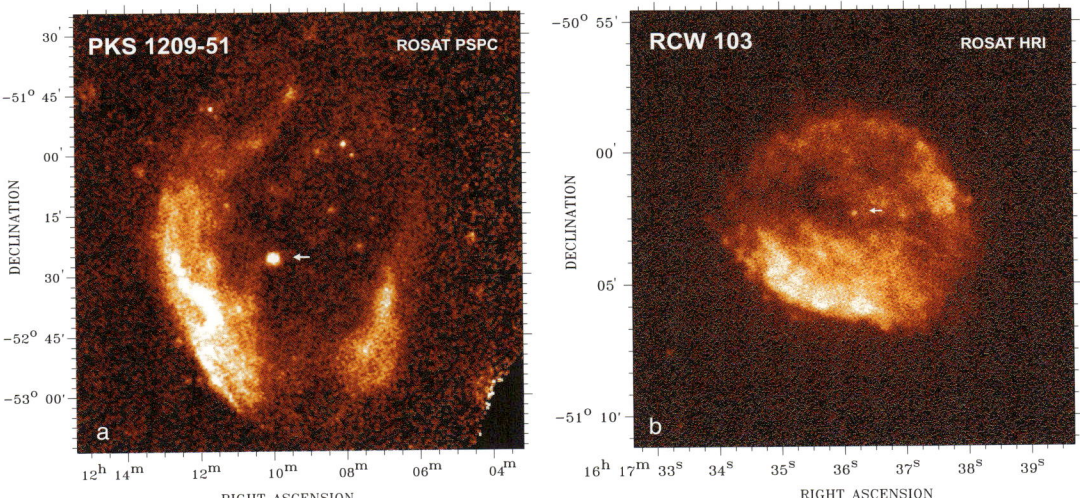

Figure 11 ROSAT images of the SNRs PKS 1209−51/52 and RCW 103. The arrows indicate the positions of the neutron star 1E 1207−5209 and the neutron star candidate 1E 161348−5055. Note the different scales of the images. PKS 1209−51/52 has an extent of about 1.5°, whereas the size of RCW 103 is about 10 arcmin. (From Becker and Aschenbach 2001.)

(plerion-like) X-ray sources with centrally peaked emission, with properties strongly resembling those observed from the plerions around Crab-like pulsars, but for which no point source has been detected. It seems very plausible that these SNRs do have active pulsars at the centers of their plerions, but an unfavorable direction of the pulsar beam precludes the detection of pulsations.

Anomalous X-ray pulsars and soft gamma-ray repeaters

In addition to the above-discussed young, X-ray-bright, and radio-quiet neutron star candidates which do not show strong pulsations, a number of apparently young neutron stars with strong X-ray pulsations, the so-called anomalous X-ray pulsars (AXPs) and soft gamma-ray repeaters (SGRs), have been discovered recently (see Table 2 for references). At least some of them are believed to be associated with SNRs. A common property of these objects is that their periods are in a narrow range of 5–12 s, substantially exceeding typical periods of radio pulsars. The AXPs and SGRs are, however, strongly different in their gamma-ray activity. While no gamma-ray emission has been detected from AXPs, SGRs emit occasional gamma-ray bursts of enormous energy, up to $10^{42}-10^{44}$ erg.

Six AXPs had been discovered by the end of the twentieth century (Table 2). They form a homogeneous class of pulsating neutron stars, clearly different from both the accreting pulsars in X-ray binaries and rotation-powered radio pulsars (Mereghetti 2000). AXPs show a relatively stable period evolution with $\dot{P} \approx (0.05-4) \times 10^{-11}$ s s^{-1}. Characteristic spin-down ages of $\tau \sim 3-100$ kyr and magnetic fields of $B \sim 10^{14}-10^{16}$ G were estimated under the assumption that the spin-down is due to magneto-dipole braking, which is not necessarily correct because these objects are not powered by their rotation. If the estimated magnetic field strengths are correct, they strongly exceed those of radio pulsars, so it has been suggested that AXPs, as well as SGRs, are magnetars – neutron stars with superstrong magnetic fields (Thompson and Duncan 1995, 1996). They have soft X-ray spectra, with characteristic black-body temperatures of $T \approx 4-7$ MK and/or power-law indices $\alpha \approx 2.5-4$, and typical luminosities of $L_x \sim 10^{34}-10^{36}$ erg s^{-1}. Typical black-body areas are one to two orders of magnitude smaller than the neutron star's surface area. At least three AXPs are associated with SNRs (Table 2).

AXPs have been studied with many X-ray observatories, but their nature remains elusive. Although it has been widely accepted that these objects are magnetars, no direct proof of their superstrong magnetic fields has been obtained. It is not clear whether their X-ray emission indeed originates from the neutron star's surface, or from the magnetosphere and/or from a synchrotron nebula, and whether it is due to "internal" radiation mechanisms (thermal or magnetospheric emission) or due to accretion from, for example, a residual disk (van Paradijs *et al.* 1995). Observations with Chandra and XMM-Newton will probably provide the answer.

Soft gamma-ray repeaters (SGRs) are among the most fascinating galactic objects. After the discovery of periods in the range $P = 5-8$ s and period derivatives of $\dot{P} \sim 10^{-10}$ s s^{-1} (in two of the four known SGRs), it has been suggested that these sources are associated with young, $\sim 1-10$ kyr, neutron stars in SNRs. The energy released during the most powerful bursts of these sources is enormous – for example, an energy of $\gtrsim 10^{44}$ erg was estimated for the 27 August 1998 outburst of SGR 1900+14 (Inan *et al.* 1999).

SGRs are not only extremely powerful sources of gamma-ray bursts, but are also bright, quiescent X-ray sources, with typical luminosities $L_x \sim 10^{34}-10^{36}$ erg s^{-1}. The origin of the quiescent radiation remains unclear. Statistically acceptable fits of the quiescent spectra can be obtained with a combined black-body plus power-law model, with typical parameters $T \sim 5$ MK, $R \sim 1$ km, and a photon index of $\alpha = 1-4$. The black-body component might be interpreted as thermal radiation from the neutron star's surface, but the area of the emitting region is two orders of magnitude smaller than the neutron star's surface area. The power-law component might be hinting that ultra-relativistic particles are involved, but no models have been put forward to explain their origin and acceleration mechanisms. However, with the poor angular resolution of the ASCA telescopes, it is difficult to separate the point source radiation from the diffuse SNR radiation, whereas the ROSAT count rates are too low for a precise spectral analysis. Chandra and XMM-Newton observations will yield much more definitive results, and will reveal the nature of the quiescent emission from SGRs and elucidate the properties of the ultra-magnetized neutron stars apparently responsible for their radiation.

Thermal emission from middle-aged pulsars

As discussed above, soft X-radiation from rotation-powered pulsars in an age interval of $\sim 10^5-10^6$ yr should be dominated by thermal emission from the neutron star's surface. These pulsars are old enough for their magnetospheric emission to become fainter than the thermal surface emission, but they are still young and hot enough to be detectable in the soft X-ray range. There are three middle-aged pulsars, Geminga, PSR B0656+14, and B1055−52, from which thermal X-radiation from the surface of the cooling neutron star has certainly been observed. Because of the similarity of their emission properties, they were dubbed the "three Musketeers" (Becker and Trümper 1997). The high-energy (IR through gammaray) spectra of these pulsars consist of two components. The thermal component dominates in the UV through soft X-ray range (up to 1–2 keV), whereas the nonthermal component with an approximately power-law (PL) spectrum prevails in IR, optical, hard X-ray, and gamma-ray ranges. It follows from the ROSAT and ASCA observations of the brightest middle-aged pulsar, B0656+14, that the thermal component cannot be described by a single temperature: that is the neutron star's surface temperature is not uniform (Greiveldinger et al. 1996; Zavlin et al. 2001). In the simplest model, the thermal component is comprised of a soft thermal component (TS) from most of the neutron star's surface (at $E \lesssim 0.5-1\,\mathrm{keV}$) and a hard thermal component (TH) from polar caps heated by relativistic particles. An example of a TS+TH+PL fit, to the IR–optical–X-ray spectrum of PSR B0656+14, is shown in Figure 12. Alternatively, the temperature nonuniformity can be due to anisotropic heat conductivity of the neutron star's crust caused by an anisotropic magnetic field – the heat flux across the field is suppressed, so that the magnetic poles are hotter than the equator (Greenstein and Hartke 1983).

The other two pulsars are not as bright as B0656+14, and their thermal components can be fitted with a single-temperature model (Figure 13 shows the X-ray spectrum of Geminga).

The existence of at least two spectral components is also confirmed by a phase-resolved analysis of the X-ray emission (Ögelman 1995). All three pulsars show a phase shift of $\sim 100°$ at an energy of 0.4–0.6 keV, accompanied by an increase in the pulsed fraction from $\sim 10-30\%$ to $\sim 20-65\%$. The X-ray pulse profiles for both the thermal and nonthermal components are found to be approximately sinusoidal. The weak modulation of the thermal soft component can be explained by the above-mentioned nonuniformity of the surface temperature due to the presence of a strong magnetic field. The surface temperatures of the three pulsars, obtained from black-body fits, are in the range $T \sim 0.3-1.2\,\mathrm{MK}$. The radii of the emitting areas cannot be found without knowing the distances to these objects. Adopting the distances estimated from the radio-pulsar dispersion measure (which may be off by a factor of ~ 2), the radii are in the range $R \sim 7-30\,\mathrm{km}$, in rough agreement with the canonical neutron star radius of 10 km. The hard X-ray spectral components, dominating at energies above $\sim 1-2\,\mathrm{keV}$, can be interpreted as magnetospheric emission (Halpern and Wang 1997, Wang et al. 1998, Greiveldinger et al. 1996, Zavlin et al. 2001).

It should be stressed that the inferred effective temperatures, and hence the radius-to-distance ratios, depend on the chosen model of the thermal component. For instance, if one assumes that the neutron star's surface is covered by a hydrogen or helium atmosphere, the effective temperatures

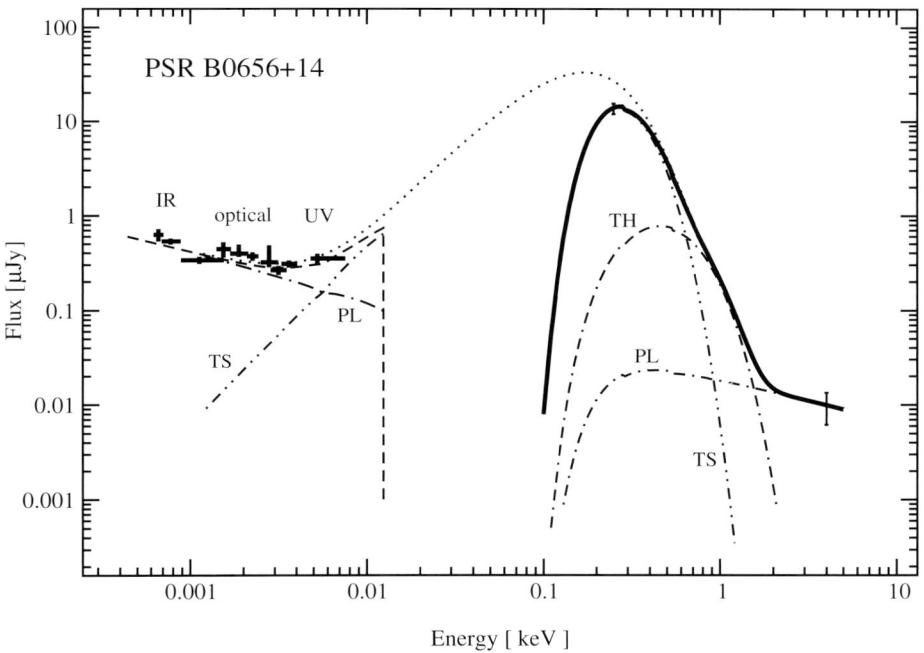

Figure 12 The energy spectrum of PSR B0656+14, a prototype middle-aged pulsar with thermal radiation dominating in soft X-rays. Shown are the X-ray (ROSAT and ASCA) spectra fitted with a model consisting of thermal soft (TS), thermal hard (TH), and power-law (PL) components, and IR–optical–UV fluxes measured with the HST and ground-based telescopes. The error bars in the X-ray range show typical uncertainties in the ROSAT and ASCA bands. The dashed and dotted lines show the continuation of the X-ray spectrum to the optical band, with and without allowance for interstellar absorption.

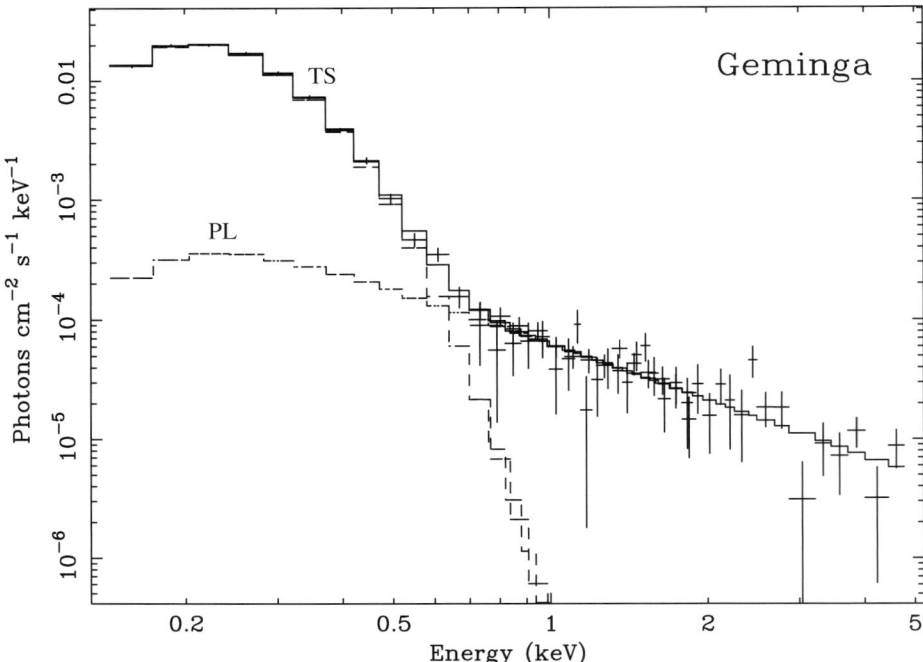

Figure 13 The X-ray spectrum of Geminga – a typical spectrum of a middle-aged pulsar, consisting of a thermal component interpreted as emission from the neutron star's surface, and a harder power-law (nonthermal) component that dominates beyond ~0.7 keV. The soft part of the spectrum was obtained with ROSAT; the harder emission was observed by ASCA. (From Halpern and Wang 1997.)

are lower than those derived from the simple black-body fits by a factor of 1.5–3 (Pavlov et al. 1995; see also the previous section "Photospheric emssion from cooling neutron stars"). An example of the difference of the temperatures inferred for the black-body and hydrogen atmosphere model fits is shown in Figure 14. The different spectral models correspond to quite different cooling scenarios, and hence to different properties of the neutron star interiors. Heavy-element atmospheres give temperatures close to the black-body temperatures. However, the heavy-element atmosphere spectra should show numerous absorption lines and photoionization edges (Rajagopal and Romani 1996, Zavlin et al. 1996). The low energy resolution of the ROSAT PSPC and the low sensitivity of the ASCA SIS in soft X-rays, has made it impossible to detect such lines and thus to determine the chemical composition of neutron star atmospheres. We hope that this problem will be solved with Chandra and XMM-Newton. Without knowing the surface chemical composition, any conclusions about the effective temperatures and radii should be treated with caution.

Important information on the emission mechanisms of middle-aged pulsars can be obtained from observations in the optical and gamma-ray ranges. PSR B0656+14 and Geminga have been observed at near-IR, optical, and near-UV frequencies (Pavlov et al. 1996a, Bignami et al. 1996, Shearer et al. 1996, Pavlov et al. 1997, Koptsevich et al. 2000), and PSR B1055−52 has been detected in a near-UV band (Mignani et al. 1997). For all three pulsars the IR–optical flux is clearly nonthermal, while the thermal component starts to dominate at UV frequencies (Figure 12). For Geminga a broad optical emission feature at ~6000 Å was reported by Bignami et al. (1996), who attributed it to proton cyclotron emission from an atmospheric plasma. This interpretation does not look plausible because it requires an artificial velocity distribution for atmospheric electrons to explain the lack of electron cyclotron lines in the hard X-ray spectrum. The nature of this feature, and the overall optical spectrum of Geminga, can hardly be understood without spectroscopic observations (Martin et al. 1998).

The two older middle-aged pulsars, Geminga and B1055−52, are bright gamma-ray pulsars in the GRO EGRET energy range, 30 MeV to 20 GeV, from which comes the main contribution to their photon luminosity (Thompson et al. 1999, and references therein). Gamma-ray emission from B0656+14 has been marginally detected, at a 3σ level (Ramanamurthy et al. 1996). The gamma-ray spectra are close to power laws, with photon indices of about 1.4–1.8 (Figure 15). A spectral turnover at about 3 GeV has been observed in the Geminga spectrum. The data can be interpreted within both the polar cap and outer gap models (see the previous section "Magnetospheric emission models"). Observations in a broader energy range

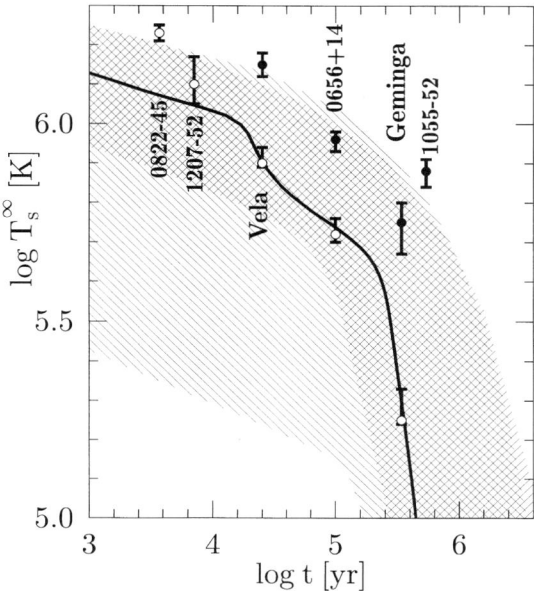

Figure 14 Surface temperatures for the "Three Musketeers" (PSR B0656+14, Geminga, and B1055−52), the Vela Pulsar, and the radio-quiet neutrons stars RX J0822−4300 and 1E 1207−52. The hatched regions indicate the possible ranges of T_s^∞ as predicted by the standard (double-hatched) and accelerated (single-hatched) cooling models for different critical temperatures of the superfluid neutrons and protons, T_{cn} and T_{cp} (∼10^6–10^{10} K). The solid line shows the standard cooling curve for a 1.30 $M_\odot$. neutron star with $T_{cn} = 200$ MK and $T_{cp} = 130$ MK. Filled and open circles indicate temperatures obtained from the black-body and hydrogen-atmosphere fits, respectively. (From Yakovlev et al. 1999.)

with more sensitive gamma-ray detectors are required to construct a detailed model of the gamma radiation.

Since all active pulsars are powerful sources of relativistic winds, one should expect them to generate pulsar-wind nebulae (PWNs), similar to those observed around the Crab-like and Vela-like pulsars. The PWN sizes should scale as $(\dot{E}/p_0)^{1/2}$, where p_0 is the pressure of the ambient medium. The existence of X-ray-bright PWNs (albeit of much larger sizes) around several pulsars, including the "Three Musketeers," was reported by Kawai and Tamura (1996) based on ASCA observations. However, the analysis of the ROSAT and BeppoSAX observations of these sources by Becker et al. (1999) did not confirm the ASCA results – the extended emission observed with ASCA was resolved in a number of unrelated background objects. In particular, Geminga and PSR B0656 + 14 are located in the Monogem Ring (Figure 16), a ∼20°-wide object which is believed to be an old and nearby SNR (Plucinsky et al. 1996). A large fraction of the sources detected by ASCA are found to be diffuse and fuzzy emission from a small part of the Monogem Ring rather than pulsar-powered nebulae.

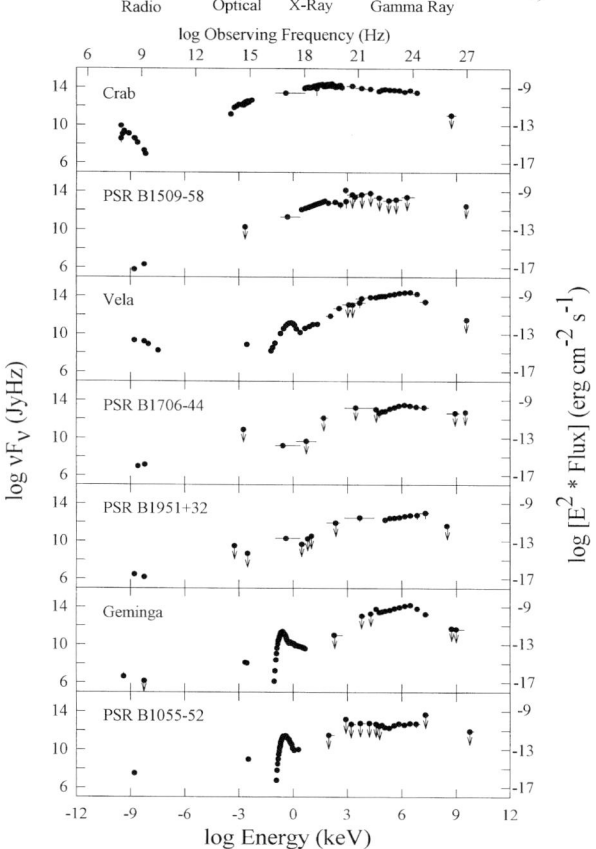

Figure 15 Multi-wavelength spectra for the known gamma-ray pulsars, showing the observed power per logarithmic energy interval. For all of the spectra the high-energy radiation power rises from the optical to the X-ray band, and the maximum observed energy output is in the gamma-ray band, which demonstrates that emitting particles are accelerated to very high energies. (From Thompson et al. 1999.)

In addition to PSR B0656+14, B1055−52, and Geminga, one could expect thermal radiation from the cooling neutron star surface to dominate in soft X-ray emission from two more middle-aged pulsars detected with ROSAT; PSR B0538+28 and B0355+54. Both have spin parameters similar to those observed for Geminga and B1055−52, and both appear to be good candidates for gamma-ray pulsars. However, these sources are approximately ten times as distant as Geminga, so the limited photon statistics has hampered a spectral or temporal analysis.

Old nearby radio pulsars

When the age of a neutron star reaches ∼10^6 yr, its temperature becomes too low for it to be observed in X-rays. At the same time, the energy loss rate $\dot{E}$, and hence the luminosity of nonthermal radiation and thermal radiation from

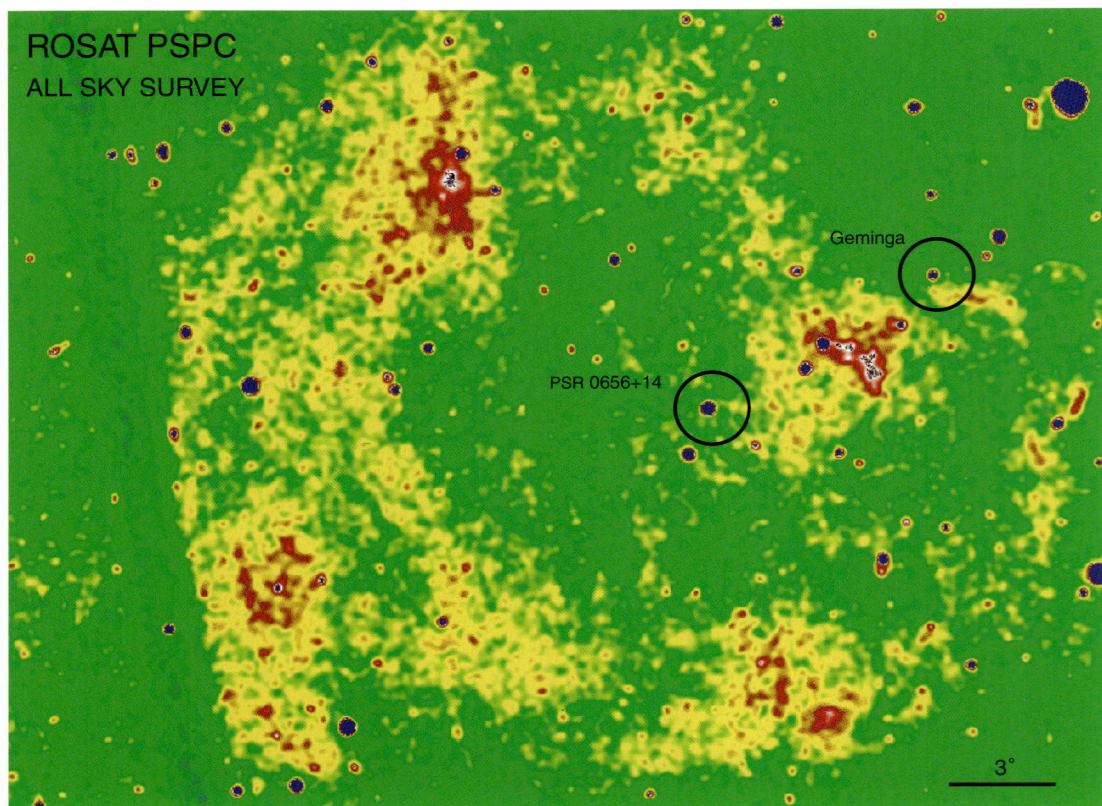

Figure 16 The 20°-wide Monogem Ring as observed in the ROSAT all-sky survey. The ROSAT PSPC full field of view during the pointed observations of Geminga and PSR 0656+14 are indicated by circles. The image demonstrates that both pulsars are located in crowded regions with patchy background emission, strongly blurred with an ASCA spatial resolution of ~3 arcmin. The image demonstrates the power of the first all-sky survey with an imaging X-ray telescope, providing X-ray images of extended celestial objects of very large sizes.

polar caps of radio pulsars, also decrease with age. Therefore, old pulsars are faint in the X-ray range and can be observed only at small distances. ROSAT and ASCA have detected X-ray emission from three old and close pulsars: PSR B1929+10, B0950+08, and B0823+26. All three are characterized by a spin-down age of 2–30 Myr and are at distances of ~0.2–0.4 kpc. Temporal and spectral information, however, is available only for PSR B1929+10 (Yancopoulos et al. 1994), whereas for the other two pulsars the sensitivity of ROSAT and ASCA was not sufficient to collect enough photons for a detailed analysis. The pulse profile of PSR B1929+10 is very broad, with a single pulse stretching across almost the entire phase cycle. Becker and Trümper (1997) and Wang and Halpern (1997) found that both the power-law and black-body models fit the observed spectrum equally well, leaving the origin of the detected X-rays unconstrained. If the observed radiation is interpreted in terms of thermal emission from hot polar caps ($T \sim 3$ MK), the caps appear to be very small ($A \sim 100$ m^2),

and their X-ray luminosity is much lower than predicted by many polar cap heating models. If this radiation is nonthermal, its luminosity is consistent with the general trend, $L_x \sim 10^{-3} \dot{E}$, found by Becker and Trümper (1997) for the nonthermal emission from those radio pulsars which are detected in the X-ray range (Figure 17). The sensitivity of XMM-Newton and Chandra is required to finally identify the emission mechanism.

Although the thermal radiation from the surface of an old neutron star no longer peaks in the X-ray band, the power of the HST allows one to observe it in the optical–UV range. Pavlov et al. (1996a) detected PSR B1929+20 at near-UV frequencies and showed that the observed flux corresponds to a temperature of about 0.2 MK, if the radiation is thermal. A candidate for the optical counterpart of PSR B0950+08 was also detected by the same authors, but its identification is less certain. Studying optical radiation from old nearby pulsars is very useful for understanding their thermal and nonthermal evolution.

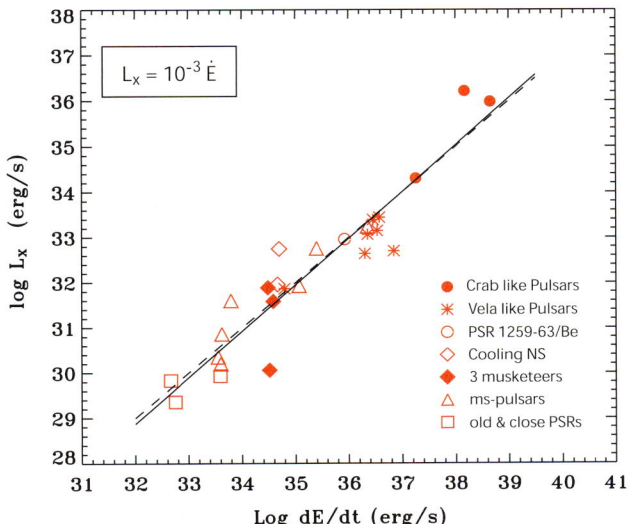

Figure 17 X-ray luminosity v. spin-down energy loss for all rotation-powered pulsars detected by ROSAT. For the "Three Musketeers" (Geminga, PSR 0656+14, and PSR 1055−52), the low-energy thermal component has been subtracted from the data. The solid line represents $L_x(\dot{E}) \propto \dot{E}^{1.03}$, the dashed line $L_x(\dot{E}) = 10^{-3}\dot{E}$. Remarkably, all the detected pulsars, from the young Crab-like to the 10^9-year-old millisecond pulsars, follow the linear trend. (From Becker and Trümper 1997.)

No gamma radiation has been detected even from the nearest old radio pulsars. Since the gamma-ray efficiency, $\epsilon_\gamma = L_\gamma/\dot{E}$, grows with age for young and middle-aged pulsars (for PSR 1055−52 it is almost 20%), it must have a break at about ~1 Myr not to exceed 100%. A reason for this break could be that the thermal surface photons are involved in the production of the observed gamma rays, via Compton up-scattering of thermal photons by ultra-relativistic particles. Since a neutron star cools as it ages, the production of gamma rays, and hence the gamma-ray efficiency, decreases.

Isolated radio-quiet neutron stars

Analyzing the ROSAT PSPC observations of a field containing the molecular cloud R CrA ($d \approx 130$ pc), Walter et al. (1996) noticed a bright point source, RX J1856−3754, projected onto the cloud. Its spectrum is very soft – the best fit with a black-body model gives a temperature of 0.66 MK and a luminosity of $\sim 5 \times 10^{31}$ erg s^{-1} for a distance of 100 pc. Based on the lack of an optical counterpart brighter than $V \sim 23$, Walter et al. (1996) suggested that the source is a nearby isolated neutron star. Like the objects described in the previous section "Radio-silent neutron stars in supernova remnants", it is radio-quiet, but unlike those objects it is not associated with any SNR (i.e., it is "truly isolated"). Since its temperature is much lower than the temperatures of the isolated neutron stars in SNRs, it is natural to assume that this object is much older – that the neutron star has cooled down. Surprisingly, no pulsations of the X-radiation, expected from a neutron star with a "typical" magnetic field and favorable orientation of the magnetic and spin axes, were found. Pavlov et al. (1996b) fitted the X-ray spectrum with neutron star atmosphere models (see the previous section "Photospheric emission from cooling neutron stars") and showed that different chemical compositions and different magnetic fields of the surface layers correspond to quite different optical magnitudes, $V = 22$–28, and distances, 5–200 pc (for a neutron star radius of 10 km).

A very faint, blue optical counterpart of RX J1856−3754 was discovered with the HST by Walter and Matthews (1997). The extremely large X-ray-to-optical flux ratio of $\sim 75{,}000$ proves unequivocally that this is indeed a neutron star. However, the optical magnitude, $V = 25.7$, is considerably different from the predictions of the four atmosphere models considered by Pavlov et al. (1996b), which means that either the atmosphere has a different chemical composition and magnetic field or that the temperature distribution is essentially nonuniform, for example because of anisotropy of heat conduction in a very strong magnetic field. The nature of RX J1856−3754 became even more puzzling after its proper motion, 0.33 arcsec/yr, was measured (Walter et al. 2000). This proper motion corresponds to a transverse velocity of 140 km s^{-1} at $d = 100$ pc, too fast for accretion from the interstellar medium to be a major heating source. This means that RX J1856−3754 is a cooling neutron star, and with its apparent surface temperature of ~ 0.7 MK it should be younger than 1 Myr, for the standard (slow) cooling models. However, projecting the proper motion backward, Walter et al. (2000) suggested that RX J1856−3754 and the well-known runaway O star ζ Oph originated from the same binary system when it was disrupted by a supernova explosion about 2 Myr ago. A neutron star of such an age should have an apparent effective temperature of <0.4 MK, in contradiction with the current data.

The lack of pulsations, which could be explained by co-alignment of the magnetic and rotational axes, or the rotational axis and the line of sight, does not allow one to measure the spin period of RX J1856−3754. Fortunately, two other objects with similar properties, for which pulsations have been measured, were discovered with ROSAT: RX J0720−3125, with $P = 8.37$ s, and RX J0420−5022, with $P = 22.7$ s (Haberl et al. 1996, 1997, 1999). Future measurements of their period derivatives will allow one to estimate their ages and elucidate the nature of these neutron stars. Particularly important will be deep optical–UV observations of these objects (a viable candidate for the optical counterpart of RX J0720−3125, with $B \approx 26.6$, has been found by Motch and Haberl (1998) and Kulkarni and van Kerkwijk (1998)).

Three more objects of apparently the same class are known at the time of writing (Table 2, and references therein). We expect the number of detected radio-silent neutron stars to grow considerably in the near future, and we shall be able to compare their properties with predictions of different models of neutron star evolution. One of the most important problems related to these objects is the source of energy which heats the radiating layers of the neutron stars up to 0.7–1.4 MK – it may be either the internal heat of relatively young cooling neutron stars, presumably with large magnetic fields, or accretion from the interstellar medium onto old neutron stars (e.g., Treves et al. 2000). Presently, we cannot exclude the possibility that the six observed sources belong to two quite different classes, young coolers and old accretors.

Recycled millisecond pulsars

In the P–$\dot{P}$ parameter space, millisecond pulsars (ms-pulsars) are distinguished from the majority of ordinary-field pulsars by their short spin periods and small period derivatives, corresponding to very old spin-down ages of typically 10^9–10^{10} years and low magnetic field strengths of $\sim 10^8$–10^{10} G (Figure 2). More than $\sim 75\%$ of the known disk ms-pulsars are in binaries with a compact companion star, compared with $\sim 1\%$ in binaries among the ordinary pulsars. This gives support to the idea that these neutron stars have been spun up by angular momentum transfer during a past mass accretion phase (Bisnovatyi-Kogan and Komberg 1974, Alpar et al. 1982, Bhattacharya and van den Heuvel 1991). Indeed, the first known accreting ms-pulsar, SAX J1808.4 – 3658, has been discovered with BeppoSAX (Chapter 33). Presumably, these pulsars were once ordinary pulsars, and would have turned off because of the loss of their rotational energy if they were not in close binaries; they are therefore often called "recycled" pulsars.

By the end of 2000 about a hundred recycled radio pulsars were known, of which 57 are located in the galactic plane (Camilo 1999, Edwards et al. 2000, Lommen et al. 2000, Lyne et al. 2000, Manchester et al. 2000). The others are in globular clusters (Kulkarni and Anderson 1996, Camilo et al. 2000), which provide a favorable environment for the recycling scenario (Rasio et al. 2000). Only 10 of the 57 ms-pulsars in the galactic plane are solitary (including PSR B1257+12, which has a planetary system); the rest are in binaries, usually with a low-mass white dwarf companion. The formation of solitary recycled pulsars is not well understood, but it is widely believed that either the pulsar's companion was evaporated or the system was tidally disrupted after the formation of the ms-pulsar.

Recycled pulsars had been studied exclusively in the radio domain until the early 1990s, when ROSAT, ASCA, EUVE, RXTE, and BeppoSAX were launched. The first ms-pulsar discovered as a pulsating X-ray source was PSR J0437−4715 (Becker and Trümper 1993), a nearby 5.75 ms-pulsar which is in a binary with a low-mass white dwarf companion. Further detections followed, and by the end of the century ms-pulsars accounted for almost a third of all X-ray detected rotation-powered pulsars (Table 1).

The available data suggest that the observed X-ray emission is likely to be generated by nonthermal processes in most ms-pulsars. This is supported by observations of the 3.05 ms-pulsar B1821−24, which is located in the globular cluster M28 (Kawai and Saito 1999), PSR B1937+21 – the fastest ms-pulsar known (Takahashi et al. 1999), and PSR J0218+4232 (Mineo et al. 2000). For these objects, power-law spectra and/or pulse profiles with narrow peaks have been measured (Figure 18). For PSR J0437 − 4715, the results of a recent Chandra observation suggest that the emission contains both the thermal component from the hot polar caps (Zavlin and Pavlov 1998) and a nonthermal component from the magnetosphere. The data on J2124−3358 do not allow one to determine unambiguously which of the two components, thermal or nonthermal, is present in the observed emission. The 4.86 ms-pulsar J0030 + 0451, which has spin parameters similar to those of J2124 − 3358, shows a high pulsed fraction of $53 \pm 3\%$. Interestingly, recent XMM observations of this ms-pulsar show that the spectrum suggests thermal polar-cap emission for the nature of the detected X-rays, with a small nonthermal contribution beyond 2 keV.

All other X-ray detected ms-pulsars (B1957+20, J1012+ 5307, B0751+18, J1744 − 1134, and J1024 − 0719) are identified only by their positional coincidence with the radio pulsar (Becker and Trümper 1999), and the low number of detected counts provide little more than flux estimates. The power of XMM-Newton and Chandra is needed to explore their emission properties in more detail. However, the fact that all ms-pulsars have roughly the same X-ray efficiency ($L_x/\dot{E} \sim 10^{-3}$) as ordinary pulsars (Figure 17), supports the conclusion that, as a rule, the non-thermal hard X-ray radiation from their magnetospheres prevails over the soft thermal radiation from their polar caps (cf. Becker and Trümper 1997).

As far as the emission of gamma rays from ms-pulsars is concerned, PSR J0218 + 4232 has been proposed as the counterpart of the EGRET source 2EG J0220 + 4228 (Verbunt et al. 1996, Kuiper et al. 2000). The final verification, however, has to await the next gamma-ray missions, GLAST (the Gamma-ray Large Area Space Telescope) and Integral, which are scheduled for the first decade of the new millennium. If J0218 + 4232 is indeed a gamma-ray pulsar, then, depending on the assumed emission model, 7–33% of

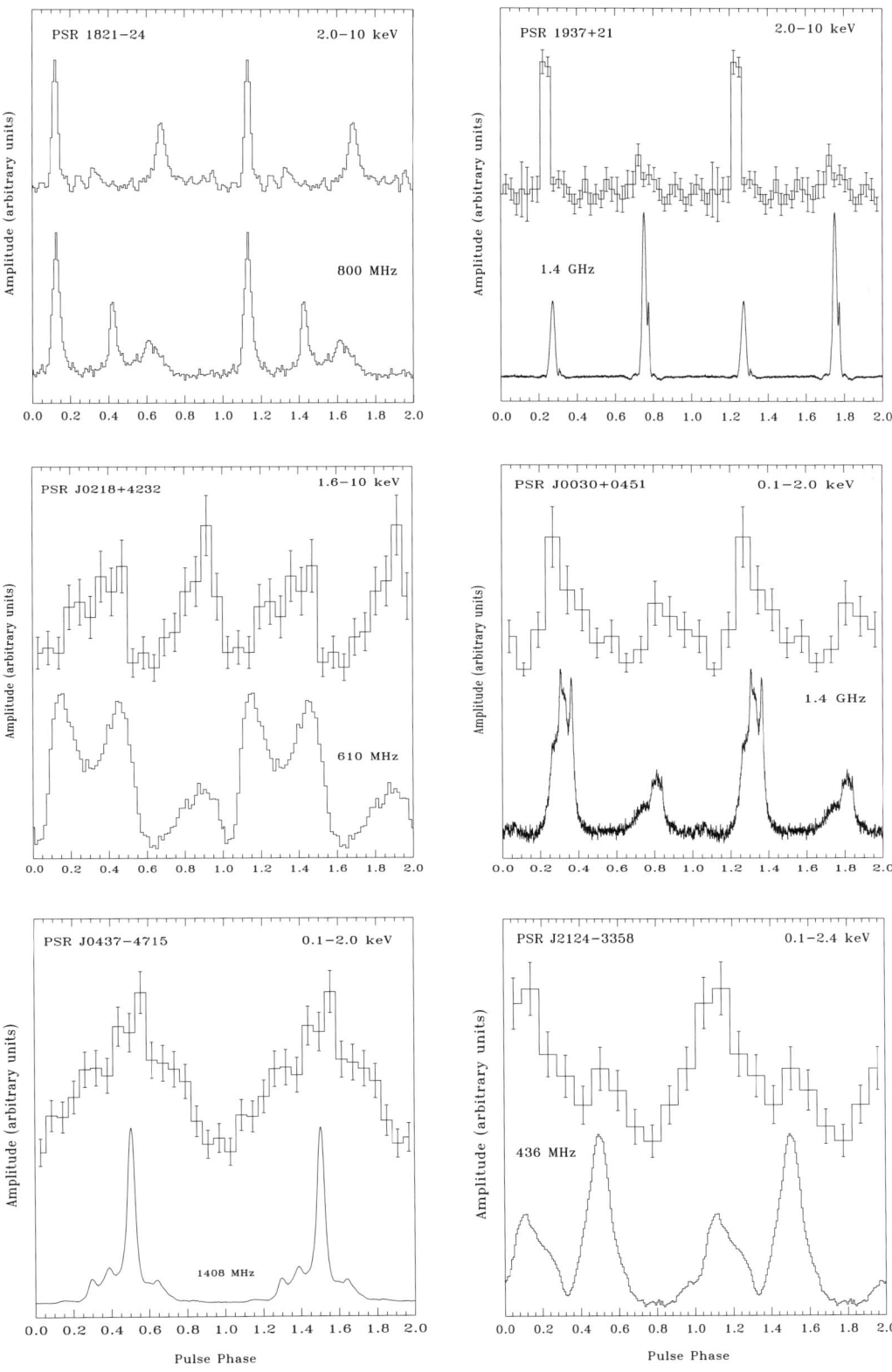

Figure 18 Integrated light-curves for all millisecond pulsars for which spin-modulated X-ray emission has been detected. The upper phase histograms show the X-ray profiles in the given energy bands. The radio light-curves are shown for comparison. Two phase cycles are shown for clarity. The relative phase between the radio and X-ray pulses is only known for PSR 1821−24, PSR B1937+21 and PSR J0437−4715. In all other cases the phase alignment is arbitrary because of the lack of accurate satellite clock calibration.

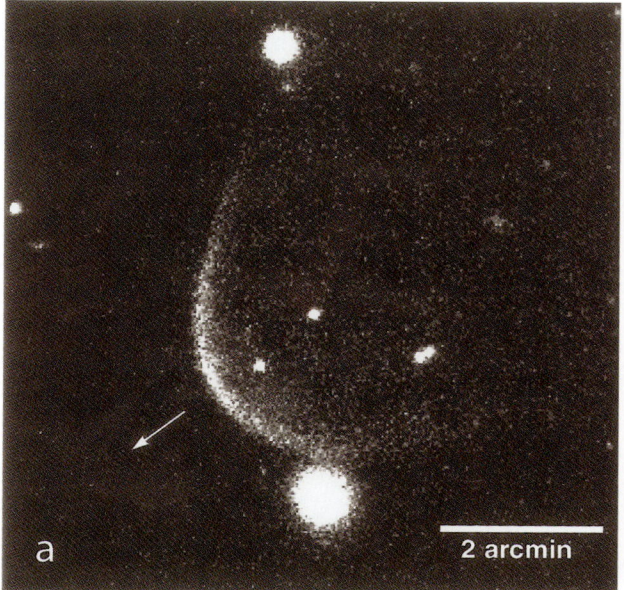 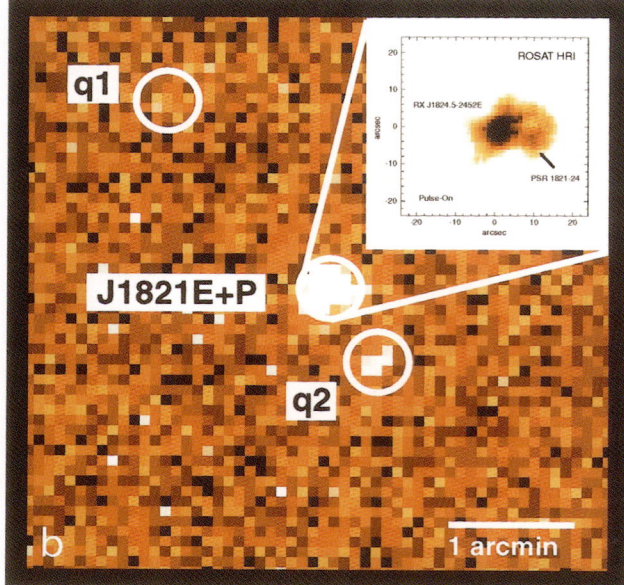

Figure 19 (a) The bow-shock nebula around PSR J0437−4715 as observed in the Hα emission (Courtesy of A. Fruchter). The arrow indicates the direction of the pulsar's proper motion. (b) 5′ × 5′ ROSAT HRI image of the globular cluster M28. RX J1824.5−2452E, RX J1824.5−2452P, q1, and q2 indicate the positions of four X-ray sources, of which q1 and q2 are globular cluster background sources. Thus upper-right inset magnifies the core encompassing J182E + P. Here, the ROSAT HRI data are oversampled at 1″ bins and temporally phased to emphasize "pulse-on" events from the millisecond pulsar B1821−14, which is the faint source indicated by the arrow.

its spin-down energy would go into the production of gamma rays. No other ms-pulsars have been identified with gamma-ray sources so far, although according to the polar cap and outer gap emission models their predicted efficiencies should be even higher than that estimated for J0218 + 4232.

An important aspect of pulsar studies is searching for PWNs. So far, bow-shock PWNs have been firmly detected in Hα emission around PSR 1957+20 (Fruchter *et al.* 1992) and PSR J0437 − 4715 (Bell *et al.* 1993). The bow-shock stand-off distance found in J0437 − 4715 is about 7 arcsec (Bell *et al.* 1995). Observations with the ROSAT HRI yielded a 3σ upper limit of $0.2 \times 10^{-3}\dot{E}$ for the X-ray emission from the nebula (Becker and Trümper 1999). Another PWN candidate is the object RX J1824.5 − 2452E, near PSR B1821 − 24 (Figure 19). However, as the pulsar is located in a globular cluster, it is quite likely that this extended X-ray source is a superposition of spatially unresolved globular cluster sources (cataclysmic variables or low-mass X-ray binaries) rather than a plerion powered by the pulsar. We expect that the true nature of RX J1824.5 − 2452E will be established from a deep Chandra observation of the globular cluster M28.

IMPRESSIVE ACHIEVEMENTS AND GREAT EXPECTATIONS

Astronomers of our generation have been truly lucky. Their collective efforts, generously supported by taxpayers of different countries, have revolutionized our understanding of the Universe and its constituents, from clusters of galaxies to neutron stars. Fifty years ago it was hard to imagine that neutron stars, very hypothetical objects at that time, not only would be discovered, but would also be studied in such detail*. The new vision of the Universe in general, and neutron stars in particular, has been made possible by opening the new windows for observing the fascinating cosmos – now we can study the Universe not only through

---

*Sachiko Tsuruta, who devoted her scientific life to studying the thermal evolution of neutron stars, recalls an episode from the mid-1960s, when she had finished her PhD thesis on the thermal evolution of neutron stars, which by that time were not expected to be observable. Says Tsuruta, "By 1965, Bahcall and Wolf published papers that a neutron star cannot be seen if there are pions in the neutron star core. Soon after that I met John Bahcall at some conference, and he urged me to bet for discovering neutron stars, while he would bet against it. To my regret, I replied that a good Japanese woman should not bet. Then in 1967 a pulsar was discovered!"

the traditional, very narrow optical window, but also in radio, X-rays, and gamma rays. Without the X-ray and gamma-ray space observatories, our understanding of neutron stars, and many other objects virtually unknown half a century ago, would be much less complete. In particular, high-energy observations of the last decade of the twentieth century allowed us to understand that the world of neutron stars is not as simple as many astronomers had believed in the 1970s and 1980s (and some of them still believe). Neutron stars are not "just dim, heavy balls of ten kilometer radius," as an expert in extragalactic astronomy claimed, explaining why proposals to observe neutron stars with the HST have been rejected so often. They are not all alike – on the contrary, their properties and observational manifestations are no less diverse than those of normal stars and galaxies.

Of the thousand-plus neutron stars that have been discovered, about 100, including 60 isolated neutron stars (Tables 1 and 2), have been observed at high energies with space observatories. These observations have firmly established that the properties of neutron stars are indeed highly unusual; in particular, their gravitational and magnetic fields are truly immense. We dare to predict that such exotic properties will never be achieved in terrestrial laboratories. Thus, neutron stars provide a unique opportunity to study matter under extreme conditions. In particular, neutron stars can be viewed as cosmic laboratories for studying nuclear interactions, general relativity, and superstrong magnetic and electric fields. This is the point where astrophysics and physics merge, and indeed cannot be separated from each other.

In spite of the impressive achievements of neutron star physics and astrophysics, a lot of work remains to be done in this recently emerged field. First, the evolution of neutron stars, starting from their violent birth in supernova explosions, is far from being well understood. Until very recently, a common prejudice had been that all neutron stars are born as active, rotation-powered pulsars, which slow down their rotation, eventually stop their activity and, after crossing a "death line," enter the "pulsar graveyard." A former pulsar remains in the graveyard for ever, cool and quiet, unless it is captured by a passing star – as may happen, for example, in a globular cluster – and forms a close binary, where accretion onto the neutron star can spin it up (recycle it) to such a short period that it again becomes an active pulsar.

Recent high-energy observations, however, show that the picture may not be so simple. In particular, it appears that many very young neutron stars are not active pulsars at all. The most recent example may be the central source of the 320-year-old Cas A supernova remnant (Figure 7) although it is still not completely clear whether it is a neutron star or a black hole. Since such objects are not seen in radio, and are extremely faint in optical, they could not be observed until the onset of the X-ray astronomy era, which means that our perception of the early evolution of neutron stars has been very strongly biased in favor of the much easily observable rotation-powered pulsars. Why are many (perhaps most) nascent neutron stars not active pulsars? Is it because they are indeed magnetars, whose superstrong magnetic field inhibits the pulsar activity? Or, on the contrary, are their magnetic fields so weak and/or rotation so slow that the pulsar does not turn on? Or is the pulsar activity quenched by the accretion of debris from the supernova explosion? Are the (apparently young) anomalous X-ray pulsars and soft gamma-ray repeaters indeed the magnetars, or do their unusual observational properties arise for quite different reasons, like the presence of a residual disk? To answer these questions, further observations with more sensitive instruments of higher angular and energy resolution are needed.

One more set of evolutionary problems is associated with the generation and evolution of neutron star magnetic fields. Although there is no doubt that very strong fields exist in many (if not all) neutron stars, there is no clear understanding of how they are generated. Why are they so different in different kinds of neutron stars (e.g., regular and recycled pulsars), what is their geometry, and do they decay during the neutron star's lifetime? It should be mentioned that the direct measurements of the magnetic field have been possible only for neutron stars in binaries. What is called the "magnetic field" in, for example, radio pulsars, is only an order-of-magnitude, model-dependent estimate. Direct measurement of magnetic fields in isolated neutron stars, for example with the aid of spectral lines formed in their photospheres or from X-ray polarimetry, is a very important goal for future observations.

One of the most important tasks in studying isolated neutron stars is to elucidate their internal composition (neutrons, quark–gluon plasma, strange matter, meson condensate?) and the properties of the superdense matter (equation of state, nucleon superfluidity). Different equations of state correspond to different mass–radius (M/R) dependences. Hence, the most direct way to determine the equation of state (which, in turn, depends on the internal composition) is to measure the masses and radii of neutron stars. One can constrain the radius from the star's bolometric flux and effective temperature, provided the thermal radiation is not strongly "contaminated" by magnetospheric radiation of relativistic particles. The effective temperature can be determined by fitting the spectrum of the thermal radiation to neutron star atmosphere models. This method requires a good knowledge of the distance to the neutron star, which can be estimated from the radio dispersion measure if the neutron star is an active radio pulsar, or, much more precisely, by measuring its parallax.

The M/R ratio can be measured directly from the gravitational redshifts of spectral lines in the X-ray range. Measuring the redshifts would require X-ray detectors with high energy resolution and reliable computations for the energies of various atoms and ions in strong magnetic fields. Since the atomic states are greatly distorted by magnetic fields that are typical of neutron stars (and, consequently, spectral lines are strongly shifted from their zero-field positions), an independent measurement of the magnetic field (e.g., from fitting the continuum radiation to neutron star atmosphere models) would be necessary. The M/R ratio can also be evaluated from the analysis of the X-ray pulse profiles. Since the temperature is not uniform over the neutron star's surface (due to the anisotropy of the heat transfer in the crust or, for active pulsars, the accretion of relativistic particles onto the polar regions), the observed X-ray flux should vary with the rotation period. Because of the gravitational bending of photon trajectories, the shape of the pulse profile depends strongly on M/R. The observations will require high detector sensitivity ($\sim 10^4 - 10^5$ counts are needed to obtain accurate pulse profiles in a few energy ranges) and good time resolution ($\sim 10^{-5}$ s for millisecond pulsars). Finally, the properties of the internal matter can be constrained by measuring the effective temperatures of neutron stars of different ages. The thermal evolution of neutron stars depends strongly on the internal composition, equation of state, and nucleon superfluidity.

Of course, no firm conclusions about the internal properties can be drawn without studying the physical conditions in the surface layers of neutron stars: magnetic fields, temperatures, densities and, in particular, the chemical composition. Elucidating the chemical composition is also important in order to understand how neutron stars interact with their environment, both in their very young age, when a fraction of the supernova ejecta can fall back to the star's surface, and during their whole life, which may include some accretion episodes. The investigation of the surface layers (atmospheres) will require the analysis of soft X-ray spectra in terms of atmosphere models (which is why the detectors need to have high sensitivity and spectral resolution). Important information about neutron stars can be obtained from their transverse velocities (proper motions) and parallaxes. These quantities have been measured for a handful of radio pulsars. X-ray telescopes with sub-arcsecond angular resolution will allow us to measure astrometric characteristics of nearby radio-quiet neutron stars.

After 32 years of radio pulsar investigations, we still lack a consistent theory of pulsar activity. New X-ray and gamma-ray data are expected to close this gap. X-radiation radiation of many pulsars is due to relativistic particles in their magnetospheres, and studying the spectra and the pulse profiles of this radiation will allow us to determine the energy spectrum and directional pattern of the relativistic particles and, consequently, conditions in the pulsar acceleration zones and their temporal evolution. Furthermore, the X-ray range is most convenient for investigating the hot polar caps of radio pulsars, inevitable components of pulsar activity.

We expect that many of these goals will be achieved with the aid of the satellite X-ray observatories Chandra and XMM-Newton, launched in 1999. The first few months of Chandra observations brought several important discoveries: the central compact object in Cas A, the unusual 6-hour period of the central source of RCW 103, and the discovery of the small-scale structure in the compact nebulae around the Crab and Vela Pulsars. Many more discoveries, from both Chandra and XMM-Newton, will have been made by the time this book is published. Furthermore, a number of new high-energy missions, GLAST, Integral, Constellation-X, and XEUS (the X-ray Evolving Universe Spectroscopy Mission) are being planned and, hopefully, will be launched within the next two decades. Particularly useful for studying isolated neutron stars will be the Constellation-X and XEUS missions. For instance, the Constellation-X mission is planned to consist of six X-ray telescopes to be launched to the Lagrangian point in 2007–2008. Each of these payloads will combine the excellent angular resolution of Chandra with the large collecting area of XMM-Newton, so that we may expect a new revolution in X-ray astronomy in the second decade of the third millennium. We are looking forward to new discoveries that will raise additional questions. Such is the nature of science.

REFERENCES

Alpar, M.A., Cheng, A.F., Ruderman, M.A. and Shaham, J. (1982). A new class of radio pulsars. *Nature*, **300**, 728–730.

Arons, J. and Scharlemann, E.T. (1979). Pair formation above pulsar polar caps: Structure of the low altitude acceleration zone. *Astrophysical Journal*, **231**, 854–879.

Arons, J. and Tavani, M. (1993). High-energy emission from the eclipsing millisecond pulsar PSR 1957+20. *Astrophysical Journal*, **403**, 249–255.

Aschenbach, B. (1999). Cassiopeia A. *IAU Circular*, 7249, 1.

Aschenbach, B. and Brinkmann, W. (1975). A model of the X-ray structure of the Crab Nebula. *Astronomy and Astrophysics*, **41**, 147–151.

Baade, W. and Zwicky, F. (1934). On super-novae. *Proceedings of the National Academy of Sciences of the USA*, **20**, 254–259.

Bhattacharya, D. and van den Heuvel, E.P.J. (1991). Formation and evolution of binary and millisecond radio pulsars. *Physics Reports*, **203**, 1–124.

Becker, W. (1995). *Untersuchung rationsgetriebener Pulsar mit ROSAT – Aur der Suche nach kühlenden Neutronensternen. MPE-Report 260.* Max-Planck-Institut für extraterrestrische Physik, Garching, Germany.

Becker, W. (1998). In R. Bandiera, E. Masini, F. Pacini, M. Salvati and L. Woltjer (eds), *The Relationship between Neutron Stars and Supernova Remnants*, Mem. S.A. It, Vol. 69, pp. 847–855.

Becker, W. (1999). In B. Aschenbach and M.J. Freyberg (eds), *Highlights in X-ray Astronomy*, MPE-Report, **272**, pp. 49–56.

Becker, W. (2000a). In P.C.H. Martens, S. Tsuruta and M.A. Weber (eds), *Highly Energetic Physical Processes and Mechanism for Emission for Astrophysical Plasmas*, IAU Symposium, Vol. 195, Astronomical Society of the Pacific, pp. 49–60.

Becker, W. (2000b). In K.S. Cheng *et al*. (eds), *Stellar Astrophysics*, Astrophysics and Space Science Library, Vol. 254, Kluwer Academic Publishers, Dordrecht, pp. 305–317.

Becker, W. and Aschenbach, B. (1995). ROSAT HRI Observations of the Crab Pulsar: An improved temperature upper limit for PSR 0531+21. In M.A. Alpar, U. Kiziloglu and J. van Paradijs (eds), *The Lives of the Neutron Stars*, Proceedings of the NATO Advanced Study Institute on the Lives of the Neutron Stars, held in Kemer, Turkey, 29 August–12 September 1993, Kluwer Academic, Dordrecht, p. 47.

Becker, W. and Aschenbach, B. (2001). Jahrbuch der Max-Planck Gesellschaft 2001, pp. 628–634.

Becker, W., Brazier, K. and Trümper, J. (1995). ROSAT observations of the radio and gamma-ray pulsar PSR 1706−44. *Astronomy and Astrophysics*, **298**, 528.

Becker, W., Kawai, N., Brinkmann, W. and Mignani, R. (1999). The putative pulsar-wind nebulae of the Three Musketeers: PSR B1055−52, B0656+14, and Geminga revisited. *Astronomy and Astrophysics*, **352**, 532–542.

Becker, W. and Trümper, J. (1993). Detection of pulsed X-rays from the binary millisecond pulsar J 0437−4715. *Nature*, **365**, 528.

Becker, W. and Trümper, J. (1997). The X-ray luminosity of rotation-powered neutron stars. *Astronomy and Astrophysics*, **326**, 682–691.

Becker, W. and Trümper, J. (1999). The X-ray emission properties of millisecond pulsars. *Astronomy and Astrophysics*, **341**, 803–817.

Becker, W., Trümper, J., Lommen, A.N. and Backer, D.C. (2000). X-rays from the nearby solitary millisecond pulsar PSR J0030+0451: The final ROSAT observations. *Astrophysical Journal*, **545**, 1015–1019.

Becker, W., Trümper, J. and Ögelman, H.B. (1993). In K.A. Van Riper, R. Epstein and C. Ho (eds), *Isolated Pulsars*, Cambridge University Press, pp. 104–109.

Bell, J.F., Bailes, M. and Bessel, M.S. (1993). Optical detection of the companion of the millisecond pulsar J0437−4715. *Nature*, **364**, 603–605.

Bell, J.F., Bailes, M., Manchester, R.N., Weisberg, J.M. and Lyne, A. (1995). The proper motion and wind nebula of the nearby millisecond pulsar J0437−4715. *Astrophysical Journal*, **440**, L81–L83.

Bell Burnell, S.J. (1977). Petit four. *Annuals of the New York Academy of Sciences*, **302**, 685–689.

Bertsch, D.L., Brazier, K.T.S., Fichtel, C.E., Hartman, R.C., Hunter, S.D., Kanbach, G., Kniffen, D.A., Kwok, P.W., Lin, Y.C. and Mattox, J.R. (1992). Pulsed high-energy gamma-radiation from Geminga (1E0630+178). *Nature*, **357**, 306–307.

Beskin, V.S., Gurevich, A.V. and Istomin, Ya.N. (1993). *Physics of the Pulsar Magnetosphere*, Cambridge University Press.

Bignami, G.F. and Caraveo, P.A. (1996). Geminga: Its phenomenology, its fraternity, and its physics. *Annual Review of Astronomy and Astrophysics*, **34**, 331–382.

Bignami, G.F., Caraveo, P.A., Mignani, R., Edelstein, J. and Bowyer, S. (1996). Multiwavelength data suggest a cyclotron feature on the hot thermal continuum of Geminga. *Astrophysical Journal*, **456**, L111.

Bisnovatyi-Kogan, G.S. and Komberg, B.V. (1974). Pulsars and close binary systems. *Soviet Astronomy*, **18**, 217.

Boyd, P.T., van Citters, G.W., Dolan, J.F., Wolinski, K.G., Percival, J.W., Bless, R.C., Elliot, J.L., Nelson, M.J. and Taylor, M.J. (1995). High-speed photometer observations of the LMC pulsar B0540−69. *Astrophysical Journal*, **448**, 365.

Bowyer, S. (1990). The Extreme Ultraviolet Explorer mission. In Y. Kondo (ed.), *Observatories in Earth Orbit and Beyond*, Kluwer Academic, Dordrecht, pp. 153–169.

Bowyer, S., Byram, E.T., Chubb, T.A. and Friedman, H. (1964). X-ray sources in the Galaxy. *Nature*, **201**, 1307–1308.

Bradt, H.V., Rappaport, S., Mayer, W., Nather, R.E., Warner, B., Macfarlane, M. and Kristian, J. (1969). X-ray and optical observations of the pulsar NP 0532 in the Crab Nebula. *Nature*, **222**, 728–730.

Bradt, H.V., Swank, J.H. and Rothschild, R.E. (1990). The X-ray Timing Explorer. *Advances in Spaces Research*, **10**, 297–310.

Brazier, K.T.S., Becker, W. (1997). High-resolution X-ray imaging of the supernova remnant MSH 15−52. *Monthly Notices of the Royal Astronomical Society*, **284**, 335–340.

Brazier, K.T.S., Johnston, S. (1999). The implications of radio-quiet neutron stars. *Monthly Notices of the Royal Astronomical Society*, **305**, 671–679.

Brinkmann, W. and Ögelman, H. (1987). Soft X-ray observations of the radio pulsar PSR 1055−52. *Astronomy and Astrophysics*, **182**, 71–74.

Bulik, T. and Pavlov, G.G. (1996). Polarization modes in a strongly magnetized hydrogen gas. *Astrophysical Journal*, **469**, 373.

Butler, R.C. and Scarsi, L. (1990). The SAX mission for X-ray astronomy. In Y. Kondo (ed.), *Observatories in Earth Orbit and Beyond*, Kluwer Academic, pp. 141–150.

Camilo, F. (1999). Pulsar searches in the northern hemisphere. In Z. Arzoumanian, F. Van der Hooft and E.P.J. van den Heuvel (eds), *Pulsar Timing, General Relativity and the Internal Structure of Neutron Stars*, Koninklijke Nederlandse Akademie van Wetenschappen, Amsterdam, p. 115.

Camilo, F., Lorimer, D.R., Freire, P., Lyne, A.G. and Manchester, R.N. (2000). Observation of 20 millisecond pulsars in 47 Tucanae at 20 centimeters. *Astrophysical Journal*, **535**, 975–990.

Caraveo, P.A., Bignami, G.F. and Trümper, J. (1996). Radio-silent isolated neutron stars as a new astronomical reality. *Astronomy and Astrophysics Review*, **7**, 209–216.

Chakrabarty, D., Pivovaroff, M.J., Hernquist, L.E., Heyl, J.S. and Narayan, R. (2001). The central X-ray point source in Cassiopeia A. *Astrophysical Journal*, **548**, 800–810.

Chandrasekhar, S. (1931). The maximum mass of ideal white dwarfs. *Astrophysical Journal*, **74**, 81.

Cheng, A.F. and Helfand, D.J. (1983). X-rays from radio pulsars: The detection of PSR 1055−52. *Astrophysical Journal*, **271**, 271–282.

Cheng, K.S., Ho, C. and Ruderman, M.A. (1986a). Energetic radiation from rapidly spinning pulsars: I. Outer magnetosphere gaps. *Astrophysical Journal*, **300**, 500–521.

Cheng, K.S., Ho, C. and Ruderman, M.A. (1986b). Energetic radiation from rapidly spinning pulsars. II. Vela and Crab. *Astrophysical Journal*, **300**, 522–539.

Chiu, H.Y. and Salpeter, E.E. (1964). Surface X-ray emission from neutron stars. *Physical Review Letters*, **12**, 413–415.

Clear, J., Bennett, K., Buccheri, R., Grenier, I.A., Hermsen, W., Mayer-Hasselwander, H.A. and Sacco, B. (1987). A detailed analysis of the high energy gamma-ray emission from the Crab pulsar and nebula. *Astronomy and Astrophysics*, **174**, 85–94.

Cocke, W.J., Disney, M.J. and Taylor, D.J. (1969). Discovery of optical signals from pulsar NP0532. *Nature*, **221**, 525–529.

Corbel, S., Chapuis, C., Dame, T.M. and Durouchoux, P. (1999). The distance to the soft gamma repeater SGR 1627−41. *Astrophysical Journal*, **526**, L29–L32.

Corbel, S., Wallyn, P., Dame, T.M., Durouchoux, P., Mahoney, W.A., Vilhu, O. and Grindlay, J.E. (1997). The distance of the soft gamma repeater SGR 1806−20. *Astrophysical Journal*, **478**, 624.

Corbet, R.H.D., Smale, A.P., Ozaki, M., Koyama, K. and Iwasawa, K. (1995). The spectrum and pulses of 1E 2259+586 from ASCA and BBXRT observations. *Astrophysical Journal*, **443**, 786–794.

Cordova, F.A., Hjellming, R.M., Mason, K.O. and Middleditch, J. (1989). Soft X-ray emission from the radio pulsar PSR 0656+14. *Astrophysical Journal*, **345**, 451–463.

Danner, R., Kulkarni, S.R. and Trümper, J. (1998). The X-ray counterpart to SGR 0526−66: Results from a monitoring campaign of N49 with ROSAT. Paper presented at AAS Meeting 192, 43.09.

Daugherty, J.K. and Harding, A.K. (1996). Gamma-ray pulsars: Emission from extended polar cap cascades. *Astrophysical Journal*, **458**, 278.

Edwards, R.T. (2000). Discovery of eight recycled pulsars: The Swinburne Intermediate Latitude Pulsar Survey. In M. Kramer, N. Wex and R. Wielebinski (eds), *Pulsar Astronomy – 2000 and Beyond*, ASP Conference Series, Vol. 202, Astronomical Society of the Pacific, San Francisco, p. 33.

Fenimore, E.E., Laros, J.G. and Ulmer, A. (1994). The X-ray spectrum of the soft gamma repeater 1806−20. *Astrophysical Journal*, **432**, 742–752.

Feroci, M., Frontera, F., Costa, E., Amati, L., Tavani, M., Rapisada, M. and Orlandini, M. (1999). A giant outburst from SGR 1900+14 observed with BeppoSAX Gamma-Ray Burst Monitor. *Astrophysical Journal*, **515**, L9–L12.

Fichtel, C.E., Hartman, R.C., Kniffen, D.A., Thompson, D.J., Bignami, G.F., Ogelman, H., Ozel, M.E. and Tumer, T. (1975). High-energy gamma-ray results from the second Small Astronomy Satellite. *Astrophysical Journal*, **198**, 163–182.

Forman, W., Jones, C., Cominsky, L., Julien, P., Murray, S., Peters, G., Tananbaum, H. and Giacconi, R. (1978). The fourth Uhuru catalog of X-ray sources. *Astrophysical Journal*, Suppl., **38**, 357–412.

Fritz, G., Henry, R.C., Meekins, J.F., Chubb, T.A. and Friedman, H. (1969). X-ray pulsar in the Crab Nebula. *Science*, **164**, 709–711.

Fruchter, A.S., Bookbinder, J., Garcia, M.R. and Bailyn, C.D. (1992). X-rays from the eclipsing pulsar 1957+20. *Nature*, **359**, 303–304.

Gaensler, B.M., Gotthelf, E.V. and Vasisht, G. (1999). A new supernova remnant coincident with the slow X-ray pulsar AX J1845−0258. *Astrophysical Journal*, **526**, L37–L40.

Garmire, G.P., Pavlov, G.G., Garmire, A.B. and Zavlin, V.E. (2000). 1E 161348−5055. *IAU Circular*, 7350, 2.

Giacconi, R. (1974). X-ray sky. In R. Giacconi and H. Gursky (eds), *X-ray Astronomy*, Reidel, Dordrecht, p. 155.

Giacconi, R., Branduardi, G., Briel, U., Epstein, A., Fabricant, D., Feigelson, E., Forman, W., Gorenstein, P., Grindlay, J., Gursky, H., Harnden, F.R., Henry, J.P., Jones, C., Kellogg, E., Koch, D., Murray, S., Schreier, E., Seward, F., Tananbaum, H., Topka, K., Van Speybroeck, L., Holt, S.S., Becker, R.H., Boldt, E.A., Serlemitsos, P.J., Clark, G., Canizares, C., Markert, T., Novick, R., Helfand, D. and Long, K. (1979). The Einstein/HEAO2/X-ray Observatory. *Astrophysical Journal*, **230**, 540–550.

Giacconi, R., Gursky, H., Paolini, F.R. and Rossi, B.B. (1962). Evidence for X-rays from sources outside the Solar System. *Physical Review Letters*, **9**, 439–443.

Giacconi, R., Kellogg, E., Gorenstein, P., Gursky, H. and Tananbaum, H. (1971). An X-ray scan of the galactic plane from Uhuru. *Astrophysical Journal*, **165**, L27.

Glen, G. and Sutherland, P. (1980). On the cooling of neutron stars. *Astrophysical Journal*, **239**, 671–684.

Glendenning, N.K. (1996). *Compact Stars*, Springer, Berlin.

Gnedin, Yu.N. and Pavlov, G.G. (1974). The transfer equations for normal waves and radiation polarization in an anisotropic medium. *Soviet Physics JETP*, **38**, 903–908.

Gold, T. (1968). Rotating neutron stars as the origin of the pulsating radio sources. *Nature*, **218**, 731–732.

Gold, T. (1969). Rotating neutron stars and the nature of pulsars. *Nature*, **221**, 25–27.

Gotthelf, E.V., Petre, R. and Hwang, U. (1997). The nature of the radio-quiet compact X-ray source in SNR RCW 103. *Astrophysical Journal*, **487**, L175.

Gotthelf, E.V., Petre, R. and Vasisht, G. (1999a). X-ray variability from the compact source in the supernova remnant RCW 103. *Astrophysical Journal*, **514**, L107–L110.

Gotthelf, E.V. and Vasisht, G. (1998). Discovery of a 7 second anomalous X-ray pulsar in the distant Milky Way. *New Astronomy*, **3**, 293–300.

Gotthelf, E.V., Vasisht, G. and Dotani, T. (1999b). On the spin history of the X-ray pulsar in Kes 73: Further evidence for an ultramagnetized neutron star. *Astrophysical Journal*, **522**, L49–L52.

Gotthelf, E.V. and Wang, Q.D. (2000). A spatially resolved plerionic X-ray nebula around PSR B0540−69. *Astrophysical Journal*, **394**, 581–585.

Gouiffes, C., Finley, J.P. and Ögelman, H. (1992). Rotational parameters of PSR 0540−69 as measured at optical wavelengths. *Astrophysical Journal*, **394**, 581–585.

Green, D.A. (1998). A Catalogue of Galactic Supernova Remnants (1998 September version), Mullard Radio Astronomy Observatory, Cambridge.

Greenstein, G. and Hartke, G.J. (1983). Pulselike character of blackbody radiation from neutron stars. *Astrophysical Journal*, **271**, 283–293.

Greiveldinger, C., Camerini, U., Fry, W., Markwardt, C.B., Oegelman, H., Safi-Harb, S., Finley, J.P., Tsuruta, S., Shibata, S., Sugawara, T., Sano, S. and Tukahara, M. (1996). Heated polar caps in PSR 0656+14 and PSR 1055−52. *Astrophysical Journal*, **465**, L35.

Gregory, P.C. and Fahlman, G.G. (1980). An extraordinary new celestial X-ray source. *Nature*, **287**, 805–806.

Gudmundsson, E.H., Pethick, C.J. and Epstein, R.I. (1983). Structure of neutron star envelopes. *Astrophysical Journal*, **272**, 286–300.

Haberl, F., Motch, C., Buckley, D.A.H., Zickgraf, F.-J. and Pietsch, W. (1997). RXJ0720.4−3125: Strong evidence for an isolated pulsating neutron star. *Astronomy and Astrophysics*, **326**, 662–668.

Haberl, F., Motch, C. and Pietsch, W. (1998). Isolated neutron stars in the ROSAT Survey. *Astronomische Nachrichten*, **319**, 97.

Haberl, F., Pietsch, W. and Motch, C. (1999). RXJ0420.0−5022: An isolated neutron star candidate with evidence for 22.7 s X-ray pulsations. *Astronomy and Astrophysics*, **351**, L53–L57.

Hailey, C.J. and Craig, W.W. (1995). Discovery of a candidate isolated neutron star in a new supernova remnant near CTB 1. *Astrophysical Journal*, **455**, L151.

Halpern, J.P. and Holt, S.S. (1992). Discovery of soft X-ray pulsations from the gamma-ray source Geminga. *Nature*, **357**, 222–224.

Halpern, J.P. and Wang, F.Y.-H. (1997). A broadband X-ray study of the Geminga Pulsar. *Astrophysical Journal*, **477**, 905.

Harding, A.K. and Muslimov, A.G. (1998). Pulsar X-ray and gamma-ray pulse profiles: Constraint on obliquity and observer angles. *Astrophysical Journal*, **500**, 862.

Harnden, F.R., Jr and Seward, F.D. (1984). Einstein observations of the Crab Nebula pulsar. *Astrophysical Journal*, **283**, 279–285.

Helfand, D.J. and Becker, R.H. (1984). Observation of stellar remnants from recent supernovae. *Nature*, **307**, 215–221.

Hill, R.J., Dolan, J.F., Bless, R.C., Boyd, P.T., Percival, J.W., Taylor, M.J. and van Citters, G.W. (1997). The spectrum of the Large Magellanic Cloud pulsar B0540−69. *Astrophysical Journal*, **486**, L99.

Hillier, R.R., Jackson, W.R., Murray, A., Redfern, R.M. and Sale, R.G. (1970). Low-energy gamma rays from NP 0532. *Astrophysical Journal*, **162**, L177.

Hoyle, R.A., Narlikar, J. and Wheeler, J.A. (1964). Electromagnetic waves from very dense stars. *Nature*, **203**, 914–916.

Hulleman, F., van Kerkwijk, M.H., Verbunt, F.V.M. and Kulkarni, S. (2000). A deep search for the optical counterpart to the anomalous X-ray pulsar 1E 2259+58.6. *Astronomy and Astrophysics*, **358**, 605–611.

Hurley, K., Kouveliotou, C., Cline, T., Mazets, E., Golenetskii, S., Frederiks, D.D. and van Paradijs, J. (1999a). Where is SGR 1806−20? *Astrophysical Journal*, **523**, L37–L40.

Hurley, K., Kouveliotou, C., Murakami, T., Ando, M., Strohmayer, T., van Paradijs, J., Vrba, F., Luginbuhl, C., Yoshida, A. and Smith, I. (1999b) ASCA discovery of an X-ray pulsar in the error box of SGR 1900+14. *Astrophysical Journal*, **510**, L111–L114.

Hurley, K., Strohmayer, T., Kouveliotou, C., Woods, P., van Paradijs, J., Murakami, T., Hartmann, D., Smith, I., Ando, M., Yoshida, A. and Sugizaki, M. (2000). ASCA observations of the quiescent X-ray counterpart to SGR 1627−41. *Astrophysical Journal*, **528**, L21–L23.

Iglesias, C.A. and Rogers, F.J. (1996). Updated Opal opacities. *Astrophysical Journal*, **464**, 943.

Inan, U.S., Lehtinen, N.G., Lev-Tov, S.J., Johnson, M.P., Bell, T.F. and Hurley, K. (1999). Ionization of the lower ionosphere by gamma-rays from a magnetar: Detection of a low-energy (3–10 keV) component. *Geophysical Research Letters*, **26**, 3357.

Israel, G.L., Covino, S., Stella, L., Campana, S., Haberl, F. and Mereghetti, S. (1999a). Further evidence that 1RXS J170849.0−400910 is an anomalous X-ray pulsar. *Astrophysical Journal*, **518**, L107–L110.

Israel, G.L., Mereghetti, S. and Stella, L. (1994). The discovery of 8.7 second pulsations from the ultrasoft X-ray source 4U 0142+61. *Astrophysical Journal*, **433**, L25–L28.

Israel, G.L., Ooostenbroek, T., Angelini, L., Campana, S., Mereghetti, S., Parmar, A.N., Segreto, A., Stella, L., van Paradijs, J. and White, N. (1999b). BeppoSAX monitoring of the 'anomalous' X-ray pulsar 4U 0142+61. *Astronomy and Astrophysics*, **346**, 929–935.

Kanbach, G., Arzoumanian, Z., Bertsch, D.L., Brazier, K.T.S., Chiang, J., Fichtel, C.E., Fierro, J.M., Hartman, R.C., Hunter, S.D., Kniffen, D.A., Lin, Y.C., Mattox, J.R., Mayer-Hasselwander, H.A., Michelson, P.F., von Montigny, C., Nel, H.I., Nice, D., Nolan, P.L., Pinkau, K., Rothermel, H., Schneid, E., Sommer, M., Sreekumar, P., Taylor, J.H. and Thompson, D.J. (1994). EGRET observations of the Vela Pulsar, PSR0833−45. *Astronomy and Astrophysics*, **289**, 855–867.

Kanbach, G., Bennett, K., Bignami, G.F., Buccheri, R., Caraveo, P., D'Amico, N., Hermsen, W., Lichti, G.G., Masnou, J.L., Mayer-Hasselwander, H.H., Paul, J.A., Sacco, B., Swanenburg, B.N. and Wills, R.D. (1980). Detailed characteristics of the high-energy gamma radiation from PSR 0833−45 measured by COS-B. *Astronomy and Astrophysics*, **90**, 163–169.

Kaspi, V.M. (2000). Neutron star/supernova remnant associations. In M. Kramer, N. Wex and R. Wielebinski (eds), *Pulsar Astronomy − 2000 and Beyond*, ASP Conference Series, Vol. 202, Astronomical Society of the Pacific, San Francisco, p. 485.

Kaspi, V.M., Chakrabarty, D. and Steinberger, J. (1999). Precision timing of two anomalous X-ray pulsars. *Astrophysical Journal*, **525**, L33–L36.

Kaspi, V.M., Johnston, S., Bell, J.F., Manchester, R.N., Bailes, M., Bessell, M., Lyne, A.G. and D'Amico, N. (1994). A massive radio pulsar binary in the Small Magellanic Cloud. *Astrophysical Journal*, **423**, L43–L45.

Kawai, N. and Saito, Y. (1999). Non-thermal X-ray emission of pulsars. *Astrophysical Letters and Communications*, **38**, 1–8.

Kawai, N. and Tamura, K. (1996). Recent X-ray observations of pulsar nebulae. In S. Johnston, M.A. Walker and M. Bailes (eds), *Pulsars: Problems and Progress*, ASP Conference Series, Vol. 105, Astronomical Society of the Pacific, San Francisco, p. 367.

Kellett, B.J., Branduardi-Raymont, G., Culhane, J.L., Mason, I.M., Mason, K.O. and Whitehouse, D.R. (1987). EXOSAT observations of the SNR PKS 1209−52. *Monthly Notices of the Royal Astronomical Society*, **225**, 199–212.

Kendziorra, E., Staubert, R., Pietsch, W., Reppin, C., Sacco, B. and Trümper, J. (1977). Hercules X-1: The 1.24 second pulsation in hard X-rays. *Astrophysical Journal*, **217**, L93–L96.

Kniffen, D.A. (1990). The Gamma-Ray Observatory. In Y. Kondo (ed.), *Observatories in Earth Orbit and Beyond*, Proceedings of the 123rd Colloquium of the International Astronomical Union, Kluwer Academic, Dordrecht, p. 63.

Kniffen, D.A., Hartman, R.C., Thompson, D.J., Bignami, G.F., Fichtel, C.E., Tümer, T. and Ögelman, H. (1974). Gamma radiation from the Crab Nebula above 35 MeV. *Nature*, **251**, 397–399.

Koptsevich, A.B., Pavlov, G.G., Shibanov, Yu.A., Sokolov, V.V., Zharikov, S.V. and Kurt, V.G. (2001). Optical photometry of the PSR B0656+14 and its neighborhood. *Astronomy and Astrophysics*, **370**, 1004–1016.

Kouveliotou, C., Dieters, S., Strohmayer, T., van Paradijs, J., Fishman G.L., Meegan, C.A., Hurley, K., Kommers, J., Smith, I., Frail, D. and Murakami, T. (1998). An X-ray pulsar with a superstrong magnetic field in the soft gamma-ray repeater SGR 1806−20. *Nature*, **393**, 235–237.

Kriss, G.A., Becker, R.H., Helfand, D.J. and Canizares, C.R. (1985). G27.4+0.0: A galactic supernova remnant with a central compact source. *Astrophysical Journal*, **288**, 703–706.

Kuiper, L., Hermsen, W., Krijger, J.M., Bennett, K., Carramiñana, A., Schönfelder, V., Bailes, M. and Manchester, R.N. (1999). COMPTEL detection of pulsed gamma-ray emission from PSR B1509−58 up to at least 10 MeV. *Astronomy and Astrophysics*, **351**, 119–132.

Kuiper, L., Hermsen, W., Verbunt, F., Thompson, D.J., Stairs, I.S., Lyne, A.G., Strickman, M.S. and Cusumano, G. (2000). The likely detection of pulsed high-energy gamma-ray emission from millisecond pulsar PSR J0218+4232. *Astronomy and Astrophysics*, **359**, 615–626.

Kulkarni, S.R. and Anderson, S.B. (1996). Pulsars in globular clusters. In P. Hut and J. Makino (eds), *Dynamical Evolution of Star Clusters − Confrontation of Theory and Observations*, IAU Symposium 174, Kluwer Academic, Dordrecht, p. 181.

Kulkarni, S.R. and Frail, D.A. (1993). Identification of a supernova remnant coincident with the soft gamma-ray repeater SGR 1806−20. *Nature*, **356**, 33–35.

Kulkarni, S.R. and van Kerkwijk, M.H. (1998). Optical observations of the isolated neutron star RX J0720.4−3125. *Astrophysical Journal*, **507**, L49–L53.

Kundt, W. and Schaaf, R. (1993). Towards solution of the pulsar problem. *Astrophysics and Space Science*, **200**, 251–270.

Landau, L. (1932). *Phys. Z. Sowjetunion*, **1**, 285.

Large, M.I., Voughan, A.E. and Mills, B.Y. (1968). A pulsar supernova association? *Nature*, **220**, 340–341.

Lattimer, J.M., Pethick, C.J., Prakash, M. and Haensel, P. (1991). Direct Urca process in neutron stars. *Physical Review Letters*, **66**, 2701–2704.

Lommen, A.N., Zepka, A., Backer, D.C., Cordes, J.M., Arzoumanian, Z., McLaughlin, M. and Xilouris, K. (2000). New pulsars from an Arecibo drift scan search. *Astrophysical Journal*, **545**, 1007–1014.

Long, K.S. and Helfand, D.J. (1979). Supernova remnants in the Large Magellanic Cloud. *Astrophysical Journal*, **234**, L77–L81.

Lyne, A.G., Camilo, F., Manchester, R.N., Bell, J.F., Kaspi, V.M., D'Amico, N., McKay, N.P.F., Crawford, F., Morris, D.J., Sheppard, D.C. and Stairs, I.H. (2000). The Parkes Multibeam Pulsar Survey: PSR J1811−1736, a pulsar in a highly eccentric binary system. *Monthly Notices of the Royal Astronomical Society*, **312**, 698–702.

Lyne, A.G, Pritchard, R.S. and Smith, F.G. (1988). Crab Pulsar timing 1982–87. *Monthly Notices of the Royal Astronomical Society*, **233**, 667–676.

Lyne, A.G, Pritchard, R.S, Graham-Smith, F. and Camilo, F. (1996). Very low braking index for the Vela Pulsar. *Nature*, **381**, 497–498.

Manchester, R.N., Lyne, A.G., Camilo, F., Kaspi, V.M., Stairs, I.H., Crawford, F., Morris, D.J., Bell, J.F. and D'Amico, N. (2000). Timing the Parkes Multibeam pulsars. In M. Kramer, N. Wex and R. Wielebinski (eds), *Pulsar Astronomy − 2000 and Beyond*, ASP Conference Series, Vol. 202, Astronomical Society of the Pacific, San Francisco, p. 49.

Manchester, R.N., Lyne, A.G., Goss, W.M., Smith, F.G., Disney, M.J., Hartley, K.F., Jones, D.H.P., Wellgate, G.B., Danziger, I.J., Murdin, P.G., Peterson, B.A. and Wallace, P.T. (1978). Optical observations of southern pulsars. *Monthly Notices of the Royal Astronomical Society*, **184**, 159–170.

Marsden, D., Rotschild, R.E. and Lingenfelter, R.E. (1999). Is SGR 1900+14 a magnetar? *Astrophysical Journal*, **520**, L107–L110.

Marshall, F.E., Gotthelf, E.V., Zhang, W., Middleditch, J. and Wang, Q.D. (1998). Discovery of an ultrafast X-ray pulsar in the supernova remnant N157B. *Astrophysical Journal*, **499**, L179.

Martin, C., Halpern, J.P. and Schiminovich, D. (1998). The optical spectrum of the Geminga pulsar. *Astrophysical Journal*, **494**, L211.

Mazets, E.P., Golenetskii, S.V. and Guryan, Yu. (1979b). Soft gamma-ray bursts from the source B1900+14. *Soviet Astronomy Letters*, **5**, 343–344.

Mazets, E.P., Golentskii, S.V., Ilinskii, V.N., Aptekar, R.L. and Guryan, Iu.A. (1979a). Observations of a flaring X-ray pulsar in Dorado. *Nature*, **282**, 587–589.

Mereghetti, S. (2000). The anomalous X-ray pulsars. In *The Neutron Star–Black Hole Connection*, NATO ASI Series, to be published.

Mereghetti, S., Bignami, G.F. and Caraveo, P.A. (1996). The X-ray source at the center of G296.5+10.0 as a young isolated neutron star. *Astrophysical Journal*, **464**, 842.

Mereghetti, S. and Stella, L. (1995). The very low mass X-ray binary pulsars: A new class of sources? *Astrophysical Journal*, **442**, L17–L20.

Michel, F.C. (1991). *Theory of Neutron Star Magnetospheres*, University of Chicago Press.

Mignani, R., Caraveo, P.A. and Bignami, G.F. (1997). Hubble Space Telescope discovers optical emission from the radio pulsar PSR 1055 − 52. *Astrophysical Journal*, **474**, L51.

Mineo, T., Cusumano, G., Kuiper, L., Hermsen, W., Massaro, E., Becker, W., Nicastro, L., Sacco, B., Verbunt, F., Lyne, A.G., Stairs, I.H. and Shibata, S. (2000). The pulse shape and spectrum of the millisecond pulsar PSR J0218+4232 in the energy band 1–10 keV observed with BeppoSAX. *Astronomy and Astrophysics*, **355**, 1053–1059.

Mineo, T., Cusumano, G., Massaro, E., Nicasto, l., Parmar, A.N. and Sacco, B. (1999). Timing and spectral properties of PSR B0540−52. Astronomy and Astrophysics, 348, 519−523.

Morrison, P. (1958). On gamma-ray astronomy. *Il Nuovo Cimento*, **7**, 858–865.

Morrison, P., Olbert, S. and Rossi, B. (1954). The origin of cosmic rays. *Physical Review*, **94**, 440–453.

Motch, C. and Haberl, F. (1998). Constraints on optical emission from the isolated neutron star candidate RXJ0720.4−3125. *Astronomy and Astrophysics*, **333**, L59–L62.

Motch, C., Haberl, F., Zickgraf, F.-J., Hasinger, G. and Schwope, A.D. (1999). The isolated neutron star candidate RX J1605.3+3249. *Astronomy and Astrophysics*, **351**, 177–184.

Murakami, T., Kubo, S., Shibazaki, N., Takeshima, T., Yoshida, A. and Kawai, N. (1999). Accurate position of SGR 1900+14 by bursts and changes in pulse period and folded pulse profile with ASCA. *Astrophysical Journal*, **510**, L119–L122.

Neuhäuser, R. and Trümper, J. (1999). On the number of accreting and cooling isolated neutron stars detectable with the ROSAT All-Sky Survey. *Astronomy and Astrophysics*, **343**, 151–156.

Ögelman, H. (1995). X-ray observations of cooling neutron stars. In A. Alpar, U. Kiziloglu and J. van Paradijs (eds), *The Lives of Neutron Stars*, Kluwer Academic, Dordrecht, p. 101.

Ögelman, H., Finley, J.P. and Zimmerman, H.U. (1993). Pulsed X-rays from the Vela Pulsar. *Nature*, **361**, 136–138.

Oosterbroek, T., Parmar, A.N., Mereghetti, S. and Israel, G.L. (1998). The two-component X-ray spectrum of the 6.4 s pulsar 1E 1048.1−5937. *Astronomy and Astrophysics*, **334**, 925–930.

Oppernheimer, J.R. and Volkoff, G.M. (1939). On massive neutron cores. *Physical Review*, **55**, 374–381.

Pacini, F. (1967). Energy emission from a neutron star. *Nature*, **216**, 567–568.

Pacini, F. (1968). Rotating neutron stars, pulsars and supernova remnants. *Nature*, **219**, 145–146.

Page, D. and Applegate, J.L. (1992). The cooling of neutron stars by the direct Urca process. *Astrophysical Journal*, **394**, L17–L20.

Page, D., Shibanov, Yu.A. and Zavlin, V.E. (1996). Temperature, distance, and cooling of the VELA pulsar. In H.U. Zimmermann, J. Trümper and H. Yorke (eds), *Röntgenstrahlung from the Universe*, MPE Report 263, pp. 173–174.

Pavlov, G.G. and Meszaros, P. (1993). Finite-velocity effects on atoms in strong magnetic fields and implications for neutron star atmospheres. *Astrophysical Journal*, **416**, 752.

Pavlov, G.G. and Potekhin, Y.A. (1995). Bound–bound transitions in strongly magnetized hydrogen plasma. *Astrophysical Journal*, **450**, 883.

Pavlov, G.G., Sanwal, D., Garmire, G.P., Zavlin, V.E., Burwitz, V. and Dodson, R. (2000a). Observations of the Vela Pulsar and its compact nebula with the Chandra High Resolution Camera. AAS Meeting 196, 37.04.

Pavlov, G.G. and Shibanov, Yu.A. (1978). Thermal emission of an optically thick plasma containing a strong magnetic field. *Soviet Astronomy*, **22**, 214–222.

Pavlov, G.G., Shibanov, Yu.A., Ventura, J. and Zavlin, V.E. (1994). Model atmospheres and radiation of magnetic neutron stars: Anisotropic thermal emission. *Astronomy and Astrophysics*, **289**, 837–845.

Pavlov, G.G., Shibanov, Y.A., Zavlin, V.E. and Meyer, R.D. (1995). Neutron star atmospheres. In A. Alpar, U. Kiziloghu and J. van Paradijs (eds), *The Lives of Neutron Stars*, Kluwer Academic, Dordrecht, p. 71.

Pavlov, G.G., Stringfellow, G.S. and Cordova, F.A. (1996a). Hubble Space Telescope observations of isolated pulsars. *Astrophysical Journal*, **467**, 370.

Pavlov, G.G., Welty, A.D. and Cordova, F.A. (1997). Hubble Space Telescope observations of the middle-aged pulsar 0656+14. *Astrophysical Journal*, **489**, L75.

Pavlov, G.G. and Zavlin, V.E. (1997). Mass-to-radius ratio for the millisecond pulsar J0437 − 4715. *Astrophysical Journal*, **490**, L91.

Pavlov, G.G. and Zavlin, V.E. (1999). Cassiopeia A. *IAU Circular*, 7270, 1.

Pavlov, G.G. and Zavlin, V.E. (2000). Polarization of thermal X-rays from isolated neutron stars. *Astrophysical Journal*, **529**, 1011–1018.

Pavlov, G.G. and Zavlin, V.E. (2001). *Astrophysical Journal*, in press.

Pavlov, G.G., Zavlin, V.E., Aschenbach, B., Trümper, J. and Sanwal, D. (2000b). The compact central object in Cassiopeia A: A neutron star with hot polar caps or a black hole? *Astrophysical Journal*, **531**, L53–L56.

Pavlov, G.G., Zavlin, V.E., Trümper, J. and Neuhäuser, R. (1996b). Multiwavelength observations of isolated neutron stars as a tool to probe the properties of their surfaces. *Astrophysical Journal*, **472**, L33.

Petre, R., Becker, C.M. and Winkler, P.F. (1996). A central stellar remnant in Puppis A. *Astrophysical Journal*, **465**, L43.

Petre, R., Kriss, G.A., Winkler, P.F. and Canizares, C.R. (1982). A high-resolution X-ray image of Puppis A: Inhomogeneities in the interstellar medium. *Astrophysical Journal*, **258**, 22–30.

Plucinsky, P.P., Snowden, S.L., Aschenbach, B., Egger, R., Edgar, R.J. and McCammon, D. (1996). ROSAT survey observations of the Monogem Ring. *Astrophysical Journal*, **463**, 224.

Rajagopal, M. and Romani, R.W. (1996). Model atmospheres for low-field neutron stars. *Astrophysical Journal*, **461**, 327.

Rajagopal, M., Romani, R.W. and Miller, M.C. (1997). Magnetized iron atmospheres for neutron stars. *Astrophysical Journal*, **479**, 347.

Ramanamurthy, P.V., Fichtel, C.E., Harding, A.K., Kniffen, D.A., Sreekumar, P. and Thompson, D.J. (1996). PSR B0656+14 in relation to other high-energy gamma-ray pulsars. *Astronomy and Astrophysics, Suppl.*, **120**, 115–116.

Rasio, F.A., Pfahl, E.D. and Rappaport, S. (2000). Formation of short-period binary pulsars in globular clusters. *Astrophysical Journal*, **532**, L47–L50.

Romani, R.W. (1987). Model atmospheres for cooling neutron stars. *Astrophysical Journal*, **313**, 718–726.

Romani, R.W. (1996). Gamma-ray pulsars: Radiation processes in the outer magnetosphere. *Astrophysical Journal*, **470**, 469.

Romani, R.W. and Yadigaroglu, I.-A. (1995). Gamma-ray pulsars: Emission zones and viewing geometries. *Astrophysical Journal*, **438**, 314–321.

Rosenfeld, L. (1974). In F. Pacini (ed.), *Astrophysics and Gravitation*, Proceedings of the 16th Solvay Conference on Physics, Editions de l'Université de Bruxelles, Brussels, p. 174.

Rothschild, R.E., Kulkarni, S.R. and Lingenfelter, R.E. (1994). Discovery of an X-ray source coincident with the soft gamma-repeater 0425 − 66. *Nature*, **368**, 432.

Ruderman, M. and Sutherland, P.G. (1975). Theory of pulsars: Polar caps, sparks, and coherent microwave radiation. *Astrophysical Journal*, **196**, 51–72.

Schwope, A.D., Hasinger, G., Schwarz, R., Haberl, F. and Schmidt, M. (1999). The isolated neutron star candidate RBS1223 (1RXS J130848.6 + 212708). *Astronomy and Astrophysics*, **341**, L51–L54.

Seward, F.D. (1990). Einstein observations of galactic supernova remnants. *Astrophysical Journal*, Suppl., **73**, 781–819.

Seward, F.D., Charles, P.A. and Smale, A.P. (1986). A 6 Second periodic X-ray source in Carina. *Astrophysical Journal*, **305**, 814–816.

Seward, F.D. and Harnden, F.R. (1982). A new, fast X-ray pulsar in the supernova remnant MSH 15 − 52. *Astrophysical Journal*, **256**, L45–L47.

Seward, F.D. and Harnden, F.R., Jr (1994). The outer shell of SNR 0540 − 69.3. *Astrophysical Journal*, **421**, 581–584.

Seward, F.D., Harnden, F.R. and Helfand, D.J. (1984). Discovery of a 50 millisecond pulsar in the Large Magellanic Cloud. *Astrophysical Journal*, **287**, L19–L22.

Seward, F.D. and Wang, Z.-R. (1988). Pulsars, X-ray synchrotron nebulae, and guest stars. *Astrophysical Journal*, **332**, 199–205.

Shearer, A., Redfern, R.M., Gorman, G., Butler, R., Golden, A., O'Kane, P., Beskin, G.M., Neizvestny, S.I., Neustroev, V.V., Plokhotnichenko, V.L. and Cullum, M. (1997). Pulsed optical emission from PSR 0656+14. *Astrophysical Journal*, **487**, L181.

Shibanov, Yu.A., Pavlov, G.G., Zavlin, V.E., Qin, L. and Tsuruta, S. (1995). Anisotropic cooling and atmospheric radiation of neutron stars with strong magnetic field. *Annals of the New York Academy of Sciences*, **759**, 291.

Shibanov, Yu.A. and Yakovlev, D.G. (1996). On cooling of magnetized neutron stars. *Astronomy and Astrophysics*, **309**, 171–178.

Shibanov, Yu.A., Zavlin, V.E., Pavlov, G.G. and Ventura, J. (1992). Model atmospheres and radiation of magnetic neutron stars: I. The fully ionized case. *Astronomy and Astrophysics*, **266**, 313–320.

Sonobe, T., Murakami, T., Kulkarni, S.R., Aoki, T. and Yoshida, A. (1994). Characteristics of the persistent emission of SGR 1806 − 20. *Astrophysical Journal*, **436**, L23–L25.

Staelin, D.H. and Reifenstein, III, E.C. (1968). Pulsating radio sources near the Crab Nebula. *Science*, **162**, 1481–1483.

Strickman, M.S., Harding, A.K. and de Jager, O.C. (1999). A Rossi X-ray Timing Explorer observation of the Vela pulsar: Filling in the X-ray gap. *Astrophysical Journal*, **524**, 373–378.

Sturner, S.J. and Dermer, C.D. (1994). On the spectra and pulse profiles of gamma-ray pulsars. *Astrophysical Journal*, **420**, L79–L82.

Sturner, S.J., Dermer, C.D., Michel, F.C. (1995). Magnetic Compton-induced pair cascade model for gamma-ray pulsars. *Astrophysical Journal*, **445**, 736–755.

Sugizaki, M., Nagase, F., Torii, K., Kunigasa, K., Asanuma, T., Matsuzaki, K., Koyama, K. and Yamauchi, S. (1997). Discovery of an 11-s X-ray pulsar in the galactic-plane section of the Scorpius constellation. *Publications of the Astronomical Society of Japan*, **49**, L25–L30.

Takahashi, M., Shibata, S., Torii, K., Saito, Y. and Kawai, N. (1998). PSR 1937 + 21. *IAU Circular*, 7030, 3.

Tanaka, Y., Inoue, H. and Holt, S.S. (1994). The X-ray astronomy satellite ASCA. *Publications of the Astronomical Society of Japan*, **46**, L37–L41.

Tananbaum, H. (1999). Cassiopeia A. *IAU Circular*, 7246, 1.

Tananbaum, H., Gursky, H., Kellogg, E.M., Levinson, R., Schreier, E., and Giacconi, R. (1972). Discovery of a periodic pulsating binary X-ray source in Hercules from Uhuru. *Astrophysical Journal*, **174**, L143.

Taylor, B.G., Andersen, R.D., Peacock, A. and Zobl, R. (1981). The EXOSAT mission. *Space Science Reviews*, **30**, 479–494.

Taylor, J.H., Manchester, R.N. and Lyne, A.G. (1993). Catalog of 558 pulsars. *Astrophysical Journal*. Suppl., **88**, 529–568.

Thompson, C. and Duncan, R.C. (1995). The soft gamma repeaters as very strongly magnetized neutron stars: I. Radiative mechanism for outbursts. *Monthly Notices of the Royal Astronomical Society*, **275**, 255–300.

Thompson, C. and Duncan, R.C. (1996). The soft gamma repeaters as very strongly magnetized neutron stars: II. Quiescent neutrino, X-ray, and Alfvén wave emission. *Astrophysical Journal*, **473**, 322.

Thomson, D.J., Bailes, M., Bertsch, D.L., Cordes, J., D'Amico, N., Esposito, J.A., Finley, J., Hartman, R.C., Hermsen, W., Kanbach, G., Kaspi, V.M., Kniffen, D.A., Kuiper, L., Lin, Y.C., Lyne, A., Manchester, R., Matz, S.M., Mayer-Hasselwander, H.A., Michelson, P.F., Nolan, P.L., Ögelman, H., Pohl, M., Ramanamurthy, P.V., Sreekumar, P., Reimer, O., Taylor, J.H. and Ulmer, M. (1999). Gamma Radiation from PSR B1055 − 52. *Astrophysical Journal*, **516**, 297–306.

Thompson, D.J., Fichtel, C.E., Kniffen, D.A. and Ögelman, H.B. (1975). SAS-2 high-energy gamma-ray observations of the Vela Pulsar. *Astrophysical Journal*, **200**, L79–L82.

Torii, K., Kunigasa, K., Katayama, K., Tsunemi, H. and Yamauchi, S. (1998). Discovery of a 7 second X-ray pulsar, AX J1845.0 − 0300. *Astrophysical Journal*, **503**, 843.

Treves, A., Turolla, R., Zane, S., Colpi, M. (2000). Isolated neutron stars: Accretors and Coolers. *Publications of the Astronomical Society of the Pacific*, **112**, 297–314.

Trümper, J. (1982). The ROSAT mission. *Advances in Space Research*, **2**, 241–249.

Trümper, J., Pietsch, W., Reppin, C., Voges, W., Staubert, R. and Kendziorra, E. (1978). Evidence for strong cyclotron line emission in the hard X-ray spectrum of Hercules X-1. *Astrophysical Journal*, **219**, L105–L110.

Tsuruta, S. (1998). Thermal properties and detectability of neutron stars: II. Thermal evolution of rotation-powered neutron stars. *Physics Reports*, **292**, 1–130.

Tucker, W. (1984). *The Star Splitters: The High Energy Astronomy Observatories*, NASA SP-466.

Tuohy, I.R. and Garmire, G.P. (1980). Discovery of a compact X-ray source at the center of the supernova remnant RCW 103. *Astrophysical Journal*, **239**, L107–L110.

van Kerkwijk, M.H., Kulkarni, S.R., Matthews, K. and Neugebauer, G. (1995). A luminous companion to SGR 1806 − 20. *Astrophysical Journal*, **444**, L33–L35.

van Paradijs, J., Taam, R.W. and van den Heuvel, E.P.J. (1995). On the nature of the 'anomalous' 6-s X-ray pulsars. *Astronomy and Astrophysics*, **299**, L41.

Vasisht, G. and Gotthelf, E.V. (1997). The discovery of an anomalous X-ray pulsar in the supernova remnant Kes 73. *Astrophysical Journal*, **444**, L33–L35.

Verbunt, F., Kuiper, L., Belloni, T., Johnston, H.M., de Bruyn, A.G., Hermsen, W. and van der Klis, M. (1996). High-energy observations of the millisecond pulsar PSR J0218 + 4232. *Astronomy and Astrophysics*, **311**, L9–L12.

Walter, F.M., An, P., Lattimer, J. and Prakash, M. (2000). The isolated neutron star RX J185635 − 3754. In P.C.H. Martens, S. Tsuruta and M.A. Weber (eds), *Highly Energetic Physical Processes and Mechanics for Emission from Astrophysical Plasmas*, IAU Symposium 195, Astronomical Society of the Pacific, San Francisco, p. 437.

Walter, F.M. and Matthews, L.D. (1997). The optical counterpart of the isolated neutron star RX J185635 − 3754. *Nature*, **389**, 358–360.

Walter, F.M., Volk, S.J. and Neuhäuser, R. (1996). Discovery of nearby isolated neutron star. *Nature*, **379**, 233–235.

Wang, F.Y.-H., Ruderman, M., Halpern, J.P. and Zhu, T. (1998). Models for X-ray emission from isolated pulsars. *Astrophysical Journal*, **498**, 373.

Wang, Q.D. and Gotthelf, E.V. (1998). ROSAT HRI detection of the 16 ms pulsar PSR J0537 − 6910 inside supernova remnant N157B. *Astrophysical Journal*, **509**, L109–L112.

Wang, Z.R. and Seward, F.D. (1984). An X-ray investigation of the unusual supernova remnant CTB 80. *Astrophysical Journal*, **285**, 607–612.

Weber, F. (1999) *Pulsars as Astrophysical Laboratories for Nuclear and Particle Physics*, Institute of Physics, Bristol.

Weisskopf, M.C., Hester, J.J., Tennant, A.F., Elsner, R.F., Schulz, N.S., Marshall, H.L., Karovska, M., Nichols, J.S., Swartz, D.A., Kolodziejczak, J.J. and O'Dell, S.L. (2000). Discovery of spatial and spectral structure in the X-ray emission from the Crab Nebula. *Astrophysical Journal*, **536**, L81–L84.

White, N.E., Angelini, L., Ebisawa, K., Tanaka, Y. and Ghosh, P. (1996). The spectrum of the 8.7s X-ray pulsar 4U0142+61. *Astrophysical Journal*, **463**, L83.

Wills, R.D., Bennett, K., Bignami, G.F., Buccheri, R., Caraveo, P.A., Hermsen, W., Kanbach, G., Masnou, J.L., Mayer-Hasselwander, H.A., Paul, J.A. and Sacco, B. (1982). High-energy gamma-ray light curve of PSR0531+21. *Nature*, **296**, 723–726.

Woods, P.M., Kouveliotou, C., van Paradijs, J., Briggs, M.S., Hurley, K., Göğüs, E., Preece, R.D., Giblin, T.W., Thompson, C. and Duncan, R.C. (1999c). Hard burst emission from the soft gamma repeater SGR 1900+14. *Astrophysical Journal*, **527**, L47–L50.

Woods, P.M., Kouveliotou, C., van Paradijs, J., Finger, M.H. and Thompson, C. (1999a). Bepposax observations of SGR 1900−14 in quiescence and during an active period. *Astrophysical Journal*, **518**, L103–L106.

Woods, P.M., Kouveliotou, C., van Paradijs, J., Finger, M.H., Thompson, C., Duncan, R.C., Hurley, K., Strohmayer, T., Swank, J. and Murakami, T. (1999b). Variable spin-down in the soft gamma repeater SGR 1900+14 and correlation with the burst activity. *Astrophysical Journal*, **524**, L55–L58.

Woods, P.M., Kouveliotou, C., van Paradijs, J., Hurley, K., Kippen, R.M., Finger, M.H., Briggs, M.S., Dieters, S. and Fishman, G.J. (1999d). Discovery of a new soft gamma repeater, SGR 1627−41. *Astrophysical Journal*, **519**, L139–L142.

Yakovlev, D.G., Levenfish, K.P. and Shibanov, Yu.A. (1999). Cooling neutron stars and superfluidity in their interiors. *Soviet Physics Uspekhi*, **169**, 825.

Yancopoulos, S., Hamilton, T.T. and Helfand, D.J. (1994). The detection of pulsed X-ray emission from a nearby radio pulsar. *Astrophysical Journal*, **429**, 832–843.

Zavlin, V.E. and Pavlov G.G. (1998). Soft X-rays from polar caps of the millisecond pulsar J0437-4715. *Astronomy and Astrophysics*, **329**, 583–598.

Zavlin, V.E., Pavlov, G.G. and Halpren, J.P. (2001). *Astrophysical Journal*, submitted.

Zavlin, V.E., Pavlov, G.G., Sanwal, D. and Trümper, J. (2000). Discovery of 424 millisecond pulsations from the radio-quiet neutron star in the supernova remnant PKS 1209−51/52. *Astrophysical Journal*, **540**, L25–L28.

Zavlin, V.E., Pavlov, G.G. and Shibanov, Yu.A. (1996). Model neutron star atmospheres with low magnetic fields: I. Atmospheres in radiative equilibrium. *Astronomy and Astrophysics*, **315**, 141–152.

Zavlin, V.E., Pavlov, G.G., Shibanov, Yu.A. and Ventura, J. (1995a). Thermal radiation from rotating neutron star: effect of the magnetic field and surface temperature distribution. *Astronomy and Astrophysics*, **297**, 441.

Zavlin, V.E., Pavlov, G.G. and Trümper, J. (1998). The neutron star in the supernova remnant PKS 1209−52. *Astronomy and Astrophysics*, **331**, 821–828.

Zavlin, V.E., Shibanov, Yu.A. and Pavlov, G.G. (1995b). Effects of the gravitational field of a neutron star on the emission from hot polar spots on the surface of radio pulsars. *Pis'ma v Astronomicheskii Zhurnal (Astronomy Letters)*, **21**, 149–158.

Zavlin, V.E., Trümper, J. and Pavlov, G.G. (1999). X-ray emission from the radio-quiet neutron star in Puppis A. *Astrophysical Journal*, **525**, 959–967.

Zhang, B. and Harding, A.K. (2000). Full polar cap cascade scenario: Gamma-ray and X-ray luminosities from spin-powered pulsars. *Astrophysical Journal*, **532**, 1150–1171.

32

ED P.J. VAN DEN HEUVEL*

Evolutionary concepts of binaries with compact objects

In 1962 the first extrasolar X-ray source, Sco X-1, was discovered (Giacconi *et al.* 1962) and by the end of the 1960s several dozen such strong point X-ray sources had been discovered with rocket and balloon experiments.

Their marked concentration in the direction of the galactic center made clear that the distances of a number of them must be of the order of 8 kpc, implying a very large energy output in the form of X-rays, of the order of 10^{37} to 10^{38} ergs s^{-1} (some 10^4 times the total energy output of the Sun). What could be the mechanism generating these enormous X-ray luminosities? Largely thanks to the Russian team of Zel'Dovitch and co-workers, since the mid-1960s the idea arose that in a binary system the process of accretion of matter flowing over from a normal companion star to a neutron star or a black hole might power these strong galactic X-ray sources. Indeed, the simple process of accretion of an amount of mass m onto a neutron star (black hole) releases some $0.15\,mc^2$ (0.06 to $0.42\,mc^2$) of gravitational binding energy which, converted into heat, is available for emission in the form of X-rays. (The process of mass accretion onto a supermassive black hole had already been suggested as the energy source for quasars and active galaxy nuclei by Salpeter (1964), Zel'Dovitch (1964), and Zel'Dovitch and Novikov (1964).)

It should be kept in mind, however, that the existence of neutron stars in nature was not known before the end of 1968 – the year in which the discovery of radio pulsars was announced (Hewish *et al.* 1968). The discovery of the Crab Nebula pulsar in November 1968 (Staelin and Reifenstein 1968) and the detection of its large spindown rate made clear that pulsars are neutron stars and that neutron stars are born in a supernova event (Gold 1969; Monaghan 1969) just as had been predicted 34 years earlier by Baade and Zwicky (1934). Before 1968 neutron stars and black holes had been purely theoretical concepts, based on theoretical insights developed in the pioneering studies by Oppenheimer and Volkoff (1938) and Oppenheimer and Snijder (1939), respectively, and subsequently studied by various groups, notably those of J.A. Wheeler in the USA and Ya.B. Zel'Dovitch in the USSR. The first to search for black holes in binary systems were Zel'Dovitch and Guseinov (Guseinov and Zel'Dovitch 1966; Zel'Dovitch and Guseinov 1966), who searched for spectroscopic binaries with massive unseen secondary stars. They did not yet, however, mention the possibility that such binaries might emit X-rays, but Novikov and Zel'Dovitch (1966) did so slightly later. Following the discovery of the faint blue optical counterpart of Sco X-1 (Sandage *et al.* 1966) this early work culminated in Shklovskii's (1967) neutron star binary model for Sco X-1. This author showed that the optical light in the system could not arise from the same source as the X-rays, and that the X-ray energy distribution is consistent with thermal bremsstrahlung from an optically thin plasma accreting onto a neutron star. Since the optical spectrum of the source resembles that of a cataclysmic variable (CV: these are binary systems consisting of an accreting white dwarf and a low-mass ordinary star; Chapter 33) and since no stellar spectrum is seen (implying that the companion star is faint) it was postulated that the neutron star is in a binary and accretes matter from a low-mass companion star. It took another nine years before this "low-mass X-ray binary" model for Sco X-1 was confirmed by the detection of its 0.86-day orbital period by Gottlieb *et al.* (1975).

Before that, however, the first X-ray satellite Uhuru (USA, 1970) had, in 1971, discovered the existence of the

*Universiteit van Amsterdam, The Netherlands

pulsating and eclipsing binary X-ray source Cen X-3, which left no doubt about the existence of neutron stars moving in orbits around massive ordinary stars (Schreier et al. 1972). Shortly before this, earlier in 1971, the Uhuru group had discovered rapid X-ray variations in Cyg X-1, with timescales down to 50 ms, with no obvious periodicity (Oda et al. 1971). This rapid variability indicated that the X-ray source in the system cannot be larger than about 10^4 km. Webster and Murdin (1972) and Bolton (1972) independently that same year identified the bright O9.7 supergiant star HD 226868 as the optical counterpart to this source. This identification was an indirect one, through the accurate arc second position of a radio source that had been discovered independently by Braes and Miley (1971) and Hjellming and Wade (1971) in the X-ray error box of Cyg X-1. Webster and Murdin and Bolton found the blue supergiant to be a 5.6-day single-lined spectroscopic binary with a large velocity amplitude (72 km s^{-1}) indicating, if the O 9.7 supergiant has a normal mass ($\geq 15 M_\odot$) for its spectral type, the presence of a companion of mass $>3 M_\odot$. As this is above the upper mass limit of a neutron star (Nauenberg and Chapline 1973; Rhoades and Ruffini 1974; Kalogera and Baym 1996) they suggested Cyg X-1 to be a black hole – a suggestion that nowadays has been fully accepted. However, at the time this was, like that for Sco X-1, a rather indirect indication for the existence of X-ray binaries, and not yet completely convincing. For example, Kristian et al. (1971) dismissed the blue supergiant as the optical counterpart of the X-ray source. On the other hand, there could be little doubt that the X-ray source and the radio source were connected, as the radio source appeared just at the time when a dramatic change in the spectrum of the X-ray source occurred (Tananbaum 1973). Since the radio source coincided within a few arc seconds with the blue supergiant star, it seemed quite likely that the star, the radio source, and the X-ray source were connected. However, the X-ray source did not eclipse and neither the X-ray source nor the radio source showed a trace of a 5.6-day period variability that might connect them with the 5.6-day blue supergiant binary (only recently such periodic variability has been established at radio wavelengths (Pooley et al. 1999)).

For these reasons it was only with the discovery of the eclipsing pulsating binary X-ray source Cen X-3 by Schreier et al. (1972) that this collection of observations found a definitive explanation. Here one observed for the first time an X-ray pulsar (neutron star) that shows a beautiful 2.087-day sinusoidal Doppler modulation of its 4.84 s pulse period and is eclipsed every orbit for about 0.5 days. From the Doppler effect a projected orbital velocity of 415.1 ± 0.4 km s^{-1} was derived leading to a minimum companion mass of about 17 solar masses. The presence of this massive companion made it immediately clear that Cyg X-1 might form a similar system with the massive blue supergiant star HD 226868. Thus, by 1972 the existence of both neutron stars and black holes in close orbits around massive companion stars appeared well established. In the same year, Uhuru discovered the second pulsating and eclipsing binary X-ray source Her X-1, in which the 1.2 s period X-ray pulsar orbits in 1.7 days a star of rather low mass, about $2.0 M_\odot$ (Tananbaum et al. 1972). In the 1970s and 1980s more and more X-ray binaries were discovered. They were found to fall into two broad categories: the high-mass X-ray binaries (HMXBs) like Cen X-3 and the low-mass X-ray binaries (LMXBs) like Her X-1 and Sco X-1 (Chapters 34–36).

The recognition that neutron stars and black holes can exist in close binary systems came at first as a surprise, as it did not seem to fit with the then current ideas about the evolution of binary systems. It was known from stellar evolution that the more massive a star is, the shorter its lifetime. Thus, the more massive component of a binary will be the first one to undergo a supernova explosion. If more than half the mass of a circular-orbit binary system is explosively ejected, the orbit of the system becomes hyperbolic and the system is disrupted (Blaauw 1961). This is a simple consequence of the virial theorem (Van den Heuvel 1994a). In a massive binary the mass of the neutron star remnant ($\sim 1.4 M_\odot$) is negligible with respect to the masses of the two components (and to the amount of mass ejected in the supernova), so at first sight one would always expect these systems to be disrupted by the first supernova explosion. Still more puzzling were the almost perfectly circular orbits of Cen X-3 and Her X-1, apparently showing no trace of the effects of the supernova (although Cyg X-1 still has a slightly eccentric orbit). For the HMXBs like Cen X-3, it was soon realized (Van den Heuvel and Heise 1972) that the survival of the systems was due to the effects of large-scale mass transfer that must have occured prior to the supernova explosion. This caused the initially more massive star in the system to have become much less massive than its companion at the time of the explosion, and prevented the system from being disrupted. Tidal effects subsequently circularized the orbits during several millions of years, before the reverse mass transfer began and the system became an X-ray source. (Several other authors somewhat later independently came to similar conclusions: Börner et al. (1972); Tutukov and Yungelson (1973)). The model of Van den Heuvel and Heise (1972) built on earlier work on close binary evolution developed primarily in the 1960s. Morton (1960) had been the first one to show that large-scale mass transfer may explain why the more evolved (sub-giant) component of a close binary system like Algol (β Persei) may be less massive than its brighter and less evolved companion star. The reason is that in the course of its evolution the envelope of a star gradually expands, and after the exhaustion of the hydrogen fuel in the stellar core, swells up to giant dimensions. However, in a close binary system there is no room for a giant star: the sizes of the stars are limited by the dimensions of their so-called Roche lobes, pear-shaped critical equipotential surfaces, surrounding the stars (Section 2). As soon as a star becomes larger than its Roche lobe, the matter outside the Roche lobe will flow over to its companion. The more massive star in a binary will be the first one to overflow its Roche lobe. Evolutionary calculations by Kippenhahn and Weigert (1967), Plavec (1967), and Paczynski (1966, 1971a) in the 1960s showed that this leads to the transfer of most of the star's hydrogen-rich envelope ($\geq 70\%$ of its mass) to its companion, reversing the mass ratio of the system.

This type of "conservative" evolution (in which mass and orbital angular momentum of the system are conserved) appears to be able to explain the formation of the HMXBs.

However, for the LMXBs it was much harder to understand how the system survived the supernova explosion. The same holds for the close double neutron stars, such as the Hulse–Taylor binary pulsar PSR 1913+16 (Hulse and Taylor 1975), which has an orbital period of only 7.75 hours and an orbital eccentricity of 0.615. These systems must have survived two supernova explosions! In the LMXBs and their close relatives the cataclysmic variables the companion of the compact object is a low-mass ordinary star like our Sun, and the orbital periods are in general very short: mostly between 11 minutes and about half a day (Chapter 34). All these systems must in the course of their lives (and, in the case of LMXBs and binary pulsars, before the last supernova in the system) have lost a very large amount of mass and orbital angular momentum. The evolution of these systems was therefore much more complicated than that of the HMXBs. The first models for the evolution of binaries with a large loss of mass and orbital angular momentum were made by Van den Heuvel and De Loore (1973) who showed that a HMXB may later in life turn into a very close binary system consisting of a helium star (the helium core of the massive star) and a compact star. They suggested that the 4.8 hour X-ray binary Cyg X-3 is such a helium-star system, which was confirmed almost 20 years later (Van Kerkwijk et al. 1992). The Hulse–Taylor binary pulsar is a logical later evolutionary product of such a system (Flannery and Van den Heuvel 1975; De Loore et al. 1975). The first model for explaining the origin of an LMXB was that of Sutantyo (1975a) for the origin of Her X-1. He showed that in order to obtain such a system one should start out with a binary with components that differ very much in mass. In this case, due to the large difference between the thermal timescales of the envelopes of the stars, the low-mass component can hardly accept any mass from its more massive companion, once that star begins to overflow its Roche lobe. As a result most of the overflowing matter is lost from the systems, together with its orbital angular momentum, and only the helium core of the more evolved star is left, together with the practically unchanged low-mass companion. Still, in order to not disrupt the system when this helium star explodes, the initial conditions must have been very fine-tuned. Therefore, the formation of a LMXB is a much rarer event than the formation of HMXBs (Van den Heuvel 1983, 1994a; Webbink and Kalogera 1994; Kalogera 1998a,b), and the same holds for double neutron stars.

An important new ingredient that has been introduced in these binary evolution models since 1975 (Flannery and Van den Heuvel 1975) is the occurrence of velocity "kicks" of the order a few hundred kilometers per second that are imparted to the neutron stars in their birth events. Ample evidence for the occurrence of such kicks has been inferred in the last decade from the space and velocity distribution of radio pulsars and from a variety of other observational facts (Dewey and Cordes 1987; Tauris et al. 2000; Hartmann 1996, 1997; Van den Heuvel and Van Paradijs 1997). Without the occurrence of these kicks a variety of properties of the X-ray binaries and the binary radio pulsars, including their birth rate in the galaxy, would be difficult to understand (Verbunt and Van den Heuvel 1995; Kalogera 1998a,b; Kalogera and Webbink 1998).

As mentioned above, in the mid-1970s a link was suggested between the massive X-ray binaries and the Hulse–Taylor binary pulsar. The magnetic field of this pulsar is relatively weak ($\sim 10^{10}$ G, some two orders of magnitude lower than the average of pulsar magnetic fields) and its spin period is very short (0.059 s). As it was believed at the time that neutron star magnetic fields decay spontaneously on a relatively short timescale ($\sim 10^7$ yr) the weak field suggested that this neutron star is old. Since in X-ray binaries with accretion disks angular momentum is fed to the neutron star, one observes the X-ray pulsars in these systems to show a gradual decrease of their pulse periods in the course of time (Chapter 35; Section 1.4), so-called "spin-up." Bisnovatyi-Kogan and Komberg (1975) suggested that if the binary system is disrupted in the second supernova, this spun-up neutron star may again become observable as a radio pulsar. It was suggested by Smarr and Blandford (1976) that PSR 1913+16 is the old "spun-up" neutron star in the system (due to its weak magnetic field it now spins down only very slowly, on a timescale of the order of 10^8 yr). The other neutron star in the system was then produced by the second supernova explosion, and must be a young, strong-field neutron star. It is not observable, either because it has already rapidly spun down (Srinivasan and Van den Heuvel 1982), or because the Earth is outside of the pulsar beam. Old neutron stars spun-up by accretion in X-ray binaries, which are now observable as radio pulsars, are called "recycled pulsars" (Radhakrishnan and

Srinivasan 1982, 1984). In order to become observable as a radio pulsar, there should no longer be gas in the system, so accretion should have terminated. The companions of recycled pulsars should therefore either be white dwarfs or neutron stars. The system may also have been disrupted, resulting in a single recycled pulsar. The discovery of the first millisecond radio pulsar in 1982 gave the recycling idea an enormous boost: Alpar *et al.* (1982) and Radhakrishnan and Srinivasan (1982) suggested that millisecond pulsars are very old neutron stars which were spun-up by accretion in LMXBs. Because of the very long duration of the accretion phase in LMXBs ($\geq 10^8$ yr), a lot of angular momentum can be fed to these neutron stars, leading to spin-up to millisecond periods. This recycling model has recently been beautifully confirmed by the discovery of the first millisecond X-ray pulsar in the LMXB system SAX 1808.4-3658 (Wijnands and Van der Klis 1998). In order to spin-up a neutron star to a millisecond period, its magnetic field should have decayed to below 10^9 G (Section 6). The general idea now is that the decay of the magnetic field in accreting neutron stars is somehow related to the accretion process (Taam and Van den Heuvel 1986), although the precise mechanisms for field decay are still being debated (Konar and Bhattacharya 1999).

This chapter is organized as follows. In Section 1 an overview is given of the various observed types of binaries with one or two compact components. In Section 2 the orbital dynamics of binary systems is described, including the main effects of mass exchange and mass loss on the orbits. In Section 3 the reasons for the existence of the two main classes of X-ray binaries, high- and low-mass systems, are discussed. In Section 4 an overview is given of the evolution of single stars, and of close binaries leading to the formation of X-ray binaries and CVs. In this section the further evolution of X-ray binaries until their final stages is also discussed. Section 5 deals with the globular cluster X-ray sources and their formation and Section 6 with the formation of binary radio pulsars as the final products of the evolution of X-ray binaries.

1 TYPES OF BINARIES WITH COMPACT OBJECTS

1.1 Introduction

The binaries with compact objects can be divided into the categories and types listed in Table 1. Basically they fall into the categories "compact star plus ordinary star" and "two compact stars." The first category consists of the X-ray binaries and the CVs, the second of the binary radio pulsars and the double white dwarfs. Each of these categories can be divided into a few main types, that can be further divided into sub-types, as indicated in the table, where examples of the different sub-types are also given. A few binaries with compact objects do not fit into the above four categories, notably the so-called "ante-deluvian" binary radio pulsars, which are young pulsars in an eccentric orbit around a massive star, which presumably are the progenitors of HMXBs. Four such systems are presently known, also indicated in the table: PSR 1259–63 ($P_{orb} = 3.4$ yr), PSR 1820-11 ($P_{orb} = 357.8$ d), PSR J1740-3052 ($P_{orb} = 231$ d), and PSR J0045-7319 ($P_{orb} = 51$ d) in the Large Magellanic Cloud. In the first system and the last-mentioned two systems the companion of the young pulsar is a B-type star, in PSR 1820-11 the companion is not known. Furthermore, there are the two peculiar X-ray binaries with relativistic jets, SS433 and Cyg X-3. In SS433 ($P_{orb} = 13$ d) the companion of the jet-producing compact object is probably an early-type hydrogen-rich star (Margon 1983). In Cyg X-3 ($P_{orb} = 4.8$ h) it is a Wolf–Rayet star (helium star; Van Kerkwijk *et al.* 1992).

I now briefly describe the main characteristics of each of the various types of neutron star and black hole binary systems listed in the table. In this review I will concentrate mainly on these systems as these are the ones that have primarily been discovered thanks to observations from space. Only where this is appropriate will I also mention the CV systems and double white dwarfs.

1.2 X-Ray Binaries

High- and low-mass X-ray binaries

The X-ray binaries can, broadly speaking, be divided into two main groups, the HMXBs and LMXBs, which differ in a number of important characteristics, listed in Table 2, and graphically depicted in Figure 1. (The characteristics listed and depicted are for systems containing neutron stars but most of them also hold for the X-ray binaries that contain black holes; also these can be divided into HMXBs and LMXBs, see Figure 3.) For further details we refer here to the reviews of these two types of systems in Chapters 34 to 36. In HMXBs the companion of the X-ray source is a luminous early-type star of spectral type O or B, like in the Cen X-3 system, with a mass typically between 10 and $40 M_\odot$. In LMXBs it is a faint star of mass $\leq M_\odot$; in most cases the stellar spectrum is not even visible, as the light of the systems is dominated by that of the accretion disk around the compact star. The orbital periods of the LMXBs are generally short (mostly ≤ 0.5 d), though on average somewhat longer than those of the CVs.

Intermediate-mass X-ray binaries

Only recently it was realized that apart from these two main groups of X-ray binaries, which each contain some 10^2 known systems in our Galaxy (Van Paradijs and McClintock 1995), there are a few X-ray binaries in which the companion

Table 1 Main categories and types of binaries with compact objects*

Category	Main types	Sub-types	Example (NS/BH)
X-ray binaries	High-mass donor ($M_d \geq 10 M_\odot$)	"Standard" HMXB	Cen X-3, $P_{orb} = 2^d.087$ (NS) Cyg X-1, $P_{orb} = 5^d.60$ (BH)
		"Be"-HMXB	A0535+26, $P_{orb} = 104^d$ (NS)
	Low-mass donor ($M_d < M_\odot$)	Galactic disk	Sco X-1, $P_{orb} = 0^d.86$ (NS) A0620-00, $P_{orb} = 7^h.75$ (BH)
		Globular cluster	X 1820-30, $P_{orb} = 11^m$ (NS)
	Intermediate-mass donor $1 \leq M_d/M_\odot < 10$		Her X-1, $P_{orb} = 1^d.7$ (NS) Cyg X-2, $P_{orb} = 9.8^d$ (NS) V 404 Cyg, $P_{orb} = 6^d.5$ (BH)
Binary radio pulsars with unevolved companion star	Main sequence star plus young pulsars in eccentric orbit	B-type companion	PSR 1259-63, $P_{orb} = 3^{yr}.4$
		Low-mass companion	PSR 1820-11, $P_{orb} = 357^d.8$
Binary radio pulsars with compact companion	"High-mass" companion ($M_c = 0.5 - 1.4 M_\odot$)	Double NS (recycled pulsar)	PSR 1913+16 ($P_{orb} = 7^h.75$)
		Massive WD companion circular orbit (recycled PSR)	PSR 0655+64 ($P_{orb} = 1.03^d$)
		Massive WD companion eccentric orbit (non-recycled PSR)	PSR 2303+46 ($P_{orb} = 12^d.34$)
	"Low-mass" companion ($M_c \leq 0.45 M_\odot$)	Circular orbits (recycled PSR)	PSR 1855+09 ($P_{orb} = 12^d.33$) PSR 0820+02 ($P_{orb} = 1232^d.4$)
CV-like binaries	Novae and related systems	$M_{donor} \leq M_{WD}$	DQ Her ($P_{orb} = 0^d.194$) SS Cyg ($P_{orb} = 0^d.275$)
	Super soft X-ray sources	$M_{donor} > M_{WD}$	CAL 83 ($P_{orb} = 1^d.04$) CAL 87 ($P_{orb} = 10^h.6$)
Double white dwarfs	AMCVn systems	He WD + CO WD (He WD fills R. Lobe)	AMCVn ($P_{orb} = 22^m$)
		CO WD + CO WD	WD1204+450 ($P_{orb} = 1^d.603$)

*NS = neutron star, BH = black hole, WD = white dwarf.

Table 2 The two main types of strong galactic binary X-ray sources

HMXB	LMXB
– Optical counterparts massive and luminous early type stars, spectrum O and early B; $L_{opt}/L_X > 1$	– Faint blue optical counterparts $L_{opt}/L_X < 0.1$
– Concentrated in space towards the galactic plane: young stellar population, age $< 10^7$ yr	– Concentrated in space towards the galactic center; fairly widespread around the galactic plane: old stellar population, age $(5-15) \times 10^9$ yr
– Type of time variability: regular X-ray pulsations; no X-ray outbursts	– Type of time variability: often X-ray outbursts; only in 3 cases regular X-ray pulsations
– Relatively hard X-ray spectra $kT \geq 15$ keV	– Softer X-ray spectra: $kT \leq 10$ keV

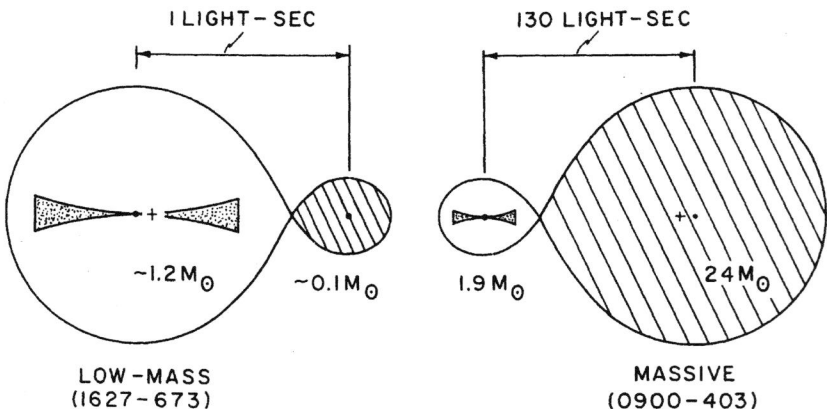

Figure 1 Comparison of the very different dimensions of a LMXB and a HMXB. The optical stars are depicted to fill their Roche lobes (hatched). The radial extent of the accretion disks (stippled) around the neutron stars is sketched nearly to scale. The dots and the crosses (+) mark the centers of mass of the individual stars and the system, respectively. (After Bradt and McClintock 1983.)

is or has been a star of "intermediate mass", that is, between $1M_\odot$ and $10M_\odot$. In fact, Her X-1 is a system of this type as well as Cyg X-2. In the latter system the companion presently has a mass $<M_\odot$, but is highly overluminous for this mass, which indicates that it is an evolved star that started out with a mass between 3 and $4M_\odot$ (Podsiadlowski and Rappaport 2000; King and Ritter 1999; Tauris et al. 2000; Kolb et al. 2000).

Spin and magnetic field in neutron star X-ray binaries

From the characteristics of the HMXBs and LMXBs listed in Table 2 and graphically depicted in Figure 1 it will be clear that the HMXBs belong to a very young stellar population, as OB stars do not live longer than about 10^7 yr. This implies that the neutron stars in these systems in general have ages of at most only a few million years. On the other hand, the characteristics of the LMXBs show that they tend to belong to a much older stellar population, with ages ranging from a few hunderd million years to over 10^{10} years. (The last-mentioned age is typical for the about a dozen known globular cluster X-ray sources.) Hence, the neutron stars in these systems tend to be old. This age difference is strikingly clear from their different galactic distribution depicted in Figure 2. A further difference is the presence of strong magnetic fields ($B \gtrsim 10^{12}$ G) in the neutron stars in HMXBs, as evidenced by their regular X-ray pulsations, and the absence of such regular pulsations and thus of strong fields in the bulk of the neutron stars in LMXBs. The occurrence of thermonuclear X-ray bursts in many LMXBs confirms that their magnetic fields must be weak, as fields stronger than 10^{10}–10^{11} G suppress such bursts (Lewin et al. 1995). This difference in magnetic field strength between the two groups is most probably due to

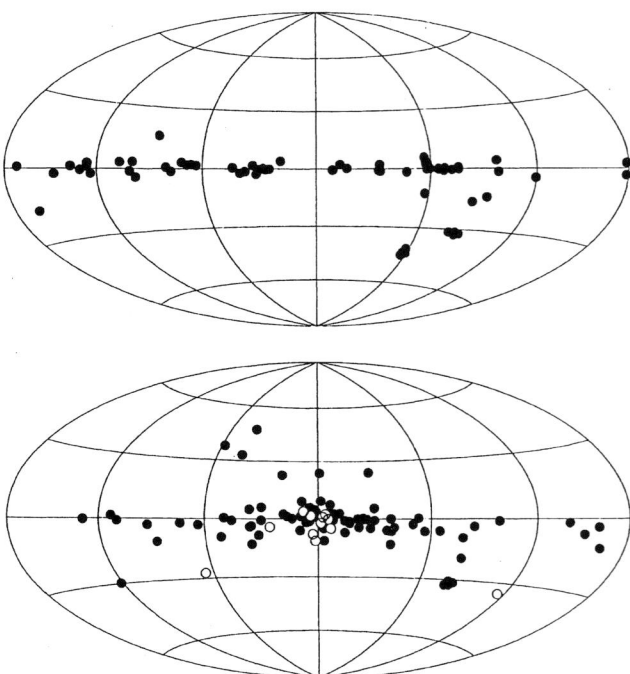

Figure 2 Sky maps (in galactic coordinates) of the HMXBs (top panel) and LMXB (bottom panel); the latter also include the globular cluster sources (indicated by open circles). The 27 LMXBs within 2° of the galactic center have not been included to avoid congestion of the map. These maps are based on the catalogue of Van Paradijs (1995).

field decay related to evolution in a binary system (Taam and Van den Heuvel 1986). The neutron stars in the LMXBs have been living next to their companions and spinning down as well as accreting on average for a much longer period (10^8–10^9 yr) than those in the HMXBs (not more than a few times 10^5 yr; Section 3). The various ways in which this large difference in timescales for evolution may affect the surface magnetic field strength are discussed in Section 6.4 (Bhattacharya and Srinivasan 1995; Ruderman 1998).

With the long-lasting accretion of matter through a disk, as occurs in the LMXBs (see Sections 3, 4.8, 5), the neutron star will be spun-up to millisecond periods, as mentioned in the Introduction. The relation between X-ray binaries and binary and millisecond radio pulsars is considered in Section 6.

The black hole X-ray binaries

A great breakthrough came with the discovery by McClintock and Remillard (1986) that the K5V companion of the "X-ray Nova Monoceros 1975" – the source A0620-00 – is a spectroscopic binary with an orbital period of 7.75 hours and a velocity amplitude >470 km s^{-1}. This large orbital velocity indicates that even if the K-dwarf would have zero mass, the compact object has a mass $>3 M_\odot$, and therefore must be a black hole. Since then eight such systems consisting of a black hole with a low-mass donor star have been discovered (Chapter 36). Figure 3 depicts the three types of black hole X-ray binaries presently known (from McClintock 1992). Also here there is an "intermediate-mass companion" system, that of LMC X-3, and there may be several more, notably V404 Cygni and X-ray Nova Sco 1994 (Tanaka and Lewin 1995; Nelemans *et al.* 1999).

HMXBs in more detail

It appears that the HMXBs fall into two sub-types which differ in a number of important characteristics (Chapter 35). (i) The "standard" HMXBs in which the massive OB-type companion of the X-ray source is close to filling its Roche lobe, as is evidenced from its double-wave optical lightcurve, that shows that the star is tidally deformed ("pear shaped"; Van Paradijs and McClintock 1995). These are the systems of the type depicted in Figure 1. They are persistent X-ray sources and except for one, they have orbital periods ≤ 11 days. (ii) The B-emission X-ray binaries, in which the B-type companion is a rapidly rotating unevolved star that is deep inside its Roche lobe. These systems have orbital periods between 15^d and over one year, and in most cases are "transient" X-ray sources that may occasionally turn on for a few weeks to several months, with long "off" periods in between (Figure 4).

Table 3 lists a number of characteristic examples of both types of systems (after Rappaport and Van den Heuvel 1982 and Van den Heuvel and Rappaport 1987, see also Apparao 1994). The Be/X-ray binaries are by far the most numerous group of HMXBs: over 50 such systems are already known in our Galaxy and a dozen of them is known in the Small

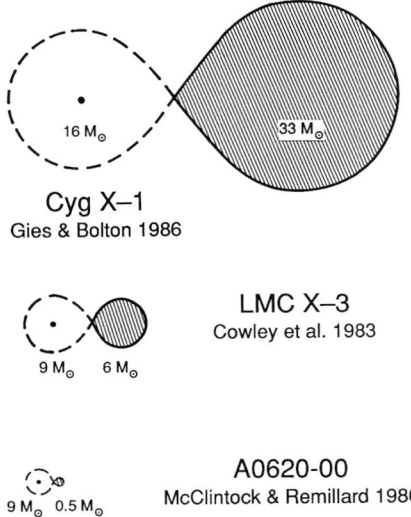

Figure 3 Schematic representation, to scale, of plausible models for three dynamical black hole candidates. The optical companions (shaded regions) are shown filling their critical Roche equipotential lobes. The value of the mass function and the masses given in the figure determine the following values of the inclination angle: $i(\text{Cyg}\,\text{X-1}) = 32°$, $i(\text{LMC}\,\text{X-3}) = 63°$, and $i(\text{A0620-00}) = 45°$. (Adapted from McClintock 1992.)

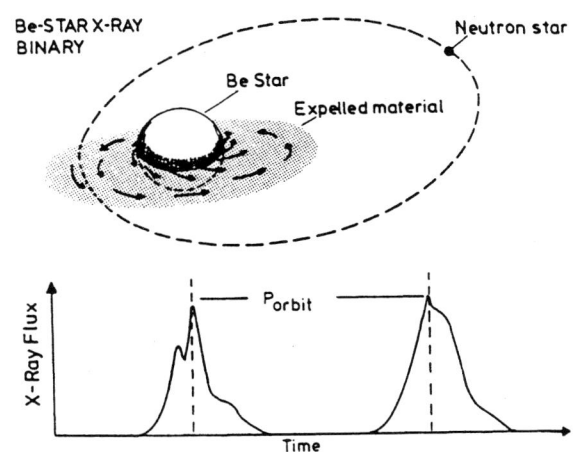

Figure 4 Schematic model of a Be star X-ray binary system such as X0535+26 or X0332+53. The neutron star moves in a moderately eccentric orbit around the Be star, which is much smaller than its own critical equipotential lobe. The rapidly rotating Be star is temporary surrounded by matter expelled in its equatorial plane. Near its periastron passage the neutron star enters this circumstellar matter and the resultant accretion produces an X-ray outburst lasting several days to weeks.

Table 3 The "standard" massive X-ray binaries (top) and some examples of Be X-ray binaries (bottom)

Source	Optical counterpart	Spectral type	Pulse period (s)	Orbital period (d)	Eccentricity
LMCX-4	Sk-Ph	O7 III–V	13.5	1.408	$e = 0.00$
Cen X-3	Krz's star	O6.5 II–III	4.84	2.087	$e = 0.00$
4U 1700-37	HD 153919	O6.5 f	—	3.412	$e \sim 0$
SMC X-1	Sk 160	B0Ib	0.717	3.89	$e = 0.00$
4U 1538-52	QV Nor	B0Iab	529	3.73	$e = 0.00$
Cyg X-1	HD 226868	O9.7Iab	—	5.60	$e = 0.05$
4U 0900-40	HD 77581	B0.5Ib	283	8.965	$e = 0.09$
GX 301-2	Wra 977	B1.5Ia	696	41.5	$e = 0.47$
4U 0115+63	John's star	Be	3.61	24.3	$e = 0.34$, transient
4U 0352+30	X Per	O9.5(III–V)e	835	250	$e \sim 0$, very weak, steady
A 0535+26	HD 245770	B0Ve	104	111	Transient
4U 1145-61	Hen 715	B1Vne	292	188	Highly variable
4U 1258-61	MMV star	B2Vne	272	133	Highly variable

Magellanic Cloud (SMC, a small satellite galaxy of our Galaxy, with a mass of only 1% of the latter). Judging from these numbers, the total number of Be/X-ray binaries in our Galaxy may easily be as large as 10^3. On the other hand, not more than 10 "standard" HMXBs are known in our Galaxy, and only one in the SMC and a few in the LMC. Extrapolating from the about 10 known "standard" HMXBs in our sector of the Galaxy, it seems likely that there are not more than some 50 to 100 such systems in the entire Galaxy.

1.3 The binary and millisecond radio pulsars

Spin and magnetic field

The binary radio pulsars are characterized by, in general, much shorter spin periods than the ordinary single radio pulsars. This can be clearly observed in the $\dot{P}$ v. P diagram of radio pulsars, where P is the pulse period and $\dot{P}$ is the period derivative. Figure 5 shows this diagram for about 1000 pulsars in the galactic disk (Camilo, private communication). Dots are single pulsars, circles are binary pulsars.

The surface dipole magnetic field strength B_s of pulsars is related to P and $\dot{P}$ by the equation (Manchester and Taylor 1977; Lyne and Smith 1990).

$$B_s = \left(\frac{3c^3 I}{8\pi^2 R^6} P \dot{P}\right) = 3.2 \times 10^{19} (P\dot{P})^{1/2} \text{ G} \qquad (1)$$

where I and R are the moment of inertia and radius of the neutron star, respectively (this follows from equating the rotational energy loss to the electromagnetic energy loss from the spinning magnetized neutron star; Manchester and Taylor 1977). In Figure 5 lines of constant B_s are indicated. The figure shows that a large fraction of the binary pulsars is millisecond pulsars, that is, has a spin period shorter than 0.01 s. In the figure lines are also drawn of

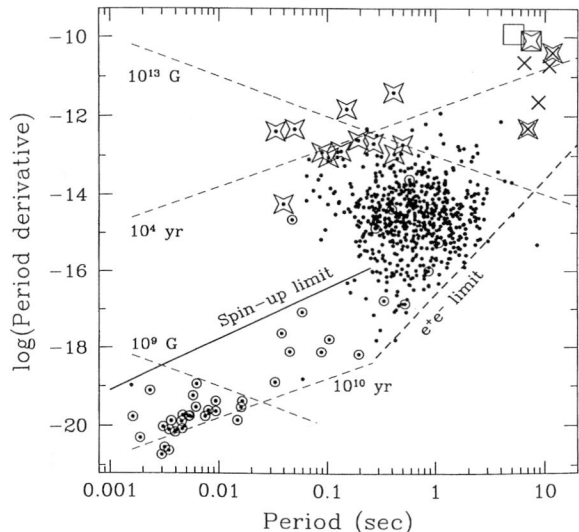

Figure 5 P–$\dot{P}$ diagram for radio pulsars. Circles indicate binary pulsars, stars indicate pulsars associated with supernova remnants, crosses are "anomalous X-ray pulsars", and squares are soft gamma-ray repeaters. (The latter two classes of objects are young, single neutron stars with extremely strong magnetic fields (10^{14} to 10^{15} G).) Lines of constant surface dipole magnetic field and of constant spin-down age are indicated, as well as the "spin-up limit" and the "death-line" (e^+e^- limit). (Diagram courtesy of F. Camilo (2000).)

constant spin-down age, as defined by (Manchester and Taylor 1977):

$$t_{sd} = P/2\dot{P} \qquad (2)$$

Figure 5 shows that the binary pulsars typically have magnetic field strengths in the range 10^8–10^{10} G and "ages" of the order of $(1-10)\times10^9$ yr, whereas the "garden variety" pulsars that make up the cloud of "ordinary pulsars" typically have magnetic field strengths of $10^{12}-10^{13}$ G and ages $\leq 10^7$ y.

A number of well-known binary radio pulsars is included in Figure 5, such as the three close double neutron stars PSR 1913+16 (Hulse–Taylor pulsar), PSR 1534+12 (Wolszczan's pulsar), and PSR 2127+11c (in the globular cluster M15), and the eccentric-orbit neutron star plus white dwarf system PSR 2303+46 (Van Kerkwijk and Kulkarni 1999). In the latter system the pulsar is an ordinary strong-magnetic-field pulsar. Table 4 lists some vital data of these binary pulsars and of some other representative binary pulsars. Many pulsars are also known in globular clusters. Most of these are millisecond pulsars and some 40% – and in some clusters, like 47 Tuc, over 60% (Camilo et al. 2000) – of them are in binaries (against some 7% of the total pulsar population).

Orbits and companion stars

With the exception of the four "ante-deluvian" systems (see Section 1.1), the companion stars in radio pulsar binaries are themselves also dead stars: neutron stars or white dwarfs. They can be divided into the following categories (see also Table 1), as depicted in Figure 6:

1. "The PSR 1913+16 class," in which the companion to the neutron star is another neutron star or a massive white dwarf.
2. "The PSR 1953+29 class," in which the companion to the neutron star is a low-mass helium white dwarf ($M < 0.45 M_\odot$). The latter systems always have circular orbits, with periods covering a large range, from less than one day to 1232 days. In systems with orbital periods shorter than ~ 150 days the pulsars tend to be millisecond pulsars, with very weak magnetic fields, in the range $10^8 - 10^9$ G.

Category (1) can be further subdivided into:

a. Systems in which the companion is itself also a neutron star. These have very eccentric orbits and their orbital periods tend to be very short. The magnetic fields of these pulsars are relatively weak, in the range 10^9-10^{10} G.
b. Systems in which the companion is a massive white dwarf ($M_{\rm wd} = 0.5 - 1.4\, M_\odot$). These fall into two groups:

   i. those with eccentric orbits, such as PSR 2303+46, in which the pulsar is a young strong-magnetic-field neutron star;
   ii. those with circular orbits, such as PSR 0655+64, in which the pulsar is an old neutron star with a relatively weak magnetic field, in the range 10^9-10^{10} G.

Table 4 Some vital data of representative binary radio pulsars with neutron star and white dwarf companions

Pulsar	$P_{\rm orb}$ (days)	$P_{\rm spin}$ (ms)	$M_{\rm p}$ ($M_\odot$)	$M_{\rm c}$ ($M_\odot$)	e	B (G)
(NS, NS)						
B1534+12	0.421	37.9	1.339	1.339	0.274	10^{10}
B1913+16	0.323	59.0	1.441	1.387	0.617	2.3×10^{10}
B2127+11C	0.335	30.5	1.349	1.363	0.681	1.2×10^{10}
(NS, CO)$_{\rm e}$						
B2303+46	12.34	1066	<1.44	>1.20	0.658	7.9×10^{11}
J114-6545	0.198	394		>0.97	0.172	
(NS, CO)$_{\rm c}$						
J2145-0750	6.839	16.1		>0.43	2.1×10^{-5}	6×10^8
J1022+1001	7.805	16.5		>0.73	9.8×10^{-5}	8.4×10^8
J0621+1002	8.319	28.9		>0.45	0.00245	1.6×10^9
B0655+64	1.029	195.7		>0.7	0.75×10^{-5}	1.26×10^{10}
(NS, He)$_{\rm c}$						
1957+20	0.38	1.6		0.02	$<10^{-3}$	1.6×10^8
1855+09	12.33	5.4		0.20	2.1×10^{-5}	3×10^8
1953+29	117.35	6.1		0.30	3.3×10^{-4}	4×10^8
0820+02	1232.4	864.9		0.30	1.2×10^{-2}	3×10^{11}

There are four categories (see also Table 1): (NS, NS)$_{\rm e}$: eccentric orbit double neutron star; (NS, CO)$_{\rm e}$: eccentric orbit neutron star +CO white dwarf; (NS, CO)$_{\rm c}$: circular orbit neutron star +CO white dwarf; (NS, He)$_{\rm c}$: circular orbit neutron star +He white dwarf. Except in the (NS, CO)$_{\rm e}$ systems the observed pulsars in these neutron star binaries are "recycled": they were the first-born compact stars in their systems. (Data partly after Brown et al. (2001) and references therein, and after van den Heuvel (1994a) and references therein.)

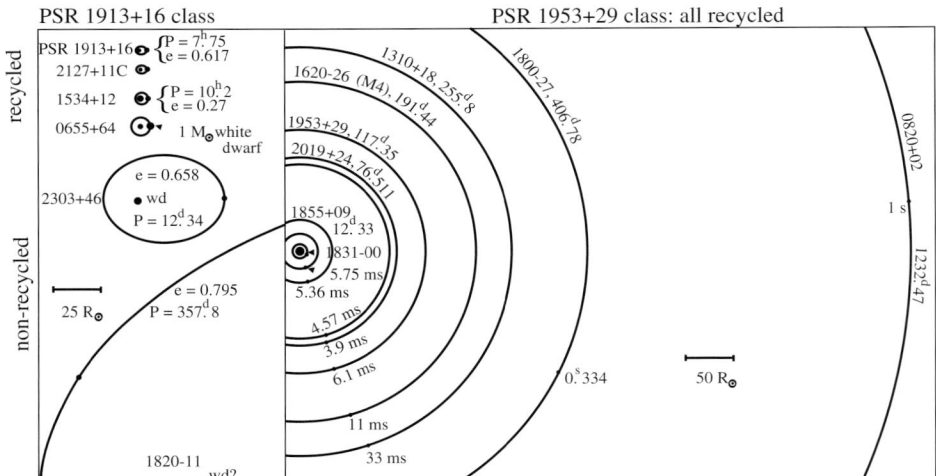

Figure 6 The main classes of binary radio pulsars (orbits drawn to scale). Left: the PSR 1913+16 class systems tend to have narrow and very eccentric orbits; the companion of the pulsar is itself a neutron star or a massive white dwarf; when, in the case of white dwarf companions, the orbit is circular, the neutron star is recycled; if the orbit is eccentric, the neutron star is not recycled – this is the case in 2303+46 (see text). Right: the PSR 1953+29 class systems tend to have wide and circular orbits; here the companion stars have a low mass, in the range 0.2–0.4 $M_\odot$, or even smaller – and in most cases probably are helium white dwarfs (see text).

1.4 Spin evolution of accreting neutron stars: The concepts of Alfven radius, magnetosphere, equilibrium spin, spin-up, and spin-down

Observed spin behavior of accreting neutron stars

Figure 7 shows the pulse period v. orbital period relation of the accreting X-ray pulsars (Liu 2001). One observes in this figure that many of these pulsars have long pulse periods between 10 and 1000 s. In persistent sources, spin periods of the order of seconds are found only in systems where from UV and optical observations we have clear evidence for the presence of an accretion disk. These are systems where (a large part of) the mass transfer is due to Roche lobe overflow: the low-mass X-ray binary Her X-1 and the high-mass systems in which the supergiant donors are just beginning to overflow their Roche lobes (Cen X-3, SMC X-1, and LMC X-4). The latter ones are expected to be powered by a combination of stellar wind and beginning Roche lobe overflow (Savonije 1978, 1983). In most of these systems the X-ray pulsars show a secular decrease of the pulse period (spin-up) on a relatively short timescale: a few thousand years in the massive systems and $\sim 10^5$ yr in the case of Her X-1 (Bildsten et al. 1997; Finger 1998). Figure 8 shows, for example, the observed spin evolution of Cen X-3. Although on short timescales episodes of spin-up and spin-down alternate, the average trend is that of spin-up. However, in persistent systems that are purely wind-fed, the HMXBs with blue supergiant companions that do not yet fill their Roche lobes, the pulse periods are very long, and they vary erratically in time showing no clear secular trends. This can be explained by the fact that the amount of angular momentum carried by the supersonic winds is negligible, and eddies form in the wind downstream of the neutron star, which alternately may feed co-rotating and counter-rotating angular momentum to it (Taam and Fryxell 1988; Fryxell and Taam 1988). The B-emission (Be) X-ray binaries show again a different spin behavior: these sources are transients, in many cases recurrent. During a transient outburst they in general show rapid spin-up, indicating that a disk has been formed. However, between outbursts they spin down, as at the beginning of the next outburst the spin period is generally observed to be much longer than at the end of the previous one. An important clue to what drives the spin-up and spin-down in the Be X-ray systems is given by the correlation between spin periods and orbital periods in these systems (Figure 7), discovered by Corbet (1984). This can be explained in terms of a slow wind that emanates from the Be star and carries off angular momentum from this rapidly rotating star (Waters and Van Kerkwijk 1989).

The concepts of Alfven radius, magnetosphere, and equilibrium spin

These concepts were introduced by Davidson and Ostriker (1973) and Lamb et al. (1973). The Alfven radius r_A is the distance from the neutron star where the kinetic energy density $0.5\rho v^2$ of the inflowing matter equals the magnetic energy density of $\mu_0 B^2/2$ of the neutron star's dipole

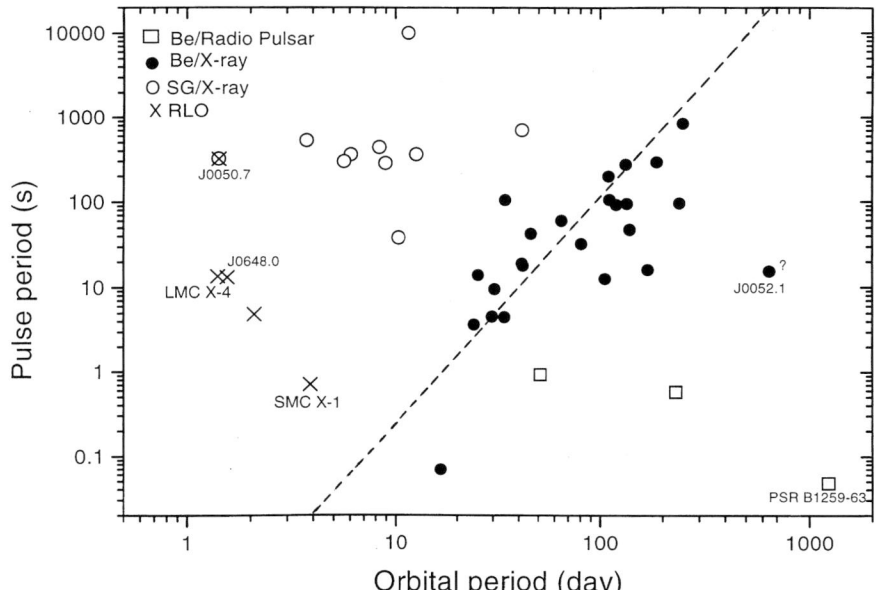

Figure 7 The pulse period v. orbital period for the accretion-powered pulsars with known orbital period. Crosses indicate persistent sources with accretion disks (Roche lobe overflow), open circles are persistent sources with blue supergiant companion (wind accretors), and filled circles are the B-emission X-ray binaries, which are mostly transient sources. Squares are radio pulsars with Be companions. The dashed line indicates Corbet's (1984) relation between pulse period and orbital period for the Be X-ray systems. (After Liu 2001.)

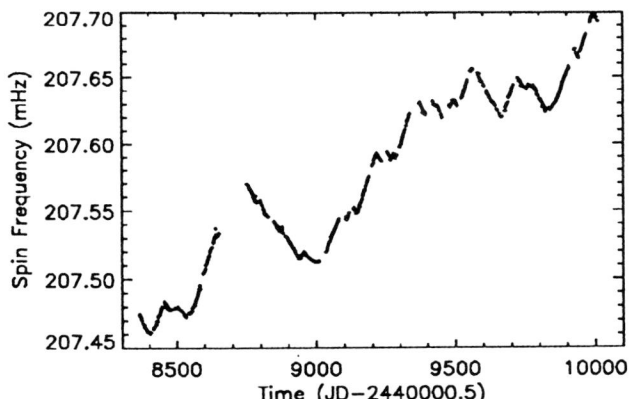

Figure 8 Cen X-3 spin frequency measurements from BATSE. (after Finger 1998).

magnetic field, so:

$$\rho v(r)^2 = \mu_0 B(r)^2 \quad (3)$$

where

$$B(r) = B_s (R/r)^3 \quad (4)$$

in which B_s is the magnetic field strength at the neutron star radius R and μ_0 is the permeability of the vacuum. Here $v(r)$ is either the free-fall velocity or the Kepler velocity in the disk (which are the same, except for a factor of $\sqrt{2}$) and $\rho(r)$ can be expressed in terms of the accretion rate $\dot{M}$, $v(r)$, and r (e.g. Bhattacharya and van den Heuvel 1991; van den Heuvel 1994a). This leads to

$$r_A = (\mu_0 B_s^2 \times R^6 / \dot{M} \sqrt{2GM})^{2/7} \quad (5)$$

For $r > r_A$ the flow of the matter is not influenced by the neutron star's magnetic field. On the other hand, for distances smaller than the Alfven radius, the magnetic field forces the matter to flow in along the field lines. The region around the neutron star closer than the Alfven radius is called the magnetosphere. For $r < r_A$ the inflowing matter is forced to co-rotate with the magnetosphere of the neutron star; for $r > r_A$ matter can freely orbit the star.

A second important radius is the co-rotation radius r_{co}. This is the distance where the Kepler velocity of matter around the neutron star just equals the rotational velocity of the magnetosphere ωr. The value of r_{co} (also indicated as R_{syn}, see Figure 9) is given by $\omega^2 \times r = GM/(r^2)$, where $\omega = 2\pi/P$, P being the rotation period of the star. Thus:

$$r_{co} = (GM \times P^2/4\pi^2)^{1/3} \quad (6)$$

There are now two possibilities for the accretion, illustrated by Figure 9:

i. If $r_A > r_{co}$, matter at the magnetospheric boundary cannot flow in: as soon as the matter enters the magnetosphere it is forced to co-rotate; however, it then rotates faster than the Keplerian velocity and is centrifuged out of the magnetosphere again. In this situation the accretion is therefore shut off.

ii. If $r_A < r_{co}$, accretion can take place. In the latter situation, matter with angular momentum will enter the magnetosphere and flow to the neutron star surface, causing the angular velocity of rotation to increase, that is, the spin period to decrease. This will go on until $r_A = r_{co}$, after which further accretion and spin-up is impossible. The star will therefore settle at a spin period for which these two radii are just equal to each other. This is called the equilibrium spin period P_{eq}, which is given by:

$$P_{eq} = (2.4 \text{ ms } (B_9)^{6/7} \times (R_6)^{16/7} \times (M)^{-5/7} \times (\dot{M}/\dot{M}_{Edd})^{-3/7} \quad (7)$$

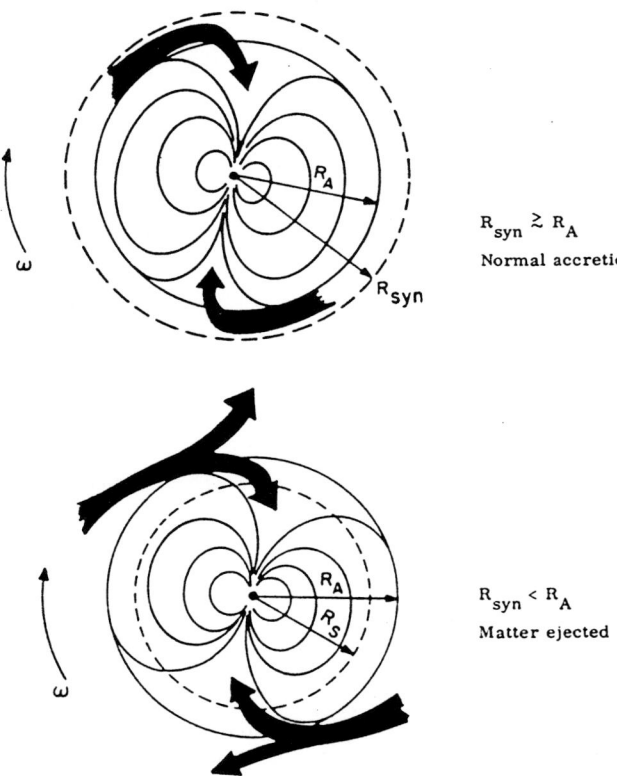

Figure 9 Schematic representation of Alfven radius R_A and co-rotation radius, indicated here as R_{syn}, of a rotating magnetized neutron star. R_A depends on accretion rate $\dot{M}$, mass M, and dipole magnetic field strength B_9 at stellar surface; R_{syn} depends on rotation period P and mass of the neutron star. When $R_A = R_{syn}$ the neutron star spins at its equilibrium spin period P_{eq}. If it rotates slower, accretion is possible; if it rotates faster, centrifugal forces on matter entering the magnetosphere will swing this matter out and accretion is impossible. (After Schreier 1977, *Annals of the New York Academy of Sciences*, **402**, 445.)

(van den Heuvel 1994a), where B_9 is the surface dipole magnetic field strength in units of 10^9G, R_6 is the neutron star radius in units 10^6cm, M is the mass of the neutron star in units of $M_\odot$, and $\dot{M}_{Edd}$ is the Eddington accretion rate (defined in Section 3, eqn (20)).

The X-ray pulsars with accretion disks all seem to be spinning near to their equilibrium spin periods. Their magnetic field strengths, inferred from X-ray cyclotron lines, are typically of the order of 10^{12}G (Trümper et al. 1978), yielding equilibrium spin periods of the order of seconds, just as observed in these sources. The fact that their spin periods still show a secularly decreasing trend (spin-up) may be due to a secular decrease of the surface dipole magnetic field strength, or a secular increase of the accretion rate, or both (Finger 1998). A decrease of the surface dipole strength might be due to temporary "burying" of the field due to the accretion process (Section 6.4).

2 PRINCIPLES OF MASS TRANSFER AND MASS LOSS IN CLOSE BINARIES

2.1 The concept of Roche lobe

Tidal forces in close binaries will, after some time, have circularized the orbits and brought the components into co-rotation. In this case one can use the co-rotating coordinate system as the reference system. The equipotential surfaces in this system are stationary and are due to a combination of the gravitational attraction of the components and the centrifugal acceleration produced by the rotation of the system. These are the so-called Langrangian or Roche equipotential surfaces, depicted in Figure 10.

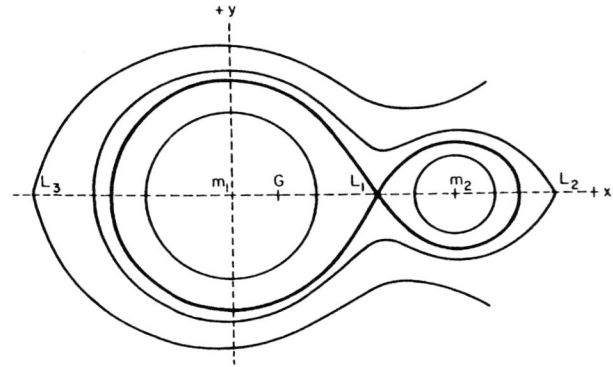

Figure 10 The cross-section with the orbital plane of some special Roche equipotentials. The surface through the inner Lagrangian point L_1 forms the two Roche lobes around the stars. The point G denotes the binary's center of mass (further explanations in Section 2). (From Plavec 1968.)

The equipotential surface passing through the first Lagrangian point L_1 is called the critical Roche surface, and the pear-shaped part of this surface around a star is called the Roche lobe. One defines the radius R_L of the Roche lobe as that of a sphere with the same volume as the lobe. R_L is a function only of the orbital radius a and the mass ratio $q = m_2/m_1$ of the binary components. Various approximation formulas for R_L exist. The most precise one is that due to Eggelton (1983):

$$R_L / a = 0.49 q^{2/3} / [0.6 q^{2/3} + \ln(1 + q^{1/3})] \quad (8)$$

which is accurate to within a few percent for all values of q. A very convenient alternative expression for $q > 1.25$ is

$$R_L / a = 0.462(m_1/(m_1 + m_2))^{1/3} \quad (9)$$

2.2 Orbital changes due to "conservative" mass transfer

When a star fills its Roche lobe, matter in its outer layers can freely flow over towards the companion star along L_1.

If one assumes that all the matter lost by the star is captured by the other star, and that the rotational angular momentum of the stars is negligible with respect to the orbital angular momentum, one will have conservation of both total mass and total orbital angular momentum of the system, which is called "conservative evolution" of the binary. In this case it is easy to calculate how the orbital radius a changes during the mass transfer, as one has the equations

$$(m_1 + m_2) = M = \text{const.} \quad (10)$$

and

$$J^2 = G \frac{m_1^2 m_2^2}{(m_1 + m_2)} a = \text{const.} \quad (11)$$

where J is the orbital angular momentum. Then, eqns (10) and (11) yield

$$a (m_1^2 m_2^2) = \text{const.} \quad (12)$$

leading to the following equation for the change in a:

$$\frac{a}{a_0} = \left(\frac{m_1^0 \times m_2^0}{m_1 \times m_2}\right)^2 \quad (13)$$

where index zero indicates the initial situation. Assuming m_1 to be the star that is losing mass along L_1, the mass of this star will decrease. It is easy to see from eqn (13) that if $m_1^0 > m_2^0$, the orbital radius a will *decrease* due to the mass transfer. On the other hand, $m_1^0 < m_2^0$ (mass transfer from the less massive to the more massive star) leads to an *increase* of a.

2.3 Orbital changes due to loss of mass and orbital angular momentum from the system

In the case of mass loss from the system we are no longer dealing with a closed "conservative" system and there is, in principle, an infinite range of possibilities for the changes of the orbit depending on the amounts of mass lost and of the angular momentum lost with this mass. For example, if a fraction α of the matter exchanged between the components leaves the system, one finds from eqn (11) that the rate of change of the orbital separation can be written as:

$$\frac{\dot{a}}{a} = -2\left[1 + (\alpha - 1)\left(\frac{m_1}{m_2}\right) - \frac{\alpha}{2}\frac{m_1}{M}\right]\left(\frac{\dot{m}_1}{m_1}\right) + 2\dot{J}_{\text{orb}}/J_{\text{orb}} \quad (14)$$

It is clear that this equation can be solved only if $\dot{J}_{\text{orb}}/J_{\text{orb}}$ is known.

In binary systems with very short orbital periods, angular momentum loss will occur due to the emission of gravitational waves (Shapiro and Teukolsky 1983):

$$\dot{J}_{\text{orb}}/J_{\text{orb}} = \frac{-32 G^3}{5 c^5} \frac{m_1 m_2 (m_1 + m_2)}{a^4} \text{s}^{-1} \quad (15)$$

For sufficiently narrow orbits, eqn (15) becomes the dominant term in eqn (14) and will cause a to decrease. Therefore, the orbits of very narrow binaries will tend to continuously shrink, forcing the components to transfer mass (if a component fills the Roche lobe). (If mass is transferred from the less massive to the more massive component the net result of gravitational radiation losses plus mass transfer may also be a widening of the orbit.) Gravitational radiation losses are therefore a major force driving the mass transfer in very narrow binaries, such as CVs and LMXBs (Faulkner 1971). For a review of a variety of "modes" of mass and angular momentum losses from binaries, see Van den Heuvel (1984a) and Soberman *et al.* (1997).

Of particular importance for CVs and LMXBs is "magnetic braking," a concept introduced by Verbunt and Zwaan (1981), which acts as follows. Stars with convective envelopes tend to have surface magnetic fields. These couple to the ejected ionized wind material and enforce co-rotation out to several stellar radii from the surface. In semi-detached binaries effective tidal interaction tends to keep the stellar rotation synchronous with the orbit, such that the spin angular momentum carried off by the magnetically coupled wind must be compensated by the system's orbital angular momentum. An effectively coupled wind can thus enforce a decrease of the orbital separation. If these wind losses can occur simultaneously with the Roche lobe overflow (RLO) the mass-transfer rate is increased by the induced shrinking of the Roche lobe, just as in the case of gravitational radiation losses. It is commonly believed that such magnetic braking occurs in systems with G- and K-type stars, for example, in CVs and LMXBs. The observationally inferred spin-down of G and K stars with age provides the quantitative information about this magnetic braking to be applied in binary stars (e.g. Rappaport et al. 1983).

2.4 Effects of sudden (explosive) mass loss on the orbits

We consider systems with initially circular orbits and we neglect the effects of the impact of the ejected shell on the companion star. We assume the explosion to be instantaneous (infinitely short duration). In this case the orbital changes can be simply expressed in terms of the ratio

$$\mu_f = \frac{m_1^f + m_2^f}{m_1^0 + m_2^0}$$

of the total mass of the system after and before the explosion. We express the orbital semi-major axis a and orbital period P in units of the initial orbital radius a_0 and initial orbital period P_0, respectively, and 0 and f indicate quantities before and after the explosion. One then obtains the following simple expressions for the orbital parameters after the explosion (Flannery and van den Heuvel 1975):

$$a_f = \mu_f/(2\mu_f - 1) \quad (16)$$

$$e_f = (a_f - 1)/a^f = (1 - \mu_f)/\mu_f \quad (17)$$

$$p_f = \mu_f/(2\mu_f - 1)^{3/2} \quad (18)$$

One thus observes that the system is disrupted (hyperbolic orbit) if $\mu_f < 0.5$. This is a consequence of the virial theorem.

The runaway velocity of the center of gravity of bound systems after star 2 (with orbital velocity V_2) ejected an amount Δm is given by

$$V_g = \Delta m \times V_2/(m_1^f + m_2^f) \quad (19)$$

Hence, the largest velocity that a post-explosion system can attain is V_2.

3 REASONS FOR THE EXISTENCE OF THE CLASSES OF HIGH- AND LOW-MASS X-RAY BINARIES: THE CONCEPTS OF EDDINGTON LIMIT, ROCHE LOBE OVERFLOW AND STELLAR WINDS

3.1 Eddington limit

The typical X-ray luminosities of the high- and low-mass X-ray binaries are in the range 10^{35}–10^{38} erg s^{-1}, corresponding to mass accretion rates onto a neutron star in the range 10^{-11} to $10^{-8} M_\odot$ yr^{-1} (Chapter 34). When the mass-transfer rate exceeds a few times $10^{-8} M_\odot$ yr^{-1}, the X-ray luminosity exceeds the so-called Eddington limit of $10^{4.5}$ $(M/M_\odot) L_\odot$ ($\sim 10^{38}$ erg s^{-1} for a $1.4 M_\odot$ neutron star), at which the radiation pressure force on the accreting matter exceeds the gravitational attraction force of the compact star (Davidson and Ostriker 1973). This places a natural upper limit to the accretion rate of

$$\dot{M}_{\text{Edd}} \simeq 1.5 R_6 \times 10^{-8} M_\odot \, yr^{-1} \quad (20)$$

where R_6 is the radius of the compact star in units of 10^6 cm. At accretion rates $> \dot{M}_{\text{Edd}}$ the excess accreting matter will pile up around the compact object and form a cloud optically thick to X-rays, thus quenching the source. Therefore, in the observed "persistent" HMXBs and LMXBs the accretion rates must be in the range 10^{-11} to $10^{-8} M_\odot$ yr^{-1}.

One may now ask the following question: assuming that neutron stars are born as companions of stars of any kind of mass, in which mass ranges do we then expect to observe (reasonably) long-lived persistent X-ray sources? That is, in what mass range do we expect companions to lose mass in such a way that the neutron star will have an accretion rate between 10^{-11} and $10^{-8} M_\odot$yr^{-1}? It appears that the answer to this question is that one then expects just the two observed groups of persistent X-ray binaries to occur: one group with companion masses $\leq M_\odot$ and another group with companion masses $\geq 15 M_\odot$. The reasons for this are simple (van den Heuvel 1975) and have to do with the ways in which companion stars can lose mass, as discussed below.

3.2 Modes of mass transfer and types of binary X-ray sources

Basically, the companion can lose mass in two ways: (1) by Roche lobe overflow: in this case the donor star fills its Roche lobe and transfers mass along the first Langranian point L_1 to its companion; (2) by a high-velocity stellar wind: in this case the star needs not fill its Roche lobe, and the neutron star captures some matter from the wind. The accretion rates predicted from these two types of mass loss are as follows.

(1) Roche lobe overflow. In this case the mass loss takes place at low velocities and practically all mass lost by the star will be accreted by the companion. The mass transfer from a donor star that is more massive than its companion takes place on the thermal timescale of the donor (Kippenhahn and Weigert 1967; Paczynski 1966, 1971a), leading to a mass-transfer rate (van den Heuvel 1994a):

$$\dot{M}_{\text{Roche}} \simeq 3.10^{-8} M^3 (M_\odot \, yr^{-1}) \quad (21)$$

where M is the donor mass in solar masses. This equation shows that only for $M \leq M_\odot$ can mass-transfer rates below the Eddington rate be obtained. However, for $M \leq M_\odot$ the mass-losing star is no longer more massive than the compact star, and eqn (21) will no longer be valid. For a donor that is less massive than the compact star, mass transfer will take place on the nuclear timescale of the donor star (10^{10} yr), or due to narrowing of the orbit by gravitational radiation losses and/or "magnetic braking" (Section 2). In both cases a mass-transfer rate of the order of $10^{-10} M_\odot$ yr^{-1} will ensue (van den Heuvel 1994a).

(2) Accretion from a stellar wind. Strong stellar winds with (total) mass loss rates $\geq 10^{-9} M_\odot$ yr^{-1} are found only in stars more massive than about $15 M_\odot$ as was first discovered in a rocket experiment by Morton (1967). Blue supergiants like the companions of Cyg X-1 and Cen X-3 typically have wind mass loss rates of a few times $10^{-6} M_\odot$ yr^{-1} with outflow velocities of the order of (1–2) $\times 10^3$ km s^{-1}. It was realized by Davidson and Ostriker (1973) that a neutron star passing through such a wind will capture only a tiny fraction of the wind matter, of the order of 10^{-11} to $10^{-10} M_\odot$ yr^{-1}, precisely enough to power a 10^{35}–10^{36} erg s^{-1} X-ray source. Figure 11 depicts this situation. The neutron star captures only the matter that passes it within its gravitational capture radius r_{acc} given by

$$r_{acc} = \sqrt{2GM_n/V_w^2} \qquad (22)$$

where V_w is the velocity of the wind and M_n is the mass of the neutron star. The amount of the wind matter captured is $\dot{M}_{acc} = (\pi r_{acc}^2/4\pi a^2)\dot{M}_w$ where a is the orbital radius and $\dot{M}_w$ is the wind mass-loss rate. For $M_n = 1.4 M_\odot$, $V_w = 10^3$ km/s, $a = 50 R_\odot$, as in a typical "standard" HMXB, one has $\dot{M}_{acc} = 5.10^{-5} \dot{M}_w$.

The conclusion from the above is that only companion stars with $M \leq M_\odot$ and $\geq 15 M_\odot$ are able to turn a compact companion star into a steady X-ray source. This explains the existence of the two main observed classes of binary X-ray sources, the LMXBs and HMXBs, and the virtual absence of persistent binary X-ray sources with donor stars with $M_\odot \leq M \leq 15 M_\odot$ (Van den Heuvel 1975).

(3) Mass accretion in transient sources. For the B-emission X-ray transients the cause for the "turn-on" as an X-ray source is clearly to be sought in the unstable outer layers of the rapidly rotating B-emission star (Maraschi *et al.* 1976). B-e stars are known to go with irregular time intervals through "emission" phases, in which, for unknown reasons, the star ejects a disk of gas in its equatorial plane. The passage of the neutron star through this disk will then cause a recurrent X-ray outburst, as depicted in Figure 4.

On the other hand, for the black-hole "X-ray novae" like NOVA Monoceros 1975 (A0620-00), see Figure 3, the cause of the outburst is probably to be sought in an instability in the accretion disk around the black hole (cf. King *et al.*, 1997; King, 1998, 2000).

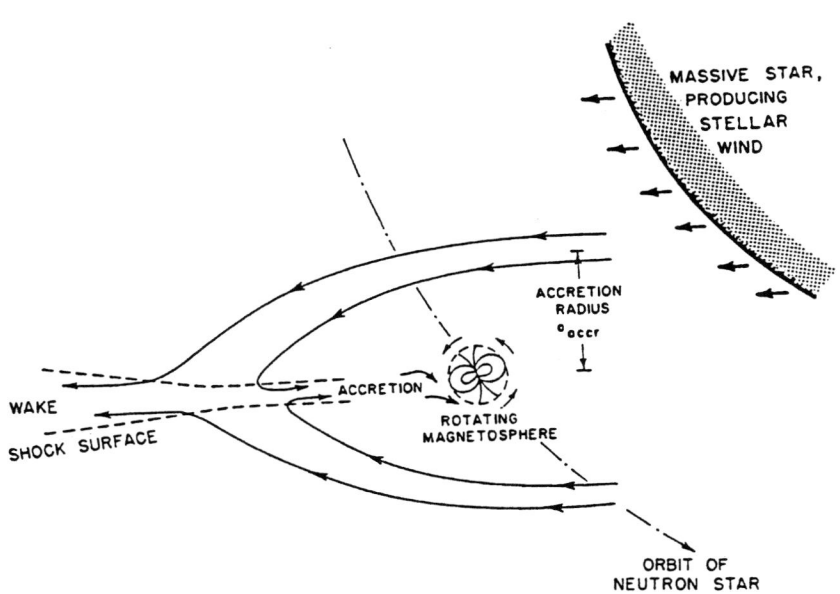

Figure 11 Streamlines of stellar-wind material, in frame of an accreting neutron star. Relative dimensions are not to scale (from Davidson and Ostriker 1973).

4 EVOLUTION OF CLOSE BINARIES AND THE FORMATION OF X-RAY BINARIES

4.1 Types of stellar remnants

Table 5 summarizes the types of remnants of single stars expected as a function of initial stellar mass. Single stars less massive than about $8\,M_\odot$ are expected to leave white dwarfs as remnants, with masses $\leqslant 1.4\,M_\odot$ and in most cases consisting of carbon, oxygen and some helium (more rarely they may also contain magnesium and neon). In single stars more massive than about $8\,M_\odot$ the burned-out core reaches a mass larger than the Chandrasekhar limit, the maximum mass possible for an electron-degenerate configuration ($\sim 1.4\,M_\odot$; Srinivasan 1998). Therefore, at the end of life the cores of such stars collapse and, if they are not too heavy, will be able to reach a stable configuration as a neutron star. During this collapse some 3×10^{53} erg of gravitational binding energy is released ($0.15 M_n c^2$, where M_n is the mass of the neutron star), far more than the binding energy of the stellar envelope, causing this envelope to be violently ejected at velocities $\sim 10^4\,\mathrm{km\,s^{-1}}$. This is observable as a supernova event.

Evidence is mounting (based on a variety of observational and theoretical considerations involving the black hole X-ray binaries and gamma-ray bursts. (Ergma and van den Heuvel 1998; Nelemans et al. 1999; Sollerman et al. 2000)) indicating that single stars more massive than about 20 to $25\,M_\odot$ produce burned-out cores too massive to leave stable neutron stars, causing them to collapse to black holes.

In binary systems the above-indicated mass ranges for progenitors of compact stars will be somewhat modified by the effects of mass transfer and mass loss, but they still can be used as a rough first approximation (van den Heuvel 1994a).

4.2 The evolution of the outer radius and luminosity of a single star

Figure 12a depicts for single stars how the observable stellar parameters – luminosity (L), radius (R), and effective surface temperature (T_e) – change during the evolution of the stellar interior. Since $L = 4\pi R^2 \sigma T_e^4$ only two of these parameters are independent of one another. The figure depicts the evolutionary tracks in the Hertzsprung–Russell diagram of a "massive" star ($15\,M_\odot$), an "intermediate-mass" star ($5\,M_\odot$), and two "low-mass" stars ($2.25\,M_\odot$ and $1.0\,M_\odot$). Figure 12b also separately depicts the evolution of the radius of the $5\,M_\odot$ star. Important evolutionary stages are indicated in the figures. Between points 1 and 2 the star is in the long-lasting phase of core hydrogen burning (nuclear timescale). At point 3 hydrogen ignites in a shell around the helium core. For stars more massive than $1.2\,M_\odot$ the entire star briefly contracts between points 2 and 3, causing its central temperature to rise. In stars more massive than about $2.3\,M_\odot$ the core continues to contract after hydrogen shell ignition, upon which the outer layers begin to expand. When the central temperature reaches $T \sim 10^8\,\mathrm{K}$, helium ignites (point 4). At this moment the star has become a red giant, with a very dense core and a very large radius. During helium burning it describes a loop in the Hertzsprung–Russell diagram. The star with $M \gtrsim 2.3\,M_\odot$ moves from point 2 to 4 on a thermal timescale (stage II in Figure 12b) and describes the helium-burning loop on a nuclear timescale (from 4 to 5). During helium shell burning the outer radius expands again and at carbon ignition the star has become a red supergiant, on the so-called asymptotic giant branch (AGB). The evolution of low-mass stars ($M \lesssim 2.3\,M_\odot$) takes a somewhat different course. After hydrogen shell ignition the helium core becomes degenerate and the hydrogen burning shell generates the entire stellar luminosity. While its core mass grows, the star gradually climbs upwards along the giant branch until it reaches helium ignition with a flash (point 4). For all stars less massive than about $2.3\,M_\odot$ the helium core at helium flash ignition has a mass of about $0.45\,M_\odot$.

The evolution described above holds for an initial chemical composition of 70% hydrogen, 28% helium, and neglecting the effects of convective overshooting (for details see Maeder and Meynet 1989).

4.3 Cases A, B, and C of close binary evolution

When a star is born as a member of a binary system (with a radius smaller than that of its Roche lobe), it may, due to

Table 5 Types of final evolutionary products expected as a function of stellar mass (see text for details)

Main sequence mass	He core mass	Final product
$3–8\,M_\odot$	$1.4–1.9\,M_\odot$	CO white dwarf
$8–10\,(12)\,M_\odot$	$1.9–2.2\,(3.0)\,M_\odot$	Degenerate O–Ne–Mg core $\to$ collapse to NS
$10\,(12)\,M_\odot–25\,M_\odot$	$2.2\,(3.0)–8\,M_\odot$	Collapsing iron core $\to$ collapse to NS
$\gtrsim 25\,M_\odot$	$8\,M_\odot$	Collapsing iron core $\to$ collapse to BH

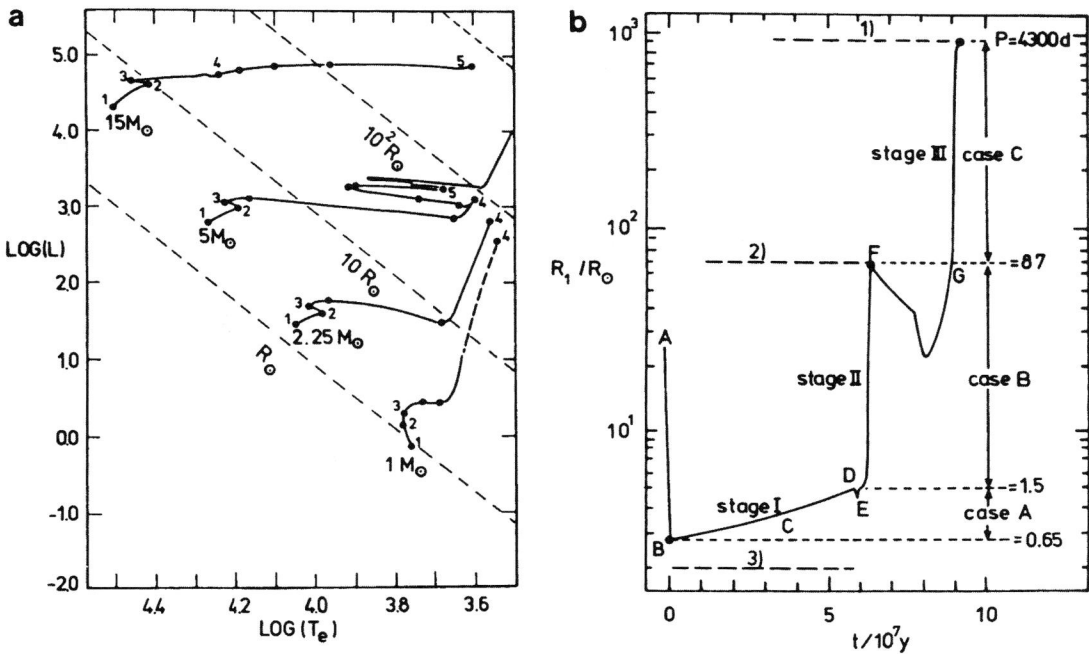

Figure 12 (a) Evolutionary tracks in the Hertzsprung–Russell diagram of stars of $1 M_\odot$, $2.25 M_\odot$, $5 M_\odot$, and $15 M_\odot$. (After Iben 1967.) (b) Secular change of the radius of a star of $5 M_\odot$, from pre-main sequence contraction (A) to carbon ignition. Also indicated are the orbital periods (in days) of close binaries in which the star just fills its Roche lobe at various evolutionary stages, for binaries with mass ratio $q = M_2/M_1 = 0.5$. The ranges of binary periods for evolution according to the cases A, B, and C (section 4) are indicated: (1) carbon ignition, (2) helium ignition, (3) main sequence model. (After Paczynski 1971a.)

evolutionary expansion of the envelope, after some time begin to overflow this lobe. When this happens, the matter flowing out along the first Langrangian point L_1 will fall towards the companion. The further evolution of the system will now depend on: (i) the evolutionary state of the star at the onset of the overflow, which is determined by the mass of the primary star, the orbital separation a, and the mass ratio of the systems; and (ii) whether the envelope of the star at the onset of the transfer is in radiative or convective equilibrium. As to the factor (i), Kippenhahn and Weigert (1967) defined three types of close binary evolution, called cases A, B, and C, depicted in Figure 12b. In case A, the system is so close that the primary star already begins to overflow its Roche lobe during core-hydrogen burning (stage 1 in Figure 12b); in case B the primary star begins to fill its Roche lobe after the end of core hydrogen burning but before helium ignition (stage II) and in this case C it overflows its Roche lobe during helium shell burning or beyond (stage III in Figure 12b). It is clear from the figure that cases B and C occur over a wide range of orbital periods, case C even up to periods of ~ 10 years. The precise orbital period ranges for cases A, B, and C depend on the initial primary mass M_1^0 and the mass ratio.

4.4 Response of the mass-losing star

Contact star with a radiative envelope

When the contact star loses an amount of mass ΔM to its companion it will restore its hydrostatic equilibrium on a dynamical timescale (free-fall timescale), that is, almost instantaneously. In stars with radiative envelopes the new hydrostatic equilibrium radius is smaller than its radius before the onset of the mass transfer, but as the star is now out of thermal equilibrium, its radius will begin to expand on a thermal timescale to restore this equilibrium. If the mass M_1 of the mass-losing star is larger than that (M_2) of its companion, the transfer will make the orbit – and with it the Roche lobe radius R_L of the mass-losing star – shrink (eqns (8), (9), and (13)). If the radius of the star after readjustment to hydrostatic equilibrium is still larger than that of its Roche lobe, the mass loss will continue until so much mass has been lost that it fits within its new Roche lobe. The subsequent expansion of the star on a thermal timescale will, however, cause the mass transfer to resume, causing the Roche lobe to shrink further, and so on. As a result the star continues to transfer mass until it has itself become the less massive component of the system. Its further expansion and mass loss will now cause, according to eqn (13), the orbit to expand such that finally the thermal equilibrium radius of the star may be able to fit within its Roche lobe. In systems evolving according to case B, practically only the helium core of the mass-losing star remains after the mass transfer, surrounded by a hydrogen-rich outer layer of only very small mass. The star may now, during its further nuclear evolution, either shrink or slowly expand on a nuclear timescale. The latter is the case in case A systems and in case B systems with original primary masses smaller than about $2.3 M_\odot$. After the mass transfer these systems are still in the phase of hydrogen burning: in case A systems hydrogen still burns in the core, in the low-mass case B systems it burns in a shell surrounding a degenerate helium core of low mass ($<0.45 M_\odot$). The radii of these stars gradually increase when hydrogen burning advances. Therefore they continue to slowly transfer mass on a nuclear timescale. The systems in this slow mass-transfer phase are the Algol-type binaries. In the more massive case B systems the helium core has a mass $\geq 0.45\ M_\odot$ and contracts further and ignites helium burning, upon which the star expels the last part of its hydrogen-rich envelope and becomes a helium-burning pure helium star. Such stars have a very small radius, and in a binary are therefore deep inside their Roche lobes. Figure 13 schematically depicts the subsequent evolutionary phases of a close binary in which the more massive star at the onset of the mass tansfer has a radiative envelope. The order of magnitude timescale of the first phase of mass transfer is simply the thermal timescale of the (initial) primary star. This leads to an order of magnitude mass-transfer rate (Van den Heuvel 1994a):

$$\dot{M}_1 = -M_1/\tau_{th} = -3 \times 10^{-8}(M_1/M_\odot)^3 (M_\odot\ yr^{-1}) \quad (23)$$

After the reversal of the mass ratio and stabilization of the system, the mass transfer takes place on a nuclear timescale, leading to a mass-transfer rate of the order of (Van den Heuvel 1994a):

$$\dot{M}_1 = 10^{-10}(M_1/M_\odot)^{3.5}(M_\odot\ yr^{-1}) \quad (24)$$

Response of the accreting companion

Being the less massive component, it has a longer thermal timescale than its companion. Therefore the rapid transfer of mass to it causes its thermal equilibrium to be strongly perturbed, especially because the accreting material carries considerable kinetic energy which is dissipated into heat when the matter accretes on the surface. The matter also carries a fraction of the orbital angular momentum which will strongly spin-up the accretion's rotation. The details of the accretion process are still poorly known, but it is likely that if the mass ratio of the system is small (≤ 0.3) the star will swell up to overflow its own Roche lobe. This occurs for radiative as well as convective envelopes of the mass-losing star. A common envelope then forms around the binary. In the next paragraph we consider the evolution which then ensues.

Contact star with a convective envelope: common envelope evolution

A star with a deep convective envelope has the characteristic that upon losing mass its radius expands on a dynamical timescale (Bhattacharya and van den Heuvel 1991), which is only of the order of hours to days. If this star is the more massive component of a close binary, its Roche lobe will shrink as a result of the mass transfer. Therefore the result is a violently unstable phase of mass transfer in which theoretically mass-transfer rates of the order of 10^{-2} to $10^{-1} M_\odot\ yr^{-1}$ could result. However, the companion will never in such a short time be able to accommodate these large amounts of mass dumped onto it. Therefore, as first suggested by Paczynski (1976) and Ostriker (1973), in this case the expanding envelope of the primary star will completely engulf the secondary star, leading to the formation of a common envelope, inside which the secondary star and the core of the primary star orbit each other. Due to the large frictional drag on this orbital motion inside the common envelope, the secondary star will rapidly spiral inwards. In this process a large amount of (orbital) potential

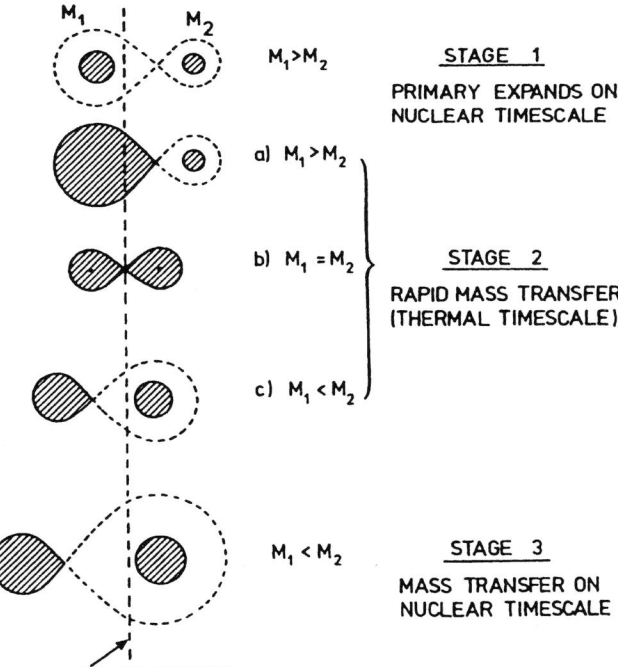

Figure 13 Subsequent evolutionary stages and orbital dimensions of a close binary system in which the primary star has a radiative envelope at the time it reaches its Roche lobe.

energy is converted into heat, causing the envelope to be expelled. As a result, one expects, after the spiral in, a very close binary to remain, consisting of the secondary and the evolved core of the primary star. Paczynski and Ostriker proposed this type of evolution to explain the existence of the CV binaries. These have extremely short orbital periods, in general of the order of only a few hours, but they often contain massive white dwarfs, which can only be produced when stars are on the asymptotic giant branch, that is, have radii of hundreds of solar radii. The original orbits of these systems, before the onset of the mass transfer, must therefore have been hundreds of days up to a number of years. An alternative outcome of the commen envelope evolution, in the case where there is not enough orbital energy available to expel the envelope, or if the core of the primary star is not sufficiently compact, is that the stars will completely merge, resulting in the formation of a rapidly rotating single star. Analytical considerations (Meyer and Meyer-Hofmeister 1978) as well as numerical hydrodynamic calculations of common envelope (CE) evolution have shown that the CE process proceeds very rapidly, on a timescale of the order of only 10^2 to 10^3 yr (Taam and Bodenheimer 1991; Taam and Sandquist 2000). CE evolution is not only important for understanding the formation of the CVs, but also for that of the LMXBs and the double compact objects with short orbital periods such as the close double neutron stars like the Hulse–Taylor binary pulsar and the close double white dwarfs such as the AM CVn systems.

Apart from mass transfer from a convective envelope or in a system of low mass ratio, formation of a common envelope may also ensue if the system is tidally unstable (Darwin 1879; Counselman 1973).

4.5 Conservative evolution and the formation of HMXBs

Formation of the "standard" HMXBs

The formation of the HMXBs can be understood in terms of the "conservative" evolution of normal massive close binary systms in the way described in Section 4.4. We consider as an example the evolution of a massive case B system as depicted in Figure 14, which started out with components of 20 and 8 solar masses and an initial orbital period of 4.7 days. The subsequent evolutionary phases are described in the figure caption and are as follows. After 6.17×10^6 yr the primary star has terminated core hydrogen burning and overflows its Roche lobe. In only 3×10^4 yr it transfers $14.66 M_\odot$ to its companion and leaves behind a $5.34 M_\odot$ helium core, which becomes a helium-burning pure helium star. The orbital period is now 10.86 days and the companion has become a 22.66 $M_\odot$ star which is still practically unevolved as it has been rejuvenated by the

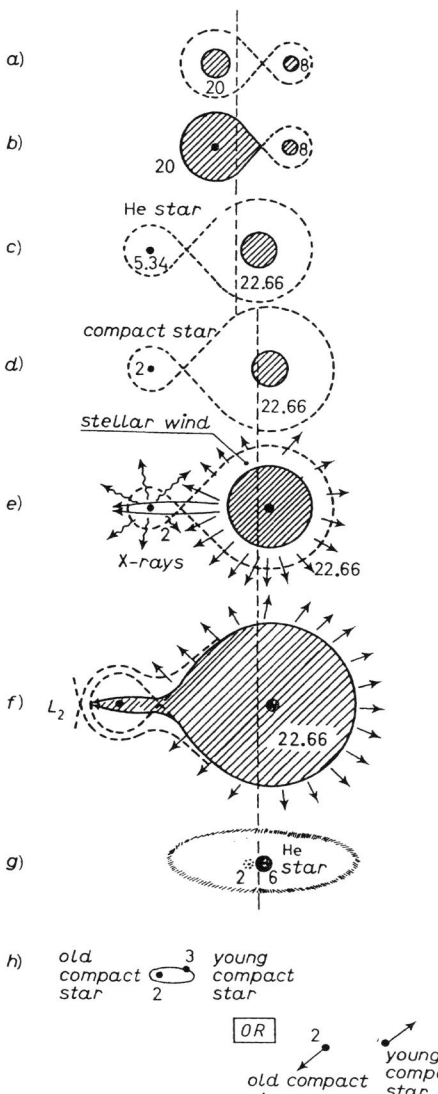

Figure 14 Subsequent stages in the evolution of a massive close binary that started out with components of $20 M_\odot$ in a circular orbit with $P = 4.70$ d (for details see van den Heuvel 1994a). It is assumed that the supernova explosion of the primary leaves a $2 M_\odot$ compact star (neutron star or black hole). (a) $t = 0$, $P = 4.70$ d, onset of the first stage of mass exchange; (b) $t = 6.17 \times 10^6$ yr, $P = 4.70$ d, onset of mass transfer; (c) $t = 6.20 \times 10^6$ yr, $P = 10.86$ d, end of mass transfer; primary has lost entire H-rich envelope and became a helium star (Wolf–Rayet star), (d) $t = 6.89 \times 10^6$ yr, $P = 12.63$ d, He star (= Wolf–Rayet star) has exploded as supernova; (e) $t = 10.41 \times 10^6$ yr, $P = 12.63$ d, the normal star becomes a supergiant, its strong stellar winds turns the compact star into powerful X-ray source; (f) $t = 10.45 \times 10^6$ yr, $P = 12.63$ d, onset of second stage of mass exchange, the X-ray source is extinguished and large mass loss from the system begins; (g) $t \sim 10.47 \times 10^6$ yr, $P \sim 4$ h, onset of second Wolf–Rayet stage; (h) $t \sim 11 \times 10^6$ yr the second He star has exploded as a supernova, survival or disruption of the system depends on the mass of the remnant. (After van den Heuvel 1977.)

14.66 $M_\odot$ of unprocessed primordial hydrogen-rich matter from its companion. As first noticed by Paczynski (1967) binary systems consisting of a helium-burning helium star plus a massive main sequence star exactly resemble the Wolf–Rayet binaries, of which hundreds are known in our Galaxy and the Magellanic Clouds. Wolf–Rayet stars are hydrogen-deficient stars of very high luminosity and relatively low mass: in the Wolf–Rayet binaries they are practically always some two to four times less massive than their O- or early B-type companion stars (van den Heuvel 1994a). Just as expected for helium-burning helium stars in this mass range, they have roughly the same luminosity as their massive main sequence companions and have a remarkable emission line spectrum indicating large mass loss by stellar winds at a rate typically of the order of $10^{-5} M_\odot \, \mathrm{yr}^{-1}$, and at velocities of the order of several thousands of kilometers per second. Some 60% of all Wolf–Rayet stars are found in close binaries of this type. The further evolution of helium stars has been studied for example by Paczynski (1971b), Arnett (1973), Nomoto (1984), Habets (1985, 1986), and Dewi *et al.* (2002). These studies have shown that the radii of helium stars more massive than 3.5 $M_\odot$ do not expand very much before the termination of their evolution with the collapse of their burnt-out core to a neutron star or a black hole. For the 5.34 $M_\odot$ helium star of Figure 14 this occurs 0.69×10^6 yr after the mass transfer. Due to the explosive mass ejection in the supernova (assuming symetric mass ejection) the orbital period increases to 12.63 days and a small orbital eccentricity is induced, and the center of gravity of the system is accelerated to a runaway velocity of about 35 km s^{-1}. At age 10.41×10^6 yr the 22.66 $M_\odot$ companion has terminated its core hydrogen burning and has become a blue supergiant with a strong stellar wind. This induces the system to become a HMXB, resembling Vela X-1. The HMXB phase, until the star overflows its Roche lobe, probably lasts only between 10^4 and 10^5 yr. After reaching its Roche lobe a common envelope will form. Assuming the system to survive this phase as a binary, the outcome will be a very close binary consisting of a helium star (Wolf–Rayet star) and the compact companion in a very narrow orbit. It is now known that the 4.8-hour X-ray binary Cyg X-3 is such a system, consisting of a helium star (Wolf–Rayet star) and a compact object (neutron star or black hole) (van Kerkwijk *et al.* 1992; Hanson *et al.* 2000; Ergma and Yungelson 1998).

Formation of the B-emission X-ray binaries

The existence of the Be/X-ray binaries can also be simply explained in terms of conservative evolution, in this case of binaries that started out with somewhat lower initial masses of the components, that is, primary stars typically in the mass range 8 to about 15 solar masses. In this case, due to the dependence of helium core mass on initial stellar mass, the mass exchange will produce systematically longer orbital periods of the post- mass-exchange binaries (van den Heuvel 1983; van den Heuvel and Rappaport, 1987; van den Heuvel *et al.* 2000). The generally high orbital eccentricities of the Be/X-ray binaries cannot be explained with symmetric supernova mass ejection (van den Heuvel 1994a; Verbunt and van den Heuvel 1995), and therefore are a clear indication that velocity "kicks" are imparted to neutron stars in their birth events (van den Heuvel and van Paradijs 1997).

4.6 Formation of double neutron stars: Descendants of Be/X-ray binaries

Calculations of common envelope evolution predict that due to their small orbital separations, the standard HMXBs are unlikely to survive the second mass-transfer phase and spiral-in as the orbits are expected to shrink by a factor 10^2 or more (Webbink 1984; Taam and Sandquist 2000). On the other hand, the wide Be/X-ray binaries such as X Per ($P_{\mathrm{orb}} = 250$ days) are expected to survive spiral-in as systems depicted in Figure 14g: short-period binaries consisting of a helium star together with a compact star. When the helium star terminates its evolution with a supernova explosion, the system is either disrupted or a double neutron star with a very eccentric orbit is formed, closely resembling the Hulse–Taylor and Wolszczan binary radio pulsars (Figure 6). It thus seems most likely that these binary pulsars are the descendants of Be/X-ray binaries (van den Heuvel 1992, 1994a). Recently some doubts have been expressed as to whether or not the neutron star can survive the spiral-in process through the envelope of a massive companion, without turning into a black hole due to highly super-Eddington accretion (Chevalier 1993; Brown 1995). However, there are various arguments indicating that the neutron star can indeed survive this spiral-in (van den Heuvel 2001). A strong argument against the neutron stars in the Be/X-ray binaries turning into black holes during spiral-in is that in that case the formation rate of close binary pulsars consisting of a black hole and a young neutron star would be of similar order of magnitude as the formation rate of the Be/X-ray binaries (as a sizeable fraction of these will survive spiral-in). (The binaries become very close if the neutron star can accrete the entire envelope of its companion.) The estimate is that there are several thousands of Be/X-ray systems in the Galaxy (van den Heuvel 1992, 1994a). These systems live shorter than 10^7 years, so their formation rate is of the order of 10^{-4} to 10^{-3} per year in the Galaxy. This implies that the formation rate of the close binaries consisting of a black hole and a young neutron star should be at least 10^{-4} per year, that is, about 1% of the pulsar birth rate. Due to the large gravitational attraction of the black hole disruption of the systems in the second supernova is unlikely, even if the neutron star receives a sizeable kick velocity when it is born. One would therefore expect some 10 close black hole/neutron star systems to be present among the about 1000 presently known young radio pulsars, while there is none. This is a strong argument against the scenario in which the neutron stars in HMXBs would not survive spiral-in as neutron stars.

Figure 15 shows the various ways in which the persistent ("standard") and Be/X-ray binaries are expected to terminate their evolution. If the Be star is less massive than 8–$10 M_\odot$ (the precise value depends on the orbital period) its burned-out core is not massive enough to leave a neutron star and becomes a CO or O–Ne–Mg white dwarf. In that case a very close binary radio pulsar with a circular orbit will remain, consisting of a neutron star and a massive white dwarf. PSR 0655+64 ($P_{\rm orb} = 1.04$) is thought to be such a system (Section 4.8).

4.7 The formation of CVs and LMXBs

The first detailed calculation of the formation of a CV binary through common envelope evolution was carried out by Meyer and Meyer-Hofmeister (1978) which started out from a system consisting of a $5 M_\odot$ red giant with a companion of one solar mass. The LMXBs must have originated through a similar type of process, the only difference being that the red giant in the progenitor system was more massive than 8–$10 M_\odot$, such that it produced a heavy-element core too massive to finish as a white dwarf. The first model put forward for such evolution was that of Sutantyo (1975a) for the formation of the Her X-1 system, which is presented as an example in Figure 16b. The progenitor was assumed to be a binary system with component masses of $15 M_\odot + 2 M_\odot$, evolving according to case B or C. The $15 M_\odot$ star produces a $4 M_\odot$ helium core and after spiral-in a close system resulted consisting of this core

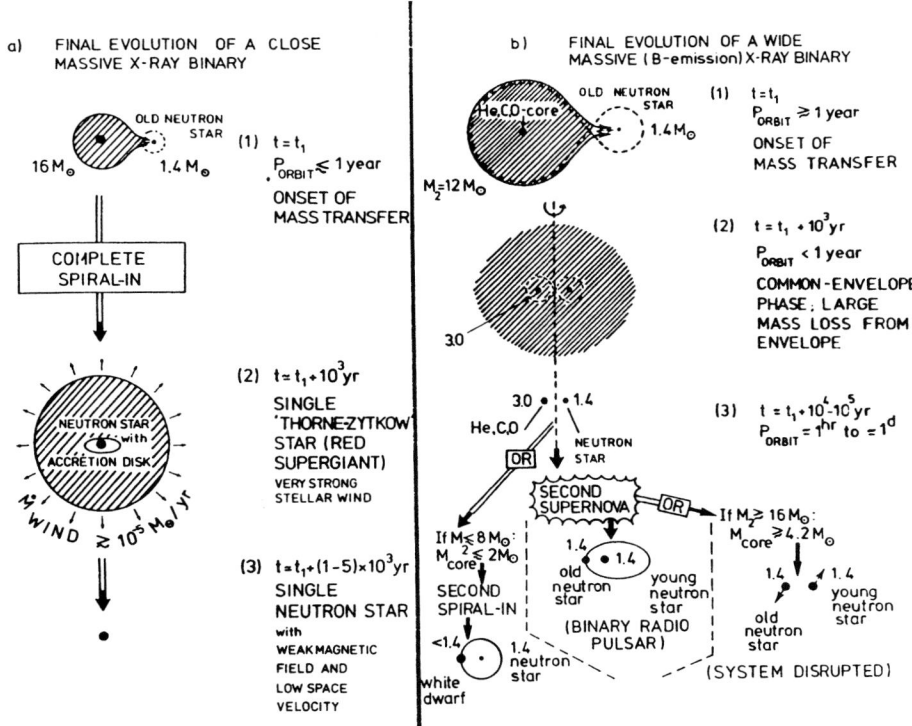

Figure 15 The various possibilities for the final evolution of a HMXB. In all cases the onset of Roche lobe overflow leads to formation of a common envelope and the occurrence of spiral-in. (a) In systems with orbital periods less than about 1 yr there is most probably not enough energy available in the orbit to eject the common envelope, and the neutron star spirals down into the core of its companion. Subsequently, the envelope is ejected by the liberated accretion energy flux, leaving a single recycled radio pulsar. (b) In systems with orbital periods longer than about 1 yr the common envelope is ejected during spiral-in, and a close binary can be left, consisting of the neutron star and the core, consisting of helium and heavier elements, of the companion. Companions initially more massive than 8–$12 M_\odot$ leave cores that will explode as a supernova, leaving an eccentric-orbit binary pulsar or two runaway pulsars. Systems with companions less massive than ~ 8–$12 M_\odot$ leave close binaries with a circular orbit and a massive white dwarf companion, similar to PSR0655+64.

together with the $2M_\odot$ companion. The evolution of the system is described in the caption to Figure 16.

In the model of Figure 16 the neutron star is formed directly from the core collapse of the evolved $4M_\odot$ helium star. This type of neutron star formation is expected for all helium stars more massive than 2.2 to $2.5M_\odot$ (Habets 1985, 1986; van den Heuvel 1994a). It is also possible that the neutron star formed by accretion-induced collapse from a massive white dwarf.

In CV-like binaries in which the white dwarf is composed of O–Ne–Mg, mass transfer can induce the white dwarf to collapse to a neutron star (Nomoto and Kondo 1991). This is most likely to occur if the mass-transfer rate is in the range $(1-4) \times 10^{-7}M_\odot$ yr^{-1}, as in this range the accreted matter will start steady hydrogen burning on the surface of the white dwarf, without much radius expansion. This leads to a continuous growth of the mass of the white dwarf. CV-like systems with mass-transfer rates in this range are the so-called super soft binary X-ray sources (SSS; Kahabka and van den Heuvel 1997) which are the most likely candidates for accretion-induced collapse. It is possible that Her X-1 formed in this way, from a SSS like Cal 83 (Kahabka and Van den Heuvel 1997).

4.8 Evolution of CVs and LMXBs driven by losses of orbital angular momentum

Evolution of CVs and compact LMXBs

For orbital periods (just after spiral-in) larger than a critical value the binary evolves slowly to a wider system when the star during hydrogen burning expands to its Roche lobe and begins to transfer mass. In the narrower systems magnetic braking of the G- or K-dwarf main sequence star is thought to dominate, causing the binary separation to decrease even when the main sequence star fills its Roche lobe and transfers matter towards the white dwarf. The enhanced mass transfer produces a hot luminous accretion disk around the white dwarf and the binary becomes a CV. The magnetic braking timescale is much shorter than both the contact star's nuclear timescale and the gravitational radiation loss timescale (e.g. Verbunt and van den Heuvel 1995, and references therein). The star therefore simply evolves down the main sequence, remaining roughly in thermal equilibrium. However, when the orbital period has decreased to about 3 hours the continuing mass loss has disturbed the star's thermal equilibrium, causing its radius to be some-what inflated. Observations of CVs indicate that there are practically no systems with orbital periods between 2 and 3 hours: the so-called "period gap" (Ritter and Kolb 1998). The period gap is believed to be caused by the sudden termination of magnetic braking when the star's mass has been reduced to $0.3M_\odot$, the point at which the star becomes completely convective (Spruit and Ritter 1983). The mass-transfer rate is then suddenly reduced and the star has time to shrink back to its thermal equilibrium radius, which is smaller than its Roche lobe at that time such that the mass transfer stops. From this moment on the orbit slowly shrinks due to gravitational radiation losses (eqn (15)). The mass transfer can resume only when the Roche lobe

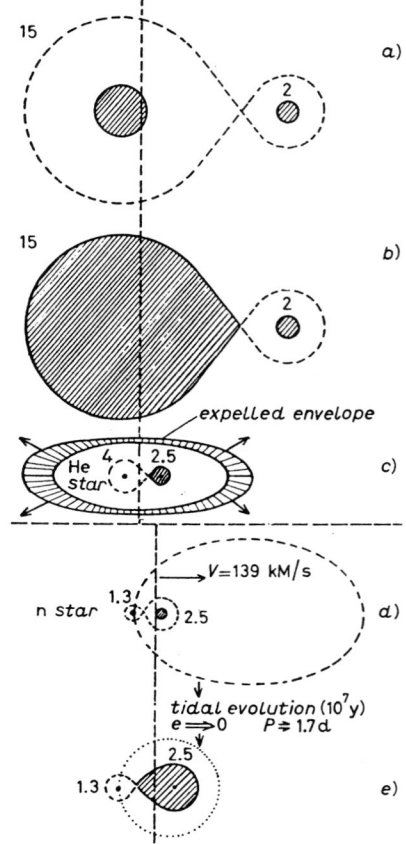

Figure 16 Scenario for the formation of the Her X-1 system (a Low-Mass X-ray Binary) out of an initially massive close binary with an extreme mass ratio. When the $15M_\odot$ primary begins to overflow its Roche lobe, the $2M_\odot$ secondary spirals in and causes the hydrogen-rich envelope of the primary to be expelled, leaving a $4M_\odot$ helium star plus the $2M_\odot$ secondary, which is expected to have accreted only $0.5M_\odot$ in this process. The supernova mass ejection and impact effects imparted a 139 km s^{-1} velocity to the center of mass of the system, and induced a large orbital eccentricity. After the supernova explosion of the helium star the system is 'quiet' for about 5×10^8 yr and the orbit is circularized by tidal friction. When the star begins to transfer mass to the neutron star the system becomes an X-ray source. (After Sutantyo 1975a; explanation in the text.)

around the star has shrunk to the thermal equilibrium radius of the $0.3 M_\odot$ star. By that time the orbital period has decreased to 2 hours. This explains the absence of CVs with periods between 2 and 3 hours. When the Roche lobe overflow has reduced the mass of the star to $0.1 M_\odot$ its thermal timescale becomes longer than the timescale for the mass loss (which is the gravitational radiation loss timescale). Further mass loss perturbs the star's thermal equilibrium causing it to become larger than its main sequence radius. At this time the nuclear fusion reactions in its interior begin to fade, and the electron gas becomes degenerate. When it has reached full degeneracy it has a reversed mass–radius relation, such that further mass loss causes it to expand. As the stellar radius equals that of the Roche lobe, it has to transfer so much mass that the expansion rate of the Roche lobe equals that of the star. This causes the orbit to expand, and determines the mass-transfer rate. The binary thus passes through a minimum orbital period which for hydrogen-rich stars is around 80 minutes (Paczynski and Sienkiewicz 1981). The observed absence of CV systems below this orbital period minimum confirms that the mass transfer in the shortest period CVs is driven by gravitational radiation (Faulkner 1971).

Evolution of compact LMXBs

The Galaxy contains about 10^2 LMXBs in which the neutron star (or black hole) receives mass from a low-mass companion star. Only for a few dozen of these systems has the orbital period been measured. Most orbital periods are shorter than 20 hours, although there are exceptions, such as Cyg X-2 ($P = 9.8$ days); (see, however, Sections 1.2 and 4.9). From the existence of wide binaries consisting of a radio pulsar with a white dwarf companion one expects the existence of LMXBs in which the X-ray source is powered by the Roche lobe overflow from the expanding (sub-)giant progenitor of the white dwarf, a phase during which the binary widens on a nuclear timescale of the (sub-)giant. This type of evolution is considered in Section 4.9 (Figure 17). Here we concentrate on the systems with short orbital periods, which closely resemble the CVs. Like in the latter systems the observationally inferred mass-transfer rates in these compact LMXBs seem an order of magnitude higher than can be driven by the nuclear expansion of the low-mass donor star or by gravitational radiation losses. Like in the CVs, the mass transfer here is apparently driven by the enhanced orbital angular momentum losses. It appears there is a bifurcation into wide and compact

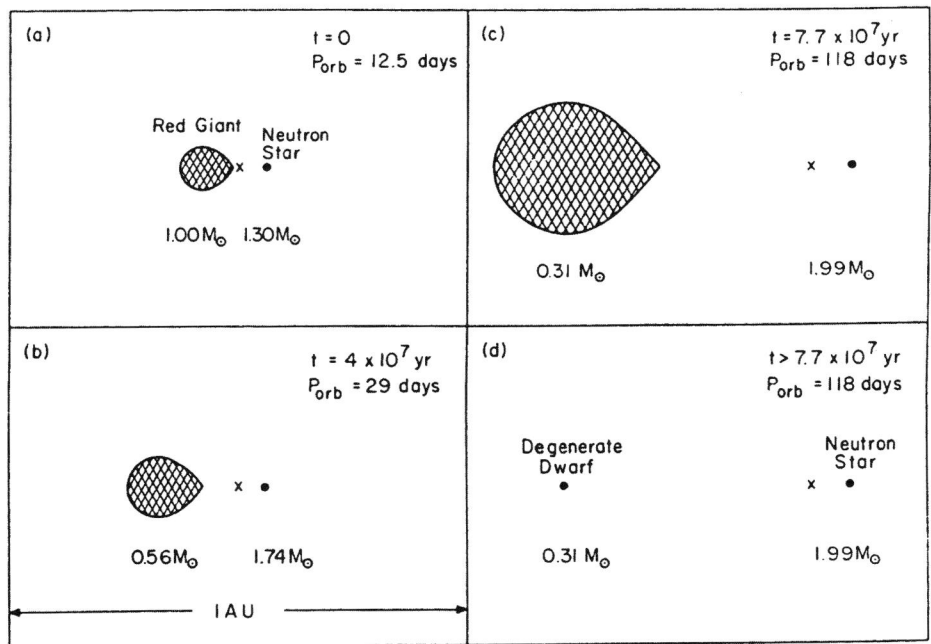

Figure 17 Evolution of a wide LMXB into a wide radio-pulsar binary with a circular orbit and a low-mass helium white dwarf companion, such as PSR 1953+29. At the onset of the mass transfer, the low-mass companion is a (sub-)giant with a degenerate helium core of $0.24 M_\odot$. Its light is generated by hydrogen fusion in a shell around the core. The mass transfer from the giant to the neutron star is due to the slow expansion of the giant, driven by this hydrogen shell burning. During the mass transfer the orbit gradually expands (due to angular momentum conservation) and after 8×10^7 yr the system terminates as a wide radio-pulsar binary. (After Joss and Rappaport 1983.)

LMXBs: short-period systems in which magnetic braking and gravitational radiation losses dominate, evolve to shorter and shorter orbital periods, while systems with periods beyond a critical initial period (of the order of a day) evolve driven largely by the nuclear evolution of the donor star, and evolve to longer and longer orbital periods (see the review by van den Heuvel 1994a). A great difference between compact LMXBs and CVs is that accretion onto the neutron star liberates far more energy which may (partly) be intercepted by the donor star and heat this star. The accreting matter can spin the neutron star up to millisecond periods (Section 6.2). It has been suggested that if the mass-transfer process is interrupted (for example, by the system entering the period gap of CVs around 3 h), the rapidly spinning neutron star becomes a millisecond radio pulsar which with its powerful relativistic pulsar wind may ultimately evaporate the low-mass donor star. That would explain the apparent absence of the (orbital) period gap between 2 and 3 hours that is observed for the CVs. Indeed several millisecond radio pulsars have been observed to be evaporating their companions, notably the "Fruchter pulsar" PSR 1957+20 (Fruchter et al. 1988). This evaporation would also explain the existence of the handful of single millisecond pulsars, such as the first discovered one PSR 1937+21. For details, see van den Heuvel (1994a).

4.9 Mass transfer in LMXBs and CVs driven by internal nuclear evolution of the donor star

Introduction

LMXBs and CVs with orbital periods longer than about 10 hours cannot contain Roche Lobe filling main sequence stars with masses $<M_\odot$. If the donor is a low-mass star, as clearly is the case in Sco X-1 ($P = 0.86$ d) or Cyg X-2 ($P = 9.8$ d, but see Section 1.2) it must be an evolved star, that is, a (sub-)giant. Recently it has been argued by various authors (King and Ritter 1999; Podsiadlowski and Rappaport 2000; Tauris et al. 2000) that the donor star in Cyg X-2, which has a mass of $0.5 M_\odot$ but a luminosity of some 30 to $50 L_\odot$, must be the evolved remnant of an "intermediate-mass" star, in this case with an initial mass of some $2 M_\odot$. (See also van den Heuvel (1981, 1994a, 1995, 1996), who suggested it to be a later evolutionary phase of a system like Her X-1.) The systems with evolved donors can therefore be divided into two groups: those with evolved "low-mass" donors, that is, stars that started out less massive than their neutron star or white dwarf companion, and "intermediate-mass" donors, that is, that started out more massive than the compact star in the system, up to about a mass ~4 to $5 M_\odot$. The latter ones produce systems consisting of a neutron star together with a massive C–O or O–Ne–Mg white dwarf ($0.5 M_\odot$ to more than $M_\odot$) in a circular orbit (Tauris et al. 2000). The former produce similar systems consisting of a neutron star and a low-mass helium white dwarf with a mass $\leq 0.45 M_\odot$, as shown by Taam (1983) and Webbink et al. (1983); see particularly Rappaport et al. (1995). Figure 17 depicts the evolution of such a low-mass system calculated by Joss and Rappaport (1983), that started out with a donor star and compact star both of $M_\odot$, and an initial orbital period of 12.5 days. After the end of the mass-transfer phase, which lasts 8.1×10^7 yr, the donor has transferred its entire hydrogen-rich envelope to the neutron star and leaves behind a $0.31 M_\odot$ degenerate helium white dwarf which is in a 117-day orbit around a $1.69 M_\odot$ neutron star. The evolution of this system was tuned to model the formation of the first discovered millisecond radio-pulsar binary PSR1953+29 (Boriakoff et al. 1983), which consist of a 6 ms pulsar and a low-mass white dwarf companion.

5 THE GLOBULAR CLUSTER X-RAY BINARIES AND THEIR FORMATION AND EVOLUTION

The discovery in 1973 by Gursky (van den Heuvel 1994b) that in three of the error boxes of Uhuru X-ray sources a globular cluster is present, came as a great surprise, since already this observation alone would indicate that X-ray sources are much more abundant (per unit mass) in globular clusters than in the Galaxy as a whole (Gursky 1973, private communication; Katz 1975). This is because the total mass of all globular clusters together is not more than 10^{-4} times the mass of the galaxy, and there are not more than some 100 strong Uhuru sources in the Galaxy. Finding three Uhuru sources in such clusters then already implies that such sources are 300 times more abundant in globular clusters than in the rest of the Galaxy as Gursky (1973) immediately realized. This point was made in writing for the first time by Katz (1975), after Clark (1975) independently discovered the globular cluster sources and had obtained much more accurate positions of these sources with the SAS-C (Small Astronomical Satellite-C, USA 1975) satellite, showing them to be inside their clusters. There are now a dozen strong globular cluster X-ray sources known, indicating a roughly thousand times higher incidence of strong sources in globular clusters than in the rest of the Galaxy (Lewin and Joss 1983). The globular cluster sources are all Type I X-ray bursters, indicating that they are neutron stars (van Paradijs 1978; Lewin et al. 1995) and binary periods have been detected in a number of them, indicating that they are definitely LMXBs. The abnormally high incidence of X-ray binaries in globular clusters indicated that here a different formation mechanism must be operating than for the X-ray sources in the galactic disk, and it was soon suggested that the formation of these LMXBs was due to close encounters of neutron stars and ordinary stars in the dense central regions of the clusters, leading to binary formation by gravitational "collision" processes. These were suggested to be either tidal capture (Fabian *et al.* 1975) or real collisions between giant stars and neutron stars (Sutantyo 1975b). Later, particularly through numerical stellar dynamics calculations it became clear that binary–binary collisions or binary–single star collisions followed by an exchange of an ordinary star by a neutron star, are the most likely processes for the formation of these systems (Hut, 1983, 1984; Hut and Bahcall 1983; Rappaport *et al.* 1989; Romani *et al.* 1987; Verbunt *et al.* 1987). It will be clear that these processes may lead to X-ray binaries that may be quite different from those that formed through the ordinary processes of close binary evolution.

Indeed, the remnants of these X-ray binaries, the globular cluster binary radio pulsars, have distributions of orbital characteristics different from those of the binary radio pulsars in the galactic disk (e.g. Camilo *et al.* 2000). For further details see Section 6.3.

6 FROM X-RAY BINARIES TO BINARY RADIO PULSARS: THE CONCEPT OF "RECYCLING"

6.1 Introduction

In the foregoing chapters we have seen that the double neutron stars and neutron star binaries with massive white dwarf companions are the remnants of HMXBs and intermediate-mass X-ray binaries, respectively, and that the binary pulsars with low-mass helium white dwarf companions (and other low-mass companions such as ablated hydrogen dwarfs) are the remnants of LMXBs. Evolutionary schemes leading to these different types of binary pulsars are given in Figure 15 for the double neutron stars and the close systems consisting of a neutron star and a massive white dwarf in a circular orbit, and Figure 17 for the neutron star plus low-mass helium white dwarf systems. The neutron stars in all these types of systems have gone through an extensive accretion phase, which particularly in the LMXBs was very long lasting ($\sim 10^8$ yr or more). As this was disk accretion the neutron stars in these systems will have acquired a large amount of angular momentum. The observed spin-up behavior of the X-ray pulsars that are accreting from a disk (Section 1.4 and Figure 8) shows indeed that one expects such neutron stars to evolve towards shorter and shorter spin periods. If at the same time the magnetic field has weakened to the values observed in the double neutron stars ($\sim 10^{10}$ G) or the binary pulsars with white dwarf companions (10^8 to 10^{10} G) (Figure 5)

then, as eqn (7) shows, these neutron stars can be spun-up to periods of the order of 20 ms in the case of double neutron stars, and 1 ms in the case of sytems with helium white dwarf companions. There is overwhelming evidence that the magnetic fields in these binary neutron stars have decayed to the values quoted above. Possible reasons for this decay are briefly discussed in Section 6.4.

6.2 Evolution of a neutron star that is born in a close binary: Recycling

The ingredients discussed in the last section allow us to describe the evolution in the surface dipole magnetic field strength (B_s) v. spin period (P) diagram of a neutron star that is born in a close binary, next to an unevolved stellar companion. This is depicted in Figure 18 (the diagram is in fact a different representation of the $\dot{P}$ v. P diagram of

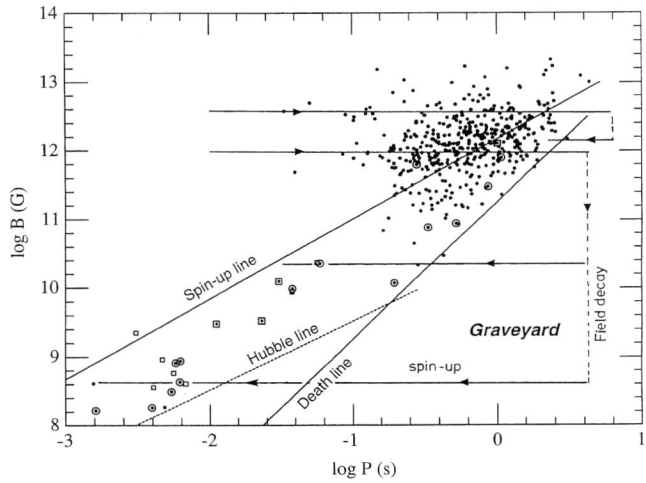

Figure 18 Magnetic field strength $\sqrt{P\dot{P}}$ v. pulse period diagram of radio pulsars with possible evolutionary tracks of pulsars in binaries indicated. Pulsars are born in the left upper part of the diagram and – if no field decay occurs – move towards the right along horizontal tracks (fully drawn). In the graveyard the field of a single pulsar probably does not decay (see text), but that of a binary neutron star does decay, presumably due to external circumstances (accretion, spin-down). Furthermore, the accretion of matter with angular momentum causes these neutron stars to be "spun-up" towards the left in the diagram along the indicated lines until they reach the spin-up line. After the companion has itself become a compact star (neutron star or white dwarf) or has disappeared, the spun-up neutron star becomes observable as a radio pulsar. It will then slowly spin-down, that is, move towards the right again in the diagram. Dots indicate normal non-recycled pulsars. Circles indicate some well-known radio pulsars in binaries, squares with dots are globular cluster binary pulsars, open squares are some single pulsars in globular clusters.

Figure 5, as $B_s = \text{const}\,(P \times \dot{P})^{1/2}$). The neutron star is born in the upper left of the diagram as a strongly magnetized rapidly spinning pulsar. It will within a few million years spin down along a horizontal track through the diagram and cross the "deathline." To the right of this line no radio pulsars can exist as the polar cap electric field has become too weak to create pairs, and no more pulsar particle wind can be produced, causing the pulsar process to stop (Chen and Ruderman 1993; Björnsson 1996). The region to the right of the deathline is the "graveyard." During its stay in the graveyard its further spin-down in the weak stellar wind of its companion (or other processes) will make its magnetic dipole moment decay (see below), causing it to move downwards in the B_s v. P diagram. When finally the companion begins to overflow its Roche lobe and begins to transfer matter through a disk, the rotation rate of the neutron star will rapidly increase, causing it to move towards shorter and shorter spin periods, that is, towards the left in the B_s v. P diagram. Thus, as shown in Figure 18, it will cross the deathline from the right and move again into the region of the "living" radio pulsars in the diagram. However, in this phase it is still an X-ray source, and will not generate radio pulses. It can at maximum be spun-up to the spin period corresponding to the "spin-up line" indicated in Figure 18, which corresponds to the shortest possible spin period which an accreting neutron star with a given dipole field strength B_s can attain, that is, for the maximum possible accretion rate, which is the Eddington rate. Thus, according to eqn (7) the equation for the spin-up line is:

$$P_{\min} = (2.4\,\text{ms})(B_9)^{6/7} \times R_6^{16/7} \times M^{-5/7} \quad (25)$$

Assuming R and M to be the same for all neutron stars, this gives the spin-up line depicted in Figure 18. The importance of the spin-up line was first realized by Radhakrishnan and Srinivasan (1982, 1984; see also Srinivasan and van den Heuvel 1982) and Alpar et al. (1982). Only after the accretion phase is terminated and the donor has itself become a compact object (neutron star or white dwarf) (or if the accretion process is otherwise interrupted, for example by the entering of the system into the period gap, Section 4.8), the rapidly spinning magnetized neutron star will become observable as a radio pulsar again. Because of their return from the graveyard, such pulsars are called "recycled." (The term "recycled" was coined by Radhakrishnan.) The recycling model has been very successful in explaining the peculiar position of the bulk of the binary and millisecond radio pulsars in the $\dot{P}$ v. P diagram (Figure 5). The model predicts that the recycled pulsars can only be found between the spin-up line and the deathline, and this is indeed exactly the region in the $\dot{P}$ v. P diagram where they are found. It also predicts that the companions should be dead stars (neutron stars or white dwarfs), which – with the exception of a few "evaporating" companions – is indeed the case. It furthermore predicts that the orbits of the double neutron stars must be eccentric (because of the second supernova in the system) whereas those of the recycled pulsars with white dwarf companions should be circular, because of the tidal circularization of the orbits during the long mass-transfer phase. All of these predictions fit excellently with the observations, confirming "recycling" to be the cause of the formation of these two classes of binary radio pulsars.

6.3 Further confirmation of the recycling model: Binary and millisecond radio pulsars in globular clusters

As noted in Section 5, the incidence of LMXBs in globular clusters is about 1000 times higher (per unit stellar mass) than in the Galaxy as a whole, indicating that the situation in globular clusters is extremely favorable for the formation of these objects. If the recycling picture is correct, one would therefore expect also many binary and millisecond pulsars to exist in globular clusters. With this idea in mind, A.G. Lyne started to survey globular clusters for such objects and found the first globular cluster millisecond pulsar: a 3 ms pulsar in the cluster M28 (Lyne et al. 1987). Radio surveys have since turned up dozens of globular cluster radio pulsars, the majority of which are millisecond pulsars. Many are in close binary systems, just as predicted by the "recycling" picture. The globular clusters richest in radio pulsars are 47 Tuc and M15, in which 20 and 8 such objects have been detected, respectively (Anderson 1993; Manchester et al. 1991; Robinson et al. 1995; Camilo et al. 2000). In 47 Tuc they are all millisecond pulsars, in M15 the population is more varied, including slowly spinning pulsars and a double neutron star (Prince et al. 1991), closely resembling the Hulse–Taylor binary pulsar. In 47 Tuc at least 13 of the 20 millisecond pulsars are in binaries. Calculations of globular cluster evolution, including dynamics predict that the total neutron star pupulation of 47 Tuc is at least of the order of 1000, many of which are expected to be recycled (Rasio et al. 2000). Detection of millisecond pulsars in close binaries in clusters is difficult, due to the Doppler variations of the pulse periods, and also since most clusters are distant and many of the pulsars are weak (Camilo et al. 2000). The observed sample is therefore likely to be just the "tip of the iceberg." Indeed, recent Chandra X-ray observation of 47 Tuc have revealed the presence of over 100 weak X-ray sources (10^{30}–$10^{33}\,\text{erg}\,\text{s}^{-1}$) among which are all of the known radio pulsars (Grindlay et al. 2000). As millisecond pulsars are known to be X-ray sources with these luminosities (Becker and Truemper 1998), it is likely that most of the over 100 faint X-ray sources in 47 Tuc are also millisecond radio pulsars. Also in several other globular clusters similar faint Chandra X-ray sources have been found: 19 in NGC 6752 and 14 in NGC 6121 (B. Gaensler, private communication). From their spectra, which are similar in hardness to those of the millisecond pulsars, the majority of them is also expected to be millisecond pulsars. Thus, the globular clusters have provided a most beautiful confirmation of the recycling piture, even though a few of the globular cluster pulsars may have had a more complex history, involving multiple stellar encounters.

6.4 Why did the magnetic fields of the recycled neutron stars decay?

We know that all the millisecond pulsars and most of the binary pulsars were recycled in binaries (the few single millisecond pulsars in the galactic disk as well as in globular clusters are also thought to have been recycled in binaries) (for a review of possible formation models see van den Heuvel 1994a). Phenomenologically one may therefore either think that the field decay is somehow related (i) to the accretion process that took place, or (ii) to the spin-down to very long spin periods that presumably took place when the companion star was still unevolved and deeply inside its Roche lobe, and had only a very weak stellar wind. For both these processes models have been proposed. Space does not permit a discussion here of the merits of these different models. I refer to the reviews by Bhattacharya and Srinivasan (1995) and Ruderman (1998).

As to field decay by accretion, models have been proposed in which the field is only confined to the neutron star crust, and there is no magnetic field in the superconducting interior; in that case, accretion can under certain conditions lead the crustal field to decay (Bhattacharya, 1995; Urpin *et al.* 1998; Konar and Bhattacharya 1999). However, the discovery of the "magnetars," neutron stars with dipole fields of the order of 10^{15} G (Kouveliotou *et al.* 1998; predicted by Thompson and Duncan 1995) makes it impossible to have the fields only in the crusts: they must thread the entire superconducting interior, as the pressures of these fields are too large to be solely sustained by the crust (Duncan, private communication). Therefore, it seems hard to make the total field decay by accretion onto the crust. If the field threads the superconducting interior the weakening of the outside field by accretion can only be achieved by "burying" the crustal field and hoping that it never comes out again. However, it seems very hard to "bury" the magnetic field by accretion for a long time (Cummings *et al.* 2001). This is because during the accretion new crust of the neutron star is continuously added, and the accreted hydrogen burns to helium, and then later, in thermonuclear flashes, to carbon and oxygen, which still later burn further to heavier elements. This creates a neutron star crust with many impurities, which causes its electrical conductivity to be low (Cummings *et al.* 2001; Bildsten, private communication). This causes any buried field to re-emerge above the stellar surface within a few million years after the termination of the accretion phase. Also, due to the relatively short timescale of the field diffusion process through the crust, one can only successfully "bury" the crustal field temporarily if the accretion rate is not too low: at rates below $10^{-10} M_\odot$ yr^{-1} one expects there will be observable surface magnetic fields (Ruderman, private communication). These considerations lead to two conclusions: (1) the internal (core) magnetic dipole moment of the millisecond pulsars must have been the internal (core) dipole moments of the neutron stars in LMXBs, and (2) the internal (core) dipole moments of these stars must in some way have been weakened by a factor of the order of 10^4 during their lifetimes.

The most promising model for achieving this is the one proposed by Ruderman (e.g. Ruderman 1998), and worked out in various ways by Bhattacharya and Srinivasan (e.g. Bhattacharya and Srinivasan 1995) which is based on the fact that in the superfluid and superconducting interior the magnetic flux is concentrated in quantized flux tubes which are strongly pinned to the quantized vortices in the rotating neutron superfluid. The density of fluxtubes per square meter is proportional to the magnetic field strength and the density of vortex tubes is proportional to the neutron star's angular velocity of rotation. When the rotation of the neutron star slows down, the density of the quantized vortex tubes in the interior decreases, forcing the quantized vortex tubes to move outwards, carrying with them the magnetic flux tubes. The outwards moving vortices disappear when they reach the crust, and the magnetic flux tubes they carried along will pile up below the inner crust. Because of the finite conductivity of the crust, the currents associated with the field will gradually dissipate, causing the field to decay. Thus, basically, the spin-down of the star is the cause of the decay of the core field. Spin-down to long rotation periods is expected in the progenitors of the LMXBs during the very long time interval before the donor star fills its Roche lobe (Bhattacharya 1995), and thus the magnetic fields of these neutron stars are expected to have decayed considerably. In the HMXBs the neutron stars had no more than about 10^7 years to spin down, such that given the diffusion timescale of several million years of the field through the crust, the field cannot have weakened much more than an order of magnitude, before spin-up occurred and the field lines were dragged in again.

Thus, in the pulsars recycled in HMXBs one does not expect the fields to have weakened to the same low values as in the LMXBs.

REFERENCES

Alpar, M.A., Cheng, A.F., Ruderman, M.A. and Shaham, J. (1982). A new class of radio pulsars. *Nature*, **300**, 728–730.

Anderson, S.B. (1993). A study of recycled pulsars in globular clusters. PhD thesis, California Institute of Technology.

Apparao, K.M.V. (1994). X-Ray emission from Be star/X-ray binaries. *Space Science Reviews*, **69**, 255.

Arnett, W.D. (1973). Advanced evolution of massive stars. III. Hydrostatic – burning nucleosynthesis and energy generation. *Astrophysical Journal*, **179**, 249–256.

Baade, W. and Zwicky, F. (1934). Supernovae and cosmic rays. *Physical Review*, **45**, 138.

Becker, W. and Truemper, J. (1998). On the X-ray emission properties of rotation powered pulsars. In R. Buccheri, J. van Paradijs, and M.A. Alpar (eds), *The Many Faces of Neutron Stars*, Kluwer, Dordrecht, pp. 525–537.

Bhattacharya, D. (1995). The evolution of the magnetic fields of neutron stars. *Journal of Astrophysics and Astronomy*, **16**, 217.

Bhattacharya, D. and Srinivasan, G. (1995). The magnetic fields of neutron stars and their evolution. In W.H.G. Lewin, J. van Paradijs, and E.P.J. van den Heuvel (eds), *X-ray Binaries*, Cambridge University Press, pp. 495–522.

Bhattacharya, D. and van den Heuvel, E.P.J. (1991). Formation and evolution of binary and millisecond radio pulsars. *Physics Reports*, **203**, 1–124.

Bildsten, L., Chakrabarty, D., Chiu, J., Finger, M.H., Koh, D.T., Nelson, R.W., Prince, T.A., Rubin, B.C., Scott, D.M., Stollberg, M., Vaughan, B.A., Wilson, C.A. and Wilson, R.B. (1997). Observations of accreting pulsars. *Astrophysics Journal Supplement*, **113**, 367.

Bisnovatyi-Kogan, G.S. and Komberg, B.V. (1974). Pulsars and close binary systems. *Astromicheskii Zhurnal*, **51**, 373 [ET: *Soviet Astronomy*, **18**, 217].

Björnsson, C.-I. (1996). Polar gap dynamics and the death line of millisecond pulsars *Astrophysical Journal*, **471**, 321.

Blaauw, A. (1961). On the origin of the O- and B-type stars with high velocities (the 'runaway' stars), and some related problems. *Bulletin of the Astronomical Institute of the Netherlands*, **15**, 265.

Bolton, C.T. (1972). Identifications of Cyg X-1 with HDE 226868. *Nature*, **235**, 271.

Boriakoff, V., Buccheri, R. and Fauci, F. (1983). Discovery of a 6.1-ms binary pulsar PSR1953 + 29. *Nature*, **304**, 417–419.

Börner, G., Meyer, F., Schmidt, H.U. and Thomas, H.C. (1972). Paper presented at the meeting of the Astronomische Gesellschaft, Wien, Austria (abstract).

Bradt, H.V.D. and McClintock, J.E. (1983). The optical counterparts of compact galactic X-ray sources. *Annual Review of Astronomy and Astrophysics*, **21**, 13–66.

Braes, L. and Miley, G. (1971). Detection of radio emission from Cygnus X-1. *Nature*, **232**, 246.

Brown, G.E. (1995). Neutron star accretion and binary pulsar formation. *Astrophysical Journal*, **440**, 270–279.

Brown, G.E., Lee, C.H., Portegies Zwart, S.F. and Bethe, H. (2001). Evolution of neutron star, carbon–oxygen white dwarf binaries. *Astrophysical Journal*, **547**, 345–354.

Camilo, F., Lorimer, D.R., Freire, R., Lyne, A.G. and Manchester, R.N. (2000). Observations of 20 millisecond pulsars in 47 Tucanae at 20 centimeters. *Astrophysical Journal*, **535**, 975–990.

Chen, K.Y. and Ruderman, M.A. (1993). Pulsar death lines and death valley. *Astrophysical Journal*, **402**, 264–270.

Chevalier, R.A. (1993). Neutron star accretion in a stellar envelope. *Astrophysical Journal*, **411**, L33–L36.

Chevalier, R.A. (1996). Neutrino-cooled accretion: Rotation and stellar equation of state. *Astrophysical Journal*, **459**, 322.

Clark, G.W. (1975). X-ray binaries in globular clusters. *Astrophysical Journal*, **199**, L143–L145.

Corbet, R.H.D. (1984). Be/neutron star binaries – A relationship between orbital period and neutron star spin period. *Astronomy and Astrophysics*, **141**, 91–93.

Counselman, C.C., III (1973). Outcomes of tidal evolution. *Astrophysical Journal*, **180**, 307–316.

Cummings, A., Zweibel, E. and Bildsten, L. (2001). Magnetic screening in accreting neutron stars. *Astrophysical Journal*, **557**, 958–966.

Darwin, G.H. (1879). *Proceedings of the Royal Society of London*, **29**, 168

Davidson, K. and Ostriker, J.P. (1973). Neutron-star accretion in a stellar wind: Model for a pulsed X-ray source. *Astrophysical Journal*, **179**, 585–598.

de Loore, C., de Greve, J.R. and de Cuyper, J.P. (1975). Evolution of massive close binaries. II – The Post X-ray binary stage: Origin of run-away and binary pulsars. *Astrophysics and Space Science*, **36**, 219–225.

Dewey, R.J. and Cordes, J.M. (1987). Monte Carlo simulations of radio pulsars and their progenitors. *Astrophysical Journal*, **321**, 780–798.

Dewi, J.D.M., Pols, O.R., Savonije, G.J. and van den Heuvel, E.P.J. (2001). The evolution of naked helium stars with a neutron star companion. *Monthly Notices Royal Astron. Soc.* (in the press).

Eggleton, P.P. (1983). Approximations to the radii of Roche lobes. *Astrophysical Journal*, **268**, 368–369.

Ergma, E. and van den Heuvel, E.P.J. (1998). On the initial progenitor masses of stellar mass black holes and neutron stars. *Astronomy and Astrophysics*, **331**, L29–L32.

Ergma, E. and Yungelson, L.R. (1998). CYG X-3: can the compact object be a black hole? *Astronomy and Astrophysics*, **333**, 151–158.

Fabian, A.C., Pringle, J.E. and Rees, M. J. (1975). Tidal capture formation of binary systems and X-ray sources in globular clusters. *Monthly Notices of the Royal Astronomical Society*, **172**, 15–18.

Faulkner, J. (1971). Ultrashort-period binaries, gravitational radiation, and mass transfer. I. The standard model, with applications to WZ Sagittae and Z Camelopardalis. *Astrophysical Journal*, **170**, L99.

Finger, M. (1998). Long-term behavior of accretion powered pulsars. In R. Buccheri, J. van Paradijs, and M.A. Alpar (eds), *The Many Faces of Neutron Stars*, Kluwer, Dordrecht, pp. 369–384.

Flannery, B.P. and van den Heuvel, E.P.J. (1975). On the origin of the binary pulsar PSR 1913+16. *Astronomy and Astrophysics*, **39**, 61–67.

Fruchter, A.S., Stinebring, D.R. and Taylor, J.H. (1988). A millisecond pulsar in an eclipsing binary. *Nature*, **333**, 237–239.

Fryxell, B. and Taam, R. E. (1988). Numerical simulation of nonaxisymmetric adiabatic accretion flow. *Astrophysical Journal*, **335**, 862–880.

Giacconi, R., Gursky, H., Paolini, F.R. and Rossi, B.B. (1962). Evidence for X-rays from sources outside the solar system. *Physical Review Letters*, **9**, 439–443.

Gold, T. (1969). Rotating neutron stars and the nature of pulsars. *Nature*, **221**, 25–27.

Gottlieb, E.W., Wright, E.L. and Liller, W. (1975). Optical studies of UHURU sources. XI. A probable period for Scorpius X-1 = V818 Sco. *Astrophysical Journal*, **195**, L33–L35.

Grindlay, J.E., Heinke, C.O., Edmonds, P.D., Murray, S.S. and Camilo, F. (2000). Deepest Globular Cluster X-ray Survey: Chandra on 47 Tuc, American Astron. Soc. Meeting 197, #126.06, Bibl. Code AAS 19712606.

Gursky, H. (1973). Talk presented at NATO Advance Study Institute on Neutron Stars and Black Holes, Cambridge, UK (unpublished).

Guseinov, O.Kh. and Zel'Dovich, Ya.B. (1966). Collapsed stars in binary systems. *Astronomicheskii Zhurnal*, **43**, 313 [ET: *Soviet Astronomy*, **10**, 251].

Habets, G.M.H.J. (1985). Advanced evolution of helium stars and massive close binaries. PhD thesis, University of Amsterdam.

Habets, G.M.H.J. (1986). Evolution of helium stars in the mass range 2.0–4.0 M. *Astronomy and Astrophysics*, **167**, 61–76.

Hanson, M.M., Still, M.D. and Fender, R.P. (2000). Orbital dynamics of Cygnus X-3. *Astrophysical Journal*, **541**, 308–313.

Hartman, J.W. (1996). The velocity distribution of young radio pulsars: simulating the observations. In S. Johnston, M.A. Walker, and M. Bailes (eds), *Pulsars, Problems and Progress*, ASP Conference Series Vol. 105, Astronomical Society of the Pacific, San Francisco, pp. 53–54.

Hartman, J.W. (1997). On the velocity distribution of radio pulsars at birth. *Astronomy and Astrophysics*, **322**, 127–130.

Hewish, A., Bell, S. 1, Pilkington, J.D., Scott, P.F. and Collins, R.A. (1968). Observation of a rapidly pulsating radio source. *Nature*, **217**, 709–713.

Hjellming, R.M. and Wade, C. (1971). Radio emission from X-ray sources. *Astrophysical Journal*, **168**, L21.

Hulse, A.R. and Taylor, J.H. (1975). Discovery of a pulsar in a binary system. *Astrophysical Journal*, **195**, L51–L53.

Hut, P. (1983). The topology of three-body scattering. *Astronomical Journal*, **88**, 1549–1559.

Hut, P. (1984). Hard binary-single star scattering cross sections for equal masses. *Astrophysical Journal Supplement*, **55**, 301–317.

Hut, P. and Bahcall, J.N. (1983). Binary-single star scattering. I – Numerical experiments for equal masses. *Astrophysical Journal*, **268**, 319–341.

Iben, I., Jr. (1967). Stellar evolution within and off the main sequence. *Annual Review of Astronomy and Astrophysics*, **5**, 571.

Joss, P.C. and Rappaport, S.A. (1983). On the origin of the 6.1-ms pulsar. *Nature*, **304**, 419–421.

Joss, P.G., Rappaport, S.A. and Lewin, B.W. (1987). The core mass-radius relation for giants – A new test of stellar evolution theory. *Astrophysical Journal*, **319**, 180–187.

Kahabka, P. and van den Heuvel, E.P.J. (1997). Luminous supersoft X-ray sources. *Annual Review of Astronomy and Astrophysics*, **35**, 69–100.

Kalogera, V. (1998a). Formation of low-mass X-ray binaries. III. A new formation mechanism: Direct supernova. *Astrophysical Journal*, **493**, 368.

Kalogera, V. (1998b). The effects of kick velocities on the formation of low-mass X-ray binaries. In R. Buccheri, J. van Paradijs, and M.A. Alpar (eds), *The Many Faces of Neutron Stars*, Kluwer, Dordrecht, pp. 505–551.

Kalogera, V. and Baym, G. (1996). The maximum mass of a neutron star. *Astrophysical Journal*, **470**, L61.

Kalogera, V. and Webbink, R.F. (1998). Formation of low-mass X-ray binaries. II. Common envelope evolution of primordial binaries with extreme mass ratios. *Astrophysical Journal*, **493**, 351.

Katz, J.I. (1975). Two kinds of stellar collapse. *Nature*, **253**, 698–699.

King, A.R. (1998). Outbursts of irradiated accretion discs. *Monthly Notices of the Royal Astronomical Society*, **296**, L45–L50.

King, A.R. (2000). Black hole transients and the Eddington limit. *Monthly Notices of the Royal Astronomical Society*, **312**, L39–L41.

King, A.R., Kolb, U. and Suszkewicz, E. (1997). Why low-mass black hole binaries are transient. *Astrophysical Journal*, **488**, 89.

King, A.R. and Ritter, H. (1999). Cygnus X-2, super-Eddington mass transfer, and pulsar binaries. *Monthly Notices of the Royal Astronomical Society*, **309**, 253–260.

Kippenhahn, R. and Weigert, A. (1966). Entwicklung in engen Doppelsternsystemen I. Massenaustausch vor und nach Beendigung des zentralen Wasserstoff-Brennens. *Zeitschrift für Astrophysik*, **65**, 251.

Kolb, U., Davies, M.B., King, A.R. and Ritter, H. (2000). The violent past of Cygnus X-2. *Monthly Notices of the Royal Astronomical Society*, **317**, 438–446.

Konar, S. and Bhattacharya, D. (1999). Magnetic field evolution of accreting neutron stars – III. *Monthly Notices of the Royal Astronomical Society*, **308**, 795–798.

Kouveliotou, C., Dieters, S., Strohmayer, T., van Paradijs, J., Fishman, G.J., Meegan, C.A., Hurley, K., Kommers, J., Smith, I., Frail, D. and Murakami, T. (1998). An X-ray pulsar with a superstrong magnetic field in the soft gamma-ray repeater SGR 1806-20. *Nature*, **393**, 235–237.

Kristian, J., Brucato, R., Visvanatan, N., Lanning, H. and Sandage, A. (1971). On the optical identification of Cygnus X-1. *Astrophysical Journal*, **168**, L91.

Lamb, F.K., Pethick, C.J. and Pines, D. (1973). A model for compact X-ray sources: Accretion by rotating magnetic stars. *Astrophysical Journal*, **184**, 271–290.

Lewin, W.H.G. and Joss, P.C. (1983). X-ray bursters and the X-ray sources of the galactic bulge. In W.H.G. Lewin and E.P.J. van den Heuvel (eds), *Accretion Driven Stellar X-ray Sources*, Cambridge University Press, pp. 41–115.

Lewin, W.H.G., van Paradijs, J.A. and Taam, R. E. (1995). X-ray bursts. In W.H.G. Lewin, J. van Paradijs, and E.P.J. van den Heuvel (eds), *X-ray Binaries*, Cambridge University Press, pp. 175–232.

Lyne, A.G., Brinklow, A., Middleditch, J., Kulkarni, S.R. and Backer, D.C. (1987). The discovery of a millisecond pulsar in the globular cluster M28. *Nature*, **328**, 399–401.

Lyne, A.G. and Graham-Smith, F. (1990). *Pulsar Astronomy*, Cambridge University Press.

Maeder, A. and Meynet, G. (1989). Grids of evolutionary models from 0.85 to 120 solar masses – Observational tests and the mass limits. *Astronomy and Astrophysics*, **210**, 155–173.

Manchester, R.N., Lyne, A.G., Robinson, C., Bailes, M. and D'Amico, N. (1991). Discovery of ten millisecond pulsars in the globular cluster 47 Tucanae. *Nature*, **352**, 219–221.

Manchester, R.N., and Taylor, J.H. (1977). *Pulsars*, Freeman, San Francisco.

Maraschi, L., Treves, A. and van den Heuvel, E.P.J. (1976). B-emission stars and X-ray sources. *Nature*, **259**, 292–293.

McClintock, J.E. (1992). Black holes in binary systems. In E.P.J. van den Heuvel and S.A. Rappaport (eds), *X-ray Binaries and Recycled Pulsars*, Kluwer, Dordrecht, p. 27–36.

McClintock, J.E. and Remillard, R.A. (1986). The black hole binary A0620-00. *Astrophysical Journal*, **308**, 110–122.

Meyer, F. and Meyer-Hofmeister, E. (1979). Formation of cataclysmic binaries through common envelope evolution. *Astronomy and Astrophysics*, **78**, 167–176.

Monaghan, J. (1969). Unpublished (manuscript submitted to *Nature*, but not placed).

Morton, D.C. (1960). Evolutionary mass exchange in close binary systems. *Astrophysical Journal*, **132**, 146.

Morton, D.C. (1967). The far-ultraviolet spectra of six stars in Orion. *Astrophysical Journal*, **147**, 1017.

Nauenberg, M. and Chapline, G. (1973). Determination of properties of cold stars in general relativity by a variational method. *Astrophysical Journal*, **179**, 277–288.

Nelemans, G., Tauris, T.M. and van den Heuvel, E.P.J. (1999). Constraints on mass ejection in black hole formation derived from black hole X-ray binaries. *Astronomy and Astrophysics*, **352**, L87–L90.

Nomoto, K. (1984). Evolution of 8–10 solar mass stars toward electron capture supernovae. I – Formation of electron-degenerate O + Ne + Mg cores. *Astrophysical Journal*, **277**, 791–805.

Nomoto, K. and Kondo, Y. (1991). Conditions for accretion-induced collapse of white dwarfs. *Astrophysical Journal*, **367**, L19–L22.

Novikov, I. D. and Zel'Dovich, Ya.B. (1966). Physics of Relativistic Collapse. *Nuova Cimento Supplement*, **4**, 810–827.

Oda, M., Gorenstein, R, Gursky, H., Kellogg, E., Schreier, E., Tananbaum, H. and Giacconi, R. (1971). X-ray pulsations from Cygnus X-1 observed from Uhuru. *Astrophysical Journal*, **166**, L1.

Oppenheimer, J. and Snyder, H. (1939). On continued gravitational contraction *Physical Review*, **56**, 455–459.

Oppenheimer, J. and Volkoff, G. (1938). On massive neutron cores. *Physical Review*, **55**, 374–381.

Ostriker, J.P. (1973). Private communication.

Paczynski, B. (1966). Evolution of close binaries. I. *Acta Astronomica*, **16**, 231

Paczynski, B. (1967). Evolution of close binaries. V. *Acta Astronomica*, **17**, 355–380.

Paczynski, B. (1971a). Evolutionary processes in close binary systems. *Annual Review of Astronomy and Astrophysics*, **9**, 183–208.

Paczynski, B. (1971b). Evolution of single stars. IV. Helium stars. *Acta Astronomica*, **21**, 1–14.

Paczynski, B. (1976). Common envelope binaries. In P. Eggleton, S. Mitton and J. Wheelan (eds), *Structure and Evolution of Close Binary Systems*, Reidel, Dordrecht, pp. 75–80.

Paczynski, B. and Sienkiewicz, R. (1981). Gravitational radiation and the evolution of cataclysmic binaries. *Astrophysical Journal*, **248**, L27–L30.

Plavec, M.J. (1967). On the problem of contact primaries in semi-detached close binaries. *Bull. Astron. Inst. Czechoslov*, **18**, 253–261.

Podsiadlowski, P. and Rappaport, S. (2000). Cygnus X-2: The descendant of an intermediate-mass X-ray binary. *Astrophysical Journal*, **529**, 946–951.

Pooley, G.G., Fender, R.R. and Brocksopp, C. (1999). Orbital modulation and longer term variability in the radio emission from Cygnus X-1. *Monthly Notices of the Royal Astronomical Society*, **302**, LI–L5.

Prince, T.A., Anderson, S.B., Kulkarni, S. and Wolszczan, A. (1991). Timing observations of the 8 hour binary pulsar 2127+11C in the globular cluster M15. *Astrophysical Journal*, **374**, L41–L44.

Radhakrishnan, V. and Srinivasan, G. (1982). On the origin of the recently discovered ultra-rapid pulsar. *Current Science*, **51**, 1096–1099.

Radhakrishnan, V. and Srinivasan, G. (1984). Are many pulsars processed in binaries? In B. Hidayat and M.W. Feast (eds), *Proceedings of the Second Asian-Pacific Regional Meeting on Astronomy*, held in Bandung, Indonesia, 24–29 August 1981, Tira Pustaka, Jakarta, p. 423.

Rappaport, S.A., Joss, P.C. and Verbunt, F. (1983). A new technique for calculations of binary stellar evolution, with application to magnetic braking. *Astrophysical Journal*, **275**, 713–731.

Rappaport, S.A., Podsiadlowski, Ph., Joss, P.C., di Stefano, R. and Han, Z. (1995). The relation between white dwarf mass and orbital period in wide binary radio pulsars. *Monthly Notices of the Royal Astronomical Society*, **273**, 731–741.

Rappaport, S.A., Putney, A. and Verbunt, F. (1989). Evolution of wide binary millisecond pulsars in globular clusters. *Astrophysical Journal*, **345**, 210–221.

Rappaport, S. and van den Heuvel, E.P.J. (1982). X-ray observations of Be stars. In *Be stars, Proceedings of the Symposium*, Munich, West Germany, 6–10 April 1981, Reidel, Dordrecht, pp. 327–344.

Rasio, F.A., Pfahl, E.D. and Rappaport, S.A. (2000). Formation of short-period binary pulsars in globular clusters. *Astrophysical Journal*, **532**, L47–L50.

Rhoades, C.E. and Ruffini, R. (1974). Maximum mass of a neutron star. *Physical Review Letters*, **32**, 324–325.

Ritter, H. and Kolb, U. (1998). Catalogue of cataclysmic binaries, low-mass X-ray binaries and related objects (Sixth edition). *Astronomy and Astrophysics Supplement*, **129**, 83–85.

Robinson, C.R., Lyne, A.G., Manchester, R.N., Bailes, M., D'Amico, N. and Johnston, S. (1995). Millisecond pulsars in the globular cluster 47 Tucanae. *Monthly Notices of the Royal Astronomical Society*, **274**, 547–554.

Romani, R.W., Kulkarni, S.R. and Blandford, R.D. (1987). Formation of a millisecond pulsar in a globular cluster. *Nature*, **329**, 309–310.

Ruderman, M. (1998). Pulsar magnetic fields and glitches. In R. Buccheri, J. van Paradijs, and M.A. Alpar (eds), *The Many Faces of Neutron Stars*, Kluwer, Dordrecht, p. 77.

Salpeter, E.E. (1964). Accretion of interstellar matter by massive objects. *Astrophysical Journal*, **140**, 796–800.

Sandage, A.R., Osmer, R, Giacconi, R., Gorenstein, P., Gursky, H., Waters, J., Bradt, H., Garmire, G., Sreekantan, B.V., Oda, M., Osawa, K. and Jugaku, J. (1966). On the optical identification of Sco X-1. *Astrophysical Journal*, **146**, 316.

Savonije, G.J. (1978). Roche-lobe overflow in X-ray binaries. *Astronomy and Astrophysics*, **62**, 317–338.

Savonije, G.J. (1983). Evolution and mass transfer in X-ray binaries. In W.H.G. Lewin and E.P.J. van den Heuvel (eds), *Accretion Driven Stellar X-ray Astronomy*, Cambridge University Press, pp. 343–366.

Schreier, E., Levinson, R., Gursky, H., Kellogg, E., Tananbaum, H. and Giacconi, R. (1972). Evidence for the binary nature of Centaurus X-3 from Uhuru X-ray observations. *Astrophysical Journal*, **172**, L79.

Shapiro, S.L. and Teukolsky, S.A. (1983). *Black Holes, White Dwarfs, and Neutron Stars: The Physics of Compact Objects*, Wiley, New York, 645 pp.

Shklovski, I. (1967). On the nature of the source of X-ray emission of Sco XR-1. *Astrophysical Journal*, **148**, L1.

Smarr, L.L. and Blandford, R. D. (1976). The binary pulsar – Physical processes, possible companions, and evolutionary histories. *Astrophysical Journal*, **207**, 574–588.

Soberman, G.E., Phinney, E. S. and van den Heuvel, E.P.J. (1997). Stability criteria for mass transfer in binary stellar evolution. *Astronomy and Astrophysics*, **327**, 620–635.

Sollerman, J., Kozma, C., Fransson, C., Leibundgut, B., Lundqvist, P., Ryde, F. and Woudt, P. (2000). SN 1998bw at late phases. *Astrophysical Journal*, **537**, L127–L130.

Spruit, H.C. and Ritter, H. (1983). Stellar activity and the period gap in cataclysmic variables. *Astronomy and Astrophysics*, **124**, 267–272.

Srinivasan, G. (1998). Neutron stars. In *Stellar Remnants*, Springer, Heidelberg, pp. 97–235.

Srinivasan, G. and van den Heuvel, E.P.J. (1982). Some constraints on the evolutionary history of the binary pulsar PSR 1913+16. *Astronomy and Astrophysics*, **108**, 143–147.

Staelin, D.H. and Reifenstein, E.C.III. (1968). Pulsating radio sources Near the Crab Nebula. *Science*, **162**, 1481–1483.

Sutantyo, W. (1975a). Supernova explosions in close binary systems. II – Runaway velocities of X-ray binaries. *Astronomy and Astrophysics*, **41**, 47–52.

Sutantyo, W. (1975b). The formation of globular cluster X-ray sources through neutron star-giant collisions. *Astronomy and Astrophysics*, **44**, 227–230.

Sutantyo, W. (1992). The evolution of Her X-1 and Her X-1 systems is very rare. In E.P.J. van den Heuvel and S.A. Rappaport (eds), *X-Ray Binaries and Recycled Pulsars*, Kluwer, Dordrecht, pp. 293–309.

Taam, R.E. (1983). The evolution of a stripped giant-neutron star binary. *Astrophysical Journal*, **270**, 694–699.

Taam, R.E. and Bodenheimer, P. (1991). Double core evolution. IV – The late stages of evolution of a 2-solar mass red giant with a 1-solar mass companion. *Astrophysical Journal*, **373**, 246–249.

Taam, R.E. and Fryxell, B. A. (1988). On nonsteady accretion in stellar wind-fed X-ray sources. *Astrophysical Journal*, **327**, L73–L76.

Taam, R.E. and Sandquist, E.L. (2000). Common envelope evolution of massive binary stars. *Annual Review of Astronomy and Astrophysics*, **38**, 113–141.

Taam, R.E. and van den Heuvel, E.P.J. (1986). Magnetic field decay and the origin of neutron star binaries. *Astrophysical Journal*, **305**, 235–245.

Tanaka, Y. and Lewin, W.H.G. (1995). Black-hole binaries. In W.H.G. Lewin, J. van Paradijs, and E.P.J. van den Heuvel (eds), *X-ray Binaries*, Cambridge University Press, pp. 126–174.

Tananbaum, H., (1973). UHURU results on galactic X-ray sources. In H. Bradt and R. Giacconi (eds), *X- and Gamma-ray Astronomy*, IAU Symposium 55, Reidel, Dordrecht, pp. 9–28.

Tananbaum, H., Gursky, H., Kellogg, E., Levinson, R., Schreier, E. and Giacconi, R. (1972). Discovery of a periodic pulsating binary X-ray source in Hercules from UHURU. *Astrophysical Journal*, **174**, L143.

Tauris, T., van den Heuvel, E.P.J. and Savonije, G.J. (2000). Formation of millisecond pulsars with heavy white dwarf companions: Extreme mass transfer on subthermal timescales. *Astrophysical Journal*, **530**, L93–L96.

Thompson, C. and Duncan, R. (1995). The soft gamma repeaters as very strongly magnetized neutron stars – I. Radiative mechanism for outbursts. *Monthly Notices of the Royal Astronomical Society*, **275**, 255–300.

Trümper, J., Pietsch, W., Reppin, C., Voges, W., Staubert, R. and Kendziorra, E. (1978). Evidence for strong cyclotron line emission in the hard X-ray spectrum of Hercules X-1. *Astrophysical Journal*, **219**, L105–L110.

Tutukov, A.V. and Yungelson, L. R. (1973). Evolution of massive close binaries. *Nauchnye Informatsii*, **27**, 70–85.

Urpin, V., Konenkov, D. and Geppert, U. (1998). Evolution of neutron stars in high-mass X-ray binaries. *Monthly Notices of the Royal Astronomical Society*, **299**, 73–77 (and references therein).

van den Heuvel, E.P.J. (1975). Modes of mass transfer and classes of binary X-ray sources. *Astrophysical Journal*, **198**, L109–L112.

van den Heuvel, E.P.J. (1977). Formation and evolution of X-ray binaries. *Annals of the New York Academy of Sciences*, **302**, 14–35.

van den Heuvel, E.P.J. (1981). The formation of compact objects in binary systems. In *Fundamental Problems in the Theory of Stellar Evolution*, IAU Symposium 93, Reidel, Dordrecht, pp. 155–175.

van den Heuvel, E. P.J. (1983). Formation and evolution of X-ray binaries. In W.H.G. Lewin and E.P.J. van den Heuvel (eds), *Accretion Driven Stellar X-ray Sources*, Cambridge University Press, pp. 303–341.

van den Heuvel, E.P.J. (1992). Formation and evolution of neutron star binaries. In E.P.J. van den Heuvel and S.A. Rappaport (eds), *X-Ray Binaries and Recycled Pulsars*, Kluwer, Dordrecht, pp. 233–56.

van den Heuvel, E.P.J. (1994a). Interacting binaries: Topics in close binary evolution. In H. Nussbaumer and A. Orr (eds), *Interacting Binaries*, Springer, Berlin, pp. 263–474.

van den Heuvel, E.P.J. (1994b). Three decades of X-ray binaries from the point of view of a theoretician. In S. Holt and C.S. Day (eds), *The Evolution of X-ray Binaries*, AIP Conference Proceedings Vol. 308, American Institute of Physics Press, New York, pp. 18–38.

van den Heuvel, E.P.J. (1995). Scenarios for the formation of binary and millisecond pulsars – A critical assessment. *Journal of Astrophysics and Astronomy*, **16**, 255.

van den Heuvel, E.P.J. (1996). Debate: The origin and evolution of millisecond pulsars. In S. Johnston, M.A. Walker and M. Bailes (eds), *Pulsars, Problems and Progress*, ASP Conference Series Vol. 105, Astronomical Society of the Pacific, San Francisco, pp. 557–559.

van den Heuvel, E.P.J. (2001). Formation and evolution of neutron stars and black holes in binary systems. In: C. Kouveliotou *et al.* (eds), *The Neutron Star – Black Hole Connection*, Kluwer, Dordrecht, pp. 173–243. Proceedings of NATO Advanced Study Institute, Kluwer, Dordrecht, pp. 173–243.

van den Heuvel, E.P.J. and de Loore, C. (1973). The nature of X-ray binaries III. Evolution of massive close binaries with one collapsed component, with a possible application to Cygnus X3. *Astronomy and Astrophysics*, **25**, 387.

van den Heuvel, E.P.J. and Heise, J. (1972). Centaurus X-3, possible reactivation of an old neutron star by mass exchange in a close binary. *Nature – Physical Science*, **239**, 67.

van den Heuvel, E.P.J., Portegies Zwart, S., Bhattacharya, D. and Kaper, L. (2000). On the origin of the difference between the runaway velocities of the OB-Supergiant X-ray binaries and the Be/X-ray binaries. *Astronomy and Astrophysics*, **364**, 563–572.

van den Heuvel, E.P.J. and Rappaport, S. (1987). X-ray observations of B-emission stars, In *Physics of Be Stars*, Proceedings of IAU Colloquium 92, Cambridge University Press, pp. 291–307.

van den Heuvel, E.P.J. and van Paradijs, J. (1988). Fate of the companion stars of ultra-rapid pulsars. *Nature*, **334**, 227–228.

van den Heuvel, E.P.J. and van Paradijs, J.A. (1997). Intrinsic kicks at birth are required to explain the observed properties of single and binary neutron stars. *Astrophysical Journal*, **483**, 399.

van Kerkwijk, M.H., Charles, P.H., Geballe, T.R., King, D.L., Miley, G.K., Molnar, L.A., van den Heuvel, E.P.J., van den Klis, M. and van Paradijs, J. (1972). Infrared helium emission lines from Cygnus X-3 suggesting a Wolf-Rayet star companion. *Nature*, **355**, 703–705.

van Kerkwijk, M.H. and Kulkarni, S.R. (1999). A massive white dwarf companion to the eccentric binary pulsar system PSR B2303+46. *Astrophysical Journal*, **516**, L25–L28.

van Paradijs, J. (1995). A catalogue of X-ray binaries. In W.H.G. Lewin, J. van Paradijs, and E.P.J. van den Heuvel (eds), *X-ray Binaries*, Cambridge University Press, pp. 536–577.

van Paradijs, J.A. (1978). Average properties of X-ray burst sources. *Nature*, **274**, 650–653.

van Paradijs, J.A. (1998). Neutron stars and black holes in X-ray binaries. In R. Buccheri, J. van Paradijs, and M.A. Alpar (eds), *The Many Faces of Neutron Stars*, Kluwer, Dordrecht, pp. 279–336.

van Paradijs, J.A. (1999). From gamma-ray bursts to supernovae. *Science*, **286**, 693–695.

van Paradijs, J.A. and McClintock, J.E. (1995). Optical and ultraviolet observations of X-ray binaries. In W.H.G. Lewin, J. van Paradijs, and E.P.J. van den Heuvel (eds), *X-ray Binaries*, Cambridge University Press, pp. 58–125.

Verbunt, F. and van den Heuvel, E.P.J. (1995). Formation and evolution of neutron stars and black holes in binaries. In W.H.G. Lewin, J. van Paradijs, and E.P.J. van den Heuvel (eds), *X-ray Binaries*, Cambridge University Press, pp. 457–494.

Verbunt, F., van den Heuvel, E.P.J, van Paradijs, J.A. and Rappaport, S.A. (1987). Formation of isolated millisecond pulsars in globular clusters. *Nature*, **329**, 312–314.

Verbunt, F. and Zwaan, C. (1981). Magnetic braking in low-mass X-ray binaries. *Astronomy and Astrophysics*, **100**, L7–L9.

Waters, L.B.F.M. and van Kerkwijk, M. H. (1989). The relation between orbital and spin periods in massive X-ray binaries. *Astronomy and Astrophysics*, **223**, 196–206.

Webbink, R.F. (1984). Double white dwarfs as progenitors of R Coronae Borealis stars and Type I supernovae. *Astrophysical Journal*, **277**, 355–360.

Webbink, R.F. and Kalogera, V. (1994). On the origin of low-mass X-ray binaries. In S. Holt and C.S. Day (eds), *The Evolution of X-ray Binaries*, AIP Conference Proceedings Vol. 308, American Institute of Physics Press, New York, pp. 321–330.

Webbink, R.F., Rappaport, S.A. and Savonije, G.J. (1983). On the evolutionary status of bright, low-mass X-ray sources. *Astrophysical Journal*, **270**, 678–693.

Webster, B.L. and Murdin, P. (1972). Cygnus X-1 – A spectroscopic binary with a heavy companion. *Nature*, **235**, 37–38.

Wijnands, R. and van der Klis, M. (1998). A millisecond pulsar in an X-ray binary system. *Nature*, **394**, 344–346.

Zel'Dovich, Y.B. (1964). The fate of a star and the evolution of gravitational energy upon accretion. *Soviet Physics Doklady*, **9**, 195–197.

Zel'Dovich, Y.B. and Guseinov, O. (1966). Collapsed stars in binaries. *Astrophysical Journal*, **144**, 840.

Zel'Dovich, Y.B. and Novikov, I.D. (1964). Estimating the mass of a super-star. *Doklady Academii Nauk SSSR*, **158**, 811–818.

33

PHIL A. CHARLES*

White dwarf binaries

The white dwarf binaries considered in this chapter are better known as cataclysmic variables (CVs) and are interacting binaries in that the white dwarf is accreting material from its (usually) cool, late-type companion star in a short (of the order of hours) orbital period. They are one of the few classes of object considered in this book that were actually known and observed prior to the twentieth century. The novae become naked-eye objects (e.g. Nova Cyg 1975 which, at a peak of 2nd magnitude, completely transformed the appearance of Cygnus for a few weeks in the summer of 1975) and hence have been observed throughout human history, and dwarf novae were first reported in the middle of the nineteenth century (Hind 1856). It was the non-periodic but continuous eruptions displayed by dwarf novae (such as U Gem and SS Cyg) that led to their cataclysmic designation, a term now widely applied to all interacting binaries where white dwarfs are accreting. However, their physical nature was not understood until the pioneering spectroscopic studies of Kraft in the late 1950s which revealed their binary signature. CVs are of considerable importance for astronomy in general because of the significance of accretion processes on virtually all scales, from star and planetary formation (the proto-star accretes material from its surrounding molecular cloud) to accretion onto supermassive black holes in the centres of active galactic nuclei. Unfortunately, in most of those circumstances, the geometry and mass accretion are very poorly constrained, there are usually multiple emitting sources that are complex to unravel and the timescales can be uncomfortably long relative to human observing lifetimes.

In CVs the visible light we see is normally dominated by the accretion process itself (the transferred material usually forms an accretion disk around the white dwarf and this disk is heated to 5000–10,000 K by internal viscous processes) as the cool, mass-losing companion is less massive than the Sun and hence very faint. The white dwarf may be hot but it is very small (comparable to the Earth in extent) and hence unable to contribute significantly to the light output except at the shortest wavelengths. CVs are also extremely common throughout our Galaxy and hence there are nearby objects for convenient study (of the several hundred CVs catalogued, the majority are estimated to be within 200–300 pc; Ritter and Kolb 1998, Downes et al. 1997). A valuable consequence of this is the low interstellar absorption for most CVs which allows their study over a wide range of wavelengths. This is particularly important for the inner part of the accretion disk and the white dwarf surface which can reach temperatures exceeding 20,000 K and hence the bulk of their radiation is in the far and extreme UV which can only be accessed from space. Furthermore the CV geometry (Figure 1) is well defined by orbital mechanics and Roche surfaces, thereby providing an outstanding astronomical 'laboratory' with which to test and improve our understanding of the physics of accretion onto compact objects, in particular the nature of the viscosity which acts within the disk. CVs thereby provide constraints on the basic physical properties of accretion disks which can then be applied in many other fields.

The typical luminosity and range of temperatures produced in CVs can be estimated in the following way. If material is being accreted by the white dwarf at a rate $\dot{M}$ then the release of gravitational potential energy will produce a luminosity $L = GM_{WD}\dot{M}/R_{WD}$. For a white dwarf of mass $M_{WD} = 1 M_\odot$ and radius $R_{WD} = 10^4$ km accreting at $\sim 10^{-10} M_\odot$ yr^{-1} (a rate expected from calculations of the evolution of close interacting binaries) then total luminosities of $\sim 10^{33}$ erg s^{-1} will result. And if this is assumed to be radiated uniformly from the white dwarf surface as blackbody radiation ($L = 4\pi R^2 \sigma T^4$) then $T \sim 30,000$ K and we would expect CVs to be copious sources of ultraviolet (UV)

*University of Southampton, United Kingdom

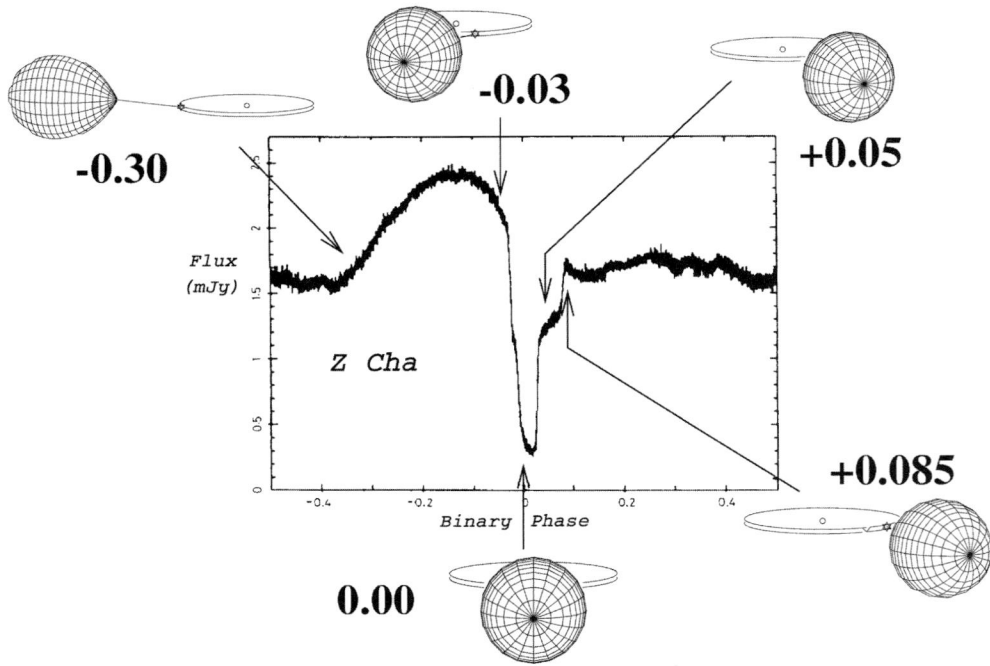

Figure 1 Basic geometry of a white dwarf interacting binary, in which a low-mass normal star transfers material into an accretion disk surrounding the white dwarf. Note that the mass-transfer stream produces a bright spot on the accretion disk, which is seen as the large hump in the light curve. The binary and resulting light curve is shown at different orbital phases using quiescent (low-state) observations of the eclipsing dwarf nova Z Cha as an example. (Adapted from Marsh 2001).

radiation. Indeed, most CVs show very blue colours rising at the UV end of the optical to imply that the bulk of the emission will lie at shorter wavelengths. However, if the accreting material is in free-fall onto the white dwarf surface then it should reach velocities of $\sim 5000\,\mathrm{km\,s^{-1}}$ which correspond to temperatures of $\sim 10^9\,\mathrm{K}$ and we would therefore expect hard X-ray emission to be produced. With these expectations both UV and X-ray satellite instruments were used to study these objects in the early days of space astronomy.

HISTORICAL SUMMARY

Dwarf novae are CVs which exhibit major optical outbursts (increasing in brightness by several magnitudes or more) at irregular intervals (usually of the order of weeks–months). The brightest are accessible with binoculars (e.g. SS Cyg and U Gem, both of which reach $V \sim 8$–9 at peak of outburst) and amateur groups have made fundamental contributions to their long-term light curves for more than a hundred years (Warner 1995). Walker's discovery (Walker 1954) that DQ Her is an eclipsing binary, combined with spectroscopic observations revealing occasional absorption and emission components, culminated in Kraft's (1962) survey of CVs that definitively established their binary nature. The presence of a white dwarf in these systems as the primary was first demonstrated observationally by Crawford and Kraft (1956). These events and others are given chronologically in Table 1, and the properties of a selection of dwarf novae are summarised in Table 2.

Both spectroscopy and photometry demonstrated that the continuum emission from CVs almost always exhibited a hot, blue, non-stellar component. With orbital periods of hours and stellar companions of usually $\sim$G-M spectral type, it became clear that the cool star (referred to as the 'secondary star' or 'companion') must be filling its Roche (equipotential) surface and hence transferring material into the hot accretion disk (Figure 1) that gives rise to the emission lines and blue continuum. For details (and equations) of the Roche geometry in interacting binaries see Warner (1995), the primary reference source for CV properties.

The accretion disk: importance of multi-wavelength studies

The material leaving the secondary's inner Lagrangian point carries its parent star's orbital angular momentum and hence cannot (in the absence of other forces) accrete

Table 1 Chronology of white dwarf binaries

Year	Discovery
Pre-history	Naked-eye observations of nova outbursts
1855	U Gem at maximum light (Hind 1856)
1896	SS Cyg found from Harvard plates (Wells 1896)
1938	Term 'dwarf novae' coined by Payne-Gaposchkin and Gaposchkin (1938)
1949	Flickering found in UX UMa (Linnell 1949, 1950)
1956	71 s pulsations in DQ Her (Walker 1956)
1956	Identification of WD as primary in CVs
	AE Aqr secondary filling Roche lobe (Crawford and Kraft 1956)
1962	All CVs are binaries undergoing mass transfer onto WD (Kraft 1962)
1971	Accretion disk responsible for outburst light (Warner and Nather 1971)
1972	Hard X-rays detected from EX Hya (Warner 1972)
1974	Novae at maximum seen in UV by OAO 2 (Gallagher and Code 1974)
1977	Large, variable polarization in soft X-ray source AM Her (Tapia 1977)
1977	Term 'polars' coined by Krzeminski and Serkowski (1977) for AM Her systems
1978	Soft X-rays from SS Cyg in outburst and quiescence (Heise *et al.* 1978)
1980	14.3 min X-ray pulsations in H2252-035 (Patterson and Price 1980)
1981	'Supersoft sources' (SSS) found by *Einstein* in LMC (Long *et al.* 1981)
1981	Term 'intermediate polars' coined by Krzeminski for asynchronous WD rotation
1983	*Einstein* X-ray survey of CVs (Córdova and Mason 1983)
1991	IUE spectra reveal cooling of U Gem WD between outbursts (Kiplinger *et al.*)
1992	Steady nuclear burning WD model for SSS (van den Heuvel *et al.* 1992)
1996	AE Aqr 'magnetic propellor' model (Eracleous and Horne 1996)

WD = white dwarf.

Table 2 Properties of selected dwarf novae

CV type	P_{orb} (h)	V			τ_{rec} (d)	τ_{decay} (d)	τ_{sup} (yr)	Spectral type
		min.	inter.	max.				
U Gem								
IP Peg	3.8	15.8		12.3	95	1.5		M4V
U Gem	4.2	14.6		9.4	101	1.2		M4.5V
SS Cyg	6.6	11.7		8.2	40	2.4		K5V
RU Peg	9.0	12.7		9.0	65	3.4		K2V
BV Cen	14.6	12.6		10.7	149	7		G5-8IV-V
Z Cam								
KT Per	3.9	16.0	12.3	11.7	26	1.3		M3
RX And	5.0	13.6	11.8	10.9	13	1.7		
Z Cam	7.0	13.6	11.7	10.4	23	2.6		K7V
EM Cyg	7.2	14.2	12.9	12.0	22	3.6		K5V
SY Cnc	9.1	13.7	12.2	11.1	27	2.3		G8V
SU UMa								
WZ Sge	1.36	14.9	—	7.1			30	
OY Car	1.51	15.6	12.4	11.5	50	—	1	
IR Gem	1.64	15.5	11.7	11.2	26	0.8	0.5	
VW Hyi	1.78	13.3	9.5	8.5	27	0.7	0.5	
Z Cha	1.79	16.0	12.4	11.9	82	—	0.6	M5.5V
SU UMa	1.83	14.8	12.2	11.2	19	1.4	0.4	
YZ Cnc	2.09	14.5	11.9	10.5	12	0.9	0.4	

(i) in the V magnitude range, "inter" corresponds to the standstill brightness of Z Cam stars, and the normal maximum of SU UMa stars.

(ii) τ_{rec} is the mean recurrence time between outbursts.
τ_{decay} is the decay time from outburst.
τ_{sup} is the mean recurrence time between super outbursts.

directly onto the white dwarf. It therefore forms an extended, rotating accretion disk which has the compact object at its centre, and some physical process acting within the disk creates an internal viscosity which allows angular momentum to be transported outward in the disk, and material to accrete inwards. The nature of this viscosity has been one of the major unsolved problems in modern astrophysics, and is being actively investigated today (e.g. Balbus and Hawley 1998). Nevertheless, by wrapping up our lack of understanding of viscosity into the parameter $\alpha = \nu/c_s H$ (ν is the effective turbulent viscosity, c_s the sound speed and H the half-thickness of the disk), Shakura and Sunyaev (1973) produced the so-called α-disks which formed the starting point for observers and theorists alike. In the steady state, the radiated spectrum of the disk does not depend on α, but not surprisingly the disk's response time to changes in accretion rate does. Consequently, it is the outbursts of dwarf novae and the intervals between them which can constrain α and hence the physical processes that must be occurring (e.g. Lasota 2001).

The original Shakura and Sunyaev formulation is summarized in the review by Pringle (1981), and can be used to demonstrate the importance of multi-wavelength observations of CVs. The circular, thin disk is divided radially into annuli of radius r, each of which is assumed to radiate as a black body at temperature $T(r)$ where

$$T(r)^4 = T_{max}^4 (r/R_{WD})^{-3}[1 - (R_{WD}/r)^{1/2}] \quad (1)$$

and T_{max} is the disk's maximum temperature given by

$$T_{max} = 41,000(\dot{M}M_{WD}/R_{WD}^3)^{0.25} \text{ K} \quad (2)$$

where the mass transfer rate $\dot{M}$ is in units of $10^{16}\,\text{g s}^{-1}$, the white dwarf mass M_{WD} is in solar masses and the radius R_{WD} is in units of 10^4 km. The total radiated emission from the disk is then the sum of all the contributions from these annuli, as demonstrated in Figure 2 (la Dous 1989), the result of which is the continuum rising into the blue and UV regions. This produces the characteristic UV excess by which many CVs have been discovered. At this range of temperatures it is necessary to observe from the near-IR into the far-UV to fully characterise the spectrum of the disk.

By application of the virial theorem it can be shown that half of the total accretion luminosity ($GM_{WD}\dot{M}/R_{WD}$) is deposited in the disk, the other half in the boundary layer with the white dwarf. An optically thick boundary layer has temperatures of $\sim 10^{5-6}$ K and thus radiates predominantly in the EUV and soft X-ray regions.

Early UV and X-ray Observations

This basic model of a CV is very similar to that of a low-mass X-ray binary (Chapter 34), apart from the difference in radius of the compact object (by a factor of $\sim 10^3$) when the neutron star is replaced by a white dwarf. Early (non-imaging) X-ray satellites of the 1970s (Uhuru, Ariel-5,

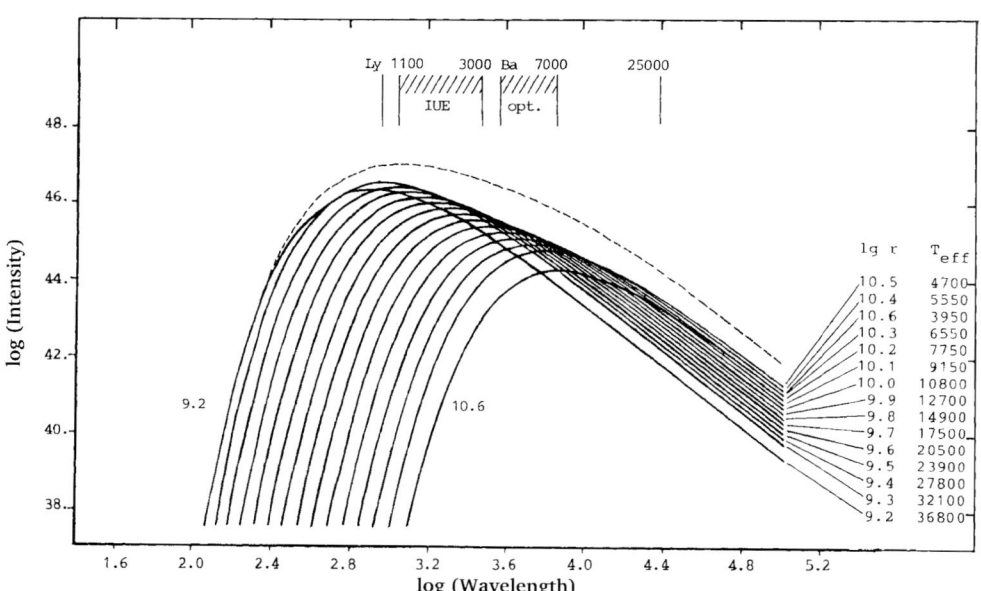

Figure 2 Simulated spectrum (dashed line) of an accretion disk computed by adding the contributions of individual rings or annuli in the disk, each of which is assumed to radiate as a black body. The contribution of each ring is labelled with its radius and temperature (coolest on the outer edge). (Adapted from La Dous 1989).

the Astronomical Netherlands Satellite (ANS), the third Small Astronomy Satellite (SAS-3) and the first High Energy Astronomy Observatory (HEAO-1)) were used to (mostly unsuccessfully) search for the hard (>1 keV) X-rays that had been expected if the accreted material were in free-fall onto the white dwarf surface. Only EX Hya and AM Her (both magnetic systems; Section on 'Magnetic CVs') are in the Uhuru X-ray catalogues, but even the latter required the discovery of strong soft X-ray emission and a more precise SAS-3 location for it to be firmly identified. In general, the field of X-ray-emitting CVs did not open up until the greatly enhanced sensitivity of the first imaging X-ray satellite, the Einstein Observatory. This allowed the detection of CVs in large numbers, with luminosities found to be in the range 10^{30-33} erg s^{-1} (Córdova & Mason 1982). However, the UV package on NASA's second Orbiting Astronomical Observatory (OAO 2) did detect Nova Ser 1970 close to maximum light and then followed it during the subsequently decline to a constant 30,000 K component from either the white dwarf or inner accretion disk, thereby establishing the nature of the UV excess.

NON-MAGNETIC CVs

The major breakthrough in CV studies came with the HEAO-1 discovery in the late 1970s that SS Cyg and U Gem were strong sources of very soft X-rays (<0.2 keV) during outburst. This corresponded to a low temperature (10–50 eV, or equivalently 100,000–500,000 K) black body which could be difficult to observe widely in other CVs due to absorption by intervening material (either intrinsic or interstellar). Such a soft component was interpreted as arising in the optically thick boundary layer between the inner accretion disk and white dwarf surface. However, it was noted that CVs exhibiting *continuously* high mass-transfer rates (the 'nova-likes') could also exhibit a hard X-ray component.

In quiescence, however, only the hard component is observed. This is compatible with what is expected at low accretion rates when most of the energy is released close to the white dwarf surface in a hot, optically thin plasma, possibly with an extended corona (due to the fact that at high temperatures, $\sim 10^8$ K, cooling can only occur by the inefficient free–free emission process and hence the plasma expands). This is shown schematically in Figure 3, and the separation between the hard X-ray (optically thin) and soft X-ray (optically thick) states is estimated to occur around mass-transfer rates of $\sim 3 \times 10^{16}$ g s^{-1} (see Warner 1995 for a more detailed discussion).

The UV delay

One of the key predictions of the early disk instability model (DIM) for dwarf nova outbursts was that the outburst would begin when the outer regions of the disk became hot enough that the material would enter the hot, high surface density state. This transition would then propagate inwards into the inner disk regions as a 'heating wave'. Since the outer regions radiate predominantly in the optical and the inner disk in the UV (Figure 2), this would imply that the dwarf nova outburst would begin earlier in the optical and later in the UV, the so-called UV delay. Such an effect was discovered in major simultaneous ground-based and International Ultraviolet Explorer (IUE) observing campaigns in the 1980s, with the UV peak being delayed by approximately half a day relative to the optical. The logistical difficulty of course in such campaigns is due to the fact that the outbursts of dwarf novae cannot be predicted accurately in advance! Consequently this is an area in which amateur observers have played a major role in providing the continuous monitoring of target dwarf novae which is crucial in catching the very early phases of outburst. Additionally, its success depends critically on the spacecraft's planners ability to respond rapidly to the information about, for example, the outbursts of dwarf novae, a category now designated Target of Opportunity or TOO observations.

However, there is a corollary to this behaviour which was anticipated from the basic physical model of the outburst and quiescent X-ray-emitting regions shown in Figure 3. Early in the outburst (before the accretion rate from the inner disk onto the white dwarf itself has changed) the hard

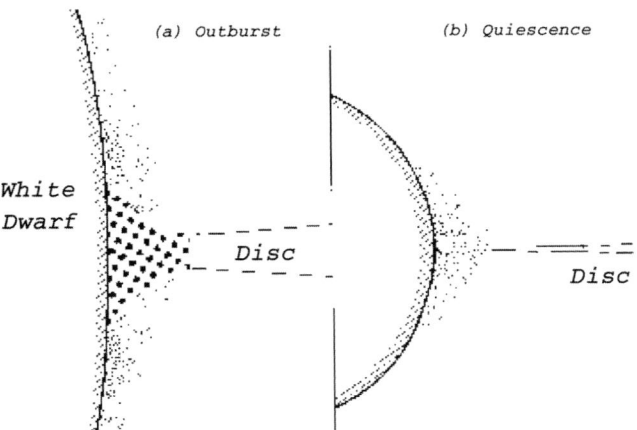

Figure 3 Schematic cross-sections of the inner disk and white dwarf surface regions of a dwarf nova in (a) outburst and (b) quiescence. In quiescence the accretion onto the white dwarf is at a low rate, but is hot and optically thin (producing mostly hard X-ray emission). In outburst the accretion rate is much higher (as the disk is hotter and more viscid) leading to an optically thick region (which is responsible for the soft X-rays observed at such times). This can still, however, be surrounded by hotter, optically thin regions giving rise to multiple spectral components. (Adapted from Patterson and Raymond 1985).

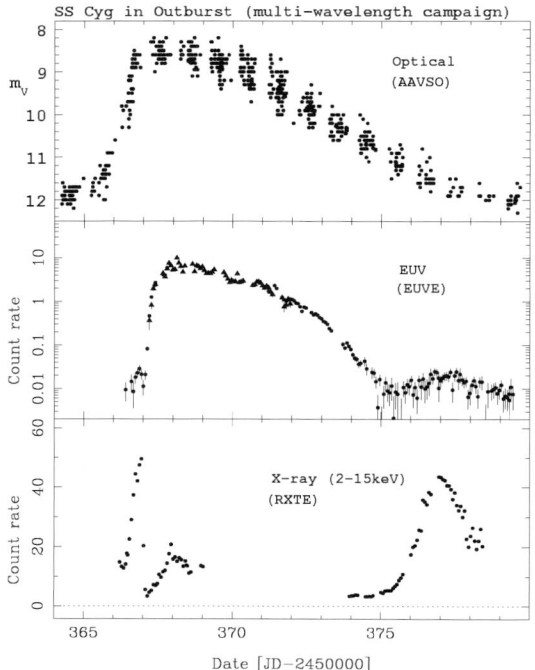

Figure 4 Optical, EUV and X-ray observations of an outburst of the dwarf nova SS Cyg showing the complex behaviour of the soft and hard X-ray components as the outburst progresses. (Adapted from Wheatley et al. 2000).

X-rays should initially be unaffected, but as the accretion rate increases they should decrease and be replaced by the much softer emission from the optically thick boundary layer. Unfortunately, it has been much harder to obtain X-ray coverage of the same quality as that offered by IUE, mostly due to the (unique for its time) almost continuous 'observatory'-class operation of IUE in its high orbit, compared to the fragmented observing available from X-ray satellites in near-Earth orbit. Early attempts were also made by Ariel 5 in the mid-1970s and EXOSAT in the early 1980s, but these were either lacking in detail at the time of the CV transitions, or had insufficient wavelength coverage. Consequently it was not until the end of the 1990s that good multi-wavelength coverage (including both EUV and hard X-rays) was added to that obtained earlier. This is shown (Figure 4) in the 1996–97 outburst of SS Cyg which was studied with excellent resolution of the different spectral components by Wheatley et al. (2000).

These observations clearly showed that the situation was more complex than had been thought. The UV delay of half a day is easily visible, but more remarkably the hard X-rays initially increased, before declining to be replaced by the EUV emission as the boundary layer between the disk and white dwarf became optically thick. This indicated that the disk must have been initially truncated in its inner regions and needed to 'fill up' before this transition could take place. It is considered to have been established that outbursts such as these are always of the outside-in variety and naturally supports the truncated disk model. A much more detailed discussion of the 'UV delay problem' can be found in Lasota (2001). (There is an amusing anecdote associated with this outburst of SS Cyg in that it began with optical observations by the amateurs of the AAVSO who telephoned their information about the start of the outburst to Wheatley in the UK. But in order for him to trigger his RXTE over-ride program a message had to be sent by e-mail from his office. In cycling in to do this his excitement got the better of him and he had an accident. But he records that he continued on to send the message successfully and trigger the program before attending to his wounds!)

Aperiodic variability

Flickering

Chaotic variability on timescales of minutes is seen at optical and X-ray wavelengths in most CVs, and is commonly referred to as 'flickering' (Córdova and Mason 1983, and references therein). Fast optical photometry of eclipsing systems such as U Gem has shown that, when the bright spot is eclipsed, the flickering level is reduced. Hence a component of this variability must be produced in the bright spot, the region where the accretion stream impacts the disk (Figure 1). However, polars (section on 'Magnetic CVs') have no accretion disk (and hence no bright spot), and yet they also flicker. This is likely to occur within shocks in the accretion column onto the white dwarf. Also, in nova-like systems (CVs in a permanently high state of mass transfer from the secondary, and hence in 'continuous' outburst) the optical flickering is found to be correlated with X-ray variability, indicating that this must be an *inner* disk phenomenon (as the outer disk is too cool to produce any X-ray emission).

Quasi-periodic oscillations

Apparently regular pulsations with periods of tens of seconds have been seen in all CV types (see Warner 1995 for a review). These have quite small amplitude ($<1\%$) in the optical, but are larger in X-rays. They are not in fact strictly periodic, but some are remarkably stable (with Q values ($=1/\dot{P}$) of $\sim 10^{11-13}$). Furthermore these pulsations can return at essentially the same values after many years; Knigge et al. (1998) detected 29 s pulsations in UV/optical spectroscopy of UX UMa with the Hubble Space Telescope (HST), at virtually the same period as first seen by Warner and Nather (1972). They were attributed to magnetically controlled accretion onto differentially rotating white dwarf surface layers (Warner 1995).

Nevertheless they are *not* completely coherent (as for the rigidly rotating components in intermediate polars and polars) and are therefore referred to as 'quasi-periodic oscillations' (QPOs). Most optical QPOs are much less stable in both period and phase. Also, X-ray QPOs seen in SS Cyg and U Gem at outburst (at $\sim$10 s and $\sim$25 s respectively) have a much higher amplitude than their optical counterparts, indicating that they are the primary source for the pulsation and that the optical QPO is then due to reprocessing in the inner disk region. The mechanism producing this effect is likely to be similar or related to that occurring in the much more luminous low-mass X-ray binaries (LMXBs; Chapter 34) and which are seen in high-time-resolution X-ray observations. These are believed to be due to beating effects between blobby material in Keplerian orbits in the inner accretion disk interacting with the magnetosphere of the rapidly spinning compact object.

UV SPECTROSCOPY FROM OUTBURST TO QUIESCENCE

UV line profiles

With components of CV binaries at temperatures of $\sim$10,000–50,000 K, UV observations are crucial to detailed studies, since the peak of their emission will occur at wavelengths $\lambda \sim 1250 \times (20000/T)$ Å. By far the major contribution to UV studies of CVs has come initially from IUE (for a detailed review of its observations of compact binaries see Córdova and Howarth 1987). UV spectra of U Gem in quiescence allowed the Lyα profile to be modelled (Panek and Holm 1984), yielding a white dwarf mass of $1.2M_\odot$ and a temperature of 30,000 K at a distance of 90 pc. Similar studies were performed on VW Hyi and WZ Sge. However, it is unclear whether the white dwarf spectrum does in fact dominate in the UV (as is assumed in this analysis), since an optically thick disk will also produce a Lyα absorption component. That the disk also has more complex structure was demonstrated by early IUE spectra of Krautter *et al.* (1981). The asymmetric and even P Cygni profiles of the emission lines (Figures 5 and 6) were indicative of outflowing material, now described as an 'accretion disk wind'.

Accretion disk winds

The velocities of this outflowing gas (a few thousand km s^{-1}) are comparable to the escape velocity from the white dwarf or inner disk regions. This was initially a surprising result as even nova-like CVs are not sufficiently luminous for radiation-driven winds to be considered likely. Such outflows would normally be expected to occur when the luminosity approached the Eddington limit, which is $\sim 10^{38}$ erg s^{-1} for a $1M_\odot$ white dwarf. However, such a limiting luminosity is calculated for pure electron scattering, whereas in an optically thin wind, the absorption in the gas would be by bound–bound opacity which is a thousand times greater. It is therefore possible to compute models (Figure 7) which achieve an explanation for these observations with mass-transfer rates of $\sim 10^{-8} M_\odot$ yr^{-1}. Currently mass loss is considered the *simplest* interpretation of the broad blue-shifted absorption that is present in these UV lines. But this could be complicated by magnetic propellor effects (see section on 'Magnetic CVs') once they are better understood (see Drew and Proga 2000 for a full discussion).

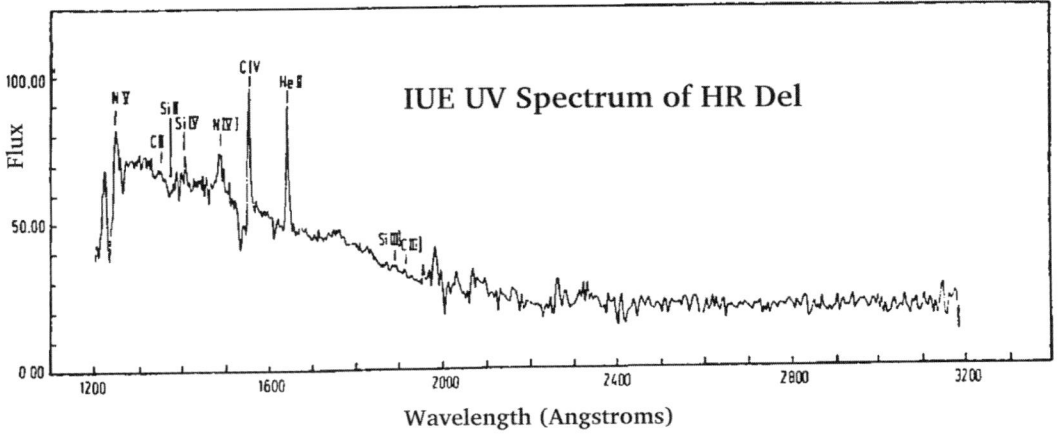

Figure 5 IUE UV spectrum of HR Del. (Adapted from Krautter *et al.* 1981).

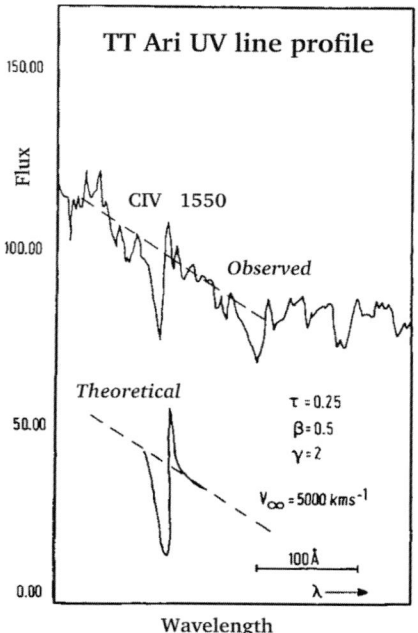

Figure 6 IUE spectrum of TT Ari showing the P Cyg profile exhibited by CIV $\lambda1550$ indicative of mass outflow from the accretion disk. Below is a theoretical line profile for an outflow velocity of 5000 km s^{-1}. (Adapted from Krautter et al. 1981).

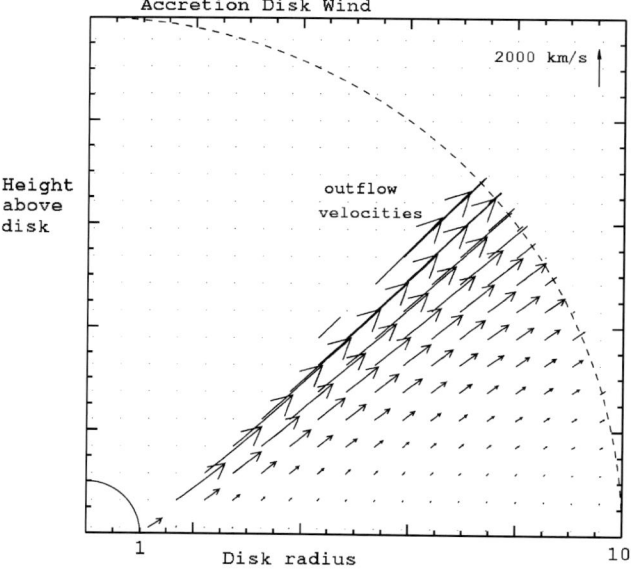

Figure 7 Theoretical calculations of radiation-driven flow of material from an accretion disk. (Adapted from Proga et al. 1999).

Heating and cooling of the white dwarf

IUE observations had also shown that the dwarf nova outburst led to the white dwarf being heated up, followed by

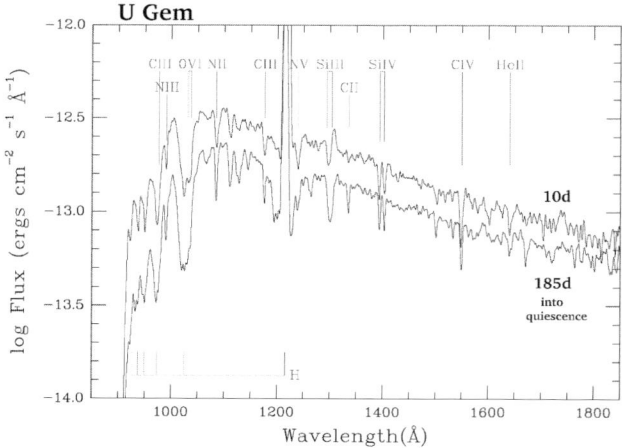

Figure 8 HUT spectra of U Gem obtained just 10 days after the end of an outburst (upper) and 185 days after (lower). The $\sim$30% fading is presumed to be due to the cooling of the white dwarf which is heated during the outburst. (Adapted from Long 2000).

its cooling during the subsequent quiescent period. This was demonstrated very nicely in UV spectra of U Gem obtained with the HUT (Hopkins Ultraviolet Telescope) which was flown on two Space Shuttle missions in the early 1990s, Astro-1 and Astro-2. U Gem had only just returned to quiescence following an outburst at the time of Astro-1, but it had been in quiescence for almost six months when Astro-2 occurred (Long 2000). The resulting UV spectra (Figure 8) show a $\sim$30% fading, which could be due to the cooling of the entire white dwarf, but is more likely to be a fading of an accretion 'belt' around the white dwarf equator (where the accretion energy had been concentrated) and the spectrum can be adequately fit with components at two temperatures.

However, the much greater collecting area of HST combined with its superior instrumentation has been providing exquisite details during the last decade of the inner disk and the accretion processes close to the white dwarf. HST's high-resolution spectrograph (the GHRS) was employed by Long and Gilliland (1999) to obtain a very high signal to noise UV spectrum of U Gem far from outburst (Figure 9). Fitting this with model white dwarf atmospheres provided evidence for CNO processing in that carbon was reduced and nitrogen enhanced relative to solar. The spectrum was of such high resolution that the rotational velocity of the white dwarf could be measured, and was found to be low enough to account for the comparable disk and boundary layer luminosities. Similar multi-temperature component spectra were required for VW Hyi (Sion et al. 1996, Long et al. 1996), but the most remarkable feature was the presence of singly ionized phosphorus at $\lambda1250$ which apparently requires an abundance enhancement of 900

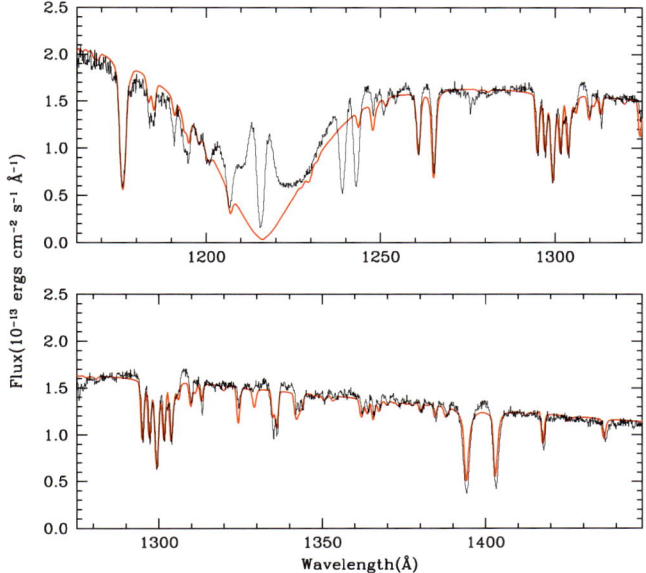

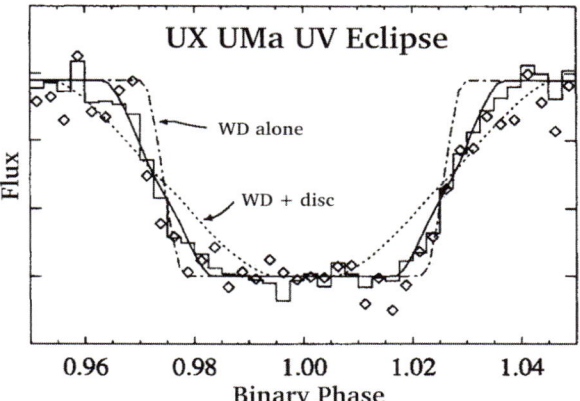

Figure 10 Mean of HST UV observations of three eclipses of the nova-like variable UX UMa. The eclipse is clearly longer than that expected from the white dwarf (WD) alone, and requires the inclusion of a disk of radius four white dwarf radii (solid line). (Adapted from Mason et al. 1995).

Figure 9 HST/GHRS spectrum of U Gem in quiescence compared with a model white dwarf spectrum (red) in which the abundances of C are reduced and N are enhanced relative to solar. (Adapted from Long and Gilliland 1999).

times solar! This could have been material that was produced during the thermonuclear runaway of a nova explosion (which all CVs are expected to eventually undergo about every 10^4 yr). To complement these models of the white dwarfs and exploit the high-quality UV spectra now available of dwarf novae, Wade and Hubeny (1998) have computed accretion disk spectra that replace the simple black-body models used in Figure 2 with stellar atmospheres.

Eclipsing systems

The nova-like variable UX UMa is an ideal target for HST to study because (a) it is a non-magnetic CV in a permanent high state (i.e. with a high mass-transfer rate), (b) it is eclipsing, thereby allowing time-resolved observations to resolve geometrical details of the emission and absorption components, and (c) it lies in HST's 'CVZ' (the continuous viewing zone of the spacecraft which refers to regions not occulted by the Earth during the HST orbit) and therefore allows the optimum opportunity to observe full eclipses. The GHRS obtained time-resolved UV spectra that confirmed the broad asymmetric emission line profile seen by IUE which had been interpreted as an outflowing wind from the accretion disk (and the eclipse profile clearly shows in Figure 10 the extent of the emitting region). However, Mason et al. (1995) also unexpectedly found absorption components that were eclipsed and therefore due to material on a scale smaller than the companion star, thereby complicating the wind geometry.

Indeed the absorption structure is almost certainly associated with azimuthal structure in the edge of the disk where it is impacted by the accretion stream, the so-called 'bright spot' (Figure 1). Extended regions of material above and below the disk at this point can obscure or absorb light from the inner disk beyond, the signature of which is absorption dips in the light curve about 0.2 in orbital phase *prior* to eclipse. These can be seen in other high-inclination CVs, such as Z Cha (with IUE), but they are best demonstrated in a soft X-ray observation of U Gem when it underwent an unusually long outburst (perhaps a superoutburst, but no superhumps were seen and the mass ratio of U Gem would seem to preclude them; see Lasota 2001, and references therein) in the mid-1980s. As shown in Figure 11 significant (both in amplitude and duration) dips were present at precisely these phases, an effect that is more dramatically demonstrated when compared with similar features in LMXBs (where the effect was first discovered; Chapter 34).

The properties and physical conditions prevailing in accretion disk material are not straightforward to determine. When a similar time-resolved study of the eclipsing CV OY Car was attempted by Horne et al. (1994) with HST, their detailed spectral analysis was complicated by the presence of a large number of additional absorption features that were identified with Fe II. Since such heavy elements would rapidly settle in the white dwarf atmosphere (and hence be unobservable), they must reside in extended material above the disk (which led to it being dubbed 'the iron curtain'). Furthermore its velocity broadening required it to be associated with turbulence in either the hot spot region or the inner disk. The presence of similar structure in UX UMa has been

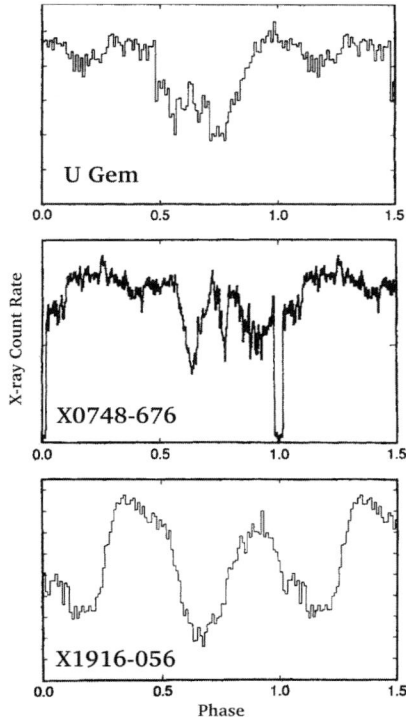

Figure 11 Comparison of soft X-ray light curve of U Gem during an extended outburst with X-ray light curves of two high-inclination neutron star X-ray binaries, X0748-676 (which totally eclipses) and X1916-056. All show the dips in X-rays around phase 0.7–0.8 when the bright spot region crosses our line of sight to the compact object. (Adapted from Mason et al. 1988).

demonstrated by Baptista et al. (1998) who used HST to produce spatially resolved spectra of the disk and gas stream. This exploited eclipse-mapping techniques, as used in many ground-based studies of CVs (Horne 1985, Marsh and Horne 1988; see Marsh 2001 for a more recent review).

MAGNETIC CVS

All of the white dwarf binaries discussed so far are dominated in their observed behaviour by the accretion disk, the inner regions of which reach right down to the white dwarf itself. However, it is known from studies of single white dwarfs that they can harbour significant magnetic fields, and there is a class of white dwarf binaries known as magnetic CVs in which the magnetic field of the white dwarf is sufficiently intense to control the accretion flow. Such systems either have a truncated disk or no disk at all, and are known as intermediate polars (IPs) and polars respectively. In the latter case the magnetic field is sufficiently strong that the rotation of the white dwarf is locked to the orbital period, whereas in IPs, where the field is generally weaker, the white dwarf spins faster. The point at which the magnetic field controls the accretion flow can be estimated by balancing the ram pressure $\sim \rho v^2$ of the accreting gas with the magnetic pressure $B^2/8\pi$ and is called the magnetospheric radius, R_{mag} (e.g. Frank et al. 1992). It can be shown that $R_{mag} \propto B^{4/7} M^{-2/7}$ and hence, for short-period binaries and sufficiently powerful magnetic fields, R_{mag} will exceed the binary separation, making it impossible for an accretion disk to form. For a typical white dwarf (0.85$M_\odot$; 0.01$R_\odot$) accreting at $\sim 2 \times 10^{-10} M_\odot$ yr^{-1} in a 2 h binary, this occurs for a magnetic field $B \sim 10^7$ G.

AM Her systems (polars)

The CV AM Her is actually in the Uhuru X-ray catalogue (Forman et al. 1978) as 4U1809+50 (and hence a source of hard X-ray emission), but the optical identification was not made until the SAS-3 satellite produced an accurate position, also showing that it was highly variable and exhibited a strong soft X-ray component. However, the true nature of this object (the prototype of the class) was not revealed until the seminal work of Tapia (1977) who discovered extremely high circular polarization which varied dramatically on a 3.09 h period. The basic model (Figure 12) has the accretion stream from the cool star immediately threading the intense magnetic field where it is channelled into a column that leads to the white dwarf polar cap. The torque in this process (and possibly via interaction with the magnetic field of the secondary) maintains the locking of the rotation and orbital periods. (Actually there are now four magnetic CVs deviating from this rule, of which the most well known is V1500 Cyg, which was discovered as Nova Cyg 1975, a spectacular naked-eye nova, and subsequently found to be a polar. It has rotation and orbital periods that differ by just 1.8% (Stockman et al. 1988), but this is almost certainly a temporary loss of synchronism arising due to the instantaneous mass loss, of possibly $\sim 0.001 M_\odot$, during the nova ejection. The periods are expected to resynchronize.)

The material in the accretion column is essentially in free-fall onto the white dwarf (very different from the buffering provided by an accretion disk), reaching velocities typically ~ 5000 km s^{-1} as it approaches the white dwarf. It is stopped in a strong shock above the surface (Figure 12) where the kinetic energy is thermalized, corresponding to energies of ~ 130 keV for protons, which is then radiated as hard X-rays by electrons. These electrons are also moving in the magnetic field and so produce cyclotron radiation (predominantly in the optical and IR) which is strongly polarized (and which was seen by Tapia). The details of the cyclotron-emitting regions and their modelling are reviewed by Wickramasinghe and Ferrario (2000), from which some of the properties of selected

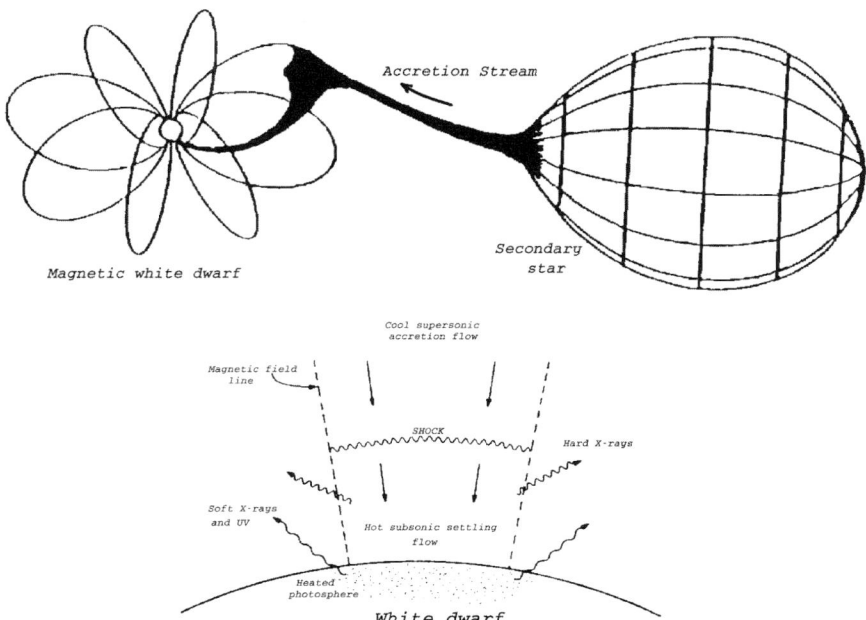

Figure 12 Schematic model of an AM Her system, or polar, in which the mass transfer from the secondary star is controlled by the intense (tens of megagauss) magnetic field of the white dwarf. The accretion stream threads onto the field which channels it in almost free-fall onto the polar cap, creating a hard X-ray-emitting shock which heats up the white dwarf surface (shown below). (Adapted from Charles and Seward 1995).

Table 3 Properties of selected polars

Star	Other designation	P_{orb} (min)	V^*	d (pc)	$kT_{BB}^\dagger$ (eV)	$B^\ddagger$ (MG)
EF Eri	2A0311−227	81	13.5	94	19	12
DP Leo	E1114+182	90	17.5	450		31
VV Pup		100	14.5	145	23–43	32
V834 Cen	E1405−451	102	14.2	86	15	24
MR Ser	PG1550+191	114	14.5	139		24
BL Hyi	H0139−68	114	14.3	128	23–40	22
ST LMi	CW1103+254	114	14.7	128	18–40	12
AN UMa	PG1101+453	115	15.5	≥270		36
AR UMa	1ES1113+432	116	15.1	88	23	230
UZ For	EXO0333−255	12.7	17	230	19–25	53
AM Her	3U1809+50	186	12	75	∼25	14
V1500 Cyg	N Cyg 75	201	17	1200		25
BY Cam	H0538+608	202	13	190		41
QQ Vul	E2003+225	226	15.3	≥320	18–29	

* Maximum brightness observed (most polars exhibit extended periods of quiescence when the mass-transfer rate is greatly reduced and their optical brightness decreases by ∼2–5 mags).
NB all spectroscopically detected companion (mass donor) stars are cool and of low mass with spectral types in the range M2–M7.
† The luminosity of this black-body (BB) component is in the range 0.4–3 × 10^{32} erg s^{-1} for all these systems.
‡ Derived from cyclotron harmonic structure in optical/infrared wide wavelength spectroscopy. Estimates of magnetic field B exist for the second accreting pole in some of these systems (Warner 1995).

polars are given in Table 3. The hard X-rays from the shock irradiate and heat up the (larger) white dwarf surface below, which then re-radiates this as an essentially black-body soft X-ray component (another key characteristic of polars).

The 'soft X-ray puzzle'

This model therefore predicts that the soft X-ray component of the white dwarf's luminosity, L_{SX}, will be comparable to that of the hard X-ray (L_{HX}) and cyclotron (L_{cyc}) components from which it must be derived. However, some of the first determinations of these fluxes (Raymond et al. 1979, Szkody et al. 1981) yielded $L_{SX} \geq 50$ ($L_{HX} + L_{cyc}$), a discrepancy that became known as the 'soft X-ray puzzle'. Later, more improved analyses reduced the factor from 50 to ~5, but it still remains. There are a variety of explanations (e.g. Warner 1995); however, it is generally accepted that the accretion flow is not smooth (as it is usually considered to be in an accretion disk) but is instead 'blobby'. This allows large blobs of material to penetrate the shock and the white dwarf surface before releasing their kinetic energy and heating up the white dwarf surface. The 'puzzle' is therefore explained as a reduction in the hard X-ray component. The X-ray spectrum is in fact more complex in that electron scattering can be important and this makes the column emission seem harder when viewed side-on, and softer at other angles (Rosen 1992). Hard X-ray flares are also seen high in the column (Cropper et al. 2000).

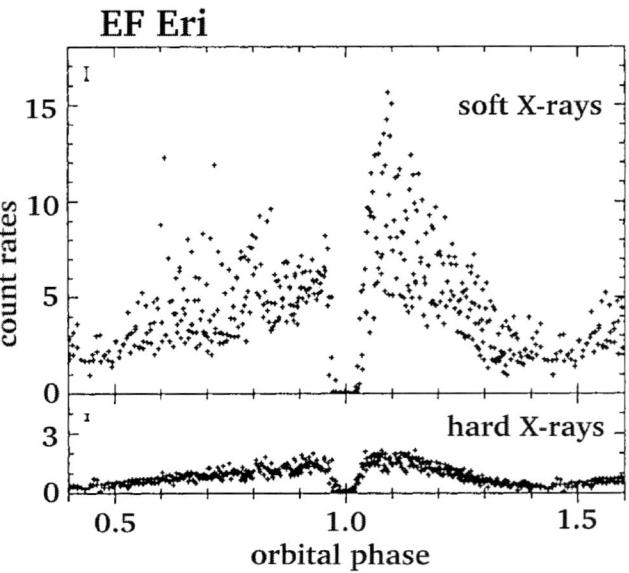

Figure 13 Soft and hard X-ray light curves of the eclipsing polar EF Eri as obtained from ROSAT observations. (Adapted from Beuermann et al. 1991).

Two-pole accretion

This simple model of accretion onto a single pole and hence of X-ray emission close to the white dwarf surface (thereby only covering a fraction of it), is used to explain the basic X-ray light curves (Figure 13). As the white dwarf rotates we see a varying fraction of the heated pole and shock region leading to a large modulation on the orbital/rotation period. However, the accretion geometry can be much more complex because there are *two* magnetic poles (Figure 12). Usually one of the poles dominates the emission, but it is possible for polars to switch between them, leading to dramatic changes in their X-ray light curves (Figure 14a). Such mode changes, as observed in E2003+225, show that the magnetic poles, already known not to be aligned with the rotation axis, are also not necessarily diametrically opposed. AM Her itself also exhibits these mode changes (Figure 14b), with soft and hard X-ray light curves that are almost in anti-phase with each other, showing that the emission is a much more complex function of viewing angle and accretion column geometry involving both poles (e.g. Cropper et al. 2000). These light curves require that even the soft X-ray-emitting region (which can only come from $\sim 10^{-4}$ to 10^{-5} of the white dwarf surface) must have vertical extent too. It is not circular, but has footprints on the surface that are extended and even double.

Not surprisingly with its soft X-ray sensitivity, the ROSAT all-sky survey in 1991 (Truemper 1993) led to a large increase in the number of polars, the period distribution of which has a peak immediately below the well-known 'period gap' of 2–3 hours (only a handful are known above the gap). It is likely that this is related to a peak in the mass-transfer rate during this evolutionary phase, since models predict substantially lower mass-transfer rates during further evolution and these will be much more difficult to discover (see Warner 1995 for more details). Indeed a notable feature of polars is that their optical brightness drops dramatically and instantly following a reduction in mass-transfer rate, as there is no disk present. This provides perhaps the only means of studying the mass-transfer process directly from the donor star itself. Hessman et al. (2000) have thereby exploited the 20 yr lightcurve of AM Her assembled by the AAVSO to show that the average mass-transfer rate is ~25% of the maximum value, and to map out the high- and low-state behaviour. These results have implications for the properties of dwarf novae, where the behaviour of the mass donor cannot be directly observed.

EUV spectroscopy

In the mid-1990s the Extreme Ultraviolet Explorer (EUVE) carried out the first spectroscopy in the optimum range (~70–180 Å) with which to study the soft X-ray/EUV excess of polars (Figure 15). In spite of EUVE's limited sensitivity, there are nine polars with good EUV spectra, and most of them peak in this range. The black-body temperatures associated with these spectra are in the range

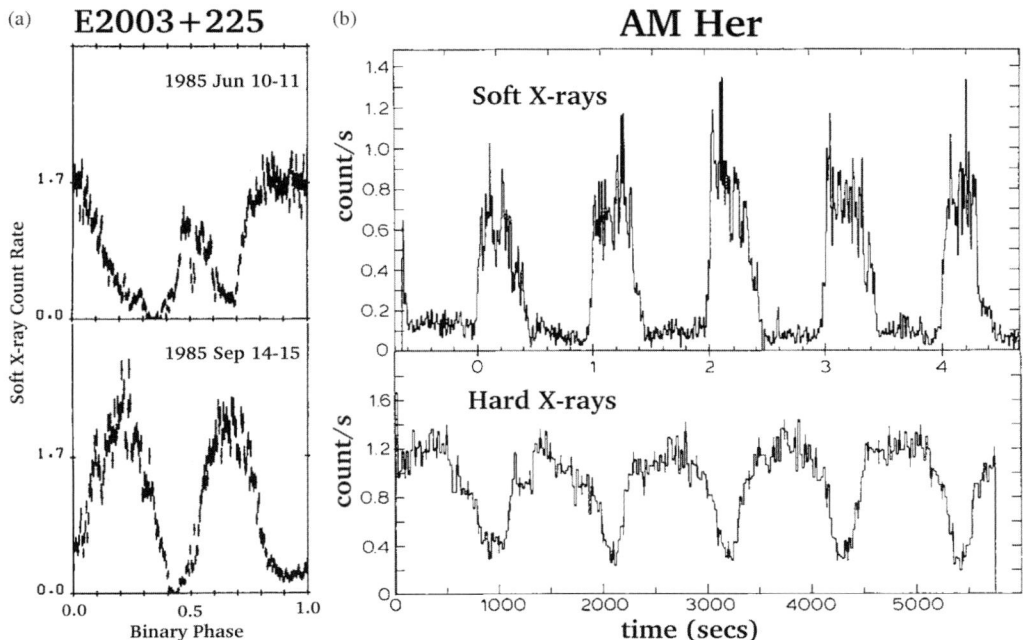

Figure 14 (a) Soft X-ray light curves of the polar E2003+225 as observed by EXOSAT on two occasions separated by three months in 1985, and plotted against orbital phase. The dominant emission on each occasion must be from completely different regions of the white dwarf, but note that they cannot be diametrically opposed. (b) Simultaneous soft and hard X-ray light curves of AM Her obtained with 1 min time resolution by EXOSAT using the LE and ME telescopes. In this new mode AM Her is emitting in both bands from both magnetic poles. (Adapted from Osborne et al. 1987, Heise et al. 1985).

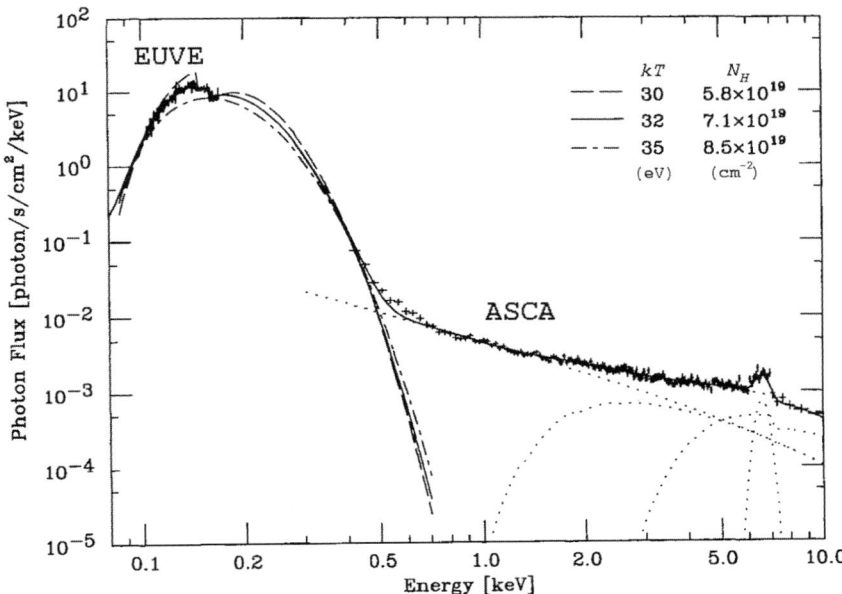

Figure 15 Simultaneous EUV and X-ray spectroscopy of AM Her obtained at rotational maximum (phases 0.3–0.7) by EUVE and ASCA. The low temperature black body ($kT = 32$ eV) is detected by ASCA in its lowest energy channels but requires EUVE for an accurate spectral measurement. (Adapted from Ishida et al. 1997).

10–30 eV but also reveal the presence of O VI and Ne VI absorption edges (likely in the hot atmospheres of the white dwarfs). However, to fit these spectra with white dwarf atmospheres would require oxygen to be underabundant (which is not expected). Hence irradiation of the white dwarf atmosphere by hard X-rays is invoked in order to produce a temperature inversion which will reduce the absorption edge strengths. As described by Mauche (1998) this explanation then requires most of the soft X-rays to be produced by hard X-ray reprocessing, and not the kinetic heating preferred as a solution of the soft X-ray puzzle.

Intermediate polars

As with AM Her, the CV EX Hya can be found in the Uhuru X-ray catalogue, but again its unusual nature was not recognized at that time. Instead the HEAO-1 X-ray instruments located the source 2A0526-328, which was subsequently identified as TV Col (Charles *et al.* 1979). However, their curious properties were mostly derived in the 1980s from optical studies (see review by Patterson 1994), followed by exploitation of EXOSAT's long satellite orbit ($\sim$4 days) to provide the first unbroken X-ray light curves of these objects lasting from many hours to days. (This property of EXOSAT, unique until the recent launch of the Chandra and XMM-Newton spacecraft, was a holdover from its original design proposal in the late 1960s as an X-ray telescope that used the Moon to occult bright X-ray sources thereby providing a precise position. By the time of its eventual launch in 1983 the technique had been rendered completely redundant by the development and launch of true X-ray imaging systems on the Einstein Observatory.) The main characteristics of these IPs can be summarized as (for more details see Buckley 2000, Warner 1995):

- X-ray pulsations representing the spin of the white dwarf (in the range $\sim$30 s to 1 h) which is *not* locked to the orbital period as in polars;
- optical pulsations at the same period, together with orbital sidebands;
- emission line variability (on the orbital, spin and sideband periods);
- a hard X-ray spectrum; and
- orbital light curves modulated by geometrical and absorption effects.

The presence of the clear spin modulation indicates that the magnetic field of the white dwarf controls the accretion flow (as in polars), but the fact that it is much faster than the orbital period suggests that the field is not sufficiently intense to control the entire flow from the secondary. Instead, a disk forms, but the inner region is truncated by the field. Secure magnetic fields (via polarization studies) are only available for three systems (PQ Gem, RX1712-24 and BG CMi), but the results do support this inference (Table 4). Hence, the basic structure of an IP is believed to be as shown schematically in Figure 16 where material from the complete inner edge of the disk threads onto the magnetic field creating a curtain of matter which can also modulate the observed signal.

Some people have differentiated between IPs and the 'DQ Her systems' which have shorter spin periods (of the order of minutes), and DQ Her itself is not an X-ray source, but they are otherwise remarkably similar. Also AE Aqr is a hard X-ray source, and the spin period is immediately visible in the light curve of GK Per (Figure 17) and so the DQ Her systems are here included with the IPs.

The basic accretion process onto the white dwarf in IPs is similar to that in polars, but the somewhat lower magnetic field leads to accretion over a larger area of the surface. There is clear evidence for a disk in some IPs (e.g. typical dwarf nova outbursts of EX Hya) but the presence of a disk has been questioned in others. It is here that the high-inclination systems have provided crucial insight, and time-resolved emission-line studies suggest that the stream collides with an outer disk (Hellier 1995, 1996, 1999), as illustrated in Figure 16. Perhaps not surprisingly, when the measured and inferred magnetic fields of polars and IPs are plotted together (Figure 18), they appear as a continuum of values, with some overlap between the classes at the high end of IPs and the low end of polars.

AE Aqr: a magnetic propellor

The behaviour of all the CVs discussed so far is consistent with most, if not all, of the matter lost from the secondary star being accreted by its compact companion. However, there is one object, AE Aqr, which appears at first sight to be an IP on the basis of its fast white dwarf spin (16.5 s pulsation in the optical, 33 s in X-rays, which is the rotation period) within a relatively long (9.9 h) orbital period. Being bright ($V \sim 11$) AE Aqr featured in the development of the basic interacting binary model for CVs of Joy (1954) and Crawford and Kraft (1956), but its strange flaring signature (only $\sim$1 mag) has been known since 1929 and is highly atypical compared to normal dwarf nova outbursts. Consequently AE Aqr has been subjected to intense multi-wavelength observing campaigns which revealed that the optical and UV flare together, as do the X-rays but with a smaller amplitude. There are radio flares too, but these are *not* associated with optical flares. But more remarkably, AE Aqr has been detected occasionally at extremely high energies (TeV γ-rays; the only CV to do so) during quiescent periods.

Table 4 Properties of selected intermediate polars

Variable	F_X	V	d (pc)	P_{orb} (h)	P_{spin} (min)	P_{syn} (min)	B (MG)
EX Hya	8.4	13.5	105	1.6	67.0		∼1
BG CMi	2.3	14.8	700	3.2	15.2	15.2	4
V1223 Sgr	5.0	13.0	600	3.4	12.4	13.2	∼1
AO Psc	4.0	13.2	420	3.6	13.4	14.3	∼1
YY Dra	4.5	16.0	155	3.9	8.8	9.2	∼1
DQ Her		15		4.6	2.4		0.14
FO Aqr	3.4	13.0	325	4.8	20.9	22.9	∼1
PQ Gem		14.5		5.2	13.9	14.5	
TV Col	4.1	13.8	≥500	5.5	31.8		∼1
TX Col	1.9	15.7	550	5.7	31.8	35.1	∼1
AE Aqr		11		9.9	0.6		0.3
GK Per	3–15	13	340	47.9	5.9	5.9	∼1

NB the distances to these objects are rather uncertain, and so only observed X-ray fluxes F_X are quoted in this table (in units of 10^{-11} erg cm^{-2} s^{-1} in the 2–10 keV band) rather than X-ray luminosities (which are believed to be in the range 10^{32-33} erg s^{-1}).

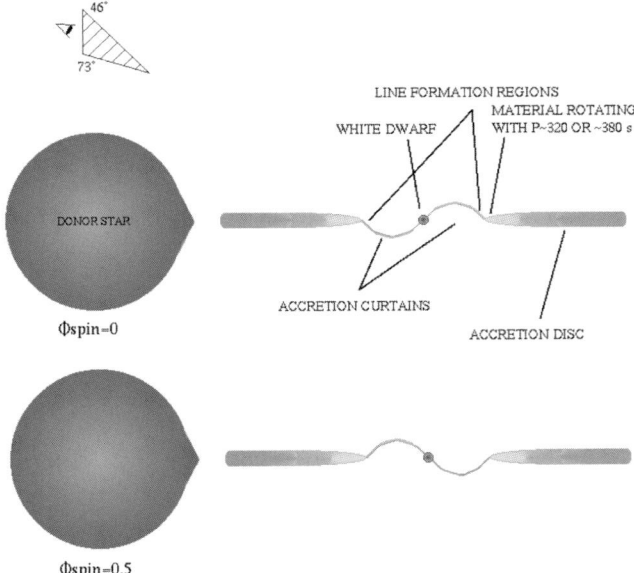

Figure 16 Schematic of the geometry of an 'intermediate polar' in which the magnetic field is not sufficiently powerful to prevent the formation of an accretion disk, but does control the flow of material onto the white dwarf itself. (Adapted from Morales-Rueda et al. 1999).

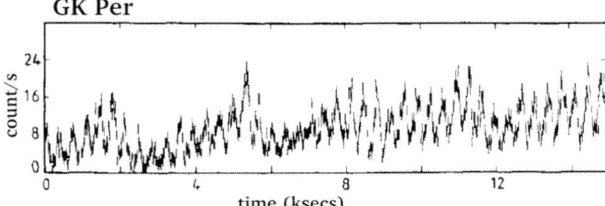

Figure 17 Hard X-ray light curve of the intermediate polar GK Per obtained with the EXOSAT observatory. The 6 min spin period of the white dwarf is clearly visible. (Adapted from Watson et al. 1985).

The pulsations were found not to be affected by the flares, indicating that the flaring is not physically associated with the magnetic cap regions of the white dwarf. Furthermore Hα (and other) emission lines do *not* follow the radial velocity curve of either the white dwarf or the companion star. And not being double peaked implies that there is no accretion disk present, even though estimates of the magnetic field would imply that there was sufficient room for one to form. The bolometric luminosity of AE Aqr (γ-, X-ray and optical) is $\leq 10^{33}$ erg s^{-1}, and yet the white dwarf's spin period is slowing at a rate that implies an energy loss of $>10^{34}$ erg s^{-1}.

This 'missing' luminosity is accounted for in the model of Wynn et al. (1997) which returns to the 'blobby' stream flow from the secondary. As the blobs approach the white dwarf they are dragged and accelerated by the magnetic field, achieving velocities that lead to their ejection from the system ($\leq 10^{-4}$ of the transferred material is accreted, the rest is ejected; Figure 19), carrying off kinetic energy which we see as the spin-down of the white dwarf. The flares are due to the interaction of the blobs with the field in the region of closest approach, producing shocks which account for the observed emission line spectrum. The presence of the pulsations indicates that there must be a white dwarf magnetosphere, and it is suggested that proton beams

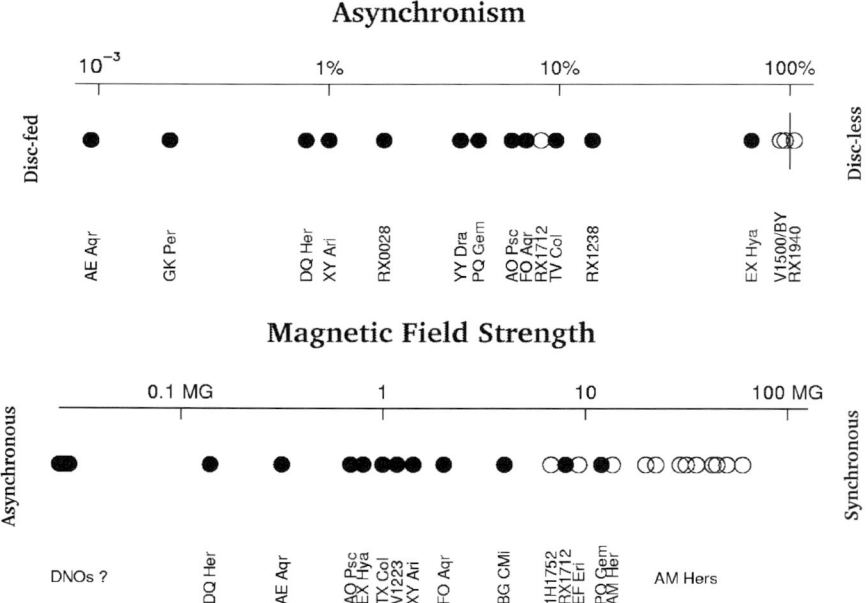

Figure 18 Distribution of periods and magnetic field strengths for polars and intermediate polars, showing that they are not distinct but form a continuum of properties. (Adapted from Hellier 1996).

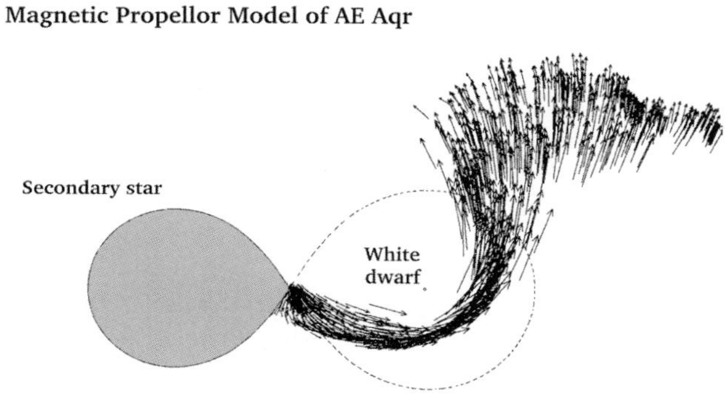

Figure 19 Schematic of the stream flow from the companion in the only γ-ray-emitting CV, AE Aqr. The blobby structure of this flow leads to it being accelerated in the white dwarf's strong magnetic field and subsequently ejected from the system. (Adapted from Wynn et al. 1997).

accelerated in this magnetosphere impact the blobs, thereby producing the observed γ-rays. Similarly electrons accelerated in this region produce the observed synchrotron radio emission.

CVs in globular clusters

It has long been known that globular clusters harbour far more LMXBs relative to the number of stars involved compared to the Galaxy as a whole (Chapter 34). Correspondingly globular clusters should provide ideal hunting grounds for cataclysmic binaries in general with anticipated enhanced populations due to the high central stellar densities involved. Comparing these with LMXBs would then provide information on the binary and dynamical evolution of globular clusters (Bailyn et al. 1990). However, such studies are very difficult from the ground due to the very severe crowding (prior to HST there was only one confirmed CV in a globular cluster: V101 in M5). Not surprisingly HST has made substantial contributions in this area. However, the great majority of those found so far, for example in NGC 6397, are IPs, indicating that search techniques hunting for CVs in outburst must be re-examined (Grindlay 1999).

Furthermore, first the Einstein Observatory, then ROSAT, demonstrated that there is a population of low-luminosity X-ray sources which it was suggested could be either quiescent soft X-ray transients or CVs, or both. The fainter luminosity range ($L_X \sim 10^{31-32}$ erg s^{-1}) is likely to contain predominantly IPs. Since almost all these objects lie within 20 arcsecs of the cluster centre, to confirm and investigate these populations in more detail will require longer-term HST observations combined with high-resolution X-ray imaging and spectroscopy, a capability that awaits the new generation X-ray observatories such as Chandra and XMM-Newton.

SUPERSOFT SOURCES

As the first X-ray observatory with direct imaging capability, the Einstein Observatory's dramatic contribution to our knowledge of the X-ray source population in the Large Magellanic Cloud (LMC) was its discovery of the very luminous ($\sim 10^{36-38}$ erg s^{-1}) but extremely soft (low temperature, $<10^6$ K) class of objects that are now known as the supersoft sources (SSS; Long et al. 1981). With such high luminosities ($>10^3$ times that of typical CVs and dwarf novae in our own Galaxy) it was not surprising that they were initially interpreted as LMXBs, possibly even containing a black hole as the compact object (Cowley et al. 1990; Chapter 34). This was due to the fact that candidate black hole systems such as LMC X-3 also exhibited soft X-ray spectra; however, the degree of 'softness' in the two cases is very different with the black hole systems producing copious X-ray emission in the 0.1–1 keV band as well as exhibiting a power-law component that extends to very high energies. Such high-energy components are never seen in the SSS. Subsequent examples of SSS were found in the Small Magellanic Cloud and M31, but very few in our Galaxy. This is exactly as expected for such extremely soft X-ray spectra, as this component is easily completely absorbed over relatively short distances (a few hundred parsecs or so) in our local interstellar medium, and hence the SSS can only readily be found and studied at high galactic latitudes using soft X-ray detectors on missions such as the Einstein, EXOSAT and ROSAT X-ray observatories.

It was then noted by van den Heuvel et al. (1992) that this combination of high luminosity and low temperature, if radiated as a black body, that is $L = 4\pi R^2 \sigma T^4$, which can be rewritten

$$R = 10^4 \text{ km} (L/10^{37} \text{ erg s}^{-1})^{0.5} (T/3 \times 10^5 \text{ K}) \quad (3)$$

had a size comparable to that of a white dwarf, not the much smaller neutron star or black hole event horizon. However, to account for the very high luminosity requires that the white dwarf be accreting material at a quite phenomenal rate ($\sim 10^{-6} M_\odot$ yr^{-1}) if this is to be accounted for entirely by release of potential energy. Such a mass-transfer rate exceeds that of a CV in outburst by at least a factor 100 and makes it essentially impossible for the soft X-rays to escape. The solution to this problem was to note that, in neutron star/black hole X-ray binaries, the energy liberated by accretion (per unit mass) exceeds that of nuclear fusion by at least a factor of 15, but in white dwarf systems the situation is reversed with fusion energy exceeding that of accretion by a factor of 30. Hence the mass-transfer rate required in the SSS could be relaxed by this same factor to the range $\sim 10^{-7}–10^{-8} M_\odot$ yr^{-1} by assuming that the accreted material be undergoing stable nuclear hydrogen burning, which then becomes the dominant emission mechanism. It would essentially be a non-ejecting nova, a possibility also suggested by Shara et al. (1977). Nevertheless even this mass-transfer rate in the SSS is a factor of at least 100 higher than that in 'normal' CVs, and van den Heuvel and co-workers attributed this to being a consequence of a binary system mass ratio of ~ 1 (i.e. a relatively massive secondary compared to normal CVs) which leads to unstable mass-transfer onto the white dwarf on a thermal timescale. The lifetime of SSS is thus expected to be of the order of 10^7 yr for a $\sim 1 M_\odot$ companion.

This model accounts for many of the SSS observed features in the long-period systems (>0.5 d), but it has still not been directly confirmed, largely due to the difficulty of obtaining accurate mass estimates for both the compact object and its companion. None have yet been obtained because no secondary star has yet been (optically) spectroscopically detected, presumably because its light has been completely overwhelmed by the steady-burning white dwarf and the surrounding inner disk. However, there is also a sub-group of the SSS which have short orbital periods (≤ 4 h, the 'SMC 13 systems'; e.g. Greiner et al. 2000) and for which this massive secondary scenario cannot be applied. This is because there is insufficient volume in the system for the Roche lobe-filling donor given that the white dwarf must exceed $\sim 0.4 M_\odot$ in order to sustain a nuclear burning layer. The nature of the SMC 13 systems is currently unclear.

The best direct evidence that the compact object in SSS is indeed a white dwarf comes from the higher resolution soft X-ray spectra available in the mid-1990s from the Japanese Advanced Satellite for Cosmology and Astrophysics (ASCA) with its CCD X-ray detectors (Einstein and ROSAT both used conventional X-ray proportional counters whose spectral resolution is $\sim E^{-0.5}$ and is extremely poor at the lowest energies). The ASCA spectrum of CAL87 (Asai et al. 1998) gives a best-fitting white dwarf model atmosphere at a temperature of 850,000 K and requires absorption edges due to O VII and O VIII.

At the end of the twentieth century, NASA's Chandra X-ray Observatory and ESA's XMM-Newton provided X-ray astronomers with true high-resolution X-ray spectroscopy (dispersive, using reflection gratings, as opposed to a CCD's intrinsic, non-dispersive spectral capability). Using XMM-Newton, Paerels et al. (2001) were able to reveal for the first time the rich and intricate absorption spectrum of the SSS component in CAL83 (Figure 20). Overplotted is a white dwarf model atmosphere with abundances of carbon, nitrogen and oxygen appropriate for the LMC. From this it is clear that the soft X-ray spectrum of CAL83 is much more complex than expected. Paerels et al. find that qualitative agreement can be reached by adding iron (M-shell absorption) together with mid-Z elements

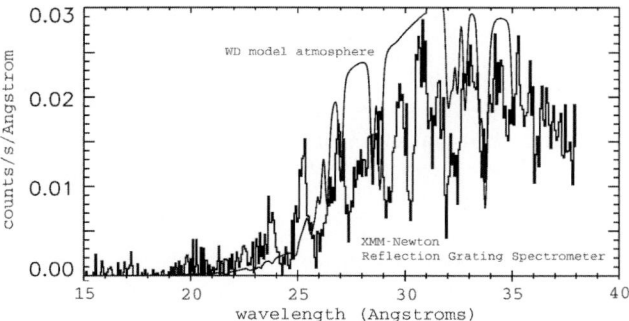

Figure 20 The soft X-ray spectrum of the SSS CAL83 obtained with the Reflection Grating Spectrometer on XMM-Newton. The continuous curve is that of a white dwarf model atmosphere with LMC abundances of CNO. Note that the absorption features of CAL83 are substantially more complex, and will require the addition of heavier elements to the model atmosphere. (Adapted from Paerels et al. 2001.)

(neon to calcium, L-shells), but more detailed analyses are now underway.

There is also excellent indirect evidence for the white dwarf nature of the compact object through ROSAT observations of the classical nova V1974 Cyg (Krautter et al. 1996). For while stable hydrogen burning does not occur in CVs, their accreted material on the white dwarf must eventually ignite in a thermonuclear runaway which we observe as a nova outburst and the ejection of a shell of the order of $10^{-4}\,M_\odot$ at velocities of $\sim 5000\,\mathrm{km\,s^{-1}}$. From a series of observations throughout the nova outburst of V1974 Cyg, ROSAT revealed the brightest SSS yet, consistent with Eddington-limited emission from the white dwarf surface at temperatures of ~ 20 eV. And since all novae have subsequently been shown to be CVs containing a white dwarf, this provides a compelling case for the basic SSS model.

Very high accretion rates can also be sustained in another well-known group of CVs, the symbiotic systems (e.g. Kenyon 1986). These are wide, very long-period ($\sim$ hundreds of days) binaries in which a white dwarf is accreting from its red giant companion. Mass transfer can be either by Roche lobe overflow or via a dense stellar wind if the red giant is on the asymptotic giant branch (AGB), and examples of both types have been detected as SSS. However, the dense winds observed ($\sim 10^{-5}\,M_\odot\,\mathrm{yr}^{-1}$) imply that such a phase would be very short lived.

The SSS are of considerable astrophysical interest because it has long been thought that neutron stars might be formed via the accretion-induced collapse of a massive white dwarf, and their 'steady' burning of accreted hydrogen provides a mechanism for their mass to grow. This is in contrast to eruptive variables where all the accreted matter is ejected in nova explosions. Furthermore, the thermonuclear detonation of a mass-accreting CO white dwarf close to the Chandrasekhar limit ($\sim 1.4\,M_\odot$) is the widely accepted explanation for Type Ia supernovae (e.g. Branch et al. 1995), and it is now considered possible for this to occur in lower mass white dwarfs via a 'double detonation' involving the large He shell (produced by burning the accreted H; Ruiz-Lapuente et al. 1995). The SSS are thus candidates for implication in all of these processes. For a detailed review of the SSS see Kahabka and van den Heuvel (1997).

Acknowledgments

This work has benefited enormously from the stimulating discussions and presentations that took place at the Warner Symposium on Cataclysmic Variables in Oxford (April 1999) and I am grateful to many colleagues for the use of figures from their reviews. I am also grateful to Danny Steeghs for a careful reading of an earlier version of this manuscript.

REFERENCES

Asai, K., Dotani, T., Nagase, F., Ebisawa, K., Mukai, K., Smale, A.P. and Kotani, T. (1998). ASCA observation of the supersoft X-ray source CAL 87. *Astrophysical Journal*, **503**, L143.

Bailyn, C.D., Grindlay, J.E. and Garcia, M.R. (1990). Does tidal capture produce cataclysmic variables? *Astrophysical Journal*, **357**, L35–L37.

Balbus, S.A. and Hawley, J.F. (1998). Instability, turbulence, and enhanced transport in accretion disks. *Review of Modern Physics*, **70**, 1–53.

Baptista, R., Horne, K., Wade, R.A., Hubeny, I., Long, K.S. and Rutten, R.G.M. (1998). HST spatially resolved spectra of the accretion disc and gas stream of the nova-like variable UX Ursae Majoris. *Monthly Notes of the Royal Astronomical Society*, **298**, 1079.

Beuermann, K., Thomas, H.-C. and Pietsch W. (1991). Short time-scale X-ray variability in the AM Her-type binary EF Eridani. *Astronomy and Astrophysics*, **246**, L36–L39.

Branch, D., Livio, M., Yungelson, L.R., Boffi, F.R. and Baron, E (1995). In search of the progenitors of Type IA supernovae. *Publications of the Astronomical Society of the Pacific*, **107**, 1019.

Buckley, D.A.H. (2000). On the power in intermediate polars. *New Astronomy Reviews*, **44**, 63–68.

Charles, P.A. and Seward F.D. (1995). *Exploring the X-ray Universe*, Cambridge University Press.

Charles, P.A., Thorstensen, J., Bowyer, S. and Middleditch, J. (1979). 2A 0526-328 – an X-ray-emitting cataclysmic variable. *Astrophysical Journal*, **231**, L131–L135.

Cordova, F.A. and Howarth, I.D. (1987). Accretion onto compact stars in binary systems. In Y. Kondo (ed), *Exploring the Universe with the IUE Satellite*, Reidel, Dordrecht, pp. 395–426.

Cordova, F.A. and Mason, K.O. (1982). High-velocity winds from a dwarf nova during outburst. *Astrophysical Journal*, **260**, 716–721.

Cordova, F.A. and Mason, K.O. (1983). Accreting degenerate dwarfs in close binary systems. In W.H.G. Lewin and E.P.J. van den Heuvel (eds), *Accretion-driven Stellar X-ray Sources*, Cambridge University Press, pp. 147–187.

Cowley, A.P., Schmidtke, P.C., Crampton, D. and Hutchings, J.B. (1990). CAL 87 – an eclipsing, black hole binary? *Astrophysical Journal*, **350**, 288.

Crawford, J.A. and Kraft, R.P. (1956). An interpretation of AE Aquarii. *Astrophysical Journal*, **123**, 44.

Cropper, M., Wu, K. and Ramsay, G. (2000). The emission from post-shock flows in mCVs. *New Astronomy Reviews*, **44**, 57–62.

Downes, R., Webbink, R.F. and Shara, M.M. (1997). A catalog and atlas of cataclysmic variables – Second edition. *Publications of the Astronomical Society of the Pacific*, **109**, 345–440.

Drew, J.E. and Proga, D. (2000). Radiation-driven accretion disk winds. *New Astronomy Reviews*, **44**, 21–26.

Eracleous, M. and Horne, K. (1996). The speedy magnetic propeller in the cataclysmic variable AE Aquarii. *Astrophysical Journal*, **471**, 427.

Forman, W., Jones, C., Cominsky, L., Julien, P., Murray, S., Peters, G., Tananbaum, H. and Giacconi, R. (1978). The fourth Uhuru catalog of X-ray sources. *Astrophysical Journal Supplement*, **38**, 357–412.

Frank, J., King, A. and Raine, D. (1992). *Accretion Power in Astrophysics*, Cambridge University Press.

Gallagher, J.S. and Code, A.D. (1974). Ultraviolet photometry from the orbiting astronomical observatory. X. Nova FH SER 1970. *Astrophysical Journal*, **189**, 303–314.

Greiner, J., Orio, M. and Schwarz, R. (2000). RX J0537.7−7034: The shortest-period supersoft X-ray source. *Astronomy and Astrophysics*, **355**, 1041.

Grindlay, J.E. (1999). Magnetic CVs in globular clusters. In C. Hellier and K. Mukai (eds), *Annapolis Workshop on Magnetic Cataclysmic Variables*, ASP Conference Series Vol. 157, Astronomical Society of the Pacific, San Francisco, p. 377.

Heise, J., Brinkman, A.C., Gronenschild, E., Watson, M., King, A.R., Stella, L. and Kieboom, K. (1985). An X-ray study of AM Herculis. I – Discovery of a new mode of soft X-ray emission. *Astronomy and Astrophysics*, **148**, L14–L16.

Heise, J., Mewe, R., Brinkman, A.C., Gronenschild, E.H.B.M., den Boggende, A.J.F., Schrijver, J., Parsignault, D.R. and Grindlay, J.E. (1978). Detection of both soft and hard X-ray emission from SS Cygni with ANS. *Astronomy and Astrophysics*, **63**, L1–L3.

Hellier, C. (1995). The accretion geometry of intermediate polars. In D.A.H. Buckley and B. Warner (eds), *Cape Workshop on Magnetic Cataclysmic Variables*, ASP Conference Series Vol. 85, Astronomical Society of the Pacific, San Francisco, p. 185.

Hellier, C. (1996). The intermediate polars. In A. Evans and J.H. Wood (eds), *Cataclysmic Variables and Related Objects*, Kluwer Academic, Dordrecht, p. 143.

Hellier, C. (1999). Recent results on intermediate polars. In C. Hellier and K. Mukai (eds), *Annapolis Workshop on Magnetic Cataclysmic Variables*, ASP Conference Series Vol. 157, Astronomical Society of the Pacific, San Francisco, p. 1.

Hessman, F.V., Gänsicke, B.T. and Mattei, J.A. (2000). The history and source of mass-transfer variations in AM Herculis. *Astronomy and Astrophysics*, **361**, 952–958.

Hind, J.R. (1856). On a new variable star. *Monthly Notes of the Royal Astronomical Society*, **16**, 56.

Horne, K. (1985). Images of accretion discs. I – The eclipse mapping method. *Monthly Notes of the Royal Astronomical Society*, **213**, 129–141.

Horne, K., Marsh, T.R., Cheng, F.H., Hubeny, I. and Lanz, T. (1994). HST eclipse mapping of dwarf nova OY Carinae in quiescence: An 'Fe II curtain' with Mach approx. = 6 velocity dispersion veils the white dwarf. *Astrophysical Journal*, **426**, 294–307.

Ishida, M., Matsuzaki, K., Fujimoto, R., Mukai, K. and Osborne, J.P. (1997). Detailed X-ray spectroscopy of AM Herculis with ASCA. *Monthly Notes of the Royal Astronomical Society*, **287**, 651–662.

Joy, A.H. (1954). Spectroscopic observations of AE Aquarii. *Astrophysical Journal*, **120**, 377.

Kahabka, P. and van den Heuvel, E.P.J. (1997). Luminous supersoft X-ray sources. *Annual Review of Astronomy and Astrophysics*, **35**, 69–100.

Kenyon, S. (1986). *The Symbiotic Stars*, Cambridge University Press.

Kiplinger, A.L., Sion, E.M. and Szkody, P. (1991). A study of the ultraviolet evolution of U Geminorum between outbursts. *Astrophysical Journal*, **366**, 569.

Knigge, C., Drake, N., Long, K.S., Wade, R.A., Horne, K. and Baptista, R. (1998). Recovery of 29 second oscillations in Hubble Space Telescope eclipse observations of the cataclysmic variable UX Ursae Majoris. *Astrophysical Journal*, **499**, 429.

Kraft, R.P. (1962). Binary stars among cataclysmic variables. I. U Geminorum stars (dwarf novae). *Astrophysical Journal*, **135**, 408.

Krautter, J., Oegelman, H., Starrfield, S., Wichmann, R. and Pfeffermann, E. (1996). ROSAT X-ray observations of nova V1974 Cygni: The rise and fall of the brightest supersoft X-ray source. *Astrophysical Journal*, **456**, 788.

Krautter, J., Vogt, N., Klare, G., Wolf, B., Wargau, W., Drechsel, H. and Rahe, J. (1981). IUE spectroscopy of cataclysmic variables. *Astronomy and Astrophysics*, **102**, 337–346.

Krzeminski, W. and Serkowski, K. (1977). Extremely high circular polarisation of AN Ursae Majoris. *Astrophysical Journal*, **216**, L45–L48.

Lasota, J.P. (2001). The disc instability model of dwarf novae and low-mass X-ray binary transients. *New Astronomy Reviews*, **45**, 449.

La Dous, C. (1989). Synthetic optical and ultraviolet spectra of stationary accretion disks. *Astronomy and Astrophysics*, **211**, 131–155.

Linnell, A.P. (1949). UX Ursae Majoris. *Sky and Telescope*, **8**, 166.

Linnell, A.P. (1950). Harv.Circ. No.455

Long, K.S. (2000). What we learn from quantitative ultraviolet spectroscopy of naked white dwarfs in cataclysmic variables. *New Astronomy Reviews*, **44**, 125–130.

Long, K.S., Blair, W.P., Hubeny, I., Raymond, J.C. (1996). Observations of the dwarf nova VW Hydri in quiescence with the Hopkins Ultraviolet Telescope. *Astrophysical Journal*, **466**, 964.

Long, K.S. and Gilliland, R.L. (1999). GHRS observations of the white dwarf in U Geminorum. *Astrophysical Journal*, **511**, 916–924.

Long, K.S., Helfand, D.J. and Grabelsky, D.A. (1981). A soft X-ray study of the Large Magellanic Cloud. *Astrophysical Journal*, **248**, 925–944.

Marsh, T.R. (2001). Observations of cataclysmic variables and double degenerate stars. In C. Lazaro and M.J. Arevalo (ed), *Binary Stars: Selected Topics on Observations and Physical Processes*, Proceedings of 12th EADN Summer School, Lecture Notes in Physics, Vol. 563 Springer-Verlag, Berlin, p. 151.

Marsh, T.R. and Horne K. (1988). Images of accretion discs. II – Doppler tomography. *Monthly Notes of the Royal Astronomical Society*, **295**, 269–286.

Mason, K.O., Cordova, F.A., Watson, M.G. and King, A.R. (1988). The discovery of orbital dips in the soft X-ray emission of U GEM during an outburst. *Monthly Notes of the Royal Astronomical Society*, **232**, 779–791.

Mason, K.O., Drew, J.E., Cordova, F.A., Horne, K., Hilditch, R., Knigge, C., Lanz, T. and Meylan, T. (1995). Eclipse observations of an accretion disc wind. *Monthly Notes of the Royal Astronomical Society*, **274**, 271–286.

Mauche, C.W. (1998). The EUV and X-ray emission of nonmagnetic cataclysmic variables. In S. Howell, E. Kuulkers, and C. Woodward (eds), *Wild Stars in the Old West: Proceedings of the 13th North American Workshop on Cataclysmic Variables and Related Objects*, ASP Conference Series Vol. 137, Astronomical Society of the Pacific, San Francisco, p. 113.

Morales-Rueda, L., Still, M.D. and Roche, P. (1999). Solving the kilosecond quasi-periodic oscillation problem of the intermediate polar GK Persei. *Monthly Notes of the Royal Astronomical Society*, **306**, 753–765.

Osborne, J.P., Beuermann, K., Charles, P., Maraschi, L., Mukai, K. and Treves, A. (1987). A new soft X-ray mode in the AM Herculis object E2003+225. *Astrophysical Journal*, **315**, L123–L127.

Paerels, F., Rasmussen, A.P., Hartmann, H.W., Heise, J., Brinkman, A.C., de Vries, C.P. and den Herder, J. W. (2001). A high resolution spectroscopic observation of CAL 83 with XMM-Newton/RGS. *Astronomy and Astrophysics*, **365**, L308–L311.

Panek, R.J. and Holm, A.V. (1984). Ultraviolet spectroscopy of the dwarf nova U Geminorum. *Astrophysical Journal*, **277**, 700–709.

Patterson, J. (1994). The DQ Herculis stars. *Publications of the Astronomical Society of the Pacific*, **106**, 209–238.

Patterson, J. and Price, C. (1980). IAUC, 3511.

Patterson, J. and Raymond, J.C. (1985). X-ray emission from cataclysmic variables with accretion disks. I – Hard X-rays. II – EUV/soft X-ray radiation. *Astrophysical Journal*, **292**, 535–558.

Payne-Gaposchkin, C. and Gaposchkin, S. (1938). *Variable Stars*, Harvard Observatory Monograph No.5, Cambridge, MA.

Pringle, J.E. (1981). Accretion discs in astrophysics. *Annual Review of Astronomy and Astrophysics*, **19**, 137–162.

Proga, D., Stone, J.M. and Drew, J.E. (1999). Line-driven disc wind models with an improved line force. *Monthly Notes of the Royal Astronomical Society*, **310**, 476.

Raymond, J.C., Davis, R.J., Hartmann, L., Matilsky, T.A., Black, J.H., Dupree, A.K. and Gursky, H. (1979). Ultraviolet observations of AM Herculis with IUE. *Astrophysical Journal*, **230**, L95–L98.

Ritter, H. and Kolb, U. (1998). Catalogue of cataclysmic binaries, low-mass X-ray binaries and related objects (Sixth edition). *Astronomy and Astrophysics Supplement*, **129**, 83–85.

Rosen, S.R. (1992). The role of electron scattering in the X-ray rotational light curves of intermediate polars. *Monthly Notes of the Royal Astronomical Society*, **254**, 493–500.

Ruiz-Lapuente, P., Burkert, A. and Canal, R. (1995). Type IA supernovae scenarios and the Hubble sequence. *Astrophysical Journal*, **447**, L69.

Shakura, N.I. and Sunyaev, R.A. (1973). Black holes in binary systems. Observational appearance. *Astronomy and Astrophysics*, **24**, 337–355.

Shara, M.M., Prialnik, D. and Shaviv, O. (1977). Non-ejecting novae as EUV sources. *Astronomy and Astrophysics*, **61**, 363.

Sion, E.M., Cheng, F.-H., Huang, M., Hubeny, I. and Szkody, P. (1996). The cooling white dwarf in VW Hydri after normal outburst and super-outburst: HST evidence of a sustained accretion belt. *Astrophysical Journal*, **471**, L41.

Stockman, H.S., Schmidt, G.D. and Lamb, D.Q. (1988). V1500 Cygni – Discovery of a magnetic nova. *Astrophysical Journal*, **332**, 282–286.

Szkody, P., Schmidt, E., Crosa, L. and Schommer, R. (1981). Simultaneous X-ray and optical observations of an Ursae Majoris during a low state. *Astrophysical Journal*, **246**, 223–230.

Tapia, S. (1977). Discovery of a magnetic compact star in the AM Herculis/3U 1809+50 system. *Astrophysical Journal*, **212**, L125–L129.

Truemper, J. (1993). ROSAT – A new look at the X-ray sky. *Science*, **260**, 1769–1771.

van den Heuvel, E.P.J., Bhattacharya, D., Nomoto, K. and Rappaport, S.A. (1992). Accreting white dwarf models for CAL 83, CAL 87 and other ultrasoft X-ray sources in the LMC. *Astronomy and Astrophysics*, **262**, 97–105.

Wade, R.A. and Hubeny, I. (1998). Detailed mid- and far-ultraviolet model spectra for accretion disks in cataclysmic binaries. *Astrophysical Journal*, **509**, 350–361.

Walker, M.F. (1954). Nova DQ Herculis (1934): An eclipsing binary with very short period. *Publications of the Astronomical Society of the Pacific*, **66**, 230.

Walker, M.F. (1956). A photometric investigation of the short-period eclipsing binary, nova DQ Herculis (1934). *Astrophysical Journal*, **123**, 68.

Warner, B. (1972). Observations of rapid blue variables – VII. EX Hydrae. *Monthly Notes of the Royal Astronomical Society*, **158**, 425–430.

Warner, B. (1995). *Cataclysmic Variable Stars*, Cambridge University Press.

Warner, B. and Nather, R.E. (1971). Observations of rapid blue variables. II. U Gem. *Monthly Notes of the Royal Astronomical Society*, **152**, 219.

Warner, B. and Nather, R.E. (1972). Observations of rapid blue variables. XIII. UX UMa. *Monthly Notes of the Royal Astronomical Society*, **159**, 429.

Watson, M.G., King, A.R. and Osborne, J.P. (1985). The old nova GK Per – Discovery of the X-ray pulse period. *Monthly Notes of the Royal Astronomical Society*, **212**, 917–930.

Wells, L.D. (1896). Harv.Coll.Obs.Circ. No.12

Wheatley, P.J., Mauche, C.W. and Mattei, J.A. (2000). RXTE, EUVE and optical observations of SS Cyg in outburst. *New Astronomy Reviews*, **44**, 33.

Wickramasinghe, D.T. and Ferrario, L. (2000). Accretion and magnetic field structure in AM Herculis systems. *New Astronomy Reviews*, **44**, 69–74.

Wynn, G.A., King, A.R. and Horne, K. (1997). A magnetic propeller in the cataclysmic variable AE Aquarii. *Monthly Notes of the Royal Astronomical Society*, **286**, 436–446.

34

JAN VAN PARADIJS[†] AND MICHIEL VAN DER KLIS[*]

Low-mass X-ray binaries

The connection between cosmic X-ray sources and compact stars is an old one: soon after the discovery of the first such source, Sco X-1 (Giacconi *et al.* 1962), it was proposed that these objects are young, hot neutron stars, formed in recent supernovae, which cool through thermal radiation from their surfaces (Chiu 1964, Chiu and Salpeter 1964, Finzi 1964). However, the finite extent of the X-ray source associated with the Crab Nebula (Bowyer *et al.* 1964), and the non-Planckian shape of the X-ray spectra of this source (Clark 1965) and of Sco X-1 (Giacconi *et al.* 1965) showed that this was not, in general, a good model for X-ray sources.

Accretion onto a compact star had meanwhile been suggested as a possible source of energy for quasars and X-ray sources (Salpeter 1964, Zel'dovich 1964, Zel'dovich and Guseinov 1966), and together with the peculiarities of the optical spectra of the counterparts of Sco X-1 (Sandage *et al.* 1966) and Cyg X-2 (Giacconi *et al.* 1967) this led to the idea that these sources are mass-exchanging binaries with a compact component (Shklovsky 1967; see also Burbidge 1972, Ginzburg 1990). The spectrum of Sco X-1 was similar to those of old novae and U Gem-type stars, which were by then known to be binary stars, in particular through the work of Crawford and Kraft in the 1950s and 1960s (Crawford and Kraft 1956; Kraft 1962, 1964). However, the single-most important characteristic of a binary star, that is, an orbital periodicity, was not found until many years later, in spite of substantial observational effort (see Hiltner and Mook (1970) and Kraft (1973) for discussions of early optical observations of X-ray sources, and references).

The discovery of the first X-ray binary, Cyg X-1, by Webster and Murdin (1972) and Bolton (1972) at the same time showed that this binary star contained an accreting object that is likely to be a black hole.

The idea that all bright galactic X-ray sources are mass-exchanging binary stars with a compact accretor became widely accepted with the observation, with the Uhuru satellite, of regular eclipses of the pulsating X-ray source Cen X-3 (Giacconi *et al.* 1971, Schreier *et al.* 1972). The variable delays of the pulse arrival times, in phase with the periodic (2.1 days) eclipses of the X-ray source, showed persuasively that in Cen X-3 the X-rays are generated by accretion onto a strongly magnetized neutron star, rotating at a 4.8 s pulse period, in orbit around a massive ($\gtrsim 10 M_\odot$) companion star. The discovery of the binary nature of Cen X-3 was soon followed by more observations of eclipsing X-ray sources, some of them pulsating, and by the identification of these X-ray sources with early-type stars. In addition, a general framework for the origin and evolution of a massive X-ray binary, as a rather normal episode in the life of a massive close binary star with successive stages of mass transfer between the two components, was readily accepted (Van den Heuvel and Heise 1972). Thus, within a few years the existence of a galactic population of high-mass X-ray binaries (HMXBs), with accreting neutron stars (or occasionally a black hole) was well established.

Already in the 1960s (e.g. Dolan 1970) it had become clear that there is a clustering of bright X-ray sources within $\sim 30°$ of the direction of the galactic center. This concentration was not accompanied by a strong background of unresolved sources, which showed that these sources are located in the central regions of the Galaxy (Ryter 1970, Setti and Woltjer 1970). It was, therefore, suspected that apart from the above-described HMXBs there is a class of low-mass X-ray binaries (LMXBs) with donor star masses of the order of a solar mass or less (e.g. Salpeter 1973), but proof for this idea was hard to obtain. Apart from the difficulty of finding orbital periods, the

---

[†] Universiteit van Amsterdam, The Netherlands. Jan Van Paradijs died on 2 November 1999
[*] Universiteit van Amsterdam, The Netherlands

apparent heterogeneity of the properties of LMXBs may have played a role. Compared to the HMXBs the first handful of known systems now classified as LMXBs (Her X-1, Sco X-1, Cir X-1) show rather more diversity than similarity in their properties. As a result, only at the end of the 1970s did it become clear that there are such objects as LMXBs, which form a group with "family traits," distinct from the HMXBs with respect to their sky distributions, X-ray spectral characteristics, optical properties, and types of X-ray variability (e.g. Lewin and Clark 1980). The LMXBs comprise the globular-cluster X-ray sources, X-ray bursters, soft X-ray transients, and the bright galactic-bulge X-ray sources.

Roughly speaking, the reason for the bi-modal distribution of the masses of donor stars in X-ray binaries is that for stars less massive than $\sim 10 M_\odot$ the stellar wind is too weak to power a strong X-ray source; however, Roche lobe overflow is unstable for stars more massive than a neutron star, and proceeds on a very short timescale, as a consequence of which the accreting neutron star is completely engulfed and X-rays cannot escape.

The differences between the X-ray properties of LMXBs and HMXBs (with accreting neutron stars) may be linked to a difference in the strength of the magnetic fields of the neutron stars they harbor. The natural assumption that the difference in donor star masses corresponds to a difference in the ages of LMXBs and HMXBs has led to the idea that the magnetic fields of neutron stars decay with time.

Within the limits of this review it would be meaningless to strive for completeness. For detailed reviews on a variety of topics related to X-ray binaries we refer the interested reader to Lewin *et al.* 1995). Further background information can be found in Shapiro and Teukolsky (1983), Frank *et al.* (1992), Ögelman and Van den Heuvel (1989), Ventura and Pines (1991), Van den Heuvel and Rappaport (1992), Alpar *et al.* (1995), and Buccheri *et al.* (1998). References on individual sources can be found in Bradt and McClintock (1983) and Van Paradijs (1995). An extensive summary of X-ray satellite missions has been given by Bradt *et al.* (1992).

1 OPTICAL COUNTERPARTS

Whereas the optical properties of HMXBs are dominated by the emission of their very luminous mass-donor stars, the optical spectra of LMXBs (e.g. Shahbaz *et al.* 1996) show only a few emission lines, particularly Hα, Hβ, He II λ4686, and C III–N III λ4630–50, superposed on a rather flat (in frequency) continuum. In a few cases the signature of a companion star can be discerned. According to Motch and Pakull (1989) the relative strength of the C III–N III emission complex relative to the λ4686 emission provides a good measure of the heavy-element abundances in the accreted matter. Spectra of cataclysmic variables (CVs), in which the accretor

Figure 1 Artist's impression of an X-ray binary. (Credit: Rob Hynes).

is not a neutron star (or black hole) but a white dwarf, bear a general resemblance to those of LMXBs, showing emission lines superposed on a continuum (see Warner 1995 for a comprehensive review of CVs). However, the equivalent widths of these lines in LMXB spectra, in particular that of Hβ, tend to be much smaller than those in CV spectra (Van Paradijs and Verbunt 1984, Shahbaz *et al.* 1996).

In LMXBs the donor star transfers mass by Roche lobe overflow (Figure 1). This mass arrives at the compact star via a relatively flat rotating configuration, the accretion disk, in which it slowly spirals inward. Gravitational potential energy is converted into heat in this process. In the inner few tens of kilometers of this disk temperatures of $\sim 10^7$ K are attained, which leads to the emission of thermal X-rays. Part of these X-rays irradiate the disk further out. The optical emission of LMXBs originates primarily from these irradiated outer regions of the accretion disk, which produce their radiation mainly through reprocessing of incident X-rays into optical and UV photons. This reprocessing dominates the internal energy generation of the disk due to conversion of gravitational potential energy into heat, generally by a large factor (Van Paradijs and McClintock 1995). For internal energy generation alone, as occurs in accretion disks in CVs, the (local) effective temperature, T, in the disk varies with radial distance, r, from the central compact star approximately as $T(r) \propto r^{-3/4}$. In LMXB disks, where reprocessing of X-rays dominates the energy budget, one would expect $T(r) \propto r^{-1/2}$. Thus, the farther out in the disk, the more reprocessing dominates internal energy generation. According to the simplified, but self-consistent calculations of Vrtilek *et al.* (1990), in which also the effect of X-ray irradiation on the height of the disk is taken into account, $T(r) \propto r^{-3/7}$.

Many LMXBs show a regular orbital variation of their optical brightness, with one maximum and minimum per orbital cycle. Minimum light occurs at the superior conjunction of the X-ray source. These variations are caused by the changing visibility of the accretion disk and the X-ray heated side of the secondary, insofar as the latter is not in the X-ray shadow of the disk (see Van Paradijs 1991 and Van Paradijs and McClintock 1995 for reviews of the optical light curves of LMXBs).

For transient LMXBs ("soft X-ray transients") in quiescence, X-ray heating of the accretion disk and the companion star is not very important, and the optical emission of the system is dominated by the Roche lobe-filling secondary star. These systems then show ellipsoidal light curves, reflecting the tidal and rotational distortion of the secondary.

The (reddening-corrected) color indices $B-V$ and $U-B$ of LMXBs have average values of -0.09 ± 0.14 and -0.97 ± 0.17, respectively (errors are 1σ standard deviations), close to those expected for a flat continuum ($F_\nu =$ constant). The distribution of the ratio of X-ray to optical fluxes is rather sharply peaked. Expressed in terms of an "optical/X-ray color index" $B_0 + 2.5 \log F_X(\mu Jy)$, the peak occurs near 21.5, corresponding to a ratio of fluxes emitted in X-rays (2–11 keV) and in optical light (3000–7000 Å) of ~ 500 (Van Paradijs and McClintock 1995).

The optical luminosities of LMXBs are, in general, much higher than those of CVs (e.g. Warner 1987, 1995); this is because the gravitational potential well of a neutron star is much deeper than that of a white dwarf, and X-ray heating of the accretion disk is not important in CVs.

Absolute visual magnitudes M_V have been estimated for LMXBs with distance estimates: (i) LMXBs in stellar systems at a known distance; (ii) X-ray burst sources (Section 5) showing bursts with Eddington-limited photospheric radius expansion; (iii) Z sources (Section 6); when these are in the "normal-branch" state their X-ray luminosity is very close to the Eddington limit (e.g. Van der Klis 1995); (iv) soft X-ray transients, whose distance can be determined in quiescence from the spectral properties of the companion star. The absolute magnitudes of active LMXBs range between -5 and $+5$ (Van Paradijs and McClintock 1994). This large range is the consequence of the large range in X-ray luminosity, L_X, of the central source, and in the size of the accretion disk. For a simple model of reprocessing of X-rays the optical luminosity, L_V, of the disk is expected to scale with L_X and orbital period P as $L_V \propto L_X^{1/2} P^{2/3}$, in agreement with the values of M_V for LMXBs with known orbital periods. This is confirmed by numerical calculations of X-ray-heated accretion disks (De Jong et al. 1996).

2 ORBITAL PERIODS

Orbital periods are known for some three dozen LMXBs, mainly from regular X-ray eclipses and X-ray "dips"
(White et al. 1995), and optical brightness variations (Van Paradijs and McClintock 1995). In several cases periodic radial velocity variations have been found for emission lines (originating from an accretion disk), or for absorption lines (secondary star of soft X-ray transients in quiescence). The known orbital periods of LMXBs range between 685 seconds and 16.6 days (Van Paradijs 1995, Liu et al. 2001). This large range indicates that LMXBs have a variety of mass donors (white dwarfs, main-sequence stars, giant stars). The orbital-period distributions of CVs and LMXBs (Figure 2) are different. Compared to the CVs a much larger fraction of LMXBs have periods above about half a day. A possible explanation for this difference is that for such long periods the companion star masses are expected to substantially exceed the mass of a typical white dwarf, but not that of a neutron star. This will make the mass transfer at long periods unstable for CVs, but not for LMXBs. The LMXB period distribution does not show the well-known period gap in the distribution for the CVs (e.g. Verbunt 1984, Spruit and Ritter 1983 for studies of the period gap); in fact, there are no LMXBs in the period range between 80 minutes and 2 hours (i.e., below the period gap), which is well populated by the CV. This may be the result of the high probability that LMXBs form at periods of about half a day or longer, whereas a large fraction of CVs form at periods below 2 hours (King and Kolb 1997). Perhaps evaporation of the LMXB secondaries plays a role after the LMXB has reached the upper edge of the period gap; mass transfer then stops, and the rapidly rotating neutron star (spun up by accretion torques) then becomes active as a millisecond radio pulsar (Ruderman

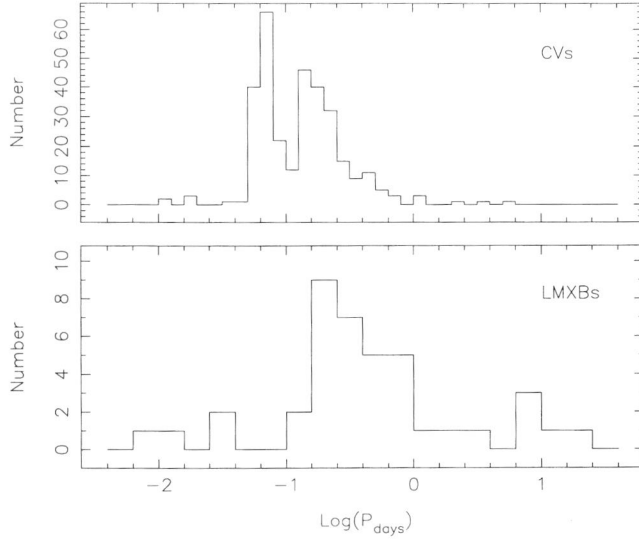

Figure 2 Distributions of orbital periods for LMXBs and CVs. Data have been taken from Van Paradijs (1995) and Ritter and Kolb (1998).

et al. 1989, Van den Heuvel and Van Paradijs 1988); however, whether or not complete evaporation of the secondary star occurs is a matter of debate.

3 GALACTIC DISTRIBUTION

The sky distributions of the HMXBs and LMXBs are quite different, as shown in Figure 3. The galactic HMXBs are distributed along the galactic plane, without an obvious concentration to the galactic center. They have an average latitude $<b^{II}> = -0.8 \pm 6.9°$; if we leave out X Per and a few other nearby high-latitude Be/X-ray systems identified by Tuohy *et al.* (1988) we find $<b^{II}> = 0.4 \pm 1.9°$. This fits the idea that HMXBs are young population I objects.

The galactic LMXBs (excluding the globular-cluster sources) have a wider latitude distribution ($<b^{II}> = 0.4 \pm 9.1°$), and are also more concentrated to the direction of the galactic center. The scale height of the (assumed exponential) distribution of distances from the galactic plane, for LMXBs with neutron stars at known distances is 900 pc (Van Paradijs and White 1995). The z distribution of LMXBs with a black hole is substantially narrower (White and Van Paradijs 1996). The wide z dispersion of LMXBs with neutron stars requires that the neutron stars in these systems formed in an asymmetric supernova explosion which gave them an extra kick velocity (Brandt and Podsiadlowski 1995, Van Paradijs and White 1995, Ramachandran and Bhattacharya 1997); the kick velocity distribution is consistent with that of single radio pulsars (Lyne and Lorimer 1994, Hansen 1996, Hansen and Phinney 1997, Hartman *et al.* 1997).

There are $\sim 10^2$ persistent luminous LMXBs ($L_X > 10^{36}$ erg s^{-1}) in the Galaxy (Van Paradijs 1995). During the last decade it has become clear that in a large fraction of the transient LMXBs the compact star is a black hole; the number of such transient black-hole binaries in the Galaxy is estimated to be between a few hundred and a few thousand (Tanaka and Lewin 1995, White and Van Paradijs 1996).

The kinematic properties of LMXBs have been studied by Cowley *et al.* (1988) and Johnston (1992). Based on the large velocity dispersion and low galactic rotation velocity, Cowley *et al.* (1988) concluded that LMXBs are among the oldest objects in the Galaxy. However, since LMXBs get a kick velocity at the formation of the neutron star, such a direct interpretation of the kinematic properties of LMXBs in terms of their ages cannot be made.

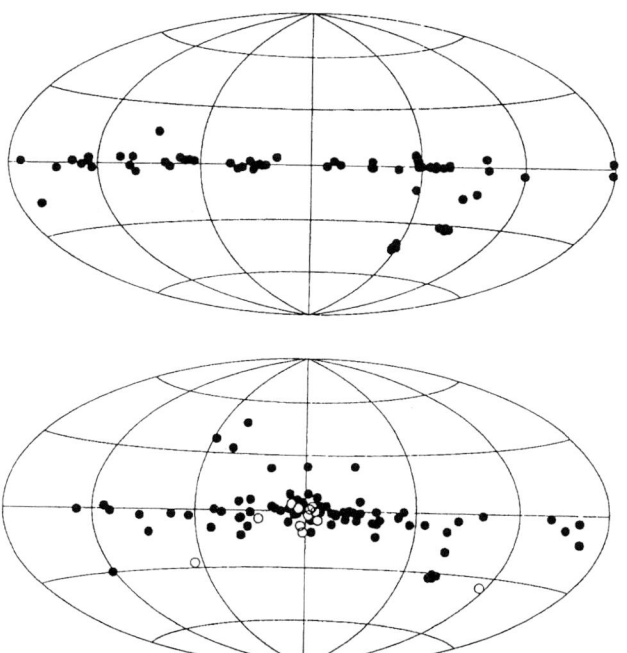

Figure 3 Sky maps (in galactic coordinates) of the HMXBs (top) and LMXBs (bottom); the latter also includes the globular-cluster sources (indicated by open circles). The 27 LMXBs within 2° of the galactic center have not been included to avoid congestion of the map. These maps are based on the catalog of Van Paradijs (1995).

4 X-RAY PULSATIONS

Almost all HMXBs show X-ray pulsations, which indicates that the accreting compact stars in these systems are strongly magnetized neutron stars (for reviews of various aspects of X-ray pulsars see, e.g. Joss and Rappaport 1984, Nagase 1989, White *et al.* 1995, Bildsten *et al.* 1997).

Pulse arrival time measurements for pulsating HMXBs, in combination with radial velocity observations of their massive companions, have provided information on the masses of accreting neutron stars.

X-ray pulsations in LMXBs are very rare. In spite of very sensitive searches (e.g. Vaughan *et al.* 1994) none of the bright LMXBs have shown the predicted millisecond pulsations, with upper limits to their amplitudes well below the 1% level. One case of a millisecond pulsar powered by accretion has recently been found. This pulsar, with a spin period of 2.5 ms, was discovered with the Rossi X-ray Timing Exporer (RXTE) in the faint transient LMXB SAX J1808.4-3658 (Wijnands and Van der Klis 1998). It has an orbital period of approximately two hours (Chakrabarty and Morgan 1998). In the last several years evidence for millisecond spin periods of the neutron stars in LMXBs has also been obtained from the "burst oscillations" seen in some X-ray bursts (Section 5); indirect evidence from aperiodic variability for millisecond spin periods also exists (Section 6).

5 X-RAY BURSTS

Discovered with the Astronomische Nederlandse Satelliet (ANS), many LMXBs emit X-ray bursts, during which the X-ray flux rises by typically at least an order of magnitude, usually within about a second (Grindlay and Gursky 1976). This is followed by a decay, generally to the pre-burst X-ray flux level, in a time interval between ~10 s and about a minute; in rare cases bursts last longer (Figure 4). A comprehensive review of X-ray bursts has been given by Lewin et al. (1993, 1995). For a recent discussion of the physics of thermonuclear burning on the surfaces of accreting neutron stars, with emphasis on the propagation of the burning, see Bildsten (1998).

Two types of X-ray bursts can be distinguished (Hoffman et al. 1978), called type I and II. The type I bursts first found with ANS show a distinct softening of the X-ray spectrum during the decay of the burst. Their recurrence times are generally of the order of hours and longer, but occasionally as short as a few minutes. The spectral development in type II bursts, discovered with the SAS 3 satellite (Lewin et al. 1976), is much less pronounced than that in the type I bursts.

Until recently, the type II bursts had only been observed from the Rapid Burster (Lewin et al. 1976). The time intervals between the type II bursts from the Rapid Burster can be as short as ~10 s, and as long as one hour (Figure 5). They come in a characteristic pattern such that the total energy in a burst is approximately proportional to the time interval to the following burst: the Rapid Burster behaves

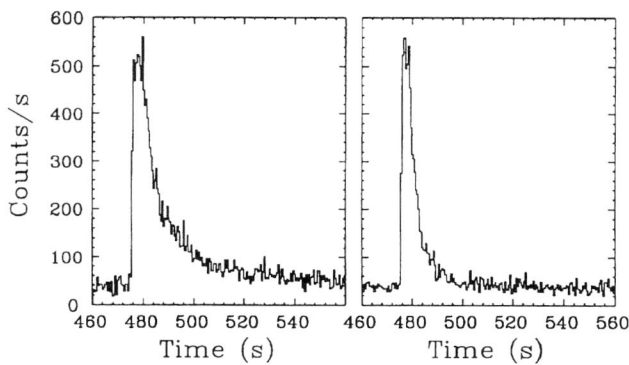

Figure 4 Type I X-ray burst from 1702–42 as observed with EXOSAT in the 1.2–5.3 keV band (left) and the 5.3–19.0 keV band (right); the softening of the X-ray burst spectrum is apparent as a longer tail in the low-energy burst profile. (Courtesy T. Oosterbroek).

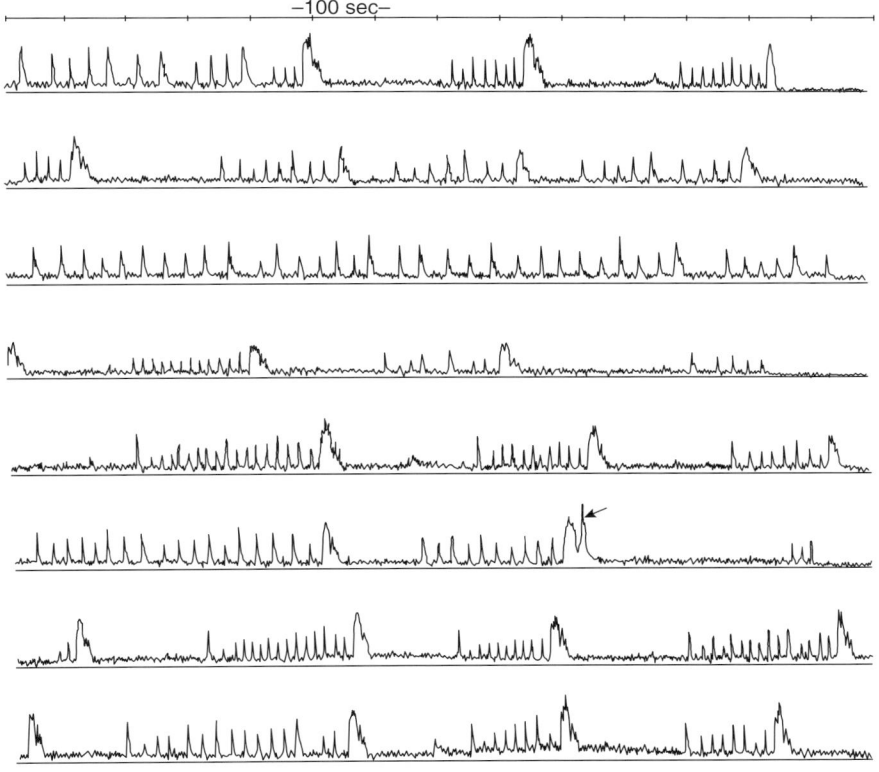

Figure 5 Train of type II bursts from the Rapid Burster as observed with the SAS 3 satellite. (From Lewin et al. 1976).

like a relaxation oscillator. This indicates that the type II bursts are the result of an accretion instability.

The sky distribution of X-ray bursters is strongly concentrated to the center of the Galaxy. Thus, X-ray bursters are located at average distances of ~8 kpc (Reid 1993), and the total energy and maximum luminosity in type I bursts are of the order of 10^{39} erg and 10^{38} erg s^{-1}, respectively. All known luminous LMXBs in globular clusters are X-ray bursters.

The ratio, α, of the total energy emitted in the persistent flux, to that emitted in bursts, is typically of the order of 10^2. This is nicely accounted for by the thermonuclear flash model of type I X-ray bursts: after a sufficient amount of matter has accreted on the neutron star surface, critical conditions may develop at the base of the accreted layer, causing unstable helium burning. The sudden release of nuclear energy gives rise to an X-ray burst. In this model α is the ratio of the gravitational binding energy to the available thermonuclear energy per gram of accreted material (i.e., of the order of $10^{2\pm0.5}$). Typical values for the rise time, decay time, and recurrence time, and for the maximum luminosity and integrated energy for type I X-ray bursts, are well reproduced by this model.

Swank et al. (1977), working with OSO 8, found that for a particular burst they observed the X-ray spectrum was best fit by a blackbody spectrum, with a temperature that decreased during the decay of the burst (Figure 6). The blackbody radius they found during burst decay (assuming a distance of 10 kpc) was ~15 km. This indicates that the

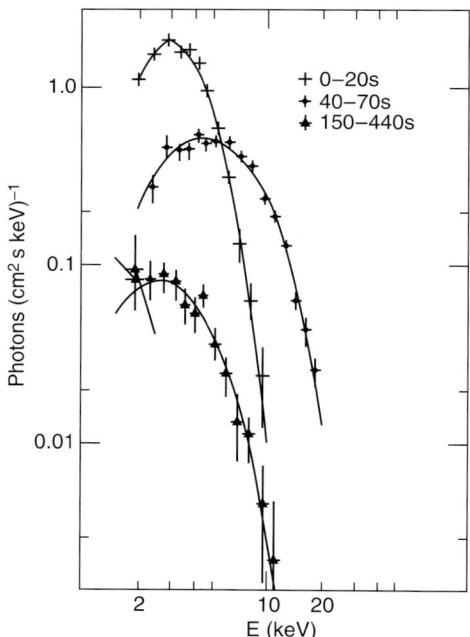

Figure 6 Blackbody spectra from a type I burst observed with OSO 8. (From Swank et al. 1977).

radiation observed during a type I X-ray burst originates directly from the surface of the neutron star. This result forms the basis for attempts to study the mass–radius relation of neutron stars from observations of the spectral evolution during X-ray bursts (see Lewin et al. 1993 for a review of these attempts).

Several properties of type I X-ray bursts show a global correlation with X-ray luminosity (i.e., mass accretion rate). Using EXOSAT, Van Paradijs et al. (1988) found that the burst duration τ and the above-mentioned ratio α are strongly correlated with the persistent X-ray luminosity L_X, as measured by the ratio γ of persistent X-ray flux to the peak flux of bursts with radius expansion (i.e., with peak luminosity equal to the Eddington limit, L_{Edd}). Above $L_X \simeq 0.3 L_{Edd}$ burst activity is extremely rare. The decrease of the burst duration with γ indicates the decreasing importance of hydrogen in the energetics of the thermonuclear flashes, as the persistent luminosity increases. The strong increase of α with γ they found, implies that independent of the accretion rate, after a given waiting time a source produces an X-ray burst with approximately the same energy. This result is not accounted for by present thermonuclear flash models. Spallation of CNO nuclei in the accretion flow has been suggested as a possible origin of the behavior of the α parameter (Bildsten et al. 1993).

Rapid, large-amplitude oscillations with frequencies between 300 and 600 Hz that drift by one or two hertz but are otherwise constant in a given source have been detected with RXTE in X-ray bursts of six different X-ray burst sources (Strohmayer et al. 1996; see Strohmayer 2001, Van der Klis 2000 for reviews). There is good evidence that these oscillations are caused by hot spots in atmospheric layers of the neutron star produced in the nuclear burning process and spinning in near corotation with the neutron star spin. The reason that the atmospheric layer is not in exact corotation is that in the X-ray burst it expands by several tens of meters and conserves its angular momentum; the observed frequency drifts of the oscillations are usually (but not always) consistent with a description where the layer initially spins slighty slower than the star, but then spins back up to the stellar spin frequency while it recontracts.

6 APERIODIC VARIABILITY AND SOURCE STATES

Pulsations and bursts are easily identifiable phenomena that allow for immediate interpretations and can be used to constrain neutron star properties such as mass, radius, and magnetic field. Irregular variability, which has a less immediate diagnostic value, has been observed on many timescales from early in the history of X-ray astronomy (Lewin et al. 1968). Such variability is expected to reflect at least in part the motions in the flow of accreting matter very close to the neutron star or the black hole, in regions of extreme space–time curvature, and is therefore of interest not only because, given the correct model, it could be used to also constrain compact object properties, but also because it provides a rare window on processes taking place in the strong-field gravity regime. The early observational information has been summarized by Bradt et al. (1982).

Our current view on fast, irregular variability of LMXBs developed from the discovery with EXOSAT of an intensity-dependent, quasi-periodic oscillation (QPO) in GX 5-1 (Van der Klis et al. 1985) and other LMXBs, and the correlation of this fast variability with the spectral properties of the X-ray source (see Lewin et al. (1988) for a review of the early developments). This has led to the recognition of two types of LMXB, with distinct correlated temporal and spectral characteristics; after the shape of the tracks that they follow in X-ray color–color diagrams (CDs) these are called the Z sources and the atoll sources (Hasinger and Van der Klis 1989; Figure 7).

Six Z sources are known. The Z-shaped track in the CDs consist of the "horizontal," "normal," and "flaring" branches (HB, NB, and FB, respectively). All the observed changes in the sources correspond to continuous movements along the track, jumps from one part to the other do not occur. On each of the branches the rapid variability, as expressed in the power spectrum, has distinct characteristics. On the HB the source intensity undergoes a quasi-periodic oscillation which in the power spectrum shows up as a peak of finite width (Figure 8), whose centroid frequency changes from ~15 Hz to ~60 Hz as the intensity increases and the source moves from the left-hand side of the HB to the top of the NB. This HB oscillation was generally interpreted as a modulation of a "clumpy" accretion flow at a magnetospheric "gate," at a frequency equal to the spin frequency of the neutron star as seen by a "clump" orbiting it with the Kepler frequency ("magnetospheric beat frequency model"; Alpar and Shaham 1985, Lamb et al. 1985). This HB oscillation (HBO) becomes much weaker when the source

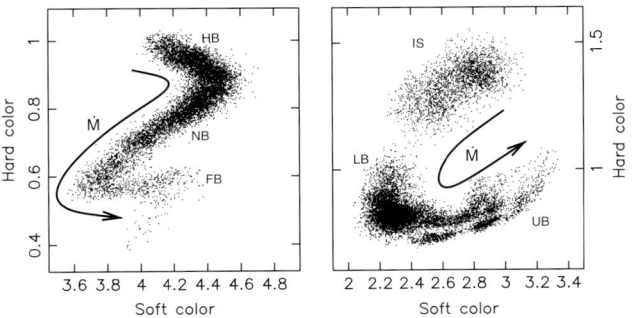

Figure 7 X-ray color–color diagrams of a representative Z (left) and atoll source (right). (From Homan 2001).

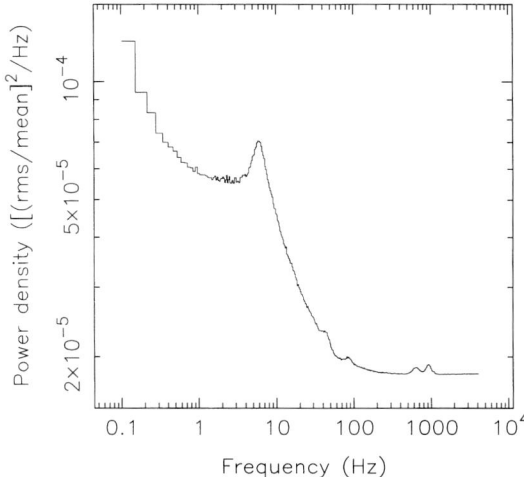

Figure 8 Power spectrum of the X-ray flux variations in the luminous LMXB Sco X-1. The NBO (near 6 Hz), HBO (near 45 Hz with a harmonic near 90 Hz), and the twin kilohertz QPOs (near 700 and 1000 Hz) are readily distinguishable superimposed on source-intrinsic broad noise components (below $\sim 10^2$ Hz) and the photon counting noise. (Data from RXTE).

reaches the upper part of the NB. The corresponding neutron star spin periods are in the millisecond range.

Near the middle of the NB the power spectrum shows a different type of QPO, with a near-constant frequency of about 6 Hz (Figure 8; the same in all Z sources). This NB oscillation is most likely the result of oscillation of the optical depth (Stollman et al. 1987) of the accretion flow, caused by radiation feedback on the flow at a luminosity near the Eddington limit (Fortner et al. 1989, Alpar et al. 1992). A strong argument for this model was provided by the observation in Cyg X-2 of an $\sim 180°$ phase shift between the oscillation above and below 5 keV (Mitsuda and Dotani 1989); this pivoting of the spectrum during the 6 Hz oscillation reflects the variation in the Compton upscattering of photons, as the scattering optical depth varies. As the Z source moves from the NB to the FB the frequency of the 6 Hz oscillation increases, and the width of the QPO peak in the power spectrum becomes much larger.

The optical and UV brightness of Z sources increases as the sources move from the HB via the NB to the FB (Vrtilek et al. 1991, Augusteijn et al. 1992). This indicates that the mass-accretion rate increases along the Z track in that order. Since along the NB the count rate decreases going from the HB to the FB this shows that count rate can be anti-correlated with mass-accretion rate.

The CDs of atoll sources, many of which are burst sources, show a "banana branch," along which the source moves on a timescale of hours, and "islands" where the source stays for much longer time intervals. In the banana state the power spectrum contains a power-law component, called the "very-low-frequency noise." In the island state the power spectrum is dominated by a broad-band noise component, called "high-frequency noise," which has a power-law shape at high frequencies, but which flattens below a cut-off frequency ν_{co}. Between the island state and the banana state the mass-accretion rate increases (Hasinger and Van der Klis 1989). Recently, a QPO similar to HBO has been detected with RXTE in a number of atoll sources; for a comparison of some of the properties of this QPO with the HBO seen in Z sources see Wijnands and van der Klis (1999).

Atoll sources are much less luminous than Z sources (Ford et al. 2000); however, it is unclear if the differences in their spectral and temporal properties can be only the result of a difference in accretion rate. Hasinger and van der Klis (1989) suggested that the neutron star magnetic fields in the atoll sources are systematically weaker than those in the Z sources.

An important recent development is the detection with RXTE of QPOs with frequencies in the range 500–1200 Hz in the power spectrum of Z and atoll sources, the kilohertz QPOs (Figure 8; Van der Klis et al. 1996, Strohmayer et al. 1996; see van der Klis 2000 for a review). These QPOs are sometimes single, but usually appear in pairs separated by several hundred hertz. Both QPOs increase in frequency when the accretion rate increases. In several cases, the frequency separation between the two peaks is close to the frequency of burst oscillations (Section 5) seen in the same source, and this has led to an interpretation in terms of a beat-frequency model, where the higher frequency kilohertz QPO occurs at the frequency of orbital motion at the inner edge of the accretion disk (at a radius of 15–30 km), whereas the lower frequency one is at the beat frequency between this frequency and the spin frequency (Miller et al. 1998). However, in several sources the two peaks move closer together while they both increase in frequency, which complicates such an interpretation. Stella and Vietri (1999) proposed that the lower frequency kilohertz QPO instead arises because the orbital motion (at the frequency of the other kilohertz QPO) is slightly eccentric and shows relativistic periastron precession at this rate. They further suggested (Stella and Vietri 1998) that the HBO is due to frame-dragging-induced nodal precession (Lense–Thirring precession) of these same orbits.

If the higher frequency kilohertz QPO indeed provides us with the frequency of orbital motion around the neutron star then important conclusions can be drawn from the value of this frequency and its variations. The orbit must be outside the star, which leads to an upper limit on the star's radius, and assuming general relativity, the orbit must also be outside the general relativistic marginally stable orbit, which provides an upper limit on the stellar mass (Miller et al. 1998). Because the observed frequencies are so high (up to more than 1300 Hz), these constraints are tight and they begin to argue against the hardest proposed equations of state for the supranuclear density matter of which neutron stars are composed. As the QPO frequency increases with increasing mass-accretion rate, the orbital radius moves in and eventually is expected to reach the stellar surface or the marginally stable orbit, whichever is larger. If the latter, then the QPO might well persist, but the frequency would not increase further while the mass accretion rate would. This effect may already have been detected in 4U 1820–30 (Zhang et al. 1998).

Components similar to several aperiodic variability components (the "very-low-frequency noise," "high-frequency noise," and the HBO) in the power spectra of the Z sources and atoll sources have also been found in the power spectra of X-ray pulsars and black hole candidates. An example is the strong similarity between the strong broad-band noise in the power spectrum of island-state atoll sources and those of accreting black holes in the low (or "hard") state. A proposal to unify the phenomenology of the power spectra of these different types of accreting compact stars based mostly on broad-band noise properties was made by Van der Klis (1994a,b; see Van der Klis 1995 for a review). Correlations between the frequencies of aperiodic variability components in Z and atoll sources and black holes have been observed (Wijnands and van der Klis 1999, Psaltis et al. 1999) which suggest that further unification including various QPOs as well may be possible.

7 X-RAY SPECTRA

X-ray spectra played an important role in the development of the view that there are different groups of compact X-ray sources; for example, Tananbaum (1973) distinguished between "X-ray binaries" and "Sco X-1-type sources." The former, which comprised systems like Cen X-3 with hard 1–10 keV X-ray spectra, are now known to be (mainly) HMXBs with strongly magnetized neutron stars: the X-ray binary pulsars. The Sco X-1-type sources had relatively soft X-ray spectra; they were suspected to belong to an old low-mass galactic population (Salpeter 1973; see introductory text).

Ostriker (1977) found that X-ray color–color diagrams provide an efficient way to separate different groups of X-ray sources: he suggested that, in addition to the above two types of X-ray binaries, accreting black holes could be distinguished this way. The value of these diagrams was shown by White and Marshall (1984) who used them as an efficient tool to distinguish accreting neutron stars with strong and weak magnetic fields (by and large pulsating HMXBs and non-pulsating LMXBs, respectively), and accreting black hole candidates. The spectra of the latter (both in HMXBs and LMXBs) occasionally showed a strong low-energy excess whose strength appears to be correlated with the mass-accretion rate.

The spectra of luminous weak-field neutron stars (e.g., Sco X-1, X-ray burst sources) can be approximately described by a thermal bremsstrahlung model, with $kT_{TB} \sim 5$ keV. The spectra of strong-field neutron stars (i.e., X-ray pulsars) are power laws (photon indices ~ 1) with a high-energy cut-off at several tens of keV. Above ~ 10 keV the spectra of black hole candidates often show a hard power-law component which in some cases can be detected up to hundreds of keV.

The difference in spectral hardness between HMXBs and LMXBs persists into the hard X-ray range, up to $\sim 10^2$ keV; it is remarkable that the few LMXBs with slow pulsations (GX 1+4, Her X-1, 1627–673) have X-ray spectra that are as hard as those of HMXBs (almost all of which are, likewise, pulsars), whereas the only millisecond pulsar among them (Section 4), SAX J1808.4-3658, has a relatively soft spectrum. For a recent review of the hard X-ray spectra of X-ray binaries see Gilfanov *et al.* (1995).

The above simple picture of the spectra of X-ray binaries was enriched by the work of the Granat/Sigma group (Barret *et al.* 1991, 1992a,b; Barret and Vedrenne 1994), who found that the spectra of several low-luminosity X-ray burst sources (i.e., atoll sources) have a power-law high-energy tail extending to the 10^2 keV range. These tails may be the high-energy extensions of the 1–20 keV power-law spectra found earlier for some low-luminosity bursters (Becker *et al.* 1977, Barret and Vedrenne 1994). Van Paradijs and Van der Klis (1994) generalized this result by showing that there is a general anti-correlation between the hardness of the 10–100 keV spectra of LMXBs and their luminosity.

8 MAGNETIC FIELD DECAY?

Many HMXBs are X-ray pulsars, but X-ray pulsations occur only rarely in LMXBs. This suggests that the magnetic fields of the neutron stars in LMXBs are generally much weaker than in HMXBs. An alternative interpretation is that the magnetic and rotational axes of the neutron stars in LMXBs are aligned. However, the observation that the X-ray spectra of LMXBs are much softer than those of HMXBs, except for the few LMXBs which do show pulsations (their X-ray spectra are as hard as those of HMXBs), strongly suggests that the division into hard and soft X-ray spectra is related to a difference in the geometry of the accretion flow. For neutron star magnetic fields of the order of 10^{12} G, and sub-Eddington accretion rates, the accretion flow is dominated by the magnetic field within a distance of $\sim 10^3$ km from the neutron star; a large fraction of the inflowing matter reaches the neutron star via an accretion column on a relatively small area (near the polar caps). For magnetic fields below 10^9 G one expects that the accreting material is distributed over a larger fraction of the neutron star surface.

It has long been true that not a single source showed both pulsations and type I bursts. (The bursts from the Bursting Pulsar GRO J1744-28 are type II bursts.) This idea of mutual exclusion of pulsations and type 1 bursts is now obsolete, as the millisecond pulsar (Section 4) SAX J1808.4-3658 is also a burst source. However, this pulsar has a *weak* (10^9–10^{10} G) magnetic field. Apparently, the presence of a strong (10^{12} G) magnetic field suppresses the instability of the nuclear reactions that gives rise to bursts (as expected from the very high rate of accretion locally in the accretion column; see e.g. Joss and Li 1980). For a recent discusion of the effect of magnetic fields on thermonuclear flashes, see Bildsten (1995). This mutual exclusion of bursts and strong magnetic fields supports the idea that it is a weaker magnetic field, and not only alignment of the field axis, which distinguishes the neutron stars in LMXBs from those in HMXBs.

There are two possible ways to understand this difference. In the first place, the magnetic fields of the (generally old) neutron stars in LMXBs may be much weaker than those of the (young) neutron stars in HMXBs, because they have always been very weak. This difference might be related to a difference in the formation mechanism of neutron stars in HMXBs and LMXBs, that is, via the normal evolution of a massive star and via the accretion-induced collapse of a white dwarf, respectively.

An alternative possibility is suggested by the observation that the neutron stars in HMXBs are all young objects, whereas those in LMXBs are typically much older: the magnetic fields of neutron stars decay. Until recently it was generally believed that all young neutron stars have strong dipolar magnetic fields ($B \sim 10^{12}$ G), which decay spontaneously on a timescale of the order of 10^7 years, to a bottom value of the order of 10^9 G. (The possibility that a substantial fraction of young neutron stars have magnetic fields much stronger than the canonical 10^{12} G has been raised with the recent discovery of "magnetars" (Thompson and Duncan 1995, 1996; Kouveliotou *et al.* 1998).) However, on the basis of a detailed analysis of the properties of the young radio pulsar population Bhattacharya *et al.* (1992) and Hartman *et al.* (1997) have shown that such spontaneous decay does not occur (with a lower limit to the exponential timescale for such decay of about 10^8 years).

It has been suggested that decay of the magnetic field of a neutron star may be caused by the accretion process, either directly (e.g. by field burial (Romani 1990)) or indirectly. In spite of much work on the relation between the neutron star magnetic field and accretion (Srinivasan *et al.* 1990, Ding *et al.* 1993, Ruderman 1991a,b,c) it is fair to say that this relation remains an unsolved issue; its clarification is central to an understanding of the evolution of neutron stars in binaries.

REFERENCES

Alpar, M.A. and Shaham, J. (1985). Is GX5-1 a millisecond pulsar? *Nature*, **316**, 239–241.

Alpar, M.A., Hasinger, G., Shaham, J. and Yancopoulos, S. (1992). 6 HZ quasiperiodic oscillations from low-mass X-ray binaries – The sound of an accretion disk? *Astronomy and Astrophysics*, **257**, 627–631.

Alpar, M.A., Kiziloglu, Ü. and Van Paradijs, J. (eds) (1995). *The Lives of the Neutron Stars*, NATO ASI Series, Vol. C450 Kluwer, Dordrecht.

Augusteijn, T., Karatasos, K., Papadakis, M., Paterakis, G., Kikuchi, S., Brosch, N., Leibowitz, E., Hertz, P., Mitsuda, K., Dotani, T., Lewin, W.H.G., van del Klis, M. and van Paradijs, J. (1992). Coordinated X-ray and optical observations of Scorpius X-1. *Astronomy and Astrophysics*, **265**, 177–182.

Barret, D., Bouchet, L., Mandrou, P., Roques, J.P., Cordier, B., Laurent, Ph., Lebrun, F., Paul, J., Sunyaev, R., Churazov, E., Gilfanov, M., Diachkov, A., Khavenson, N., Novikov, B., Chulkov, I. and Kuznetsov, A. (1992b). Sigma detection of hard X-ray emission from the soft transient type I X-ray burster KS 1731–260. *Astrophysical Journal*, **394**, 615–618.

Barret, D., Mereghetti, S., Roques, J.P., Mandrou, P., Salotti, L., Lebrun, F., Laurent, Ph., Ballet, J., Churazov, E., Gifanov, M., Sunyaev, R., Khavenson, N., Chulkov, I., Novikov, B., Kuznetzov, A. and Dyachkov, A (1991). Discovery with the Sigma telescope of hard X-rays from the globular cluster Terzan 2. *Astrophysical Journal*, **379**, L21–L24.

Barret, D., Roques, J.P., Mandrou, P., Bouchet, L., Lebrun, F., Goldwurm, A., Laurent, P., Paul, J., Sunyaev, R., Churazov, E., Gilfanov, M., Diachkov, A., Kavenson, N., Novikov, B., Chulkov, I. and Kuznetsov, A. (1992a). SIGMA observation of hard X-ray emission from the ultrasoft X-ray transient Triangulum Australe X-1 (A1524-62). *Astrophysical Journal*, **392**, L19–L22.

Barret, D. and Vedrenne, G. (1994). Hard X-ray emission from weakly magnetized neutron stars. *Astrophysical Journal Supplement*, **92**, 505–510.

Becker, R.H., Smith, B.W., Swank, J.H., Boldt, E.A., Holt, S.S., Serlemitsos, P.J. and Pravdo, S.H. (1977). Spectral characteristics of 3U 1915-05, a burst source candidate. *Astrophysical Journal*, **216**, L101–L104.

Bhattacharya, D., Wijers, R.A.M.J., Hartman, J.W. and Verbunt, F. (1992). On the decay of the magnetic fields of single radio pulsars. *Astronomy and Astrophysics*, **254**, 198–212.

Bildsten, L. (1995). Propagation of nuclear burning fronts on accreting neutron stars: X-ray bursts and sub-hertz noise. *Astrophysical Journal*, **438**, 852–875.

Bildsten, L., Salpeter, E.E. and Wasserman, I. (1993). Helium destruction and gamma-ray line emission in accreting neutron stars. *Astrophysical Journal*, **408**, 615–636.

Bildsten, L., Chakrabarty, D., Chiu, J., Finger, M.H., Koh, D.T., Nelson, R.W., Prince, T.A., Rubin, B.C., Scott, D.M., Stollberg, M., Vaughan, B.A., Wilson, C.A. and Wilson, R.B. (1997). Observations of accreting pulsars. *Astrophysical Journal Supplement*, **113**, 367.

Bildsten, L. (1998). Thermonuclear burning on rapidly accreting neutron stars. In R. Buccheri, J. van Paradijs, and M.A. Alpar (eds), *The Many Faces of Neutron Stars*, Kluwer, Dordrecht, pp. 419–449.

Bolton, C.T. (1972). Identifications of CYG X-1 with HDE 226868. *Nature*, **235**, 271.

Bowyer, S., Byram, E.T., Chubb, T.A. and Friedman, H. (1964). Lunar occultation of X-ray emission from the Crab Nebula. *Science*, **146**, 912–916.

Bradt, H.V., Kelley, R.L. and Petro, L.D. (1982). Rapid aperiodic X-ray variability. In P.W. Sanford, P. Laskarides and J. Salton (eds), *Galactic X-ray Sources*, Wiley, Chichester, UK, pp. 89–112.

Bradt, H.V. and McClintock, J.E. (1983). The optical counterparts of compact galactic X-ray sources. *Annual Review of Astronomy and Astrophysics*, **21**, 13–66.

Bradt, H.V.D., Ohashi, T. and Pounds, K.A. (1992). X-ray astronomy missions. *Annual Review of Astronomy and Astrophysics*, **30**, 391–427.

Brandt, N. and Podsiadlowski, Ph. (1995). The effects of high-velocity supernova kicks on the orbital properties and sky distributions of neutron-star binaries. *Monthly Notices of the Royal Astronomical Society*, **274**, 461–484.

Buccheri, R., Alpar, M.A. and Van Paradijs, J. (eds) (1998). *The Many Faces of Neutron Stars*, NATO ASI Series C, Vol. C515, Kluwer, Dordrecht.

Burbidge, G. (1972). *Comments Astrophysics and Space Physics*, **4**, 105

Chakrabarty, D. and Morgan, E.H. (1998). The two-hour orbit of a binary millisecond X-ray pulsar. *Nature*, **394**, 346–348.

Chiu, H.Y. (1964). Supernovae, neutrinos, and neutron stars. *Annals of Physics*, **26**, 364–410.

Chiu, H.Y. and Salpeter, E.E. (1964). Surface X-ray emission from neutron stars. *Physical Review Letters*, **12**, 413–415.

Clark, G.W. (1965). Balloon observation of the X-ray spectrum of the Crab Nebula above 15 keV. *Physical Review Letters*, **14**, 91–94.

Cowley, A.P., Hutchings, J.B. and Crampton, D. (1988). Spectroscopy and kinematics of the low-mass X-ray binaries. *Astrophysical Journal*, **333**, 906–916.

Crawford, J.A. and Kraft, R.P. (1956). An interpretation of AE Aquarii. *Astrophysical Journal*, **123**, 44

De Jong, J.A., Augusteijn, T. and Van Paradijs, J. (1996). Reprocessing of X rays in low-mass X-ray binaries. *Astronomy and Astrophysics*, **314**, 484–490.

Ding, K.Y., Cheng, K.S. and Chan, H.F. (1993). Magnetic field decay from the core of neutron stars – Effects of interpinning of 3P2 neutron superfluid and 1S0 proton superconducting fluid. *Astrophysical Journal*, **408**, 167–178.

Dolan, J.F. (1970). A catalogue of discrete celestial X-ray sources. *Astronomical Journal*, **75**, 223.

Finzi, A. (1964). Neutron stars as a possible source of X-rays from outside the Solar System. *Astrophysical Journal*, **139**, 1398.

Ford, E.C., van der Klis, M., Méndez, M., Wijnands, R., Homan, J., Jonker, P.G. and van Paradijs, J. (2000). Simultaneous measurements of X-ray luminosity and kilohertz quasi-periodic oscillations in low-mass X-ray binaries. *Astrophysical Journal*, **537**, 368–373.

Fortner, B., Lamb, F.K. and Miller, G.S. (1989). Origin of 'normal-branch' quasiperiodic oscillations in low-mass X-ray binary systems. *Nature*, **342**, 775–777.

Frank, J., King, A.R. and Raine, D.J. (1992). *Accretion Power in Astrophysics*, 2nd edition, Cambridge University Press.

Giacconi, R., Gorenstein, P., Gursky, H., Usher, P.D., Waters, J.R., Sandage, A., Osmer, P. and Peach, J.V. (1967). On the optical search for the X-ray sources CYG X-1 and CYG X-2. *Astrophysical Journal*, **148**, L129.

Giacconi, R., Gursky, H., Kellogg, E., Schreier, E. and Tananbaum, H. (1971). Discovery of periodic X-ray pulsations in Centaurus X-3 from UHURU. *Astrophysical Journal*, **167**, L67.

Giacconi, R., Gursky, H., Paolini, F. and Rossi, B. (1962). Evidence for X-rays from sources outside the Solar System. *Physical Review Letters*, **9**, 439–443.

Giacconi, R., Gursky, H. and Waters, J.R. (1965). Spectral data from the cosmic X-ray sources in Scorpius and the near the galactic centre. *Nature*, **207**, 572–575.

Gilfanov, M., Churazov, E., Sunyaev, R., Vikhlinin, A., Finogenov, A., Sitdikov, A., Dyachkov, A., Khavenson, N., Laurent, P., Ballet, J., Claret, A., Goldwurm, A., Roques, J.P., Mandrou, P., Niel, M. and Vedrenne, G. (1995). Hard X-ray observations of black-hole candidates. In Alpar, M.A., Kiziloglu, Ü. and Van Paradijs, J. (eds) (1995). *The Lives of the Neutron Stars*, NATO ASI Series, Vol. C450 Kluwer, Dordrecht, pp. 331–354.

Ginzburg, V.L. (1990). Notes of an amateur astrophysicist. *Annual Review of Astronomy and Astrophysics*, **28**, 1.

Grindlay, J. and Gursky, H. (1976). Scattering model for X-ray bursts – Massive black holes in globular clusters. *Astrophysical Journal*, **205**, L131–L133.

Hansen, B.M.S. (1996). The Ages, Speeds and Offsprings of Pulsars. Ph.D. Thesis, California Institute of Technology.

Hansen, M.M.S. and Phinney, S.R. (1997). *Monthly Notices of the Royal Astronomical Society*, **291**, 569–577.

Hartman, J.W., Bhattacharya, D., Wijers, R.A.M.J. and Verbunt, F. (1997). A study of the evolution of radio pulsars through improved population synthesis. *Astronomy and Astrophysics*, **322**, 477–488.

Hasinger, G. and Van der Klis, M. (1989). Two patterns of correlated X-ray timing and spectral behaviour in low-mass X-ray binaries. *Astronomy and Astrophysics*, **225**, 79–96.

Hiltner, W.A. and Mook, D.E. (1970). Optical observations of extrasolar X-ray sources. *Annual Review of Astronomy and Astrophysics*, **8**, 139.

Hoffman, J.A., Marshall, H.L. and Lewin, W.H.G. (1978). Dual character of the rapid burster and a classification of X-ray bursts. *Nature*, **271**, 630–633.

Homan, J. (2001). X-ray timing studies of low-mass X-ray binaries. PhD thesis, University of Amsterdam.

Johnston, H. (1992). Compact Objects in the Disk and Globular Clusters. PhD Thesis, California Institute of Technology.

Joss, P.C. and Li, F.K. (1980). Helium-burning flashes on accreting neutron stars – Effects of stellar mass, radius, and magnetic field. *Astrophysical Journal*, **238**, 287–295.

Joss, P.C. and Rappaport, S.A. (1984). Neutron stars in interacting binary systems. *Annual Review of Astronomy and Astrophysics*, **22**, 537–592.

King, A.R. and Kolb, U. (1997). The formation of low-mass transient X-ray binaries. *Astrophysical Journal*, **481**, 918.

Kouveliotou, C., Dieters, S., Strohmayer, T., van Paradijs, J., Fishman, G.J., Meegan, C.A., Hurley, K., Kommers, J., Smith, I., Frail, D. and Murakami, T. (1998). An X-ray pulsar with a superstrong magnetic field in the soft gamma-ray repeater SGR 1806-20. *Nature*, **393**, 235–237.

Kraft, R.P. (1962). Binary stars among cataclysmic variables. I. U Geminorum Stars (dwarf novae). *Astrophysical Journal*, **135**, 408.

Kraft, R.P. (1964). Binary stars among cataclysmic variables. III. Ten old novae. *Astrophysical Journal*, **139**, 457.

Kraft, R.P. (1973). Binary systems as X-ray sources: A review. In H. Bradt and R. Giacconi (eds), *X- and Gamma-Ray Astronomy*, IAU Symposium 55, Reidel, Dordecht, pp. 36–50.

Lamb, F.K., Shibazaki, N., Alpar, M.A. and Shaham, J. (1985). Quasi-periodic oscillations in bright galactic-bulge X-ray sources. *Nature*, **317**, 681–687.

Lewin, W.H.G., Clark, G.W. and Smith, W.B. (1968). Observation of an X-ray flare from SCO X-1. *Astrophysical Journal*, **152**, L55.

Lewin, W.H.G., Doty, J., Clark, G.W., Rappaport, S.A., Bradt, H.V.D., Doxsey, R., Hearn, D.R., Hoffman, J.A., Jernigan, J.G., Li, F.K., Mayer, W., McClintock, J., Primini, F. and Richardson, J. (1976). The discovery of rapidly repetitive X-ray bursts from a new source in Scorpius. *Astrophysical Journal*, **207**, L95–L99.

Lewin, W.H.G. and Clark, G.W. (1980). Galactic bulge sources – What are they? X-ray sources in globular clusters and burst sources. *Annals of the New York Academy of Sciences*, **336**, 451–478.

Lewin, W.H.G., Van Paradijs, J. and Van der Klis, M. (1988). A review of quasi-periodic oscillations in low-mass X-ray binaries. *Space Science Reviews*, **46**, 273–378.

Lewin, W.H.G., Van Paradijs, J. and Taam, R.E. (1993). X-ray bursts. *Space Science Reviews*, **62**, 223.

Liu, Q.Z., Van Paradijs, J., Van den Heuvel, E.P.J. (2001). *Astronomy and Astrophysics*, **368**, 1021–1054.

Lyne, A.G., Lorimer, D.R. (1994). High birth velocities of radio pulsars. *Nature*, **369**, 127.

Miller, M.C., Lamb, F.K., Psaltis, D., (1998). Sonic-point model of kilohertz quasi-periodic brightness oscillations in low-mass X-ray binaries. *Astrophysical Journal*, **508**, 791–830.

Mitsuda, K. and Dotani, T. (1989). Energy-dependent time lags in QPO from Cygnus X-2. *Publications of the Astronomical Society of Japan*, **41**, 557–575.

Motch, C. and Pakull, M.W. (1989). The strength of N III-C III complex emission in low-mass X-ray binaries as a possible indicator of metallicity. *Astronomy and Astrophysics*, **214**, L1–L4.

Nagase, F. (1989). Accretion-powered X-ray pulsars. *Publications of the Astronomical Society of Japan*, **41**, 1–79.

Ögelman, H. and Van den Heuvel, E.P.J. (Eds), (1989). *Timing Neutron Stars*, NATO ASI Series, Vol. C262 Kluwer, Dordrecht.

Ostriker, J.P. (1977). The astronomy of X-ray sources and the phsyics of accretion. *Annals of the New York Academy of Sciences*, **302**, 229–243.

Psaltis, D., Belloni, T., Van der Klis, M., (1999). Correlations in quasi-periodic oscillation and noise frequencies among neutron star and black hole X-ray binaries. *Astrophysical Journal*, **520**, 262–270.

Ramachandran, R. and Bhattacharya, D. (1997). Kinematics of low-mass X-ray binaries and millisecond pulsars. *Monthly Notices of the Royal Astronomical Society*, **288**, 565–571.

Reid, M. (1993). The distance to the center of the Galaxy. *Annual Review of Astronomy and Astrophysics*, **31**, 345–372.

Ritter, H. and Kolb, U. (1998). Catalogue of cataclysmic binaries, low-mass X-ray binaries and related objects (Sixth edition). *Astronomy and Astrophysics Supplement*, **129**, 83–85.

Romani, R. (1990). A unified model of neutron-star magnetic fields. *Nature*, **347**, 741–743.

Ruderman, M.A. (1991a). Neutron star crustal plate tectonics. I – Magnetic dipole evolution in millisecond pulsars and low-mass X-ray binaries. *Astrophysical Journal*, **366**, 261–269.

Ruderman, M.A. (1991b). Neutron star crustal plate tectonics. II – Evolution of radio pulsar magnetic fields. *Astrophysical Journal*, **382**, 576–586.

Ruderman, M.A. (1991c). Neutron star crustal plate tectonics. III – Cracking, glitches, and gamma-ray bursts. *Astrophysical Journal*, **382**, 587–593.

Ruderman, M.A., Shaham, J. and Tavani, M. (1989). Accretion turnoff and rapid evaporation of very light secondaries in low-mass X-ray binaries. *Astrophysical Journal*, **336**, 507–518.

Ryter, C. (1970). The location of X-ray sources and the galactic background. *Astronomy and Astrophysics*, **9**, 288.

Salpeter, E.E. (1964). Accretion of interstellar matter by massive objects. *Astrophysical Journal*, **140**, 796–800.

Salpeter, E.E. (1973). Models for compact X-ray sources. In H. Bradt and R. Giacconi (eds), *X- and Gamma-Ray Astronomy*, IAU Symposium 55, Reidel, Dordrecht, pp. 135–142.

Sandage, A.R., Osmer, P., Giacconi, R., Gorenstein, P., Gursky, H., Waters, J., Bradt, H., Garmire, G., Sreekantan, B.V., Oda, M., Osawa, K. and Jugaku, J. (1966). On the optical identification of SCO X-1. *Astrophysical Journal*, **146**, 316.

Schreier, E., Levinson, R., Gursky, H., Kellog, E., Tananbaum, H., Giacconi, R. (1972). Evidence for the binary nature of Centaurus X-3 from UHURU X-ray observations. *Astrophysical Journal*, **172**, L79–L89.

Setti, G. and Woltjer, L. (1970). X-ray sources and the diffuse background. *Astrophysics and Space Science*, **9**, 185.

Shahbaz, T., Smale, A.P., Naylor, T., Charles, P.A., van Paradijs, J., Hassall, B.J.M. and Callanan, P. (1996). An atlas of optical continuum and line emission from low-mass X-ray binaries. *Monthly Notices of the Royal Astronomical Society*, **282**, 1437–1453.

Shapiro, S.L. and Teukolsky, S.A. (1983). *Black Holes, White Dwarfs and Neutron Stars*, Wiley, Chichester, UK.

Shklovsky, I.S. (1967). On the nature of the source of X-ray emission of SCO XR-1. *Astrophysical Journal*, **148**, L1.

Spruit, H.C. and Ritter, H. (1983). Stellar activity and the period gap in cataclysmic variables. *Astronomy and Astrophysics*, **124**, 267–272.

Srinivasan, G., Bhattacharya, D., Muslimov, A.G. and Tsygan, A.I. (1990). A novel mechanism for the decay of neutron star magnetic fields. *Current Science*, **59**, 31–38.

Stella, L. and Vietri, M. (1998). Lense–Thirring precession and quasi-periodic oscillations in low-mass X-ray binaries. *Astrophysical Journal*, **492**, L59.

Stella, L. and Vietri, M. (1999). kHz Quasiperiodic oscillations in low-mass X-ray binaries as probes of general relativity in the strong-field regime. *Physical Review Letters*, **82**, 17–20.

Stollman, G.M., van Paradijs, J., Hasinger, G., Lewin, W.H.G. and van der Klis, M. (1987). Time lag between hard and soft X-ray photons in QPO. *Monthly Notices of the Royal Astronomical Society*, **227**, 7P–12P.

Strohmayer, T.E. (2001). Oscillations during thermonuclear X-ray bursts: A new probe of neutron stars. AIP Conference Proceedings, **599**, pp. 377–386.

Strohmayer, T.E., Zhang, W., Swank, J.H., Smale, A., Titarchuk, L., Day, C. and Lee, U. (1996). Millisecond X-ray variability from an accreting neutron star system. *Astrophysical Journal*, **469**, L9.

Swank, J.H., Becker, R.H., Boldt, E.A., Holt, S.S., Pravdo, S.H. and Serlemitsos, P.J. (1977). Spectral evolution of a long X-ray burst. *Astrophysical Journal*, **212**, L73–L76.

Tanaka, Y. and Lewin, W.H.G. (1995). Black-hole binaries. In Lewin, W.H.G., Van Paradijs, J. and Van den Heuvel, E.P.J. (eds), *X-ray Binaries*, Cambridge University Press, pp. 126–174.

Tananbaum, H. (1973). UHURU Results on galactic X-ray sources. In H. Bradt and R. Giacconi (eds), *X- and Gamma-Ray Astronomy*, IAU Symposium 55, Reidel, Dordrecht, pp. 9–28.

Thompson, C. and Duncan, R.C. (1995). The soft gamma repeaters as very strongly magnetized neutron stars – I. Radiative mechanism for outbursts. *Monthly Notices of the Royal Astronomical Society*, **275**, 255–300.

Thompson, C. and Duncan, R.C. (1996). The soft gamma repeaters as very strongly magnetized neutron stars. II. Quiescent neutrino, X-ray, and Alfven wave emission. *Astrophysical Journal*, **473**, 322.

Tuohy, I.R., Buckley, D.A.H., Remillard, R.A., Bradt, H.V. and Schwartz, D.A. (1988). Twelve X-ray emitting Be candidates from HEAO-1. In Y. Tanaka (ed), *Physics of Neutron Stars and Black Holes*, Universal Academic Press, Tokyo, pp. 93–96.

Van den Heuvel, E.P.J. and Heise, J. (1972). Centaurus X-3, possible reactivation of an old neutron star by mass exchange in a close binary. *Nature – Physical Science*, **239**, 67.

Van den Heuvel, E.P.J. and Van Paradijs, J. (1988). Fate of the companion stars of ultra-rapid pulsars. *Nature*, **334**, 227–228.

Van den Heuvel, E.P.J. and Rappaport, S.A. (eds) (1992). *X-ray Binaries and Recycled Pulsars*, NATO ASI Series, Vol. C377, Kluwer, Dordrecht.

Van der Klis, M. (1994b). A comparison of the power spectra of Z and atoll sources, pulsars and black hole candidates. *Astronomy and Astrophysics*, **283**, 469–474.

Van der Klis, M. (1994a). Similarities in neutron star and black hole accretion. *Astrophysical Journal Supplement*, **92**, 511–519.

Van der Klis, M. (1995). Rapid aperiodic variability in X-ray binaries. In Lewin, W.H.G., Van Paradijs, J. and Van den Heuvel, E.P.J. (eds), *X-ray Binaries*, Cambridge University Press, pp. 252–307.

Van der Klis, M. (2000). Millisecond oscillations in X-ray binaries. *Annual Review of Astronomy and Astrophysics*, **38**, 717–760.

Van der Klis, M., Jansen, F., van Paradijs, J., Lewin, W.H.G., van den Heuvel, E.P.J., Trumper, J.E. and Szatjno, M. (1985). Intensity-dependent quasi-periodic oscillations in the X-ray flux of GX5 – 1. *Nature*, **316**, 225–230.

Van der Klis, M., Swank, J.H., Zhang, W., Jahoda, K., Morgan, E.H., Lewin, W.H.G., Vaughan, B. and van Paradijs, J. (1996). Discovery of submillisecond quasi-periodic oscillations in the X-ray flux of Scorpius X-1. *Astrophysical Journal*, **469**, L1.

Van Paradijs, J. (1991). Optical light curves of X-ray binaries. In J. Ventura and D. Pines (eds), *Neutron Stars, Theory and Observations*, NATO ASI Series, Vol. C344 Kluwer, Dordrecht, pp. 289–317.

Van Paradijs, J. (1995). A catalogue of X-ray binaries. In W.H.G. Lewin, J. van Paradijs, and E.P.J. van den Heuvel (eds), *X-ray Binaries*, Cambridge University Press, pp. 536–577.

Van Paradijs, J. and McClintock, J.E. (1994). Absolute visual magnitudes of low-mass X-ray binaries. *Astronomy and Astrophysics*, **290**, 133–136.

Van Paradijs, J. and McClintock, J.E. (1995). Optical and ultraviolet observations of X-ray binaries. In W.H.G. Lewin, J. van Paradijs, and E.P.J. van den Heuvel (eds), *X-ray Binaries*, Cambridge University Press, pp. 58–125.

Van Paradijs, J., Penninx, W. and Lewin, W.H.G. (1988). On the relation between X-ray burst properties and the persistent X-ray luminosity. *Monthly Notices of the Royal Astronomical Society*, **233**, 437–450.

Van Paradijs, J. and Van der Klis, M. (1994). Luminosity dependence of the hardness of the 13–80–keV X-ray spectra of low-mass X-ray binaries. *Astronomy and Astrophysics*, **281**, L17–L20.

Van Paradijs, J. and Verbunt, F. (1984). A comparison of soft X-ray transients and dwarf novae. In S.E. Woosley (ed), *High Energy Transients in Astrophysics*, AIP Proceedings Vol. 115, American Institute of Physics, New York, p. 49.

Van Paradijs, J. and White, N.E. (1995). The galactic distribution of low-mass X-ray binaries. *Astrophysical Journal*, **447**, L33.

Vaughan, B.A., van der Klis, M., Wood, K.S., Norris, J.P., Hertz, P., Michelson, P.F., van Paradijs, J., Lewin, W.H.G., Mitsuda, K. and Penninx, W. (1994). Searches for millisecond pulsations in low-mass X-ray binaries, 2. *Astrophysical Journal*, **435**, 362–371.

Ventura, J. and Pines, D. (eds), (1991). *Neutron Stars, Theory and Observations*, NATO ASI Series, Vol. C344 Kluwer, Dordrecht.

Verbunt, F. (1984). Mass transfer and the period gap of cataclysmic variables. *Monthly Notices of the Royal Astronomical Society*, **209**, 227–240.

Vrtilek, S., Penninx, W., Raymond, J.C., Verbunt, F., Hertz, P., Wood, K., Lewin, W.H.G. and Mitsuda, K. (1991). Observations of Scorpius X-1 with IUE – Ultraviolet results from a multiwavelength campaign. *Astrophysical Journal*, **376**, 278–288.

Vrtilek, S., Raymond, J.C., Garcia, M.R., Verbunt, F., Hasinger, G. and Kurster, M. (1990). Observations of Cygnus X-2 with IUE – Ultraviolet results from a multiwavelength campaign. *Astronomy and Astrophysics*, **235**, 162–173.

Warner, B, (1987). Absolute magnitudes of cataclysmic variables. *Monthly Notices of the Royal Astronomical Society*, **227**, 23–73.

Warner, B. (1995). *Cataclysmic Variables*, Cambridge University Press.

Webster, B.L. and Murdin, P. (1972). Cygnus X-1 – A spectroscopic binary with a heavy companion? *Nature*, **235**, 37–38.

White, N.E. and Marshall, F.E. (1984). The unusually soft X-ray spectrum of LMC X-3. *Astrophysical Journal*, **281**, 354–359.

White, N.E. and Van Paradijs, J. (1996). The galactic distribution of black hole candidates in low mass X-ray binary systems. *Astrophysical Journal*, **473**, L25.

White, N.E., Nagase, N., Parmar, A.N. (1995). The properties of X-ray binaries. In Lewin, W.H.G., Van Paradijs, J. and Van den Heuvel, E.P.J. (eds), *X-ray Binaries*, Cambridge University Press, pp. 1–57.

Wijnands, R. and Van der Klis, M. (1998). A millisecond pulsar in an X-ray binary system. *Nature*, **394**, 344–357.

Wijnands, R. and Van der Klis, M., (1999). The broadband power spectra of X-ray binaries. *Astrophysical Journal*, **514**, 939–944.

Zel'dovich, Ya.B. (1964). The radiation of gravity waves by bodies moving in the field of a collapsing star. *Soviet Physics Doklady*, **9**, 246–249.

Zel'dovich, Ya.B. and Guseinov, O.H. (1996). Collapsed stars in binaries. *Astrophysical Journal*, **144**, 840.

Zhang, W., Smale, A.P., Strohmayer, T.E. and Swank, J.H. (1998). Correlation between energy spectral states and fast time variability and further evidence for the marginally stable orbit in 4U 1820–30. *Astrophysical Journal*, **500**, L171.

35

NICHOLAS E. WHITE*

High-mass X-ray binaries

High-mass X-ray binary (HMXB) systems were first identified in the early 1970s amongst the handful of mysterious bright X-ray sources found in the 1960s from brief sounding rocket and balloon flights. The launch of NASA's Uhuru (SAS 1) satellite, which made the first all-sky X-ray survey, established the class. One of the first results from this mission was the discovery by Giacconi et al. (1971) of 4.8 s X-ray pulsation from Cen X-3 (Figure 1). Precise pulse timing and the observation of an eclipse of the pulsar showed that Cen X-3 is in a 2.1-day orbit around a massive early-type star (Schreier et al. 1972). Within the Uhuru error box of Cen X-3 Krzeminski (1974) identified the optical counterpart as a 13th magnitude highly reddened early OB-type star. It is ironic to note that if the radio astronomers had not discovered radio pulsars in 1967, then X-ray astronomy would surely have found them a few years later. Around the time of the launch of Uhuru, Bolton (1972) and Webster and Murdin (1972) identified the optical counterpart to the then anonymous X-ray source Cyg X-1 (discovered by Bowyer et al. 1965) with an OB supergiant HD226868. For a brief time it was thought that Cyg X-1 was also a pulsar (Oda et al. 1971), but it turned out the pulses were not coherent but rather a manifestation of "shot noise" (Terrell 1972), caused by chaotic energy release in the accretion disk around the black hole. The mass function of the HD226868/Cyg X-1 system suggested that the X-ray source was too heavy to be a neutron star, and might be a black hole with a mass of 10 or more solar masses. This caused headline news at the time, as it provided the first direct evidence that these enigmatic objects might exist in nature. Over the course of time the black hole candidacy of Cyg X-1 has strengthened and it is now widely accepted. While this was one of the first HMXBs to be found, it is a relatively unique system in containing a black hole.

Following these initial discoveries several more HMXBs similar to Cen X-3 were found, primarily from the identification of other Uhuru sources. These included Vela X-1 (Ulmer et al. 1972; Forman et al. 1973), SMC X-1 (Webster and Murdin 1972; Schreier et al. 1972), 4U1700-37 (Jones et al. 1973), and GX 301-2 (Vidal 1973; Jones et al. 1974).These discoveries stimulated a flood of theoretical papers. Pringle and Rees (1972), Davidson and Ostriker (1973), Lamb et al. (1973), and Shakura and Sunyaev (1973) all proposed and discussed variants on the basic model that successfully explained the broad observational properties of Cyg X-1 and Cen X-3. The X-rays are driven by material captured from a relatively normal star, falling into the deep gravitational potential well of a neutron star in the case of Cen X-3 and a black hole in Cyg X-1 (Figure 2). The strong wind of the OB star provides a natural source of material to be captured by the pulsar or black hole. The wind may be enhanced due to the fact that the OB star is close to filling its critical Roche potential lobe, so that material spills through the inner Lagrangian point onto the compact object.

These early results provided the first direct confirmation of the earlier suggestion by Shklovsky (1967) and Pendergast and Burbidge (1968) that the mysterious bright X-ray sources were binary systems containing a compact object. These earlier models were based on the identification of Sco X-1 and Cyg X-2 with nova-like objects. These eventually became the class of low-mass X-ray binaries (LMXBs) where the mass donor is a late-type star. But it was the HMXB systems that provided the first direct proof

---
* NASA – Goddard Space Flight Center, Greenbelt, MD, USA

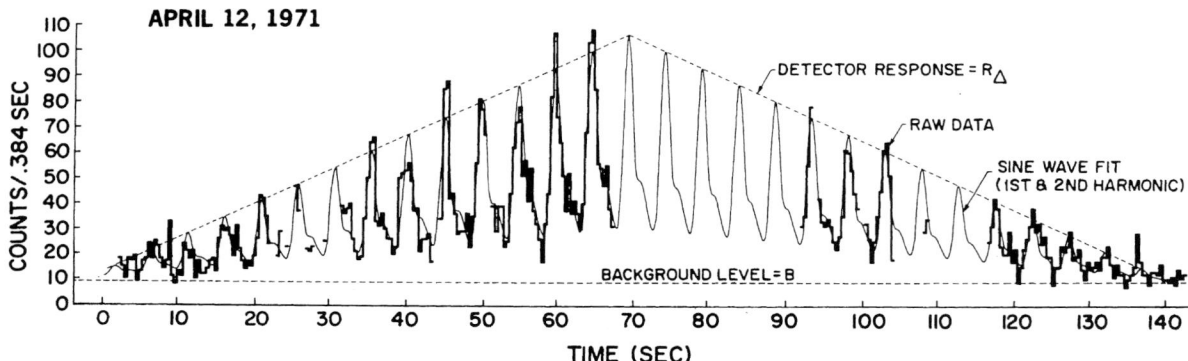

Figure 1 The discovery of X-ray pulsations from Cen X-3 made by Giacconi et al. (1971). The 4.8 s pulsations are seen as the Uhuru satellite scanned across the source.

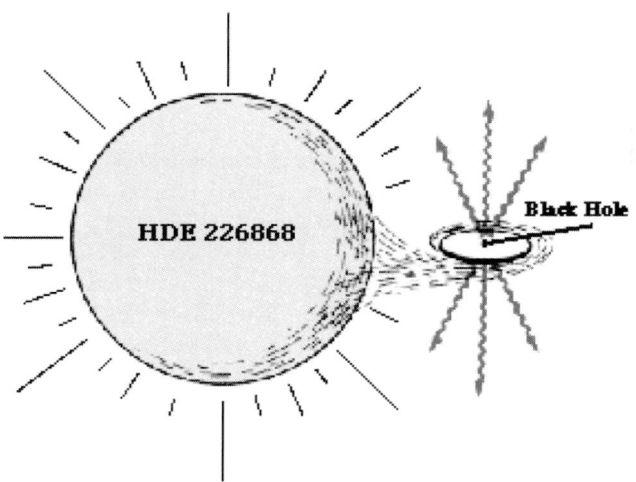

Figure 2 An artist's impression of the HMXB Cyg X-1, which has an orbital period of 5.6 days. This system contains a black hole accreting from the wind of a massive OB star.

for both the presence of accretion onto a neutron star and for the binary orbit.

The OB star companion has a substantial stellar wind, removing between 10^{-6} and 10^{-10} solar masses per year with a terminal velocity up to $2000\,\mathrm{km\,s^{-1}}$. Roche lobe overflow has also been discussed as a supplement to the mass transfer rate in some of the supergiant HMXBs. But full Roche lobe overflow is not likely to be the case because when the mass ratio of the compact object to its companion is greater than unity, then mass transfer via Roche lobe will become unstable $\sim 10^5$ yr after it starts (Savonije 1983). As the supergiant approaches its Roche lobe, the reduced gravity and/or X-ray illumination can cause a focusing of the wind towards the compact object (Friend and Castor 1982; Day and Stevens 1993).

The newly launched Ariel V observatory detected transient X-ray pulsars that brightened quickly, and then faded away over several tens of day (Ives et al. 1975). Several were identified with Be star optical counterparts. Maraschi et al. (1976) suggested they formed a distinct class of HMXB object, with the X-ray outburst driven by mass-loss episodes from the Be star. This idea was subsequently confirmed with the determination by Rappaport et al. (1978) using the SAS 3 satellite of a 24-day orbital period for one of these transients, 4U0115+63. The Be X-ray binaries are much more widely separated than the supergiant systems. Their typical X-ray luminosity is many orders of magnitudes lower, both because the wind is less strong and the neutron star is further out. The Be stars are well known as a class of star that is rotating close to breakup velocity. They are thought to have a disk in the equatorial plane and to undergo mass-ejection episodes. Most are only seen as X-ray sources when the Be star undergoes such a mass-ejection episode, so that the X-ray luminosity increases dramatically (Stella et al. 1986).

Not all the Be systems are transients. The Uhuru error box of the persistent, relatively faint source 4U0352+52 contained X Per, a bright Be star that up until then had been assumed to be a single star. White et al. (1976a) using the small X-ray telescope on the Copernicus observatory discovered a coherent 13.9-minute pulsation from the X-ray source associated with X Per. The luminosity of this system is $10^{33}\,\mathrm{erg\,s^{-1}}$, 1000 to 1 million times fainter than the other members of the HMXB class. The Be star systems are by the far the most populous members of the HMXB class. They tend to be nearer by (by virtue of their higher space density) and have a lower luminosity.

The term "high-mass X-ray binary" did not enter the vernacular until the early 1980s. By then the systematic studies using the X-ray observatories of the 1970s (Uhuru, Ariel V, SAS 3 and HEAO 1) had resulted in the identification of ~ 25 systems associated with OB stars (Bradt and McClintock 1983). By that time it was clear that the overall properties of an X-ray binary were dictated by the nature of the mass donor. The spectral type of the mass donor in an

LMXB is K or later and does not have a strong wind – Roche lobe overflow drives the X-ray emission. In an HMXB the wind of the OB star provides a ready source of fuel for the X-ray source, and also attenuates the X-ray source. This natural division also means that HMXBs are younger systems, where the neutron star has been relatively recently formed.

PULSE AND ORBITAL PERIODS

Orbital periods

The orbital period and eccentricity of a representative selection of HMXB systems are listed in Table 1. The orbital periods range between 4.8 h and 250 d. The supergiant systems typically are eclipsing and show extreme intensity and absorption variability on all timescales. The longer orbital period systems are more eccentric. This eccentricity is to be expected from the "kick" imparted during the supernova explosion that created the neutron star.

Tidal effects will tend to circularize the orbit, and will be more important in the shorter orbital period systems. All but one of the supergiant systems have orbital periods less than 15 d, and dominate this part of the period distribution. Most Be star systems have longer periods of several tens or hundreds of days. The orbital periods of the Be X-ray binaries are not so well known, except that they must be long to avoid any tidal modulation of the optical star or variations of the pulsar arrival times. For these systems it can take many years of observations to reveal the orbital period. For example the orbital period of X Per remained elusive for many years until very recently the Rossi XTE, by monitoring the pulse timing over several years, revealed a 250-day period (Delgado-Martí et al. 2001).

As the baseline of the pulsar timing measurements increased it became possible to search for changes in HMXB orbital periods. Kelley et al. (1983) determined that the orbital period of Cen X-3 is decreasing on a timescale of ~500,000 yr. This decrease is the result of tidal torque between the neutron star and the outer layers of the

Table 1 The parameters of some well-known HMXBs

Name	Alternative name	Orbital period (day)	Pulse period[*] (s)	Eccentricity	Type[†]
Gamma Cas	4U 0053+604	?	?		Be
A 1118-61	He3-640	?	405		Be transient
Cyg X-3	V1521 Cyg	0.20	—		WR
LMC X-4	Sk-Ph	1.4	13.5		SG
LMC X-3	—	1.7	BHC		Be
Cen X-3	Kra	2.1	4.8	0.0008	SG
4U 1700-37	HD 153919	3.41	—		SG
4U 1538-522	QV Nor	3.73	529		SG
SMC X-1	Sk 160	3.89	0.71	<0.0007	SG
Cyg X-1	HD226868	5.60	BHC		SG
4U 1907+097		8.38	438	0.22	SG
Vela X-1	HD77581	8.96	283	0.092	SG
OAO 1657-415		10.4	38		SG
4U 0115+63	LS I+65 010	11.6	850	0.34	Be
SS433	4U 1909+048	13.1	—		SG
1E 1145.1-6141	V830 Cen	?	298		SG
A 0535-668		16.7	0.069		Be transient
4U 0115+634	V662 Cas	24.3	3.6		Be transient
4U 1553-542		30.6	9.3		Be?
V 0332+53	BQ Cam	34.25	4.4	0.31	Be transient
GX 301-2	Wra 977	41.5	696		SG
EXO 2030+375		46.0	41.8		Be
A 0535+262	HD 245770	111	104		Be transient
GX 304-1	V850 Cen	133	272	0.47	Be
4U 1145-619	Hen 715	187.5	292		Be
X Per	4U 0352+30	250	835		Be

[*]BHC, black hole candidate.
[†]SG, supergiant; WR, Wolf Rayet.

supergiant. A similar decay in the orbit of SMC X-1 has been measured by Levine *et al.* (1993). There are several other HMXBs where marginal evidence for a change in period has been reported, including LMC X-4, Vela X-1, and X0115+634 (Nagase 1992, and references therein).

Neutron star orbit and mass determination

The X-ray pulsar can be used as a precise clock to determine the orbital parameters of the neutron star and to obtain a mass function (Figure 3). A mass function can also be obtained for the OB star using optical spectroscopy. These can be combined with other constraints to determine the orbital inclination, such as the duration of the eclipse, to yield the mass of the neutron star. These measurements also give the overall system parameters.

The neutron star mass has now been determined in seven X-ray binary systems and three radio pulsar binaries. The uncertainties for the X-ray pulsars are dominated by those in the optical mass function (which can be difficult to measure) and the system inclination. Within the uncertainties, the neutron star masses obtained are consistent with 1.4 solar masses, the canonical Chandrasekhar mass.

Pulse periods

HMXB pulse periods are also given in Table 1. They range from 60 ms to 850 s. Many of the HMXBs have long pulse periods of the order of minutes, the slowest rotating neutron stars known. During the mid-1970s many of these long-period X-ray pulsars were found using a combination of Ariel V, SAS 3, and Copernicus. These missions provided the first pointed capability where observations lasting many hours to days were possible (as opposed to Uhuru which scanned a narrow field of view across the sky). Many of these long-period pulsars were discovered in quick succession (e.g. McClintock *et al.* 1976, 1977; White *et al.* 1976b). A striking example of how common these long-period pulsars are was the discovery of the "twin pulsars" by White *et al.* (1978) using Ariel V. The field of view of the instrument being used for the search was 3.75°. In one field containing the Uhuru source 4U1145-61 a power spectrum of the time series revealed two peaks at 292 and 298 s. This suggested there were two pulsars within 1° of each other, with very similar pulse periods. This result was confirmed a year later when the Einstein Observatory took the first images of the region to reveal two separate sources, pulsing at the two periods found by Ariel V (Lamb *et al.* 1980). The original 4U1145-61 source is in a widely separated Be system with an orbital period of 292 days (Watson *et al.* 1982).

The longer pulse periods were an unexpected discovery. While there was no reason to exclude such a population, the radio pulsars cuts off at about 10 s because they cross a "death line" where the dipole radiation is no longer efficient. These newly found long-period X-ray pulsars were dubbed "slow rotators" (e.g. Fabian 1975; Lea 1976) and provided a new view on the evolution of neutron star spin periods.

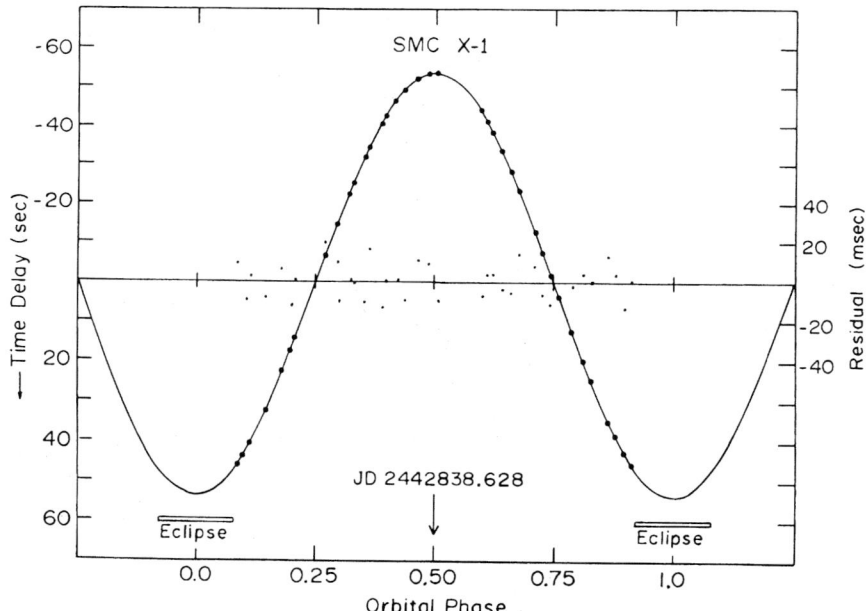

Figure 3 The orbit of SMC X-1 determined from an arrival time analysis of the 0.7 s pulsations. (Taken from Primini *et al.* 1977).

Spin-up and spin-down

Initially there was some debate as to whether the newly discovered long-period X-ray pulsars were neutron stars or white dwarfs. There were models proposed that could explain the observed luminosity and spectra from accretion onto white dwarfs (e.g. Cameron and Mock 1967). The issue was quickly resolved by monitoring the pulse periods and searching for a change in the spin period. This gives a measure of the underlying moment of inertia of the compact object. Because a neutron star is 100 times smaller than a white dwarf, the former is expected to have a two orders of magnitude lower moment of intertia.

Long-term monitoring of the pulse period of the X-ray pulsars quickly revealed that the periods were changing. Radio pulsars are spinning down at a rate that is consistent with the loss of rotational energy via dipole radiation. An accreting neutron star or white dwarf is expected to spin-up, i.e., its period will decrease, because of the angular momentum gained from the accretion flow (Pringle and Rees 1972; Lamb et al. 1973; Davidson and Ostriker 1973). The spin-up rate is determined by the radius where the accretion flow is threaded onto the magnetic field lines, the so called magnetosphere radius, and the angular momentum of the accretion flow given to the compact object. The measured rate of period change is crucial in determining the nature of the underlying object. As the pulsar is accreting material from the companion star, it also will gain (or lose) angular momentum at a rate that depends on the mass accretion rate, the direction of the accretion flow, the magnetic field strength of the object, and its moment of inertia. Mason (1977) and Rappaport and Joss (1977) independently pointed out that the observed spin-up (and spin-down) of these pulsars are too large to be explained by a white dwarf, which simply could not respond quickly enough to the accretion torque. Ghosh and Lamb (1978, 1979a,b) went on to develop the accretion torque model in great detail and provided new insight into the interaction of the accretion flow with the neutron star magnetosphere.

There is an empirical relationship between the spin-up rate, $-\dot{P} = -dP/dt$, and the observed luminosity of $-\dot{P} \propto (P.L^{3/7})^2$, with the normalization determined by the magnetic dipole moment and the moment of inertia of the neutron star. The measured values of these parameters for the pulsars that exhibit spin-up are in good agreement with this simple model (Rappaport and Joss 1977; Mason 1977). The observed spin-up timescale ranges from 100 to 100,000 yr. Figure 4 compares the more detailed theoretical relationship determined by Ghosh and Lamb (1979a,b) with the observed values for pulsars showing strong spin-up. The agreement is good. In the X-ray transient system EXO 2030+375, this relationship has been confirmed by directly measuring $\dot{P}$ over a wide range of luminosity (Parmar et al. 1989).

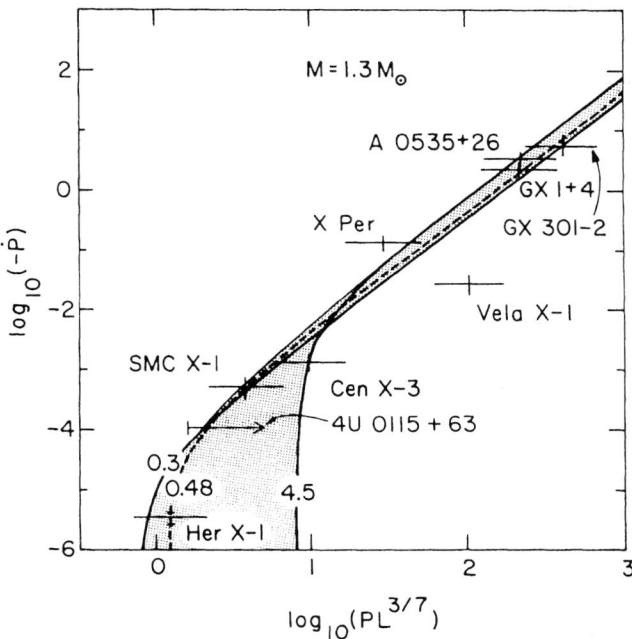

Figure 4 Log$(-\dot{P})$ v. log$(P \cdot L^{3/7})$ for nine X-ray pulsars, superposed on the theoretical relation for the spin-up of an accreting magnetized neutron star. The theoretical curves are labeled by the corresponding value of the magnetic moment (μ_{30} in units of 10^{30} G cm^3. (From Ghosh and Lamb 1979b).

An equilibrium is reached when the neutron star magnetosphere co-rotates with the inner edge of the accretion disk, i.e., when the magnetosphere radius equals the co-rotation radius. Elsner and Lamb (1976) divided pulsars up into two types: slow rotators where the magnetosphere radius is larger than the co-rotation radius and fast rotators where they are similar. The boundary layer between the unperturbed disk and the magnetosphere is complex, and there will be a transition region where the magnetic field begins to thread the disk. As the neutron star reaches co-rotation the spin-up torque diminishes. Elsner and Lamb (1977) introduced a dimensionless fastness parameter that is the ratio of the angular velocity of the neutron star and the Keplerian angular velocity at the magnetosphere radius. Close to co-rotation, when this parameter approaches unity, the field lines in the transition region are swept backwards, and a negative torque is exerted on the neutron star. The neutron star will be spun down, even though accretion continues. This causes a turnover in the predicted values at low luminosity (Figure 4).

Period fluctuations

The best monitoring of the pulse periods of X-ray pulsars has come from the BATSE experiment on the Compton

Gamma Ray Observatory (CGRO). This instrument continuously monitored the brightest X-ray pulsars for 9 yrs (Bildsten *et al.* 1997). The lower luminosity ($<10^{37}$ erg s^{-1}) HMXB pulsars show no overall trend in the pulse period. There are large fluctuations in pulse period that persist down to timescales at least as short as a few days. For the transient Be systems there is a large spin-up seen during the outburst, due to the large amount of material accreted.

The pulse timing noise from Vela X-1 can be described as white noise in the angular acceleration of the neutron star (Boynton *et al.* 1984; Deeter *et al.* 1987). A torque internal to the neutron star, e.g., caused by the coupling of the superfluid interior and the solid outer crust, would give red noise. Boynton *et al.* (1984) conclude that variations in the accretion flow cause the period fluctuations. Baykal and Ogelman (1993) went on to show that the pulse period fluctuations are most likely caused by episodic mass transfer with positive and negative torque.

Accretion flow

The amount of angular momentum captured by a neutron star from a stellar wind determines whether or not an accretion disk forms around the neutron star. The existence of an accretion disk has been a controversial subject. Illarionov and Sunyaev (1975) and Shapiro and Lightman (1976) show that sufficient angular momentum to form a disk can be captured from the wind if a density or velocity gradient exists across the accretion radius. Shapiro and Lightman (1976) also found that the accretion flow could reverse around a black hole if there are fluctuations in the density across the accretion radius. However, Davies and Pringle (1980) pointed out that for accretion to take place at all, the transverse component of the momentum must be lost, and the only way to do this in an ideal situation is for the matter passing either side to collide. This results in a net of zero angular momentum being captured. Wang (1981) found instead that turbulence in the accretion process allows the bulk of the angular momentum to be captured. The problem is that this is a very complex situation, with the wind–neutron star interaction far from simple.

Full hydrodynamical simulations of accretion from a stellar wind are necessary to understand the real situation. The first such numerical simulations by Livio *et al.* (1986) and Soker *et al.* (1986) showed a steady-state situation, similar to that envisioned by Davies and Pringle. But more detailed calculations using a much finer grid reveal a non-steady situation (Matsuda *et al.* 1987; Taam and Fryxell 1988; Taam *et al.* 1991). These models show that in the presence of inhomogeneities in the wind, there is a tendency for instabilities in the accretion flow associated with oscillatory motion in the accretion wake. The accretion flow shows flip-flop behavior where an accretion disk briefly forms, disperses, then forms again but rotating in the opposite sense. The models seem to be in qualitative agreement with the observed period behavior from the wind-driven pulsars. The simulations also reproduce the unusual periodic flaring behavior seen from the Be transient EXO 2030+375 (Taam *et al.* 1988).

The Corbet diagram

The pulse periods are distributed between 0.069 and 835 s, with no evidence for a clustering at any particular period. In Figure 5 the pulse periods are shown *v.* orbital period for those systems where both are known. This plot was first made by Corbet (1984) and has since then become known as the "Corbet diagram." For the Be star systems there is a strong correlation between orbital period and pulse period. The supergiant systems show no obvious dependence. For new Be star X-ray transients the correlation between pulse period and orbital period has been quite a good predictor of the orbital period.

The correlation between pulse period and orbital period for the Be star X-ray binaries is not understood. One suggestion is that these pulsars have reached equilibrium where the co-rotation period is equal to the magnetosphere radius. Combining this with Kepler's law and assuming a wind that is flowing at a constant velocity gives $P_p \propto P_o^{4/7}$ (Corbet 1984; Stella *et al.* 1986; Van den Heuvel and Rappaport 1987; Waters and Van Kerkwijk 1989). However, this is not consistent with the observed linear relationship for the Be systems. Another puzzle is that the observed luminosity of many Be X-ray binaries is 1–2 orders of

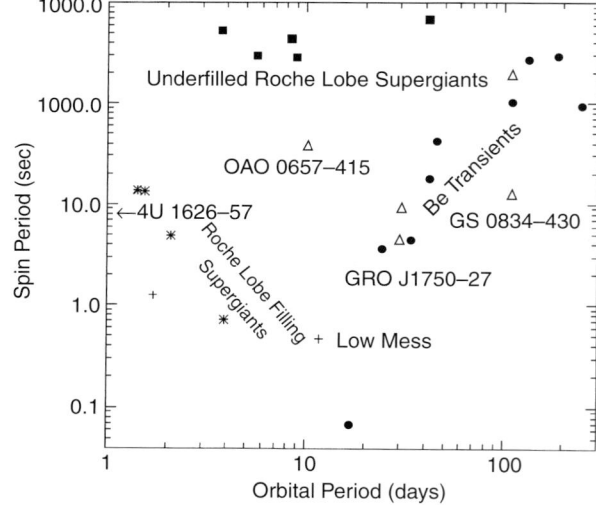

Figure 5 The "Corbet diagram" showing the distribution of pulse period *v.* orbital period of HMXBs. (Taken from Bildsten *et al.* 1997).

magnitude higher than predicted for stellar wind accretion (White *et al.* 1982). Be stars are rapidly rotating, close to breakup, and exhibit an infrared excess that comes from a dense equatorial ring. The material in this ring flows outwards with a much more gradual velocity law than the radially expanding wind from the polar regions. The observed X-ray luminosities of the Be systems are consistent with accretion from an enhanced equatorial flow (Waters *et al.* 1988).

The pulse period–orbital period distribution found for the longer period pulsars in supergiant systems is similar to that predicted, but the normalization requires a mass accretion rate two orders of magnitude lower than is observed (Van den Heuvel and Rappaport 1987; Stella *et al.* 1986). This could be explained if it reflects an earlier evolutionary phase when the companion was on the main sequence and its wind much weaker (Waters and Van Kerkwijk 1989). The pulsar has simply become "stuck" at the longer period.

A centrifugal barrier

Many of the Be star HMXBs are transient systems, i.e., they undergo outbursts that last a few days to weeks, and are then not detected for long intervals. For such a transient system, as the outburst decays the magnetosphere radius will increase because the decreasing accretion rate reduces the pressure on the magnetosphere. At some point the magnetosphere radius may exceed the co-rotation radius. If this happens centrifugal forces will throw material away as soon as it becomes attached to the magnetic field lines. Accretion onto the neutron star surface will cease. The X-ray source will abruptly turn off at a critical luminosity that primarily depends on the magnetic field of the neutron star. Stella *et al.* (1986) realized that if the luminosity can be measured just before the transient turns off it gives an estimate of the magnetic dipole moment. They also pointed out that this effect can give rise to much more dramatic outbursts. In particular as a pulsar orbits the companion the change in accretion rate expected around the eccentric orbit can, for a particular mass-loss rate from the Be star, cause the pulsar to cross the centrifugal limit. This will result in a large orbital modulation, much larger than expected simply from the expected varition in stellar wind accretion. This effect was first observed from the Be transient V 0332+530 (Stella *et al.* 1986).

The centrifugal barrier means that an X-ray outburst may be much shorter and more dramatic than any associated optical phenomena, which may reflect the mass-ejection episode from the OB star. This has been seen for 4U 0115+634 where an outburst in December 1980 was preceded some months before by an optical brightening of the Be star (Kriss *et al.* 1983).

An extreme Be X-ray transient system is the 69 ms pulsar A 0538-66, which is located in the Large Magellanic Cloud (LMC) (White and Carpenter 1978; Skinner *et al.* 1982). This is a highly eccentric system with an orbital period of 16.7 d (Johnston *et al.* 1980; Charles *et al.* 1983). This transient system can be dormant for long intervals. When it does become active, it has a super-Eddington luminosity. A 69 ms rotation period requires a magnetic field of $\sim 10^{11}$ G for the observed luminsoity, a factor of 10 less than typical, simply to overcome the centrifugal barrier.

THE STELLAR WIND LABORATORY

Accretion from the wind

The neutron star or black hole is deeply embedded in the wind of the OB star. It was quickly realized that the HMXB systems provide an exciting new laboratory for the study of the properties of OB star winds and the effects of X-ray photoionisation (Pringle 1973; Buff and McCray 1974). The neutron star will capture a fraction of the wind via Bondi–Hoyle accretion (Bondi and Hoyle 1944; Davidson and Ostriker 1973). The Bondi–Hoyle accretion radius is defined by equating the kinetic energy of the wind to the gravitational potential energy surrounding the neutron star. The accretion radius is inversely proportional to the fourth power of the wind velocity, i.e., as the wind velocity increases, the accretion radius decreases and a smaller fraction of the wind is captured.

An OB star wind is driven by the transfer of momentum from the star's radiation field to the wind by scattering of radiation in UV spectral lines (Lucy and Solomon 1970). The wind velocity rapidly accelerates within a couple of stellar radii to reach a terminal velocity of several thousand kilometers per second (Castor *et al.* 1975). IUE and Copernicus UV spectra obtained in the 1970s revealed strong P-Cygni profiles, indicating mass loss. These line profiles can be used to derive the mass-loss rate, terminal velocity, and wind acceleration profile. The X-ray source usually lies at the point where the wind has reached 90% or more of the terminal velocity. It is straightforward to calculate the expected X-ray luminosity for the wind parameters measured for the OB star companion. Lamers *et al.* (1976) and later Conti (1978) made this calculation and found that the luminosity of the lower luminosity systems like Vela X-1 were consistent with the observed X-ray luminosity. In contrast, for the highest luminosity systems such as Cen X-3, the luminosity is one or two orders of magnitude higher than can be accounted for by stellar wind capture. This suggests the presence of some form of enhanced flow to the neutron star. This may be quasi-Roche lobe overflow, from focusing of the stellar wind towards the neutron star.

As discussed in the next section, the X-ray source in the wind will ionise a cavity within which the radiative acceleration process is inhibited because the UV line transitions which drive the wind acceleration are destroyed. This adds a complication since the magnitude of the disruption of the acceleration process must be included in any model. There may well be a feedback imposed, because the disruption of the wind acceleration will reduce the wind velocity and increase the mass captured by the neutron star. For wind-driven systems this is a relatively small effect, although it must be included in the modeling to obtain consistency (Haberl *et al.* 1989).

The lower luminosity systems ($<10^{37}\,\text{erg s}^{-1}$) show erratic flaring activity with in some cases luminosity variations up to a factor of 100 on a timescale of tens of minutes. The observed flaring may reflect inhomogeneities in the wind (White *et al.* 1983a). These inhomogeneities may be a common feature of all radiation-driven stellar winds, or be found only in HMXBs because of the disrupting effect of the X-ray source (Haberl *et al.* 1989). The more luminous systems ($>10^{37}\,\text{erg s}^{-1}$) are much less variable. This flow may carry sufficient angular momentum to cause a stable accretion disk to form which mediates the flow and smooths out any inhomogeneities.

Absorption by the wind

Buff and McCray (1973) pointed out that the X-ray spectrum of an HMXB is expected to show substantial absorption in the soft X-ray band, caused by the medium-Z elements such as iron, oxygen, and carbon. Absorption edges from the K and L shells are expected, as well as emission lines from fluorescence and recombination. They point out that the absorption will vary as a function of orbital phase, because the depth of material in the line of sight to the X-ray source will vary.

A variable iron K absorption edge at 7.1 keV was first resolved from the HMXB GX301-2 by Swank *et al.* (1976) using OSO 8 proportional counter data (Figure 6). Shortly thereafter, iron K emission line features, as well as edges, were detected with OSO 8 and HEAO 1 from many other HMXBs (Becker *et al.* 1978; Pravdo *et al.* 1979; Rose *et al.* 1979). The GSPCs on the Japanese Tenma satellite and the ESA EXOSAT observatory gave a factor of two improvement in spectral resolution. While this may seem a small improvement, it did allow the line and edge energy to be measured with sufficient precision to identify the ionisation state of the emitting material. This constrained the iron emission line ionisation state to be close to neutral, and the absorption edge energy "warm" with an ionization state of V–X (Nagase 1989). The advent of CCD spectrometers on the ASCA mission showed strong fluorescence lines from many of the medium-Z elements, with a spectacular

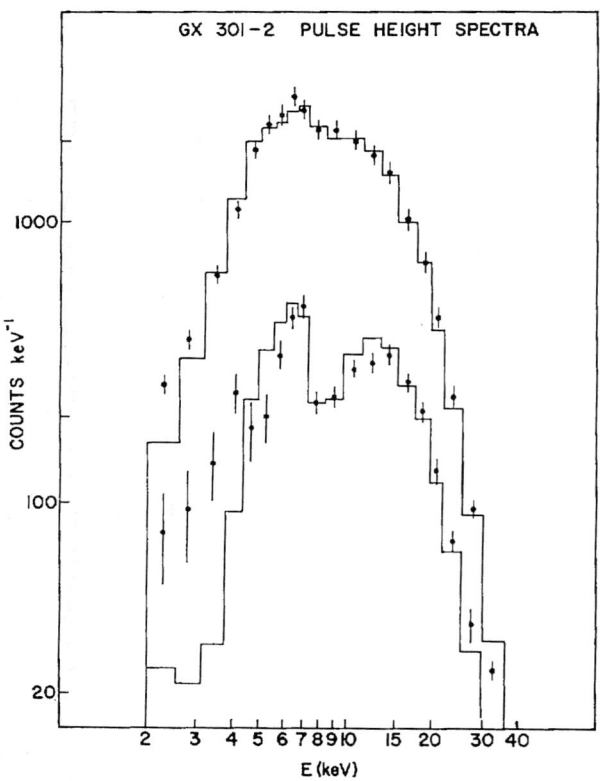

Figure 6 The first evidence for an iron K edge in a cosmic X-ray source, seen in the spectrum of the HMXB GX301-2 by the OSO 8 satellite. (Taken from Swank *et al.* 1976).

line-rich spectrum (Figure 7) during the eclipse of Vela X-1 (Nagase *et al.* 1994).

The absorption of soft X-rays by circumstellar matter is most prominent in the spectra of wind-fed X-ray pulsars, such as Vela X-1 and GX301-2. The amount of the absorption column density is variable and can range up to 10^{23} to $10^{24}\,\text{H cm}^{-2}$. These systems show a smooth decrease in absorption after eclipse egress, to a minimum around phase 0.5. During eclipse, when the X-ray source is hidden behind the OB star, X-rays are still observed, scattered around the primary by the OB star wind. The eclipse spectrum is particularly line rich.

There is a dependence of the iron-line intensity on orbital phase in Vela X-1 (Sato *et al.* 1986). Part of this can be explained as due to expected variations in the intensity of the scattered/fluorescent emission with orbital phase. But some of the line emission appears to be eclipsed, which indicates an additional component of fluorescent emission in the vicinity of the pulsar (Inoue 1985). The absence of a pulsed modulation of the iron-line intensity gives another clue to the distribution of the reprocessing material (Ohashi *et al.* 1984; Leahy and Matsuoka 1990).

For a purely wind-driven system the observed and predicted X-ray luminosity should be consistent with the

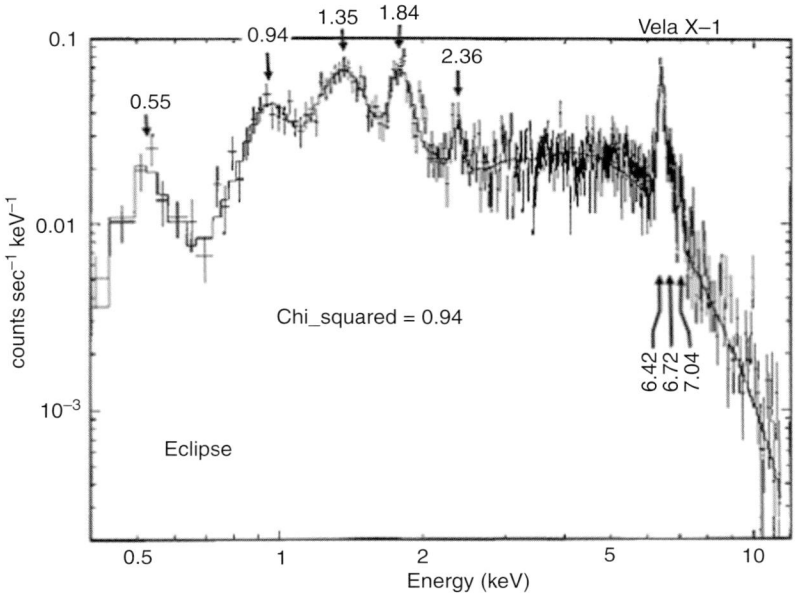

Figure 7 The spectrum of Vela X-1 during eclipse as seen by the ASCA satellite using a CCD spectrometer. This detector has an energy resolution a factor of 10 better than that of the proportional counter flown on OSO 8 15 years earlier. (Taken from Nagase et al. 1994).

observed and predicted absorption. A simultaneous fit to the flare-averaged luminosity and the X-ray absorption of Vela X-1 does indeed give good agreement (Haberl et al. 1989; Haberl and White 1990). An additional absorption component is required to account for the sudden absorption increase at orbital phase 0.6, which is discussed in the next section.

The absorption of the X-ray source from eclipse egress to phase 0.5 is dominated by the relatively undisturbed OB star wind. Observations of the X-ray pulsar can be used to determine the density distribution of the wind and the outer atmosphere of the supergiant (Clark et al. 1988; Sato et al. 1986). These indicate the radial density profile of the wind is divided into two zones. In an inner zone with a radius less than 1.5 stellar radii the atmosphere can be well modeled as an exponential atmosphere with a scale height of the order of one-tenth the radius of the underlying star. Above this a radiation-driven wind forms. The scale height of the atmosphere is an order of magnitude larger than that expected for a supergiant. X-ray irradiation of the atmosphere may form a hot, gravitationally unbound coronal region, giving rise to a thermally driven stellar wind from the X-ray heated face of the companion (Day and Stevens 1993).

Gas streams and accretion wakes

The early observations of supergiant HMXBs revealed a sharp increase in absorption of the X-ray source after phase 0.5 (when the X-ray source passes in front of the OB star). This was seen from 4U 1700-37 (Mason et al. 1976), Cen X-3 (e.g. Pounds et al. 1976), and Vela X-1 (Watson and Griffiths 1977). There followed an extensive discussion as to its cause. Suggestions included an accretion wake from the supersonic bow shock of the neutron star in the wind (Jackson 1975), a gas stream from the supergiant trailing the neutron star (Petterson 1978), and disruption of the radiation-driven wind by the X-ray source illumination (Fransson and Fabian 1980).

Two-dimensional gas-dynamical simulations by Blondin et al. (1991) favor the gas-stream interpretation. The simulations show that a tidal stream will be drawn off the primary if it is close to filling its Roche lobe. This tidal stream will evolve into Roche lobe overflow when the surface of the primary approaches its critical radius. The tidal stream is deflected behind the X-ray source, and will cause the observed sharp increase in absorption at orbital phase 0.6. Modeling of the absorption profiles (Figure 8) over a single orbital cycle also suggests a gas stream from the supergiant trailing behind the X-ray source (Haberl et al. 1989; Kallman 1989; Haberl and White 1990).

Photoionization of the wind by the X-ray source

The presence of the luminous X-ray source in the wind will photoionize material nearby. Building on early work by Tarter et al. (1969), Pringle (1973) made the first estimates of the size and shape of the fully ionized region surrounding the X-ray source. The luminosity of Cen X-3 in the high state is such that the wind not eclipsed by the supergiant will be photoionized. But low-state spectra reported by Schreier et al. (1976) indicate higher absorption and much

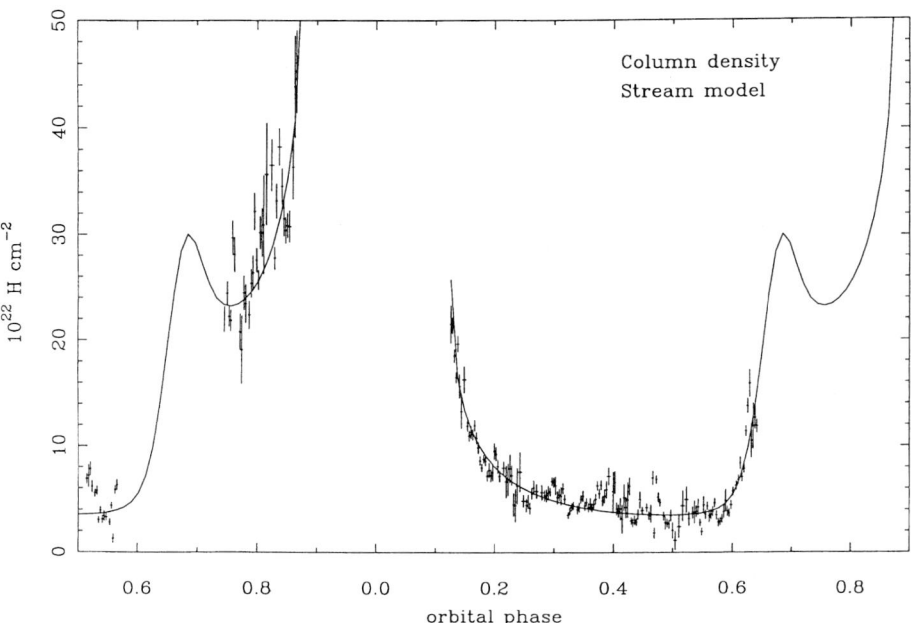

Figure 8 The observed absorption from 4U 1700-37 as a function of orbital phase recorded by the EXOSAT observatory. The solid line shows the absorption expected through the wind, including the effects of a gas stream trailing the X-ray source. (Taken from Haberl et al. 1989).

lower-amplitude X-ray pulsation. The low–high-state transitions of Cen X-3 revealed a "spike" of emission centered around phase 0.5 that became progressively broader as the transition to the high state progressed. This was explained by Schreier et al. (1976) in terms of the Pringle (1973) model of ionised spheres that gradually increased in radius as the source luminosity increased.

Hatchett and McCray (1977) set out to explain in detail the Cen X-3 "low" to "high" state behavior and to provide a framework in which to predict the UV and X-ray properties of the HMXB supergiant systems. They predicted the "spherical surfaces" surrounding the X-ray source that have constant ionisation parameter (Figure 9). These predict that particular lines such as NV will show variations as a function of orbital phase. These were later confirmed by observations with IUE by Dupree et al. (1980) and modeled in detail by McCray et al. (1984). More recently observations using HST of Vela X-1 have detected the X-ray pulsation in the UV lines, which reflect the ionization of the wind as the X-ray beam sweeps through the wind (Boroson et al. 1996).

The Wolf–Rayet system Cyg X-3

Cyg X-3 is a 4.8 h period system that shows an asymmetric sinusoidal-like modulation. It is in a reddened region of the galaxy and no optical counterpart had been identified. Up until recently Cyg X-3 had been assumed to be an LMXB,

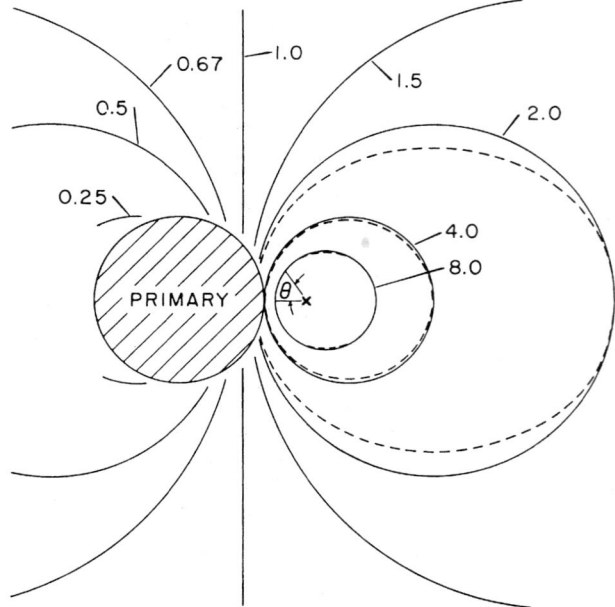

Figure 9 Lines of constant ionization (solid) and column density from the X-ray source (dashed) for the wind of an OB star that is illuminated by a bright X-ray source. (Taken from Hatchett and McCray 1977).

because its orbital period is similar to that of other LMXBs. A big surprise came when infrared spectroscopy revealed that the companion is a Wolf–Rayet star (Van Kerkwijk et al. 1992). This was consistent with an earlier model for Cyg X-3 proposed by Davidson and Ostriker (1974) and Pringle (1974) where the X-ray source is immersed in the optically thick stellar wind of a companion star. Cyg X-3 shows giant radio outbursts, with evidence for jet-like emission expanding at one-third the speed of light (Gregory et al. 1972; Geldzahler et al. 1983), that suggests periods of super-Eddington accretion.

The 4.8 h orbital period of Cyg X-3 is increasing on a relatively short timescale of 4.5×10^5 years (Kitamoto et al. 1987; Molnar 1988; Van der Klis and Bonnet-Bidaud 1989). This is the opposite of what might be expected if the mass transfer is driven by Roche lobe overflow (unless the companion is degenerate (Molnar 1988)). The timescale is consistent with a wind that has a mass loss rate of $\sim 10^{-6}\,\mathrm{yr}^{-1}$ and carries away orbital angular momentum (Van Kerkwijk et al. 1992).

Cyg X-3 was the first X-ray binary to reveal strong iron K line emission (Serlemitsos et al. 1975). An absorption edge at around 9 keV was reported from EXOSAT observations by Willingale et al. (1985), and confirmed by Tenma and Ginga observations (Kitamoto et al. 1987; Nakamura et al. 1993). The X-ray spectrum of Cygnus X-3 observed by the CCD spectrometers on ASCA showed the first clear spectroscopic evidence for recombination in an X-ray photoionized medium, presumably the stellar wind from the Wolf–Rayet companion star (Liedahl and Paerels 1996). The ASCA spectrum revealed the distinctive signature of a radiative recombination continuum from H-like S, Mg, and Si. These narrow continua are an unambiguous indicator of excitation by recombination in an X-ray photoionized gas. Thus Cygnus X-3 bears some strong similarities to the more traditional HMXB counterparts.

SS433

The X-ray source A1909+04 was identified with SS433 by Clark and Murdin (1978). SS433 is an unusual X-ray binary, and not typical of the class. It is famous for moving lines in the optical spectrum which indicate the ejection of jets of material at a velocity of 0.26 times the speed of light. The movement of the optical emission lines from the jets gives a period of 162 d, which is thought to represent a precession period either of the disk or the underlying compact object (Margon 1984). The binary period is 13.1 d and the companion is a massive early-type star. The nature of the compact object is unknown. EXOSAT observations of SS433 revealed a 6.7 keV iron K emission line whose energy varies with time, in concert with the optical lines (Watson et al. 1986). The optical lines from both the receding and advancing jets are always seen, but for most of the precession cycle only the X-ray line from the advancing jet is seen. This means the iron emission comes from very close to the compact object, such that the X-ray line from the receding jet is usually hidden behind the disk. Ginga observations by Kawai et al. (1989) show a partial eclipse of the X-ray emission by the companion star, and a spectral softening during the eclipse. This is further evidence for a temperature gradient along the jet. The SS433 system is probably a short-lived system that is undergoing critical Roche lobe overflow, causing super-Eddington accretion onto a neutron star or black hole. It is illustrative that HMXB systems go through a number of different phases.

THE NEUTRON STAR LABORATORY

The X-ray pulsars in HMXBs provide a unique laboratory for the study of matter in the vicinity of one of the most extreme environments known. One of the most exciting discoveries in the study of accreting neutron stars was a cyclotron line from Her X-1 by Trümper et al. (1978). This provided a direct measure of the magnetic field at the surface of a neutron star and showed that fields of $\sim 10^{12}\,\mathrm{G}$ are involved. In the late 1980s improved spectral measurements with Ginga over a broader energy band (2–30 keV) with better statistical precision revealed cyclotron lines in many HMXB X-ray spectra (Makishima et al. 1990).

The basic accretion model for X-ray pulsars was first proposed by Pringle and Rees (1972), Davidson and Ostriker (1973), and Lamb et al. (1973). The inflowing material from the companion is threaded onto the neutron star magnetic field lines, and channeled onto the magnetic poles. The accretion energy is released as X-rays in a shock near or at the surface (Figure 10). This model was successfully

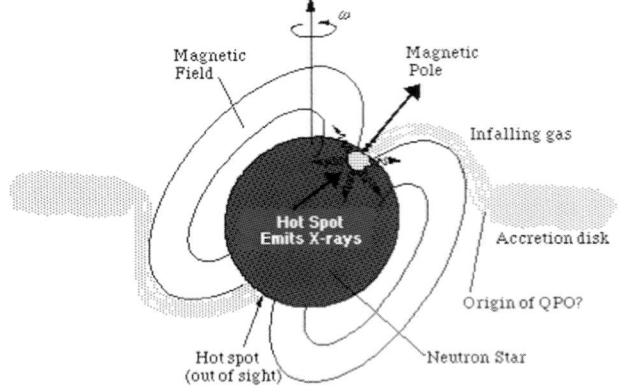

Figure 10 An artist's impression of the accretion flow onto a highly magnetized neutron star. Material flowing in via an accretion disk is disrupted by the neutron star's magnetic field and is channeled to the magnetic pole. X-rays are released in a shock at the magnetic pole.

extended during the late 1970s and 1980s (e.g. Meszaros and Nagel 1985) to account qualitatively for the overall properties of both the X-ray spectra and pulse profiles.

The first comprehensive review to classify the properties of X-ray pulsars was made by White *et al.* (1983b). This showed that there are only very subtle changes with the underlying parameters. The spectra are in general hard, with a tendency to be steeper in the lower luminosity cases. The pulse profiles show great variety, and tend to be simpler for the lower luminosity systems. The observed pulse profiles can be reproduced by an offset dipole model, with a beaming of the X-rays either as a pencil or fan beam, where the emission is beamed respectively parallel or perpendicular to the magnetic field lines. The variety of profiles can be explained by employing either a pencil or fan beam, and varying the geometrical aspect of the rotation axis the magnetic axis, and the line of sight (e.g. Wang and Welter 1981). In some cases an offset of the magnetic axis (Parmar *et al.* 1989c) and/or two polar-cap emission regions with different size are required to give an asymmetric pulse profile (Leahy 1991).

The dependence of the pulsar properties, in particular the pulse profile, on X-ray luminosity is most dramatically seen from observations of X-ray transients. The best example is the 42 s X-ray pulsar X2030+375 where Parmar *et al.* (1989) made observations through the decay phase of an outburst over two orders of magnitude in luminosity. The pulse profile changed dramatically with a reversal in the strength of an interpulse with the main pulse, indicating the beam pattern changed from a fan beam at high luminosity to a pencil beam at low luminosity. This is explained by a shock front developing above the neutron star surface at high luminosity.

CONCLUSION

As we enter the twenty-first century, HMXBs will provide an important laboratory for future studies of a wide range of phenomena. The improved spectral resolution provided by the grating spectrometers on the newly launched Chandra and XMM-Newton observatories will enable the line features to be exploited. The photoionized wind will provide a rich source of lines (Sako *et al.* 1999) that can be used to probe the structure of the stellar wind of an OB star and the effects of the X-ray source on its structure. The detection of line features from the photosphere of the neutron star itself remains a long-sought, but elusive goal – but would, if found, provide a direct measure of the equation of state of a neutron star. The potential for discovery of unexpected phenomena from HMXBs remains large and the new generation of telescopes and instruments will continue this story.

Acknowledgment

The author thanks Fumiaki Nagase for helpful discussions during the writing of this chapter.

REFERENCES

Baykal, A. and Oegelman, H. (1993). An empirical torque noise and spin-up model for accretion-powered X-ray pulsars. *Astronomy and Astrophysics*, **267**, 119–125.

Becker, R.H., Pravdo, S.H., Rothschild, R.E., Boldt, E.A., Holt, S.S., Serlemitsos, P.J. and Swank, J.H. (1978). Extended observations of Vela X-1 by OSO 8. *Astrophysical Journal*, **221**, 912–916.

Bildsten, L., Chakrabarty, D., Chiu, J., Finger, M.H., Koh, D.T., Nelson, R.W., Prince, T.A., Rubin, B.C., Scott, D.M., Stollberg, M., Vaughan, B.A., Wilson, C.A. and Wilson, R.B. (1997). Observations of accreting pulsars. *Astrophysical Journal Supplement*, **113**, 367.

Blondin, J.M., Stevens, I.R. and Kallman, T.R. (1991). Enhanced winds and tidal streams in massive X-ray binaries. *Astrophysical Journal*, **371**, 684–695.

Bolton, C.T. (1972). Identifications of Cyg X-1 with HDE 226868. *Nature*, **235**, 271.

Bondi, H. and Hoyle, F. (1944). On the mechanism of accretion by stars. *Monthly Notices of the Royal Astronomical Society*, **104**, 273.

Boroson, B., McCray, R., Kallman, T. and Nagase, F. (1996). Pulsations in the ultraviolet and X-ray spectra of Vela X-1. *Astrophysical Journal*, **465**, 940.

Bowyer, S., Byram, E.T., Chubb, T.A., Friedman, H. (1965). Cosmic X-ray sources. *Science*, **147**, 394–398.

Boynton, P.E., Deeter, J.E., Lamb, F.K., Zylstra, G., Pravdo, S.H., White, N.E., Wood, K.S. and Yentis, D.J. (1984). New evidence on the nature of the neutron star and accretion flow in Vela X-1 from pulse timing observations. *Astrophysical Journal*, **283**, L53–L56.

Bradt, H.V.D. and McClintock, J.E. (1983). The optical counterparts of compact galactic X-ray sources. *Annual Review of Astronomy and Astrophysics*, **21**, 13–66.

Buff, J. and McCray, R. (1974). Soft X-ray variability of binary X-ray stars. *Astrophysical Journal*, **188**, L37.

Cameron, A. and Mock, M. (1967). Stellar accretion and X-ray emission. *Nature*, **215**, 464–466.

Castor, J.I., Abbott D.C. and Klein R.I. (1975). Radiation-driven winds in of stars. *Astrophysical Journal*, **195**, 157–174.

Charles, P.A., Booth, L., Densham, R.H., Bath, G.T., Howarth, I.D., Willis, A.J., Skinner, G.K., Thorstensen, J.R. and Olszewski, E. (1983). Extreme variability in the Be-type, periodic recurrent X-ray transient A0538-66 – A highly eccentric interacting binary. *Monthly Notices of the Royal Astronomical Society*, **202**, 657–682.

Clark, G.W., Minato, J.R. and Mi, G. (1988). The atmospheric structure of the O-type supergiant Krzeminski's star and the mass of its companion neutron star Centaurus X-3. *Astrophysical Journal*, **324**, 974–994.

Clark, D.H. and Murdin, P. (1978). An unusual emission-line star/X-ray source/radio star, possibly associated with an SNR. *Nature*, **276**, 44–45.

Conti, P.S. (1978). Stellar parameters of five early type companions of X-ray sources. *Astronomy and Astrophysics*, **63**, 225–235.

Corbet, R.H.D. (1984). Be/neutron star binaries – A relationship between orbital period and neutron star spin period. *Astronomy and Astrophysics*, **141**, 91–93.

Davidsen, A. and Ostriker, J.P. (1974). The nature of Cygnus X-3: A prototype for old-population binary X-ray sources. *Astrophysical Journal*, **189**, 331–338.

Davidson, K. and Ostriker, J.P. (1973). Neutron-star accretion in a stellar wind: Model for a pulsed X-ray source. *Astrophysical Journal*, **179**, 585–598.

Davies, R.E. and Pringle, J.E. (1980). On accretion from an inhomogeneous medium. *Monthly Notices of the Royal Astronomical Society*, **191**, 599–604.

Day, C.S.R. and Stevens, I.R. (1993). An X-ray excited wind in Centaurus X-3. *Astrophysical Journal*, **403**, 322–331.

Deeter, J.E., Boynton, P.E., Lamb, F.K. and Zylstra, G. (1987). Apsidal advance in Vela X-1. *Astrophysical Journal*, **314**, 634–640.

Delgado-Martí, H., Levine, A.M., Pfahl, E. and Rappaport, S.A. (2001). The orbit of X Persei and its neutron star companion. *Astrophysical Journal*, **546**, 455–468.

Dupree, A., Gursky, H., Black, J.H., Davis, R.J., Hartmann, L., Matilsky, T., Raymond, J.C., Hammerschlag-Hensberge, G., van den Heuvel, E.P.J., Burger, M., Lamers, H.J.G.L.M., vanden Bout, P.A., Morton, D.C., de Loore, C., van Dessel, E.L., Menzies, J.W., Whitelock, P.A., Watson, M., Sanford, P.W. and Pollard, G.S.G. (1980). Simultaneous ultraviolet, optical, and X-ray observations of the X-ray source Vela X-1 (HD 77581). *Astrophysical Journal*, **238**, 969–981.

Elsner, R.F. and Lamb, F.K. (1976). Accretion flows in the magnetospheres of Vela X-1, AO535+26 and Her X-1. *Nature*, **262**, 356–360.

Elsner, R.F. and Lamb, F.K. (1977). Accretion by magnetic neutron stars. I – Magnetospheric structure and stability. *Astrophysical Journal*, **215**, 897–913.

Fabian, A.C. (1975). Slowly rotating neutron stars and transient X-ray sources. *Monthly Notices of the Royal Astronomical Society*, **173**, 161–165.

Forman, W., Jones, C., Tananbaum, H., Gursky, H., Kellogg, E. and Giacconi, R. (1973). Uhuru observations of the binary X-ray source 2u 0900-40. *Astrophysical Journal*, **182**, L103.

Fransson, C. and Fabian, A.C. (1980). X-ray induced shocks in stellar winds. *Astronomy and Astrophysics*, **87**, 102–108.

Friend, D.B. and Castor, J.I. (1982). Radiation-driven winds in X-ray binaries. *Astrophysical Journal*, **261**, 293–300.

Geldzahler, B.J., Johnston, K.J., Spencer, J.H., Klepczynski, W.J., Josties, F.J., Angerhofer, P.E., Florkowski, D.R., McCarthy, D.D., Matsakis, D.N. and Hjellming, R.M. (1983). The 1982 September radio outburst of Cygnus X-3 – Evidence for jetlike emission expanding at not less than about 0.35 C. *Astrophysical Journal*, **273**, L65–L69.

Ghosh, P. and Lamb, F.K. (1978). Disk accretion by magnetic neutron stars. *Astrophysical Journal*, **223**, L83–L87.

Ghosh, P. and Lamb, F.K. (1979a). Accretion by rotating magnetic neutron stars. II – Radial and vertical structure of the transition zone in disk accretion. *Astrophysical Journal*, **232**, 259–276.

Ghosh, P. and Lamb, F.K. (1979b). Accretion by rotating magnetic neutron stars. III – Accretion torques and period changes in pulsating X-ray sources. *Astrophysical Journal*, **234**, 296–316.

Giacconi, R., Gursky, H., Kellogg, E., Schreier, E. and Tananbaum, H. (1971). Discovery of periodic X-ray pulsations in Centaurus X-3 from Uhuru. *Astrophysical Journal*, **167**, L67.

Gregory, P.C., Kronberg, P.P., Seaquist, E.R., Hughes, E.R., Hughes, V.A., Woodsworth, A., Viner, M.R., Retallack, P., Hjellming, R.M. and Balick, B. (1972). Nature of the first outburst. *Nature Phys Sci*, **239**, 114.

Haberl, F. and White, N.E. (1990). The X-ray absorption spectrum of Vela X-1. *Astrophysical Journal*, **361**, 225–234.

Haberl, F., White, N.E. and Kallman, T.R. (1989). An EXOSAT X-ray observation of one orbital cycle of 4U 1700-37/HD 153919. *Astrophysical Journal*, **343**, 409–425.

Hatchett, S. and McCray, R. (1977). X-ray sources in stellar winds. *Astrophysical Journal*, **211**, 552–561.

Illarionov, A.F. and Sunyaev, R.A. (1975). Why the number of galactic X-ray stars is so small. *Astronomy and Astrophysics*, **39**, 185.

Inoue, H. (1985). Tenma observations of bright binary X-ray sources. *Space Science Reviews*, **40**, 317–338.

Ives, J.C., Sanford, P.W. and Bell-Burnell, S.J. (1975). Observations of a transient X-ray source with regular periodicity of 6.75 min. *Nature*, **254**, 578–580.

Jackson, J.C. (1975). Parameters of Cen X-3 deduced from observations of its accretion wake. *Monthly Notices of the Royal Astronomical Society*, **172**, 483–492.

Johnston, M.D., Griffiths, R.E. and Ward, M.J. (1980). Improved position and new optical candidate for A0538-66. *Nature*, **285**, 26–27.

Jones, C.A., Chetin, T. and Liller, W. (1974). Optical studies of Uhuru sources. VIII. Observations of 92 possible counterparts of X-ray sources. *Astrophysical Journal*, **190**, L1.

Jones, C.A., Forman, W. and Liller, W. (1973). Optical studies of Uhuru sources. IV. The long-term behavior of HZ Herculis=Hercules X-1. *Astrophysical Journal*, **182**, L109.

Kawai, N., Matsuoka, M., Pan, H.-C. and Stewart, G.C. (1989). Ginga observations of the X-ray eclipse of SS 433. *Publications of the Astronomical Society of Japan*, **41**, 491–507.

Kelley, R.L., Rappaport, S., Clark, G.W. and Petro, L.D. (1983). Orbital period changes in Centaurus X-3. *Astrophysical Journal*, **268**, 790–799.

Kitamoto, S., Miyamoto, S., Matsui, W. and Inoue, H. (1987). 4.8-hour modulation of X-rays from Cygnus X-3. *Publications of the Astronomical Society of Japan*, **39**, 259–285.

Kriss, G.A., Cominsky, L.R., Remillard, R.A., Williams, G. and Thorstensen, J.R. (1983). The 1980 outburst of 4U 0115+63 (V635 Cassiopeiae). *Astrophysical Journal*, **266**, 806–813.

Krzeminski, W. (1974). The identification and UBV photometry of the visible component of the Centaurus X-3 binary system. *Astrophysical Journal*, **192**, L135–L138.

Lamb, R.C., Markert, T., Hartman, R., Thompson, D. and Bignami, G.F. (1980). Two X-ray pulsars – 2S 1145-619 and 1E 1145.1-6141. *Astrophysical Journal*, **239**, 651–654.

Lamb, F.K., Pethick, C.J. and Pines, D. (1973). A model for compact X-ray sources: Accretion by rotating magnetic stars. *Astrophysical Journal*, **184**, 271–290.

Lamers, H., van den Heuvel, E., and Petterson, J.A. (1976). Stellar winds and accretion in massive X-ray binaries. *Astronomy and Astrophysics*, **49**, 327–335.

Lea, S. (1976). Pulsating X-ray sources – Slowly rotating neutron stars. *Astrophysical Journal*, **209**, L69–L72.

Leahy, D.A. (1991). Modelling observed X-ray pulsar profiles. *Monthly Notices of the Royal Astronomical Society*, **251**, 203–212.

Leahy, D.A. and Matsuoka, M. (1990). The pulse-phase dependence of the spectrum of GX 301-2. *Astrophysical Journal*, **355**, 627–634.

Levine, A., Rappaport, S., Deeter, J.E., Boynton, P.E. and Nagase, F. (1993). Discovery of orbital decay in SMC X-1. *Astrophysical Journal*, **410**, 328–341.

Liedahl, D.A. and Paerels, F. (1996). Photoionization-driven X-ray line emission in Cygnus X-3. *Astrophysical Journal*, **468**, L33–L36.

Livio, M., Soker, N., de Kool, M. and Savonije, G.J. (1986). On accretion of angular momentum from an inhomogeneous medium. *Monthly Notices of the Royal Astronomical Society*, **218**, 593–604.

Lucy, L.B. and Solomon, P. (1970). Mass loss by hot stars. *Astrophysical Journal*, **159**, 879.

Makishima, K., Ohashi, T., Kawai, N., Matsuoka, M., Koyama, K., Kunieda, H., Tawara, Y., Ushimaru, N., Corbet, R.H.D., Inoue, H., Kii, T., Makino, F., Mitsuda, K., Murakami, T., Nagase, F., Ogawara, Y., Tanaka, Y., Kitamoto, S., Miyamoto, S., Tsunemi, H. and Yamashita, K. (1990). Observations of the peculiar hard X-ray transient X0331+53 (V0332+53). *Publications of the Astronomical Society of Japan*, **42**, 295–315.

Maraschi, L., Treves, A. and van den Heuvel, E.P.J. (1976). B-emission stars and X-ray sources. *Nature*, **259**, 292–293.

Margon, B. (1984). Observations of SS 433. *Annual Review of Astronomy and Astrophysics*, **22**, 507–536.

Mason, K.O. (1977). Secular period changes in X-ray pulsators. *Monthly Notices of the Royal Astronomical Society*, **178**, 81P–85P.

Mason, K.O., Branduardi, G. and Sanford, P.W. (1976). The X-ray behavior of 3U 1700-37. *Astrophysical Journal*, **203**, L29–L33.

Matsuda, T., Inoue, M. and Sawada, K. (1987). Spin-up and spin-down of an accreting compact object. *Monthly Notices of the Royal Astronomical Society*, **226**, 785–811.

McClintock, J.E., Rappaport, S., Joss, P.C., Bradt, H., Buff, J., Clark, G.W., Hearn, D., Lewin, W.H.G., Matilsky, T., Mayer, W. and Primini, F. (1976). Discovery of a 283-second periodic variation in the X-ray source 3U 0900-40. *Astrophysical Journal*, **206**, L99–L102.

McClintock, J.E., Rappaport, S., Nugent, J. and Li, F. (1977). Discovery of a 272 second periodic variation in the X-ray source GX 304-1. *Astrophysical Journal*, **216**, L15–L18.

McCray, R.A., Kallman, T.R., Castor, J.I. and Olson, G.L. (1984). Spectral variability in early-type binary X-ray systems. *Astrophysical Journal*, **282**, 245–255.

Meszaros, P. and Nagel. W. (1985). X-ray pulsar models. I – Angle-dependent cyclotron line formation and comptonization. *Astrophysical Journal*, **298**, 147–160.

Molnar, L.A. (1988). Interpretation of the period derivative of Cygnus X-3. *Astrophysical Journal*, **331**, L25–L28.

Nagase, F. (1989). Accretion-powered X-ray pulsars. *Publications of the Astronomical Society of Japan*, **41**, 1–79.

Nagase, F. (1992). Ginga Memorial Symposium. F. Makino and F. Nagase (eds), Institute of Space and Astronautical Research, p. 1.

Nagase, F., Zylstra, G., Sonobe, T., Kotani, T., Inoue, H. and Woo, J. (1994). Line-dominated eclipse spectrum of Vela X-1. *Astrophysical Journal*, **436**, L1–L4.

Nakamura, H., Matsuoka, M., Kawai, N., Yoshida, A., Miyoshi, S., Kitamoto, S. and Yamashita, K. (1993). Unified model fitting to variable X-ray spectra of Cygnus X-3. *Monthly Notices of the Royal Astronomical Society*, **261**, 353–365.

Oda, M., Gorenstein, P., Gursky, H., Kellogg, E., Schreier, E., Tananbaum, H. and Giacconi, R. (1971). X-ray pulsations from Cygnus X-1 observed from Uhuru. *Astrophysical Journal*, **166**, L1.

Ohashi, T., Inoue, H., Koyama, K., Makino, F., Matsuoka, M., Suzuki, K., Tanaka, Y., Hayakawa, S., Tsunemi, H. and Yamashita, K. (1984). Properties of the iron line from Vela X-1. *Publications of the Astronomical Society of Japan*, **36**, 699–707.

Parmar, A.N., White, N.E., and Stella, L. (1989). The transient 42 second X-ray pulsar EXO 2030+375. II – The luminosity dependence of the pulse profile. *Astrophysical Journal*, **338**, 373–380.

Pendergast, K.H. and Burbidge, G.R. (1968). On the nature of some galactic X-ray sources. *Astrophysical Journal*, **151**, L83.

Petterson, J.A. (1978). On the occurrence of streams and disks in massive X-ray binary systems. *Astrophysical Journal*, **224**, 625–630.

Pounds, K.A., Cooke, B.A., Ricketts, M.J., Turner, M.J. and Elvis, M. (1975). An extended observation of Cen X-3 with the Ariel-5 Sky Survey. *Monthly Notices of the Royal Astronomical Society*, **172**, 473–481.

Pravdo, S.H., White, N.E., Szymkowiak, A.E., Boldt, E.A., Holt, S.S., Serlemitsos, P.J., Swank, J.H., Tuohy, I. and Garmire, G. (1979). HEAO 1 observations of the X-ray pulsar 4U 1626-67. *Astrophysical Journal*, **231**, 912–918.

Primini, F., Rappaport, S. and Joss, P. (1977). Pulse profile and refined orbital elements for SMC X-1. *Astrophysical Journal*, **217**, 543–548.

Pringle, J.E. (1973). X-ray sources – optical appearance and evolution. *Nature Phys Sci*, **243**, 90.

Pringle, J.E. (1974). Model for Cygnus X-3. *Nature*, **247**, 21–22.

Pringle, J.E. and Rees, M.J. (1972). Accretion disc models for compact X-ray sources. *Astronomy and Astrophysics*, **21**, 1.

Rappaport, S., Clark, G.W., Cominsky, L., Li, F. and Joss, P.C. (1978). Orbital elements of 4U 0115+63 and the nature of the hard X-ray transients. *Astrophysical Journal*, **224**, L1–L4.

Rappaport, S. and Joss, P.C. (1977). Accretion torques in X-ray pulsars. *Nature*, **266**, 683–685.

Rose, L.A., Marshall, F.E., Holt, S.S., Boldt, E.A., Rothschild, R.E., Serlemitsos, P.J., Pravdo, S.H. and Kaluzienski, L.J. (1979). Observations of the transient X-ray source 4U 0115+63. *Astrophysical Journal*, **231**, 919–926.

Sako, M., Liedahl, D.A., Kahn, S.M. and Paerels, F. (1999). The X-ray spectrum and global structure of the stellar wind in Vela X-1. *Astrophysical Journal*, **525**, 921–934.

Sato, N., Hayakawa, S., Nagase, F., Masai, K., Dotani, T., Inoue, H., Makino, F., Makishima, K. and Ohashi, T. (1986). X-ray probing of the circumstellar matter in the Vela X-1 system from observations over an eclipse phase. *Publications of the Astronomical Society of Japan*, **38**, 731–750.

Savonije, G.J. (1983). Evolution and mass transfer in X-ray binaries. In W.H.G. Lewin and E.P.J. van den Heuvel (eds), *Accretion Driven Stellar X-ray Sources*, Cambridge University Press, pp. 343–366.

Schreier, E., Levinson, R., Gursky, H., Kellogg, E., Tananbaum, H. and Giacconi, R. (1972). Evidence for the binary nature of Centaurus X-3 from Uhuru X-ray observations. *Astrophysical Journal*, **172**, L79.

Schreier, E., Swartz, K., Giaconni, R., Fabbiano, G. and Morin, J. (1976). The long-term intensity behavior of Centaurus X-3. *Astrophysical Journal*, **204**, 539–547.

Serlemitsos, P.J., Boldt, E.A., Holt, S.S., Rothschild, R.E. and Saba, J.L.R. (1975). Spectral variability of Cygnus X-3. *Astrophysical Journal*, **201**, L9–L13.

Shakura, N.I. and Sunyaev, R.A. (1973). Black holes in binary systems. Observational appearance. *Astronomy and Astrophysics*, **24**, 337–355.

Shapiro, S.L. and Lightman, A.P. (1976). Black holes in X-ray binaries – Marginal existence and rotation reversals of accretion disks. *Astrophysical Journal*, **204**, 555–560.

Shklovsky, I.S. (1967). On the nature of the source of X-ray emission of Sco XR-1. *Astrophysical Journal*, **148**, L1.

Skinner, G.K., Bedford, D.K., Elsner, R.F., Leahy, D., Weisskopf, M.C. and Grindlay, J. (1982). Discovery of 69 ms periodic X-ray pulsations in A0538-66. *Nature*, **297**, 568–570.

Soker, N., Livio, M., de Kool, M. and Savonije, G.J. (1986). Accretion of angular momentum from an inhomogeneous medium. II – Isothermal flow. *Monthly Notices of the Royal Astronomical Society*, **221**, 445–452.

Stella, L., White, N.E. and Rosner, R. (1986). Intermittent stellar wind accretion and the long-term activity of Population I binary systems containing an X-ray pulsar. *Astrophysical Journal*, **308**, 669–679.

Swank, J.H., Becker, R.H., Boldt, E.A., Holt, S.S., Pravdo, S.H., Rothschild, R.E. and Serlemitsos, P.J. (1976). Variations in the spectra and pulse period of GX 301-2. *Astrophysical Journal*, **209**, L57–L60.

Taam, R.E. and Fryxell, B.A. (1988). On nonsteady accretion in stellar wind-fed X-ray sources. *Astrophysical Journal*, **327**, L73–L76.

Taam, R.E., Fryxell, B.A. and Brown, D.A. (1988). A model for the recurrent flares in EXO 2030+375. *Astrophysical Journal*, **331**, L117–L120.

Taam, R.E., Fu, A. and Fryxell, B.A. (1991). Accretion in wind-driven X-ray sources. *Astrophysical Journal*, **371**, 696–707.

Tarter, C.B., Tucker, W.H. and Salpeter, E.E. (1969). The interaction of X-ray sources with optically thin environments. *Astrophysical Journal*, **156**, 943.

Terrell, N.J., Jr. (1972). Shot-noise character of Cygnus X-1 pulsations. *Astrophysical Journal*, **174**, L35.

Trümper, J., Pietsch, W., Reppin, C., Voges, W., Staubert, R. and Kendziorra, E. (1978). Evidence for strong cyclotron line emission in the hard X-ray spectrum of Hercules X-1. *Astrophysical Journal*, **219**, L105–L110.

Ulmer, M., Baity, W.A., Wheaton, W.A. and Peterson, L.E. (1972). Observations of the Vela XR-1 by the UCSD X-Ray Telescope on OSO-7. *Astrophysical Journal*, **178**, L121.

Van den Heuvel, E.P.J. and Rappaport, S. (1987). X-ray observations of B-emission stars. In A. Slettebak and T.D. Snow (eds), *Physics of Be Stars*, Cambridge University Press, pp. 291–307.

Van der Klis, M. and Bonnet-Bidaud, J.M. (1989). The X-ray ephemeris of Cygnus X-3. *Astronomy and Astrophysics*, **214**, 203–208.

Van Kerkwijk, M.H., Charles, P.A., Geballe, T.R., King, D.L., Miley, G.K., Molnar, L.A., van den Heuvel, E.P.J., van der Klis, M. and van Paradijs, J. (1992). Infrared helium emission lines from Cygnus X-3 suggesting a Wolf–Rayet star companion. *Nature*, **355**, 703–705.

Vidal, N.V. (1973). The association of a high-luminosity B star with 3u 1223–62. *Astrophysical Journal*, **186**, L81.

Wang, Y.-M. (1981). Spin-reversed accretion as the cause of intermittent spindown in slow X-ray pulsars. *Astronomy and Astrophysics*, **102**, 36–44.

Wang, Y.-M. and Welter, G.L. (1981). An analysis of the pulse profiles of the binary X-ray pulsars. *Astronomy and Astrophysics*, **102**, 97–108.

Waters, L.B.F.M., van den Heuvel, E.P.J., Taylor, A.R., Habets, G.M.H.J. and Persi, P. (1988). Evidence for low-velocity winds in Be/X-ray binaries. *Astronomy and Astrophysics*, **198**, 200–210.

Waters, L.B.F.M. and Van Kerkwijk, M.H. (1989). The relation between orbital and spin periods in massive X-ray binaries. *Astronomy and Astrophysics*, **223**, 196–206.

Watson, M.G. and Griffiths, R.E. (1977). Ariel V Sky Survey Instrument – Extended observations of 3U 0900-40. *Monthly Notices of the Royal Astronomical Society*, **178**, 513–524.

Watson, M.G., Stewart, G.C., Brinkmann, W. and King, A.R. (1986). Doppler-shifted X-ray line emission from SS433. *Monthly Notices of the Royal Astronomical Society*, **222**, 261–271.

Watson, M.G., Warwick, R. and Corbet, R. (1982). The orbital period of 2S 1223–624 (GX301-2). *Monthly Notices of the Royal Astronomical Society*, **199**, 915–924.

Webster, B.L. and Murdin, P. (1972). Cygnus X-1 – A spectroscopic binary with a heavy companion? *Nature*, **235**, 37–38.

White, N.E. and Carpenter, G.F. (1978). The recurrent X-ray transient A0538-66. *Monthly Notices of the Royal Astronomical Society*, **183**, 11P–15P.

White, N.E., Kallman, T.R. and Swank, J.H. (1983a). The X-ray absorption spectrum of 4U 1700-37 and its implications for the stellar wind of the companion HD 153919. *Astrophysical Journal*, **269**, 264–272.

White, N.E., Mason, K.O., Huckle, H.E., Charles, P.A. and Sanford, P.W. (1976b). Periodic modulation of three galactic X-ray sources. *Astrophysical Journal*, **209**, L119–L124.

White, N.E., Mason, K.O., Sanford, P.W. and Murdin, P. (1976a). The X-ray behaviour of 3U 0352+30 (X Per). *Monthly Notices of the Royal Astronomical Society*, **176**, 201–215.

White, N.E., Parkes, G., Sanford, P.W., Mason, K. and Murdin, P.G. (1978). Two X-ray periodicities from the vicinity of 4U1145-61. *Nature*, **274**, 664–666.

White, N.E., Swank, J.H. and Holt, S.S. (1983b). Accretion powered X-ray pulsars. *Astrophysical Journal*, **270**, 711–734.

White, N.E., Swank, J.H., Holt, S.S. and Parmar, A.N. (1982). A comparison of the X-ray properties of X Persei and Gamma Cassiopeiae. *Astrophysical Journal*, **263**, 277–288.

Willingale, R., King, A.R. and Pounds, K.A. (1985). EXOSAT MEDA observations of Cygnus X-3. *Monthly Notices of the Royal Astronomical Society*, **215**, 295–314.

36

YASUO TANAKA*

Black-hole binaries

1 INTRODUCTION

The black hole is the most exotic object in the universe predicted as a direct consequence of general relativity. Research of stellar mass black holes dates back to 1939, when Oppenheimer and Snyder (1939) predicted the possible presence of black holes. They discovered that, based on general relativity, a sufficiently massive star would collapse indefinitely when all thermonuclear energy was exhausted and disappear inside a sphere of a limiting radius from which even photons could not come out (the "event horizon"). This limiting radius, called the Schwarzschild radius, equals twice the gravitational radius, r_g ($= MG/c^2$, where M, G and c are the mass, the gravitational constant and the light velocity, respectively). Such an object is what is later called a black hole. However, black holes remained merely theoretical objects for a long time.

The actual existence of stellar-mass black holes came to light for the first time in the early seventies, as described below. At present, the presence of stellar mass black holes in a certain type of X-ray binaries has become beyond doubt. The "discovery" of black holes was a great victory for general relativity theory, and was certainly one of the highlights of astronomy in the twentieth century. However, the presence of stellar mass black holes in our Galaxy has not been established by a single indisputable discovery (unlike the discovery of neutron stars with radio pulsars). In actuality, their real existence has become convincing as a result of steadily growing evidence accumulated over a length of time.

It is not straightforward to identify black holes observationally. The most direct proof for a black hole would be

1. to demonstrate the presence of an event horizon in a compact (gravitationally collapsed) object.

However, this demonstration is extremely difficult. The next best thing is

2. to find the general relativistic effects (particle motion near the speed of light, a large gravitational redshift, etc.) that are events unique to close vicinities of event horizons.

For the reasons discussed later, attempts along this line have not been as yet successful (although successful for some Seyfert galaxies). So far, the most reliable evidence for a black hole has come from mass determination:

3. showing the mass of a compact object exceeds $3M_\odot$.

To the best of our current knowledge, any compact object more massive than $3M_\odot$, the mass upper limit of a stable neutron star, is believed to have no other fate than to collapse into a black hole.

As of the end of 1999, the compact objects in twelve X-ray binaries have been shown to have a mass greater than $3M_\odot$ (see Table 1, Section 4.1). On this theoretical basis, they are considered to be "reliable" black holes. (It is to be noted that black holes can in principle have any mass, hence black holes of $<3M_\odot$, if they exist, are not recognized using this criterion.) Yet, genuine general relativistic tests such as the above-mentioned (1) and/or (2) are still lacking. In this sense, one can say that the presence of stellar mass black holes in some X-ray binaries is virtually certain, but rigorously speaking the observational proof is not yet perfect.

A wealth of observational results on X-ray binaries has become available in the last few decades. Accordingly, studies of the X-ray properties of these sources have made great progress, and we have acquired a fair understanding

---

*Max-Planck-Institut für Extraterrestrische Physik, Garching bei München, Germany and Institute of Space and Astronautical Science, Sagamihara Kanagawa-ken, Japan

of the nature of X-ray emission. In particular, we find that in certain circumstances, X-ray binaries containing a black hole (hereafter referred to as "black-hole X-ray binaries") show X-ray properties distinctly different from those containing a neutron star ("neutron-star X-ray binaries"), a point to which we will return in a later section. Based on these properties, one can identify the most probable black holes with the X-ray observations alone. As a result, it has become certain that many more black-hole binaries exist in our Galaxy. Interestingly enough, most of the black-hole binaries are not bright in X-rays all the time, but they are transient sources. The current observational facts on these transients make us suspect that there exist as many as several hundred or even more black-hole binaries (though most of them are X-ray quiet) in our Galaxy.

It is worth emphasizing that X-ray observations, which are only possible from space, have played a unique role in the investigation of black holes. The currently known stellar-mass black holes have all been discovered among bright X-ray binaries. There are good reasons for this, since, explained later, X-ray observation is practically the only means to discover stellar-mass black holes. Thus, one can say that without the development of observations from space, this fundamentally important field could not have been opened.

The above summarizes the present status of observational research into black-hole binaries *as of the end of 1999*. We shall discuss these topics in detail in the rest of this chapter. For previous reviews, see e.g. Tanaka and Lewin (1995), Tanaka and Shibazaki (1996), and references therein. The Research in this field has been long and is expanding rapidly. A great many investigators have been involved in the course of development. However, it is beyond the capability of the author to cover all the important contributions and giving them their credit due. Readers must be aware that this article is not intended to be a complete review of the subject and that the coverage of topics and the assessment of the results and interpretations may well be subject to the author's personal bias.

Following this introduction, we begin with a brief history of the early developments that led to the discovery of stellar-mass black holes.

2 EARLY DEVELOPMENTS

Soon after the discovery of X-ray stars in 1962 (Giacconi *et al.* 1962), the concept of an accreting compact object emerged in order to account for the large X-ray luminosity. If a compact object is in a binary system and accretes matter from a companion star, the matter falling into a deep gravitational potential well of the compact object would be heated to a very high temperature and efficiently emit X-rays. Shklovsky (1967) argued that the compact object of Sco X-1 (the first discovered and the brightest X-ray source in the sky) must, on observational grounds, be a neutron star, which later turned out to be the case. This speculation occurred even before the discovery of neutron stars. Neutron stars were discovered in 1967 by Hewish and Bell (Hewish *et al.* 1968) as radio pulsars. Soon, a rapidly spinning neutron star was found in the center of the Crab Nebula. This discovery provided direct evidence for the birth of a gravitationally collapsed object as a consequence of supernova explosions, a phenomenon predicted earlier by Baade and Zwicky in 1934. These discoveries re-excited interest in black holes, another class of gravitationally collapsed object, that might exist in reality. X-ray sources were considered to be the best locations to look for black holes because the deep gravitational potential of the latter make them strong X-ray emitters. However, in early days of X-ray astronomy there was no observational clue to identify black holes.

An epoch-making development concerning black holes took place in 1972. This was with Cyg X-1. It is worth a brief account here. Cygnus X-1 was one of the bright X-ray sources known from the earliest X-ray observations in sixties. From the early days, this source drew much attention because of its distinct characteristics among then known X-ray sources. It showed a hard X-ray spectrum much harder than others, and also irregular time variabilities. (For those who are not familiar with the expression "hard" or "soft", see explanation in Section 4.3.) The first X-ray astronomy satellite, *UHURU* launched in 1969, found that Cyg X-1 had a bimodal behavior switching between a low-intensity hard state and a high-intensity soft state (see Section 4.3). This added another peculiarity to the source. Various efforts had been made to localize the position of the source, but they were not yet precise enough to allow optical identification. In 1971, a new radio source was detected within this error region by Braes and Miley (1971) and Hjellming and Wade (1971). It became convincing that this radio source was Cyg X-1 itself, since the radio source had emerged in coincidence with the epoch when Cyg X-1 had changed from a high/soft state into a low/hard state. The accurate position of the radio source immediately led to an identification of the optical counterpart of Cyg X-1 with the O-type supergiant (O9.7I_{ab}) HDE226868.

A big surprise soon followed, when Webster and Murdin (1972) and Bolton (1972) discovered a sinusoidal variation with a 5.6-day period in the optical Doppler curve of HD 226868 (Figure 1), clearly indicating that it is a binary system. This was the first discovery of the binary nature of Galactic X-ray sources, and was prior to the discoveries of binary X-ray pulsars. The invisible companion must be a compact object accreting matter from the supergiant and emitting strong X-rays. Not only that, but the mass function (see equation below) obtained by them indicated that the mass of the compact object probably exceeded $3M_\odot$ (more recent results listed in Table 1). They independently considered that this binary system contained a black hole, Webster and Murdin wrote, "it is inevitable that we should also speculate that it might be a black hole". Bolton wrote, "this raises the distinct possibility that the secondary is a black hole".

Because of the great impact on the astronomy community, this discovery excited various critical discussions. However, none of the suggested alternatives to exclude the black hole hypothesis (e.g. invoking an equation of state with strange matter, or a three-body system, etc.) survived. Thus, Cyg X-1 remained as a strong candidate for a black-hole binary. Details of these accounts on Cyg X-1 are found in a review by Oda (1977).

It is to be cautioned, however, that the above-mentioned X-ray properties of Cyg X-1 are no longer unique, nor are they signatures of black holes. Similar properties are observed from many other X-ray binaries regardless of whether the compact object is a black hole or a neutron star.

3 BLACK HOLES IDENTIFIED FROM MASS FUNCTIONS

During the following ten years, Cyg X-1 was the only black hole candidate until two bright X-ray sources in the Large Magellanic Cloud, LMC X-3 and LMC X-1, were discovered. Both of them were optically identified as early-type stars (Cowley et al. 1983 for LMC X-3, and Hutchings et al. 1983 for LMC X-1). The mass functions obtained indicated that the masses of the compact objects in these sources were also larger than $3M_\odot$. (The result for LMC X-1 in Table 1 was taken from the follow-up work by Hutchings et al. 1987).

The mass function $f(M)$ is given by the following equation:

$$f(M) = \frac{M_x^3 \sin^3 i}{(M_x + M_c)^2} = \frac{PK^3}{2\pi G}$$

where M_x and M_c are the masses of the X-ray emitting compact object and of the companion star, respectively, and i is the inclination angle of the binary orbit. P is the orbital period, and K denotes the amplitude of the Doppler curve (giving the line-of-sight component of the radial velocity) of the companion, which are both optically measurable quantities.

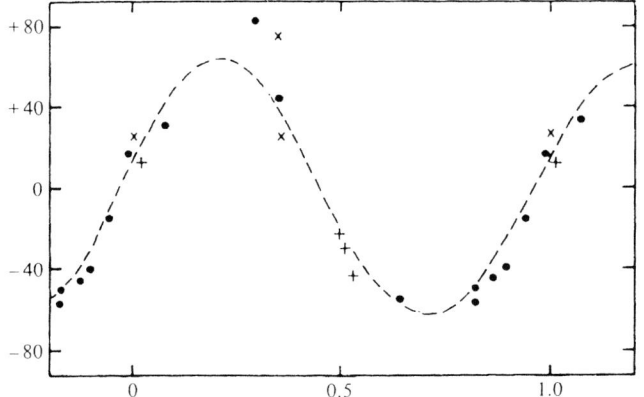

Figure 1 The radial velocity curve of HD 226868 plotted against the phase of 5.6-day orbital period. The vertical scale is in units of km s^{-1}. (From Webster and Murdin 1972.)

Table 1 Black-hole binaries established from the mass functions

Source name	Spectrum[a]	Companion	F(M) ($M_\odot$)	BH mass ($M_\odot$)	Ref.[b]
Cyg X-1	S+PL	O9.7I$_{ab}$	0.241 ± 0.013	~16 (>7)	1
LMC X-3	S+PL	B 3 V	2.3 ± 0.3	>7	2
LMC X-1	S+PL	O 7–9 III	0.14 ± 0.05	~6	3
J0422+32 XNova Per '92	PL	M 2 V	1.21 ± 0.06	>3.2	4
0620−003 XNova Mon '75	S+PL	K 5 V	3.18 ± 0.16	>7.3	5
1009−45 XNova Vel '93	S+PL	K 7–8	3.17 ± 0.12	~4.4	6
1124−684 XNova Mus '91	S+PL	K 0–4 V	3.1 ± 0.4	~6	7
1543−475 XNova '71, '83, '92	S+PL	A 2 V	0.22 ± 0.02	2.7–7.5	8
J1655−40 XNova Sco '94	S+PL	F 3–6	3.24 ± 0.09	7.02±0.22	9
1705−250 XNova Oph '77	S+PL	K ~3 V	4.0 ± 0.8	~6	10
2000+251 XNova Vul '88	S+PL	early K	4.97 ± 0.10	6–7.5	11
2023+338 XNova Cyg '89	PL	K 0 IV	6.26 ± 0.31	8–15.5	12

[a] X-ray spectrum at high luminosities, S+PL: soft + power-law, PL: power law.
[b] References:
1. Gies and Bolton 1982
2. Cowley et al. 1983
3. Hutchings et al. 1987
4. Filippenko et al. 1995a
5. McClintock and Remillard 1986
6. Filippenko et al. 1999
7. McClintock, Bailyn and Remillard 1992
8. Orosz et al. 1998
9. Orosz and Bailyn 1997
10. Remillard et al. 1996
11. Filippenko et al. 1995b
12. Casares et al. 1992

It is clear from the equation that $f(M)$ gives an absolute lower limit of the mass of the compact object. Once $f(M)$ is obtained, the actual mass of the compact object can be estimated if i and M_c are known by some means. These quantities are however subject to a fair amount of uncertainty. In particular, the possible systematic effects in the early type systems (large M_c), where the $f(M)$ values are usually much smaller than M_x (see Table 1), complicates the setting of a firm mass lower limit for the compact object. For instance, the companion mass M_c is usually estimated from the optical spectral type and our knowledge of the masses of those stars of the same spectral type. However, such an estimate might not be correct for the stars in close binary systems such as X-ray binaries that transferred a large amount of mass and might have experienced an unusual evolutionary process. According to the critical discussions, e.g. McClintock et al. (1992), $M_x > 3M_\odot$ is secure for Cyg X-1 and LMC X-3, but less secure for LMC X-1. (Yet, we shall see later that the X-ray spectrum of LMC X-1 supports the proposition that it is a black-hole binary.)

Up until 1986, these three were the only known black hole binaries, and they were all high-mass ($M_c \gg M_\odot$) systems. Since 1986, research on black hole binaries entered into a new era of rapid development. It began with

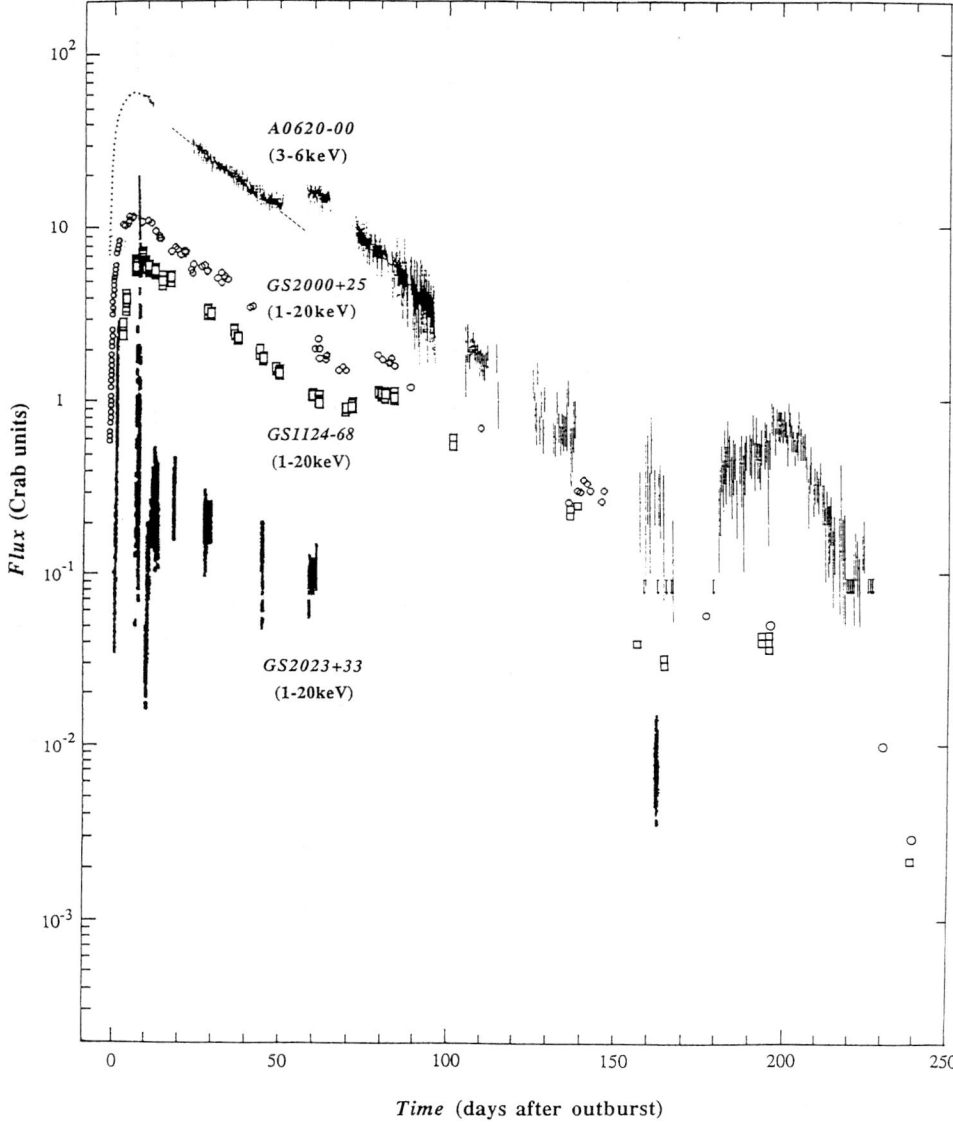

Figure 2 X-ray light curves of the transient outbursts of four black-hole binaries. The observed fluxes are shown in units of the Crab Nebula flux in an energy band indicated for each source. (Tanaka and Shibazaki 1996.)

the discovery by McClintock and Remillard (1986) that the compact object in a low-mass X-ray binary system (for which $M_c \lesssim M_\odot$) was definitely more massive than $3M_\odot$. This was A 0620-00, which underwent a transient X-ray outburst in 1975 and was detected with the X-ray satellites *Ariel V* and *SAS*-3 (Figure 2). At its peak, it became the brightest X-ray source in the sky, three times as bright as Sco X-1. Since the measured mass function itself was very close to $3M_\odot$, there was little doubt that $M_x > 3M_\odot$. This discovery was earmarked not only because it presented the most convincing case for a black hole, but also because it was the first in low-mass binary systems and furthermore the source was a transient.

In the following years, X-ray outbursts of three low-mass binaries were successively detected with the X-ray satellite *Ginga* in 1988, 1989, and 1991. Surprisingly, the mass functions of all of them indicated that they contained black holes (see Table 1). Of these three, the mass functions of GS 2023+33 and GS 2000+25 were $6.3M_\odot$ (Casares *et al.* 1992) and $5.0M_\odot$ (Filippenko *et al.* 1995b), respectively, the largest values ever found so far. These cases, where the mass functions are not just marginally more but far more than $3M_\odot$, completely refuted the still surviving suspicion against the existence of black holes, i.e. a possibility that the masses of some neutron stars may be close to $3M_\odot$ and are somewhat overestimated by systematic uncertainties. As a matter of fact, for those binaries for which the compact objects are known to be neutron stars, the masses estimated from the mass functions have always been close to $1.4M_\odot$ (the Chandrasekhar limit) within the uncertainties, and no convincing cases of significantly more massive neutron stars are known.

Further in the nineties, successive progress has been made in identification of the optical counterpart and determination of the mass function for more X-ray binaries. As a result, "reliable" black-hole binaries that satisfy $M_x > 3M_\odot$ have been discovered one after another. Thus far, we have twelve reliable black hole binaries in the list, as indicated in Table 1.

Remarkably, except only three (Cyg X-1, LMC X-1, and LMC X-3), they are all low-mass binaries and are all transients detected during X-ray outbursts (see Table 1). Since these low-mass transient sources are X-ray quiet for most of the time, this fact indicates that there must be many more low-mass binaries containing a black hole that have not yet been detected. Those low-mass X-ray binaries that undergo transient outbursts belong to a class called soft X-ray transients, because they show a distinctly softer (steeper) X-ray spectrum than that of the transient X-ray pulsars (strongly magnetized neutron stars). We shall come back to this subject in a later section (see Section 4.2).

So far, the identification of black holes relied on the mass lower limit of the compact object in X-ray binaries.

It is true that the criterion $M_x > 3M_\odot$ based on the optical measurements is still the most reliable black hole discriminator. On the other hand, continuous efforts have been made to find out any specific signatures of black holes in the observed properties, in particular the X-ray properties.

In preparation for going into these subjects, we shall summarize the current concept of the mass accretion process which determines the X-ray properties of X-ray binaries.

4 X-RAY PROPERTIES

4.1 Mass accretion

X-ray binaries are powered by mass accretion. In high-mass X-ray binaries of which the companion is a massive early-type star, the stellar wind feeds mass to the compact object. On the other hand, low-mass X-ray binaries are contact binary systems in which the companion fills its own Roche-lobe. As the binary separation tends to become smaller (hence the Roche lobe tends to shrink), the mass of the companion overflows into the gravitational potential well of the compact object (Roche-lobe overflow). These are the basic mechanisms of feeding mass to the compact object.

Matter from the companion star forms a disk-like structure called an accretion disk circulating around the compact object, and gradually moves inward into the deep gravitational potential well. In this process, half the gravitational energy released is converted to rotational energy of the Keplerian motion, and the other half to thermal energy through viscous heating. This thermal energy is the source of X-ray radiation. The accretion process and the accretion disk structure have been studied extensively over the past decades. Among others, Shakura and Sunyaev (1973) have given a most comprehensive description of the disk accretion, which is often referred to as the standard accretion disk model.

In the case of an accreting black hole, the accretion disk can extend inward to 3 Schwartzschild radii ($6r_g$; the radius of the innermost stable Keplerian orbit) for a non-spinning black hole (Schwartzschild black hole), or even closer if the black hole is spinning (Kerr black hole) in the same direction as the accretion disk rotation. Inside the innermost disk, matter falls freely into the black hole. In the case of a non-spinning black hole, the gravitational energy converted to thermal energy amounts to as much as 6% of the rest mass energy of the accreting matter (and even more for a spinning black hole). According to the standard disk model, the accretion disk becomes geometrically thin and optically thick when the accretion rate is sufficiently high (more quantitative discussion follows). In such a disk, the thermal energy is efficiently radiated away in the form of blackbody radiation. Because the temperature of the inner disk becomes so high, the radiation is predominantly in the X-ray band. This accounts for the large X-ray luminosities of black hole X-ray binaries that can go up to $10^{38}\,\mathrm{erg\,s^{-1}}$ or even higher.

For neutron-star X-ray binaries, accretion depends on the magnetic fields of the neutron stars. If the neutron stars are strongly magnetized, typically $B \sim 10^{12}$ Gauss, the accretion disk stops at the magnetospheric boundary where the accretion pressure balances with the magnetic pressure, inside of which matter flows along the magnetic field lines, funnels onto the magnetic poles and emits X-rays there. Such stars manifest themselves as X-ray pulsars, an unambiguous signal that they are neutron stars.

On the other hand, it is known that neutron stars in low-mass binaries have much weaker magnetic fields, estimated to be of the order $B \sim 10^8$ Gauss. Such weak magnetic fields do not disturb the accretion flow unless the accretion rate is very low, and the accretion disk can extend close to the neutron star surface. Since the radius of a canonical $1.4\,M_\odot$ neutron star is of the order of 10 km, the innermost accretion disk may approach 3 Schwartzschild radii (12 km). This situation is not much different from an accreting black hole. Since the accreting matter does not know whether the central object is a neutron star or a black hole, the structure of the accretion disk and its radiation is expected to be similar in both cases. For a given mass accretion rate, the X-ray luminosity of the accretion disk will also be little different between them, since the potential depth in a given Schwartzschild radius is the same regardless of the mass of the compact object. Hence, it is not so straightforward to distinguish black holes from weakly-magnetized neutron stars by just looking at their appearances. Detailed study of their X-ray properties is necessary.

A fundamental difference between a neutron star and a black hole is the presence or absence of a solid surface. In the case of a neutron star, accreting matter eventually lands on the neutron star surface and releases the remaining kinetic energy into the dense neutron star atmosphere. Part of this energy will be emitted as additional blackbody radiation. On the other hand, the matter accreting around a black hole simply disappears beyond the event horizon. This difference can provide an important means of distinguishing a black hole from a neutron star, as we shall see in Section 4.3.

The accretion disk structure may change with the accretion rate. While the standard model predicts a geometrically-thin optically-thick disk at high accretion rates from which thermal energy is efficiently radiated, the disk structure at low accretion rates remains unclear. This subject has attracted much attention in recent years.

It is generally considered that the inner part of the disk will become optically thin (geometrically-thick) below a certain accretion rate. The efficiency of radiation in such a disk is very low, and most of the thermal energy will be carried away (advected) by hot accreting matter into the compact object. Such a flow is called an advection-dominated accretion flow (ADAF). A similar idea had once been presented early on by Ichimaru (1977) in an attempt to explain the bimodal behavior of Cyg X-1. Since the mid-nineties, the ADAF has been studied extensively. Narayan, McClintock and Yi (1996) proposed an ADAF disk model consisting of two distinct zones: a standard thin disk at large radii and a hot quasi-spherical disk in the inner region. This model was applied to soft X-ray transients at very low luminosity levels (quiescent state; see Section 4.2), and also to explain X-ray quiescence of the supermassive black hole at the center of our Galaxy (Sgr A*) and other galactic nuclei. At present, the concept of ADAF at low accretion rates is generally accepted. Yet, the current ADAF models still contain problems to be solved (Lasota 1999). For a review of ADAF, see e.g. Narayan, Mahadevan and Quataert (1999).

4.2 Soft X-ray transients

As shown in Section 3, most of the reliable black-hole binaries are soft X-ray transients. For this distinct nature, it is worthwhile to outline what we have learned about soft X-ray transients so far.

Among the bright X-ray binaries known to date, a significant fraction of them are not persistently visible and brightened only occasionally. These sources are categorized as transient sources (or "transients"; see catalogue by van Paradijs 1995). These transients have been discovered by various X-ray satellites, in particular those with large field-of-view detectors on board, e.g. *Ariel V, Tenma, Ginga, GRANAT, CGRO (BATSE)* and more recently, *RXTE, BeppoSAX*.

The transients are further divided into two different classes; the high-mass X-ray transients, and the low-mass X-ray transients. The high-mass X-ray transients are systems with an O or B star companion ($M_c \gg M_\odot$), while the low-mass X-ray transients have primarily K or M dwarf companions ($M_c \lesssim M_\odot$).

Almost all high-mass X-ray transients are recurrent X-ray pulsars, in which a strongly magnetized neutron star in an eccentric orbit periodically encounters a stellar wind zone around a massive companion. Because their X-ray spectra are hard, typical of accretion-powered X-ray pulsars, they are called "hard X-ray transients". They are outside the scope of this chapter.

Of about 120 low-mass X-ray binaries known to date, one third are transients (see Section 6). These low-mass X-ray transients are characterized by episodic X-ray outbursts without a fixed periodicity. They are in a low-luminosity quiescent state for most of the time, and occasionally undergo dramatic X-ray outbursts. During an outburst, they show a soft X-ray spectrum that is characteristic of high-luminosity low-mass X-ray binaries, as explained in Section 4.3. These are the "soft X-ray transients", or sometimes called "X-ray novae". The soft X-ray transients are a subset of low-mass X-ray binaries containing either a weakly-magnetized neutron star or a black hole. Many soft X-ray transients exhibit recurrent outbursts with intervals ranging from a few years to tens of years or even longer. Perhaps all of the soft X-ray transients are recurrent. X-ray outbursts are accompanied by optical outbursts that allow identification of the optical counterparts. For reviews, see Tanaka and Shibazaki (1996) and also van Paradijs and McClintock (1995).

Figure 2 shows the X-ray light curves of the outbursts from four black-hole binaries which exhibit monotonous exponential decays except for intermediate increases. However, note that many other outbursts show much more complex light curves (see a compilation of light curves by Chen et al. 1997). In general, the source brightens to an X-ray luminosity, L_x, as high as 10^{38} erg s^{-1} or sometimes even $\sim 10^{39}$ erg s^{-1} at the outburst peak, and becomes quiescent after a few months of decay.

An X-ray outburst is triggered by a sudden onset of accretion flow on to the compact object, followed by the decay due to gradually diminishing accretion rate. Because the accretion rate changes over orders of magnitude through the decay, soft X-ray transients are extremely useful for studying the physics of mass accretion and X-ray properties as a function of accretion rate. In fact, much of our current knowledge described in the following sections has come from the studies of soft X-ray transients.

Whether a low-mass binary remains persistently X-ray active or not seems to depend on the accretion rate. Apparently, there is a threshold accretion rate below which X-ray emission is practically turned off. (For instance, notice the abrupt fall in the last part of the decay recorded in Figure 2.) The available data indicate that the threshold rate is around $\sim 10^{16}$ g s^{-1} corresponding to an X-ray luminosity L_x of 10^{36} erg s^{-1}. In fact, there exist few persistent sources of L_x below 10^{36} erg s^{-1}. Thus, soft X-ray transients are those for which the accretion rate from the companion is below the threshold required to sustain persistent X-ray emission.

What causes transient outbursts? An interesting fact is found in the data of the recurrent outbursts of several soft X-ray transients. The outburst is bigger when the interval since the last outburst is longer. Regardless of the source, the time-averaged outburst luminosity, i.e. the total energy radiated in an outburst divided by the interval from the last event, is of the same order of magnitude. This time-averaged luminosity is converted to an average accretion rate on the order of 10^{15}–10^{16} g s^{-1} (see Tanaka and Shibazaki 1996). These values are close to or somewhat less than the above-mentioned threshold rate. This fact suggests the following picture: (1) The mass transfer rate from the companion is lower than the threshold rate necessary for being a persistent source. (2) A reservoir exists in the system, in which the mass transferred from the companion is stored during the quiescent period. The reservoir occasionally empties all the accumulated mass that floods into the compact object (an outburst), and the system returns to quiescence.

This behavior of soft X-ray transients is currently explained by the disk instability models originally developed for dwarf novae, in which the outer part of an accretion disk serves as a mass reservoir. The concept of the disk instability is summarized as follows: An accretion disk has two stable states: a cool, neutral state and a hot, fully-ionized state. In a quiescent state, the outer accretion disk is considered to be in the cool state. As matter accumulates in the disk, the surface density and temperature increase gradually. When the surface density reaches a certain critical value, a thermal instability sets in and the disk jumps to the hot state. As a consequence, it gives rise to a high accretion rate, causing rapid infall of matter into the inner part of the disk and hence an X-ray outburst. When the surface density drops below another critical value, the disk returns to the cool state.

Understanding the inner accretion flow in the quiescent state is an important issue. The X-ray luminosity in quiescence is typically $\sim 10^{32}$ erg s^{-1} or lower. This luminosity is, however, orders of magnitude lower than expected when much of the gravitational energy released for an accretion rate of 10^{15}–10^{16} g s^{-1} goes into radiation. The question is whether the inner flow during quiescence is quenched or in the form of ADAF (Section 4.1) of very low efficiency of radiation. At present, the interpretation involving ADAF is generally supported. It considers that a part of the accreting matter is stored in the cool outer disk while the rest (a comparable amount) goes into the inner disk in ADAF. If that is the case, the ADAF is expected to produce a significantly higher luminosity for a neutron star than a black hole for the following reason. For a black hole, the bulk of the energy of accreting matter is advected into the event horizon with little radiation. On the other hand, for a neutron star, the advected energy is deposited on the solid surface and eventually radiated. Several workers have reported that the quiescent luminosities of neutron-star transients observed with *ROSAT* and *ASCA* are higher than those of the reliable ($M_x > 3 M_\odot$) black-hole transients, although the available sample is still small (also see Section 4.3). The lower luminosity of the black-hole transients is consistent with the presence of the event horizon. See e.g. Narayan et al. (1999); Menou et al. (1999).

4.3 X-ray spectra

From the late seventies through the eighties, the quality of the observed X-ray spectra was dramatically improved by

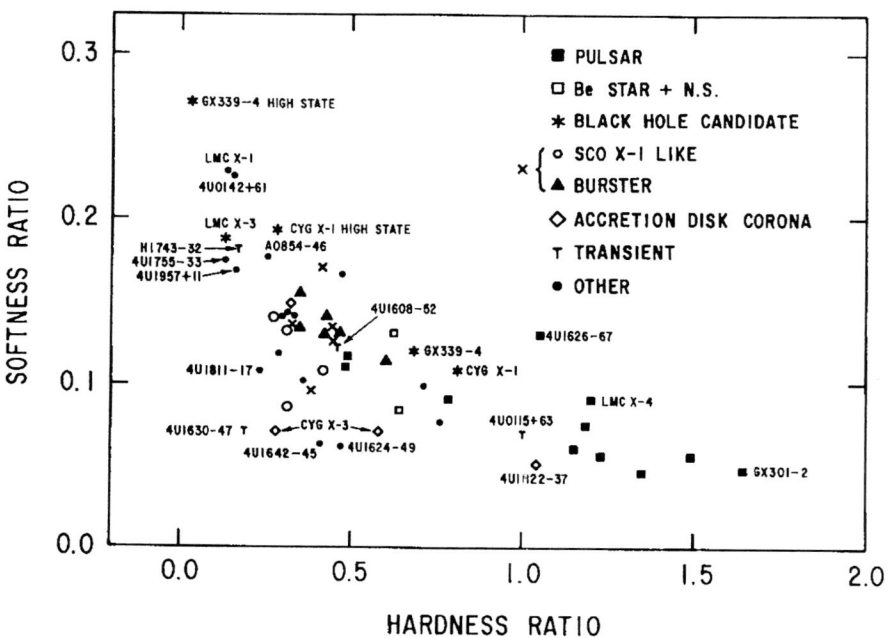

Figure 3 An X-ray color–color diagram. They measured fluxes C1, C2, C3, and C4, in four energy bands 7–30 keV, 3–7 keV, 1.5–3 keV, and 3–10 keV, respectively, with *HEAO-1*. A hardness ratio C1/C2 is plotted against a softness ratio C3/C4 for each source. (From White and Marshall 1984b.)

new X-ray satellites, such as *HEAO 1, Tenma, EXOSAT, Ginga* and *GRANAT*. Thus obtained high-quality X-ray spectra allowed detailed investigation of X-ray spectra of various sources. This prompted the study of the X-ray spectrum of black hole binaries.

In most X-ray binaries, the X-ray spectrum is dominated by a featureless continuum. In earlier times, when the theoretical models of X-ray spectrum had not been well developed, it was a practical exercise to characterize the spectral shape by means of the "softness" (or "hardness") ratio, expressed by the ratio of fluxes in two separate energy bands. The softness ratio is the ratio of the flux in a lower energy band to that in a higher energy band, and the hardness ratio is the inverse. This is a similar approach to one characterizing the color by *UVB* photometry in the optical band. In comparing two spectra of different shapes, when one contains relatively larger flux in the lower energy bands, and the softness ratio is hence larger, the spectrum is said to be softer. The softness (or hardness) ratio defines the average slope of a spectrum, i.e. steepness (or flatness). For thermal emission, it qualitatively represents the temperature, i.e. the softer the spectrum, the lower the temperature.

White and Marshall (1984b) (also White, Kaluzienski and Swank 1984a) were the first to propose, based on the comparison of the softness of the observed X-ray spectra from X-ray binaries, that those with the softest spectra should be regarded as "potential" black hole candidates. At that time, only Cyg X-1 and LMC X-3 were known as probable black hole binaries, and as shown in Figure 3, both of them were among those showing the softest spectra, which White and Marshall called "ultrasoft". At that time, this looked like a bold speculation based on only a few samples of black-hole binaries. Yet, this was a pioneering work in the development of the spectral study of black-hole binaries.

Since then, the spectral study of X-ray binaries has greatly advanced. In what follows, we shall describe the general picture of the X-ray spectra of black-hole binaries in comparison with those of neutron-star low-mass binaries (of which neutron stars are weakly magnetized). For a review, see e.g. Tanaka and Shibazaki (1996) and Tanaka (1997). It is well known by now that the spectral characteristics of X-ray binaries at high X-ray luminosities and low X-ray luminosities are quite different. We deal with them separately.

The X-ray spectrum at high luminosities

When the X-ray luminosity is well above $10^{37}\,\mathrm{erg\,s^{-1}}$, accreting neutron stars in low-mass binaries show a common spectral shape, as shown in Figure 4. It is well known

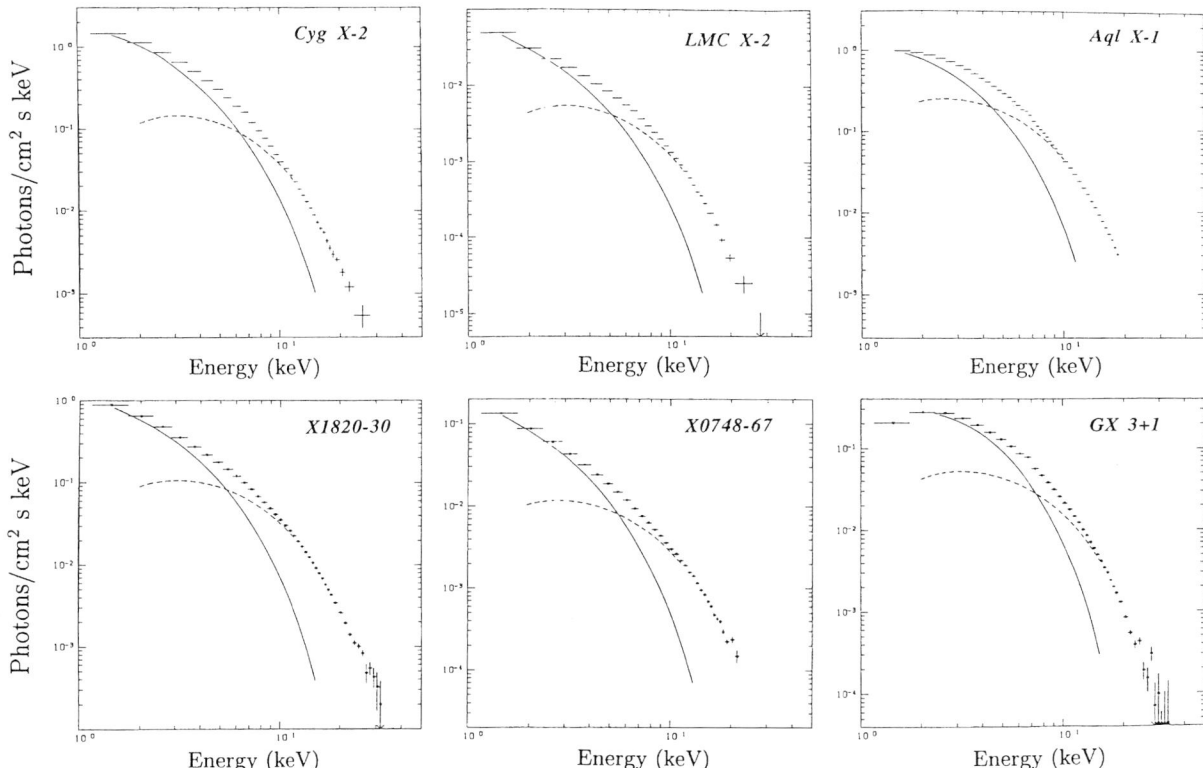

Figure 4 X-ray photon spectra of neutron-star low-mass X-ray binaries at high luminosities obtained with *Ginga*, each consisting of a soft component (solid curve) and a blackbody component (dashed curve).

that the spectrum actually consists of two separate components: a soft component and a harder component. It is important to note that these two components are real, and not those introduced in a mathematical model for the purpose of reproducing the observed spectrum. Since they change in intensity over time independently of each other, these two components can be identified and their spectra determined separately (see Tanaka 1997).

The soft component shows the characteristics of thermal emission. It is well expressed by the "multicolor blackbody disk" model developed by Mitsuda *et al.* (1984) for the emission from an optically-thick accretion disk. Such a disk emits predominantly black-body radiation. The blackbody temperature increases toward the center, and reaches its maximum near the innermost disk; hence the emission is "multi-colored". The multi-color blackbody disk model is a formulation of such a spectrum based on the standard disk model. Based on the good fit of this model, the soft component is identified to be the emission from such an accretion disk. Some other model spectra that fit the observed data were also proposed (White *et al.* 1986, 1988), but the multicolor blackbody disk model on the physical basis has been widely accepted.

The simplest multicolor blackbody disk model includes only two free parameters, i.e. r_{in} and kT_{in}, where r_{in} represents the innermost disk radius, and kT_{in} is the color temperature (note that it is not the effective temperature) at r_{in}. The observed color temperature kT_{in} is typically 1.4–1.5 keV at X-ray luminosities $L_x \sim 10^{38}\,\mathrm{erg\,s^{-1}}$ and goes down as the luminosity decreases. The color temperature is substantially higher than the effective temperature due to the electron scattering effect that dominates at such high temperatures. It is also important to note that this model is based on a non-relativistic approximation. The effects of relativity, special as well as general, become significant in the innermost part of the disk. Correct models should take into account these effects, which gives a slight modification of the spectral shape (see e.g. Hanawa 1989).

The harder component shows approximately a blackbody spectrum with a color temperature of ~2 keV. This blackbody component is most probably the emission from the neutron star surface, where the kinetic energy of accreting matter is eventually thermalized. The fact that this spectrum is very similar to that of X-ray bursts (thermonuclear flash on the neutron star surface, which is a unique signature of a neutron star) also supports this interpretation. The intensity of this blackbody component varies irregularly with time by a large factor (its maximum luminosity being comparable to the soft component) without changing shape. The reason for this time variation is still unknown.

On the other hand, the spectrum of black-hole X-ray binaries at high X-ray luminosities, well above 10^{37} erg s^{-1}, appears distinctly different from that of the neutron-star low-mass X-ray binaries described above. Of the twelve reliable black-hole X-ray binaries known so far (Table 1), ten show X-ray spectra of a common characteristic shape, as shown in Figure 5. They consist of a soft thermal component and a hard tail.

The soft component of these black-hole X-ray binaries is also well expressed by a multicolor blackbody disk model of the same functional form as that of neutron-star low-mass X-ray binaries, hence identified to be the emission from an optically-thick disk. However, the observed color temperature kT_{in} is significantly lower than that for the neutron-star low-mass X-ray binaries at the same luminosity level, and is typically ~1.2 keV when the X-ray luminosity is ~10^{38} erg s^{-1}.

Remarkably, none of them displays the presence of a blackbody component that is characteristic of the spectra of luminous ($L_x > 10^{37}$ erg s^{-1}) neutron-star low-mass X-ray binaries. This is a clear difference. For the reason discussed in the previous section, the absence of the blackbody component argues for the absence of a solid surface. This should be taken as a strong indication of a black hole.

Furthermore, the low kT_{in} values for black holes are also understood in terms of the standard disk model. Suppose the blackbody-emitting disk extends to the innermost stable orbit, the actual radius is proportional to the mass of the central object (see below). For a fixed luminosity, a larger innermost disk radius gives a lower blackbody temperature (kT_{in} scales as $\propto M_x^{-1/4}$ for a given luminosity). Hence, the X-ray spectrum from the disk around a black hole, much more massive than a neutron star, is expected to appear softer (lower temperature) than that around a neutron star. This indeed confirms the earlier suggestion by White and Marshall (1984) that the X-ray spectra of black-hole binaries appear "ultrasoft". They were right!

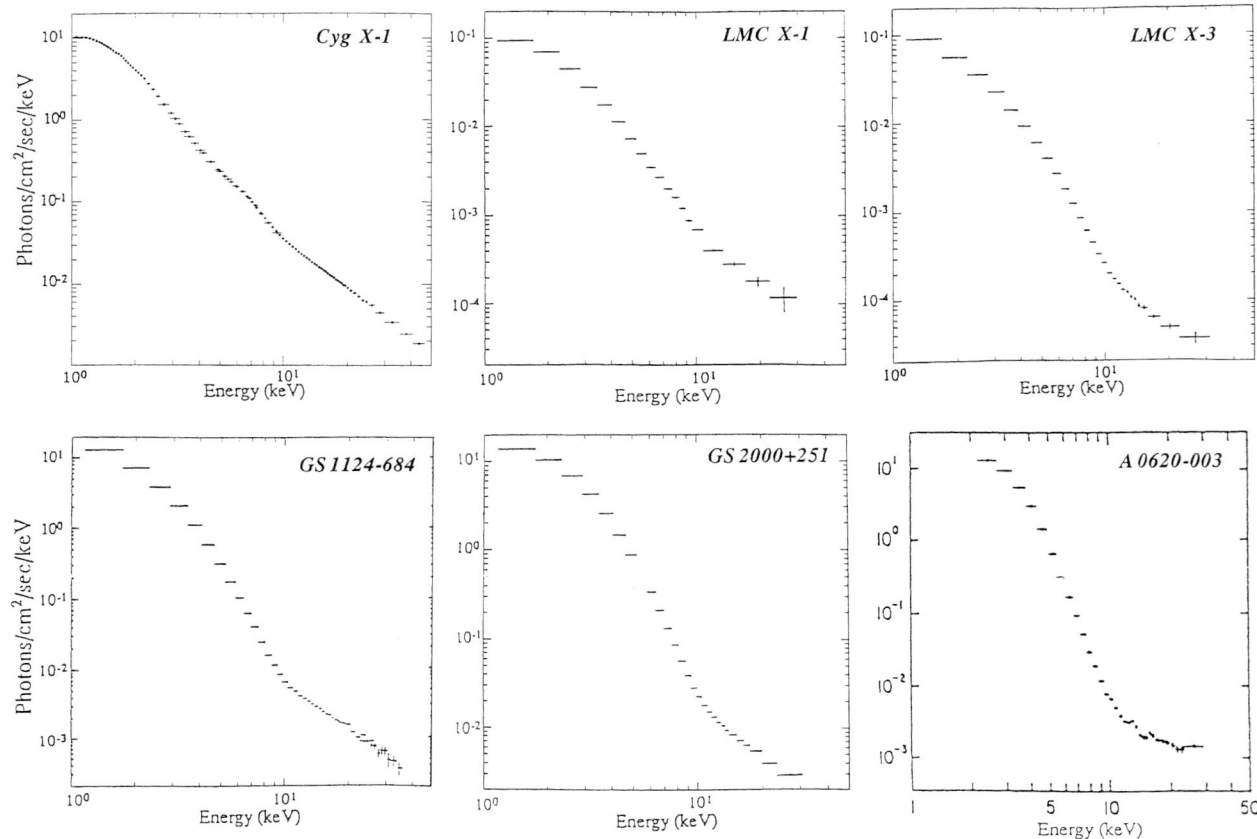

Figure 5 X-ray photon spectra of reliable black-hole X-ray binaries at high luminosities. The spectrum of A 0620-00 is from the OSO-8 data (White et al. 1984a), and that of Cyg X-1 is from the data of ASCA and RXTE (Gierliński et al. 1999). Others are obtained with Ginga. They are commonly characterized by a soft component and a hard tail.

For at least three long-observed black-hole binaries, the observed value of r_{in} remained constant despite large-factor luminosity changes (Tanaka and Lewin 1995). This fact gives support to the interpretation that r_{in} indeed represents the radius of the innermost stable orbit. For a non- or slowly spinning black hole, the innermost stable orbit is 3 Schwartzschild radii ($6 GM_x/c^2$), hence proportional to the compact object mass. If the source distance is known, r_{in} can be estimated from the observed L_x and kT_{in} of the soft component. The estimated values of r_{in} for black-hole X-ray binaries has turned out to be larger by a factor of 3 to 4 than those for neutron-star low-mass X-ray binaries, implying that the compact objects are more massive by this factor than a neutron star (see Tanaka and Lewin 1995; Tanaka 1997 for more detail). Note, however, that the r_{in}-values should not be taken as the real innermost disk radius, because the model is a non-relativistic approximation. The estimation of the actual mass should take into account the general relativistic effect and the difference between the color temperature and the effective temperature (see e.g. Hanawa 1989; Ebisawa, Mitsuda and Hanawa 1991; Zhang, Cui and Chen 1997). In fact, thus-estimated mass values are generally consistent with those obtained from the mass functions (except for the case of GRO J1655−40, as explained below).

The above considerations make it convincing that such a "soft + hard-tail" spectrum is a signature of an accreting black hole. (Note that this holds only when $L_x \gg 10^{37}\,\mathrm{erg\,s^{-1}}$ for the reasons discussed below.) In addition to ten among the twelve reliable black-hole X-ray binaries, fourteen more soft X-ray transients are so far known to show this characteristic soft + hard-tail spectrum when $L_x \gg 10^{37}\,\mathrm{erg\,s^{-1}}$. Based on this property, they are also considered to be "probable" black-holes binaries. As further support, X-ray bursts (a definitive signature of an accreting neutron star) have never been detected from any of them at any luminosity level. Hopefully, these probable black-holes would be upgraded to reliable black holes by future mass function measurements. (Cir X-1 had once been suspected of being a black-hole binary because of its ultrasoft spectrum during flares. However, this source turned out to be a neutron-star binary, as X-ray bursts were detected. Its ultrasoft spectrum is considered to be caused by the expansion of the photosphere as a result of super-critical accretion, since the r_{in}-value increased with luminosity. Also, the ultrasoft X-ray spectrum of Cir X-1 is not accompanied by a hard tail).

It is of interest that the reliable black-hole binary GRO J1655−40 and the probable black-hole binary GRS 1915+105 (showing a "soft + hard-tail" spectrum), both super-luminal jet sources (see Section 5), show significantly higher kT_{in} than other black-hole X-ray binaries (hence smaller r_{in}). Zhang, Cui and Chen (1997) suggest that the black holes in these systems are rapidly spinning

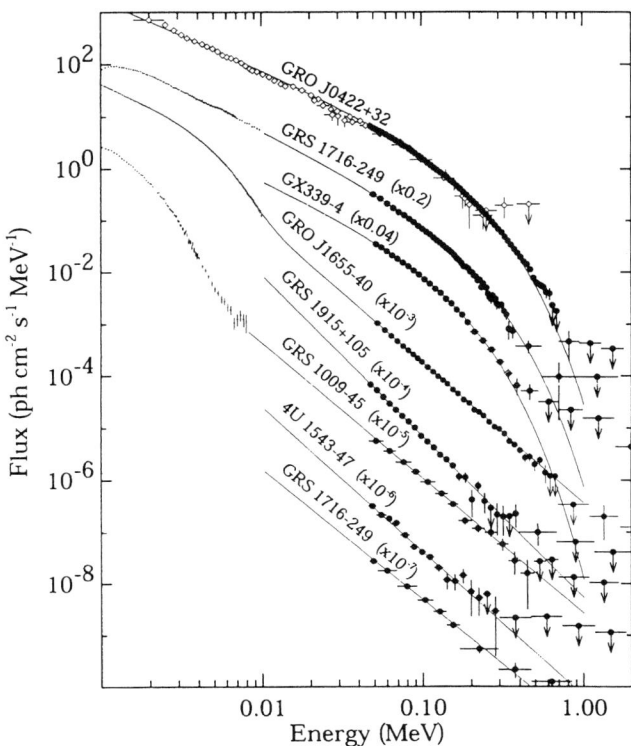

Figure 6 Hard X-ray spectra of reliable and probable black-hole soft X-ray transients observed with *CGRO* OSSE. Note that the upper three are those when the sources are in the hard state, whereas the lower five are those in the soft state. (From Grove *et al.* 1998.)

(Kerr hole) in the same direction as the disk rotation. In this case, the maximum disk temperature can be substantially higher. It is suspected that the high black hole spin is related to the formation of the highly relativistic jets from these sources.

The hard tail that characterizes the black-hole spectrum has a power-law form, and extends to well over 100 keV, sometimes observed up to ~1 MeV without a cut-off, according to *GRANAT* and *CGRO OSSE* results, as shown in Figure 6. The luminosity of the hard component relative to the soft component varies irregularly by a large factor, (see Figure 7). The photon index α of the power-law tail (for a photon number spectrum of the form $E^{-\alpha}$) is ~2.5, and remains remarkably constant against changes in intensity. The power-law spectrum has been considered to be produced as a result of multiple scattering of soft photons with high-energy electrons, by which photons gain energy by the inverse Compton effect, the process called Comptonization (Sunyaev and Titarchuk 1980). Yet, the origin of the power-law tail is still unclear. How to accelerate electrons to such high energies and how to maintain the electron energy against Compton cooling remain serious problems.

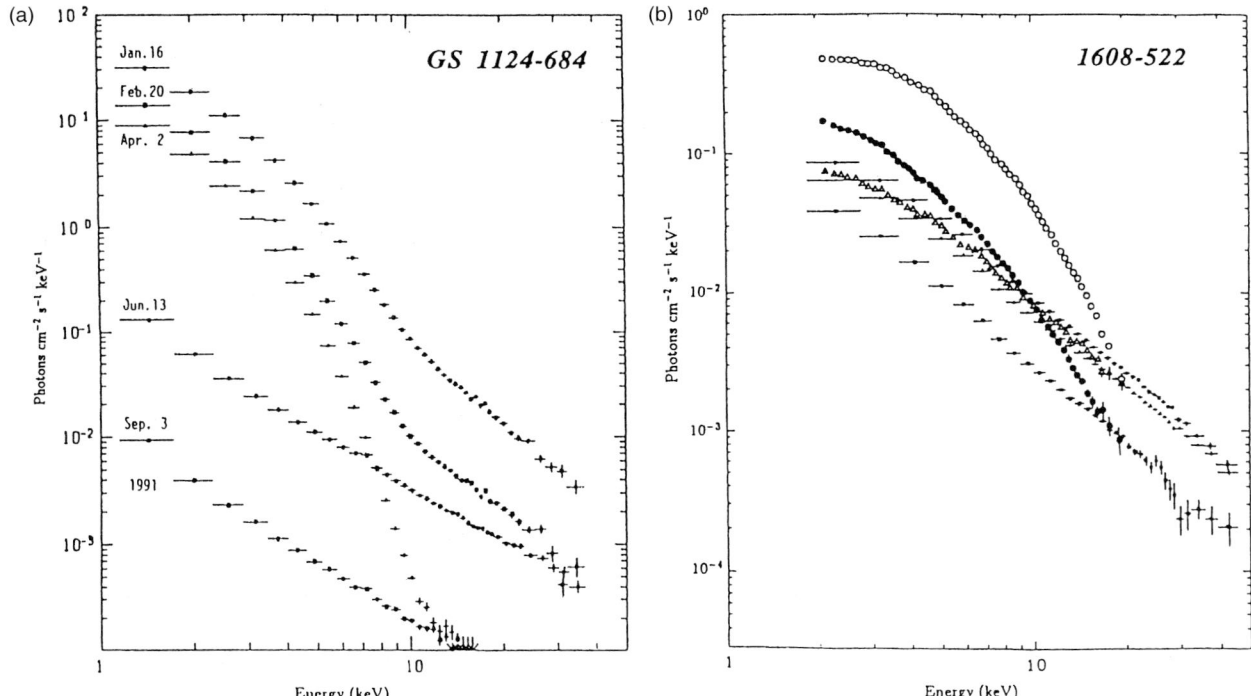

Figure 7 Changes in the spectral shape with luminosity: (a) the spectra of the black-hole X-ray binary GS 1124-684 (*Ginga* data), and (b) the spectra of the neutron-star low-mass X-ray binary 4U 1608-522 (*Tenma* and *Ginga*).

The fact that such a power-law tail is absent in the luminous neutron-star low-mass binaries might suggest that it is formed in the gap between the innermost stable orbit and the Schwartzschild radius. (For the case of a neutron star, the gap is presumably small or absent.) Such a model has been proposed (see e.g. Laurent and Titarchuk 1999). However, the present model still contains such problems as those mentioned above.

A remark is necessary here that among the twelve reliable black-hole X-ray binaries, GS 2023+338 and GRO J0422+32, are exceptions. Both sources showed an approximately single power-law spectrum (typical form at lower luminosities, see below) even at high luminosities. The reason why they did not show the soft + hard-tail spectrum is still unknown. Either an optically-thick thin disk was not formed, which is rather unlikely, or the thin disk was covered by an optically-thin hot disk in which soft photons from the thin disk are Comptonized. This puzzle remains to be understood.

The X-ray spectrum at lower luminosities

The X-ray spectrum at low luminosities is distinctly different from that at high luminosities. Both neutron-star low-mass X-ray binaries and black-hole X-ray binaries exhibit a dramatic change in the spectral shape around a certain luminosity level. The spectral shape changes between a soft thermal form (at high L_x as described above) and a hard power-law form (at low L_x), as shown in Figure 7. Available data indicate that the transition occurs at around an X-ray luminosity $L_x \sim 10^{37}\,\mathrm{erg\,s^{-1}}$ (or a mass accretion rate around $\sim 10^{17}\,\mathrm{g\,s^{-1}}$), but it may vary from source to source and even one transition to another (see Tanaka and Shibazaki 1996). Another outstanding change associated with the spectral state transition is the properties related to time variability. When sources enter into the hard state, rapid large-amplitude intensity fluctuations (flickering) build up in all time scales down to ms. Such transition between the two spectral states has been observed in several soft X-ray transients during the decay phase and also in Cyg X-1. Since this bimodal behavior is seen regardless of whether the compact object is a neutron star or a black hole, it is believed to be a fundamental property of an accretion disk, dependent on the accretion rate.

The power-law spectrum in the hard state is clearly different from the hard tail of black-hole X-ray binaries in the soft state. It is substantially harder with the observed photon indices α in the range 1.7–1.9. Also, unlike the hard tails of black holes in the soft state, the power law spectrum shows a clear fall-off above several tens keV (high-energy cut-off), as seen in a few examples in Figure 6.

Remarkably, these properties are essentially the same for both black-hole binaries and neutron-star low-mass binaries. Therefore, once they go into the hard state, one can no longer distinguish between black holes and neutron stars as far as the spectral shape is concerned.

The power-law spectrum with a cut-off observed in the hard state can well be reproduced by Comptonization of soft photons with hot thermal electrons (thermal Comptonization, see Sunyaev and Titarchuk 1980), which was first applied successfully to the Cyg X-1 spectrum in the hard state by Sunyaev and Trümper (1979).

Transition between the soft state and the hard state has been considered to be due to a change in the disk structure. There is evidence that an optically-thin hot plasma quickly builds up when a source goes into the hard state. For instance, when X 1608 − 522 (a neutron-star low-mass binary) was about to go into the hard state, the spectra of X-ray bursts (blackbody emission from the neutron star surface) from the source also began to show a significant hard tail due to Comptonization (Nakamura et al. 1989). Note that, for this effect, the spectrum of a neutron-star low-mass binary may mimic the "soft + hard-tail" shape near the transition of the spectral state. Therefore, it is important that the spectral distinction between a black hole and a neutron star holds only when L_x is well above $10^{37}\,\mathrm{erg\,s^{-1}}$. Disappearance of a blackbody component in the neutron-star systems (expected anyway from the neutron star surface) also indicates a wide-angle coverage with a hot plasma in the hard state that turns the blackbody photons into a power-law form by Comptonization.

The debate still goes on as to whether the thin disk still extends to the innermost stable orbit in the hard state, embedded in a hot plasma, or whether it recedes out to a larger radius. It is noteworthy that, despite a radical change in the spectral shape, the total luminosity remains essentially the same before and after transition (see e.g. Zhang et al. 1997 for the 1996 transition of Cyg X-1). Hence, the spectral state transition is not due to a change between a radiation-dominated accretion flow and an advection-dominated accretion flow (ADAF) of low radiation efficiency. The physical mechanism for the transition and other unique properties of the hard state are yet to be understood.

Incidentally, the properties of X-ray binaries in the hard state are strikingly similar to those of many active galactic nuclei, i.e. the same power-law index, and high time-variabilities on time scales down to the shortest Keplerian periods. These similarities suggest that, despite huge differences in the scale and power, the basic process of accretion is essentially the same in both systems. On this basis, many properties of active galactic nuclei are interpreted in the light of the results of low-luminosity X-ray binaries. Most of active galactic nuclei are considered to be accreting supermassive black holes in a state corresponding to the hard state of low-luminosity X-ray binaries.

The X-ray spectrum in quiescent state

Soft X-ray transients spend most of their life-times in a quiescent state. Since the X-ray luminosity in quiescence is below $10^{33}\,\mathrm{erg\,s^{-1}}$, observation of them requires high sensitivity. Most of the observations so far were made in the nineties with the imaging instruments of *ROSAT*, *ASCA* and *BeppoSAX* (see Tanaka 1999).

As mentioned in Section 4.2, the observed results of soft X-ray transients in the quiescent state seem to show a systematic difference in luminosity between neutron-star binaries and black-hole binaries, i.e. neutron-star binaries are significantly more luminous than black-hole binaries. In addition, there is a good possibility that the spectral shapes of these systems reveal an important difference, although it is still too early to arrive at a definitive conclusion. Figure 8

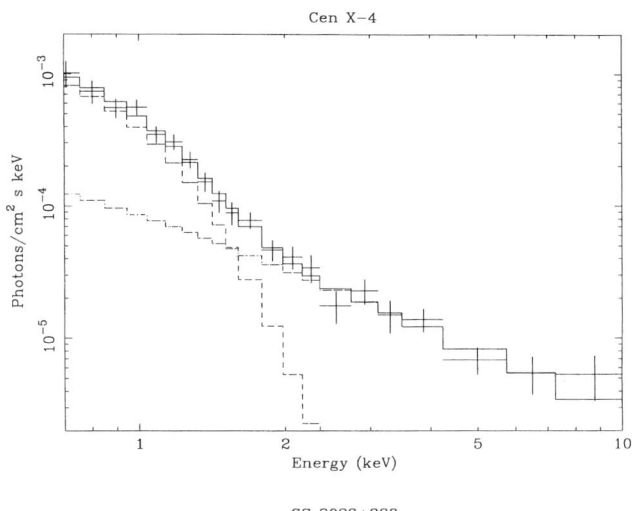

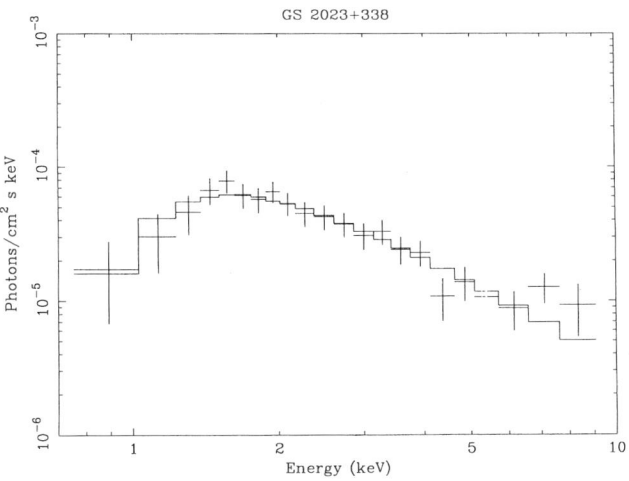

Figure 8 X-ray spectra of two soft X-ray transients in quiescence, the neutron-star binary Cen X-4 and the black-hole binary GS 2023+338, observed with *ASCA*. (Asai *et al.* 1998.)

shows some examples of the observed spectra in the quiescent state. Three neutron-star binaries for which the spectrum was measured commonly show a blackbody component of a low temperature (much lower than in a luminous state), and a hard tail (significantly in at least two of them; Asai et al. 1998). Rutledge et al. (1999) show that the blackbody component is consistent with the thermal emission from the surface of the neutron stars. In contrast, the available spectra of black-hole binaries (yet only two with meaningful statistical accuracy) do not have a blackbody component but show a hard power-law like spectrum (Narayan et al. 1997; Asai et al. 1998). For other reliable black-hole binaries, only luminosity upper limits have so far been obtained, except for a positive detection of A 0620-00 at $\sim 10^{31}$ erg s^{-1}. If these differences are confirmed by future high-sensitivity observations, there could be a chance of finding black-hole binaries even in the quiescent state.

Reflection component

This is another topic of interest related to the X-ray spectra of black-hole binaries. We return here to the subject of luminous X-ray binaries. When an optically-thick disk is illuminated with X-rays, part of the X-rays are reflected by Thomson scattering. This reflection component, predicted by Lightman and White (1988), was first discovered from active galactic nuclei (AGN) by Pounds et al. (1990). Since photoelectric absorption dominates Thomson scattering at low energies, the reflected component is characterized by a hard continuum, much harder than the incident spectrum, a K-absorption edge (7.1 keV), and a fluorescent emission line (6.4 keV) of iron. The iron lines from some Seyfert galaxies are found to be relativistically broadened as predicted by Fabian et al. (1989), which provides clear evidence for the general relativistic effect unique to the close vicinity of massive black holes (see e.g. Tanaka et al. 1995; Nandra et al. 1997).

The reflection component of X-ray binaries has recently been investigated by many people (see e.g. a review by Zdziarski 2000), because it provides useful diagnostics of accretion disks and also has a potential of finding the relativistic effect near the compact object. Done and Życki (1999) claim to have detected relativistic broadening in the reflection spectrum of Cyg X-1. (Also, see Gierliński et al. 1999.) However, it is a subtle feature and may not be conclusive yet for the reasons stated below.

The situation in X-ray binaries seems to be more complicated than in Seyfert galaxies. Because of a smaller disk size by several orders of magnitude, the effect of photoionization becomes much more important. Ross, Fabian and Young (1999) demonstrated a pronounced dependence of the line intensity and relative contribution of the reflected component on the ionization degree of the disk, and showed that Compton scattering in the ionized disk also broaden the line and the absorption edge of iron.

5 RELATIVISTIC JETS

The sources that show relativistic jets in our Galaxy are listed in Table 2 and were taken from the recent review by Mirabel and Rodríguez (1999). In particular, three soft X-ray transients, GRS 1915+105, GRO J1655−40 and XTE J1748−288, have shown superluminal radio jets

Table 2 Sources of relativistic jets in the Galaxy[1]

Source	Compact object	V_{app}[2]	V_{int}[3]	Θ[4]
GRS 1915+105	black hole	1.2c–1.7c	0.92c–0.98c	66°–70°
GRO J1655−40	black hole	1.1c	0.92c	72°–85°
XTE J1748−288	black hole	1.3c	>0.9c	
SS 433	neutron star?	0.26c	0.26c	79°
Cyg X-3	neutron star?	~0.3c	~0.3c	>70°
CI Cam	neutron star?	~0.15c	~0.15c	>70°
Sco X-1	neutron star	~0.5c		
Cir X-1	neutron star	≥0.1c	≥0.1c	>70°
1E 1740.7−294	black hole[?]			
GRS 1758−258	black hole[?]			

[1] Sources reported as of December 1998.
[2] V_{app} is the apparent speed of the highest velocity component of the ejecta.
[3] V_{int} is the intrinsic velocity of the ejecta.
[4] Θ is the angle between the direction of motion of the ejecta with the line of sight.
[?] Note by the present author: Other than the radio jets, no firm evidence for a black hole as yet (see text).

during outbursts with an intrinsic velocity exceeding 0.9c (Mirabel and Rodríguez 1994 for GRS 1915+105; Tingay et al. 1995 and Hjellming and Rupen 1995 for GRO J1655−40; Hjellming et al. 1998 for XTE J1748−288). GRO J1655−40 is a reliable black-hole binary (see Table 1), and the other two are probably also black-hole binaries, as indicated by their spectral signatures discussed in Section 4.3.

On the other hand, Sco X-1 and Cir X-1 are neutron-star binaries. In addition to those listed in Table 1, radio outbursts were detected in association with X-ray outbursts from several other soft X-ray transients, including those of neutron-star low-mass binaries (see Hjellming and Han 1995). Even if radio images were not obtained in earlier observations, Hjellming and Rupen (1995) suggest that these radio outbursts may have been relativistic jets as well, based on the observed properties showing relativistic expansion of plasmas. These facts indicate that relativistic jets alone cannot be evidence for a black hole, although there could be possible systematic differences in the speed and power of jets between black-hole and neutron-star systems. 1E 1740.7−294 and GRS 1758−258 are associated with well-collimated radio structures, most probably relativistic jets. These sources are sometimes suspected to be black-hole binaries. However, they show X-ray spectra of a single power-law form typical of low-luminosity X-ray binaries (see Section 4.3), and no other evidence for being black holes has yet been obtained.

SS433 is a unique source that generates a stationary pair of relativistic jets. Despite extensive studies at various wavelengths, whether the compact object is a black hole or a neutron star is not as yet known. Some people suspect that SS433 and also Cyg X-3 may be black-hole binaries that are accreting matter at a super-critical rate.

6 POPULATION OF BLACK-HOLE BINARIES

Table 3 lists the numbers of persistent and transient low-mass X-ray binaries taken from the catalogue by van Paradijs (1995). The sources of transient outbursts detected later are also added, but previously unknown faint sources that have been discovered more recently, particularly with *BeppoSAX WFC*, are not included. The number of black-hole low-mass binaries includes the "probable" ones classified as such based on the spectral signature described in Section 4.3, in addition to nine reliable ($M_x > 3M_\odot$) ones. The number of low-mass neutron-star binaries includes the uncertain cases that might possibly be black-hole binaries. The classification is more secure for those detected during outbursts (the numbers in parentheses). One can readily notice the following remarkable facts from the table.

1. The majority of neutron-star low-mass X-ray binaries is persistent, whereas none of black-hole low-mass X-ray binaries is persistent. The reason for this outstanding difference is not completely understood as yet. A key may be found in the large difference in the primary to secondary mass ratio between the two systems.

2. The majority of transients that underwent X-ray outbursts are (either reliable or probable) black-hole binaries. This clearly indicates that the best place to look for black holes is the soft X-ray transient outbursts. Since soft X-ray transients spend almost all the life in quiescence, those discovered during outbursts must be a tiny fraction of all black-hole binaries existing in our Galaxy.

Attempts have been made to estimate the total number of black-hole binaries in our Galaxy, based on the current statistics of X-ray outbursts and applying corrections for the detection biases (see Tanaka and Shibazaki 1996; Romani 1998; White and van Paradijs 1996). Outbursts of black-hole binaries are detected roughly twice a year according to the statistics in the last several years. This rate is subject to the detection sensitivity and the efficiency of sky coverage. The true rate of outbursts in the entire Galaxy could be significantly higher. The biggest unknowns that influence the estimation are the average recurrence time of outbursts and the galactic distribution of low-mass black-hole binaries in the Galaxy. Both of them are not too well known at present.

Nonetheless, modest estimations show that at least several hundred and possibly more than one thousand black-hole binaries are present in the whole Galaxy. Note that this is the number of systems in which the companion already fills its Roche-lobe and the mass transfer to the black hole is going on. If one includes those binaries that are still detached, the total number of black-hole binaries would be significantly larger. Since neutron-star low-mass binaries are the minority among soft X-ray transients, this would imply that the total population of black-hole binaries is no smaller, possibly even larger, than that of neutron-star low-mass binaries in our Galaxy.

Table 3 Persistent and transient low-mass X-ray binaries

Class	Persistent	Transient	Transient fraction
Neutron-star binaries	~80	~16[a] (7)[b]	<20%
Black-hole binaries	0	~23 (all)[b]	100%

[a] Includes those for which either a neutron-star or a black-hole binary is uncertain.
[b] Those detected in outburst.

7 FORMATION OF BLACK-HOLE BINARIES

Stellar-mass black holes are generally thought to be produced in supernova explosions of very massive stars, probably of $\geq 30 M_\odot$. However, unlike the case of neutron stars, association of black holes with supernova remnants has never been seen. Nor exists convincing theoretical demonstration of black hole formation in supernovae of very massive stars as yet. Furthermore, formation and evolution of low-mass binaries, containing either a black hole or a neutron star, are not yet fully explained, and are still the subjects of current study. (For a review, see Verbunt and van den Heuvel 1995.) This remains to be one of the key issues regarding black-hole binaries.

Almost all black-hole X-ray binaries known are low-mass systems, except a few high-mass binaries (only one, Cyg X-1, in our Galaxy). This implies that they are predominantly old systems, 10^{8-9} years old, similar to neutron-star low-mass binaries. As far as the optical properties are concerned, there is no noticeable difference in the types of companion stars or in the orbital period distribution between black-hole low-mass binaries and neutron-star low-mass binaries (van Paradijs and McClintock 1995). No systematic trend that suggests that these two systems are in different stages of evolution is apparent.

Yet, there seem to be some differences between the two systems. White and van Paradijs (1996) suggest a difference in the Galactic distribution between neutron-star low-mass binaries and black-hole low-mass binaries, though not conclusive yet and requiring further investigation. Another contrasting fact is that, while as many as 10% of neutron-star X-ray binaries that we know are located in globular clusters (van Paradijs 1995), none of black-hole X-ray binaries has been found to be associated with globular clusters.

On the other hand, the number of black-hole low-mass binaries is at least as large as, or even larger than, that of neutron-star low-mass binaries. There must be reasons that favour formation of black-hole binaries. These facts may present important hints for understanding the formation and evolution of black-hole low-mass binaries.

We have so far presented the results on the black-holes in X-ray binaries in our own galaxy, for which the estimated black-hole masses are around $10 M_\odot$. In the end, it is worth noting that there is increasing X-ray evidence for the presence of more massive compact objects in other galaxies (see e.g., Fabbiano 1989; Colbert and Mushotzky 1999; Makishima et al. 2000). They are ultraluminous X-ray sources ($L_x \sim 10^{40}\,\mathrm{erg\,s^{-1}}$) located off-center of the galaxies (hence not AGN), and are believed to be accreting black holes with masses ranging up to as high as $100 M_\odot$. Such high mass black holes could not be produced by supernovae. These possible black holes of intermediate mass, between stellar mass and that of supermassive black holes in AGN, will undoubtedly become a new important subject.

In conclusion, the presence of stellar mass black holes has been well established, and they are known to be abundant in our Galaxy. However, many fundamental problems, in particular concerning formation and evolution, have not been solved and are left for continued study in the new century.

REFERENCES

Asai, K., Dotani, T., Hoshi, R., Tanaka, Y., Robinson, C.R. and Terada, K. (1998). ASCA observations of transient X-ray sources in quiescence. *Publications of the Astronomical Society of Japan*, **50**, 611–619.

Bolton, C.T. (1972). Identifications of CYG X-1 with HDE 226868. *Nature*, **235**, 271.

Braes, L. and Miley, G.K. (1971). Detection of radio emission from Cygnus X-1. *Nature*, **232**, 246.

Casares, J., Charles, P.A. and Naylor, T. (1992). A 6.5-day periodicity in the recurrent nova V404 Cygni implying the presence of a black hole. *Nature*, **355**, 614–617.

Chen, W., Shrader, C.R. and Livio, M. (1997). The properties of X-ray and optical light curves of X-ray novae. *Astrophysical Journal*, **491**, 312.

Colbert, E.J.M., Mushotzky, R.F. (1999). The nature of accreting black holes in nearby galaxy nuclei. *Astrophysical Journal*, **519**, 89–107.

Cowley, A.P., Crampton, D., Hutchings, J.B., Remillard, R. and Penford, J.E. (1983). Discovery of a massive unseen star in LMC X-3. *Astrophysical Journal*, **272**, 118–122.

Done, C. and Życki, P.T. (1999). Relativistic distortions in the X-ray spectrum of CYG X-1. *Monthly Notices of the Royal Astronomical Society*, **305**, 457–468.

Ebisawa, K., Mitsuda, K. and Hanawa, T. (1991). Application of a general relativistic accretion disk model to LMC X-1, LMC X-3, X1608 - 522, and X1636-536. *Astrophysical Journal*, **367**, 213–220.

Fabbiano, G. (1989). X rays from normal galaxies. *Annual Review of Astronomy and Astrophysics*, **27**, 87–138.

Fabian, A.C., Rees, M.J., Stella, L. and White, N.E. (1989). X-ray fluorescence from the inner disc in Cygnus X-1. *Monthly Notices of the Royal Astronomical Society*, **238**, 729–736.

Filippenko, A.V., Matheson, T. and Barth, A.J. (1995b). A black hole in the X-ray nova GS 2000+25. *Astrophysical Journal Letters*, **455**, L139.

Filippenko, A.V., Matheson, T. and Ho, L.C. (1995a). The mass of the probable black hole in the X-ray nova GRO J0422+32. *Astrophysical Journal*, **455**, 614.

Filippenko, A.V., Leonard, DC., Matheson, T., Li, W., Moran, E.C. and Riess, A.G. (1999). A black hole in the X-ray Nova Velorum 1993. *Publications of the Astronomical Society of the Pacific*, **111**, 969–979.

Giacconi, R., Gursky, H., Paolini, F. and Rossi, B. (1962). Evidence for X-rays from sources outside the solar system. *Physical Review Letters*, **9**, 439–443.

Gies, D.R. and Bolton, C.T. (1982). The optical spectrum of HDE 226868 = Cygnus X-1. I – Radial velocities and orbital elements. *Astrophysical Journal*, **260**, 240–248.

Grove, J.E., Johnson, W.N., Kroeger, R.A., McNaron-Brown, K. and Skibo, J.G. (1998). Gamma-ray spectral states of galactic black hole candidates. *Astrophysical Journal*, **500**, 899.

Gierliński, M., Zdziarski, A.A., Poutanen, J., Coppi, P.S., Ebisawa, K. and Johnson, W.N. (1999). Radiation mechanisms and geometry of Cygnus

X-1 in the soft state. *Monthly Notices of the Royal Astronomical Society*, **309**, 496–512.

Hanawa, T. (1989). X-ray emission from accretion disks in low-mass X-ray binaries. *Astrophysical Journal*, **341**, 948–954.

Hewish, A., Bell, S.J., Pilkington, J.D.H., Scott, P.F. and Collins, R.A. (1968). Observation of a rapidly pulsating radio source. *Nature*, **217**, 709–713.

Hjellming, R.M. and Wade, C.M. (1971). Radio emission from X-ray sources. *Astrophysical Journal Letters*, **168**, L21

Hjellming, R.M. and Han, X. (1995). Radio properties of X-ray binaries. In W.H.G. Lewin, J. van Paradijs and E.P.J. van den Heuvel (eds), *X-Ray Binaries*, Cambridge University Press, Cambridge, pp. 308–330.

Hjellming, R.M. and Rupen, M.P. (1995). Episodic ejection of relativistic jets by the X-ray transient GRO:J1655-40. *Nature*, **375**, 464.

Hjellming, R.M., Rupen, M.P., Mioduszewski, A.J., Smith, D.A., Harmon, B.A., Waltman, E.B., Ghigo, F.D. and Pooley, G.G. (1998). Radio and X-ray observations of the new relativistic jet X-ray transient XTE Jl748-288. *American Astronomical Society Meeting*, 193, 103.08.

Hutchings, J.B., Crampton, D. and Cowley, A.P. (1983). A spectrographic orbit for LMC X-l – Another massive X-ray source? *Astrophysical Journal Letters*, **275**, L43–L47.

Hutchings, J.B., Crampton, D., Cowley, A.P., Bianchi, L. and Thompson, I.B. (1987). Optical and UV spectroscopy of the black hole binary candidate LMC X-1. *Astronomical Journal*, **94**, 340–344.

Ichimaru, S. (1977). Bimodal behavior of accretion disks – Theory and application to Cygnus X-l transitions. *Astrophysical Journal*, **214**, 840–855.

Lasota, J.-P. (1999). ADAFs – models, observations and problems. *Physics Reports*, **311**, 247–258.

Laurent, P. and Titarchuk, L. (1999). The converging inflow spectrum is an intrinsic signature for a black hole: Monte Carlo simulations of Comptonization on free-falling electrons. *Astrophysical Journal*, **511**, 289–297.

Lightman, A.P. and White, T.R. (1988). Effects of cold matter in active galactic nuclei – A broad hump in the X-ray spectra. *Astrophysical Journal*, **335**, 57–66.

Makishima, K., Kubota, A., Mizuno, T., Ohnishi, T, Tashiro, M, Aruga, Y., Asai, K., Dotani, T., Mitsuda, K., Ueda, Y., Uno, S., Yamaoka, K. and Ebisawa, K. (2000). The nature of ultraluminous compact X-ray sources in nearby spiral galaxies. *Astrophysical Journal*, **535**, 632–643.

McClintock, J., Bailyn, C. and Remillard, R. (1992). MUSCAE 1991, IAU Circ., 5499, 1.

McClintock, J.E. and Remillard, R.A. (1986). The black hole binary A0620-00. *Astrophysical Journal*, **308**, 110–122.

Menou, K., Esin, A.A., Narayan, R., Garcia, M.R., Lasota, J.-P. and McClintock, J.E. (1999). Black hole and neutron star transients in quiescence. *Astrophysical Journal*, **520**, 276–291.

Mirabel, I.F. and Rodríguez, L.F. (1994). A superluminal source in the Galaxy. *Nature*, **371**, 46.

Mirabel, I.F. and Rodríguez, L.F. (1999). Sources of relativistic jets in the Galaxy. *Annual Review of Astronomy and Astrophysics*, **37**, 409–443.

Mitsuda, K., Inoue, H., Koyama, K., Makishima, K., Matsuoka, M., Ogawara, Y., Shibazaki, N., Suzuki, K., Tanaka, Y. and Hirano, T. (1984). Energy spectra of low-mass binary X-ray sources observed from TENMA. *Publications of the Astronomical Society of Japan*, **36**, 741–759.

Nakamura, N., Dotani, T., Inoue, H., Mitsuda, K., Tanaka, Y. and Matsuoka, M. (1989). TENMA observation of X-ray bursts from X1608-52. *Publications of the Astronomical Society of Japan*, **41**, 617–639.

Nandra, K., George, I.M., Mushotzky, R.F., Turner, T.J. and Yaqoob, T. (1997). *Astrophysical Journal*, **477**, 602.

Narayan, R., Barret, D. and McClintock, J.E. (1997). Advection-dominated accretion model of the black hole V404 Cygni in quiescence. *Astro-physical Journal*, **482**, 448.

Narayan, R., Mahadevan, R. and Quataert, E. (1999). Advection-dominated accretion around black holes. In M.A. Abramowicz, G. Bjornsson and J.E. Pringle (eds), *The Theory of Black Hole Accretion Discs*, Cambridge University Press, Cambridge, pp. 148–182.

Narayan, R., McClintock, J.E. and Yi, I. (1996). A new model for black hole soft X-ray transients in quiescence. *Astrophysical Journal*, **457**, 821.

Oda, M. (1977). CYG X-l – A candidate of the black hole. *Space Science Reviews*, **20**, 757–813.

Oppenheimer, J.R. and Snyder, H. (1939). On continued gravitational contraction. *Physical Review*, **56**, 455–459.

Orosz, J.A. and Bailyn, C.D. (1997). Optical observations of GRO J1655-40 in quiescence. I. A precise mass for the black hole primary. *Astrophysical Journal*, **477**, 876.

Orosz, J.A., Jain, R.K., Bailyn, C.D., McClintock, J.E. and Remillard, R.A. (1998). Orbital parameters for the soft X-ray transient 4U 1543-47: Evidence for a black hole. *Astrophysical Journal*, **499**, 375.

Pounds, K.A., Nandra, K., Stewart, G.C., George, I.M. and Fabian, A.C. (1990). X-ray reflection from cold matter in the nuclei of active galaxies. *Nature*, **344**, 132–133.

Remillard, R.A., Orosz, J.A., McClintock, J.E. and Bailyn, C.D. (1996). Dynamical evidence for a black hole in X-ray nova Ophiuchi 1977. *Astrophysical Journal*, **459**, 226.

Romani, R.W. (1998). A census of low mass black hole binaries. *Astronomy and Astrophysics*, **333**, 583–590.

Ross, R.R., Fabian, A.C. and Young, A.J. (1999). X-ray reflection spectra from ionized slabs. *Monthly Notices of the Royal Astronomical Society*, **306**, 461–466.

Rutledge, R.E., Bildsten, L., Brown, E.F., Pavlov, G.G. and Zavlin, V.E. (1999). The Thermal X-ray spectra of Centaurus X-4, Aquila X-1, and 4U 1608-522 in quiescence. *Astrophysical Journal*, **514**, 945–951.

Shakura, N.I. and Sunyaev, R.A. (1973). Black holes in binary systems. Observational appearance. *Astronomy and Astrophysics*, **24**, 337–355.

Shklovsky, I.S. (1967). On the nature of the source of X-ray emission of SCO XR-1. *Astrophysical Journal*, **148**, L1.

Sunyaev, R.A. and Titarchuk, L.G. (1980). Comptonization of X-rays in plasma clouds – Typical radiation spectra. *Astronomy and Astrophysics*, **86**, 121–138.

Sunyaev, R.A. and Trümper, J. (1979). Hard X-ray spectrum of CYG X-1 *Nature*, **279**, 506–508.

Tanaka, Y. (1997). X-ray spectrum of low-mass X-ray binaries. In E. Meyer-Hofmeister and H. Spruit (eds), *Accretion Disk – New Aspects*, Lecture Notices in Physics 487, Springer-Verlag, Berlin, p. 1.

Tanaka, Y. (1999). X-ray observations of X-ray novae. In S. Mineshige and J.C. Wheeler (eds), *Disk Instabilities in Close Binary Systems. 25 Years of the Disk-instability Model*, Universal Academy Press, Tokyo, p. 21.

Tanaka, Y. and Lewin, W.H.G. (1995). Black-hole binaries. In W.H.G. Lewin, J. van Paradijs and E.P.J. van den Heuvel (eds), *X-Ray Binaries*, Cambridge University Press, Cambridge, pp. 126–174.

Tanaka, Y., Nandra, K., Fabian, A.C., Inoue, H., Otani, C., Dotani, T., Hayashida, K., Iwasawa, K., Kii, T., Kunieda, H., Makino, F. and Matsuoka, M. (1995). *Nature*, **375**, 659.

Tanaka, Y. and Shibazaki, N. (1996). X-ray novae. *Annual Review of Astronomy and Astrophysics*, **34**, 607–644.

Tingay, S.J., Jauncey, D.L., Preston, R.A., Reynolds, J.E., Meier, D.L., Murphy, D.W., Tzioumis, A.K., McKay, D.J., Kesteven, M.J., Lovell, J.E.J., Campbell-Wilson, D., Ellingsen, S.P., Gough, R., Hunstead, R W., Jones, D.L., McCulloch, P.M., Migenes, V., Quick, J., Sinclair, M.W. and Smits, D. (1995). Relativistic motion in a nearby bright X-ray source. *Nature*, **374**, 141.

van Paradijs, J. (1995). A catalogue of X-ray binaries. In W.H.G. Lewin, J. van Paradijs and E.P.J. van den Heuvel (eds), *X-Ray Binaries*, Cambridge University Press, Cambridge, pp. 536–577.

van Paradijs, J. and McClintock, J.E. (1995). Optical and ultraviolet observations of X-ray binaries. In W.H.G. Lewin, J. van Paradijs and

E.P.J. van den Heuvel (eds), *X-Ray Binaries*, Cambridge University Press, Cambridge, pp. 58–125.

Verbunt, F. and van den Heuvel, E.P.J. (1995). Formation and evolution of neutron stras and black holes in binaries. In W.H.G. Lewin, J. van Paradijs and E.P.J. van den Heuvel (eds), *X-Ray Binaries*, Cambridge University Press, Cambridge, pp. 457–494.

Webster, B.L. and Murdin, P. (1972). Cygnus X-1 – A spectroscopic binary with a heavy companion? *Nature*, **235**, 37–38.

White, N.E., Kaluzienski, J.L. and Swank, J.H. (1984a). The spectra of X-ray transients. In S.E. Woosley (ed), *High Energy Transients in Astrophysics*, AIP, New York, p. 31.

White, N.E. and Marshall, F.E. (1984b). The unusually soft X-ray spectrum of LMC X-3. *Astrophysical Journal*, **281**, 354–359.

White, N.E., Peacock, A., Hasinger, G., Mason, K.O., Manzo, G., Taylor, B.G., Branduardi-Raymont, G. (1986). A study of the continuum and iron K line emission from low-mass X-ray binaries. *Monthly Notices of the Royal Astronomical Society*, **218**, 129–138.

White, N.E., Stella, L. and Parmar, A.N. (1988). The X-ray spectral properties of accretion discs in X-ray binaries. *Astrophysical Journal*, **324**, 363–378.

White, N.E. and van Paradijs, J. (1996). The galactic distribution of black hole candidates in low mass X-ray binary systems. *Astrophysical Journal Letters*, **473**, L25.

Zdziarski, A.A. (2000). Radiative processes and geometry of spectral states of black-hole binaries. In P.C.H. Martens, S. Tsuruta and M.A. Weber (eds), *Highly Energetic Physical Processes and Mechanisms for Emission from Astrophysical Plasmas*, IAU Symposium 195, Astronomical Society of the Pacific, San Francisco, pp. 153–170.

Zhang, S.N., Cui, W. and Chen, W. (1997). Black hole spin in X-ray binaries: Observational consequences. *Astrophysical Journal Letters*, **482**, L155.

Zhang, S.N., Cui, W., Harmon, B.A., Paciesas, W.S., Remillard, R.E. and van Paradijs, J. (1997). The 1996 soft state transition of Cygnus X-1. *Astrophysical Journal Letters*, **477**, L95.

37

CHRISTOFFEL WAELKENS*

The formation of stars and protoplanetary disks

The discovery that stars and galaxies evolve, and with them the Universe, is probably the major legacy of twentieth-century astronomy. On the threshold of the new millennium, questions about the origins of the Universe and its constituent galaxies, stars and planets are the main drivers of astronomical research. Answers to these big, fundamental questions will require a timely interplay between ground-based and space-borne observations.

Stars are the main energy sources in the Universe, and their study has always been a central theme in astronomy. In the quest for our origins, stars will continue to play a crucial role. 'Galaxy formation cannot be understood without incorporating a detailed theory of star formation' (Silk 1997). Such a theory also is a prerequisite for understanding the formation of planetary systems in the circumstellar disks with which most stars appear to be born.

While the basic insights into the gravitational instability of molecular clouds date back to Jeans's work (Jeans 1902), the theory of star formation is still far from complete. The formation of structure in the Universe turns out to be a complex process which needs sustained observational efforts to be unravelled. Our present database on star formation is still restricted, since many relevant phenomena occur in regions that are shielded from our view at optical wavelengths. Ground-based and space-borne facilities that are able to probe newly formed stars have been developed only recently. While optical pictures of star-forming regions are often spectacular (Figure 1), one must realize that the spectral energy distribution (SED) of young stellar objects (YSOs) peaks at far-infrared wavelengths, where the Earth's atmosphere is a strong emitter. Consequently,

infrared and submillimeter telescopes in space are crucial for unravelling the star formation process observationally and for detecting YSOs which can be observed in adjacent wavelength intervals from the ground. A second window suitable for the detection of YSOs from space is the X-ray window: low-mass YSOs develop convective envelopes, the activity of which generates X-rays that pass relatively unaffected through the molecular cloud material.

Observations from space have, moreover, revealed the existence of circumstellar disks not only of YSOs, but also of stars which already have significantly evolved on the main sequence. Such disks were first discovered from their thermal infrared radiation, and subsequently directly observed in scattered light, in particular by the Hubble Space Telescope. The study of such disks, and the realization that their occurrence is fairly widespread, have triggered very active research on the evolution of such disks towards planetary systems. Only from space can the spectra of these disks and their observational links to Solar System objects be studied in detail. The Infrared Space Observatory (ISO) has highlighted the rich chemistry and mineralogy of the dusty environments of YSOs. Observations from above the Earth's atmosphere can also provide the high angular resolution that is necessary for probing the structure of such disks, with possible links to the presence of planets.

STAR FORMATION: BASIC SCENARIOS AND TERMINOLOGY

Following Shu *et al.* (1987), star formation can be described as consisting of four stages. The first stage is the formation of dense cores in molecular clouds; these

* Katholieke Universiteit Leuven, Belgium

Figure 1 Hubble Space Telescope image of the star-forming region in the Eagle Nebula. (Courtesy of NASA.)

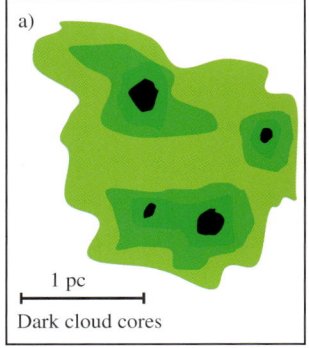
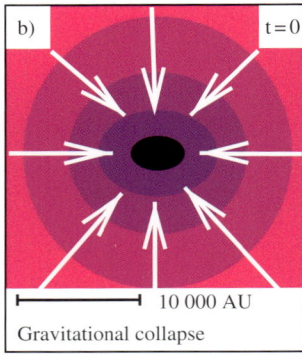
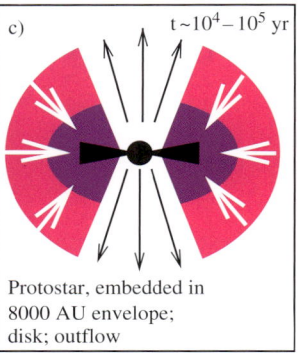
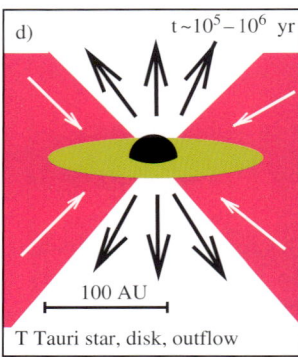
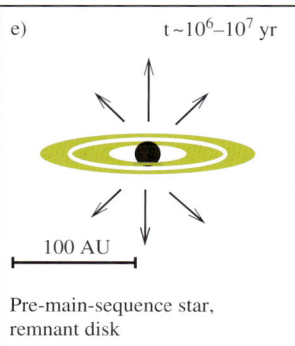
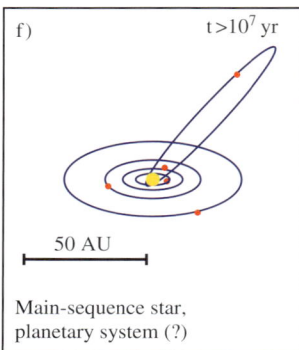

Figure 2 Schematic representation of stellar formation and early stellar evolution (Hogerheijde 1997). With respect to the original scheme, the third (outflow) phase has been split up in two subphases before and after the evaporation of the envelope, and a sixth phase, where planets have formed in the disk, has been added. (After Shu *et al*. 1987.)

cores are supported by turbulent and magnetic pressure, which gradually decrease due to ambipolar diffusion (Mouschovias 1991). Magnetic pressure is not able to support the more massive cores, which are thought to collapse more rapidly. In the second stage, a central protostar already occurs: it collapses rapidly in its centre, and less so in its outer layers. Matter originating far from the rotation axis has too large an angular momentum to fall onto the protostar and settles in a circumstellar disk; the mass of the latter itself, may, in fact, be higher than that of the central object. The third stage is characterized by a bipolar outflow aligned with the rotation axis. The confinement of the outflow to a narrow cone has led to the suggestion that it is related to magnetic activity, but the occurrence of powerful outflows in massive objects and in the accretion-powered UX Orionis stars is consistent with the hypothesis that accretion energy feeds the outflows. During the fourth stage both infall and outflow are terminated, and a newly formed star with a circumstellar disk emerges. A schematic view of this evolution process was pictured by Hogerheijde (1997) and is shown in Figure 2.

From an observational point of view, YSOs emerge as optically visible, but still heavily obscured, objects during the third stage in the Shu *et al*. (1987) scenario. Younger, still completely embedded, objects have SEDs with a positive spectral index in the far-infrared and are called 'class I' sources in the observational classification scheme proposed by Lada and Wilking (1984; see also Lada 1987, Lada and Shu 1990). This scheme is illustrated in Figure 3. YSOs in stages 3 and 4 are also called 'class II' sources; their SEDs are characterized by a more or less reddened stellar component and an infrared excess. This classification scheme has been extended towards 'class III' YSOs, the infrared excess of which has essentially disappeared, but for which circumstellar gas causes atomic emission lines, and also towards 'class 0' sources, which emit the bulk of their energy in the sub millimetre and millimetre domains (André *et al*. 1993), and even towards 'class-I' sources (Boss and Yorke 1995), where the collapse has just been initiated and which still await detection.

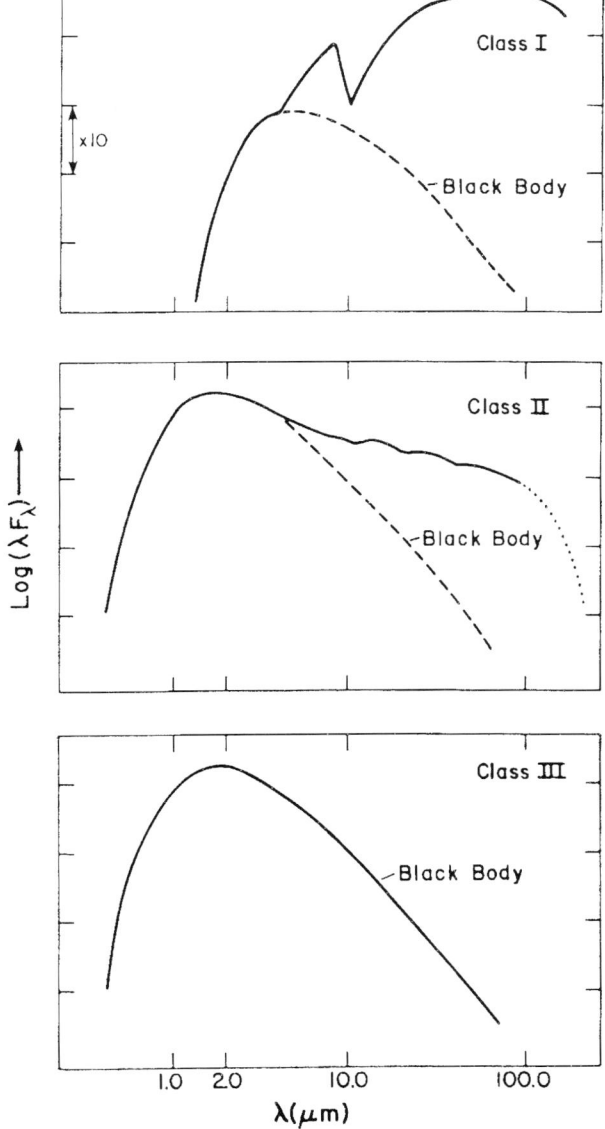

Figure 3 Observational classification scheme of YSOs. The 'black bodies' in the plots incorporate a correction factor to account for circumstellar reddening. (After Lada and Shu 1990.)

Recent results have provided evidence for the fact that many stars, if not all, are formed as binaries. Inventories of binary objects in star-forming regions (SFRs) indicate a larger fraction of binaries than for main-sequence stars in the field and in evolved clusters (Mathieu 1994). Some studies suggest that in low-mass SFRs nearly all stars are born as binaries: for the Taurus SFR, Leinert et al. (1993) and Ghez et al. (1993) find that some 85% of the objects are binaries, that is, much more than the 53% found among G dwarfs in the solar neighbourhood (Duquennoy and Mayor 1991). In high-mass SFRs an excess of binaries is still observed, but it is less marked. Prosser et al. (1994) derive a 'normal' binary frequency for the Orion Trapezium cluster, but current data are felt to be insufficient for firm conclusions. The two main scenarios for binary-star formation are fragmentation (where the cloud cores fragment before rapid collapse is initiated), and gravitational instability in the massive circumstellar disk. It is possible that the inventories of binaries in SFRs teach us that binary-star formation is the natural outcome, and that single stars are either disrupted binaries, following dynamical interactions which indeed are more common in the denser high-mass SFRs, or objects whose disks have been rapidly photo-evaporated.

Dynamical interaction in SFRs clearly also influences the evolution of the circumstellar disks, as does the interaction with the ionized winds of already formed stars. In various clusters it has been noted that for stars which occur close to each other in the Hertzsprung–Russell diagram, and thus are of about the same mass and age, the infrared excesses – which directly probe the mass of the disk – widely differ (for example, NGC2244; Pérez et al., 1987). The highest fraction of infrared excesses are found for low-mass SFRs (e.g. Hillenbrand et al. 1995), where dynamical interactions are less frequent and the ionizing flux is small. The dynamical evolution of the disks and their ablation by the ionizing radiation may thus indeed contribute to the lower binary fraction in high-mass SFRs. In addition, infrared excesses as well as X-ray surveys, as tools for detecting YSOs, have revealed a surprisingly high number of putative YSOs that occur outside known SFRs, suggesting the possibility of almost isolated star formation, with relatively long-lived disks.

In the Hertzsprung–Russell diagram, the 'stellar birth-line' (Stahler 1983) is defined as the locus of points at which a protostar no longer accretes mass, and thus initiates its pre-main-sequence phase. For galactic stars more massive than about eight solar masses, the circumstellar environment is still optically thick when the central object initiates core hydrogen burning and thus arrives on the main sequence; in more metal-deficient and therefore less dusty galaxies, this mass limit is higher. High-mass pre-main-sequence stars are thus not observable at optical wavelengths. YSOs with masses between two and eight solar masses, which emerge before having reached the main sequence, are called Herbig-Ae/Be stars, hereafter called Haebe stars. During their quasi-static contraction towards the main sequence, they have radiative envelopes. YSOs of lower mass are called T Tauri stars and have convective envelopes, with associated activity which causes intense emission of X-rays. Class-II T Tauri stars are also called 'classical T Tauri stars' (CTTs), and class-III T Tauri stars correspond to 'weak-lined T Tauri stars' (wTTs) or 'naked

T Tauri stars' for which the circumstellar material has almost disappeared.

DETECTING THE YOUNGEST STELLAR POPULATIONS

The cool and embedded nature of YSOs implies that infrared, sub millimetre, millimetre and radio techniques, from the ground and from space, are needed to detect and study these objects. All techniques are plagued with their own selection effects, involving sensitivity as well as angular resolution, so that a suitable combination is required in order to obtain a complete picture. The all-sky survey in the 12 to 100 μm region by the Infrared Astronomical Satellite (IRAS) led to the detection of a large number of YSOs of classes I and II, but was hampered by both relatively low sensitivity and low spatial resolution. A significant improvement has been obtained with ESA's Infrared Space Observatory (ISO), especially from the ISOCAM surveys. Nevertheless, all space surveys so far are confusion limited except for the few nearest SFRs, so that an unbiased picture will probably not emerge before the surveys foreseen by the Herschel satellite, to be launched in 2007, are available. Overcoming the confusion limit in surveys is essential for a proper understanding of the initial mass function (IMF) in various environments.

Large-scale, ground-based continuum surveys are sensitive to the small amounts of ionizing gas occurring in YSOs of several classes, but often miss the low-mass YSOs for which such radiation is faint. However, X-ray surveys have proved rather efficient in detecting the population of young, often naked, objects in which vigorous stellar activity takes place. After initial successes in this field with the Einstein observatory (Vaiana et al. 1981), substantial progress has been made with the ROSAT all-sky survey. Optical follow-up of the ROSAT survey in various SFRs has impressively increased the amount of – mostly weak-lined – known T Tauri stars (for a review, see Neuhauser 1997). In the Taurus region, Neuhauser et al. (1995) detected with ROSAT 43 out of 65 wTTs, but only 9 out of 79 CTTs; they found that the larger detection rate in weak-lined objects was intrinsic, being related to, among other parameters, a higher rotation rate. Interestingly, the X-ray technique has also proved useful for detecting objects in the brown dwarf mass range (Neuhauser and Comeron 1998).

A somewhat surprising result of the IRAS and ROSAT all-sky surveys – that is, surveys not limited to known SFRs – and their follow-up, has been the detection of several YSOs which appear isolated in the sky or as small associations (Gregorio-Hetem et al. 1992, Kastner et al. 1997). The relative importance of such a mode of small-scale star formation is currently not very clear, one possible bias being the fact that the isolated nature of an object may imply a less disturbed and therefore longer evolution for its circumstellar structures. The so-called Bok globules, which are small, dark clouds and are easily detected from the obscuration they cause to the background, appear to be the natural sites for isolated star formation (Bok and Reilly 1947).

Some 10 or even 25% of the known giant molecular clouds appear devoid of current star formation (Blitz 1993). Though for distant clouds this apparent absence of star formation may be due to the limited sensitivity of current surveys, Mooney and Solomon (1998) argue that the case for quiescent molecular clouds is genuine. Detailed mapping of these clouds with future instrumentation should reveal whether they represent an early stage in molecular cloud evolution or whether star formation is inhibited in them. It is clear that we are currently still lacking a definitive theory for understanding the processes of fragmentation and core formation in molecular clouds (André 1997).

CIRCUMSTELLAR DISKS REVEALED

Disks appear to be a natural byproduct of star formation, but direct confirmation of the disk-like structure of the matter close to young stars has had to await the recent advances in high-angular-resolution observing techniques. With the Wide Field and Planetary Camera on the Hubble Space Telescope (HST), Burrows et al. (1996) for the first time were able to resolve the vertical structure of a YSO disk, for the Herbig-Haro object HH30 (Figure 4). The object, viewed nearly edge-on, appears as a circumstellar disk, the equatorial part of which is optically thick, and which shows a vertical decrease in density so that scattered light from the central object can emerge. Perpendicular to the disk, highly collimated bipolar jets are observed. ISOCAM and ISOPHOT observations of this object (Stapelfeldt and Moneti 1999) reveal a double-peaked SED, where the scattered light peaks around 2 μm and the thermal emission from the disk peaks in the far-infrared.

In the youngest stellar objects, circumstellar disks are strongly affected in their appearance by their surroundings. The so-called 'proplyds' (Figure 5) are flattened circumstellar clouds of gas and dust which are rendered visible as a result of the ionizing radiation of hot stars in a H II region to which they belong, or close to which they are located. In some cases (Figure 5) the optically thick dusty inner disk can be seen in absorption against the background, but the optical appearance of proplyds is most often dominated by recombination radiation from their outer parts. Such objects were detected in the central parts of the Orion Nebula as emission-line sources (Laques and Vidal, 1979) and as compact radio continuum sources (Churchwell et al. 1987, Garay et al. 1987), and were resolved spatially with

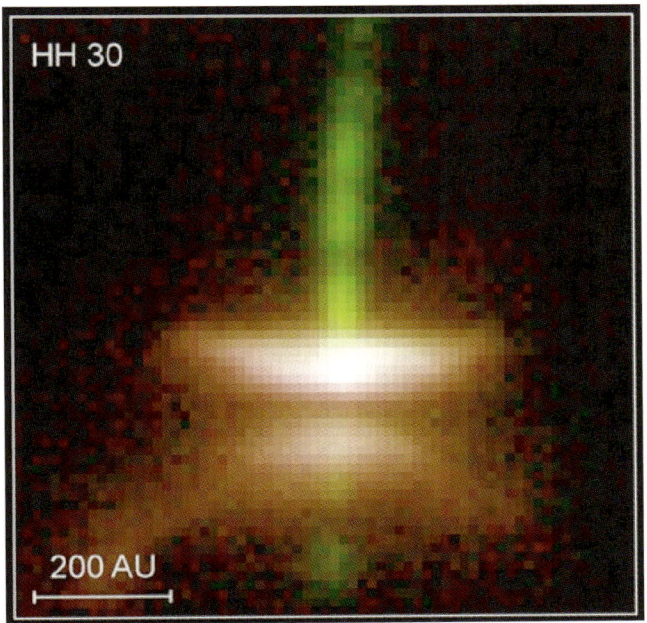

Figure 4 Hubble Space Telescope image of the YSO disk of the Herbig-Haro object HH30 (Burrows *et al.* 1996). The optically thin upper layers dominate in this optical picture, where a bipolar outflow perpendicular to the disk is also apparent. (Courtesy of NASA.)

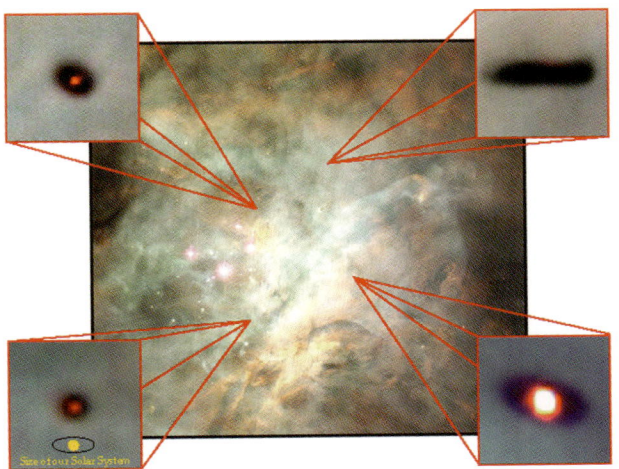

Figure 5 HST image of proplyds in the Orion Nebula. (Courtesy of NASA.)

HST by O'Dell *et al.* (1993), O'Dell and Wen (1994), McCaughrean and Stauffer (1994) and O'Dell and Wong (1996).

The mere existence of proplyds in large numbers indicates that they cannot be short-lived objects, an inference which is consistent with their compact structure, revealing a brightness distribution which is steeper than that of a freely expanding gas (O'Dell 1998). Outflow velocities as observed with the Keck telescope (Henney and O'Dell 1999) imply, however, mass loss rates of the order of some 4×10^{-6} solar masses per year, which are consistent with the inferred lifetimes only if the disk masses are of the order of one solar mass, that is, much larger than initially thought.

ACTIVITY OF YOUNG STELLAR OBJECTS

In the early days of ultraviolet astronomy it was doubted whether the ultraviolet window would ever be useful for exploring the cool and rather faint T Tauri stars. The advent of the International Ultraviolet Explorer (IUE) has dramatically altered this picture, since this satellite revealed important continuum ultraviolet excesses for such objects as well as numerous emission lines which became powerful diagnostics for an apparently vigorous activity (for a review, see Imhoff and Appenzeller 1987). IUE was launched on 26 January 1978 as a collaborative project between NASA, ESA and the UK, and was operated until 30 September 1996, when it was turned off from the ground (Kondo 1998). Because of the exceptionally long lifetime of IUE, some 150 YSOs could be studied, and many of them frequently so; the large extent of this database did not result in simple interpretations, since a large variety of behaviour was observed to occur. For stellar astronomers who are aware that the time dimension is essential to many cosmic phenomena, IUE proved to be an extraordinarily rich mission.

The ultraviolet continuum excess of T Tauri stars was initially thought to originate from hydrogen free–free and free–bound emission from a dense chromosphere (Kuhi 1966, 1974; Calvet *et al.* 1984), but later studies (see below) showed that accretion energy also contributes. Model fits yield electron temperatures between 10000 and 50000 K (Lago *et al.* 1984, Herbig and Goodrich 1986). Not surprisingly, in the framework of both hypotheses, the strength of the continuum excess is observed to increase from naked over weak-lined objects towards classical T Tauri stars.

The strong ultraviolet Mg II recombination lines are popular diagnostics for the study of the chromospheric activity of T Tauri stars. For the lower-mass objects, these lines correlate well with the surface fluxes diagnosed by the optical Ca II lines, but in more massive objects it appears that the Mg II emission is also affected by rather strong stellar winds. Hot chromospheric gas also results in a rich spectrum of far-ultraviolet (120–200 nm) emission lines. It seems, however, that when compared with chromospherically active main-sequence stars (Imhoff and Giampapa 1980, 1981), pre-main-sequence stars show relatively weaker emission in the highest-temperature (2×10^5 K) lines, hence suggesting a less developed coronal activity.

Since chromospheric activity is usually linked to the occurrence of convective envelopes in low-mass stars, it came as a surprise that for the more massive Haebe stars substantial amounts of warm gas are also present. The continuum ultraviolet excess of Haebe stars is relatively rare and rather weak (Malfait et al. 1998a), but strong and variable emission is often observed for Fe II and Mg II resonance lines (Talavera et al. 1982, Praderie et al. 1982; for an overview, see Talavera 1998). These observations led to the hypothesis that Haebe stars also have chromospheres (Catala et al. 1986, Praderie et al. 1986, Catala 1988). An alternative hypothesis is that the warm gas surrounding Haebe stars derives its energy from accretion of the inner parts of the circumstellar disk (Blondel and Tjin A Djie 1994), and this interpretation is strengthened by the observation of redshifted Ly α emission (Hu et al. 1991, Blondel et al. 1993, Talavera et al. 1994).

Since accretion contributes to the heating of circumstellar gas of T Tauri stars as well as of Haebe stars (Bertout et al. 1988, Simon et al. 1990, Blondel et al. 1993, Edwards et al. 1993), and the outflows observed for both kinds of objects are similar (Mundt and Ray 1994), it appears that accretion may also provide the energy source for the outflows of T Tauri stars. This hypothesis is consistent with the correlations of the strengths of the winds and various indicators of the importance of the circumstellar disks (Bertout 1989, Cabrit et al. 1990, Blondel et al. 1993) and also with the occurrence of massive jets for the UX Orionis stars (Hartmann et al. 1989, Poetzel et al. 1989), which are characterized by episodes of intense accretion activity.

An important characteristic of the ultraviolet line spectra of YSOs is their strong variability. Variable absorption features have a convincing explanation in terms of clumpy accretion of circumstellar matter (Graham 1992), presumably attesting to the infall of comet-like bodies onto the star (Grinin et al. 1994, Grady et al. 1996), a phenomenon which will be further discussed below for more evolved objects such as β Pictoris. Periodicities for the Mg II emission from particular objects have been found and interpreted by Praderie et al. (1986) and Catala et al. (1989), but in most cases the apparently irregular nature of the variability still defies a definitive interpretation.

ISO RESULTS ON STAR FORMATION

The ideal instrument for unravelling the star-formation process is an infrared and sub millimetre space observatory which combines sensitivity and high spatial and spectral resolution. The ISO satellite (Kessler et al. 1996), launched in November 1995, was equipped with four instruments: the near-infrared camera ISOCAM (Cesarsky et al. 1996), the two medium-to high-resolution spectrometers SWS (2.4–48 μm; de Graauw et al. 1996a) and LWS (47–187 μm; Clegg et al. 1996) and the broad-band photometer ISOPHOT (Lemke et al. 1996) with filters in the range 5–270 μm. The size of the aperture of the telescope (60 cm) implied that an adequate combination of both spatial and spectral resolution could be achieved only for nearby bright objects, but from the richness of the results it can be anticipated that the ISO database will serve as a major source for follow-up studies with powerful ground-based instruments and that the ISO legacy will serve as a guideline for larger future space projects.

The spectral coverage, sensitivity and spatial resolution of ISOCAM proved adequate for achieving a fairly complete census of the class-I and class-II source content of nearby SFRs such as the ρ Ophiuchus cloud and the Chamaeleon complex (Nordh et al. 1996, Bontemps et al. 1999, Olofsson et al. 1999, Prusti 1999). Typically twice as many sources as known from previous surveys were found, and confusion did not limit the detection even for sources in the substellar mass range. Excellent correlations between ISOCAM and ROSAT surveys were found, suggesting that virtually all T Tauri stars are X-ray emitters, and that it is obscuration which prevents the X-ray detection of the most embedded sources.

Converting the luminosity function of a SFR to its initial mass function (IMF) is a model-dependent exercise. The mass dependence of the luminosity of a collapsing object implies that substellar objects are most easily detected because they are brightest during the initial collapse, hence increasing the confidence that the source counts from ISOCAM maps deliver a robust picture. In this picture it is found that the IMF rises slowly towards the substellar mass domain, thus confirming and extending the Scalo (1998) IMF, but also implying that brown dwarfs do not dominantly contribute to the local mass density.

Pre-stellar cores, that is, class-0 and class-'-I' sources, do not radiate significantly in the ISOCAM spectral domain. However, by studying the obscuration of class-0 sources against the diffuse mid-infrared background, Bacmann et al. (1999) were able to detect the density profiles of pre-stellar cores, which steepen sharply at typically 15–30 $\times 10^3$ AU, thus quantifying the typical sizes where cores decouple from their parent cloud. The thermal emission of pre-stellar cores, which were not even detectable with IRAS, has been detected at the longer (200–270 μm) wavelengths covered by ISOPHOT (Ward-Thompson and André, 1999). The SEDs of these sources, and also those of class-0 sources detected by IRAS and LWS, are well reproduced with black bodies affected by extinction. It is clear that further study of the earliest stages of stellar evolution will require future far-infrared and sub millimetre observatories in space as well as on the ground, with the latter also providing millimetre mapping and spectroscopy.

ISO spectroscopy of YSOs with SWS and LWS has revealed an impressive amount of atomic, molecular and solid-state features (Figure 6). These spectra contain a rich potential: many diagnostic lines have been uncovered by ISO for the first time, the broad wavelength coverage allows a complete picture involving both the gas and solid phases and sensitive observations could be obtained for a large number of objects that cover the evolutionary phases from dark cores to circumstellar disks. A drawback, however, is the limited spatial resolution of the data. This implies that the full exploitation of these rich data will require ground-based follow-up with infrared cameras using large telescopes and far-infrared follow-up with missions such as SOFIA and Herschel. The study of interstellar and circumstellar ices, in particular those which are radio-quiet such as CO_2 and CH_4, has received an important boost from the ISO-SWS observations (van Dishoeck et al. 1999, d'Hendecourt et al. 1999). Interstellar CO_2 absorption at 15.2 μm was discovered with IRAS observations by d'Hendecourt and Jourdain de Muizon (1989). Its much stronger mode at 4.27 μm was first observed using the low-resolution ISOPHOT-S spectrometer (Gürtler et al. 1996) and has since also been seen in many sources at the higher SWS resolution (e.g. de Graauw et al. 1996b). The abundance of CO_2 ice with respect to H_2O ice proves to be remarkably constant at about 18%. The detection of gas-phase CO_2 in highly obscured protostars (van Dishoeck et al. 1996) opens new avenues for the study of the interaction between grain mantles and gas-phase chemistry.

Of similar interest is the detection around YSOs of H_2O in both the gas and solid phases. An early surprise was the detection of abundant, hot ($T_{ex} \approx 300$ K) water vapour towards a number of YSOs (Helmich et al. 1996, van Dishoeck and Helmich 1996, van Dishoeck 1998). This molecule is predicted to be the dominant oxygen-bearing species in warm, dense gas, and thus a rich diagnostic for both dense cores and shocks. Other important tracers of warm, dense gas are the pure rotational lines of H_2, which could be observed by ISO-SWS up to high J levels. The surprisingly high excitation temperatures observed for H_2 are interpreted in terms of photodissociation regions ($T \approx 500$–1000 K; Draine and Bertoldi 1999) or shocks, and the broad spectral range in which such lines are observed offers a tool for discriminating between both models (van den Ancker et al. 1999). The latter authors interpreted the H_2 emission from T Tauri in terms of a dissociative shock, also responsible for the atomic fine-structure lines, and a second, much slower non-dissociative shock.

THE VEGA PHENOMENON

The opening of new windows on the Universe by new space facilities is always an exciting adventure, but unfortunately also one which may still require extrapolation by use of models of celestial sources to explore unknown domains. This is true particularly for the infrared spectral range, where proper procedures that compare celestial standards directly to laboratory standards are lacking. When the rare circumstance occurs that during such an in-orbit calibration a significant scientific result emerges, the opportunity should not be missed to pay tribute to the teams, the technical expertise of which enables a large community to address their science with new facilities.

Such a circumstance occurred during the early phases of the IRAS mission, when the fluxes in the four infrared bands at 12, 25, 60 and 100 μm were calibrated with the help of observations of the 'well-understood' stellar candle α Lyrae or Vega. This rapidly turned out to display an important excess of radiation at 60 and 100 μm, implying the presence of cool circumstellar material (Aumann et al. 1984). The discovery of dusty circumstellar material around Vega was then rapidly followed by similar ones for the stars β Pictoris and Fomalhaut (Gillett 1986), ε Eridani, and later for other main-sequence objects. At the limited angular resolution of IRAS, only the excesses of Vega, Fomalhaut and β Pictoris could be spatially resolved, but these observations were sufficient to point out the disk-like structure of the circumstellar material and the fact that relatively large grains were involved.

It rapidly became clear that the evolved nature of Vega and Fomalhaut implies ages much older than the typical timescales on which the circumstellar disks associated with the star-formation process should disappear, thus implying the need for a continuous replenishment of the dust debris

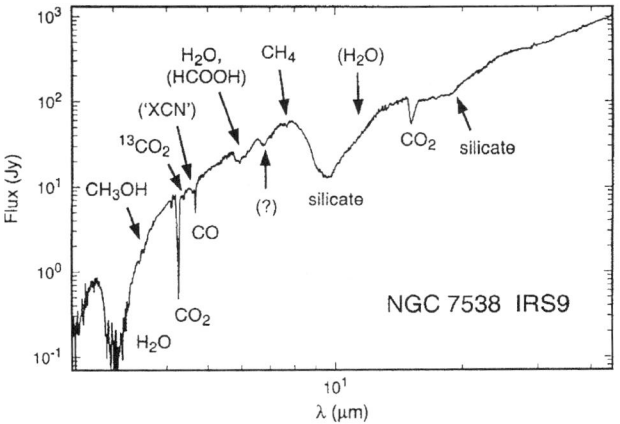

Figure 6 ISO-SWS spectrum of the star-formation region NGC 7538, an object with a particularly rich solid-state infrared spectrum. (From Whittet et al. 1996.)

disks, presumably through collisions of larger bodies. It is not an exaggeration to stress that the Vega phenomenon has opened a new avenue for the observational study of the formation and evolution of planetary systems. Since then extrasolar planets have been discovered (Mayor and Queloz 1995, Marcy and Butler 1998), and one system (ρ Cancri) is known for which both a planet and an optically resolved dust debris disk occur (Dominik et al. 1998, Trilling and Brown 1998, Jayawardhana et al. 2000).

Due to the limited sensitivity of the IRAS survey, a statistically sound census of the occurrence of debris disks around main-sequence stars has been difficult to achieve, and initial estimates diverged widely. The most robust estimate of the occurrence of the phenomenon from IRAS data is probably the one by Plets and Vinckier (1999), who performed a statistical analysis on the data. The result of this study is that about 13% of main-sequence stars display the phenomenon; interestingly, a similar fraction was found for GK giants, of which the stars in the mass range for which the Vega phenomenon could be studied are the natural progenitors.

A more sensitive census of the Vega phenomenon could be achieved with the PHOT instrument on board ISO (Habing et al. 1999). A somewhat larger fraction of, on average, 21%, but slightly dependent on spectral type, emerged from this study. Habing et al. were also able to show that the occurrence of the Vega phenomenon decreases with stellar age on the main sequence, the decline starting after some 400 Myr. This timescale corresponds very well to that after which the heavy impacts ceased in our Solar System. However, a study of the ISOPHOT excesses of G-type dwarfs by Decin et al. (2000) suggests that debris disks are much longer-lived in solar-type stars.

A Vega-like star of particular interest is β Pictoris. Its disk is among those with the highest optical depth, and has been successfully resolved at optical wavelengths (Smith and Terrile 1984). HST (Figure 7) as well as coronographic ground-based observations have since revealed a warped structure of the β Pictoris disk, which has been interpreted as dynamical evidence for a planet. Beta Pictoris is considered to be younger than typical Vega-like stars, though its Hipparcos age of some 9 Myr (Crifo et al. 1997) is still substantially larger than the estimated dissipation timescale of the disk, which thus also requires a replenishment mechanism.

The presence of large bodies in the β Pictoris disk can be directly inferred from the frequent, most often redshifted, absorption events detected with the IUE satellite (e.g. Lagrange et al. 1989) as well as in optical spectra (e.g. Ferlet et al. 1993). The interpretation of these events requires infalling bodies with a filling factor which is a substantial fraction of the stellar disk, most likely comets (Beust et al. 1990, 1991). An independent argument for the presence of a vast cloud of cometary material around β Pictoris is the appearance of the 10 μm silicate feature (Telesco and Knacke 1991, Knacke et al. 1993), which is more reminiscent of the crystalline feature observed in some Solar System comets than of the amorphous feature which is commonly observed in circumstellar envelopes and in the interstellar medium. A comet dust model for the β Pictoris disk which fits the available data has been proposed by Li and Greenberg (1998).

MAPPING CIRCUMSTELLAR DEBRIS DISKS

The rapid initial success with the optical detection of the β Pictoris disk has triggered many efforts to detect, through scattered optical and near-IR light, similar disks around other main-sequence stars with infrared excesses or with intermittent absorption features. The success rate of these studies has been low, however. The disks surrounding Vega, Fomalhaut and τ Eridani appear to be too faint for optical detection with present instrumentation. It thus appears that β Pictoris is rather exceptional, and that it is a fortunate

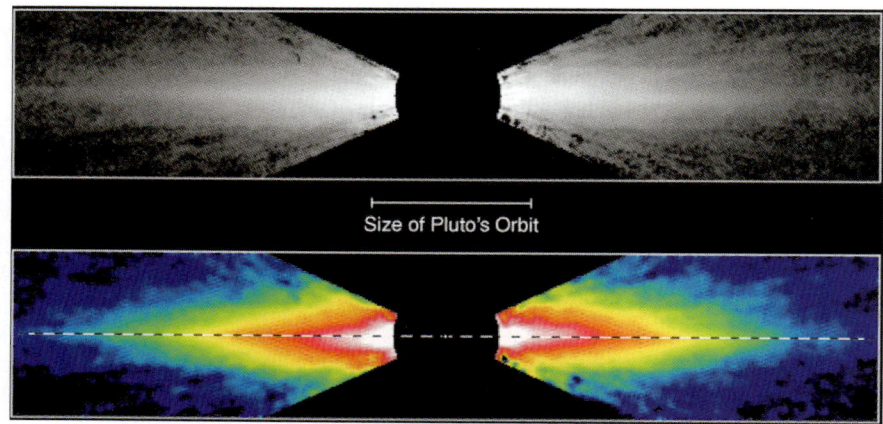

Figure 7 Hubble Space Telescope image of the β Pictoris disk (Burrows et al. 1995). (Courtesy of NASA.)

circumstance that the main-sequence object with one of the larger excesses discovered so far is located close to the Solar System. Beta Pictoris is also one of the younger Vega-like stars, though most age estimates, including the ones obtained from its Hipparcos parallax and stellar evolution models (Crifo et al. 1997, Barrado y Navascués et al. 1999), are higher than the expected disappearance timescale for its disk, implying that the large IR excess is linked to the particular replenishment history of the object.

Recent searches for β Pictoris analogues focus on young objects in the relevant mass range, which may still be evolving towards the main sequence or arrived on it recently, but appear as isolated stars in the sky. Such objects are isolated or 'old' Haebe stars: their hydrogen recombination-line spectra and substantial infrared excesses prove that they are still surrounded by significant amounts of circumstellar matter, but no obvious link with a SFR is found. With the advent of the NICMOS camera on HST and its coronographic mode it has become possible to resolve spatially the disks of HR 4796A (Schneider et al. 1999, Augereau et al. 1999) and of HD 141569 (Weinberger et al. 1999), as shown in Figure 8. The age of the former object could be estimated rather accurately at 8 ± 2 Myr, due to the presence of a companion which is an active M star (Stauffer et al. 1995), while van den Ancker et al. (1998b) derived an age in excess of 7 Myr for the latter object, from the fitting of stellar evolution models to a Hipparcos-based HR diagram.

For HR 4796A it is found that the bulk of the circumstellar material is located in a ring some 17 AU wide at a distance of about 70 AU from the central star. The confinement of the ring also calls for the presence of 'shepherding' planetary bodies, while its long lifetime suggests, as for β Pictoris, that collisions of smaller objects, at least larger than 10 μm but presumably planetesimals, continuously occur. Similar conclusions apply to the HD 141569 circumstellar disk, the flux density of which peaks at 185 AU from the star. In order to explain the near-infrared SED of HR 4796A, Augereau et al. (1999) invoke the presence of a second component of warmer grains closer to the star, and suggest that these grains may be replenished by cometary infall. In both cases the albedo of the grains is substantially larger than that of the zodiacal dust in our Solar System (i.e. 0.4 rather than 0.1), suggesting that the grains are covered with ices (Jura et al. 1998).

The evolutionary link between β Pictoris and Haebe stars has in the meantime been further confirmed for AB Aurigae, which is a Haebe star associated with a SFR, with an age between 2 and 4 Myr: its disk could be spatially resolved and studied with HST-NICMOS (Grady et al. 1999), after having been resolved at millimetre wavelengths by Mannings and Sargent (1997). For AB Aurigae, for which an inner gap of the disk has been reported from ground-based millimetre observations (Marsh et al. 1995), the gaps seen for the older stars do not appear, but instead a clumpy structure of the disk is observed, maybe suggesting that giant planets are still in the process of forming.

ISO SPECTROSCOPY OF THE DISKS OF HERBIG AE/BE (HAEBE) STARS

A distinctive characteristic of β Pictoris is the peculiar appearance of its 10 μm silicate feature (Knacke et al.

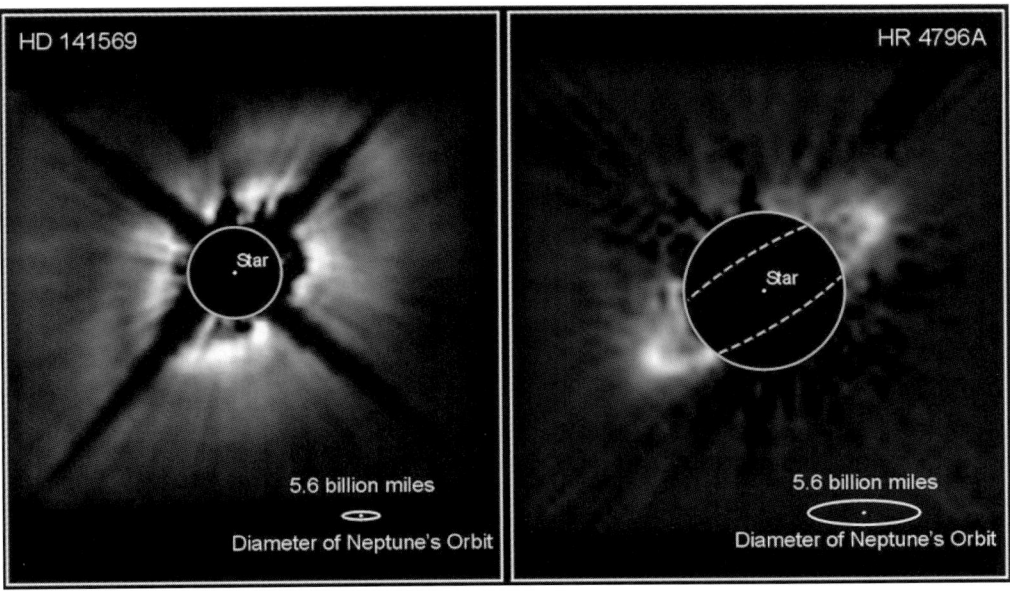

Figure 8 HST-NICMOS images of the circumstellar disks surrounding HR 4796A and HD 141569. (Courtesy of NASA.)

1993), which is also seen in comets such as P-Halley (Campins and Ryan 1989, Herter *et al.* 1987), Bradfield 1987 (Hanner *et al.* 1990) and Levy 1990 (Lynch *et al.* 1992), and points towards the presence of a crystalline variety of silicates. The 'cometary' nature of its silicate feature thus confirms the hypothesis that cometary objects orbit β Pictoris in huge quantities (Ferlet *et al.* 1993). Peculiar 10 μm features are now well documented in ground-based IR spectra of several young stars (e.g. Fajardo-Acosta *et al.* 1993, Hanner *et al.* 1998).

Before ISO, it was thought that 'astronomical' silicates were essentially amorphous, producing distinctive broad features around 10 and 18 μm, both being accessible to ground-based observations (Draine and Lee 1984). The full richness of the crystalline silicate spectra could only be disclosed by ISO, since it appears in the 20 to 80 μm region (Waters *et al.* 1996, Waelkens *et al.* 1996) which is inaccessible from the ground. These results have triggered new efforts in laboratory astrophysics (Jäger *et al.* 1998, Koike *et al.* 2000) from which it appears that a reliable identification in terms of different crystalline olivines ($[Mg/Fe]_2SiO_4$) and pyroxenes ($[Mg/Fe]SiO_3$) is possible, but only if a broad spectral range is observed. Jäger *et al.* (1998) coined the term 'the crystalline revolution' to describe the spectacular advance in astronomical silicate mineralogy made possible by ISO. That it was a space-induced revolution can be illustrated by a comment on a project introduced by the present author in 1994, where the referee still thought that 'some of the ISO observing goals, for example the study of silicate emission, can be achieved with large new telescopes on good ground-based sites'.

A remarkable object in this respect is the isolated Haebe star HD 100546, the Hipparcos-based age of which is consistent with it being a 5 Myr-old main-sequence star (van den Ancker *et al.* 1998). The full ISO-SWS-LWS spectrum of HD 100546 has been discussed by Malfait *et al.* (1998b), who pointed out the striking similarity of its silicate features with those of comet Hale-Bopp (Crovisier *et al.* 1997). The modelling presented in Figure 9 is the result of a radiative-transfer calculation with the MODUST programme (Bouwman *et al.* 2000). From the presence at 69 μm of the longest-wavelength silicate feature, which is very sensitive to the Mg/Fe ratio, it follows that forsterite, that is, pure Mg olivine, is present around this star, and that this particular mineral is responsible for nearly all sharp emission features in the spectrum. From the modelling it also follows that the amorphous silicate component is Mg rich, and that iron preferentially occurs in its metallic form and in FeO grains.

In the ISO-SWS-LWS spectrum of the younger Haebe star HD 142527 (Figure 10; Malfait *et al.* 1999), the dominant component of the IR flux stems from crystalline water ice (50% in mass), from cold amorphous silicates (35%)

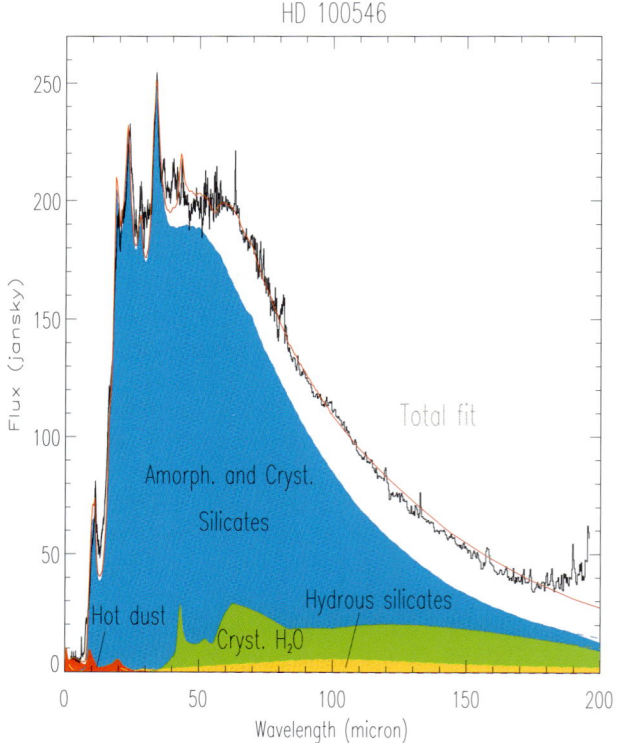

Figure 9 ISO-SWS-LWS spectrum of the isolated Herbig Ae/Be star HD 100546 (modelled with the MODUST code). The sharp peaks in the spectrum point to the presence of crystalline silicates and water ice.

and from layered hydrous silicates (15%), which were modelled with the available optical constants of a particular mineral called montmorillonite (Koike and Shibai 1990). While the attribution to a particular mineral is uncertain, and was only done on the basis of the availability of laboratory data, the relevant result to note is that the broad shoulder present around 100 μm points to a silicate with water inclusions.

Hydrous silicates are found in carbonaceous chondrites of CI-type (Wilkening 1978) and in interplanetary dust particles (IDPs), which are thought to be of cometary origin (Sandford and Walker 1985), and may occur in comet spectra (e.g. Hanner *et al.* 1990), but it is currently not clear whether they arrived there unprocessed from the early solar nebula. Some controversy exists as to whether the presence of crystalline water in comets reflects their original composition or is rather due to processing during earlier passages close to the Sun. The presence of both hydrous silicates and crystalline water ice in the cold and young disk of HD 142527 appears to be of value in the context of the study of the early stages of our Solar System.

The striking similarity between the silicate spectra of HD 100546 and that of of comet Hale-Bopp suggests that

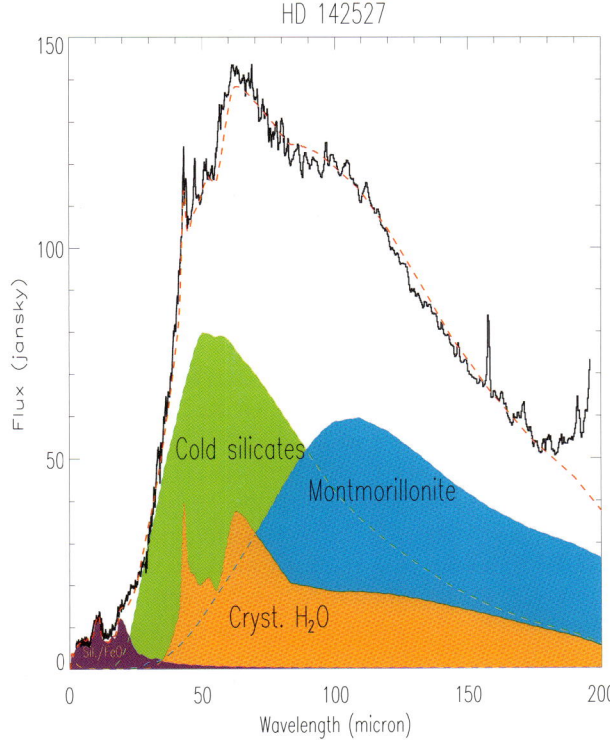

Figure 10 ISO-SWS-LWS spectrum of the isolated Herbig Ae/Be star HD 142527 (modelled with the MODUST code). The circumstellar disk of this object is characterized by huge amounts of crystalline water ice. The broad shoulder around 100 μm points to the presence of hydrous silicates.

this star is caught in an early stage when comets are abundant. It does appear, however, that crystalline silicate spectral fingerprints are not visible in the spectra taken from the youngest objects. The silicate spectra of the youngest, still partially embedded sources consistently appear amorphous or with at most modest crystalline contributions (e.g. AB Aurigae and HD162396; van den Ancker et al. 2000), and sources such as HD100546, for which crystalline silicates are abundant, are definitely evolved. Moreover, the occurrence of crystalline forsterite in HD100546 over the large temperature range from 50 to 220 K suggests that the processes dispersing the comets to large distances from the star are presently operating in this system, thus indirectly suggesting the presence of the giant planets which are thought to be dynamically responsible for the formation of an Oort cloud (Fernandez 1978).

PLANET FORMATION AROUND OLD STARS?

Crystalline silicate features in ISO spectra were first detected for AGB and post-AGB stars (Waters et al. 1996),

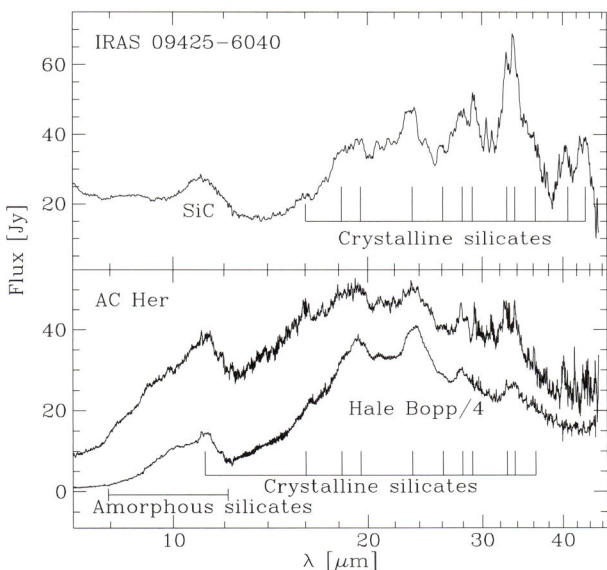

Figure 11 Comparison of the ISO-SWS spectra of two evolved (post-AGB) objects and that of comet Hale-Bopp. The infrared emission for the stars originates in disks which are supposed to enclose a binary system. The similarity of the stellar and cometary spectra suggests that the formation of larger bodies in long-lived circumstellar disks is a widespread process.

that is, objects which are close to the end of their evolution as stars. Subsequent studies have revealed that around some evolved objects these features are fairly strong. It now appears that, while weak features are observed in the dense outflows of some luminous evolved stars for which no evidence for binarity exists (Waters et al. 1996), crystalline silicates are most prominent around binary objects, and occur there in long-lived circum-binary disks.

A remarkable example is the famous Red Rectangle nebula, which is the prototype of a class of carbon-rich reflection nebulae surrounding low-mass stars in their final stages of evolution. While the near-infrared spectrum of this object is dominated by features of polycyclic aromatic hydrocarbons (PAHs) in the extended nebula, its 20–40 μm spectrum (Waters et al. 1998) is characterized by crystalline oxygen-rich emission bands which originate in the disk known to surround the central binary (Roddier et al. 1995, Van Winckel et al. 1995, Jura et al. 1995). A second example is the more massive binary system AFGL4106 (Molster et al. 1999a).

Figure 11 shows the ISO-SWS spectra of the carbon-rich giant star IRAS09425-6040 and of the RV Tauri-type variable post-AGB star AC Herculis (Molster et al. 1999b). Direct evidence for binarity only exists so far for the latter object (Van Winckel et al. 1998), but the resemblance of the spectrum of the former to that of the Red Rectangle, and in particular the coexistence of C-rich and O-rich dust, strongly argues for the binarity of IRAS09425. It certainly is surprising that the object whose ISO spectrum is most alike that of comet Hale-Bopp is the evolved star AC Herculis!

It is tempting to conclude that the ISO spectra of these evolved objects suggest that the processes which initiate planet formation also occur in the disks surrounding these evolved stars, thus implying that planet formation is a most natural process in circumstellar disks. While the spectral diagnostics are indicative of comet and/or planetesimal formation only, the presence of a density contrast in the disk of the Red Rectangle has been interpreted as evidence for the occurrence of a genuine giant planet (Jura and Turner 1998).

The observations that crystalline silicates are not seen in the disks of the youngest stars and, for evolved stars, are most prominent in the long-lived disks surrounding binary systems, suggest that time is a critical parameter for the crystallization of silicates to occur. If crystallization is the result of thermal annealing, as laboratory experiments suggest (Hallenbeck and Nuth 1998), the typical temperatures prevailing in the disks would, however, imply timescales for annealing which are prohibitively large. While it may be possible that the processing in the disks of young stars involves outwards migration of grains after a passage in the warm environment close to the star, such a hypothesis seems most unlikely for the evolved binary systems, so that it may be concluded that the crystallization process itself is currently not understood.

SUMMARIZING AND ANTICIPATING

Astronomy is a multidisciplinary science, where astronomers, chemists, geologists and physicists need to collaborate. In the field of star formation, as well as for the fields described in other contributions, progress has involved a continuous interplay between observations from space and from the ground, but it has often been the case that the opening of new windows from space has been the main trigger for spectacular new developments.

In a most healthy way, these new developments force us to do away with the sometimes artificial walls we tend to erect between various aspects of science. For an astronomer who, as many of his kind, was first attracted to our science by the appeal of the Solar System as our home and by the spectacular aspects of the first crewed missions to explore it, but then professionally turned to the then much more fashionable stellar astrophysics, it is a wonderful experience to witness how nowadays stellar astrophysics and Solar System research are becoming interlinked. The analysis of the composition and structure of meteorites entails rich constraints for our understanding of the formation of dust in the outflows of evolved stars, and the study of circumstellar disks has opened completely new perspectives for the study of the origin of our own planetary system.

Stars are born in dust and gas and at the end of their lives return much of their mass in dust and gas to the interstellar medium, allowing new stars to be formed in a richer world. Infrared spectroscopy from space is now revealing impressive detail about the circumstellar environments of both young and old stars, and surprising similarities between both suggest that researchers on stellar evolution and star formation will have to pay attention to each others' results in order to understand the dusty cosmic cycle of evolution.

Not only is the study of star formation and circumstellar disks situated at the crossroads between stellar and planetary science, it also appears obvious that a good understanding of star formation will be essential for a correct interpretation of the spectacular observations that are now within reach concerning the distant extragalactic Universe. In order to understand the formation of structure in the early Universe, we will have to understand the vigorous star formation which is observed in high-redshift galaxies in more detail, and to measure accurately the role of dust in the appearance of these galaxies to us.

At the turn of the century, the star formation community is eagerly awaiting the promised new facilities which will explore with much more detail the infrared, submillimetre and millimetre windows, that is, the large ground-based telescopes equipped with mature instrumentation, and in particular the planned space missions such as SIRTF, Herschel and NGST. Moreover, learning from the unanticipated results which were obtained early on from X-ray-satellites and IUE, we should remain aware that a multiwavelength approach is most likely to yield the most complete answers to our questions about cosmic origins.

REFERENCES

André, P. (1997). The potential of FIRST for the earliest stages of star formation. In A. Wilson (ed), *The Far InfraRed and Submillimetre Universe*, ESA SP-401, p. 53.

André, P., Ward-Thompson, D. and Barsony, M. (1993). Submillimeter continuum observations of Rho Ophiuchi A – The candidate protostar VLA 1623 and prestellar clumps. *Astrophysical Journal*, **406**, 122–141.

Augereau, J.C., Lagrange, A.M., Mouillet, D., Papaloizou, J.C.B. and Grorod, P.A. (1999). On the HR 4796 A circumstellar disk. *Astronomy and Astrophysics*, **348**, 557–569.

Aumann, H.H., Beichman, C.A., Gillett, F.C., de Jong, T., Houck, J.R., Low, F.J., Neugebauer, G., Walker, R.G. and Wesselius, P.R. (1984). Discovery of a shell around Alpha Lyrae. *Astrophysical Journal*, **278**, L23–L26.

Bacmann, E., André, P., Abergel, A., Puget, J.L., Bontemps, S., Ward-Thompson, D. and Bernard, J.P. (1999). Density structure of pre-stellar cores seen in absorption with ISOCAM. In P. Cox and M.F. Kessler (eds), *The Universe as Seen by ISO*, ESA SP-427, pp. 467–470.

Barrado y Navascués, D., Stauffer, J.R., Song, I. and Caillault, J.P. (1999). The age of Beta Pictoris. *Astrophysical Journal*, **520**, L123–L126.

Bertout, C. (1989). T Tauri stars – Wild as dust. *Annual Review of Astronomy and Astrophysics*, **27**, 351–395.

Bertout, C., Basri, G. and Bouvier, J. (1988). Accretion disks around T Tauri stars. *Astrophysical Journal*, **330**, 350–373.

Beust, H., Vidal-Madjar, A., Ferlet, R. and Lagrange-Henri, A.M. (1990). The Beta Pictoris circumstellar disk. X – Numerical simulations of infalling evaporating bodies. *Astronomy and Astrophysics*, **236**, 202–216.

Beust, H., Vidal-Madjar, A., Ferlet, R. and Lagrange-Henri, A.M. (1991). The Beta Pictoris circumstellar disk. XI – New Ca-II absorption features reproduced numerically. *Astronomy and Astrophysics*, **241**, 488–492.

Blitz, L. (1993). Giant molecular clouds. In E.H. Levy and J.I. Lunine (eds), *Protostars and Planets III*, University of Arizona Press, Tucson, AZ, pp. 125–161.

Blondel, P.F.C., Talavera, A. and Tjin A Djie, H.R.E. (1993). Lyman alpha emission in spectra of Herbig Ae stars – An indication of accretion? *Astronomy and Astrophysics*, **268**, 624–640.

Blondel, P.F.C. and Tjin A Djie, H.R.E. (1994). Detection of the accretion disk boundary layer of Herbig Ae/Be stars. In P.S. Thé, M.R. Pérez and E.P.J. van den Heuvel (eds), *The Nature and Evolutionary Stage of Herbig Ae/Be Stars*, ASP Conference Series, Vol. **62**, pp. 211–214.

Bok, B.J. and Reilly, E.F. (1947). Small dark nebulae. *Astrophysical Journal*, **105**, 255–257.

Bontemps, S., Nordh, L., Olofsson, G., André, P., Huldtgren, M., Kaas, A.A., Abergel, A., Blommaert, J.A.D.L., Boulanger, F., Burgdorf, M., Cesarsky, C.J., Cesarsky, D., Copet, E., Davies, J., Falgarone, E., Lagache, G., Montmerle, T., Perault, M., Persi, P., Prusti, T., Puget, J.L. and Sibille, F. (1999). An ISOCAM deep census of low-mass stars in ρ Ophiuchi. In P. Cox and M.F. Kessler (eds), *The Universe as Seen by ISO*, ESA SP-427, pp. 475–478.

Boss, A.P. and Yorke, H.W. (1995). Spectral energy of first protostellar cores: Detecting 'class -I' protostars with ISO and SIRTF. *Astrophysical Journal*, **439**, L55–L58.

Bouwman, J., de Koter, A., van den Anker, M.E. and Waters, L.B.F.M. (2000). The composition of the circumstellar dust around the Herbig Ae stars AB Aur and HD 163296. *Astronomy and Astrophysics*, **360**, 213–226.

Burrows, C.J., Krist, J.E., Stapelfeldt, K.R. and WFPC2 Investigation Definition Team. (1995). HST observations of the beta Pictoris circumstellar disk. *Bulletin of the American Astronomical Society*, **187**, 32.05.

Burrows, C.J., Stapelfeldt, K.R., Watson, A.M., Krist, J.E., Ballester, G.E., Clarke, J.T., Crisp, D., Gallagher, J.S., III, Griffiths, R.E., Hester, J.J., Hoessel, J.G., Holtzman, J.A., Mould, J.R., Scowen, P.A., Trauger, J.T. and Westphal, J.A. (1996). Hubble Space Telescope observations of the disk and jet of HH 30. *Astrophysical Journal*, **473**, 437–451.

Cabrit, S., Edwards, S., Strom, S.E. and Strom, K.M. (1990). Forbidden-line emission and infrared excesses in T Tauri stars – Evidence for accretion-driven mass loss? *Astrophysical Journal*, **354**, 687–700.

Calvet, N., Basri, G. and Kuhi, L.V. (1984). The chromospheric hypothesis for the T Tauri phenomenon. *Astrophysical Journal*, **277**, 725–737.

Campins, H. and Ryan, E.V. (1989). The identification of crystalline olivine in cometary silicates. *Astrophysical Journal*, **341**, 1059–1066.

Catala, C. (1988). Line formation in the winds of Herbig Ae/Be stars – The CIV resonance lines. *Astronomy and Astrophysics*, **193**, 222–228.

Catala, C., Czarny, J., Felenbok, P. and Praderie, F. (1986). Spectral similarities in the visible and UV spectrum of Herbig Ae/Be stars. *Astronomy and Astrophysics*, **154**, 103–114.

Catala, C., Simon, T., Praderie, F., Talavera, A., Thé, P.S. and Tjin A Djie, H.R.E. (1989). Active phenomena in the pre-main sequence Herbig Ae star HD 163296. *Astronomy and Astrophysics*, **221**, 273–286.

Cesarsky, C.J., Abergel, A., Agnese, P., Altieri, B., Augueres, J.L., Aussel, H., Biviano, A., Blommaert, J., Bonnal, J.F., Bortoletto, F., Boulade, O., Boulanger, F., Cazes, S., Cesarsky, D.A., Chedin, A., Claret, A., Combes, M., Cretolle, J., Davies, J.K., Désert, F.X., Elbaz, D., Engelmann, J.J., Epstein, G., Franceschini, A., Gallais, P., Gastaud, R., Gorisse, M., Guest, S., Hawarden, T., Imbault, D., Kleczewski, M., Lacombe, F., Landriu, D., Lapègue, J., Léna, P., Longair, M.S., Mandolesi, R., Metcalfe, L., Mosquet, N., Nordh, L., Okumura, K., Ott, S., Pérault, M., Perrier, F., Persi, P., Puget, P., Purkins, T., Rio, Y., Robert, T., Rouan, D., Roy, A., Saint-Pe, O., Sam Lone, J., Sargent, A., Sauvage, M., Sibille, F., Siebenmorgen, R., Sirou, F., Soufflot, A., Starck, J.L., Tiphene, D., Tran, D., Ventura, G., Vigroux, L., Vivares, F. and Wade, R. (1996). ISOCAM in flight. *Astronomy and Astrophysics*, **315**, L32–L37.

Churchwell, E., Wood, D.O.S., Felli, M. and Massi, M. (1987). Solar System-sized condensations in the Orion Nebula. *Astrophysical Journal*, **321**, 516–519.

Clegg, P.E., Ade, P.A.R., Armand, C., Baluteau, J.-P., Barlow, M.J., Buckley, M.A., Berges, J.-C., Burgdorf, M., Caux, E., Ceccarelli, C., Cerulli, R., Church, S.E., Cotin, F., Cox, P., Cruvellier, P., Culhane, J.L., Davis, G.R., di Giorgio, A., Diplock, B.R., Drummond, D.L., Emery, R.J., Ewart, J.D., Fischer, J., Furniss, I., Glencross, W.M., Greenhouse, M.A., Griffin, M.J., Gry, C., Harwood, A.S., Hazell, A.S., Joubert, M., King, K.J., Lim, T., Liseau, R., Long, J.A., Lorenzetti, D., Molinari, S., Murray, A.G., Naylor, D.A., Nisini, B., Norman, K., Omont, A., Orfei, R., Patrick, T.J., Pequignot, D., Pouliquen, D., Price, M.C., Nguyen-Q-Rieu, Rogers, A.J., Robinson, F.D., Saisse, M., Saraceno, P., Serra, G., Sidher, S.D., Smith, A.F., Smith, H.A., Spinoglio, L., Swinyard, B.M., Texier, D., Towlson, W.A., Trams, N.R., Unger, S.J. and White, G.J. (1996). The ISO Long-Wavelength Spectrometer. *Astronomy and Astrophysics*, **315**, L38–L42.

Crifo, F., Vidal-Madjar, A., Lallement, R., Ferlet, R. and Gerbaldi, M. (1997). β Pictoris revisited by Hipparcos. Star properties. *Astronomy and Astrophysics*, **320**, L29–L32.

Crovisier, J., Leech, K., Bockelee-Morvan, D., Brooke, T.Y., Hanner, M.S., Altieri, B., Keller, H.U. and Lellouch, E. (1997). The spectrum of Comet Hale-Bopp (C/1995 O1) observed with the Infrared Space Observatory at 2.9 AU from the Sun. *Science*, **275**, 1904–1907.

Decin, G., Dominik, C., Malfait, K., Mayor, M. and Waelkens, C. (2000). The Vega phenomenon around G dwarfs. *Astronomy and Astrophysics*, **357**, 533–542.

de Graauw, T., Haser, L.N., Beintema, D.A., Roelfsema, P.R., van Agthoven, H., Barl, L., Bauer, O.H., Bekenkamp, H.E.G., Boonstra, A.-J., Boxhoorn, D.R., Coté, J., de Groene, P., van Dijkhuizen, C., Drapatz, S., Evers, J., Feuchtgruber, H., Frericks, M., Genzel, R., Haerendel, G., Heras, A.M., van der Hucht, K.A., van der Hulst, T., Huygen, R., Jacobs, H., Jakob, G., Kamperman, T., Katterloher, R.O., Kester, D.J.M., Kunze, D., Kussendrager, D., Lahuis, F., Lamers, H.J.G.L.M., Leech, K., van der Lei, S., van der Linden, R., Luinge, W., Lutz, D., Melzner, F., Morris, P.W., van Nguyen, D., Ploeger, G., Price, S., Salama, A., Schaeidt, S.G., Sijm, N., Smoorenburg, C., Spakman, J., Spoon, H., Steinmayer, M., Stoecker, J., Valentijn, E.A., Vandenbussche, B., Visser, H., Waelkens, C., Waters, L.B.F.M., Wensink, J., Wesselius, P.R., Wiezorrek, E., Wieprecht, E., Wijnbergen, J.J., Wildeman, K.J. and Young, E. (1996a). Observing with the ISO Short-Wavelength Spectrometer. *Astronomy and Astrophysics*, **315**, L49–L54.

de Graauw, T., Whittet, D.C.B., Gerakines, P.A., Bauer, O.H., Beintema, D.A., Boogert, A.C.A., Boxhoorn, D.R., Chiar, J.E., Ehrenfreund, P., Feuchtgruber, H., Helmich, F.P., Heras, A.M., Huygen, R., Kester, D.J.M., Kunze, D., Lahuis, F., Leech, K.J., Lutz, D., Morris, P.W., Prusti, T., Roelfsema, P.R., Salama, A., Schaeidt, S.G., Schutte, W.A., Spoon, H.W.W., Tielens, A.G.G.M., Valentijn, E.A., Vandenbussche, B., van Dishoeck, E.F., Wesselius, P.R., Wieprecht, E. and Wright, C.M. (1996b). SWS observations of solid CO_2 in molecular clouds. *Astronomy and Astrophysics*, **315**, L345–L348.

d'Hendecourt, L. and Jourdain de Muizon, M. (1989). The discovery of interstellar carbon dioxide. *Astronomy and Astrophysics*, **223**, L5–L8.

d'Hendecourt, L., Jourdain de Muizon, M., Dartois, E., Dmyk, K., Ehrenfreund, P. and Heras, A.M. (1999). SWS observations of protostars: The relevance of laboratory ices simulations. In P. Cox and M.F. Kessler (eds), *The Universe as Seen by ISO*, ESA SP-427, pp. 589–597.

Dominik, C., Laureijs, R.J., Jourdain de Muizon, M. and Habing, H.J. (1998). A Vega-like disk associated with the planetary system of ρ(1) Cnc. *Astronomy and Astrophysics*, **329**, L53–L56.

Draine, B.T. and Bertoldi, F. (1999). Heating the gas in photodissociation regions. In P. Cox and M.F. Kessler (eds), *The Universe as Seen by ISO*, ESA SP-427, pp. 553–559.

Draine, B.T. and Lee, H.M. (1984). Optical properties of interstellar graphite and silicate grains. *Astrophysical Journal*, **285**, 89–108.

Duquennoy, A. and Mayor, M. (1991). Multiplicity among solar-type stars in the solar neighbourhood. II – Distribution of the orbital elements in an unbiased sample. *Astronomy and Astrophysics*, **248**, 485–524.

Edwards, S., Ray, T.P. and Mundt, R. (1993). Energetic mass outflows from young stars. In E.H. Levy and J.I. Lunine (eds), *Protostars and Planets III*, University of Arizona Press, Tucson, AZ, pp. 567–602.

Fajardo-Acosta, S.B., Telesco, C.M. and Knacke, R.F. (1993). Detection of silicates in the 51 Ophiuchi system. *Astrophysical Journal*, **417**, L33–L36.

Ferlet, R., Lagrange-Henri, A.M., Beust, H., Vitry, R., Zimmermann, J.-P., Martin, M., Char, S., Belmahdi, M., Clavier, J.-P., Coupiac, P., Foing, B.H., Sevre, F. and Vidal-Madjar (1993). The beta Pictoris protoplanetary system. XIV – Simultaneous observations of the Ca-II H and K lines; evidence for diffuse and broad absorption features. *Astronomy Astrophysics*, **267**, 137–144.

Fernandez, J.A. (1978). Mass removed by the outer planets in the early Solar System. *Icarus*, **34**, 173–181.

Garay, G., Moran, J.M. and Ried, M.J. (1987). Compact continuum radio sources in the Orion Nebula. *Astrophysical Journal*, **314**, 535–550.

Ghez, A.M., Neugebauer, G. and Matthews, K. (1993). The multiplicity of T Tauri stars in the star forming regions Taurus-Auriga and Ophiuchus-Scorpius: A 2.2 micron speckle imaging survey. *Astronomical Journal*, **106**, 2005–2023.

Gillett, F.C. (1986). IRAS observations of cool excess around main sequence stars. In F.P. Israel (ed), *Light on Dark Matter*, Reidel, Dordrecht, pp. 61–69.

Grady, C., Perez, M.R., Talavera, A., Bjorkman, K.S., de Winter, D., Thé, P.S., Molster, F.J., van den Ancker, M.E., Sitko, M.L., Morrison, N.D., Beaver, M.L., McCollum, B. and Castelaz, M.W. (1996). The β Pictoris phenomenon among Herbig Ae/Be stars. UV and optical high dispersion spectra. *Astronomy and Astrophysics Supplement Series*, **120**, 157–177.

Grady, C., Woodgate, B., Bruhweiler, F.C., Boggess, A., Plait, P., Lindler, D.J., Clampin, M. and Kalas, P. (1999). Hubble Space Telescope Space Telescope Imaging Spectrograph Coronagraphic Imaging of the Herbig Ae Star AB Aurigae. *Astrophysical Journal*, **523**, L151–L154.

Graham, J.A. (1992). Clumpy accretion onto pre-main-sequence stars. *Publ. Astron. Soc. Pacific*, **104**, 479–488.

Gregorio-Hetem, J., Lepine, J.R.D., Quast, G.R., Torres, C.A.O. and de La Reza, R. (1992). A search for T Tauri stars based on the IRAS point source catalog. *Astronomical Journal*, **103**, 549–563.

Grinin, V.P., Thé, P.S., de Winter, D., Giampapa, M., Rostopchina, A.N., Tambovtseva, L.V. and van den Ancker, M.E. (1994). The beta Pictoris phenomenon among young stars. 1: The case of the Herbig Ae star UX Orionis. *Astronomy and Astrophysics*, **292**, 165–174.

Gürtler, J., Henning, T., Koempe, C., Pfau, W., Kraetschmer, W. and Lemke, D. (1996). Detection of an absorption feature at the position of the 4.27-μm band of solid CO_2. *Astronomy and Astrophysics*, **315**, L189–L192.

Habing, H.J., Dominik, C., Jourdain de Muizon, M., Kessler, M.F., Laureijs, R.J., Leech, K., Metcalfe, L., Salama, A., Siebenmorgen, R. and Trams, N. (1999). Disappearance of stellar debris disks around main-sequence stars after 400 million years. *Nature*, **401**, 456–458.

Hallenbeck, S. and Nuth, J. (1998). Infrared observations of the transition from chaotic to crystalline silicates via thermal annealing in the laboratory. *Astrophysics and Space Science*, **255**, 427–433.

Hanner, M.S., Brooke, T.Y. and Tokunaga, A.T. (1998). 8–13 Micron spectroscopy of young stars. *Astrophysical Journal*, **502**, 871–882.

Hanner, M.S., Newburn, R.L. and Gehrz, R.D. (1990). The infrared spectrum of Comet Bradfield (1987s) and the silicate emission feature. *Astrophysical Journal*, **348**, 312–321.

Hartmann, L., Kenyon, S.J., Hewett, R., Edwards, S., Strom, K.M., Strom, S.E. and Stauffer, J.R. (1989). Pre-main-sequence disk accretion in Z Canis Majoris. *Astrophysical Journal*, **338**, 1001–1010.

Helmich, F.P., van Dishoeck, E.F., Black, J.H., de Graauw, T., Beintema, D.A., Heras, A.M., Lahuis, F., Morris, P.W. and Valentijn, E.A. (1996). Detection of hot, abundant water toward AFGL 2591. *Astronomy and Astrophysics*, **315**, L173–L176.

Henney, W.J. and O'Dell, C.R. (1999). A Keck high-resolution spectroscopic study of the Orion Nebula proplyds. *Astronomical Journal*, **118**, 2350–2368.

Herbig, G.H. and Goodrich, R.W. (1986). Near-simultaneous ultraviolet and optical spectrophotometry of T Tauri stars. *Astrophysical Journal*, **309**, 294–305.

Herter, T., Campins, H. and Gull, G.E. (1987). Airborne spectrophotometry of p/Halley from 16-microns to 30-microns. *Astronomy and Astrophysics*, **187**, 629–631.

Hillenbrand, L.A., Meyer, M.R., Strom, S.E. and Skrutskie, M.F. (1995). Isolated star-forming regions containing Herbig Ae/Be stars. 1: The young stellar aggregate associated with BD +40°4124. *Astronomical Journal*, **109**, 280–297.

Hogerheijde, M.R. (1997). The molecular environment of low-mass protostars. PhD Thesis, Leiden University.

Hu, J.Y., Blondel, P.F.C. and Thé, P.S. (1991). IUE observations of the bright Herbig Ae star HD 104237. *Astronomy and Astrophysics*, **248**, 150–154.

Imhoff, C.L. and Appenzeller, I. (1987). Pre-main sequence stars. In Y. Kondo (ed), *Scientific Accomplishments of the IUE*, Astrophysics and Space Science Library, Reidel, Dordrecht, Vol. 129, pp. 295–319.

Imhoff, C.L. and Giampapa, M.S. (1980). The ultraviolet spectrum of the T Tauri star RW Aurigae. *Astrophysical Journal*, **239**, L115–L119.

Imhoff, C.L. and Giampapa, M.S. (1981). The ultraviolet variability of the T Tauri star RW Aurigae. In *The Universe at Ultraviolet Wavelengths*, NASA CP-2171, pp. 185–191.

Jäger, C., Molster, F.J., Dörschner, J., Henning, T., Mutschke, H. and Waters, L.B.F.M. (1998). Steps toward interstellar silicate mineralogy. IV. The crystalline revolution. *Astronomy and Astrophysics*, **339**, 904–916.

Jayawardhana, R., Holland, W.S., Greaves, J.S., Dent, William R.F., Marcy, G.W., Hartmann, L.W. and Fazio, G.G. (2000). Dust in the 55 Cancri planetary system. *Astrophysical Journal*, **536**, 425–428.

Jeans, J.H. (1902). *Phil. Trans.*, **199A**, 49.

Jura, M., Balm, S.P. and Kahane, C. (1995). A long-lived disk around the Red Rectangle? *Astrophysical Journal*, **453**, 721–726.

Jura, M., Malkan, M., White, R., Telesco, C., Pina, R. and Fisher, R. (1998). A proto-cometary cloud around HR 4796A? *Astrophysical Journal*, **505**, 897–902.

Jura, M. and Turner, J. (1998). A mysterious dust clump in a disk around an evolved binary star system. *Nature*, **395**, 144–145.

Kastner, J.H., Zuckerman, B., Weintraub, D.A. and Forveille, T. (1997). X-ray and molecular emission from the nearest region of recent star formation. *Science*, **277**, 67–71.

Kessler, M.F., Steinz, J.A., Anderegg, M.E., Clavel, J., Drechsel, G., Estaria, P., Faelker, J., Riedinger, J.R., Robson, A., Taylor, B.G. and Ximenez de Ferran, S. (1996). The Infrared Space Observatory (ISO) mission. *Astronomy and Astrophysics*, **315**, L27–L31.

Knacke, R.F., Fajardo-Acosta, S.B., Telesco, C.M., Hackwell, J.A., Lynch, D.K. and Russell, R.W. (1993). The silicates in the disk of beta Pictoris. *Astrophysical Journal*, **418**, 440–450.

Koike, C. and Shibai, H. (1990). Optical constants of hydrous silicates from 7 μm to 400 μm. *Monthly Notices of the Royal Astronomical Society*, **246**, 332–336.

Koike, C., Tsuchiyama, A., Shibai, H., Suto, H., Tanabé, T., Chihara, H., Sogawa, H., Mouri, H. and Okada, K. (2000). Absorption spectra of Mg-rich Mg–Fe and Ca pyroxenes in the mid- and far-infrared regions. *Astronomy and Astrophysics*, **363**, 1115–1122.

Kondo, Y. (1998). Space astronomy and IUE. In W. Wamsteker and R. González Riestra (eds), *Ultraviolet Astrophysics Beyond the Final IUE Archive*, ESA SP-413, pp. 5–6.

Kuhi, L.V. (1966). Ultraviolet continuous emission in T Tauri stars. *Publications of the Astronomical Society of the pacific*, **78**, 430–433.

Kuhi, L.V. (1974). Spectral energy distributions of T Tau stars. *Astronomy and Astrophysics*, **15**, 47–89.

Lada, C.J. (1987). Star formation – From OB associations to protostars. In M. Peimbert and J. Jugaku (eds), *Star Forming Regions*, International Astronomical Union Symposium 115, Reidel, Dordrecht, pp. 1–17.

Lada, C.J. and Shu, F.H. (1990). The formation of sunlike stars. *Science*, **248**, 564–572.

Lada, C.J. and Wilking, B.A. (1984). The nature of the embedded population in the Rho Ophiuchi dark could – Mid-infrared observations. *Astrophysical Journal*, **287**, 610–621.

Lago, M.T.V.T., Penston, M.V. and Johnstone, R. (1984). A UV glimpse of T Tauri stars. In *The 4th European IUE Conference*, ESA SP-218, pp. 233–237.

Lagrange, A.M., Beust, H., Ferlet, R. and Vidal-Madjar, A. (1989). The circumstellar gas around Beta Pictoris. VIII – Evidence for a clumpy structure of the infalling gas. *Astronomy and Astrophysics*, **215**, L5–L8.

Laques, P. and Vidal, J.L. (1979). Detection of a new type of condensations in the center of the Orion Nebula by means of S20 photocathode cells associated with a Lallemand electronic camera. *Astronomy and Astrophysics*, **73**, 97–106.

Leinert, Ch., Zinnecker, H., Weitzel, N., Christou, J., Ridgway, S.T., Jameson, R., Hass, M. and Lenzen, R. (1993). A systematic approach for young binaries in Taurus. *Astronomy and Astrophysics*, **278**, 129–149.

Lemke, D., Klaas, U., Abolins, J., Abraham, P., Acosta-Pulido, J., Bogun, S., Castaneda, H., Cornwall, L., Drury, L., Gabriel, C., Garzon, F., Gemuend, H.P., Groezinger, U., Gruen, E., Haas, M., Hajduk, C., Hall, G., Heinrichsen, I., Herbstmeier, U., Hirth, G., Joseph, R., Kinkel, U., Kirches, S., Koempe, C., Kraetschmer, W., Kreysa, E., Krueger, H., Kunkel, M., Laureijs, R., Luetzow-Wentzky, P., Mattila, K., Mueller, T., Pacher, T., Pelz, G., Popow, E., Rasmussen, I., Rodriguez Espinosa, J., Richards, P., Russell, S., Schnopper, H., Schubert, J., Schulz, B., Telesco, C., Tilgner, C., Tuffs, R., Voelk, H., Walker, H., Wells, M. and Wolf, J. (1996). ISOPHOT – capabilities and performance. *Astronomy and Astrophysics*, **315**, L64–L70.

Li, A. and Greenberg, J.M. (1998). A comet dust model for the beta Pictoris disk. *Astronomy and Astrophysics*, **331**, 291–313.

Lynch, D.K., Russell, R.W., Hackwell, J.A., Hanner, M.S. and Hammel, H.B. (1992). 8- to 13-micron spectroscopy of Comet Levy 1990 XX. *Icarus*, **100**, 197–202.

Malfait, K., Bogaert, E. and Waelkens, C. (1998a). An ultraviolet, optical and infrared study of Herbig Ae/Be stars. *Astronomy and Astrophysics*, **331**, 211–223.

Malfait, K., Waelkens, C., Bouwman, J., de Koter, A. and Waters, L.B.F.M. (1999). The ISO spectrum of the young star HD 142527. *Astronomy and Astrophysics*, **345**, 181–186.

Malfait, K., Waelkens, C., Waters, L.B.F.M., Vandenbussche, B., Huygen, E. and de Graauw, M.S. (1998b). The spectrum of the young star HD 100546 observed with the Infrared Space Observatory. *Astronomy and Astrophysics*, **332**, L25–L28.

Mannings, V. and Sargent, A.I. (1997). A high-resolution study of gas and dust around young intermediate-mass stars: Evidence for circumstellar disks in Herbig Ae systems. *Astrophysical Journal*, **490**, 792–802.

Marcy, G.W. and Butler, R.P. (1998). Detection of extrasolar giant planets. *Annual Review of Astronomy and Astrophysics*, **36**, 57–98.

Marsh, K.A., van Cleve, J.E., Mahoney, M.J., Hayward, T.L. and Houck, J.R. (1995). Spatially resolved mid-infrared observations of circumstellar dust around AB Aurigae. *Astrophysical Journal*, **451**, 777–783.

Mathieu, R. (1994). Pre-main-sequence binary stars. *Annual Review of Astronomy and Astrophysics*, **32**, 465–530.

Mayor, M. and Queloz, D. (1995). A Jupiter-mass companion to a solar-type star. *Nature*, **378**, 355–359.

McCaughrean, M.J. and Stauffer, J. (1994). High resolution near-infrared imaging of the trapezium: A stellar census. *Astronomical Journal*, **108**, 1382–1397.

Molster, F.J., Waters, L.B.F.M., Trams, N.R., Van Winckel, H., Decin, L., van Loon, J.T., Jger, C., Henning, T., Kufl, H.-U., de Koter, A. and Bouwman, J. (1999a). The composition and nature of the dust shell surrounding the binary AFGL 4106. *Astronomy and Astrophysics*, **350**, 163–180.

Molster, F.J., Yamamura, I., Waters, L.B.F.M., Tielens, A.G.G.M., de Graauw, T., de Jong, T., de Koter, A., Malfait, K., van den Ancker, M.E., Van Winckel, H., Voors, R.H.M. and Waelkens, C. (1999b). Low-temperature crystallization of silicate dust in circumstellar disks. *Nature*, **401**, 563–565.

Mooney, T.J. and Solomon, P.M. (1988). Star formation rates and the fair-infrared luminosity of galactic molecular clouds. *Astrophysical Journal*, **334**, L51–L54.

Mouschovias, T.Ch. (1991). Magnetic braking, ambipolar diffusion, cloud cores, and star formation – Natural length scales and protostellar masses. *Astrophysical Journal*, **373**, 169–186.

Mundt, R. and Ray, T.R. (1994). Optical outflows from Herbig Ae/Be stars and other high luminosity young stellar objects. In P.S. Thé, M.R. Pérez and E.P.J. van den Heuvel (eds), *The Nature and Evolutionary Stage of Herbig Ae/Be Stars*, ASP Conference Series, Vol. 62, pp. 237–252.

Neuhauser, R. (1997). Low-mass pro-main sequence stars and their X-ray emission. *Science*, **276**, 1363–1370.

Neuhauser, R. and Comeron, F. (1998). ROSAT X-ray detection of a young brown dwarf in the Chamaeleon I dark cloud. *Science*, **282**, 83–85.

Neuhauser, R., Sterzik, M.F., Schmitt, J.H.M.M., Wichmann, R. and Krautter, J. (1995). ROSAT survey observation of T Tauri stars in Taurus. *Astronomy and Astrophysics*, **297**, 391–417.

Nordh, L., Olofsson, G., Abergel, A., Andre, P., Blommaert, J., Bontemps, S., Boulanger, F., Burgdorf, M., Cesarsky, C.J., Cesarsky, D., Copet, E., Davies, J., Falgarone, E., Huldtgren, M., Kaas, A.A., Lagache, G., Montmerle, T., Perault, M., Persi, P., Prusti, T., Puget, J.L. and Sibille, F. (1996). ISOCAM observations of low-luminosity young stellar objects in the Chamaeleon dark clouds. *Astronomy and Astrophysics*, **315**, L185–L188.

O'Dell, C.R. (1998). Observational properties of the Orion Nebula proplyds. *Astronomical Journal*, **115**, 263–273.

O'Dell, C.R. and Wen, Z. (1994). Postrefurbishment mission Hubble Space Telescope images of the core of the Orion Nebula: Proplyds, Herbig-Haro objects, and measurements of a circumstellar disk. *Astrophysical Journal*, **436**, 194–202.

O'Dell, C.R., Wen, Z. and Hu, X. (1993). Discovery of new objects in the Orion nebula on HST images – Shocks, compact sources, and proto-planetary disks. *Astrophysical Journal*, **410**, 696–700.

O'Dell, C.R. and Wong, S.K. (1996). Hubble Space Telescope mapping of the Orion Nebula. I. A survey of stars and compact objects. *Astronomical Journal*, **111**, 846–855.

Olofsson, G., Kaas, A.A., Bontemps, S., Nordh, L., Huldtgren, M., Abergel, A., André, P., Boulanger, F., Burgdorf, M., Casali, M.M., Cesarsky, C.J., Copet, E., Falgarone, E., Montmerle, T., Perault, M., Persi, P., Prusti, T., Puget, J.L. and Sibille, F. (1999). An ISOCAM survey of nearby star formation regions. In P. Cox and M.F. Kessler (eds), *The Universe as Seen by ISO*, ESA SP-427, pp. 459–462.

Pérez, M.R., Thé, P.S. and Westerlund, B.E. (1987). On the distances to the young open clusters NGC 2244 and NGC 2264. *Publications of the Astronomical Society of the Pacific*, **99**, 1050–1066.

Plets, H. and Vynckier, C. (1999). An analysis of the incidence of the Vega phenomenon among main-sequence and post-main-sequence stars. *Astronomy and Astrophysics*, **343**, 496–506.

Poetzel, R., Mundt, R. and Ray, T.P. (1989). Z CMa – A large-scale high velocity bipolar outflow traced by Herbig-Haro objects and a jet. *Astronomy and Astrophysics*, **224**, L13–L16.

Praderie, F., Catala, C., Simon, T. and Boesgaard, A.M. (1986). Short-term spectral variability in AB Aurigae – clues for activity in Herbig Ae stars. I – The ultraviolet lines of Mg-II and Fe-II. *Astrophysical Journal*, **303**, 311–326.

Praderie, F., Felenbok, P., Czarny, J., Talavera, A. and Boesgaard, A.M. (1982). The chromosphere and wind of the Herbig Ae star, AB Aurigae. *Astrophysical Journal*, **254**, 658–662.

Prosser, C.F., Stauffer, J.R., Hartmann, L., Soderblom, D.R., Jones, Burton F., Werner, M.W. and McCaughrean, M. (1994). HST photometry of the trapezium cluster. *Astrophysical Journal*, **421**, 517–541.

Prusti, T. (1999). Luminosity distribution of young stars. In P. Cox and M.F. Kessler (eds), *The Universe as Seen by ISO*, ESA SP-427, pp. 453–457.

Roddier, F., Roddier, C., Graves, J.E. and Northcott, M.J. (1995). Adaptive optics imaging of proto-planetary nebulae: Frosty Leo and the Red Rectangle. *Astrophysical Journal*, **443**, 249–260.

Sandford, S.A. and Walker, R.M. (1985). Laboratory infrared transmission spectra of individual interplanetary dust particles from 2.5 to 25 microns. *Astrophysical Journal*, **291**, 838–851.

Scalo, J. (1998). The IMF revisited: A case for variations. In D. Gilmore, I. Parry and S. Ryan (eds), *The Stellar Initial Mass Function*, ASP Conference Series 142, pp. 201–236.

Schneider, G., Smith, B.A., Becklin, E.E., Koerner, D.W., Meier, R., Hines, D.C., Lowrance, P.J., Terrile, R.J., Thompson, R.I. and Rieke, M. (1999). NICMOS imaging of the HR 4796A circumstellar disk. *Astrophysical Journal*, **513**, L127–L130.

Shu, F., Adams, F.C. and Lizano, S. (1987). Star formation in molecular clouds – Observation and theory. *Annual Review of Astronomy and Astrophysics*, **25**, 23–81.

Silk, J. (1997). Current issues in star formation. In A. Wilson (ed), *The Far Infrared and Submillimetre Universe*, ESA SP-401, p. 158.

Simon, T., Vrba, F.J. and Herbst, W. (1990). The ultraviolet and visible light variability of BP Tauri – Possible clues for the origin of T Tauri star activity. *Astronomical Journal*, **100**, 1957–1967.

Smith, B.A. and Terrile, R.J. (1984). A circumstellar disk arund Beta Pictoris. *Science*, **226**, 1421–1424.

Stahler, S.W. (1983). The birthline for low-mass stars. *Astrophysical Journal*, **274**, 822–829.

Stapelfeldt, K. and Moneti, A. (1999). The spectral energy distribution of HH 30 measured with ISO. In P. Cox and M.F. Kessler (eds), *The Universe as Seen by ISO*, ESA SP-427, pp. 521–524.

Stauffer, J.R., Hartmann, L.W. and Barrado y Navascués, D. (1995). An age estimate for the beta Pictoris analog HR 4796A. *Astrophysical Journal*, **454**, 910–916.

Talavera, A. (1998). Herbig Ae/be stars: Mass flows and disks. In W. Wamsteker and R. González Riestra (eds), *Ultraviolet Astrophysics Beyond the Final IUE Archive*, ESA SP-413, pp. 69–73.

Talavera, A., Blondel, P.F.C. and Tjin A Djie, H.R.E. (1994). Lyman α emission in the sepctra of two young B9e-type stars. In P.S. Thé, M.R. Pérez and E.P.J. van den Heuvel (eds), *The Nature and Evolutionary Stage of Herbig Ae/Be Stars*, ASP Conference Series, Vol. 62, pp. 115–118.

Talavera, A., Catala, C., Crivellari, L., Czarny, J., Felenbok, P. and Praderie, F. (1982). On the Mg-II and Fe-II resonance lines in Herbig Ae stars: Preliminary results. In *The 3rd European IUE Conference*, ESA SP-176, pp. 99–105.

Telesco, C.M. and Knacke, R.F. (1991). Detection of silicates in the Beta Pictoris disk. *Astrophysical Journal*, **372**, L29–L31.

Trilling, D.E. and Brown, R.H. (1998). A circumstellar dust disk around a star with a known planetary companion. *Nature*, **395**, 775–777.

Vaiana, G.S., Fabbiano, G., Giacconi, R., Golub, L., Gorenstein, P., Harnden, F.R., Jr, Cassinelli, J.P., Haisch, B.M., Johnson, H.M., Linsky, J.L., Maxson, C.W., Mewe, R., Rosner, R., Seward, F., Topka, K. and Zwaan, C. (1981). Results from an extensive Einstein stellar survey. *Astrophysical Journal*, **245**, 163–182.

van den Ancker, M.E., Bouwman, J., Wesselius, P.R., Waters, L.B.F.M. and Dougherty, S.M. (2000). ISO spectroscopy of circumstellar dust in the Herbig Ae systems AB Aur and HD 163296. *Astronomy and Astrophysics*, **357**, 325–329.

van den Ancker, M.E., Wesselius, P.R., Tielens, A.G.G.M., van Dishoeck, E.F. and Spinoglio, L. (1999). ISO spectroscopy of shocked gas in the vicinity of T Tauri. *Astronomy and Astrophysics*, **348**, 877–887.

van den Ancker, M.E., Wesselius, P.R., Tielens, A.G.G.M. and Waters, L.B.F.M. (1998a). PDRS and shocks in S106 IR and Cep A East. *Astrophysics and Space Science*, **255**, 69–75.

Van den Ancker, M.E., de Winter, D. and Tjin A Djie, H.R.E. (1998b). Hipparcos photometry of Herbig Ae/Be stars. *Astronomy and Astrophysics*, **330**, 145–154.

van Dishoeck, E.F. (1998). What can ISO tell us about gas-grain chemistry? In *Chemistry and Physics of Molecules and Grains in Space*, Faraday Discussions No. 109, Royal Society of Chemistry, London, p. 31.

van Dishoeck, E.F., Black, J.H., Boogert, A.C.A., Boonman, A.M.S., Ehrenfreund, P., Gerakines, P.A., de Graauw, T., Helmich, F.P., Keane, J.V., Lahuis, F., Schutte, W.A., Tielens, A.G.G.M., Whittet, D.C.B., Wright, C.M., van den Ancker, M.E., Blake, G.A., Creech-Eakman, M., Waters, L.B.F.M. and Wesselius, P.R. (1999). ISO spectroscopy of young stellar objects. In P. Cox and M.F. Kessler (eds), *The Universe as Seen by ISO*, ESA SP-427, pp. 437–448.

van Dishoeck, E.F. and Helmich, F.P. (1996). Infrared absorption of H_2O toward massive young stars. *Astronomy and Astrophysics*, **315**, L177–L180.

van Dishoeck, E.F., Helmich, F.P., de Graauw, T., Black, J.H., Boogert, A.C.A., Ehrenfreund, P., Gerakines, P.A., Lacy, J.H., Millar, T.J., Schutte, W.A., Tielens, A.G.G.M., Whittet, D.C.B., Boxhoorn, D.R., Kester, D.J.M., Leech, K., Roelfsema, P.R., Salama, A. and Vandenbussche, B. (1996). A search for interstellar gas-phase CO_2. Gas: solid state abundance ratios. *Astronomy and Astrophysics*, **315**, L349–L352.

Van Winckel, H., Waelkens, C. and Waters, L.B.F.M. (1995). The extremely iron-deficient 'Post-AGB' stars are binaries. *Astronomy and Astrophysics*, **293**, L25–L28.

Van Winckel, H., Waelkens, C., Waters, L.B.F.M., Molster, F.J., Udry, S. and Bakker, E.J. (1998). The binary RV Tauri star AC Her and evidence for a long-lived dust-disc. *Astronomy and Astrophysics*, **336**, L17–L20.

Waelkens, C., Waters, L.B.F.M., de Graauw, M.S., Huygen, E., Malfait, K., Plets, H., Vandenbussche, B., Beintema, D.A., Boxhoorn, D.R., Habing, H.J., Heras, A.M., Kester, D.J.M., Lahuis, F., Morris, P.W., Roelfsema,

P.R., Salama, A., Siebenmorgen, R., Trams, N.R., van der Bliek, N.R., Valentijn, E.A. and Wesselius, P.R. (1996). SWS observations of young main-sequence stars with dusty circumstellar disks. *Astronomy and Astrophysics*, **315**, L245–L248.

Ward-Thompson, D. and André, P. (1999). ISO and the initial conditions of star formation. In P. Cox and M.F. Kessler (eds), *The Universe as Seen by ISO*, ESA SP-427, pp. 463–466.

Waters, L.B.F.M., Cami, J., de Jong, T., Molster, F.J., van Loon, J.T., Bouwman, J., de Koter, A., Waelkens, C., Van Winckel, H. and Morris, P.W. (1998). An oxygen-rich dust disk surrounding an evolved star in the Red Rectangle. *Nature*, **391**, 868–970.

Waters, L.B.F.M., Molster, F.J., de Jong, T., Beintema, D.A., Waelkens, C., Boogert, A.C.A., Boxhoorn, D.R., de Graauw, T., Drapatz, S., Feuchtgruber, H., Genzel, R., Helmich, F.P., Heras, A.M., Huygen, R., Izumiura, H., Justtanont, K., Kester, D.J.M., Kunze, D., Lahuis, F., Lamers, H.J.G.L.M., Leech, K.J., Loup, C., Lutz, D., Morris, P.W., Price, S.D., Roelfsema, P.R., Salama, A., Schaeidt, S.G., Tielens, A.G.G.M., Trams, N.R., Valentijn, E.A., Vandenbussche, B., van den Ancker, M.E., van Dishoeck, E.F., Van Winckel, H., Wesselius, P.R. and Young, E.T. (1996). Mineralogy of oxygen-rich dust shells. *Astronomy and Astrophysics*, **315**, L361–L364.

Weinberger, A.J., Becklin, E.E., Schneider, G., Smith, B.A., Lowrance, P.J., Silverstone, M.D., Zuckerman, B. and Terrile, R.J. (1999). The circumstellar disk of HD 141569 imaged with NICMOS. *Astrophysical Journal*, **525**, L53–L56.

Whittet, D.C.B., Schutte, W.A., Tielens, A.G.G.M. and Boogert, A.C.A. (1996). An ISO SWS view of interstellar ices: First results. *Astronomy and Astrophysics*, **315**, L357–360.

Wilkening, L.L. (1978). Carbonaceous chondritic material in the Solar System. *Naturwissenschaften*, **65**, 73–79.

38

ROBERTO PALLAVICINI*

High-energy radiation from outer stellar atmospheres

Although the Sun had been known to be a source of X-ray and ultraviolet radiation since the late 1940s, it was only in the late 1970s that normal stars of nearly all spectral types and luminosity classes were recognized to be sources of high-energy radiation detectable from space (see Mewe 1996 for a historical perspective). The origin of this high-energy emission from normal (i.e. non-accreting) stars constitutes a fundamental and yet unsolved problem in stellar astrophysics, which space observations have helped to elucidate over the past 20 years. It was quite obvious from the very first observations that the generation of high-energy photons by thermal processes, such as the ones usually observed in normal stars, requires temperatures far in excess to those responsible for the stellar optical radiation. In particular, it requires that the temperature profile in stellar atmospheres, after reaching a minimum in the upper photosphere, rises again, attaining values of one to several million degrees in the corona. These outer atmospheric layers, where the temperature increases outwards through the chromosphere, transition region and corona, had been known and widely studied for a long time in the case of the Sun, but it was only much later with the advent of grazing incidence X-ray telescopes and sensitive X-ray detectors that their existence for the vast majority of stars was definitely proved.

The study of high-energy radiation from normal stars is a powerful diagnostic tool for the physical conditions in outer stellar atmospheres and for the processes in the stellar interiors that are at the origin of coronal heating. X-ray and UV radiation in fact provides information on non-radiative heating mechanisms as well as on the processes that lead to wind acceleration and mass loss. In the case of cool stars – which possess subphotospheric convective zones – both the high coronal temperatures and the various manifestations of stellar activity appear to be related to the generation of magnetic fields through a dynamo process that involves rotation and convection. Thus, observations of stellar chromospheres, transition regions and coronae, albeit limited to the outer atmospheric layers, allows us to put constraints also on the physical processes that occur deep in the interior of stars, with important implications for our understanding of stellar structure and evolution. Moreover, observations of stars in different evolutionary stages, from the early phases of contraction to the main sequence to the late stages of evolution out of it, allow for the investigation of the effects of magnetic braking and angular momentum loss on the generation of magnetic fields and on the heating processes.

The traditional view, first developed in the case of the Sun, attributed the temperature inversion that occurs in the outer stellar atmospheres to shock dissipation of acoustic waves generated in the subphotospheric convective zone. Although this theory is probably still valid for the Sun's lower chromospheric layers and for non-magnetic regions, it fails badly for solar magnetic regions, for active stars and in general for stars, both hotter and cooler than the Sun, with an internal structure significantly different from the solar one. For cool stars, magneto-acoustic and Alfvén waves and/or resistive dissipation of electric currents are more likely to be responsible for the observed high-energy radiation, thus stressing the importance of magnetic fields in the outer atmospheres of these stars. Hot stars, which do not have outer convective zones, are also observed to be sources of X-rays, but high-energy emission in this case is more likely due to shock dissipation in their massive radiatively driven winds rather than to the presence of a high-temperature corona.

*Osservatorio Astronomico di Palermo, Italy

In this chapter, I will review X-ray and UV emission from outer stellar atmospheres with emphasis on X-ray emission from the coronae of late-type stars, that is stars that possess subphotospheric convective zones. According to the usual classification based on their optical spectra, these stars cover the spectral types from F to M, and all luminosity classes from dwarfs to supergiants. Along the main sequence, the effective temperature decreases from about 7000 K to less than 3000 K, the mass decreases from about 1.7 solar masses to one-tenth of a solar mass and the radius decreases from about 1.3 to a few tenths of the solar radius. On the contrary, the convection zone depth increases from a negligible fraction of the stellar radius in early F-type stars to the whole stellar radius in late M dwarfs. The wide range of physical parameters covered by F to M stars along the main sequence, not to mention the wider range provided by the inclusion of giants, offers a variety of different conditions for the generation of magnetic fields and for the heating of the outer atmospheric layers. In addition, the broad range of ages and evolutionary stages covered by cool stars implies also different values of rotation rates and hence of the efficiency of magnetic field generation by the dynamo process. All this offers to our physical speculations a much more complex picture that the one made possible only by observations of the Sun. The X-ray and UV observations carried out from space during the past 20 years have largely unravelled this complex picture taking us to the point where we can finally start to understand the physical processes at the origin of coronal activity in different types of stars.

EARLY RESULTS ON STELLAR X-RAY EMISSION

The beginning of stellar X-ray astronomy

Before the launch of the International Ultraviolet Explorer (IUE) and of the Einstein Observatory (formerly HEAO 2) in 1978, there was only a handful of stellar coronal sources, besides the Sun, known to emit X-ray and UV radiation (Mewe 1979). They were either very close objects (e.g. α Centauri) or very active stars with coronal emission many orders of magnitude higher than that of the Sun. Most of the latter objects were close binary systems of the RS Canum Venaticorum (RS CVn) type, a class of binary stars known for their extreme activity at optical and radio wavelengths. One of these objects, Capella, was indeed the first non-solar coronal X-ray source to be detected in 1974 during a rocket flight and is still the brightest X-ray source among normal stars, except for the Sun. This early discovery was soon followed by the detection with the first High Energy Astronomical Observatory (HEAO 1) of a number of other RS CVn stars with typical X-ray luminosities in the range of 10^{30}–10^{31} erg s^{-1} in the soft X-ray band. This is 1,000 to 10,000 times more than the average X-ray luminosity of the Sun or of quiet solar-type stars like α Cen. At any rate, X-ray emission at appreciable levels was thought at that time to be limited to somewhat peculiar objects and not to be a general characteristic of all types of stars.

Neither was there much expectation from a theoretical point of view of detecting stellar coronal emission (Mewe 1979). The mechanism of coronal heating generally accepted at that time was shock dissipation of acoustic waves generated in subphotospheric convective zones. In spite of large uncertainties present in available estimates of the power dissipated by acoustic waves, the theory invariably predicted that maximum dissipation should occur around spectral types F and G, decreasing rapidly towards later spectral types along the main sequence. It was not expected therefore that X-ray emission could be detected in K and M dwarfs, nor in early-type stars that do not possess outer convective zones. Even more importantly, the acoustic heating theory predicted that stars at a given position in the HR diagram should all have the same level of chromospheric and coronal emission, which in turn should depend only on the properties of their convective zone and on the efficiency of generation of acoustic noise. Spatially resolved observations of the solar chromosphere and corona (which showed the presence of localized regions of enhanced emission associated with regions of stronger magnetic fields) were clearly at variance with this simplified picture; however, the effect of magnetic fields was considered only as a second-order perturbation which could hardly change the overall picture. As for early-type stars, acoustic waves could not be driven by convection but it was suggested that waves could be amplified by radiation pressure leading some authors to believe that early-type stars might eventually be detected in X-rays.

The early observations by IUE and Einstein showed that these expectations were fundamentally wrong. In particular, the Einstein Observatory showed that virtually all stars, with the exception of A-type dwarfs and of late K and M giants, were X-ray emitters (Vaiana et al. 1981). For late-type stars, there was little or no dependence of the X-ray emission on spectral type, in contrast with the predictions of the acoustic theory. Only for early-type stars was there a clear dependence of the X-ray emission on the stellar bolometric luminosity. For late-type dwarfs of spectral types F to M a broad range (more than three orders of magnitude) of coronal emission levels existed at each effective temperature, indicating that parameters other than the position of a star in the HR diagram (i.e. effective temperature and luminosity) were responsible for the activity level of stellar outer atmospheres. These early observations clearly demonstrated that the acoustic theory was basically untenable as far as coronal heating is concerned and that other mechanisms, possibly involving magnetic fields, were to be

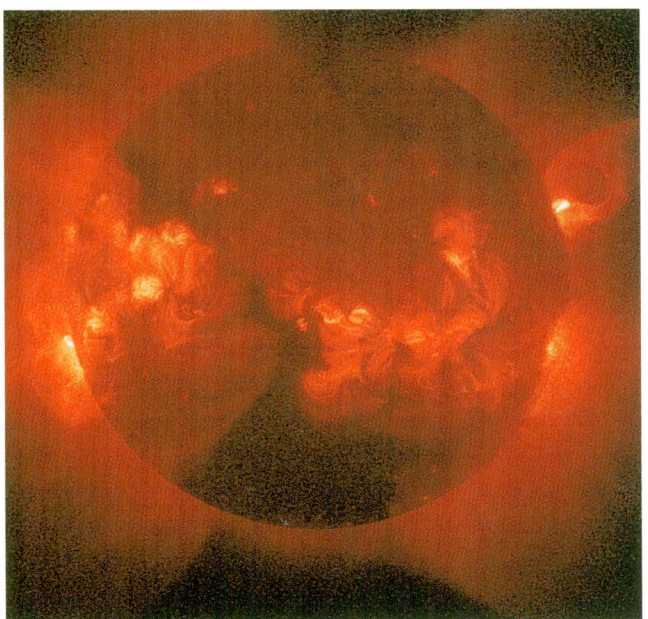

Figure 1 Soft X-ray image of the solar corona obtained on 27 January 1992 with the Soft X-ray Telescope (SXT) on the Japanese satellite Yohkoh. (Courtesy of the Institute of Space and Astronomical Sciences.)

explored. These results for stars were in line with increasing evidence from solar observations that X-ray-emitting coronae are dominated by magnetic field effects (Rosner et al. 1985), and that surface magnetic fields are perhaps the dominant factor in determining the structure and variability of late-type coronae and in providing for their heating (Figure 1).

It is interesting to note that this interplay of solar and stellar results, all pointing in the same direction, is itself a consequence of space observations at X-ray and UV wavelengths, being carried out at about the same time, but with different emphasis and different objectives, for the Sun and stars. In the case of the Sun, sensitivity is not a problem, since the Sun, in spite of being an intrinsically weak X-ray and UV source, appears to us very strong because of its proximity. The Sun is the only star which appears to us as an extended source; it is therefore not surprising that the emphasis of solar research was, and still is, in obtaining high-spatial-resolution observations which allow for the study of fine details in the solar atmosphere. The X-ray and UV observations obtained in 1973–74 by *Skylab* and in 1980 by SMM, as well as the solar images obtained more recently by Yohkoh, SOHO and TRACE, revealed a highly structured outer atmosphere, where the coronal plasma appears confined by the magnetic field in loop-like structures tracing the field lines. The same observations also revealed that the solar corona is highly dynamic in response to fluctuations in the heating rate. All this suggested that magnetic fields not only confine the coronal plasma (since the magnetic pressure exceeds the gas pressure in the corona), but are also responsible for the heating of the plasma to temperatures far in excess of a star's effective temperature.

Coronal X-ray emission in late-type stars

The broad range of X-ray emission levels observed for late-type stars of similar effective temperatures was soon found to be related to stellar rotation and age, thus enforcing the notion that dynamo-generated magnetic fields are responsible for coronal heating in stars of late spectral types (Pallavicini et al. 1981). Although a fully consistent dynamo theory does not yet exist, not to mention a generally accepted coronal heating theory, a qualitative expectation is that the efficiency of dynamo action should increase with stellar rotation rate in stars that possess subphotospheric convective zones. We may expect, therefore, that the activity level of a star, as it results from the cumulative effects of the multitude of magnetically confined coronal structures that form its outer atmosphere, should depend on the rotation rate, or better on a combination of rotation and convection zone properties. This combination of dynamo factors is often parameterized by means of a Rossby number (Noyes et al. 1984) defined as the ratio of the star rotation period P_{rot} to a colour-dependent convective turnover time τ_{conv}, computed at the base of the convective zone (where most of the dynamo action is believed to occur). The observed dependence of stellar chromospheric and coronal emission on either rotation rate or Rossby number is a strong, albeit only qualitative, support of this interpretation. The details should depend on the still poorly understood properties of the dynamo process and of the conversion of magnetic fields into plasma heating.

Since coronal emission among late-type stars appears to depend on both rotation and age, a relevant question is whether these parameters are related to each other and, if so, which of the two is the primary one for coronal heating. The existence of a dependence on age was demonstrated by observations of open clusters, that is by homogeneous samples of stars with approximately the same age and chemical composition. It was evident from early IUE and Einstein observations of the Hyades and the Pleiades, and from the comparison of solar-type stars in these clusters with the much older Sun, that there was a dependence of chromospheric, transition region and coronal emission upon age (Stern et al. 1981, Caillault and Helfand 1985, Micela et al. 1990). On the other hand, it is well known that late-type stars with outer convective zones suffer angular momentum loss during their main sequence lifetime, owing to the braking action of magnetized stellar winds. As stars get older, they rotate more slowly, and their chromospheric and coronal emissions (which have been found by Einstein and IUE to be related one to the other, e.g. Ayres *et al.* 1981) both decline with age. Rotation, rather than age, appears therefore to be the primary factor, at least for stars that have already reached the main sequence (the situation may be more complex for pre-main sequence stars, owing to the interaction of the young star with the surrounding disk). That rotation is indeed the primary factor is confirmed by the high activity and strong coronal emission of RS CVn binaries which are relatively old stars (a few Gyr) that still rotate rapidly because of the tidal interaction of the two components in a close binary system.

The above picture was developed in the early 1980s at the Center for Astrophysics in Cambridge, MA, by the group responsible for the early analysis of the stellar data from the Einstein Observatory. This group, led by Giuseppe S. Vaiana (an Italian scientist who, not surprisingly, had worked previously on X-ray observations of the solar corona, and who had been responsible for one of the main instruments on ATM-Skylab), greatly benefited from the interaction with a number of Guest Investigators from all parts of the world (Jeff Linsky, Tom Ayres, Fred Walter, Rick Harnden, Joe Cassinelli, Rolf Mewe, Bob Stern and Thierry Montmerle, just to mention a few) and included at that time a number of young scientists (Bob Rosner, Leon Golub, Roberto Pallavicini, Jürgen Schmitt) who continued for many years to work in the field of stellar X-ray astronomy. Later, it also included a number of young researchers from the Astronomical Observatory of Palermo, where Vaiana, upon his return to Italy, established an active centre of stellar corona research. The interpretation of stellar X-ray activity in terms of dynamo-generated magnetic fields and solar analogues was largely suggested by previous work on chromospheric Ca II H and K emission and on its dependence on rotation and age, as well as by the lesson learned from spatially resolved observations of the solar chromosphere, transition region and corona. It received further support from the wealth of ultraviolet data provided at about the same time by IUE on stellar chromospheres and transition regions up to temperatures of about 10^5 K (see Jordan and Linsky 1987 for a review).

Late-type giants

A further surprise in these early days came from the observations of late-type giants and supergiants. Whereas yellow giants are typically X-ray emitters, sometimes at very strong levels, this was not the case for red giants later than spectral type K2 for luminosity III giants and of spectral type G0 for supergiants. On the basis of UV and X-ray observations from IUE and Einstein, the presence of a dividing line (DL) in the HR diagram was inferred (Linsky and Haisch 1979, Maggio *et al.* 1990), separating stars with high-temperature coronae and low mass losses (to the left of the DL) from stars with high mass losses and the absence of high-temperature plasma (to the right of the DL; note that the cool, low-speed winds of late-type giants are fundamentally different from the hot, high-temperature winds of early-type stars). A class of stars was also found (the so called 'hybrid' stars) which have both large mass losses and plasma in excess of 10^5 K in their outer atmospheres, suggesting that the transition between the two classes of evolved stars (with and without coronae) may not be as abrupt as originally thought. For some late giants (for instance Arcturus) extremely low upper limits to their X-ray luminosity (as low as 10^{25} erg s^{-1}, Ayres *et al.* 1991) have now been established, which indicates that the X-ray surface fluxes in these evolved low-gravity stars must be many orders of magnitude lower than in solar coronal holes. The onset of strong mass losses nearly coincident with the disappearance of hot coronae strongly suggests a causal relationship between the two phenomena. By analogy with the Sun, the presence of a high-temperature corona in a late-type star implies the existence of magnetically confined structures which are denser and hotter than the surrounding regions. On the contrary, areas of open field lines in the Sun (i.e. coronal holes) are somewhat cooler and much less dense than the emitting loops: apparently the energy deposited in these regions (by either magneto-acoustic waves or electric currents) goes into accelerating the wind and in enhancing mass losses rather than into plasma heating. By analogy, the transition across the DL could be interpreted as a change of magnetic topology in the outer atmospheres of cool stars, from a corona dominated by closed magnetic structures (similar to solar coronal loops) to an essentially open corona, with most field lines being open to the interplanetary space. Whether this picture is correct, and how the supposed change of magnetic topology relates to the dynamo process and to the internal changes experienced by a star during its evolution out of the main sequence, remain questions to be solved.

X-rays from stellar winds in early-type stars

The situation with early-type (O-B) stars is completely different. Their X-ray luminosity, ranging from 10^{29} to 10^{34} erg s^{-1}, is virtually independent of rotation, while it depends on bolometric luminosity (albeit with some scatter). This suggested that the X-ray emission might be related to the strong, high-speed radiatively driven winds known to exist in these massive stars. Early models of X-ray emission from hot stars assumed the presence of a thin, high-temperature corona at the base of the wind, possibly confined by fossil magnetic fields (Cassinelli and Olson 1979). This model, however, was not confirmed by spectral observations with the Einstein Solid State Spectrometer (SSS) which did not show the presence of the predicted low-energy absorption (the X-ray radiation emitted by a corona at the base of the wind would be strongly absorbed by the overlying dense material). On the contrary, other models attributed the X-ray emission of early-type stars to shock heating throughout the wind (Lucy and White 1980, Lucy 1982), thus overcoming the difficulty of the lack of absorption at low energies. The winds of early-type stars are in fact radiatively unstable and form high-density blobs against which the outflowing wind can shock, thus raising the plasma temperature to a few million degrees. Although the wind model remains still phenomenological, it is in reasonable agreement with the observations and is the only one to have resisted the test of time after almost 20 years since it was first proposed. Alternative models assumed wind collisions with the surrounding interstellar medium or colliding winds in binary systems. The latter possibility is indeed a likely one in binary systems as those formed by an O-type star and by a Wolf–Rayet companion. Observations of single O stars, and of binary systems at different orbital phases, can provide crucial tests of the wind model for early-type stars: this is an active area of investigation at the present time.

Time variability and eclipse mapping

The decade from the launch of the Einstein Observatory until the launch of ROSAT in 1990 was largely utilized to refine the above picture for both early- and late-type stars, capitalizing on the large body of stellar data provided by Einstein (and by IUE at UV wavelengths). The same decade, however, also witnessed the launch of the European satellite EXOSAT (in 1983) and of the Japanese X-ray satellites TENMA (in 1983) and GINGA (in 1987). Although the results of these missions in the area of stellar coronae are not comparable in quantity with those obtained with Einstein and later with ROSAT, these missions provided some new interesting results which complemented those obtained previously by Einstein. In particular, the role played by EXOSAT in stellar coronal studies, both for stellar spectroscopy and variability, has often been overlooked. EXOSAT had a new important feature, a highly eccentric orbit, that allowed continuous uninterrupted observations for periods of up to 3 days. This was a very important capability to study stellar flares or to monitor rotational modulation and eclipses. Moreover, the Transmission Grating Spectrometer (TGS) on board EXOSAT (developed by the SRON group in Holland), albeit used only for a few observations due to a hardware problem with the grating insertion mechanism, provided coronal spectra that remained unsurpassed in terms of resolution until the launch of AXAF and XMM 15 years later. Most of the Einstein observations were in fact obtained with the Imaging Proportional Counter (IPC) which had only moderate energy resolution ($E/\Delta E$ of the order of 1). Except for a high-resolution spectrum of Capella obtained with the Objective Grating Spectrometer (OGS) on Einstein, moderate-resolution spectroscopy at resolution significantly better than that of the IPC was possible with Einstein using the SSS (Swank et al. 1981), which, however, suffered from calibration problems due to ice formation on the detector. Yet the spectra of early- and late-type stars obtained by the SSS were sufficient to show the inadequacy of the base coronal model for hot stars proposed by Cassinelli and Olson (1979) as well as for demonstrating the multi-temperature nature of the coronae of late-type stars. With the TGS on EXOSAT it was possible to go a step further. The TGS spectra of Capella, σ CrB and Procyon obtained by EXOSAT allowed the separation of line complexes (but not yet of individual lines) and a much better determination of the temperature structure in stellar coronae (see Mewe 1991 for a review). At least two isothermal components (one at a few million degrees and the other at about 20 million degrees) were necessary to fit the TGS spectrum of Capella, whereas much cooler temperatures were required to fit Procyon. Evidence for a two-temperature structure came also from Einstein SSS and IPC spectra of active stars (e.g. Schmitt et al. 1990), although it remained unclear whether these two temperatures (which appeared to be somewhat detector dependent) referred to physically distinct coronal regions (e.g. to two different families of loop structures) or were instead the result of a continuous distribution of coronal temperatures in either a single type of coronal loops or in a multitude of different loop structures.

In order to disentangle the temperature and spatial structure of coronae, and to map the distribution of active regions in the coronae of stars, observations of eclipsing binary systems were carried out with EXOSAT, by means of long continuous observations. Algol, AR Lac and TY Pyx were observed for a full orbital period or at crucial phases centred on the eclipses. The EXOSAT observation of AR Lac showed the presence of eclipses in the low-energy detector (0.04–2 keV), but not at high energies (2–10 keV), which suggested that the high-temperature component indicated by the spectral fits was to be large, comparable to or larger than the stellar radius (White et al. 1990). However, subsequent observations of the same binary system with ROSAT and ASCA have not confirmed the absence of eclipses at high energies, suggesting either that the corona of AR Lac has significantly changed since the time of the EXOSAT observation or that the EXOSAT data were of insufficient S/N to detect the eclipse at high-energies (Siarkowski et al. 1996). At any rate, these early EXOSAT observations pioneered by Nick White, who was leading at that time the EXOSAT Observatory team, clearly showed the potential of eclipse monitoring for mapping the structure of (otherwise unresolved) stellar coronae. It must be noted anyhow that the inversion of eclipse data to reconstruct the spatial structure of a corona is not a trivial task, and the results are often ambiguous due to the non-uniqueness of the solution (e.g. see Schmitt 1998 for a review). Continuous monitoring, together with high S/N and possibly large bandwidth, are essential for this type of study. Coverage of more than one orbital period is also desirable, to disentangle orbital modulation from intrinsic variability of the two components.

The continuous viewing capability of EXOSAT also proved essential to study stellar flares and to obtain a statistical characterization of their properties in terms of duration, physical parameters and energy release (Pallavicini et al. 1990). Both long- and short-lived flares were found to exist, similarly to the two classes of long-duration and compact flares known to exist on the Sun (and which are attributed to different physical mechanisms of conversion of magnetic energy into particle acceleration and plasma heating). Although flares had been observed and studied with Einstein, the large flare on Algol detected by EXOSAT (White et al. 1986) remained for many years the prototype flare to study the energy release in transient stellar events.

ADVANCES IN STELLAR CORONAL RESEARCH

Over the past 10 years substantial progress has been made in our understanding of stellar coronal emission thanks to the launch of ROSAT (in 1990), EUVE (in 1992) and ASCA (in 1993). These new missions have confirmed and extended the results of previous space missions (particularly those of Einstein and EXOSAT). At the same time, they have provided many new, sometimes unexpected, results that have significantly increased our knowledge of high-energy phenomena in stellar atmospheres. Among these new results, I mention a much better characterization of coronal emission across the HR diagram, a more complete coverage (with regard to both age and area coverage) of open clusters and star-forming regions, and the availability of better quality spectral data that for the first time have allowed us to address the question of coronal elemental abundances in addition to determinations of coronal temperatures and emission measures. In this section I will review some of these new findings, with emphasis again on coronal emission of late-type stars. I will also mention some recent results by the Italian–Dutch satellite BeppoSAX (launched in 1996) particularly in the area of stellar spectroscopy and for the study of stellar flares. A unique feature of BeppoSAX, unmatched by any previous X-ray satellite, is the very broad simultaneous spectral coverage (from 0.1 to more than 200 keV) which has allowed for the first time the detection of hard (>20 keV) X-ray emission from stars other than the Sun.

X-ray emission across the HR diagram

During the first six months of its lifetime, the German–US–British satellite ROSAT carried out an All-Sky Survey (RASS) at a typical sensitivity of about 2×10^{-13} erg cm^{-2} s^{-1}, which was later supplemented by pointed observations. The sensitivity of the RASS is comparable to the typical sensitivity of the previous pointed Einstein observations, whereas the ROSAT pointed observations were typically a factor of 10 deeper. Also the area coverage was nearly a factor of 10 larger than for Einstein and extended to the whole sky. This combination of higher sensitivity and larger area coverage has resulted in larger unbiased samples of stars, often volume limited (i.e. including all stars of a given type within a given volume), as opposed to the typical flux-limited samples of Einstein (i.e. including only stars brighter in X-rays than a given flux limit) and/or to samples largely dominated by upper limits, which required with Einstein a statistical analysis using survival techniques to properly take into account both detections and upper limits. Moreover, the method used in scanning the sky (which involved the repetition of very short observations of the source over a period of up to 2 days) also provided information of stellar variability over these timescales (see Schmitt 1993 for a review of the early ROSAT results). As a drawback, the spectral resolution of the ROSAT PSPC data was only slightly better than that of the IPC and virtually no spectral resolution was provided by the HRI detectors on both ROSAT and Einstein. Thus ROSAT observations provided only a crude temperature determination for the observed sources, and only for those sources sufficiently strong to allow accumulation of enough counts (at least a few thousands) for a proper spectral analysis. Better spectral data were provided by EUVE and ASCA, and these will be discussed later, together with the lower resolution PSPC data provided by ROSAT.

The ROSAT satellite (whose stellar X-ray group was led at MPE by Jürgen Schmitt) confirmed that X-ray emission is common to stars of nearly all spectral types and luminosity classes, with the exception of A-type dwarfs and of late-type giants to the right of the coronal DL previously established by IUE and Einstein. They have also confirmed the dichotomy that appears to exist among X-ray-emitting stars between early-type stars (whose X-ray luminosity is proportional to bolometric luminosity and independent of rotation) and late-type stars (whose X-ray luminosity is independent of bolometric luminosity but strongly dependent upon rotation and age). This dichotomy parallels the fundamental difference in the internal structure of stars that exists between hot stars with no outer convective zones (and hence with no dynamo-generated magnetic fields) and cool stars with outer convective zones of increasing depth towards later spectral types. Another fundamental difference between hot and cool stars with regard to their coronal emission is time variability which is ubiquitous and often of large amplitude in cool stars (especially in active stars like RS CVn binaries and in flare stars), whereas it is much weaker and rarer in hot stars. The current understanding of this phenomenology is the same as that suggested by the Einstein Observatory, that is the X-ray emission of hot stars originates from shock heating in radiatively driven winds (Hillier *et al.* 1993, Cohen *et al.* 1996), while coronal emission in cool stars is basically of solar type and results from the dissipation of dynamo-generated magnetic fields (e.g. Schmitt 1997). The heating mechanism (either magneto-acoustic waves or electric currents) remains elusive (as it is for the Sun), but there seems to be no question as to the primary importance of magnetic fields in heating and structuring the outer atmospheres of cool stars. Albeit we cannot observe directly spatially resolved details on the surface of other stars, it seems plausible to assume that the coronae of cool stars are formed by a multitude of magnetically confined loop-like structures similar to the ones we observe directly in the case of the Sun. As for the Sun, we expect that fluctuations in the heating rate (due to the emergence of new magnetic flux and/or to the motions of the plasma that carry with it the magnetic field lines) lead to a highly dynamic state of the corona, with chaotic variability on all timescales, from seconds, hours and days to weeks, months and even years. At times, this variability becomes extremely violent and results in intense, short-lived flares which in the most active stars may be many orders of magnitude stronger than on the Sun.

As an example of the kind of study that has been possible to carry out with ROSAT using large unbiased samples of stars, I mention the analysis of complete volume-limited samples of F and G stars (within 13 pc) and K and M stars (within 6 pc) carried out using both the All-Sky Survey data and pointed ROSAT observations (Schmitt 1997). A very large range of emission levels (from about 10^{26} to nearly $10^{30}\,\mathrm{erg\,s^{-1}}$) exists for stars of otherwise similar colour, whereas there is very little or no variation of the median emission level with the effective temperature (Figure 2). Even more interesting is the fact that the lower emission level of these stars (determined by actual detections, not by upper limits!) is comparable in terms of surface flux to that of solar coronal holes, suggesting that the lowest activity stars are nearly deprived of magnetically confined regions. On the contrary, the most active stars must be completely covered by active regions, or the active regions on these stars must be much denser than typical solar active regions (Güdel *et al.* 1997). The X-ray luminosity of the Sun as a star at different times during the solar cycle has been recently studied by Peres *et al.* (2000) from solar Yohkoh/XST data, substantiating the above conclusions. For stars, there is also a correlation between X-ray luminosity and average coronal temperature, suggesting that the most

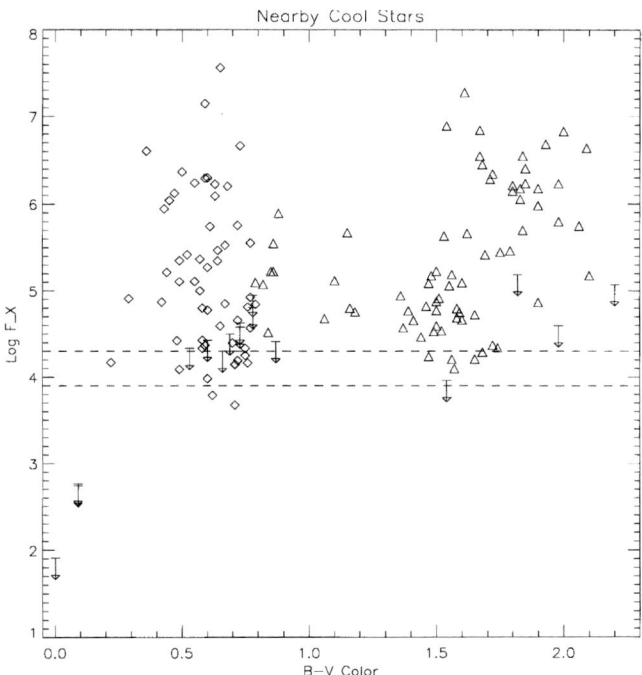

Figure 2 Soft X-ray surface fluxes v. $B-V$ colour for nearby cool dwarfs of spectral types F and G (diamonds) and K and M (triangles). The horizontal dashed lines represent the typical range of soft X-ray surface fluxes for solar coronal holes. The upper limits to the left refer to two single A-type stars. (From Schmitt 1997, Springer-Verlag: reproduced with permission of ESO.)

active stars are indeed heated more efficiently and that their coronae must be formed by magnetically confined structures that are both hotter and denser than typical solar coronal loops. At the extreme of this activity scale, we find RS CVn binaries, which have quiescent coronal temperatures of the order of 10^7 K (as opposed to the order of 10^6 K for the Sun and low-activity solar-type stars) and coronal surface fluxes orders of magnitude higher than the average of the Sun.

If coronal emission in late-type stars is produced by dynamo-generated magnetic fields, it should be possible to use coronal emission as a diagnostic tool of the dynamo process, similarly to what has been done over the past 30 years at Mt. Wilson for the systematic monitoring of chromospheric Ca II H and K emission from stars (Baliunas and Vaughan 1985). Unfortunately, this is much more difficult to do from space than from the ground, unless one can use a dedicated mission to carry out a systematic monitoring programme at regular intervals and for a sufficiently large sample of stars. No such space mission has been available so far and indeed our knowledge of the variability of stars remains rather limited. Eventually, it should be possible in the future to search for X-ray activity cycles in stars, and to determine in a more quantitative way the amplitude of variability and its timescales for stars of different levels of activity. On the Sun, the variability amplitude is much larger in X-rays than in Ca II H and K emission suggesting that space observations could be much better diagnostics of stellar dynamos than ground-based Ca II observations. This remains a programme for the future. In the meanwhile, anyway, we can use coronal emission to put at least some constraints on the dynamo process.

An interesting question in this respect is how late in spectral type coronal emission survives or, equivalently, whether hot coronal plasma exists in very late fully convective stars. According to some dynamo models, the efficiency of magnetic field generation should be higher at the bottom of the convection zone or better in the overshoot region at the interface between the convective zone and the underlying radiative interior. If this is the case, one should observe a significant drop-off of coronal emission in very late M dwarfs that are fully convective. No such drop-off has been convincingly observed by ROSAT and there is no indication that the efficiency of coronal heating (as measured by the ratio of X-ray to bolometric luminosity) significantly decreases for fully convective M dwarfs except possibly for very late spectral types (Fleming *et al.* 1993, Schmitt *et al.* 1995). This favours a distributed dynamo, which operates throughout the convection zone, with respect to a shallow dynamo which operates only at the interface of the outer convective zone and of the radiative interior. Similar evidence comes also from the strong coronal emission of young pre-main sequence stars along the Hayashi track as well, more recently, from the reported detection by ROSAT of X-ray emission from fully convective brown dwarfs (Neuhäuser and Comeron 1998). Constraints on dynamo models come also from the study of the coronal DL among giant stars (Haisch *et al.* 1992), if the transition from yellow giants with hot coronae to red giants with no detectable coronal plasma but large mass losses is indeed due to a change of magnetic topology from predominantly closed magnetic structures to essentially open field lines (Rosner *et al.* 1995). It is interesting to note in this respect that ROSAT has shown that all hybrid stars indeed posses hot coronal plasma, and that the coronal DL appears to vanish among supergiants (Reimers *et al.* 1996).

A strong argument in favour of the magnetic nature of coronal emission in late-type stars is the absence of coronae in late B and early A-type stars, that is stars which have neither outer convective zones nor strong winds and which therefore should be unable to generate hot plasmas in their outer atmospheres by either magnetic heating or wind shocks (Simon *et al.* 1995). The earliest single A-type star that has been detected in X-rays by both Einstein and ROSAT is Altair at spectral type A7V. ROSAT has confirmed the previous finding by Einstein (which, however, was somewhat debated owing to the possibility of ultraviolet leak in the detector) that the nearby A0V star Vega is not an X-ray source at a level as low as $10^{26}\,\mathrm{erg\,s^{-1}}$. Even more convincing is the case of the eclipsing binary α CrB formed by a G- and an A-type star. When the A star eclipses the G star there is no detectable X-ray emission, indicating that the A star is dark in X-rays (Schmitt and Kürster 1993). Incidentally, the ROSAT observation of this eclipsing system has been used to map the distribution of active regions over the surface of the X-ray-emitting G star. In spite of the above evidence, X-rays have been reported from several A-type stars at levels comparable to those of later spectral type stars (a well-known example is Castor A+B which was also observed to flare by EXOSAT): the possible inference is that, in these cases, coronal emission arises from previously undiscovered late-type companions. An exception to this rule is represented by magnetic Ap stars, in which X-ray emission is possibly generated by magnetic reconnection of the strong fossil fields of these stars, and by Herbig Ae-Be stars, for which the origin of coronal emission remains elusive (Zinnecker and Preibisch 1994).

X-ray emission from stellar clusters

An area in which ROSAT has provided very significant advances is X-ray emission from open clusters. Prior to ROSAT, only two clusters (the Hyades and the Pleiades) had been observed in sufficient detail at X-ray wavelengths. ROSAT has observed more than two dozen nearby open clusters covering the age interval from about 30 Myr to about 700 Myr (Figure 3). These observations have allowed

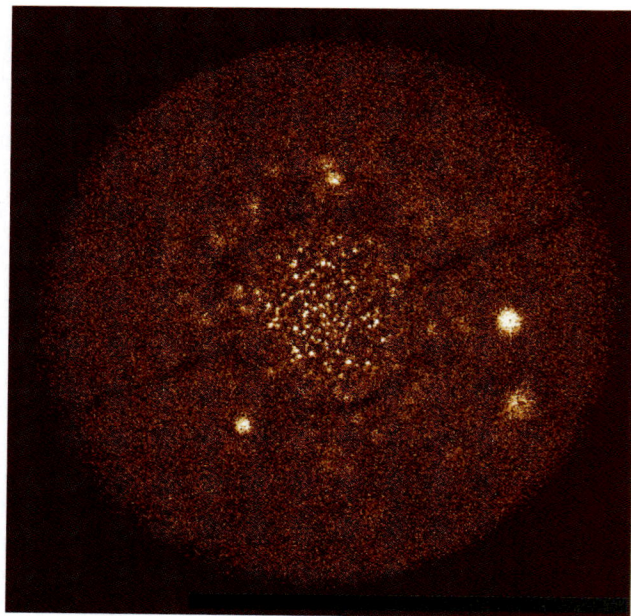

Figure 3 Soft X-ray image of the open cluster NGC 2516 obtained with the PSPC detector on board ROSAT. (Retrieved from the ROSAT archive by courtesy of the Max-Planck Institut für Extraterrestrische Physik.)

the investigation in detail of the decline of coronal emission with age as it is determined in turn by the evolution of stellar angular momentum on the main sequence (Pye et al. 1994, Stern et al. 1994, Stauffer et al. 1994, Micela et al. 1996, Jeffries et al. 1997; see also Randich 1997, 2000; Jeffries 1999 for reviews).

A comparison of the X-ray luminosity functions of different clusters shows that the median X-ray emission steadily decreases from very young clusters (like IC 2602 and IC 2391 at ≈ 30 Myr or α Persei at ≈ 50 Myr) to intermediate-age clusters (like the Pleiades and NGC 2516 at ≈ 100 Myr), to older clusters (like the Hyades, Coma and Praesepe at an age of ≈ 500–700 Myr). Whereas this is true in general, there are significant differences for stars of different spectral types. For instance, while G and K stars in α Persei (age ≈ 50 Myr) are typically brighter in X-rays than stars of the same spectral type in the Pleiades (age ≈ 100 Myr), this is not true for M stars, which show virtually the same X-ray luminosity function in α Persei and in the Pleiades (Randich et al. 1996). In the framework of stellar spindown by magnetized winds, this can be understood as due to the different timescale of the braking which is primarily determined by the depth of the convection zone. For G and K stars, the convective zone is sufficiently shallow to allow a significant braking in the time interval from the age of α Per to that of the Pleiades, whereas there is no significant braking in the same time interval for M stars which have a much deeper convective zone.

The dependence of coronal emission upon rotation in cluster stars had been a matter of debate for a long time, for example the observed existence (first pointed out with Einstein; Caillault and Helfand 1985, Micela et al. 1990) of rapidly rotating, low-luminosity K stars in the Pleiades that seemed to contradict the dynamo model. X-ray observations by ROSAT, and more precise rotational data from the ground, have allowed clarification of this issue. Coronal activity saturates quite rapidly with increasing rotation rate at a saturation value L_x/L_{bol} of about 10^{-3} (Vilhu 1987): the dependence of X-ray activity upon rotation in the Pleiades is evident only for rotation rates lower than a threshold value of ≈ 15 km s^{-1} (Stauffer et al. 1994). For higher velocities, the X-ray luminosity is independent of rotation, as shown also by observations of other clusters. This saturation phenomenon is not well understood: it could be due to a saturation of the dynamo itself, or to coverage of the whole corona by active regions with a given maximum temperature and density. Even more difficult to understand is the phenomenon of 'supersaturation' which has been observed in some clusters, that is the X-ray luminosity seems to turn over at still higher rotation rates (Randich 1998). Several explanations have been suggested for this behaviour, but none of them appears to be completely satisfactory. At any rate, the existence of saturation (and supersaturation), together with the availability of more

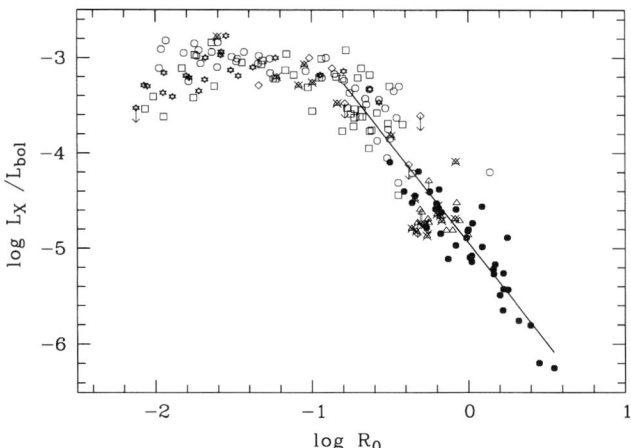

Figure 4 X-ray to bolometric luminosity ratio L_x/L_{bol} v. Rossby number $R_0 = P_{rot}/\tau_{conv}$ for cluster stars (open symbols) and field stars (filled symbols). Symbols for cluster stars are as follows: circles (Pleiades), squares (IC 2602 and IC 2391), stars (α Persei), triangles (Hyades), diamonds (IC 4665). The line represents the regression fit of the data points with $\log R_0 > -0.8$. Note the effect of saturation at low Rossby numbers, that is for high rotation rates. (From Randich 2000, Astronomical Society of the Pacific: reproduced with permission.)

accurate X-ray and rotational data, has allowed a fairly successful interpretation of cluster data (and hence of coronal activity in late-type stars in general) in terms of the Rossby number, that is in terms of both rotation and convection, as expected from the dynamo mechanism. A diagram of X-ray luminosity v. Rossby number for stars of different clusters and different masses (Patten and Simon 1996, Randich 2000) is in fact an evolutionary diagram, where the coronal emission of any given star depends on the combined effect of rotation (age) and spectral type (i.e. convection zone depth), taking proper account of the saturation effect (Figure 4).

Although the general picture is quite consistent, there are still a number of disturbing facts, indicating that the current scenario is at best only a first-order approximation. For instance, late-type stars in the Praesepe cluster (which has the same age and chemical composition as the Hyades) are on average much weaker in X-rays than similar stars in the Hyades (Randich and Schmitt 1995). On the contrary, F and G stars in the Coma cluster (at an age of ≈ 500 Myr) are in good agreement with the Hyades. Similar discrepancies have also been found between NGC 2516 (a southern 'twin' of the Pleiades) and the pattern shown by the Pleiades, NGC 6475 (age ≈ 200 Myr) and the Hyades (Jeffries et al. 1997). Possible reasons for these discrepancies could be different cluster metallicities (which could affect angular momentum evolution through their effects on the convection zone depth) or different rotational histories.

It is interesting to note that X-ray observations are also a powerful tool to determine cluster membership for low-mass stars in clusters with no (or poor) membership determination from proper motion surveys. The strong dependence of X-ray coronal activity on age via rotation and its effects on magnetic field generation by dynamo action means that members of young clusters are typically strong X-ray sources. A tentative list of candidate cluster members can thus be obtained from X-ray selected samples by combining X-ray observations with optical photometry and high-resolution spectroscopy. In fact, the photometric colours of the candidate members must be consistent with the theoretical isochrone at the age of the cluster, while radial velocity determinations and other spectroscopic indicators (such as lithium abundance and chromospheric activity) provide additional evidence for membership.

Although the emphasis on cluster research has been so far on relatively young clusters (younger than the Hyades), some work has also been carried out for older clusters, with ages of a few to several Gyr. These clusters are typically too far away and too old for detecting single solar-type and later spectral-type stars at the current sensitivity levels; yet, they allow the detection of active binary systems like RS CVn and W UMa binaries. Observations of a few intermediate-age clusters (like NGC 752, NGC 3680 and NGC 4651 at about 2 Gyr) and of older clusters (like M 67 at about 4.5 Gyr) have indeed been obtained with ROSAT, detecting a number of coronal sources (Belloni 1997, Verbunt 2000). These observations provide important clues for understanding the evolution of binary systems.

X-ray emission from star-forming regions

The dependence of coronal emission upon age, which is quite evident in open cluster data, extends to very young stars that are still in the pre-main sequence (PMS) evolutionary phases. The early observations by Einstein showed in fact vigorous coronal emission from PMS objects in regions of on-going star formation such as Orion, Taurus–Auriga, ρ Ophiuchi and Chamaeleon (Feigelson and De Campli 1981, Montmerle et al. 1983, Feigelson and Kriss 1989). These findings have been confirmed by extensive ROSAT observations which have allowed a complete mapping of virtually all star-forming regions (SFRs) at distances up to about 150 pc and of a few other regions at even greater distances (e.g. Orion, see Figure 5). The ROSAT all-sky survey has further allowed the exploration of regions that are outside molecular clouds, with the unexpected discovery of young stars up to tens of degrees away from known sites of star formation (Neuhäuser 1997 and references therein).

One of the most important finding of the Einstein Observatory was that X-ray emission among PMS stars is not limited to T-Tauri stars previously known from optical

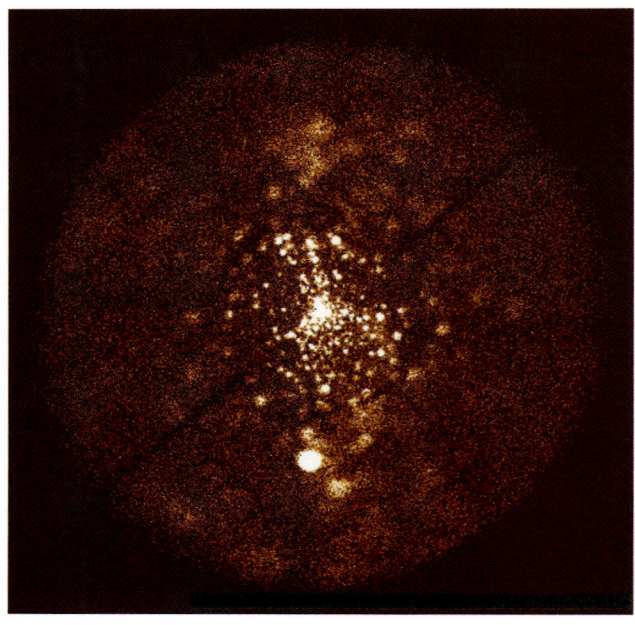

Figure 5 Soft X-ray image of the cental region of the Orion nebula obtained with the PSPC detector on board ROSAT. (Retrieved from the ROSAT archive by courtesy of the Max-Planck Institut für Extraterrestrische Physik.)

and infrared surveys; instead, a large population of previously unidentified PMS stars were discovered to be vigorous X-ray emitters (Walter 1993, Montmerle 1996). X-ray surveys in fact have proved to be the most efficient way to identify PMS stars, allowing a complete census of young stars in SFRs in all cases in which the sensitivity of the X-ray survey was sufficiently high. This has important consequences for the determination of the initial mass function (IMF) in SFRs and for understanding the processes of star formation and early stellar evolution (e.g. Preibisch and Zinnecker 1999).

We now know, thanks to the extensive imaging observations carried out by ROSAT that have confirmed and extended previous findings by Einstein, that X-ray coronal emission is common for both classical T-Tauri (CTT) and weak-lined T-Tauri (WTT) stars. The latter stars are characterized by small equivalent widths of the Hα line and by little or absent infrared excess. This indicates that the surrounding disk, which is a characteristic of CTT stars, has largely disappeared in WTT stars. The high level of X-ray coronal emission, even in the absence of a disk, and the extreme time variability of this emission, shows that the X-ray activity originates predominantly from the star itself and is likely to be of magnetic origin, as for late-type stars in general. The extreme variability, with the frequent occurrence of large, long-duration events, suggests that coronal activity is due, at least in part, to magnetic reconnection, as believed to occur for solar flares. Quiescent emission, likely arising from magnetically confined structures, is also present, although the possibility cannot be excluded that quiescent emission is the result of continuous low-level flaring activity, as could also be the case for other active stars like, for example, M dwarfs (Güdel 1997).

One might think that WTT stars represent a later phase than CTT stars in the PMS evolution of low-mass stars. Although this might be true for a fraction of the X-ray detected PMS stars, this does not seem to be true in general. When located precisely on the HR diagram, WTTs and CTTs often share the same position indicating that many WTTs are indeed coeval with CTTs in the same star-forming region. Either they originated with little or no circumstellar material or they dissipated the disk on timescales shorter than for CTT stars. The second possibility is more in line with current ideas about the evolution of stellar rotation during PMS and main sequence evolution (Bouvier 1994, 1997 and reference therein). Very young open clusters, with ages of a few tens of Myr (e.g. α Persei), show in fact a very large range of rotational velocities, from less than $10\,\mathrm{km\,s^{-1}}$ up to about $200\,\mathrm{km\,s^{-1}}$. This range of rotational velocities is much larger than observed in typical PMS stars at ages of 1 to 10 Myr (both CTTs and WTTs). These stars have rotational velocities that range from the minimum detectable value up to about $50\,\mathrm{km\,s^{-1}}$. The higher rotational velocities of some late-type stars in the younger clusters is due to the spin-up that occurs as a star contracts towards the ZAMS while conserving angular momentum. The large spread in rotational velocities in the young clusters can thus be interpreted as due to the different timescales for the dissipation of the disk: the longer is this timescale, the greater is the braking action exerted by the disk, and the smaller is the rotational velocity eventually attained by the star on the ZAMS.

PMS stars in the CTT and WTT phases have typical ages of 1 to 10 Myr. Younger objects, with ages as low as 0.1 Myr, have been detected recently in ρ Oph and R CrA by ROSAT and ASCA (Koyama et al. 1994, 1996). These very young objects are still in the late stages of accretion and are deeply embedded in the parent molecular cloud. They are usually invisible at optical wavelengths, but are associated with infrared sources. The discovery of these X-ray sources (facilitated by the harder spectral response of ASCA which allows penetration deeper into the clouds) is important in many respects. It shows that coronal activity originates very early in the lifetime of a star and that X-rays can be used effectively to investigate star formation and early stellar evolution in conjunction with the more traditional methods of IR and millimetre astronomy. It also shows that the emission is most likely to be magnetic in nature, as indicated by the powerful flares that are often observed, and by the typical temperatures (about 10^7 K) and quiescent emission levels (about $10^{31}\,\mathrm{erg\,s^{-1}}$) characteristic of these sources. The activity could originate from reconnection events in magnetic fields connecting the star to the surrounding accreting disk, rather than from magnetic structures rooted to the star, as it is more likely for CTT and WTT stars. Finally, strong X-ray activity in the early phase of evolution of a star can produce significant ionization of the accreting matter and hence can affect the accretion process itself. This has important consequences for the evolution of proto-planetary disks and the formation of planetary systems (Feigelson and Montmerle 1999).

Observations by ROSAT have allowed us to get a better understanding of young stars not only in molecular clouds and in regions of on-going star formation, but also in the surrounding regions, often at distances of tens of degrees from the cloud core. The All-Sky Survey found hundreds of X-ray sources widespread outside SFRs, which were tentatively identified with PMS stars on the basis of low- and medium-resolution lithium observations (Krautter et al. 1994, Neuhäuser 1997 and references therein). If these are really PMS stars, their existence at great distances from the cloud core is difficult to understand (Feigelson 1996). They could be born in the cloud and later be ejected by three-body encounters, or they could be born locally in small cloudlets that rapidly dissipated after star formation. Alternatively, these X-ray sources could not be PMS stars, but rather a population of ZAMS stars with typical ages as old as about 100 Myr (Briceño et al. 1997). If these stars are older than typical PMS stars, they might have had enough time to move out of their parent cloud and the latter may have already dissipated, thus alleviating the problem raised by their widespread distribution. A cross-correlation of the ROSAT All-Sky Survey data with the Tycho catalogue (Guillout et al. 1998a,b) has revealed an asymmetry in the distribution of the X-ray-emitting stars in galactic coordinates, showing a density enhancement in agreement with the expected position of the Gould Belt, an alignment of O-B stellar associations and young stellar clusters that surrounds the Sun at a distance of a few hundred parsecs (Figure 6). Taking all factors into account, it is likely that the widespread population of young stars detected by ROSAT is a combination of different populations, including both PMS and Pleiades-like stars, with a significant fraction coming from the Gould Belt. At any rate, these observations show that X-ray surveys over large sky areas offer a unique opportunity to investigate stellar populations in the Galaxy and to trace the star formation history in the solar neighbourhood in the past billion years.

X-ray spectroscopy of stellar sources

Stellar coronae have thermal spectra due to optically thin line plus continuum emission. One-temperature (1-T) and two-temperature (2-T) models and solar abundances were usually assumed in the analysis of early observations with low spectral resolution, such as those obtained with the IPC detector on Einstein and the PSPC detector on ROSAT (Schmitt et al.

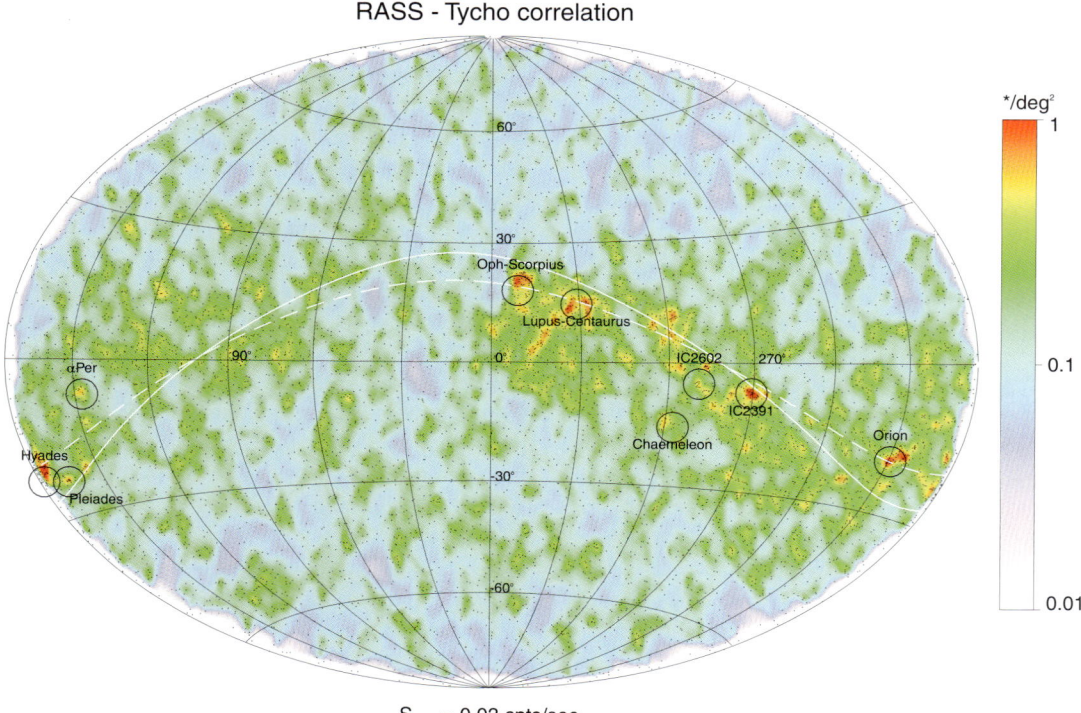

Figure 6 Cross-correlation of the ROSAT All-Sky Survey and Tycho catalogues in galactic coordinates showing an excess (full line) of young stars along the Gould Belt (dashed line). The position of star-forming regions and young, open clusters is also indicated by circles (from Guillout et al. 1998a, Springer-Verlag: reproduced with permission of ESO.)

1990, Dempsey et al. 1993). Only recently with EUVE, ASCA and BeppoSAX, has it become possible to start exploiting the diagnostic potential of X-ray coronal spectra, to derive differential emission measure (DEM) distributions and elemental abundances. This is a field that is expected to blossom with the new generation of X-ray satellites launched at the turn of the century (AXAF, XMM, ASTRO-E) or those which are under study for the first or second decade of the twenty-first century (Constellation and XEUS).

Somewhat surprisingly, most (albeit not all) coronal sources observed with ASCA (and EUVE as well) showed reduced metal abundances (typically by factors of 3 to 5) with respect to *solar* abundances. In many cases, the best fit to the ASCA CCD spectra over the range 0.5 to 10 keV was obtained with a 2-T thermal model with variable non-solar abundances, that is by leaving all abundant elements free to vary (Antunes et al. 1994, Singh et al. 1996, White 1996; Figure 7). In other cases, when either the statistics or the resolution were not high enough, a sufficiently good fit was obtained by leaving only the overall metallicity Z free to vary, that is by varying the individual abundances in the same proportion with respect to solar abundances. When the individual abundances are left free to vary, they show little evidence for a first ionization potential (FIP) effect, although evidence for a FIP effect has been reported for some of the stars observed with EUVE (Drake et al. 1996). The FIP effect, which consists of different coronal to photospheric abundance ratios depending on the FIP of the ion emitting the line, has been observed in the case of the Sun and has been attributed to fractionation due to diffusive processes across magnetic field lines. If a FIP effect is present in stellar coronae, it is the opposite to that observed in the solar corona, where ions with low FIP values like iron are enhanced with respect to ions with high FIP values like oxygen. Stars instead show invariably a metal abundance deficiency (MAD) effect.

The fact that the coronal abundances measured by ASCA and EUVE in a given star are typically lower than the solar photospheric abundances does not imply that they are also lower than the photospheric abundances of the same star (Drake 1996a). Most stellar coronal sources for which medium-resolution spectroscopy has been possible so far with ASCA are bright RS CVn and Algol-type binaries, which are known often to have subsolar photospheric abundances, albeit it is still unclear whether the observed weakness of the optical absorption lines is due to lower metallicity or rather to the effect of surface activity. For instance, the RS CVn binary λ And observed by ASCA was found to have a coronal abundance a factor of five lower than solar but perfectly consistent with its measured photospheric abundance. Unfortunately, there are still large uncertainties

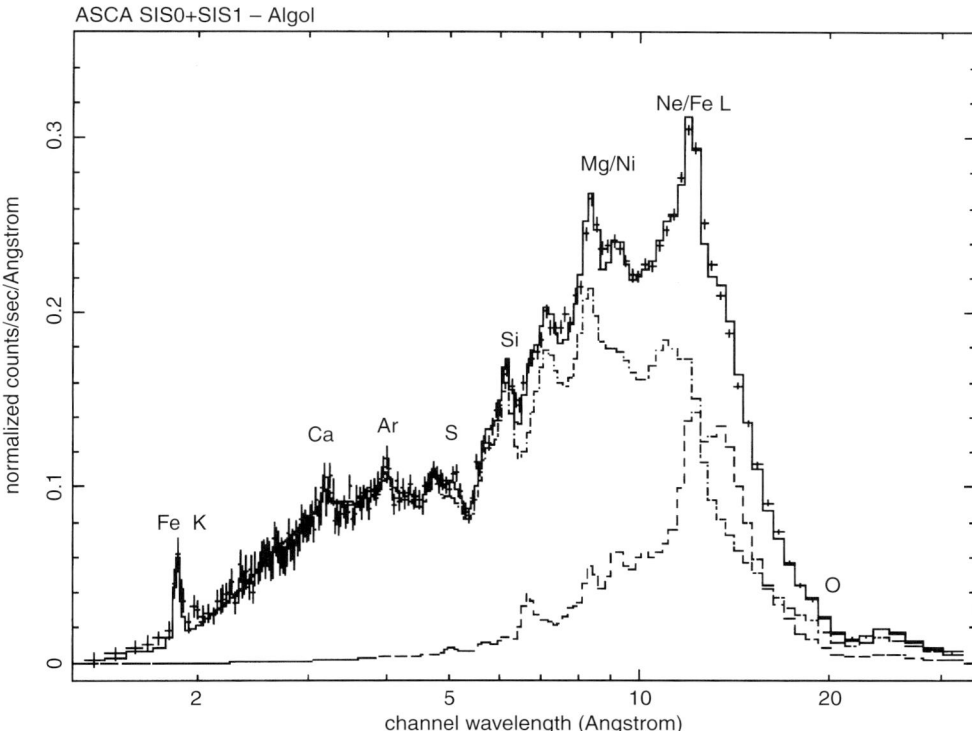

Figure 7 The X-ray spectrum of Algol as observed with the SIS detector on ASCA. The 2-T model fit is also shown. The positions of the most prominent spectral features are indicated. (Courtesy of the High Energy Astrophysics Archive Research Center at NASA/GSFC.)

in the (optical) determination of the photospheric abundances of active stars, and in many cases accurate photospheric abundances are totally lacking. This leaves the issue of coronal v. photospheric abundance differences largely open.

At any rate, while the subsolar coronal abundances found by ASCA (and BeppoSAX as well) in several RS CVn and Algol-type binaries may not be inconsistent with the photospheric abundances of these stars as a class, there are at least a few cases in which there is an obvious discrepancy between the measured coronal and photospheric abundances. One case is the young, rapidly rotating star AB Dor observed simultaneously by ASCA and EUVE, and later by BeppoSAX. The coronal metallicity of this star is [Fe/H] ≈ 0.3 (Mewe et al. 1996) in spite of the fact that AB Dor is a ZAMS star with a measured photospheric abundance that is solar. On the contrary, the active giant β Ceti observed by both ASCA and BeppoSAX shows a nearly solar photospheric abundance and the analysis of a recent BeppoSAX observation of Capella gives a coronal metallicity [Fe/H] of about 0.8 consistent with both the EUVE value and the photospheric abundance. Other recent BeppoSAX observations of active binaries indicate instead low coronal abundances (e.g. UX Ari, HR 1099 and II Peg), consistent with the results of ASCA (e.g. Covino et al. 2000; see Pallavicini 2001a for a review).

In conclusion, the issue of low coronal abundances in stars, and of possible discrepancies between coronal and photospheric abundances, is far from clear. While in several cases (particularly for RS CVn binaries), the reported subsolar abundances may not be inconsistent with the photospheric metallicity of the star, in other cases there is evidence for coronal abundances lower than photospheric ones. Future higher resolution observations with AXAF and XMM should clarify this issue, determining whether this is a genuine physical effect or whether it depends at least partly on uncertainties in the atomic physics and/or on instrument calibration.

The 2-T models usually assumed to fit stellar coronal spectra have not necessarily a physical meaning and may be simple parameterizations of more complex situations involving a continuous DEM distribution. A more physical approach is to fit the observed spectra with fully consistent loop models in hydrostatic equilibrium and energy balance. Attempts in this direction have been made by assuming the corona to be formed by an ensemble of unresolved loops all equal one to the other or belonging to a limited number of different loop families (two families of loops are usually sufficient at the resolution and S/N of the available spectra). Fits of coronal spectra with loop models provide a more sound physical description of stellar coronae and allow putting constraints on pressure, temperature and surface coverage by coronal structures (e.g. Maggio and Peres 1997). However, the assumption of only one or two families of loops, all equal one to the other, may be a fundamental limitation of this type of model in interpreting spatially unresolved stellar observations (the Sun shows in fact a much more complex situation with a virtually continuous spectrum of loops with different temperatures and pressures).

X-ray emission from stellar flares

Stellar flares observed at X-ray wavelengths are the most spectacular manifestations of coronal activity in stars. Large stellar flares (with total radiative outputs many orders of magnitude larger than the most powerful solar flares) have been detected frequently from RS CVn and Algol-type binaries, from M dwarf flare stars and from PMS objects (see Schmitt 1994, Pallavicini 1996 for reviews). I have already mentioned the observations of stellar flares by EXOSAT and Einstein. Large flares were also detected with ROSAT, GINGA and ASCA and these observations have been used to study the energy release process and the evolution of plasma parameters in the course of the flare (e.g. Tsuru et al. 1989, Stern et al. 1992). Temperatures as high as 10^8 K (much higher than typically observed in solar flares) have been observed sometimes, particularly from flares on RS CVn stars. Spectroscopic observations of large flares with ASCA and BeppoSAX have revealed in some cases variations of elemental abundances in the course of the flare, with typically lower Fe abundances at the peak of the flare (Favata and Schmitt 1999). In a few cases (as in a flare on Proxima Cen observed with Einstein and in a flare on Algol observed with BeppoSAX) variations of the hydrogen column density N_H were also observed, and interpreted as due to a prominence eruption in the early phases of a flare, analogous to what sometimes is observed on the Sun. Spectacular flares from Algol, AB Dor and UX Ari have been observed recently with BeppoSAX (Pallavicini 2001, Schmitt and Favata 1999; Figure 8), and a very large flare on EV Lac has been observed with ASCA. Most surprisingly, a very large flare was also detected from an embedded protostellar object (Grosso et al. 1997), and attributed to possible

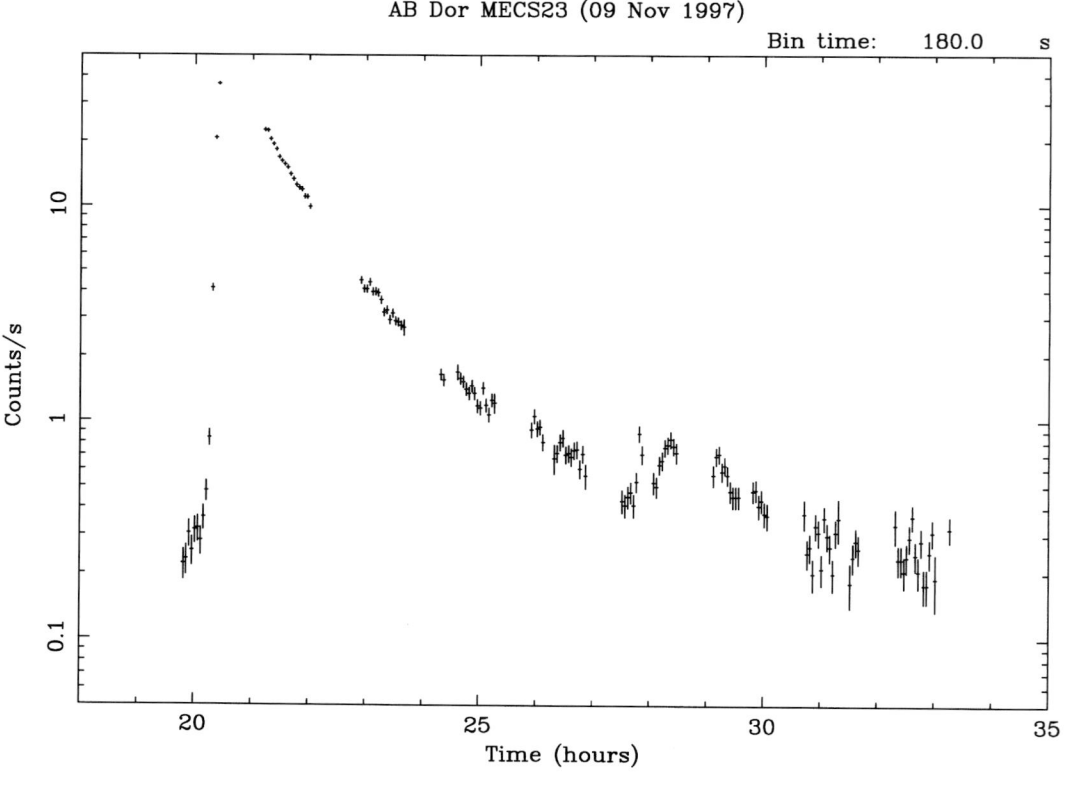

Figure 8 Light curve of an intense, long-duration X-ray flare on the young star AB Doradus observed on 9 November 1998 with the MECS detectors on the Italian–Dutch satellite BeppoSAX. Another similar flare was detected by BeppoSAX from the same star on 29 November 1998. (Unpublished data from the author.)

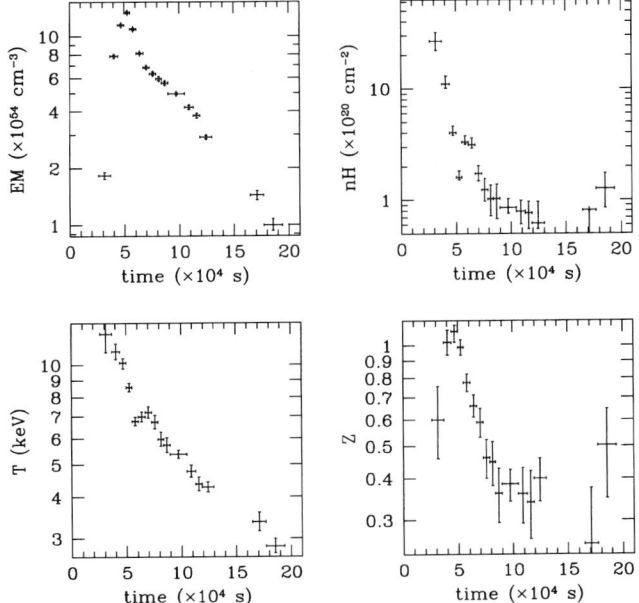

Figure 9 Temporal evolution of temperature T, emission measure EM, metal abundance Z and hydrogen column density N_H for an intense flare on Algol observed on 30 August 1997 by BeppoSAX. (From Favata and Schmitt 1999, Springer-Verlag: reproduced with permission of ESO.)

reconnection between field lines connecting the central star with the surrounding protostellar disk.

Observations of stellar flares from space, particularly the large, long-duration flares which allow better statistics and a finer temporal sampling of the variations of their physical parameters (Figure 9), allow application of flare models originally developed for the Sun. Although the mechanisms of energy release in flares remain elusive, even in the case of solar flares, the application of solar-type flare models to soft X-ray observations of stellar flares allows us to get insights into the flare phenomenon and to derive important physical parameters (in particular the size of the flaring region) that would remain otherwise inaccessible in the case of stars. Recent applications of this technique to large flares observed on stars from ASCA and BeppoSAX have led to the conclusion that the time evolution of long-duration flares is determined primarily by the time profile of the energy release (which appears to operate for a significant fraction of the flare decay) rather than by the natural decay due to energy losses by radiation and conduction (e.g. Favata and Schmitt 1999, Maggio et al. 2000). As a consequence, the estimated flare loop sizes turn out to be significantly smaller (by a factor of 3–4) than inferred by assuming negligible heating during the decay. These sizes are also much smaller than those derived by assuming a quasi-static model, as was previously done in interpreting some EXOSAT and ROSAT flares. The flare loop sizes determined in this way come out to be comparable to or smaller than the stellar radius, and certainly not as large as several stellar radii as sometimes claimed in the past. Although more physically consistent, these estimates remain, however, model dependent and rely on the assumption that stellar flares are scaled-up versions of solar flares. In the case of the large flare observed by BeppoSAX on Algol, the flare region on the K0 IV active star was completely eclipsed by the X-ray dark B8 V primary and this led to an estimate of the size of the flaring region on purely geometrical grounds (Schmitt and Favata 1999).

An important result obtained recently by BeppoSAX is the first detection from stars other than the Sun of hard (>20 keV) X-ray emission during large flares (Pallavicini 2001 and references therein). This emission, extending up to ≈ 50 keV, has been detected by the Photoswitch Detector System (PDS) on board BeppoSAX during the rise phase and at the peak of large flares on Algol, UX Ari and AB Dor (Figure 10). The emission appears to be thermal and due to a plasma at $\approx 10^8$ K. These high temperatures are already required to fit the low-energy part of the spectrum, between 0.1 and 10 keV, observed with the LECS and MECS instruments. If this thermal emission is extrapolated to higher energies it accounts fairly well for the hard tail observed by the PDS. There is therefore no need to invoke a non-thermal power-law component as the one observed during the impulsive phase of solar flares. Moreover, while in solar flares the hard X-ray emission is much smaller than the soft one (typically by a factor $\approx 10^5$), the hard tail seen by the PDS is about one-tenth of the soft X-ray emission observed at the peak of the flare, making its identification with a non-thermal component even more inplausible. Thus, even if non-thermal X-ray emission has been observed for decades in solar flares, its detection in stellar flares still remains elusive owing to the insufficient sensitivity of current high-energy detectors.

A LOOK AT THE FUTURE

The twenty-first century looks very promising for stellar coronal physics and for X-ray astronomy in general. New powerful missions (AXAF now renamed Chandra and XMM now renamed Newton) were launched in July and December 1999 (unfortunately the Japanese satellite ASTRO-E that followed had a failed launch). Both Chandra and Newton carry on board CCD arrays for medium-resolution spectroscopy and grating arrays for high-resolution spectroscopy. On ASTRO-E microcalorimeter detectors were expected to fly for the first time to allow high-resolution spectroscopy at energies of 1 to 10 keV, and high-energy detectors, sensitive to harder X-rays, were also aboard. The subarcsecond imaging capability of the

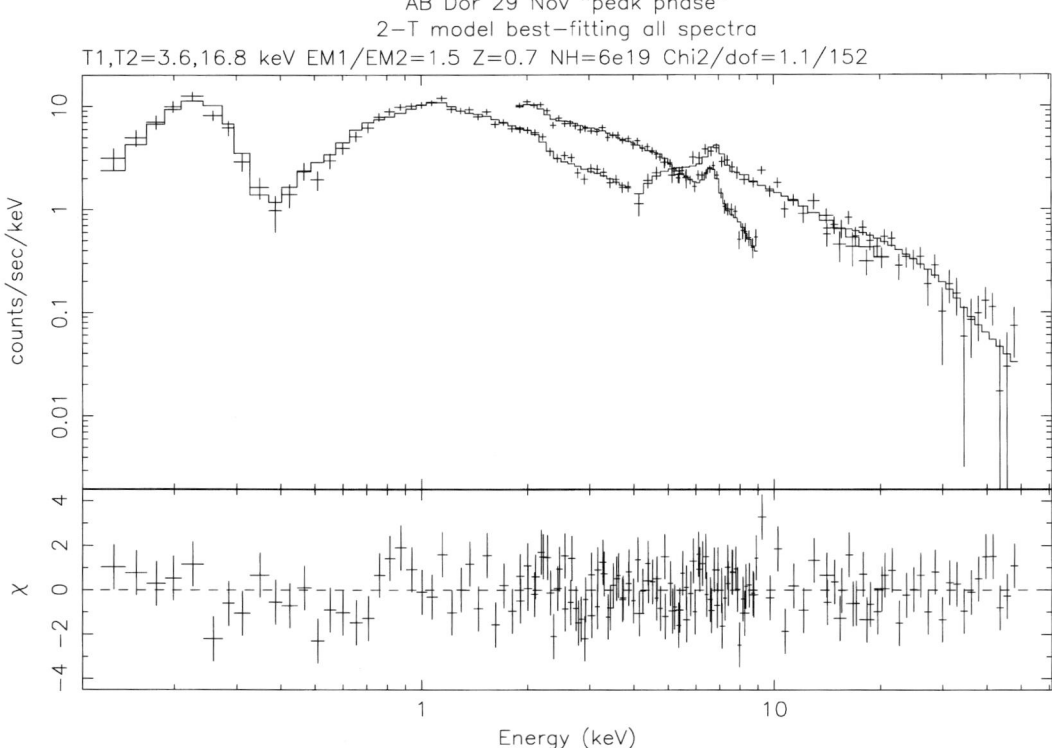

Figure 10 The X-ray spectrum at the peak of a large flare on AB Doradus observed by BeppoSAX on 29 November 1998 over the full energy range from 0.1 to 50 keV. X-ray photons were detected by all narrow-field instruments (LECS, MECS, HPGSPC and PDS). A 2-T spectral fit to the spectrum is also shown. (Unpublished data from the author.)

Chandra mirrors will be invaluable for resolving crowded stellar regions like open clusters and star-forming regions, as the images obtained by Chandra for Orion have already confirmed. The larger collecting area of Newton will be invaluable for medium- and high-resolution spectroscopy of large samples of stars, providing the required sensitivity and spectral resolution. Plans are already under way, both in the USA and in Europe, for the next generation of X-ray missions (like Constellation and XEUS) with improved collecting area and advanced spectroscopic devices.

Looking back to the study of stellar coronae from space, it is astonishing that all this occurred in just two decades. Twenty years ago, the new discipline of stellar X-ray astronomy was just at its starting point and the situation was not much different for ultraviolet studies of stellar chromospheres and transition regions. It is significant in this respect that IUE and Einstein (which represented the true beginning of stellar astronomy from space at UV and X-ray wavelengths) were both launched in the same year, at the beginning and at the end of 1978. A few non-accreting stellar sources, besides the Sun, had been detected at X-ray and UV wavelengths before IUE and Einstein; yet 20 years ago only a negligible fraction of the observing time of any major X-ray and UV mission was devoted to stellar studies. Now, the situation has changed completely, thanks to the success of the stellar observations of IUE and Einstein and later of ROSAT. Stellar studies represent a significant fraction of the observing programmes in space missions like Chandra and Newton, and it will be even more so in the future.

There are several reasons for this. First, stars are ideal sources for exploiting the full spectroscopic capabilities of current and future X-ray missions. They are thermal sources (except for possible non-thermal flaring events which, however, have not yet been detected) rich in emission lines. Optically thin thermal spectra are in fact dominated by lines, rather than by continuum emission, at temperatures of up to 10^7 K, that is at temperatures typical of stellar coronae in quiescence (albeit the temperature distributions can extend to higher values in the most active stars and during flares). These lines have great diagnostics potential, for temperature, density, elemental abundances and flow velocities. So far, these potentials have been exploited only partially, if at all. With the current (Chandra, Newton) and future (Constellation, XEUS) sensitive high-resolution spectroscopic instruments, it will be possible to apply to the study of stars the same diagnostic tools currently in use for the study of the Sun.

Secondly, the results obtained so far clearly indicate that high-energy emission from stellar atmospheres is indeed a physically interesting topic, and that not all stars are the same (as might be thought superficially by those who do not work in the field!). I hope I have demonstrated in the course of this chapter that the study of high-energy radiation from stars allows one to address a number of fundamental problems in stellar astrophysics. These include the heating of stellar chromospheres and coronae, the acceleration of stellar winds and the mechanisms of stellar mass loss, the instability of radiatively driven winds and the formation of shocks, both in the wind itself and in colliding winds, the generation of magnetic fields by dynamo action in different types of stars, the process of magnetic braking and angular momentum loss in the course of stellar evolution, the physics of magnetically confined coronal structures, both quiescent and flaring, the formation and early evolution of stars and protostellar objects, the distribution of different stellar populations in the Galaxy, and more.

Many, perhaps most, of these topics have received only partial answers from the data collected from space over the past 20 years. In many cases, the answers were only qualitative, and the picture assembled so far remains largely phenomenological. We have many ideas on how things should possibly work, but we lack sound physical models capable of explaining the bulk of the available data. With the better quality data that will be available in the coming years, we should be able to progress from the pioneering phase in which we have been up to now to a stage where we should be able to obtain a full physical understanding of the processes at work. This is the right time to enter this new, exciting field.

REFERENCES

Antunes, A., Nagase, F. and White, N.E. (1994). ASCA observations of the coronal X-ray emission of Algol. *Astrophysical Journal*, **436**, L83–L86.

Ayres, T.R., Fleming, T.A. and Schmitt, J.H.M.M. (1991). Digging in the coronal graveyard: A ROSAT observation of the red giant Arcturus. *Astrophysical Journal*, **376**, L45–L48.

Ayres, T.R., Marstad, N.C. and Linsky, J.L. (1981). Outer atmospheres of cool stars. IX. A survey of ultraviolet emission from F-K dwarfs and giants with IUE. *Astrophysical Journal*, **247**, 545–559.

Baliunas, S.L. and Vaughan, A.H. (1985). Stellar activity cycles. *Annual Review of Astronomy and Astrophysics*, **23**, 379–412.

Belloni, T. (1997). X-ray sources in intermediate-age and old open clusters. In G. Micela, R. Pallavicini and S. Sciortino (eds), *Cool Stars in Clusters and Associations: Magnetic Activity and Age Indicators*, Memorie della Societa Astronomia Italia Vol. 68, pp. 993–1000.

Bouvier, J. (1994). The rotational evolution of low-mass pre-main sequence stars. In J.-P. Caillault (ed), *Cool Stars, Stellar Systems, and the Sun*, ASP Conference Series Vol. 64, Astronomical Society of the Pacific, San Francisco, pp. 151–162.

Bouvier, J. (1997). The surface and internal rotation of low-mass stars in young clusters. In G. Micela, R. Pallavicini and S. Sciortino (eds), *Cool Stars in Clusters and Associations: Magnetic Activity and Age Indicators*, Memorie della Societa Astronomia Italia Vol. 68, pp. 881–894.

Briceño, C., Hartmann, L.W., Stauffer, J.R., Gagné, M., Stern, R.A. and Caillault, J.-P. (1997). X-ray surveys and the post-T Tauri problem. *Astronomical Journal*, **113**, 740–752.

Caillault, J.-P. and Helfand, D.J. (1985). The Einstein soft X-ray survey of the Pleiades. *Astrophysical Journal*, **289**, 279–299.

Cassinelli, J.P. and Olson, G.L. (1979). The effects of coronal regions on the soft X-ray flux and ionization conditions in the winds of OB supergiants and Of stars. *Astrophysical Journal*, **229**, 304–317.

Cohen, D.H., Cooper, R.G., MacFarlane, J.J., Owocki, S.P., Cassinelli, J.P. and Wang, P. (1996). Evidence for wind attenuation and a multitemperature plasma in the combined EUVE and ROSAT observations of ε Canis Majoris (B2 II). *Astrophysical Journal*, **460**, 506–521.

Covino, S., Tagliaferri, G., Pallavicini, R., Mewe, R. and Poretti, E. (2000). The active binary star II Pegasi with BeppoSAX. *Astronomy and Astrophysics*, **355**, 681–687.

Dempsey, R.C., Linsky, J.L., Schmitt, J.H.M.M., Fleming, T.A. (1993). The ROSAT All-Sky Survey of active binary coronae. II. Coronal temperatures of the RS Canum Venaticorum Systems. *Astrophysical Journal*, **413**, 333–338.

Drake, J.J. (1996a). Stellar spectroscopy with the Extreme Ultraviolet Explorer. In R. Pallavicini and A.K. Dupree (eds), *Cool Stars, Stellar Systems, and the Sun*, ASP Conference Series Vol. 109, Astronomical Society of the Pacific, pp. 203–214.

Drake, J.J., Laming, J.M. and Widing, K.G. (1996). The FIP effect and abundance anomalies in late-type stellar coronae. In S. Bowyer and R.F. Malina (eds), *Astrophysics in the Extreme Ultraviolet*, Kluwer, Dordrecht, pp. 97–104.

Favata, F. and Schmitt, J.H.M.M. (1999). Spectroscopic analysis of a super-hot giant flare observed on Algol by BeppoSAX on 30 August 1997. *Astronomy and Astrophysics* **350**, 900–916.

Feigelson, E.D. (1996). Dispersed T Tauri stars and galactic star formation. *Astrophysical Journal*, **468**, 306.

Feigelson, E.D. and De Campli, W.M. (1981). Observations of X-ray emission from T Tauri stars. *Astrophysical Journal*, **243**, L89-L93.

Feigelson, E.D. and Kriss, G.A. (1989). Soft X-ray observations of pre-main sequence stars in the Chamaeleon dark cloud. *Astrophysical Journal*, **338**, 262–276.

Feigelson, E.D. and Montmerle, T. (1999). High-energy processes in young stellar objects. *Annual Review of Astronomy and Astrophysics*, **37**, 363–408.

Fleming, T.A., Giampapa, M.S., Schmitt, J.H.M.M. and Bookbinder, J.A. (1993). Stellar coronae at the end of the main-sequence: A ROSAT survey of the late M dwarfs. *Astrophysical Journal*, **410**, 387–392.

Grosso, N., Montmerle, T., Feigelson, E.D., André, P., Casanova, S. and Gregorio-Hetem, J. (1997). An X-ray superflare from an infrared protostar. *Nature*, **387**, 56–58.

Güdel, M. (1997). Are coronae of magnetically active stars heated by flares? *Astrophysical Journal*, **480**, L121–L124.

Güdel, M., Guinan, E.F. and Skinner, S.L. (1997). The X-ray Sun in time: A study of the long-term evolution of coronae of solar-type stars. *Astrophysical Journal*, **483**, 947–960.

Guillout, P., Sterzik, M.F., Schmitt, J.H.M.M., Motch, C., Egret, D., Voges, W. and Neuhäuser, R. (1998b). The large-scale distribution of active stars. *Astronomy and Astrophysics*, **334**, 540–544.

Guillout, P., Sterzik, M.F., Schmitt, J.H.M.M., Motch, C. and Neuhäuser, R. (1998a). Discovery of a late-type stellar population associated with the Gould Belt. *Astronomy and Astrophysics*, **337**, 113–124.

Haisch, B., Schmitt, J.H.M.M. and Fabian, A.C. (1992). Disappearance of coronal X-ray emission in stars with cool dense winds. *Nature*, **360**, 239–241.

Hillier, D.J., Kudritzki, R.P., Pauldrach, A.W., Baade, D., Cassinelli, J.P., Puls, J. and Schmitt, J.H.M.M. (1993). The 0.1–2.5 keV X-ray spectrum of the O4f star ζ Puppis. *Astronomy and Astrophysics*, **276**, 117–128.

Jeffries, R.D. (1999). X-rays from open clusters. In C.J. Butler and J.G. Doyle (eds), *Solar and Stellar Activity: Similarities and Differences*, ASP Conference Series Vol. 158, Astronomical Society of the Pacific, San Francisco, pp. 75–86.

Jeffries, R.D., Thurston, M.R. and Pye, J.P. (1997). An X-ray survey of the young open cluster NGC 2516. *Monthly Notices of the Royal Astronomical Society*, **287**, 350–380.

Jordan, C. and Linsky, J.L. (1987). Chromospheres and Transition Regions. In Y. Kondo (ed), *Exploring the Universe with the IUE Satellite*, Kluwer, Dordrecht, pp. 259–293.

Koyama, K., Hamaguchi, K., Ueno, S., Kobayashi, N. and Feigelson, E.D. (1996). Discovery of hard X-rays from a cluster of protostars. *Publications of the Astronomical Society of Japan*, **48**, L87–L92.

Koyama, K., Maeda, Y., Ozaki, M., Ueno, S., Kamata, Y., Tawara, Y., Skinner, S. and Yamauchi, S. (1994). ASCA observations of hard X-ray emission from the ρ Ophiuchi dark cloud. *Publications of the Astronomical Society of Japan*, **46**, L125–L129.

Krautter, J., Alcalà, J.M., Wichmann, R., Neuhäuser, R. and Schmitt, J.H.M.M. (1994). ROSAT observations of star forming regions. *Revista Mexicana de Astronomía y Astrofísica*, **29**, 41–53.

Linsky, J.L. and Haisch, B.M. (1979). Outer atmospheres of cool stars. I. The sharp division into solar-type and non-solar-type stars. *Astrophysical Journal*, **229**, L27–L32.

Lucy, L.B. (1982). X-ray emission from the winds of hot stars. II. *Astrophysical Journal*, **255**, 286–292.

Lucy, L.B. and White, R.L. (1980). X-ray emission from the winds of hot stars. *Astrophysical Journal*, **241**, 300–305.

Maggio, A., Pallavicini, R., Reale, F. and Tagliaferri, G. (2000). Twin X-ray flares and the active corona of AB Dor observed with BeppoSAX. *Astronomy and Astrophysics*, **356**, 627–642.

Maggio, A. and Peres, G. (1997). Loop modeling of coronal X-ray spectra. IV. One- or two-loop model fitting of ROSAT PSPC spectra: Three test cases. *Astronomy and Astrophysics*, **325**, 237–248.

Maggio, A., Vaiana, G.S., Haisch, B.M., Stern, R.A., Bookbinder, J.A., Harnden, F.R., Jr. and Rosner, R. (1990). Einstein Observatory magnitude-limited X-ray survey of late-type giant and supergiant stars. *Astrophysical Journal*, **348**, 253–278.

Mewe, R. (1979). Stellar coronae: evidence for their existence from X and UV observations. *Space Science Reviews*, **24**, 101–144.

Mewe, R. (1991). X-ray spectroscopy of stellar coronae. *Astronomy and Astrophysics Review*, **3**, 127–168.

Mewe, R. (1996). X-ray spectroscopy of stellar coronae: History, present, future. *Solar Physics*, **169**, 335–348.

Mewe, R., Kaastra, J.S., White, S.M. and Pallavicini, R. (1996). Simultaneous EUVE and ASCA observations of AB Doradus: Temperature structure and abundances of the quiescent corona. *Astronomy and Astrophysics*, **315**, 170–178.

Micela, G., Sciortino, S., Kashyap, V., Harnden, F.R., Jr. and Rosner, R. (1996). ROSAT observations of the Pleiades. I. X-ray characteristics of a coeval stellar population. *Astrophysical Journal Supplement*, **102**, 75–103.

Micela, G., Sciortino, S., Vaiana, G.S., Harnden, F.R., Jr., Rosner, R. and Schmitt, J.H.M.M. (1990). X-ray studies of coeval star samples. II. The Pleiades cluster as observed with the Einstein Observatory. *Astrophysical Journal*, **348**, 557–579.

Montmerle, T. (1996). X-rays from young stellar objects: from T Tauri stars to protostars. In R. Pallavicini and A.K. Dupree (eds), *Cool Stars, Stellar Systems, and the Sun*, ASP Conference Series Vol. 109, Astronomical Society of the Pacific, San Francisco, pp. 405–416.

Montmerle, T., Koch-Miramond, L., Falgarone, E. and Gridlay, J. (1983). Einstein observations of the ρ Ophiuchi dark cloud – An X-ray Christmas tree. *Astrophysical Journal*, **269**, 182–201.

Neuhäuser, R. (1997). Low-mass pre-main sequence stars and their X-ray emission. *Science*, **276**, 1363–1370.

Neuhäuser, R. and Comeron, F. (1998). ROSAT X-ray detection of a young brown dwarf in the Chamaeleon I dark cloud. *Science*, **282**, 83–85.

Noyes, R.W., Hartmann, L.W., Baliunas, S.L., Duncan, D.K. and Vaughan, A.H. (1984). Rotation, convection, and magnetic activity in lower main-sequence stars. *Astrophysical Journal*, **279**, 763–777.

Pallavicini, R. (1996). Flares in late-type stars: X-rays. In J. Greiner, H.W. Dürbeck and R.E. Gershberg (eds), *Flares and Flashes*, Lecture Notes in Physics 454, Springer, Berlin, pp. 148–157.

Pallavicini, R. (2001). Studies of late-type Stars with BeppoSAX. In R.J. García Lopez, R. Rebolo and M.R. Zapaterio Osorio (eds), *Cool Stars, Stellar Systems, and the Sun*, ASP Conference Series, Vol. 223, Astronomical Society of the Pacific, San Francisco, pp. 337–387.

Pallavicini, R., Golub, L., Rosner, R., Vaiana, G.S., Ayres, T. and Linsky, J.L. (1981). Relations among stellar X-ray emission observed from Einstein, stellar rotation and bolometric luminosity. *Astrophysical Journal*, **248**, 279–290.

Pallavicini, R., Tagliaferri, G. and Stella, L. (1990). X-ray emission from solar neighbourhood flare stars: A comprehensive survey of EXOSAT results. *Astronomy and Astrophysics*, **228**, 403–425.

Patten, B.M. and Simon, T. (1996). The evolution of rotation and activity in young open clusters: IC 2391. *Astrophysical Journal Supplement*, **106**, 489–531.

Peres, G., Orlando, S., Reale, F., Rosner, R. and Hudson, H. (2000). The Sun as an X-ray star. II. Using the Yohkoh/SXT-derived solar emission measure vs. temperature to interpret stellar X-ray observations. *Astrophysical Journal*, **528**, 537–551.

Preibisch, T. and Zinnecker, H. (1999).The history of low-mass star formation in the Upper Scorpius OB association. *Astronomical Journal*, **117**, 2381–2397.

Pye, J.P., Hodgkin, S.T., Stern, R.A. and Stauffer, J.R. (1994). ROSAT X-ray luminosity functions of the Hyades dK and dM stars. *Monthly Notices of the Royal Astronomical Society*, **266**, 798–804.

Randich, S. (1997). X-ray emission from young open clusters. In G. Micela, R. Pallavicini and S. Sciortino (eds), *Cool Stars in Clusters and Associations: Magnetic Activity and Age Indicators*, eds., Memorie della Societa Astronomia Italia Vol. 68, pp. 971–984.

Randich, S. (1998). Supersaturation in X-ray emission of cluster stars. In R.A. Donahue and J.A. Bookbinder (eds), *Cool Stars, Stellar Systems, and the Sun*, ASP Conference Series Vol. 154, Astronomical Society of the Pacific, San Francisco, pp. 501–510.

Randich, S. (2000). Coronal activity among open cluster stars. In R. Pallavicini, G. Micela and S. Sciortino (eds), *Stellar Clusters and Associations: Convection, Rotation, and Dynamos*, ASP Conference Series Vol. 198, Astronomical Society of the Pacific, San Francisco, pp. 401–410.

Randich, S. and Schmitt, J.H.M.M. (1995). A ROSAT X-ray study of the Praesepe cluster. *Astronomy and Astrophysics*, **298**, 115–132.

Randich, S., Schmitt, J.H.M.M., Prosser, C.F. and Stauffer, J.R. (1996). The X-ray properties of the young open cluster around α Persei. *Astronomy and Astrophysics*, **305**, 785–805.

Reimers, D., Huensch, M., Schmitt, J.H.M.M. and Toussaint, F. (1996). Hybrid stars and the reality of 'dividing lines' among G to K bright giants and supergiants. *Astronomy and Astrophysics*, **310**, 813–824.

Rosner, R., Golub, L. and Vaiana, G.S. (1985). On stellar X-ray emission. *Annual Review of Astronomy and Astrophysics*, **23**, 413–452.

Rosner, R., Musielak, Z.E., Cattaneo, F., Moore, R.L. and Suess, S.T. (1995). On the origin of "dividing lines" for late-type giants and supergiants. *Astrophysical Journal*, **442**, L25–L28.

Schmitt, J.H.M.M. (1993). ROSAT observations of late-type stars. In J.L. Linsky and S. Serio (eds), *Physics of Solar and Stellar Coronae*, Kluwer, Dordrecht, pp. 327–336.

Schmitt, J.H.M.M. (1994). ROSAT observations of stellar flares. *Astrophysical Journal Supplement*, **90**, 735–742.

Schmitt, J.H.M.M. (1997). Coronae on solar-like stars. *Astronomy and Astrophysics*, **318**, 215–230.

Schmitt, J.H.M.M. (1998). Inference of stellar coronal structure. In R.A. Donahue and J.A. Bookbinder (eds), *Cool Stars, Stellar Systems, and the Sun*, ASP Conference Series Vol. 154, Astronomical Society of the Pacific, San Francisco, pp. 463–486.

Schmitt, J.H.M.M., Collura, A., Sciortino, S., Vaiana, G.S., Harnden, F.R., Jr. and Rosner, R. (1990). Einstein Observatory coronal temperatures of late-type stars. *Astrophysical Journal*, **365**, 704–728.

Schmitt, J.H.M.M. and Favata, F. (1999). Continuous heating of a giant X-ray flare on Algol. *Nature*, **401**, 44–46.

Schmitt, J.H.M.M., Fleming, T.A. and Giampapa, M.S. (1995). The X-ray view of the low-mass stars in the solar neighborhood. *Astrophysical Journal*, **450**, 392–400.

Schmitt, J.H.M.M. and Kürster, M. (1993). A spatially resolved X-ray image of a star like the Sun. *Science*, **262**, 215–218.

Siarkowski, M., Pres, P., Drake, S.A., White, N.E. and Singh, K.P. (1996). Corona(e) of AR Lacertae. II. The spatial structure. *Astrophysical Journal*, **473**, 470–482.

Simon, T., Drake, S.A. and Kim, P.D. (1995). The X-ray emission of A-type stars. *Publications of the Astronomical Society of the Pacific*, **107**, 1034–1041.

Singh, K.P., White, N.E. and Drake, S.A. (1996). Corona(e) of AR Lacertae. I. The temperature and abundance distribution. *Astrophysical Journal*, **456**, 766–776.

Stauffer, J.R., Caillault, J.-P., Gagné, M., Prosser, C.F. and Hartmann, L.W. (1994). A deep imaging survey of the Pleiades with ROSAT. *Astrophysical Journal Supplement*, **91**, 625–657.

Stern, R.A., Schmitt, J.H.M.M., Pye, J.P., Hodgkin, S.T., Stauffer, J.R. and Simon, T. (1994). Coronal X-ray sources in the Hyades: A 40 kilosecond ROSAT pointing. *Astrophysical Journal*, **427**, 808–821.

Stern, R.A., Uchida, Y., Tsuneta, S. and Nagase, F. (1992). GINGA observations of X-ray flares on Algol. *Astrophysical Journal*, **400**, 321–329.

Stern, R.A., Zolcinski, M.-C., Antiochos, S.K. and Underwood, J.H. (1981). Stellar coronae in the Hyades. A soft X-ray survey with the Einstein Observatory. *Astrophysical Journal*, **249**, 647–661.

Swank, J.H., White, N.E., Holt, S.S. and Becker, R.H. (1981). Two-component X-ray emission from RS Canum Venaticorum binaries. *Astrophysical Journal*, **246**, 208–214.

Tsuru, T., Makishima, K., Ohashi, T., Inoue, H., Koyama, K., Turner, M.J.L., Barstow, M.A., McHardy, I.M., Pye, J.P., Tsunemi, H., Kitamoto, S., Taylor, A.R. and Nelson, R.F. (1989). X-ray and radio observations of flares from the RS Canum Venaticorum system UX Arietis. *Publications of the Astronomical Society of Japan*, **41**, 679–695.

Vaiana, G.S., Fabbiano, G., Giacconi, R., Golub, L., Gorenstein, P., Harnden, F.R., Jr., Cassinelli, J.P., Haisch, B.M., Johnson, H.M., Linsky, J.L., Maxson, C.W., Mewe, R., Rosner, R., Seward. F., Topka, K. and Zwaan, C. (1981). Results from an extensive Einstein stellar survey. *Astrophysical Journal*, **245**, 163–182.

Verbunt, F. (2000). X-rays from old open clusters. In R. Pallavicini, G. Micela and S. Sciortino (eds), *Stellar Clusters and Associations: Convection, Rotation, and Dynamos*, ASP Conference Series Vol. 198, Astronomical Society of the Pacific, San Francisco, pp. 421–430.

Vilhu, O. (1987). Heating of stellar chromospheres and coronae: Observational constraints and evidence for saturation. In J.L. Linsky and R.E. Stencel (eds), *Cool Stars, Stellar Systems, and the Sun*, Lecture Notes in Physics 291, Springer-Verlag, Berlin, pp. 110–122.

Walter, F.M. (1993). X-rays and activity in pre-main sequence stars. In J.L. Linsky and S. Serio (eds), *Physics of Solar and Stellar Coronae*, Kluwer, Dordrecht, p. 249.

White, N.E. (1996). Stellar X-ray spectroscopy with ASCA. In R. Pallavicini and A.K. Dupree (eds), *Cool Stars, Stellar Systems, and the Sun*, ASP Conference Series Vol. 109, Astronomical Society of the Pacific, San Francisco, pp. 193–202.

White, N.E., Culhane, J.L., Parmar, A.N., Kellett, B.J., Kahn, S., van den Oord, G.H.J. and Kuijpers, J. (1986). An EXOSAT observation of quiescent and flare coronal X-ray emission from Algol. *Astrophysical Journal*, **301**, 262–274.

White, N.E., Shafer, R.A., Parmar, A.N., Horne, K. and Culhane, J.L. (1990). X-ray eclipse mapping of AR Lacertae. *Astrophysical Journal*, **350**, 776–795.

Zinnecker, H. and Preibisch, T. (1994). X-ray emission from Herbig Ae/Be stars: A ROSAT survey. *Astronomy and Astrophysics*, **292**, 152–164.

FURTHER READING

Caillault, J.-P. (1996). ROSAT observations of stellar clusters. In R. Pallavicini and A.K. Dupree (eds), *Cool Stars, Stellar Systems, and the Sun*, ASP Conference Series Vol. 109, Astronomical Society of the Pacific, San Francisco, pp. 325–334.

Caillault, J.-C., Briceño, C., Martín, E.L., Palla, F. and Wichmann, R. (1998). The ROSAT field sources: What are they? In R.A. Donahue and J.A. Bookbinder (eds), *Cool Stars, Stellar Systems, and the Sun*, ASP Conference Series Vol. 154, Astronomical Society of the Pacific, San Francisco, pp. 237–256.

Drake, S.A. (1996b). X-ray measurements of coronal abundances. In S.S. Holt and G. Sonneborn (eds), *Cosmic Abundances*, ASP Conference Series Vol. 99, Astronomical Society of the Pacific, San Francisco, pp. 215–226.

Feigelson, E.D. (1997). X-rays, star formation and the solar nebula. In G. Micela, R. Pallavicini and S. Sciortino (eds), *Cool Stars in Clusters and Associations: Magnetic Activity and Age Indicators*, Memorie della Societa Astronomia Italia Vol. 68, pp. 1007–1016.

Jordan, C., Doscheck, G.A., Drake, J.J., Galvin, A.B. and Raymond, J.C. (1998). Coronal abundances: What are they? In R.A. Donahue and J.A. Bookbinder (eds), *Cool Stars, Stellar Systems, and the Sun*, ASP Conference Series Vol. 154, Astronomical Society of the Pacific, San Francisco, pp. 91–110.

Krautter, J. (1996). The impact of ROSAT observations on our understanding of star forming regions. In R. Pallavicini and A.K. Dupree (eds), *Cool Stars, Stellar Systems, and the Sun*, ASP Conference Series Vol. 109, Astronomical Society of the Pacific, San Francisco, pp. 395–404.

Laming, J.M. (1998). Spectroscopy of stellar coronae. In R.A. Donahue and J.A. Bookbinder (eds), *Cool Stars, Stellar Systems, and the Sun*, ASP Conference Series Vol. 154, Astronomical Society of the Pacific, San Francisco, pp. 447–462.

Linsky, J.L. (1985). Nonradiative activity across the HR diagram. Which types of stars are solar-like? *Solar Physics*, **100**, 333–362.

Linsky, J.L. (1990). Einstein and stellar sources. In M. Elvis (ed), *Imaging X-ray Astronomy*, Cambridge University Press, Cambridge, pp. 39–59.

Linsky, J.L. and Gagné, M. (1998). Studying new problems in stellar coronal physics with AXAF and XMM. In R.A. Donahue and J.A. Bookbinder (eds), *Cool Stars, Stellar Systems, and the Sun*, ASP Conference Series Vol. 154, Astronomical Society of the Pacific, San Francisco, pp. 527–543.

Mewe, R. (1992). X-ray spectroscopy of solar and stellar coronae. In R. Pallavicini (ed), *Solar and Stellar Coronae*, Memorie della Società Astronomica Italiana Vol. 63, pp. 681–696.

Mewe, R. (1993). Stellar coronal X-ray spectroscopy. In J.L. Linsky and S. Serio (eds), *Physics of Solar and Stellar Coronae*, Kluwer, Dordrecht, pp. 225–236.

Montmerle, T. (1997). Magnetic activity in young stellar objects: from T Tauri stars to protostars. In G. Micela, R. Pallavicini and S. Sciortino (eds), *Cool Stars in Clusters and Associations: Magnetic Activity and Age Indicators*, Memorie della Società Astronomica Italiana Vol. 68, pp. 1017–1028.

Pallavicini, R. (1986). Stellar coronae: The EXOSAT picture. In R. Pallavicini and N.E. White (eds), *X-ray Astronomy with EXOSAT*, Memorie della Società Astronomica Italiana Vol. 59, pp. 71–94.

Pallavicini, R. (1989). X-ray emission from stellar coronae. *Astronomy and Astrophysics Review*, **1**, 177–207.

Pallavicini, R. (1993). Time variability of stellar X-ray emission. In J.L. Linsky and S. Serio (eds), *Physics of Solar and Stellar Coronae*, Kluwer, Dordrecht, pp. 237–248.

Pallavicini, R. (1999). X-ray coronae of stars: Some theoretical questions. In C.J. Butler and J.G. Doyle (eds), *Solar and Stellar Activity: Similarities and Differences*, ASP Conference Series Vol. 158, Astronomical Society of the Pacific, San Francisco, pp. 334–347.

Schmitt, J.H.M.M. (1990a). Stellar X-ray astronomy. *Advances in Space Research*, **10**, 115–124.

Schmitt, J.H.M.M. (1990b). X-ray spectroscopy across the HR diagram. In P. Gorenstein and M. Zombeck (eds), *High Resolution X-ray Spectroscopy of Cosmic Plasmas*, Cambridge University Press, Cambridge, pp. 110–121.

Schmitt, J.H.M.M. (1992). First stellar results from the ROSAT XRT. In R. Pallavicini (ed), *Solar and Stellar Coronae*, Memorie della Società Astronomica Italiana Vol. 63, pp. 563–576.

Schmitt, J.H.M.M. (1996a). Eclipse mapping at X-ray wavelengths. In K.G. Strassmeier and J.L. Linsky (eds), *Stellar Surface Structure*, Kluwer, Dordrecht, pp. 85–94.

Schmitt, J.H.M.M. (1996b). Stellar coronae. In J.P. De Greve, R. Blommme and H. Hensberge (eds), *Stellar Atmospheres: Theory and Observations*, Lecture Notes in Physics 497, Springer, Berlin, pp. 277–315.

Sciortino, S. (1993). Stellar coronal emission: What we have learned from pre-ROSAT observations. In J.L. Linsky and S. Serio (eds), *Physics of Solar and Stellar Coronae*, Kluwer, Dordrecht, pp. 211–224.

Stern, R.A. (1998). Long-term X-ray variability in cool stars. In R.A. Donahue and J.A. Bookbinder (eds), *Cool Stars, Stellar Systems, and the Sun*, ASP Conference Series Vol. 154, Astronomical Society of the Pacific, San Francisco, pp. 223–234.

Stern, R.A. (1999). X-ray magnetic activity and variability in stellar coronae: What are the limits of the standard picture? In C.J. Butler and J.G. Doyle (eds), *Solar and Stellar Activity: Similarities and Differences*, ASP Conference Series Vol. 158, Astronomical Society of the Pacific, San Francisco, pp. 47–56.

Rosner, R. (1991). Implications for coronal heating theories from stellar X-ray observations. In P. Ulmschneider, E.R. Priest and R. Rosner (eds), *Mechanisms of Chromospheric and Coronal Heating*, Springer-Verlag, Heidelberg, pp. 287–299.

Vaiana, G.S. (1990). X-ray emission from stars: A sharper and deeper view of our Galaxy. In M. Elvis (ed), *Imaging X-ray Astronomy*, Cambridge University Press, Cambridge, pp. 61–88.

Vaiana, G.S., Maggio, A., Micela, G., Sciortino, S. (1992). Coronal emission and stellar evolution. In R. Pallavicini (ed), *Solar and Stellar Coronae*, Memorie della Società Astronomica Italiana Vol. 63, pp. 545–562.

39

JOSEPH P. CASSINELLI*

Mass loss from stars

Stars return mass to the interstellar medium throughout their lives. A low-luminosity star such as the Sun is losing mass by a stellar wind at a rate of 10^{-14} solar masses per year ($M_\odot \text{yr}^{-1}$). Very luminous stars, whether they are early-type hot stars or cool giants, lose mass at a rate of up to about $10^{-5} M_\odot \text{yr}^{-1}$, and such rates have major consequences. The evolutionary track of the star on the Hertzsprung–Russell diagram is modified, the terminal state can be changed, and the winds affect the surrounding interstellar medium by depositing momentum, energy, and chemically enriched material.

This chapter reviews selected highlights of the origin and development of our understanding of stellar winds and mass loss. Much of what we currently know about stellar winds and the effects of mass loss comes from space observations. The space era in this field started with the rocket UV research by Donald Morton and his colleagues at Princeton in the mid-1960s (Morton 1967, Morton et al. 1968). The results were unexpected and led to a strong increase in activity regarding the subject of mass loss from hot stars. Figure 1 (top) shows the spectrum of one of the belt stars of Orion. On it were discovered strong and broad features at several UV resonance lines. Some of the line profiles were found to have longward shifted emission and shortward shifted absorption, and these lines are called P-Cygni profiles after the B supergiant which shows many such lines in the visible part of the spectrum. An early study of P-Cygni profiles and their significance as indicators of outflows was carried out in the 1930s by Beals (1929, 1931). He compiled catalogs showing spectra of Wolf-Rayet (WR) stars and η Carinae, and some of the novae and central stars of planetary nebulae which show the profiles. Chandrasekhar (1934a,b) introduced the now often used iso-velocity contour diagnostic for finding outflow properties from line profiles. He also introduced ways for interpreting the flattened continua of WR stars in terms of an extended atmosphere. Kosirev (1934) used these diagnostics to derive a rate of mass loss and outflow velocity ($10^{-5} M_\odot \text{yr}^{-1}$ and 1000 km s^{-1}) of a WR star, and his values are within a factor of three of current estimates. Most of the initial work on profiles had assumed that the outflow was transparent to line radiation. Sobolev (1947) developed an escape probability method for treating optically thick lines, and his method served as the basis for essentially all of the subsequent work on line profile analysis. Underhill (1949) computed continuum model atmospheres for hot stars and arrived at the important conclusion that luminous stars with $T_{\text{eff}} > 20,000$ cannot have stable hydrostatic atmospheres. In spite of this early activity, the understanding of mass loss from hot stars progressed very slowly in the several decades before Morton's rocket observations.

During that interval, the attention of stellar astronomers was on mass loss from cool stars. This began with the discovery by Adams and McCormack (1935) of spectral evidence for slow ($\sim 5 \text{ km s}^{-1}$), outward, mass motion. Lyman Spitzer (1939), who would later lead the development of space astronomy, carried out a PhD thesis on a study of the supergiants α^1 Her (M5 II) and α Ori (M2 Iab). In his thesis he developed the idea of "fountains" to explain the fact that the optical spectra of cool stars show evidence for expansion, but only at a speed well below the escape speed. He postulated that there would occur a change in the ionization to an unobservable stage, followed by a fountain-like infall. Many decades later this concept of fountain flow was found useful for modeling the infalling gas trajectories in our Galaxy. The first strong evidence that cool giants involve matter actually leaving the stars was developed by Deutsch (1956) from his observations of α Her (M5 II + G0 III). His analysis showed that the expansion of the primary star's

*University of Wisconsin, Madison, WI, USA

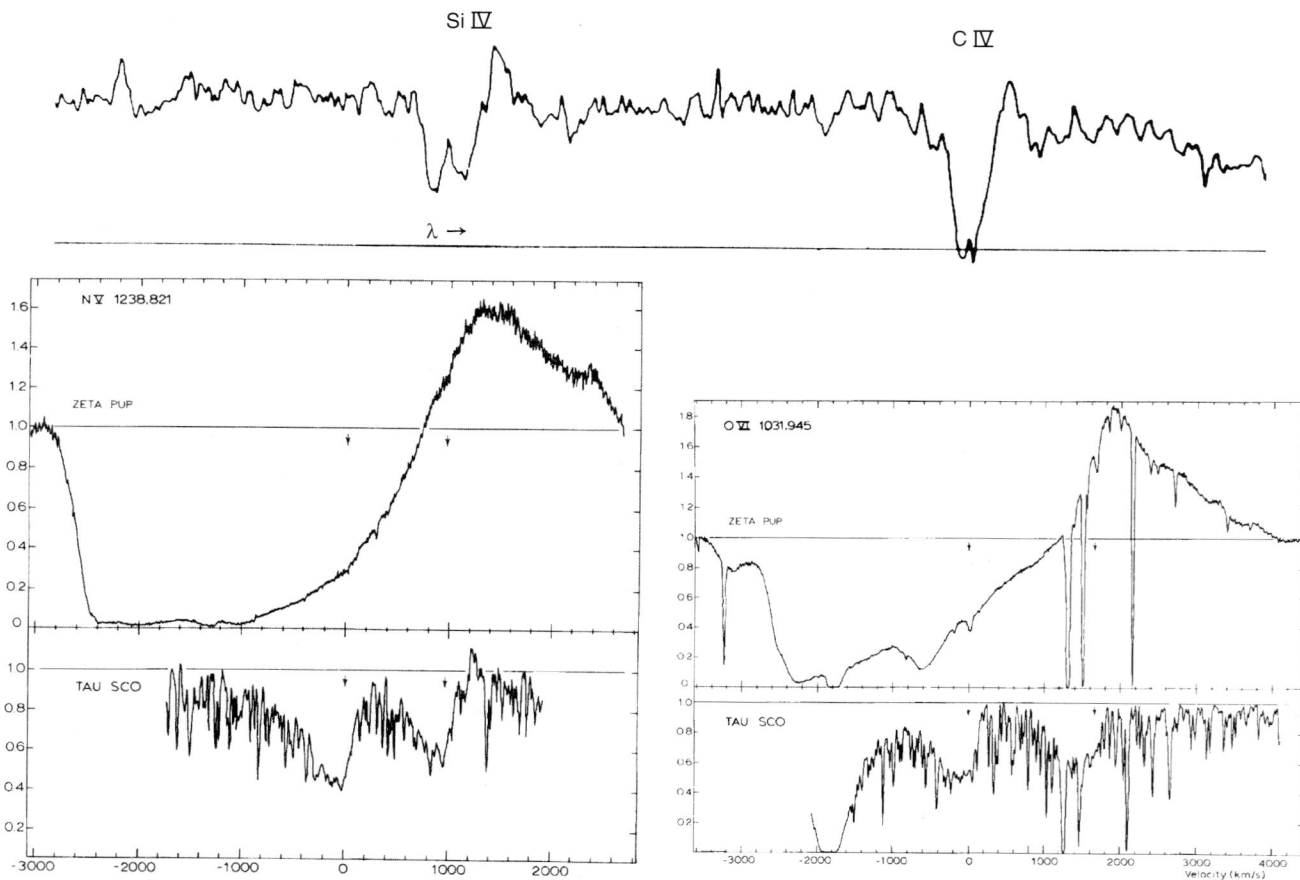

Figure 1 The upper panel shows Morton's (1967) rocket observation of the ultraviolet spectrum of ζ Ori (O 9.5 Ia). The wavelength increases to the right from 1140 to 1630 Å. Note the P-Cygni profiles at C IV and Si IV, and the shortward displaced absorption lines of C III, N V, and Si III. The lower panels show the Copernicus satellite spectra of the doublets of N V and O VI in the O4f star ζ Pup and the B0 V star τ Sco. The arrows indicate the location of the rest wavelength of the lines. The sharp lines in the O VI spectrum are interstellar lines and the strong line at -1900 km s^{-1} is Ly-β. (Adapted from Lamers and Cassinelli 1999.)

outer atmosphere extended beyond the orbit of its distant ($360 R_*$) G0 companion. At such a large distance the flow had reached local escape speed leading Deutsch to infer that there is an outflow with a mass loss rate of $\sim 10^{-7} M_\odot$ yr^{-1}.

Ludwig Biermann (1951) recognized from the directions of the ion tails of comets that there must be a fast and continuous "corpuscular radiation" from the Sun. Eugene Parker (1958) developed an explanation of this mass outflow, which he called solar wind theory. This theory formed the basis for all the subsequent work on stellar winds and stellar mass loss (Chapters 9 and 47). The high speed or "wind" outflow that he predicted was verified by the *Mariner II* interplanetary space mission (Neugebauer and Snyder 1962). Weymann (1960, 1963) pioneered the application of wind theory to other cool stars, initially using extensions of coronal or thermally driven wind theory, and then considering other forces such as radiation pressure on resonance line and dust grain opacities.

The following reviews the development of the subject of mass loss by stellar winds in the space age. As a unifying theme, we give special attention to the various dividing lines and boundaries on the Hertzsprung–Russell (HR) diagram that mark sharp changes in outflow properties associated with slight changes in stellar parameters. These boundaries were discovered mostly through space satellite surveys. Their locations on the HR diagram led to insight regarding the driving mechanisms, and the dependence of the winds on basic stellar properties such as effective temperature, surface gravity, and stellar rotation.

MASS LOSS FROM EARLY-TYPE STARS

In the late 1960s research was being carried out on stellar atmospheres theory regarding stars with a high luminosity-to-mass ratio, and hence near the stars' Eddington limit (Böhm and Deinzer 1965, Cassinelli 1970). Proximity to the Eddington limit was seen as a requirement for producing stars with extended density distributions which have scale heights comparable to the stellar radius. The rocket observations had begun to make it clear that the outer atmospheres of hot luminous stars are not hydrostatic, but are expanding at high velocities.

Lucy and Solomon (1970) wrote a ground-breaking paper that explained the high speed of the winds from hot stars. They showed that the same P-Cygni lines that had been used to recognize the outflow could be responsible for accelerating the winds to speeds of $v_\infty \sim 3000 \, \text{km s}^{-1}$, and that each strong line could drive a mass loss rate of $\dot{M} = L/c^2$, which corresponds to a total mass-loss rate of OB supergiants of about $\dot{M} \sim 10^{-8} M_\odot \, \text{yr}^{-1}$. Furthermore, they showed that coronal wind theory could not explain both the high speeds and the low-ion stages observed through P-Cygni profiles. The basic theory of outflows driven by *selective absorption* of the continuum radiation in the shortward wing of the line opacity in an expanding atmosphere was originally developed in a sequence of papers by Saha (1919), Milne (1924, 1926), and Johnson (1925). This early interest in the acceleration of atoms was in part to find a mechanism to explain the heating of the solar chromosphere and corona.

Castor, Abbott, and Klein (CAK) (Castor *et al.* 1975, 1976) developed a powerful method to treat the effects of driving of an outflow by very large numbers of lines, which have a statistical distribution in line strengths. CAK theory led to predictions of mass-loss rates and terminal velocities as a function of stellar properties and the line statistics parameters. With the modifications by Friend and Abbott (1986), Pauldrach *et al.* (1986), and Kudritzki *et al.* (1989), CAK multi-line theory gives good agreement with observationally derived values of $\dot{M}$ and v_∞. In particular, strong winds could be driven by radiation alone, with no extra driving associated with mechanical energy deposition. Thus it was the expectation of the CAK theory that the wind temperatures would be cool, that is, about 80 to 90% of the star's effective temperature, as would be the case for an extended atmosphere that is in radiative equilibrium with the photospheric radiation field. Space observations showed evidence for a quite different temperature structure.

Superionization of hot star winds

The Copernicus satellite, launched by NASA in 1973, provided the first extensive set of data regarding winds from bright, hot stars. The spectra covered the far-UV region in the 1000 to 1450 Å band with very high resolution ($\lambda/\Delta\lambda = 10^4$). Among the most surprising discoveries was the presence of anomalously high ionization stages with broad, strong P-Cygni profiles. Lamers and Morton (1976) called this "superionization" in the winds, and two superionization lines are shown in Figure 1 (bottom). Doublets of O VI near 1040 Å and N V near 1240 Å were seen in the spectra of stars across the entire O star spectral sequence, and resonance lines of C IV and Si IV extended into the B spectral range (Snow and Jenkins 1977). The extent of these superionization zones with their well defined *dividing lines* on the HR diagram are shown in Figure 2. These ions are anomalous because they were not expected to be present in a gas photoionized by the stellar radiation field. Their presence indicated that that there was a problem regarding the temperature structure assumed in line-driven wind theory, and that an extra source of heating was present and a source for that was needed.

In the Copernicus observation of ζ Pup (O4 f) shown in Figure 1 (bottom), the profiles of the superionization stages are fully developed as saturated P-Cygni profiles. The nature of the profiles could be taken as good evidence that the ions O^{5+} and N^{4+} are present at all levels in the wind, not just near the base nor just far from the star where high speeds are reached. Initially the superionization was interpreted by Lamers and Morton (1976) by the "warm wind model." In this model superionization would be produced collisionally in a gas with a temperature above 10^5 K. Collisional ionization required that stars with O VI should have winds with temperatures of about 200,000 K. Lamers and Snow (1978) extended the warm wind idea to B supergiants which showed a lower stage of superionization (N V and C IV and Si IV), by using progressively lower wind temperatures such that $T_{\text{wind}} \sim 80{,}000$ K for the later B supergiants.

Producing the warm wind temperatures required some sort of mechanical heating or wave-energy deposition. The amount needed was troubling, however. In the case of ζ Pup, the required mechanical luminosity was estimated to be about 10% of the radiative luminosity of the star. Such a large proportion of the luminosity is required because gases with temperatures near 2×10^5 K are near maximal efficiency for radiating away thermal energy. The mechanical flux required for ζ Pup was considered to be unacceptably large given that hot stars have no obvious source of wave energy such as an outer convection zone, so alternative solutions were sought.

At a workshop at JILA, Cassinelli, Castor and Lamers (1978) critically analyzed three models for explaining the superionization: Lamers' warm wind model, Cassinelli's "corona plus cool wind" model, and Castor's non-LTE model with a wind temperature of ~60,000 K, which was dubbed the "tepid wind" model. In the corona plus cool wind idea, a geometrically thin corona with a temperature of a few million kelvins, at the base of the wind was postulated. The idea that a hot star could have a thin corona had been proposed by Hearn (1975). Some mechanical flux

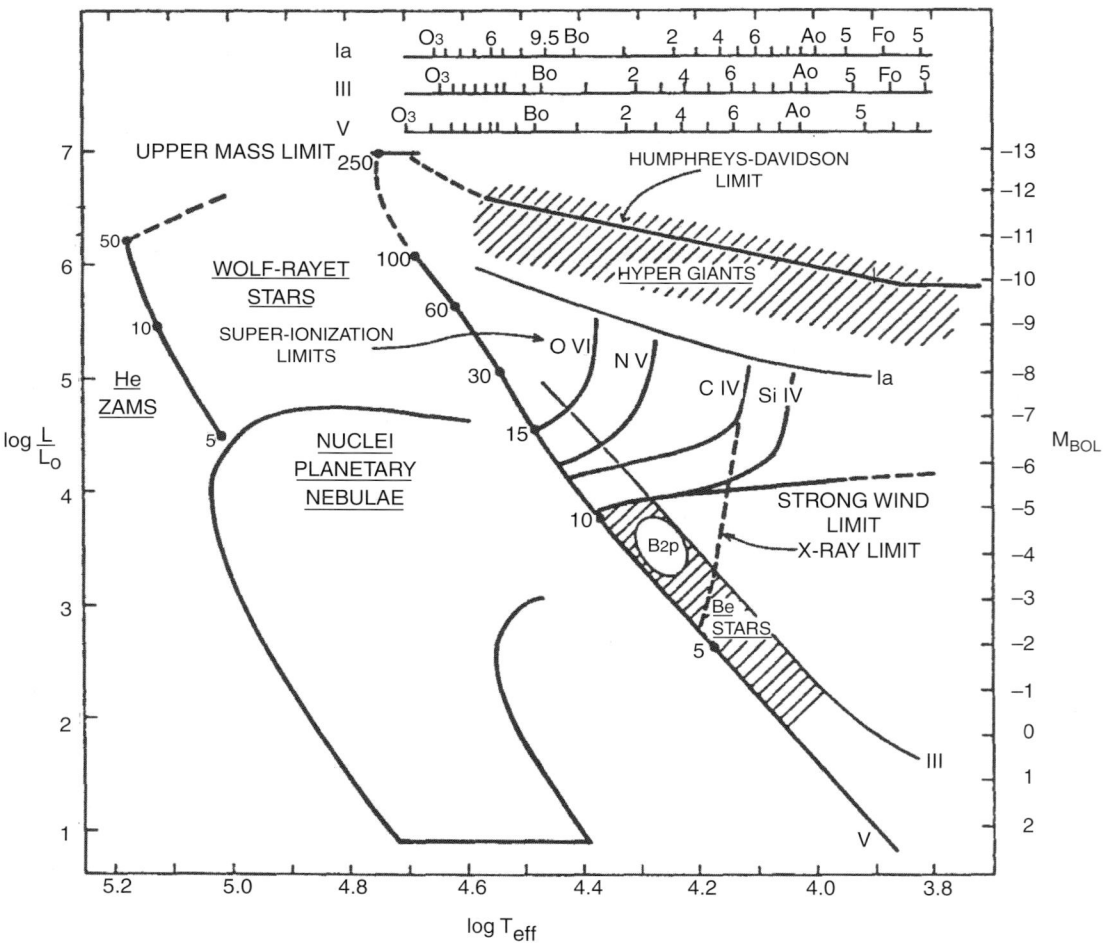

Figure 2 Regions and boundaries for hot stars in the HR diagram. The locations of various classes of stars with interesting mass-loss properties are shown. Of particular interest are the "superionization boundaries" of O VI, N V, C IV, and Si IV, which demarcate the range in effective temperatures of stars that show these high-ion stages. (From Cassinelli and Lamers 1987.)

from the star was invoked to produce the thin coronal region. However, the amount of mechanical flux would be orders of magnitude below that required by the warm wind model, because gas at million-degree coronal temperatures has a much lower emissivity than does warm gas. A corona would produce X-rays at energies beyond the K-shell edges of carbon, nitrogen, and oxygen, and the absorption of these X-rays in the cool wind above the corona could produce superionization. Castor showed that the photospheric radiation field of the O4f star ζ Pup could produce some O VI ions. However, the O VI line persists through the O spectral range to τ Sco at B0.5 V. This star has a much lower luminosity and effective temperature than ζ Pup (30,000 K v. 42,000 K). The non-LTE effect could not explain the O VI superionization at this low stellar temperature.

Cassinelli and Olson (1979) made use of the boundaries of the superionization on the HR diagram (Figure 2) to argue that the superionization is produced by the Auger effect, following X-ray photoionization by the star. If an X-ray is absorbed by an ion with fewer than 10 electrons, then *two* electrons are essentially always ejected from the ion (Daltabuit and Cox 1972). The X-ray photon ejects a K-shell electron, and then in the transition of an electron from the L shell to the K shell, a second electron is ejected. As a simple numerical consequence, if a wind has stage "n" as its dominant ion, Auger ionization will lead to an overabundance of the "n + 2" stage. Cassinelli and Olson (1979) argued that the O VI line should be present in winds for which O^{3+} is the dominant ion. The ion O^{3+} persists as the dominant stage for all of the O stars and extending as late as B0.5. Similarly, the UV line of N V should persist out to spectral type B2 I for which N^{2+} is the dominant ion stage, and C IV and Si IV can be produced to B8 I (Odegard and Cassinelli 1982). The mechanism involving X-rays thus simply and clearly explained the observed superionization boundaries shown in Figure 2. The coronal plus cool wind model further led to the prediction that OB stars would be X-ray sources at a well-defined flux level. This prediction was confirmed by Harnden et al. (1979) and Seward et al. (1979) with the very first observations from the Einstein satellite launched by NASA in 1978.

X-ray emission and shocks in hot-star winds

As is usual in astronomy, the observations did not totally conform to theoretical predictions. In particular, the base corona plus cool wind model predicted that there should be significant attenuation of coronal soft X-ray emission by the extended cool wind region above the coronal zone. The absorption should have been especially large at energies at and beyond the K-shell edges of C, N, and O (located respectively at 0.2, 0.4, and 0.6 keV). Since the opacities decrease as ν^{-3} beyond the edges, the winds would become optically thin for energies above about 1 keV. The spectral energy distributions were relatively flat below 1 keV, and did not show the predicted deficit shortward of the K-shell edges. The idea that X-rays cause the superionization via the Auger effect was confirmed by the Einstein satellite observations, but the auxiliary idea that X-rays originate only from a thin base corona was not supported because the soft X-rays were not absorbed as predicted in the corona cool wind model (Long and White 1980, Cassinelli and Swank 1983). The superionization occurs over the whole range in the wind from the low- to the high-velocity outflow, so the star can neither have X-rays produced just at the base of the wind, nor just in zones far out in the wind.

Lucy and White (1980) proposed that shocks in the winds produce the X-rays. They argued that the instability of line-driven winds to velocity perturbations should cause density enhancements in the wind to grow and form clumps. As radiation pressure gradients accelerated these outward, bow shocks would form and lead to temperatures of the order of 10^6 K. Lucy (1982a) developed a model with fewer free parameters called the "periodic shock model." However, in a comparison of the predicted energy distributions with the rather high spectral resolution Solid State Spectrometer (SSS) data from the Einstein satellite, Cassinelli and Swank (1983) showed that these early shock models could not explain the observed hardness nor the large X-ray luminosities of the three OB supergiant belt stars of Orion. In fact one of the stars, ζ Ori (O9.5 Ia), showed evidence for line emission from the ions Si XIII (1.8 keV) and S XV (2.2 keV). These ions indicated the presence of temperatures of 1.5×10^7 K. Such temperatures are hotter than maximal coronal temperatures, and led them to propose that the hot gas is magnetically confined at the base of the wind. Accounting for such very high temperatures observed in this and other hot stars has continued to be a problem for models in which X-rays arise solely from shocks embedded in winds.

The Einstein satellite and ROSAT (Röntgensatellite) showed the following general properties regarding hot-star X-ray emission. The most widely known property is the "10^{-7} law" that related the X-ray luminosity to the stellar bolometric luminosity through the O spectral region to B1. The relation $L_x/L_{bol} \approx 10^{-6.5 \pm 1.0}$ was discovered even in the earliest surveys (Long and White 1980, Cassinelli et al. 1981, Seward and Chlebowski 1982). There is a sharp dividing line at B1 Ia in that supergiants later than this do not emit X-rays above $10^{-8} L_{bol}$. The observation that the X-ray luminosity obeys the 10^{-7} law was not completely understood even near the end of the twentieth century (Owocki and Cohen 1999). There is also a departure from the L_x/L_{bol} relation for the early B main sequence stars. A sharp decrease in the ratio occurs because the stars have too little wind to account for the emission measure needed to keep the X-ray emission at the $10^{-7} L_{bol}$ level (Cassinelli et al. 1994). The Of stars and OB supergiants have winds that are thick to soft X-rays (below about 1 keV) so that the soft X-rays arise only from the outer regions of a wind. Thus, the actual production of X-rays in optically thick winds can correspond to a much greater rate than implied by the 10^{-7} law (Waldron 1984).

Another major property of the X-rays from early-type stars is that the X-ray flux is nearly constant in time. This is not what would be expected from propagating shock models. Therefore, Cassinelli and Swank (1983) argued that the shocks are not really in the form of spherically symmetric expanding shells, but rather must be in the form of "shock fragments." In this picture, there must be a roughly constant number of X-ray-producing regions distributed throughout the accelerating wind.

Ultraviolet evidence for the presence of shocks and the associated non-monotonic velocity structure of the winds was provided by Lucy (1982b). He showed that shocks lead to enhanced backscattering of radiation and hence to extended black absorption troughs in resonance line profiles, as is observed in stars with strong winds such as the WR stars.

The origin of the shocks

An explanation for how shocks are produced in hot-star winds was developed by Owocki et al. (1988). The line-driven winds are unstable because small-scale perturbations to the velocity law rapidly grow to form shocks. This initial explanation assumed an isothermal wind and so it was not specifically aimed at explaining X-rays. MacFarlane and Cassinelli (1989) used a simpler kinematic shock production picture in which a fast wind collides with a slower sector of the wind in front of it. This collision produces a driven wave with a structure that is simple enough to allow

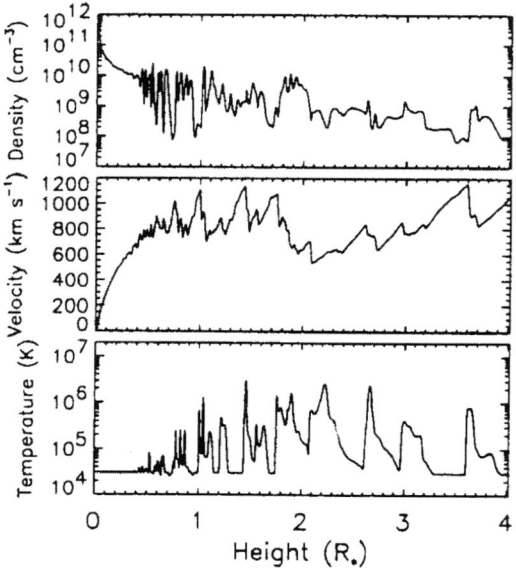

Figure 3 The results of a model in which the line-driven wind instabilities produce shocks in the wind. This shows the non-monotonic distributions v. radius of the density, velocity, and temperature in the flow. (From Cooper and Owocki 1994.)

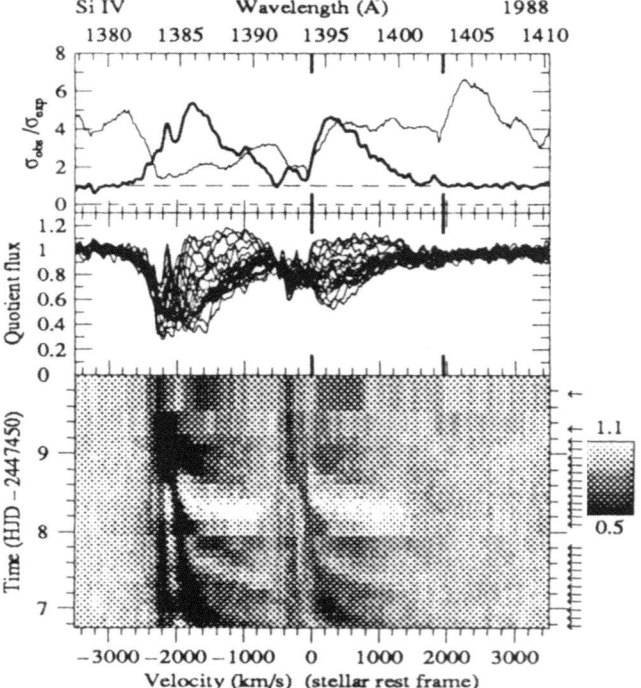

Figure 4 Time series spectra of ξ Per, observed regularly with IUE. The DACs, which appear here as dark zones, show a repeatability related to the rotational period of the star, indicating a connection between the wind features and the occurrence of phenomena at the surface of the star. (From Henrichs et al. 1994.)

the inclusion of an energy equation. The model showed that the X-rays and Auger ionization in τ Sco could be understood in the context of winds with embedded shocks.

A useful way to picture the temperature and velocity structure in the winds of O stars is provided by radiation hydrodynamical calculations (Cooper and Owocki 1994, Feldmeier 1995). An example of the complicated velocity structure associated with shocks in winds is shown in Figure 3. Clearly the properties of line-driven winds can be quite different from the smooth winds that were derived from basic CAK wind theory: the velocity law is not monotonic, and segments of the wind can even have a velocity larger than the terminal velocity. This was important in the interpretation of the wind momentum properties of WR stars (Willis 1991). X-rays from the hot regions can lead to superionization in the cooler layers by way of the Auger process. Also the compressions associated with shocks could lead to enhanced absorption in observed line profiles.

Time variability of hot-star outflows

The IUE (International Ultraviolet Explorer) satellite, a joint project of NASA, the UK, and ESA was launched in 1978. Its long operational lifetime of 18 years allowed for statistical studies and lengthy investigations of the wind properties of a large number of stars. Of particular interest was the time variability of discrete absorption features that were seen in non-saturated P-Cygni line profiles. It became clear that wind line profile variations are not chaotic but occur in well-defined patterns. Absorption components with widths of the order of $200\,\mathrm{km\,s^{-1}}$ appeared episodically at low but supersonic speeds. These components were seen in line profiles to accelerate up to a displacement near the terminal wind speed. During the acceleration phase, lasting of the order of days, the width of the absorption feature decreases. Because it is possible to follow the time development in the line of a specific absorption feature, they are called "discrete absorption components" (DACs). In their survey of 203 O stars Howarth and Prinja (1989) found that DACs were present in 80% of stars. The DACs also contained information about the cause of the wind variability, which is rooted near the base of the wind. The primary tool for the study of the DACs is a stack of gray-scale spectra with time running upwards along the stack and the height of each spectrum proportional to the time between consecutive observations. Figure 4 shows the gray-scale representation of the variability in the lines of ξ Per. Prinja (1988) and Henrichs et al. (1988) found that DAC patterns repeated on a rotational timescale. Furthermore, Kaper et al. (1997) found a correlation between variability in Hα emission and the DACs indicating a photospheric connection to the shock formation process, since Hα emission is formed near the base of the wind.

A model that seems to explain the features is based on the co-rotating interaction region (CIR) model by Mullan (1986). The CIR phenomenon is taken from the analogous behavior occurring in the solar wind, in which a longitudinal sector of fast-streaming material catches up with a neighboring sector of slowly moving wind. This collision of the two longitudinal zones of wind can occur because of the rotation of the star. Many properties of the observed DACs were explained by Cranmer and Owocki (1996) using a CIR model. The cause of the wind-velocity sectors is still a matter of debate, as they could be due to non-radial pulsations or to magnetic sectoring of stellar surface properties which affect the overlying wind speed.

The bi-stability limit for B supergiants

The Copernicus satellite spectra of B supergiants showed that the winds tend to get slower with later spectral type. This could be due to the decrease in the escape speed of stars as they evolve through the B supergiant phase. However, there is a interesting boundary line with a rather sharp separation in mass-loss rates and terminal velocities, which occurs for stars with effective temperatures of about 17,000 K. This observational boundary is called the "bi-stability limit." It was initially discovered from theoretical modeling of line-driven wind theory by Pauldrach and Puls (1990). They noted in their modified CAK models of the B1 Ia^+ supergiant P-Cygni that there was a significant change in $\dot{M}$ and v_∞ predicted by the line-driving theory for a rather modest change in stellar properties. A series of P-Cygni models v. T_{eff} would undergo a major change when the winds became optically thick in the Lyman continuum. The ionization in the wind shifts to a lower stage, and the dominant driving lines shift from being the strong EUV resonance lines of C, N, and O ions to being the weaker but more numerous iron lines located in the Balmer continuum region. This caused the cool states of P-Cygni to have a lower wind speed and a higher mass-loss rate. The location on the HR diagram of the bi-stability jump is at about B1 I. In going from hot to cooler stars there is steep decrease on the terminal wind speed from $v_\infty = 2.6v_{\text{esc}}$ to $v_\infty = 1.3v_{\text{esc}}$, (Lamers et al. 1995). There is an accompanying increase in mass-loss rate by about a factor of five (Vink et al. 1999). The latter authors attributed the change to an increase in the line acceleration of Fe III below the sonic point. The bi-stability jump provides another example of how phenomena involving hot-star winds were predicted and finer details explained using the CAK theory.

Rotating winds

The IUE satellite observations also led to the recognition that the UV resonance lines of emission-line B stars (called Be stars) are anomalously strong for their near-main-sequence spectral types. Another dividing line shown in Figure 2 for near-main-sequence hot stars is associated with luminosity and stellar rotation. The only hot stars with luminosities below $10^4 L_\odot$ that also have significant wind lines are the Be stars. These objects show Hα in emission with widths of a few hundred km s^{-1}. From their absorption line profiles the Be stars are deduced to have larger rotational speeds than normal B stars, and it is commonly believed that the rapid stellar rotation causes asymmetry in their extended atmospheres.

The energy distribution of Be stars in the infrared was measured by IRAS (Infrared Astronomical Satellite), which was a joint project of the Netherlands, the UK, and NASA launched in 1983. The Be stars were found by Waters (1986) to show large IR excesses. These IR distributions could be used to estimate the density distribution in the disks. Assuming a disk geometry with a specific opening angle, and assuming a density distribution of the form $\rho(r) = \rho_0(r/r_0)^{-n}$, Waters et al. (1987) could also derive $\rho(r)$ and the mass-loss properties of Be stars. By the end of the 1980s, empirical models based on infrared, Hα, UV lines, and optical polarization observations led to a "two-component picture" for the outer envelope. This was a piecewise spherical model, in which the stars have a slow outflow in a dense equatorial disk zone that gives rise to the IR and Hα emission. The second component is the polar cap regions, from which a fast wind consistent with line-driven wind theory emerges.

The UV evidence for winds of Be stars (Grady et al. 1987, 1989), comes mostly from the UV superionization lines of C IV 1550 Å and Si IV 1400 Å, or, for the early Be stars, N V. The Be stars show high-speed expansion with speeds in the range 500 to 1500 km s^{-1}, but not the strong longward-shifted emission features that are a characteristic of O star P-Cygni profiles. This indicates that the optical depth of the material in the high-speed outflow is relatively small compared to that of the O stars.

Based on our discussion above regarding O-star X-rays, one would think that the superionization of Be stars would be caused by the Auger mechanism, whereby X-rays lead to the ejection of two electrons. However, the X-ray luminosity and the hardness of the X-rays from both B stars and Be stars has been found to be nearly identical as a function of spectral type (Cohen et al. 1997). Thus the X-rays are produced in the winds of both types of star in the same way, perhaps by wind shocks. The X-ray production appears to occur whether the star is rotating or not. Therefore, the X-ray-producing shocks cannot be used to explain the excess level of superionization that is present in the Be stars v. the normal B stars.

Grady et al. (1987) found that the excess C IV equivalent width in Be stars may be attributed entirely to the presence of the DACs. Thus it appears that the underlying C IV profile is produced by a typical B-star wind, while the DACs are associated with some feature of Be stars that is not present in ordinary B stars: an equatorial disk. The C IV line varies with changes in the relative strength of the longward (R) and shortward (V) emission peaks of the Hα line (described as the V/R ratio). Since the Hα emission and the V/R variations are commonly thought to be produced in the equatorial disk, this indicates that the superionization and the equatorial disk are somehow linked. This was puzzling because it implied that the high-velocity superionized region exists at the same latitude sector as the low-ionization stage, low-velocity equatorial disk. This picture would not be expected in the simple two-component model with the high-speed polar wind and the low-speed disk wind which do not interact with one another.

New types of models had to be developed for the Be stars. Extensions of CAK theory were explored by Friend and Abbott (1986) and Pauldrach et al. (1986). The rotation would lead to an enhancement of the equatorial mass-loss rate and to a reduction of the wind speed there, but the net effects were far too small to explain the emission excesses and polarization. Friend and MacGregor (1984) and Poe and Friend (1986) examined the effect of combining magnetic rotator forces, as developed for the Sun by Weber and Davis (1967), with the CAK line-driving mechanism. This combination failed because the fields led to an increase of the equatorial speeds to $1000\,\mathrm{km\,s^{-1}}$ while observations of Be/X-ray binary stars indicated that disk speeds were below $100\,\mathrm{km\,s^{-1}}$ (Waters et al. 1988).

The bi-stability idea of Pauldrach and Puls (1990) described above, was considered by Lamers and Pauldrach (1991). They noted that bi-stability could be operating in latitudinal sectors of rapidly rotating stars that have temperatures near that of P-Cygni, and such is the case for the Be and the supergiant emission-line B[e] stars. In the polar region of a rapidly rotating star, the temperature and the ionizing flux tend to be larger than at the equator owing to the von Zeipel effect (Clayton 1968). According to bi-stability theory these conditions should lead to a fast, lower density polar flow. The equatorial region should have a lower surface gravity and a lower flux of ionizing photons, so bi-stability would lead to a slower, more massive outflow because there is a greater optical thickness in the Lyman continuum. This idea could explain the supergiant B[e] stars, but not the classical Be stars, because the equatorial outflows are just not optically thick enough to produce the zone of high mass loss and low speed.

It became necessary to develop two-dimensional models instead of piecewise spherical ones discussed thus far. Bjorkman and Cassinelli (1993) developed approximations allowing for a solution of the two-dimensional supersonic flow. They argued that material leaving the star at a given latitude should experience only gravity and outward line radiation forces after it has reached supersonic speeds. As the two forces are radial, a wind parcel can be pictured as having a trajectory that remains in a plane defined by the center of the star and the initial velocity vector. The material remains in this plane as it flows away from the star. This plane crosses the equator, so if the speed of the outflow is reduced because the radial force is weaker (as expected from the line-driven wind parameters of B stars), then the trajectory will cross the equatorial plane. The crossing of flows at the equator cannot occur, because a collision of flows from the northern and southern hemispheres occurs instead. The collision leads to a compression shock and to density enhancements by a factor of 100 or so and thus an increase in Hα emissivity by a factor of about 10^4. This model is called the "wind compressed disk" (WCD) model. The disk is very thin, bounded from both top and bottom by a shock that can have temperatures of the order of 10^5 K. This provides an extra region for the production of the superionization that is seen in the DACs in the UV spectra of Be stars.

Owocki et al. (1996) found, however, that a fundamental assumption of the WCD model is not correct. The radiation forces are not truly radial, both because the star is rotationally distorted, and an absorption asymmetry in the line transfer produces a force away from the equator. These effects impede the formation of wind-compressed disks. Nevertheless, from an observational point of view the disks were deduced to be present and seemed to have the properties predicted by WCD theory. Wood et al. (1997) modeled the Be star polarization obtained with the WUPPE (Wisconsin Ultraviolet Photo Polarimeter Experiment) on the Space Shuttle Astro Missions, launched by NASA in 1990 and 1995. The models showed that the equatorial region was very thin, as expected, and the presence of significant absorption of near-UV light by a confluence in wavelength of Fe II absorption lines in the disk. So the cause of the disk structure is uncertain.

The most serious problem regarding the WCD disks was pointed out by Telting et al. (1994). The Hα lines in Be stars show two emission peaks – shortward, V, and longward, R, of line center. The slow periodic variations of the V/R line strength ratio in the stars imply that the equatorial disk material is nearly in Keplerian orbital motion. Hence, there is not only a transfer of matter from the star to the disk, but also a transfer of angular momentum. The latter appears to require a magnetic field. For this and several other reasons, magnetic fields in early-type stars became a topic of significant interest toward the end of the twentieth century. Unfortunately, due to their broad absorption lines, magnetic fields are difficult to measure in early-type stars. Nevertheless progress has been made by Donati et al. (2001) who have used a multi-line Zeeman method to measure β Cep (B1 IIIe) to have a oblique rotator field with a 360 G strength. This star is classified as both a Be star and a pulsating variable.

WR stars

The WR stars were among the first of the early-type stars for which it was clear that fast, massive outflows were present (Beals 1931). Early observational work was summarized by Smith (1973). From a theoretical point of view, the WR stars also posed some of the most perplexing problems, the most notable of which is the so-called "momentum problem." In radiatively driven winds it is the transfer of photon momentum to the gas motions that is responsible for the outflow. Thus, at maximum efficiency, one would expect that the photon momentum L/c could lead to a wind momentum equal to $\dot{M} v_\infty$. The wind momentum is most conveniently discussed in terms of the ratio

$$\eta = \frac{\dot{M} v_\infty}{L/c}$$

Barlow et al. (1981) derived values for η in the range 4 to 50 well beyond values of <0.5 for luminous OB stars (Barlow and Cohen 1977) and beyond the maximum of $\eta = 1$ derived from the early radiation-driven wind model by Cassinelli and Castor (1973). The problem persisted through the 1980s. Using IUE observations, radio free–free observations and expanding atmosphere models, Willis (1991) derived WR momentum ratios in the range $\eta = 1$ to 72. The high value of η was the major problem for WR wind theory and various attempts to explain it were made. Friend and Castor (1983) and Abbott and Lucy (1985) considered the increased momentum deposition owing to multiple scattering of line radiation, and derived values of an $\eta \approx 5$. Wavelength gaps between spectral lines that were needed for the multiple scatterings limited the number of scatterings and the η enhancement that was achievable. Poe et al. (1989) developed the luminous magnetic rotator. In this model the wind momentum problem is solved by having a large equatorial mass-loss produce the large radio fluxes by which WR mass-loss rates are measured, and by having the a lower density polar wind form a fast wind. Thus the combination of the two sectors produces a large value for η for WR stars observed at inclinations other than equatorial. However, polarization observations show that only a few of the WR stars have geometrically distorted winds, and hence it is unlikely that the magnetic rotator model provides an explanation for the WR momentum problem in general.

During the 1990s it became widely accepted that the WR winds can be driven by radiation in spite of the momentum problem, by way of multiple scattering of the stellar light as the photons progress through the outer atmosphere and wind. Monte Carlo analyses showing that η values of 9.2 could be explained were carried out by Lucy and Abbott (1993). Gayley et al. (1995) also developed a clever way to illustrate how multiple scattering could lead to an enhanced wind momentum. The momentum problem shifts to the need for a large amount of ovelapping line opacity in the wind. The opacity intercepting the flow of radiation is significantly enhanced if there is a large range of temperatures in the WR winds from high values of the order of 10^5 K near the base of the wind to rather cool values of 10^4 K in the outer layers. When this occurs there can be a large range in ionization stages and thus the opacity remains high at the wavelengths at which the radiative flux is large. Models of the ionization structure show that helium can be just singly ionized in the outer regions of the winds (Hillier 1991) and that the electron temperature in the even more distant radio continuum formation region can have rather low temperatures ($\approx 10^4$ K).

In concluding this section we should mention other topics of great interest with regards to hot-star mass loss. These include (a) the very luminous star, η Car, as imaged and studied spectroscopically with the Hubble Space Telescope (Davidson 1999), (b) the colliding winds of hot stars, and the unexpected associated phenomena such as dust formation occuring in these most unlikely of environments (Williams and van der Hucht 1987), and (c) observations with EUVE (Extreme Ultraviolet Explorer) by Cassinelli et al. (1995, 1996), which showed that the EUV energy distributions of B giant stars disagree by an order of magnitude from photospheric model atmosphere predictions. High-spectral-resolution observations from X-ray to IR wavelengths in the next decades will provide solutions to such problems and hopefully even more insight into the winds of many other hot stars.

MASS LOSS FROM COOL STARS

There are significant differences between the winds of cool stars and those of the hot stars discussed above. Most notably, the wind speeds are slower, instead of the roughly $10^3 \, \text{km s}^{-1}$ speeds associated with hot stars, the cool giant stars have expansion speeds in the range of 10 to 200 km s^{-1}. The low speeds are also well below the surface escape speed of the stars, and since escape must be reached at some radius, the wind acceleration mechanisms must operate over geometrically extended regions. Another difference from the hot stars is the lack of a strong correlation between mass-loss rate and location on the HR diagram. Two cool stars at nearly the same location can have very different mass-loss properties (Dupree 1986). Thus, properties other than luminosity and effective temperature play a role in the mass loss. Over the years, it was found that stellar pulsations, magnetic field geometries, and different evolutionary states are also important. The pre-main sequence T-Tauri stars have been a major topic of study since the early work by Kuhi (1964). No comparable objects exist in the case of the massive stars, because the formation phases are shrouded by the star-forming clouds, and only the ultra-compact H II regions can be observed directly (Wood and Churchwell 1989).

Given the general triangular shape in the HR diagram of the zone of stellar evolution (Figure 5), there is a much greater range of luminosity and a richer variety of evolutionary phases in the cool star part of the HR diagram. With regards to mass loss, there are two broad classes of outflows that occur for the cool stars: those for which dust grain formation appears to play a role in the mass loss, and others for which there is no evidence for dust. The stars with dusty winds are late M supergiants and stars on the asymptotic giant branch. These can have mass-loss rates as large as those of the hot, luminous stars and values for $\dot{M}$ range from 10^{-6} to more than $10^{-5} M_\odot \, \mathrm{yr}^{-1}$. Radiation driving combined with stellar pulsation are important for the stars with dust. Various aspects of pulsation plus dust-driven mass loss have been reviewed by Bowen (1988), Lewis (1991), Willson and Bowen (1998), and Arndt et al. (1997). For stars other than the luminous late M stars, grain formation does not occur, either because the outer atmospheric temperatures are too high, or the ambient densities are too low. These stars with purely gaseous winds have properties that to first order are explainable by the effects of Alfvén waves.

The accumulative post-main sequence mass loss can determine the terminal state of evolution of intermediate-mass stars. Wiedemann and Koester (1983) showed that essentially all stars with initial masses below about $8 M_\odot$ lose sufficient mass in winds to make possible the evolution to the white dwarf state. If stars with initial masses in the range 2 to $8 M_\odot$ did not reduce to masses below the Chandrasekhar limit of about $1.4 M_\odot$, there would be far more supernovae than in fact occur.

In the remainder of this chapter we will focus on the stars with purely gaseous winds, and especially on the important sector boundary called the "dividing line" (or DL) in the HR diagram. The discovery and exploration of the cool-star DL region was a major accomplishment of space astronomy. Figure 5 shows the cool portion of the HR diagram with several prominent stars identified and the location of the DL at about K1 III that marks a transition in outflow properties.

The cool-star dividing line

Stencel (1978) and Mullan (1978) were the first to draw attention to the fact that there are boundaries in the HR diagram associated with asymmetries in the cores of the strong resonance doublet of Ca II. Linsky and Haisch (1979) coined the phrase "dividing line," after an analysis of their early IUE survey showed the apparent bifurcation in the HR diagram regarding the presence of hot gas with temperatures of about 10^5 K. To the "blue-ward" side of the DL, the stars showed UV emission lines such as C IV $\lambda 1550$. Stars on the "red-ward" side of the DL showed only low ion emission features such as the Mg II doublet. A nearly coincident boundary was also seen in the Einstein X-ray surveys of cool stars by Ayres et al. (1981). Stars blue-ward of the DL showed X-rays, stars to red-ward for the most part did not. The evidence for mass loss in the asymmetric profiles at the center of strong resonance lines was described by Judge and Stencel (1991) (discussed further below). An intermediate class called the hybrid stars was discovered (Hartmann et al. 1980) that showed both strong emission lines and mass-loss properties.

For the stars with purely gaseous winds, the useful conceptual framework of the "solar–stellar connection" was developed, and it provided a guide for studying the magnetic activity and winds of cool giants. Above the Sun's photosphere the outward temperature decrease is reversed in

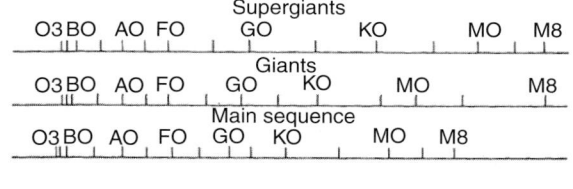

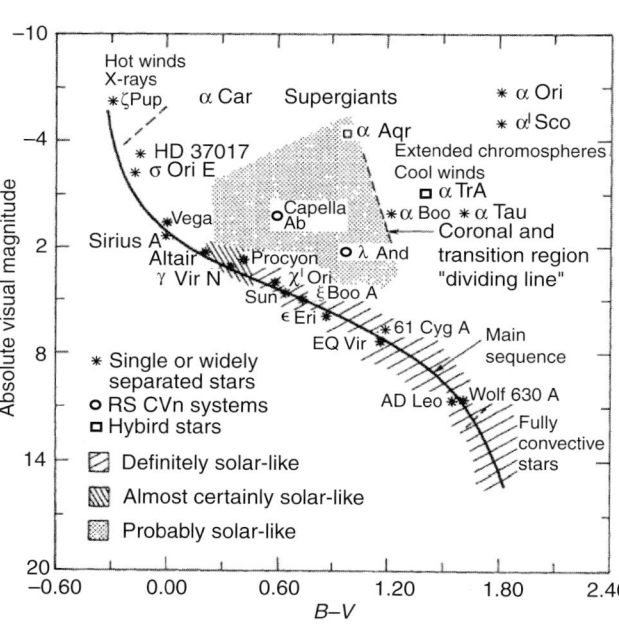

Figure 5 An HR diagram showing various well-studied cool stars and the dividing line (DL) corresponding to an abrupt change from stars with solar-like transition regions to those with extended chromospheres and cool winds. (From Linsky 1985.)

the chromosphere and gives rise to emission in the cores of the resonance lines of Ca II and Mg II. Then above the solar chromosphere, in the transition region, the temperature rises rapidly from about 10^4 K to values above 10^6 K at which X-ray emission is produced. The magnetic field emergent from the Sun threads through all of these layers and gives rise to phenomena generally referred to as "solar magnetic activity." The Sun also has coronal hole regions, which are open field line regions with lower levels of X-ray and transition region emission, from which the fast solar wind originates. Other cool main sequence stars were also discovered to have X-ray emission by the Einstein satellite (Vaiana et al. 1981) and extreme ultraviolet line emission by the EUVE satellite (Mewe et al. 1995). However, observations from neither of these satellites could be considered as evidence for actual mass loss. Skylab, launched in 1973 as NASA's first space station, contained a major solar observatory, the so-called Apollo Telescope Mount. Observations from this mission (Orrall 1981) showed that most of the X-ray luminosity and line emission from the transition region comes from material confined in solar magnetic structures and not from the outflowing wind. Thus, the Sun and presumably other cool main sequence stars have extended closed-field regions, and these confine most of the X-ray-emitting gas.

Even before space telescopes permitted direct spectroscopic studies of winds, there was good indirect evidence for mass loss deduced from ground-based observations. Skumanich (1972) compared of stars in different clusters, and found that during the main sequence evolution of stars with $M_* < 1.3 M_\odot$, the rotation rates decrease with age, and angular momentum loss was known to be a consequence of mass loss as had been shown in the magnetic rotator theory of Weber and Davis (1967).

The nature and extent of the DL in the HR diagram was most thoroughly studied in the survey of X-ray and UV emission lines carried out by Ayres et al. (1995) in their "ROSAT and IUE all sky survey" or RIASS. Nearly 100 stars were examined for the presence of X-ray or emission lines from high stages. ROSAT was sensitive to the presence of gas at temperatures greater than about 10^6 K. The IUE was sensitive to transition region temperatures of the order of 10^5 K. For the main sequence stars, the C IV emission line was found to remain present in all stars from F5 V and later. For single giant stars C IV emission was found to be present out to about K0 III, and for single supergiants C IV persists out to G0 Ib. In regards to X-rays, there is a disappearance of coronal X-ray flux red-ward of the DL that was found in the RIASS by Haisch et al. (1992), as is illustrated in Figure 6. The DL is not as abrupt as implied by the word "line," but there is a fairly narrow region which separates stars of two contrasting behaviors: high temperatures and no clear evidence for mass loss blue-ward of the

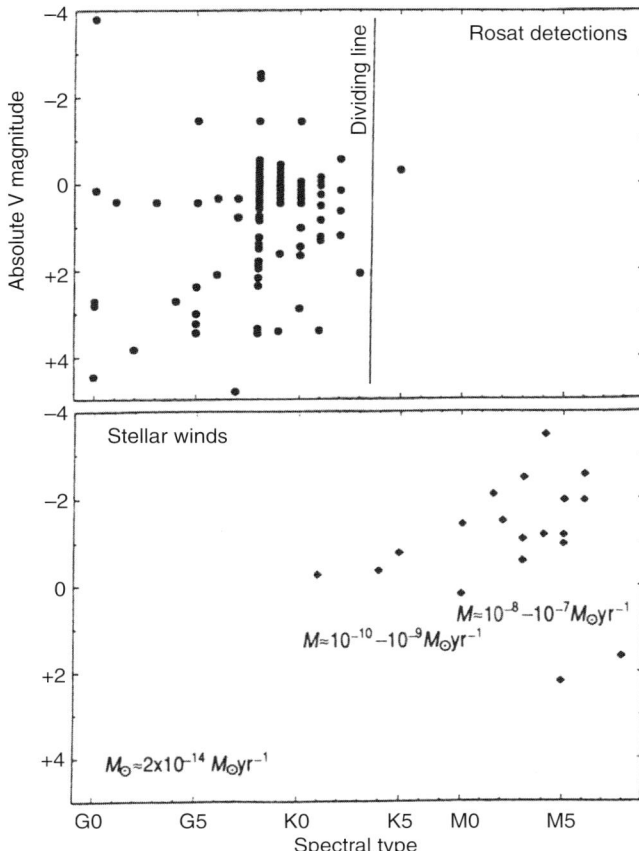

Figure 6 The HR diagram of late-type non-binary evolved stars from the Bright Star catalog. The top panel shows the cutoff of X-ray emission at the DL, and the points in the bottom panel denote stars with measurable stellar winds. (From Haisch et al. 1992.)

DL, and relatively low temperatures plus mass loss redward of the DL.

For binary stars the RIASS results were more complex in that the RS CVn binary systems containing close companions could show vigorous coronal activity even redward of the DL. This was taken to be evidence that the maintenance of rapid rotation by binary tidal effects can sustain the presence of high-temperature phenomena even where single stars would not show it.

Wind properties derived from cool giant stars in binary systems

Some of the best information about cool-star mass loss has been derived from observations of giants and supergiants that have a hotter, less luminous star embedded in the outflowing wind. It was through the analysis of the binary star α Her that Deutsch (1956, 1960) was first able to prove that there is a wind from cool stars. The techniques for studying cool-star winds through the analysis of space observations of binary stars were developed by Reimers (1977). Two especially interesting classes of binary systems were intensely studied. The ζ Aur binaries which consist of a K supergiant and a B-type main sequence star at separations of about 5 to 10 AU, and the VV Cep stars, which are luminous M stars with hot-star companions at much larger separations.

Figure 7 illustrates the advantages of binary stars for studying the winds. The wind of the cool star can be observed in the UV spectra of the hot star because the light from the hot star passes through the wind. Space observations of the ultraviolet flux allow for a clean separation of the continuum of the hotter star from that of the cooler one. The binary systems proved to be especially useful for determining mass-loss rates for cool stars, because we observe the circumstellar line of a dominant stage of ionization in the cool wind seen in absorption in the hot-star spectrum. IUE observations of cool stars with hot binary companions were summarized by Dupree and Reimers (1987).

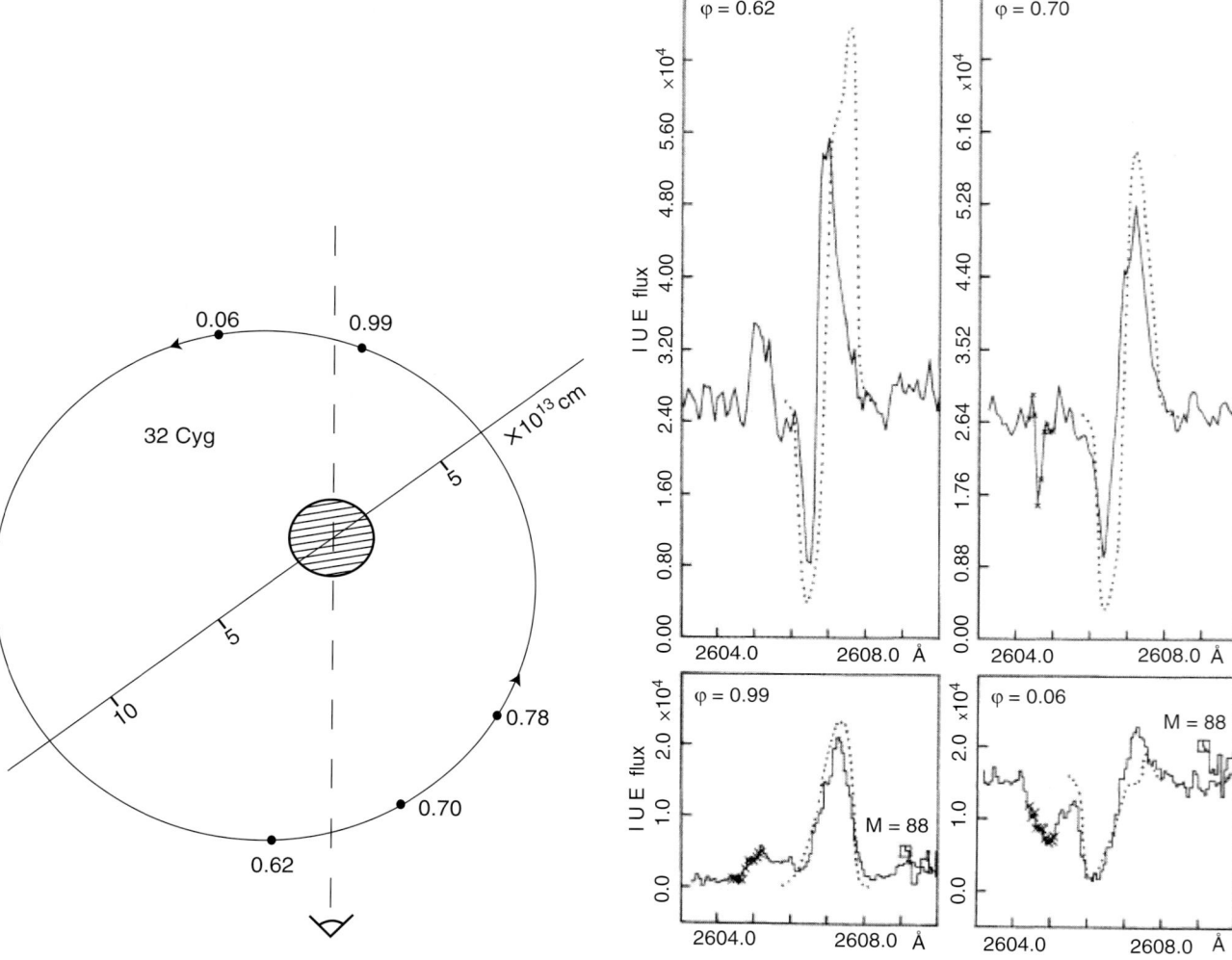

Figure 7 The binary phase dependence of the wind lines in the ζ Aur star 32 Cyg. The left panel shows the location of the B star in the orbit around the K5 Iab star 32 Cyg at the times of four spectroscopic observations. Note that the intervening column density of cool wind material changes with orbital phase. The B star emits a much larger flux of UV radiation, so that absorption by material in the cool-star wind can be detected in the observed spectrum. The right panel shows the profile of the resonance line of Fe II at 2607 Å as seen in the spectrum of the B star through the wind of the K star. The solid lines are the observations. The dotted lines are predictions for a wind model of the K5 Iab star with a mass loss rate of $2.8 \times 10^{-8} M_\odot \text{yr}^{-1}$ and a wind velocity of 60 km s^{-1}. The panel on the right shows that there are large variations in the line profiles depending on the column density of material along the line of sight to the B star. (From Che et al. 1983.)

Mass-loss properties of non-binary cool stars

For the single stars the prime indicators of mass loss are the emission features at the core of the Ca II K and H lines (3933.7 Å and 3968.5 Å) and the analogous Mg II k and h lines in the space UV at 2795.5 Å and 2802.7 Å, respectively. Both doublets form in the chromospheres of these stars. Magnesium has a greater abundance and the Mg II lines have larger optical depths in the chromospheres than the Ca II lines, so the boundary line on the HR diagram for visibility of Mg II was observed to be shifted to slightly lower luminosities and higher temperatures (Linsky 1981). All of the cool giants and supergiants show these emission features, but red-ward of the DL, the emission is stronger and broader, and the profiles can show asymmetries, indicating mass loss.

In contrast with the thin solar chromosphere, the chromospheres in the red giant stars have large thicknesses, corresponding to several density scale heights. Wilson and Bappu (1957) showed that the width of the Ca II core emission is wider for stars with increasingly large luminosities. An analogous correlation was found for the ultraviolet Mg II lines by Kondo et al. (1976). The core emission is symmetric for stars that are blue-ward of the Mg II DL in the HR diagram, but for stars red-ward of the DL the Mg II core emission shows significant asymmetry. Hummer and Rybicki (1968) showed that differential expansion of the line formation region shifts line opacity to shorter wavelengths. This is analogous to having a P-Cygni profile superposed on the emission core of the Mg II lines. The central emission feature thus serves as a pseudo-continuum background on which the outflow absorption feature can appear. Lines from other ions that have such P-Cygni profiles include Fe II, Si II, Al II, Mg II, C II, and O I. For many years there was a large discrepancy from binary star mass-loss rates, but the problem was resolved after it was realized that the higher mass-loss values were based on the resonance lines and these are contaminated by interstellar absorption. Figure 8 illustrates the line profile formation process and the effects of interstellar matter.

Understanding the observational boundaries that comprise the DL line region was a major research goal regarding cool-star winds for many years. A unifying picture was developed by Rosner et al. (1991, 1995). They noted that to explain the DL region, it is not necessary to appeal to discontinuous changes in the mechanical energy generated in the subphotospheric regions of stars as crossing the DL region. Reimers (1989) had shown that the wind luminosities of stars on the mass-loss side of the DL region are comparable to the X-ray plus transition region luminosities on the solar-like side of the DL region. In addition, Goldberg (1986) had found that the mass-loss rates on the red-ward side of the DL region are proportional to the surface area. Hence, it appears that there is no significant change in the mass-driving energy *flux*, that is, energy flow per unit area along the cool star evolutionary track for stars with purely gaseous winds.

Instead of mechanical flux changes, Rosner et al. (1995) proposed that the stellar magnetic fields may be responsible for the DL region changes. Magnetic confinement is required for very hot gas to be present in the atmospheres of late-type giants, because the hot gas cannot be confined by gravity because of the low escape speed of these large stars. The generation of magnetic Alfvén waves can be responsible for behavior on both sides of the DL, as the waves can either heat gas or drive a relatively massive cool outflow. Rosner et al. (1995) suggested that at the DL there is a switch in the magnetic field topology from being

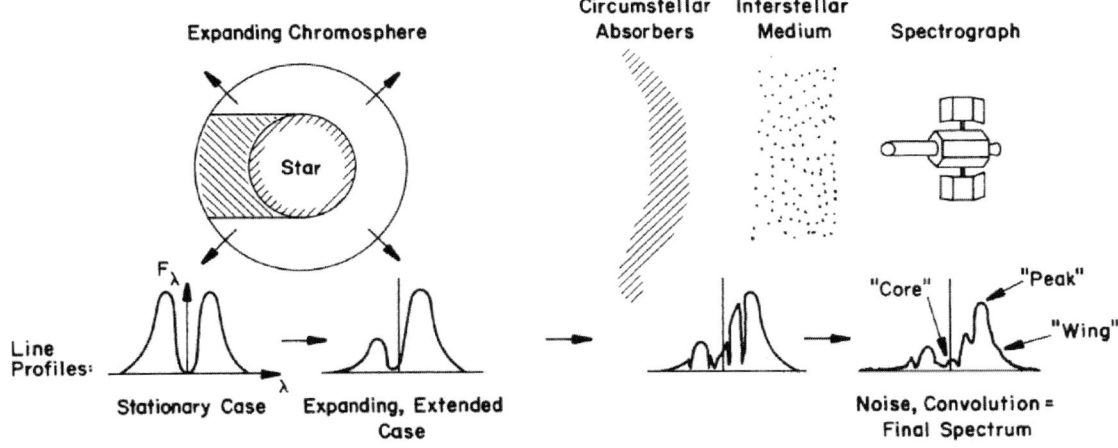

Figure 8 The physical and instrumental effects that influence the observed chromospheric wind features of cool stars. (From Judge et al. 1993.)

predominantly "closed" to one that is predominantly "open." A closed topology can lead to the magnetic confinement of the high-temperature gas, with the heating provided by Alfvén wave energy. An open topology allows for massive outflow to be driven by Alfvén waves.

As the star evolves, the scale of the magnetic field structures at the photosphere goes from large to small values because of changes in the nature of the magnetic dynamo that is operating. Large-scale magnetic structures can be generated by the "α–ω" dynamos, in which the required helicity of the fluid is produced by the combination of convection and differential rotation. As a star evolves to the right on the HR diagram, its rotation rate decreases, making this dynamo mechanism ineffective, and thus to a disappearance of large-scale magnetic fields. Another dynamo mechanism driven by turbulence in the convection zones of stars then becomes dominant. The Sun shows both structures. At times of solar maximum, the solar magnetic fields have field structures with dimensions comparable to the solar convection zone. Whereas at times of minimum magnetic activity, the size of the surface magnetic fields are small and characterized as being "salt and pepper" fields. The magnitudes of the absolute value of the magnetic flux is comparable in both magnetic configurations. The transition across the DL region is thus accompanied by a change in the nature of the surface magnetic field; from being organized on a large scale, to having a less organized magnetosphere that is "open" on large scales.

The study of winds and mass loss has been possible mostly because of observations from space. By the end of the twentieth century astronomers had developed a reasonably good understanding of the basic properties of the winds. Particularly useful have been observational studies of transitions that occur in winds corresponding to a slight displacement on the HR diagram. Examples so far include the superionization and X-ray boundaries and the bi-stability limits for the O and B stars, and the dividing line for cool giant stars. Major questions regarding the roles of magnetic fields, rotation, and pulsation still remain to be answered. However, during the twentieth century we were able to formulate theories that can explain how mass outflows from both the hot and cool classes can be driven.

REFERENCES

Abbott, D.C. and Lucy, L.B. (1985). Multiline transfer and the dynamics of stellar winds. *Astrophysical Journal*, **288**, 679–693.

Adams, W.S. and MacCormack, E. (1935). Systematic displacements of lines in the spectra of certain bright stars. *Astrophysical Journal*, **81**, 119.

Arndt, T.U., Fleischer, A.J. and Sedlmayer, E. (1997). *Astronomy and Astrophysics*, **327**, 614.

Ayres, T.R., Linsky, J.L., Vaiana, G.S., Golub, L. and Rosner, R. (1981). The cool half of the H-R diagram in soft X-rays. *Astrophysical Journal*, **250**, 293–299.

Ayres, T.R., Fleming, T.A., Simon, T., Haisch B.M., Brown, A., Lenz, D., Wamsteker, W., de Martino, D., Gonzalez, C., Bonnell, J., Mas-Hesse, J.M., Rosso, C., Schmitt, J.H.M.M., Truemper, J., Voges, W., Pye, J., Dempsey, R.C., Linsky, J.L., Guinan, E.F., Harper, G.M., Jordan, C., Montesinos, B.M., Pagano, I. and Rodono, M. (1995). The RIASS coronathon: Joint X-ray and ultraviolet observations of normal F-K stars. *Astrophysical Journal Supplement*, **96**, 223–259.

Barlow, M.J. and Cohen, M. (1977). Infrared photometry and mass loss rates for OBA supergiants and Of stars. *Astrophysical Journal*, **213**, 737–755.

Barlow, M.J., Smith, L.J. and Willis, A.J. (1981). Mass-loss rates for 21 Wolf-Rayet stars. *Monthly Notices of the Royal Astronomical Society*, **196**, 101–110.

Beals, C.S. (1929). On the nature of Wolf-Rayet emission. *Monthly Notices of the Royal Astronomical Society*, **90**, 202.

Beals, C.S. (1931). *Publications of Dominion Astrophysical Observatory*, Vol. 4, 271.

Biermann, L. (1951). Kometenschweife und solare Korpuskularstrahlung. *Zeitschrift für Astrophysik*, **29**, 274.

Bjorkman, J.E. and Cassinelli, J.P. (1993). Equatorial disk formation around rotating stars due to Ram pressure confinement by the stellar wind. *Astrophysical Journal*, **409**, 429–449.

Böhm, K.H. and Deinzer, W. (1965). Atmosphären von Zentralsternen planetarischer Nebel. II. Modellatmosphären für die O'Dell- und die Harman-Seaton-Sequenz. *Zeitschrift für Astrophysik*, **63**, 177.

Bowen, G.H. (1988). Dynamical modeling of long-period variable star atmospheres. *Astrophysical Journal*, **329**, 299–317.

Cassinelli, J.P. (1970). Extended model atmospheres for the central stars of planetary nebulae. *Astrophysical Journal*, **165**, 265.

Cassinelli, J.P. and Castor, J.I. (1973). Optically thin stellar winds in early-type stars. *Astrophysical Journal*, **179**, 189–208.

Cassinelli, J.P., Castor, J.I. and Lamers, H.J.G.L.M. (1978). Expanding envelopes of early-type stars – Current status. *Publications of the Astronomical Society of the Pacific*, **90**, 496–505.

Cassinelli, J.P., Cohen, D.H., MacFarlane, J.J., Drew, J.E., Lynas Gray, A.E., Hoare, M.G., Vallerga, J.V., Welsh, B.Y., Vedder, P.W., Hubeny, I. and Lanz, T., (1995). EUVE spectroscopy of epsilon Canis Majoris (B2 II) from 70 to 730 Å. *Astrophysical Journal*, **438**, 932–949.

Cassinelli, J.P., Cohen, D.H., MacFarlane, J.J., Drew, J.E., Lynas Gray, A.E., Hubeny, I., Vallerga, J.V., Welsh, B.Y. and Hoare, M.G. (1996). EUVE Spectroscopy of beta Canis Majoris (B1 II–III) from 500 angstrom to 700 angstrom. *Astrophysical Journal*, **460**, 949.

Cassinelli, J.P., Cohen, D.H., MacFarlane, J.J., Welsh, B.Y. and Sanders, W.T. (1994). X-ray emission from near-main-sequence B stars. *Astrophysical Journal*, **421**, 705–717.

Cassinelli, J.P. and Lamers, H.J.G.L.M. (1987). Winds from hot young stars. In Y. Kondo (ed), *Exploring the Universe with the IUE Satellite*, Reidel, Dordrecht, pp. 139–155.

Cassinelli, J.P. and Olson, G.L. (1979). The effects of coronal regions on the X-ray flux and ionization conditions in the winds of OB supergiants and Of stars. *Astrophysical Journal*, **229**, 304–317.

Cassinelli, J.P. and Swank, J.H. (1983). X-ray spectra of Orion OB supergiants. *Astrophysical Journal*, **271**, 681–690.

Cassinelli, J.P., Waldron, W.L., Sanders, W.T., Harnden, F.R., Jr., Rosner, R. and Valana, G.S. (1981). X-ray emission from Of stars and OB supergiants. *Astrophysical Journal*, **250**, 677–686.

Castor, J.I., Abbott, D.C. and Klein, R.I. (1975). Radiation driven winds in Of stars. *Astrophysical Journal*, **195**, 157–174.

Castor, J.I., Abbott, D.C. and Klein, R.I (1975). Radiation driven stellar winds. In R. Cayrel and M. Steinberg (eds), *Physique des Mouvements dans les Atmosphères Stellaires*, CNRS, Paris, pp. 363–393.

Chandrasekhar, S. (1934b). On the hypothesis of the radial ejection of high-speed atoms for the Wolf-Rayet stars and the novæ. *Monthly Notices of the Royal Astronomical Society*, **94**, 522.

Chandrasekhar, S. (1934a). The radiative equilibrium of extended stellar atmospheres. *Monthly Notices of the Royal Astronomical Society*, **94**, 444.

Che, A., Hempe, K. and Reimers, D. (1983). A study of ultraviolet spectra of Zeta Aur/VV CEP systems. II – Mass loss of supergiants in Zeta Aur, 32 CYG and 31 CYG. *Astronomy and Astrophysics*, **126**, 225–239.

Clayton, D.D. (1968). *Principles of Stellar Evolution and Nucleosynthesis*, University of Chicago Press.

Cohen, D.H., Cassinelli, J.P. and MacFarlane, J.J. (1997). ROSAT PSPC observations of 27 near-main-sequence B stars. *Astrophysical Journal*, **487**, 867.

Cranmer, S.R. and Owocki, S.P. (1996). Hydrodynamical simulations of corotating interaction regions and discrete absorption components in rotating O-star winds. *Astrophysical Journal*, **462**, 469.

Cooper, R.G. and Owocki, S.P. (1994). X-ray emission in wind instability simulations. In A.F.J. Moffat, S.P. Owocki, A.W. Fullerton and N. St-Louis (eds), *Instability and Variability of Hot Star Winds*, Kluwer, Dordrecht, pp. 427–436.

Daltabuit, E. and Cox, D.E. (1972). K-shell photoionization cross-sections. *Astrophysical Journal*, **177**, 855.

Davidson, K. (1999). Basic parameters of η Carinae. In J.A. Morse, R.A. Humphreys, and A. Daminelli (eds), η *Car at the Millenium*, ASP Conference Series 179, Astronomical Society of the Pacific, San Francisco, pp. 6–11.

Deutsch, A.J. (1956). The circumstellar envelope of Alpha Herculis. *Astrophysical Journal*, **123**, 210.

Deutsch, A.J. (1960). The mass loss from Red Giant Stars. In J.L. Greenstein (ed), *Stellar Atmospheres*, Stars and Stellar Systems Series Vol. VI, University of Chicago Press, pp. 543–568.

Donati, J.F., Wade, G.A., Babel, J., Henrichs, H.F., de Jong, J.A. and Harries, T.J. (2001). The magnetic field and wind confinement in beta Cephei: New clues for interpreting the Be phenomenon. *Monthly Notices of the Royal Astronomical Society*, **326**, 1265–1279.

Dupree, A.K. (1986). Mass loss from cool stars. *Annual Review of Astronomy and Astrophysics*, **24**, 377–420.

Dupree, A.K. and Reimers, D. (1987). Mass loss from cool stars. In Y. Kondo (ed), *Exploring the Universe with the IUE Satellite*, Reidel, Dordrecht, pp. 321–353.

Feldmeier, A. (1995). Time-dependent structure and energy transfer in hot star winds. *Astronomy and Astrophysics*, **299**, 523.

Friend, D.B. and Abbott, D.C. (1986). The theory of radiatively driven stellar winds. III – Wind models with finite disk correction and rotation. *Astrophysical Journal*, **311**, 701–707.

Friend, D.B. and Castor, J.I. (1983). Stellar winds driven by multiline scattering. *Astrophysical Journal*, **272**, 259–272.

Friend, D.B. and MacGregor, K.B. (1984). Winds from rotating, magnetic, hot stars. I – General model results. *Astrophysical Journal*, **282**, 591–602.

Gayley, K.G., Owocki, S.P. and Cranmer, S.R. (1995). Momentum deposition on Wolf-Rayet winds: Nonisotropic diffusion with effective gray opacity. *Astrophysical Journal*, **442**, 296–310.

Goldberg, L. (1986). Mass loss. In H.R. Johnson and F. Querci (eds), *The M-type Stars*, CNRS/NASA Monograph Series, Washington, DC, pp. 245–289.

Grady, C.A., Bjorkman, K.S. and Snow, T.P. (1987). Highly ionized stellar winds in Be stars – The evidence for aspect dependence. *Astrophysical Journal*, **320**, 376–397.

Grady, C.A., Bjorkman, K.S., Snow, T.P., Sonneborn, G., Shore, S.N. and Barker, P.K. (1989). Highly ionized stellar winds in Be stars. II – Winds in B6-B9.5e stars. *Astrophysical Journal*, **339**, 403–419.

Haisch, B.M., Schmitt, J.H.M.M. and Fabian, A.C. (1992). Disappearance of coronal X-ray emission in stars with cool dense winds. *Nature*, **360**, 239–241.

Harnden, F.R., Jr., Branduardi, G., Gorenstein, P., Grindlay, J., Rosner, R., Topka, K., Elvis, M., Pye, J.P. and Vaiana, G.S. (1979). Discovery of an X-ray star association in VI Cygni/Cyg OB2/. *Astrophysical Journal*, **234**, L51–L54.

Hartmann, L.W., Dupree, A.K. and Raymond, J.C. (1980). Hybrid atmospheres and winds in supergiant stars. *Astrophysical Journal*, **236**, L143–L147.

Hearn, A.G. (1975). The mass loss from the O 9.5 Ib supergiant, zeta Orionis, derived from the H-alpha profile. *Astronomy and Astrophysics*, **40**, 277–283.

Henrichs, H.F., Kaper, J.A. and Nichols, J.S., (1994). Wind variability in O-type stars. In L.A. Balona, H.F. Henrichs and J.M. Le Contel (eds), *Pulsation, Rotation and Mass Loss in Early Type Stars*, IAU Symposium 162, Kluwer, Dordrecht, pp. 517–529.

Henrichs, H.F., Kaper, L. and Zwarthoed, G.A.A. (1988). Rapid variability in O star winds. In E.J. Rolfe (ed), *A Decade of UV Astronomy with the IUE Satellite*, Proceedings of 10th IUE Conference, ESA SP-281, ESA, Noordwijk, pp. 145–149.

Hillier, D.J. (1991). Theory of Wolf-Rayet atmospheres (review). In K.A. van der Hucht and B. Hidayat (eds), *Wolf-Rayet Stars and Interrelations with Other Massive Stars in Galaxies*, IAU Symposium 143, Kluwer, Dordrecht, pp. 59–73.

Howarth, I.D. and Prinja, R.K. (1989). The stellar winds of 203 Galactic O stars – A quantitative ultraviolet survey. *Astrophysical Journal Supplement*, **69**, 527–592.

Hummer, D. and Rybicki, G. (1968). Redshifted line profiles from differentially expanding atmospheres. *Astrophysical Journal*, **153**, L107.

Johnson, M.C. (1925). The emission of hydrogen and helium from a star by radiation pressure, and its effect in the ultra-violet continuous spectrum. *Monthly Notices of the Royal Astronomical Society*, **85**, 813.

Judge, P.G., Luttermoser, D.G., Neff, D.H., Cuntz, M. and Stencel, R.E. (1993). Line profile variations in M giants – Clues to mass-loss and chromospheric heating mechanisms. *Astronomical Journal*, **105**, 1973–1986.

Judge, P.G. and Stencel, R.E. (1991). Evolution of the chromospheres and winds of low- and intermediate-mass giant stars. *Astrophysical Journal*, **371**, 357–379.

Kaper, L., Henrichs, H.F., Fullerton, A.W., Ando, H., Bjorkman, K.S., Gies, D.R., Hirata, R., Kambe, E., McDavid, D. and Nichols, J.S. (1997). Coordinated ultraviolet and Hα spectroscopy of bright O-type stars. *Astronomy and Astrophysics*, **327**, 281–298.

Kondo, Y., Morgan, T.H. and Modisette, J.L. (1976). The behavior of the MG II doublet emissions near 2800 Å observed in late-type giants. *Astrophysical Journal*, **207**, 167–173.

Kosirev, N.A. (1934). Radiative equilibrium of the extended photosphere. *Monthly Notices of the Royal Astronomical Society*, **94**, 430.

Kudritzki, R.P., Pauldrach, A.W.A., Puls, J. and Abbott, D.C. (1989). Radiation-driven winds of hot stars. VI – Analytical solutions for wind models including the finite cone angle effect. *Astronomy and Astrophysics*, **219**, 205–218.

Kuhi, L.V. (1964). Mass loss from T Tauri stars. *Astrophysical Journal*, **140**, 1409.

Lamers, H.J.G.L.M. and Cassinelli, J.P. (1999). *Introduction to Stellar Winds*, Cambridge University Press.

Lamers, H.J.G.L.M. and Snow, T.P. (1978). Ionization conditions in the expanding envelopes of O and B stars. *Astrophysical Journal*, **219**, 504–514.

Lamers, H.J.G.L.M., Snow, T.P. and Lindholm (1995). Terminal velocities and the bistability of stellar winds. *Astrophysical Journal*, **455**, 269.

Lamers, H.J.G.L.M. and Morton, D.C. (1976). Mass ejection from the O4f star Zeta Puppis. *Astrophysical Journal Supplement*, **32**, 715–736.

Lamers, H.J.G.L.M. and Pauldrach, A.W.A. (1991). The formation of outflowing disks around early-type stars by bi-stable radiation-driven winds. *Astronomy and Astrophysics*, **244**, L5–L8.

Lewis, B.M. (1991). Some effects of dust formation on M and C stars. II – Expansion velocities. *Astronomical Journal*, **101**, 254–267.

Linsky, J.L. and Haisch, B.M. (1979). Outer atmospheres of cool stars. I – The sharp division into solar-type and non-solar-type stars. *Astrophysical Journal*, **229**, L27–L32.

Linsky, J.L. (1985). Nonradiative activity across the H-R diagram – Which types of stars are solar-like? *Solar Physics*, **100**, 333–362.

Linsky, J.L. (1981). Outer atmospheres of late-type stars. In I. Iben and A. Renzihi (eds), *Physical Processes in Red Giants*, Reidel, Dordrecht, pp. 247–261.

Long, K.S. and White, R.L. (1980). A survey of soft X-ray emission from hot stars. *Astrophysical Journal*, **239**, L65–L68.

Lucy, L.B. (1982b). The formation of resonance lines in locally nonmonotonic winds. *Astrophysical Journal*, **255**, 278–285.

Lucy, L.B. (1982a). X-ray emission from the winds of hot stars. II. *Astrophysical Journal*, **255**, 286–292.

Lucy, L.B. and Abbott, D.C. (1993). Multiline transfer and the dynamics of Wolf-Rayet winds. *Astrophysical Journal*, **405**, 738–746.

Lucy, L.B. and Solomon, P.M. (1970). Mass loss by hot stars. *Astrophysical Journal*, **159**, 879.

Lucy, L.B. and White, R.L. (1980). X-ray emission from the winds of hot stars. *Astrophysical Journal*, **241**, 300–305.

MacFarlane, J.J. and Cassinelli, J.P. (1989). Shock-generated X-ray emission in radiatively driven winds – A model for Tau Scorpii. *Astrophysical Journal*, **347**, 1090–1099.

Mewe, R., Kaastra, J.S., Schrijver, C.J., van den Oord, G.H.J. and Alkemade, F.J.M. (1995). EUV spectroscopy of cool stars. I. The corona of α Centauri observed with EUVE. R. *Astronomy and Astrophysics*, **296**, 477.

Milne, E.A. (1924). An astrophysical determination of the average life of an excited calcium atom. *Monthly Notices of the Royal Astronomical Society*, **84**, 354.

Milne, E.A. (1926). On the possibility of the emission of high-speed atoms from the sun and stars. *Monthly Notices of the Royal Astronomical Society*, **86**, 459.

Morton, D.C. (1967). The far-ultraviolet spectra of six stars in Orion. *Astrophysical Journal*, **147**, 1017.

Morton, D.C., Jenkins, E.B. and Bohlin, R.C. (1968). Rocket observations of Orion stars with an all-reflective ultraviolet spectrograph. *Astrophysical Journal*, **154**, 661.

Mullan, D.J. (1978). Supersonic stellar winds and rapid mass loss in cool stars. *Astrophysical Journal*, **226**, 151–166.

Mullan, D.J. (1986). Corotating interaction regions in stellar winds. *Astrophysical Journal*, **283**, 303–312.

Neugebauer, M. and Snyder, C.W. (1962). The mission of Mariner II: Preliminary observations. *Science*, **138**, 1095–1097.

Odegard, N.P. and Cassinelli, J.P. (1982). X-ray luminosities of B supergiants estimated from ultraviolet resonance lines. *Astrophysical Journal*, **256**, 568–577.

Orrall, F.Q. (ed) (1981). *Solar Active Regions: A Monograph from Skylab Solar Workshop III*, Colorado Associated University Press, Boulder, CO.

Owocki, S.P., Castor, J.I. and Rybicki, G.B. (1988). Time-dependent models of radiatively driven stellar winds. I – Nonlinear evolution of instabilities for a pure absorption model. *Astrophysical Journal*, **335**, 914–930.

Owocki, S.P. and Cohen, D.H. (1999). A simple scaling analysis of X-ray emission and absorption in hot-star winds. *Astrophysical Journal*, **520**, 833–840.

Owocki, S.P., Cranmer, S.R. and Gayley, K.G. (1996). Inhibition FO wind compressed disk formation by nonradial line-forces in rotating hot-star winds. *Astrophysical Journal*, **472**, L115.

Parker, E.N. (1958). Dynamics of the interplanetary gas and magnetic fields. *Astrophysical Journal*, **128**, 664.

Pauldrach, A.W.A. and Puls, J. (1990). Radiation-driven winds of hot stars. VIII – The bistable wind of the luminous blue variable P Cygni (B1 Ia/+/). *Astronomy and Astrophysics*, **237**, 409–424.

Pauldrach, A.W.A., Puls, J. and Kudritzki, R.P. (1986). Radiation-driven winds of hot luminous stars – Improvements of the theory and first results. *Astronomy and Astrophysics*, **164**, 86–100.

Poe, C.H. and Friend, D.B. (1986). A rotating, magnetic, radiation-driven wind model applied to Be stars. *Astrophysical Journal*, **311**, 317–325.

Poe, C.H., Friend, D.B. and Cassinelli, J.P. (1989). A rotating, magnetic, radiation-driven wind model for Wolf-Rayet stars. *Astrophysical Journal*, **337**, 888–902.

Prinja, R.K. (1988). Evidence for rotationally modulated variability in O star winds. *Monthly Notices of the Royal Astronomical Society*, **231**, 21P–24P.

Reimers, D. (1977). Observational evidence for mass-loss from K giants, G and K supergiants. *Astronomy and Astrophysics*, **57**, 395–400.

Reimers, D. (1989). Observations of the chromospheres, coronae, and winds of F, G, and K stars. In L.E. Cram and L.V. Kuhi (eds), *FGK Stars and T Tauri Stars*, CNRS/NASA Monograph Series, CNRS, Paris, pp. 53–98.

Rosner, R., An, C.H., Musielak, Z.E., Moore, R.L. and Suess, S.T. (1991). Magnetic confinement, Alfven wave reflection, and the origins of X-ray and mass-loss 'dividing lines' for late-type giants and supergiants. *Astrophysical Journal*, **372**, L91–L94.

Rosner, R., Musielak, Z.E., Cattaneo, F., Moore, R.L. and Suess, S.T. (1995). On the origin of 'dividing lines' for late-type giants and supergiants. *Astrophysical Journal*, **442**, L25–L28.

Saha, M.N. (1919). On radiation-pressure and the quantum theory. *Astrophysical Journal*, **50**, 220.

Seward, F.D., Forman, W.R., Giacconi, R., Griffith, R.B., Harnden, F.R., Jones, C. and Pye, J.P. (1979). X-rays from Eta Carinae and the surrounding nebula. *Astrophysical Journal*, **234**, L55–L58.

Seward, F.D. and Chlebowski, T. (1982). X-ray emission from the Carina Nebula and the associated early stars. *Astrophysical Journal*, **256**, 530–542.

Skumanich, A. (1972). Time scales for CA II emission decay, rotational braking, and lithium depletion. *Astrophysical Journal*, **171**, 565.

Smith, L.F. (1973). Classification and distribution of WR stars and an interpretation of the WN sequence. In M.K.V. Bappu and J. Sahade (eds), *Wolf-Rayet and High Temperature Stars*, IAU Symposium 49, Reidel, Dordrecht, pp. 15–41.

Snow, T.P. and Jenkins, E.B. (1977). A catalog of 0.2 Å resolution far-ultraviolet stellar spectra measured with Copernicus. *Astrophysical Journal Supplement*, **33**, 269–360.

Sobolev, V.V. (1947). *Moving Envelopes of Stars*, Harvard University Press, Cambridge, MA (1960). [Russian Edition, 1947]

Spitzer, L. (1939). Spectra of M supergiant stars, *Astrophysical Journal*, **90**, 494.

Stencel, R.E. (1978). The CA II V/R ratio and mass loss. *Astrophysical Journal*, **223**, L37–L39.

Telting, J.H., Heemskerk, M.H.M., Henrichs, H.F. and Savonije, G.J. (1994). Observational evidence for a prograde one-armed density structure in the equatorial disc of a Be star. *Astronomy and Astrophysics*, **288**, 558–560.

Underhill, A.B. (1949). On the effect of radiation pressure in the atmospheres of early-type stars. *Monthly Notices of the Royal Astronomical Society*, **109**, 562.

Vaiana, G.S., Fabbiano, G., Giacconi, R., Golub, L., Gorenstein, P., Harnden, F.R., Jr., Cassinelli, J.P., Haisch, B.M., Johnson, H.M., Linsky, J.L., Maxson, C.W., Mewe, R., Rosner, R., Seward, F., Topka, K. and Zwaan, C. (1981). Results from an extensive Einstein stellar survey. *Astrophysical Journal*, **245**, 163–182.

Vink, J.S., deKoter, A. and Lamers, H.J.G.L.M. (1999). On the nature of the bi-stability jump in the winds of early-type supergiants. *Astronomy and Astrophysics*, **350**, 181–196.

Waldron, W.L. (1984). Recombination stellar wind model for the coronae of early-type stars. *Astrophysical Journal*, **282**, 256–266.

Waters, L.B.F.M. (1986). The density structure of discs around Be stars derived from IRAS observations. *Astronomy and Astrophysics*, **162**, 121–139.

Waters, L.B.F.M., Coté, J. and Lamers, H. (1987). IRAS observations of Be stars. II – Far-IR characteristics and mass loss rates. *Astronomy and Astrophysics*, **185**, 206–224.

Waters, L.B.F.M., Taylor, A.R., van den Heuvel, E., Habets, G. and Persi, P. (1988). Evidence for low-velocity winds in Be/X-ray binaries. *Astronomy and Astrophysics*, **198**, 200–210.

Weber, E.J. and Davis, L., Jr. (1967). The angular momentum of the solar wind. *Astrophysical Journal*, **148**, 217.

Weymann, R. (1960). Coronal evaporation as a possible mechanism for mass loss in red giants. *Astrophysical Journal*, **132**, 380.

Weymann, R. (1963). Mass loss from stars. *Annual Review of Astronomy and Astrophysics*, **1**, 97.

Wiedemann, V. and Koester, D. (1983). The upper mass limit for white dwarf progenitors and the initial–final mass relation for low and intermediate mass stars. *Astronomy and Astrophysics*, **121**, 77–84.

Williams, P.M., van der Hucht, K. and The, P. (1987). Infrared photometry of late-type Wolf-Rayet stars. *Astronomy and Astrophysics*, **182**, 91–106.

Willis, A.J. (1991). Observations of Wolf-Rayet mass loss (review). In K.A. van der Hucht and B. Hidayat (eds), *Wolf-Rayet Stars and Interrelations with Other Massive Stars in Galaxies*, IAU Symposium 143, Kluwer, Dordrecht, pp. 265–280.

Willson, L.A. and Bowen, G.H. (1998). Pulsation and stellar winds: Some lessons learned from dynamical models of the atmospheres of cool stars. In L. Kaper and A.W. Fullerton (eds), *Cyclic Variability in Stellar Winds*, Springer-Verlag, Berlin, pp. 294–305.

Wilson, O.C. and Bappu, M.K.V. (1957). H and K emission in late-type stars: Dependence of line width on luminosity and related topics. *Astrophysical Journal*, **125**, 661.

Wood, D.O.S. and Churchwell, E. (1989). The morphologies and physical properties of ultracompact H II regions. *Astrophysical Journal Supplement*, **69**, 831–895.

Wood, K., Bjorkman, K.S. and Bjorkman, J.E. (1997). Deriving the geometry of Be star circumstellar envelopes from continuum spectropolarimetry. I. The case of zeta Tauri. *Astrophysical Journal*, **477**, 926.

40

STUART R. POTTASCH*

Planetary nebulae

HISTORY

Before 1917

Several hundred years ago it became apparent to astronomers that other objects were present in the sky besides stars, planets and an occasional comet. These objects had a hazy or nebulous appearance, and it was for this reason that they were called nebulae. The nature of these objects was not clear and it took many years to discover that these objects do not form a homogeneous group, but that several very different classes of objects were being grouped together.

In the eighteenth century telescopes were small and imperfect; the images were not sharp and photography had not yet been discovered. This made it difficult to study the properties of nebulae. A controversy arose as to their nature: did the nebulae consist of many faint stars that were close together or did they consist of a luminous fluid? The regular appearance of the outer regions of globular clusters provided strong evidence for the former point of view, but the wide variety of shapes which were seen implied that some of the nebulae consisted of fluid or gas.

The first attempt to catalogue these objects was made by Charles Messier (1730–1817). He was a comet seeker and his motivation was to avoid confusion between the nebulous objects and comets. His catalogue, published in 1784, contained 103 entries. Four of these are now known to be planetary nebulae. A copy of this catalogue was given to the German-born English astronomer William Herschel (1738–1822), who immediately set to work to observe all these objects with his 30 cm and 48 cm telescopes. He concluded that most of the nebulae could be resolved into stars, and that the Milky Way could also be resolved into individual stars. He attempted to enlarge Messier's list: Herschel discovered 2000 new nebulae in seven years.

In a paper published in 1785, Herschel distinguished a class of nebulae that he considered to be distinct from the rest. He called them 'planetary nebulae' because they vaguely resembled the greenish disk of a planet. He found these objects intriguing. In a paper in 1791, he reported on an observation made the previous year: 'A most singular phenomenon! A star of about the 8th magnitude with a faint luminous atmosphere, of circular form. The star is perfectly in the centre, and the atmosphere is so diluted, faint and equal throughout that there can be no surmise of its consisting of stars; nor can there be a doubt of the evident connection between the atmosphere and the star.' Several examples are shown in Figure 1.

Herschel's argument that the nebulae did not consist of stars was simple. He was certain that the star at the centre and the nebula were associated because a chance coincidence of such a bright star so perfectly centred on the nebula was highly improbable. Thus, the star and the nebula are at the same distance. If the nebulosity is composed of stars they must either be very faint (assuming the central star to be ordinary) or, if they are normal stars, the central star must be of 'enormous size'. Herschel rejected both of these possibilities from which it followed that the nebula was not composed of stars.

Figure 1 shows photographs of a sample of planetary nebulae obtained by the Hubble Space Telescope. While all nebulae are different in detail, all have a generally similar morphology. They are usually symmetric, at least about one axis, and there is always a star at a centrally located position, clearly indicating a physical connection. Occasionally this star is so faint that it cannot be seen above the nebular background and an observer may conclude that it is absent. But careful observations, which suppress background light, will

---

*Rijksuniversiteit Groningen, The Netherlands

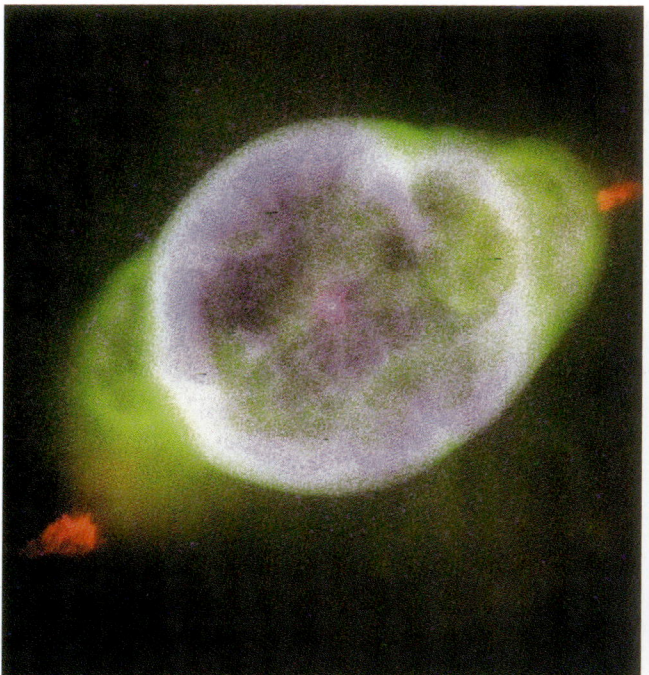

Figure 1 (a) The planetary nebula NGC 3242. The colours indicate the strength of certain lines as the image is a composite of three images of the nebula in different spectral lines. Note the strong red (N II) ansae or fliers at each end of the nebula. The central star is clearly visible. The matter appears to be dispersed in the form of a shell. This type of morphology is quite common.

Figure 1 (b) The planetary nebula MyCnl 8. The morphology of this nebula is unusual. In addition, the central star appears to be offset from the centre. The reasons for this behaviour are unknown.

almost always render the star visible. Thus Herschel's conclusion that nebulae do not consist of stars (other than the central star) applies to the whole class of nebulae, and implies that all are in a similar stage of development or evolution.

By the middle of the nineteenth century further evidence became available which confirmed these nebulae as a separate class. The spectroscope had become available and was being used with telescopes to observe the Sun and stars. Joseph Fraunhofer (1787–1826) had discovered that the Sun emitted a continuous spectrum interspersed with sharp absorption lines. The planets showed many of the same features as the solar spectrum. The stars also showed a continuous spectrum, but each had its own set of absorption lines. In 1859 Gustav Kirchhoff (1824–1887), working in the laboratory of Bunsen in Heidelberg, discovered that certain elements in gaseous form emit lines at just the wavelengths of the solar absorption lines. In this way over 25 elements were identified in the atmosphere of the Sun.

William Huggins (1824–1910) was the first to examine a planetary nebula with a spectroscope. In 1864 he observed the bright nebula in Draco, NGC 6543. Huggins had been observing star spectra for over a year at that time, so that the spectrum he observed was completely unexpected. He found 'a single bright line only'. This bright line provided a means of distinguishing between starlight and gaseous light. The Great Nebula in Andromeda (as an example of 'spiral nebulae') was observed and showed a continuous spectrum indicating starlight. It was clearly a very different object than a planetary nebula. It was thus possible to distinguish gaseous nebulae from nebulae consisting of stars by using a spectroscope.

In 1865 Huggins used a spectroscope with higher resolving power, and was able to resolve the 'single' bright line into three individual lines. One line could be identified with a Balmer line of hydrogen (Hβ), while the other two stronger lines to the red remained unidentified. When it became clear that no element known in the laboratory would produce these lines, they were ascribed to a new element, nebulium. This was not the first new element named in this way. An unidentified line observed in the solar chromosphere during the eclipse of 1859 was ascribed to the then unknown element helium, while a line found in the solar corona during an eclipse 10 years later was ascribed to the element coronium. Helium was identified in the laboratory in 1895. The other two 'new elements' were identified only much later: nebulium in 1927 ([O III]) and coronium in 1939 ([Fe XIV]).

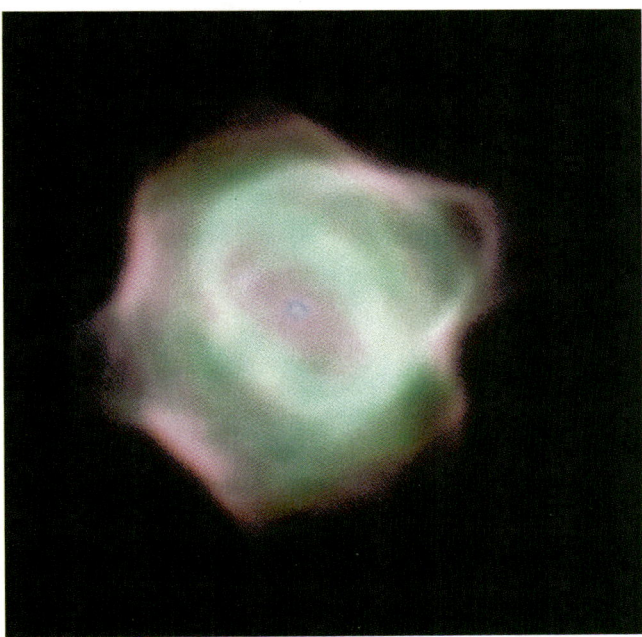

Figure 1 (c) The so-called Stingray nebula. This is a very young nebula in which the central star has been found to increase in temperature over a period of several years. Note the companion star, which is probably physically related.

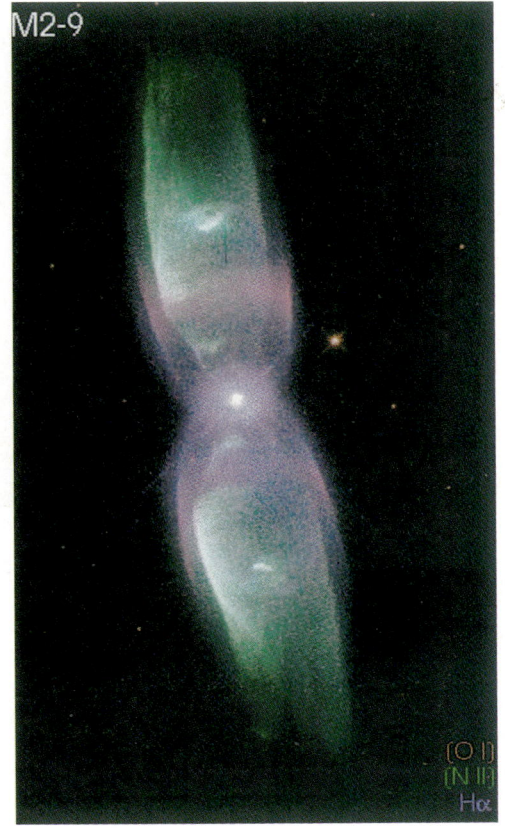

Figure 1 (e) The planetary nebula M2-9. The morphology of this nebula suggests that it is an extreme form of the 'bipolar' type, but the normal nitrogen and helium nebular abundance suggests that this is not true.

Figure 1 (d) Hubble 5. This is an example of a 'bipolar' planetary nebula. About 15% of all nebulae have this form. It is thought that the central stars of these nebulae are of higher mass than the average nebula, and that the nebulae have a higher nitrogen and helium abundance.

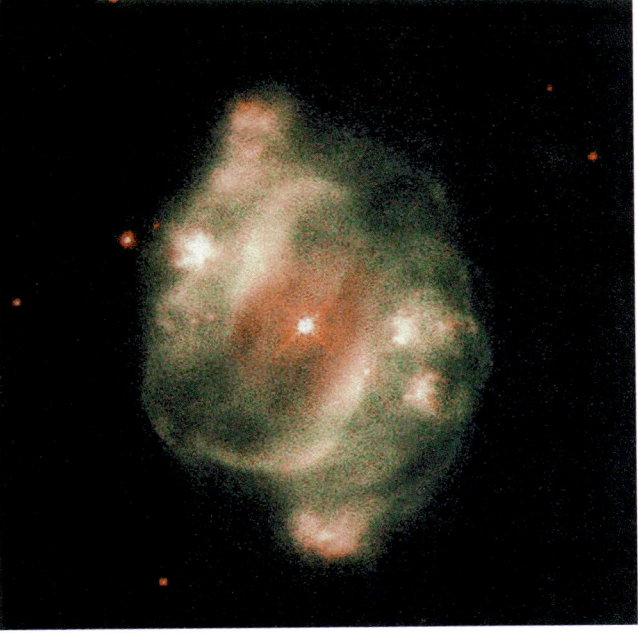

Figure 1 (f) The planetary nebula NGC 5307. An example of a nebula that does not show axial symmetry.

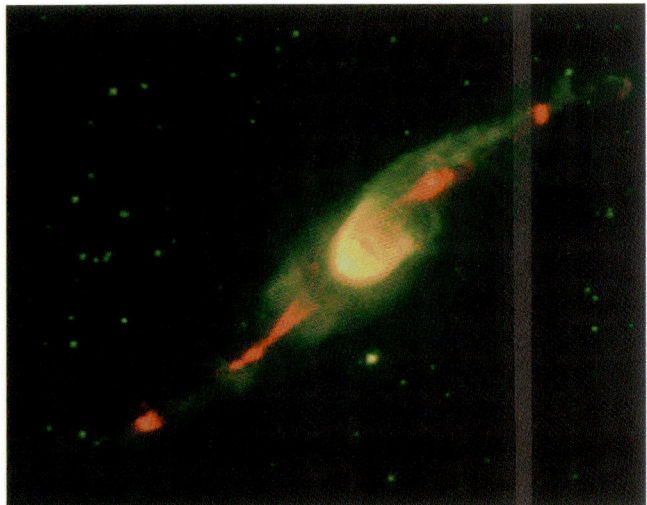

Figure 1 (g) He3-1475 is another very young nebula whose central star temperature has increased by a significant amount in the past 10 years. The origin of its unusual shape is not known.

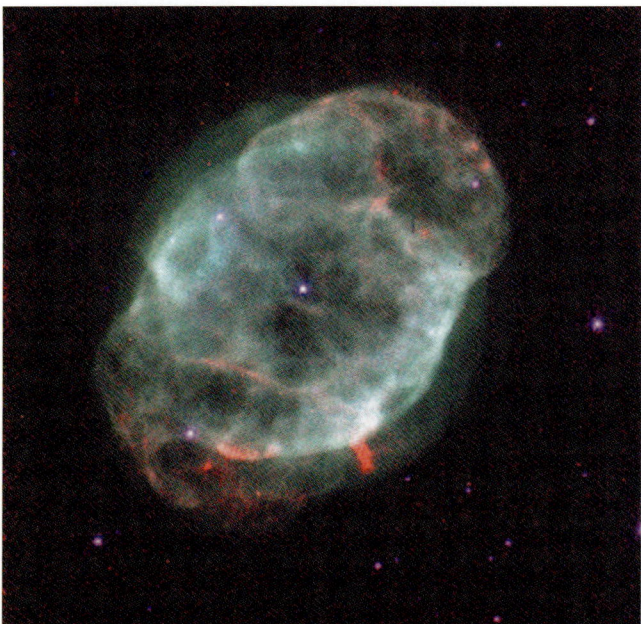

Figure 1 (h) The nebula 1C 4663. The elliptical morphology is quite common, as is the small-scale structure seen in the nebula. Only the central star is related to the nebula; the other stars in the field are foreground or background stars.

The work of Curtis and beyond

In 1918 Heber Curtis published an important paper in the *Lick Observatory Publications*. This is the first publication that attempts to define the status of planetary nebulae in terms of stellar evolution. The paper begins with a thorough summary of all the observational data available, including photographs and drawings of all 78 planetary nebulae then known and observable from the Lick Observatory. Curtis plotted the position of diffuse and planetary nebulae in the plane of the sky and showed that both were to be found close to the galactic plane. However, the 'spiral nebulae' were distributed uniformly except for a 'zone of avoidance' toward the galactic plane. When coupled with the spectroscopic information it was clear to Curtis that both diffuse and planetary nebulae were 'an integral part of our own galactic system' while spiral nebulae were 'very clearly a class apart – not only unconnected with our Galaxy but perhaps individual galaxies'.

On the basis of this information, it was possible to speculate on the place of planetary nebulae in stellar evolution. Planetary nebulae were rare objects: 'fewer than 150 are known in the entire sky. The relative proportion of planetaries to the total number of stars must be of the order of 10^{-5} or less. This minute percentage would seem to stamp the planetary nebula as an exceptional case, a sporadic manifestation of a path which has been rarely followed in stellar evolution.' The only alternative to this conclusion was to regard 'the planetary stage of existence as one of relatively very brief duration, through which the great majority of stars have long since passed.' Adopting the latter hypothesis, the lifetime in the planetary stage was calculable; less than 10,000 years. 'The very short life which must be presumed for the planetary stage of existence ... does not seem inherently probable; it is as yet unsupported by any direct evidence.'

It is somewhat ironic that the 'direct evidence' actually existed. In the same volume in which the work of Curtis appeared was an article by Campbell and Moore which presented observations which are now used to show that planetary nebulae are actually about 10,000 years old. These astronomers obtained spectra at a higher resolution than was previously possible. They found evidence for important line broadening in 23 nebulae and in four cases the broadening was so large that the line was split. It is now known that this splitting is due to an expansion of the nebula with a velocity of about 20 to 30 km s^{-1}. Coupled with the size of the nebula this leads to an age of 10,000 years. But Campbell and Moore did not recognize the correct reason for the line splitting; they attributed it to a rotation of the nebula. The splitting was thought to be caused by matter on the outside of the nebula, which was supposed to be rotating more slowly than the emitting matter and absorbing the central part of the line.

Ten years later, in 1928, this interpretation was shown to incorrect when the nebulium lines were identified with forbidden line radiation. The absorption coefficient for these transitions is very low so that absorption as a cause of the central dark portions of the nebulium lines was out of the question. First Perrine, then Zanstra, argued convincingly that the observed splitting could be better explained by expansion of the nebula. Using rough estimates of the distance to the nebula, it was possible to calculate that the velocity of expansion is much greater than the velocity of escape from the gravitational forces, even if the planetary nuclei have masses as great as 100 times that of the Sun. This has as a consequence, as discussed above, that the planetary nebula stage is very short lived, and that many, and perhaps most, stars go through this stage.

Curtis had additional observational evidence concerning the nature of the planetary nebulae. He measured the average space velocity (from spectral radial velocity measurements) of the nebulae and compared it with the average space velocity of the various stellar spectral classes and the diffuse nebulae. The results were very interesting. The diffuse nebulae have the lowest space velocity, close to zero. The B stars have a somewhat higher space velocity, followed by the A stars, then the F stars and the G, K and M stars. The planetary nebulae have the highest space velocity of all. Curtis noted that the lowest value of space velocity occurs for the hottest stars and increases as the stars become cooler. Looked at in this way the planetary nebulae are an anomaly, since their central stars have the highest temperature and should therefore have the lowest space velocity. Curtis used this information to conclude that planetary nebulae are 'merely sporadic cases of stellar evolution, presumably of cataclysmic origin'. But there is a better explanation for the behaviour of the space velocities.

It took more than 40 years to gain sufficient insight into stellar evolution in order to understand that the sequence of space velocity is only incidentally related to the stellar temperature. It is actually a sequence of stellar ages, in the sense that the diffuse nebulae are the youngest objects, the B stars on the average somewhat older, followed by the A stars, and so on. The planetary nebulae are on average the oldest objects, and are the furthest developed form of stellar evolution. We are witnessing the beginning of the last stage of stellar evolution.

Imaging of planetary nebulae

In Curtis's article about 80 planetary nebulae (PN) were pictured by drawings made at the telescope. The first impression is that they are round or elliptically shaped, but a more detailed study of the morphology shows many differences of form which are not yet well understood. There are many competing phenomena that may contribute to the observed shape, and each type of phenomenon contributes in a different way to the final shape. Some of the nebulae have multiple shells, which are most likely produced by episodic mass-loss events during the final phases of stellar evolution. The deviations from symmetry during the mass ejection, and the (non-)symmetry of the stellar wind after ejection also contribute to the nebular shape. Several mechanisms have been identified which are also capable of producing asymmetry in nebulae: (1) the presence of another star close to the central star which is ejecting matter, (2) the presence of a magnetic field of sufficient strength surrounding the central star and (3) rotation of the central star. The subject is of great interest to many astronomers; so much so that two international conferences with the title 'Asymmetric Planetary Nebulae' have been organized in recent years to discuss the subject. The first took place in Haifa (Israel) in 1994 and the second in Cambridge, Massachusetts, in 1999. In the following we summarize the images which have become available in the past 15 years.

In 1987 Balick published a catalogue of images of about 50 PN. They were usually made in the light of a particular nebular line: $H\beta$, He II, O III, N II and O I were mostly used. Balick discussed these images and constructed an empirical evolutionary sequence of three different types: round, elliptical and butterfly (which later was called bipolar). In the same year Chu *et al.* published a catalogue of nebulae with more than one shell. They discussed the types and morphologies of these PN, with particular attention to the relative distances between the inner and outer shell(s). In 1992 Schwarz *et al.* published a catalogue containing images of more than 250 PN. They used the ESO telescopes in Chile, with filters so that the images were mainly due to either $H\alpha$ or O III. By their geographic position they were limited to the southern sky. A complementary catalogue in the northern sky was published by Manchado *et al.* in 1996, using the telescopes located at Tenerife and La Palma in the Canary Islands. In addition to the filters used by Schwarz *et al.*, they also made images in the light of the N II line. In total they measured 243 nebulae. Together, most of the known PN whose diameters are greater than 4 arcsec are contained in one or another of these catalogues.

The most spectacular images have been made with the Hubble Space Telescope (HST). This 2.4 m telescope was launched on 25 April 1990 in a low-Earth orbit (about 600 km). It was intended to achieve a spatial resolution of 0.1 arcsec, which is an order of magnitude better than a telescope on the ground, whose resolution is limited by atmospheric turbulence. Due to spherical aberration, however, this resolution was not achieved until December 1993 when a service mission placed a corrective optics package on the Faint Object Camera, which obtains the images. (HST was built and launched by NASA, but the European Space Agency built the Faint Object Camera.)

The HST has obtained images of at least 60 PN, and the number continues to increase. Very small nebulae such as NGC 6741 and IC 4997 have been imaged for the first time.

The limitation is on large PN, which cannot fit into the limited field of view. Figure 1 shows eight examples of these images; they are chosen to represent the various morphologies seen. The images are composites of several images made in the light of different spectral lines. The presence of different ions in different structures does not represent an abundance change in different positions; rather it represents different ionization conditions, probably mainly caused by different densities.

Possible causes of the different morphologies have been summarized above, but the origins are not fully understood. Frank (2000) has given a recent summary of the theoretical work.

SPECTRA

Spectra of the nebula

As noted above, it was found that the strongest lines in the nebular spectrum have no counterpart in the laboratory; they are so-called 'forbidden' lines. After the identification by Ira Bowen of the 'nebulium' lines as originating in the lowest energy levels of doubly ionized oxygen (O^{2+}) many other forbidden lines were identified. In spectra taken from the ground lines of O I, O II, O III, N II, S II, S III, Ar IV and Ne III were found to be strong. Besides the forbidden lines, recombination lines from ionized hydrogen and from singly and doubly ionized helium are observed. The forbidden lines are formed by a different process from that of the recombination lines, which accounts for their great strength in gaseous nebulae. They are formed by excitation of the lowest excited levels of the atom or ion by the collision of an electron with it, followed by a spontaneous transition or cascade to the lowest level. The energy of many of these low-lying levels is such that at the electron temperature existing in the nebula, about 10,000 K, a large fraction of the electrons are able to excite these levels. This is necessary to account for the great strength of these lines. At the same time it is necessary that the excited level decays to the ground state radiatively, that is, with the emission of one or more photons. The alternative is that the excited level will be de-excited by an electron collision, in which case no line radiation will be emitted. Which of these processes dominates depends on the ratio of the collisional de-excitation rate to the spontaneous radiative rate. Since the former depends on the electron density, there will always be a value of the density when the two rates are equal. This is called the 'critical density' and its value will vary from line to line. For most lines, however, this density is 10^5 electrons/cm^3 or somewhat greater. Most PN have densities lower than this value, which is the reason that 'forbidden lines' are so strong. It is also the reason that the lines are not seen in the laboratory: even in the best laboratory vacuums the density is still so much higher than the 'critical density' that almost all the de-excitations occur collisionally. Consequently extremely little 'forbidden line' radiation is seen in the laboratory.

The lines of hydrogen and helium seen in nebulae are formed by recombination from the ionized state. Although the recombination process is several orders of magnitude slower than the collisional excitation rate discussed above, this is compensated for by the much larger abundance of hydrogen and helium, so that the line intensities are in general comparable. Recombination processes also take place in ions other than hydrogen and helium. But because the heavier elements are much less abundant, their recombination lines are much weaker. However, modern detector systems are linear, even when intensity ratios of 10^4 are measured, and it should also be possible to obtain abundances from these weak recombination lines. First attempts give somewhat different abundances from those obtained from the 'forbidden lines'. We shall come back to this problem when discussing the chemical composition of the nebulae.

Spectra as measured from the ground

Spectra have a double function. They may be used for studying the kinematics of the gas in the nebula, and they may be used for studying the chemical composition of the gas. For the study of the kinematics a high spectral resolution is required, and it is usually not critical which line is being observed. Ground-based measurements are ideal for this purpose. This is because high-resolution spectrographs are much easier to obtain using ground-based telescopes and the observing time necessary is more readily available. In addition all instruments and detectors are optimized for ground-based instruments, and the exact position at which the measurement is made easier to determine.

For determining the chemical composition, ground-based spectra have limitations. The most important of these is the fact that the number of emission lines that can be observed from the ground is limited. Carbon is not represented by a forbidden line in ground-based spectra so that its abundance cannot be determined. Nitrogen is represented by several lines of N II, which allows the determination of the abundance (relative to hydrogen) of the N^+ ion. But because in many nebulae most of the nitrogen is in the form of N^{2+}, N^{3+} and N^{4+}, a (large) correction factor must be used. This 'ionization correction factor' can be obtained in several ways. For example, the ratio N^+/N^{2+} is often set equal to O^+/O^{2+}, because the ionization potentials are very similar. This can be very uncertain, more so for some ions than for others. Another method is to calculate a model of the nebula. This involves changing the abundance of a particular element until the correct line intensity is produced. But models involve the temperature of the exciting star, the density in the nebula, possible nebular asymmetries, the distance of the nebula and so on, all of which may be uncertain. Thus the theoretical ways of determining the 'ionization correction factor' are all rather uncertain. The best method is the direct measurement of a line of the 'missing' stage of ionization. This usually involves measurements in a spectral region that is only available from space.

Before considering measurements from space, it is useful to mention two shortcomings sometimes found in ground-based observations. First, the observations are made with a slit, and thus refer to a particular region of the nebula. This region is often a bright region, but it may have a spectrum that is not typical of the nebula as a whole. The second shortcoming is that the position at which the spectrum is taken is not given. Both of these difficulties could be resolved by giving the integrated spectrum of the nebula. The only disadvantage of this approach in the determination of the chemical composition is that possible variations in composition cannot be found.

Spectra as measured from space

Ultraviolet spectra

The Earth's atmosphere becomes opaque at wavelengths shorter than about 3100 Å, although most observations from the ground do not extend this far into the ultraviolet. The interesting lines of Ne V, especially the line at 3425 Å, can be measured from the ground. However, the important lines of nitrogen and carbon discussed above can only be measured from space. The first ultraviolet spectrum of a planetary nebula was obtained by the TD satellite, which was launched by ESRO on 12 March 1972. This was a survey instrument that catalogued the important ultraviolet sources with their positions and magnitudes. The measurements were made with a spectrograph, but the signal to noise ratio was so low that integration over a wide spectral region was necessary to obtain a meaningful result. This precluded obtaining line spectra. An exception was made for NGC 6543. This is because NGC 6543 is located close to the ecliptic pole, and the satellite made a measurement of it on each orbit. Thus more than 2000 spectra were measured and could be added. The noise was therefore reduced by a factor of 50, and the strongest lines in the spectrum could be observed. A second set ultraviolet measurements were made in 1974 by the Astronomische Nederlandse Satelliet (ANS), a Dutch/US collaboration. The satellite was built in the Netherlands and launched in the USA. It had three experiments on board: two X-ray instruments (one Dutch and one US) and an ultraviolet instrument built in Groningen. It is only this last experiment which is of interest for planetary nebula research. The instrument could be pointed, and integrated emission was measured using a diaphragm of several arc minutes square and 5 different filters. For one of the wavelengths used, 1550 Å, both a wide- and narrow-band filter was used, so that combining the two measurement gave the strength of the C IV emission line at 1548 and 1550 Å. The emission in the other filter bands was usually dominated by the central star, so that for the first time the ultraviolet continuum of the stars could be measured. Because one of the filters was centred on 2200 Å, the peak of the interstellar extinction, a value for the interstellar extinction in the direction of the planetary nebula could also be deduced. About 40 nebulae were measured with this instrument. The results are reported in two papers (Pottasch et al. 1977, 1978). The most important ultraviolet spectra have been obtained with the IUE (International Ultraviolet Explorer). This satellite was begun by ESRO, but launched by ESA in 1978, in collaboration with the USA and UK. The spectrograph was built in the UK under the direction of R. Wilson, and continued to work for more than 15 years after launch. Spectra were recorded in two wavelength ranges: from about 1170 to nearly 2000 Å (the SWP camera), and from 1900 to about 3250 Å (the LWR and later LWP cameras). There were two possibilities for spectral resolution: low resolution (about 6 Å) and high resolution (about 0.1 to 0.2 Å). The low resolution option was more reliably calibrated and it was possible to use it for much fainter objects. However, quite a few lines were not resolved at low resolution (e.g. the C III lines at 1906 and 1909 Å, whose ratio is an important indicator of electron density). High-resolution spectra are also desirable when the nebula contains a central star that also radiates line emission. Because the stellar lines are usually considerably broader than the nebular lines, the use of high-resolution spectra can determine whether a given line is of nebular or stellar origin. This is of special importance for the helium, carbon and nitrogen lines, which can be emitted both by the stars and the nebulae. High resolution is also necessary in measuring interstellar absorption lines in the spectra of nebulae.

There were two possible diaphragm sizes available. The first was an ellipse roughly 10 by 23 arcsec, while the second was a circle of 3 arcsec in diameter. This latter diaphragm turned out to be too small for most purposes. Jitter prevented it from being useful for isolating the central star from the nebula, and it was too small for use in measuring nebular emission. The spectrograph was of the echelle type, and in each wavelength range an image of the star and nebula was available. When the nebula is larger than several arcseconds, it is possible to separate stellar lines from nebular lines simply from the image. When the nebular is much larger than the diaphragm size considerably longer exposure times are necessary. For a large nebula such as NGC 246 with a diameter of 100 arcsec, it was impossible to obtain a good spectrum of the nebula, while a high-resolution spectrum of the central star was easy to obtain.

Recently, new and more accurate reductions of the IUE spectra have been made and are available to all astronomers via the Internet. These IUE spectra are still being used in all phases of current research, in particular for abundance determinations. Carbon and nitrogen abundances are very uncertain without ultraviolet spectra. A replacement for these spectra is not yet in sight, although the FUSE satellite should obtain spectra in the wavelength range between 900 and 1200 Å.

Infrared spectra

On the red side of the visual spectrum the atmosphere again becomes opaque. This region begins at about $2\,\mu$m and extends to about $600\,\mu$m. The opacity of the atmosphere varies in this region and there are wavelength ranges that can be observed from the ground. An example of such a region is between 8 and $13\,\mu$m. It is in this wavelength range that the first measurements of a planetary nebula (NGC 7027; Gillett *et al.* 1967) showed that an unexpectedly high emission flux exists. In addition, in some spectral regions the opacity occurs low in the atmosphere, so that measurements can be made from highflying aircraft, usually specially equipped for astronomical observations. Such an aircraft is the Kuiper Airborne Observatory, which flies at a height of at least 10 km. At these heights the spectral regions $5-8\,\mu$m and $16-100\,\mu$m become largely available, although there are still small regions that are opaque.

The first infrared spectra measured from a satellite were obtained by IRAS, the Infrared Astronomical Satellite, developed by the Netherlands, the USA and the UK, which was launched in 1984. Proper operation of the focal plane detectors required cryogenic cooling to liquid helium temperature (3 K). This limited the lifetime of the satellite to less than a year. There was a low-resolution spectrograph on board that surveyed the entire sky. Spectra were obtained in the wavelength range $7.6-23\,\mu$m for every object for which a broadband photometric measurement was made. Most of these objects were too faint to yield useable spectra, but 5000 good spectra were obtained, among which were about 60 PN. But because of the low resolution and high noise, only the strongest lines could be observed.

Data from a much more advanced satellite (ISO) are now available. The highest resolution spectrographs on board are the SWS (short-wave spectra) and LWS (long-wave spectra). The SWS measures the spectral region $2.4-43\,\mu$m at a resolution $\lambda/\Delta\lambda$ of about 3000. The LWS measures the spectral region $46-195\,\mu$m with a resolution of about 200. The diaphragm of the SWS varies slightly with wavelength, and is about 15×20 arcsec. This is very similar to that of the IUE. The LWS has a much large diaphragm, about 80 arcsec in diameter.

There are three important advantages of using the infrared forbidden lines. First, these lines originate from energy levels very close to the ground level, and the population of these levels is not sensitive to the electron temperature of the gas in the nebula. Thus the abundances derived from these lines are not temperature sensitive, an important advantage compared to the ultraviolet lines. Second, there are many ions represented in the infrared spectra that have not been seen in other spectral regions, lowering the uncertainty of the 'ionization correction factor' discussed earlier. Third, the extinction in the infrared is small to negligible, which could be of great importance if (a part of) the extinction occurs within the nebula. In addition, the emission for most of the ions observed is not sensitive to the electronic density for the range of densities expected in nebulae. However, for some ions the density does play a role, and these cases must be treated with care.

Spectra of the central star

It has been known for a long time that the central stars are hot. The evidence for this is varied. The intrinsic colours of the stars are very blue; thus very similar to O stars. The spectra contain lines of hydrogen and helium, again indicating that they are hot. Furthermore, they must be hot to be able to ionize the nebula. In spite of this, classifying the stars has proved to be a difficult task. This is partly because the central stars are faint, and partly because nebular emission often contaminates the stellar spectrum. The latter is made more difficult due to the fact that the stellar spectrum often contains emission lines as well as absorption lines.

The earliest classifications looked for similarities of the central star spectra with the spectra of main sequence stars. This led to a classification dominated by O and Of-type stars, Wolf–Rayet stars and subdwarf O stars. To this were added the O VI-type stars (no main sequence stars show O VI). Finally many stars were placed in the category 'continuous spectra' because low-resolution spectra were not able to distinguish the absorption lines which were actually present in the spectra. With the rapid improvement of observations in recent years, insight into the spectra has changed considerably. Roberto Mendez was especially influential in developing this insight. It is clear that the spectra are most influenced by the surface temperature, the surface gravity and the chemical composition of the surface layers. A division is first made according to chemical composition: the hydrogen-rich central star atmospheres are treated separately from the hydrogen-poor (and consequently helium- and carbon-rich) central stars. There are about three times as many central stars in the first category than in the second. The stars in the first group are usually designated as O(H) or Of(H) stars, depending on the presence of emission lines in the star. It has been found recently that some very large (and therefore old) nebulae contain central stars that are already hot white dwarfs. When their spectra contain hydrogen lines they are listed in the category DA or DAO. These stars are at present thought to be the most evolved stars in the hydrogen-rich group.

Most of the members of the helium-rich group have spectra of the Wolf–Rayet type. These stars have spectra almost identical to the high-mass, high-luminosity early-type counterparts. From their spectra it is known that they form a sequence in surface 'temperature', ranging from about 25,000 K (spectral type WC 11) to at least 100,000 K (WC 2). Analysis of Wolf–Rayet spectra are difficult because the emission lines originate in a very extended atmosphere, which is further complicated by the existence of high-velocity mass motions. The other members of the helium-rich group are the so-called PG 1159 stars, which often show non-radial pulsations. Broad, shallow lines of He II and C IV dominate their optical spectra; hydrogen lines are not detected. Roughly half of the stars in this class are variable stars, and these variations are known to be due to gravity-mode non-radial pulsations. Both pulsating and non-pulsating PG 1159 stars are planetary nebula central stars, but not all PG 1159 stars have known nebulae. The reason for this is not clear; perhaps they have had nebulae in the past that have already dissipated. The PN surrounding PG 1159 stars are very large-diameter objects, indicating that they have evolved recently from other spectral types. Because it is likely that the previous stage was also a hydrogen-poor star, it is generally thought that the PG 1159 stars evolve directly from Wolf–Rayet central stars.

The analysis of their spectra plays a very important part in the study of the evolution of nebulae, since it can lead to the determination of the surface temperature, gravity and composition. These quantities are discussed in later sections. It should be borne in mind, however, that the analysis of spectra is continually improving so that the present results may not be the final ones.

SPATIAL DISTRIBUTION

At present about 1500 PN are known to exist in the Galaxy. This is probably between 2 and 5% of the total number which actually exist. Various methods are used to discover new nebulae. These methods use data at many wavelengths, so that sometimes they are limited by extinction (visual observations) and sometimes they are independent of extinction (radio and some infrared observations). Often surveys are undertaken in particular regions of the sky to look for nebulae, while sometimes a nebula is found accidentally while investigating other objects. Figure 2 shows the distribution of PN found by visual observations in galactic coordinates. Notice that most of the nebulae are found within about 10° of the galactic plane (a fact already known to Curtis). Notice also the 'holes' or lack of nebulae very close to the galactic plane, especially near to the galactic centre. This is caused by extinction; infrared and radio observations show that there are indeed many nebulae in these directions. The density of nebulae is seen to increase strongly towards the galactic bulge. Several conclusions can be drawn from these simple observations. First, PN are sometimes bright enough to be seen at the galactic centre. Second, the concentration towards the galactic centre

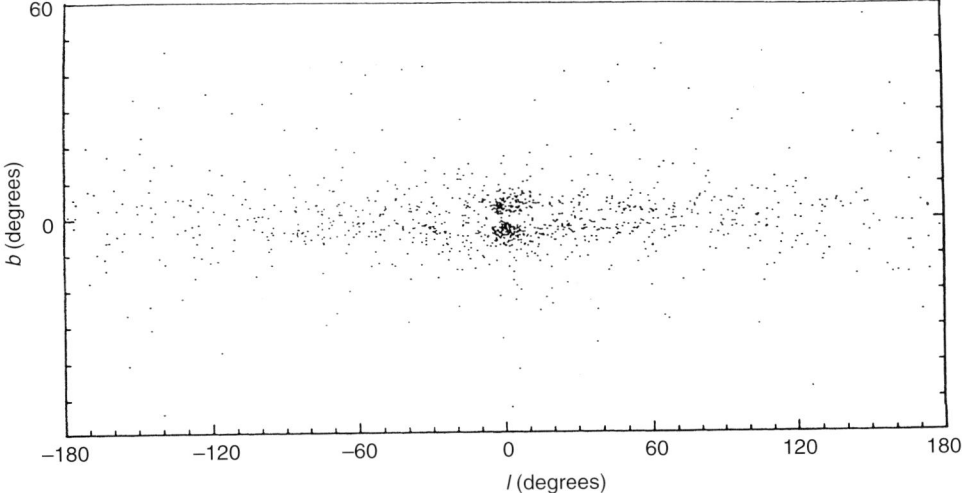

Figure 2 The distribution of planetary nebulae in galactic coordinates found by visual observations. The longitude *l* and the latitude *b* are in degrees. Note the concentration toward the galactic plane and toward the galactic centre. The lack of nebulae very close to the centre is not a real effect, but is caused by the very large extinction near the galactic centre.

indicates an older population, that is, similar to the globular clusters. However, the concentration towards the galactic plane indicates a younger population, that is, similar to the early-type stars. This apparent contradiction is unique for PN, being young and old at the same time. As will be seen, some of the PN are descendants of young, high-mass stars, while others are formed from older, low-mass stars.

Figure 3 shows two histograms of the number of nebulae as a function of the galactic latitude. The upper diagram is for PN which have a diameter of less than 20 arcsec and shows a very pronounced concentration toward the galactic centre. The lower diagram contains those PN with diameters greater than 20 arcsec; it is clear that these are all local objects, since the distribution with galactic longitude is roughly uniform. All PN that belong to the galactic bulge have diameters of less than 10 arcsec.

Reliable radial velocities are known for about 800 PN. Figure 4 shows the radial velocities of a sample of the nebulae as a function of galactic longitude. The same nebulae have been used as in the upper diagram in Figure 3; that is, the smaller nebulae close to the galactic plane. Also plotted are predictions of the expected radial velocities assuming circular orbits at three different distances from the Sun. It can be seen that the observed radial velocities are compatible with circular motion but with a very large dispersion. Thus within about 10° of longitude of the galactic centre there appears to be a rotation of the system of PN with a velocity of about $10.2\,\mathrm{km\,s^{-1}}$ per degree longitude. Superimposed on this rotation is a very high dispersion, much higher than is found in directions other than the galactic bulge. This presumably indicates that the galactic bulge PN are on average considerably older than those found in the spiral arms.

DISTANCES

Obtaining accurate distances to PN has proved to be very difficult. This is because the usual methods for obtaining distances to astronomical objects are not applicable to more

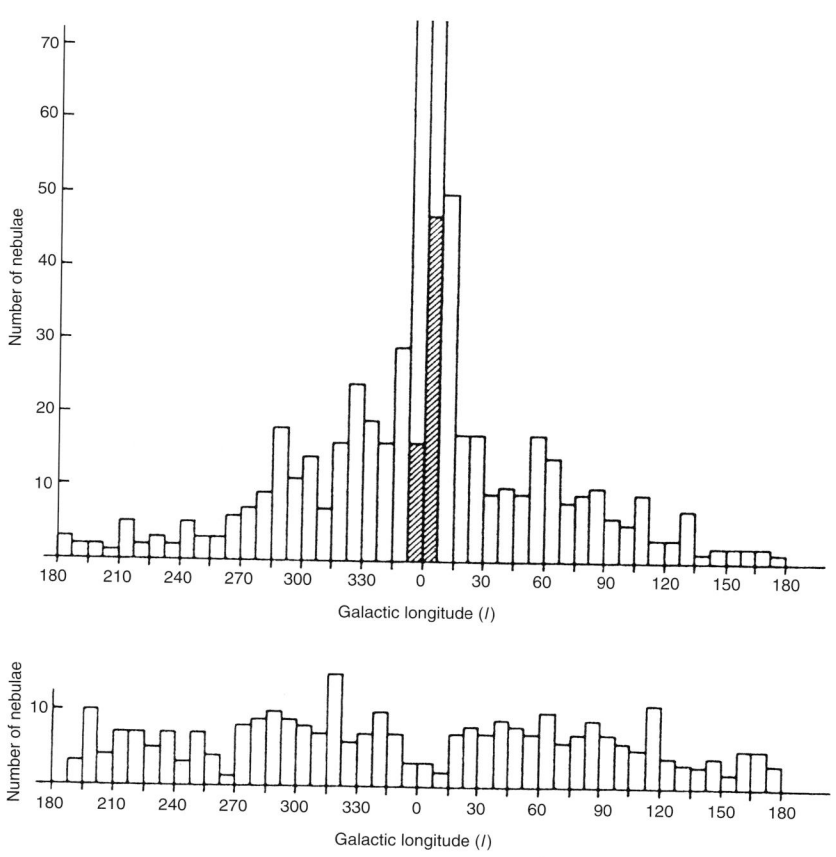

Figure 3 Two histograms of the number of nebulae as a function of galactic longitude. The upper diagram includes only nebulae whose diameters are smaller than 20 arcsec and the concentration toward the galactic centre is clearly seen (the two shaded parts close to the centre should be added to the lighter ones). The lower diagram includes nebulae larger than 20 arcsec. No concentration toward the galactic centre is seen, so that these are all 'local' nebulae.

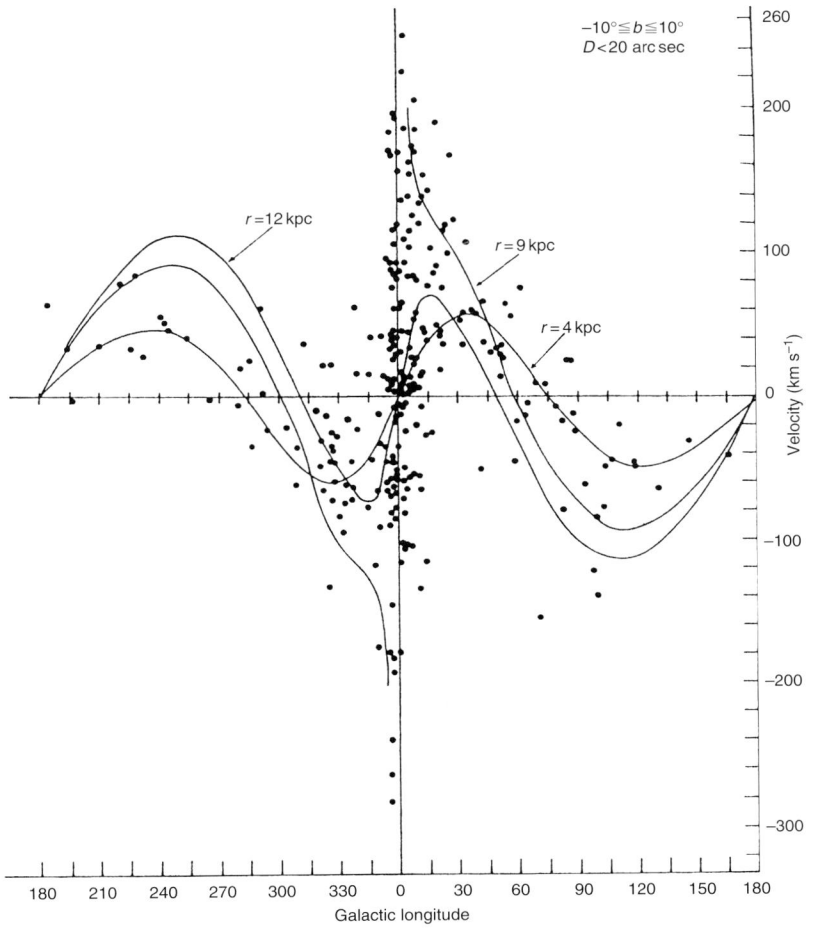

Figure 4 The radial velocities of the nebulae of the upper diagram of Figure 3. The lines show the predicted radial velocities assuming circular orbits at three different distances from the Sun.

than a handful of PN. For the rest, various methods have been found, but it has been difficult to assess their accuracy. The first methods used were statistical methods: it is assumed that all PN have one property in common. An example is that all PN have the same radius, or that they all have the same ionized mass. Then this radius, or the ionized mass, is found from a nebula whose distance is known accurately from some independent method. The difficulty with this method is that it is impossible to find a property of nebulae that does not change as the nebulae evolve. This is obvious for the radius, which increases as the nebula expands. This is not as obvious for the ionized mass, which without doubt increases as the stellar temperature increases, but could remain constant after the entire nebula has been ionized. This is a method that has been in use for more than 40 years: the 'Shklovskii method', named after the Russian astronomer who first suggested it. It assumes that all PN are fully ionized, and all have the same mass. Many who question these assumptions continue to use the method because, as they point out, the distance obtained is only dependent on the ionized mass to a small power (0.4). While this is true, it is a valid argument only if the mass change from one nebula to the other is not very large.

Twenty years ago I computed the ionized mass of those nebulae for which a distance was known by an independent method (Pottasch 1980). The various independent methods are described below. It was found that the ionized mass varied strongly, from a value of about $10^{-3} M_\odot$ for nebulae of small radius (about 2×10^{16} cm) to almost $1 M_\odot$ for PN with radii of 10^{18} cm. Furthermore the correlation between ionized mass and nebular radius is very good. This is what can be expected if the number of ionizing photons remains constant and the nebula is optically thick. This is because the number of recombinations is

$$N_e^2 \Delta V = K$$

where N_e is the electron density, ΔV is the ionized volume and K is a constant if the number of ionizing photons is constant. The ionized mass M_i can be written as

$$M_i \sim N_e \Delta V = K/N_e$$

Since the nebula is expanding and the N_e is therefore decreasing, the ionized mass is increasing. In this way the conclusion was reached that only a small fraction of the total mass is usually ionized, except for the larger nebulae. The smaller nebulae are optically thick for ionizing radiation. The Shklovskii method therefore gives a substantial overestimate of the distance for the smaller nebulae. Although the Shklovskii method is still being used, it is gradually being replaced by more reliable methods. These methods are now discussed.

The most direct method is that of trigonometric parallaxes. Because of the comparatively large distances to PN, rather accurate measurements are required. The group at the US Naval Observatory has now achieved an accuracy of 0.5×10^{-3} arcsec for a single observation, which is an order of magnitude better than that achieved 15 years ago. To date, about 10 central stars have been measured, and about 2 or 3 times this number are close enough that they may be measured in the near future. Results are given by Harris et al. (1997). In addition to ground-based observations, the Hipparcos satellite has obtained fairly accurate measurements of about 5 additional PN (Acker et al. 1998), but these should be checked by ground-based observations.

A second method of obtaining distances can be used for central stars that have physical companions. By measuring the spectral type of the companion, its distance can be determined, assuming that it is a normal star. If the stars are actually physically related, this is also the distance to the nebula. This method is difficult to effectuate from the ground since the companion is often so close to the central star that it is hard to measure an uncontaminated spectrum. There has been a great deal of progress recently by using the HST, since the much improved seeing above the atmosphere extends significantly the number of cases which can be measured. Results are given by Ciardullo et al. (1999).

Distances may also be determined from the 'expansion' method. As a nebula expands its rate of motion appears as a proper motion measurable in arcseconds per year. Spectra obtained at the same time give the expansion velocity directly in kilometres per second. If the expansion is uniform, then the proper motion corresponds to this velocity. This method is an old one, and when measurements were made on photographic plates it was considered an unreliable method. This is because plates taken at two epochs are needed, with a spread in time as long as possible. But the emulsions on the old plate may have shrunk by an unknown amount in the course of time. This is probably not large, but one is looking for a very small effect, of the order of a few milli-arcseconds per year. Determination of the angular expansion rates has become more reliable in recent years due to the use of high-resolution radio frequency measurements. If a very careful comparison of maps made with the same telescope using a baseline of as short as 3 years is done then a reasonably reliable angular expansion rate can be obtained. The further interpretation is not straightforward, however. The increase of the size of the nebula between the two epochs is not only because of the proper motion expansion of the nebula. As the nebula expands its density decreases; consequently, the same number of ionizing photons can now ionize a larger region of space. Thus the apparent increase in the nebular size is due to both the proper motion of the nebular and the increased size of the ionized region. To separate the two effects a model of the nebula must be made. This increases the uncertainty of the result. A discussion of this method, including the latest results, can be found in Terzian (1997).

The fourth method makes use of the spectrum of the central star. An analysis of the spectrum, and in particular of the profiles of the absorption lines of hydrogen and helium, is able to give the properties of the central star's atmosphere: effective temperature, gravity and chemical composition (in this case the helium to hydrogen ratio). Modelling of the atmosphere is not simple, however. The populations of the absorbing atoms and ions are not in local thermal equilibrium (LTE). The non-LTE calculations are difficult because all physical processes must be taken into account. The first systematic discussion of O-type central star atmospheres (Mendez et al. 1988) derived temperatures and gravities using the $H\gamma$ line of hydrogen. Shortly thereafter Napiwotzki (1992) found that no consistent fit to the first four Balmer lines was possible: the higher Balmer lines yielded higher temperatures. For several years thereafter various possible reasons for this were discussed in the literature: effects of stellar winds, magnetic fields, pressure ionization, defects in the line broadening theory and modification of the atmospheric structure due to line blanketing of the heavier elements. It is this last effect which now seems to be the most important: Werner (1996) demonstrated that Stark broadening of C, N and O lines can strongly influence the atmospheric structure. Thus caution must be used in interpreting the results obtained for this method as well as for the other methods.

In Table 1 distances are compared for those nebulae for which the distance has been determined by more than one of the four methods discussed above. There is very little overlap between PN with expansion distances and distances from the other methods so that it is not possible judge the reliability of this method. Most of the overlap is between the model atmosphere distances and the other values. The model atmosphere values for NGC 3242 and 6210 should be ignored because these are early determinations using

Table 1 Distance determinations (parsecs)

Nelula	Trig parallax	Spectro binary	Expansion	Model atmosphere
NGC 246	630	490	570	750
NGC 1360	350			420
NGC 1535		2300		2000
NGC 3242			440	1800
NGC 6210			1570	4800
NGC 6720	700			1090
NGC 6853	380			436
NGC 7293	210			290
A 31	210	440		990
A 33		1160		1260
A 36	240			600
S216	130			185
PW 1	430			695
PHL 932	110			235
K1-27		470		1200

the HB line profile that is likely to be affected by the problem discussed above. For the remainder of the objects there appears to be reasonable agreement, considering that the individual errors are usually about 50%. However, there appears to be a systematic trend of the model atmosphere distances being larger than those determined by other methods. This is probably significant and should be further investigated.

There are other methods for determining distance. One relates to the membership of a planetary nebula of a stellar group, that is, a globular cluster, an open cluster or the galactic bulge. The first two lead to accurate distances but are limited to a very small number of objects. Membership of the galactic bulge can be uncertain. A method that is applicable to those nebulae lying close to the galactic plane is the 'extinction distance'. In this method the extinction is measured as a function of distance in the line of sight to a particular nebula. This is usually done by measuring the extinction and distance to many stars located close to the nebula in the plane of the sky. Then the extinction to the nebula is measured. If it is assumed that this extinction is all in the line of sight (and not local to the nebula) the distance can be determined. The difficulties with this method are twofold. First, it requires many observations, because the spectral type and the extinction to many foreground and background stars must be measured. It is necessary to select the stars so that a wide range of distances is represented but this cannot be done *a priori*. Second, the results show a rather large amount of scatter in the extinction–distance diagram. It is not clear whether this is due to poor measurement of the spectral type and extinction of the stars used, or is due to a very non-uniform distribution of the interstellar dust. It is worth doing a good deal more work on this method, however, since it has the potential of being useful for obtaining distances to many nebulae.

TEMPERATURE OF THE CENTRAL STAR

The effective temperature, defined in terms of the total flux, integrated over all wavelengths, per unit area of the star surface, is extremely difficult to measure. This is mostly because the stellar energy is in the inaccessible ultraviolet, since the stars are hot. Thus it is necessary to specify the spectral distribution of the emergent radiation of the central star (or of a model atmosphere which represents the central star) and then determine the temperature by comparing a specific observation with what is predicted by the model atmosphere. It should be noted that the temperature is independent of the distance to the star. In this section we will summarize the various methods which can be used to estimate the temperature.

Zanstra temperatures

This rather ingenious method for determining the temperature of any hot star surrounded by a tenuous nebula was first discussed by Zanstra in 1931 and applied by him to determine the temperature of the central stars of PN. In his later years, Zanstra referred to this method as a 'poor man's space research', because it enables one to count the number of photons which can ionize hydrogen. These ultraviolet photons cannot penetrate the Earth's atmosphere, hence the reference to space research. However, many of these ultraviolet photons will not even leave the nebula so that space measurements will be of limited value.

The necessary condition for this method to work is that the nebula absorbs all the photons that are able to ionize hydrogen, that is, the optical depth in the Lyman continuum is greater than unity. Thus each photon ionizes one hydrogen atom. This atom then recombines, and in the cascade of the electron one, and only one, Balmer photon (line or continuum) is emitted. By counting the number of Balmer photons one is therefore counting the number of photons emitted shortward of the Lyman limit at 912 Å. A temperature can then be determined by comparing this number with the number of photons emitted in the visual, as found from the magnitude of the star or by some other method. This ratio is a unique measure of the temperature for a specified type of model atmosphere. A blackbody spectral distribution is usually used, and seems to give a reasonable first approximation for the temperature, but any type of model atmosphere may be used. The resulting temperature is usually written as $T_Z(H)$, the hydrogen Zanstra temperature. In practice it is not necessary to measure and add up all the Balmer photons. A single Balmer line is enough, as theory can supply the ratio of a single Balmer line to the total Balmer emission.

The same reasoning, in combination with a He II emission line (4686 Å) instead of a Balmer line, yields the number of photons shortward of 228 Å, if it is assumed that all these photons are absorbed in the nebula. As in the discussion of hydrogen, it is necessary to assume a model atmosphere to interpret the ratio of the number of ultraviolet photons to visual photons in terms of a temperature. The ionized helium temperature is usually denoted as T_Z(He II). The hydrogen and helium temperatures should be the same if the correct model atmosphere has been used. The most often used model is a blackbody. This usually gives the same value for the hydrogen and helium temperatures when a high temperature, about 10^5 K, is found. This indicates that a blackbody appears to be a reasonable approximation for the high-temperature stars.

For the lower temperature central stars the ionized helium Zanstra temperature is often considerably higher than the hydrogen Zanstra temperature. This could indicate that a blackbody is not the correct model for these stars, but it could also indicate that the basic assumption in deriving the temperature, that the nebula is absorbing all the photons emitted by the star, is not being fulfilled. It is not always clear which of these possibilities is the correct one, which has led to a controversy in the literature. Probably no single answer can be given. We shall return to this problem later.

Stoy or energy balance temperatures

This method of determining temperature is named after the South African astronomer who first discussed it. 'Energy balance' is a more descriptive term. In this method the observational quantity measured is the ratio of the sum of the energy in the 'forbidden' lines to the energy emitted in one of the hydrogen lines. In this context 'forbidden' line is used to designate all transitions, both forbidden and permitted, formed in the nebula by electron collisions. These transitions are usually optically thin so that they escape from the nebula, and are the most important source of energy loss. Energy input in the nebula takes place by ionization, primarily of hydrogen and helium. The ionizing photon gives up that part of its energy required for ionization; the energy which it had in excess of this amount goes into the motion of the ejected electron. This electron then shares its energy with the other particles by elastic collisions. This is how the electrons are heated.

The ratio of the 'forbidden lines', F(FL), to Hβ is therefore the average value of the excess energy of an ionizing photon. The higher the temperature of the star, the higher is the value of this ratio. The precise value will depend somewhat on the spectral distribution in the star and, for the higher temperatures, on the amount of helium present and its ionization. Also the optical depth in the nebula shortward of 912 Å will play a role, but it is not of such critical importance as it is in the Zanstra method, where not registering a photon affects only the inferred number of ionizing photons, leaving the number of visual photons unchanged. In the Stoy method, not registering a photon affects both the 'forbidden' line flux and the Hβ, in only slightly different ways.

In order to apply this method the entire spectrum must be measured. For high-temperature stars most of the energy loss will occur in the ultraviolet, and this part of the spectrum must be well measured. For low-temperature stars an important fraction of the energy is lost in the infrared so that measurement of this part of the spectrum, especially the Ne II line at 12.8 μm, is important. This method is therefore only applicable to the best-observed nebulae. When applied to nebulae for which the complete spectrum has not been observed, the results are quite uncertain.

Temperatures from model atmosphere analysis

Because the model atmosphere analysis yields not only temperature but gravity and chemical composition as well, this method and its uncertainties is discussed in the section concerning distances. The method fits the predicted line profile to the observed one. Thus the method is at present limited to those central stars which have absorption line spectra.

Comparison of the various methods

Table 2 summarizes the resultant temperatures for a selection of central stars; these were chosen to illustrate a range of temperatures. There are not many more central stars with reliable determinations of temperature by several methods. Several things may be deduced from the table. First, the high-temperature central stars, the last five items in the

Table 2 Central star temperatures (K)

Nebula	T_Z(H)	T_Z(He II)	T(spect)	T(Stoy)
IC 418	36,000		36,000	36,000
IC 3568	34,000	50,000	50,000	52,000
NGC 6826	34,000		50,000	40,000
NGC 6891	30,000		50,000	51,900
NGC 3242	60,000	82,000	75,000	62,000
NGC 1535	38,000	65,000	70,000	87,000
NGC 2392	27,000	67,000	47,000	109,000
NGC 6720	110,000	121,000	101,000	80,000
NGC 6853	112,000	134,000	109,000	95,000
NGC 7293	116,000	102,000	104,000	
NGC 246	40,000	86,000	140,000	
NGC 6445	18,0000	180,000		180,000

table, have values from the different methods that agree with each other (with one exception) within the uncertainties. This indicates that (1) the nebulae are absorbing all the ionizing photons and (2) a blackbody representation of the energy distribution is adequate for these stars. The single exception is NGC 246 for which it appears that the nebula is too diffuse to absorb all the ionizing photons, so that the Zanstra method gives only a lower limit to the temperature. Second, the low-temperature central star of IC 418 appears to be well represented by a blackbody. Third, the other five central stars, with the possible exception of NGC 3242, have lower Zanstra hydrogen temperatures than would be expected on the basis of the other temperatures. This is quite common for stars in this temperature range. It is not clear whether this is caused by departures from a blackbody energy distribution or whether some of the photons that could ionize hydrogen are not absorbed by the nebula. NGC 2392 is an exception; in this case there is other evidence that the temperature variations are caused by departures from blackbody radiation. The continuum radiation over a large spectral range (1500–7000 Å) is known for this star and it indicates a temperature of not more than 35,000 K. The ionization in the nebula is quite high (Ne V lines are seen) indicating the presence of 100 eV photons in substantial numbers. The central star of NGC 2392 must be seen as an extreme example, however. Nevertheless it is likely that departures from blackbody energy distribution play a substantial role in stars of 'temperature' below 10^5 K.

DUST IN PLANETARY NEBULAE

In 1967, 10 μm continuum emission from NGC 7027 was first measured. The radiation was considerably stronger than had been expected; the nebular continuum is expected to increase slowly at longer wavelengths due to increased free–free emission, but the measured increase was substantially in excess of this prediction.

The continuous spectrum is now known past 100 μm for at least several hundred nebulae. An example is shown in Figure 5 for the nebula NGC 6572. It can be seen that a strong emission peak occurs near 60 μm that is about 10^3 times stronger than the expected free-free nebular emission. This emission can be approximately fit by a blackbody with a temperature of 180 K. This far-infrared emission is found in every planetary nebula; the peak wavelength of the emission varies somewhat so that the blackbody fit varies from 60 to 250 K. The total energy in this far-infrared radiation is substantial: it may amount to 10% of the total energy emitted by the central star. It appears to originate in the same region of space as the ionized gas. As Figure 5 shows, atomic processes in the nebula cannot explain the infrared radiation.

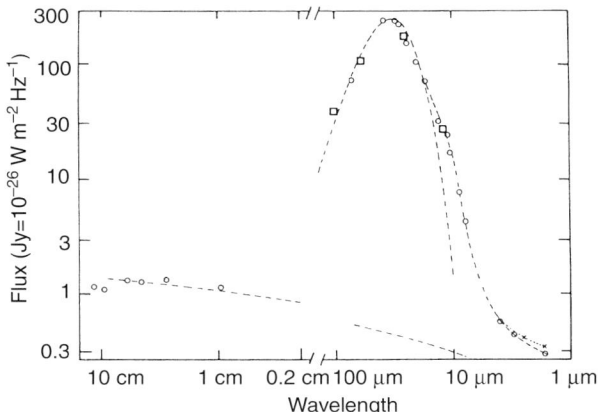

Figure 5 Flux density of the nebular continuum emission as a function of wavelength for the nebula NGC 6572. The circles, squares and crosses are the measurements (the data represented by squares were measured with the IRAS satellite). The lower dashed line is the continuous emission expected from atomic processes, mainly free–free emission. The observed 60 μm emission is seen to be almost three orders of magnitude higher. The emission at radio wavelengths, however, is that expected from atomic processes.

Dust

There are several reasons for believing that the emission is produced by small particles which, without defining them in any more detail, will be referred to as 'dust'. Dust is known to be a component of the interstellar medium, where it is identified by the extinction and polarization that it causes. To be an efficient cause of extinction, or an efficient emitter, the particle size must be of the same order as the wavelength. Thus the particle size must be 10^{-3}–10^{-4} cm. The particles cannot be very much bigger because in that case the volume to area ratio becomes very large. This will lead to there having to be an impossibly large amount of material to produce the emission.

Composition of the dust

Various features are present in the infrared spectra. These same features are also found in the spectra of other astronomical objects such as young stars, H II regions, the interstellar medium and in other galaxies as well. The features are recognizable since they are considerably broader than atomic or ionic lines. The most important of these bands in the spectral region to 20 μm are at 3.3, 5.2, 6.2, 7.7, 8.6, 11.2, 12.6 and 17.0 μm. These features are ascribed to C—C and C—H stretching and bending vibrations in aromatic hydrocarbon material. Emission features are also present at 9.7 and 18 μm that are attributed to silicates. In addition features have been found further in the infrared such as those at 32.8, 33.6, 40.4 and 43.1 μm. Some of these features have been tentatively identified with crystalline silicates, such as olivine and pyroxenes, with various mixtures of iron and magnesium. The study of these features is continuing both in the laboratory and in the nebulae. While the final word has not yet been said, it is clear that a substantial amount of silicon, iron, magnesium and perhaps other metals, together with oxygen and carbon are present in the dust.

CHEMICAL COMPOSITION OF THE NEBULAE AND THE STARS

The determination of the chemical composition has implications that extend beyond the evolution of PN themselves. For example, it is probable that the production of nitrogen in the Galaxy occurs in low- or intermediate-mass stars on their way to becoming PN. In the nebulae the nitrogen enrichment can be measured. The chemical composition of the nebulae is known in much more detail than that of the central star atmosphere. The composition below the surface of the star can only be inferred, either with the help of non-radial pulsation measurements, or by theoretical considerations. Even the composition of the nebula presents some problems, however, which have been alluded to above. Each of these subjects will be discussed separately.

Composition of the nebulae

It is at least 60 years since the first calculations of the composition of nebulae, and even for the best-observed nebulae problems still exist. The observed forbidden lines are used to determine the abundance of the ion from which they are emitted. The line is formed by an electron collision with the ion, which raises it to an excited level, followed by a radiative transition. The following problems arise:

1. There are uncertainties in the atomic parameters, especially the cross-section for electron collisional excitation.
2. There are uncertainties in the correction for extinction, or the extinction occurs within the nebula.
3. There are uncertainties in the electron temperature, or temperature fluctuations in the nebula.
4. There are uncertainties in the density, or density fluctuations in the nebula.
5. Not all ions of a given element are measured. Sometimes the most populous ions are missing.

Concerning point 1. New values of the collision cross-sections have become available in the past few years (the IRON project). It is estimated that the cross-sections that have been calculated are accurate to within 25%. From nebular observations these cross-sections usually give consistent results: only in the case of the Ne III lines at 3868 and 3967 Å are there indications that the error is considerably larger.

Concerning point 2. Extinction values are usually obtained from either the ratio of the radio continuum flux density to the Hβ emission or from the Balmer decrement. Sometimes the additional absorption at 2200 Å is used to find the extinction. All these methods require the assumption of a standard extinction law in all directions and the assumption that none of the extinction occurs within the nebula itself. When any two of these methods give a different extinction value (which occurs not infrequently) one should question whether this difference is really outside the observational error before interpreting the results using a non-standard extinction law.

Concerning point 3. The use of (far) infrared lines can minimize this problem, because the upper level of these lines is very close to the ground level, and therefore not temperature sensitive. The use of these lines minimizes the extinction problem as well.

Concerning point 4. The density usually does not enter into the abundance determination with respect to hydrogen because the hydrogen line has the same density dependence as the forbidden line. The only exceptions to this are lines where the transition probability is very low so that collisional de-excitation can occur. This will occur for a few of the far-infrared lines of nitrogen and oxygen, and is more important in high-density nebulae.

Concerning point 5. The correction for unseen ionization stages has always been a problem. The usual methods of making the correction make use of similarities in ionization potentials of various ionization stages of different elements (e.g. $N^{2+}/N^+ = O^{2+}/O^+$). Another possibility is to construct a model of the nebula, assuming reasonable values for the central star temperature, the nebular density and the nebular geometry, and predict what the ratios of the various ionization stages of a given element are expected to be. But this method is only as good as the assumptions that are used in the model. Sometimes this is unavoidable. The best solution is to measure the spectrum over as wide a spectral range as possible, so as to include the maximum number of ionization stages. Thus inclusion of ultraviolet and infrared measurements can substantially improve the accuracy of the abundance determination. In this case the accuracy of the abundance is estimated to be about 50%, when at least several ionization stages have been measured. The helium abundance is more accurately known, probably to within 10%. Results for four bright, well-studied PN are presented in Table 3, and compared to the solar abundance and the abundance in the interstellar medium (ISM). The ISM abundance is the average of the abundance in the well-studied cloud in the line of sight to ρ Oph and the Orion Nebula.

Several things may be noted from Table 3. There are elements whose abundance in PN is substantially lower than in the Sun (which presumably has an abundance representative of stars formed 5 billion years ago). These elements are Al, Ca and Fe. The reason for this is probably that these elements are present in the dust and are only found in small amounts in the gas. They are also depleted in the ISM, which also has an important dust component. Two other elements that are expected to be present in substantial amounts in the dust are Si and Mg. These two elements are only marginally depleted in PN, by a factor of two or three. In contrast their depletion is much higher in the ISM. The elements Na and K are also only marginally depleted in PN relative to the Sun, while they are significantly depleted in the ISM. The dust in the ISM must have a different equilibrium with the surrounding gas than in PN. S and Cl apparently are not constituents of the dust, judging by the good agreement of their abundance in the Sun and the ISM. Their abundance in PN is in reasonable agreement with the solar abundance indicating that these elements are not produced in the course of the evolution of the central stars.

It is expected that the compounds that form the dust contain O, C and perhaps N in addition to the elements discussed above. The amount of these elements attached to the dust is limited by the abundance of Si, Mg, and so on, and therefore it is not expected that more than about 20% of these elements will be present in the dust. This seems to be confirmed by the reasonable agreement of the abundance of these elements in the Sun and the ISM. The oxygen abundance in PN seems to agree roughly with its abundance in the Sun and ISM (with the possible exception of NGC 6302) indicating that its abundance has not changed in the course of its evolution. However, the nitrogen abundance is clearly higher in most of the nebulae, and it is clear that the N/O ratio has increased in the course of evolution. This is especially clear in the case of NGC 6302, which is one of a small number of such nebulae where there is more nitrogen than oxygen. For the nebulae listed, carbon appears to have the solar value except in NGC 6302 where it is substantially lower and may have been destroyed in the nuclear processes which have occurred. The C/O ratio appears to be somewhat higher than solar in NGC 7027, but it is not clear whether this is an initial condition or is caused by evolution. The somewhat higher neon abundance in these PN is marginal and awaits confirmation. The argon abundance seems to be solar.

Composition of the central star atmospheres

The analysis of the atmospheres of central stars is difficult for at least two reasons. First, the atmospheres are not in LTE and thus non-LTE analysis must be applied. This means that all physical processes that are important must be taken into account, and their rates must be correct. Second, these stars often have extensive winds that contribute to the line profiles (either in absorption or in emission) and which must also be taken into account. Therefore precise values of atmospheric abundances are difficult to determine, but approximate values are sufficient for drawing interesting conclusions.

There appears to be a contrast in the atmospheric abundance of helium and hydrogen. There are stars whose atmosphere consists mainly of hydrogen, and there are stars

Table 3 Abundances in planetary nebulae

Element	NGC 7027	BD+30	NGC 6302	NGC 6445	Sun	ISM
He/H	0.11		0.15	0.14	0.1	
O/H	4.1 (−4)	3.8 (−4)	2.3 (−4)	7.4 (−4)	7.4 (−4)	3.5 (−4)
N/H	1.5 (−4)	1.2 (−4)	2.9 (−4)	2.4 (−4)	0.93 (−4)	0.75 (−4)
C/H	4.5 (−4)	2.0 (−4)	0.60 (−4)	4.8 (−4)	3.6 (−4)	2.1 (−4)
Ne/H	1.1 (−4)	1.9 (−4)	2.2 (−4)	2.1 (−4)	1.2 (−4)	0.81 (−4)
Na/H	1.2 (−6)		2.6 (−6)		2.0 (−6)	0.23 (−6)
Mg/H	2.2 (−5)		1.3 (−5)	1.7 (−5)	3.8 (−5)	0.11 (−5)
Al/H			1.0 (−8)		300 (−8)	
Si/H	1.1 (−5)		2.3 (−5)	0.7 (−5)	3.6 (−5)	0.28 (−5)
S/H	9.5 (−6)	5.5 (−6)	7.8 (−6)	8.0 (−6)	18 (−6)	16 (−6)
Cl/H	1.1 (−7)	1.4 (−7)	3.4 (−7)		1.9 (−7)	1.2 (−7)
Ar/H	0.69 (−6)	0.6 (−6)	6.0 (−6)	3.8 (−6)	3.6 (−6)	1.8 (−6)
K/H	5.2 (−8)		8.5 (−8)		14 (−8)	1.1 (−8)
Ca/H	1.0 (−8)		7.4 (−8)		219 (−8)	0.04 (−8)
Fe/H	3.0 (27)		8.7 (27)		323 (27)	6.7 (27)

The numbers in parentheses indicate the power of 10 by which the previous number should be multiplied.

whose atmosphere consists mainly of helium, with little or no measurable hydrogen. These different sequences are usually identifiable from their spectral type: O-type stars are in general hydrogen rich, while stars with a Wolf–Rayet-type spectrum, or the PG 1159 stars, almost always consist mainly of helium. This is reminiscent of white dwarf stars, which show the same contrast. Apparently this difference in abundance begins early in the lives of PN, because there are low-temperature stars of both classes. The ratio of hydrogen-rich to helium-rich central stars is about 4:1, which is the same as that of hydrogen-rich to helium-rich white dwarfs. It is likely that the hydrogen-rich central stars evolve into hydrogen-rich white dwarfs, and never become helium-rich objects. Among the hydrogen-rich objects the amount of helium is found to vary. It can be from 10 to 30% for low-gravity, younger central stars, and it slowly decreases as the star evolves and the gravity increases. This is probably due to gravitational settling. The helium-rich sequence probably evolves from the Wolf–Rayet spectral type through the higher temperature PG 1159-type stars to helium-rich white dwarfs.

It is interesting to note that the atmospheric abundances are not related to the nebular abundances. This is most strikingly seen in the case of the Wolf–Rayet stars that have less than 1% hydrogen in their atmospheres. The surrounding nebulae, without exception, are hydrogen rich, containing roughly 10% helium. In fact, the nebular abundances are roughly the same for hydrogen-rich and hydrogen-poor central stars. An example is shown in Table 3, BD +30 3639. The central star of this planetary nebula is a low-temperature Wolf–Rayet star with no measurable hydrogen in its atmosphere. The nebula, which has the form of an elliptical shell, is seen to have a normal composition. The shell was ejected from the star about 1000 years ago. Apparently the ejection stopped at precisely the moment when the hydrogen-rich material was entirely used. With the help of HST observations, it has been shown that the mass loss continued at a much lower rate after the hydrogen-rich material was gone, and that this ejecta no longer contains hydrogen.

The question as to why one-quarter of the central stars eject all their hydrogen-rich material while in the remainder the ejection stops while a substantial amount of hydrogen-rich material remains on the surface is unanswered. There does not appear to be a difference in the stellar mass or the luminosity in the two cases, but it is possible that the uncertainty in the distance could mask such an effect. The stellar mass has been directly measured for only one Wolf–Rayet central star, that of NGC 1501. Its mass, $M = 0.55 M_\odot$, is not very high. There is also no evidence that the nebular composition is different in the two cases. This problem must be further investigated.

EVOLUTION

The study of stellar evolution is generally accomplished in the following way. A model star is constructed and allowed to evolve via computer calculations. The results of the computer model are then plotted on a Hertzsprung–Russell (HR) diagram. This is a plot of the luminosity of the star against its temperature. Agreement of the position of the model on this diagram with the observed position of the star is taken as a confirmation of the correctness of the model. The model is, of course, not perfect. For this discussion it will be assumed that the computations are always correctly done, with small enough time steps to reveal all the details. The physical situation may still not be well enough known and important physical processes may possibly be neglected. Rates of known physical processes may not be correctly given. Thus when there is a lack of agreement between the model prediction and the observed position on the HR diagram, the model must be reconsidered and adjusted. The situation is complicated by the fact that the observations are far from perfect. The temperature is usually known to within 10%, but the luminosity can be uncertain by a much larger factor because the distance can be very uncertain.

The asymptotic giant branch

By the method described above it is known that PN are a late stage of stellar evolution. They originate from stars on the asymptotic giant branch (AGB). The structure of an AGB star is rather complicated. It has a hydrogen exhausted core, M_{core}, which contains two nuclear-burning shells. The hydrogen-burning shell is at the surface of the core, while the helium-burning shell is deeper in the core. The helium-exhausted inner part of the core consists of carbon and oxygen. The core contains most of the stellar mass, usually more than 99%, and is comparatively small: the stellar radius exceeds the core radius by factors of up to 10^4. This comparatively large, almost fully convective envelope, M_{env}, contains the unprocessed stellar matter. The core may be thought of as a very hot white dwarf that increases in mass by accreting nuclear-processed matter from the envelope. The luminosity of the star depends essentially only on the mass of the stellar core, that is, the mass within the hydrogen-burning shell.

The helium-burning shell is thermally unstable: a small increase in temperature results in an excess of energy release that cannot immediately be removed. This leads to a thermonuclear runaway, and is called the helium shell flash. Ultimately expansion and cooling halt this thermonuclear runaway, but not before a number of nuclear reactions take place, which are discussed below. After the thermonuclear runaway ends, the helium-burning shell continues burning for a short time. Expansion and cooling then extinguish this shell. The hydrogen-burning shell, which had been extinguished by the helium flash, resumes burning. It provides more than 90% of the total stellar luminosity. When the mass of the helium-rich region left behind by the advancing hydrogen-burning shell reaches a critical value, the helium-burning shell is again ignited and the whole process repeats itself. This is called a thermal pulse and is periodic. The interval between pulses depends on the mass of the core, varying from about 30 years for high-mass cores to several thousand for those of lower mass. At the time of the helium flash the luminosity in the helium-burning shell is substantially increased, but because this energy does not immediately reach the surface the luminosity of the star may not significantly change, and an observer may not detect these flashes.

Changes in the composition

Changes occur in the surface composition of the star before the planetary nebula is ejected. These changes are important, because it is the surface which is ejected and whose composition can be measured in the nebula. Of particular concern here are possible changes in the composition of nitrogen, carbon and helium. It is not so much the production of these elements that determines the surface composition, but how and when they are brought to the surface. Bringing the matter to the surface involves convective mixing, and for this reason the various mixing processes are known as 'dredge-up' events.

The first dredge-up event occurs in all low- and intermediate-mass stars on their first ascent of the red giant branch. Envelope convection reaches down into the interior of the star, bringing material processed in the CN cycle to the surface. Carbon is convected inward and nitrogen is convected outward. By the time convection has extended inward almost to the hydrogen-burning shell, the surface abundance of carbon has dropped by about 30%. At the same time the surface abundance of nitrogen has roughly doubled, assuming an initial solar abundance.

The second dredge-up event occurs only in intermediate-mass stars, roughly in the range 2.5–8 $M_\odot$. Following the exhaustion of helium at the centre of the star and the increasing strength of the helium-burning shell, the hydrogen burning temporarily stops. A convection zone then forms which extends from the atmosphere almost to the outward advancing helium-burning shell. Thus matter that has previously been completely processed by hydrogen burning is mixed with relatively unprocessed envelope material. The result is an increase in the surface helium and nitrogen abundance, and a slight reduction in the amount of surface carbon and oxygen. The surface nitrogen abundance may ultimately exceed that of carbon abundance, and the amount of helium may be increased by about 40%.

The third dredge-up event occurs just after a thermal pulse (helium shell flash). The hydrogen-burning shell is extinguished and a convection zone may penetrate deeply into the region where helium burning has taken place. Mainly carbon is then brought to the surface and the C/O ratio may increase from smaller than unity to larger than unity. At the same time s-process elements, particularly Ba, Zr, Y, La and Tc, are brought to the surface. Technetium has no stable isotope, and yet it is often observed in red giants. Since the observed isotope has a half-life of about 10^5 years, it must have been formed and brought to the surface within this time. Its existence in the stellar atmosphere is direct evidence that the products of nuclear burning in the interior of the star have been carried to the surface.

As the luminosity increases along the AGB, envelope convection extends further and further downward during the quiescent hydrogen-burning phase. For models using masses greater than $5M_\odot$, the envelope can penetrate into the hydrogen-burning region, with temperatures at the bottom in excess of 5×10^7 K, thus mixing newly processed matter throughout the stellar envelope. This is often referred to as hot bottom burning (HBB). In combination with the third dredge-up, this has important consequences for the surface abundance. For instance, lithium is produced and brought to the surface by this mechanism. Furthermore carbon in the envelope can be converted into nitrogen. This is the opposite to what occurs in the third dredge-up. Thus the surface composition depends on the competition between HBB and dredge-up. Both effects depend sensitively on the envelope mass, which in turn depends on the mass loss. Since the mass loss is not yet understood, it is clearly difficult to predict the surface abundance.

This is not the only difficulty in predicting the surface composition. The physics of convection is not completely understood, and the present theory used to describe it is probably inadequate. Furthermore, mass loss is probably an important factor in calculations of evolution, but the mechanism of mass loss is not completely understood, so that it enters the calculations more or less as an unknown parameter. Other mixing processes that are expected from rotationally driven instabilities are not yet included in the theory. In spite of these difficulties, there have been some successes. The second dredge-up, for example, is shown by the observed correlation of N/O and He/H (Figure 6). Further, the low carbon abundance coupled with the very high nitrogen abundance in NGC 6302 (Table 3) indicates that it is likely that HBB has taken place. Comparison of the composition of an individual nebula with a theoretical result must assume that the original composition is solar, which may not always be the case; it is also possible that the solar composition is in error.

This is the present status of the theory of composition changes in PN. There is another parameter which is coupled

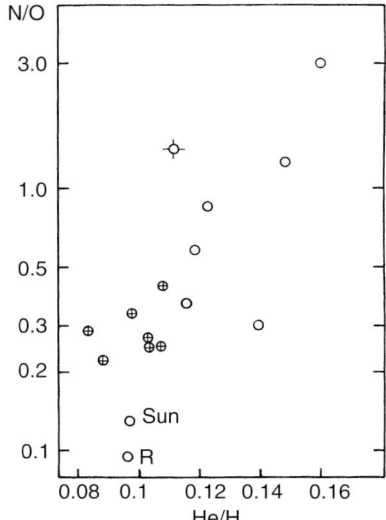

Figure 6 The correlation between the nitrogen to oxygen ratio and the helium to hydrogen ratio. The symbols are related to the nebular morphology: the open circles indicate 'bipolar' nebulae and the crosses indicate elliptical or circular nebulae.

to the composition, and which plays an important part in the evolution. This is the luminosity.

Luminosity

The luminosity can be obtained if the distance of the planetary nebula, the temperature of the central star and its radius (or alternatively its brightness) are known. While this is only strictly true for a blackbody, it is likely that this is a good approximation for the actual central star. In addition, if the gravity is known, the mass of the central star can also be found. Alternatively, one can assume the mass of the central star (taken from theory) and obtain the distance from the temperature and gravity. This is discussed in the section on distances. In practice, the most uncertain quantity in determining the luminosity is the distance. There are less than 30 central stars that meet the above requirements: it is these central stars which must be used to check the calculations of the evolution. Not only is this a small number, but there are important selection effects which enter into the choice of objects for which the distance can be obtained. These central stars are listed in Table 4 along with their distances. Also listed is the visual magnitude of each central star after

Table 4 Luminosity, temperature and gravity of central stars

Nebula	Distance (kpc)	Stellar magnitude, visual (corrected for extinction)	Stellar temperature (K)	Gravity (log (g/cm s^{-2}))	Kinetic age (yr)	Luminosity (solar)
IC 4637	0.50	10.56	50,000		1,100	750
K1-22	1.3	16.58	141,000		21,000	420
K1-27	0.47	15.54	105,000		2,700	57
NGC 246	0.5	11.91	140,000		8,000	4,100
NGC 1535	2.3	11.86	70,000		6,000	12,000
NGC 3132	0.77	15.26	110,000		2,700	230
NGC 3242	0.42	12.07	75,000		1,200	410
NGC 6572	1.2	12.27	60,000		1,700	1,500
NGC 6720	0.7	15.47	101,000		5,000	120
NGC 6853	0.38	13.87	109,000		12,000	190
NGC 7008	0.37	12.51	100,000		2,000	500
NGC 7027	0.8	13.68	109.000		800	2,200
NGC 6210	1.57	12.75	50,000		2,700	990
NGC 7293	0.21	13.48	104,000		25,000	72
NGC 7662	0.8	13.63	100,000		1,200	830
A 74	0.75	17.07	108,000	6.82	240,000	38
BD+30 3639	1.5	9.18	30,000		1,000	5,800
EGB 1			147,000	7.34	20,000	310
HDW 4	0.25		48,000	7.93	3,000	0.95
HW 5	0.42		38,000	7.58	1,700	0.81
K2-2	0.63		67,000	6.09	65,000	250
PHL 932	0.11	12.06	35,000	5.93	5,000	3.5
PW 1	0.43	15.1	94,000		43,000	52
Sh2-174			69,000	6.71	40,000	17
Sh2-216	0.13	12.37	83,000		300,000	41

correction for extinction, the stellar temperature, the gravity when it can be measured directly from the line profile, the kinetic age and the luminosity in units of solar luminosity.

The kinetic age is obtained by dividing the radius of the nebula by the measured expansion velocity of the nebula. This assumes that the expansion velocity has been constant since the expulsion of the nebula, which, considering the small range of the velocities observed in nebulae, is a reasonable assumption. If the ejection of the nebula took place directly following a thermal pulse, this is the time since the star began to move away from the AGB.

In Figure 7 the luminosity of these objects has been plotted against the temperature, the HR diagram. Also plotted on the diagram are the predictions of theoretical models which have been calculated for various values of the mass of the central star (the core mass, but the envelope mass is almost certainly very small). The central stars appear to coincide reasonably well with the theoretical predictions. However, if a comparison is made of the predicted age of the nebula with the observed kinetic age, the agreement is less good. For this purpose dashed lines have been drawn on the diagram, which represent equal ages since the star left the AGB (usually when the star had a temperature of 5000 K, thus somewhat later than the thermal pulse). One of the problems lies in the group of nebulae NGC 3242, 6210, 6572, 7008, 7662, IC 4637 and K2-2. These are all predicted to be greater than 10^5 years old and sometimes substantially more. However, with the exception of K2-2, all these nebulae are about 1000–3000 years old. While it is possible that the distance to an individual object is in error, it is unlikely to be systematically underestimated for the entire group. Furthermore inconsistencies are also found on the lower part of the diagram. The position of PHL932 would indicate a very low mass and extremely old age, compared to its kinetic age of 5000 years. Another problem is the central star of Sh2-174. Its position on the HR diagram would indicate an extremely old age, certainly more than an order of magnitude greater than the kinetic age. Even more extreme are the central stars of HW5 and HDW4. These two stars are not shown on the diagram because their luminosity is quite low: slightly less than one solar luminosity. Although they can be placed on an extension to lower luminosity of the curves shown, the time predicted to get to this position is of the order of 5×10^6 years, while the kinetic age is only 3000 years. Many of these cases have been discussed in the literature, where it is concluded that their evolution does not follow the standard post-AGB evolution theory. There are suggestions that in these

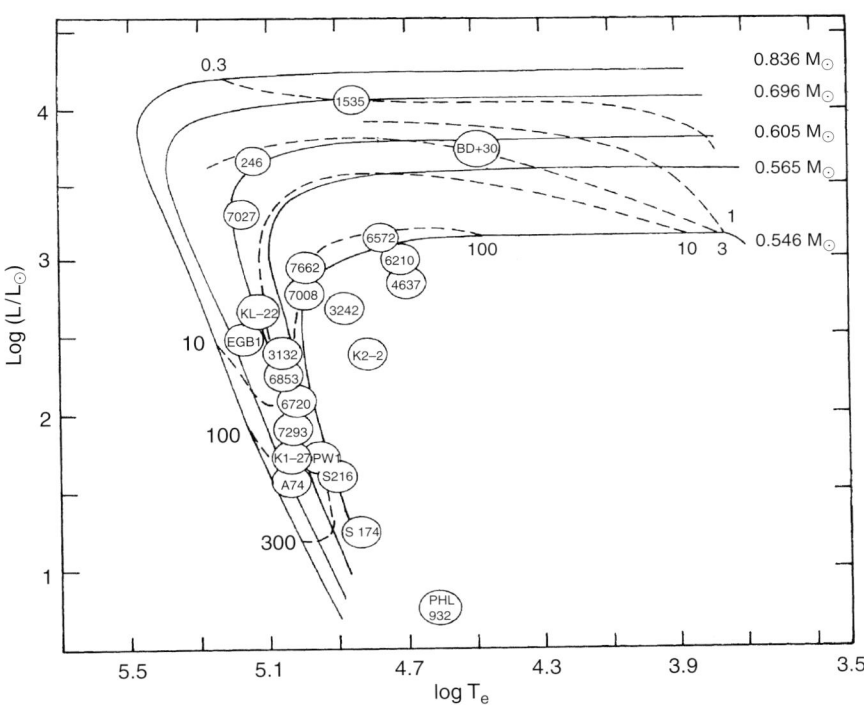

Figure 7 The Hertzsprung–Russell diagram in the region of planetary nebula central stars. The stars are indicated by circles that contain the name of the nebula in which they are situated. Only nebulae with reasonably reliable distances are shown. The solid lines are the theoretical predictions of Schonberner (for the lower mass stars) and Blocker (for the two higher mass stars). The dashed lines indicate the time interval elapsed since the last thermal pulse. The time is given in units of 1000 years.

cases the formation of the PN occurs at an earlier stage in the evolution, just after the first red giant branch. There are also suggestions that the presence of a close companion may cause important modifications of the evolution theory. However, since almost one-half of the PN with reasonably well-determined luminosities do not agree with the prediction, it may be that the theory itself should be modified. One of the most uncertain aspects of the theory is that the mass loss, which plays an essential role, is not well understood. It is included as a parameter in the calculations, and may actually differ considerably in individual cases from the general assumption. This is a subject for future work.

Evolution after nuclear burning has stopped

As already described, the central star consists of a carbon–oxygen core surrounded by two nuclear-burning shells. The luminosity is dictated by the mass of the core and by the nuclear-burning shells. The burning continuously depletes the envelope, the products of which fall into the core. The envelope is also depleted by a continuous mass loss. While the nuclear burning is active, the decrease of the envelope size is compensated for by an increase in temperature so as to maintain the luminosity at a constant value. At some point there is not sufficient material in the envelope to maintain the nuclear burning, and the burning stops. The main source of energy is now gone; the remaining energy source is gravitational contraction. The contraction continues, but the energy is not sufficient to increase or even maintain the surface temperature. The star begins to cool and its luminosity begins to decrease. This is shown in Figure 7. The rate at which the luminosity decreases depends on how quickly the energy is lost. There are two main sources of energy loss at this stage. First there is radiation from the surface, which is a measurable quantity. Second there is neutrino emission from the core, which is not measurable and must be determined theoretically.

The star is now essentially a white dwarf. This is confirmed by the spectrum. Of the stars listed in Table 4, NGC 7293, 6853, 6720, PW1, A74, Sh2-216, Sh2-174, EGB1, HDW4 and HW5 all have white dwarf spectra. All of these central stars are seen to be on the cooling paths in Figure 7. All of these stars have hydrogen-rich atmospheres, that is, they have a spectral type designated by DA or DAO. Other hydrogen-rich white dwarfs are known that are not surrounded by nebulae, which join this sequence at lower temperatures and luminosity. Thus there is no doubt that many, if not all, DA white dwarfs have passed through the planetary nebula stage.

There are comparatively few helium-rich central stars listed in Table 4. Only the central stars of BD +30 3639 (a Wolf–Rayet star), NGC 246 (PG 1159 star) and K1-27 have helium-rich atmospheres. Only the latter has reached the cooling track; its position is indistinguishable from that of the hydrogen-rich stars although its kinetic age is rather young for this position.

FUTURE RESEARCH

One of the important problems is the determination of the distances to nebulae. Eventually this problem will be solved by some kind of instrument capable of measuring stars an order of magnitude fainter than Hipparcos could measure, with an accuracy 100 times greater. Until this is done only slow progress will be made. Ground-based parallax measurements will probably give the best results. But studies of systems with well-established distances, such as the Magellanic Clouds, may be important in circumventing the distance problem. This work can only be done from space, because the extra spatial resolution is absolutely necessary. Studies of the galactic bulge PN could be accomplished from the ground, but the extra extinction in this direction will probably require that this work be done in the infrared. Theoretical work is needed in two respects. First, the interpretation of the spectra of these stars is still in an early stage and much needs to be done. Second, the investigation of mass loss in these stars, and the integration of this aspect in the modelling of than, is indispensable in obtaining adequate results.

There are many further points that will require much thought. The role of companion stars and their influence on the evolution of the system is not well understood. The theory of the chemical composition, and how elements are produced and brought to the surface, is not yet completely satisfactory. The study of the morphology of the nebula still has problems, in spite of the important progress made in recent years. It is difficult to predict in which directions progress will be made in the near future, but it is clear that observations taken from satellites will be one of the aspects for further progress.

REFERENCES AND FURTHER READING

History of the measurements of nebulae in the eighteenth and nineteenth centuries:
Whitney, C.A. (1971). *The Discovery of our Galaxy*, Knopf, New York.
Status of the subject in the early twentieth century:
Campbell, W.W. and Moore, J.H. (1918). The spectrographic velocities of the bright-line nebulae. *Lick Observatory Publications*, **13**, 77.
Curtis, H.B. (1918). The planetary nebulae. *Lick Observatory Publications*, **13**, 57.
Modern monographs:
Kwok S. (2000) *The Origin and Evolution of Planetary Nebulae*, Cambridge University Press, Cambridge.
Pottasch, S.R. (1984). *Planetary Nebulae*, Reidel, Dordrecht.

Catalogues containing images of PN:

Balick, B. (1987). The evolution of planetary nebulae: I. Structures, ionizations, and morphological sequences. *Astronomical Journal*, **94**, 671–678 (about 50 objects in 4 colours).

Chu, Y.-H., Jacoby, G.H. and Arendt, R. (1987). Multiple-shell planetary nebulae: I. Morphologies and frequency of occurrence. *Astrophysical Journal Supplement*, **64**, 529–544 (multiple shell PN).

Corradi, R.L.M. and Schwarz, H.E. (1995). Morphological populations of planetary nebulae: Which progenitors? I. Comparative properties of bipolar nebulae. *Astronomy and Astrophysics*, **293**, 871–888 (46 bipolar nebulae).

Gorny, S.K., Schwarz, H.E., Corradi, R.L.M. and von Winckel, H. (1999). An atlas of images of planetary nebulae. *Astronomy and Astrophysics Supplement*, **136**, 145–171.

Manchado, A., Guerrero, M.A., Sanghellini, L. and Serra-Ricart, M. (1996). *IAC Morphological Catalog of Northern Galactic Planetary Nebulae*, Instituto de Astrofísica de Canarias, La Lagune.

Schwarz, H.E., Corradi, R.L.M. and Melnik, J. (1992). A catalogue of narrow band images of planetary nebulae. *Astronomy and Astrophysics Supplement*, **96**, 23–113 (southern PN).

Atlas of many colour images as a CD ROM:

Hajian, A.R. and Terzian, Y. (1999). *Images of Galactic Planetary Nebulae Obtained with the Hubble Space Telescope Wide Field Planetary Camera 2*, USNOHSTPNE, US Naval Observatory (Figure 1 is a selection of eight of these images).

Hydrodynamic problems in forming nebulae:

Frank, A. (2000) A paradigm lost: New theories for aspherical planetary nebulae. In J.H. Kastner, N. Sokar and S. Rappaport (eds), *Asymmetrical Planetary Nebulae II: From Origins to Microstructures*, ASP Conference Series, Vol. 199, Astronomical Society of the Pacific, San Francisco, pp. 225–234.

Kastner, J.H., Soker, N. and Rappaport, S. (eds) (2000). *Asymmetrical Planetary Nebulae II: From Origins to Microstructures*, ASP Conference Series, Vol. 199, Astronomical Society of the Pacific, San Francisco.

Spatial distribution:

Acker, A., Ochsenbein, F., Stenholm, B., Tylenda, R., Marcout, J. and Schohn, C. (1992). *The Strasbourg-ESO Catalogue of Galactic Planetary Nebulae*. Published by the European Southern Observatory.

Pottasch, S.R. (1984). *Planetary Nebulae*, Reidel, Dordrecht.

Results concerning PN from the ANS satellite:

Pottasch, S.R., Wesselius, P.R., Wu, C.-C. and von Duinen, R.J. (1977). Ultraviolet observations of planetary nebulae: I. Determination of extinction. *Astronomy and Astrophysics*, **54**, 435–442.

Pottasch, S.R., Wesselius, P.R., Wu, C.-C., Fieten, H. and van Duinen, R.J. (1978). Ultraviolet radiation from planetary nebulae. *Astronomy and Astrophysics*, **62**, 95–110.

IUE results:

Wamsteker, W. and Gonzalez-Riestra, R. (eds) (1998). *Ultraviolet Astrophysics Beyond the IUE Final Archive*, ESA SP-413, ESA, Noordwijk.

Infrared emission from NGC 7027:

Gillett, F.C., Low, F.J. and Stein, W.A. (1967). Infrared observations of the planetary nebula NGC 7027. *Astrophysical Journal*, **149**, L97.

Broadband IRAS emission:

Pottasch, S.R., Baud, B., Beintema, D., Emerson, J., Habing, H.J., Harris, S., Houck, J., Jennings, R. and Marsden, P. (1984). IRAS measurements of planetary nebulae. *Astronomy and Astrophysics*, **138**, 10–18.

IRAS spectra of PN:

Pottasch, S.R., Preite-Martinez, A., Olnon, F.M., Mo, J.-E. and Kingma, S. (1986). IRAS spectra planetary nebulae: III. *Astronomy and Astrophysics*, **161**, 363–375.

Discussions of distances:

Acker, A., Fresneau, A., Pottasch, S.R. and Jasniewicz, G. (1998). A sample of planetary nebulae observed by Hipparcos. *Astronomy and Astrophysics*, **337**, 253–260.

Ciardullo, R., Bond, H.E., Sipior, M.S., Fullton, L.K., Zhang, C.-Y. and Schaefer, K.G. (1999). A Hubble Space Telescope survey for resolved companions of planetary nebula nuclei. *Astronomical Journal*, **118**, 488–508.

Habing, H.J. and Lamers, H.J.G.L.M. (eds) (1997) *Planetary Nebulae*, IAU Symposium 180, Kluwer, Dordrecht, pp. 29 and 40.

Pottasch, S.R. (1980). Masses of planetary nebulae. *Astronomy and Astrophysics*, **89**, 336–341.

Shklovskii, I.S. (1956). The nature of planetary nebulae and their nuclei (*Russian*). *Ast. Zh.*, **33**, 315.

Zanstra temperature:

Zanstra, H. (1931). Luminosity of planetary nebulae and stellar temperature. *Publications of the Dominion Astrophysical Observatory*, **4**, 209.

Stoy temperature:

Kaler, J.B. (1976). The exciting stars of low-excitation planetary and diffuse nebulae. *Astrophysical Journal*, **210**, 843.

Preite-Martinez, A. and Pottasch, S.R. (1983). The temperature of central stars of planetary nebulae: The energy-balance method. *Astronomy and Astrophysics*, **126**, 31–44.

Stoy, R.H. (1933). Nebulæ, planetary: The temperatures of the nuclei of planetary nebulæ. *Monthly Notices of the Royal Astronomical Society*, **93**, 588.

Model atmosphere analysis to determine stellar temperatures and gravities:

Napiwotzki, R. (1999). Spectroscopic investigation of old planetaries: IV. Model atmosphere analysis. *Astronomy and Astrophysics*, **350**, 101–119.

Formation of dust in evolved stars:

Beintema, D.A. (1997). ISO-SWS observations of planetary nebulae (invited review). In H.J. Habing and H.J.G.L.M. Lamers (eds), *Planetary Nebulae*, IAU Symposium 180, Kluwer, Dordrecht, p. 145.

Winters, J.M. (1997). Dust formation and mass loss in evolved stars. *Astrophysics and Space Science*, **255**, 257–266.

Chemical composition of planetary nebulae:

De Marco, O., Barlow, M.J. and Storey, P.J. (1996). The winds of WC10 central stars of planetary nebulae. *Astrophysics and Space Science*, **238**, 91–95.

Hamann, W.R. (1996). Spectral analysis and model atmospheres of WR central stars (invited paper). *Astrophysics and Space Science*, **238**, 31–42.

Leuenhagen, U. (1996). Spectral analyses of [WCL]-type central stars. *Astrophysics and Space Science*, **238**, 75–78.

Marigo, P., Bressan, A. and Chiosi, C. (1996). The TP-AGB phase: A new model. *Astronomy and Astrophysics*, **313**, 545–564.

Napiwotzki, R. (1999). Spectroscopic investigation of old planetaries: IV. Model atmosphere analysis. *Astronomy and Astrophysics*, **350**, 101–119.

Pottasch, S.R. (2000). Abundances and morphology in planetary nebulae. In J.H. Kastner, N. Soker, and S. Rappaport (eds), *Asymmetrical Planetary Nebulae II: From Origins to Microstructures*, ASP Conference Series Vol. 199, Astronomical Society of the Pacific, San Francisco, pp. 289–296.

Theoretical evolution:

Blöcker, T. (1997). Theory of AGB evolution. *Astrophysics and Space Science*, **255**, 267–277.

Iben, I., Jr. (1995). Planetary nebulae and their central stars: Origin and evolution. *Physics Reports*, **250**, 2–94.

Schönberner, D. (1998). Structure and evolution of central stars of planetary nebulae (invited review). In H.J. Habing and H.J.G.L.M. Lamers (eds), *Planetary Nebulae*, IAU Symposium 180, Kluwer, Dordrecht, pp. 379–388.

Wood, P. (1997). Final stages of AGB evolution (invited review). In H.J. Habing and H.J.G.L.M. Lamers (eds), *Planetary Nebulae*, IAU Symposium 180, Kluwer, Dordrecht, pp. 297–302.

41

HIROSHI TSUNEMI*

Supernovae and supernova remnants

1 GENERAL INTRODUCTION TO SUPERNOVAE AND SUPERNOVA REMNANTS

A supernova explosion is among the most dramatic events that can be seen. The term '*supernova*' is somewhat misleading, as such an event represents not a new star (that is, a '*nova*'), but instead the end of a star's life. Nuclear fusion, the energy source of the stars, creates heavier elements from lighter elements. In this way, almost all of the elements that make up the universe, with the exception of hydrogen and helium, are created inside a star through this process. These elements accumulate inside a star over its lifetime and are dispersed into space through a supernova explosion. We may therefore say that all of the material of the Earth (excepting hydrogen and helium) were probably created inside some star many eons ago, and that they were ejected by a supernova, soon to become the primordial material of the Earth. Nuclear fusion inside a star can generate heavy elements ranging from helium up to the most stable nucleus, iron. It may therefore be said that the iron that plays such a vital role in the hemoglobin of our blood must have been generated inside a massive star that soon went supernova. This must have happened at least five billion years ago, since this is the age of our solar system.

Based on the statistical study of the occurrence rate of supernovae in other galaxies, we believe that a supernova should occur every few tens of years in our Galaxy. Further, the energy released by a supernova is of the order of 10^{53} erg, which is two orders of magnitude larger than the energy radiated by our Sun over its entire 10-billion-year lifetime. Although 99% of this energy is carried away by neutrinos that go nearly completely undetected, the remaining 1% still represents an enormous amount of energy released over a very short time. The maximum absolute magnitude of the supernova is about -19^M. This means that if a supernova occurred at a distance of 10 parsecs (pc: $1\,\text{pc} = 3.1 \times 10^{18}\,\text{cm}$) from the Earth, it would appear to be about 1,000 times as bright as a full moon. If it were to appear at the outer edge of the Galaxy, it would be as bright as Venus, if it did not suffer from extinction due to interstellar matter. The Galaxy, however, does contain a great deal of diffuse matter so that there is a great deal of extinction, particularly at optical wavelengths. This is especially true in the Galactic plane. Since most of the stars of the Galaxy are in the Galactic plane, only those supernovae that occur nearby are visible. On the contrary, supernovae appearing in external galaxies lying outside of the Galactic plane can be easily seen. We can therefore study the supernova rates in these galaxies.

The appearance of several supernovae has been documented over the past 2,000 years. Most of these were recorded in the Eastern as opposed to the Western world, for reasons that appear to be socio-political. In the Western worldview, God created the universe perfect and forever fixed. It is thus likely that the appearance of a '*new star*', a phenomenon at variance with a perfect, unchanging universe, would go unrecorded. Such a '*mistake*' could not be accepted. On the contrary, in the Eastern worldview, changes in the celestial sphere could result from God's '*mistake*', a forecast for an impending disaster. In China, in fact, it became an important concern for the emperor to be able to predict a drought, a flood, or other such disasters. It was thus a regular job for astronomers to search the sky for any omens of such events. Furthermore, a nearby supernova would be bright enough to be easily visible even during the

*Osaka University, Japan

day, thus not only professional astronomers, but even common people would have been able to observe such events.

During a supernova explosion, the heavy elements synthesized inside the star expand at very high velocity, up to 10,000 km sec^{-1}. Afterwards the supernova slowly fades through adiabatic expansion. The major heat source during this phase comes from the following series of radioactive decays: ^{56}Ni (7 days) → ^{56}Co (77 days) → ^{56}Fe. This decay series will keep the supernova bright for several months. The ejecta, rich in heavier elements, will eventually form another generation of stars and planets. The ejecta first expand at high velocity with no deceleration. Eventually, it will sweep up enough interstellar material so that deceleration becomes effective. The deceleration will heat up both the ejecta and the interstellar material and the temperatures will become high enough so that the majority of the radiation will be not in the visible but in the X-ray regime. The swept-up material will form a shell-like structure, leaving a low density, high temperature interior.

The expanding shell contains a mixture of interstellar matter and the heavy elements from the supernova progenitor and the supernova itself. Once the temperature becomes high enough to ionize most of the elements, the shell barely cools down. The dominant emission mechanism is thermal bremsstrahlung (also referred to as free-free emission), which results from free electrons colliding with the nuclei within the shell, losing energy by emitting radiation. The result is an X-ray continuum spectrum. When the temperature has cooled enough so that some heavier atoms, such as iron, capture electrons, much more efficient cooling processes take over. When an iron nucleus, for example, captures a free electron, free-bound radiation is emitted. When a free electron collides with a bound electron, the bound electron is excited to a higher energy level. Almost immediately, this electron returns to its ground state, emitting a photon in a bound-bound transition (line emission). Once the matter cools down to the line-emitting temperature, the cooling process accelerates. The lower the temperature, the more efficient the cooling process. The temperature range for which the iron family (the most abundant species among the heavy elements) captures electrons is thermally unstable. When the temperature reduces further, the more abundant light elements such as oxygen begin to capture electrons that again accelerate the cooling process. The cooling region is compressed by its surroundings and further increases in density, further accelerating the cooling process. Sheets that are visible in the optical are formed as a result of the thermal instability. Figure 1 shows a portion of the Veil nebula in the Cygnus Loop, that results from the radiative instability of the hot gas.

The supernova leaves behind a high-temperature low-density cavity a few tens of pc in diameter. It can survive for a few million years due to its low density. Taking into

Figure 1 Close up of the Veil Nebula (the Cygnus Loop) obtained by the HST (see http://oposite.stsci.edu/pubinfo/gif/CygnusLoop.gif) (courtesy of J. Hester, Arizona State Univ. and NASA). A high density region forms a filamentary structure.

account the occurrence rate of supernovae, we find that the hot gas, with temperatures of a few million degrees and only visible in X-rays, potentially occupies a large volume of our Galaxy.

The fraction of heavy elements in our Galaxy is gradually increasing. Moreover, apart from this in-flux of material, there is also a flow of thermal and kinetic energy. The heavy elements are generated inside the stars as a by-product of the nucleosynthesis in the stars. The energy released by this nucleosynthesis is continuously radiated away by the stars as visible and ultraviolet light, while the heavy elements continue to accumulate within the star. When the star becomes a supernova, a large amount of the star's gravitational energy is released along with the heavy elements. These elements then disperse through the Galaxy and form the building blocks for the next generation of stars. The supernovae and supernova remnants (SNRs) provide the most dramatic examples of this cycle of life-and-death in the Galaxy.

A star, such as the Sun, is a large sphere of hot plasma with nucleosynthesis taking place only in its innermost regions. The energy is transported away from the center via emission and, in the outer layers, by convection. The photons emitted from the center (excluding the neutrinos) are scattered, absorbed and re-emitted by the surrounding material. Therefore, photons that are created in the center of the star cannot be directly detected. The photons that we detect from stars come from only a very shallow layer near to the star's surface. This region is known as the photosphere.

The supernova remnant (SNR) is a blow-up of the star, where the matter is at very low density. Because of this, we may directly detect emission from any portion of the plasma. That is, we do not need to worry so much about intervening material in the remnant absorbing photons. Such a plasma is referred to as being optically thin. In this circumstance, we can measure directly the abundances of the heavy elements in the remnant. Because of the remnant's great size and low density, self-absorption and scattering do not play an important role in the radiation process. We can therefore directly see emission from the heavy elements irrespective of their position in the remnant. When heavy elements such as O, Ne, Mg, Si, S, A, Ca, and Fe emit photons at their characteristic energies (line emission) the photons will reach us undisturbed (thermal broadening of the lines is very small, in practice). Thus we are provided with a unique opportunity to directly see these heavy elements before they redistribute themselves through the Galaxy. Further, the morphology of the remnants is also intricate, making them interesting in their own right.

Because so many remnants have already been observed, it is impossible to review all of them. In this chapter, we will discuss a sample of SNRs, ranging from young remnants to old remnants. This will help us to understand how the heavy elements and the energy of the supernova are distributed throughout the Galaxy.

1.1 Nuclear fusion inside a star

Atomic nuclei consist primarily of baryons (protons and neutrons) which determine the mass of the nucleus. Figure 2 shows the average baryon mass inside the atomic nucleus as a function of mass number, A. The figure gives the energy level of each nucleus: the lower the energy level, the more stable the nucleus. The isolated proton, hydrogen (^{1}H, A=1) has the highest energy level among the various nuclei. The baryons inside the helium atom (^{4}He, A=4) form a local minimum in Figure 2, being lower than lithium (^{6}Li, ^{7}Li), beryllium (^{9}Be) and boron (^{10}B, ^{11}B) that follow it. ^{12}C becomes lower than ^{4}He. Beyond this, the data points fluctuate up and down. Generally, however, the energy level gradually decreases with increasing mass number until ^{56}Fe is reached. This nucleus (A=56) has the lowest energy level among all nuclei and is therefore the most stable atomic nucleus. Beyond this nucleus, the energy level of nuclei increases. This is a fundamental result of nuclear mass measurements.

After the Big Bang, the matter of the universe consisted mostly of hydrogen along with a little helium. Because of the gap between helium and lithium seen in Figure 2, very little high-Z elements (i.e. elements beyond helium, where

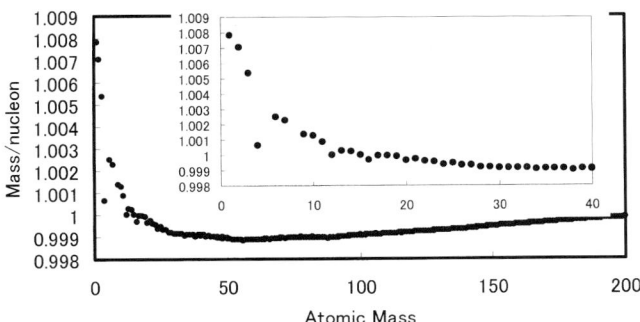

Figure 2 Atomic mass per nucleon as a function of mass number. Hydrogen has the highest mass per nucleon while the nucleons in ^{56}Fe have the lowest ones.

Z is the atomic number) existed. This energy gap hindered the creation of elements beyond helium in the first 3-minute after the Big Bang. In contrast, the universe today is abundant in high-Z elements. In the first epoch of the universe, matter was partly clustered and formed into stars. Once the stars began to form, the nuclear fusion of hydrogen into helium began in their cores. This phase of a star's life is called the 'main sequence phase', and it is the most stable phase of a star's life. The Sun, for example, has been in the main sequence phase for the last five billion years and will remain in this phase for another five billion years.

Once the star has exhausted the hydrogen in its core, the main sequence phase ends, and the star intermittently expands and contracts. During this phase, a new class of fusion processes proceeds, creating heavier elements such as C, N, O, Ne, etc. Each element is created by a process that occurs at its own particular temperature. In general, the higher the temperature, the heavier the element created. Because the temperature increases as one goes deeper into the star and because the temperature in the core of a star is slowly increasing during this phase, an 'onion-skin structure' results. That is, the elements are created in layers so that, as one proceeds from the outer layers to the core of the star, different layers of elements are passed through. Elements with higher Z are found deeper in the star. Elements with a lower energy level are more stable, so that the fusion of two or more lighter nuclei into a heavier one creates a nucleus with a lower energy level, that is, a more stable nucleus. The most stable nucleus is ^{56}Fe. It therefore represents a barrier beyond which nuclei cannot be easily created through the fusion process. Since a certain minimum temperature is required for fusion and since the temperature in the core of a star depends primarily on its mass, iron is created only in the most massive stars. It may thus be said that iron represents the ashes of the nuclear burning that took place in the star. Through this process the high Z elements are created and accumulate in a massive star.

Meanwhile, the energy released in this process is ultimately released through the star's surface, making the star shine.

A star is stable when the gravity of the star, which tends toward collapsing the star, is balanced by the pressure in the star's interior, which tends toward expanding the star. This outward pressure is maintained by the thermal energy of the star which is in turn regulated by the nuclear reactions. Therefore, once the star has exhausted its source of nuclear energy, it can no longer support itself against its own gravity. If the star has less than a certain mass, another source of pressure will ultimately halt the collapse. However, for stars beyond this lower mass, the collapse cannot be stopped and a supernova will result. This supernova will shine as brightly as the entire Galaxy. During the explosion, more nucleosynthesis takes place, creating more high-Z elements. In some supernova, the entire star is detonated while in others a compact object – a black hole or a neutron star – is left behind.

The metal abundance in the universe is referred to as the '*cosmic abundance*', which is essentially the same as the abundances of the elements within our solar system. These abundances are determined through studies of the sun as well as of meteorites (Anders and Grevesse 1989). There are many elements above iron. The Earth, for example, has Ni, Cu, Ag, Au, Pt, U, etc., although their abundances are relatively small. These higher-Z elements (elements beyond Fe) are created by a variety of processes. One of these processes, known as the rapid-process, occurs only in the presence of a large flux of neutrons. A supernova has the necessary high flux so that this rapid-process occurs during a supernova explosion. The process is able to operate because neutrons do not carry an electric charge, so that the Coulomb barrier in a nucleus is ignored by them. The elements thus created have large numbers of neutrons, some of which migrate to higher-Z elements via β–decay. In this way, the high-Z elements up to uranium are created. These high-Z elements have a higher energy level than iron, as is seen in Figure 2. Therefore, excess energy can be released when they are split into two or more lower-Z elements. This process is called nuclear fission. We extract energy from these nuclei via the fission process in nuclear power stations and in nuclear weapons.

1.2 The supernova explosion

When the supernova occurs, the high-temperature plasma, the ejecta, expands outward and mixes with the interstellar matter, giving rise to an increase in its metal abundance. Early in the explosion, the ejecta expand adiabatically and cool very rapidly. In general, there are two types of supernova: Type I and Type II supernovae. The supernova type is phenomenological, depending on whether or not the H–β absorption line (in the visible region) is seen in its spectrum near the maximum brightness. If the line is present, the progenitor must have had a hydrogen envelope, an indication of a massive star. The internal structure of the supernova depends on the mass of the progenitor star.

If the mass of the progenitor star is less than a certain critical mass, a few times $M_\odot$ ($M_\odot$ represents the mass of the sun, 2×10^{33} g), it will not become a supernova. It will evolve instead into a white dwarf. Our sun will therefore not explode after it has exhausted its hydrogen about five billion years in future. A star whose mass is over the critical mass accumulates high-Z elements in its center forming an onion-skin structure. In the deepest layer of the star, we will see that the matter consists mainly of elements from the iron family (Fe, Ni, etc.), which represent the ashes of the nuclear fusion. Above this layer, lighter elements form layers one by one. The outermost layer consists of hydrogen, if the star has not lost its hydrogen envelope. Otherwise, it consists primarily of relatively low-Z elements such as carbon, oxygen etc.

If the mass of the progenitor star is in some range (up to several times $M_\odot$), it is believed that the entire star will be detonated in a supernova. All of the nucleo-synthesized matter is ejected, creating the building blocks for the next generation of stars. If the mass is greater than this critical mass, some inner region of the star is not ejected in the explosion but instead forms a compact object, i.e. a neutron star or a black hole. Whether a neutron star or a black hole is left behind thus depends on the mass of the progenitor. In this way, we can say that a Type I supernova supplies a large amount of iron since the entire star is blown up, whereas a Type II supernova leaves a compact source behind: a neutron star or a black hole.

1.3 The supernova remnant

The supernova releases both an enormous amount of energy and a large mass of high-Z elements into space. The effects of these two components are long-lasting and are seen as a supernova remnant (SNR).

As the ejecta expand into the interstellar medium, they sweep up the surrounding interstellar matter. As long as the material swept up is less than the mass of the ejecta, the expansion is considered to be a free expansion. This phase lasts for a few hundred years, depending on the density of the interstellar matter around the supernova. As the ejecta expand into the interstellar matter, a violent collision occurs between the ejecta and the interstellar matter. The former is expanding at a velocity of a few thousand km sec^{-1} while the latter is stationary. At the boundary of the collision, a shock front is formed, resulting in shock heating of the matter up to tens of million degree. The shock front represents a boundary at which physical conditions, such as temperature and density, are discontinuous. After the collision, the ejecta and the interstellar matter gradually mix.

We may consider a very simple case: an inelastic collision of two particles having the same mass, m. One particle has a velocity v, while the other is at rest. After the inelastic collision, they move together with a velocity of $v/2$ (momentum is conserved). The kinetic energy in this system decreases from $\frac{1}{2}mv^2$ to $2 \times \frac{1}{2}m(v/2)^2$. In this case, half of the kinetic energy is converted into an increase in the internal energy, creating a loud sound or a bright flash and a high temperature. This is a schematic representation of shock heating, which efficiently converts kinetic energy into thermal energy in a very short time. The formation of the shock front usually heats the matter up to a temperature where it starts to emit radiation.

Once the swept-up material becomes more massive than the ejecta, the evolution of the remnant changes from the free-expansion phase. At the boundary between the ejecta and the swept-up material, a bi-layered structure exists. On the inner side of the boundary is the ejecta, while on the outer side is the swept-up matter. This boundary is known as a contact discontinuity. The shock heating starts at the contact discontinuity and expands in both directions. Gradually, the shock-heated region expands. The front portion of the shock propagating into the interstellar matter is called the fore shock, while the shock propagating back into the ejecta is called the reverse shock. Figure 3 shows a very typical composite image (Gaetz et al. 2000) of a young SNR, E0102-72, seen in the Small Magellanic Cloud. A reddish image shows a fore shock seen in radio wavelength and a bluish image shows a reverse shock in X-ray. Green filamentary structures are optical filaments.

During this phase, the ejecta expand into the interstellar medium, which is stationary in the rest frame. The shock heated layer between the fore and reverse shocks is also expanding outward with a velocity slower than that of the unshocked ejecta. Therefore, when we examine the situation from a reference frame that is at rest with respect to the shock-heated material, the interstellar material is rushing in forward. After they enter the shock, they lose their kinetic energy and are heated. When both the ejecta and the interstellar matter are heated, the increase in temperature results in thermal emission. This emission is primarily in the X-ray region rather than in the optical region.

2 THERMAL EMISSION

As the material is shock-heated, it becomes bright by virtue of its increased temperature. The material may be characterized by three temperatures in this instance. In other words, there are three heat reservoirs. Equipartition within a reservoir takes place quickly, whereas between reservoirs it takes a long time for equipartition. The three temperatures are the electron temperature, T_e, the ion (primarily hydrogen) temperature, T_{ion}, and the ionization temperature, T_z. The ionization temperature, T_z, determines the ionization level of the (primarily high-Z) ions.

The shock heating is a conversion process from kinetic energy into thermal energy. Before the occurrence of the shock heating, the electrons and ions have the same velocity, since they are expanding together. Furthermore almost all of the ions were in a low ionization state. We therefore expect that, just after the shock heating, the temperatures have the following relation: $T_z < T_e < T_{ion}$. Most of the energy is stored in the ion (hydrogen in this case) thermal energy, while the ion emits little radiation because of its large mass. On the contrary, the electrons emit thermal bremsstrahlung through collisions with the ions. Since the line emission is the most efficient radiation process in this case, the high-Z ions in ionized states can radiate more if they are ionized through collisions with electrons. Thus, there is a flow of energy from the ion thermal energy to ionization of high-Z elements through the electrons. Due to the relatively low density, the timescale for thermal equilibrium between these components is long (thousands of years or longer). At high density, like that in laboratory, the timescale is so short that they are almost always in a state of thermal equilibrium.

In practice, the initial ionization state can be considered to be neutral. The ions are gradually losing electrons

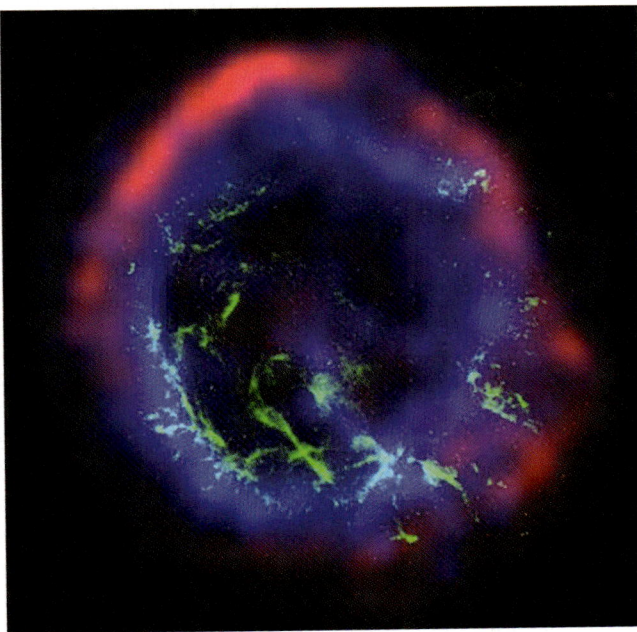

Figure 3 Composite image of a young SNR, E0102-72, in the Small Magellanic Cloud (NASA/CXC/SAO). The reddish part shows a forward shock in radio and the bluish part shows a reverse shock in X-rays. Green structures are the optical filaments.

one-by-one through collisions with thermal electrons until their ionization states are in equilibrium with the thermal electrons. The ionization degree is roughly proportional to the number of collisions with thermal electrons, which is proportional to the product of the electron density, the electron velocity (or temperature) and the elapsed time after the shock-heating. Therefore, the plasma condition can be characterized by T_e and τ, which is the product of the electron density and the elapsed time after the shock-heating. τ is called the '*ionization parameter*'. In the shock heating process, thermal equilibrium is achieved when τ is about $10^{12\sim13}\,\mathrm{cm}^{-3}\,\mathrm{sec}$.

As τ increases, the ionization state of the heavy elements increases. The line energy of the $K\alpha$ emission (the transition of an electron from the L-shell to the K-shell) depends on the ionization state of the ion. Because K-shell emission from heavy elements provides crucial information on the plasma conditions, we need to know the precise line energy for the various ionic states.

Figure 4 shows the energy of the $K\alpha$ emission line of iron as a function of τ and T_e. It is calculated based on models developed by Masai (1984). When τ is very small, almost all of the iron is in the lowest ionization state (26 electrons remaining) whose $K\alpha$ emission has the energy of the neutral iron $K\alpha$ line around 6.4 keV. As τ increases, the thermal electrons strip the bound electrons from iron, resulting in an increase in its ionization state. As the ionization state proceeds, the line center energy of the $K\alpha$ emission increases gradually. It is at about 6.7 keV that the iron becomes a helium-like ion (two electrons remaining). Therefore, the line center energy of the $K\alpha$ emission of iron gradually increases from 6.4 keV to 6.7 keV. However, this figure does not include the emission of iron in the hydrogen-like state (one electron remaining) whose $K\alpha$ line is around 6.9 keV. This is because there is a relatively large gap between the emission line from the helium-like ion and that from the hydrogen-like ion. We can therefore easily derive the plasma conditions by measuring the line energy of the $K\alpha$ emission.

Because the interstellar particle density is of the order $1\,\mathrm{cm}^{-3}$, it takes about $3\times10^{4\sim5}$ years to reach thermal equilibrium. Therefore, most of the X-ray emitting SNRs are in a non-equilibrium ionization (NEI) state. In the NEI condition, the emission from free electrons produces a continuum spectrum, representing T_e, while the discrete emission lines from various ionization states represent T_z. However, it is very difficult to measure T_{ion}, since it appears only in the line broadening of the discrete emission lines whereas the major part of the energy is stored in the ion thermal energy.

2.1 Simple model of the evolution of an SNR (Sedov model)

When the matter swept up by the ejecta exceeds the mass of the ejecta itself, the evolution of the SNR is determined by the swept-up interstellar matter. Since the energy loss through radiation is negligibly small, the evolution of the SNR is governed by a few parameters: the density of the surrounding interstellar matter (assumed to be uniform), the explosion energy, and the remnant age. The expansion velocity and the ion temperature gradually decrease as the SNR expands. This scenario is referred to as the '*Sedov phase*' after the pioneering studies of Sedov (1959).

In this phase, the pressure of the surrounding medium is negligibly small compared with that inside the SNR. Under these conditions, the density at the shock front is $(\gamma+1)/(\gamma-1)$ times that of the surrounding medium, where γ is the specific heat of the matter. In general, the interstellar matter consists primarily of protons (hydrogen) for which $\gamma=5/3$. In other words, the strong shock compresses the interstellar matter by a factor of 4. Assuming spherical expansion, the interstellar material inside the SNR cavity is compressed towards the outer part of the sphere with a thickness about 1/12 of the radius, forming a shell structure. The inner part of the SNR is left almost vacant, probably occupied by a thin, hot gas consisting of ejecta material.

2.2 Radiative phase and dissolving phase

When the temperature of the shock region decreases, the ions gradually capture electrons and become partially ionized. The ions then efficiently emit line radiation, which has the effect of quickly reducing the temperature. Because of radiative instability, the cooling process speeds up. The pressure decreases with the temperature. The shell is therefore compressed by the surrounding medium, which increases its

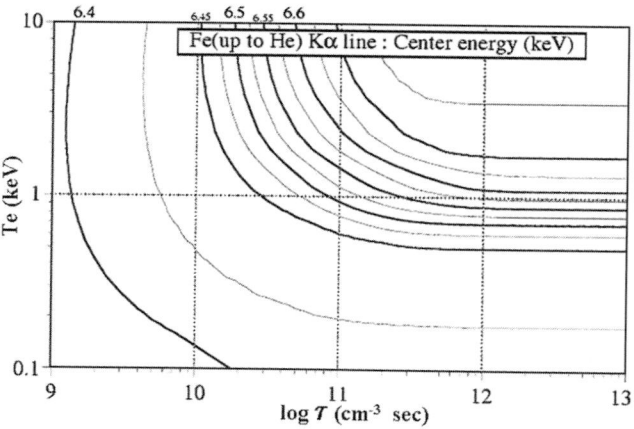

Figure 4 Apparent line energy of the iron $K\alpha$ emission line as a function of T_e and τ. In the low ionization condition (τ is small), the line center energy is around 6.4 keV, whereas the high ionization stage (τ is large) depends only on T_e. (From Tsunemi *et al*. 2000.)

density. The radiation power is proportional to the square of the density (the number of collisions between the particles increases as the square of the density). This further accelerates the cooling. Furthermore, the line emission becomes more effective at relatively low temperature, which also accelerates the thermal instability. In this way, a high-density region is formed that is at a low temperature and is optically bright. This high-density region propagates into interstellar space, continuing to accumulate the interstellar matter in front, acting as a snowplow.

The shell will ultimately slow down to the proper motion of the interstellar cloud in which it resides. At that stage, the shell region is indistinguishable from the interstellar cloud. We can then consider the expansion of the SNR to have ceased. The hot cavity still survives and will continue to exist for a long time, however. The cavity crossing time for the typical proper motion of an interstellar cloud is typically about one million years.

3 SN 1987A: A BIG BANG IN THE NEIGHBORHOOD

Astronomy is full of examples of dramatic events: flares, bursts, novae, supernovae, etc. Observation of these phenomena is obviously time-critical. Because of this need for immediate notification of such events, the astronomical community has its own electronic telegram: the International Astronomical Union Circular (*IAUC*). Urgent information is distributed electronically, via *e-mail*. In this way, the first naked-eye visible supernova in the telescope-age was reported through the *IAUC* in 1987 by Ian Shelton. Shelton, at the University of Toronto Las Campanas Station, discovered a supernova in the Large Magellanic Cloud (LMC, the galaxy nearest to our own Galaxy) on February 24.23 (UT) 1987 at an apparent magnitude of 5 (*IAUC* 4316). R. H. McNaught, Siding Spring Observatory, found no difference between two separate films of this region of the sky, taken on January 24 and on February 22. However, on a film taken on February 23.443 an object appeared at a magnitude of 6.0.

This event is the now famous SN 1987A, the first naked-eye supernova since the invention of the telescope in the 17th century. The observation of hydrogen lines immediately placed the supernova as a Type II supernova. SN 1987A is a unique supernova in many respects. It appeared in the LMC, whose distance is well known to be 55,000 pc. Furthermore, its progenitor, Sanduleak-69 202, is clearly identified. This was the first time that a supernova progenitor was observed. Sanduleak is an American astronomer who has studied and catalogued hot stars in the LMC. The value -69 indicates the latitude while 202 is the number in the catalog of the star. This star was a blue giant, whereas stellar evolution theory had predicted that Type II supernova resulted from the explosion of a red giant. The maximum light of SN 1987A was only 1/100 of what was expected from a typical Type II supernova. We now have an understanding of why a blue giant in the LMC could have gone supernova. The LMC has a metal abundance that is only a few tens of percent of cosmic values. This metal deficiency affects the star's evolution in such a way that the mass-loss rate during the red giant phase is increased. The theory of stellar evolution can therefore account for the explosion of a blue giant. It is interesting to note that the theory did not predict this result beforehand.

Another important observational aspect of SN 1987A was the first ever detection of a neutrino flux from outside of the solar system. M. Koshiba reported that the neutrino telescope KAMIOKANDE detected a neutrino burst on February 23.316, just at the beginning of the brightening phase of SN 1987A. KAMIOKANDE is a nickname derived from KAMIOKA, the location of the detector, and the name of the experiment, Nucleon Decay Experiment. The experiment consists of a large tank containing 2,140 tons of water along with a series of large phototubes surrounding the tank. Its original intent was to search for evidence of proton decay. KAMIOKANDE is sensitive to the study of the solar neutrino problem but is also expected to be able to detect a burst of neutrinos associated with a nearby supernova. Thus a great deal of attention was focused on the KAMIOKANDE experiment soon after SN 1987A exploded. The result, shown in Figure 5 (Hirata *et al.* 1987), was a detection of 11 events from SN 1987A, recorded on 07:35:35, February 23, 1987, with a timing accuracy of 1 minute. Because the original intent of KAMIOKANDE was the study of proton decay and the

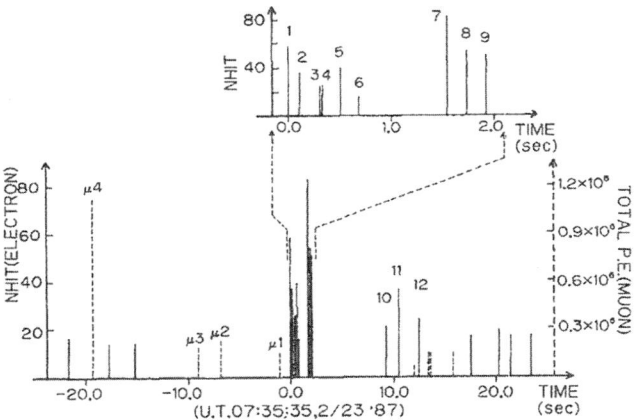

Figure 5 Time history of the neutrino burst detection at KAMIOKANDE (Hirata *et al.* 1987). Vertical lines show the detection epoch and its relative energy. Eleven events are clustered about 10 seconds.

solar neutrino problem, timing accuracy was not a concern, thus the 1 minute uncertainty. Since SN 1987A, a more precise clocking system has been installed.

Type II supernovae are believed to result from a progenitor star that has mass in excess of several $M_\odot$. Such a supernova is expected to leave behind either a neutron star or a black hole. With the gravitational collapse that occurred in the center of SN 1987A, a compact star heated to a very high temperature must have formed. The gravitational energy released must have been of the order 10^{53} ergs. The compact star, apart from its high temperature, has a radius of a few tens of kilometers. Most of the gravitational energy is carried away by neutrinos on a timescale of about 10 seconds. A temperature of roughly 10^{10} K will give rise to neutrinos with an energy of about 10 MeV. The neutrino flux at the Earth was about $10^{18}\,\mathrm{cm}^{-2}$ and lasted for about 10 seconds. The cross-section for the interaction of neutrinos with matter (primarily electrons) is very small, being of the order $10^{-44}\,\mathrm{cm}^2$. In this way, only 11 neutrinos were detected in a 2,140 ton volume of water in KAMIOKANDE. The detection of this neutrino burst is a direct confirmation that gravitational collapse is the real mechanism that drives Type II supernovae.

On February 5, 1987, the third Japanese X-ray astronomy satellite, GINGA, was launched. It carried a large area gas proportional counter with a sensitive area of $\sim 4,000\,\mathrm{cm}^2$. Once a satellite has been placed into orbit, it typically enters a commissioning phase, i.e. a warming up phase. In the first two weeks, all of the pointing systems are checked to see if they are functioning properly. Following this, observations used to calibrate the detectors are carried out. It was at the time that the calibration plan was finalized that we were informed of the appearance of SN 1987A. The calibration plan was immediately abandoned and the satellite quickly pointed in the direction of the supernova. From this time on, GINGA intermittently observed SN 1987A and detected it to be brightening in X-rays only 130 days after the explosion (Dotani *et al.* 1987), a time that is much earlier than we expected. A similar detection in X-rays up to 200 keV (Sunyaev *et al.* 1987) was confirmed by the Mir-Kvant Space Station.

The Hubble Space Telescope (HST) also intermittently observed SN 1987A. It clearly detected the expanding ring, as shown in Figure 6. Finally, it detected a three-fold ring structure of the ejecta colliding with circumstellar matter. Because there is a region of dense circumstellar matter, we expect X-ray brightening once the collision with the ejecta occurs. This is expected to occur within a few tens of years after the explosion. There are currently several recently launched X-ray satellites that will be able to detect any X-ray brightening when this collision occurs. Many studies have been done to look for the pulsations from a newly-born rapidly rotating neutron star, a pulsar. As yet, no success has been reported.

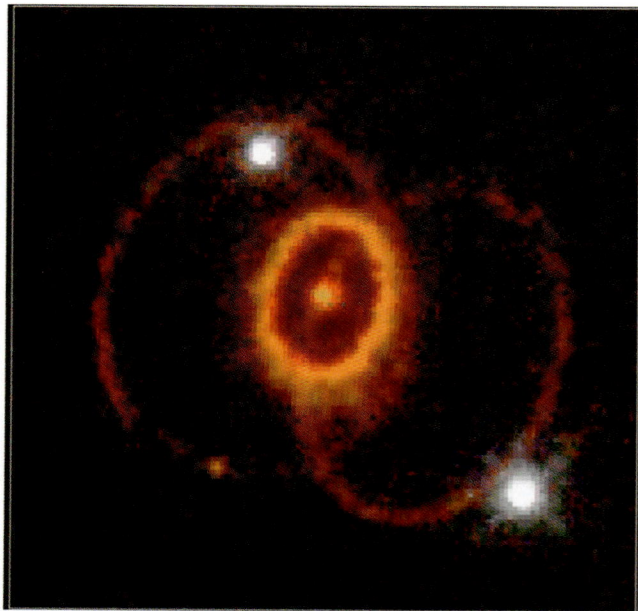

Figure 6 HST obtained the image of SN 1987A 7 years after the explosion (courtesy of Dr. C. Burrows, ESA/STScI and NASA). Two ring structures are clearly seen with some brighten spots on them.

4 YOUNG SNRS SHOWING THIN THERMAL EMISSION

Table 1 lists the supernovae which have been witnessed in our Galaxy over the last 2,000 years. Four SNRs show a shell-like structure in X-rays, while the others display a filled-center morphology. A typical shell structure is seen in the last three remnants listed in the table. These remnants display thin thermal emission in X-rays.

4.1 Cassiopeia-A

The X-ray spectrum of a thin hot plasma provides information on the conditions within the hot gas. If the object is a point-source, such as a star, a dispersive instrument, for example a grating or a crystal spectrometer, is useful in the same way as in the laboratory case. The energy resolution of such a wavelength-dispersive instrument depends on the image quality of the source. If the image quality is high enough, a grating or crystal spectrometer will have a very good energy resolution. On the other hand, if the image quality is poor or if the image is not a point source, a dispersive instrument is of limited use. In order to use such an instrument on an extended source, we would need to restrict the region observed so that it was small enough to

Table 1 Historical supernova explosion in our Galaxy

Year	Constellation	Name	X-ray morphology
185	Centaurus	G314.4-2.3 (?)	?
386	Sagittarius	G11.2-0.3	filled center
1006	Lupus	SN1006	shell
1054	Taurus	Crab nebula	filled center
1181	Cassiopeia	3C58	filled center
1572	Cassiopeia	Tycho	shell
1604	Ophiuchus	Kepler	shell
17th C.	Cassiopeia	Cassiopeia-A	shell

act like a point-source. We therefore must wait until energy dispersive detectors have improved enough to allow us to observe extended sources with a high energy resolution.

Cassiopeia-A is $3' \times 4'$ in size and shows the strongest thin, thermal X-ray emission from an extended source. As X-ray detection techniques were improved, Cassiopeia-A was always among the objects observed with the new technology. As such, a sketch of the X-ray studies of this object can be made. Studies made with gas proportional counters, a widely used device in the early days of X-ray astronomy, revealed the existence of Fe–K line emission around $6\sim7\,\mathrm{keV}$. The existence of this emission line is strong evidence for thin, thermal emission. Figure 7a shows the X-ray spectrum obtained with the ME (a gas proportional counter) onboard the European EXOSAT satellite (1983), whose energy resolution at 6 keV was about 5. The energy resolution is given by $E/\Delta E = \lambda/\Delta\lambda$, where E and λ are the incident X-ray photon energy and wavelength while ΔE and $\Delta\lambda$ are the narrowest possible widths (in energy and wavelength space) at which a perfectly narrow emission line would be detected.

A more detailed spectrum was obtained with the gas scintillation proportional counters employed by the Japanese TENMA satellite (1983) and the EXOSAT satellite. The energy resolution at 6 keV is about twice that of a gas proportional counter. Figure 7b shows a spectrum obtained with such a detector onboard the Japanese ASCA observatory (1993). This spectrum clearly shows that the X-ray spectrum below 6 keV contains emission lines from Si (1.8 keV) and S (2.5 keV). The TENMA and EXOSAT satellites found that the thin hot plasma in Cassiopeia-A did not reach the condition of collisional ionization equilibrium, since the energy of the Fe–K line center was not high enough when compared to what was expected from the electron temperature. The German–US–UK satellite ROSAT (1990) observed Cassiopeia-A and obtained an image using the PSPC (Position Sensitive gas Proportional Counter) as shown in Figure 8. A clear shell-like structure is seen, suggesting an expanding, strong shock. The energy range of

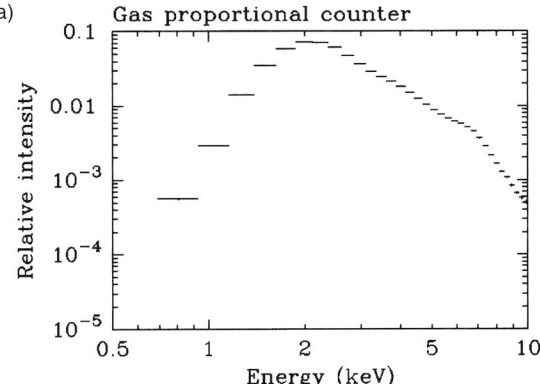

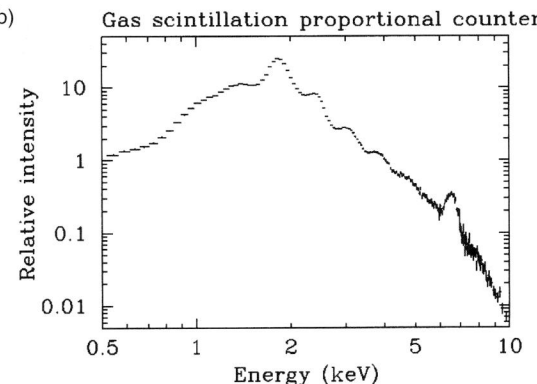

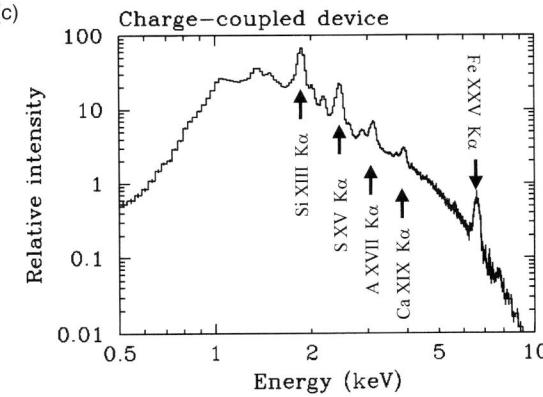

Figure 7 X-ray spectra of Cassiopeia-A with various X-ray detectors: (a) a gas proportional counter whose energy resolution, $E/\Delta E$, is about 5; (b) a gas scintillation proportional counter whose energy resolution is about 12 and; (c) an X-ray CCD whose energy resolution is about 50.

ROSAT was about $0.1\sim2.0\,\mathrm{keV}$ and the PSPC energy resolution was comparable to other gas proportional counters. The image therefore shows the X-ray photon intensity.

The Solid State Spectrometer (SSS) onboard the Einstein observatory (US, 1978) revealed a detailed spectrum. The SSS had an energy resolution comparable to that of the solid-state detectors (SSD) widely used in laboratory

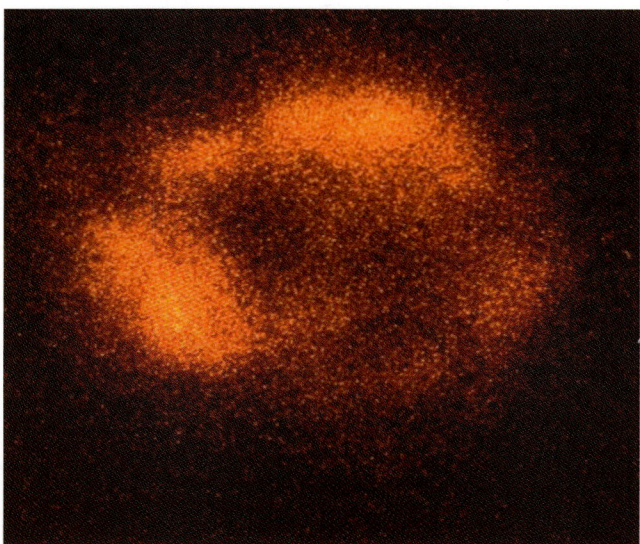

Figure 8 Cassiopeia-A X-ray image (see http://heasarc.gsfc.nasa.gov/docs/rosat/gallery/snrs.html and also see ROSAT Mission (http://www.xray.mpe.mpg.de/) and the MPE (http://www.mpe.mpg.de/)) obtained with ROSAT, whose effective energy range is 0.1~2.0 keV. Limb brightening is evident. The spatial resolution is about 30″.

experiments: the energy resolution is about $E/\Delta E = 40$ at 6 keV. It had no spatial resolution. The spectrum represents an integrated spectrum over the entire remnant. Clear emission lines from Si, S, A and Ca were seen (Becker et al. 1979). However, the effective energy range was only up to 4 keV, so that any emission from Fe–K was missed.

The Focal Plane Crystal Spectrometer (FPCS) onboard the EINSTEIN observatory also observed Cassiopeia-A (Markert et al. 1983). The field of view was restricted so that the intrinsic resolution of the detector could be met. Because the field of view was rectangular, a narrow slice was selected by the energy resolution. The observations entailed two pointings, one to the northwest portion of the remnant and the other to the southeast portion. Both of these regions form a portion of the bright shell of the remnant. The energy resolution of the FPCS could be as much as several hundred, which was good enough to resolve the Si–K emission in detail. The field of view was increased, however, in order to obtain good photon counting statistics. Three components of the Si–K line were observed: a helium-like Kα line, a helium-like Kβ line and a hydrogen-like Kα line. An analysis of the line intensity ratios of these lines enabled an ionization temperature, T_z, to be determined, which was found to be substantially lower than the electron temperature, T_e. Furthermore, they found the line-center energy of the helium-like Kα line differed between the two pointings. Strictly speaking, the helium-like Kα line has 3 components: a forbidden line a resonance line, and an intercombination line. The FPCS did not have a high enough energy resolution to be able to resolve these lines individually, so that the apparent line center of the helium-like Kα line depends on the relative strengths of these three components. These three components are also affected by the plasma conditions. The interpretation is that the energy difference between the two pointings did not originate from differences in the ionization state, but rather it resulted from a difference in Doppler shifts. The northwest region was expanding away along the line of sight, while the southeast was expanding towards us along the line of sight.

The X-ray spectrum obtained with the ASCA CCDs showed many emission lines, including Mg, Si, S, A, Ca and Fe, as shown in Figure 7c. The effective energy range is from 0.5 keV to 10 keV. The spatial resolution was about 2′.5. The energy resolution of the ASCA CCDs was high enough to be able to resolve different lines from the same species. This was the first time that spatially resolved X-ray spectra up to 10 keV were obtained for Cassiopeia-A. Three emission lines were seen from Si: helium-like Kα, helium-like Kβ and hydrogen-like Kα. The hydrogen-like Kβ line of Si is contaminated by the helium-like Kα line of S. A similar set of emission lines was seen for S as well.

There is a clear difference in the energy of the line center from position to position in the remnant (Holt et al. 1994). Due to its low spatial resolution, the results obtained from the ASCA data were similar to those obtained by EINSTEIN. A Doppler map of the Si–K line reveals the remnant to be axially symmetric with the north part running away from us, while the south part is running towards us. A simple interpretation is that the plasma is expanding in an axially symmetric way. In this model the axis of symmetry is almost perpendicular to the line of sight, pointing to the northwest. The plasma is expanding in the plane perpendicular to the axis of symmetry. The shell structure is, therefore, seen to be slightly distorted, showing an elongation from east to west.

The AXAF satellite, renamed Chandra just prior to its July 1999 launch, observed Cassiopeia-A with its X-ray CCDs. The spatial resolution is about 0″.5, which is comparable to ground-based optical telescopes. A detailed structure is revealed for the first time, as shown in Figure 9. The CCDs used in the observation had an energy resolution comparable to that of the ASCA CCDs. Most interesting is the detection of a point source in the center of the remnant, perhaps the neutron star generated by the explosion. A detailed study of the Chandra data is still under way. Figure 10 (Hwang et al. 2000) shows the images by using emission lines of Si, Ca and Fe; they clearly indicate the distribution of each element. In the eastern part, Si and Ca lines are clearly seen, while the Fe line is quite dim, etc. We see how the heavy elements nucleo-synthesized in the supernova explosion are distributed into the interstellar space and mixed with the interstellar matter.

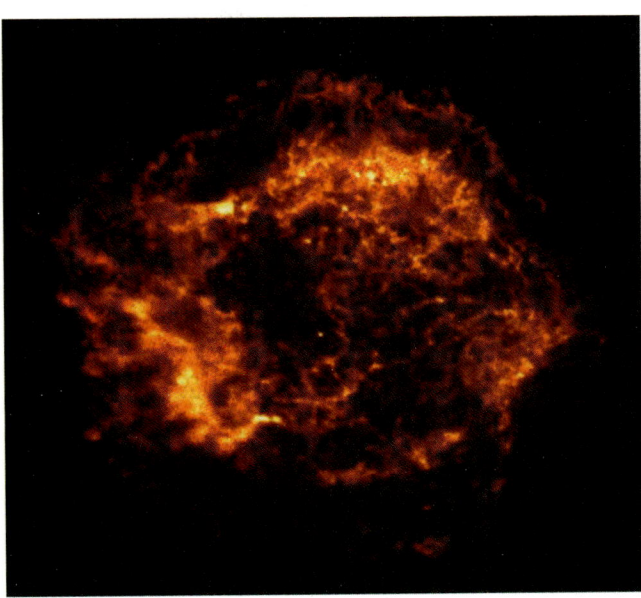

Figure 9 X-ray image of Cassiopeia-A obtained with Chandra whose energy range is up to 10 keV with a spatial resolution of about 0".5 (see http://Chandra.harvard.edu/photo/) (NASA/CXC/SAO). A point source is seen in the center.

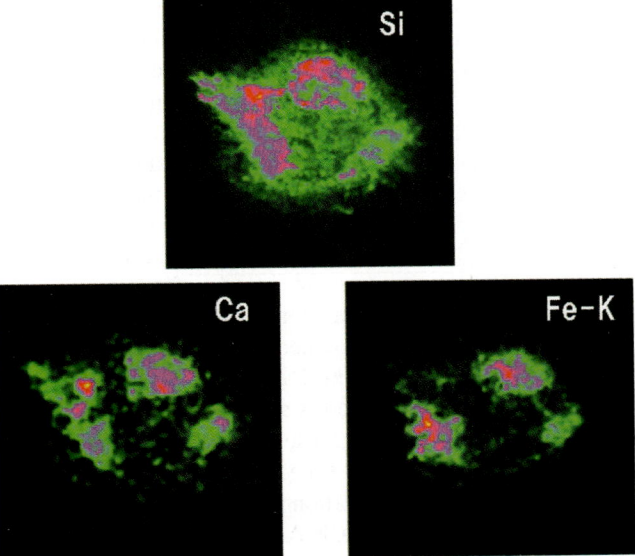

Figure 10 Intensity maps for Si, Ca and Fe–K emission lines of Cassiopeia-A (NASA/CXC/SAO). The relative line intensity is not uniform over the remnant.

The intensity ratio between the emission line and the continuum allows us to estimate the abundance of the element, if the continuum comes from the collision of electrons with ions. The detailed spatially resolved spectroscopy provided by Chandra reveals that some fraction of the continuum comes from a non-thermal origin, probably synchrotron emission. Synchrotron emission, which will be explained later, is characterized by a power law type spectrum, while the continuum of the thermal emission is characterized by an exponential type spectrum. The biggest observational difference is whether or not it extends to a higher energy range. BeppoSAX (an Italian–Dutch satellite, 1996) detected the X-ray spectrum of Cassiopeia-A to be extending up to several tens of keV (Favata et al. 1997), a factor that cannot be explained by a simple extrapolation of the low energy portion of the spectrum. Therefore, non-thermal emission may play an important role in the SNR, showing a shell-like structure with thin thermal emission.

4.2 Kepler's SNR

Kepler's supernova is so-named because of the intense study that it received from Johannes Kepler in 1604. Based on his light curve, it must have been a Type I supernova. This SNR is also a typical shell-like young SNR, showing thin thermal emission. Its apparent size is a little smaller, however, than Cassiopeia-A. Figure 11 shows the ROSAT image. The bright shell structure is clearly seen.

The GINGA satellite observed Kepler's SNR (Hatsukade et al. 1990) and found strong Fe–K emission, indicating an overabundance of iron. The interesting point is that the energy of the line center for Fe–K is around 6.5 keV. This clearly means that the iron has not yet reached collisional ionization equilibrium. Figure 12 (Kinugasa and Tsunemi 1999) shows the X-ray spectrum obtained with ASCA, displaying many emission lines of Si, S, A, Ca and Fe, much like Cassiopeia-A.

Detailed analysis of the spectrum indicates that the plasma in Kepler's SNR is in various physical conditions. This is an indication that the spatial structure is consistent with what is expected from a reverse shock propagating into the ejecta, rather than with what is expected from a fore shock propagating into interstellar matter. The metal abundance is certainly higher than the cosmic abundance. The total iron, however, is much less than what is expected from calculations of theoretical models of Type I supernovae. This situation is similar to that for Tycho's SNR.

4.3 Tycho's SNR

Tycho's SNR is the largest in apparent size of the three most recently observed supernovae in our Galaxy. It derives its name in virtue of its having been intensively studied by Tycho Brahe in 1572. It is about 8' in diameter at present, and it clearly shows a bright shell structure originating from the Type I supernova. Its spectrum is also similar to the spectra of Cassiopeia-A and Kepler's SNR. The apparent line center energies were measured to be 6.4 keV for Tycho and 6.6 keV for Cassiopeia-A (Tsunemi et al. 1986).

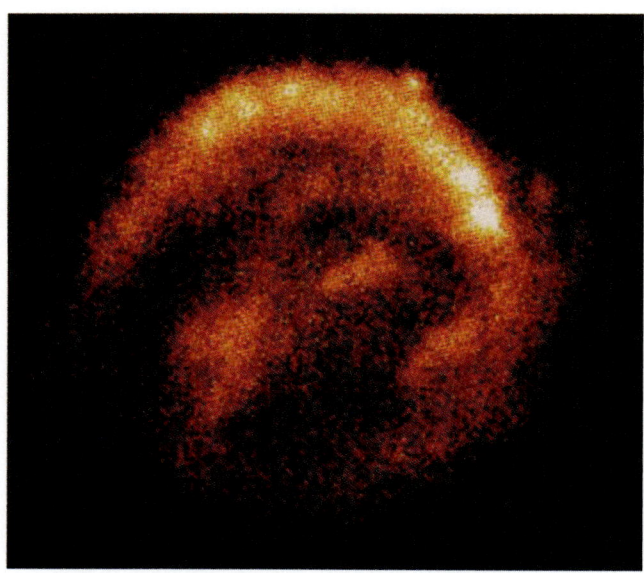

Figure 11 X-ray image of Kepler's SNR obtained with ROSAT (see http://heasarc.gsfc.nasa.gov/docs/rosat/gallery/snrs.html and also see ROSAT Mission (http://www.xray.mpe.mpg.de/) and the MPE (http://www.mpe.mpg.de/)). It shows a circular shape with complicated structure.

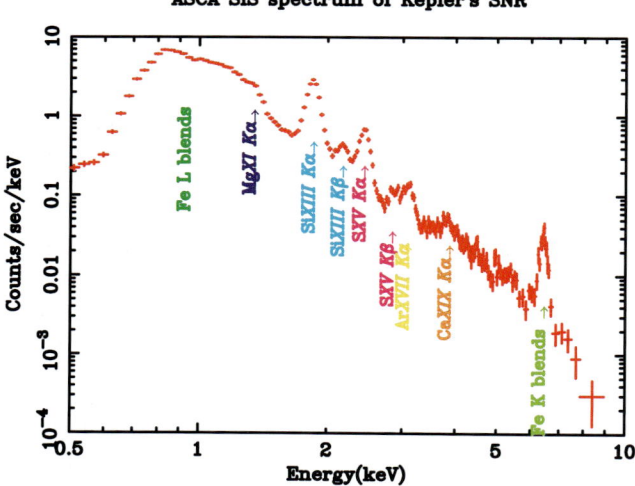

Figure 12 X-ray spectrum of Kepler's SNR (Kinugasa and Tsunemi 1999) obtained with the ASCA CCD.

The 6.4 keV emission line coming from Fe–K clearly shows that the iron atom is in a very low ionization state. This is an indication that the iron has just begun to heat up. ASCA observations revealed many emission lines along with their spatial distribution (Hwang and Gotthelf 1997). The XMM-Newton satellite, launched in 1999, clearly shows images of Tycho's SNR in various energy ranges, as shown in Figure 13 (Decourchelle et al. 2001). These images show that the metal distribution in the shell region depends on the azimuth. This is an indication that the metals nucleosynthesized inside the star are not uniformly distributed but instead display a substantial non-uniformity in abundances. The spectrum also indicates that the iron is in a state of very low ionization. Furthermore, the total amount of iron is found to be much less than what is expected from theoretical models for Type I supernovae. We therefore expect that a large amount of iron remains in the core region of Tycho's SNR. The inner most region of the star contains a large amount of iron, whereas the reverse shock does not reach the entire ejecta leaving the iron still unshocked.

4.4 W49B

Another SNR, W49B, displays very strong emission lines from various metals. Figure 14 shows the spectrum of W49B obtained with ASCA (Fujimoto et al. 1995). Emission lines from Si, S, A, Ca and Fe are seen. There are, in fact, three emission lines from each element: helium-like $K\alpha$, helium-like $K\beta$ and hydrogen-like $K\alpha$. These clearly show the plasma conditions for each element. The interesting point is that the heavier elements are in a lower state of ionization, suggesting that they are not well mixed and that they have been recently shock-heated. The imaging capability of ASCA is not good enough to reveal the distribution of the heavy elements. However, there is a clear indication that the iron resides inside the remnant, while silicon and sulfur are, relatively speaking, more outwardly distributed. This finding may suggest an onion-skin structure in the ejecta of this SNR. Observations with substantially higher spatial resolution (i.e. Chandra and XMM-Newton observations) will reveal how the metals are ejected by the explosion.

5 THE CRAB NEBULA: A HISTORICAL RECORD AND DETECTION OF A POWER-LAW SPECTRUM

In 1054, a bright star abruptly appeared in the constellation Taurus. It was bright enough that it must have been witnessed around the world. However, historical records of the event are restricted to the Eastern world. The 'guest star', which is how the Chinese referred to a new star, was one of the brightest objects in the sky and faded away within a year. Later, Charles Messier, an 18th century comet hunter who also catalogued objects which might be confused for comets by a naked-eye astronomer, assigned a nebulous object in Taurus to be the first object in his catalog of 110 objects. It is therefore known as M1. Lord Rosse (William P. Rosse), another 18th century astronomer, observed M1 with a large

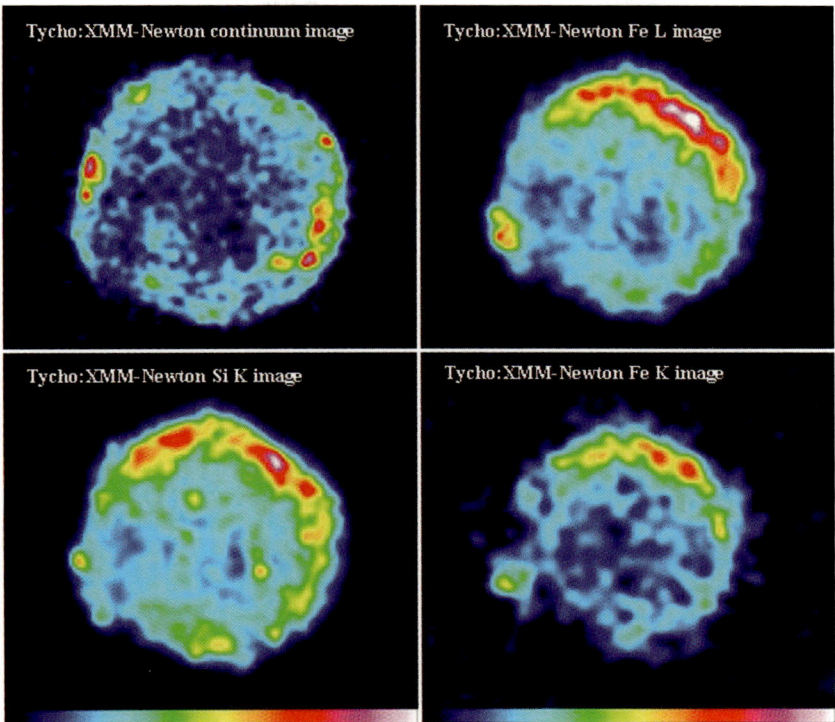

Figure 13 X-ray images of Tycho SNR for various energy ranges (Decourchelle et al. 2001) obtained with the XMM-Newton satellite.

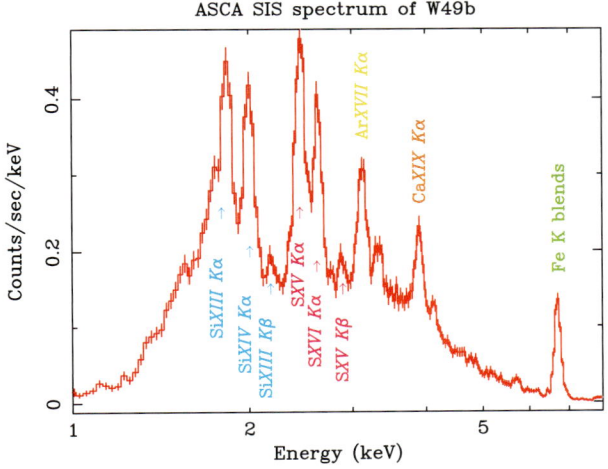

Figure 14 Thermal emission from W49B showing many emission lines. Each element is different plasma condition. (Fujimoto et al. 1995.)

telescope and called it the '*Crab Nebula*', given its similarity in shape to a crab (it is admittedly difficult to imagine a crab, however, when looking at a photographic image of it). Images in various wavelengths are shown in Figure 15.

In the 1960s, a strong X-ray source was detected from the direction of the Crab nebula (M1). It is now known that most of the emission comes from the extended nebula, while there is an X-ray point source at its center, as shown in Figure 15. This is a pulsating neutron star, which has a 33 msec rotation period. The pulse period of the pulsar is slowing down at a rate which indicates 'a characteristic age' of about 1,200 years (if the slow-down rate is constant through the electro-magnetic radiation process, the pulse period should have been zero about 1,200 years ago). The energy loss rate predicted by the slow down rate of the neutron star is almost equal to that of the energy radiated by the Crab nebula. It is therefore concluded that emission from the Crab nebula is powered by the rotational energy loss of the pulsar. The radiation is generated by synchrotron emission, which is generated by relativistic electrons spiraling in a magnetic field. The spectrum is characterized by a continuum with no emission lines. It is well characterized by a power law function (that is, the spectrum $\sim E^{-\alpha}$, where E is the photon energy and α is the so-called power-law index), which arises from a distribution of electrons whose energy spectrum is also described by a power law.

The bluish star at the center of the Crab nebula, seen as an optical image in Figure 15, was well known by astronomers before its pulsation was detected. It is called Minkowski's

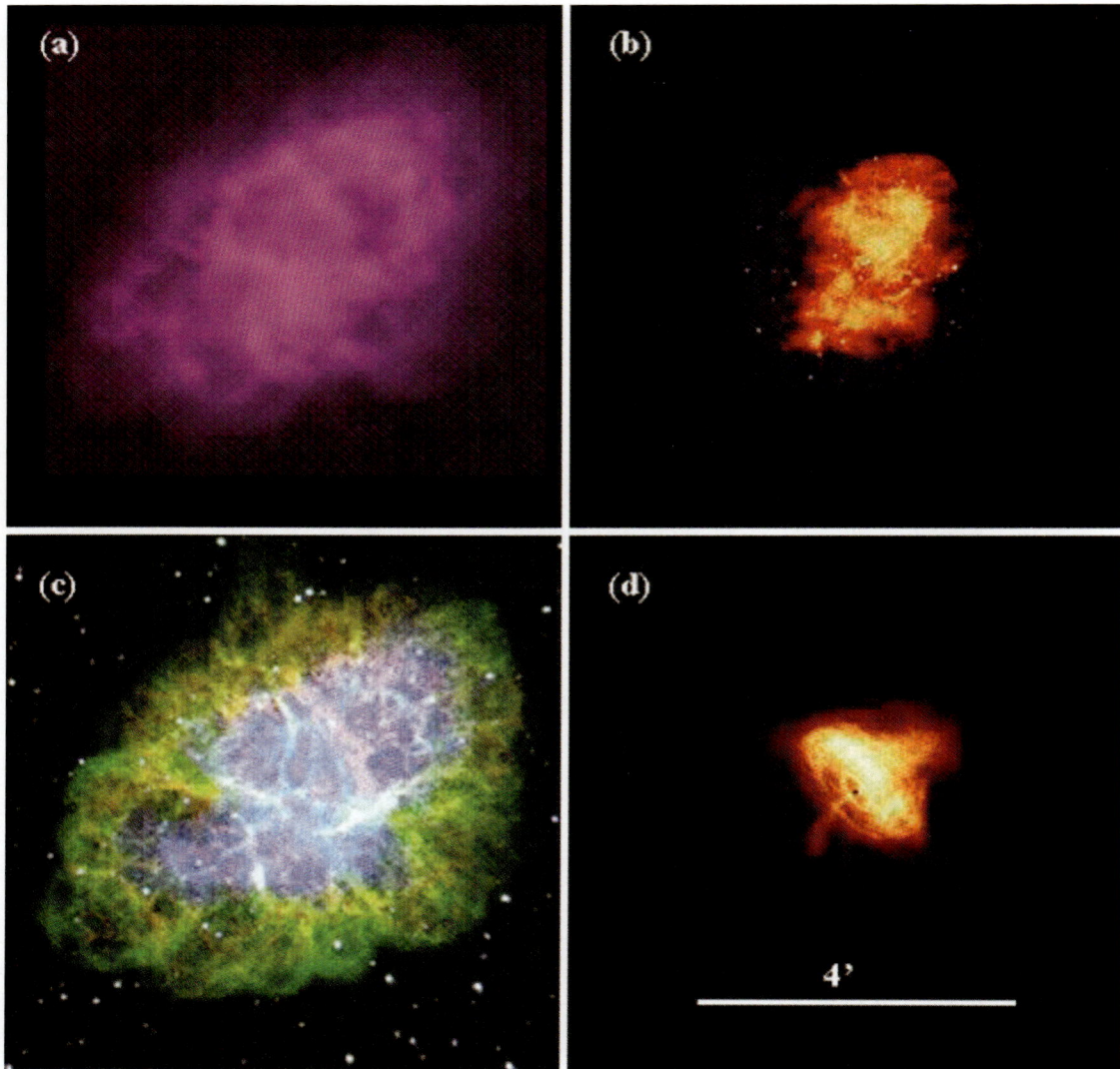

Figure 15 Crab nebula images for various wavelengths (see http://chandra.harvard.edu/photo/): (a) radio (VLA/NRAO), (b) infrared (W M Keck Observatory), (c) optical (Palomar Observatory) and (d) X-ray (NASA/CXC/SAO). The Crab pulsar can be seen in the center. All are in the same scale centered on the Crab pulsar.

star, and it is clearly bluer than the surrounding stars. This star now ranks as one of the most famous pulsars, even though it was not among the first discovered pulsars. The first discovered pulsar was found by J. Bell at radio wavelengths. She discovered radio pulsars with pulse periods on the order of one second, while the Crab pulsar has a period of 33 msec (that is, it 'flashes' 30 times per second).

The story about the discovery of pulsars has many interesting facets. The following example comes from the conference lecture, 'Neutron Stars and Pulsars' held in Rikkyo University, Tokyo, Japan in November, 1997. J. Bell gives the anecdote (Bell 1997) from the Yerkes Observatory, which houses the world's largest refractor (40-inch diameter):

This story, which Roger Romani reminded me of, also involves the Crab nebula. It concerns an occasion in the late 50s or early 60s when there was a public viewing evening at Yerkes Observatory. The telescope was set on Minkowski's star – now better known as the Crab pulsar, which pulses in the optical as well as in the radio. A female member of the public, viewing this object through the Yerkes telescope, claimed that it was flashing. The astronomer on duty described how stars twinkle or scintillate, but the woman maintained that she could appreciate the difference, and that this object was regularly flashing!

This episode may provide an interesting warning to the professional. Once people have a prejudice, it is very difficult to wipe out. At present, the Crab pulsar is one of the best-studied pulsars and will be discussed elsewhere in this book. Figure 15 shows the appearance of the Crab nebula at various wavelengths: radio, infrared, optical and X-rays. The appearance of the Crab nebula depends on the wavelength. It elongates from the south-east to the north-west at relatively long wavelengths (radio, infrared, optical), while it elongates perpendicular to this direction in X-ray images. This feature is also seen in the hard X-ray region (up to 100 keV) (Pelling *et al.* 1987). Furthermore, we see jet-like streams in the X-ray region along the elongation direction of long wavelengths. The jet-like streams follow the rotation axis of the X-ray pulsar.

6 A COSMIC RAY BIRTHPLACE: SN1006

The most brilliant supernova that has been observed was first noticed in May, AD 1006 and is now known as SN1006. Within a day, it became the brightest star. According to historical records in China and Arabia, it was as bright as the crescent moon and was sighted in the daytime. When it appeared in the Southern hemisphere, Mars was near Antares in Scorpio, which rose from the East just after sunset. The new star was south of Mars and Antares, about 25° away. Due to the location of the new star (15 h, $-42°$), the northern observation limit of SN1006 ran north of Paris. It took several years before it faded below the naked-eye detection limit. The radio and X-ray appearances show, at present, a clear shell-like structure, as shown in Figure 16 (Koyama *et al.* 1995). The apparent size at present is as large as the full moon. There are two bright regions in the shell that look like two crescents. At any rate, it was believed by many that this must be a typical shape for a shock wave.

In the early phase of X-ray astronomy, SN1006 was a puzzling object. The solid-state spectrometer (SSS) on-board the EINSTEIN observatory detected many emission lines from Si, S, A, Ca and Fe from young SNRs that showed a clear shell-like structure, such as Cassiopeia-A (Becker *et al.* 1979), Tycho (Becker *et al.* 1980), and Kepler (Becker *et al.* 1980). But no clear emission lines were detected from SN1006 (Becker *et al.* 1980). The SSS had good enough energy resolution to detect such emission lines, even though it had no spatial resolution. Other instruments on board EINSTEIN had either poor energy resolution or low detection efficiency for extended sources. A

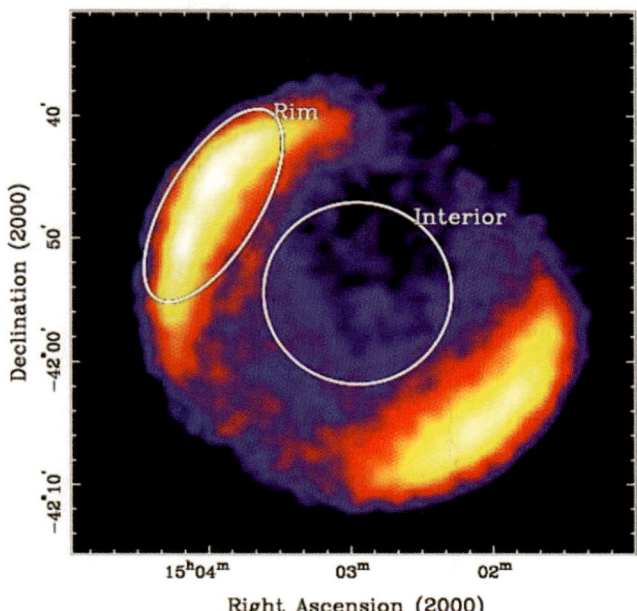

Figure 16 The full image of SN1006 obtained by ASCA. SN1006 is about 30′ in diameter showing a brightened shell. (Koyama *et al.* 1995.)

sounding rocket experiment, carrying a gas scintillation proportional counter, reported the detection of an O-K line (Vartanian *et al.* 1985), which was just below the energy detection threshold of the SSS.

The ASCA satellite, the first mission to carry an X-ray CCD, provided a spatially-resolved X-ray spectrum. This spectrum is shown in Figure 17 (Koyama *et al.* 1995). The spectrum of the interior shows several emission lines from O, Mg, Si and S. These lines must therefore originate from an optically thin thermal emission. However, the spectra from the bright shell regions are featureless. These spectra are power laws, such as those that would originate from a synchrotron source. The intensity distribution shows that the synchrotron emission comes from the shell region, while the thin thermal emission comes from the interior of the remnant. The spectrum from the direction of the interior is mainly optically thin thermal emission, along with some synchrotron emission. The spectrum at the edge of the remnant is from a synchrotron emission. Because of the relatively long path-length through the synchrotron emission region along the line of sight, limb brightening is seen, as shown in Figure 16.

The synchrotron emission comes from high energy electrons spiraling through a magnetic field. The electrons are accelerated to high energies via Fermi-acceleration within the shock. According to this mechanism, an electron is gradually accelerated as it bounces back and forth through

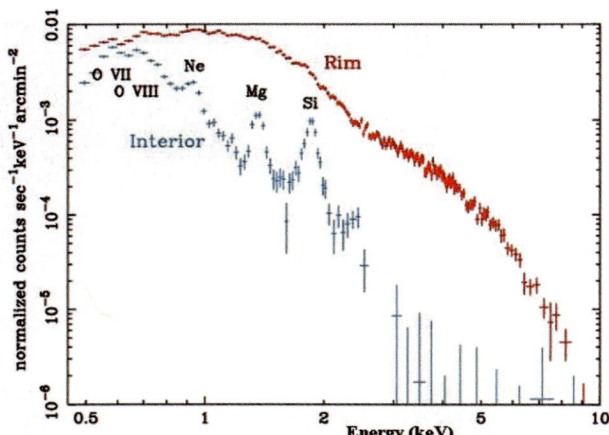

Figure 17 X-ray spectra for the shell region and the interior of SN1006 that is quite contrast to each other. (Koyama *et al.* 1995.)

the shock front. Because the X-rays originate from synchrotron radiation, there must be both high-energy electrons and a magnetic field in the shell region. Considering that the magnetic field, of the order of a few tens μ-gauss, results from the compression of the interstellar matter (which has a magnetic field), X-ray emission of photons with energies of a few keV suggests the existence of electrons with energies up to 100 TeV (1 TeV = 10^{12} eV).

The thermal electrons in the interior of SN1006 have energies of a few keV, and must act as 'seed' electrons in the shell region. Are the electrons really acclerated up to energies of 100 TeV?

The universe today is filled with the 2.7 K cosmic microwave background radiation, a fossil of the Big Bang. We expect that these photons will collide with the high-energy electrons and thus have their energies increased from 2.7 K up to TeV energies. This is the well-known inverse Compton Scattering mechanism. The 2.7 K photons are the most abundant in the universe, and it is therefore these photons that we expect to be primarily increased up to TeV energies. Because the electrons have such a high energy, the 2.7 K photons are seen in the electron frame of rest as having a very high energy. The energetic photon hits the electron at rest in its own frame of reference and Compton scattering results. In the laboratory frame, the opposite is observed: the 2.7 K photon is raised to a 100 TeV photon through a collision with the high-energy electron. Photons at these energies are γ rays, in particular, TeV γ rays. We therefore expect to observe 100 TeV γ rays from SN1006, if 100 TeV electrons are really present.

A TeV photon that enters the atmosphere of the Earth will encounter an atmospheric nucleus, producing charged particles that have velocities close to the vacuum-velocity of light. This velocity is greater than the velocity of light within the atmosphere (recall that the velocity of light in a medium is reduced from the velocity of light in vacuum by a factor equal to the index of refraction for that medium). Under this circumstance, the medium emits Cerenkov radiation in the visible portion of the spectrum. This radiation can be detected with phototubes. A TeV γ ray telescope consists of a very large reflector and an array of phototubes at the focal point that are able to detect the Cerenkov radiation and to determine its incoming direction. In 1997, such a TeV γ ray telescope detected a signal from the shell region of SN1006 (Tanimori *et al.* 1998). This provides us with direct evidence of the presence of electrons with up to TeV energies in the shell of SN1006.

One of the biggest current puzzles in Astrophysics is the origin of the cosmic rays, whose energy ranges up to very large values. The energy spectrum is characterized by a power law up to about 10^{20} eV, with a 'knee' around $10^{14\sim15}$ eV. The cosmic-ray energy density is about $1\,\mathrm{eV\,cm^{-3}}$. The energy loss rate of cosmic rays from our Galaxy is believed to be about $10^{40\sim41}\,\mathrm{ergs\,sec^{-1}}$. The source of the cosmic rays must therefore supply both this total energy and the high-energy particles. A supernova will deposit an energy of the order of 10^{51} ergs, and will occur on average every few tens of years in our Galaxy. Thus, if only $1\sim10\%$ of the supernova energy is incorporated into cosmic rays, they could provide the necessary energy. Furthermore, the SNRs can be a birthplace for the very high energy particles.

Do all young SNRs emit synchrotron radiation? When the thin thermal emission is mixed with the synchrotron emission, it is very difficult to distinguish the one from the other. Cassiopeia-A, which is believed to be the youngest SNR in our Galaxy, is one of the strongest X-ray emitters among those sources that show thin thermal emission. The X-ray spectrum shows many emission lines, which makes detecting any synchrotron emission very difficult. The synchrotron spectrum would extend up to high energies, whereas thin thermal emission drops off quickly at high energies. A power-law component in the broad-band spectrum of Cassiopeia-A has now been firmly detected at several tens of keV by BeppoSAX (Farata *et al.* 1997). We therefore expect synchrotron emission from SNRs that show predominantly thin thermal emission. In this way, we may expect that the SNRs in our Galaxy are potential candidates for cosmic-ray acceleration in our Galaxy.

7 ASSOCIATION OF COMPACT OBJECTS WITH SNRS

When a star has a mass greater than some critical mass, it will end its life in a supernova, leaving behind a compact object. Depending on its mass, the compact object will be either a neutron star or a black hole. The critical mass for the star to become a compact object is around $8\sim10\,M_\odot$. We therefore expect a compact star to be born after a Type II supernova. If the compact object is a neutron star and if it has a large enough magnetic field (it is generally expected that new born neutron stars have such a magnetic field), it will be a pulsar. The pulsar embedded in the Crab nebula is one of the most famous compact objects that was created in a supernova. Moreover, it is a very important pulsar, since we know with certainty the date of its birth, AD1054.

The X-ray emission from an X-ray pulsar extends to high energy, which is called to be a very hard spectrum. It is characterized by a power-law spectrum. Synchrotron nebulae also show such hard spectra. The Crab nebula is a typical example of this class. We therefore expect that when we find such a hard spectrum, synchrotron radiation must be responsible. In contrast, the X-ray emission from a black hole is rather soft. The difference between neutron star and black hole spectra results from the fact that the neutron star has a rigid surface, unlike the black hole. The X-ray emission from a shock-heated plasma shows thin thermal emission, which is rather soft. We can therefore infer the existence of a neutron star from an object showing such hard emission.

The Crab nebula is not the only SNR that is associated with a compact object. Such an association is confirmed when a compact source is found very near to the SNR, in most cases inside the remnant. Figure 18 shows the image of the Puppis-A SNR obtained with ROSAT. The inset shows a neutron star inside the remnant. Similarly, Figure 9 shows a point-source inside Cassiopeia-A that was discovered in 1999 by Chandra. This source is thought to be a newly born compact object. The presence of a neutron star would be consistent with the current assumption that Cassiopeia-A originated from a Type II supernova. No compact object has yet been found in either the Tycho or the Kepler SNRs, which is consistent with the belief that both of these remnants originated from Type I supernovae.

The remnant 3C58, which is believed to have originated from a supernova that occurred in AD1181, shows a similarity to the Crab nebula in both its morphology and in its X-ray spectrum. Though a compact object is expected in its center, pulsations from the expected pulsar have not been detected.

The remnant CTB109 observed with ROSAT (Rho and Petre 1997) is shown in Figure 19. Both its apparent shape and size are similar to those of the half moon. An expanding shell forms the circular structure, so that it is puzzling that half of the disk is missing. There is a pulsar, 1E 2259+586, at its center appearing as a very bright point source. The pulsar has a period of 6.98 seconds. This remnant shows an association of an accretion-powered pulsar and an expanding shell with a high inhomogeneous structure. It is quite interesting that the pulsar is in a binary

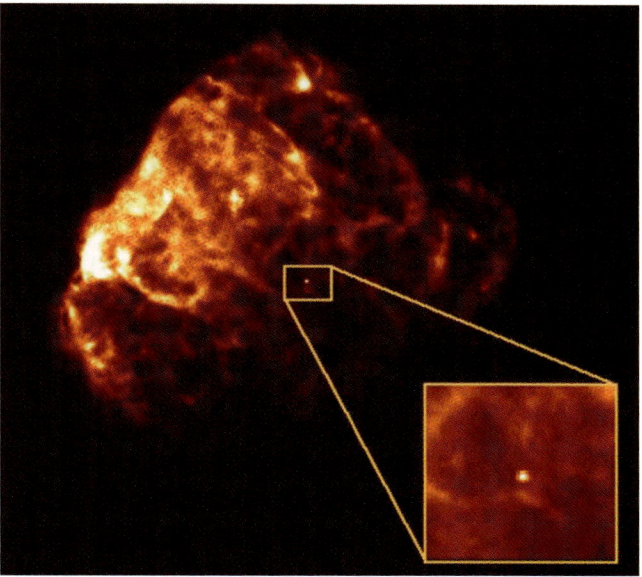

Figure 18 ROSAT clearly shows a complicated structure of the Puppis-A SNR (see http://heasarc.gsfc.nasa.gov/docs/rosat/gallery/snrs.html and also see ROSAT Mission (http://www.xray.mpe.mpg.de/) and the MPE (http://www.mpe.mpg.de/)). There is a point source in its center, a neutron star generated by the supernova explosion.

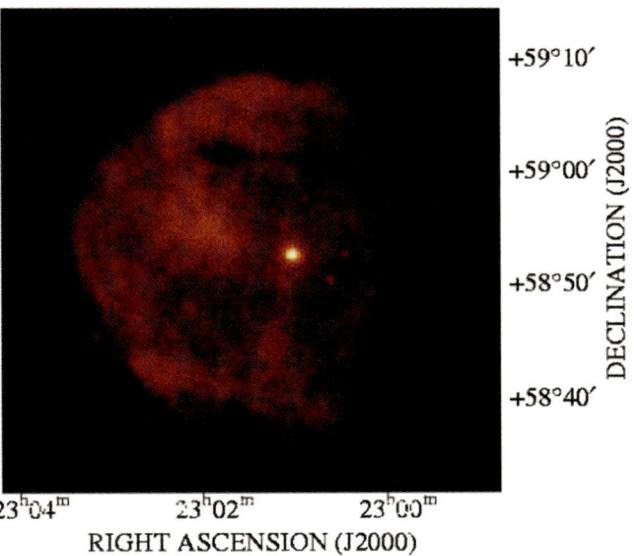

Figure 19 X-ray image of CTB109 obtained with the ROSAT PSPC (see http://heasarc.gsfc.nasa.gov/docs/rosat/gallery/ snrs.html and also see ROSAT Mission (http://www.xray.mpe.mpg.de/) and the MPE (http://www.mpe.mpg.de/)). It shows an expanding half disk with an X-ray pulsar at its center.

system with a faint optical counterpart. This is an example of the binary system that is not disrupted by the supernova explosion.

The remnant RCW103 also displays a shell-like morphology in X-rays, with a diameter of about 8'. It also has a point source at its center, 1E 161348-5055. In 1992 a pulse-period of 69.319 msec was detected by the GINGA satellite (Aoki *et al.* 1992). The FOV of GINGA was 1° × 3°, which is much larger that the extent of RCW103. The energy range of GINGA was 2~60 keV. The energy range of imaging satellites (EINSTEIN ranged up to 4 keV, ROSAT ranged up to 2 keV) available at that time indicated that RCW103 was the only X-ray source in the FOV of GINGA, as shown in Figure 9. The detected pulsation, therefore, had to originate from the point source within the remnant. Since the pulsar usually shows a hard X-ray spectrum, the pulsed component is very weak at low energies, where the thermal emission dominates. Thus, imaging at hard X-ray energies was desired. The ASCA satellite observed RCW103 with imaging detectors that were sensitive up to 10 keV. Even though a substantial search for the pulsations was carried out for RCW103, no pulsation was found for any energy in the range 0.5~10 keV. The pulsations reported for the GINGA observations were therefore suspected to be the result of a statistical effect.

ASCA obtained images for RCW103 below and above 3 keV. Although ASCA images have poorer spatial resolution than ROSAT, the low-energy image is consistent with what was found with ROSAT. What is interesting is that in the high-energy image two sources are seen, while in the low-energy image the northern source is not seen. Torii *et al.* (1998) detected pulsations from the northern source, AXS J161730-505505, which is located about 10' away from the center of RCW103, outside of the shell. Figure 20 shows the images of pulse-on and pulse-off that clearly indicates the northern source is pulsating while the southern source is not. The pulse period was 69.338 msec, slightly longer than that reported by GINGA. The difference indicates the spin down rate. Interestingly, these data were obtained in 1993. At this time, many attempts were made to find these pulsations. The data became public – everyone had access to them – in 1994, with no pulsations having been found. The pulsations were reported in 1997. Prior to this discovery, there had been a strong prejudice that there was a point-source at the center of RCW103 that must be responsible for the pulsation. The FOV of ASCA is about 25', which easily covers RCW103 and its northern source. These two sources are always observed simultaneously, while the pulsar was always missing from the pulsation searches.

The two sources around RCW103 are located at its center (1E 161348-5055) and a little outside of its shell (AXS J161730-505505). The northern source (lying outside of the shell) has a spectrum that is harder than the central source. In particular, the northern source is not seen below 3 keV because of interstellar absorption. A hard spectrum is an indication of a neutron star, while a soft spectrum is an indication of a black hole. The nature of the central source is not yet understood, while it must be related to RCW103. The characteristic age of the pulsar derived from the pulse period change is about 8,000 years while the thermal emission from the shell gives an age estimate of about 4,000 years. The association between the pulsar and the SNR is therefore not yet established. If the supernova occurred in a binary system that contained a compact object one would remain at the center while the other would be ejected from the system. Further study of these two objects is thus being vigorously pursued.

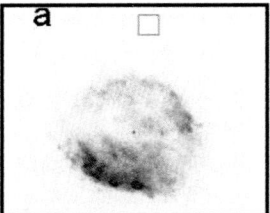
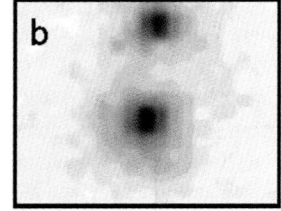
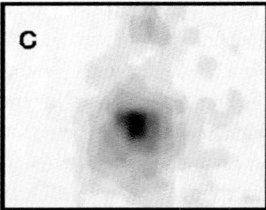

Figure 20 (a) ROSAT (0.1 keV~2.0 keV) clearly shows a shell like structure of RCW103 with a point source at its center (see http://heasarc.gsfc.nasa.gov/docs/rosat/gallery/snrs.html and also see ROSAT Mission (http://www.xray.mpe.mpg.de/) and the MPE (http://www.mpe.mpg.de/)). The upper square denotes the position of the X-ray pulsar. ASCA showed a relatively high energy X-ray image (3 keV~10keV) around RCW103 for the pulse-on phase (b) and the pulse-off phase (c) of the pulsar. The center source shows constant intensity while the northern source is pulsating.

8 THE VELA SNR: A MIDDLE-AGED SNR

The Vela SNR is well known since the discovery of the pulsar, PSR 0833-45 (89 msec pulsation period), inside the remnant. PSR 0833-45 and the Crab pulsar are the two most well-studied pulsars. This SNR is in contrast to the Crab nebula in many ways. It is only 250 pc away from us so that there are few absorption features seen in X-rays. The estimated age is 10,000 years, resulting in low-temperatures in X-rays. The ROSAT All Sky Survey revealed the entire appearance of this source for the first time, shown in Figure 21 (Aschenbach *et al.* 1995). It shows a circular structure with a diameter of about 8°. Another bright SNR, Puppis A, is seen on the upper right edge of the Vela SNR. The pulsar is also seen as a point source near the geometric center. Because there is a pulsar within the remnant, it must have originated from a Type II supernova.

There are a few interesting features seen outside of the Vela SNR. Several 'boomerang' like structures are shown in Figure 21a. These structures give the appearance of supersonic motion in a tenuous plasma, as though they were shooting out from the center of the remnant. They may be shrapnel from the supernova event, and would thus represent 'fossils' of the progenitor. The opening angle of the boomerang-like structures gives a Mach number, i.e. the ratio between the local sound velocity and the supersonic velocity. The local sound velocity is derived from the measured X-ray temperature. In this way, the velocity of these shrapnels is estimated to be of the order of a few hundred $km\,sec^{-1}$. If they were shrapnel from the supernova, their initial velocity would have been a few thousand $km\,sec^{-1}$ and would therefore have suffered from a severe deceleration. In this deceleration, the shrapnels apparently did not dissolve into the surrounding matter.

X-ray spectra of one of the shrapnel (called shrapnel A) were acquired with ASCA. The spectra revealed a strong emission line from Si–K, suggesting an overabundance of Si (Tsunemi *et al.* 1999). The absolute abundance measured depends on assumptions made for the abundances of other elements. Therefore, a relative abundance between Si and O was made and was found to be very different from the cosmic value. This would indicate, then, that shrapnel A is silicon-rich, and could have originated in a silicon-rich layer in the progenitor star. Even in old SNRs such as the Vela SNR, complete mixing of the ejecta with the interstellar medium has not yet taken place.

Figure 21b shows an image of the same region as Figure 21a, for relatively higher energies (Aschenbach 1998). In this map, a major portion of the Vela SNR has disappeared, while Puppis-A is still visible in the upper right. Apart from this, we can see a circular structure with a diameter of about 2° at the lower left of the Vela SNR. This source, RXJ0852-4622, seems to be on the near side of the Vela SNR. The size of the SNR is large only because it occurred nearby. It must therefore be very young.

γ ray line emission has been detected from RXJ0852-4622 by COMPTEL (Iyudin *et al.* 1998). This is the second SNR showing γ ray line emission: the first is Cassiopeia-A, a very young SNR (Iyudin *et al.* 1994). The γ ray emission originates from ^{44}Ti, which decays into ^{44}Sc with a half life of 60 years. The ^{44}Sc in turn decays into ^{44}Ca with a half-life of only several hours. During this decay chain, three γ ray lines are produced: the first decay results in a line at 68 keV and another at 78 keV, while the second decay results in a line at 1157 keV. Due to the relatively short half-life, we can assume that the supernova event must have occurred in the recent past. ^{44}Ti is a fusion product associated with the ^{56}Ni that is nucleo-synthesized during the

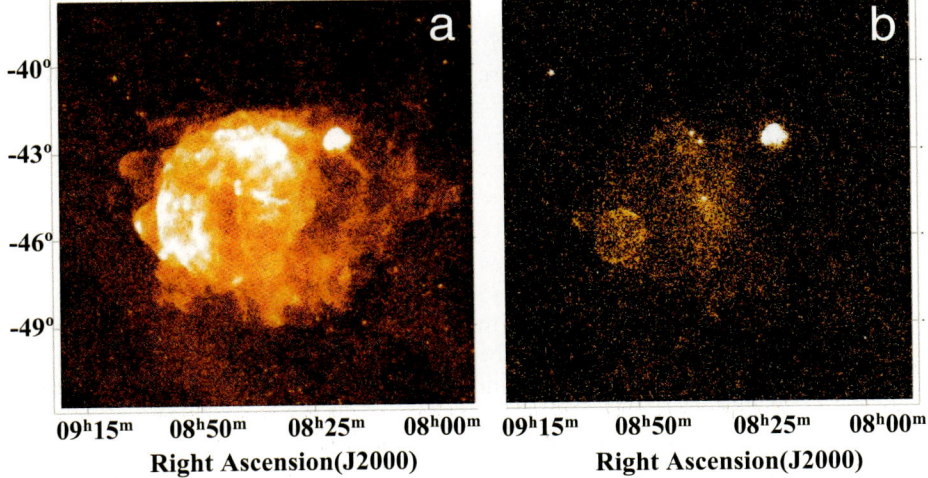

Figure 21 A full image of the Vela SNR obtained with ROSAT (see http://heasarc.gsfc.nasa.gov/docs/rosat/gallery/snrs.html and also see ROSAT Mission (http://www.xray.mpe.mpg.de/) and the MPE (http://www.mpe.mpg.de/)) below 1.5 keV. (a) Several boomerang structures are seen around the main shell. Above 1.5 keV a new circular structure in the lower left appears (b), that coincides in position with the γ ray line emitter from ^{44}Ti (Iyudin *et al.* 1998.)

supernova explosion. Therefore, the detection of the 1157 keV γ rays provides a tool useful in determining whether or not a SNR is young.

9 THE CYGNUS LOOP: A MIDDLE-AGED SNR

The Cygnus Loop, one of the most heavily studied SNRs, resulted from a supernova that occurred about 20,000 years ago. Its diameter is about 3° and it shines brightly over a wide range of energies. It shows clear limb brightening due to a shock front expanding into the interstellar medium. Figure 22 shows the highest quality X-ray image available at the time this article is being written. It was obtained by the ROSAT HRI. We can see very fine details in the shell region where veil-like structures are seen.

In the early stages of X-ray astronomy, it was uncertain as to whether the emission came from a thin thermal spectrum or from a power-law spectrum. At the time, the energy resolution was not high enough to distinguish between the two using the gas proportional counters that were widely used in X-ray observations. Gas scintillation proportional counters, having improved energy resolution, were introduced in 1977 for space use. Using such a detector, emission lines of O VII and O VIII were detected (Inoue et al. 1980). The Einstein observatory was equipped with a focal plane crystal spectrometer (FPCS) that revealed detailed emission line information for a small portion of the northern portion of the Loop (Vedder et al. 1986). The FOV of the FPCS was very limited, while the energy resolution was far superior to that for gas scintillation proportional counters. Three emission lines were detected from helium-like oxygen: the resonance line, the forbidden line and the intercombination line. Further, it was found that the emission-line structure from oxygen was not in collisional ionization equilibrium.

9.1 ASCA observations of the Cygnus Loop: Type II SN

ASCA observed the Cygnus Loop with its high energy resolution CCDs. It detected emission lines from O, Mg, Si, S and Fe with a spatial resolution of a few arcminutes (Miyata et al. 1994). In particular, one sees strong emission lines from oxygen, including the $K\alpha$ lines from the helium-like ion (O VII) and the hydrogen-like ion (O VIII). The line intensity ratio of these two lines gives a relation between T_e and τ. The result found was that the plasma in the shell region showed a radially dependent plasma condition, far from the CIE condition. Even in a well-aged SNR, the shock front is expanding into the interstellar matter so that it is continually supplied with fresh material. This results in a radial dependence in τ, with τ increasing rapidly at the shock front, then reaching a maximum just behind the shock and then decreasing. ASCA found that the NEI condition is universal throughout the Galaxy. The emission from the shell region is thus a superposition of different hot plasmas, each possessing a different NEI condition. The metal abundance in the shell region is below cosmic values, particularly for oxygen, which has an abundance of only 0.2~0.3 of that of cosmic values. Since interstellar matter is swept up in the shell region, the spectrum of the shell region reflects the abundances in the interstellar matter. For the Cygnus Loop, this consists of a volume of a sphere with a radius of about 20 pc. Oxygen may thus be under-abundant on this scale.

A simple model of SNR evolution indicates that there is a hot, thin plasma left in the center of the remnant. It remains at a high temperature for a long time due to its low density, which results in a low rate of energy loss via emission. ASCA detected strong emission lines from Si and S in the inner region of the Loop, as shown in Figure 23 (Miyata et al. 1998). The abundances of Fe, Si and S are at least several times higher than those of the cosmic values, a few tens of times higher than those in the shell region. This is interpreted as the fossil ejecta of the supernova. In this way, we found the Cygnus Loop to have a two-fold structure.

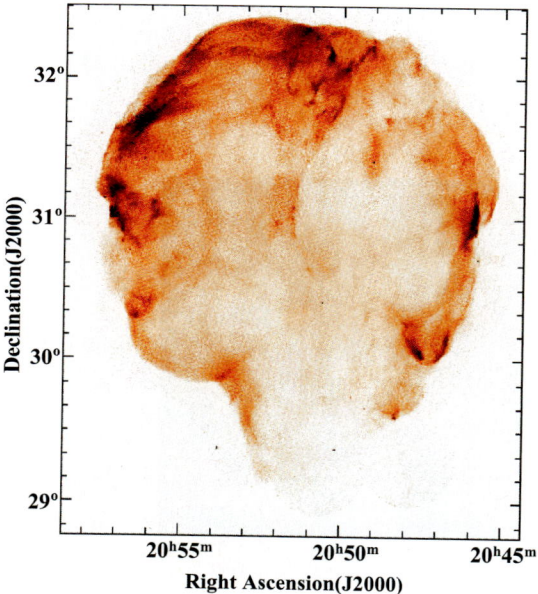

Figure 22 The Cygnus Loop image obtained with the ROSAT HRI (Levenson et al. 1998; see http://heasarc.gsfc.nasa.gov/docs/rosat/gallery/snrs.html and also see ROSAT Mission (http://www.xray.mpe.mpg.de/) and the MPE (http://www.mpe.mpg.de/)). The effective energy range is below 1 keV. A brightened shell is clearly visible.

A metal rich plasma is left in the inner part of the remnant that is a fossil of the ejecta. The interstellar matter swept up by the shock wave surrounds the ejecta as the white of an egg surrounds the yolk. They are not yet mixed together but are still separate, resulting from the fact that the Cygnus Loop is still 'fresh'.

The metal abundance of the ejecta reflects the metal abundance near to the core of the supernova progenitor. A careful study of the relative abundance between Fe, Si and S indicates that the progenitor mass was $20 \sim 25 M_\odot$ and that the supernova responsible for the Cygnus Loop must therefore have been a Type II event. We would expect a compact object to have been left behind.

The distribution of the emission from silicon, as measured by its equivalent width, is shown in Figure 24. The emission is well concentrated inside the Loop. The large equivalent width of the line is a strong indication that it is fossil ejecta from the progenitor. We thus expect that silicon ought to be a good tracer in the Loop of ejecta from the progenitor. The larger the equivalent width of the line, the greater the fraction that came from the ejecta. The distribution of the ejecta is concentrated inside the Loop, but it is not point-symmetric in shape. It is well elongated southward from the center where there is a break out, as shown in Figure 22. The X-ray intensity drops off in the break out region, where the shock front is not so clear. This situation also holds for the high-energy map obtained with ASCA, as shown in Figure 24. It is interesting to note that the intensity dims in the southward break-out, where the ejecta fraction is quite high.

There is a point source at the southern edge of the fossil region, as shown in Figure 24 (Miyata *et al.* 1998), as if the ejecta were pulled out by this source. This source becomes clearer at higher energies. A careful study of the ROSAT PSPC image revealed a point source in the same location as found by ASCA. The X-ray spectrum of this source is a power law with an interstellar absorption feature that makes it difficult to detect at lower energies. No pulsation has been detected from this source.

What is the nature of the point source? Is it related to the Cygnus Loop and was, therefore, created in the supernova? If it is related to the supernova explosion, it has to be a compact star (neutron star or black hole) that has been kicked out from the explosion center. Based on the X-ray spectrum, there are two candidate sources: a magnetized neutron star or an active galactic nucleus (AGN). AGNs are distributed uniformly in the sky and are detected more frequently than are neutron stars. If it is an AGN, the optical counterpart will show the emission lines that distinguish AGN from field stars. Figure 25 shows an optical image acquired with the 4-meter telescope at Kitt Peak National Observatory, where the circle denotes the position uncertainty of the X-ray point source (Miyata *et al.* 2001). As seen in Figure 25, there are several candidates for the

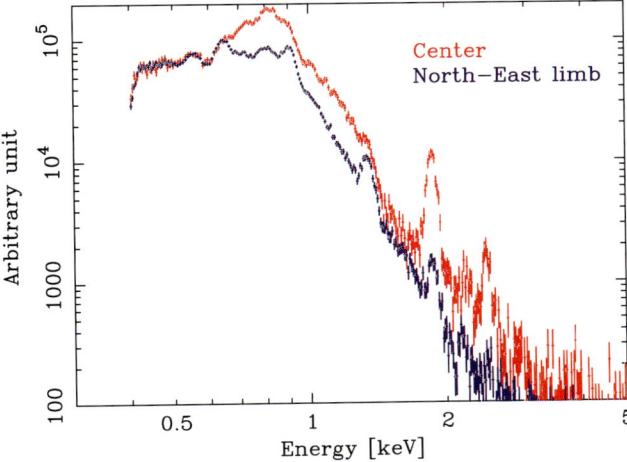

Figure 23 X-ray spectra of the Cygnus Loop: a shell region in the north-east limb and a central region of the Loop. Si–K (1.8 keV) and S–K (2.4 keV) in the center are much stronger than in the shell region. (Miyata *et al.* 1998a.)

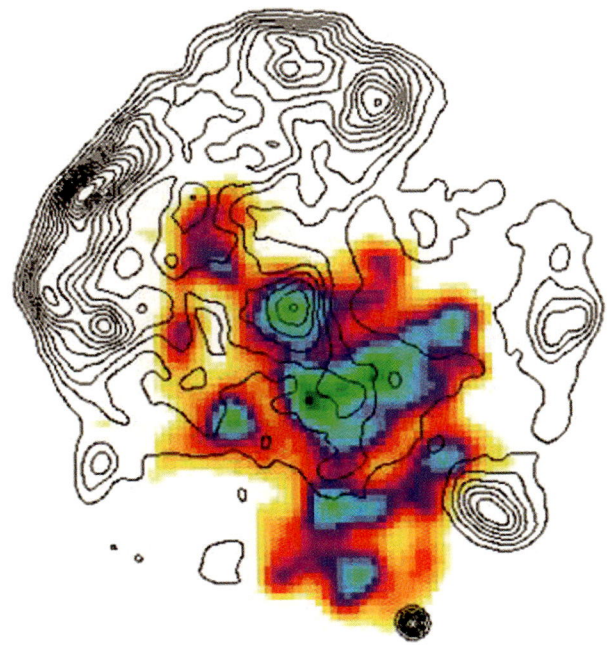

Figure 24 Distribution of the Si abundance inside the Cygnus Loop is shown by gray scale while the X-ray intensity above 0.7 keV is superimposed by contours.

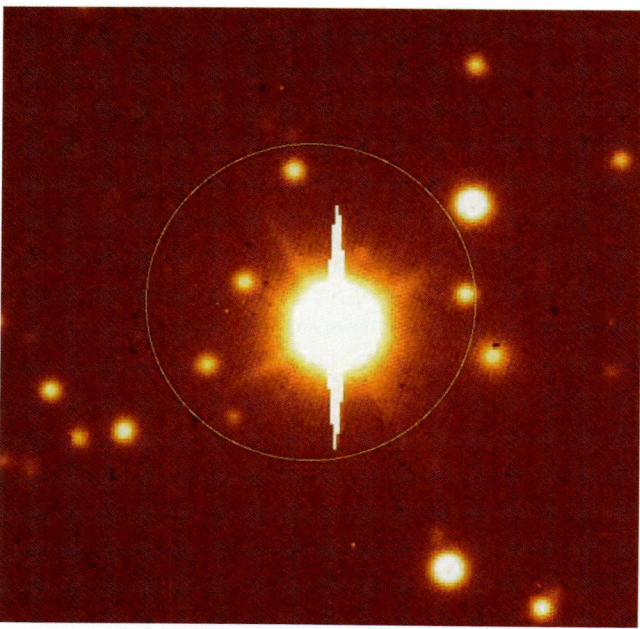

Figure 25 X-ray point source location is shown by a circle with a radius of 10″ on the optical image obtained with Kitt Peak National Observatory. No counterpart is established. (Miyata *et al.* 2001.)

optical counterpart, that have been carefully studied. In this way we have confirmed that there is no AGN among the optical counterpart candidates. At this stage, it is not yet clear whether or not the point source is associated with the remnant. If it has been kicked out from the remnant center by the explosion, the velocity perpendicular to the line-of-sight has to be about 1,000 km sec^{-1}, which would make it a high velocity compact star. A careful study in the radio band may reveal its proper motion.

10 SUMMARY OF THIS CHAPTER AND A RELATED EPISODE

Observations of SNRs are conducted primarily at radio and X-ray wavelengths, since most of the emission appears in these wavelength bands. In particular, a great deal of progress on SNR research has been made through X-ray observations. In the 1970's, most of these observations were carried out via sounding rockets. Such flights allowed for observations that extended over several minutes. Since that time, however, several X-ray observatory satellites have been launched. These observatories have allowed for much longer observations.

The apparent size of SNRs ranges from point-like to large extended sources. SN 1987A is now expanding from having a point-like structure to having a noticeable size, as observed by the HST. Young SNRs, such as Cassiopeia-A, Tycho and Kepler are several arc minutes in size. SN1006, Puppis-A and CTB109 are about one half-degree in size, similar to the full moon. The Cygnus Loop is about 3° in diameter, while the Vela SNR is more than twice that size.

The linear extent of SNRs ranges from point-like to a few tens of pc. The younger objects are relatively small since they are still expanding rapidly, i.e. they have not yet swept up enough material to have been decelerated appreciably. The older objects are quite large. These remnants have had their expansions appreciably decelerated. When the expansion of the shell is comparable to the proper motion of the circumstellar environment, expansion is no longer traceable. Even then, a hot cavity still remains inside the shell where all of the material is highly ionized due to the high temperature. The remnant expansion will reach this stationary phase within about 100,000 years.

There are two flows in the Galaxy that govern its internal dynamics. One is an energy flow: the energy transfers from hot matter to cold matter. Our Galaxy, and probably our universe as well, is so violent that it does not reach equilibrium. The other is a flow of metals: the high-Z elements transfer from the core of stars to the interstellar space through the supernova explosions. Our Galaxy is not in a static phase but instead is in a dynamic phase all the time. These activities are mainly maintained by the supernovae. We have studied the SNRs through X-ray observations, which allow direct measurements of the high temperature, metal-rich plasma to be made. We have only glimpsed the initial phase of the SNR, where the metal rich plasma is still separated from the surrounding interstellar matter. After this, a hot cavity will be left that is very difficult to study in detail.

The hot cavity that is left behind will remain hot for several hundreds of thousands of years. Taking into account the occurrence rate of supernovae in our Galaxy, many such hot cavities ought to remain today. We can easily guess that a substantial fraction of the Galaxy is occupied by these hot cavities, where no neutral matter is left.

The sounding rocket observations of the 1970's revealed that there was very little neutral matter around the solar system. In other words, the solar system is surrounded by thin hot gas, with a temperature of about one-million degrees (Tanaka and Bleeker 1977). In the observations of early-type stars, we can measure the interstellar absorption feature by highly ionized O, allowing us to estimate how much hot gas is present along the line of sight to the star. Such an analysis also indicates that the solar system is surrounded by hot gas (Jenkins and Meloy 1974). This is not surprising if we assume that a substantial fraction of the Galaxy is filled by hot gas originating from supernova explosions. The solar system happened to slide into the hot cavity of a nearby supernova or was instead engulfed by a nearby supernova.

What would have happened if the solar system encountered an SNR? The solar system should have passed the shell region where there is a region of enhanced density. If we were there, we would look through some regions in the sky with a relatively high density of interstellar matter. These regions would be in a direction tangential to the cavity surface. In the winter sky, there is a winter triangle formed by three bright stars: Sirius in Canis Major, Procyon in Canis Minor and Betelgeuse in Orion. We can see them at night and notice that Betelgeuse appears very reddish in color while the other two are bluish. The great Greek astronomer of the 2nd century, Claudius Ptolemy, recorded interesting events in the Almagest. He reported that Sirius was reddish, which is sharply in contrast with the situation at present (see for example Whittet 1999 and references therein). Is it possible for Sirius to change color in a few thousand years? From the perspective of stellar evolution theory, it is not possible. If it really was reddish, it may have been due to extra interstellar absorption between us and Sirius. This is, of course, merely speculation. However, supernova explosions certainly play a crucial role in the evolution of matter and energy in our Galaxy.

REFERENCES

see http://oposite.stsci.edu/pubinfo/gif/CygnusLoop.gif

Anders, E. and Grevesse, N. (1989). Abundances of the elements – meteoritic and solar. *Geochimica et Cosmochimica Acta*, **53**, 197–214.

Aoki, T., Dotani, T. and Mitsuda, K. (1992). IAU Circ., 5588.

Aschenbach, B., Egger, R. and Trümper, J. (1995). Discovery of Explosion Fragments Outside the VELA Supernova Remnant Shock-Wave Boundary. *Nature*, **373**, 587.

Aschenbach, B. (1998). Discovery of a young nearby supernova remnant. *Nature*, **396**, 141–142.

Becker, R.H., Smith, B.W., White, N.E., Holt, S.S., Boldt, E.A., Mushotzky, R.F. and Serlemitsos, P.J. (1979). X-ray spectrum of Cassiopeia A measured with the Einstein SSS. *Astrophysical Journal Letters*, **234**, L73–L76.

Becker, R.H., Smith, B.W., White, N.E., Mushotzky, R.F., Holt, S.S., Boldt, E.A. and Serlemitsos, P.J. (1980). Elemental abundances in a Type I supernova remnant. *Astrophysical Journal Letters*, **235**, L5–L8.

Becker, R.H., White, N.E., Boldt, E.A., Holt, S.S. and Serlemitsos, P.J. (1980). X-ray spectrum of Kepler's supernova remnant. *Astrophysical Journal Letters*, **237**, L77–L79.

Becker, R.H., Szymkowiak, A.E., Boldt, E.A., Holt, S.S. and Serlemitsos, P.J. (1980). Is the remnant of SN 1006 Crablike? *Astrophysical Journal Letters*, **240**, L33–L35.

Bell, J. (1997). In: N. Shibazaki, N. Kawai, S. Shibata and T. Kifune (eds), *Neutron Stars and Pulsars*. Universal Academy press inc., Tokyo, Japan, p. 2.

Decourchelle, A., Sauvageot, J.L., Audard, M., Aschenbach, B., Sembay, S., Rothenflug, R., Ballet, J., Stadlbauer, T. and West, R.G. (2001). XMM-Newton observation of the Tycho supernova remnant. *Astronomy and Astrophysics Letters*, **365**, L218–224.

Dotani, T., Hayashida, K., Inoue, H., Itoh, M., Koyama, K., Makino, F., Mitsuda, K., Murakami, T., Oda, M., Ogawara, Y., Takano, S., Tanaka, Y., Yoshida, A., Makishima, K., Ohashi, T., Kawai, N., Matsuoka, M., Hoshi, R., Hayakawa, S., Kii, T., Kunieda, H., Nagase, F., Tawara, Y., Hatsukade, I., Kitamoto, K., Miyamoto, S., Tsunemi, H., Yamashita, K., Nakagawa, M., Yamauchi, M., Turner, M.J.L., Pounds, C.A., Thomas, HD, Stewart, G.M.C., Cruise, A.M., Patchett, B.E. and Reading, D.H. (1987). Discovery of an unusual hard X-ray source in the region of supernova 1987A. *Nature*, **330**, 230–231.

Favata, F., Vink, J., dal Fiume, D., Parmar, A.N., Santangelo, A., Mineo, T., Preite-Martinez, A., Kaastra, J.S. and Bleeker, J.A.M. (1997). The broad-band X-ray spectrum of the CAS A supernova remnant as seen by the BeppoSAX observatory. *Astronomy and Astrophysics*, **324**, L49–L52.

Fujimoto, R., Tanaka, Y., In oue, H., Ishida, M., Itoh, M., Mitsuda, K., Sonobe, T., Tsunemi, H., Murakami, H., Koyama, K., Hayashi, I., Ikeda, K., Rasmussen, A., Ricker, G. and Becker, C.M. (1995). ASCA observation of the supernova remnant W49B. *Publications of the Astronomical Society of Japan*, **47**, L31–L35.

Gaetz, T.J., Butt, Y.M., Edgar, R.J., Eriksen, K.A., Plucinsky, P.P., Schlegel, E.M. and Smith, R.K. (2000). Chandra X-Ray observatory arcsecond imaging of the young, oxygen-rich supernova remnant 1E0102.2-7219. *Astrophysical Journal Letters*, **534**, L47–L50.

Hatsukade, I., Tsunemi, H., Yamashita, K., Koyama, K., Asaoka, Y. and Asaoka, I. (1990). The X-ray spectrum of Kepler's supernova remnant. *Publications of the Astronomical Society of Japan*, **42**, 279–286.

Hirata, K., Kajita, T., Koshiba, M., Nakahata, M., Oyama, Y., Sato, N., Suzuki, A., Takita, M., Totsuka, Y., Kifune, T., Suda, T., Takahashi, K., Tanimori, T., Miyano, K., Yamada, M., Beier, E.W., Feldscher, L.R., Kim, S.B., Mann, A.K., Newcomer, F.M., Van, R., Zhang, W. and Cortez, B.G. (1987). Observation of a neutrino burst from the supernova SN1987A. *Physical Review Letters*, **58**, 1490–1493.

Holt, S.S., Gotthelf, E.V., Tsunemi, H. and Negoro, H. (1994). ASCA observations of Cassiopeia A. *Publications of the Astronomical Society of Japan*, **46**, L151–L154.

Hwang, U. and Gotthelf, E.V. (1997). X-ray emission-line imaging and spectroscopy of Tycho's supernova remnant. *Astrophysical Journal*, **475**, 665–682.

Hwang, U., Holt, S.S. and Petre, R. (2000). Mapping the X-ray-emitting ejecta in Cassiopeia A with Chandra. *Astrophysical Journal Letters*, **537**, L119–L123.

Inoue, H., Koyama, K., Matsuoka, M., Ohashi, T., Tanaka, Y. and Tsunemi, H. (1980). Emission line features in the soft X-ray spectra of the North Polar Spur and the Cygnus Loop. *Astrophysical Journal*, **238**, 886–891.

Iyudin, A.F., Diehl, R., Bloemen, H., Hermsen, W., Lichti, G.G., Morris, D., Ryan, J., Schoenfelder, V., Steinle, H., Varendorff, M., de Vries, C. and Winkler, C. (1994). COMPTEL observations of Ti-44 gamma-ray line emission from CAS A. *Astronomy and Astrophysics*, **284**, L1–L4.

Iyudin, A.F., Schönfelder, V., Bennett, K., Bloemen, H., Diehl, R., Hermsen, W., Lichti, G.G., van der Meulen, R.D., Ryan, J. and Winkler, C. (1998). Emission from 44Ti associated with a previously unknown Galactic supernova. *Nature*, **396**, 142–144.

Jenkins, E.B. and Meloy, D.A. (1974). A survey with Copernicus of interstellar O VI absorption. *Astrophysical Journal Letters*, **193**, L121–L125.

Kinugasa, K. and Tsunemi, H. (1999). ASCA observation of Kepler's supernova remnant. *Publications of the Astronomical Society of Japan*, **51**, 239–252.

Koyama, K., Petre, R., Gotthelf, E.V., Hwang, U., Matsuura, M., Ozaki, M. and Holt, S.S. (1995). Evidence for shock acceleration of high-energy electrons in the supernova remnant SN:1006. *Nature*, **378**, 255.

Levenson, N.A., Graham, J.R., Keller, L.D. and Richter, M.J. (1998). Panoramic views of the Cygnus Loop. *Astrophysical Journal Supplement*, **118**, 541–561.

Markert, T.H., Clark, G.W., Winkler, P.F. and Canizares, C.R. (1983). High-velocity, asymmetric Doppler shifts of the X-ray emission lines of Cassiopeia A. *Astrophysical Journal*, **268**, 134–144.

Masai, K. (1984). X-ray emission spectra from ionizing plasmas. *Astrophysics and Space Science*, **98**, 367–395.

Miyata, E., Tsunemi, H., Pisarski, R. and Kissel, S.E. (1994). The plasma structure of the north-east rim of the Cygnus Loop as observed with ASCA. *Publications of the Astronomical Society of Japan*, **46**, L101–L104.

Miyata, E., Tsunemi, H., Kohmura, T., Suzuki, S. and Kumagai, S. (1998a). Metal-rich plasma at the center portion of the Cygnus Loop. *Publications of the Astronomical Society of Japan*, **50**, 257–269.

Miyata, E., Tsunemi, H., Torii, K., Hashimotodani, K., Tsuru, T., Koyama, K., Ayani, K., Ohta, K. and Yoshida, M. (1998b). Discovery of a compact X-ray source inside the Cygnus Loop. *Publications of the Astronomical Society of Japan*, **50**, 475–481.

Miyata, E., Ohta, K., Torii, K., Takeshima, T., Tsunemi, H., Hasegawa, T. and Hashimoto, Y. (2001). On the Nature of AX J2049.6+2939/AX J2050.0+2914. *Astrophysical Journal*, **550**, 1023–1029.

Pelling, R.M., Paciesas, W.S., Peterson, L.E., Makishima, K., Oda, M., Ogawara, Y. and Miyamoto, S. (1987). A scanning modulation collimator observation of the high-energy X-ray source in the Crab Nebula. *Astrophysical Journal*, **319**, 416–425.

Petre, R., Becker, C.M. and Winkler, P.F. (1996). A central stellar remnant in Puppis A. *Astrophysical Journal*. **465**, L43–L46.

Rho, J. and Petre, R. (1997). X-ray imaging and spectroscopy of the supernova remnant CTB 109 and its associated pulsar 1E 2259+586. *Astrophysical Journal*, **484**, 828–843.

Sedov, L.I. (1959). *Similarity and Dimensional Methods in Mechanics*, New York: Academic Press.

Sunyaev, R., Kaniovsky, A., Efremov, V., Gilfanov, M., Churazov, E., Brebenev, S., Kuznetsov, A., Melioranskiy, A., Yamburenko, N., Yunin, S., Stepanov, D., Chulkov, I., Pappe, N., Boyarskiy, M., Gavrilova, E., Loznikov, V. and Prudkoglyad, A. (1987). Discovery of hard X-ray emission from supernova 1987A. *Nature*, **330**, 230.

Tanaka, Y. and Bleeker, J.A.M. (1977). The diffuse soft X-ray sky – Astrophysics related to cosmic soft X-rays in the energy range 0.1-2.0 keV. *Space Science Reviews*, **20**, 815–888.

Tanimori, T., Hayami, Y., Kamei, S., Dazeley, S. A., Edwards, P. G., Gunji, S., Hara, S., Hara, T., Holder, J., Kawachi, A., Kifune, T., Kita, R., Konishi, T., Masaike, A., Matsubara, Y., Matsuoka, T., Mizumoto, Y., Mori, M., Moriya, M., Muraishi, H., Muraki, Y., Naito, T., Nishijima, K., Oda, S., Ogio, S., Patterson, J.R., Roberts, M.D., Rowell, G.P., Sakurazawa, K., Sako, T., Sato, Y., Susukita, R., Suzuki, A., Suzuki, R., Tamura, T., Thornton, G. J., Yanagita, S., Yoshida, T. and Yoshikoshi, T. (1998). Discovery of TeV gamma rays from SN 1006: Further evidence for the supernova remnant origin of cosmic rays. *Astrophysical Journal Letters*, **497**, L25–L28.

Thielemann, F., Nomoto, K. and Hashimoto, K. (1996). Core-collapse supernovae and their ejecta. *Astrophysical Journal*, **460**, 408–436.

Torii, K., Kinugasa, K., Toneri, T., Asanuma, T., Tsunemi, H., Dotani, T., Mitsuda, K., Gotthelf, E.V. and Petre, R. (1998). Discovery of a 69 millisecond X-ray pulsar: A compact source in the vicinity of the supernova remnant RCW 103. *Astrophysical Journal Letters*, **494**, L207–L210.

Tsunemi, H., Yamashita, K., Masai, K., Hayakawa, S. and Koyama, K. (1986). X-ray spectra of the Cassiopeia A and TYCHO supernova remnants and their element abundances. *Astrophysical Journal*, **306**, 248–254.

Tsunemi, H., Miyata, E. and Aschenbach, B. (1999). Spectroscopic study of the Vela-Shrapnel. *Publications of the Astronomical Society of Japan*, **51**, 711–717.

Tsunemi, H., Miyata, E., Kinugasa, K. and Hiraga, J. (2000). Thermal emission from SNRs and plasma diagnostics. *Advances in Space Research*, **25**, 539–548.

Vartanian, M.H., Lum, K.S.K. and Ku, W.H. (1985). Imaging X-ray spectrophotometric observation of SN 1006. *Astrophysical Journal Letters*, **288**, L5–L9.

Vedder, P.W., Canizares, C.R., Markert, T.H. and Pradhan, A.K. (1986). High-resolution X-ray spectroscopic evidence of nonequilibrium conditions in the Cygnus Loop. *Astrophysical Journal*, **307**, 269–274.

Whittet, D.C.B. (1999). A physical interpretation of the 'red Sirius' anomaly. *Monthly Notices of the Royal Astronomical Society*, **310**, 355–359.